The Anatomical Basis of Medical Practice

The Anatomical Basis of Medical Practice

R. FREDERICK BECKER, Ph.D.
Michigan State University;
Formerly: Duke University Medical Center

JAMES W. WILSON, Ph.D., M.D.
Duke University Medical Center

JOHN A. GEHWEILER, M.D.
Duke University Medical Center

1971
The Williams & Wilkins Company
Baltimore

Made in the United States of America

Library of Congress Catalog Card Number 71-167736
SBN 683-00494-8

Composed and printed at the
WAVERLY PRESS, INC.
Mt. Royal and Guilford Aves.
Baltimore, Md. 21202, U.S.A.

DEDICATION

This book is dedicated to YOU:

You may be a medical student beginning anatomy in your first year; You may be taking an in depth elective course in anatomy in a later year in the curriculum; or You may be reviewing anatomy in a Residency Training Program. Perhaps, even You, the continuing student out in practice, might find reading it a refreshing experience.

We certainly did not write it to impress our colleagues with our erudition. Now that we have reread it, we are convinced of that. Nevertheless, we hope our colleagues will recommend the book to you as one which will be helpful in your clinical career.

We agree with Dr. Robert Salter of Toronto, author of a superb "Textbook of Disorders and Injuries of the Musculo-skeletal System," that a textbook is one "containing the principles of a subject, used as a basis for instruction." This is why we chose "The Anatomical Basis of Medical Practice" as a title. The book is designed to *instruct* and *familiarize* you with the concepts of anatomy useful to you as a clinician, not as a professional anatomist.

As a consequence, we have not given you a complete reference book in anatomy. There are many of these useful, large, reference compendia available. As good students you ought to consult them from time to time, as you would any complete reference work, to clarify certain details. Neither do we claim to have presented you with a Synopsis, an Outline nor a Handbook. Such are useful tools, to be sure, but more often as review media after one has had a reasonable amount of prior instruction in an unfamiliar area. Even then, such quick reviews may be more factual than conceptual.

One of the most difficult things to get across to the beginning student is that memorization of facts is not the goal of anatomy. It is to relate these facts to some meaningful whole that is important. Probably it is because anatomy, often a first course in medical schools, introduces the neophyte to such a vast new vocabulary of medical terms, that learning this vocabulary has been conducive to memorizing. You must try hard to overcome this tendency, for memorization alone does not necessarily lead to real understanding. We would prefer that you get a sound idea of how the human body is put together and how it functions, even at the expense of your having forgotten the specific names of some of the multitude of structures which comprise it. These names you can look up again in a hurry.

If the truth were known, this started out to be a "core text" which grew beyond bounds. Everyone we talked to about the "core content of anatomy" had particular and different opinions about what could be pared from the whole anatomical apple and about how much flesh should be left to constitute a useful "core." Had we attempted to satisfy everyone in this regard, we would have been engaged in preparing another voluminous reference text.

In the end, we decided to satisfy You, the medical student, who wanted to understand enough helpful anatomy to contend with the problems of the patient. Thus, we left enough flesh around the core so that you could take a healthy, palatable bite.

One parting bit of advice to You, the modern medical student:

In this age of instant everything, we still have not found a way of imparting instant knowledge despite all the gadgetry of modern pedagogy. Knowledge in any area which is being approached by the student for the first time grows by gradual accretion and use. Reading good books is still one of the best ways to acquire knowledge through self-learning. The pace may be a bit slower than by some other methods, but the incubation period it affords makes it worthwhile. We hope we have offered You one of the good books to help You on your way.

Sincerely,

R. Frederick Becker
James Wilson
John Gehweiler

FOREWORD

Medical education is currently in the midst of curricular experimentation and course evaluations. In these changing times dedicated teachers strive constantly to correct course deficiencies, as well as update and narrow the gaps in our medical knowledge. This is often a formidable task inasmuch as academic traditions can form a rigid mold whereby "established hours" become the yardstick, and relevance to medicine often receives only secondary consideration. Without a strong foundation of basic scientific knowledge, the art of medicine indeed rests upon tenuous pillars placed in quick-sand.

A vast array of excellent teaching and audiovisual innovations have been devised in recent years by anatomists to facilitate programmed instruction and promote intellectual absorption by the student. Other teachers have accepted the challenge of doing better teaching in less time by writing short core-type monographs. The voluminous amount of old and new information in the several aspects of anatomy are sifted for relevance, resynthesized and condensed into brief, palatable scientific packets. Sheer brevity necessitates the exclusion of desirable explanations, original descriptions when necessary and the use of key analogies. Such additions often promote clarity of communication, reader interest and student comprehension. The present unique volume represents an attempt by three highly competent, experienced and energetic teachers to bridge the gap midway between a synopsis and the more encyclopedic textbook. In this endeavor they have accepted the greatest challenge in all the medical and dental disciplines—namely, to simplify the tremendous storehouse of actual and applied knowledge that is embodied in Gross Anatomy.

"Human anatomy is a new experience for most students . . . and it is an exercise in logic. . . . It requires three dimensional thinking; it is invested with a new language which, like all languages, needs to be learned. . . . Anatomy has some sense and order and can quite easily be made an interesting and exciting subject." These are the published words by one of the authors. They give prophetic insight to the constructive thoughts and ultimate objectives of the authors who conceived this very practical book. Early embryology and organogenesis provide a logical starting point to explain the derivation and structural arrangement of the adult organ systems. This and the orderly sequence of subsequent chapters are all written in a lucid, interesting and refreshing scientific style. The rationale for important terms, the distillate of essential basic and applied anatomy, and the profuse use of excellent illustrations are all expertly blended. The result is a smooth, readable, accurate and useful student textbook of Gross Anatomy. Inclusion of pertinent anatomic details with their clinical significance merits special commendation, for the presentation of such material is often deferred to later years or deleted entirely from many anatomy texts. One is tempted to quote passages and give examples from each of the chapters of this fascinating book. However, to do so would detract from a basic prerogative of the reader. It should be his pleasure to discover interesting and important details sprinkled with a myriad of choice tid-bits.

The Anatomical Basis of Medical Practice will be appreciated by all those who desire to learn Anatomy either for the first time, or the busy physician who wishes to review the subject. The character of the textbook is molded by the authors while its usefulness will be determined by the intended reader. In the present attractive book the undergraduate, as well as the professional student, will find that the acquisition of sound anatomic knowledge can be an interesting and truly exciting experience. Throughout the 892 pages of the text the authors have successfully proved that Anatomy is, indeed, alive.

Raymond C. Truex, Ph.D.
Professor of Anatomy
Temple University School of Medicine

PREFACE

The title page makes it apparent that the authors are both allopathically and osteopathically oriented. In morphological outlook, they have no special axes to grind. Even in clinical outlook, they are thinking first of the whole patient and his welfare as a healthy, living being. Fortunately, anatomists, radiologists and pathologists make compatible bed-fellows in both fields. Perhaps the day is dawning when all of our colleagues will cooperate in the common cause of total health care, each in his own way, and become less concerned with private and often petty philosophical differences.

In giving this text a clinical orientation, we have placed strong emphasis upon Surface and Normal Radiographic Anatomy.

We emphasized surface anatomy because the clinical examination begins there, and because we wanted the student to develop an "x-ray sense" of his own, by looking and feeling, so that he attains some three-dimensional ideas of what is under the surface. Perhaps we should have included photographs of garden-variety, American males and females who have let their physiques go to pot. Instead, we used female models as model females. The student will see the ordinary specimen every day. Only on rare occasions will the attractive, well-turned specimen appear before him for consultation. He should be prepared for this pleasant shock. For the growing ranks of female medics, we included the body beautiful of a robust, healthy male. We are sorry that we cannot make available the addresses of the young ladies who grace our pages. Our wives burned our little address books at our last barbecue get-together.

The use of normal radiographs helps to provide an early clinical introduction. The sooner the student learns to identify normal anatomy on x-ray film, the easier it will be for him later to understand the radiological changes seen in diseased states. We feel strongly that gross and microscopic anatomy ought to provide a solid base for radiology and pathology as well as for clinical medicine in general. It is not our intention to proselytize students to become anatomists, radiologists, or pathologists, any more than we would push them to become biochemists or physiologists. Rather, we want them to use anatomy as a useful stepping stone to all of medicine. Our main aim is to inculcate an appreciation for the so-called "normal" morphological constitution of human beings. Strangely, despite our anatomically-oriented backgrounds, we are far less concerned with cadavers than we are with living, breathing individuals, the future patients of our students.

We must admit that we did have in mind providing anatomical correlation for the radiological resident as well. His knowledge of anatomy may have become a bit rusty, or he might still be wondering what some of the fuzzy lines on roentgenograms actually signify.

Medical curricula are in such a state of flux these days, that it has been difficult to plan a text in anatomy for the modern scene. If the first offering in anatomy is more of an introductory nature, the early chapters may provide adequate material. However, all chapters were written in an easy-going, literary style so that any student could read ahead on his own without difficulty, fitting the other sections to the needs of his particular course structure. We have not felt that anatomy needs to be written in dead seriousness; neither have we felt that its presentation needs to be deadly. So we have avoided cadaverous overtones as much as possible.

Where students still encounter considerable dissection as a requirement sometime in their curriculum, they will find the main chapters of this book arranged by bodily regions rather than by systems. The thorax and the abdomen have been given particular attention since they are going to be the "bread and butter" areas of most future physicians whether they approach these areas directly or through visceral manifestations reflected in the soma. Still, head, neck, pelvis and perineum have not been given short shrift. If we have slighted anywhere, it may be in the section on the limbs. Even here, as throughout, our major concern has been to develop concepts of organization, leaving much detail for later specialty consideration. No doubt we have played upon our own preferences in regard to certain bodily regions, and more detail creeps into these sections than elsewhere. If this is not a satisfactory arrangement, we will await the flood of constructive criticism that usually attends first efforts in publishing works of this kind. We are prepared for invective as well, though we may pay less attention to it.

A thread of development is woven throughout. Indeed, the first chapter opens upon a developmental note, and the theme is picked up again and again where it seems most appropriate in later chapters.

We could not possibly have touched on all clinical bases and still have kept the book within reasonable compass. But we feel we did dangle the clinical carrot in plain enough view to tempt the student along a forward path. We ordered the chapters to follow the sequence of the clinical examination from the surface out to the limbs. We left the chapter on central nervous system and special senses to the last in adhering to this order. But we also wanted that chapter to build a bridge over to the neural sciences. Thus, its removal from proximity to head and neck was arbitrary. Each chapter contains enough of the fundamentals of the nervous system to be relevant in these places.

If one dissects, circulation is also encountered regionally, and we adhered to this plan in the main. But we did add a special chapter which treats circulation systematically for overall review purposes.

Believing thoroughly that a picture is worth, ofttimes, a thousand words, we were most liberal with illustrations. Still, we were vain enough to be unwilling to sacrifice the thousand words. These tactics drove the bulk of the text to greater limits than we had originally intended it to be. We hope that this has not been too unhappy an amalgamation and that the words and pictures will reinforce one another to leave a satisfactory and lasting impression.

Pontiac, Michigan
July, 1971

ACKNOWLEDGMENTS

Textbooks rarely spring full blown from authors' heads. Neither do authors labor entirely alone in assembling it for publication. A glance over our shoulders at this time amazes us at the long file of concerned individuals standing there who, in one way or another, had a hand in bringing this book to fruition. Here we thought we were closeted in our own isolated ivory towers all these months! Certainly our wives have harped often enough on this latter point to make it seem entirely believable. Without their acceptance of this whole project and without their constant prodding, bless their mercenary little hearts, we might not be finished yet. They have spent long ago all of the nebulous royalties in fancy, if not in fact. But who could have been more patient, considerate and encouraging in our periods of blue funk or even, at times, of utter cantankerousness? Among them all we are particularly indebted to the spirit of *Faith* Gehweiler who more than once buoyed up our *hopes* and succored us with *charity* by devoting long hours to editorial assistance.

We are deeply indebted to our mentors, past and present, whose brains we picked. The senior author looks back, in particular, to his early training under Dr. William F. Windle and Dr. Leslie B. Arey, and hopes he has not disgraced them altogether. He regards with warm memories his long association with the late Dr. Joseph E. Markee of Duke University for an outpouring of encouragement up to the last. He is grateful for the strong vote of confidence from Dean Myron S. Magen and from this author's close associates here at Michigan State University College of Osteopathic Medicine.

In like manner the younger members of the team express thanks to their chiefs for encouragement and for the furnishing of certain materials: Dr. Richard G. Lester, Chairman and Professor of Radiology; and Dr. Thomas D. Kinney, Chairman and Professor of Pathology and Director of Medical Education, at Duke University.

A number of other people made direct contributions of materials used in the text:

Mr. Charles R. Kilgore, Kilgore International Importers of Coldwater, Michigan donated the use of much of the skeletal materials depicted in the photographs.

Several other contributed illustrations of one kind or another: Dr. Stephan Ryan, St. Luke's Hospital, New York City, photographs and drawings of lung capillaries; Dr. Joseph M. Lauweryns and his associate Dr. Luke Bosseau, Department of Pathology, University of Leuven, Belgium, actually wrote and illustrated the section on pulmonary lymphatics; Dr. Donald Hackel and Dr. Leona Ayers, Department of Pathology, Duke University contributed material on the conducting system of the heart and the paraffin-constructed model of this system; Mr. Leon Schlossberg of the Medical Illustration Department, Harvard University provided the American Heart Association drawings; Dr. Erwin Johnsrude, Department of Radiology, Duke University, assisted in labelling many of the vascular radiographs presented; The General Electric and Eastman Kodak Companies provided materials for Chapter 4; last, but by no means, least in these regards, we thank Mr. Peter Gowland, internationally famous women's photographer of California, for his generosity in providing the striking photos of female models used throughout the text. We give special thanks to our male model who also went unpaid because he is the senior author's son.

Authors of other outstanding textbooks have been most generous in allowing us to use certain figures from their own works. For this we are especially indebted to Dr. Jan Langman; Dr. Raymond C. Truex and Dr. Malcolm B. Carpenter; Dr. Wilfred M. Copenhaver; and Drs. John C. B. Grant and John V. Basmajian. In addition, Robert N. Cooley, M.D.; Melvin Schreiber, M.D. and Christian V. Cimmino, M.D. loaned us selective radiographs from their excellent books. The use of these figures, in some measure, took the drain off our own overworked medical illustrators: Mr. Wayne Williams and Mr. Bob Blake of Duke University; Mr. William Loechel of Wayne State University; and Ralph Johnson, D.O. of Flint, Michigan. Dr. Barry J. Anson, Department of Otolaryngology, University of Iowa, is to be thanked for allowing us to use his full-body sections of x-ray cadaver material.

In certain other areas, technical assistance was required. Mr. Phillip Pickett and Mrs. Betty Schlitz, Histology Preparations Laboratory, Pathology Department, Duke University, provided the sections of endocrines, kidneys, uterus and large organs. We appreciate the work of Miss Barbara Miller and Miss Kathy Horn who lightened the work in the photographic darkroom. Physicist Alice Mc Rae at Duke graciously consented to read Chapter 4 for content and accuracy. Technical assistance in preparing many of the radiographs was provided especially by John B. Cahoon, R.T.; Mary Stump, R.T.; Ron Cromartie, R.T; Karen Schurm, R.T.; and Frances Apple, R.T.—all of Duke University.

Besides a patient wife, there stands by every author a harassed secretary willing to burn the midnight oil when necessary. We could not have gotten along at all without the help of Eleanor Hart, Faith L. Gehweiler, Annette Grant and Doris Klausmeyer.

All of us owe a great deal to those who have complete faith in our credit ratings. Special kudos in this department must go to Mr. Bill Harrison of Foister's Camera Store, Inc. of Chapel Hill, North Carolina who virtually fitted our darkroom on tick and at wholesale prices.

One of the nicer things about a professional career are the warm friendships that develop between members in the same spheres of scientific interest. Among the many fine and accomplished gentlemen of the American Association of Anatomists, we have been most fortunate in being counted as one of the friends of Dr. Raymond C. Truex. We are most honored by his willingness to peruse this text and by his kind comments in the Foreword.

In the final analysis, we owe the most to certain men who looked upon our efforts with high regard, were willing to take extreme capital risks in our behalf, and who backed us and encouraged us every step of the way. We could not have found two nicer gentlemen anywhere in the field of medical publishing than Mr. Charles Reville, President, and Mr. Earle V. Hart, Editor, of The Williams &Wilkins Company, Baltimore, Maryland. Our advice to anyone with a saleable medical book in the works is to get in touch with them at once. You will never be in better hands, not even with Allstate.

Pontiac, Michigan and
Durham, North Carolina,
July, 1971

CONTENTS

ONE

Early Beginnings of Human Form

INTRODUCTION

Developmental anatomy, the study of the changes undergone by a living organism, begins at the time of union of the male and female sex cells. Often this study is pursued only until the time of birth, but it can be carried beyond through infancy, childhood, and into adulthood, for some bodily organs mature late in life whereas others begin to regress after birth. If we are going to tell the story of human anatomy in its essentials in this book, we should start at the beginning for a number of reasons.

First, the beginning is always a good and sensible place to start any story. Secondly, form and relationships of certain adult structures often first begin to make sense to the student when certain aspects of their developmental history are understood. Finally, morphological abnormalities are of clinical significance and they frequently are understood best as deviations from the normal course of developmental events. Clinicians in virtually every branch of medicine constantly face problems that require a fundamental knowledge of the anatomy and physiology of reproduction and embryogenesis.

We shall begin the story of development in this chapter, but we shall not finish it here. Rather, throughout the text, we shall reintroduce the topic where it seems pertinent for a better understanding of adult structures. We hope that the developmental point of view will serve as a drawstring to pull things together in a meaningful way.

THE GERM CELLS

Any new individual begins as the fusion of two cells, one from each parent, the male germ cell or **spermatozoon,** and the female germ cell or **ovum.** The process of this union is known as fertilization, and the new organism formed by the union is the **zygote.** From the zygote, the **embryo** develops. At the end of the second month after fertilization, when the embryo begins to look reasonably like a human being, the as yet unborn organism is called a **fetus.**

Before launching into a discussion of the zygote and its development, let us consider the components of the male and female reproductive system.

Male Reproductive System

The male germ cells are produced in the **testes,** paired glands housed in a skin pouch, the **scrotum.** The spermatozoa are conducted to the exterior by a system of genital ducts associated with special accessory glands (Fig. I-1A).

The testes contain a number of compartmentalized, convoluted, **seminiferous tubules** within the walls of which male germ cells develop. In the surrounding connective tissue, between the tubules, lie nests of **interstitial cells** responsible for the secretion of the male sex hormone, **testosterone.** The latter cells, lying in close proximity to capillaries as they do, serve as the **endocrine** portion of the testes. Their product is transported *via* the blood stream. The tubules proper serve as the **exocrine** portion of the sex gland since their products, the male sex cells, are expelled through ducts. The terminal ends of the coiled tubules straighten to end in a network, the **rete testis.** This network system of epithelially-lined channels leads by a system of **efferent ductules** into the **epididymis,** a structure applied cranially and posteriorly to the testis. Sperm are stored and undergo partial maturation in the epididymis, and it is erroneous to attribute a sperm storage function to the seminal vesicles, glandular organs remote from the testes.

The tubules in the head of the epididymis unite to form a single, coiled duct which, after descending along the posterior aspect of the testis, straightens to become the thick-walled, main excretory duct of the testis-epididymal complex, the **ductus deferens** (vas deferens). The latter ascends in the scrotum to become one of the structures within the **spermatic cord,** a complex of ducts, vessels, and nerve fibers wrapped in layers of connective tissue. The cord contains vessels and nerves for both the ductus deferens itself and for the testis. The coverings of the cord also enclose some skeletal muscle fibers, the **Cremasteric muscle,** which reflexly pulls the testes toward the abdomen in cold weather or when the skin on the inner aspect of the thigh is stroked. Clinically, a positive cremasteric reflex signifies that the motor nerve fibers emanating from the first lumbar cord segment are intact, at least those that are conveyed by the genitofemoral nerve.

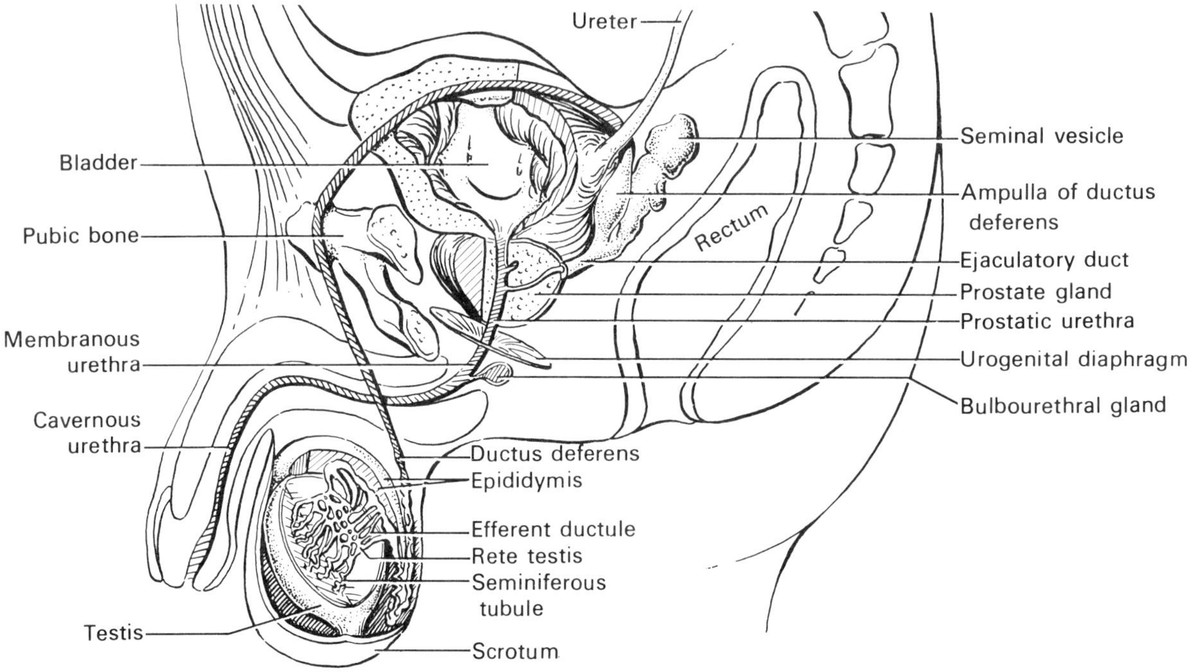

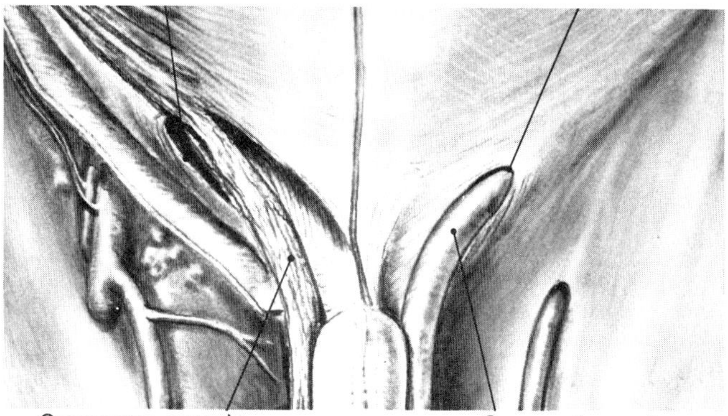

Fig. I-1A. Midsagittal schema of parts of male reproductive system. (The spermatic cord associated with the ductus deferens is not shown nor is a second lamina of the urogenital diaphragm below the bulbourethral gland. These were deleted to show schema of duct system to advantage.)

B. Spermatic cord with Cremasteric muscle passing through inguinal canal.

After passing with the cord structures through the **inguinal canal** (Fig. I-1B), the ductus deferens finds itself under the **peritoneal lining** of the pelvic cavity where it remains. It crosses in turn the iliac vessels and the ureter and courses medially downward in the pelvic cavity toward the **prostate gland** upon which the urinary bladder rests in the male. Before reaching the prostate, the ductus widens into an **ampulla** or dilated portion and, at this point, joins with the duct of the **seminal vesicle,** the common union becoming the **ejaculatory duct.** The seminal vesicle is, actually, a diverticulum or outgrowth of the ductus deferens, and it contributes a secretion to the composite of seminal fluid in which spermatozoa will be suspended at the time of ejaculation. The ejaculatory duct now pierces the prostate gland from behind and empties into the **urethra,** the urinary outlet of the bladder, which has pierced the prostate from above. It should be clear that the reproductive duct system just described is a paired system, one set of ducts coming from each testis, and that both pierce the prostate to open into a common, single urethra. The **prostatic urethra** is the most proximal portion of the long male urethra which will pierce the urogenital diaphragm to course as a channel to the exterior through the penis. The prostate supplies most of the secretory component of the **semen,** but the secretions from the seminal vesicles and from two small **bulbourethral glands** (of Cowper) add to it.

Thus, the reproductive system of the male (Fig. I-1) consists of: (1) two sperm-producing glands, the testes; (2) a system of paired ducts to convey spermatozoa to a single-channeled urethra; (3) some paired accessory glands (seminal vesicles and bulbourethral) and a single, midline gland, the prostate, all of which contribute to the seminal fluid in which spermatozoa are suspended in the ejaculate; and (4) finally, there is the copulatory organ itself which contains the greater portion of the urethra, the final common duct for the expulsion of both the urinary wastes and the reproductive cells in the male.

Spermatogenesis: Figure I-2 illustrates certain

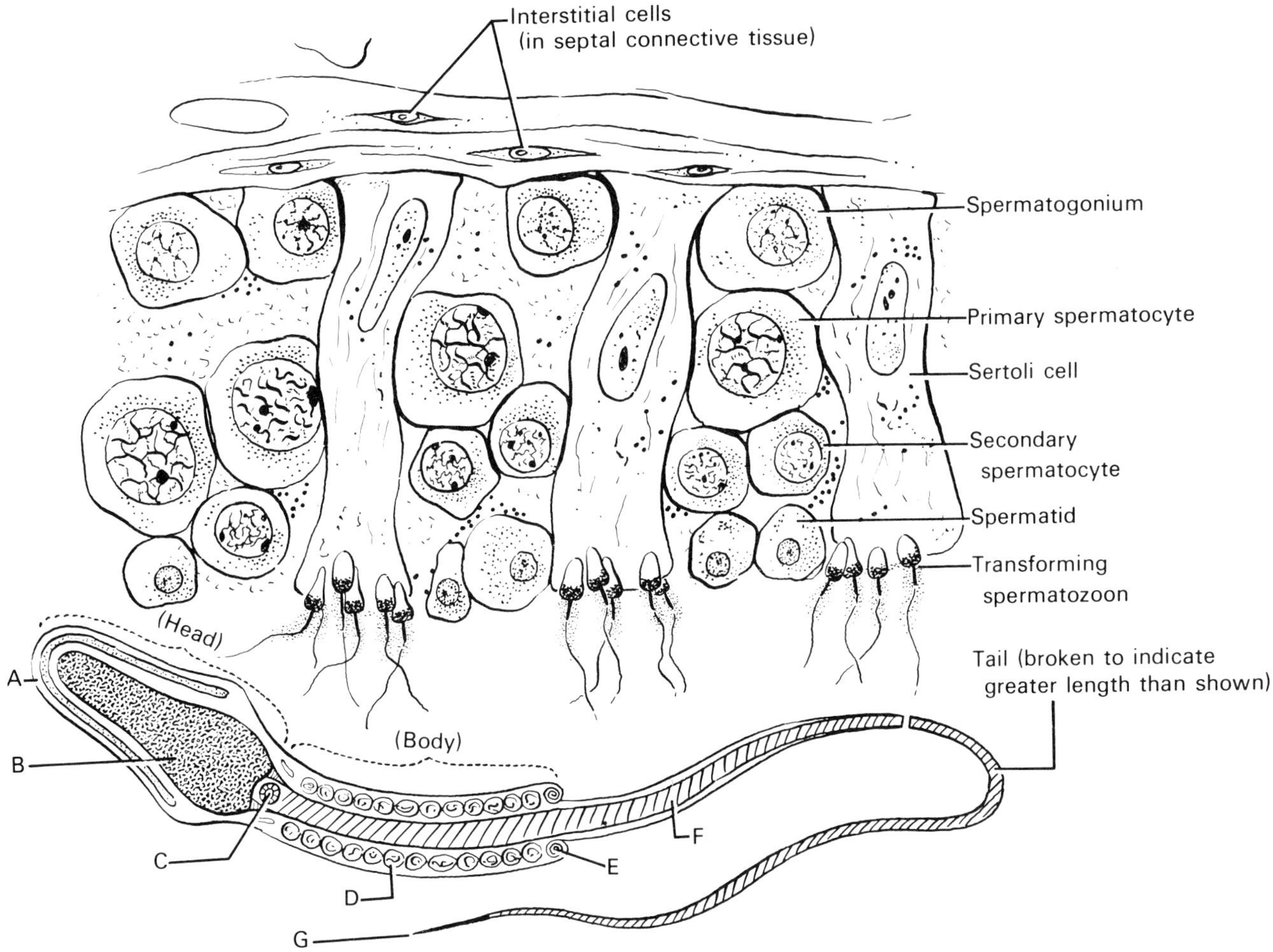

Fig. I-2. Schema of a portion of the wall of a seminiferous tubule showing types of male sex cells undergoing development. *Inset:* A mature spermatozoon. A, acromosal cap; B, nuclear material of sperm head; C, proximal centriole; D, spiral packing of mitochondria in body; E, distal ring centriole; F, axial filament of flagellum; G, terminal filament.

stages of the formation of spermatozoa within the walls of the seminiferous tubules. Two types[1, 2] of cells rest upon the basal lamina of the tubule: (1) **spermatogonia,** block-shaped cells of fairly large size—the most primitive of the cells involved in sperm production; and, (2) tall, columnar, **Sertoli cells,** believed to act as a kind of nurse cell to later generations of immature germ cells. Spermatogonia contain the **diploid number** of chromosomes of the species within their nuclei.[3-6] In man[7, 8] this is 44 + x + y; x and y represent female and male sex-determining chromosomes, respectively. By ordinary **mitotic** division, spermatogonia give rise to daughter cells larger than the parent, the **primary spermatocytes.** These cells have moved closer to the tubule lumen where they, too, divide, but by a process of **meiosis** where each daughter cell, the smaller **secondary spermatocyte,** receives half the number of chromosomes of the parent (**haploid number**); *e.g.,* 22 + x or 22 + y. The mitotic division of the secondary spermatocytes gives rise to still smaller cells, the **spermatids,** which, of course, carry the same haploid chromosomal complement. Spermatids divide no further. Instead, they attach in clusters to the luminal surface of the Sertoli cells and slowly metamorphose into spermatozoa. In this process much of the cytoplasm of the spermatid is lost by extrusion. What becomes of this residual material is not well known. The remainder is wrapped as a cytoplasmic **acrosomal cap** about the cell nucleus, about a cylindrical middle piece or body, and, except for the terminal end, about the **axial filament** of the tail or **flagellum.**[9] The **centrioles** of the spermatid which give rise to the axial filament reside in the proximal and distal ends of the sperm body, the distal one being ring-shaped and pierced by the axial filament. **Mitochondria** are packed spiral-fashion around the axial filament in the central body portion. Thus, spermatozoa transform from spermatids which they in no way resemble. Overall, the tiny spermatozoon which can only be visualized under the microscope measures about 55 to 60 **microns** (μ). Recall that a micron equals $\frac{1}{1000}$ of a millimeter, the smallest division on your centimeter rule.

While spermatozoa remain within the seminiferous tubules they are nonmotile though they may become motile gradually within the epididymis.[10] The excess fluid and extraneous material of the seminiferous tubules appear to be absorbed by the lining epithelium of the epipidymis.[11, 12] This resorption, apparently, creates a negative pressure which could aid in the transport of spermatozoa from the seminiferous tubules into the epididymis. Prior to ejaculation spermatozoa are moved along the ductus deferens by muscular peristalsis. When they mix with the secretions of the accessory glands, they become very active, lashing their flagella and moving forward at the rate of 15 to 20 μ per second.

Thus, in the male gonad, contrary to what we shall find in the female gonad, everything is directed toward the production of millions of tiny, active germ cells whose potential for survival after ejaculation is short. They have sacrificed cytoplasm which could store food material. Within the testis and epididymis they can survive much longer, but, even so, they have been relatively remote from an adequate oxygen supply during most of their lives. Apparently, they obtain their energy by fermentative metabolism rather than by oxidative metabolism and can live anaerobically. Carbohydrates are supplied by the tubules through which they pass. The epididymis, for example, is rich in glycogen, and fructose is a product of the seminal vesicles. Spermatozoa thrive in the alkaline environment of the male reproductive tract and are extremely sensitive to changes in pH, especially if it is an acid one such as they will encounter in the vagina where great numbers of them die rather quickly. For this reason, many spermatozoa are suspended in a single ejaculate. In healthy males, the concentration is about 100 million per cc. of semen. If the concentration falls below 20 million per cc., the male is usually considered sterile.[13]

Spermatozoa begin to lose their motile power and, hence, their ability to reach and fertilize an ovum about 24 to 36 hours after ejaculation. They must meet the ovum in the upper reaches of the uterine tube if fertilization is to be successful. Having survived the acidity of the vagina, the trip through the uterus and uterine tube, a distance of some 7 inches in the human, can be made with luck in about an hour by hardy swimmers. Even so, the traversing of the tube itself poses problems. Many spermatozoa get trapped in deep mucosal folds and swim about in blind pockets until exhausted. The main function of the uterine tube is to propel the egg downward into the uterus. This is accomplished by muscular peristaltic waves directed toward the uterus, and the beat of cilia, which beset the uterine tube mucosa, is likewise directed downward. Both of these factors thwart the upstream advance of spermatozoa. Not many of the total number are ever found in the upper end of the tube. In rabbits, for example, only 500 (one out of every 200,000) were recovered.[13] That human eggs do get fertilized and that much of our world is in a state of overpopulation today can be attributed not so much to the relentless amoratory propensity of the human male the world over but to the ardent, single-purposed heroics of certain vigorous spermatozoa in the face of all obstacles. The brief interlude of survival after expulsion during the act of copulation stands in sharp contrast with the relative longevity within the male reproductive tract. One cycle from spermatogonium to spermatid occupies, roughly, 16 days, and it takes about 58 days more to mature in the seminiferous tubules and epididymis.[14] It is not known exactly how long fully-formed spermatozoa can survive in the epididymis prior to ejaculation, but, obviously for many days at times. If not ejaculated, they degenerate and eventually liquify.

Female Reproductive System

The paired female sex glands are the **ovaries;** the paired ducts, the **uterine tubes.** The **uterus** and **vagina** are single, midline structures (Fig. I-3). A pair of accessory glands, the **vestibular glands** (of Bartholin) have ducts opening into the vaginal vestibule. They are comparable to the mucin-producing bulbourethral glands of the male. The lateral suspension of the uterus as it lies between the urinary bladder and rectum is achieved by a pair of **broad ligaments** which are reflections of peritoneum from the pelvic body wall. The ovaries and uterine tubes are attached to the broad ligament by short peritoneal folds emanating from that structure, and these folds are designated as the **mesovarium** and the **mesosalpinx** (salpinx, Greek for "trumpet" because of the flared shape of the open ends of the uterine tubes). The finger-like **fimbriae** at the open ends of the tubes lie in close proximity to each ovary; indeed, one or more of these fingers may attach to the ovary. Thus, at ovulation the egg or ovum generally finds its way into the upper end of the tube where fertilization must occur before the zygote is moved into the uterus for **implantation.** Delivery of a formed fetus from the uterus is, of course, by way of the vagina. The much shorter urinary tract, the **female urethra,** does not join with these reproductive tracts as it did in the male but opens on its own into the vaginal vestibule.

Oogenesis: The ovaries (Fig. I-4) are the site of development of the female sex cells, or ova, and of certain female sex hormones. In this sense they are comparable to the male testes, but only their central portion or medulla can be considered to have come from the same origin as the testes. Like all organs in the abdomen and pelvis, they lie behind a covering of **peritoneum.** Over the ovary this covering differs from that elsewhere in these body cavities. Generally, the epithelial component of peritoneum (as well as of pleura and pericardium) consists of a single layer of thin flat cells (**simple squamous epithelium** or **mesothelium**). Over the ovary it is still a single-layered lining, but the cells are **cuboidal,** as tall as they are wide, and they constitute the so-called **germinal epithelium** of the ovary. The function of this germinal epithelium has long puzzled investigators. Two main schools of thought have evolved about it:

1. A current view[15-24] holds that it is a non-functional covering in the adult and that, perhaps, it may never have functioned at any time as a germinal epi-

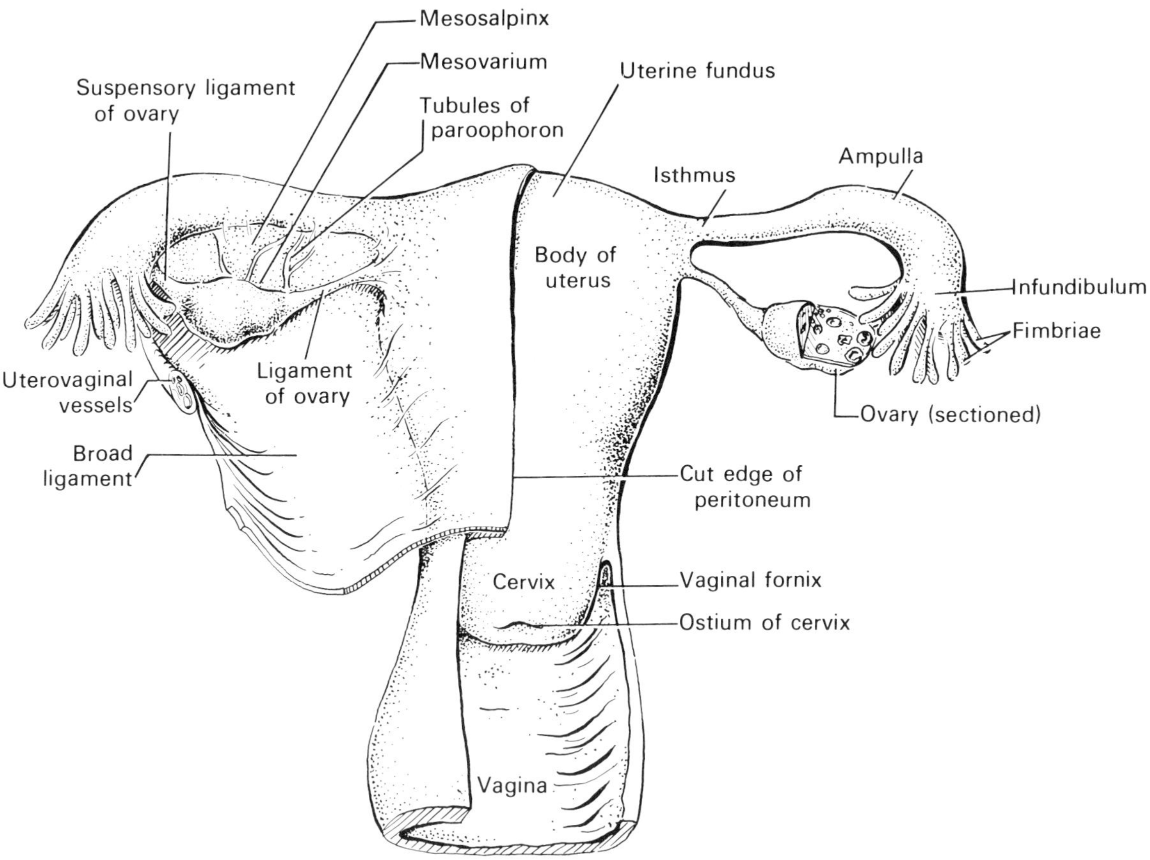

Fig. I-3. The female reproductive organs.

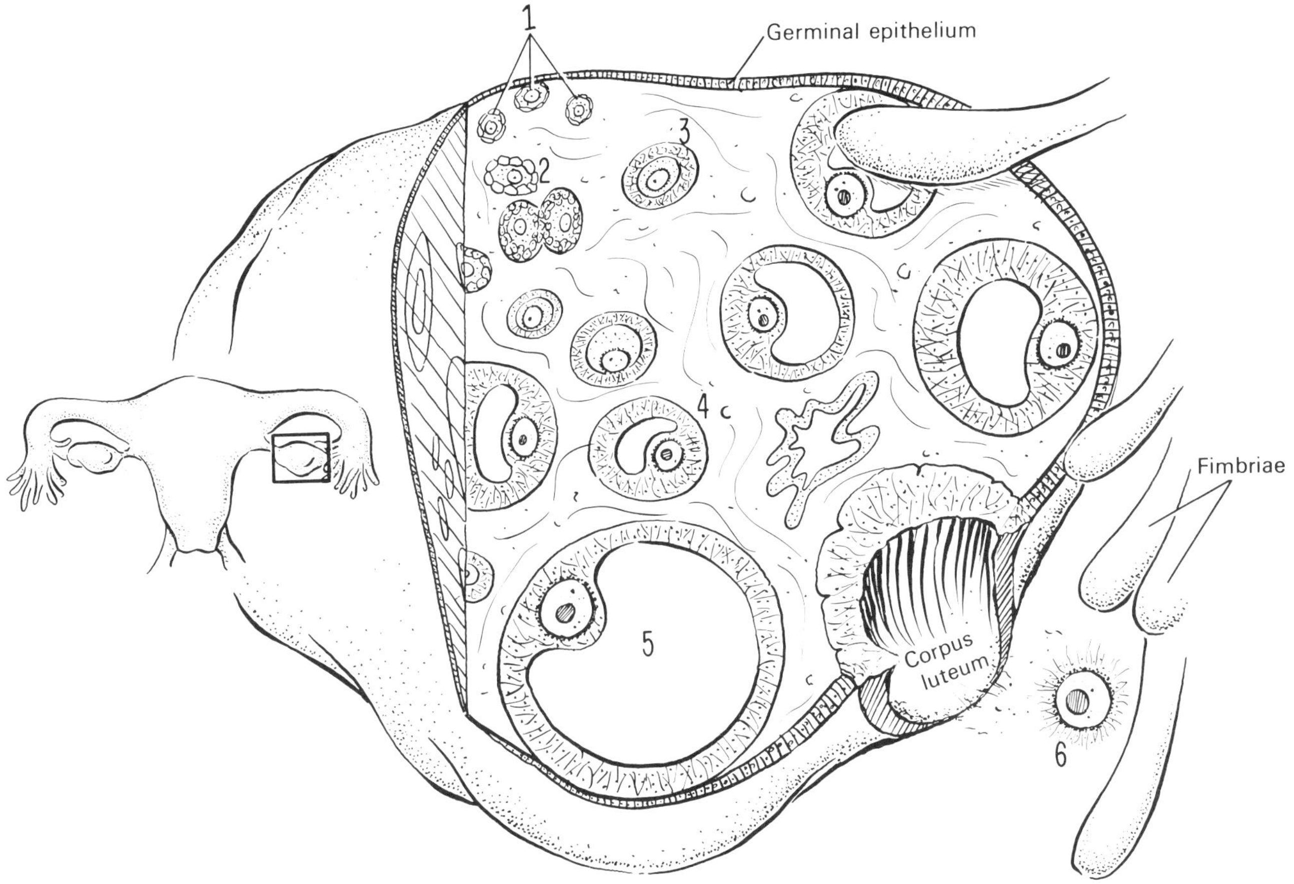

Fig. I-4. Schematic section through an ovary to show stages in oogenesis. 1, Primordial follicles; 2 and 3, early stages of growing follicles; 4, antrum formation; 5, Graafian follicle about to rupture; 6, ovulated egg (secondary spermatocyte) about to enter uterine tube. See Figure I-6 for structural details.

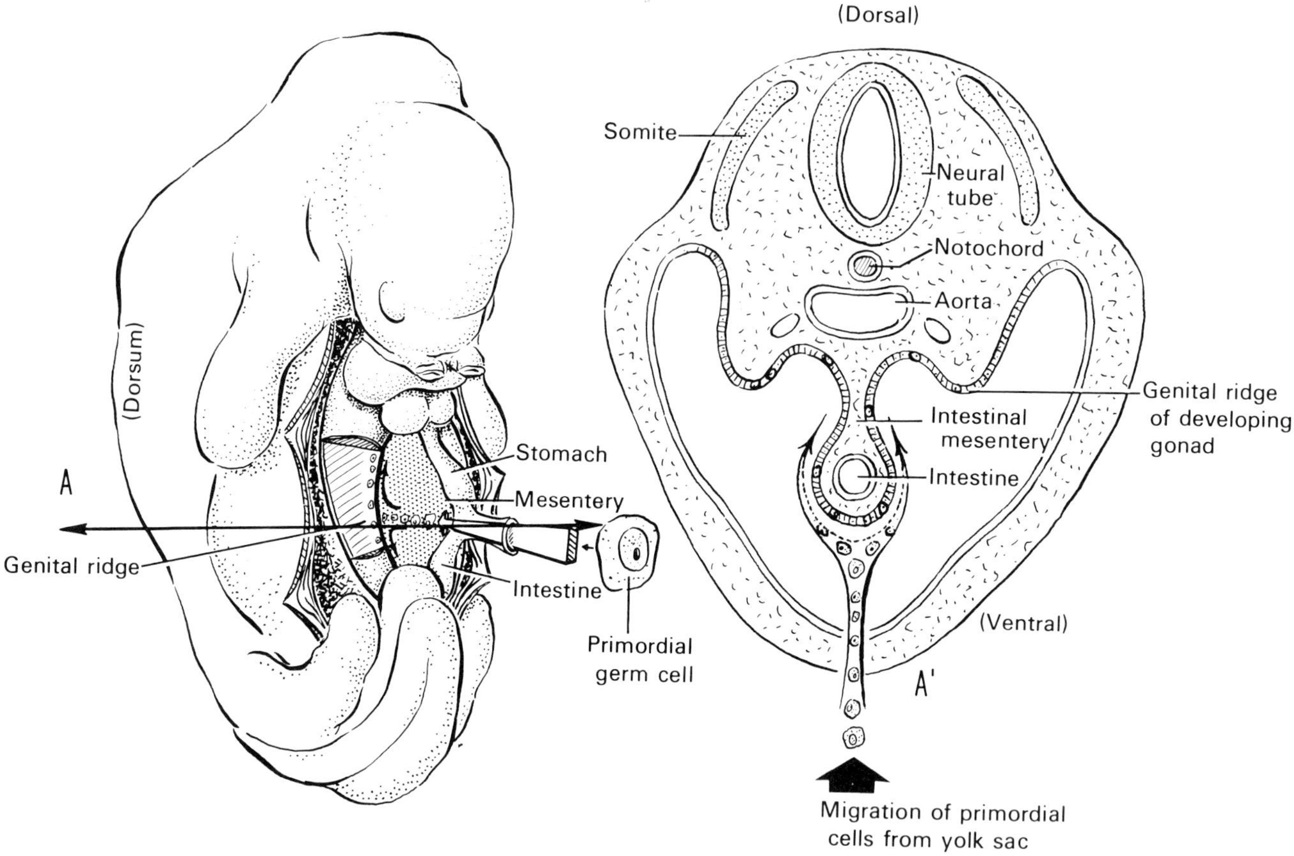

Fig. I-5. Origin of primordial germ cells schematically illustrated. A′ is cross-section through embryo in plane A.

thelium which produced new ova. This idea is not new but rests upon an old and almost irrefutable observation (Fig. I-5), that the original germ cells do not arise in the gonads at all but arise in the embryonic yolk sac and migrate upward along the embryonic intestinal mesenteries to invade the tissues at the higher site of embryonic gonad formation. They are said to migrate directly into the ovarian cortex, not into the germinal epithelium. By mitotic division of the primordial cells of the cortex of the gland, all of the 40,000 to 300,000 primary oocytes that could be called upon to mature from time to time during the 30- to 40-odd years of useful reproductive life of the female have, thus, been produced during embryonic life. They have not come from the germinal epithelium, and no more are produced after birth.

2. A second view[25-29] holds that by puberty all of this original group of primary oocytes have long degenerated (become **atretic**). Beginning at puberty, under hormonal influence, the germinal epithelium is induced monthly to proliferate cords of cells into the ovarian cortex. These give rise to primary oocytes and their follicular nurse cells. This would presuppose that some of the primitive oogonia came to rest in the germinal layer and are capable of cyclic reproduction.

Both groups are somewhat adamant in adhering to their points of view so that the student is likely to encounter controversy as he meets different clinicians during his medical school career. He will have to sift the evidence for himself if he feels it is essential to adopt one hypothesis over the other. Both rest upon inferences, and neither answers all the questions that might be raised. The bibliography at the end of this chapter supplies some of the key references on both sides of the argument for the interested student.[15-29]

If the first view is correct, why, then, at puberty do not all primary oocytes mature at once under a driving hormonal influence since they are all, presumably, in the same dormant state of development? What causes some few to mature each month and others to lie dormant 40 years or more?

If the second view is correct, how can all the primary oocytes in the prepubertal ovary be accounted for? It could be argued that the number seen in the ovary at any time represents the current crop arising from the germinal epithelium. Prior to puberty, when none of these primary oocytes are capable of maturing to the point of ovulation because the pituitary gland may not yet be producing its proper titer of follicle-stimulating hormone (FSH) and the ovary is not yet producing adequate estrogen, maturation may be attempted but not progress very far before many oocytes suffer atresia. These could be replaced by a new shower of ingrowths from the germinal epithelium in subsequent months or over longer cycling periods prior to puberty.

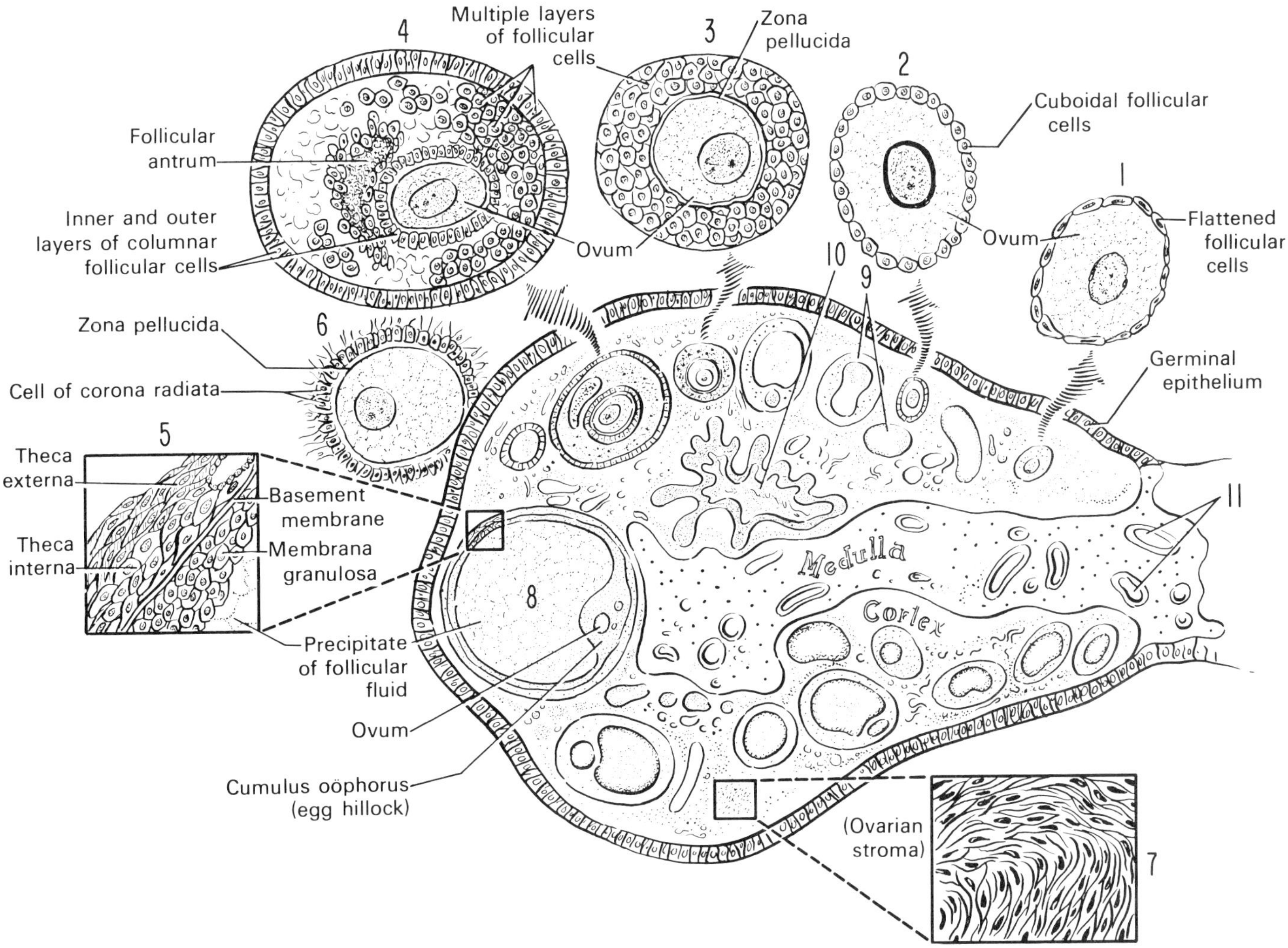

Fig. I-6. Cross-section of an ovary showing cortex and medulla. Compare with structures in Figure I-4. 1, Primary follicle; 2 and 3, early growing follicles; 4, follicle developing an antrum (not all the follicular cells are drawn); 7, cortical stromal connective tissue; 8, Graafian follicle with details of its wall (5) and its ovum (6); 9, follicles undergoing atresia; 10, old degenerating corpus luteum; 11, blood vessels.

Still, why are there so many more primary oocytes in the prepubertal ovary when they are of little use than in the late-functioning adult ovary?

Another common argument advanced against the second hypothesis has been the fact that the germinal epithelium lies upon a dense, fibrous, connective tissue, **tunica albuginea,** which would prevent downgrowth of cellular cords into the looser ovarian cortex below. Still, close to the time of ovulation, such a connective tissue barrier presents no obstacle to the growing follicle which must make its way to the surface and rupture the surface.

In some ways, all the fervor engendered by these arguments reminds one of the old theological debates about how many angels were able to sit on the head of a pin. No agreement can be reached until we know more about eggs and angels. Whichever viewpoint you, as a student, adopt, it is basic to realize that the ovary, at best, appears to be a wasteful gland, producing thousands of prospective eggs that will never mature fully.

On the other hand, there is a point of view which holds that continuous production of ova, even if apparently wasteful, is the very thing which maintains the ovary through life as a functional organ and is an essential event. Nevertheless, there is constant regression and death of eggs in one state of partial maturity or another. Only a select few survive to the time of ovulation, and even these do not survive long if they are not fertilized after ovulation.

We are on safer ground if we pick up the story of how any primary oocyte, no matter what its origin, changes and grows as it matures under the hormonal influences that prevail in an adult cycling ovary. An examination of an ovary under the microscope shows that it has a **cortex** and a **medulla** (Fig. I-6). (In any organ so arranged, the cortex is its "outer rind"; the medulla is its central core or, figuratively, its "marrow part.") The ovarian cortex, below the tunica albuginea, contains **primary oocytes** in various stages of development surrounded by their **follicle cells** and some loosely-arranged, connective-tissue **stroma** infiltrated with capillaries. The medulla, into which ova never advance, contains a more robust connective tissue, large blood vessels and nerves that enter and leave it *via* the mesovarium, and some vestigial remnants of the embryonic male duct system.

The ova are fairly large spherical cells, the most immature of them measuring about 20 μ in diameter; those fully grown measure about six times this size. They have large vesicular nuclei with prominent nucleoli. The cytoplasm is somewhat granular, the larger particles within it being yolk granules. Young eggs have a delicate plasma membrane. As they mature, a thicker, tougher, refractible membrane is secreted about them. This **zona pellucida** is thought to be produced by the follicular cells which surround the ovum.[30-32] Electron microscopy shows it to be double layered, so that part of it might be contributed by the ovum, itself.

Primordial follicles contain an immature egg surrounded by a single layer of thin, flat cells (Fig. I-6). They lack a zona pellucida. As the follicle matures, follicular cells divide rapidly so that they ensheathe the egg in multiple strata. Most of these cells are now cuboidal in shape, though those closest to the developing zona pellucida become columnar, as do those at the periphery of the follicle. In follicles of about 0.2 mm. in diameter, pools of fluid begin to collect between the cells. These pools coalesce to form a central cavity in the follicle, the **antrum** (Latin, cave). This fluid is thought to be secreted by the follicular cells themselves. Follicular cells have been held by some to play a role in the secretion of estrogenic hormones. Recent studies[33] seem to assign this role more directly to other stromal cells that surround the follicle. Within the follicle, the expanding fluid crowds the ovary and some of its nurse cells into a mound-like **egg hillock** (cumulus oophorus) at one pole of the cavity. The remaining follicular cells are pressed by the fluid as a peripheral shell of several layers in depth about the cavity.

The follicle is now a fluid-filled ball with the egg burrowed in a mound of cells at one pole inside the hollowed ball. The follicular cells are now called the **membrana (stratum) granulosa** of the follicle. Very large units of this kind are called **Graafian follicles,** and they are found only in the ovaries of mammals. They were first described in a monograph in 1672 by Reijnier de Graaf (1641–1673), a Dutch anatomist. Figure I-6 shows the progressive stages undertaken by a primary follicle to reach this size (about 1 cm. in diameter) just before it is ready to be ovulated. By this time, stromal cells have compacted about the follicle to form two outermost coats. The inner one of these is the highly cellular coat, the **theca interna;** the outermost is a more fibrous connective tissue ensheathment, the **theca externa.** The cells of the theca interna are believed to produce estrogen, and it is this hormone, induced by **FSH** of the **pituitary gland,** which is responsible for stimulating follicular growth. A wedge-shaped cap of thecal cells at one pole of the Graafian follicle aids in pushing other ovarian stromal and cellular elements aside as the enlarged follicle begins to snow-plow its way toward the germinal epithelium.

The follicle started its growth as a tiny structure in the ovarian cortex just below the tunica albuginea. As it increased in size it sank progressively deeper into the cortex and encroached upon the medulla. It is now ready to make its way back toward the germinal epithelium which it will thin and raise into a surface blister. Some 10 to 14 days after growth first begins and with the added impetus of a second hormone of the pituitary, the **luteinizing hormone (LH),** the follicle ruptures through the surface and the egg is released in the **process of ovulation** (Fig. I-4). That is, it detaches from the egg hillock and slowly oozes into the peritoneal cavity where, as a rule, it is captured by the fimbriae of the uterine tube and guided into the tube. The released egg is still surrounded by the zona pellucida and a number of columnar follicular cells forming a **corona radiata** about it (Fig. I-7). Under the influence of LH, the granulosa and thecal elements of the ruptured follicle remain behind in the ovary and organize into an endocrine structure, the **corpus luteum.** The secretory product, **progesterone,** is a hormone that will maintain the mucosa of the uterus in a *highly secretory state,* receptive for a fertilized ovum that may now reach it to *implant* within it and to transform into an embryo. This hormone will maintain the uterus in this state throughout pregnancy, for, after a successful implantation, the corpus luteum persists within the ovary until pregnancy terminates. Should the egg fail to be fertilized, the corpus luteum regresses, and the uterine mucosa sloughs in the cyclic process of menstruation.

It is apparent that the ovary is an active gland with both exocrine and endocrine functions. Cells of various types are constantly marshaled to form a variety of structures, all in preparation for an anticipated event, only to regress and die away if the blessed event fails to occur. All that is required of any ovum in this process is that it grow robust, rotund, and lazy. No attempt is made to sacrifice the cytoplasm of this germ cell; rather, all efforts are directed toward cytoplasmic hypertrophy. The ovary is not concerned with matters of cell number and cell motility as was the testis. Its sole objective is to produce one large target (or, at best, a few) as its exocrine product which, hopefully, spermatozoa can hardly fail to miss. Nothing is wasted, even in cell division. Until just prior to ovulation, the ovum has remained a primary oocyte containing the diploid complement of chromosomes, 44 + x + x in man. While still within the Graafian follicle, it undergoes meiosis, giving rise to a very tiny, insignificant daughter cell which remains adherent between the ovum and the zona pellucida (Fig. I-7). This **first polar body,** like the ovum, contains the haploid num-

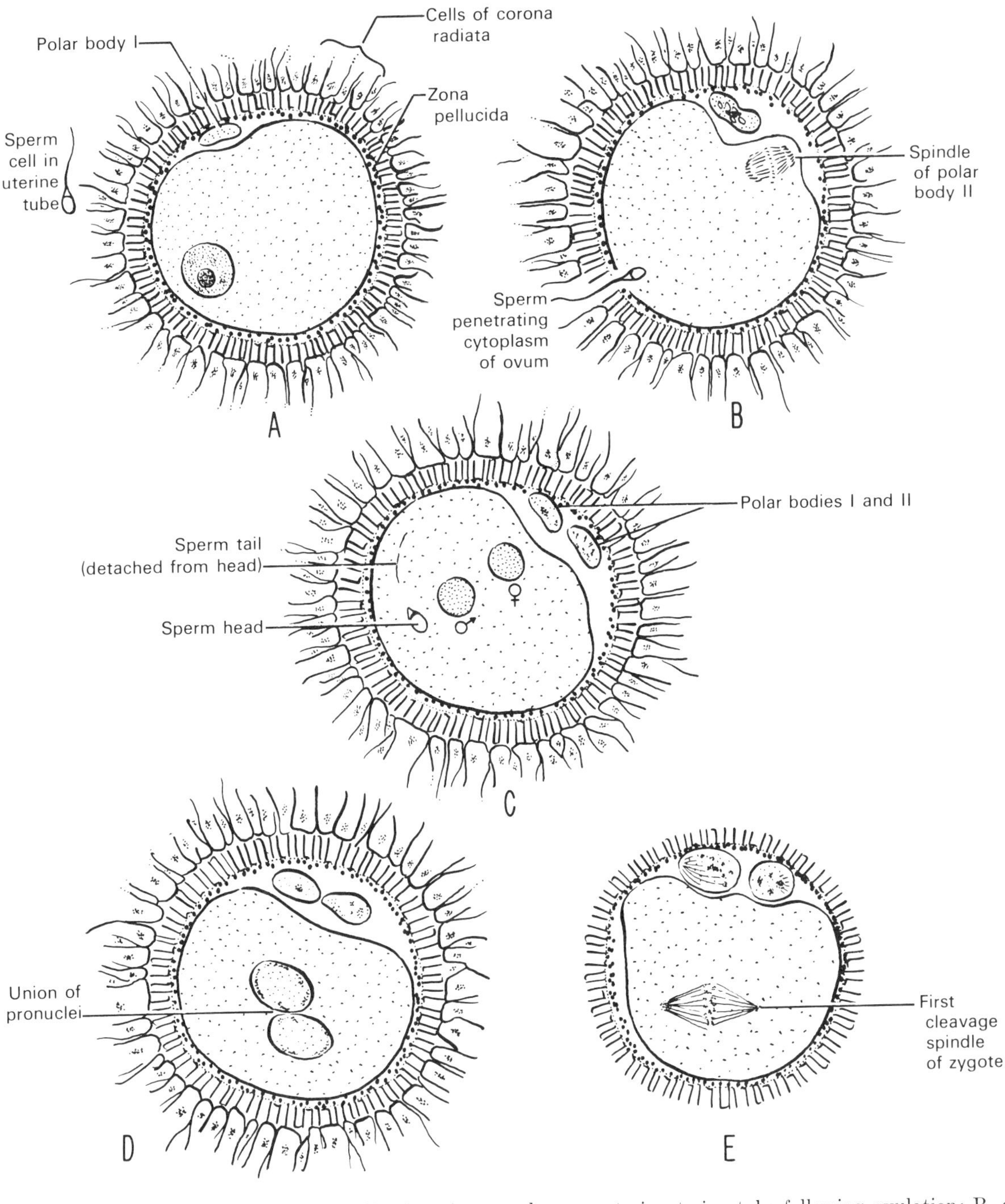

Fig. I-7. Schema of ovum and events of fertilization. A, secondary oocyte in uterine tube following ovulation; B, entrance of spermatozoon and cutting off of second polar body; C, approach of male (♂) and female (♀) pronuclei and disintegration of sperm cell; D, merging of pronuclei; E, spindle for first cleavage stage of zygote.

ber of chromosomes, 22 + x (x, the female sex chromosome). Recall that different sperm bear either 22 + x or 22 + y chromosomes in their nuclei (y, the male sex chromosome). In fertilization of the ovum, the zygote will receive once more the diploid number of chromosomes for the species, half from each parent germ cell. In man this will be 46 in all, 44 + x + y or 44 + x + x. The first combination determines that the zygote will develop into a male; the second, into a female.

Thus, each month in the life of a fertile woman a number of primary follicles start on their way to become the "Egg of the Month." Among primates, only one lucky candidate is chosen, as a rule, to be crowned in any month. Rarely are there ties in this contest. The unlucky aspirants to this rather uncertain title languish and die within the ovarian cortex, not all of them having had talent enough to become even a Graafian follicle. More often than not, even the winner is disappointed, left waiting in the upper end of the uterine tube for a bridegroom that fails to get to the church on time. Jilted at the altar, such prospective brides also quickly waste away and are absorbed in the uterine tube without ever inspecting their newly-furnished quarters in the exclusive section of Uterine Heights. The uterus then cleans house through menstruation and rebuilds anew in anticipation of a possible new tenant in the following month. The corpus luteum of the ruptured unfertilized egg also begins to regress in the ovary as progesterone titer drops and estrogen titer begins to increase. A new crop of primary follicles gets under way once more.

Not all eggs get into the uterine tube. We do not know how many eggs fall into the peritoneal cavity and are absorbed there because of faulty fielding accuracy on the part of the uterine tube fimbriae. But we know that some do because of the rare instances when eggs get fertilized in the peritoneal cavity and start to develop there. Such an egg generally falls into the retrouterine pouch (of Douglas), a cul-de-sac between uterus and rectum. It implants and establishes a placenta upon the mesentery of the large bowel or upon the mesentery of the coils of the small bowel that occupies the pelvic basin. Such an abdominal pregnancy may progress to full term. Delivery, of course, must be achieved by cesarean section and great care must be taken not to disturb the placental attachment lest intraperitoneal hemorrhage and intestinal rupture result. At times, too, eggs are fertilized at the site of ovarian rupture, and these attempt to implant upon the ovary. Ovarian pregnancies rarely progress very far, and they are apt to result in teratomatous malformations of small size within the ovary. The fertilized ovum within the uterine tube may fail to move down it to the uterus and may implant upon the walls of the tube. Such tubal pregnancies often rupture into the peritoneal cavity and may account secondarily for the rare incidence of so-called abdominal pregnancy. All of these places are abnormal sites for a zygote to begin to grow and develop into an embryo; hence, they are termed **ectopic pregnancies.** The normal site for implantation, placenta formation, growth, and development is within the uterus, either on its back, front or side walls. Sometimes implantation occurs at the uterine cervix close to where the uterus opens into the vagina. Placental formation obstructing the cervical orifice (placental praevia) also interferes with delivery of the fetus. Thus, this is an ectopic site within the uterus.

FERTILIZATION

For fertilization to occur at all, both male and female sex cells must be properly mature, neither under-ripe nor over-ripe. The time at which an egg is ready to receive a sperm varies in different species. In man, apparently, the first polar body must be extruded and the second polar spindle must be formed but in a stage of arrest (Fig. I-7A). While both male and female germ cells must have spent some time in residence within the female genital tract before they are completely capacitated for union,[34] it cannot be for too long a time, for fertilization must occur in the upper third of the tube. Unfertilized eggs 2 or 3 days old, recovered in the lower end of the tube, already display signs of degeneration.

Only one spermatozoon of the millions released during ejaculation is essential for the fertilization of the ovum. Additional ones may be found attached to the corona radiata or even to have penetrated into the space between the zona pellucida and the egg membrane.[35] But it is the first successful arrival to penetrate the egg cytoplasm, tail and all, that initiates the events which follow penetration (Fig. I-7B). An enzyme, **hyaluronidase,** thought to be present in the acrosome, seems to aid in penetration. The dispersal of the corona radiata takes place somewhat later through some as yet unknown tubal factor.[36] Fertilization is also species-specific in that, were cross-breeding attempted experimentally, dogs would not produce cats, nor human beings, apes. This may depend upon immunological reactions similar to those between antigens and antibodies. Such an immune response against foreign invaders has been considered as a possible reason why certain individuals within the same species may remain sterile.

After the head of the spermatozoon enters the ovum, it breaks from its body, swells and becomes the **male pronucleus** of the zygote; the nucleus of the ovum, on the other hand, becomes the **female pronucleus.** The second polar body is given off (Fig. I-7B), and the pronuclei advance toward each other to fuse (Fig. I-7 C and D). The sperm cell body and tail parts are absorbed, liberating the two centrioles from the body part. These form the mitotic spindle (Fig. I-7E) which will aid in the distribution of the diploid number of chromosomes to each daughter cell at the first division of the zygote.

DEVELOPMENT OF THE EMBRYO

Formation of Morula

Mitosis of the zygote continues until, after four sets of divisions, a compact ball of 16 cells, the **morula** (Latin, mulberry), has been produced. Since there is little yolk to interfere with cell division in human eggs, all of the cells of the morula are about the same size (Fig. I-8A). Some may be slightly larger than others, however, and at times, the larger cells may divide slightly in advance of the smaller ones. Hence, **cleavage** in the human zygote does not proceed necessarily in nice, geometric progression; *i.e.*, 2, 4, 8, and then 16 cells. At various times after fertilization, in studies on the development of the zygote, 3-, 9-, and 10-cell stages have been flushed from the human uterine tube.[37] The zona pellucida is lost some time after morula formation; the corona radiata, even earlier. The quantity of yolk that was stored in the ovum is used up in energy production during cleavage; hence, the morula is not much larger than the original zygote. If anything, there has been a reduction in total cytoplasmic mass with nuclear material having increased

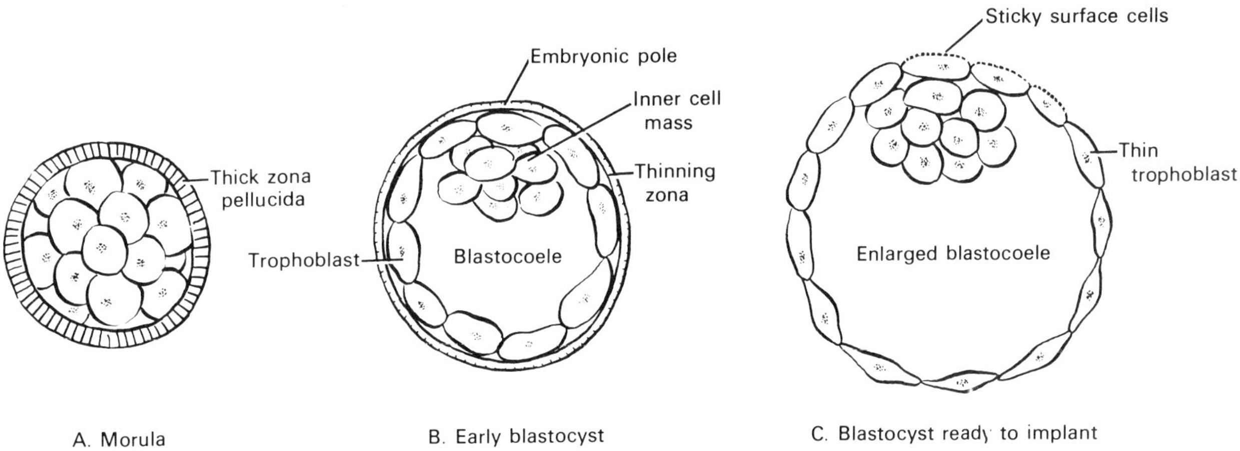

Fig. I-8. Morula and blastocyst.

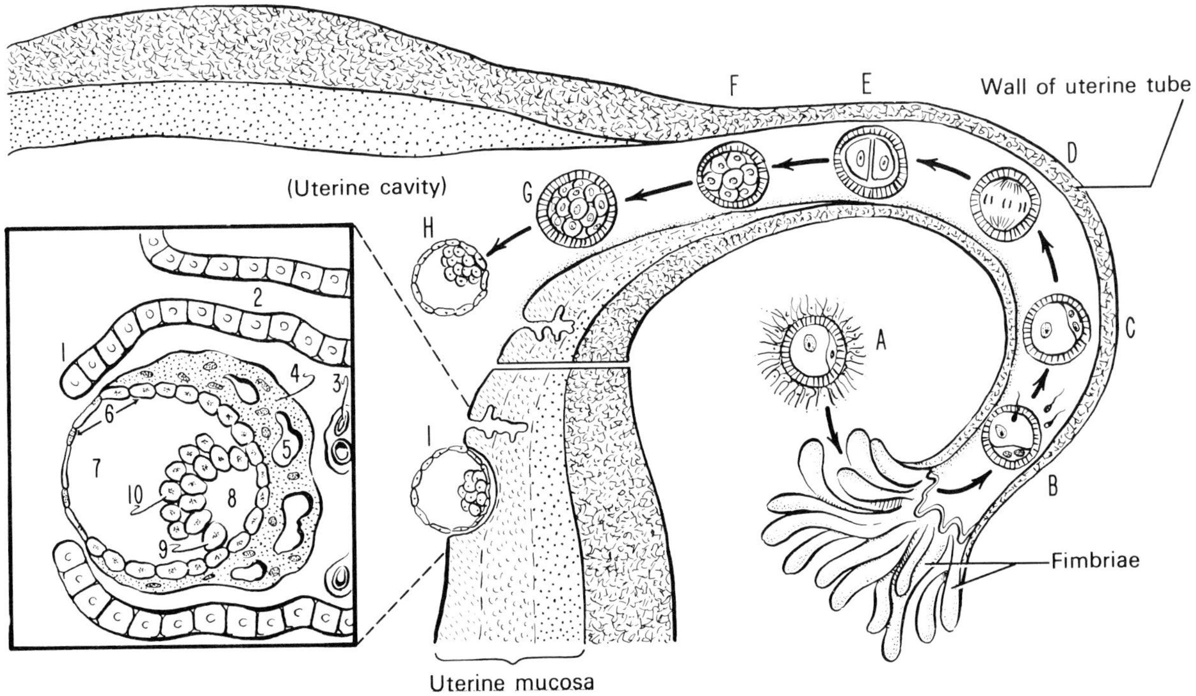

Fig. I-9. Schema to show passage of ovum into uterine tube; its fertilization; subsequent cleavage and eventual implantation of zygote. A, ovulated egg with first polar body; B, fertilization and second polar body; C, eccentrically placed pronuclei and polar bodies; D, first mitotic division of zygote; E, two-cell stage; F, 7- to 8-cell stage; G, morula; H, blastocyst free in uterine cavity; I, early implantation into uterine mucosa 7 to 8 days after ovulation.

Inset: Schema of blastocyst almost fully implanted. 1, Uterine epithelium; 2, lumen of uterine gland; 3, uterine blood vessels (X-sections); 4, syntrophoblast; 5, lacuna in syntrophoblast; 6, cytotrophoblast; 7, blastocystic cavity; 8, amniotic cavity; 9, ectodermal layer of inner cell mass; 10, entodermal layer of inner cell mass. Edges of uterine epithelium will seal over to repair implantation site.

at the expense of cytoplasmic loss. During its cleavage stages, the zygote has been moved along the uterine tube, and it enters the uterus about 3 to 4 days after fertilization. It spends another 3 to 4 days within the uterus seeking a suitable site in which to embed (Fig. I-9).

Formation of Blastocyst

The cells of the morula continue to divide. They derive their nourishment and oxygen supply from the uterine secretions. So much fluid is absorbed that it forms in pools between cells, much as it did between the follicular cells in the ovary. And, as in the latter instance, the pools coalesce to hollow out a central cavity within the morula. This cavity (or **blastocoele**) is surrounded by a single layer of flat cells, and a hillock of cells accumulates as an **inner cell mass** at one pole. The latter protrudes into the cavity much as the egg hillock did in the Graafian follicle. The whole structure is now called the **blastocyst** (Fig. I-8 B and C). The outer shell of cells is known as the **trophoblast** (the nourishment-former), and where these cells cover the inner cell mass is the area of the **embryonic pole.** The outer cells at the pole are more sticky than the rest, and they are the first to adhere against the uterine mucosa. This process allows the region of the inner cell mass, from which the embryo will arise, to become the first part of the blastocyst to burrow beneath the uterine mucosa (Fig. I-9). Once the sticky cells make contact, they become very active. They proliferate and split into an inner and an outer trophoblastic layer. The cells of the inner layer, **cytotrophoblast,** are a single layer of cuboidal cells. The outer lamina of cells, in direct contact with the uterine mucosa, divide so rapidly that they form a **syncytium** of cells with no recognizable borders between cells. Hence, the outermost layer of trophoblast is called the **syntrophoblast** (Fig. I-10).

Among these cells and those of the inner cell mass, metabolic rates accelerate. Bicarbonates form and are passed to the uterine mucosa. Under the influence of progesterone from the corpus luteum of the ovary, **carbonic anhydrase** is also formed in the cells of the mucosa. This enzyme acts as the catalyst in producing an alkaline environment at the mucosal surface. Such an environment favors the softening of the intercellular matrix between the maternal mucosal cells so that they can be pushed aside as the blastocyst implants. In its proliferating syncytial form, the trophoblast is able to continue to spread and invade, moving the blastocyst ever deeper below the uterine surface. Finally, the blastocyst encroaches deeply enough so that the uterine mucosa is able to seal over the implantation site as it repairs the break in its surface wall. Were you to examine the uterine surface now, you would see only a slightly domed spot, the only visible sign of a recent implantation. The rapidly proliferating syntrophoblast becomes a spongy mass of **lacunae** filled with parts of destroyed uterine tissue. The large **decidual cells** of the maternal uterus are rich in glycogen, and these, along with blood from eroded vessels, fill the lacunae. This autolysed material is the **embryotrophe,** the fodder for embryonic growth at a time when this minute organism has not established a circulation of its own.

The Two-Layered Embryo

The inner cell mass also segregates into two major lamina (Fig. I-10), a sheet of tall columnar cells that lie upon a thinner sheet of flat cells. These are the first **germ layers** of the new individual, the columnar **ectoderm** (outer skin) and the thinner **entoderm** (inner skin), respectively. A fluid-filled space, the **amniotic cavity,** appears above the ectoderm of the inner cell mass, between it and the cuboidal cells of the cytotrophoblast. This cavity will become the "bag of waters" in which the growing embryo will come to float until the time of birth.

Ectodermal Derivatives: The **ectodermal layer** of cells of the inner cell mass will give rise to the following bodily structures:

1. Outer skin and skin derivatives of the new organism, including hair, nails, and teeth, as well as skin glands.
2. The lens of the eye.
3. The epithelium of the sense organs.
4. The lining of the nasal cavity, of the air sinuses, of the skull, of the mouth, and of the anal canal.
5. The oral glands.
6. The nervous system, pituitary gland (hypophysis), and the adrenal medulla.

Entodermal Dervivatives: The **entoderm** will provide:

1. Lining material for the pharynx, including that covering the root of the tongue, auditory (Eustachian) tube, tonsils, thyroid, and thymus.
2. Lining of larnyx, trachea, and lungs.
3. Lining of the digestive tube from esophagus to rectum.
4. The secretory cells of liver and pancreas.
5. The bladder lining and that of the vagina, at least in part, as well as the epithelium of the urethra and its associated glands.

No wonder these sheets of cells are called "germ layers," since they contribute so many of the bodily parts as they fold and develop. Soon a third germ layer, the **mesoderm proper,** will make its appearance. When it does, we shall list its contributions to body structure.

Extraembryonic (Nonembryonic) Mesoderm

In Figure I-10A, we see the appearance of something called **primary mesoderm** which is not contributing to the embryo at all. It is depicted as a middle or inner group of stellate-shaped cells that, in all probability, have come from cytotrophoblast, though, at these very early stages, it is difficult to determine whether they may not have come, in part at least, from the inner cell mass. Next to the amniotic cavity, this primary mesoderm transforms into cuboidal cells forming **amniotic ectoderm** continuous with the columnar **embryonic ectoderm;** *i.e.*, the ectoderm of the uppermost germ layer (Fig. I-10B). But most of this primary mesoderm remains as stellate-shaped cells whose processes contact as if holding hands. A jelly-like matrix forms between these cells to form a kind of packing tissue within the blastocyst. This is a type of connective tissue called **early mesenchyme.** Some of these mesenchymal cells below the entoderm form a flattened lining layer for the blastocoele (major cavity of the blastocyst). In this sense, they form an **exocoelomic membrane.** It is, perhaps, this mesenchymal-lined coelomic cavity beneath the early entoderm that some writers wish to call the **primary yolk sac,** simply because of its early association with the entoderm. This cavity appears to be merely what is left of a blastocoele once the primary mesoderm begins to pro-

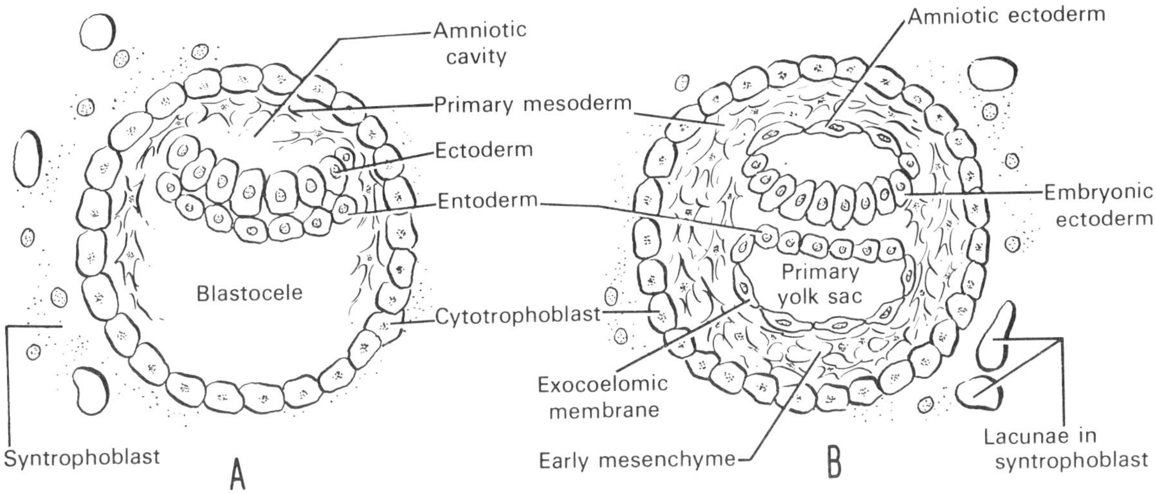

Fig. I-10. Formation of amniotic cavity and exocoelom (primary yolk sac).

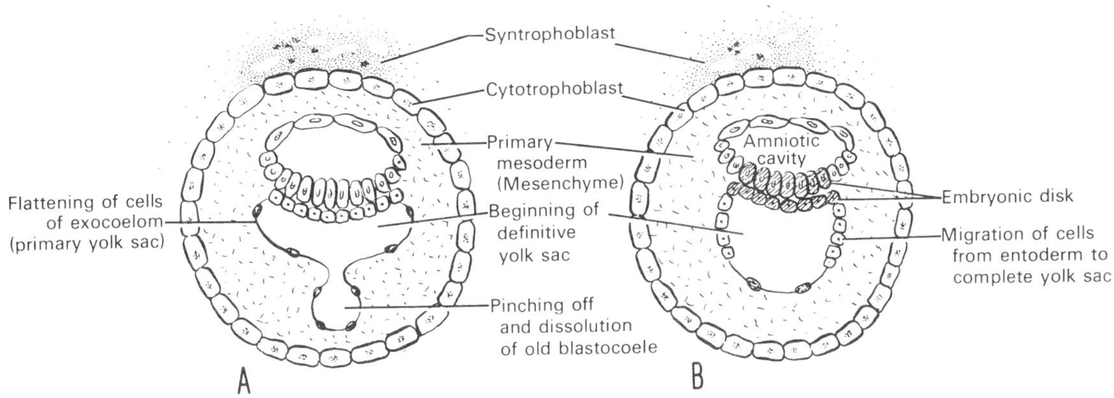

Fig. I-11. Formation of a two-layered embryonic disk and the definitive yolk sac.

liferate. A true or **definitive yolk sac** appears later as entodermal cells, proper, migrate out to line the reduced blastocoele (Fig. I-11 A and B). At this stage, the double-layered trophoblast (cyto- and syntrophoblast) encloses a mass of primary mesoderm or mesenchymal tissue in which two cavities or vesicles exist: (1) the ectodermal vesicle or **amniotic cavity;** (2) the entodermal vesicle or **definitive yolk sac.** The ectodermal vesicle is lined by two kinds of cells, amniotic and embryonic (those of the ectodermal germ layer); the entodermic vesicle is lined by entoderm throughout, part of which constitutes the entodermal germ layer.

To gain a *surface view* of the ectodermal germ layer at the stage of Figure I-11B, we should have to stand atop the amnion, which is placed over the embryonic ectoderm like the transparent, domed cover of a Pyrex roasting pan, and peer down through the clear amniotic fluid. We would see below a rounded-to-oval **plate of cells** (Fig. I-13). In Figure I-11B, this plate is cut through in the midsagittal plane and is presented schematically, edge-on. It is evident that the plate is one layer of cells thick, that the cells are columnar, and that they rest upon another layer of thinner cells, the entoderm. If we stood in the bottom of the yolk sac and peered upward, the embryonic entoderm would also appear as a rounded-to-oval plate. This plate is not quite as extensive as the ectodermal plate.

You must learn early to think and reconstruct in your mind in three dimensions that which you see in flat, two-dimensional views, and to take into account the plane of section through three-dimensional structures. In Figure I-11B, the two layers of cells that constitute the **embryonic germinal disk** have been shaded to impress upon you that they represent all there is of the **two-layered embryo** at this stage. All the rest of the cells in the figure are ancillary and will not become a part of the embryo.

We must turn our attention once more to the primary mesoderm. We have seen it proliferate as a packing tissue to the point where it reduced the blastocoele to a small yolk sac, and we saw it give rise to the amniotic ectoderm. Now it begins to disintegrate. Clefts appear in it which coalesce until, once more, a large cavity appears, into which the ectodermal and entodermal vesi-

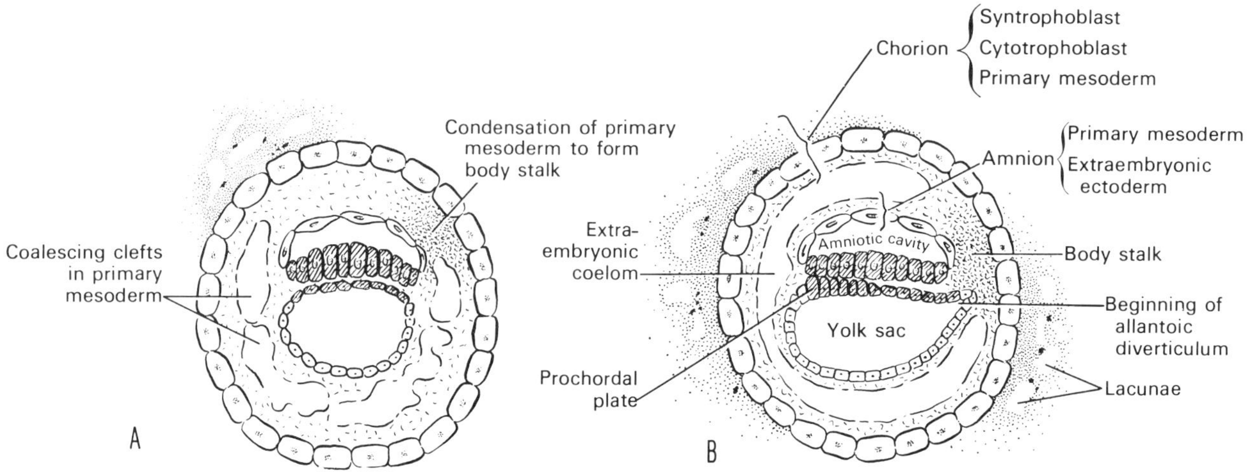

Fig. I-12. Formation of extraembryonic coelom, body stalk, allantois, and prochordal plate.

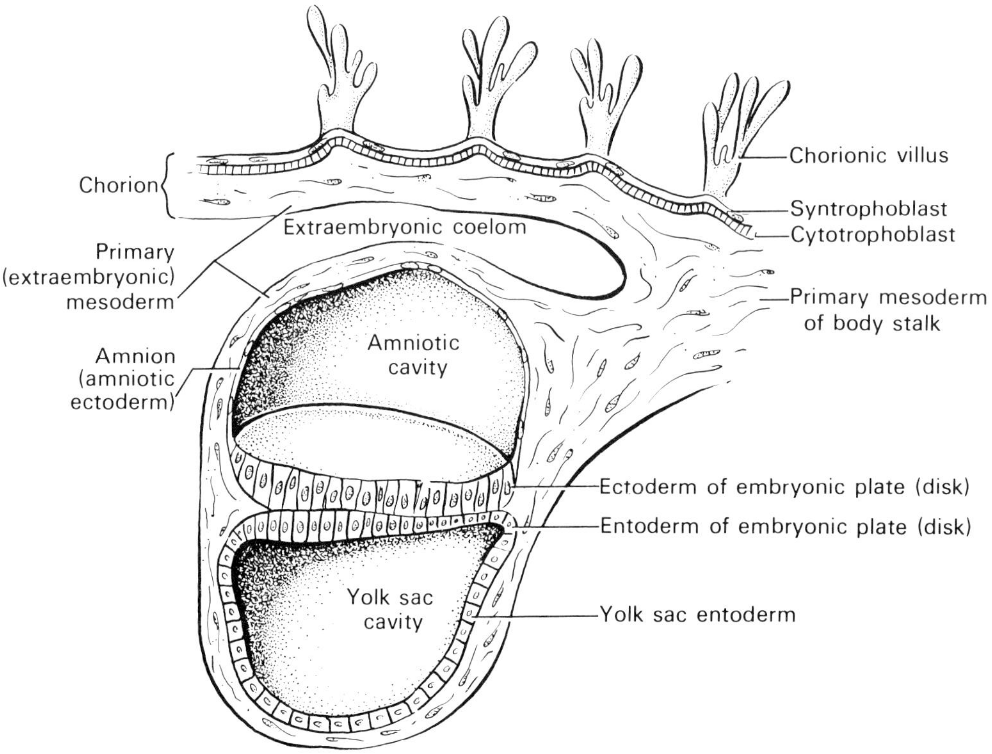

Fig. I-13. Sectioned, three-dimensional schema of the two-layered embryo hanging by body stalk in extraembryonic coelom (not shown in entirety).

cles hang suspended from a broad stalk of primary mesoderm (Fig. I-12 A and B). This large cavity is the **extraembryonic coelom.** It is the cavity or spatial environment outside the embryo proper. The embryo, itself, has no body cavities as yet, but when it begins to develop them, they will, for a time, open into this space. Note that vestiges of the primary mesenchyme do remain to coat the inner aspects of the cytotrophoblast, to clothe the amniotic ectoderm and the yolk sac entoderm, and, finally, to form a broad suspensory stalk for the germinal disk and its associated vesicles. This **body stalk** eventually will become the **umbilical cord** that will attach the developing embryo to a structure called the **placenta.**

The Allantois

Figure I-12B shows a diverticulum of the yolk sac already beginning to invade the body stalk as the **allantois.** The area of the germinal disk in this region will become the **caudal** or tail end of the organism. At the opposite pole of the disk, the **cranial** or **cephalic** end, other entodermal cells thicken in the midline to

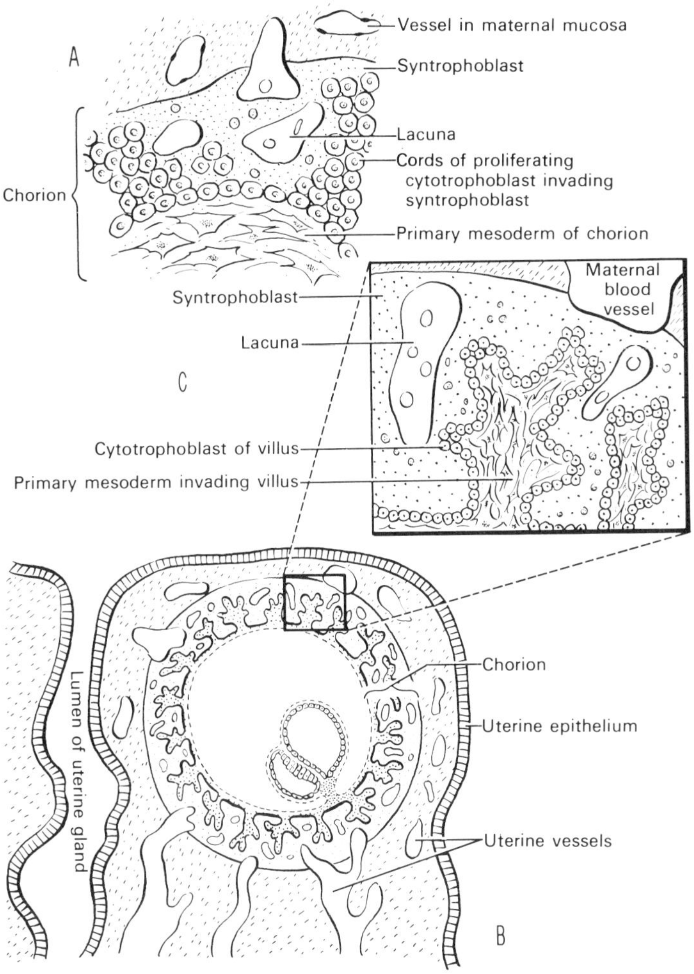

Fig. I-14. Formation of chorionic villi by implanted embryo. A, early stage of proliferating villi from cytotrophoblast; B, later stage showing chorion beset with villi; C, enlarged detail of a villus at a stage before capillary networks form in mesoderm of villus core.

form a **prochordal plate.** This will become the **stomodeal membrane** of the future mouth region. In the human embryo, the allantois appears to be useful only as an advance guide for blood vessels which will later make their way from the embryo into the body stalk on their way to the placenta. The allantois, then, is a rudimentary structure which remains in the stalk to waste away, never to become the large, membranous, extraembryonic, urinary reservoir which invades the extraembryonic coelom in lower forms.

Amnion and Chorion

The stage of development reached in Figure I-12B, in an embryo of less than 15 days since fertilization, allows us now to define two important membranes, namely, the **chorion** and the **amnion.** The amnion is the membrane enclosing the amniotic cavity, and, as we have noted, it is two-layered, consisting of primary mesoderm on the outside and amniotic ectoderm on the inside. The chorion, in contrast, is three-layered, surrounding the extraembryonic coelom. Outermost is the syntrophoblast; its middle layer, the cytotrophoblast; and its inner layer is, again, primary mesoderm. So long as there is an extraembryonic coelom, and so long as the embryo remains small, these two membranes, amnion and chorion, remain separated from each other. But as the embryo enlarges, it encroaches progressively upon the space of the extraembryonic coelom until the space is obliterated. Then the amnion is pushed against the chorion in the process, and these membranes fuse loosely.

Chorionic Villi

At this stage, too (Fig. I-12B), the syntrophoblast has eroded about as much of the uterine mucosa as is required for temporary nutritive purposes. Small uterine blood vessels now empty into the lacunae of the syntrophoblast, the lacune begin to interconnect with each other, and maternal blood oozes into these spaces. As embryonic growth begins to accelerate, cellular projections of the cytotrophoblast push into the syntrophoblast (Fig. I-14A). Primary mesoderm, in turn, invades the cytotrophoblastic cell cords as a central core of loose connective tissue (Fig. I-14B and C). All around the outer surface of the blastocyst these series of chorionic fingers, or **villi,** sprout out (Fig. I-13). The whole structure will eventually look like a small ball beset by fringed projections (Fig. I-15). In between the projecting villi, in the **intervillous space,** lie the lacunae filled with maternal blood (Fig. I-14C). It is not difficult to imagine that once blood vessels develop in the embryonic region and migrate to invade the body stalk, they will follow the primary mesoderm of the chorion, enter the villous projections, and form capillary nets in the core connective tissue. When this happens, the embryonic circulation will be brought close to the maternal blood pools within the lacunae. The exchange of blood gases and nutrients between embryo and mother can then take place more readily across the tissues of the chorionic villi. This area of exchange will become limited eventually to a discoidal region of the chorion where it is more closely associated with the body stalk. Here the chorionic villi will lengthen greatly and enter into relationship with the uterine mucosa to form a placenta. Elsewhere around the blastocystic ball the early chorionic projections will disappear gradually as localized placental exchange is established. But for the nonce, there is, as yet, no placenta, nor are there any embryonic blood vessels within the chorionic villi. Before blood vessels can emerge from the area of the embryonic disk, the disk must develop its third germ layer, the **embryonic mesoderm.**

The Three-Layered Embryo

Around 16 days after fertilization, certain changes take place in the ectodermal cells in the *midline* of the

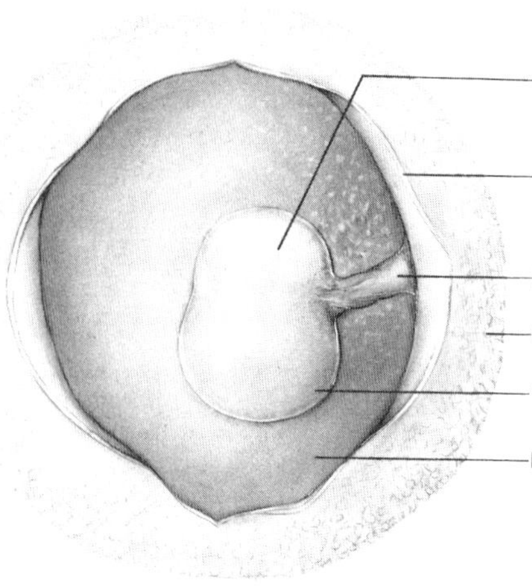

Fig. I-15. Opened chorionic sac with fringed villi at about 35 days (embryo not shown inside amnion). The plush surface area of villi is known as the **chorionic frondosum.** × 4.

germinal disk near its *caudal end* (near the body stalk). These cells divide rapidly, lose their columnar shape, and make a thickened, midline, longitudinal streak, the so-called **primitive streak.** Cells from this thickened, longitudinal mass migrate *downward* and then spread *laterally* between ectoderm and entoderm to form a middle lamina of cells in the disk, the **intraembryonic mesoderm** or the third **primitive germ layer** (Fig. I-16 A′ and B). At its anterior end the primitive streak bulges into a localized *knot of cells*, or **node of Henson.** At the posterior end of the streak, the ectoderm thickens and fuses with the entoderm in the region of the allantois to form a **cloacal membrane** (Fig. I-16A).

If the overlying amnion (and chorion) were cut off and the amnion were lifted away, as one might remove a casserole cover, the plate-like surface of the embryonic disk could be viewed as in Figure I-17A. With X-ray vision, imagine peering *through* the ectoderm to look *at* the mesoderm, just below it. At 16 days it would lie as a midline streak of cells with a rounded knot of cells anteriorly, and it would lie between ectoderm and entoderm. Actually, we would see in surface view what was seen in midsagittal section in Fig. I-16. Just caudal to the streak, the ectoderm and entoderm would be in contact with each other at the cloacal membrane. Similarly, quite cephalad in the disk, they would be in contact again in the midline as the prochordal plate. In these two areas, mesoderm will not insinuate itself between ectoderm and entoderm.

A bit later in time (Fig. I-17B), mesoderm migrates away laterally from the primitive streak, moving in wave-like progression everywhere between ectoderm and entoderm except in the prochordal and cloacal regions. There also grows forward, from the node toward the prochordial plate, a solid midline rod of mesodermal cells, the **notochord.** The latter delineates the future midaxial line of the embryo, above which a **neural tube** will develop from ectoderm and around both of which the **bones** of the **vertebral column** will form from mesoderm. The notochord will be absorbed, in part, into the **bodies** of the **vertebrae,** while those portions which remain between vertebrae will form the central parts of the **intervertebral disks.** The actual formation of a solid notochord is more complicated than we have stated here, but the changes which precede this stage are of such short duration in the human embryo, almost as abrupt as the appearance and disappearance of the primitive yolk sac, that the details are of little consequence for the beginning student. If, for purposes of comparing these events in man with those occurring in lower forms, the student desires to know all about the notochordal canal, primitive pit or blastopore, and the short-lived neurenteric canal, he should consult the larger texts on development listed under general references at the end of this chapter.

Somites, Lateral Plate Mesoderm, and Embryonic Coelom

As mesoderm migrates lateralward, it condenses into a rather thick plate of cells where it flanks the notochord (Fig. I-18). This material is designated the **paraxial mesoderm** and will soon break up into segmental masses called **somites.** Farther laterad, toward the perimeter of the disk, the mesoderm thins into a lamina known as the **lateral plate mesoderm.** Clefts soon appear in the lateral plate mesoderm, and these spaces fuse until, eventually, they become a U-shaped cavity between ectoderm and entoderm. This cavity is the **intraembryonic coelom,** or **body cavity.** The two limbs of this U-shaped coelom flank the paraxial mesoderm, while the bend lies rostral to the prochordal plate. This bent part will become the **pericardial cavity** into which a heart will bulge; the straight-sided parts will become the **pleural and peritoneal cavities.**

At the lateral edges of the disk, the intraembryonic lateral plate mesoderm will meet and fuse with the primary mesoderm which clothes amnion and yolk sac (Fig. I-18A′). As the intraembryonic coelom forms and pushes laterad toward this junction point, it will, for a time, be in open communication with the larger extraembryonic coelom (Fig. I-18B′). This will be true until the lateral edges of the disk fold ventrally to meet in the midline, ectoderm of one side meeting ectoderm of the other. Then the lateral edges of the straight limbs of the intraembryonic coelom will also meet in the midline and a body cavity will come to lie once more inside the embryo lateral and ventral to the notochord. There will also be some later ventral bending of the head and the tail ends of the disk but this will not be so drastic as the ventral curling under of the lateral margins of the disk. As a result of this tucking under, a tubular body form will emerge.

For the present, however, at about 16 days after fertilization, we have a flat, three-layered disk with mesoderm sandwiched between ectoderm and entoderm. The relative thickness and the density of the cell population in this middle layer varies regionally in the plate. It is thick and heavy in the midline notochord, node, and streak, and quite compact and dense in the

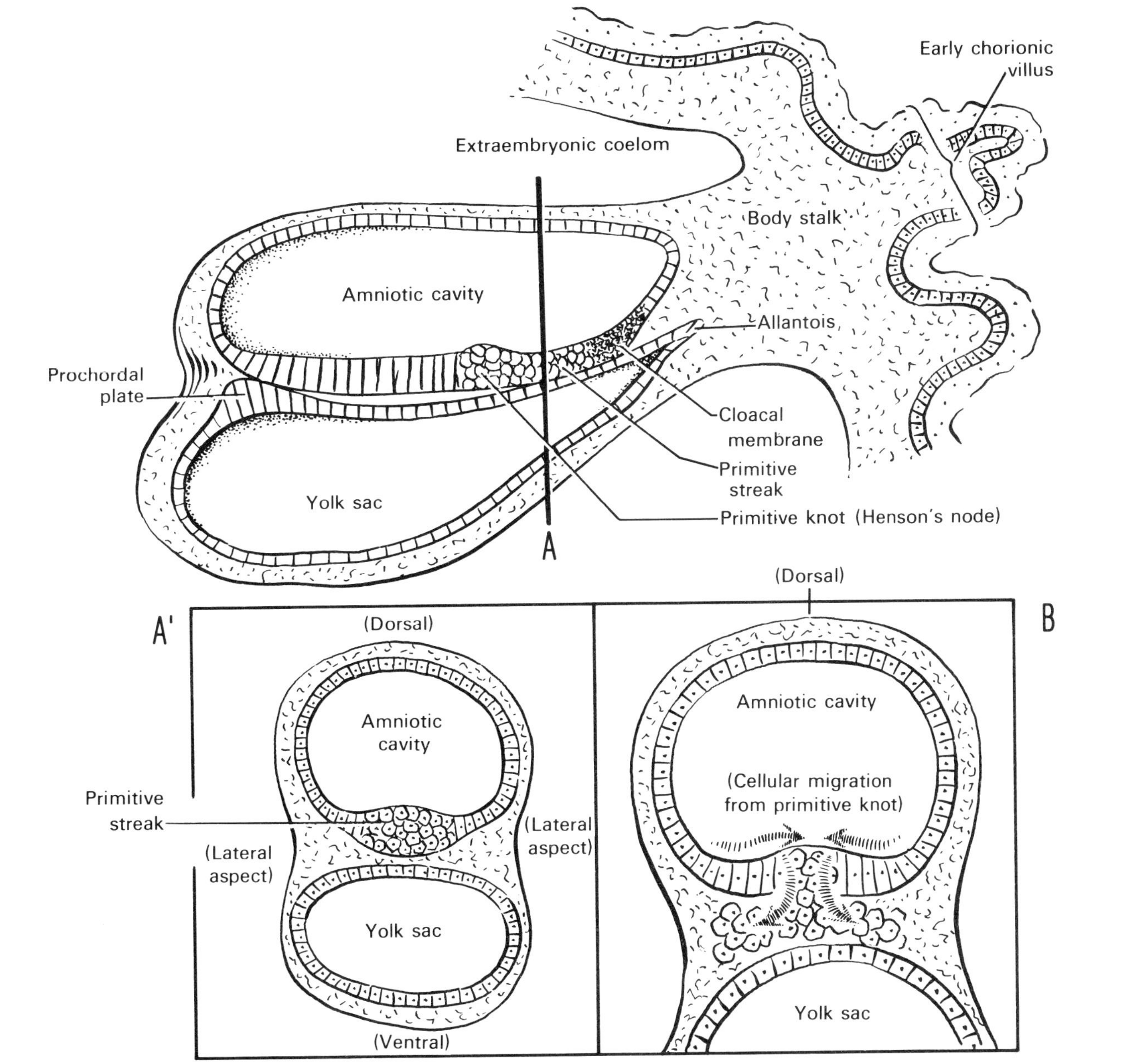

Fig. I-16. Early beginnings of mesoderm (third germinal layer) by migration of cells from primitive streak area. A′ is a cross-section taken at plane A above; B, a later stage showing mesodermal cells migrating to form a layer between ectoderm and entoderm of germinal disk. (Stippled area into which these cells migrate is primary mesoderm, not embryonic mesoderm, proper.) Migration is ventralward and laterad in the disk. Top drawing is a midsagittal section through the disk.

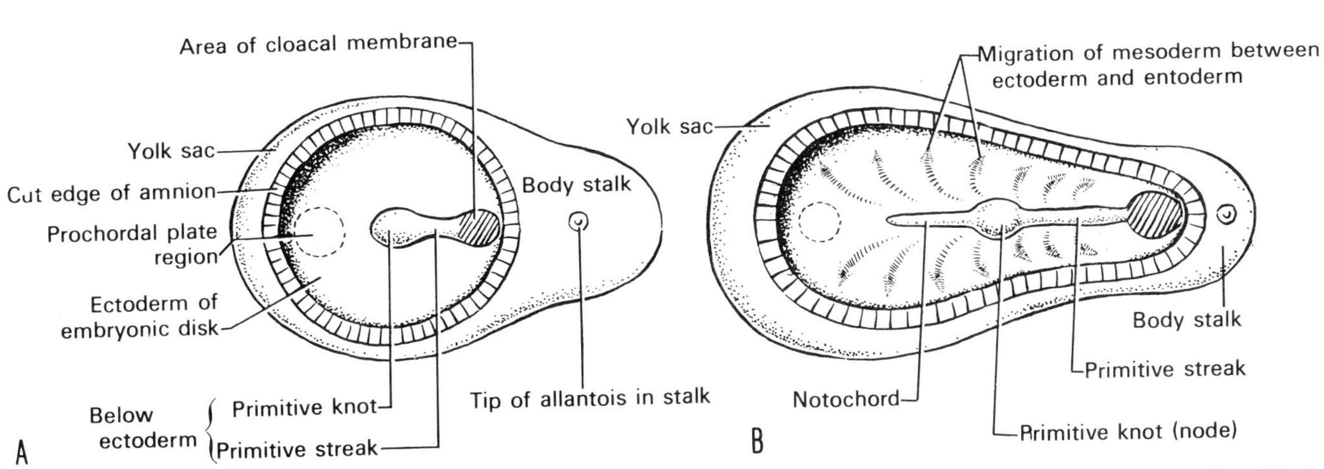

Fig. I-17. Primitive streak, primitive knot, and notochord. Surface view of embryonic disk with amnion removed and looking through ectoderm at mesoderm. (After Haines and Mohiuddin, *Handbook of Human Embryology*, 3rd ed., Williams & Wilkins, 1965.)

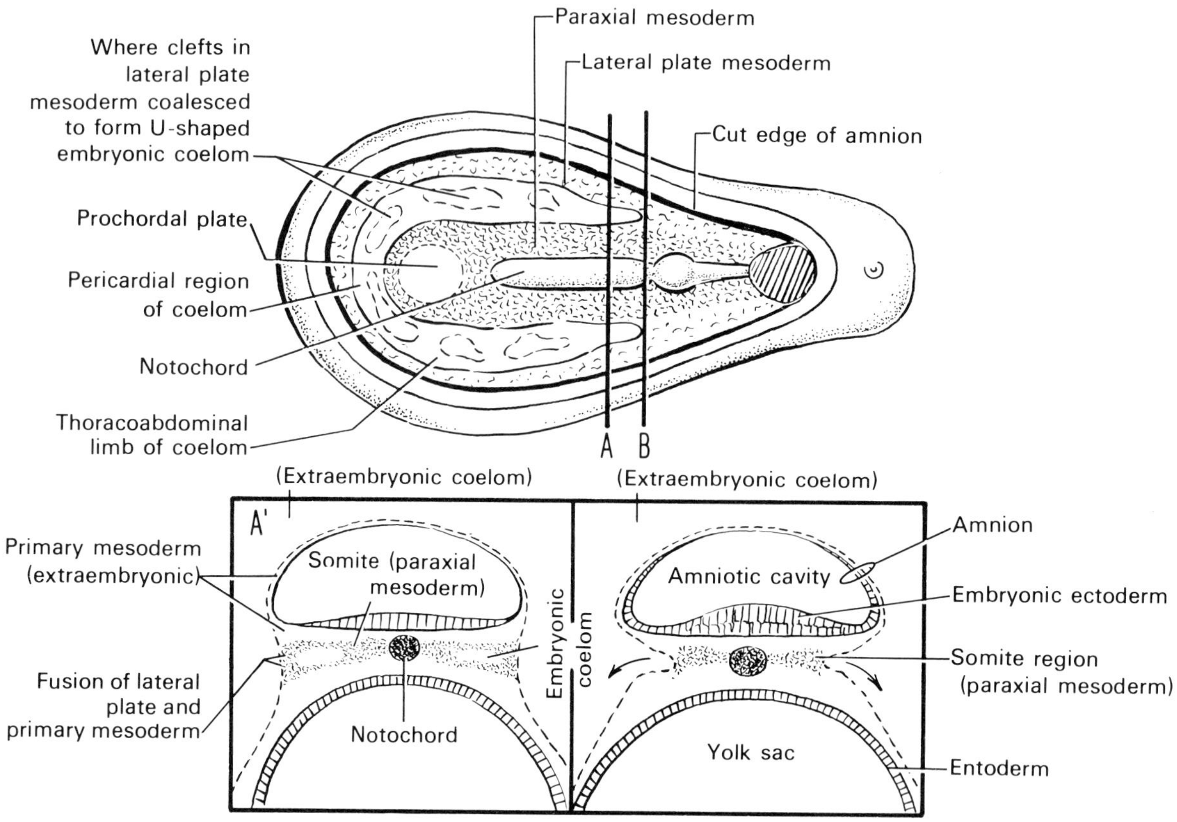

Fig. I-18. Regional differentiation of mesoderm and the formation of embryonic coelom (body cavity). A′ and B′ are cross-sections in planes of A and B above. Arrows in B′ show where intraembryonic coelom communicates with extraembryonic coelom. (After Haines and Mohiuddin, *Handbook of Human Embryology*, 3rd ed., Williams & Wilkins, 1965.)

paraxial region and much thinner in the lateral plate area. It is absent entirely in the prochordial and cloacal regions where ectoderm and entoderm remain in contact. Between paraxial and lateral plate areas, it has spun out so thinly as to allow clefts to unite and open up a U-shaped cavity within the substance of the mesodermal layer itself. Eventually, node and streak will wear themselves out and disappear as the notochord increases in length and more and more cells migrate outward as filler tissue elements between ectoderm and entoderm.

Mesodermal Derivatives

The intraembryonic mesoderm will give rise to the following bodily structures:

1. All types of muscle.
2. All types of connective and supportive tissues, including bone and cartilage. The notochord is the first of these supportive structures to appear, and the axial skeleton will form around it.
3. Blood, bone marrow, and lymphoid tissue.
4. The epithelium of: blood vessels and lymphatics; the body cavities; kidneys; ureters; gonads and genital ducts; and the joint cavities.
5. The cortex of the suprarenal (adrenal) gland.

The Neural Tube

As the mesoderm of the embryo proliferates, a midline thickening of cells appears in the overlying ectoderm. This is the **neural plate,** the first sign of a nervous system (Fig. I-19A′). The neural plate extends from a point slightly in advance of the anterior tip of the notochord to the primitive node and to a part of the primitive streak which, eventually, it will overlie. Its lateral extent is partly over the paraxial mesoderm, but it does not spread over the lateral plate mesoderm (Fig. I-20A). It is wider at the cranial end of the disk where the brain will form from it than at the posterior end where it will roll up to form the spinal cord. Cell proliferation is more rapid on the undersurface of the ectodermal neural plate so that a depression or **neural groove** forms here (Fig. I-19B′), raising the lateral margins into progressively taller **neural folds.** Eventually, the tips of the folds meet in the midline and a tubular structure with a hollow central bore is produced by this rolling-up process (Fig. I-19). The **neural tube** is completed in its middle portion first, and closure progresses slowly toward both ends. For some little time openings persist at the **anterior** and **posterior neuropores,** but these seal off in time so that the **neural canal,** the cavity of the tube, becomes completely enclosed. The anterior neuropore marks the site of the **lamina terminalis,** the cranial-most extent of the future brainstem. Thus, two important hallmarks of the chordate body plan have made their appearance, namely, a more or less rigid, primitive "backbone" in the form of the mesodermal notochord and a dorsally-placed, ectodermal, hollow nerve cord in the

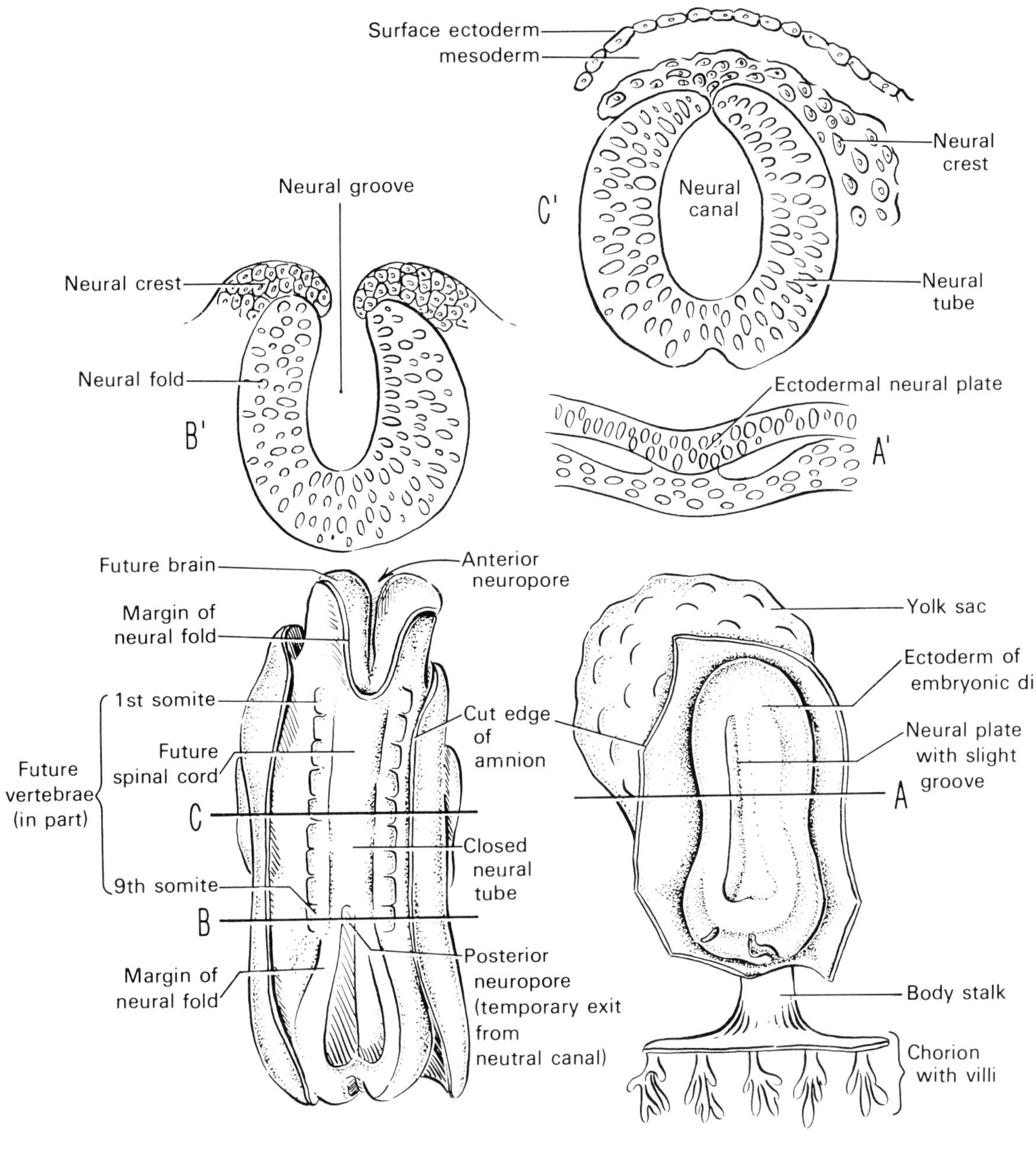

Fig. I-19. Rolling up of a neural tube from an ectodermal neural plate. Early stage (lower right) and later stage (lower left). A′, B′, and C′ are cross-sections taken at planes A, B, and C, below.

form of the neural tube above the notochord. For the moment, the organism wears its nervous system outermost on its back, exposed to the prevailing currents of the amniotic cavity. Soon it will sink below the surface ectoderm of the disk which will cover it over as an outer skin (Fig. I-19C′). But, even then, it will be unprotected, devoid of its bony encasement for a while.

Even before the tube rolls up completely (Fig. I-20), lateral bulges at the anterior end foreshadow the developmental site of the **forebrain** or **cerebral hemispheres. Optic grooves** and slight thickenings of the **otic placodes** mark areas of eye and ear development. Slight bulges behind the forebrain mark the **tectal** or roof area of the **midbrain,** and a slight constriction, the **isthmus,** separates midbrain from **hindbrain.** The latter will become **pons, medulla,** and **cerebellum.** Except for the optic groove, the **diencephalic** ("between brain") or **thalamic** region is not too evident as yet. Thalamus, midbrain, and hindbrain will constitute the **brainstem** as distinct from the forebrain or cerebral hemispheres. Forebrain and brainstem constitute the bulk of the neural tube. The caudal **spinal cord** region of the tube is still quite short. It is best to hold the rest of the story of notochord and neural tube development in abeyance at this point. It will make more sense to discuss it further as the plot thickens in subsequent chapters.

Meanwhile, the embryonic disk has undergone a gradual change in shape. First it was oval, then pear-shaped, and now it is fiddle-shaped with a pinched-in

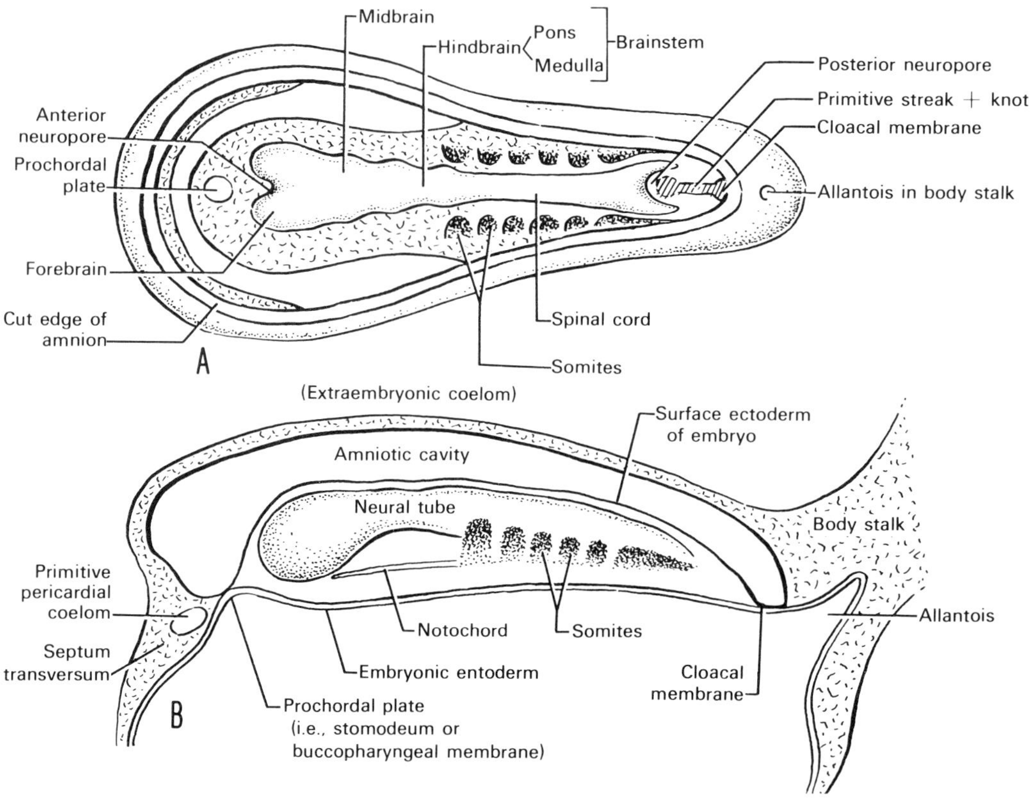

Fig. I-20. Schema of neural tube and somites. Surface (A) and lateral (B) views. (After Haines and Mohiuddin, *Handbook of Human Embryology*, 3rd ed., Williams & Wilkins, 1965.)

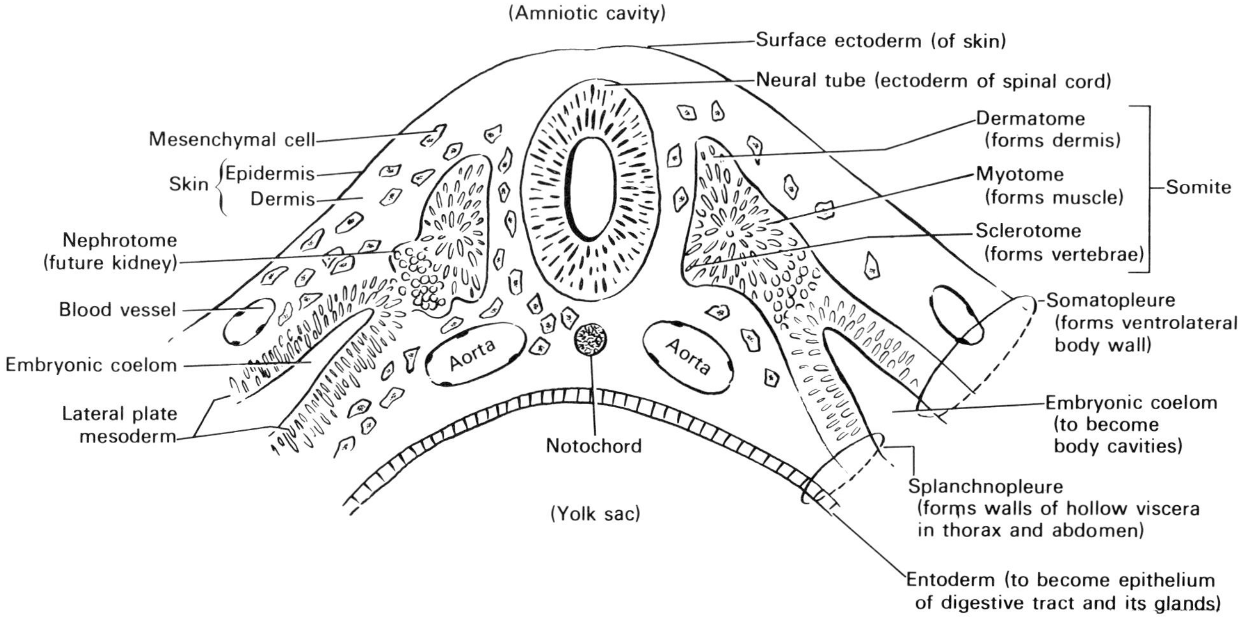

Fig. I-21. Cross-section of germinal disk through spinal cord and flanking somites. On the right, the parts into which a somite will differentiate are shown. On the left, the nephrotome is seen as an intermediate mass of condensed mesoderm connecting somite and lateral plate mesoderm which surrounds the coelom. Loose mesodermal connective tissue fills in the rest of the area between surface ectoderm and entoderm (labeled mesenchymal cell). Note lateral plate mesoderm contributing to somatopleure and splanchnopleure on each side of embryonic coelom.

waist. It bulges upward slightly into the amniotic cavity as the growing organism attempts to raise itself off the flat plate. There is a very slight attempt at undercutting at both ends where more marked head and tail folds will appear later.

Sclerotome, Myotome, Dermatome, and Nephrotome

At about the 20th day after fertilization, the paraxial mesoderm begins to segment into block-shaped chunks, the **somites,** which flank the notochord on each side. The first pair appears near the hindbrain (Fig. I-20), and subsequent pairs put in an appearance over the next week and one-half until, in caudalward progression, there are some 42 to 44 pairs of somites (4 occipital, 8 cervical, 12 thoracic, 5 lumbar, 5 sacral, and 8 to 10 coccygeal). The significance of this enumeration will become clear when we discuss the vertebrae and the arbitrary segmentation of the spinal cord later. Each somite is separated from its neighbors fore and aft, but it maintains a lateral contact with lateral plate mesoderm (Fig. I-21). In fact, at these lateral points of contact, small intermediate masses of mesoderm will soon form the **nephrotome.** Urinary and reproductive ducts will develop from the intermediate nephrotomic masses. A portion of each somite is designated as the **sclerotome.** Sclerotomal mesoderm differentiates into the segmentally arranged vertebrae which will incorporate the notochord in part and which will enclose and protect the spinal cord. Another portion of each somite, the **myotome,** contributes to the segmental musculature of the body which will be supplied by segmentally-emerging peripheral nerves from the adjacent neural tube. In time, a longitudinally-oriented aorta, developing ventral to the vertebral bodies, will contribute paired segmental arteries to the segmental muscle masses. A final portion of the somites, the **dermatome,** contributes a portion of the connective tissue **dermis** to the overlying skin. Sensory branches of the segmental nerves will follow this migrating connective tissue into the skin. This information should help you later in accounting for the pattern of sensory distribution in the body.

This kind of segmentation with bilateral symmetry is another feature which man has inherited from his chordate ancestry. The somites set a pattern for the organization of the axial skeleton and the trunk musculature and its blood and nerve supply. Some reasonable segmentation will hold for head, neck, and extremities, but it will not be as nice in these regions as it is in the trunk.

Formation of Digestive Tube

At the prochordal plate where ectoderm and entoderm are in contact, ectodermal cells thin first, followed by an entodermal thinning to produce a **stomodeal** or **buccopharyngeal membrane.** We have indicated that an eventual break-through here will produce a mouth, opening into an oral cavity. This area still lies cephalad to the most forward extent of the neural tube (Fig. I-23 A–C).

Between the 22nd and 25th days, the embryo begins to bend C-shaped, humping up somewhat in the trunk region to bring head and tail ends closer toward each other. Concomitantly, growth continues at both of these ends. In this process, it could be said that the embryo "sucks in its gut" at the expense of the yolk sac (Fig. I-22). The sac is constricted into an open neck or **yolk stalk** so that the forming gut and the reduced sac remain in communication for some time. The cul-de-sacs that are pushed into the head and tail regions by this process are the **foregut** and **hindgut.** The **midgut** lies above the communicating yolk stalk. The beginning of an entodermally-derived digestive tube with localized regions for future differentiation has folded into the embryo.

A glance at Figure I-23 should help you to understand what is occurring at the head region when the embryo begins to bend. It is apparent that the forebrain vesicle bulges in rapid forward growth and is involved in the bending. Whereas the prochordal plate once lay in the horizontal plane with entoderm below and ectoderm above (Fig. I-23A), a C-shaped bending of the head now moves it progressively into a vertical plane (Fig. I-23B) and then into a reversed horizontal plane where entoderm is the uppermost lamina (Fig. I-23C). In this folding-under process, the stomodeal membrane is brought into a more appropriate position.

Rathke's Pouch and the Pituitary Gland

In the roof of this potential mouth region, the ectoderm dimples into a cranial depression known as Rathke's pouch (Fig. I-23C). The anterior wall of this depression will thicken to become a mass of secretory cells, the **anterior lobe** of the **hypophysis** (pituitary). The posterior wall will become the important **pars intermedia** of this gland. The cavity of the depression will close to become the residual lumen in the adult gland. In man, such a cleft in the hypophysis degenerates into cysts or obliterates entirely. The **neural lobe** of this important endocrine organ does not arise from this pharyngeal diverticulum at all but arises as a downgrowth from the diencephalic (hypothalamic) portion of the neural tube in this region. It will apply itself to the pars intermedia developing from the Rathke's pouch.

Septum Transversum and Cardiogenic Plate

We need to turn our attention now to the mesoderm of the head region. In Figure I-23A there is a thickened partition developing, the **septum transversum.** It is continuous with the primary mesoderm that covers the amnion and yolk sac at the cephalic region of the germinal disk. Immediately adjacent (Fig. I-23B) is another mesodermal region, the **cardiogenic plate** from

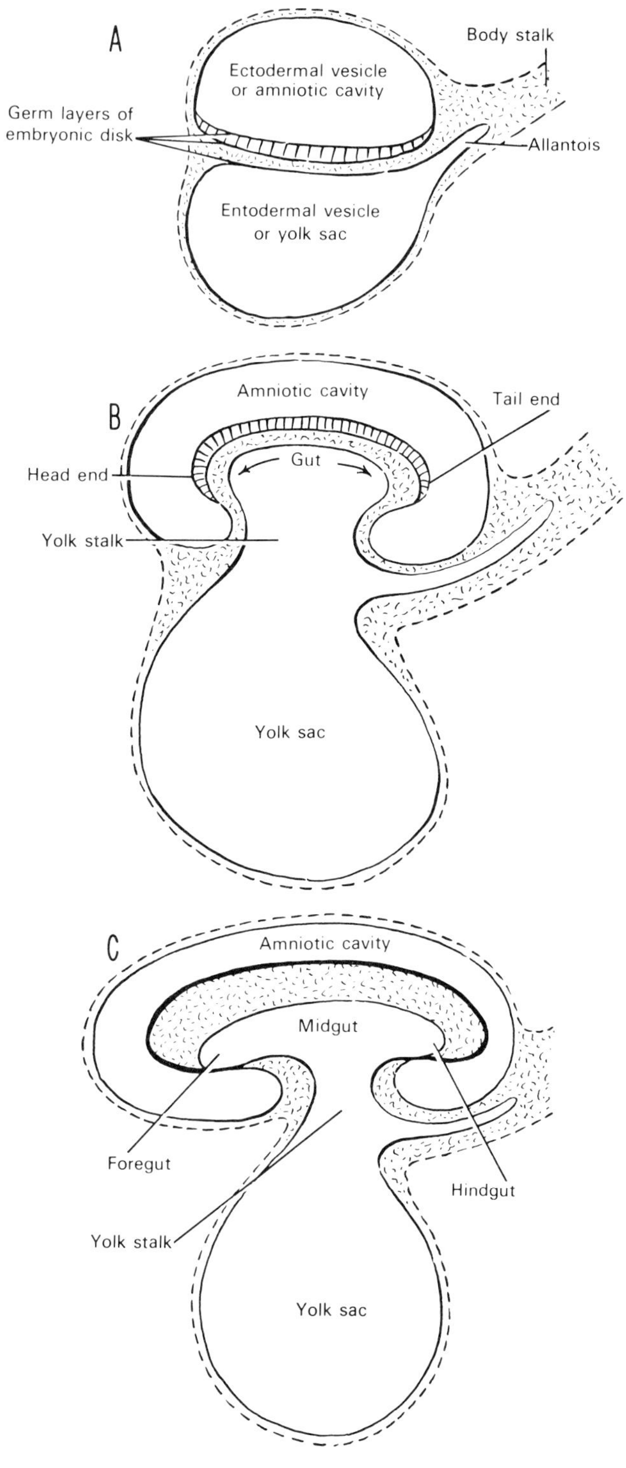

Fig. I-22. Progressive stages of head and tail folding of embryo to form a gut tube. (After Haines and Mohiuddin, *Handbook of Human Embryology,* 3rd ed., Williams & Wilkins, 1965.)

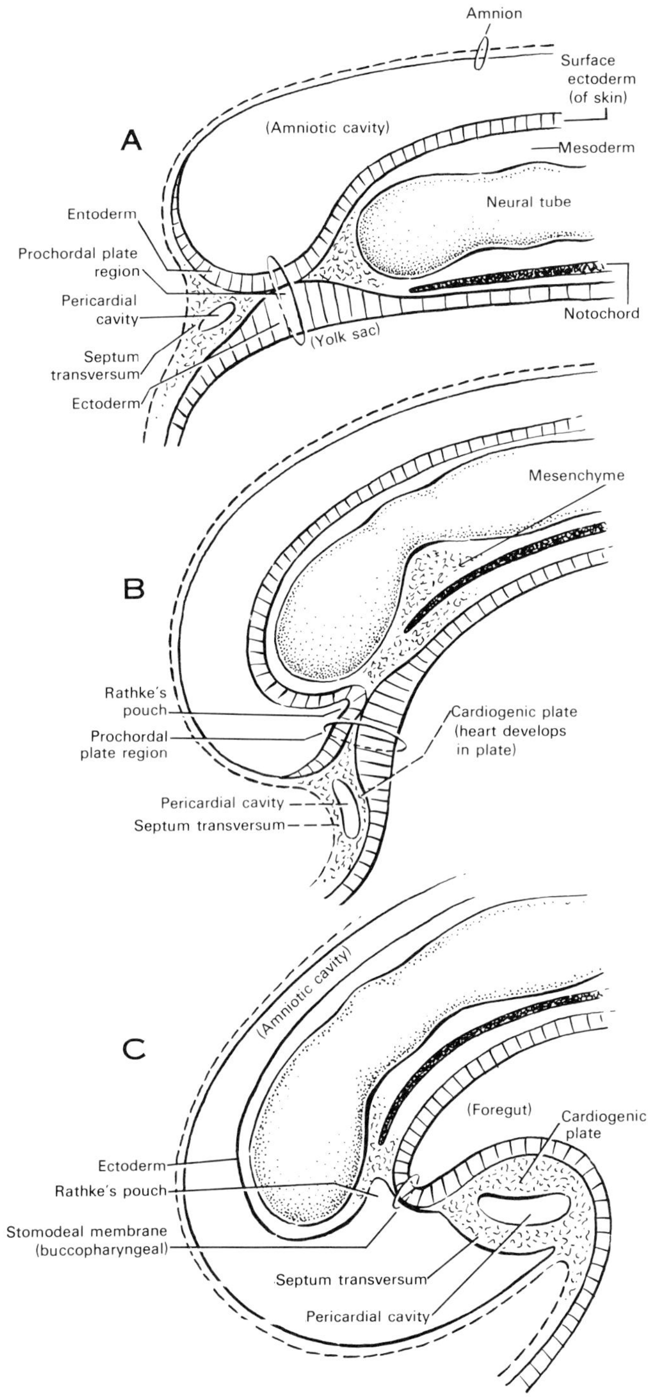

Fig. I-23. Details in the region of head folding. Observe changes in position of: prochordal plate and its transformation into a stomodeal membrane; the pericardial coelom and cardiogenic plate; the septum transversum or future diaphragm. Rathke's pouch marks the early beginning of the adenohypophysis (anterior lobe of pituitary). (After Haines and Mohiuddin, *Handbook of Human Embryology,* 3rd ed., Williams & Wilkins, 1965.)

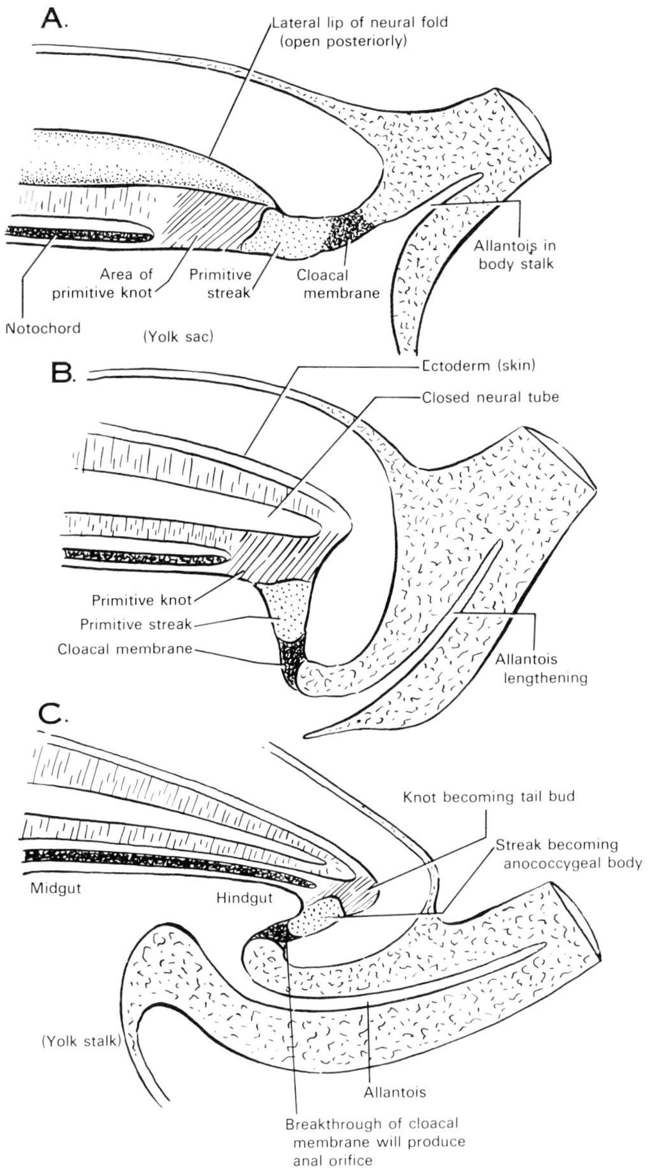

Fig. I-24. Progressive changes in tail region with caudal bending of germinal disk. (After Haines and Mohiuddin, *Handbook of Human Embryology,* 3rd ed., Williams & Wilkins, 1965.)

which the heart will develop. Associated with the cardiogenic plate is the bent portion of the horseshoe-shaped intraembryonic coelom into which the developing heart will bulge. This part of the coelom, cut in cross-section in Figure I-23, is the future pericardial cavity. At first, the newly-developing organism appears to be starting out with its heart virtually in its mouth. One wonders if it will ever fully recuperate from this initial state of apprehension in the long, uncertain journey through life. But as the head bending occurs, the heart is moved from this strange anatomical location and is tucked more properly into position in the ventral thorax with the foregut (**future esophagus**) behind it as it ought to be. The base of the heart will rest within the pericardium on the septum transversum, the **future diaphragm,** which also has folded into proper position (Fig. I-23C).

Further Changes in Tail Region

As bending occurs in the tail end (Fig. I-24), the primitive streak is also moved into a ventral position. Once it has finished giving rise to migrating mesodermal cells, what is left of it becomes the connective tissue of the **anococcygeal body.** The primitive node (of Henson), which is pulled around the bend with the streak, becomes a tail bud. In man, in whom no tail develops even though the human embryo appears at first glance to be bent on developing a rather substantial one, only a few **coccygeal vertebrae** arise in this region. A breakthrough of the cloacal membrane will eventually contribute to an **anal orifice** entering the hindgut. As the allantois enters the primary mesoderm of the body stalk, note that it is applied to the ventral surface of the hindgut (Fig. I-24C).

Blood Islands

Even though many of the bodily systems are in a state of construction, the embryo still exists upon embryotrophe. It has not begun to construct a circulatory system of its own which can come in contact with that of the maternal host. Only a pericardial cavity exists to house the future circulatory pump.

As the gut forms, a series of nodular areas breaks out on the surface of the yolk sac within the primary mesoderm that clothes its outer aspect (Fig. I-25). They are the primitive **blood islands** organized as clusters of irregularly-shaped mesodermal cells. Some of these cells flatten to form short tubes. These are the primitive **capillaries,** and the lining cells are **endothelium,** that is, simple squamous epithelium derived from mesoderm. Within the tubes, other cells of the mesoderm round up and begin the manufacture of **hemoglobin** within their cytoplasm. Soon they float in accumulating fluid **plasma.** The first **red cells** of the blood have formed in localized spots over the surface of the yolk sac. The isolated islands extend processes toward each other, and as they link up, the short capillaries that form build up an irregular, canalized network. This new **capillary bed** forming over the yolk sac does not link up for some little time with the similar networks that are now beginning to appear in the mesodermal layer of the embryo itself.

Multipotentiality of Early Connective Tissue (Mesenchyme)

Actually, such dense, structured, mesodermal masses as the notochord, somites, and lateral plate mesoderm are not surrounded by air or fluid as our black and white schematic diagrams would lead you to believe, nor is the neural tube suspended in space. As mesoderm wandered from the primitive streak it condensed into these larger-structured masses, but myriads of mesodermal cells also formed a loose packing tissue, the **early mesenchyme,** to fill in all the other empty spaces and chinks, like excelsior in a crate of eggs (See Fig. I-21B, mesenchyme). These are similar to the stellate-shaped cells we mentioned earlier that appear to

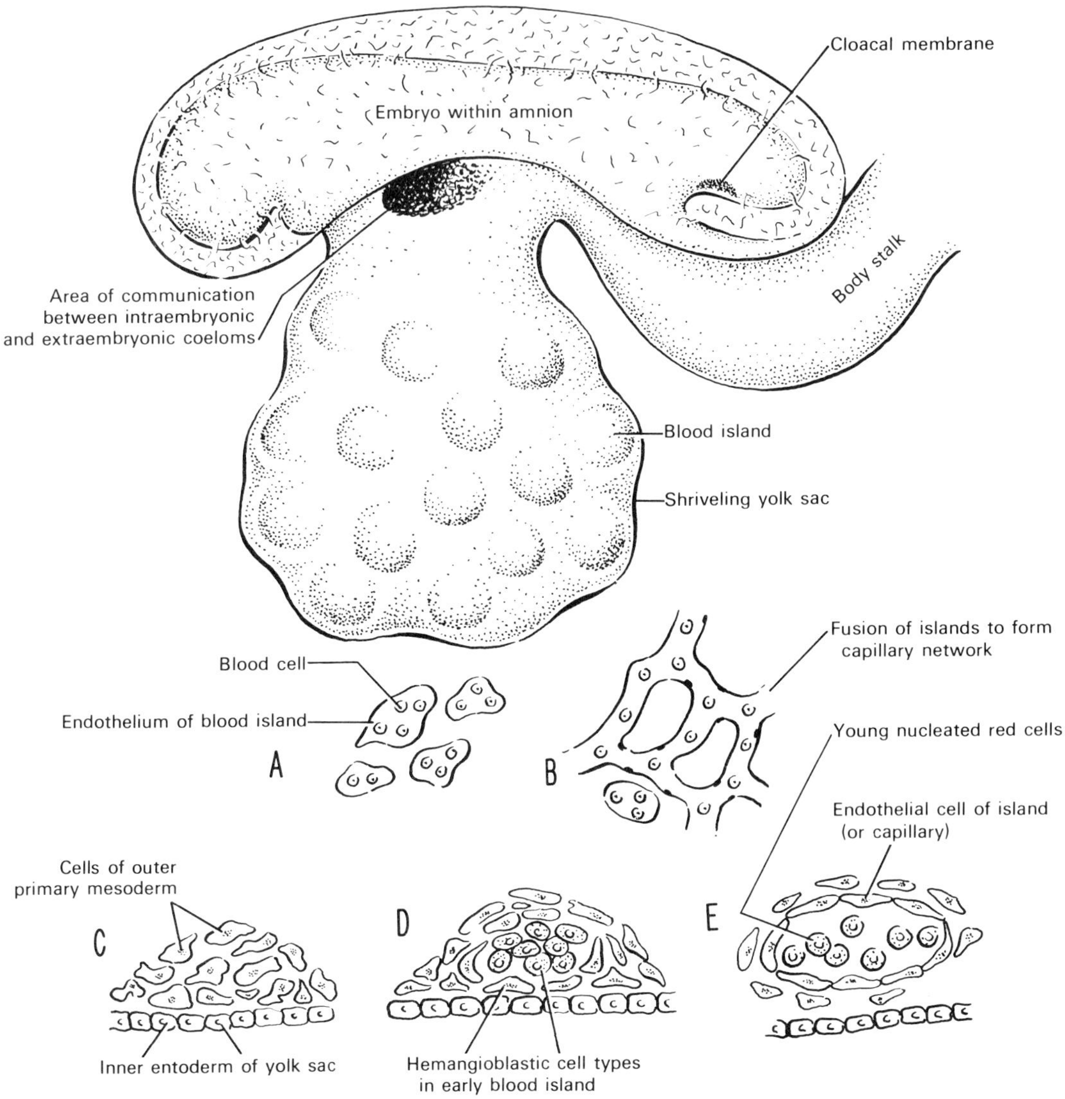

Fig. I-25. Blood islands of yolk sac. A, low-power sketch of isolated islands; B, fusion of islands into capillaries; C, primary mesoderm of sac before island formation; D, differentiation of hemangioblasts into lining endothelium and blood elements; E, high-power view of completed island (or capillary) in cross-section. (After Haines and Mohiuddin, *Handbook of Human Embryology*, 3rd ed., Williams & Wilkins, 1965.)

hold hands with each other in the jelly-like matrix of the primary blastocoelic mesoderm. They represent a loose kind of embryonic connective tissue. At any time, some cells can retract their processes, alter their internal chemistry, and begin to differentiate into stem cells which will give rise to more adult types of connective tissue cells. Differentiation follows varied and specialized lines. They can become **fibroblasts** (protein fiber formers); **chondroblasts** (cartilage formers); **steatoblasts** (fat formers or, better, fat storers); **osteoblasts** (bone formers); or **myoblasts** (muscle formers). It should be no surprise, therefore, to find some of them becoming **angioblasts** or blood formers.

Figure I-26A is the now familiar cross-section through the neural tube, notochord, and flanking somites (cf. Fig. I-21). But now we see that all the blank space between these structures and the ectoderm and entoderm is occupied by a loose connective tissue, **embryonic mesenchyme.** Some of the cells in it have freed themselves from the general network of cells to become **hemangioblasts** and will soon form blood islands that will anastomose in capillary nets and even larger channels, the beginnings of arteries and veins of the embryo proper. All this occurs in the embryonic mesoderm where other cells (of the somites and notochord, for example) will become cartilage, bone, and muscle.

The dorsal limb of the lateral plate mesoderm will

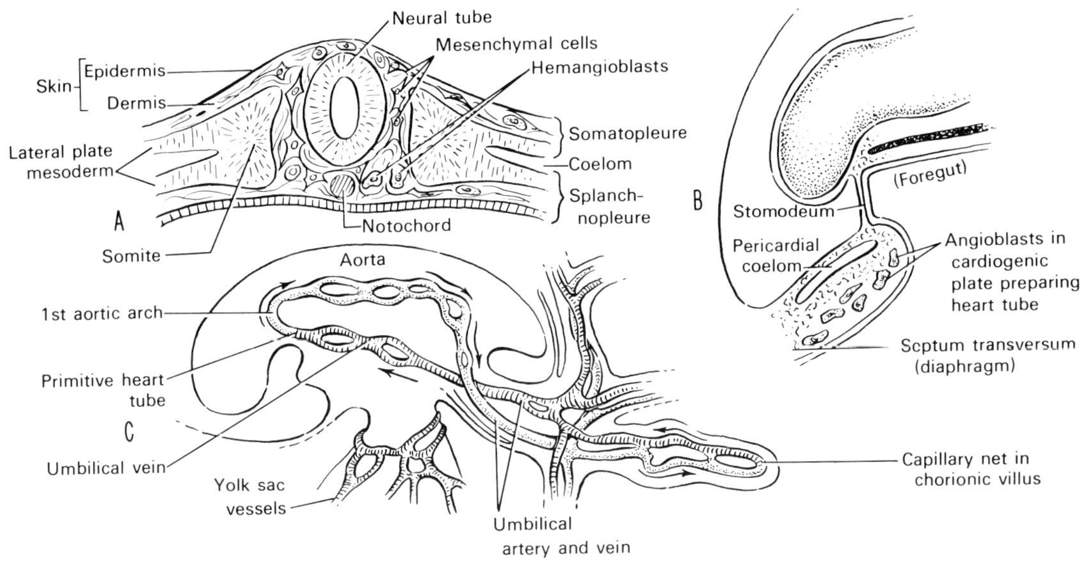

Fig. I-26. Angioblasts and the formation of early blood vessels. A, cross-section through neural tube and somites showing some mesenchymal cells free of reticular net transforming into angioblasts of embryo; B, angioblasts in cardiogenic plate preparatory to formation of heart tubes; C, pattern of early blood supply in a 22-day embryo. Vitelline vessels from aorta and umbilical vein have not yet linked with developing capillary bed on yolk sac. Connections have been made with placental region through body stalk and chorionic villi.

develop into skeletal muscle and into the inner lining cells (**mesothelium**) of the thoracic, abdominal, and pelvic body walls. The alliance of this mesoderm with the dermis and surface ectoderm above forms the **somatopleure** (see Fig. I-21). **Dermis** is the connective tissue layer of the skin of mesodermal origin; the outer **epidermis** is of ectodermal origin.

The ventral limb of the lateral plate mesoderm will join with the embryonic entoderm and the rolled-in yolk sac entoderm to form the walls of the hollow digestive tube. The muscle in these walls will be **smooth,** or **involuntary muscle** in contrast to **skeletal,** or **voluntary muscle** forming in the somatopleure. The ventral limb of the lateral plate mesoderm is, thus, **splanchnopleuric** in nature since it will form the walls of visceral organs (see Fig. I-21). The intraembryonic coelom, at this stage of development, lies between somatopleure and splanchnopleure.

DEVELOPMENT OF THE CIRCULATORY SYSTEM

The Heart Tubes

In Figure I-26B, the front end of the disk is seen in midsagittal view. Note the angioblasts forming in the cardiogenic plate between pericardial coelom of the embryo and the septum transversum. These are forming the heart tubes, for the heart will form from two capillary-like tubes that lie parallel, side by side, in this region. Very quickly the side walls will meet and fuse. Later other mesenchymal cells will form **cardiac muscle** around the outside of the fused tubes. But for the moment, only endothelially-lined tubes are forming with blood cells and plasma inside, much as occurred in the blood islands of the yolk sac.

Dorsal Aortae and Descending Aorta

Elsewhere in the embryonic mesenchyme, in the body stalk close to the allantois and in the mesoderm of the chorionic villi, other areas of angioblastic proliferation appear. These form complex, interconnecting, tubular channels at first. Then the larger ones fuse to form more regular and lengthy channels (Fig. I-26C). Actually, the building up of vascular channels proceeds more rapidly in the periphery (chorionic villi) and in the caudal end of the embryo at first than in the more cephalic heart region. In the midline, below the somites and flanking the notochord, a pair of channels form which represent the **dorsal aortae.** In Figure I-21, a cross-section through the embryo, these paired tubes appear to be ventral in position only because the anterior body wall below the entodermal layer shown has not closed off to form an abdominal cavity. Their position is really dorsal, adjacent to the area where the vertebral bodies will form. Later, they will fuse in all but their most cephalic portions, and a single **descending aorta** will lie ventral to the bodies of the vertebral column as it does in the adult. There is nothing in the human embryo that could be designated as a ventral aorta as in the lower vertebrates.[38]

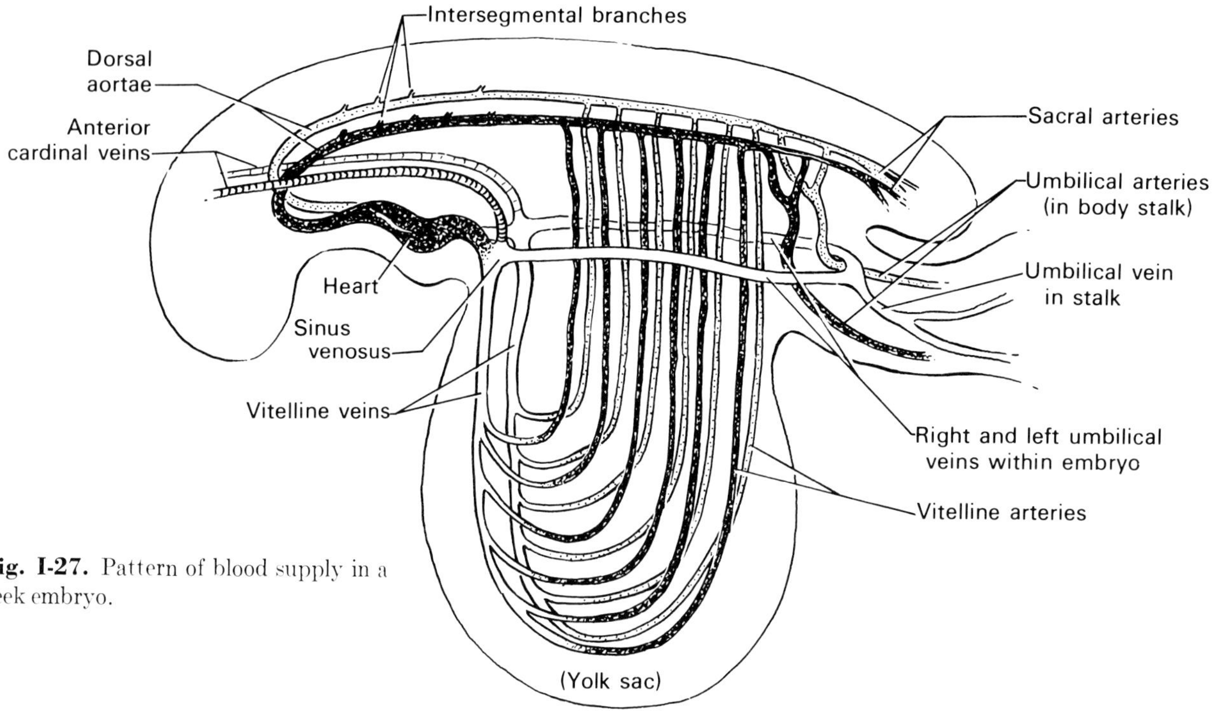

Fig. I-27. Pattern of blood supply in a 3-week embryo.

Aortic Branches

Extending caudally from the heart region, a second pair of channels will develop along the borders of the embryonic disk. These contain portions that will link with the yolk sac vessels and with the umbilical vein developing in the body stalk region near the allantois. Thus, in embryos of 22 to 23 days, parts of the arterial and venous sides of the **embryonic systemic circulation** are developing, though completed portions have not as yet linked. In embryos of 24 to 25 days (12 to 14 somites), a single heart tube communicates by a pair of **aortic arches** with the paired dorsal aortae (Fig. I-27). The dorsal aortae give off **intersegmental arteries** to somites in the trunk region, extend a series of ventral branches to link up with yolk sac vessels (**the vitelline arteries**), and form a pair of ventral **umbilical arteries** caudally which follow the allantois into the body stalk to link up with forming capillary beds in the mesoderm of the chorionic villi. The aortae end blindly in the embryo proper as paired sacral arteries in the region of the hindgut. Fusion of this pair will form the definitive **median sacral artery.**

Umbilical, Cardinal, and Vitelline Veins

On the venous side, a single **umbilical vein** leaves the body stalk. Peripherally it communicates with the venous portion of the chorionic capillary networks in the villi. At the embryonic disk it joins the paired venous channels that are extending to it from the heart. Thus, in the embryo proper, the entering single umbilical vein can be said to bifurcate into the right and left umbilical veins (Fig. I-27). Later on, as the liver develops within the septum transversum, the right umbilical channel will disappear. But for the moment, both pierce the septum transversum to enter the heart tube at its base in the developing **sinus venosus** region. Into this region, too, empty a pair of venous channels that drain the capillary bed of the yolk sac, the **vitelline veins.** Likewise, a smaller pair of veins draining the head region, the **anterior cardinals,** empty into the sinus venosus. Paired channels extending cephalad from the region of the developing kidneys (nephrotome territory), the **posterior cardinal veins,** will fuse into a common trunk with the anterior cardinals, and this will be the common **cardinal trunk** which returns blood from the body of the embryo itself to the sinus venosus.

Of the three pairs of venous channels emptying into the sinus venosus; *i.e.,* umbilical, common cardinal, and vitelline, the umbilicals are of the utmost importance throughout embryonic life. Remember that the left umbilical channel will persist as the fetal life line in contact with the placenta; the right one disappears shortly. The vitellines are the least important. At no time will the embryo derive much nourishment from its short-lived yolk sac. They will transform into the **hepatic, portal,** and **superior mesenteric veins** of the growing organism. The **vitelline arteries** will transform into the three unpaired branches of the abdominal aorta, namely, the **coeliac** and **superior** and **inferior mesenteric arteries** that serve the gastrointestinal tract and its associated glands, the liver and pancreas.

If the umbilical vein serves to bring oxygen and nourishment from the maternal blood pools of a developing placenta, the umbilical arteries will carry waste products of fetal metabolism back to the placenta. Here is a distinct reversal of function in the general circulatory plan. A student generally assumes that arteries carry oxygen and nutriment, whereas veins carry CO_2 and waste products of metabolism. The reverse is true of the umbilical vessels. Indeed, in respect to blood-gases they function like the pulmonary vessels of the adult—veins transporting oxygen, and arteries, CO_2. In terms of a respiratory exchange mechanism, the lung is to the adult as the placenta is to the fetus, for under ordinary circumstances the growing organism will not use its own developing lungs until birth. At birth, remains of the umbilical vein within the infant will obliterate, but parts of its umbilical arteries will be rescued for its own use. They will form a connection with the **internal iliac arteries.** Their proximal parts below this connection will continue to supply the walls and viscera of the pelvic region with arterial blood. Their distal parts, in the ventral body wall, will become obliterated, solid cords to the level of the umbilicus, the **lateral umbilical ligaments** of the adult. These structures, of course, lie internally in the abdominopelvic cavity, not on the external aspects of the ventral body wall.

What of the common cardinals? The right will contribute to the structure of the superior vena cava. The left becomes a very small and inconstant **oblique vein** (of Marshall) on the posterior aspect of the left atrium of the heart. The once large sinus venosus region, itself, gets almost completely incorporated into the right atrium of the developing heart. Only a small segment of it is left in the posterior atrioventricular sulcus of the completed heart as the **coronary sinus.** Into the latter will drain most of the cardiac veins along with the remnant of the left common cardinal vein, the now tiny left oblique cardiac vein mentioned above. There is no need to understand these details thoroughly at this time. We shall bring them up when we discuss the development of the heart and circulatory plan in later chapters. Realize, mainly, that the umbilical vessels serve throughout **intrauterine life** and then they become pretty much functionless, while vitelline and cardinal vessels last, as such, for a brief time in the embryo and are then transformed into other useful parts of a circulatory system that will serve the organism throughout its **extrauterine life.**

The Heart Beat

The story of the organization of the body plan is almost complete. We need survey only some major changes that take place between the 3rd and 8th weeks of embryonic life. Before we do, we had best consider the heart beat and the formation of a placenta, for growth has reached a stage where a better means of obtaining ready nourishment and oxygenation is becoming essential. A heart beat has been established, and a reasonable circulatory system is present. The heart started beating as early as the three-somite stage[39] before the heart tubes fused into a single chamber, with the left tube beginning to beat earlier and at a faster rate than the right tube. Upon fusion, the faster beat dominated in the single tube. Thus, in the 3-week embryo, some blood is sloshing to and fro in this circulatory system in a desultory fashion. With continued rapid growth of the heart and improvement of major vascular channels in capillary beds, it will soon be possible to move blood in a complete circuit from heart to chorionic villi and back again.

FORMATION OF THE PLACENTA

We have seen the fetus implant and bury itself below the endometrium of the uterus. We have noted the partial erosion of the endometrium and the inclusion of decidual cells and blood from opened maternal vessels in the lacunar spaces of the syntrophoblast. Only that portion of the endometrium below the body stalk region of the implanted blastocyst will enter into placental formation (see Fig. I-28). This portion of the uterine endometrium is called the **decidua basalis.** That which grows over the sides and surfaces of the implant, into which the shorter chorionic villi push and later degenerate, is the **decidua capsularis.** The endometrium lining the rest of the uterine cavity is the **decidua vera.** After the 8th week, there is rapid growth and elongation of villi in the region of the decidua basalis, while regression takes place in those related to the capsularis. This leaves the chorion next to the decidua capsularis smooth and relatively avascular. By the 4th to 5th month, the decidua capsularis thins to such an extent that the bare, smooth chorion abuts against the decidua vera of the opposite wall, virtually obliterating the uterine cavity (Fig. I-28). In the process, the amnion has fused loosely to the inner aspect of the chorion, completely obliterating the extraembryonic coelom, and both have been reflected as a sheath to clothe the body stalk. The latter is now a large, thin structure, the definitive, jelly-like **umbilical cord** which contains umbilical vessels and a degenerating allantois. The degenerated yolk sac has been squeezed between amnion and chorion after the ventral body wall has sealed off in the midgut region where it once opened into the yolk stalk. A large fetus, anchored by the umbilical cord to the placenta, now floats in an abundance of amniotic fluid (Fig. I-28). The disk-shaped placenta, the combined product of fetal and maternal tissues, is broken up into 15 to 20 smaller subregions or **cotyledons** by connective tissue septae that project upward from the decidual plate. The cytotrophoblast of the villi gets used up by proliferating and transforming into syntrophoblast. Some cytotrophoblast persists in parts of the basal plate along the septal walls and on the upper stems of some of the larger chorionic villi.

Coiled uterine end-arteries of the maternal endometrium open into the intervillous spaces (Fig. I-29). Since these vessels have muscular walls that can constrict intermittently, the maternal blood is projected into the intervillous spaces in a pulsatile fashion. The

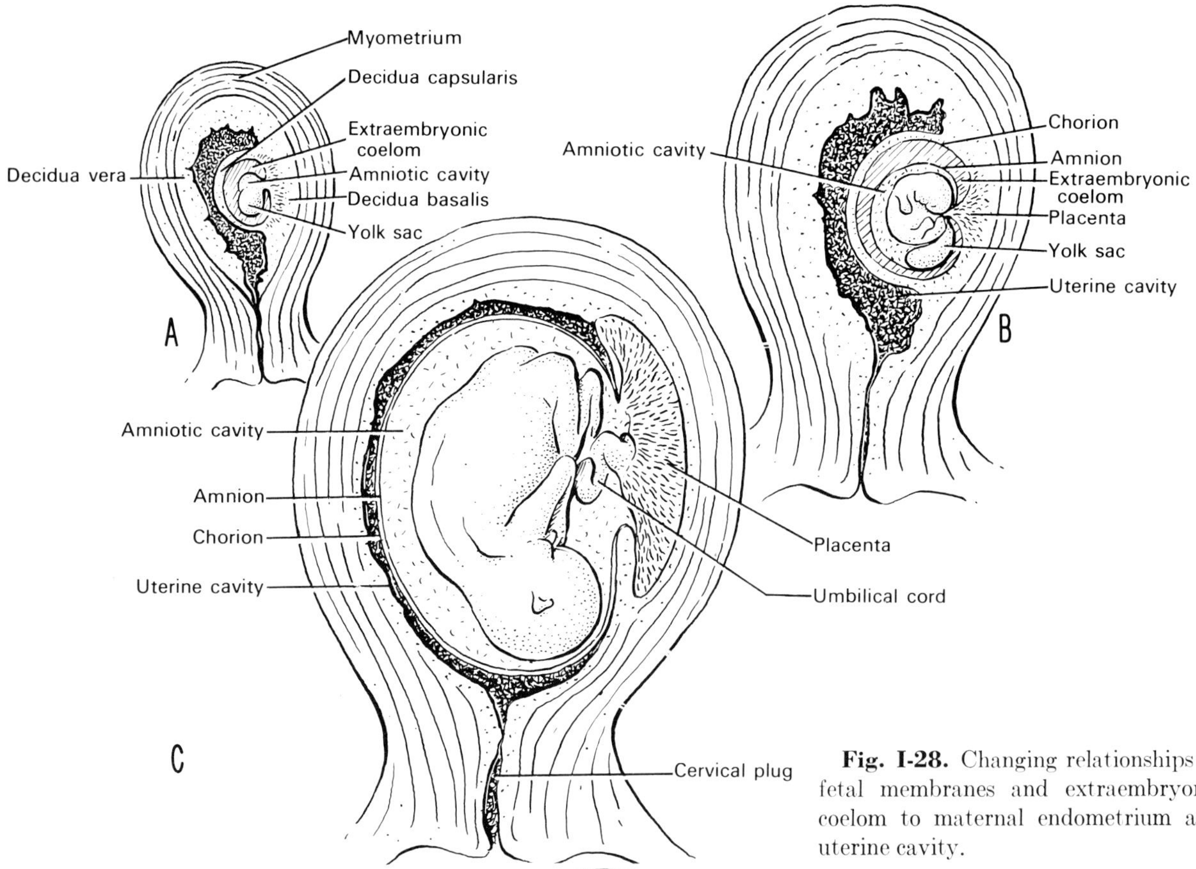

Fig. I-28. Changing relationships of fetal membranes and extraembryonic coelom to maternal endometrium and uterine cavity.

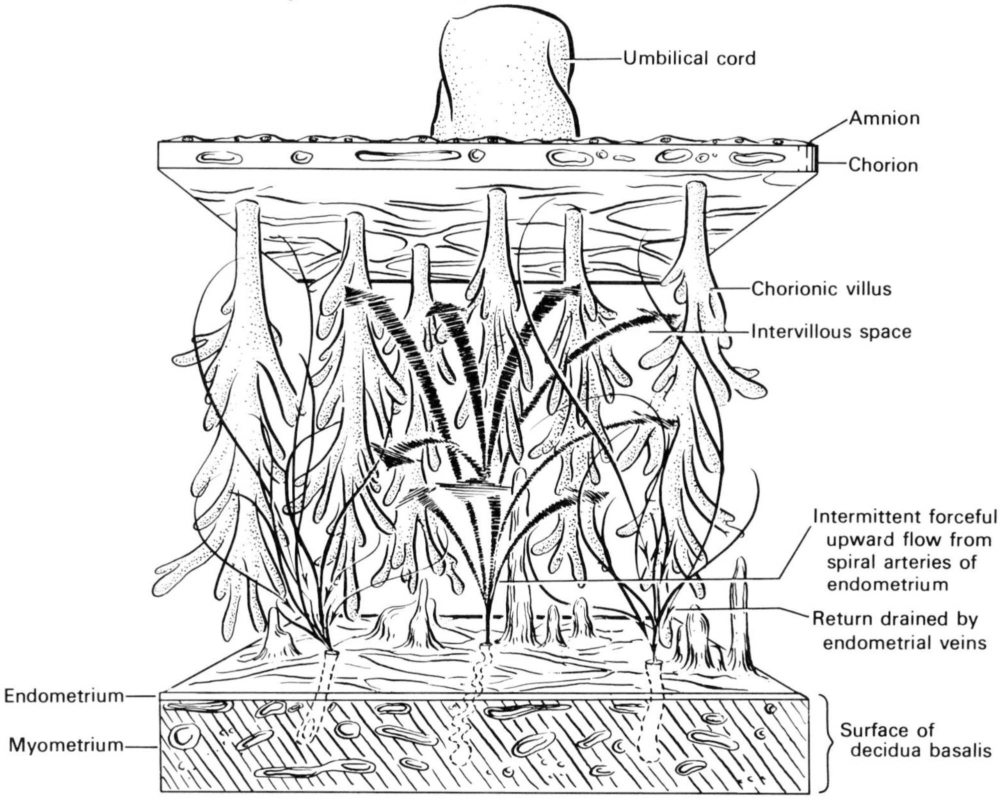

Fig. I-29. A schema of relationships of chorionic villi to pulsatile blood flow from maternal spiral arteries in the decidua basalis of the placenta. Arterial blood, spurting upward in intervillous spaces, forces circulation of venous return as a down flow into open collecting veins of the maternal endometrium.

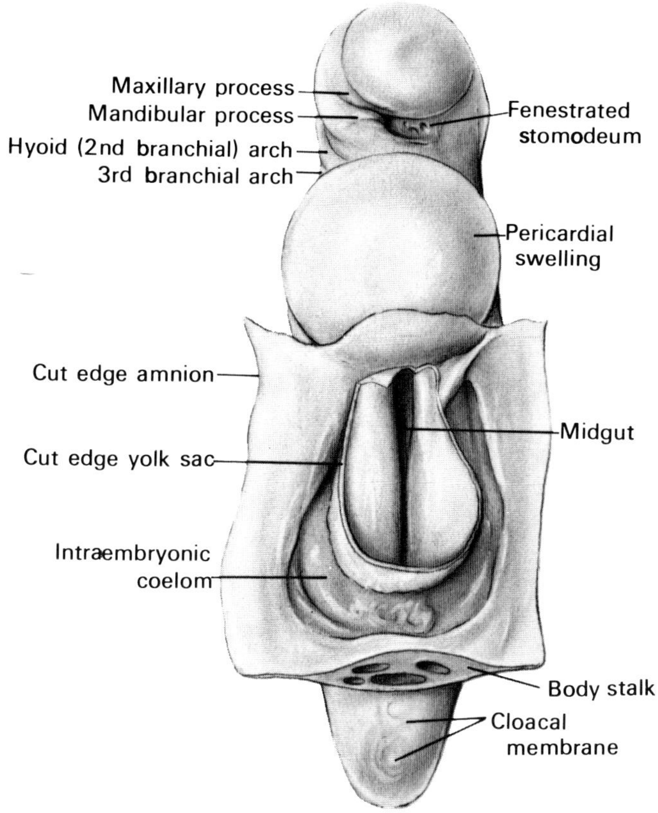

Fig. I-30. Reconstruction of a 26-day embryo, ventral view. (After Hamilton, Boyd and Mossman, *Human Embryology*, 3rd ed., Williams & Wilkins, 1962.)

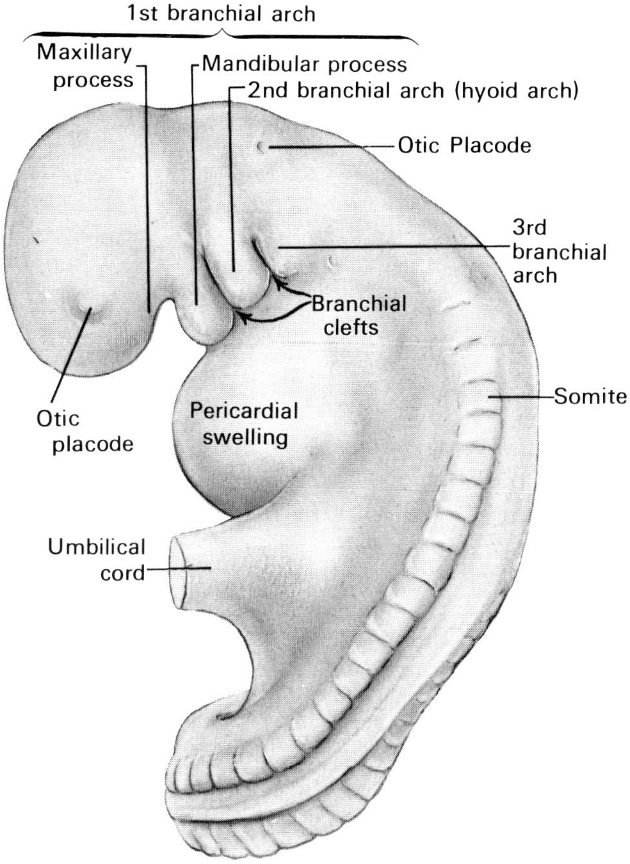

Fig. I-31. Left side of a 28-day embryo. (After Hamilton, Boyd and Mossman, *Human Embryology*, Williams & Wilkins, 1962.)

pressure in these vessels is also greater than in the intervillous pools. Thus, a jet stream of blood fountains upward to bathe the dangling villi.[40-41] Oxygen and metabolites diffuse readily in this **hemochorial placenta,** for diffusion is through the thin syntrophoblast, a basal lamina of connective tissue, and the endothelium of the capillaries that lie in the center of the villi. The head of arterial pressure, bursting into the intervillous space, forces blood downward toward the basal plate where it is drained away by maternal uterine veins that open into the bottom of these spaces (Fig. I-29). In these blood-bathed cotyledonous pools of the placenta, the capillaries of the embryonic villi can exchange carbon dioxide and waste products of metabolism from the fetus for a fresh supply of oxygen and nutrients from the maternal blood. The waste products are removed by the maternal lungs, kidneys, and skin. The oxygen and nutrients are conveyed to the fetal heart by the umbilical vein. This blood is pumped to the actively-growing cells of the organism by the fetal heart. Waste products of cellular activity are returned from fetal capillary beds by the umbilical arteries.

Figures I-30 to 38 are a well-labeled series to show the salient features of embryonic growth from the 26th day to the 60th day after fertilization.

FEATURES OF DEVELOPING EMBRYO

The 26-day Embryo

In Figure I-30, the chief features to note are the open midgut (yolk sac cut off), the nature of the intraembryonic coelom which opens outward toward you into the extraembryonic coelom, and the forming of maxillary and mandibular processes around the stomodeal membrane. The latter is partially fenestrated (Latin, *fenestra*—window), opening into the pharynx or foregut. If you pushed your index finger forward into the midgut, you could peep at its end through these little windows. If you pushed the finger backward it would be in the blind-ending hindgut where the cloacal membrane has not yet become patent.

The 28-day Embryo

Figure I-31 shows a lateral view at 28 days. Somites have increased from 20 to 25 since the stage shown in Figure I-30. Optic and otic (auditory) placodes are evident. The heart swelling is prominent. The mandibular process of the mandibular arch (**1st branchial arch**) and the 2nd hyoid arch (**2nd branchial arch**) are prominent. There will be six of these **branchial arches** in all, with **branchial clefts** between. Out-

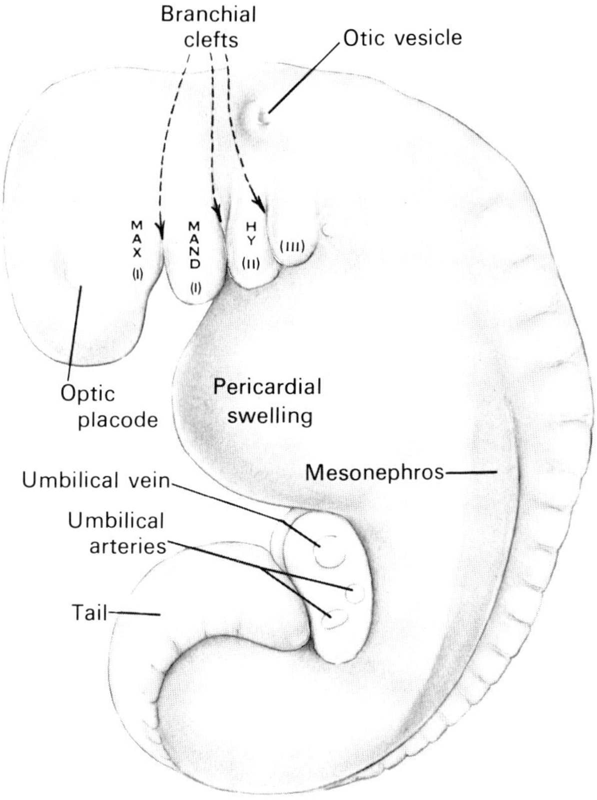

Fig. I-32. Left side of a 30-day embryo. Branchial arches are labeled I, II, and III, and branchial clefts or grooves between are pointed out by arrows. Arch I is split into maxillary and mandibular processes. (After Hamilton, Boyd and Mossman, *Human Embryology*, 3rd ed., Williams & Wilkins, 1962.)

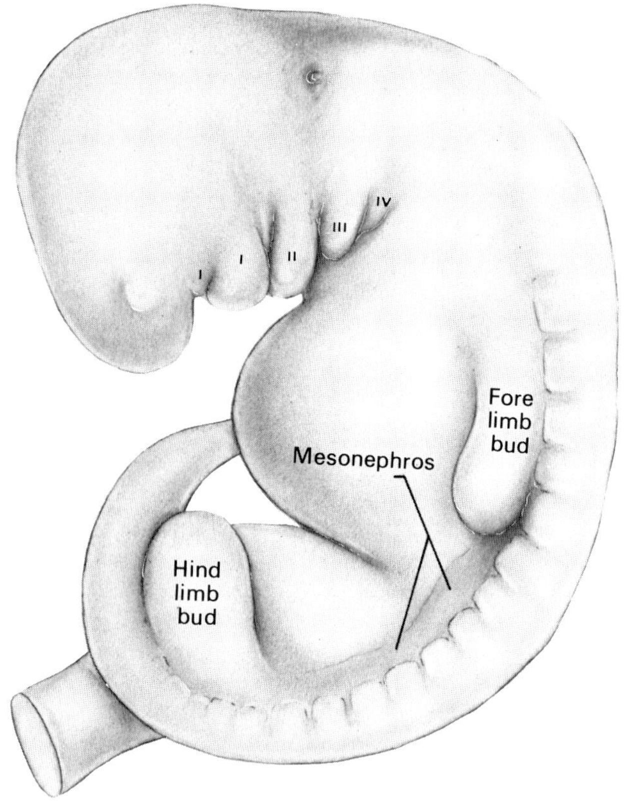

Fig. I-33. Left side of a 32-day embryo. Fore- and hindlimbs appear. Arch IV present. Mesonephros more prominent than heretofore. (After Hamilton, Boyd and Mossman, *Human Embryology*, 3rd ed., Williams & Wilkins, 1962.)

pouchings from the pharynx (**pharyngeal pouches**) will originate in the foregut region to meet the invaginations of the branchial clefts or grooves. These areas in fish would develop into gills. But man puts them to other uses in developing bones, cartilages, muscles, and glands of the middle ear, face, pharynx, and larynx. Each branchial arch will have a specific cranial nerve associated with it to innervate muscles developing from it. You might note early that the following cranial nerves relate to these specific arches:

Arch I—Trigeminal, or 5th cranial nerve.

Arch II—Facial, or 7th cranial nerve.

Arch III—Glossopharyngeal, or 9th cranial nerve.

Arches IV and V—Vagus, or 10th cranial nerve.

Rudimentary Arch VI—Spinal accessory, or 11th cranial nerve.

The 30-day Embryo

Externally, nothing spectacular has happened further in the 30-day-old embryo except for a faint lateral swelling which marks the region of the mesonephros, an early type of kidney (Fig. I-32). We will worry about internal changes in subsequent chapters. There are now 28 somites, and the 3rd branchial arch is well formed. The pericardial swelling has increased. Part of the bulge in the caudal part of this region is due to rapid growth of the liver in the underside of the septum transversum. A tail region is prominent. The otic placode has a pit and has become an **otic vesicle.** This structure is below the skin, and we look through the thin skin at it. It will become the **inner ear.**

The 32-day Embryo

By 32 days (Fig. I-33), the upper and lower limb buds appear and a 4th branchial arch is evident. A ridge in the lateral body wall, the **mesonephros,** is more prominent than it has been previously. Kidney and gonads are developing internally in this region. Most of the somites are present.

The 34-day Embryo

At 34 days (Fig. I-34) the C-shaped bending has intensified, and the head end is becoming massive. The smaller remaining branchial arches are lost in the cervical (Latin, *cervix*—neck) region. Olfactory pits are prominent. The tip of the tail is regressing. The

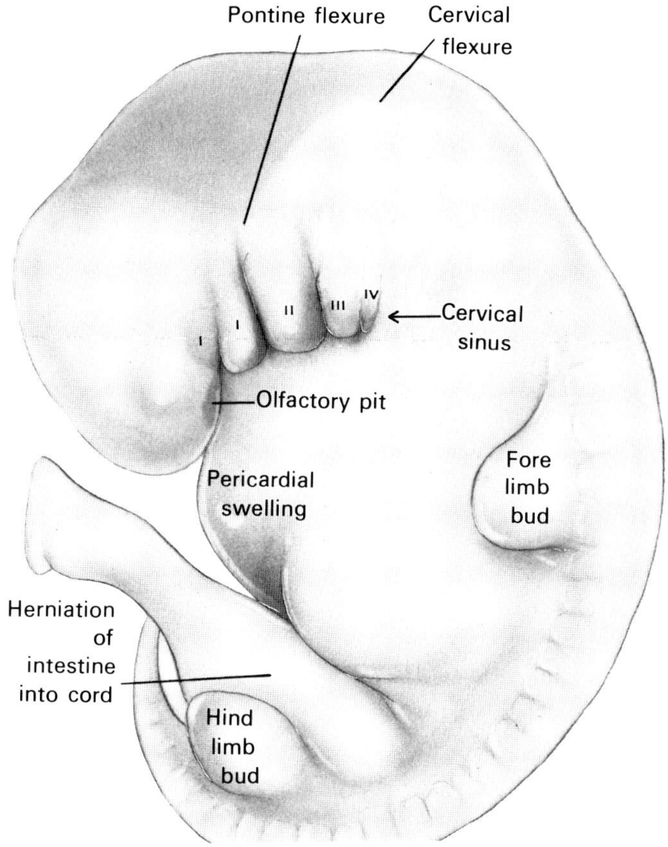

Fig. I-34. Left side of a 34-day embryo. Increase in head size. Regression of arches III and IV. Olfactory pit appears. Beginning herniation of intestine into umbilical cord. (After Hamilton, Boyd and Mossman, *Human Embryology*, 3rd ed., Williams & Wilkins, 1962.)

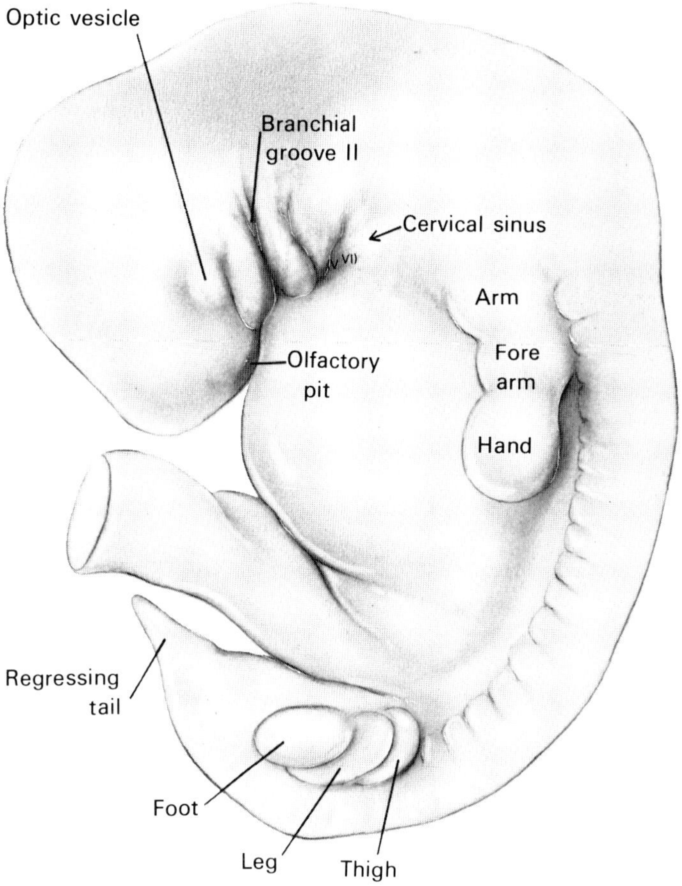

Fig. I-35. Left side of a 37-day embryo. Arches III and IV have virtually disappeared. Branchial groove II has deepened to mark locus of external auditory meatus. Limbs show differentiation into arm, forearm, and hand and into thigh, leg, and foot regions. Tail is regressing. Optic vesicle is present for the first time. (After Hamilton, Boyd and Mossman, *Human Embryology*, 3rd ed., Williams & Wilkins, 1962.)

intestinal tract has developed so rapidly that the ventral body wall has not kept pace to contain it. Small intestine and parts of the large bowel have herniated into the umbilical cord. They will remain there for some time until the abdominal cavity can retain them, then they will be withdrawn. Thus, an umbilical hernia is a natural phenomenon during embryonic growth. If still present at birth, it is an untoward event requiring surgical repair.

The 37-day Embryo

By 37 days (Fig. I-35), the cartilaginous patterns of many of the long bones have formed. The upper limb has lengthened and appears to have shoulder and elbow areas. The hand is a ping-pong paddle disk. There is a depression in the eye region, the optic vesicle.

The 40-day Embryo

At 40 days (Fig. I-36), the head is quite enlarged. The growing brain is largely brainstem, but cerebral vesicles are bulging laterally and growing forward. Eventually they will curve upward and backward in an arching course to cover over and hide the brainstem beneath them. There is an optic cup and a lens placode in the eye region. The external ear is a puckering region with a cleft, the **external auditory meatus.** It lies inferiorly in position to the otic vesicle which, itself, becomes the inner ear. The placode is dimpling into a pit, pinched off from the skin and sunken below it to develop into the auditory sense organ. The hand plate is ridged where fingers will form. The lower limb has increased in length and has a semblance of a hip, knee, and ankle joint.

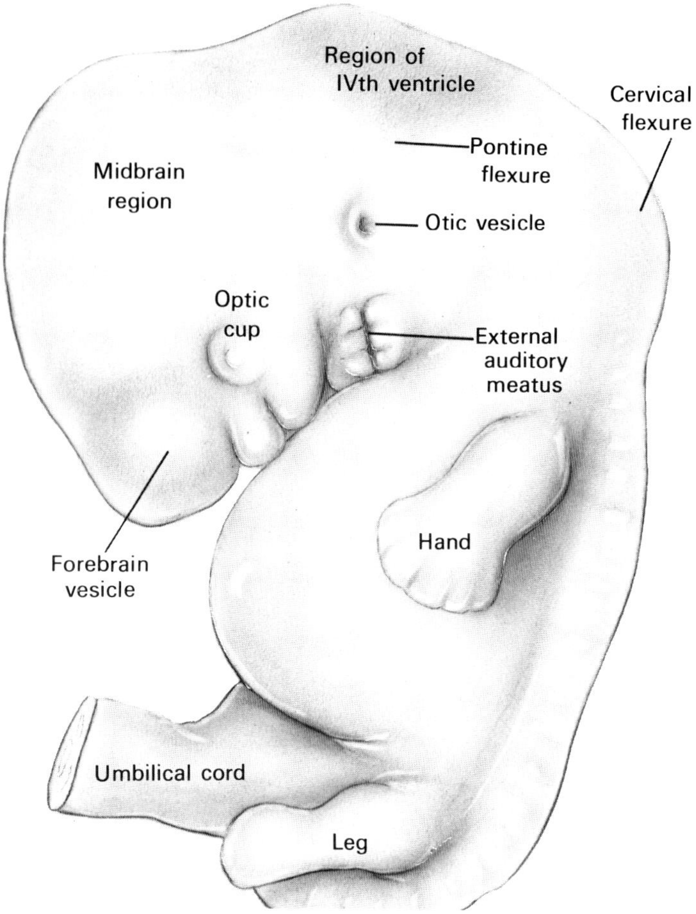

Fig. I-36. Left side of a 40-day embryo. Swelling of ventral body wall due to rapid growth of liver; pericardium lies superior to this bulge. Fingers are being cut in the hand plate. (After Hamilton, Boyd and Mossman, *Human Embryology*, 3rd ed., Williams & Wilkins, 1962.)

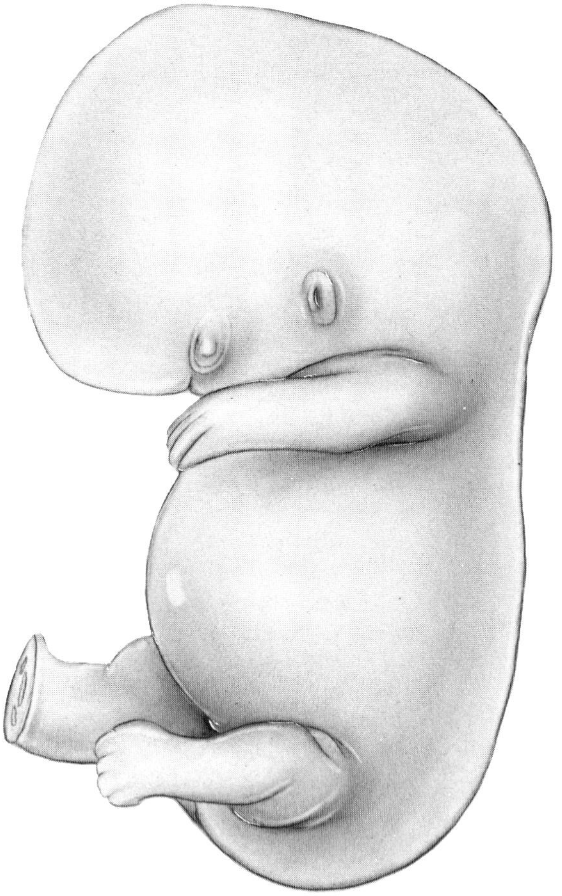

Fig. I-37. The human fetus at about 2 months, a fair replica of future human form. Crown-rump length, c., 17 mm. (After Hamilton, Boyd and Mossman, *Human Embryology*, 3rd ed., Williams & Wilkins, 1962.)

THE FETUS (2 MONTHS)

Finally (Fig. I-37), by 2 months the embryo has become a fetus. It has some recognizable characteristics that qualify it as a probable human being, though we might wish to reserve judgement and classify it as something from outer space. Continued overgrowth of the cerebrum gives it a high-browed and massive-domed appearance, yet its vacuous visage seems to belie any capacity for intellectual achievement. It is somewhat pot-bellied, like your author, and sits like a little Buddha with the soles of its paddle-feet opposed in the reflective posture, while it gravely contemplates its navel. But it has two eyes, two ears, a button nose, and a mouth; five fingers on its hands, and it has lost most of its long tail. It has long since given up the idea of becoming a fish, and it has converted its gills to more worthwhile purposes. It has a four-chambered heart and some rudimentary lungs in its thorax. It has separated thorax from abdomen by a diaphragm. It has a liver, pancreas, spleen, incomplete kidneys, and gonads that can't make up their minds as to sex. It has a well-developed tongue and pharynx and a long digestive tube that still is partially herniated into the umbilical cord. On the whole, this little creature, some 17 mm. in length, has become a pretty complete package and a fair replica of the human race at 2 months of age, a time when some women are not as yet certain that they are pregnant. By 60 days (Fig. I-38), it already looks like a fairly cuddly infant. All that remains is for it to grow and develop in perfection in the uterus for another 220 days until it becomes the pride and joy of a pair of fawning parents.

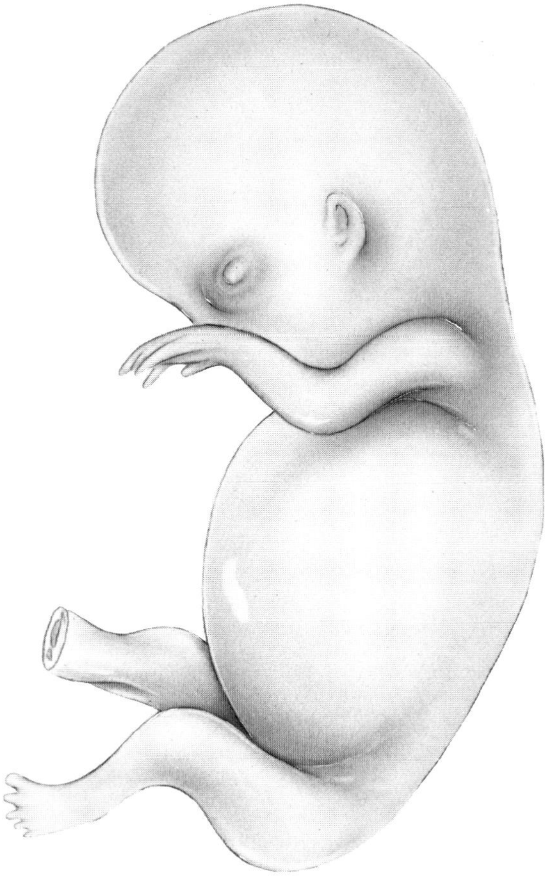

Fig. I-38. The human fetus at about 60 days. Crown-rump length about 31 mm. (After Hamilton, Boyd and Mossman, *Human Embryology*, 3rd ed., Williams & Wilkins, 1962.)

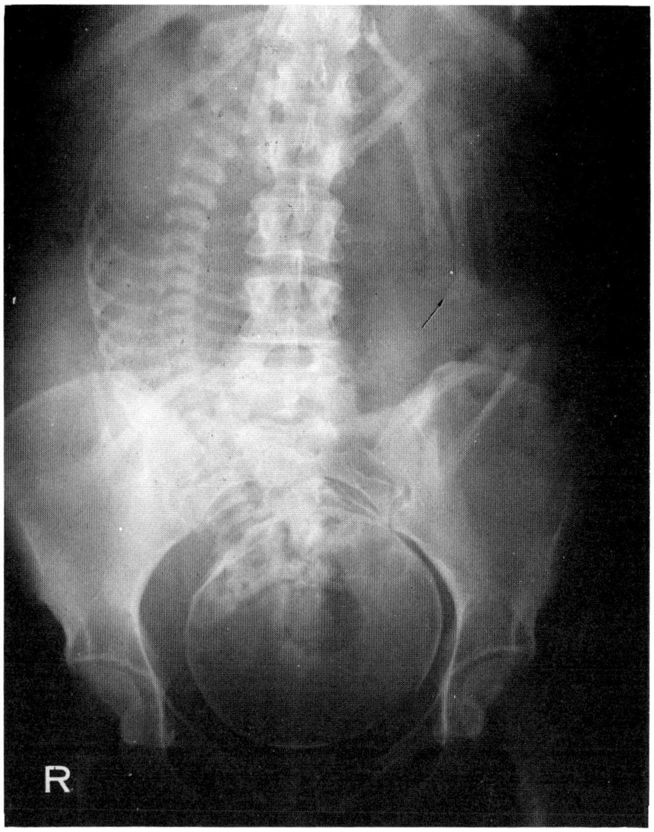

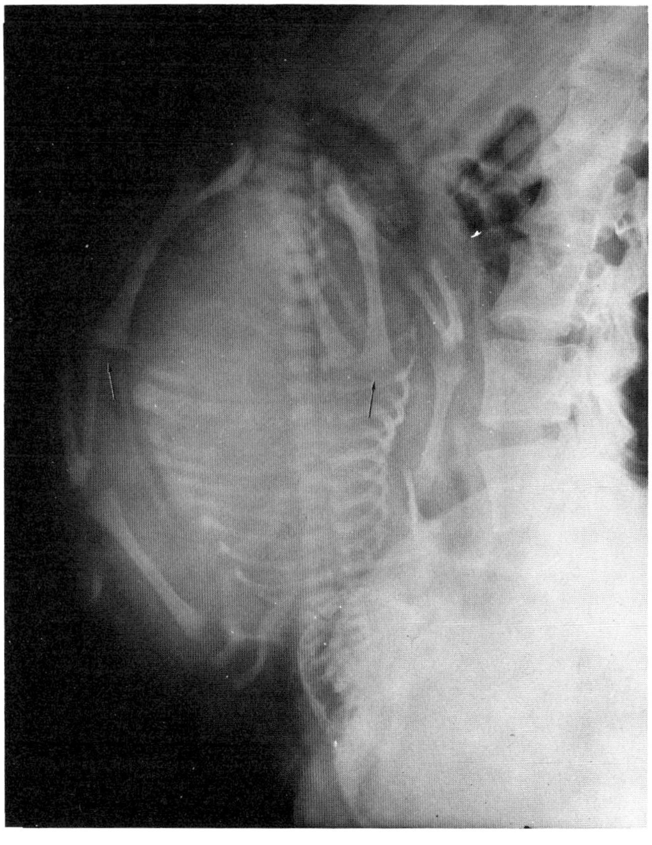

Fig. I-39. The end of the beginning. Radiographs of the fetus *in utero* at term in the vertex presentation. Fetal bones may first be detected by radiographic means about 14 to 15 weeks from conception. The arrow in each view points to the distal femoral epiphysis which is calcified in some 90% of term fetuses. This valuable sign is one of many used to determine fetal age.

A. anterior-posterior (AP) view of the mother's abdomen.

B. lateral view.

REFERENCES

1. Clermont, Y., and Perry, B.: Quantitative study of the cell population of the seminiferous tubules in mature rats. Amer. J. Anat., *100:* 241, 1957.
2. Mancini, R. E., Narbaitz, R., and Lavieri, J. C.: Microscopic anatomy of the sex glands and seminiferous tubules in growing adult and male albino rats. Amer. J. Anat., *108:* 79, 1961.
3. Clermont, Y., and Leblond, C. P.: Renewal of spermatogonia in the rat. Amer. J. Anat., *93:* 39, 1953.
4. Clermont, Y., and Leblond, C. P.: Differentiation and renewal of spermatogonia in the monkey. Amer. J. Anat., *104:* 237, 1959.
5. Oakberg, E. F.: Degeneration of spermatogonia in the mouse following exposure to x-rays, and stages in the mitotic cycle at which death occurs. J. Morph., *97:* 39, 1955.
6. Oakberg, E. F.: A description of spermiogenesis in the mouse and its use in the analysis of the cycle of seminiferous epithelium and germ cell renewal. Amer. J. Anat., *99:* 167, 1952.
7. Ford, C. E., and Hamerton, J. L.: The chromosomes of man. Nature (London), *178:* 1020, 1956.
8. Tjio, J. H., and Levan, A.: The chromosome number in man, Hereditas, *42:* 1, 1956.
9. Leblond, C. P., and Clermont, Y.: Spermiogenesis of man, monkey, ram, and other mammals as shown by the periodic acid-Schiff technique. Amer. J. Anat., *90:* 167, 1952.
10. Blandau, R. J., and Rumery, R. E.: Fertilizing capacity of rat spermatozoa recovered from various segments of the epididymis. Anat. Rec., *139:* 209, 1961.
11. Mason, R. E., and Shaver, S. L.: Some functions of the caput epididymis. Ann. N. Y. Acad. Sci., *55:* 585, 1952.
12. Clermont, Y.: Contractile elements in the limiting membrane of the seminiferous tubules. Exp. Cell. Res., *15:* 438, 1958.
13. Hotchkiss, R. S.: *Fertility in Man.* Lippincott, Philadelphia, 1944.
14. Heller, C. G., and Clermont, Y.: Kinetics of the germinal epithelium in man. Rec. Progr. Hormone Res., *20:* 545, 1964.
15. Mintz, B., and Russel, E. S.: Gene-induced embryological modifications of primordial germ cells in the mouse. J. Exp. Zool., *134:* 207, 1957.
16. Mintz, B.: Embryological development of primordial germ cells in the mouse. J. Embryol. Exp. Morph., *5:* 396, 1957.
17. Mintz, B.: Embryological phases of mammalian gametogenesis. J. Cell. Comp. Physiol., *56:* 31, 1960.
18. Mintz, B.: Formation and early development of germ cells. In *Germ Cells and Development, Symp. Inst. Internat.* d'Embry., pp. 1–24, Fondoz A. Baselli, 1960.
19. Witschi, E.: *Development of Vertebrates.* Saunders, Philadelphia, 1956.
20. Rudkin, R. T., and Griech, H. A.: On the persistence of oocyte nuclei from fetus to maturity in the laboratory mouse. J. Cell. Biol., *12:* 169, 1962.
21. Sirlin, J. L., and Edward, R. G.: Timing of DNA synthesis in ovarian oocyte nuclei and pronuclei of the mouse. Exp. Cell. Res., *18:* 190, 1959.
22. Witschi, E.: Migration of the germ cells of the human embryo from the yolk sac to the primitive gonadal folds. Contrib. Embryol., *32:* 67, 1948.
23. Everett, N. B.: Observation and experimental evidence relating to origin and differentiation of the definitive germ cells in mice. J. Exp. Zool., *92:* 49, 1943.
24. Chiquoine, A. D.: The identification, origin, and migration of the primordial germ cells in the mouse embryo. Anat. Rec., *118:* 135, 1954.
25. Evans, H. M., and Swezy, O.: Oogenesis and the normal follicular cycle in adult animals. Univ. Calif. Mem., *9:* 119, 1931.
26. Latta, J. S., and Pedersen, E. S.: The origin of ova and follicle cells from the germinal epithelium of the ovary of the albino rat as demonstrated by selective intravital staining with India ink. Anat. Rec., *90:* 23, 1944.
27. Zuckerman, S.: The number of oocytes in the mature ovary. Rec. Progr. Hormone Res., *6:* 63, 1951.
28. Brambell, F. W. R.: Ovarian changes. In *Marshall's Physiology of Reproduction,* 3rd Ed., edited by A. S. Parkes. London, Longmans-Green, London, 1956.
29. Gaillard, P. J.: Sex cell formation in explants of foetal human ovarian cortex. I. and II. Konink Nederl. Akad. Wetenschap, *53:* 1300, 1950.
30. Chiquoine, A. D.: The development of the zona pellucida of the mammalian ovum. Amer. J. Anat., *106:* 149, 1960.
31. Kemp, N. E.: Protoplasmic bridges between oocytes and follicle cells in vertebrates. Anat. Rec., *139:* 324, 1958.
32. Shettles, L. G.: The nourishment of the human ovum. Bull. Sloane Hosp. Wom., *4:* 34, 1958.
33. Ryan, K. J., and Short, R. V.: Formation of estradiol by granulosa and thecal cells of equine ovarian follicle. Endocrinology, *76:* 108, 1965.
34. Bishop, D. W.: Section D 13. In *Sex and Internal Secretions,* edited by Young. Williams & Wilkins, Baltimore, 1961.
35. Pincus, G.: *The Eggs of Mammals.* Macmillan, New York, 1936.
36. Blandau, R. F.: Section D 14. In *Sex and Internal Secretions,* edited by Young. Williams & Wilkins, Baltimore, 1961.
37. Hertig, A. T., Adams, E. C., and Milligan, W. J.: On the pre-implantation stages of the human ovum. Contrib. Embryol., *35:* 199, 1954.
38. Congdon, E. D.: Transformation of the human aortic-arch system during the development of the human embryo. Contrib. Embryol., *14:* 47, 1922.
39. Goss, C. M.: The first contractions of the heart in rat embryos. Anat. Rec., *70:* 505, 1938.
40. Ramsey, E. M.: Circulation in the intervillous space of the primate placenta. Amer. J. Obstet. Gynec., *84:* 1649, 1962.
41. Ramsey, E. M., and Harris, J. W. S.: Comparison of uteroplacental vasculature and circulation in Rhesus monkey and man. Contrib. Embryol., *38:* 61, 1966.

GENERAL REFERENCES

1. Arey, Leslie Brainerd: *Developmental Anatomy—A Textbook and Laboratory Manual of Embryology,* 7th Ed. W. B. Saunders Co., Philadelphia, 1965.
2. Hamilton, W. J., Boyd, J. D., and Mossman, H. W.: *Human Embryology,* 3rd Ed. The Williams & Wilkins Co., Baltimore, 1962.
3. Langman, Jan: *Medical Embryology—Human Development—Normal and Abnormal,* 2nd Ed. The Williams & Wilkins Co., Baltimore, 1969.
4. Patten, Bradley M.: *Foundations of Embryology,* 2nd Ed. McGraw-Hill Book Co., New York, 1964.

TWO

Anatomical Terminology and General Concepts

PART I. TERMINOLOGY

Introduction

Anatomy still furnishes the lion's share of the terms that make up the "language of medicine." Many of these terms came to us from Latin and Greek, the "dead tongues," which most students no longer study in the course of their premedical training. It is no wonder that the acquisition of an extensive vocabulary of medical English during the student's first year in medical school almost becomes a matter of learning a foreign language. To make matters even more difficult, time is of the essence. The rapid pace of modern curricula makes it impossible to discuss terms as they are met anew, or to go into their background history. The few courses that exist in medical terminology by-pass etymology. These courses are designed to familiarize medical secretaries with words commonly encountered in report typing and to teach accurate spelling.

Like all good teachers everywhere, we can only exhort the student to make frequent use of a good medical dictionary.[1, 2] However, even this tool is limited when it comes to pointing up word derivatives. Dictionaries are concerned chiefly with word definitions. For the sheer enjoyment of discovery, we would recommend a text like H. A. Skinner's, *The Origin of Medical Terms*[3] which, happily, happens to be published by our publishers. It is alphabetized in dictionary style, giving a full and an entertaining account of word ancestry. It is a boon reference companion. For a cold winter night's entertainment at the fireside, we strongly recommend a brief little book by A. M. Lassek, *Human Dissection, Its Drama and Struggle*,[4] unfortunately not published by our publisher.

The very early Greeks, our forebears in medical sciences, coined a part of our medical vocabulary, but not many of their words came down to us directly. Some of the few that have survived were rather prosaic household terms in ancient Greece. The shape of the socket of the hip joint resembled a vinegar cruet, so it was named the **acetabulum,** while one of the fetal membranes looked "like a sausage" so it was called the **allantois.** But the pure originals from the Hippocratic era are a mere handful compared to those that first stopped over in Rome to become Latinized as part of the vocabulary of the Roman physician; *e.g.,* the Greek **perikardion** became **pericardium.** In Rome some of these Greek terms might exist in two forms, one with a Greek ending and one with a Latin ending; *e.g.,* **pelvis-pelvic; dermal-dermic;** and we continue this practice even today, sometimes using the adjective forms interchangeably. Even more confusing in present day practice is to find Latin nouns used as names for an anatomical structure while the pathologic processes associated with this same structure are derived from the Greek; *e.g.,* **renis,** the kidney, but **nephritis** instead of **renitis; oculus,** the eye, but **ophthalmitis** instead of **oculitis.**

But our heritage of Romanized Greek terms may not even be as great as the number of Greek **neoclassicisms** that have been minted over the past 200 years as scientists have continually coined new terms. Greek has been favored for this purpose because of its wealth of prefixes and suffixes.

Neither have many pure Latin anatomical terms survived the times aside from some of the particularly earthy ones like **penis** and **testis.** The former came from **pendere,** "to hang down." But, like the Elizabethans, the Romans had other choice appellations for this organ; *e.g.,* cauda (tail); radix (root); ramus (branch); gladius (sword); clava (club); and vomer (the plow). Testis very readily translates into "a witness," or one who testifies or gives testimony. Under Roman law no one was acceptable as a witness unless his testicles were present. Thus, testis also came to mean one who was present, or one who voted. Among other things, "testa" meant a shell, and shells were used for voting. And, so, in a roundabout way the English "blackball" came into being as a means for voting "nay."

We must not suppose that the classical Greeks of Hippocrates' time were true anatomists. Human dissection was frowned upon and not permitted on the homefront. It was not until 332 B.C., after the widespread conquests of Alexander the Great, that the first medical school permitting human dissection was established at Alexandria in Egypt under Greek dominion. The Greeks apparently had no qualms about dissecting abroad, for a number of them flocked to Alexandria. Among them was Herophilus, an avid investigator of the human body by dissection and a teacher at the school. If anyone deserves the title of "Father of Anatomy," it is probably he. But by 152 A.D. when Galen, another great Greek physician, arrived at Alexandria, anatomy there was quite defunct. Galen was already famous as a comparative anatomist, having secretly dissected many forms of animals on his father's farm in Greece, and he had hoped to round out his studies by dissecting the human. All he found at Alexandria were two old human skeletons with which he improved his knowledge of osteology. Later he constantly sent his students in Rome and Greece to Alexandria just to see these two rare specimens, the only fully-assembled human skeletons in existence in the world. At no time in his life did Galen dissect a human body. By 200 A.D. the libraries of Alexandria were ordered burned by Julius Caesar. While the Romans did not seem to mind the gory shedding of blood in the gladiatorial arenas, they were quite squeamish about the discipline of human anatomy which they regarded as outright sacrilegious.

Shortly thereafter, with the overthrow of Rome by the barbarians, medicine and anatomy went into limbo during the Middle Ages. Whatever specific terminology in these disciplines survived was to be found in the writings of the monks, mere transcribers of the world's knowledge who were more concerned with magic, herbs, and drugs than with morphology.

The reawakening in medicine occurred between 800 and 1000 B.C. when the varied Arab tribes finally merged into a unified nation and began establishing new seats of learning. They were hardly the original investigators after the mold of Herophilus or Galen, but they were avid restorers of former Greek and Roman writings and teachings. In time, they became the teachers in the new universities that were springing up toward the close of the Middle Ages in Spain and Italy from 1200 to 1350 A.D. It was not until 1315 at the University of Bologna that Mondino, an Italian, resurrected the practice of dissecting the cadaver as a prescribed training in a medical school after a lapse of some 1700 years. During the Renaissance from 1350 to 1600, anatomy made its great comeback under such men as da Vinci, Sylvius, and Versalius in Italy and France. Medical schools gradually fluorished in England, Holland, and Germany, as well, and each of these European nationals influenced our anatomical nomenclature in their own inimitable ways.

Having been put through these varied historical wringers, it is not difficult to see why our present-day terminology has become so complex. There are a few surviving Arabic terms such as **nucha** for the spine or spinal cord. When we meet the term as **ligamentum nuchae,** the supraspinous ligament in the neck, we are apt to mistake it for a genitive form of Latin origin. Probably more mathematical and chemical terms than purely medical ones hang over from the Arabic period; *e.g.,* **algebra, alkali,** and **aniline.** With the Norman invasion of England, Anglo-Saxon vocabulary became sprinkled with terms of French origin and the French, of course, had altered Latin terminology as much as the Romans had altered the Greek. We see the French influence in terms like **cretin** and **migraine.** The first is a pure French concoction from Chretien, meaning "a Christian," but in a most derogatory way. It described a group of persecuted Christians who had settled in the remote valleys of the Pyrennes to avoid harassment. Because of iodine-deficient diets, isolation, and intermarriage, they had also raised a crop of deficient children, stunted in body growth and dimwitted. Migraine, on the other hand, seems to be Frenchified Latin for hemicranium and referred to a type of one-sided headache.

Fortunately, we inherited few words from the German, or we would all find ourselves jumping into one side of the Atlantic and coming up on the other with an umlaut in our mouths. But we did latch on to **anlage** as a developmental precursor or the beginning site of some future structure. Moreover, within the memories and experiences of your authors, almost every anatomy department in this country had its **Diener,** or servant. His status within the department could be high or low depending upon whether he merely lugged bodies up from the morgue and mopped the dissection room floors or whether he was a curator and preparator of exquisite anatomical preparations and models and frequently a better dissector than any of the professors.

Despite all this, it is still not difficult to locate words in common usage whose exact parentage is questionable. Strangely, **anus** is one of them. At first blush, it would appear to be almost straight-forward Latin. But it is argued that it comes from the Sanskrit, "as", meaning "to sit." Add another "s," and it all begins to make sense. Others of equally dubious derivation include: **abdomen, clitoris, coccyx, ileum, omentum, retina, saphenous,** and **thymus,** even though we know very well what *we* mean when we use these terms.

Probably one factor that has plagued anatomical and clinical jargon more than anything else has been the prolific use of **eponyms** (Greek, *eponymos*—named after; from *epi*—upon, and *onoma*—a name). These refer to the use of a man's name, usually the discoverer or first describer, for the name of a structure or a phenomenon; *e.g.,* Köllicker's columns; Babinski sign; Broca's area; Sylvian fissure. Taken out of historical context, they have little immediate meaning. International committees on nomenclature have striven to remove most of them as acceptable forms of usage in scholarly anatomical reporting. Still, they hang upon the lips of teachers, clinicians, and students everywhere. Like old soldiers, some of them never die, and the ones that do eventually pass away are frequently replaced in the rank and file by others. A few, like

Caesarean (cesarean) and Roentgen, are here to stay. The purists among us have a point in decrying their use.

On the other hand there is something to be said in their favor. They honor our forefathers and make us aware of our anatomical and clinical heritage. They are the bas-reliefs on the timeworn sculptoring of anatomical and medical history. To recognize some of these figures, eroded by the centuries as they may be, is often essential at times to be able to understand what is being discussed in the clinic or even in some journal reports. If nothing else, an occasional bit of familiarity with these figureheads as their names are dropped from time to time makes us aware that there were giants in our fields of interest in a former day who made amazingly shrewd deductions with far less technology and implementation than we have at our command today.

We should caution that there is some danger in knowing them too well, for familiarity may breed contempt. They were all human beings first and scientists secondarily. They were not infallible giants, though a number of them would have liked posterity to have regarded them in this light. There were many plodders who kept their noses to a single grindstone until they fashioned something reasonably monumental. Others were of the order of the genius, making brilliant discoveries in diverse fields. The contributions of still others were almost serendipitous; some, of minor consequence.

They varied in personality, as well. Some were so kind, humble, and generous that they were beloved by students, colleagues, and peers. But there were unsavory, bombastic, egotistical types who would brook no criticism as they attempted to ram their single-tracked ideas down the throats of all opposition. There were cunning, conniving, political schemers and big operators who robbed underlings of the fruits of their labor and who climbed over the backs of everyone to reach posts of esteem. There were even frank charlatans who sold a bill of goods woven out of whole cloth. Unfortunately, we still peddle some of their ancient and questionable contraband under one guise or another even today.

Indeed, for all the fuss about the nuisance of eponyms, they hold a mirror to ourselves. The individuals so honored differed not one whit in the worthiness and tactics of achievement from our current polyglot crop of scientific workers in our anatomical and medical vineyards. We can only profit by emulating the most honorable of them and by shunning the ways of the transgressors against human decency, for they all labored in the name of science and discovery.

Specific Terms

Terminology has two functions: (1) to give names to structures so that they may be identified readily and distinguished from all others; and, (2) to describe relationships of structures clearly by means of terms of location, position, and direction (or plane). It is important to use this kind of language accurately in history taking or in writing medical reports, for those who examine the same patient a few days or even a few months later will be interested in looking at the same site in which you reported seeing a lesion, wound, or scar. If no adequate reference points were provided by the original reporter, such observations are often lost and useless for further documentation.

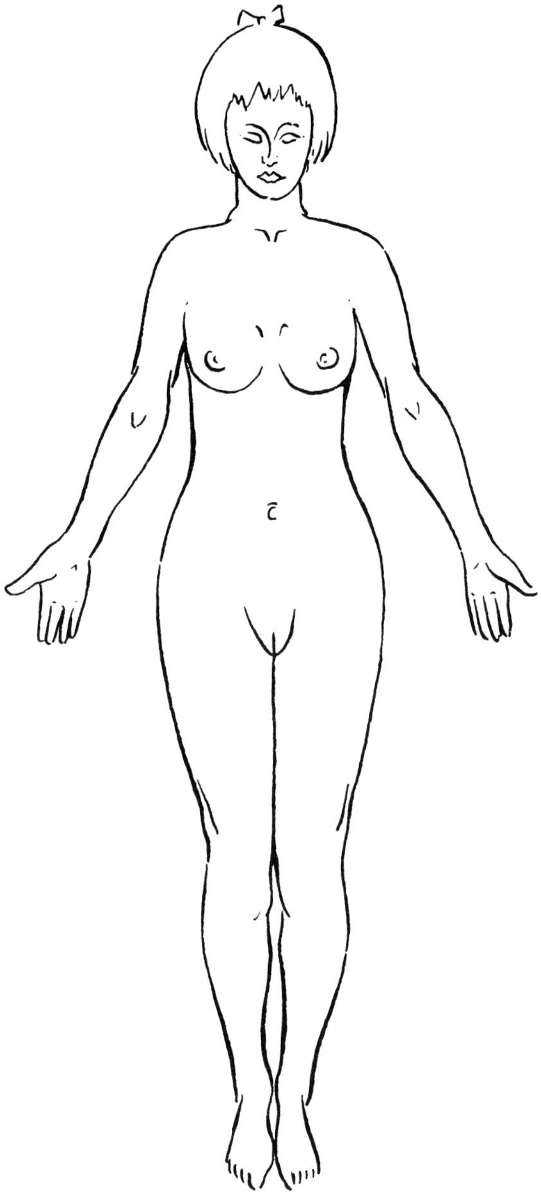

Fig. II-1. The body in the anatomical position.

A. Terms Referring to Location or Position: The following terms assume the body to be in the "anatomical position" (Fig. II-1), a standard viewpoint of reference with the individual standing erect, face forward, arms at the sides, and palms forward (supinated):

Anterior. The front of the body; the belly-button side; the same as **ventral.**

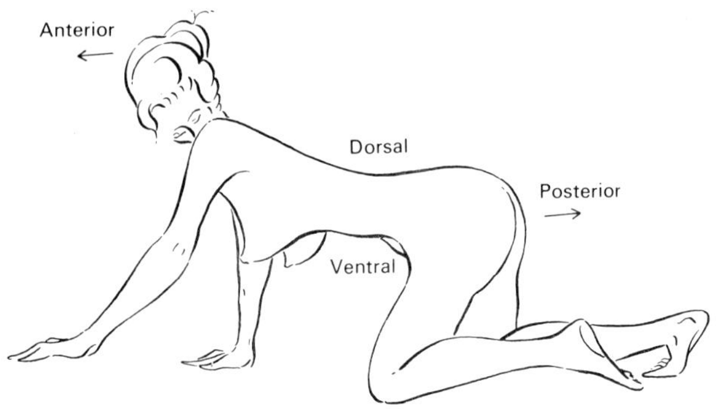

Fig. II-2. The body in the quadrupedal position. Terms shown are applicable to quadrupedal animals only, so that **anterior** and **posterior** are **not** used in this sense with man and other bipeds.

Posterior. The back of the body; the same as **dorsal.** (Were the young lady on her hands and knees in Fig. II-2 a **quadruped,** then **anterior** would refer to the head end; **posterior,** the tail or caudal end; **ventral** would become the underneath or belly side; and **dorsal,** the back or the uppermost side. Anterior and posterior are not used in this sense with man and other bipeds.)

Dorsal. Always means the back side or situated near the back.

Ventral. Always means the belly side or situated near that side.

Superior. Higher; situated higher than some other point or structure (*e.g.*, the thoracic cavity lies superior to the abdominal cavity).

Inferior. Situated lower or below (*e.g.*, the ankle joint lies inferior to the knee joint), though in the limbs the terms **proximal** and **distal** will be used more than "superior" and "inferior").

Superficial. On or near the surface.

Deep. Remote from the surface; or, more remote than some other structure that lies nearer the surface than it (*e.g.*, the Soleus muscle lies deep to the Gastrocnemius).

Internal. Within, inside.

External. Without, outside.

As a rule, however, **inside** or **interior,** and **outside** or **exterior** are used when referring to the body cavities or the hollow organs (*e.g.*, the stomach is lined on its inside or interior by a glandular mucosa and clothed on its outside or exterior by visceral peritoneum).

When **internal** and **external** refer to the rotation of the limbs, the terms denote a turning movement toward or away from the midline of the body; toward the midsagittal plane (**internal rotation**) or away from it (**external rotation**).

Proximal. Nearest the trunk for limbs or nearest the point of origin for a structure (*e.g.*, the renal arteries arise from the aorta proximal to the ovarian arteries).

Distal. Farthest from the trunk for the limbs or farthest from the point of origin of a structure (*e.g.*, the phalanges lie distal to the metacarpals).

Medial. Relating to the middle or center; nearer the midsagittal plane (*e.g.*, the thoracic duct lies medial to the azygos vein in the right inferior part of the thorax).

Lateral. Situated to the side; farther from the midsagittal plane (*e.g.*, the lateral malleolus is on the little-toe side of the ankle).

B. Terms Referring to Direction:

Rostral (or Rostrad). Toward the nose or beak end.

Cranial (or Craniad). Toward the cranium.

Cephalic (or Cephalad). Toward the head end of the body.

Caudal (or Caudad). Toward the tail end; away from the head end.

Mesad (Mesiad). Toward the midline of the body or a structure.

Laterad. Away from the midline of the body or a structure.

C. Terms Referring to Body Planes: For the purposes of study or description, a body or an organ may be cut into **sections** that lie in various **planes.** Basically, there are four planes commonly mentioned:

Midsagittal. A plane that sections the body into equal or perfect right and left halves; a midline section along any longitudinal axis (like the young lady being attacked a lá Lizzie Borden in Fig. II-3).

Parasagittal (or simply, sagittal). A vertical plane sectioning the body parallel to the midline but obviously dividing it into unequal right and left parts. Many such planes are possible.

Transverse (or cross-sectional). A horizontal plane at right angles to the long axis of the body, dividing it into superior and inferior portions. Often simply called a horizontal plane (Fig. II-4).

Coronal or **Frontal Plane.** A vertical plane at right angles to the midsagittal, dividing the body into anterior and posterior portions (Fig. II-4). The young lady in this figure *was* the wife of an infuriated, circus knife-thrower.

Of course, organ structures can also be sectioned in these planes, generally with the long axis constituting the sagittal. The term **longitudinal** often replaces sagittal and, hence, **median longitudinal** substitutes for midsagittal. Any slanting cut is called an **oblique section.**

Medial and **median** are often misused as if they were interchangeable. Median signifies "in the middle of," while medial means that a structure is situated nearer the middle than one farther from the midline.

D. Terms Referring to Specific Areas of the Body: Body structures are related to certain regions or divisions of the body within definitive boundaries such as the **head, neck, thorax, abdomen, pelvis,** and **the limbs.** The thorax and abdomen together constitute the **trunk** or **torso.** The following are some terms used in connection with these regions:

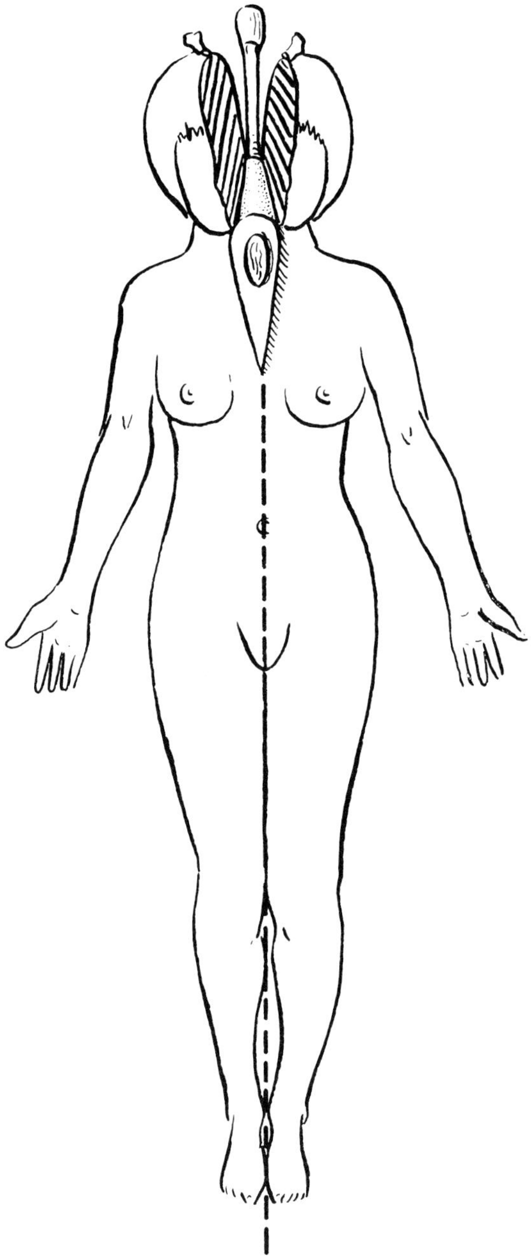

Fig. II-3. Body sectioned in midsagittal plane.

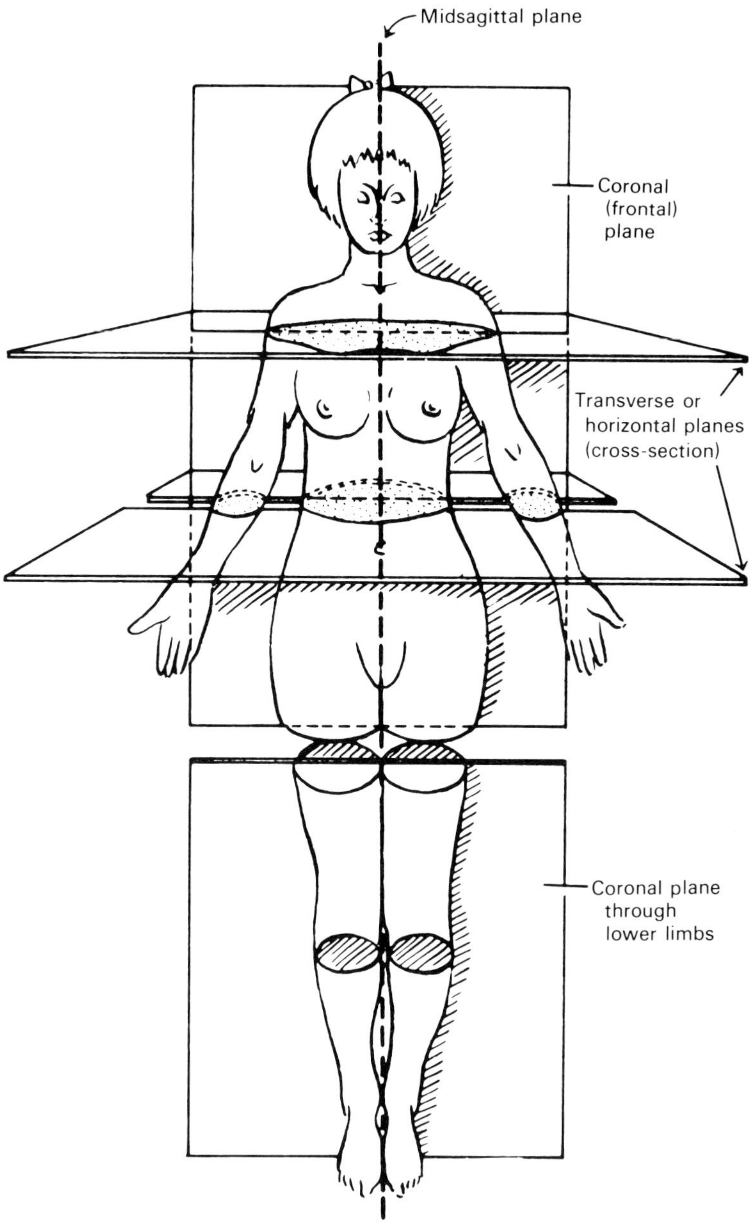

Fig. II-4. The body planes.

Axial Skeleton. The skull and vertebral column, sternum and ribs.

Skull. The bony skeleton of the head, situated above the cervical vertebrae.

Face. Visage; that part of the head facing forward and lying below the orbital ridges. Fourteen bones of the skull contribute to the skeleton of the face.

Cranium. Those bones of the skull which house and protect the brain (of which there are eight).

Calvaria (Calvarium). The superior aspect of the skull, forming the skull cap (*calva*—hairless head).

Neck (Column). This may be defined as the region extending from the lower border of the mandible to the clavicle and from the base of the skull to the 7th cervical vertebra, or vertebra prominens. The neck contains, among other things, the esophagus, larynx, trachea and thyroid and parathyroid glands.

Appendicular Skeleton. Shoulder and hip girdles and the attached bones of the upper and lower limbs (members; extremities).

Shoulder Girdle. Clavicle and scapula.

Pelvic (Hip) **Girdle.** The two innominate bones (os coxae).

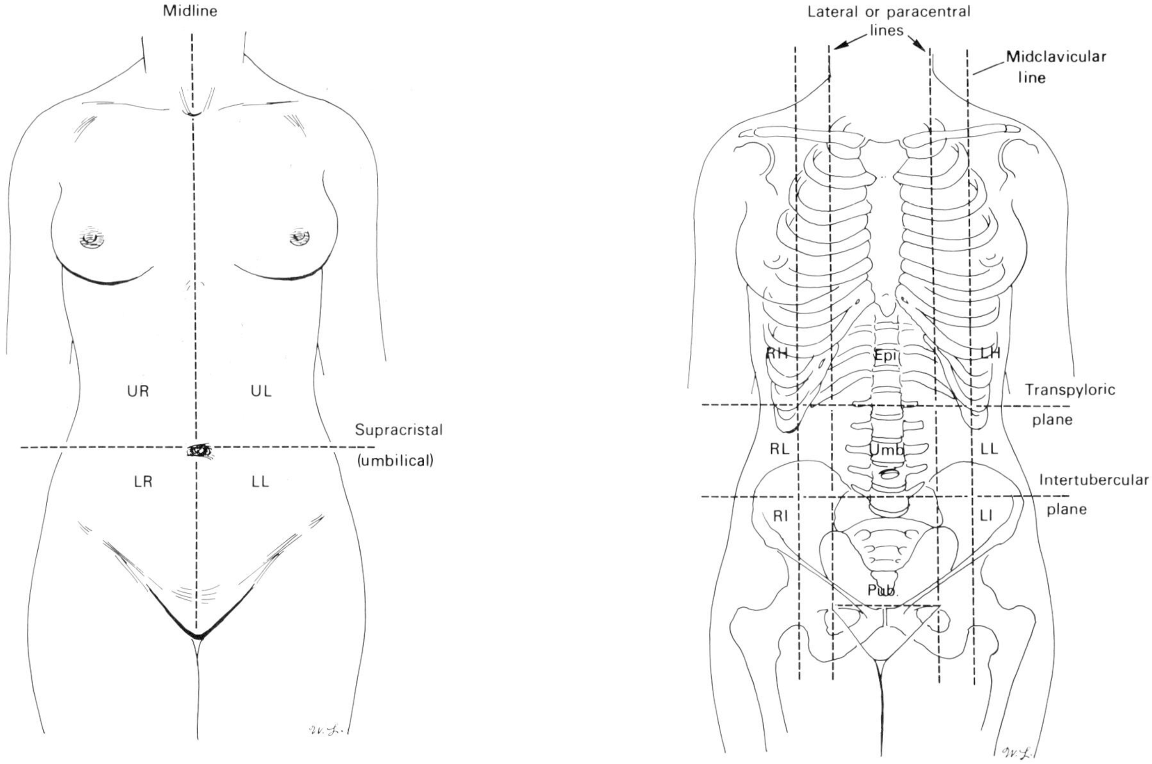

Fig. II-5. Planes, lines, and zones of the abdomen. **A.** The abdomen in quadrants: UR, upper right quadrant; UL, upper left quadrant; LR, lower right quadrant LL, lower left quadrant. **B.** Conventional zones of the abdomen: RH, right hypochondriac zone; LH, left hypochondriac zone; RL, right lateral (formerly lumbar) zone; LL, left lateral zone; RI, right inguinal (formerly iliac) zone; LI, left inguinal zone; Epi, epigastric region; Umb., umbilical; Pub., pubic (formerly hypogastric) region.

Arm (Brachium). Upper limb extending from shoulder to elbow.

Forearm (Antebrachium or Antibrachium). That part of the upper limb extending from elbow to wrist.

Volar. Of or pertaining to the palmar (flexor) surface of the forearm, wrist, or hand; the anterior surface of these regions when the body is in the anatomical position. Rarely, it is used pertaining to the sole of the foot.

Thigh. Part of the lower limb extending from hip to knee.

Leg. Part of the lower limb extending from knee to ankle.

Plantar. Sole of the foot.

The abdominal region is often subdivided for convenience of localizing structures related to it, either into quadrants or into nine regions by special planes and vertical lines (See Chapter III, p. 136). The regions are drawn and the abdomen quartered for you in Figure II-5.

E. Terms Referring to Movements of the Limbs or Parts of the Body: Since anatomy is the study of the living and not the dead, it is important to think in terms of the living body and its movements. All movement, of course, takes place at joints, and the relation of structures at these joints is the clue to understanding movements that are possible at these joints. We shall consider some of these relationships in future chapters. For the moment, we are interested only in understanding the terms that describe types of movements.

Flexion is bending, often more strictly, ventral bending, which descreases the angle between parts of the body. It can also mean the condition of being bent. The young lady in Figure II-6A sits with her neck flexed and her trunk flexed upon the thigh, while her right thigh is flexed upon the trunk. Flexion at her elbow is reversed from the position of flexion at her knee. The flexor muscle groups clothe the anterior aspect of the upper limb but, because of the rotation of the lower limb, in the process of development, they clothe the posterior aspect of the lower extremity. Her fingers are flexed about her right leg, but the entire left lower limb, including ankle and toes, is held in a position of **extension.**

Extension is straightening or stretching out, increasing the angle between parts of the body (Fig. II-6B). This is a muscle movement by which the flexed part is made straight or nearly straight. A notable exception to this is **dorsiflexion** as applied to the foot (Fig. II-7). In this case the ankle is actually flexed with the toes in extension as in rocking back on the heels. Extension of the foot, as in standing on tip-toe, is called **plantar flexion** (Fig. II-6B). Curling the toes, of course, is flexing them.

Abduction is a movement away from the midsagittal plane, as the right arm and leg are moved in the ballet routine shown in Figure II-8. Raising the arm laterally abducts the shoulder joint (Latin, *ab*—away from); returning it to the side is **adduction** (Latin, *ad*—toward); *i.e.*, bringing it back toward the midaxis of the body (Fig. II-9A). Abduction and adduction of the fingers and toes are movements away from and toward the median planes of the hand and foot; *i.e.*, the middle finger of the hand and the second toe of the foot (Fig. II-9B).

Circumduction is a movement in which the distal end of a part is free to describe a circle and the part, itself, describes the sides of a cone as it moves about its joint. Such a movement is possible at the ball and socket joints of the shoulder and hip so that the entire upper or lower limb can engage in circumduction. The movement can occur as well at the rather freely movable saddle joint of the thumb; *i.e.*, at the first carpometacarpal joint between metacarpal I and the trapezium (Fig. II-10D). In the process, the movement combines flexion, extension, abduction, and

Fig. II-6A. The body in a flexed posture with the exception of the left lower limb which is in extension.

B. A beautiful example of extension and plantar flexion. There is extension of the neck, entire trunk and the upper and lower limb.

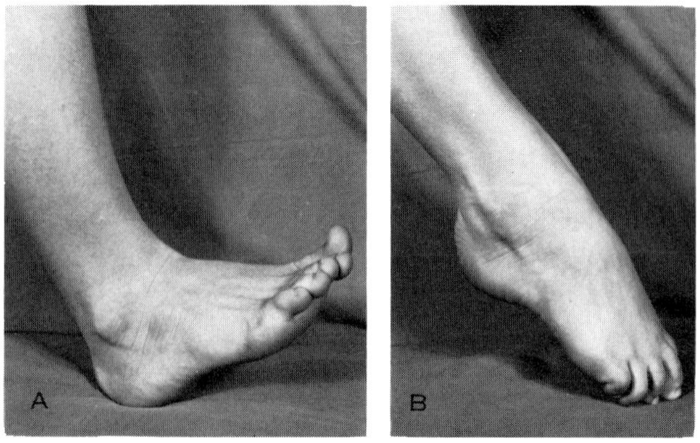

Fig. II-7. Movements at ankle and toes.
A. Dorsiflexion: ankle flexed, toes extended.
B. Ankle in extension, toes flexed (plantar flexion).

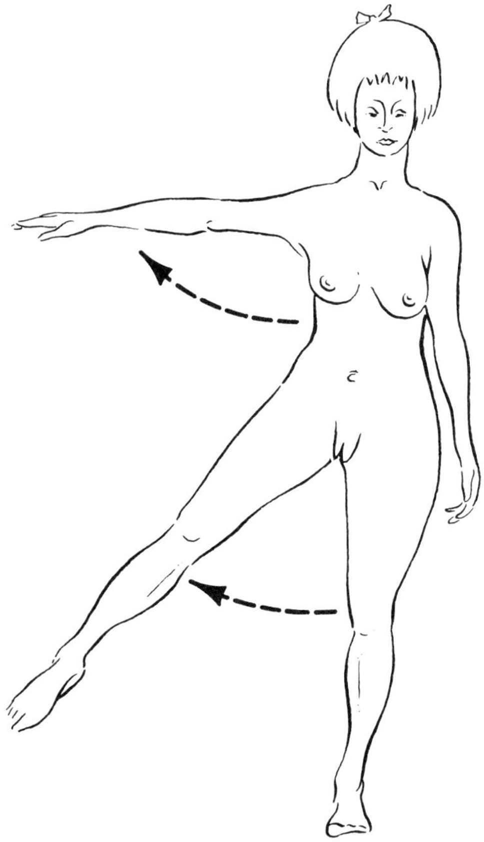

Fig. II-8. Abduction of the right shoulder and hip joint.

Fig. II-9. Abduction and adduction.
A. Arrows show movements of extremity in abduction (AB) and adduction (AD) of the shoulder joint.
B. In abduction and adduction of the fingers *arrows* show movement away from or toward the middle finger (median plane of hand).

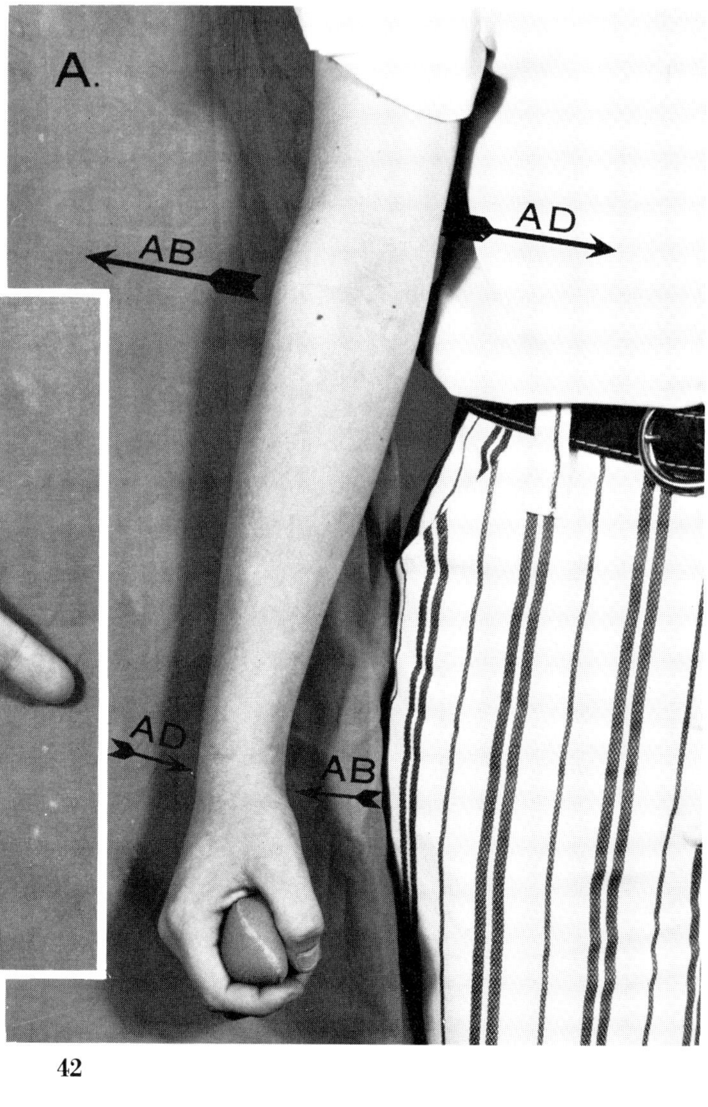

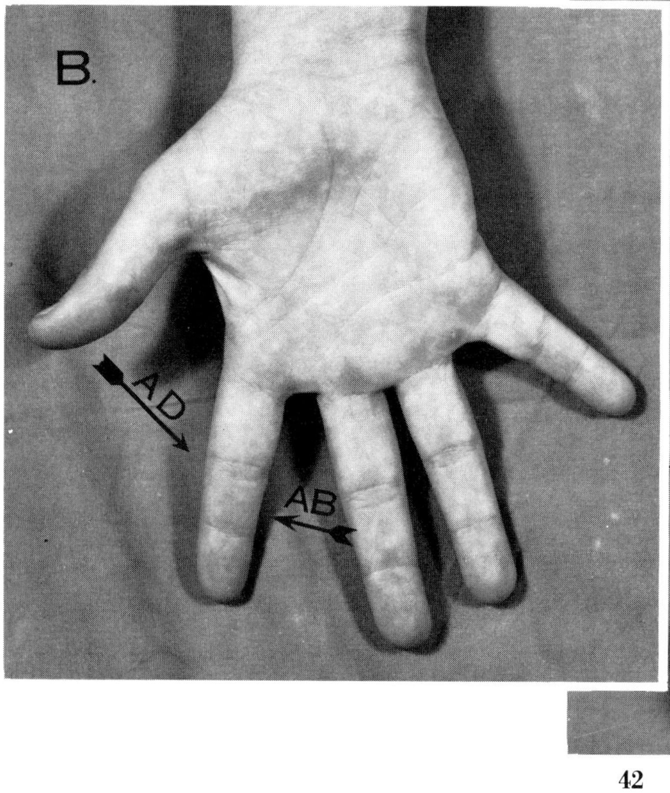

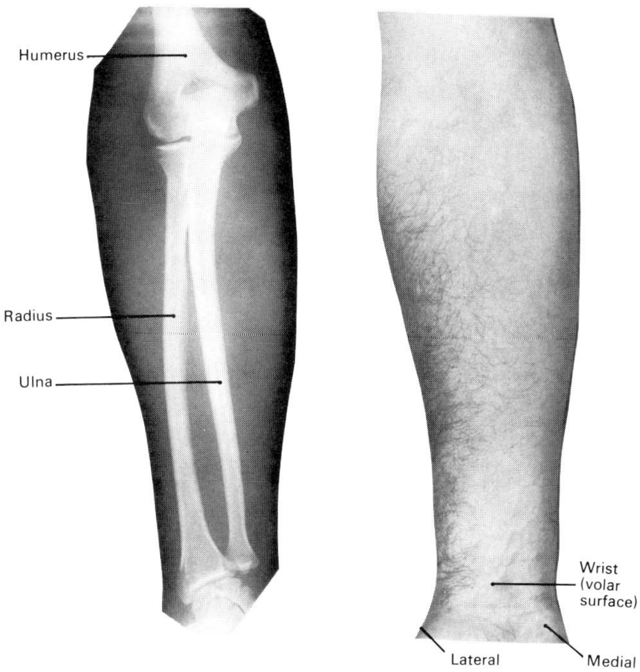

Fig. II-10A. Supination. Radiograph on the left with a matching photograph of the subject's arm on the right.

adduction in a smooth continuum. Waggle your own thumb about in circumduction to verify this.

Rotation is also a movement of a part about an axis but without displacement, as the atlas (1st cervical vertebra) turns upon the **axis** or **epistropheus** (2nd cervical vertebra). We have mentioned above the **internal** and **external rotation** of the limbs. Rotation of the foot which turns the plantar surface outward or moves it laterad is called **eversion;** the opposite movement, facing this surface inward toward the midline of the body, is **inversion** (Fig. II-11).

Supination is a movement which orients the palm and the hand cranialward when the elbow joint is flexed, or faces the palm anteriorly when the body is in the anatomical position (Fig. II-10A, C). Turning the palm dorsally from the anatomical position, or facing the palm caudalward with the elbow flexed, as when placing the palms flat upon a table, constitutes **pronation.** In this movement the radius of the forearm crosses the ulna diagonally, moving the thumb medialward (Fig. II-10B, D). When the hands of the Christian are raised in supplication, the forearm and hands are in supination. When the Moslem prostrates himself before his God, the forearms and hand are in pronation. The body lying on its back is **supine;** on its belly, **prone.**

Opposition is a movement by which the thumb pad is brought to a finger pad and held there. This is a most useful movement in man and one in which the ape with his unopposed thumb and man with a median nerve injury cannot engage. It is employed in holding the pen in writing, buttoning shirts, raising cups by their handles, and, in France and Italy, for tweaking the derrière of females *en passage*. Thus, most delicate

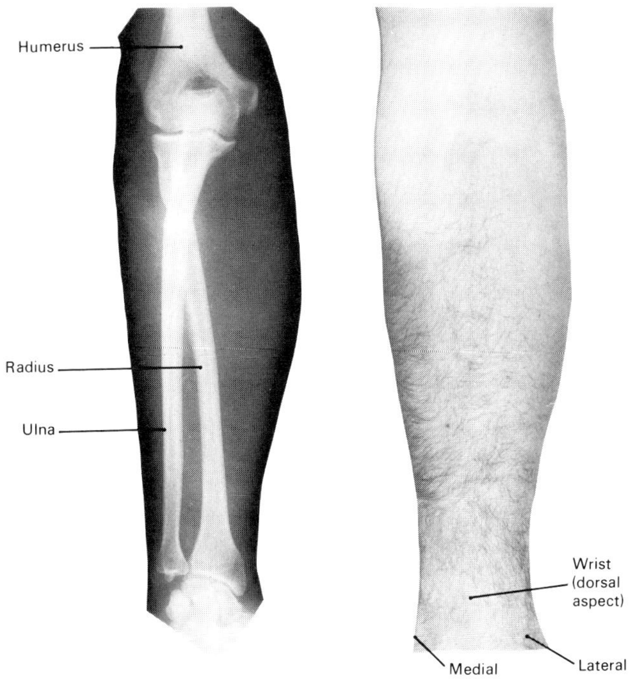

B. Pronation. The hand is in pronation with the radius crossing the ulna as it does in this movement. Radiograph and photograph are matching views as in part **A**.

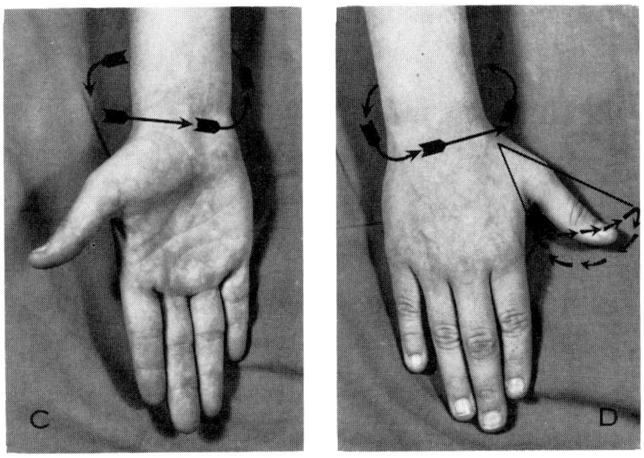

C. Supination of the hand shows the palm facing forward in the anatomical position. The arrows indicate the direction of movement of the radius as it crosses the ulna in going from supination to pronation.

D. Pronation of the hand is accomplished by partial rotation at the wrist. The carpal bones and the hand will follow the movement of the radius since the ulna does not articulate directly with the carpal bones. **Circumduction** of the thumb is accomplished by fixing the interphalangeal joint and rotating the thumb about the metacarpo-carpal joint (in the direction of the arrows.)

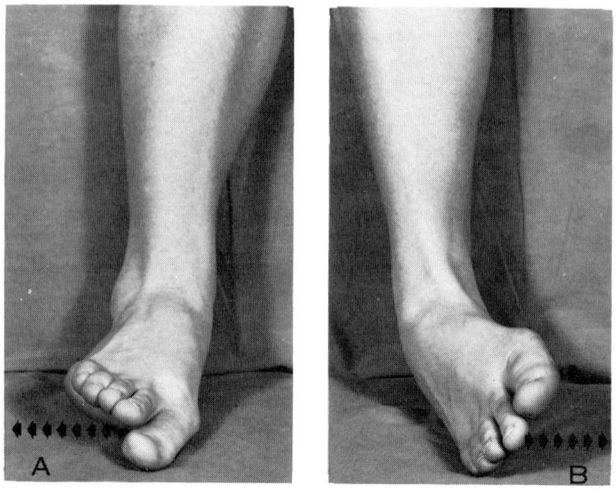

Fig. II-11. The right foot in eversion (**A**) and inversion (**B**).

as well as indelicate acts of the thumb involve opposition along with adduction.

F. Possible Movements at Certain Joints:

1. **Spine.** Rotation, flexion, extension.
2. **Shoulder.** Flexion of arm at shoulder; extension of humerus (as when pulling the handle of an oar toward you in rowing); rotation of the humerus internally or externally (as in using a screwdriver); abduction and adduction (as in raising or lowering the arm laterally; circumduction (as in describing a circle with the hand while the arm is extended and abducted to the horizontal).
3. **Scapula.** Rotation with arm abduction to raise the arm over the head; elevation and depression; protraction and retraction. In **protraction,** or in drawing the shoulder forward, the vertebral borders of the scapulae are drawn away from each other; the inferior angle is turned obliquely lateral as the scapula itself slides laterad along the posterior rib cage; and the glenoid cavity is tipped upward. In **retraction,** the shoulders are drawn backward or "squared" in military fashion. The vertebral borders of the scapulae approximate the midline.
4. **Elbow.** Flexion and extension.
5. **Wrist.** Flexion and extension; ulnar and radial abduction. Abduction and adduction of the wrist relate to the ulna or radius as a reference point. Thus, in the anatomical position, **ulnar abduction** means movement of the wrist away from the ulnar side.
6. **Fingers.** Flexion and extension; abduction and adduction from the medial finger. The thumb and index finger are capable of circumduction; the others not as easily. We have mentioned the role of the thumb in opposition.
7. **Hip.** Flexion and extension; rotation internally and externally of the femur; abduction and adduction of the thigh; and partial circumduction (as in the gait of the unilateral paraplegic who swings his stiffly-extended lower limb in a partial arc in locomotion—circumducted gait).
8. **Knee.** Flexion and extension; partial rotation (as the femur rotates medialward on the tibia when the knee is bent).
9. **Ankle.** Dorsiflexion and plantar flexion; eversion (lateral rotation, abduction, fibular adduction); and inversion (medial rotation, adduction, or tibial adduction).
10. **Toes.** Flexion and extension; abduction and adduction with the second toe as a median axis.

G. Other Terms: Invaginations or **evaginations** (Latin, *vagina*—a sheath or scabbard) are inward or outward bulgings or pocketings, respectively, of the wall of a cavity or from a surface.

Ipsilateral or **homolateral** refer to the same side of the body (as the right arm and leg are ipsilateral structures). **Contralateral** refers to the opposite side of the body (as the left arm is contralateral to the right leg).

Terms such as **over, under, above, below,** and **on** can get us into some trouble if we are not careful in their use. They serve best in a general sense without specific regard to anatomical position. But they can be used with some judiciousness and whenever it is quite obvious what we mean. Just as a flea may be *on* a dog's back or leg, so can a tuberosity be *on* any bone or a gyrus *on* the surface of the brain. A nerve can pass *under* or *over* or *through* some other structure, or lie *beneath* a particular muscle or *under* the deep fascia. In this sense these terms have some specificity and meaning. But what do you conclude when you say the transverse colon lies *under, below,* or *beneath* the stomach? Do you mean *inferior* to it, or *dorsal* to, *deep* to, or *behind* the stomach?

Muscle Attachments: Since muscles attach at both ends it is important to establish which end you wish to describe. The proximal attachment of a limb muscle, for example, is its **origin;** its distal attachment, its **insertion.** Another logical approach would be to refer to the fixed end as the origin; the moving end, the insertion if they were constant. But, they are not always constant and can be reversible. When we chin ourselves on a limb, the Latissimus Dorsi draws the humerus toward the body, but when lifting weights it draws the body toward the humerus. In walking, the right foot is stationary on the ground while the left is advancing. The condition is reversed when the next step is taken. Thus, origins and insertions of the lower limb muscles alternately are fixed and moving.

Blood Vessels: Anatomists compare arteries to the **branches** of a tree; veins, with the **rivers and their tributaries.** Before the time of William Harvey and his thesis that the blood is propelled in a circuit through the body, both arteries and veins were described as having branches. It is still done today and constitutes only a minor offense against which no penalty is ever assessed. The terms **proximal** and **distal** refer to the heart when we talk about arteries, but while these same terms use the heart as a reference point, when we speak of veins, direction is reversed. The more distal one drains into the venous system, the closer to he heart he gets.

It becomes increasingly evident from the foregoing that to describe the relationship of one structure to another, accepted anatomical terminology should be used. Only when you make yourself precisely clear can you avoid being misunderstood. Put the terms to frequent use and in proper context. Only by practice will you realize how much rapid progress you have made with the language of anatomy. Your skill in its use begins with your introduction to some 10,000 to 15,000 new terms in the first 2 years of medical training alone, and the importance of acquiring skill in their use cannot be overemphasized. This is precisely the reason your authors chose to include this chapter in this particular portion of the text.

Nomina Anatomica: Systems of anatomical nomenclature have come and gone since the first attempt was made in 1895 to construct a universal language for anatomists the world over. At that time, the German Anatomical Society met at Basel, Switzerland, to arrive at a gentlemen's agreement called **BNA** (Basle Nomina Anatomica). Terminology was Latinized because that seemed to be the best approach toward a universal tongue. The system was accepted by most countries, but many of them, as we were prone to do in the United States, soon translated the Latin into their own vernacular. The French would have nothing to do with this system at all, preferring their own version of nomenclature.

Since then, the shorthand designations for systems of nomenclature have been almost as numerous and varied as the "alphabet soup" of our own governmental agencies. There have been **BR,** British Revision, in 1933; **NK,** German Nomenklature Kommission, in 1935; and the **JNA,** Jena Nomina Anatomica, of 1937. In general, British systems held to Anglicized terminology and referred to the human body in the anatomical position. The Germans preferred a universal, pure, scientific language (Latin) and wished to accommodate the comparative anatomist as much as the human anatomist.

The latest revision occurred in Paris in 1955 when the International Congress of Anatomists approved a revision of anatomical terminology and a classification of parts. The list has been revised twice since, once in 1960 and again in 1965. This compromise agreement, known as the Nomina Anatomica, or **NA,** is supposed to be the current tongue of all official, respectable, union anatomists.

Someone is sure to notice that we do not adhere to it wholeheartedly. It is not that we are not dues-paying members that we depart from the party line. Basically, we are conformists and Milquetoasts on every major issue—social, political, and anatomical. But this book was born in the South where the strains of "Dixie" can still arouse a rebel attitude in the mildest of human beings. Despite miniskirts, it sounds indecent to call the limbs "the members," and we can't quite bring ourselves to call the auditory meatus "the porus." Neither can we visualize bones like the nasal, vomer, lacrimal, and inferior conchae as being parts of the cranium since they have so little to do with the constitution of the brain case. Other bits of the new terminology stick in our craw. The terms sound frankly foreign to our true-blue American ears. If it makes our dialogue saltier and our meaning more facile, we have no compunction against slipping into **BA,** a kind of Bastard Anatomica which is a concoction of the past and the present. In all of American anatomical history there seems to be a precedent to do so, pulled as we have been between our British and German colleagues.

PART II—GENERAL CONCEPTS

General Osteology

The **skeletal system** consists of the **bones** of the body and their related **cartilages.** Isolated bones are useless structures unless there is some provision to tie them together in meaningful relationship. Hence, we offer no objections if anyone cares to include the **joints** or **articulations** as integral parts of the skeletal system. Osteology is the study of bones; arthology, the study of joints (Greek, *osteon*—bone; *arthron*—joint). Sometimes the Latin **os** is used in the names of bones as **os coxae,** the hip bone. It should be obvious from whence the term **osseous** was derived.

Unlike the Arthropods, we do not wear an **exoskeleton** like a tough, outer skin which will not grow as our soft parts grow and which must be shed periodically. Instead, we clothe our **endoskeleton** with our soft parts. Hence, bone must be an adaptive structure, capable of growing as we grow, able to change with stress and to display regenerative potential when injured.

The functions of the skeletal system are: (1) to provide support and protection; (2) to afford a mechanical basis for movement; and, (3) to play a role in blood formation. As muscles attach to bone and cross joints, they work as levers or act to fix parts involved in movement. Also, muscles chiefly round out the body contours so that, along with the skeleton, they are responsible for body form. Certain bones, such as those forming the brain case and vertebral column, the rib cage and the bones of the pelvis, furnish protection for many of the soft organs. Bone marrow at the ends of long bones and in the diploe of flat bones are sites of development of red cells and the granulocytic white cells of blood. It could be questioned as to whether structural support and protection are the main functions of bone. They might well take second place to that of providing a mineral reservoir. While it would be difficult to imagine a jellyfish of a man without his skeleton, it is more difficult still to conceive of functional metabolic activity in man without extra supplies of calcium and phosphorus on hand at all times.

Classification of Bones According to shape:

1. **Long Bones,** such as those of the limbs. *Examples:* Femur (thigh); humerus (upper arm). Oddly,

such short-appearing bones as the **phalanges** are considered as miniature long bones according to shape.

2. **Short bones** which are six sided, as a rule. *Examples:* Carpal (wrist); tarsal (ankle).

3. **Flat bones.** *Examples:* Bones of cranial vault; ribs and sternum.

4. **Irregular bones.** *Examples:* Vertebrae; scapulae; the hip bones; *i.e.,* os coxae or innominate bones.

5. **Sesamoid bones,** shaped "like a sesame seed." These lack a periosteum. *Examples:* Certain tendons bear flattish or nodular pieces of bone especially where they cross joints. The patella can be considered to be the largest of this type of bone. Its shape is hardly that of a sesame seed. It was named because of its resemblance to a flat frying pan (without the handle). It is, however, embedded in tendon. Figure II-12 illustrates an example of each of these types.

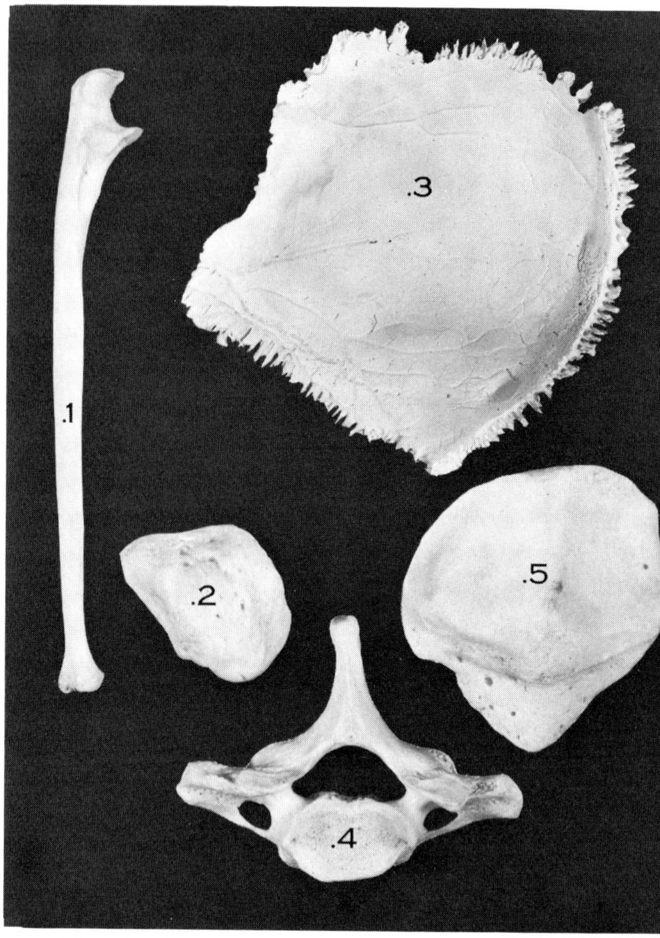

Fig. II-12. Bones classified by shape. 1, Long bone (ulna); 2, short bone (carpal bone—pisiform); 3, flat bone (parietal); 4, irregular bone (7th cervical vertebra); 5, sesamoid bone (patella). (Magnification relative and not to scale.)

Vertebrae are irregular bones. We will deal with specific examples of them later. Here we should consider the general make-up of a typical vertebra. There are 33 vertebrae, though when we list them under the axial skeleton later they will appear to tally only to 26. This is because the five sacral vertebrae are fused into the single sacrum. This fusion occurs by the 23rd year, but a gap may exist between the 1st and 2nd sacral vertebrae until the 32nd year.

Parts of a Typical Vertebra: The typical vertebra consists of the following parts (Fig. II-13A and B):

1. A weight-bearing **body.**
2. A **vertebral arch** that protects the spinal cord.
3. Three levers for muscle attachment; one **spinous process,** posteriorly; and two lateral **transverse processes.**
4. Two **articular processes**—each with one superior and one inferior articular facet.

The term **pedicle** refers to the part of the arch between body and transverse process.

There will be variations in shape of the vertebrae, length and slope of spines, nature of transverse processes, and locus of articular facets and their shapes. These differences can be considered later when specific vertebrae are encountered.

Glossary of Bony Parts: Bones are by no means smooth and even in contour. They display a variety of bumps, depressions, and holes. Such surface features have attracted attention, and many of them have been given names, sometimes quite fanciful ones. The following alphabetized glossary may help you to distinguish between them, a task that always cannot be accomplished distinctly.

Condyle. The exact meaning is a knob. The exact spelling in English is without the final "e," which is Frenchified. It is used to mean a rounded, articular surface.

Crest. A prominent ridge.

Epicondyle. A projection just above a condyle (*epi*—upon).

Facet ("little face," from French, *facette*). A term applied to any smooth surface of a bone, especially where it articulates with another bone. It seems to be applied more particularly where the articulating surface is flat, or nearly so. The term comes from the gem cutters who speak of a small, cut, polished surface of a precious stone as a facet.

Fissure. A narrow, cleft-like passageway (from the Latin verb *findere*—to cleave).

Foramen. An aperture or hole (from the Latin verb *forare*—to bore or pierce).

Fossa. A rather deep depression, but don't bet on it. Some named fossae are quite shallow. The term comes from the Latin verb *fodere*—to dig.

Fovea. A term derived from the same Latin verb as fossa, apparently by a less ardent digger. It signifies a small pit.

Hamulus. A little, hook-shaped projection.

Head. A large rounded articular end of a bone often set off from the main shaft by a neck of bone.

Line. A slight ridge, less prominent than a crest.

Meatus. A canal (from the Latin *meare*—to go or to pass).

Process. A rather widely-used term for a projecting part (from the Latin verb *procedere*—to go forward or advance).

Spine. A pointed process (from the Latin *spina*—a

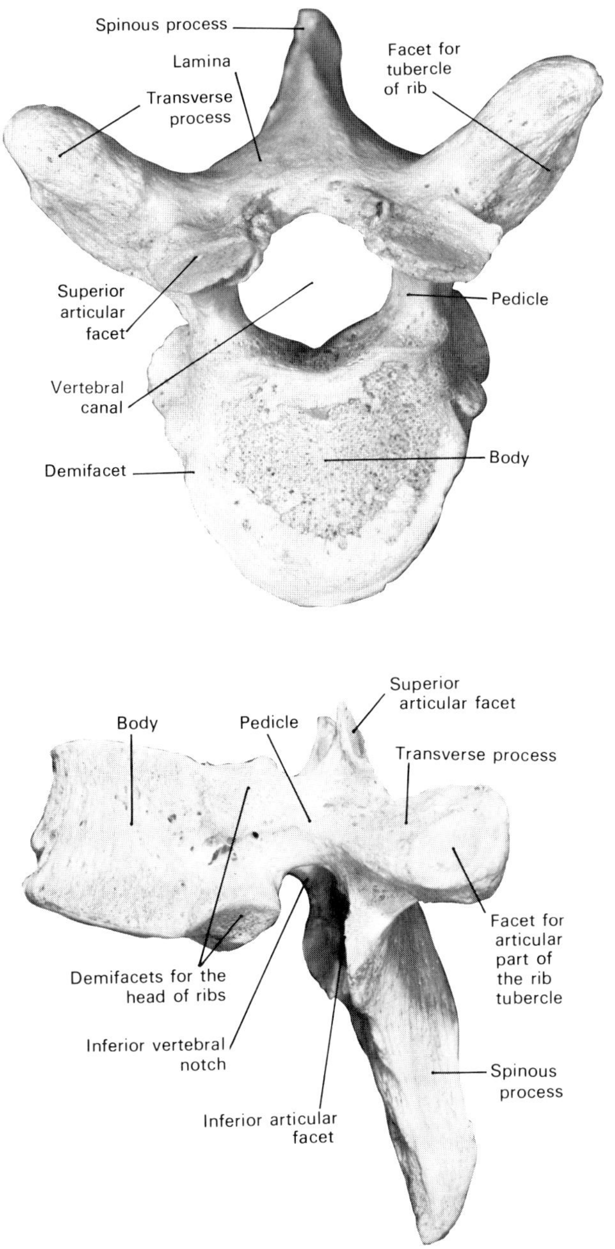

A.

B.

Fig. II-13. A typical vertebra.
A. Viewed from above.
B. Lateral view.

thorn). Not always is this term used with precision. Some named spines are pretty blunt. The term "spine" for the spinal column seems to have even a different conotation, having been named from its fanciful resemblance to the dividing wall of the Roman circus. The top of this wall was decorated with projecting altars and statues, giving it a bristling appearance. The spinal column, which separates the back into two equal parts, has its similar series of irregular, projecting processes.

Trochanter. A large, blunt **process.** There has been a peculiar change in the strict use of this term. It comes from the Greek, meaning "a runner, or wheel." It was first used for the head of the femur which turned like a wheel in the socket of the hip joint. Then it began to be used for **processes** of the femur to which muscles attached, because these *processes* acted as levers to turn the wheel.

Tubercle. A **small** swelling, bump, or protuberance, usually smooth.

Tuberosity. A **large** bump or protuberance, often rough.

Structure of Adult Bone (Fig. II-14A): A longitudinal section of a long dried bone, as shown here, demonstrates that the bony material can be very dense (**compact bone**) or fine, light, and spongy (**cancellous bone**). Where these two kinds of tissue are distributed and in what proportions in any bone will depend upon the kind of stress that is put on it in growth and function. You will note that the compact bone is thin over the articular ends; *i.e.*, at the **epiphysis,** but thick in the wall of the tubular shaft, or **diaphysis.** The epiphyseal end is always capped and strengthened by a glassy layer of **hyaline cartilage** which serves as a tough, smooth gliding surface where bones are likely to rub in joint cavities. Undoubtedly, you have seen this milky-white cartilage at the end of any knuckle bone which you obtained from the butcher shop for your best friend, the dog. In a young, growing bone there is considerable cartilage at the epiphysis. There is also an **epiphyseal plate** that acts as a **growth zone** between diaphysis and epiphysis. Any increase in the length of long bone is added at the expense of the epiphyseal plate. When growth ceases, the cartilage of the plate disappears, and we see a **synostosis.** This is the final fused joint of the epiphysis to the diaphysis, the plate having behaved as a kind of temporary disk in a temporary joint cavity between these two parts of the developing bone. Perhaps this is why epiphyseal fractures are more common among the young than in older individuals where tight fusion of these parts has finally occurred.

The region just distal to the epiphyseal plate nearest the diaphysis of a long bone has been called the **metaphysis.** It is an area of active bone modeling as various stresses are placed upon growing bone. Its extent distalward is rather indeterminate depending upon whether you are conferring with an anatomist or a clinician. Each has his own peculiar ideas about the extent of this region. Yet, the terms **epiphysis, metaphysis,** and **diaphysis** are not to be dismissed lightly, for these regions serve as valuable identification markers in various pathological processes involving bone. Generally, where tendons attach to bone, tendon cells give way to fibrocartilage. Near puberty, certain of these cartilaginous areas, especially where tendons of heavy muscles attach, begin to undergo ossification. Thus, **traction epiphyses** are formed, such as the trochanters of the femur (Fig. II-14A) or the tubercles of the humerus. These gradually become fused to the bony shaft.

Spongy bone, as you can see in Figure II-14A, is

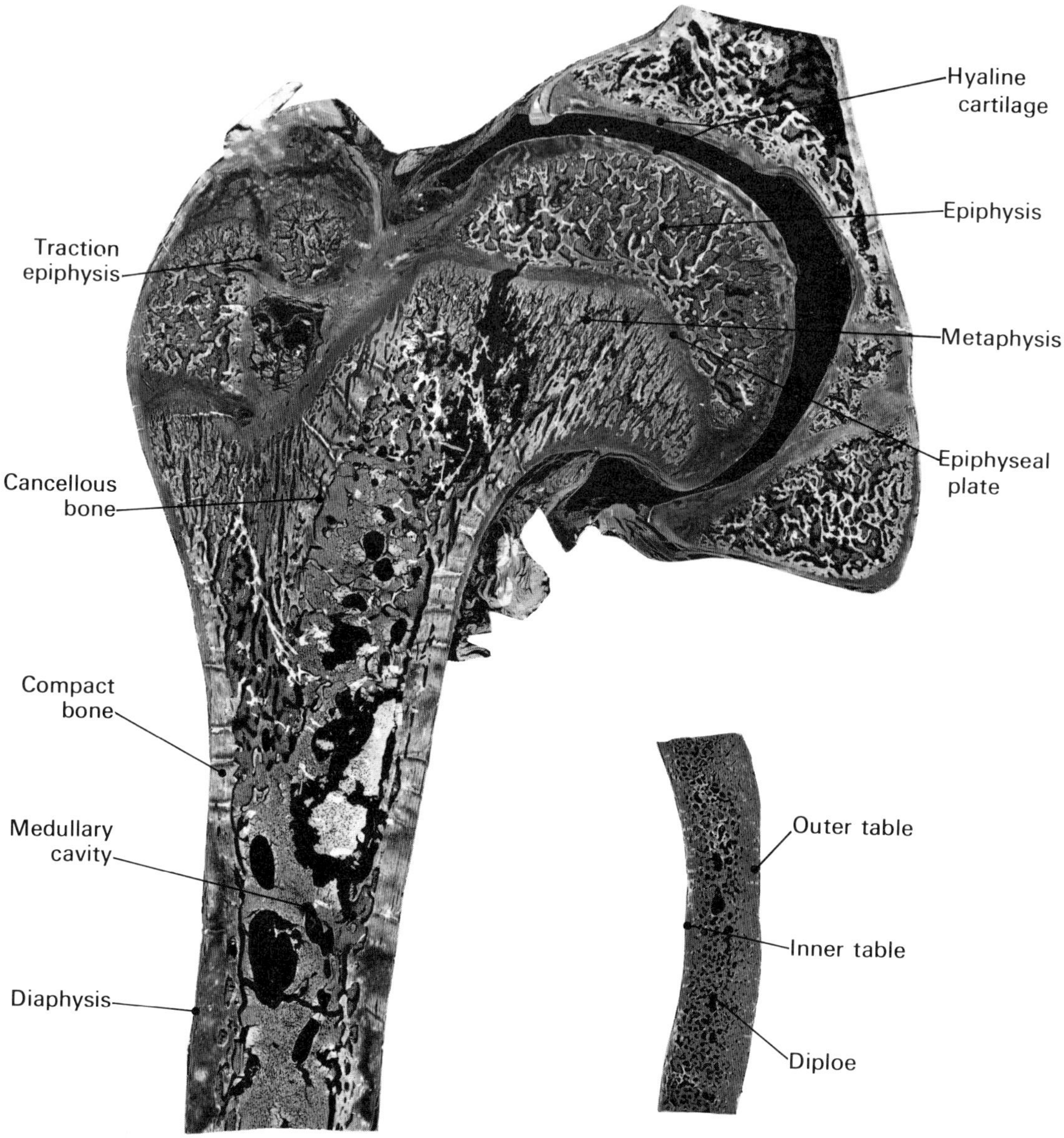

common at the ends of long bones and around the marrow cavity. As any structural engineer knows, the compact bone of the shaft, forming a hollow cylinder, provides greater strength, architecturally, for weight bearing than any solid shaft of equal length. You will also find that the diaphysis bears crests and ridges as strengthening buttresses wherever heavy muscle attaches. Moreover, long bones tend to curve along their length, and the ridges and grooves curve likewise. These features provide some elasticity and shock absorption against the vicissitudes of outrageous fortune. Remember this when next you receive a sharp kick in the shins under the bridge table and walk away without a fracture.

In flat bones of the skull, spongy bone forms **diploe** (inset Fig. II-14B) while compact bone forms the **inner** and **outer table bone.**

Bone as a tissue consists of cells, collagenous fibers, and a mucoalbuminoid ground substance. It is hard and rigid because of the calcareous material deposited in this matrix. The compact portion of a long bone, especially at the diaphysis, repeats the hollow cylinder effect many times in the longitudinal channels, the **osteones** or **Haversian canals,** that run through it (Fig. II-15). The canals are surrounded by **concentric lamellae;** *i.e.,* fibers and matrix are laid down in alternating longitudinal and circular or spiral layers around these longitudinal canals. Bone cells are embedded in **lacunae** in or between lamellae. Figure II-15 shows that the longitudinal Haversian canals are cross connected with each other by oblique or horizontal channels that are nonlamellated. These communicating channels are known as **Volkmann's canals.** Blood vessels occupy both Haversian and Volkmann's canals. Note that it is only the central area of the compact shaft that contains Haversian canals. Outermost, directly under the fibroelastic periosteum, there are **circumferential** bone lamellae, several layers of which encircle the shaft. Where the marrow cavity is finished off more smoothly than shown in Figure II-14, and where there is not so much spongy bone, a few layers of **endosteal lamellae,** also devoid of Haversian canals, line the marrow cavity. In bone remodeling, many Haversian canals are torn down as fast as they are built up, since stress and strain on bone varies with growth changes. The chinks between the final osteones or Haversian systems are filled with the partially concentric lamellae of old Haversian systems (Fig. II-15). These incomplete areas of lamellation are known as **interstitial lamellae.** Thus, the diaphysis is characterized as a lamellated structure capable of great strength. Its hollow central marrow cavity contains reticular connective tissue where blood formation in the embryo is prominent. Blood vessels of the marrow cavity, as well as those in the periosteum, are reached by Volkmann's canals. In the adult, this central cavity becomes filled with fat—yellow marrow instead of red marrow—and blood formation is relegated to the spongy epiphyses of long bones and to the diploic spaces in flat bones.

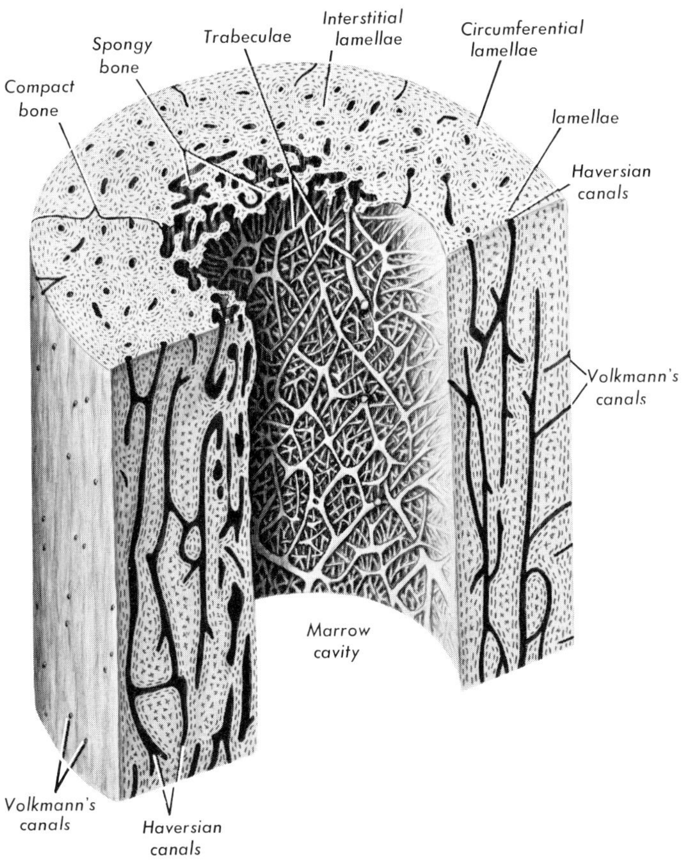

Fig. II-15. Section from a shaft of long bone after maceration; undecalcified. (Courtesy of W. M. Copenhaver; from *Bailey's Textbook of Histology,* 15th ed., Williams & Wilkins, 1965.)

Bones of the Skeleton: Next, to help you satisfy the curiosity of laymen of your acquaintance who will invariably inquire as to the number of bones in the human body, we present you with a partial enumeration of them. This should not become an exercise in memorization. It is presented as reference material only.

The human skeleton (Fig. II-16), not counting certain variable sesamoid and Wormian bones, contains 206 bones in all. Look up the definition of Wormian bones in a good dictionary. We will mention them again in Chapter VI where we will also tally the skull bones in detail. For convenience, the skeleton is divided into **axial** and **appendicular parts.** To the axial skeleton belongs the **skull,** along with the **hyoid bone, vertebrae, sternum,** and **ribs**—all bones that form the axis of the body. The appendicular skeleton consists of the bones of the appendages; *i.e.,* the shoulder and hip girdles (scapulae, clavicles, and the innominate bone) and the bones of the upper and lower limbs (Fig. II-16).

A. *Axial Skeleton* (80 bones)

Skull	28	(See Chapter VI)
Hyoid	1	
Vertebrae:		
Cervical	7	
Thoracic	12	
Lumbar	5	
Sacrum	1	(Fusion of 5 vertebrae into one bone)
Coccyx	1	
Sternum	1	
Ribs:		
True ribs	14	
False ribs	10	
Subtotal	80	

B. *Appendicular Skeleton* (126 bones)

Pectoral (shoulder) girdle	4	(2 clavicles, 2 scapulae)
Bones of upper limb	60	
Pelvic (hip) girdle	2	(right and left hip bones)
Bones of lower limb	60	
Subtotal	126	
Grand total	206	

Note that the first seven pairs of ribs that attach directly to the sternum of the axial skeleton are called "true ribs." The next upper three pair anchor as "false ribs" to the costal cartilages of the 7th ribs and so gain only an indirect attachment to the sternum. The lower two pair of "false ribs" do not attach to any bony or cartilaginous structures anteriorly and are often designated as "floating ribs" (see Chapter VIII, Thorax).

About 1 person in 20 has a 13th pair of ribs, and mongoloid children may have only 11 pair. The infant in whom parts of bones have not yet fused might be said to have 305 bones in all. In some occupations it may be possible to acquire subsidiary bones. For example, the hard core cavalryman of yore occasionally had extra bones in his buttocks and thighs. Whether these were due to fragmentation or to constant friction is a moot point. It is quite possible that such subsidiary "bones" are examples of myositis ossificans. Hemorrhagic areas in muscle may undergo calcification and eventual ossification. But man, unlike his ape cousins, does not have an os penis, for all the friction it may have endured in a lifetime.

When one considers mammals, from the largest to the smallest, the bony skeleton remains fairly consistent despite certain differences. We sport seven cervical vertebrae, and so does the whale without a neck and the giraffe with too much neck. The horse that carries us has one less bone *in toto* than we have, but he gains in ribs (18 pair to 12 pair) and in backbone (54 vertebrae to 26 vertebrae). Since he trots around in this present era on his middle fingers and toes, he has suffered evolutionary short shrift in the extremities (20 bones per limb to our 31 to 32).

Bone Formation: We should consider the topic of bone formation, for this is an area that most beginning students have some difficulty comprehending. We hope that the following will help to clarify the problem. Bones are formed as either **intramembranous bones** or as **endochondral (intracartilaginous) bones.** Except for the **clavicle,** all of the bones of the appendicular skeleton are endochondral bones, as are the vertebrae, sternum, and ribs of the axial skeleton. Most of the skull bones, and certainly all of those in the face, are "membrane" bones. There are exceptions in the skull. The ethmoid is fashioned entirely from cartilage, while the occipital, temporal, and sphenoid bones are partly cartilaginous (endochondral) and partly membranous. The hyoid is also cartilaginous but of branchial arch (gill cleft) origin, as are the ear ossicles. The clavicle is odd. Not only is it the first bone in the body to ossify, as early as in the 15-mm. embryo, but it starts out as a membrane bone which soon undergoes an atypical kind of endochondral development.

In intramembranous bone formation, bone develops in a primitive field of fibrous connective tissue. It is never preceded by cartilage formation or the eventual replacement of this cartilage by bone. In endochondral bone formation, a cartilaginous dressmaker's pattern of the whole bone, or of a part of it, develops first. The cartilage gradually calcifies and breaks down. Bone-forming cells enter the disintegrating field and transform the cartilaginous pattern to bone. Most textbooks of histology limit the discussion of endochondral bone formation to the tubular long bones with their specialized osteones or Haversian canal systems—as if this type of bone were typical of all endochondral bone. As a matter of fact, it is the exception rather than the rule, since so many endochondral bones are flat or irregular in shape and have no Haversian canal systems. We advise you to consult the textbooks of histology for the story of long bone formation and recommend Ham[5] or Bailey[6] as your most readable and reliable sources in this particular regard. When we discuss endochondral bone formation below, it will not pertain to long bones, but to their second cousins about whom little is written, the endochondral flat and irregular bones.

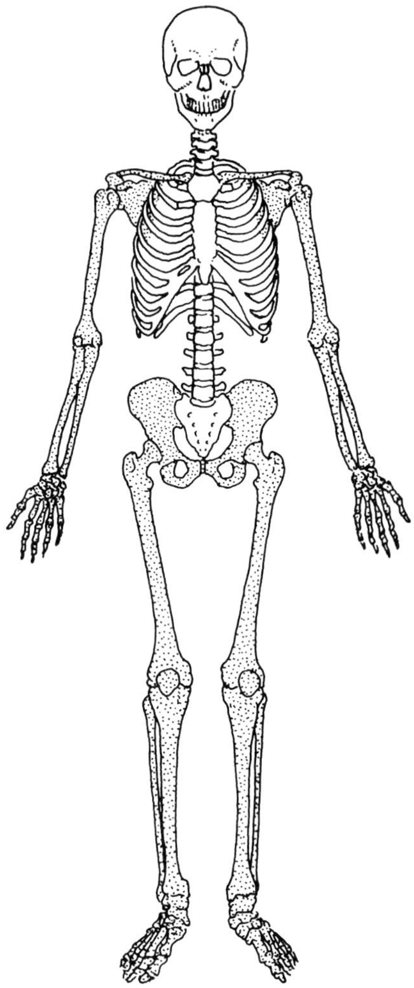

Fig. II-16. Axial and appendicular skeleton. Axial = white; appendicular = dotted.

Superficially, the two methods of bone formation look quite different, but the fundamental process of bony matrix formation and its deposition is the same in both cases. This matrix seems to have to be applied by apposition to some structural framework in the field—to collagenous fibers, to calcified cartilage, or to bone, itself, once the initial trabeculae or lamellae of bone begin to form. The manufacture and secretion of the bony matrix is carried out by special cells called **osteoblasts.**

Intramembranous Bone Formation: Where intramembranous bone is to form in the embryo, certain events precede the appearance of osteoblasts upon the scene[5-26]:

1. Mesenchymally-derived angioblasts have proliferated and organized a rich vascular bed. No similar vascularization occurs in areas where cartilage formation precedes bone formation; *e.g.*, as in endochondral bone formation.

2. Mesenchymally-derived fibroblasts have laid down a network of fine collagenous fibers (Fig. II-

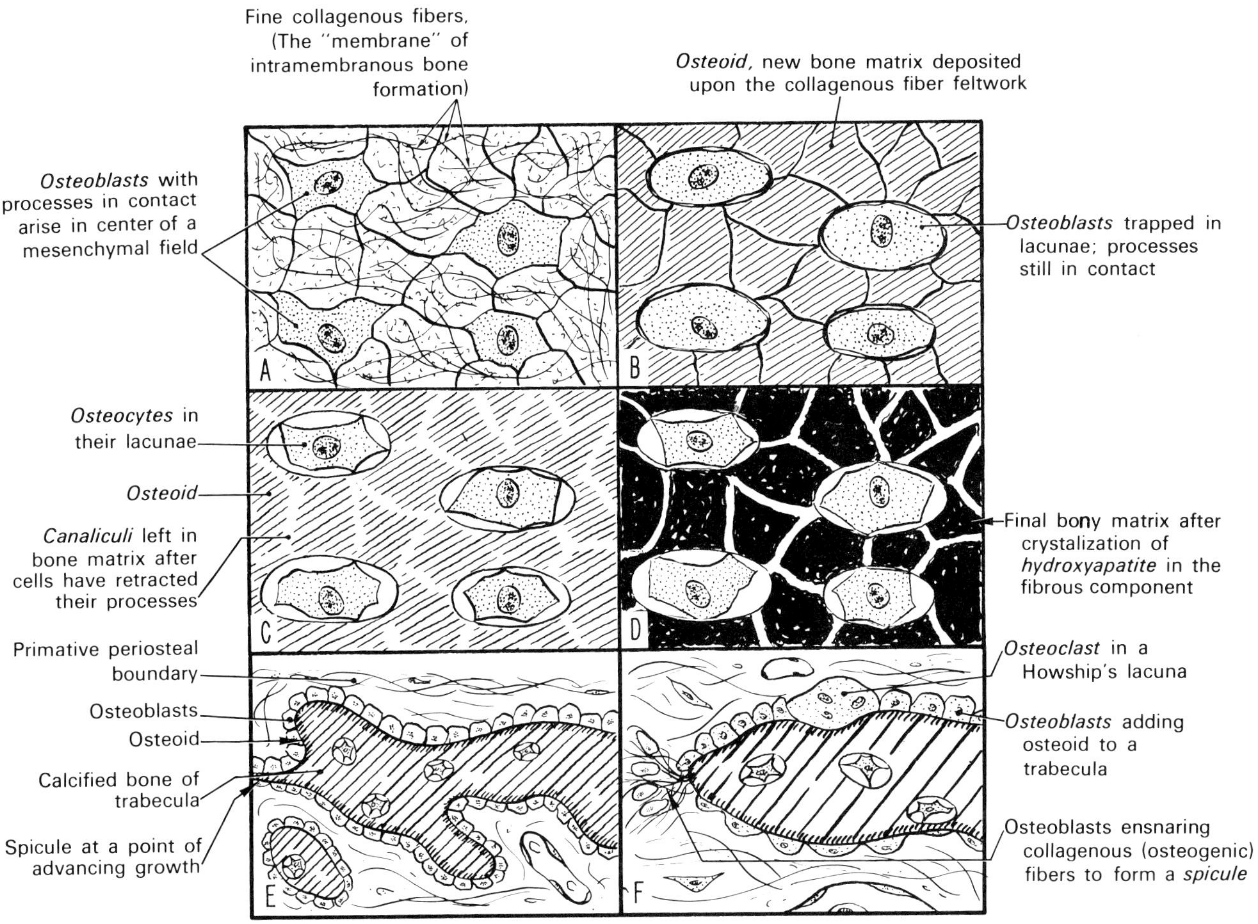

Fig. II-17. Early stages of intramembranous bone formation.
A. Membrane field in mesenchyme.
B. Osteoid deposition.
C. Canalicular formation.
D. Calcification of matrix.
B–D are interior portions, greatly enlarged, of the bony trabeculae seen in **E** and **F**. (Modified from Ham's *Histology*, 5th ed., Lippincott, 1965.)

17A). Indeed, this fiber meshwork is the "membrane" of intramembranous bone formation. It is these fibers which provide the early scaffolding upon which osteoblasts can begin to secrete **osteoid.** The latter is a sticky, semisolid, primitive bone matrix which is free, at first, of mineral salts. When mineral salts in the form of **hydroxyapatite** are deposited in the matrix, the salts align along the fibers and, eventually, crystallize within the fibers at sites of ultrastructural periodicity in collagen.

3. Coincident with the appearance of osteoblasts in this field, and often even earlier, a heavy fibrous condensation of mesenchyme delineates the perimeter boundaries of the more loosely organized central field (Fig. II-17E). Early bone (or cartilage) formation begins in the central area and progresses outward toward the boundary area of primitive **periosteum** (or of primitive **perichondrium** in the case of endochondral bone formation).

Thus, most of the props and the supporting players are already on stage when the stellar performers, the osteoblasts, make their entrance. These bone formers arise from mesenchyme as did the angioblasts and fibroblasts before them. They appear among the fibers in the semifluid matrix of mesenchyme as somewhat stellate cells with radiating processes that remain in contact with each other (Fig. II-17A). In the usual light microscope preparations (Fig. II-17E and F), the processes are not seen readily, probably due to shrinkage artifacts or fixation, and the cells tend to appear more block shaped. From the standpoint of electron microscopy, the cells have an extensive endoplasmic reticulum which accounts for their basophilia in stained preparations, and the Golgi apparatus is prominent and has large vesicles. Thus, the cell is organized for protein synthesis and has the means for transporting these formed products to the exterior for export into the tissue matrix. Without question, one chief product is the primitive, noncalcified, bone-matrix material which we have called osteoid. This seems to be a mucopolysaccharide containing, among other chondroitins, chondroitin sulfate C, all bound to a long-chained protein.

In intramembranous bone formation, the cells secrete the sticky, semisolid osteoid not only about themselves but about the collagen fibers along which they align themselves in the primitive mesenchymal field. It is doubtful in these early stages of bone formation that the osteoblasts are concerned directly with the synthesis of collagen. There are abundant collagen fibrils already present in the adjacent mesenchymal field. New membranous bone begins as fine **spicules** as osteoid is deposited upon wisp-like bundles of neighboring collagenous fibers. Figure II-17F shows such a spicule in an advanced stage where a bar of bone is having a point of advancement added to it. But the whole bar began from such a point in the beginning. As secretory activity continued, spicules were widened into irregularly-shaped bars of bone or **trabeculae,** such as we see in Figure II-17E and F. The rest of Figure II-17A–D, simply shows at high magnification what goes on in the **interior** of such a trabecula, and we will discuss this shortly. Each trabecula starts as a separate, isolated, jig-saw puzzle piece which, through extension along advancing spicules, eventually is joined to other trabeculae in the three-dimensional volume field. Each broad-beamed trabecula, as it widens, encroaches closer and closer upon the neighboring blood vessels in mesenchyme without quite touching them. In this way, the spongy diploe of the flat bones of the skull, for example, are formed "in membrane." In the intertrabecular spaces, mesenchyme organizes into a reticular connective tissue network to support vessels and to give rise to hemopoietic cells. So arises the primitive marrow of the diploic spaces of flatbones. As yet no compact bone has formed (Fig. II-17E and F).

In this early process of cementing a sticky matrix about themselves, the early generation of osteoblasts soon become entrapped in tiny, hollow spaces or **lacunae** within the forming trabeculae (Fig. II-17B). Once mineral salts deposit in this surrounding matrix, such cells are isolated as permanent indwellers in the bone. They cease their secretory activity and become **osteocytes,** or the cells of living bone (Fig. II-17D). But by virtue of having kept their cytoplasmic processes in contact and by having become sealed in trabeculae that maintained close contact with a neighboring blood supply, osteocytes (unlike chondrocytes) do not cut themselves off from a ready means of procuring nourishment. As matrix traps them in lacunae, the cells can withdraw their processes when the matrix solidifies, leaving fine, three-dimensional, interlinking channels or **canaliculi** to interconnect lacunar spaces (Figs. II-17C and D). These channels reach out to the edges of the trabeculae, as well. Tissue fluids can percolate through the hard bony matrix to serve the inhabitants trapped within.

However, if subsequent osteoblasts did not follow upon the heels of these entombed original settlers, spicules would never widen into trabeculae; neither would trabeculae anastamose to form spongy bone. Wherever early bars of bone are seen, their outer edges are beset by a swarm of new osteoblasts, the replacement crew for those already entombed (Figs. II-17E and F). They must have arisen from adjacent mesenchymal cells, for osteoblasts rarely engage in mitotic division. The newcomers add new matrix to the already-formed trabeculae to increase them in width and length or to change their shape by adding projecting sidearms to them. Eventually these cells, too, share the same fate as those before them and become ensnared in their own handiwork to become indwelling bone cells.

When the anastamosing trabeculae of the central mesenchymal field have spread out to contact the peripheral, periosteal, boundary zone (Fig. II-18A), new generations of osteoblasts arise in the periosteum. These commence to lay down layer upon layer of parallel bone lamellae under the periosteum, pushing the latter farther to the periphery as they form the outermost, compact, **table bone** (Fig. II-18B). The spongy, trabecular diploe are, at last, sandwiched between limiting layers of compact bone derived from the periosteum that surrounded the central field on all sides.

But nothing in development is all that simple. Sooner or later in every area of growing bone, the processes of bone resorption and remodeling come into play. Neighboring developing structures, such as a growing brain which impinges upon the bone, or muscles and ligaments which begin to attach and pull upon it, impose stresses upon growing bone to which bone must conform as it shapes up. Parts of the growing bone must be torn down and remodeled in the very process of its building up. Even the blood vessels and nerves which will run through bone demand smooth, not rough, channels. So bone constantly has to change its architecture to meet the demands of other structures that seek its protection or that care to anchor upon it.

Wherever such processes of reconstruction are underway, large, multinucleated giant cells, the **osteoclasts** or "bone breakers," appear in the field (Fig. II-17F). That they actually are involved in bone resorption seems to be largely a matter of guilt by association, since most evidence to prove that they are has been rather circumstantial to date. They are seen snuggled in little coves or depressions along the sides of trabeculae. Or, they are found at the tips of spicules. It appears as if osteoclasts were wearing away bony edges or blunting spearheads of bony advance. Some have said that these cells are simply fusions of osteoblasts.[19, 20] Some have argued that they are, probably, defunct cells. Osteoclasts could, of course, have arisen from a fusion of mesenchymal cells and could have very special functions. It is even possible that they arose through endomitosis (a process of nuclear division without cytoplasmic division), though this phenomenon is uncommon among mammalian cells. The concavities which osteoclasts occupy long have been known as Howship's lacunae after the English surgeon, John Howship, who first noticed them in 1841. It was he, perhaps, who first assigned the uncertain role of bone resorption to osteoclasts and who enshrouded these cells in a garment of mystery that present investigators are still trying to unravel.

Figure II-18A–D, taken through the microscope, reviews the salient features of intramembranous bone formation.

Much has been reported of late on the ability of osteoblasts, the bone-forming cells, themselves, to synthesize collagen. This has been said to be true of chondroblasts, or cartilage formers, as well. The protein is synthesized in the rough endoplasmic reticulum as tropocollagen, and these primitive fibrils have been identified outside the cell close to its plasma membrane. The formation, extrusion, and organization of tropocollagen strands into the coiled helices of collagen fibrils with quarter-staggered linkage and a 640 Å, periodic cross-banding has been similar to that described for collagen formation by fibroblasts. Tropocollagen, composed of one-third glycine and one-fifth proline and hydroxyproline, is an unusual substance, limited in its appearance in the body. Hence, fibroblasts, osteoblasts, and chondroblasts are, apparently, close cousins in respect to their ability to manufacture this fibrous protein. We have pointed out that it need not be manufactured by osteoblasts at all in the early stages of intramembranous bone formation where collagenous fibers are as ubiquitous as the osteoblasts themselves and even present in the field before osteoblasts appear there. Simply because tropocollagen appears in the close vicinity of the osteoblastic border could, likewise, mean very little. The field contains many fibroblasts or undifferentiated mesenchymal cells that could be equally responsible for its presence wherever the early osteoblasts abound.

The evidence that collagen is produced by osteoblasts has not come so much from the study of intramembranous bone formation but from what has happened in the vicinity of fracture and repair of **adult long bones** (endochondral bones). Even here there still remains the distinct possibility for much of the collagen in the area of repair to have been produced by fibroblasts and undifferentiated cells of the periosteum and endosteum. It becomes less likely that this would be the case in late stages of **endochondral bone formation,** especially in the tubular shafts of long bones when osteoblasts must enter an area of cartilaginous breakdown, invade tubular channels of calcified cartilage, and proceed to lay down concentric lamellae of bone against the interior walls of these cartilaginous tunnels (Haversian canals). In each lamella, collagen fibers become oriented in opposite planes as the bony lamellae are added. Here, certainly it would seem to be desirable, if not absolutely essential, for the osteoblasts to be able to synthesize their own collagen as they go about constructing **osteones;** *i.e.,* **Haversian canals.**

So, too, in the later stages of intramembranous bone formation, as resorption and remodeling progress rapidly, collagen in the area may be at a premium so that osteoblasts are now required to secrete their own for incorporation into the new bony design. Baker,[17] Pritchard,[11] and Weinmann and Sicher[18] all have pointed out that the content of bone matrix, in respect to the number of bone cells and the amount of collagen and mucopolysaccharide, varies widely at different stages of bone formation. It seems reasonable that collagen synthesis is in greater demand at certain times than at others. The osteoblast, always on the spot, is in a good field position to supply collagen as the need for it arises and appears capable of doing so.

The chondroitin sulphate C which osteoblasts secrete into the matrix also seems to serve as a "calcium bank," for it tends to form a spherical "domain." Its macromolecule apparently folds into a sphere that is negatively charged on its outer surface so that calcium ions are attracted to it. As the protein backbone of the mucopolysaccharide is degraded through hydrolyses by proteinases, the "domain" is broken up with the release of calcium in the matrix. The quarter-staggered collagen fibrils embedded in the matrix then act as nucleating sites for the crystallization of calcium salts, probably as simple calcium phosphates at first. These, in turn, may serve as nucleating centers for the deposition of more complex hydroxyapatite. In this way the minerals crystallize upon and within the fibers.

Endochondral Bone Formation: *This kind of bone formation, as it occurs in certain skull bones, in ribs and sternum, and in vertebrae and other irregularly-shaped bones, is more akin to the events that occur at the* ***epiphyseal ends*** *of long bones than to those events that take place in the* ***diaphysis*** *or shaft of long bone.*

Fortunately, the **very early events,** those dealing with the formation of the precartilagenous model, are the same for all cases of endochondral bone development. As in intramembranous bone, the story opens in a primitive field of mesenchyme (Fig. II-19A). Fibroblasts have contributed a meshwork of fine fibers, but the activity of angioblasts has been conspicuous by its absence, even after chondroblasts (cartilage formers) have appeared in the area. Already, too, a stout mesenchymal condensation in the form of an outer limiting **perichondrium** is present when the large, rounded chondroblasts begin to congregate in the more loosely-organized center of the field. These cells, derived from mesenchyme, huddle closely together as if grouped about some chondroblastic quarterback, awaiting preplay instruction. As if on signal, they move slowly apart to blanket the entire fiber field with their mucopolysaccharide secretory product which contains chondroitin sulfates. Like osteoid, this substance is somewhat fluid before it "sets up" as a semisolid mass. It is, actually, the laying down of this substance that moves the cells apart. Unlike osteoblasts, chondroblasts retain the ability to divide as they secrete. In this way new workers are born and marshaled to help out *within* the field. Cartilage, then, unlike bone, expands internally by this kind of **interstitial growth.** Bone, on the other hand, can only grow by **apposition;** *i.e.,* by adding on to the exterior of already-formed matrix. Interstitial growth of cartilage is relatively short-lived, however, for the central mass soon reaches the perichondrial boundaries, and in time, too, matrix begins to harden. Then chondroblasts, like osteoblasts,

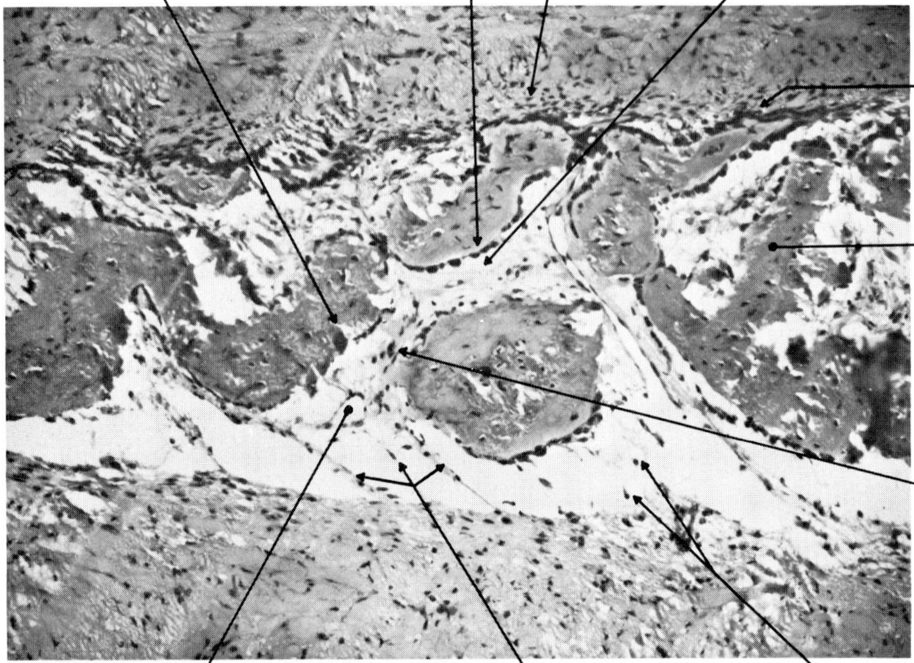

Fig. II-18. Photomicrographs of intramembranous bone formation.

A. Early spicules and trabeculae forming in mesenchyme. Between rows of osteoblasts and the larger trabeculae, new bone (osteoid) stains lighter than that formed earlier. Same light staining seen in area of spicules. There is as yet little trabecular anastomosis, but some have extended to contact the periosteal boundary.

B. Later stage. Parallel lamellae of table bone being added from periosteum.

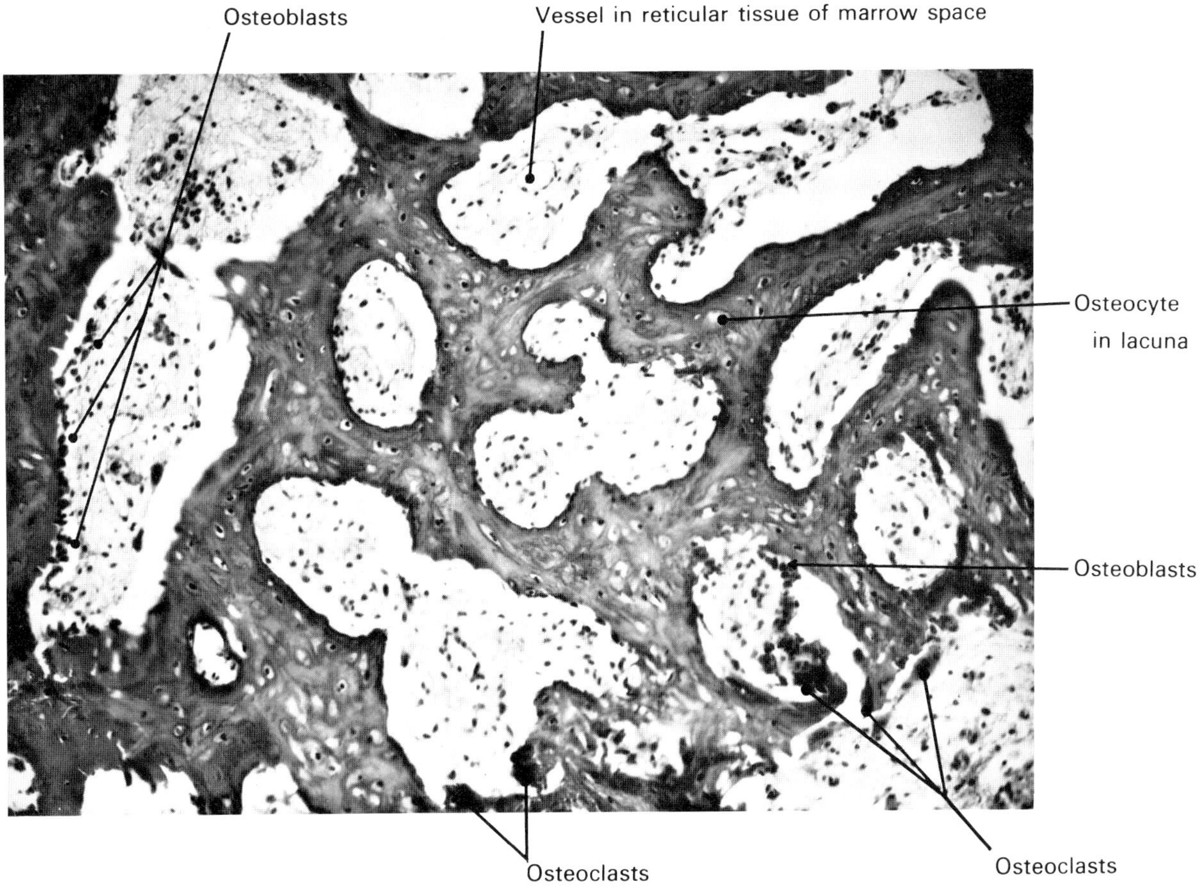

C. Later stage. Considerable trabecular anastomosis in the spongy bone of the diploic region. Few osteoblasts seen, but **osteoclasts** are present in regions of trabecular erosion and remodeling.

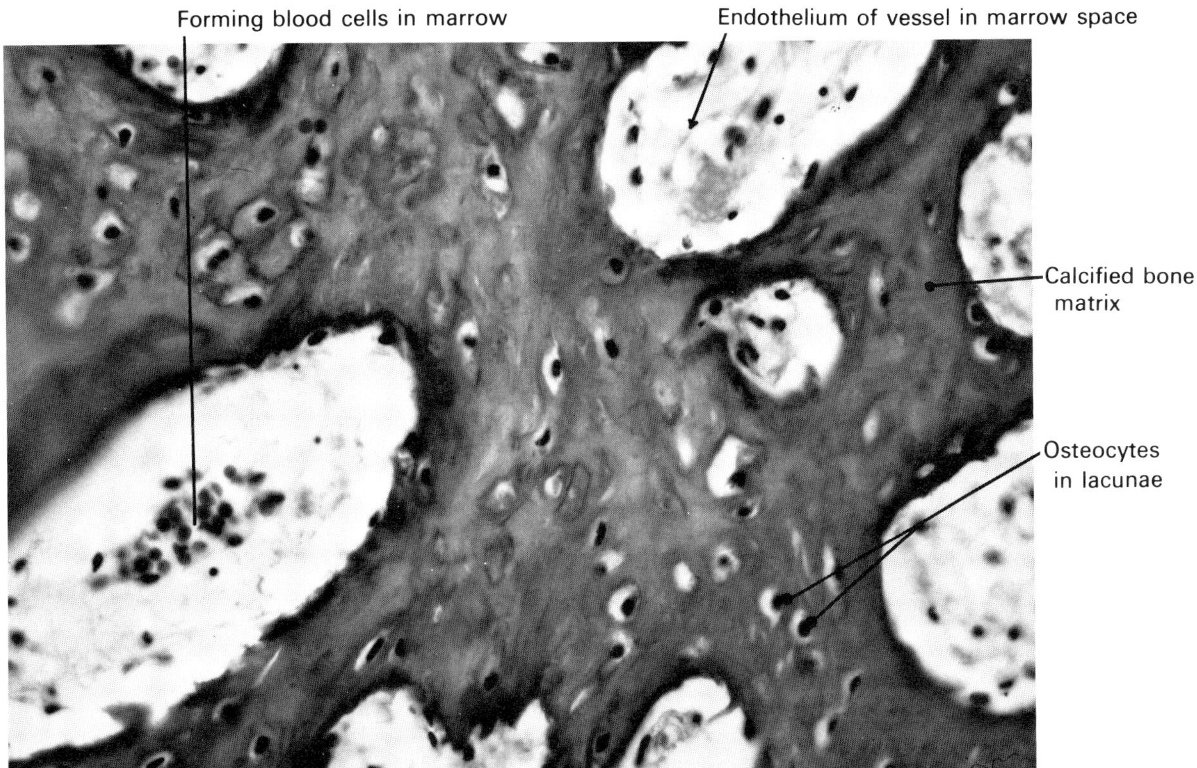

D. Higher power of **C**. Note blood formation in the endothelially-lined channels within the marrow spaces of spongy diploe. Bone matrix now well calcified.

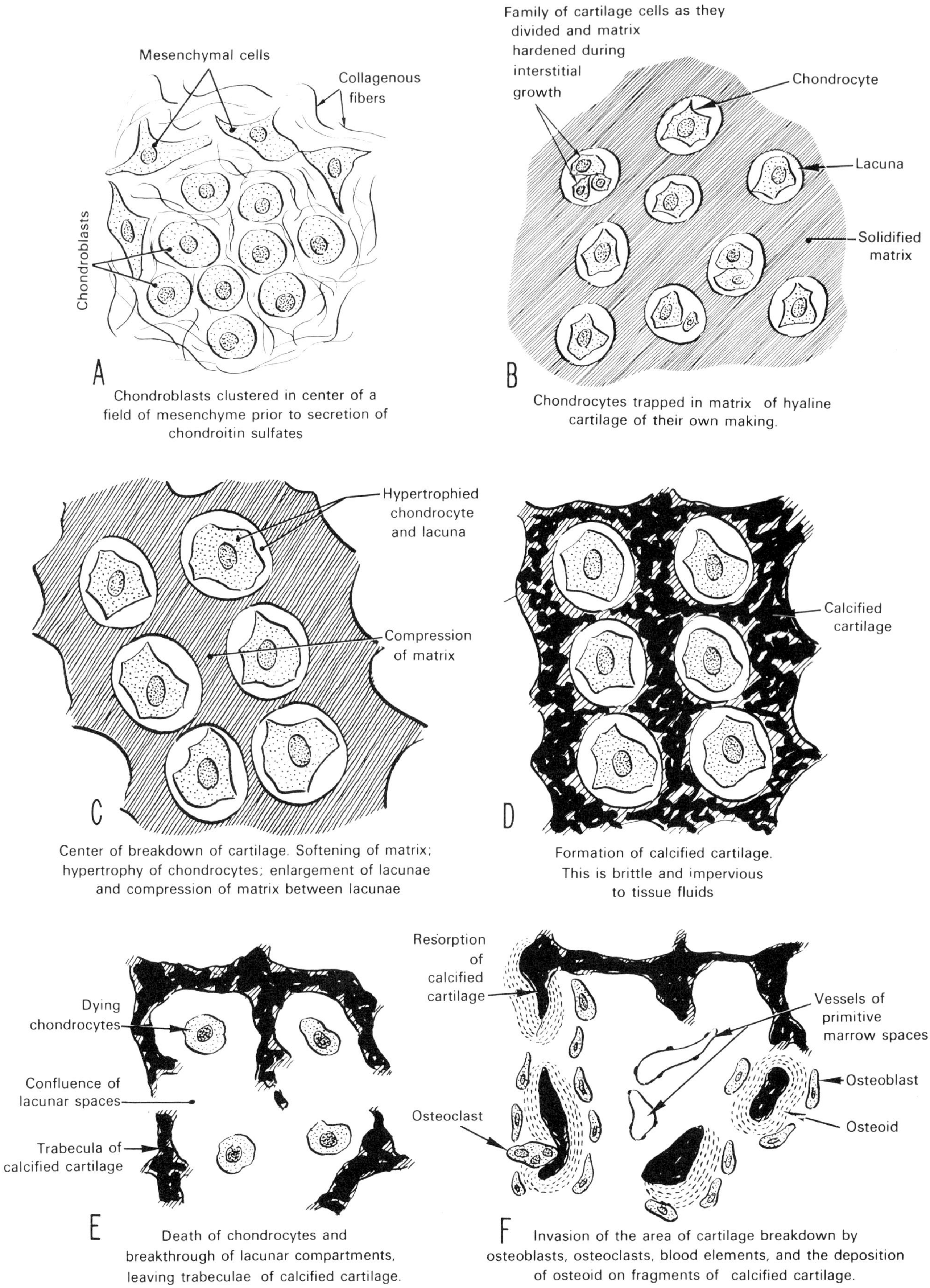

Fig. II-19. Early stages in endochondral bone formation. Stages **A–F** are self-explanatory in figure. (Modified from Ham's *Histology*, 5th ed., Lippincott, 1965.)

find themselves trapped in lacunae and faced with becoming indwelling **chondrocytes** (Fig. II-19B). Sometimes a whole family of cells, when mitosis was underway as matrix solidified, finds itself entombed in the same lacuna (Fig. II-19B). Chondroblasts, unlike osteoblasts, have lacked the foresight to keep in touch with each other by means of cytoplasmic processes. As a result, cartilage matrix is devoid of canaliculi to serve as interlacunar connecting channels. Cells trapped in lacunae farthest from the meagerly-vascularized perichondrium must obtain their thinly-rationed nourishment as it oozes toward them through the semisolid matrix by osmosis. For this reason, only thin cartilaginous structures survive very well throughout life.

But, the cartilaginous patterns of early bone generally are fairly chunky masses. Once the primary center of chondrification reaches perichondrial borders, new cartilage is added to the mass by appositional growth from chondroblasts arising in the inner layer of perichondrium. Thus, these masses continue to grow in thickness, width, and length. Because of their size, such masses, as a rule, are not very stable. They tend to break down locally in one or more areas designated as **centers of ossification** (Fig. II-19C) where cartilage cells hypertrophy and where matrix between lacunae becomes compressed as well as increasingly brittle as it undergoes a process of calcification (Fig. II-19D). It becomes even less pervious to the passage of tissue fluids as it calcifies. The cells die, lacunar spaces break open into each other, and remaining brittle matrix begins to crumble and undergo dissolution (Fig. II-19E). The matrix, perhaps, would disappear entirely and the whole pattern would break down and become nonfunctional were not the processes of bone formation initiated to replace the cartilage by degrees (Fig. II-19F). Apparently the cartilaginous model would not, by itself, be strong enough to meet the skeletal requirements of a large and heavy organism even if it were to survive as cartilage. Indeed, calcification of the matrix could be regarded as the first attempt to strengthen the cartilage mass, except for the fact that it is a self-defeating effort. Cartilage cells, themselves, play a role in initiating calcification. But we have seen that in so doing they render matrix more impervious to the passage of tissue fluid and so seal their own doom. When the cells die the brittle matrix between enlarged lacunae disintegrates in short order.

The exact events that bring about calcification, the deposition of mineral salts in cartilage matrix, are still not well known. Somehow calcium and phosphorus have to be brought to the matrix to form calcium phosphate. This salt can hardly be conveyed by the blood stream as such, since it is a relatively insoluble compound. Nor is there much phosphate in the blood in free ionic form in plasma. Most of it is in the form of a phosphoric phosphate ester. Fortunately, this ester can pass through endothelium and arrive in the matrix of cartilage. But something there has to release phosphate from the ester.

We can hardly expect much calcium in the blood in free ionic form, else it would combine with phosphate ions there as an insoluble precipitate. There is some ionic calcium, however, and enough to diffuse through endothelium and into the matrix of cartilage, but not enough in terms of mass action to combine with the free phosphate ions in blood. Most of the calcium *in blood* is protein-bound and does not leave the plasma. Thus, *in the matrix of cartilage* the minerals are present as ionic calcium and as a phosphate ester. Probably even the calcium is first stored in the "calcium bank," the spherical domain of chondroitin sulphate C, to be released by gradual hydrolysis. If there were something at the matrix site to release phosphate from its ester, calcification could proceed.

The hypertrophied cartilage cells in the area of disintegration (*i.e.*, in the centers of ossification) are able to synthesize an enzyme, **alkaline phosphatase,** which is capable of releasing phosphate from its ester, but only in an environment of alkaline pH. If this is so, some other reactions must occur first, for calcification does not occur in normal cartilage in the presence of chondroitin sulfuric acids. The acid content of matrix must be buffered or altered in some way. Certainly, it must occur if the matrix is to soften so that it can be compressed by swelling cells in enlarged lacunae since these cells take on water, and the matrix does soften and become compressed before calcium salts deposit in it.

As hypertrophied chondrocytes appear in centers of ossification, or in centers of cartilage breakdown, if that connotation makes better sense, vascular channels begin to proliferate in the perichondrium (Fig. II-20A). The channels invade inward into the area of breakdown much as periosteal buds erode through the forming shafts of developing long bones. Osteoblasts and marrow tissue are brought into the crumbling cartilaginous region along with the vessels, for with vascular proliferation the original perichondrium becomes transformed into a periosteum (Fig. II-20B). The osteoblasts use the splinters of calcified cartilage as a framework upon which to lay down bony matrix (first as osteoid) by apposition. In this way, through the routines of trabecular formation, a spongy network of bone gradually replaces cartilage in these "centers of ossification." Osteoblasts in the transformed periosteum proceed to lay down parallel lamellae of compact table bone outermost about the central area of diploe formation. In this roundabout way, for example, **endochondral bones of the skull** end up fashioned very much like those that developed in membrane, directly. Since osteoclasts also invade the interior and since they lie out in the periosteum, as well, the usual processes of bone resorption and remodeling proceed hand-in-hand with bone construction. On the whole, the process is much more simple than that which takes place in the formation of the tubular shafts in the long bones. Still, it is, undoubtedly, more complicated than we have described here. (For information on long bone formation see references 5 and 6 at end of chapter.)

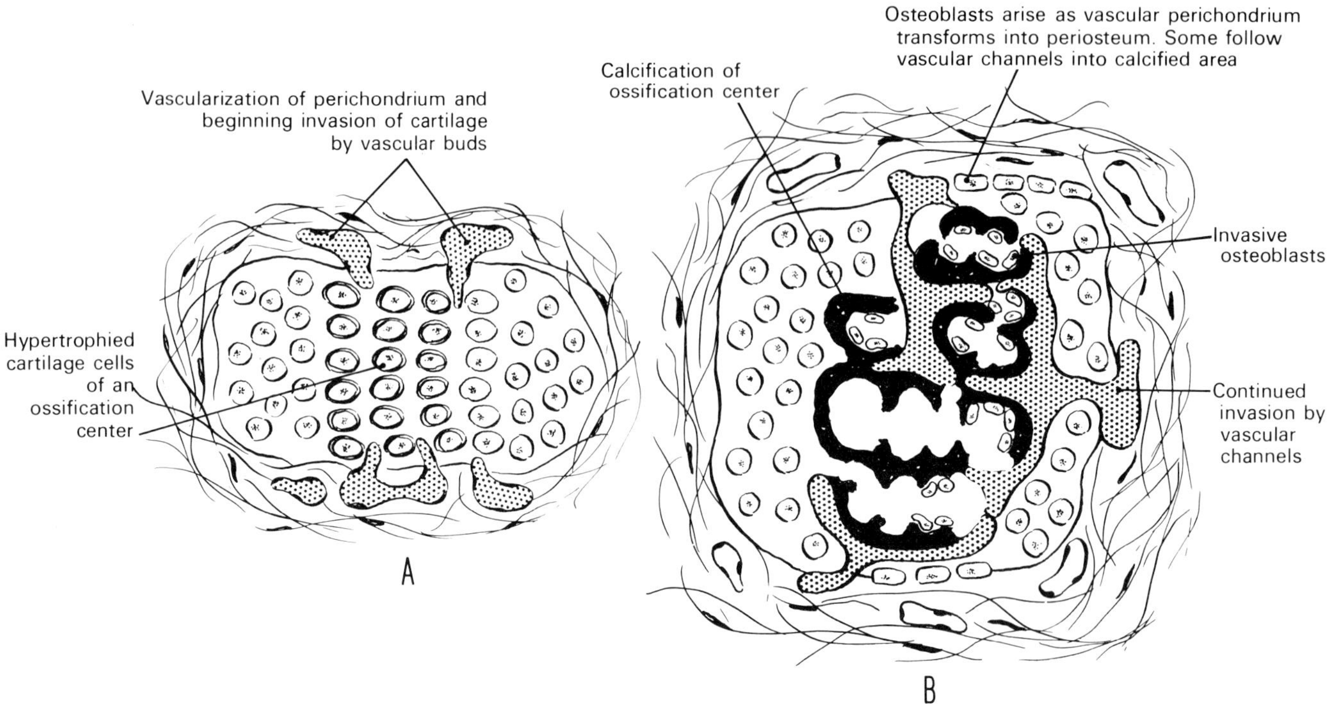

Fig. II-20. Stages in the osification of an irregular mass of hyaline cartilage, as in the case of endochondral bones of the skull.

Joints

Articulations between bones are classified as synarthroses, amphiarthroses, or diarthroses; that is, they are either **immovable, slightly movable,** or **freely movable joints.** Such functional differences are reflected in morphological structure, and much depends upon whether fibrous connective tissue, cartilage, or a synovial cavity is interposed between the ends of bones.

In the fibrous joints, if very little collagenous material holds adjacent bones together we have a **suture.** In the infant, a suture, as in skull bones, may allow for some slight bending or buckling of the joined plates of bone, but with aging, fibrous tissue is replaced by bone. Thus, bone fused to bone is a **synostosis** and is immovable. The final epiphyseal lines in long bones are synostoses. The fibrous joint, itself, is slightly movable. It is a syndesmotic amphiarthrosis, or a **syndesmosis.** When greater amounts of fibrous tissue hold bones together by means of ligaments, slight-to-considerable movement can be achieved.

In **synchondroses** some slight movement is also achieved when bone is joined to bone by cartilage. Hyaline cartilage is generally involved as at the epiphyseal plate of growing bone and where the costal cartilages join the ends of ribs. Both of these areas can transform into rigid synostoses. Where the first costal cartilage joins the sternum, there is also a synchondrosis. But where the other costal cartilages join the sternum there are synovial cavities, and the joints here are freely movable. A **symphysis** involves fibrocartilage held in place by ligaments. Symphyses are slightly movable joints and are limited in locus as at the **pubic symphysis** between the rami of the pubic bones and the **vertebral symphyses** at the vertebral disks. The "symphysis menti" of the mandible is a misnomer, being a synostosis.

Table II-1. Joint Types

Joint	Example
Fibrous Joints—no joint cavity.	
A. Suture.	
1. Syndesmosis—slightly movable.	Bones of infant skull.
2. Synostosis—immovable.	Bones of adult skull (except mandible); fusion of epiphysis and diaphysis in adults.
B. Ordinary ligamentous syndesmoses (slightly movable).	Laminae of vertebrae; distal ends of tibia and fibula.
Cartilaginous Joints—no joint cavity.	
A. Synchondrosis—slightly movable.	Epiphysis and diaphysis where epiphyseal plate is present; Costal cartilage and ribs; 1st costal cartilage and sternum.
B. Symphysis—slightly movable.	Between pubic bones and bodies of vertebrae (fibrocartilaginous disks interposed).
Synovial Joints—freely movable. Always a joint cavity present.	Temporomandibular joint (Fig. III-12); rest of costal cartilages with sternum; joints of limbs; articulating processes of vertebrae.

All movable joints are encased in fibrous capsules lined by mesenchymal epithelium which secretes a synovial fluid into the smooth-lined capsule. There may or may not be an articular disk separating the ends of opposing bones and converting the single cavity into two synovial cavities (see Fig. III-13). From the standpoint of action, these diarthroses have been classified as follows:

1. **Gliding joints** (arthroidia), such as those between carpal and tarsal bones.
2. **Hinge joint** (ginglymus), where movement is in one axis at right angles to the bones involved, as at elbow and finger joints.
3. **Pivot joint** (trochoid), allowing rotation around a long axis, as the atlas around the axis (epistropheus), or the radius about the ulnar.
4. **Condyloid joint** (ellipsoid), allowing movement in two directions; *e.g.*, flexion and extension as in the hinge joint, as well as abduction and adduction. The distal joints at the wrist are good examples, or the occipital condyles where we can nod "Yes."
5. **Saddle joint,** where the opposing surfaces are reciprocally concave and convex. These allow movement in several directions. The carpometacarpal joint at the base of the thumb would be an example.
6. **Ball and socket joints,** where universal movements are permitted, such as in the case of the head of the humerus in the glenoid fossa of the scapula or the head of the femur in the acetabulum.

Table II-1 may help summarize the above.

General Myology

The ancients thought some muscles looked like mice, the tendons, probably, reminding them of tails. Hence the Latin derivation for muscle from *mus*—a mouse (Gr. *myos*—muscle). Table II-2 classifies the three types of muscle in the body.

Muscles are built to contract and relax. When skeletal muscle contracts it generally moves bones by virtue of crossing joints, having an **origin** on one bone and an **insertion** on another. Cardiac and smooth muscle tend to encircle tubes or cavities and to express the contents of these structures as the muscle contracts.

We generally think of skeletal muscle as being under the control of our wills; the other two as being involuntary or unable to be moved willfully. Still the Diaphragm is a skeletal muscle which operates involuntarily during ordinary breathing yet it can be brought under voluntary control in deep breathing or in momentarily holding our breath. Some individuals have learned to contract certain smooth muscles by a process of conditioning, and even to alter the heart rate.

Individual skeletal muscle fibres are really multinucleated, elongated cells in the form of a true syncytium. The common plasma membrane is the **sarcolemma.** In a large muscle bundle or fascicle, the individual muscle fibers are surrounded by some loose fibroelastic tissue, the **endomysium.** The whole fascicle is bandaged in a tougher fibroelastic tissue, the **perimysium.** If a muscle contains several fascicles in a common sheath so that individual fascicles will separate away later to make their own special attachments, **perimysial septae** separate them within the bundle. The loose areolar tissue that separates one adjacent muscle from another is called **epimysium.** In gross anatomy, perimysium is called variously **deep fascia, muscle fascia,** or **innominate fascia.** Sometimes it bears the name of a particular muscle; *e.g.*, Psoas fascia. The fleshy part of a muscle is its **belly.** Some muscles are fleshy throughout, but most are fibrous at one or both ends. The dense, white fibrous connective tissue (collagen) which attaches muscle to bone is its **tendon.** The same tissue in the form of a flat sheet anchoring one muscle to another is an **aponeurosis.**

Table II-2. Types of Muscle

Histological Appearance	Name	Type of Neural Control
I. Striated.	A. **Skeletal** Muscle. (Branchial muscle is skeletal muscle of branchial arch origin.)	Voluntary.
	B. **Cardiac** Muscle.	Involuntary (visceral).
II. Non-striated.	**Smooth** Muscle.	Involuntary (visceral).

The orientation of fibers in any particular muscle varies. The fibers may parallel the long axis of the muscle, may run obliquely, or pass radially as in a fan. The "strap" muscles of the neck and many flat and fusiform-shaped muscles have a parallel fiber arrangement; *e.g.*, Sternomastoid, Rectus Abdominis, Biceps Brachii.

When fibers are obliquely arranged they look feather-like and are called **pennate.** The muscle fibers, themselves, appear like the barbs of the feather, and the tendon into which they peg corresponds to the shaft of the feather (Fig. II-21). In **unipennate** muscles the arrangement is like that of half a feather, and the fleshy fibers have a narrow origin. This is the arrangement in the Peroneus Tertius, for example. The Peroneus Longus, on the other hand, is a **bipennate** muscle (whole feather), and its fibers have a broad origin. If the tendon runs through the central part of the muscle so that fibers converge upon it in several planes, as in the Tibialis Anterior, it is a **circumpennate** muscle. Pennate muscles are very powerful. Triangular or fan-shaped muscles start from a wide origin and converge upon a narrow point (the apex of the triangle, or the handle of the fan); *e.g.*, the Temporalis or the Pectoralis Minor.

Muscles are well supplied by blood vessels. These enter either *with* the nerves, separately, or come in as a series of multiple entries. Capillaries tend to form rather rectangular networks. The veins in muscles generally have valves; thus, exercise and massage will foster circulation. The so-called **motor nerve** to skeletal muscle is not purely motor in the strictest sense. It contains sensory fibers as well, carrying back **proprioceptive** information from muscle and tendon spindles. Proprioception is "muscle and joint sensibility," giving us an awareness of the position of our limbs in space. There are visceral motor fibers in this nerve (sympathetic) for the smooth muscle of arterioles. A single axon of any somatic motor nerve fiber in the

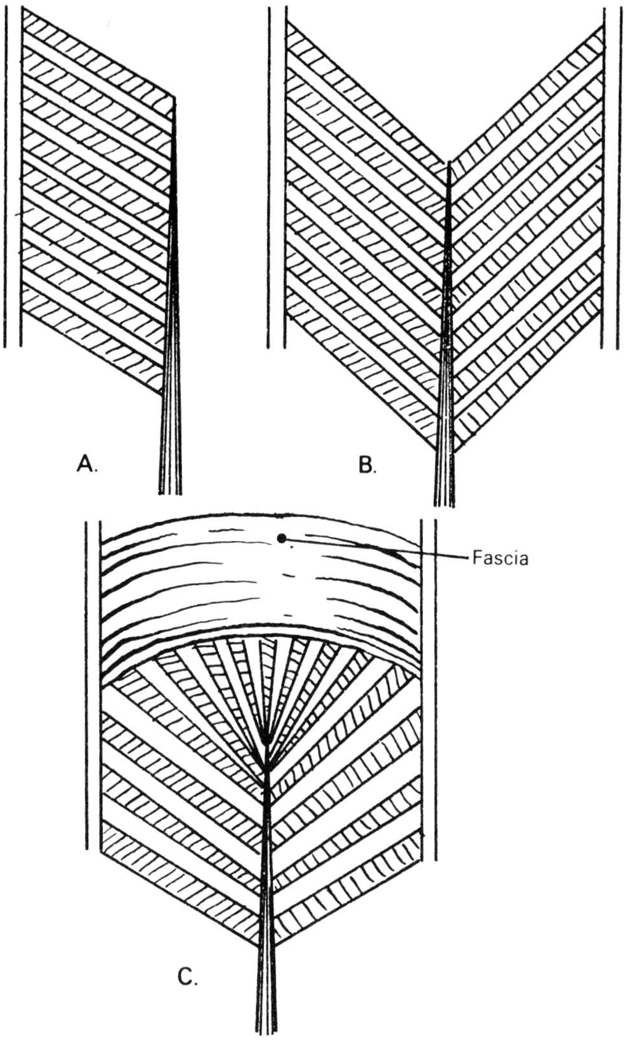

Fig. II-21. Diagrams showing muscle fiber arrangement in some of the powerful pennate muscles.
A. Unipennate; *e.g.*, Peroneus Tertius.
B. Bipennate; *e.g.*, Peroneus Longus.
C. Circumpennate; *e.g.*, Tibialis Anterior. (After Grant's *Method of Anatomy*, 7th ed., Williams & Wilkins, 1965.)

nerve bundle may supply from 1 to 100 muscle fibers with **motor end plates,** one plate per fiber, depending upon how finely or how grossly the muscle is to be controlled in purposeful individualistic movements. This kind of relationship between one nerve cell process and a number of muscle fibers in a muscle bundle constitutes a **motor unit** (Fig. II-22). The **motor point** is something different. Physical therapists attempt to get weak muscles to contract by placing electrodes close to the point where a nerve fiber enters a muscle. This is its motor point.

Synovial Bursae

A small sac of connective tissue containing synovial fluid is a bursa. The fluid in the tiny bag allows free movement in this area and a lessening of frictional effects. Bursae can be subtendinous so that tendons slide

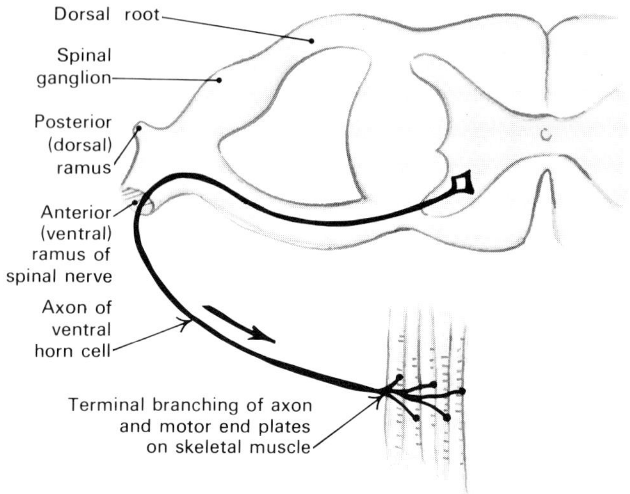

Fig. II-22. A motor unit: The number of muscle fibers in any muscle that are supplied by the single axon of a ventral horn cell. In large muscles, a single axon may supply 50–100 skeletal muscle fibers in the bundle. Hence, the larger the muscle, the more motor units required for its innervation.

over them in areas where tendons otherwise would rub against bone, cartilage, or against each other. Bursae can be subcutaneous at the convex surface of joints where skin has to flex acutely or they can be articular as extensions of synovial capsules around certain joints.

Synovial Sheaths

Long tendons in the hand and foot often run through tubular bursae filled with fluid. The nature of these synovial sheaths with their "visceral" and "parietal layers" and their "mesotendon" carrying in a blood supply is sketched in Figure II-23.

The Nervous System

General Principles: The central nervous system and special senses will be covered briefly in the last chapter of this book. Here we simply want to introduce you to the general make-up of the nervous system. For convenience it is subdivided into a **central portion** encased in bone and a **peripheral portion** lying outside the bony shelter. For convenience, once more, we can consider that each division contains three parts:

A. Central Nervous System (CNS) (Fig. II-24)
   1. Brain (cerebral hemispheres)
   2. Brain Stem and Cerebellum
   3. Spinal Cord
B. Peripheral Nervous System (PNS)
   1. 12 pairs of Cranial Nerves
   2. 31 pairs of Spinal Nerves
   3. The Autonomic Nervous System (ANS)
      a. Sympathetic Division
      b. Parasympathetic Division

The chief function of the central nervous system is to maintain voluntary control over muscles of the body and to be concerned with all information coming to it from any source, sensory or otherwise, whether at or

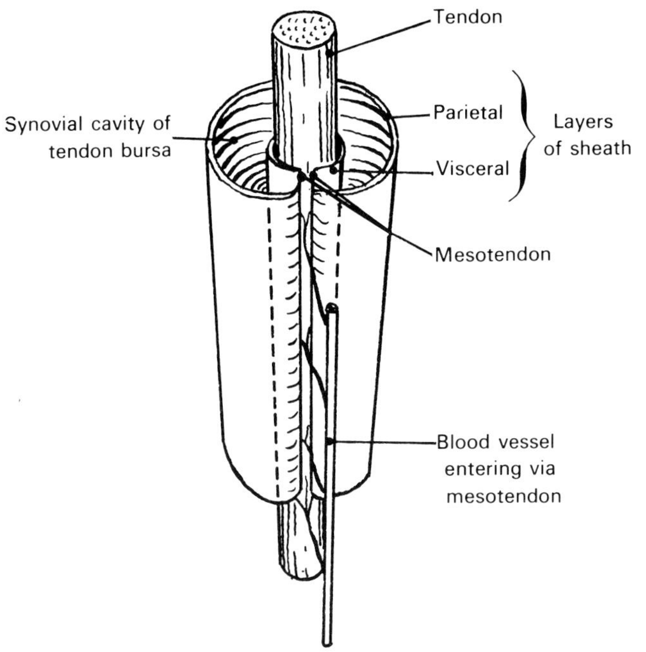

Fig. II-23. Diagram of the synovial sheath about a tendon. Most tendons do not have such tubular bursae about them, but they are common in the flexor and extensor tendons of the hand and foot where many bones are crossed by the tendon on its way to its insertion. Sheaths extend about a half inch above and below the sites of friction.

below the level of conscious awareness. With it we perceive, formulate concepts, dream, act, and even exert control over involuntary bodily mechanisms that we rarely notice in our daily living.

We can delay discussion of the brain and brainstem until a later chapter, but we can hardly neglect a consideration of the spinal cord and its relation to spinal nerves and the autonomic nervous system at this time. Even as the skin is reflected in dissection, one meets with the peripheral endings of certain spinal nerves. Most of these cutaneous nerve fibers are sensory in function, but those to smooth muscle and glands in the skin are visceral motor. Somatic motor nerves to skeletal muscle just below the skin also will be encountered. It will be well to know from whence such fibers come and how they are organized.

The spinal cord is the part of the CNS that is well protected by the bones of the vertebral column (Fig. II-25B). The vertebral canal of any single vertebra (except, perhaps, the atlas) hardly slips over a finger. Hence, the diameter of the spinal cord is less than that of a finger and is not uniform, being narrow in its thoracic segments and wider at its cervical and lumbar enlargements (Fig. II-25A). The cord does not fill the entire length of the vertebral canal in the adult as it does in the fetus. Its terminal conical tip **(conus)** *ends at the upper border of the 2nd lumbar vertebra* (*VL-2*) (Fig. II-25B and Fig. II-27B).

As in the case of the brain, the spinal cord has three **meningeal coverings** (Fig. II-26). Closest to it, like a snug, thin surgeon's glove, is the delicate **pia mater,**

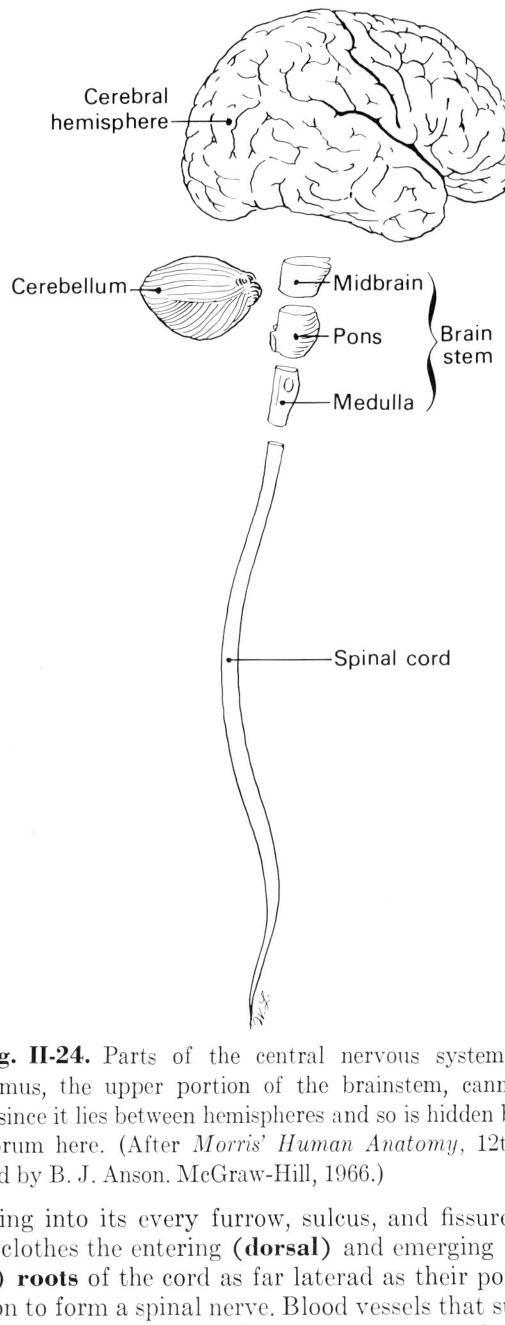

Fig. II-24. Parts of the central nervous system. The thalamus, the upper portion of the brainstem, cannot be seen since it lies between hemispheres and so is hidden by the cerebrum here. (After *Morris' Human Anatomy*, 12th ed., edited by B. J. Anson. McGraw-Hill, 1966.)

dipping into its every furrow, sulcus, and fissure. Pia also clothes the entering **(dorsal)** and emerging **(ventral) roots** of the cord as far laterad as their point of fusion to form a spinal nerve. Blood vessels that supply cord and roots run upon this coat and are carried into the fissures with the pia.

The second investment of the cord is the filmy, spider-webby **arachnoid.** This coat is separated from the pia by a space filled with **cerebrospinal fluid,** the **subarachnoid space.** The arachnoid fastens to pia by means of thread-like **arachnoidal trabeculae** which bridge the fluid-filled space (Fig. II-26A).

The fluid pushes the arachnoid tightly against the third and outermost meningeal covering, the tough **dura mater,** thus eliminating a potential subdural space. The dura around the cord is single-layered, not double as it is over the brain. Dura and arachnoid (and, hence, the subarachnoid space) descend to a lower level in the vertebral canal than the spinal cord, itself—to the level of the 2nd sacral vertebra (VS-2)

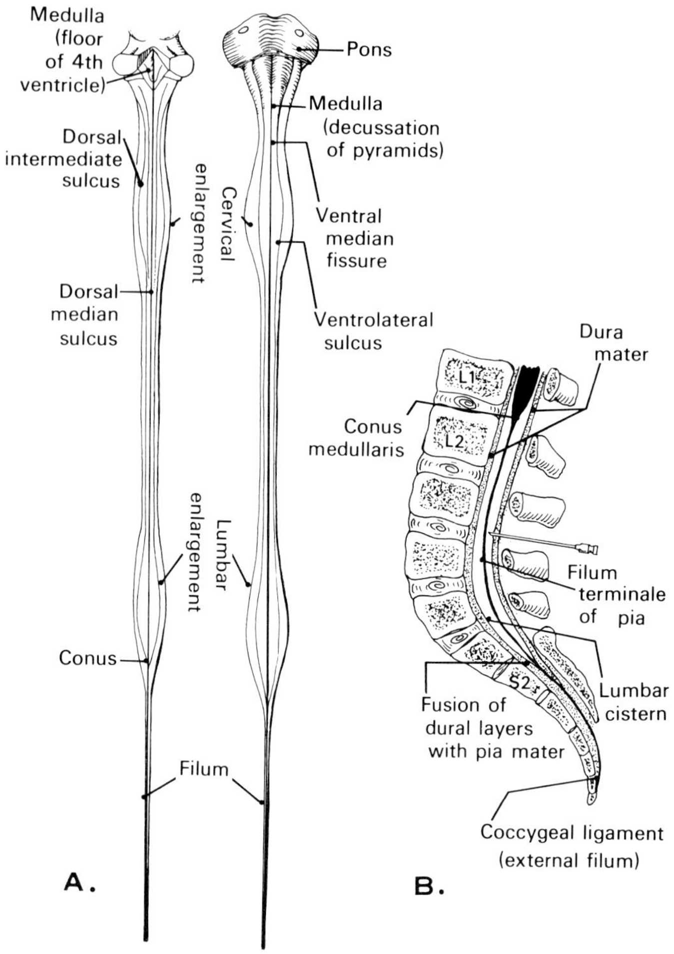

Fig. II-25. The spinal cord.

A. In ventral view on the right and in dorsal view on the left. It blends insensibly with the medulla of the brain stem above.

B. Extent of cord in the lower part of the vertebral canal. Needle of a syringe is shown in subarachnoid space below cord level. This is the area of the lumbar cistern which is tapped for cerebrospinal fluid samples in a lumbar puncture.

(Fig. II-25B). Meanwhile, the pia, having invested the cord, spins out at the conus as a glistening pial thread, the **internal filum terminale.** At the level of VS-2 pia and dura join and the filum continues in the vertebral canal to emerge, finally, through the sacral canal. It anchors on the coccyx as the **filum terminal externum,** or **coccygeal ligament,** by spreading out in a series of strands. In this way, the cord is anchored in the vertebral canal caudally. The continuity of the cord with the brainstem through the foramen magnum of the occipital bone furnishes stability at that end. Pia flattens into two lateral bands that run much of the length of the cord. A toothed process from this longitudinal band projects laterally between dorsal and ventral roots to anchor to dura. This lateral anchoring structure of pia mater is called the **denticulate ligament** (Fig. II-26A and B). The first tooth of this band goes through the foramen magnum to anchor on dura of posterior fossa; the last tooth comes off at about VT-12.

The dura, while free in the vertebral canal, anchors laterally at the inner margins of the **intervertebral foramina.** In so doing, it forms dural sleeves over the dorsal and ventral roots, forcing them to combine at these foramina of exit from the bony vertebral canal as **spinal nerves.** The dural sleeve encloses a swelling along the dorsal root which houses the sensory nerve cells in the **dorsal root ganglion** or **spinal ganglion** (Fig. II-26A and B). The dura of the cord, cranially, is continuous with the outer layer of dura over the brain, but it does not rest directly against the bone of the vertebral canal in the way the dura of the brain rests against the brain case. Fat and a plexus of veins fill an epidural space, keeping the spinal dura away from bone except in the regions of the intervertebral foramina.

The subarachnoidal sac below the level of cord termination (Fig. II-25B) is known as the **lumbar cistern.** A syringe needle entering the cistern below VL-2 and above VS-2 can draw off cerebrospinal fluid for clinical sampling without running the risk of damage to central nervous tissue; *i.e.,* to the spinal cord. Figure II-25B might lead you to believe that the vertebral canal here is empty of other structures. Figure II-27 indicates otherwise. It is filled, as well, with the rootlets of the last 11 spinal nerves from cord segments L1–L5, S1–S5, and CO1, all of which must seek intervertebral foramina of exit below the level of cord termination in the vertebral canal. These lowest rootlets descend around the filum terminale in the fashion of the hairs of a horse's tail, so that the complex has been called the **cauda equina.**

Dorsal and ventral roots do not spring spigot-like from the cord but emerge gradually over a distance along the cord as a series of filaments or radicals that gather together into a single root (Fig. II-26B). The distance between the top filament of one root and the highest filament of the next root below constitutes the length of a spinal cord segment. This length or thickness varies locally along the cord; the last 11 segments, in particular, must pack very closely at the conus (Fig. II-27).

Since the cord does not fill the vertebral canal, the cord segments do not correspond well with the vertebral segments. Yet every spinal nerve must exit at its proper intervertebral foramen. Hence, roots below VL-2, in particular, must become quite long as they run downward to reach their foramina of exit. They are anchored to the rims of their respective foramina as they receive dural sheath, you will recall. In addition, there are 8 cervical cord segments and only 7 cervical vertebrae. This means that the 1st cervical nerve, which has no sensory ganglion, must exit between the base of the cranium and *above* the level of the body (if we can say the atlas has a body) of the 1st cervical vertebra (Fig. II-27). The next 7 cervical nerves all exit in intervertebral foramina *above* the level of their corresponding vertebral body; *e.g.,* C4 *above* VC-4, *etc.,* C8 being

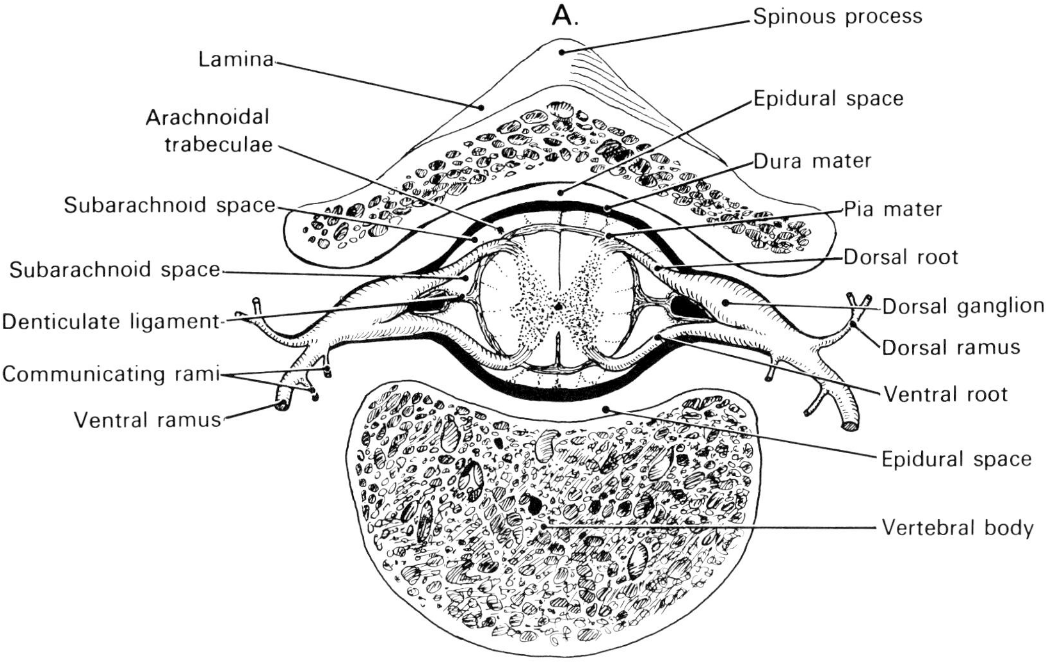

Fig. II-26. The spinal cord in relation to meningeal investments within the vertebral canal. **A.** Cross-sectional view at midthoracic cord level.

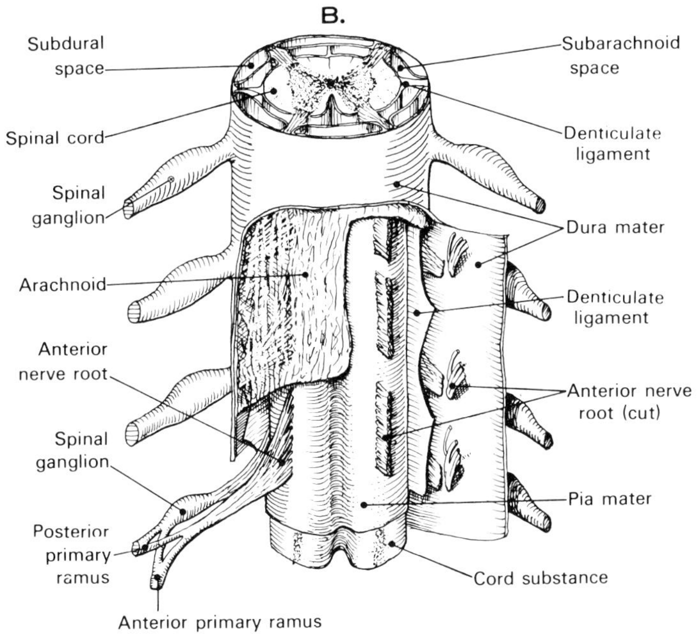

B. Longitudinal view with meninges cut away in stages to expose filamentous rootlets and the pial denticulate ligament.

an exception. It emerges above the body of VT-1. As a consequence, *all* other spinal nerves must, therefore, exit in intervertebral foramina *below* the level of their respective vertebral bodies; *e.g.,* T6 *below* VT-6; L3 *below* VL-3; and S2 *below* VS-2, *etc.* (Fig. II-27).

Typical cross sections at the thoracic cord level are sketched in Figures II-26A and II-29. There is an H-shaped core of **gray matter** surrounded by a mass of **white matter.** The pinpoint canal in the gray is the **spinal canal,** a remnant of the neural canal of the fetal neural tube. (Do not use the terms "spinal canal" and "vertebral canal" interchangeably.) **Nerve cells (neurons)** dwell in the **gray matter** of the cord as **motor neurons** and **interneurons** (connector, internucial, or associational neurons). **Primary sensory neurons** live in the **dorsal root ganglia.** The **white matter** consists of the **axonal processes** of interneurons; the incoming axonal processes of some dorsal root ganglion cells; and axonal processes of neurons of the brain and brainstem which descend in the white matter to enter the gray matter of the cord at various levels. Thus, white matter consists of ascending and descending **fiber tract systems** formed by the axonal processes of nerve cells. This distinction between white and gray matter will hold for brain and brainstem, as well, but in these regions the location of gray matter and its patterning will differ from the H-shaped central core of the spinal cord.

Even the configuration of the H-shaped gray and the

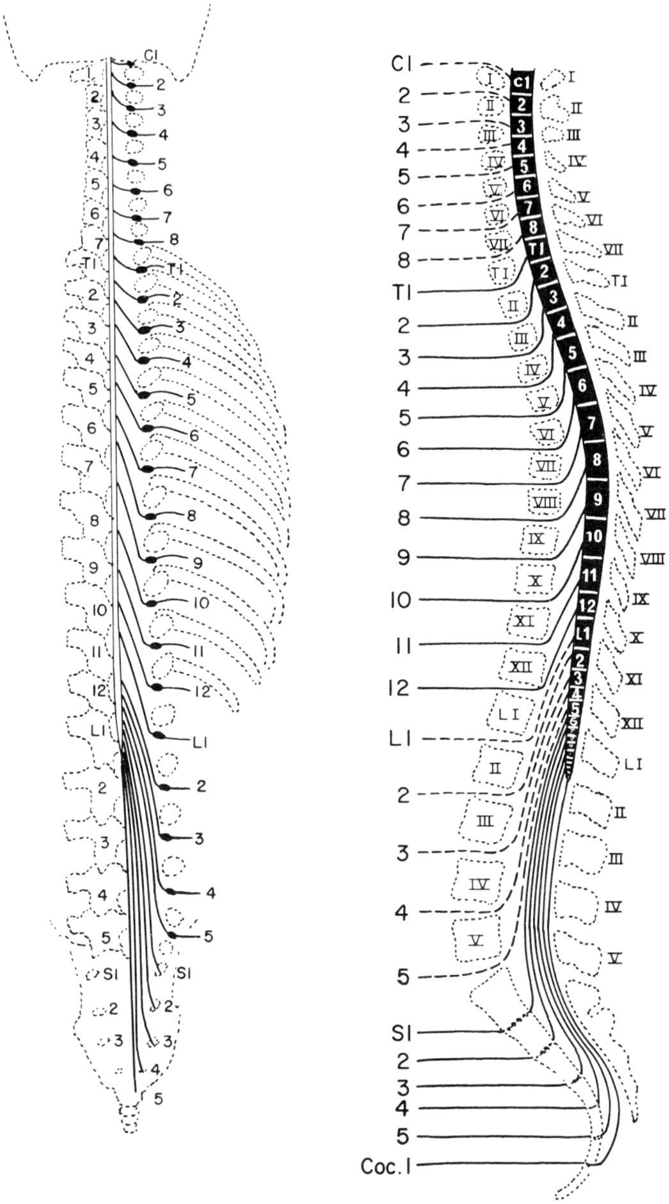

Fig. II-27. Relation of cord segments to vertebral segments. These relationships affect the emergence of spinal nerves relative to the bodies of their corresponding vertebrae. (See text for discussion.)

A. As viewed from the back so that intervertebral foramina can be identified.

B. Schematic lateral view so that vertebral bodies can be identified.

relative amount of white matter will vary at different cord levels. The dorsal and ventral horns will be bulkier at cervical and lumbosacral levels where there are many more neurons to serve the heavier limb muscles. Where, chiefly, trunk muscles must be served, gray columns in the thoracic regions can be much more slender. Also, at cord segments T1 through L2 there exits an intermediate horn of gray matter where *visceral motor neurons of the sympathetic division of the ANS reside.* Such an **intermediolateral cell column** of gray exists nowhere else in the cord. Thus, thoracic cord segments contain sensory related interneurons in the dorsal horn of gray matter; visceral motor neurons for smooth muscle, cardiac muscle, and glands in the intermediate gray; and somatic motor neurons for skeletal muscle in the ventral horns.

Incoming axons from cells in the dorsal root ganglia (primary sensory neurons) synapse, as a rule, on interneurons. Only the subserving muscle synapse directly on ventral horn cells. The role of the interneuron is twofold. Its axonal process can remain in the gray matter to synapse with a motor nerve cell (in the intermediolateral *or* in ventral gray at thoracic cord levels) and so form a **reflex arc** between sensory and motor neurons. Or, its axonal process can leave the gray matter to enter the white matter as a fiber tract carrying sensory information to higher or lower levels.

Thus, it is the **nerve cell or neuron** that is the **structural unit** of the nervous system and *its interlinkages* (by contact) with other neurons that *make up functional neural mechanisms.* The nerve cell consists of a **cell body** variable in size which contains its nucleus; one or more cytoplasmic extensions from this body; called its **dendrites;** and one extension, called its **axon.** Neurons will vary considerably in shape in respect to the configuration of these processes (Fig. II-28). Dendrites may be long or short, few or multiple and branching, but they do not leave the gray matter. The exception, of course, is the peripheral process of the primary sensory neuron. This process has to be classed as a dendrite on a functional basis, for in the *intact* nervous system the direction of neural conduction of an impulse proceeds from dendrites to cell body to axon. *Thus, dendrites conduct toward the cell body; axons, away from it.*

Axons are always **single** processes as they leave the cell body. Farther away from it they may give off collaterals, and they, generally, branch at their terminal ends. They can be short or extremely long; those to the leg in the sciatic nerve, for example, may be as much as 3 feet in length.

Primary sensory neurons in dorsal root ganglia and most primary sensory neurons in cranial ganglia are **unipolar** (Fig. II-28A). One process *only* leaves the cell body to bifurcate secondarily into a peripheral process (the dendrite) and into a central process directed toward the cord or brainstem (the axon). Such neurons have also been designated as **pseudounipolar** on this account. Because the peripheral dendritic proc-

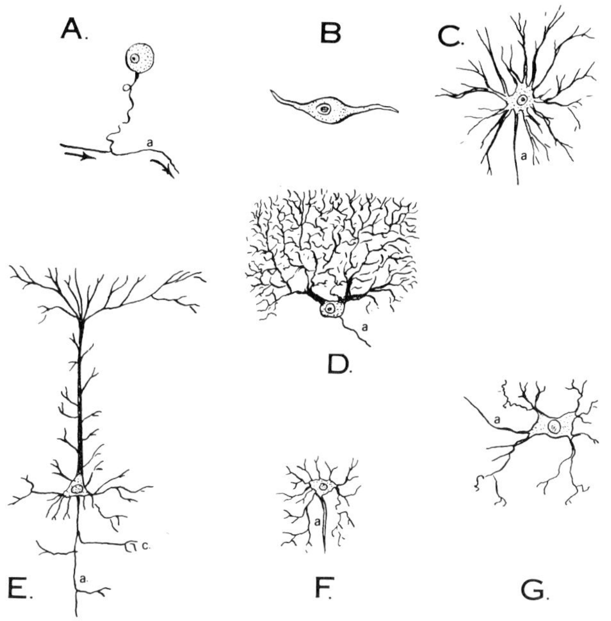

Fig. II-28. Types and variations in nerve cell configuration.

A. Unipolar neuron of the dorsal root ganglion (primary sensory neuron).

B. Bipolar neuron. Special sensory as in spiral ganglion of cochlea.

C–G are all multipolar neurons; a = axon; c = collateral.

C. Ventral horn cell (motor neuron). Note the many dendrites and the single, fine axon.

D. Purkinje cell of cerebellum (motor neuron). Bushy dendritic tree and very fine axon (with recurrent collaterals to other Purkinje cells).

E. Pyramidal (voluntary motor) cell of cerebral cortex. Apical and lateral basal dendrites. Axon is fine and leaves base, later giving off collaterals.

F. Small interneuron of cord. Difficult to locate axon.

G. Autonomic ganglion cell.

ess has much the appearance of an axon, you will find physiologists calling it an axon, and to confuse matters further, they will talk about fanciful "axonal reflexes" in the periphery. Fence sitters may call this sensory dendrite the dendraxon. But as anatomists, we will call it a dendrite and realize that it does *not* give off peripheral collaterals to smooth muscle in the periphery, simply because it is *not* a motor neuron. Whatever an "axon reflex" is, other than a schematic sketch on a blackboard or a scheme in someone's mind, it has, as yet, no morphological basis. That antidromic conduction over this process produces an effect of some kind is one thing. What its explanation is, is quite another; and even at the present moment, it is imaginative and conceptual, rather than factual. The event is factual; how it is produced is conjectural.

Only the more phylogenetically-older sensory neurons are **bipolar** (Fig. II-28B); *i.e.*, have one process, an axon and a dendrite, at opposite poles of the cell body. The most ancient of these are the cells in the olfactory mucosa. Others are found in the retina, the spiral ganglion of the cochlear, and in the vestibular ganglia.

All motor neurons and interneurons are **multipolar** (Fig. II-28C–G), having several dendrites and one axon. Motor neurons of the ventral horn are called **somatic motor neurons** or **ventral horn cells.** They are somatic motor neurons because they supply skeletal muscle derived from embryonic somites. (Branchial motor neurons supply **skeletal muscle** derived from branchial arch mesoderm of the embryo. Most textbooks call them special visceral motor neurons, a poor name, since they do not subserve the viscera. Most branchial motor neurons arise at brainstem levels.) Ventral horn cells are also called the **lower motor neurons** since they are the "final common path" to skeletal muscle of the neck, trunk, and limbs (with a few exceptions where brainstem branchial motor neurons reach some neck and shoulder muscles). They are "lower motor neurons" because they are the last nerve cells to make contact with muscle. They are the "final common path" in the sense that neural impulses of many varieties impinge upon them to discharge or inhibit them—direct sensory impulses, or indirect impulses through interneurons or the desecending axons from brain and brainstem levels. Even the latter generally contact an interneuron first. The left side of Figure II-29 shows that the axons of ventral horn cells leave by the ventral roots to join spinal nerves and are distributed to skeletal muscle.

The Autonomic Nervous System (ANS): This is our visceral or involuntary nervous system which innervates smooth muscle, cardiac muscle, and glands. The ANS is under the control of the CNS and is regulated, to a degree, hormonally, but if its connections with the CNS are severed, certain of its organs continue to function; *e.g.*, the heart and gastrointestinal tract. The activity of these organs simply is not as easily accelerated or slowed under such conditions.

Visceral motor neurons of the intermediolateral gray of cord segment T1–L2 are lower motor neurons, also, and while their axons exit over ventral roots in the spinal nerves T1–L2, they make no contact with smooth muscle, cardiac muscle, or glands. Instead, they leave these particular spinal nerves through **white rami communicantes** to synapse with other visceral motor neurons in outlying **autonomic ganglia.** The right side of Figure II-29 shows that its takes *two* visceral motor neurons to reach visceral structures. The first, whose cell body lies in the cord, is called the **preganglionic neuron;** the second, in a ganglion, is called the **postganglionic neuron.** The axon of the postganglionic cells then terminate adjacent to smooth muscle, cardiac muscle, or glands.

Autonomic ganglia of the **thoracolumbar outflow** (Sympathetic Division) of the ANS have their outlying ganglia in one of two places. They lie as somewhat segmental, ganglionated clusters in **chain** (paraverte-

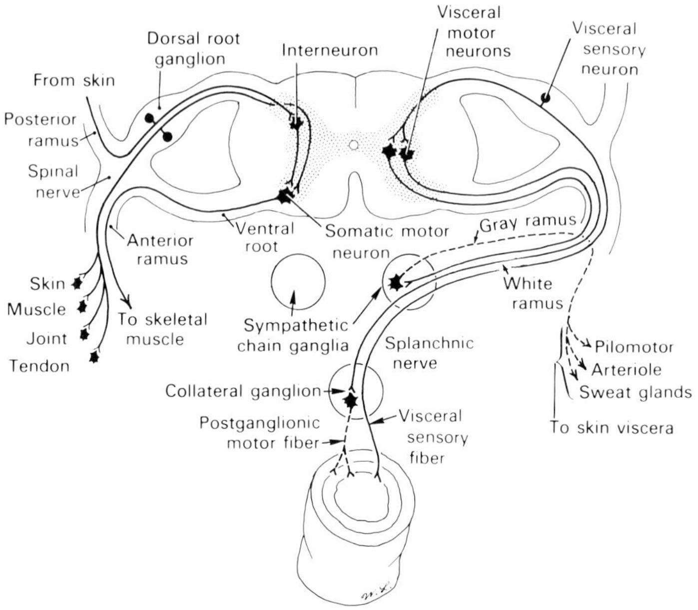

Fig. II-29. Peripheral nerve composition at thoracic cord level and a comparison between somatic motor and visceral moto r(sympathetic) innervation (see text for discussion).

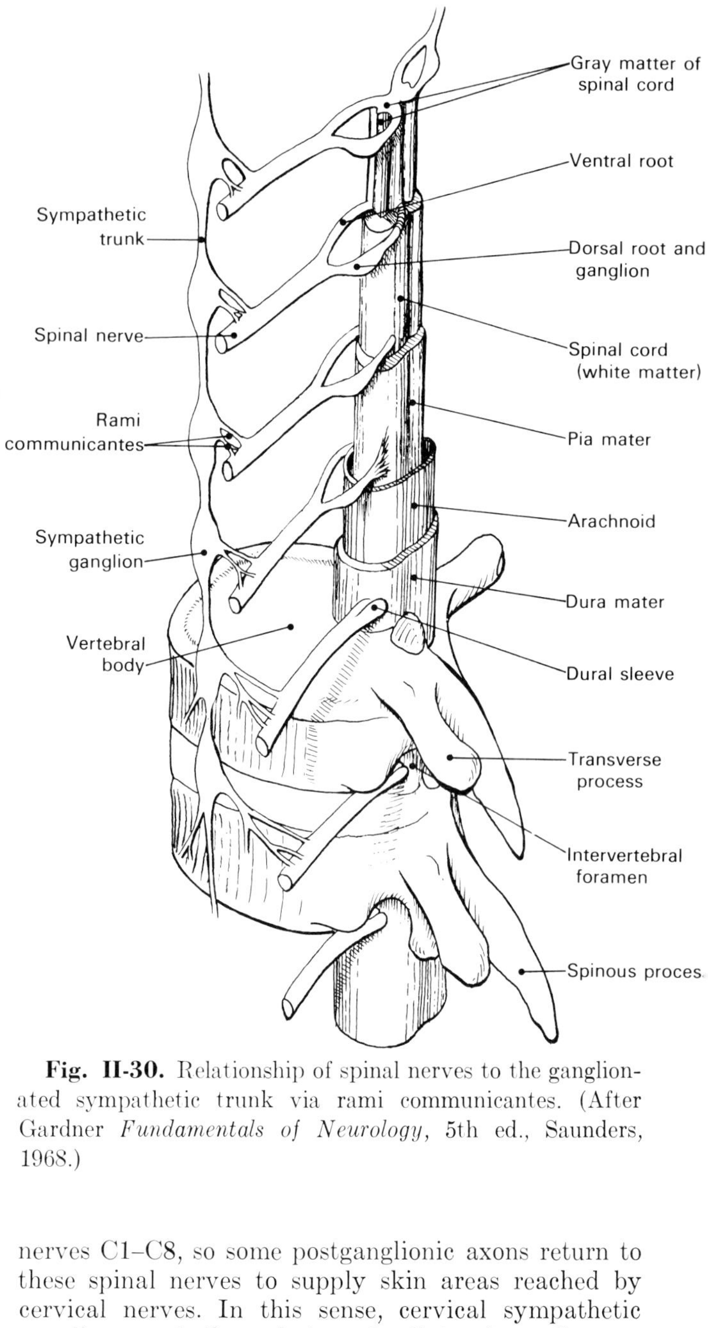

Fig. II-30. Relationship of spinal nerves to the ganglionated sympathetic trunk via rami communicantes. (After Gardner *Fundamentals of Neurology*, 5th ed., Saunders, 1968.)

bral) **ganglia** of the **sympathetic trunk** which descend upon the vertebral bodies, (Fig. II-30) or they lie farther removed in **collateral ganglia.** Collateral ganglia lie in one of two places. They lie in the neck as the **cervical sympathetic ganglia** somewhat isolated from the rest of the ganglionated chain. Or they lie along the base of the three unpaired branches of the abdominal aorta, at the roots of the coeliac artery and superior and inferior mesenteric arteries. If a preganglionic neuron synapses in a chain ganglion (reaching it through a white ramus), the postganglionic axon returns to the spinal nerve through a **gray ramus communicantes** and is distributed by the spinal nerve to **skin viscera;** *i.e.*, to pilomotor muscle of hairs, to smooth muscle of arterioles, and to sweat glands. However, a preganglionic axon reaching a chain ganglion need not synapse there. It can pass *through* the ganglion and traverse the sympathetic trunk to reach the cervical sympathetic ganglia in the neck; or it can pass *through* the chain ganglia and *leave* the sympathetic trunk as a **splanchnic nerve** to reach collateral ganglia of the abdomen. If it reaches the **superior cervical sympathetic ganglion** to synapse, postganglionic axons generally follow the internal and external carotid arteries and their branches to **head viscera.** If the synapse is in *any* of the cervical ganglia (which vary in number but rarely are more than four), **cervical sympathetic nerves** leave these ganglia to supply **viscera of the thorax.** Thus, the sympathetic ganglia of the neck, at the upper end of the sympathetic trunk, supply viscera of the head and thorax. However, all these cervical ganglia have **gray rami,** as well, to spinal nerves C1–C8, so some postganglionic axons return to these spinal nerves to supply skin areas reached by cervical nerves. In this sense, cervical sympathetic ganglia serve both as chain and collateral ganglia.

If the preganglionic neuron sends its axon as a splanchnic nerve to the coeliac, superior, or inferior mesenteric ganglia, postganglionic neurons of these ganglia send their axons over the branches of the aorta to reach all viscera within the abdomen and pelvis. Sympathetic chain and collateral ganglia lie some distance from the visceral organs supplied, so their postganglionic axons are sometimes quite long. But all smooth muscle, cardiac muscle, and glands get a sympathetic motor supply in the manner we have indicated above.

The ANS has a **parasympathetic** or **craniosacral**

division, as well. Preganglionic parasynaptic neurons of this division lie in the gray matter of the brainstem and at sacral cord levels S2–S4. Their axons have nothing to do with spinal nerves or rami communicantes. They leave in cranial nerves **III, VII, IX, and X** (oculomotor, facial, glossopharyngeal, and vagus), or from the sacral cord as the **pelvic nerve.** They all end in outlying ganglia, also, but in small, **terminal ganglia** close to or within the walls or capsules of the organs innervated. Hence, parasympathetic preganglionic axons are very long; axons from the postganglionic cells in terminal ganglia can be quite short since they lie in or close to smooth or cardiac muscle or the glands of the visceral organs. *Parasympathetic fibers do not reach skin viscera, nor do they innervate the adrenal gland.* But they supply every other viscus supplied by the sympathetic or thoracolumbar outflow of the ANS. Cranial nerves III, VII, and IX limit their distribution to the head viscera. The vagus, or cranial nerve X is, indeed, a "vagrant" or wanderer. It supplies thoracic viscera and abdominal viscera as far as the middle of the transverse colon. The sacral pelvic nerve supplies the rest of the colon and all the pelvic viscera. Thus, visceral organs, except skin and adrenal gland, get a dual autonomic supply, sympathetic and parasympathetic, and in each case by two neurons, a pre- and postganglionic neuron. The latter lies either in sympathetic chain or collateral ganglion, or in a parasympathetic terminal ganglion.

The two systems tend to act antagonistically in furnishing this dual supply to viscera. One system or the other will either facilitate motility, contractility, or secretion of a given viscus while the other inhibits or arrests such action. It is not always easy to remember which system does what to which organ. As a fairly reliable generality we can say that the sympathetic division of the ANS arouses us to respond to meet an emergency. The parasympathetic system returns the organism to the calm, steady, homeostatic state once the emergency has passed.

We need then only recall what would logically happen to our bodily organs in the emergency situations of "Fright, Fight, or Flight" to understand the effect of sympathetic influence upon them and realize that the parasympathetic innervation of these same viscera will exert the opposite effect.

Take the case in which the organism is aroused to fever pitch in fight. The pupils are dilated in order to provide a wide field of vision; it's nice to know from whence the fists or missiles are coming in order to avoid them. The mouth is dry; saliva is thick and ropy, not watery. Mucous rather than serous glands are activated. This is no time to eat a sandwich. Peripheral skin arterioles constrict. Blood is shunted to feed the muscles that are working hard. Moreover, the sight of blood from a superficial wound in the heat of the fray might send some erstwhile warrior running home. Naturally, the heart rate accelerates to move blood faster for tissue oxygenation and to remove metabolic wastes. Its own coronary vessels must, therefore, dilate since the heart, too, is working harder than normal, and its tissues must be fed first. The bronchial tubes dilate to increase the size of the airways as oxygen demands rise. As the body overheats, sweat glands are activated to cool the skin by evaporation of moisture. Peristalsis and gastric secretion come to a standstill. There is no time for dinner now. The secretions of the adrenal medulla pour into the blood stream to keep these sympathetic discharges aroused as long as the emergency situation exists. When it has passed, a reversal of these visceral behavior patterns ensues gradually as the parasympathetic system once more assumes control.

You have seen how cartoonists express fright in their characters. The hairs of the head stand straight on end. Our own hairs do not act in this exaggerated way, though those of lower animals do bristle in fright or fight. We do get "goose pimples" on the arms and back of the neck in creepy situations as pilomotor muscles partially erect hairs. This is another sympathetic response of the ANS, since parasympathetic fibers do not supply skin.

The ANS is not entirely a motor system. It has its sensory side, as well, for we can feel the pain of a diseased viscus and we are aware of distention in many of our hollow organs. Figure II-29 shows the route of return of sensory messages from abdominal viscera supplied by sympathetic splanchnic nerves, for example. The sensory fibers follow blood vessels back to a collateral ganglion through which they pass to enter a splanchnic nerve. The return route continues *through* a chain ganglion, the sympathetic trunk and *via* a **white ramus** to a spinal nerve. The final leg of the journey is over the **dorsal root** to a visceral sensory cell in a **dorsal root ganglion.** The central process (axon) of such a cell can enter the gray matter of the cord to synapse directly, or indirectly through an interneuron, upon a visceral motor neuron in the intermediate gray, or upon a somatic motor neuron of the ventral horn. Thus, **viscerovisceral reflex arcs** may be established; *e.g.,* distention of the gut by a bolus initiates reflex contraction behind it to move it along. Or, **viscerosomatic reflex arcs** can be effected, as well; *e.g.,* pain in the appendix leading to reflex contraction of abdominal wall musculature. **Somatovisceral** reflex arcs are also possible; *e.g.,* cooling of the skin eliciting "goose pimples." Incoming visceral sensory neurons can also contact interneurons that build up pathways to conscious levels, for we are aware of uncomfortable distention and pain from the viscera.

Somatic sensory information such as pain, touch, temperature, and proprioception from skin, muscle tendons, and joints is conveyed to somatic sensory cells of the dorsal root ganglia by spinal nerves. The central processes (axons) of these cells enter the cord and by way of direct or collateral branches are able to do a number of things, separately or simultaneously.

1. Muscle afferents can effect reflex arcs by ending directly upon motor neurons, or other afferents can use

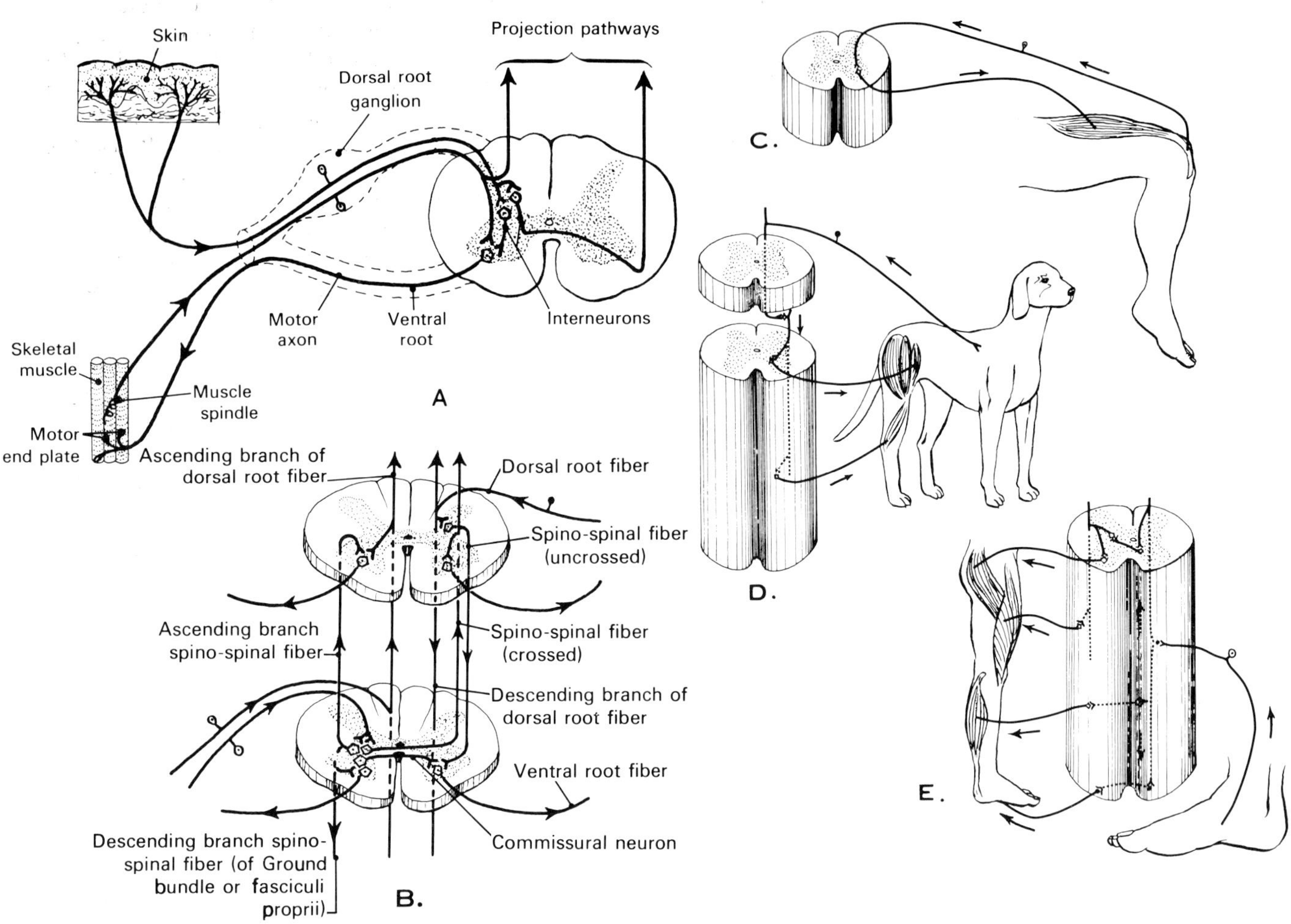

Fig. II-31. Some types of spinal cord reflexes.
A and **B.** Two- and three-neuron intrasegmental reflex arcs and projection tracts to higher levels.
C. The knee-jerk, a simple, two-neuron **intersegmental** reflex arc.
D. The scratch reflex, a more complicated multineuronal, homolateral, **intersegmental** reflex arc.
E. The crossed-extensor reflex, a multineuronal, contralateral reflex arc. (Descriptions of each in text.)

an interneuron to reach the motor cells in the gray matter of the cord indirectly (Fig. II-31A).

2. They can end upon interneurons whose axons, in turn, enter white matter to effect reflex arcs at a higher or lower level (Fig II-31B).

3. The interneuron upon which the sensory fiber ends may build up a projection tract to levels of conscious awareness so that we can appreciate the particular sensory modality that is being transmitted (Fig. II-31A).

4. Or, the sensory fiber can enter the dorsal white matter to build up its own projection tract to the lower brainstem (medulla) before it synapses upon an interneuron (Fig. II-31A).

5. Or, by means of a number of collateral branches, the sensory axon can do a number of these things simultaneously as it enters the cord (Fig. II-31A).

Reflex arcs can be **intrasegmental;** *i.e.*, effecting a connection at the immediate site of entry of a sensory axon; or they can be **intersegmental,** the interneuron connecting at a level higher or lower than the level of the entering sensory axon. They can be **homolateral** (ipsilateral) on the side of entry; or by way of a **commissural interneuron** whose axon crosses the midline, they can be **contralateral,** effecting a movement on the side of the body opposite the side of entry. Generally, three or more neurons are involved in reflex arcs; the sensory ganglion cell, one or more interneurons, and a motor neuron. **Tendon reflexes,** however, are examples of simple, **two-neuron,** intrasegmental, reflex arcs. For example, when, in the freely-suspended knee, the patellar tendon is tapped, proprioceptive, neurotendinal spindle receptors are stimulated as the tendon is stretched. Impulses travel over spinal nerve L3 to cells of its dorsal root ganglion. The axons of these sensory neurons synapse directly on ventral horn cells at cord level L3. The motor axons enter the same spinal nerve to innervate the Quadriceps Femoris which extends the leg. Such a "knee jerk" is a quick reflex re-

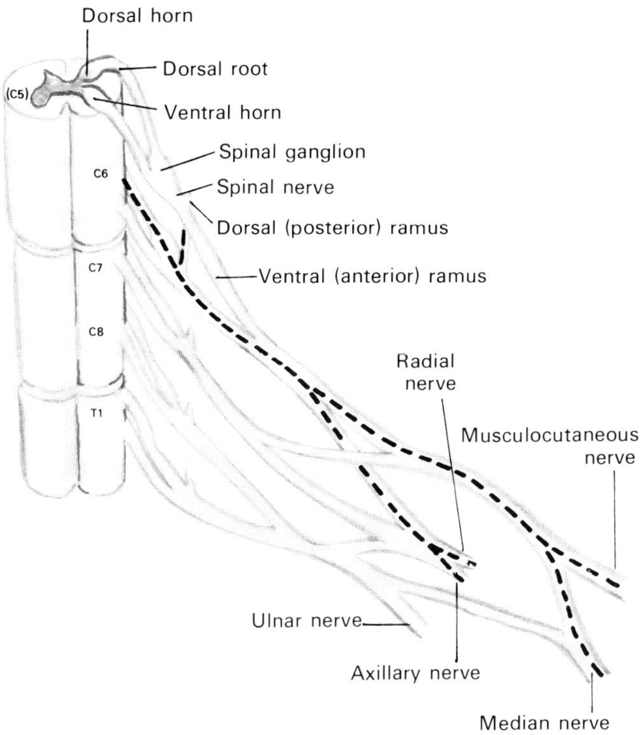

Fig. II-32. Spinal nerves forming the brachial plexus. Not all the branches of the plexus are shown. The sensory fibers of C-6 (dotted) have been traced through the plexus to their final distribution by four peripheral nerves. They also enter the posterior ramus of the spinal nerve which takes no part in plexus formation.

sponse of the muscle to tendon stretching, the leg kicking out almost as soon as the tendon is tapped (Fig. II-31C). If the range of movement is normal, we know that the arc connections are intact; if they are hyperactive or hypoactive, we suspect central nervous system malfunction at this cord level.

The homolateral, **intersegmental** arc illustrated in Figure II-31D is familiar to all of you who own household pets. If the animal is scratched in the upper back region, the hind leg comes into play to scratch the spot you are scratching, as if to attend to a noxious flea. The sensory impulse enters the cord at the thoracic level to synapse with an interneuron at that level. The axon of the interneuron enters white matter to descend to ventral horn cells at the lumbosacral levels of the same side. Such a "scratch reflex" is purely automatic. If the animal's spinal cord were severed, say at C7, it would still occur without conscious awareness on the part of the animal.

The crossed extensor reflex is an example of a **contralateral reflex arc** (Fig. II-31E). Every soldier standing at "Parade Rest" has, probably, experienced this without being too aware of it, since we all pay so little attention to our postural reflexes. After standing with most of the weight distributed on the left leg and foot for some time, the leg gets weary and we shift weight to the opposite limb. The muscles of the right thigh, leg, and foot now hold that leg in extension with the joints fixed. The sensory stimulus to make the shift comes from the constant bombardment of proprioceptive impulses entering the cord on the left side, at the level of S1. Finally, an interneuron is discharged whose axon bifurcates to traverse up and down in the white matter on the left side. At various levels, axonal collaterals re-enter gray matter to synapse on **commissural interneurons** whose axons, in turn, cross the midline to synapse on ventral horn cells. Thus, ventral horn cells on the right side, from L4–S2, send motor impulses to extensor groups of muscles in the right thigh, leg, and foot. Except for the vague sensations of weariness and relief, we are hardly aware of the changing of duties by the muscles themselves. We can begin to appreciate now how important interneurons are at cord levels by the complicated roles they play in postural reflex arcs.

One fundamental and peculiar fact, for which there is no good explanation except that we are built that way, is that *any sensation that is perceived as coming from the right side of the body is registered in the left cerebral cortex, and vice versa.* So all **projection interneurons** eventually cross to the contralateral side either at cord or brainstem level. Similarly, voluntary motor impulses (as when you willfully raise your right hand) initiated in the left cerebral hemisphere ultimately reach motor neurons of the ventral horns on the right side, and *vice versa.*

Nerve Plexuses: The spinal nerve as it leaves the intervertebral foramen at once divides into an **anterior and a posterior primary division, or ramus.** The posterior ramus supplies skin and muscles of the back. It never forms plexuses. The anterior division or ramus supplies skin and muscles of the ventrolateral body wall, but it also supplies neck and extremities. To do the latter, it enters into plexus formations (recall, too, that it was the anterior ramus of the spinal nerves that interconnected with the ANS through white and gray rami communicantes). The anterior rami of C1–C4 intercommunicate with each other to form the **cervical plexus** for neck musculature. Anterior rami of C5–T1 form the **brachial plexus** for upper limb musculature (Fig. II-32). Several major peripheral nerves emerge from such a plexus (more than are shown in the figure). By this means, sensory or motor fibers from a given set of dorsal or ventral roots can be distributed to several peripheral nerves. Anterior rami of spinal nerves L1–L4 form a **lumbar plexus;** L4–S3, a **sacral plexus.** These distribute a series of peripheral nerves to the lower limb. Anterior rami of spinal nerves S4-CO1 form **pudendal** and **coccygeal plexuses** to supply pelvic and perineal structures and the skin over the coccyx.

The anterior divisions of the 12 thoracic nerves (except for T1) do not form plexuses. They become the relatively-simple, segmental, intercostal nerves (first 11 of them) and the subcostal nerve (T12) to supply the skin and muscles of the thoracic and abdominal walls.

General Principles of Circulation

The total concept of a cardiovascular system, the concept that blood flows from the heart, which acts as a pump, through arteries and back into the heart *via* the veins, was first made known by the noted English physician, William Harvey (1578–1657) in 1616. His brilliant concise treatise, *Exercitatio Anatomica de Motu Cordis et Sanguinis in Animalebus* appeared in 1628. For 1500 years prior to this date physicians had adhered to the teachings of Galen (130 A.D.–200 A.D.). Galen's concepts now seem mighty strange. He had the venous blood of the liver endowed with "natural spirits" receive nutritive materials from the digestive tract. This blood then entered the right ventricle of the heart through the inferior vena cava. Here its impurities were removed and expelled *via* the pulmonary arteries into the lungs where the impurities were exhaled. The purified blood remaining in the right ventricle passed back into the venous channels, but some of it went through tiny pores in the interventricular septum to mix with arterial blood in the left ventricle.

The left ventricle, Galen taught, was filled with blood containing "spirits of pneuma" which had reached it from the lungs and pulmonary veins. When some of the purified blood from the right ventricle mixed with that in the left by seeping through the interventricular system, "vital spirits" were added to left ventricular blood. This blood coursed through arteries to the brain where the "vital spirit" became transformed to "animal spirit," which, in turn, was distributed to bodily organs by the nerves.

In a sense, Galen had some kind of a circuit from right heart to lungs by pulmonary arteries and from lungs to left heart by pulmonary veins. He also recognized the right heart as a venous receptacle and the left heart as an arterial receptacle. He got oxygen ("spirits of pneuma") into arterial blood. But chiefly, he regarded the heart as a distillation apparatus for "spirits" (of the nonalcoholic kind), and he got his nerves mixed up with blood vessels as conduits for fluids.

Harvey's experiments on animals clearly showed the circular course of blood throughout the body as it was propelled by a beating heart. He observed the ventricles contracting in systole to force blood into the great arteries (pulmonary and aorta) from which reflux was prevented by cardiac valves. He measured cardiac output and reasoned that there must be connections between arteries and veins to return the same volume of blood to the heart *via* the veins to complete the cycle. He showed that arteries carried blood away from the heart, while valves in veins prevented any other course of flow in them other than toward the heart. **De Mortu Cordis** was a revolution in medical conceptualization of that era.

Today, of course, we realize that a vast **capillary plexus** of thin-walled vessels in the tissues interconnect arteries and veins (Fig. II-33). Arteries enter plexuses after they have branched and reduced to small-calibered **arterioles,** the last arterial twigs with muscle in their walls. The capillary plexuses are drained by small venules, tributaries of veins that increase in caliber as they join and course to the heart. We also realize that the heart is a two-chambered pump, a right-sided venous pump for the **pulmonary** or **lesser circulation** and a left-sided arterial pump for the **systemic** or **greater circulation,** and that it has, in reality, four chambers, a right and left **atrium** and a right and left **ventricle,** to serve these purposes (Fig. II-34).

More than that, tissues are also drained by a series of blind-ending **lymphatic capillaries** that begin in the connective tissue spaces (matrix) alongside blood capillaries (Fig. II-33). They drain into larger lymphatic vessels which empty eventually as the right or left lymphatic ducts at the junction of the jugular and subclavian veins in the neck. In this way, like blood, lymph is returned to the right heart by venous channels. Lymph returns some of the water and crystalloids of plasma which have left the blood capillaries for the tissue spaces; the rest of this tissue juice, along with cellular metabolites, re-enters the venous side of the blood capillary plexus (in health). In this sense, lymphatics furnish accessory venous drainage for connective tissue matrix. Lymph channels pass through lymph nodes along their course. Here debris in lymph is filtered out by phagocytic cells, and here the nongranular leucocytes and monocytes that are manufactured in lymph nodes are constantly added anew to the circulation. Lymphatic capillaries of the digestive tract may also absorb considerable fat (chyle) after particularly fatty meals.

Figure II-34 illustrates the plan of circulation through an adult heart. This plan differs considerably from circulation through the fetal heart when the unborn organism is not breathing air.

Venous blood from levels above the Diaphragm are returned to the **right atrium** by the **superior vena cava** and azygos system, while that from below the Diaphragm enters the right atrium by the **inferior vena cava.** During right ventricular **diastole,** when the right ventricle is at rest, blood from the right atrium rushes into the right ventricle through the right **atrioventricular (tricuspid) valve.** The entire right heart is, therefore, a venous receptacle containing **deoxygenated** blood that has been returned after tissue utilization. Moreover, Galen was correct in indicating that absorbed nutritive materials from the digestive tract (*via* the portal veins) returned to the liver and then to the right side of the heart (*via* liver sinusoids, hepatic veins, and the inferior vena cava). Only Galen thought the inferior vena cava emptied directly into the right ventricle instead of the right atrium.

In ventricular **systole** (contraction) the flaps of the right A-V (tricuspid) valves are closed, preventing reflux of blood into the right atrium. The **semilunar**

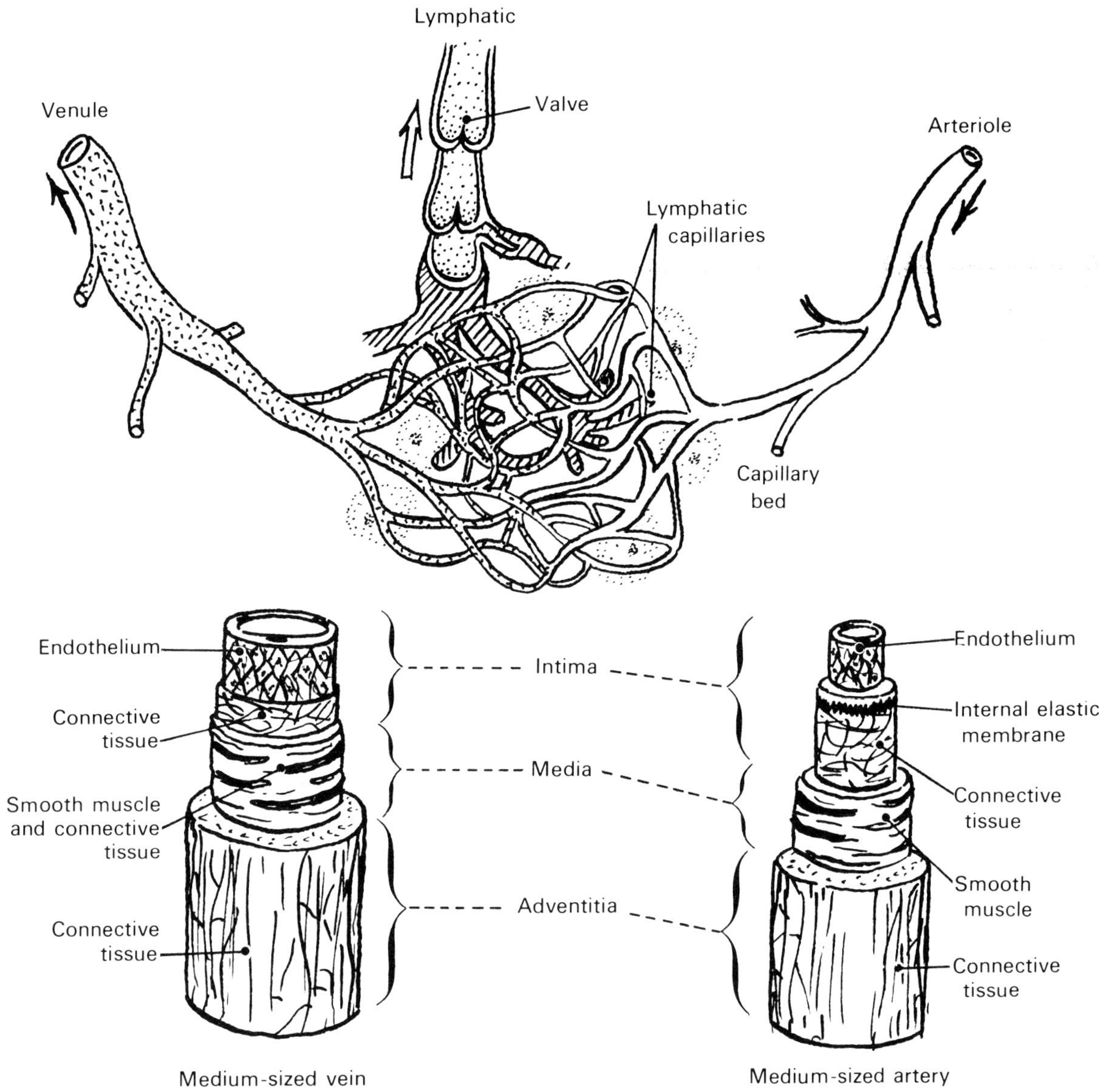

Fig. II-33. The arterial supply of a tissue capillary bed and its venous and lymphatic drainage (central). On the right, the coats of a typical muscular artery are shown and on the left, the coats of its companion vein.

valves of the **pulmonary trunk** are forced open under pressure, and the deoxygenated blood (with its added nutrients from the digestive tract) is sent to the lung *via* the **pulmonary arteries** to be aerated in the capillary beds of the lung. Here CO_2 is given up from the blood and O_2 enters the plasma to be transported by the hemoglobin of the red cells. The pulmonary arteries and umbilical arteries are exceptions to the rule that arteries convey oxygenated blood, for these contain venous or deoxygenated blood. The **pulmonary veins** return oxygenated blood from the lungs to the left atrium. This circuit from right atrium to left atrium completes the **pulmonary** or **lesser circulation.** In this sense, too, pulmonary veins and umbilical veins are odd, since they carry oxygenated blood.

The left side of the heart is concerned with the **systemic** or **greater circulation.** It must pump the oxygenated blood with its nutrients throughout the entire body with enough force to move it through the resistances of narrowed arteriolar beds and the veins, to be returned, with its loss in oxygen and increase in carbon dioxide, to the right heart, once more. This is a greater workload than merely moving blood from right to left side of the heart *via* a pulmonary circuit. The left ventricle has, therefore, much heavier cardiac muscle in its walls than the right ventricle.

In left ventricular diastole, blood from the left atrium enters the ventricle through the **bicuspid** or **mitral valve** (left A-V valve). In ventricular systole, this valve is closed and the semilunar valves of the

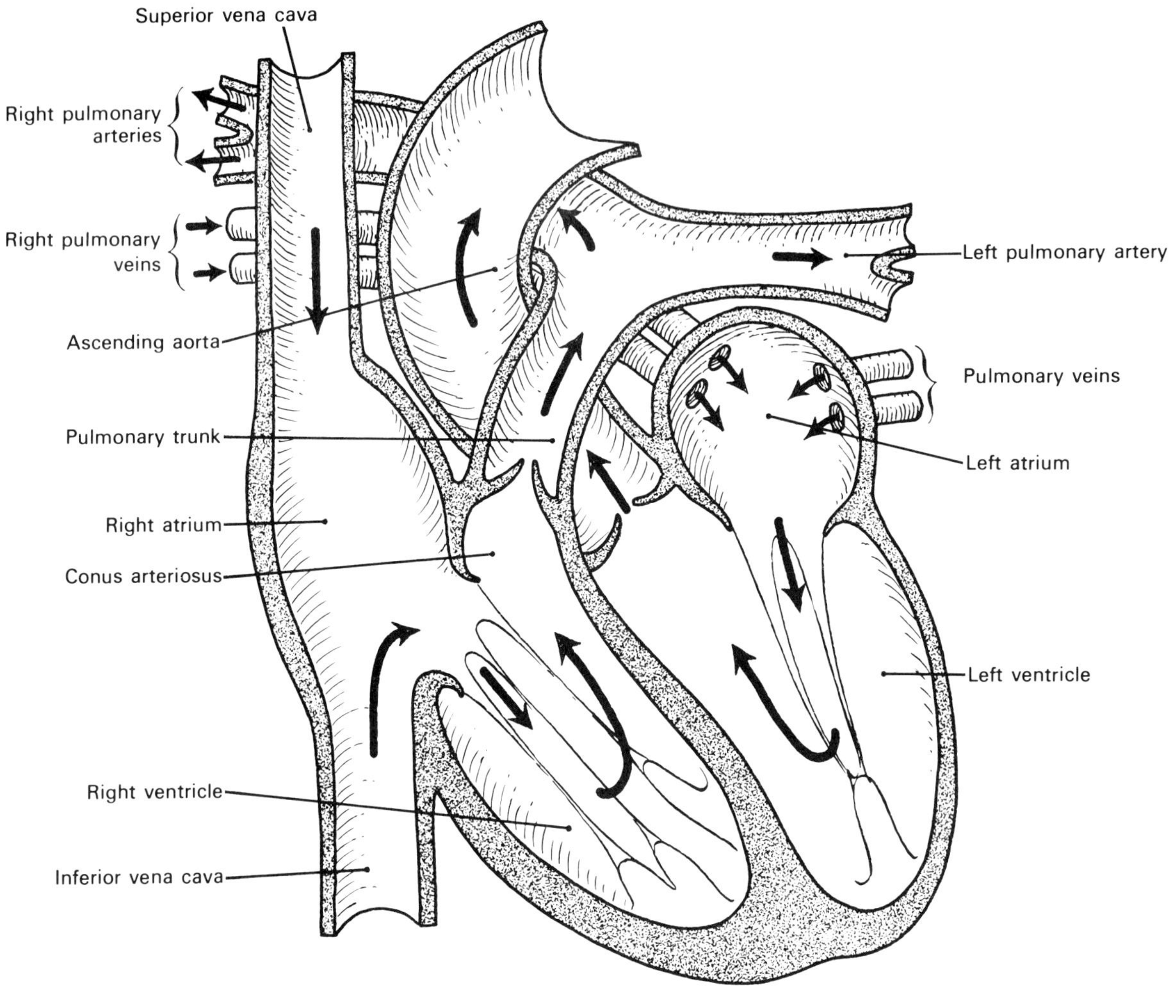

Fig. II-34. Schema of circulation through adult heart.

aorta are pushed open by the surge of blood from ventricle into this great distributing channel. The aorta *ascends* slightly to the right behind the pulmonary trunk, forms its **arch** over the right pulmonary artery, and becomes the **descending aorta** anterior to the vertebral bodies in the chest and abdomen. It must, of course, pierce the Diaphragm to enter the abdomen. Its terminal bifurcation is to the left of vertebra L4.

As seems fitting, the first branches of the aorta, the **right and left coronary arteries,** are to the very pump, itself, which must work constantly. These branches to the heart spring from the ascending segment of the aorta just above two of the pocket-like semilunar valves at dilatations of the vessel known as the **aortic sinuses** (of Valsalva). The coronary arteries can be regarded as **vasa vasorum** (vessels that supply vessels) if we regard the heart as a very muscular blood vessel. They are also **end arteries.** While there are rich anastamoses between their own smaller arterioles, this supply is not always sufficient nor readily functional should there be a sudden occlusion of a major coronary artery branch. Cardiac tissue in the region supplied by the occluded branch will die. We say we have an infarction in this region.

How is coronary blood returned to the heart? Chiefly through capillary beds in its own tissues which are drained by the cardiac veins. All but a few of these empty into the common **coronary sinus** which opens into the right atrium. The **anterior cardiac veins** make their own entrance into this chamber. There are also several small veins, the **vena cordis mimimae** or Thebesian veins, which open into all heart chambers. So some arterial blood of the left heart suffers some slight degree of venous contamination from its own return. In the lung a few bronchial veins also contaminate the pulmonary venous return to the left atrium so that blood leaving the left heart is not exactly 100% "pure" in terms of oxygen content.

The great arteries stemming from the aorta and the great veins returning blood to the right heart are depicted and named in Figure II-35, though not all branches from them are shown. You should begin gradually to learn the names of these vessels and their chief branches as you dissect or see pictures in texts

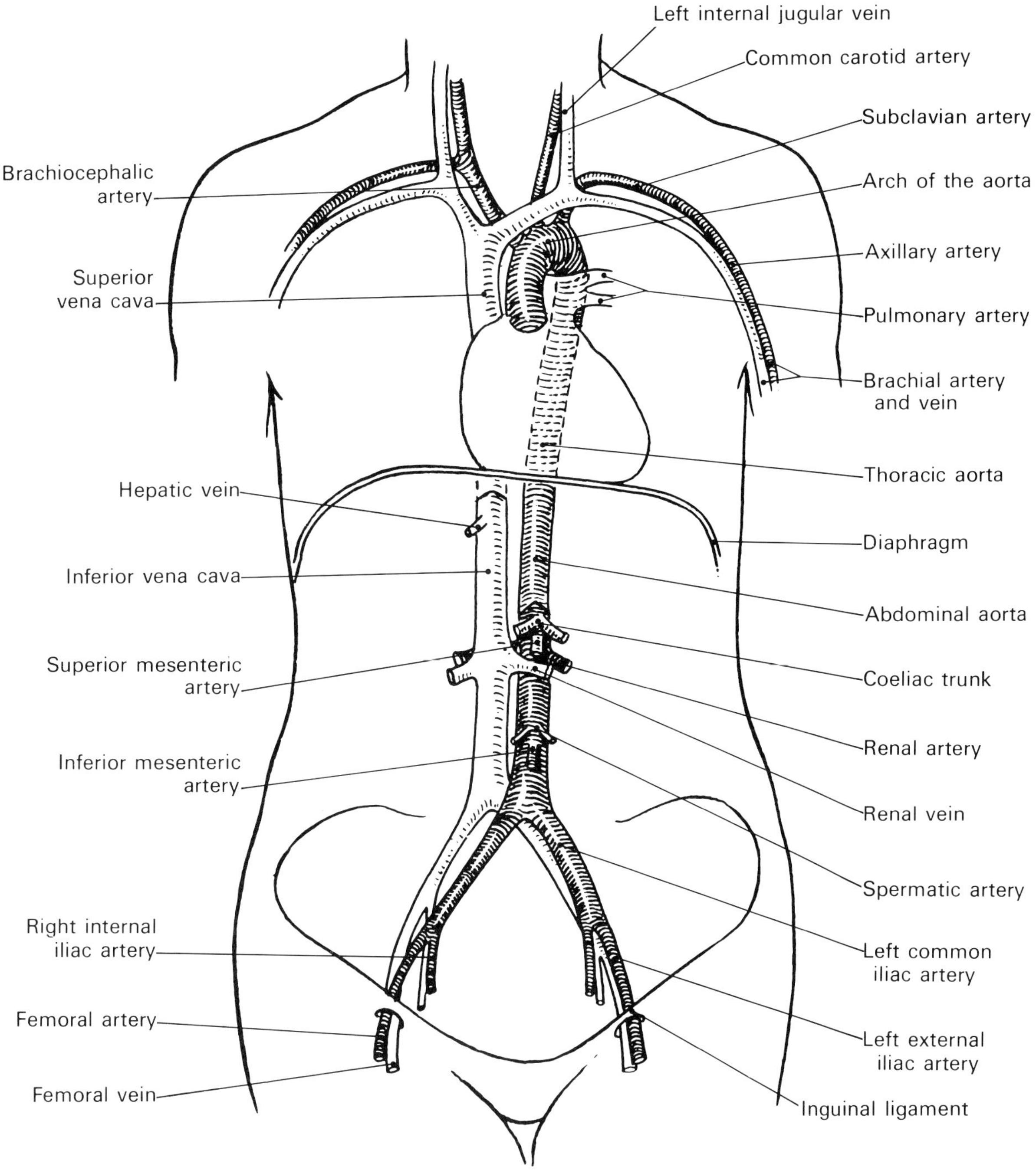

Fig. II-35. Major arteries and veins of the greater or systemic circulation (not all branches of these vessels are shown).

and atlases. It is enough to know these main pipelines at this time.

Currently, "brachiocephlic" is a term that is replacing "innominate" in usage. There are *two* innominate or brachiocephalic **veins,** a right and a left. There is only *one* innominate or brachiocephalic **artery,** the one on the right side in normally-developed individuals. Note that the heart in its pericardial sac sits upon the upper surface of the Diaphragm in the chest.

The Nature of Blood Channels: Both arteries and veins come in small, medium, and large sizes. Arteries regress in size as they run from heart to capillary plexuses; veins increase in size as they course from capillary plexuses to the heart. Lymphatic vessels likewise increase in this direction. All are constructed of three essential coats of tissue, designated as **intima, media,** and **adventitia** (Fig. II-33). The innermost or luminal coat is the *intima.* In most arteries it is a complex trilaminar layer of variable thickness; in veins it is bilaminar. Next to the lumen is the thin lining layer of **simple squamous epithelium,** called **endothelium** when it lines blood vessels. Both blood and lymphatic capillaries consist *only* of this **endothelial lining** with no other external layer. In the intima of veins the endothelium rests upon a layer of fibroelastic connective tissue variable in thickness. In arteries, even in small arterioles, an **internal elastic membrane** lies next outermost in the intima. It stains well with elastic tissue stains. In the medium-sized arteries, two or three interlacing, circumferential, elastic laminae may form this internal elastic membrane. This membrane is not present in veins.

Except in the largest arteries (aorta and its mainstem branches), the *middle coat* or *media* is quite muscular (smooth muscle, generally circularly disposed). Even small arterioles have a hefty muscular media for their size. The largest arteries are **elastic vessels;** that is, elastic fibers predominate in their stout media to such an extent that the internal elastic membrane of the intima blends into the media. A few smooth muscle fibers scatter in the media.

The muscular media in venules and small veins is almost nil, but a definite media appears prominent, though never as heavy as in companion arteries, as veins increase in size. Yet, the muscle fibers intermingle with some collagenous and elastic fibers and do not present the neat compactness witnessed in the media of arteries. Large lymphatic vessels present this appearance, also, except the muscular weave is even looser. Large veins and some arteries may have scattered, longitudinally-oriented, smooth muscle bundles in the media. In arteries and some large veins an external elastic membrane may separate media from the outermost coat, or adventitia.

The *adventitia* is the heaviest coat in veins and is never quite as robust as the media in arteries, though sometimes it almost equals that of the media. It consists of a dense fibroelastic connective tissue which gradually blends with the looser surrounding connective tissue through which vessels run. In it will be found the *vasa vasorum* that feed the thick walls of the larger vessels. These penetrate almost to the intima. The adventitia contains nerve bundles, as well (sympathetic fibers for the smooth muscle present in the walls). Figure II-33 illustrates the vascular coats in a medium-sized muscular artery and its companion vein. Certain veins and lymphatics have internal valves along their luminal wall. We shall consider these below.

The heart has the same three coats, but in different proportion. The heaviest is the media of cardiac muscle, or the **myocardium.** Its intima, or **endocardium** consists of the endothelium which lines its chambers and surfaces its valves. This thin coat rests upon fibroelastic connective tissue through which, in the ventricles, run some of the specialized cardiac muscle fibers (Purkinje fibers) that make up the conducting system of the heart. (Purkinje fibers of the heart are cardiac muscle fibers; Purkinje fibers of the cerebellum are nerve fibers.) The adventitia of the heart is its **epicardium,** a fibroelastic connective tissue of variable thickness. In it run the coronary and lymphatic vessels and autonomic nerve fibers, and, posteriorly, over the atria and in the atrioventricular sulcus, parasympathetic terminal ganglia are found in the epicardium. In some individuals it contains a great deal of fat, much to their detriment.

The aorta and pulmonary trunk have diameters of about 30 mm. The aorta is an extremely elastic vessel with little smooth muscle in its media. So are early large vessels that branch from it. Thus, they can expand and compress during systole and diastole of the ventricles, serving as subsidiary pulsating pumps. They can do this with no great expenditure of energy, for elastic tissue requires little oxygen to keep it contractile compared with the oxygen requirements of the muscular arteries farther out in the vascular tree. Muscular arteries are also under neural control, so that the amount of blood going into particular capillary beds can be regulated as needs dictate. Arterioles which feed capillary beds can cut flow off and on as if it were coming through a spigot. Arterioles have been trimmed down to about 0.3 mm. in diameter, and they lead to capillaries that are 7 to 7.5 μ in diameter, just large enough to admit red cells in single file, though red cells have enough elasticity to squeeze through tighter fits. Small as capillaries are, in diameter, the cross-sectional area of the entire network is about 800 times that of the aorta, so a considerable volume of blood can be stored in capillary pools.

The pulmonary artery is constructed much more like a medium-sized muscular artery but with some elastic tissue mixed in its media. The pulmonary veins are thinner walled than the pulmonary arteries, but they do have a respectable muscular media. As they approach the heart they may have cardiac muscle in the adventitia.

Upon their return routes, most veins pretty much accompany the course of their named companion arteries, flowing into larger and larger channels like streams into a river. In some places, as below the elbow and the knee, veins pair on each side of the companion artery and cross-link with each other around it. Such **venae comitantes** may be aided in moving blood toward the heart by the pulsating of the companion artery against them. In the limbs, a system of **superficial veins** under the skin will course quite independently of any accompanying artery. Most medium-sized and small veins have cusps or valves at intervals along their inner walls. These modifications of the intima are constructed much like the cusps of the semilunar valves at the entrance to the aorta and pulmonary trunk except that they are paired, generally, rather than being triplicate, and face each other along opposing walls of the vessel. These flap valves are all directed toward the heart so that backflow fills the "pockets" behind the flaps and their meeting edges prevent much reflux, constantly lifting blood toward the heart against gravity. Each new pulse from the arterial side, forcing blood ahead of it, pushes the flaps open against the venous walls, inexorably moving the flow on to the heart. The positions of the closest valves

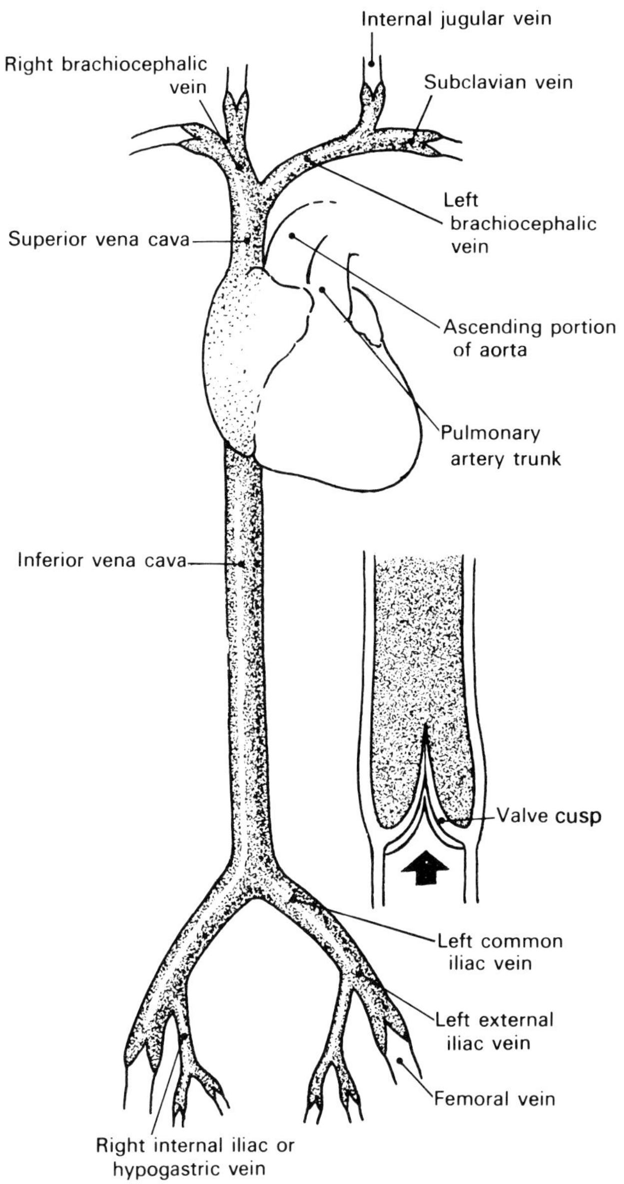

Fig. II-36. Locus of valves in great veins to prevent reflux into limbs and neck when increases in intrathoracic and intra-abdominal pressures occur.

to the heart in the large veins are shown in Figure II-36. These are able to trap blood in the trunk during periods of increased intrathoracic or intra-abdominal pressure (expiration, defecation, and the like) and so prevent it from moving back into the head or out into the limbs. Valves are much more prevalent distally in the limbs where lifting against gravity is essential. There are few valves in the cardiac veins and these valves are rarely functional; there are none in the portal system of veins to the liver.

It is possible to demonstrate valve action in the veins of your own arm as Harvey did (Fig. II-37). Place a tourniquet on the arm midway between shoulder and elbow. With the arm held in a dependent position, open and close your hand vigorously to cause the veins to distend and become visible. Muscular action helps neighboring veins to move blood along. Select a prominent distended superficial vein of the forearm and place the middle and index fingers of the other hand on it to compress it. Now stroke the proximally-placed index finger upward to empty a short segment of the vein of its blood (Fig. II-37B). Keep the middle finger in place, meanwhile. Now lift the index finger off the proximal part of the vein (Fig. II-37C). The collapsed segment does not fill completely by reflux from above but only as far as the nearest valve, leaving a segment below the valve still collapsed and unfilled. Now remove the middle finger and the whole collapsed segment will refill from below.

The following terms will be encountered from time to time as one reads about circulation in specific regions of the body. It is well to have some early idea of their meaning:

Portal System: One thinks at once of the liver with blood coming to it from the digestive tract through the portal veins, their emptying into sinusoids (wide capillaries) of the liver, and the sinusoids being drained by the hepatic veins. The system can be generalized as one in which a particular capillary bed (*e.g.,* that of the intestinal tract) is drained by venules and enters a second capillary bed (*e.g.,* sinusoids of liver) which is, in turn, drained by veins. Normally, capillary beds are fed from the arterial side and drained by the venous side. Here both the feeding and drainage of the second capillary bed are from the veins. This is true only in part, for hepatic arteries convey oxygenated blood into these *same* liver sinusoids.

The **hypophyseal-portal system** is similarly constructed. The superior hypophyseal artery supplies a capillary bed at the upper end of the pituitary stalk. Its venous drainage empties into sinusoids of the anterior lobe of the gland which, in turn, are drained by the inferior hypophyseal vein. As in the case of the liver, however, an oxygenated supply is delivered to the sinusoids of the anterior lobe by *direct* branches to it from the superior hypophyseal arteries, as well.

Sinusoids: These are specialized wide-bore capillaries (15–30 μ in diameter) found in liver, spleen, bone marrow, and glands like the pituitary (hypophysis) and adrenal. They may be a little different in construction in some of these places, especially in the spleen. The sinusoids of lymph nodes may be mere non-endothelially-lined channels washed through the reticular tissue network by incoming lymph, channels varying in diameter, shape, and location as lymph washes lymphocytes along the connective tissue network in irregular fashion.

Venous Sinuses: These are simply wide collecting venous channels principally in the double-layered dura mater over the brain and in the placenta.

Arteriovenous Anastomoses: These are direct connections between arteries and veins without benefit of

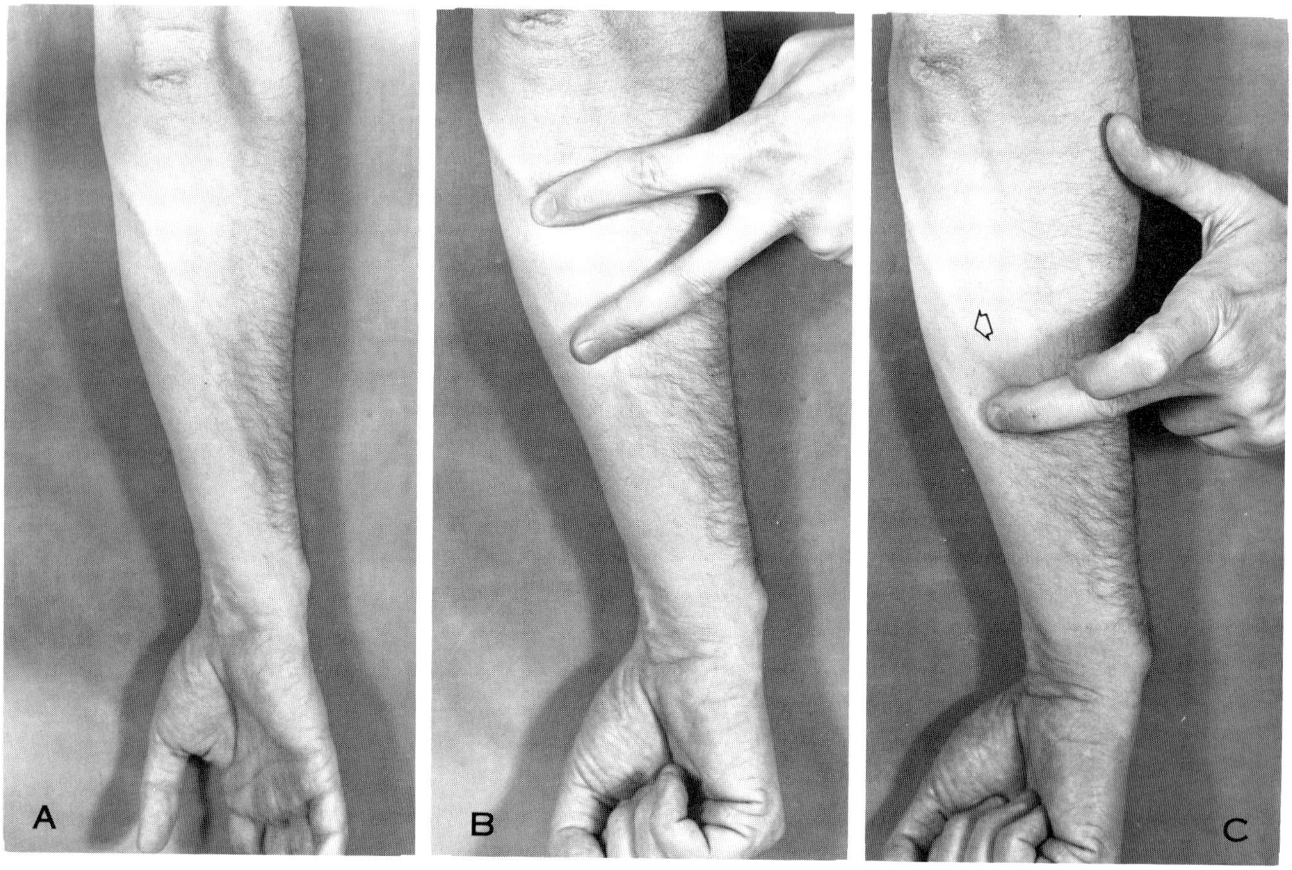

Fig. II-37. Demonstration of a valve in a superficial vein of the forearm.
A. Distended superficial vein.
B. Collapse of a segment by compression.
C. Backflow to next distal valve (*arrow*) after removal of the index finger. Segment distal to the valve remains unfilled until the lower finger is removed from the arm.

capillary beds and have been described in skin, lungs, alimentary tract, eyelids, tip of the tongue, ears, nail-beds, and lips. These same regions will have ordinary arterial-capillary-venous connections. The exact function of A-V anastomoses is still unclear, and they seem to be more prevalent in the infant than in adults.

Rete Mirabile: This is a capillary network fed by a small artery and drained by a small artery, as in the case of the glomeruli of the kidney.

Cavernous Tissue: There are large spaces lined with endothelium and crossed by fibrous septae which contain smooth muscle. Such smooth-walled venous spaces are found in the male penis, the female clitoris, and in the mucous membranes of the nose. Erectile tissue of this type is capable of immense dilatation in fortunate individuals. In the nose, of course, its filling brings blood close to the surface of the respiratory mucosa and warms inspired air, aiding the sense of smell. Smell plays an important role in sexual behavior among both lower animals and man; thus, it may not seem farfetched to find cavernous tissue in the nose as well as in the genital parts.

REFERENCES

1. *Dorland's Illustrated Medical Dictionary,* 24th ed. Saunders, Philadelphia, 1965. (If an older copy, vintage 1948, can be obtained there is the added advantage of having all the portraits of famous men in anatomy and medicine who were chiefly responsible for the eponyms that clutter our jargon. Sometimes it is nice to know what a particular culprit looked like. This rogue's gallery will be found missing after the 21st edition.)
2. *Stedman's Medical Dictionary,* 21st ed. Williams & Wilkins, Baltimore, 1966.
3. Skinner, H. A.: *The Origin of Medical Terms.* Williams & Wilkins, Baltimore, 1961.
4. Lassek, A. M.: *Human Dissection, Its Drama and Struggle.* Thomas, Springfield, Ill., 1958.
5. Ham, A. W.: *Histology,* 5th ed., Ch. 18. Lippincott, Philadelphia, 1965.
6. Copenhaver, W. M., and Johnson, D. D.: *Bailey's Textbook of Histology,* 15th ed., Ch. 6. Williams & Wilkins, Baltimore, 1965.
7. Bardeen, C. R.: Studies of the development of the human skeleton. Amer. J. Anat., *4:* 265, 1905.
8. Bardeen, C. R.: The development of the skeleton and the connective tissue. In *Manual of Human Anatomy,* edited by Kiebel and Mall. Lippincott, Philadelphia, 1910.
9. Ham, A. W.: Cartilage and bone. In *Special Cytology,* edited by Cowdry, vol. 2., p. 979. Hoeber, New York, 1932.
10. de Beer, G. R.: *The Development of the Vertebrate Skull.* Oxford, London, 1937.
11. Pritchard, J. J.: A cytological and histochemical study of bone and cartilage formation in the rat. J. Anat., *86:* 259, 1952.
12. Bourne, G. H.: *The Biochemistry and Physiology of Bone.* Academic Press, New York, 1956.
13. Ascenzi, A., and Benedetti, L.: An electron microscope (Basel), *37:* 370, 1962.
14. Bassett, A. L.: Current concepts of bone formation. J. Bone Joint Surg., *44-A:* 1217, 1962.
15. Arey, L. B.: *Developmental Anatomy,* 7th ed., Ch. 21. Saunders, Philadelphia, 1965.
16. Bloom, W., and Fawcett, D. W.: *A Textbook of Histology,* 9th ed., Ch. 10. Saunders, Philadelphia, 1968.
17. Baker, S. L.: Introduction to the pathology of bone. In *X-Ray Diagnosis,* 2nd ed., Saunders, Philadelphia, 1952.
18. Weinmann, J. P., and Sicher, H.: *Bone and Bones: Fundamentals of Bone Biology,* 2nd ed., Mosby, St. Louis, 1965.
19. Arey, L. B.: The origin, growth and fate of osteoclasts and their relation to bone resorption. Amer. J. Anat., *26:* 315, 1920.
20. Bloom, M. A., Domm, L. V., Nalbandov, A. V., and Bloom, W.: Medullary bone of laying chickens. Amer. J. Anat., *102:* 411, 1958.
21. Hancox, N. M.: The osteoclast. Biol. Rev. *24:* 448, 1949.
22. Hancox, N. M.: The osteoclast. In *The Biochemistry and Physiology of Bone,* edited by Bourne. Academic Press, New York, 1956.
23. Hancox, N. M., and Boothroyd, B.: Structure and function relationships in the osteoclast. In *Mechanisms of Hard Tissue Destruction,* edited by Sagnaes, p. 497. Amer. Assoc. Adv. Sci., Washington, D. C., 1963.
24. Robinson, R. A.: *The Significance of Phosphoric Esters in Metabolism.* N. Y. Univ. Press, New York, 1932.
25. Wells, H. G.: *Chemical Pathology.* Saunders, Philadelphia, 1925.
26. Cooper, R. R., Milgram, J. W., and Robinson, R. A.: Morphology of the osteon, an electron microscopic study. J. Bone Joint Surgery, 48-A: 1239–1271, 1966.

THREE
Surface Anatomy

INTRODUCTION

A prominent paint producer once advertised: "Save the surface and you save all." In anatomy and medicine, this slogan could be paraphrased: "Observe the surface and you learn a lot." A knowledge of surface anatomy can save you much useless memorization of facts, for you can always refer to your own body, once you learn to read it, as a kind of animated textbook. Even before you attempt to examine a colleague or a patient, you can always profit considerably by admiring your own physique in the privacy of your household mirror. Once you get the hang of localizing structures on yourself, it is well to try to identify them on someone with an entirely different body build. In our own student days, we discovered that studying surface anatomy with a wife or girl friend proved to be not only instructive, but highly entertaining.

Since the majority of medical students still tend to be males, we have liberalized this text by making use of the female form. But, more to the point, we have done so because a large portion of your future patients will be women and few texts have pointed out surface landmarks on the female. If female students feel put out on this account, and if it affords any solace to them, we point out that most of the skeletal material depicted came from male skeletons chosen as prime examples of virile underpinnings.

As you study your own body, you will note prominent bony landmarks by sight and feel, and will note contours of certain muscle masses. Changes in posture can bring some of these muscles into such bold relief that their very tendons become visible. Application of slight pressure makes it possible to roll certain vessels, nerves, and ducts under the fingers so that their course, in part at least, can be localized. Pulse can be taken in several places. Once you learn to use the stethoscope, ophthalmoscope, otoscope, and reflex hammer, you can inquire deeper into the nature of some of the structures below the surface. Work with patients will make you observant of posture, gait, range of movement at joints, skin conditions, speech, nutritive states, and mental states. Thus, as you continue through medical school you constantly draw upon surface anatomy to aid in narrowing down a diagnosis of illness.

In this respect you study the living body, and the practice of medicine consists of and exists for the living. They are your patients. You will have very little recourse to the cadaver after gross anatomy is over as a prescribed course unless you choose to become a professional anatomist, a staff pathologist, or a coroner. You may, of course, wish to redissect from time to time to refresh yourself on certain details that become vague after one has been away from the dissecting room for any length of time. We all live with forgetfulness which is no crime so long as we make an attempt to review and relearn that which we hold to be important.

In studying the living body, we not only look to see what is at the surface, but we feel for what lies beneath the surface. Even more, we learn to project the hidden structures upon the surface and to see them in the mind's eye or, at least, to localize them with some reasonable degree of accuracy. In the process, the voluminous terminology of anatomy builds up so that we are ready to approach the cadaver with meaning, not merely to dissect willy-nilly with little idea of what we expect to find there. If, for example, you have felt the pulse of the superficial temporal artery, as an anesthetist does daily, you have some idea already where to locate this branch of the external carotid artery when you begin to dissect. So it will be with many other structures.

It is also possible to draw certain imaginary planes upon the body by erecting them upon prominent bony landmarks. These planes will transect certain structures and from these points of reference still others can be located by measurement. If you have no other measuring tools, you do have your fingers at hand. You can, if you wish, premeasure their widths, but a general rule of thumb (rule of fingers?) is that each one approximates 2 cm. in breadth, or ¾ of an inch for those who have a decimetric phobia.

In all of this undertaking, realize the approximate nature of these measurements and localizations. Body build varies from individual to individual, and structures that are bilaterally represented in the same individual will often vary in size, shape, and position on the two sides. On the whole, you can expect to find most variability in veins, next arteries, with nerves, perhaps, showing more relative stability in locus and degree of branching. Muscles will vary in size, in their origins and insertion, and even in their presence or absence among individuals and races. Probably the skeleton will afford the most reasonable base for stability, yet even it changes in growth, differs in the sexes and throughout the races of man. Pathological processes may also alter these structures.

SURFACE ANATOMY OF THE HEAD

Cranium and Face

For this study you shall have to use your head and any other loose skull available. If the latter happens to be that of a colleague, handle it gently and please do not detach it from his neck. If it comes from your bone box, treat it gently, also. Do not poke or probe foramina or fissures with a pencil or any hard instrument. To sound foramina or canals use nothing harder than a stiff, waxed string. If you have access to a so-called "dissectable skull," one with hinges and latches so that you can take it apart to examine recesses, remember that each separate part is more fragile than the whole. Lay the pieces down on cotton wool rather than on a hard table top, and be sure they reassemble easily and without pressure. Skull material is expensive to replace when broken. Other bones are easier to come by, but even they last for years if handled carefully. Never mark on any bone with crayon or pencil. Do not add colored modeling clay to bleached skeletal material in an attempt to reconstruct muscle masses. The coloring matter stains the bone and will not clean off. If you must determine exactly where each muscle has its origin or insertion on the skeleton, use special skeletons that are already prepainted for this purpose or consult the figures in standard anatomical reference texts.[1-10]

To orient a skull so that it approximates the anatomical position, place a 2-inch block under the large hole or foramen, the **foramen magnum** of the **occipital bone.** Now the *lower margin* of the **orbital cavity** will lie in a direct line with the *upper margin* of the **external auditory meatus.** This line will parallel the table top, and the skull is said to be in the Frankfort Plane. This is a standard reference plane adopted by a convention of anthropologists in Frankfort in 1882. At this convention a number of reference terms were adopted for points on the skull from which measurements could be made. These terms, such as bregma, lambda, nasion, inion, *etc.*, have become common parlance in anatomy and anthropology, and some of them serve well to localize underlying structures *within* the cranium.

Structures beneath the Nasion-Inion Line: The **nasion** is the point where the midline suture between the two, tiny, external **nasal bones** meets the horizontal suture line between **frontal** and nasal bones. On the surface, it is the dimpled depression where you wear the bridge of your eyeglasses. A prominent posterior bump in the midline base of the skull is the **inion** or **external occipital protuberance** (Figs. III-1 and 2). An imaginary line drawn over the **calvaria,** or skull cap, from nasion to inion forms a midsagittal cleavage plane that would divide the skull into two equal halves.

Significance: Under this line or landmark lie three midline structures:

1. The **sagittal suture** of the cranium between the two **parietal bones** (Fig. III-3).
2. The course of the **superior sagittal sinus,** a venous channel in the **dura mater** covering the brain. This channel drains the veins of the brain, veins of the scalp, and veins of the spongy parts (**diploe**) of the cranial bones. It also drains the cerebrospinal fluid channels (Figs. III-4B and 5).
3. The **superior longitudinal fissure,** a midline cleft between the cerebral hemispheres and in whose depth lie the transverse fibers of the **corpus callosum,** stout nerve fiber tracts which connect both hemispheres (Fig. III-4A).

The relationship of these three structures can be seen in Figure III-4A. Such a section cuts through the two parietal bones in a coronal plane with the layers of the scalp above and the brain with its three meningeal coverings below. Note the **sagittal suture** of the cranium between the parietal bones. It lies under the nasion-inion line. Below it, in the midline and within the tough dural covering of the brain, is a wide venous channel with some lateral extensions. This is the **superior sagittal sinus** with its **lateral lacunae.** It lies in the upper part of a dural fold (falx cerebri) which dips down in a cleft between the brain hemispheres, the **superior longitudinal fissure.** There is another sinus, the **inferior sagittal sinus,** in the bottom of this dural fold. The fold rests upon some transversely-running, white fiber tracts, the corpus callosum, which interconnect the hemispheres. The full extent of the superior and inferior sagittal sinus can be seen in Figure III-5 where several other dural sinuses are also depicted. All such sinuses are simply wide venous channels which run between the inner and outer layers of dura mater. The **outer layer of dura** forms the **inner periosteum** of the cranial bones, and its **inner surface,** presented toward the brain, is its smooth **meningeal layer** in contact with a filmy, spider-webby, meningeal covering, the **arachnoid.** A third, delicate, meningeal covering, the **pia mater,** fits the brain like a thin glove and dips into its every crevice. Between pia and arachnoid is a space filled with **cerebrospinal fluid.** Thus, the brain is kept away from the cranial bones above by a water cushion as well as by the three protective **meninges.** We shall discuss these relationships further in another chapter. The point here is to note the structures that lie between brain and the surface along the nasion-inion line and to realize that the meninges and the subarachnoid space with its cerebrospinal fluid cushion are always interposed between cortical gray matter and the cranial bones.

The scalp is certainly the only surface feature that can be seen and felt in this region. It is a multilayered

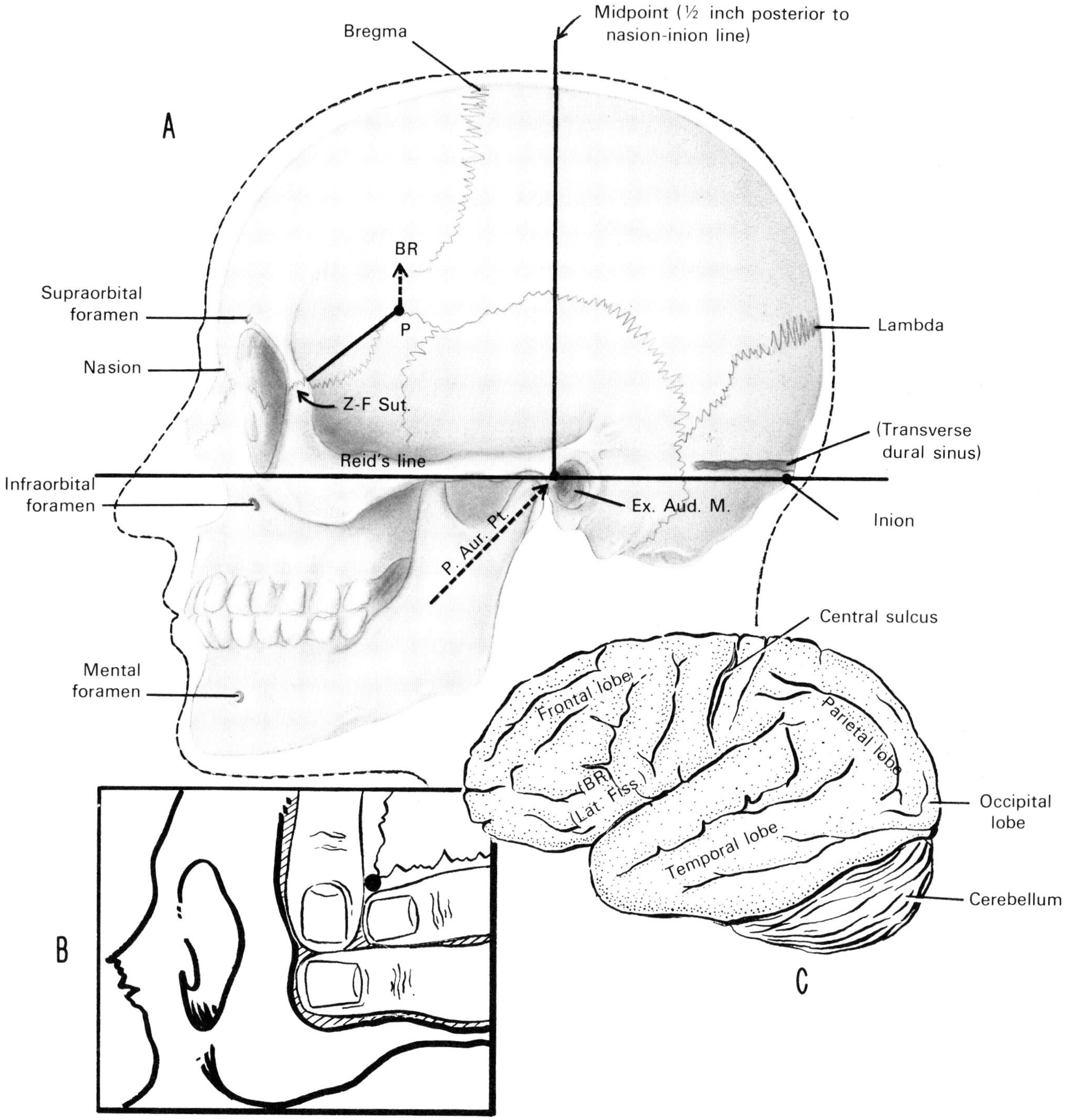

Fig. III-1A. Lateral aspect of skull with landmarks. Br., Broca's motor speech area; Ex. Aud. M., external auditory meatus; P., pterion; P. aur. pt., preauricular point; Z.-F. sut., zygomaticofacial suture. **B.** Inset to localize pterion shown at black dot **C.** Inset to localize structures in brain. Br., Broca's motor speech center in frontal lobe; Lat. fiss., lateral fissure (of Sylvius). (Modified from Brantigan, *Clinical Anatomy*, McGraw-Hill, 1963.)

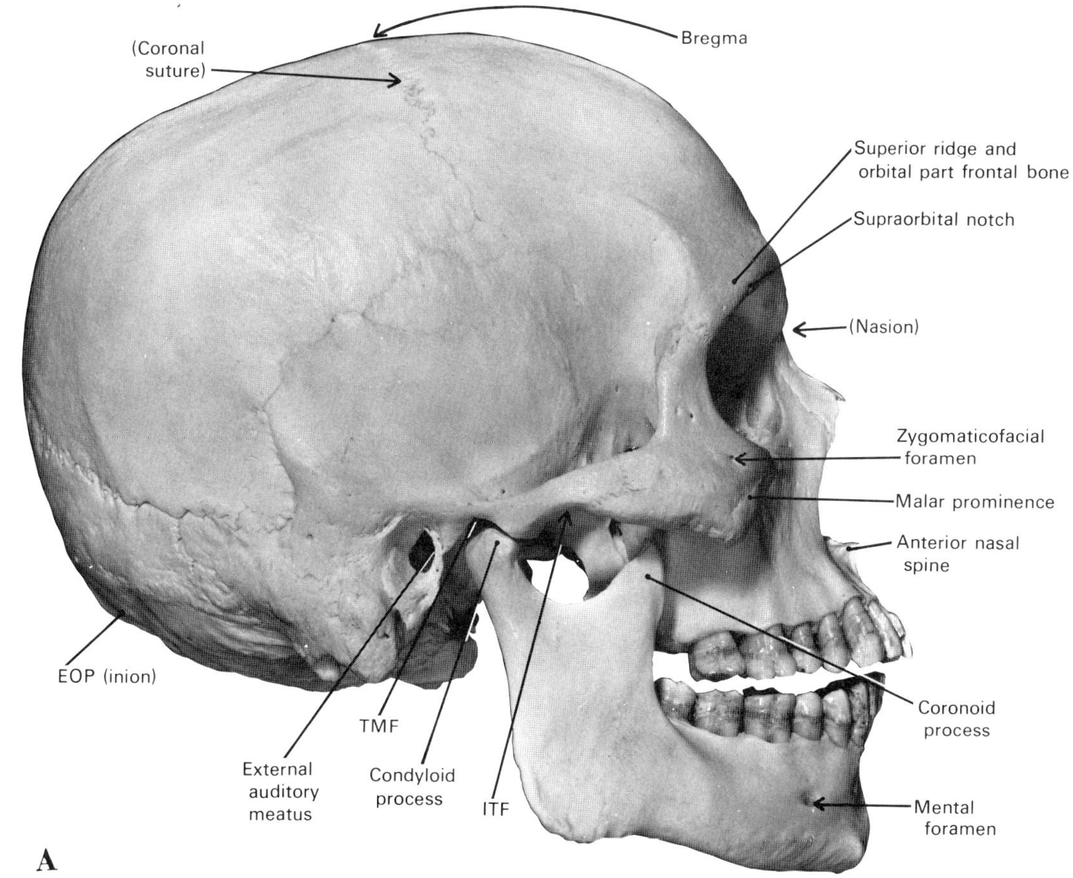

A

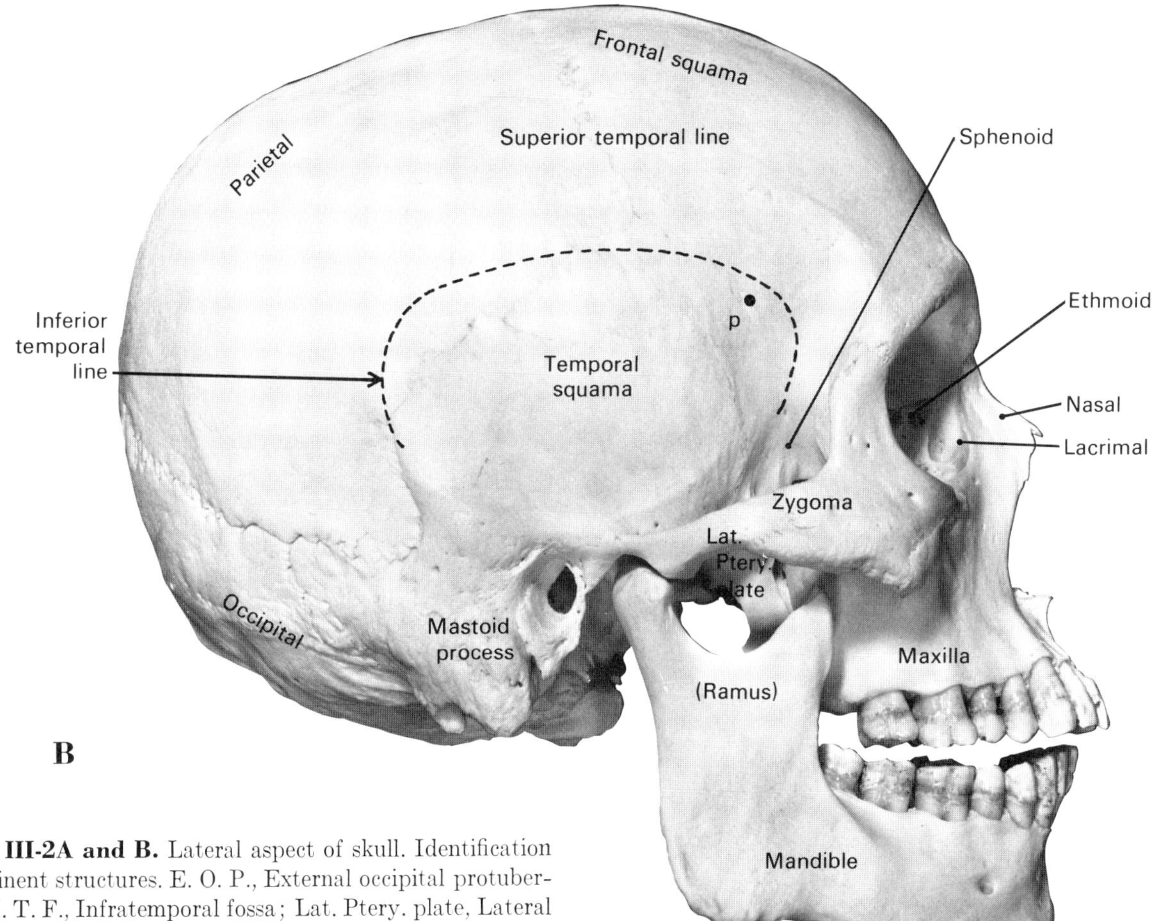

B

Fig. III-2A and B. Lateral aspect of skull. Identification of pertinent structures. E. O. P., External occipital protuberance; I. T. F., Infratemporal fossa; Lat. Ptery. plate, Lateral pterygoid plate; T. M. F., Temporomandibular fossa; ZYG., Zygoma; P., Pterion.

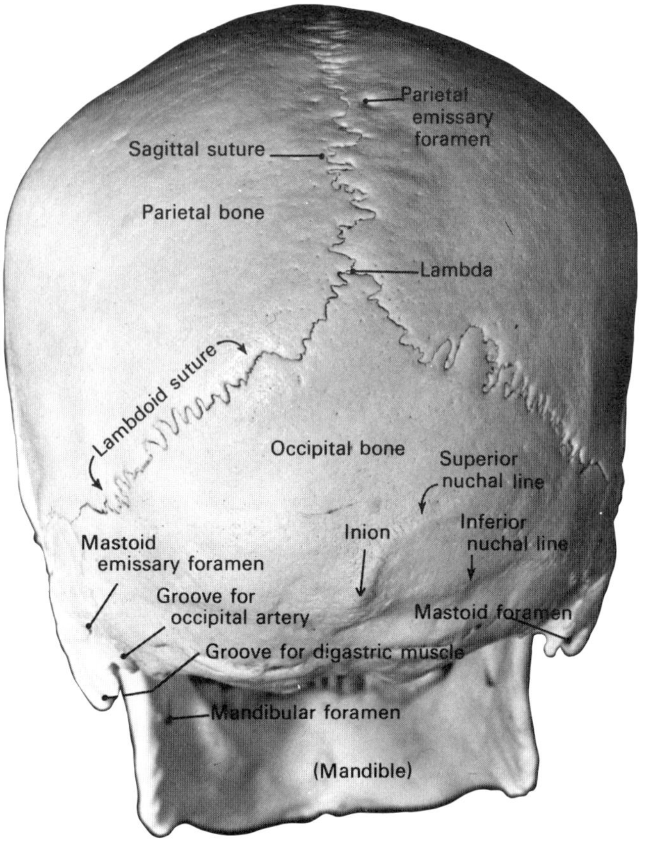

Fig. III-3A. Skull, posterior aspect.

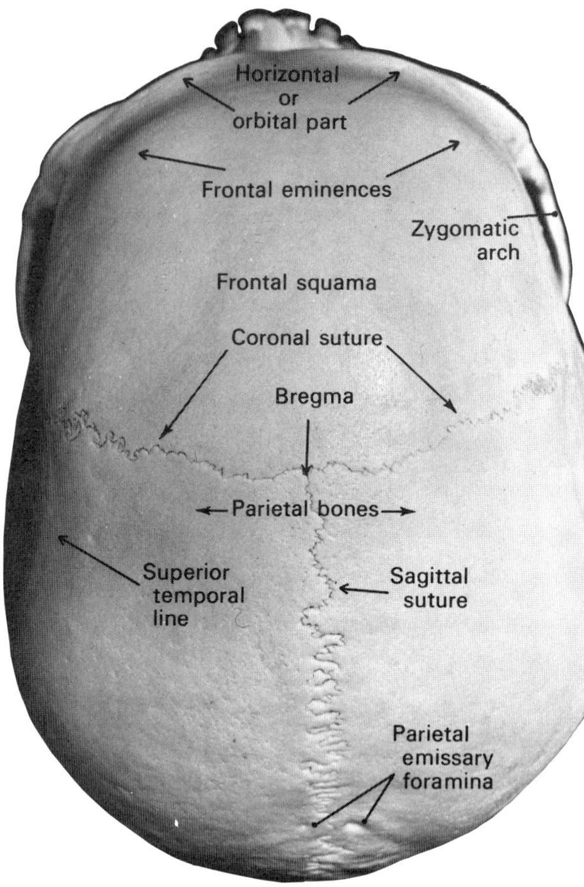

B. Calvaria from above.

structure as Figure III-4A indicates. Like skin anywhere, its outermost layer, the **epidermis,** consists of stratified squamous epithelium which rests upon a fibroelastic **dermis** of connective tissue. The latter contains numerous hair follicles and considerable fat as well, so that the term "fat head" is not exactly a misnomer or necessarily an epithet. It is also one of the most richly vascularized cutaneous areas of the body.

Below the skin of the scalp is its **aponeurotic layer.** This consists of a tendinous sheet of dense collagenous tissue covering much of the central area of the bony skull cap. It is called the **galea aponeurotica** ("the helmet of nerves"). This fanciful term was coined long ago by Galen when he first saw it. The tough flattened tendon to which the **Occipitalis** and **Frontalis** muscles are attached appeared to him to resemble a leathery helmet (galea), and he mistook the collagenous fibers in this sheet for nerve fibers. Nevertheless, wherever a *flattened tendon* attaches *muscle to muscle* we still call it an **aponeurosis.** Generally, we conceive of *tendons* as ropelike collagenous strands anchoring *muscle to bone.* Similar ropelike strands connecting *bone to bone* we call *ligaments.* The student should be warned that the word "ligament" receives considerable loose usage in anatomy, for many mesenteric visceral folds in the abdomen, quite unrelated to bony structures, masquerade as "ligaments." Note that the galea and the muscles attached to it are *within* the scalp as one of its layers.

The **subaponeurotic layer** of the scalp is of a looser fibroelastic weave than that above, making the scalp a fairly mobile structure. Test this by massaging the scalp under the fingers, or by alternately contracting the Occipitalis and Frontalis muscles if you can (some of us can't). It was to the depth of this layer that the Indians sliced to lift scalps as they whooped their war cries in the wild and woolly West.

Finally, the deepest layer serves as the **periosteum** for the outer table of the flat bones of the skull.

Significance: The fat of the scalp is tightly packaged in lobules separated by stout connective tissue septae, much as it is in palms and soles. Hence, while occasional **lipomata,** *i.e.,* **benign fatty tumors,** are encountered on the scalp, they are rare. Neither is the scalp used as a routine depot for fat storage by the obese.

Because of its extensive blood supply, scalp wounds bleed severely and the ends of cut vessels retract from each other readily. Bleeding must be stopped by compression and suturing. Still, large areas of the scalp can be ripped loose in accidents, but so long as a small patch is left in contact with underlying tissues, repair

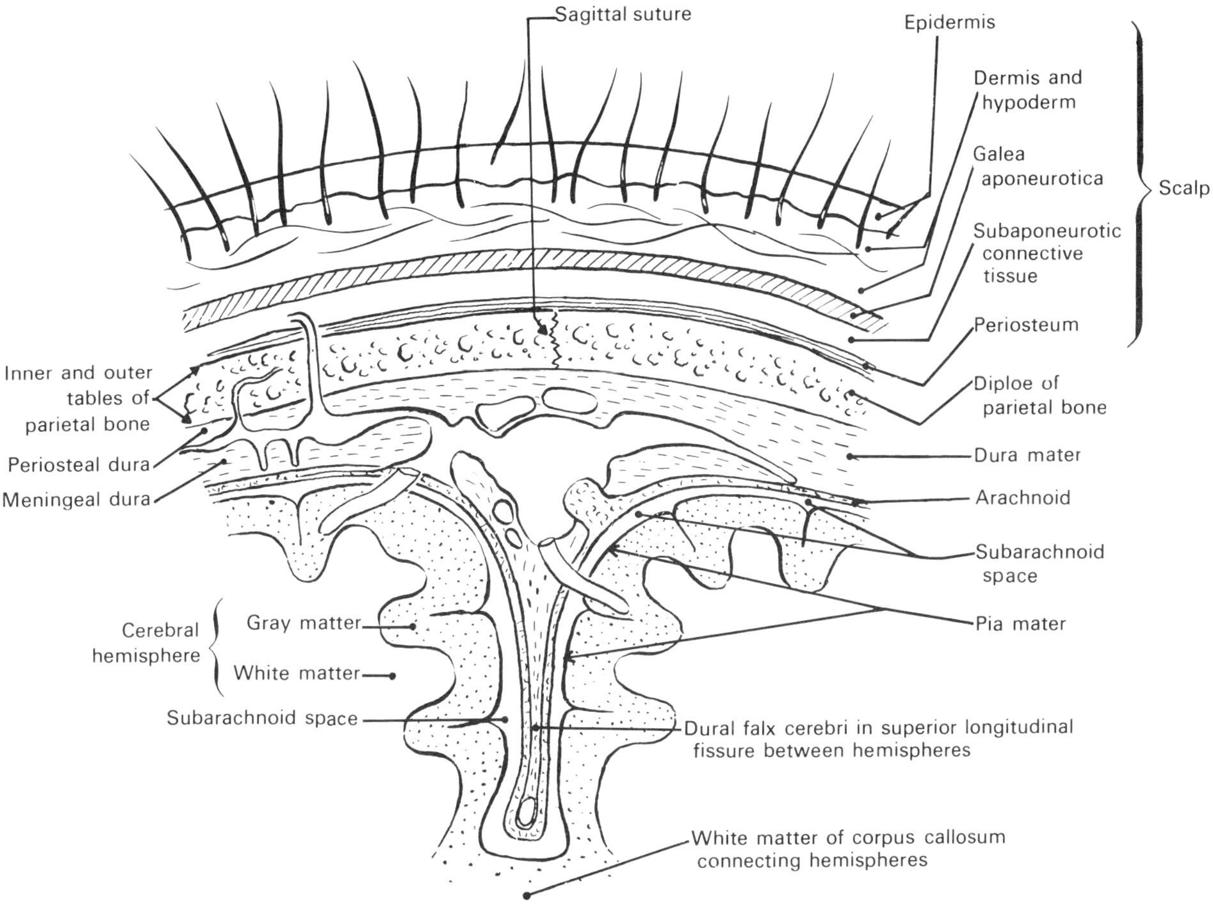

A

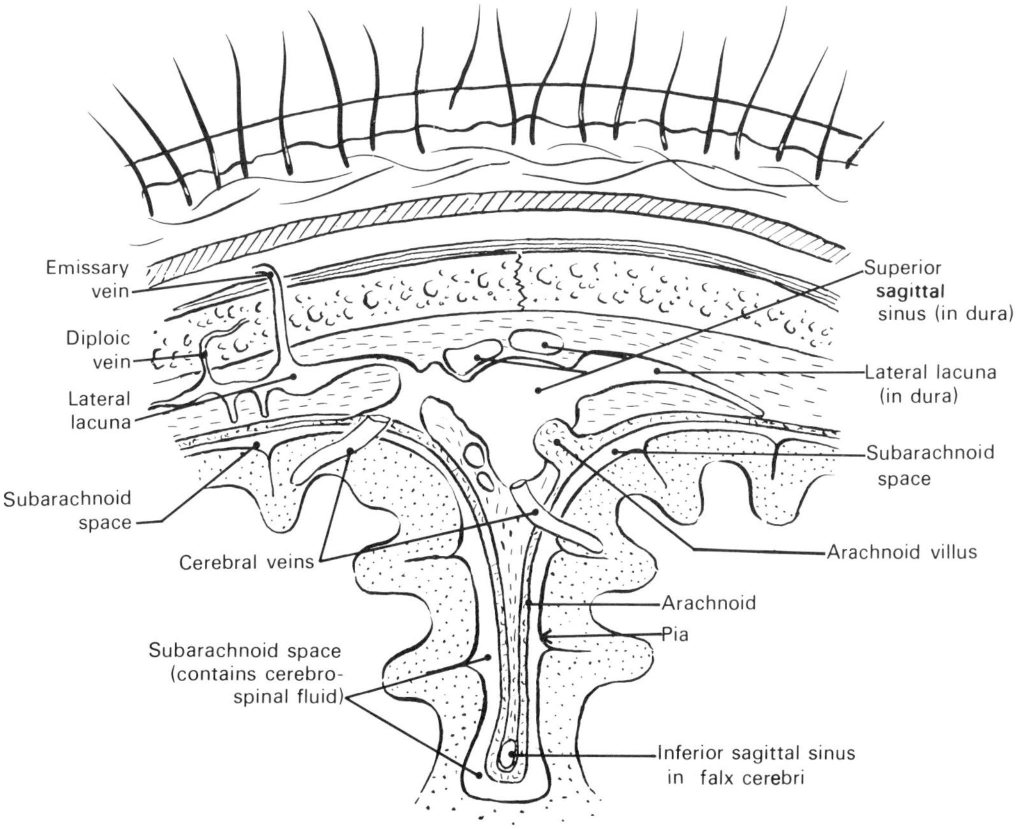

B

Fig. III-4. Coronal section through scalp, parietal bones, meninges, and cerebral hemispheres. **A.** Relationship of scalp, bone, and meninges to underlying brain tissue. **B.** Vascular relationships with superior longitudinal sinus of dura mater.

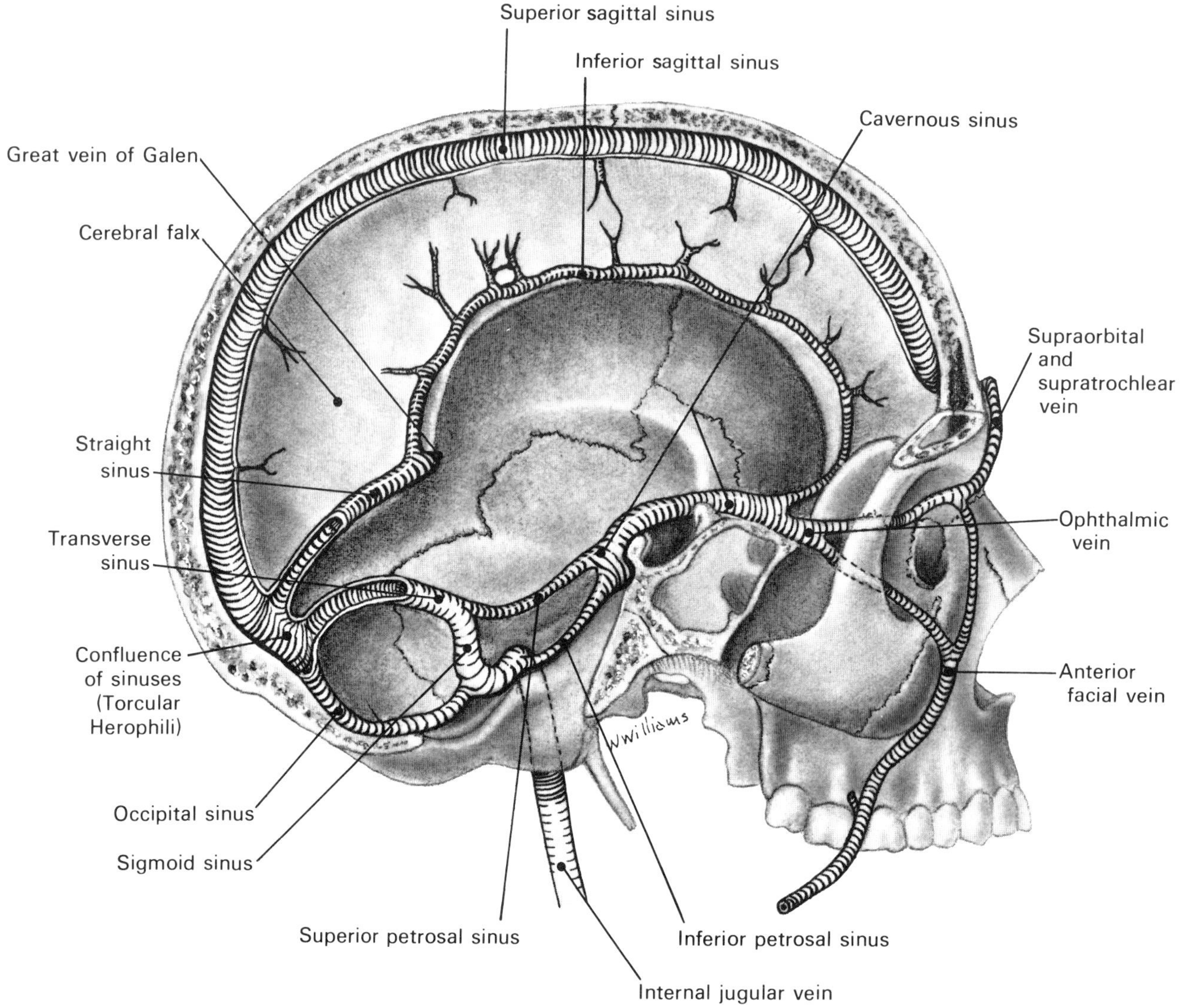

Fig. III-5. A midsagittal section of the skull with cerebral hemispheres removed to illustrate some of the dural sinuses.

of such extensive wounds is amazing. Revascularization of the damaged area extends outward from the intact patch.

The Forehead and Supraorbital Region: The **frontal bone** associated with the forehead region consists of a flat, vertical or **squamous part** and a horizontal or **orbital part** (Fig. III-2A, 3B, 6 and 7). The rounded elevations in the squamous part, more conspicuous in some individuals than others, are the **frontal eminences.** They can be felt to bulge about two fingerbreadths above the eyebrows in your own head.

Along the line of the eyebrows feel the **superciliary ridges or arches** in the orbital part of the bone (Figs. III-2, 6 and 7). These mask the loci of the **air sinuses** within the frontal bone (Figs. III-6 and 8). Such sinuses, as distinct from the dural sinuses, are irregular, rarely symmetrical, air-containing spaces lined by the same kind of a respiratory mucous membrane that lines the nasal cavity. Indeed, these frontal air sinuses, like those found in certain other bones of the skull and face, communicate with the nasal cavity by narrow channels. Vascular engorgement of the mucosal lining can easily obstruct these channels as we are often well aware when we have "a cold in the head." If you can obtain a "dissectable skull," open the latch in the frontal bone and peer into the frontal sinus. Pass a **waxed string** into the **frontonasal duct** and note where it emerges in the nasal cavity.

Significance: 1. The superciliary ridges are sharper in the male than in the female. Consequently, a blow, delivered accidentally or on purpose in boxing or other sports, will lacerate the skin over the ridge and result in bleeding. Contusion or bruising of the overlying skin causes fluids and blood to accumulate in the loose connective tissue in the area and in the upper eyelid, resulting in the familiar "black eye."

2. Just as the meninges intervened between parietal

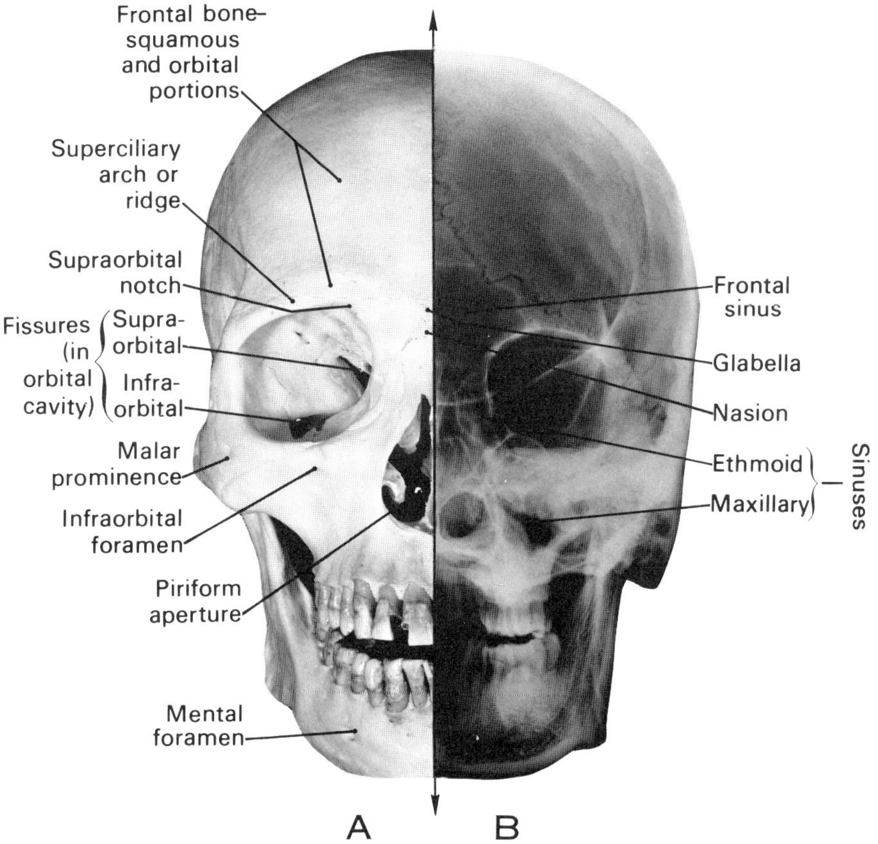

Fig. III-6. Frontal aspect of the face and skull. Composite of (A), gross skull photograph, and (B), matching radiograph of the same dry skull. (Note some slight distortion due to magnification on the radiograph.)

bones and brain, so do they between frontal bone and the brain. This brings the frontal air sinuses with their fragile walls in close relationship to the frontal lobe of the brain, separated from it by the meninges only. A fracture of the frontal sinus which tears dura and arachnoid will allow cerebrospinal fluid to enter the air sinus, traverse the frontonasal duct, and trickle from the nose.

3. Bacterial infection, breaking through the wall of the sinuses can lead to **meningitis** or even **frontal lobe abscess.**

4. Obviously, if the neurosurgeon wishes to use a frontal approach to the brain, he should first consult a roentgenogram (see Fig. III-8) to ascertain the variable topography of the frontal sinus. He can then plan his skin and bone flap without opening into the sinus and, thus, avoid the risk of carrying infection into the brain.

The smooth elevation, felt between the two brows in the midline just above the nasion, is the **glabella,** the glabrous or hairless region (*cf.* Figs. III-6 and 7). Look at a number of skulls in the laboratory and examine the region of the glabella. See if you can find remnants of a **frontal sagittal suture** near the glabella (Fig. III-7). If you do, it signifies that what appears to be one frontal bone in the adult was, developmentally, two bones. Should this suture be present in its entirety, a rare occurrence in adults, it would be found to be continuous with the sagittal suture separating the parietal bones. It would then be known as a **metopic suture,** one normally out of place. Still, fragments of it exist in enough skulls near the glabella to make us cognizant of its normal existence in infants.

Supraorbital Notch, Foramen, and Nerve: Just below the superciliary ridge, feel with your finger tip the sharp upper margin of the orbital cavity. At the junction of the medial and intermediate thirds of this margin, can you feel the **supraorbital notch** (Figs. III-1, 2, 6, and 7). If you do not feel it, it may have been converted into a tiny foramen. Check a number of skulls in the laboratory to see how variable supraorbital notches and foramina can be. A notch may be present on one side and a foramen on the other in the same skull.

Significance: The nerve emerging through here, as you might suspect, is the **supraorbital nerve** (Fig. III-9). It is a **sensory twig** of the uppermost or **ophthalmic division** of the **trigeminal** or **5th cranial nerve.**

Infraorbital and Mental Foramina: Once you

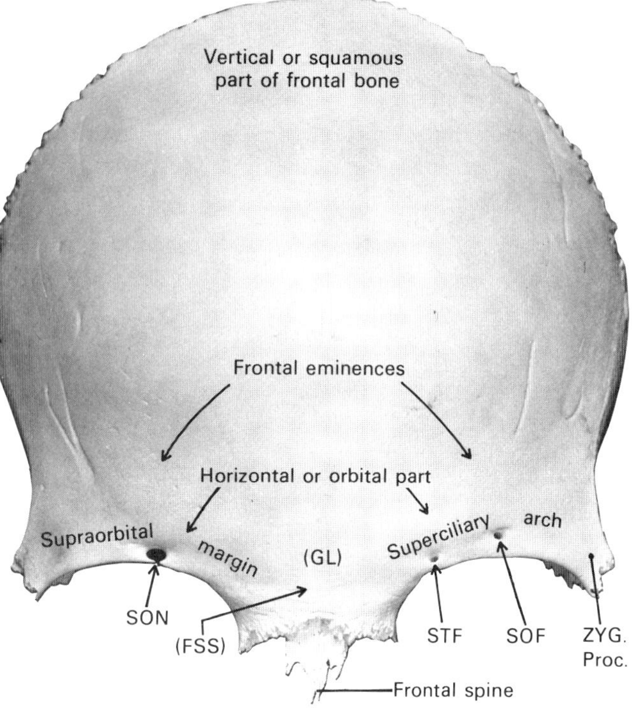

Fig. III-7. Frontal view of disarticulated frontal bone. F. S. S., Remnants of frontal sagittal suture; Gl., glabella; S. O. F., Supraorbital foramen; S. O. N., Supraorbital notch; S. T. F., Supratrochlear foramen; Zyg. Proc., Zygomatic process of frontal bone.

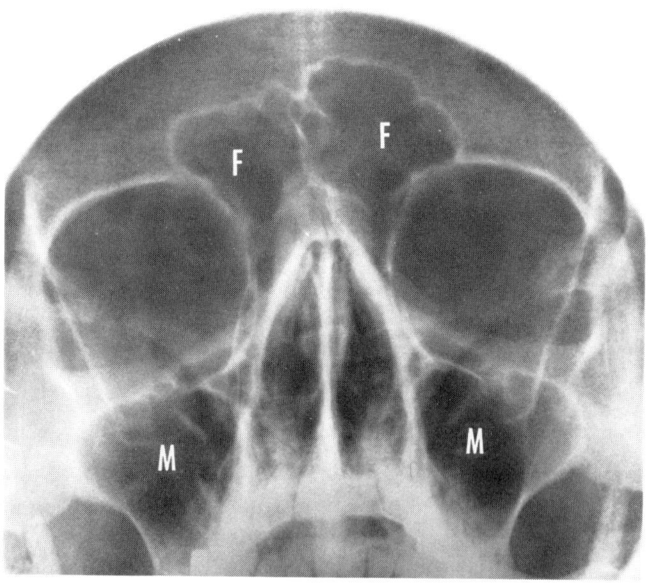

Fig. III-8. Air sinuses, X-ray, frontal view. F., Frontal sinus; M., Maxillary sinus.

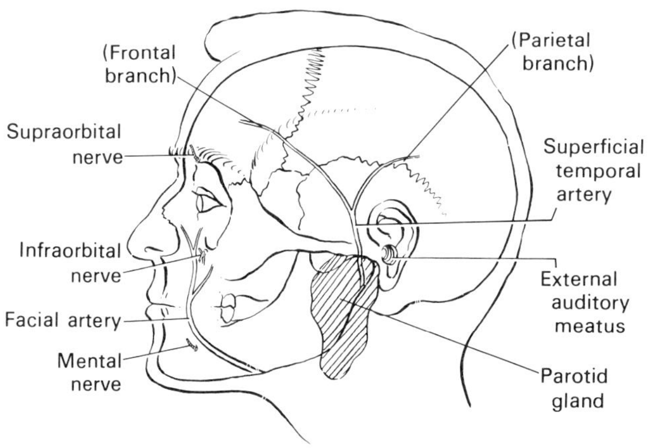

Fig. III-9. Laterial view, face and skull, localizing pertinent structures.

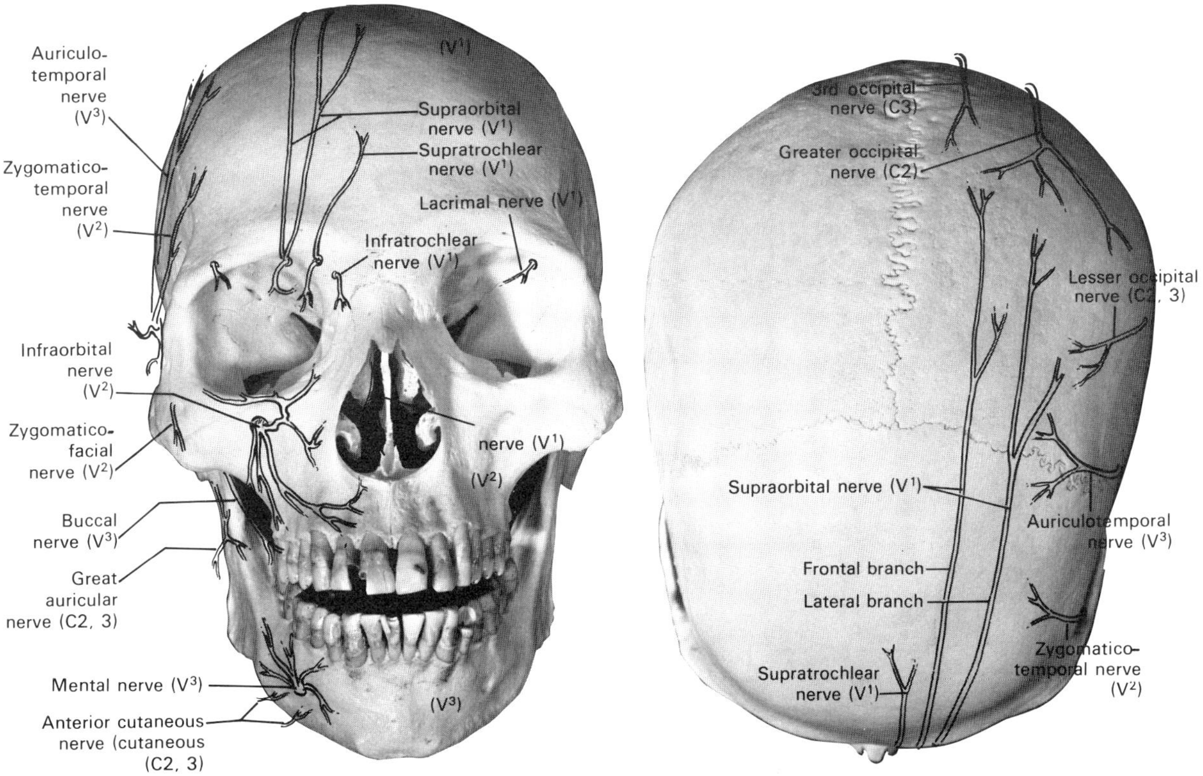

Fig. III-10. Sensory nerves of face and scalp. **A.** Frontal view. **B.** View from above. V_1, from ophthalmic division of trigeminal; V_2, from maxillary div.; V_3, from mandibular div.; C, cervical cord levels of named cervical nerves.

have located the supraorbital notch or foramen on yourself and on a skull, use the skull to locate the prominent **infraorbital foramen** below the orbital cavity in the **maxillary bone** and the **mental foramen** in the chin area of the mandible. A vertical line could be drawn which would almost connect these three foramina (Fig. III-6). Through them some of the **sensory branches** of the **three chief divisions** (V1, V2, and V3) of the *trigeminal nerve* emerge to supply the skin of the face and the scalp.

1. The **supraorbital nerve** is the largest of the five branches of the **ophthalmic division (V1)** to supply the skin above the level of the eyes. It is sensory for touch, pain, the temperature to the upper lid and to the forehead and scalp at least as far back as the intra-auricular line (imaginary line drawn from ear to ear across the scalp). The rest of the scalp is supplied by cervical spinal nerves (Fig. III-10). There are four other smaller twigs of the ophthalmic division to the skin in this area (Fig. III-10), but these details will be covered in another chapter.

2. The **infraorbital nerve** is the largest of three branches of the **maxillary division (V2)** of the trigeminal nerve to supply the skin around the nose and mouth. It is sensory to the lower lid and conjunctiva, the inner and outer aspects of the upper lip, a part of the mucous membrane of the cheek, and to the dorsum and alae (wings) of the nose. Two other tiny sensory branches of the maxillary division to skin of the face are seen in Figure III-10.

3. The **mental nerve** is one of three sensory branches of the **mandibular division (V3)** of the trigeminal nerve. It distributes to the skin of the chin and to the skin and mucous membrane of the lower lip. Figure III-10 shows two other sensory nerves from the mandibular division which do not utilize the mental foramen, the auriculotemporal and the buccal. We will consider the complexity of the trigeminal nerve in another chapter. For the moment, attend only to the three nerves emerging from the three foramina discussed above. One general concept about the trigeminal nerve, as a surface feature, can be developed at this time. Place your finger just in front of your right ear; trace over the scalp in the interauricular line to in front of the left ear, then down the angle of the jaw to the tip of the chin (symphysis menti) and along the right mandible to the front of the ear. You have outlined within these boundaries the *entire* territory of **sensory supply** of the *trigeminal (5th cranial) nerve* to the *skin* of *scalp* and *face*. No other nerve gives a sensory innervation to this area. The **10th** cranial nerve (2 × **5**) supplies a patch of skin behind the ear.

Deep to the infraorbital foramen **within** the maxil-

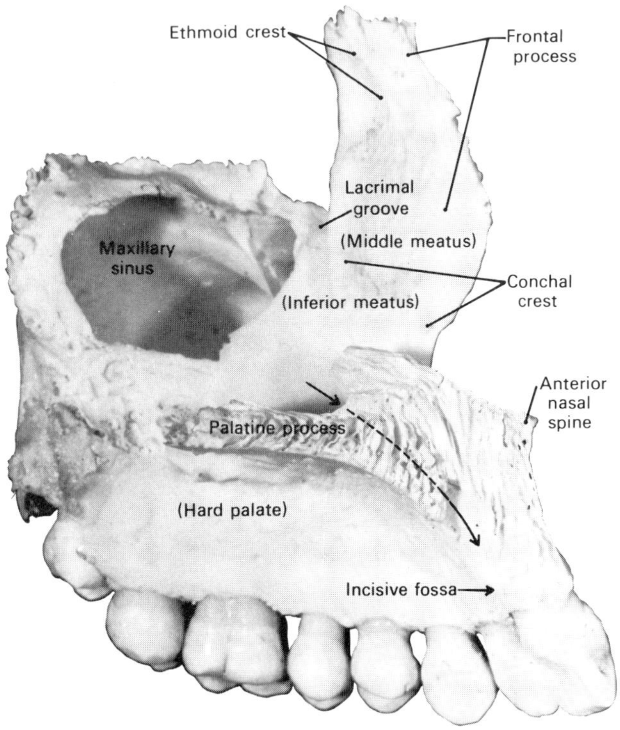

Fig. III-11. Left maxilla, medial view looking into maxillary sinus. Dotted arrows show course of incisive canal (in part) and, hence, course of nasopalatine nerve.

lary bone will lie the largest air sinus of any bone in the skull, the **maxillary sinus,** or antrum of Highmore (Figs. III-6, 8, and 11). In a dissectable skull, open the trap door in the maxilla and locate the **ostium** of the sinus high up on its medial wall, poorly placed for free drainage. Pass a waxed string into it to determine where it opens in the nasal cavity. It does not do so directly, but opens into a deep groove, the narrow ethmoidal infundibulum, into which some of the ethmoidal air sinuses also drain. Inflammation of the nasal mucosa of this narrow groove also interferes with ready drainage of the maxillary sinus at times.

Superficial Temporal and Middle Meningeal Arteries: Run a finger along the lateral (temporal) margin of your orbital cavity. You are feeling the **frontal process** of your **zygomatic bone** whose **malar prominence** can be felt as the cheek bone (Figs. III-2 and 6). Trace your finger back with heavy pressure over its **zygomatic process** which joins with the zygomatic process of the **temporal bone** to form the **zygomatic arch.** Your finger will lead directly to the **external auditory meatus** of the temporal bone (Figs. III-1, 2, and 9).

Approximately two fingerbreadths anterior to the meatus, place your middle finger on the zygomatic arch to feel the pulse of the **superficial temporal artery** (Fig. III-9). Obviously, this branch of the external carotid artery *crosses* the arch on its way to the temple and scalp above.

About three fingerbreadths anterior to the meatus the **middle meningeal artery,** the largest and most constant arterial supply to the dural covering of the brain, *passes beneath* the arch. Hence, its pulse cannot be felt. We shall pick up its course again later.

The Mastoid Process: Just behind the ear feel the prominent anterior part of the **mastoid** ("titlike"; an "oid" ending always signifies "like" **process** of the temporal bone (Figs. III-2 and 3A). Its posterior position and tip are difficult to feel since they lie deep to the insertion of a large neck muscle, the **Sternomastoid.** The bony process contains many small air cells lined with the same kind of a mucous membrane that clothes the nasopharynx (Fig. II-14). Indeed, these air spaces communicate with the nasopharynx by a devious route through the middle ear cavity and the auditory (Eustachian) tube.

The Temporomandibular Joint: The whole surface of the U-shaped **mandible,** or lower jaw, can be palpated except for its **coronoid process** which is hidden behind the zygomatic arch (Figs. III-1, 2, and 12). Place a finger in front of the external auditory meatus and open and close your mouth. The **condyloid process** then can be felt moving as it articulates in the **mandibular fossa** (depression) of the temporal bone (Fig. III-2). This articulation is the *only freely movable joint* (diarthrosis) in the head. All the others are *immovable joints* (synarthroses) of the *suture* type (Fig. III-2).

If you love to dig out the meaning of medical terms, the last paragraph is an intriguing goldmine. "Mandible" comes from the Latin "mandibula" which means lower jaw and is derived from the verb, "mandire", "to chew." Both jaws were once called "maxilla, superior and inferior." Hence, the odd term "submaxillary" for the salivary gland lying below the mandible and now often and more correctly called "submandibular." "Maxilla" came from the Greek as a diminitive form of "mala," the cheek (*cf.* **malar prominence**). Both terms are related to the verb "macerare" which again means "to chew." And if you think of cheeks as round and rosy, you are not far off, for "mala" is Greek also for fruits like apples and peaches. "Temporo" and "temple" relate to the Latin "tempus," a word meaning both "temple" and "time." But this goes back to the Greek word, "to cut," since time can be considered as "a slice off eternity," and a blow to the "temple of the head" can cut one off from life. It might be reasoned that "coronoid" meant like a corona, or crown. But there is a similar Greek word for the notched end of a bow to hold the bowstring, and the coronoid processes of the mandible and the ulna bones are both notched structures. Condyloid again is Greek,

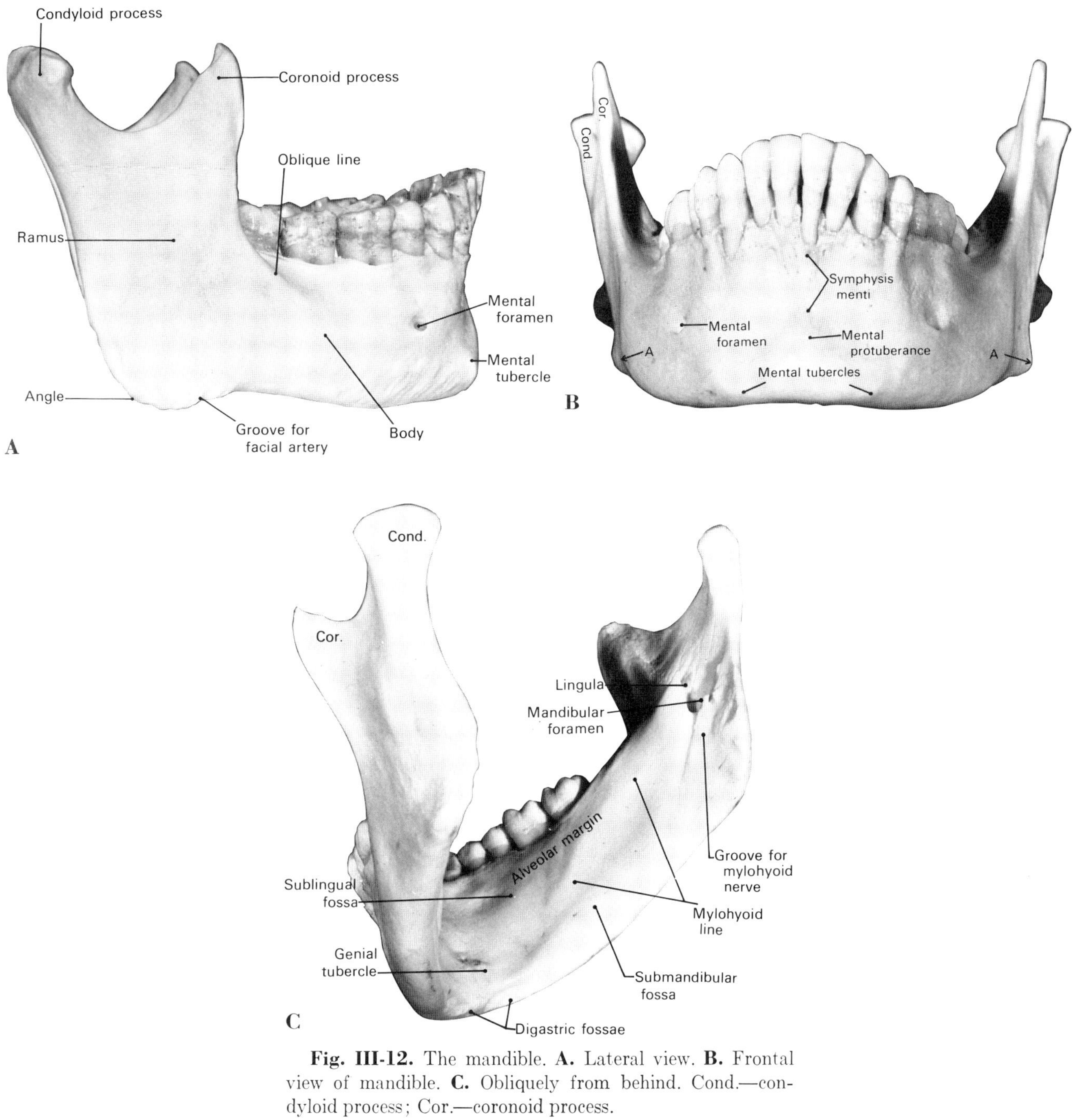

Fig. III-12. The mandible. **A.** Lateral view. **B.** Frontal view of mandible. **C.** Obliquely from behind. Cond.—condyloid process; Cor.—coronoid process.

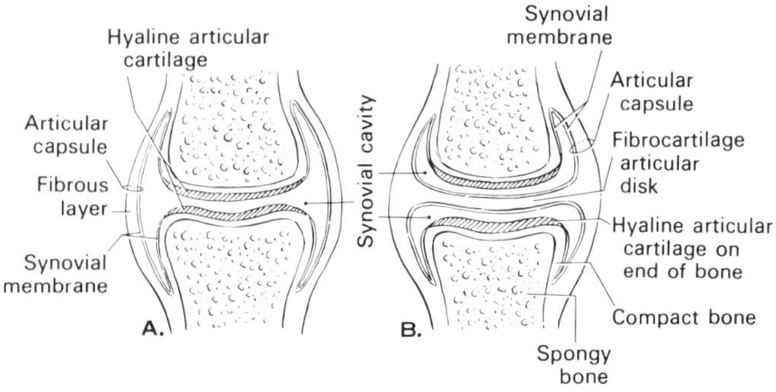

Fig. III-13A. Diagram of usual type of synovial (diarthrotic) joint. **B.** Synovial joint with an articular disk. (Schematic).

"knoblike," as the knob of a knuckle; "knuckle," a Dutch diminutive form for "bone." Diarthrosis comes from two Greek words meaning "throughout" and "a joint," so it means "a thorough or complete, or freely movable" articulation. Synarthrosis comes from the Greek "bound or bonded together" plus "a joint"; hence, it signifies an immovable joint. Suture is from the Latin "suturer" or "seam," or "a sewing together" from the verb "suo," "to sew." A third large class of joints is the *amphiarthroses,* Greek for "both" and "joints"; hence, inbetween or incomplete, or partly movable joints. There is, seemingly, no end to the useful trivia you can glean once you start to cultivate medical terms. Try it as often as you can if it helps you toward a better understanding. If you don't particularly care for this kind of fun, you are simply stuck with the task of rote memorization of each term you acquire which, in the interest of conservation of time and energy, is quite all right so long as you use (and, hopefully, spell) the terms properly.

Generally, in diarthrotic joints the articulating ends of opposing bones are encased in a ligamentous capsule whose inner lining layer rests upon well-vascularized connective tissue. This lining membrane secretes a **synovial fluid** into the capsule. The fluid, consisting chiefly of **hyaluronic acid,** serves as a lubricant bathing the cartilaginous-covered ends of bones as they slide against each other in the joint. Figure III-13A schematically illustrates this point, and Figure III-13B shows how such a joint can be separated into two synovial cavities by the interpositioning of a fibrocartilaginous **articular disk.** The temporomandibular joint is modified according to this latter schema (Fig III-14C). Its capsule is lax between disk and temporal bone but strong between disk and condyloid process, and these actual relationships are depicted in Figure III-14C and D.

Significance: The disk moves forward when the joint is **protracted** (pulled forward) as in a yawn (Fig. III-14A and B). Protraction can be so forcible as to slip the condyloid process out of the mandibular fossa anteriorly over its **articular tubercle.** It will then fall into a deeper depression behind the zygomatic arch, the **infratemporal fossa,** and the mouth will be left open. This dislocation can only be reset by forcing the condyle downward and backward to ease it back into place behind its articular tubercle.

To feel the tenseness in the **Masseter muscle** over the **ramus** of the mandible and in the **Temporalis muscle** in front of the ear and above the zygomatic arch, clench your teeth. Both of these *muscles of mastication* (hence, innervated by the 5th cranial nerve) serve in closing the mouth.

Significance: Most of this closing pressure is borne by the molar teeth and by the synovial membrane of the temporomandibular joint. Besides having a robust blood supply, the membrane is well innervated by sensory nerve fibers. Malocclusion of the teeth or spastic contraction of Masseter and Temporalis muscles (as in pipe-stem clenchers) can, in time, cause pain and lead to degenerative changes in the joint.

The Parotid Gland and Duct: Mark out a triangular region by running your finger from the jaw angle approximately to the middle of the zygomatic arch and then posteriorly to the mastoid process and back to the angle of the jaw (Fig. III-15). The **parotid gland,** the one whose swelling within its tight connective tissue capsule causes the pain of mumps, somewhat overflows these approximate boundaries (Fig. III-9). Place a finger along the lower border of the zygomatic arch and parallel with it. The **parotid duct** (of Stenson) will course along the under border of your finger, and the fingertip will end at about the second upper molar tooth (Figs. III-15 and 16). Here the duct pierces deep to empty into the **buccal cavity** (Fig. III-16B). In a fortunate photograph of a subject's open mouth, our camera incidentally caught the ejection of a drop of saliva from the parotid duct (see Fig. III-27). Tense your Masseter muscle by clenching your teeth while you place your fingers on the outside of the cheek and try to roll this stout duct beneath the fingertips. Peer into the open mouth with a flashlight to discover the duct orifice opposite the second upper molar tooth.

The Facial Artery: Three fingerbreadths anterior to the angle of the jaw, the **facial artery,** another branch of the external carotid, curls upward around the lower border of the mandible and crosses it superficially. It angles across the corner of the mouth and up along the side of the nose (Fig. III-9). *Feel its pulse:*

1. Three fingerbreadths from the jaw angle.
2. Alongside the nose.
3. By grasping the cheek between thumb and forefinger at the angle of the mouth.
4. By grasping the upper lip between thumb and fingers to feel the pulse in its **labial** branches.

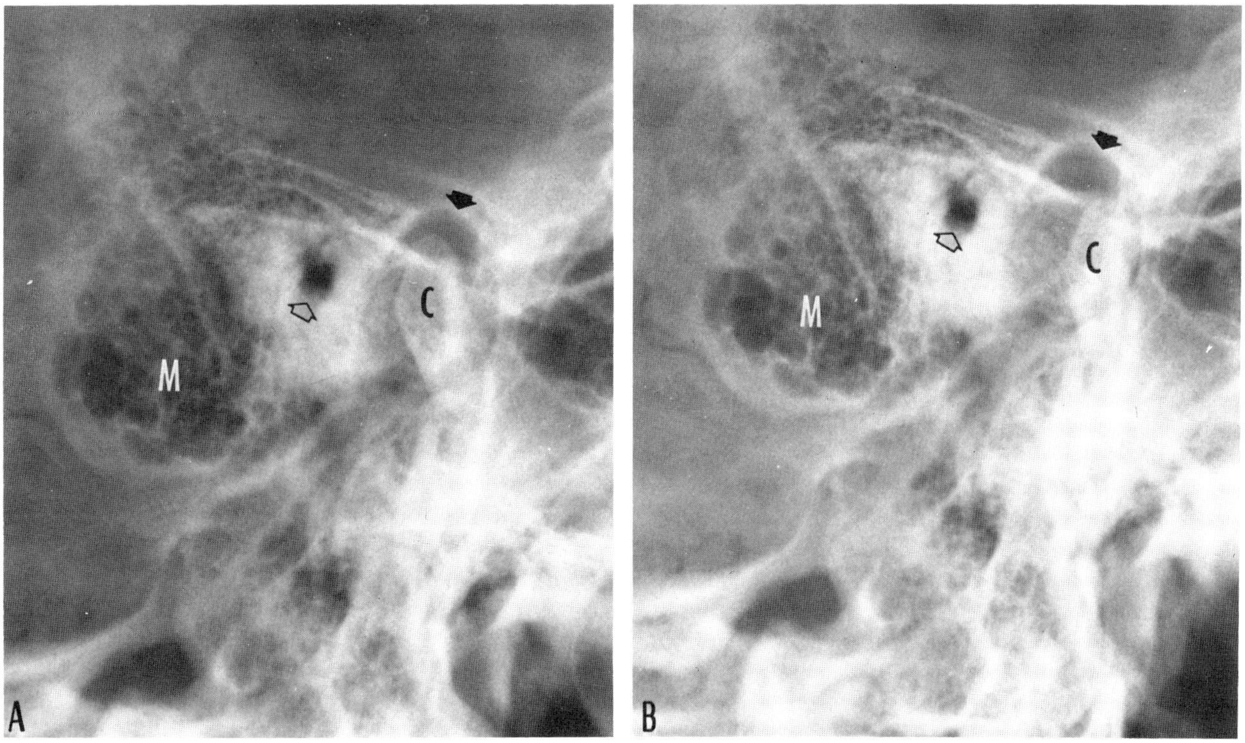

Fig. III-14. Temporomandibular joint. **A.** Lateral oblique X-ray of joint with mouth closed. C, condyloid process in temporomandibular fossa (at black arrow); M, mastoid air cells; external auditory meatus at open arrow. **B.** X-ray of same joint with mouth open. Note change of position of C. The dark region beyond C at the right margin of both fissures is the infratemporal fossa.

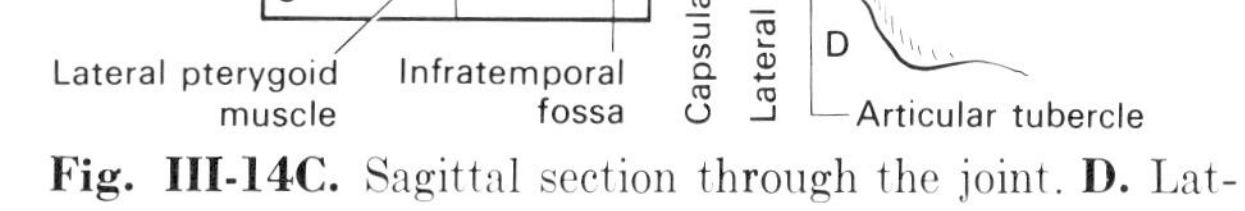

Fig. III-14C. Sagittal section through the joint. **D.** Lateral aspect of joint with capsular ligaments in place.

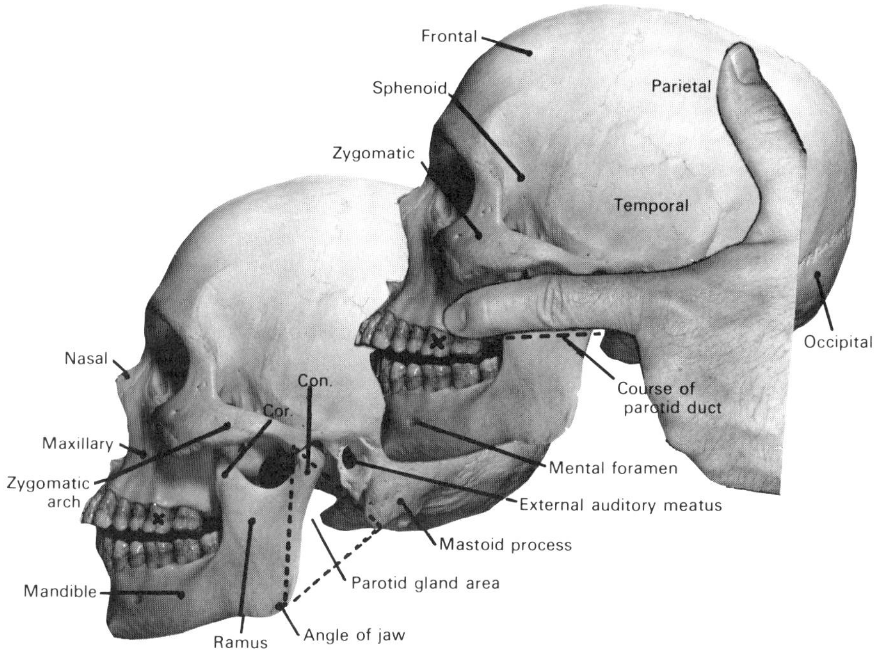

Fig. III-15. Method for localizing parotid gland and duct. Cor. and Con., coronoid and condyloid processes of mandible; x, 2nd molar tooth.

The 7th Cranial or Facial Nerve: Place your *left* hand with fingers extended and thumb down upon the *right side* of a colleague's face (Fig. III-17). Let the little finger lie along the upper lateral margin of the orbital rim; the ring finger, alongside the nose; the middle finger, along the line of the upper teeth; the index finger, along the mandible; and let the thumb point obliquely downward toward the anterior aspect of the neck. You will have outlined roughly the course of the *five major branches of the facial nerve* as its main stem emerges from the **stylomastoid foramen** in front of the mastoid process. Most of these branches wlll have had to pierce the substance of the parotid gland before they reached the **muscles of facial expression** (Fig. III-18) which they innervate.

It is not important that you should know all of the muscles of facial expression shown in the figure. Many are insignificant slips whose very names are soon forgotten. Realize merely, that some arise from bone and insert in skin; others are skin residents entirely, originating and inserting there locally; and that a few wander farther afield into the scalp, ears, and neck region.

What is important is that you distinguish between the muscles in the face that are supplied by the 7th cranial nerve, and those by the 5th cranial nerve.

The 7th cranial or facial nerve supplies all those skeletal muscles dervied from the mesoderm of the

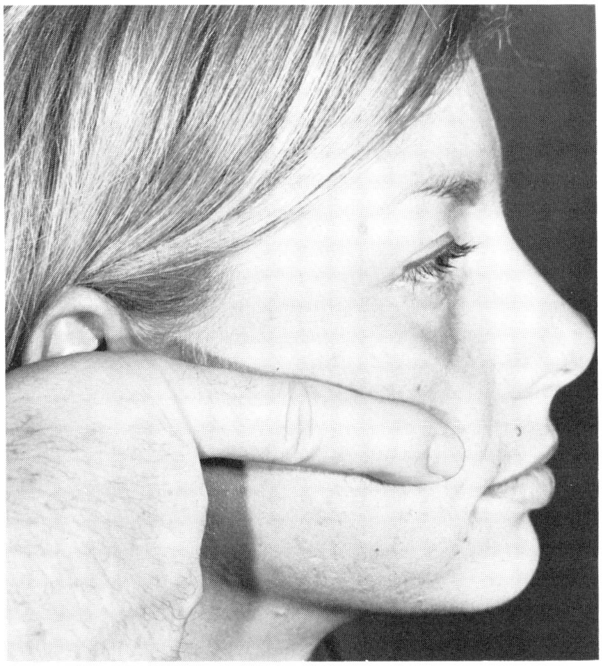

Fig. III-16A. Localizing parotid duct on a subject.

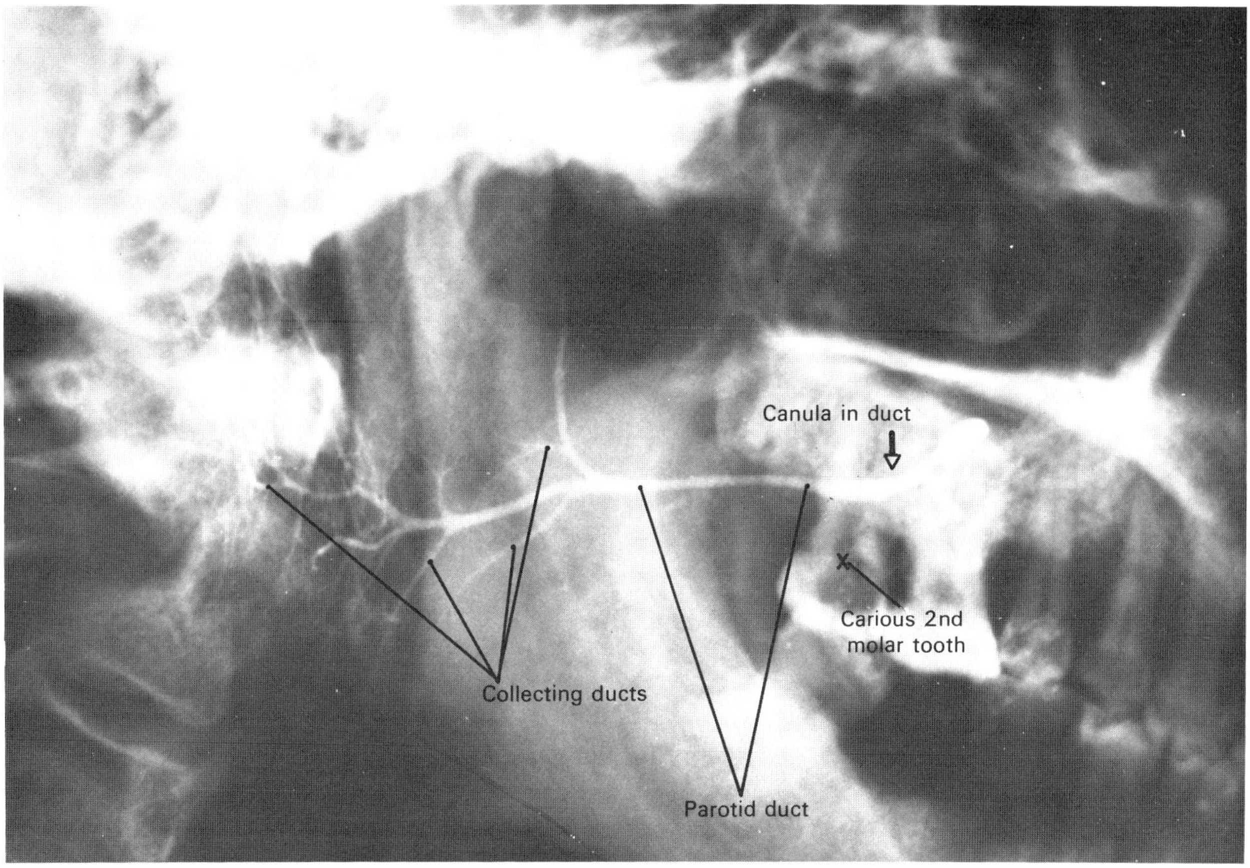

Fig. III-16B. X-ray showing contrast media filling parotid duct and collecting ducts of the gland (lateral view) (*Sialogram*).

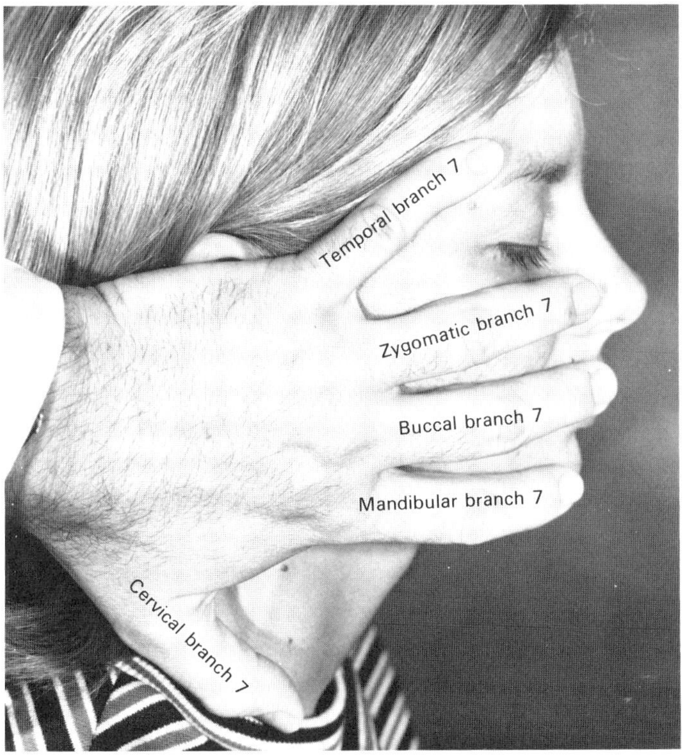

Fig. III-17. Localization of the branches of the 7th cranial nerve in a subject.

2nd branchial arch region of the embryo, most of which happen to be muscles which alter facial expression (Fig. III-18). The nerve does innervate a few other muscles that are not so directly related to this task; *e.g.,* the **Stapedius** of the middle ear cavity, the **Stylohyoid,** and the **posterior belly** of the **Digastric** in the neck.

The 5th cranial or trigeminal nerve, *through its mandibular division only,* supplies those muscles derived from the mesoderm of the *1st branchial arch region* of the embryo, chief of which are the *muscles of mastication* (**Temporalis, Masseter,** and **Pterygoids**). It also supplies a few other odd muscles that are not as concerned with jaw movement; *e.g.,* the **anterior belly** of the **Digastric,** the **Mylohyoid,** and the **Tensors Tympani** and **Palatini.**

For the present, you might consider remembering the names of only a few **muscles of facial expression:** Certainly, the two sphincteric ones which close the mouth and the eyes, the **Orbicularis Oris,** and the **Orbicularis Oculi;** and those that you are not likely to associate with facial expression at all, the **Frontalis, Occipitalis,** and the **Auricular muscles** (which may be uselessly atavistic in some of us); the **Platysma** in the neck region, and the **Buccinator** of the cheek pouches. There is a tendency to want to think of this

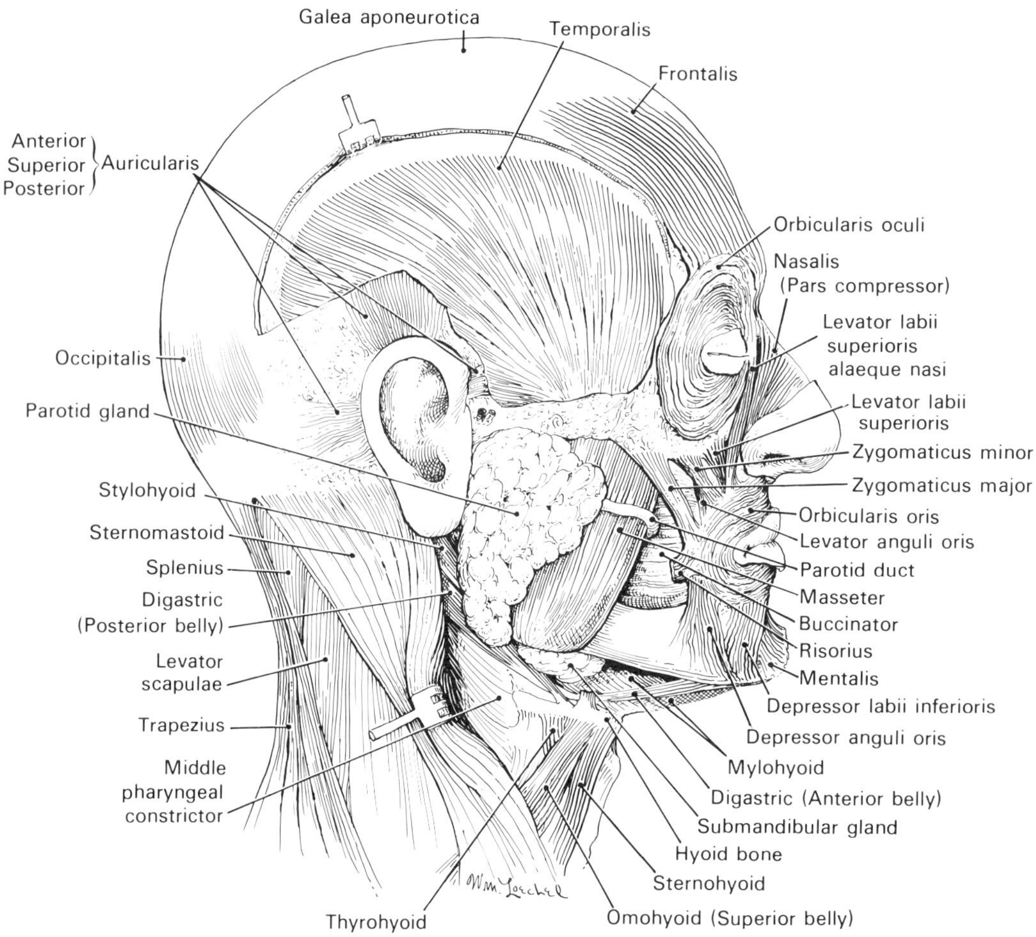

Fig. III-18. Muscles of facial expression: Muscles without leaders in this sketch are *not* muscles of facial expression. The following muscles of facial expression are *not* shown: Depressor septi; Levator anguli oris; Risorius; Dilator naris; Transversus menti; Platysma.

last muscle as a muscle of mastication. Its Latin name means "trumpeter"; hence, it is a muscle of facial expression, functioning to puff out the cheeks as in blowing.

The Pterion as a Reference Point: Observe the inset in Figure III-1 where the middle finger, which is slightly longer than the index finger, is laid along the upper border of the zygomatic arch. The thumb of the other hand is placed vertically along the lateral orbital ridge until its tip meets the tip of the index finger. This is easier to do on a skull, or the head of a colleague, than upon yourself. In the angle between thumb and index finger will be the **pterion,** or **Sylvian point.** On the skull, it represents the junction point of **parietal** and **sphenoidal** suture lines on the lateral aspect of the brain case. In the living, it marks the point where the **anterior branch** of the **middle meningeal artery** courses obliquely upward and backward over the surface of the dura. Its **posterior branch** courses backward over the dura surface two fingerbreadths above the zygomatic arch (Fig. III-19).

If you erect an oblique line from the suture between frontal and zygomatic bones to the pterion (Fig. III-1), the **frontal lobe** of the **brain** will lie above this line and the **temporal lobe** below it (Fig. III-19). This line will also mark the position of the stem of the **lateral cerebral fissure,** or **Sylvian fissure,** named in 1641 after Francis de la Boe, a French anatomist, who was known as Sylvius by his Latin-speaking colleagues of that era. This deep groove on the lateral hemispheric surface separates frontal from temporal lobes of the brain.

Moreover, one fingerbreadth above the pterion, in the frontal lobe of the dominant hemisphere will lie the **motor-speech center** (Broca's area). In the right-

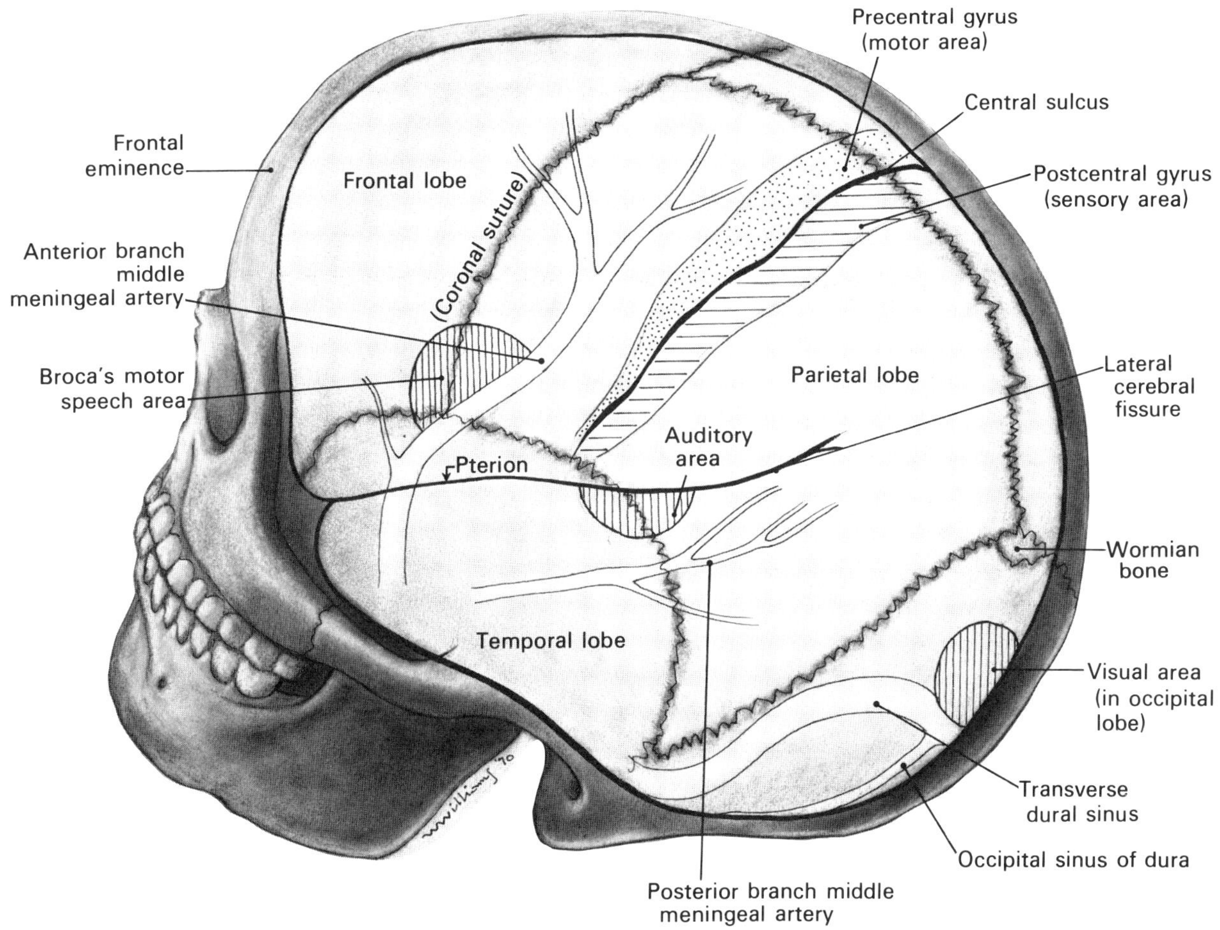

Fig. III-19. Diagram—looking through "transparent" skull to orient landmarks and structures on dura and brain surface.

handed individual, the left hemisphere is considered to be the physiologically dominant one in respect to language function; *i.e.*, in respect to the symbolic formulation and the understanding of the spoken and written word. The reverse is thought to be true in the left-handed individual where control of these functions is in the right hemisphere.

Significance: With a brain lesion in Broca's area there is a disturbance in *using* words, alone or in combination, even though the patient can understand what is said and can follow instructions. In severe cases the patient is almost entirely speechless. Such a condition if it exists as a pure motor deficit (and it rarely does exist in pure form) is known as **motor or verbal aphasia.** It is interesting to note that this area of the brain is one of the last to develop, and that speech is the last important motor faculty to develop in the child.

Relationships around the Preauricular Point: The exact center of the external auditory meatus is the **auricular point.** The anterior point of a horizontal diameter drawn through this center constitutes the **preauricular point,** at the anterior rim of the meatus (Fig. III-1). A line projected from the lower border of the orbital margin through the preauricular point and through the inion is known as **Reid's line.** A fingerbreadth above this line from meatus to inion will run the **transverse dural sinus** (Figs. III-1, 6, and 19). Like the superior longitudinal dural sinus, it is one of the major channels for collecting venous drainage from the brain, conducting this venous blood to the jugular vein *via* an S-shaped **sigmoid sinus,** the connecting link between transverse sinus and vein.

If a perpendicular is erected upon Reid's line at the pre-auricular point, it will almost bisect the nasion-inion line (Fig. III-1). It falls short by crossing this line ½ inch posterior to its midpoint. Three fingerbreadths along this perpendicular above the pre-auricular point will lie the *lower end* of the **central sulcus of the brain** (Figs. III-1C and 19). **Sulci** are the shallow fissures of the brain; convolutions or **gyri** of the brain lie between sulci or fissures. The prominent central sulcus extends upward in each hemisphere to the superior longitudinal fissure from our measured low point. It not only separates **frontal** from the **parietal** lobe of the brain, but the **main motor control area** of the brain (voluntary motor area) from its **general somatic sensory receptive center** (Figs. III-1C and 19), where pain, touch, and proprioception reach consciousness so that we become aware of these sensory modalities.

Fontanelles: Two landmarks of minor significance on the adult skull, but of some relative importance on the infant skull, are the **bregma** and the **lambda** (Figs. III-1–3). To the ancient Greeks, bregma simply meant the vertex of the head. Actually, the vertex, as the highest point, lies posterior to the bregma in most heads. The bregma is the point where coronal and sagittal sutures of the cranium meet in the midline (Fig. III-1). The lambda is a similar midline point posteriorly where the sagittal suture between the parietal bone meets the transverse **lambdoid suture** that separates parietal and occipital bones. The Greeks used the term, "lambda," for this suture junction because it was shaped like *their* letter "L" (Fig. III-3).

Significance: In the adult, all these suture lines are tight, bone interlacing with bone in serrated seams, to form immovable joints (*synarthroses*). These seams are wider apart, extremely wide at the bregma and lambda, in infant skulls (Fig. III-20). The gaps between the bones of the cranium are filled by connective tissue which later becomes bone by intramembranous ossification. As a consequence, the suturelike areas in the infant skull are partially movable joints (*amphiarthroses*) of the *syndesmotic* type (joined with connective tissue), while the immovable adult suture joints are of the *synostotic type* (joined with bone).

There are other wide gaps like this in the infant skull. Witness those between parietal and sphenoid and parietal and temporal bones on the lateral aspect of the infant skull (Fig. III-20). These flexible sutures allow the skull bones to compress and override each other somewhat as the infant head moves through the pelvic birth canal. Therefore, fathers should be revived and encouraged when they first take a look at some of the peaked, "Denny Dimwit" crania of their newborn offspring. In time, the head will remold into a reasonably esthetic contour.

These "soft spots" at the wide gaps are also the areas of the infant **fontanelles.** You may recall that in the ancient days of medicine, barbers and surgeons were one and the same—blood brothers, as it were. The barber pole was the symbol of their guild—red for bloodletting and white for bandaging. To effect a cure of any number of diseases, it was deemed proper procedure to purge the blood of its putrid humors (fluids) through bloodletting. In the case of a sick infant, the barber-surgeon found it convenient to knick with his razor the vessels in these "soft spots" of the infant head. Blood spurted forth "as from little fountains"; hence the term, *fontanelles.*

Take an adult skull and search along the suture lines in the areas of these former wide gaps. You may find small, separate, pieces of bone in these sutures that were formed as the connective tissue underwent intramembranous ossification with growth of the cranium. These little chunks of bone are the so-called **Wormian bones,** not because they look like worms, but because they were first described in the 1600's by Claus Worm, a Danish anatomist who later became both Professor

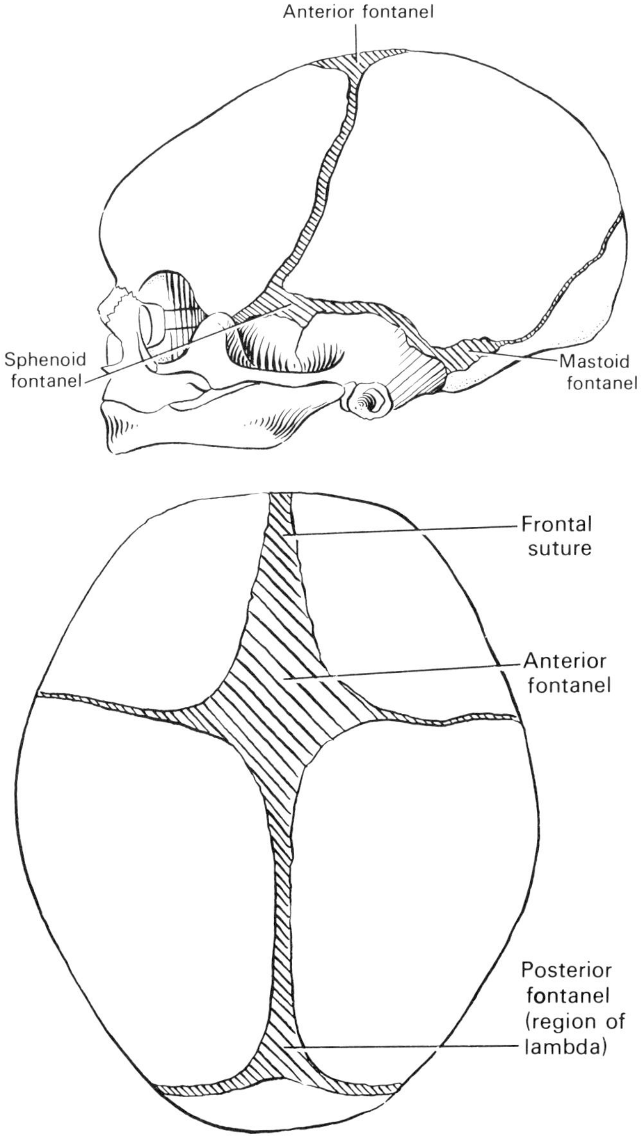

Fig. III-20. The fontanelles of the infant skull.

of Physiology and Medicine at the University of Copenhagen. As a matter of fact, it was Thomas Bartholin (of Bartholin's glands in the vaginal vestibule) who first named these insignificant bones in honor of his uncle, Claus. Things were kept in the family in those days. Thomas became Dean of the College of Medicine at Copenhagen, and his father and son were Professors of Anatomy there.

Area of the Eye, Lid, and Lacrimal Apparatus

The Cornea: Look at your own eye in a mirror and then later examine a colleague's eye. Most of the globular eyeball lies almost completely within the bony orbital cavity. But, if you look at it from the side (Fig. III-21A), you will find its anterior curvature protruding between the cleft of the lids; *i.e.,* through the **palpebral fissure.** The most prominent anterior bulging is produced by the transparent **cornea** which becomes continuous at its margins with the white **sclera.**

Significance: The greatest refraction of light rays coming into the eye occurs at this outer, curved, corneal surface, greater even than that which occurs when the lens of the eye bulges to increase its own curvature, as it does to accommodate for near vision. The greater the degree of curvature of any convex lens, the greater its refraction power; *i.e.,* its power to bend and converge parallel rays of light coming from an object and to bring them to focus close behind the lens. Both the cornea and the lens of the eyeball are convex lenses.

The Pupil and Iris Diaphragm: Examine the eye straight on (Figs. III-21C and 22A). Note the **pupil** which is merely a hole in the surrounding colored **iris.** Realize that you are looking through the transparent cornea to see both of these structures and that a chamber of aqueous fluid, the **anterior chamber** of the eyeball, lies behind the cornea, between its concave inner surface and the colored iris. Realize, too, that the iris is a variable diaphragm which sits in front of the lens. It is shaped like a shallow, truncated cone, having a narrower diameter at the pupillary margin than posteriorly where it spreads over the lens (which, of course, you can't see, since the lens lies behind this shallow funnel of an iris diaphragm).

Significance: Because the iris contains an anular, smooth muscle sphincter at its pupillary margin and radially-disposed strands of smooth muscle directed toward its lens margin, the pupillary orifice can be decreased in diameter or widened. You can test this in several ways. Flash a bright light (pencil flashlight) in the eye as you look in the mirror. Bright light falling on the photosensitive retina of the eyeball (focused there by cornea and lens) reflexly causes the pupil to constrict to pinpoint diameter.

The pupillosphincter has been goaded to constrict in a purely involuntary and protective manner. Try looking in the mirror while you willfully command it to constrict and see what success you have!

While you watch his eyes, ask a colleague to look off into the distance and then to focus on his fingertip closely. As he does so, both eyes will turn toward his nose, *i.e., the pupils will constrict consensually* (together) in order to cut out distracting peripheral light in the act of scrutinizing a near object. The lens will bulge also, but you will not be able to detect this action. In a sense, while this is called an **accommodation reflex,** there is here an act of will on his part, for he is purposefully focusing his attention upon a near object. The neural pathway for this type of "reflex" response of the pupil goes through the cerebral cortex. The reflex pathway in response to bright light does not.

Next, grasp him firmly by the nape of the neck and squeeze suddenly. His pupils will **dilate** reflexly. Like the **pupillolight reflex,** this **ciliospinal reflex** does not involve cortical pathways, and, in this case, the radial dilator fibers are activated rather than the sphincteric constrictors Pupilloconstriction is mediated by the parasympathetic system; pupillodilatation, by the sympathetic.

The Eyelids: As you look in the mirror, note that the entire lower half of the iris is normally uncovered by the lower lid while the upper lid crosses the iris superiorly (Fig. III-21C and 22A). Notice also the deep groove above the upper lid and beneath which most of the upper lid seems to be recessed. This is the **orbital groove** coursing between the eyebrows above and the eyelashes below (Fig. III-21C). The area of skin above the groove up to the brows is the region of the **orbital septum;** that below the groove is the **tarsal plate region** within the lid proper. Both of these structures are, actually, connective tissue layers lying beneath the skin, but their areal extent can be seen as surface features. If you will examine Figure III-22B, you will see that the *Orbicularis Oculi* below the skin consists of two parts. An **outer circle of muscle** runs above and below the upper and lower lids. This is its **orbital portion.** The thin, fascial, connective tissue *covering* over this part is the *orbital septum.* There is an **inner circle of muscle** running *through* each lid. This is the **palpebral portion,** and it rests *upon* a thick connective tissue lamina which forms the central skeleton of the lid, the **tarsal plate** (*cf.* Figs. III-22B and C).

In the lower lid (Fig. III-21C) where the tarsal plate region is quite prominently displayed but the orbital groove is shallow and the orbital septal region below it (or the lower orbital portion of the Orbicularis Oculi) is more difficult to circumscribe as a surface marking. The septal region blends off into the skin over the maxillary bone. If the skin here is "baggy" or somewhat pigmented, the lower orbital septum region is more readily visualized.

Significance: We have noted that the normal pattern is for the upper lid to cover part of the iris, but this will change when nerves that control lid movement or tone are injured:

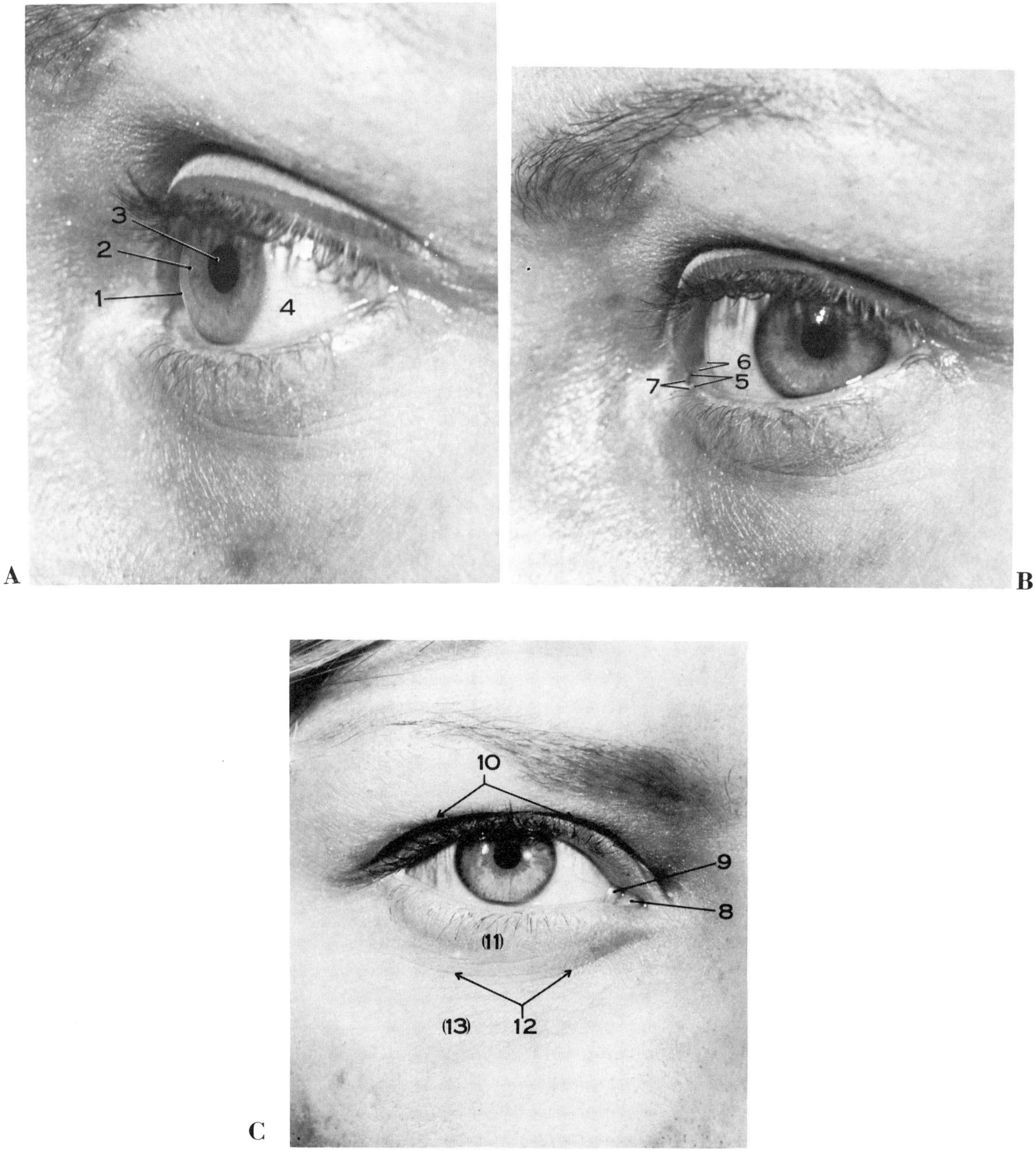

Fig. III-21. Anatomical relationships of the eye, lid, and lacrimal apparatus. **A.** Lateral view, eye looking straight forward. **B.** Lateral view with lateral gaze. **C.** Frontal view. 1, Anterior curvature of cornea; 2, iris; 3, pupil; 4, sclera, note blood vessels; 5, lacrimal papilla; 6, lacrimal lake; 7, lacrimal canaliculi; 8, caruncle; 9, plica semilunaris (8 and 9 in lacrimal lake); 10, superior orbital groove; 11, area of lower tarsal plate; 12, inferior orbital groove; 13, area of lower orbital septum.

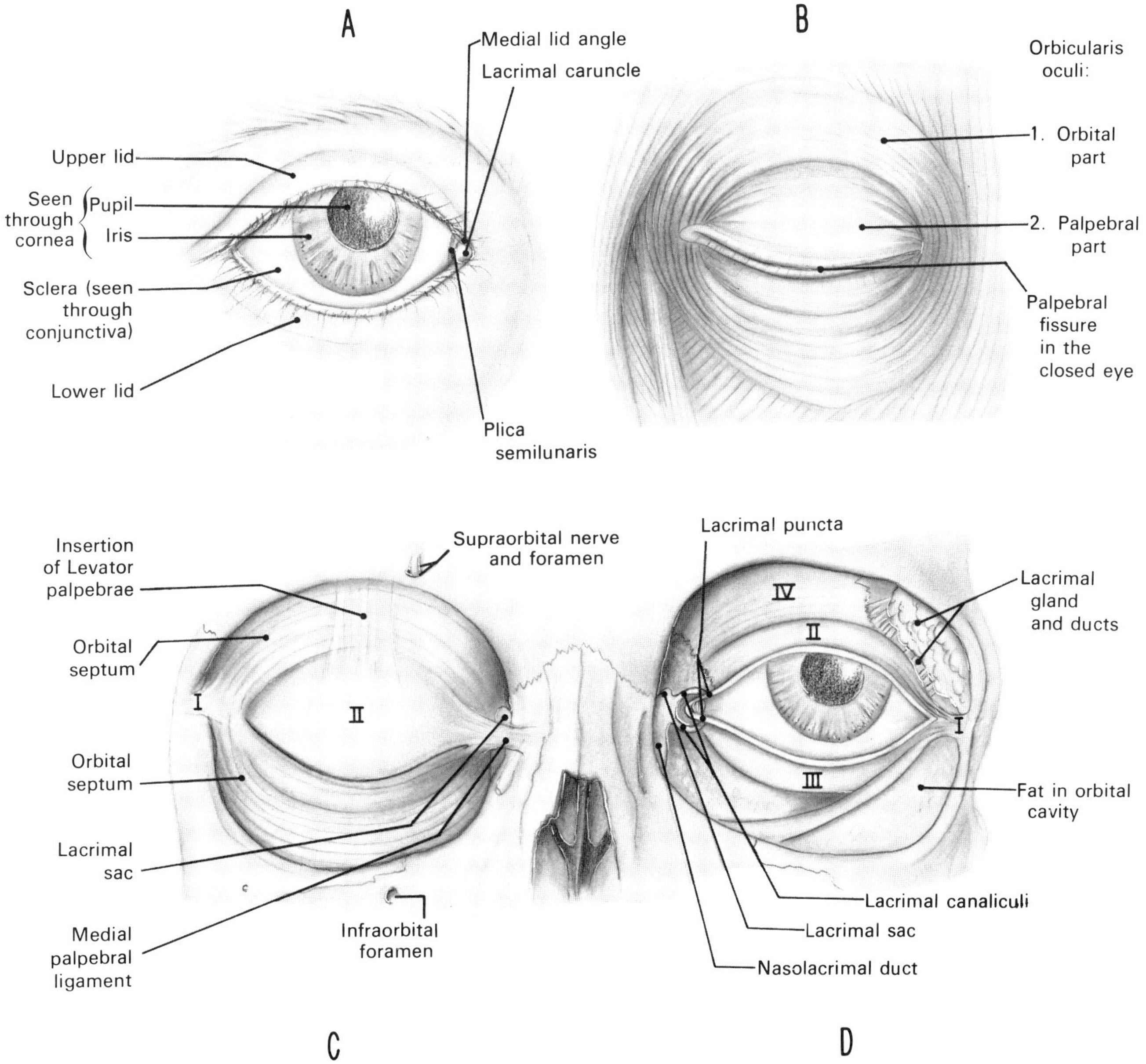

Fig. III-22. Progressive dissection of eyelids and lacrimal apparatus. **A.** Right eye. **B.** Left eye. Eye is closed with skin removed to the level of Orbicularis oculi muscle in the lower stratum of the lid dermis. **C.** Right eye. Obicularis oculi removed to show orbital septum and tarsal plate. **D.** Left eye. Further dissection to show parts of lacrimal apparatus. In **C** and **D:** I, lateral palpebral ligament; II, superior tarsal plate; III, inferior tarsal plate; IV, orbital septum (in part). (After Craft.)

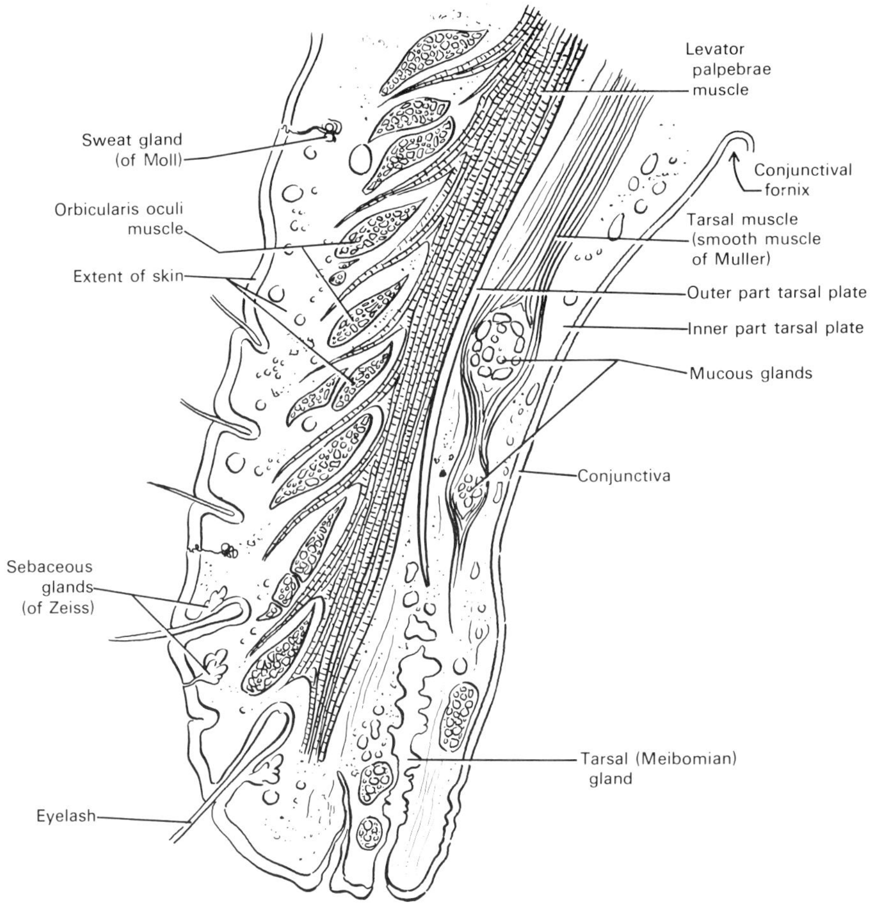

Fig. III-23. Cross-section of upper lid.

1. In a 3rd cranial nerve injury, the levator muscle (*Levator Palpebrae*) of the upper lid (Fig. III-22C) is involved, and the lid will be in complete **ptosis;** *i.e.*, the eye will be entirely closed, and its upper lid cannot be voluntarily raised.

2. Injury to the cervical sympathetic trunk causes degeneration of the *smooth muscle* of the lids (*Tarsal muscles of Müller*), the lids lose tone, and there is a narrowing of the palpebral fissure resulting in a partial lid droop, or partial ptosis. More of the iris will be covered, both from above and below.

3. If the 7th nerve is injured, the lids cannot be closed because of paralysis of the Orbicularis Oculi which is the lid closer and one of the muscles of facial expression. Consider winking and recall the "crow's feet" that develop in the lateral margins. The same occurs in smiling or when the eyes close tightly as in wincing with pain. The orbital part of the muscle closes most forcibly and is responsible for the "crow's feet" phenomenon. In paralysis of the 7th nerve, the upper lid retracts so as to uncover the iris completely.

The Glands of Zeiss: Evert the upper lid by rolling it up over a match stick. Note the lashes aligned in two or three irregular rows. These hairs have their own **sebaceous glands** (of Zeiss) to keep them oiled. Some of these glands open directly upon the surface instead of into the sides of the hair follicles (Fig. III-23).

Significance: If the ducts of these glands become obstructed or inflamed, a **sty** will protrude from the front of the lid.

Tarsal Glands: On the posterior aspect of the rolled lid note a series of yellow streaks that shine through the moist, thin, **conjunctival covering** of the inner lid surface. These are the **tarsal glands** (Meibomian glands)—*large, modified, sebaceous glands* whose ducts open along the posterior lid margin (Fig. III-23).

Significance: The oil from these glands seals the lid margins of the closed eye and decreases the rate of evaporation from the moist inner conjunctival surfaces of the lid and eyeball. An obstruction of a tarsal gland produces a **chalazion** that will protrude toward the orbit and rub against it as the lids blink. Such an inflammation can be more annoying than a sty bulging from the outer surface of the lid.

Glands of Moll: Of less importance are the **sweat glands** (of Moll) which empty upon the lid surface. The locus of all three of the gland types just mentioned can best be appreciated in a cross-sectional view of the upper eyelid (Fig. III-23). In this same figure, cross-sectioned bundles of the palpebral part of the Orbicularis Oculi can be seen just beneath the outer skin. These *skeletal muscle* fibers pass in front of the tarsal plate and get their motor supply from the 7th cranial (facial) nerve. Also seen are the longitudinal strands of *skeletal muscle* belonging to the **Levator Palpebrae** (Levator of the upper lid; there is no comparable muscle or, for that matter, no depressor muscle, in the lower lid). The Levator also lies in front of the tarsal plate and gets its motor supply from the 3rd cranial (oculomotor) nerve. Its fibers course between and over the fascicles of the Orbicularis Oculi to anchor into the skin of the lid. Within the connective tissue of the tarsal plate, proper, and under the conjunctiva that surfaces the inner lid can be found longitudinal strands of *smooth muscle* (Tarsal muscle of Müller). This smooth muscle gets a motor supply from sympathetic nerve fibers. The tarsal glands (of Meibom) are also residents of the tarsal plate connective tissue.

The Lacrimal Apparatus: Only parts of the lacrimal apparatus can be visualized by surface inspection, but you should understand the basic idea of its constituent parts. The **lacrimal** (tear) **gland,** itself, lies hidden from view in a tiny fossa on the superolateral aspect of the bony orbit (Fig. III-22D). Thus, the glands lie on the temporal aspects of each eye. The gland has several tiny **ducts** opening into the area where the conjunctiva of the orbit reflects onto that of the upper lid; *i.e.,* at the superior **fornix** of the **conjunctiva.** The conjunctiva is the thin outer layer of cells which covers sclera and cornea and which reflects from the eyeball onto the inner surface of the lids. The gland receives its secretory motor supply from the **facial nerve** (7th), but these twigs use the sensory fibers of the trigeminal nerve as a highway to reach the gland. Therefore, just because a branch of the trigeminal is called the lacrimal nerve, it *cannot* be concluded that the 5th cranial nerve is secretory motor to the gland. The thin film of lacrimal secretion is wiped across the eyeball toward the nose by the automatic and intermittent blinking of the lids. In this way, particulate matter which might irritate the sensitive cornea, a structure richly endowed with pain receptors, is constantly washed toward the medial or nasal corner of the eye.

Obviously, this fluid must drain away, else it would constantly trickle down the cheeks in overflowing. Look at the medial angle of your eye and note a reddish-yellow elevation, the **caruncle,** and a thin, whitish, **semilunar fold** (plica semilunaris) just lateral to the caruncle. Both of these structures make up the **lacrimal lake** where small quantities of lacrimal fluid pool before they are drained away by other structures (Fig. III-21B and C). Evert the lower lid slightly by pulling it down with the forefinger. Note a little white papilla on its medial margin close to the lacrimal lake. This is the **lacrimal papilla** and the orifice is its **punctum.** There is a similar papilla on the medial margin of the upper lid. In Figure III-21B, the elevated papilla is best seen on the lower lid, and the punctum is a distinct pit on the upper lid. In Figure III-21C, the caruncle is prominent and the plica shows as a wide band. These papillae with their puncta drain tears from the lacrimal lake into **lacrimal canaliculi** (Figs. III-21B, 22C and D) which lead to the **nasolacrimal duct.** The blind upper end of the duct forms the **lacrimal sac** behind the medial palpebral ligament. Of course, these latter structures cannot be seen by surface study. When the lower lid is everted, notice also the series of orifices along it lateral to the papilla. As in the upper lid, these are the duct openings of the tarsal (Meibomian) glands.

Significance: In paralysis of the 7th (facial) nerve, the Orbicularis Oculi is functionless on the affected side. The lid cannot be closed. Protective blinking is lost, and the lacrimal fluid is not moved to the medial angle of the eye. Irritation of the unprotected eyeball leads to excessive lacrimation and "crocodile tears" roll down the patient's cheeks.

The Retinal Fundus: Have an instructor help you examine the eye of a colleague with an **ophthalmoscope.** You will be seeing the **fundus oculi,** the back part of the inner aspect of the globe, the photosensitive inner **retinal coat** of the eyeball. In the handle of the examining instrument, a prism guides light through the lower part of the subject's pupil. Returning light is *via* the upper part of his pupil, through the top of the prism to your own retina.

The fundus will appear red even though our retinal photographs in Figure III-24A and B are black, white, and shades of gray. The color is due chiefly to the great many vessels which run through the middle or **choroid coat** of the eyeball and, in part, to the pigmented cells of the choroid. The **sclera** is the outer coat, the retina is the inner coat, and the choroid is the middle coat of the eyeball. Most blood vessels reach the eyeball through this middle coat. But the retina has special vessels of its own, and it will be these you will see.

If you look toward the nasal side of center of the eye which you are examining, retinal vessels seem to radiate out of a white spot, and, chiefly, from the nasal side of this spot. This is the **optic disk.** All **nerve fibers** of the retina run toward this disk to leave the eyeball in forming the **optic nerve** behind the globe. Retinal veins leave and retinal arteries enter to spread out over the retina at this point so that all of these vessels travel for a time within the optic nerve.

Look carefully at Figure III-24A at about "11 o'clock" and at "5 o'clock" where you can see light striae, the nerve fibers, moving over the retina toward the disk. The veins are thick and dark; the arteries are light and narrow, and, in the figure, the arteries seem to have a wider white streak along them than the veins.

Since the optic disk is filled with nerve fibers and vessels and devoid of photosensitive rod and cone cells of the retina, an object whose image falls upon the disk area is invisible. Such a **blind spot** in the retina and, hence, in your **visual field,** should not worry you, for if the image falls here in one eye, it falls upon a photo-

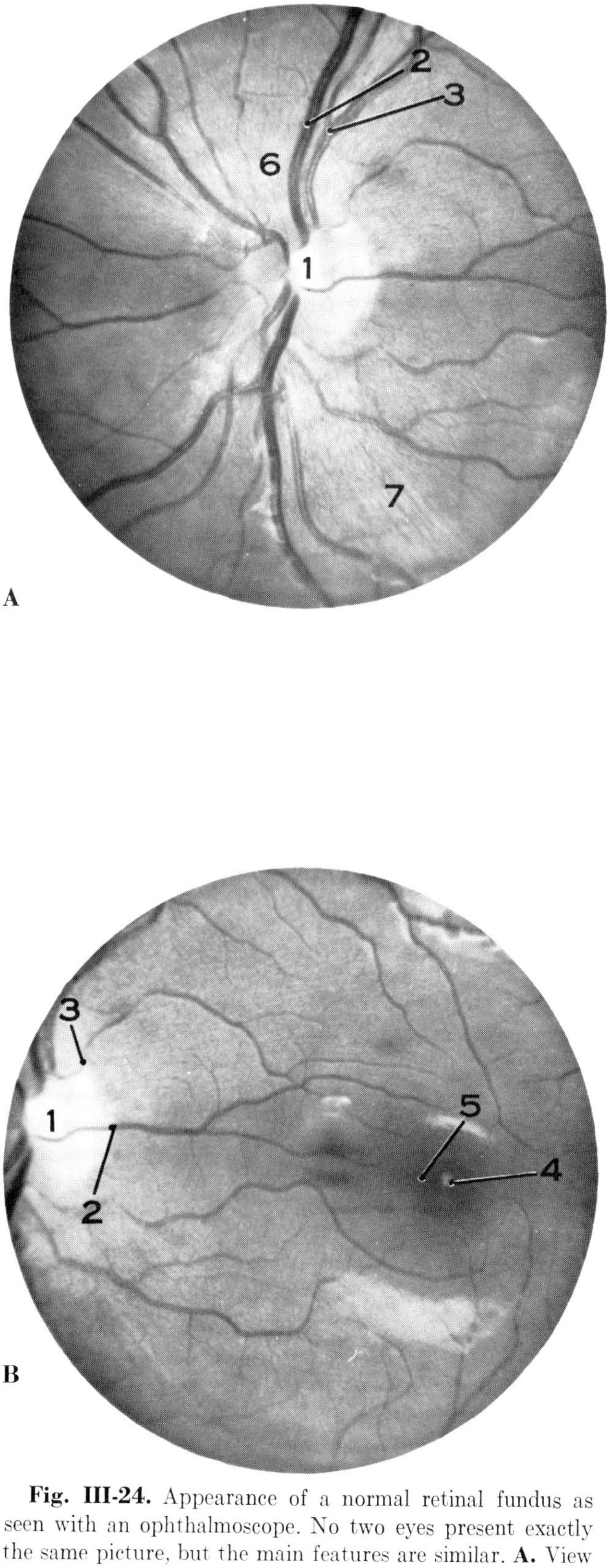

Fig. III-24. Appearance of a normal retinal fundus as seen with an ophthalmoscope. No two eyes present exactly the same picture, but the main features are similar. **A.** View of optic disk, nasal side of left eye. 1, Optic disk; 2, retinal vein; 3, retinal artery. Nerve fibers coursing toward disk 6 and 7. **B.** Macula, 5; and fovea centralis, 4; seen to the temporal side of the disk in the same eye.

sensitive area in the other. Even if you have only one eye functional or if you close one, very fine, involuntary ocular movements keep shifting the fixation point of the open eye, just enough, though slight in the extreme, so that you are completely unaware that you have a photo-insensitive region.

Now look toward the temporal side of the eye you are examining. Can you see a dark, red, oval spot, the **macula**? In Figure III-24B, there is a tiny white spot in the macula. This is the **fovea centralis,** an area occupied exclusively by cone cells of the retina.

Significance: Cone cells are responsive to wave lengths of colored light in the visible spectrum. Elsewhere in the retina rod cells and cone cells are arranged in a mixture. Rod cells are infinitely more sensitive to minimal intensities of white light than are cone cells, but they do not respond to colored light. As one moves from macula toward the peripheral boundaries of the retina, cone cells decrease in number and rod cells increase until they predominate.

At night, therefore, the eyeball shifts to let light fall nearer the periphery to achieve clearer vision. Images are in shades of gray and lack color. There is an old German expression which, when translated, illustrates this point nicely: "All cows look black at night." You can test rod and cone vision on yourself in the daylight, if you don't cheat. Get a number of pencils of different colors, say, blue, red, and yellow. Look straight ahead with both eyes. Hold the blue pencil in your right hand about a foot opposite your ear. If you are looking straight ahead you won't see it. Slowly move it in an arc until it is straight in front of your eye. Until it is almost dead center, it will only appear as a black or gray stick. When the blue light reflected from it falls on the macula you will see it as blue. As you move the red pencil through the same arc you will see the change from black to red occur sooner, and even much sooner still as you bring a yellow pencil through this arc. But all three as first perceived by the peripheral rods of the retina are black-to-gray shapes. Of course, if you cheat and shift both eyes to the right as you bring the pencils forward, you will see their specific hues much earlier, and the yellow pencil may never appear gray at all.

In conditions where intracranial pressure (cerebrospinal fluid pressure) is increased, *e.g.*, as in growth of brain tumors, hydrocephalus, and uremia, the optic disk becomes red and swollen. The central vein of the retina, venules, and capillaries become engorged. There may be small hemorrhages. Fluid collects between nerve fibers where these make their way to pierce the sclera and to form the optic nerve. The disk elevates. The whole entity is known as "choked disk" or **papilledema.** It is easily recognized by ophthalmoscopy, and it is frequently the *first indication* that the patient may have a brain tumor. In this sense, as in others, we can regard the eyes as the "windows of the brain."

Why are the retinal veins dilated in this manner? Recall two points: (1) the nature of meningeal investments; and (2) that retinal veins and arteries are special vessels which leave and enter the retina *via* a short course within the optic nerve. The optic nerve is a continuation of the brain itself, and so carries a sleeve of dura, arachnoid, and pia out upon itself. Between pia and dura is the subarachnoid space filled with cerebrospinal fluid. At the back of the globe all of these investments except dura terminate. Dura modifies as sclera to form the tough outer coat of the eyeball. Increases in cerebrospinal fluid pressure arising elsewhere in the brain are transmitted through the watery system as far as the back of the eyeball.

The walls of the retinal veins are thinner than arterial walls and the pressure within them is less than arterial pressures. An increase in cerebrospinal fluid pressure in the meningeal cuff about the optic nerve can either compress the vein within the nerve or bend it as it leaves the nerve to pierce the dural sleeve. Thus, venous drainage from the retina is obstructed, but arterial input under higher pressure is not. In the retina, capillaries, venules, and the central vein fill to overcapacity up to the point of venous obstruction in the nerve behind the eyeball. Capillaries leak plasma and continued retinal edema ensues as the connective tissue spaces fill with fluid about the nerve fibers at the disk. Nerve fibers are compressed as they try to squeeze through the sievelike holes in the scleral lamina cribosa. Pressure on them leads to retrograde degeneration in the ganglion cells of the retina, neurons whose axonal processes are these very fibers that are gathering to form the optic nerve. If there is no relief of the cause, eventual visual dysfunction follows, for the ganglion cells are the final common pathway from the retina by means of which visual impulses are transmitted to the brain.

The External Ear

We have used the external auditory meatus several times as a reference point, but we have never really examined the ear. Actually, there isn't much point in doing so, for the human auricle or pinna is pretty much of a useless embellishment. In dogs, which can move these structures quite provocatively, the external ear serves as a mobile funnel to trap and focus sound waves. In man, this auricular appendage is more of a useful adjunct to vision than it is to hearing, since it keeps our hats from falling down over our eyes. Still, it is well to see the anatomical names applied to it in print so that if you ever hear anyone using the terms later on you will know they are not talking about the elbow.

The **auricle** is an odd-shaped piece of cartilage covered with skin and it contains a number of named depressions and elevations (Fig. III-25A). The **helix** is its outer rim and this ends inferiorly in the fleshy **lobe** which females may or may not pierce for earrings. Similar ornamentation also can be dangled from the nose. The shell-like structure centrally above the meatus is aptly named the **concha.** Anterior to the meatus is an elevation called the **tragus,** which means "goat,"

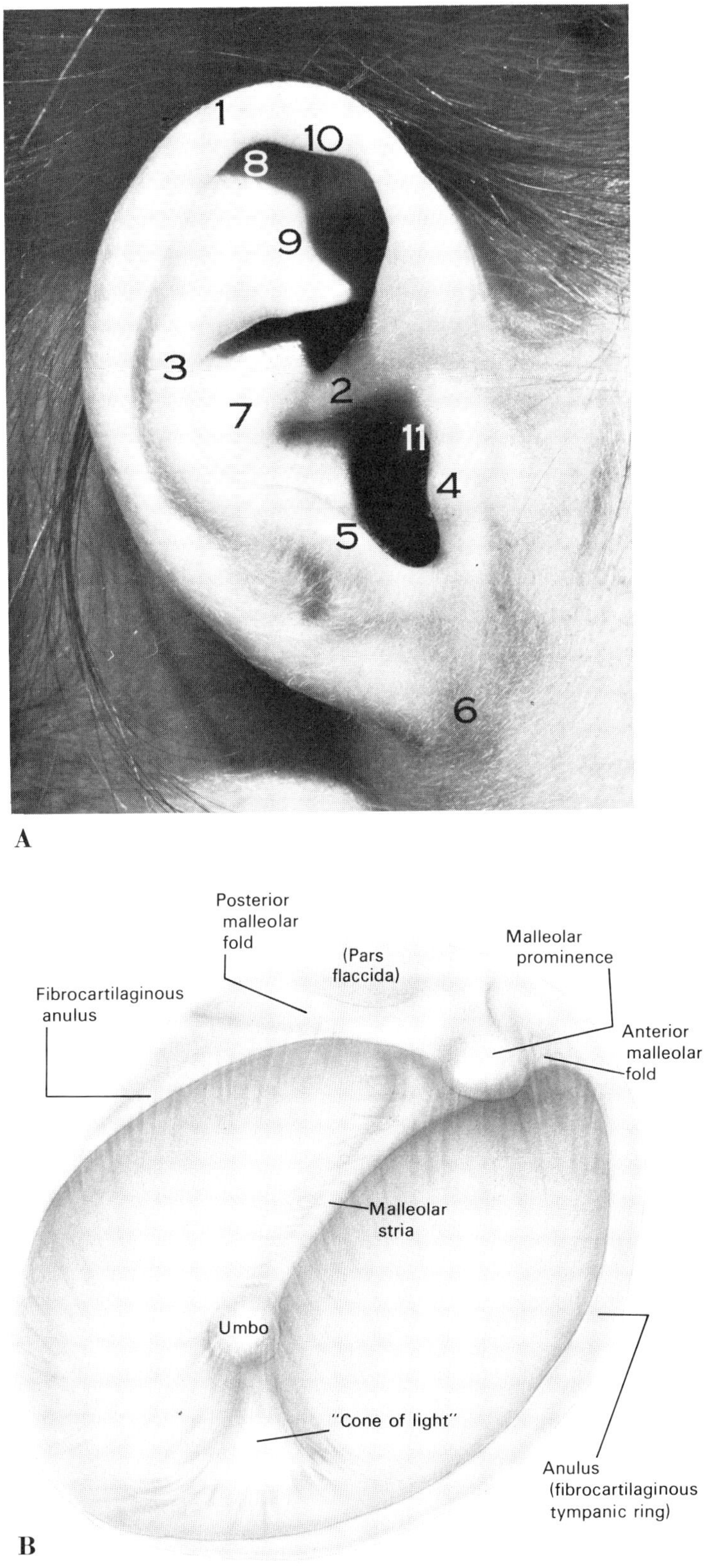

Fig. III-25. External ear and tympanic membrane. **A.** External ear (auricle; pinna). 1, helix; 2, crus of helix; 3, antihelix; 4, tragus; 5, antitragus; 6, lobule; 7, concha; 8, scaphoid fossa; 9, triangular fossa; 10, Darwin's tubercle; 11, external auditory meatus. **B.** Tympanic membrane (eardrum) as viewed through an otoscope.

probably because it tends to grow hair in the aged, like a billy-goat's whiskers. Posterior to the meatus is the **antitragus,** a structure whose name might lead one to think that it is opposed to hair growth; yet, it does not always remain glabrous. If you run your finger upward from the antitragus and inside the helix, you are tracing the **antihelix** and you may end up in a tight cul-de-sac from which, if you are lucky, you can extract some accumulated dirt. What the antihelix is protesting against is anybody's guess.

There are also three extrinsic muscles which are supposed to move this ear. They are named the **Posterior, Superior, and Anterior Auricular muscles.** In some of us in which they are not merely atavistic vestiges, they prove useful for amusing grandchildren on a rainy afternoon. Obviously, if they can be moved at all, they change facial expression, as witnessed by the gale of children's laughter. Thus, you can expect them to be innervated by the 7th cranial nerve.

Sensory innervation, anteriorly, is *via* the auriculotemporal branch of the trigeminal. Posteriorly, it is by the 2nd and 3rd cervical nerves (greater and lesser auriculars), save for a small patch of skin on the back of the ear supplied by the vagus or 10th cranial nerve.

Aside from these facts, not much further can be said for the external ear except to note that, like noses, ear shapes tend somewhat to highlight family resemblances. But like so many other anatomical and psychological traits, they also earmark people as being extraordinarily individualistic. None of us is an average statistic, anatomically, or otherwise, in this wide, wide world. We are not struck from a common mold, and this should give us courage to keep from conforming to one. If anything, an awareness of this fact ought to keep us from joining and becoming cultists of any kind. We do not have to be led by the nose or the ear by any group in order to find our identity.

The External Auditory Meatus

This structure is a somewhat tortuous, skin-lined canal which leads to the **tympanic membrane** or **eardrum,** an important structure which separates the meatus from the **middle ear cavity** of the temporal bone. A prominent textbook of physiology[11] gives the following directional description of the meatus: "It is directed first medially upwards and forwards, then medially and backwards, and finally, medially forwards and downwards." At this point, one visualizes the author as stopping and asking: "Are you following us?" We can assume, then, that its meandering is tricky, but it can be straightened, as it must be if one is to insert an otoscope. To accomplish this, the ear is pulled up and back. Have an instructor assist you in examining a **tympanic membrane.**

Through the otoscope the membrane will be seen as a smooth, glistening, gray-pink curtain placed at a slant so that its inferior border lies farther medialward than its superior border. Notice a faint white streak running down from above to almost the center. This is the **malleolar stria,** the handle of the malleolus (hammer), the outermost of the three, tiny **ossicles** (ear bones) of the middle ear cavity. The handle is attached to the inner surface of the eardrum (tympanic membrane), and we observe it shining through (Fig. III-25B). Where the stria terminates centrally, the membrane is drawn inward, this point being called the **umbo.** Thus, the eardrum not only sits at an oblique slant, it is also drawn inward as a shallow funnel, the deepest point of which is the umbo. If you observe the lower anterior segment of the membrane, just below the stria, you will see a triangular "cone of light." This also indicates that you are looking into the floor of a shallow funnel. At the upper end of the stria, one sees another point of light reflecting from the head of the malleolus, the **malleolar prominence.** Lateral lines diverging from this point are the **malleolar folds,** and the stretch of membrane between them and above the malleolar prominence is the flaccid portion. The rest of the membrane is quite taut.

The membrane is held in place by a fibrocartilaginous ring. The membrane is a fibrous connective tissue sheet sandwiched between two epithelia. The surface observed is covered by stratified squamous epithelium, while that facing the middle ear cavity is covered with pseudostratified columnar epithelium. Its chief arterial supply descends from above, crossing the stria. Smaller vascular twigs radiate inward from the sides. The chief nerve supply follows the main vessel. Hence, the favorite site for piercing the drum in middle ear infections is in its posteroinferior quadrant. Certain prominent middle ear structures lie behind each quadrant. The jugular bulb of the jugular vein lies behind the posteroinferior quadrant, but the danger of wounding it is slight for the most part. The carotid canal is more in danger by piercing the anteroinferior compartment in the area of the "cone of light." In the upper quadrants, the opening of the Eustachian tube, which communicates with the nasal pharynx, lies anteriorly; posteriorly lie the long process of the incus (2nd ear bone), the stapes (3rd ear bone), and the oval window of the inner ear vestibule.

The Mouth and Teeth

The Lips: Examine the lower lip from the outside inward as you turn it down while looking in the mirror. Each lip at its midline is joined to the corresponding **gum** by a mucosal fold, the **frenulum.** Note how the gums rise higher *between* the teeth than they do on the oral and labial aspects of the teeth (Fig. III-26). In the mirror, identify the three zones of the lip:

1. The outer **cutaneous zone** covered by typical dry skin which contains hair follicles, sebaceous glands, and sweat glands.
2. The **red zone,** an area of translucent skin where blood capillaries come close to the surface. This is the zone which females artificially color to suit their moods or their decor. It lacks glands of any kind, as a rule.
3. The inner, moist, **mucosal zone** is continuous with gums and inner cheeks. Its glands are of the mixed mucoserous secreting variety.

A

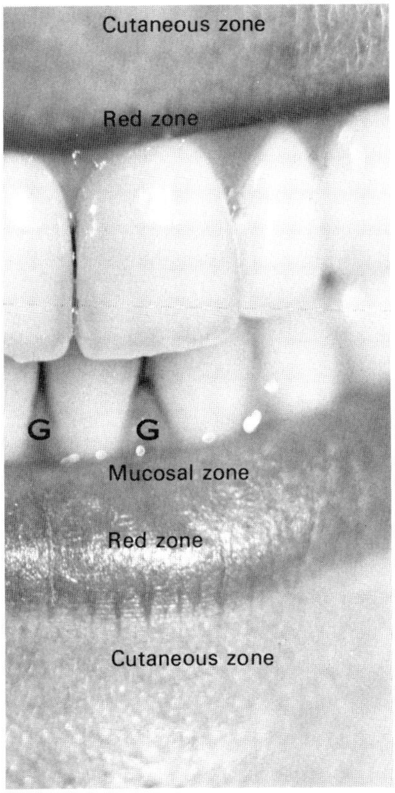

B

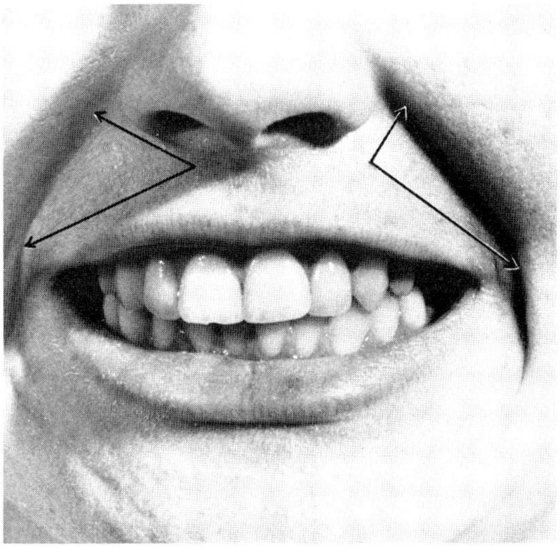

Fig. III-26. Lips and teeth. **A.** Zones of the lips. G, gingiva of gums rising between teeth. **B.** Nasolabial fold in smiling (at arrows).

When the subject in Figure III-26B was asked to show her teeth, both **nasolabial folds** were rendered prominent, much as they would be in smiling. Obviously, muscles of facial expression under these folds were acting. When the 7th nerve suffers unilateral injury, the nasolabial fold is flattened on the injured side, and the mouth angle on that side does not rise when the patient smiles or attempts to show the teeth. The teeth, then, will be only partly uncovered on the intact, nonparalyzed side.

The Oral Cavity: When the mouth is opened (Fig. III-27A), it is obvious that this cavity is limited in the front and on the sides by the lips and cheeks. Behind, the **palatine arches** on each side and the midline **uvula** mark the boundary of the **oropharyngeal isthmus** where the **oral cavity** and **oropharynx** *communicate*. Inspect the cavity with a flashlight. The side wall pillars or arches each consist of two muscular mucosal-covered folds, one behind the other, the **palatoglossal folds** anteriorly, and the **palatopharyngeal folds** posteriorly. In the fossa between them nestles the **palatine tonsil,** one on each side. Its medial surface may be in view if this lymphoid organ is somewhat enlarged.

The roof of the oral cavity is the **hard** and **soft palate,** the partition between the nose. The posteriorly-placed soft palate terminates in an engaging midline structure, the uvula, which dangles down between the palatine arches and which bobbles about when you open your mouth to say, "Ah!" (Fig. III-27A).

Significance: When the mouth is closed the uvula rests upon the tongue so that when you breathe through the nose the whole soft palate is tipped obliquely downward and backward. Air can pass from the nasopharynx, above the hard palate, into the oropharynx, behind the soft palate, and, thence, into the larynx, trachea, bronchi, and lungs. In swallowing, the soft palate is brought up against the back wall of the oropharynx, so that the soup you sip will not regurgitate through the nose but will enter the pharynx and esophagus where it will do you the most good.

There is nothing really outstanding to observe about the hard palate (Fig. III-27A). It has a faint median ridge which terminates behind the upper incisor teeth as the **incisive papilla.** Radiating transversely from the papilla on each side are about six laterally-directed ridges. These tend to smooth out and disappear in the aged.

The space between teeth and cheeks is known as the **vestibule** of the mouth which ends behind the last erupted tooth. Place a finger in the vestibule behind the lower teeth and press backwards and laterally to feel the **ramus** of the mandible. If you press a little medial and behind the 3rd upper molar, you will feel the **hamular** (hooked) **process** of the **sphenoid bone.** The tendon of the **Tensor Palatini** muscle swings around this hook as the muscle continues to its insertion into the palatine aponeurosis.

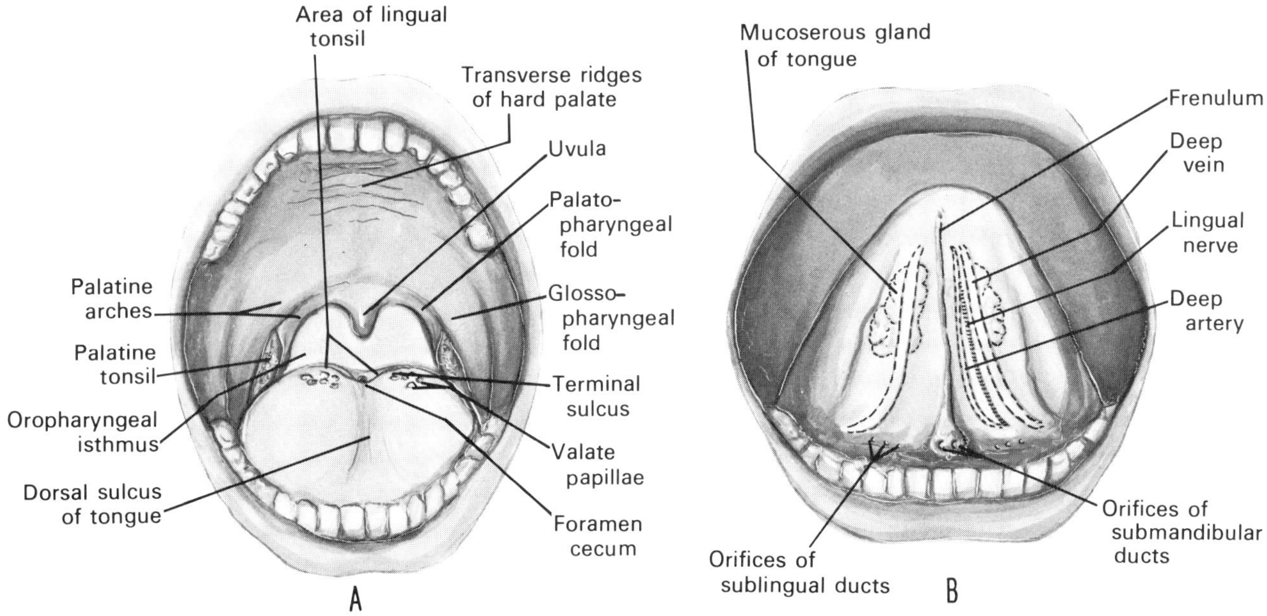

Fig. III-27. The oral cavity. **A.** Mouth open, presenting dorsum of tongue. **B.** Inferior surface of tongue with mucous membrane partially removed. (After Appleton, Hamilton, and Simon, Surface and Radiological Anatomy, Williams & Wilkins, 1958).

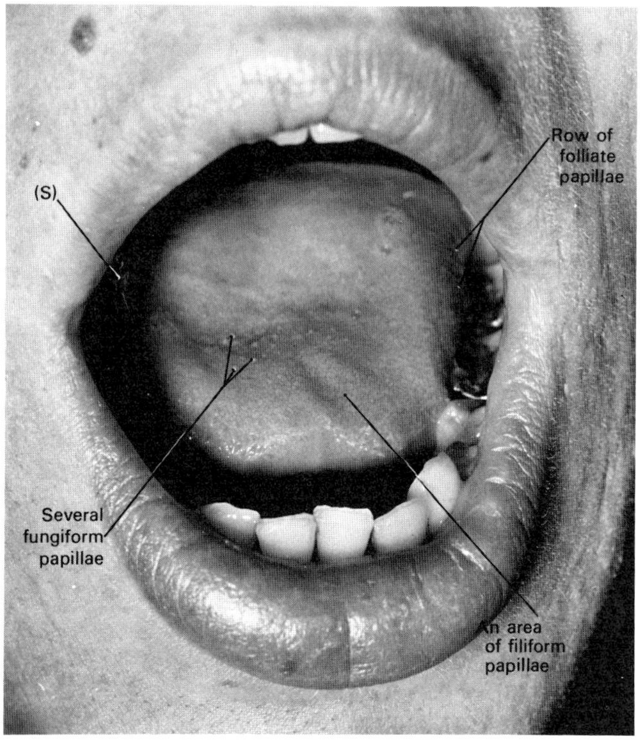

Fig. III-27C. Papillae of tongue. (S), drop of saliva ejected from parotid duct.

The Tongue: In the floor of the mouth, the tongue lies partly in the oral cavity and partly in the oral pharynx. Some morning when you are feeling particularly healthy, stick it out and examine it in the mirror. Discovery of its topography can enliven an otherwise dull session of shaving. It is separated into two halves by a faint longitudinal furrow, the **dorsal sulcus.** You may want to use a magnifying lens to examine its shaggy dorsal surface. This surface is mostly beset with many fine, pointed, **filiform papillae** (Fig. III-27C) which generally do not bear taste buds. Mixed in among them are papillae that are larger, redder, more rounded elevations, the **fungiform papillae** (Fig. III-27C), many of which do bear taste buds. Taste buds, of course, are visible only under a microscope. Do not expect wonders from your mirror, hand magnifier, or naked eye. Along the posterolateral margins of the tongue look for some vertical ridges, the **folliate papillae** (Fig. III-27C), which bear many taste buds in the walls of their lateral troughs. These particular papillae are more numerous in the infant tongue and tend to regress in aging. Pull the tongue as far forward as you can (Fig. III-27A). Perhaps you can just begin to see a V-shaped row of large, rounded, **vallate** papillae. These have circumvallate trenches containing many taste buds.

Significance: There seem to be four primary modalities of taste sensation: sweet, salt, bitter, and sour, elicited by sugars, salts, bases, and acids in solution. A mixture of these chemostimuli plus olfactory and general somatosensory impressions account for our variety of taste experiences. Taste buds at the tip of the tongue appear to be more sensitive to sweet, though they do respond to the other modalities, as well. The infant tongue is particularly sensitive to sweet tastes. At the margins the taste buds respond well to sour and also to salt tastes. Those in the posterior region give the best response to bitter taste substances. Buds in some papillae seem to respond only to one modality; other taste buds respond to several of them.[12]

Foramen Cecum and Its Significance: Behind the row of vallate papillae there are no taste buds. There is a V-shaped sulcus, the **sulcus terminalis,** immediately behind them, and it terminates in a pit at the point of the "V," the **foramen cecum** (Fig. III-27A). Here in the embryo, the **thyroid gland** and its **thyroglossal duct** began their downward migration into the neck region. The duct once opened into the foramen, but the duct disappears in development, cutting the gland off from its former connection with the digestive tube. Occasionally, remnants exist as cysts along the route of its old course; only very rarely does the duct persist and maintain communication with the pit of the foramen cecum.

Lingual Tonsils: Behind the sulcus terminalis the tongue has a nodular appearance because of the 100 or more small lymphoid follicles with their single crypts that make up the **lingual tonsils.** These, along with the palatine tonsils and the pharyngeal tonsils (adenoids), have been said to constitute a guardian lymphoid ring (of Waldeyer) around the oral end of the digestive tract, protecting it against bacterial invasion. Now that we know more about the role of lymphocytes than was known in 1884 when Waldeyer, a German anatomist, proposed this theory, the "lymphoid ring" hypothesis seems more fanciful than factual. Lymphocytes, the products of these tonsilar tissues, rarely behave as phagocytic cells capable of engulfing bacteria. Under certain circumstances, they can transform into phagocytes, but this process is altogether too slow for them to be regarded as a ready, first line of defense. Moreover, palatine tonsils and adenoids are often removed in a diseased state even early in childhood, having succumbed rather easily to bacterial infection.

Touch the tip of the tongue to the hard palate so that you can examine its ventral surface in the mirror. A prominent median anchoring fold, the **frenulum,** is seen. The Latin meaning for this structure is "bridle," so there appears to be some anatomical basis for the admonishment to "Hold your tongue!". In the midline posteriorly at the root of the frenulum are two orifices for the right and left **ducts** of the **submandibular** (submaxillary) **glands.** The **sublingual glands** lie under the base of the tongue below the **sublingual folds** which run transversely on either side of the frenulum. The folds are beset with numerous small papillae at whose summits are the orifices of the many **sublingual ducts** (Fig. III-27B). Do not mistake the gland seen in this figure for the sublingual gland. It is one of the many **mucous** or **serous glands** under the mucosa of the tongue whose ducts open directly upon the mucosal surface or indirectly into the troughs of folliate or vallate papillae.

Observe the **deep vein** (profunda linguae) shining through the mucosa a short distance on each side of the frenulum and parallel with it (Fig. III-27B). The **lingual nerve** lies medial to the vein, and the **deep artery,** medial to the nerve, closest to the frenulum. The nerve is a branch of the mandibular division (V3) of the trigeminal and is sensory for pain, temperature, and touch to the mucosal sac of the tongue.

The Teeth: Each tooth consists of a **root, neck,** and **crown,** the latter being the only visible part. The roots are embedded in separate alveolar sockets in the mandible or maxilla, while the necks are covered by the gums. Necks and roots also are covered by a **cement;** crowns are covered by a hard, epithelially-derived **enamel.** Figure III-28 illustrates tooth anatomy as seen in X-ray.

Adults with a full set have 32 teeth in all; 2 **incisors,** 1 **canine,** 2 **premolars** (bicuspids), and 3 **molars,** counting from the midline laterally on each side of each jaw (Fig. III-29). The 3rd molar may be late in erupting, fail to erupt at all, or fail to erupt to the same level as the others. Sometimes there may even be a 4th molar. Incisors and canines have only one root; premolars generally have one but there can be two; molars of the lower jaw have two, and those of the upper jaw have three.

Generally, teeth are enumerated as shown in Figure III-29, *letters* used for the nonpermanent, **deciduous** or **milk** teeth and numbers for the **permanent** teeth. Only incisors, canines, and molars make up the milk teeth, young molars tending to lie above the crowns of the future premolars forming in the jaw below them. As the permanent molars develop, there must, of necessity, be considerable reconstruction of the jaw. There is displacement of existing teeth, build-up and lengthening of the bone to accommodate new and larger teeth, and, of course, much bony resorption in the region of the alveolar sockets. Because of these growth changes, the spaces in children's teeth, as they lose one after another of the milk teeth and get paid off by the "Good Tooth Fairy," tend to close up. In the adult, where there is no pay-off but only a shelling out of hard cold cash, there is only resorption around alveolar margins of lost teeth and bridgework to anchor in artificial ones for those who can afford this luxury in the name of esthetics and a proper bite.

Figure III-30 gives some information about expected ages of eruption of teeth, but for the benefit of curious fathers in each class who want to check on their offspring by occasionally "looking their gift horses in the mouth," we have prepared the following table after that of Appleton, Hamilton, and Simon[8]:

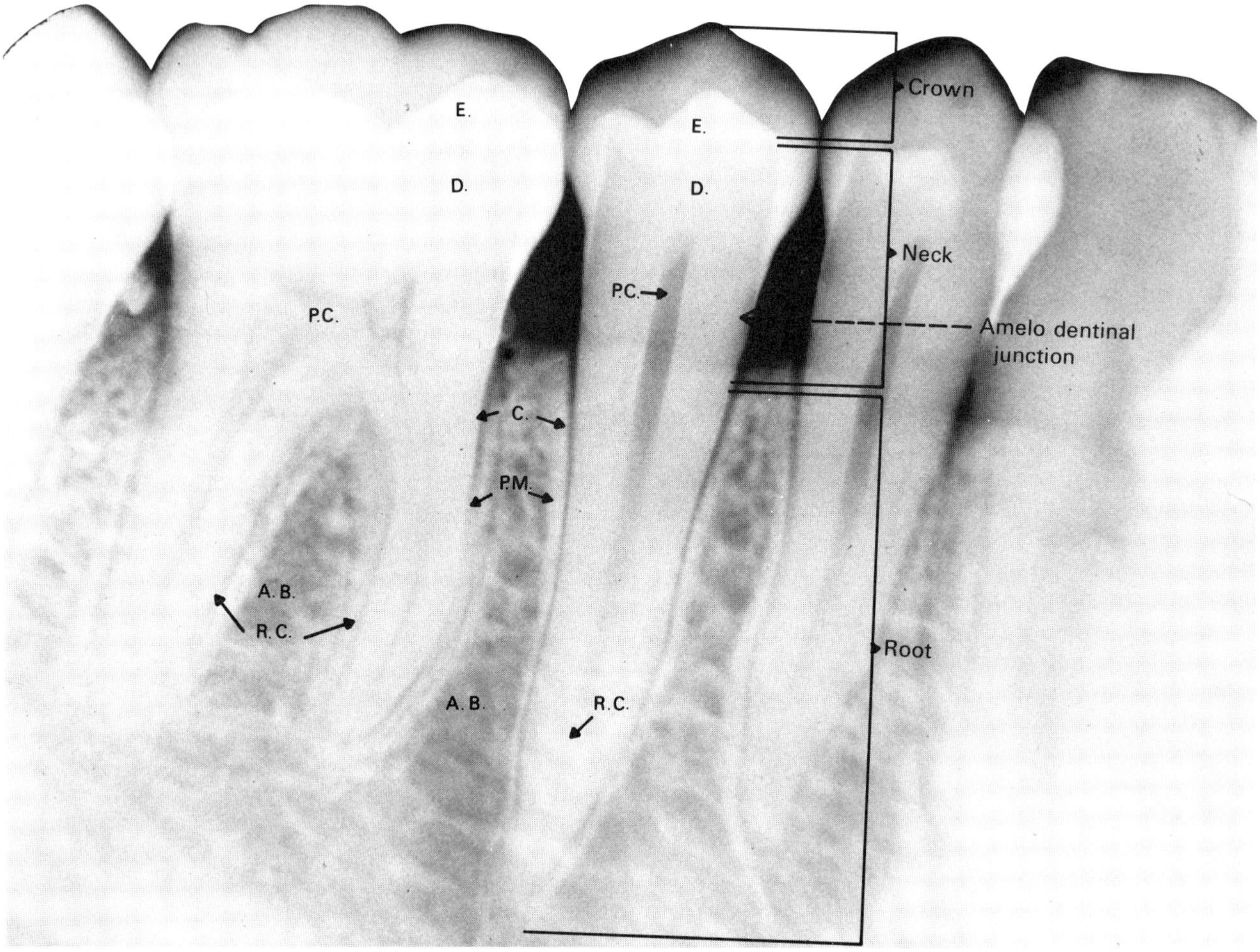

Fig. III-28. Anatomy of teeth as seen in X-ray. A.B., alveolar bone; C, cementum; D, dentin; E, enamel; G, gingiva (gum); P.C., pulp chamber; P.M., peridontal membrane; R.C., root canal.

A. Deciduous or milk teeth
- 1st yr.
  - a. Central lower incisors
  - b. All 4 upper incisors
- 2nd yr.
  - a. Lateral lower incisors
  - b. 1st milk molars
  - c. Canines
- 3rd yr.
  - 2nd milk molars

B. Permanent teeth
- **6th yr., 1st permanent molars**
- 7th yr., Central incisors
- 8th yr., Lateral incisors
- 9th yr., 1st Premolars
- 10th yr., 2nd Premolars
- 11th yr., Canines
- 12th yr., 2nd Molars
- Young adult, 3rd Molars.

It is not easy to find a young adult with a full set of 32 perfect teeth, despite our purported state of reasonably good nutrition in this country and despite widespread advertising and propaganda directed toward oral hygiene. We were fortunate to capture such a set on film in a living head and present it to you in Figure III-31. The details of their discovery are set down in another book of ours entitled: *Head-Hunting on the Eno River*.[13]

SURFACE ANATOMY OF THE NECK

The "Adam's Apple," or **laryngeal prominence** of the **thyroid cartilage,** is generally visible in the midline of the neck anteriorly. It is less visible in women and children whose "voice boxes" are smaller. But it can be palpated easily and felt to **recede** upon swallowing. It marks the level of the 4th cervical vertebra (VC_4).

Above the thyroid cartilage and about ½ inch below the lower border of the mandible the two greater horns of the **hyoid bone** can be grasped and the bone can be moved from side to side. Its horse-shoe shape can be appreciated. If you can't find it easily because of too much fat under the chin, swallow and it will move *forward* to meet the fingers. It lies at the level of VC_3.

You may have to open your collar to palpate the

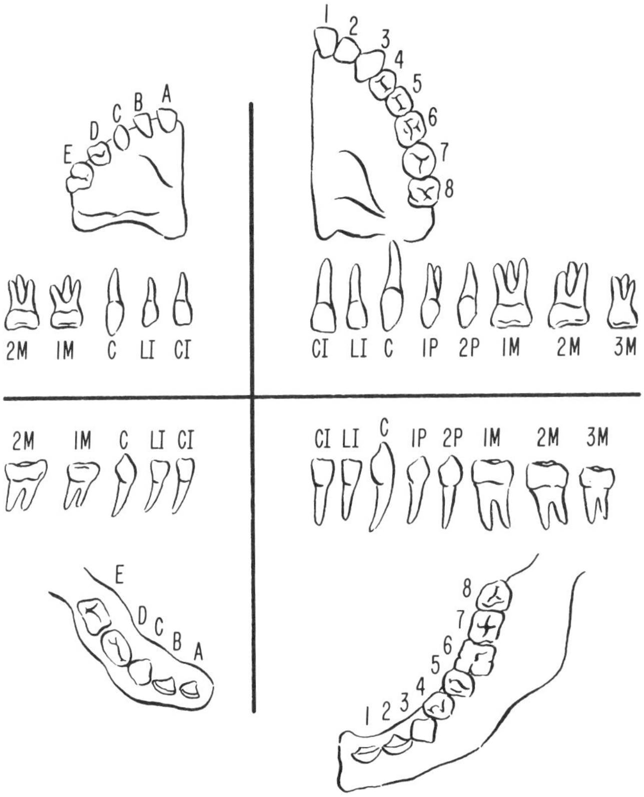

Fig. III-29. Chart of deciduous (left) and permanent (right) teeth; A–E, deciduous; 1–8, permanent. CI, central incisor; LI, lateral incisor; C, canine. 1 and 2 P, 1st and 2nd premolars. 1, 2, and 3 M, 1st, 2nd, and 3rd molars. (After Appleton, Hamilton, and Simon, *Surface and Radiological Anatomy*, Williams & Wilkins, 1958.)

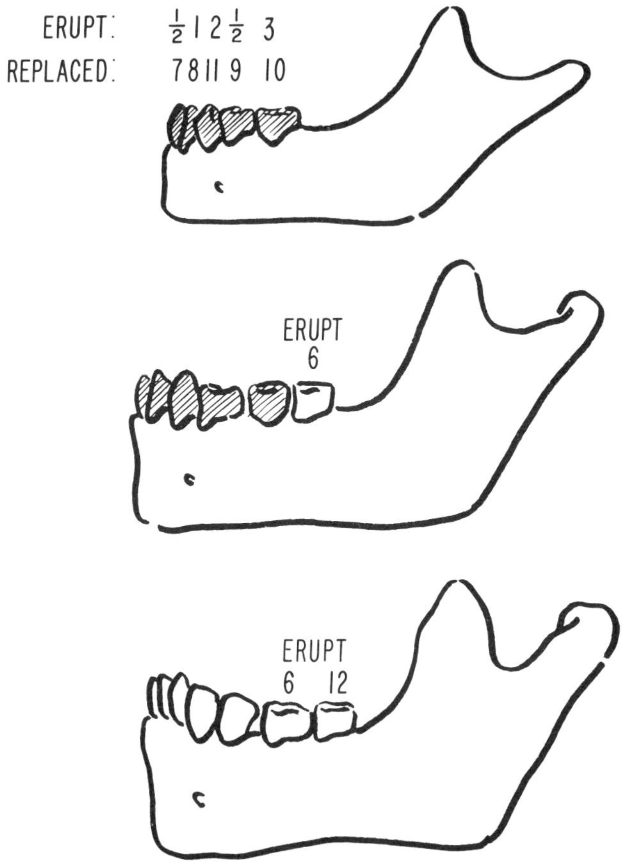

Fig. III-30. Dates of eruption of deciduous teeth (shaded) and permanent 1st and 2nd molars. Dates of replacement of deciduous (milk) teeth also given. (After Appleton, Hamilton, and Simon, *Surface and Radiological Anatomy*, Williams & Wilkins, 1958.)

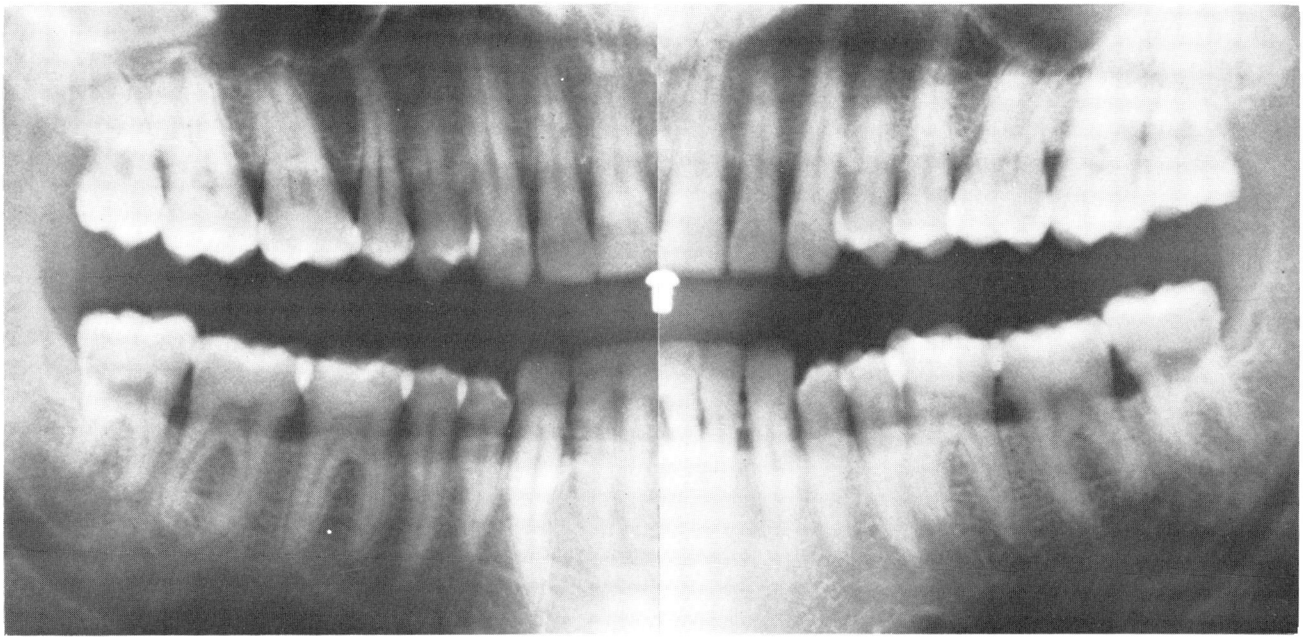

Fig. III-31. Complete and perfect dentition in a young adult seen in panoramic X-ray view (count them and name them). White marker shows midline.

cricoid cartilage of the larynx. It can best be palpated by an up-and-down movement of the finger in the midline and felt as a narrow horizontal bar just below the thyroid cartilage. Laterally, it is covered by the thyroid gland and some musculature. Make a real effort to palpate this rather small structure, for it lies at the *level of VC^6*, a most important landmark in the neck.

The cricoid cartilage at the level of VC_6 can serve as a guide to the following:

1. The junction point of **pharynx** with **esophagus.**
2. The junction point of **larynx** with **trachea.**
3. The point at which the **recurrent laryngeal nerve,** an all-important branch of the vagus, enters the larynx.
4. The **isthmus** of the **thyroid gland** lies ½ inch below the cricoid. The upper poles of the gland flank it on either side.
5. It is the level of the **inferior thyroid artery** and **middle thyroid vein.**
6. Here the tendon of the Omohyoid muscle crosses the **carotid sheath** which contains the **common carotid artery, internal jugular vein,** and the **vagus nerve.**
7. The carotid can be compressed laterally against the transverse process of VC_6, and its pulse can be felt at this level.
8. A pecularity of the cervical vertebrae is that their transverse processes contain foramina. At this level the **vertebral artery,** one of the two vessels that supply blood to the entire brain, enters the lower of these foramina through all of which it will course to enter the foramen magnum of the occipital bone eventually. (The other vessel to the brain is the **internal carotid.**)
9. A **middle cervical sympathetic ganglion** lies at this level, if there is one. Contrary to general textbook descriptions, this ganglion is most frequently absent. But if present, it lies at VC_6 along the sympathetic ganglionic chain which runs behind the carotid sheath.
10. In some individuals the **apices of the lungs** rise almost this high in the neck under cover of the Sternomastoid muscle. In most, they fall short of this point by a good fingerbreadth.

The Triangles of the Neck: It has been fashionable in anatomy to divide the neck arbitrarily into triangular regions for the sake of convenient localization of structures. The geographic peculiarities of certain muscles in the neck have helped to circumscribe these triangular boundaries. Unfortunately, students tend to memorize the muscular boundaries as if this were the important thing to do, when it is more important to know the general area of the triangle and most important of all to know its contents. Part of the misdirection of effort is the fault of the anatomist who has meticulously surveyed these areas, naming muscular landmarks to the north, south, east, and west, and, at times, even to the north-northeast. Moreover, he has tagged the triangles with virtually useless anatomical names when he should have emphasized their importance by naming them for their chief contents. We will rename them in what we feel to be a more useful way, but, unfortunately, we will have to give their old names also so that the student can find them on the ancient maps of the neck region.

They should be located in a general way in the study of surface anatomy, though they can best be seen in the course of dissection after the skin and platysma are removed. Unfortunately, too, only two large muscles can be localized well from the surface; the other smaller ones, which are so important in subdividing large triangles into smaller, three-cornered plots, remain almost undetectable from the surface. So we shall have to look at maps in which a bit of dissection has been done in order to better appreciate the lie of the land.

First of all, two large triangles, one in the front of the neck and one in the back, need to be mapped out. The **Sternomastoid** muscle is the key landmark, running obliquely from **sternum** and **clavicle** to the **mastoid process.** Its entire anterior border can be felt and the slight depression between its **sternal** and **clavicular heads of origin** can be seen. This is nicely demonstrated in Figure III-32 if you focus your attention upon the right place. The anterior border of the *right* Sternomastoid in this shapely young lady is sharply outlined. The lower border of the *left* muscle is bold, and its narrow sternal and broader clavicular heads of origin are separated by a small pronounced fossa, called by some the **lesser supraclavicular fossa,** beneath which lies the lowest part of the **internal jugular vein** (see also Fig. III-68A). Between the two sternal heads is the larger **jugular fossa** with the jugular notch of the **manubrium** of the sternum forming its lower boundary. In it, the **inferior thyroid veins** descend on their way to right and left **brachiocephalic veins.** Some transverse connections between the **anterior jugular veins** may also cross in this fossa, though major anastamoses between these vessels generally occur a bit higher up in the neck (see also Fig. III-68A).

The Sternomastoid at once divides the neck into large **anterior** and **posterior** triangles. The **anterior triangle** lies between the midline of the neck and the anterior border of this landmark muscle. It has its apex at the sternoclavicular joint and its base at the lower border of the mandible. Compare these relationships in Figure III-32 with the similar but far less glamorous representations in Figures III-33 and 34.

A second large muscle of importance in the neck is the capelike **Trapezius** in the posterior aspect, running from the occiput and attaching, in part, to the **scapula** and the lateral part of the *clavicle* (Fig. III-35). Between its anterior margin and the posterior margin of the Sternomastoid lies the **posterior triangle** with its apex at the occipital bone and its base at the clavicle (Figs. III-33 and 34).

Minor Vascular or Subclavian Triangle: Both of

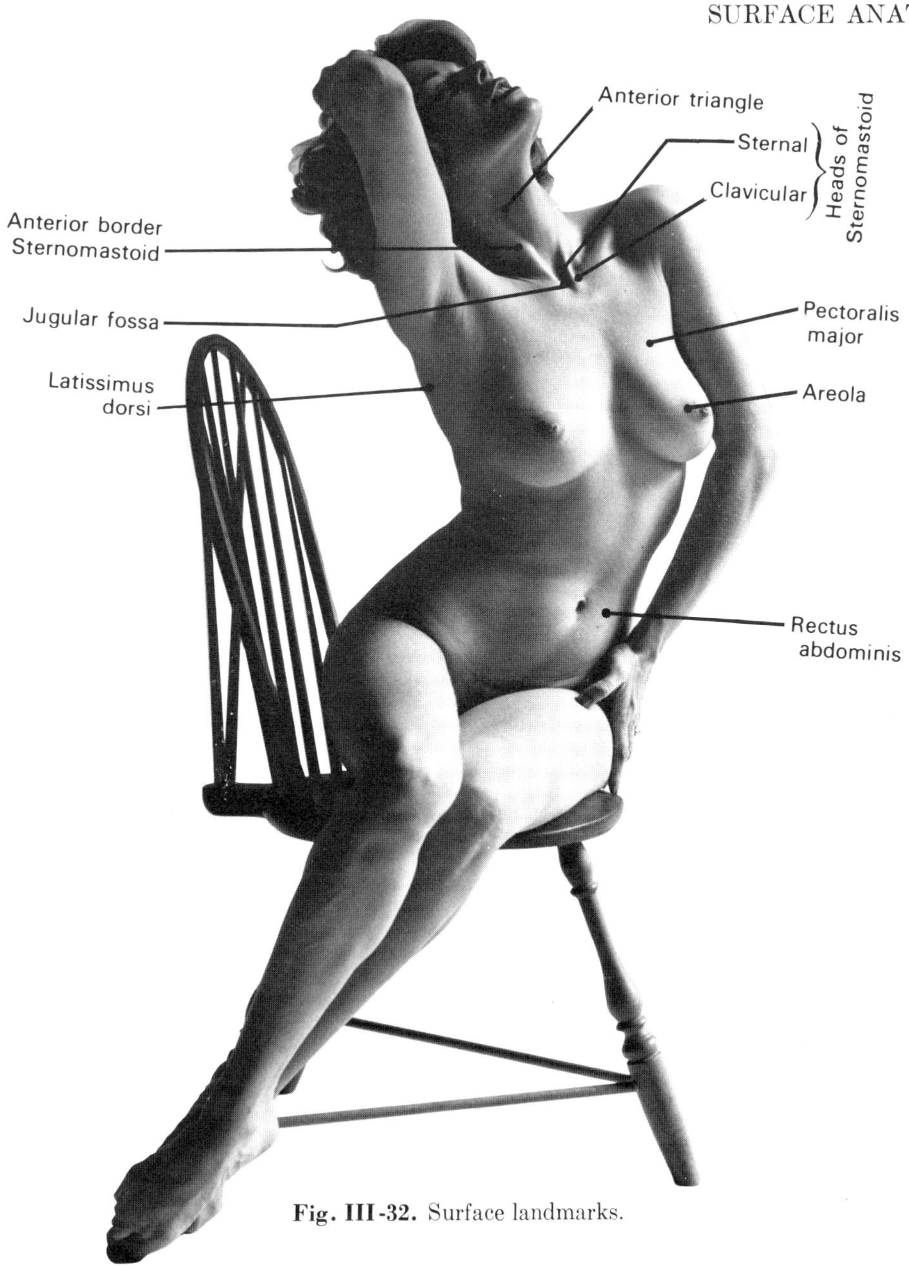

Fig. III-32. Surface landmarks.

the large triangles are cut into smaller ones by two small muscles of lesser importance, neither of which is easily seen from the surface as a rule. These are the two-bellied **Digastric** and **Omohyoid muscles** (Fig. III-34). Omohyoid derives from two Greek words, "omos," meaning "shoulder," and "hyoid," "like the Greek letter 'U.' " Thus, we know that it attaches to the shoulder bone, or scapula, and to the U-shaped hyoid bone in the neck. The inferior belly of this muscle cuts the posterior triangle into a large, superiorly-placed triangle and a small, inferiorly-placed one. This can be seen very nicely in Figure III-35 where the head is turned to one side. The **inferior belly** of the **Omohyoid** can just barely be seen under the skin, crossing the posterior triangle obliquely. In the fork between it and the Sternomastoid, just above the clavicle, can be seen a triangular depression, the **supraclavicular fossa.** The fossa corresponds to the **minor vascular triangle** of the neck or, anatomically stated, the **subclavian triangle.** This is once when the anatomical designation hits the nail on the head, for the chief content of it is the **subclavian artery.** If you place your thumb on this triangle as shown in Figure III-36 and press downward toward the 1st rib, you can shut off the arterial flow from this vessel to the entire upper extremity. Besides some important lymph nodes, this triangle also contains part of the **external jugular vein** which crosses it superficially as it descends from above where it is also superficial to the Sternomastoid muscle (see also Fig. III-58A).

The Neural or Occipital Triangle: Anatomically, the superior subdivision of the posterior triangle lying

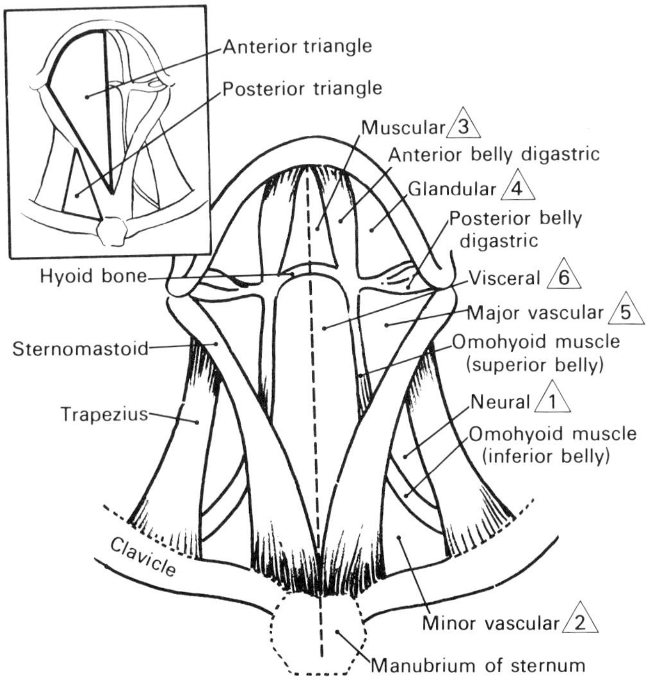

Fig. III-33. A mapping of the triangles of the neck as seen from in front with chin tilted upward. Inset locates the large anterior and posterior triangles which are subsequently subdivided into smaller triangles as shown in enlarged sketch. (Modified from Brantigan, *Clinical Anatomy,* McGraw-Hill, 1963.)

above the inferior belly of the Omohyoid has been called the occipital triangle because a minuscule portion of the occipital bone is included at its apex. We prefer to call it a nervous or **neural triangle** because its chief contents are peripheral nerves (Fig. III-37). This area has even been called a muscular triangle by some writers since its floor is carpeted by bands of muscle, some broad and some narrow. But its chief contents are the nerves which cross it. They emerge from a point midway along the posterior border of the Sternomastoid. The most important of these to cross the muscular floor is the **spinal accessory** or **11th cranial nerve.** It emerges at the *nerve point* from under the Sternomastoid and crosses the triangle to disappear under the Trapezius one thumb's breadth above the upper lateral border of the clavicle. It is the proper nerve supply for both Sternomastoid and Trapezius, since these muscles were derived from the mesenchyme surrounding the 6th cartilaginous branchial arch of the embryo. These muscles quickly migrated to their present positions in the neck and the back.

The other nerves seen in Figure III-37 are all *cutaneous* twigs of **cervical nerves C_2, C_3, and C_4** (C_1 has no sensory branches). These are the **spinal nerves** of the **cervical plexus** which supply overlying skin areas (dermatomes). Many curl around the Sternomastoid to cross it or to run along it in their course.

In the triangle, but not coming *out* of it, are also motor branches of the cervical plexus to the Levator of the scapula, the Rhomboid muscles and the Serratus Anterior—all muscles that assist in shoulder movement, as does the Trapezius, itself. An important nerve carrying both motor and sensory fibers from cord segments C_3–C_5 is the **phrenic nerve** which descends vertically upon the Anterior Scalene muscle in the lowest part of the neural triangle. It enters the chest behind the internal jugular vein as it makes its way to the **Diaphragm,** that most important skeletal muscle of respiration.

We can now turn our attention to the large anterior triangle of the neck which is subdivided by the bellies of the Digastric and Omohyoid muscles into four smaller triangles on each side (Figs. III-33 and 34). Two lie above the hyoid bone and two below it. While these have been variously named, we shall call them the **muscular, glandular, major vascular,** and **visceral triangles** after their chief contents.

The Muscular (Submental, Suprahyoid) Triangle: This is the least important of the anterior triangles. It lies under the mandible between the midline of the neck and the anterior belly of the Digastric (Fig. III-33). If any of these triangles deserves the name "muscular triangle," this is it, for, aside from a few lymph nodes, the **Mylohyoid muscle** fills it, serving as a diaphragm for the oral cavity above. The Mylohyoid muscles arise on each side from a line along the medial aspect of the mandible and insert onto the body of the hyoid bone and along the **median raphe** which is a connective tissue anchoring strand extending from the hyoid to the symphysis menti. As a consequence, these muscles also serve as the floor for the glandular triangles (Fig. III-34).

The Glandular (Digastric) Triangle: This triangle lies under the mandible between the two bellies (anterior and posterior) of the Digastric muscle (Figs. III-33 and 34). It houses the major portion of the **submandibular** (submaxillary) **gland.** Besides, it contains:

a. Hypoglossal nerve, in part
b. Mylohyoid nerve
c. Facial (external maxillary) artery, in part
d. Lingual artery, in part
e. Facial vein, in part
f. A few lymph nodes.

The Major Vascular (Carotid, Superior Carotid) Triangle: The superior belly of the Omohyoid muscle divides the subhyoid region of the neck into a lateral, large vascular triangle and a median visceral triangle. These are both important areas. The boundaries of the major vascular triangle are the posterior belly of the Digastric above, the Sternomastoid posteriorly, and the superior belly of the Omohyoid anteriorly (Figs. III-33

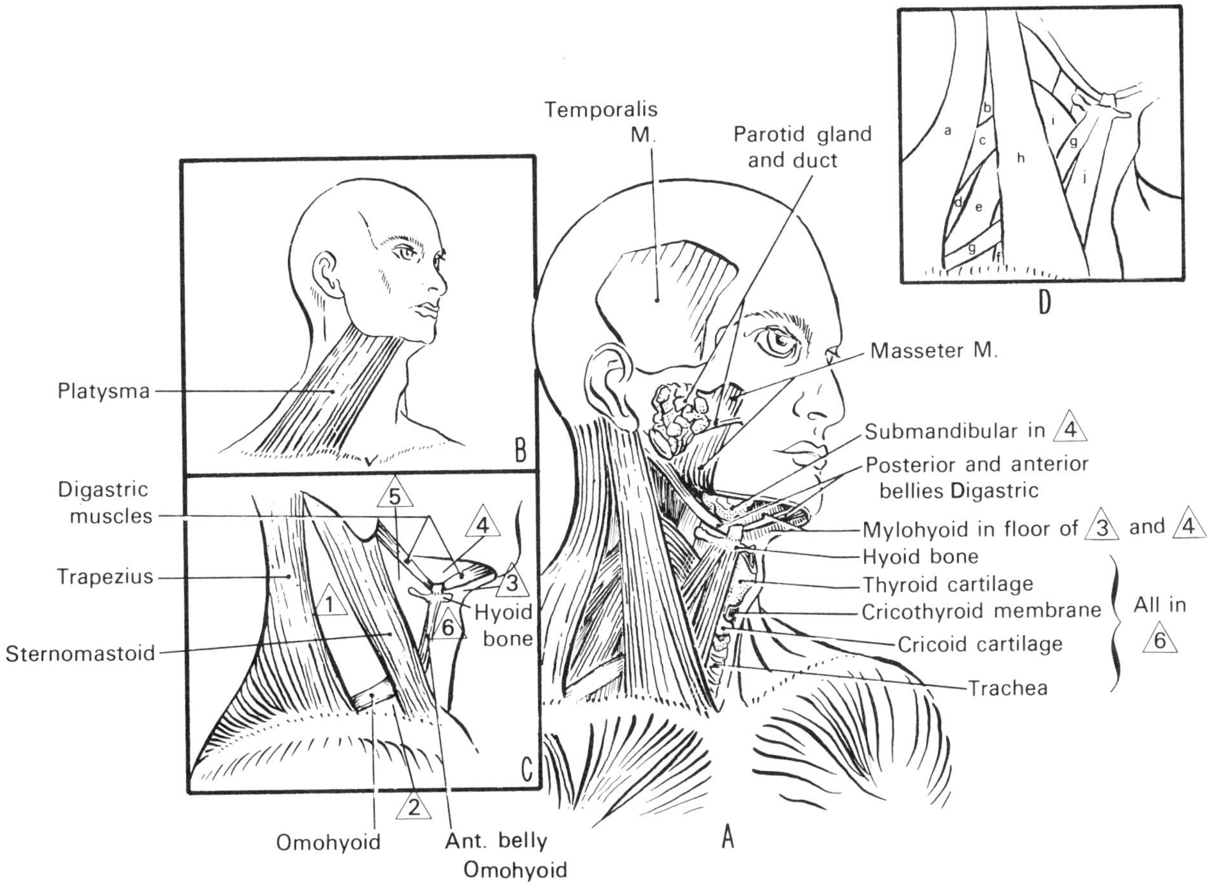

Fig. III-34A. Dissection to show muscular relations and some associated structures in triangles of neck. **B.** Platysma. **C.** Triangles viewed from the side. Numbers correspond to named and numbered triangles in Figure III-33. **D.** Key to muscles seen in **A:** a, Trapezius; b, Splenius capitis; c, Levator scapulae; d, Scalenus posterior; e, Scalenus medius; f, Scalenus anterior; g and g′, inferior and superior bellies of Omohyoid; h, Sternomastoid; i, Inferior constrictor of pharynx (with esophagus showing above); j, Sternohyoid (one of the "strap" muscles). Muscles "b–f" are in the posterior triangle, all except "g" forming its floor. Muscle "i" is in major vascular triangle, while "j" is in visceral triangle. (Modified from Appleton, Hamilton, and Simon, *Surface and Radiological Anatomy,* Williams & Wilkins, 1958.)

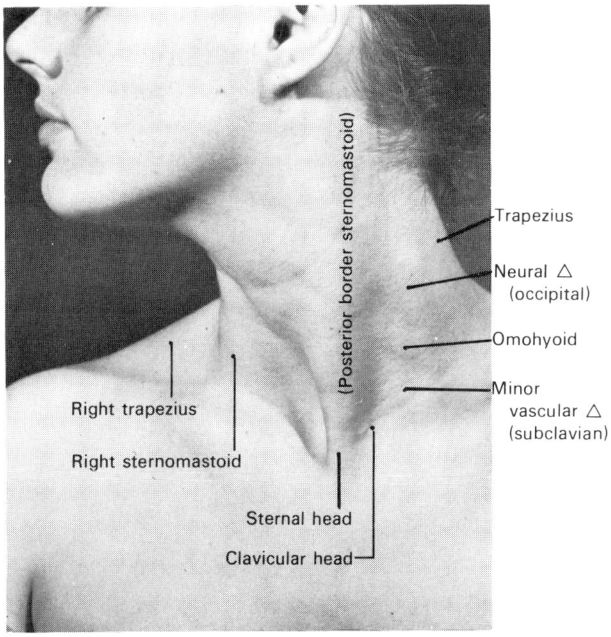

Fig. III-35. Surface landmarks related to posterior triangle of the neck.

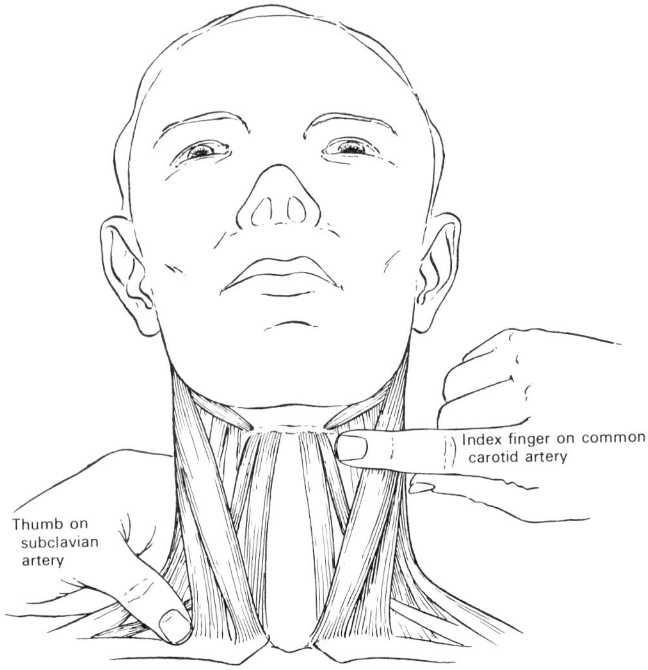

Fig. III-36. The major vascular (carotid) and minor vascular (subclavian) triangles. Pressure points for occluding these vessels are shown.

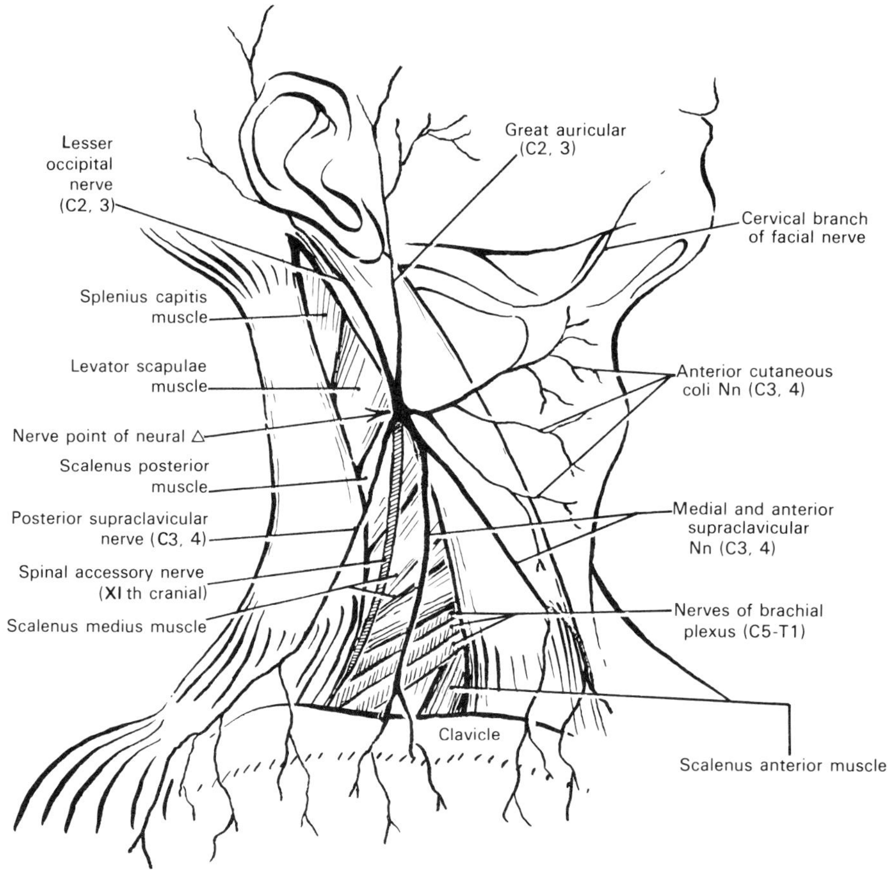

Fig. III-37. The major peripheral nerves of the neural (or occipital) triangle.

and 34). As you might suspect, it contains the **common carotid artery** whose pulse can be felt by placing a finger in the triangle, as shown in Figure III-36, compressing the vessel against the transverse processes of the cervical vertebrae. A common, stout, connective tissue sheath wraps the *artery*, the **internal jugular vein,** and the **vagus,** or **10th cranial nerve,** into a common tubular package, the vein lateral to the artery and the nerve posterior between the two vessels.

In the upper end of the triangle (Fig. III-38), the common carotid bifurcates into its internal and external branches. The **internal carotid** gives off no further branches, but the **external carotid** gives off its five major branches to the face and head. The **internal jugular vein** receives tributaries named after the corresponding branches of the artery.

Anterior to the sheath, the *nerves of the cervical plexus form a loop* to supply the Infrahyoid muscles as well as the Omohyoid. (The Mylohyoid muscles and the anterior belly of the Digastrics get their motor supply from the mandibular division of the trigeminal; the posterior belly of the Digastric, from the facial nerve.)

Posterior to the carotid sheath runs the **cervical part** of the **ganglionated sympathetic trunk.**

Certain nerves, some common in part to other triangles, also, cross the major vascular triangle; *e.g.,* **hypoglossal** in the upper part and the spinal accessory in the uppermost end. The **internal branch** of the **superior laryngeal nerve** traverses the triangle (Fig. III-38). It is an offshoot of the vagus and it supplies the laryngeal muscles and mucosa of the larynx above the level of the vocal cords.

The Visceral (Muscular) Triangle: This midline, infrahyoid triangle is the one which anatomists tend to call the muscular triangle because the "strap" muscles of the neck form its floor beneath its fascial roof. We prefer to call it the **visceral triangle** because of its deep contents. It houses the neck viscera—larynx, upper pharynx, trachea, esophagus, and thyroid gland. Its boundaries are the midline medially, the Sternomastoid and Omohyoid muscles laterally, and the hyoid bone serves as its base (Fig. III-33). The important *recurrent laryngeal nerves* course on either side in the groove between trachea and esophagus, and similarly situated on the left at this level is the upper part of the *thoracic duct*, the major lymph-collecting channel of the body. Tracheotomy, when needed, is done in the lower aspect of this region, in the midline between the opposing Sternohyoid and Sternothyroid muscles.

Some idea of the location of these triangles in the anterior neck can be gained from Figure III-39. The location of the **infraclavicular fossa** below the clavicle and between the Deltoid and Pectoralis Major muscles also is well delineated here (see also Fig. III-68A). In this groove and fossa the **cephalic vein** courses upward from the arm on its way to the **axillary vein** of which it is a tributary. Nor can we help noticing here the extent to which well-developed mammae cover the ribs and parts of the axillary region.

The Platysma: By now, you must have realized that the contents of the triangles of the neck cannot be well displayed at best without reflecting the skin. But even below the skin most of the contents of the anterior triangles and the lowest parts of the posterior triangle are covered as well by a broad, thin sheet of skeletal muscle, the **Platysma,** one of the muscles of facial expression (Fig. III-34B). It runs from the lower borders of the mandible as far down as the 2nd rib and lies in the subcutaneous tissues. It can be removed so easily along with skin reflection that most students miss seeing it. It is the remains, in man, of a larger **panniculus carnosus,** a muscular layer well developed under the hairy coat of some animals such as the horse and cow. They use it to shake off flies; we use it to ease a tight collar and to relieve the pressure on underlying superficial veins. You may wonder why this broad muscle in the neck is called a muscle of facial expression. First, it comes from the mesenchyme of the 2nd branchial arch from whence all muscles of facial expression are derived; second, it is innervated, therefore, by the facial nerve; and, third, you use it to pull the skin taut and, therefore, to contort the face every morning that you shave.

Major Lymph Nodes of the Neck: In doing a physical examination you will learn to palpate the neck for enlarged lymph nodes. Until they enlarge, they are not easy to feel, so it is well to have some idea where they lie. Fortunately, they string out in a horizontal chain at the junction of face and neck and orient in vertical chains within the neck proper. The horizontally-oriented, superficial nodes form a high, circular necklace. You can follow them around from just under the point of the chin:

1. Submental nodes in the muscular (submental) triangle.
2. Submandibular nodes in the glandular (digastric) triangle.
3. Preauricular nodes in the parotid region.
4. Mastoid nodes.
5. Suboccipital nodes near the inion.

Thus, you have come half circle between head and neck and need only enumerate them in reverse order (5 to 1, above) on the other side to come full circle.

In the neck, itself, the most important nodes begin in the minor vascular (subclavian) triangle and course upward vertically along the *internal jugular vein.* These are called **deep cervical nodes,** and it is into them that all the other nodes in the neck drain. Their efferent lymphatic vessels in turn join the **lymphatic duct** on the **right side** and the **thoracic duct** on the **left side.** Both of these main lymphatic channels empty into the venous system at the junction of subclavian and internal jugular veins at the root of the neck.

Other nodes are vertically arranged along the *external jugular vein,* and since this lies upon the Sternomastoid muscle, these are called **superficial cervical nodes.** They, too, drain into the deep nodes. There is an **anterior vertical group** on the membrane

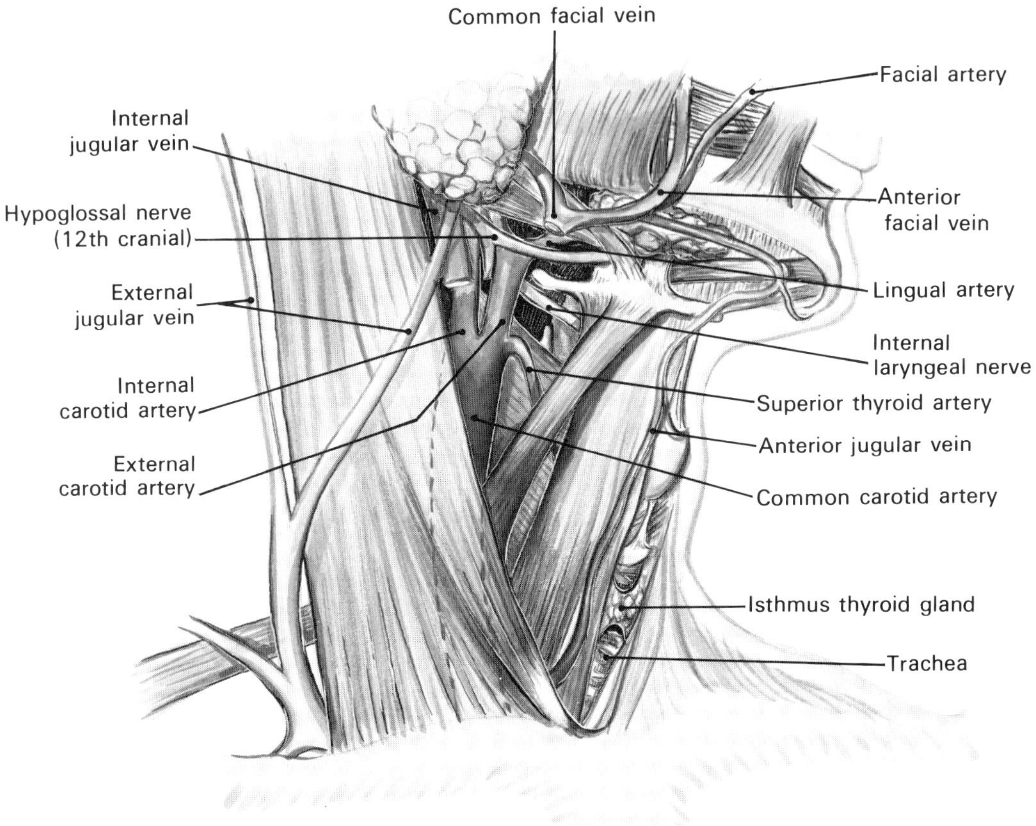

A

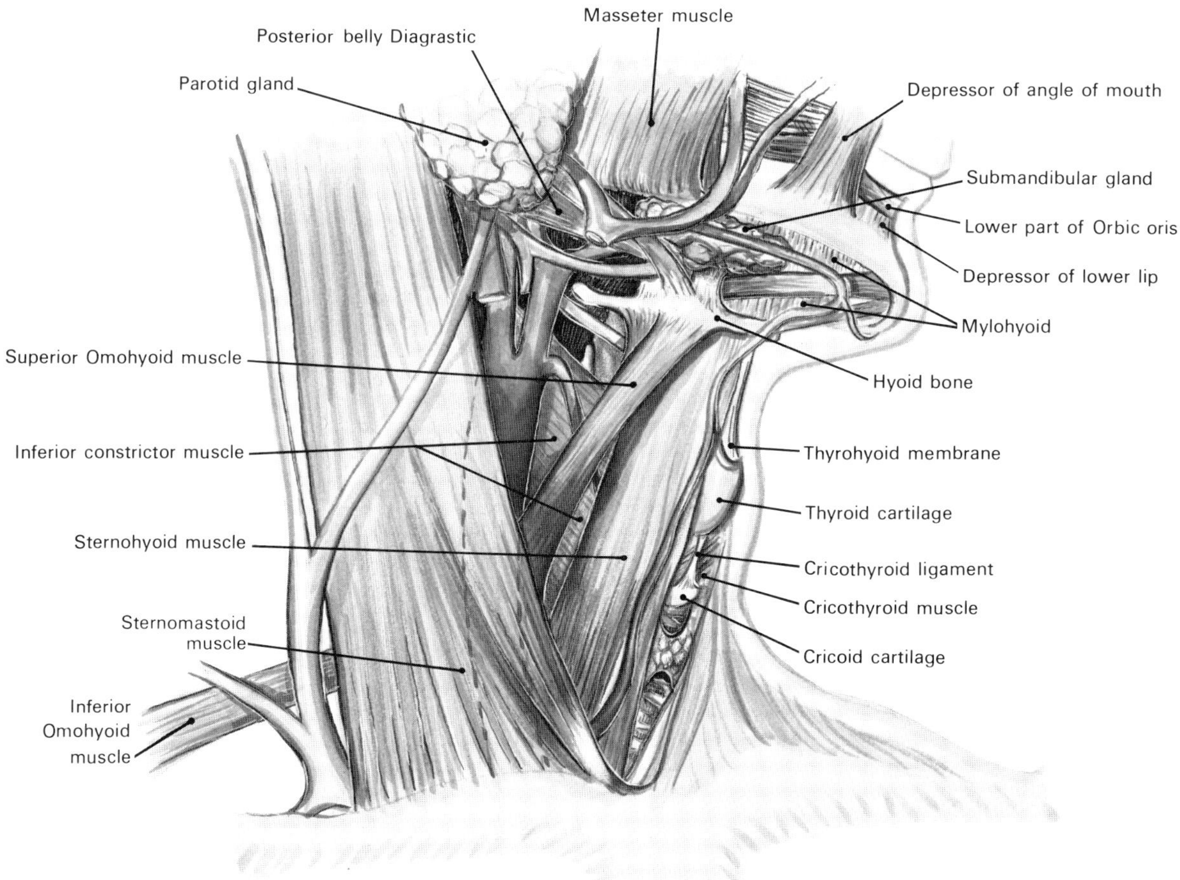

B

Fig. III-38. Dissection in the region of the anterior triangles of the neck. **A.** Major vessels and nerves. **B.** Major muscles. The dotted line continues the course of the internal jugular vein in the carotid sheath. (After Appleton, Hamilton, and Simon, *Surface and Radiological Anatomy*, Williams & Wilkins, 1958.)

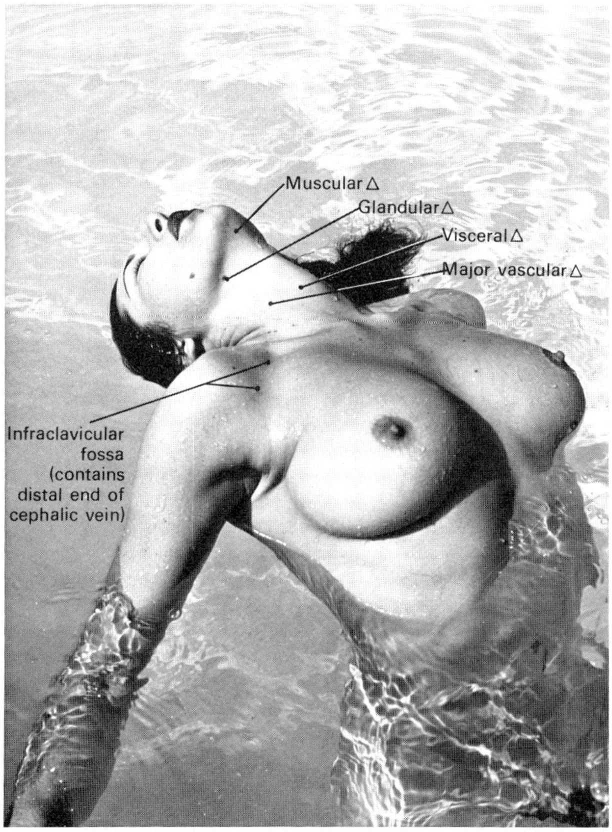

Fig. III-39. Surface localization of anterior triangles of neck and region of infraclavicular fossa.

that stretches from thyroid cartilage to hyoid bone. These drain the thyroid gland, larynx, trachea, and part of the pharynx. A number of posterior superficial nodes lie in the posterior triangle where they form vertical alignments. Deeper retropharyngeal nodes drain the pharynx, posterior nose, and the auditory (Eustachian) tube.

While most of these nodes are barely palpable in the healthy individual, they do become enlarged and firm in many diseased states. In some instances they become so strikingly swollen as to become visible without palpation. While the superficial cervical group beginning at the minor vascular triangle (Supraclavicular node) and following along the Sternomastoid may start out as discrete swellings, some the size of baseballs, in Hodgkin's disease and tuberculous lymphadenitis, they soon mat together to form a veritable horse collar about the neck. This may be unilateral, only, in some individuals.

The horizontal ring of superficial nodes and the vertical group in the posterior triangle form massive discrete swellings in lymphatic leukemia. Similar swellings in the posterior triangle used to be common as a result of bites from head lice (pediculosis capitis), but this is now rare among us. Perhaps we should look for its return among our unwashed and hirsute Hippie generation of the 70's. Some forms of protest among the younger generation seem downright unhygienic at times.

Vertebral Levels in the Neck: We use the backbone in anatomy like a rosary to count the stations of the body. We refer to structures as lying at some vertebral level. As a rule, it is to the **bodies** of the **vertebrae** that we refer, not to their **spines,** since the obliquity of the latter varies so from region to region that they are not useful as markers. Let us begin at the top of the neck and work downward. You are already familiar with the inion, or external occipital protuberance as a landmark. If, next, you projected a straight line along the midpoint of the hard palate, it would bisect the foramen magnum of the occipital bone, below which the neck begins posteriorly. The **3rd cervical vertebra** lies in line with the *body* of the *hyoid* bone. The *top* of the **5th cervical body** is in line with the *top* of the **thyroid cartilage;** the *top* of the **6th,** with the *bottom* of that cartilage. In the *beginning* of this section on the neck you learned that the **6th cervical vertebra** was the most important landmark in the neck, corresponding with the **cricoid cartilage** and many anatomical structures at that level. *Turn back to review these points* (page 110). At the level of the **7th cervical vertebra,** the *thoracic duct* arcs downward to empty into the junction of the subclavian and jugular veins on the left side of the neck. The **isthmus** of the **thyroid gland** lies at this level. In some few individuals the **apices of the lungs** rise this high under cover of the Sternomastoid. The **spine** of the **7th cervical vertebra** may be the most prominent landmark on the back of the neck, and the **ligamentum nuchae,** one of the few elastic ligaments in the body and the longest, will stretch from it to the external occipital protuberance (inion). However, in some instances, the 1st thoracic spine may protrude more prominantly than the 7th cervical.

SURFACE ANATOMY OF THE THORAX

The thorax can mean different things to different people. It can mean the chest containing the pleural cavities. It can mean the bony rib cage in its entirety. But, because the first rib is so obliquely placed anteriorly, there is at once an overlap into the neck region. The lungs in the pleural cavity and the heart in the pericardial cavity rest their bases upon the Diaphragm, the partition which separates thorax and abdomen anatomically. Again, the Diaphragm is so obliquely placed that, anteriorly, abdominal contents lie within the bony rib cage, making another area of overlap. Hence, we must think of the thorax in regions; *e.g.,* thorax and neck, thorax proper, and thorax and abdomen (Fig. III-40A). At least we know that, posteriorly, the thorax will contain the 12 thoracic vertebrae. Once more we can use them as references for the levels of other structures.

As a rule, the *apices of the lungs* tend to lie more at the level of vertebra T_1 than at vertebra C_7.

The *jugular notch* at the upper border of the

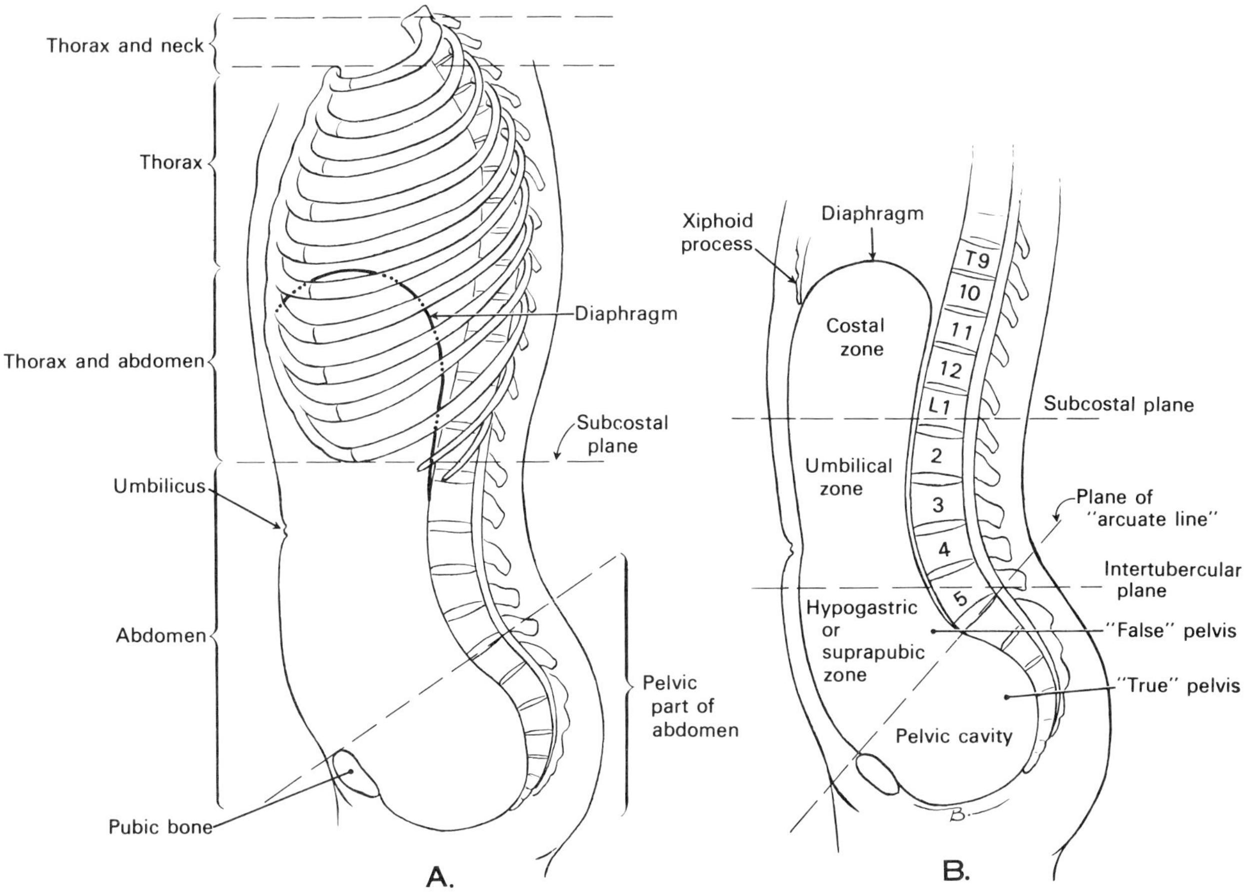

Fig. III-40. Regions of the thorax and abdomen. **A.** Illustrates how the bony thoracic cage trespasses upon territory of the neck and abdomen. **B.** Zones of the abdominal cavity.

manubrium of the sternum lies at the level of VT_2. (Were we to use the shorthand symbol "T_2" alone, we would be speaking of spinal cord segments, not vertebral levels. Cord segment levels do not necessarily correspond to vertebral body levels since the **spinal cord terminates** at the lower border of VL_1.)

A line can be drawn laterally through the middle of **VT_3** to the **base** of the **spine of the scapula.** A similar line through the middle of VT_7 will just knick the inferior tip (**inferior angle**) of the *scapula* (Fig. III-41A). Of course, the arms must be held at the sides for these relationships to remain true. Abduction of the arms to the horizontal or over the head will progressively rotate the inferior angle outward and upward. Ask a colleague to abduct the upper extremity while you watch his scapula. Even hunching of the shoulders pulls the inferior angle well above the level of VT_7 (*cf.* Fig. III-41B).

If VC_6 was the most important vertebral level in the neck, VT_4 is its counterpart in the thoracic region. Here, anteriorly, is the **sternal angle (angle of Louis,** pronounced "Looie") or the angle that the manubrium of the sternum makes with the body of the sternum (Fig. III-42C). Feel it on your own breast bone. As a rule, these two parts of the sternum are joined at a slight angle with fibrocartilage. In about one-third of the cases, this articulation may even have a synovial cavity and be quite movable. In the elderly, in disease, and even in some children (*e.g.,* progeria), the joint here may ossify. The **costal cartilages of the 2nd ribs** articulate with the manubrium and body of the sternum at this angle (Fig. III-42). Hence, one can slide a finger off the angle onto the 2nd rib and start counting interspaces and ribs from this point.

Despite what most textbooks may say, the *arch (i.e., its concavity) of the aorta hardly attains the level of the angle of Louis in the living adult,* so that the ascending and descending aorta do not end and begin here but below this level. Only in children and the very aged does the arch of the aorta rise and ride above this level so that the vessel can be palpated in the jugular notch. This can be confirmed in X-ray findings. The difference in opinion comes from reading anatomists' impressions from the dead, embalmed, and supine cadaver rather than from the living body. But the right and left pleura, which line the thoracic cavity, meet in the midline slightly above the level of VT_4. The left line of pleural reflection will deviate from the right line lower down near the heart, as we shall see later. The trachea ends at VT_4 and the mainstem bronchi begin.

From **VT_5 to VT_7** the *thoracic duct,* the main

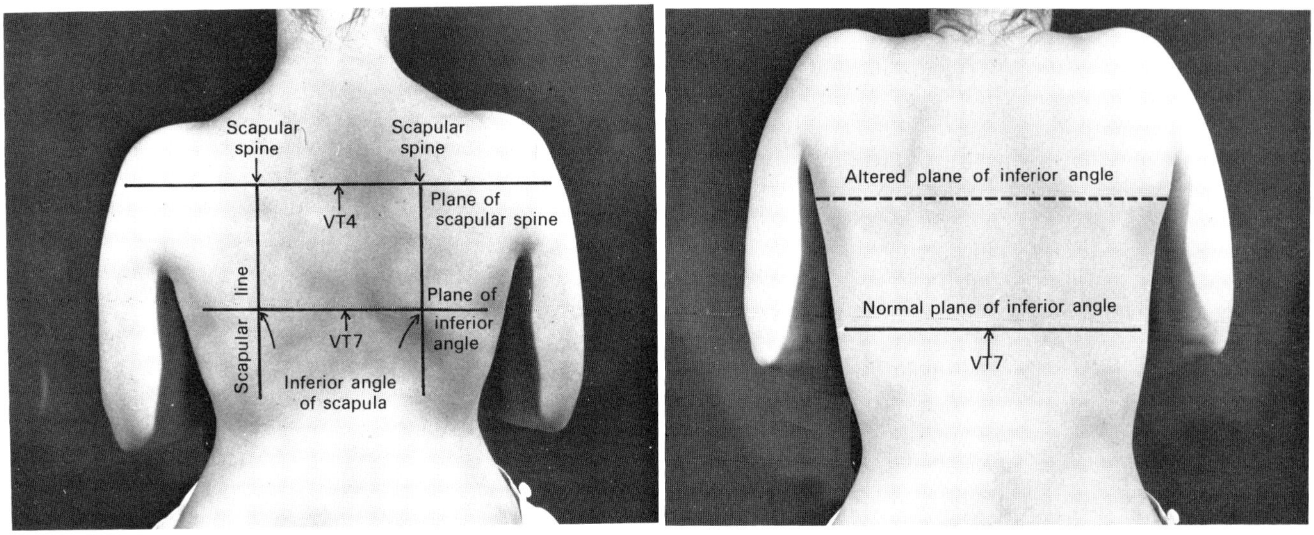

Fig. III-41. Hunching of shoulders changes relationship of scapular angle to vertebral level. **A.** Back in normal sitting posture. **B.** Back with shoulders elevated.

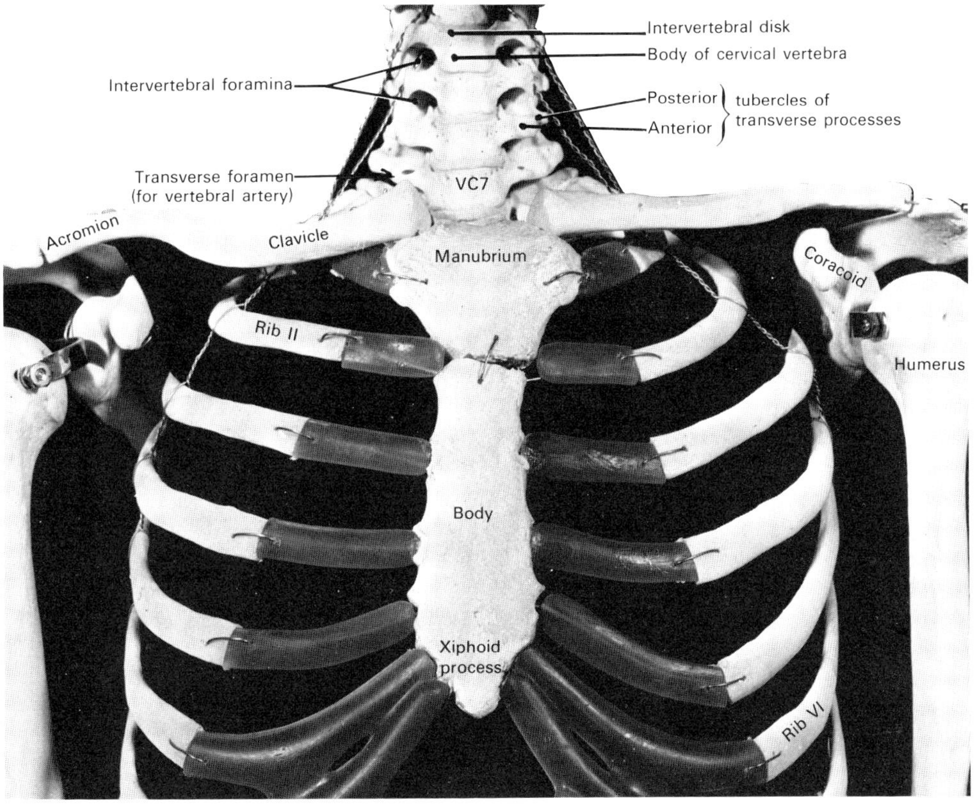

Fig. III-42A.

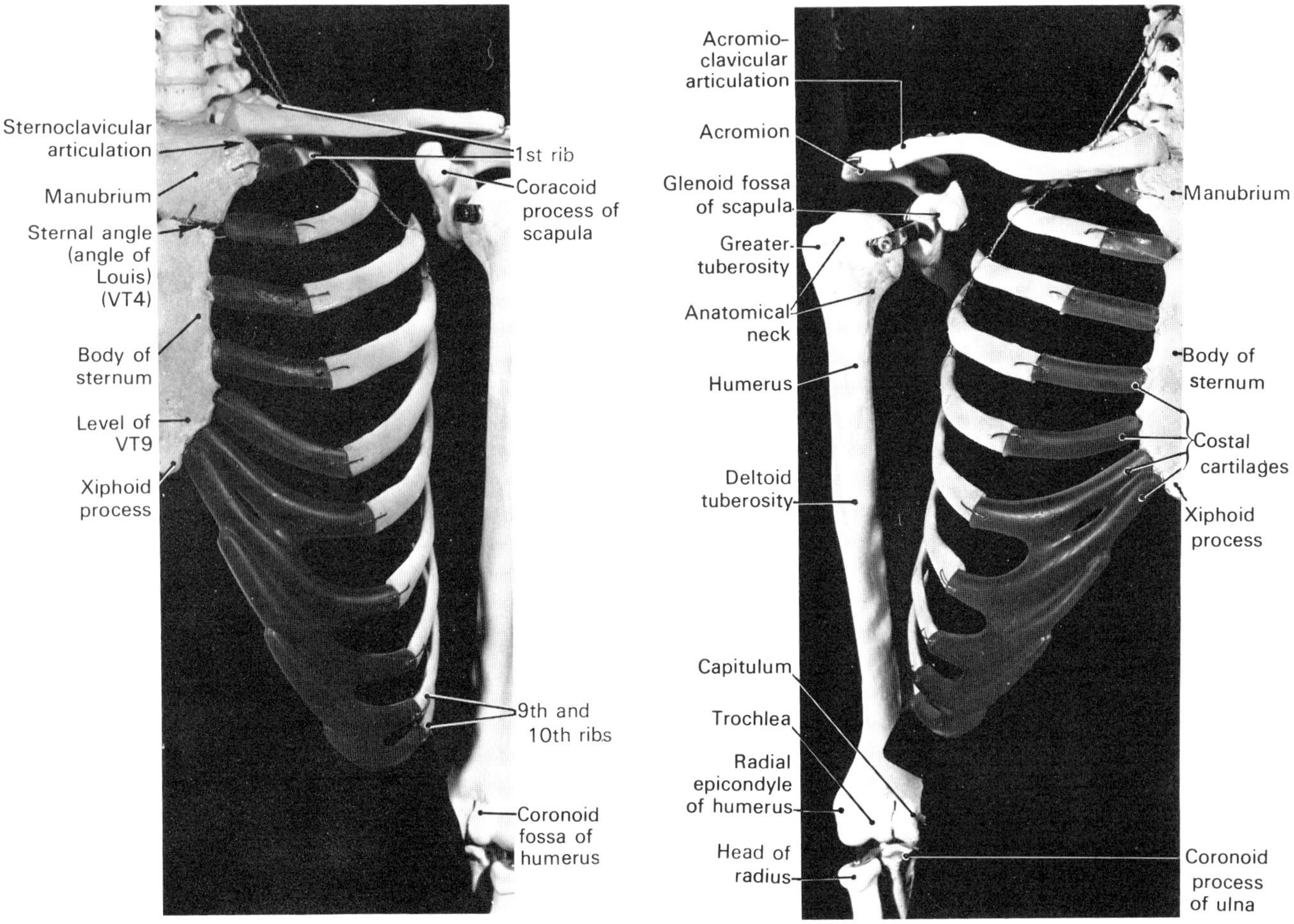

Fig. III-42. Ventral views of the thoracic cage of the skeleton. **A.** Relations with the shoulder girdle and neck; **B.** Right half and **C.** left half, showing some features of the bones of the upper arm and forearm. The xiphoid process in this specimen is broad and blunt, and it articulates with the sternum as a synostosis. (The costal cartilages and the intervertebral disks are artificial Black velvet lines the inner aspect of the thoracic cage to screen out posterior features during photography.)

lymph collecting channel in the body, *crosses behind the esophagus.* Above VT_5 it lies to the left of the esophagus; below VT_7, to the right of it. We have already indicated that VT_7 is the level of the inferior angle of the scapula.

At VT_8, the **inferior vena cava** *passes up* from the abdomen *through the Diaphragm.*

The **xyphoid** (swordlike) **process** articulates with the body of the sternum at the level of $\mathbf{VT_9}$ (Fig. III-42). This process is anchored to the main body of the breast bone by cartilage which generally ossifies by the 15th year. Try palpating it on yourself at the chondrocostal angle. It tends to dip inward, and deep palpation of it may be somewhat painful. Remember this if you need to feel for it in other people.

The esophagus goes through the Diaphragm more posteriorly than does the vena cava since it lies nearer the bodies of the vertebrae and since the Diaphragm slants obliquely from front to back. The **diaphragmatic-esophageal hiatus** is at $\mathbf{VT_1}$. The aorta, which lies on the vertebral bodies with the esophagus in front of it, *passes through* even lower at $\mathbf{VT_{12}}$. The thoracic duct and azygos vein go through with it.

Vertical Lines: Early in this chapter we intimated that certain useful lines could be drawn on the body to help in the localization of structures. The most useful and most stable, since body builds vary so, is, obviously, the **midline,** whether you draw it on the back or the front. At least, in any given body, the most objective measurements can be made from the midline.

It is good practice to draw some of these lines using a colored chalk (red is pretty) and employing the services of a willing colleague stripped to the waist. A tape measure helps if you want to determine midpoints.

1. **Midclavicular line:** A perpendicular dropped from the middle of the clavicle (Fig. III-43A) runs about 3½ inches from the midline and will lie medial to the nipple in the male. It is equivalent to a point midway between midline and the lateral border of the **acromion** (tip of the shoulder). If it is extended below the chest, it will not quite correspond to the so-called **lateral or paracentral line on the abdomen.** This means that it will not cross the inguinal ligament midway between midline and anterior superior iliac spine, but will cross it about an inch lateral to this point. It will, thus, also miss cutting across the tip of the 9th costal cartilage by about an inch. From this we can conclude that the **lateral line** parallels the midline but

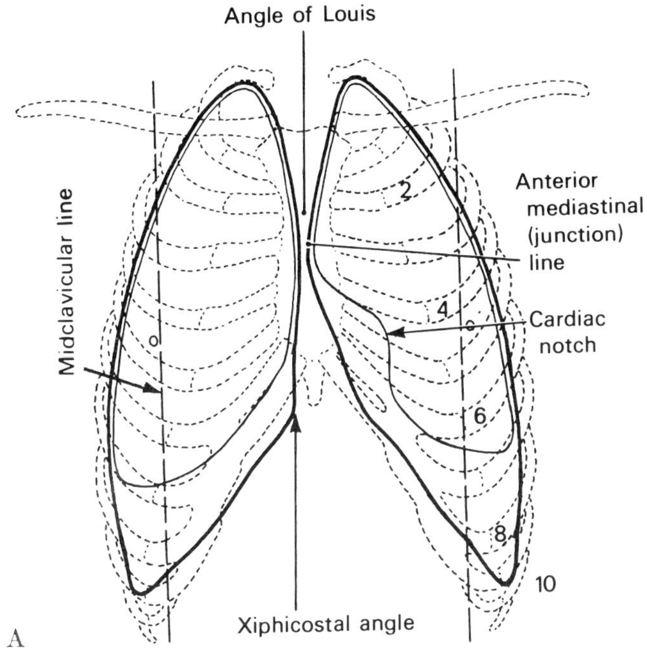

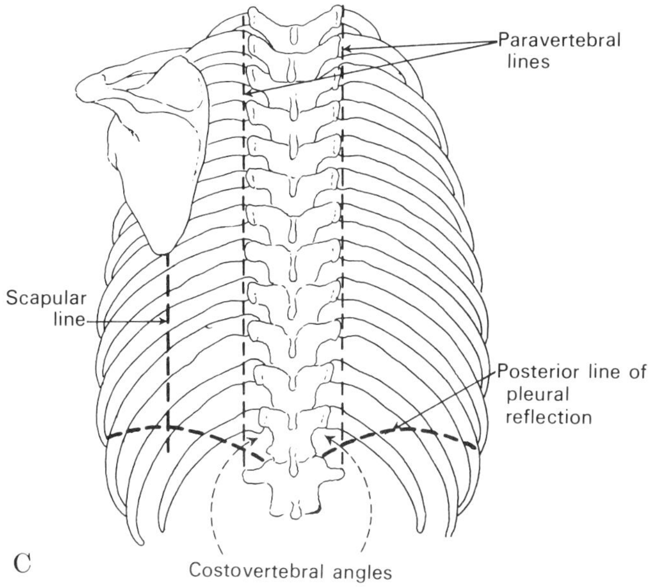

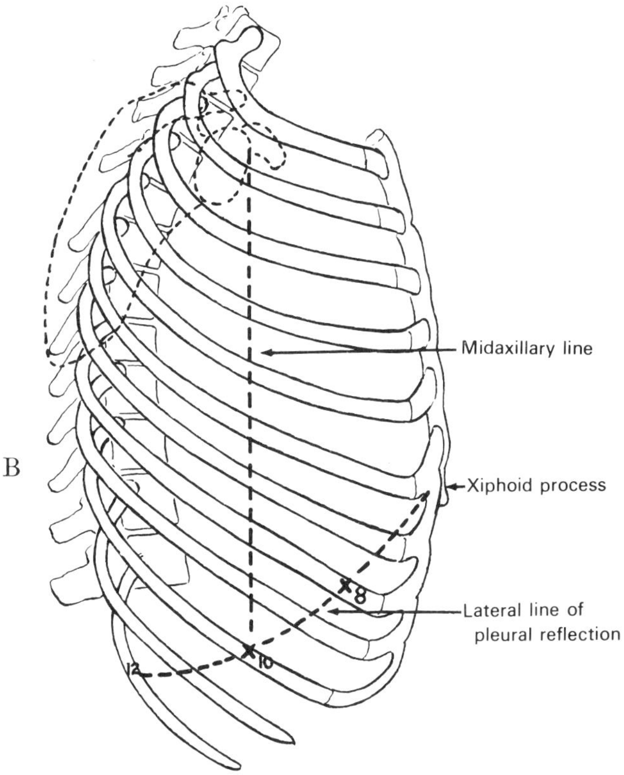

Fig. III-43. Lines of pleural reflection. **A.** Anterior. Note cardiac notch on left and that midclavicular line meets pleural reflection at 9th rib; misses tip of 9th costal cartilage and lies medial to nipple. **B.** Lateral. Line of pleural reflection cuts 10th rib at midaxillary line. **C.** Posterior. Line of reflection crosses 11th rib in scapular line and 12th at neck to meet paravertebral line.

lies about 2½ inches on either side of it in the abdominal region. Extended, the lateral line passes through the "femoral point" of the thigh just below the inguinal ligament.

2. The **midaxillary line** is one dropped perpendicularly so as to bisect the armpit when the arm is raised over the head (Fig. III-43B). This means that anterior and posterior axillary lines could be drawn, as well. They would run along the anterior and posterior axillary folds, the former as a vertical line along the lateral chest at the lateral border of the Pectoralis Major muscle and the latter a similar vertical line along the anterior border of the Latissimus Dorsi.

3. The **scapular line** is perpendicular, dropped from the inferior angle of the scapula with the arms at the sides (Figs. III-41A and 43C).

4. **Paravertebral lines** connect most of the tips of the transverse processes of the vertebrae. These, of course, cannot be drawn accurately except on an X-ray film (Fig. III-43C).

Occasionally other lines are mentioned in respect to the thorax but are of lesser importance:

1. **Lateral sternal line,** a line drawn along the lateral margins of the sternum.

2. **Parasternal line,** a line lying midway between sternal and midclavicular lines.

3. **Mammary line,** a perpendicular drawn through the nipple. It is of little use as a marker on females since their breasts vary. It lies about 4 inches from the midline and, hence, is lateral to the midclavicular line in males.

Horizontal Planes: These we have discussed, in a sense, when we noted vertebral levels. The important ones are:

1. **Plane of the sternal angle:** Cutting through the manubriosternal joint at VT_4–VT_5 interspace.

2. **Xiphisternal plane:** Cutting through the junction of the xyphoid process with the sternum at VT_9.

3. **Plane of scapular spine** (Fig. III-41A).

4. **Plane of inferior angle** (Fig. III-41A).

The Sternoclavicular Joint: This is a useful landmark because of the vessels behind it. The **internal jugular** and **subclavian veins** empty into the **right** and **left brachiocephalic veins** behind each joint. On the right side alone, the arterial **brachiocephalic trunk** (innominate artery) just begins to branch into the **right subclavian** and **right common carotid arteries.** The trunk, itself, may give off the inconstant **thyroidea ima artery** and sometimes a **thymic** and a **bronchial** branch. Often it gives off no small branches at all. Moreover, the lines of pleural reflection cross obliquely behind each joint.

This brings us to a consideration of the **lines of pleural reflection.** While you are in the process of marking up a partner with chalk, try mapping the extent of the pleural cavities on his chest. The lines you will draw follow the boundaries of the simple squamous epithelium **(parietal pleura)** which lines the inside of each pleural cavity like wallpaper upon the walls of a cone-shaped room.

Lines of Pleural Reflection: Start on the *left side* since the left cavity departs from the pattern of the right side because of the position of the heart within the thorax. Burden your memory only to the extent of recalling even numbers from **2** to **12** (Fig. III-43 A–C).

1. From a point one fingerbreadth above the clavicle, draw oblique lines over each sternoclavicular joint and let them meet just above the angle of Louis. This puts you at the level of the **2nd costal cartilage.**

2. Continue a single line down the middle of the sternum to the level of the **4th costal cartilage.** Angle the line slightly to the left as far as the left sternal margin.

3. Continue down the left sternal margin to **costal cartilage 6.**

4. Carry the line obliquely to the left so that it cuts the **8th rib** in the midclavicular line.

5. Continue laterad obliquely and cut the **10th rib** at the midaxillary line.

6. Continue posteriorly to cut the **12th rib** at its neck at the *paravertebral line.* This should carry across the 11th rib about at the scapular line. Since you cannot see transverse processes very well without X-ray vision, how will you know how to erect a *paravertebral line?* Bring your finger rule into play. Carry the line across the 12th rib **2** fingerbreadths from the midline drawn upon the back. You can then erect a perpendicular from VT_{12} up to VT_1, always **2** fingerbreadths from the midline.

How do you know where VT_1 lies? The prominent 7th cervical spine ends at the upper border of VT_1 because of its slight obliquity. Or you could draw the **plane of the scapular spine** and measure up **4** fingerbreadths along the midline from it to arrive at approximately the same place (Fig. III-43C).

To outline the lines of pleural reflection on the *right side,* begin once more at the angle of Louis:

1. Continue down the midsternum to the xiphisternal joint.

2. Angle the line to the right to cut **8th, 10th,** and **12th** ribs at the midclavicular, midaxillary, and paravertebral lines respectively.

3. Finally at VT_1, connect right and left paravertebral lines to those started on the anterior chest wall by drawing oblique lines around the base of the neck.

4. Your partner's chest, sides, and front should look something like Figure III-43A–C, provided he had the foresight to remove shirt, tie, and undershirt before you got busy with the chalk.

Significance: With landmarks and certain vertical lines, you have projected upon the surface the outlines of the two **pleural cavities** which, of course, lie entirely within the bony cage formed by the ribs and sternum, except at three small angles and at the root of the neck.

They lie below the margins of the cage *posteriorly* (Fig. III-43C) between the neck of the rib and the body of the 12th thoracic vertebra at the **costovertebral angles.** *Anteriorly,* on the right (Fig. III-43A), they lie below the costal margin at the **xiphocostal** angle. The apices or cupolas of these somewhat cone-shaped cavities extend above the first rib into the neck for a distance of about 1½ inches (Fig. III-44C). Sharp instruments piercing at any of these places (as well as between the ribs, of course) will enter the pleural cavity to produce a **pneumothorax** (collapse of the lung).

How are the lungs related to these cavities? Strange as the thought may seem, the lungs lie outside of these cavities, having only invaginated its walls to the extent that the cavities have become virtually obliterated and all that fills them normally is a small amount of serous fluid (Fig. III-44A). Think of these two cavities as rooms whose costal, diaphragmatic, and mediastinal (medial) walls are "papered" with a thin epithelial pleura, called **parietal pleura** (parieties, "walls"; *cf.* parietal bones, the sidewalls of the cranial vault) (Fig. III-44A, left side). The **mediastinum** is a central connective tissue partition whose thickness varies locally and which forms a common wall between these pleural "rooms" (Fig. III-44B). It carries the "plumbing" between them to upper and lower stories, to and from neck and abdomen. Included as "plumbing" are such ducts and conduits as trachea, esophagus, major blood vessels, and even the heart which can be considered as an overly muscular blood vessel. You can fancy the important nerves as the "wiring" in the partition, if you wish.

The lungs develop within the mediastinal partition as buds at the ends of the endotracheal bifurcation (Fig. III-44A). In growth, they push laterad against

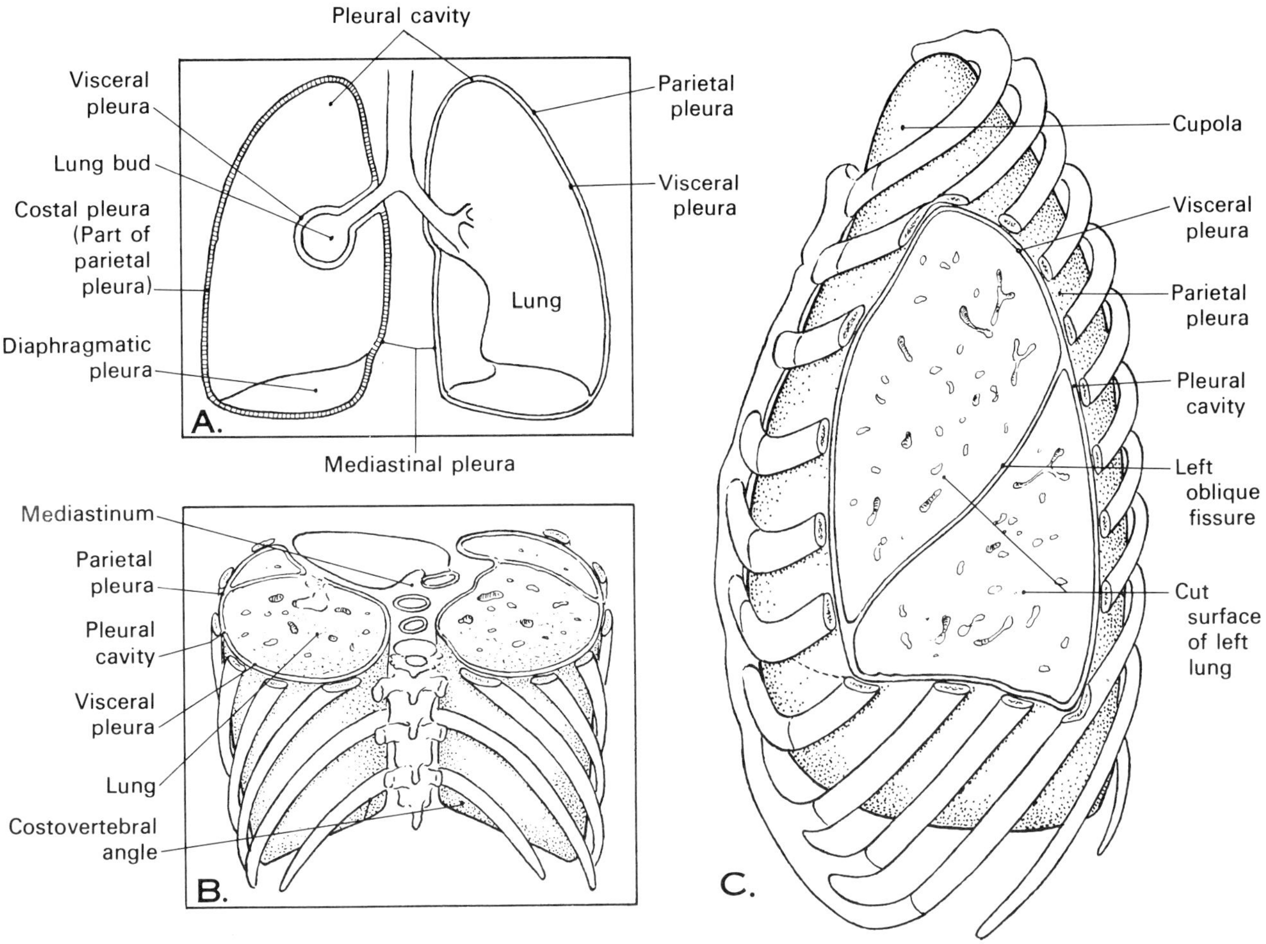

Fig. III-44. The parietal and visceral pleurae. **A.** Lung invaginating pleural cavity to gain its visceral pleura and to reduce the space of the cavity. **B.** The costovertebral recesses (angles). Note position of mediastinum between pleural cavities in the cross-section. **C.** The extent of the parietal pleura and its cupola. Parasagittal section shows lung, visceral pleura, and reduced pleural cavity.

the medial walls of each room (pleural cavities), bulge into them, lift the "wall paper" off these medial walls, and push it before them. The growing lungs conform to the shape of the rooms and encroach upon and obliterate the space therein as they wrap the loosened, mediastinal pleura about themselves as an outer skin. Since the lung is a **viscus** containing smooth muscle, this reflected outer skin, now lying upon it, is called the **visceral pleura.** In time visceral pleura almost touches parietal pleura on the walls of the rest of the cavity (Fig. III-44A, right side). It is separated from parietal pleura only by a thin film of serous fluid. This lubricating fluid is all that is contained *within* the pleural cavity, or what is left of it. The lungs are still *outside* of the cavity *behind* their covering of visceral pleura which was once a portion of the parietal pleura of the medial or mediastinal wall of the cavity (Fig. III-44B and C).

While the foregoing may seem to be a matter of semantics and definition and while this concept is a very simple one, it is because students so often only begin to realize it so late in their studies that they have difficulty with the concepts of respiratory physiology. An old Chinese proverb says: "The simplest things are keys to the wonders of complexity." While this may be true, so is our own occidental proverb which says: "The simplest things are most wonderously complex." For no matter how simple morphological and physiological concepts about the lung may be stated, they simply are not that simple.

One thing, at least, is certain. The lung is not a simple football bladder suspended in a rigid chamber whose floor is a rubbery diaphragm and whose neck is a one-holed rubber stopper. In such a system, pulling down on the diaphragm will inflate the bladder somewhat. Let up on the diaphragm and the bladder shrivels and collapses as air escapes from the neck of the chamber. True, like this bladder, the lung has its own, intrinsic, elastic recoil propensities and it always seeks to collapse and reduce its size as much as possible. But what actually contracts most in the elastic recoil phase of expiration is the diameters of its internal alveoli (air pockets) and the diameters of those bronchiolar tubes that lack cartilaginous walls. Its overall *outside* dimensions decrease only as much as the pleural cavity, itself, decreases in volume during expiration, which is rather minimal. The serous fluid between parietal and visceral pleurae exerts surface tensions that maintain a *negative intrapleural pressure*. This pressure is more negative during inspiration than during expiration so that air rushes in to expand alveoli and bronchiolar diameters. The visceral surface of the lung also expands somewhat, creeping along the fluid interface as the body walls expand. Surface tension factors make the visceral pleura follow parietal pleura in thoracic cage expansion and keep it from pulling away from parietal pleura during the expiratory phase.

The lung in its invagination does not encroach entirely upon all the space within the cavity. At least, at the **costodiaphragmatic recess** (anterolaterally), the lower margins of the lung are thin and sharp and do not reach the lower lines of parietal pleural reflection upon the ribs and diaphragm in this region (Fig. III-45C). But they do encroach upon this space in inflation, only partly, and they withdraw from it in expiration. When they withdraw, the two opposing parietal surfaces come together, as much as they can with the serous fluid interposed. On every inspiration, the lung margins redissect these parietal surfaces apart. Posteriorly, in all probability, the posterior lung margins occupy the **costovertebral recesses** almost fully, though it is not as easy to determine whether this is so by means of percussion as it is anteriorly because of overlying heavy muscle masses. You have also drawn a **costomediastinal recess** or **cardiac notch** region to the left of the sternum (Fig. III-43A). It is generally thought that the lung does not extend into this area, and that here the heart will drum a tattoo against the anterior chest wall with its every beat. Such a recess may not be as marked in adults as in children, and lung tissue may encroach nearer the midline upon inflation than we normally presume. The relationships of pleural reflection here are variable. It is again difficult to determine whether there is a parietal pleural flap here that the lung invades in inspiration because the pectoral muscles make determination by percussion difficult.

It does no good to open the chest to find out how much space under its visceral pleura the lung actually occupies in inspiration. Even tiny holes that penetrate parietal pleura lead to **pneumothorax.** Atmospheric pressures break the surface tension seal between pleurae, negative intrathoracic pressure is overcome, and visceral pleura is pulled away from parietal pleura as the lung now does what it always wants to do—contracts into a tiny mass. Even if you were to clamp the trachea of an animal when the lung was in full inflation, a hypodermic needle inserted through the base of the Diaphragm would cause the lung to contract to about one-fifth of its inflated volume at once. The extent of collapse is amazing when one measures the increase in intratracheal pressure in the clamped trachea only to find it increasing by as little as 2 cm. of water! Obviously most of the air is still within the lung tissues, and this means that a considerable volume of blood must have moved out for the lung to have collapsed to this extent. To bring this lung back to where it once again fills the pleural cavity requires a filling pressure of 30 cm. H_2O! These are most abnormal, aphysiological pressures which overextend and even burst alveolar walls.

These simple facts have never been taken into proper account by most people who try to fix lung tissue for histological study. Either they fix collapsed lung, or they attempt to fix the reinflated lung under abnormal pressures. For this reason probably no *normal* histological pictures of lung tissue have appeared in our textbooks of histology, and we can probably say as well that no electron photomicrographs to date show *normal alveolar morphology* as it exists in the closed pleural cavity. Pouring fluids down the trachea or vascular perfusion of the lung in the unopened chest leads to other artifacts of fixation equally as damaging to normal alveolar relationships. This is why we say that though concepts about lung morphology and function are basically rather simple ones, the overall picture is quite complex, and there is much that we still have to learn about internal lung structure and function. We have digressed again from pure surface anatomy because we feel that these concepts should be borne in mind when you think about pleural relationships as applied to respiratory function and to the problem of pneumothorax. The anatomy of this region will be considered in more detail in the chapter on the thorax (page 448).

The Lobes and Fissures of the Lungs: There are three major lobes in the **right lung: upper, middle,** and **lower.** An **oblique fissure** separates the right upper and middle lobes from the lower lobe. The **horizontal (minor) fissure,** which passes forward from the oblique fissure, separates the right middle and upper lobes (Fig. III-45A).

The **left lung** has two major lobes: **upper** and **lower.** These lobes are separated by an **oblique fissure.** This fissure can be mapped by beginning posteriorly at the 3rd thoracic spine (its tip lies at the upper level of VT_4). To get the proper angle of obliquity over the back, have your laboratory partner place the palm of his hand on his head (Fig. III-45B). The vertebral border of the scapula will now slant in proper orientation so that you can draw an approximate line along it that will cut the **5th rib** in the midaxillary line. Carry the line over the anterior chest wall

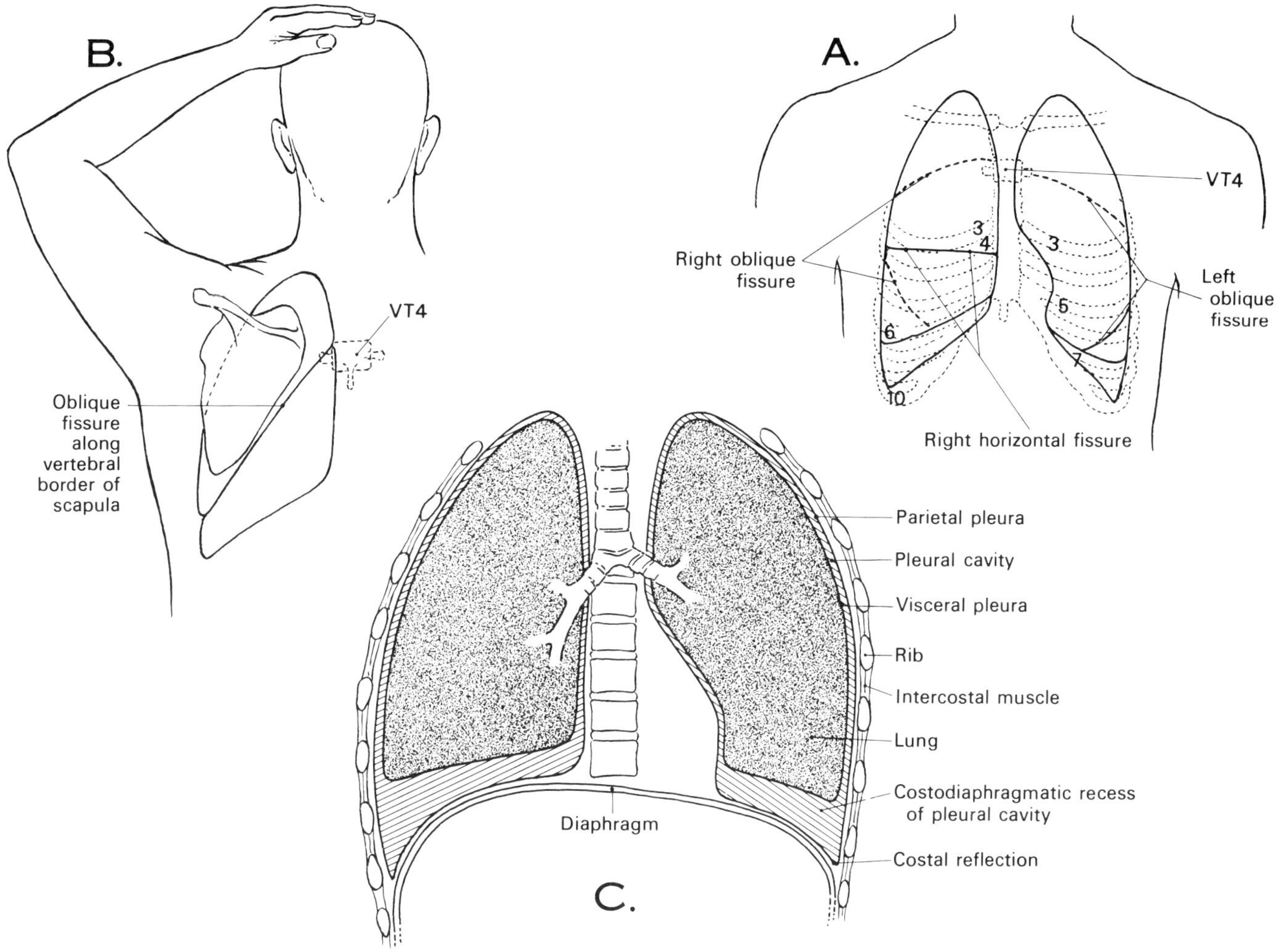

Fig. III-45. Fissures of the lung and costodiaphragmatic recess. **A.** Localization of the fissures of the lung on the surface of the trunk. **B.** Method for localizing oblique fissure posteriorly. **C.** Coronal section of thorax showing costodiaphragmatic recess.

so that it cuts the **6th rib** at its costochondral junction.

In life the lungs will be variable in shape, and this shape is dependent upon the space within the thoracic cavity as well as the influence from adjacent structures including the Diaphragm. Then, too, the fissures of the lungs can only be discussed in the broadest terms for they are inconstant not only in shape, but in extent and position (part or all may be absent). The illustrations shown (Fig. III-45A) are based upon the work of Brock.[15] These reveal the position of the fissures relative to the costal pleura. The student must realize that these pleural-lined clefts do not follow a horizontal course medially toward the hilus of the lung, but that they are often more obliquely oriented. In some individuals, the fissures bend cranially or caudally as they pass medially. In others they may bend both ways in somewhat of a lazy-S configuration.

Realize that the oblique fissure on the right begins at a lower level (about one intercostal space) than the same fissure on the left. Thus, the right oblique fissure begins, within the thoracic cage relative to the costal pleura, at about the level of the vertebral end of the 5th rib or the 5th intercostal space. Such a level, when projected cranially and posteriorly through the rib cage and overlying soft tissues, would place it to the right of the 3rd thoracic spine (its tip lies at the upper level of VT_4) (Fig. III-45A). From this point the right oblique fissure curves obliquely downward along a line paralleling the 6th rib, and finally ends near the 6th costochondral junction, some 5 to 10 cm. from the midline.

The horizontal (minor) fissure begins near the lower border of the 4th costal cartilage and extends posteriorly in a horizontal plane to join the oblique fissure in the midaxillary line (Fig. III-45A).

The posterior beginning of the left oblique fissure is rather an indefinite point. It may leave the vertebral column on a level with the inner margin of the 3rd to 5th ribs. In passing downward, it is more steeply inclined than the right oblique fissure. In the midaxillary line, it is at the level of the 5th rib, and it ends to the left of midline at the 7th costochondral junction.

The Location of the Heart: Projecting the anterior outline of the heart upon the chest wall can be approximated well enough by using three landmarks and your fingers. Landmarks are the angle of Louis, the xiphisternal junction, and the **5th** anterior costal interspace.

1. Starting at the midpoint of the angle of Louis, draw a gently curving line to a point one finger breadth beyond the right xiphisternal junction (Fig. III-46).
2. Draw a similar curved line from the midpoint of the sternal angle to a point 4 fingerbreadths left of the midline in the 5th anterior costal interspace.
3. Connect the bottom of these lines with a straight line which angles slightly up toward the left.

Most of the area outlined on the right side, about the lateral two-thirds, will represent chiefly **right ventricle.** Near the right sternal angle will lie a bit of the **right atrium (its auricular appendage).** The left one-third of the outline will represent **left ventricle** only. This means that the heart lies in the chest with its base upon the Diaphragm and with, chiefly, right ventricle and a part of left ventricle oriented anteriorly. Both atria and most of the left ventricle are posteriorly oriented. The *left apex beat* can be felt in the 5th interspace 3½ inches (four fingers) from the midline. How does this relate to the left nipple in the male? The latter lies in the 4th interspace lateral to the apex beat.

Suppose you wished to draw this outline on the back of your partner to determine where piercing wounds from the back would reach the heart. The heart lies between the upper border of VT_5 and the lower border of VT_8, the middle four of the 12 thoracic vertebrae. The upper border of VT_5 lies one fingerbreadth *above* the tip of spine VT_4; the lower border of VT_8 lies one fingerbreadth *below* the tip of spine VT_7. Place upper and lower midline dots at these places. From the lower dot measure 4 fingers to the left and 2 to the right. Place a dot at each of these places. Connect them with gentle curving lines to the upper midpoint dot and draw the base line between them.

You may have noticed that the level of the tips of the vertebral spines varies in respect to the levels of the vertebral bodies next below. Is there any rule about this that you can evoke since spines seem to vary in their obliquity? From VC_2 to VT_3 spines reach to the *top* of the vertebrae next below. From VT_4 to VT_7 they reach lower still to the *midpoint* of the vertebral body next below. From VT_8 to VT_{12} the angulation is still steeper, and the tips of the spines reach to the *bottom* of the vertebral body next below. Lumbar spines revert to the first rule, reaching only to the *top* of the vertebra next below. Consider a vertebral body to be 2 fingerbreadths in height, approximately.

The right posterior aspect of the heart is one great venous channel including **superior vena cava, right atrium,** and **inferior vena cava.** These lie virtually

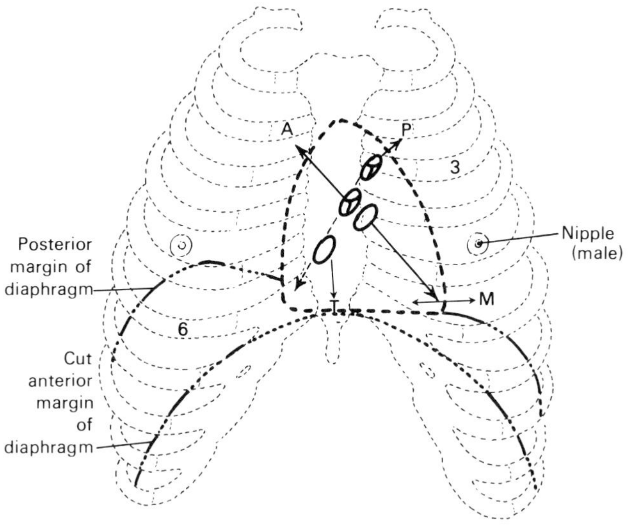

Fig. III-46. Surface localization of heart, heart valves, and heart sounds. The valves are drawn along an oblique line crossing the sternum from the 3rd left costal cartilage to the 6th right. This is approximately the line of the atrioventricular groove. A, P, M, and T show where heart sounds are best heard from the aortic, pulmonary, mitral, and tricuspid valves, respectively. The arrow in the 5th costal interspace indicates that there is some latitude in placing the stethoscope here to pick up sound of the mitral valve.

in a straight line. Draw a line through the sternoclavicular joint and run it straight down along the right lateral border of the sternum to the 3rd costal interspace. The superior vena cava will lie behind this line and will enter the right atrium at the level of the 3rd intercostal space. Continue the line downward almost to the xiphisternal junction, or to about 4 fingerbreadths from the 3rd costal interspace. At this point, the inferior vena cava pierces the diaphragm from below to enter the right atrium.

If you draw another straight line from the cricoid cartilage to the midpoint of the sternal angle and beyond, the trachea will lie behind this line and will bifurcate into its primary bronchi at the angle (VT_6) (anatomists, viewing cadavers, will say VT_4).[15]

The Position of the Heart Valves: The **atrioventricular (coronary) groove** of the heart can be indicated approximately by a line drawn from the sternal end of the 3rd left costal cartilage obliquely across the sternum to the sternal end of the 6th right cartilage. The four *large* orifices of the heart and, hence, the position of the heart valves lie slightly to the left and a bit below this line (see dotted line, Fig. III-46). The pulmonary orifice with its tripocketed, semilunar, **pulmonary valves** lies at the upper left of the line at the level of the 3rd left costal cartilage. The aortic orifice and its similarly fashioned **aortic valves** lie a little lower, slightly to the right behind the left margin of the sternum at the level of the 3rd intercostal space. The **mitral** or **left atrioventricular valve** is below the aortic valve at the level of the 4th costal cartilage.

and its counterpart on the right side of the heart, the **tricuspid valve** lies opposite the midline of the sternum at the level of the 4th intercostal space.

Students often forget whether the **bicuspid** or **mitral valve** is the right or the left atrioventricular valve. It is called mitral because its two cusps resemble the prongs on a bishop's mitre (crown). Long ago David Hume, a worldly philosopher, had heated debates with Bishop Berkley of England over religious doctrines. Invariably the arguments ended with Hume exclaiming in exaspiration: "The Bishop is *never* in the right!"

Heart Sounds: When the heart valves close they give rise to sounds that can be heard with a stethoscope. The **first heart sound** in the cardiac cycle represents the closure of the atrioventricular valves (bicuspid and tricuspid). The **second sound** represents the closure of the two sets of semilunar valves (pulmonary and aortic). The two sounds together give the "lupp-dupp" of the heart beat. Sometimes a **third sound** is heard following closely upon the second. This is due to the opening of the A-V valves and the vibrations of the ventricular chamber walls as they fill with blood.[14]

While you might think that you would place your stethoscope directly over the anatomical position of these valves (along the oblique line which you have drawn) in order to hear them best, this is not so. It is better to place the stethoscope at some distance from each valve along the direction of blood flow after it leaves the valve. This gets placements far enough apart so that you are not in areas of anatomical overlap, and, thus, you avoid confused information from the valves. The areas of maximal audibility from each of the separate valves are shown at the end of the black arrows in Figure III-46:

1. To listen to the *pulmonary valve* place the stethoscope to the left of the sternum in the 2nd costal interspace.
2. To listen to the *aortic valve*, place the stethoscope in the same interspace but *to the right of the sternum.*
3. The *bicuspid (mitral) valve* of the left ventricle is best heard at the apex of the heart in the 5th left anterior costal interspace. The arrow in Figure III-46 indicates that there is some latitude within this interspace for maximal audibility of this valve.
4. The *tricuspid valve* is best heard near the xiphisternal junction in the midline.

Heart sounds will be louder in children than in adults and duller in the obese. The second sound at the pulmonary valve will be louder in children than the second sound at the aortic valve; the reverse is true in adults.

Aside from some odd sounds from pericardial and pleural friction, you may occasionally hear heart "murmurs" from the great vessels or at the valves, and these may be due to differences in the size of the orifice in respect to the size of the heart chamber into which they empty or from which they drain. Such murmurs are classed generally according to the phase of the cardiac cycle in which they occur and according to the valve involved; *e.g.*, "tricuspid systolic murmur." Have your instructor help you locate and listen to the heart sounds and then practice on your wife, children, or sweetheart. As they say, "Practice makes perfect," so take your stethoscope along on all blind dates as the perfect icebreaker. If you find toting a stethoscope a bit awkward, tuck a reflex hammer in your back pocket. There is nothing like tapping a miniskirted patellar tendon to get a kick out of life!

Muscular Contours of the Back and Chest: Consider the back muscles first. Those in the shoulder region are a superficial group that can be localized rather easily by sight. They are not truly back muscles at all, but muscles of the shoulder girdle and upper limb. In a similar category we can place the rounded Glutei along the lower back and buttocks which are primarily muscles of the hip joint. The true intrinsic muscles of the back that may be somewhat visible from the surface are the **Erector Spinae** group. The other intrinsic back muscles are small, deep, and hidden beneath the Erector Spinae.

Figures III-47 and 48 illustrate a number of the superficial muscles in male and female subjects. Both subjects were asked to "square the shoulders." This approximates the vertebral borders of the scapulae and accounts for the deep midline furrow seen in these photographs. You can expect the bony prominences (acromion and spine of scapula) to stand out less in the well-padded female and to see the muscular outlines better in the more athletic male. We simply call your attention to the muscles named in these plates at this time. Details and function will be covered more fully in a later chapter (p. 418).

Both subjects (Figs. III-47 and 48) demonstrate certain well-developed muscles in common: the Trapezius, Rhomboideus Major, and Teres Major. The latter shows especially well in the female subject in this particular posture. The Supraspinatus and the Infraspinatus stand out better in the male, since he has been engaged in a weight-lifting regime. All of these muscles are concerned with the movement of the scapula, and the last four lie deeper than the Trapezius which covers them almost completely.

The **Latissimus Dorsi,** more concerned with extension and adduction of the upper arm, is broad upon the lower back and narrows on its way to its insertion on the **humerus** so that it becomes the prominent **posterior axillary fold.** In some individuals the tendon of the Teres Major can be seen giving a contribution to this fold, as well.

Generally, the Erector Spinae group, in its lower reaches at least, will be more evident in males. This is a complex group of muscles, as you will discover when you dissect, and they run on each side of the vertebral column from sacrum to occiput. See them stand out in Figure III-49 where back and neck are in virtual *opisthotonus.* The latter term comes from two Greek

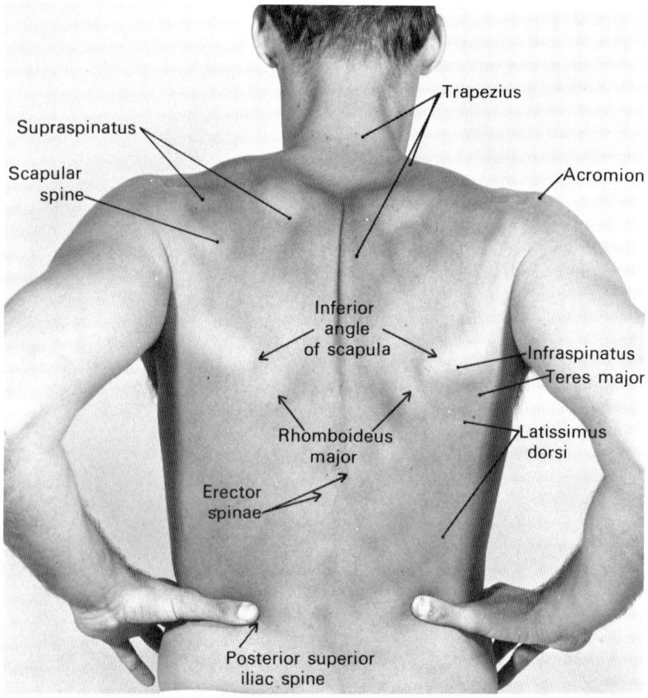

Fig. III-47. The male back.

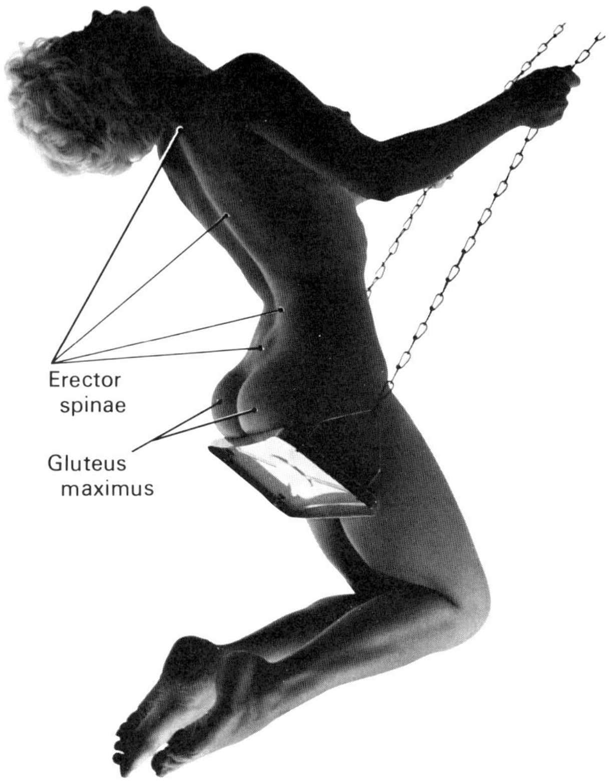

Fig. III-49. Back in opisthotonus. To illustrate Erector spinae.

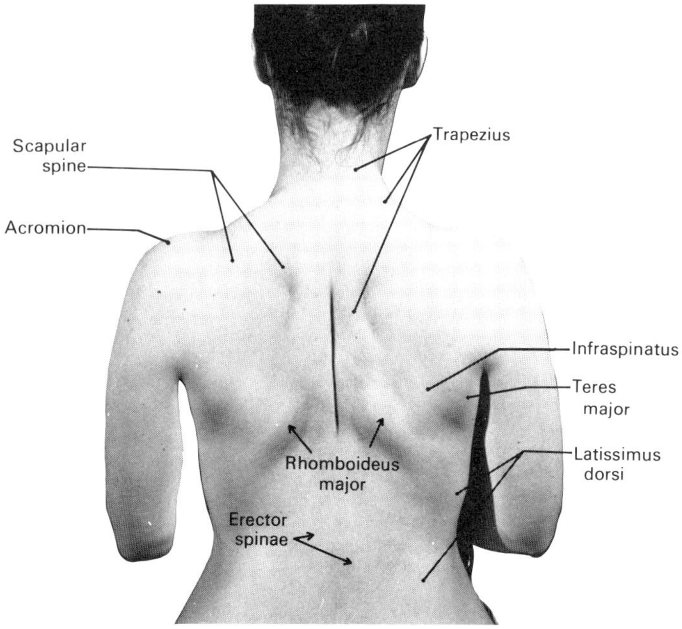

Fig. III-48. The female back.

words meaning "to be drawn backwards" and "a stretching." In certain neurological disorders the muscles are thrown in spasm in this way, and the term is used to describe this abnormal state of posture.

If you think that once you have seen the backside of one female, you have seen them all, then you haven't sat in a sidewalk cafe in Italy where girl watching is a cultivated art. Your authors, whose zeal in this regard never flags, refer you to Figures III-50 and 53 as proof that female backs can keep an interest in anatomy alive. In Figures III-50 and 52, the scapula spine, several vertebral spines, and the general outline of the sacrum can be identified. The *vertebra prominens* (VC_7) is particularly well demonstrated in Figures III-51 and 53. The young lady admiring herself in the mirror without glass (Fig. III-50) is an intriguing study in proportions. Compare her with her more mature sisters (Figs. III-51 to 53) who have much to be said in their favor, also.

The **Pectoralis Major** occupies much of the anterior rib cage below the clavicle (Fig. III-54). But like the superficial muscles of the back, it is a mover of the shoulder girdle and upper limb. However, certain females with tassels attached to "pasties" have been known to do other amazing feats with it. Just as the Latissimus Dorsi formed the posterior axillary fold

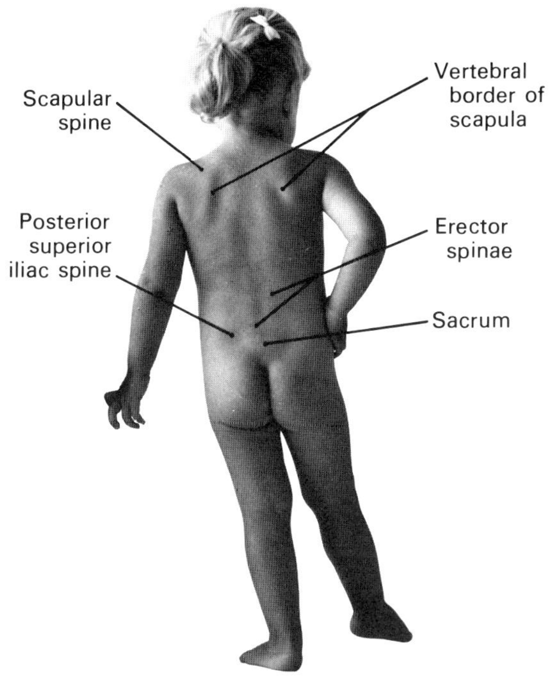

Fig. III-50. The back of the child. Compare proportions of trunk to limbs here with Figures III-51 to 53 of the mature form.

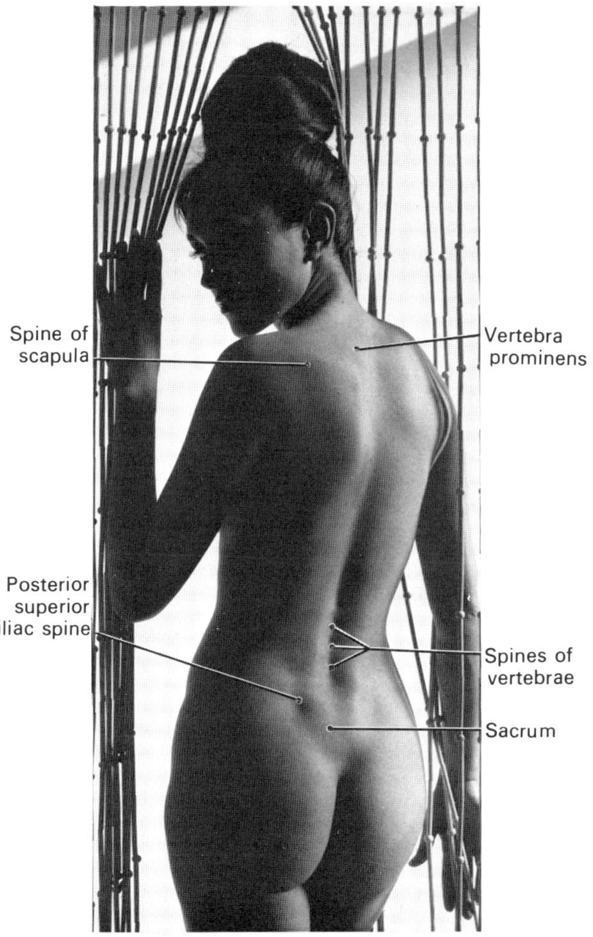

Fig. III-51. Back of adult female. Note particularly the vertebra prominens, posterior superior iliac spine, and sacrum.

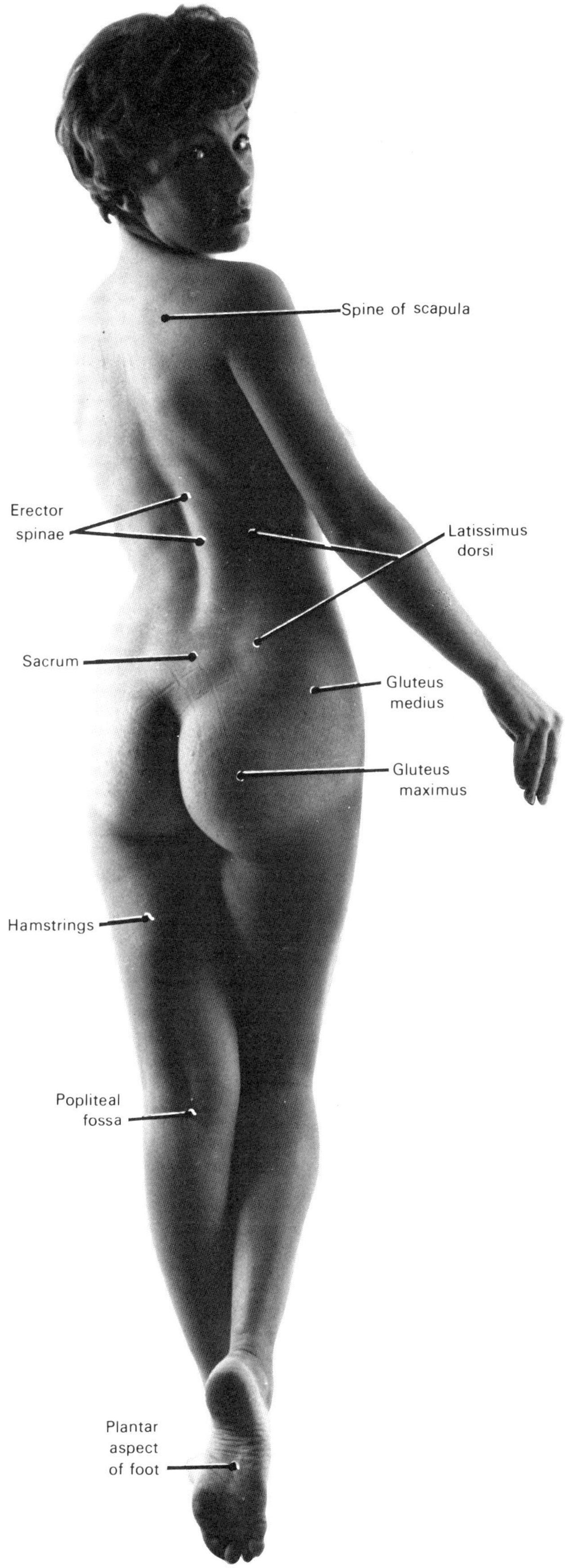

Fig. III-52. Back of adult female, highlighting sacrum and Erector spinae.

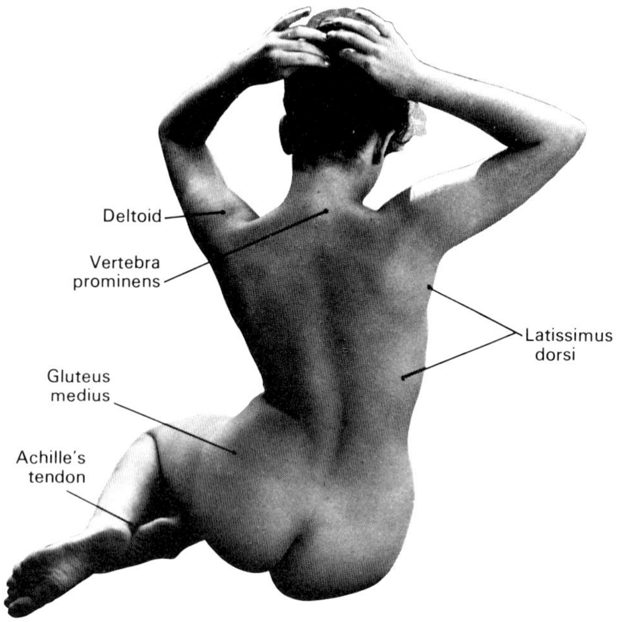

Fig. III-53. Back of adult female showing vertebra prominens and Latissimus dorsi.

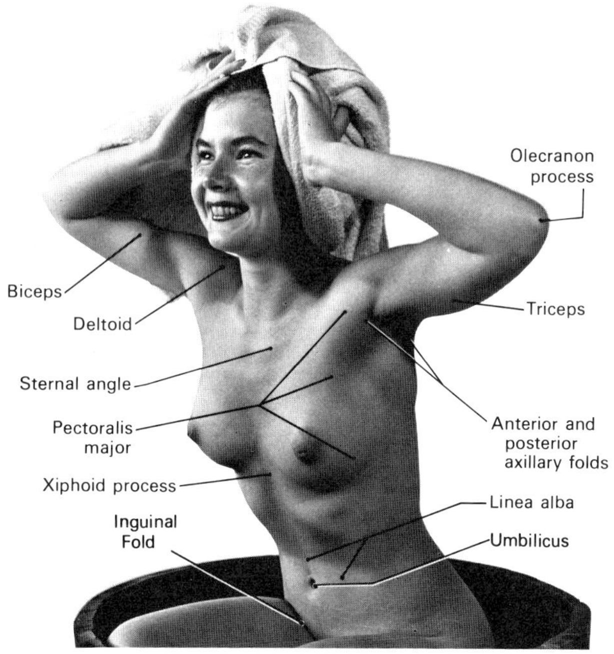

Fig. III-54. Adult female, pectoral region.

so does the Pectoralis Major form the **anterior axillary fold** (Figs. III-54 to 57). While these two muscles have other functions of their own, they both act as adductors and medial rotators of the **humerus.** The latter, despite its peculiar name, is not the so-called "funny bone."' This latter is a common term applied to the **olecranon process** of the **ulna,** seen in our "Lady of the Bath" (Fig. III-54). Since the **ulnar nerve** crosses the medial epicondyle of the humerus in the vicinity of the process, it is vulnerable to being bumped occasionally, leading to a peculiar tingling sensation in the little finger and half the ring finger.

There is no question that the anatomy of the Pectoralis Major generally looks better on the female than on the male. All that can be said in favor of the male (Fig. III-58), unless subject to gynecomastia, is that the nipple sits upon its lower margin in a stable way at about the 4th anterior costal interspace. In the female, the breast overlies the greater part of the muscle, extending from the 2nd to the 6th rib and to points south with parturition and aging. Laterally, it extends from sternum to midaxillary line and may even reach the border of the Latissimus Dorsi. As a result, it overlaps the fleshy slips of the **Serratus Anterior** and of the **External Oblique** muscle of the abdomen. These slips are seen very nicely in Figure III-58 and in the "Surfbather," Figure III-59. Glandular mammary tissue may even extend into the axilla in some instances. The burgeoning of the virginal breast is a boon to all mankind, young and old, to Playtex, and to the sweater industry.

Other features, just as important, to notice on the anterior chest wall are the costal angle and xiphoid process, the sternal angle, and the infraclavicular fossa between **Pectoralis Major** and **Deltoid** (Figs. III-54, 56–58). The lateral rib cage stands out in such bold relief that the **Intercostal muscles** can be visualized in Figure III-59.

The musculature between the ribs forms a threeplied layer. Outermost are the **External Intercostal muscles** with their fiber direction oriented obliquely downward and ventralward. Deep to them the fibers of the **Internal Intercostals** run at right angles so that their obliquity is posterior in direction. This muscle layer splits off a still deeper lamina, the **Innermost Intercostal muscles.** The intercostal vessels and nerves run between Internal and Innermost Intercostal laminae. We mention this now for it will also be the basic plan of muscular arrangement in the anterolateral abdominal wall below the costal margin. The **Recti of the abdominal wall,** on the other hand, continue the plan of a midline group of short, vertically-oriented, muscle strands which began in the neck but, for the most part, disappeared over the sternum.

Rarely, therefore, and more often in Japanese males, are there midline muscles on the anterior chest wall. When present, they are called the **Sternalis muscles.**

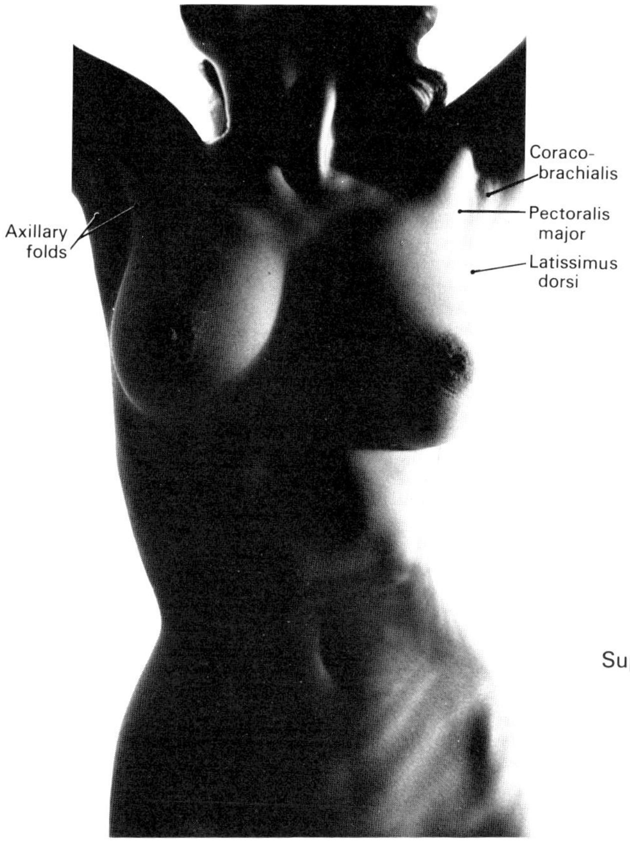

Fig. III-55. Adult female, pectoral region. Note especially the axillary folds and Coracobrachialis.

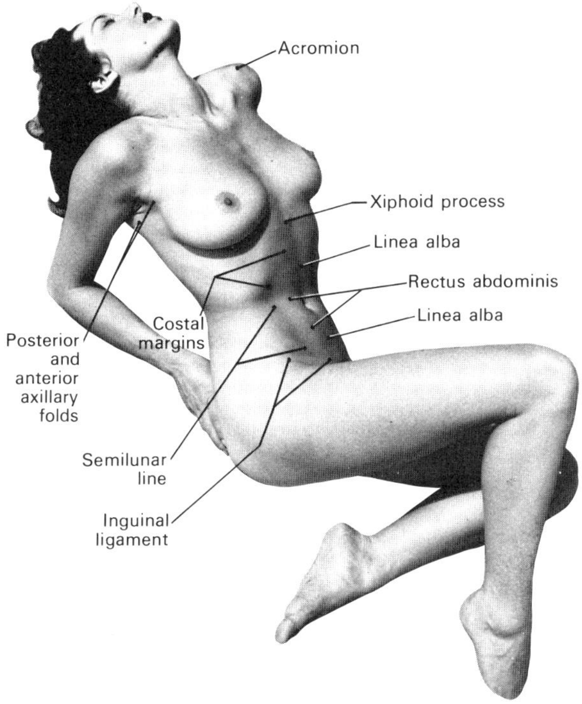

Fig. III-56. Adult female, surface anatomy of the abdomen.

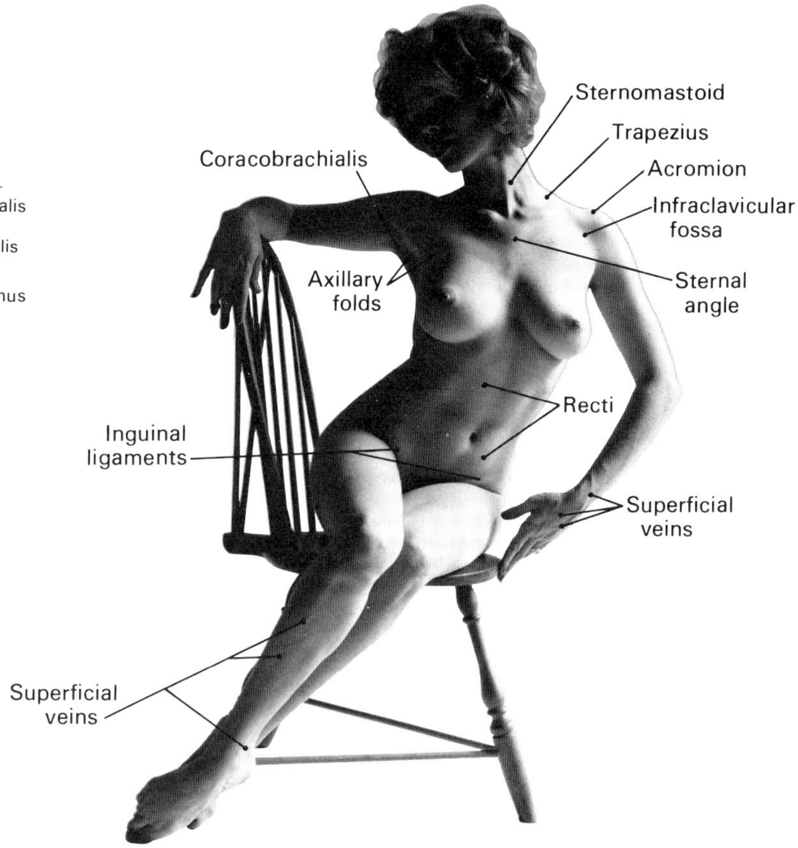

Fig. III-57. Adult female, surface landmarks in the pectoral and abdominal regions.

These lie upon the sternum, and they are representatives of a group that run in the midline from symphysis menti to symphysis pubis. In the neck from symphysis menti to hyoid bone these would be: **Geniohyoid** and the anterior belly of the **Digastric;** from hyoid bone to sternum there is the infrahyoid or "strap" muscle group: **Sternohyoid, Sternothyroid, Thyrohyoid** (a tongue-twisting group), and the **Omohyoid.** In the abdomen, the **Rectus Abdominis** runs from sternum to symphysis pubis. At first consideration, this last, long, straight muscle seems to violate the rule of a midline, segmental chain of short muscles. In actuality, the Rectus, itself, is broken into short stretches of muscle bellies separated by tendinous inscriptions (Fig. III-58).

We have not mentioned all the muscles, external or internal, that clothe the rib cage, either ventrally or dorsally. Nor have we mentioned all the muscles in the neck region. These you will meet up with in subsequent chapters on head, neck, thorax, and abdomen. But we have developed a basic plan for the thoracic wall which will carry over into the abdominal region.

SURFACE ANATOMY RELATED TO THE ABDOMEN

Boundaries: If the Diaphragm is the floor of the thoracic cage, so is it the roof of the abdominal cavity. Where the abdominal cavity becomes continuous with the pelvic cavity, there is no floor. The floor lies in a

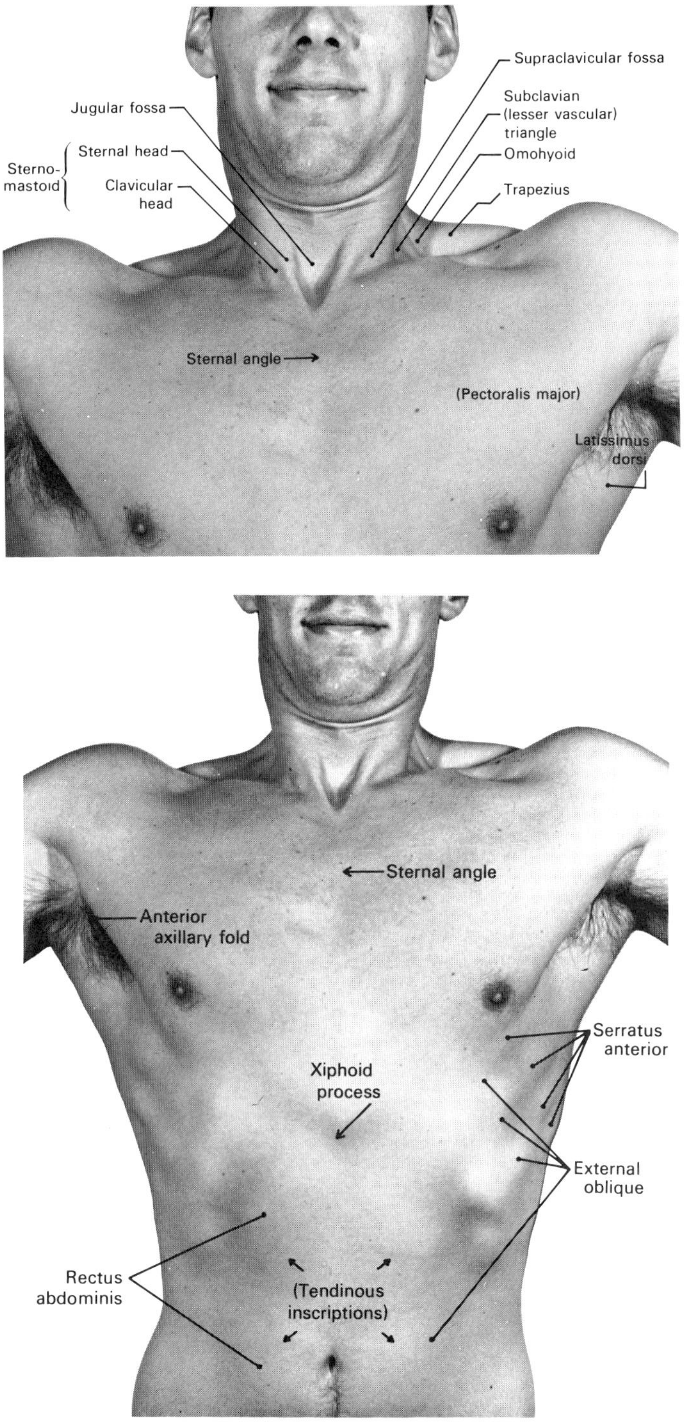

Fig. III-58. Male torso. **A.** Chest and relationships at the neck. **B.** Chest and abdomen.

Fig. III-59. Female torso, surface anatomy of the abdominal region.

different plane, as the **pelvic diaphragm,** in what is spoken of as the **lesser or true pelvis.** However, the pelvis, in reality, is part of the abdominal cavity, separated from the latter in a midsagittal plane only by an imaginary, oblique line extending from the top of the **pubic symphysis** to the **sacral promontory** (see Fig. III-40). This places the pelvic cavity not in direct line with the abdominal cavity (proper), but somewhat behind as well as below it (Fig. III-40B).

For convenience only, the abdominal cavity can be divided from above downward into three zones—**costal, umbilical,** and **hypogastric** (suprapubic) (Fig. III-40B). The first two zones are separated from each other by the **subcostal plane,** an imaginary girdle line drawn around the trunk at the lowest level of the costal margin that can be *felt from the front.* That would place the plane at the lower margin of the 10th costal cartilage. The last two zones are separated by a still lower plane, the **intertubercular,** which, in encircling the trunk, crosses through the two highest points of the **iliac crests** that you can see *from the front; i.e.,* through the **iliac tubercles.** You can feel the full extent of the upper margin of the fan-shaped ilium laterally at the belt line of your waist.

Thus delineated, this leaves the hypogastric zone in continuity with the pelvic cavity below. Certain structures are common to both abdominal and pelvic cavi-

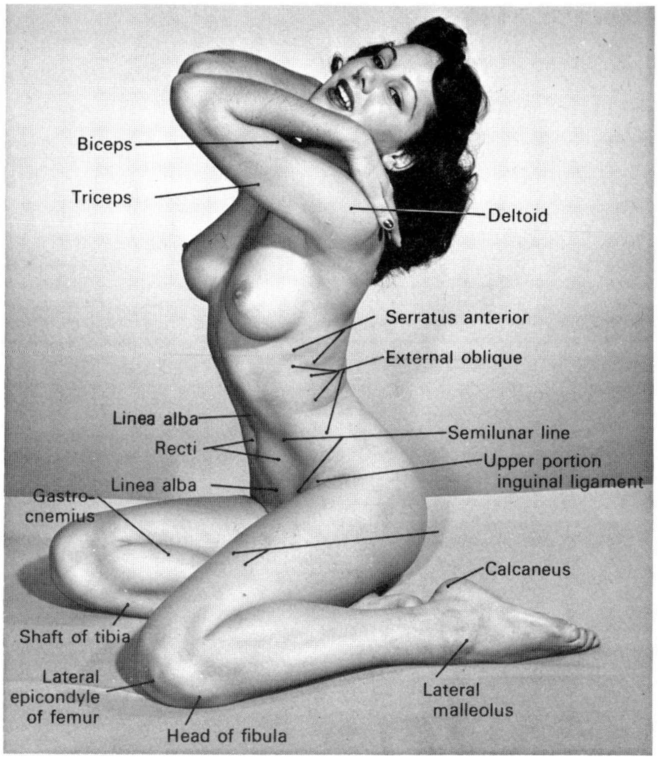

Fig. III-60. Adult female, landmarks of the abdomen.

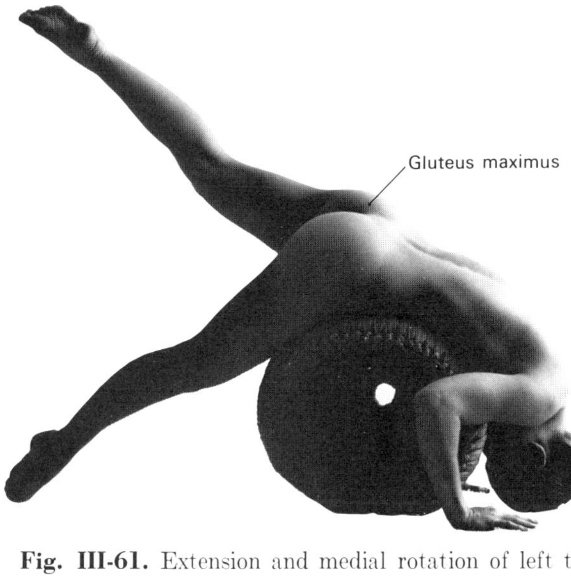

Fig. III-61. Extension and medial rotation of left thigh by Gluteus maximus.

ties; *e.g.*, some vessels and nerves that dip over the pelvic brim; the lower part of the large bowel which enters the pelvis from above to reach the anal orifice through the pelvic diaphragm; parts of the small bowel which merely trespass by hanging into the pelvic cavity; and the ureters on their way to the bladder. However, the urinary bladder and the internal reproductive organs are local residents of the pelvic cavity, while the stomach, liver, pancreas, spleen, kidneys, adrenals, and the bulk of large and small intestines inhabit the three zones we have classed as abdominal cavity. Again, like the lung in its relation with the pleural cavity, or the heart with the pericardial cavity, none of these organs really lie *within* the abdominopelvic cavity. They developed in the connective tissues of the body wall deep to the muscles and pushed inward in their growth, lifting away the thin **parietal peritoneum** that lined the empty cavity and draped it over and around themselves as an outer visceral garment, the **visceral peritoneum.** Only a serous fluid occupies the space between parietal and visceral peritoneum, the space that *is* the abdominopelvic cavity. Only in the female (*via* uterine tube, uterus, and vagina) does this cavity communicate with the exterior.

Muscles of the Posterior Abdominal Wall: We have already seen how the Latissimus Dorsi covers the lower back as a broad muscular sheet, attaching as it does to the lumbodorsal fascia and the lower four ribs. It is the most superficial muscle of the lower back, even if its main function is to move the upper limb (Figs. III-47, 48 and 53, 54). In the midline between spinous and transverse processes lie the lower portions of the Erector Spinae. The only other superficial group over the buttocks and hip are the Glutei (Figs. III-49 and 52–53). These are actually muscles of the hip joint. The Gluteus Maximus extends and rotates the thigh laterally as demonstrated in a rather intriguing fashion in Figure III-61. The Gluteus Medius and the smaller, deeper, hidden Gluteus Minimus are abductors and medial rotators of the thigh.

All the other muscles in the lower back are deep and do not give surface markings. We shall mention them briefly here to round out the picture. In the midline, under the Erector Spinae group, are all the other small short muscles that move the vertebral column—the **Mutifides, Rotatores,** and **Intertransversarii.** These, of course, are found in the thoracic and neck region, as well. Laterally, under the Latissimus, arising from the lower thoracic and upper lumbar transverse processes and inserting upon the lower four ribs, is the **Serratus Posterior Inferior.** Its counterpart in the upper thorax is the **Serratus Posterior Superior** (Fig. III-62A). These should not be confused with the Serratus Anterior, a mover of the scapula. The posterior Serrati can be regarded as subsidiary muscles of respiration that lengthen the thoracic cage. The lower ones in the posterior costal zone of the abdominal region draw the ribs downward and outward, resisting the pull of the Diaphragm; the upper ones in the thoracic region raise the upper ribs.

The innermost muscles in the lower back are both quite thick and lie in the troughs formed between bodies and transverse processes of the lumbar vertebrae. These are the **Psoas** (Major and Minor) and the **Quadratus Lumborum** (Fig. III-62B). The Psoas anchors upon the lumbar bodies and their transverse processes and passes anteriorly to emerge below the inguinal ligament and inserts upon the **lesser trochanter** of the **femur.** It is, thus, less a mover of the trunk than of the thigh. Up to this point, the Erector Spinae

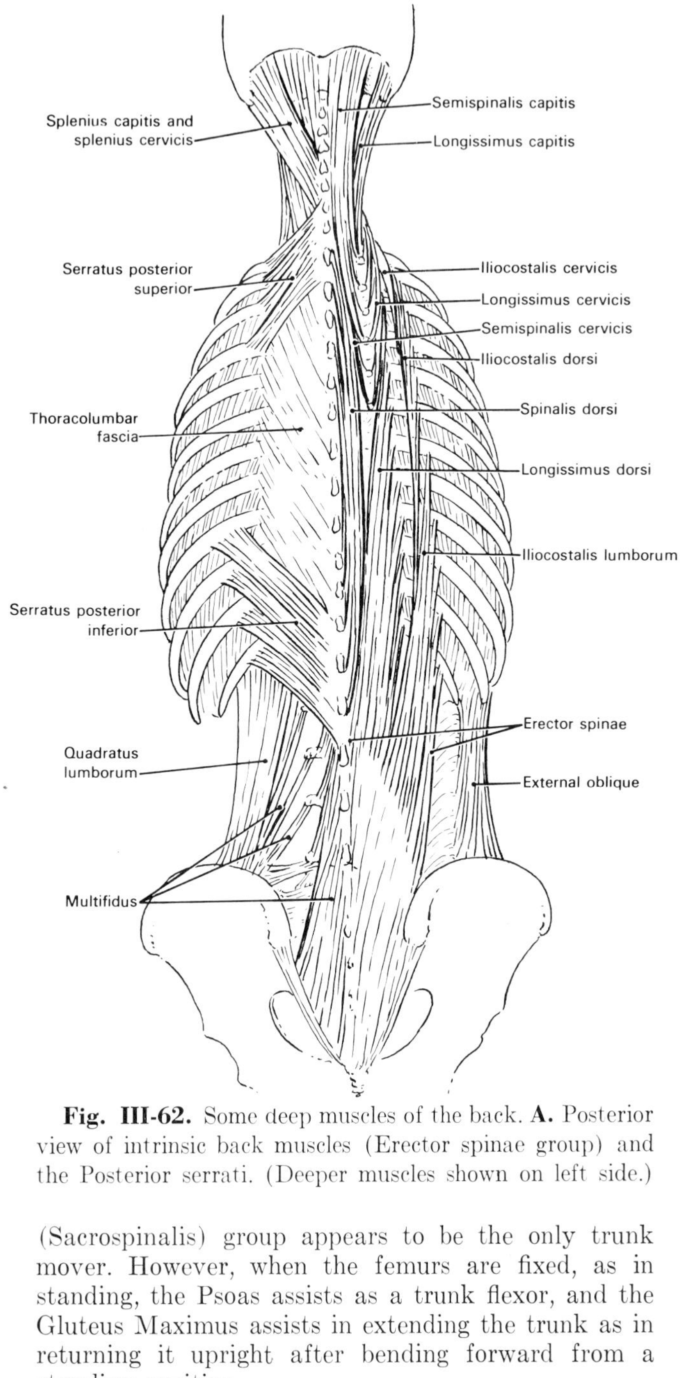

Fig. III-62. Some deep muscles of the back. **A.** Posterior view of intrinsic back muscles (Erector spinae group) and the Posterior serrati. (Deeper muscles shown on left side.)

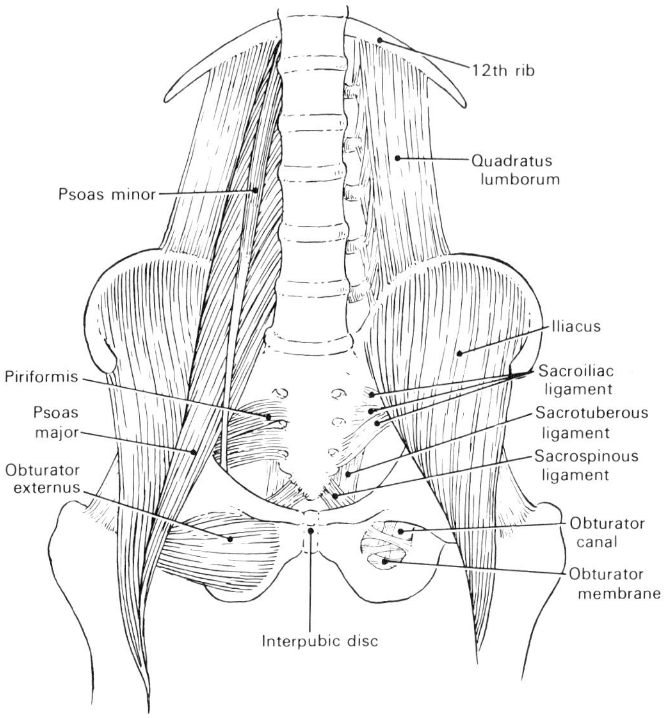

B. Anterior views, Iliopsoas and Quadratus lumborum.

(Sacrospinalis) group appears to be the only trunk mover. However, when the femurs are fixed, as in standing, the Psoas assists as a trunk flexor, and the Gluteus Maximus assists in extending the trunk as in returning it upright after bending forward from a standing position.

The major trunk flexor and chief antagonist of the Erector Spinae at the lower trunk is the Quadratus Lumborum, the thickest back muscle of all, arising from the posterior iliac crest to insert upon the four upper lumbar transverse processes. It flexes the spine at the lumbar vertebrae.

The Rectus Abdominis and the Umbilical Region: The soft outer contours of the abdomen are formed by a ventrolateral muscle group along with the midline Recti and accumulations of fat (adipose tissue) in the subcutaneous layers of the skin and the underlying fascia. A midline furrow, interrupted as it encircles the **umbilicus,** separates one Rectus Abdominis muscle from the other. This is called the **linea alba** (Figs. III-56, 57 and 59, 60), but it is not really seen as a white line until the skin is reflected in dissection (Figs. III–63 and 64). It is the midline aponeurotic junction of the various flattened tendons of the ventrolateral muscle groups from each side as these tendon sheets pass over or behind the Recti to interdigitate their fibers in the midline. As these flat, aponeurotic tendons of the **External Oblique,** the **Internal Oblique,** and the **Transversus** muscles of the abdomen come into relationship with the Recti, they form a rectus sheath for these longitudinal muscles (Fig. III-64).

As in the case of California oranges, the *umbilicus,* or navel, protrudes in ungainly fashion in some individuals or dimples in neatly in others, depending upon the former Boy Scout training of the obstetrician who tied the cord when the placenta was separated at birth. Since the fetal intestines once protruded into the umbilical cord and then withdrew into the abdomen again, this area in the belly wall is one of potential weakness, and it is an area where loops of bowel, fat, or remnants of the vitelline vessels may herniate, especially in some infants at birth, more rarely in adults, and, in the latter case, more often among Negroes than in whites. Sometimes, too, the **urachus** may remain patent, connecting bladder to umbilicus. After the 2nd month of fetal life, the urinary bladder expands into a sac whose apex tapers into a long tube, and it is this extension that is called the urachus. It, in turn, is continuous at

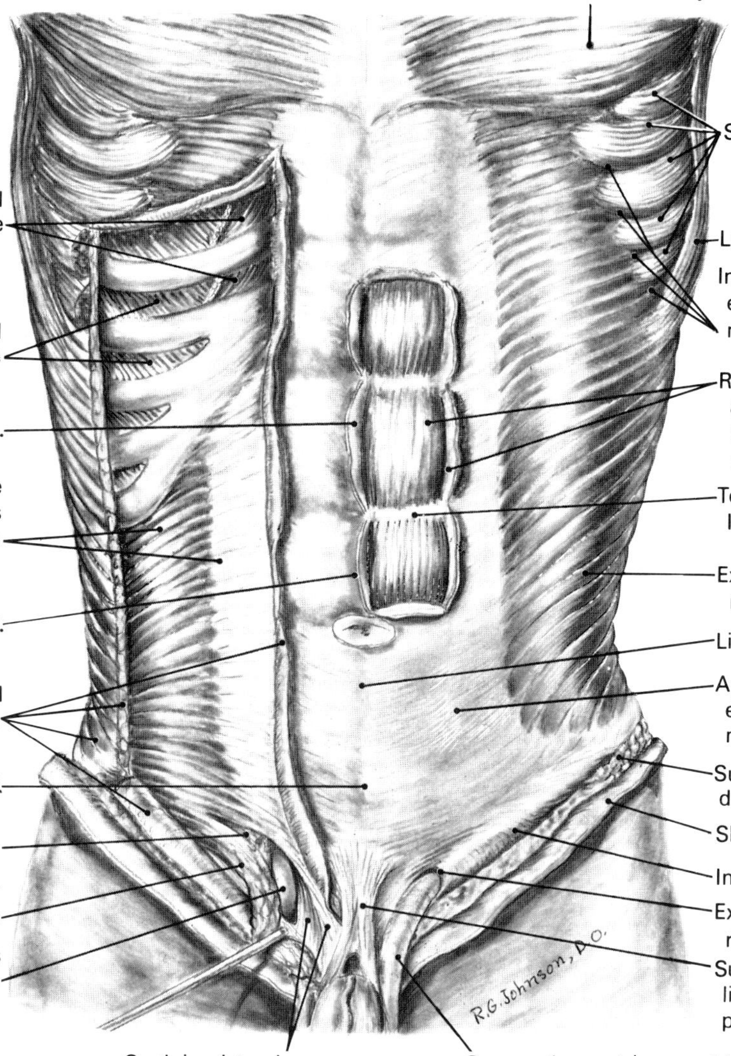

Fig. III-63. Muscles of ventrolateral abdominal wall. **A.** On left, anterior part of rectus sheath and the External oblique muscle have been removed. **B.** Transversus abdominis; most of Rectus abdominis removed. b and b′ are cross sections of ventral abdominal wall in planes shown in **B** above.

the umbilicus with the proximal remnant of the allantoic stalk, but it is believed that the allantois contributes nothing to bladder or urachus. In very rare instances, instead of degenerating into a midline peritoneal fold (**medial umbilical ligament**), the urachus persists as an open connection between bladder and umbilicus and urine leaks from the navel. But in most of us, fortunately, the navel persists as a cute but useless "belly button," not even serviceable to anchor down a shirt tail. In contemplating their own navels, your philosophic authors have found them to be pragmatic receptacles for salt while eating celery in bed once accumulated fuzz has been removed with a vacuum cleaner or a stiff-bristled toothbrush.

Each Rectus Abdominis is about 3 inches wide, parallels the midline, and extends from the xiphoid process and the 5th–7th costal cartilage to the crest of the pubis. In the well-developed male (Fig. III-58), when the muscle is tensed, each short stretch of muscle between the *tendinous inscriptions* bulges. When the rectus sheath is opened in dissection, such tendinous intersections can be seen (Figs. III-63 and 64). These transverse tendon bands which attach to the anterior rectus sheath are not too evident on the posterior aspect of the muscle. Each of them is pierced by twigs of the **superior epigastric vessels** which supply the muscle (Fig. III-64). In the female, the Rectus Abdominis is more smoothly contoured in contraction than in the male (Figs. III-56, 57 and 59, 60). Laterally in some individuals, the Rectus Abdominis (or its sheath) forms a gentle curved line, the **semilunar line** (Figs. III-56 to 59). Do not confuse this with the **semicircu-**

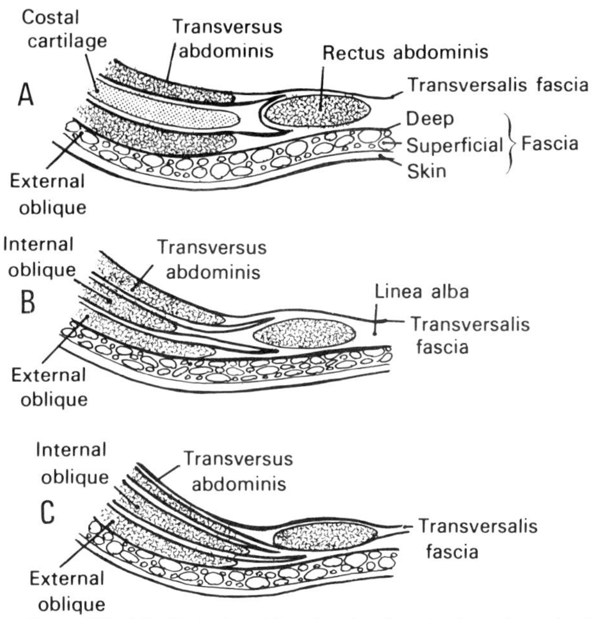

Fig. III-64. Relationships in the inguinal region. **A.** External inguinal ring and spermatic cord. **B.** Deep structures related to inguinal canal.

lar or arcuate line (of Douglas) which is not a surface marking but lies behind the Rectus muscle in the posterior wall of its sheath and below the level of the umbilicus. The **inferior epigastric vessels** cross the arcuate line as they enter the sheath from behind (Fig. III-64).

The Ventrolateral Abdominal Wall: The ventrolateral muscles form a three-plied corset for the abdominal wall. The **External Oblique** is the only one of the three that can be seen from the surface, and then only by means of its fleshy interdigitations where it takes origin from the ribs laterally. These are seen only in muscular or quite skinny individuals, and these fleshy "fingers" interlock with those of the **Serratus Anterior** (Figs. III-58 to 60). From these slips of origin, fiber direction is downward and medialward as it was in the External Intercostal muscles above. The lower, wide, fleshy belly of the muscle stops well short of the semilunar line. A vast aponeurotic sheet continues to anchor the muscle to the midline where it interdigitates its tendinous fibers with those of the aponeurosis of the opposite side, becoming partly responsible for the linea alba (Fig. III-63). The aponeurosis of the External Oblique anchors inferiorly to the **anterior superior iliac spine** and to the **pubic tubercle.** Its thick, rolled-under, free, edge between these points of attachment is called the **inguinal (Poupart's) ligament.** We can see here how the flat tendinous sheet that anchors one muscle to another at the linea alba has been designated the **aponeurosis,** while its inferior free edge, attaching bone to bone, has been called a **ligament** (Fig. III-63). Were these flat sheets ropelike, instead, and were they to anchor fleshy muscle to bone, they would have been called **tendons.** From the histological point of view *most* ligaments, and *all* aponeuroses and tendons are structurally similar in that they consist of **dense white fibrous (collagenous) connective tissue.**

Bony Points to Palpate:

1. The **tip of the 9th costal cartilage** lies about half way along the costal margin.

2. Feel for **the point of the 12th rib** where it just emerges along the lateral border of the Sacrospinalis (Erector Spinae). It may be much shorter and not reach this far laterad in some individuals, and it may even be absent in others. Those in whom it is absent are not necessarily direct descendants of Adam, not even if they sport a prominent "Adam's apple."

3. The **anterior superior spine** of the ilium at the anterior crest is quite prominent in some individuals (Fig. III-65).

4. The **tubercle of the ilium** lies behind the anterior spine by some 2 inches along the iliac crest. It marks the highest point of the crest as viewed on the anterior aspect of the body, but it is not the highest point of the iliac crest.

5. The **highest part of the iliac crest** can be seen from the dorsal aspect (Figs. III-65 and 66), and a line drawn through both points of the crest constitutes the **supracristal plane.**

6. The **posterior superior iliac spines** are the easiest prominences to detect in the lower back. They lie $1\frac{1}{2}$ inches from the midline and tend to dimple the skin (Figs. III-65 and 66).

7. The two **pubic rami** meet in the midline in a **symphysis** with a fibrocartilagenous disk, held in place by ligaments anteriorly, posteriorly, superiorly, and inferiorly. One inch from each side of the symphysis, along the pubic crest can be felt the **pubic tubercles.** They are hard to palpate on the female but can be felt in the male by invaginating the scrotum with a finger.

8. The **pubic arch** below the symphysis can again be felt by invaginating the scrotum of the male. In the female it can be felt below the **mons pubis** in front of the **clitoris.**

Planes and Lines of the Abdomen: Such a variety of lines and planes have been drawn upon the abdomen that it appears as though surgeons play tick-tack-toe while waiting for the anesthetist to bring the patient under. Figure III-67A and B shows the commonest of these.

The abdomen can be "drawn" and quartered by erecting a midsagittal and a supracristal plane. As you might suspect, the quadrants are named upper and lower right (UR and LR) and upper and lower left (UL and LL). It is still popular to lay out nine surface regions by erecting a pair of **lateral (paracentral) lines** and drawing the subcostal and intertubercular planes (Fig. III-67B). The areas have been named **right and left hyochondriac;** the **right and left lateral** (formerly lumbar); the **right and left inguinal** (formerly iliac); the **epigastric, umbilical,**

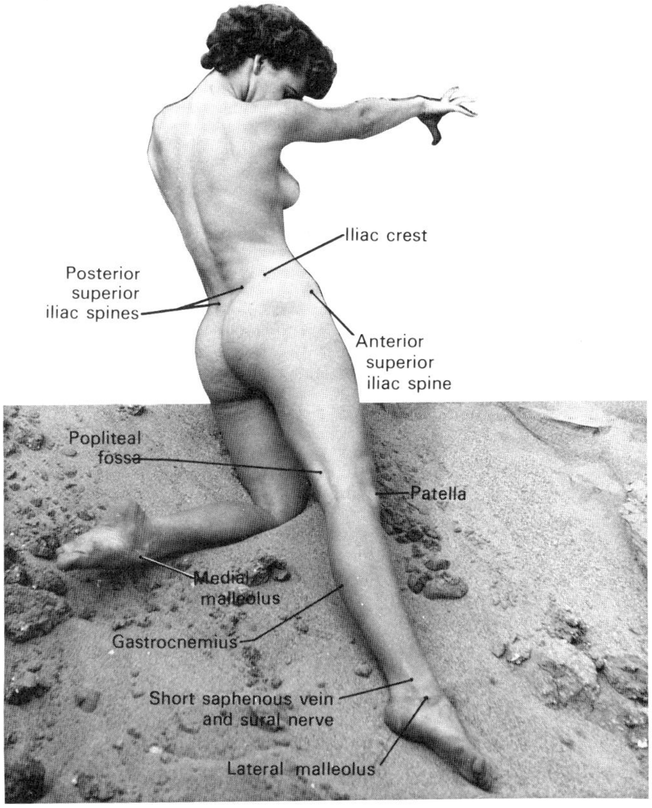

Fig. III-65. Adult female, surface anatomy of pelvic girdle and lower limb.

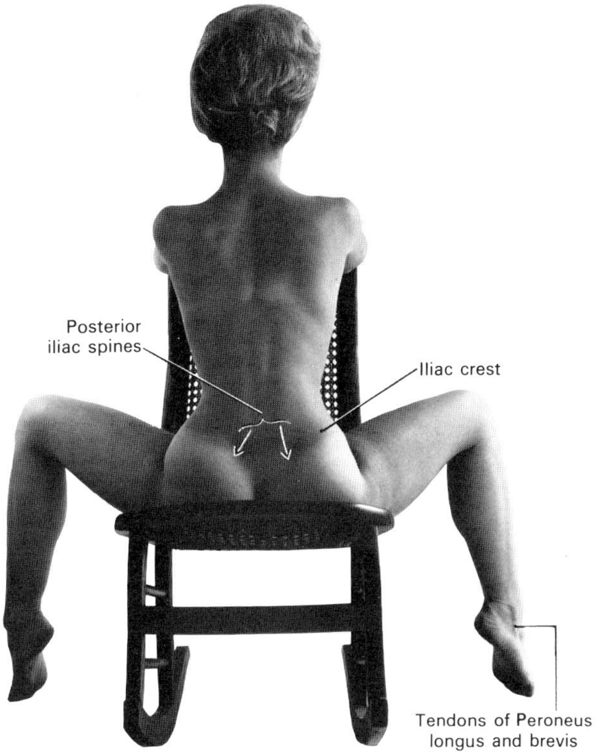

Fig. III-66. Adult female, iliac crest and posterior iliac spines.

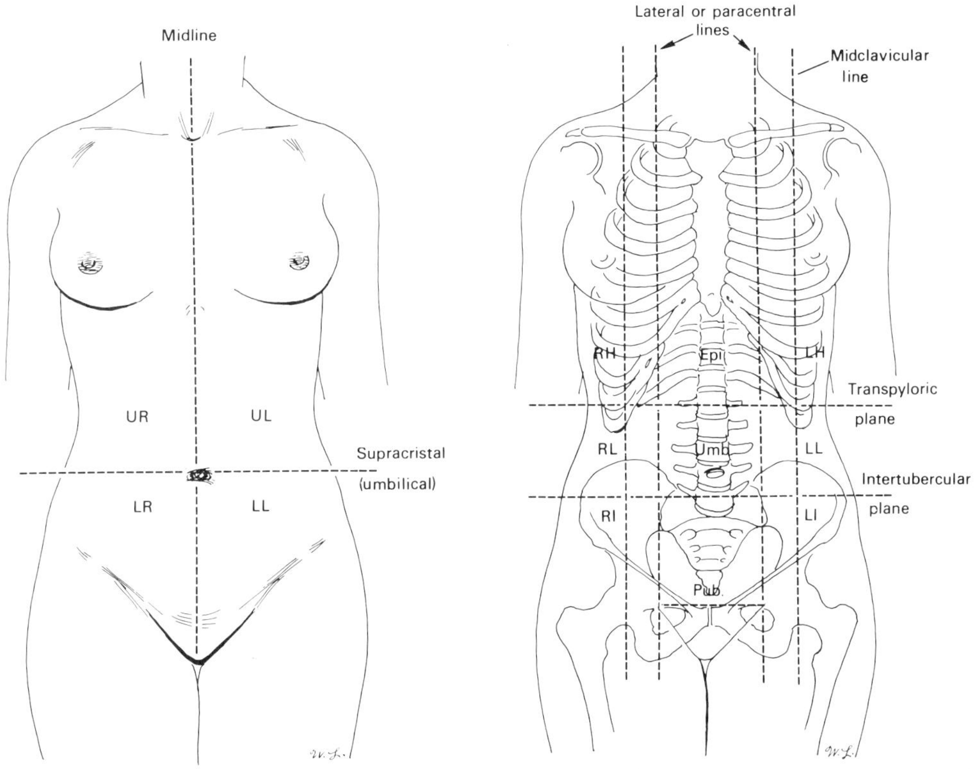

Fig. III-67. Planes, lines, and zones of the abdomen. **A.** The abdomen in quadrants: UR, upper right quadrant; UL, upper left quadrant; LR, lower right quadrant; LL, lower left quadrant. **B.** Conventional zones of the abdomen: RH, right hypochondriac zone; LH, left hypochondriac zone; RL, right lateral (formerly lumbar) zone; LL, left lateral zone; RI, right inguinal (formerly iliac) zone; LI, left inguinal zone; Epi, epigastric region; Umb., umbilical; Pub., pubic (formerly hypogastric) region.

and pubic (formerly hypogastric). The lateral lines are 2½ inches from the midline and cut the tip of the 9th costal cartilage about where it is touched by the upper extent of the semilunar line. It also passes through the *femoral point* of the thigh midway between anterior superior spine of ilium and symphysis pubis, missing the midinguinal point which lies ½ inch laterad. There are certain drawbacks to this method of subdivision. The subcostal plane tends to ride high in the elderly and in the short, stout (pyknic) subject so that it may lie anywhere from 1½ to 4 inches above the intertubercular line in different individuals. Le Clerc felt it was better to measure from the xiphisternal point to symphysis pubis and then to draw three horizontal lines equidistant apart, one through the joint and two below it.

The **transpyloric plane** has also been a favorite horizontal plane with many. It is a plane drawn midway between symphysis pubic and the *upper margin* of the manubrium of the sternum. The **pylorus** of **the stomach** is generally found at this level *in the supine cadaver*. In the living, it may lie 1 to 3 inches below this plane. *Near* this plane the following *approximate:*

1. The 12th rib at the lateral margin of the Sacrospinalis muscle.
2. Tip of the 9th costal cartilage at the costal margin.
3. Midpoint of the costal margin.
4. Spine of VL_1.
5. Upper extent of the semilunar line.
6. Pyloric junction of the stomach; duodenojejunal flexure of the small intestine; neck of the pancreas and hila of the kidneys (points of entrance and emergence of the renal artery, vein, and ureter).

Vertebral Levels:

1. VT_9, the level of the xiphoid
2. VL_1, the transpyloric plane
3. VL_3, the subcostal plane
4. VL_4, the supracristal plane
5. VL_5, the intertubercular plane

Location of Certain Organs:

1. The **liver** extends from the tip of the right 10th costal cartilage across to just below the left nipple as a lower border. Thus, it occupies the right and left hypochondriac regions as well as the epigastrium. Its **gall bladder** will lie at the tip of the 9th costal cartilage where the semilunar line ends superiorly (or about at the middle of the right costal margin).
2. Since the transpyloric plane cuts the neck of the **pancreas** at about the midsagittal plane, this gland will be found lying on the stomach bed (*i.e.*, behind the stomach) with its head nestled in the arms of the duodenum to the right of the midline.
3. The **spleen** lies against the 9th to 11th ribs posteriorly on the left side.
4. The **kidneys** have their hila on the transpyloric plane about 4 fingerbreadths from the midline. Posteriorly, the upper pole of the left kidney lies deep to the 12th rib; the upper pole of the right one, about 1 inch lower.
5. The aorta terminates inferiorly just left of the midline at the supracristal plane.

SURFACE ANATOMY OF THE UPPER LIMB

Parts and Regions: Parts constituting the upper limb are:

1. The **acromion,** or tip of the shoulder (Fig. III-68A) along with the **clavicle** and the **spine** of the scapula with which it merges insensibly, forms a continuous bony arch visible over most of its length. The clavicle is seen as slightly S-shaped, convex anteriorly in its medial part and concave anteriorly in its lateral part. It forms a rounded elevation at the sternum lateral to the **suprasternal fossa (notch)** and it may rise higher than the acromion laterally, forming another elevation at the acromioclavicular joint (Fig. III-68A).
2. The **axilla,** or arm pit, is the space we have already bounded by its anterior fold (Pectoralis Major) and its posterior fold (Latissimus Dorsi and the tendon of the Teres Major) (Figs. III-54 to 57). Its medial wall is the Serratus Anterior and its lateral wall, the head and neck of the humerus. The head can be felt indistinctly by upward pressure in the armpit, especially if the limb is rotated to and fro as the arm hangs by the side. Crossing the roof of the axilla to enter the medial aspect of the arm is a muscular elevation often readily visible when the arm is abducted to the horizontal. The elevation is due to the **Coracobrachialis muscle** and the **short head of the Biceps** (Figs. III-57 and 59). In the groove posterior to this elevation runs the **neurovascular bundle** bringing vessels and nerves into the arm. It contains the **axillary artery** (the continuation of the **subclavian** which will be named **brachial** in the upper arm) and its companion vein (which becomes **basilic** in the arm); and the trunks and cords of the **brachial plexus.** These latter robust nerve bundles, arising from cord levels C_5–T_1, will provide the major motor and sensory supply to the arm as the **musculocutaneous, axillary, median, radial, ulnar,** and the **medial cutaneous nerves of the arm and forearm.** The plexus gives twigs, as well, to the pectoral muscles, the shoulder muscles (except the Trapezius and Omohyoid), the Latissimus Dorsi, the Serratus Anterior, and the Subclavius which is a small muscle deep to the clavicle. The whole neurovascular bundle is enveloped in the stout connective tissue of the **axillary sheath.**
3. **The brachium** is the upper arm containing the shaft of the humerus and the muscles that clothe it. The rounded contour of the **Deltoid** is readily seen and even its point of insertion upon the lateral aspect of the humerus can be appreciated in the muscular

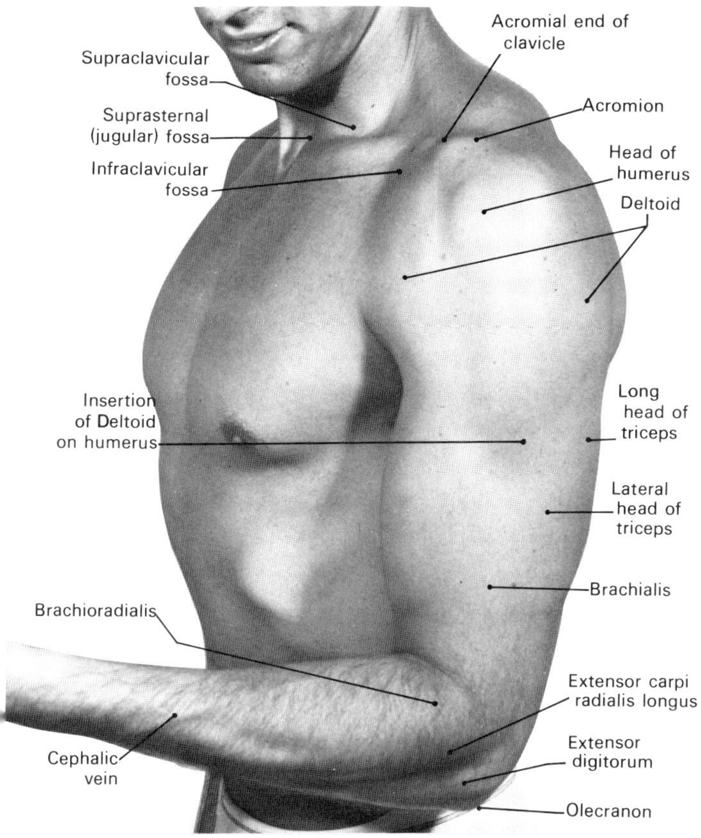

Fig. III-68. Surface anatomy of the upper limb. **A.** Lateral aspect of the left arm and forearm.

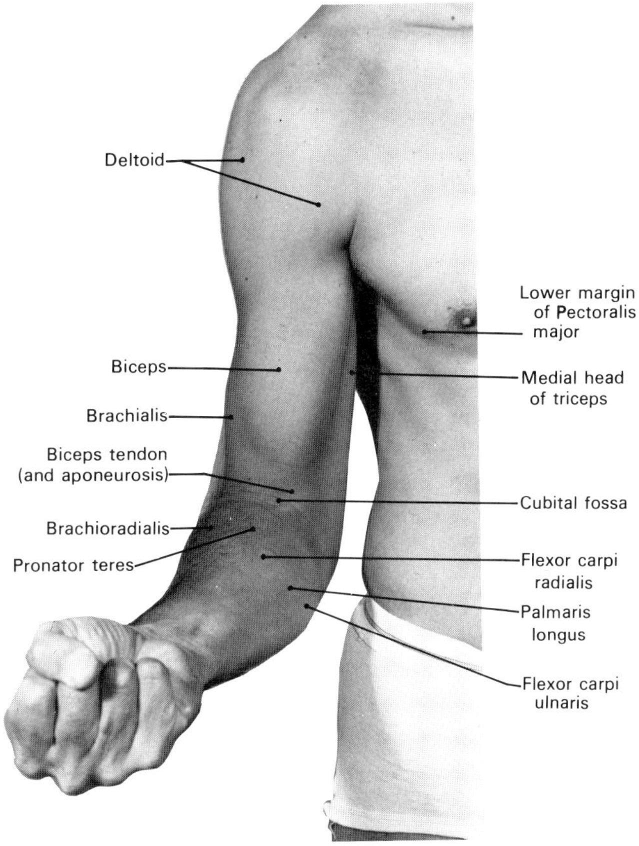

Fig. III-68B. Anterior aspect of the right arm and forearm.

male (Fig. III-68A). It is the *chief abductor* of the arm (assisted in part by the Supraspinatus). Its anterior part is also the *chief flexor* of the upper arm. Its innervation is by the **axillary nerve.** (The only other muscle this nerve supplies is the small Teres Minor, an **external rotator.**)

The **Biceps Brachii** (so named because it arises by two heads, a short and a long) accounts for the familiar bulge you raise on the ventral aspect of the humerus when someone says "Let's feel your muscle." It is commonly thought that this muscle (Figs. III-68 and 69) is chiefly responsible for "bending an elbow," as in raising a beer mug, but this is erroneous. It is a weak flexor of the forearm and flexes best only when the forearm is supinated. It has a rounded tendon that crosses the elbow joint to attach to the **tuberosity of the radius,** but it has a much stouter **aponeurotic band** passing medially to reach the **ulna.** Hence, its *chief action* is as a supinator, but only when the forearm is already flexed.

The *chief flexor* of the forearm is the **Brachialis** (Figs. III-68 and 69) muscle upon which the Biceps lies. Hence, this muscle is not as readily seen as a surface feature. The only other muscle on the ventral aspect of the humerus is the **Coracobrachialis** which was responsible for the elevation noted in the axilla and which cannot be seen otherwise from the surface. It serves as an *accessory adductor*, the chief adductors being the Pectoralis Major, Latissimus Dorsi, and Teres Major.

The nerve supply of the ventral group of the arm (Coracobrachialis, Biceps, and Brachialis) is the *musculocutaneous.* This nerve also supples the *skin of* **the forearm** on its lateral (radial) aspect from the elbow to the base of the thumb, both dorsally and ventrally. The posterior aspect of the humerus is clothed by the large **Triceps Brachii** with its **long, medial,** and **lateral heads** (Figs. III-68 and 69). Its nerve supply is the **radial nerve.** The triceps is the *major extensor* of the elbow joint, the medial head being most active in this respect. The long head serves also as an accessory extensor and adductor of the arm.

Other surface features to note on the brachium are

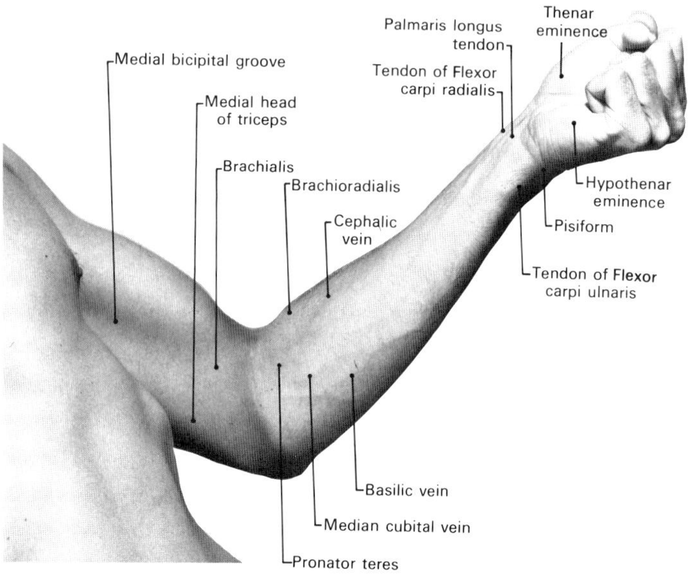

Fig. III-69. Forearm of the male.

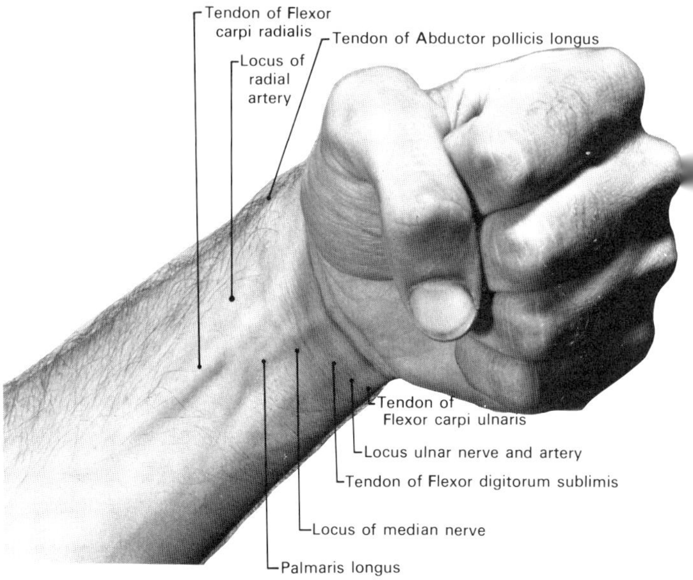

Fig. III-70. Male wrist.

the rounded eminence produced by the **greater tubercle** on the head of the humerus in the deltoid region and the shaft of the humerus which can be felt close to the surface at the point of insertion of the Deltoid laterally and along the medial bicipital groove (Figs. III-68 and 69). The **medial bicipital groove** is apt to be more prominent than the lateral bicipital groove. The pulse of the **brachial artery** (Fig. III-78) is sometimes even visible at the distal end of the groove and certainly can be felt here by compressing it against the shaft of the humerus. The **basilic vein,** a superficial vein, is commonly visible at the distal end of the groove (Fig. III-79). The **lateral bicipital groove** lies slightly posterior to the course of another superficial vein, the **cephalic,** which is also commonly visible (Fig. III-79). The cephalic vein runs up the lateral side of the limb from the palmar venous arches and lies in the infraclavicular fossa between Deltoid and Pectoralis Major where it makes its way to enter the axillary vein by passing deeply below the clavicle. The basilic vein does not course as far along the medial aspect of the arm but joins the brachial vein at the border of the axilla after having pierced deeply about midway along the medial bicipital groove.

4. At **the cubitus, or elbow,** three bony landmarks can be easily felt: the medial and lateral **epicondyles of the humerus;** and the **olecranon process** of the **ulna.** In the hollow produced posterolaterally when the elbow is extended, feel the **head of the radius** distal to the lateral epicondyle. Supinate and pronate the arm to feel it rotate. Note that you can follow the posterior border of the ulna from the olecranon process along the entire length of the bone. It is completely subcutaneous and no named vessel or nerve crosses it. The ulnar nerve does cross posteriorly in the groove behind the medial epicondyle and can be set atingling when the olecranon is bumped sharply.

Anteriorly is the **cubital fossa,** a depression bounded medially by the elevation of the **Pronator Teres** and the *superficial flexors of the forearm* that arise from the medial epicondyle. The lateral boundary of the depression is the **Brachioradialis muscle** (Fig. III-69). The tendon of the Biceps and its aponeurosis can be felt crossing the fossa. In the fossa superficially is the **median cubital vein,** a cross connection between cephalic and basilic veins (Fig. III-79). It is often visible when all other superficial veins are collapsed or buried in fat. It is convenient for the drawing of a sample of venous blood, but not a good site for the injection of irritating drugs because of the danger of injecting into the nearby brachial artery or an ulnar artery that may branch high in this region.

5. **The antebrachium (antibrachium) or forearm** (Figs. III-68 to 70) contains the radius on the thumb side and the ulna on the little finger side as its skeletal components. We have already palpated the *head* of the radius proximally just below the lateral epicondyle. The *head of the ulna* is at its distal, not proximal, end, and it can be seen as a rounded prominence at the wrist when the hand is pronated. Its **styloid process** can be palpated just distal to the head. The **styloid process** of the **radius** can be palpated anterolaterally at the wrist with the hand in supination, and it extends further distally than that of the ulna. Pronate the hand once more and feel for the **dorsal tubercle** (of Lister) on the radius approximately midway between its styloid process and the rounded head of the ulna.

As in the upper arm, the posterior aspect is the territory of the *extensor muscles* (extensors of the wrist and hand), and as in the upper arm (Triceps) these extensors are supplied by the radial nerve. Indeed, the radial nerve, emanating from the **posterior cord** of the **brachial plexus,** supplies all the *extensor-supinator muscles* on the posterior (and posterolateral) aspect of the upper limb. (Recall that the Biceps is a supinator, also, but attaches to the medial aspect of the limb and has its own nerve, the musculocutaneous, which arises from the lateral cord of the brachial plexus. Another exception to this general rule is the nerve supply to the Brachioradialis. The nerve supply to this muscle is the radial, but muscle action is, chiefly, forearm flexion, not extension.) All nerves arising from the lateral and medial cords of the plexus which innervate muscles of the upper limb supply the anterior muscles of the limb, the *flexor-pronator groups, in particular*. These nerves are the musculocutaneous for muscles of the brachium and the median and ulnar nerves for the antebrachium and hand. The median nerve supplies the bulk of the muscles in the forearm but few in the hand; the ulnar, only 1½ muscles in the forearm and the bulk of the muscles in the hand. This, then, is the basic plan of the distribution of muscles in the upper limb and their motor nerve supply. This was why the brachial plexus was organized: (1) to distribute to the anterior chest wall (medial and lateral pectoral nerves); (2) to the lateral chest wall (long thoracic nerve); (3) to the anterior aspect of the limb (musculocutaneous proximally; and, distally, the median nerve to the anterolateral aspect; the ulnar to the anteromedial aspect); (4) the posterior muscles of the shoulder (nerves to the Deltoid and Teres Minor, Latissimus and Teres Major, Rhomboids, Supra- and Infraspinatus, and Subscapularis); and, (5) to the posterolateral aspect of the limb (radial nerve).

6. The skeleton of **the carpus, or wrist** contains eight carpal bones in two rows of four each. Unfortunately, some of them have been given new names so that a mildly pornographic jingle formerly used by medical students as a memonic device to recall them no longer applies. Undoubtedly, new versions are making the rounds since medical students are always highly inventive in this regard. At any rate, from radial to ulnar side these are now named: **scaphoid, lunate, triquetral, pisiform** for the proximal row, and **trapezium, trapezoid, capitate,** and **hammate** for the distal row. If you will just remember the three that can be felt, that will be sufficient as a first exposure to the carpal bones. The *scaphoid* (navicular) has a **tuberosity** that can be felt by pressing at the base of the **thenar eminence** (Figs. III-70 and 74A) at the radial side of the distal crease of the wrist. The *pisiform* can be felt at the ulnar end of this same skin crease. The *hook of the hammate* can be located by deep pressure distal to the pisiform in the **hypothenar eminence** more toward the center of the palm. No muscles have their origins or insertions on the dorsal aspect of these bones. On the ventral or palmar aspect, three muscles for the little finger and four for the thumb take origin from them and one muscle in the forearm (flexor carpi ulnaris) inserts on the pisiform. Otherwise, the bones are crossed ventrally and dorsally by the **long flexor and extensor tendons** of the forearm muscles. The palmar aspect of the carpal bones forms a concave arch bridged over by a stout connective tissue band or ligament, the **flexor retinaculum** (Fig. III-71A) passing from scaphoid laterally to pisiform and hook of the hammate medially. The flexor tendons pass below the retinaculum in the "carpal tunnel" where they course through three synovial sheaths, actually modified **bursae.** There is a common sheath ("ulnar bursa") for the tendons of the Flexors Digitorum Superficialis and Profundus; a radial sheath or "bursa" for the Flexor Pollicis Longus; and a separate third sheath for the Flexor Carpi Radialis.

There is an **extensor retinaculum** on the dorsum of the wrist attaching to various points on the radius, to the styloid process of the ulna and to the hammate. Six synovial compartments house the extensor tendons as they pass under the extensor retinaculum to cross the carpal bones (Fig. III-71B).

7. **The manus, or hand** (*cf.* with the term "manual"), consists of five metacarpals which can be palpated dorsally. They look much alike and are named by Roman numerals, counting from the thumb side. The **styloid process** on the radial side of the **base of metacarpal III** can be felt about 1½ inches from the radial tubercle (of Lister). The **heads** of these bones form the knuckles when a fist is made. The digits include the fingers and thumb, and they have both names and numbers: The thumb or pollex is digit I; index finger, II; middle or long finger, III; ring finger, IV; and little finger (digitus minimus), V. Each digit consists of **phalanges,** two for the thumb, a prominal and distal one; three for each finger, proximal, middle, and distal.

If the dorsum of the hand is examined with the wrist extended (dorsiflexed) and with the fingers abducted, the **extensor tendons** generally stand out boldly in most hands, but, perhaps, not as well in the female as in the male (Fig. III-72). They can be traced to the base of the phalanges but no farther. Beyond the knuckles they are bound down tightly on the fingers and can be neither seen nor felt.

If the thumb is extended in the "thumbs-up" position (Fig. III-73) rather than as it is in Figure III-72A, the so-called "anatomical snuffbox" is displayed as a matter of historical interest. It was used in former days by elegant gentleman as a depository for a bit of snuff from their ornamental snuff boxes. The wrist was then lifted to the nose and the individual engaged in a hearty sneeze that cleared all of the nasal sinuses more effectively than Dristan. Whereupon the nose was wiped gently upon his freshly-laundered lace wristlets. It is of little consequence that we no longer apply this little bit of anatomy to utilitarian purposes. What is important is that the tendons of the long abductor and the short extensor of the thumb lie lateral to the depression, while the long extensor of the thumb lies

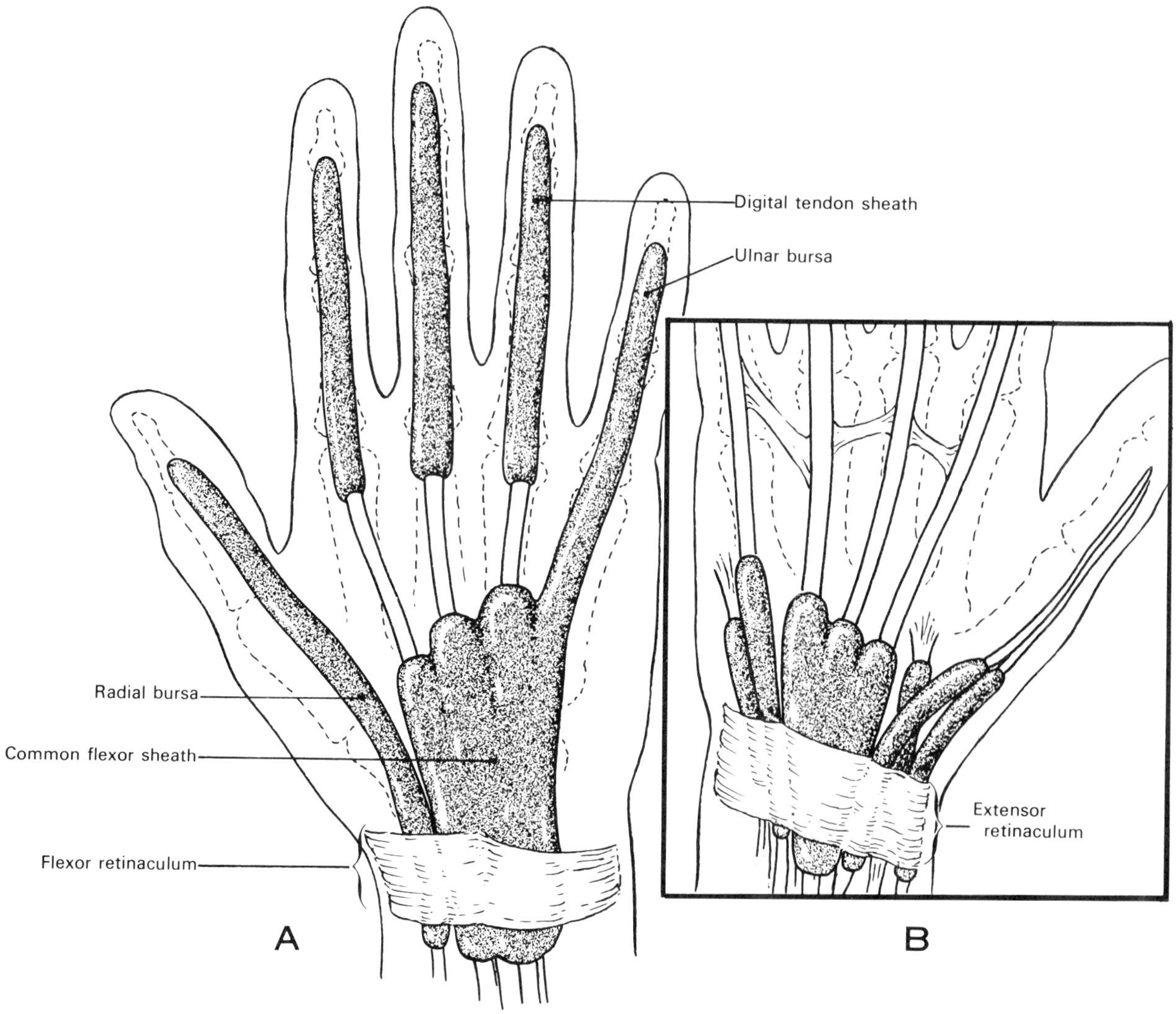

Fig. III-71. Retinaculi and tendon sheaths. **A.** Flexor retinaculum. **B.** Extensor retinaculum.

medial to it (remember the "anatomical position"). More important, the **radial pulse** can be felt in the depression of the "anatomical snuffbox." Some of the other **extensor tendons** are named for you in Figure III-72A so that you can read the back of your own hand. The positions of the Dorsal Interossei, the finger-spreading muscles, are also shown.

The **flexor tendons** on the anterior aspect of the wrist are not always as easily delineated as the extensor tendons on the dorsum of the hand. But, if you will tighten the forearm muscles as you flex your wrist, the **Palmaris Longus** tendon generally will stand out sharply about dead center and in direct line with the midpalmar skin crease of the hand (Figs. III-70 and 74A). There is always the possibility that this tendon will be absent in some individuals, in which case, the *median nerve* will lie superficially in its place. As a rule, the nerve lies deep to this tendon, or partly under its ulnar border near the distal skin crease of the wrist.

To the radial side of the Palmaris tendon lies the **tendon** of the **Flexor Carpi Radialis,** and this, too, is generally prominent in most wrists (Fig. III-70). Lateral to it, in turn, lies the *radial artery* in the spot where its pulse is generally taken. It can be felt in the "anatomical snuffbox," as well. Satisfy yourself in both regards.

When we move to the ulnar side of the Palmaris Longus we encounter, in turn, (if we are lucky) the cluster of tendons of the **Flexor Digitorum Superficialis** and that of the **Flexor Carpi Ulnaris** most medialward of all where it inserts upon the pisiform (Fig. III-70). The **ulnar pulse** can be felt to the radial side of this last tendon, and the **ulnar nerve** will pass into the hand at this point, also.

The Skin Creases: As an ace in the hole for those harboring doubts about getting through medical school we include the skin creases on the hand (Fig. III-74B). Palmistry, especially on the West coast, pays almost as well as psychiatry and the overhead is slight. All it

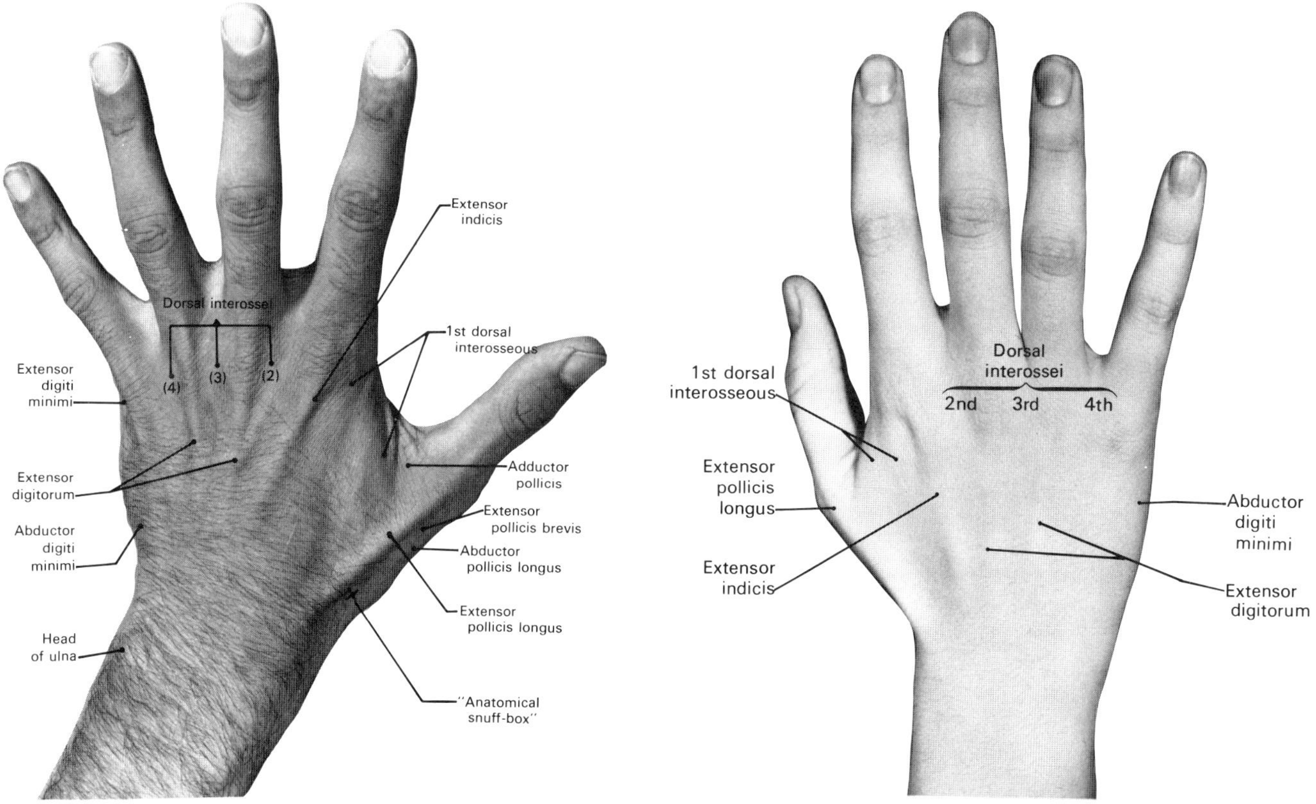

Fig. III-72. Dorsum of the hand. **A.** Male. **B.** Female.

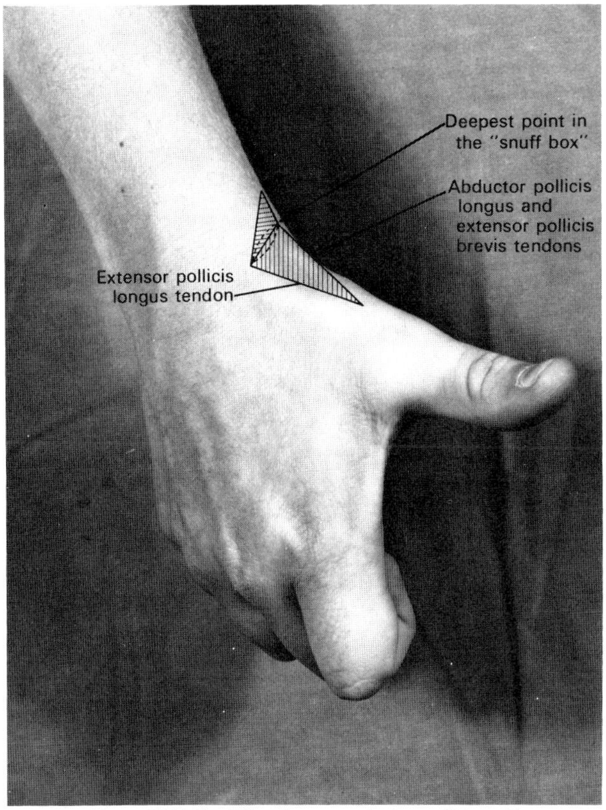

Fig. III-73. The "anatomical snuff-box."

takes is a leak-proof tent to set up business. Two creases, distal and proximal, run roughly transversely across the palm, the proximal one joining the longitudinal palmar crease at the distal end of the thumb-webbing. The distal transverse crease begins at the ulnar edge of the palm, and, with luck, meets the base of the middle digit. In the center of the hand a midpalmar crease runs more or less longitudinally from middle finger to the Palmaris Longus tendon. The middle metacarpal lies under it. There are three creases at the wrist: distal, middle, and proximal. The scaphoid on the radial side is just distal to the distal wrist crease. The middle one at the wrist is about at the ends of the styloid processes of the radius and ulna and at the limit of the synovial sheath for the common flexor tendons (Superficialis and Profundus). There are three digital creases, as well, and the metacarpophalangeal joints lie ½ inch proximal to the proximal digital crease and distal to the transverse creases of the palm.

There is enough variety in the hand creases to make fortune telling interesting. The proximal transverse crease of the palm is the "head line" and if it doesn't make it to the web of the thumb you can advise your client that he doesn't have the intellectual prowess to put his grandiose schemes into effect. The distal transverse crease is the "heart line." If it is segmented or falls short of the middle finger, your client had better cool his romantic notions, but if the midpalmar crease

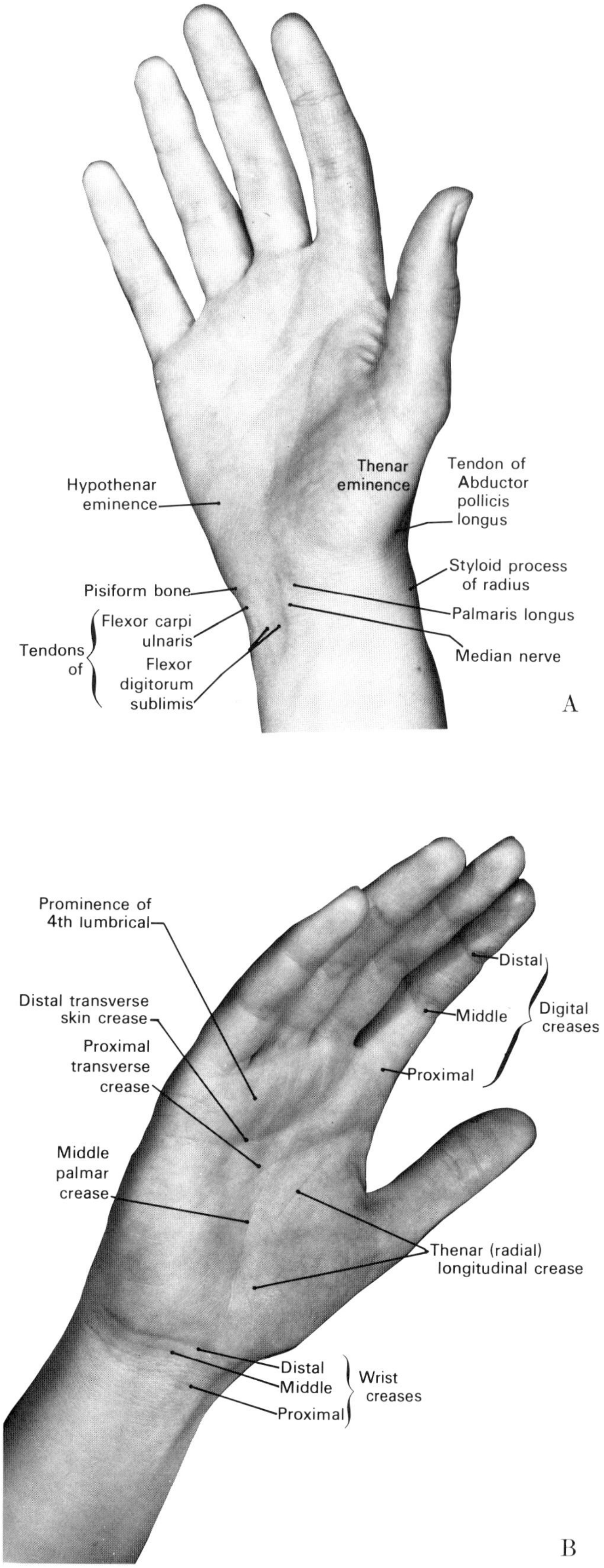

Fig. III-74. The female hand. **A.** Palmar surface and wrist. **B.** Skin creases of the wrist, palm, and digits.

crosses the two transverse creases so as to form a letter "M," you can console him with: "Lucky at cards; unlucky in love." The midpalmar crease is the "fate line," and with ingenuity you can come up with all sorts of good or ill omens for his fortune. Having provided you with this cushion for *your* future—just in case—we wish we could read your palm to tell you how well you would fare on your next anatomy exam.

Nerves to the Upper Limb (Figs. III-75 to 78A): The skin over the upper parts of the pectoral region and shoulder derive their sensory supply from C_3 and C_4. T_2 generally supplies the skin on the medial aspect of the upper arm. Otherwise, all other motor *and* sensory supply comes from the brachial plexus (C_5–T_1). Several of the nerves to the upper limb can be palpated. It is easier to do this in thin individuals. Those that can be felt superficially at times by rolling them under the fingers are: the supraclaviculars as they cross the clavicle; the cords of the plexus in the axilla when the arm is raised to the horizontal; the ulna behind the medial epicondyle of the humerus; the radial at the snuffbox where it is superficial to the long extensor of the thumb. The median may be superficial at times when it crosses the brachial artery from lateral to medial side about midway in the upper arm, but it is not always this accessible to touch.

It is also possible to zero in roughly on some of these nerves in spots where they can't be felt. Palpate the arcomion. Four fingerbreadths from it along the humerus should place you where the axillary nerve winds round the surgical shaft. We have already located the ulnar nerve at the wrist just medial to where you feel the ulnar pulse and we noted that the median nerve generally lay beneath the Palmaris Longus tendon. The radial nerve winds posteriorly around the shaft of the humerus in the spiral radial groove on the bone midway along its shaft.

General Plan of Blood Supply to the Upper Limb: This is basically simple (Fig. III-78B). All the named arteries run deep and they have companion veins of the same name that also run deep to return blood toward the heart. Often veins are double. Besides this, there are separate **superficial veins** which we have already discussed; there are no comparable superficial arteries. The **subclavian artery** lies behind the clavicle. It gives off branches to the thoracic wall, neck, and shoulder before it becomes the **axillary artery.** The latter also gives branches to the shoulder, clavicular region, and thoracic region before it enters the arm as the **brachial artery.** The brachial artery ends at the elbow after it has given off a deep branch that follows the radial nerve behind the shaft of the humerus. It terminates as a lateral **radial artery** and a medial **ulnar artery** which course anterolaterally along their respective bones. Both give **superior** and **inferior collateral branches** that anastomose around the elbow joint. The ulnar sends a **deep branch** to pierce the intermuscular connective tissue septum between the bones. This bifurcates to run on the anterior and posterior aspects of the septum between the planes

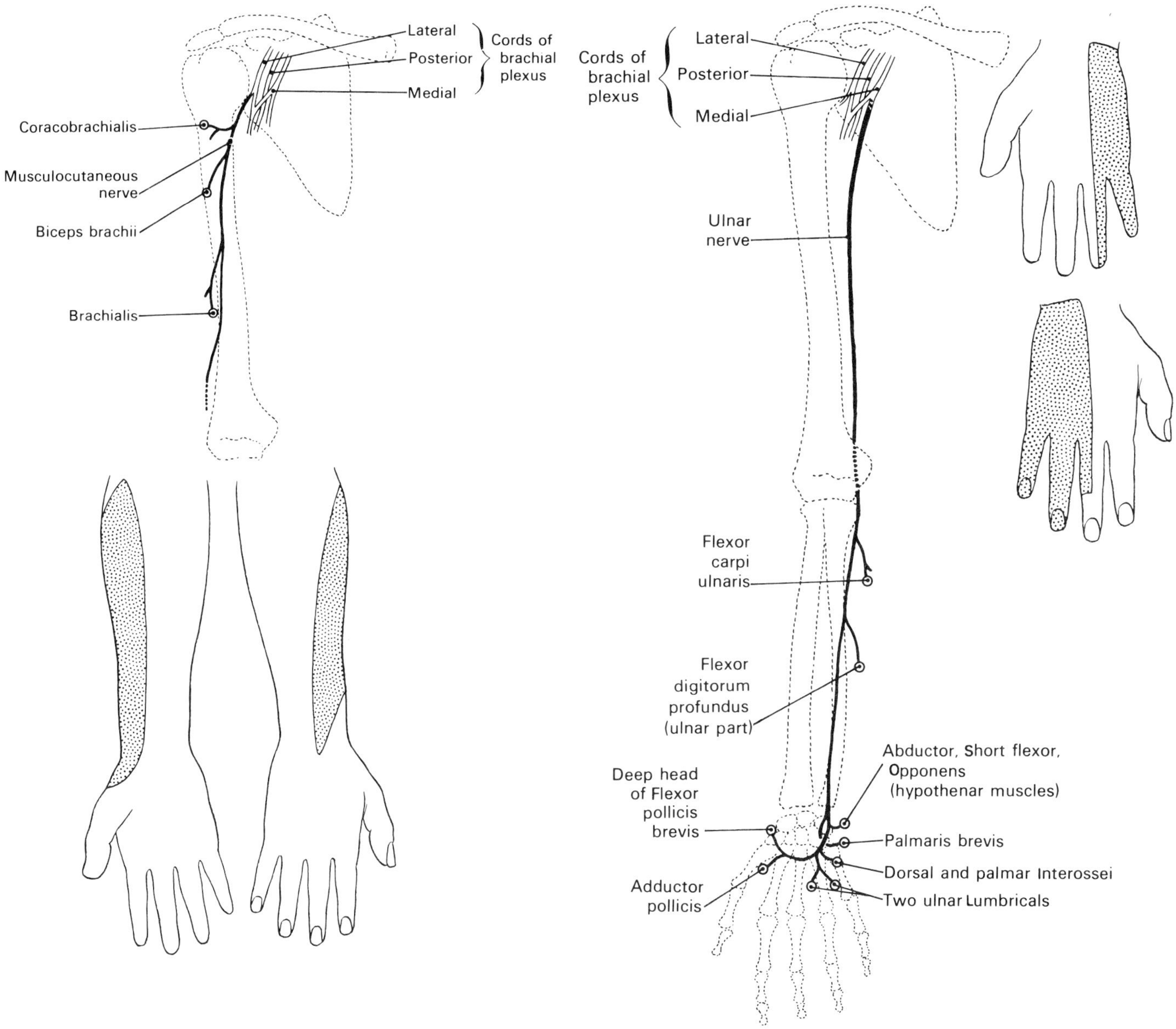

Fig. III-75. Distribution of musculocutaneous and ulnar nerves to muscles and skin (shaded areas in insets). **A.** Musculocutaneous distribution. **B.** Ulnar distribution. The lateral part of the Flexor digitorum profundus is not labeled. It gets its supply from the median nerve; see Figure III-76.

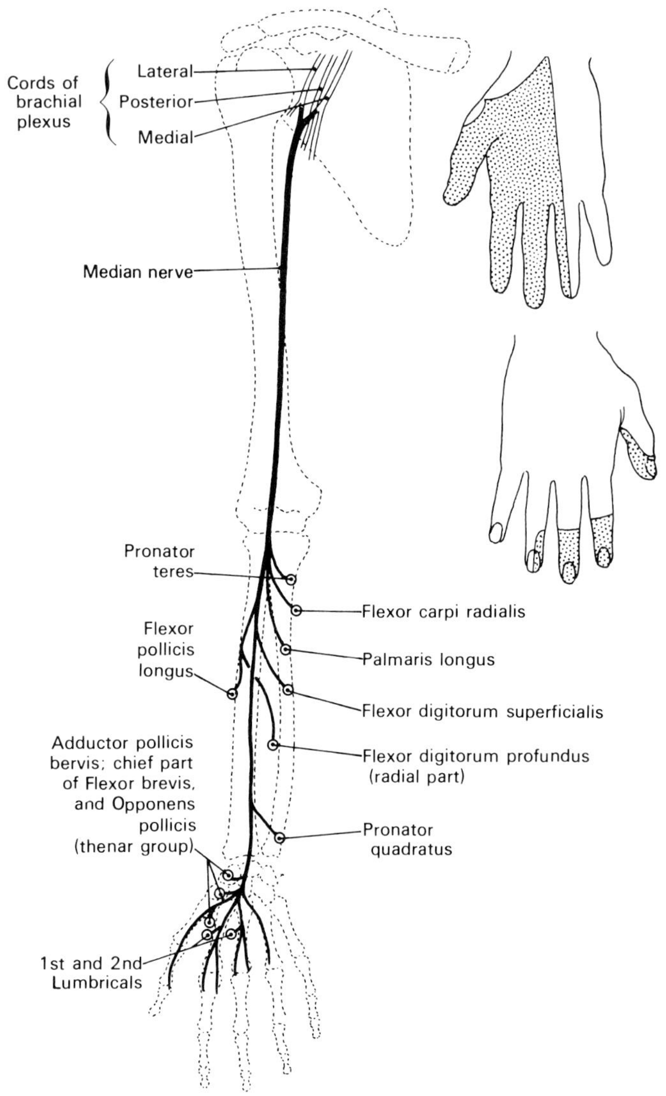

Fig. III-76. Distribution of the median nerve to muscles and skin (shaded areas of inset). The medial part of the Flexor digitorum profundus is not labeled. It gets its supply from the ulnar nerve; see Figure III-75.

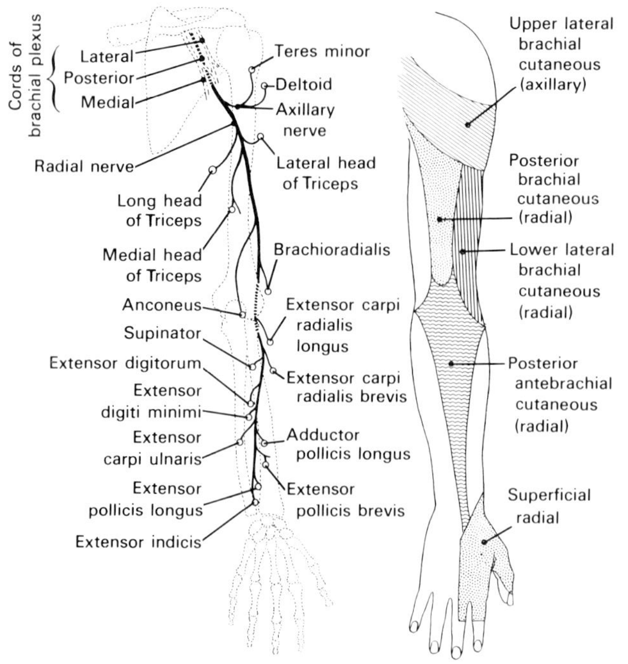

Fig. III-77. Distribution of axillary and radial nerves to muscles and skin (large shaded areas). Posterior view of upper limb.

of the longitudinally-arranged flexor and extensor groups of muscles of the forearm. At the styloid processes of each bone, **deep and superficial palmar arches** are formed between the radial and ulnar arteries. Digital branches to the fingers leave these arches. The deep veins follow the same route to the subclavian veins in reverse. If this requires extreme memorization on anyone's part, it's time to throw in the towel.

Bursae of the Upper Limb: A bursa (Latin, meaning a purselike structure) is simply a connective tissue space, nonepithelially lined, in an area of friction between tendon and bone, between muscles, and a few, simply, within fascia. There may be 50 or more of them named in man and they all contain a small amount of fluid until they become inflamed, when they may contain much fluid and become swollen enough to be visible or painful enough to call attention to their location. Some are found close to joint cavities that contain synovial fluid, and they may even be continuous with these cavities. We mentioned tendon sheaths as a kind of bursae. They are more like fluid-filled tubes that run longitudinally along a tendon for some distance, especially where tendons have to cross a stretch of bone as in the wrist (Fig. III-71). It is as if a bursa developed on one side of a tendon as a flattened meniscus and then almost enveloped it completely, leaving a space, the **mesotendon,** between the ends of the double-layered bursa that almost met, but not quite (Fig. III-80B). Again, these elongated tubes (or partial sausage-cased tubes) contain fluid. The outer layer is the parietal layer; that against the tendon, its visceral layer. The mesotendinous space makes it possible for blood vessels to reach the tendon much as the connective tissue of the mesentery of the gut allows blood vessels and nerves to reach visceral organs (Fig. III-80B).

There is a **subacromial bursa** where the Supraspinatus tendon lies close upon the acromion (Fig. III-80). Since this muscle assists the Deltoid in abducting the arm, if the tendon is inflamed, great pain is experienced as the arm is raised beyond 60°. A similar bursa lies under the Deltoid as it reaches for attachment to the acromion. Another, the subscapular bursa, lies near the coracoid process of the scapula and opens into the capsule enclosing the humerus and glenoid cavity.

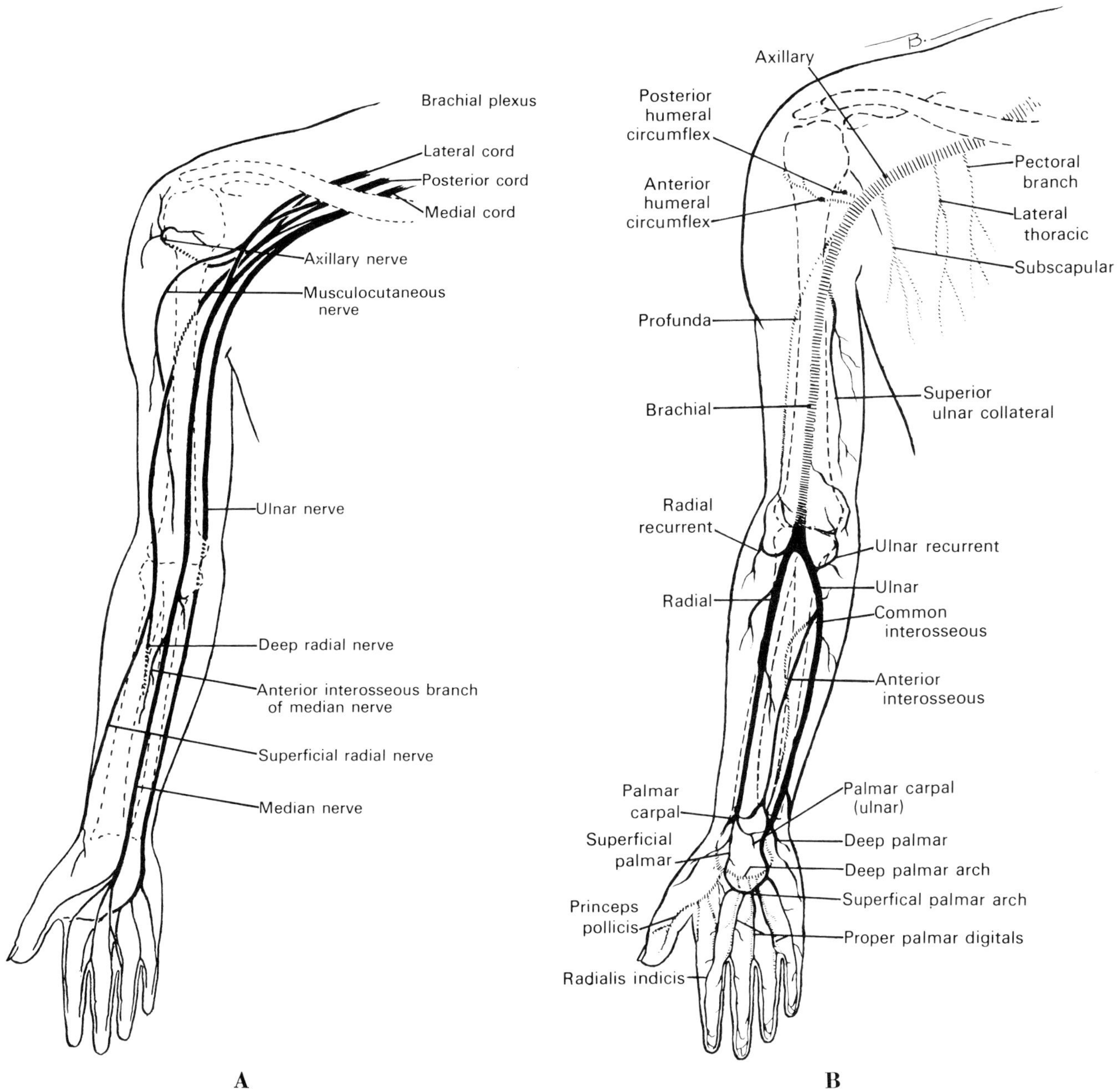

Fig. III-78. Major nerves **(A)** and arteries **(B)** to upper limb. (Anterior aspect.)

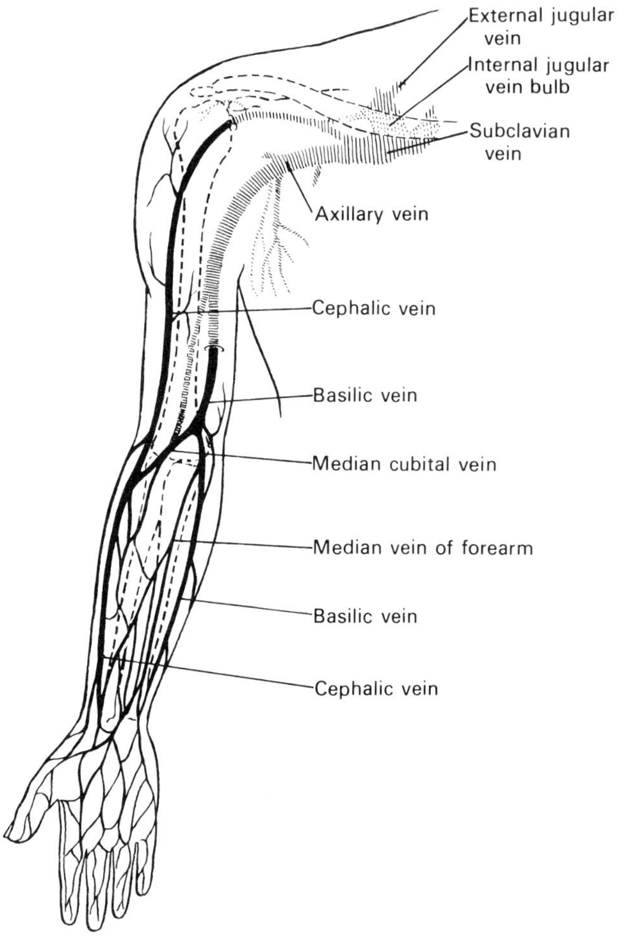

Fig. III-79. Superficial veins of the upper limb. (Anterior aspect.)

You can feel the coracoid process ("like a crow's beak") but not the bursa when you push your finger deep and laterally in the infraclavicular fossa between Pectoralis Major and Deltoid. The long head of the Biceps arises from a **supraglenoid tubercle** and, hence, has to pass as a tendon through the capsule of the shoulder joint. It carries a reflection of the synovial membrane with it as the **intertubercular synovial sheath** that surrounds the tendon in the groove between the **greater and lesser tubercles** of the head of the humerus (Fig. III-80A). As the Triceps tendon inserts upon the olecranon process there is a bursa beneath it. Inflammation results in the so-called "tennis elbow." Tennis, anyone? Or would you prefer "house-maid's knee"? There is a bursa in front of the patella and extensions of the synovial cavity of the knee joint above and below the patella posteriorly as supra- and infrapatellar bursal pouches. We have already mentioned the specialized "bursae" or tendon sheaths about the flexor and extensor tendons as they enter the wrist.

Lymphatics of the Upper Limb: There are few lymph **nodes** in the upper arm and those inconstant ones as exist follow deep lymphatic channels which, in turn, follow the deep arteries and veins. The nodes lie principally deep in the region of the elbow. There are many more superficial **lymphatic vessels** than there are deep ones. They begin along the digits and run to follow the palmar arches around each side of the hand to the dorsum. Hence, infection in the palm causes fluid accumulation on the back of the hand. In the forearm the channels converge on the anteromedial aspect and run upward toward the axillary nodes where the deep lymphatic channels that have plied the deep vascular route also join them.

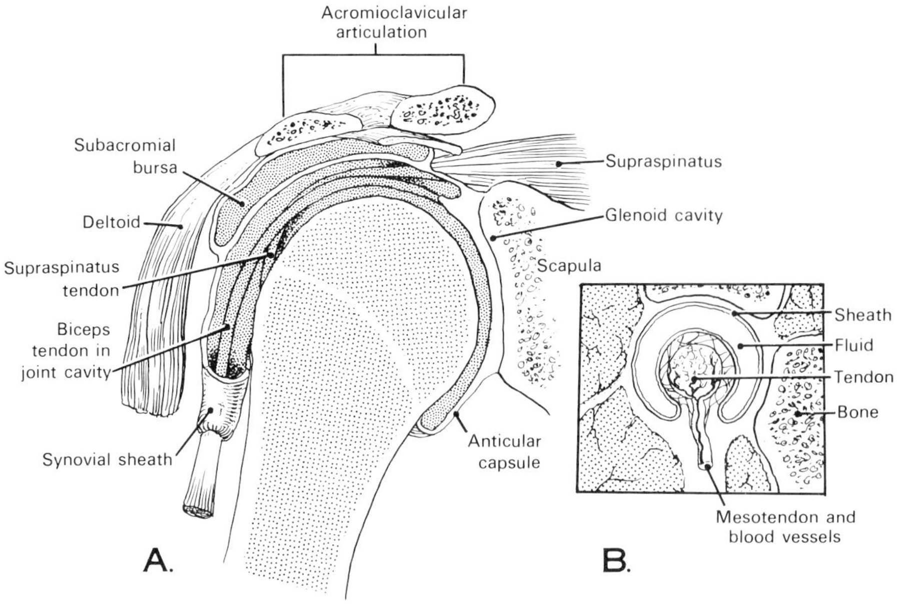

Fig. III-80. Bursae and tendon sheaths. **A.** Subacromial bursa and intertubercular synovial sheath of Biceps tendon. **B.** Structure of a tendon sheath.

SURFACE ANATOMY OF THE LOWER LIMB

By now we can probably agree that anatomy is something that generally looks better on the female than on the male. The same would appear to apply to the lower limb where most males sport calves that only a motherly cow could love. Still, in both sexes, the fundamental plan of construction of the lower limbs is much the same as the upper ones, with, of course, some salient differences. Most Primates abandoned the quadrupedal posture, though a few still use the knuckles of the forelimb in locomotion. Once we, as members of this clan, sacrificed our tail and prehensile toes to stand squarely on our own two feet, the upper limb was greatly freed for a high degree of mobility, while the lower limb specialized in the direction of stability, sacrificing some of the finer movement which some of our more lowly cousins can still perform with it.

Embryology of the Limbs: For a while in the embryo, it seemed as if legs and feet were going to be carbon copies of arms and hands. But by the 8th week the lower limbs began to rotate in an opposite direction from the upper ones, reversing the original dorsal and ventral surfaces. Limb buds first appear at the end of the somite stage as faint, short, longitudinal thickenings in the somatopleure of the body wall. The upper ones make this advent first just ventral to the lowest cervical somites, the lower ridge appearing a bit later at the level of the upper sacral somites. The underlying mesenchyme pushes a cylindrical bud of ectoderm before it with the thickened ridge at its apical end. The distal end flattens into a paddlelike plate from which fingers or toes will be cut, and constrictions delineate the regions of the elbow and wrist, knee, and ankle.

A capillary plexus develops in the mesenchyme, connected at first with several intersegmental arteries coming to the basal portion of the limb bud from the dorsal aorta. The plexus eventually develops into a **midline axial artery** of the cylindrical limb, connecting to the 7th cervical intersegmental branch in the upper limb region and to the 5th lumbar artery in the lower limb (Fig. III-82A). Distally in the bud, just deep to the apical ridge, a **marginal vein** develops which drains superficially along the **preaxial (cephalic) border** and the **postaxial or caudal border** of each limb, for both project about at right angles from the body. The marginal vein and the epithelial apical ridge both appear to be essential if a whole limb is to develop. Interference with one or the other at this early stage of development can lead to a lack of limb (**amelia**); a lack of the distal part beyond the elbow or knee (**hemimelia**); or to the absence of a hand, foot, or digit. The preaxial and postaxial veins that drain the marginal one become the *cephalic* and *basilic*, respectively, in the upper limb and the **greater** and **smaller saphenous,** respectively, in the lower limb.

The preaxial (cephalic) border also marks the radial or thumb side of the upper limb and the tibial or great toe side of the lower. The ulna and little finger and the fibula and little toe make up the postaxial or caudal side of these early limbs once the cartilaginous patterns of these bones form in the mesenchyme. Thus, both limbs begin with a similar orientation. In both, the dorsal surfaces are the *extensor* surfaces and the ventral surfaces the *flexor* surfaces, just as in the case of the trunk.

Beginning prenatally but completed postnatally, the lower limb (which had its ventral surface flexed medially across the body just as did the upper limb in the fetal position) began a medial rotation, an adduction and an extension. These combined movements rotated the limb almost 180° and in a direction opposite to that of the upper limb. In the process the original dorsal surface of the lower limb shifted ventrally and somewhat laterally; the original ventral surfaces, dorsally and somewhat medially. Thus, flexor and extensor surfaces became reversed from what they were in the upper limb. In the anatomical position, the lateral rotation of the upper limb about its longitudinal axis faces its flexor surface ventralward. This surface faces posteriorly in the lower limb. If we wish to place the arm and hand in the same position as our leg and foot we have to pronate the forearm and assume a quadrupedal posture (Fig. III-81C). Then the thumb is medial like the great toe, the little finger lateral like the little toe.

We pointed out in Chapter I that segmental somites influenced the segmental arrangement of muscles developing from myotomes of the somite, the arrangement of segmental nerves that follow the migration of these muscle masses and the segmentation of skin dermatomes. A dermatome is a segmental band of skin supplied by sensory nerve fibers that are associated with a specific dorsal root ganglion. Dermatomes are distributed in circular bands about the trunk (Fig. III-81C). They also run out over the limbs, but not as complete girdles. Sensory fibers of the 5th cervical spinal nerve supply the proximal preaxial dermatome in the upper limb, while those of the 1st thoracic supply the proximal postaxial dermatome (Fig. III-81A). The nerves to the limb dermatomes supply both dorsal and ventral aspects, but only rarely do they cross the axial line set down by the midaxial artery of the limb bud. That is, sensory fibers from C_5 do not lap over into the postaxial territory of T_1; nor C_6 into C_8; C_4 into T_2. (C_4 and T_2 actually supply trunk dermatomes which encroach upon the shoulder, axilla, and upper medial aspect of the arm as if they were pulled out into the limb as the limb bud lengthened.) You will not find C_5–T_1 supplying the skin of the trunk, however. When the adult upper limb is in the anatomical position (Fig. III-81B), you can trace the limb dermatomes progressively down the lateral aspect of the arm and back up its medial aspect. In the leg (Fig. III-81C), the progression is down over the ventral aspect and up the dorsal aspect from L_3–S_2, but not as precisely as in the upper limb because developmental rotation gave a semitwist to some of these skin areas.

It should be made clear that while the sensory supply of a preaxial limb dermatome does not overlap into the territory of a postaxial dermatome, the supply to each dermatome is overlapped by one-half from the two dermatomes next adjacent above and below it. For example, while the sensory fibers of C_5 are *specific* for a given patch of skin, one-half of the area of this patch receives sensory fibers from spinal nerve C_4 and the other one-half of its area from spinal nerve C_6. Thus, each dermatome is supplied by sensory fibers from three adjacent dorsal root ganglia. You could not sever the dorsal root of spinal nerve C_5 and render the dermatome it supplies insensitive. It would still receive a sensory innervation in part from C_4 and in part from C_6. To render this skin patch totally anesthetic you would have to sever all three dorsal roots (C_4–C_6). Then you could stick pins into this area, pinch it, or hold a lighted match to it and elicit no responses to pain, touch, or temperature sensations.

What becomes of the original axial artery that supplied the early limb bud? It is quite apparent (Fig. III-82A) that the axial artery of the bud lies between extensor and flexor territories where muscles will develop in the mesenchyme as myotomal parts of the somites migrate to these regions. In the upper limb this primitive axial artery becomes the **axillary, brachial,** and **anterior interosseus arteries,** regional differentiations of one main stem. Next, a **median** and a **posterior interosseus artery** branch from this main stem. The median artery supplies the hand at first, but it is replaced in this function as radial and ulnar branches come off the main trunk later to form the palmar arches (Fig. III-82B). In the leg, the main axial artery is short lived, and only short pieces of it survive in the adult as the **inferior gluteal** and short pieces of the **popliteal** and **peroneal arteries.** With the rotation of the leg, the **femoral** and its **deep (profunda) branches** become the chief vessels in the thigh. It connects with the popliteal artery *via* Hunter's adductor canal. Other branches develop from the popliteal to supply the leg as **anterior** and **posterior tibial arteries** and the proximal part of the **peroneal artery.**

Thus far we have been concerned with the sensory supply to the skin dermatomes of the limbs. The myotomes which form muscle groups in the limb receive not only a motor supply but a sensory one as well, since there are proprioceptive endings in muscles and tendons (neuromuscular and neurotendinal spindles). A segmental spinal nerve follows each segmentally-derived myotome into the limb. Both the myotomes and their nerves divide into anterior and posterior portions. The posterior division of the myotomes will form the muscles of the axial column in the back and the *posterior primary division of the spinal nerves supply these back muscles.* The anterior divisions of the myotomes become ventral and lateral trunk musculature and some of them migrate into the limbs. The **anterior primary divisions of the spinal nerves**

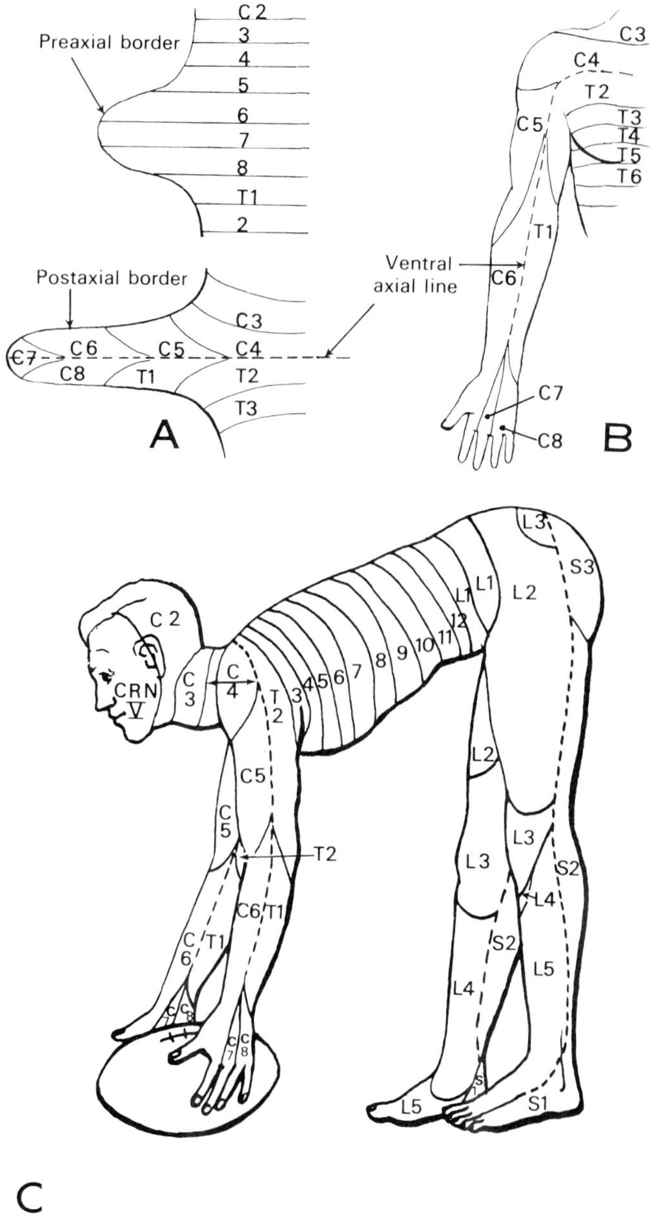

Fig. III-81. Dermatomes of the upper and lower limb. **Fig. III-81A.** Development of dermatomal supply to the upper limb; early and late stages. **B.** Dermatomes of the adult upper limb in the anatomical position. **C.** Comparison of dermatomal supply in upper and lower limbs. The body is in the quadrupedal position.

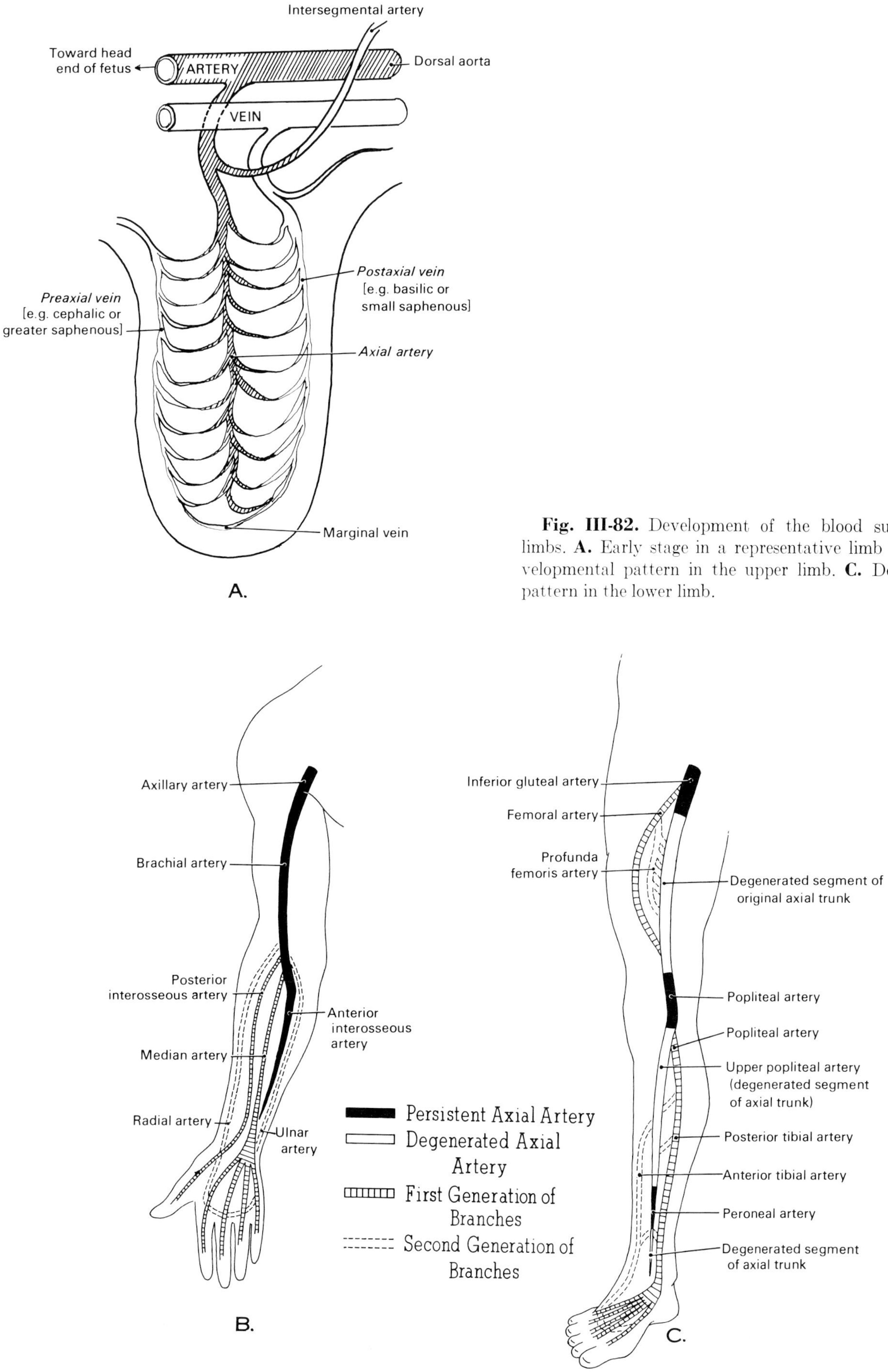

Fig. III-82. Development of the blood supply to the limbs. **A.** Early stage in a representative limb bud. **B.** Developmental pattern in the upper limb. **C.** Developmental pattern in the lower limb.

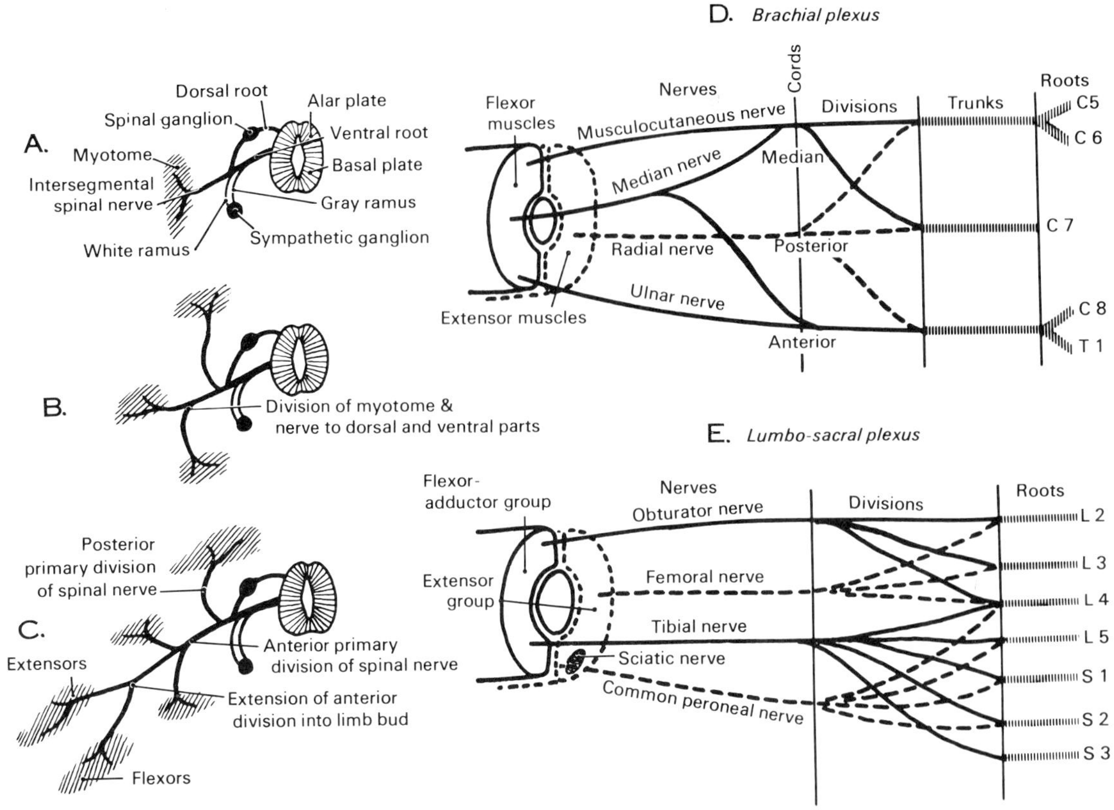

Fig. III-83. Motor supply of limb myotomes and plexus formation. **A.** Parts of segmental spinal nerves and their relation to developing muscle groups in the limbs. **B.** Major peripheral nerves emerging from the brachial and lumbosacral plexuses and the functional muscle groups of the limbs which they supply.

innervate these muscles. All of the so-called nerve plexuses (**cervical, brachial, lumbosacral,** and **pudendal**) come from the anterior primary divisions of the spinal nerves. In a plexus, nerve fibers which have first organized to form segmental spinal nerves regroup for distribution as certain named peripheral nerves; *e.g.*, radial and ulnar in the upper limb; femoral, tibial, in the lower limb. Each of these peripheral nerves may convey the fibers of more than one segmental spinal nerve to the muscles in the periphery (Fig. III-83). The *brachial plexus,* for example, gathers fibers from spinal nerves C_5–T_1 and delivers them to the upper limb through five important peripheral nerves. (Actually, there are more than these five that arise from the plexus, but these are the most important ones for motor supply to the upper limb.) The plexus combines the spinal nerves into three divisions: medial, lateral, and posterior. The *medial* and *lateral divisions* form three nerves destined for the *anterior aspect of the limb* to provide *flexor-pronator groups* of muscles. These nerves are the **musculocutaneous,** the **median,** and the **ulnar nerve.** The *posterior division* supplies the *extensor-supinator muscles* on the *dorsal aspect of the limb* by means of the **radial nerve** (Fig. III-83). (The **axillary nerve** to the Deltoid is also derived from the posterior division.)

In the leg, something similar takes place as spinal nerves L_2–S_3 form the lumbosacral plexus. The *anterior division of the plexus* supplies the *flexor-adductor* groups of muscles by way of the **obturator and tibial nerves.** These muscles are on the posteromedial aspect of the lower limb in the adult, but they were anterior in the limb bud before rotation; hence, they still are supplied by the anterior parts of the plexus. The *extensor group* or muscles in the lower limb are supplied from the *posterior division of the plexus* by means of the **femoral** and **common peroneal nerves.** In the adult they lie anterolaterally, yet they were dorsal in the early limb bud; hence, the posterior division of the plexus reached them early before rotation of the limb occurred (Fig. III-83). These are the big generalities in respect to the brachial and lumbosacral plexuses. Other muscles are supplied by them, and in the leg, especially, a few muscles such as the Biceps Femoris, Pectineus, and part of the Adductor Magnus, may get supplied either by the anterior or the posterior division of the plexus.

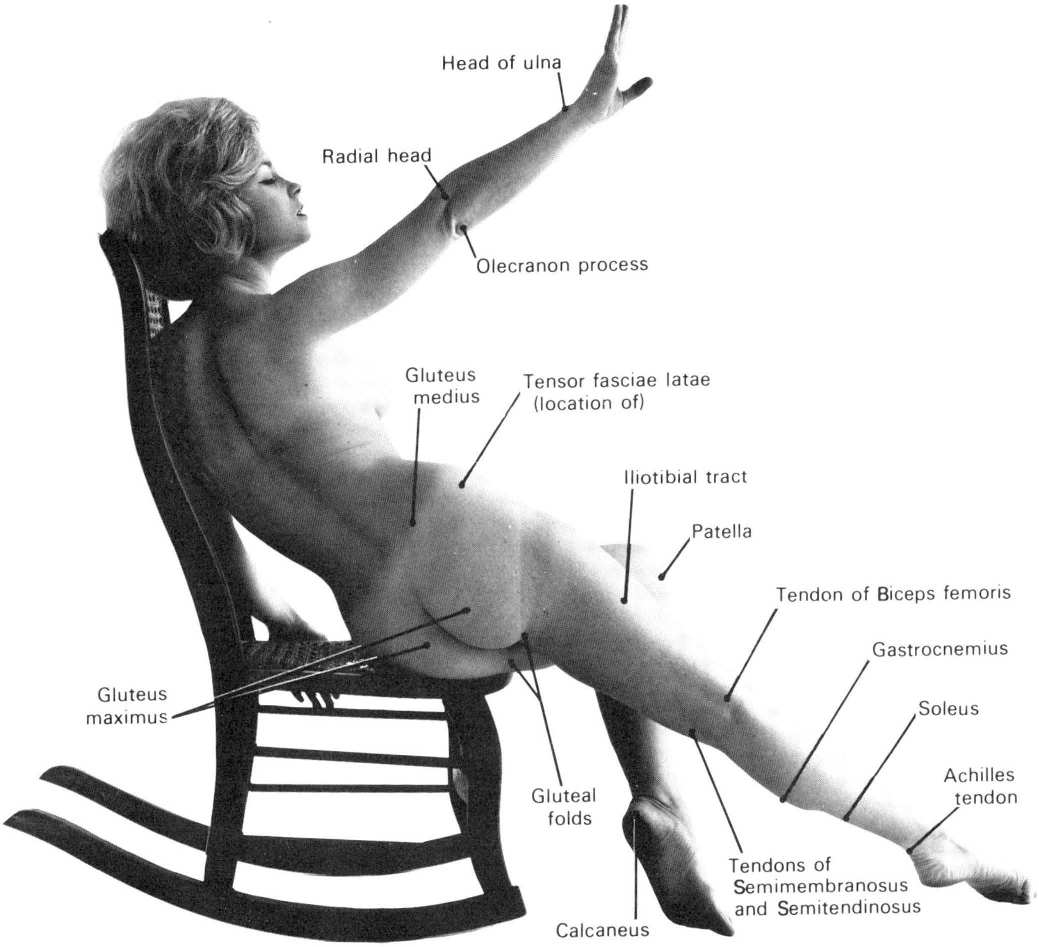

Fig. III-84. Surface anatomy of the female lower limb.

Parts and Regions of the Lower Limb:

1. Only two muscles can actually be seen superficially in the **gluteal region** (sometimes referred to as natis, clunis, or buttock). The large **Gluteus Maximus** covers most of the structures in this region, but a part of the Gluteus Medius peeks out above and somewhat lateral to it (Fig. III-84). The gluteal fold delineates buttock from posterior thigh. We often have the mistaken idea that we sit upon our Gluteus Maximus. Feel for the **ischial tuberosity** of the innominate bone as you bend to sit, and notice that this big muscle of the buttock rides up to uncover the tuberosity; it is this upon which we sit. The Gluteus Maximus arises from the outer posterior surface of the ilium, the sacrum and coccyx, and from the sacrotuberous ligament. Most of it inserts into a stout fascial band, the **iliotibial band** or **tract,** which runs from iliac crest to the **lateral tibial condyle,** helping to form a lateral fibrous capsule of the knee joint (Figs. III-84 and 85B). A smaller muscle, the **Tensor Fasciae Latae,** also pegs into this tract, so that, in a sense, the stout band could be regarded as the major tendon for this small muscle and the Gluteus Maximus, though the latter inserts in part on the **gluteal tuberosity of the femur** as well. The Gluteus Maximus is innervated from the sacral plexus (posterior division) by the **inferior gluteal nerve** and is supplied by the artery of the same name. It is a chief extensor of the thigh and a good lateral rotator of the hip joint.

The **Gluteus Medius** arises from the wing of the ilium and inserts posteriorly on the upper surface of the **great trochanter of the femur.** It is supplied by the **superior gluteal** nerve and vessels as is the **Gluteus Minimus** which lies deep to it. The *chief action* of both of these muscles is abduction of the thigh, though they can act in part to flex and medially rotate the thigh or to assist the Gluteus Maximus in lateral rotation. The Gluteus Maximus and Medius also cover a group of smaller muscles of the buttock (Piriformis, Obturator Internus, Quardratus Femoris, and the Superior and Inferior Gemelli), all external (lateral) rotators of the thigh. The **Tensor Fasciae Latae** is primarily a hip flexor and a partial internal (medial) rotator as it flexes. It lies too far forward to be a useful abductor.

2. **The Thigh:** Officially the term for this region is **femur,** though usage restricts this noun to the bone of the thigh. The adjective, femoral, *does* apply to structures in this region, however. The thigh has anterior, medial, and posterior compartments, each with a separate nerve supply and each containing special functional muscle groups; *e.g.*, **anterior, the extensors; medial, the adductors; posterior, the flexors.** The nerves to these function groups are the **femoral,** the **obturator,** and the **sciatic,** respectively (Figs. III-89 to 91).

There is not too much surface anatomy to be seen in the thigh. Structures can only be localized chiefly by realizing what muscle groups are covered by skin in given regions. The only bony landmarks to be palpated are pubic tubercle, the anterior superior iliac spine, and the ischial tuberosity. The skin crease between thigh and anterior abdominal wall marks the course of the inguinal ligament. We have already observed the gluteal fold posteriorly separating thigh from buttock. Except for the Iliopsoas muscles which arise from the vertebral column and inner surface of the ilium, all the other muscles arise from the pelvic girdle and, hence, are intrinsic thigh muscles. In the fairly muscular individual (Fig. III-85), the location of some of the anterior thigh muscles can be approximated. The **Quadriceps Femoris** constitutes the major anterolateral mass forming the extensor group which acts on the knee joint to extend the leg. Three parts of this four-part muscle are shown in Figure III-85A, the **Rectus Femoris** and the **Vastus Lateralis and Medialis.** The fourth part, the **Vastus Intermedius,** as you might expect, lies deep. All of the parts attach to the upper surface of the **patella,** and the patella, in turn, attaches to the **tibial tuberosity** by means of the **patellar ligament.** Both these supra- and infrapatellar attachments constitute the **quadriceps tendon** which, when tapped with a reflux hammer, gives rise to the knee jerk.

Only the lowermost part of the long, straplike **Sartorius** muscle can be localized, as it reaches its insertion medially on the body of the tibia close to where the Gracilis and Semitendinosis will also insert (Fig. III-85A). Actually, the origin of this muscle is quite high and lateral on the anterior superior iliac spine so that in crosses the thigh diagonally. Its superior border, therefore, forms the lateral margin of the **femoral triangle,** the medial border of which is the Adductor Longus. When some individuals sit cross-legged, like the tailors of old (hence, the name, Sartorius), the diagonal course of this muscle stands out. Thus, you can expect it to assist in crossing your legs by simultaneous flexion of the hip and knee.

Both parts of the **Iliopsoas** arise within the abdomen

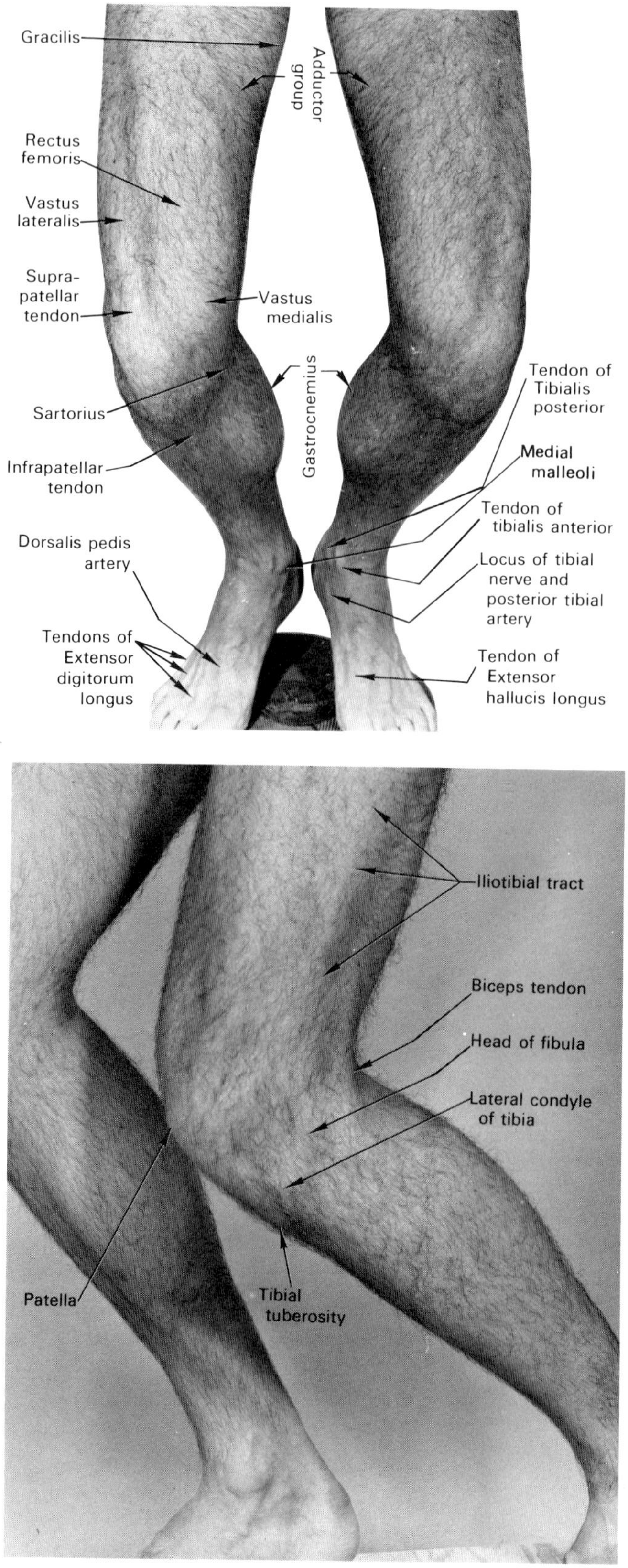

Fig. III-85. The lower limb of the male. **A.** Anterior aspect. **B.** Lateral aspect.

and emerge under the lateral part of the inguinal ligament to become part of the floor of the femoral triangle. The femoral nerve lies in the junctional groove between these two conjoined muscles and supplies the Iliacus portion. The Psoas portion receives direct twigs from the lumber plexus (Fig. III-89B). We have pointed out in the section on the abdomen that this strong muscle flexes the thigh on the trunk, or in sitting up from a supine position, flexes the trunk upon the thigh. It would seem, probably because of its internal origin, that this muscle remained as an anterior flexor, unable to take much part in the rotation of the limb. Each portion of the Quadriceps Femoris and the Sartorius is also supplied by the femoral nerve.

The **Pectineus** which lies just medial to the Iliopsoas in the floor of the femoral triangle is usually considered one of the muscles of the anterior compartment because it is sometimes supplied by the femoral nerve (Fig. III-89A). At other times it receives an innervation from the obturator or from an accessory obturator nerve. This could qualify it as an upper member of the medial adductor group. It is, however, a hip flexor and a medial rotator of the thigh. If it belongs to the adductor group, it is, functionally, a most weak adductor.

The medial or adductor compartment consists of the **Adductors Longus, Brevis, and Magnus,** the **Gracilis,** the **Obturator Externus,** and, sometimes, the **Pectineus.** Like the Pectineus, the Adductor Magnus is somewhat of an oddball in this group. Part of it, its *posterior portion,* does not arise from the pubis but from the ischial tuberosity which would put this portion of the muscle into the hamstring group of the posterior compartment. While the bulk of the muscle is supplied by the obturator nerve (Fig. III-89A), as are all the others in this group, the posterior portion of the Adductor Magnus is innervated by the tibial division of the sciatic nerve, the same nerve that supplies the hamstrings (Fig. III-91). All of these muscles of the anteromedial aspect of the thigh lie in the floor of the femoral triangle. Except for the Pectineus which is a flexor and internal rotator of the thigh and the Obturator Externus which is an external rotator, all the rest are strong adductors. They can also act to flex the thigh but they do not do this as well as the Pectineus. The Gracilis, for example, can only assist in thigh flexion when the knee is extended. It is also an internal rotator of the thigh. It can also internally rotate the leg when the knee is flexed.

The Femoral Triangle: Before we consider the posterior compartment of the thigh (which contains the *leg flexors*), we should consider the important femoral triangle in the upper anterior aspect of the thigh. It is a large triangular area only recognizable after skin reflection, hence, not a prominent surface feature but important for the structures within it. Its boundaries are: (1) the inguinal ligament superiorly; (2) medially, the medial border of the Adductor Longus; and (3) laterally, the superior or medial border of the Sartorius. Its floor contains the fleshy fibers of the Iliacus, Pectineus, and Adductor Longus from lateral to medial, with the tendon of the Psoas between Iliacus and Pectineus. The other Adductors and the Obturator Externus lie deep. Its roof is fascia, containing inguinal lymph nodes and the upper end of the saphenous vein with its tributaries. Its contents, emerging from under the inguinal ligament above the Psoas tendon and in order laterally to medially, are the **femoral nerve, femoral artery, femoral vein,** and **femoral canal.** The latter is a space which just about admits the tip of the little finger. It is larger in the female; hence, herniations of the bowel into the thigh through this canal (femoral hernia) are more common in females than in males. The reverse is true of inguinal hernias, especially the indirect type through the inguinal canal where the predilection is greatest in males. The femoral canal contains fat and a constant lymph node (of Cloquet). The superficial **saphenous vein** joins the femoral vein in the triangle (Fig. III-93). The **femoral artery** can be felt pulsating at the *midinguinal point; i.e.,* half way between the anterior superior iliac spine and symphysis pubis. When you have your finger on the pulse, the **vein** is just medial, the **femoral nerve** about a fingerbreadth lateral and you are right above the **head of the femur** against which the artery is compressed (Fig. III-92). The femoral artery runs a 4-inch superficial course in the triangle before it gains cover by muscle masses. This makes it vulnerable to any number of occupational hazards, so bullfighters, take care! The vessel can be easily tapped for arterial blood samples and is available for arteriography of the leg vessels. Its deep branch, the **profunda femoris,** arises high in the triangle and is directed posterolaterally. It is the chief supply to the thigh—the main femoral trunk continuing into the leg to become the popliteal artery. The perforating branches of the profunda femoris supply the adductor compartment as well as the anterior compartment and pass through the tendons of insertion of the adductor muscles to supply the muscles in the posterior compartment of the thigh too.

The Adductor (Hunter's) Canal: If a perpendicular is dropped from the midinguinal point (when the hip is slightly flexed and externally rotated) it will meet the **adductor tubercle** on the distal and medial part of the femur (Fig. III-86A). The Adductor Magnus attaches to the tubercle. The line marks the course of the femoral artery to the hiatus in the tendon of the Adductor Magnus which the artery uses to reach the popliteal fossa. We can say that the apex of the femoral triangle empties into the Adductor canal which leads to the tendinous hiatus. Thus, the Adductor Longus and Magnus form the posterior boundaries of this canal; the Vastus Medialis, the medial boundary; and the Sartorius, the roof and slanting lateral boundary of the Adductor canal. The **saphenous nerve** goes through the canal with

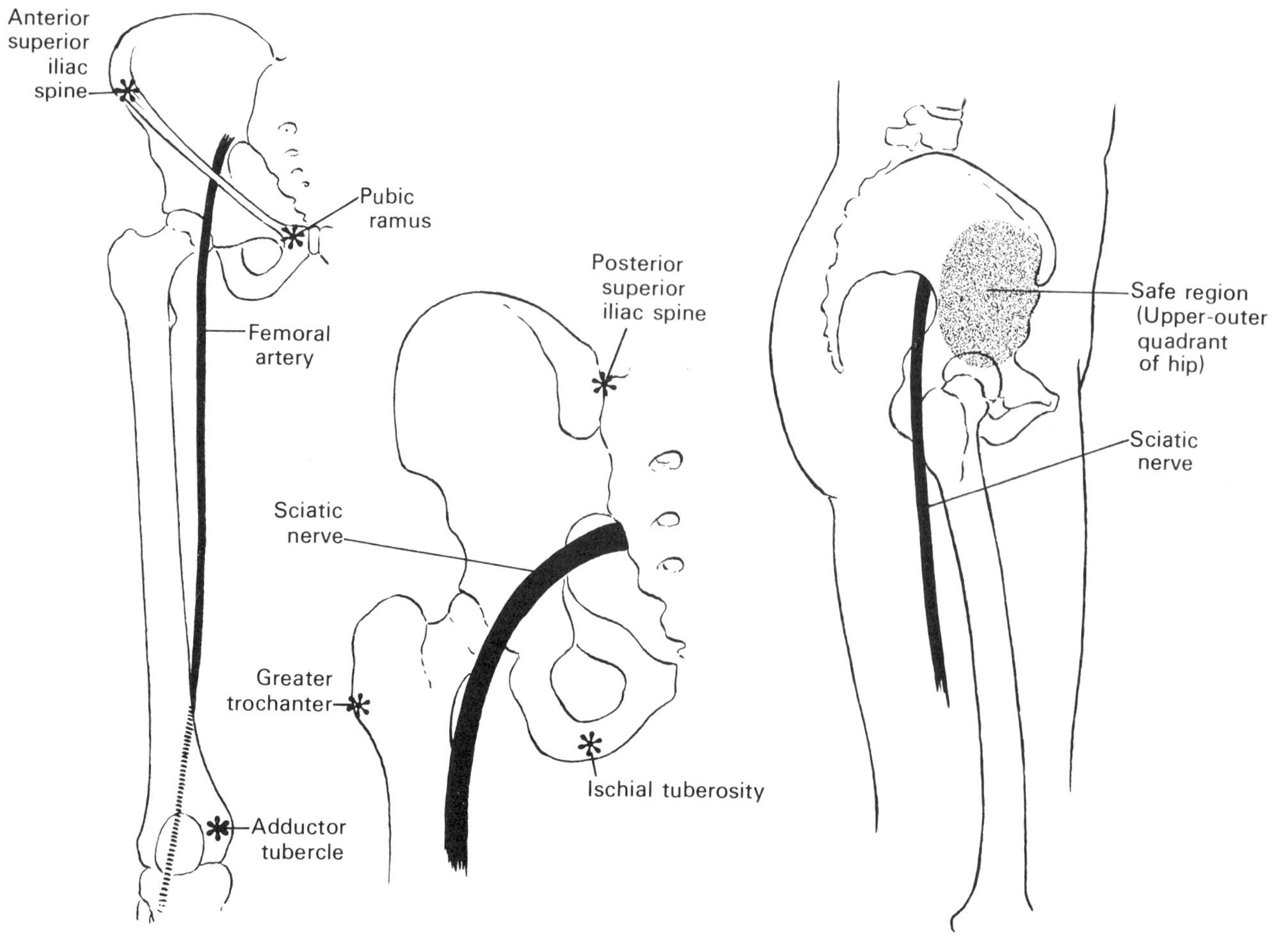

Fig. III-86. Course of the femoral artery and sciatic nerve in the thigh. **A.** Landmarks localizing the path of the femoral artery. **B.** Landmarks localizing the path of the sciatic nerve. **C.** Area for safe intramuscular injection in the hip.

the artery, and before the artery enters the tendinous hiatus to reach the popliteal fossa it gives off its **genicular branch.**

The Posterior Compartment of the Thigh: Inferior to the Gluteus Maximus, the *flexor muscles of the leg* arise from the ischial tuberosity and form the posterior group or **hamstring muscles** of the thigh (Fig. III-87A and B). These consist of the **Biceps Femoris** laterally, the **Semimembranosus** and **Semitendinosus** medially, and the **posterior** part of the **Adductor Magnus** anteriorly. All except the latter cross the knee joint. All except the short head of the Biceps arise from the tuberosity. It has its origin on the lateral lip of the **linea aspera** of the femur. All except this short head of the Biceps lie medial to the midline of the posterior thigh. As a consequence, all except the short head of the Biceps are supplied by the medial division or **tibial division of the sciatic nerve** (Fig. III-91). The short head is supplied from the lateral or **common peroneal division** of the **sciatic nerve** (Fig. III-90).

All of these muscles help the Gluteus Maximus extend the hip except the short head of the Biceps which does not cross the hip joint. All flex the leg on the knee except the posterior part of the Adductor Magnus, but it does help in extending the thigh without flexing the knee every time we get up out of a chair. Actually, because of the degree of tone engendered by the hamstrings, they are fine thigh extensors as we maintain a standing posture. When the hamstrings are paralyzed, the Gluteus Maximus cannot maintain this degree of tone alone and the individual tends to fall forward.

The sciatic nerve is such a large structure in this region that it ought to be localized accurately to prevent damage to it when intramuscular injections are made in the buttocks. Locate the posterior iliac spine and the ischial tuberosity. Midway between these points the nerve curves downward through a point that lies midway between the ischial tuberosity and the greater trochanter of the femur (Fig. III-86B). From here it runs straight down the posterior midline of the thigh to the superior border of the popliteal fossa where it divides into tibial and common peroneal trunks. The general instructions call for

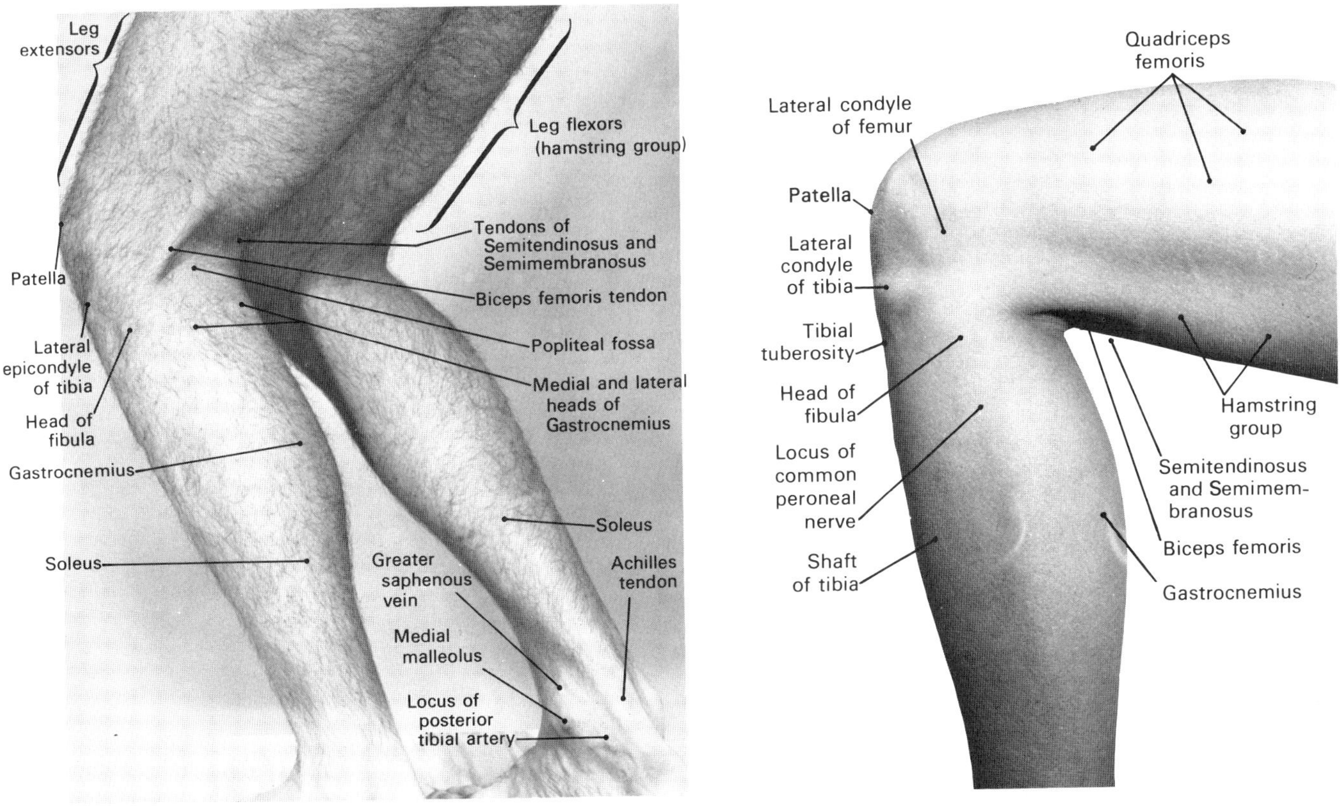

Fig. III-87. The knee and leg. **A.** Surface anatomy in the male. **B.** Surface anatomy in the female.

muscular injections to be made in the upper quadrant of the buttock. The uninitiated have a rather poor conception of the extent of the buttock. They see the cute, rounded, natal hillock, assume this to be buttock, and plunge into the upper quadrant of this area, whereupon they can hardly miss driving the needle into the nerve. To play it perfectly safe, place the tip of the thumb on the anterior superior iliac spine and spread the hand out along the iliac crest. It will cover an area which affords ample musculature well away from the course of the nerve (Fig. III-86C).

3. **The Knee:** All good Catholics and Royalists everywhere who genuflect know that this is the **genu** or bend in the lower limb. Few are too familiar with the **poples,** however, or the back of the knee, the area of the **popliteal fossa.** The term, poples, comes from the Latin, meaning "the ham." And the "hamstrings" are the tendons of the posterior thigh muscles that form the prominent superior popliteal folds. In Roman days when warriors wore short, kilted skirts and went bare kneed, it was great sport to deliver a cut behind the knee and so "hamstring" one's opponent. A fellow can't run very far after that, especially when the common peroneal and tibial nerves are severed in the fossa, as well. Nor have we accounted for the exsanguination resulting from the severance of the popliteal artery, the direct continuation of the femoral artery into the leg. This blow was probably one of the unkindest cuts of all.

As you sit in your chair with knee flexed, it is easy to feel the lateral stout cord of the Biceps tendon (Fig. III-87A and B). On the medial side of the fossa the tendons of the Semitendinosus and Semimembranosus can also be easily grasped between the fingers (Fig. III-87A and B). The nerves in the fossa are the most superficial structures. The *common peroneal* leaves the compartment laterally to wind about the neck of the fibula. It is the only nerve in the whole lower limb that you can feel by rolling it under the fingers against the bone, and, of course, it is quite vulnerable to injury at this point. The **popliteal vein** is next most superficial and it receives the **small saphenous** vein in the fossa. The artery lies deepest. The **tibial nerve** is lateral to the vessels in the fossa, but as it moves down toward the back of the leg it crosses them medially.

Certain bony structures around the knee joint can be seen and felt. The *patella,* of course, is prominent, as is the anterior bump on the tibia, the *tibial tuberosity,* to which the lowest part of the patellar tendon (ligament, or Quadriceps tendon) attaches. The medial and lateral condyles of the femur can be felt if you stretch your hand across the upper aspect of the flexed joint as you sit in your chair. Run the hand slightly

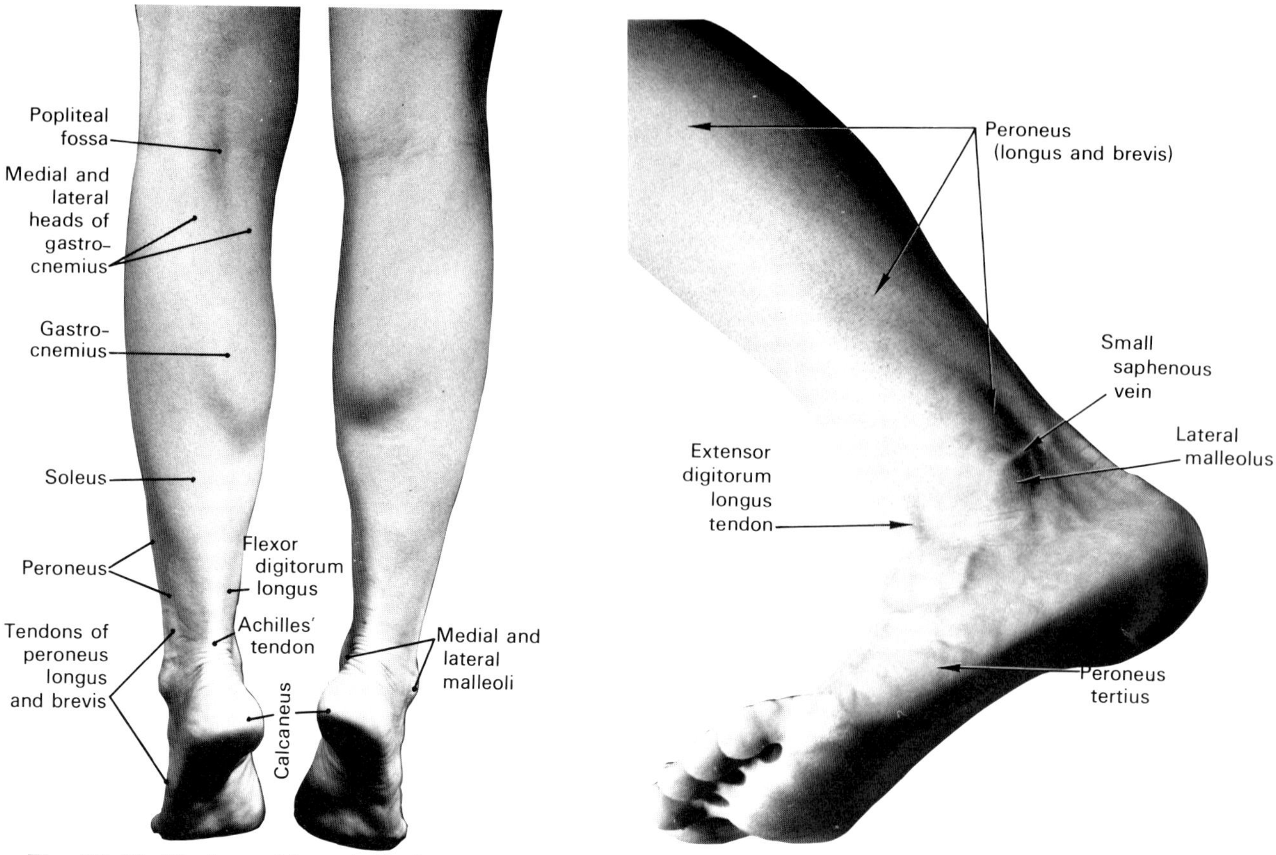

Fig. III-88. The leg and foot. **A.** Surface anatomy of the sural region. **B.** Surface anatomy of the peroneal region.

upward along the thigh and feel the large **medial epicondyle** of the femur under your thumb, the uppermost margin of which constitutes the *adductor tubercle.* Laterally the head of the fibula can be palpated. The fibula will not be subcutaneous again until it approaches the lateral malleolus above the ankle joint where its terminal 3 inches can be felt. But the shaft of the tibia, as the shin bone, is palpable along its entire length (Fig. III-87A and B).

Even with extreme strengthening by stout ligaments, the knee joint is a poor one at best. The large, rounded, rockerlike condyles of the femur rest on the flat table of the tibia with nothing much to prevent them from sliding off in anteroposterior directions, and with very little bony molding to keep them from sliding sideways. Yet this is the main weight-bearing surface. The fibula contributes not at all to the knee joint and bears none of the weight. It articulates with the lateral proximal end of the tibia and the lateral side of the ankle joint but transmits no weight across the ankle joint.

4. **The Leg (Crus):** As in the thigh, muscles in the leg are in three compartments separated by intermuscular septae, and each compartment will have its own special nerve supply. The **sural** region is the calf and, hence, the area of the posterior compartment. The **medial** and **lateral malleoli** are bony projections of the tibia and fibula respectively at the ankle joint.

Rotate your arm internally at the shoulder; pronate the forearm and dorsiflex the wrist and hand. You can now compare upper and lower limbs. The leg and foot are in the same position as the forearm and hand, so that the tibial side of the leg corresponds to the radial side of the forearm; the fibula side, to the ulna side. The extensor surface of the forearm faces forward, and the anterior aspect of the leg is its extensor surface; the posterior aspect, its flexor surface. The sole of the foot (plantar surface) faces down as does the palm of the hand, and the dorsum of both face up. Thumb and great toe, little finger, and little toe are on corresponding sides.

But there are, of course, differences. The great toe lacks the mobility of the thumb, and the digits are not as dexterous as those in the hand. While the wrist can be abducted and adducted in the usual way, the ankle does it in a peculiar fashion in that the sole turns out in abduction (eversion) and inward and slightly upward upon adduction (inversion). The elbow allows for considerable rotation in pronation and supination, but the knee is far less flexible in this regard.

Muscular Compartments of the Leg and Their Nerves: two muscles only form the *lateral compartment.* These are the **Peroneus Longus and Brevis** (Figs. III-88 and 90). As the **common peroneal nerve** winds around the neck of the fibula it divides

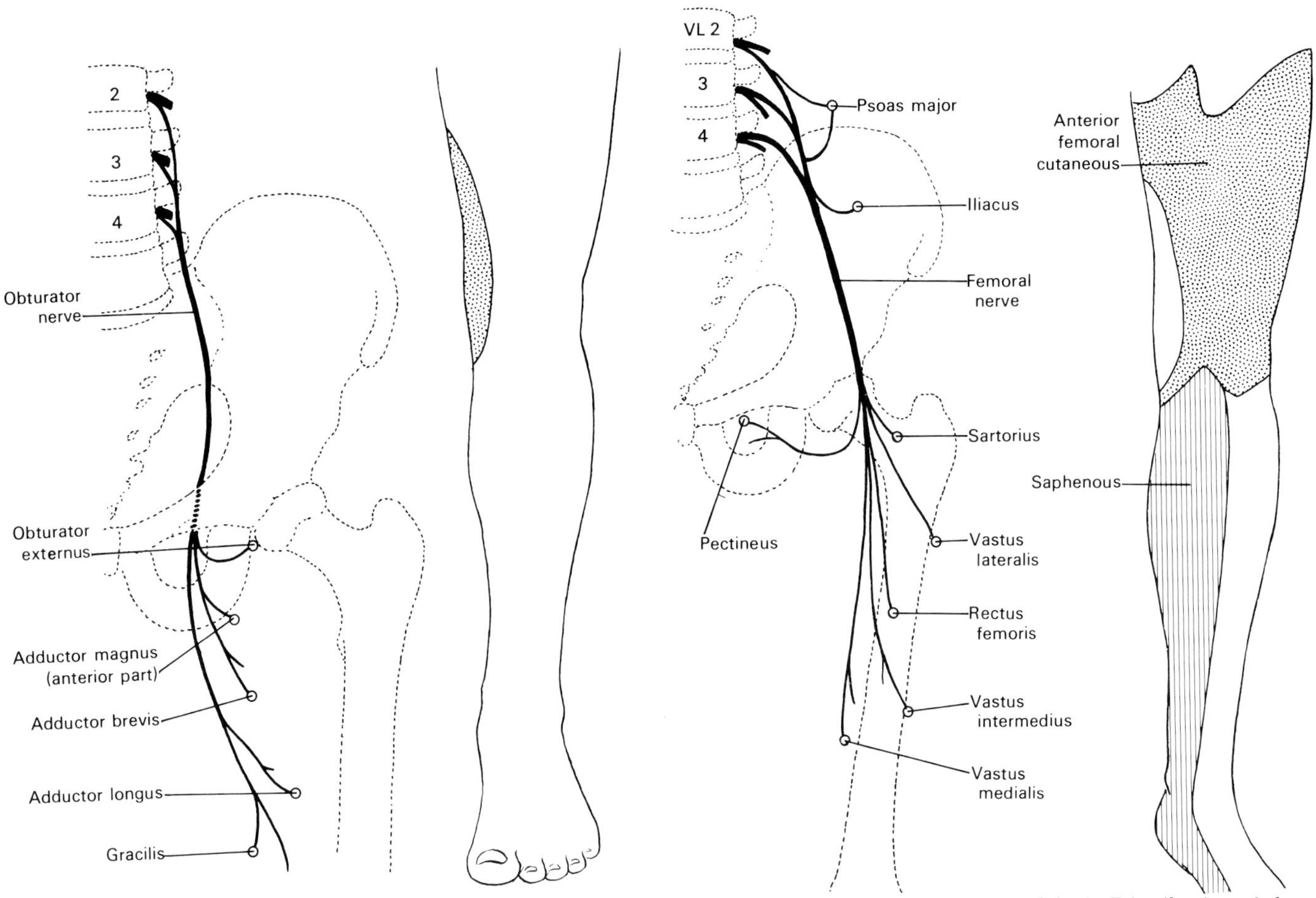

Fig. III-89. The obturator and femoral nerves (shaded areas indicate the region of cutaneous supply). **A.** Distribution of the obturator nerve. Occasionally the Pectineus muscle gets its supply from the obturator nerve also. **B.** Distribution of the femoral nerve.

into superficial and deep branches. As you might suspect, the **superficial peroneal nerve** supplies these two superficial muscles that take their origin from the fibula. The **deep peroneal nerve** leaves the lateral compartment to enter the anterior compartment and will innervate the muscles there. The tendons of these two peroneal muscles wind behind the lateral malleolus to enter the foot, and we will note these relationships later.

The *anterior compartment* is composed of four muscles in the leg. The **Tibialis Anterior** arises from the lateral surface of the tibia. The other three, **Extensor Hallucis Longus, Extensor Digitorum Longus,** and **Peroneus Tertius** (Fig. III-90), arise from the fibula and the intermuscular septum between the leg bones. Their long tendons pass under an extensor retinaculum on the dorsum of the ankle just as did the extensor tendons at the dorsum of the wrist. The Tibialis Anterior and the extensor of the big toe have their own separate tendon sheaths beneath the retinaculum; the other two muscles share a common sheath. The tendon of the Tibialis Anterior only reaches the medial cuneiform and the base of the 1st metatarsal. The Peroneus Tertius, the dorsal part of the 5th metatarsal and the longer extensors to the toes reach the distal phalanges. The **anterior tibial artery** is the chief blood supply to this compartment and parallels the course of the **deep peroneal nerve** which supplies these muscles. The nerve will also supply two muscles on the dorsum of the foot, the **Extensor Hallucis Brevis** and the **Extensor Digitorum Brevis.**

The Peronei of the lateral compartment evert the foot. The muscles of the anterior compartment are chiefly dorsiflexors (extensors) of the foot, though the Tibialis Anterior inverts it.

The *posterior compartment* contains superficial and deep muscles. The superficial group are the large **Gastrocnemius** and **Soleus** and the almost infinitesimal **Plantaris** which is virtually all tendon and which is the equivalent of the Palmaris Longus in the hand. The first two together constitute the **Triceps Surae,** and their main function is plantar flexion at the ankle joint. When they are paralyzed the individual cannot stand on tiptoe, and when they are in marked contracture the foot is held in plantar flexion and we say the individual is "clubfooted." The calcaneus tendon

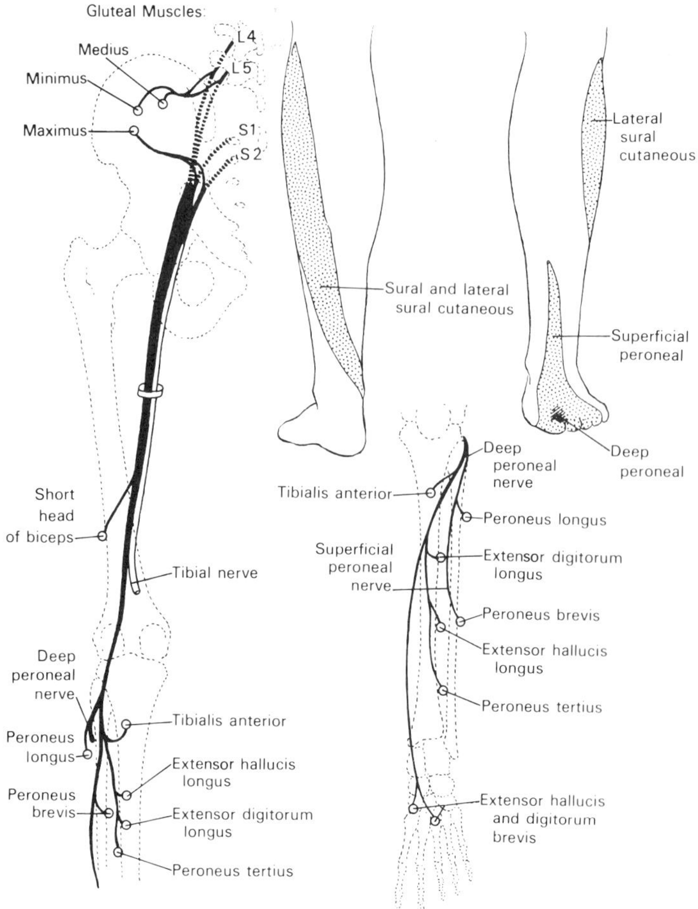

Fig. III-90. The distribution of the sciatic nerve through its common peroneal division. (Shaded areas indicate sensory supply.)

(Achilles' tendon) is the common tendon of the Triceps Surae. The Plantaris seems to be an ativistic muscle in most of us but capable of some plantar flexion in others. Its tendon is often easily ruptured as in the case of sprinters and ballet dancers. Surprisingly, the pain of rupture is often so acute as to render it impossible to bear weight upon the limb which seems to collapse at once. A 2-inch wad of cotton wool placed in the heel of the shoe works wonders, and locomotion is almost instantly possible.

The deep groups are formed by the **Popliteus,** the **Tibialis Posterior,** the **Flexor Hallucis Longus,** and the **Flexor Digitorum Longus.** The Popliteus, from its name, can be expected to arise highest in the back of the leg. It arises from the lateral femoral condyle and crosses the knee joint diagonally in the popliteal fossa to insert on the upper medial aspect of the tibia. It is an internal rotator of the knee joint when the leg is free, but in standing, it externally rotates the femur. The Tibialis Posterior rises from the posterior aspect of the tibia and fibula and the intermuscular septum between them and runs down the midline of the leg with its tendon gradually trending medialward. It is flanked and overlapped by the Flexor Digitorum Longus arising from the tibia and the Flexor Hallucis Longus arising from the fibula. The tendons of these two muscles must obviously have to cross if the latter is to reach the great toe and the former the other toes.

The tibial nerve and popliteal vessels have to cross the Popliteus in the fossa, and at the lower border of this muscle, the popliteal artery branches into the anterior tibial and posterior tibial arteries. The former we found in the anterior compartment; the latter accompanies the tibial nerve, the nerve of the posterior compartment, down the leg. The Tibialis Posterior is an inverter of the foot. The other muscles plantar flex the toes.

5. **The Ankle and Foot:** We have already noted the medial and lateral malleoli as subcutaneous bony prominences raised by the distal portions of the tibia

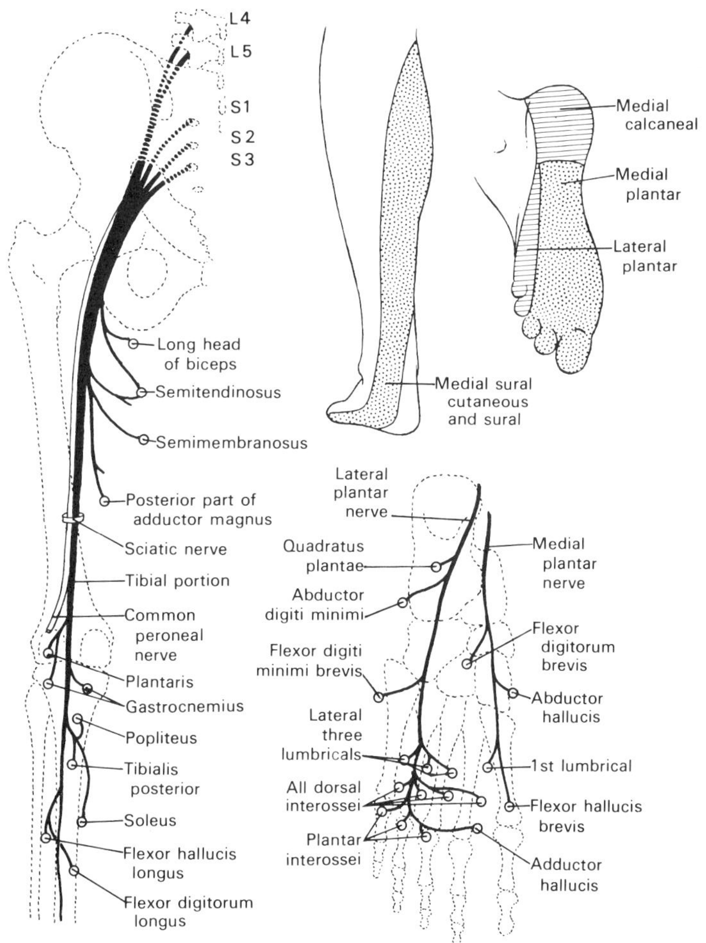

Fig. III-91. Distribution of the sciatic nerve through its tibial division. (Shaded areas indicate sensory supply.)

and fibula at the ankle (Figs. III-85A, 87A, and 88). They act as pullies for certain tendons and their sheaths as these make their way into the foot. Behind the lateral malleolus, bound down by peroneal retinacula, the tendons of the Peroneus Brevis and Longus pass in that order from anterior to posterior. The order of progression, anterior to posterior behind the medial malleolus, is the tendon of the Tibialis Posterior, tendon of the Flexor Digitorum Longus, posterior tibial vein, artery, tibial nerve, and, finally, the tendon of the Flexor Hallucis Longus, all bound down by the flexor retinaculum. Anteriorly the great saphenous vein courses in front of the medial malleolus; the lesser saphenous, in front or over the lateral malleolus (Figs. III-87A and 88B). The calcaneus (Achilles') tendon is a prominent structure posteriorly where the Triceps Surae insert upon the heel (Calx) (Figs. III-84, 87A, and 88A).

There are seven tarsal bones forming the ankle. The largest of these is the **calcaneus** which projects posteriorly to form the heel. A large block-shaped **talus** sits upon the calcaneus and transmits all of the body weight from the tibia to all of the other bones of the foot. The **navicular** articulates proximally with the head of the talus and distally with **three cuneiforms,** small wedge-shaped bones. Finally, a **cuboid** articulates with the calcaneus proximally and with metatarsals IV and V distally. Each cuneiform articulates separately with **metatarsals** I, II, and III. The **phalangeal structure** is similar to that in the hand.

Inversion and eversion take place at the talocalcaneal joint and at the midtarsal joints (calcaneocuboid and talonavicular). These rotating movements can be impaired in arthritis, making it very difficult to walk over rough ground or down slopes. Inversion, recall, is the function of the anterior and posterior tibial muscles; eversion, that of the peronei.

When you stand, most of the weight is on the heel and the distal heads of the metatarsals, but a wet footprint will also show markings of the tips of the toes and the whole lateral aspect of the sole. The articulated foot in your bone box makes it clear that the foot has two longitudinal arches: the medial arch consisting of calcaneus, talus, navicular, three cuneiforms, and three metatarsals; and a lower lateral arch consisting of the calcaneus, cuboid, and the lateral two metatarsals. These give rigid support to the body in standing, for then the bones interlock and the many ligaments of the foot are in maximum tension. But in walking the bones unlock, and the ligaments and the tendons of certain muscles act as mobile levers, giving a spring-board action to locomotion.

There are many small muscles in the foot itself upon its plantar aspect which we have not discussed here (see Fig. III-91). As the posterior tibial nerve enters the foot after winding behind the medial malleolus it divides into medial and lateral plantar nerves. The **lateral plantar nerve** behaves like the ulnar in the hand, supplying most of these intrinsic muscles; the **medial plantar nerve,** like the median nerve of the hand, supplies fewer of them.

The course of the vessels in the leg is shown, at least as principal branches, in Figure III-92. In Figure III-93 are shown the superficial veins of the leg which correspond to the cephalic and basilic veins of the arm. You should attempt to feel the pulse of the following vessels:

1. *Femoral artery.*
2. *Popliteal artery.* This is not always easy to do, particularly on yourself. Have a partner lie down, flex his knee, and relax his muscles by resting the leg on your arm. Press firmly in the popliteal fossa to compress the vessel against the femur.
3. Feel for the pulse of the *dorsal pedis artery* (Fig. III-85A) between the tendons of the Extensors Hallucis and Digitorum Longus. Do not be too surprised of you can't feel it. It may be normally absent in about 15% of individuals.
4. Try to feel the pulse of the *posterior tibial artery* about a fingerbreadth behind the medial malleolus. Sometimes this vessel is replaced by a peroneal artery rounding the lateral malleolus. Even if you can't feel either dorsal pedal or posterior tibial pulses, it does not necessarily mean that anything is drastically wrong with your subject.
5. You should look for both *greater* and *small saphenous veins,* especially in the lower part of the leg near the ankle in normal subjects. You should see some sign of them, for the Arabs saw them long ago and gave them their name which means "visible veins." The greater is almost certain to be found coursing anterior to the medial malleolus. It runs up along the medial side of the leg to the knee and then follows the course of the Sartorius muscle diagonally across the thigh to empty into the femoral vein about an inch below the inguinal ligament in the region of the femoral triangle. The small saphenous crosses the lateral malleolus to come up the back of the leg to empty into the popliteal vein in the fossa.

The cutaneous nerves: The distribution of these nerves to skin of the lower extremity is shown by the shaded areas in Figures III-89 to 92. More will be said about their course and distribution in a subsequent chapter. Figures III-75 to 77 illustrate comparable distributions to the skin of the upper extremity.

The Rule of 5: As a conclusion to this chapter on the salient aspects of surface anatomy, let us next attend to Figure III-94 which depicts an adagio movement of the ballet terminating in a split. Its purpose is to demonstrate *a simple rule of "5,"* where the number "5" becomes the key for localizing the general distribution of a sensory supply to the skin areas of the body. Recall the distribution of the *5th* cranial nerve to the skin of the face. This area lies in front of the interaural line drawn from in front of one ear over the cranium to the other and down the jaws from their angles to the symphysis menti. No other nerve supplies the skin of the face. But *behind* the ear, a small patch of skin gets its sensory supply from *2 × 5*, or from the *10th* cranial nerve.

The rest of the skin of the scalp posterior to the coronal suture and the skin over the neck to the shoulders is supplied by the sensory twigs of two cervical spinal nerves, $\mathbf{C_2}$ and $\mathbf{C_3}$ (which add to *5*). (Spinal nerve C_1 has no *sensory twigs,* only *motor* ones to certain small, deep muscles in the back of the neck (suboccipital triangle), and nerves C_5–T_1 will not give a sensory supply to the trunk.)

The *5th thoracic nerve* supplies a brassierelike belt of skin around or just below the nipple, and *2 × 5*, or the *10th thoracic* nerve, a beltlike zone around the umbilicus.

You have *5* fingers, generally, at the end of each hand. The upper extremity, as we have already noted, gets its sensory supply from *5* spinal nerves, beginning with C_5 and ending with T_1. (Some supply from trunk dermatomes C_4 and T_2 does encroach upon the upper extremity in the shoulder area and along the upper medial aspect of the arm.)

You have *5* toes on the end of your foot (count them). The lower extremity derives its sensory supply from the middle *5* of *10* spinal nerves (*5* lumbar and *5* sacral nerves of the lumbosacral plexus). Fortunately, these *5* begin with L_3 and end with S_2 in progressive order (which, once more, add together to make *5*). (Like C_4 and T_2, the trunk dermatome L_2 does encroach upon the anterior thigh.)

Finally, the last of the sacral dermatomes, beginning with S_3 and ending with S_5, form smaller and smaller sensory, concentric rings around the anus. The latter supply is reminiscent of the flight pattern of that strange Australian bird, the Kiwi, which is said to soar in closer and closer circles until, with a cry of utter anguish and despair, it disappears from sight up its own cloacal canal.

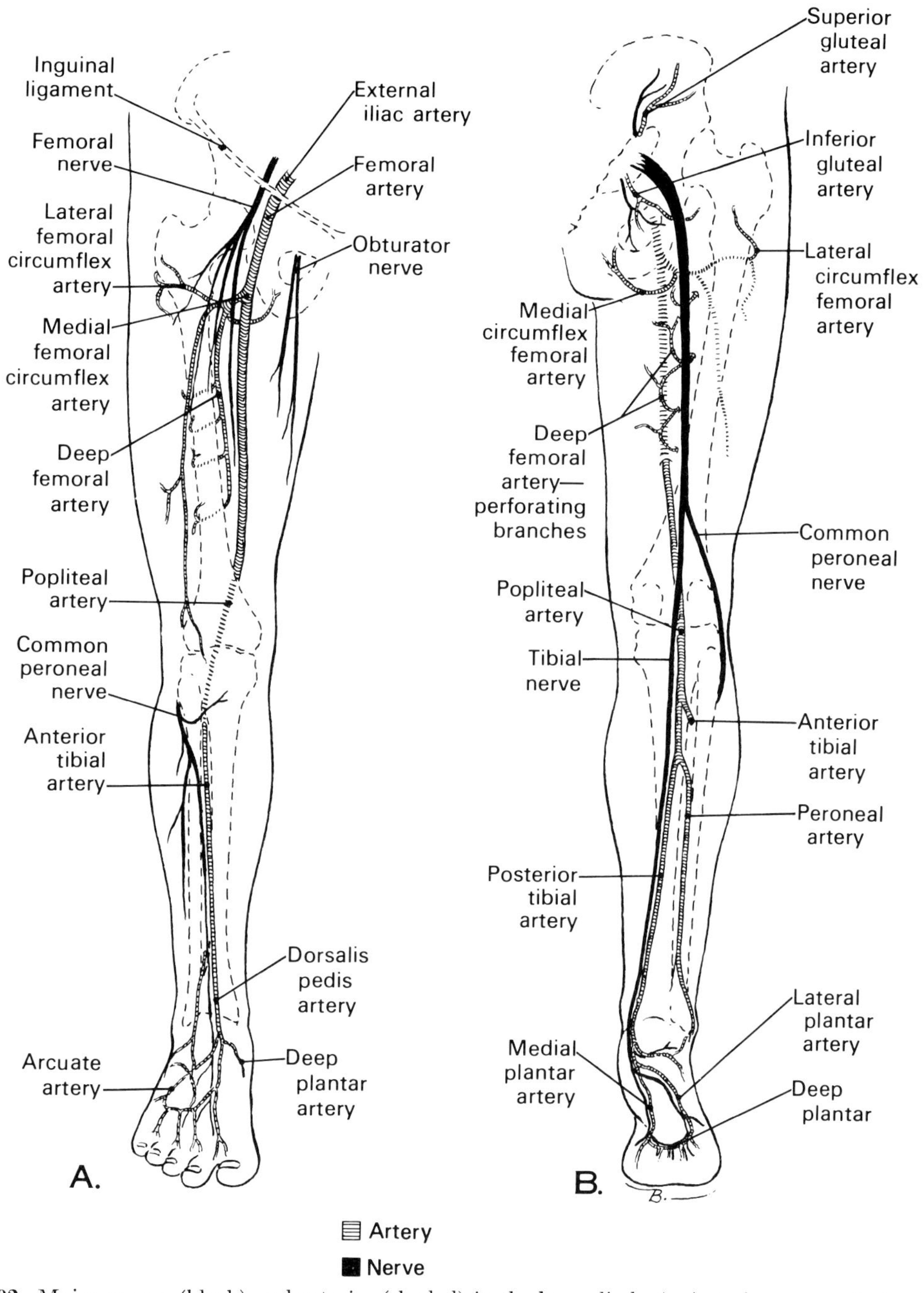

Fig. III-92. Major nerves (black) and arteries (shaded) in the lower limb. **A.** Anterior aspect; **B.** Posterior aspect.

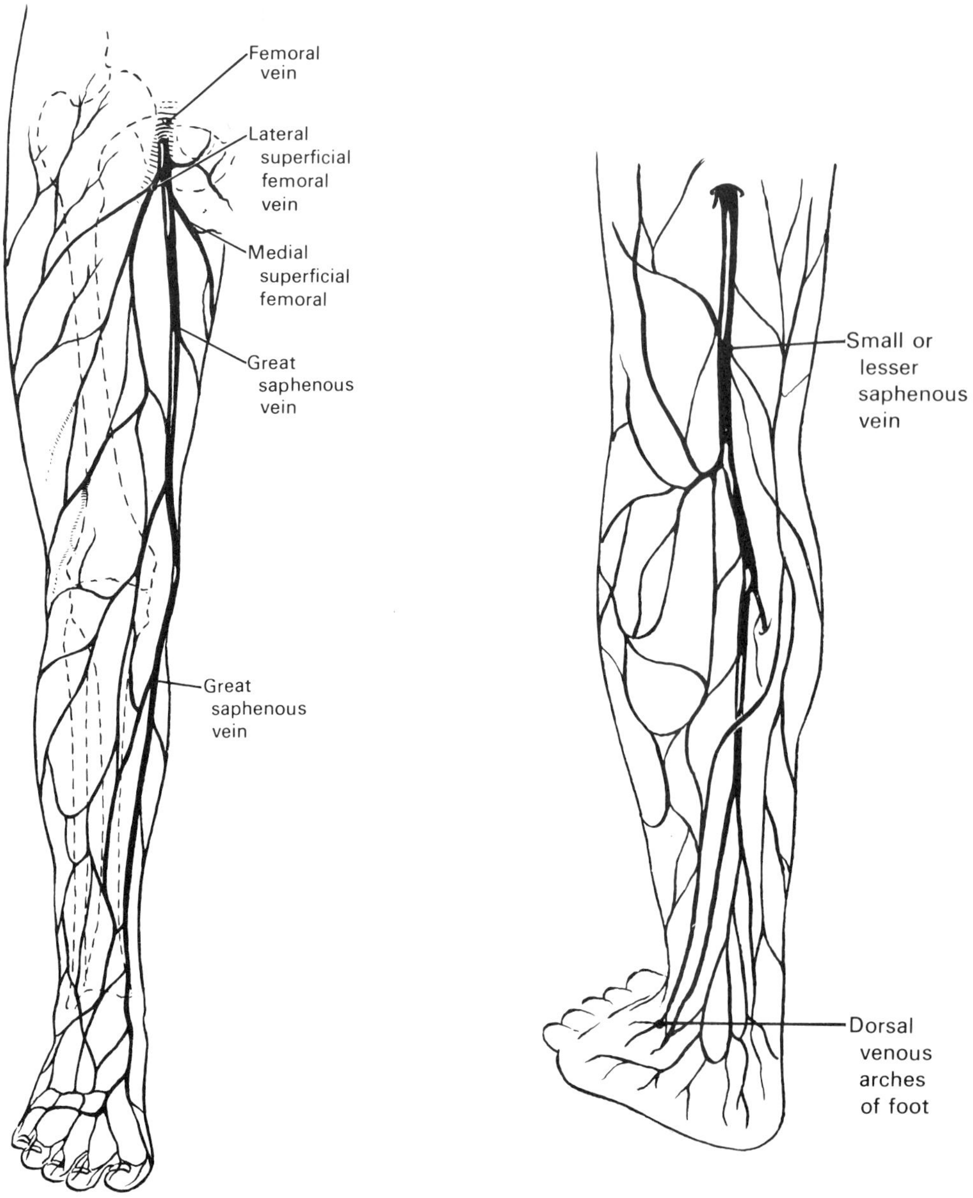

Fig. III-93. Superficial veins of the lower limb.

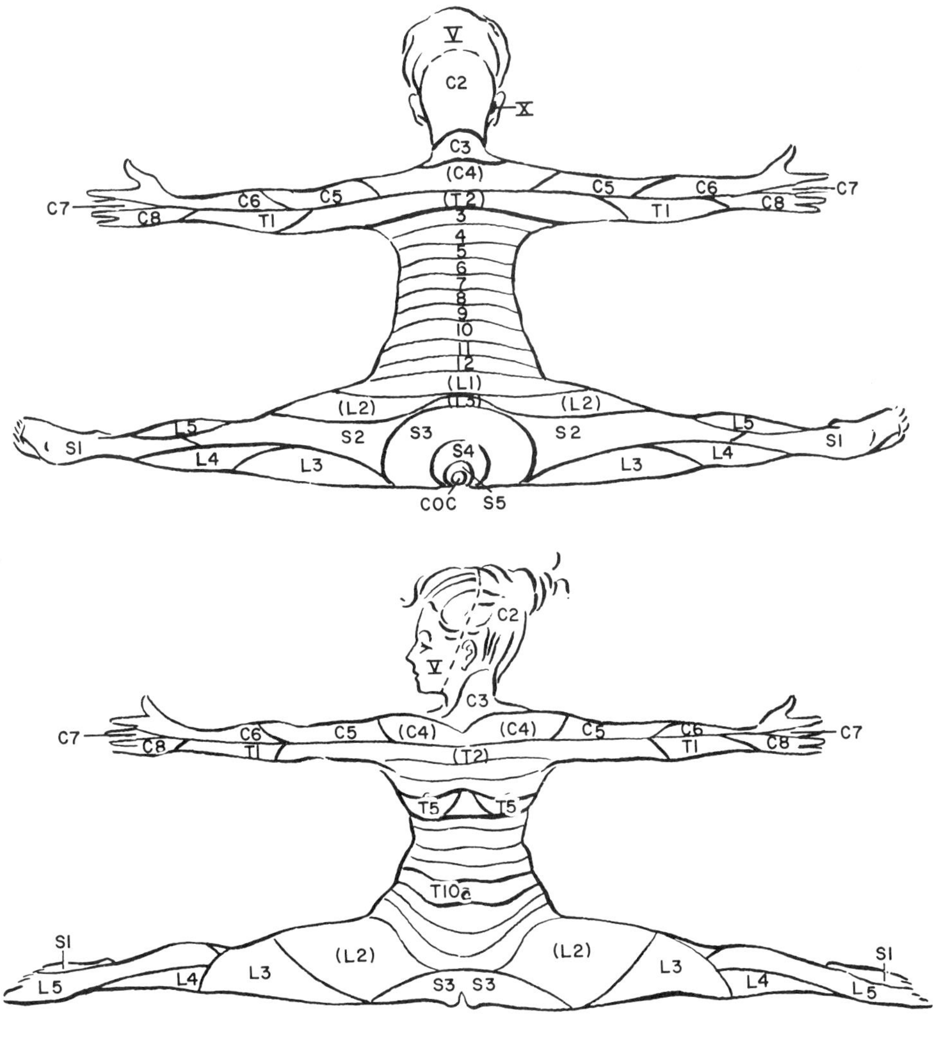

Fig. III-94. The Dermatomes. Above, posterior view; below, anterior view. Note absence of cutaneous innervation by the first cervical segment.

REFERENCES

A. General Systemic Reference Texts

These are useful for minute details of anatomy and for their illustrations. They are arranged by systems, *e.g.,* osteology, arthrology, myology, blood vessels, the nervous system, digestive system, *etc.* They are best used as you would a reference encyclopedia; hence, use the index freely in locating specific topics.

1. Goss, C. M. (Ed.): *Gray's Anatomy,* 28th ed., Lea & Febiger, Philadelphia, 1967.
2. Anson, B. J. (Ed.): *Morris' Human Anatomy,* 12th ed., McGraw-Hill, New York, 1966.
3. Romanes, G. S. (Ed.): *Cunningham's Textbook of Anatomy,* 10th Ed. Oxford, London, 1964.

B. Regional Texts

These are more useful texts to consult while dissecting in a given region than are the systemic texts.

4. Hollinshead, W. H.: *Textbook of Anatomy,* 2nd ed. Hoeber (Harper-Row), New York, 1967.
5. Gardner, E., Gray, D. J. and O'Rahilly, R.: *Anatomy.* Saunders, Philadelphia, 1967. (Considerable clinical information.)
6. Grant, J. C. B. and Basmajian, J. V.: *Grant's Method of Anatomy.* Williams & Wilkins, Baltimore, 1965. (An excellent deductive approach to anatomy.)
7. Lockhart, R. D., Hamilton, G. F., and Fyfe, F. W.: *Anatomy of the Human Body.* Lippincott, Philadelphia, 1965. (Well illustrated; clinical details available that are not likely to be presented anywhere else.)

C. Clinical Anatomy

8. Appleton, A. B., Hamilton, W. J., and Simon, G.: *Surface and Radiological Anatomy.* Williams & Wilkins, Baltimore, 1958.
9. Ellis, H.: *Clinical Anatomy.* Davis, Philadelphia, 1966. (A short readable gem full of clinical pearls. Best read after you know some anatomy.)
10. Brantigan, O.: *Clinical Anatomy.* McGraw-Hill, New York, 1963.

D. Specific References in this Chapter

11. Best, H. C., and Taylor, N. B.: *The Physiological Basis of Medical Practice,* Ch. 77, p. 1185. Williams & Wilkins, Baltimore, 1955.
12. Pfaffman, C.: Gustatory afferent impulses. J. Cell. Comp. Physiol. *17:* 243, 1941.
13. Gehweiler, J., Becker, R. F. and Wilson, J.: *Head-Hunting on the Eno River.* F. J. Ludlow, Ltd., Pelham, England, 1890. (Out of print.)
14. White, P. D.: *Heart Disease.* Macmillan, New York, 1931.
15. Brock, R. C., *The Anatomy of the Bronchial Tree,* Ch. IV. Oxford, London 1954.

FOUR

An Introduction to Roentgenology

INTRODUCTION

The history of medicine abounds in great scientific achievements. Not all of these monumental accomplishments are credited to physicians, but rather to the talents of men in other, seemingly unrelated, fields. Such is the case in the discovery of X-rays by a physicist in the year 1895. To many people this discovery ranks with the investigations of Harvey on the circulation of blood. In 1896 when scientists throughout the world learned of the X-ray, an entirely new era in medical knowledge was born. Progress has been rapid during the years in devising and perfecting new techniques in roentgen diagnosis and therapy. Much is yet to be achieved, and many of you reading this book will be in the forefront of the next round of great medical accomplishments.

To aid in illustrating the anatomy under discussion, we have incorporated many radiographs (X-ray films, roentgenograms) throughout the following chapters of this book. Experience has taught us that the use of normal radiographs readily complements the study of human anatomy as well as pathology, physiology, and the clinical disciplines. Exposure to these normal radiographs will help train the eye to see and to observe carefully and to reason through every problem presented you. Each shadow cast on the radiograph and every detail of your laboratory specimens should be observed, and their significance should be fully comprehended, not only for today and the approaching examination, but in terms of their practical application to the everyday practice of medicine. The structures that we present radiographically have a definite significance to one or more of the specialty fields of medicine. We have purposely excluded the minutia and have emphasized the "core" of information which you must know before your clinical years begin.

HISTORY

In order to study normal radiographic anatomy we must first learn about X-rays (röntgen or roentgen rays) and how the radiograph is produced. In this manner you will have an appreciation of why the images cast on the film appear as they do. This chapter, then, will introduce you to the art and science of roentgenology.

In November 1895, Wilhelm Conrad Röntgen, Professor of Physics at the University of Würzburg in Germany (Fig. IV-1), accidentally discovered the new form of radiant energy which he later termed X-rays. Röntgen had been totally absorbed for several years in his experiments on **cathode rays** before his monumental discovery occurred. Cathode rays, which are accelerated **electrons,** can be generated within a partially evacuated tube which has been electrically energized (Figs. IV-2 and 3). Previous investigators had shown that cathode rays could cause darkening of photographic plates and fluorescence of certain chemical substances. They were known to make air electrically conductive, and they were also known to be absorbed after a few centimeters' passage through air.

During one of his experiments, Röntgen covered a Crookes tube with heavy black cardboard. While energizing the tube within his darkened laboratory, he noticed a faint luminescence some distance away from the tube. The luminescence emanated from a barium platinocyanide screen lying on a nearby bench. Röntgen realized, and soon proved, that cathode rays could not penetrate air for such a distance to induce the fluorescence. Apparently a new form of radiant energy was produced as the stream of cathode rays struck the thick-walled glass tube (Fig. IV-3). Over the next few weeks, Röntgen found that the unknown ray had remarkable penetrating ability and was capable of passing through paper, wood, and thin sheets of metal. While moving his hand between the energized tube and the barium platinocyanide screen one day, he first distinguished the bones within his hand. With the substitution of photographic plates for the crude fluorescent screen to document his findings, the science of roentgenology was born.[1] Figure IV-4 is an early radiograph made by Professor Röntgen of the hand of the famous anatomist and the father of histology, Albert von Kölliker.

Röntgen published his first paper[2] on the new form of radiant energy ("Ueber Eine Neue Art von Strahlen") which he called X-rays within two months of the discovery. Two other classical papers[3, 4] soon followed in which Röntgen accurately described many of the physical properties of X-rays. The scientific community was astounded by the discovery, and its vast significance, particularly to medicine, became immedi-

Fig. IV-1. Wihelm Conrad Röntgen—1845–1923. Awarded the first Nobel Prize in Physics in 1901 for his discovery of X-rays. (Courtesy of the General Electric Company, X-ray Department.)

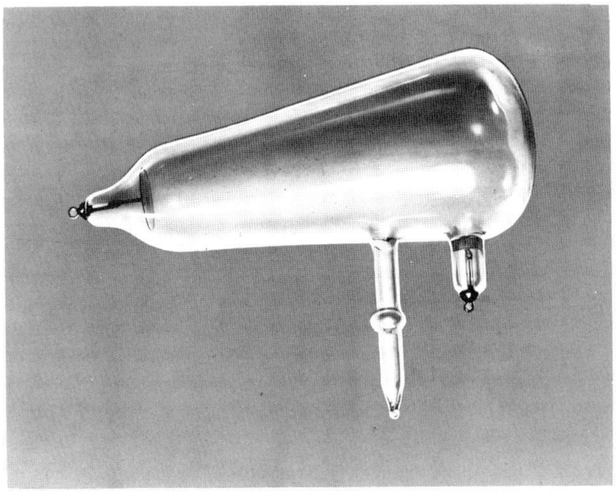

Fig. IV-2. Crookes tube, typical of the type used by Röntgen to produce X-rays. (Courtesy of the General Electric Company, X-ray Department.)

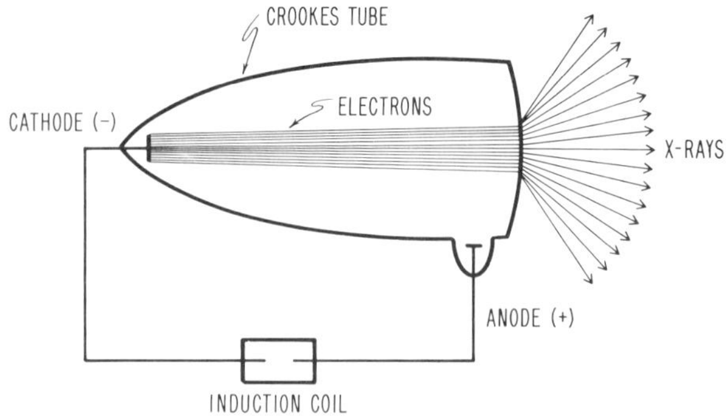

Fig IV-3. Schematic diagram illustrating the generation of X-rays in a Crookes tube. The induction coil maintained a high voltage across the tube. Electrons produced at the cathode were accelerated across the tube to strike the thick-walled glass end. The interaction of these high speed electrons with atoms in the glass resulted in the production of X-rays and heat.

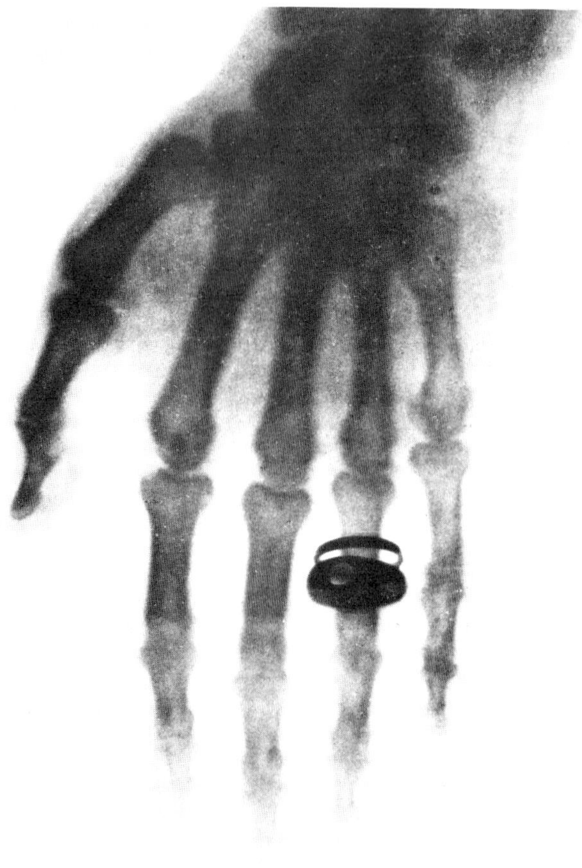

Fig. IV-4. Radiograph of the hand of the father of histology, Professor Albert von Kölliker. This radiograph was made on January 23, 1896, by Wilhelm Conrad Röntgen. (Courtesy of the General Electric Company, X-ray Department.)

ately obvious. Within a few years hundreds of papers about Röntgen's rays were written, many of which described new developments in the generation of these rays. For his discovery of X-rays, Röntgen was awarded the first Nobel Prize in Physics in 1901.

GENERATION OF X-RAYS

Radar, radio, light, and X-rays are all **electromagnectic waves,** and these make up a part of the large electromagnetic **spectrum** illustrated in Figure IV-5. All electromagnetic waves have certain properties in common:

1. They are without mass.
2. They travel in straight lines unless deflected by an object; however, their new direction is also linear.
3. They travel at the speed of light in a vacuum (3×10^{10} cm. per second or 186,000 miles per second).
4. They do not require a material medium in which to travel.
5. They are termed "electromagnetic" since their effects are due to their electrical and magnetic forces on charged particles (usually electrons).
6. All are produced by the acceleration of charged particles.

Many of the physical properties of electromagnetic radiation can be explained if these radiations are thought of as having a dual nature; that is, behaving as both waves and particles.[7] X-rays have such short wave lengths that they are measured in terms of **Angstrom units** (1 Angstrom = 10^{-8} cm.). It is the extremely short wave length of X-rays that permits their penetration of matter, and it is this distinction that makes them useful in the study of the anatomy of the body.

In order to generate X-rays, electrons traveling at high speed must be stopped suddenly by some material object, called a **target.** In so doing, neither the target atoms nor the electrons are permanently affected. The X-rays produced are a manifestation of the loss in kinetic energy by the electrons when stopped suddenly by the target atoms.

X-ray tubes have been markedly improved since Röntgen's time. Here we will examine two types of diagnostic tubes, the **stationary anode** and the **rotating anode tube.** Figure IV-6 is a diagram of a stationary anode X-ray tube, which is an evacuated glass envelope containing two electrodes: the **anode** (+) and the **cathode** (−). At the cathode end there is a thin tungsten wire filament which is inserted into a shallow holder referred to as a **focusing cup.** When the spiral wire filament is heated to incandescence, electrons literally "boil off." The hotter the filament temperature becomes the greater the number of electrons that are available. These electrons are given high speed when a high voltage is applied between the cathode and the target anode. The negatively-charged electrons are repelled by the negatively-charged cathode and they are attracted to the positively-charged anode. The electrons are directed by the focusing cup to strike a small rectangular area of the target known

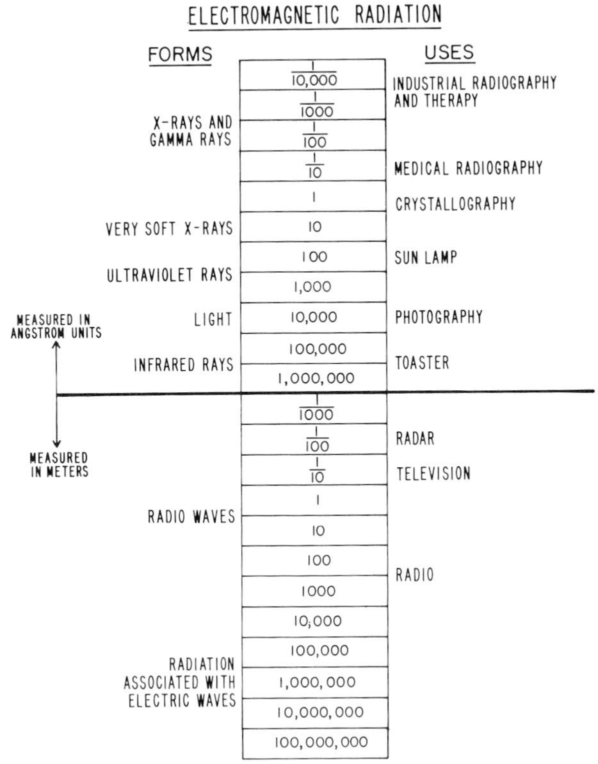

Fig. IV-5. The spectrum of electromagnetic radiation. (Modified from *The Fundamentals of Radiography,* 10th ed. The Eastman Kodak Company, 1960.)

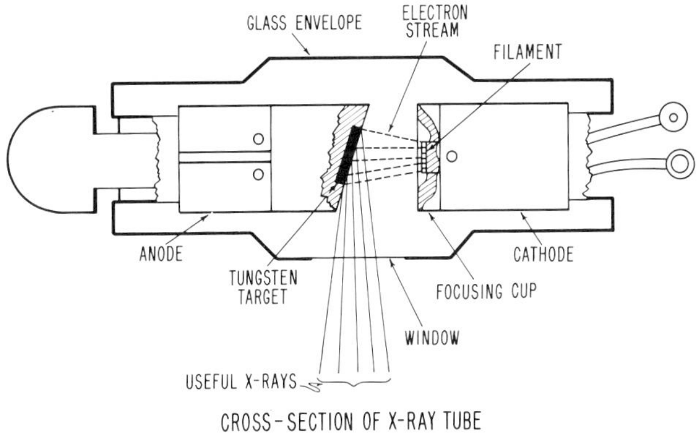

Fig. IV-6. Diagram of a stationary anode diagnostic X-ray tube. (Courtesy of the General Electric Company, X-ray Department.)

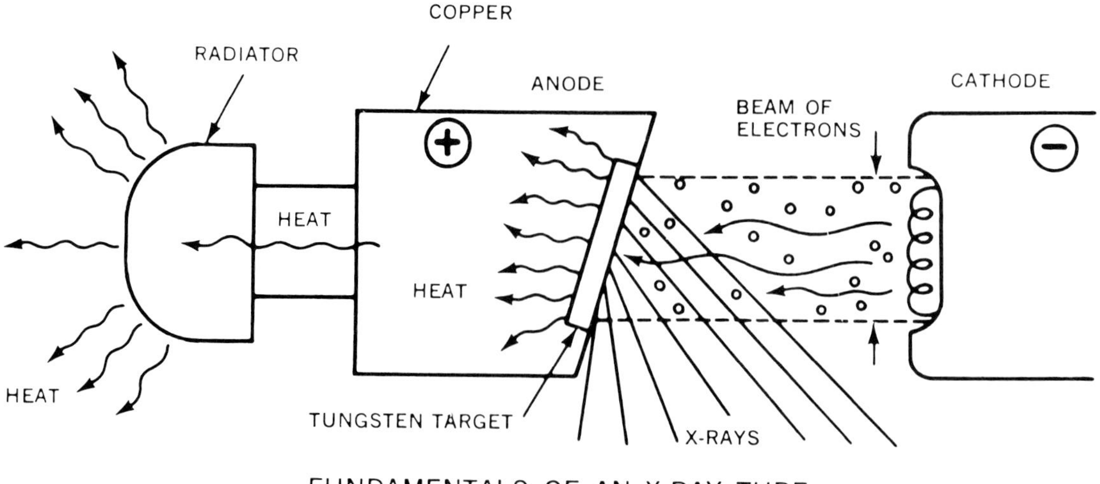

Fig. IV-7. Composite close-up of the anode and cathode in a stationary anode X-ray tube. (Courtesy of the General Electric Company, X-ray Department.)

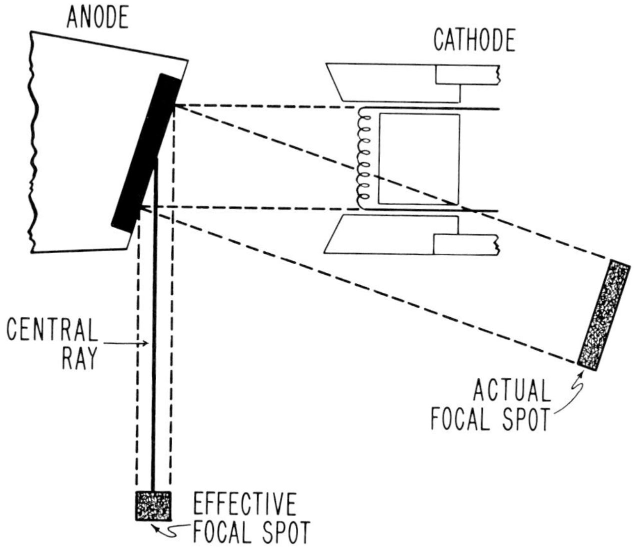

Fig. IV-8. Diagram showing the relative size of the actual focal spot of the tungsten target, and the smaller projected or effective focal spot. This effect is made possible by angling the tungsten target 16 to 20° from the vertical plane of the tube. The smaller effective focal spot permits greater definition of images on the radiograph. (Modified from *The Fundamentals of Radiography,* 10th ed. The Eastman Kodak Company, 1960.)

as the **focal spot.** The sudden stoppage of the high-speed electrons by atoms in the target results in the loss of energy by the electrons and the production of X-rays and heat (Figs. IV-7 and 8).

The **target** usually consists of a small tungsten button imbedded in the copper anode. Tungsten is used as the target material since its high melting point permits it to withstand the high temperatures to which it is subjected, and its high atomic number permits an increased efficiency of X-ray production. Note that the target face is set at an angle of approximately 16 to 20° which permits a larger surface area **(actual focal spot)** for the electron beam to strike (Fig. IV-8). The **projected** or **effective focal spot** size as seen from a position below the tube window is much smaller, and it actually appears to be a small square. By utilizing X-rays emerging from such a small effective focal spot, radiographic definition is markedly improved. The anode heat capacity is increased by spreading the electron beam over the larger area of the actual focal spot.

Thus far we have considered only the operation of a stationary anode tube. With the greater demands for tubes with smaller focal spots and greater heat capacity, manufacturers have developed the rotating anode tube (Fig. IV-9). As you can see in the diagram, the anode is disk-shaped with a beveled edge. The anode disk rotates at very high speed on an axis through the center of the tube. The cathode filament is offset so that the electron beam is directed against the beveled edge. This permits the electrons to strike a cool surface of the target at all times, while keeping the target focal spot small. The heat produced is distributed over the broad area of the disk rather than one small area.

It is important to remember that the higher the voltage applied across the tube, the greater will be the energy of the electrons interacting with the target

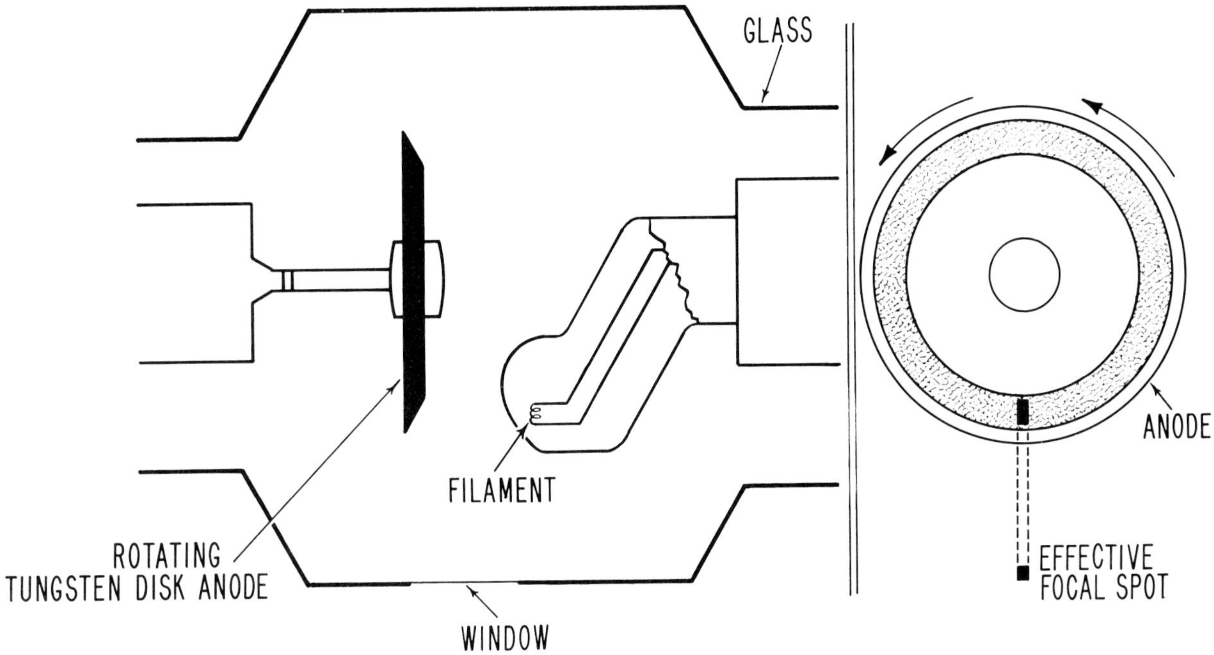

Fig. IV-9. The anode-cathode complex of a rotating anode X-ray tube (left). The edge of the anode is beveled to provide a small effective focal spot. At right, the anode face-on showing the actual focal spot and the projected or effective focal spot. (Modified from *The Fundamentals of Radiography,* 10th ed. The Eastman Kodak Company, 1960.)

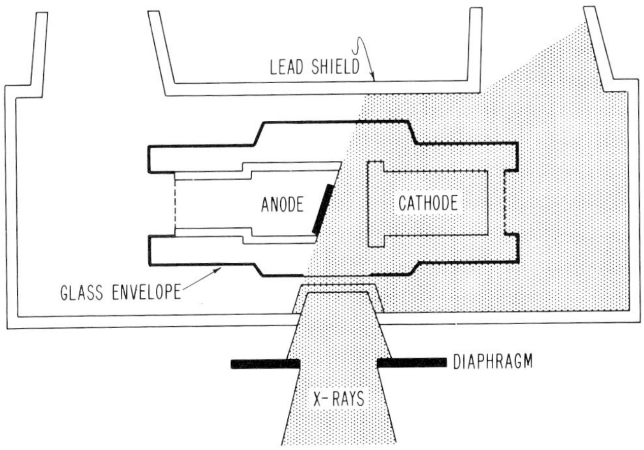

Fig. IV-10. Diagram showing that the X-rays produced in the tungsten target radiate from the target surface in straight lines through a 180° arc. The useful X-ray beam is seen emerging through the tube and lead shield windows. The remaining X-rays are absorbed by the components of the tube, the oil bathing the tube, and the lead protective shield. (Modified from *The Fundamentals of Radiography,* 10th ed. The Eastman Kodak Company, 1960.)

atoms, and, therefore, the higher the energy of the resultant X-rays and the shorter their wave lengths. The majority of the high-speed electrons striking the target do not give up all of their energy in a single encounter with the target atoms, but rather, through a series of collisions, contribute to a spectrum of X-rays of varying energies and wave lengths. Some of these X-rays are produced below the target surface and are absorbed in the target material itself, while others are radiated from the surface of the target through a 180° arc. Unfortunately, X-rays cannot be focused, and therefore, only a small portion of the *beam* is *useful*—that part emerging through the window provided (Fig. IV-10) in the tube and its surrounding metal housing. The remaining X-rays are absorbed in the glass envelope of the tube, in oil surrounding the tube, and in the metal tube housing. The efficiency of X-ray production is very low, less than 1% in the diagnostic energy range. It is beyond the scope of this book to consider the circuitry of an X-ray machine. Several excellent references on this subject are given at the end of this chapter.[5-7]

CHARACTERISTICS OF X-RAYS

Photographic Effect

Now that we have learned how X-rays are produced, let us examine some of the properties and effects of X-rays as they apply in the production of a radiograph. Those of you familiar with photography will remember that a **latent image** is formed in a photographic emulsion after its exposure to light. The image

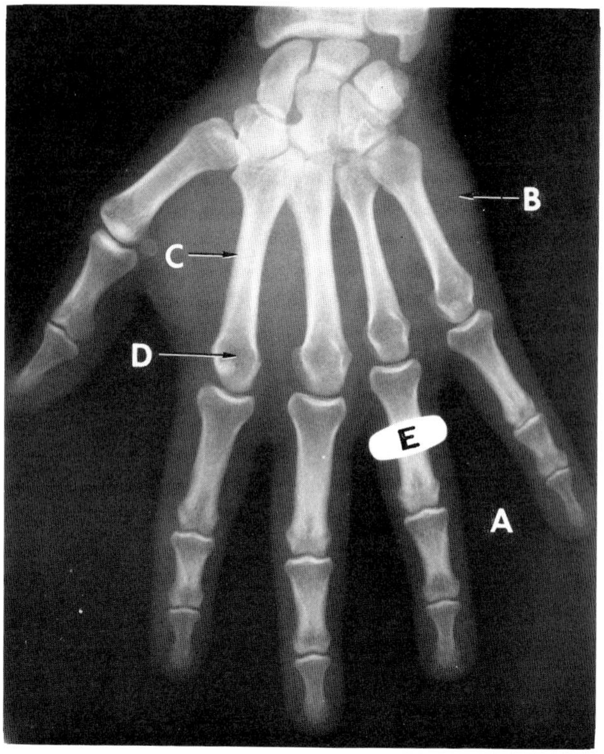

Fig. IV-11. Posterior-anterior (PA) view of the right hand. The relative absorption of X-rays in: A, air; B, soft tissue: C, dense cortical bone; D, less dense cancellous bone; and E, gold ring is shown.

is brought out by appropriate chemical development and fixation. X-rays also have a direct **photochemical effect** on photographic film; however, this direct effect is much less rapid than that of light. The radiograph of a hand shown in Figure IV-11 is a negative comparable to the negatives you produce with your camera. Those areas of the film which have been exposed to X-rays are shades of black, depending upon the differential X-ray **absorption** in the various tissues in the part under examination. We will explain the absorption of X-rays in greater detail shortly. Those areas of the film which have been completely shielded from the radiation will appear white.

In order to enhance the small direct photochemical effect of X-rays on photographic film, special techniques have been developed. In radiography, a film which has been coated on both sides with sensitive silver bromide emulsion is placed in a holder called a **cassette.** The film is closely sandwiched in the cassette between two **screens** coated with calcium tungstate crystals (Fig. IV-12). Upon striking these crystals, X-rays cause **fluorescence** or the production of visible light. The radiographic film used is extremely sensitive to the blue visible light emitted by each screen. Thus we can see that when the X-rays strike the cassette, there is not only a small direct photochemical effect by the X-rays, but also an image registered from the light-producing screens. By using screens, the radiation exposure to the patient can be markedly reduced. The time of exposure may also be shortened, and this prevents blurring of the image by motion of the patient.

Absorption

If an object is placed between the X-ray tube and the film cassette, some of the X-ray beam will be **attenuated** by either **absorption** or **scattering** before striking the film. It is the difference in X-ray attenuation between adjacent structures which permits their differentiation on the radiograph. Absorption is dependent upon several interconnected factors, including the wave length of the X-rays, chemical composition, density, and thickness of the object encountered.

X-rays produced at lower kilovoltage have a longer **wave length** and are therefore more readily absorbed than X-rays produced at higher kilovoltage and having shorter wave length. Tissues of similar thickness show differences in absorption based upon their **effective atomic number** and **density** as we will illustrate in Figure IV-11. The greater the effective atomic number and density, the greater the degree of absorption. As you can see in Figure IV-11, bone absorbs more radiation than soft tissues, and therefore, fewer X-rays penetrate to the film below. The effective atomic number of bone is 11.6, while that of soft tissue and water is 7.4. Now notice that the gold ring does not permit any penetration of the X-ray beam. (Such an area of whiteness on the radiograph, where little or no radiation was permitted to strike the film, is termed **radiopaque.**) If you will consult a periodic table of the elements you will see that gold has an atomic number of 79, and therefore, you would expect it to be a better absorbant than bone or soft tissue. Where would you place the absorption of X-rays in fat in this scheme, knowing that fat has an effective atomic number of 6.3? You would expect that fat would not absorb quite as effectively as soft tissue or water and thereby appear slightly darker gray than those images cast by soft tissue.

Now note that the areas outside of the soft tissues are black, or **radiolucent.** In these areas the X-ray beam passed through the air surrounding the hand. There is no significant difference in the effective atomic numbers of air (7.64) and soft tissue (7.4) to account for the difference in their appearance on the radiograph. Therefore, the differential absorption must be due to differences in **density** as measured in grams per cubic centimeter. The density of soft tissues is 1.0 gm. per cm.^{3}, while that of air is 0.0013 gm. per cm.^{3}; or in other words, air is one $\frac{1}{1000}$ as dense as soft tissue.[5] This enables the X-ray beam to penetrate air readily. If you now re-examine the bones you will see that the denser compact cortical bone along the shaft is a greater absorber of the X-rays than the less dense, cancellous bone at the ends of the shaft.

Thickness also affects absorption as illustrated in Figure IV-13. The X-ray beam has been passed through an aluminum step wedge consisting of pro-

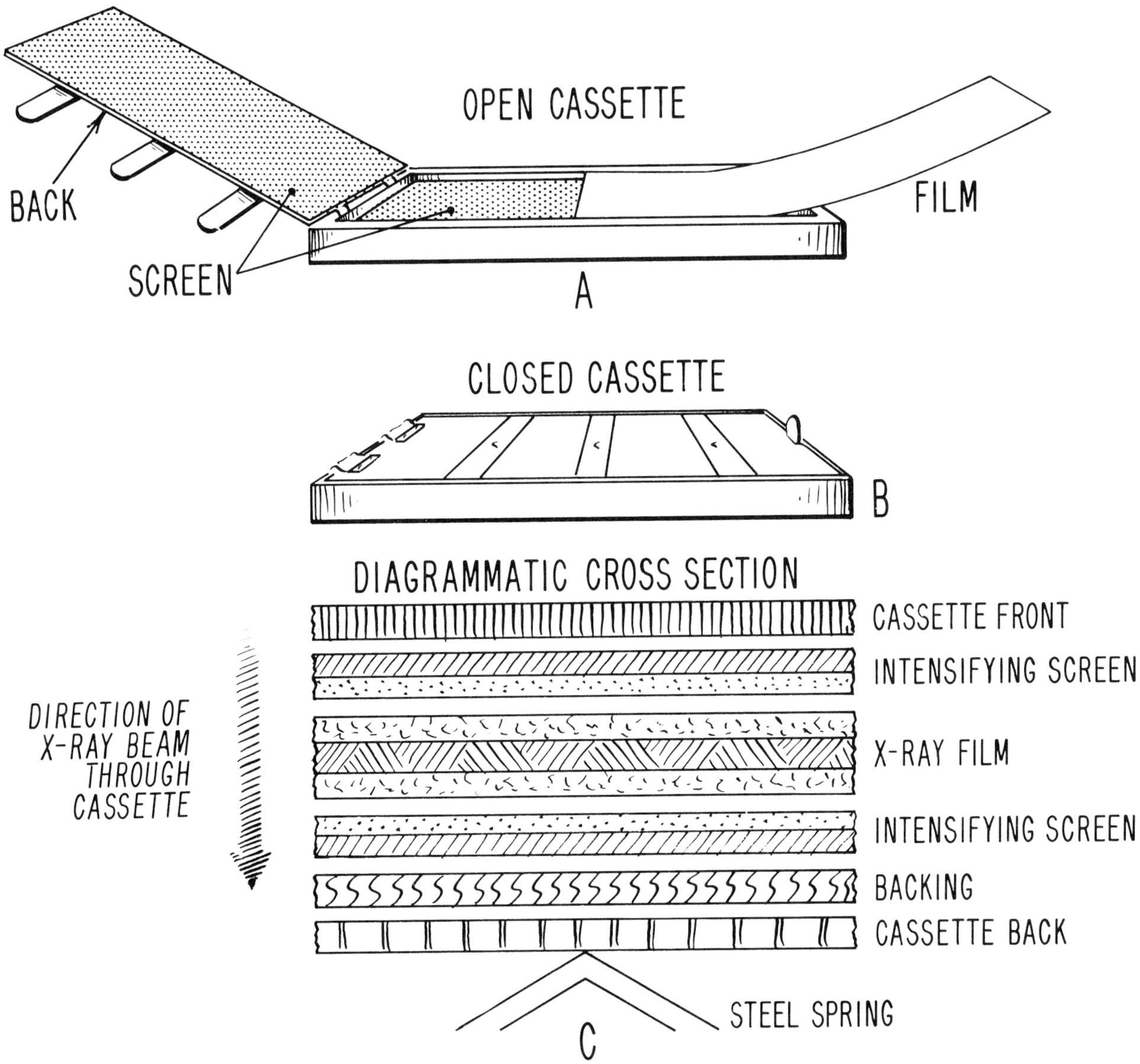

Fig. IV-12. The radiographic cassette. **A.** Open cassette placed face down, ready for loading with unexposed X-ray film. **B.** Cassette closed after loading. **C.** Cross section of a cassette in position for use. Note the layers that the X-ray beam passes through. X-ray films have a double emulsion, and therefore, a screen is placed on each side of the film in order to intensify the image. The use of screens permits less radiation to be used in order to achieve desired results.

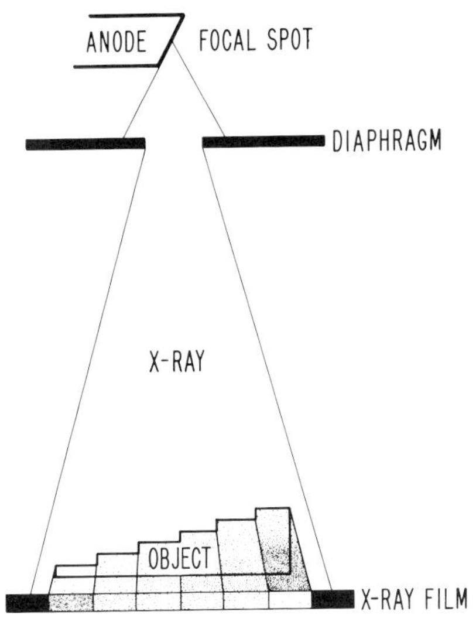

Fig. IV-13. Schematic diagram illustrating the effect of thickness on the absorption of the X-ray beam. The object is a stepped wedge of aluminum. With progressively increasing thicknesses of the metal, there is greater and greater absorption of the X-ray beam resulting in less blackening of the X-ray film pictured below. (Modified from *The Fundamentals of Radiography*, 10th ed. The Eastman Kodak Company, 1960.)

gressively increasing thicknesses of the same metal. The resultant film is shown below the step wedge, and as you can see, more radiation penetrated the thinner layers to produce more blackening of the X-ray film.

Scattering

As X-rays pass through tissues, not only are some absorbed, but many are scattered and deflected. Some new radiation is also produced by interactions within the tissue. This scattered radiation may be projected in any direction from the object. It reduces the detail and sharpness of the image by producing a uniform blackening **(fogging)** of the radiograph. Since scattered radiation is of no value, every effort is made to eliminate it, particularly when examining thicker parts of the body. One ingenious method of decreasing scattered radiation was devised in 1913 by Dr. Gustave Bucky. His invention, called a **grid,** consists of thin linear strips of lead set on edge, with a strip of some radiolucent material interposed between each lead strip. The grid is rectangular in shape and only 2 to 3 mm. in thickness. It is placed between the object being examined and the film cassette. If you will examine the cross section presented in Figure IV-14, you will see that the strips are inclined slightly from the vertical the further they are from the center strip. It is as if they were arranged along the radius of a circle with the circle's center represented by the focal spot of the tube. In this manner, only X-rays traveling in straight lines from the focal spot can pass through, while most of the scattered radiation traveling in oblique lines is absorbed by the lead.

Grids may be **stationary** or **moving** in type. When stationary grids are used fine lines due to the lead strips will be visible on the finished radiograph. By keeping the grid constantly in motion, these lead lines are blurred out. The grid moves at right angles to the grid lines at a speed sufficient to eliminate the lead lines. Radiographs made with these moving grids are referred to as **Bucky films.**

Another method of decreasing scattered radiation is to restrict or cone down the useful X-ray beam emerging from the tube to as small an area as possible consistent with the part under examination. This coning may be accomplished by the use of lead extension tubes or a lead diaphragm which has been placed as close as possible to the X-ray tube window. Figure IV-15 illustrates this principle.

BASIC PRINCIPLES OF LEARNING ROENTGENOLOGY[10, 11]

Now that you are familiar with **absorption,** you should have no problem predicting how an object with known size, shape, and composition would appear on a radiograph. For instance, what would you predict the appearance of a light bulb to be?; an egg with its contents intact and another with contents removed?; a starfish? Knowing about an object enables you to perceive it in your mind's eye before it is shown to you (Fig. IV-16).

What is represented on the radiograph in Figure IV-17? Such a collection of junk could only be found in a lady's handbag. However, unless you are familiar with the individual contents of the bag, you will not be able to name all of them. Identify: hairpins, compact, perfume vial, keys, pill box, *etc.* Now what do you think is represented on the radiograph in Figure IV-18? It is a human foot still in its shoe. Did you deduce this by recognizing the shoe and assuming that it contained a foot, or did you recognize the bones of the foot and conclude that the object surrounding it must be a shoe? In any case, you can see how **familiarity with an object or an anatomical structure** permits a logical analysis of its appearance in various projections. Thus, in the following chapters, we have adhered to the plan of thoroughly discussing the gross anatomy of a region or structure before presenting the radiographs and the anatomy to be gleaned from them.

Structures of uniform composition lying **parallel** to the X-ray film cast a **uniform shadow;** however, that same structure turned **perpendicular** to the film plane will appear much *more dense* and white. Examine the egg shell (empty) in Figure IV-16B and note how densely white the oval periphery is compared with the center although the shell is of uniform thickness. This same principle may be seen in any of the long bones of the hand (Fig. IV-11) where that portion of cortical bone lying tangential to the X-ray beam appears more dense than the cortical bone superimposed on the less dense marrow cavity.

Whenever objects or structures overlap there is a **summation effect** produced on the radiographs as each shadow is added to the one above or below. Figure IV-19 shows this principle nicely. Note the composite picture resulting from some marbles overlapping. The patterns produced do not look like round marbles at all. This illustration leads us to the last but probably the most important principle in learning roentgenology. You must begin to **think three-dimensionally** (3-D) whenever you view a two-dimensional radiograph. Look again at the bag of marbles (Fig. IV-19) and imagine them stacked one upon the other in some areas, and in other areas appearing in a single layer. With your newly acquired knowledge of gross anatomy, you will have little difficulty in visually dissecting your patient's radiographs layer by layer. To aid you in thinking three-dimensionally, we recommend that you make a visit to the department of radiology and ask to be shown some stereoscopic X-rays. Practice viewing through the stereoscope, and then looking at them singly while thinking 3-D.

CONTRAST MEDIA

Many important anatomical structures cannot be differentiated from those adjacent to them due to insufficient differences in X-ray attenuation. In many instances, however, the structures may be made visible by introducing contrast media into or around them. These media may be either **radiopaque** or **radiolucent.**

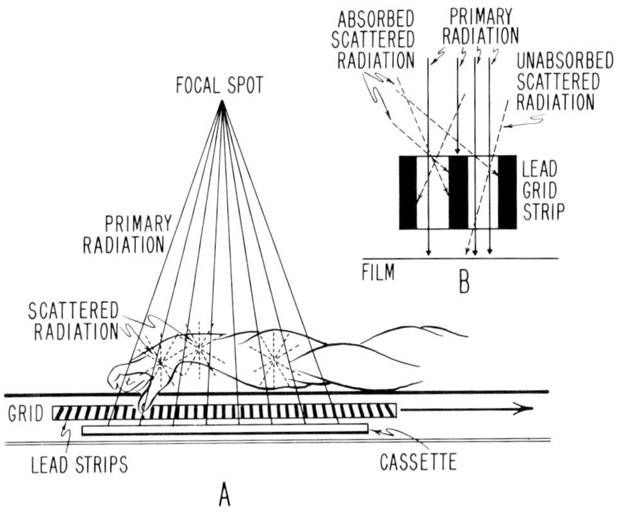

Fig. IV-14A. Cross section of a grid which is interposed between the object to be examined and the film below. Note that the lead strips are inclined slightly from the vertical the further they are from the center strip. The arrow indicates that the grid moves during the exposure, thereby blurring the image cast by the lead strips. Moving grid radiographs are called Bucky films. **B.** Cross-sectional enlargement of the lead strips comprising the grid. Much of the scattered radiation produced is absorbed by the lead strips before striking the film below.

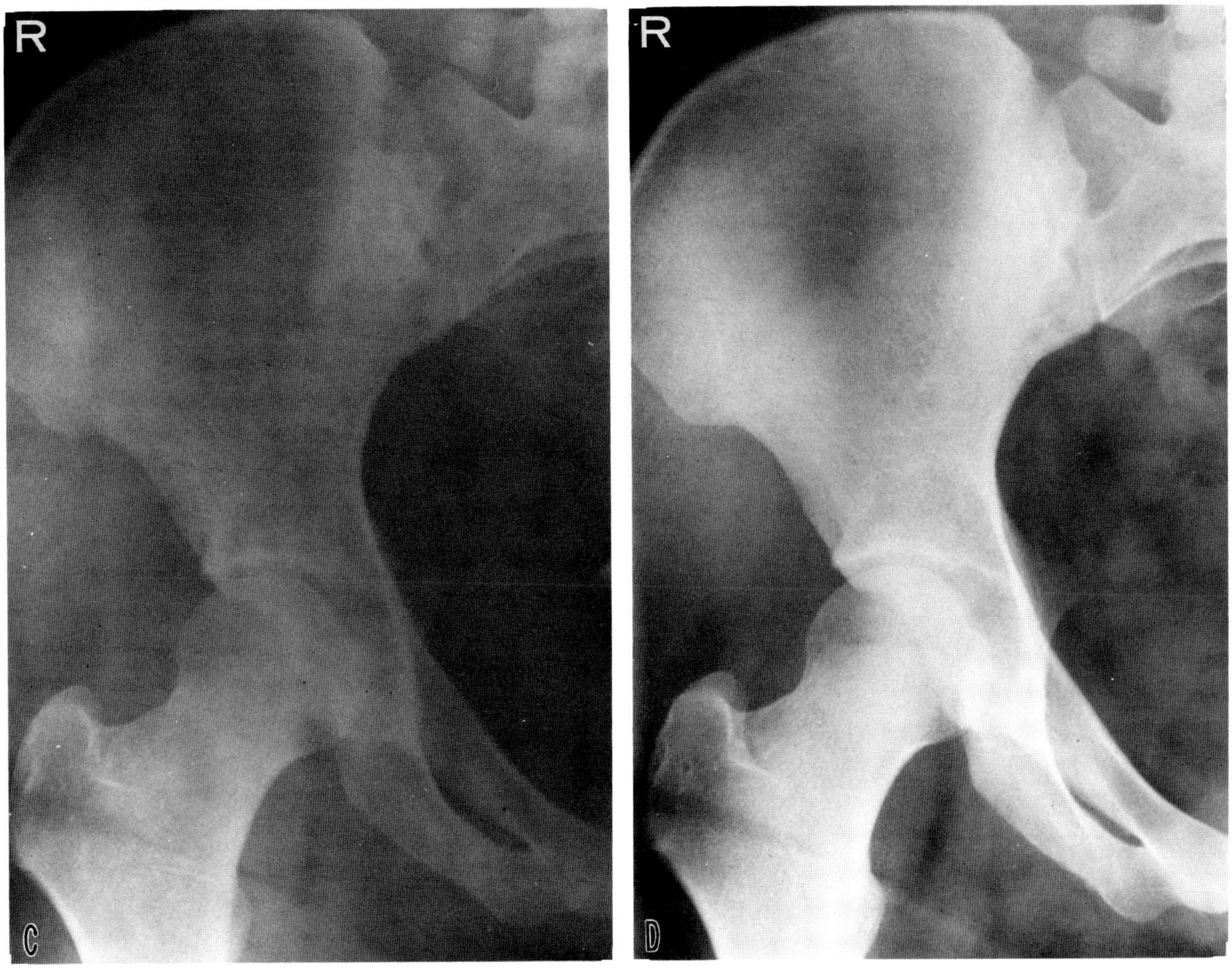

Fig. IV-14C. Anterior-posterior (AP) view of the right side of the pelvis, taken *without* using a Bucky. **D.** Same view as **C;** however, a Bucky has been used. Note that the bone detail is markedly improved.

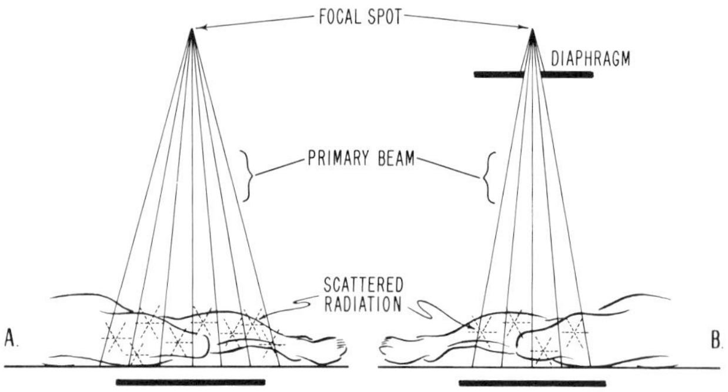

Fig. IV-15. The importance of coning the X-ray beam to decrease scattered radiation. **A.** No restriction of the useful X-ray beam. **B.** Coning the useful beam by means of a lead diaphragm results in less scattered radiation.

Fig. IV-16A. Radiograph of light bulb. **B.** Radiograph of eggs—one has had its contents removed.

Fig. IV-16C. Radiograph of starfish.

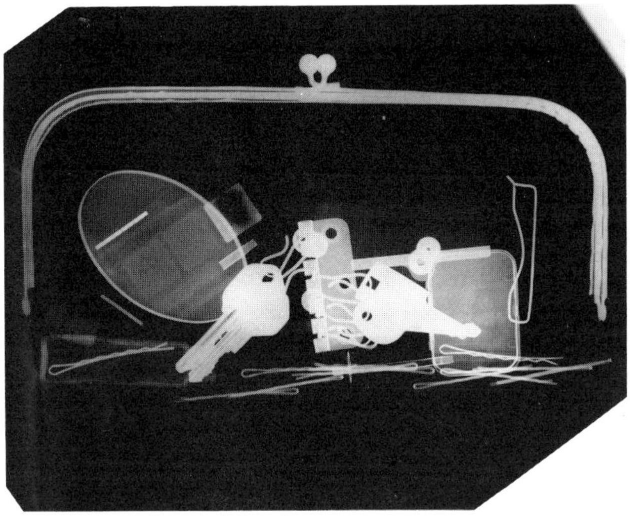

Fig. IV-17. Unknown—see text.

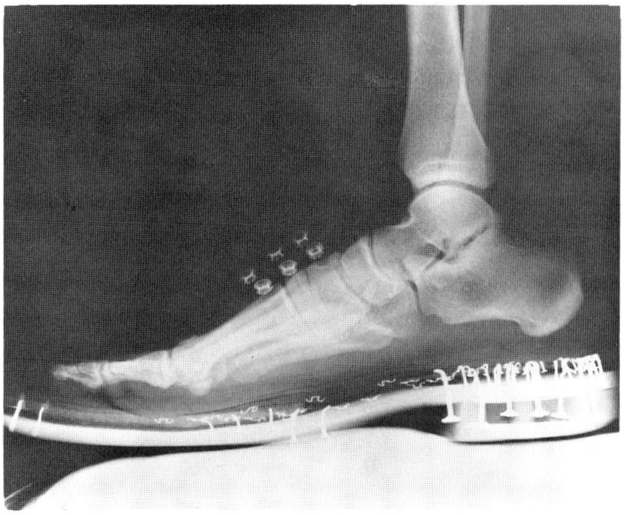

Fig. IV-18. Unknown—see text.

Fig. IV-19. Radiograph of a bag of marbles (cat's eyes).

Radiopaque contrast media absorb more of the X-ray beam than the surrounding structures and, therefore, appear white. The most frequently used radiopaque contrast media are barium sulfate and iodine-containing fluids. Barium sulfate (suspended in water) is primarily used to study the gastrointestinal tract. It passes through the bowel without being absorbed. Many iodine-containing fluids are used to study blood vessels, heart chambers, the urinary tract, main biliary ducts, gall bladder, and the bronchial segments of the lung.

Radiolucent contrast media, such as air, oxygen, and carbon dioxide, are less dense than the surrounding tissues. These appear black on the radiograph and are used to study the ventricles of the brain, body cavities, the gastrointestinal tract, and, occasionally, joints. Many examples of the uses of contrast media will be shown you throughout this book.

MAGNIFICATION AND DISTORTION

In every radiographic examination, there will be some magnification and distortion of the object on the X-ray film. However, these effects may be minimized by following certain principles which we will now illustrate by substituting light as our source of radiant energy. The smaller the source (focal spot) of X-rays or light, the sharper the resulting image. In Figures IV-20A and C the hand has been interposed between a source of light and a flat recording plane, in this instance a wall. In both illustrations the hand has been placed close to the wall. In Figure IV-20A, a small-sized light bulb has been used, while in Figure IV-20C a larger-sized bulb has been substituted. Note how much sharper the shadow of the hand is in Figure IV-20A.

Magnification and vagueness are increased the further the object is from the recording plane. These effects are lessened if a small source of X-rays or light is used. In Figure IV-20B and D, the hand has been placed further from the reflecting wall. Both illustrations show magnification of the hand shadow, but it is much sharper in outline in Figure IV-20B where a small light source has been used. We have illustrated these principles diagrammatically in Figures IV-21C to E.

X-rays and light diverge from their source along straight lines, gradually losing intensity as they travel outward. The radiation intensity at any given point from this source is inversely proportional to the square of the distance from the source.[7] Figure IV-22 shows that at distance D, the light or X-ray beam covers an area on the grid (C-1) of four squares. At twice the distance 2D, the beam covers an area on the grid (C-2) of 16 squares or four times the area. It holds, therefore, that the intensity of the beam at 2D is one-quarter of the intensity at distance D. In performing any radiographic examination, therefore, the distance of the object from the X-ray source must always

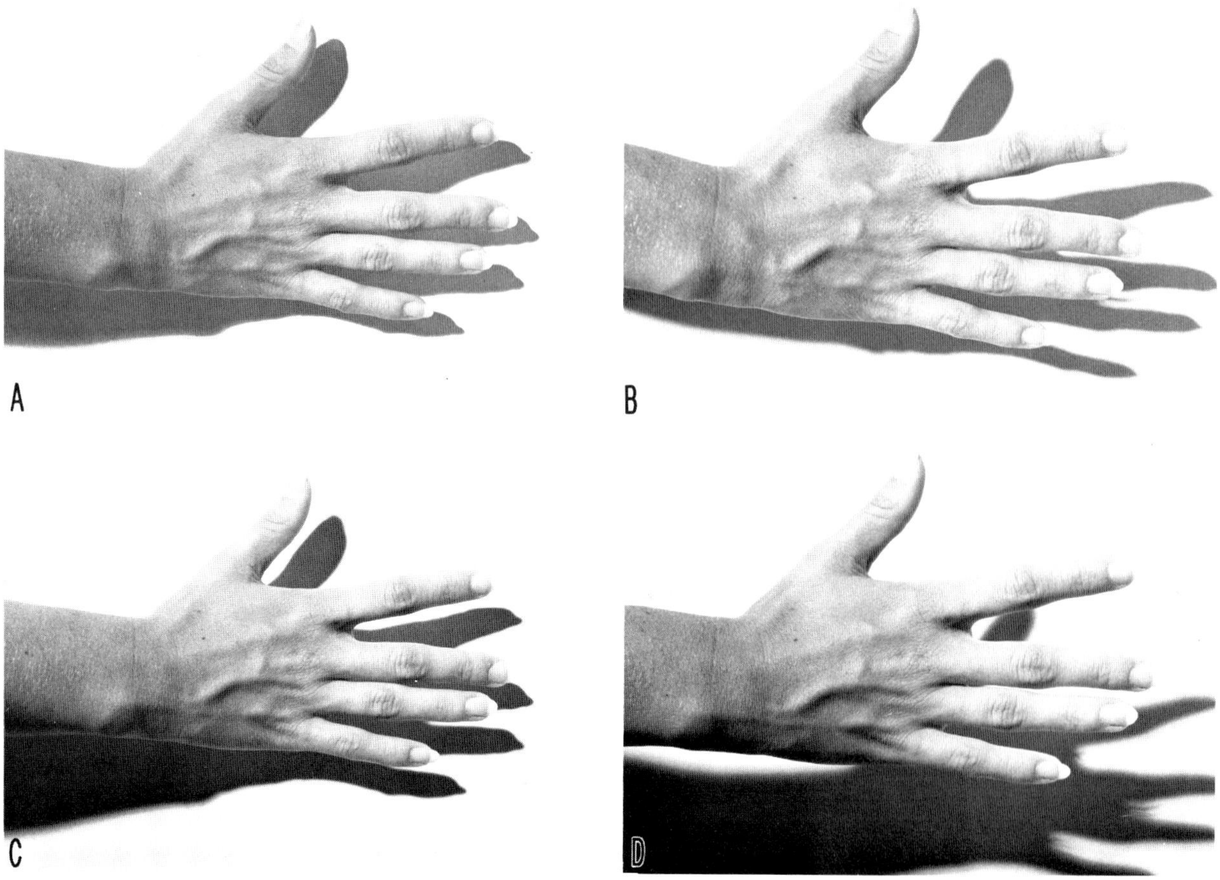

Fig. IV-20. Magnification and distortion illustrated, substituting light for X-rays as the source of radiant energy. A small light source (bulb) is used in **A** and **B,** while a larger light source is used in **C** and **D**. In **A** and **C,** the hand is held close to the reflecting surface. Note that the shadow cast is unsharp when the large light source is used (**C**). The hand is moved closer to the camera or further from the reflecting surface in **B** and **D**. Although there is some magnification of the shadow in these views, the shadow is still sharper when a small light source is used (**B**).

be considered. If we double the distance of an object from the X-ray tube, we must increase the beam intensity fourfold to achieve the same degree of film blackening. You will find that most examinations are performed at a 36 to 40 inches tube target-film distance. Therefore, the possibility of patient movement is minimized because a shorter exposure time may be utilized.

Image sharpness on the radiograph may also be improved by increasing the target-film distance, and magnification and distortion are reduced. This principle is best utilized in chest radiography, where a distance of 6 to 10 feet is used between the tube and the film. Since only a relatively small amount of radiation is required to penetrate the air-filled lungs, this can easily be accomplished with present-day X-ray tubes. In other areas of the body such as the abdomen where large radiation intensity is needed to penetrate, these 6- to 10-foot distances become impractical.

You will note in Figures IV-21A, C, D, and E that the object being examined is parallel with the recording plane. Although magnification and vagueness may be present, there is no distortion of the shadow case by the object. However, in Figure IV-21B when the object is not parallel to the recording plane, distortion results. That portion of the object farthest from the angled recording plane is magnified, and its true shape is thereby distorted.

BODY SECTION RADIOGRAPHY (TOMOGRAPHY)

Tomography is among the many specialized techniques in radiography to delineate superimposed structures. This simply means that a detailed radiograph of a preselected plane or section of the body is obtained with all anatomical details above and below this plane blurred out. In order to understand the principle involved let us briefly examine Figure IV-23A and B. The X-ray tube-housing and the film cassette tray beneath the table top are joined by a metal arm, and these are made to move or pivot about a fulcrum (Fig. IV-23A). The height of this fulcrum may be

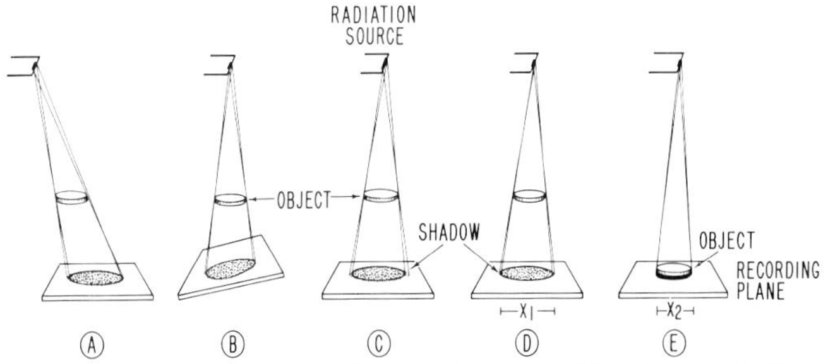

Fig. IV-21. Diagram illustrating magnification and distortion. In **A, B, C,** and **D** there is magnification of the object shadow since the object-recording plane distance is long. The magnified shadow diameter (X_1) can be reduced in size (X_2) by getting the object as close as possible to the recording plane. **B.** By keeping the object parallel with the recording plane (**A**), there is no distortion of the shadow as in (**B**) where the object and recording plane are not parallel. **C.** A large source (focal spot) is used which causes unsharpness of the shadow or image. **D.** Unsharpness is minimized by decreasing focal spot size. **E.** For the best method of image recording, use, whenever possible, a small focal spot and keep the object to be examined close to the recording plane. (Modified from *The Fundamentals of Radiography,* 10th ed. The Eastman Kodak Company, 1960.)

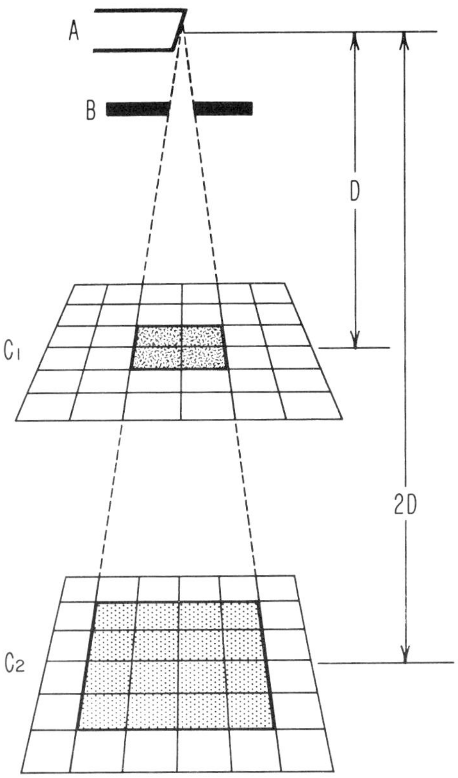

Fig. IV-22. Diagram illustrating the "inverse square law." A, tube focal spot; B, lead diaphragm. Radiation intensity at any given point from the source is inversely proportional to the square of the distance from the source. (Modified from *The Fundamentals of Radiography,* 10th ed. The Eastman Kodak Company, 1960.)

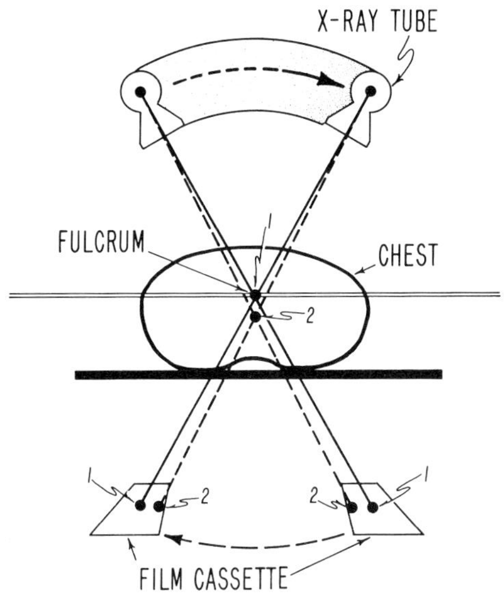

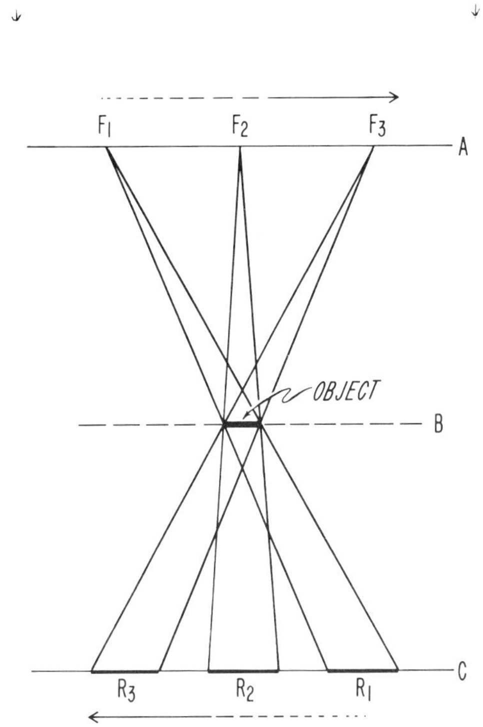

Fig. IV-23. The principles of body-section radiography (tomography). Both **A** and **B** illustrate the basis of the technique in slightly different fashion. Arrows indicate that the tube and film cassette move simultaneously, but in opposite directions. **A.** The dot at 1 is in the plane in focus. Note that it maintains the same position on the film cassette throughout the examination while the dot at no. 2, in an out of focal plane, moves across the film and, thus, is blurred out.

changed as desired by the radiologist to that level of the body plane that is to be examined (Plane B) (Fig. IV-23B). The X-ray tube in Plane A moves from F1 through F3, while the film in Plane C is moving in the opposite direction from R1 to R3. Throughout this examination, the object under study remains constantly in focus on the same area of the film, while structures above and below Plane B are blurred out and therefore obscured on the radiograph[7] (Fig. IV-23A).

FLUOROSCOPY

Earlier in this chapter we mentioned that when X-rays interact with certain inorganic crystals, visible light is emitted by the crystals. This principle is used in both radiography and fluoroscopy. In our discussion of the intensifying screens in the film cassette, we noted that calcium tungstate crystals within the screen emit blue light to which the X-ray film is extremely sensitive. In fluoroscopy, screens of crystalline zinc cadmium sulphate are used and these emit a yellow-green fluorescence when struck by the X-ray beam. This type of screen is used in fluoroscopy since the **rods in the retina** of the eye are more sensitive to yellow-green than to blue light.[7]

The patient is placed between the table which houses the X-ray tube and the fluoroscopic screen. An arm connects the screen support with the X-ray tube-housing so that both move synchronously. The X-rays pass in turn through the radiolucent table top and the patient, and then strike the fluoroscopic screen. In order to protect the radiologist from any direct radiation penetrating through the fluoroscopic screen, a sheet of glass which is impregnated with lead is placed over the screen. The lead within the glass absorbs any direct radiation while allowing the passage of the visible light emitted by the screen's crystals. Since the intensity of light is extremely low in conventional fluoroscopy, the radiologist must dark-adapt his eyes before beginning the procedure. This dark-adaptation is achieved by wearing red goggles for approximately 20 minutes.

The **conventional** fluoroscope has, for the most part, been replaced in radiology departments by **image intensification** equipment. Image intensifiers electronically amplify the intensity of the light produced in the fluorescent screen several hundred to several thousandfold. Fluoroscopy, therefore, may be performed in a lighted room without prior dark-adaptation by the radiologist, and this permits the more sensitive **cones** of the eye to be used.[6] With this equipment and with the addition of a motion picture camera to the system, extremely fine detail can be visualized and permanently recorded on film. This is of the utmost importance in examining such rapidly occurring phenomena as deglutition, flow of contrast media through the chambers of the heart and associated blood vessels, and peristalsis of bowel, since the movie films may be studied at leisure in slow motion.

Another method currently in use combines a television system with image intensification. The image is then viewed on a television screen, and the images produced may also be recorded on tape. The "instant replay" is now as much a reality in medicine as in televised sporting events.

During fluoroscopy the images produced are shades of greenish-yellow when conventional and image amplifier systems are used, whereas with television systems, all of the images are in shades of black and white (Fig. IV-24A). Note that in fluoroscopy the patient's bones, heart, and blood vessels appear dark in color, while the lungs appear light. This, of course, is the reverse of their appearance on the radiograph (Fig. IV-24B). The reasons for this difference should be obvious to you from our preceding discussions of X-ray attenuation.

INTERPRETATION OF RADIOGRAPHS

Placement of Radiographs on the Viewing Box

Before examining any radiograph, you must learn how it is placed on the viewing box. By tradition, all radiographs are viewed as if you were looking at the patient face to face, with the patient standing in the normal *anatomical position; i.e.*, with the palms of his hands forward and his feet together. The technician will always mark each radiograph with a "right" or "left" lead metal marker according to the patient's anatomical right or left side. This lead "R" or "L" will appear white on the finished radiograph so that it is readily discernible. In Figure IV-24B you will see that the patient's left side has been clearly marked with an "L," and since you view a radiograph as if you were looking at the patient, the patient's left side will go on your right as we have shown. One note of caution: when you read the radiologist's report describing your patient's radiographs, "right" or "left" refers to the patient's anatomical side and not your own.

Terminology of Positioning

Figure IV-24B is a PA (posterior-anterior) view of the chest, which is the standard view of the chest taken whenever the patient can stand and cooperate for the examination. Roentgen terminology is based, for the most part, on that anatomical surface first penetrated by the X-ray beam; *i.e.*, in examining the chest, the patient (Fig. IV-25A) stands with his anterior thorax against the film cassette, with the X-ray tube placed horizontally at his back. The X-ray beam strikes the posterior aspect of the thorax and exits at the anterior surface before reaching the cassette. Therefore, the term PA (**posterior-anterior**) view is used. It follows then that an AP (**anterior-posterior**) view of the chest is made when the patient's back is against the film cassette. The horizontal X-ray beam enters the anterior surface of the thorax and exits at the posterior surface.

Certain notable exceptions to the above rule of terminology should also be considered. In examining any part of the body **laterally** (Fig. IV-25B), the film

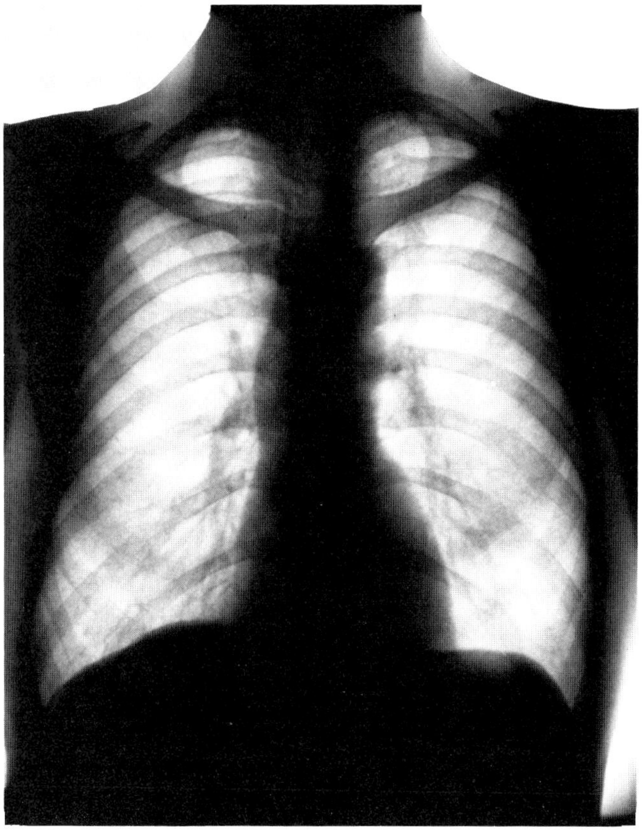

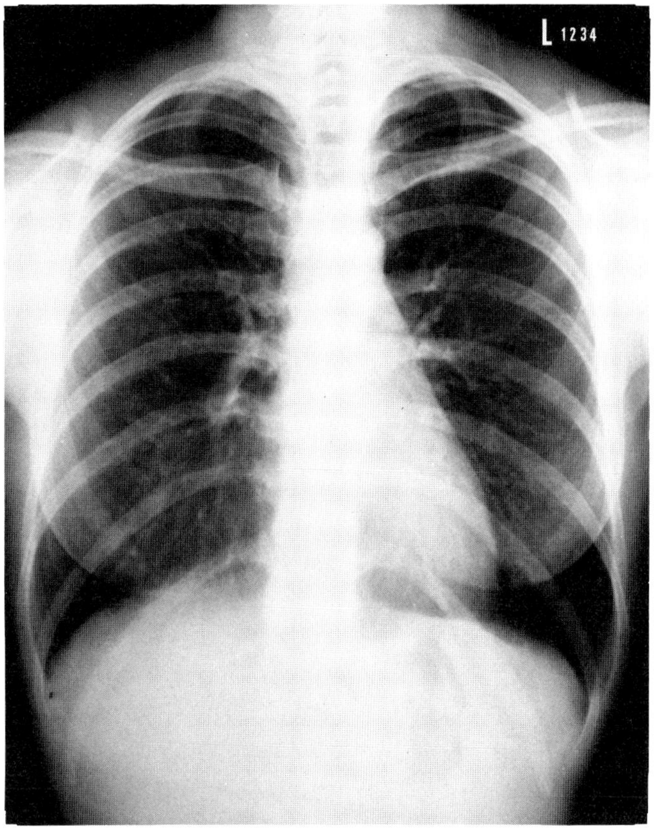

Fig. IV-24A. Appearance of the chest as you would see it during fluoroscopy with a television monitor. Contrast this with **B** which is a posterior-anterior (PA) *radiograph* of the chest. These examples are not of the same patient. **B.** Posterior-anterior PA) *radiograph* of chest.

(is marked with a "right" or "left" according to the side closest to the film. Thus, we can see that a left lateral view of the chest is made by placing the patient's left side against the film cassette. The horizontal X-ray beam first enters the right side of the thorax and exits on the left side. **Oblique** films are also named according to the patient's anatomy closest to the film. An RAO (**right anterior oblique**) view of the chest (Fig. IV-25C) is made by turning the patient approximately 45° so that the right anterior aspect of the thorax touches the film cassette. The X-ray beam enters the posterior thorax and exits anteriorly, while the beam remains perpendicular to the film cassette. You should now be able to reason out for yourself the patient's position in the following views: left anterior oblique (Fig. IV-25D), right posterior oblique, and left posterior oblique.

Recumbent means that the patient is lying down during the examination, either on his back (**supine**) or on his abdomen (**prone**). In a **decubitus** view the patient lies on his right (**right lateral decubitus**) or left (**left lateral decubitus**) side, and the horizontal X-ray beam may enter the chest or abdomen from anteriorly or posteriorly. When the patient stands up for the examination the term **erect** is used, and when the patient is approximately 45° upright the term **semierect** is used. Every radiograph throughout the following pages will be labeled for you as to the position used in making the film.

Standard Positions

Most X-ray examinations are performed in at least two projections at right angles to each other—frontal and lateral views. Since each radiograph represents a composite picture of the tissues passed through by the X-ray beam, many structures will overlap each other. Therefore, more than one view is usually required to localize an abnormality accurately. A simple illustration of this principle is given in Figure IV-26. In Figure IV-26A, you see the radiograph of an ordinary chicken egg end on. In the center of the egg there is a faint, round, gray, radiolucency. From this single projection it is impossible to explain or accurately localize this radiolucency. However, with the egg now in a lateral position (rotated 90°) (Fig. IV-22B), it is immediately obvious that the radiolucency represents an air cap localized to the broad end of the egg. In similar fashion in clinical practice, you will find that having more than one projection, preferably at right angles to each other, will permit you to localize accurately pneumonias, tumors, and foreign bodies, to name only a few. In subsequent chapters we will present multiple

Fig. IV-25. Terminology of positioning in radiography. **A.** Posterior-anterior (PA) view. **B.** Left lateral view. **C.** Right anterior oblique (RAO) view. **D.** Left anterior oblique (LAO) view.

projections of each part under consideration so that you will be able to practice this important lesson.

Film Reading Techniques

In interpreting any radiograph there are certain techniques which should always be followed. First, examine the radiograph to confirm that this is, in fact, your patient, since occasionally another patient's X-ray films will find their way into your patient's film envelope. **Identification** is made by checking the name or hospital number recorded in the corner of the film. Note the **dates** recorded on the films so that each examination of a part may be examined and compared in a logical sequence. Remember to determine which side of the radiograph is right or left so that you will place the film correctly on the viewing box.

For many years radiology residents were taught to examine each radiograph in an orderly manner, adhering to a rigid search pattern by systems. However, Tuddenham[8] has found that most experienced radiolo-

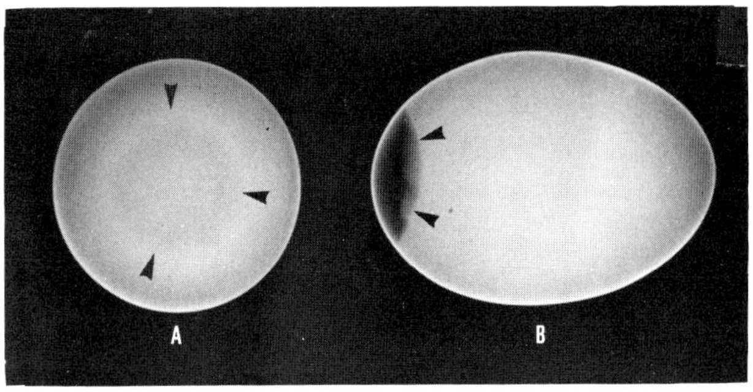

Fig. IV-26. Radiographs of a chicken egg illustrating the value of having more than one projection, preferably at right angles, in order to accurately localize structures. The arrows indicate the air cap. **A.** Egg end-on. **B.** Lateral view.

gists do not search in any systematic fashion but rather in a random pattern. He believes that once an abnormality has been found the film reader should be free to search out other clues to confirm or amplify the initial finding. Remember not to stop looking once you have detected an abnormality, but rather, to force yourself to search the entire radiograph completely.

Tuddenham also found that most errors in detection occurred when more than one abnormality was present on the X-ray film. He found that experienced radiologists will overlook 20 to 30% of the significant abnormalities in any series of radiographs. In order to improve our abilities to retrieve information on the radiographs several steps should always be followed:

1. Double read every radiograph—the more times that the X-ray film is viewed the greater the chance of detecting abnormalities. Initially examine each radiograph without benefit of clinical history, findings of physical examination, or laboratory data. After studying the pertinent clinical data, re-examine the X-ray films for specific abnormalities that you know may be seen in the suspected disease entities. Throughout this search your knowledge of normal anatomy will be put to the test, since it is not possible to see the abnormal without first thoroughly knowing what is normal and what constitutes the variations from the normal. In many institutions at least two radiologists read every radiographic examination in order to avoid errors of omission.

2. Tuddenham has recommended the use of a reducing lens in order to increase your ability to perceive large, vague, low-contrast shadows on the radiographs. Every X-ray film should be examined with a reducing lens after the initial search with the naked eye. A magnifying lens is of great value in resolving closely approximated high-contrast shadows. This is particularly true when examining bones for small fractures. Your author has learned that both types of lenses are of inestimable value to the radiologist and are a vital part of his armamentarium.

3. In any examination you will probably find that while the radiograph has been optimally exposed for one portion of the part, other areas will be overexposed. In order to study these blackened areas, a "hot light" or high-intensity light will be needed.

CONCLUSION

With this brief introduction to the fundamentals of roentgenology, you are now ready to use the radiographs included in this book. The knowledge of anatomy that you will acquire will greatly aid your thinking and enable you to recognize abnormalities and to accurately place them in their proper perspective in the body. This is of the utmost importance to you in future years of medical study, for you will not be able to comprehend the radiologist's report unless you have a thorough knowledge of gross anatomy and an understanding of roentgen anatomy and terminology. You will find that certain diseases and tumors have a high incidence of occurrence in certain sites. The accurate anatomical placement of a lesion, along with the knowledge of pathology and the other medical disciplines that you will acquire, will give you a high degree of accuracy in making correct diagnoses. Throughout the ensuing chapters we will stress many of the "pearls" used by radiologists to aid in the accuracy of placing lesions. It is, of course, not our intention in this book to stress pathological conditions. However, once you have mastered the normal radiographic anatomy in this text, we would urge you to read a classical work entitled *Fundamentals of Roentgenology*.[10] This book was written by one of the foremost *teachers* of radiology, Dr. Lucy Frank Squire. It is unique in its content for it takes the student through an exciting exercise in deductive reasoning relating pathology and radiology. Many of the basic principles for learning roentgenology that we have presented are based upon Dr. Squire's methods of teaching medical students.

REFERENCES

1. Glasser, O.: *Dr. W. C. Röntgen,* 2nd ed. Thomas, Springfield, Ill., 1958.
2. Röntgen, W. C.: Ueber eine neue Art von Strahlen—(Vorläufige Mittheilung) Sitzber. Physik.-Med. Ges., Würzburg, #9, 1895.
3. Röntgen, W. C.: Ueber eine neue Art von Strahlen. II. Mittheilung. Sitzber. Physik.-Med. Ges., Würzburg, #1 and 2, 1896.
4. Röntgen, W. C.: Weitere Beobachtungen über die Eigenschaften der X-Strahlen. Sitzber. K. Preuss. Akad. Wiss., Kl. Physik.-Math., 1897.
5. Ter-Pogossian, M. M.: *Physical Aspects of Diagnostic Radiology,* Hoeber (Harper-Row), New York, 1967.
6. Johns, H. E.: *The Physics of Radiology,* 2nd ed. Thomas, Springfield, Ill., 1961.
7. Glasser, O., Quimby, E., Taylor, L., Weatherwax, J., and Morgan, R.: *Physical Foundations of Radiology,* 3rd ed., New York, Hoeber (Harper-Row), 1961.
8. Tuddenham, W.: Problems of perception in chest roentgenology: facts and fallacies. Radiol. Clin. N. Amer., *1:* 277–289, 1963.
9. *Fundamentals of Radiology,* 10th ed. Radiography Markets Division, Eastman Kodak Company, Rochester, N. Y.
10. Squire, L. F.; *Fundamentals of Roentgenology.* Harvard Univ. Press, Cambridge, Mass., 1966.
11. Squire, L. F.: Perception related to learning radiology in medical school. Radiol. Clin. N. Amer., *7:* 485–497, 1969.

FIVE

Skin

The Integument and Its Derivatives

INTRODUCTION

The skin you love to touch may be regarded either as an organ or an organ system, depending upon how you wish to view it.

Organs comprise a number of tissues organized to carry out one or more common functions. The skin, composed as it is of an epithelium, connective tissue, a small amount of muscle, and a wealth of nerves all organized for a common purpose, does, indeed, fit the concept of an organ.

But when we consider its extent as a surface covering for the body, its multiplicity of functions, and the specialization of its components such as glands, hair, nails, sensory receptors, and fatty depots, we often find ourselves speaking of it as the integumentary system, one of the largest and most complex of the body systems. A body system consists of several organs associated to carry out one or more functions. For example, the oral cavity with its glands, the pharynx, esophagus, stomach, and intestines with their own intrinsic glands as well as their associated glands, the liver and pancreas, all comprise the digestive system. Each of the parts is engaged in some special way in the processing of the food we eat for assimilation and final elimination. It is more difficult to think of the skin as such a body system since anything about it that we would like to regard as an organ has microscopic dimensions. Yet, this outer investment or integument of ours has macroscopic dimensions as well, and the tiny organs within work together in specialized ways to carry out the function of the whole skin. Macroscopically the skin covers a surface area of 19.4 square feet in the average adult male and 17.2 square feet in the average adult female, and that's a lot of skin! Perhaps, because it is at a slight premium in the average female, the average male values it more when it is draped over her. When things become somewhat scarce they become slightly more precious.

FUNCTIONS OF THE SKIN

The principal functions of the skin are: (1) to protect the underlying tissues from injurious assault which may be physical, chemical, electrical, thermal, or biological; (2) to prevent excessive loss of water or drying out of the tissues; (3) to act as a temperature control device, preventing heat loss in cold environments and facilitating heat loss in warm ones; (4) to serve as a reservoir for water and some nutrients; (5) to assist in excretion (through sweating) by eliminating water, salts, and, to a limited extent, urea; (6) to act as a sense organ in that it is the most extensive sense organ of the body for the reception of tactile, painful, and thermal stimuli, while, at the same time, it is the most generalized and least elaborate of our sense organs; (7) to synthesize and store the antirachitic vitamin (vitamin D); and, (8) to prevent, in large measure, the entrance of foreign bodies.

The skin not only clothes the outer aspects of the entire body but dips into the body orifices where it becomes continuous with the mucous membranes of the respiratory, digestive, urinary, and reproductive tracts.

NATURE OF THE SKIN

In a sense, the integument is more than the skin, since it includes certain accessory organs which are derivatives of skin such as nails, hair, teeth, and glands of a variety of types as set forth in the following table:

- A. Skin
  - Epidermis
    - Stratum corneum
    - Stratum lucidum
    - Stratum granulosum
    - Stratum germinativum
  - Dermis (corium)
    - Papillary layer
    - Reticular layer

B. Skin derivatives — Sudoriferous glands; Sebaceous glands; Ceruminous glands; Mammary glands; Hair; Nails; Teeth

The skin, itself, consists of an **epidermis,** its **epithelial** part, and a **dermis,** its underlying **connective tissue.** Embryologically, epidermis was derived from ectoderm; dermis from mesoderm. Below the dermis the skin blends with the **superficial fascia,** a subcutaneous connective tissue which attaches skin to underlying structures. The subcutaneous fascia may contain considerable fat in some areas of the body as is shown schematically in Figure V-1A. This fat often invades the dermis as well. The deeper layers of superficial fascia are usually fat free and are composed of more compact collagenous fibers than in its superficial lamina (*cf.* Camper's and Scarpa's fasciae in the abdominal wall).

We sometimes describe some individuals as being "thin skinned," meaning that the petty irritations of life "get under their skin" very easily. Of others, we say that their concern about matters we feel to be important runs only "skin deep." Yet, many of us have little idea of how thin or how deep skin is. It varies considerably over the body from 0.5 mm. to about 4 mm. or more. While this variation is great, that is still pretty thin even at the maximum range when you consult your millimeter rule. Generally, it is thicker over the extensor and dorsal body surfaces than in flexor regions or ventrally. The reverse is true in hands and feet.

The Epidermis

The epithelium of the epidermis is of the **stratified squamous** variety. This means that it is made up of many layers of cells in depth, but that no matter what the shape of the cells may be in the lower layers, those at the free surface are thin and flat, like pavement stones. The typical four-layered epithelium depicted in Figure V-1A and in the upper part of Figure V-1B holds only for the thickest skin, as in the palms and soles. In these places four strata of cells can be identified microscopically in skin sections—**corneum, lucidum, granulosum,** and **germinativum**—from free surface to the basement membrane where the lowest cells rest upon the dermis. The stratum germinativum, the deepest stratum and the one that is actively **mitotic,** can be divided further on the basis of cell shape into "stratum cylindricum" and "stratum spinosum." This means that the cells that actually rest upon the basement membrane form a single lamina of **columnar cells.** Their daughter cells in the stratum spinosum next above are fat, irregularly-shaped, 14-sided individuals (tetrakaidecahedrons) because they push and crowd each other so. They got their nickname of "spiny" or "prickle" cells before the days of electron microscopy, for when they shrunk and pulled away from each other in routine tissue preparations, delicate "spines" or processes beset them, hedgehog fashion, wherever they were buttoned together by desmosomes. It was once thought that these processes were protoplasmic "intercellular bridges" linking these cells together in a kind of syncytium. But we now feel much wiser, having examined them at infinitely higher power with the electron microscope. There is no longer any doubt that intercellular bridges are desmosomal cell junctions.

At any rate, whether cells in the stratum germinativum are columnar, 14-sided, prickly, or not, they all appear to be healthy and very busily engaged in the process of cell division. They become so engrossed in the population explosion that some of these very cells are continually and progressively raised toward the surface where they undergo the chemical changes of aging, senescence, and gradual death. In the germinativum, they are also busy synthesizing a precursor of a special protein, **keratin,** which remains in the cell as filamentous **tonofibrils.** These seem to have a high sulfhydryl content.

As the cells move toward the surface, chemical changes leading to the production of keratin continue. But the cells themselves degenerate in the process. Their nuclei degenerate and are lost, the cells become spindle shaped, and finally thin, flattened, horny plates at the surface.

The cells in the stratum granulosum are already spindle shaped. Most seem to have viable nuclei, though some nuclei are becoming pyknotic. The cytoplasm contains granules, called "keratohyalin" (for want of knowledge as to its exact nature). The granules are not seen in the living cell as colored elements, but only in dead fixed preparations stained with basic dyes. Sulfhydryl groups seem to have disappeared.

In the next stratum, the lucidum, the cells are flat and so refractile that they appear as a homogeneous band of translucence. Instead of granules, the cytoplasm is filled with semifluid droplets called "eleidin" (again, for want of specific information). All scientists seem to be driven to make up fancy names for things about which they know very little, even anatomists. The cells in this layer seem to have lost their nuclei once and for all.

Finally, at the surface, the cells appear as very thin, tough, horny, flat plates piled up as a thick stratum corneum. These are the squamous cells for which the whole stratified epithelium was named. They have no nuclei. Their cytoplasm is filled with filaments composed of the fibrous protein called **keratin.** The fibrils contain disulphide groups which leads us to believe they are the end result of a chemical manufacturing process that began in the deep stratum of the germinativum. We say, therefore, that the surface cells have become **keratinized** or **cornified.** They are, of course, dead cells and are constantly being sloughed from the

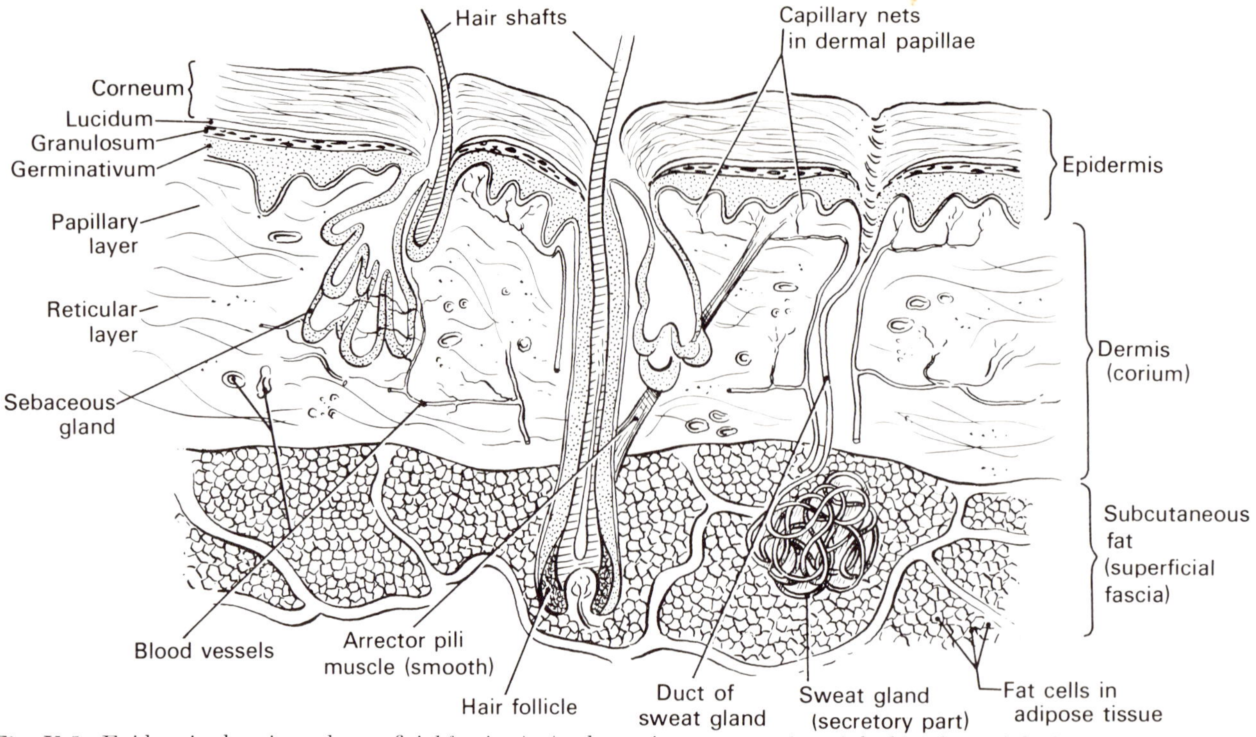

Fig. V-1. Epidermis, dermis, and superficial fascia. **A.** A schematic representation of the histology of thick skin in cross section, including the upper stratum of the superficial fascia.

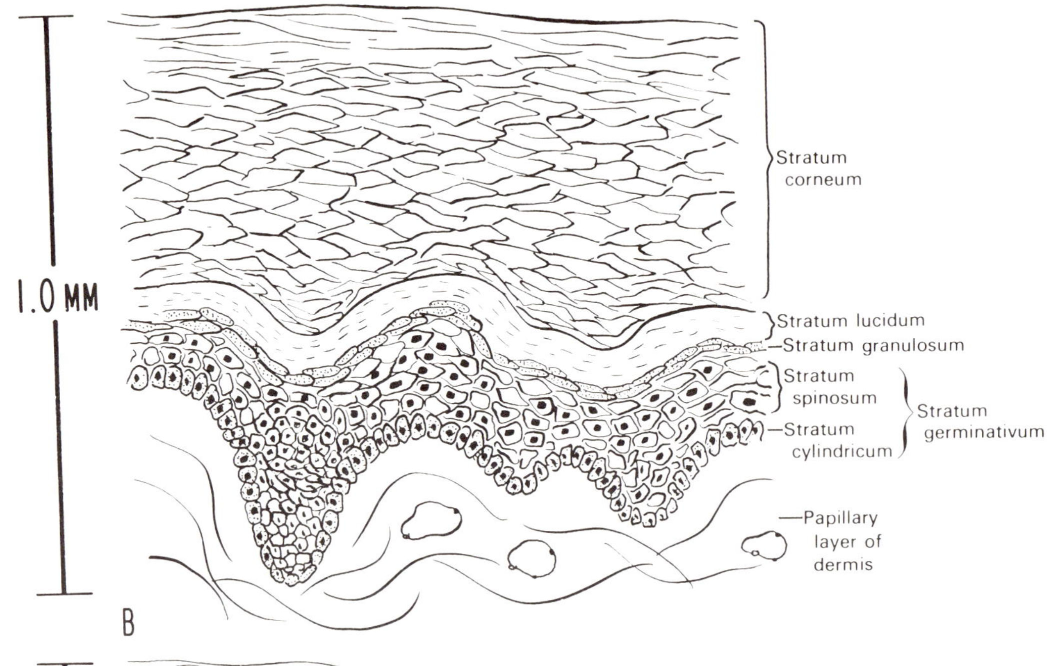

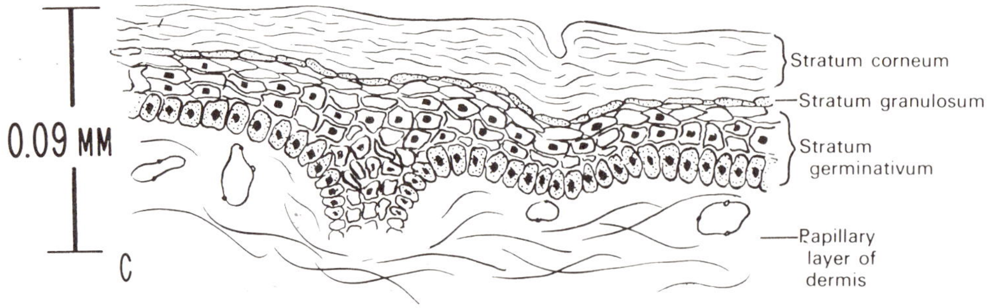

Fig. V-1B. Strata in the epidermis of thick skin of palms and soles. **C.** Strata in the epidermis of thin skin of the abdomen.

surface. This accounts in part, along with soap and dirt, for the daily ring around the bath tub. On our bald pates and among our mange, we call this debris dandruff and whisk-broom it off the collar of our threadbare blue serge.

The thickness of the stratum corneum will vary depending upon the amount of friction to which the surface is exposed. Hence, the thickness of the epidermis, itself, will vary from 0.07 mm. to 0.12 mm. over most of the body, but it may range from 0.8 mm. to 1.4 mm. on the palms and soles (Fig. V-1B). When people habitually go barefoot, the epidermis may thicken to 4 mm. or more in **callus formation** (Fig. V-2). It thickens also over areas of **bunion formation.** A bunion (from the Latin, meaning a "turnip") is a swelling of the bursa over the base of the great toe with a thickening of the overlying skin. A **corn** (clavus) on the other hand, is a hard thickening of the skin produced by pressure and friction. It forms a conical mass which invades the dermis to produce pain. The skin over the little toe is often a favored spot for corn formation.

In areas other than palm and sole, the skin may be quite thin, as over the abdomen (Fig. V-1C). In such areas it may consist basically of two to three layers, the stratum lucidum being absent or not recognizable as a rule, and the stratum granulosum being reduced to a single row of cells, if it is present at all (Fig. V-1B, lower).

You will note that no blood vessels enter the epidermis, but remain below in the dermis (Fig. V-1A and B). They are brought quite close to the epithelium by means of elevated **dermal papillae** over which the epithelium is draped. Indeed, the generic term, **epithelium,** for any membrane covering a free surface or lining a body cavity, a tubular duct or a hollow interior of a body organ, was derived from the relationship of epidermis to dermis. This obviously was one of the most available surface membranes for study and one of the first to be investigated. It was named "epithelium" ("epi," upon; "thelia," the nipples) because it draped over the dermal, nipplelike papillae as they appeared in cross sections of skin. Unfortunately, the term applies more readily to the surface covering of the outer body since few other epithelia are invaginated by the underlying connective tissue in this way. Moreover, the papillae are not so much a series of isolated nipples under the epidermis as they are longitudinal ridges (Fig. V-4). The fact that blood vessels remain below in the dermis means that the epithelial cells must obtain their nutrients and oxygen by diffusion. These substances must leave the thin-walled capillaries in the dermal papillae for the tissue fluid in the matrix of connective tissue and then pass up between the cell borders of the epithelial cells. This accounts for the gradual death of the cells in the higher strata since they are so far removed from the main supply depot and since the lower cells are also fouling the water ways by dumping their metabolic wastes.

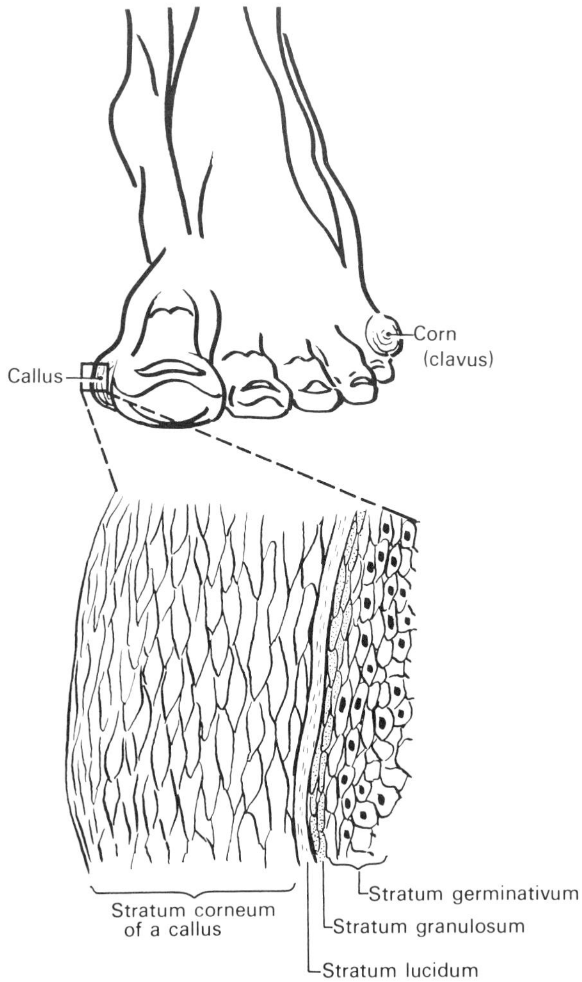

Fig. V-2. Thickening of the stratum corneum in callus and corn (clavus) formation.

The Dermis

The dermis (or corium) is a tough, flexible mixture of fibroelastic connective tissue in which the fibers are irregularly oriented. It varies in thickness from 2 to 6 mm. on the palms and soles to a very thin lamina on the eyelids, scrotum, and penis. It is thinnest over the outer drumhead of the ear, the mammary areola, lids, and prepuce, in that order. Closest to the epidermis, we have noted already its **papillary layer.** It consists of closely intermeshed elastic and collagenous fibers, and the papillae bear numerous nerve endings, as well as being profusely vascularized. The deeper **reticular layer** is a looser felt-work weave of collagenous and elastic fibers. It may contain considerable adipose tissue as well in certain regions of the body so that its transition over into subcutaneous fascia below is not always sharply demarcated.

The entire corium is well supplied with blood vessels, lymphatics, and nerves. Sensory (afferent) nerve fibers carry sensory messages to the spinal cord from specialized receptor organs located throughout the dermis, and visceral motor (efferent) fibers bring in impulses to regulate contraction of smooth muscles in the blood vessels and the smooth muscles which erect hairs, as well as for regulation of the sweat glands. Smooth muscle is also found in loose strands in localized regions of the corium; *e.g.*, as the Dartos muscle of the scrotum, in the prepuce, in the perianal region, and around the nipples. Contraction of these fibers causes a wrinkling. In other areas, skeletal muscle occupies the corium, either originating or inserting into it, or remaining attached at both ends within it. This is true over the face where many of the muscles are "skin" muscles. The broad **Platysma** is another example, coursing in the skin from mandible down the neck and over the clavicle to almost the second rib.

Certain connective tissue fibers in the dermis tend to orient parallel to the direction of pull upon the skin and at right angles to any pressure to which skin may be subjected. These form lines in the dermis which are enforced upon the epidermis as well. Such lines of pull or force are called **Langer's lines** (Fig. V-3). As a result, stab wounds made by a round awl or an ice pick become slitlike rather than rounded in the skin due to the splitting apart of these oriented connective tissue strands. The direction of the cleft indicates fiber direction. Skin incisions made parallel to Langer's lines usually heal well with little scarring because the lines of force pull the two cut surfaces into approximation much better than suture material put in by a physician.

We are all familiar with finger prints or **dermatoglyphic patterns,** those latent imprints of epidermal ridges and whorls that we leave behind on glass tumblers, door knobs, rear vision mirrors, revolver butts and in the F.B.I. files when we are imprinted. What is not generally known are the facts reported by Markee and Wray[1] that the vascular bed of the underlying dermis takes on a similar pattern (Fig. V-4). You are not looking at the outer skin of the fingers in Figure V-4A, though it appears very much as if you were. Here the vascular bed of the dermis had been injected with a liquid plastic. When it hardened, all the soft tissues—epithelium, connective tissue, sweat glands, and other dermal structures—were removed with a strong alkali solution. This was done in fresh specimens obtained at accidental amputations. All that is left in Figure V-4A is the **dermal pattern of blood supply** to give shape to these "fingers," and these patterns are forming the **cristae cutis** or dermal whorls that you see here. Each of these cristae or ridges when observed at higher magnification (Fig. V-4B) consists of a double line of longitudinally-oriented fine vessels, some of which branch. At still higher magnification (Fig. V-4C) we note that the longitudinal vessels give rise to vertical ones, the central capillaries of the dermal papillae. Obviously, sweat glands must come up in the troughs between each double row of vascular ridges, for here the overlapping epithelium would dip down as it drapes the vascularized dermal papillae.

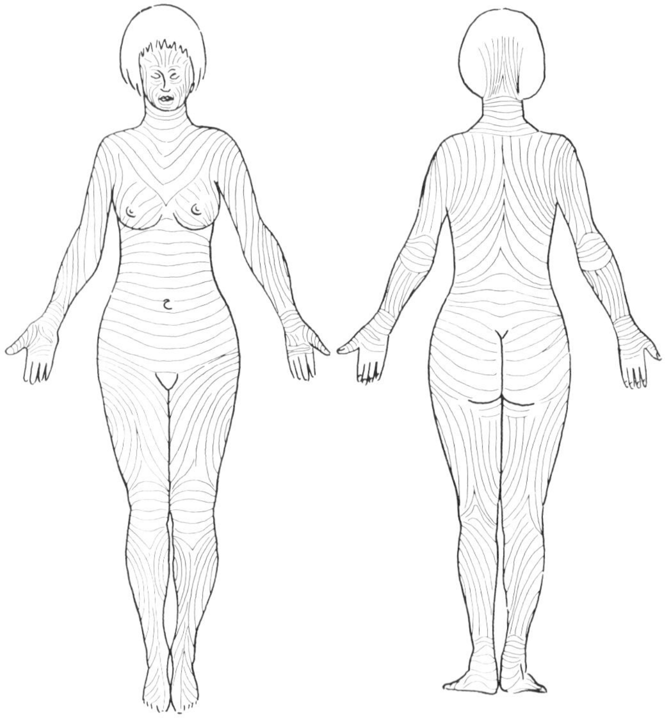

Fig. V-3. Langer's lines. **A.** General distribution over the body.

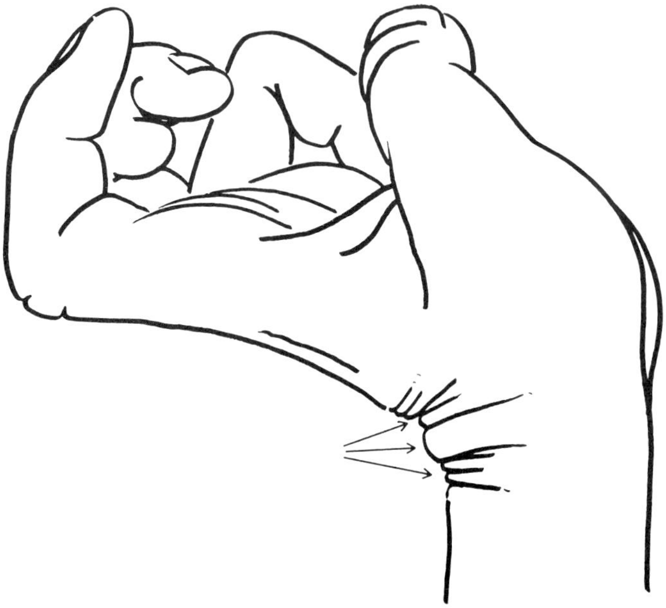

Fig. V-3B. Seen in wrist extension.

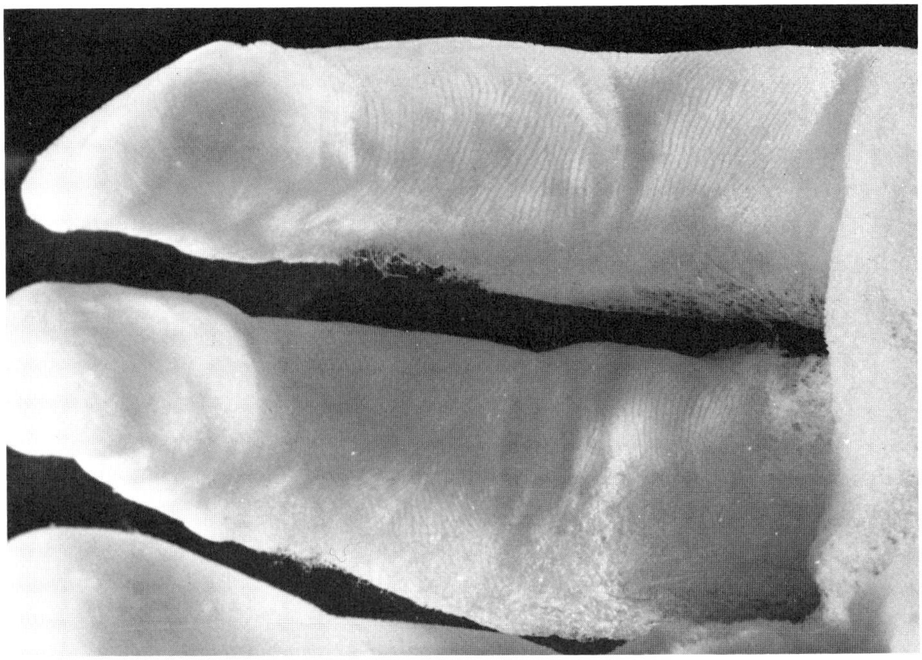

A.

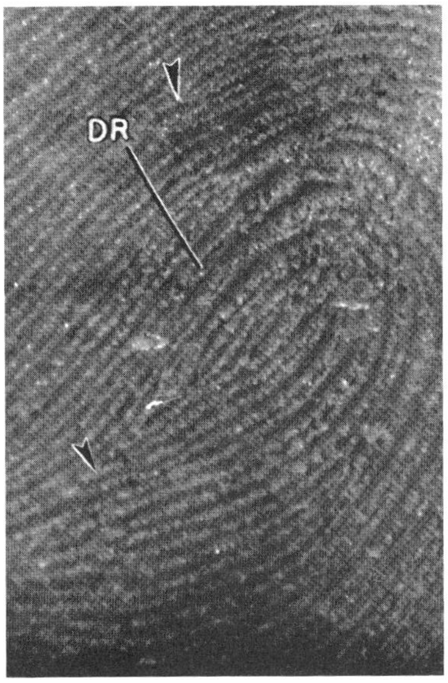

B.

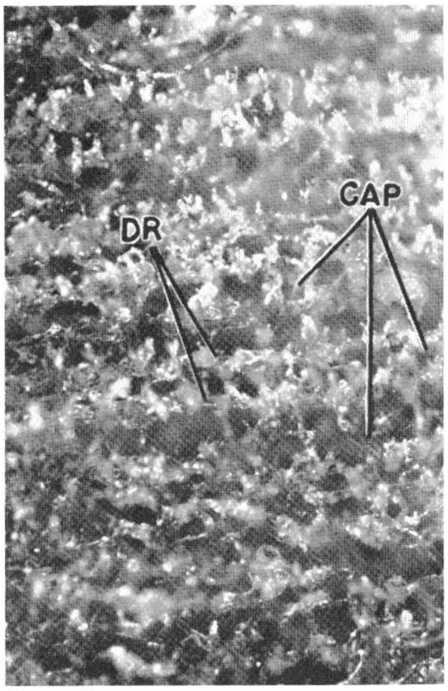

C.

Fig. V-4. Vascular patterns in the dermis in injected-corrosion specimens of the fingers. **A.** The vascular bed of the fingers after injection with a liquid plastic and the removal of **all** soft tissues by strong alkali corrosion. The cristae cutis that look like epidermal whorls are longitudinally-oriented, plastic-impregnated blood vessels giving shape to these "fingers." **B.** Higher power view of these cristae cutis or dermal ridges (DR) as double lines of blood vessels in each ridge, some of which branch (arrows). **C.** Still higher power view of these same double rows of vessels in each ridge (DR) giving rise to vertical vessels, the central capillaries of the dermal papillae (CAP). (Courtesy of J. E. Markee.[1])

Unless the skin is bound down firmly, as it is in some places (scalp, external ear, palms, and soles), the skin is relatively mobile. It exhibits the properties of extensibility and marked elasticity because of the great number of elastic fibers in the dermis. Any temporary displacement produced by joint and muscle movement and stretching is overcome, and the smoothness of skin, so conspicuous in early life, is maintained. But with advancing age, elasticity is impaired and such folds are no longer erased. This accounts for the permanent wrinkles of the elderly and the accentuation of palmar creases and folds seen in the neighborhood of joints (Figs. V-5 and 6).

The Subcutaneous Fascia

The main points of its composition have been discussed above. This layer is always a potential fat depot, especially in certain places such as the back of the arm, the midriff, abdomen, hips, thighs, and buttocks of the female. Most women round out to shaplier proportions when fat accumulates in these areas than do males when it accumulates in the same places. In this layer lie the larger blood vessels to the skin, the heavier nerve trunks, and the specialized deep pressure receptors, the **Pacinian corpuscles,** though structures similar to the latter may be found in the dermis proper, in joint capsules, in the mesenteries of the gut, and in the connective tissue of such glands as the parotid and pancreas. The superficial layer of this fascia seems to retain the characteristics of the part of the body to which it originally belonged. A full-thickness graft, if it is transferred from the abdominal wall to the hand, becomes obese at the same time that the abdominal wall becomes obese.

Skin Color and Pigmentation

Skin coloration results from the blending of the tint of the tissues with that of the blood in the superficial vessels. Lack of blood in these vessels, as in death, causes the skin to assume its characteristic pallor and ashen hue. Where the capillaries are numerous and the overlying strata thin, the skin exhibits a rosy color as in the lips, cheeks, ears, and parts of the hand. The beaconlike effect in Rudolf, the Red-nosed Boozer, is due to increased capillarity.

In certain areas, *e.g.*, the mammary areola after pregnancy, axillae, external genitalia, and around the anus, the skin has a brownish color because of the quantity of melanin pigment in the epidermis. Pigmentation differs among races and individuals (blondes, brunettes, and redheads), and with age and exposure to the sun, as witnessed by the rosy tint of the infant and the bronzed tan of the weather-beaten, surfing teenager.

A number of pigments are known to alter skin color such as melanin, melanoid, carotene, and the blood pigments (oxy- and reduced hemoglobin).

Probably the first and the last, melanin and blood pigments, are the most ubiquitous. With special histological staining techniques, pigment-forming cells known as **melanocytes** can be located scattered through the basal layers of the epidermis. They are present in about the ratio of one to every dozen or so regular epithelial cells of the germinativum, but they may extend a dendritic process well up into the more superficial layers of the stratum germinativum. The yellow, brown, or black **melanin** granules which they secrete are taken up by the neighboring ordinary epithelial cells and may appear in them in greater concentration than in the melanocytes themselves. Obviously, melanin is present in greater concentration in Negro skin than in white, yet it is not lacking in the latter except in degree, since its exposure to sunlight results both in a darkening of the pigment present and an increase in new pigment formation. The contrast between exposed and unexposed parts of the epidermis is quite stark when the bathing suit is removed (Fig. V-7). Hormonal factors also increase pigment formation, as in the case of darkening of areolae around nipples during pregnancy. Melanin, of course, is present in the matrix of hairs, giving them their characteristic shades.

When the two forms of hemoglobin are in imbalance so that reduced hemoglobin predominates, the skin takes on a bluish hue. We say the patient appears **cyanotic,** as in shock and heart failure cases. When the total hemoglobin is low, as in anemias, the skin has a distinct pallor. We have pointed out the ruddiness of the thin-skinned (literally, not figuratively) individual whose capillaries lie close to the surface. You are all familiar with the effects of a blush when many superficial vessels fill with arterial blood, and perhaps you have noted, too, that for no apparent reason someone seems to have turned on the heat so that the ambient temperature has suddenly become warmer during life's embarrassing moments. Light-scattering properties also play a role in affecting skin color, but again, this has more to do with the deep, rather than the superficial, pigments. The superficial layers of epidermis scatter little light because of their relative transparency. The deeper layers, containing melanin granules, tend to scatter blue light more easily than red; hence, when heavily pigmented, they may give almost a blue cast to skin color.

Melanoid is still somewhat of an enigma. It is thought by some to be an altered, degraded form of melanin; others hold it to be an altered, oxidized form of keratin.

If you eat your carrots and other yellow vegetables like a good little man, you may not necessarily see better in the dark, but their **carotene** content will be transformed to vitamin A for storage in body fat and to some degree, in the stratum corneum. The effect is a yellow cast to the skin seen best in the thick palm and sole pads.

Some Special Considerations

As a rule, little absorption occurs through the skin. It is pretty impervious to water and aqueous solutions

Fig. V-5. Skin of the child and of the aged.

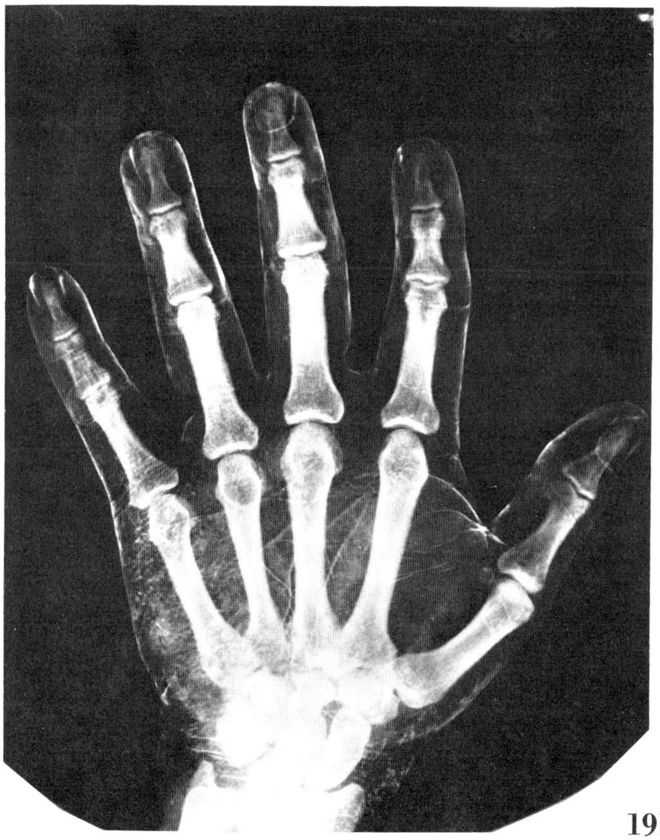

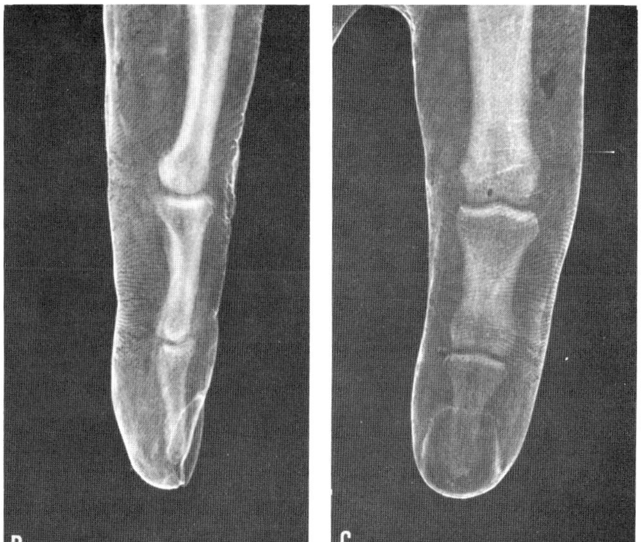

Fig. V-6. Radiographs of the hand (**A**) and index finger (**B** and **C**) after coating the skin with a thick barium paste. The skin folds and ridges are now visible, and their relationship to the underlying bones may be seen. A feeling of depth is also achieved with this technique. **A.** Posterior-anterior (PA) view of the right hand. **B.** Lateral view of the index finger. **C.** PA view of the index finger.

Fig. V-7. Effect of ultraviolet on exposed skin. Sunlight darkens melanin already present in skin and activates melanocytes to increase pigment production.

of salts and sugars. Fatty substances in oils and ointments may be absorbed to a limited extent. A few substances such as lead and mercuric salts and some analine dyes are taken up readily. Infectious organisms rarely pass through unbroken skin, but bacteria can enter **pores** (openings of sweat glands and hair follicles) as infections of these structures indicate.

SKIN DERIVATIVES

Sudoriferous (sweat) and **sebaceous** glands (collectively called the **cutaneous** glands), **ceruminous** glands, **mammary** glands, **hair** and **hair follicles,** and **nails** are all structures derived from the epidermis. Parts of them may lie in the dermis, and the dermis may contribute tight capsules, sheaths, or papillae to some of them.

Sudoriferous or Sweat Glands

Glands of this type are found over the entire body except on the prepuce, glans penis, margin of the lips, and the deeper portion of the auditory meatus. Each gland is a simple, coiled, tubular structure with a tortuous excretory duct. When we say the gland is simple, we mean it is a direct invagination of the epithelial surface as a simple tube would invaginate. Its secretory end is coiled complexly like a skein of yarn (Fig. V-8). Actually, the gland developed as a cord of cells which invaded the dermis from the stratum germinativum and channelization occurred secondarily.

Sweat glands were described as early as 1833 by Purkinje, a Hungarian anatomist who has left his name in all of our hearts as well as in our cerebelli. Later, in 1922, Schiefferdecker[2] appears to be responsible for pointing out that these glands are of two types, **eccrine** and **apocrine.** This simply means that there are small and large ones and that the large apocrine glands are located in special places. The small eccrine glands have a widespread distribution over the body surface except in those areas mentioned above. They inhabit the deeper parts of the dermis. Their ducts wind snakelike toward the surface and they are rarely cut in full longitudinal section as we have drawn them in our figures. They always meet the epidermis at an **epidermal invagination,** never in the dermal papillae. Once the duct reaches the stratum germinativum it ceases to have the structure of a tube lined with cuboidal cells. Sweat simply erodes a coiled channel through the epithelial cells, pushing them aside irregularly in its course to the surface. The opening at the surface is called a **pore.** It has no structure and no muscular sphincter. Hence, the old notion that hot baths "open" pores and cold baths "close" them has no basis in anatomical fact. A hot environment will make you sweat; a cold one won't.

While all eccrine sweat glands look alike, there appear to be some differences among them in the way they behave. Some appear to be more readily responsive to thermal stimulation, and others to psychic or emotional stimulation. All receive their final visceral motor supply from the chain ganglia of the autonomic nervous system which parallels the vertebral column and which are in contact with visceral motor preganglionic neurons that emanate from spinal cord segments T_1 to L_2. It is the source of regulatory command messages from higher neural levels to these visceral motor neurons in the cord that seems to make a difference in the way certain sweat glands function.

Those in the palms and soles seem to respond readily when we are frightened or emotionally upset. Police make use of this fact when they explore the "psychogalvanic skin reflex" by means of the "lie detector" or polygraph. The regulation of these glands in this instance would appear to be cortical, coming, probably, from "limbic lobe" though there is no real evidence that such a lobe exists anatomically or even functionally) areas through connections with our premotor and motor centers in the frontal lobe. And, so, we give our innermost "feelings" away through a skin response.

On the other hand, metabolic rate increases with hard labor, and the body temperature rises. Sweat beads upon the brow, trickles down the chest and the back to soak our shirts, and pours over the arms and

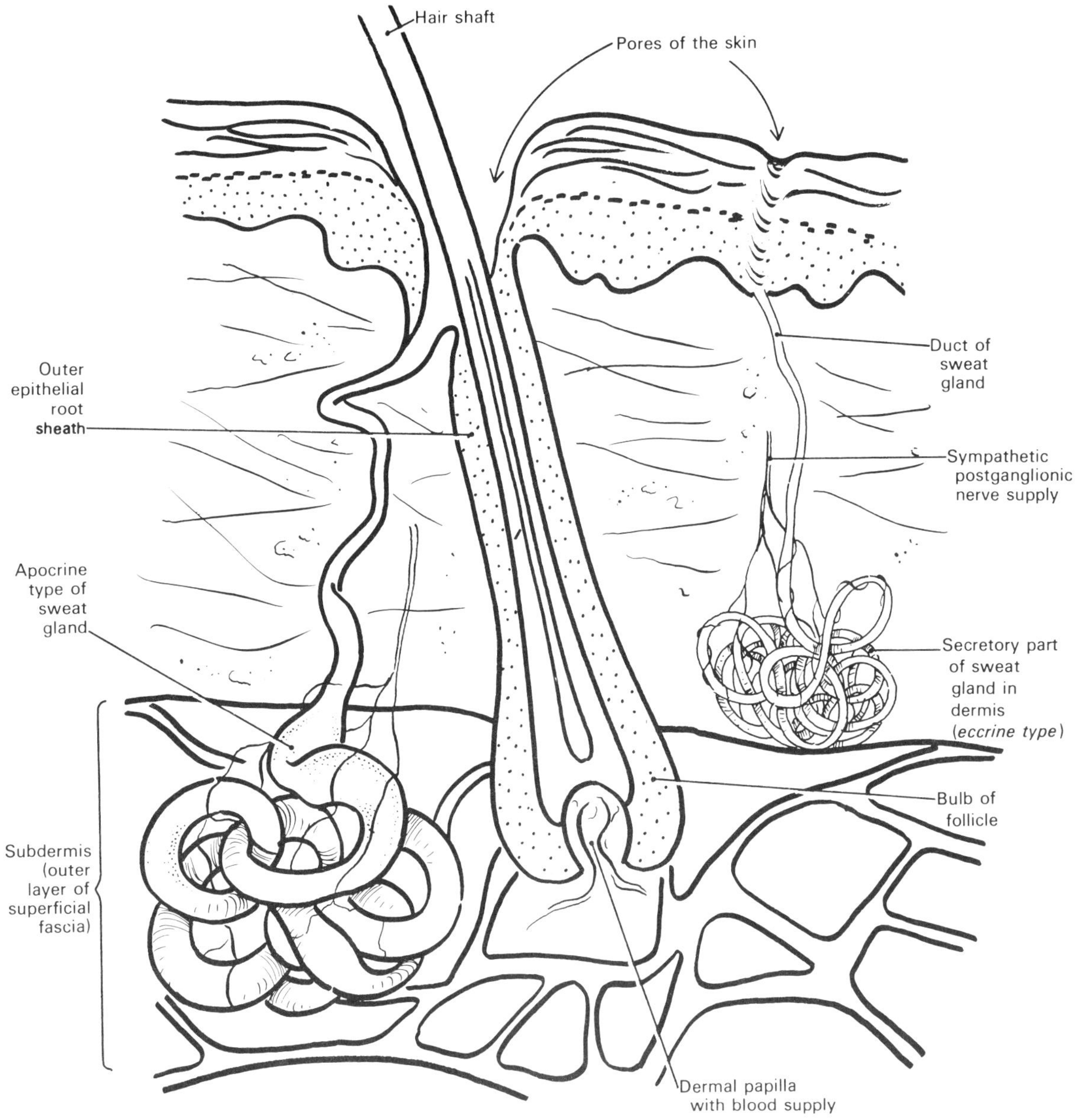

Fig. V-8. Eccrine and "apocrine" sweat glands.

dorsum of the hand. Sweat glands in these areas are responding reflexly to thermal stimulation, and they appear to do so more readily than they do to psychic stimulation. Their control would appear to be directed more by a subcortical heat loss center in the preoptic region of the hypothalamus than by the cerebral cortex. This is a reflex response by means of which the body seeks to regulate temperature through cooling of the surface by evaporation of moisture.

While it makes a nice story, this dichotomy in eccrine gland behavior may not be all this distinct. Even our palms sweat when we engage in heavy labor. They simply may not react to thermal stimuli as rapidly or as intensely. The purpose of palmar sweating under these circumstances may not even be for thermal regulation, but to increase the grip on work tools by increasing palmar surface tension. We have all witnessed (even if only in movies) the scrub nurse as she constantly mopped the surgeon's brow when the operative procedure became difficult and critical. In this same situation, sweat from the forearm and hands may begin to overrun the cuffs of surgical gloves, and there is danger of its contaminating the operative field. Hence, these other glands can respond to psychic or emotional stimuli, but perhaps not as rapidly as palmar glands. In all probability, both higher control mechanisms, cortical and hypothalamic, can assert their influence upon the visceral motor neurons at cord level, and **all** sweat glands can be influenced **by either** of these regulatory channels depending upon the total stimulus situation. This means there must be inhibitory neuronal circuits somewhere to allow one control system full sway while the other is cut off from its influence upon the lower visceral motor neurons. This inhibition could reside at cord level, but it is also well known that the cerebral cortex effects direct connections with the hypothalamus.

We do not seem to be faced with solving such a dilemma when we come to consider the larger apocrine sweat glands. Our major problem with them is to devise ways and means of keeping them from functioning at all so that we become socially acceptable to our neighbors. These inhabit the hypoderm; *i.e.*, the superficial fascia, and, thus, they lie deeper than eccrine glands. They also inhabit our smelliest places, *e.g.*, axillae, perigenital and perianal regions, around the umbilicus, and around the nipples of the breast. Their thicker, goaty-smelling secretions encourage continued research for new deodorants and perfumes. Many of these regions are also hairy ones, and the ducts of these glands utilize the canals of hair follicles to reach the surface (Fig. V-8). Some few reach the surface directly. The chemical composition of their secretions differs from that of eccrine glands, the latter rarely contributing to body odors. We all may have remarked at one time or another that "we have sweat blood." Of course, we haven't, but if you care to enrage a hippopotamus, it will appear to sweat blood. Its apocrine secretions have a red color.[3] We do not advise this as an extracurricular experiment, however, either at the zoo or on the veldt.

The large, malodorous sweat glands were termed apocrine by Schiefferdecker because he thought the apical ends of the cells in the coiled portion pinched off when distended with secretion. This relatively gross fragmentation which he observed with the light microscope may have been an artifact of tissue fixation. However, electron microscopy reveals that these cells have microvilli of variable length at their free surface bordering the gland lumen. Secretion could occur through the plasma membrane of the microvilli; or these microvilli could pinch off. Apocrine secretion would still hold in the latter event, but entirely at a submicroscopic level which could not have been foreseen by Schiefferdecker in 1922. If apocrine secretion *via* microvilli proves to be the rule, we can only credit him with extreme serendipity.

The number of sweat glands varies from 60 to 80 per $cm.^2$ on the back to 400 per $cm.^2$ on the palms, so that about two million have been estimated for the whole body. On the average they produce about 500 to 1,000 cc. of sweat in 24 hours under normal conditions. After strenuous exercise the amount will greatly exceed these figures and 8 to 10 gm. of sodium chloride may be lost in 1 hour. Sweat consists of water (99%), salts (mainly sodium chloride at 0.2 to 0.5%), and traces of urea (0.08%). The amount of urea lost in sweat represents about 1 or 2% of the total urea excreted by the body.

Like the Arrector Pili muscle and smooth muscle of arterioles, sweat glands receive only a sympathetic supply from the autonomic nervous system. There is evidence for a heat loss center in the hypothalamus which promotes sweating. The center seems to be affected directly by a rise in blood temperature in the capillaries surrounding the hypothalamic neurons, or reflexly by stimulation of heat receptors in the skin. It discharges through pathways that end upon visceral motor neurons in cord segments T_1 to L_2, the centers of preganglionic outflow of the sympathetic nervous system. We have mentioned psychic or "cold sweat," most noticeable on the forehead, palms, and hands, which results from fright, embarrassment, or nervousness. Sweating may also accompany nausea, asphyxiation, severe pain, ingestion of highly-spiced foods, and, of course, fever.

Ceruminous Glands

These are a peculiar kind of sweat gland (apocrine?) found in the eyelid (glands of Moll) and in the external auditory meatus. In the lid they form apical coils. In the ear the gland is large with a short duct. The secretory gland cells contain a brown pigment and oily fat droplets. Their secretion, **plus** that of the regular sebaceous glands in the ear, **plus** desquamated epithelium, constitutes the brownish "ear wax" we dig out of our ears. Technically, this is called **cerumen.** It is not a wax.

Sebaceous Glands

Sebaceous glands, with some few exceptions, are connected with hair follicles since they develop from the follicular epithelium of the hair. They are either simple or multilobed structures, variable in size, and their short ducts empty directly into the hair root canal (Fig. V-9). The size of the hair bears little relation to the size of the gland, some of the smallest hairs being associated with the largest glands. Sebaceous glands unconnected with hairs but with short ducts open directly on the surface occur along the margin of the lips, in the nipples, on the eyelids, in the glans and prepuce of the penis, and in the labia minora.

The glands produce an oily secretion called **sebum** which forms the greater part of the lipid component of the surface film over the skin and which oils the hairs. Cuboidal cells in the periphery of the gland seem to be direct descendants of the cells of the stratum cylindricum of the epidermis and are actively mitotic. Their daughter cells fill the lumen of the saclike gland and become impregnated with fat droplets as they swell and degenerate. It is this product of disintegration that constitutes sebum. In this sense, the gland is a **holocrine** type; *i.e.*, the whole cell becomes the main secretory product. Only the ovaries and testes behave in a somewhat similar fashion, though it may be somewhat farfetched to regard the gonads as a type of exocrine gland whose reproductive cells are its secretory product.

Sebaceous glands are odd in the sense that they are the only organs in the body to manufacture a true wax, *sebum*, as a secretory product. While cholesterols and free fatty acids are constant constituents of this product, its other ingredients vary depending upon the species of animal studied. For example, esterified fatty acids are present in human sebum, and only in man and the horse is squalene, an acyclic triterpene alcohol, formed.[4] Why we should have formed this close biological partnership with the horse when St. Bernards are man's best friend as alcohol purveyors, is difficult to fathom. You need only mop your forehead with a clean handkerchief at almost any time to note that you wipe off an oily (and sometimes dirty) surface film which seems to be there as a protective layer to limit the evaporation and absorption of water. Since this surface lipid is quite hydrophilic, its spread over the skin surface is facilitated by the moisture produced by sweat glands.

Your authors know of no good anatomical evidence that sebaceous glands receive a nerve supply, though its existence has been claimed by others.[4] Hormonal regulation appears to be an accepted fact, and testosterone and progesterone may account for glandular enlargement and prepubertal acne in males and females respectively. It should not be overlooked that androgen levels are also fairly high in the normal female. If the pituitary gland is removed, sebaceous glands atrophy. This, of course, would be a most radical procedure for the cure of our pimply youth.

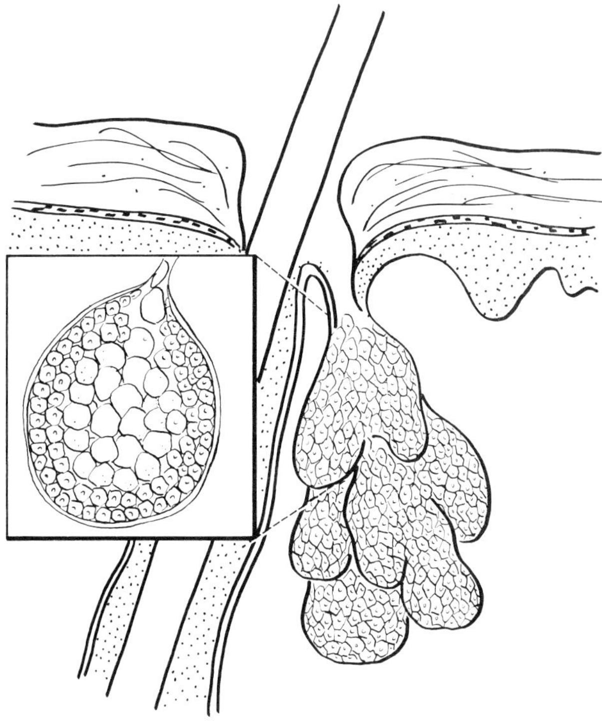

Fig. V-9. Sebaceous glands. Inset is a section of a part of the gland. Outer layer of cuboidal cells in the saccule is the germinal layer. Their daughter cells fill the sac, altering their synthetic processes as they enlarge and degenerate. The end product of their degeneration is a semisolid sebum which oils hairs and forms a lipid surface film on skin.

The elongated tarsal (Meibomian) glands of the upper and lower eyelids are a modified kind of sebaceous gland discharging their oily secretions on the posterior edge of the lids. Retained secretions and swelling of the gland may result in a cyst (chalazion).

Mammary Glands

These glands are said to be modified sweat glands, but their primordia (anlagen) appear early in the sixth week of fetal life. This precedes the appearance of typical sweat glands by several weeks. The first sign of development takes the form of a thickened strip of ectoderm located bilaterally and extending from axilla to the inguinal region as a **"milk ridge"** or **"milk line."** It is, therefore, possible for more than a pair of nipples to develop anywhere along this line. These nipples may or may not persist. Such **supernumerary nipples,** if they do persist, often secrete in the female, become tender with the menstrual cycle, and are generally found bilaterally. They can occur elsewhere; *e.g.*, on the back, buttocks, arm, or thigh, and then they represent independent differentiation. It is said that at least 1% of the population (even some males) show traces of them.[5] Mammary glands develop as a solid ingrowth from the germinal epithelium, much like the sweat gland, except that they are rounder and broader. These soon form 16 to 24 sprouts as solid outbuddings from the main mass which canalize by the end of prenatal life to become **lactiferous ducts** in the female. These, in turn, give rise to a number of solid buds which will transform into **shorter ducts** and **glandular alveoli.** While the larger lactiferous ducts at first open into a pit, shortly after birth the pit becomes a nipple. Thus, the breast is already formed at birth and it may even evidence pseudosecretory activity, producing a thin fluid, **colostrum** (witch's milk), due to the influence of maternal hormones having reached the fetus *via* the placenta. Completion of differentiation awaits changes coincident with puberty and pregnancy. The anatomy of these glands will be discussed in more detail in Chapter VIII.

Hair and Hair Follicles

A hair is an elastic, horny, threadlike structure also derived from the epidermis. It consists of a **shaft,** that part which projects above the epidermis and extends slightly below the surface; and a **root,** the part imbedded in the dermis, the lower portion of which expands to form the **hair bulb** (Fig. V-10). The dermis indents the bulb just as dermal papillae indent the epithelium above, bringing a blood supply close to the living epithelial cells of the hair follicle.

Hair makes its appearance in the embryo at the end of the second month over the upper lip and chin and the brows, as if we were all destined to have moustaches and beards. By the seventh month, fine hairs (lanugo hairs) cover the entire human body. Each hair starts as a thickening of the stratum germinativum which grows down obliquely into the dermis as a solid cord and which becomes evaginated by a **dermal papilla.** The dermis also condenses about the cord as a **connective tissue sheath.** The epithelial cells about the papilla form the **bulb** and project upward as a conical mass. The outer layer of these epithelial cells differentiate into the **outer root sheath.** Above the apex of the cone, the central cells begin to cornify and disintegrate. The channel they form penetrates the surface epithelium, and it is this channel that the hair shaft, itself a modified corneum, will follow to the surface.

Besides the outer root sheath, the cells of the bulb form several layers of specialized cells about the forming hair follicle. Even the shaft, itself, is a trilaminar structure with cells concentrically arranged as a **cuticle,** a **cortex,** and a **medulla.** All the other laminae of the follicle are named for you in Figure V-11 for what they may be worth to you. Our advice is not to try to commit them to memory at any time. If you do, you will have more useless information about hair at your fingertips than your barber. Best, let him talk to you about hair while you talk to him about politics, unless, of course, you expressly wish to impress him with your erudition. Should that be the case, we have added a few more paragraphs about hair below. If you do nothing else with these facts, you can always enliven a dinner conversation when you find a hair in the soup.

The day will come all too soon when it will be nice to know that the intracellular spaces of the hair shaft may contain air, which accounts for the gray or silvery hue when hair pigment is lost. The pigment is located in the medulla or central axis, but even the medulla is absent in fine hairs, and it may not extend the full length of the shaft in others.

We have already mentioned the association of sebaceous glands with hair. Sebum annoints hair and keeps it from becoming dry and brittle. Since it tends to solidify at surface temperatures, it appears on the hair in a semisolid state which leaves greasy marks on the wallpaper behind the sofa. Also associated and attached to the hair follicle in the dermis is a thin band of smooth muscle, the **Arrector Pili,** usually located on the "windward side" in the obtuse angle between follicle and the surface. When things go "Bump!" in the night, the muscle contracts and the hair "stands up." The epidermis dimples in except regionally about the point of hair emergence, where it elevates slightly, and we experience a case of "goose flesh" or "goose pimples." The muscle is innervated by sympathetic fibers of the autonomic nervous system. The hair follicle, itself, is surrounded by a **peritrichal network** of sensory fibers which discharge when the hairs are stroked, as with a wisp of cotton, resulting in the sensation of *light touch.*

Hair on each part of the body has its own definite period of growth, after which time it is shed and replaced. In some animals; *e.g.,* your best friend, the dog, this is a periodic process; in man, growth and loss is continuous, and among your authors it has become a losing battle. The life span of hair on the scalp is from 2 to 5 years; on the eyebrows and lashes, about 3 to 5 months. As scalp hair, lashes, and eyebrows are replaced in children, they become progressively larger and coarser. If you were to measure hair growth in the mirror every day, you might be startled to learn that hair grows at the rate of 1.5 to 3.0 mm. per week. If you do not see any signs of growth, do not be alarmed; you may have a case of baldness or **alopecia.** This can be pathological in some instances, but in most of us it is hereditary. There is nothing your barber can do about it. To have avoided it, you should have castrated your grandfather before you were born. Be content with the fact that sex hormones at puberty influence the growth of hair in the axillary and pubic regions in both sexes, and on the face and trunk in males, and these last a lifetime. Admire these luxurious growths, instead. Baldheaded men can adopt no other philosophy but that of making their lemon sweet.

Nails

Finger and toenails are flat, horny, scalelike modifications of the epidermis which correspond to hooves and claws in lower animals. They serve to protect the end of the digit and to scratch itches. The **body** of the nail is its attached, uncovered portion; the **free edge** is the distal uncovered part that you snip off or paint to suit your fancy; and the **nail root** is the part embedded under a fold of skin, the **proximal nail fold** (Fig. V-11). Since fingers were made before toes, fingernails developed earlier in the embryo than toenails.

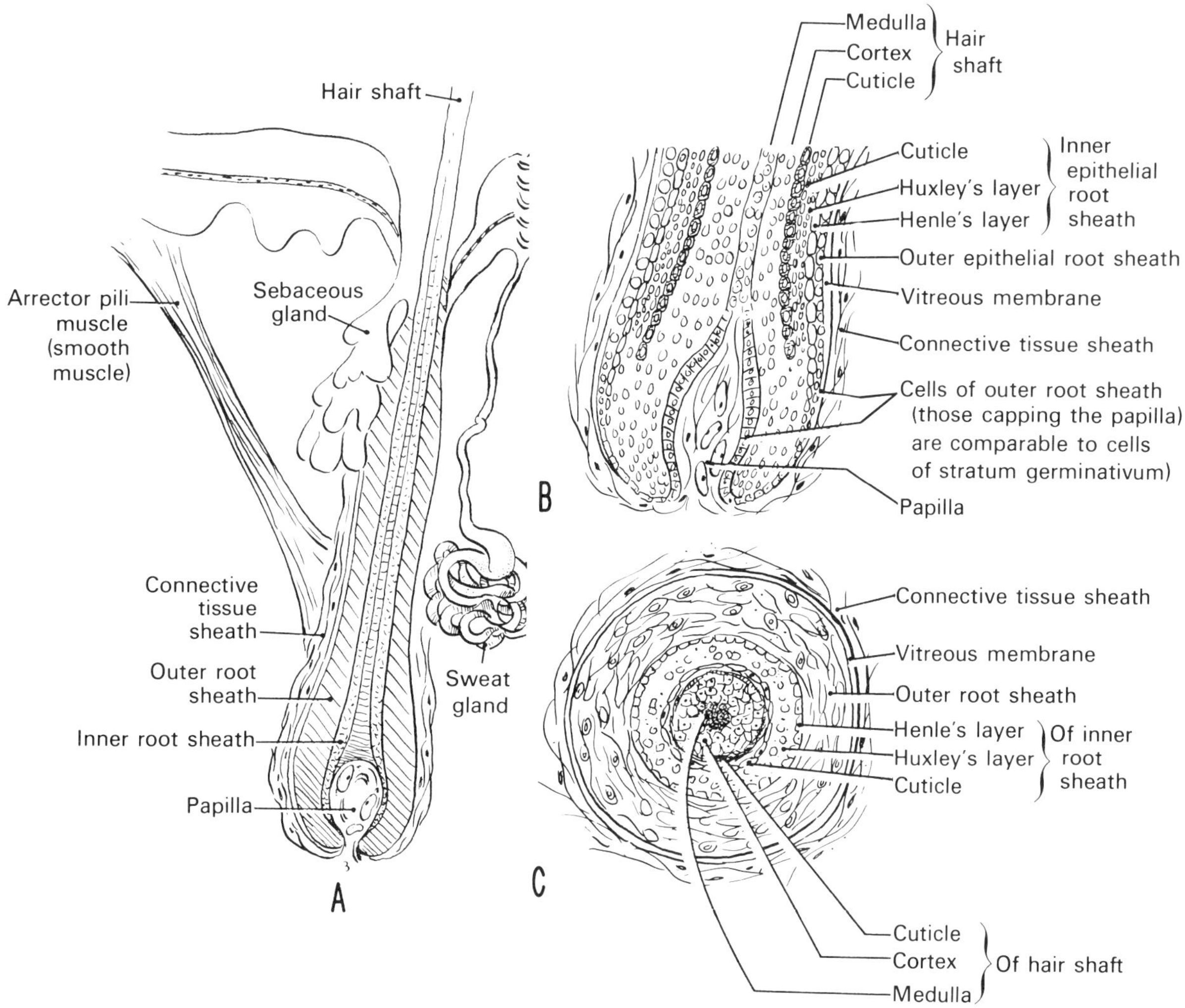

Fig. V-10. Hair, **A.** Schematic longitudinal section through hair, showing basic parts and associated structures. **B.** Detailed longitudinal section of hair at lower end of root, showing composition of follicle and the connective tissue papilla. **C.** Cross section through root of hair and hair follicle with various investing sheaths shown.

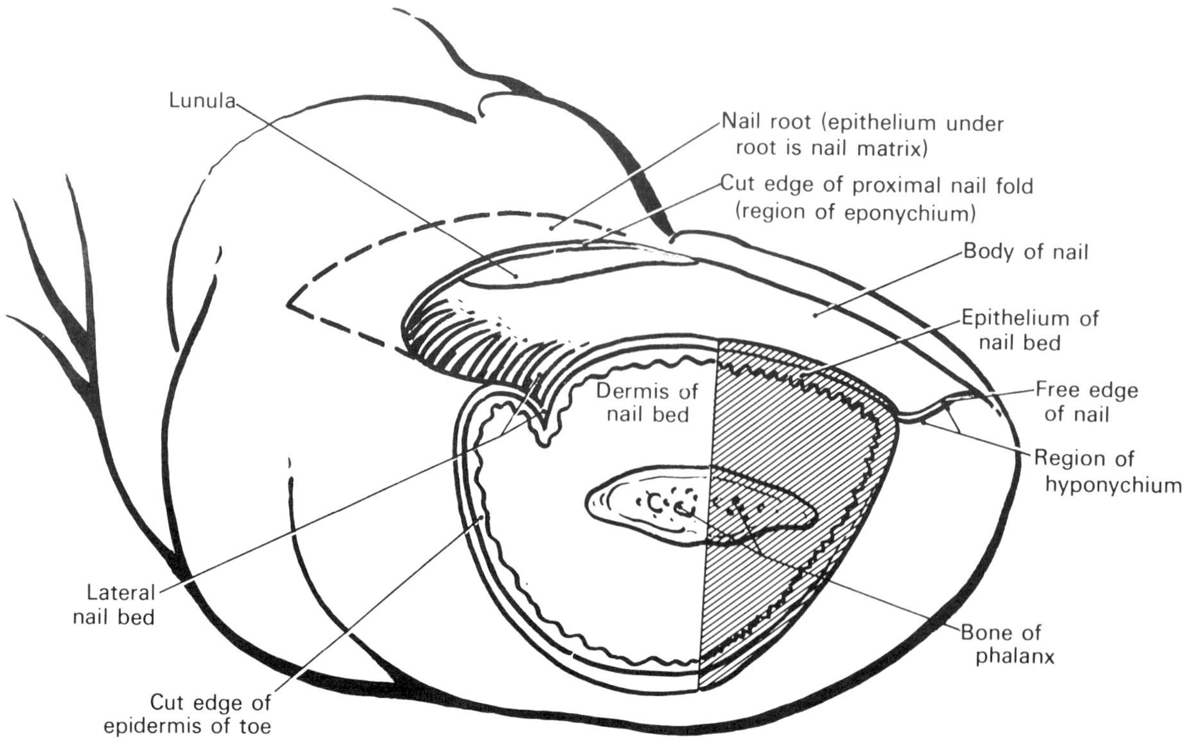

Fig. V-11. Nail of great toe.

The first indication, at 10 weeks, is a thickened area of epidermis on the dorsal aspect of each finger, the **nail field.** The epidermis on the sides and proximal end of this thick area form shallow lateral nail folds or grooves and a deeper proximal nail fold. Like the shaft of the hair, nails are a type of modified corneum. While some local cornification does take place at the surface of the nail field, the real nail is a later development from the underlayer of the proximal nail fold, the area of **nail matrix.** Here, at the fifth month, keratin forms from the matrix directly without going through the stages of keratohyalin and eleidin formation typical of epidermal corneum. If you examine a finished nail you will see some stratum corneum at the surface of the proximal nail fold that extends down onto the nail. This is the **eponychium** which females push back with a cuticle stick so that the pale, half-moon **lunula** at the base of the nail will show to advantage. Under the free edge of the nail, epithelial corneum is also heavy. This is the **hyponychium** which drys, splits, and gets rough if you use the wrong kind of dishwashing detergent. Brand X is always to be eschewed.

Nails grow in length and thickness by sliding forward over the **nail bed** (original nail field) as they grow out from the nail matrix under the proximal nail fold. Some nail may be added from the epithelial cells of the bed. Lost or torn nails will regenerate if the stratum germinativum of the buried matrix is not damaged.

The appearance of nails changes with age and health status. In the healthy young they are smooth, free of ridges and grooves, and are neither too soft nor too brittle. In illness they become ridged and easily broken or torn. In the aged they are likely to be brittle, thick, rather opaque, and quite rough.

The Teeth

Teeth form early in embryonic life sometime during the seventh week. They develop from ectoderm (oral epithelium) and from a mesodermal primordium, but they do not erupt until the sixth or seventh month after birth. We have discussed and named the temporary milk or deciduous teeth and the later permanent ones in Chapter III (p. 107). The **cusp** of each tooth is its raised chewing portion; the **crown,** the part above the gum; the **root,** the portion embedded in the bony socket or alveolus of the mandible or maxilla; and the **neck,** a narrowed region between crown and root. The neck generally has gum, or gingiva, flanking it on the sides. Figure V-12 presents details of tooth structure seen when it is cut in longitudinal section.

The **enamel,** the only true ectodermal derivative, is the material covering the crown, and it is composed of thin prisms or rods, twisted in form, standing upright upon the **dentin** below. Enamel is the hardest substance in the body, even harder than bone. It contains more mineral than bone (96%, as compared to 46%). The bulk of the tooth is made up of dentin, a hard, yellow-

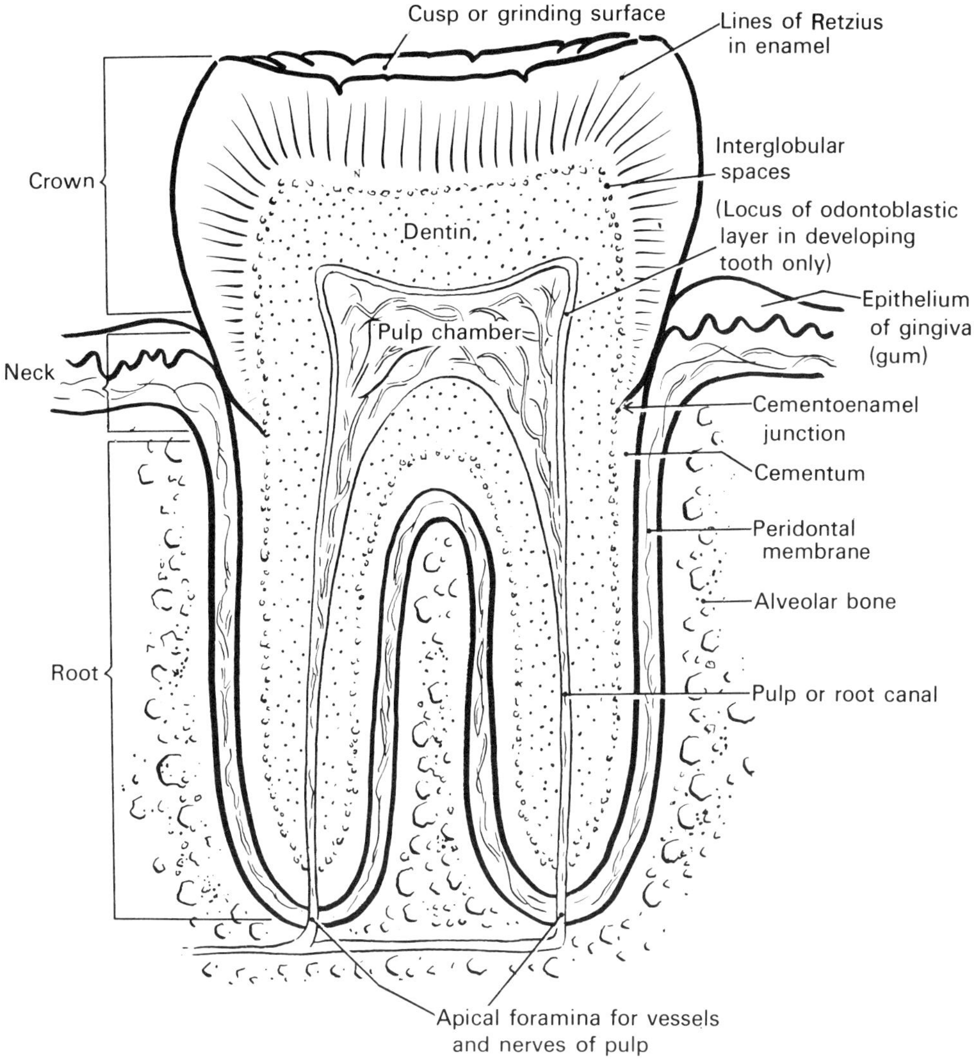

Fig. V-12. Details of a tooth (schematic).

ish, but yet somewhat elastic, material. It also contains more mineral than does bone (69%). In a way it resembles bone in that it contains collagenous fibers within a calcified matrix, but no cells live in the matrix as they do in bone. Only the processes of cells, the **odontoblasts,** project into the matrix of dentin, while their cell bodies lie adjacent in the **pulp cavity.** The **cementum** covers the dentin of the root, and it is continuous with the enamel at the neck of the tooth. It looks much like bone and may be cellular or acellular. It acts like a thin, bony sheath around the tooth, ending at the **cemento-enamel junction,** and it can be disturbed by moving the teeth in orthodontic procedures without undergoing resorption the way alveolar bone does when it is disturbed or damaged in any way.

The soft parts of the tooth are its **pulp** and **peridontal membrane.** The dental pulp, consisting of a fine connective tissue, has the shape of a miniature tooth where it fills the crown as the **pulp chamber,** and the roots as the **root canals.** It brings in the blood and nerve supply to the tooth, much as dermal papillae invaginate surface epithelium or the bulbs of hair follicles. It has draped over its surface the **odontoblastic** layer of cells that give rise to dentin. These, of course, are absent or nonfunctional in adult teeth. The peridontal membrane is a connective tissue sheath that surrounds the root, attaching to it, the wall of the bony socket, and to the gingiva. It holds the tooth firmly in the socket. Its impairment may be responsible for advancing gum disease which sometimes leads to decay and deterioration of teeth. All the parts mentioned, except enamel, are of mesodermal origin.

SKIN AS A SENSE ORGAN

Probably the simplest level of human awareness resides in the general senses, and these, for the most part, reside in the skin. Truly, the general senses are more widespread than this, however. General sensory receptors in our muscle tendons and joints provide us with an awareness of the orientation of our limbs in space, and this portion of our experience we call **proprioception.** There is, in addition, **vibratory sensibility,** tested by placing a tuning fork upon bony prominences. We experience **visceral sensations** (interoception) of a general nature as a sense of fullness or distention in some of our hollow organs and as pain, often ill referred from them, only when they become diseased. When the body cavities are opened in surgery in the lightly-anesthetized individual, the viscera can be pinched, probed, and even cut with impunity without arousing sensations from them unless their mesenteric attachments are pulled up, twisted, or stretched. Yet, any contact with the inner surfaces of the parities will be felt and quite accurately localized. We also have some odd **chemical senses** which can be grouped under the heading of general sensations. Such are the chemoreceptors and baroreceptors in the carotid and aortic sinuses and bodies that are responsive to pH changes and pressure alterations of the circulation and whose effects we appreciate only secondarily in changes in the rates of heart beat and respiratory efforts. More readily experienced are the responses of the nose and other mucous membranes of the eyes and mouth to the acrid qualities of smoke and the irritating, noxious vapors of other chemical substances. Thus, these mucous membranes have a generalized chemical sensitivity, also, and this kind of stimulus applied to the nose is often confused with smell, though it is trigeminal nerve endings that are being stimulated, rather than the olfactory endings within the nasal mucosa. Yet, by and large, the host of general sensory receptors are found in the skin which is virtually peppered with them.

It is strange, upon reflection, that our general sensory receptors so often fail to arouse our attention. We can turn our thoughts to other things and never pay them heed. For one thing, when they are under constant stimulation, their rates of discharge diminish. Responding rapidly at first, tactile receptors, for example, gradually discharge more and more slowly and finally cease to transmit neural impulses entirely, though they are still subject to stimulation. We say they adapt rapidly. The shirt collar that felt a bit too snug when we first buttoned it soon does not bother us at all. For this, we can give thanks, else belts, ties, garters, and the like, not to mention everyday shoes and clothing, would become a nuisance and a burden to us. Even so, one of your authors, who is nursing a slipped intervertebral disk, can sympathize fully with the girls when he is able to retire to the privacy of his boudoir to slip out of that damned girdle!

Still, these old sensory skin receptors of ours have provided us with some of our most pleasant experiences —the feel of cool grass underfoot, the sun's warmth and the breeze's caress, the languor of a tepid bath and the luxury of clean sheets. All these, and many more, affect us deeply and conjure up some of our most precious memories. Perhaps, we are so overwhelmed by the greater flood and wider range of experiences provided by our special senses of vision and hearing and even taste and smell that we quite overlook the joys of our general senses, for even pain is exquisite at times, and we assign to our general senses the role of carrying the minor theme in the complete symphony of the senses.

Enough of fancy and memories. It is time for facts and morphology. **Pain receptors** have the widest distribution of any in the skin. They are widespread, freely-arborizing, **naked terminals** of lightly-myelinated to unmyelinated sensory nerve fibers. For the most part, they branch profusely subepidermally in the papillary layer of the dermis. A few venturesome twiglets may penetrate into the stratum germinativum but no higher (Fig. V-13).

Sensory endings within the epidermis are rare at best. Some fine terminals with small concave disks which embrace the underside of some epidermal cells have been described as epithelial tactile end organs, though none of your authors have encountered them in their lifetime except as schematic doodlings in certain textbook illustrations. These are the so-called **Merckle's tactile disks** (Fig. V-13). They have been touted as inhabiting the snouts of hogs and the penis and clitoris of the guinea pig, if, indeed, the female of the latter species is endowed with a reasonable organ of this kind. With the vaginal membrane closed for 13 days out of every 15 in the estrus cycle, she can hardly keep a clitoris gainfully employed. But that is her private affair.

The bulk of pain receptors, as well as all other types of sensory endings, lie at some level in the dermis. This was discovered to be true by a brave group of British investigators who talked a laboratory flunky, a much braver Mr. McAdoe, into allowing them to take progressive skin slices off his hide with a razor blade.[6] First they tested his skin surface for the locus of pain, touch, warmth, and cold spots, and then proceeded to slice, staining each slice with neurohistological techniques to see how well it matched their surface map for specific receptor-type locus, and to determine the depth of residence level. Mr. McAdoe did not flinch much until they began cutting across dermal papillae, where, of course, they encountered considerable bleeding, as well as the main welter of pain terminals. By that time, they had to strong-arm Mr. McAdoe down on the table to collect a few more deeper slices.

Fortunately, they weren't exploring parts where Merckle's disks might be encountered, but they did pinpoint the locus of some other more widespread tactile receptor organs, namely, **Meisner's tactile corpuscles** (Fig. V-13). These are endings of a moderately-myelinated nerve fiber which branches within the horizontal compartments of a fairly stout connective tissue capsule. In the fingers these reside as solitary units in certain dermal papillae. Elsewhere they lie deeper in the papillary layer in clusters of four or five, with the clusters fairly widely separated. These receptors account for the bulk of our tactile sensibility and their

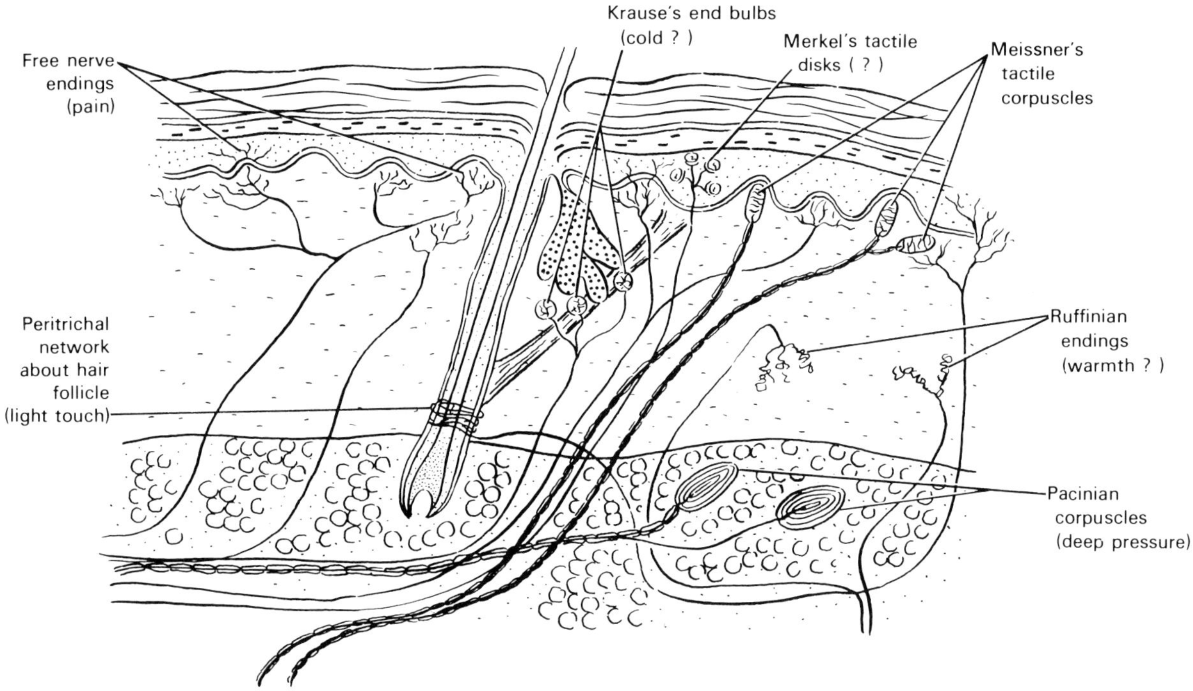

Fig. V-13. Sensory receptors in the skin (schematic).

varied distribution determines our ability to distinguish between one-point and two-point tactile stimuli.

We have already mentioned the *peritrichal ringed network at the base of hairs* which gives us a sense of *light touch* (Fig. V-13). These are naked, non-encapsulated terminals. The British workers did not slice down this far on Mr. McAdoe, for which he, of course, was extremely grateful.

But they did determine that warmth receptors lay deeper in the dermis than cold receptors. In this they were supported by a still more Spartan group of American investigators.[7] The latter (behind locked doors, we presume) anchored *their own* prepuces to cork boards with fishhooks and lowered thermocouples to various depths in the skin thereof in exploring for warmth and cold receptors. While we are still not quite certain of the nature of these receptors, **Krause's end bulbs** have commonly been designated as *cold receptors* and **Rufinian end organs** as *warmth receptors* (Fig. V-13). The former is a rounded encapsulated ending in the papillary layer, while the latter is a naked, tangled skein in the reticular layer. The nerve fibers are of the diameter of fibers that conduct pain and of the myelinated, as well as the unmyelinated, variety. Pain and temperature fibers are always of much smaller diameter than tactile fibers. One problem is that the Rufinian type of ending is also found in lips and in joint capsules where it does not appear to be functioning particularly as a warmth receptor.

The largest of the sensory endings for deep pressure sensibility has, correspondingly, the heaviest amount of myelinization and, consequently, the greatest fiber diameter of any of the sensory nerve fibers in the skin. It also lies deepest, as a rule, in the subcutaneous fascia, and its terminal end organ is large enough to see with routine tissue-staining techniques. It has a heavy, concentrically-lamellated capsule so that it looks like a slice through a small onion. This is the *Pacinian corpuscle* (Fig. V-13).

The general rule is that the more heavily myelinated the nerve fiber, the faster its rate of impulse conduction. Tactile and pressure fibers relay messages more rapidly than pain temperature fibers. Moreover, the faster the rate of conduction, the lower the threshold of sensitivity of the receptor organ. This simply means tactile endings can respond to less intense stimuli than pain endings. Perhaps you can attest to these facts, yourself. If you cut yourself while shaving, you are likely to burst forth with an expletive before you feel the pain. You react to the tactile breaking of the skin first and heed the smarting later.

While it is no longer certain that the receptor organ types we have named are entirely as specific in their separate functions as we may have made them appear to be, at least all of these types have been described and reasonably localized in the skin. We have labeled them, therefore, in Figure V-13 and placed them roughly at their proper depths.

CLINICAL NOTES

The skin is of great importance to the physician in evaluating changes in body function during illness. What the observer sees in or on the skin is frequently a reflection of an underlying disorder and may lead to early diagnosis if properly assessed. Increased redness (**erythema**) results when capillaries are dilated. It may indicate hyperemia, hypertension, infection, fever, allergy, or an emotional disturbance. **Pallor,** which may be due to constriction of the vessels as in the case of cold exposure or fright, may also be due to a reduced number of red blood cells, or anemia. **Cyanosis,** a bluish appearance which results from incomplete oxygenation of the blood as in pneumonia, asphyxiation, heart failure, or congenital heart anomalies, is an important sign to the physician and is easy to detect. **Jaundice** (a yellow pigmentation) results from liver disorders in which bile pigments are absorbed by the blood and circulated through the body.

Increased pigmentation, such as that found in Addison's disease; **excessive dryness,** as found with vitamin deficiency states or heat stroke; **extreme moisture,** which can be a sign of hyperthyroidism; and many other subtle changes in the skin may provide the primary clue to systemic disorders.

Examples of common skin ailments are: **abscess,** a collection of pus formed by localized disintegration of tissue; **acne,** caused by inflammation of the sebaceous glands and accompanied by small pimples or pustules, most often found on the face; **boils,** a localized swelling in the skin caused by infection of hair follicles; **eczema,** an inflammatory disease characterized by vesiculation, watery discharge, and the development of scales or crusts; and **urticaria,** characterized by severe itching, skin patches, and rash—known commonly as hives.

The skin is exposed to a greater variety of irritations than other organs because of its location, and for the same reason, finer variations in its pathology can be more readily observed. The essential changes are similar. The majority of common skin diseases come under the category of inflammation and are produced by such irritants as heat, cold, light, and friction, from which the internal organs are relatively remote. From the clinical standpoint the skin is capable of showing 11 main types of pathological change, and these are referred to as the **primary skin lesions:**

> Erythema, macula, wheal, papule, vesicle (bulla), pustule, furuncle, carbuncle, nodule, tumor, and gumma.

In addition, there are 5 types of secondary lesions which depend for their occurrence upon the evolution of an existing primary lesion. These **secondary lesions** are:

> Scale, crusts, fissure, ulcer, and scar.

In spite of the fact that the clinical types of lesions are limited in number, there are a great variety of easily distinguishable clinical entities. This is possible because of modifications in the structure of the primary lesion, in their arrangement with regard to one another, and in their distribution on the body surface. It must be noted that each disease is characterized clinically by a single, predominant primary lesion.

REFERENCES

For general references to skin consult the chapters devoted to it in:

Bloom, W., and Fawcett, D. W.: *A Textbook of Histology.* Saunders, Philadelphia, 1968.

Greep, R. O.: *Histology.* McGraw-Hill, New York, 1966.

Grollman, S.: *The Human Body, Its Structure and Physiology.* Macmillan, New York, 1964.

Specific references made in this chapter:

1. Markee, J. E., and Wray, J. B.: Circulation of the hand, injection-corrosion studies. J. Bone Joint Surg. (Brit.), *41A:* 673, 1959.
2. Bunting, H., Wislocki, G., and Dempsey, W.: The chemical histology of human eccrine and apocrine sweat glands. Anat. Rec., *100:* 61, 1948.
3. Szodoray, L.: Hetero-topic apocrine glands. Orv. Hetil., *4:* 360, 1948.
4. Grollman, S.: *The Human Body, Its Structure and Physiology,* Ch. 6. Macmillan, New York, 1964.
5. Speert, H.: Supernumerary mammae, with special reference to Rhesus monkey. Quart. Rev. Biol., *17:* 59, 1942.
6. Woolard, H. H.: Observation on the terminations of cutaneous nerves. Brain, *58:* 352, 1935.
7. Bazett, H. C.: *Temperature: Its Measurement and Control in Science and Industry,* pp. 489–501. Reinhold, New York, 1941.

SIX

The Head

INTRODUCTION

The detail involved in any full discussion of the head and neck is voluminous. We have already presented an introduction to the region in Chapter III, pages 79–108. It is, perhaps, best that that portion be reread at this time to gain a fresh overview. Some things will be taken up in more detail than they were in the previous chapter; others, adequately discussed there, may be mentioned only briefly in passing, if at all. Our aim here is to present enough facts and concepts to paint a broad canvas for the student who is approaching the region for the first time. The brain and sense organs will be discussed in a separate chapter later.

THE SCALP

Having treated this integumentary covering of the dome rather fully already in Chapter III, we need only review it quickly here in more or less outline form.

Layers of the Scalp (Fig. VI-1A)

If we spell out S C A L P, each letter serves as the key to one of its layers: **S**kin, **C**onnective tissue, **A**poneurotica, **L**oose connective tissue, **P**eriosteum.

1. The **skin** is quite thick and replete with numerous hairs and sweat glands. When, alas, the hair goes, as it sometimes does in males, the shiny, glabrous pate attests mutely to the presence of the glands, both in "the front rows" and in every church pew of the nation.

2. The **connective tissue beneath the dermis** of the skin, the hypoderm, is a dense, irregular, fibro-elastic weave. It is richly vascularized and surrounds well-lobulated areas of fat.

3. The **aponeurotic layer** and the associated **epicranial musculature:**

The **Occipitofrontalis** is a two-bellied muscle which attaches to each end of the central tendinous **galea aponeurotica.**

I. **Frontalis:** Consists of two long, wide, but united muscles over the frontal bone.
   A. Origin: Galea aponeurotica.
   B. Insertion: Skin of brow and nose.
   C. Nerve supply: Facial (cranial nerve VII); its temporal branch.
   D. Action: Raises eyebrows and wrinkles forehead.

II. **Occipitalis:** Consists of two small, separate muscle masses over the occipital bone.
   A. Origin: Lateral two-thirds of superior nuchal line.
   B. Insertion: Galea aponeurotica.
   C. Nerve Supply: Facial; its posterior auricular branch.
   D. Action: Draws scalp posteriorly.

III. **Galea Aponeurotica:** The flattened common tendon for the above muscles is as integral a part of the scalp as the muscles themselves.
   A. Attachments: *Posteriorly,* to the external occipital protuberance (inion); *laterally,* to the temporal line of the brain case, but it creeps farther laterad as a thin fascial membrane to anchor upon the zygomatic arch.

4. **Loose fibro-elastic tissue** makes up the subaponeurotic layer and allows for mobility when the muscles contract. Here lie the emissary veins which pierce some of the bones of the skull to drain parts of the scalp into the dural venous sinuses (Fig. VI-1B).

5. The **periosteum** on the external aspect of the skull cap is a part of the scalp and consists of dense, irregular, fibro-elastic, connective tissue. In the main, it attaches the scalp loosely to the skull except along suture lines. At these places, a firm attachment is achieved as fibers insinuate between bony serrations to become continuous with the dura mater below. The dura, as it lines the inner surfaces of the skull bones, forms the **internal periosteum** (Fig. VI-1C).

Blood and Nerve Supply of Scalp

See Tables VI-1 and 2 for arterial and nerve supply. Figure VI-2A depicts the motor and sensory innervations; Figure VI-2B shows the arterial and venous distribution of the scalp. Note that the scalp veins bear the same names as their arteries. Two of the nerves have names in common with the vessels. This relationship should pare down the burden of recall.

Briefly, ***the veins*** leave a venous network in the subaponeurotic layer of the scalp. Their external courses and important connections are as follows:

1. **Supraorbital and supratrochlear veins:** These

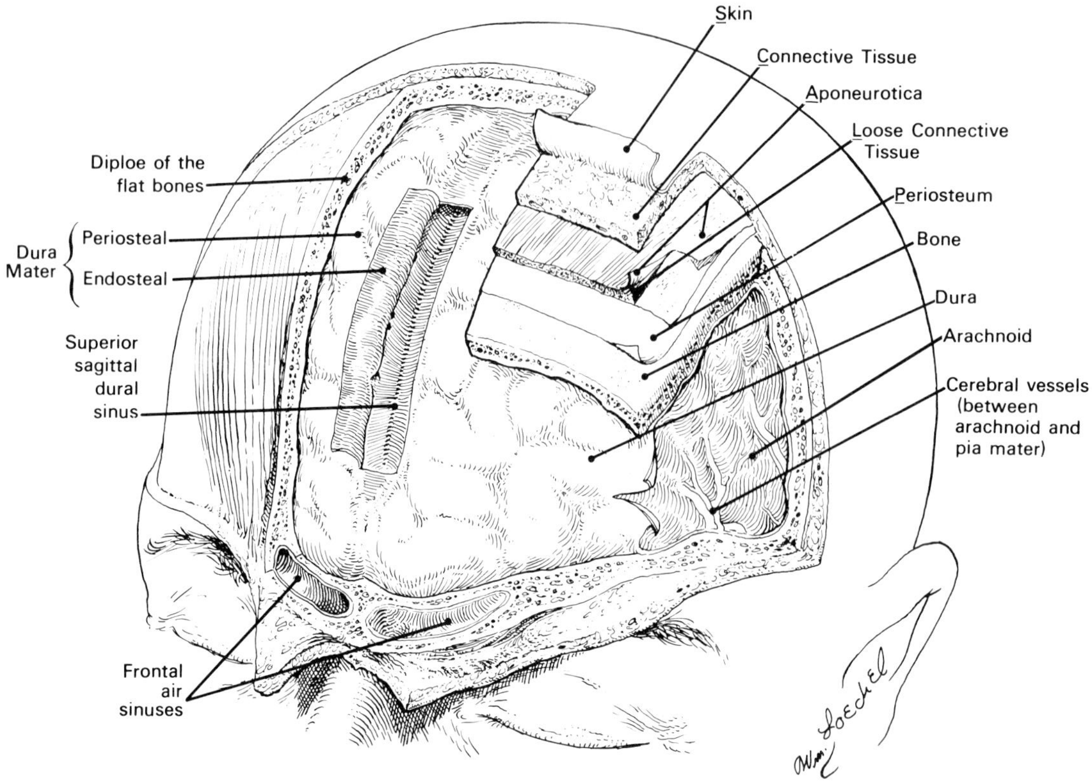

Fig. VI-1. The layers of the scalp and the meninges. **A.** A dissection from the surface.

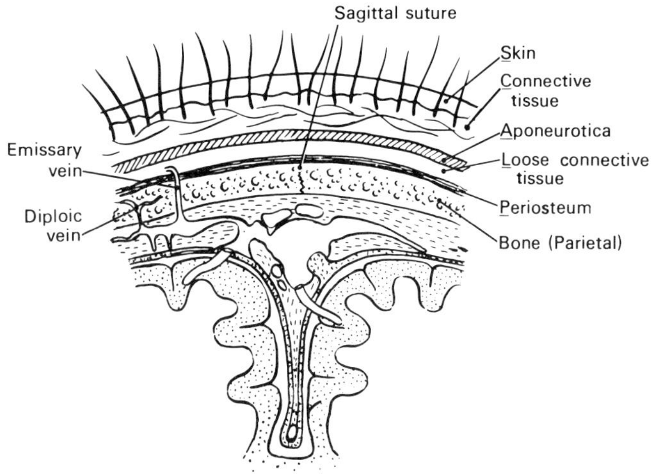

Fig. VI-1B. A schematic coronal section to show layers of scalp; parietal bones joined at sagittal suture; and the relation of emissary and diploic veins to scalp, bone, and dural sinuses.

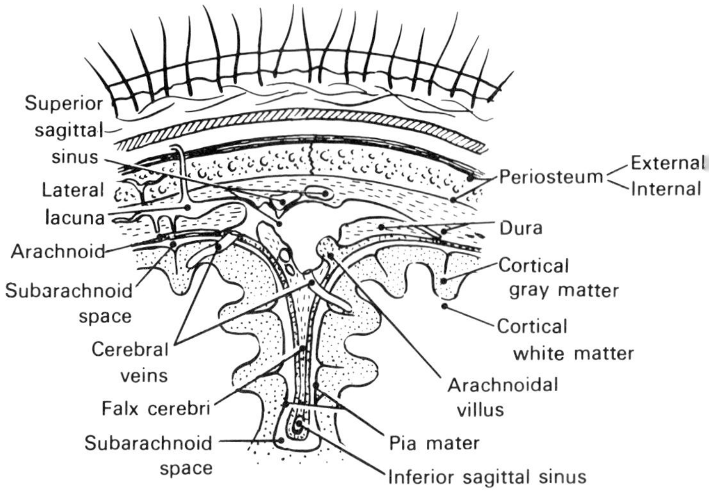

Fig. VI-1C. Same showing relationship of meninges to bone and brain, and to dural sinuses and cerebral vessels. Note where cerebral vessels lie in subarachnoid space. Veins must pierce arachnoid and subdural space before emptying into sinuses, *e.g.*, here into superior sagittal sinus.

Table VI-1. Arterial Supply of Scalp

Artery	Parent Stem	Course	Region Served
Supratrochlear	Ophthalmic artery	Emerges in medial angle of orbital cavity	Forehead
Supra-orbital	Ophthalmic artery	Emerges in supraorbital foramen or notch	Forehead; vault of skull
Superficial temporal	External carotid artery	Terminal branch in front of auricle	Frontal and parietal surfaces
Posterior auricular	External carotid artery	Branch near mastoid process	Scalp areas above ear, and behind ear
Occipital	External carotid artery	Emerges deep from groove in mastoid	Scalp over occiput

Table VI-2. Nerves of Scalp

Nerve	Course	Region Innervated
A. Sensory		
1. Supratrochlear	From orbital cavity at medial angle of upper lid to pierce Frontalis muscle (but not to innervate it); branch of ophthalmic division of trigeminal (cranial V) nerve	Skin of medial aspect of lid and forehead
2. Supraorbital	From supraorbital foramen or notch ***through*** Frontalis muscle; branch of ophthalmic division of trigeminal	Skin of scalp from forehead to vertex
3. Zygomaticotemporal	From a foramen on the temporal or posteromedial aspect of zygomatic bone; branch of maxillary division of trigeminal	Skin of anterior part of temple
4. Auriculotemporal	Out of infratemporal fossa to emerge in front of ear after piercing parotid gland; branch of mandibular division of trigeminal	Skin over side of head, outer aspect of ear, meatus, and the tympanic membrane on its external aspect
5. Lesser occipital	From anterior division of spinal nerve C2 as part of cervical plexus. Emerges midway behind posterior border of Sternomastoid and ascends along that border of the muscle	Neck and skin of scalp behind ear
6. Greater occipital	From posterior divisions of spinal nerves C2 and C3; pierces Trapezius close to its attachment to skull.	Skin of posterior aspect of scalp
7. Third occipital	From posterior division of spinal nerve C3; pierces Trapezius medial to greater occipital nerve	Lower posterior aspect of scalp
B. Motor		
1. Temporal	Branch of facial nerve from stylomastoid foramen to emerge at superior border of parotid gland in front of ear	Motor to Frontalis, Anterior and Superior auricular muscles
2. Posterior auricular	Branch of facial nerve from stylomastoid foramen that crosses mastoid process behind ear	Motor to Occipitalis and Posterior auricular muscles

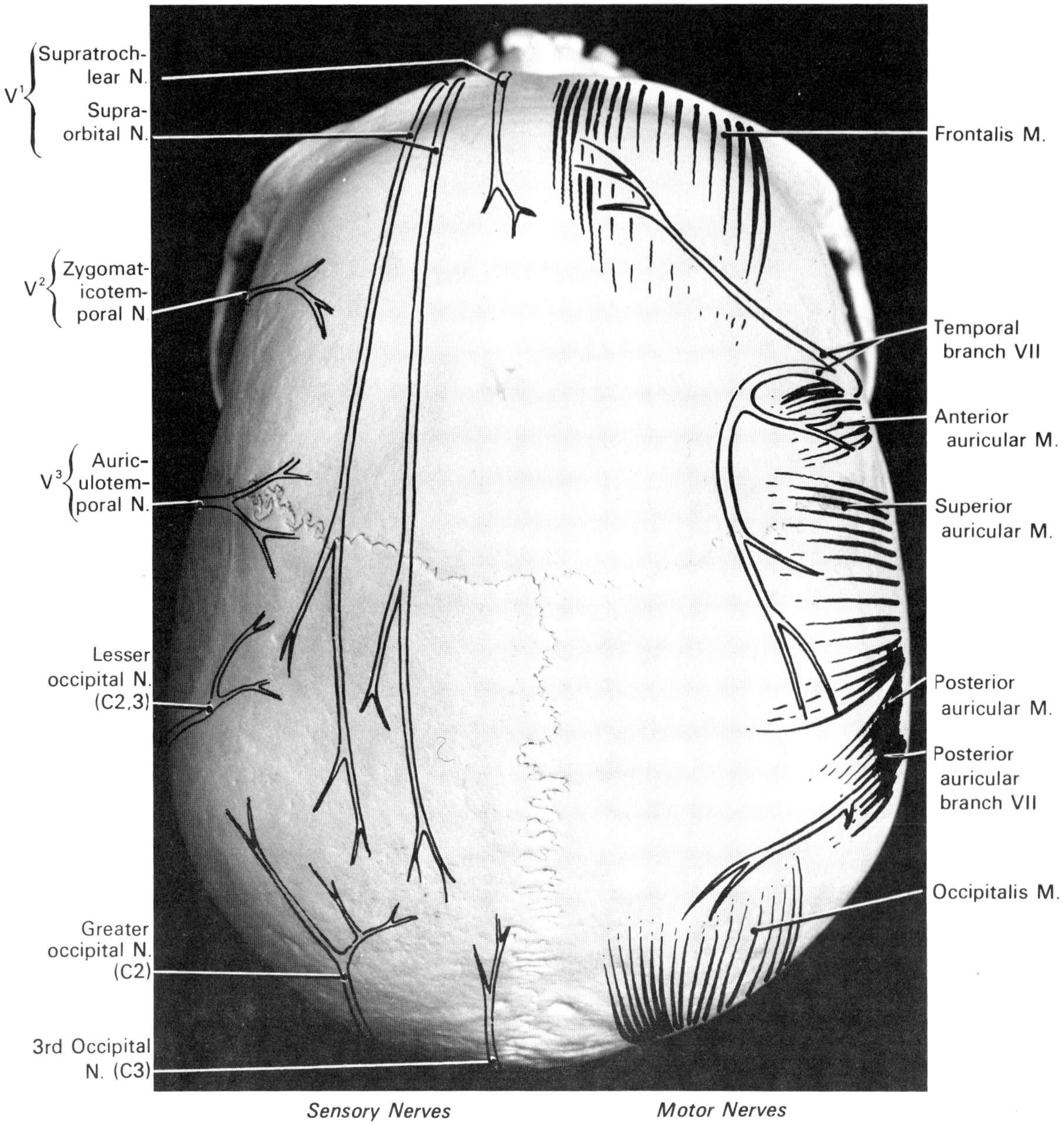

Fig. VI-2A. Sensory (cutaneous) and motor nerve supply of the scalp.

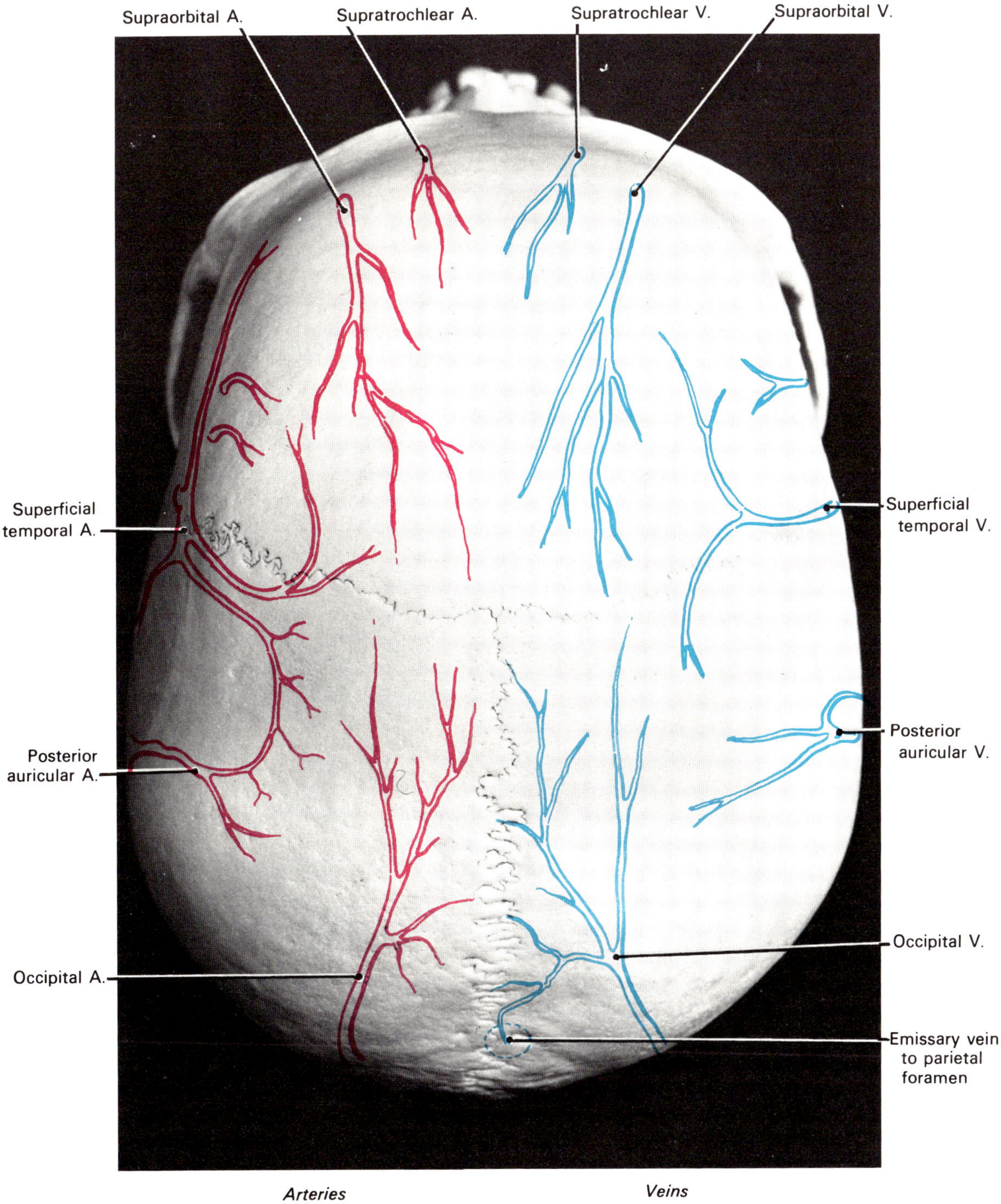

Fig. VI-2B. Arterial and venous supply of the scalp.

drain the forehead and merge to form the **facial vein** at the medial eye angle (Fig. VI-3A). They send twigs through their respective foramina to communicate with the **ophthalmic vein** which drains into the **cavernous sinus** of the dura (Fig. VI-3B), a circular venous channel around the stalk of the pituitary gland. Here, then, is a route where scalp infections could be carried internally.

In Figure VI-3A, note that the facial vein also gives off a deep communicating twig which reaches both the **pterygoid plexus of veins** (between the Pterygoid muscles) and, eventually, the cavernous sinus. The danger of infections to these deep areas from face and scalp wounds is, of course, no longer as serious as it was before the advent of antibiotics, but clotting of blood in the sinus (cavernous sinus thrombosis) is still a clinical problem. There is backflow and dilatation of veins within the bony orbital cavity and edema in the tissues there as a result. The edema may occur more often than we think and as often as we dream. The daily change to a nightly recumbent posture accounts for our puffy lids in the morning light.

2. Superficial Temporal Vein: It descends along the route of its companion artery until it reaches the upper border of the parotid gland. There it joins with the **maxillary vein** to form the **retromandibular vein.** The latter (Fig. VI-3A) continues a direct descent through the gland to terminate below it in two branches, one of which continues into the **external jugular vein** and the other, by joining the common facial vein, empties into the **internal jugular vein.**

3. Posterior auricular vein (Fig. VI-3A): It descends along its companion artery to a point below the ear where it joins the branch of the retromandibular vein that leads to the external jugular.

4. Occipital vein (Fig. VI-3A): It follows its artery to pierce the Trapezius where it terminates in a **suboccipital plexus of veins** among the small muscles of the suboccipital triangle. Normally the plexus drains into the **vertebral or the deep cervical veins.** The occipital vein can, occasionally, end in the internal jugular or join the posterior auricular vein to reach the external jugular. In the scalp, it is its twigs which become the emissary veins that traverse the **parietal emissary foramen** to connect with the **superior sagittal sinus** (Fig. VI-1B); or which traverse the **mastoid emissary foramen** to join the transverse dural sinus. It often receives a **diploic vein** from the marrow spaces of the occipital bone. Here, then, are other routes for scalp infections to reach the **meninges** or into the cranial bones. The dura mater with its venous channels forms the outermost of the three **meningeal investments** of the brain (dura, arachnoid, and pia).

THE EXTRADURAL SPACE

Beneath the skull cap, between the inner aspect of the skull bones and the dura mater, is a *potential space,* the **extradural space,** which becomes important in the case of hemorrhage (Fig. VI-4A). Here the middle meningeal arteries on the surface of the dura, or the superior sagittal venous sinus with its associated lateral lacunar extensions within the dura, can bleed into the space as a result of a blow to the head. Hemorrhage within the rigid confines of the cranium soon provokes an increase in intracranial pressure within the cerebrospinal fluid system (*i.e.,* within the ventricles and the subarachnoid space). A prolonged rise in these pressures soon leads to death; hence, treatment of extradural hemorrhage constitutes an urgent emergency. Generally a burr hole opening is made midway along a line joining the orbital cavity with the external auditory meatus, and the exposure is enlarged until the bleeding vessels can be reached and cauterized. The early flow of blood from the opening itself helps to bring intracranial pressure under control as well. Figure VI-1B does not depict this space and need not, since it exists only under these abnormal conditions.

THE SUBDURAL SPACE

Figure VI-1B shows that the inner surface of dura rests snugly upon a filmy, arachnoidal membrane, the second investment of the brain. Between these two opposed sheets of tissue is a second ***potential space*** that can be dissected by hemorrhage. This is the **subdural space,** and it has not been shown in the figure. However, the relationships here are nicely illustrated in Figure VI-4B, where a dural flap has been incised. The venous sinuses are more likely to bleed into this space, and so can the cerebral veins. The latter, as do the cerebral arteries, course over the brain within the **subarachnoid space,** the fluid-filled space between arachnoid and pia which *is* shown in Figure VI-1B. Normally, arachnoidal villi pierce the dura at intervals along the sagittal sinus and protrude into the sinus as microscopic, endothelial-lined structures (see Fig. VI-1B). Through such villi, cerebrospinal fluid of the subarachnoid space is returned into the venous circulation. To enter the dural sinuses, the cerebral veins must pierce the arachnoid and so they find themselves for a short span within the potential subdural space (Fig. VI-4B). At this point, when injured, they can produce subdural hematomas. The cerebral arteries, of course, leave the subarachnoid space in the reverse direction by piercing the innermost meningeal coat, the pia, in order to enter the brain substance, proper. We will return to the matter of cerebral vessels and the subarachnoidal fluid system in another chapter. The student should note now, however, exactly where these vessels lie as they run over the brain surface, realizing that they are *between* arachnoid and pia. He should also note the relationship of cerebral veins to the dural sinuses at this time, for it is by emptying into the dural sinuses that venous drainage of the brain eventually reaches the internal jugular for return to the heart *via* the superior vena cava.

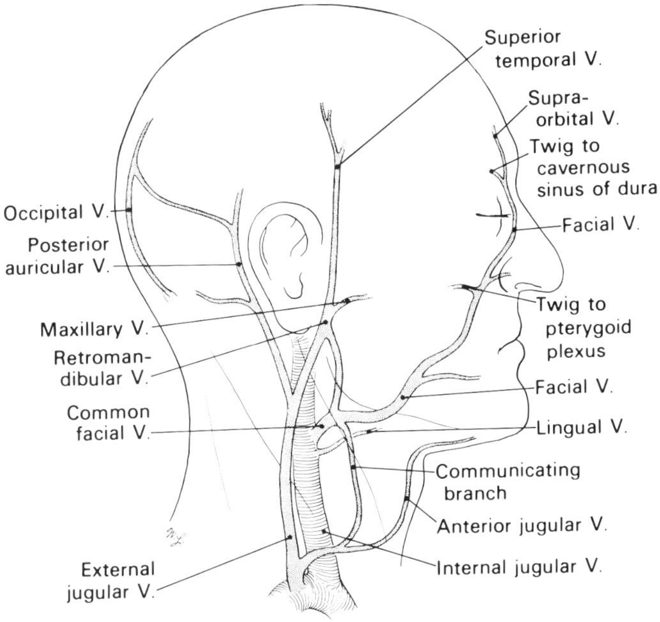

A

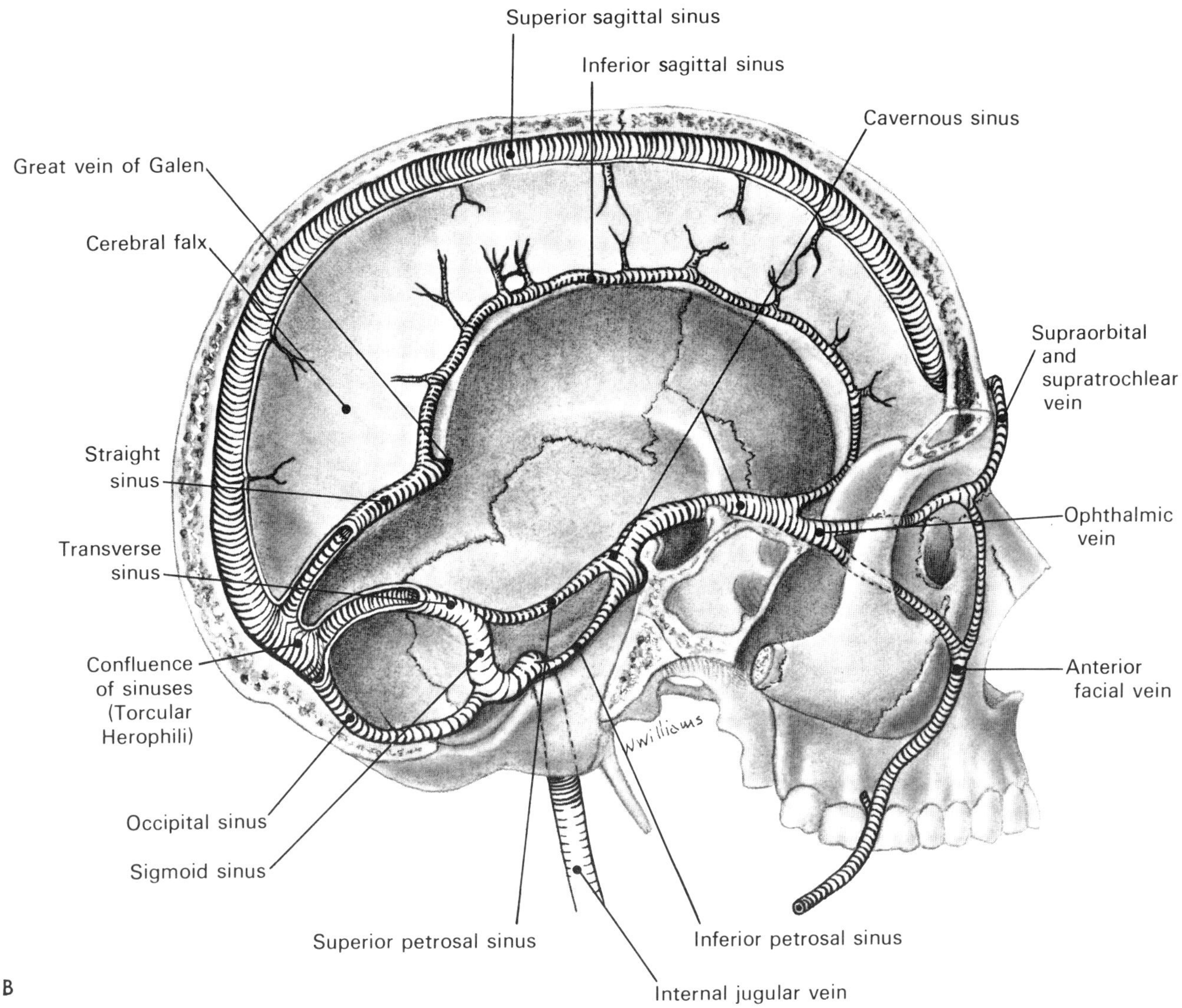

B

Fig. VI-3A. Venous drainage of the scalp and face. **B.** Midsagittal view of cranium showing venous dural sinuses.

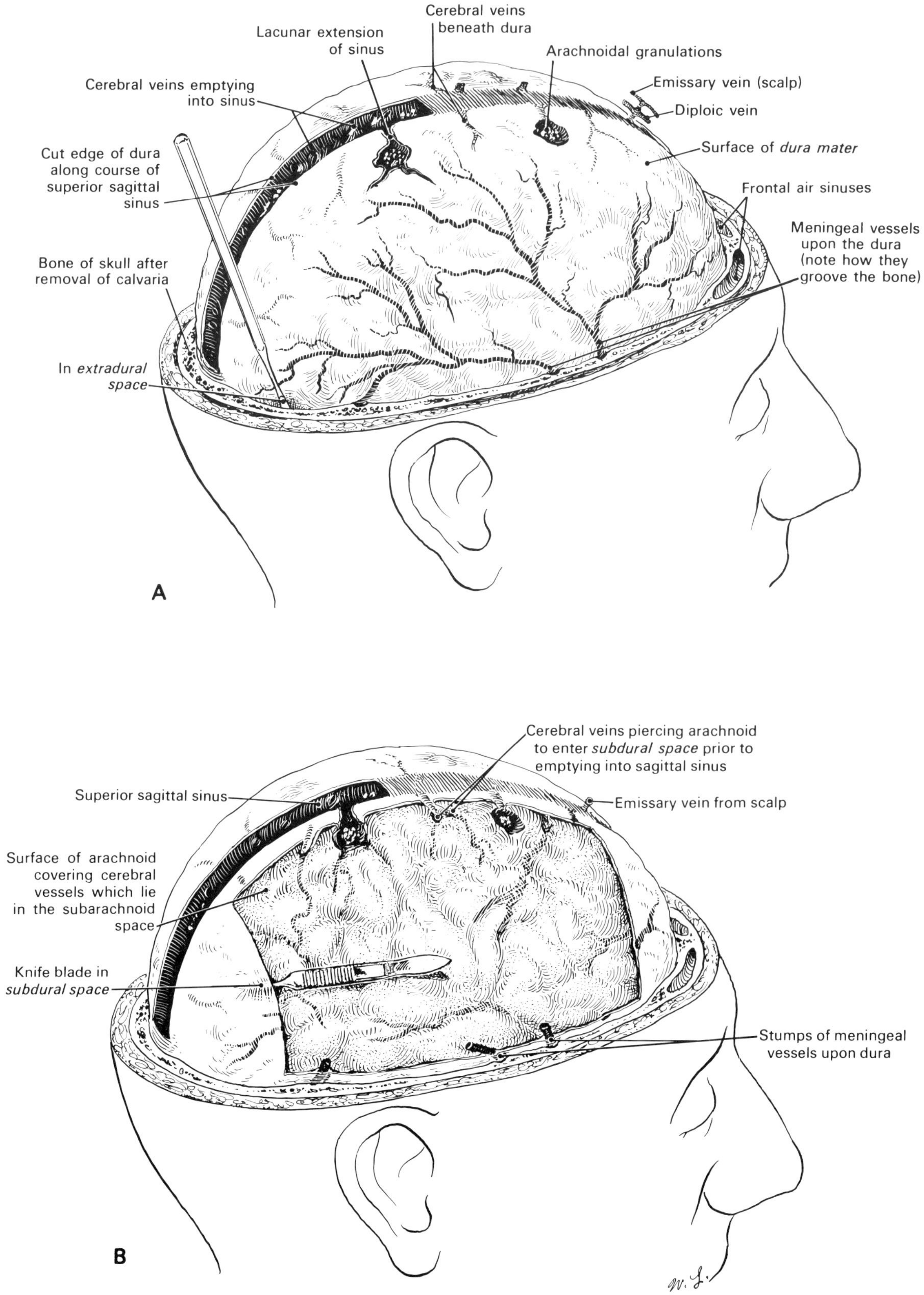

Fig. VI-4 A. The potential extradural space. Notice the location of the arachnoidal (Pacchionian) granulations alongside the superior sagittal sinus. **B.** Cerebral vessels in the subdural space.

THE SKULL

The Bones of the Skull

It is time next for a bit of "skullduggery." Structures in the head are protected by a bony framework, **the skull,** which is a bit more intricately contrived than the rest of the skeleton. The aim of this section is not to learn the details of all of the bones of the skull but to appreciate the skull as a whole, as well as some of its important holes. For it is through certain holes or **foramina** that important nerves and vessels make their way in and out of the skull.

The skull has 28 irregularly-shaped bones, 11 of them paired and 6 single. Of these 28, six (3 pair) are the tiny bones of the middle ear; 14 are bones of the face; and 8 make up the neurocranium, or that part of the skull which protects and houses the brain.

Bones of the Skull

	No.	*Total*
A. Cranial bones		
I. Paired		
a. Parietal	2	
b. Temporal	2	
II. Single		
a. Frontal	1	
b. Occipital	1	
c. Sphenoid	1	
d. Ethmoid	1	
Total		8
B. Facial bones		
I. Paired		
a. Nasal	2	
b. Maxillary	2	
c. Zygomatic	2	
d. Lacrimal	2	
e. Palatine	2	
f. Inferior conchae	2	
II. Single		
a. Mandible	1	
b. Vomer	1	
Total		14
C. Middle ear bones (paired, only)		
a. Malleus	2	
b. Incus	2	
c. Stapes	2	
Total		6
Total for skull		28

Development of the Bones of the Skull: During embryonic life bones develop along two major pathways, either as **intramembranous** bones or as **endochondral (intracartilaginous)** bones. The bulk of the bones of the cranium and the face are formed by the first method and often are called "membrane bones." The ethmoid, hyoid, middle ear ossicles, and all the bones of the appendicular skeleton arise through endochondral bone formation. Others, like the occipital, temporal, and sphenoid, because they represent fusions of once separate bones in lower forms, arise partly as membrane bones and partly from cartilage (Fig. VI-5).

Branchial Arch Derivatives: The cartilaginous bars in the gill cleft region of the embryo may form some small bones (middle ear ossicles and the hyoid); may form parts of other bones (styloid process of temporal); may remain as cartilage (thyroid and cricoid); may degenerate into ligaments; or may degenerate completely; or, in some anomalous situations, they may remain in part or even completely (Table VI-3 and Fig. VI-6).

Table VI-3. Development of Bones of Skull

Type of Development	Bones (Cartilage and Ligaments) so Derived
A. Endochondral bone formation	Most of occipital, temporal and sphenoid; all of ethmoid
B. Intramembranous bone formation	Interparietal part of occipital; squamous and tympanic parts of temporal, orbital and temporal parts of greater wing and the pterygoid plates of the sphenoid; vomer and inferior conchae; nasal bones; lacrimal bones; frontal; parietal; maxillary; palatine; zygomatic; and the mandible
C. Derivation of branchial arches	*Arch I* 1. Malleus and incus 2. Tip of mandible at chin only (Resorbed parts become mandibular canal and mental foramen) 3. Anterior malleolar and sphenomandibular ligaments *Arch II* 1. Styloid process of temporal bone 2. Stapes 3. Lesser horns and cranial part of hyoid bone 4. Stylohyoid ligament *Arch III* 1. Greater horns and caudal part of hyoid bone *Arch IV* 1. Cuneiform and thyroid cartilages *Arch V* 1. Corniculate, arytenoid and cricoid cartilages (perhaps some thyroid cartilage) *Arch VI* 1. Disappears entirely

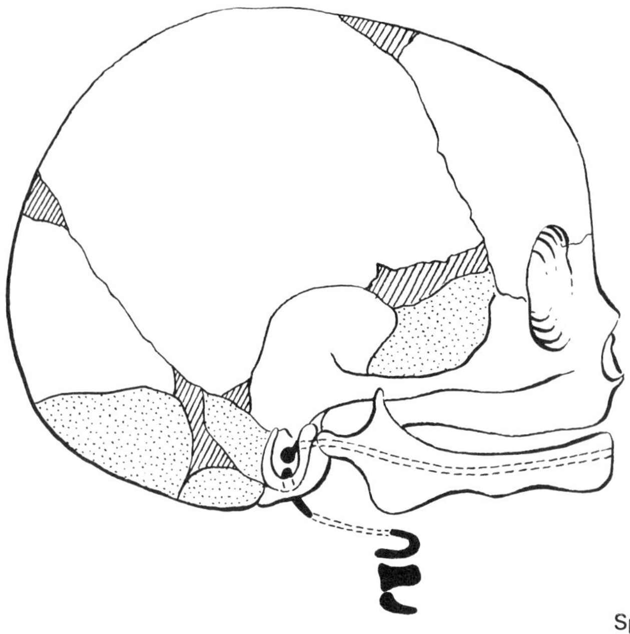

Fig. VI-5. Bony composition of the human skull at birth. White areas, membrane bone; stippled areas, endochondral bone (cartilage): black areas, branchial arch derivatives; striped areas, fontanelles.

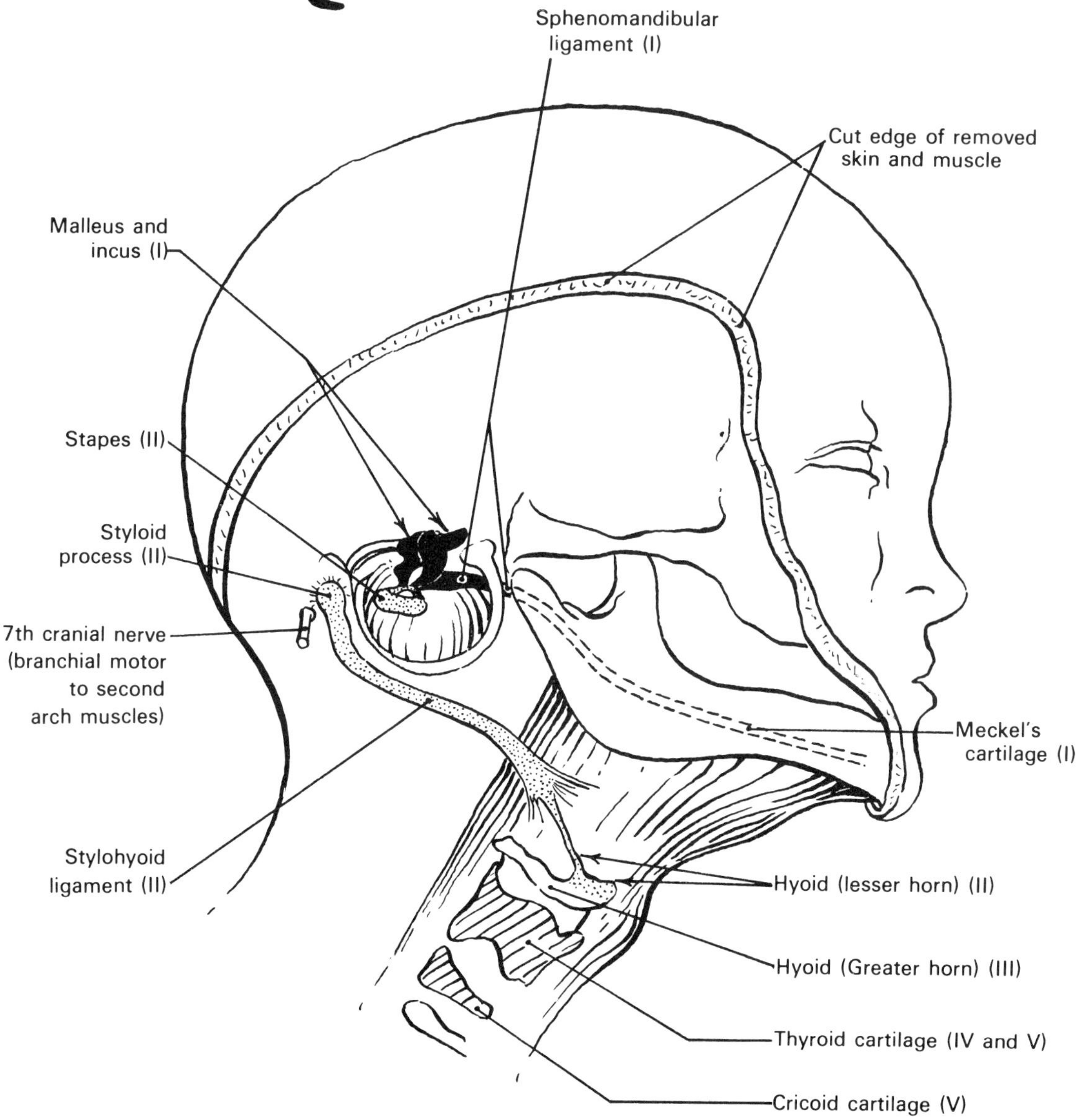

Fig. VI-6. Lateral dissection of a fetal head to illustrate derivatives of the human branchial arches. Black and dashed lines, arch I; stippled, arch II; white (inferior part and horn of hyoid), arch III; striped, arches IV and V.

An understanding of the branchial arch cartilage helps to understand development in parts of the face and neck. Since each arch is associated with a specific cranial nerve, some knowledge of muscular innervation in the head and neck is gained, as well. The specific muscles involved are derived from mesoderm which surrounds the cartilaginous gill cleft bars. Specific blood vessels are also associated with the bars and these will become meaningful when we study the vascular supply in this region.

The cartilaginous bars are bilateral structures. The first pair forming arch I (Fig. VI-6) are rather narrow pieces that are rounded at their cranial ends. Distally they extend into the territory of a future chin. This cartilage has been called Meckel's cartilage after Johann Friederich Meckel, II, one of three generations of anatomists and the surgeon who first described the diverticulum of the ileum. He discussed the fate of this cartilaginous arch in 1850. Its rounded cranial end develops into two tiny ear ossicles, the **malleus** and the **incus,** and into the **anterior malleolar ligament** which attaches the malleus to the tympanic wall of the middle ear cavity. The fibers of this ligament traverse the petrotympanic fissure of the skull to gain secondary attachment to the **spine of the sphenoid.** Some minor contribution to the malleus and incus may be made by the second arch.[1]

The stretch of cartilage from the spine of the sphenoid to the **lingula,** a spur of bone on the inner aspect of the **ramus of the mandible,** also degenerates into a ligamentous structure, the **sphenomandibular ligament** (Fig. VI-6). The rest of the cartilaginous bar extends distalward and appears as if it might become the mandible or lower jaw bone. Instead, the mandible develops from the adjacent mesoderm to *surround* the distal part of this cartilaginous bar (Fig. VI-6), incorporating it into its very substance. The cartilage degenerates, leaving a canal which traverses the mandible from *lingula* to the **mental foramen** as the **mandibular canal.** The canal conveys the **inferior alveolar nerve** as a sensory supply to the lower teeth and **mental nerve** which is sensory to the skin of the chin and lower lip. A bit of the mandible near the chin may arise as endochondral bone from this cartilage (as did the malleus and incus at its proximal end). The bulk of the mandible, however, is "membrane bone," like all other facial bones.

Karl Bogislaus Reichert, another German anatomist, described the fate of the second branchial arch in 1836. It is a much thinner bar than the first arch cartilage but, again, its nodular cranial end becomes an ear ossicle, the **stapes** (Fig. VI-6). It forms, as well, the **styloid process of the temporal bone,** and the stretch of cartilage from this process to below the base of the tongue converts into the **stylohyoid ligament.** Its distal end becomes the **lesser horn** and **the cranial portion of the body of the hyoid bone.**

Reichert, wrote the autobiography of the third arch, also. It completes the hyoid bone by contributing the **greater horn** and the **lower part of the body of the hyoid.**

The fourth and fifth arches remain as cartilage to give shape to the skeleton of the larynx; *e.g.,* as **thyroid, cricoid, cuneiform, corniculate,** and **arytenoid cartilages** (see Chapter VII, p. 390, for locations of these smaller cartilages).

The cartilage of the rudimentary sixth arch disappears completely.

From the mesoderm around these cartilages, muscle masses arise. These from the first arch region are, chiefly, the *muscles of mastication* (Fig. VI-7A). Most of them are associated with the mandible (Temporalis, Masseter, the Pterygoids, Mylohyoid, and the anterior belly of the Digastric); some few are not (the Tensors Tympani and Palatini). The proper fifth arch nerve for these muscles is the **trigeminal or cranial nerve V.**

The muscles related to the second or hyoid arch are, chiefly, the *muscles of facial expression* (Fig. VI-7B). These include the Fronto occipitalis, the Auriculars, and Platysma, the Buccinator, and the host of smaller named muscles in the skin of the face. In addition, they include the Stapedius (the tiny muscle attached to the stapes), the Stylohyoid, and the posterior belly of the Digastric. As you might suspect, the proper arch nerve for these muscles is the **facial nerve or cranial nerve VII.**

Only one small muscle, **the Stylopharyngeus,** develops from mesoderm associated with the third branchial arch cartilage (Fig. VI-7C). Its proper arch nerve is the **glossopharyngeal or cranial nerve IX.**

The **vagus nerve or cranial nerve X** supplies all the named muscles arising from both fourth and fifth branchial arch territories (Fig. VI-7C). These are the rest of the *muscles of the palate* (except the Tensor Palatini which the cranial nerve V supplies); the rest of the *pharyngeal muscles* (except the Stylopharyngeus); and all the *laryngeal muscles,* both extrinsic and intrinsic. (These will be discussed in more detail in the next chapter.)

Despite the complete disappearance of the sixth arch cartilage, its associated muscular derivatives survive and migrate into the neck as the **Sternomastoid** and **Trapezius.** The **spinal accessory or cranial nerve XI** is their proper arch nerve.

Thus, we see that certain fiber components of cranial nerves V, VII, IX, X, and XI subserve *skeletal muscle* derived from mesoderm associated with the branchial arches. Such muscle is designated as *branchial musculature* to differentiate it from *skeletal muscle* derived from the segmental, mesodermal somites of the embryo and which is often referred to as *somatic musculature.* The nerve fibers which supply these two types of skeletal or voluntary muscle are, therefore, called **branchial motor** and **somatic motor nerves.** Most of the somatic motor nerves are **spinal nerves.** Only **cranial nerves III, IV, and VI** to extra-occular muscles (Fig. VI-8A) and cranial nerve **XII** to tongue musculature innervate skeletal muscle derived from an odd group of "head somites" (Fig. VI-8B). *All branchial motor nerves are cranial nerves.*

The branchial arches, related as they are to the

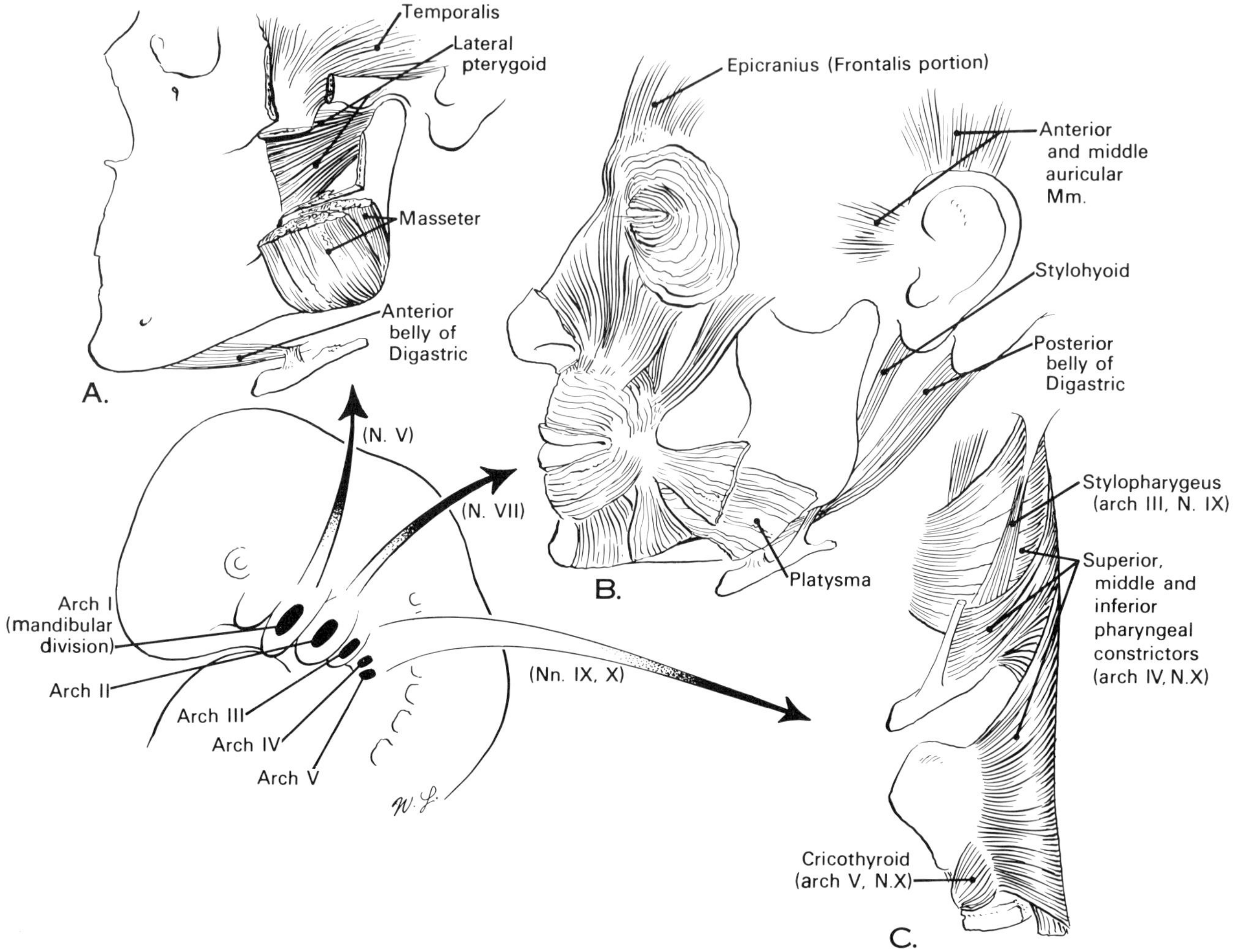

Fig. VI-7. Branchial arch derivations of certain muscles in the head and neck. **A.** The muscles of mastication from the first arch. The Medial pterygoid, Mylohyoid, and the Tensors palatini and tympani lie too deeply to be seen in this view. **B.** The muscles of facial expression from the second arch. Not all are shown. **C.** Muscle derivatives from the mesoderm around the third, fourth, and fifth arches. Not all of the palatine and pharyngeal muscles and none of the intrinsic laryngeal muscles are shown. Cranial nerves V, VII, and IX are proper to the first, second, and third arch muscles, respectively. Cranial nerve X supplies all muscles derived from the fourth and fifth arches.

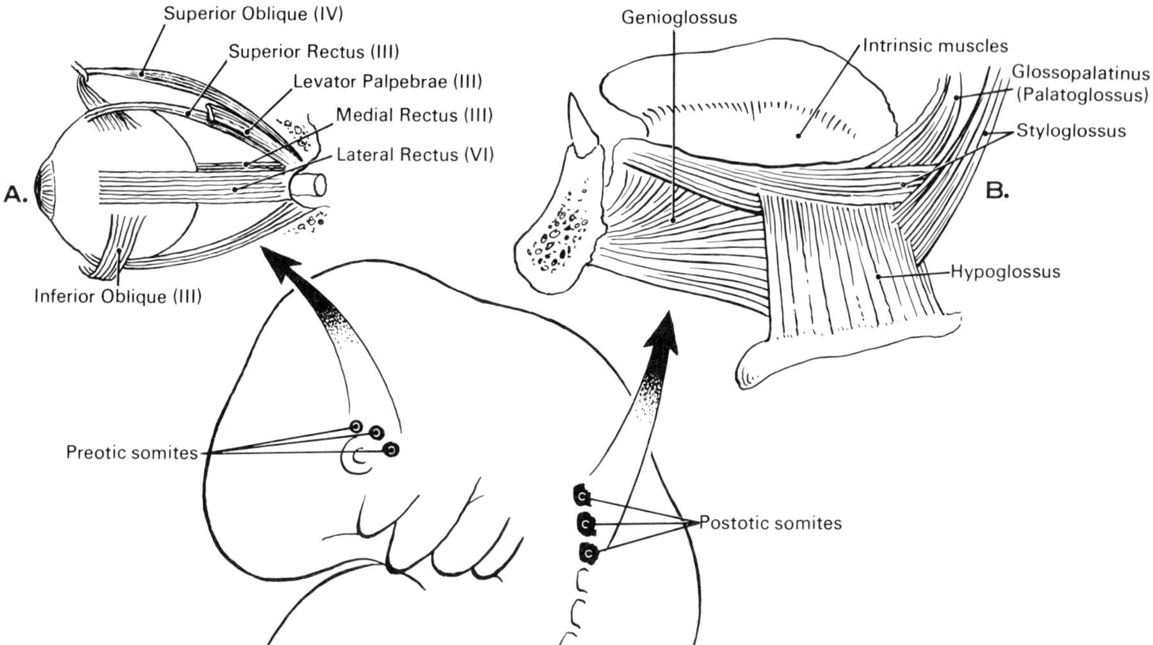

Fig. VI-8. Muscular derivatives of the head somites. **A.** Extraocular muscles from preotic somites. The cranial never to each is given. **B.** Skeletal muscle of the tonguef rom postotic somites innervated by cranial nerve XII (hypoglossal).

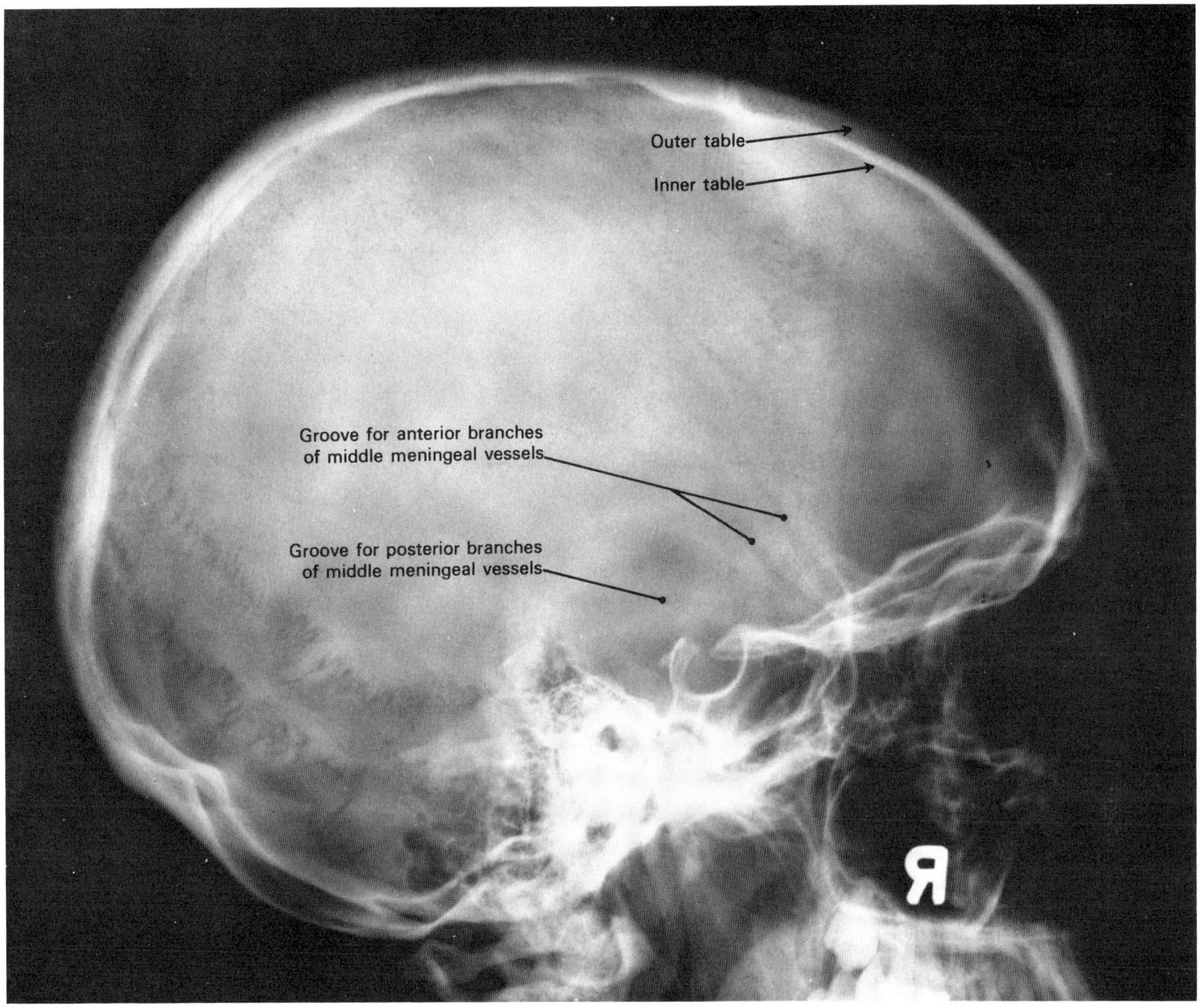

Fig. VI-9. Lateral radiographs of the skulls of living adults. **A.** Twenty-year-old patient. The meningeal grooves are clearly seen since they are etched on the inner table of the skull. Diploic veins are poorly defined in the young; however, they may become more prominent with increasing age.

region of the upper digestive and respiratory tubes and to certain glands (thyroid, parathyroid, and thymus), sometimes have been called the "visceral arches," an unfortunate and confusing bit of terminology. Its use leads the student to think that the muscular derivatives must be smooth or visceral musculature, and we have just noted that this is not the case.

The Calvaria—Structure and Vascular Supply

The flat bones of the skull cap or calvaria, as we attempted to point out in the section on bone development, consist of two plates, laminae, or tables of compact, cortical bone which hold between them a spongy bone rich in marrow. This spongy filling is called the **diploe.** The tables and diploe are well demonstrated in Figure VI-9. In scanning the curvilinear calvaria in this radiograph, notice that the bone is not everywhere of uniform thickness. Definite thinning is seen in the frontoparietal area, in the region of the coronal suture. The temporal and sphenoidal squama and the suboccipital part of the occipital bone, which are covered by heavy muscle, do not have much, if any, diploic space. They are areas of thin, compact bone that, as a consequence, transmit light easily (Fig. VI-10).

Four *diploic veins* are present on each side of the calvaria (Fig. VI-9B). No arteries accompany these veins. That function is fulfilled by meningeal and pericranial arteries. The diploic veins commonly encountered are:

1. The *frontal diploic vein* coursing through a small foramen in the superciliary region to join with the supraorbital vein or superior sagittal sinus.
2. The **anterior temporal diploic vein** which empties into the sphenoparietal sinus through a tiny foramen on the inner aspect of the greater wing of the sphenoid.
3. The **posterior temporal diploic vein** which empties into the transverse sinus from an internal opening near the postero-inferior angle of the parietal bone.
4. The **occipital diploic vein** which also joins the

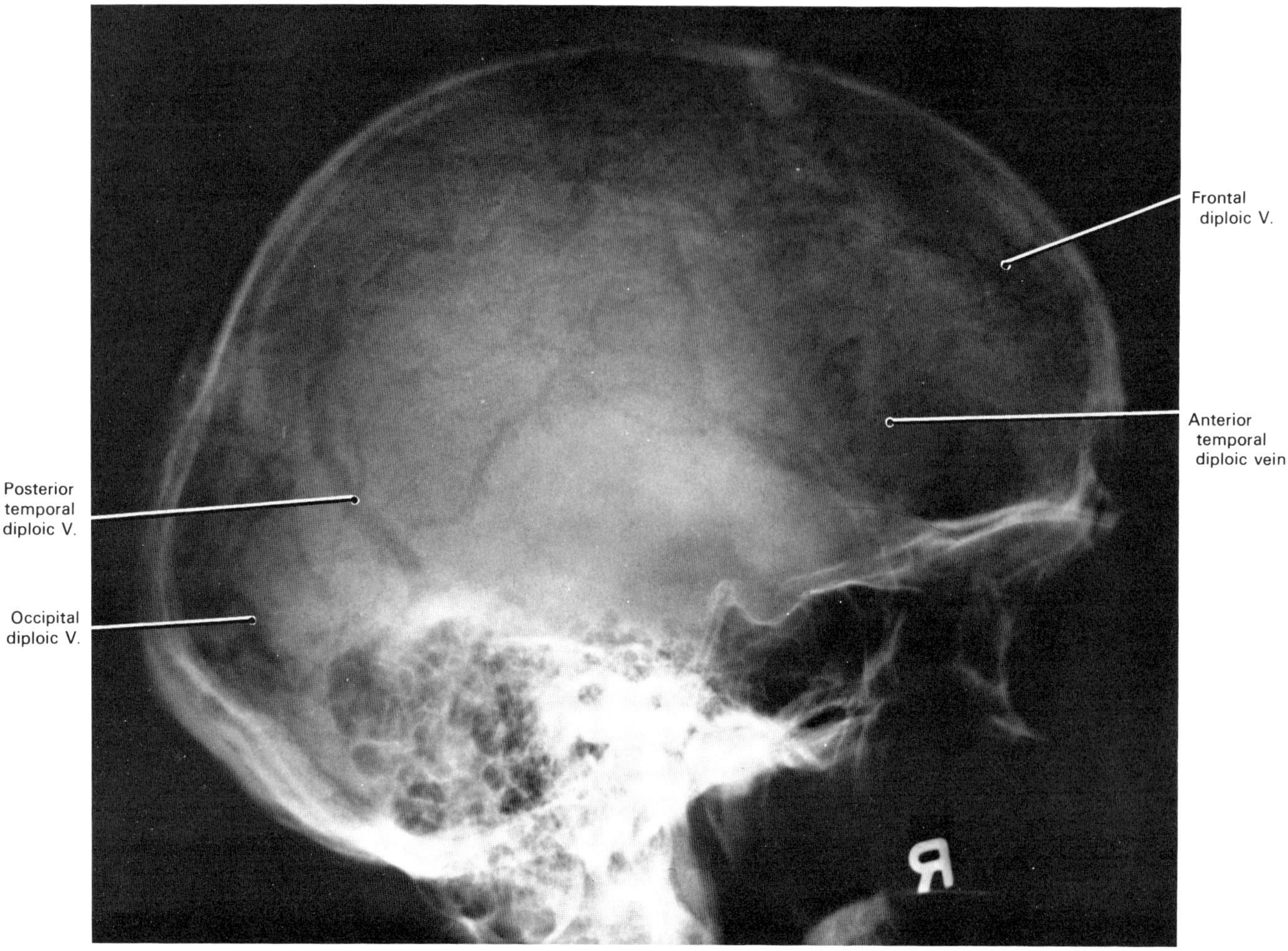

Fig. VI-9B. Sixty-year-old patient. Prominent venous channels in the diploe are seen.

transverse sinuses by leaving the occipital bone near the internal occipital protuberance.

Compare Figures VI-9 A and B. The former X-ray is from a 20-year-old patient; the other is from a patient of some 60-odd years. See the marked difference in the prominence of the diploic vessels in these heads. Indeed, it is difficult to be entirely certain in the X-ray of the younger patient which markings truly represent diploic veins. Certain individuals of advanced age, on the other hand, will readily show the symmetrical diploic channels due to relative osteoporosis of the overlying table bone. So long as these markedly, prominent, diploic venous channels are not associated with concomitant enlargement of meningeal arterial channels, this is of no clinical significance.

The Superior Aspect of the Skull

Four, not five, bones make up the calvaria or skull cap. The student nurse who listed these bones as "a frontal, a backal, two sidals, and a topal" was not quite correct. We see them as **a frontal, two parietals, and an occipital** (Fig. VI-11). There is nothing which corresponds to the apocryphal "topal." The student should now work with a skull to locate the **coronal** and **sagittal** sutures and as many of the landmarks as we may mention below. He would do well to review the early figures and related text in Chapter III and especially Figure III-1A. Realize that the **bregma** is the meeting point of the coronal with the sagittal suture, just as the **lambda** is the junction of the sagittal and lambdoidal sutures. The **vertex,** on the other hand, is merely the highest point along the arch of the sagittal suture and it rarely corresponds to the bregma. In a like manner, the **parietal eminences** are the most convex points laterally of each parietal bone. Generally, one or both parietal bones show a prominent **parietal foramen** through which an emissary vein, a branch of the occipital vein, communicates with the superior sagittal sinus.

In Figure VI-11 the central surface view of the calvaria brings only a snipet of the **interparietal portion** of the **occipital bone** into view posteriorly. As a consequence, neither the lambda nor the parietal foramina are displayed too well. The skull would have to be

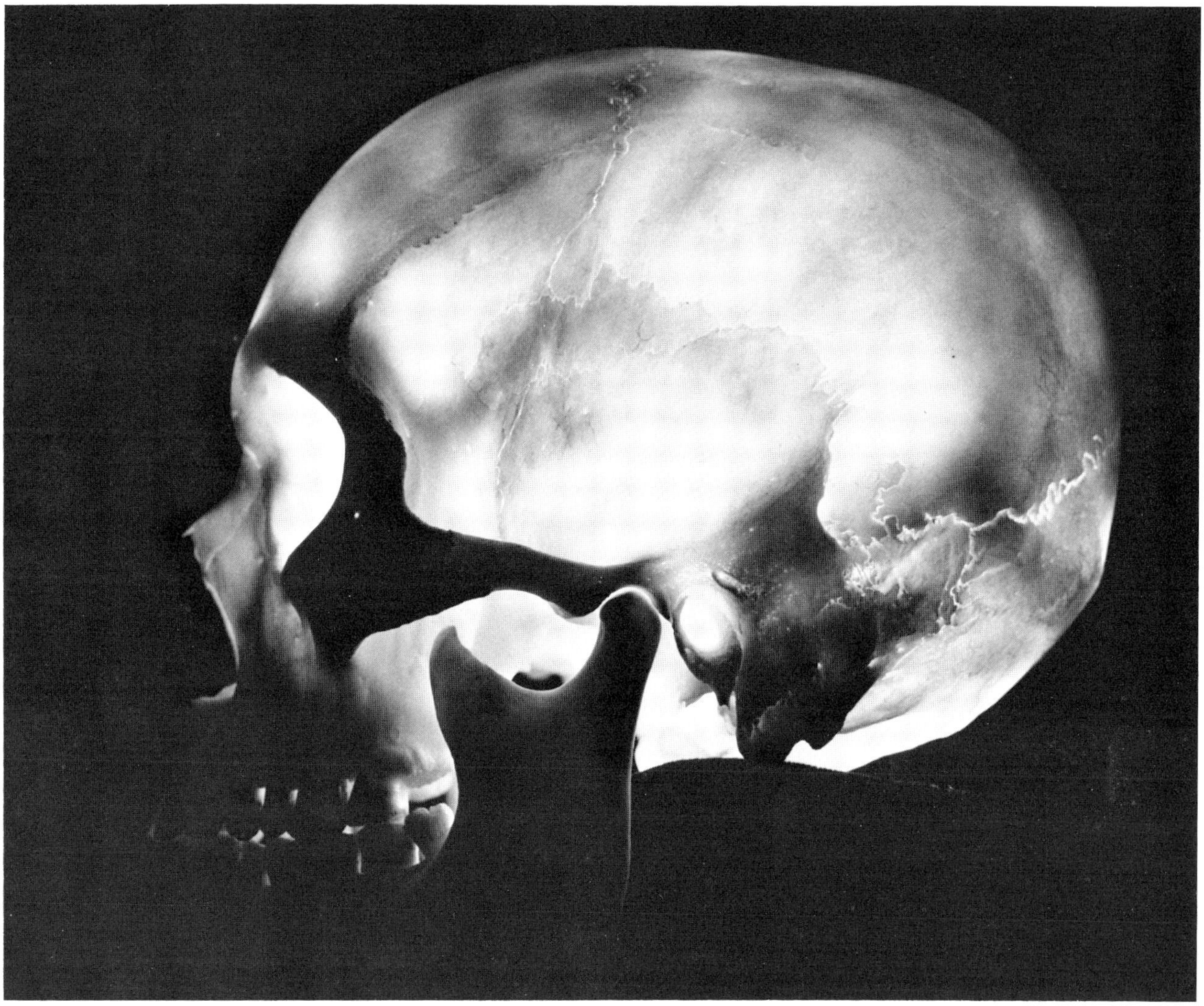

Fig. VI-10. Lateral view of an internally illuminated skull. Note areas of thinning in the parietal bones, sphenoidal and temporal squama, and in the suboccipital region. These areas are relatively free of diploe.

tipped forward for better viewing of these landmarks. Around the central figure, frontal, parietal, and occipital bones are shown as if they had been hinged back so that their inner aspects become available for examination. Both brain and dura have left their marks on the insides of these bones.

The frontal bone (Fig. VI-11A) presents a concave inner surface. In the upper part of the midline is a faint and pock-marked groove, the **sagittal sulcus** which lodges the superior sagittal sinus. Anteriorly, the edges of the groove sharpen and unite as a **frontal crest** to give attachment to the **falx cerebri,** a duplicated fold of dura that separates the cerebral hemispheres. The crest ends in a pit, the **foramen cecum,** which may be merely a closed depression. But when it is a patent canal, it transmits an emissary vein from the nose to the superior sagittal sinus. On either side of the midline the convolutions of the frontal lobe of the brain make shallow depressions in the bone, often more prominent in some skulls than in this one. Quite distinct, however, are many thin, small furrows which are the impressions left by the anterior branches of the **middle meningeal vessels.** Larger pits scattered about, but especially near the midline, once housed **arachnoidal granulations (Pacchionian bodies).** If we could examine the dura mater in these corresponding regions, several rough structures, appearing like coarsely granular mushrooms, would be encountered upon its surface. These manage to embed themselves partway in the bony vault (see Fig. VI-4A). They represent senescent forms of arachnoidal villi, now, presumably, non-functional. The large, plugged, non-functional, granulated villi can be seen grossly with the naked eye. They are hardly ever noticeable before the third year of life, increase by the 10th year, and become more numerous and still larger in old age.

Note, particularly, in Figure VI-11A the location of the **frontal air sinuses,** the **supraorbital notches or**

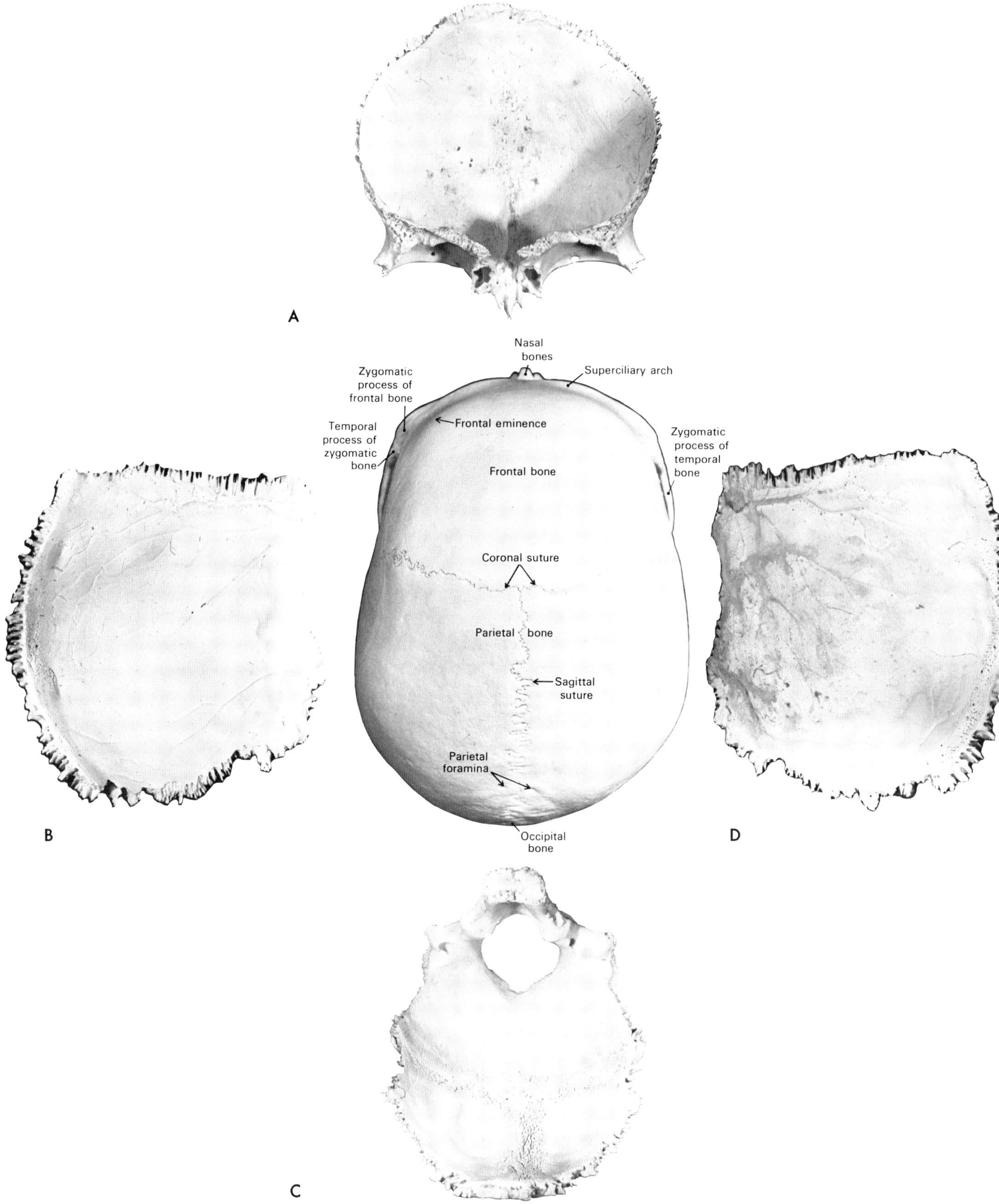

Fig. VI-11. The bones of the calvaria. Center view looks down on top of skull from above. **A.** The inner aspect of the frontal bone as if unhinged from central figure and opened forward with nasal spine as a fulcrum. **B.** Inner aspect of left parietal bone as if swung open laterally from central figure. **C.** Inner part of occipital bone with lambdoid suture swung posteriorly with basilar part of the bone as a fulcrum or hinge. **D.** Inner aspect of right parietal bone similarly swung open to the right. On the opposite page **A–D** have been enlarged so that their inner markings can be seen to better advantage.

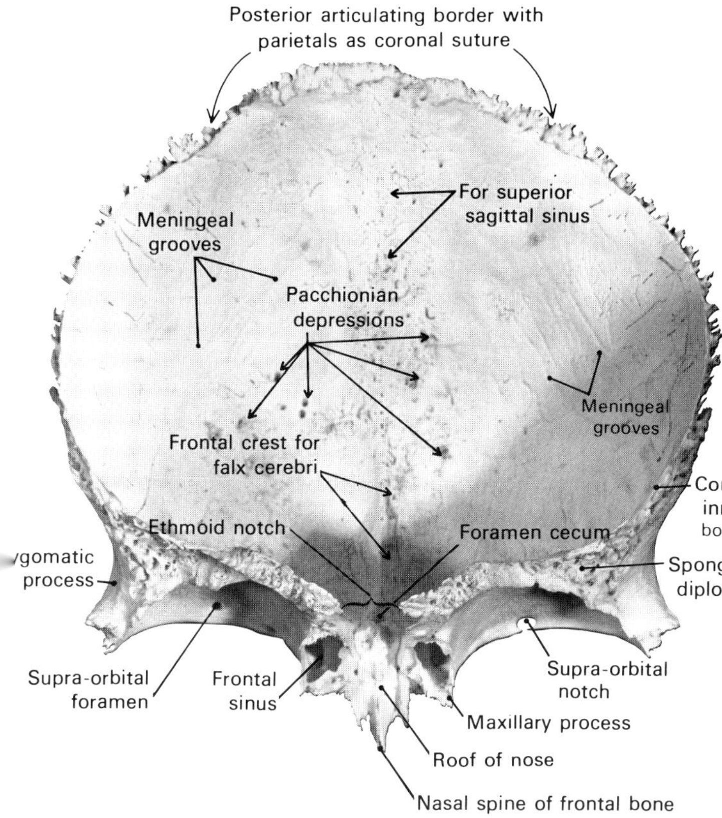

Fig. VI-11A

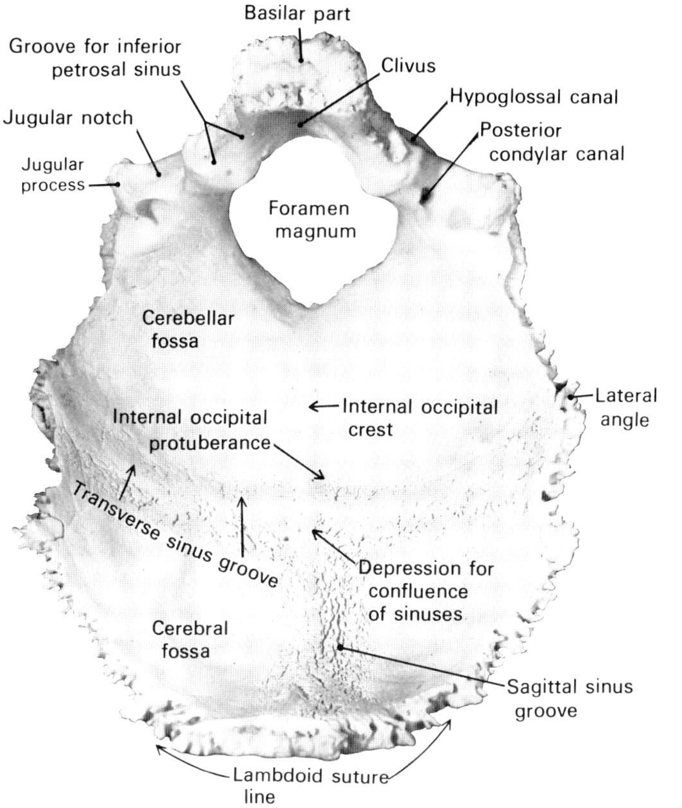

Fig. VI-11C

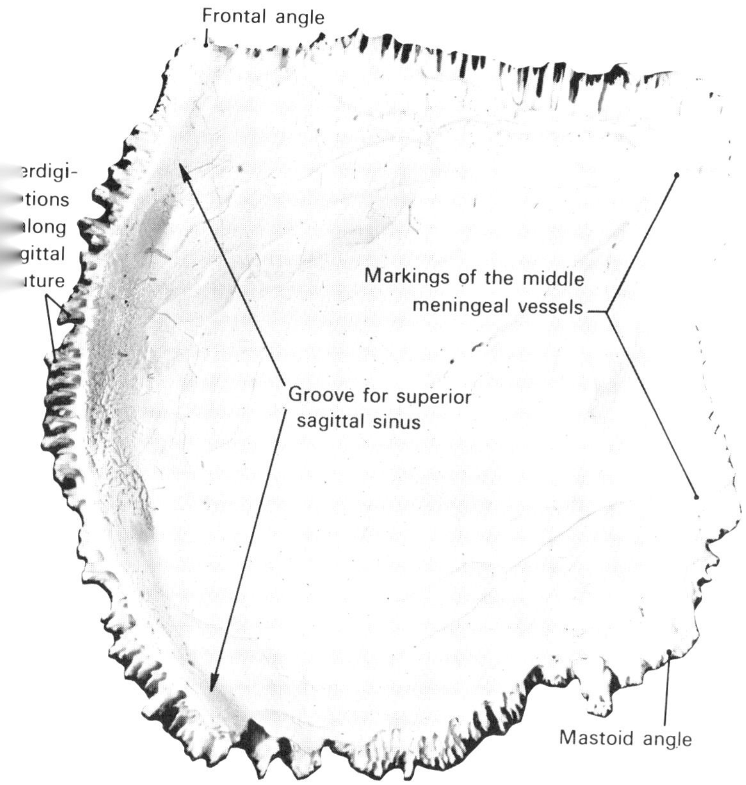

Fig. VI-11B

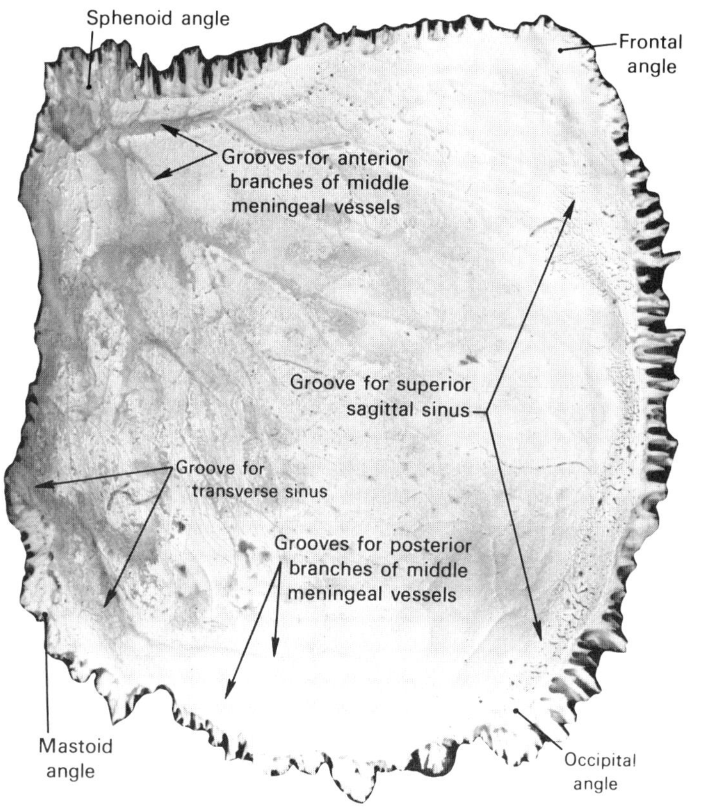

Fig. VI-11D

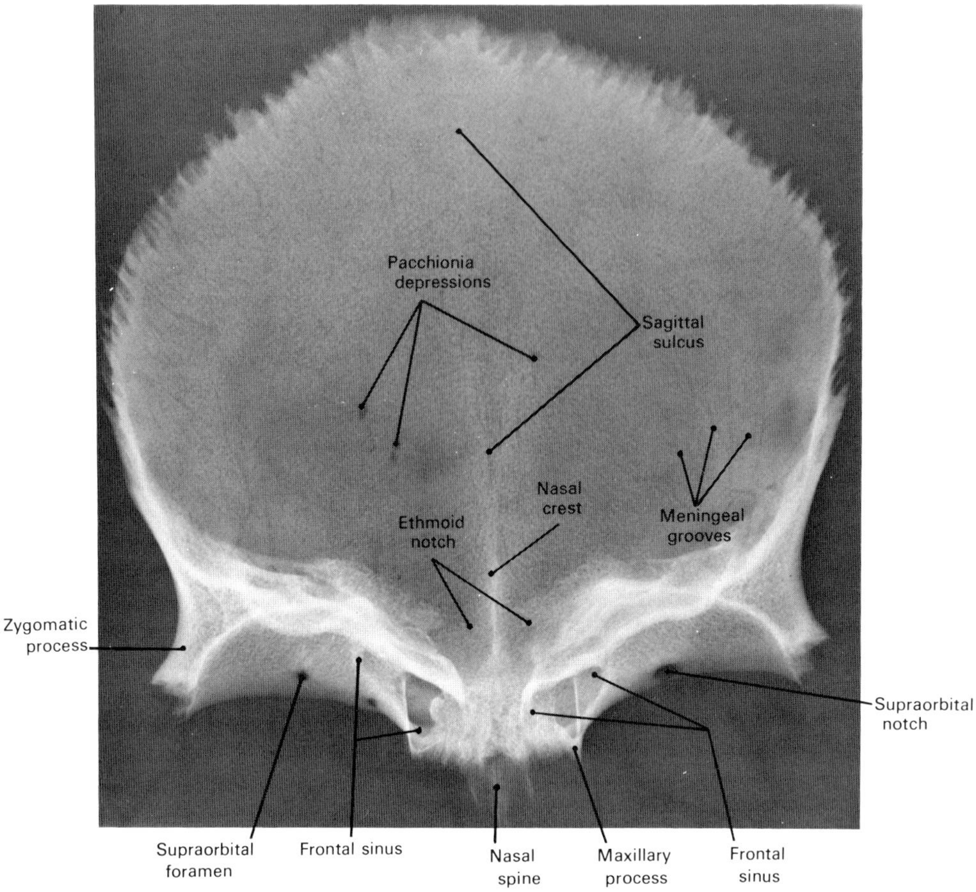

Fig. VI-12. X-rays of the bones comprising the calvaria or skull cap. **A.** Frontal bone.

foramina, the **zygomatic and maxillary processes** of the frontal bone, and note, too, how the frontal bone is constructed of spongy diploe and compact table bone. In the area of the foramen cecum, the orbital plates of the bone are separated by a cleft. This is the **ethmoid notch** which is occupied by the **cribiform plates** of the ethmoid in the intact skull. The part of the frontal bone anterior to the notch is its **nasal part,** and, centrally, here the **nasal spine** projects forward. It will insert into the **nasal septum** in the articulated skull.

The very same frontal bone is seen on X-ray views in Figure VI-12A. The similarities of detail in the X-ray are amazing. Certainly, the midline sagittal sulcus stands out more prominently than in the original bone, itself. Now the lateral extent of the frontal air sinuses are apparent. Even certain Pacchionian depressions and the fine meningeal grooves show up. Note how radiopaque the areas of heaviest compact bone appear in contrast to the massive gray regions which consist of thinner compact bone and the thinner structural diploe. Compact bone in these thin regions absorbs little of the roentgen rays. Hence, most of the bone appears as a mottled gray.

The **parietal bones** form a great part of the cranial vault. Their inner surfaces are concave (Fig. VI-11 B and D). The middle meningeal vessels make deeper grooves on them than they did upon the frontal bone. A long groove, following the sagittal suture line, marks the continued course posteriorly of the superior sagittal sinus. A smaller groove, cutting the **mastoid angle,** indicates a small area where the **transverse dural sinus** left its mark. Some pits for arachnoidal granulations can be identified here and there, especially near the sagittal sulcus. It is difficult to pinpoint the locus of the parietal foramina. They vary in size in different skulls and may be absent in some. Usually they lie closer to the lambdoid suture lines than anywhere else. When present, they correspond to the position of the (parietal) **sagittal fontanelle** of the infant skull. If ossification of this fontanelle is incomplete a fissure rather than a foramen may form. Wormian bones may be seen in this suture line near the lambda at times.

X-ray views of parietal bone (Fig. VI-12B) are not

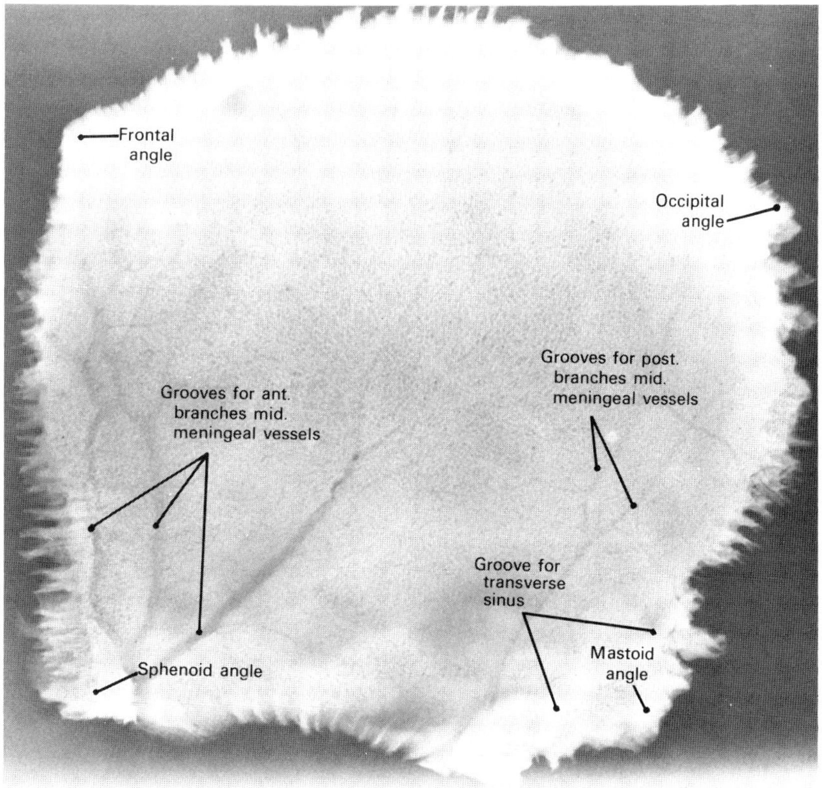

Fig. VI-12B. X-ray view of right parietal bone.

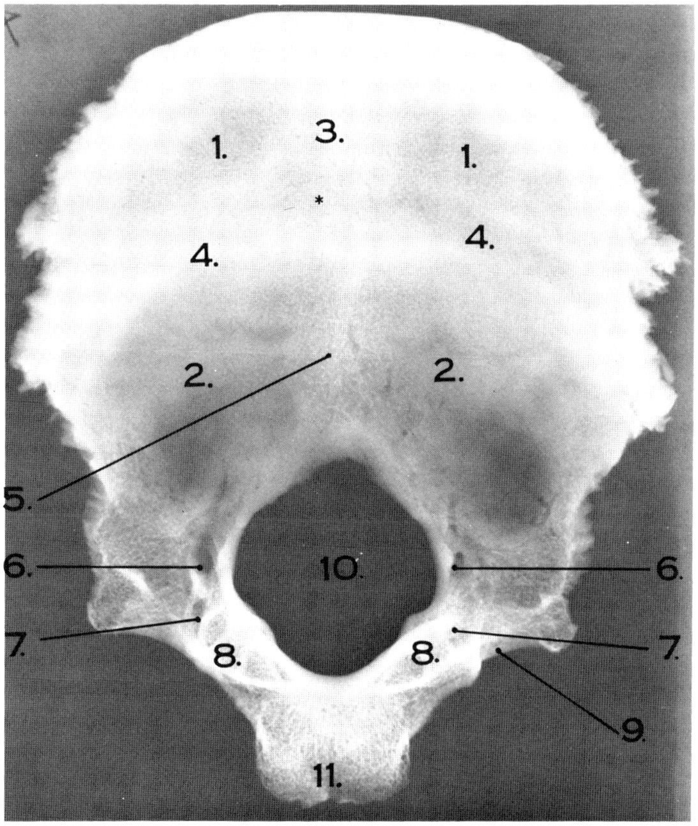

Fig. VI-12C. X-ray view of occipital bone. In C: 1. cerebellar fossae; 2. cerebral fossae; 3. groove for the superior sagittal sinus on the *right side;* 4. grooves for transverse sinus; 5. internal occipital crest for falx cerebelli; 6. posterior condylar canal; 7. hypoglossal canal; 8. outlines of condyles showing through from below; 9. jugular notch; 10. foramen magnum; 11. basilar part of the bone; *, confluence of the sinuses (torcular Herophili). (Note: C in this view has been rotated 180° from view C in Fig. VI-11 so that the lambdoid suture line is superiorly oriented as it should be.)

too impressive save for the markings of the meningeal vessels which can be easily identified.

While the frontal and parietal bones are completely formed in membrane, only the **squamous** or **interparietal part** of the **occipital bone** is formed in this way. The rest of the bone contributes to the base of the skull and is part of the **chondrocranium** and, hence, is intracartilaginous in its development. We shall discuss the chondrocranium after we look at the occipital bone as a unit.

The large hole in the occipital bone is the **foramen magnum** (Fig. VI-11C), through which the brainstem insensibly blends into the spinal cord. The foramen conveniently divides the bone into its major parts. The expanded **squamous part** lies superior to the foramen (below it in Fig. VI-11C, since the bone is bent back toward you as on a hinge). The thick, rod-like part (uppermost in Fig. VI-11C) is the **basilar part.** A pair of condylar parts lies at the sides of the foramen.

The concave internal surface of the squamous part is divided into four **fossae** or depressions by a cruciate pattern of grooves and ridges which meet at the **internal occipital protuberance.** The **cerebellar hemispheres** occupy the pair of fossae nearest the foramen magnum, while the **occipital lobes** of the **cerebral hemispheres** nestle in the superior pair of fossae in the intact head. The midline ridge from the protuberance to the foramen is the **internal occipital crest** to which the **falx cerebelli** attaches. In this latter dural fold runs the **occipital sinus,** sometimes as a double venous channel. The **transverse sinus** is responsible for the grooves that pass sideways from the protuberance to form the horizontal arms of this cross. The shallow groove completing the upper part of the cross (upper, when the bone is hinged back into the skull; lower in the Fig. VI-11C) is the impression of the posterior extremity of the superior sagittal sinus. It generally empties, as we see it here, into the right transverse sinus. Where these two sinuses meet (superior and transverse) at an angled union is the area known as the confluence of sinuses, or the torcular Herophili. Here at this junction, the **straight sinus,** as well as the occipital sinus, join with the two major large sinuses. Torcular as defined by Herophilus, meant a twisted or tortuous place where the blood was wrung through "a wine press." (Refer to Fig. VI-3B to locate venous sinuses in lateral view.)

The condylar parts of the occipital bone are so named because they bear a pair of "rockers" on their outer surface (not seen in Fig. VI-11C) which articulate with the atlas or first cervical vertebra. At this articulation we can waggle our heads in silent assent. In each condyle there is a short anterior tunnel, the **hypoglossal canal,** for the transmission of cranial nerve XII or the hypoglossal nerve which innervates the somatic musculature of the tongue. Sometimes there is also a posterior condylar canal through which an emissary vein can reach the transverse sinus. Laterally, each condylar part extends as the **jugular process,** part of which is excavated as a smooth notch, the **jugular notch.** In the intact skull, the notch takes part in forming the **jugular foramen.** The internal jugular vein draining the dural sinuses passes through the foramen, and cranial nerves IX, X, and XI manage to squeeze through along with the vein.

The basilar part of the occipital bone fuses with the sphenoid after the 21st year. The site of union is seen as a roughened blunt end in Figure 11C. Clinically, this slanting floor of the cranium that gives support to the brainstem is known as the **clivus** ("slope of a hill").

The X-ray view (Fig. VI-12C) of this bone shows the cruciate pattern of grooves and ridges better than one often sees it in the skull, itself. The fossae stand out sharply, but the posterior condylar canal is seen to better advantage than the hypoglossal canal even in the X-ray when the bone is oriented in this plane. The oval outlines of the condyles show through from below.

The Chondrocranium (Fig. VI-13)

We have just seen how the frontal and parietal bones and the occipital squama, all membrane bones, form the **roof of the neurocranium** which houses the brain. The **floor of the neurocranium,** upon which the brain sits, is the **chondrocranium** since most of it is preformed in cartilage. Involved are the basilar and condylar parts of the occipital bone; all of the temporal bone except its squamous portion, much of the sphenoid; and all of the ethmoid.

In the construction of the floor, the notochord figures in, as well, for in the embryo it extended forward into the head to terminate behind the **sella turcica,** the part of the sphenoid that resembles a Turkish saddle in whose depression the pituitary gland (hypophysis) rests. Chondrification begins in the mesenchyme which surrounds this cephalic extension of the notochord so that a **basal** or **parachordal plate** of cartilage is formed here (Fig. VI-13B). The plate extends from the sella turcica back to four rather typical somites in the head, the **occipital somites,** the most cephalad of which seems to disappear.[2] The sclerotomal parts of the other three occipital somites fuse with the parachordal plate to form the base of the occipital bone. In the fusion, of course, somitic segmentation is lost, but signs of it sometimes persist in the hypoglossal canal where a bar of bone may divide the canal into two channels.

(The myotomic parts of these head somites, by the way, have been said to migrate to the tongue region to account for its intrinsic and extrinsic skeletal musculature. Cranial nerve XII or the hypoglossal nerve has been classed, therefore, as a *somatic motor nerve,* and it has been assumed that its course follows the

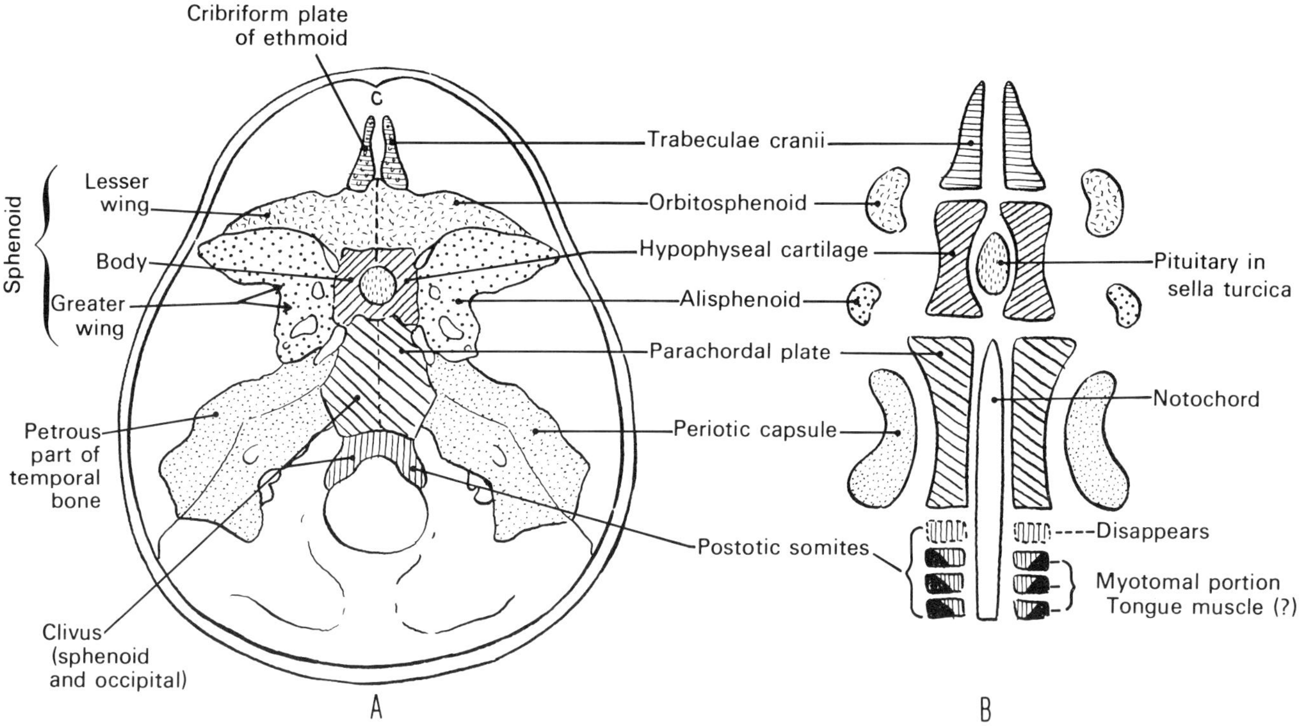

Fig. VI-13. The chondrocranium. **A.** Derivatives of the chondrocranium in the floor of the skull. **B.** Cartilagenous anlage of the bones of the cranial floor oriented around the notochord in the early embryo.

migration of these occipital myotomic masses. No cervical nerves enter the tongue, though C1 does travel with the hypoglossal to reach infrahyoid muscles.)

Other plates of cartilage contribute to the chondrocranium. Two such, the **hypophyseal cartilages,** flank the sella turcica, and rostral to them arise the cartilaginous **trabeculae cranii.** These four plates fuse and give rise to the **ethmoid bone** and **body of the sphenoid.** Along with the parachordal plate and the occipital somites just mentioned, all these cartilages merge to form the *midline floor* or the *embryonic skull* from the nasal region to the foramen magnum (Fig. VI-13A).

A number of lateral masses begin to chondrify on either side of this midline group of cartilages. The most rostral of the lateral masses are the **orbitosphenoids** (bilaterally) from which the **lesser wings** of the **sphenoid bone** develop. Just caudal to these masses are the **alisphenoids,** the source of the **greater wings** of this keystone, bat-shaped, sphenoid bone in the base of the skull. As these four cartilaginous masses fuse with the midline mass, holes are left in the floor of the skull for the passage of **cranial nerves.**

The most caudal of the lateral masses are the **periotic capsules** which surround the embryonic ear vesicle and become, as well, **petrous** and **mastoid parts** of the **temporal bones.**[3] The squamous part of the temporal bone is of membrane origin, not cartilage. The styloid process of this bone is from Reichert's second arch cartilage. The bones of the middle ear are first and second branchial arch derivatives, as we also have noted above. When the periotic capsule unites with alisphenoid and parachordal plate, the jugular notch, which we observed in the disarticulated occipital bone, becomes converted to a foramen for cranial nerves IX, X, and XI and for the internal jugular vein.

The petrous part of the temporal bone sits obliquely wedge-shaped in the floor of the skull. Recall that of all the Twelve Disciples, Peter (Petrus) was called "The Rock." The petrous bone is purported to be the hardest bone in the head. It is probably well that it is such a solid structure, since it houses the delicate receptor organs for hearing and equilibrium within the cochlea and the semicircular canals. It is pierced medially by a foramen which is extended as a canal, the **internal auditory meatus and canal,** for the passage of cranial nerves VII and VIII.

Rostrally, the ethmoid forms a pair of sieve-like plates in the floor of the skull, the **cribiform plates,** derived from the trabeculae cranii. Through its many tiny holes, filaments of the **olfactory or cranial nerve I** pass upward from the nasal mucosa to terminate on the **olfactory bulbs** of the brain. The latter (in their dural investments) lie directly over each cribiform plate when the brain sits in the skull. Most caudalward, we noted the hypoglossal canal in the condylar parts of the occipital bone for the emergence of the last cranial nerve (XII).

We have just seen how the neurocranium has developed in part from membrane bone and in part from cartilage (the chondrocranium). We have noted how foramina have been left in the floor for passage of all of the cranial nerves. Once we have become a bit more familiar with the outside of the adult skull, we will come back to consider these bones and foramina in its floor.

The Anterior Aspect of the Skull

Viewed from in front, the skull presents a forehead region, orbital cavities, the bony parts of the cheek, the dorsum of the nose, a large piriform aperture, and the bones of the upper and lower jaw (Fig. VI-14). Most of the area seen, therefore, relates to the **visceral or facial cranium,** though the major part of the frontal bone relates more to neurocranium than to the forehead region.

Throughout this chapter we have resorted frequently to photographs of the skull as well as to an admixture of radiographs of these same skulls and radiographs of the living subject. It is possible to see much more anatomical detail in radiographs of the dry skull than in X-rays of the living head as a rule. Such material is useful before trying to demonstrate the same anatomy in radiographs of patients. In using X-rays of the dry skull, radiographic anatomy becomes more meaningful to the beginning student and to the resident physician in Radiology. It challenges them both to search harder to find the structures in their patients' films that they known to be present in dry skull films. The left half (A) of Figure VI-14, for example, is such a dry skull radiograph. Note how well it matches the skull (B) from which it was taken if you allow for slight differences in magnification which you can expect in radiographs made at short target-film distances (36″–40″).

To render A and B in Figure IX-14 more meaningful, the skull was photographed from the position of the film cassette rather than from the position of the X-ray tube. This is, in radiographic terms, an inclined posterior-anterior projection. The camera was angled to complement the angle of the X-ray tube. The inset (Fig. VI-14C) shows how the radiograph was made. In skull radiographs the reference base line which is usually employed is the **orbitomeatal (OM) line.** The line extends from the outer canthus of the eye to the center of the external auditory meatus. It differs from the anthropological Frankfurt plane referred to in Chapter III, page 79. The latter passes from the infraorbital point to the upper margin of the external auditory meatus. Thus, these lines would cross at an angle of approximately 10°. (This last statement has nothing to do with the 15° angle of inclination in Fig. VI-14C.)

One thing must be realized when you look at these radiographs. As in any radiograph, all three-dimensional structures are reduced to two dimensions. Thus, structures located posteriorly or midway in the skull will overlap those located anteriorly. You must learn to think in depth as you "look through" Figure VI-14A from anterior to posterior in this particular instance.

The *numerals* in the list of structures mentioned below correspond to the comparable numerals on both sides of Figure VI-14 so that you can compare similar features in the skull and the radiograph of the skull.

1. **Mental protuberance.** (Note its triangular shape in Fig. VI-15B.)
2. **Mental foramen.**
3. **Angle of mandible.**
4. **Ramus of mandible.**
5. **Maxilla**—with air-containing **maxillary sinus** in (A).
6. Coronoid process of mandible, obscured in (A) by overlying **occiput** (6A) and **mastoid process** (6B) with its air spaces.
7. **Hard palate** seen on edge view.
8. **Vomer** seen joining the **perpendicular plate of the ethmoid** (11) to complete the **nasal septum** and to bisect the **piriform aperture** whose outline is not clear in (A).
9. **Malar prominence** of the zygomatic bone seen as a dense, white, lateral bulge in (A).
10. **Inferior concha.** Other conchae in (A) are obscured by superimposition of deeper-lying structures.
11. Perpendicular plate of ethmoid (nasal septum).
12. **Zygomatic process** of **frontal bone** forming superolateral part of orbital wall.
13. **Orbital rim.** Note its position (13A) in the radiograph. The dense, white, curvilinear line (13B) above it is the highest point of the orbital roof *within* the anterior cranial fossa (p. 225). It (13B) does not correspond to (13) on the right side of Figure VI-14, but 13A does correspond.
14. Frontal air-sinus (obviously not seen in (B)).

15, 16, 17 are, respectively, the **lambdoid suture,** the **lambda** and the **sagittal** suture. The lambdoid suture projects from the posterior aspect of the skull where it separates the occipital from the two parietal bones. The lambda is the point of junction posteriorly of the two lambdoid sutures with the sagittal suture. The latter separates the two parietal bones in the midsagittal plane. Obviously none of these more posteriorly-lying structures can be seen in (B), the frontal view of the solid skull, but "X-ray vision" now brings them into view. The uninitiated might wish to consider them sutures of the frontal bone, which of course, they are not.

Enlarged views of the skull (Fig. VI-15 A and B) show certain other features. The frontal bone is smoothly convex over the forehead region except for two lateral bulges, the **frontal eminences.** These may be more (or less) prominent in different individuals. Heavy ridges, the **superciliary arches,** appear over each orbital cavity. Each arch bears a **supraorbital foramen (or notch)** at its inferomedial angle through which the supraorbital nerve emerges (Fig. VI-15A).

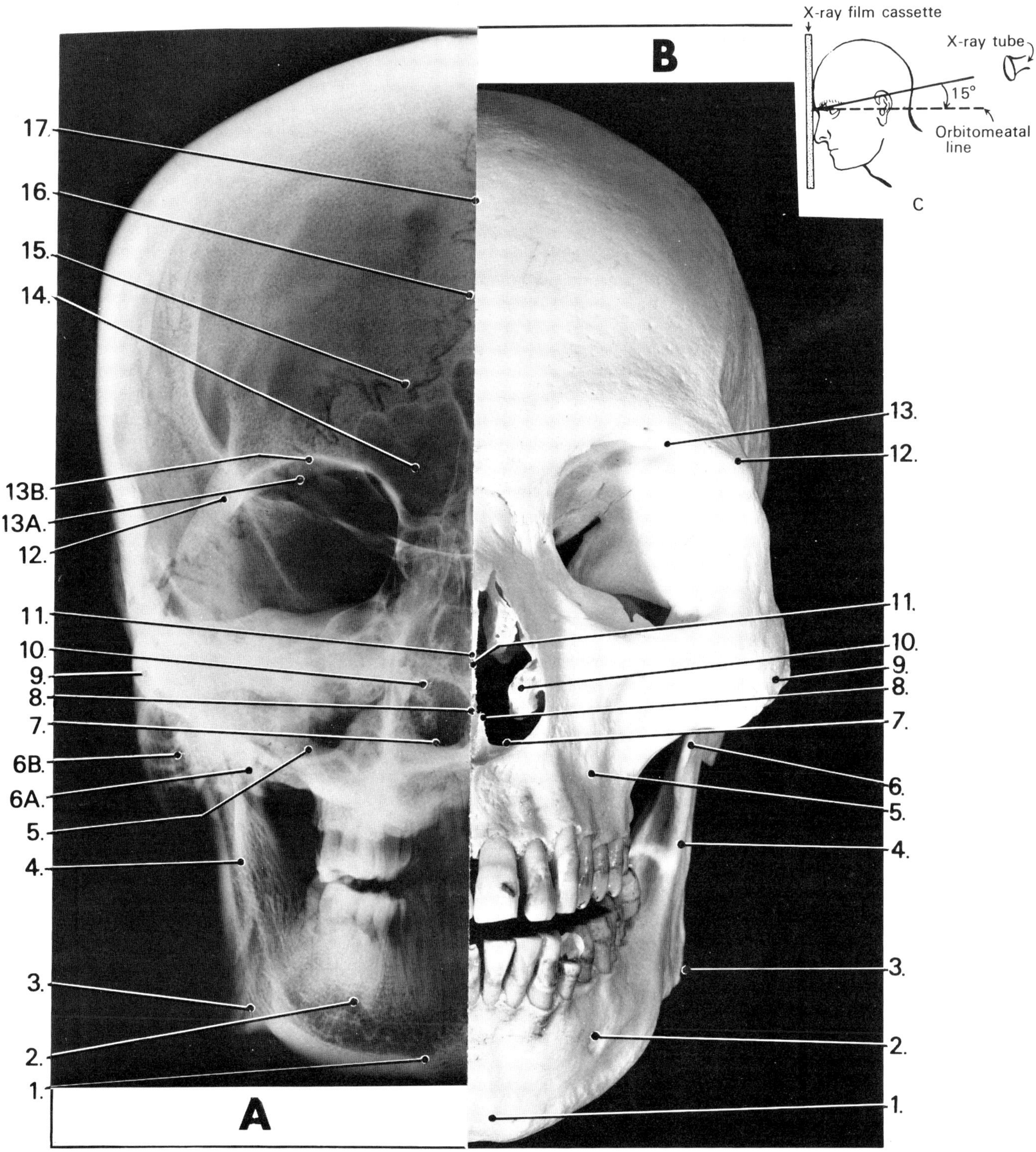

Fig. VI-14. The frontal aspect of the skull (**B**) and an inclined posterior-anterior projection of the skull (**A**). **C.** The inset C shows the angle of inclination to be 15°. The numerals on each side display corresponding features. See text (p. 224) for descriptive enumeration.

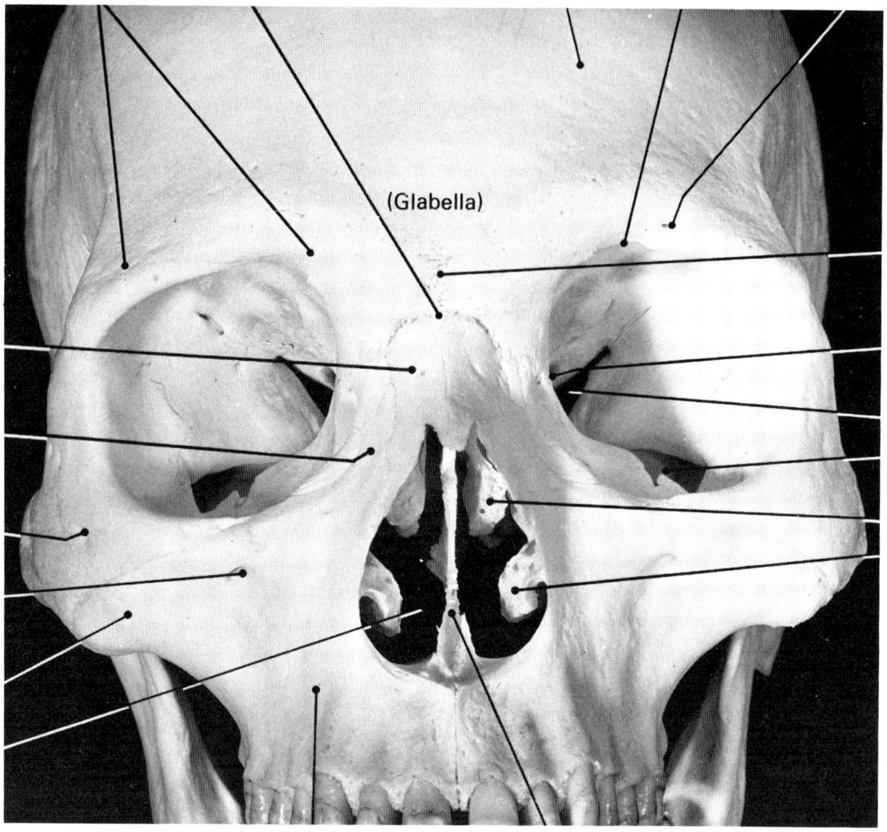

Fig. VI-15A. Enlarged view of upper part of skull in frontal aspect.

The nerve is one of the terminal twigs of the **ophthalmic divsion** (V_1) of the trigeminal. There is usually a small foramen above the margin of the arch for the *frontal diploic vein* (Fig. VI-15A). The **glabella** resides as a smooth, depressed area centered between the arches. Below it, or often coursing through the center of the glabella may be a remnant of the old frontal suture (**metopic suture**). When present in its entirety it indicates that the frontal bone is a fusion of two separate bones, as in the infant skull. Fusion is complete and most of the suture is wiped out by the fifth or sixth year. In Figure VI-15A, a bit of the suture is seen near the **nasion,** *i.e.,* at the frontonasal suture line.

The cheek region shows a *malar prominence* formed in part by the *zygomatic bone* itself and by the **zygomatic process of the maxilla.** A **frontal process** of the cheekbone extends upward, and a **temporal process** extends backward. The latter takes part in the formation of the **zygomatic arch** (**zygoma**) in the region below the temple (Fig. VI-17A). On the surface of the zygomatic bone is a foramen for the **zygomaticofacial nerve.** This is a branch of the larger *infraorbital* nerve which emerges in the **infraorbital foramen** of the maxilla. Look in the floor of the orbital cavity (Fig. VI-15A) to see the large **inferior orbital fissure.** The infraorbital nerve, as a branch of the **maxillary division** (**V_2**) (Fig VI-16) of the trigeminal, leaves the fissure to enter the foramen. Along its course it gives off the zygomaticofacial twig as a side branch.

The **nasal bones** are two short, separate, slanting bones that meet in a midline suture to form the dorsum of the nose. They are separate bones and are not parts of the ethmoid which contributes in large measure to the interior of the nose. The more wobbly skeleton of the nose is cartilaginous distally and is not preserved in dried skulls. Thus, we peer into a wide, pear-shaped, **piriform aperture.** In the internally illuminated skull (Fig. VI-16) this aperture stands out sharply below the orbital cavities. It is partitioned by a midline **nasal septum.** The upper part of the septum consists of the **perpendicular plate of the ethmoid** which meets the smaller **vomer** at the *hard palate.* The rest of the septum, distally in the nose, is cartilage, and so it is absent. From the upper lateral walls of the piriform area (Fig. VI-15A), the **middle**

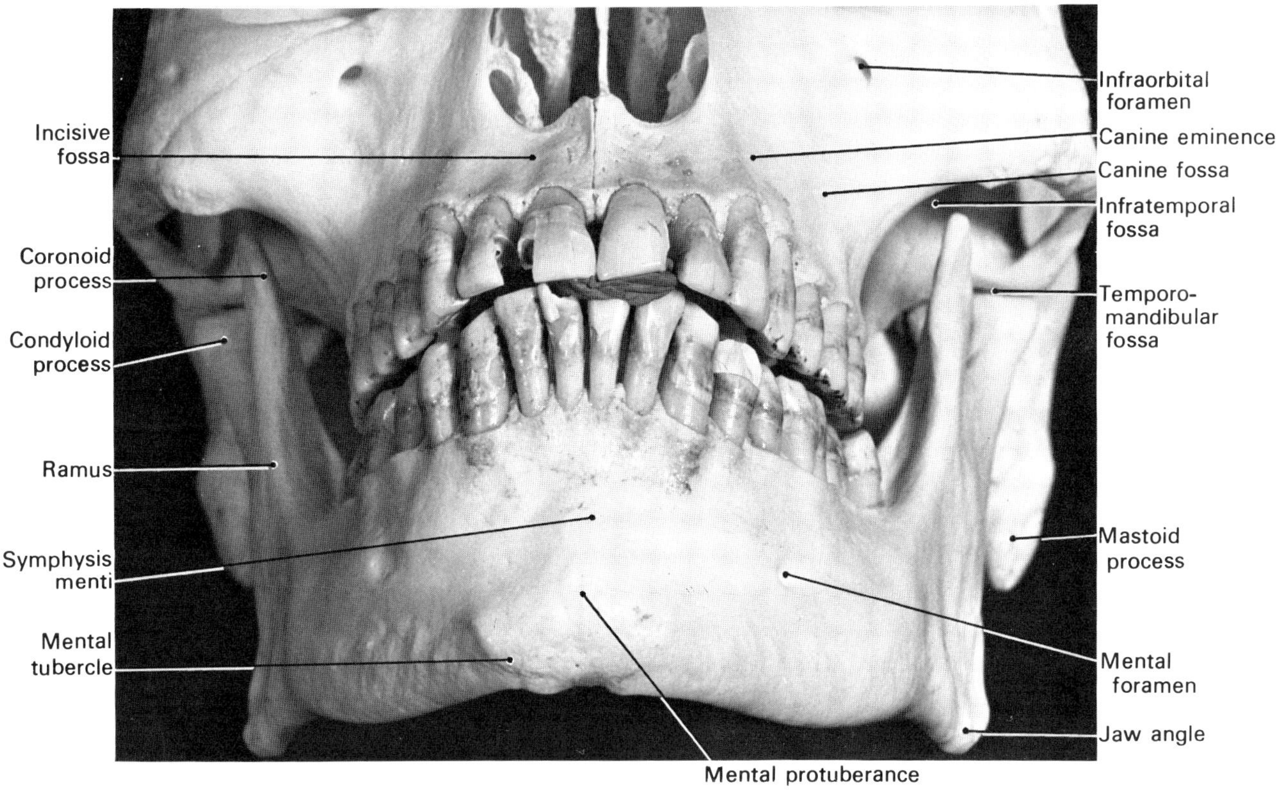

Fig. VI-15B. Enlarged view of lower part of the skull in frontal aspect.

concha can be seen to bulge medialward toward the septum. Superior conchae (turbinates) are hidden from sight below the nasal bones but are seen in Figure 17A. Middle and superior conchae are also parts of the ethmoid bone. The inferior conchae below are not but are separate skull bones in their own right.

We shall say more about the orbital cavities later on. At this point, however, note the **supraorbital fissure** in the roof of the orbit (Fig. VI-15A). Medial to it you can just barely see the rounded **optic canal** for the optic nerve or cranial nerve II.

In the Jack-O'-Lantern skull of Figure VI-16, notice that regions other than the orbit and the piriform area are lighted. Streaks of light flood from beneath the zygomatic arches in the **infratemporal fossa** to brighten portions of the mandible. Light shines through some pin-point holes, as through the two tiny foramina in the nasal bones and in the foramina of the superciliary arches where frontal diploic veins emerge. Several bones of the cranium display translucence where they thin markedly. Above the frontonasal suture is such a large area marking the region of the *frontal air sinuses,* the left sinus in this skull being enlarged more than the right one. Other thin spots are seen in the frontal bones. The squamous parts of the temporal bone are so paper-thin that they appear as bright sidewalls of the brain case. This should indicate the ease with which a blow to the temple area could splinter bone fragments into the brain.

The two **maxillae** form the upper jaw which bear the upper teeth in their **alveolar process.** From the body of the maxilla, a **frontal process** projects upward along the side of the nose; a **zygomatic process** projects laterad to the cheek, and the alveolar process projects downward. Over each canine tooth (third from midline) a prominent bulge the **canine eminence** separates an **incisive fossa** medially from a **canine fossa** laterally (Fig. VI-15B). The infraorbital foramen, recall, delivers a branch of cranial nerve V. This is part of the maxillary division (V_2) of the trigeminal.

It should be obvious that the lower jaw or **mandible** bears the lower teeth in its alveolar process. The two halves of this jawbone have fused in the midline at the symphysis menti (a synostosis, not a true symphysis). Here a triangular eminence, the *mental protuberance,* (Fig. VI-15B) projects forward as a chin. Neither gibbon, gorilla, orangutan, or chimpanzee can boast of chins as outstanding as ours which reach their epitome in Dick Tracy and Fearless Fosdick. Even in primitive man, while the entire jawbone may be heavier, this prominence is not so characteristically marked as in modern man. The bilateral bulges at the base of the protuberance are the **mental tubercles.** Lateral to them lie the **mental foramina.** If the supraorbital fis-

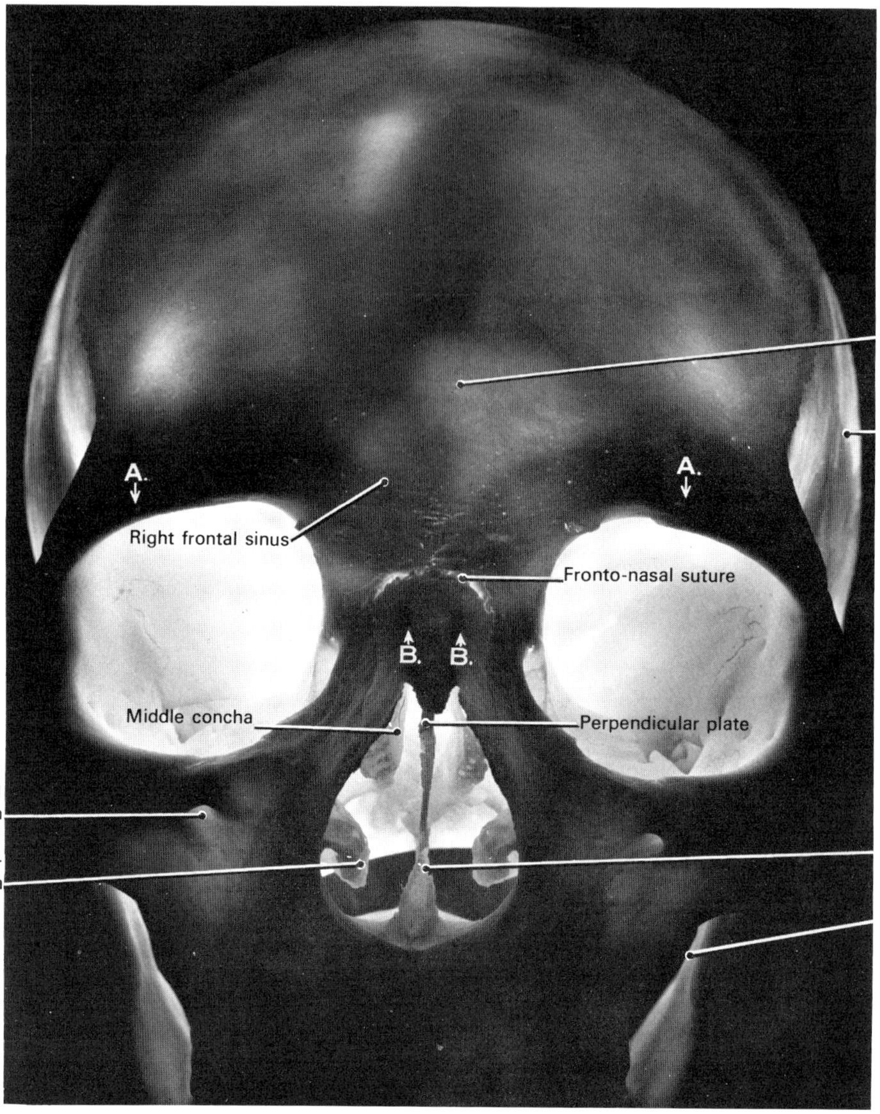

A. Foramina for frontal diploic veins
B. Foramina for blood vessels in nasal bones

Fig. VI-16. The internally illuminated skull.

sure and notch (foramen) transmit terminal branches of V_1 to the face (ophthalmic branches of cranial nerve V or trigeminal nerve) and if the infraorbital foramen transmits maxillary branches of V_2, then the mental foramen transmits **mandibular branches** of the trigeminal (V_3) to the skin of the face. These are all cutaneous (sensory) nerves.

The body of the mandible supports a pair of lateral **rami** which extend upward into the *infratemporal fossa* (Fig. VI-15B). The anterior, triangular-shaped process of each ramus, the *coronoid process,* gives attachment to the Temporalis muscle, one of the muscles of mastication (Coronoid—Greek—"like the heel-end of a bow" where the string is attached to form an arch). The other process of the mandible, the **condylar,** is not seen well in these frontal views of the skull. It has a constricted **neck** and a **head** that articulates with a disk within the capsule of the **temporomandibular joint.** This single, movable joint, the only movable joint in the head, functions when we speak, chew, and yawn. The condyloid process sits in a slight depression of the temporal bone, in the **temporomandibular fossa.**

If the skull is tipped up slightly (Fig. VI-17A) certain other features are rendered prominent. The superciliary arches now stand out sharply as do the

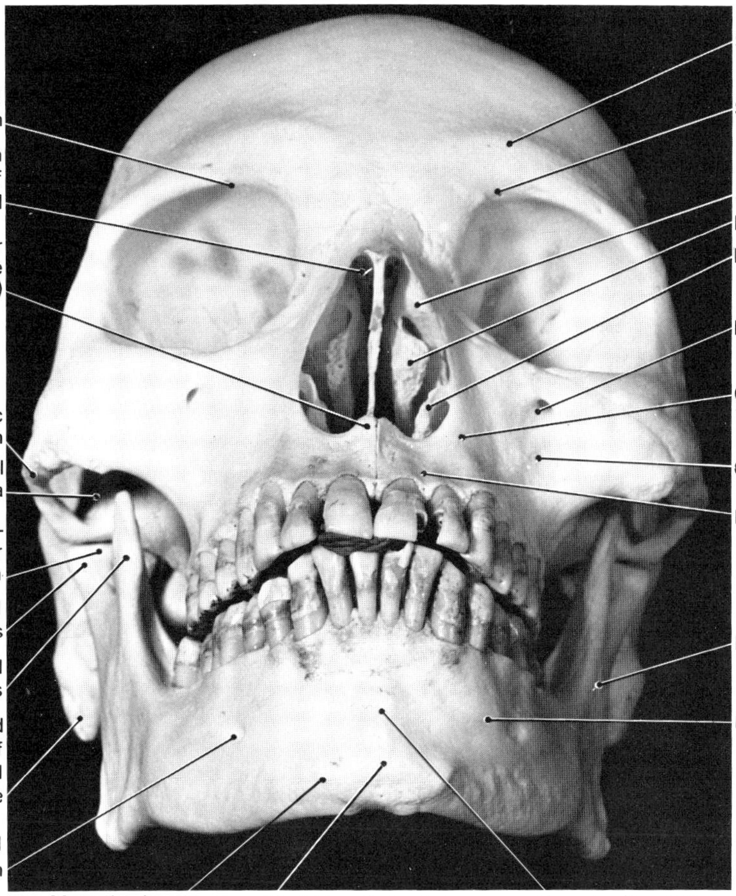

Fig. VI-17. The skull is tipped slightly and viewed frontally. **A.** With mandible in place.

mental tubercles. In this skull a supraorbital notch is present on the right while the notch has been converted into a foramen on the left side.

The **superior conchae** are now in view, and it is possible to see, as well, the perforated *cribriform plate,* a portion of the ethmoid bone in the roof of the nasal cavities.

The anterior **nasal spine of the maxilla** and the infraorbital foramina are clearly seen. For the first time a definite *zygomatic arch* can be appreciated, displaying the *infratemporal fossa* below it. The coronoid process of the mandible projects into the fossa and the condyloid process is seen to articulate with the temporal bone in the fossa. Structures within the orbit are rendered less visible however, with the skull oriented in this plane.

In Figure VI-17B the mandible of the tilted skull has been removed to allow us to see the hard palate and some of the underparts of the sphenoid, temporal, and occipital bones. The *hard palate* is formed by the **palatine processes of the maxillae** and by the **horizontal plates** of the **palatine bone.** These will be seen better in full basal views of the skull. In the posterior midline, where the horizontal plates meet, a slight projection, the **posterior nasal spine,** is found. The **uvula** takes its origin from the spine. Posterolaterally the **greater palatine foramina** can be seen. A waxed string passed into one of these enters a canal, the **greater palatine canal,** which lies between the maxilla and the **perpendicular plates** of the **palatine bone.** The palatine bones are "L"-shaped. Their horizontal plates form the foot of the "L," face each other, and fuse to complete the back part of the hard palate. The canal contains the greater palatine vessels and nerve. These supply the mucosa of the palate. A tinier **lesser palatine foramen** communicates with the canal and bears twigs from the main vessels and nerve to supply soft palate, uvula and tonsils. These nerves, greater and lesser palatine, are all sensory components of V_2, and their tonsilar twigs join branches of cranial nerve IX in a plexus upon the tonsil.

Inferior to the lesser palatine foramen is a hook-shaped piece of bone, the **hamulus** (Fig. VI-17B). This is a projection from the **medial pterygoid plate of the sphenoid.** The tendon of the Tensor Palatini muscle bends around this hook to anchor to the side of

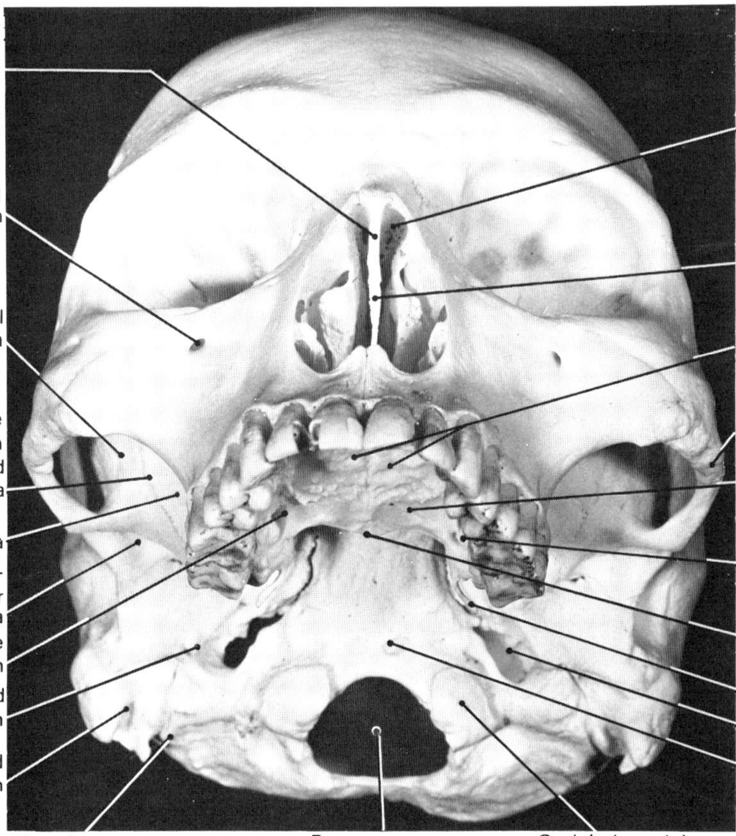

Fig. VI-17B. With mandible removed.

the **soft palate.** In this manner, during swallowing, the soft palate can be pulled taught and moved backward to seal off the nasopharynx.

In the region of the infratemporal fossa an **oblique suture line** (* in Fig. VI-17B) can be seen separating **sphenoidal squama** from **temporal squama.** Inferior and lateral to this suture at the base of the zygomatic process of the temporal bone is a depressed area, the **temporomandibular fossa,** for the head of the condyloid process of the mandible. Below this fossa a horizontal suture line runs medialward seeming to end in two tiny foramina. The medial one of these holes is for the **tympanic artery** to gain entrance into the middle ear cavity. The lateral hole is for emergence of the **chorda tympani branch of cranial nerve VII.** The nerve is moving out of the middle ear cavity into the infratemporal fossa at this point.

Superior and laterad of the great *foramen magnum* are the irregular contours of the **jugular foramen.** Lateral to the lowest point of this foramen it is possible to identify the **stylomastoid foramen.** Normally, the stylomastoid foramina are hidden behind a downward-projecting, **styloid process.** The processes break off rather easily in prepared skulls and, as in this case (Fig. VI-17B), they are blunted and virtually missing.

Lateral to the foramen magnum, only the anterior

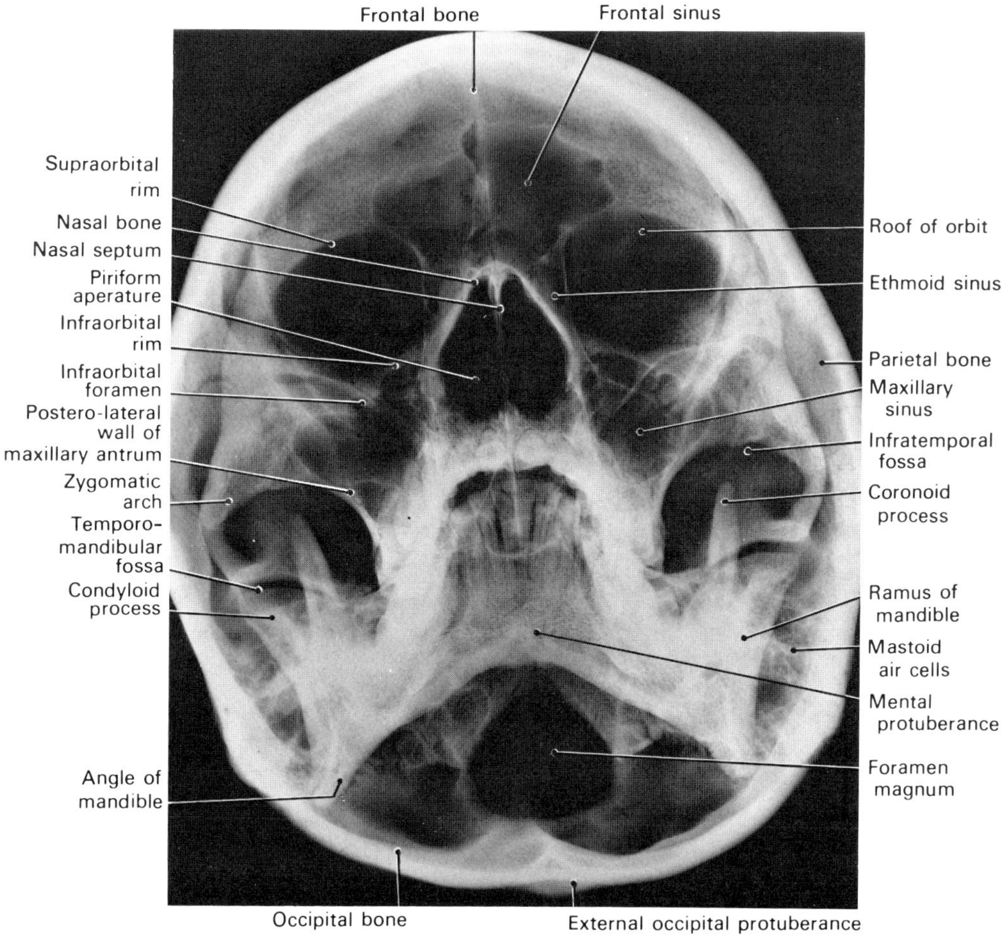

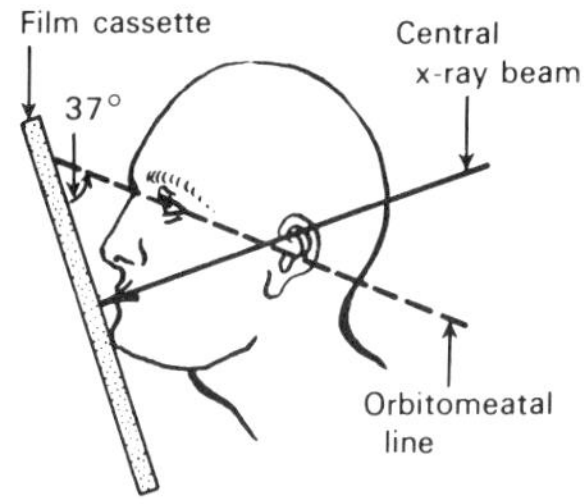

Fig. VI-17C. Radiograph of A in the occipitomental or Water's projection. **D.** Inset shows how radiograph was made. *, Oblique suture line between sphenoid and temporal squama.

portions of the **occipital condyles** show. On the inferior aspect of the basilar part of the occipital bone a **pharyngeal tubercle** comes into view. Here the uppermost of the three pharyngeal constrictor muscles inserts.

Figure VI-17C is a radiograph of the skull in a position comparable to that of Figure VI-17A with mandible attached. Its coronoid and condyloid processes can now be seen in the infratemporal fossa. The articular space between condyloid process and temporal bone is quite noticeable. The infraorbital foramen is also more prominent than it was in the inclined PA view (Fig. VI-14A). A number of air sinuses are now visible: frontal, ethmoidal, and maxillary. The orbital margins are more clearly defined, and both internal and external occipital protuberances can be identified. Such an occipitomental projection of the skull with an angle of inclination of 37° is referred to as a Water's view. Try to identify as many of the other structures we have mentioned in the skull as you are able.

Do not, however, try to memorize structures as mere words. Instead, gradually begin to see them as familiar landmarks, associating minor structures as parts of a larger whole. That is, associate given terms with parts of a larger bone or with geographic locations. Wherever possible gradually tie the term to something significant—a vessel, a nerve, a muscle, or to some facet of clinical significance that may be mentioned. Some terms which we introduce are bound to have less significance than others. We use them as markers to zero-in on something more important. It is not our intention to burden you with the unimportant. Yet, always in teaching and in learning some useless information is bound to be conveyed and to be picked up. You can always jettison useless baggage later if it does not suit your particular fancy. Fortunately, much may drain away with disuse and only the larger concepts may remain. But do not be too alarmed if even some useful information takes this route in time. It can be reviewed at a more rapid rate if needed and if some overlearning has occurred in the first place. In actuality, no fact is really unimportant. It may have a place

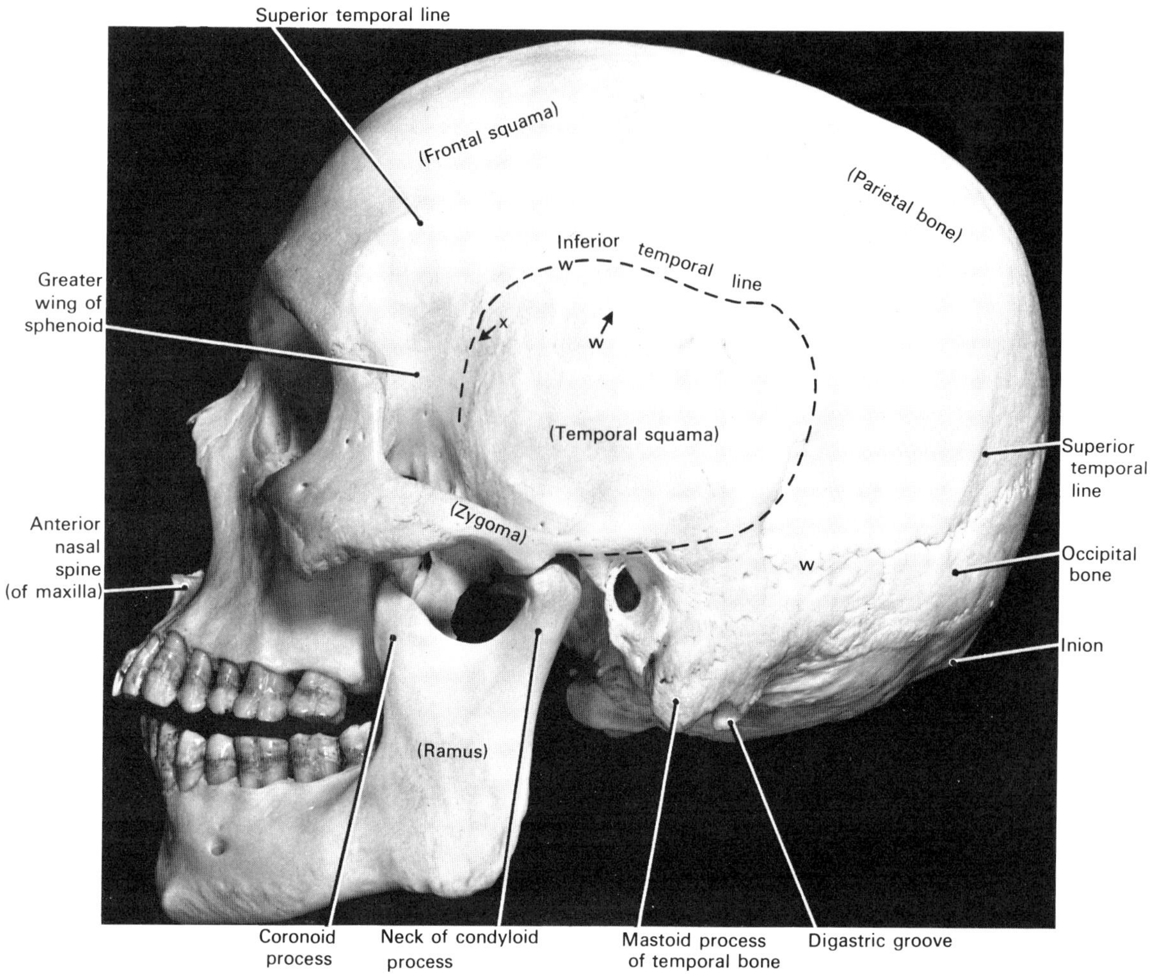

Fig. VI-18A. Norma lateralis of skull. In A, x = pteryion; W = Wormian bones.

in the scheme of things later on. You will become amazed at the plethora of seemingly unimportant facts that the minds of our most brilliant scholars and researchers tote around like a spare bag of nuts and bolts. You never know when some particular item may come in handy. A little extra toting does little harm in the long run.

Lateral Aspect of the Skull (Fig. VI-18 A and B)

This aspect includes: (1) the temporal fossa; (2) the infratemporal fossa; (3) the lateral aspects of the maxilla and mandible; (4) the region of the external auditory (acoustic) meatus; and (5) the mastoid region.

The Temporal Fossa: This region is a large, shallow, oval depression over much of the side of the skull above the level of the *zygoma* (zygomatic arch). The fossa is crossed by the **superior and inferior temporal lines,** each of which describes a faint oval region upon its surface, one within the other. Often the superior line can be a double one. It arches backward from the zygomatic process of the frontal bone (Fig. VI-18A and B). Postero-inferiorly, it blends with the parietotemporal suture lines to end, finally, in the **suprameatal crest** and **posterior root of the zygoma** (Fig. VI-18B). The inferior line is often much more faintly demarcated. It was not prominent in Figure VI-18A and B, so we dotted it in. It starts across the **greater wing of the sphenoid** anteriorly and terminates posteriorly by cutting across the temporal squama to become continuous, finally, with the superior margin of the zygoma.

The medial wall of the temporal fossa consists, therefore, of parts of the frontal and parietal bones, the greater wing of the sphenoid (**sphenoidal squama** at this lateral aspect), and the **squamous part of the temporal bone.** In the skull shown in Figure VI-18A, the fossa contains two fairly large *Wormian bones,* marked "W." They fill in an area once occupied by the anterolateral fontanelle in the infant head, an area usually filled by parietal bone and temporal squama in most skulls. Here they reduce the borders and size of

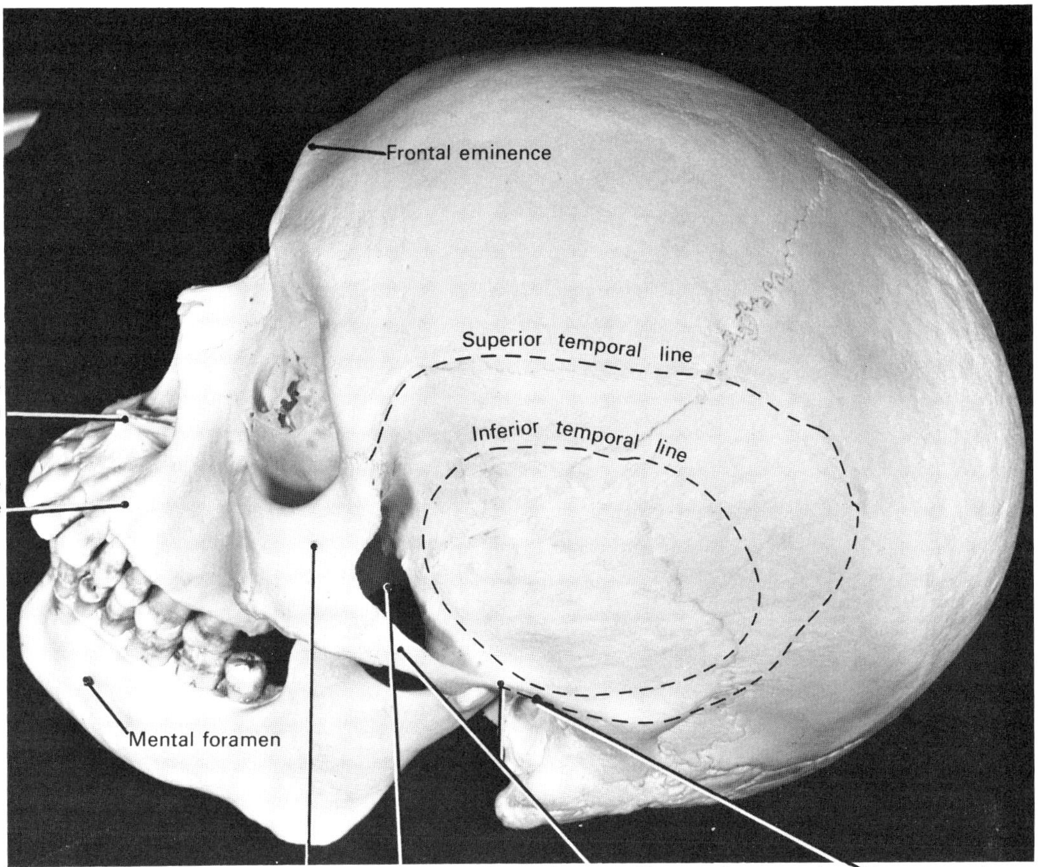

Fig. VI-18B. The skull has been rotated slightly so that the superior opening into the infratemporal fossa can be shown.

the latter two bones. This makes it difficult to mark the **pterion,** a useful landmark on the lateral aspect and the common meeting point of the greater wing of the sphenoid with the temporal squama and parietal bone. We have marked the location of the pterion with an "X" in Figure VI-18A. See Chapter III (p. 93) for a review of its significance.

The Infratemporal Fossa: The lateral wall of the temporal fossa has to be the inner or medial surface of the zygoma. As this bony arch moves away from the side of the skull at its posterior root (Fig. VI-18B), it swings laterad and anteriorly, leaving a considerable gap through which certain structures (principally, the tendon of the Temporalis muscle) can pass into the **infratemporal fossa** below the zygoma. This region has the inner aspect of the ramus of the mandible as its lateral boundary (Fig. VI-18A and B). If the skull is rotated and examined from below as in Figure VI-19, it becomes apparent that the infratemporal fossa is roofed over by the inferior aspect of the greater wing of the sphenoid. Two important holes pierce this roof, one of them very tiny. Through the larger, oval-shaped one (foramen ovale), the mandibular division (V_3) of the trigeminal nerve drops down into the fossa to reach the major muscles of mastication associated with the mandible. The Temporalis is one of these muscles of mastication. Its tendon entered the fossa in the manner noted above. Its motor nerve will have to come up from the fossa to innervate it. The nerve will be a branch of V_3. Through the tiny hole, which is the **foramen spinosum** (Fig. VI-19), pass the important **middle meningeal arteries.** These are twigs of the **maxillary artery.** The maxillary artery is the largest terminal branch of the external carotid, the superficial temporal being the smaller, end-branch of this great vessel. The maxillary artery arises at the neck of the condyloid process within the substance of the parotid gland which covers this region. It passes medial to the mandible through the heads of the Lateral Pterygoid muscle to find itself headed deep into the **pterygopalatine fossa** beyond (Fig. VI-20). The middle meningeal artery is one of the earliest of the many branches of the maxillary artery, and it escapes from the infratemporal fossa almost immediately through the tiny foramen spinosum in the roof.

If we look through a window frame in the skull formed by the zygoma and arch of the mandibular processes (Figs. VI-18A and 20), we can see in this oval frame some of the lateral structures that reside in the

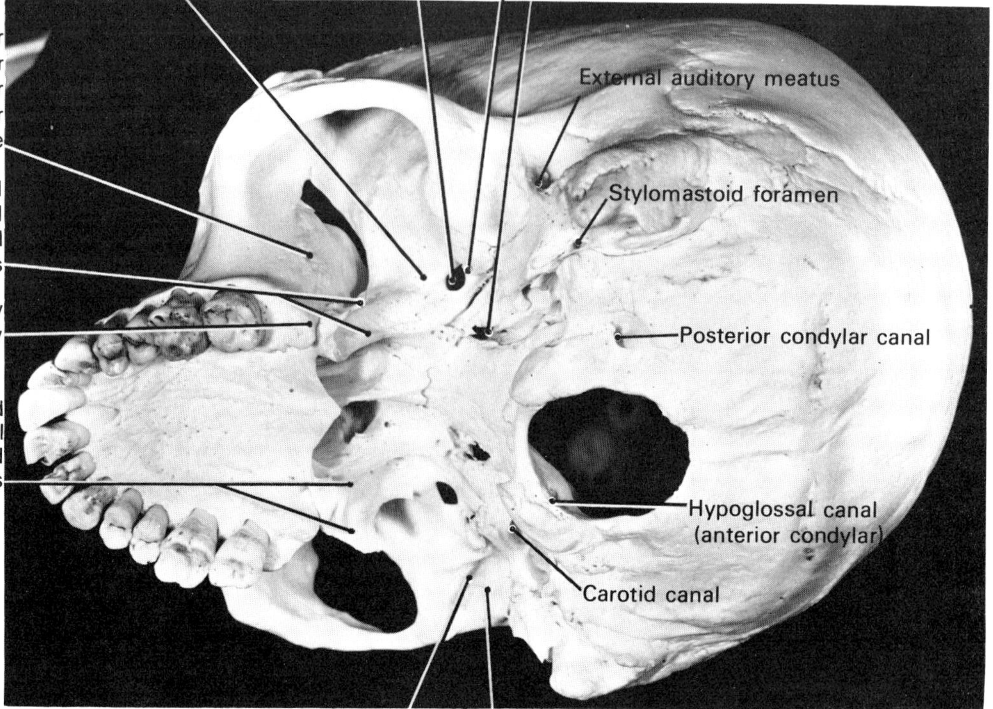

Fig. VI-19. Skull rotated in order to examine inferior or basal surface.

infratemporal fossa. From rostral to caudal these are:

1. A tiny bit of the **posterior wall of the maxilla.**
2. A part of the **palatine bone.**
3. A slit-like **pterygopalatine fossa** with the sphenopalatine foramen deep within it.
4. The **lateral pterygoid plate** of the sphenoid.

The **medial and lateral pterygoid plates** or **laminae** can be identified in Figure VI-19 where the posterior margins of these plates, rather than their sidewalls, come into view. A deep fossa appears between them as the *pterygoid fossa.*

This somewhat buried region of the infratemporal fossa is fraught with clinical significance for both the doctor and the dentist. We have mentioned the fact that the parotid gland hides it and that the maxillary division of cranial nerve V enters it while the middle meningeal vessels leave it. Later we will discuss the muscles, nerves, and vessels related to it. The superior and inferior alveolar nerves to upper and lower teeth arise in it and are accessible for anesthetization. A sphenoidal electrode pushed against its roof can pick up electroencephalograms from the temporal lobe, since this lobe abuts against the anterior portion of the wing of the sphenoid of the cranial floor. The nerves and vessels of the fossa become of extreme concern in any reconstructive surgery of the mandible, and posteriorly lies the temporomandibular joint, the only movable joint in the head.

The Maxilla and Mandible: In a lateral view, the **anterior nasal spine** of the maxilla comes into prominence (Fig. VI-18A and B). Posterior to the last molar tooth, the maxilla presents the **maxillary tuberosity** as a rounded eminence (Figs. VI-19 and 20). Its roughened surface provides partial attachment for the **Medial Pterygoid** muscle. Superior to the tuberosity much of the posterior wall of the maxilla in the infratemporal region is obscured by the position of the mandibular ramus. Through the posterior maxillary wall, foramina open into **alveolar canals** for the conduction of the **posterior superior alveolar nerves and vessels** (Fig. VI-19).

In a dissected skull such as that shown in Figure VI-21, the posterior superior alveolar nerves and vessels can be followed to the **upper molars,** while the **middle superior alveolar nerve** is seen to supply the **upper premolars.** As a rule, the nerve to the premolars descends vertically to them in a separate canal as a branch that leaves the infra-orbital nerve. In this skull, it is a terminal twig of the posterior superior alveolar nerve. However, we can see the **anterior superior alveolar nerve** in its own vertical canal coursing to the **upper canine and incisors** as a direct

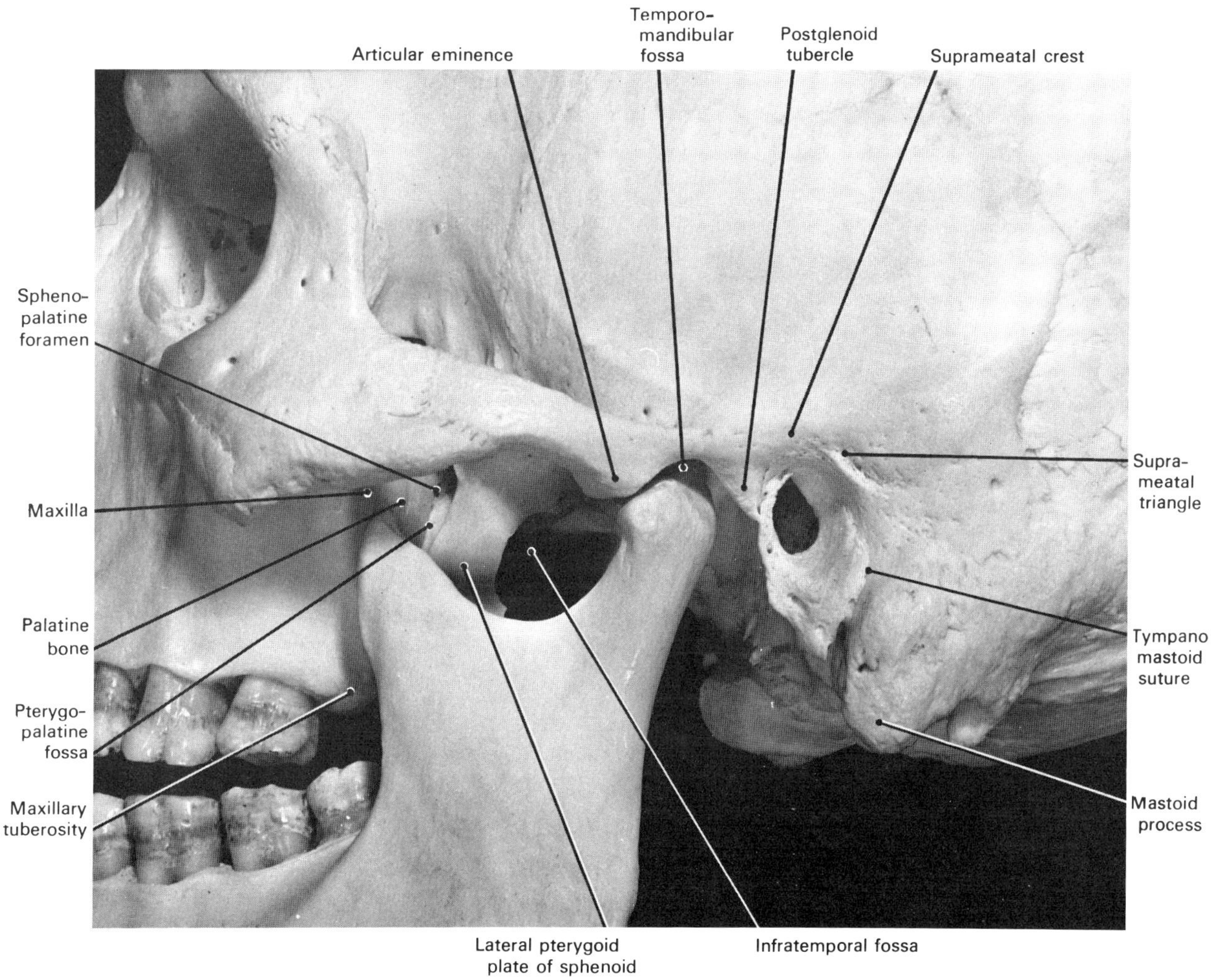

Fig. VI-20. Enlarged view of lateral aspect of skull showing pterygoid plates and fossa and the region of the temporomandibular articulation.

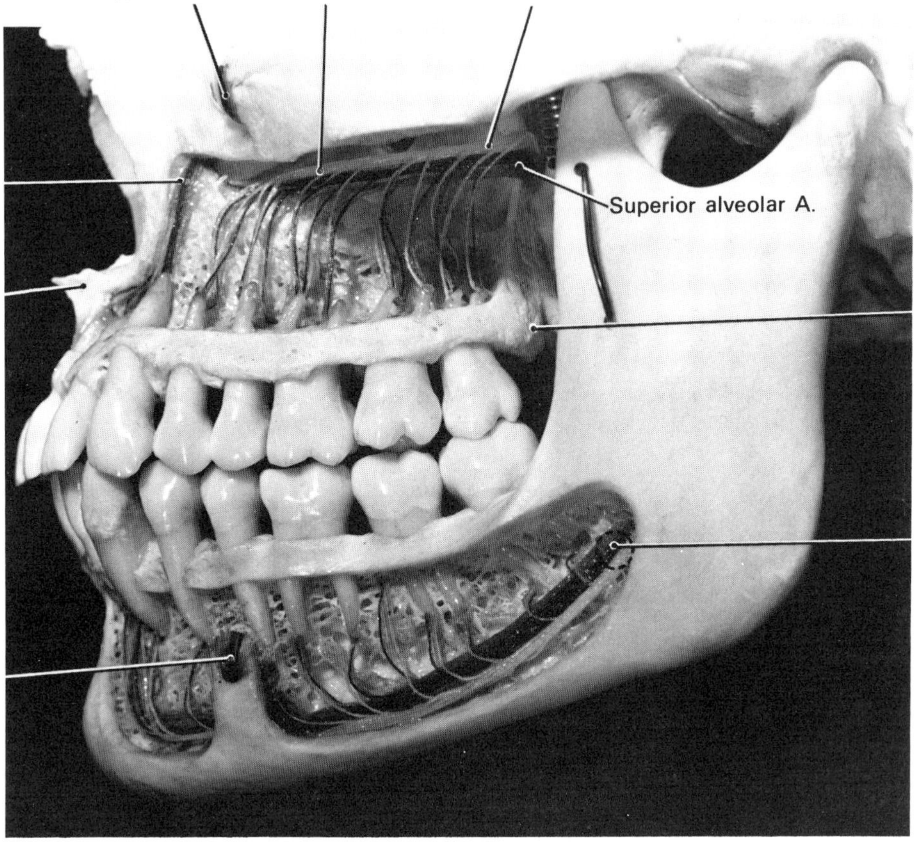

Fig. VI-21. Dissected maxilla and mandible showing the nerve and vascular supply to the teeth.

branch of the infraorbital. Note how separate filaments enter a foramen in the dental root of each tooth. The point to remember is that like the infra-orbital nerve, all of these nerves to the upper teeth are branches of V_2, ***the maxillary division of cranial nerve V.***

In the lower jaw the picture is somewhat different (Fig. VI-21). All the teeth are supplied by *one nerve,* the **inferior alveolar,** which is a branch of V_3, *the mandibular division of cranial nerve V.* The nerve and vessels to the lower teeth entered the mandibular canal in the infratemporal fossa by passing beneath a spur of bone, *the lingula,* on the medial aspect of the ramus (Fig. VI-22A). Note in the dissected skull (Fig. VI-21) that the **mental nerve** to the skin of the chin and lower lip leaves its own foramen anteriorly in the mandible to reach these superficial structures.

An X-ray of this region in the oblique lateral projection (Fig. VI-22B) shows the mandibular canal and mental foramen, as well as a tremendous cavitation in the maxilla, the **maxillary air sinus.** The superior alveolar nerves run in the thin lateral wall of the sinus and can be expected to provide sensory twigs to its mucosal lining and to the mucous membranes of the inner cheek and gums. Note the root canals and entering foramina in the canine, premolar, and molar teeth of the lower jaw in this X-ray.

The parts of the mandible can be reviewed in Figure VI-15B and in Figure VI-22A from in front and from behind. When it is in place in the skull, the coronoid process is in the infratemporal fossa, while the condyloid process is in the temporomandibular fossa. An **articular eminence** of the zygoma helps to restrain the latter process in the temporomandibular fossa (Fig. VI-20). It prevents dislocation of the process into the infratemporal fossa during excessive yawning, during anesthesia, and when one attempts to pass a laryngoscope or to extract teeth. Warning: In attempting to reset such a dislocation, keep fingers out of the open mouth when the process slips back into place and the jaws snap shut!

Region of the External Auditory Meatus (Fig. VI-20)**:** It is apparent that the posterior root of the zygoma takes part in fashioning the region superior to the meatus, for it is continued as the **suprameatal crest.** The inferior spur of bone below the posterior zygomatic root, just anterior to the meatus, is known as the **postglenoid tubercle.** The other three bony

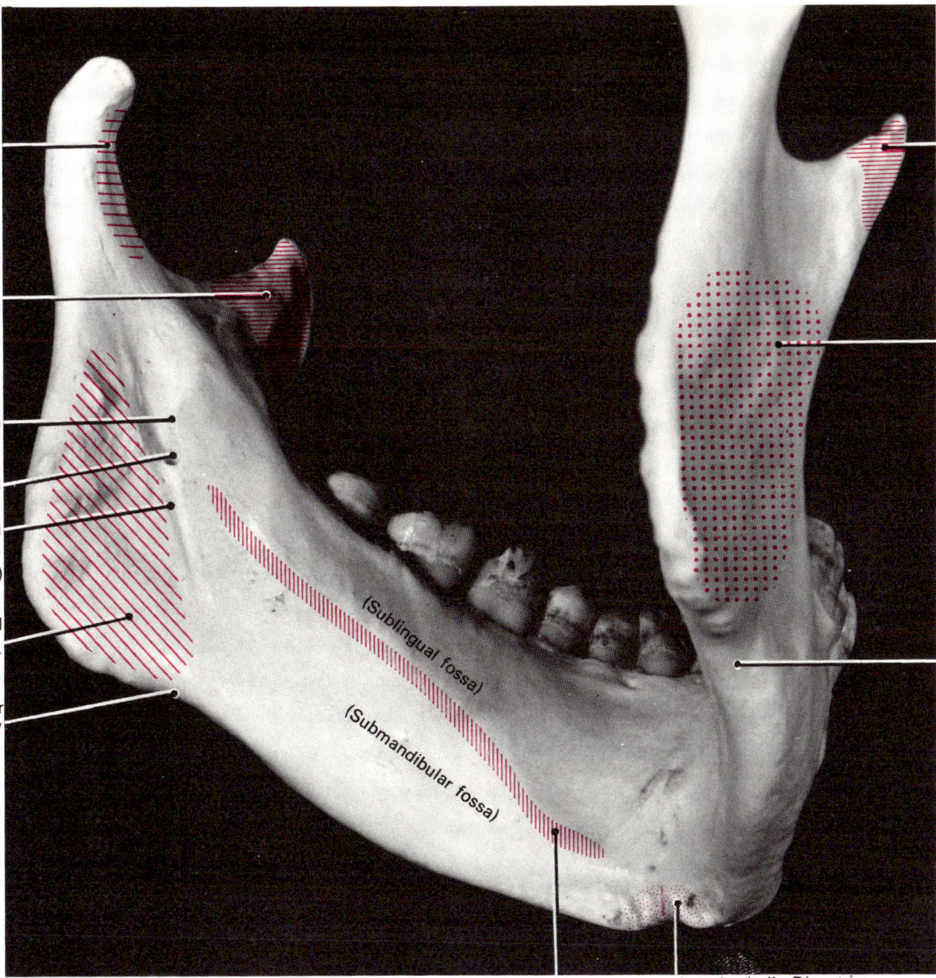

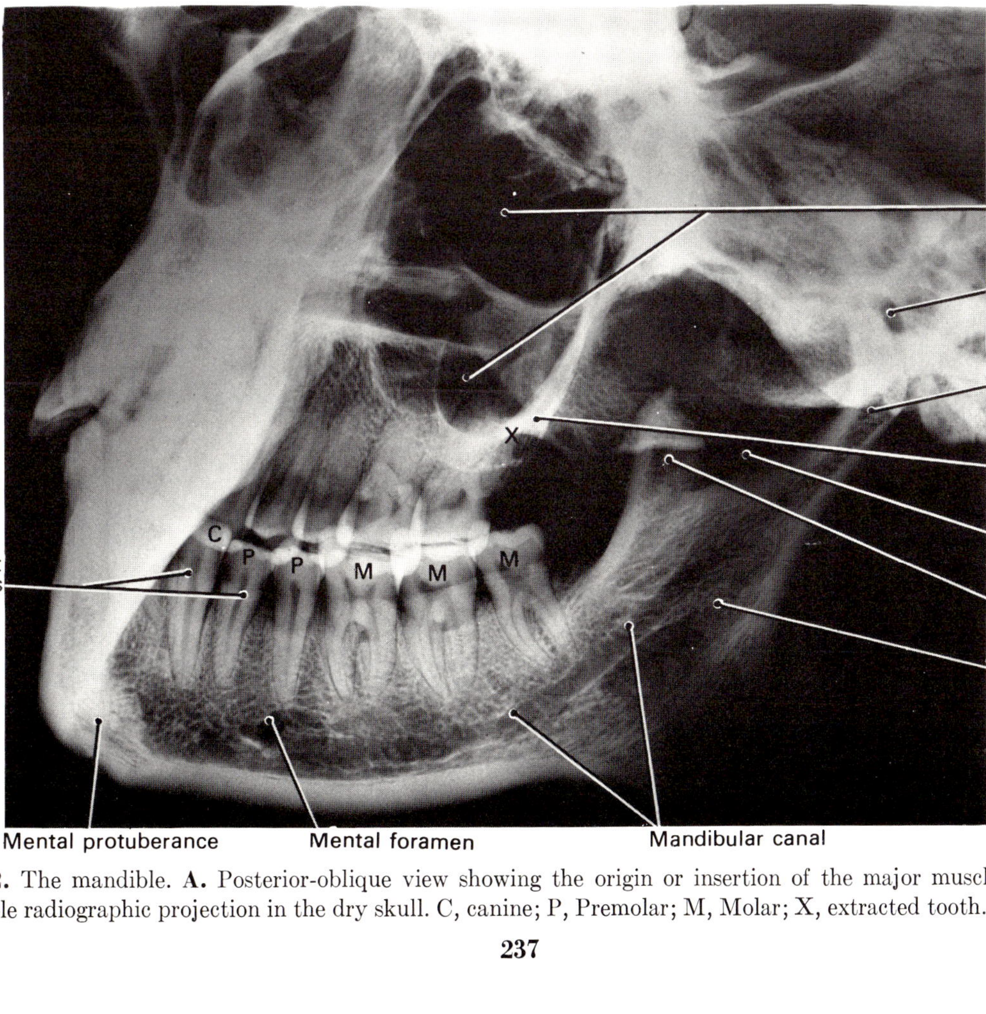

Fig. VI-22. The mandible. **A.** Posterior-oblique view showing the origin or insertion of the major muscles of mastication. **B.** Comparable radiographic projection in the dry skull. C, canine; P, Premolar; M, Molar; X, extracted tooth.

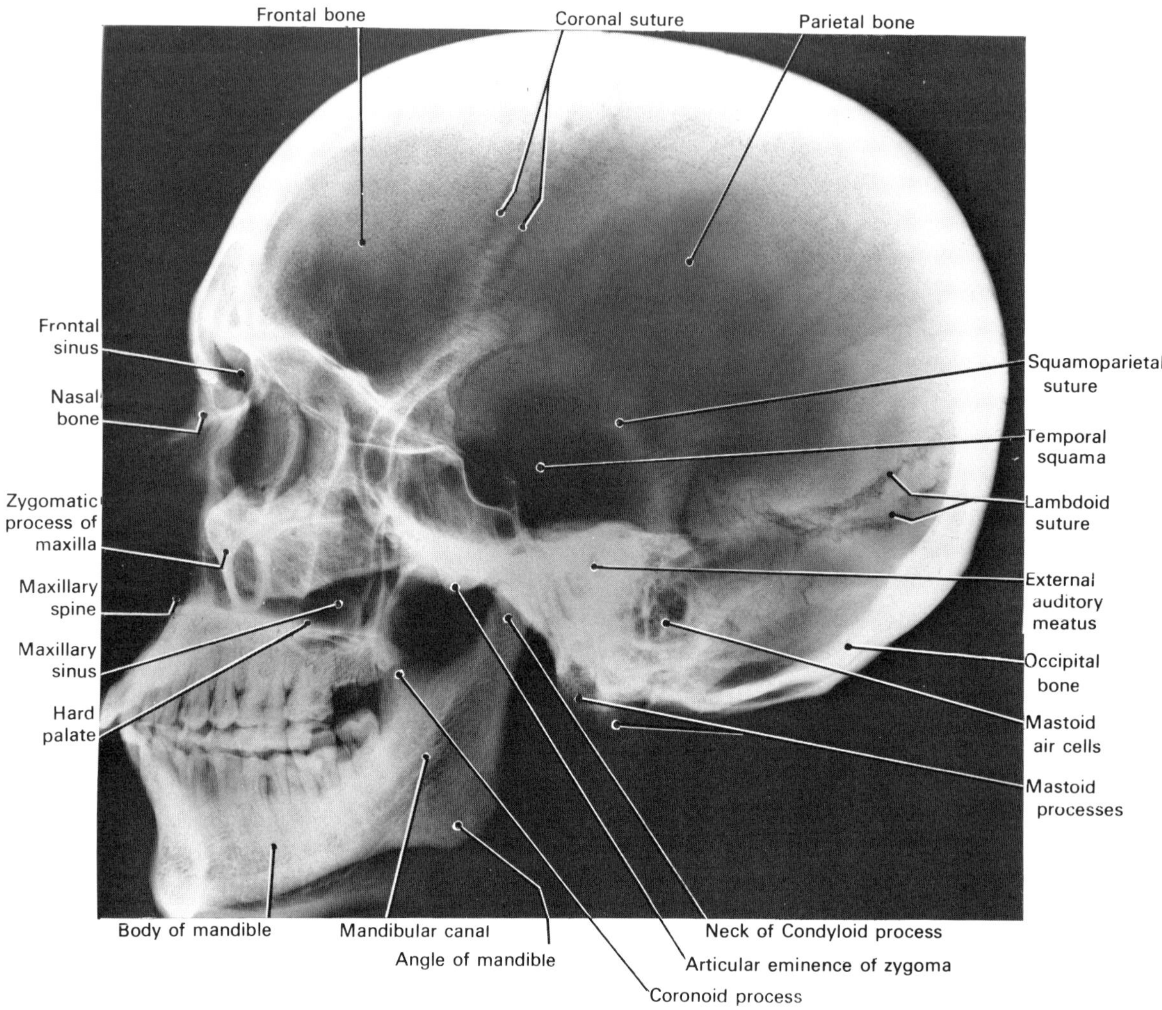

Fig. VI-23. Lateral radiograph of the dry skull. Compare with Figure VI-18A.

sides of the meatus are the roughened margins of the **tympanic part** of the temporal bone. Slightly superior and posterior to the meatus is a small depressed oval, the **suprameatal triangle.** Formerly this was a popular site of exposure in mastoid operations. By using such a postaural approach, however, parts of the dural sigmoid sinus are placed in danger, along with the facial nerve as it descends in its canal through the temporal bone toward the *stylomastoid foramen* (Fig. VI-19) where its branchial motor branches emerge. It has become more fashionable, therefore, to approach the mastoid air cells endauraly; *i.e.*, through a skin incision between tragus and helix (see Fig. III-25A) of the auricle which will follow along the bony meatus and, by going through its wall, will enter the mastoid region. The tympanic ring will serve as a landmark of depth to the facial nerve in this procedure.

The Mastoid Region (Fig. VI-20): The **mastoid process** is a tit-like, down-hanging mass of the temporal bone behind the ear. Generally it is pneumatized by many air cells variable in size, some of which communicate with the posterior part of the middle ear cavity (Fig. VI-23). The mastoid process can also be sclerotic; *i.e.*, consist of very dense bone; or it can be diploic; *i.e.*, it can consist of a mixture of diploe and air cells. Usually, though, most mastoids are well pneumatized by age three or four, but the infant skull does not contain air cells in this region. The mastoid process is separated from the tympanic ring by the tympanomastoid suture (Fig. VI-20). We have mentioned the mastoid foramen for an emissary vein before, as well as the groove or notch for the posterior belly of the Digastric muscle.

Now examine Figure VI-23, which is a lateral radiograph of the dry skull that we have been discussing. This view matches Figure VI-18A. We are concerned only with the superficial structures that are clearly visible, and we shall return to study the deeper structures in this same Figure at a later time. Note that overlap in the mastoid region obscures much of the anatomy there. The following features can be readily identified:

1. Neck of condyloid process
2. Articular eminence of zygoma anterior to
3. Coronoid process
4. Mandibular canal
5. Angle of mandible
6. Body of mandible
7. Shelf of hard palate
8. Maxillary sinus
9. Anterior nasal spine of maxilla
10. Edge view (almost as if in cross-section) of the zygomatic process of maxilla.
11. Nasal bones
12. Frontal sinuses—one virtually inside the other as seen in this lateral view. We know from the frontal view that the left one is the larger of the two
13. The frontal bone, itself
14. Coronal suture
15. Parietal bones
16. Squamoparietal suture
17. Temporal squama
18. Lambdoid sutures
19. Region of external auditory meatus, for the most part obscured by overlapping temporal structures
20. Occipital bone
21. Mastoid processes. These are asymmetric in this skull, and, therefore, they do not overlap
22. Mastoid air cells

The Muscles of Mastication and Relationships in the Infratemporal Fossa: This seems to be a reasonable time and place to introduce the major muscles that move the mandible. Recall (Fig. VI-7A) that they are derivatives of the first branchial arch mesoderm and, as such, are innervated by V_3, the mandibular division of the trigeminal nerve. This is the division that entered the infratemporal fossa through the oval hole in its roof (foramen ovale). While V_3 supplies the motor innervation to these muscles, it also contains many sensory fibers, as we shall see, but it is the only one of the three divisions of the trigeminal nerve (V) to convey motor fibers to skeletal muscle. *All divisions of the trigeminal also serve as transport conduits for visceral motor fibers of certain other cranial nerves which join some of the specific branches of the trigeminal. The trigeminal nerve, itself, does not contribute any motor supply to smooth muscle or glands in the head.*

The **Temporalis** is a fan-shaped muscle (Fig. VI-24) which takes its origin from the margins of the inferior temporal line, though its muscle fascia spreads to the superior temporal line. Its tendon passes, as the handle of the fan, behind the zygoma to enter the infratemporal fossa and to insert upon the coronoid process of the mandible within the fossa. Its points of attachment upon the mandible, as well as those for the other muscles of mastication, are shown cross-hatched in Figure VI-22A. In general, the muscle draws the mandible upward, but its posterior fibers can retract the lower jaw. Two or three temporal nerves, branches of V_3, reach the muscle by crossing between the bony roof of the fossa and the superior border of the Lateral Pterygoid (Fig. VI-26B). They end in the Temporalis on its deep surface behind the zygoma.

The Lateral Pterygoid is a key structure in the fossa since structures entering or leaving must pass over or under it or between its two heads of origin. The maxillary artery, for example, in moving medially into the fossa, passed between its heads, and we have just seen the temporal nerves pass laterad above it to leave the fossa.

The Masseter (Fig. VI-24) is another superficial muscle in the region, rhomboid in shape, and arising in part from the inferior border and medial surface of the zygoma. It covers most of the lateral aspect of the mandibular ramus to insert at the jaw angle (Fig. VI-22A). It, too, draws the mandible upward. Its masseteric nerve (Fig. VI-26B), a branch of V_3, must enter its deep surface by passing between the upper and lower heads of the Lateral Pterygoid, and through the mandibular arch which is formed by the condyloid and coronoid processes. This nerve conveys some sensory fibers, as well, which it returns from the temporomandibular joint. These are, undoubtedly, both pain and proprioceptive fibers, the latter affording us a sense of awareness of joint positioning and movement.

Deep to these two massive superficial muscles which we have just mentioned, and truly arising from within the infratemporal fossa, are two smaller muscles. The key **Lateral Pterygoid** is two-headed (Fig. VI-24), the upper head coming from the underportion of the greater sphenoidal wing and the lower from the lateral surface of the lateral pterygoid plate or lamina of the sphenoid. The insertion is upon the condyloid neck (Fig. VI-22A), onto the joint capsule and even upon the articular disk of the temporomandibular joint. It can help its partner, the Medial Pterygoid, to protrude the mandible, but alone, it is able to move the jaw from side to side in grinding movements of the molars. Its nerve, also from V_3, is short and enters its deep surface at once.

The **Medial Pterygoid** lies deepest of all in the infratemporal fossa as a quadrilateral muscle which originates from the pterygoid fossa and the medial side of the lateral pterygoid plate (Fig. VI-24). It angles medialward and downward to attach to the medial side of the jaw angle; *i.e.*, at the opposite side of the bone from the Masseter (Fig. VI-22A). Working with the latter, it draws the mandible upward; with its partner, it protrudes the jaw.

All of these muscles have one common action. They work to close the mouth. When they relax, gravity alone opens it, as you may have observed in your snoring roommate. When the mouth opens against resistance, however, the infrahyoid muscles ("strap" muscles of the anterior neck) tense to fix the hyoid bone. With the bone so immobilized, the suprahyoid muscles (anterior belly of the Digastric, Mylohyoid,

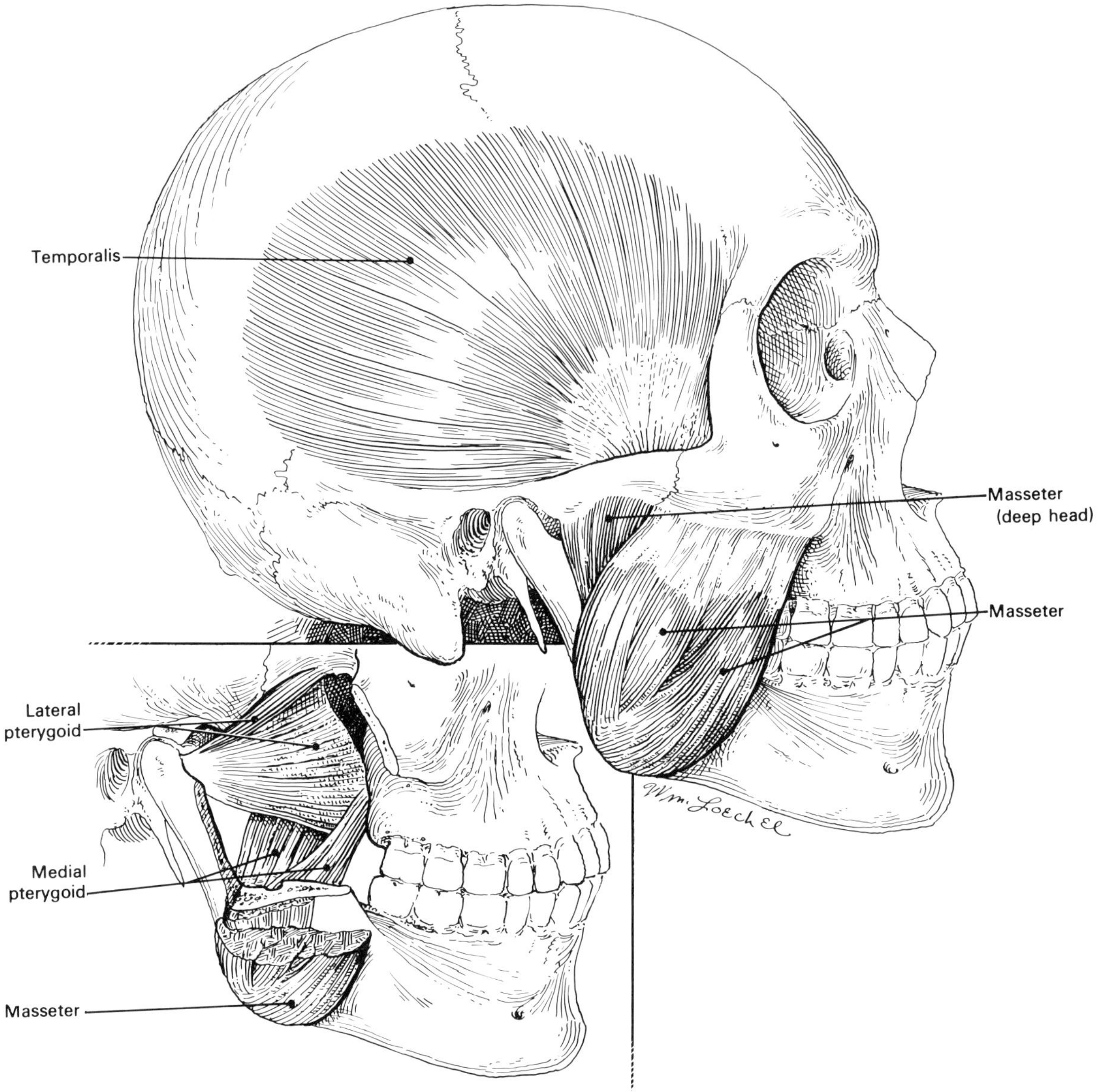

Fig. VI-24. The muscles of mastication and their relationship to the temporal and infratemporal fossae.

and Geniohyoid) can come into play to pull the mandible downward.

Figure VI-22A shows also where the **Mylohyoid** and the **anterior belly of the Digastric** insert upon the mandible. These muscles get branches from V_3 through the **mylohyoid nerve** which leaves a groove on the inner aspect of the mandible medial to and outside of the mandibular canal. The Mylohyoid muscle, in forming a diaphragmatic floor for the mouth, has a long line of attachment which separates the body of the mandible on its inner aspect into two fossae. The **sublingual gland** nestles against the bone in the upper one, and, as you might expect, the **supramandibular** (submaxillary) **gland,** in the lower one (Fig. VI-22A).

We must not forget that V_3 also supplies motor fibers to the **Tensor Tympani,** a tiny muscle of the middle ear cavity which it reaches by penetrating the cartilage of the auditory tube in the neighborhood of the foramen lacerum (Fig. VI-19). So this muscle can be reached from the infratemporal fossa after V_3 enters the fossa through the foramen ovale. In a like manner, V_3 innervates the **Tensor Vili Palatini** which arises from a depression in the pterygoid fossa, the **scaphoid fossa,** and inserts into the velum or soft palate by a tendon that swings about the hamulus, a projection of the medial pterygoid plate (Fig. VI-17B). Both of these Tensors, like the Mylohyoid and the anterior belly of the Digastric, are first branchial arch derivatives.

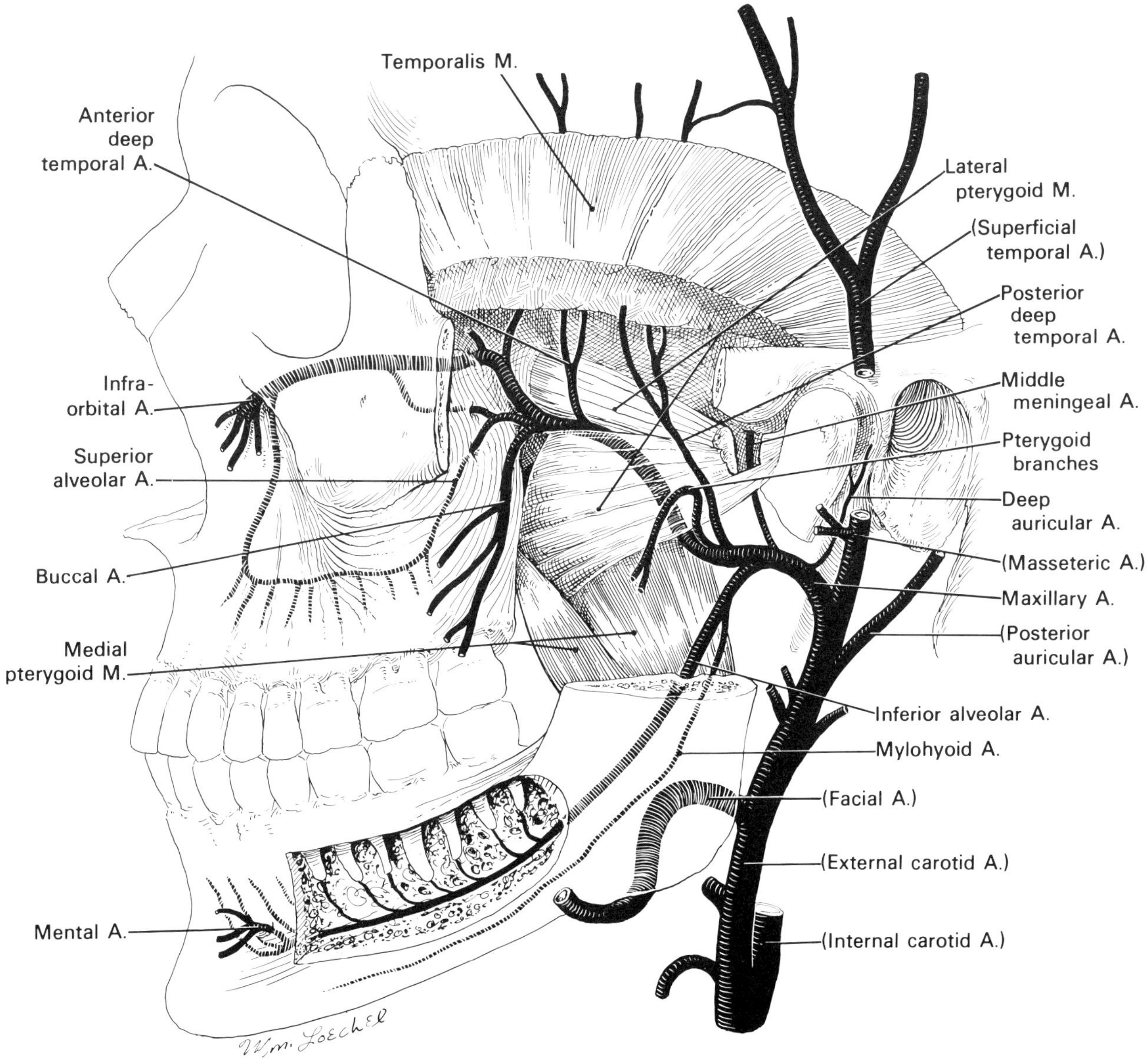

Fig. VI-25. Branches of the maxillary artery and their relationship to the pterygoid muscles. (Vessels in parentheses are not branches of the maxillary artery.)

Vascular and Neural Relationships in Infratemporal Fossa: The vascular relationships at the jaw and in the crowded infratemporal fossa are shown in Figure VI-25. It is obvious that only *one* artery, the **maxillary,** is the main vascular supply to the region. The figure is virtually self-explanatory and the distribution of its named branches should be studied.

Neural relationships are a bit more complicated. Still, *two* divisions of the trigeminal, $\mathbf{V_2}$ **and** $\mathbf{V_3}$, furnish the main neural input in the fossa, discounting any autonomic (visceral motor) supply from other cranial nerves. Even so, the fibers of the trigeminal (cranial nerve V) will serve to distribute these visceral motor fibers. The trigeminal will also convey all general sensory information from this area. But V_3 alone, the mandibular division of the nerve, will carry all the motor supply to the muscles of mastication or to any other muscles derived from the mesoderm of the first branchial arch.

The **branchial motor fibers** of the trigeminal leave the brainstem in the motor root of the nerve (Fig. VI-26A) to be distributed entirely by V_3. In this Figure the branches to the Masseter, Pterygoids, Temporalis, and the Tensors Tympani and Palatini are cut off short, but fibers can be traced to the Mylohyoid and to the anterior belly of the Digastric by the mylohyoid nerve. The Temporalis, Masseter, and Lateral Pterygoid are seen being supplied in Figure VI-26B, however.

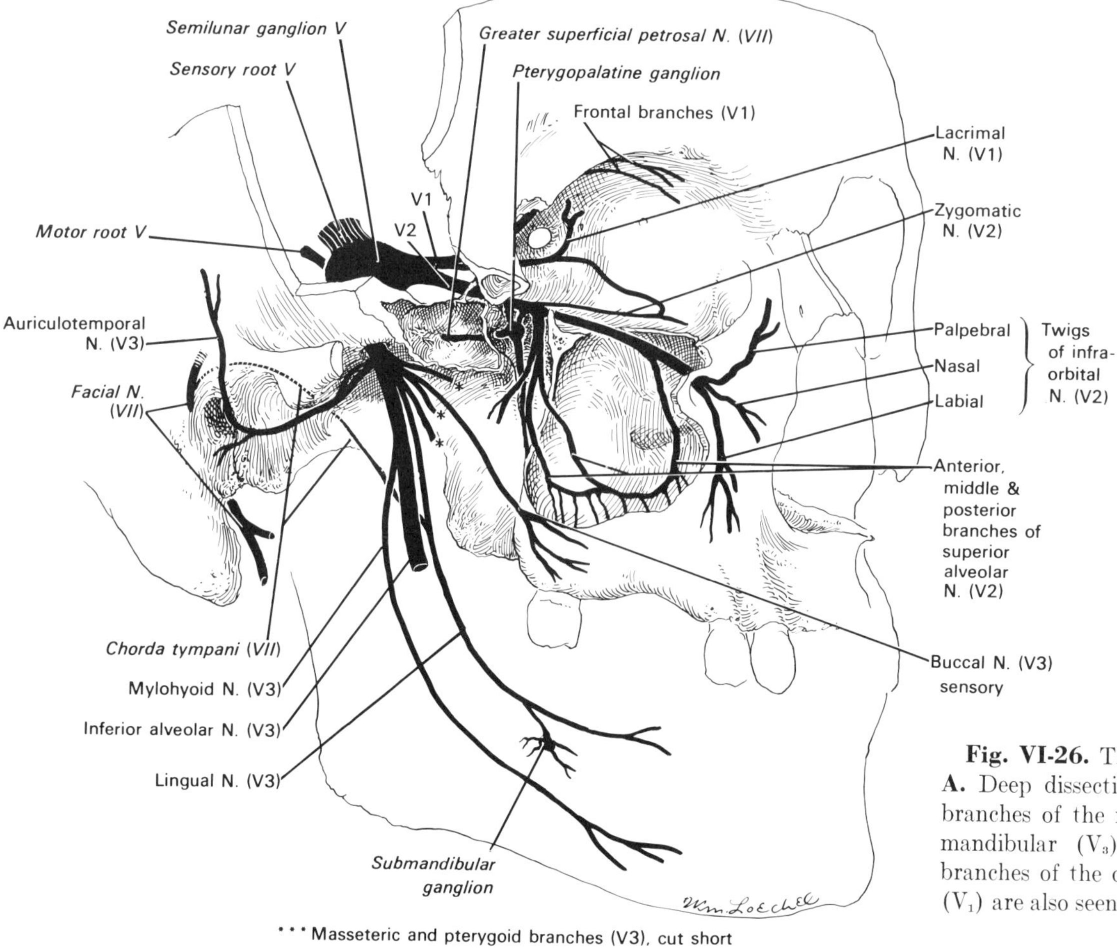

Fig. VI-26. The trigeminal nerve. **A.** Deep dissection to show chiefly branches of the maxillary (V_2) and mandibular (V_3) divisions. A few branches of the ophthalmic division (V_1) are also seen.

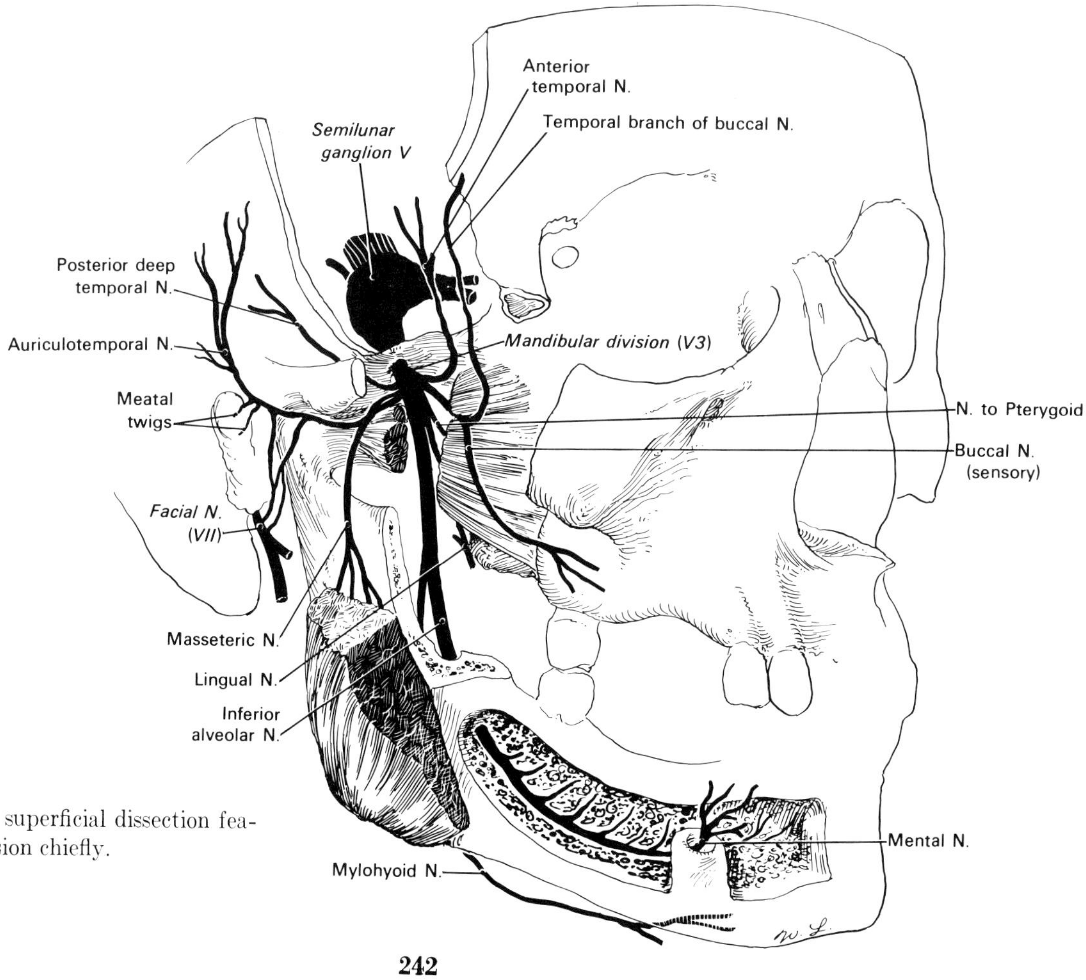

Fig. VI-26B. More superficial dissection featuring mandibular division chiefly.

All the other branches of V_3 depicted in Figure VI-26 are *sensory nerves*. These latter account for the **auriculotemporal** nerve from skin around the ear, meatus, and temple; the *inferior alveolar nerve* (cut short) from the lower teeth and gums; the *lingual nerve* from the mucous membrane of the tongue (but not from its taste buds); the *buccinator nerve* from the mucous membranes of the cheek; and the *mental nerve* from skin of chin and from the lower lip. All of these sensory fibers have their cell bodies in the **semilunar (Gasserian) ganglion** of the trigeminal (Fig. VI-26A). This ganglion sits in the floor of the middle cranial fossa at the anterior tip of the petrous bone. It is covered over by dura mater.

Parasympathetic and sympathetic motor fibers (axons of nerve cells) are also distributed with the trigeminal nerves. But, as we have indicated, this cranial nerve has absolutely nothing to do with these functions other than to serve as a convenient highway to carry such visceral motor fibers to where they wish to go. The parasympathetic supply comes principally from cranial nerves VII (facial) and IX (glossopharyngeal) which mingle their postganglionic axons with the fibers of V_2 and V_3. We will find later that the parasympathetic outflow of cranial nerve III (oculomotor) uses V_1 as the same kind of a convenient highway. Thus, the facial nerve uses V_2 to reach nasal, lacrimal, and palatine glands. Its **chorda tympani** branch, on the other hand, joins with V_3 by way of the **lingual nerve** to reach submaxillary and sublingual glands. In a similar manner visceral motor fibers of the glossopharyngeal nerve (IX) hook a ride with the auriculotemporal branches of V_3 to reach the parotid gland. *Let it be made clear for once and all that visceral motor fibers serve glands, smooth (and cardiac) muscle.* The trigeminal nerve is never, itself, concerned with these structures. It just obliges neighbors that are concerned by piggy-backing them to their destination.

As a rule, **sympathetic visceral motor fibers** shun cranial nerve V. They prefer to reach structures in the head over perivascular routes. In the areas of the infratemporal fossa they would use the branchings of the maxillary artery as their highway.

The branches of the trigeminal are also utilized to conduct *special sensation from the taste buds* of the tongue. These travel back in the lingual nerve (a branch of V_3) but transfer to the *chorda tympani* in the infratemporal fossa. The latter is a branch of VII and it, not V, is concerned with taste sensation from the anterior two-thirds of the tongue. Hence, the cell bodies of these taste fibers are buried in the geniculate ganglion within the depths of the petrous part of the temporal bone. They are not in the semilunar ganglion of V.

It is for these reasons that Figure 26A includes the **facial nerve,** the *chorda tympani,* and two parasympathetic terminal ganglia, the **pterygo(spheno)-palatine ganglion** and the **submaxillary (submandibular) ganglion.** These are where preganglionic visceral motor axons of VII terminate. Postganglionic axons from cells of these ganglia then use the fibers of V_2 or V_3 to reach nasal, palatine, lacrimal, submandibular, and sublingual glands. There is a third terminal parasympathetic ganglion, the **otic ganglion,** in the infratemporal region just below the foramen ovale. This is a terminal parasympathetic ganglion for IX. Postganglionics from it join the auriculotemporal twigs of V_3 (mandibular division of V) to run to the parotid gland since this nerve passes through this territory.

Another point of common confusion relates to the ***buccinator nerve,*** a branch of V_3. It does *not,* as you might think, supply the Buccinator muscle. That muscle is derived from brachial arch II mesoderm and that is the territory of VII since that cranial nerve supplies skeletal muscles of facial expression. The Buccinator is a muscle of facial expression not a muscle of mastication. While it might play a minor role in keeping food compressed in the mouth, it is chiefly used in blowing and so puffs out the cheeks and alters facial expression. The muscle has to get its motor supply, therefore, from a branch of VII. The **buccinator branch of the trigeminal nerve** is *purely sensory* to mucous membranes and skin in the cheek and lip areas.

Some branches of V_2 are shown in part in Figure VI-26A, but not those to the nose and palate. The ones to the upper teeth are shown forming loops within the walls of the maxilla. Their pattern is variable.

A sensory branch of V_1 is also shown in Figure VI-26A, the **lacrimal nerve,** a trigeminal branch. It is sensory to the region of the conjunctiva. *It has nothing to do with the innervation of the lacrimal gland* except in being a gracious host to fibers of VII which reach it by a devious route from the pterygopalatine ganglion. Fibers of VII also steal a ride with some fibers of V_2 *en route* to the gland. VII is motor for the lacrimal gland, as we have mentioned above, not V.

We hope this step by step clarification of the branches of the trigeminal and their association with the visceral motor fibers of VII and IX will help you overcome some confusion among the welter of nerve bundles in this region.

The Posterior Aspect of the Skull (Fig. VI-27)

The skull from behind presents a convex surface with the parietal bones above; the occipital squama (interparietal part of the occipital bone) below; and the mastoid process projecting laterally as part of the temporal bone. The mandible is in place so we view it from behind. Even a few teeth in its alveolar process peek into view.

Note the position of the parietal foramina in this skull and of the lambda and lambdoid sutures. The *superior nuchal lines* of the inferior aspect of the occipital bone are unequally represented bilaterally, with the left one less well developed. They give attach-

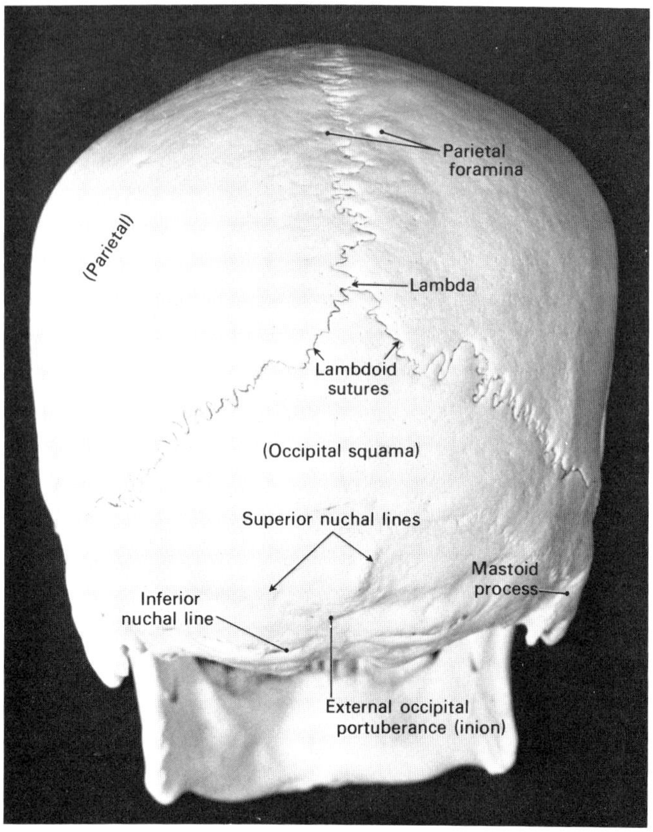

Fig. VI-27. Posterior aspect of the skull.

ment to the posterior neck muscles, and they meet in the midline at the external occipital protuberance, or inion. The *inferior nuchal lines* appear quite normal in this skull.

The Cranial Fossae

When the calvaria is removed, and when the skull is tipped in various ways (Fig. VI-28A–C), it is possible to look down into a bowl-shaped area that housed the brain and brainstem. Most of this region is made up of the bones of the endocranium (discussed above), but parts of the bones of the calvaria form its walls. This inner cranial floor can be divided into three large, depressed areas, the **anterior, middle,** and **posterior cranial fossae,** each one at a slightly lower level than the one rostral to it. The boundaries of each are seen in Figure VI-28A and B. The separate bones involved are outlined in Figure VI-28C. Many foramina perforate the cranial floor. Most of these are for the passage of the 12 cranial nerves, but some are for vessels.

Through the largest of these foramina, the *foramen magnum* (Fig. VI-28B), pass:

1. The spinal cord, joining with the brainstem.
2. The vertebral arteries that have traversed the foramina within the cervical vertebrae. They ascend through the base of the skull here to join as the single **basilar artery** on the ventral aspect of the **pons.** The latter structure is a part of the brainstem. Along with the **medulla oblongata** caudal to it and the cerebellum dorsal to it, it is part of the so-called "hindbrain."
3. The last toothed process of the **denticulate ligament,** a lateral pial anchoring of the spinal cord in the vertebral canal, barely ascends over the lip of the foramen to attach to the bone on the inner aspect of the dura.
4. The motor fibers of the **spinal portion of the spinal accessory nerve.** These leave spinal cord segments C1 to C4, traverse between the dorsal and ventral rootlets of these nerves, and utilize the posterior surface of the denticulate ligament as a route to enter the skull *via* the foramen magnum, only to leave the skull at once through the jugular foramen in company with the **vagus nerve.**
5. **Anterior and posterior spinal arteries,** arising inside the skull from the upper parts of the vertebral arteries, leave the foramen in the reverse direction to supply the spinal cord.

The Anterior Cranial Fossa (Fig. VI-29A)**:** This uppermost level of the cranial floor houses the frontal lobes of the brain. The **orbital plates** of the frontal bone roof over the orbits. Some of the **orbital gyri** of the brain leave depressions and ridges on the orbital roof. Parts of the ethmoid bone and the lesser wings and forepart of the body of the sphenoid complete the fossa.

The *frontal crest,* to which the falx cerebri attaches, is a stout midline structure anteriorly. We see it end in its blind pit, the *foramen cecum,* which, of course, is a foramen only when it is patent and able to transmit an emissary vein from the nose.

Just posterior to this pit, two thick alae of the *crista galli* converge to form a midline crest. The whole structure, alae and crest, fancifully resemble a cock's comb that projects up from the anterior cranial floor to furnish additional attachment for the falx cerebri. The crista galli is the upper terminal end of the perpendicular plate of the ethmoid. On either side of it lie the perforated *cribiform plates* of the ethmoid upon which the olfactory bulbs would rest.

The lesser wings of the sphenoid have sharp posterior margins that overhang the middle cranial fossa. The concavity formed by the greater wings of the sphenoid makes anterolateral recesses in the middle fossa to house the temporal lobes of the brain. The lesser wings end in two projections near the midline as the **anterior clinoid processes,** the anterior pommels of a fanciful Turkish saddle in which the pituitary sits, or the clinoids can be regarded as the front pair of Vesalius's bedposts, if you would rather think of this area as a bed for the pituitary.

The Middle Cranial Fossa (Fig. VI-29A–C)**:** The floor in this part of the cranium consists of the poste-

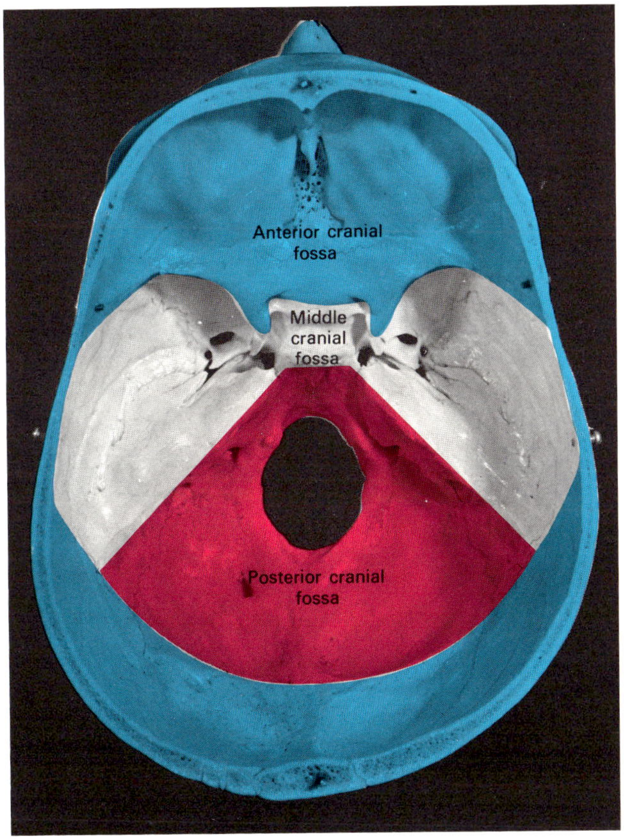

Fig. VI-28A

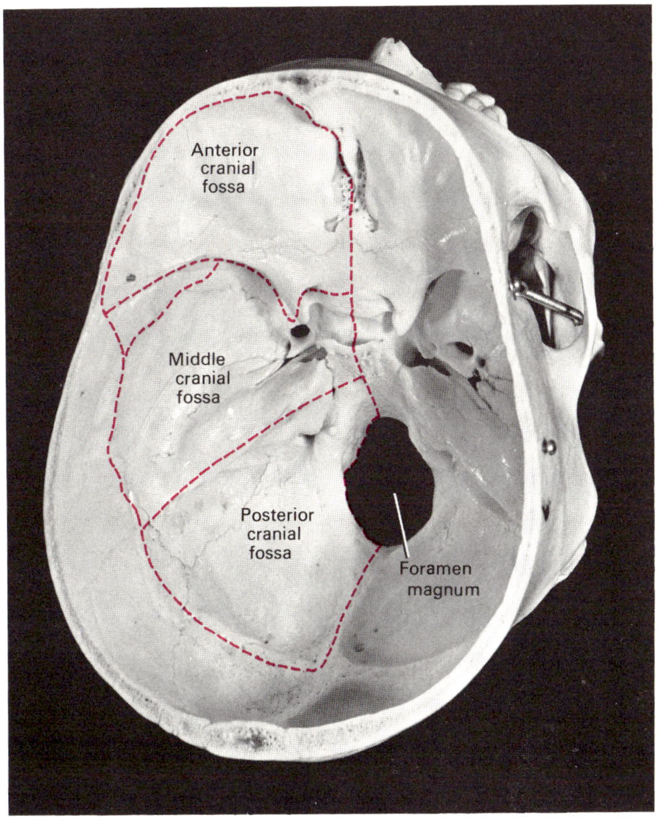

Fig. VI-28B

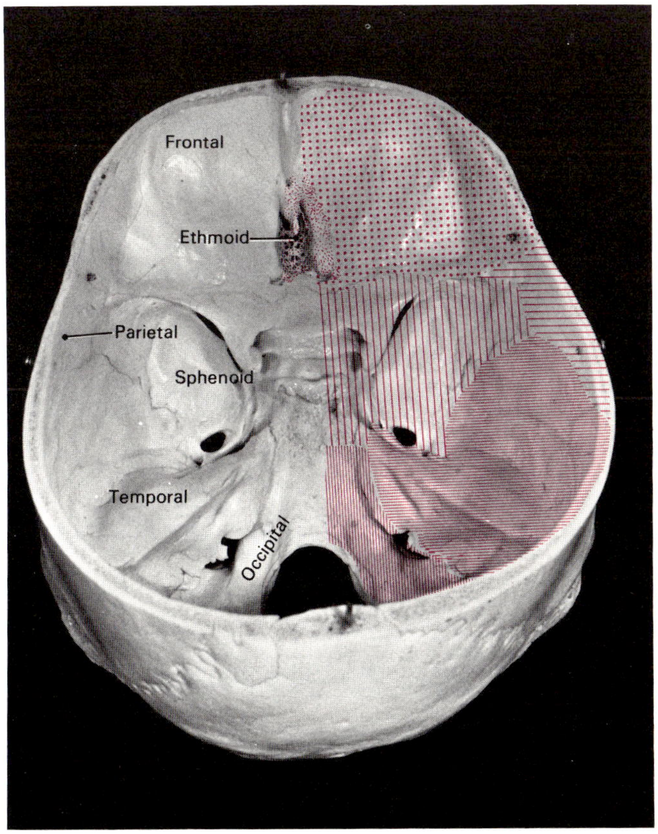

Fig. VI-28C

Fig. VI-28. The cranial fossae. **A.** Boundaries of the anterior, middle, and posterior cranial fossae. **B.** Oblique view of the cranial fossae. **C.** Bones comprising the fossae. Half of the frontal, sphenoid, and occipital bones have been shaded on the right side. All of the ethmoid, temporal, and part of the parietal bones are also shaded differentially.

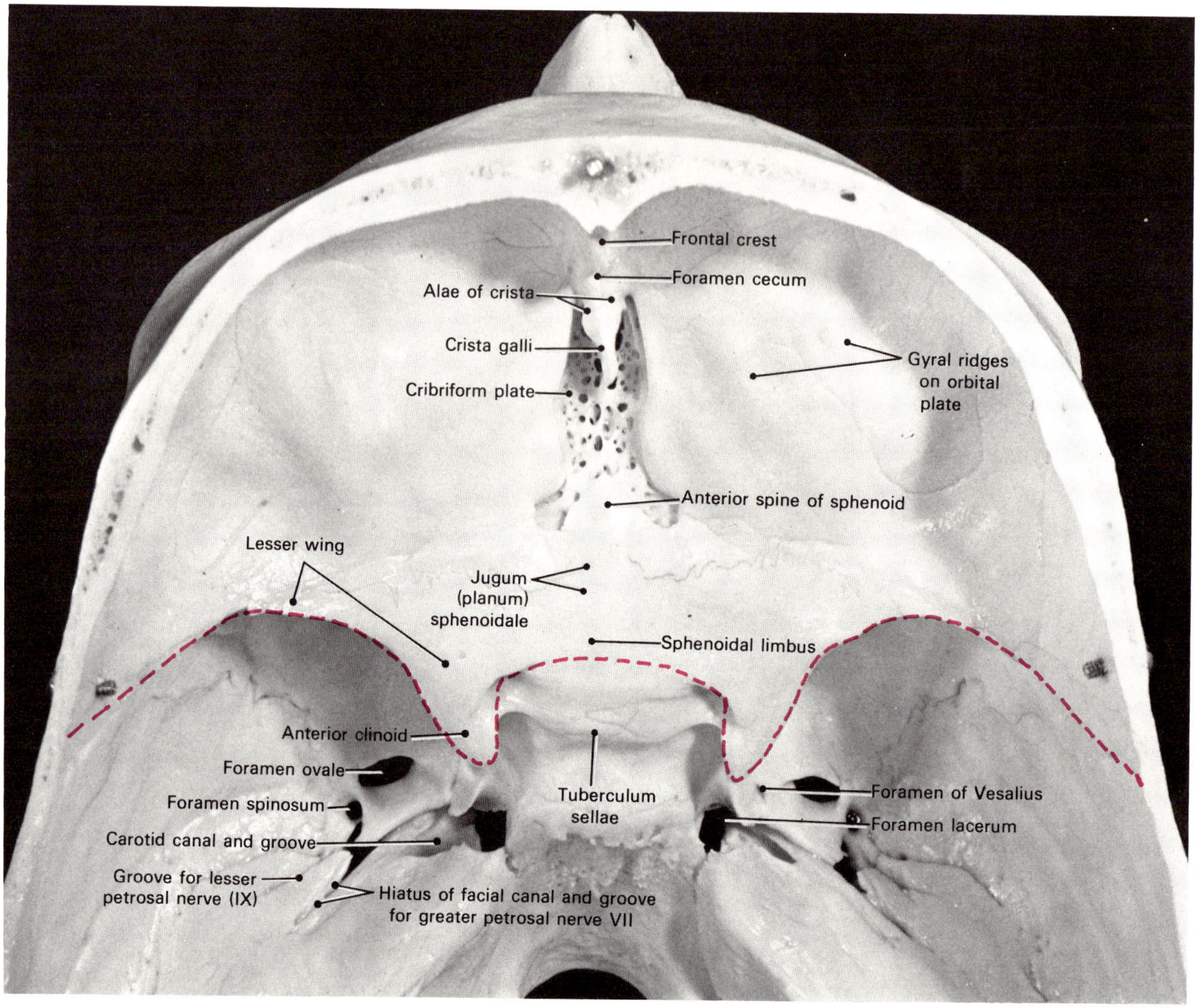

Fig. VI-29. Details of the anterior and middle cranial fossae. **A.** Division of anterior and middle cranial fossae.

rior part of the body of the sphenoid and its greater wing; the anterolateral half of the petrous part of the temporal bone, its crest forming a dividing line between middle and posterior fossae; and, finally, an antero-inferior corner of the parietal bone.

A transverse sulcus lies between the anterior clinoids just below the **sphenoidal limbus** (Fig. VI-29A and B). This is the **optic chiasmatic sulcus** in which the anterior margin of the **optic chiasma** would rest in the living head. This is where the **optic or cranial nerve II** would decussate, in part, as each nerve emerged from the orbital cavities in the **optic canals** at the extreme ends of the optic sulcus. The canals can be seen (Fig. VI-29B) just rostral to the anterior clinoids, leading away under the edges of the lesser sphenoidal wing into the orbital cavity. The **ophthalmic artery** will use the canal, along with the optic nerve. Caught up in the adventitial coat of the artery will run *sympathetic nerve fibers* for the ***Ciliary muscle of* the lens** and the **Dilator muscle of the pupil.** These fibers have traversed the vascular branchings of the **internal carotid,** joining it in the neck from the superior cervical sympathetic ganglion.

Just medial to the tip of each anterior clinoid is a deep depression, the **carotid sulcus** (Fig. VI-29B), where the internal carotid gives off the ophthalmic artery just before it makes an upward bend to send its terminal branches to the brain.

Posteromedial to the carotid sulci are the **posterior clinoids,** the much shorter horns on the back of the "Turkish saddle" (Fig. VI-29C). The depressed seat of the "saddle," between anterior and posterior clinoids, is the **sella turcica,** itself, which contains the **pituitary gland (hypophysis)** [Fig. VI-29B]. A tiny pit may exist in its depth to mark the course of the old **pharyngeal diverticulum,** the site of dimpling-in

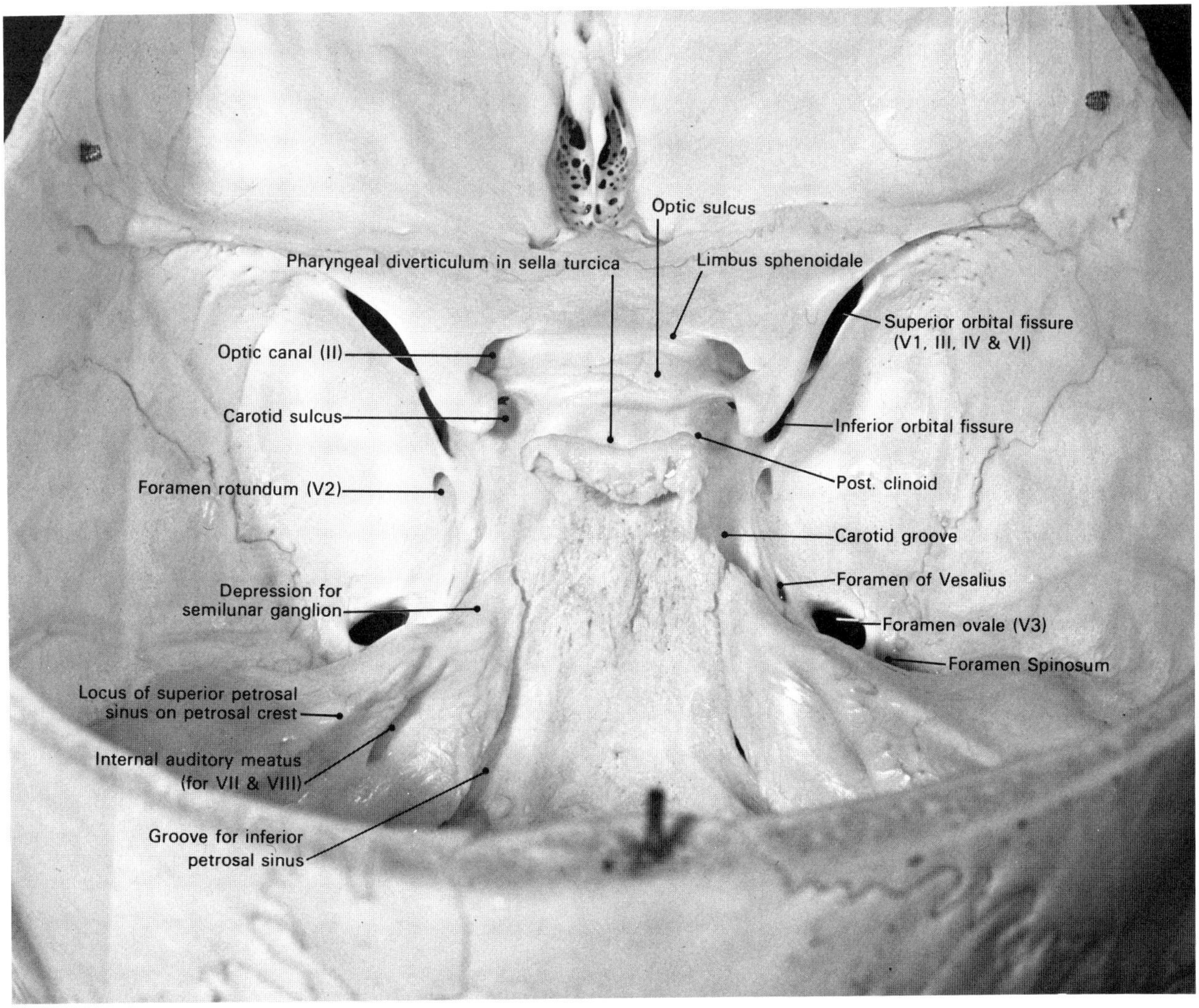

Fig. VI-29B. Features of middle fossae.

of the pharyngeal roof as it gave rise, embryologically, to the **anterior lobe of the pituitary.** The pituitary, snuggled in the sella, is roofed over by a diaphragm of dura that attaches to the four clinoid processes and to the margin of the tuberculum sellae (olivary eminence) (Fig. VI-29A).

In the margin of this diaphragm the two layers of endosteal and periosteal dura are separated by a circular sinus, the **cavernous sinus** (Figs. VI-49B; 50B). The venous channel and the gland, itself, come into close relationship anteriorly with the optic chiasma. It is easy to visualize that pituitary tumors, meningeal infections, or vascular accidents in this area would affect visual function through pressure upon, or irritation of, optic nerve fibers as well as upon the oculomotor fibers of cranial nerves III, IV, and VI which course nearby.

Inclined lateral grooves (Fig. VI-29A and B) on either side of the posterior clinoids lead *across* (not through) a ragged foramen, the **foramen lacerum,** into the **carotid canals.** By means of the carotid canal the internal carotid artery enters the skull. Since these canals move posterolaterally, we can expect them to come into relationship with the middle ear cavity. The "lacerated" foramen is in this sad shape because it is part of a cartilaginous floor of the skull which does not hold up well. While the foramen lacerum may *contain* a number of important structures in the sense that they rest upon it as they cross over it in the cranial floor, it *transmits* no important structures *through* it. Passing *through* are a meningeal twig of the ascending pharyngeal artery, some meningeal lymph vessels, and some small veins that connect the cavernous sinus with the pharyngeal plexus of veins. These structures you will probably never remember. But every student feels that some key structures *ought* to go through this big hole (Fig. VI-29C). All important structures merely *cross over* it on the way to

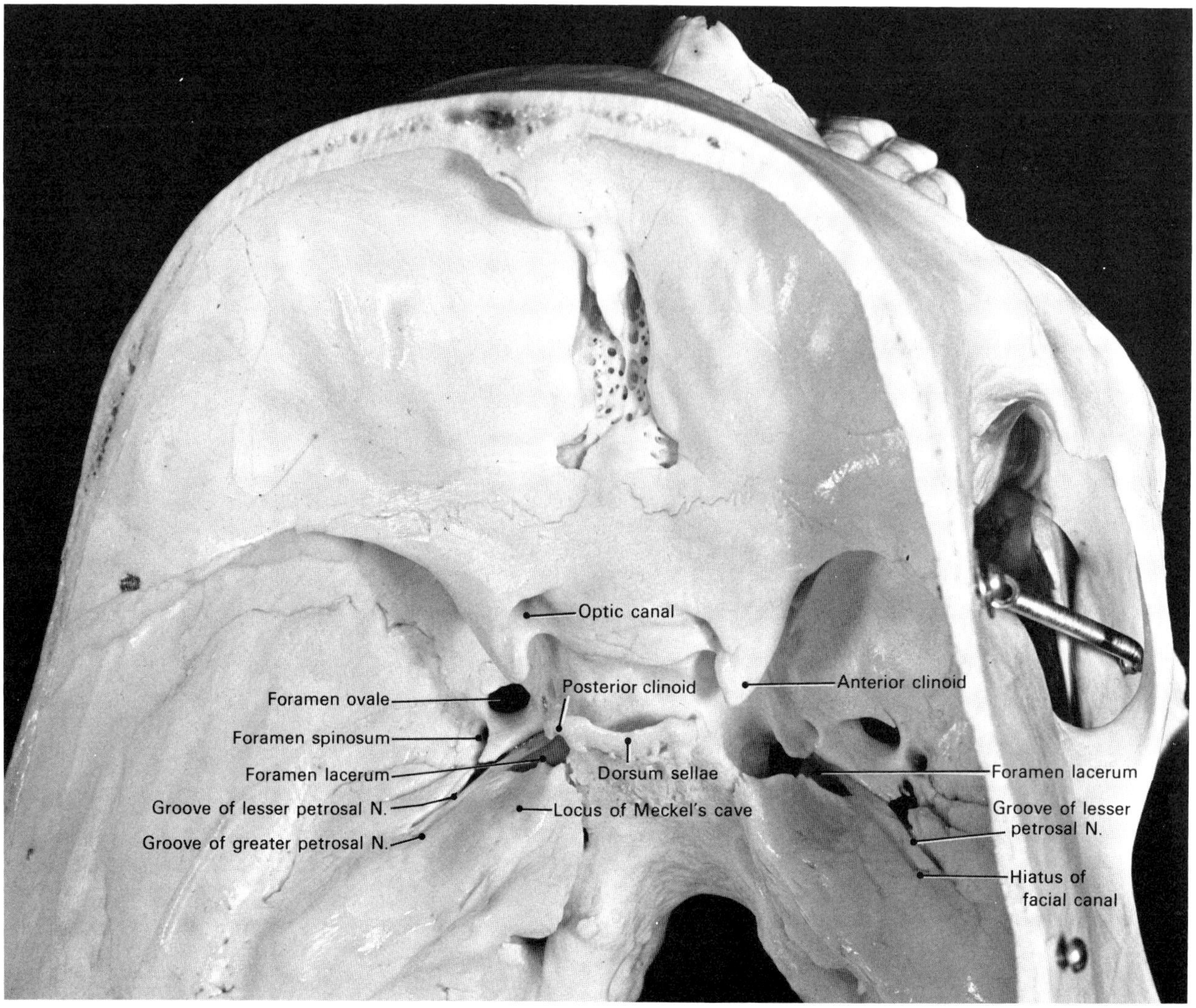

Fig. VI-29C. The middle fossa. C is the same as B with skull tipped obliquely.

someplace else. These are: (1) the **internal carotid artery;** (2) with the internal carotid **plexus of sympathetic fibers** in its adventitia; (3) the **deep petrosal nerve** (simply twigs branching off this plexus); (4) the terminal end of the **greater** (superficial) **petrosal nerve;** and (5) the beginning of the **nerve of the pterygoid** (Vidian) **canal.** The latter nerve is formed by the junction of the deep petrosal nerve (sympathetic twigs) with the superficial petrosal nerve (parasympathetic fibers of cranial nerve VII). It heads for the pterygoid canal in the sphenoid just at the rostralmost border of the foramen lacerum. None of these structures go *through* the foramen in life because cartilage fills it in for the most part, and not much of a foramen exists. There are simply chinks for the transmission of the less important items mentioned above.

If the region around the clinoid processes bears some resemblance to a Turkish saddle, then on either side of this midline zone the crescent of the Turkish flag is emblazoned on the cranial floor of the middle fossa in the form of a series of foramina. If you are not enough of an old philatelist to recognize this crescent, you will have to look at the Turkish flag in your Britannica. The foramina line up in an arc, one behind the other (Fig. VI-29B).

The first of these foramina is an elongate slit between the greater and lesser wings of the sphenoid. Through this *superior orbital fissure* a number of structures enter the orbital cavity. If the optic canal is the royal entrance to this important orbital chamber, the fissure is a side door for important personages that serve royalty. Among these peers are: (1) the oculomotor nerve (III); (2) the trochlear nerve (IV); (3) the abducens nerve (VI); (4) the ophthalmic branch of nerve V with its frontal, nasociliary, and lacrimal twigs; (5) a recurrent branch of the lacrimal artery; and, (6) the ophthalmic branch of the middle meningeal artery.

A spur of bone separates the superior orbital fissure from the *inferior orbital fissure* (Fig. VI-29B). This fissure in the *floor* of the orbit is not usually seen by looking in the floor of the skull unless the skull is tipped in just the right plane. The inferior orbital fissure can be regarded as a kind of tradesman's entrance to the orbital chamber through which personages on business elsewhere scurry. Among such are the **inferior orbital vein** and the *branch of* V_2 *that makes its way to the infraorbital foramen* on the face. Actually, neither V_2 nor this branch leave the skull through the inferior orbital fissure. They make their exit through the next posterior foramen in the crescent, the **foramen rotundum.** The maxillary division (V_2) uses this rounded exit to reach the pterygopalatine and infratemporal fossae where it gives off its branches. Its *infraorbital nerve* flits back through the inferior orbital fissure briefly simply because the infra-orbital foramen is in direct line beyond the fissure.

The next foramen along this crescenteric arc is a tiny and inconstant one. It is the foramen of Vesalius which conducts an emissary vein. It can be dismissed from your mind as unimportant. We point it out simply because it shows up in the skull illustrated in Figure VI-29A and B. We would not want you to think you had made an important discovery should you come across it in your specimen, for Vesalius spotted it long ago.

The next large, oval foramen beyond, along the lower curve of the crescent (Fig. VI-29A–C), ***is*** important. Here, in this **foramen ovale,** V_3 drops down into the infratemporal fossa where we have seen most of its branches already. The parasympathetic **lesser petrosal nerve of cranial nerve IX** may often reach the fossa through this same foramen on its way to its terminal **otic ganglion** which lies just below the foramen.

Thus, we see that the three divisions of the trigeminal exit through three separate holes in the middle cranial fossa: V_1 (ophthalmic division) uses the superior orbital fissure; V_2 (maxillary division), the foramen rotundum; and V_3 (mandibular division), the foramen ovale.

The last important hole in the posterior tail of the crescent is the *foramen spinosum.* It is tiny and is best seen in Figure VI-29A. It transmits the middle meningeal artery and vein.

A very inconsistent hole sometimes appears between the foramen ovale and spinosum as the **innominate** ("no name") **canal.** When present, the lesser petrosal branch of IX prefers to use this route to the infratemporal fossa over that of the foramen ovale. Sometimes this nerve passes through a chink in the petrosquamous suture line. Because of these variations, it is easiest to think of the lesser petrosal as sharing the foramen ovale most of the time.

At the apex of the petrous bone, anteriorly, is a fingertip depression (Fig. VI-29B). Snuggled in it lies the **semilunar ganglion of the trigeminal nerve.** This sensory root ganglion, like the pituitary in its sella, is roofed over by dura which bridges from the petrous to sphenoid bone. So the ganglion hides in a dural cave, or **Meckel's cave.** The details of the ganglion and of the trigeminal nerve were first written up in 1748 by Johann Frederich Meckel, Senior, the first of three famous German anatomists. All were named Johann Frederich. The second did not remain in Anatomy long, but moved from the University of Berlin to the University of Halle to become a surgeon. We hear little of his accomplishments. It is Johann Frederich, III, grandson of Meckel, Senior, that we associate with the first branchial arch cartilage and the diverticulum of the ileum (1781–1833). Some anatomists, as you must have begun to suspect, also carry all sorts of historical trivia in their bag of spare bolts and washers in the hope that a little of it may wash off on a few students.

The Posterior Cranial Fossa (Fig. VI-30A and B)**:** The bones involved are the occipital and the parts of the petrous bone posteromedial to its crest. The anterior limits are nebulous and, apparently, arbitrary. Some authors include as an anterior boundary the *dorsum sellae* (the part of the sphenoid between the posterior clinoids) and the *clivus* (the body of the sphenoid below the posterior clinoids to its junction with the occipital bone (basilar part)). We are excluding the sphenoid entirely, limiting it to the middle fossa. A clearer division is achieved this way. Earlier we may have led you to believe that the entire incline in the midline of this fossa, upon which the brainstem rests, was the clivus. Strictly speaking, the *clivus* is only the sharply inclined part of the sphenoid that fuses with the basilar part of the occipital bone. Again, the use of this term by clinicians and authors of anatomy texts is somewhat loose and nebulous and, on the whole, unimportant. It is a localizing or "zeroing-in" kind of term. The broad, shallow grooves of the transverse sinus (Fig. VI-30B) limit the fossa posteriorly and superiorly. Thus, in the fossa are contained, chiefly, the cerebellum, medulla, and pons, all parts of the hindbrain. It is roofed over by a dural fold placed transversely, the **tentorium cerebelli** (Fig. VI-49B).

The foramen magnum is the lowest point of the fossa, narrower anteriorly in the condylar region than posteriorly (Fig. VI-30 A and B). As one moves up and laterad out of its depth, the **hypoglossal canal** for cranial nerve XII is encountered. The canal pierces the **jugular tuberculum,** a stout bar of bone which IX, X, and XI must cross on their way to the irregularly-shaped *jugular foramen.* Higher and more laterally still, in the medial wall of the petrous bone, is the *internal auditory meatus* (Fig. VI-30A). The facial nerve (VII) enters it, and the auditory nerve with its vestibular component (VIII) emerges from it.

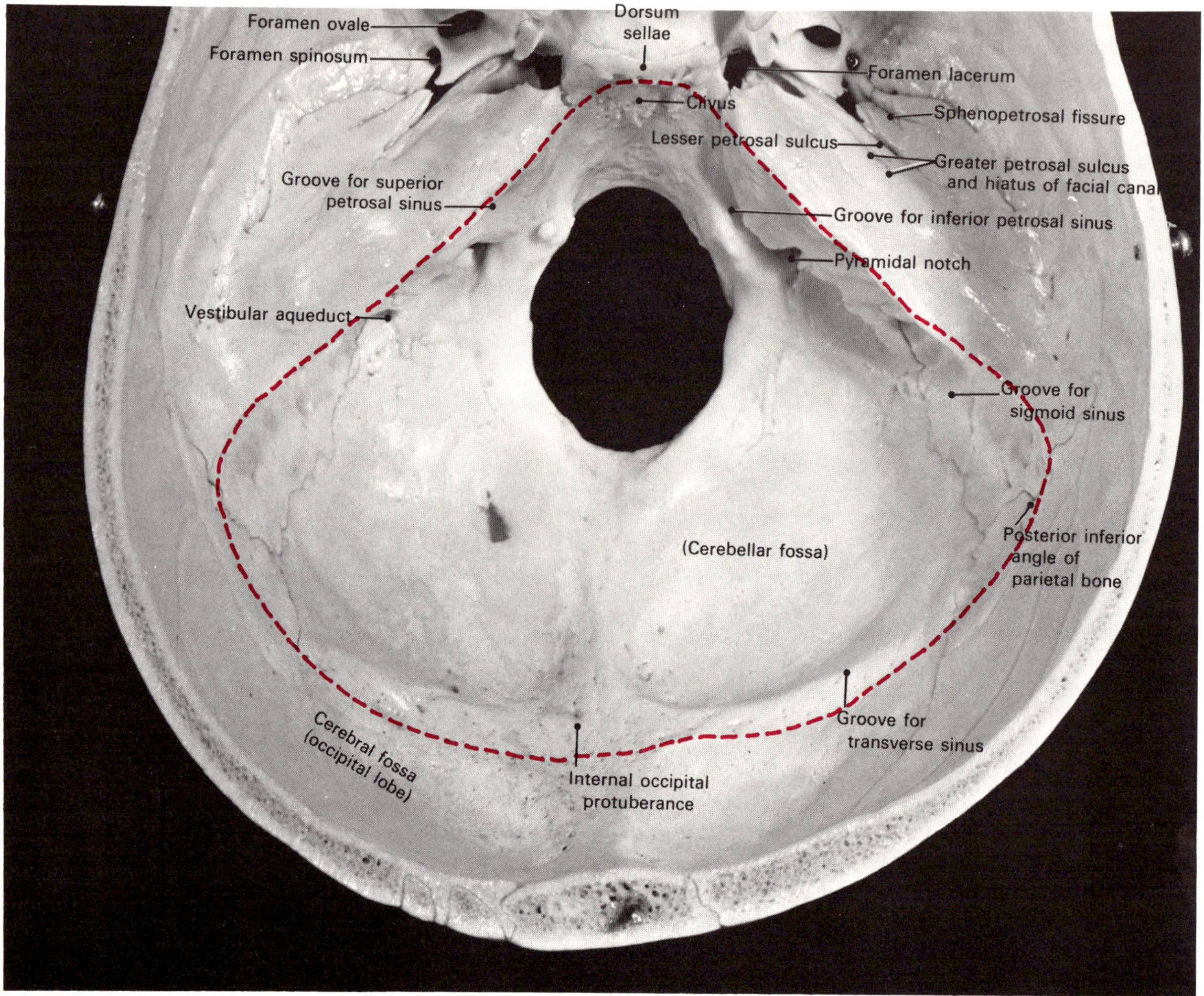

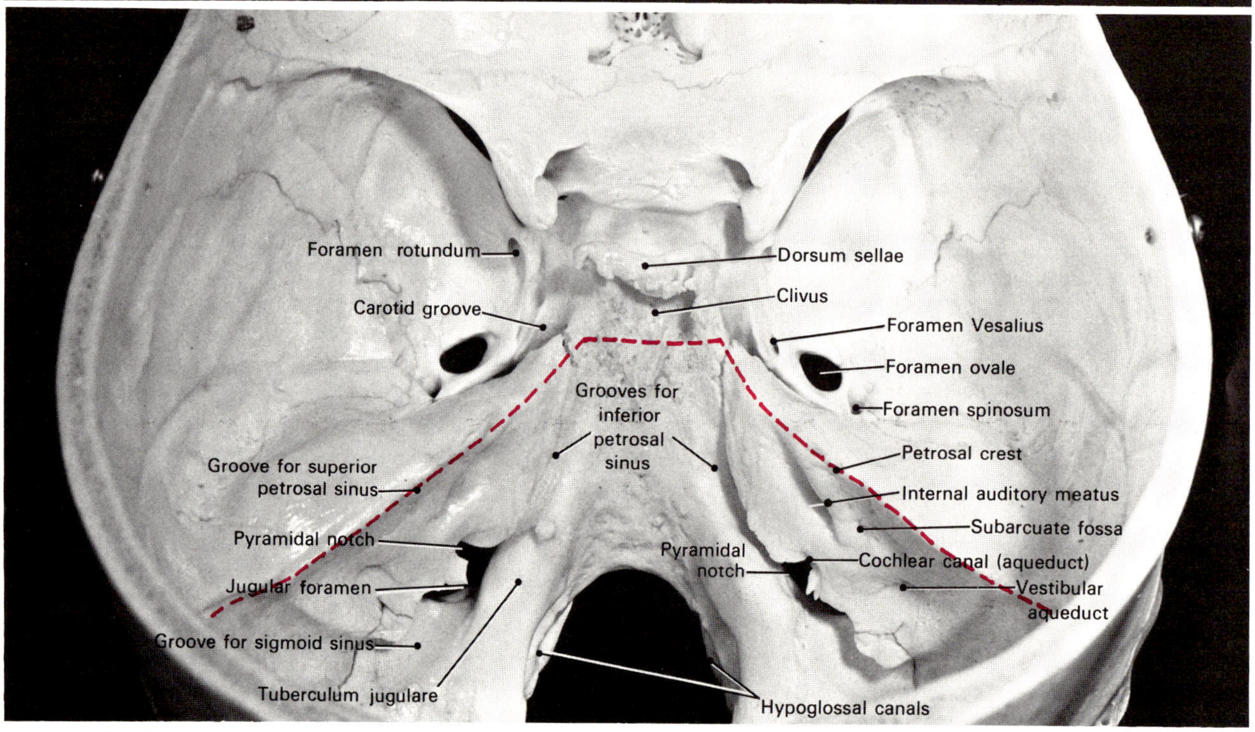

Fig. VI-30 A. Details of the posterior cranial fossa. **B.** An enlargement of **A** with skull tipped upward slightly.

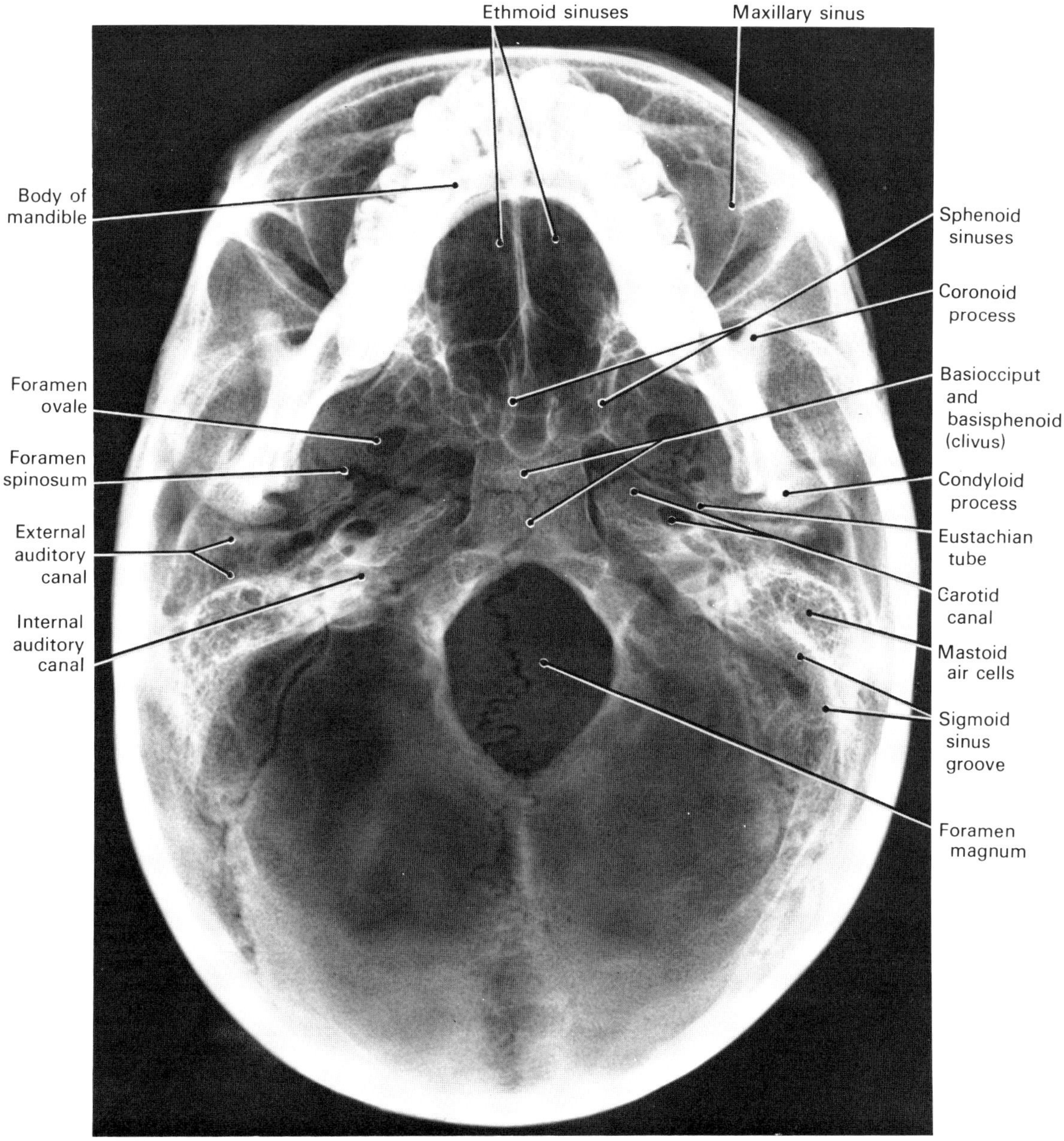

Fig. VI-31. Basal (submentovertex) projection of the dry skull. Note particularly the position of the external and internal auditory canals. The positioning for this radiograph is similar to but not identical with the photograph of the dry skull in Figure VI-32A. The basal projection is discussed in greater detail on p. 282.

The latter dual nerve could be called the stato-acoustic nerve since it has to do with hearing as well as position sense and balance. The internal auditory meatus lies almost in line with the external auditory meatus. This alignment is well demonstrated in a basal radiograph (Fig. VI-31). As a result, all of the last six cranial nerves (VII–XII) come off the brainstem to enter or leave the skull at or behind the line that connects these two auditory openings.

The jugular foramen, between the petrous, temporal, and occipital bones, lies between two channels that descend toward it from in front and behind. Descending to it anteriorly is a groove which follows the anterior tip of the wedge-shaped petrous bone along the petro-occipital suture line. The **inferior petrosal sinus** will follow this course in the dura to reach the internal jugular vein. *It is the only dural sinus to enter the vein directly.* All the rest have to reach the transverse sinus or meet at the confluence of sinuses to be conveyed to the vein *via* the **sigmoid sinus.** It is the broad groove of the latter that we see coursing toward the jugular foramen posteriorly (Fig. VI-30A). We can regard the transverse and sigmoid sinuses as constituting a very large channel, the **lateral sinus.** If we do, the groove for the lateral sinus begins at the internal occipital protuberance (Fig. VI-30B) and runs horizontally laterad on the occipital squama to the postero-inferior angle of the parietal bone. Here it turns medialward in a downward curve between the mastoid and petrous parts of the temporal bone and comes once more in relationship with occipital bone at its jugular process to descend in front of it to the jugu-

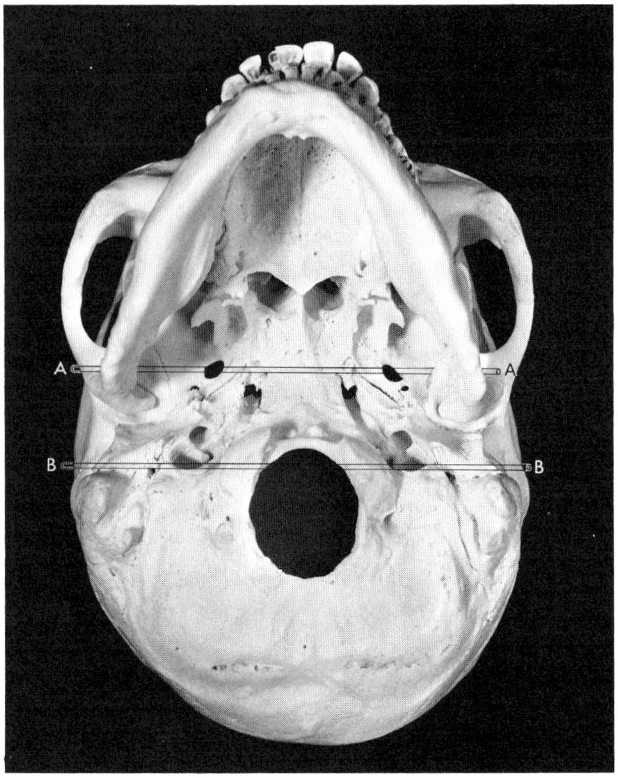

Fig. VI-32. The base or inferior surface of the skull. **A.** Divisions into anterior, middle, and posterior parts by lines shown.

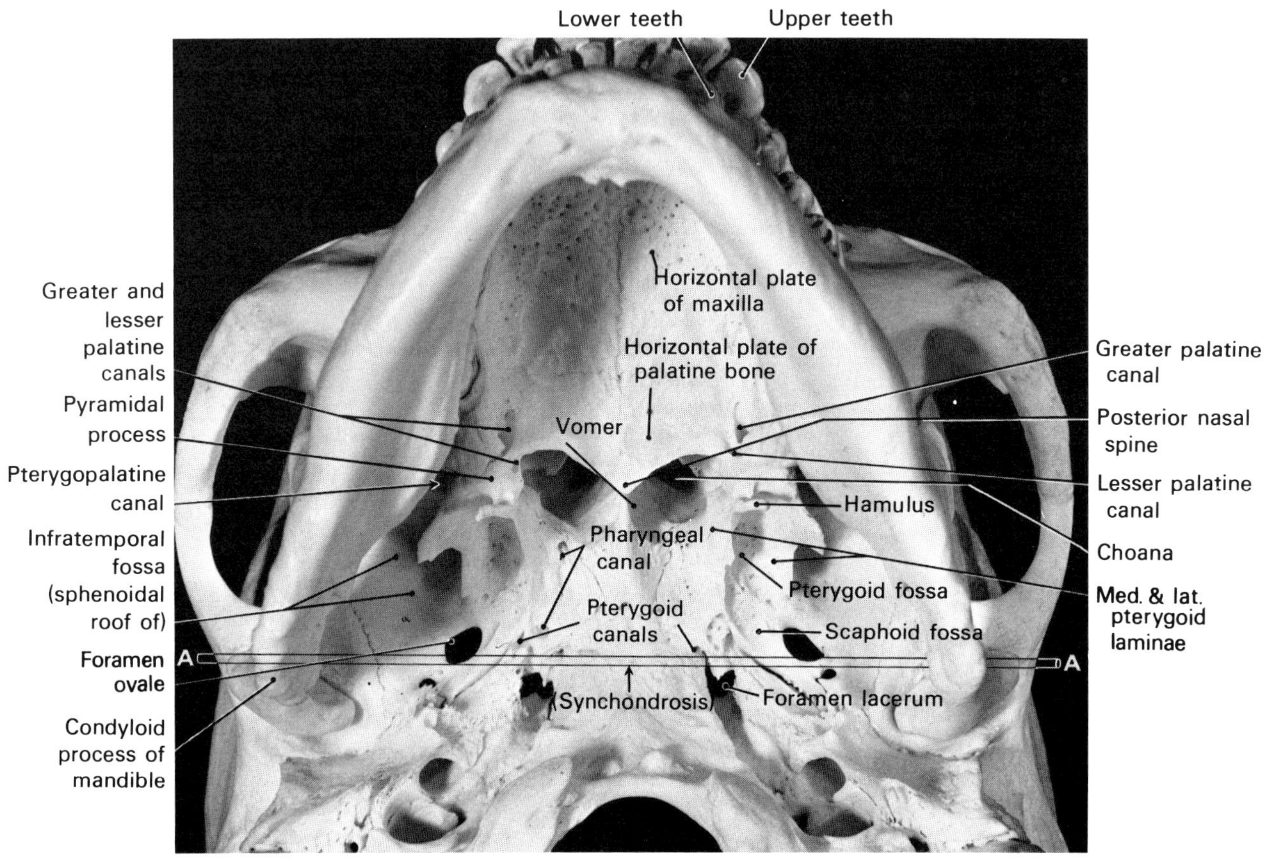

Fig. VI-32B. Details of anterior part.

Spine of sphenoid
Orifice of auditory tube
External auditory meatus

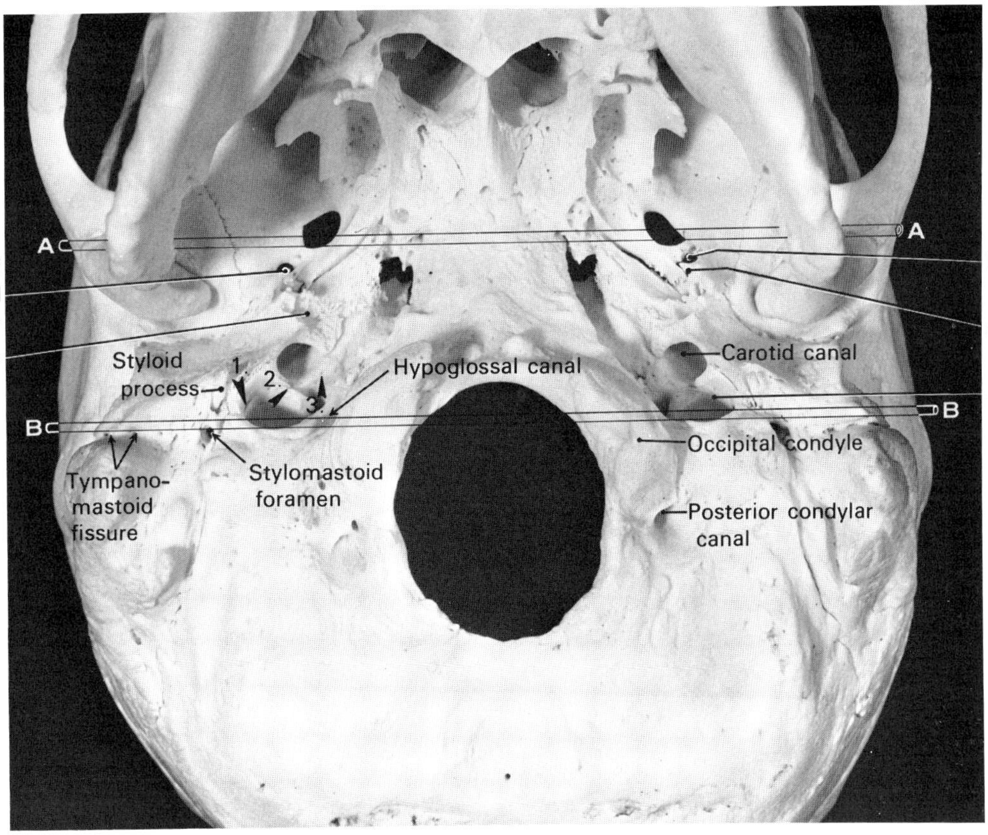

Fig. VI-32C. Details of middle and posterior part. In C, 1 marks mastoid canaliculus or Arnold's canal; 2 marks tympanic canaliculus or Jacobson's canal; 3 marks locus of caroticotympanic canals.

lar foramen. As this venous return from the brain *via* the various dural sinuses passes through the jugular foramen, it finds itself in the internal jugular vein on its way to the superior vena cava and the right heart chambers.

The whole posterolateral surface of the petrous bone is flat. The bone, itself, is triangular by virtue of its being bordered by grooves for three dural sinuses. We have mentioned the inferior petrosal and sigmoid sinuses (lower portion of the lateral sinus). Along the petrous crest runs the **superior petrosal sinus** marking a groove (Fig. VI-30) as it courses to a point of union posteriorly with the sigmoid sinus. Posterosuperior to the internal auditory meatus is a small depression in the petrous bone. It is all that is left of a once-deeper depression over which, at birth, the superior semicircular canal arched. Therefore, it is called the **subarcuate fossa** (Fig. VI-30A). Posterior and inferior to it is a tiny slit, an opening of the **vestibular aqueduct** (Fig. VI-30B) which houses the **endolymphatic sac,** a storage reservoir of endolymph for the membranous labyrinths of the inner ear buried within the petrous bone. Below the meatus is a prominent **pyramidal notch** in the petrous bone. The base of the notch opens into the jugular foramen. Here the **glossopharyngeal nerve (IX)** makes its exit through the foramen, and here will lie the **superior sensory ganglion** of this nerve. Its **petrous sensory ganglion** lies below the foramen. At the apex of the notch is a tiny opening, the **aqueduct of the cochlea, or cochlear canal** (Fig. VI-30A). It has been held by some that the cerebrospinal fluid in the subarachnoid cuff around cranial nerve IX enters the osseous labyrinths of the inner ear through this canal, and that perilymph in these channels and cerebrospinal fluid are identical on this account. Others claim that a membrane blocks the orifice of the aqueduct. This is based on experiments showing that mercury does not leave the osseous labyrinth under pressure through this tiny hole. One wonders why such a heavy fluid metal was used to proove the point in the first place. Moreover, a Wood's metal cast of the osseous labyrinthus chambers shows no communication through this hole with the posterior fossa. This, in itself, also seems to be weak proof, since the injection mass soon solidifies and adheres to the bony walls. It can only be stated that not too much is really known about the function of this pinpoint cochlear canal to date. It either has no function, or it may be important. Contrasting points of view about it run from the 1800's up to the present time.[4-10]

The Inferior Aspect of the Skull (Figs. VI-32–34)

By far the most difficult area of the skull to comprehend is its inferior or basal surface. Some kind of an arbitrary division of it into sections has to be made to gain some reasonable comprehension. If two imagi-

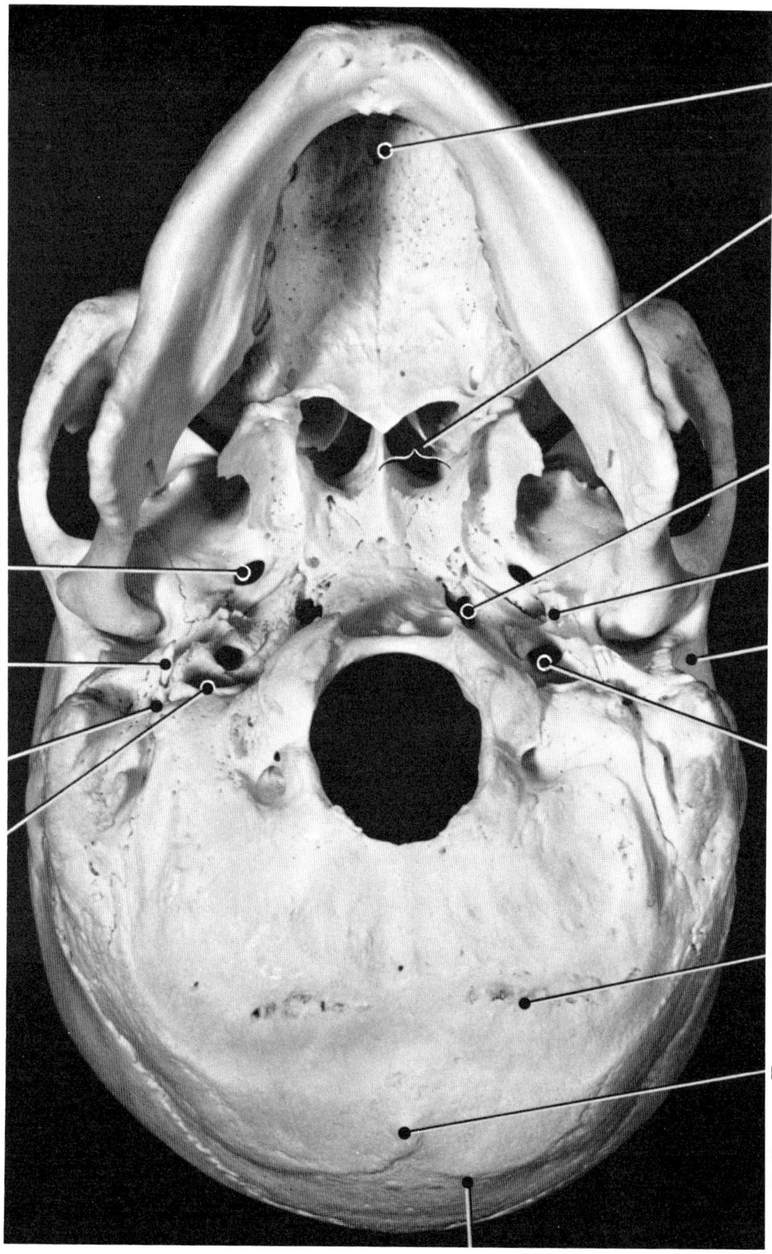

Fig. VI-33. Base of the skull tipped more anteriorly than in Figure VI-32. **A.** Full view.

nary lines could be made to cross most of the foramina at the base, or at least come close to doing so, it would be possible to consider an anterior, middle, and posterior portion of the base (Fig. VI-32A).

The Anterior Area: If a waxed string or a long pipe cleaner is pushed through the mandibular arch so that it rests close to the neck of the condyloid process, the anterior imaginary line (Fig. VI-32B, plane A-A) *should come close* to cutting across the following (but does not actually do so in the figure) from lateral to medial:

1. The posterior part of the foramen ovale seen from below in the infratemporal fossa.
2. The foramen lacerum, or close to it.
3. And the junction where the basi-occiput joins the basi-sphenoid at its point of fusion as a synchondrosis. We have indicated previously that this fusion should be complete by the 21st year.

The anterior area lies in front of this line. How well all of the structures will be seen here depends upon how much the skull is tipped. This is why we have added Figures VI-33A–C, and 34. You will note, for example, that in Figure VI-32C the foramen spinosum

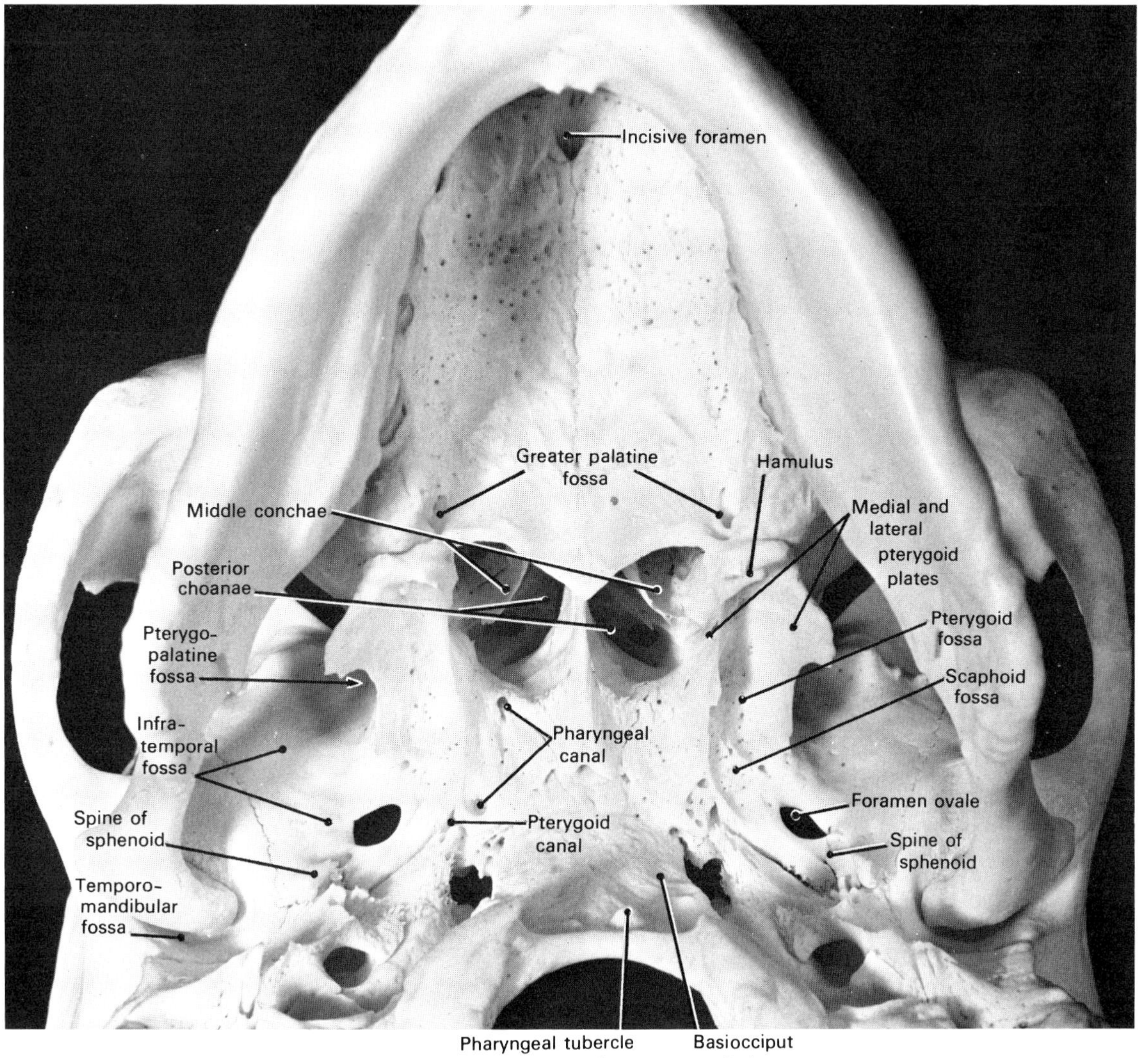

Fig. VI-33B. Details of the anterior division.

is nicely in view behind the foramen ovale; in Figure VI-33, it disappears; whereas in Figure VI-34, almost all of the foramen ovale has been lost from sight. But other structures that were not seen come into view, and others that were in view in the first figure are gradually seen in better perspective in the succeeding ones. Even if these illustrations duplicate somewhat, they should all be studied carefully.

It is obvious that the anterior boundary of the bony, hard palate stops at the upper teeth (Fig. VI-34). Its posterior border is free, scalloped, and has a midline point, the *posterior nasal spine*. Cross-shaped suture lines separate the palate into unequal quadrants. The larger anterior quadrants are the **horizontal palatine processes of the maxilla;** the smaller posterior ones, meeting in the spine, are the **horizontal plates of the palatine bones,** proper. Anteriorly (Figs. VI-33, 34), the long vertical limb of the cross ends in the *incisive foramen*. Here the midline suture diverges to run to the lateral margin of each incisor tooth, enclosing a triangular piece of bone, the **premaxilla** (incisive bone).

In the posterior part of the palate, opposite the third molar, is the **greater palatine canal** (Fig. VI-32B). Notice that in this region the horizontal bar of the cross-shaped suture line does not cut all the way to the teeth but stops short so that the canal lies between the palatine bone medially, and the alveolar process of the maxilla laterally. A **pyramidal process** of the **palatine bone** projects still farther posteriorly and laterad to contact the pterygoid laminae of the sphenoid. This process bears the **lesser palatine canals.** Below the pyramidal process in Figure VI-32B, the *hamulus* of the medial pterygoid lamina protrudes almost directly laterad as a short, stout rod of bone.

On either side of the posterior nasal spine we can peer into the paired, wide, posterior, nasal apertures or **choanae.** The roof over each is the palatine bone; the

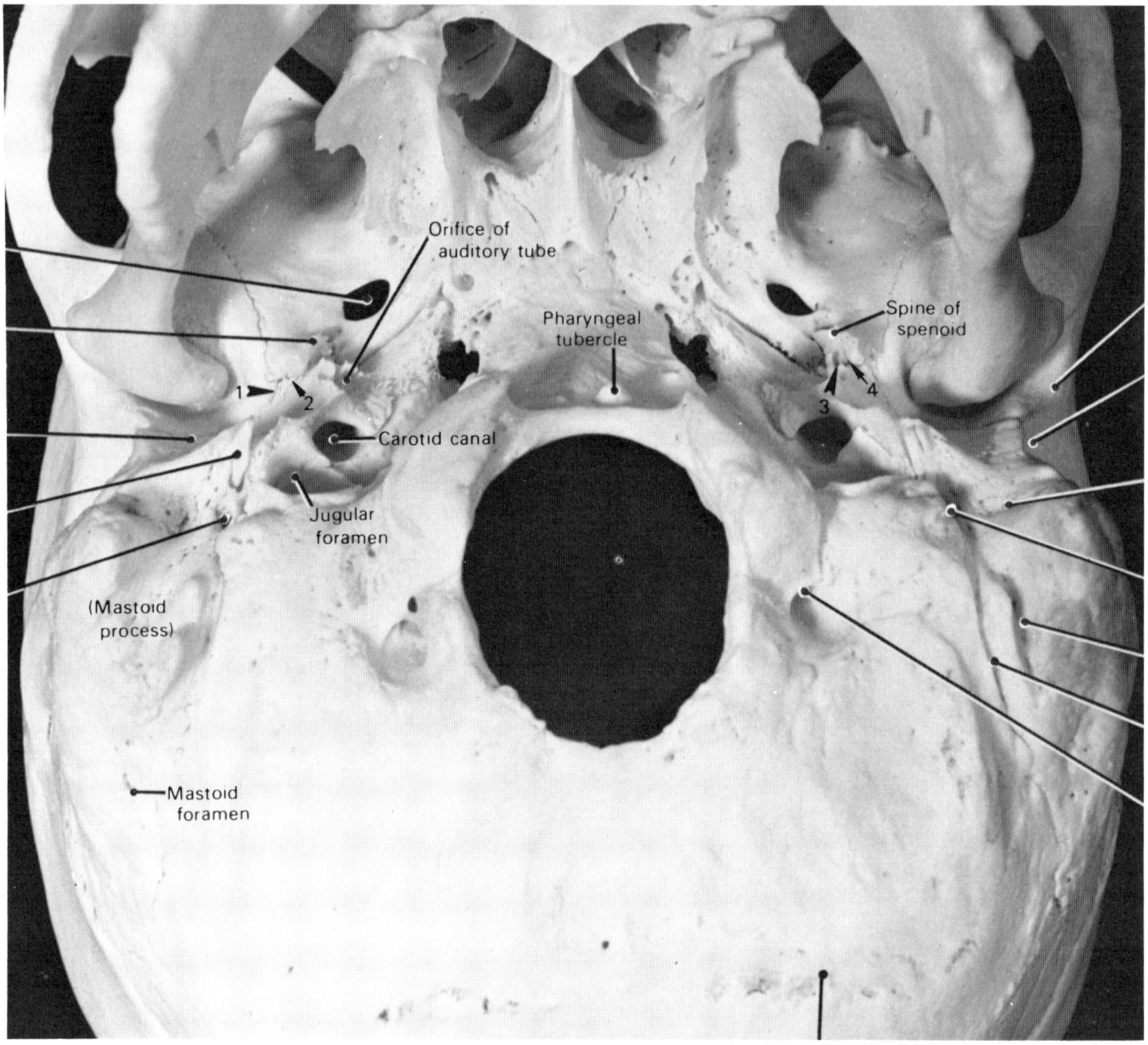

Fig. VI-33C. Details of the middle and posterior divisions. Small arrows on left side indicate: (1) continuation of squamo-tympanic suture; (2) pterygotympanic suture. On right side arrows indicate: (3) fissure for anterior temporal artery and (4) foramina for emergence of chorda tympani in infratemporal fossa.

vomer is the floor and part of their medial wall; and the medial pterygoid lamina furnishes the lateral wall (that is, with the skull viewed in the plane of Fig. VI-32B).

Posterolateral to the choanae are the *medial* and *lateral pterygoid plates* with the *pterygoid fossa* between the plates. As the edge of the medial plate is traced posterolaterally, a canoe-shaped depression is seen, the **scaphoid fossa** (Fig. VI-32B).

Lateral to the scaphoid fossa is the foramen ovale; medial to it is the foramen lacerum. Between the foramen lacerum and the scaphoid fossa, just at the terminal edge of the medial pterygoid lamina, is a tiny opening into the **pterygoid (Vidian) canal** (Fig. VI-32B and 33B) for the joined deep and greater superficial petrosal nerves. This canal leads forward through bone to open anteriorly, hidden from view here by the pterygoid laminae and fossa, into the **pterygopalatine** (sphenopalatine) **fossa.** At its anterior exit, a parasympathetic visceral motor ganglion of VII will lie in the fossa. Fibers from this spheno- or pterygopalatine ganglion serve nasal, palatine, and lacrimal glands.

Just above the posterior entrance of the *pterygoid* (*Vidian*) *canal,* in the vaginal part of the sphenoid bone, lies the **pharyngeal (palatovaginal) canal** (Fig. VI-32B). The term "vagina," whenever it is used, signifies "a sheath," and, in this sense, a bony one for an artery and nerves. This particular canal is fragile and often broken into in frequently-handled skulls. It transmits a branch of the maxillary artery and fibers from the **pterygopalatine ganglion.**

The Middle Region: If we lay a long pipe cleaner across the base of the skull as shown in plane B-B′ of Figures VI-32A and C, the middle region becomes the narrow zone between the first and this second imaginary line. Our new, posterior, transverse line will come close to cutting across the following, depending upon minor variabilities in skull anatomy:

1. The **tympanomastoid fissure** between the tip of the mastoid and the external auditory meatus.
2. The stylomastoid foramen between mastoid and styloid processes.
3. Part of the jugular foramen.
4. The hypoglossal canal.
5. The occipital condyle, such that about a half of it lies posterior to the line.
6. The anterior portion of the foramen magnum.

Again, you see, we are treading upon very "holey ground" in this region of the skull.

If you hold the skull in the basal view so that the maxilla points rostrad, and if you tilt it downward and away from you, you are able to look deeper still into the nasal choanae (Fig. VI-33A and B). The pterygoid plates and fossae stand out in relief, but the *spine of the sphenoid* hides the foramen spinosum from view. The basi-occiput is broad posteriorly, and the pharyngeal tubercle is prominent.

Most of the structures that comprise the bulk of the middle region are seen best in Figure VI-33C. However tiny and seemingly insignificant the spine of the sphenoid may appear to be, it is the major orienting key to the region. The foramen spinosum lies at its anterior base (hidden from view unless the anterior part of the skull is tipped upward again). But, farther rostrad, the foramen ovale remains in clear view. Directly laterad of the spine is the temporomandibular fossa (Fig. VI-33B). Posteromedially, the *bony orifice* of the **auditory tube (tympanic or Eustachian canal)** can be seen (Figs. VI-32C and 33C). The auditory canal is continued in an anteromedial direction, almost to the foramen lacerum, as a cartilaginous tube, most of which does not preserve well in the dry skull. Here its *mucosal-lined orifice* opens into the posterolateral wall of the nasopharynx lying above the level of the soft palate. The auditory tube or canal connects the nasopharynx with the middle ear cavity. When the auditory tube constricts during rapid ascents and descents in planes, it is difficult to regulate air pressure on both sides of the ear drum. We try, sometimes with little success, to overcome pain and discomfort and to equalize pressure by forcing the auditory canal open, either through swallowing, yawning, or blowing through pinched nostrils.

Against the *lateral* side of the spine of the sphenoid lies the **auriculotemporal branch of V_3;** against its *medial* aspect lies the *chorda tympani branch of VII.* A blow to the side of the jaw could drive the mandibular condyle medialward to fracture the spine of the sphenoid and to injure these two nerves. A unilateral functional deficit, as a result, is of minor consequence. The greater danger would be rupture of the middle meningeal vessels coming through the foramen spinosum. But, it is interesting to speculate upon the neural deficits, since they review the composition of these nerve bundles. The auriculotemporal branch of V_3 carries sensory fibers from the skin of the temple and conveys visceral motor fibers of IX to the parotid gland. The chorda tympani conveys visceral motor fibers of VII to sublingual and submaxillary glands (*via* lingual branch of V_3), and it also conducts taste sensations from the anterior two-thirds of the tongue. While one might expect a dry mouth and some loss in taste sensation as a result of damage to these nerves, only one side of the tongue would be insensitive and only the glands on the same side of the face would be involved. With function intact on the non-injured side, the patient might not notice much of a loss of function in the oral cavity at all. He might be much more likely to complain of a lack of pain and touch sensation from the skin over the temple on the injured side because of a feeling of dullness or numbness in this area (contribution of auriculotemporal branch of V_3).

If we examine, next, the bar of bone between jugular foramen and carotid canal (left side of Fig. VI-32C) it is possible to locate some tiny foramina here. The first of these, marked "1," comes to view just inside the jugular foramen. This is the **mastoid canaliculus,** frequently referred to as Arnold's canal. As the vagus (cranial nerve X) leaves the jugular foramen, it sends a recurrent twig back into the canal. The twig escapes from the skull again in the *tympanomastoid fissure.* It becomes the *auricular branch of the vagus* to supply a patch of skin behind the ear.

A second tiny foramen is found ***on*** the ridge between jugular foramen and carotid canal (marked "2" in Fig. VI-32C). This has been confusingly named the **tympanic canaliculus.** The main auditory tube is often called the **tympanic canal.** Neither of these structures have much in common. Here is one case then where an old eponym for the tiny canaliculus may save confusion for the tympanic canaliculus was earlier known as Jacobson's canal. We can say, then, that as cranial nerve IX (glossopharyngeal) emerges in the jugular foramen, it, too, gives off a recurrent branch, and a far more important one than the vagus gives off. The recurrent twig enters Jacobson's canal to reach the middle ear cavity where it breaks up into a plexus upon the medial wall of this cavity. It is from this plexus that the **lesser superficial petrosal nerve** (of IX) arises.

On the rim of the carotid canal (marked "3" in Fig. VI-32C) are one or two other very tiny foramina. These are the caroticotympanic canals to carry vascular twigs into the middle ear cavity. Of all three of

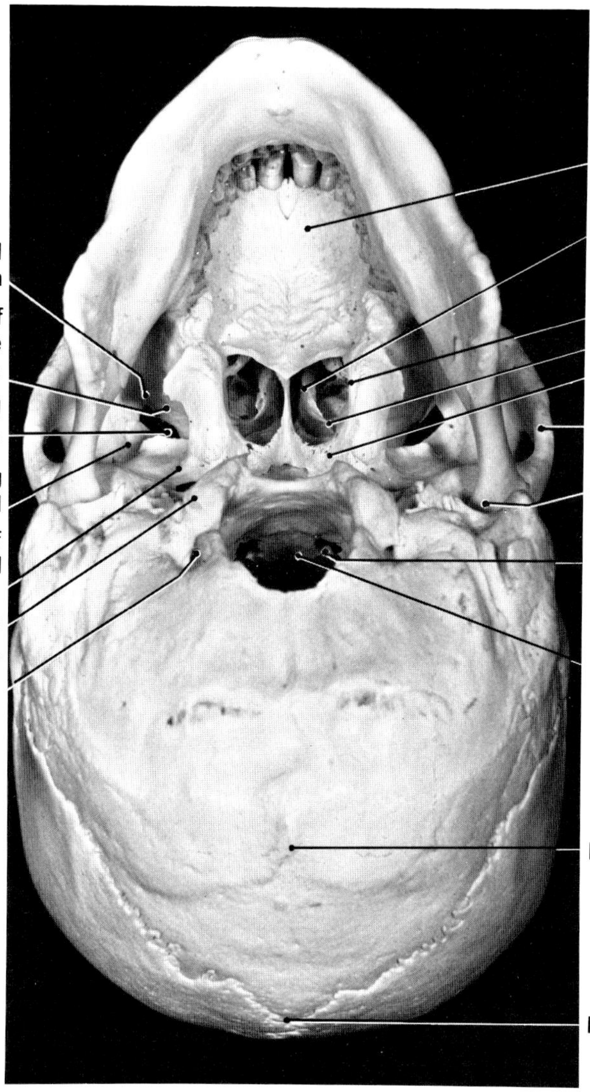

Fig. VI-34. Base of the skull tipped farther anteriorly than in Figures VI-32 and 33.

these minute foramina probably Jacobson's canal (tympanic canaliculus) is the only one worth remembering.

Figure VI-33C makes it clear that the condylar process of the mandible is restrained by a vertical wall or plate of bone posteriorly. This back wall of the temporomandibular fossa is the **tympanic plate** which is not too strong. A blow to the chin could fracture it, also. The suture line along the anterior border of this plate is the **squamotympanic suture.** As this suture moves medially, it forks into two tiny sutures, the anterior one (marked "1" in Fig. VI-33C) is a continuation of the **squamotympanic suture;** the posterior fork (marked "2") is the **petrotympanic suture.** We see it continue as a groove between the spine of the sphenoid and the auditory tube (on the left side of the figure). At the terminal end of the petrosquamous suture or fissure lie two tiny foramina (marked "3" and "4" on the right side of Fig. VI-33C). From the lateral of these foramina emerges the chorda tympani branch of VII from its **iter** or canal in the petrous bone ("4" in Fig. VI-33C). It marks a groove on the medial aspect of the spine of the sphenoid. The medial foramen in the petrotympanic fissure transmits the anterior tympanic artery (marked "3") which will enter the middle ear cavity. The vascular twig is a branch of the maxillary artery.

If the orifice of the auditory tube is located once more, the rounded opening of the *carotid canal* lies posterior to it (Fig. VI-33C). At this point on the

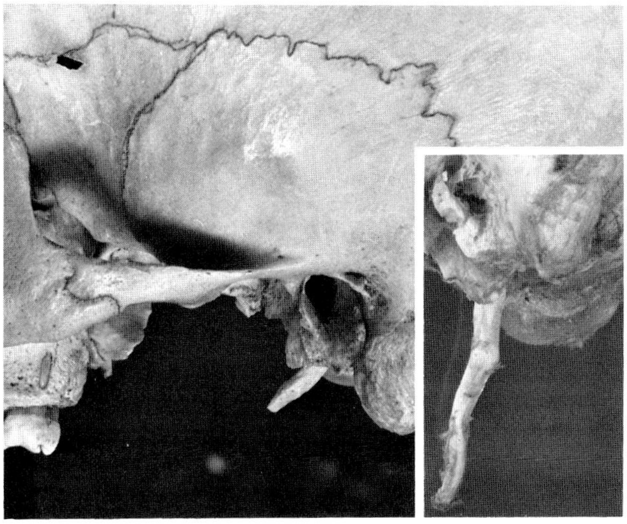

Fig. VI-35. The styloid process. **A.** Lateral view of the normal process. **B.** Abnormally long styloid process (inset).

basal surface, the internal carotid artery enters the skull. Only a narrow partition of bone separates this canal from the jugular foramen next posterior. Medial to the jugular foramen, the hypoglossal nerve (XII) exits from its condylar canal in the occipital bone (Fig. VI-32C). It will run for a short time, as it leaves the base of the skull, in company with IX, X, and XI. Lateral to the jugular foramen is the *stylomastoid foramen* for the emergence of the branchial motor components of VII. Just anterior to the latter foramen the *styloid process* can be seen as a fair remnant in the left side of Figure VI-33C. Even this one has been broken off short. The normal size of a styloid process is illustrated in a lateral view of the skull in Figure VI-35. The inset shows how abnormally long this process can get if a portion of the stylohyoid ligament should remain calcified or become ossified. Developmentally, process and ligament were common parts of the second branchial arch cartilage, the former changing to bone and the latter becoming a tough, noncalcified, fibrous strand, as a rule.

In Figure VI-34 we have given the skull its last tip forward so that the lambda comes into view posteriorly. If we trace from back to forward (posteroanteriorly) the following prominent foramina are seen:

1. Paired posterior condylar canals.
2. Foramen magnum, *through* which we see the dorsum sellae and posterior clinoids.
3. On the left, a bit of the foramen ovale.
4. In the midline, the paired choanae with all three conchae and the nasal septum viewed from behind.
5. Well forward, the incisive foramen, and a wide gap between the middle incisor teeth in this skull.

As to the fossae we have mentioned, Figure VI-34 shows the pterygopalatine (sphenopalatine) fossa to lie between the posterior walls of the maxillae and the anterior aspects of the pterygoid plates (laminae). Still farther posteriorly is the territory of the infratemporal fossa. It lies under shelter of the greater wing of the sphenoid medial to the zygoma and it includes the pterygoid plates, the posterior aspects of which we see, and the fossae between these plates (pterygoid and scaphoid). The infratemporal fossa ends at the foramen spinosum just anterior to the spine of the sphenoid.

Development of Face and Palate

The palate which separates nasal from oral cavity develops as the face is formed. Its premaxilla is contributed in the embryo by the **median nasal processes** as these form the nostrils and the median portion of the upper lip (Fig. VI-36, B, C, and E). The first branchial arches bifurcate to give rise to the mandibular and maxillary processes (Fig. VI-36A). The fusion of the two mandibular processes from each side completes the lower jaw at the symphysis menti. The upper jaw construction is more complicated (Fig. VI-36 A–E), since it involves a fusion of maxillary and medial nasal processes (Fig. VI-36B). Each of these on its undersurface develops shelf-like projections that grow toward the midline (Fig. VI-36E). Thus, the maxillary processes develop **lateral palatine processes** that move toward each other to complete the anterior part of the hard palate. Their midline fusion accounts for the long vertical arm of the cross-shaped suture line in the palate. The median nasal processes do not in any real sense contribute to the palate but form **median palatine processes** that are, actually, the premaxillary parts of the upper jaw. Their fusion is always incomplete, hence the gap or incisive foramen for the **greater palatine arteries and the nasopalatine nerve.** The foramen has lateral canals opening into it behind the incisor teeth (Fig. VI-36F), but these get covered by mucous membrane as a rule. Sometimes the canals may stay uncovered as in some animal forms. It is easy to imagine from Figure VI-36 how incomplete fusions could account for cleft palate or harelip. The etiology of these anomalies has never been well accounted for in the human, but much is known about their genetic frequency of occurrence.[11-14]

When we consider palatine and pterygoid muscles we find them all related to some bony prominences within the infratemporal fossa on the antero-interior aspect of the skull. The *Lateral Pterygoid* takes origin from the lateral side of the lateral pterygoid lamina or plate. The *Medial Pterygoid* arises from the medial wall of the same plate as well as from the pterygoid fossa between the laminae, the maxillary tubercle and the pyramidal process of the palatine bone. The **Tensor Veli Palatini** arises from the scaphoid fossa and

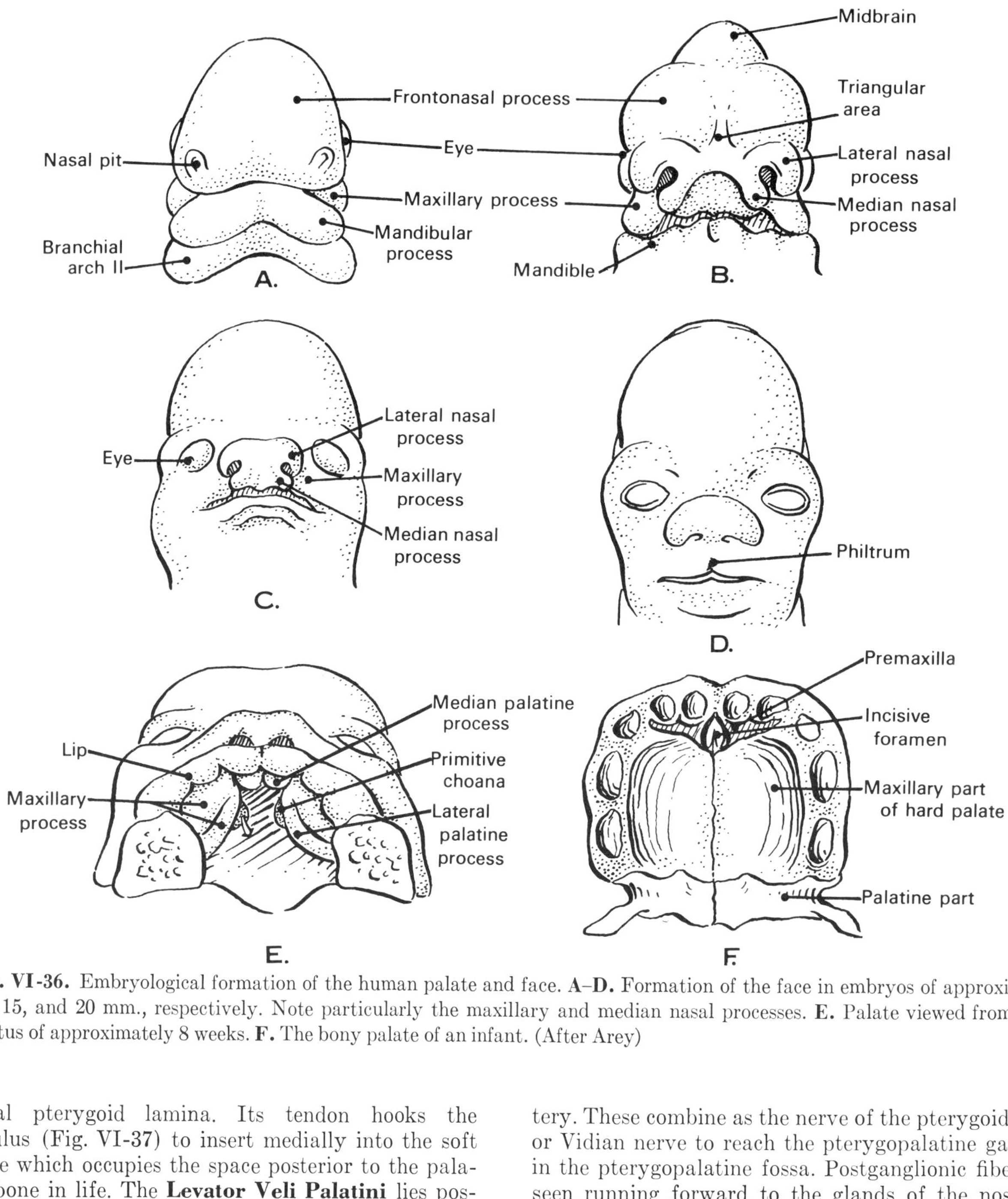

Fig. VI-36. Embryological formation of the human palate and face. **A–D.** Formation of the face in embryos of approximately 6, 10, 15, and 20 mm., respectively. Note particularly the maxillary and median nasal processes. **E.** Palate viewed from below in a fetus of approximately 8 weeks. **F.** The bony palate of an infant. (After Arey)

medial pterygoid lamina. Its tendon hooks the hamulus (Fig. VI-37) to insert medially into the soft palate which occupies the space posterior to the palatine bone in life. The **Levator Veli Palatini** lies posterolateral to the choanae and arises from the apex of petrous bone and the cartilage of the auditory tube. Just above the concave margin of the Superior Pharyngeal Constrictor (Fig. VI-37), its fibers slip medialward and obliquely into the soft palate. The posterior upper part of the Superior Pharyngeal constrictor takes origin from the pharyngeal tubercle.

In a midsagittal view through the nose, palate, and nasopharynx (Fig. VI-38), preganglionic fibers from the facial nerve (as greater superficial petrosal fibers) are seen joining deep petrosal fibers on the carotid artery. These combine as the nerve of the pterygoid canal or Vidian nerve to reach the pterygopalatine ganglion in the pterygopalatine fossa. Postganglionic fibers are seen running forward to the glands of the nose and downward in the sphenopalatine canal to glands in the mucous membranes of the hard and soft palate. Still other postganglionic fibers are seen taking a backward course in the pharyngeal (vaginal) canal to reach structures in the pharynx, auditory tube, and sphenoid air sinuses which lie posterior to the palate. The structures innervated are glands or smooth muscle of arterioles. Note the proximity of the semilunar ganglion of V and even the olfactory bulb on top of the cribiform plate to the region of these bony canals. The pharyngeal ostium of the auditory tube is also shown.

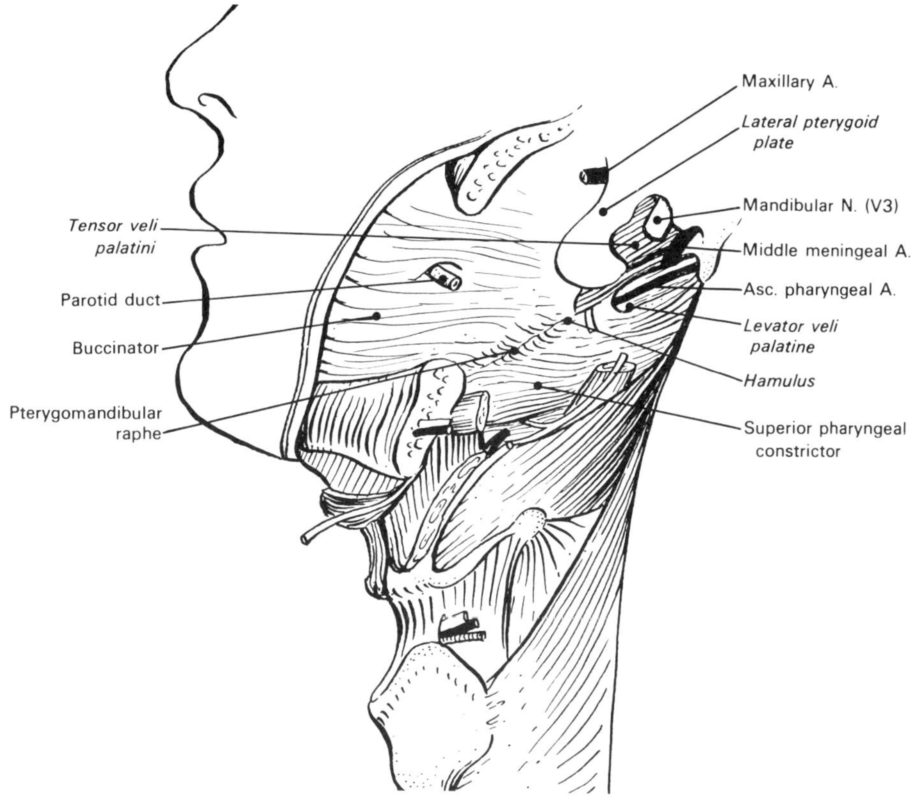

Fig. VI-37. Relation of palatine muscles to pharynx and pterygoid lamina.

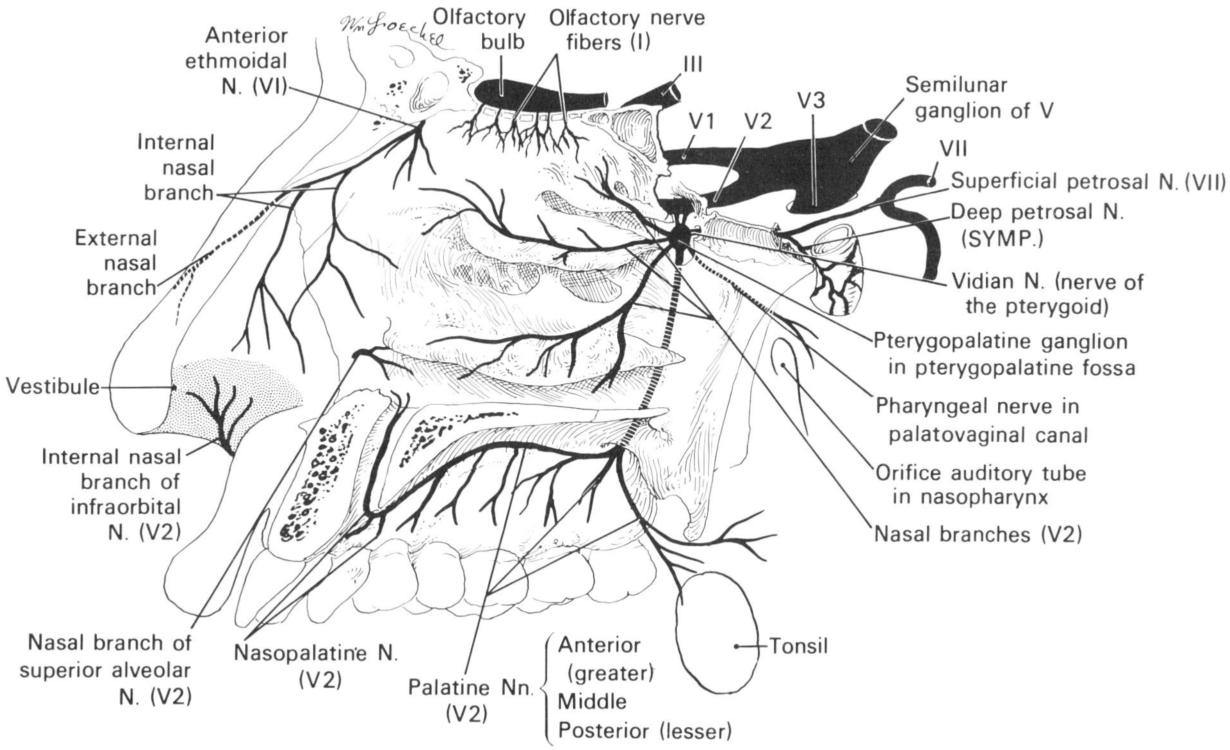

Fig. VI-38. Locus of the pterygopalatine ganglion and some of its relationships in a midsagittal section through nose, palate, and nasopharynx.

Table VI-4 lists many foramina, canals and fissures. Some we have mentioned, others we have not. The structures that pass through them are indicated. The Table is to be used for reference only and is not to be committed to memory. Memorization for memorization's sake serves no useful purpose (unless we are truthful and say it helps in passing some exams). It is far more important to grasp relationships, develop concepts and strive to understand. The finer details themselves, as memorized facts are not that important. Still there is no gain saying that without some firm grasp on facts our concepts are often vague ones. The good student marshals facts as he constructs sound concepts, and there is no instant way to do this. Instant knowledge is not dispensed from jars. Most of us gain a fair idea of relationships as we read and try to visualize from illustrations. Fortunately books, pictures, the dissection, and other media serve as reference tools for recall of the many small details that might otherwise be forgotten. If a good initial attempt is made to localize any new structure that is mentioned, memory attrition is much less pronounced.

RADIOGRAPHIC ANATOMY OF THE SKULL[15-17]

In the following series of illustrations, you will have an opportunity to review much of the anatomy that has just been covered; however, the emphasis in this section will be to examine structures as they appear in radiographs. We will consider five basic radiographic projections used in *clinical practice* to survey the skull: (1) **lateral;** (2) **inclined posterior-anterior;** (3) **occipitomental** (Water's); (4) **basal** (axial or submentovertex); (5) **anterior-posterior half-axial** (Towne's). Although the radiologist does, at times, use other projections of the skull to examine certain structures not seen to advantage on these five views, these are the pertinent projections with which you should become familiar. A thorough understanding of these projections will afford an excellent foundation to build upon during the young physician's clinical years.

Each of the basic radiographs has been placed near a matching photograph or photographs of the dry skull in the hope that some better understanding can be gleaned. In the case of the inclined PA and occipitomental projections, the photographs of the dry skull were made from the position of the X-ray film cassette and not from the position of the X-ray tube. The reason for this will become obvious shortly. However, in the basal and AP half-axial projections, the dry skull photographs were made from ***both*** the position of the X-ray tube and the film cassette. In all instances, the photographs were made to take into account the appropriate angulation of the tube needed to produce the particular projection.

1. Basic Radiographic Lines and Planes

We will not dwell upon the complexities of positioning the patient for projections to be discussed, but rather we will quickly review the basic lines and planes, and let the inserts, which detail the positioning, speak for themselves.

a. Orbitomeatal line (the OM line): This line joins the center of the orbit with the center of the external auditory meatus.

b. Infraorbitomeatal (anthropological) base line: This line joins the infraorbital point to the upper margin of the external auditory meatus. These two lines, the infraorbitomeatal and orbitomeatal, cross at approximately a 10° angle.

c. Median sagittal plane: This divides the skull into symmetrical halves.

Although there are many other lines and planes, these will suffice for our purposes. For the budding radiologists in your group, we can find a few more such bits of information when you arrive in the clinical year.

2. General Overview of the Skull:

A quick survey of your fellow students will immediately reveal that skulls come in all shapes and sizes. Anatomists prefer intricate terminology in describing the skull, relative to its length and breadth, and, of course, many of these terms are carried over into clinical practice. For instance, the average head is termed **Mesocephalic,** and it has a **cephalic index** of 75 to 80. (The cephalic index is calculated from the formula: 100 × breadth/length.) Individuals with skulls that are longer than average, **Dolichocephalic,** have a cephalic index of less than 75, while **Brachycephalic** skulls (cephalic index greater than 80) are shorter and broader than average. These same terms may be applied clinically to describe the shape of the head resulting from premature closure of the cranial sutures (craniostenosis). Although the sutures normally remain open into adult life, on occasion one or more of them will close early and the clinical problem will be evident at or shortly after birth. As you expect, the suture that closes early will establish the shape of the head. Thus, with premature closure of the sagittal suture, the head assumes a Dolichocephalic shape; with premature closure of the coronal or lambdoid sutures, the head is of the Brachycephalic type. Other terms are also used to describe the oft-times weird shapes; however, we must defer these to your clinical mentors.

From the foregoing, then, it becomes obvious that in viewing radiographs of the skull you should evaluate the shape carefully. Size is also important, for many pathological entities cause the head to enlarge (hydrocephalus, acromegaly, Paget's disease) or to remain quite small (microcephaly). You will find that as a general rule in adults, the facial area is approximately

Table VI-4. An Alphabetized List of Passages in the Skull and Structures That Pass through Them

Passage	Structures	Passage	Structures
Anterior ethmoidal foramen	Anterior ethmoidal artery, vein, and nerve	Mandibular canal	Inferior alveolar artery, nerve, and vein; mental nerve
Carotid canal	Internal carotid artery and sympathetic carotid plexus	Mastoid canal	Auricular branch of nerve X.
Cochlear aqueduct	Nothing? Cerebrospinal fluid to perilymphatic channels of inner ear?	Mastoid foramen	Emissary vein
Condylar canal (posterior)	Emissary vein	Mental foramen	Mental nerve
Foramen lacerum	Meningeal branch of ascending pharyngeal artery; meningeal lymph vessels; small veins from cavernous sinus to pharyngeal plexus of veins	Nasal slit	Anterior ethmoidal artery and nerve
Foramen magnum	Junction cord and brainstem; vertebral aa.; anterior and posterior spinal aa.; last tooth of pial denticulate ligament	Optic canal	Nerve II; ophthalmic artery; sympathetic nn. from internal carotid plexus for ciliary muscle; dilator of pupil; smooth muscle of blood vessels
Foramen ovale	Mandibular nerve (V_3); lesser petrosal nerve; accessary meningeal artery	Petrosquamous fissure	Tympanic branch middle meningeal artery
Foramen rotundum	Maxillary nerve (V_2).	Petrotympanic fissure	Chorda tympani branch of nerve VII; anterior tympanic artery; tympanic branch of maxillary artery
Foramen of Vesalius (inconstant)	Emissary vein	Pharyngeal canal	Pharyngeal artery and vein; pharyngeal branch from pterygopalatine ganglion (parasympathetic nerve VII)
Greater palatine canal	Greater palatine artery, vein, and nerve	Posterior ethmoidal foramen	Posterior ethmoidal artery, vein, and nerve
Hiatus of facial canal	Greater petrosal nerve; petrosal branch, middle meningeal artery	Posterior superior alveolar foramen	Posterior superior alveolar artery, nerve, vein (sometimes middle alveolar nerve)
Hypoglossal canal	Hypoglossal nerve	Pterygoid canal	Artery, vein and nerve of the canal (Vidian nerve—greater and deep petrosal nerve)
Incisive foramen and canal	Greater palatine aa; nasopalatine nerve	Pterygomaxillary fissure	Maxillary artery and vein
Inferior orbital fissure	V_2, infraorbital artery and vein; zygomatic nerve; orbital branch from pterygopalatine ganglion; veins from orbit to pterygoid plexus; ophthalmic vein	Stylomastoid foramen	Nerve VII; stylomastoid branch of postauricular artery
Infraorbital foramen	Infraorbital nerve	Superior orbital fissure	Nerves III, IV, VI; V_1 (ophthalmic nerve); frontal nerve; lacrimal nerve; recurrent branch of lacrimal artery; orbital branch of middle meningeal artery
Innominate canal (inconstant)	Lesser petrosal nerve	Supraorbital foramen	Supraorbital artery, vein, and nerve
Jugular foramen	Nerves IX, X, and XI; internal jugular vein as formed by sigmoid sinus and inferior petrosal sinus	Vestibular aqueduct	Endolymphatic duct
Lesser palatine foramina	Middle and posterior palatine nn.	Zygomaticofacial foramen	Zygomaticofacial artery, vein, and nerve
		Zygomaticotemporal foramen	Zygomaticotemporal nerve

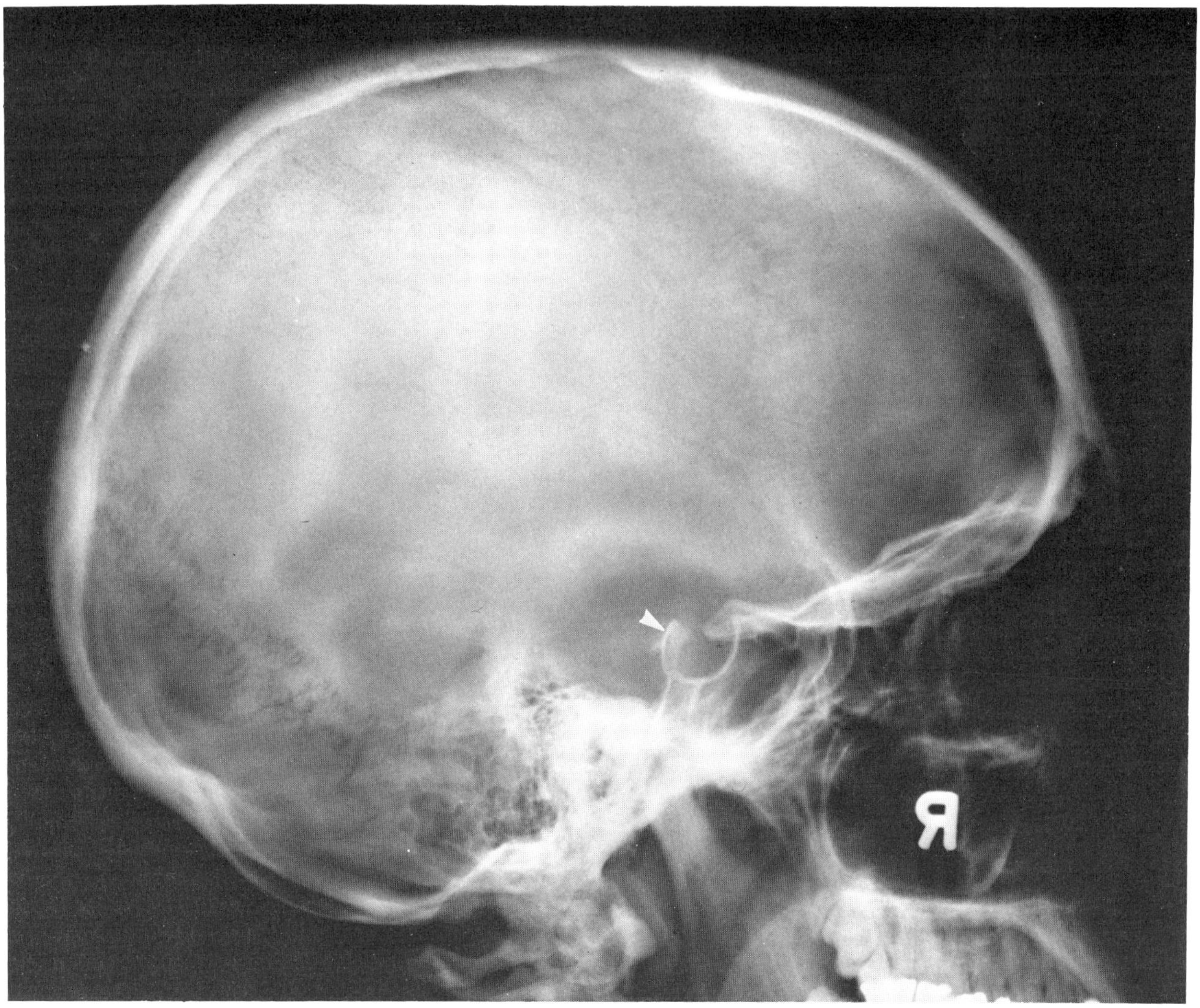

Fig. VI-39A. Roentgen appearance of the dorsum sellae in a patient aged 20 years.

half that of the cranium, whereas in children, the facial area is proportionately smaller.

When examining frontal views of the skull on radiographs, you will have the advantage of comparing each side of the skull for symmetry of structures, as well as for density of bone. Most skulls will show some slight degree of asymmetry. The density of the bones of the cranium is at times difficult to evaluate unless the technical factors used in making the radiograph are known. A common cause of decreased bone density is senile osteoporosis. This becomes particularly noticeable in the dorsum sellae which tends to demineralize with advancing age but is not thinned or eroded. If you compare the dorsum sellae in the lateral skull films of Figure VI-39A and B this will become obvious. Other entities, such as hyperparathyroidism, also cause demineralization of the vault. Still other clinical conditions, such as osteopetrosis, cause increased bone density.

In the discussion to follow on each projection, no attempt has been made to discuss every minute structure or even those that are readily apparent. Only those details that are bound to produce confusion will be explained.

3. Basic Projections

a. Lateral projection (Fig. VI-40): *Superficial structures:* Review these structures in Figure VI-40. Note particularly that the cranial sutures in the adult are serrated and often surrounded by a zone of sclerosis (perisutural sclerosis). This point may be of aid to you in differentiating a suture from a fracture line—the latter usually appear quite a bit more radiolucent and, if not depressed, will not show surrounding sclerosis. Be particularly careful not to confuse the occipitomastoid suture for a fracture. This suture probably causes more confusion to the tyro than any other one.

The calvaria varies from about 3 to 8 mm. in its

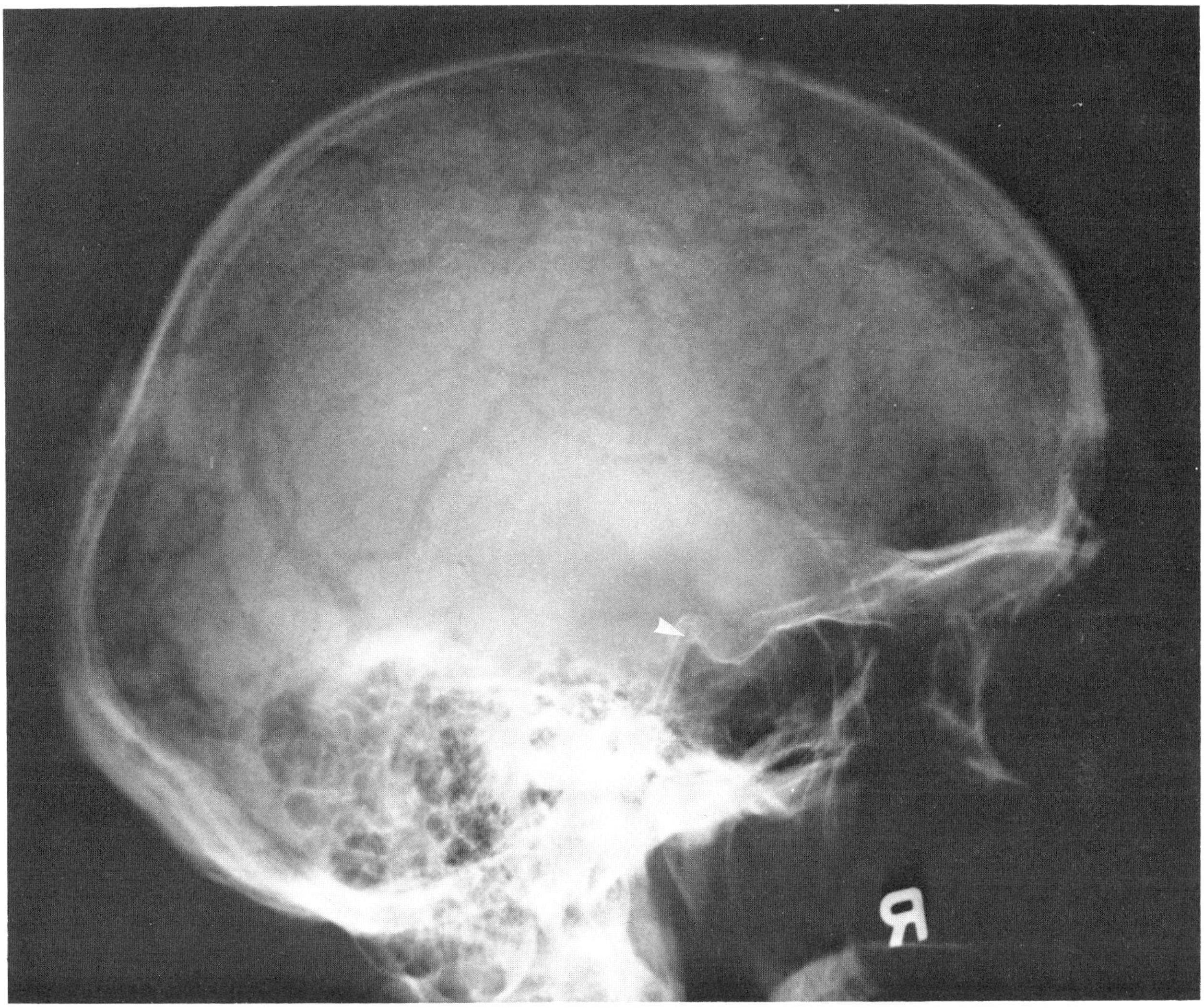

Fig. VI-39B. Roentgen appearance of the dorsum sellae in a patient aged 60 years. The arrow indicates the dorsum sellae. Follow the bony cortex (lamina dura) of the sella around and note that it is intact in each instance, although somewhat less mineralized in B. All of the normal anatomic structures are present with no sign of bony erosion or destruction.

thickness; however, this is typically increased over the occipital protuberance. Areas of particular thinness occur beneath the Temporalis muscle, especially the temporal squama, and in the subnuchal portion of the occipital bone.

Deep structures (Fig. VI-41): Although the lateral projection is made to demonstrate the calvaria, it is extremely valuable in evaluating the base of the skull. Let us examine the structures visible in each of the three cranial fossae.

(1) Anterior cranial fossa: Along the central area of the base of the skull, find the two overlapping anterior clinoid processes in Figure VI-41A and the right anterior clinoid in the sagittally-sectioned skull (Fig. VI-41B). The **anterior clinoids** may be of varying size. This is normal unless evidence of erosion is present. Trace the clinoids forward in Figure VI-41 to two heavy white lines. The uppermost of these two lines is produced by the superimposed orbital roofs. About one-fourth of the way forward, this line crosses two concave anteriorly-directed lines which are formed by the **greater wings of the sphenoid bone** (Fig. VI-41C). These curved lines represent the most anterior margin of the middle cranial fossa wherein are nestled the tips of the temporal lobes of the brain. The two curved lines of the greater wing of the sphenoid ex-

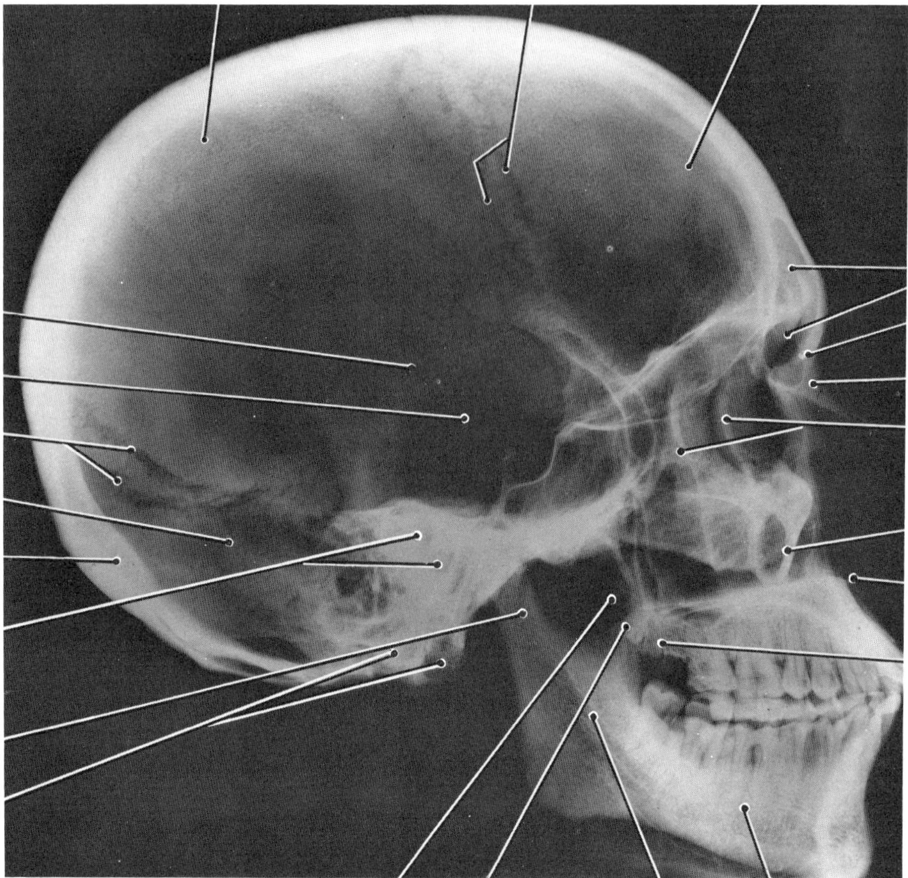

Fig. VI-40A. The lateral aspect of the skull as seen in matching (**A**) radiograph and in (**B**) the skull itself. Use these to review pertinent superficial landmarks.

tend above the orbital line and form an area of density which is rather triangular in shape. This ridge of bone is well seen in Figure VI-41A and C and is formed by the junction of the horizontal plate of the frontal bone and the lateral aspect of the greater sphenoid wing. Note the middle meningeal groove just behind this ridge (Fig. VI-41A and C). The meningeal grooves are usually seen on the inner table of the calvaria. These grooves are sharply defined and their parallel walls remain of the same caliber until they branch. At that point, the caliber decreases. Venous diploic channels are more commonly uneven in their caliber.

As we continue to trace the roofs of the orbits anteriorly, it is apparent that they are irregular as, indeed, this floor of the anterior fossa is. At its highest point, the roof of the orbit is actually higher than the palpable supraorbital rim, and we will see this again illustrated in the inclined posterior-anterior projection.

Now again find the anterior clinoid process and trace it forward to the lower of the two lines mentioned earlier. This lower line is formed by the **planum (jugum) sphenoidale,** and its most posterior margin is the **limbus sphenoidale** (Fig. VI-41A and B). The planum sphenoidale is continuous with the **cribiform plate** which may actually be slightly lower in position. The cribiform plate is not seen in its entirety, since it is quite thin. The crista galli, although well demonstrated in Figure VI-41B, is completely obscured in the radiograph (Fig. VI-41A) because of overlapping structures.

Between the lines formed by the roof of the orbits and the planum sphenoidale, we occasionally see a third line which is the roof of the **fronto-ethmoid air cells** (Fig. VI-41A).

(*2*) *Middle cranial fossa:* As we have just observed, the most anterior margin of the middle fossa is seen as two separate anteriorly-directed concave lines formed by the greater wings of the sphenoid bone. Al-

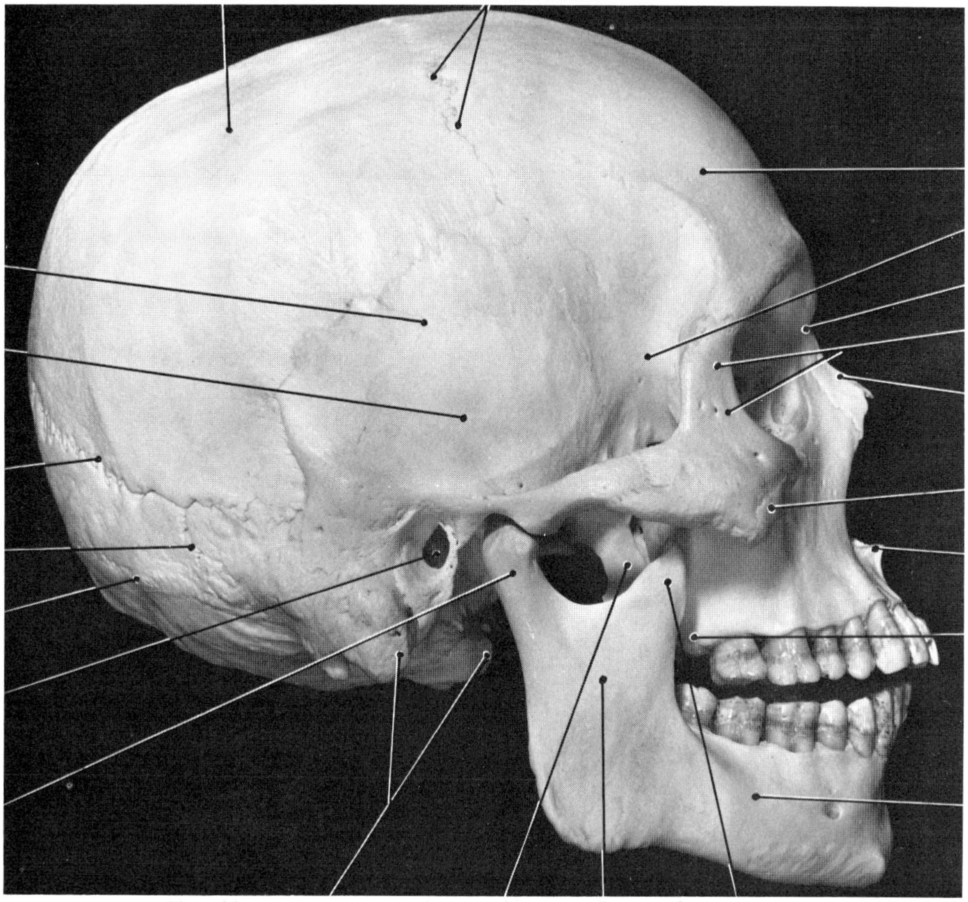

Fig. VI-40B.

though these two lines appear to be widely separated on the radiograph (Fig. VI-41A), this is actually due to projection and slight rotation of the skull. The line farthest forward is actually formed by the greater sphenoid wing that is farthest from the X-ray film. As the greater wings of the sphenoid are traced posteriorly they become lost in the superimposed petrous pyramids and mastoid air cells.

Now identify the **floor of the sella turcica** (Fig. VI-41A and B) and note the **sphenoid sinus** which is extremely variable in its size, immediately below and often anterior to the floor. Follow the anterior wall of the sella upward until you encounter a small projection, the **tuberculum sellae** (olivary eminence). Between the tuberculum sellae and the limbus sphenoidale is the **chiasmatic sulcus.** This sulcus is inconstant in its depth (Fig. VI-42A, B, and C) and in some instances may not be discernible at all. The posterior wall of the sella turcica is the **dorsum sellae** which is crowned laterally by the **posterior clinoid processes.**

Before leaving the middle cranial fossa, we should say a few words about the size and shape of the sella turcica. The sella is protean in both its size and shape. The small sella is usually of no clinical import; however, enlargement of the sella may have grave significance. Figure VI-42D shows the simple measurements that are used to determine sella size in the lateral adult skull radiograph made at a 40-inch target-film distance. The greatest AP diameter should not exceed 17 mm. The depth, which is determined by first constructing a line from the tuberculum sellae to the top of the dorsum sellae and dropping a perpendicular from this line to the deepest portion of the floor, should not exceed 13 mm.

(*3*) *Posterior cranial fossa:* As we have mentioned earlier, the clivus is that portion of the sphenoid bone sloping downward from the dorsum to the synchondrosis of the sphenoid with the basi-occiput (Fig. VI-41B). Clinicians, however, consider the inner sloping portion of the basiocciput as part of the clivus, also. Follow the basiocciput posteriorly and downward

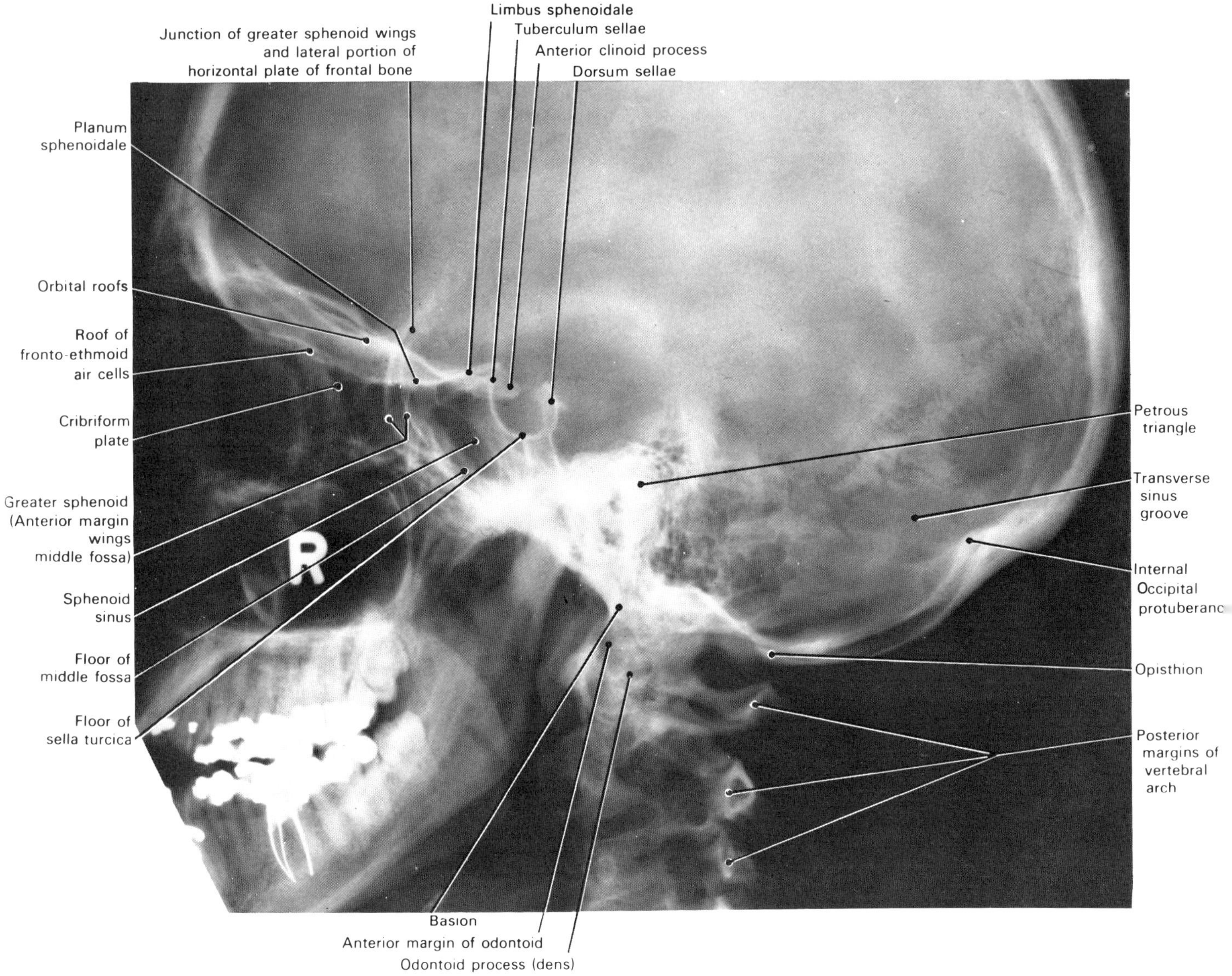

Fig. VI-41. Deeper structures of the skull in lateral view. **A.** Lateral radiograph in the living subject.

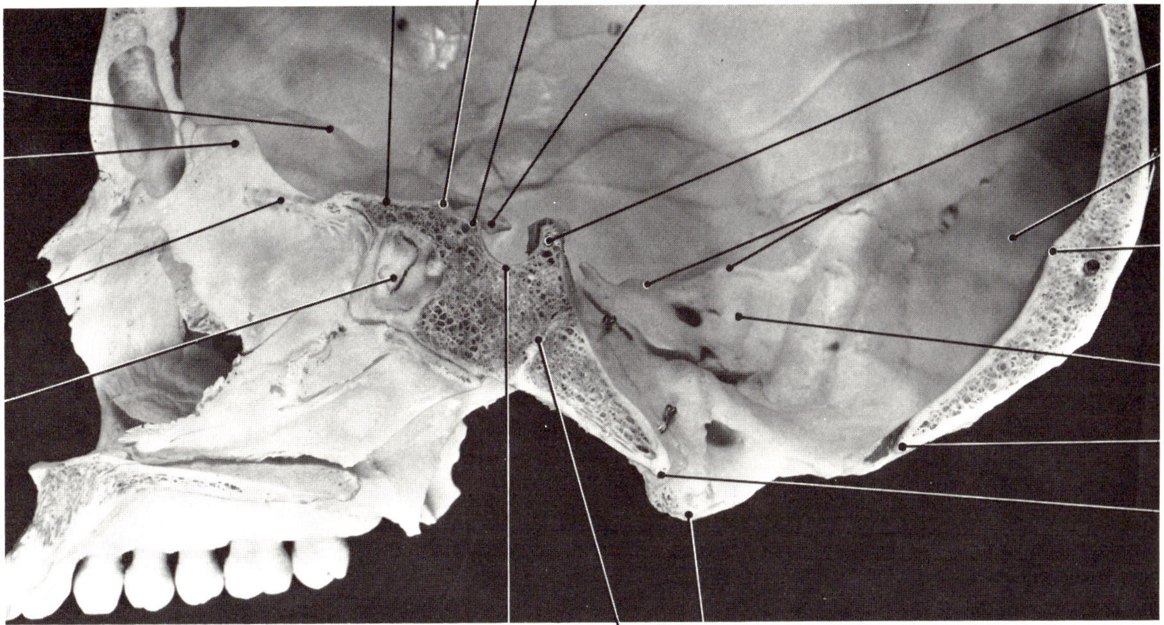

Fig. VI-41B. Sagittal section of the dry skull.

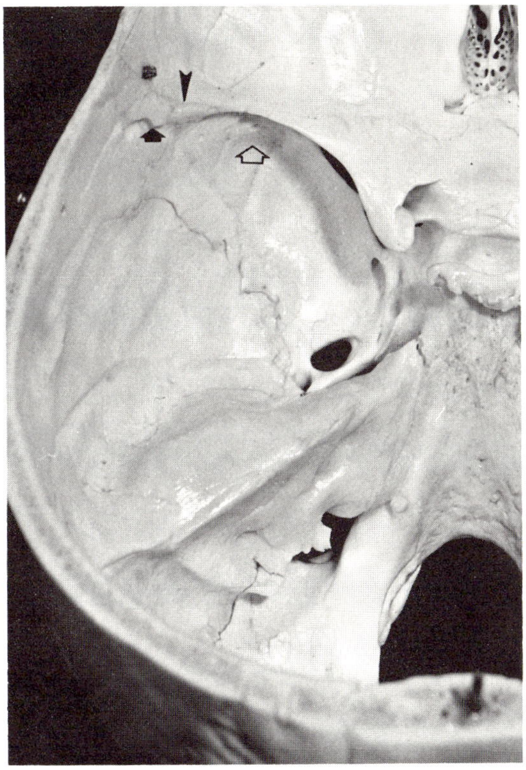

Fig. VI-41C. Interior view of the skull illustrating the junction of the greater sphenoid wings and the lateral portion of the horizontal plate of the frontal bone (*black arrow head*); greater wing of the sphenoid (*open arrow*)—the most anterior portion of the middle cranial fossa; and the groove for the anterior branches of the meningeal vessels on the inner table of the parietal bone (*black arrow*).

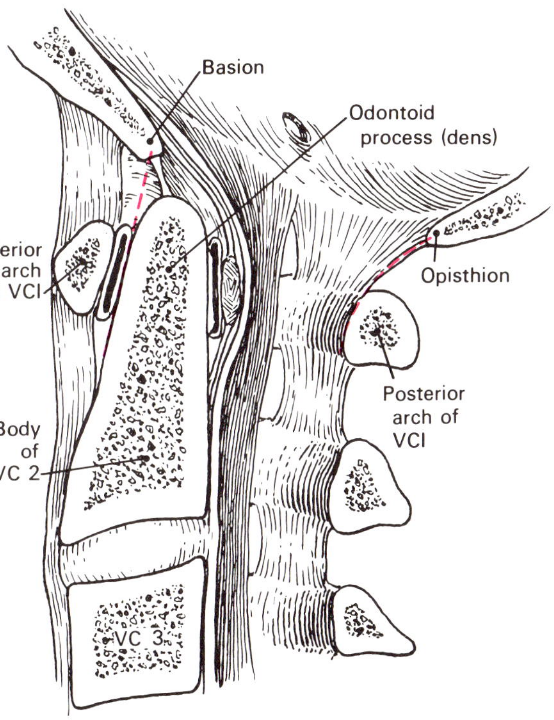

Fig. VI-41D. Schematic drawing showing location of the basion and opisthion. See text for two methods of finding these anatomic points in lateral skull radiographs.

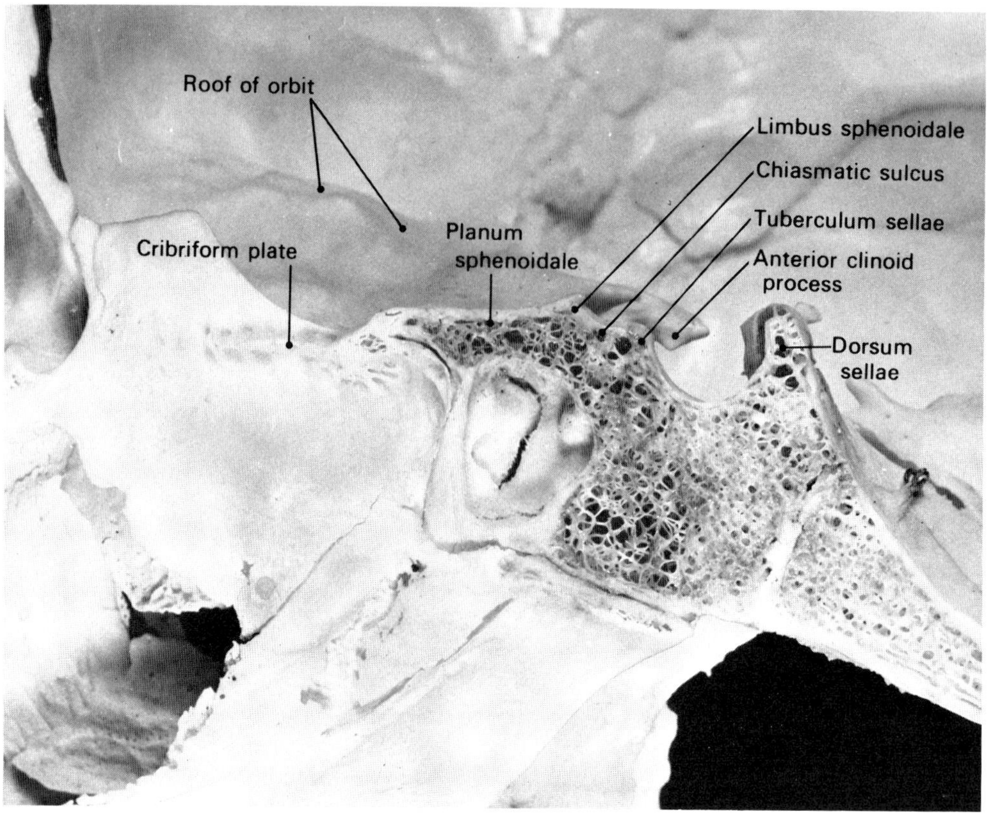

A

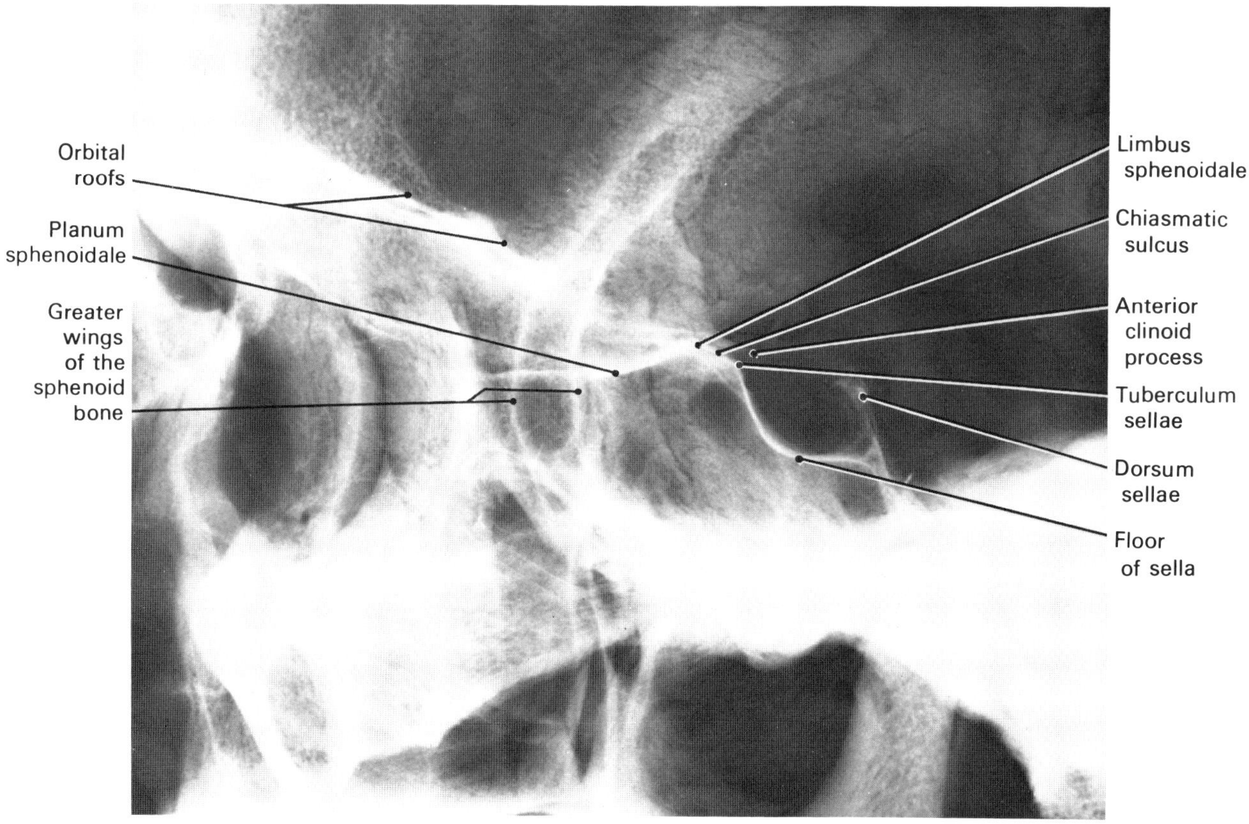

B

Fig. VI-42. Detailed anatomy of the anterior cranial fossa. **A.** Sagittal section of the dry skull. **B and C.** Lateral radiographs. Note particularly the difference in appearance of the chiasmatic sulcus in Figs. VI 42B and C. (Dry skull preparations.)

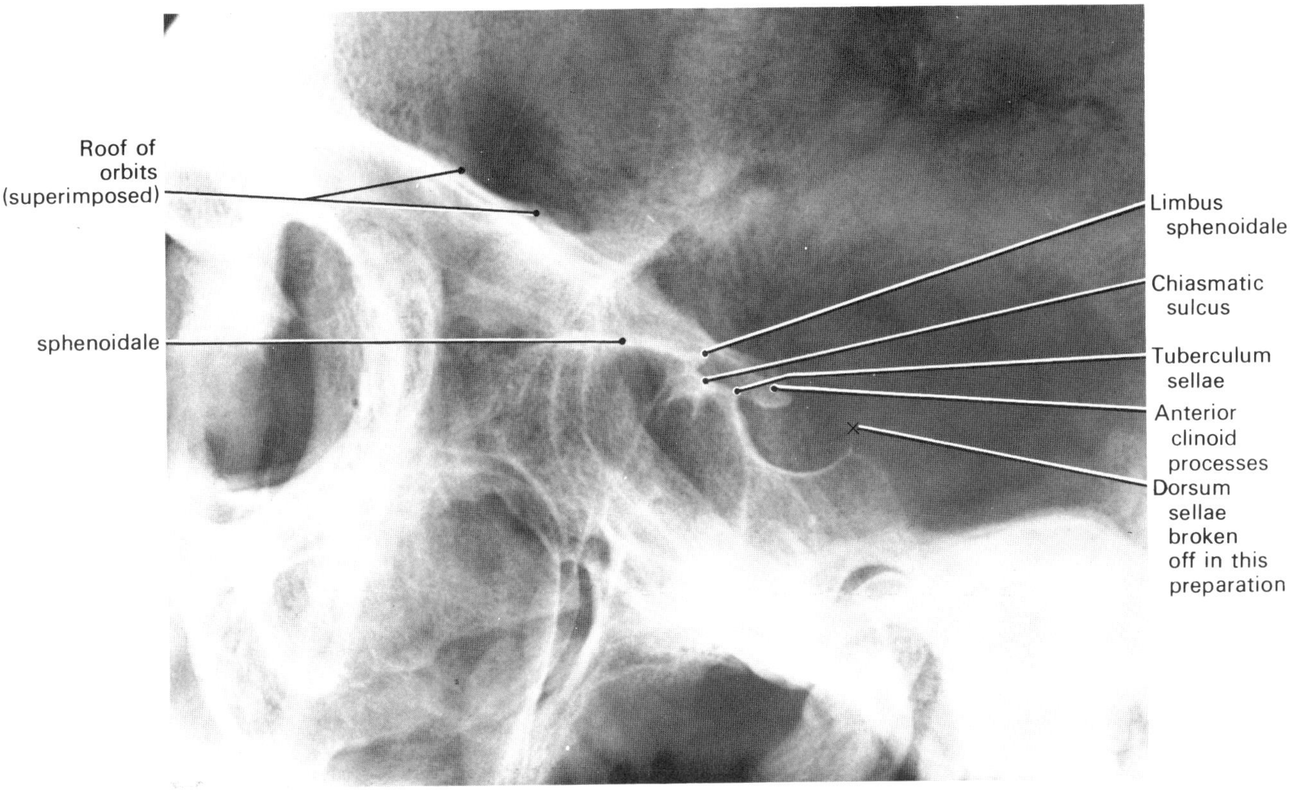

Fig. VI-42C.

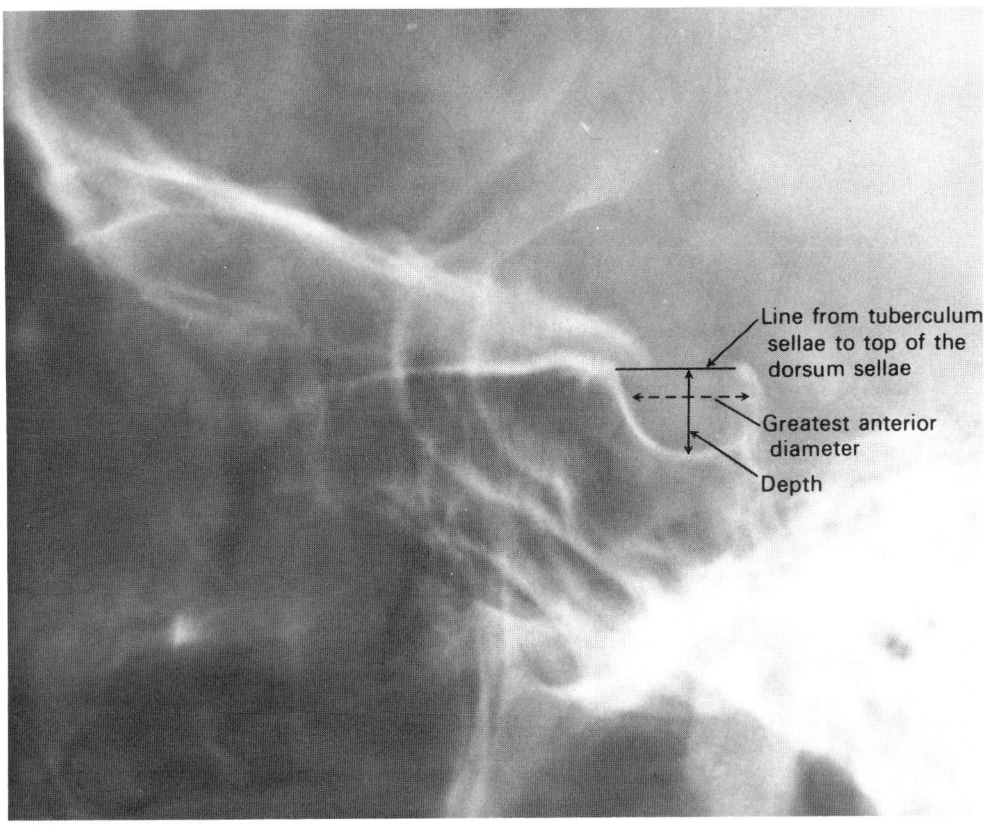

Fig. VI-42D. Technique of measuring the depth and the AP diameter of the sella turcica (see text). (Living subject.)

where it soon joins its own outer table to round off at the midline **anterior margin of the foramen magnum (basion).** This is well demonstrated in the sagittal section of the skull, Figure VI-41B, and it may also be seen in the lateral radiograph, Figure VI-41A. The anterior margin of the foramen magnum may also be found in the radiograph by directing an imaginary line upward from the anterior margin of the odontoid process (dens) of the second cervical vertebra (Fig. VI-41D).

It is important to recognize the midline **posterior margin of the foramen magnum (opisthion)** in the lateral radiograph. This may be found in most cases by tracing the inner table of the occipital bone forward and downward to the point where the inner and outer tables merge (Fig. VI-41A, B, and D), or by following an imaginary line which curves upward along the posterior wall of the vertebral canal.

The posterior border of the posterior cranial fossa ends at the internal occipital protuberance as shown in Figure VI-41A and B. The petrous ridge forming the anterior boundary of this fossa runs obliquely posteriorly, and although it is well seen in Figure VI-41B, it is partially obscured in detail due to superimposition of the two sides in the radiograph (Fig. VI-41A). Note the transverse sinus groove forming the lateral posterior boundary of the fossa.

Review all of these structures in Figure VI-43.

b. Inclined posterior-anterior projection (Fig. VI-44)**:** *Superficial structures:* Review the superficial structures as labeled in Fig. VI-44A and B. Because of the angulation of the X-ray tube, the petrous pyramids have been projected at the lower margin of the orbits.

Whenever examining the skull in the frontal view, you will have the opportunity to compare each side of the calvaria for symmetry, density, and thickness of the vault. When visible, also check the vascular channels for symmetry. If marked degrees of asymmetry exist, a vascular malformation or tumor must be considered.

Deep structures: (*1*) *Anterior cranial fossa:* The highest point of the **floor of the anterior fossa** is projected above the supraorbital rim (Fig. VI-44C and D). The roof of the orbit slopes posteriorly to meet the lesser sphenoid wing. At the posterior aspect of the fossa, we can see that each of the lesser wings of the sphenoid slope medially to be joined by the planum (jugum) sphenoidale. Note the anterior clinoid process at the medial aspect of the lesser wing of the sphenoid (Fig. VI-44D). At times, the lateral margin of the **optic canal** may be seen. The superior orbital fissure, situated between the lesser and greater wings of the sphenoid bone (Fig. VI-44C and D), shows well. Between the optic canal and the superior orbital fissure is a bar of bone, the **sphenoid strut** (see p. 317). Perpendicular to the planum sphenoidale is the **crista galli,** which we could not see in the lateral projection (Fig. VI-44C).

(*2*) *Middle and posterior cranial fossae:* When the sphenoid air cells extend beneath the floor of the **sella turcica,** the floor becomes readily visible in this projection (Fig. VI-44C and D). The dorsum sellae and floor of the middle fossa are not well seen due to overlapping structures.

The most inferior aspect of the **floor of the posterior fossa** may be seen in Figure VI-44C.

(*3*) *The oblique orbital line* (Figure VI-44C): In this projection, the lower two-thirds of the line is made up of the cross-section of the squamozygomatic portion of the greater wing of the sphenoid bone. The upper one-third is produced by a tangential section of the temporal surface of the frontal bone (immediately behind the zygomatic process).

(*4*) *Straight PA Projection:* At times it is useful to direct the central X-ray beam directly down the OM line in a straight PA projection (Fig. VI-44E). Thus the petrous pyramids are projected into the center of the orbits. This affords an excellent look at the internal auditory canals. Note, too, in this dry skull how clearly defined the cochlea and semicircular canals (superior, lateral) are. The highest point of the roof of the orbit is now for the most part superimposed upon the lesser sphenoid wing. Thrown into view are the anterior clinoid processes, the dorsum sellae and the posterior clinoid processes, crista galli, and curving upward across the sella turcica, the shadow of the tuberculum sellae.

c. Occipito-mental (Water's) projection: *Superficial structures:* These may be reviewed by studying labeled parts of Figure VI-45A to C. Note the great clarity of the facial bones. This frontal projection again affords the opportunity to compare symmetry of structures. Thus, it is of great aid particularly when you are concerned with a case involving trauma.

Deep structures: The Water's projection is one of a series used to study the paranasal sinuses (others include the inclined PA and the lateral). The normal paranasal sinuses have a bony cortical margin, which is lost particularly in the presence of infection or by tumor invasion. Always compare the two sides of the same sinus for changes in density or, when the films are made with the patient erect, for air-fluid levels. The latter may of course be seen in an acute sinusitis. Following trauma it may signify the presence of blood within the sinus. In allergic or infectious sinusitis, the normal mucosal lining will thicken and become readily visible, whereas normally it is not seen to advantage. To visualize the sphenoid sinuses, the angulation is exaggerated to about 45°, and the patient is told to open the mouth. The sphenoid sinuses are visualized through the open mouth (Fig. VI-45D).

The floor of the anterior cranial fossa, which is composed of the orbital roofs, is extremely thin, but is seen superimposed upon the calvaria posteriorly. A destructive process of the orbital roofs is usually well seen in this view.

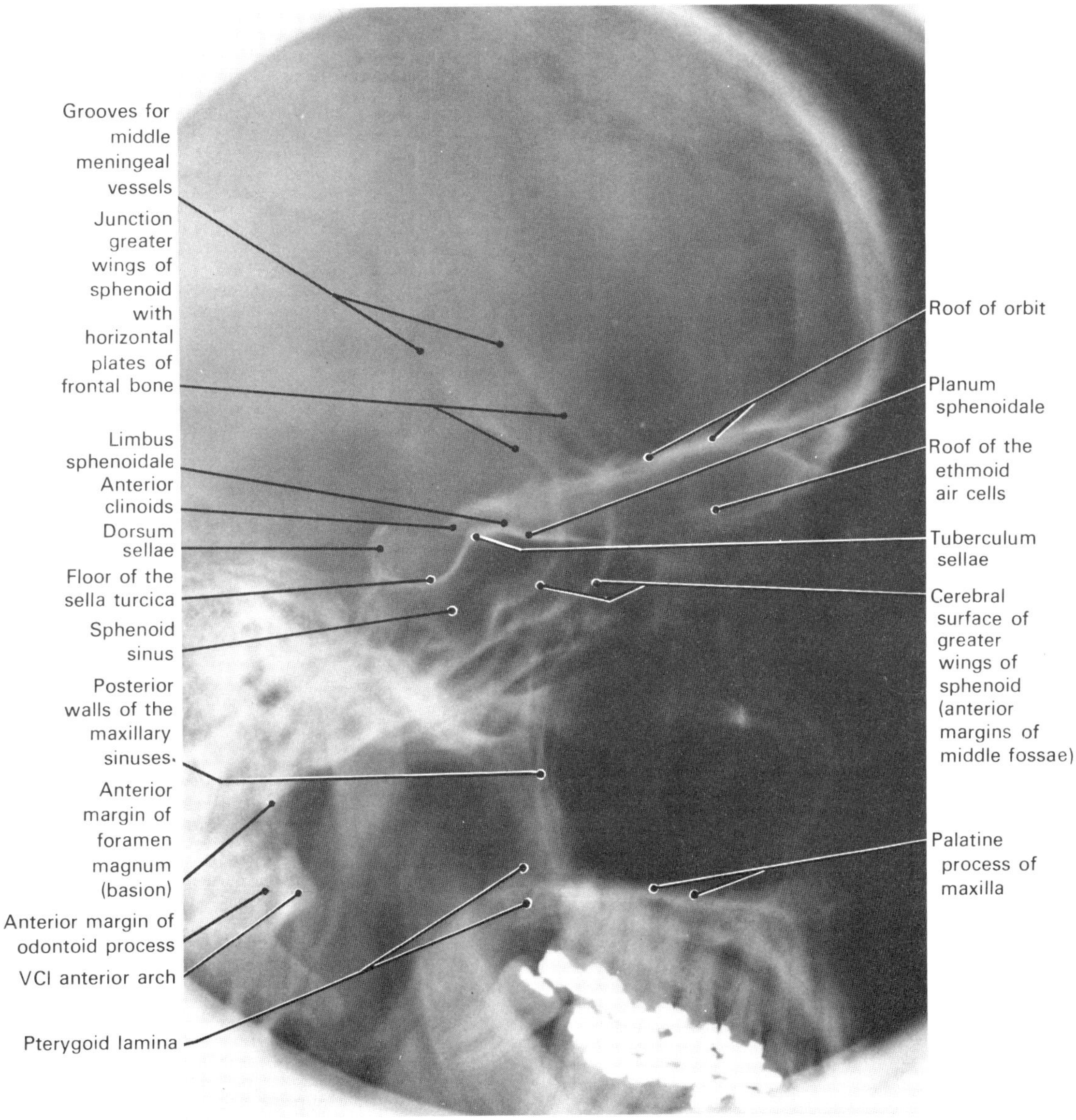

Fig. VI-43. Coned lateral radiograph of the skull in a living subject.

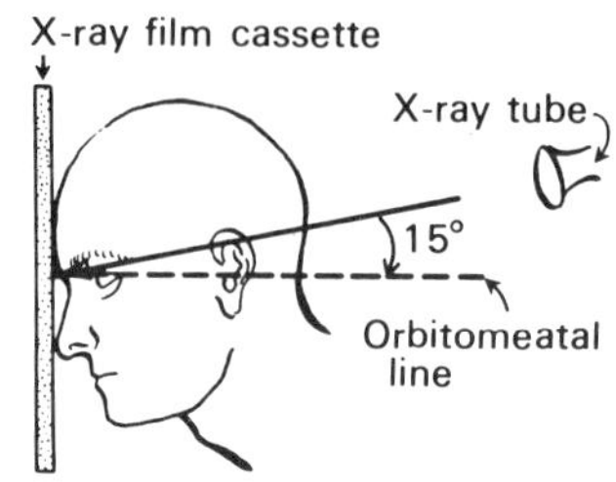

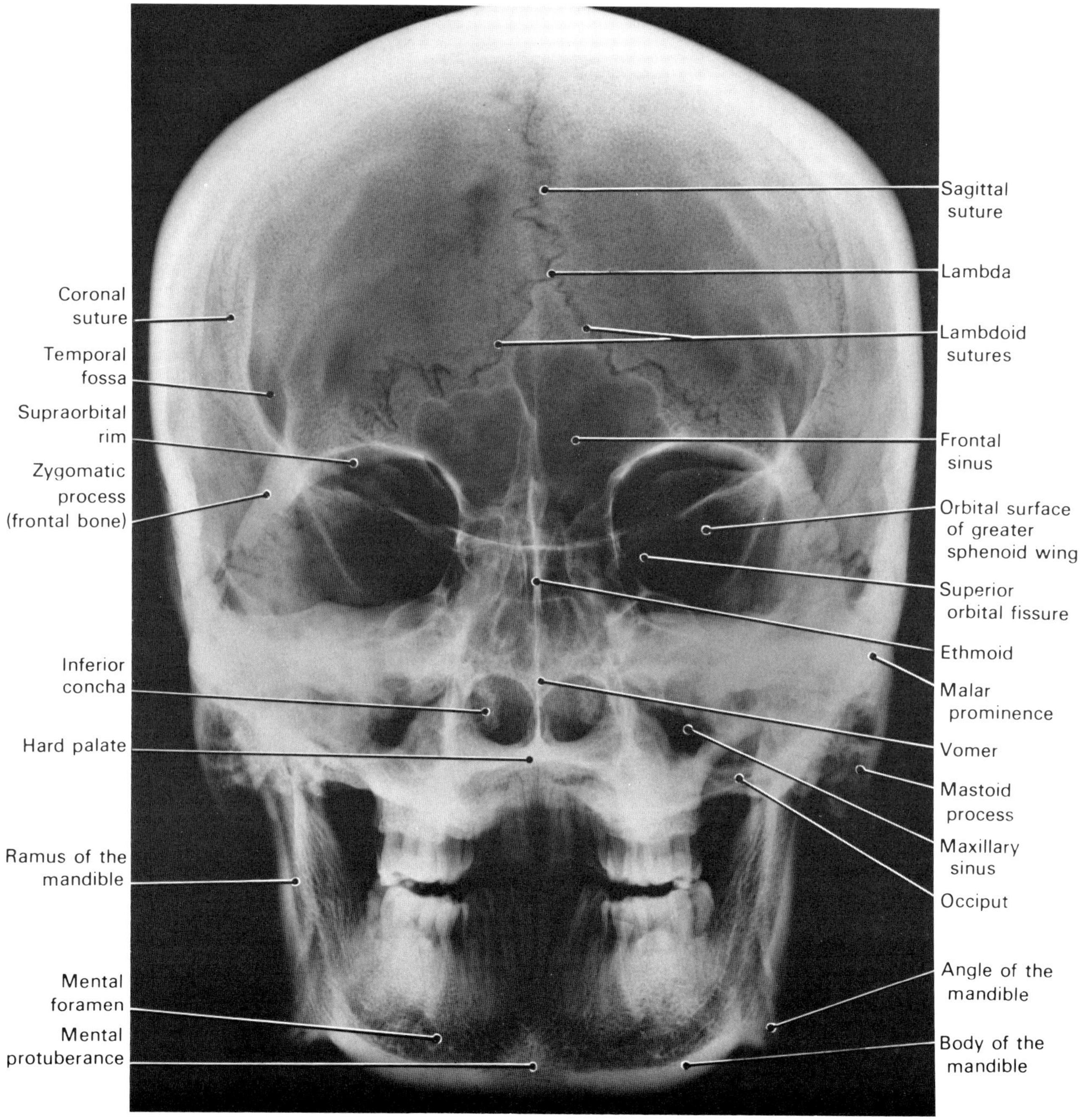

Fig. VI-44. The inclined posterior-anterior projection of the skull. **A.** Radiograph of the dry skull. Superficial structures labeled.

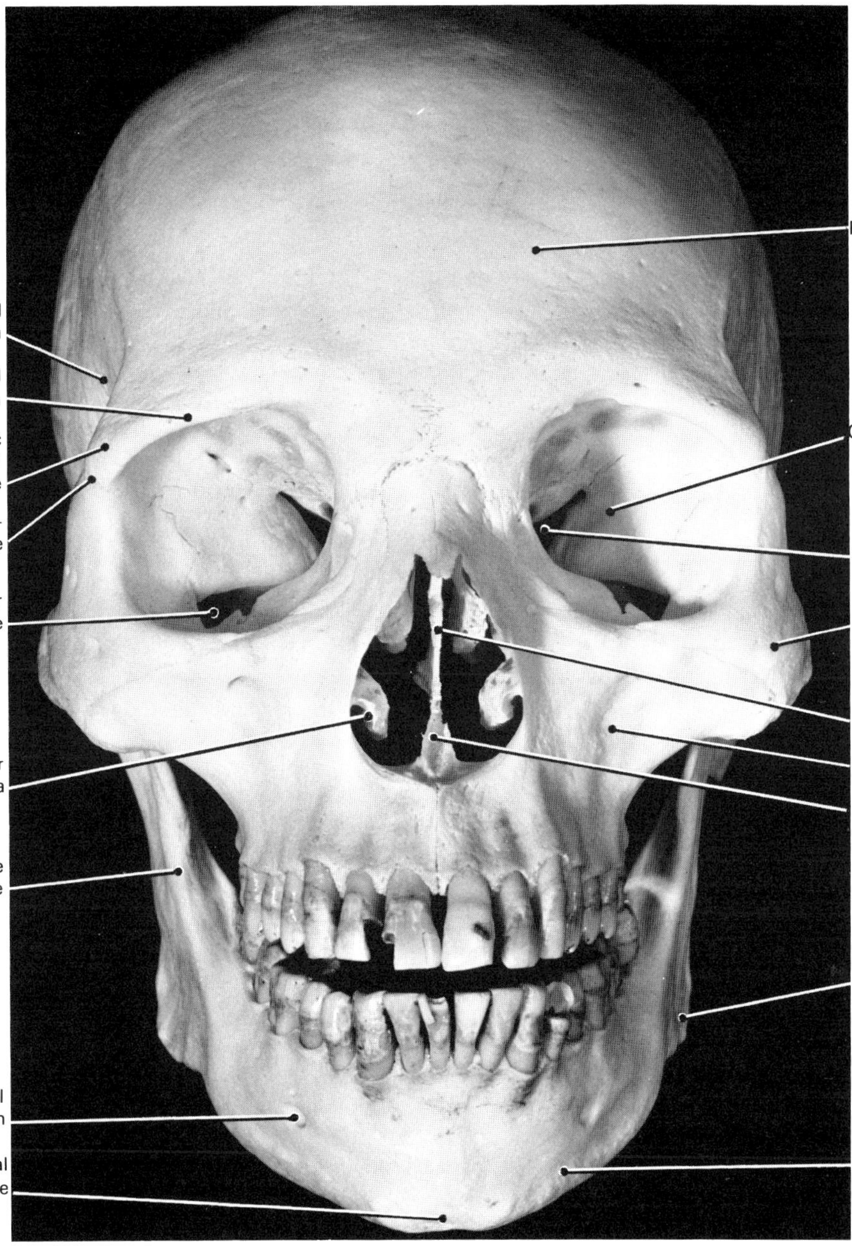

Fig. VI-44B. Photograph of the dry skull from the position of the film cassette.

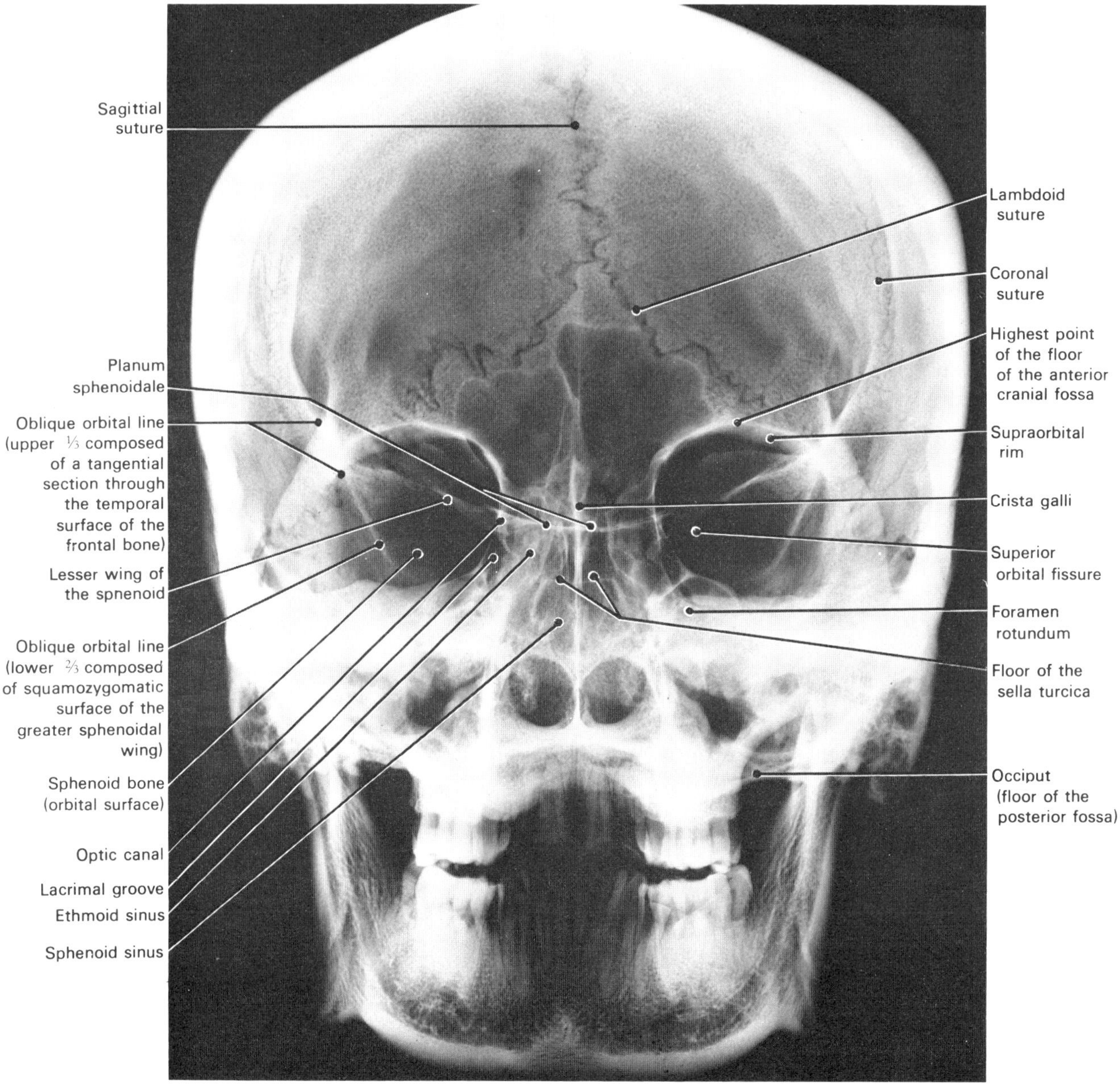

Fig. VI-44C. Same projection with the deep structures labeled.

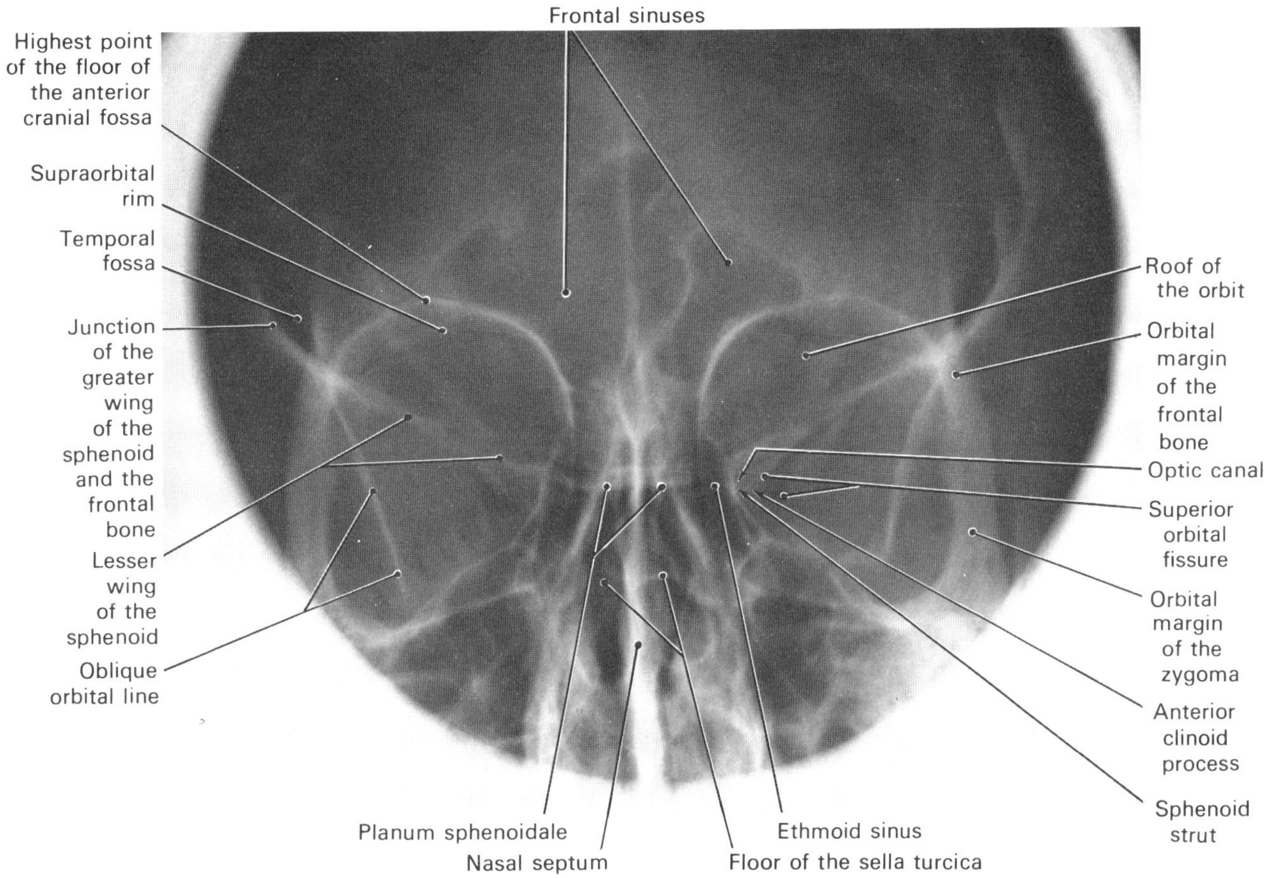

Fig. VI-44D. Coned inclined PA view in the living subject.

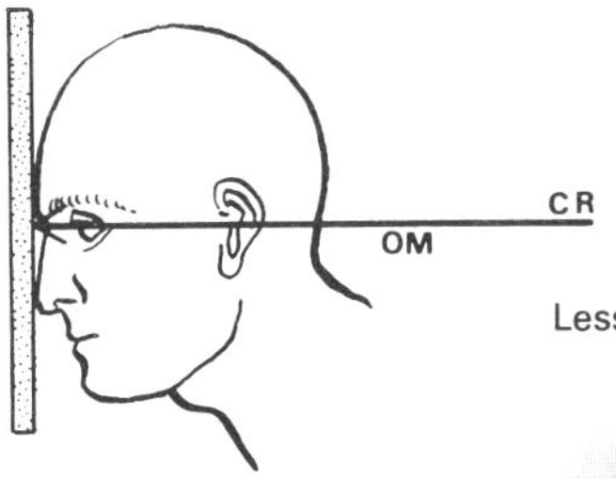

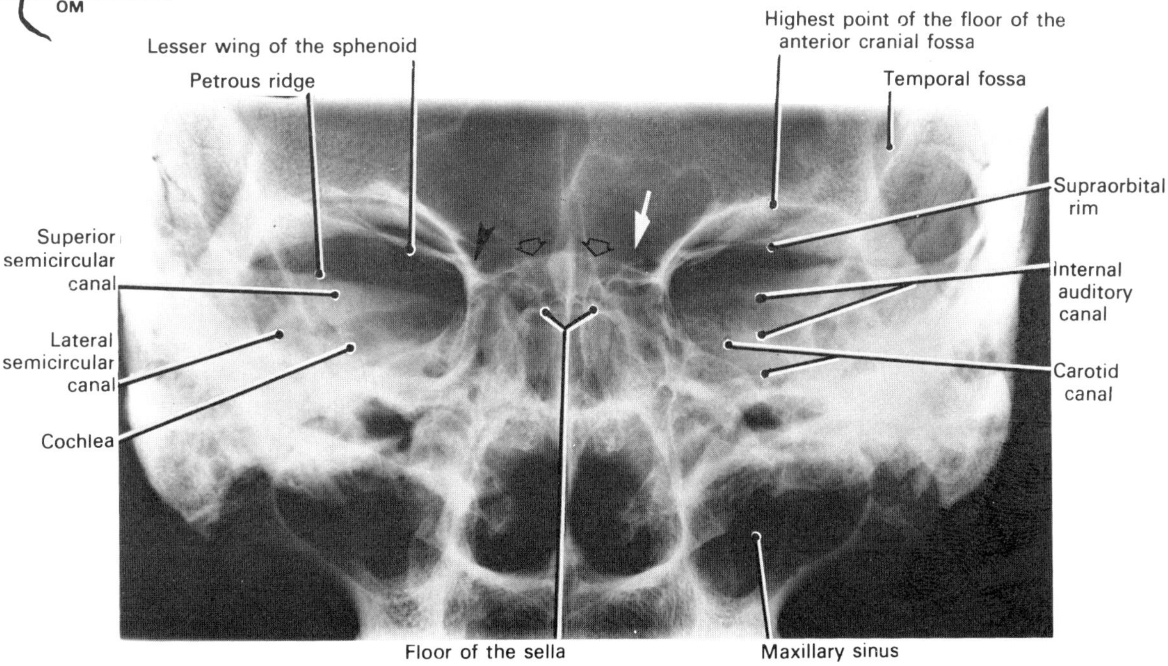

Fig. VI-44E. Straight PA projection of the dry skull. The central X-ray beam is directed along the OM line in order to project the petrous ridges in the orbits. The *open arrows* indicate the line cast by the tuberculum sellae; the *black arrows head* points to the anterior clinoid process, while the *white arrow* shows the posterior clinoid process.

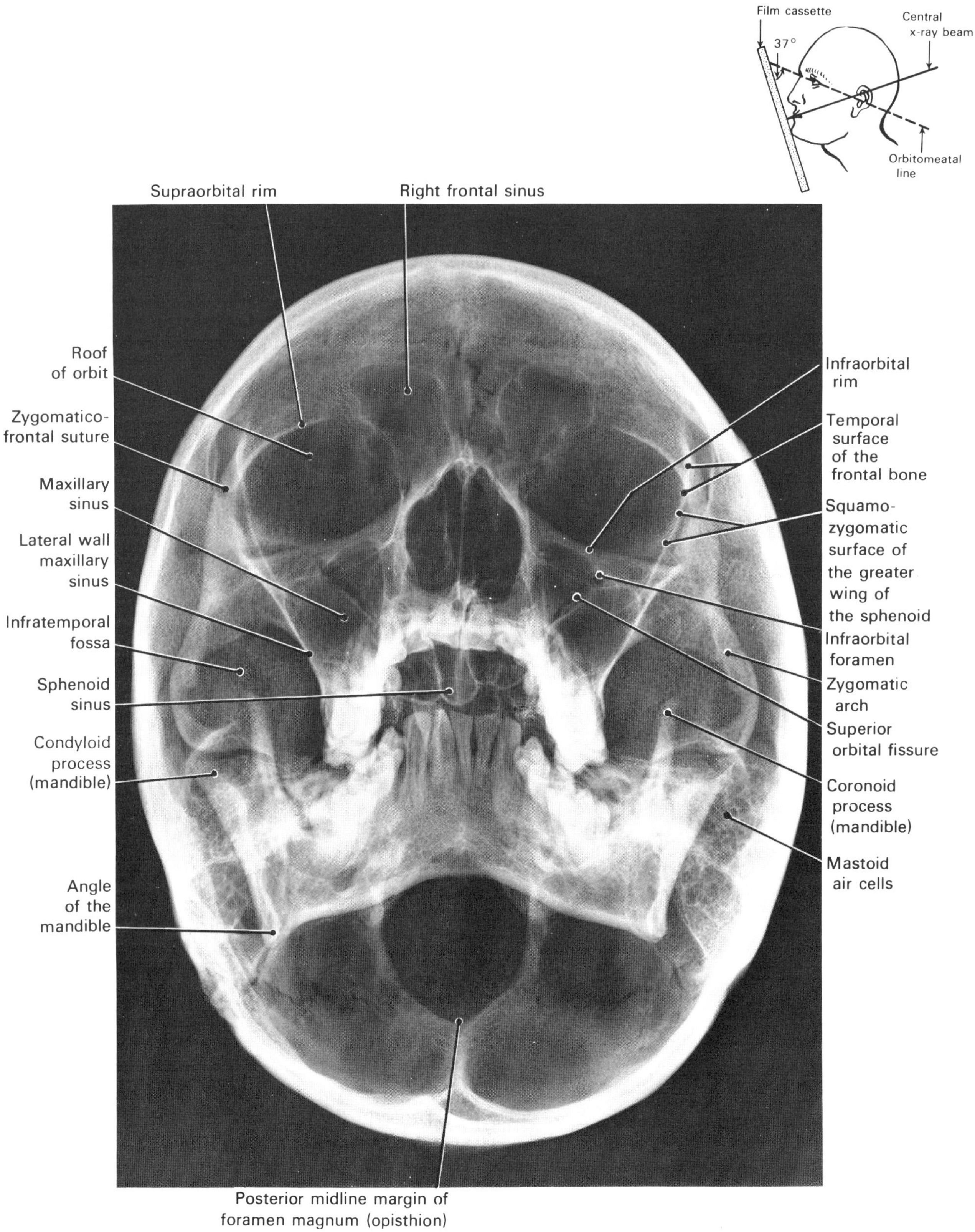

Fig. VI-45. The skull in the occipitomental projection. **A.** Radiograph of the dry skull.

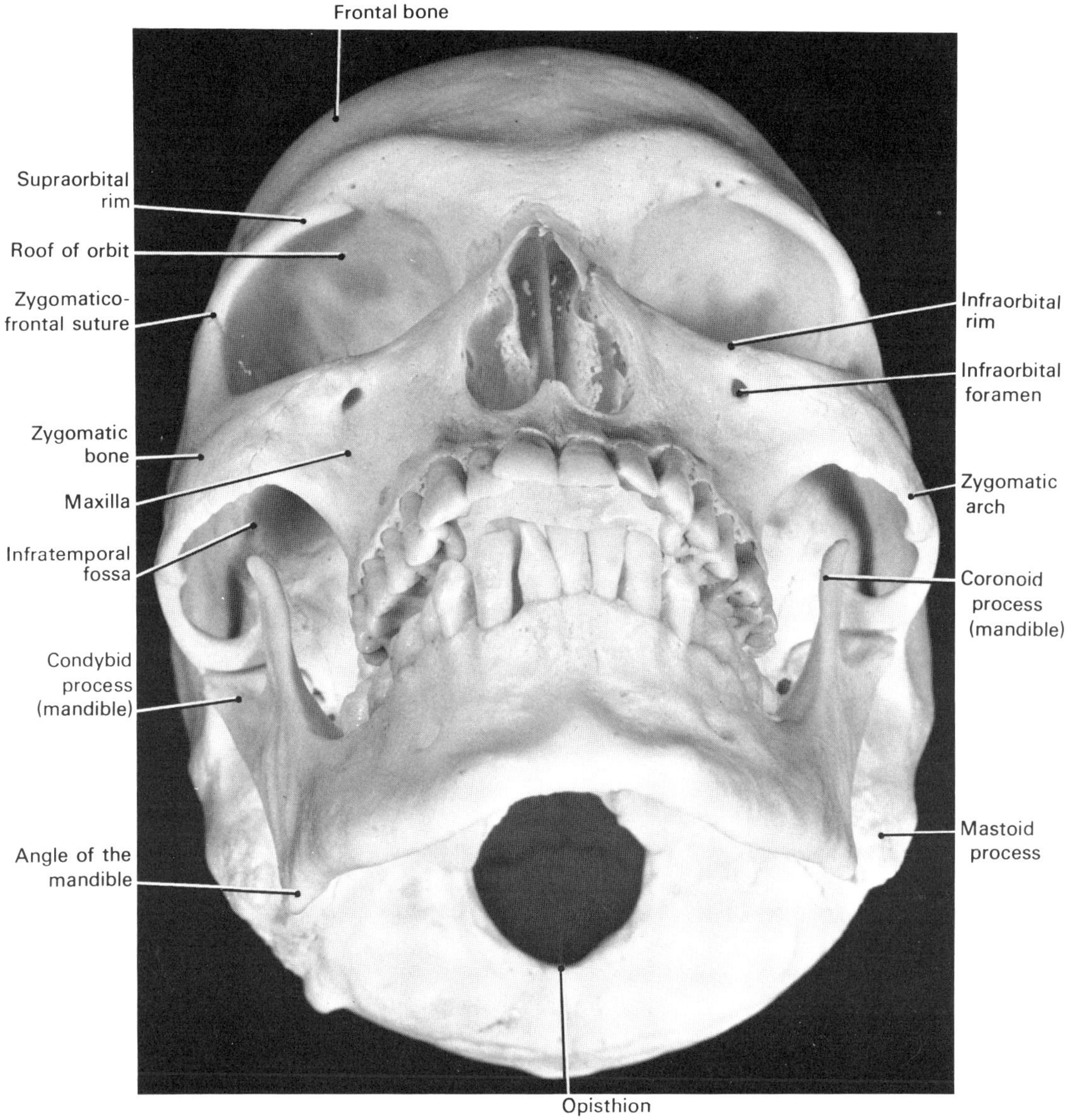

Fig. VI-45B. Photograph of the same dry skull as Figure VI-45A from the position of the film cassette.

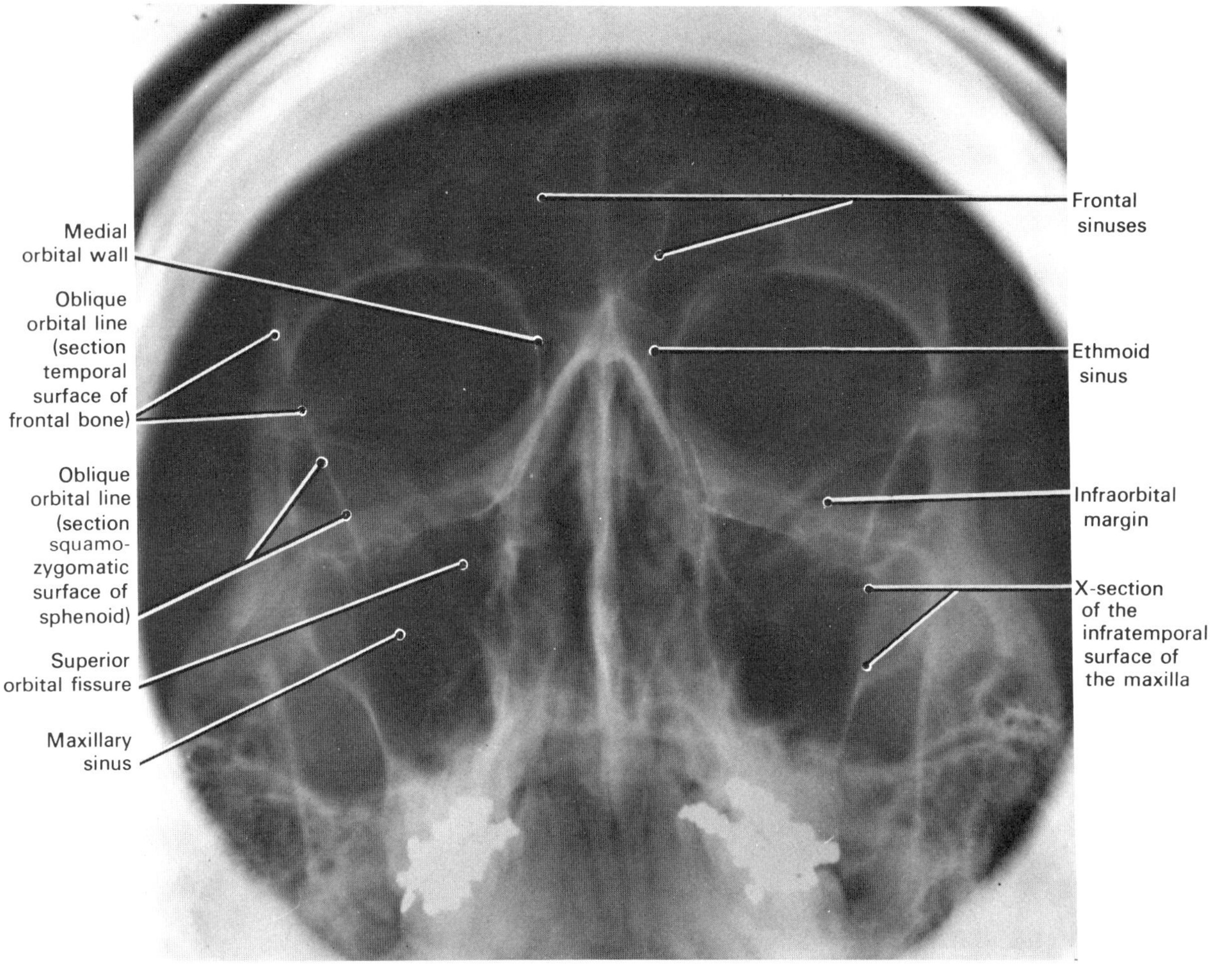

Fig. VI-45C. Coned radiograph in the living subject—closed mouth.

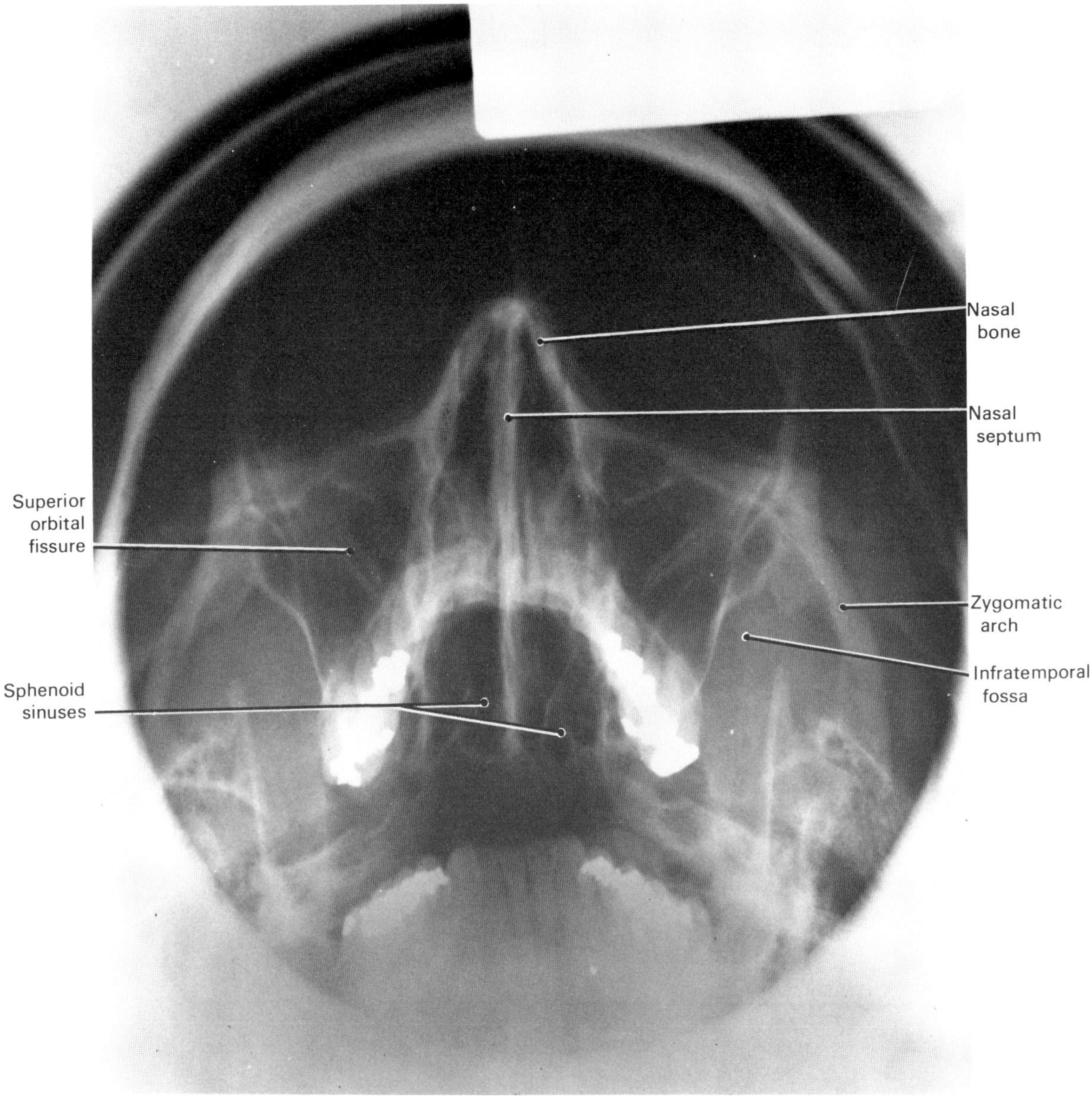

Fig. VI-45D. Coned radiograph in the living subject with the mouth opened to show the sphenoid air cells. This view is more exaggerated than that in C.

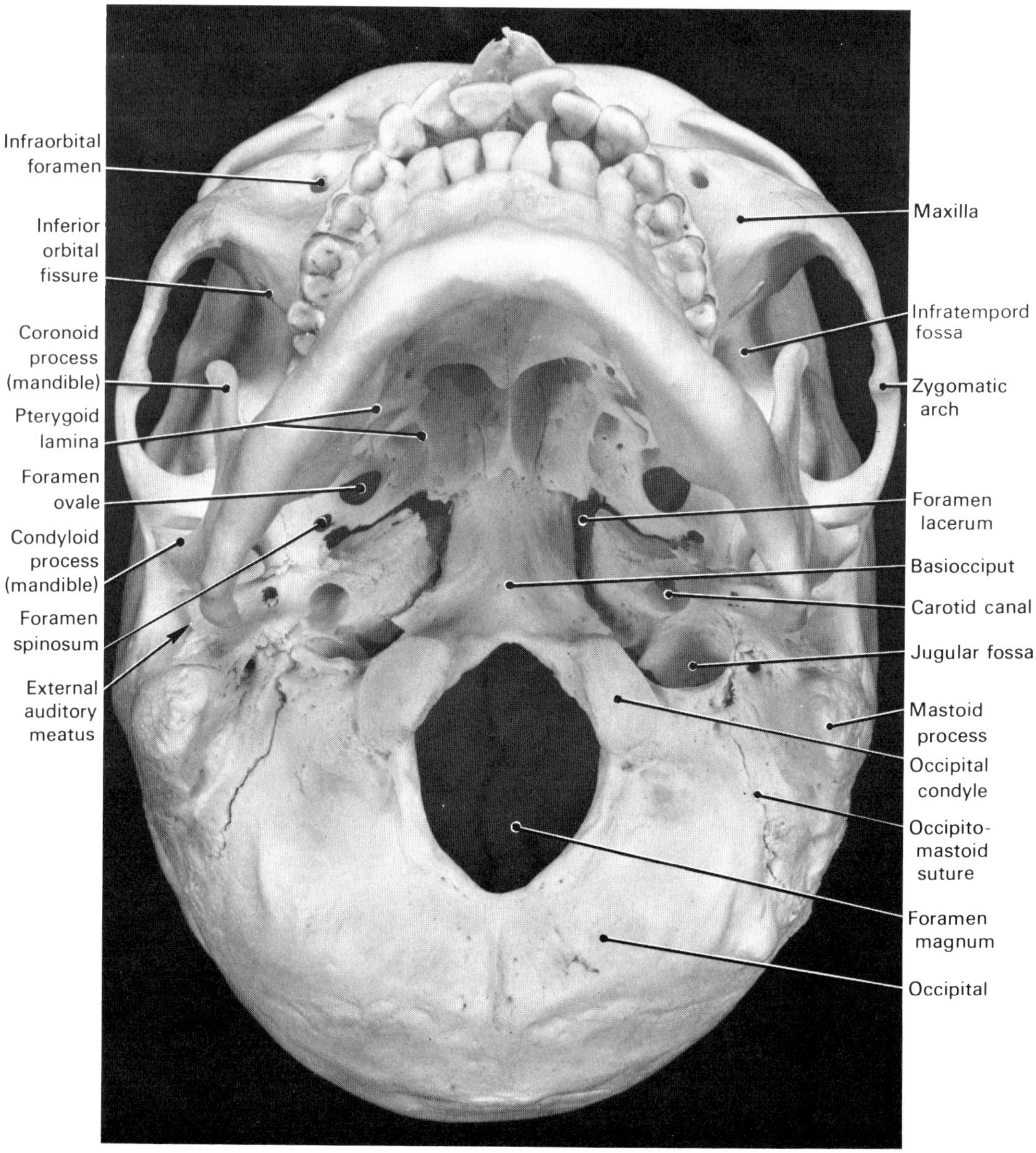

Fig. VI-46. Basal projection of the skull. **A.** Photograph of the dry skull taken from the position of the X-ray tube.

d. Basal (axial or submentovertex) projection: *General:* In order to give you a better understanding of how structures appear when a three-dimensional skull is reduced to two dimensions, we have presented exterior and interior photographs (Fig. VI-46A and B) of the dry skull which correspond with the radiograph in Figure VI-46C. Please do not expect to find every structure that you can see clearly in the dry skull photographs; there is simply too much overlap of structures for that.

Discussion: The ethmoid, maxillary and sphenoid sinuses are all visible in this projection. Note that the sphenoid sinus is divided by a bony septum which is not midline. The sphenoid air cells are surrounded by a bony cortical margin normally. As we have commented earlier, the sinuses may be inconstant in size and extent.

Between the curving line of the posterior wall of the maxillary sinus and the posterior orbital surface of the greater sphenoidal wing is the inferior orbital fissure (top of Fig. VI-46D).

Identify the pterygoid lamina (Fig. VI-46D) which

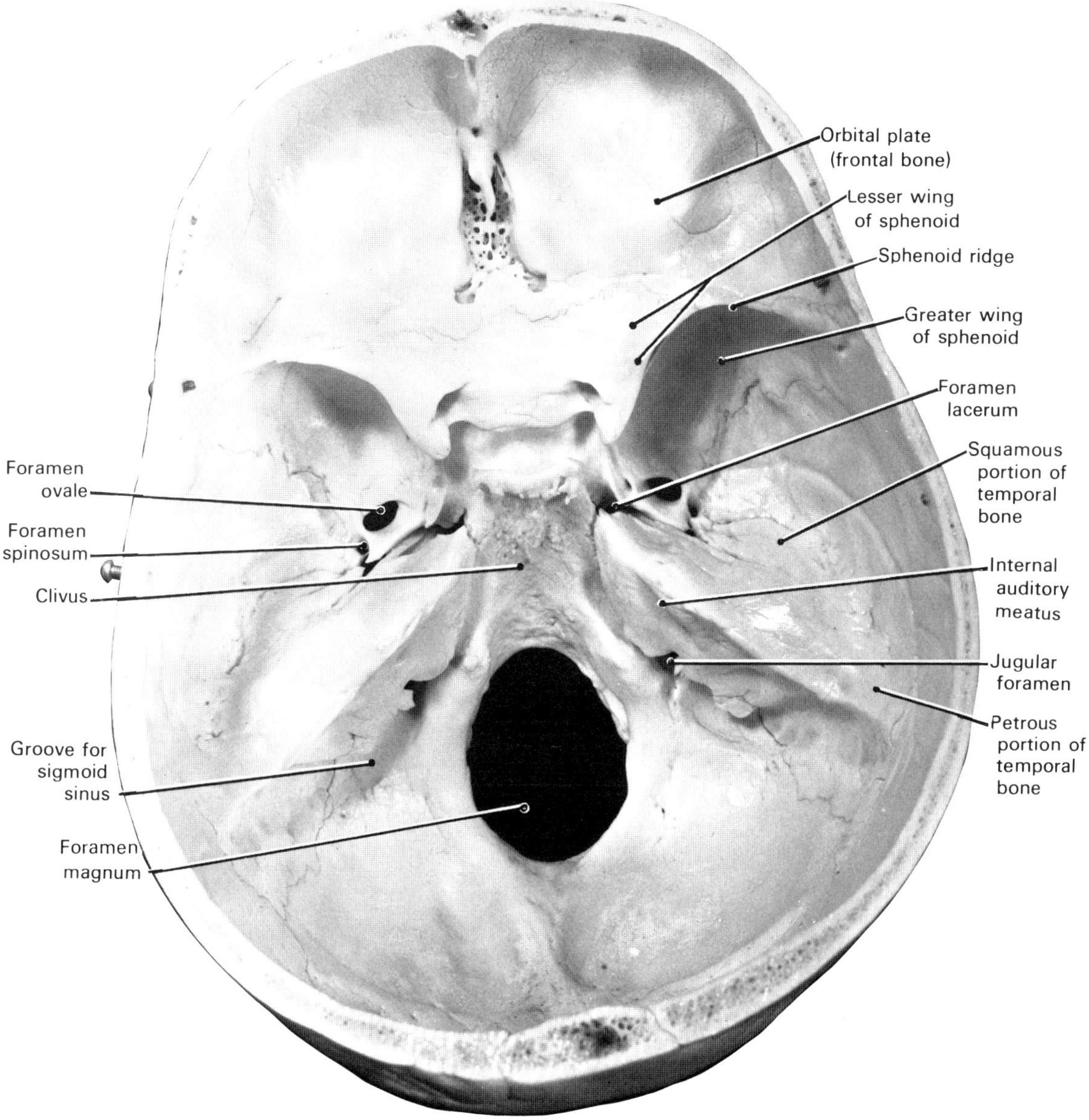

Fig. VI-46B. Photograph of the dry skull taken from the position of the film cassette.

are projected over the most medial margin of the **sphenoid ridge.** The hamulus is rarely identified due to superimiposition of other structures on the small hook.

Now carefully examine Figure VI-46D and note the dense anteriorly directed convexity of the greater sphenoidal wing which is the anterior aspect of the middle cranial fossa. Immediately posterior to this dense white line is an accompanying thinner line which is in fact the sphenoid ridge lying slightly more posterior. The sphenoid ridge is formed by the greater wing of the sphenoid in its outer one-third, and by the lesser wing of the sphenoid in its medial two-thirds.

Identify the foramen ovale (Fig. VI-46D) which, as in this case, may be somewhat asymmetric and still be considered normal. Notice that the foramen ovale on the left is rather indistinct laterally because its wall is quite thinned but not eroded. Rarely, this foramen is divided into two parts as the pterygo-alar ligament calcifies. Immediately behind the foramen spinosum a small spur of dense bone is visualized. This is the spine of the sphenoid.

The eustachian tube passes through the eustachian canal, and the canal may be visualized behind and medial to the foramen spinosum. Note the position of the carotid canal as it courses toward the foramen lacerum. Recall that the carotid artery crosses the fo-

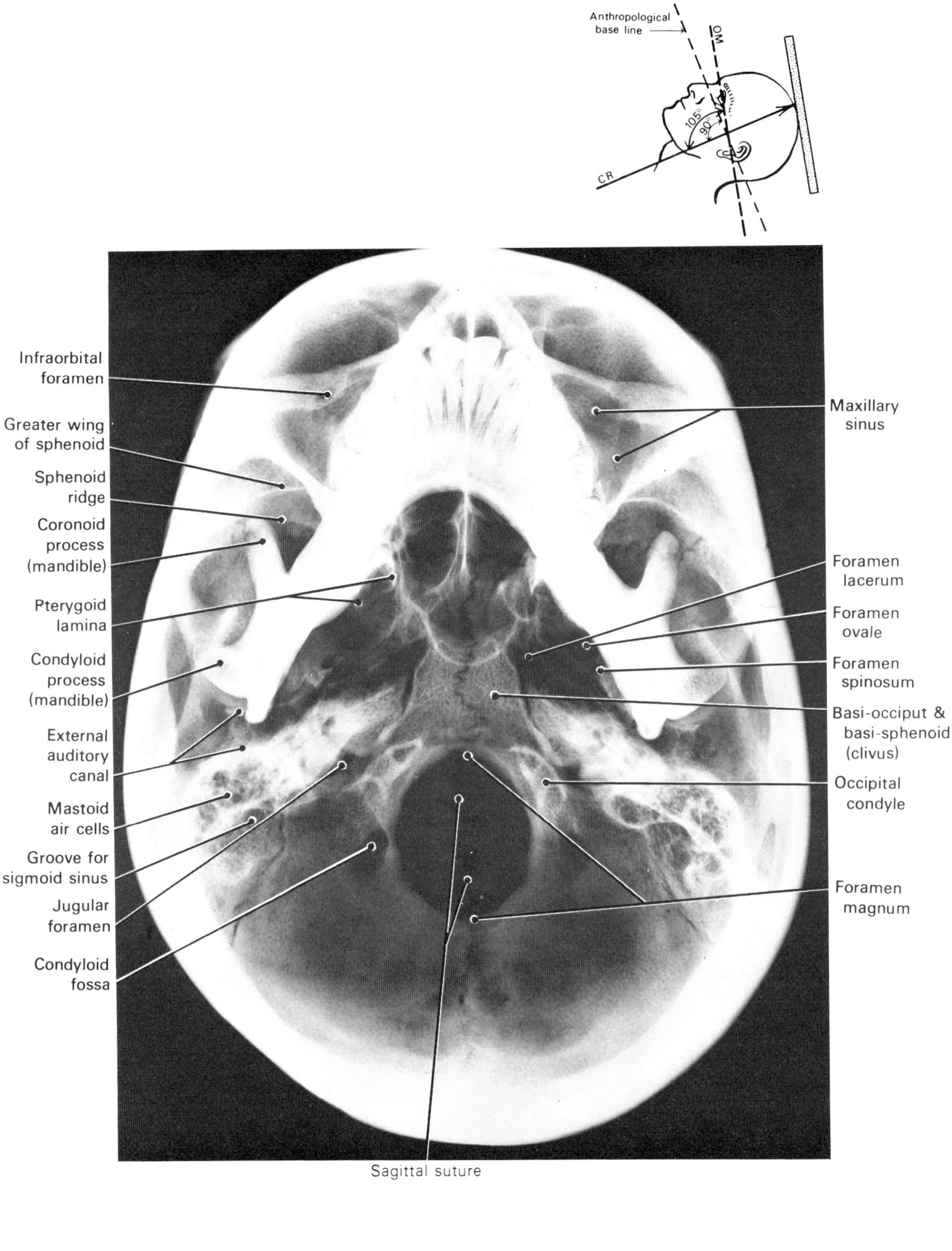

Fig. VI-46C. Radiograph of the dry skull—a composite of A and B.

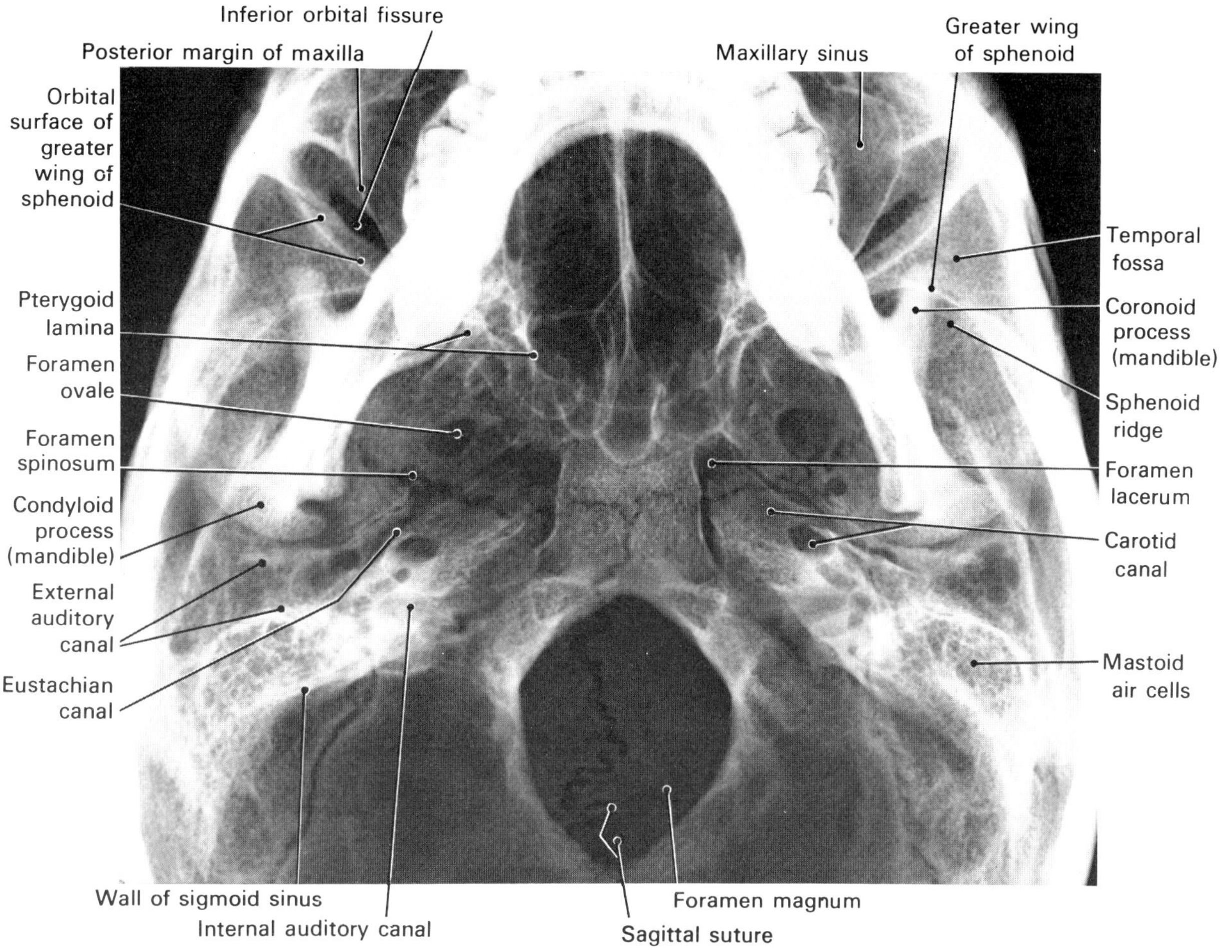

Fig. VI-46D. Enlargement of significant structures of Figure VI-46C.

ramen lacerum and does not pass through it. In life this foramen is closed by cartilage, and is not usually seen to advantage.

The jugular fossae are typically asymmetric, and care must be exercised to avoid overinterpreting the presence of tumor in these instances. Special views and tomography are usually required to rule on tumors in this site.

e. Anterior-posterior half-axial (Towne's) projection (Fig. VI-47): *General:* the combination of photographs in Figure VI-47A and B when reduced to two dimensions results in the radiograph of Figure VI-47C.

Discussion: This projection is useful in evaluating the occipital bone, and many of the structures of the anterior, middle, and posterior fossae. Identify the sagittal, coronal, lambdoid, and occipitomastoid sutures. The latter may give trouble at times because of its straight course, and it can simulate a fracture. Its serrated border, perisutural sclerosis and, of course, your knowing where it should be, will help to avoid this pitfall.

Identify the grooves on the occipital bone along the course of the transverse and superior sagittal sinuses which join at the torcula Herophili. The point of insertion of the falx on the internal occipital crest may also be seen.

In the midline the impressive opening of the foramen magnum may be seen. Framed within it is the dorsum sellae, crowned laterally by its posterior clinoid processes. Immediately beneath the dorsum is the thickened bony shadow of the superimposed basiocciput and basisphenoid (the clivus).

The dense petrous bone is shown to good advantage in this view. Within it one may see the external and internal auditory canals, the semicircular canals and cochlea, and the mastoid antrum.

Note the anterior margin of the middle cranial fossa. This shadow is cast by the greater wing of the sphenoid bone which is curved and directed convexly down-

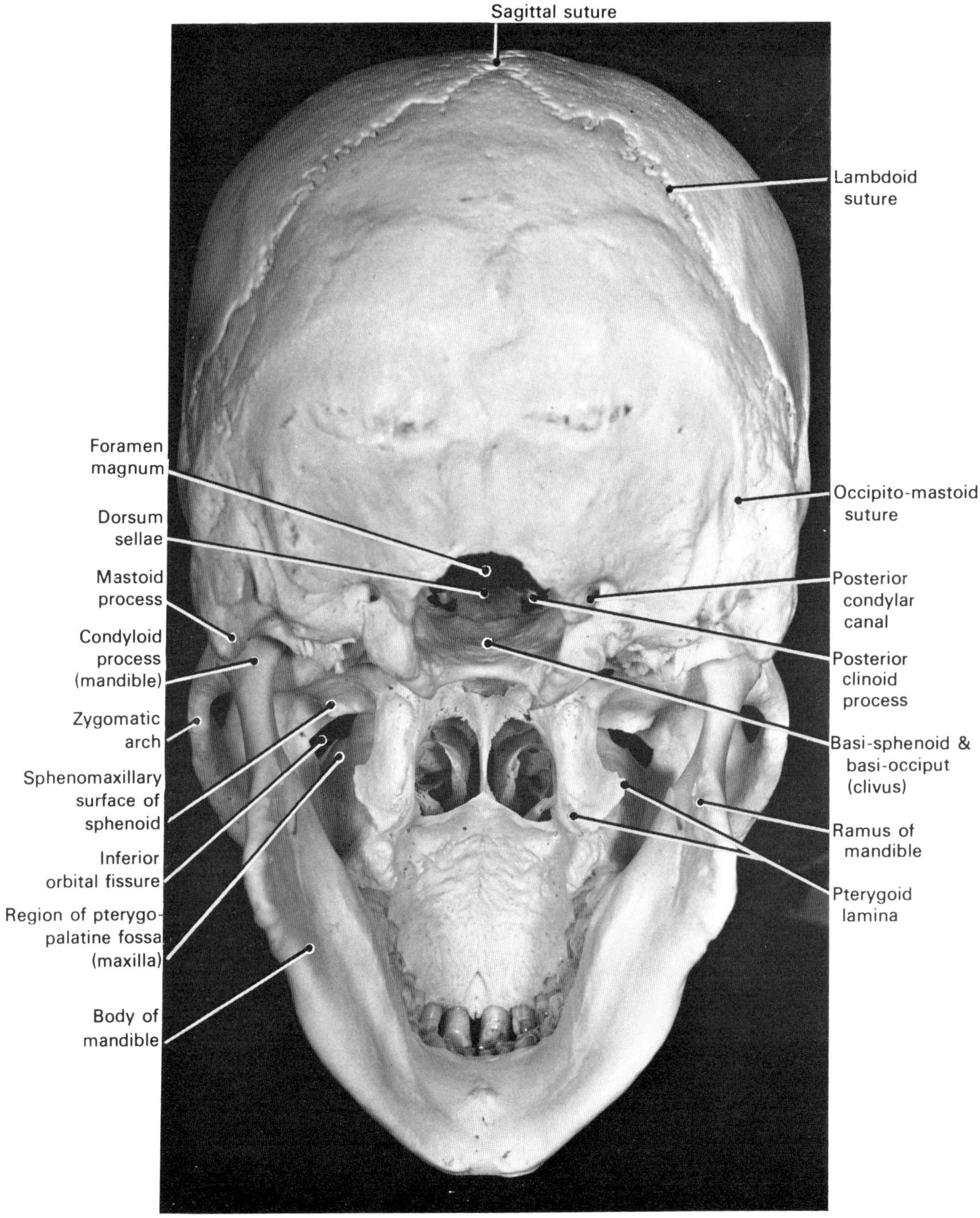

Fig. VI-47. Anterior-posterior half-axial projection. **A.** Photograph of the dry skull taken from the position of the film cassette.

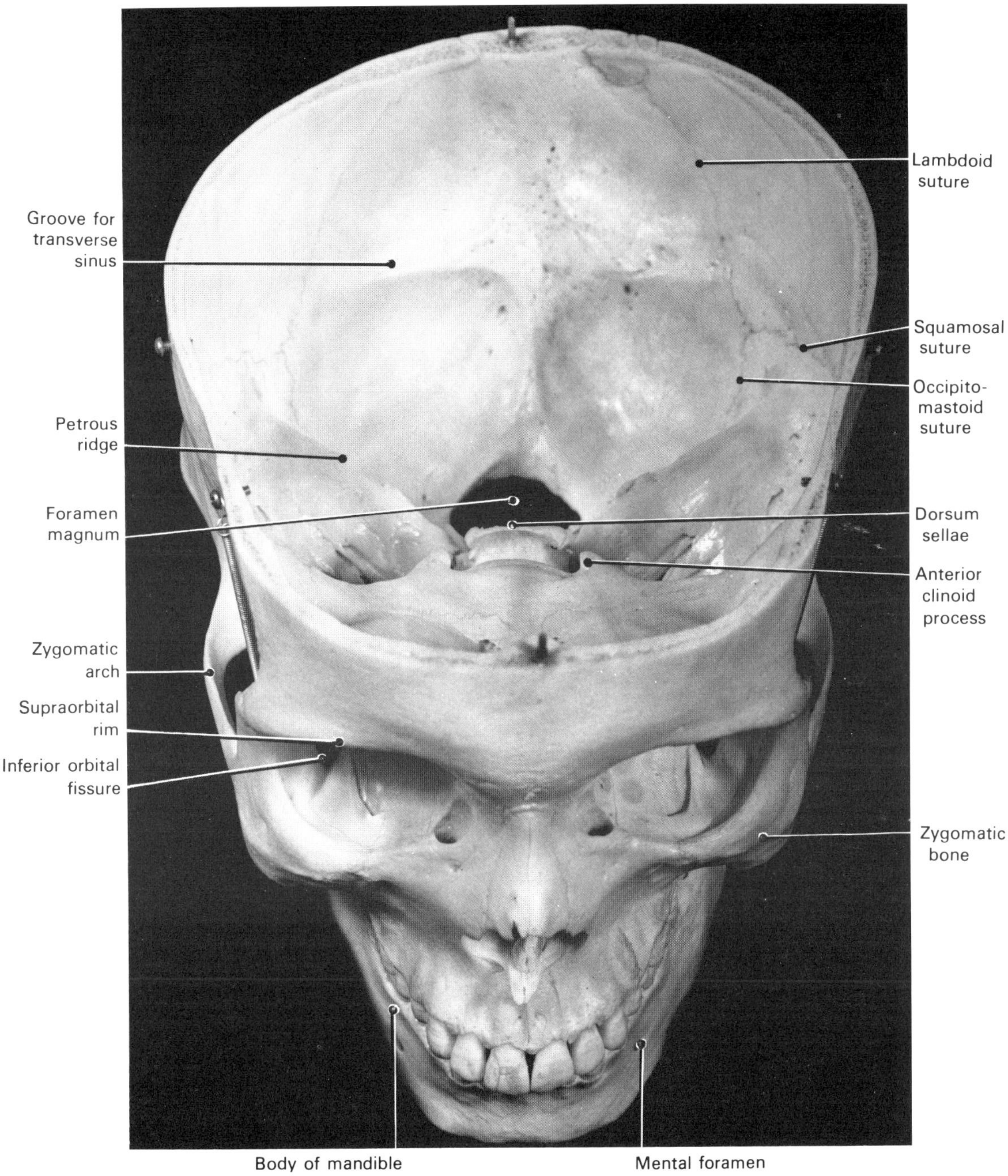

Fig. VI-47B. Photograph of the dry skull taken from the position of the X-ray tube.

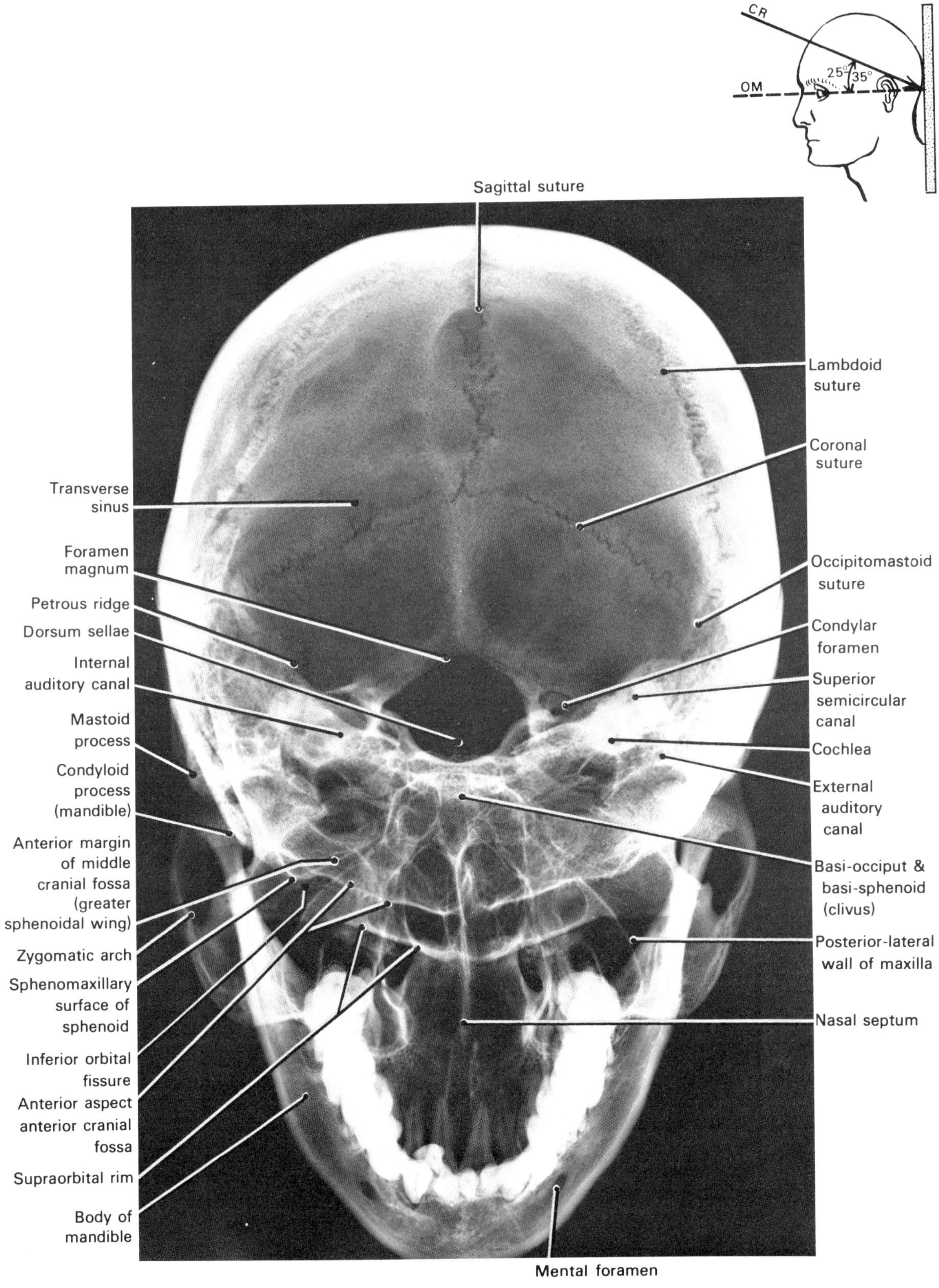

Fig. VI-47C. Radiograph of the dry skull. A composite of Figures VI-47A and B.

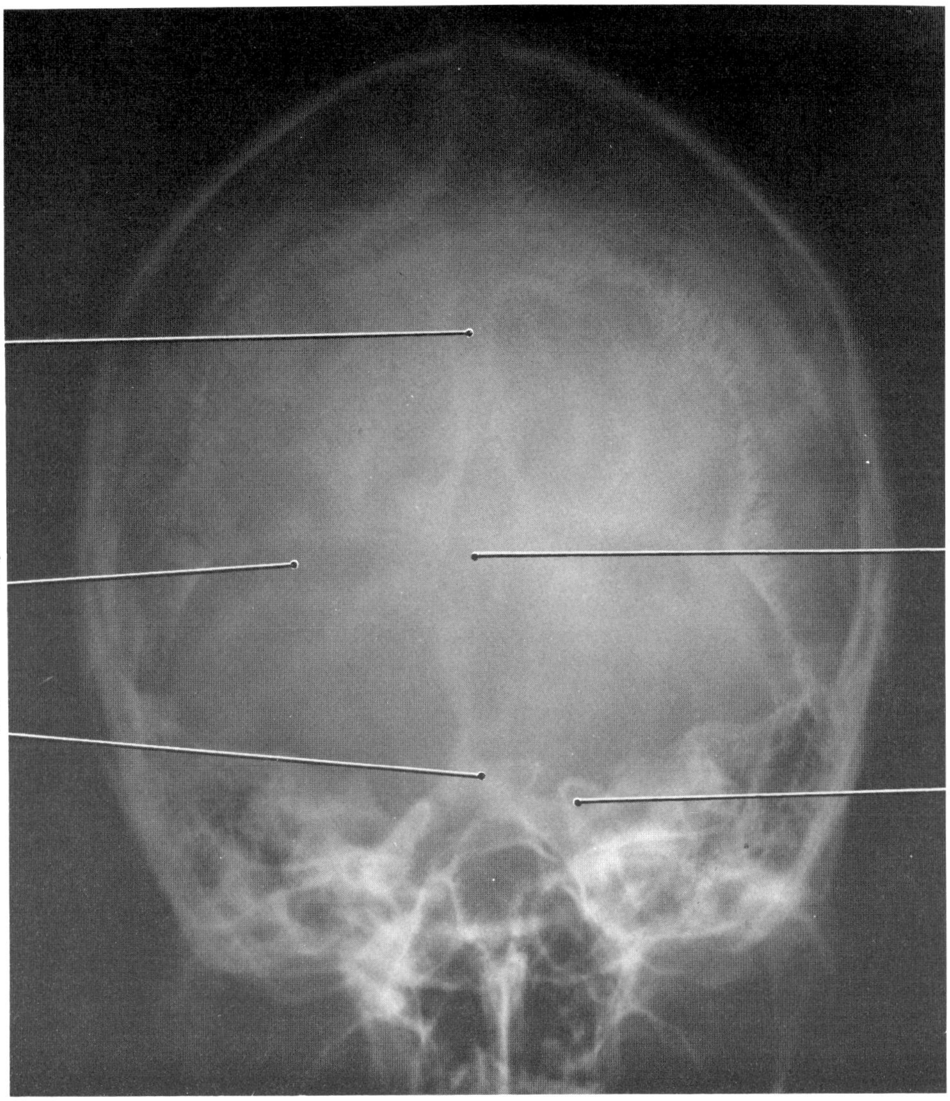

Fig. VI-47D. Modified half-axial projection. As demonstrated here, when angulation is decreased to about 25°, the foramen magnum can not be seen.

ward in this view. Any asymmetry of the middle fossa could readily be appreciated here and the pathology sought out.

Compare the anatomic details in Figure VI-47A and C, and identify the inferior orbital fissure peeking out between the sphenomaxillary portions of the sphenoid bone and the posterior lateral wall of the maxillary antrum. One can even identify the pterygoid lamina quite well although they are in part superimposed by the palpable supraorbital rim. Immediately above the supraorbital rim and, seemingly, paralleling it, may be seen the most anterior and inferior part of the anterior fossa.

Notice what an excellent view this is when examining a patient with facial trauma. Beautifully seen are the zygomatic arches and the condyles of the mandible as well as the ramus and body of the mandible. Again, this frontal projection affords an opportunity to compare sides for symmetry of structure.

Modified half-axial projection: Many radiologists prefer a modified half-axial view to the standard projection. Instead of angling the X-ray tube 35° caudally, one angulates it only 25° caudally. The resultant radiograph (Fig. VI-47D) shows that the foramen magnum is now obscured. However, the petrous pyramids are better visualized, and in addition, the anterior clinoid processes can be seen.

4. Radiographic Anatomy of the Newborn Skull (Fig. VI-48)

In order to prepare you better for a later course in pediatrics, we have included here the oft-times con-

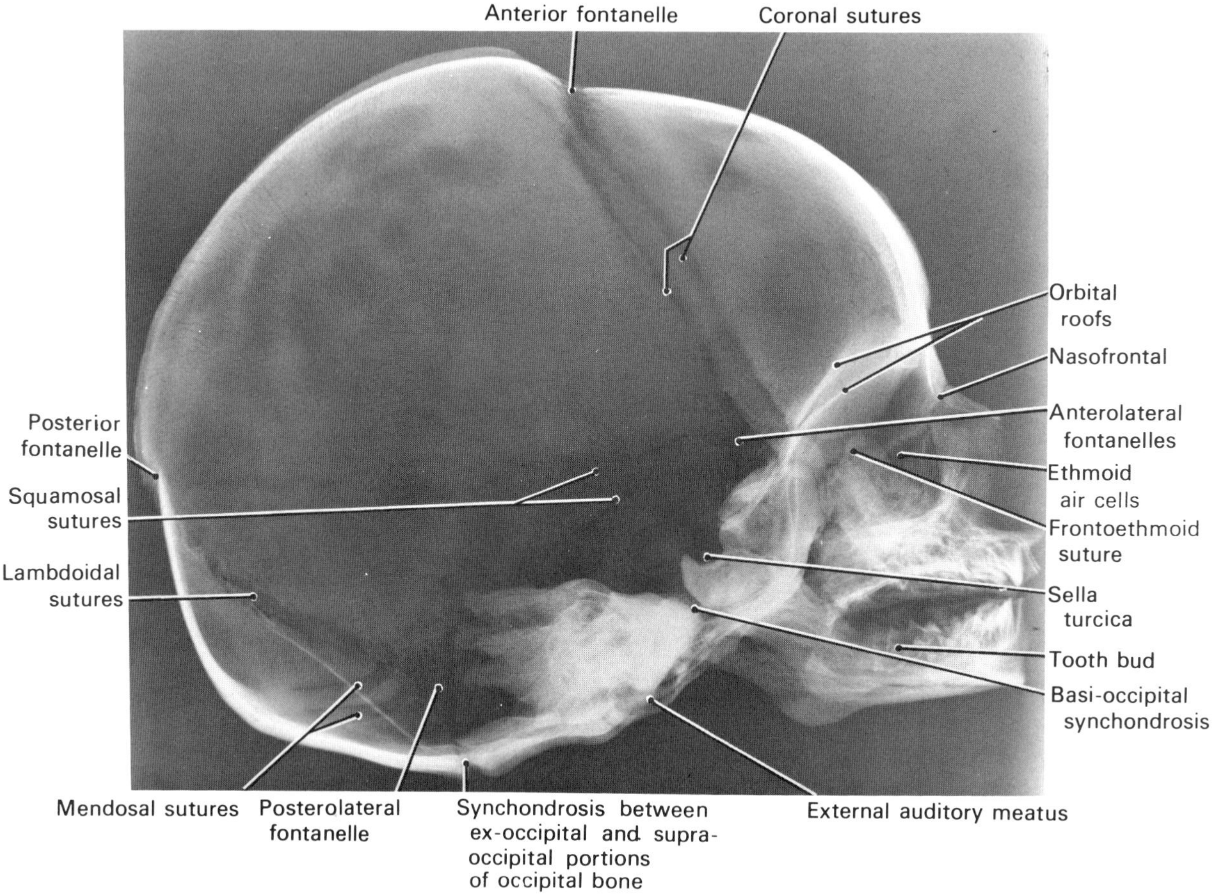

Fig. VI-48. Normal radiographic anatomy of the newborn skull. **A.** Lateral projection.

fusing anatomy of the newborn skull. The skull of these little people contains a myriad of sutures, fontanelles, and synchondroses which are not found in the adult. Fortunately, the anatomy that we have just reviewed applies quite well in the newborn, and we will not dwell upon those structures which by now should be well in mind.

In Figure VI-48A it is immediately apparent that the normal facial area is relatively small when compared with the great bulk of the cranium. The bones of the skull are quite thin, since they have not yet formed separate inner and outer tables with diploe interposed. Therefore, the calvaria casts a roentgen shadow of uniform density. Many of the vascular channels etched into the calvaria are visible this early and should not be mistaken for a fracture. The parietal bone, notoriously, has many, tiny, linear fissures radiating down from the sagittal suture (Fig. VI-48A) which simulate fractures to the unwary. At times, Wormian bones may also be confused as abnormalities. These are small accessory bones which may be found in the sutures.

The frontal and parietal eminences are extremely convex and therefore prominent. Notice that the edges of the bones are separated and have not yet become serrated as in the adult.

The paranasal sinuses are virtually non-existent. The frontal sinuses, that are present, are to be found amid the anterior group of ethmoid air cells (Fig. VI-48A and B), and it is not possible to separate them. The maxillary sinuses are elongated from front to back, but appear slit-like in frontal views. They are seen medial to the infraorbital foramina. The ethmoid sinuses are sufficiently formed to be readily visible (Fig. VI-48A and B). The sphenoid sinuses are rudimentary at this age. During the first decade of life the paranasal sinuses gradually increase in dimension, and by the end of this time, closely approximate those of the adult.

At intervals between the bones of the skull are membranous spaces which are called *fontanelles.* These are usually six fontanelles: the unpaired **anterior**

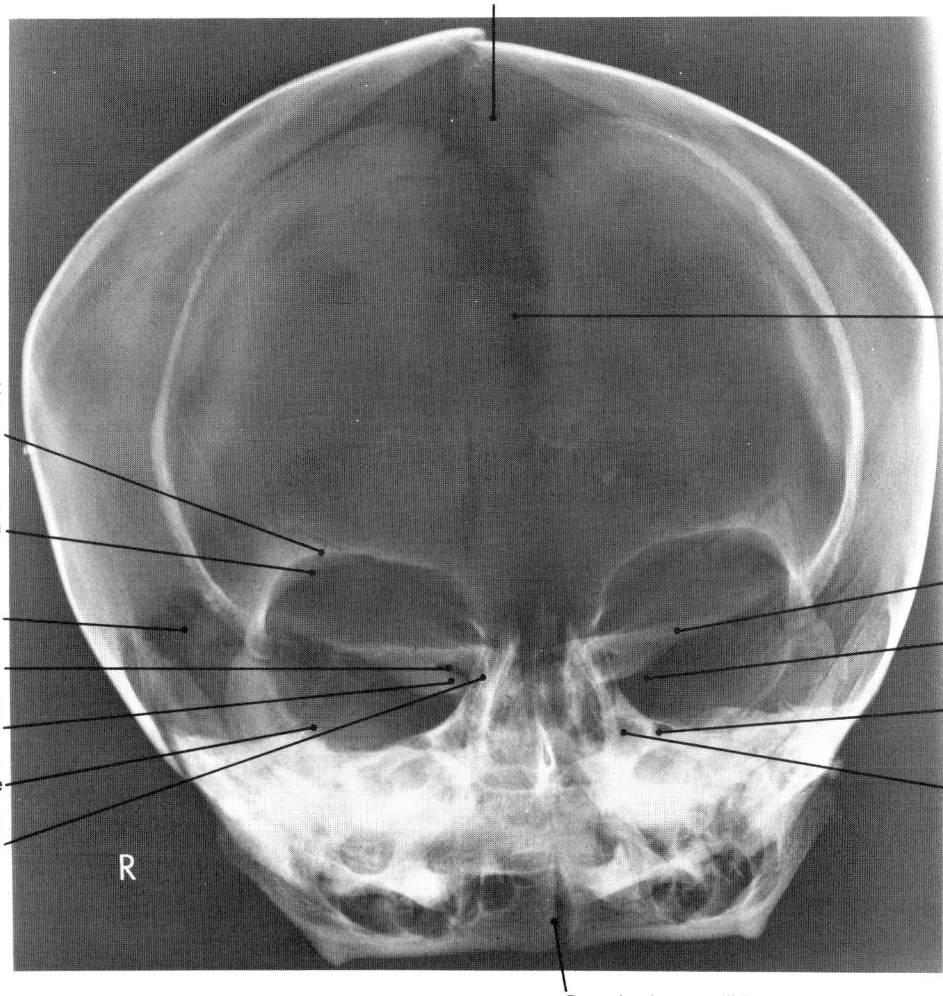

Fig. VI-48B. Inclined PA projection of newborn skull.

(frontal, bregmatic) and **posterior** (occipital); and the paired **anterolateral** (sphenoid) and **posterolateral** (mastoid) fontanelles. Identify these in Figures VI-48A to E. It is apparent from the radiographs, that the lozenge-shaped anterior fontanelle is the largest of the six fontanelles. It is situated at the junction of the metopic (between the unfused frontal bones), coronal and sagittal sutures. The small triangular-shaped posterior fontanelle may be closed at birth, but when present is located at the junction of the lambdoidal and sagittal sutures. In obstetrics, palpation of the anterior (lozenge-shaped) and posterior (triangular-shaped) fontanelles through the mother's rectum permits the physician to determine the position of the fetal skull. At the sphenoid angle of the parietal bone, the irregularly-shaped, anterolateral fontanelles are found. These are bounded by the frontal, parietal, and temporal bones, and by the greater wing of the sphenoid (Fig. VI-48A). Posterolateral fontanelles are present at the mastoid angle of the parietal bone, or between the parietal, temporal, and occipital bones (Fig. VI-48A).

Within a few months of birth, the anterolateral and posterolateral fontanelles are closed. As we have mentioned before, the posterior fontanelle may have closed before birth, or it, too, may close within the first 2 months of life. During the second year of life, the anterior fontanelle usually closes. Because the exact time of closure of the fontanelles is variable, their presence or absence (closure) is not of much value in judging the age of an infant.

In the various illustrations identify the following **sutures:** nasofrontal (Fig. VI-48A); coronal (Fig. VI-48A and C); squamosal (Fig. VI-48A); lambdoidal

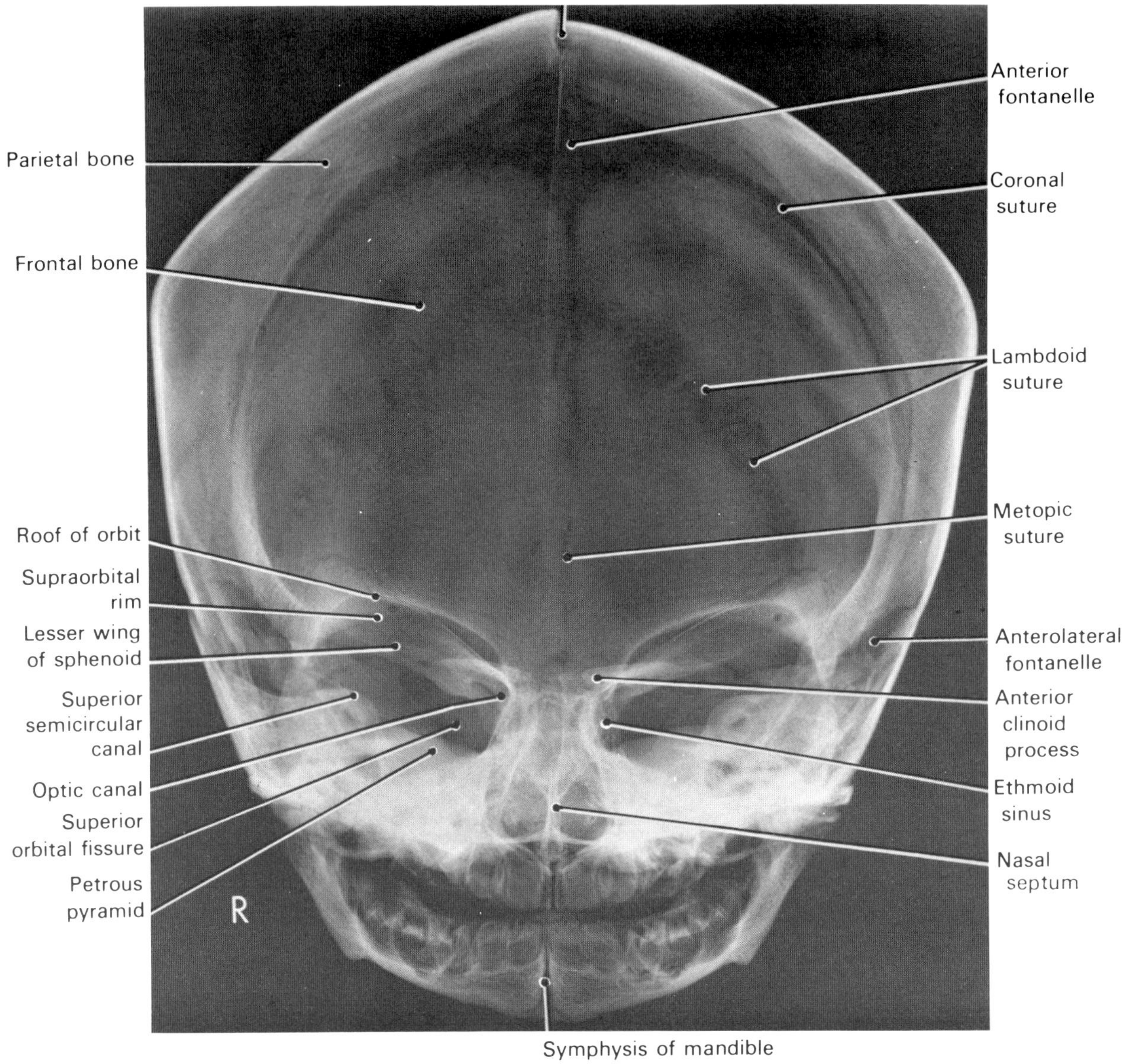

Fig. VI-48C. Straight PA projection of newborn skull.

(Fig. VI-48A); mendosal (Fig. VI-48A, E, and F); frontoethmoid (Fig. VI-48A); and metopic (Fig. VI-48B).

Several **synchondroses** are also visible in the illustrations, namely: spheno-occipital; the one between the ex-occipital and supra-occipital portions of the occipital bone; and that between the basi-occipital and the ex-occipital portions of the occipital bone.

The sutures and synchondroses also close at variable times, but the mendosal suture usually closes within a few weeks of birth. The coronal, sagittal, and lambdoidal sutures begin to close late in the third decade of life. This progresses very slowly with marked variations. The spheno-occipital synchondrosis unites at about age 21. The synchondrosis between the ex-occipital and supra-occipital portions of the occipital bone close between the second and third years of life.

THE CRANIAL NERVES

The 12 pairs of nerves associated with the brainstem and brain have been mentioned from time to time. We have given certain details about some of them. More will be said about others of them as we encounter them. They are designated, as a rule, either by name or by Roman numeral. Name and number ought to be quickly associated. They are listed, and their chief functions are given in Table VI-5. Locate them in Figures VI-49A and B, and 50A and B, and, at the same time, notice their relation to dural folds, dural sinuses, and other vascular structures.

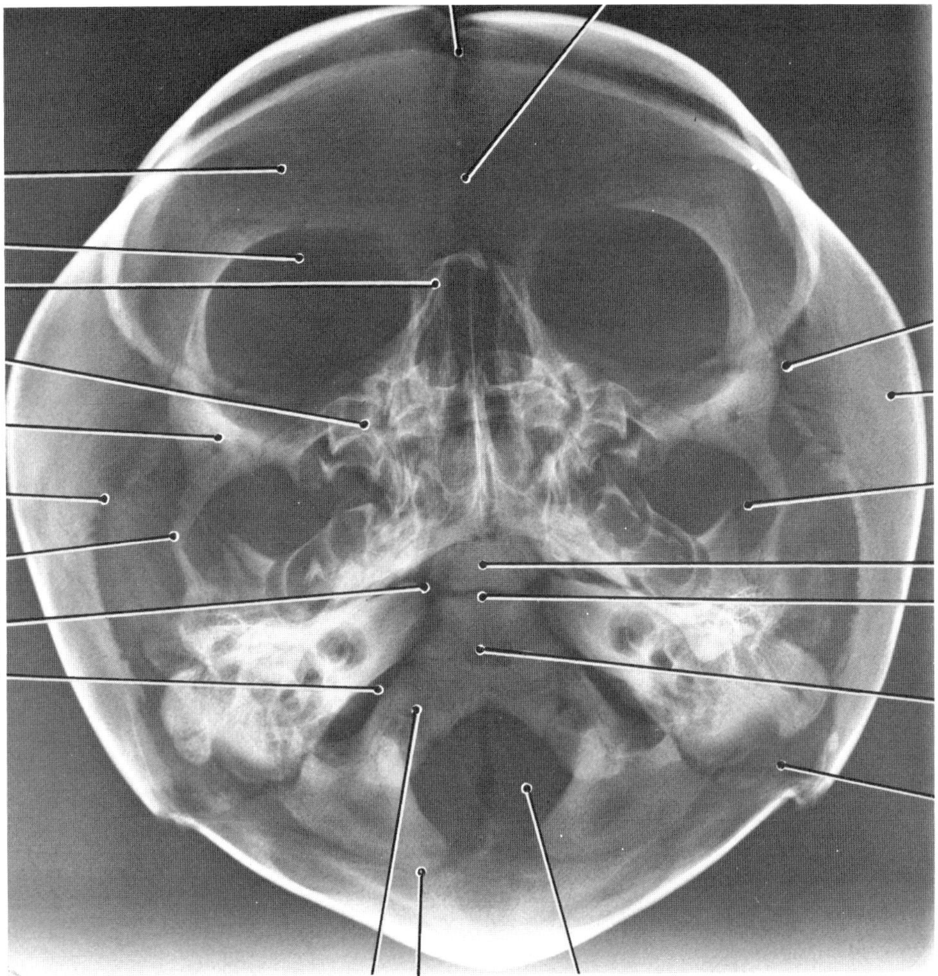

Fig. VI-48D. Occipitomental (Water's) projection of newborn skull.

If you pay attention to the "Components" section of Table VI-5, you will note that some cranial nerves convey functional fibers of more than one type and that, at times, there may even be different types of **motor and sensory** fibers in the same nerve. The following *functional groupings* evolve from the table:

A. Motor function:

1. Somatic motor: Nerve fibers that supply *skeletal* muscle derived from embryonic mesodermal somites: III, IV, VI, and XII (all factors of 12, mathematically)

2. Branchial motor: Nerve fibers that supply *skeletal* muscle derived from mesoderm of the branchial arch region: V, VII, IX, X, and XI

3. Visceral motor: Nerve fibers that supply **glands** and muscle derived from splanchnic mesoderm; *i.e.*, *smooth* and *cardiac* muscle: III, VII, IX, and X. (This functional group forms the *cranial outflow* of the **parasympathetic nervous system.**)

B. Sensory Function:

1. Special sensory: For smell, vision, hearing, and equilibrium: I, II, and VIII

2. Somatic sensory: Pain, temperature, touch, proprioception: V and X (the Woolworth twins)

3. Visceral sensory: General pain or feelings of distention from mucous membranes or walls of viscera: VII, IX, and X

4. Gustation (taste): VII, IX, and X

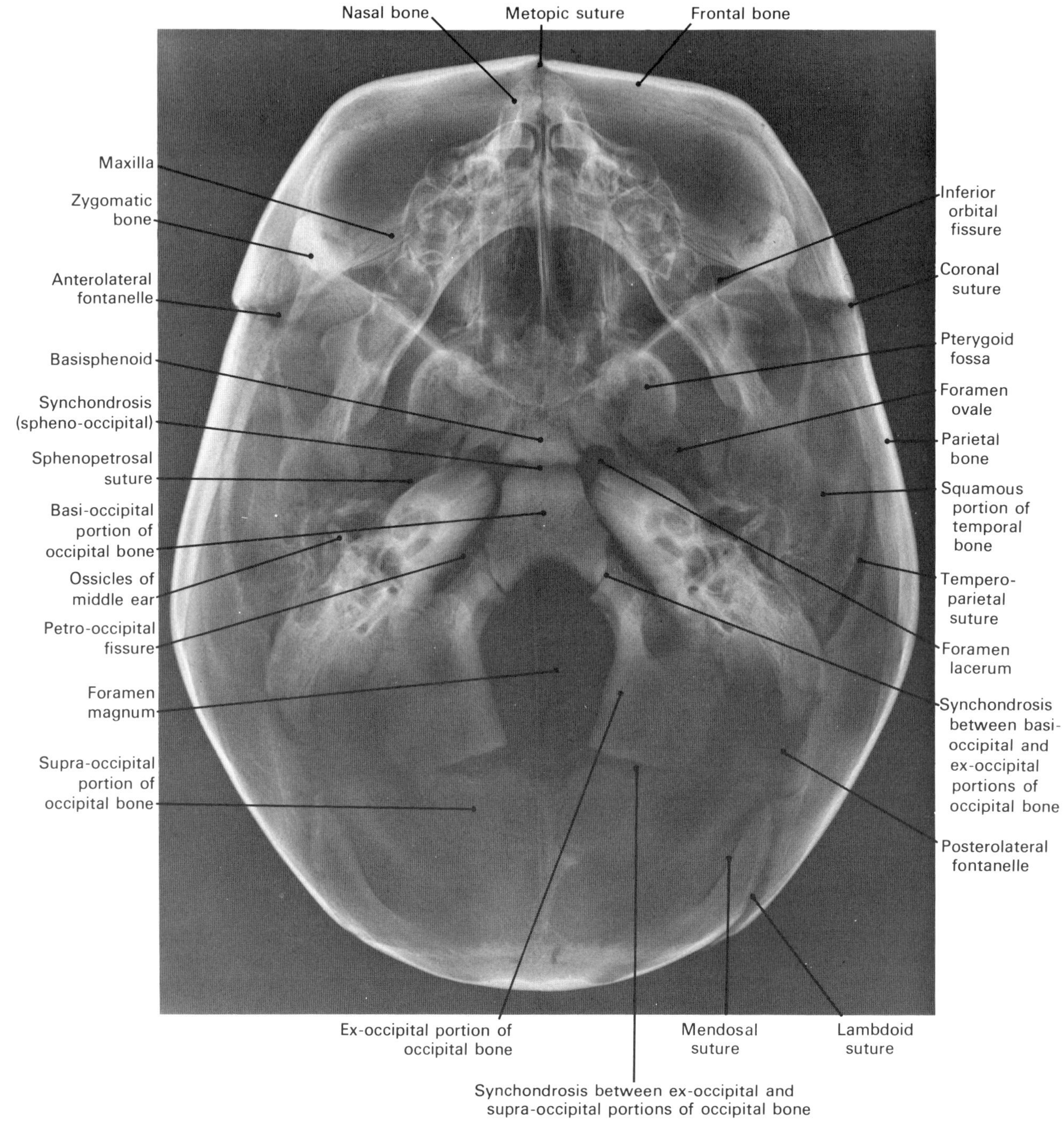

Fig. VI-48E. Basal projection of newborn skull.

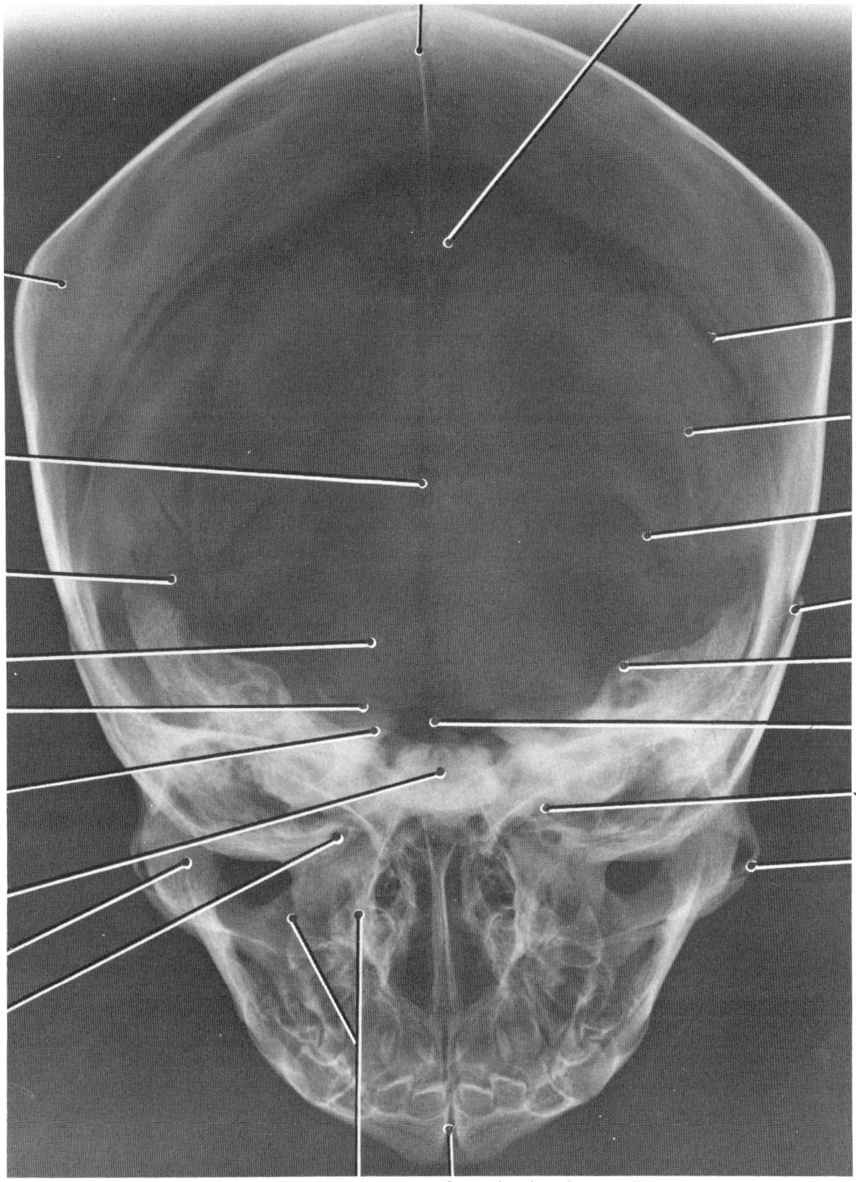

Fig. VI-48F. AP half-axial (Towne's) projection of newborn skull.

Table VI-5. Cranial Nerves

No.	Name	Component	Function	Cranial Foramen (or special nerve branches)
I	Olfactory	Special sensory	Smell	Cribriform foramina
II	Optic	Special sensory	Vision	Optic canal
III	Oculomotor	A. Somatic motor	Ocular movement	Superior orbital fissure
		B. Visceral motor	Pupiloconstriction and accommodation	Superior orbital fissure
IV	Trochlear	Somatic motor	Ocular movement	Superior orbital fissure
V	Trigeminal	A. Branchial motor	Muscles of mastication (and other Arch I muscles)	Foramen ovale
		B. Somatic sensory: (pain, temperature, and touch)		
		V_3	Skin, teeth (lower), mucous membranes of mandibular region, and tongue	Foramen ovale (to semilunar ganglion)
		V_2	Skin, upper teeth, and mucous membranes of maxillary region	Foramen rotundum (to semilunar ganglion)
VI	Abducens	Somatic motor	Ocular movement	Superior orbital fissure
VII	Facial	A. Branchial motor	Muscles of facial expression (and all Arch II muscles)	Internal auditory meatus → facial canal → stylomastoid foramen
		B. Visceral motor	1. Lacrimal, nasal, and palatine glands	Internal auditory meatus → greater petrosal nerve → hiatus of facial canal → pterygoid canal → pterygopalatine ganglion → branches of V_1 and V_2 to glands
			2. Sublingual and submaxillary glands	Internal auditory meatus → chorda tympani nerve → iter chorda tympani → lingual branch V_3 → glands
		C. Visceral sensory (pain, temperature, touch)	Mucous membranes, nose, and palate	*Via* greater petrosal branch to geniculate ganglion
		D. Gustation	Taste buds of anterior ⅔ of tongue	*Via* lingual nerve (V_3) → chorda tympani → geniculate ganglion
		E. Somatic sensory (?) (pain, temperature, touch)	Skin, ear, internal meatus, and mastoid region	Connecting twig with auricular branch of X → geniculate ganglion
VIII	Auditory (acoustic or stato-acoustic)	Special sensory	1. Hearing	Internal auditory meatus → spiral ganglion of cochlea

Table VI-5. Continued

No.	Name	Component	Function	Cranial Foramen (or special nerve branches)
			2. Equilibrium	Internal auditory meatus → vestibular ganglia
IX	Glossopharyngeal	A. Branchial motor	Stylopharyngeus (Arch III)	Jugular foramen
		B. Visceral motor	Parotid gland	Jugular foramen → Arnold's canal → middle ear plexus → lesser petrosal nerve → foramen ovale → otic ganglion → auriculo-temporal branch V_3 → gland
		C. Visceral sensory (pain, touch, chemoreception)	1. Mucous membranes; tympanic cavity	Arnold's canal → petrous and superior ganglia → jugular foramen
			2. Pharynx; carotid body	Petrous ganglion → jugular foramen
		D. Gustation	Taste buds of posterior ⅓ of tongue	Petrous ganglion → jugular foramen
X	Vagus	A. Branchial motor	Muscles, palate, pharynx, and larynx (Arches IV and V [except muscles supplied by nerves V and IX])	Jugular foramen
		B. Visceral motor	Smooth muscles and glands of neck, thorax, and abdomen to midtransverse colon	Jugular foramen → terminal ganglia
		C. Visceral sensory (pain and distension)	1. Thoracic and abdominal viscera supplied	Nodose ganglion → jugular foramen
			2. Barroreception from aortic and carotid sinuses	Nodose ganglion → jugular foramen
		D. Gustation	Taste buds of epiglottis	Nodose ganglion → jugular foramen
		E. Somatic sensory (pain, temperature, and touch)	Skin back of ear (meatus ?)	Auricular branch of X → Jacobson's (mastoid canal) → jugular ganglion → jugular foramen
XI	Spinal accessory	A. Branchial motor	Move head and shoulder (Sternomastoid and Trapezius as Arch VI muscles)	Vertebral canal → foramen magnum → jugular foramen
		B. Visceral motor	Swallowing and phonation (?)	Jugular foramen in company with vagus
XII	Hypoglossal	Somatic motor	Extrinsic and intrinsic muscles of tongue	Hypoglossal (anterior condylar) canal

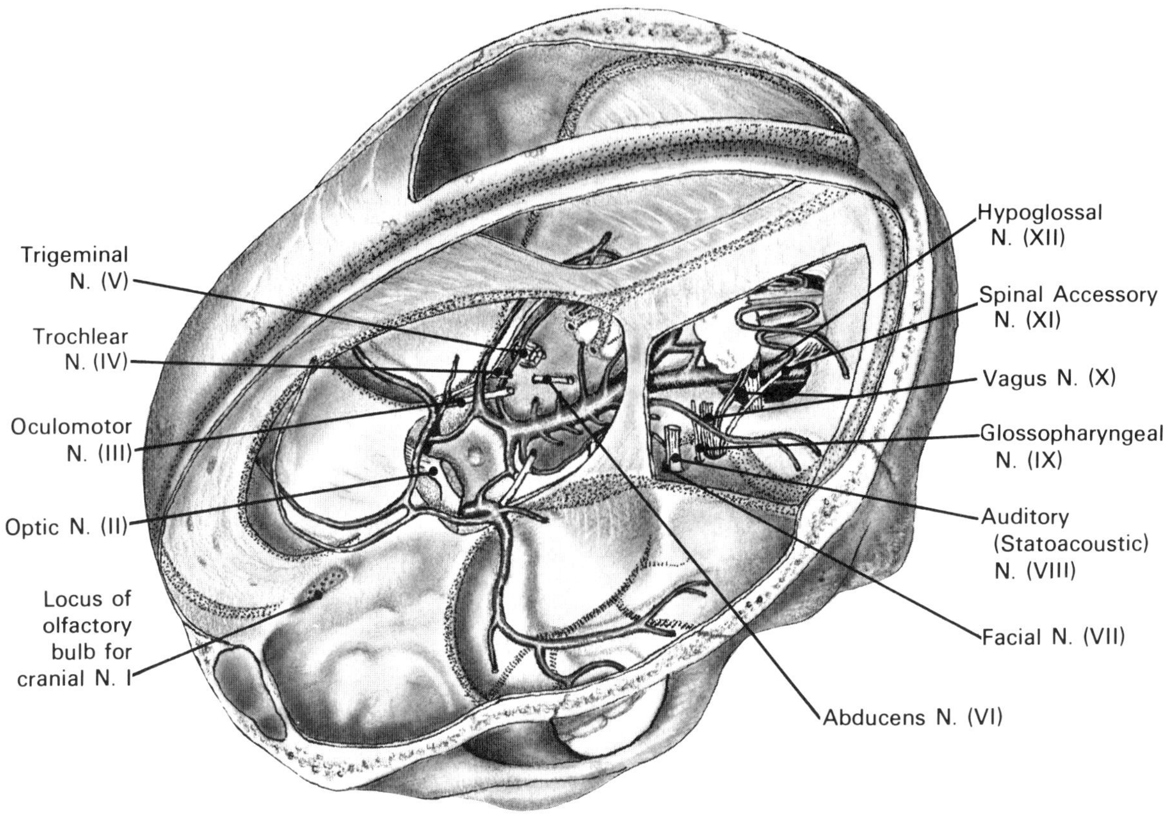

Fig. VI-49A. Dura and floor of cranium showing cranial nerves.

Recall that both the glossopharyngeal and the vagus have two sensory ganglia each along their course. These are located just outside the jugular foramen. For IX these are the **superior** and **petrous** ganglia. As a rule the superior ganglia subserves sensation from middle ear cavity and the auditory tube; the petrous ganglion subserves carotid sinus and taste from the posterior one-third of tongue. In a similar manner the cell bodies for somatic sensation (from back of ear) are located in the **jugular ganglion of X** whereas sensation from the baroreceptors and taste buds of the epiglottis have their cell bodies in the **nodose ganglion of X.** Visceral sensation is also subserved by the nodose ganglion. The latter would involve pain from thoracic and certain abdominal viscera, though most abdominal visceral pain is conducted to the spinal cord over sympathetic fibers.

A number of cranial nerves may play a role in proprioception from skeletal muscles of the face, *e.g.*, III, IV, VI, VII, and XII. It is not always certain where the cells for proprioception are located. Proprioception from the eye muscles may be served by sensory cells *within* the midbrain (near mesencephalic tract of V). Cranial nerve VII may transfer twigs to V so that proprioception from the face may have cell bodies in the semilunar ganglion. There may be small clusters of ganglion cells along the motor rootlets of XII. In a sense, the vestibular fibers of VIII may be said to play a proprioceptive role since they are concerned with equilibratory reflexes. In this sense, we are aware of our bodily orientation in space.

In general, we can say that of all the cranial nerves, three of them are purely somatic motor (IV, VI, and XII); three are purely special sensory (I, II, and VIII); and the remaining six are *mixed nerves.* The mixed motor nerves are III and XI; the mixed sensory and motor nerves are V, VII, IX, and X.

Our best advice is for you to understand these nerves as *functional groupings* rather than by their Roman numeral sequence from I to XII. The former method is more useful, too, than trying to recall their names by means of an ancient mnemonic device that has no logic at all. Any enterprising Finn or German would be busy brewing his own hop in his own basement. You would not find him traipsing around on old Olympus's tiny top trying to view some. He'd find little hop there—grape vines and olive trees in the valleys, maybe, but no hop. If these geographic observations make no sense to you, you are fortunate and have been spared some useless memorization, something your authors should have avoided in medical school. The only mnemonic devices they can recall

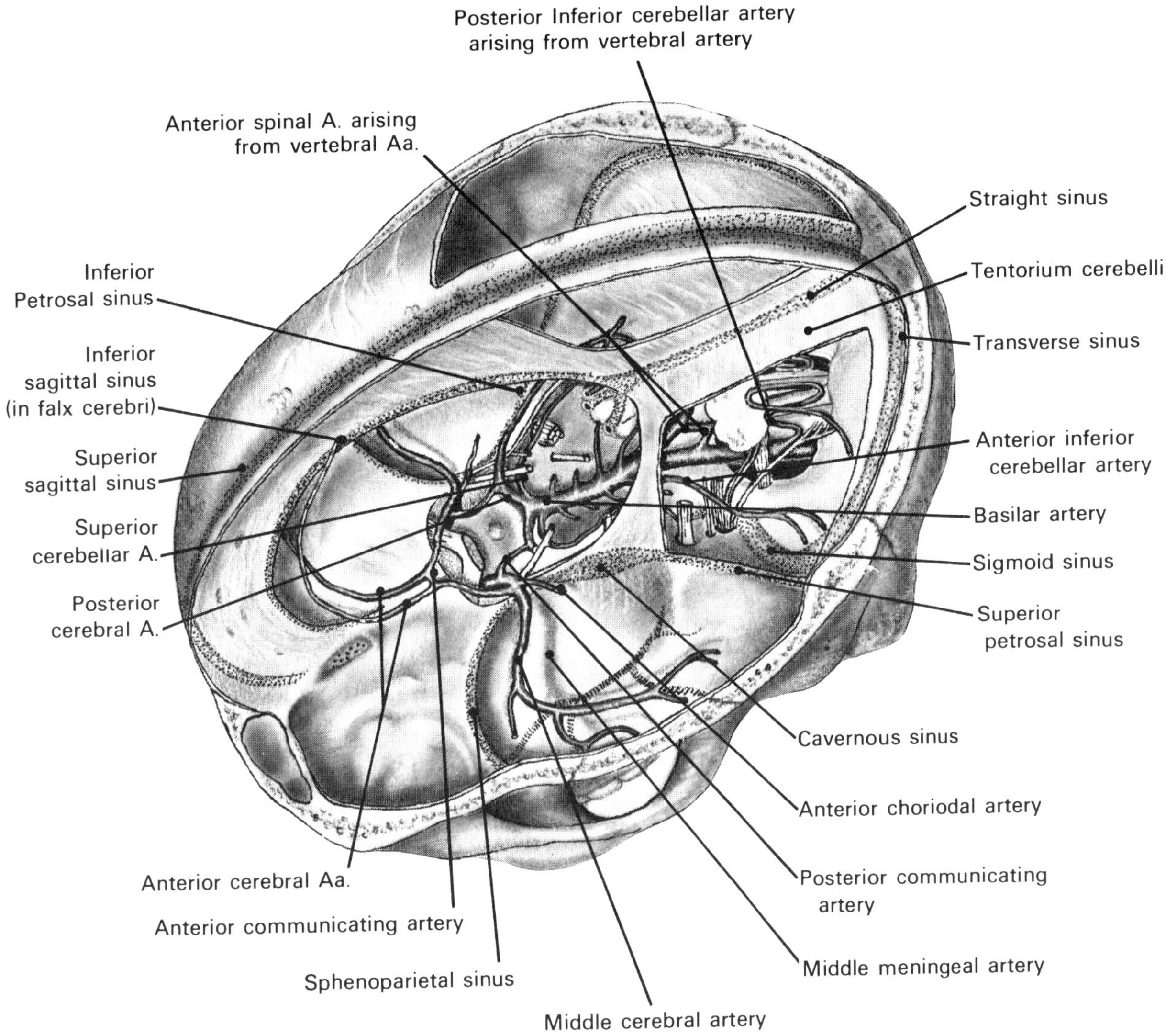

Fig. VI-49B. The relations of these nerves to dural sinuses and blood supply of brain and brainstem. (Modified from Lockhart.)

now are not fit to print. At least they weren't until the Supreme Court took a liberal attitude toward pornography.

THE FACE

The face has been described as the area that extends from ear to ear and from symphysis menti to the hairline. In that case, the senior author's face extends to the inion and is smoother from that point to the brows than from the brows to the chin.

The skin of the face is of average thickness, 1 to 2 mm. or less. The larger-calibered vessels of the well-vascularized face lie in the superficial fascia. As a rule, facial muscles are skin muscles in that they, too, occupy the superficial fascia and do not lie deep to it. Elsewhere in the body muscles lie deep to superficial fascia invested in their own, deep, innominate fascia (perimysium). The facial muscles originate either from bone or the superficial fascia, itself, and insert into the dermal portion of the skin.

Figure VI-51 illustrates the general location of most of the muscles of facial expression. None of these need be studied or even recalled by specific names. Note, simply, whether they lie superficially or deep in Figure VI-51. Use Table VI-6 as a regional reference index if you need to look up one or more of these muscles for any reason in the future.

The Buccinator Muscle

The Buccinator (E.2. of Table VI-6) does not conform as a typical facial muscle. It lies deep in the region of the cheek, filling in the gap between the maxilla and mandible, lateral to the teeth. Only anterior to the Masseter does it come close to contacting

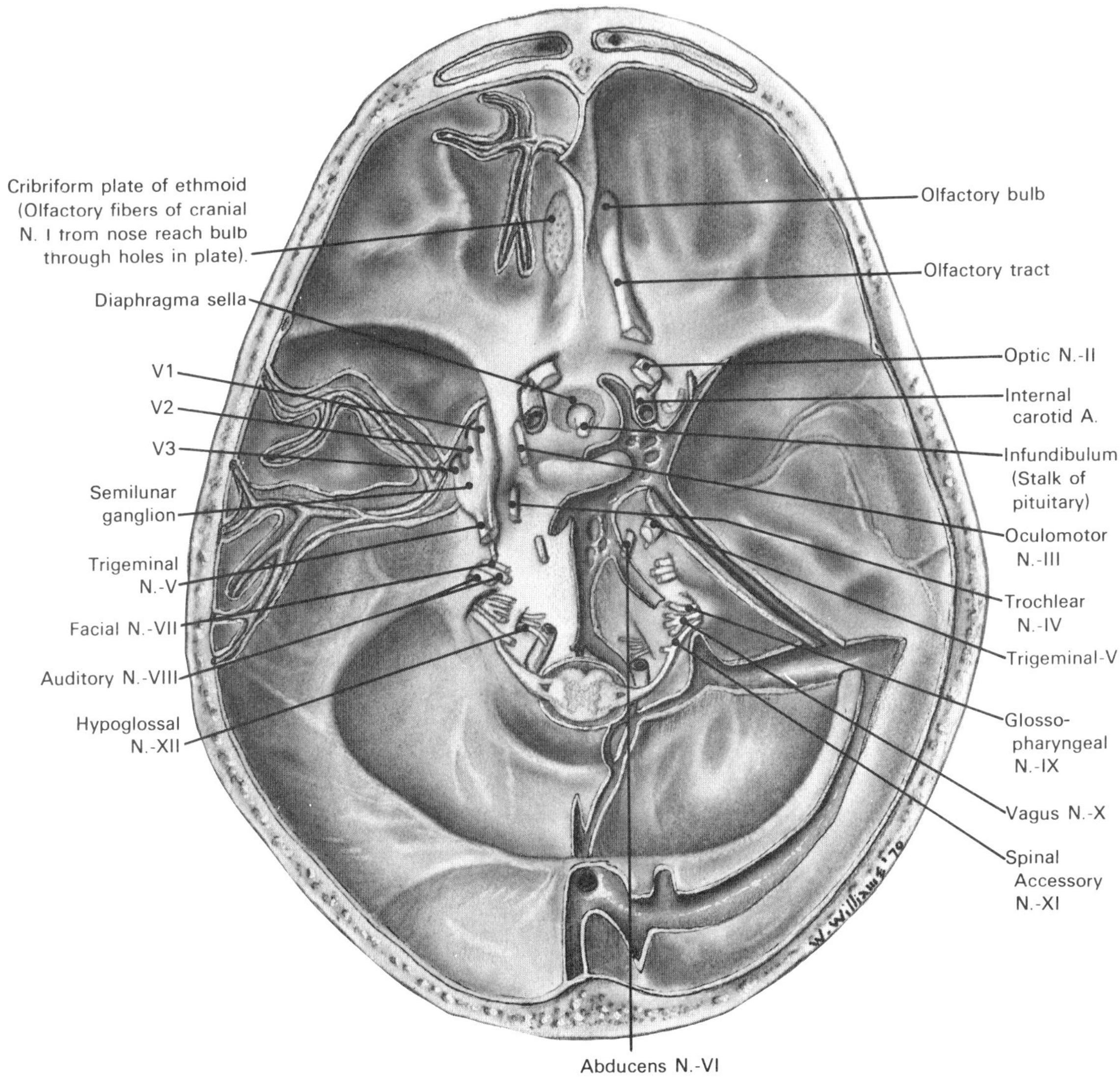

Fig. VI-50A. Base of skull, interior view showing cranial nerves as they exit.

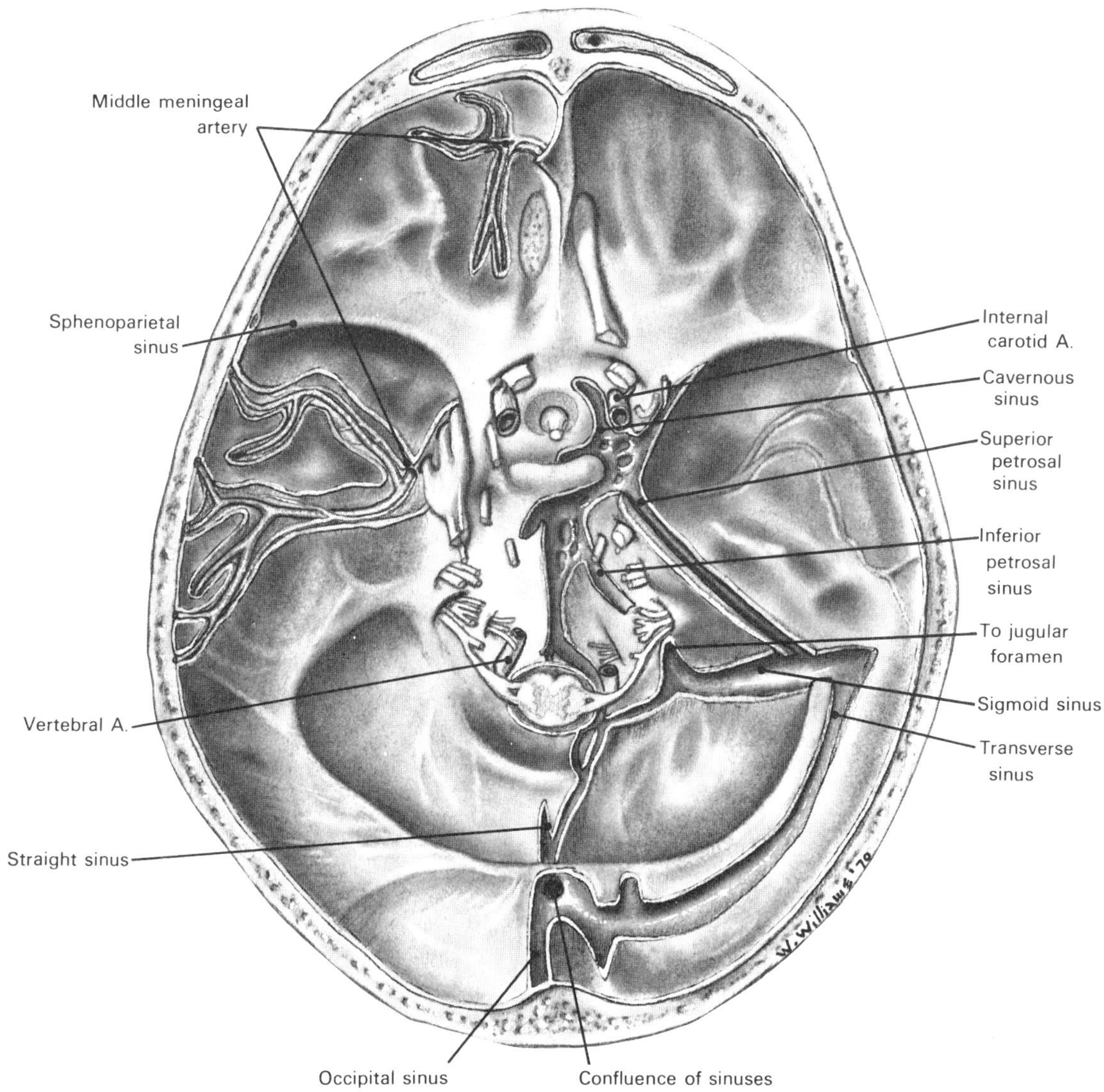

Fig. VI-50B. Certain dural sinuses at the base of the skull.

Table VI-6. Muscles of Facial Expression

Muscle	Origin	Insertion	Action
A. Scalp muscles			
1. **Occipitofrontalis**			
a. Frontal bellies	Epicranial aponeurosis	Skin of brows	Raises brows
b. Occipital bellies	Superior nuchal line	Epicranial aponeurosis	Retracts scalp
2. **Temporoparietalis**	Superficial temporal fascia	Skin above ear	Retracts scalp
B. Ear muscles			
1. **Auricularis**			
a. Superior	Epicranial aponeurosis	Root of auricle	Wiggle ears (?); generally none
b. Anterior			
c. Posterior			
C. Eye muscles			
1. **Orbicularis oculi**			
a. Palpebral part	Medial palpebral ligament	Lateral palpebral raphe	Closes lid gently
b. Orbital part	Medial and anterior orbital margin	Forms an ellipse around orbit and over temple and cheek	Closes lid forcibly
c. Lacrimal part	Orbital surface of lacrimal bone	Superior and inferior tarsal plates	Draws lids and lacrimal canal close to globe to favor tear collection; compresses lacrimal gland
2. **Corrigator**	Medial part of superciliary arch	Skin of brow	Draws brow down and medialward; wrinkles brow to produce a frown
D. Nose muscles			
1. **Procerus**	Nasal bone	Between brows	Pulls down medial brow angle to make a transverse crease over nose
2. **Compressor naris**	Canine eminence	Bridge (dorsum) of nose	Compresses nostril
3. **Dilator naris**	Skin above lateral incisor	Alar cartilage	Widens aperture of nostril
4. **Depressor septi**	Incisive fossa	Lower part of nasal septum	Constricts nostril by drawing ala downward
E. Mouth muscles			
1. **Orbicularis oris**	From many muscles that converge on mouth	Forms an ellipse around lips	Closes lips
2. **Buccinator**	Pterygomandibular ligament	On Orbicularis and in upper and lower lip	Flattens cheek in sucking and chewing; when cheeks are distended with air, compresses them to force air through lips in blowing
3. **Risorius**	Masseteric and parotid fascia	Mouth angle	Retracts mouth as in grinning
4. **Depressor anguli oris**	Oblique line of mandible	Mouth angle	Depresses mouth angle
5. **Depressor labii inferioris**	Between symphysis menti and mental foramen	Lower lip	Draws lip down and laterad as in expressing irony.
6. **Levator labii superioris**	Lower margin, orbit, ***above*** infra-orbital foramen	Mouth angle	Elevates and everts upper lip (see 8, below)
7. **Levator anguli oris** (Caninus)	Canine fossa ***below*** infra-orbital foramen	Mouth angle	Acts with other muscles to deepen nasolabial furrow as in contempt or disdain
8. **Levator labii superioris alaeque nasi**	Narrow slip from frontal process of maxilla	Alar cartilage and upper lip	Dilates nares; along with zygomaticus minor and Levator labii superioris, it forms nasolabial fold in expressing sadness
9. **Zygomaticus major**	Zygomatic bone	Mouth angle	Draws mouth upward and out in laughing
10. **Zygomaticus minor**	Zygomatic bone	Upper lip	See 8, above
11. **Mentalis**	Incisive fossa	Chin	Raises and protrudes lower lip as it wrinkles chin as in pouting, doubting, or expressing disdain
12. **Platysma**	Fascia of Pectoralis major of 2nd rib and fascia of Deltoid	Lower jaw, mouth angle, and lower part of mouth	Widens mouth aperture at corners as in horror; pulls skin taut as in shaving or in relieving pressure of a tight collar

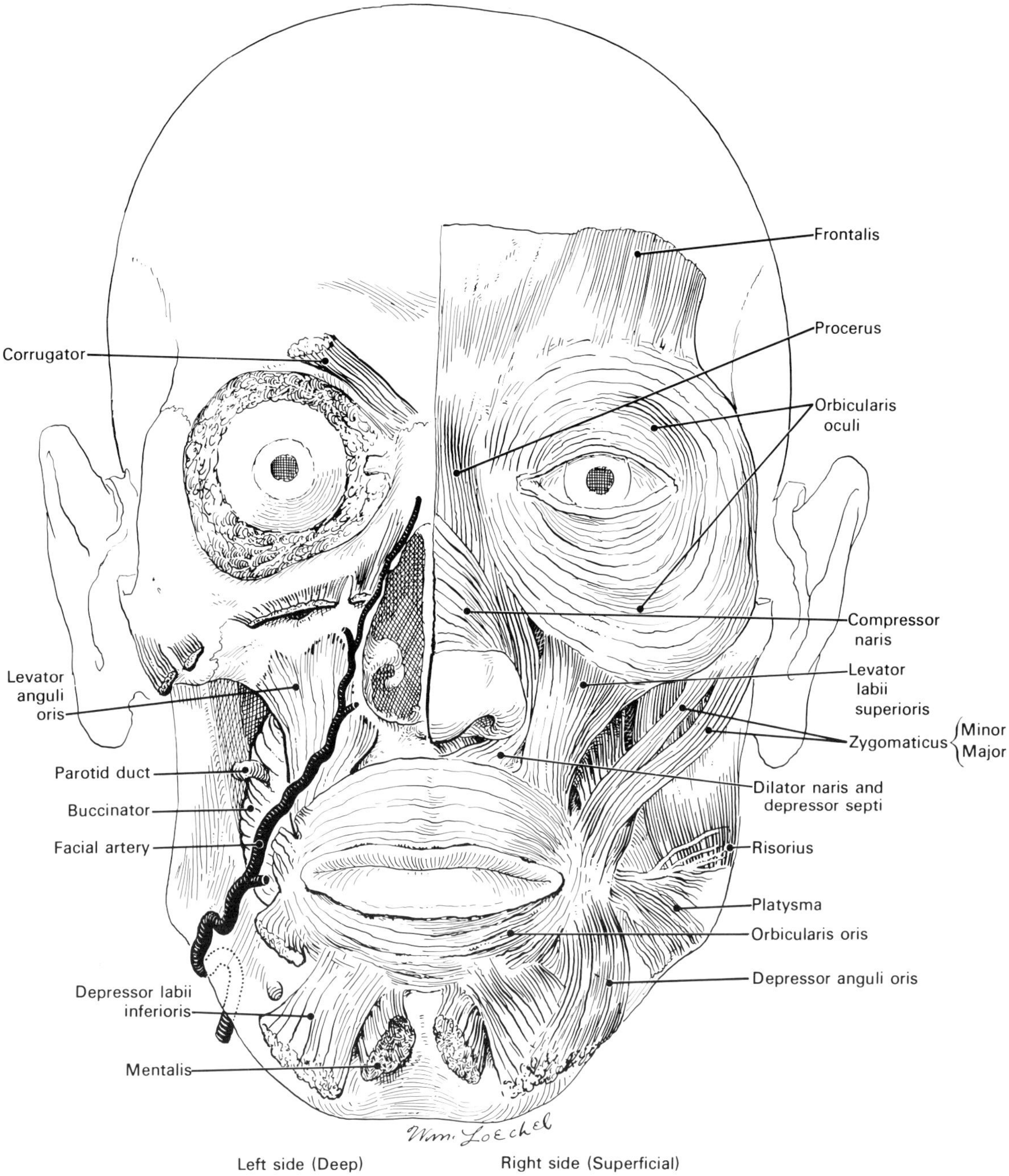

Fig. VI-51. Deep and superficial muscles of facial expression.

superficial fascia in the mouth region. Between the Buccinator and the fascia is an encapsulated **buccal fat pad** in the cheek region. The pad is most prominent in the infant where it is supposed to aid in sucking. Finger-like extentions of this fat creep between the Masseter and Temporalis muscles, and some of it infiltrates the infratemporal fossa to separate Temporalis from the Lateral Pterygoid. Superficially, the Buccinator and the **Superior Pharyngeal constrictor,** with which it is continuous, are surfaced by a tough **buccopharyngeal fascia.** The fascia lies deep to the fat pad and serves as the perimysium for these two muscles. Both muscles are joined to the **pterygomandibular raphe,** a kind of common tendinous inscription. The raphe runs from hamulus to the posterior end of the mylohyoid line on the inner aspect of the mandible (Fig. VI-52). In the figure, observe where the **parotid duct** passes through the Buccinator.

The Parotid Area of the Face

The largest of the salivary glands, weighing 15 to 30 gm. and from 6 cm. in length to 3 cm. in width, extends from zygomatic arch to the angle of the mandible. It reaches upward to and even behind the external ear. It is invested in a complete sheath derived from the splitting of the superficial layer of the cervical fascia.[18] The deepest aspect of this sheath has the consistency of areolar tissue but it is of tougher consistency and thicker externally over the gland. Projections pass into the gland from the outer part of this sheath.[19] Caudally and medially the cervical fascia splits again to invest the submandibular gland. The stylohyoid ligament plays a role in separating parotid from submandibular gland.

The lateral surface of the gland is covered by subcutaneous fascia and skin, and, in its caudalmost extent, even by a bit of Platysma. This is the surface that yields as the gland emlarges in mumps. Its anterior border embraces the posterior aspect of the mandible, the Masseter and Medial Pterygoid. Thus, there may be a pterygoid lobe and a retromandibular portion of the gland. Most of it can be seen lying upon the Masseter in Figure VI-53. Its posterior surface is in contact with the mastoid process, the Sternomastoid, the posterior belly of the Digastric and Stylohyoid muscles, the base of the styloid process and the tympanic part of the temporal bone. Still more caudally, the posterior surface relates to the Styloglossus and Stylopharyngeus muscles and, most caudalward of all, to the structures in the carotid sheath (internal carotid artery, internal jugular vein and vagus nerve).

If you wish to reconstruct the dissection shown in Figure VI-53 you would have to realize in your minds eye that the Buccinator attaches to the Superior Pharyngeal constrictor at the Pterygomandibular raphe; that the Pterygoids lie superficial to the constrictor; and that the Buccinator will be hidden in part by the Masseter. The mandibular ramus would have to be put in place. The parotid duct then crosses the Masseter and at the anterior margin of that muscle (Fig. VI-53) the duct pierces the buccal fat pad, the buccopharyngeal fascia, the Buccinator, and the mucous membrane of the inner cheek. It opens then into the **vestibule of the mouth,** *i.e.,* the space between cheek and teeth, at the level of the second upper molar.

Blood Supply of the Face

The two main arteries are the **facial** and the **transverse facial** (Fig. VI-54). The facial artery is one of the eight chief branches of the external carotid. It appears on the face at the base of the jaw immediately anterior to the jaw angle and superficial to the Masseter at that point. The pulse of the facial artery can be felt there when it is compressed against the mandible (see Chapter III, p. 92, for other points where its pulse can be felt). Just posterior to the artery will run the **facial vein** and anteriorly, the **cervical branch** of the **facial nerve.** The facial artery slants across the lower face almost to the corner of the mouth and then turns upward along the side of the nose. Its terminal **angular branch** plays out at the medial corner of the eye. At the mouth it gives off branches to the upper and lower lips.

The **transverse facial artery** is a branch of the **superficial temporal artery** which, itself, is one of the two terminals of the external carotid (recall that the maxillary artery is the other end branch of this main trunk line). The transverse facial follows the zygomatic arch onto the cheek prominence. Figure VI-54 illustrates the arteries to the face, and Table VI-7 gives the principal branches of the facial artery.

Some supply comes through anastamoses with the ophthalmic and maxillary arteries, as well. The veins have been considered earlier. The venous supply is reviewed in Figure VI-55.

Nerve Supply to the Face

Three nerves are involved. The chief of these are the **trigeminal** (cranial nerve V) and the **facial** (cranial nerve VII). Of lesser importance is the ascending branch from the cervical plexus of nerves, the **greater auricular** nerve. The latter is sensory to the skin over the parotid and Masseter in front of the ear.

The **trigeminal** is the main sensory nerve to the face. It reaches as high as the interaural line, and remains to supply an area in front of the ear and along the jaw angle to the symphysis menti (Fig. VI-56). We have located the supra-orbital notch or foramen, the infra-orbital and mental foramina. Through these orifices in the skull the major branches of V_1, V_2, and V_3 to the skin emerge. Four nerves constitute the main branches of $\mathbf{V_1}$, or the **ophthalmic division** of the trigeminal: **frontal, lacrimal, nasociliary,** and **tentorial** nerves. The first three of these send some branches to the skin, but they will supply deeper structures, as well, such as the iris and cornea; the mucous

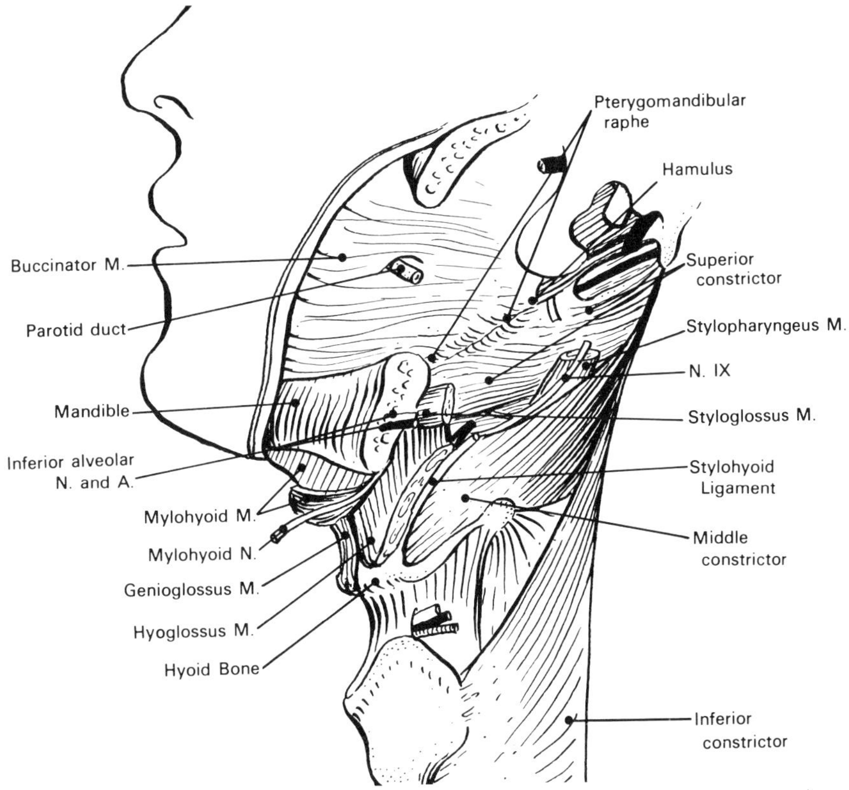

Fig. VI-52. Relationships at the pterygomandibular raphe. Note Buccinator, Superior constrictor, and parotid duct.

membranes of the frontal, ethmoidal, and sphenoid air sinuses; and the internal nose. The last, the tentorial, is a recurrent branch to the dura and never reaches the skin.

The *supraorbital nerve* is the largest of the branches from V_1 to reach the skin, and it can be considered to be a direct continuation of the frontal nerve (Fig. VI-58A). The rest are all small twigs and they are listed in Table VI-8.

The *maxillary division of V_2* has roughly seven main nerves: **zygomatic, infraorbital, middle meningeal, pterygopalatine,** and **three alveolar nerves** (anterior, posterior superior, and middle superior). All seven nerves are sensory nerves, but twigs from the first two are the only ones to reach the face (Figs. VI-56 and 58 and Table VI-8) as the zygomaticotemporal, zygomaticofacial, and infraorbital nerves.

We have come to look upon *V_3 or the mandibular division* of the trigeminal as mainly supplying the muscles of mastication, but it has several sensory branches, as well: a **meningeal branch to the dura,** the **lingual nerve,** the **inferior alveolar** nerve, the **auriculotemporal, buccal,** and **mental nerves.** Only the last three reach the skin of the face (Figs. VI-56 and 58 and Table VI-8). As a rule, the mandibular nerve distributes as an anterior and posterior division (Fig. VI-57). The only sensory component of the anterior division is the buccal nerve. The only motor nerves of consequence in the posterior division are those to the Mylohyoid and anterior belly of the Digastric. The meningeal branch to the dura comes off before the anterior and posterior divisions of the mandibular nerve form. Thus, all the other sensory nerves of V_3 listed above run with its posterior division (Fig. VI-57).

Two points must be repeated, for they are often sources of confusion:

1. The lacrimal nerve *does not supply* visceral motor fibers to the lacrimal gland. *Cranial nerve VII provides* these fibers, and they are *conveyed* by the lacrimal nerve to the gland. The lacrimal nerve (V_1) (Fig. VI-57) receives these fibers *via* a communicating twig from the zygomaticotemporal nerve (V_2). Hence, two sensory divisions of the trigeminal piggyback the visceral motor fibers of the facial nerve to the gland.

2. The buccal nerve has nothing to do with the Buccinator muscle. It is a sensory branch of V_3 to the cheek. The muscle is one of facial expression, and it gets its motor supply directly from VII (buccal branch of VII, Fig. VI-58B).

It is also interesting to observe the way certain

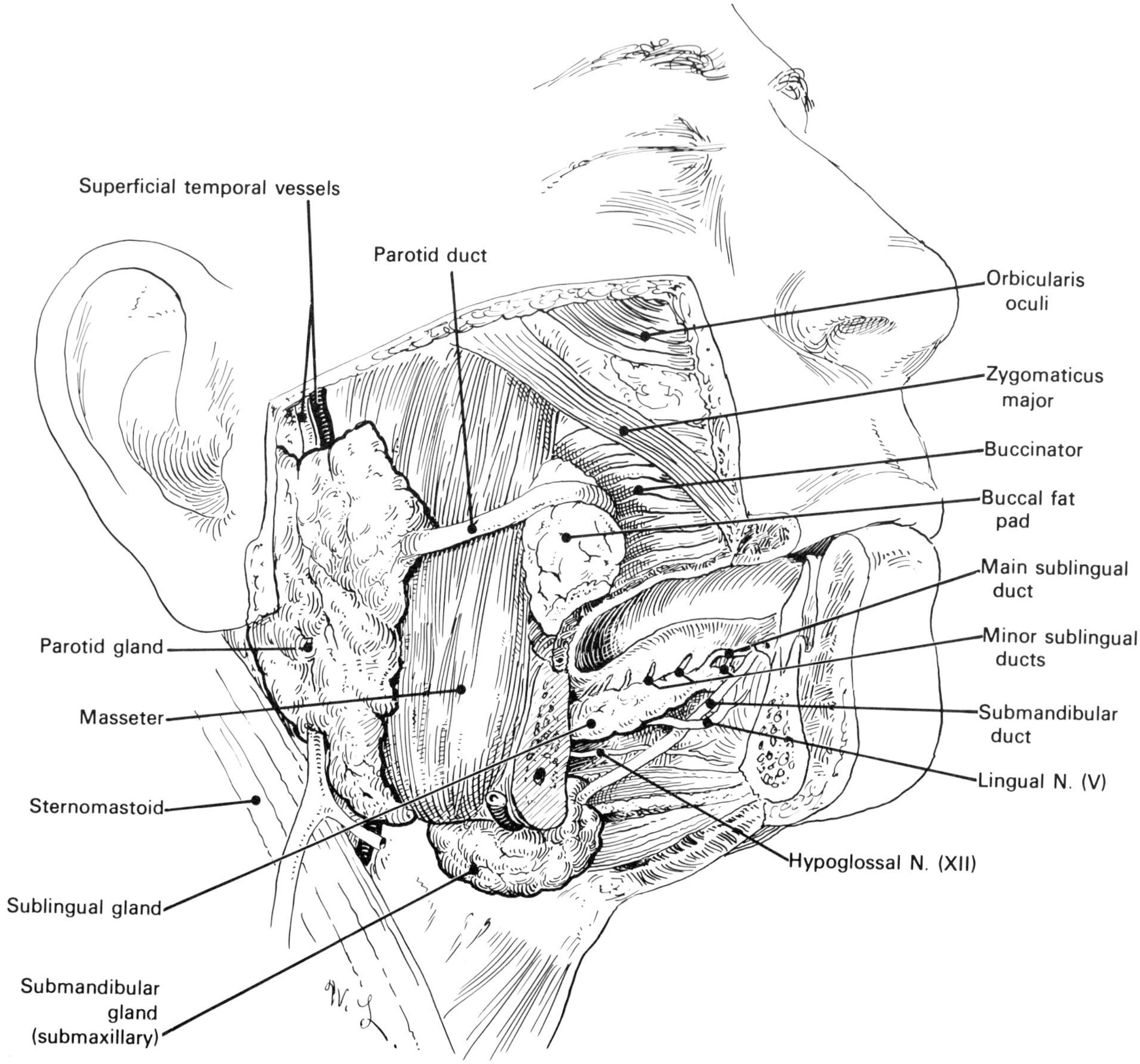

Fig. VI-53. Superficial dissection of the face to show relationships in parotid region.

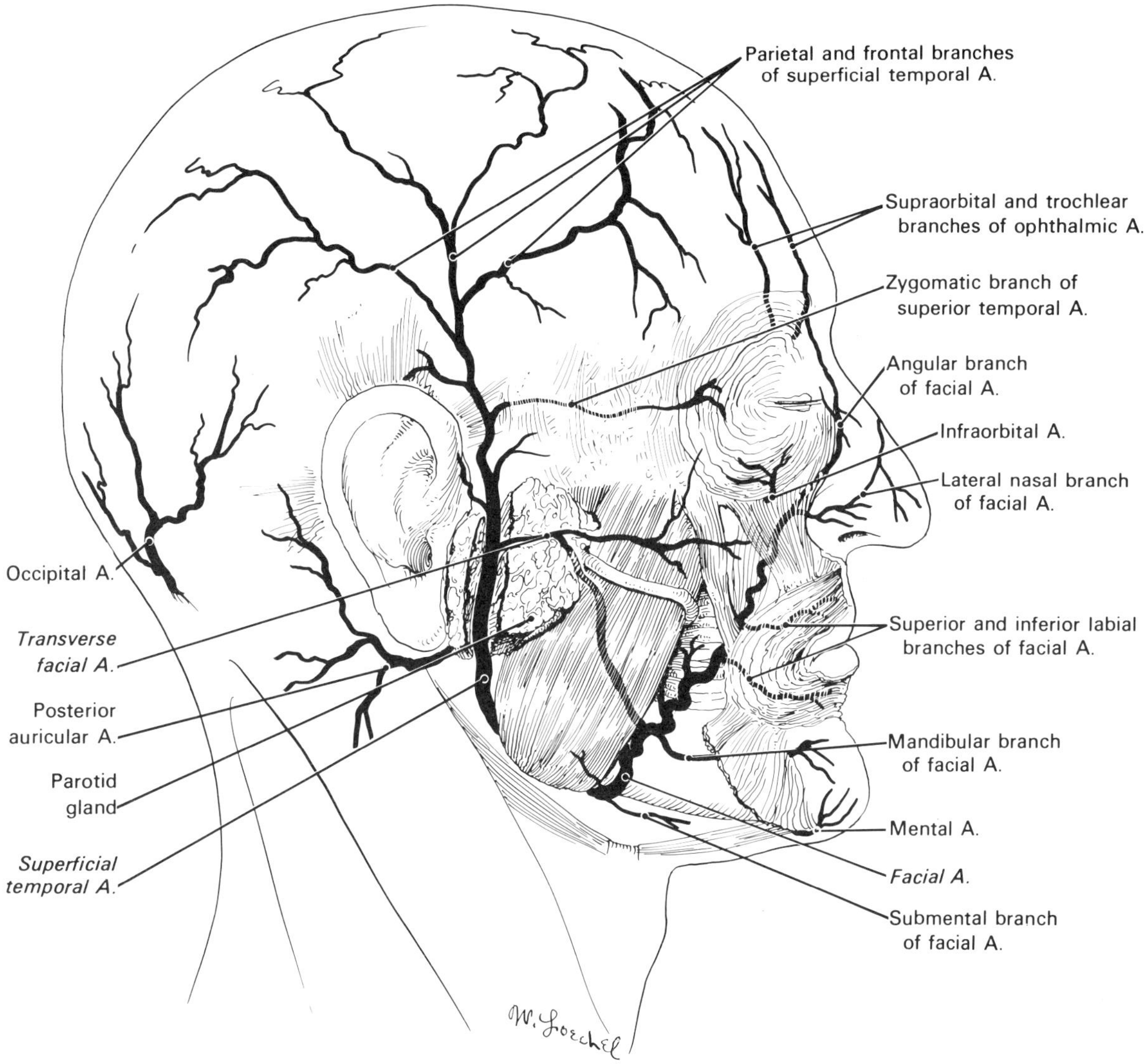

Fig. VI-54. Arterial supply of the face.

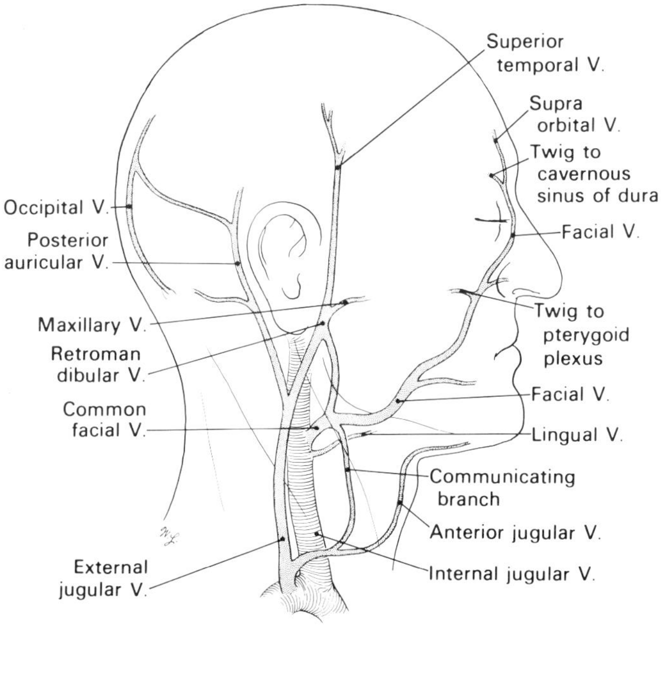

Fig. VI-55. Venous drainage of the face.

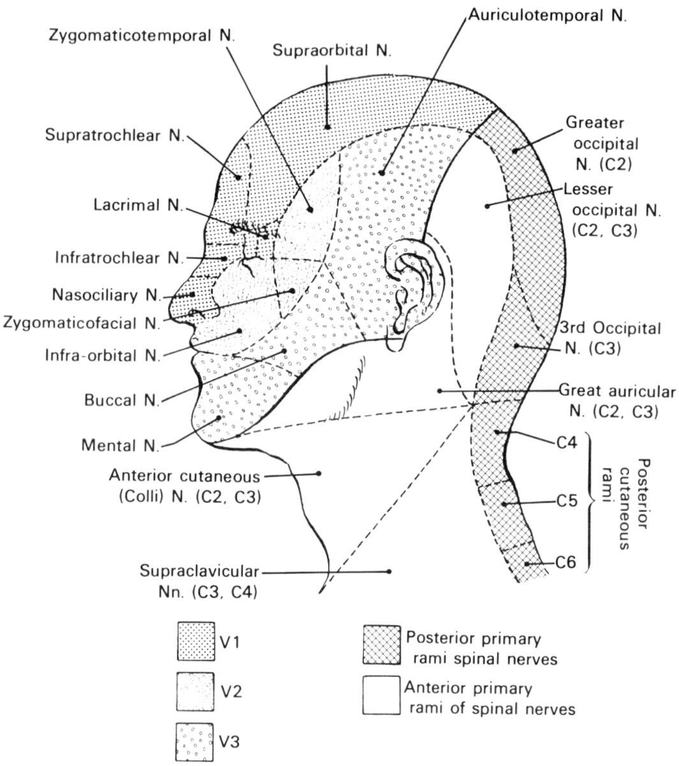

Fig. VI-56. Distribution of trigeminal and sensory cervical branches to skin of face and neck.

Table VI-7. Branches of the Facial Artery

Branch	Origin	Area Supplied
1. Inferior labial	Mouth angle	Labial glands and lower lip, skin and muscle.
2. Superior labial	Just superior to 1, above	Labial glands and upper lip, skin and muscles.
a. Septal branch		Nasal septum, antero-inferior part
3. Lateral nasal	Superior to alae alongside nose	Alae and dorsum of nose
4. Angular	Terminal part of facial artery	Lacrimal sac and Orbicularis oculi; skin in region

Table VI-8. Sensory Branches of Trigeminal Nerve to the Face

Division	Nerve	Branch of	Area Supplied
A. Ophthalmic (V_1)	1. Supraorbital	Frontal nerve	Scalp and central part of upper eyelid
	2. Supratrochlear	Frontal nerve	Forehead and medial part of upper lid
	3. Lateral palpebral	Lacrimal nerve	Lateral part of upper lid; conjunctiva
	4. Infratrochlear	Nasociliary nerve	Lower lid; lateral nose; conjunctiva; lacrimal sac caruncle
	5. External nasal	Anterior ethmoidal nerve (itself a branch of nasociliary)	Alae and tip of nose
B. Maxillary (V_2)	1. Zygomaticotemporal	Zygomatic nerve	Side of forehead (before it leaves orbit it sends a twig to lacrimal nerve, which conveys parasympathetic fibers of VII to lacrimal gland)
	2. Zygomaticofacial	Zygomatic nerve	Malar prominence
	3. Infraorbital	Maxillary nerve	Face, nose, lower lid, and upper lip
C. Mandibular (V_3)	1. Auriculotemporal	Mandibular nerve	Auricle, meatus, and internal surface of tympanic membrane
	2. Buccal	Mandibular nerve	Skin of cheek
	3. Mental	Superior alveolar nerve	Skin of chin and lower lip

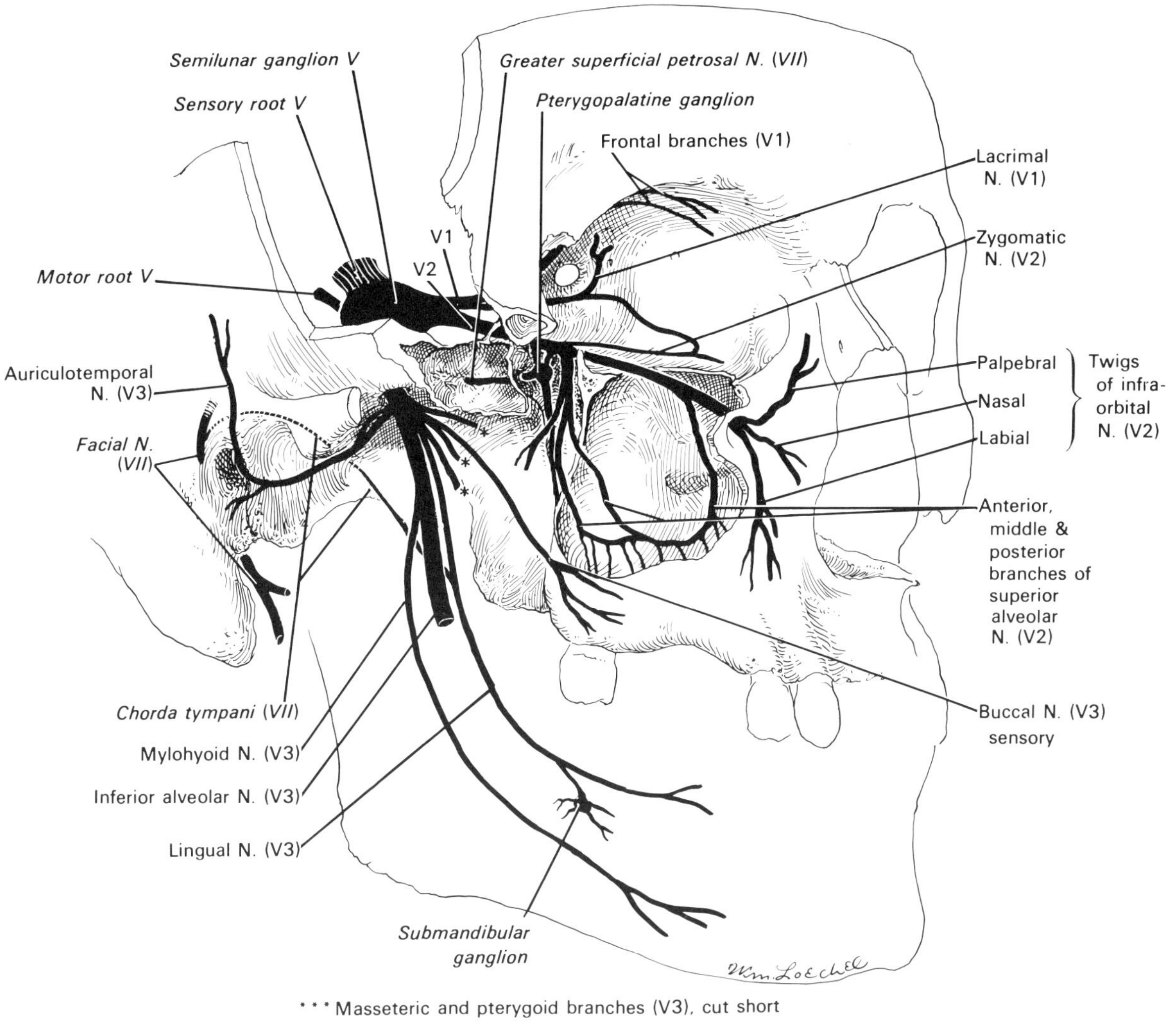

Fig. VI-57. Distribution of maxillary and mandibular divisions of the trigeminal nerve.

arterial branches follow the branches of the trigeminal nerve on the face (see Table VI-9).

The *facial nerve*, insofar as it gives a supply to the muscles of facial expression, leaves the cranial cavity *via* channels in the petrous bone; *i.e.*, internal auditory meatus and facial canal, to emerge through the stylomastoid foramen. The terminal branches of the nerve appear at the margins of the parotid and then they spread out like the ribs of a fan over and through the gland to reach their more forward destination (Fig. VI-58). Its **temporal branches** cross the zygoma to supply facial muscles above the zygoma. Its **zygomatic branches** tend to run to the infra-orbital musculature, remaining above the level of the parotid duct. The **buccal branch of the facial nerve** tends to follow the parotid duct across the Masseter. The **mandibular branch** slopes along the lower jaw toward the muscles of the lip and chin, while the **cervical branch** descends even farther to join with the anterior cutaneous branches of the cervical nerves and, finally, distributes to the Platysma. A **posterior auricular branch** of little consequence departs from these major routes to curl behind the ear, giving a supply to the Occipitalis and Auricularis Posterior and Superior. Figure VI-58 shows these branches of VII in relation to the sensory twigs of the trigeminal to the face.

The part of the facial nerve to reach facial muscles is easy to understand. It is the part in the head which runs through the bony canals of the skull that seems to bother students. It lies too deep to be seen readily. Figure VI-59 attempts to show these deeper relationships. Its relation to the terminal ganglia in the head (pterygopalatine and submandibular) is seen in Figure VI-57. We see the main branches of the facial nerve emerging from the **facial canal** in the mastoid region (Fig. VI-59). The canal in the petrous bone is essentially "C"-shaped. The stylomastoid foramen is an exit for the main branches posteriorly. The **hiatus of the facial canal** is the anterior exit for the **greater (superficial) petrosal branch of VII.** The latter nerve soon joins with the **deep petrosal nerve** (sympathetic fibers from the internal carotid plexus). The two, together, form the **nerve of the pterygoid canal** (Vidian nerve). The parasympathetic visceral motor fibers of VII in this nerve terminate on nerve cells of the **pterygopalatine (sphenopalatine) ganglion** (see Figs. VI-57 and 59). From the ganglion, postganglionic axons course to nasal, lacrimal, and palatine glands *via* the fibers of $\mathbf{V_1}$ and $\mathbf{V_2}$ which they join for distribution to these glands (Fig. VI-60).

Table VI-9. Arteries Which Follow Branches of Trigeminal Nerve on Face*

Artery	Branch of
Supraorbital	Ophthalmic artery
Supratrochlear	Ophthalmic artery
Lateral palpebral	Lacrimal artery
Medial palpebral	Ophthalmic artery
Dorsal nasal	Ophthalmic artery
Zygomaticotemporal	Lacrimal artery
Zygomaticofacial	Lacrimal artery
Infra-orbital	3rd part, maxillary artery†
Superficial temporal	One of the terminal branches of external carotid artery
Buccal	2nd part, maxillary artery†
Mental	1st part, maxillary artery†

* See Table VI-8 for area of supply.

† First part, maxillary artery—neck of mandible to lower border Lateral pterygoid.

Second part, maxillary artery—in the infratemporal fossa.

Third part, maxillary artery—in pterygopalatine fossa.

In Figure VI-59 we see the internal auditory meatus burrowing into the left petrous bone. It leads eventually to the C-shaped facial canal which traverses the bone longitudinally. The **geniculate or sensory ganglion of VII** sits in the bone at the junction point of meatus and canal. The meatus conveys sensory axons of VII toward the brainstem and motor axons of VII from the brainstem to be distributed *via* the canal to motor end-organs (branchial musculature and certain glands). We see also the auditory nerve (VIII) in the canal. The sensory axons of VII in the meatus form its *sensory root* or **nervus intermedius** (nerve of Wrisberg).

As the main part of the facial canal moves postero-inferiorly (toward the right in Fig. VI-59) it conveys chiefly motor but a few sensory fibers toward an exit from the skull at the stylomastoid foramen. On the way through this portion of the facial canal, the nerve first gives off a small twig which enters the middle ear cavity to supply the small **Stapedius muscle.** Next, before emerging from the stylomastoid foramen, it gives off its *chorda tympani branch.* This branch crosses in the lateral wall of the middle ear cavity, re-enters the petrous bone and emerges near the spine of the sphenoid at the base of the skull. It finds itself in the infratemporal fossa close to the *lingual branch of* V_3 as the mandibular division of V drops through the foramen ovale into the fossa. The chorda tympani then join with the lingual nerve (V) to be distributed to its terminal ganglion, the submandibular (submaxillary) ganglion. It, thus, carries *parasympathetic preganglionic fibers of VII* to the ganglion. Postganglionic axons from the ganglia distribute with V to the submandibular and sublingual glands. Sensory fibers from the *taste buds of the anterior two-thirds of the tongue* are conveyed back over this complicated routing to the geniculate ganglion. Thus, the chorda tympani serves two functions, taste and visceral motor to salivary glands (but not to the parotid gland).

Sometimes a third tiny twig of VII pierces the petrous bone from the facial canal to join with the **auricular branch of X in Arnold's canal.** This facial twig assists X, along with a variable contribution from V and IX, to supply the skin of the external auditory

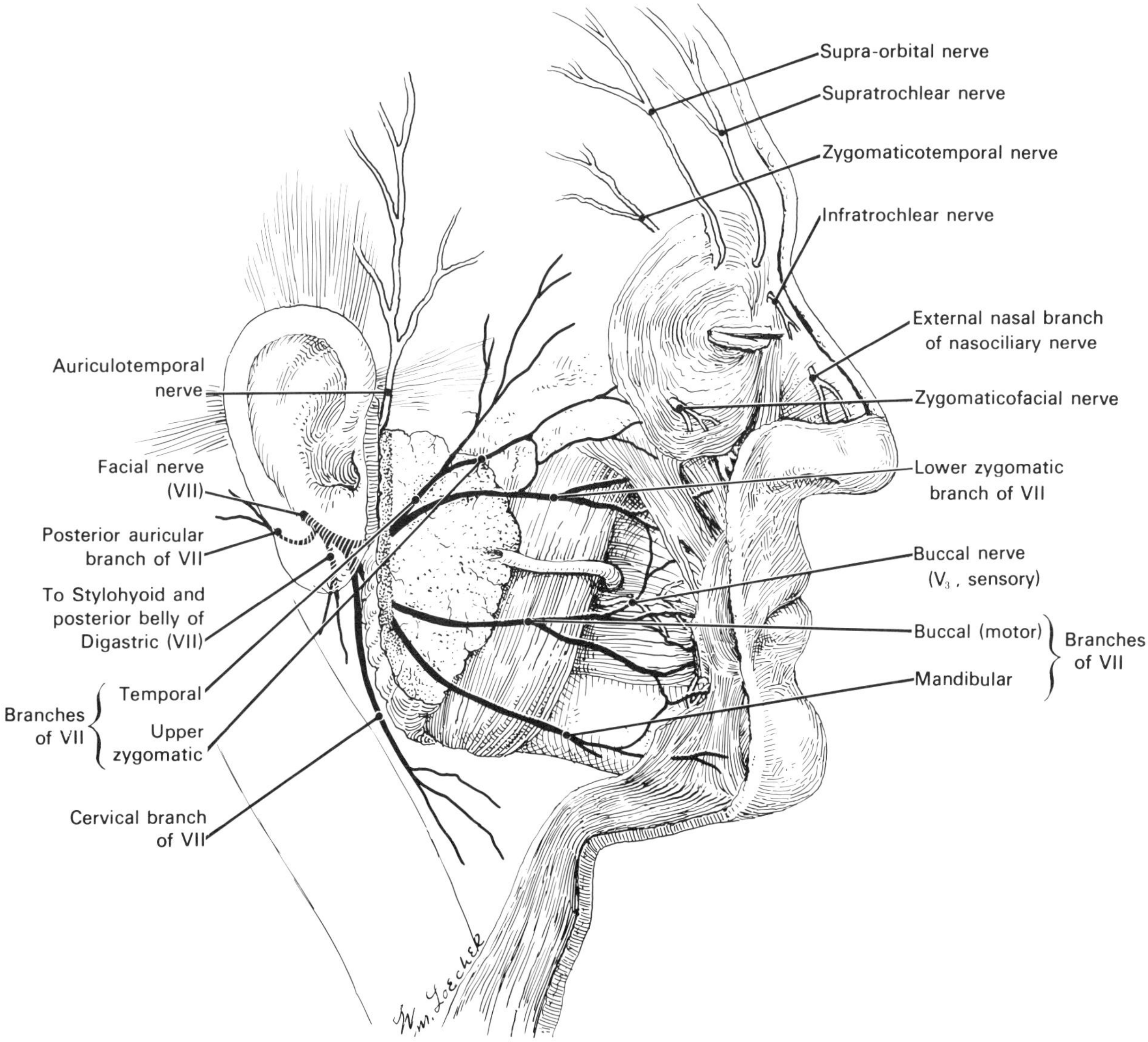

Fig. VI-58. Branches of trigeminal nerve (white) and facial nerve (black) to face.

meatus and the skin of the mastoid region behind the ear.

Finally, the branchial motor fibers of VII emerge from the stylomastoid foramen to supply the muscles of facial expression (as shown in Fig. VI-58B), and also the Stylohyoid and the posterior belly of the Digastric.

We have already noted how the anteriorly-directed limb of the facial canal tunnels through the petrous bone toward its hiatus in the middle cranial fossa and conveys visceral motor fibers of VII *via* the greater petrosal nerve to nasal and palatine glands (Figs. VI-59 and 60). The route to the lacrimal gland *via* zygomatic twigs of V_2 and lacrimal twigs of V_1 is shown in Figure VI-57.

Figure VI-59 also depicts the route of visceral motor fibers of IX to the parotid gland. A recurrent branch of IX is seen to re-enter the skull in Jacobson's canal. It burrows through the petrous bone to reach a plexus on the medial wall of the middle ear cavity. An anterior branch from the tympanic plexus continues forward through the petrous bone to emerge in the floor of the middle cranial fossa as the *lesser petrosal nerve* (IX). It finds the foramen ovale (or the innominate canal) almost dead-on near its point of emergence from the petrous bone. With one or the other of these foramina it leaves the skull for the infratemporal fossa where it at once finds its terminal parasympathetic ganglion, the *otic ganglion*. So close is the ganglion to the mandibular division of V in the infratemporal fossa that the postganglionic axons of the ganglion cells can attach themselves immediately upon the *auriculotemporal branch of V* which conveys them to the parotid gland.

Refer back to the descriptions of the middle cranial fossa and infratemporal fossa to fix relationships. Realize, once again, that both IX and X leave the skull

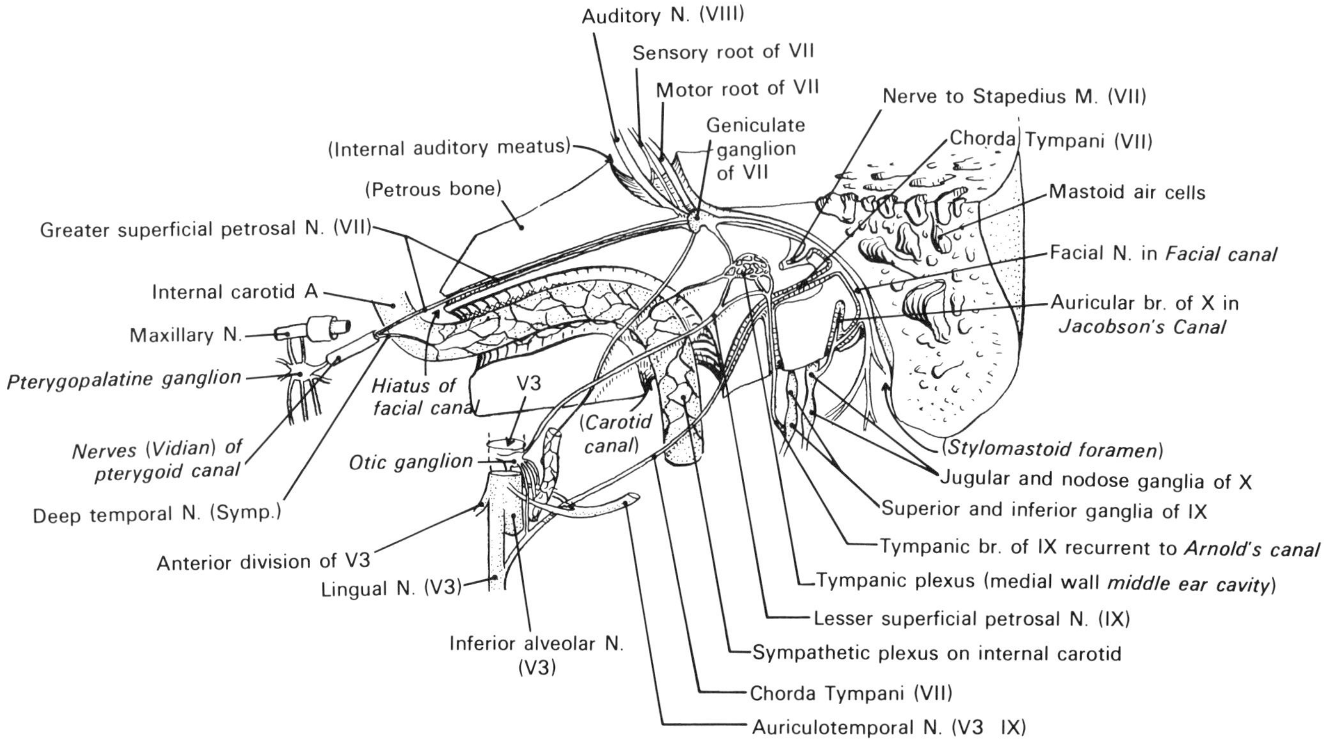

Fig. VI-59. Course of the seventh and ninth nerves through bony channels of temporal bone.

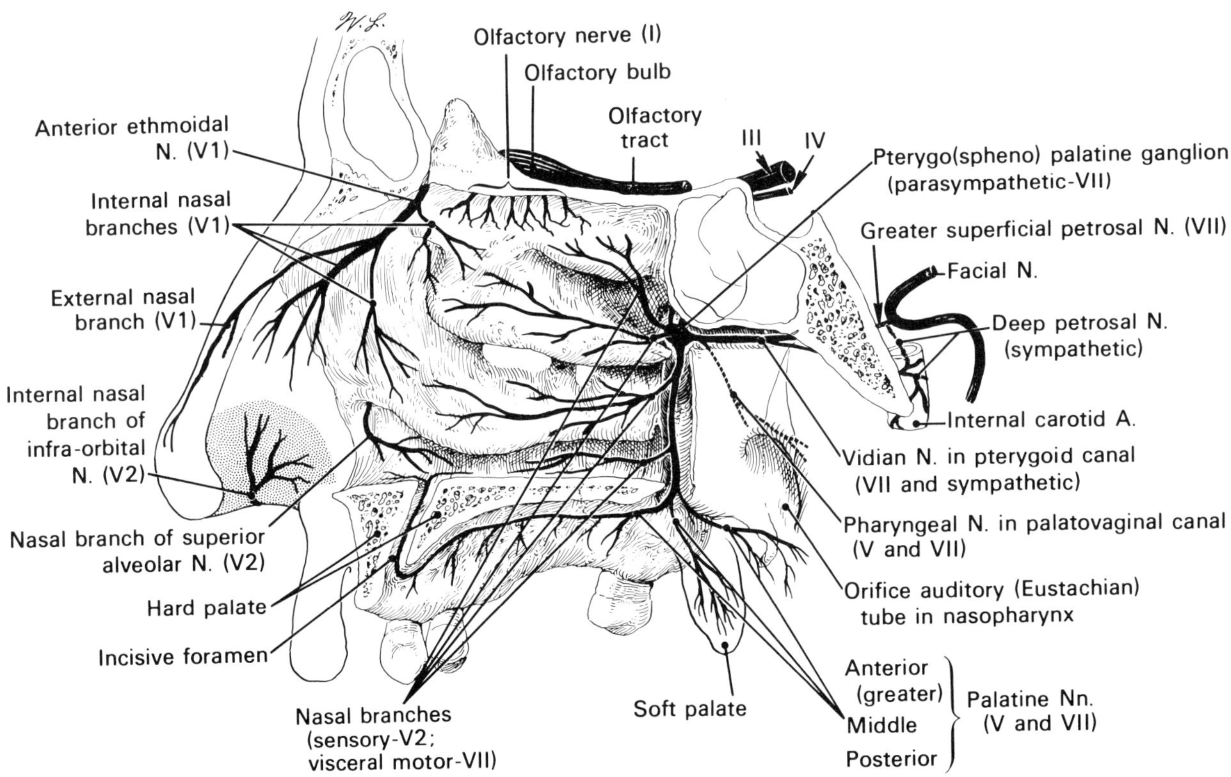

Fig. VI-60. The nasal and palatine nerves and their relationship to cranial nerves V and VII. The trigeminal distribution is sensory to the mucous membranes while the facial postganglionic fibers from the pterygopalatine ganglion are parasympathetic, visceral motor to glands of the mucosa and are distributed by the trigeminal nerve branches.

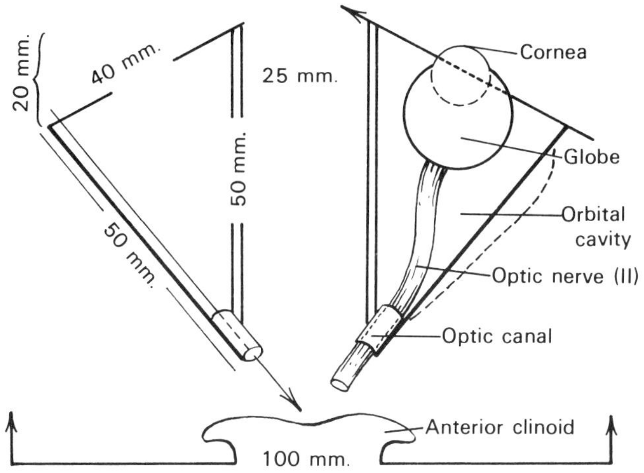

Fig. VI-61. Schematic horizontal section through orbital cavities to show their dimensions. (After Grant.)

via the jugular foramen before they send recurrent twigs back into the skull *via* Arnold's and Jacobson's canals (tympanic and mastoid canals), respectively.

THE ORBITAL CAVITY AND ITS CONTENTS

The orbital cavity often is approached in the cadaver after the calvaria, brain, and the thin roof of frontal bone in the anterior cranial fossa have been removed. Taking away part of the lateral orbital wall also helps to gain dissection room in this rather limited space. But before such destruction occurs, it is well to gain a concept of the empty orbit as seen in an intact skull (Figs. VI-61 and 63).

Each orbit (or orbital cavity) is pyramidal in shape, with the *optic canal* as its apex and the margins as a base; each has a roof, floor, and medial and lateral walls. The medial walls of each cavity are parallel to each other with the upper parts of the nasal cavities separating them (Fig. VI-61). The lateral walls stand at right angles to one another. This can be tested by passing stiff-waxed strings along each lateral wall and through each optic canal. The strings will meet at right angles near the dorsum sellae. The orbital margins are recessed about an inch behind the root of the nose. The *vertical diameter* at the margin approximates 35 mm., 5 mm. less than the *horizontal diameter,* and this latter is 10 mm. less than the depth or length from margin to optic canal. In the caucasian female the shape of the margin is more circular than in the male; it is more circular in yellow races than in white and in white races than in black, where the shape tends toward the oblong.

That the long axis of the orbital cavity is directed laterad as well as anterior will become important when we have to consider the *visual axis* of the globe or eyeball and have to judge the effective action of two extra-ocular muscles (Superior and Inferior Recti) that parallel the long axis of the orbit. These muscles work best as respective elevators and depressers of the globe when the visual axis of the globe is in line with the long axis of the orbit. (The visual axis is that passing from midcornea through the center of the lens to the posterior wall of the globe.) When the globe is in its primary position (looking straight ahead), these axes are *not* in line, and the action of the two muscles named above is not as effective as when the eye is abducted (looks laterad), for only then is the axis of the globe in line with the long axis of the orbital cavity. Students often overlook such facts when they attempt to study eye movements produced by the extra-ocular muscles.

What bones enter into the composition of the orbital cavity? These are illustrated in Figure VI-63A to D. Three bones each contribute about one-third to the orbital margin—the *frontal, zygomatic,* and *maxillary* (Fig. VI-63A). The first forms the upper margin and roof; the last two, the lower margin and sides, meeting midway at the lower margin in a tubercle. The zygomatic contributes to the lateral inner side wall; the maxillary contributes to the floor. If you trace the lower margin from tubercle medially, it terminates at the **anterior lacrimal crest** to which the **medial palpebral ligament** of the eyelids attach. If you trace the upper margin medialward, it does not join the lower medial margin, but spirals inward, instead, as the **posterior lacrimal crest** to which the **orbital septae (palpebral fascia)** of the lids attach. In the depression of the **lacrimal bone** between these crests lies the upper end of the tear duct, the **lacrimal sac** (see Fig. VI-63A and B). Thus, it is not essential to enter the orbit at all or disturb the globe in operating upon the tear sac. Do not confuse this sac for the lacrimal **gland** which you can see lies superolaterad to the globe in Figure VI-62D and under the orbital roof. See Chapter III, page 101 for a discussion of the movement of tears from gland to nasolacrimal sac and duct.

The radiologist is frequently called upon to evaluate the optic canal and its surrounding structures for abnormalities. We use two special projections, the optic canal and orbital views, to complement other radiographs of the skull. Both the optic canal and orbital views are basically posterior-anterior oblique projections with some differences in angulation of the patient and tube. Let us examine the structures visualized in these projections by comparing photographs of the dry skull taken from the position of the cassette with the radiographs.

In the optic canal view (Fig. VI-64A and B), note that the optic canal is projected into the inferolateral quadrant of the orbit. The optic canal is therefore seen end-on. Find the optic canal (Fig. VI-64B) by tracing the curving slope of the lesser sphenoid wing laterally. The first, and in most instances, the only foramen encountered is the optic canal. At times, however, the anterior clinoid process is pneumatized and it therefore simulates a foramen. The pneumatized clinoid would lie lateral to the real optic canal. In some individuals, the **caroticoclinoid ligament,** which runs from the tip of the anterior clinoid to the middle clinoid process, becomes ossified (Fig. VI-64E). (At times you will be able to see the **middle clinoid processes** in dry

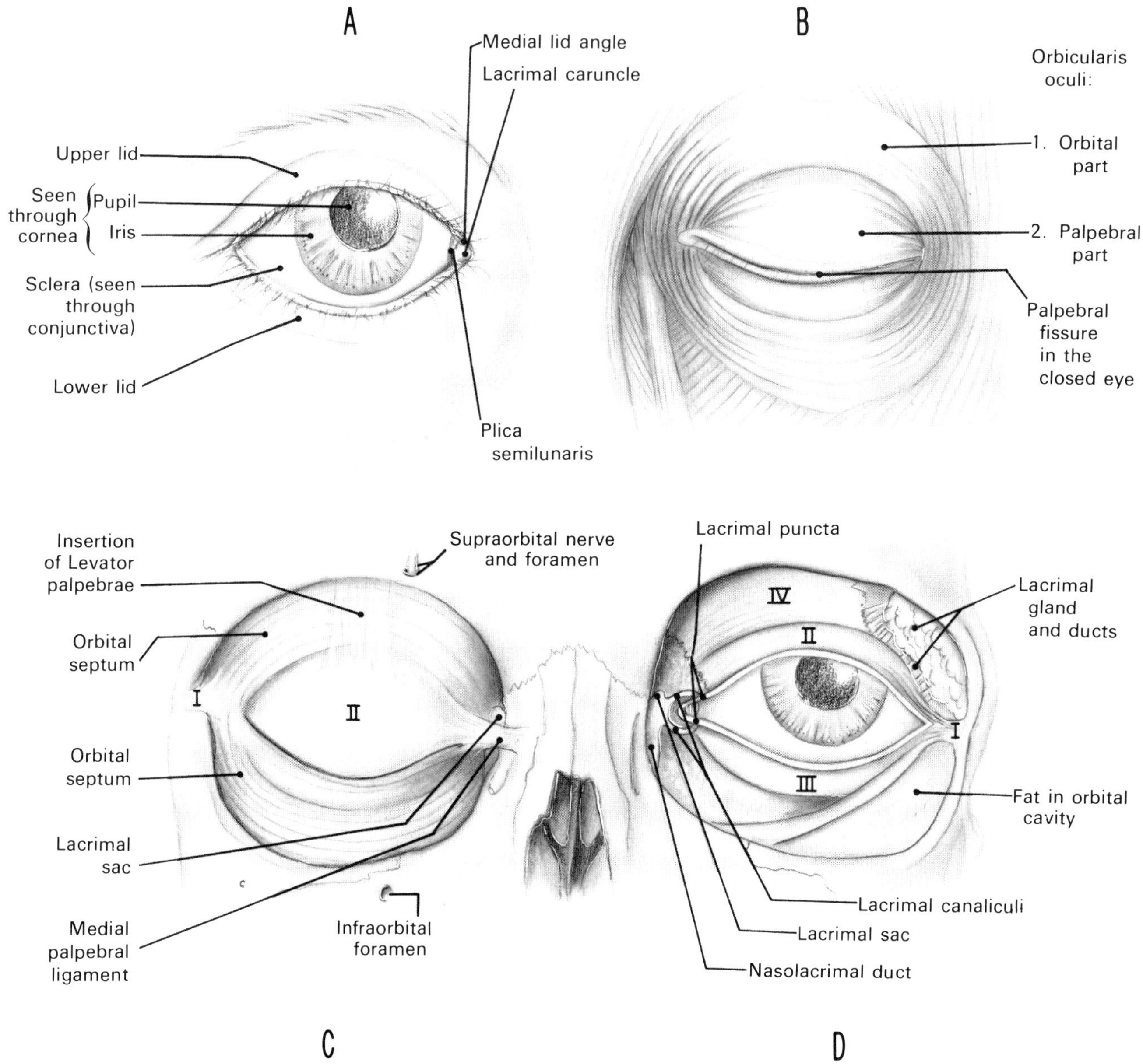

Fig. VI-62. Progressive dissection of the eyelids and lacrimal apparatus. **A.** Right eye. **B.** Left eye. Eye is closed with skin removed to the level of the Orbicularis oculi muscle in the lower stratum of the lid dermis. **C.** Right eye. Orbicularis oculi removed to show orbital septum and tarsal plate. **D.** Left eye. Further dissection to show parts of the lacrimal apparatus. In C and D: I, lateral palpebral ligament; II, Superior tarsal plate; III, Inferior tarsal plate; IV, orbital septum (in part). (After Craft.)

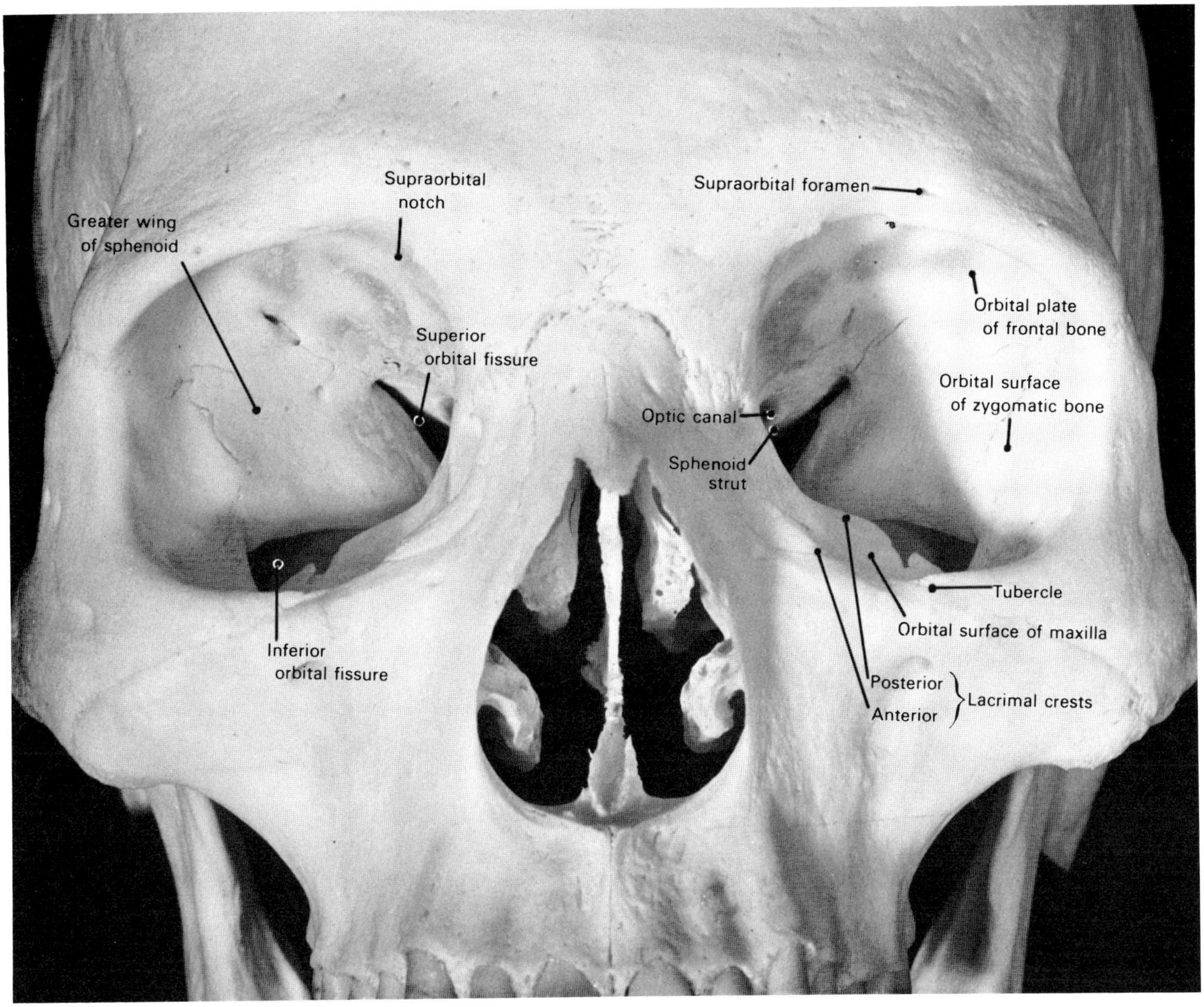

Fig. VI-63. Bony walls of the orbit. **A.** Frontal view.

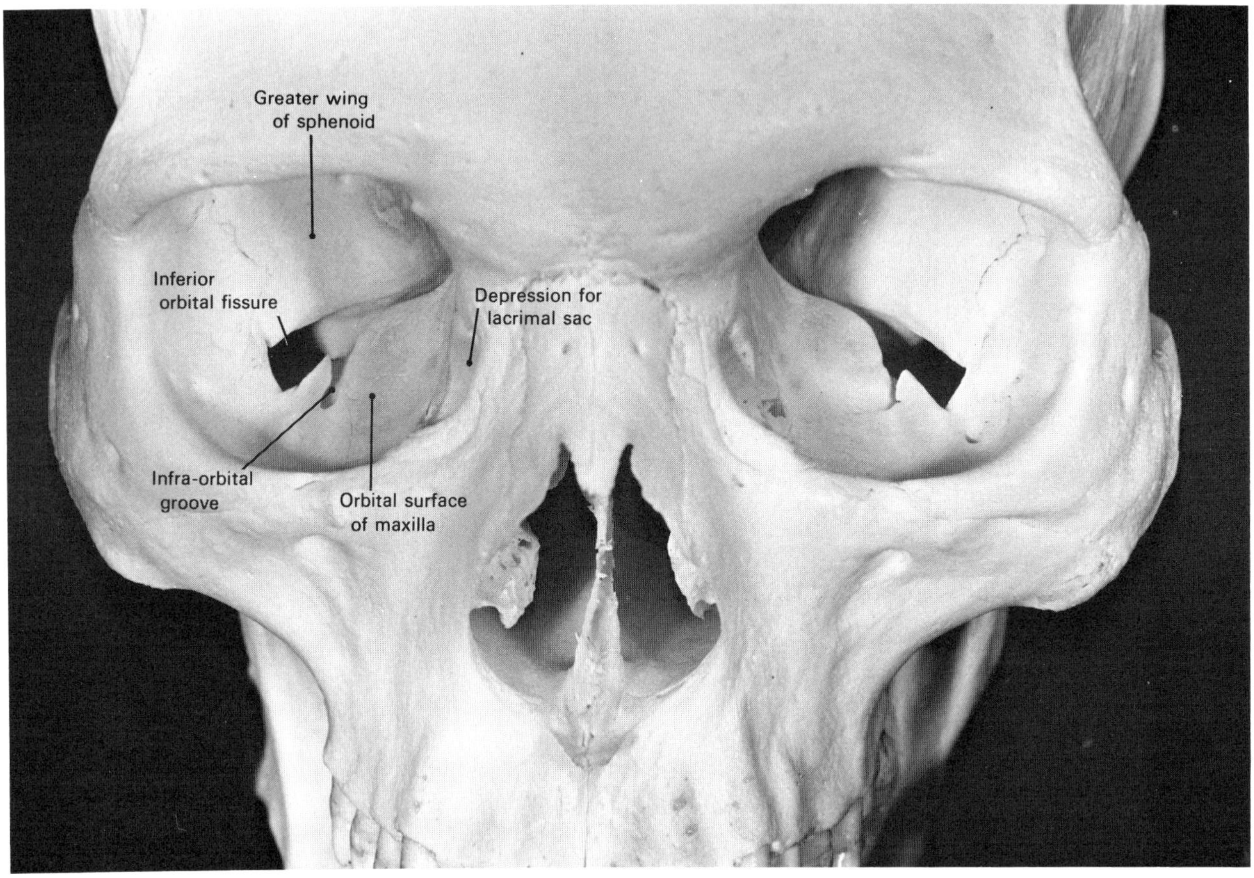

Fig. VI-63B. Skull tilted downward to view the orbital floor.

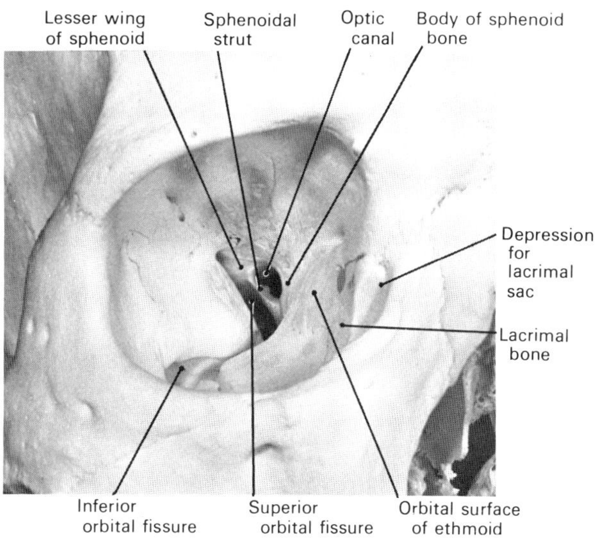

Fig. VI-63C. Frontal view of the right orbit with the skull turned slightly toward the left.

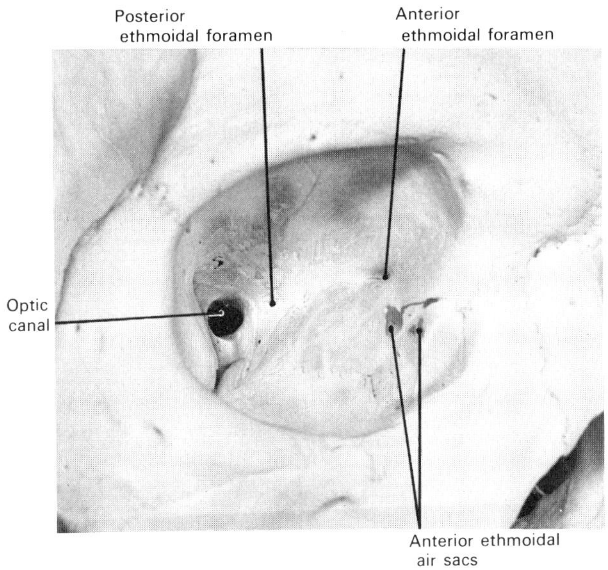

Fig. VI-63D. Frontal view of the right orbit with the skull turned farther to the left.

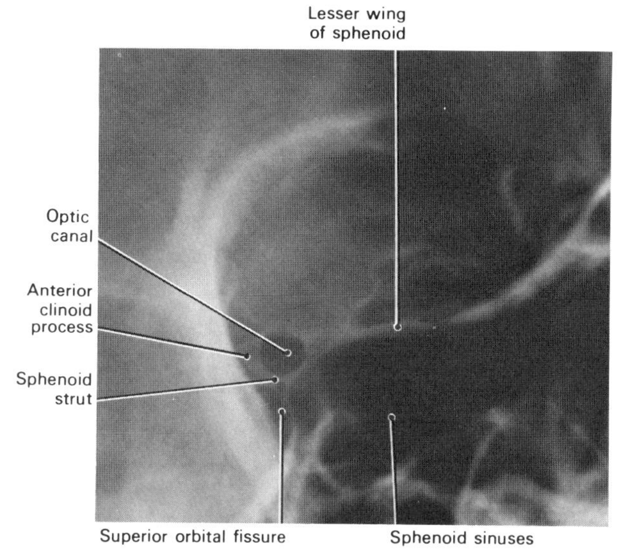

Fig. VI-64. Normal roentgen anatomy of the orbit. **A.** Optic canal projection in the living patient.

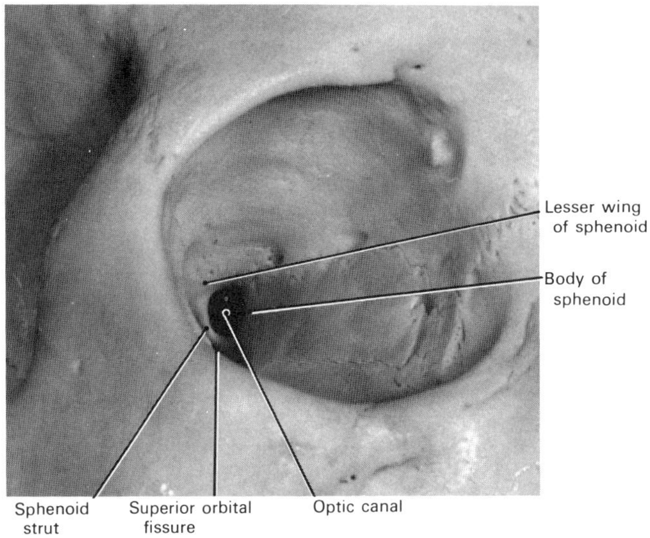

Fig. VI-64B. Optic canal view as seen from the position of the film cassette in a dry skull.

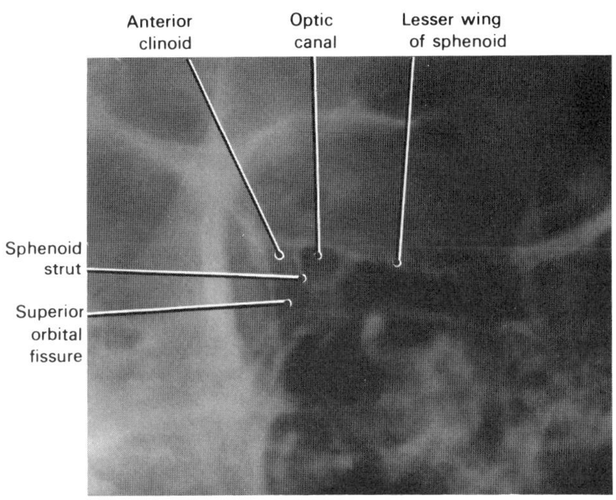

Fig. VI-64C. Orbital projection in the living subject.

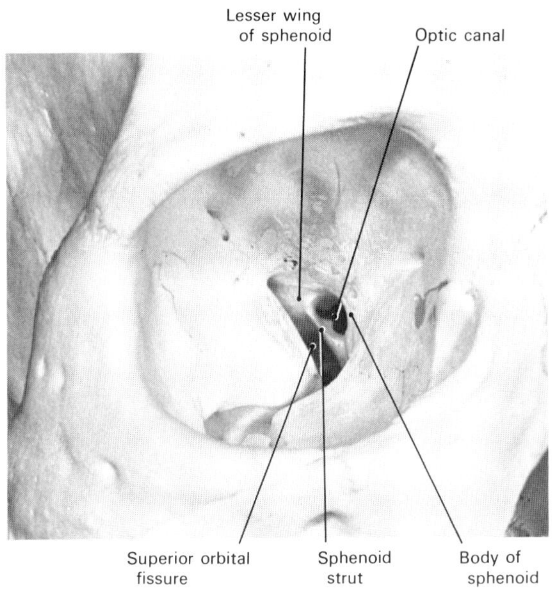

Fig. VI-64D. Orbital view taken from the position of the film cassette.

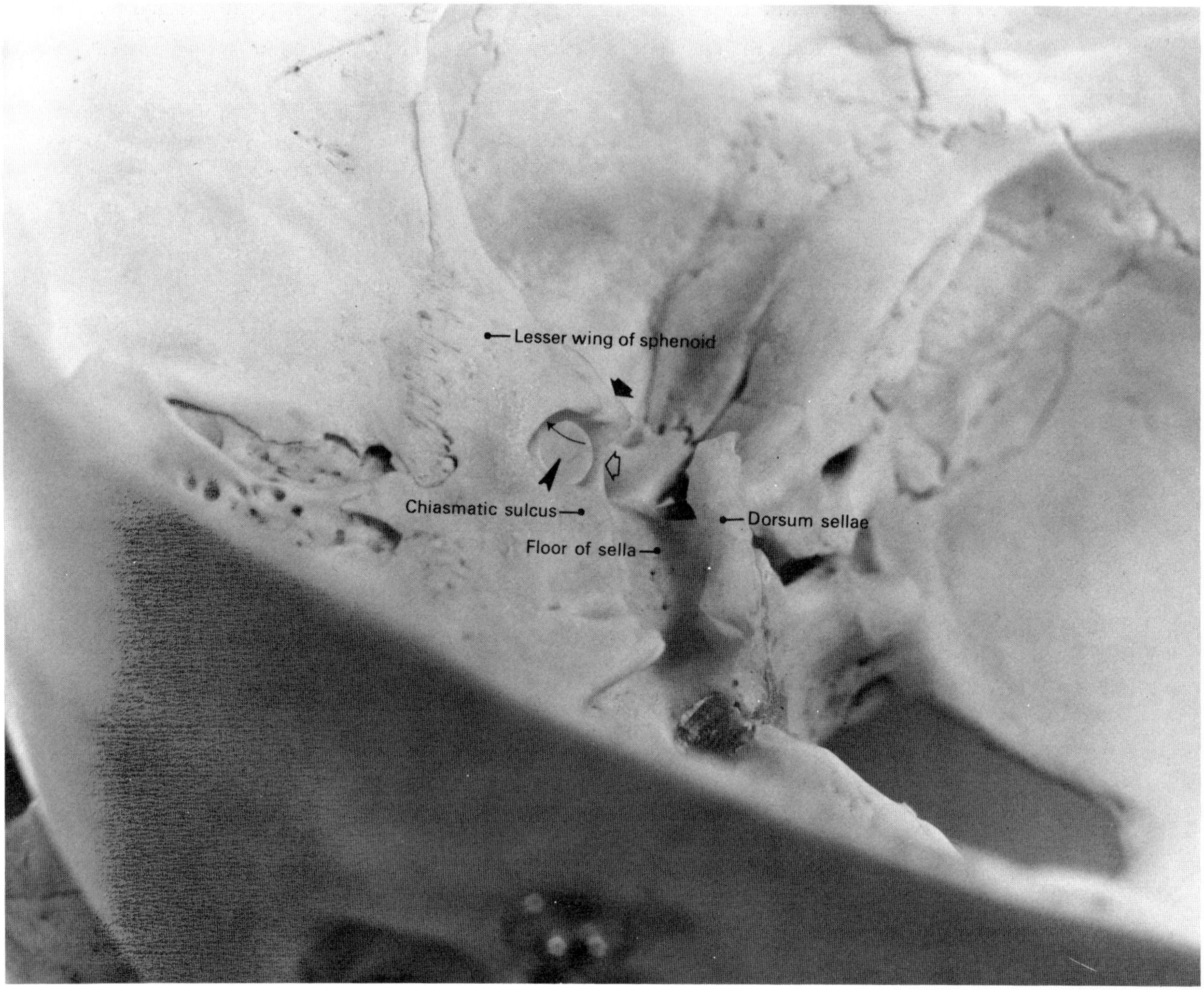

Fig. VI-64E. Caroticoclinoid foramen as viewed obliquely from above. *Black arrow,* anterior clinoid process; *black arrow head,* caroticoclinoid foramen; *open arrow,* ossified ligament running from the anterior clinoid process to the middle clinoid process; *long fine arrow,* indicates direction into the optic canal.

skull preparations. These tiny eminences arise near each side of the tuberculum sellae). The foramen formed is called the **caroticoclinoid foramen** and through it passes the internal carotid artery. This foramen is typically seen inferior and lateral to the optic canal, and it overlaps the superior orbital fissure.

Since the optic canal projection is made so that the central X-ray beam passes along the central axis of the canal there is no distortion. Therefore it affords an excellent means of measuring the canal. Realize that the optic canal reaches its ultimate size by the third year of life. On radiographs obtained at 36″ to 40″ target-film distance the optic canal measures in its greatest diameter no more than 6.5 mm. Both sides should be compared, and although there is frequently some disparity in size this should not exceed 2.0 mm.[15]

In the orbital view, the optic canal is projected into the center of the orbit (Fig. VI-64C and D). Note that the optic canal is distorted and appears oval. The structures surrounding the optic canal now are minimally distorted. The sphenoid strut may be quite variable in its appearance, and occasionally is even pneumatized or congenitally absent.

Other views used to examine the orbits adequately include: inclined PA, lateral, basal, and occipitomental projections. All of these have been discussed earlier.

The lower limb of the V-shaped fissure (Fig. VI-63B) is, obviously, the *inferior orbital fissure.* We see it separating the *greater wing of the sphenoid* in the lateral wall from the maxillary bone in the floor of the orbit, and it encroaches upon the zygomatic bone.

You are already familiar with the foramina in the first of these three bones—supraorbital, zygomaticofacial, and infraorbital foramina. Figure VI-63B shows that the infraorbital foramen becomes a canal which opens in the floor of the orbit as the **infra-**

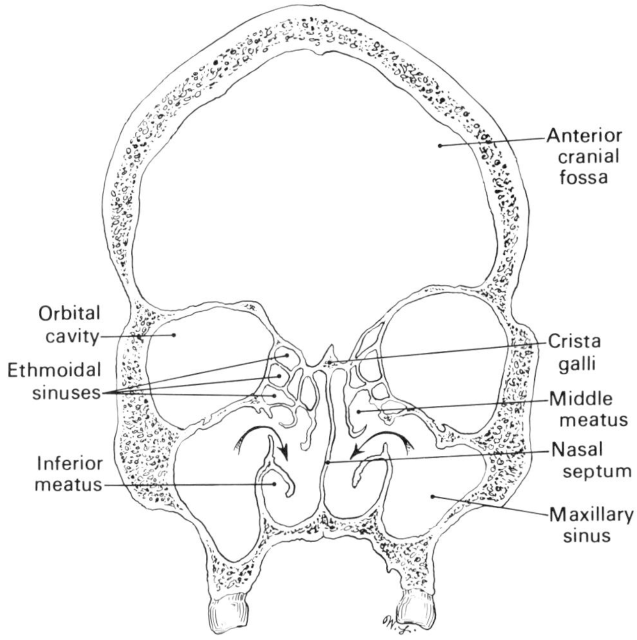

Fig. VI-65. Relation of the orbital cavity to the maxillary, frontal, and ethmoidal sinuses. The apparent increased thickness of the cranial vault is caused by the bone sloping obliquely at this point in the coronal sectioning of the skull.

orbital groove. The groove meets the inferior orbital fissure about at its middle.

Medial to the optic canal, between it and the lacrimal bone, a small flat patch of the *ethmoid* appears as part of the medial orbital wall (Fig. VI-63D). Thus, *frontal, zygomatic, maxillary, greater and lesser wings of the sphenoid,* the *ethmoid,* and the *lacrimal* bones all contribute to the orbital cavity. Along the fronto-ethmoidal suture line are the **anterior and posterior ethmoidal foramina.** There has been a breaking through of the medial wall above the lacrimal bone so that we look into some of the **anterior ethmoidal air sacs** (Fig. VI-63D) which range well forward, medial to the lacrimal fossa. This is a good point for future ophthalmic surgeons to note. While the surface of the ethmoid in the orbit looks innocent enough, behind its blank face lies a most complex honey-comb of ethmoidal sinuses which communicate with the nasal cavity.

Except for the foramina and fissures, the whole orbit appears stable enough, yet it is most vulnerable, since its walls are all quite thin. A simple sketch like Figure VI-65 should demonstrate this fact at once. In such a frontal or coronal section through the orbit, we come to realize that its floor is the roof of the maxillary sinus; its roof is the floor of the anterior cranial fossa and, in part, of the frontal sinus; its medial wall is a thin partition separating the orbit from the ethmoidal sinuses. Medial to the optic canal lies the sphenoidal sinus. Students generally fail to realize how close the optic nerve in its canal lies to the sphenoidal and posterior ethmoidal sinuses. Pus from any of the sinuses can erode through the thin bone. Then, only the periosteum lies between pus and the orbital contents. The lateral wall in man is the strong wall. In non-primates, there is no lateral wall; Müller's smooth muscle tends to fill the gap. In some mammals, *e.g.,* the family cat, even the lateral orbital margin is missing.

If you dissect the orbital contents, you will, of course, follow the directions of your instructor or of some dissection manual. Methods of approach vary. Once the exposure is made, the following structures will be found occupying the confines of the pyramidal orbital space:

1. The extra-ocular muscles, seven of them (if we include the levator of the upper lid) forming a muscular cone about
2. The optic nerve.
3. The nerves supplying the muscles (III, IV and VI).
4. The ophthalmic artery and veins.
5. The ciliary nerves and vessels.
6. The rest of the branches of V, *en route* to their peripheral distribution.
7. The lacrimal gland.

If bone is removed carefully, the periosteum; *i.e.,* the **peri-orbita,** will tend to remain behind, fairly well intact, since it is only loosely adherent to the bony walls. All the contents are within the peri-orbita, and they are well packed with **peri-orbital fat** (Fig. VI-67), a bit of a nuisance since it must be removed to get at the fine structures packed in this small space. In life, of course, the fat is semifluid, not semisolid as you find it in the cadaver. The ophthalmic surgeon tries to disturb it as little as possible so that it will not leak away, else the exquisite mechanisms controlling eye movement become impaired from want of the essential support this fat affords.

Structures will not necessarily be encountered in the order in which we listed them above, and only after as careful a dissection as you can perform can you begin to put the pieces of the following description in place, for each of the separate structures must be cleaned and teased apart before you can sort them out as you proceed.

The **bulbar fascia** (capsule of Tenon) (Fig. VI-67) is the fibrous tunic that forms a sheath about all but the corneal part of the eyeball. It plays out thinly over the optic nerve, keeps the eyeball aloof from fat and other contents in the orbit, and provides a socket for limited movement. The extra-ocular muscles pierce it to gain anchorage upon the sclera beyond the equator of the globe (Fig. VI-66A and B).

The extrinsic muscles of the eyeball comprise four straight muscles or recti and two obliques (Fig. VI-

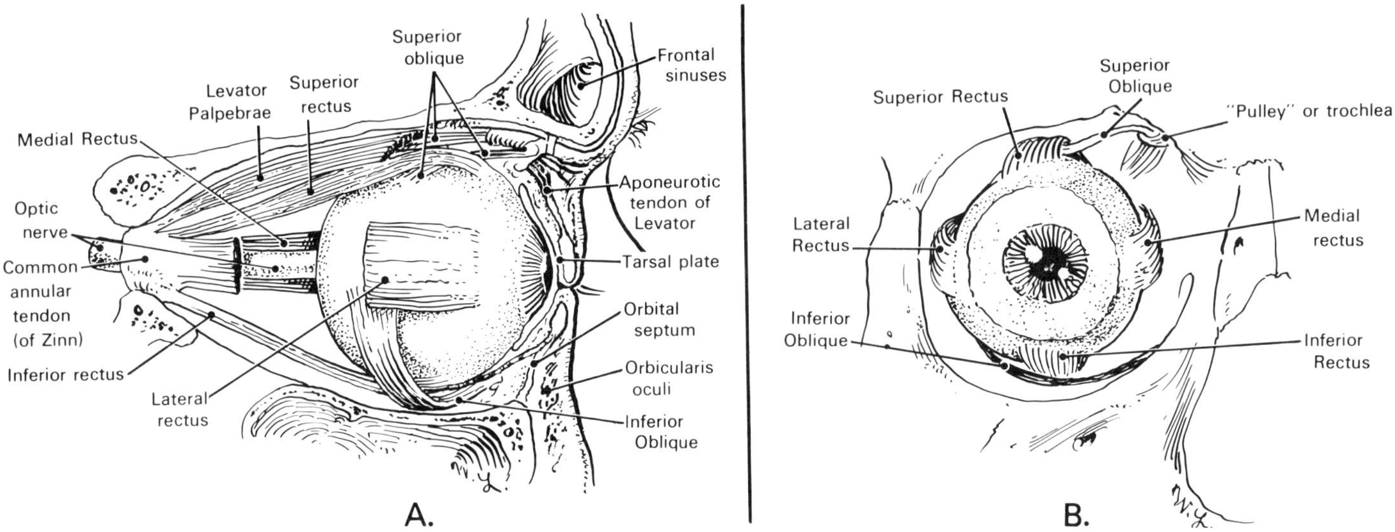

Fig. VI-66. The muscular cone in the right orbit. **A.** Lateral view. **B.** Frontal view of the globe and muscular attachments.

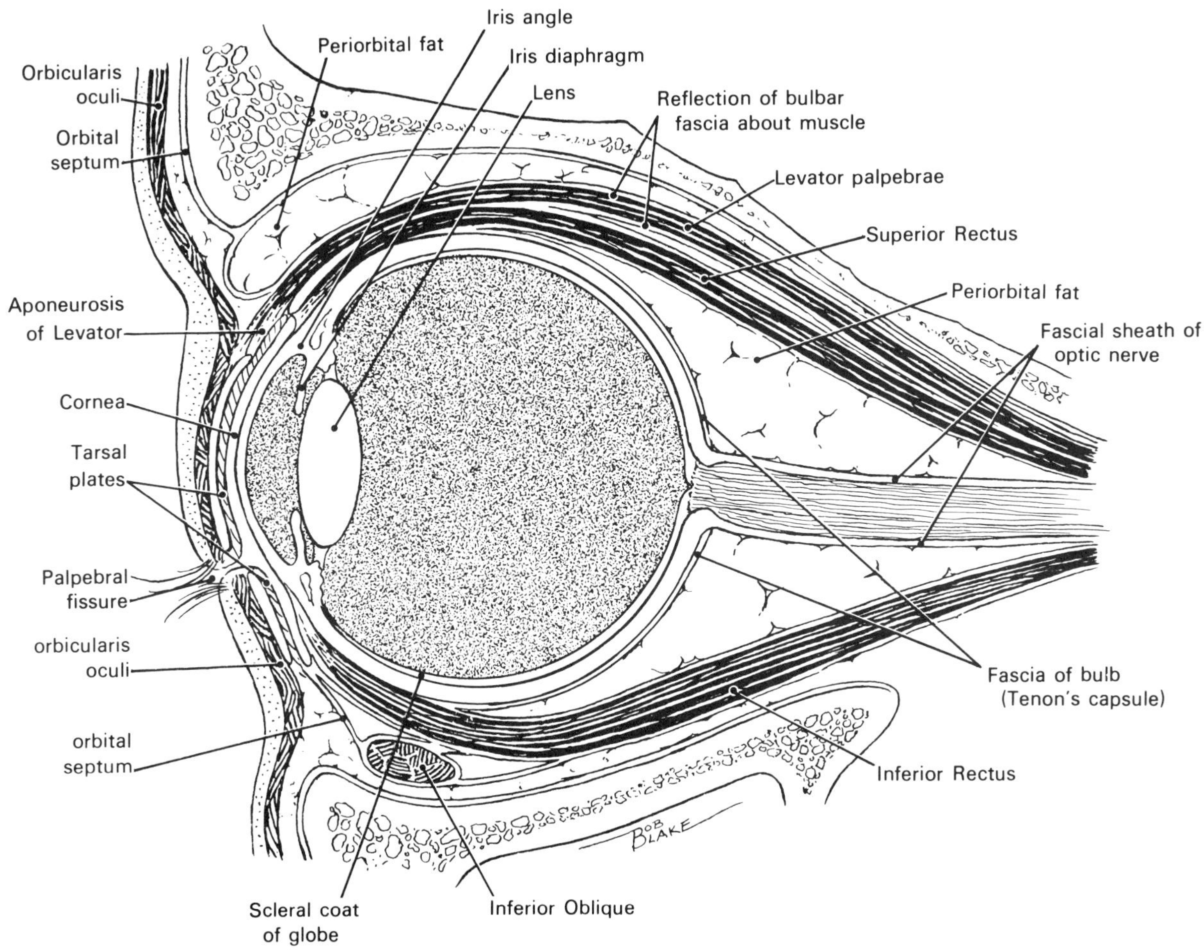

Fig. VI-67. Schematic vertical section through the orbit to show orbital fasciae and periorbital fat.

66). The seventh, the **Levator Palpebrae,** is the elevator of the upper lid and does not move the globe. There is no comparable muscle to move the lower lid. In essence, the Levator is a delaminated portion of the **Superior Rectus** and shares the same nerve (III). The Recti—**Superior, Inferior, Medial,** and **Lateral** (the first three innervated by III; the last, by VI) all arise from a fibrous cuff or tendinous ring that is attached behind at the optic canal, to the posteromedial part of the superior orbital fissure, and to the dural sleeve of the optic nerve. From the ring the muscles spread forward in barrel-stave fashion as a muscular cone (Fig. VI-66A) to envelope the globe, piercing the bulbar fascia as noted above. Where they attach to the globe, the bulbar fascia thickens, and it is reflected back over each muscle as a tight sleeve. Between the sleeves, the alveolar tissue spreads thinly from muscle to muscle. Orbital fat impinges upon structures both within and without the muscular cone. We shall find the optic nerve and ophthalmic artery entering the "cone" through the optic canal from the middle cranial fossa. Also coming into the cone through the superior orbital fissure will be nerves VI and III and the nasociliary branches of V_1. The other branches of V_1 (frontal and lacrimal) and cranial nerve IV will remain outside the cone area to course under the orbital roof even though they, too, enter the orbital cavity *via* the superior orbital fissure (Fig. VI-68 A–C).

Before we consider the oblique muscles, we should note that extensions of the fascial sleeves from the Medial and Lateral Recti run as **check ligaments** to anchor medially on the posterior lacrimal crest and laterally upon the orbital tubercle of the zygomatic bone. The bulbar fascia also thickens under the globe as a sling or hammock between these points. This **suspensory ligament** of the globe is strong enough to hold the eyeball up even if the maxilla is removed, and it will serve as a socket to support an artificial eye.

The *Superior Oblique* is the longest and thinnest of the extra-ocular muscles (Fig. VI-66). It arises from the sphenoid in front of the optic canal. The muscle then runs forward under the medial margin of the orbital roof, until it reaches its pulley or **trochlea** which is a fibrous loop located forward along the frontal bone. Here, in riding through the "pulley," it naturally becomes tendinous, bends about 50° to take a posterolateral course, and crosses under the Superior Rectus muscle for its insertion onto the globe (Fig. VI-68B). Because the muscle goes through a "trochlea" you can expect that cranial nerve IV or trochlear nerve supplies its innervation. If you can recall the sulphate radical, SO_4, you can easily associate the Superior Oblique with cranial nerve IV. It is the only extrinsic eye muscle innervated by this nerve.

The *Inferior Oblique* is the shortest of all of these voluntary skeletal muscles that move the globe. Recall that they were all derived from pre-otic somites; hence, their nerves are somatic motor. The Inferior Oblique is the only one to arise well forward in the orbit; the rest have posterior origins. It arises from the orbital floor anterolateral to the beginning of the lacrimal fossa. The muscle then courses posterolaterad under the globe, crossing the Inferior Rectus on its way to insert upon the globe (Fig. VI-66B). The oculomotor nerve III supplies the Inferior Oblique just as this nerve supplies the majority of the extra-ocular muscles (Superior, Inferior, and Medial Recti and the Levator Palpebrae). Cranial nerve IV we saw innervating the Superior Oblique. This leaves unaccounted only the nerve supply of the Lateral Rectus, and since this muscle moves the eyeball laterally, or *abducts* it, you ought to conclude that the Abducens (Abducent) or VI will innervate it. The Lateral Rectus is the only extra-ocular muscle that cranial nerve VI supplies.

The *Levator Palpebrae* originates from the tendinous ring above the Superior Rectus (Figs. VI-66A and 68A). It broadens as it moves forward under the roof of the orbit and above the Superior Rectus. It descends anteriorly through peri-orbital fat to insert into the root of the upper lid. There its fibers cross perpendicular to those of the Orbicularis Oculi and through them to reach dermis of the lid. In entering the lid, its lower aponeurotic portion (Fig. VI-67) contains the smooth **Tarsal muscle** (Müller's muscle) of the lid which receives a cervical sympathetic supply and is responsible for lid tonicity. Similar involuntary Tarsal muscle enters the lower lid from the advancing aponeurotic strands of the Inferior Rectus. As noted above, the palpebral portion of the Levator is supplied by III since it is voluntary skeletal muscle.

If we take out all of the orbital contents except parts of the extra-ocular muscles and then peer into the orbit from in front (Fig. VI-69), we can visualize the muscle origins from the tendinous ring and their nerve supply coming to them. Only the Inferior Oblique lies well away from the rest in its origin. Only the optic nerve and ophthalmic artery enter the muscular cone through the optic canal. All the motor nerves enter the orbital cavity *via* the superior orbital fissure, and only IV fails to enter the muscular cone. This nerve crosses medialward under the orbital roof (over the cone) to reach the Superior Oblique. The Oculomotor nerve (III) breaks up on entry into an upper and lower division, the upper to the Superior Rectus and Levator; the lower passing below the optic nerve to Medial and Inferior Recti and the Inferior Oblique. The Abducens passes solely to the Lateral Rectus.

It is the branch of III to the Inferior Oblique that conveys the visceral motor fibers (parasympathetic preganglionics) to the **ciliary ganglion** (Figs. VI-68 to 70). About a dozen, *short* **ciliary nerves** leave the ganglion, some above and others coursing below the optic nerve, to pierce sclera of the globe and to enter its vascular, choroidal coat. Exactly to which cranial nerve these short ciliary branches belong has never

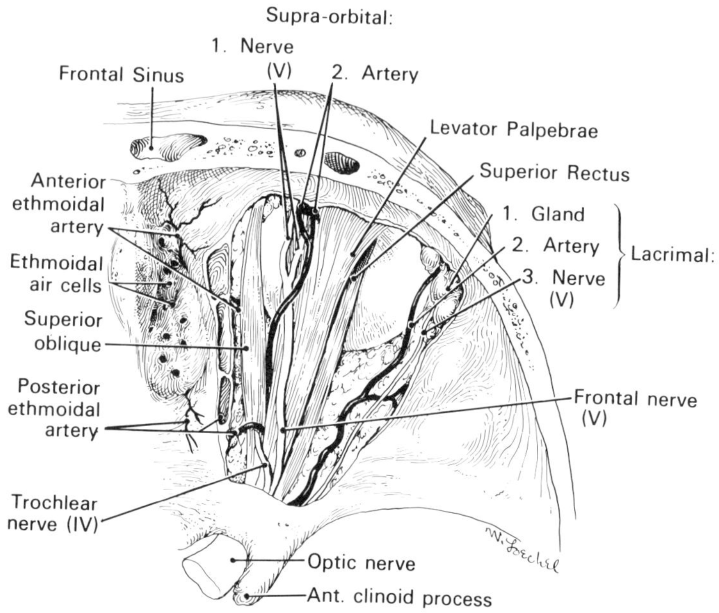

Fig. VI-68. Dissection of the right orbital cavity. **A.** After removal of orbital plate of frontal bone.

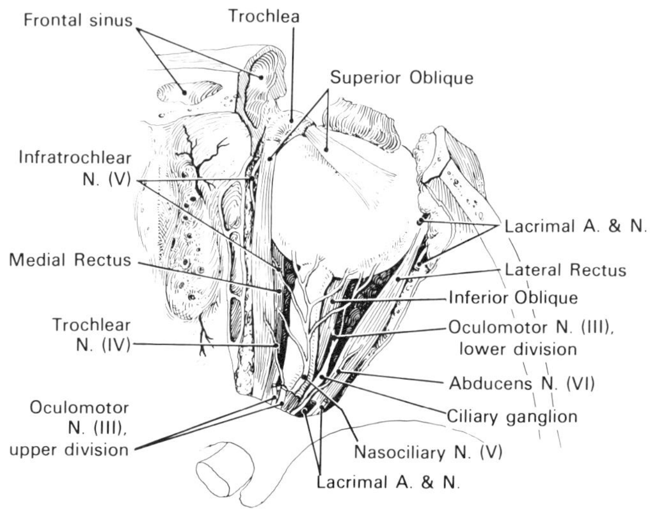

Fig. VI-68B. After removal of Superior rectus and Levator palpebrae.

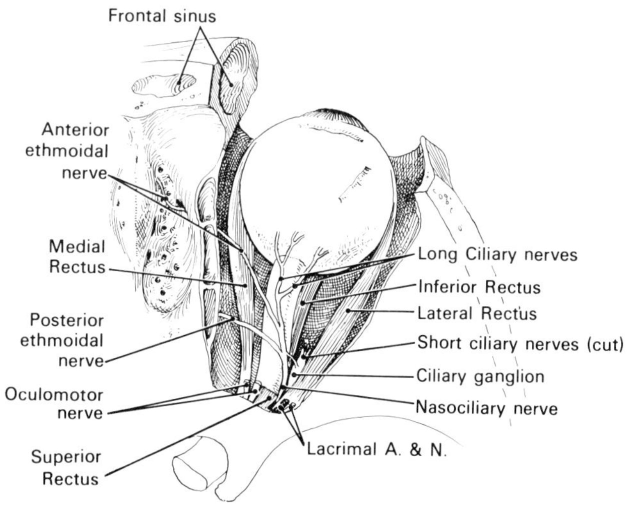

Fig. VI-68C. The lacrimal gland and Superior oblique has also been removed.

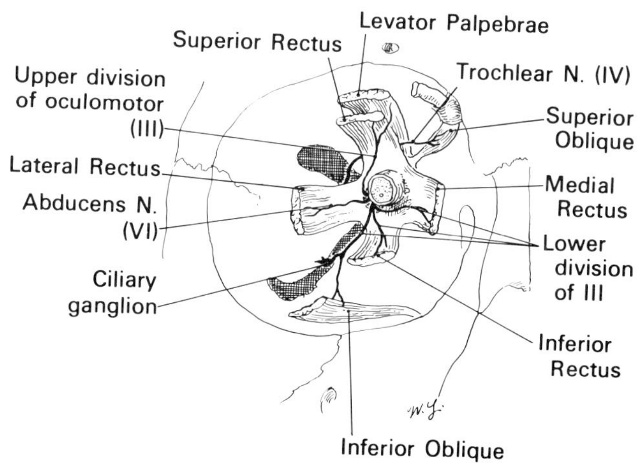

Fig. VI-69. Motor innervation of extraocular muscles of the right orbit. The globe has been removed.

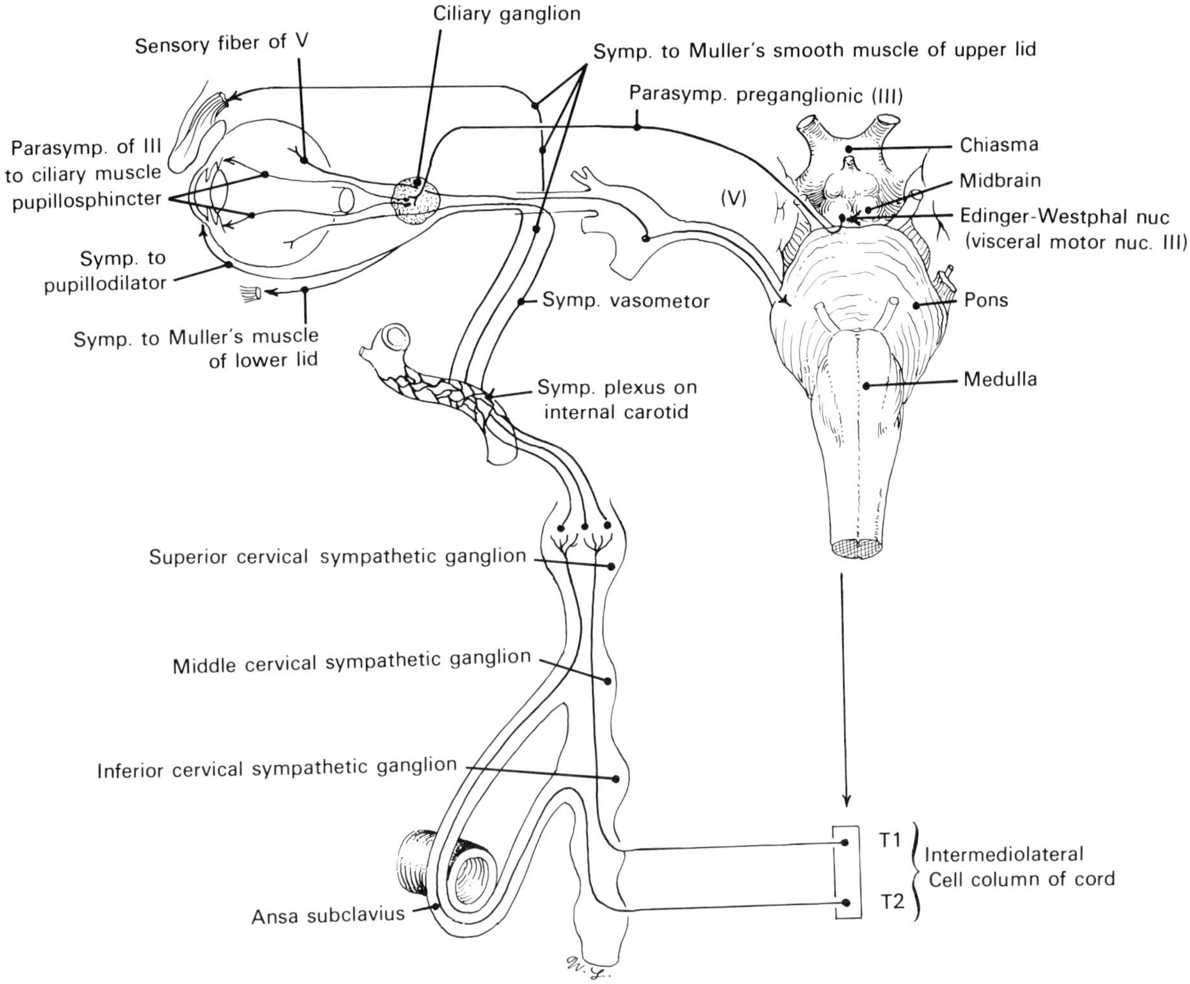

Fig. VI-70. Sympathetic and parasympathetic innervation of smooth muscle structures within the globe.

been made clear. They are described as branches of the ganglion and, generally, are discussed as if they were part of the oculomotor nerve. But several types of fibers run *through* the ganglion without interruption; *e.g.,* sensory fibers from the cornea and other parts of the eye and *sympathetic* fibers from the superior cervical sympathetic ganglion that have taken the vascular route *via* the internal carotid and its ophthalmic artery. Only the parasympathetic fibers of III *synapse* in the ganglion. Postganglionic axons from the ganglion enter the short ciliary nerves. The sensory fibers in these nerves go *through* the ganglion to join the **nasociliary nerve branch** of $\mathbf{V_1}$. It would make more sense, then, as it does elsewhere in the head, *to consider the short ciliary nerves to be twigs of the trigeminal which, as usual, are piggy-backing visceral motor fibers to their terminal depots in the head.* In this case, twigs of V_1 are carrying third nerve fibers to smooth muscle of the **Pupillosphincter** in the iris and to the smooth **Ciliary** muscle of the **ciliary body,** and *sympathetic fibers* to smooth muscles of the choroidal coat of the eye ball. The iris, containing the sphincteric muscle (Pupilloconstrictor), and the ciliary body are simply anterior modifications of the choroid, the middle vascular coat of the eyeball.

But the pudding gets stiffer, for the nasociliary nerve gives off two, *long* **ciliary branches,** also, which contain sensory and visceral motor components. These by-pass the ciliary ganglion and pierce the sclera farther forward than the short ciliary nerves. They convey pain from the cornea; carry *sympathetic fibers* to the smooth, radially-arranged *Dilator Pupillae* of the iris; and *sympathetic fibers to the Tarsal muscle of the lower lid. Sympathetic twigs* to the *Tarsal muscle of the upper lid* appear to join the *upper division of cranial nerve III to the Levator Palpebrae.* The diagram in Figure VI-70 should help to make these relationships clear.

As a result of this kind of distribution, you could expect the nasociliary branch of V_1 to lie within the muscular cone, and, indeed, it enters the orbit by piercing the heads of origin of the Lateral Rectus so

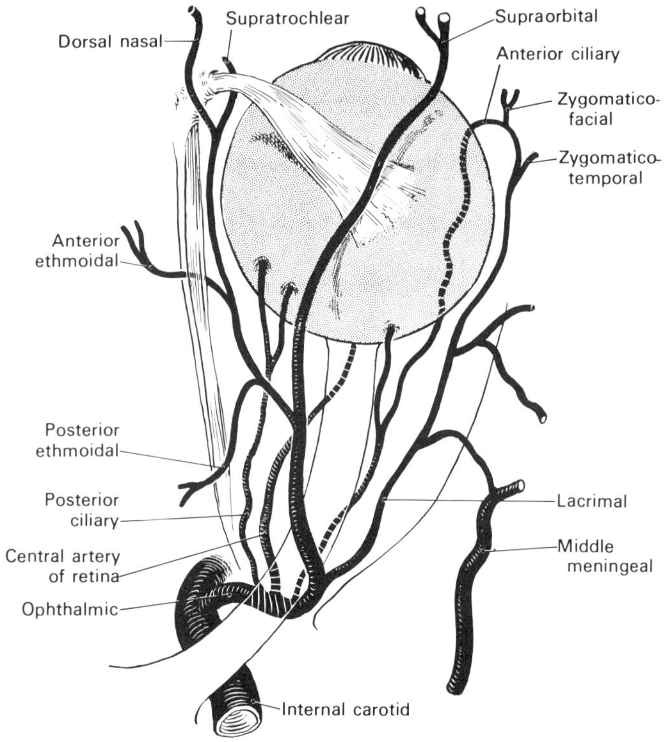

Fig. VI-71. Schema of the arterial supply to the right orbit as viewed from above.

that it lies, at first, lateral to the optic nerve. But its *nasal component* (Fig. VI-68C) runs obliquely forward and medialward toward the medial wall of the orbit, giving off, in turn, a **posterior ethmoidal nerve** through the posterior ethmoidal foramen for the mucous membranes of the posterior ethmoidal air cells and sphenoid sinus; the **anterior ethmoidal nerve** through the anterior ethmoidal foramen; and the **infratrochlear nerve** which passes under the pulley of the Superior Oblique to supply the eyelids with sensory fibers, as well as the skin on the side of the nose. The anterior ethmoidal nerve wanders much more deviously through the anterior ethmoidal air cells to enter the anterior cranial fossa once more. It leaves the fossa again through the **nasal slit** at the side of the crista galli to enter the nose as the **internal nasal nerve.** It finally appears in the skin of the nose as the **external nasal nerve** by piercing the junction between the nasal bone and upper lateral cartilage.

The two other branches of V_1, the **frontal** and **lacrimal** nerves (Fig. VI-68A), remain outside of the cone (like IV). The frontal crosses the orbit just under its roof above the Levator Palpebrae. We know its terminal cutaneous branches as the *supraorbital* and *supratrochlear* nerves. Often the frontal continues as a branch of the supraorbital well back over the scalp to the vertex. This nerve is the stoutest branch of V_1.

The *lacrimal nerve* (Fig. VI-68A and B) is the smallest branch of V_1. It courses outside the cone along the lateral wall of the orbital cavity just above the upper border of the Lateral Rectus muscle. It supplies the conjunctiva and skin of the lateral eye angle (not the gland).

The major arterial supply in the orbit is the *ophthalmic artery* (Fig. VI-71). Its terminal branches, as you might guess from most of their names, follow trigeminal nerve branches; *e.g.,* dorsal nasal (follows infratrochlear nerve), supratrochlear, supraorbital, lacrimal, and anterior ethmoidal. All of these end in the skin of the forehead or upper face. The posterior ethmoidal branch supplies the mucous membranes of air cells and nose.

The *important veins* are the **superior** and **inferior ophthalmic.** The superior vein anastomoses with the anterior facial vein. It crosses *above* the optic nerve and dives through the superior orbital fissure to enter the cavernous sinus. The inferior ophthalmic vein enters from its communcation with the pterygoid plexus of veins *via* the inferior orbital fissure; hence, it runs below the optic nerve. It either joins with the superior vein or goes directly to the cavernous sinus on its own.

The *lacrimal gland* rests under the superolateral orbital margin (Figs. VI-62; 68A). It is indented by the lateral edge of the Levator Palpebrae so that it has upper and lower divisions much in the manner of the submandibular gland which is bent around the border of the Mylohyoid. The larger upper orbital part anchors by fibrous strands to the periosteum of the orbital wall. Its smaller inferior palpebral part lies below on the conjunctival fornix and pushes part way into the upper lid. The 14 or so fine ducts drain into the fornix at the lateral eye margin, the upper ducts piercing the inferior gland and joining with ducts from that portion. However, some ducts from the inferior division open independently. Tears are wiped by lid action medialward across the globe to the lacrimal lake and drained by puncta and canaliculi into the nasolacrimal ducts (Fig. VI-62).

EYE MUSCLE ACTION

If one simply considers the direction of pull of the extra-ocular muscle fibers and their anatomical attachment to gain a clue to the direction of eye movements, the problem seems simple enough. Indeed, it is overly simplified, and the eyes do not work exactly in this way even in monocular vision, much less in binocular vision. It would seem that the diagram in Figure VI-72A should tell the story for direction of movement and, roughly, for range of movement for a patient's right eye as you look at it. That is, the Lateral Rectus moves the eye away from the nose or *abducts* it; the Medial Rectus, toward the nose or *adducts* it; the Superior Rectus elevates it; the Inferior Rectus depresses it; and just the opposite of what you think will occur in the case of the obliques actually takes place—namely, the Inferior Oblique makes the anterior pole look *up* and out; the Superior Oblique is the "*down* and out" muscle. But we must dismiss Figure

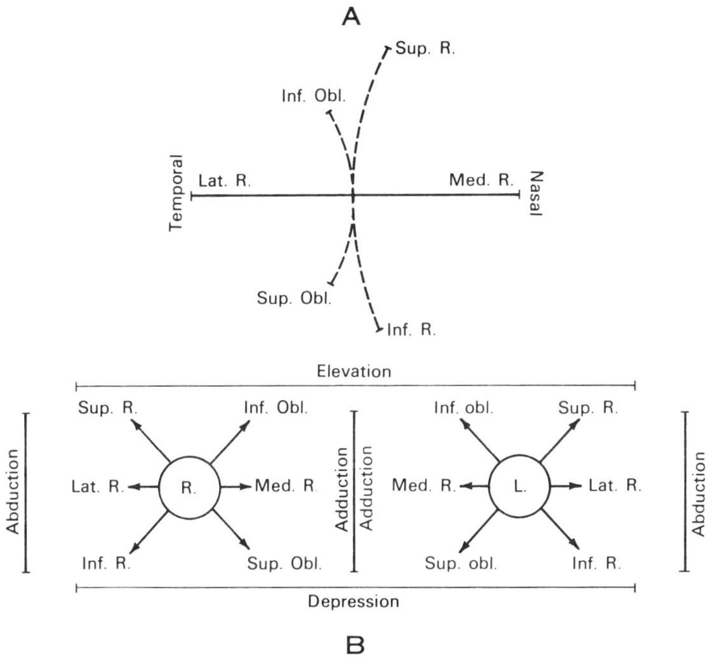

Fig. VI-72. Charts illustrating eye movements. **A.** Anatomical movements possible in isolated, single muscle contraction in the right eye. **B.** Actual movements of a patient's right (R) and left (L) eye during binocular vision. The reader looks into the patient's eyes.

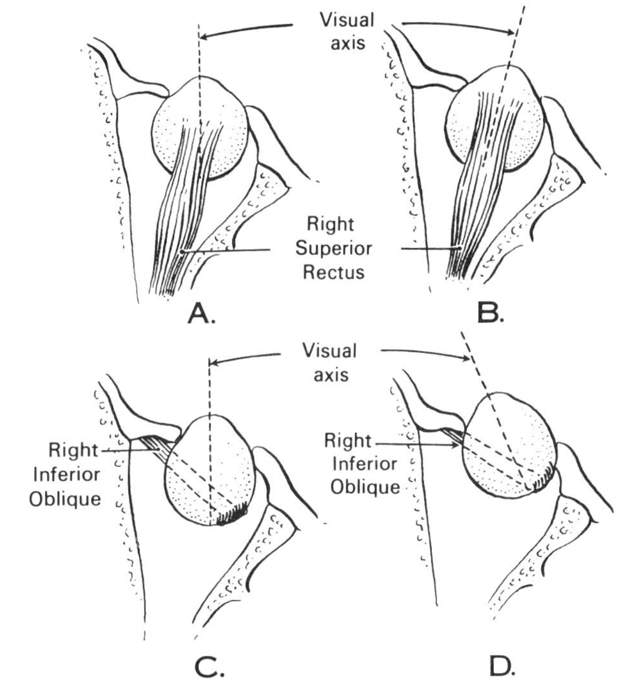

Fig. VI-73. Action of the Superior rectus and Inferior oblique muscles as elevators during abduction and adduction, respectively. **A** and **C.** Eye looks straight ahead. **B.** Eye in abduction. **D.** Eye in adduction. (After Laurenson.)

VI-72A as being too anatomical and too simple and even downright false when we come to study the very complicated and very precise eye movements of patients in the clinic.

In the first place, when the Medial Rectus moves the right eye toward the nose, the right Lateral Rectus, as its antagonist, must relax and, *vice-versa;* when the right eye looks toward the patient's right, the Medial Rectus must relax. But, of course, when one looks to the right (without head-turning) the right eye does not move alone. The left Medial Rectus contracts, as well, and the left Lateral Rectus relaxes. This is easy enough to understand.

It is in vertical movements where problems begin to arise, for each eye has two elevators and two depressors. If we attend again only to the single right eye, its elevators are the right Superior Rectus and the right Inferior Oblique. Its depressors are the right Inferior Rectus and the right Superior Oblique. So we always have a rectus muscle and an oblique muscle working in combination (and the two Superiors and the two Inferiors do not work as units). Obviously, again, in elevation both elevators of the right eye would contract and both depressors, as antagonists, would have to relax. Once more, Figure VI-72A (a very standard, if misleading, diagram, by the way) would have both elevators of the right eye pulling in opposite directions, the Superior Rectus up and *in;* the Inferior Oblique up and *out,* and this would make for a bit of a problem. These "in" and "out" vectors have to be "ironed out" to achieve some happy medium as we look straight ahead and roll our eyeball up in a movement of elevation.

Normally, we think of the Obliques as the great rollers of the eyeball because of their seemingly aberrant muscular attachments which would cause rotation as their muscle bellies contracted. But all eye muscles have some subsidiary action that will cause the eyeball to rotate somewhat about its anteroposterior axis. Of course, the eyeball rotates on a horizontal axis and about a vertical axis, as well.

The degree of abduction or adduction will also determine which of the two elevators will work most efficiently. We have pointed out before that when the eye looks straight ahead, its visual axis (AP pole) is not in line with the long axis of the orbital cavity (Fig. VI-73A). Neither is it in line with the direction of the fiber pull of the Superior Rectus which lines up with the orbital axis. The Superior Rectus is less efficient as an elevator with the eye in this position than it is when the eyeball is *abducted* (Fig. VI-73B). In the latter instance, visual axis, orbital axis, and fiber direction all coincide, and the Superior Rectus can act to maximum advantage as an elevator.

Similarly, when the eyeball is *adducted,* turned toward the nose (Fig. VI-73D), the visual axis of the

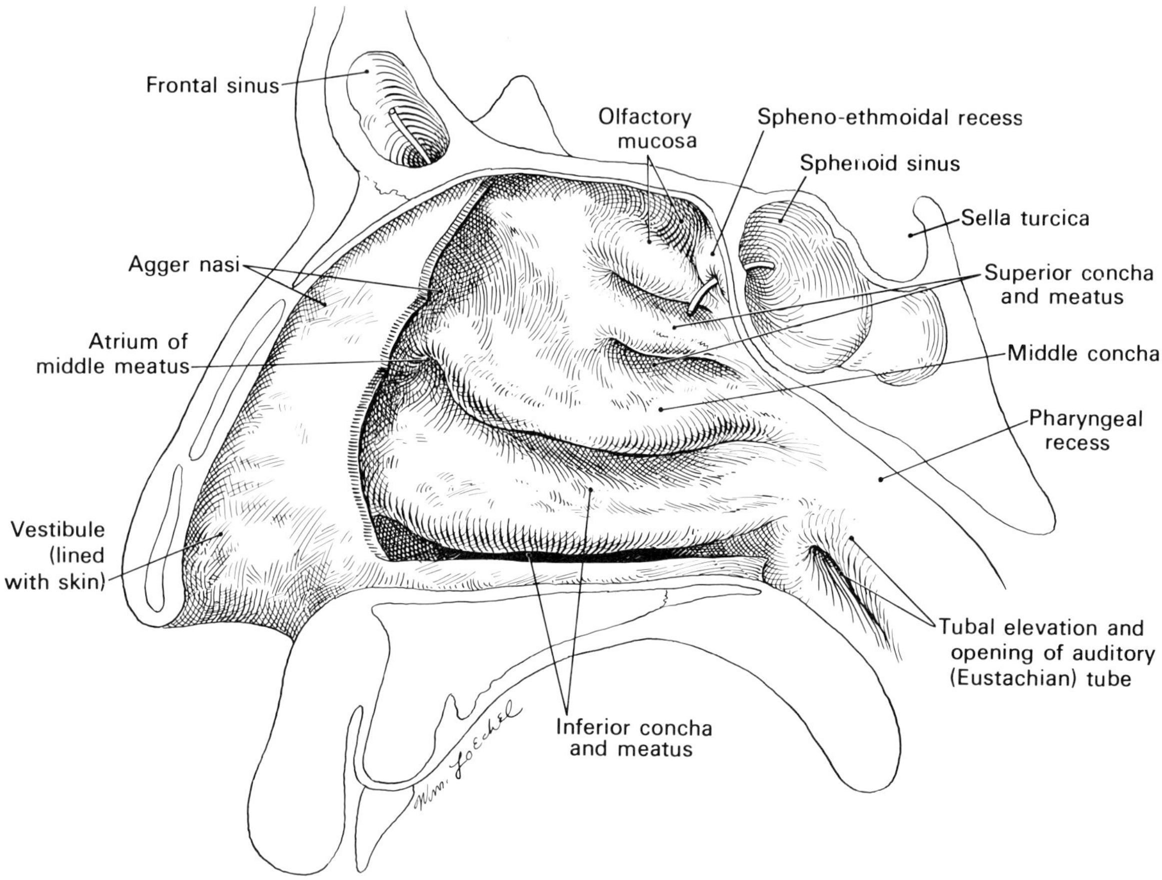

Fig. VI-74A. Nasal cavity and its relations to paranasal sinuses. Lateral nasal wall intact.

globe is much more in line with fiber direction of the Inferior Oblique, and that muscle is now a stronger elevator than when the eye looks straight ahead (Fig. VI-73C).

If the picture seems to get a bit complicated in the case of monocular vision alone, it becomes more so in the case of binocular vision where both eyes must move in nicely-coordinated fashion if we are to benefit from stereoscopic vision. Figure VI-72B presents an entirely different concept of eye movement than did Figure VI-72A where movements were deduced only from the anatomical findings on isolated muscle action. You will find yourself consulting such a chart as Figure VI-72B when you become involved in ophthalmology. Consider, by looking at this chart, what muscles would be involved if you asked a patient to look up and to the left without moving his head. The *left* Superior Rectus and the *right* Inferior Oblique would contract, while their antagonists, the *left* Inferior Rectus and the *right* Superior Oblique, would have to relax. But the other elevators and depressors which would normally make the eye look *up* and to the *right* would also have to make compensatory adjustments.

You should always be aware of the head-movers in the clinic, for they may be compensating for some kind of an eye movement deficiency. For example, a patient with a right Lateral Rectus involvement prefers to look to the left. When asked to look straight ahead, he turns his head to the right, so the direction of the head-turning serves as a clue to the side of eye muscle involvement. Why does this patient prefer looking to the left? Because paralysis or weakness of any eye muscle leads to **diplopia,** or double vision. This is the common complaint of all patients with injuries of cranial nerves III, IV, or VI. If the eyes are turned in the direction opposite to the weak or paralyzed muscle, diplopia tends to be reduced or overcome. So this patient with right Lateral Rectus involvement has learned that double images fall closer together or may even superimpose as one if he looks to the left instead of straight ahead. This may make him appear somewhat shifty-eyed, but, at any rate, he has learned to live as comfortably as possible with his functional deficit. If you now ask the patient to look straight ahead, and then turn his eyes to the right *without moving his head,* the right eye will not move beyond the midline. The left one, of course, will move all the way to the right; *i.e.,* toward the nose.

The above account has been limited to the *primary* actions of the extra-ocular muscles. These muscles are capable of secondary and tertiary movements, as well.

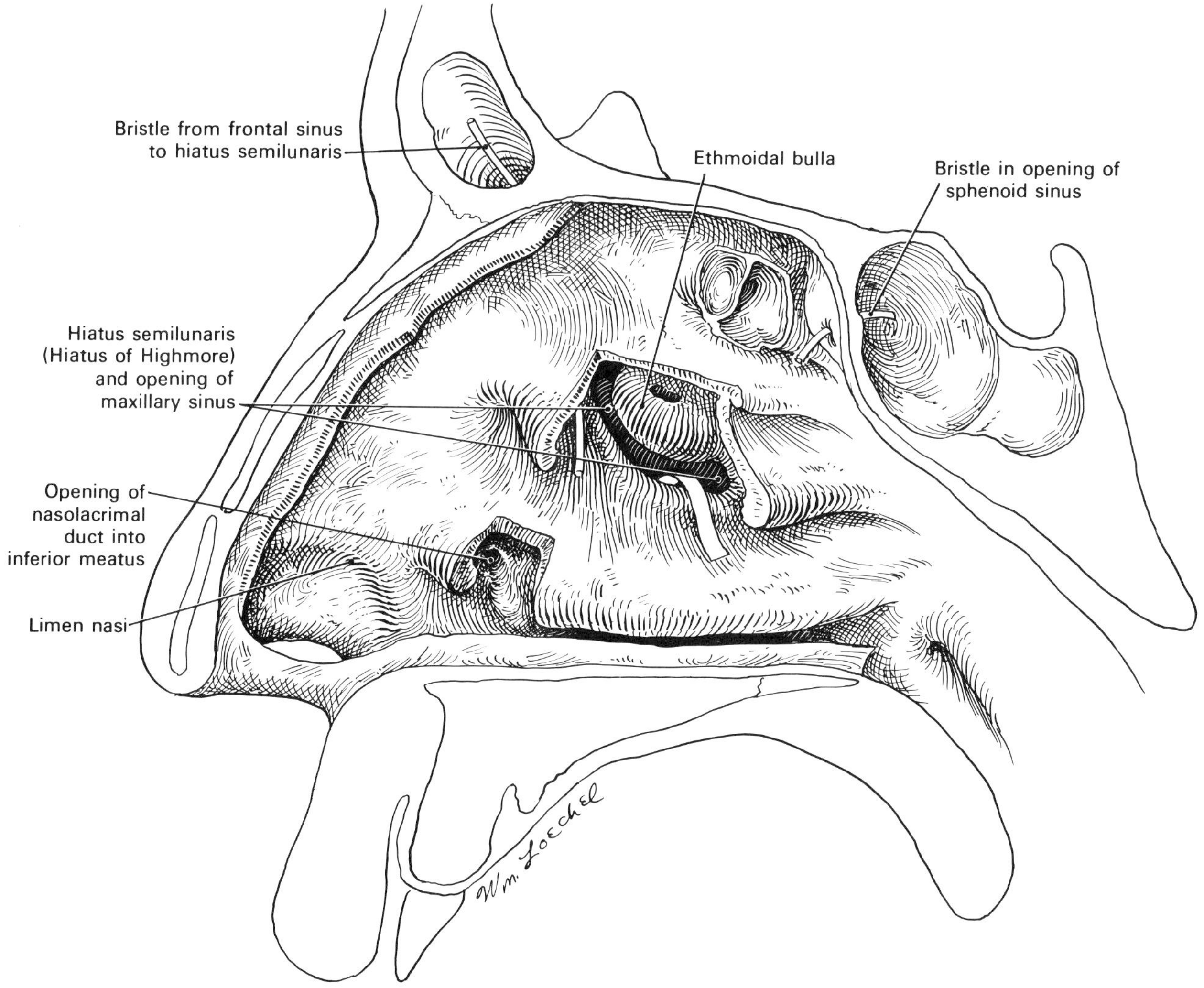

Fig. VI-74B. Conchae removed to expose meatal communication with sinuses.

An understanding of the latter is beyond the needs of the beginning medical student and will be pertinent only to those who will make ophthalmology a specialty. Should any student become interested in the correction and treatment of eye muscle deficits, such details are grist for the mill in later years.

THE NOSE

Noses come in a variety of shapes and sizes. The senior author sports one of magnificient proportions. His mother went around twice when noses were being handed out for male infants. But his father and grandfather had not been exactly short-changed in this respect, either. Even nostrils differ in that they are narrow in Europeans and may widen and diverge as much as 90° in Negroes. The nostrils are more circular and less elliptical in children than in adults.

The nose is well fixed to the skull at its root by the two nasal bones and the frontal process of the maxilla (Fig. VI-74A and B). More distally and especially at its apex, the nose is quite mobile because of its framework of hyaline cartilage which is more pliable than bone (Fig. VI-75). There is *one* septal cartilage which completes the separation of the nasal cavities anteriorly; there are *two* large, triangular lateral cartilages; *two* large alar cartilages around the nostrils; *three* smaller alar cartilages inferolaterally; and a *few* accessory cartilages at the anterior end of the septal cartilage, between it and the large alar cartilages.

The medial walls of the nasal cavities are the bony and cartilaginous **septum** which can deviate or buckle to one side or the other, narrowing one of the cavities. If it does, the conchae of the larger cavity tend to enlarge to fill up the space toward the concave side of the septum. The **septal cartilage** which finishes off the partition anteriorly is a quadrilateral plate that meets the perpendicular plate of the ethmoid posterosuperiorly; the vomer, maxillary crest, and anterior nasal spine of the maxilla postero-inferiorly; and the septal processes of the large alar cartilages antero-inferiorly. Thus, the tip-end of the septum is not septal cartilage, *per sé*, but part of the alar cartilages, a most mobile part of the nose.

The **nares** are the anterior apertures kept open by

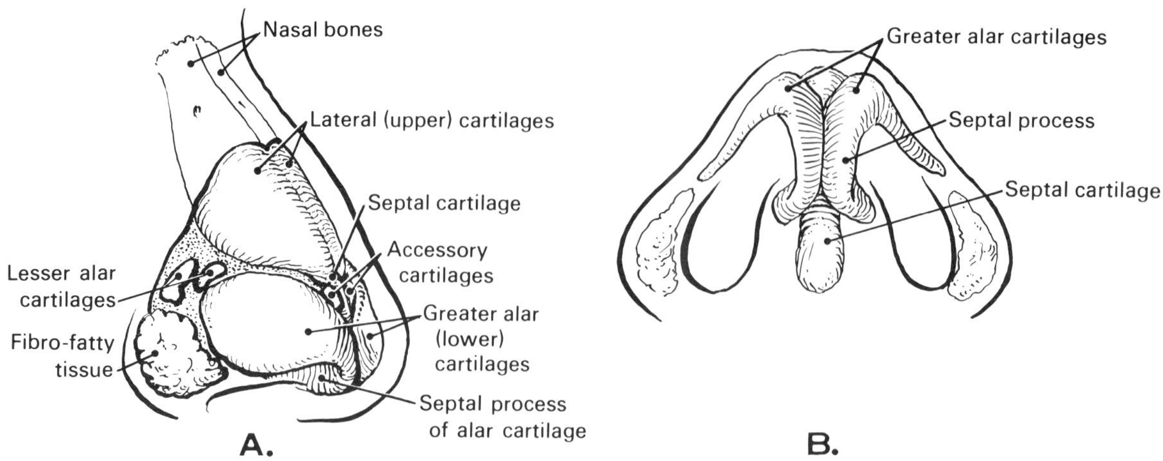

Fig. VI-75. Cartilages of the nose. **A.** Lateral aspect. **B.** Inferior aspect.

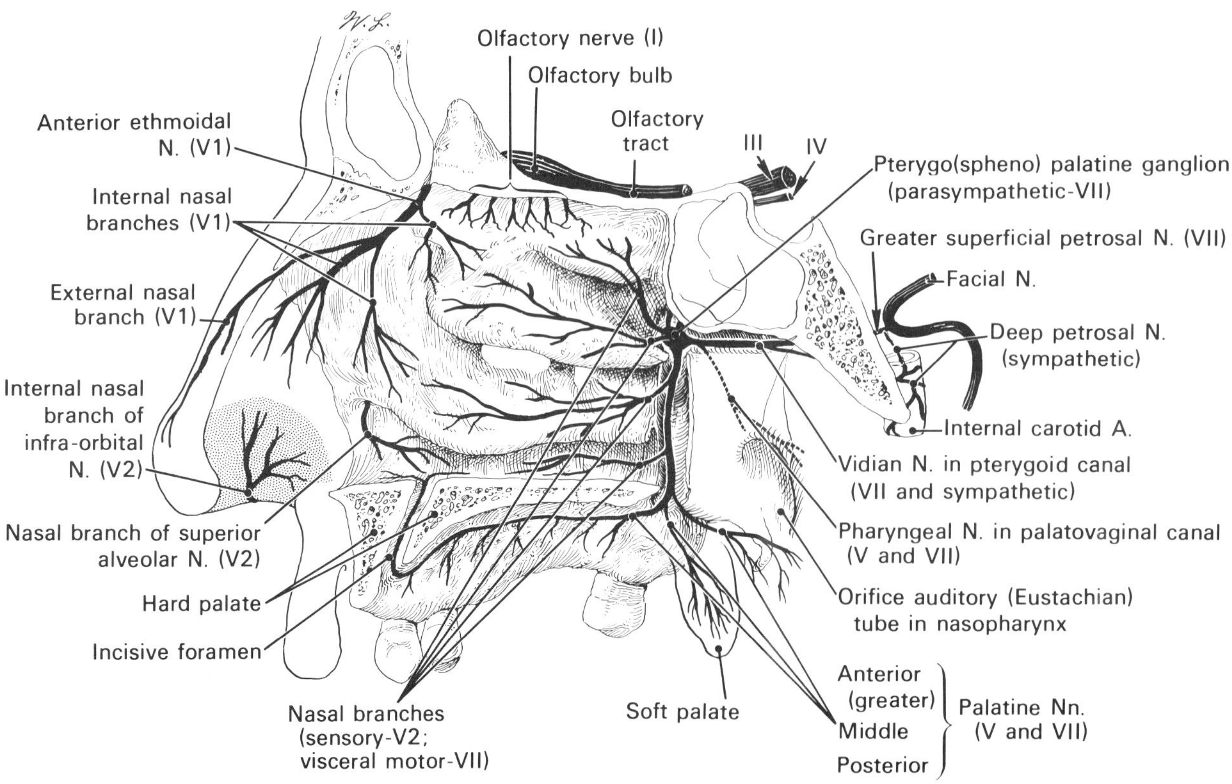

Fig. VI-76. Relation of trigeminal and facial nerve supply to mucosa of nose, palate, and nasal pharynx.

alar cartilage. They lead into a skin-lined, somewhat-dilated **vestibule** beset with hairs (Fig. VI-74A). The posterior openings are the oblong choanae leading from nose into nasopharynx.

The floor is the superior part of the hard palate; the roof, the upper nasal cartilages and bone anteriorly; the antero-inferior surfaces of the body of the sphenoid, posteriorly; with the cribiform plate of the ethmoid between.

The lateral walls are complex. They include contributions from the maxilla (its frontal process); the lacrimal bone; vertical plate of the palatine; medial pterygoid plate of the sphenoid; the ethmoidal conchae (superior and middle); and the lowermost concha, the inferior, a separate nasal bone in its own right. Septum and walls are lined for the most part by pseudostratified respiratory mucosa with goblet cells and cilia that beat toward the nasopharynx. There are mucous and serous glands that empty on the surface though they are embedded in the connective tissue below the respiratory epithelium.

Only high in the nasal vault, over superior conchae and along the upper portion of the septum, does one encounter olfactory epithelium with its specialized bipolar cells in the surface epithelium. This is the only place in man where sensory cells lie as they did in lower vertebrates—out in surface epithelium. The bipolar sensory cell is also of ancient lineage. We have examples of it in three other areas: in the retina; in the spiral ganglion of the cochlea; and in the vestibular ganglion of the internal auditory meatus. Elsewhere ganglion cells are unipolar neurons, and except in the nose, they lie deep in the body in well-protected areas. Thus, the olfactory cells are easily impaired by infection and by certain chemical agents. Their sensory hairs are stimulated in still some unknown way (though theories of how this might be accomplished are legion) only if odiferous materials are in lipid solution. One must be careful to rule out chemical agents as odiferous materials if they simply stimulate the rich trigeminal nerve supply to the nasal mucosa. Axons from the bipolar cells, we have learned, traverse the holes in the sieve-like cribiform plate to terminate in the olfactory bulb. Man does not use his nose and relatively small olfactory bulb to sample his sensory environment as much as lower animals, so we admit his sense of smell is not so keen. Yet, in extremely well-controlled tests with specific odors; *e.g.*, oil of clove or wintergreen, that have little biological significance for dogs, we found human individuals with a threshold of sensitivity close to that of the dog.[20-22] These were medical students selected at random, not odor panel experts or perfume testers. But for odors that were particularly meaningful for dogs, including human odor trace, the humans were no match for the dog in olfactory acuity. Since the olfactory mucosa sits so high up in the nose, it is not surprising that we sniff when we wish to appreciate certain odors best, especially the faint, delicate ones. In quiet breathing only the more concentrated, intense odor stimuli reach these upper regions almost incidentally, wafted about by eddy currents as we inhale and exhale. We do not depend too much upon olfaction in our ordinary daily life, but depend much more upon our eyes and ears. It is not surprising, then, that the pathways for olfaction in our brains are much more concerned with emotional affect than with sampling the environment for information or detective purposes.

The **meatuses** are the curved spaces under each concha or turbinate; *i.e.*, they are the air passageways from nostrils to choanae. They and some of the conchae can be seen in Figure VI-74A. The spaces in life are more constricted, since the bones are clothed with respiratory mucosa save in the upper olfactory portion. The mucosa is well vascularized so that the air is warmed, and moisture from the mucosal surface is added before the air reaches the lungs. This warming should also aid in smelling, since warm air rises. However, in inspiration most of the air stream is directed under the inferior meatus directly to the posterior chonae. In exhalation, the main stream flows under the middle meatus. Only in active sniffing is the main stream directed significantly upward. But we have indicated that eddy currents do move some small portions of the air mass through all passageways, even during quiet breathing. The direction of air streams in the nose has been solved by various ingenious techniques in man,[23-26] rabbit, and other forms,[27-28] as well as in the dog.[29]

The meatuses and conchae are best visualized in the lateral walls of the parasagittally-sectioned head (Fig. VI-74A). If the turbinates or conchal shelves are clipped back, openings under them for the various air sinuses can be demonstrated (Fig. VI-74B). Notice that the **nasolacrimal duct** empties into the **inferior meatus.** Beneath the middle concha is a curved fissure, the **semilunar hiatus.** It is bounded above by a rounded swelling, the **ethmoidal bulla.** The *frontal sinus* opens by a very narrow passage, the **infundibulum,** into the upper part of the hiatus; the *anterior and middle ethmoidal air cells* open in front of or on the bulla; the **maxillary sinus** opens into the posterior part of the hiatus. Thus, *several sinus areas* open into the fissured region below the **middle meatus.** The *posterior ethmoidal air cells* open below the **superior meatus.** The space above the superior concha is the **spheno-ethmoidal recess,** and here the *sphenoid sinus* opens into the nasal cavity. In this way the *paranasal* sinuses, into which the respiratory mucosa invaginates, communicate with the nasal cavity. Indeed, they are diverticulae of it.

Nerve Supply of the Nose

Aside from the olfactory, the rich nerve supply to the nasal mucosa is from various branches of the trigeminal (Fig. VI-76).

1. We have discussed the odd route that the *anterior ethmoidal nerve,* a terminal branch of the

nasociliary nerve, took into the orbit, out through the anterior ethmoidal foramen, into the anterior cranial fossa, and out the nasal slit into the nose, the **internal nasal branches.** Finally, it is out in the superficial skin at the tip of the nose as the **external nasal nerve.** Try tracing a tactile sensation from a finger placed on the tip of your nose to the semilunar ganglion.

2. Three nerves pass through the pterygopalatine (sphenopalatine) foramen:

a. Posterior superior nasal nerves: They go to the upper side walls and to roof septum and ethmoidal air cells.

b. The Nasopalatine (or long sphenopalatine) nerve: It crosses the undersurface of the body of the sphenoid to reach the septum, grooves the vomer, and passes through the incisive foramen. It supplies the septum in its passage, the anterior under-surface of the hard palate, and the gums.

c. Pharyngeal nerve: This is the one passing back in the pharyngeal canal to innervate the roof and back wall of the nasopharynx as far as the auditory tube.

All of these branches are off-shoots of the maxillary division of V.

3. Two other branches pass down in the pterygopalatine (sphenopalatine) canal.

a. Greater palatine nerve: It emerges through the greater palatine foramen of the hard palate, supplying its under-surface, while in the canal it sends **posterior inferior nasal nerves** to the middle and inferior conchae.

b. The lesser palatine nerves: These pass through the lesser palatine foramina to innervate the soft palate and its uvula, as well as the palatine tonsils.

All these, too, are branches of V_2. Another branch from V_2, the **internal nasal branch of the infraorbital,** supplies the skin of the vestibule, and still another from the **anterior superior alveolar nerve** gives a piercing nasal branch to the lateral wall under the inferior concha. Only the **anterior ethmoidal** twigs represent V_1 in the nose.

Blood Supply of the Nose

Arteries: As you might suspect, branches from the *ophthalmic artery* supply the area innervated by the anterior ethmoidal twigs, while branches of the *maxillary artery* tend to follow the nerve supply from the branches of V_2.

Veins: There is a rich venous network beneath the nasal mucosa. It is drained as follows:

1. Ethmoidal veins through the cribiform plate to the superior sagittal sinus of the dura.
2. Nasal veins to the ophthalmic veins to cavernous sinus.
3. Pterygopalatine veins through the formen of the same name to the pterygoid plexus of veins.
4. All other venous twigs drain into the facial vein.

Radiographic Anatomy of the Nose

Figure VI-77A is a lateral radiograph of the nose of a patient which not only shows the bony parts, but some of the soft tissue structures in the area. Figure VI-77B is a lateral photograph of the same area, and Figure VI-77C is a sagittal section, both in the dry skull. In the lateral radiograph, take careful note of the prominent nasomaxillary suture, and the finer grooves for the nasociliary nerves which you can see in Figure VI-77C are actually on the inner surface of the nasal bones. These nasociliary grooves and the nasomaxillary suture are not uncommonly mistaken for fractures. Fractures of the nasal bone, however, are most commonly transverse in direction, rather than longitudinal. If, as a result of a direct blow, the perpendicular plate of the ethmoid bone is also fractured, the nasal fragments are usually depressed. Patients may also have fractures of the maxillary spines, and these should be looked for carefully. In this type of fracture, there is commonly marked soft tissue swelling anterior to the spines.

In Figure VI-77A, you may also observe the contours of the upper and the lower lids and the cornea. Even the orbital groove of the upper lid may be seen. Most cartilages of the nose do not show well in radiographs made for bone detail; however, in this instance, we can see the septal process of the lower nasal cartilage in the region of the nares.

We shall consider the eye and the entire ear (external, middle, and inner) in another chapter when we discuss the nervous system.

REFERENCES

1. Hanson, J. R., Anson, B. J., and Strickland, E. M.: Branchial sources of the auditory ossicles in man. Observations of embryonic stages from 7 mm. to 28 mm. CR length. Arch. Otolaryng. (Chicago), *76:* 200, 1962.
2. Arey, L. B.: The history of the first somite in human embryos. Contrib. Embryol., *27:* 235, 1938.
3. Bast, T. H., and Anson, B. J.: *The Temporal Bone and Ear.* Thomas, Springfield, Ill., 1949.
4. Meurman, Y.: Zur Anatomie des Aqueductus Cochleae nebst einigen Bermerkungen über dessen Physiologie. Acta Soc. Fenn. Duodecim, ser. B., *13:* 1, 1930.
5. Young, M. W.: Preparations of casts of the bony labyrinth. Science, *86:* 619, 1937.
6. Perlman, H. B., and Lindsey, J. R.: Relations of the internal ear spaces to the meninges. Arch. Otolaryng. (Chicago), *29:* 12, 1939.
7. Altman, F., and Waltner, J. G.: The circulation of the labyrinthian fluids. Ann. Otol., *56:* 648, 1947.
8. Waltner, J. G.: Barrier membrane of the cochlear aqueduct: Histological studies on the patency of the cochlear aqueduct. Arch. Otolaryng. (Chicago), *47:* 656, 1948.
9. Altman, F., and Waltner, J. G.: New investigations of the labyrinthian fluids. Laryngoscope, *60:* 727, 1950.
10. Young, M. W.: Mercury as a casting and contrast medium. Science, *116:* 207, 1952.
11. Fogh-Anderson, P.: *Inheritance of Harelip and Cleft Palate.* Arnold Busck, Copenhagen, 1942.

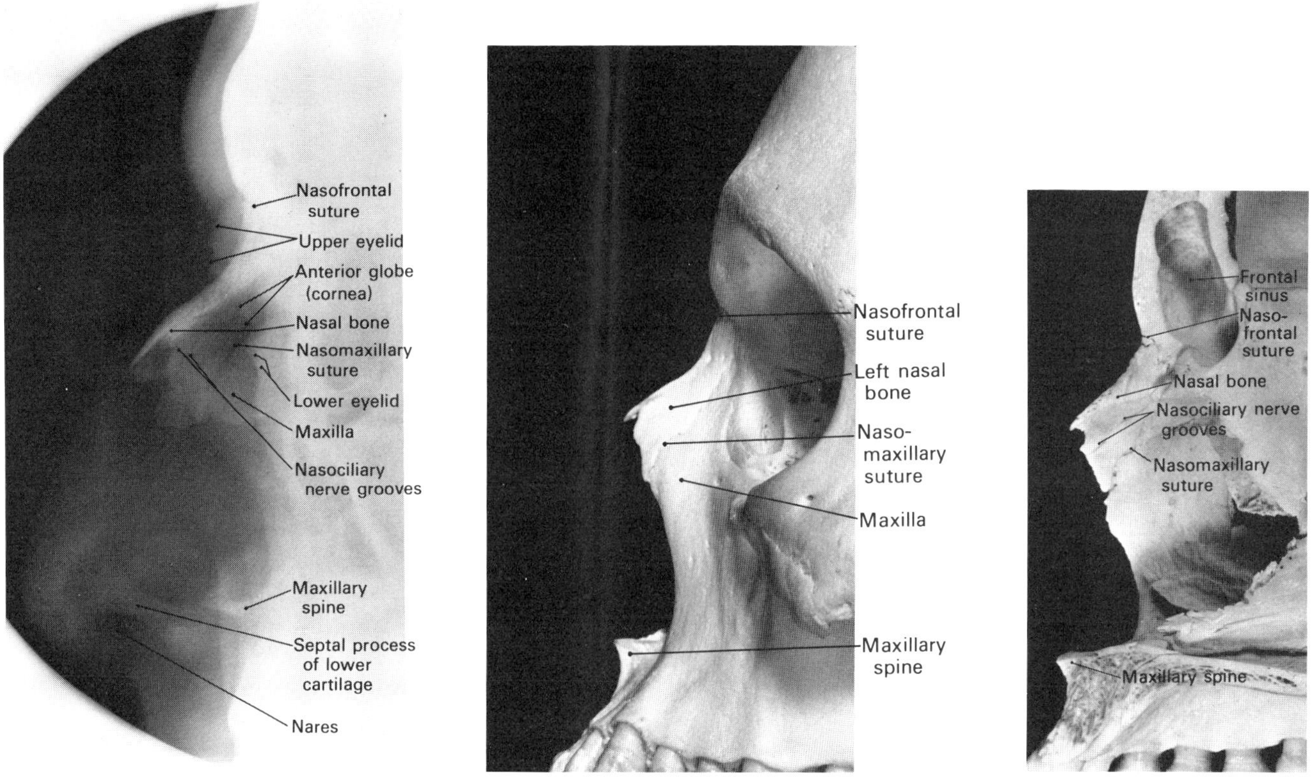

Fig. VI-77. The nose. Comparable views in: **A.** Lateral radiograph of the nose in a normal adult; **B.** Lateral photograph of the intact dry skull; **C.** Sagittally sectioned dry skull.

12. MacMahon, B., and McKeown, T.: The incidence of harelip and cleft palate related to birth rank and maternal age. Amer. J. Hum. Genet., *5:* 176, 1953.
13. Fraser, F. C.: Thoughts on the etiology of clefts of the palate and lip. Acta Genet. (Basel), *5:* 358, 1935.
14. Curtis, E. J., Fraser, F. C., and Warburton, D.: Congenital cleft lip and palate. Amer. J. Dis. Child., *102:* 853, 1961.
15. Tavaras, J. M., and Wood, E. H.: *Diagnostic Neuradiology.* Williams & Wilkins, Baltimore, 1964.
16. Etter, L. E.: *Atlas of Roentgen Anatomy of the Skull.* Thomas, Springfield, Ill., 1955.
17. Pendergrass, E. P., Schaeffer, J. P., and Hodes, P. J.: *The Head and Neck in Roentgen Diagnosis.* Thomas, Springfield, Ill., 1956.
18. Grodinsky, M., and Holyoke, E. A.: The fasciae and fascial spaces of the head, neck, and adjacent regions. Amer. J. Anat., *63:* 367, 1938.
19. Patey, D. H., and Ranger, I.: Some points in the surgical anatomy of the parotid gland. Brit. J. Surg., *45:* 250, 1957.
20. Becker, R. F., Markee, J. E., and King, J. E.: Studies on olfactory acuity in dogs. I. Discriminatory behavior in problem box situation. Anim. Behav., *5:* 94, 1957.
21. Becker, R. F., King, J. E., and Markee, J. E.: Studies on olfactory discrimination in dogs. II. Discriminatory behavior in a free environment. J. Comp. and Physiol. Psychol., *55:* 773, 1962.
22. King, J. E., Becker, R. F., and Markee, J. E.: Studies on olfactory discrimination in dogs. III. Ability to detect human odor trace. Anim. Behav., *12:* 311, 1964.
23. Paulsen, E.: Experimentelle Untersuchungen Über die Strömungen der Luft in der Nasenhöhle. Sitzungsbericht der. K. Akad. Wissensch. Wien. Math. u. Naturwissensch. Kl. 85 (Heft III, Abt. III), 352, 1882.
24. Franke, G.: Experimentelle Utersuchungen Über Luftdruck, Luftbewegung und Luftwechsel in der Nase und irhen Nebenhöhlen. Arch. Laryng. u. Rhin., *1:* 230, 1894.
25. Parker, C. A.: Some observations and remarks on air currents in nasal respiration. J. Laryng., *16:* 345, 1901.
26. Proetz, A. W.: *Essays on the Applied Physiology of the Nose.* Annals Publ. Co., St. Louis, 1953.
27. Lucas, A. M., and Douglas, L. C.: Principles underlying ciliary activity in the respiratory tract. II. A comparison of nasal clearance in man, monkey, and other mammals. Arch. Otolaryng. (Chicago), *20:* 518, 1934.
28. Dawes, J. D. K.: The course of the nasal air streams. J. Laryng., *66:* 583, 1952.
29. Becker, R. F., and King, J. E.: Delineation of the nasal air streams in the dog. A.M.A. Arch. Otolaryng. (Chicago), *65:* 428, 1957.

SEVEN

The Neck

INTRODUCTION

The surface anatomy of the neck was emphasized in Chapter III. To aid in your immediate review and recall, we present a few neckable subjects in Figures VII-2 through 5. Similar features in the less esthetic male are shown in Figures VII-6 to 8.

Locate at once the cricoid cartilage in your own neck and realize that it lies at the level of the sixth or seventh cervical vertebrae. It marks the junction point of pharynx with esophagus and of larynx with trachea. The recurrent laryngeal branch of the vagus enters the larynx at this level, and here lie the inferior thyroid artery and the middle thyroid vein. If there is a middle cervical sympathetic ganglion in the sympathetic trunk, it will lie at the level of the cricoid cartilage, and the vertebral arteries will enter the foramina of the transverse processes of VC_6 as they make their way through similar foramina of the vertebrae above them on their way to supply the brain and brainstem (Fig. VII-1). At this level, the Omohyoid muscle crosses the carotid sheath which contains in it the common carotid artery, the internal jugular vein, and the vagus nerve.

Except for the heavy muscles posteriorly, only the Sternomastoid gives much protection to the neck. Otherwise, the neck is covered with skin that is freely movable, superficial fascia, and a variable amount of fat. The neck, in fact, is not well protected when compared with some other areas of the body.

THE CERVICAL VERTEBRAE

It is the cervical part of the vertebral column which, through its rigidity, lends support to the head and, through its limited flexibility, allows for some movement. The column also provides protection for the spinal cord in the neck region and aids in transmitting one of the two main sources of blood supply to the brain *via* the **vertebral arteries.** Like all parts of the vertebral column, that in the neck is flexible because it consists of several slightly-movable blocks of bone separated by intervertebral disks. Its stability depends more upon muscular and ligamentous attachments. Though there is reasonable variability in the size of cervical vertebral bodies, in general, bodies of vertebrae increase in size progressively in a caudal direction, and this also provides a measure of stability.

In Chapter II we discussed the nature of a typical vertebra, choosing one in the thoracic region for our model. The third to sixth cervical vertebrae depart from this basic plan in minor ways, yet they can be regarded as typical for the cervical region. The bodies of cervical vertebrae VC_3 through VC_7 have a slightly concave anterior surface, and their inferior anterior margin is lipped, while the corresponding superior anterior margin of the vertebrae below is beveled (Fig. VII-9A and B. If you will examine these vertebrae in the radiographic sagittal plane (Fig. VII-10A, you will find that the superior intervertebral surface is slightly convex, while the inferior intervertebral surface is concave. In the coronal radiographic plane (Fig. VII-

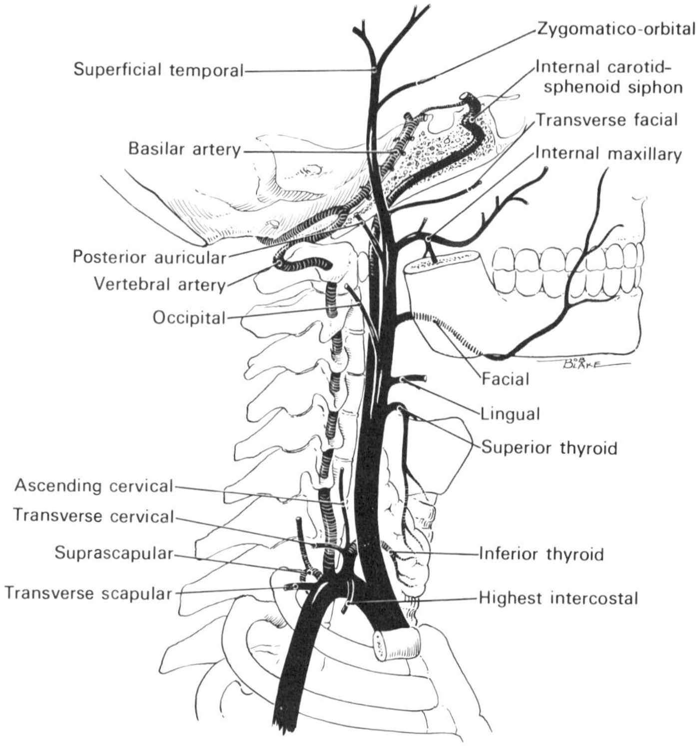

Fig. VII-1. Branches of external carotid and subclavian arteries. Note course of vertebral artery.

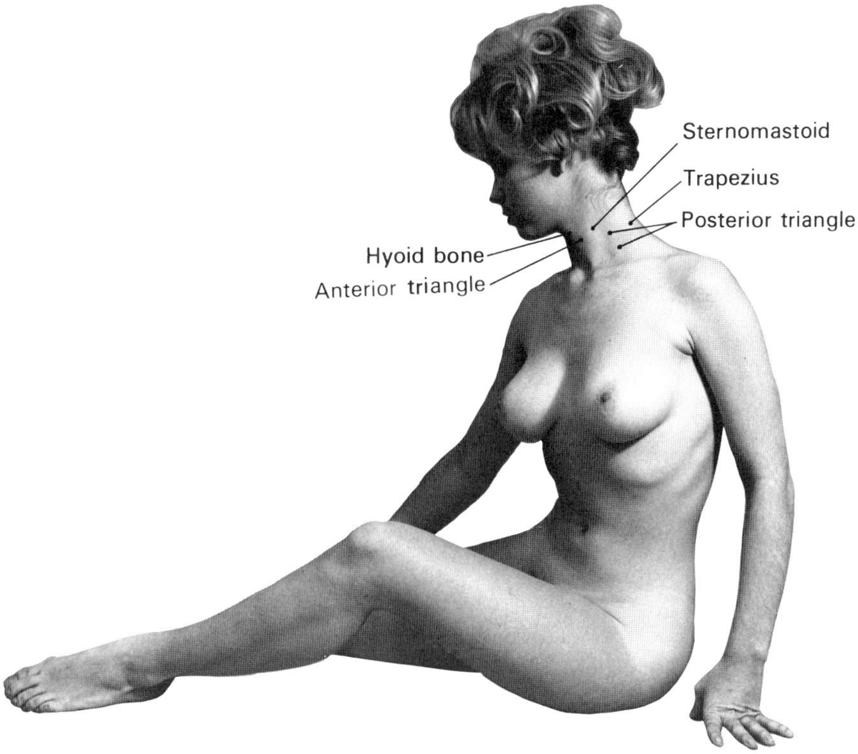

Fig. VII-2. Surface anatomy of the female neck.

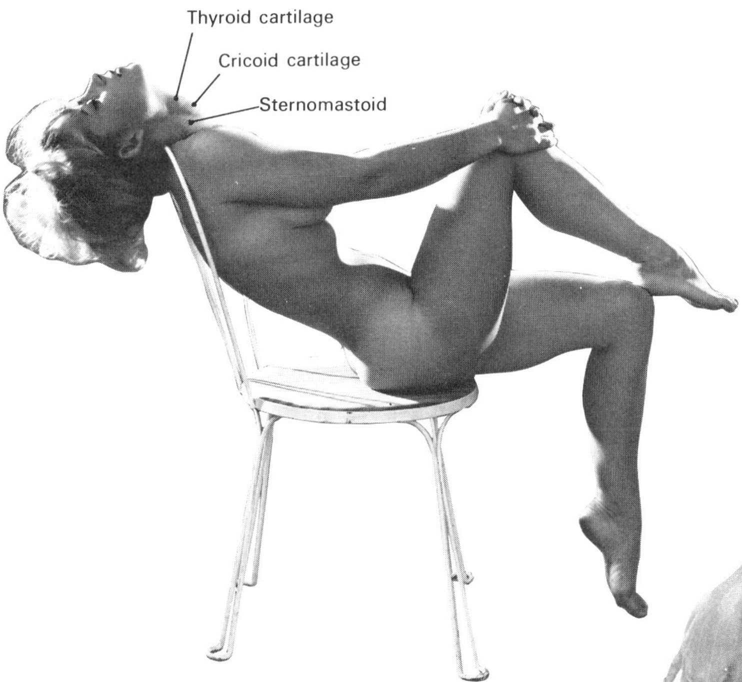

Fig. VII-3. Surface anatomy of the female neck.

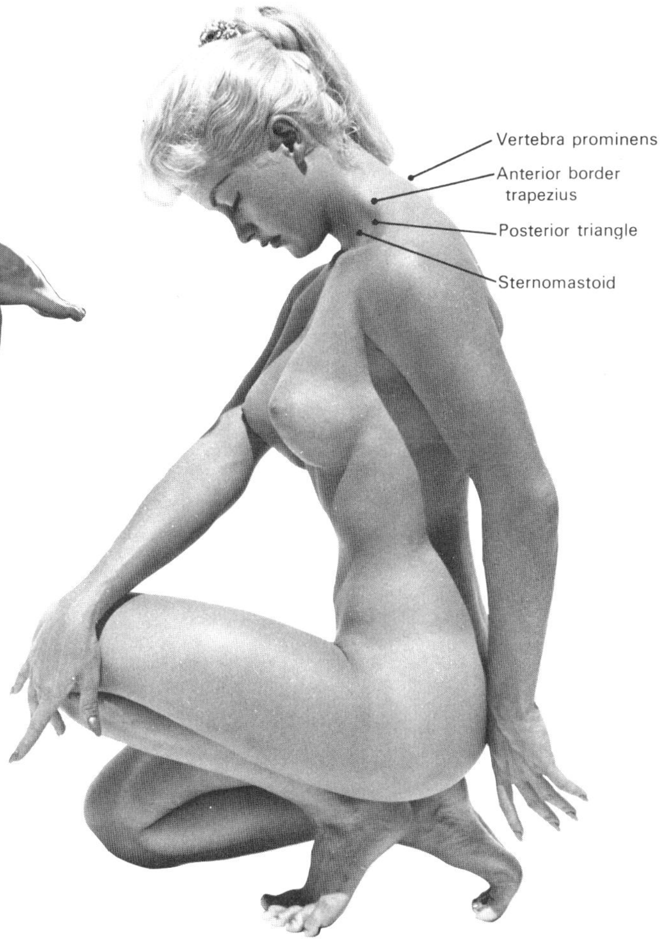

Fig. VII-4. Landmarks of the posterior (neural) triangle.

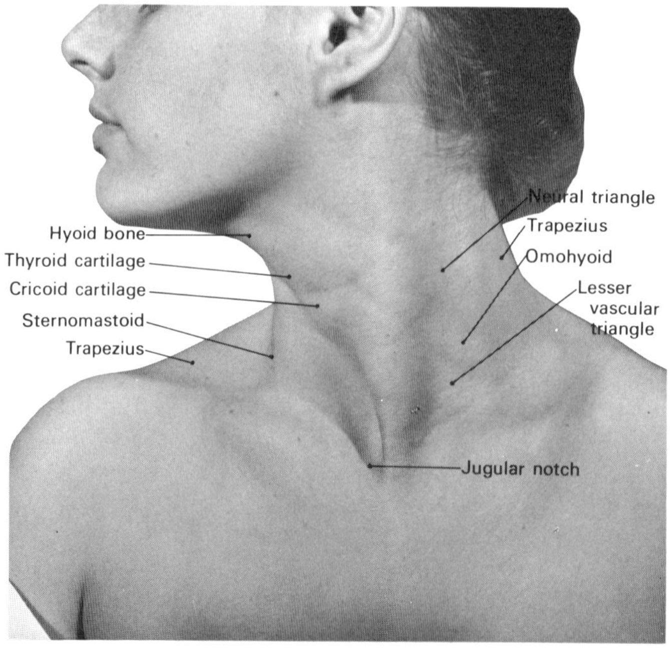

Fig. VII-5. Neural and lesser vascular triangle of the neck.

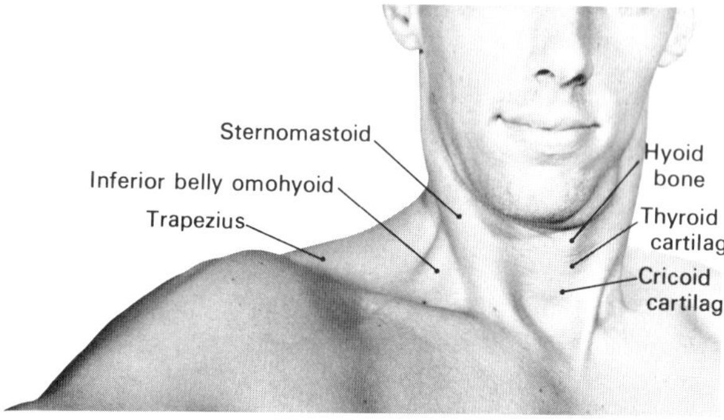

Fig. VII-6. Surface landmarks—neck of male.

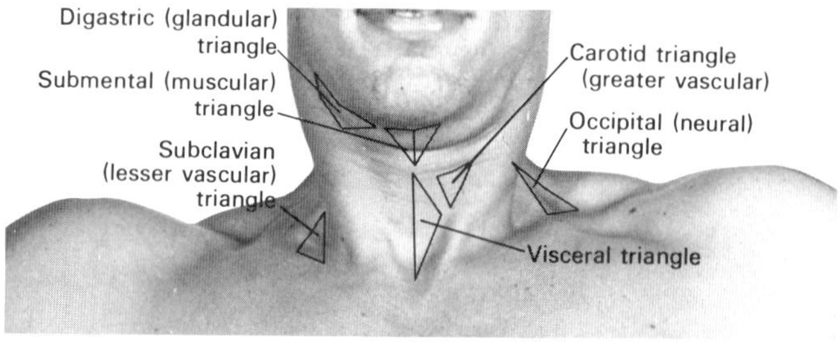

Fig. VII-7. Surface localization of triangles of neck.

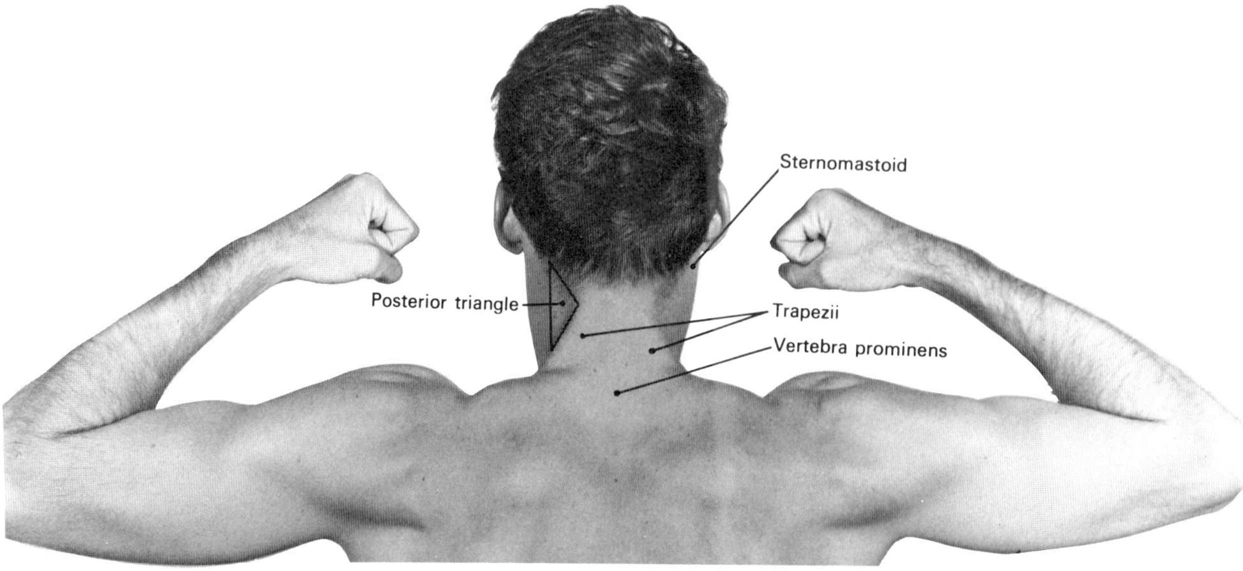

Fig. VII-8. Posterior aspect of neck in surface view.

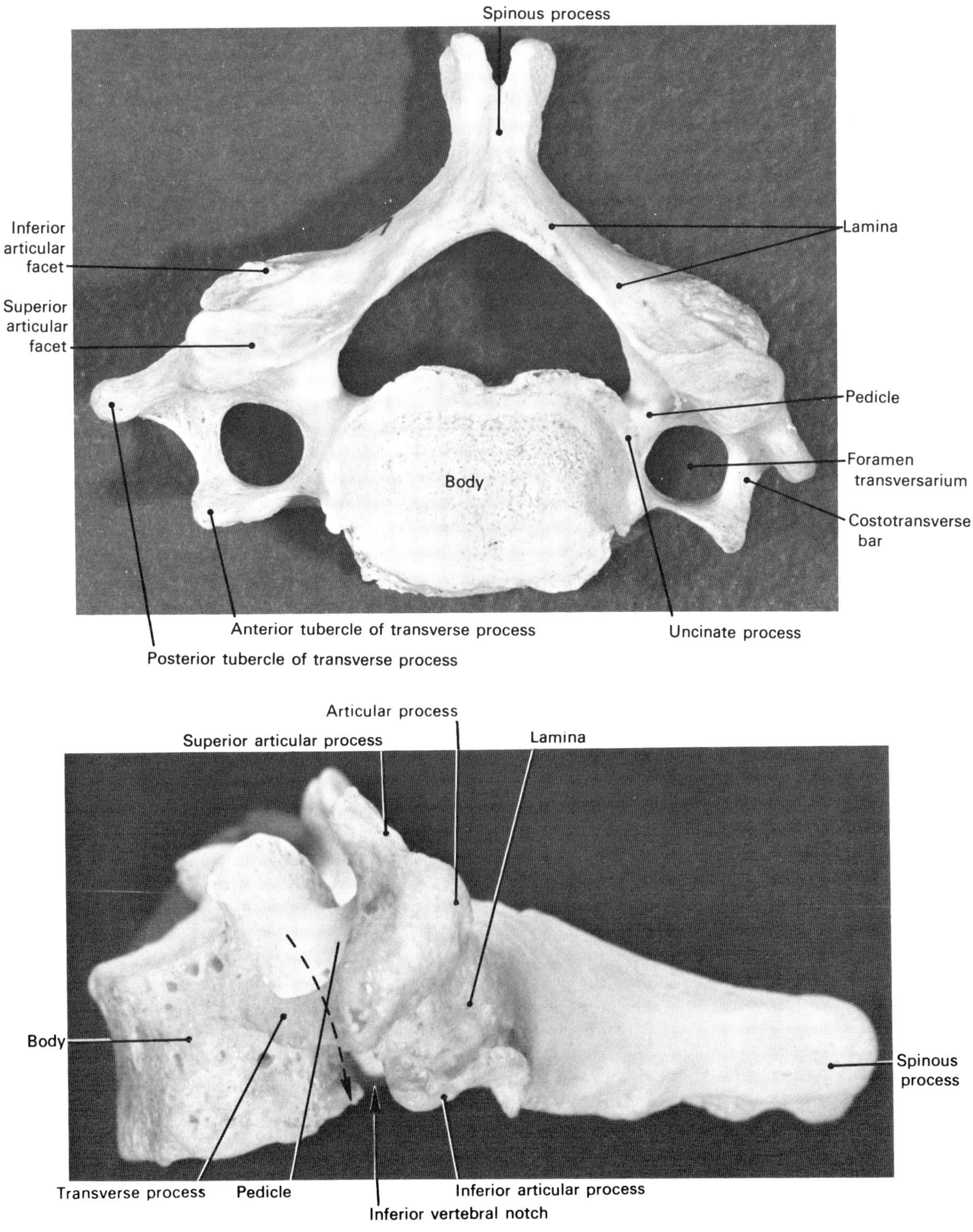

Fig. VII-9. Sixth cervical vertebra. **A.** From above. **B.** Lateral view.

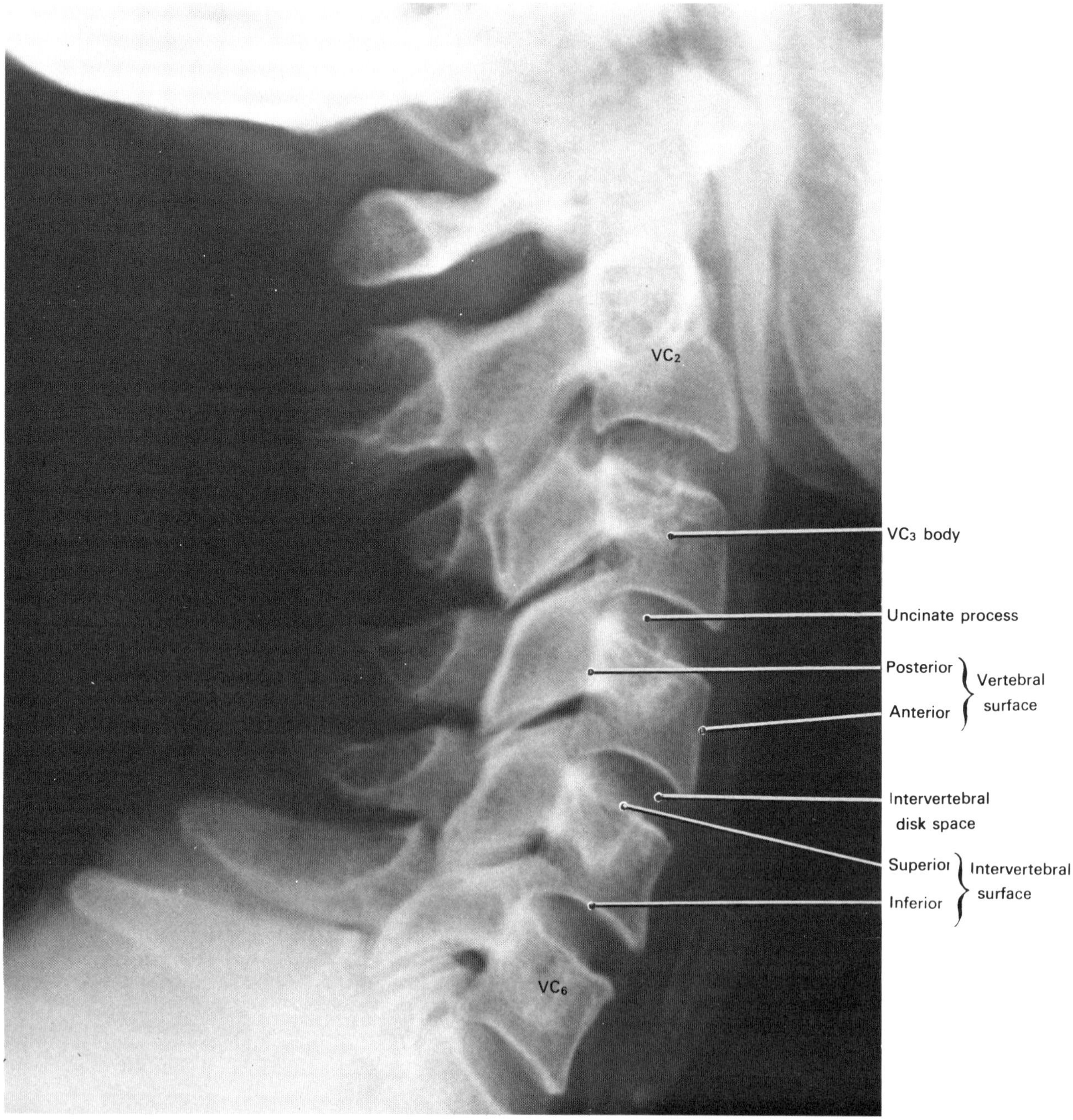

Fig. VII-10. Radiographs of the cervical spine. **A.** Lateral view. Note shape of vertebrae.

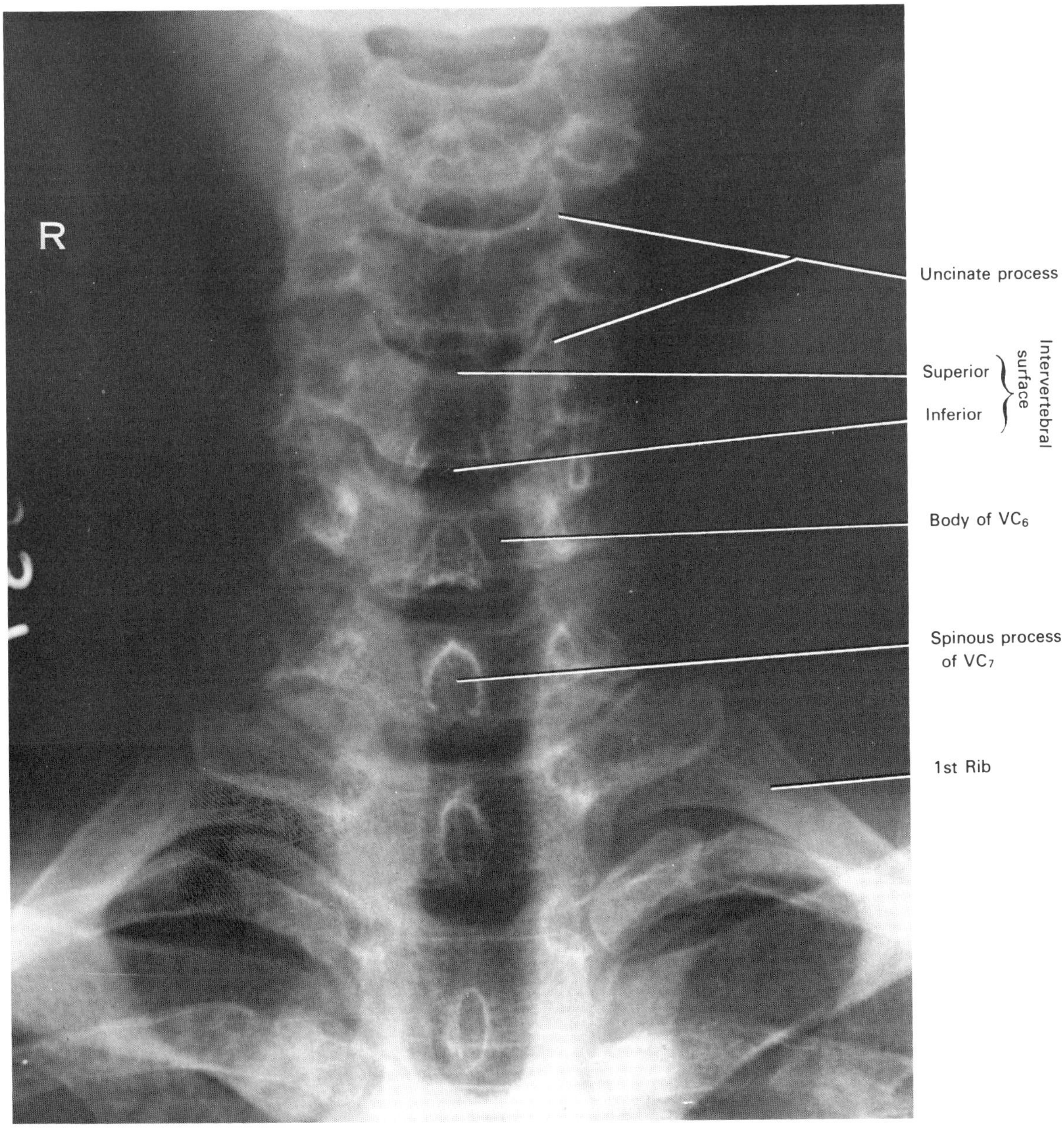

Fig. VII-10B. Anterior-posterior view of cervical spine to demonstrate the uncinate process.

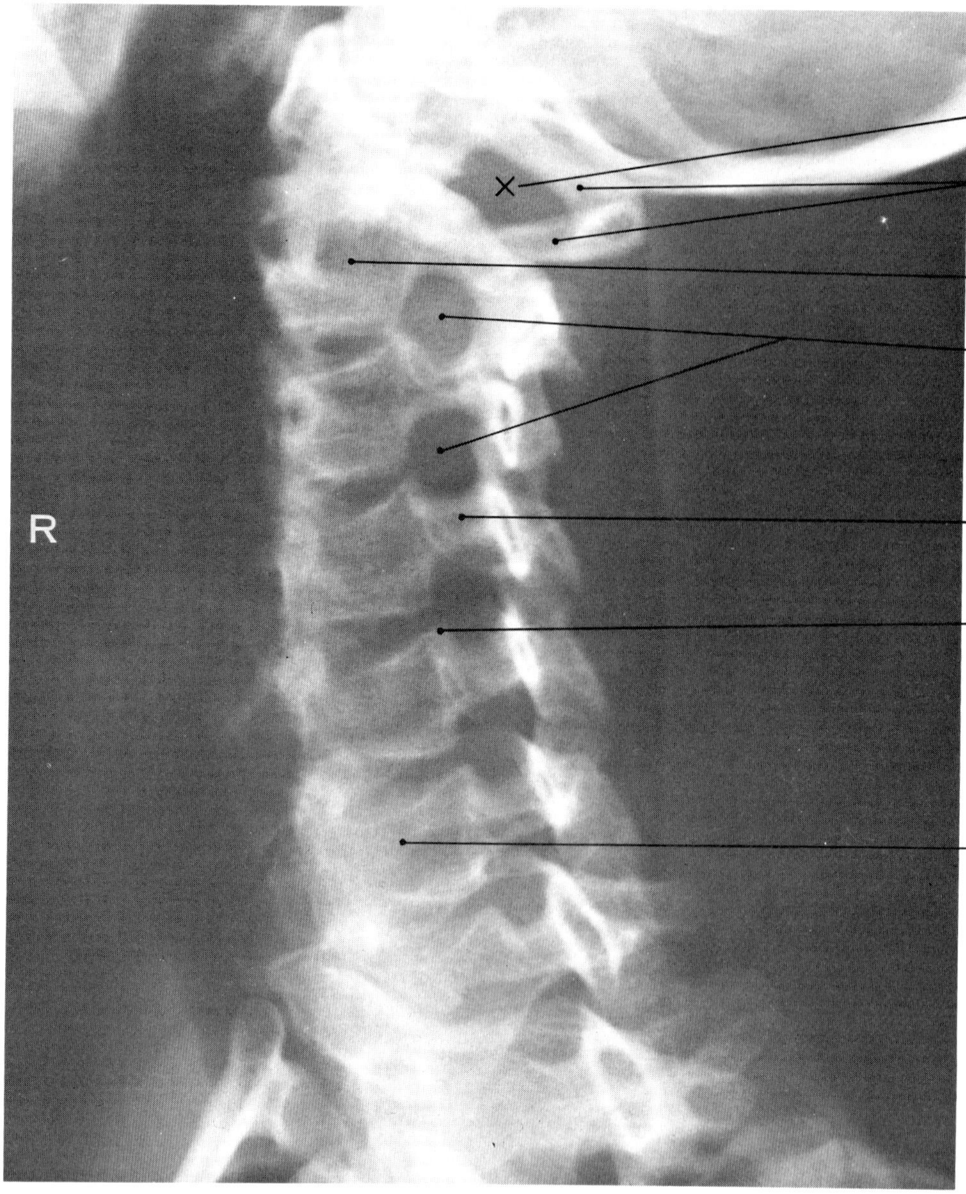

Fig. VII-10C. Oblique view of cervical spine to show intervertebral foramina.

10B), the superior intervertebral surface is concave, and the inferior intervertebral surface is convex. The posterior surface of these vertebrae is flat. Notice also that these vertebrae are considerably wider in their transverse diameters than in their anterior-posterior diameters.

The **uncinate processes** (uncus, lunate process) (Figs. VII-9A, 10A, B, and C) are present only on the superior lateral margins of VC_3 to VC_6 and on the superior dorsolateral margins of VC_7 to VT_1, but not below this level. These processes are virtually flat during the first decade of life; however, they attain their full height by about 18 years of age. The corresponding inferior surfaces of the vertebrae above each process become beveled. The uncinate processes are thought to function as a guide rail for the vertebral body above to prevent any lateral displacement of the vertebrae during flexion—extension movements of the cervical spine. Between the uncinate processes and the beveled undersurface of the vertebrae above, the **uncovertebral (Luschka, covertebral, lateral interbody) "joints" or fissures** are formed. These are considered further on pages 353–354 of this chapter.

The **intervertebral foramina** (Fig. VII-10C) are short bony canals found at the VC_2–VC_3 through the VC_6–VC_7 levels in the cervical spine. They are bounded caudally and cranially by the pedicles, laterally by the superior articular process, and medially by the uncinate process and uncovertebral "joint." They become progressively smaller from the VC_2–VC_3 level to the VC_6–VC_7 level. They are somewhat ovoid in shape and have a greater vertical than horizontal diameter. The intervertebral foramina open obliquely in an anterolateral axis which varies at different levels. Thus, the upper VC_2–VC_3 foramina are angled more to the coronal plane than the lower foramina at VC_6–VC_7.

Passing through these intervertebral foramina are both the ventral and dorsal nerve roots, spinal arteries, intervertebral veins, and some fatty tissue. The **cervical vertebral foramen or canal,** housing the spinal cord, is large and triangular. Cervical **pedicles** are short, and they project laterally as well as posteriorly. Cervical **laminae** are thinner and longer horizontally than in most other vertebrae (Fig. VII-9A).

The outstanding features, the hallmarks of cervical vertebrae, are their bifid **spinous processes** (though they can be irregularly non-bifid in some individuals) and their **transverse processes which possess a foramen for the transmission of the vertebral artery.** At the ends of the transverse processes, a **costotransverse bar** separates **anterior** and **posterior tubercles** (Fig. VII-9A).

It is the first two and the last of the seven cervical vertebrae that are odd (Figs. VII-11, through 13). Vertebra C_1 lacks a body and a spinous process. From the body of VC_2 anteriorly, a stout **odontoid process** or **dens** projects upward. While the spines of VC_2 to VC_6 are usually bifid, that of VC_7 is not (Fig. VII-11A and B). This spine is long, and it ends in a single tubercle to which the **ligamentum nuchae** attaches. Because of its length, VC_7 often protrudes in the flexed neck as the visible **vertebra prominens.** However, the spines of VC_6 and VT_1 can often be seen as well, and in some individuals the spine of VT_1 is more prominent than that of VC_7. The transverse processes of VC_7 may lack foramina entirely. If they are present, they are often small and may transmit the vertebral **vein,** but not the vertebral artery except on rare occasions when the artery passes through on the left side of the neck. Generally, both vein and artery tend to pass in front of the seventh cervical transverse process. The anterior tubercle of the transverse process of VC_7 often is missing; yet, at other times, it can enlarge as a cervical rib.

Developmentally, all the anterior tubercles of cervical transverse processes correspond to ribs. The posterior tubercle corresponds to the true transverse process. When a cervical spinal nerve exits from its intervertebral foramen, the anterior division or ramus of the nerve hugs the posterior tubercle. Thus, it comes to pass behind the vertebral artery which comes up through the foramina of the transverse processes. The posterior ramus of the nerve passes immediately dorsalward, wrapping around the articular process.

Joints between the first and second vertebrae allow rotation, and that between the first and the skull permit a rocking or nodding movement. The rest of the cervical joints allow flexion, extension, and a moderate degree of lateral bending, but permit little or no rotation. This is because their superior articular facets face backward while the inferior facets face forward. When adjacent vertebrae articulate, these faces tend to interlock (Fig. VII-10A).

The First Cervical Vertebra (Atlas)

The first cervical vertebra, lacking a body, is ring-shaped (Fig. VII-12A and B). Viewed from above, we see its lateral masses and the broad articular facets for the condyles of the occipital bone. Since the skull articulates here, this first vertebra has been named the **atlas** after the Greek god who bore the world upon his shoulders. Terminology was more confusing originally in respect to this bone. Strangely, the Greeks, themselves, named it the epistropheus, meaning "the turner," from the verb "to turn," and "epi," upon. They regarded this vertebra as turning upon the second. Later, the term, "epistropheus," was transferred to the second cervical vertebra, implying that it was the pivot upon which the first turned. Such a situation could lead to confusion; hence, the first is now known as the **atlas** and the second as the **axis.**

Instead of a spinous process, the atlas has a posterior tubercle for the ligamentum nuchae. Its anterior arch also bears a tubercle to which the **anterior longitudinal ligament** attaches. Its posterior arch represents the laminae seen in other vertebrae. A groove crosses the posterior arch anterolaterally. The vertebral arteries curves *medialward* in this groove to enter the foramen magnum (Fig. VII-1) after they pierce the transverse processes of the atlas. Once inside the skull, the right and left arteries unite on the ventral aspect of the medulla and pons as the **basilar artery.** Cranialward the last branches of this united vessel diverge as the **posterior cerebral arteries.** The first cervical spinal nerve shares the groove on the posterior arch with the vertebral artery. The nerve exits *laterally* between the skull and atlas.

Seen from below (Fig. VII-12B), the lateral masses present a smaller inferior articulating surface for the axis. Notice how obliquely the transverse foramina are directed. Behind the anterior tubercle is a smooth, somewhat-grooved facet for the projecting dens of the axis. That this ring of an atlas is, after all, a fairly chunky bone can be appreciated from a lateral view (Fig. VII-12C). Its anterior tubercle is by no means a mere nubbin.

The Second Cervical Vertebra (Axis)

Except at its superior surface, the axis (VC_2) begins to resemble a more normal cervical vertebra (compare Fig. VII-13A and B). Its **odontoid** ("tooth-like") **process** or **dens** does not look as formidable in surface view as it does in lateral view (compare Fig. VII-13A and C). Notice the apical terminations of the dens

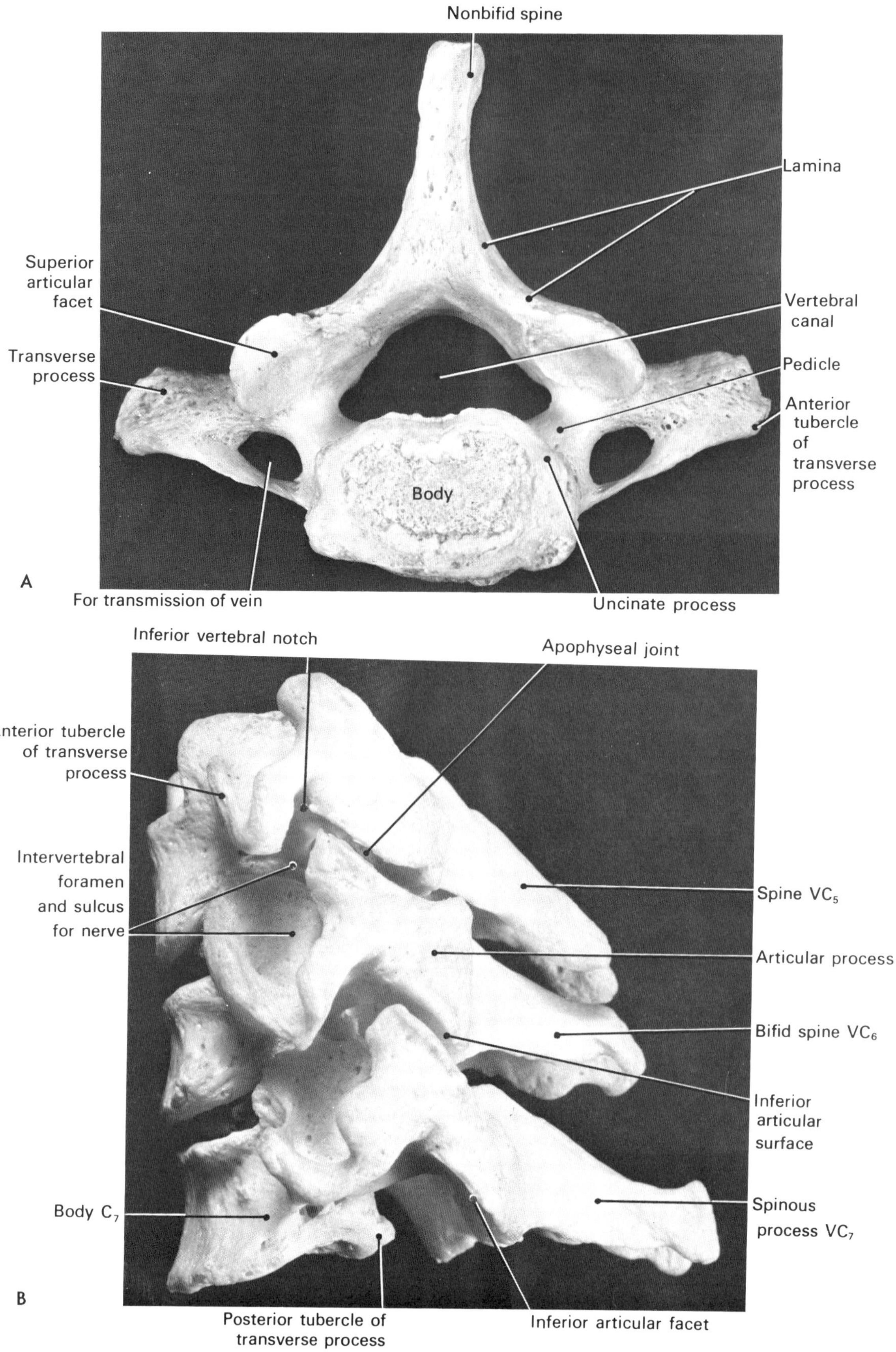

Fig. VII-11. Seventh cervical vertebra. **A.** From above. **B.** Lateral view. Relationship of cervical vertebrae C_5 and C_6 is shown. All three are aligned in their normal anatomical relationship in this true lateral view.

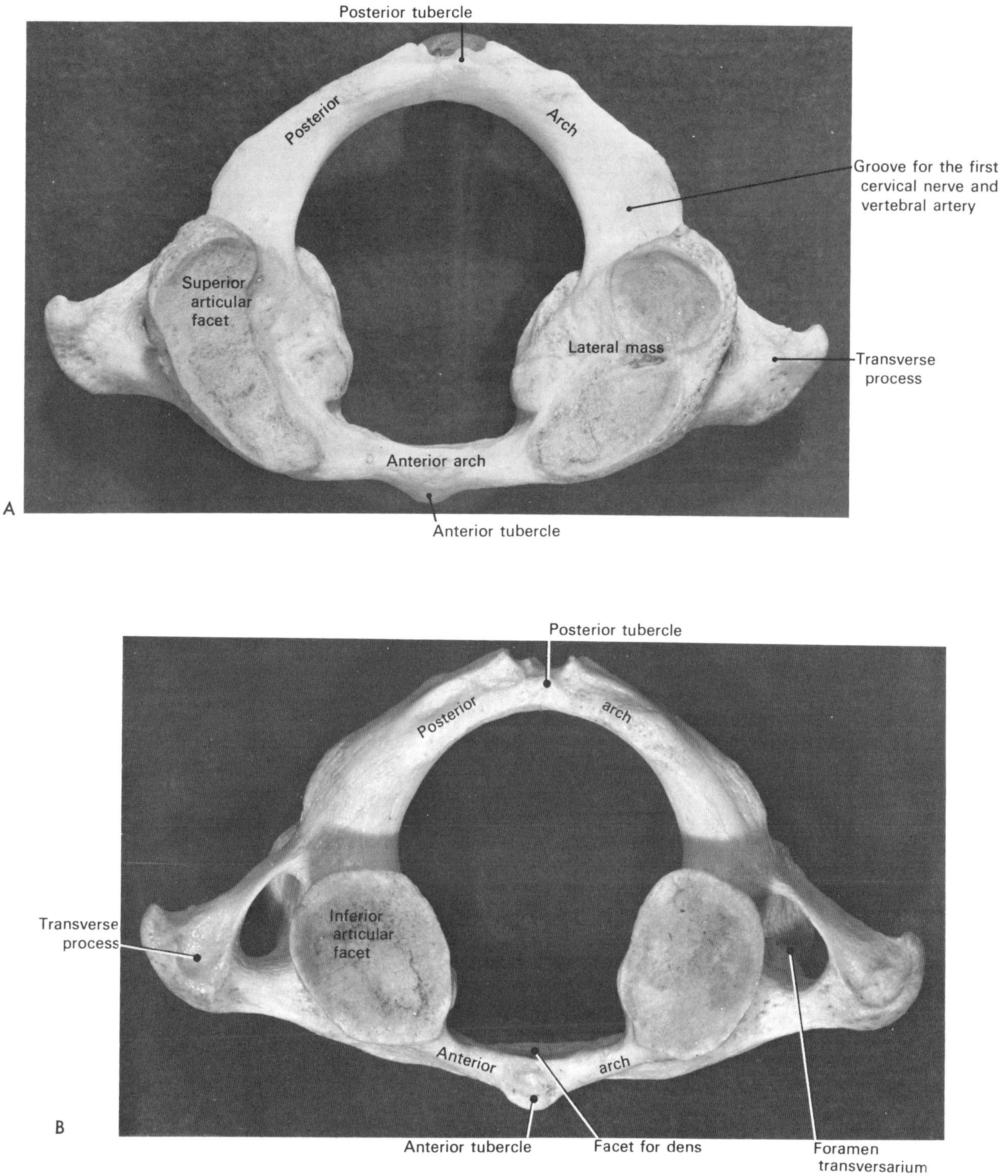

Fig. VII-12. The Atlas (VC_1). **A.** From above. **B.** From below.

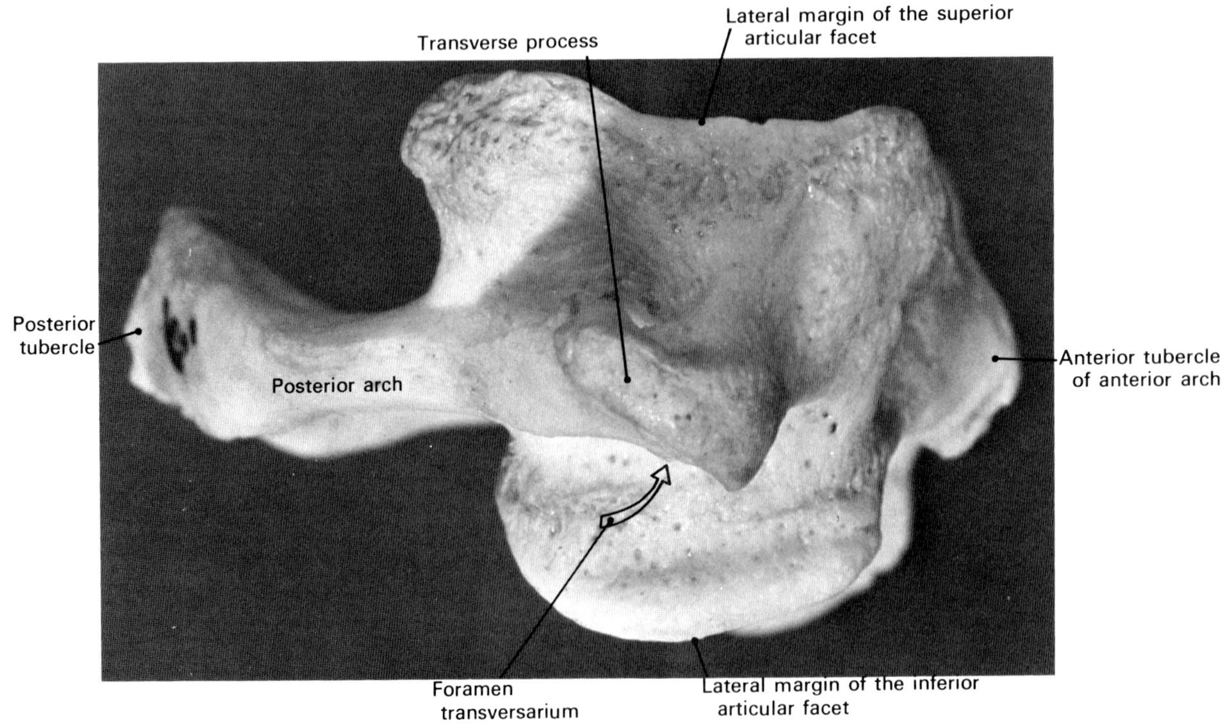

Fig. VII-12C. The Atlas (VC_1). Lateral view.

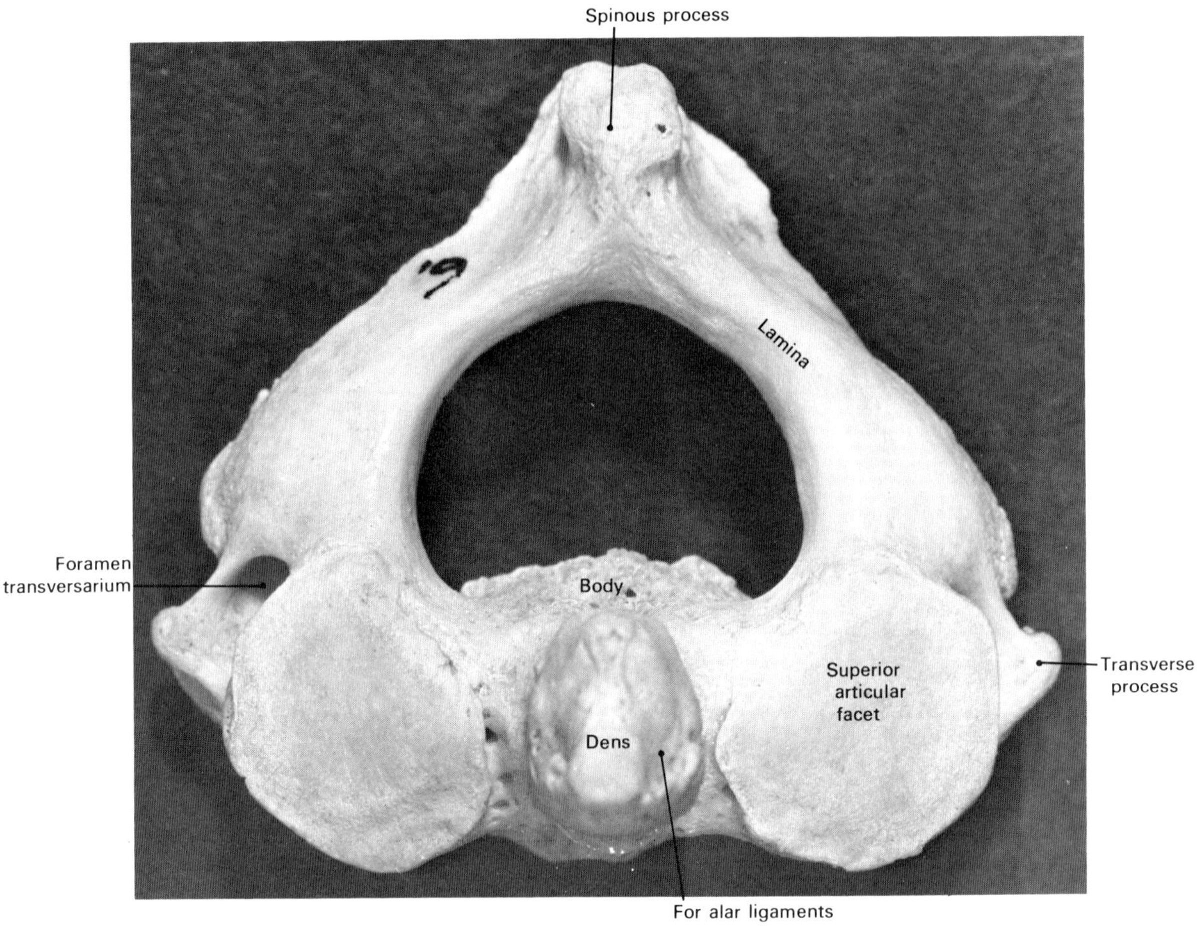

Fig. VII-13a. The Axis (VC_2). From above.

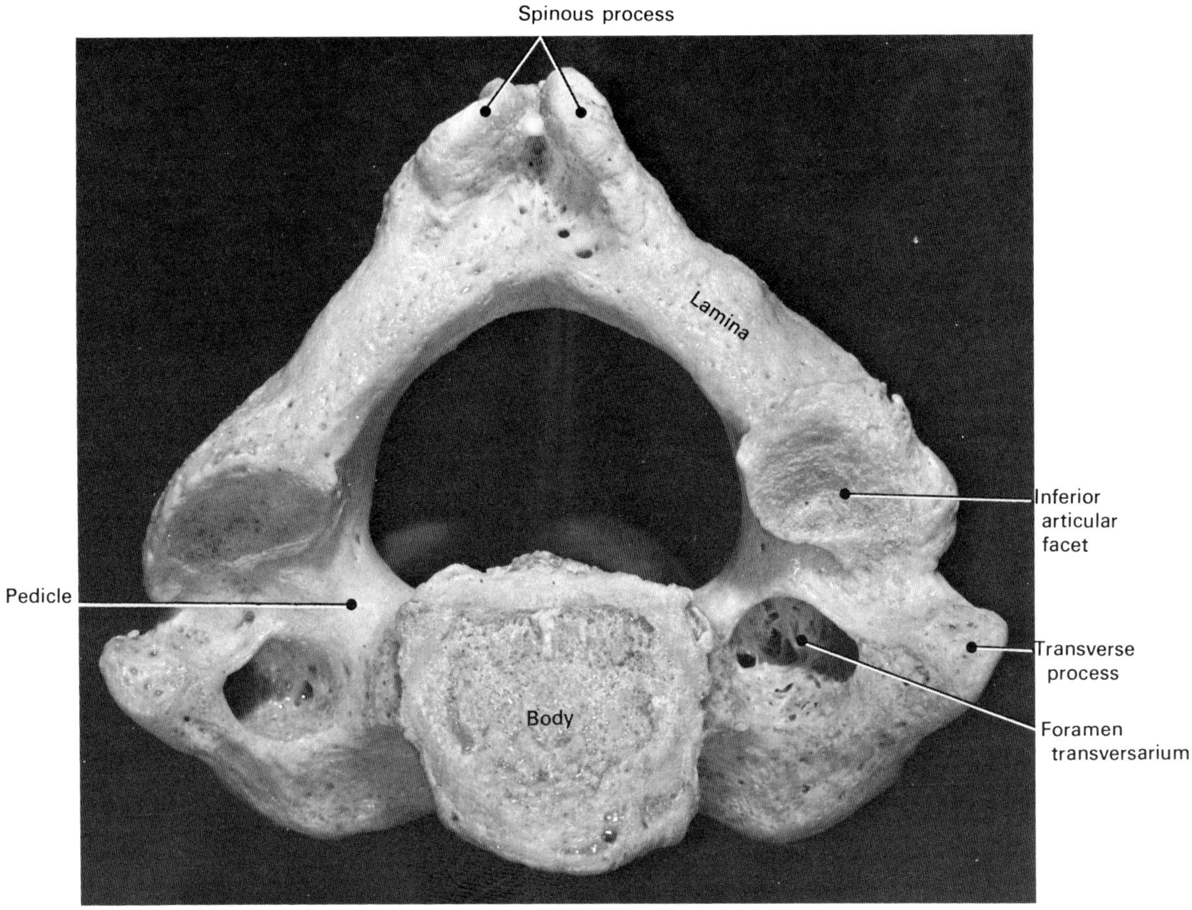

Fig. VII-13B. The Axis (VC_2). From below.

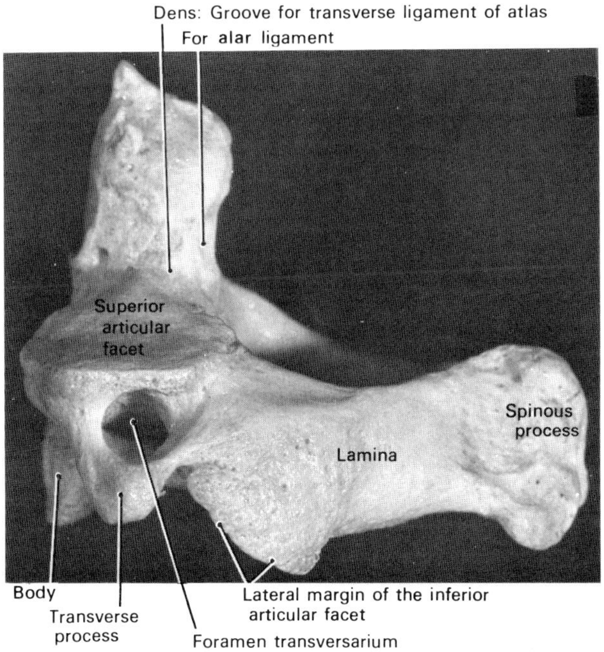

Fig. VII-13C. The Axis (VC_2). Lateral view.

(Fig. VII-13C) to which a weak **apical ligament** will attach. The grooves along the sides of the dens (Fig. VII-13A) are for the **alar ligaments.**

The manner in which atlas and axis fit together is shown from above and in lateral view in Figure VII-14A and B. The dens snuggles against the anterior arch of the atlas, separated from bone in front and from a restraining **transverse ligament** behind by synovial capsules (Fig. VII-14A). Rotation of the atlas, thus, can occur without the odontoid process slipping posteriorly. The rotation takes place at the respective superior and inferior facets of these articulating bones.

INTERVERTEBRAL DISKS AND VERTEBRAL SYMPHYSES

The last six cervical vertebrae have fibrocartilaginous disks interposed between their bodies. In this respect they do not differ from the rest of the vertebrae, except for the sacrum and coccyx. In the sacrum the disks have ossified and fused with the sacral bones to form a single structure. There is a disk between sacrum and coccyx, but the single coccyx represents a fusion

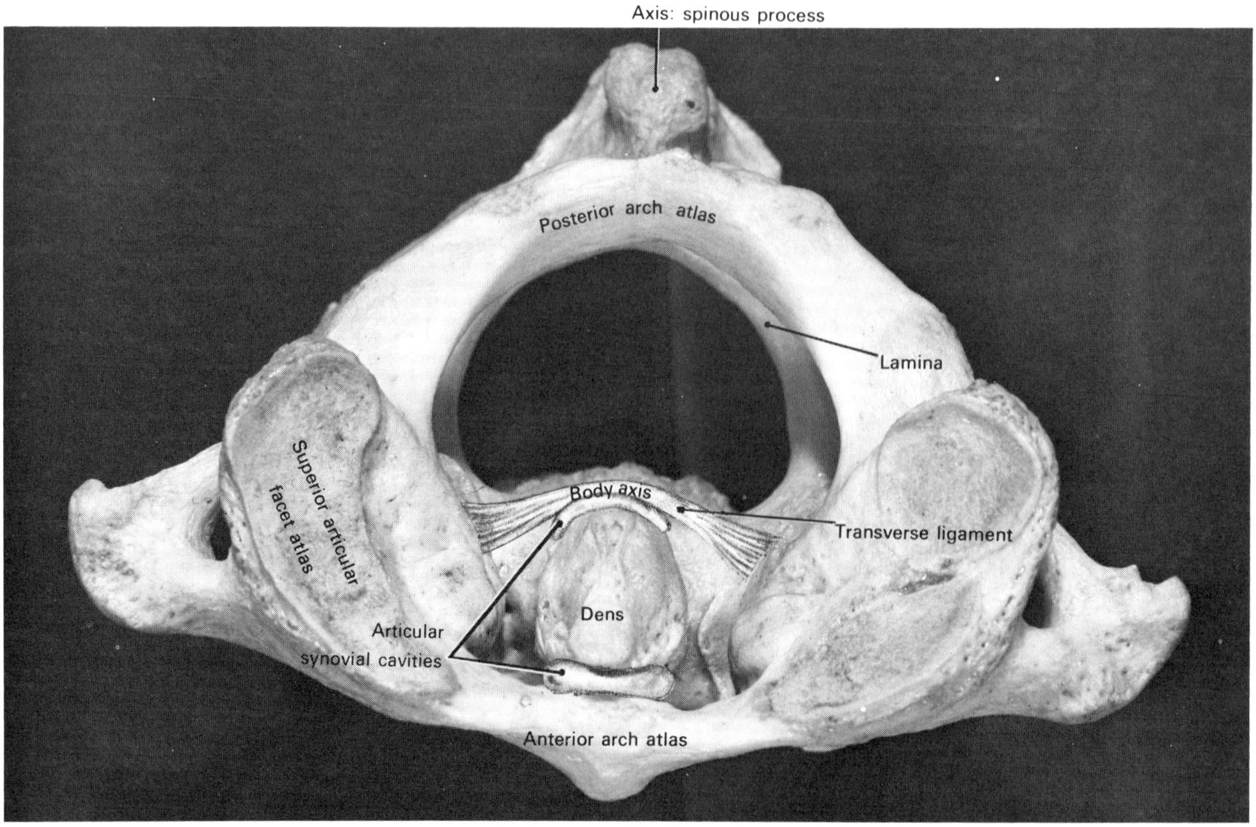

Fig. VII-14. First and second cervical vertebrae. **A.** From above showing transverse ligament and synovial joints about dens.

of small, vestigial tail bones. In the cervical region the disks are slightly thicker anteriorly than posteriorly and this wedging contributes to cervical lordosis, a secondary curvature which is normal.

The intervertebral disks allow the blocks of bone to align one upon the other, and they provide a resilient cushion which allows some slight rocking between vertebrae. The disks compress as muscle pulls on bones and they withstand the weight of gravity which alone could erode and grind the bones of the column were they to move directly against each other. The peripheral laminae of the disks consist of concentric rings of fibrocartilage wrapped around an eccentric central core material. The peripheral **anulus fibrosus** is interwoven laterally by obliquely-directed collagenous fibers within the matrix. These fibers cross each other at right angles. The anulus is not well developed posteriorly. The central core material or **nucleus pulposus** is more cartilaginous than fibrous and has some elastic qualities, especially when the disk is compressed. The core sits more posteriorly than centrally. In forward bending (flexion), because of its position in the disk, the nucleus pulposus tends to be pushed even more posteriorly under compression. If there is wearing or damage of the fibrocartilaginous anulus, either posteriorly or laterally, the nucleus pulposus can be protruded or extruded to press upon the cord or upon the emerging spinal nerves. We speak of this as a "slipped disk." The consequences can be excruciatingly painful, and, depending upon the locus of compression of the neural structures, can lead to sensory and/or motor paralysis and hypoactive tendon reflexes.

Common sites of occurrence for slipped disks are in the cervical and lumbar regions. In the neck the most frequent sites of involvement are the disks at the VC_5–VC_6 and VC_6–VC_7 interspaces. Recall that the first disk lies between vertebrae C_2 and C_3. Recall that except for spinal nerve C_8, all cervical spinal nerves emerge *above* their correspondingly-numbered vertebral body (C_8 emerging above VT_1). Protrusion of disks at the VC_5–VC_6 interspace or at the VC_6–VC_7 interspace will involve **spinal nerves C_6 and C_7,** respectively.

If the disk at VC_5–VC_6 compresses spinal nerve C_6, the general symptoms are pain over the shoulder, extending into the arm; vague, abnormal burning or prickling sensations (paresthesia) in the thumb and index fingers; and loss or weakness of the Biceps tendon reflex.

If spinal nerve C_7 is compressed by disk VC_6–VC_7, the sensations extend into the middle finger, as well, but the thumb is spared. There is absence of the Tri-

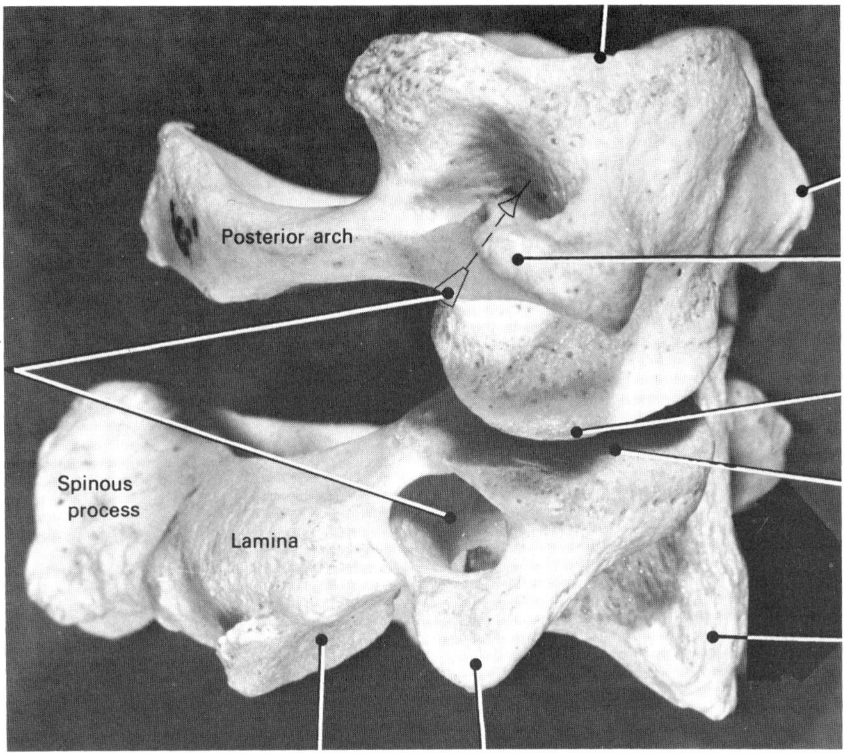

Fig. VII-14B. Lateral view showing articulations.

ceps tendon reflex, as well. Motor paralysis is a rare finding in disk disorders, in general, and the sensory findings are more reliable than the reflex symptoms.

The most common sites in the lumbar region are at the VL_4–L_5 and VL_5–S_1 interspaces. Here it will be spinal nerves L_5 and S_1 that will be involved. This may seem strange when you recall that below VT_1 spinal nerves emerge *below,* not above, the level of their correspondingly-numbered vertebral body. But in the lumbar region spinal nerve L_5 exits *above* disk VL_4–L_5 to seek its correspondingly-numbered intervertebral foramen of exit; hence, it will be the nerve to be compressed by this disk. Disk VL_5–S_1 compresses spinal nerve S_1 for similar reasons.

Symptoms are lumbar rigidity with scoliosis (patient bent forward and in pain if he attempts to straighten up); pain along the distribution of the sciatic nerve, especially if you squeeze his popliteal fossa; and weakening of the knee and ankle jerk (Quadriceps tendon and Achille's tendon reflexes).

Disks between vertebrae would be useless if they were not held in place by ligaments. These two tissues, ligaments and fibrocartilage, fashion a **symphysis,** in this case, the vertebral symphysis (*cf.,* pubic symphysis). Symphyses are **amphiarthrotic** or slightly movable joints. (Recall that articulations between **vertebral articular facets** are **diarthrotic;** *i.e.,* synovial and freely movable. These are the ones variously called intervertebral, apophyseal, zygapophyseal, or posterior joints.)

The disks of vertebral symphyses are sandwiched between two longitudinal ligaments which extend practically along the entire length of the vertebral column, covering both disks and bodies. The outer, ventral one is the **anterior longitudinal ligament;** the other, the **posterior longitudinal ligament,** invests the inner, posterior aspects of bodies and disks *within* the vertebral canal. Both are bands of dense fibro-elastic connective tissue. Lateral ligaments along the sides of bodies and disks have been described, but they may simply be the spreading fascicles of the anterior one (Fig. VII-15).

LIGAMENTOUS SYNDESMOSES

The **ligamenta flava** are short bands of elastic tissue (yellow ligaments) that pass between adjacent vertebral **laminae.** They help maintain erect alignment and stretch and retract somewhat in flexion and extension. Although the amount of movement between any two vertebrae may be slight, the sum of a number of such movements over a stretch of the column is additive and allows for considerable flexibility. Ligaments be-

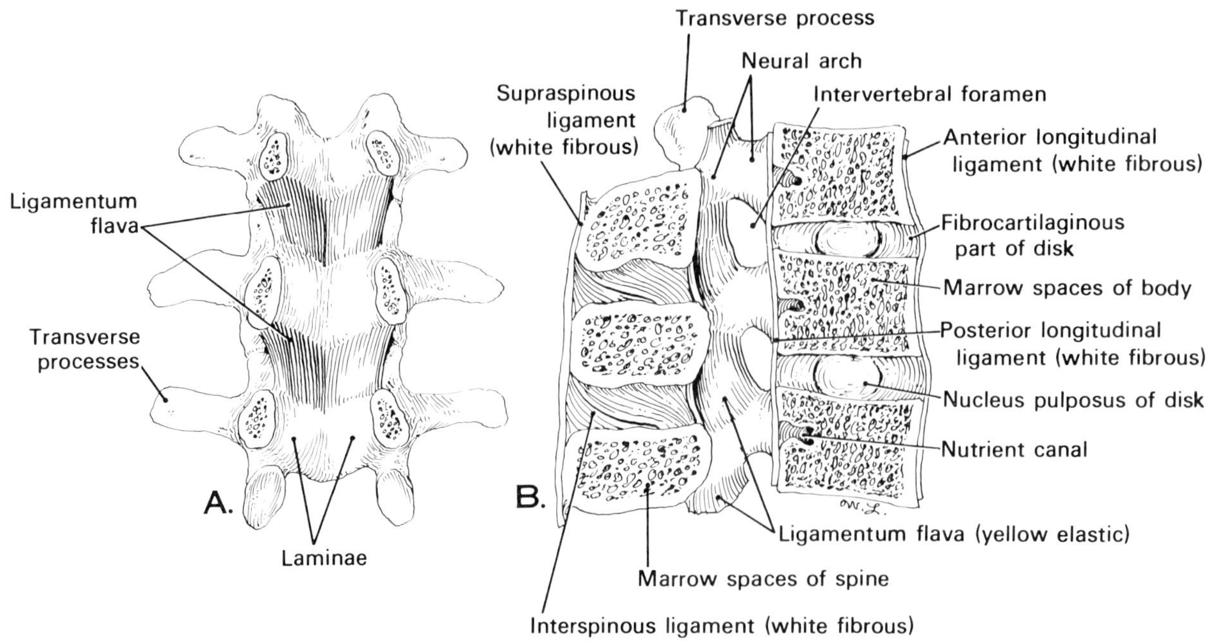

Fig. VII-15. Articulating vertebrae and disks. **A.** Sagittal section through vertebral column. **B.** Ligaments associated with vertebral column.

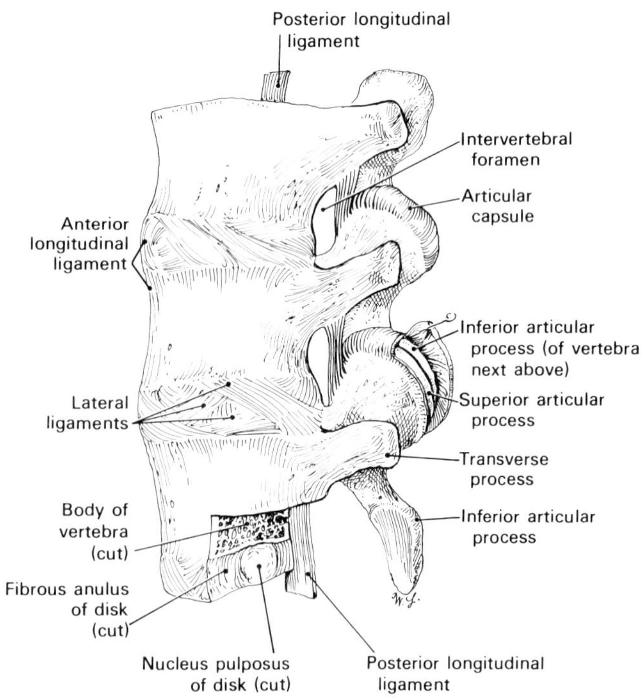

Fig. VII-15C. Lateral view of column, slightly rotated.

tween **spines** (interspinous) are weak in the cervical region and they blend with the **supraspinous ligament** running from spine to spine longitudinally at their extremities (Fig. VII-15). The interspinous and supraspinous ligaments of the cervical region, in turn, blend with the **ligamentum nuchae,** a strong median cord to which muscles from each side of the back attach. The ligamentum nuchae begins at the external occipital protuberance, attaches to the posterior tubercle of the atlas, and continues over the ends of the spines to attach to the tubercle of the seventh cervical spine. In man, it may contain more fibrous tissue than elastic, but it is a heavy elastic cord in grazing animals and, in them, it helps bear the weight of the dependent head. The ligamentum nuchae and the ligamenta flava are the only elastic ligaments associated with the vertebral column. The rest are ordinary collagenous (white fibrous) ligaments known for their tensile strength rather than for their elasticity.

ATLANTO-OCCIPITAL JOINTS

At VC_1 and VC_2 the ligamentous attachments become more complicated (Fig. VII-16A and B). At the upper end of the cervical column where occipital condyles articulate with the superior facets of the atlas laterally, these joints are typically encapsulated synovial joints. These joints allow only two types of movement: anteroposterior flexion and extension and some slight lateral tipping. Rotation is impossible because of the oblique orientation of the condyles and the medial slant of the facets of the atlas. In this area, the anterior longitudinal ligament modifies slightly and

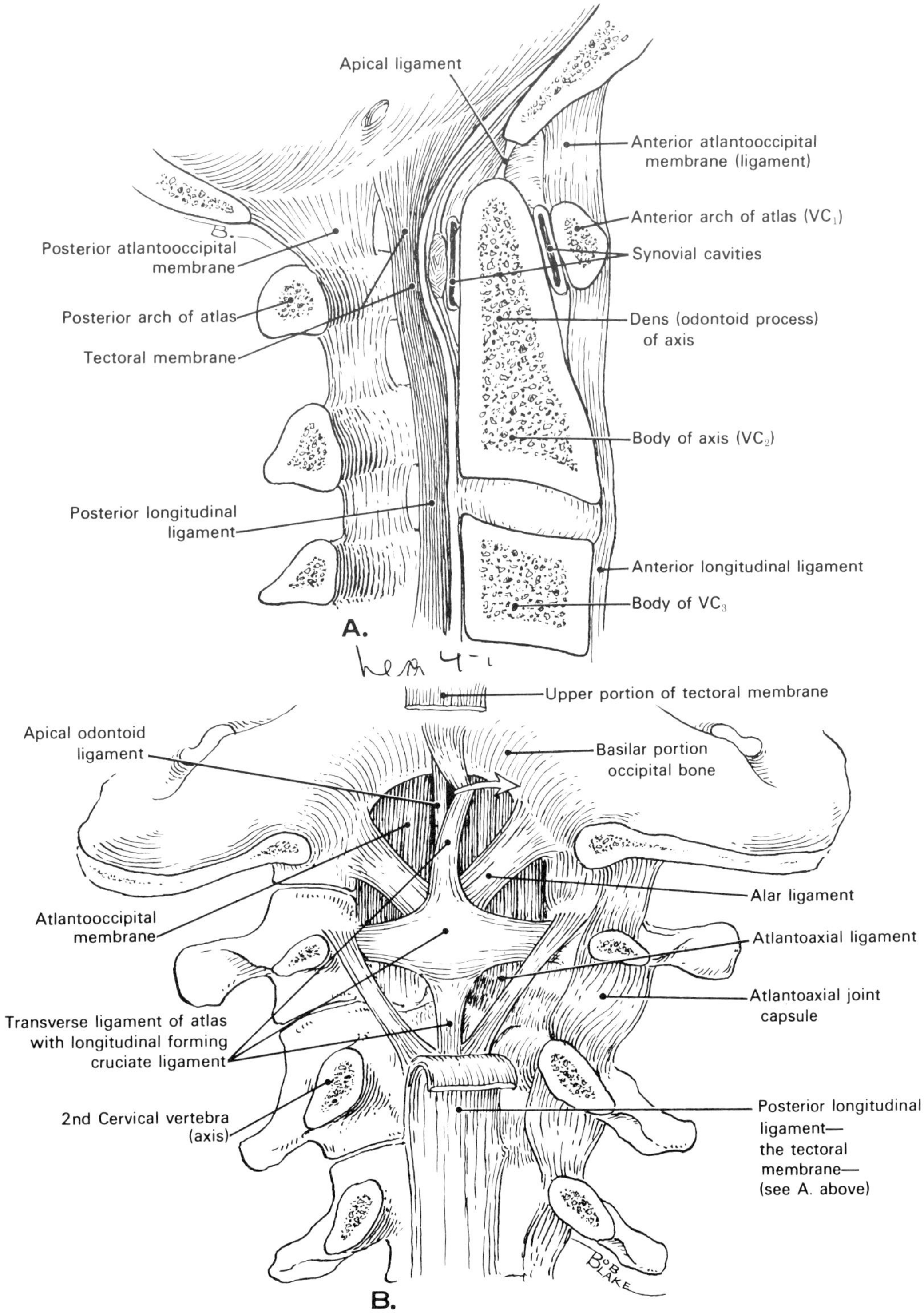

Fig. VII-16. Atlanto-occipital (A) and atlanto-axial (B) joints.

continues over the anterior arch of the atlas to the base of the skull as the **anterior atlanto-occipital membrane** (Fig. VII-16A). It is impossible for the posterior longitudinal ligament to attach to the body of the atlas since there is no body, so it bridges the inner aspect of the posterior arch of the atlas as the very weak **tectorial membrane,** attaching inside the floor of the skull. From the posterior arches of VC_1 to the base of the skull, the ligamenta flava must now serve to complete the atlanto-occipital membrane. These elastic laminal ligaments from each side meet and fuse in this region as the *posterior atlanto-occipital membrane*. Thus, the atlanto-occipital joint is not so complex after all. Old friends have changed names as they moved into the cervical region, much as if the Goldsteins suddenly preferred to be known as the Stoners when they moved into a swanky neck of the woods.

ATLANTO-AXIAL JOINTS

These joints look a bit more complex because of the dens, but they are relatively easy to comprehend. If cervical laminae and spines are removed and the cord structures, fat, and veins are cleared away, these joints can be examined from behind (Fig. VII-16B). One looks down upon the thin tectorial membrane which must be cut and turned back to detect the stouter ligaments. We now see that the **transverse ligament** is really cruciate or cross-shaped. Its transverse bands hold the axis to the anterior arch of the axis, as we noted in Figure VII-14A. Its rostral and caudal extensions attach to the cranial surface of the occipital bone and to the body of the axis, respectively. Deep to the rostral extension are the **alar** or **check ligaments,** responsible for the lateral grooves we observed on the odontoid process or dens (Fig. VII-13A). The alar ligaments run obliquely laterad from dens to occipital condyles, limiting or checking rotation between axis and atlas. They also hold the skull to the axis and tighten when the atlas rotates on the axis, carrying the skull with it. Laterally, the articular facets, as usual, are encapsulated synovial joints. There is a weak but insignificant **apical ligament** (Fig. VII-16A) that holds the tip of the dens to the anterior lip of the foramen magnum.

RADIOGRAPHIC ANATOMY OF THE CERVICAL SPINE

Now that you have become familiar with the individual members of the cervical vertebral column, we will consider them all together as they appear in the living subject. The standard radiographic examination of the cervical spine consists of: (1) anterior-posterior; (2) lateral; and, (3) oblique projections.

1. Anterior-Posterior Projection

In this radiographic projection, the mandible will normally obscure the upper two cervical vertebrae unless special techniques are employed. The simplest method to visualize the atlanto-axial relationship is by extending the neck slightly so that the occiput and the upper teeth are in the same plane. The central X-ray beam is then directed perpendicular to the film cassette through the open mouth (Fig. VII-17). Another method often used requires the patient to move his jaw constantly during the exposure which results in blurring out the mandibular shadow (Fig. VII-18). To visualize the lower cervical vertebrae (VC_3–VC_7), the patient, ideally sitting erect, slightly hyperextends the head, and the central X-ray beam is directed about 10° to 15° cephalad in order to better visualize the intervertebral joint spaces (Fig. VII-19).

Figures VII-17 and 18 show the relationship of the occipital condyles to the atlas, and the atlas to the axis. The convex articular surface of the occipital condyles may be seen in close apposition to the concave superior facets of the atlas. The contents of the atlanto-occipital joint are radiolucent and, therefore, do not cast a shadow; hence, the use of the term joint "space."

The lateral masses of the atlas have a triangular shape, and these are readily discernible, while the thinner anterior and posterior arches of the atlas are only faintly seen. The long transverse processes of VC_1 are directed laterally from the lateral masses. In this projection, the foramen transversarium cannot be detected. Between the lateral masses of the atlas, the odontoid process is seen projecting cranially from the body of the axis. On occasion, the air gap between the front upper incisor teeth may overlie the odontoid, causing it to appear bifid. The atlanto-axial joint space is seen sloping inferolaterally in Figures VII-17 and 18. It has smoothly concave articular margins. Remember that there are no uncovertebral "joints" (fissures) at this level.

Now examine the Figure VII-19 to study the relationship between the lower five cervical vertebrae and the upper two thoracic vertebrae. The intervertebral disk spaces are, of course, filled with the radiolucent anulus fibrosis and the nucleus pulposus. At times the nucleus pulposus may be seen when it is calcified. This may occur in some systemic diseases or when the disk degenerates. The uncovertebral "joints" are located between the uncinate process and the beveled undersurface of the vertebrae above. Osteophytes presenting off these joints could be readily detected in this projection when they beak laterally in certain individuals. Osteophytes are not seen in Figure VII-19.

The pedicles of the cervical vertebrae are not well seen in the AP projection since they are small and emerge from the body dorsolaterally. The distance between the inner margins of the pedicles is the *width of the vertebral canal* in the coronal plane. The intervertebral foramina are partially visualized. The pedicles and the intervertebral foramina are best evaluated in the oblique views. Since the intervertebral (apophyseal) joints leave the coronal plane of the vertebrae at an acute angle, they are obscured for the most part. The spinous processes of the vertebrae usu-

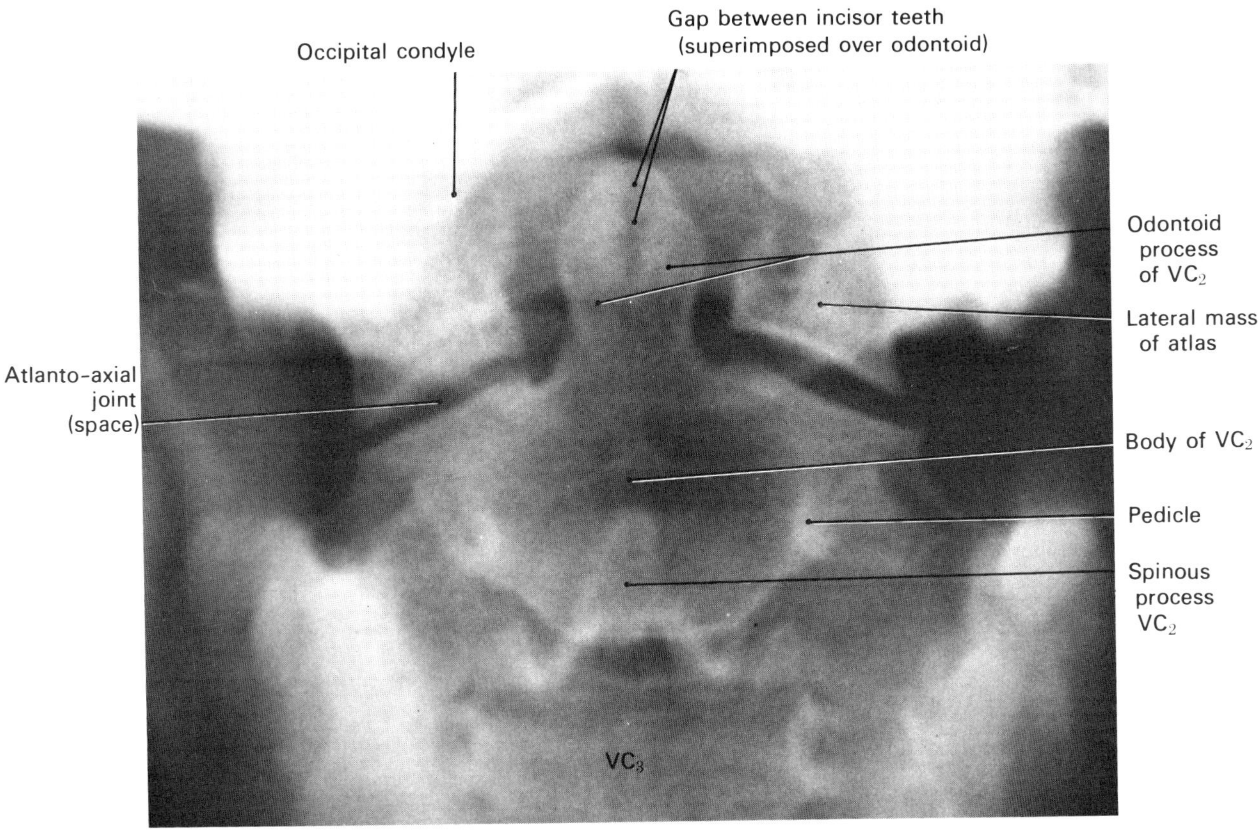

Fig. VII-17. Radiograph of atlanto-occipital joint and odontoid process; anterior-posterior view taken through the open mouth.

ally line up in a vertically-directed line. Note that those spinous processes from VC_3 to VC_6 are bifid. The second most prominent *transverse process* in the cervical spine is at VC_7. Cervical transverse processes are directed laterally and inferiorly, as opposed to the transverse processes of the upper thoracic spine which are directed laterally and cranially. Occasionally a rudimentary cervical rib may articulate with the transverse process of VC_7.

At times the overlapping shadows of the **thyroid** (laryngeal) **cartilages** are confused with abnormalities of the cervical vertebrae. Their normal location is shown in Figure VII-19. Also notice the **tracheal airway** superimposed upon the spinal column in Figure VII-19.

2. Lateral Projection

For this projection the patient is usually sitting erect, with the head slightly hyperextended so that the anatomic base line (infra-orbito-meatal line) is parallel with the floor. The central X-ray beam is directed perpendicular to the X-ray film cassette at the level of the thyroid cartilage.

The cervical spine normally has a smooth lordotic curvature which has its apex at the VC_4–VC_5 level. Let your eyes follow the anterior and then the posterior margins of the vertebral bodies in Figure VII-20A and observe how these bodies are aligned. Any deviation from the smooth curvature in this neutral position should immediately alert you to an abnormality such as a fracture with luxation, tearing of ligaments secondary to trauma, or pathological abnormalities of the apophyseal joints and intervertebral disks.

Figure VII-20A and B shows the extremely close approximation of the anterior margin of the odontoid process and the anterior arch of the atlas, the latter here seen end-on. The distance between the anterior aspect of the odontoid and the posterior margin of the atlas should not normally exceed 2.5 to 3.0 mm. (4.0–5.0 mm. in children). Anything beyond this value is usually abnormal and intense search must be made as to its etiology.

As we mentioned earlier there is no intervertebral disk at the VC_1–VC_2 level. The intervertebral disks are wedge-shaped, and, therefore, the disk spaces are wider anteriorly than posteriorly. As the disks degenerate, or if the interspace becomes infected, this vertical height will decrease. Be certain to check the disk spaces, then, for even a moderate loss in height may have very real clinical significance.

The body of the axis appears massive compared with the other lower five cervical vertebrae. The pos-

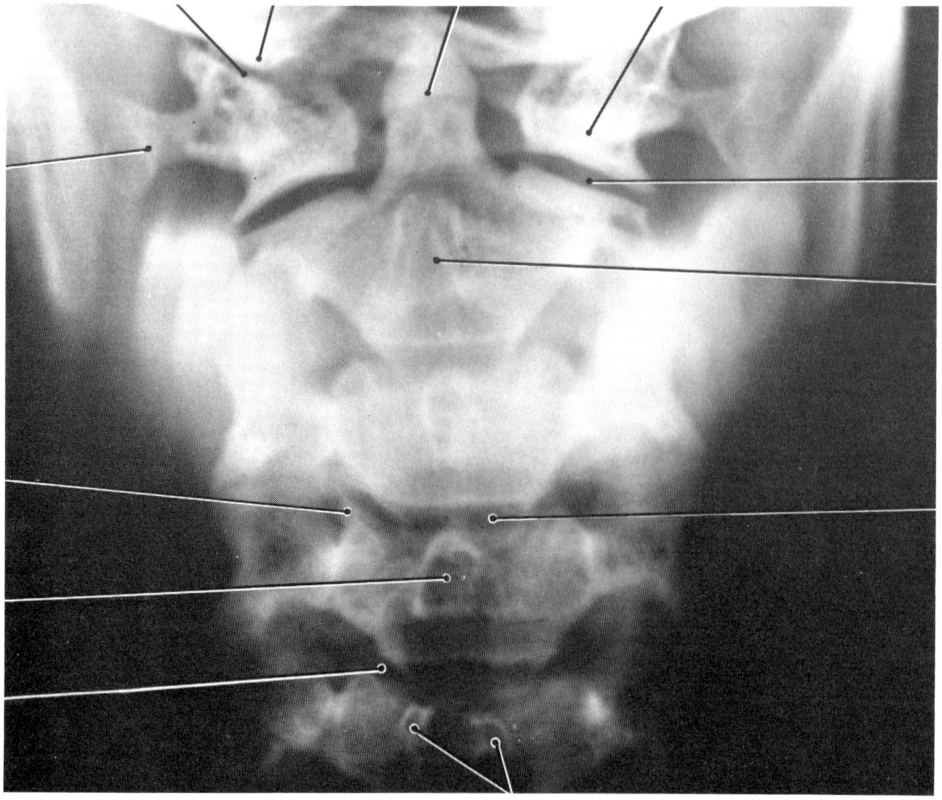

Fig. VII-18. Anterior-posterior projection of cervical spine ("wagging jaw" technique).

terior margins of the bodies of cervical vertebrae are relatively flat. These *margins, of course, mark the anterior aspect of the vertebral canal* (Fig. VII-20A). The anterior inferior margins of the bodies are lipped, while the superior anterior margins are beveled. The transverse processes are superimposed on the bodies.

Because of their anatomic position, little, if any, portion of the pedicles and intervertebral foramina is seen in the lateral projection. The rhomboid-shaped articular processes are seen posterior to the bodies. Between the inferior articular facet of VC_4 and the superior articular facet of VC_5 is the **intervertebral (apophyseal) joint.** Where apophyseal joints of opposite sides are superimposed, a clear joint "space" is readily visualized, such as at the VC_3–VC_4 level. At VC_6–VC_7 the apophyseal joints are overlapping and a joint space is not seen to good advantage. Thus, you can readily see why the lateral view is not the best projection for demonstrating pathological processes in the apophyseal joints. The modified oblique of Buetti-Bäuml is far superior and will be considered later. One further point is in order before we leave the apophyseal joints. In some people those joints at VC_2–VC_3 may slope inferolaterally 10° to 20° and, therefore, will not be sharply defined since you will not be looking directly through the joint.

The vertebral arches and spinous processes are posterior to the articular processes. The *dense white line separating these two structures* (Fig. VII-20A) *represents the most posterior margin of the vertebral canal.* You now know both the anterior and posterior limits of the vertebral canal in the coronal and sagittal planes. This is a useful piece of information for your clinical years because certain pathological processes may cause this canal to be abnormally large or small. Many charts are available which give the range of normal measurements of the vertebral canal.

It is to your advantage to learn the anatomy of the normal cervical spine, particularly as seen in the lateral projection. In many instances it may be the only

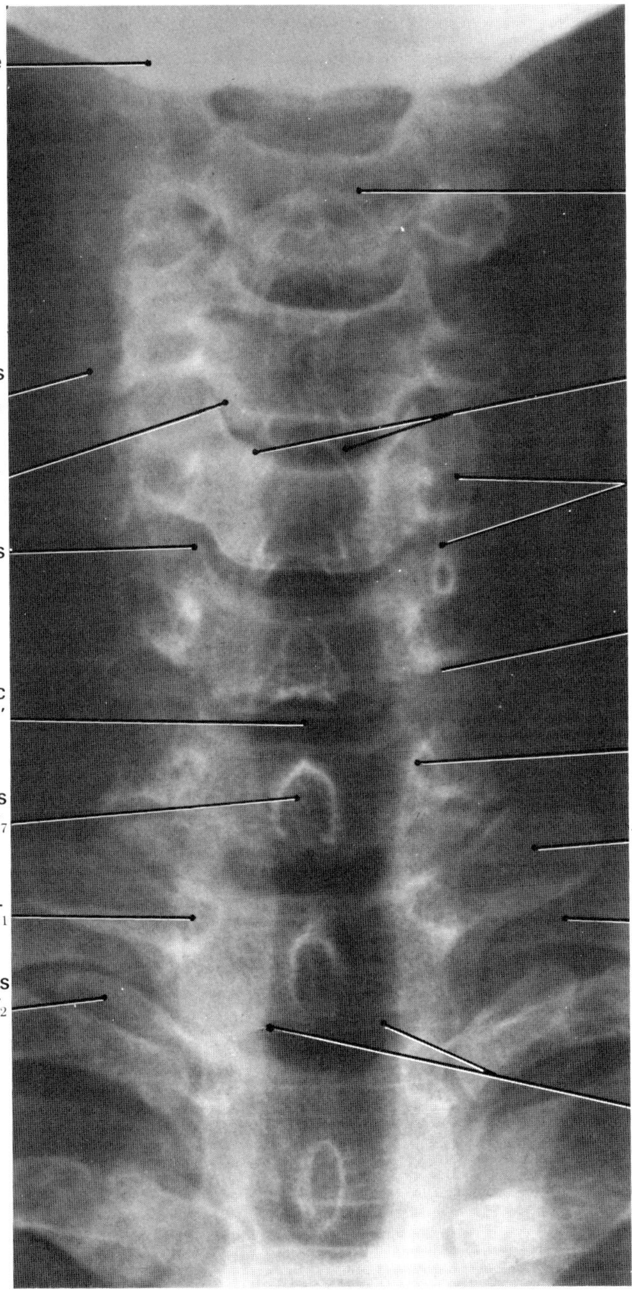

Fig. VII-19. Anterior-posterior view, VC_3 to VT_2.

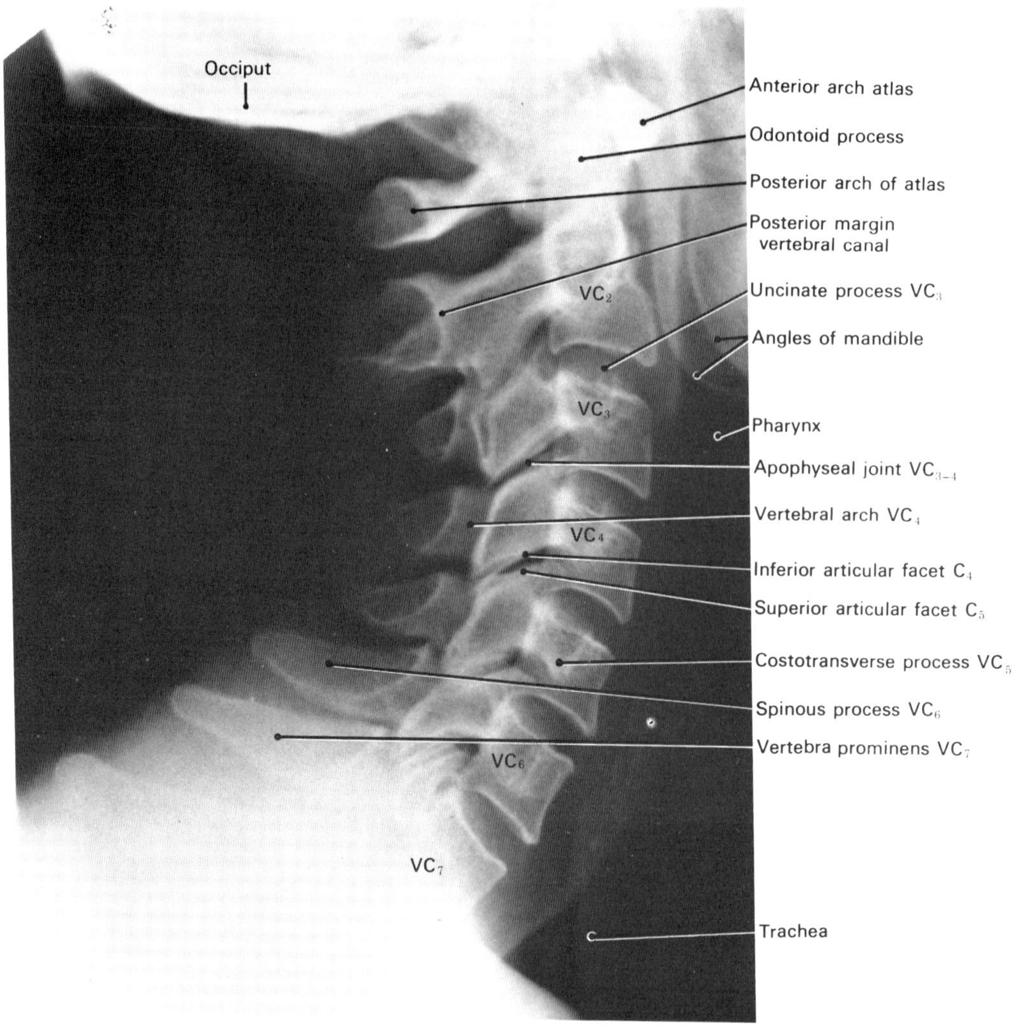

Fig. VII-20. Lateral radiographs of cervical spine. **A.** Neutral-lateral position.

projection available to you in evaluating the neck of a severely-injured patient. Where fracture is suspected clinically, a single, cross-table, lateral view is usually made and evaluated before the patient is moved for further projections. Obviously, if you have not mastered the normal, you cannot be expected to detect the abnormal.

3. Oblique Projections

Two different degrees of obliquity are necessary to completely study the cervical spine. The first is the standard 45° to 55° right and left anterior oblique (Fig. VII-21A), made with the patient erect. The central X-ray beam is directed 10° caudally. The second projection is the modified oblique of Buetti-Bäuml, where the erect patient is rotated 20° from the lateral, and again the X-ray beam is directed 10° caudally (Fig. VII-21B).

Each of these projections offers valuable information to the clinician and to the radiologist with whom he is consulting. In the standard 45° oblique (Fig. VII-21A), the intervertebral foramina are seen well as are the uncovertebral "joint" regions. Bony encroachment such as osteophytes off the uncovertebral joints or from the apophyseal (intervertebral) joints could be readily detected in such a view. Such compromise of the foramina may be symptomatic in your patient. In this left anterior oblique projection (Fig. VII-21A), the pedicles on the right side are seen end-on and, therefore, are quite dense. The pedicles on the patient's left side are seen in profile.

Observe the normal shingle effect (imbrication) of the articular facets on the patient's left side. For example, the superior facet of VC_6 is situated below and ventral to the inferior articular facet of VC_5. Occasionally in severe hyperflexion injuries to the cervical spine (whiplash), unilateral or bilateral facet jumping or locking may occur; that is, the superior facet is luxated and is situated below and dorsal to the inferior

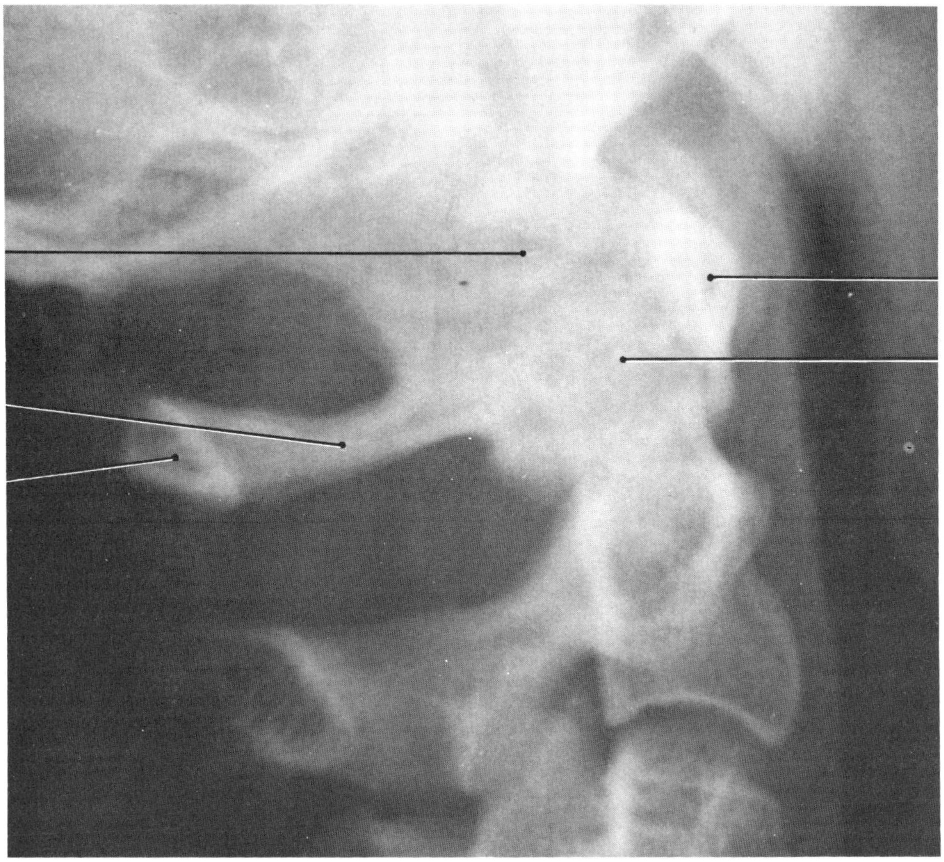

Fig. VII-20B. Enlargement of relationship of odontoid (dens) to atlas.

facet. An anterior oblique projection, therefore, is needed to confirm the diagnosis. Such locking of facets may also be detected in the lateral projection, but not as well as in this oblique view.

This standard oblique projection affords yet another opportunity to evaluate the *width* of the vertebral canal; that is, *from the inner margin of the right pedicle to the inner aspect of the vertebral arch.*

The modified oblique of Buetti–Bäuml (Fig. VII-21B) provides us with an excellent view of the intervertebral (apophyseal) joints and the articular processes. Again a shingle effect is seen as the articular processes are layered one upon the other. Arthritic processes involving the apophyseal joints can be detected easily in this view, as can be fractures through the facets and pedicles of the vertebrae.

Throughout our discussion we have mentioned only briefly the functional anatomy of the cervical spine. The subject is extremely complex and beyond the scope of this book, and, therefore, we refer you to the excellent monograph of Penning for a review of functional anatomy.

THE UNCOVERTEBRAL "JOINTS" OR FISSURES

Although most of the textbooks of anatomy ignore the subject of the uncovertebral "joints" or fissures, we cannot overlook this controversial subject, since it has clinical importance. In 1858, von Luschka[8] described synovial joints in that portion of the disk between the uncinate processes and the corresponding beveled undersurface of the cervical vertebral body above. Luschka termed these "joints" the lateral intervertebral half-joints. Others have given other names to these "joints," and, thus, you will find them designated uncovertebral joints,[13] lateral interbody joints,[7] covertebral joints,[4] or the joints of Luschka[10] in the literature.

Rathke[9] disputed the concept of a true synovial joint in 1933, and since then an argument has raged as to the significance of these uncinate structures. Some authors support the original concept of von Luschke,[1, 2] but most authors disagree.[3, 5, 6, 11, 12] It is now believed that these are not synovial joints, but rather that there are fissures formed in the lateral and posterolateral margins of the intervertebral disks of the cer-

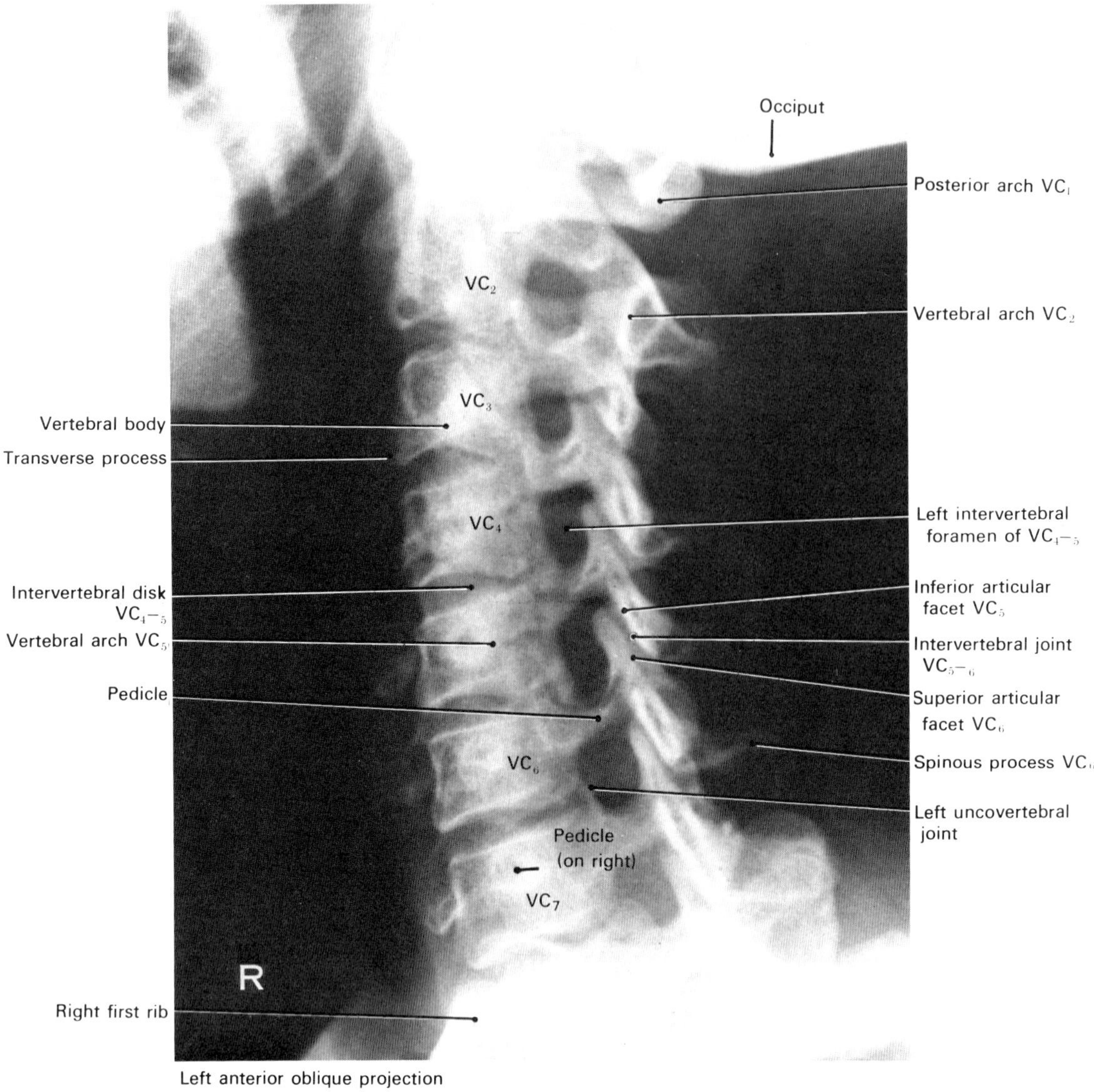

Fig. VII-21. Oblique radiographs of the neck. **A.** A 45° left anterior oblique projection.

vical spine. These fissures are not present at birth, and, indeed, they are first detected at from 4 to 7 years of age. They begin to develop in the anulus fibrosis near the growing uncinate processes and gradually enlarge in both size and shape until they eventually reach the nucleus pulposus. Although the cavities formed certainly resemble a joint, most investigators have failed to find a synovial membrane lining them. It is thought that the uncovertebral "joints" or fissures represent a functional adaptation by the disks to the increased mobility of the spine at the time that the uncinate processes are reaching full height.

The intervertebral disks are incapable of regeneration, so that as we grow older, the disks begin simply to wear out and become thinned. This causes the uncinate processes to approach the beveled underbelly of the cervical vertebrae above. Osteophytes eventually develop at the uncovertebral "joints" (an osteochondrosis), and as these lips of bone turn lateralward they encroach upon the intervertebral foramina. Thus, the mobility of the cervical spine may be decreased and the adjacent nerve roots, and even the vertebral artery, may be compromised. These uncovertebral spurs are readily demonstrable in both oblique and anterior-posterior projections of the cervical spine. The location of the uncovertebral "joints" is shown in the normal radiographs, Figures VII-18, 19, and 21A.

TRIANGLES OF THE NECK

The so-called triangles of the neck were outlined previously. They exist because anatomists exist and because the latter must organize material for student comprehension. Figures VII-4 and 6 demonstrate them

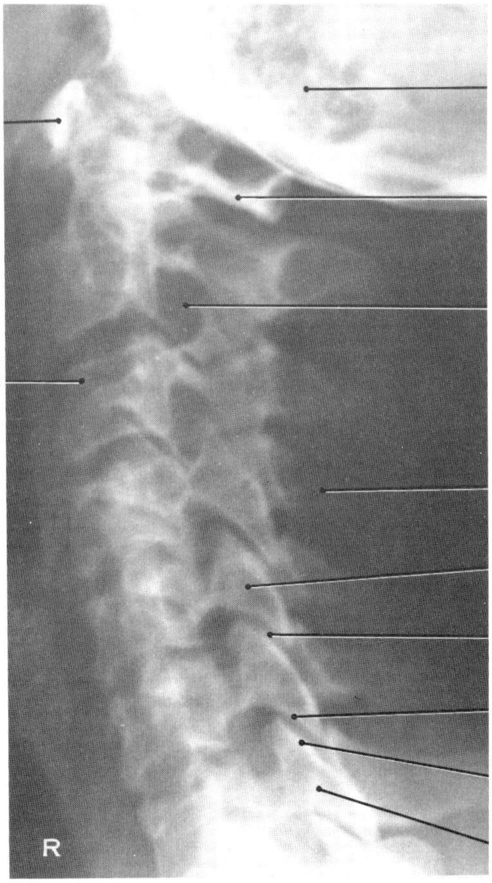

Fig. VII-21B. A 20° left anterior oblique (Buetti-Bäuml) projection.

best in the living subject. Their muscular borders and position in respect to the midline of the neck and the hyoid bone are reviewed in Figure VII-22. The **subclavian artery** can be compressed against the first rib in the subclavian (lesser vascular) triangle, and the common carotid can be compressed against the transverse process of VC_6 in the carotid (greater vascular) triangle, or its pulse can be felt there as shown in Figure III-36. You should review the *contents* of each of these triangles (see Chapter III, p. 109).

Posterior Neck and Suboccipital Region

As glasses are raised in English pubs there is a toast which goes: 'Here's hoping you never see the back of your neck." What this means is an enigma. Perhaps, while we are in the neighborhood of the atlanto-occipital and atlanto-axial joints, we can solve this enigma in part if we look under the superficial Trapezius and the origin of the Sternomastoid. With these in place (Fig. VII-8), the back of the neck does not look like much.

At the apex of the posterior triangle, deep to it and medialward to the midline, extending from skull to axis, is the **suboccipital region.** Among its contents are four small muscles which can move the head relatively independent of neck movements. But before we discuss these little muscles, it should be realized that there are still other muscles in the back of the neck and along its posterolateral slope. Some of the muscles of the back, itself, reach up to the spine of the axis (Semispinalis Cervicis, Multifides, Rotatores, and Interspinalis). Others of the back muscles have to reach the head since the atlas has no spine (Semispinalis Capitis, Splenius Capitis, and Longissimus Capitis) (see Fig. VII-23). The floor of the posterior triangle, especially its large upper neural triangle, is carpeted by a number of muscles [Levator Scapulae, Rhomboids (Major and Minor), Scalenus Anterior, and the Scaleni (Posterior and Medius)] (see Fig. VII-24).

The **Splenii,** for the most part, lie under the Trapezius. They run as an oblique cover across the ascending muscle slips of the Erector Spinae group: hence, their Latin name, splenius, "a bandage." The **Splenius Capitis** extends from the ligamentum nuchae and the spines of VT_1 to VC_3 to insert upon the head in the mastoid region and along the lateral one-third of the superior nuchal line (Figs. VII-23B and 24). Lower muscle slips from spines VT_3 to VC_5 arise as the **Splenius Cervicis** to insert on transverse processes VC_1 to VC_3. These muscle fibers course lateral to the Splenius Capitis (Fig. VII-23B). Both muscles, in their attachment, represent a **spinotransverse group.** The Splenii of one side can turn the head and tilt the face

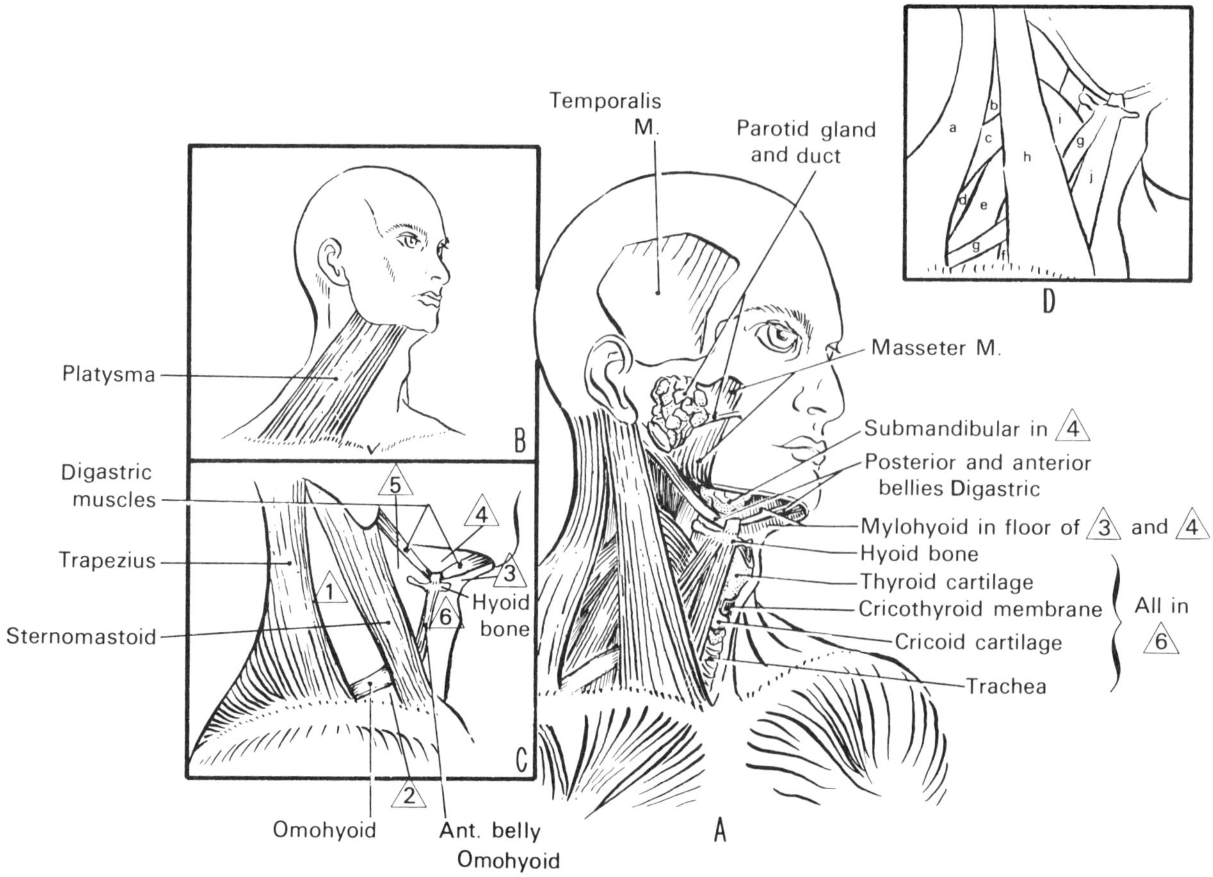

Fig. VII-22. Review of triangles of neck. **A.** Dissection to show muscular relations and some associated structures in triangles of neck. **B.** Platysma. **C.** Triangles viewed from the side. Numbers correspond to named and numbered triangles in Figure III-33. **D.** Key to muscles seen in view **A:** a, Trapezius; b, Splenius capitis; c, Levator scapulae; d, Scalenus posterior; e, Scalenus medius; f, Scalenus anterior; g and g′, inferior and superior bellies of Omohyoid; h, Sternomastoid; i, Inferior constrictor of pharynx (with esophagus showing above); j, Sternohyoid (one of the "strap" muscles). Muscles b–f are in the posterior triangle, all except g forming its floor. Muscle i is in the major vascular triangle, while j is in the visceral triangle. (Modified from Appleton, Hamilton, and Simon.)

toward their same side. Acting together, they extend the head. In this sense they oppose the Sternomastoids. The latter together, flex the neck, pulling the chin toward the chest; or, in acting alone, each Sternomastoid turns and tilts the head so that the face looks toward the opposite side. Which way would the face look if the spinal accessory nerve to the right Sternomastoid were severed?

The **Semispinales** belong to a **transverospinal** group of muscles that originate on transverse processes and insert on spines (Fig. VII-23C). The **Semispinalis Capitis** may peek through in the upper part of the posterior triangle, but it is largely under the cover of the Splenius, and it conceals the suboccipital region as well as the **Semispinalis Cervicis.** The Capitis arises from transverse processes VC_3 to VT_7 to insert on the head along the medial two-thirds of the nuchal lines of the occipital bone. The cervical slips of this muscle-complex arise from processes VT_2 to VT_{11} to reach spines VC_2 to VC_4.

An inconstant **spinospinal muscle,** the **Spinalis,** extends from spine VC_7 to spine VC_2 and may extend to the head by blending with the Semispinalis Capitis. The only other muscle of the back to reach the head is the **Longissimus Capitis** (Fig. VII-23B). It arises from transverse processes VC_4 to VT_7 to insert posteriorly on the mastoid process. The cervical part of this muscle arises from transverse processes VT_1 to VT_5 to insert on transverse processes VC_3 to VC_6. Another back muscle, the **Iliocostalis Cervicis,** reaches the neck, only. It arises laterally from the lower ribs to reach transverse processes VC_4 to VC_7 (see Fig. VII-23A).

Deep to the Semispinales are short oblique muscles which form the **Multifides.** In the cervical region their fleshy fibers arise from articular processes. They cross about three vertebral levels where they narrow their insertions upon a spinous process above them. Akin to them are the **Rotatores** which bridge only one vertebral segment from transverse process to the

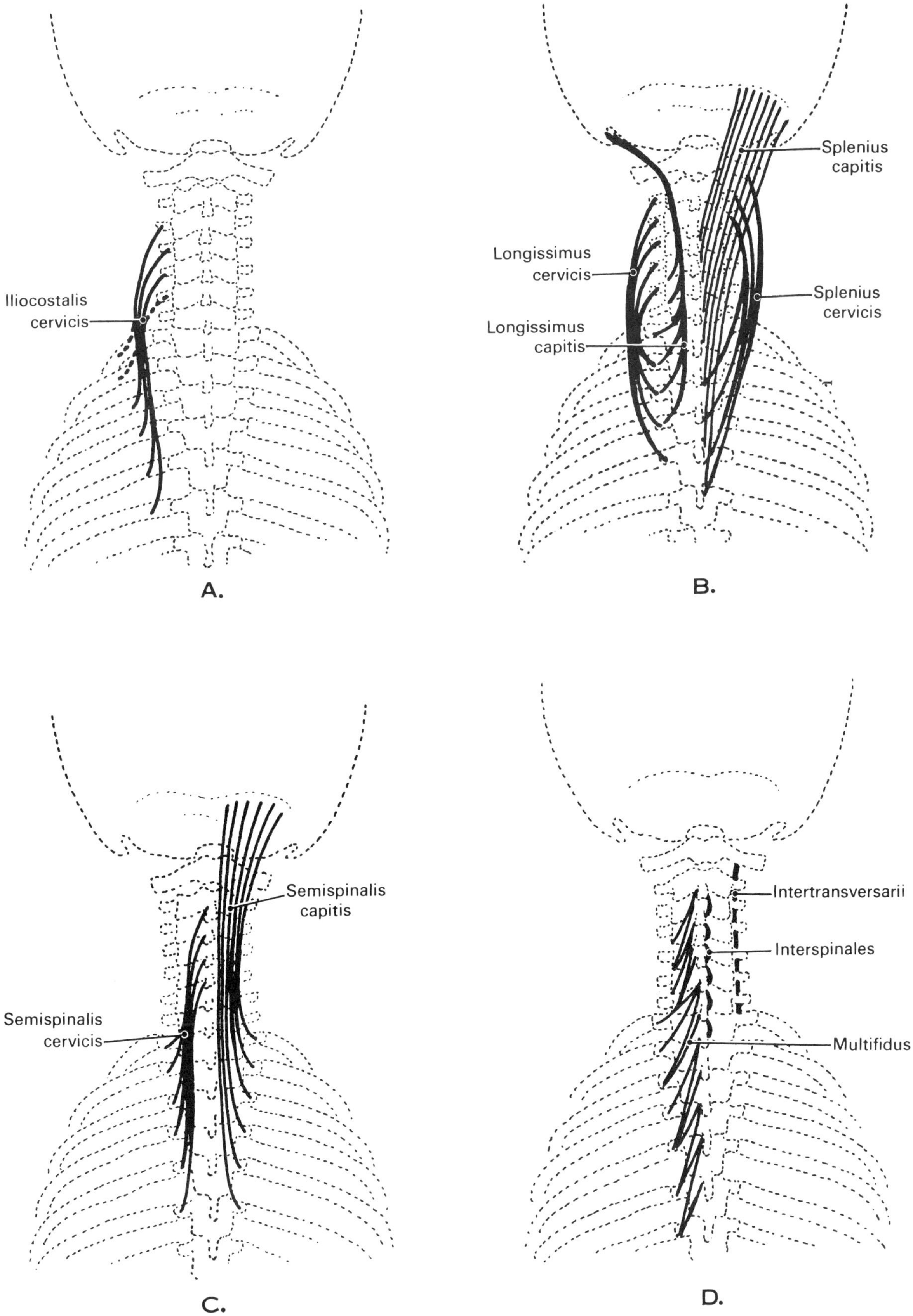

Fig. VII-23. Schematic views of the Erector spinae and Splenii at neck level.

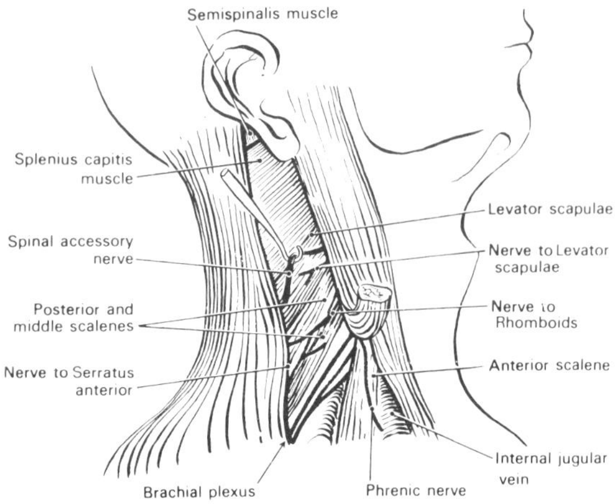

Fig. VII-24. The floor of the posterior (neural) and lesser vascular (subclavian) triangles and their contents.

next lamina above. The **Interspinales** and **Intertransversarii** are also well developed in the cervical region. The first of these unite bifid tubercles of adjacent spines; the second unite anterior and posterior tubercles of adjacent transverse processes (Fig. VII-23D). The Longissimus, the Iliocostalis, and all these short muscles lie deep to the Semispinales. All, when they act with their own partners of the opposite side, extend the neck. The Multifides and Rotatores, acting unilaterally, twist the neck toward the opposite side. Unilaterally, the Semispinalis, Longissimus, Iliocostalis, and Intertransversarii bend the neck laterad to the same side. Action, of course, is not of single muscles so much as it is of groups in concert.

The **suboccipital region or triangle** contains four small muscles, two straight ones (Recti), and two obliques; two vessels; and two nerves of importance in the area (Fig. VII-25). One of the muscles is thick and fleshy; the others are flat and triangular. The thick one (Obliqus Capitis Inferior) courses from the spine of the axis to the transverse process of the atlas and, as its name implies, marks the inferior border of the triangular area. The other oblique one (Obliqus Capitis Superior) runs from the transverse process of the atlas to the base of the skull between the superior and inferior nuchal lines marking the lateral boundary of the triangle. The two Recti (Rectus Capitis Posterior Major and Minor) lie medially, the Major arising from the posterior tubercle of the atlas; the Minor, lower down from the spine of the axis. Both insert in the base of the skull between the foramen magnum and the inferior nuchal line. The Minors, alone, are the only muscles attaching to the posterior tubercle of the atlas. Five muscles attach to the spine of the axis; *i.e.*, the Rectus Capitis Posterior Major and the Obliqus Capitis Inferior of the suboccipital group, and the Semispinalis, Multifidus, and Interspinalis.

Posterior Divisions of Spinal Nerves

Since this is in the back of the neck, only the **posterior primary divisions of spinal nerves** can be expected to supply the area. Anterior primary divisions of cervical nerves form the **cervical plexus** for supply of lateral and anterior parts. It is a general principle that when a posterior primary ramus of a spinal nerve reaches the back it divides into medial and lateral branches. In the neck, both branches supply muscle, but only the **medial branch supplies skin.** But C_1 rarely has a dorsal root, so its chief job is to supply the four suboccipital muscles. Hence, it is called the **suboccipital nerve.** This nerve is the one that occupies the same groove on the posterior arch of the atlas that the vertebral artery uses in its bend toward the foramen magnum (Fig. VII-25).

The posterior primary ramus of C_2 appears in the triangle below the Obliqus Capitis Inferior (Fig. VII-25). Its lateral branch supplies adjoining muscles but its medial branch ranges farther, ascending and piercing the Semispinalis Capitis and Trapezius to follow the occipital artery to the skin of the scalp as far as the vertex of the head. This sensory branch is known as the **greater occipital nerve.**

Table VII-1 gives the general distribution of the posterior primary rami in the neck for all cervical nerves.

There is a small **third occipital nerve,** the medial branch of the posterior primary ramus of C_3, which pierces the Trapezius to reach the skin of the back of the head.

The two arteries in the triangle are the **vertebral,** about which we already know, and the **occipital artery.** The latter is a branch of the external carotid, and so it must be hidden over much of its course if it first comes to light here. Its *anterior part* arises in front of the Sternomastoid in the carotid triangle of the neck. Its second, or *deep part,* crosses under the

Table VII-1. The Posterior Primary Divisions of the Cervical Nerves

Medial Branches	Lateral Branches
C_2. **Great Occipital Nerve:** Sensory to posterior parts of neck and scalp.	C_2. **Motor:** Obliqus capitis inferior; Splenius; Longissimus capitis
C_3. **Third Occipital Nerve:** Sensory to small part of posterior scalp and neck.	C_3. **Motor:** Semispinalis capitis
C_4–C_5. Sensory to skin, lower posterior parts of neck. C_6–C_8. **Motor only:** Multifides and neighboring muscles.	C_4–C_8. **Motor:** Longissimus capitis and cervicis; Semispinalis capitis and cervicis; Iliocostalis cervicis.

C_1. No branches. **Motor** to suboccipital muscles.

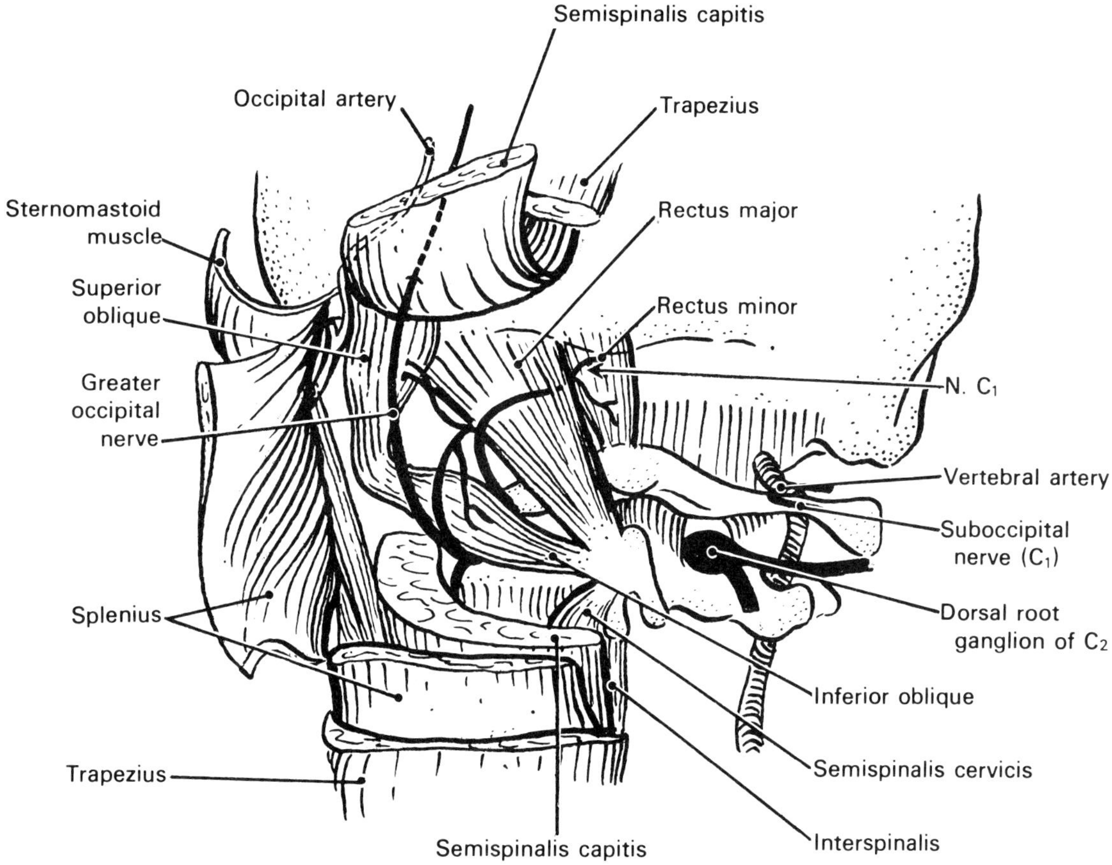

Fig. VII-25. The suboccipital triangle and contents.

Sternomastoid to reach the occipital groove on the mastoid process of the temporal bone. Here it has to be covered not only by the Sternomastoid but by the Splenius Capitis, Longissimus Capitis, and the posterior belly of the Digastric. Its *posterior part* then comes to lie upon the Obliqus Capitis Superior and the Semispinalis; it pierces the Trapezius and divides into its terminal branches to supply the skin of the scalp. One twig sometimes enters the mastoid foramen to supply the air cells of the mastoid process. Other twigs supply the Sternomastoid muscle and, sometimes, the medial aspect of the ear.

Superficial Structures of the Neck

When we think of superficial structures of the neck we think first of the two big muscles, the Sternomastoid and Trapezius, which divide the neck into its main, large, anterior and posterior triangles; the nerve to these muscles (spinal accessory); the infrahyoid muscles, especially those that divide the larger triangles into smaller triangles; the Platysma, which forms the roof of the anterior triangle; some of the superficial nerves that come to view in the posterior triangle; and the external jugular vein. Most of these were discussed in Chapter III, and we will only add some details here.

The Sternomastoid can serve as a key muscle since, in its oblique course in the neck, it presents its anterior border to the anterior triangle and its posterior border to the posterior triangle. The rest of the posterior triangle is demarcated by the anterior margin of the Trapezius and by the clavicle as a base. The mandible and the midline of the neck ventrally delimit the rest of the anterior triangle.

The **Sternomastoid** has two heads of origin—the manubrium of the sternum and the medial one-third of the clavicle. Its insertion overrides the mastoid process to reach around to the lateral aspects of the superior nuchal line.

The **Trapezius** occupies the rest of the superior nuchal line not usurped by the Sternomastoid, and it attaches as far medialward as the external occipital protuberance. It has a *long* origin in the midline of the back from the ligamentum nuchae and from *all* of the thoracic spines, as well. It inserts on the lateral one-third of the clavicle and upon the acromion and spine of the scapula. So it is cape-like and triangular. It is difficult to see why it resembles a trapezoid.

Both the Sternomastoid and Trapezius developed from the sixth branchial arch mesoderm in the embryo. As a consequence, they are innervated by the cranial nerve proper for this arch, the **spinal accessory (XI).**

We mentioned the role of the Sternomastoid as a neck flexor and rotator (see above). Electromyographic recording indicates that it also becomes active in neck extension when the individual is recumbent, probably by increasing the holding tension as the normal extensors pull the head backward.[14] The Sternomastoid can also be an important respiratory muscle when regular respiratory muscles malfunction. The Trapezius is hardly a neck muscle in the true sense but, rather a rotator and elevator of the shoulder which simply has some of its attachments in the neck region. It is, probably, the only muscle in the neck that can be relaxed.[15]

The **spinal accessory nerve** (Fig. VII-26) has its neurons located in two places—in the spinal cord and in the medulla. In both of these places the neurons lie in the ventrolateral gray matter, a position which all branchiomotor neurons seek in the neural tube. At cervical cord levels C_1 to C_4, and sometimes as far as C_6, the cells lie in the ventral horn in its lateral upper portion. Their axons do not emerge in the ventral rootlets but exit directly laterad between the dorsal and ventral roots. They use the posterior aspect of the pial denticulate ligament as a guide path to the foramen magnum (Fig. VII-26). Once inside the skull, these fibers appear to be more at home, for they at once join with the emerging fibers of their brethren whose cell bodies lie in the **nucleus ambiguous** of the medulla where branchial motor neurons of IX, X, and XI have a common home. The joined fibers of XI (spinal and medullary) then leave the skull once more through the jugular foramen in company with those of cranial nerve X (Fig. VII-26).

In the neck, the spinal accessory nerve parts company from the vagus to take up a more posteriorly-directed course. It crosses posterior to the transverse process of the atlas, runs deep to the styloid process and the posterior belly of the Digastric and then appears under the Sternomastoid which it innervates. We do not see it until it crosses the floor of the posterior triangle obliquely (Fig. VII-26B), appearing from under the midpoint of the posterior border of the Sternomastoid. It has no business to transact in the triangle, so it dives under the anterior border of the Trapezius which it also innervates on its undersurface. Along its course it may be joined by sensory twigs from C_2 to C_4. These are probably proprioceptive fibers from muscle spindles of both of these muscles, at least in most of us. Some of us, however, still have anthropoid hangovers and still retain function in the lower half of the Trapezius when the spinal accessory nerve is severed, indicating that these cervical nerves (C_3 and C_4, in particular) can supply a motor innervation to part of this muscle. If they do, then this lower half of the muscle is of cervical somite origin in some individuals and only the upper part is of branchiomeric origin in such instances.

Figure VII-26A shows the rest of the fibers of the spinal accessory being distributed along with the vagus. This part is thought to contain the fibers that arose from the medulla rather than those of spinal cord origin. If this is so, then the name, "spinal accessory," is not quite accurate. The part with the vagus ought to be called a bulbar or vagal accessory nerve. Your authors defy you to satisfy your curiosity as to where these particular fibers of XI are going. No two standard textbooks seem to tell quite the same story, so that the very existence of these fibers may be more Galenic than factual, in the sense that writers have copied and transcribed the words of some other previous author, who copied it from someone else. Original sources may have come from comparative rather than from human anatomy.

The muscles forming the *floor of the posterior triangle* are shown in Figure VII-24 as are the important nerves that lie under *their prevertebral* fascia. These are the nerves to the Levator Scapulae, the Rhomboids, and the Scalenus Anterior; the phrenic nerve; and the beginnings of the brachial plexus. The **third part of**

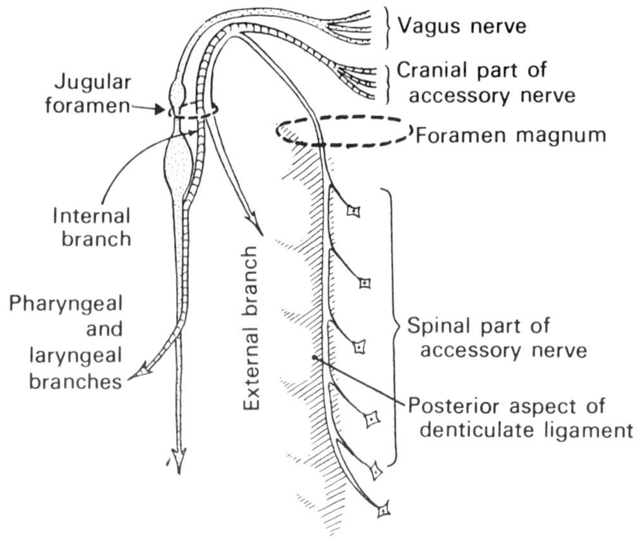

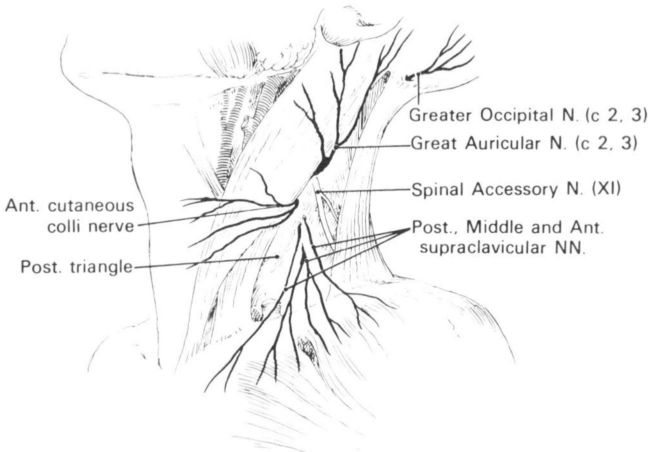

Fig. VII-26. Schema of course of spinal accessory nerve.

the subclavian artery lies here in the floor, also, but it is in the lesser vascular rather than the neural part of the posterior triangle. The structures that *cross the roof; i.e.,* cross the deep investing fascia, are: **the accessory nerve** overlying the Levator Scapulae (shown hooked in Fig. VII-24); and the **superficial cutaneous branches of the cervical plexus.** These are seen in Figure VII-26B, emerging at the posterior midpoint along the Sternomastoid in company with the accessory nerve. They spread as sensory branches upward, along and over the surface of the Sternomastoid—the **supraclavicular branches** even crossing the clavicle below. The other named sensory branches of this plexus are the **lesser or small occipital, greater auricular,** and the **transverse cervical** (anterior cutaneous colli) nerves. We shall discuss the complete makeup of the cervical plexus below. The rest of the nerves of the plexus are motor and run deep to reach muscles in the subdivisions of the anterior triangles. Otherwise, beside the *one vessel* and its transverse cervical branch and the *nerves* mentioned, the posterior triangle in its entirety contains only lymph glands. The part of it above the inferior belly of the Omohyoid is predominantly neural in content; below this muscle (subclavian triangle) it is predominantly vascular in content.

The course of the superficial veins in the neck are tributaries of the external jugular for the most part, and they are shown and named in Figure VII-27.

The boundaries of the lesser triangles within the greater anterior triangle of the neck are reviewed for you in Figure VII-22. Their contents were discussed in Chapter III. The **Platysma** is shown in this figure as covering parts of both the anterior and posterior triangles of the neck. It is a very variable muscle, indeed, of second branchial arch origin. It may form an entire roof over the large anterior triangle, reaching to the anterior midline and lying between the skin and deep investing fascia of the neck. Since it is branchiomeric it is supplied by the **cervical branch of the facial nerve,** the nerve for the second branchial arch muscle. This is not a cervical (spinal) nerve, but a *branch of cranial nerve VII in the cervical region.* So the Platysma is another skeletal muscle in the neck that gets a cranial nerve supply.

The **infrahyoid muscles** can be considered as superficial structures within the anterior triangle of the neck. The suprahyoid group of muscles lie in an area where head and neck blend. Suprahyoid muscles will be considered later in a discussion of the submandibular region. The infrahyoid group of the "strap" muscles are ribbonlike bands, four in number, on each side of the midline. These are the **Sternohyoid** and **Omohyoid** which lie superficial to the **Sternothyroid** and the **Thyrohyoid.** If you know that "omos" means shoulder and that the "thyroid" in these names refers to the thyroid cartilage of the larynx, you have an idea of the points between which these four muscles run (Fig. VII-28), their only other points of anchorage, obviously,

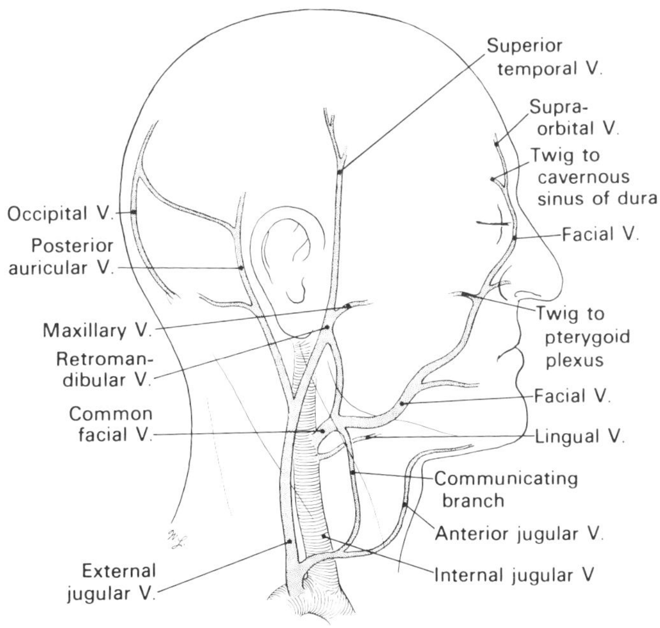

Fig. VII-27. Superficial veins of the face.

being the sternum and hyoid bone. It is these muscles which hide the contents of the **visceral triangle** of the neck for the most part. We will concern ourselves with these visceral contents later under the heading of **deep structures in the neck.**

The **Sternohyoid** takes origin from the manubrium of the sternum and even from a bit of the clavicle to insert upon the lower border of the hyoid bone. The **Omohyoid** is slightly more complicated since it has two bellies that meet in a common tendon. The **inferior belly** runs from a point near the scapular notch on the upper border of the shoulder bone to the tendon. It must pass under the Sternomastoid to do so. The **superior belly** continues from the tendon to the lower border of the hyoid. A sling of deep cervical fascia anchors the tendon to the clavicle and the first rib.

Like the Sternohyoid, the **Sternothyroid** arises from the manubrium, but deep to the former muscle. It fails to reach the hyoid bone but inserts, instead, upon the **oblique line** of the thyroid cartilage. The small, deep muscle which continues from the oblique line to the hyoid bone is the Thyrohyoid. It may be regarded as an interrupted part of the Sternothyroid. Despite reports to the contrary, *none* of these muscles is supplied by *any* cranial nerve. They all receive their innervation from the deep part of the cervical plexus. All act to depress the floor of the mouth, the hyoid bone, and the larynx, or to resist the elevation of these structures. Along with the suprahyoid group of muscles, the infrahyoids steady the hyoid bone which rides at anchor, fore and aft, like a ship at sea. The bone drifts upward and downward as we chew, swallow, or speak, these two opposing hyoid muscle groups (Supra- and Infra-) acting as guy ropes to steady it.

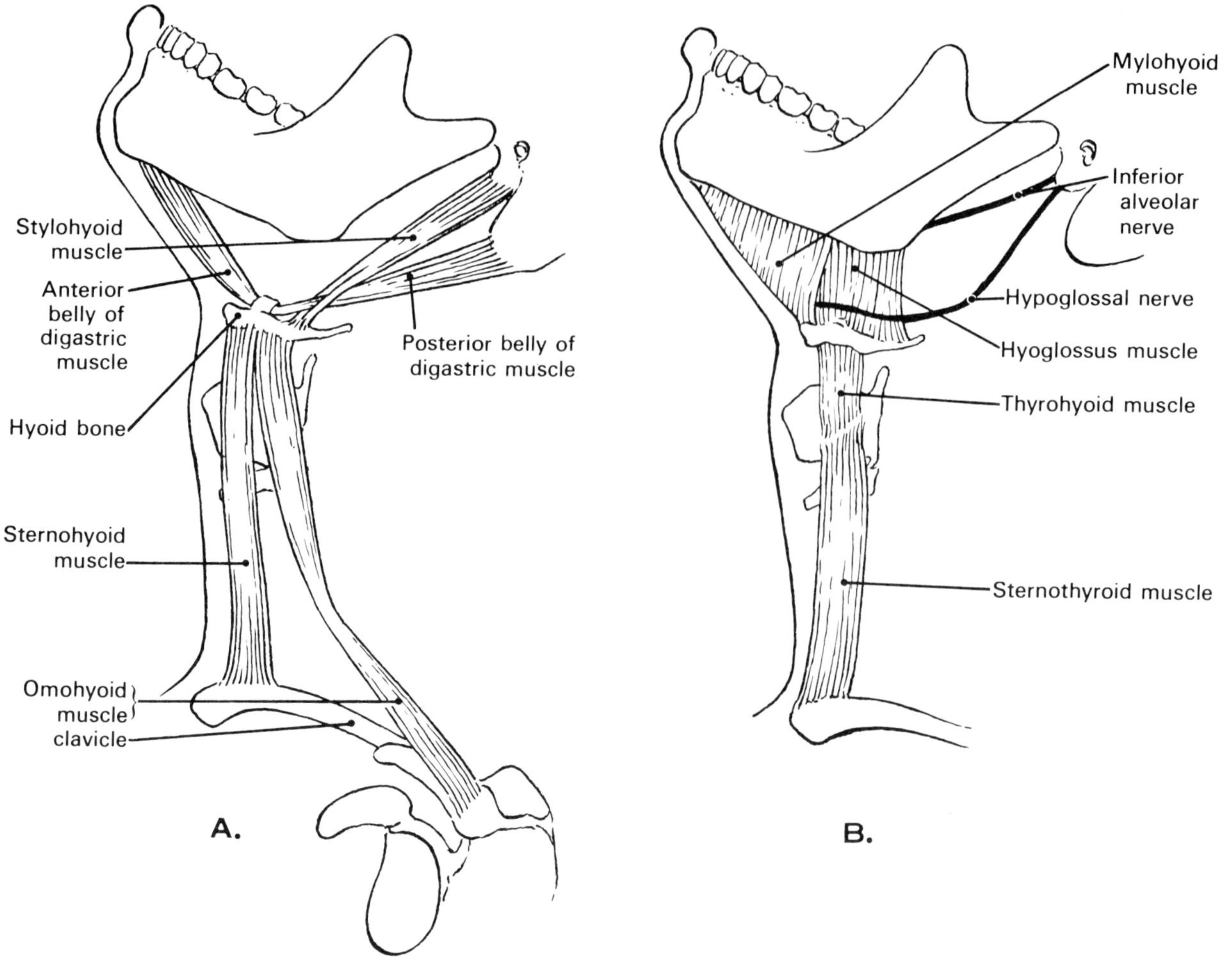

Fig. VII-28. Supra- and infrahyoid muscles. **A.** Superficial group. **B.** Deep group. (Geniohyoid not shown.)

The **Cervical Plexus** is formed by the **anterior primary divisions or rami** of the **first four cervical nerves.** Recall that the last four, along with T_1, will enter into the formation of the brachial plexus. Figure VII-29 illustrates the nature of the cervical plexus. Its **superficial sensory branches** (greater auricular, small occipital, cervical cutaneous, and supraclavicular nerves) we noted above in the posterior triangle. Figure VII-32 shows other sensory fibers returning from muscle spindles of the Trapezius and Sternomastoid, and these can be followed back to C_2 to C_4. The contribution of C_5 to the brachial plexus is shown. Some communicating twigs to or from other nerves should be noted. The plexus receives twigs from the gray rami of the sympathetic trunk as visceral motor input to skin viscera. C_1 and sometimes C_2 send motor twigs along with the hypoglossal nerve (XII). These cervical fibers do not innervate the tongue, however. That organ receives its motor supply from the hypoglossal nerve, alone. The hypoglossal innervates *all* the extrinsic and intrinsic tongue muscles and *no others.* The fibers of C_1 and C_2, traveling with it, break away from the hypoglossal nerve to innervate some of the suprahyoid and infrahyoid muscles.

Indeed, the innervation of infrahyoid and suprahyoid muscles is one of the chief motor functions of the *deep part* of the *cervical plexus.* The twigs of C_1 to C_2 descending with XII, caught up in the perineurium of this cranial nerve as it were, form the superior branch of a loop system (**ansa cervicalis,** "ansa" meaning loop). Anastomosing twigs of C_2 to C_4 form the inferior limb of this loop, and off-shoots from it distribute to infrahyoid muscles (see Fig. VII-29). The superior limb has been called the "descendens hypoglossi," and the inferior limb, the "descendens cervicalis." We mention this to acquaint you with other terms commonly employed in discussing the deep part of the cervical plexus and to emphasize once more that the descendens hypoglossi contains *no hypoglossal nerve fibers.*

In addition, fibers from C_3 to C_5, both sensory and motor, unite to form the **phrenic nerve,** the princi-

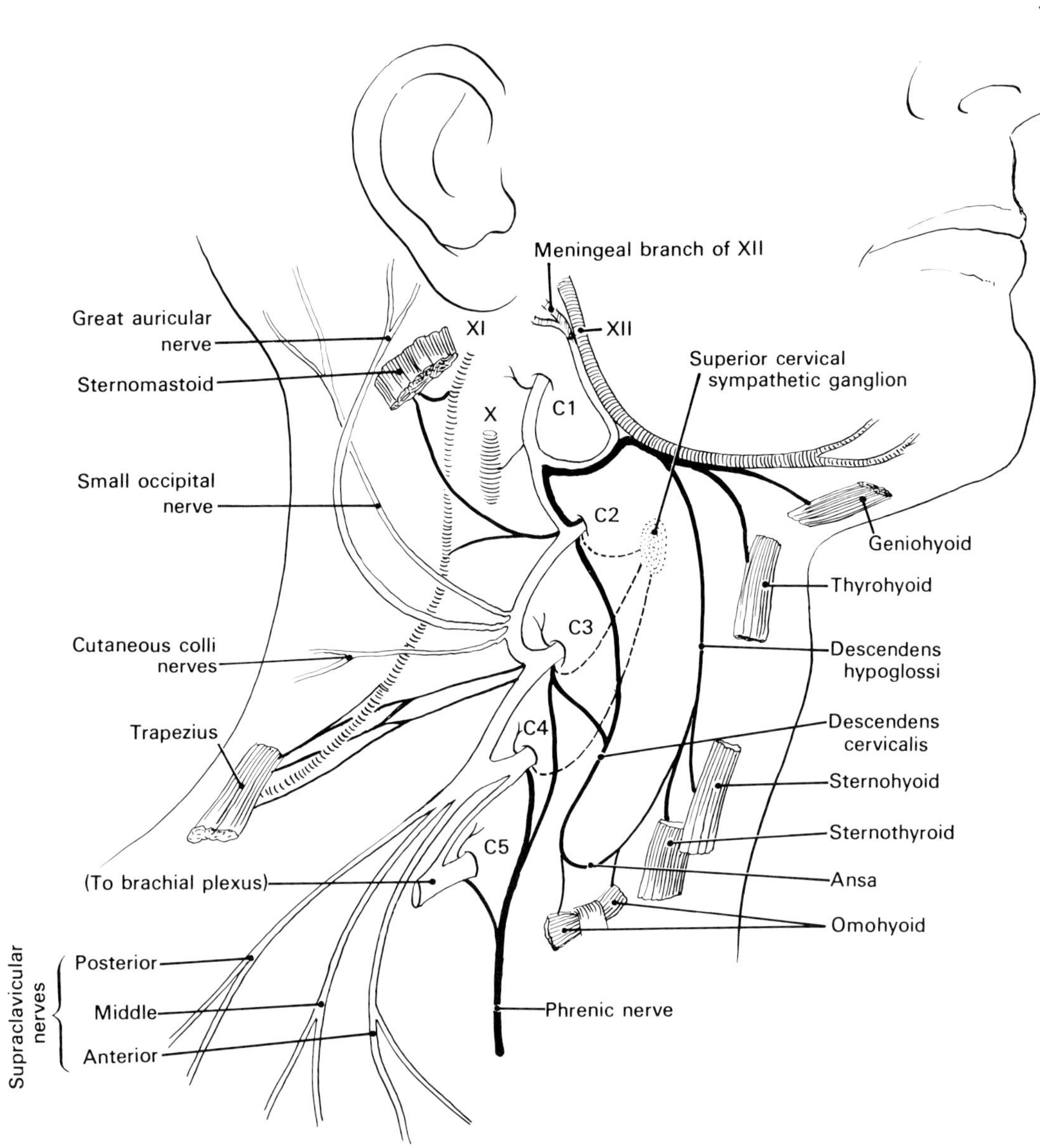

Fig. VII-29. The cervical plexus. Cranial nerves are shown striped. Motor nerves of plexus are black; sensory nerves of plexus are white. Note, however, that the fine black fibers from plexus to Sternomastoid and Trapezius are more likely to be proprioceptive sensory fibers than they are to be a motor supply in the majority of humans. Gray rami from superior cervical sympathetic ganglion to first four cervical nerves are shown as dashed lines. (After Chusid and McDonald.)

pal motor nerve to the Diaphragm. Its sensory branches return pain sensations from the diaphragmatic surface, from costal and mediastinal pleurae, and from the pericardial sac. Clinically, injury to the phrenic nerve is the most important lesion of the cervical plexus. If it is involved unilaterally, the diaphragmatic paralysis causes few symptoms. The fluoroscope may indicate relative immobility that is one-sided, and indicate that the liver or spleen seems to lie at a higher level than normal. It is bilateral paralysis which leads to severe dyspnea. The accessory muscles of respiration are brought into play so that heaving of the chest and shoulders accompany respiration. The abdomen does not protrude in expiration. Obviously, sneezing and coughing become difficult. Other irritations of the phrenic nerve can lead to hiccups (singultus) or to neuralgic pain, the latter induced by tumors in the neck, by aneurysms of the aorta, and by pericardial or mediastinal infections. The pain can be felt as high as the chin or may extend down the arm; it can be felt at the far border of the ribs or under the clavicle, or deep in the neck and mostly on the left side. Respiration is short and rapid as if the patient acts to protect himself from pain were he to breathe normally. Rigidity of the neck, as a protective reflex, is common in neuralgias involving cervical nerves, as well as in meningitis.

In the neck, the phrenic nerve descends vertically

along the Scalenus Anterior muscle under the Sternomastoid and deep to the internal jugular vein. It passes between the subclavian artery and vein with the vein in front. The left nerve will cross the aortic arch anterior and lateral to the vagus nerve. Its course in the thorax will be discussed later.

The Submandibular Region

This region lies largely under cover of the mandibular body and between it and the hyoid bone. Its important contents are glandular and muscular; hence, the area involved accounts for the **muscular** (submental) and **glandular (digastric) triangles** of the neck, subdivisions of the greater anterior triangle. It will include the **submandibular** (submaxillary) and **sublingual glands;** their terminal parasympathetic ganglion associated with them; two nerves, lingual and hypoglossal; two arteries, lingual and facial; and the suprahyoid muscles.

The 10- to 20-gm. **submandibular gland** is, mainly, a serous-type salivary gland. Much as the lacrimal gland is cleft by the Levator Palpebrae, so the submandibular gland is bent U-shaped about the posterior border of the **Mylohyoid** (Fig. VII-30). Hence, it has a large superficial part and small deep part. Its *inferior surface* would be covered by skin, Platysma within the dermis, and fascia (Fig. VII-31B). Some lymph nodes and the **facial vein** also relate to this inferior surface. Laterally, it snuggles against the submandibular fossa of the mandible, and it comes in contact with the Medial Pterygoid. Its *medial surface* relates to the Hyoglossus, Mylohyoid, and the anterior belly of the Digastric (Fig. VII-30). The *deep part of the gland* slips between the Mylohyoid and the Hyoglossus. The lingual nerve lies above it and the hypoglossal nerve below it (Fig. VII-31A).

The **submandibular duct (of Wharton)** is short, about some 5 cm. in length, and it emerges from the anterior tip of the deep part of the gland to continue forward between the Hyoglossus and Mylohyoid. It opens by one to three orifices on the sublingual papillae along the frenulum of the tongue. As an old but censored jingle states, the **lingual nerve** takes a swerve across the Hyoglossus and crosses the duct on its lateral aspect (Fig. VII-31A). This branch of V carries with it secretory motor fibers of VII, the parasympathetic preganglionic supply to both submandibular and sublingual glands. Recall that this supply reached the lingual nerve in the infratemporal fossa *via* the chorda tympani branch of VII. These preganglionics distribute to the **submandibular ganglion** lying medial to the gland (Fig. VII-31A). This is one of the **terminal parasympathetic ganglia** within the head. Postganglionics from this ganglion run to the submandibular and sublingual glands (Fig. VII-32A). A sympathetic supply comes as postganglionic fibers from the superior cervical sympathetic ganglion by following the external carotid and the branches of the facial artery. The sympathetic supply, of course, does not synapse in the submandibular ganglion.

The disposition of the facial and lingual arteries, both branches of the external carotid, is shown in Figure VII-32B.

Of the three paired salivary glands in the head, the parotid, submandibular, and sublingual—the latter is the smallest. While the parotid is entirely serous secreting, the submandibular is mostly so despite some mucous acini. The **sublingual gland** is mostly mucous-secreting, though it does contain some few serous demilunes. When the tip of the tongue is touched against the hard palate in the open mouth, the **sublingual folds** and their papillae can be seen posteriorly in the floor of the mouth on each side of the frenulum. Under these folds lie the sublingual glands. Some of their 10 to 30 fine ducts open through the papillae directly into the oral cavity. Others of them open into the submandibular duct. The lateral surfaces of these glands rest in the sublingual fossae of the mandible. Anteromedially, they come into relationship with the Genioglossus and even with part of the Hyoglossus posteromedially. They are separated from muscle by the submandibular duct and lingual nerve (Fig. VII-31A and B).

In the discussion of the submandibular region, we have casually mentioned most of the suprahyoid muscles by name except, possibly, the **Stylohyoid.** The **Digastric,** like the Omohyoid, is two-bellied. Its *posterior belly* arises in the mastoid notch (digastric notch) of the temporal bone and it runs to a common tendon. The *anterior belly* attaches on the inner lower border of the mandible near the symphysis at the digastric fossa. It is directed posteriorly and inferiorly to the common tendon which has a fibrous attachment to the greater horn and body of the hyoid. This tendon sometimes pierces the Stylohyoid. The facial nerve (VII) supplies both the posterior belly and the Stylohyoid. The anterior belly of the Digastric, obviously, was a derivative of the first branchial arch, not the second, for it is innervated by the mylohyoid nerve, a branch of V.

The **Stylohyoid,** from styloid process to the posterior base of the greater horn of the hyoid bone, is a slender muscle that runs along the upper border of the posterior belly of the Digastric. In pulling the hyoid bone posteriorly, it elongates the floor of the mouth. The posterior belly of the Digastric pulls the chin down and backward to assist in mouth-opening, while the anterior belly of this muscle supports the hyoid bone.

The two **Mylohyoids** form a kind of floor beneath the front of the mouth (diaphragma oris). They arise from the mylohyoid line on the inner aspect of the mandible, running roughly from last molar tooth to symphysis. The muscles from each side insert into a midline raphe, though the most posterior fibers insert on the body of the mandible (Fig. VII-30). The

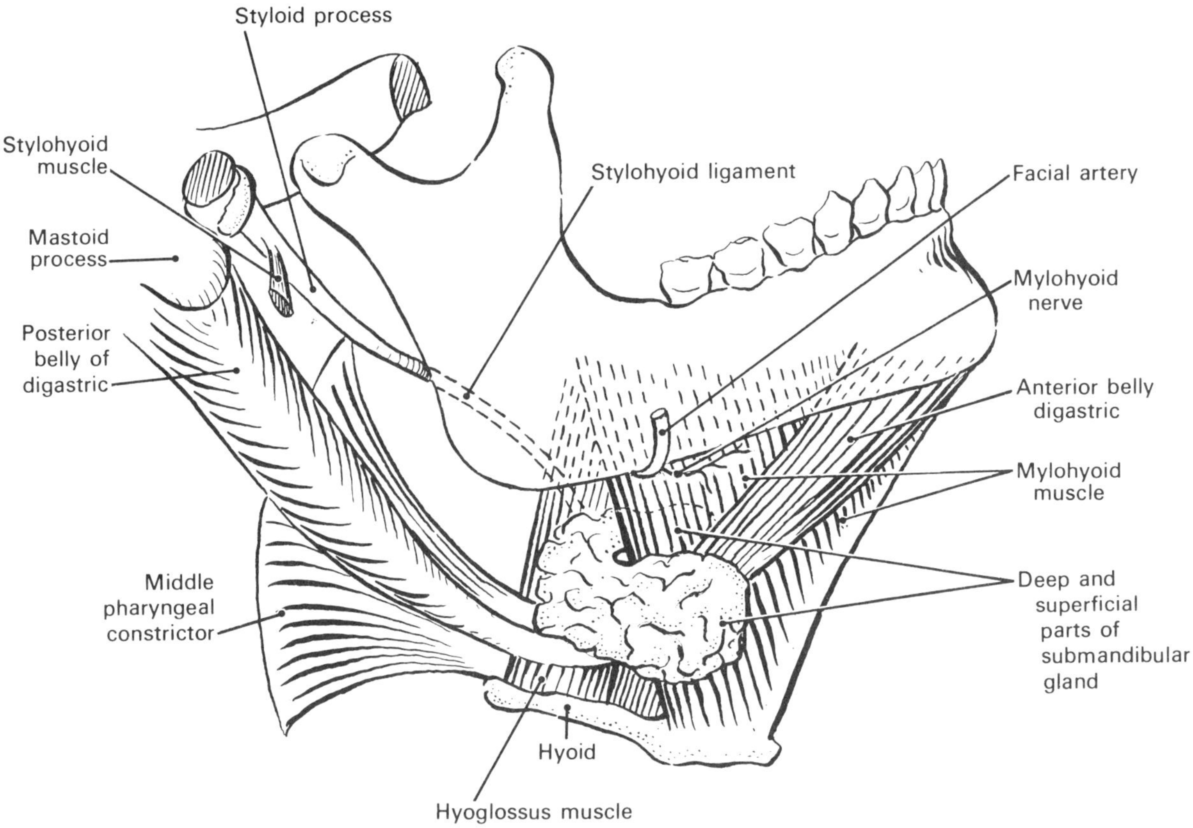

Fig. VII-30. Submandibular region.

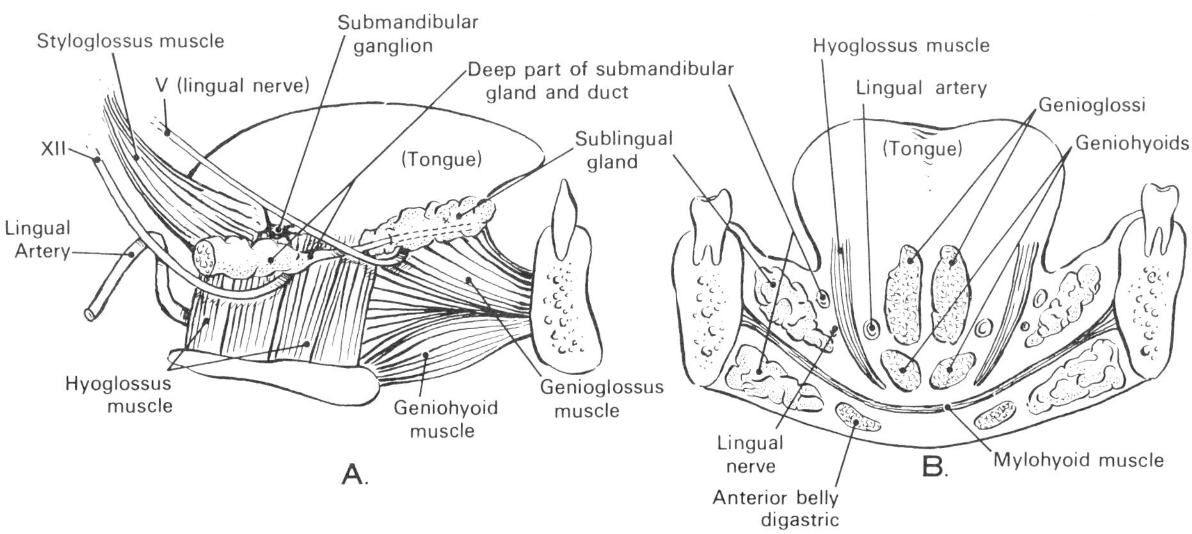

Fig. VII-31. The tongue. **A.** Extrinsic muscles and nerves. **B.** Tongue and floor of mouth in coronal section.

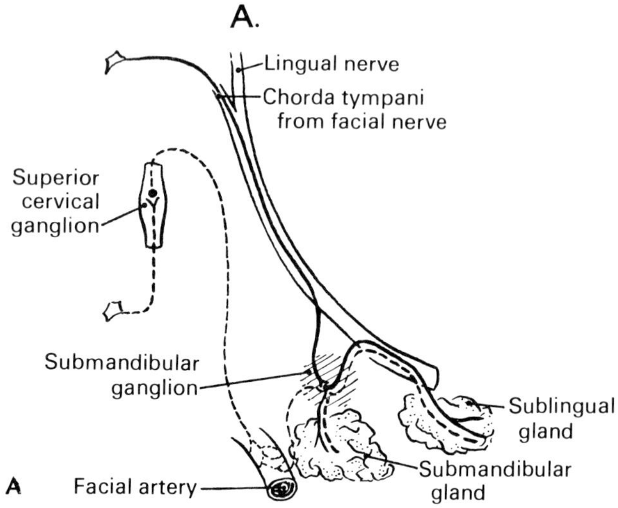

Fig. VII-32. Lingual nerve and submandibular ganglion (A), and the lingual artery (B). In A, dotted lines indicate sympathetic fibers; black line, parasympathetic fibers.

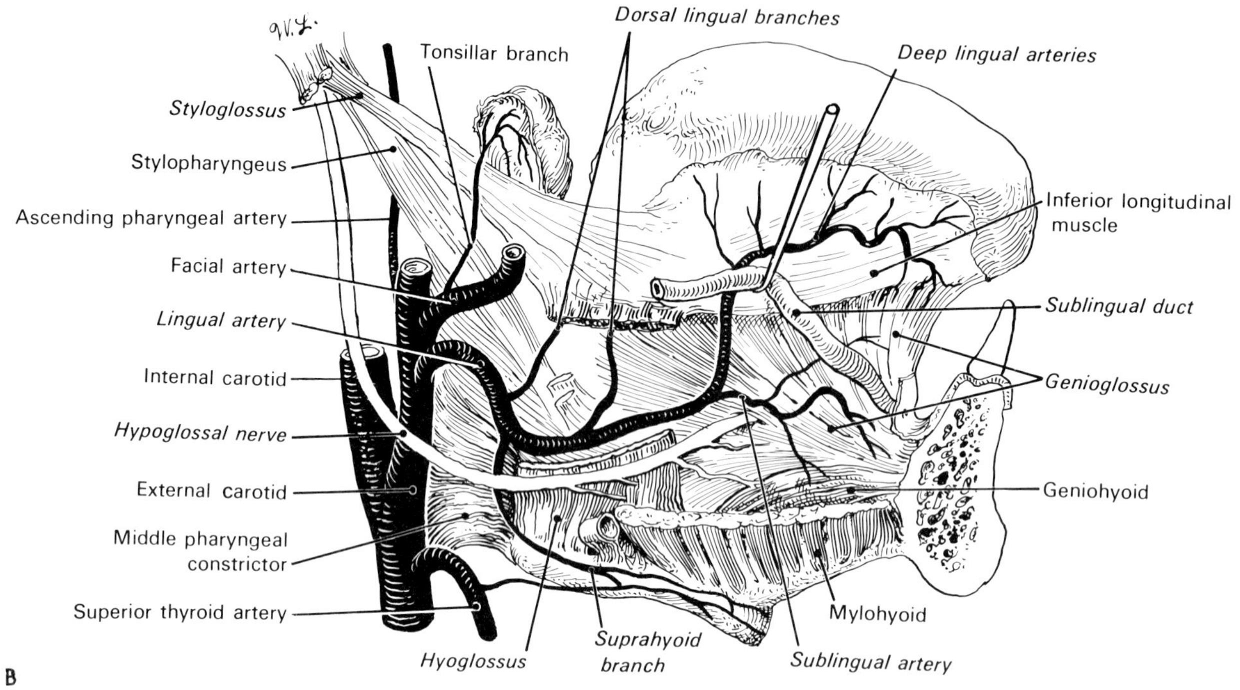

Mylohyoids hide a goodly stretch of the Hyoglossi, which are extrinsic tongue muscles even though they appear in the suprahyoid region.

The last of the suprahyoid group, proper, are the **Geniohyoids** (Fig. VII-31A). These lie above and under cover of the Mylohyoids, arising from the genial tubercles on the inner aspect of the mandible near the symphysis to insert anteriorly on the body of the hyoid bone. The Mylohyoid, like the anterior belly of the Digastric, is supplied by the same nerve, a branch of V. This nerve, the mylohyoid nerve, is given off as a motor twig just before the inferior alveolar nerve enters the mandibular canal. The Geniohyoid, on the other hand, is supplied by C_1 (and sometimes C_2) from the deep part of the cervical plexus—indeed, from the cervical twigs in the descendens hypoglossi (*but not by the hypoglossal nerve!*).

The Mylohyoids, in elevating the floor of the mouth upon contraction, force the tongue backward and so aid in accomplishing swallowing movements, especially when the teeth are occluded. Indeed, all four suprahyoid muscles elevate the hyoid bone and aid in swallowing. The Stylohyoid and posterior belly of the Digastric draw the bone posteriorly; the anterior belly and the Geniohyoid draw it ventrally and, in so doing, shorten the floor of the mouth. The Digastric, Geniohyoid, and Mylohyoid can depress the mandible when the hyoid bone is fixed by the infrahyoid muscles. The Digastrics, acting unilaterally, can rotate the jaw to the same side; acting together, they can retract the jaw. The posterior bellies may even be able to bend the head backward slightly.

We have discussed the **tongue** in relation to the oral cavity in Chapter III, page 106. Review this material there at this time. The mucous sac of the tongue is stuffed with intrinsic skeletal muscles, while extrinsic muscles arise nearby to peg into the intrinsic musculature. All muscles are bilaterally represented, those on one side being separated from those on the other by a peculiar septum fashioned by transverse, crosslinking muscle fascicles. The **intrinsic muscles** run in all planes and are simply named longitudinal, transverse, or vertical according to fiber direction.

The five pairs of **extrinsic muscles** all have the Greek term for tongue, "glossa," as part of their name.

The fan-shaped **Genioglossus** arises from the superior genial tubercle behind the mandibular symphysis. It inserts partly on the hyoid bone and into the inferior aspect of the tongue (Figs. VII-31A and 32).

We have seen that the **Hyoglossus** is partly hidden by the Mylohyoid (Fig. VII-30), but when the latter muscle is stripped away (Fig. VII-31A), the Hyoglossus is revealed as a quadrilateral muscular plate running from superior horn and body of the hyoid bone into the underside of the tongue. It is crossed laterally by the lingual nerve and the hypoglossal nerve, while on its deep side pass the glossopharyngeal nerve, the lingual artery, and the stylohyoid ligament.

The **Styloglossus** arises from the styloid process and from the **stylomandibular ligament.** The latter is a modification of deep cervical fascia (parotid and submandibular fascia) running between the styloid process and the jaw angle; *i.e.*, the posterior border of the mandibular ramus. The muscle pegs into the tongue posterolaterally and postero-inferiorly (Fig. VII-31A).

The **Palatoglossal** occupies the palatoglossal fold, arising from palatine aponeuroses to insert posterolaterally into the tongue.

The **Chondroglossus** (a poor name) is an inconstant muscular slip that runs from the lesser horn of the hyoid bone to the dorsum of the tongue. (Obviously, the greater horn and body of the hyoid bone are every bit as much derivatives of branchial arch cartilage as the lesser horn, so that the Hyoglossus could just as well have been named "Chondroglossus.")

The tongue is shaped by its intrinsic musculature and is moved by the extrinsic muscles. The lower posterior parts of the two Genioglossi, acting together, protrude the tongue; *i.e.*, stick it out. But, mainly the Genioglossi are tongue depressors, and their anterior attachment prevents the tongue from falling backward in the oral cavity and obstructing respiration. Along with the Styloglossus and Hyoglossus, parts of the Genioglossi can also assist in retracting the tongue. All of the extrinsic, as well as the intrinsic muscles, except the Palatoglossus, are innervated by cranial nerve XII. The Palatoglossus draws the sides of the tongue upward and backward, and it receives its innervation from the pharyngeal plexus of the vagus (X). It also will draw the sides of the soft palate downward and so constrict the faucial isthmus. This is one of several muscles in the body where origin and insertion can reverse depending upon the function attributed to it. Generally, one considers it as much a part of the pharyngeal musculature as of tongue musculature, proper.

The two Genioglossi tend to roll the tongue forward when you stick it out, the right-hand muscle pushing toward the left; the left-hand one, toward the right. If, in a patient who has been asked to protrude his tongue, it deviates to one side of the midline, the hypoglossal nerve *on the side toward which the tongue points,* can be suspected of being injured *only if the musculature on that side of the dorsal septum shows signs of atrophy.* If atrophied, the mucosa of the tongue will be thrown up in rugged folds on the affected side. But the tongue can also deviate to one side or the other upon protrusion if the entire mucosa appears perfectly smooth. In this instance, the deviation signals a lesion in the corticobulbar tract coming from the motor hemisphere *opposite to the side of tongue pointing,* or even a lesion in the opposite motor cortex, itself; *e.g.*, lower part of the precentral gyrus.

So far as innervation of the tongue is concerned, you should now recall that cranial nerve XII is its somatic motor supply; V is somatic sensory to the mucosa (pain, touch, temperature, chemo-irritation); IX may play a similar role in the posterior one-third of the tongue but it definitely supplies the taste buds of the circumvallate papillae; and VII supplies the taste buds of the fungiform papillae in the anterior two-thirds of the tongue. The vagus innervates taste buds of the epiglottis.

CERVICAL FASCIAE

Here in the neck, as elsewhere in the body, layers of fasciae sometimes are difficult to distinguish because they are, principally, condensations of fibro-elastic connective tissue of variable thicknesses laid down in laminar fashion and formed by the stresses of structures which move within their confines. What degree of condensation is necessary before a connective tissue is called a fascia often depends upon the anatomist who views it and upon his methods of studying it. Does it separate away easily upon dissection? Will fluids injected into the cadaver run between certain layers? Do infections in the living spread between definite sheets which tend to isolate spread into other regions or tend to direct spread toward given locales? Is the given sheet tough enough to hold a surgical suture in repair? If so, these fascial sheets are apt to be given a name, either that of an individual or some anatomical appellation appropriate to their location or to the structures they invest.

Indeed, these connective tissue laminae invest muscles as well as other structures, so that each structure is not only enclosed in a layer, but the layers are fastened to each other by loose areolar tissue. Thus, movement between sheaths are permitted along with

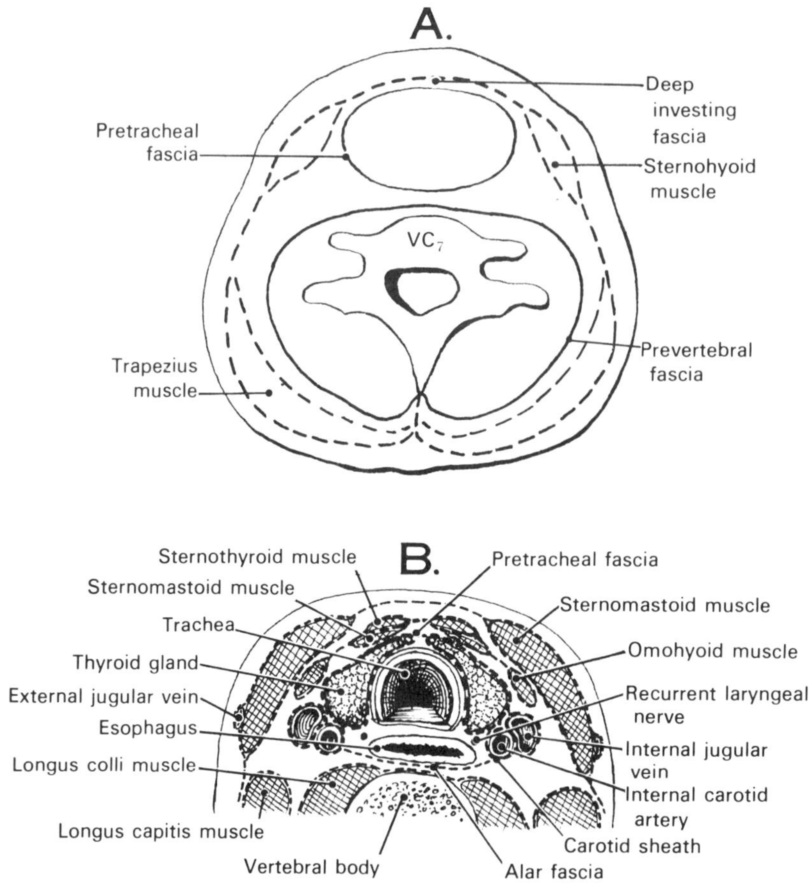

Fig. VII-33. Facial planes of the neck. **A.** Schematic plan of the three layers of the cervical fascia. **Fig. B.** Schematic section at level of CV_7 showing pretracheal fascia. Note particularly alar fascia interconnecting carotid sheaths.

movements of structures within the sheaths. In the neck it is customary to consider three fascial layers: deep investing, pretracheal, and prevertebral.

In its encirclement of the neck, the **deep investing fascia** (Fig. VII-33) forms the thin roofs over the anterior and posterior triangles, splitting to invest the Sternomastoid, Trapezius, and the inferior belly of the Omohyoid, including its intermediate tendon. At the sternum it splits to anchor to the front and back of the manubrium to enclose a **suprasternal space.** This space contains the sternal heads of the Sternomastoids, a few lymph nodes, and the jugular venous arch. Laterally, it forms sheaths for the parotid and submandibular glands. The fascia gains anchorage to various bony structures in the area: sternum, clavicle, and acromion; the hyoid bone, the mandible at its inferior border and along the mylohyoid line, the zygomatic arch, the styloid process *via* the stylomandibular ligament, mastoid process, superior nuchal line, ligamentum nuchae and the external occipital protuberance.

The **pretracheal fascia,** while somewhat limited, still covers considerable territory. It covers the sides and front of the trachea and invests infrahyoid muscles, as well (Fig. VII-33A and B). Behind the trachea it envelopes the esophagus. It divides to enclose the thyroid gland and attaches to the oblique line of the thyroid cartilage and the anterior arch of the cricoid. As a result, the thyroid gland rises and falls with the larynx during swallowing. Proceeding inferiorly, the pretracheal fascia surrounds the thyroid veins to terminate in a fusion with the posterior surface of the pericardium. Infections in the head and neck can spread in front of the trachea or behind the esophagus in this fascial sleeve to reach the superior mediastinum. Superiorly, but posterior to the esophagus, it becomes continuous with the buccopharyngeal fascia which clothes the Superior Constrictor muscle of the pharynx and the Buccinator.

The **prevertebral fascia** (Figs. VII-33A and 43) anchors to the base of the skull and the transverse cervical processes. In so doing it invests the prevertebral muscles (Longus Capitis and Longus Colli), the Scaleni, phrenic nerve, the back muscles that we mentioned as reaching up to anchor to the skull and cervical vertebrae, and all the muscles lying in the floor of the posterior triangle. Anterior to the subclavian artery it is prolonged as a sleeve, the **axillary sheath,** which contains the subclavian vessels as well as the emerging nerves of the brachial plexus.

In front of the vertebral bodies, between pretracheal and prevertebral fascia, stretching from the transverse processes of one side to those of the opposite side, is an isolated fascial lamina, called the **alar**

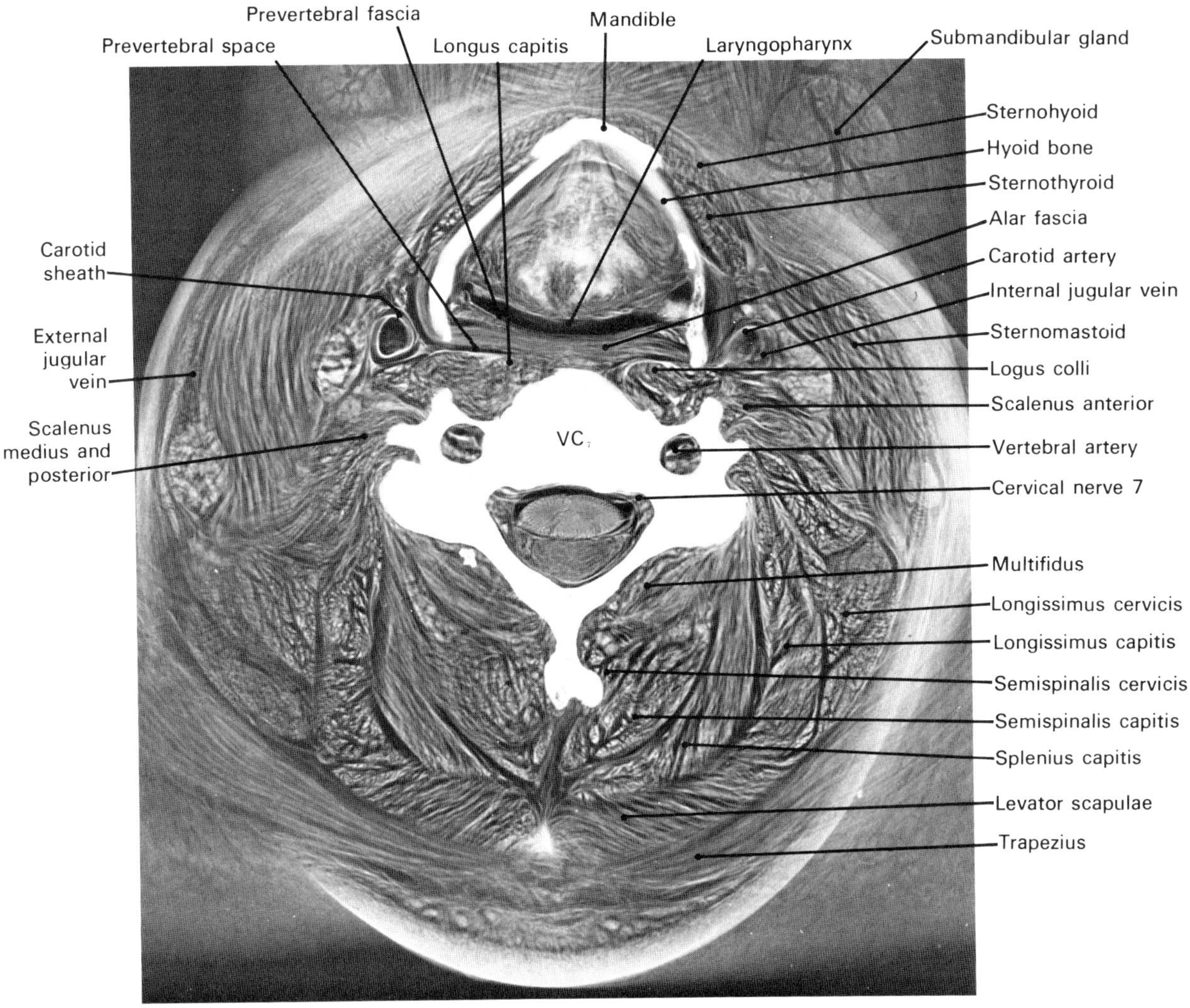

Fig. VII-33C. Radiograph of thick section through neck at level of VC_6 (Fig. VII-33C, courtesy of Dr. B. J. Anson).

fascia.[16] Thus, there is a space between alar fascia and prevertebral fascia filled by extremely loose connective tissue (Fig. VII-33B and C).

A condensation of fascia surrounds the common and internal carotid arteries, jugular vein, and vagus nerve. This condensation is derived from all three layers of cervical fascia and is commonly called the **carotid sheath,** as if it were a tubular structure. The term may be a misnomer since its original description was that of a "sheaf" such as is used to bind bundles of straw. Sheath or sheaf, this interwoven connective tissue around these structures extends from the base of the skull to the root of the neck where it is tightly enclosed between pretracheal and prevertebral fascia. Posteriorly, outside of the carotid sheath and under the prevertebral fascia, runs the ganglionated, cervical, sympathetic trunk.

It is important to recognize that these fascial structures connect with each other. A few may be traced as continuities into the anterior face, and, most important, is the connection of prevertebral and retropharyngeal fascia with a mediastinal partition to be described in the chapter on the thorax, page 440.

DEVELOPMENTAL PROCESSES IN THE OROPHARYNX AND LARYNGEAL REGIONS

Before we say much about the deep structures in the neck, it is well to consider the developmental history of some of them. Recall that as the embryo raised itself up off the yolk sac it pinched off an entodermal foregut and hindgut, leaving the midgut still open to the yolk sac (Chapter I, p. 21). The foregut (or pharyngeal gut) was in direct communication with the ectoderm at the stomodeum, below which lay the pericardium in the early embryo (A, in Fig. VII-34).

During the fourth and fifth weeks of development, recall that a series of **branchial grooves and arches** appears on the surface of the embryo (Fig. VII-34B). Previously, for didactic reasons only, we have talked about six branchial arches and their cartilaginous or muscular derivatives. Actually, the human embryo, unlike lower vertebrate forms, develops only four branchial grooves and five branchial arches, even the fifth arch being poorly defined. We postulated a possible sixth arch merely to account for the development of the Trapezius and Sternocleidomastoid, and if such an

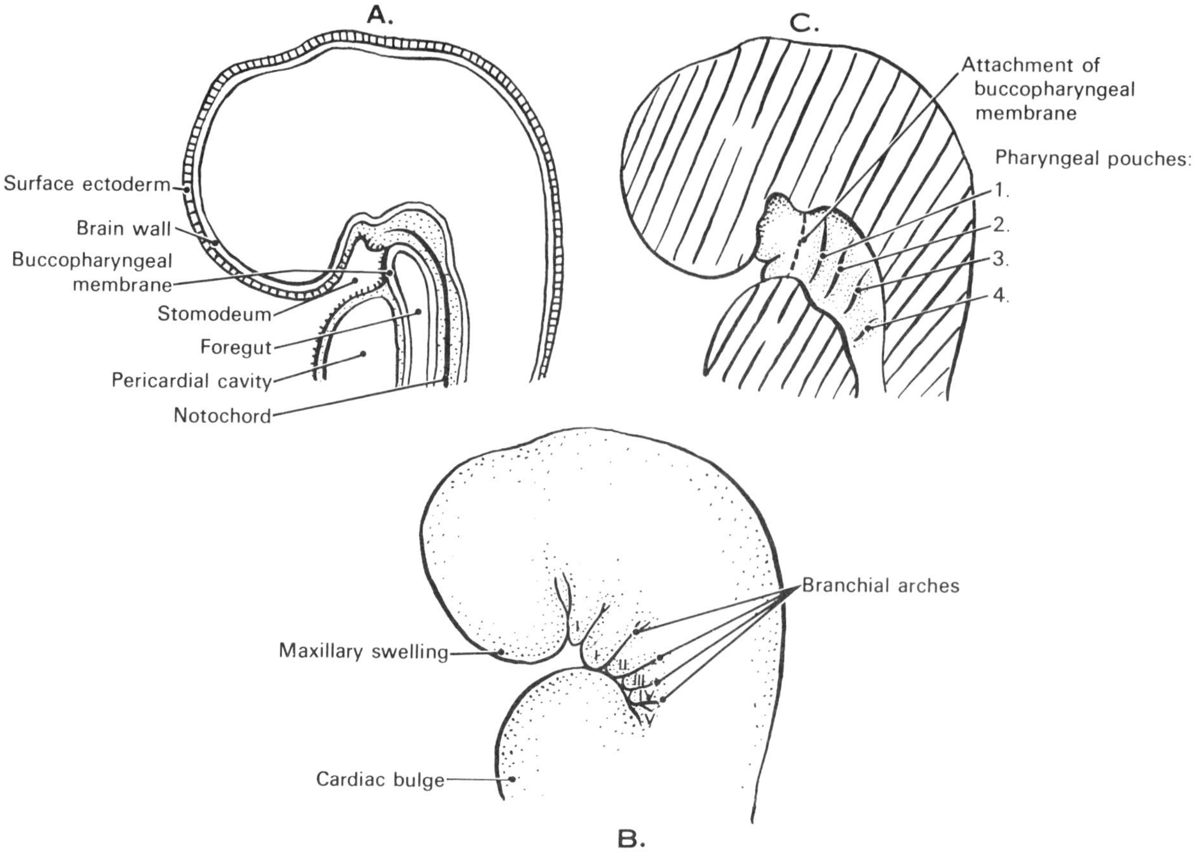

Fig. VII-34. Pharyngeal clefts (grooves) and pouches (schematic). **A.** Sagittal section through head of 25-day-old embryo. **B.** Branchial (pharyngeal) arches, external view, with grooves or clefts between in 5-week-old embryo. **C.** Locus of pharyngeal pouches internally in pharynx of 5-week-old embryo. (After Langman.)

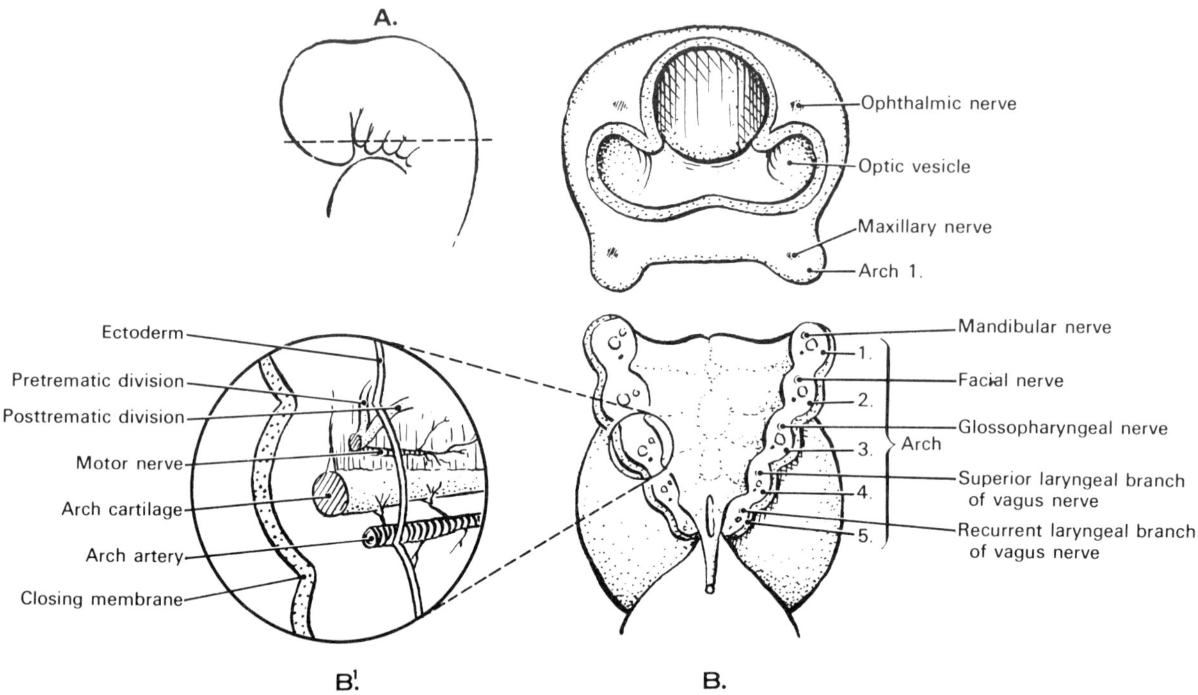

Fig. VII-35. Branchial grooves and pharyngeal pouches. **A.** Showing level of section B. **B.** Neural innervation of branchial arches. **B′.** Enlargement of arch 3 showing cartilage, muscle, vessels and its proper nerve. (After Williams-Wendell Smith.)

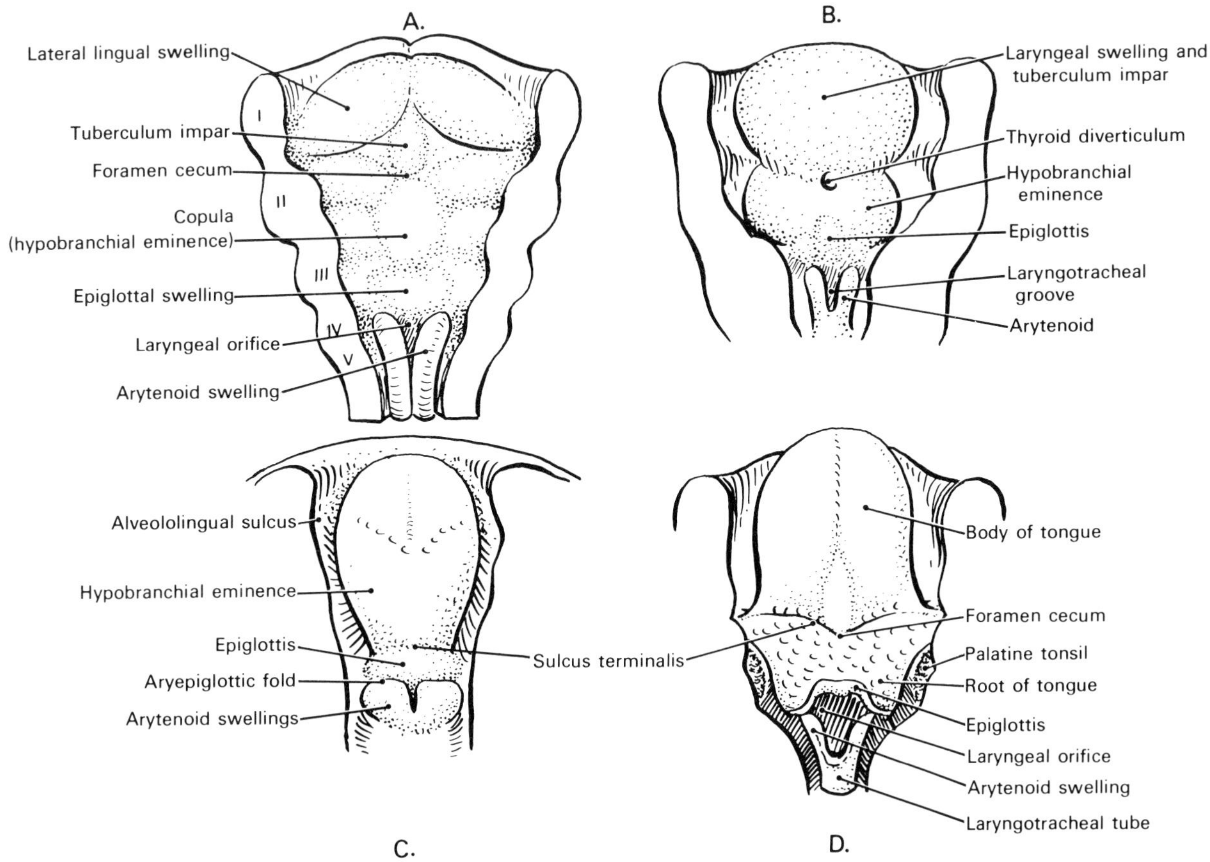

Fig. VII-36. Stages of tongue development from fifth week to fifth month in human fetus.

arch were present it must have disappeared in a twinkling so as to be virtually unnoticeable in the course of human development.

The distal ends of the paired arches grow ventralward from their lateral position so that those of the opposite sides tend to meet toward the midline (Fig. VII-35B). In doing so, they progressively dissect the pericardium away from the stomodeum or buccopharyngeal membrane. The first and second arches, as the most cranial ones, develop earliest and remain larger than the rest. The first arch splits off a maxillary and a mandibular process so that the stomodeum will break through as a mouth between them. The second or hyoid arch grows rapidly caudalward, as well, so that, in time, it covers the succeeding arches from view, much like the operculum of a fish covers the more caudal gill clefts (Fig. VII-38A). Indeed, the ectodermal grooves or clefts invaginate deeply to approach corresponding, inner, entodermal evaginations or pouches so that contiguous arches are almost separated by epithelial closing plates. The ectoderm and entoderm have not quite met in Figure VII-35B, but they have indented and approached each other so that separation of the arches is evident in this section through the pharyngeal region. The structures associated with any given arch, namely, cartilage, branchial musculature, a vessel, and a nerve, are shown schematically in Figure VII-35B′, while the cranial nerves proper to each arch are given in Figure VII-35B.

We now want to examine carefully the ventral floor of this pharyngeal gut on the inside where a tongue is developing (Fig. VII-36A–D). As the ends of the arches tend to approach the ventral midline, some may be blocked by elevations of subjacent mesoderm which raises the entoderm up from the floor of the mouth region. One small such median elevation, the **tuberculum impar** (Fig. VII-36A and B), appears between the mandibular processes of the first arch. **Lingual swellings** appear at the ends of the first arches, and these join with the tuberculum to form the anterior mass; *i.e.*, anterior portion, of the tongue.

The ventral ends of the second arch fuse only transiently in the midline behind the tuberculum impar adjacent to a midline depression just behind the tuberculum. This point is the anlage of the thyroid gland. It is the **thyroid diverticulum** (Fig. VII-36B) from which point the gland will begin its downward migration into the neck. In the adult tongue, this spot will remain as a blind dimple, the **foramen cecum,** between the anterior and posterior parts of the tongue.

The succeeding arches never quite meet in the midline, for they are kept from it by a second median swelling, the **hypobranchial eminence** or **copula** (Fig. VII-36A). The anterior part of this swelling from the third arch territory migrates forward to fuse with the enlarging tuberculum impar. In this way, the posterior mass of the tongue is formed and the ventral ends of the second arch appear to be obliterated in this mass. It is difficult not to think that the second arch is involved in tongue formation since it lies in the way of the over-growing copula, but how much it is involved is questionable. The junction of anterior and posterior parts of the adult tongue is marked by a V-shaped furrow, the **sulcus terminalis.** In front of the sulcus lie the large vallate papillae with taste buds in their circular troughs; behind it, the numerous small lingual tonsils; and at the apex of the V, the dimpled pit—all that remains of the original thyroid diverticulum (Fig. VII-36C and D).

The caudal part of the hypobranchial eminence, cleft from the main portion by a transverse sulcus, develops into the epiglottis in fourth arch territory in the posterior floor of the pharyngeal gut. We should also note that caudal to this part of the hypobranchial eminence a groove appears whose lips round up to form the **laryngotracheal tube** (Fig. VIII-36D). Thus, the respiratory passageway also buds off from the ventral part of the foregut. This is why you expect to see the trachea lying anterior to the esophagus (Fig. VII-37). While the groove begins as a median slit it soon becomes T-shaped by the additions of the **arytenoid swellings** added from the ends of the fifth arches (Fig. VII-36C). So now the larynx begins to appear with epiglottis and cunieform cartilages derived from the fourth arch and the rest from the fifth arch, in all probability. **Epiglottic valleculae** appear between tongue and epiglottis, and the **piriform fossae** appear lateral to the **aryepiglottic folds** (Fig. VII-36C and D).

Below the level of the larynx the endotracheal tube lengthens and finally bifurcates (Fig. VII-37 A–E). The original respiratory diverticulum makes its appearance in the third week of life. The two lateral outpocketings at the birfucation of the trachea are the **lung buds.** The right lung bud branches into **three main bronchi;** the left into two, thus forecasting the development of three lobes in the right lung and two in the left (Fig. VII-37 D–E). During further development, bronchi continue to subdivide by dichotomous branching. By the seventh month there have been 17 generations of these, the last and smallest known as the **respiratory bronchioles.** Some six more divisions will occur during postnatal life. The final sac-like structures are, of course, the thin, epithelially-lined sacs or **alveoli** in contact with surrounding capillaries for air exchange.

We must now turn our attention to the *lateral walls* of the **pharyngeal gut** where **pharyngeal pouches** evaginate (Fig. VII-34C). Each of these pouches has a peculiar fate of its own. Except for the first and second, each of these pharyngeal pouches tends to have its terminal ends split into **dorsal** and **ventral diverticulae.** Probably development of the tongue ventrally prevents the ventral diverticulae from forming in the first two pouches, so that only their dorsal portions become definitive structures.

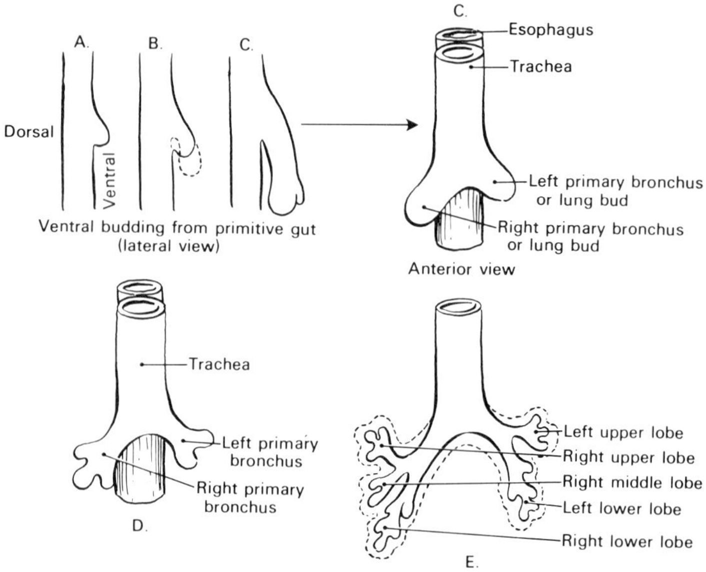

Fig. VII-37. Stages in development of trachea and lungs (schematic) from 3 weeks to 6 months in human fetus.

The dorsal part of the **first pharyngeal pouch** forms a canal with a sac-like terminus (Fig. VII-38A). The saccular dilatation becomes the **tympanic or middle ear cavity** and the proximal canal, the **pharyngotympanic or Eustachian tube.** The epithelial closing plate between the branchial groove and the pharyngeal pouch converts to the **tympanic membrane or ear drum,** the branchial groove, itself, becoming the external auditory meatus.

The history of the **second pharyngeal pouch** is simple. Lacking a ventral diverticulum, its dorsal one becomes the primordium of the **palatine tonsil** (Fig. VII-38 B and C) between the faucial pillars of the lateral oropharynx. From the third to the sixth month of gestation, the structure becomes infiltrated by lymphocytes.

The history of the **third and fourth pouches** is a bit more complicated since they have dorsal and ventral diverticulae. The dorsal part of the third pouch differentiates into **parathyroid tissue,** while its ventral part forms the anlage of the **thymus** (Fig. VII-38B and C). The luminal canal to these pouches obliterates, and by the sixth week of gestation[17] the differentiating dorsal and ventral ends of the pouches lose all connection with the pharyngeal wall. The thymus begins migrating caudomedially, dragging the parathyroid along with it. In its hasty migration, the main bulk of the thymus moves to its final resting place in the superior mediastinum rather rapidly. There it joins in the midline with its partner of the opposite side. But, it drags behind a slower moving tail of thymic tissue. This elongate strand of lagging material eventually breaks up into little fragments and becomes obliterated. However, occasionally small chunks of it fuse with the developing thyroid gland or persist as

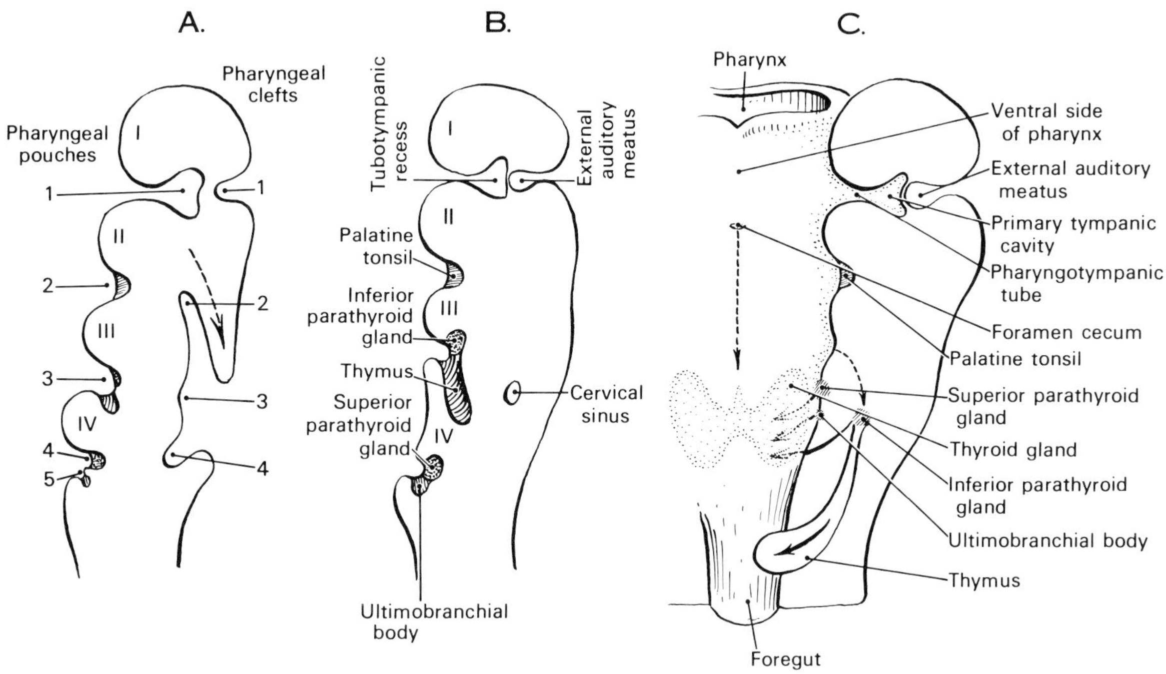

Fig. VII-38. Schema of development of pharyngeal clefts and pouches. (After Langman.)

isolated remnants along the migratory path.[18, 19] Because the parathyroid anlage was pulled downward by the migrating thymic tissue, it becomes, paradoxically, the **inferior parathyroid** of man applied to the posterior inferior pole of the thyroid, even though it arose more cranially in development than the superior parathyroid.

The dorsal part of the fourth pharyngeal pouch likewise becomes parathyroid tissue, ending up as the **superior parathyroid** by attaching itself to the craniodorsal part of the migrating thyroid gland as it passes the fourth pouch region in its descent into the neck (Fig. VII-38 B and C). Thus, both parathyroids hook rides to reach their final location: the superior parathyroids by migrating directly with passing thyroid tissue; the inferior parathyroids from the third pouch by being dragged along unwillingly to a lower level by rapidly-migrating thymic tissue. It is as if the inferior parathyroid jumped the wrong freight train. Whether the ventral part of the fourth pharyngeal pouch differentiates into thymic tissue or not is debatable. If it does it may disappear quickly, contributing little or nothing to the gland.

Since the fifth arch is rather rudimentary in man, the **fifth pharyngeal pouch** develops close to the fourth, and, by some, it is described as part of the fourth pouch. It gives rise to large, pale cells which, for want of a better name, have been called the **ultimobranchial body.** Whatever the nature or function of these cells, they are incorporated laterally into the thyroid gland.[20-22]

We noted earlier the anlage of the **thyroid gland** in the floor of the pharynx behind the tubercum impar of the developing tongue. The primordium appears on the 17th day of development,[23] and by the 19th, it has begun to push through the underlying mesoderm to descend in front of the foregut as a bilobed diverticulum (Fig. VII-38C). It pulls out a tubular canal behind it as the **thyroglossal duct** which later solidifies into a strand of tissue and subsequently disappears, leaving the gland detached from the foramen cecum. Eventually the gland reaches its definitive location in front of the hyoid bone and laryngeal cartilages by the seventh week of gestation. It already has its two lobes and, at times, a connecting midline isthmus. How much the ventral part of the fourth pharyngeal pouch and the ultimobranchial body contribute to it is questionable.

The overgrowth caudally of the large second branchial arch covers the second, third, and fourth branchial grooves or clefts and creates a **cervical sinus** which is normally absorbed (Fig. VII-38A and B). But if this area fails in being covered, **cervical cysts** can open to the outside of the neck through narrow channels. Such communications to the exterior are called **cervical fistulas,** and they can occur anywhere along the anterior margin of the Sternomastoid[23] (see Fig. VII-39B). Most often they occur at the jaw angle, and they may not be noticed until some time after birth when they have enlarged.

The obliterating thyroglossal duct can also become cystic in various places along its original course. The locations of these anomalies are shown in Figure VII-39C. About half of them occur close to or even behind the hyoid bone. They can also occur at the base of the tongue, as you might expect, and alongside the thyroid cartilage. It is possible for such cysts to open to the surface as fistulas but only secondarily as they enlarge

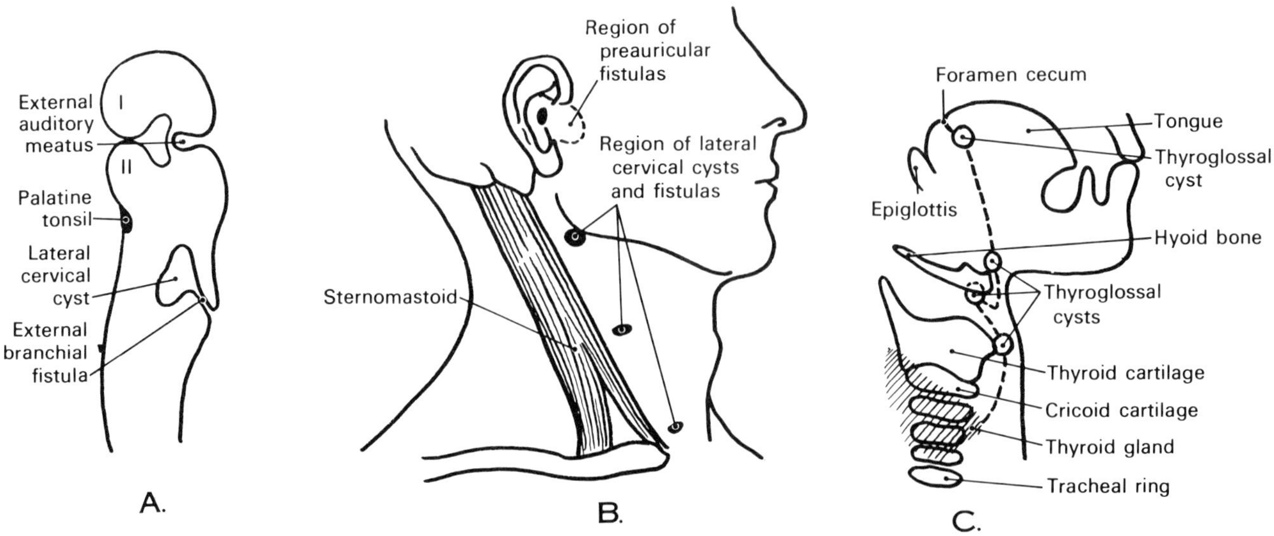

Fig. VII-39. Cervical and thyroglossal cysts. **A.** Schema of development of a branchial fistula (opening of embryonic cyst in lateral part of neck). **B.** Common locus of cervical cysts when present abnormally in adult. **C.** Common locus of thyroglossal cysts when present abnormally in adult. Broken line shows course of old thyroglossal duct. (After Langman.)

and rupture.[24] Since the floor of the mouth develops so close to the heart and its pericardial sac in the early embryo, it should not seem strange on rare occasions to find thyroid tissue adhering to the pericardium in the "old classical" middle mediastinum or anterior mediastinum.

DEEP STRUCTURES IN THE NECK

We are now ready to consider the deeper-lying viscera of the neck and related blood vessels and nerves.

The Thyroid Gland

This organ is an endocrine gland which is highly vascular. It cradles the upper part of the trachea, extending from the fifth tracheal ring inferiorly to the sides of the thyroid cartilage superiorly (Fig. VII-40). While the gland is invested by the pretracheal fascia, its own capsule lies internal to this tissue, and the vessels and nerves that supply it lie between these two investments. The gland varies in size and is larger in women and children than in men. It is bilobed and may or may not be joined by an isthmus across the midline.

The convex, cone-shaped superficial surfaces are covered by the strap muscles, and they are overlapped by the anterior border of the Sternomastoid. The medial surface relates to the Cricothyroid muscle, trachea, Inferior Pharyngeal Constrictor, esophagus, and the external and recurrent laryngeal nerves. The latter course laterally between the esophagus and trachea to reach the larynx. The posterior surface is applied to the prevertebral musculature and over parts of the sympathetic trunk. Laterally, it overlaps the carotid sheath. Embedded in the posterior surface are the parathyroids.

The isthmus is variable in its width and in its presence, generally lying over the second to fourth tracheal rings under the skin and fascia. It is in the way of being cut in performing a tracheostomy. A paramedial **pyramidal lobe,** long and slender, frequently arises from the upper border of the isthmus, chiefly on the left side. It may be attached to the hyoid bone by a fibrous strand which is a remnant of the thyroglossal duct.

The gland gets its arterial supply from two major sources and sometimes from a third (Fig. VII-40). The **superior thyroid artery,** as a branch of the external carotid, divides into a spray of two or three branches as it reaches down from above to supply the apex of each lobe. The **inferior thyroid artery,** a branch of the **thyrocervical trunk** (a stump arising from the subclavian), reaches the base and posterior aspects of the gland. Occasionally the **thyrodea ima** arises as a small artery from the brachiocephalic trunk, left common carotid, or even from the arch of the aorta. It ascends as far as the isthmus over the anterior surface of the trachea. These arteries anastomose freely over the gland, particularly on the posterior surface, but there is little anastomosis across the midline. What cross-connections there are in this area are affected chiefly by twigs of the superior thyroid coursing along the upper border of the isthmus.

Veins leave the deep substance of the gland to merge into a vast venous plexus over its surface. The plexus is drained by three pairs of veins. The **superior thyroid veins** arise from the upper poles of each lobe, cross the common carotids, and join the internal jugular vein. Sometimes the superior thyroid vein ascends with its companion artery to empty into the facial vein. The **middle thyroid vein** leaves the mid or lower

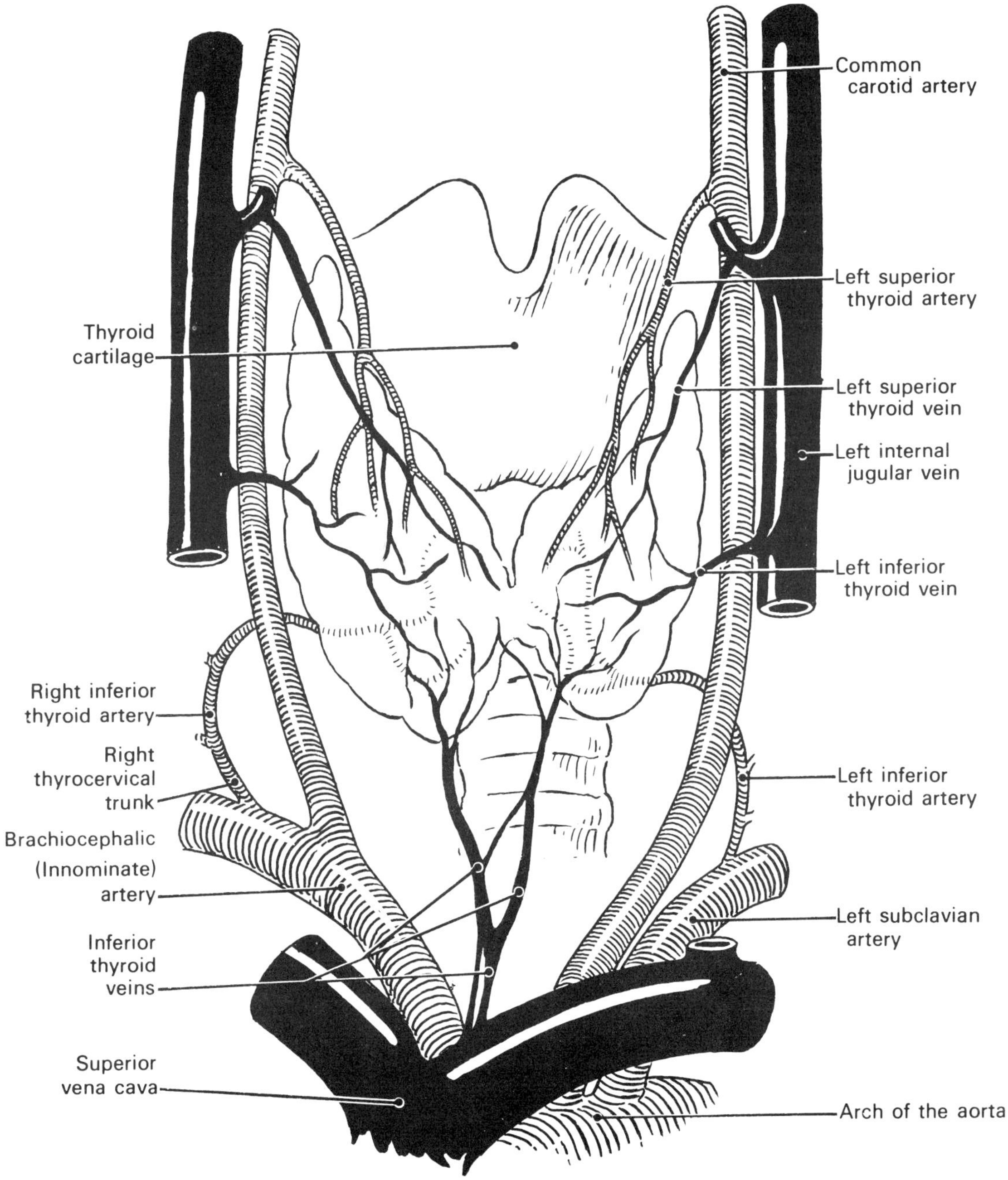

Fig. VII-40. Blood supply of thyroid gland. Veins, black; arteries, hatched.

part of the gland, crossing the common carotid to empty into the jugular vein. The **inferior thyroid vein** comes from the network on the isthmus and lower part of the gland as a single stem which splits into two tributaries that descend into the thorax on each side of the anterior trachea. One way or another, these empty into the brachiocephalic veins, either into the right and left brachiocephalic as separate vessels or by way of a joined trunk, the single **vena thyroidea ima.**

Both sympathetic fibers from the superior cervical sympathetic ganglion and from the vagus (superior laryngeal branch) have been described as entering the gland. Functionally, these nerves relate to the blood vessels and not to the glandular parenchyma, for severing them does not affect glandular function to any significant extent. Vagal contributions may represent sensory fibers from the blood vessels but certainly not visceral motor if, indeed, they do exist in the substance of the gland.

The Parathyroids

These tiny bodies are two pairs of yellow-brown endocrine glands embedded in the posterior surface of the thyroid capsule. The superior pair are more constant in position than the inferior pair and easier to find. They lie in the middle of the posterior surface of the lobes, the inferior ones lying closer to the inferior surface. The most reliable guide to the parathyroids is the small individual twig which the inferior thyroid artery gives to each. It should not seem strange from a developmental viewpoint to find parathyorid tissue occasionally embedded in the thymus in the superior mediastinum. Which parathyroid would you expect to find in such a location?

The Thymus

While the **thymus** is really a thoracic structure, we will consider it here since it started out in the neck. The thymus of the fetus and the newborn is remarkable in that it is nearly as large as the heart. It grows until puberty and then involutes. In the adult it is usually reduced to two narrow bands of lymphoid and fibrous connective tissue well infiltrated with fat. There is considerable variation from individual to individual depending upon the rate of involution. Its superior portion may extend into the neck above the superior mediastinum where the bulk of it resides. It can sometimes be a confusing mass in radiographs of the infant neck and thorax (Fig. VII-41). Its blood supply is from the **internal thoracic** (internal mammary) artery and occasionally from the aortic arch directly. Thymic veins run superiorly on the deep surface of the gland to enter the left brachiocephalic vein posterior to the manubrium of the sternum. The function of these endocrine glands will be considered briefly in Chapter 12.

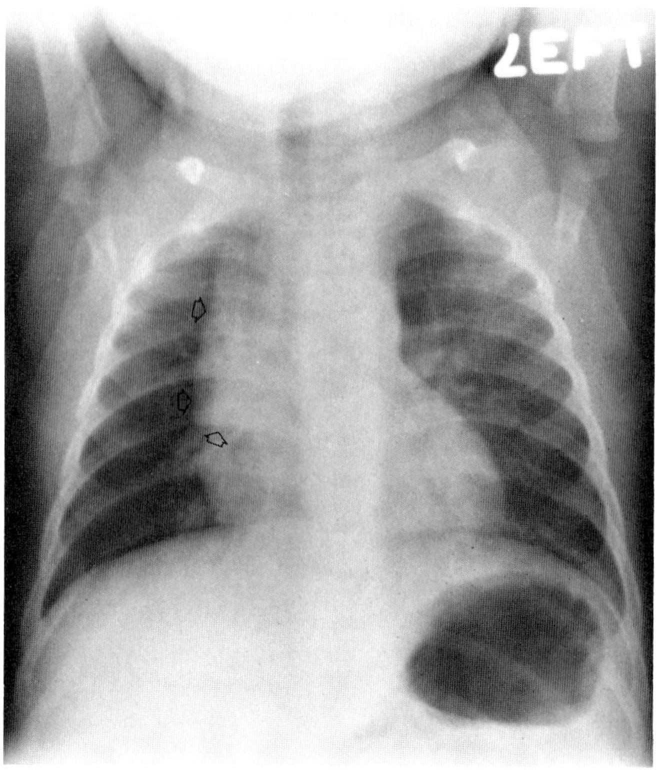

Fig. VII-41. Radiograph of infant chest in AP projection. Arrows demarcate right border of normal thymus. This normal "sail-shaped" shadow to right of heart should not be mistaken for an abnormal mediastinal mass.

The Trachea

This tubular organ, consisting of a series of C-shaped, hyaline cartilage sections closed behind by transverse muscle, lies both in the neck and thorax (Figs. VII-42 and 43). In life it is about 15 cm. long and 12 mm. in diameter. In the newborn this diameter is much smaller, 1 to 7 mm.[25] Probably, its most important relationships are to the thyroid gland in front, the esophagus against its muscular back wall, and the recurrent laryngeal nerve laterally between the esophagus and trachea. Not too far laterad lies the common carotid artery and internal jugular vein. The trachea gets its blood supply mainly from the inferior thyroid vessels. Its nerve supply comes from the cervical sympathetic ganglia and from the recurrent laryngeal branch of the vagus, the latter going to terminal ganglion cells within its walls. One can understand why its posterior wall consists of transverse bands of smooth muscle rather than cartilage. As a bolus of food traverses the esophagus and bulges it, the tracheal wall in front of the esophagus must indent to conform and to allow passage downward of the bolus. A rigid tracheal wall would tend to interfere with esophageal peristalsis.

Tracheostomy may be required for laryngeal obstruction for any number of reasons—inhaled foreign

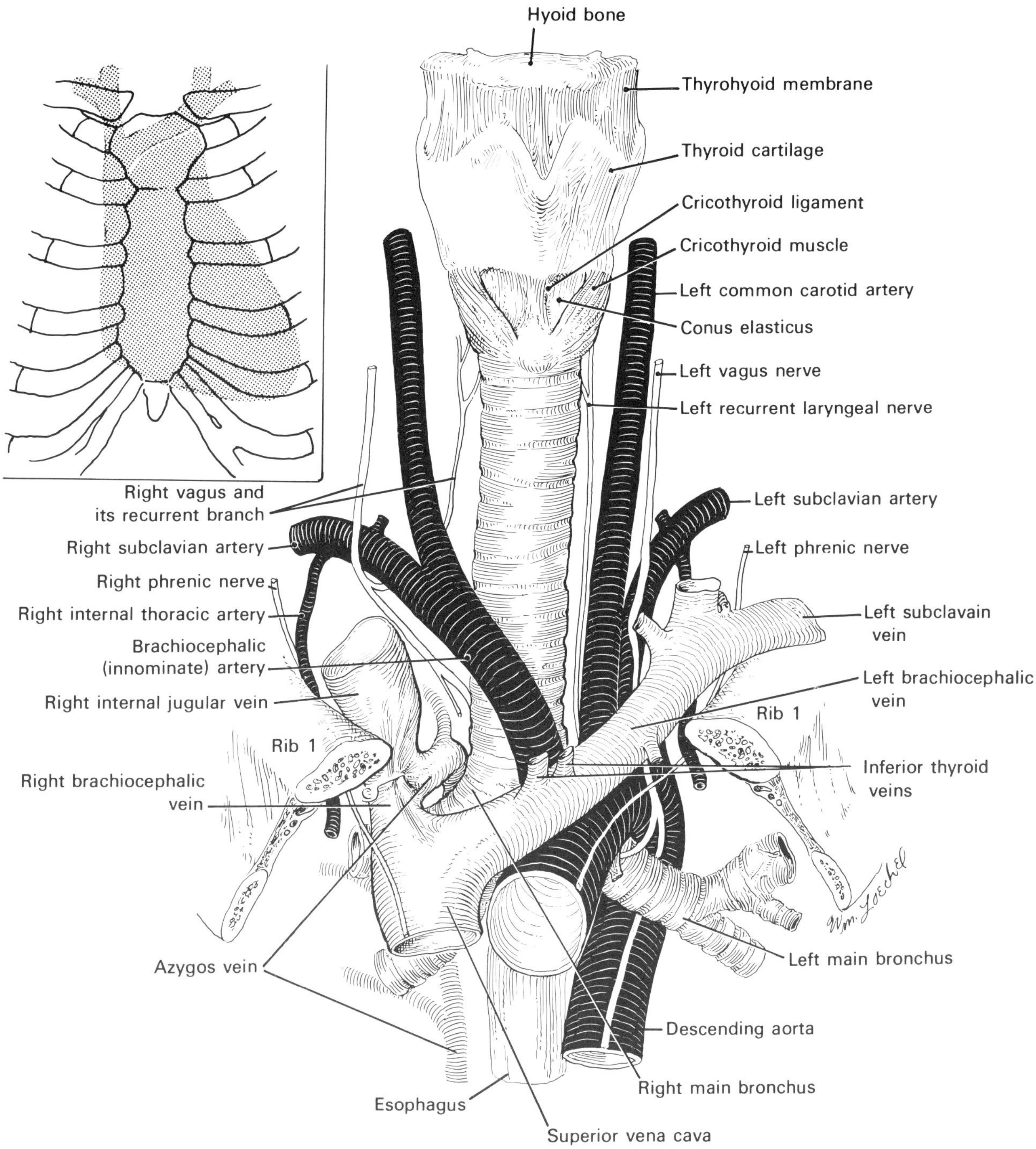

Fig. VII-42. Relationship of trachea and bronchi to great vessels as seen from in front.

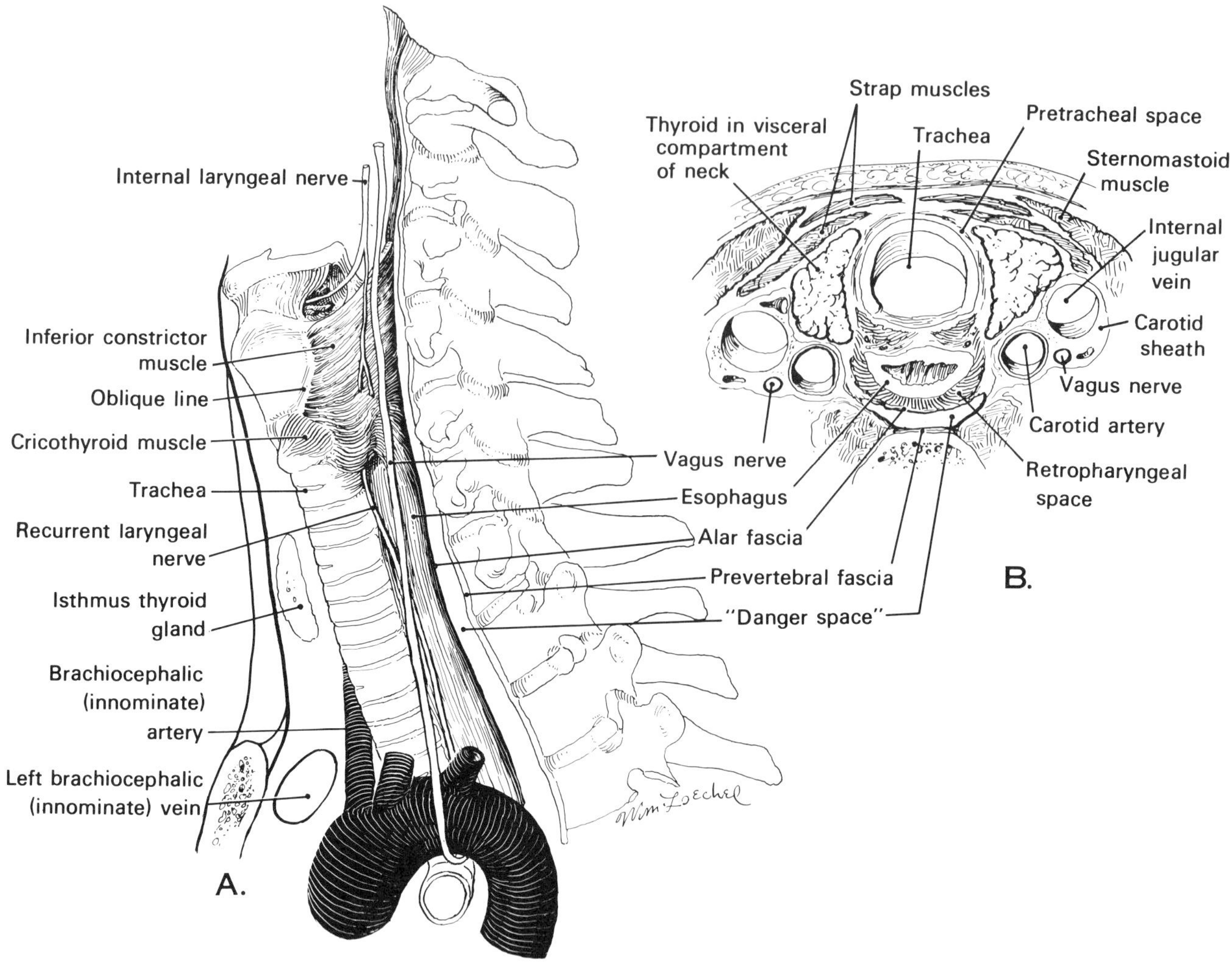

Fig. VII-43. Relationship of trachea to esophagus. **A.** Seen from left side. **B.** In cross-section.

bodies, tumors, diphtheria, drainage of postoperative secretions, or to render long-standing artificial respiration through a tracheal cannula as in bulbar poliomyelitis where respiratory control centers in the medulla are damaged. The main rule in making an incision is to *stick to the midline* so that major vessels will not be jeopardized. The neck is extended and held perfectly straight by an assistant. A vertical slit is made through the skin and underlying soft tissues from the notch of the thyroid cartilage to just above the jugular notch of the manubrium and the third and fourth rings of the trachea are incised. This is more difficult to do in children where the trachea is soft and mobile. In the ossified tracheas of some elderly individuals, even a small bone snips may be needed to cut the tracheal rings.

The Esophagus

This muscular tube is obviously longer than the trachea since it must pierce the diaphragm to reach the stomach (25–30 cm.). It projects beyond the left side of the trachea more than on the right, and about one-fifth of its length occupies the neck. Its muscular wall contains skeletal muscle in the cervical region. This gains anchorage to the posterior part of the cricoid cartilage by a tendon. It is constricted in the neck where the Inferior Pharyngeal Constrictor narrows upon it to serve as a sphincter. Its blood and nerve supply in the neck region are similar to that of the trachea. The skeletal muscle in the upper parts of the esophageal wall receives branchiomotor fibers from the vagus, while vagal visceromotor nerves supply mucosal and submucosal glands, the smooth muscle of the muscularis mucosae, and the smooth muscle in the lower walls of the tube as it begins to approach its junction with the stomach.

GREAT VESSELS AND NERVES

To approach deep structures of the neck best, a considerable amount of prior dissection has had to occur. The parotid and infratemporal regions should have been dissected. The posterior belly of the Digastric with its associated vessels, the occipital and posterior auricular arteries, needs to be cut and laid back. The three muscles arising from the styloid process (Stylohyoid-glossus-pharyngeus) must have been severed. One can then get at the structures contained within or related to, the carotid sheath, namely the **internal jugular vein, internal carotid artery, the last four cranial nerves,** and **the sympathetic trunk.**

All of these structures pass through the skull by way of only three openings: (1) the **jugular foramen,** containing the vein and nerves IX, X, and XI; (2) the **hypoglossal canal** for the cranial nerve XII; and, (3) the **carotid canal** for the artery and extensions of the sympathetic postganglionics along it. Actually, the jugular foramen, itself, is a three-compartmented structure. The vein sits in a posterolateral compartment, the three nerves in a middle one, and the inferior petrosal sinus in a ventromedial compartment. There would be a fourth compartment for the hypoglossal nerve if a bar of bone failed to separate its canal from the jugular foramen. It is no wonder, then, that these four cranial nerves emerge from the skull crowded together between internal jugular vein and internal carotid artery (Fig. VII-44A). The carotid canal, as a channel through petrous bone, lies in front of the jugular foramen. Hence, upon appearing in the neck, the artery will lie in front of the vein and somewhat medial to it. The last four nerves and the sympathetic trunk will be in close relation to the vessels, the sympathetic trunk actually posterior to the carotid sheath.

All four nerves will descend for a short distance between artery and vein before they diverge from each other. The vagus (X) continues straight down through the neck between artery and vein. At the sternoclavicular joint it crosses *behind* the junction of the brachiocephalic veins (Fig. VII-45B). To reach tongue and pharynx, the glossopharyngeal nerve (IX) must pass between internal and external carotid arteries (Fig. VII-44A). It follows the posterior border of the Stylopharyngeus muscle to which, of course, it supplies the sole innervation. The spinal accessory (XI) we have noted taking a posterior route under the Sternomastoid. It may cross in front of or behind the internal jugular vein or form a loop around it on its way (Fig. VII-45B). The hypoglossal nerve (XII), to reach the tongue, must cross the internal and external carotid arteries, but it lies deep to the posterior belly of the Digastric as it makes its way into the submandibular region (Fig. VII-44A). Figure VII-45 shows that the facial nerve (VII) in its bony canal relates to the jugular bulb; that the ansa cervicalis encircles the internal jugular at the levels of the superior and middle thyroid veins; and that the phrenic nerve, at first to the right of the vessel, emerges behind and to the left of the brachiocephalic junction lateral to the vagus.

The **Common carotid artery** appears from under cover of the Omohyoid in the carotid triangle where it extends to the upper border of the thyroid cartilage be-

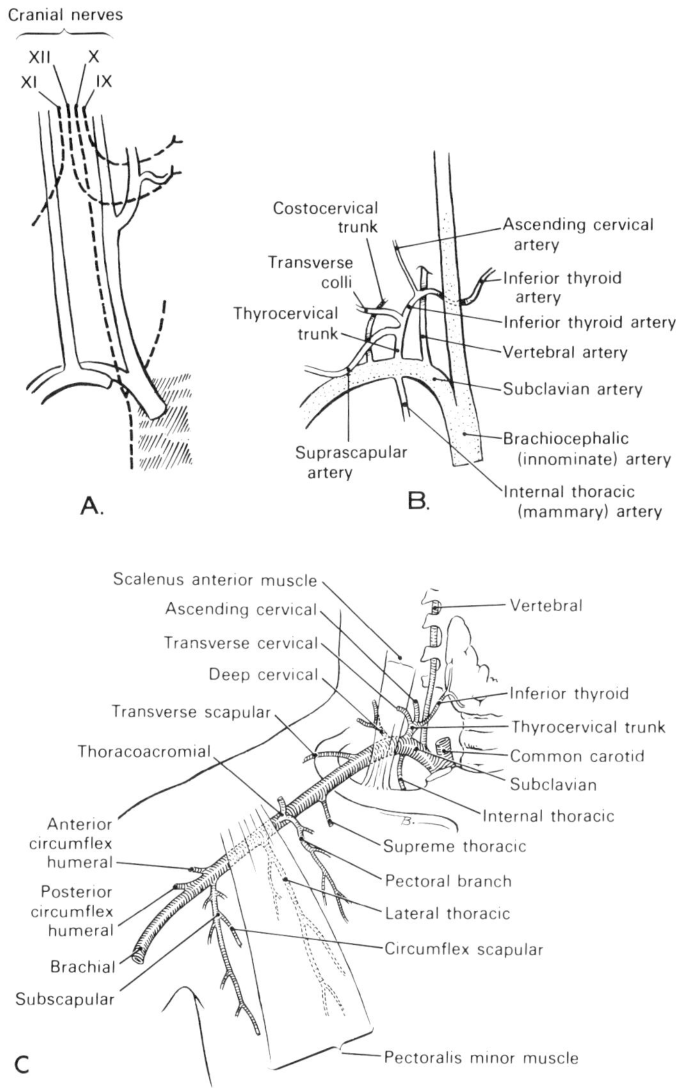

Fig. VII-44. The great vessels of the neck. **A.** Relationship of last four cranial nerves to vessels. **B.** Branches of subclavian artery at root of neck. **C.** Subclavian and axillary arteries. (A and B Modified from Grant.)

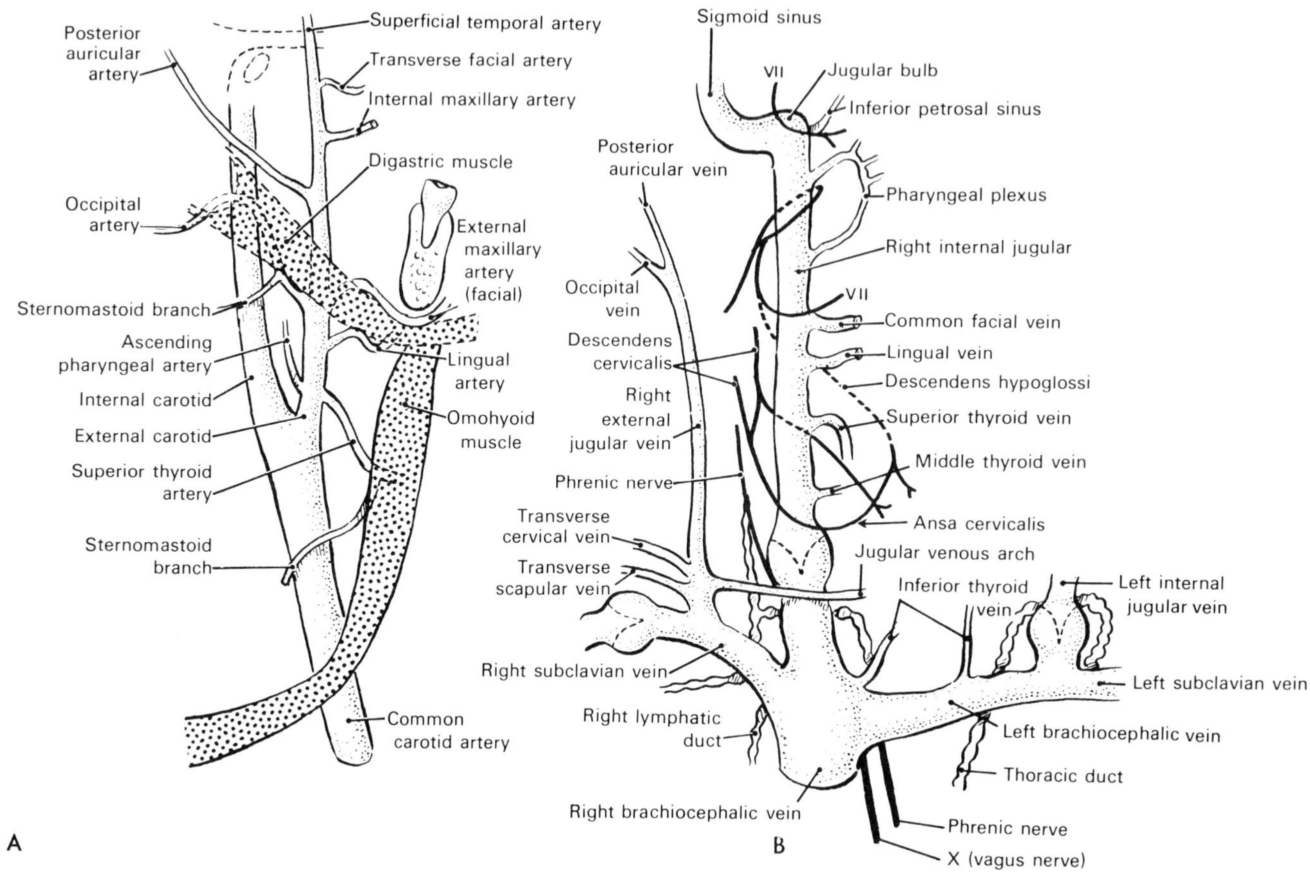

Fig. VII-45. Carotid arteries (**A**) and internal jugular vein (**B**). Phrenic nerve, certain branches of the cervical plexus, and some of the cranial nerves are shown related to the vein. (After Grant.)

fore bifurcating into its main internal and external stems. The internal carotid will distribute inside the skull, the external outside; the internal passing deep to the parotid gland, the external entering the gland. Neither common nor internal carotid has branches in the neck, only the external (Fig. VII-45A).

The **external carotid artery** terminates as the **superficial temporal** and **internal maxillary arteries.** The largest branch of the superficial temporal is a **transverse facial artery** that sometimes arises from the external carotid as a common trunk with the temporal vessel (Fig. VII-45A). The common carotid gives off six radiating branches near the posterior belly of the Digastric. Three of these arise together from below the Digastric and all take a meandering course in this most mobile part of the face and neck to get to where they are going. These are, in order from below craniad, the **superior thyroid, lingual,** and **facial** (external maxillary) **arteries.**

We have observed the **superior thyroid artery** (Fig. VII-45A) arising above the level of the gland and noted its descent to the apical poles of the lateral thyroid masses. This vessel uses the oblique (line of the thyroid cartilage) as a guide-path as does its companion, the **external laryngeal nerve.** The branches of the artery are:

1. An **infrahyoid artery** below the bone.
2. A **superior laryngeal** branch which pierces the thyrohyoid membrane to enter the larynx.
3. A **cricothyroid** branch which crosses the Cricothyroid muscle and ligament and enters the larynx by piercing the ligament.
4. **Three twigs to the thyroid gland.**
5. A **sternomastoid branch** which follows the Omohyoid across the carotid sheath to reach the Sternomastoid inferiorly.

Recall that the hairpin arch of the **lingual artery** was crossed by the hypoglossal nerve before the artery passed from sight deep to the Hyoglossus. The vessel passes forward to the tip of the tongue, applied to the Genioglossus. Its terminal part, the **profunda artery,** anastomoses with its partner only at the tongue tip, so that a bisection of the tongue is almost a bloodless procedure. The branches of the **lingual artery** are:

1. A **suprahyoid twig.**
2. The **dorsales linguae** to the posterior part of the tongue which anastomose in the tonsilar bed.

3. The vessel provides other muscular twigs to the tongue and then gives off the **sublingual artery** to the gland. The latter can anastomose with the submental branches of the facial artery.

We have referred previously to the sinuous course of the **facial** or **external maxillary artery** (Fig. VII-45A). The submandibular gland, Digastric, and Stylohyoid muscles keep the facial artery separated from the facial vein, and they force the artery upward between the **Medial Pterygoid** and the tonsilar bed. As a consequence, the following vessels arise from the lower or cervical portion of the vessel:

1. An **ascending palatine artery** to the soft palate.
2. A **tonsilar branch** to the palatine tonsil and root of the tongue.
3. Branches **to the submandibular gland.**
4. A **submental artery** to the anterior belly of the Digastric, Mylohyoid, Platysma, and adjacent skin.

Three relatively long, ascending branches also arise from the **external carotid** (Fig. VII-45A):

1. The **ascending pharyngeal** begins near the origin of the external carotid and its ascends deep upon the pharyngeal constrictor muscles to the base of the skull. On its way it supplies the pharynx, pharyngeotympanic tube, and soft palate. Its final meningeal twigs seek out a variety of foramina to reach dura—hypoglossal canal, jugular and lacerum foramina, and one twig follows the tympanic branch of IX.

2. The **occipital artery** is another vessel we have seen before. Its chief branches are one to the **Sternomastoid** and a **descending branch** that courses on both the superficial and deep aspects of the Seminspinalis Capitis. This branch anastomoses with the **transverse cervical artery.** The terminal meningeal twigs of the occipital artery utilize both mastoid and parietal foramina.

3. In addition to supplying the skin behind the ear, the **posterior auricular artery** gives off a **stylomastoid twig** which enters the facial canal to anastomose with the petrosal branch of the middle meningeal artery. The **auricular branches** supply the medial aspect of the ear but lap over on its superficial or lateral aspect enough to anastomose with branches of the superficial temporal artery.

The **subclavian artery** is generally described as having three parts: Part 1 lies medial to the Anterior Scalene muscle; Part 2 is crossed by the muscle and, thus, separated from the subclavian vein which lies anterior to the muscle; and Part 3 extends from the lateral border of the Anterior Scalene to the lateral border of the first rib where it insensibly becomes the axillary artery (Fig. VII-44C).

The right common carotid and right subclavian arise from a brachiocephalic (innominate) trunk behind the right sternoclavicular joint. These two vessels on the left side arise as separate vessels from the arch of the aorta.

As a rule all the branches arise from Part 1 so that their origins hide behind the internal jugular vein.

In outline, these are the major branches: (Fig VII-44C)

1. Vertebral artery
2. Thyrocervical trunk—its branches are:
    - A. Inferior thyroid artery—its branches are:
        - a) Ascending cervical artery
        - b) Esophageal and tracheal twigs
        - c) Pharyngeal and laryngeal twigs
        - d) Thyroid supply
    - B. Transverse cervical artery
    - C. Transverse (supra) scapular artery
3. Costocervical trunk—its branches are:
    - A. Deep cervical artery
    - B. Highest intercostal artery—its branches are:
        - a) First posterior intercostal artery
        - b) Second posterior intercostal artery
4. Internal thoracic (mammary) artery

The **vertebral artery** we have met before. It ascends between the Anterior Scalene and Longus Colli to enter the foramen transversarium of the sixth cervical vertebra. We have followed it craniad from here. We emphasize again that only the two vertebral arteries and the two internal carotids supply blood to the brain.

The **thyrocervical trunk** is virtually non-existent, as three vessels branch away from its short stump almost at once. Both the **transverse cervical** and **transverse scapular** arteries cross the posterior triangle to disappear into the shoulder region under the Trapezius, crossing the Anterior Scalene on the way and clamping the phrenic nerve to it. The third branch is already familiar to us as the **inferior thyroid artery** which gives off the various visceral twigs mentioned above. Its laryngeal branch follows the recurrent laryngeal nerve (X) into the larynx. It gives off the **ascending cervical artery** as a major branch which ascends on the Anterior Scalene medial to the phrenic nerve, and this vessel is thin enough to be mistaken for the nerve. This artery supplies the deep neck muscles.

The **costocervical trunk** is also a short stem and can be said almost to arise from Part 2 of the subclavian, just behind the medial border of the Anterior Scalene. Its two branches come off quickly, the **deep cervical** crossing back between the neck of the first rib and the transverse process of VC_7 to anastomose with the occipital artery. The **highest (superior, supreme) intercostal** descends in front of the necks of the first two ribs, giving arteries to the first and second intercostal spaces.

The **internal thoracic artery** arises from the lower border of the subclavian opposite the thyrocervical trunk. It descends under the clavicle and behind the cartilage of the first rib. It crosses deep to succeeding costal cartilages to the sixth interspace where it divides

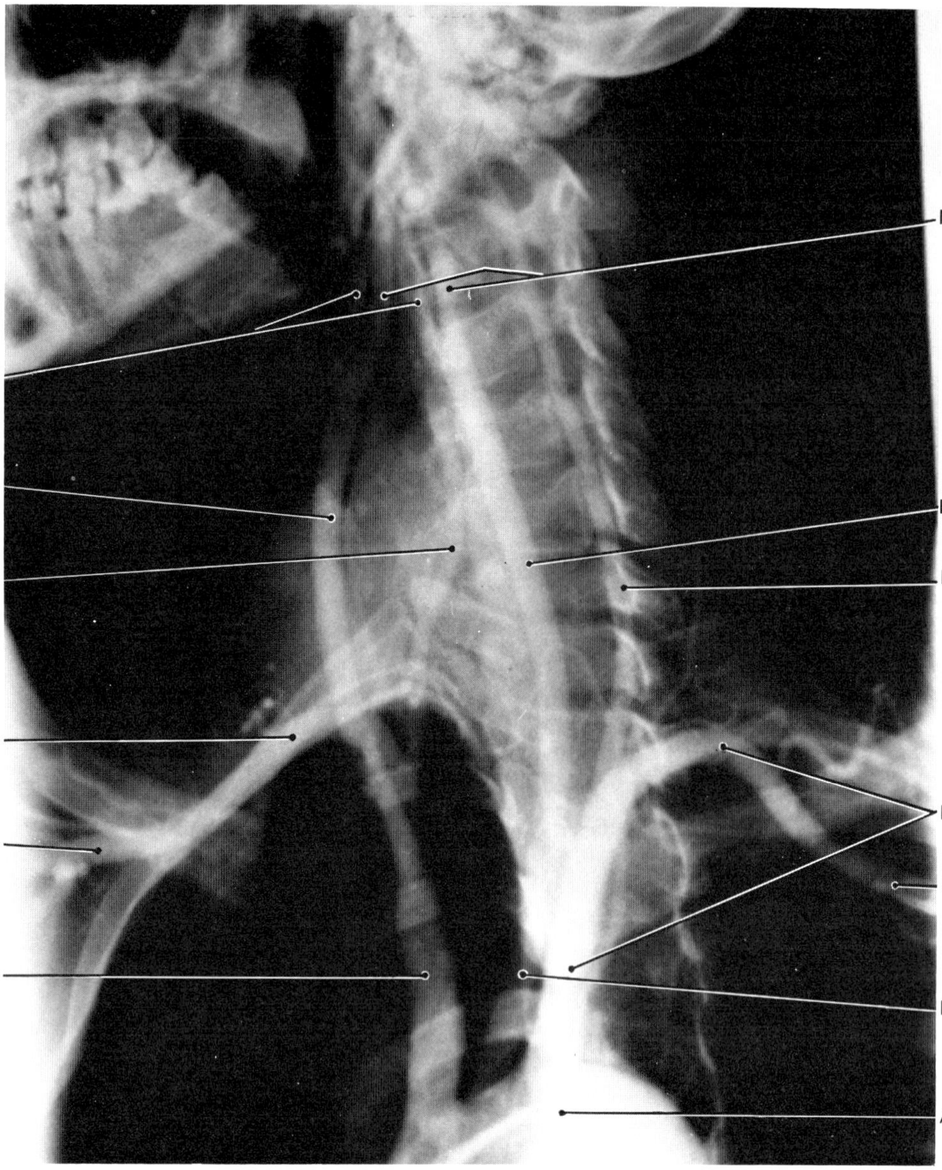

Fig. VII-46. Arteriograms showing certain vessels of the neck and head. **A.** Injection of contrast media into aortic arch (right posterior oblique projection).

into the **superior epigastric** and **musculophrenic arteries.** Just as it enters the chest it gives off the **pericardiacophrenic artery** which runs with the phrenic nerve over the lateral pericardium to the Diaphragm. Some of these major arteries can be seen on radiographs where the vessels have been injected with contrast medium (Fig. VII-46A–D).

The Veins of the Neck

In contrast to the plan of the arteries where internal carotid vessels have no branches in the neck and the external gives off all of them, the **internal jugular vein** has the most branches in the neck and the **external jugular** the least. Most of these venous branches are companion vessels of the carotid artery branches, and they need not be named (see Fig. VII-45B). There are few new veins to attend to, then: the right and left brachiocephalics; the pharyngeal plexus; and three low branches from the external jugular.

The **brachiocephalic (innominate) veins** drain blood from the head, neck, and limbs into the superior vena cava. They are formed by the junction of the internal jugular and subclavian veins and terminate in a confluence behind the first costal cartilage on the right side. The right one is short, about 1 to 1.5 inches; the left is longer, about 2.5 to 3 inches. On their upper surface they also receive the **vertebral veins,** the **deep**

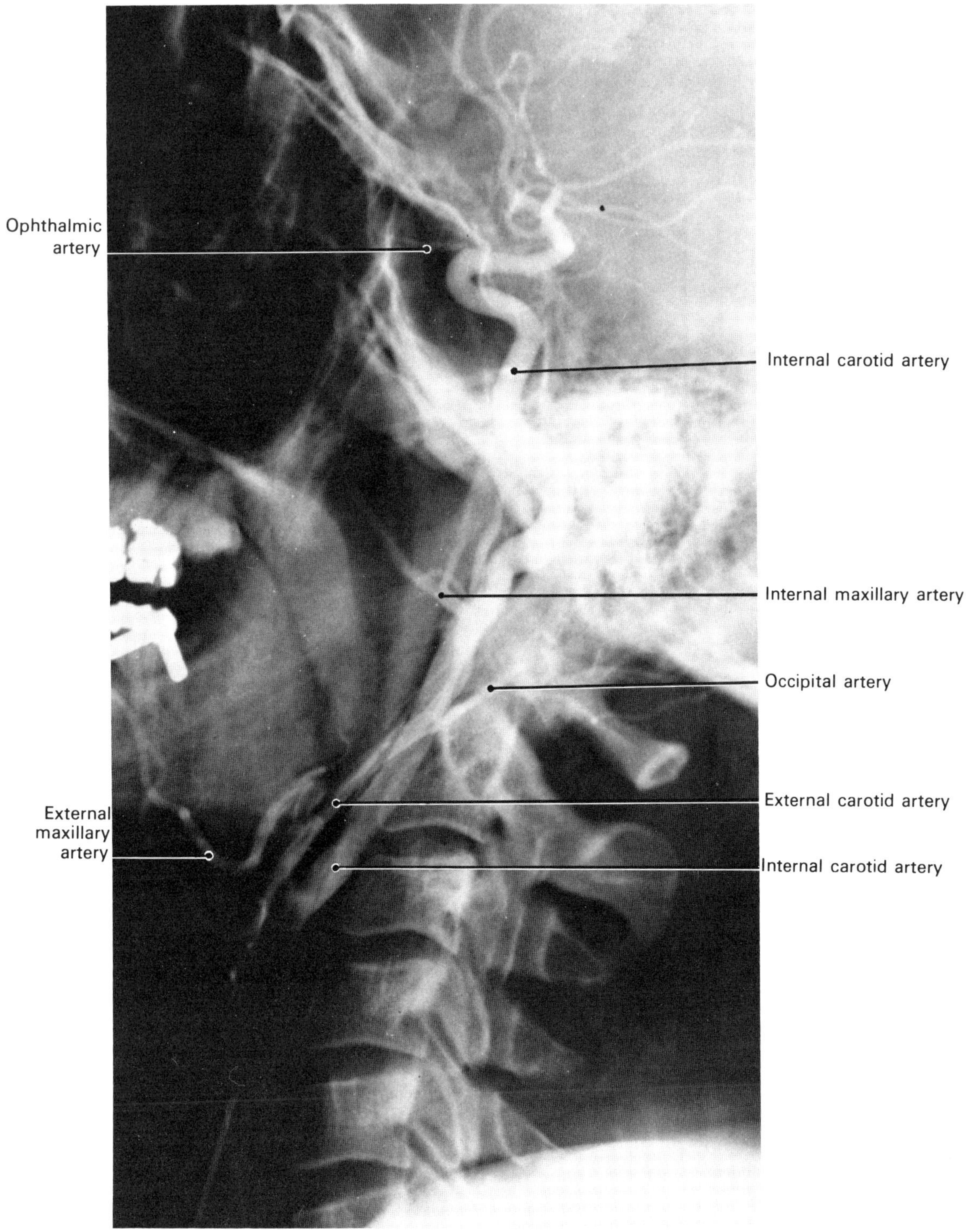

Fig. VII-46B. Injection of contrast media into common carotid artery.

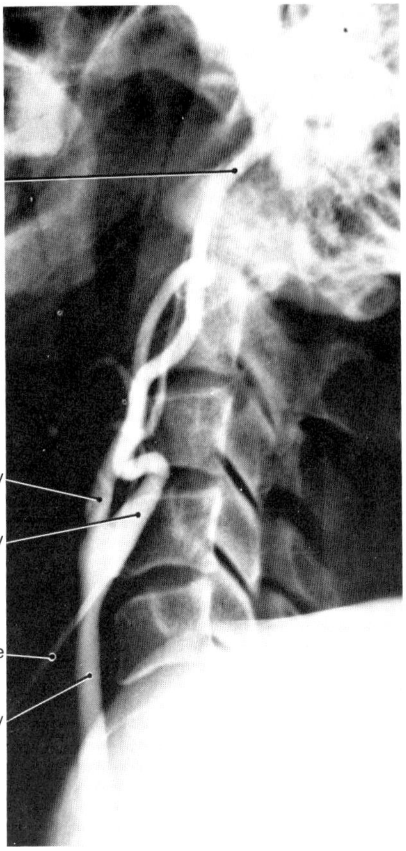

Fig. VII-46C. Injection of contrast media into common carotid artery.

cervical veins, and the **inferior thyroid veins.** The **internal thoracic veins** empty into their lower border. At the junction of the jugular and subclavian, the **thoracic duct** empties its accessory venous drainage as the major lymphatic return from the body, while the **right lymphatic duct,** draining less of the body area, empties similarly on the right side (Fig. VII-45B).

The **veins of the pharyngeal plexus** lie between the three pharyngeal constrictor muscles and the prevertebral fascia. This plexus drains the mucosa of the pharynx, soft palate, auditory tube, pterygoid canal, and the Anterior Recti and Longus Colli muscles. It communicates above with the pterygoid plexus of veins and drains below into the internal jugular.

The **transverse (suprascapular) vein** drains the upper part of the shoulder supplied by the companion artery. If double, it generally combines into a single trunk before emptying into the **external jugular vein** which, in turn, joins the subclavian vein.

Oftentimes the **transverse cervical veins,** which emerge under the Trapezius to cross the posterior triangle in company with the Transverse scapular veins, emtpy into the external jugular. They may communicate directly with the subclavian vein, however.

The **anterior jugular vein** begins at the chin, draining mental, submental, inferior labial, and inferior hyoid veins. It descends near the midline, occasionally

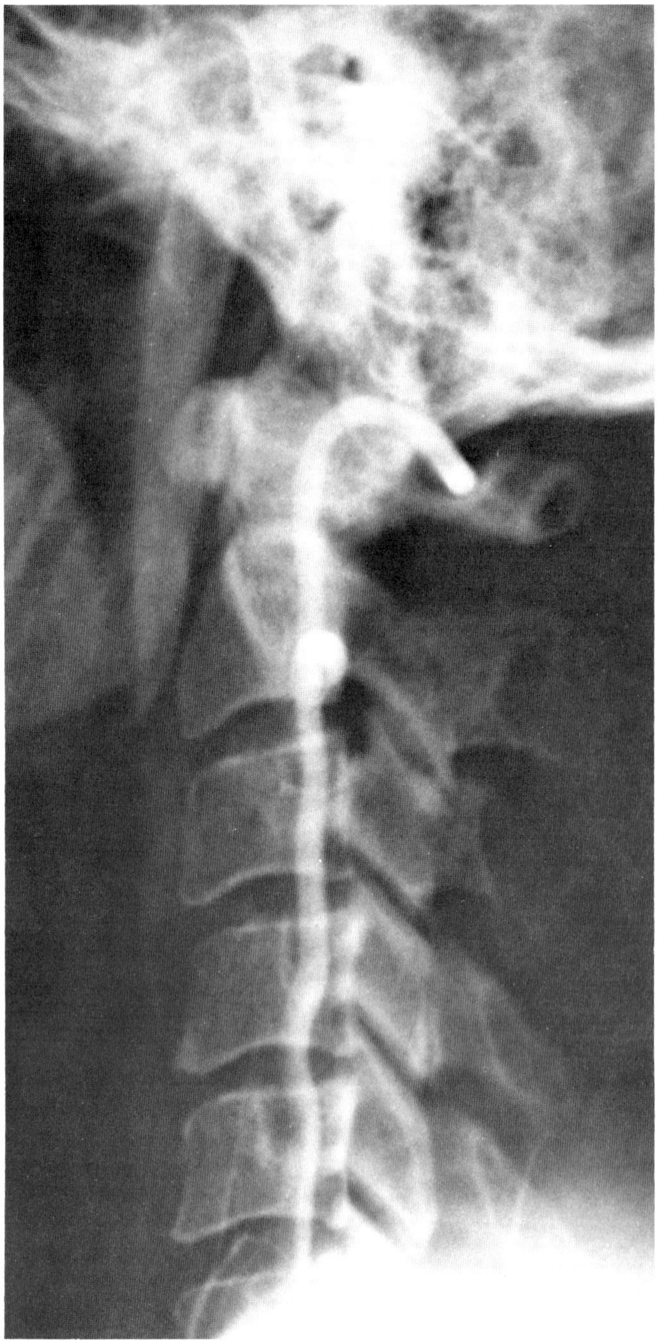

Fig. VII-46D. Selective injection of vertebral artery. Note course through foramina transversaria.

getting branches from the larynx and thyroid gland. Just above the clavicle the anterior jugulars turn laterally to join the external jugulars. Sometimes the anterior jugular veins join the subclavian directly. At the point of lateral turning, a transverse connection, the **jugular venous arch,** may connect the two anterior jugular veins across the trachea.

The **common facial vein** is a short, stout stem, on the whole, draining anterior and posterior facial veins into the internal jugular.

Carotid Sinus and Carotid and Aortic Bodies (Fig. VII-47)

There is a dilatation in the wall of the internal carotid artery just as it bifurcates from the common carotid which is known as the **carotid sinus.** Its walls contain an elaborate network of nerve filaments and terminal receptor endings, proprioceptive in nature. These are reached by the glossopharyngeal nerve (IX) by a branch known as the **sinus nerve** (or nerve of Hering). The receptors are sensitive to changes in blood pressure. These **baroceptors** play an important role in the reflex regulation of circulation.

The **carotid body** (glomus caroticum) has been variously named carotid gland or intercarotid body. The last name, not often used, locates it well in the bifurcation of the internal and external carotid vessels where it sits as a small, 5-mm., oval mass. Its fibrous capsule contains rounded clumps of polyhedral cells well surrounded by sinusoidal capillaries which present a glandular appearance. Some of the cells give a chromaffin reaction (stain brown with chromic acid), but it does not contain adrenaline as this reaction might imply, nor is it an endocrine gland. It receives nerve fibers and sensory terminals from the vagus (X) and serves as a **chemoreceptor** sensitive to changes in oxygen tension of the blood and, hence, as a reflex **regulator of respiration.** Some physiologists deny that it plays any role in the regulation of circulation, but others of equal importance as investigators claim that activation of the carotid and aortic bodies by hypoxia or by CO_2 excess can cause reflex vasoconstriction and a rise in blood pressure.

The **aortic bodies** (glomera aortica) are similar to the carotid bodies. They are known to exist in mammals and are thought to occur in man. The right ones are usually found high at the junction of the right common carotid and subclavian arteries; the left, near the arch of the aorta. They are supplied by the cardiac branches of the vagus and, apparently, also serve as chemoreceptors.

The Glossopharyngeal Nerve (Fig. VII-48)

This nerve leaves the jugular foramen in its own dural sleeve, descends between the jugular and internal carotid vessels to turn around the posterior border of the Stylopharyngeus where it moves forward along the Middle Pharyngeal Constrictor and beneath the Hyoglossus to reach the pharynx and tongue.

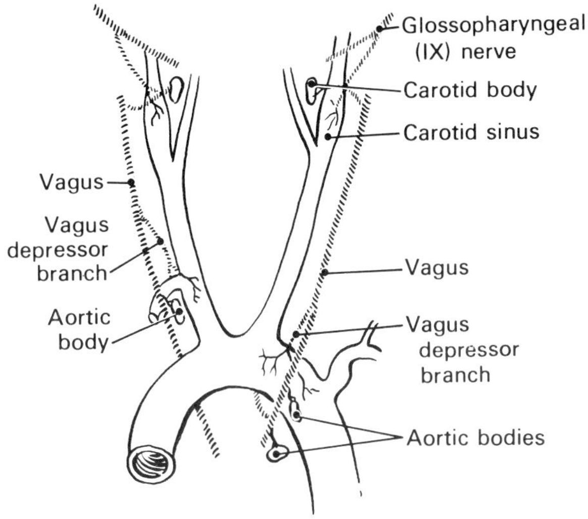

Fig. VII-47. Schema of innervation to carotid and aortic sinuses and bodies.

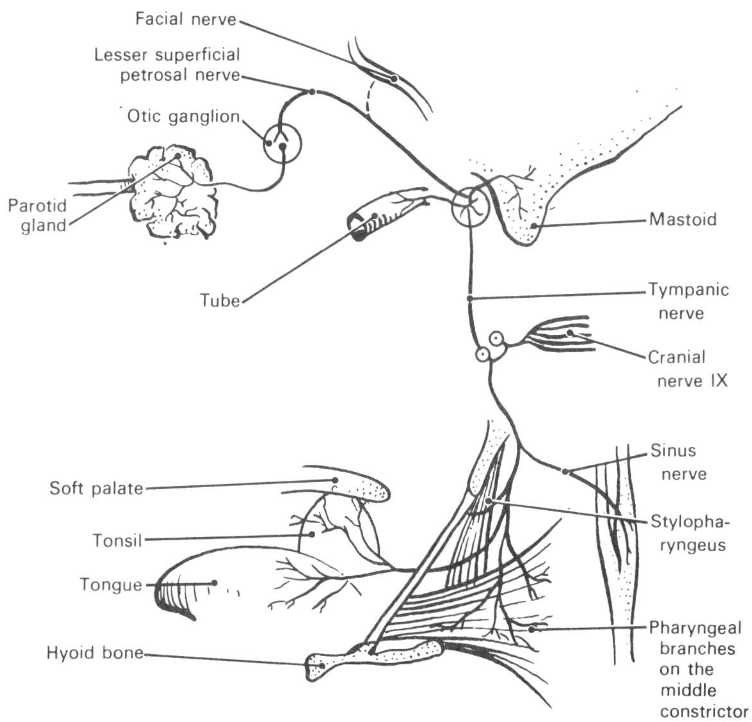

Fig. VII-48. Schema of distribution of glossopharyngeal nerve. (After Grant.)

It has one branch to **skeletal muscle** only, the twig to the third branchial arch derivative, the **Stylopharyngeus,** itself. Its **sensory branches** are to the pharyngeal mucosa, in part, and to the palatine arches, soft palate, tonsilar mucosa, and taste buds of the posterior part of the tongue. It also innervates the barroreceptors of the carotid sinus *via* its **sinus nerve.** It is **visceral motor** for the parotid gland. Recall that this latter parasympathetic supply takes a peculiar recurrent course after leaving the jugular foramen. It returns to the skull *via* **Jacobson's Canal** or the **tympanic canaliculus** to reach the middle ear cavity. Here it supplies **sensory branches** to the pharyngotympanic (auditory) tube, mucosa of the medial wall of the tympanium, and the mastoid air cells. The visceral motor branch leaves the tympanic plexus by piercing the petrous bone as the **lesser superficial petrosal nerve.** This nerve makes its final emergence from the skull in the foramen ovale. Here it reaches its terminal parasympathetic ganglion, the **otic ganglion,** in the infratemporal fossa. Postganglionic axons leave the ganglion to join the **auriculotemporal branch of V_3** which piggy-backs them to the parotid gland.

The Vagus Nerve (Fig. VII-49)

This nerve courses through the neck in the carotid sheath between the internal jugular and carotid vessels as far as the sternoclavicular joint. It gives off certain branches along the way before it crosses behind the brachiocephalic vein to enter the superior mediastinum.

Close to the jugular foramen it sends a recurrent meningeal twig back to the dura of the posterior cranial fossa (sensory). It sends a somatosensory twig back into the skull *via* **Arnold's canal** which leads to the **tympanomastoid fissure** where the nerve supplies the outer aspect of the ear drum, part of the external auditory canal, and the skin back of the ear.

In the neck it gives off **pharyngeal branches,** the **superior laryngeal nerve, carotid body branches, cardiac branches,** and the **right recurrent laryngeal nerve.** The pharyngeal twigs are *branchiomotor* and supply all pharyngeal and palatal muscles (except the Stylopharyngeus (IX), 2nd Tensor Palatini (V)) and the Inferior Constrictor. The vagus supplies the last muscle by lower branches not related to the pharyngeal plexus.

The **superior laryngeal nerve** crosses behind the internal and external carotid vessels to course medialward before breaking into its external and internal branches. The **external branch** is *branchiomotor only*. It supplies the **Inferior Pharyngeal Constrictor** and the one and only external muscle of the larynx, the **Cricothyroid.** The **internal branch** is *sensory only*. It pierces the thyrohyoid membrane to enter the larynx where it innervates the mucosa above the level of the vocal cords.

The **right recurrent laryngeal nerve** descends to the level of the right subclavian artery where it loops over, under, and behind it to run a recurrent course

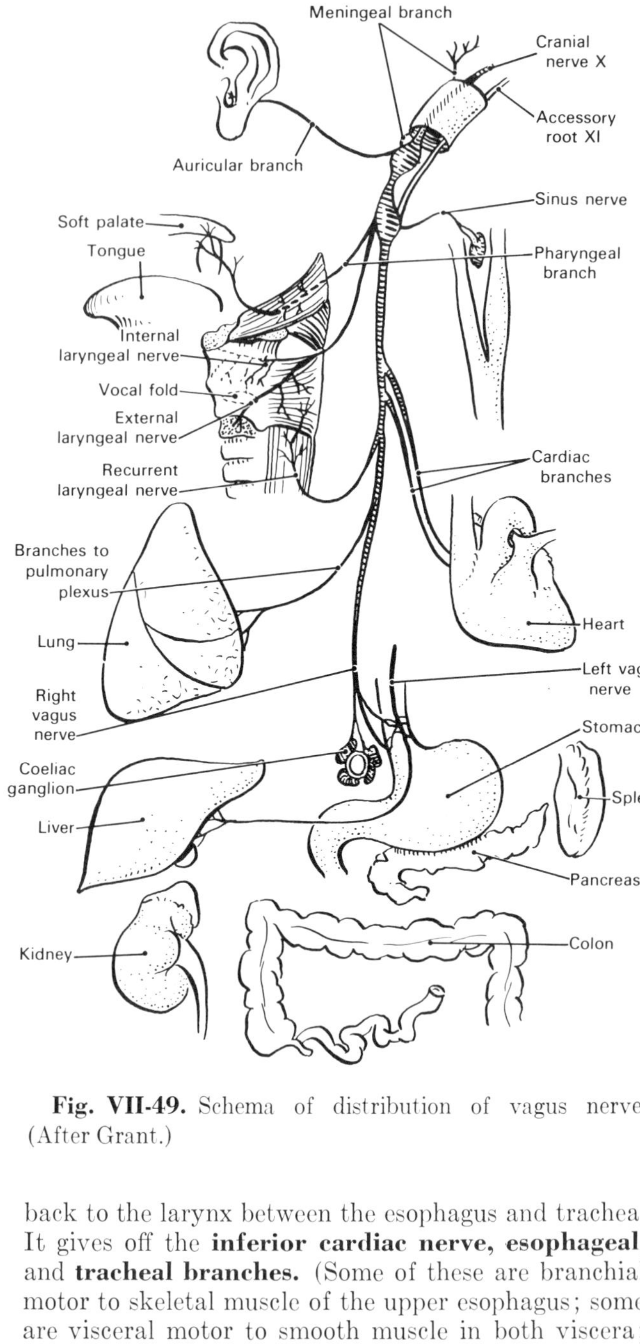

Fig. VII-49. Schema of distribution of vagus nerve. (After Grant.)

back to the larynx between the esophagus and trachea. It gives off the **inferior cardiac nerve, esophageal,** and **tracheal branches.** (Some of these are branchial motor to skeletal muscle of the upper esophagus; some are visceral motor to smooth muscle in both viscera; and others are visceral sensory to their mucosae.) Its main concern is to supply *all internal laryngeal muscles* on its side and to give sensory branches to the laryngeal mucosa below the level of the vocal cords.

The **left recurrent laryngeal nerve** functions likewise, but since the sixth aortic arch persists on the left side (as the ductus arteriosus) and disappears on the right side during development, the left recurrent nerve hooks the old sixth aortic arch instead of riding up around the subclavian artery (fourth aortic arch derivative). Hence, the left nerve makes its recurrent bend below the level of the arch of the aorta in the adult.

The vagus gives branches to the chemoreceptors of the carotid and aortic bodies as it passes by them.

Branches to the Cardiac and Pulmonary Plexuses

The exact details of the way autonomic cardiac nerves enter this plexus and the differences on each side of the neck relative to the supply of the deep and superficial parts of the cardiac plexus and to the pulmonary plexus are not worth worrying about by the beginning student. Textbook descriptions are stereotyped and may bear little relation to the case in any individual body.

Suffice it to realize at this time that the vagus gives off three cardiac branches to these plexuses from each side of the neck and trunk. The **first vagal cardiac nerve** comes off high in the neck. The **second vagal cardiac nerves** come from the recurrent laryngeals as they make their loops around the right subclavian or around the ductus arteriosus on the left side. The **third vagal cardiac branches** arise from the nerve as it passes through the thorax. These branches enter the cardiac and pulmonary plexuses and course through them to reach terminal ganglia in the posterior atrial walls of the heart or in the walls of bronchi. Postganglionics supply the cardiac and smooth muscle as well as bronchial glands.

The vagal contributions in the plexuses are joined by three sympathetic cardiac nerves which arise already as postganglionic fibers from the ganglia of the cervical sympathetic chain. Their origins are as variable as the nature of the cervical ganglionated trunk. But one, at least, the **superior cervical sympathetic cardiac nerve,** invariably arises from the superior cervical sympathetic ganglion. These nerves run directly to cardiac and smooth muscle by traversing the plexuses. Sensory fibers from the heart and from alveolar sacs and bronchi traverse the plexuses in reverse direction. Thus, in essence, these plexuses consist of: (a) preganglionic, visceral motor, vagal fibers; (b) postganglionic, visceral motor, sympathetic fibers; and (c) visceral sensory fibers returning from the same visci to which an autonomic motor supply is being delivered. The three types of fibers intermingle and criss-cross in plexiform fashion to get to where they are going. This is standard practice for most of the autonomic plexuses you will encounter. Sometimes sympathetic preganglionics will also enter plexuses, but they are not involved in the cardiac and pulmonary plexus. Since parasympathetic postganglionics lie in, or close to, the walls of the organs they innervate, they rarely get involved in traversing the plexuses. Plexuses form along blood vessels as a rule, but sometimes along visceral organs; *e.g.*, esophageal plexus.

Figure VII-49 demonstrates schematically that, after entering a plexus upon the esophagus, both right and left vagi follow that organ through the Diaphragm to supply the liver, stomach, spleen, gall bladder, pancreas, small intestine, ascending colon, and at least half of the transverse colon. These preganglionic fibers always terminate in ganglia within the walls or capsules of the structures they innervate.

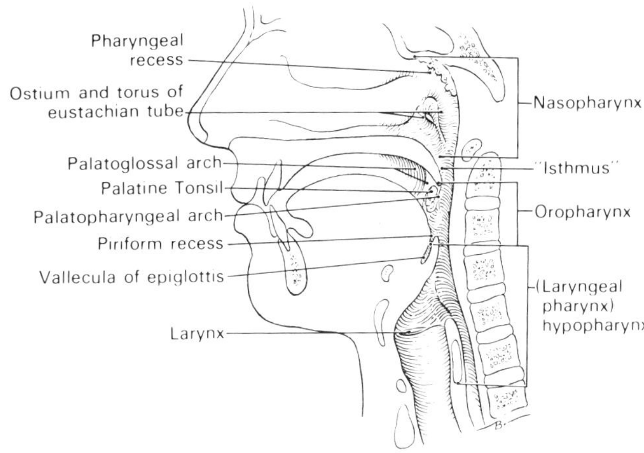

Fig. VII-50. Hemisected head showing divisions of pharynx.

THE PHARYNX

Food introduced into the mouth, if all goes well, moves into that portion of the digestive tract known as the **pharynx,** or throat. The part of the pharynx above the level of the soft palate is the **nasal pharynx,** for the choanae open into it posteriorly. The short stretch between the soft palate and the laryngeal opening is the **oral pharynx** (Fig. VII-50). The section posterior to the larynx, between the epiglottis and esophagus, is known as the **laryngeal** or **hypopharynx.** Because of all of these various anterior openings into it, its muscular wall exists primarily laterally and posteriorly.

The muscular composition of the pharynx can then be approached and seen best from behind, after parts of the occipital bone and vertebral column have been cleared away (Fig. VII-51A and B). The muscles will be covered by the thin coating of buccopharyngeal fascia continuing down over them from the Buccinator muscle. At the upper end of the pharynx our old friends which traverse the carotid sheath—the jugular vein, vagus nerve, and internal carotid artery, can be found **(Fig. VII-51B).** We have noted already that IX supplies the Stylopharyngeus muscles. The vagus supplies the rest of this funnel-shaped pharyngeal tube, its external laryngeal branch going to the lowest of the three overlapping constrictor muscles which form most of its bulk (Fig. VII-51A). The pharynx receives its blood supply from the ascending pharyngeal artery, some small twigs from the facial and lingual arteries, and from the superior and inferior thyroids which serve its lower parts (Fig. VII-51B).

The three constrictors and one somewhat longitudinally-oriented muscle slip form a rather simply-constructed tubular pharynx. The lower end of the longitudinal **Stylopharyngeus** disappears between the

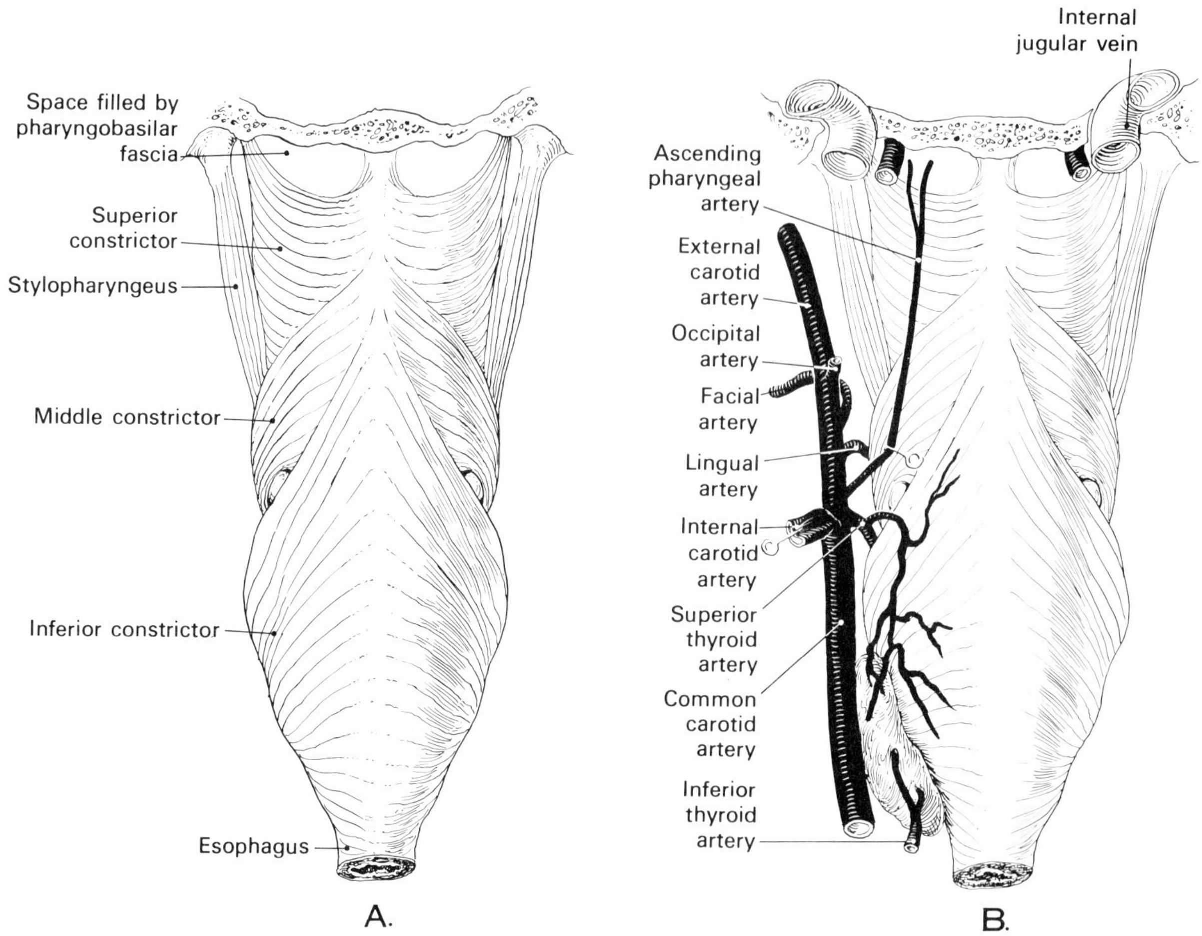

Fig. VII-51. Muscles and arteries of pharynx (seen from behind).

Superior and Middle Constrictors (Fig. VII-51A). The muscles, even the constrictors, are bilaterally represented, the constrictors from each side meeting posteriorly in a midline raphe.

The upper border of the Superior Constrictor does not quite fill in the posterior pharyngeal wall. This border is scalloped, looping on each side from the **medial pterygoid plate** to the midline **pharyngeal tubercle** on the basi-occiput. Through the gap passes the **pharyngeotympanic (auditory, Eustachian) tube,** packed around by stout **pharyngeobasilar fascia.** The ascending palatine artery also goes through this gap. This fascia supports the mucous membrane of most of the nasopharynx with the Tensor Veli Palatini external to it and the Levator Veli Palatini internal to it (Fig. VII-52). If the fascia is dissected away, these two palatine muscles tend to cover the gaps above the Superior Constrictor. Other origins of the Superior Constrictor become, progressively caudad, the pterygomandibular raphe; the mylohyoid line of the mandible; and the lateral tongue margins. Its lower border is under cover of the Middle Constrictor, and we note an anterolateral gap for the passage of the Stylopharyngeus between these two constrictors (Fig. VII-51). Coming through this second gap with the Stylopharyngeus will be the cranial nerve IX.

The Middle Constrictor has a narrow origin anteriorly and widens out posteriorly to reach the midline raphe. It comes from the greater and lesser horns of the hyoid bone and from the stylohyoid ligament (Fig. VII-52). It is overlapped by the Inferior Constrictor. In the anterolateral gap between the Middle and Inferior Constrictors will pass the internal laryngeal nerve and the superior laryngeal vessels. They run beneath the Thyrohyoid muscle and enter the larynx by piercing the thyrohyoid membrane (Fig. VII-52).

The Inferior Constrictor arises from the region of the oblique line on the thyroid cartilage and from the cricoid cartilage (Fig. VII-52). The lower cricopharyngeal part of this muscle is sphincteric in action. It maintains tonus until swallowing begins, and then it serves as a sphincter between the pharynx and esophagus which it encircles. This normally stops regurgitation of material from the stomach back to the laryngeal level unless

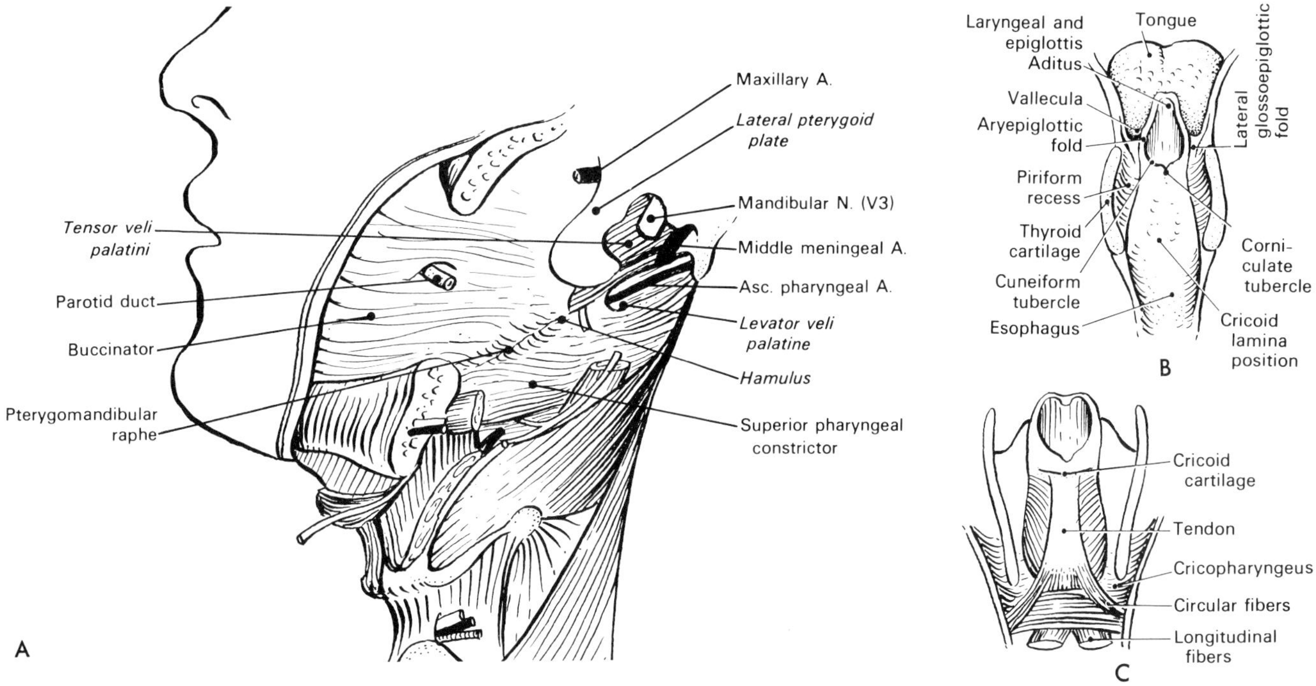

Fig. VII-52. The laryngopharyno (hypopharynx). **A.** Relation of palatine muscles to pharynx and pterygoid lamina. **B.** From behind with posterior wall removed. **C.** Showing attachments of esophagus to larynx. (**A** and B after Hollinshead.)

vomiting is in active progress. The more longitudinally-oriented of its fibers in this region sometimes go into spasm to cause esophageal obstruction and, in time, produce caudally-directed outpouchings, or **hypopharyngeal (Zenker's) diverticulae.** When these become so large as to interfere with the normal propulsion of food toward the stomach, they require surgical intervention. Through the anterolateral gap between the Inferior Constrictor and esophagus pass the recurrent laryngeal nerve and the inferior laryngeal artery (Fig. VII-52).

The Nasopharynx (Fig. VII-50)

This portion of the pharynx remains constantly open. It has, essentially, no anterior wall since the **choanae** open into it here. It has a roof, or **fornix,** and lateral and posterior walls. A pseudostratified epithelium, non-ciliated, forms the surface layer of the mucosa which is applied to the basal parts of sphenoid and occipital bones in the roof and to the **pharyngobasilar fascia** that lines the Superior Constrictor muscle internally in the lateral and posterior walls. If there is any mobility at all in this upper portion of the pharynx, it is limited to the soft palate in its anterior floor, or partial floor, since the nasopharynx is a passageway for air directed toward the larynx.

Above the level of the soft palate in the lateral wall is the opening (ostium) of the **auditory tube** which leads to the middle ear cavity. The **tubal torus** forms an elevation posterior to the ostium. Between this elevation and the occipital bone in the roof is a narrow projection of the pharynx upward as the **pharyngeal recess** (Fig. VII-50).

In the posterior wall and roof, about opposite the point where the inferior meatus of the nose empties into the choana, lies the **pharyngeal tonsil.** There may be a **tubal tonsil** at the posterior lip of the opening into the auditory tube. These, together, are masses of lymphoid tissue that constitute the **adenoids** of childhood. When they become infected and enlarged they may obstruct the nasopharynx and interfere with breathing. They can also obstruct the auditory tube in such a way as to impair hearing. Hence, they are often removed in children though they dwindle and disappear by adulthood if they are left intact. Along with the lingual tonsils at the base of the tongue and the palatine tonsils between the faucial pillars at the entrance of the oral pharynx, these masses make up the "lymphoid ring of Waldeyer" at the upper end of the digestive tract. Wilhelm Gottfried Waldeyer proposed in 1884 that this ring of tonsilar tissue served as a line of defense against bacteria entering the mouth. Others postulated a similar role for the aggregates of lymphoid nodules in the walls of the appendix and in the ileum (Peyer's patches) at the lower end of the digestive tract. Perhaps the "lymphoid ring" hypothesis has always been more fanciful than factual since tonsils, adenoids, and appendices are so readily overcome by bacterial invasion and are commonly removed in inflamed states. It is unfortu-

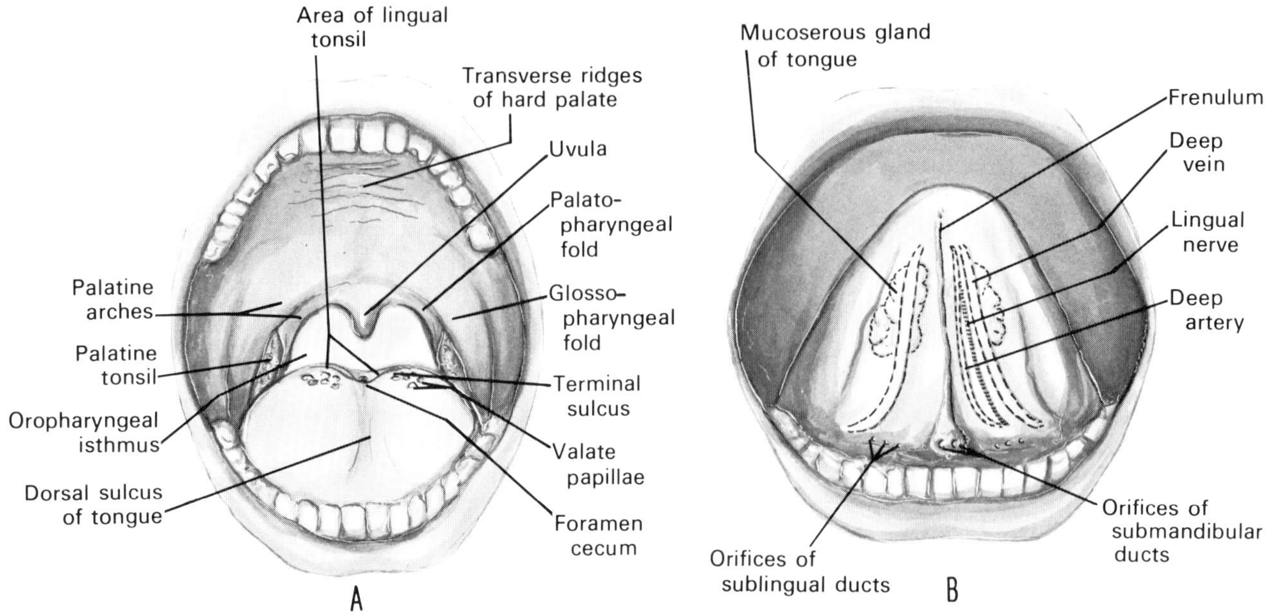

Fig. VII-53. The oral cavity. **A.** Tongue depressed, note uvula. **B.** Tongue elevated. (After Hamilton, Appleton, and Simon.)

nate that most clinicians associate Waldeyer's name with an outmoded concept. He really should be remembered for his outstanding contributions to the understanding of nerve structure and function, and as the father of the "Neuron Doctrine." This stressed the structural uniqueness of the single nerve cell, and the lack of continuity of cytoplasm at the synaptic junction at a time when there was grave controversy and confusion about the nature of nerve cells.

The **oral pharynx** can be regarded as extending from the tip of the soft palate to the upward protecting tip of the **epiglottis** (Fig. VII-50). The mouth opens into it. The lateral walls of this opening are bounded by the **palatine arches.** These are two folds that curve from the soft palate to the tongue, the **palatoglossal arch** anteriorly, the **palatopharyngeal arch** posteriorly (Fig. VII-50). These mucosal folds cover the muscles of the same name, and between them in a fossa nestles the **palatine tonsil** that we know as the garden variety "tonsil," the one we have removed at times when we complain of sore throat. The arches are also called faucial pillars, and the term "fauces" means throat. Below the palatine arches the posterior part of the tongue projects somewhat into the oral pharynx. Behind the line of vallate papillae many small, single-crypted, lymphoid masses are found at the base of the tongue. These are the lingual tonsils.

The tonsil is well supplied with blood. Its major supply enters its lower pole from the tonsilar branch of the facial (external maxillary) artery. It gets twigs from the ascending pharyngeal, descending palatine, ascending palatine branch of the facial, and from the lingual artery, as well.

The epiglottis lies in a kind of no-man's-land between the oral and laryngeal pharynx (Fig. VII-50). It is anchored to the tongue by mucosal folds, one in the midline and two laterally (**median and lateral glosso-epiglottic folds**). The paired depressions between the medial and lateral folds are the **epiglottic valleculae** (Fig. VII-52). Swallowed foreign bodies often get trapped in these "little valleys" or pockets on each side of the epiglottis. Chiefly, the epiglottis stands sentinel, guarding the orifice to the larynx, but occasionally bits of food or drink are misguided past this guard into "the wrong pipe." A fit of reflex coughing often sets things aright by forcibly rejecting these offensive intruders.

The tip of the soft palate or **uvula** hangs down between the palatine arches (Fig. VII-53). If you ask a patient to say "Ah" and you notice a deviation of the uvula from the midline, you can suspect damage to the vagus nerve on the side opposite that to which the uvula points. The muscles of the intact side are pulling the uvula toward their side.

The **laryngopharynx** is the lower stretch of pharynx from epiglottis to esophagus (Fig. VII-50). The pharynx rides down laterally around the larynx to make another pair of blind pockets, the **piriform recesses or sinuses** (Fig. VII-52B). The superior laryngeal nerve may raise a fold as it courses through the floor of the recess toward the larynx.

Junction of Pharynx and Esophagus

The internal mucosae of these two tubes is continuous. The outer muscular laminae are only partly in continuity. The esophagus attaches to the posterior aspect of the cricoid cartilage for the most part. Posteriorly the external longitudinal muscular coat of the esophagus becomes concentrated near the cricoid into two diverging longitudinal bands. This leaves the inner

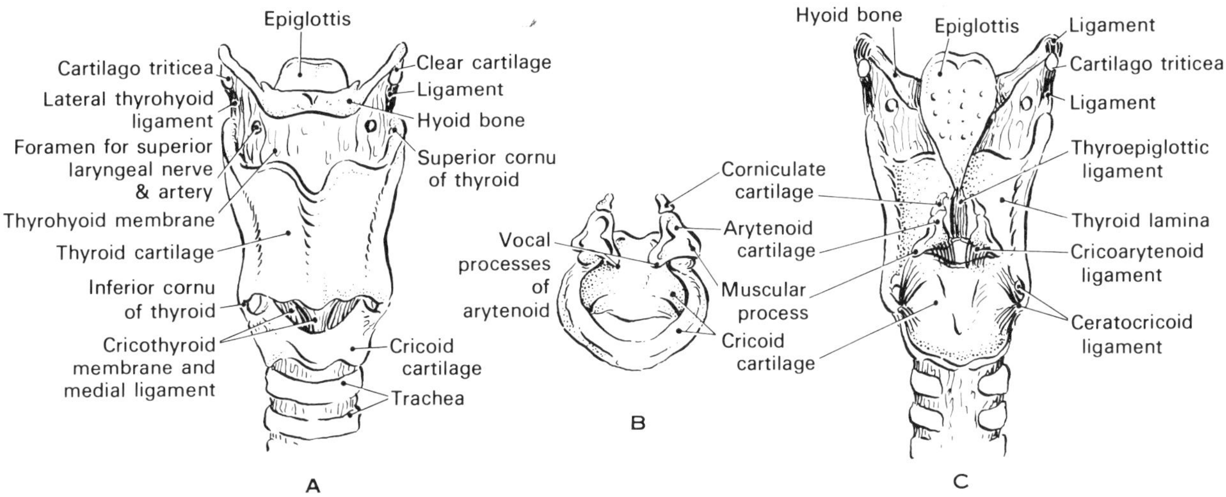

Fig. VII-54. The larynx. **A.** Frontal view. **B.** Relation of cricoid to arytenoid cartilages. **C.** Larynx from behind.

circular coat of muscle exposed in this region (Fig. VII-52C). The longitudinal bands unite to form the **cricoesophageal tendon** attaching to the lower posterior border of the cricoid. Here the circular muscle of the esophagus blends with the lowermost sphincteric fibers of the Inferior Constrictor (the so-called Cricopharyngeal muscle). Some of the circular muscle also blends into the tendon at this junction. Recall that laterally in this region there is a slight gap between the Inferior Constrictor and esophageal wall through which recurrent laryngeal nerve and inferior laryngeal vessels pass to reach the larynx.

THE LARYNX

The larynx is the first part of the respiratory system which opens into the oral pharynx and continues into the trachea. We noted its early development in the back part of the foregut, and observed how the development of the epiglottis was bound up with that of the tongue.

The larynx is best understood by constructing first its cartilaginous and membranous framework, hanging the muscles upon this framework, and then considering muscular action.

The chief cartilages are the **thyroid, cricoid,** and **epiglottis** as single, unpaired structures; and the smaller **arytenoid, corniculate,** and **cuneiform** cartilages as paired structures (Fig. VII-54). There are, in addition, two minor nodules, the **triticeal cartilages** (Fig. VII-54A and C), which are little pieces of the thyroid cartilage that became detached in development. These, along with corniculate (Fig. VII-54B) and cuneiform cartilages, are of minor significance. All of the cartilages are of the hyaline variety except the epiglottis, corniculate, cuneiform, and the vocal process of the arytenoid, which are elastic cartilage. The thyroid, cricoid, and major portion of the arytenoid can ossify with age. The epiglottis can become infiltrated with fat as the cartilage deteriorates with age.

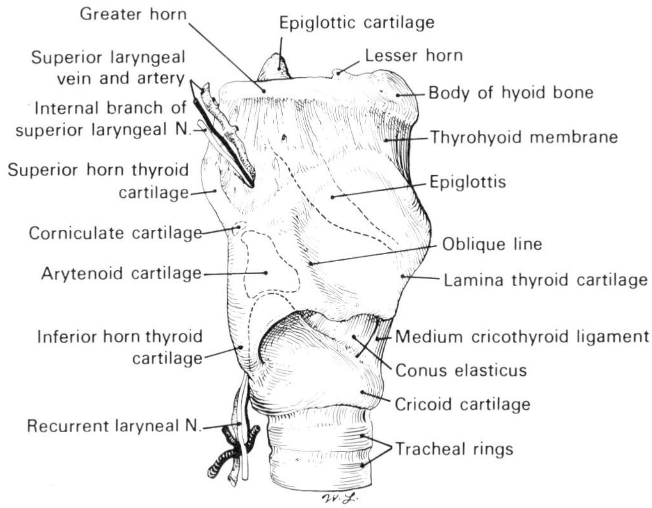

Fig. VII-55. Cartilages of the larynx viewed from the side.

The **thyroid cartilage** is the largest, and it sits anteriorly like a bowed shield, open behind (Fig. VII-54A and C). Its wings flare back from a permanent midline prominence, the Adam's apple. Supposedly, this is the level where the forbidden fruit stuck in Adam's craw, and this is one fanciful way to account for the greater size of this structure in the male than the female. The posterior borders of this cartilage are prolonged into **upper** and **lower horns.** Along the surface of each lateral plate is an **oblique line** (Fig. VII-55) to which three muscles attach—the Inferior Pharyngeal Constrictor and two of the strap muscles, the Sternothyroid and Thyrohyoid. A membrane, the **thyrohyoid membrane,** stretches from the upper horn

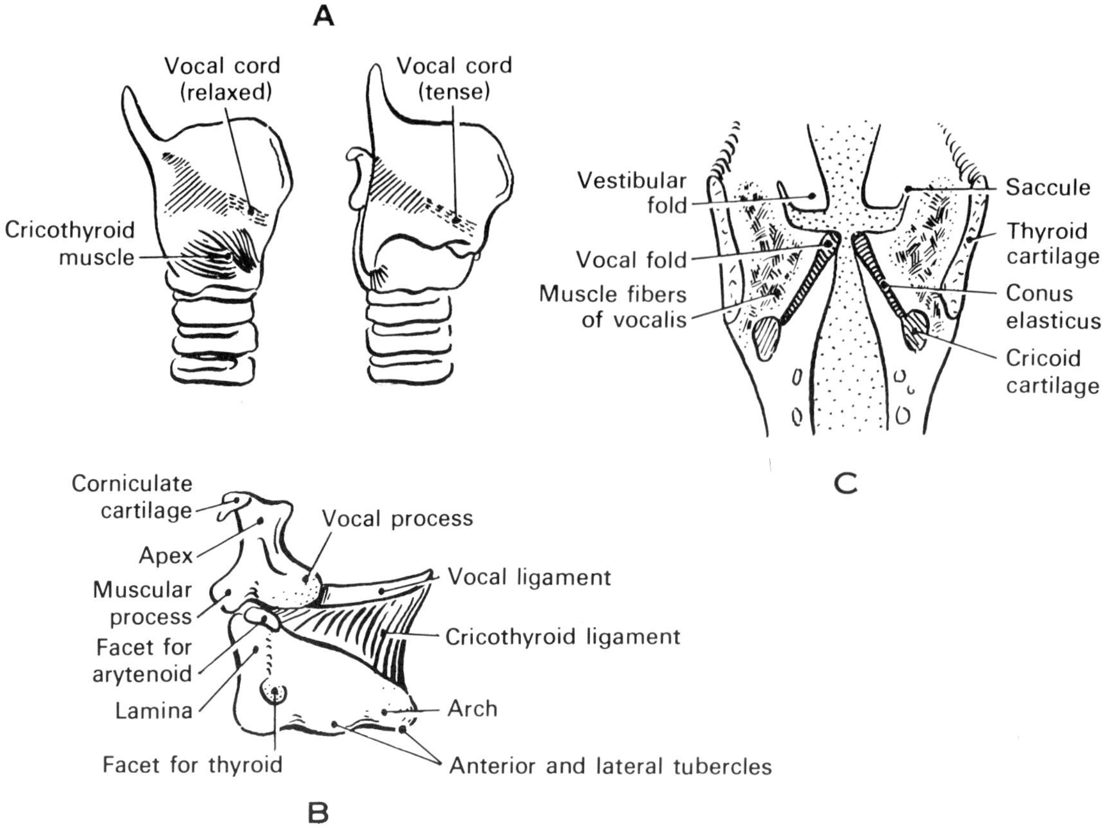

Fig. VII-56. Larynx. **A.** Movement of vocal cords by Cricothyroid muscle. **B.** Lateral view, thyroid cartilage removed to show cricothyroid membrane (conus elasticus). **C.** Vestibular folds and conus elasticus in sagittal view. (A and B after Grant.)

and upper border of the thyroid cartilage to the lower border and greater horn of the hyoid bone, closing in the gap between these structures anteriorly. Its thickened lateral edge (**lateral thyrohyoid ligament**) contains the displaced triticial cartilage (Fig. VII-54A). Laterally on each side of the membrane there is a tiny aperture through which the internal branch of the superior laryngeal nerve passes to enter the layrnx.

The **cricoid cartilage** (Fig. VII-54B) is ring-shaped but taller behind than in front. Actually, it is the only complete ring of cartilage in the respiratory tubes, those of the trachea tending to be somewhat C-shaped and open behind like the thyroid cartilage, and those of the bronchial tree to be disconnected and overlapping plaques. There is a circular facet on the lateral wall of the cricoid (Fig. VII-56B) with which the inferior horn of the thyroid cartilage articulates. The thyroid cartilage rotates at this point (Fig. VII-56A), tliting forward and backward like a visor on a knight's helmet. When the Cricothyroid muscle pulls the thyroid cartilage forward and downward, it tenses the vocal cords, relaxing them on the backswing. How this can be accomplished we shall note later when we understand what the cords are and how they attach from the vocal processes of the arytenoids to the midline at the inner angle between the diverging thyroid laminae (wings, plates). Stretching between cricoid and thyroid cartilages is the **cricothyroid membrane** (Fig. VII-54A), a tough elastic membrane. When the larynx is viewed laterally, we only see the flat base of this membrane. Its apex projects inward and upward into the interior of the larynx (Fig. VII-56C) to anchor to the vocal process of the arytenoid and to the thyroid cartilage at its inner angle. Thus, it is the free upper edge of the membrane stretching between these two points that thickens as the **vocal ligament.** Cover this "ligament" with a lining mucosa and you have constructed a **vocal cord** (or **vocal fold**) (Fig. VII-56B and C). Another name for the odd-shaped cricothyroid membrane is the **conus elasticus** (Figs. VII-55 and 56C). Its anterior, outer, midline thickening is called the **cricothyroid ligament** (Fig. VII-55).

The paired arytenoid cartilages sit as pyramids upon the posterosuperior border of the cricoid (Fig. VII-54B). They are really shaped like the quadrant of a cone whose surfaces have been molded somewhat so that the base and posterior surfaces are concave, the medial surface flat and the anterolateral surface more or less rounded. They also have three projections—an anterior one that is sharp to which the vocal ligament attaches; a blunt lateral one to which muscles attach; and an apical one to which a detached piece, the **corniculate cartilage,** is fastened by perichondrium.

At the **crico-arytenoid joints,** the arytenoids are

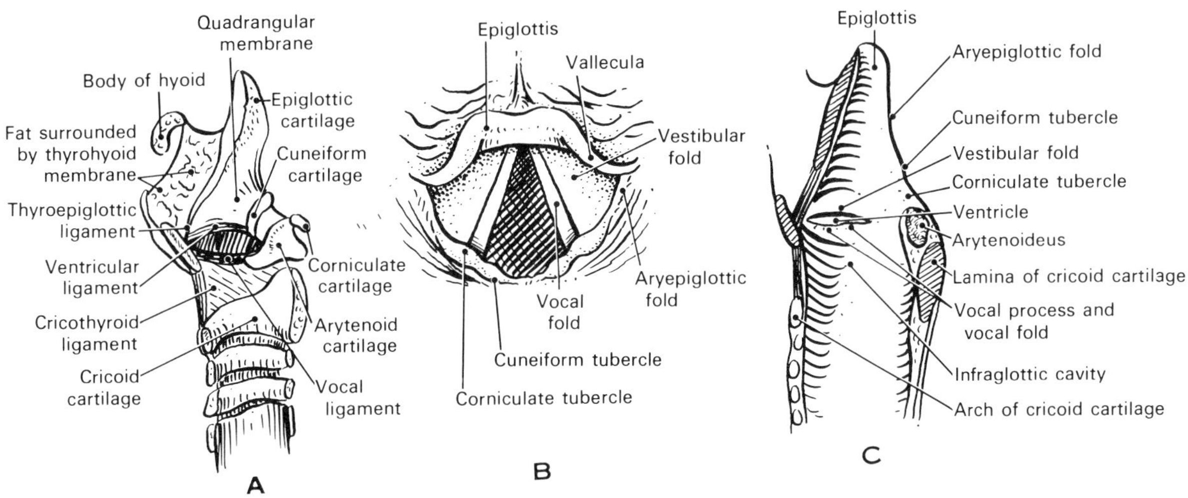

Fig. VII-57. Larynx. **A.** Cartilaginous and membranous skeleton, midsagittal section. **B.** Vocal cords abducted, viewed from above. **C.** Interior of larynx, midsagittal view with mucous membranes intact.

quite movable. They can slide medially and laterally along the cricoid, tilt forward and backward, or rotate so that their muscular processes can be turned medialward or laterad. This is most important, for this is the way vocal cords will be approximated, tensed or relaxed, and the way the glottis will be opened. The arytenoids are, probably, the most important cartilages in the larynx on this account, since most of the intrinsic muculature activates them and the vocal cords attached to them.

The **epiglottis** stands up in the pharynx behind the tongue. It is broad above and narrows to a stalk below (Fig. VII-54C) which is attached to the inner midline angle of the thyroid cartilage, above the level of the vocal cords, by the **thyro-epiglottic ligament** (Fig. VII-54C). Extending from the arytenoid cartilage and its surmounted corniculate cartilage, upward to the lateral border of the epiglottis, is the **quadrangular membrane** (Fig. VII-57A), a connective tissue sheet which, when covered with a mucous membrane, constitutes the **aryepiglottic fold.** This fold separates the piriform recess of the pharynx from the laryngeal entrance. In its free border is buried the last of the cartilages, the club-shaped **cuneiform cartilage.** This is really a developmentally-detached part of the epiglottis.

We have now constructed the framework of the larynx and viewed it mostly from the outside devoid of its muscles. Before we add these, it might be best to view the inside. This can best be appreciated by a sagittal hemisection. Its **aditus or entrance** is bounded above by the epiglottis, laterally by the aryepiglottic folds, and by the Arytenoid muscles below covered by mucous membranes. The cavity of the larynx proper can be divided into three sections from above and below (Fig. VII-57C): the **vestibule** extends from the aditus to the **vestibular (ventricular) folds** which are prominent lateral mucosal folds (false vocal cords); the **glottis,** lying between ventricular and **vocal folds** (true vocal cords); and the area below to the cricoid cartilage, the **infraglottic cavity.**

The vestibular folds are rather bulky mucosal folds compared to the shaper-edged, more medial-lying vocal folds below them (Fig. VII-57B). Muscle fibers underlie the vestibular folds, and their edges can be brought together in the midline to form a narrow slit, the **rima vestibuli.** This apposition occurs when the breath is held against build-up of thoracic pressure as in weight lifting. Between vestibular and vocal folds, the glottis expands laterally as the **laryngeal ventricle.** Near the anterior end of the ventricle a small blind sac, the **laryngeal saccule,** protrudes. Only rarely in man does it protrude through the thyroid membrane to lie deep to the strap muscles. This condition is reminiscent of the great saccule of some of the howling monkeys and, especially, of the gorilla where it may extend to the axilla (arm pit).

The slit between the vocal folds is the **rima glottidis.** The anterior part of each vocal fold is membranous in that the mucous membrane covers the vocal ligament which, you recall, is the thickened, apical end of the triangular, cricothyroid membrane (conus elasticus) (Fig. VII-56B and C; 57A). The posterior ends of the vocal folds, as you might suspect, are cartilaginous. Indeed, the cartilage represents the vocal processes of the arytenoids to which the mucosal vocal folds attach.

Some of these structures related to the larynx can be seen in the soft tissue radiographs (Fig. VII-58A and B).

The two vocal folds or cords must be brought together for phonation to allow for vibration as air passes between them. Since they can swing apart much like the blades of shears, by muscle action, a wide opening aids in maximum breathing.

There is nothing much to be said about the lower

part of the laryngeal cavity except that it is wider fore-and-aft than laterally and narrows down as the trachea is approached.

Thus, a mucosal membrane lines the interior of the cartilagino-membranous framework (Fig. VII-57C) and *all* but *one* of the muscles that activate the larynx lie beneath this mucsoa. The epithelium of the mucosa is stratified squamous where attrition and friction are most prevalent; *e.g.*, over the epiglottis as far as its tubercular swelling and in the area of the vocal cords, themselves. Elsewhere it is typically respiratory; *i.e.*, ciliated and pseudostratified. The lamina propria upon which the epithelium rests is rich in elastic fibers, mixed glands, and lymphoid nodules which are prevalent except in the vocal folds. They are especially numerous in the vestibular folds.

The laryngeal muscles are divided into two functional groups: (1) those serving to protect the larynx and to act sphincterically upon the opening and the vestibule; (2) those serving to adjust larynx and vocal cords during phonation and breathing. Of these latter, only the **Posterior Crico-arytenoids** abduct the vocal cords; that is, open the rima glottidis (Figs. VII-59 and 60). So these are very important muscles. The rest in this second group act to adduct or approximate the cords in one way or another.

The first group of muscles consist of the Aryepiglotticus, the Thyro-epiglotticus, the Thyro-arytenoideus, and the Oblique Arytenoideus.

The **Thyro-arytenoideus** lies under the mucosa on the inner side of the larynx as a thin sheet that is heaviest in the region of the vestibular fold. It arises from the lamina of the thyroid near its inner angle and inserts on the lateral border of the arytenoid (Fig. VII-59A; 60A and B). One might think that it would draw thyroid and arytenoid cartilages toward each other when it contracted. But it does not contract alone and it serves as one of the general sphincters of the vestibule, acting mainly to approximate vestibular folds.

The **Thyro-epiglotticus** (Fig. VII-59A; 60A) consists of rather thin wisps of muscle arising from the upper part of the Thyro-arytenoideus. These sweep upward over the quadrangular membrane to blend with the Aryepiglotticus (Fig. VII-59A; 60A and C). The latter runs in the free margin of the quadrangular membrane. The two together act to constrict the entrance to the larynx. The Aryepiglotticus can also be regarded as an upward continuation of the Oblique Arytenoideus to be described below.

The muscles primarily concerned with effecting the vocal cords are: **Lateral** and **Posterior Crico-arytenoids,** the **Vocalis,** the **Transverse Arytenoideus,** and the **Cricothyroideus.** The latter is the only one on the outer aspect of the thyroid cartilage, all other laryngeal muscles being truly intrinsic rather than extrinsic, in the sense that they are hidden behind the thyroid shield and the mucosa and cannot be seen from the anterolateral aspect. Of course, all laryngeal muscles are "intrinsic" in the sense that they are confined to the larynx.

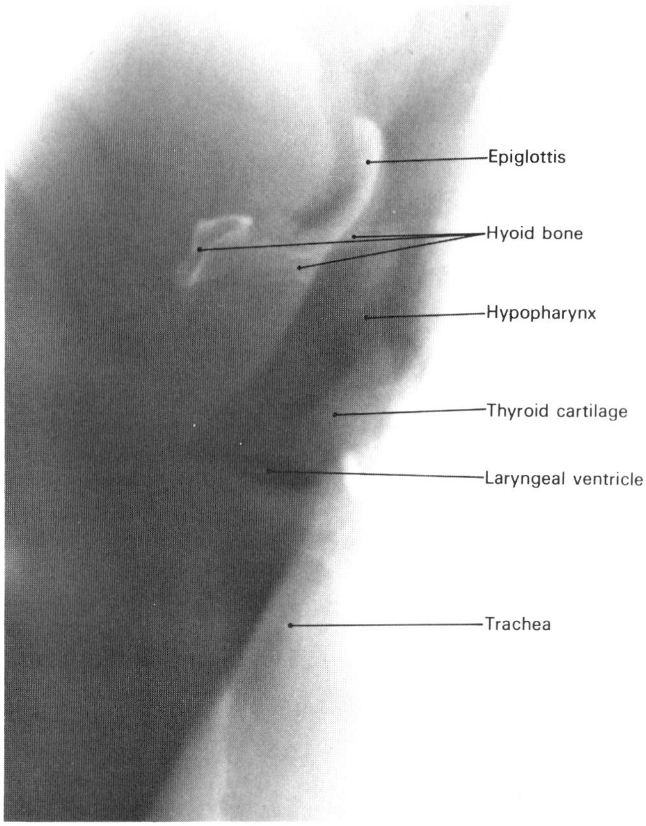

Fig. VII-58. Larynx. **A.** Soft tissue radiogram, lateral view, of hypopharynx, larynx, and trachea.

We have described the **Cricothyroid muscle** (Fig. VII-56A) as the external one that rotates the whole thyroid cartilage forward and backward visor-like, lengthening or thinning the vocal cords in its forward tilt; shortening and relaxing them in the backward swing to normal position. If one of the Cricothyroids becomes paralyzed, the cord on that side is relaxed. The other intact muscle twists the thyroid cartilage toward the paralyzed side.

The **Lateral Crico-arytenoid** (Fig. VII-59A; 60A and B) is applied to the cricothyroid membrane (conus elasticus) below the level of the Thyro-arytenoideus. It arises from the posterolateral arch of the cricoid cartilage to insert in the muscular process of the arytenoid cartilage. In contracting, it rotates the arytenoid medialward and so adducts (Fig. VII-61C) the vocal cord attached to the vocal process of this cartilage.

The **Posterior Crico-arytenoid** (Fig. VII-59A and 60) lies on the posterior surface of the cricoid cartilage, its upper fibers passing almost horizontally and its lowermost fibers, almost vertically to the muscular process of the arytenoid. The horizontal fibers will ro-

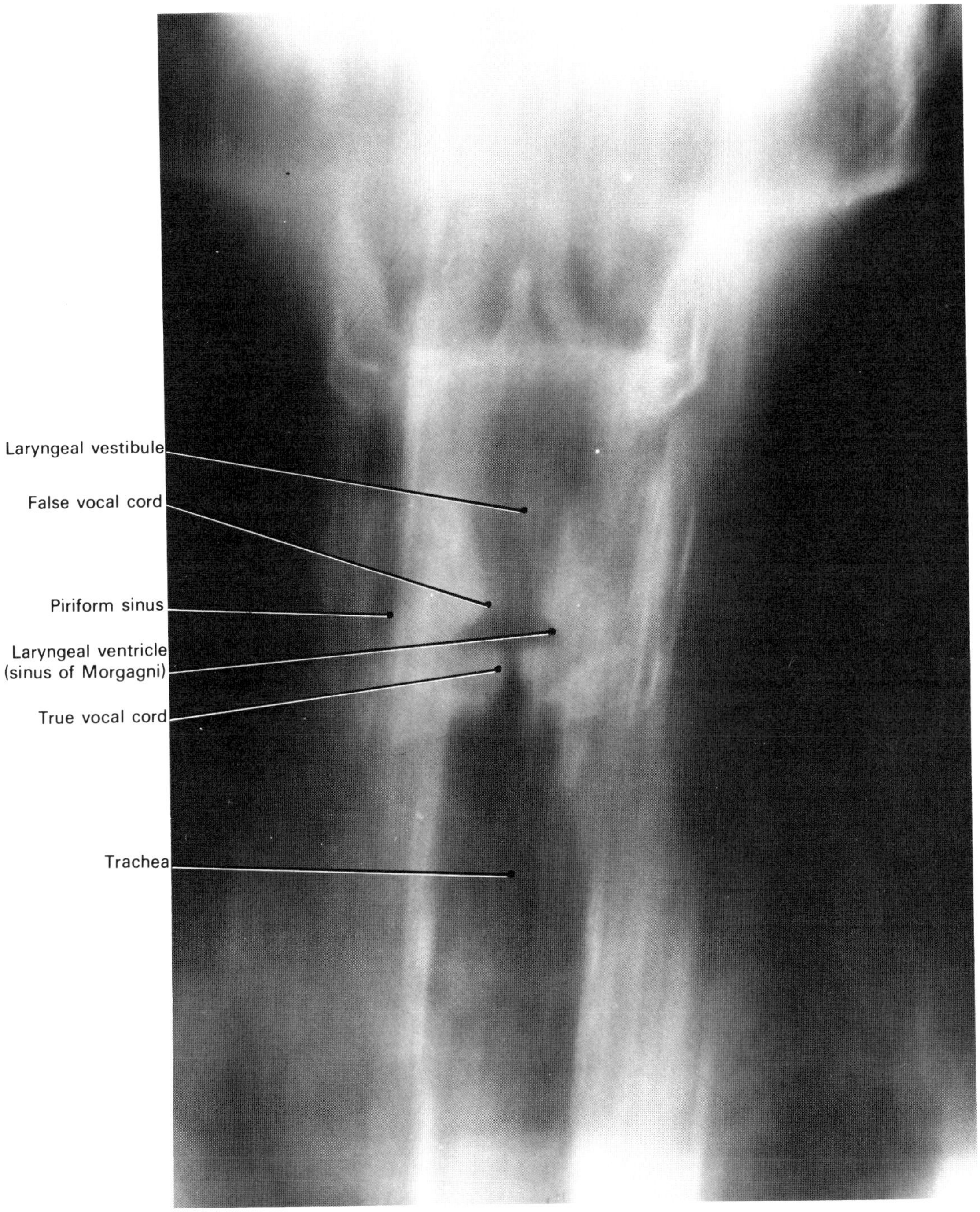

Fig. VII-58B. Tomogram of larynx and trachea.

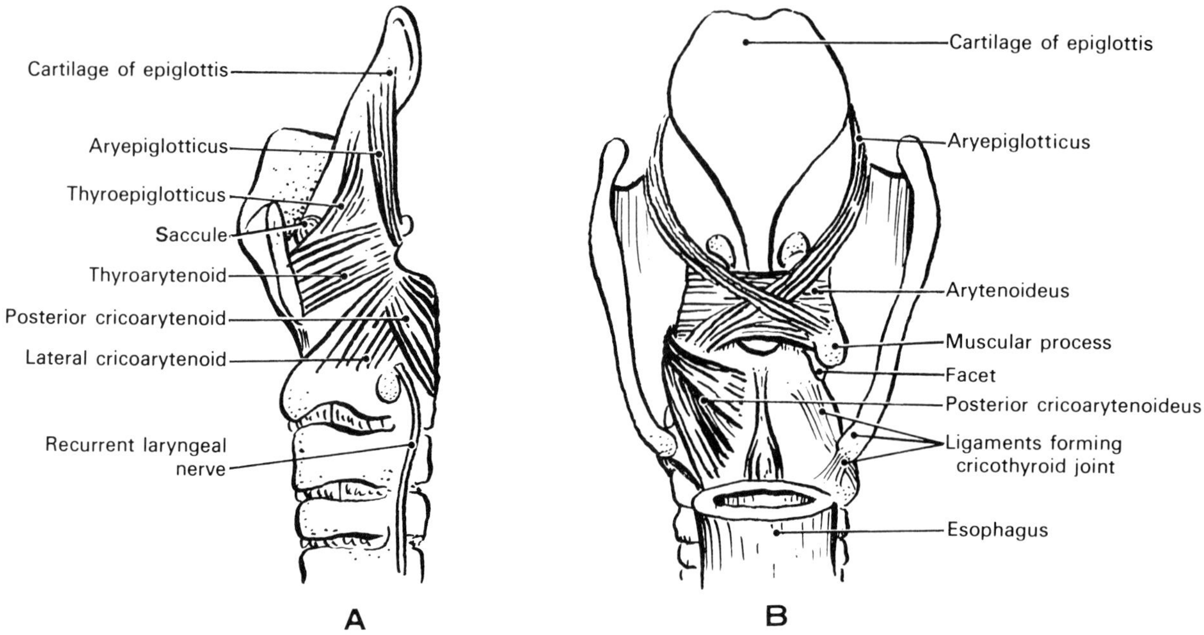

Fig. VII-59. Intrinsic muscles of larynx (schematic). **A.** From side. **B.** From behind. (After Grant.)

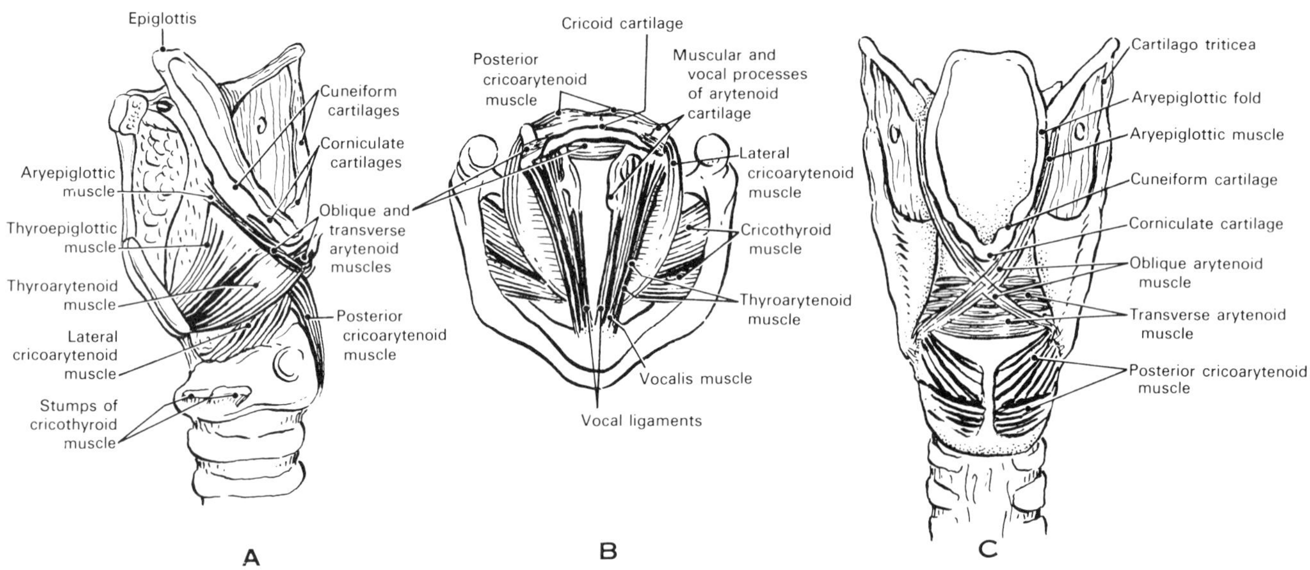

Fig. VII-60. Intrinsic muscles of larynx. **A.** Seen in midsagittal view. **B.** From above, with mucosa stripped away. **C.** From behind.

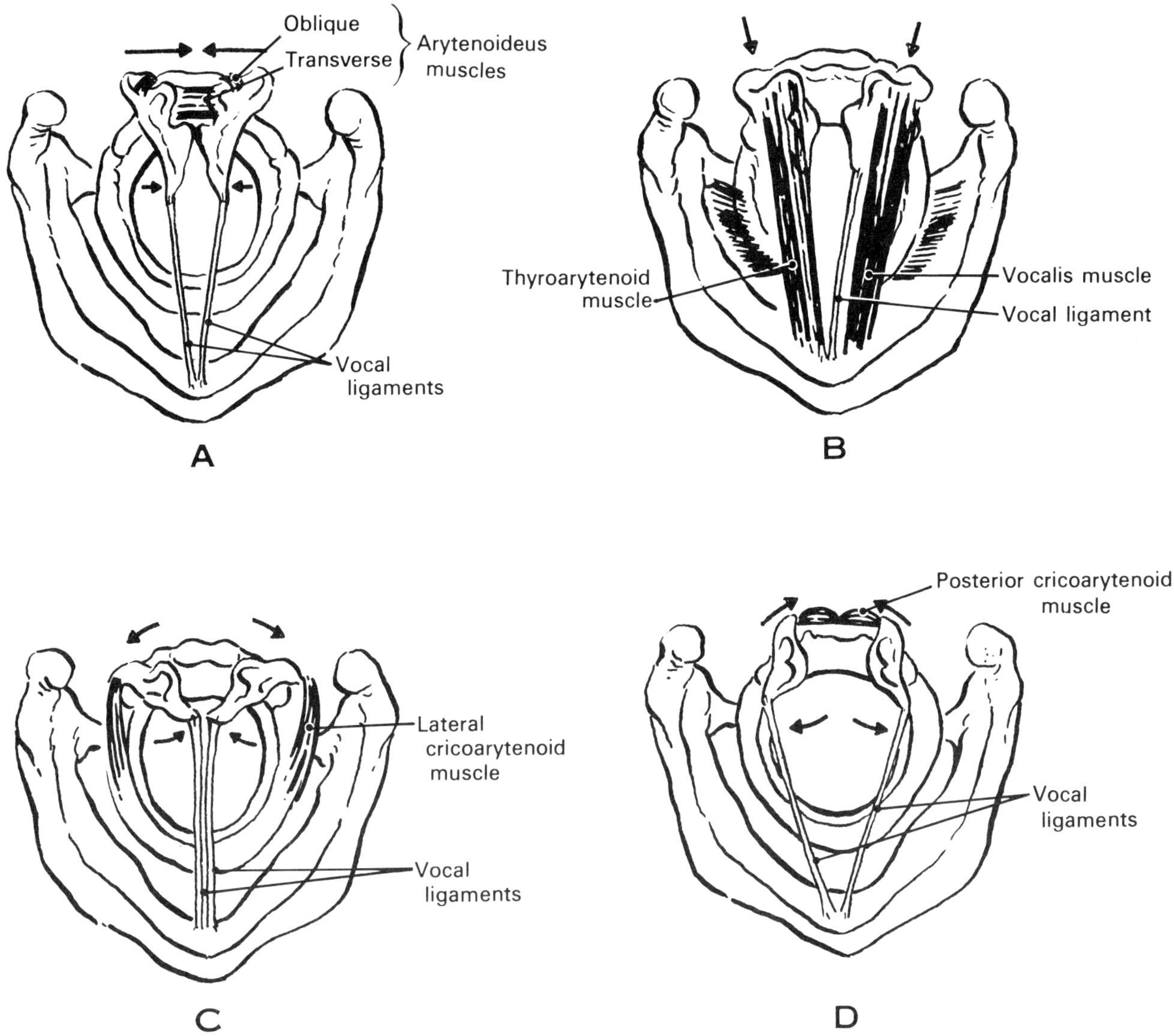

Fig. VII-61. Schema to illustrate movements of the larynx.

tate the arytenoids (Fig. VII-61D) laterally, thus abducting the vocal cords. The vertical fibers steady the arytenoids and may even tilt them backward slightly. As we have said, this is the safety muscle of the group, the only one to open the rima glottidis. In bilateral paralysis, suffocation ensues since all the other muscles close the rima glottidis.

The **Arytenoideus** is a bipartite muscle whose thick transverse fibers attach on the lateral aspect of each arytenoid cartilage (Fig. VII-59B; 60B) and cover them posteriorly except for their exposed tips. In contracting, the transverse fibers slide the arytenoids along the cricoid rim, toward each other and, hence, toward the midline (Fig. VII-61A), again approximating the vocal cords. Oblique strands (the **Oblique Arytenoideus**) arise from each muscular process of the arytenoids to cross their fellow of the opposite side and continue into the aryepiglottic folds, and blend with the Aryepiglotticus in the lateral margins (Fig. VII-60C). These strands help in narrowing the laryngeal opening.

The **Vocalis** is the fine adjuster of tension in the vocal cords (Fig. VII-60B and 61B). It is, actually, the deep medial strands of the Thyro-arytenoideus that peg into the undersurface of the vocal ligaments. When the Cricothyroids act to tense the cords by tilting the thyroid cartilage forward and the Posterior Cricothyroids (vertical bundles) steady or pull back on the arytenoids, the Vocalis can act effectively and with fine finesse to achieve the proper degree of tension upon the vocal ligaments for fine changes in pitch and the nuances of speech.

REFERENCES

A. General:

1. Penning, L.: *Functional Pathology of the Cervical Spine.* Williams & Wilkins, Baltimore, 1968.

B. On the Uncovertebral "Joints" or Fissures:

1. Boreadis, A. G., and Gershon-Cohen, J.: Luschke joints of the cervical spine. Radiology *66:* 181, 1956.
2. Compere, E. L., Tacholjian, M. O., and Kernahan, W. T.: The Luschke joints: Their anatomy, physiology and pathology. Orthopedics *1:* 159, 1959.
3. Frykholm, R.: Lower cervical vertebrae and intervertebral discs: Surgical anatomy and pathology. Acta Chir. Scand. *101:* 345, 1951.
4. Hadley, L. A.: The covertebral articulations and cervical foramen encroachment. J. Bone Joint Surg., Am. Vol. *39-A:* 910, 1957.
5. Hall, M. C.: *Luschke's Joint.* Thomas, Springfield, Ill., 1965.
6. Hirsch, C., Schajowicz, F., and Galante, J.: Structural changes in the cervical spine. Acta Orthoped. Scand. (Suppl.) *109,* 1967.
7. Jackson, R.: *The Cervical Syndrome,* 3rd Ed. Thomas, Springfield, Ill., 1966.
8. Luschke, H.: *Die Halbgelenke des menschlichen Körpers.* Reimer, Berlin, 1858.
9. Rathke, L.: Zur normalen und pathologischen Anatomie der Halswirbelsäule. Deutsch. Z. Chir. *242:* 122, 1933.
10. Töndury, G.: Entwicklungsgeschichte und Fehlbildungen der Wirbelsäule. In *Die Wirbelsäule in Forschung und Proxis,* Bd. 7. Hippokrates, Stuttgart, 1958.
11. Töndury, G.: The cervical spine—its development and changes during life. Acta Orthop. Belg. *25:* 602, 1959.
13. Schmorl, G. and Junghanns, H.: *The Human Spine in Health and Disease.* 1st Amer. Ed., translated by S. P. Wilk and L. S. Goin, Grune & Stratton, New York, 1959.

C. Other References:

14. Weddell, G., Feinstein, B., and Pattle, R. H.: The electrical activity of voluntary muscle in man under normal and pathological conditions. Brain *67:* 178, 1944.
15. Causey, G., and Slome, D.: Noted in Anat. Soc. G. B. and Ireland, Feb. 1954. (No title available.)
16. Grodinsky, N., and Holyoke, E. A.: The fasciae and fascial spaces of the head, neck, and adjacent regions. Amer. J. Anat. *63:* 367, 1938.
17. Norris, E. H.: The parathyroid glands and the lateral thyroid in man; their morphogenesis, histogenesis, topographic anatomy and prenatal growth. Contrib. Embryol. *26:* 247, 1937.
18. Norris, E. H.: The morphogenesis and histogenesis of the thymus gland in man; in which the origin of the Hassall's corpuscle's of the human thymus is discovered. Contrib. Embryol. *27:* 193, 1938.
19. Gilmour, J. R.: The embryology of the parathyroid glands, the thymus, and certain associated rudiments. J. Path. Bact. *45:* 507, 1937.
20. Kingsbury, B. F.: The question of a lateral thyroid in mammals, with special reference to man. Amer. J. Anat. *65:* 333, 1939.
21. Van Dyke, J. H.: Behavior of the ultimobranchial tissue in the postnatal thyroid gland. Amer. J. Path. *76:* 201, 1945.
22. Boyd, J. D.: Development of thyroid, parathyroid glands, and the thymus. Ann. Roy. Coll. Surg. Eng. *7:* 455, 1950.
23. Neel, H. B., and Pemberton, J. De J.: Lateral cervical (branchial) cysts and fistulas. Surgery *18:* 267, 1945.
24. Marshall, S. F., and Becker, W. F.: Thyroglossal cysts and sinuses. Ann. Surg. *129:* 642, 1949.
25. Donaldson, S. W., and Tompsett, A. C.: Tracheal diameter in the normal newborn infant. Amer. J. Roentgen. *67:* 875, 1952.

EIGHT

The Thorax

INTRODUCTION

The heart and lungs are the most important viscera within the thorax. Since both of these organs are in constant motion during respiration and during the repeated cardiac cycle, it soon becomes apparent that this is one of the most dynamic areas of the body.

Some important concepts relating to the thorax have already been discussed in Chapter III. This introductory information should be reviewed at this time (pp. 117–131).

PART I. THORAX

The Chest Wall

The thoracic wall consists of skin, underlying fascia and muscles, the ribs and their cartilages, the sternum and, dorsally, the 12 thoracic vertebrae. The shape of the thorax is roughly that of a cone flattened in front and behind (Fig. VIII-1). A cross-section taken through the upper portion of the chest is flatter than one from the middle or from the lower parts. The vertebrae and the dorsal convexity of the ribs indent and project so far into the central portion of the thorax that two wide dorsal "gutters" flank the vertebrae along the back (Fig. VIII-2). These gutters are occupied mainly by the musculature of the Erector Spinae group. The inward projection of these bony parts is so great that at the level of VT_8 the esophagus, situated just in front of the aorta, occupies the center of the body as viewed in a cross-section.

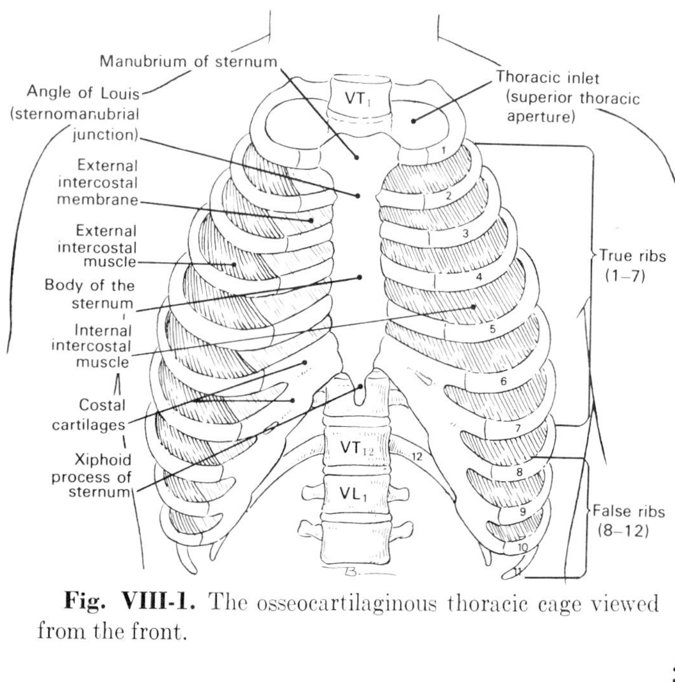

Fig. VIII-1. The osseocartilaginous thoracic cage viewed from the front.

The thoracic cone is not quite complete above, and the superior opening or **thoracic inlet** is filled with structures entering or leaving the bony cage. The trachea, esophagus, various vessels and nerves, muscles, and connective tissue fill the inlet. The cupola of the pleural cavity also rises through this opening, and this may be seen in the lateral radiograph of the cervical spine, Figure VIII-3, and in the coronal radiograph of the body section, Figure VIII-5.

Surprisingly, the inlet is small in its anteroposterior diameter, measuring about 2¼ inches (Fig. VIII-2). Transversely it measures about 4⅜ inches. The inlet is bounded by the sternum in front, the body of VT_1 behind and laterally by the first pair of ribs and their cartilages (Fig. VIII-4).

Inferiorly, the **thoracic outlet** is bounded by the vertebra VT_{12}, the 12th rib, the lowermost costal cartilages, and the xiphoid process of the sternum. The 12th vertebra is joined posteriorly by the 12th ribs which project downward and end rather abruptly. The free edges of the 10th ribs join the costal cartilages of the lower six pairs of ribs. The conjoined cartilaginous bar swings upward to articulate at the xiphisternal junction, forming an articulation with the sternum alone at this point if the xiphoid process is missing (Fig. VIII-1). The thoracic outlet is closed by the Diaphragm with its openings for esophagus, blood vessels and nerves, and for the thoracic duct as a major lymph channel. The Diaphragm domes up and attaches higher in the thorax anteriorly than it does posteriorly and higher by an interspace or so on the right side than on the left (Fig. VIII-5). This permits some of the abdominal viscera (liver, stomach, and spleen) to seek shelter under the ribbed cage. The cage, in this respect, behaves much like the grill of an automobile that offers some protection to the radiator provided that the grill, itself, suffers no grave damage. Entering the ab-

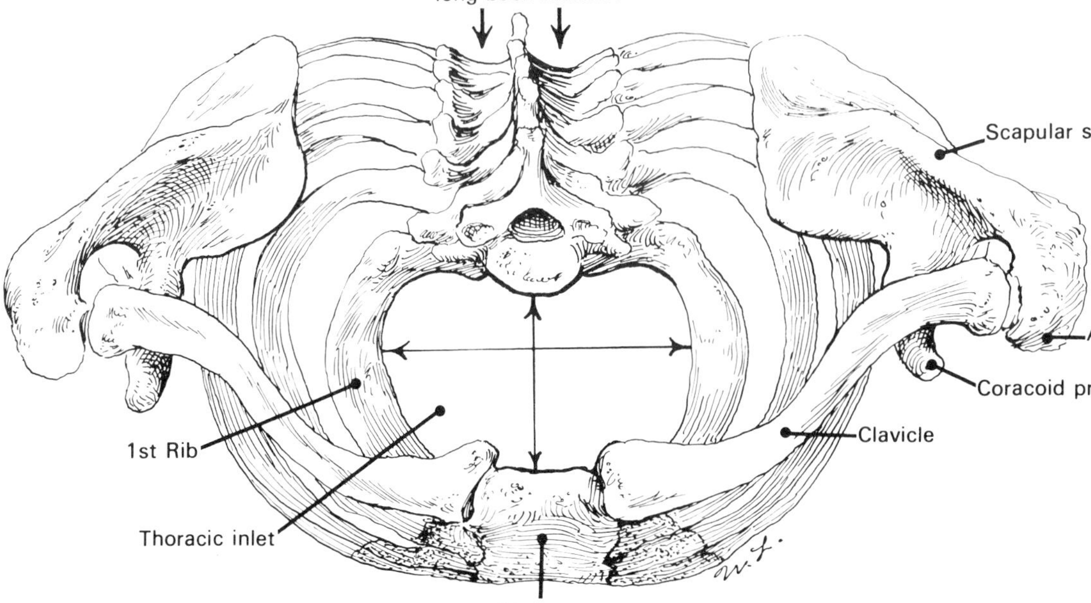

Fig. VIII-2. Thoracic inlet.

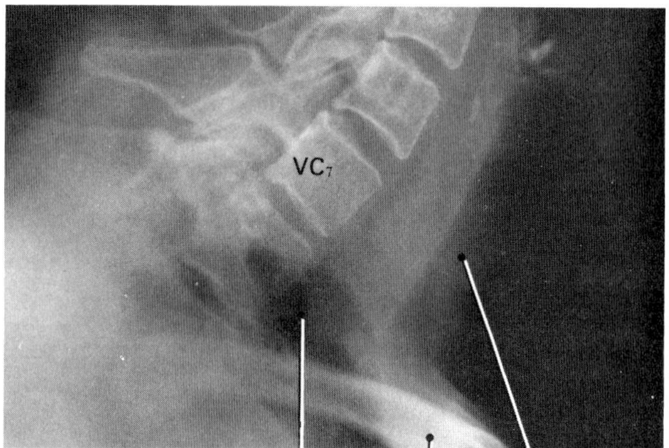

Fig. VIII-3. Lateral radiograph of the neck. Note the apex of the lung at the cupola extending above the clavicle.

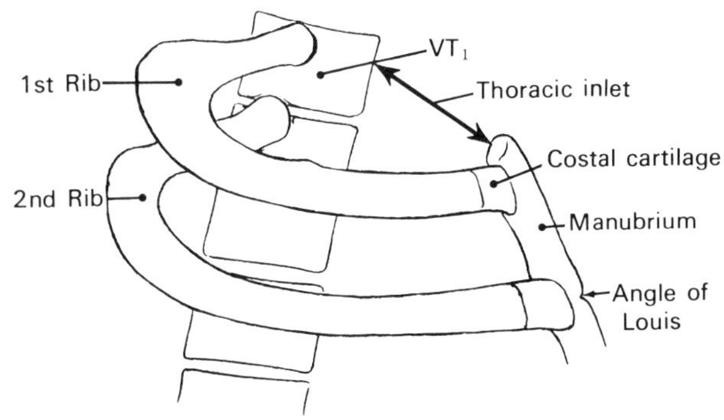

Fig. VIII-4. The thoracic inlet in lateral view (schematic).

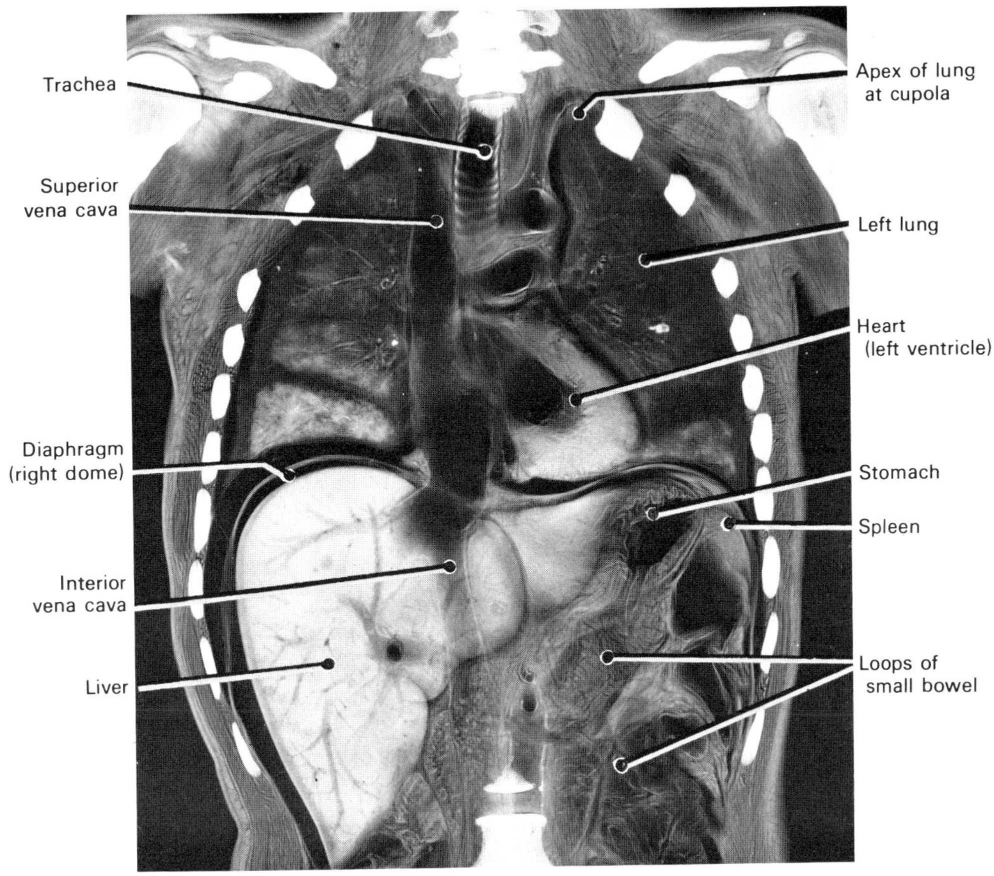

Fig. VIII-5. Radiograph of a thick coronal section of the thorax in the cadaver. (Courtesy of B. J. Anson.)

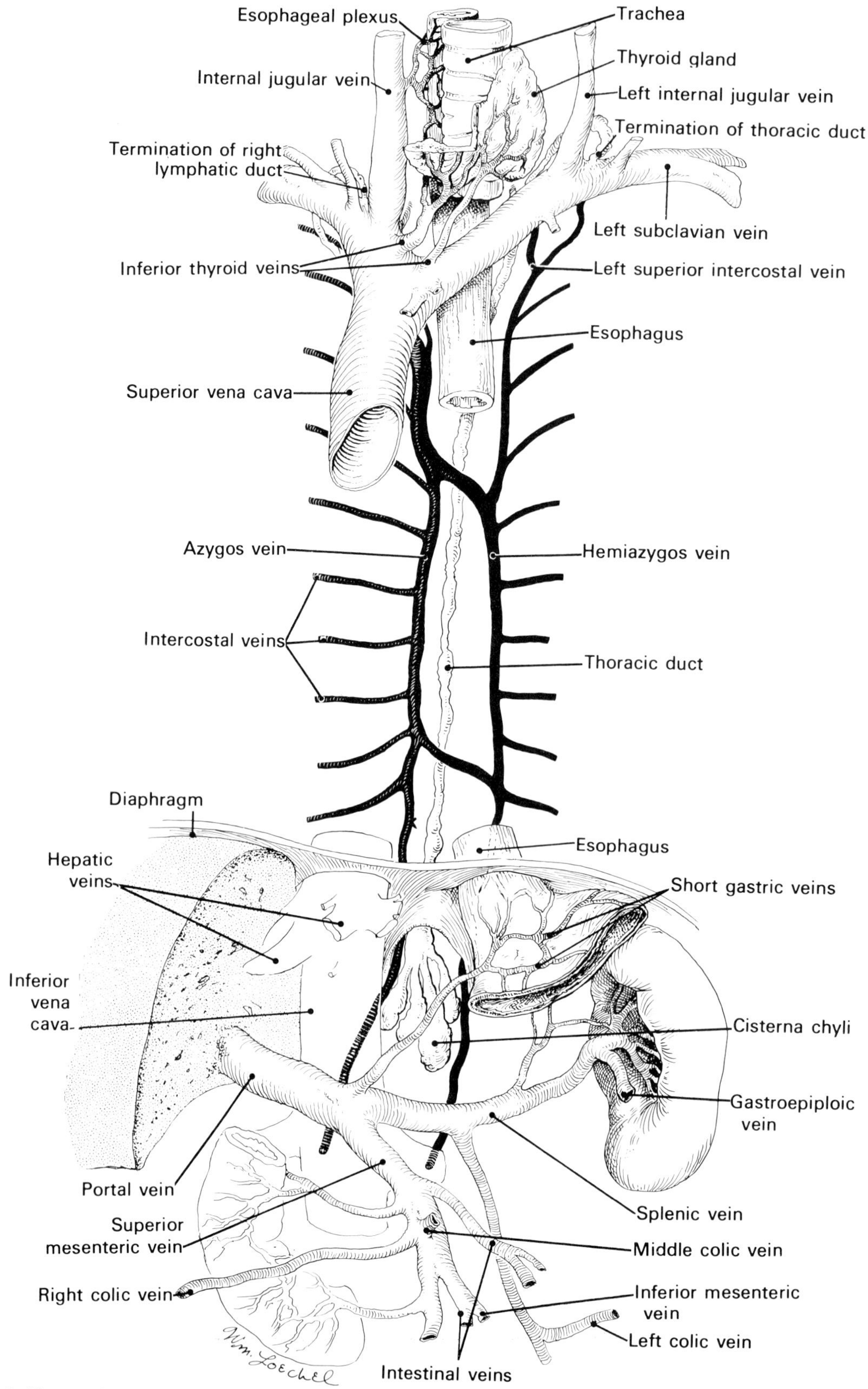

Fig. VIII-6. Venous drainage of thorax and abdomen, the azygos system of veins, and chief lymphatic trunks. (Abdominal portions of azygos system frequently arise from ascending lumbar veins.)

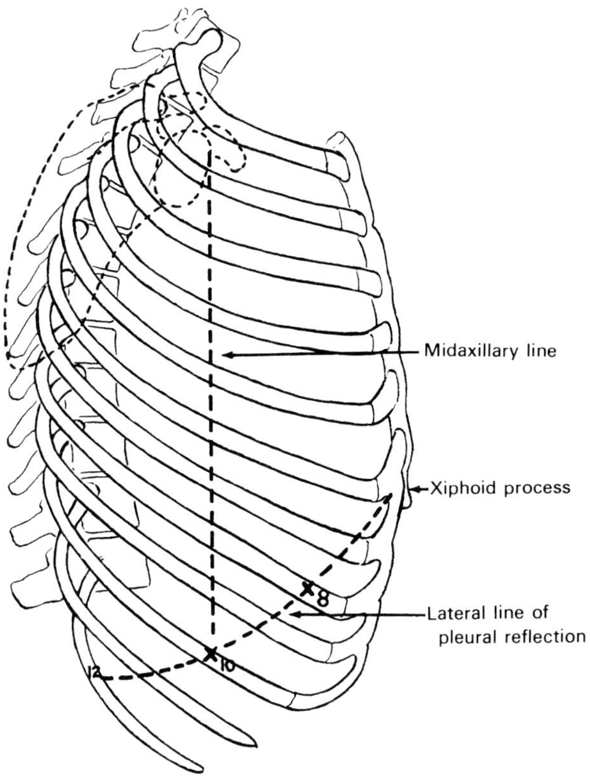

Fig. VIII-7. Lateral view of the thoracic cage.

domen through the Diaphragm are the esophagus, vagus nerve and sympathetic trunk, internal mammary (internal thoracic) artery, and the aorta. Traveling in the reverse direction into the thorax are the inferior vena cava, thoracic duct, and the subcostal veins which anastomose with the azygos and hemiazygos system of veins (Fig. VIII-6).

The Bony Thorax

Ribs: From the second to the eighth, the outer convex surfaces of the ribs tend to slant outward while that of the first faces upward. Below the eighth rib, the external surfaces tend to slant inward slightly. The upper intercostal spaces are wider than the lower spaces, and the widths of the spaces are greater in the front of the thorax than laterally and posteriorly. The ribs, themselves, slope downward and forward with an obliquity that increases from the first to the ninth, the latter being the most obliquely pitched of all the 12 ribs. This means that the thorax will also be widest around the eighth to ninth rib and it will decrease slightly in width below that point. The obliquity is greater in women than in men and in the old than in the young. It can best be appreciated in a lateral view of the bony thoracic cage (Fig. VIII-7). Actually, the female thorax also tends to be rounder and shorter than in the male, and the total thoracic capacity is less. However, the upper ribs of the female are often quite mobile in contrast to the more rigidly fixed upper ribs of the male.

The longest rib is the seventh while the shortest is the first, but in breadth the ribs decrease from the first to the 12th. While the 12 pairs of ribs (costae) are actually long, thin, curved bones, they are classified as flat bones that attach posteriorly to the 12 thoracic vertebrae. Each consists of two parts, the rib bone itself, and its anterior extension as a costal cartilage (hyaline). The osseocartilaginous cage is best viewed from in front (Fig. VIII-1). The first seven ribs are the **true** or **vertebrosternal** ribs because they articulate with the sternum through their costal cartilages and with the vertebrae posteriorly. The remaining five pairs are **false** or **vertebrochondral** ribs since their costal cartilages do not reach the sternum directly. The eighth to 10th ribs reach the sternum only through secondary fusion with the costal cartilage of the seventh rib which articulates at the xiphisternal junction. Sometimes the eighth costal cartilage reaches the sternum, and it does so more often on the right side than on the left. The pattern of costal fusion among the lower ribs is variable and often is not the same on both sides. However, the 11th and 12th ribs, as false ribs, are truly **vertebral ribs.** They attach posteriorly only and bury their costal cartilage in muscles. In this sense, they are often called **floating ribs.** The 12th pair of ribs may on occasion be absent or rudimentary.

The isolated ribs from one side are shown in Figure VIII-8A. Because of the individual variation in shape only ribs five and six and seven to 12 can be grouped into nests that fit together, one rib within the concavity of the next adjacent rib above (Fig. VIII-8B). If we look at ribs from the middle of the series (fifth or sixth ribs) as typical examples, we see that ribs have a vertebral end divided into two facets (Fig. VIII-9, A,B), and a sternal end with a porous depression into which the costal cartilage fitted during life. The sternal ends are oriented toward your left side in Figure VIII-8. The **head** of the rib at the vertebral end articulates with the lower and upper **demifacets** of two adjacent **vertebral bodies** as a rule, the intervertebral disk lying at a point between these articulations (Fig. VIII-10). The **neck** of the rib extends laterad from the head, and bears a **tubercle** (Fig. VIII-9B) which is divided into two parts. The medial part of the tubercle articulates with the transverse process of the lower of the two vertebrae with which the head articulates. The lateral part of the tubercle is non-articular, giving rise to a ligamentous attachment.

Beyond the tubercle is a short line directed downward and laterally which is more prominent in some ribs than in others. The line marks the **rib angle** and a tendon of the Iliocostalis, one of the muscles of the Erector Spinae group, generally attaches along this line. The distance between the tubercle and angle gets progressively greater from the second to the 10th rib, and the section of rib between these points serves for the attachment of the Longissimus, another of the Erector Spinae muscles. Thus, the space between rib angle and vertebral spine is a veritable deep trough to house the long intrinsic muscles of the back. The internal surface of a rib is concave and smooth and tapers to a sharp edge. A strong ridge of bone begins at the rib angle and runs forward along the upper, inner as-

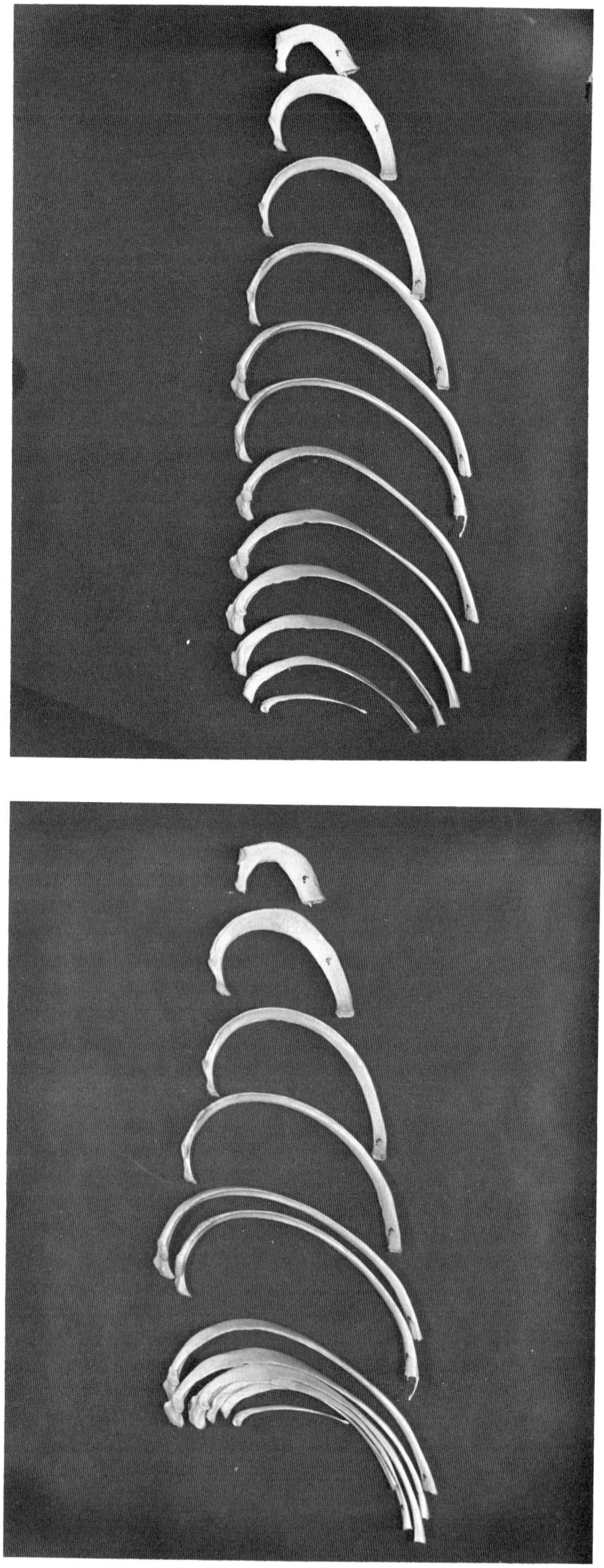

Fig. VIII-8. Ribs of the right side of the thorax. **A.** Isolated ribs. **B.** Grouping into nests of ribs 5 and 6, and 7 to 12.

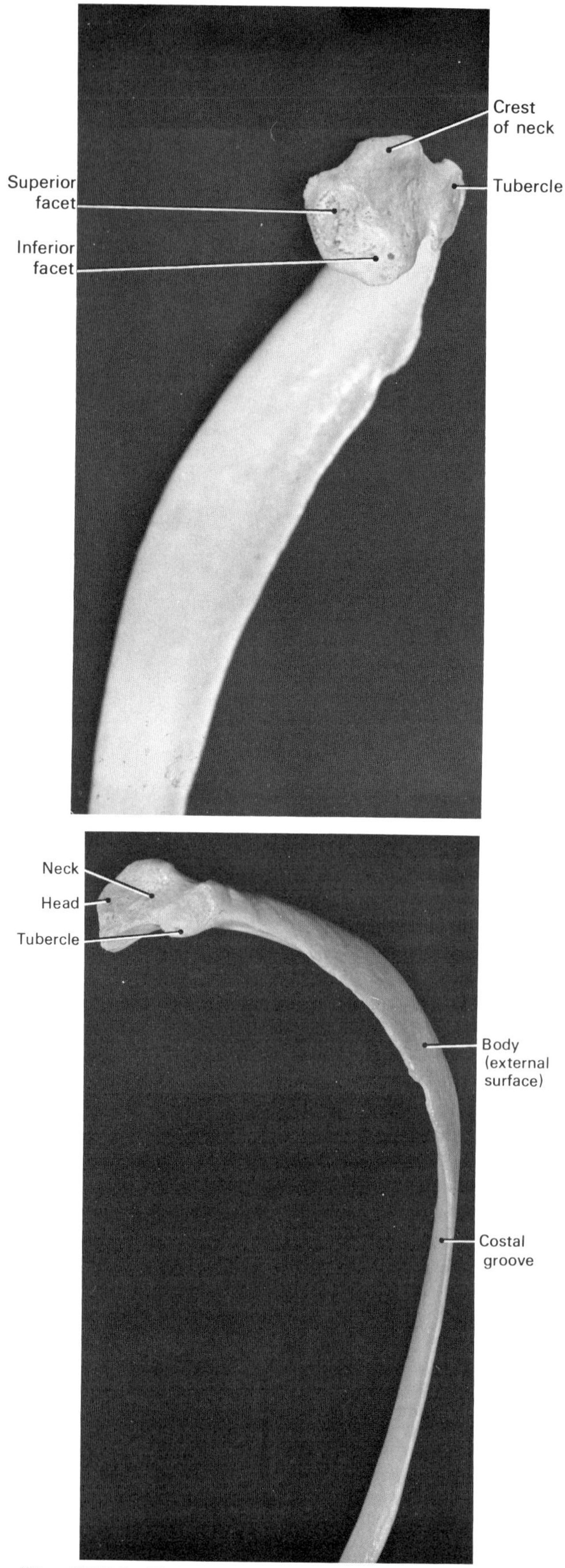

Fig. VIII-9. Photograph of the fifth rib. **A.** Medial aspect. **B.** From below.

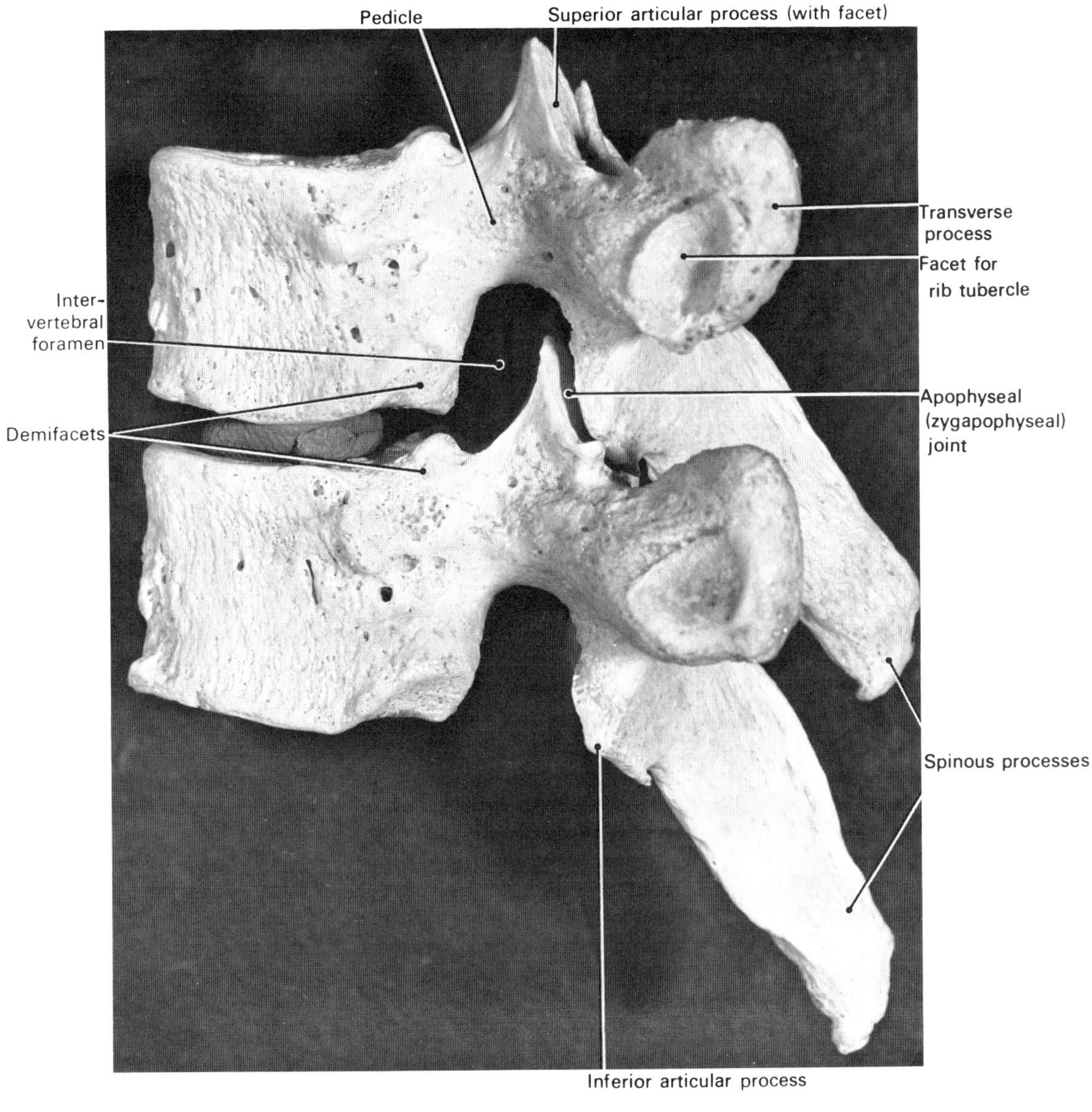

Fig. VIII-10. Lateral view of two adjacent "typical thoracic vertebrae."

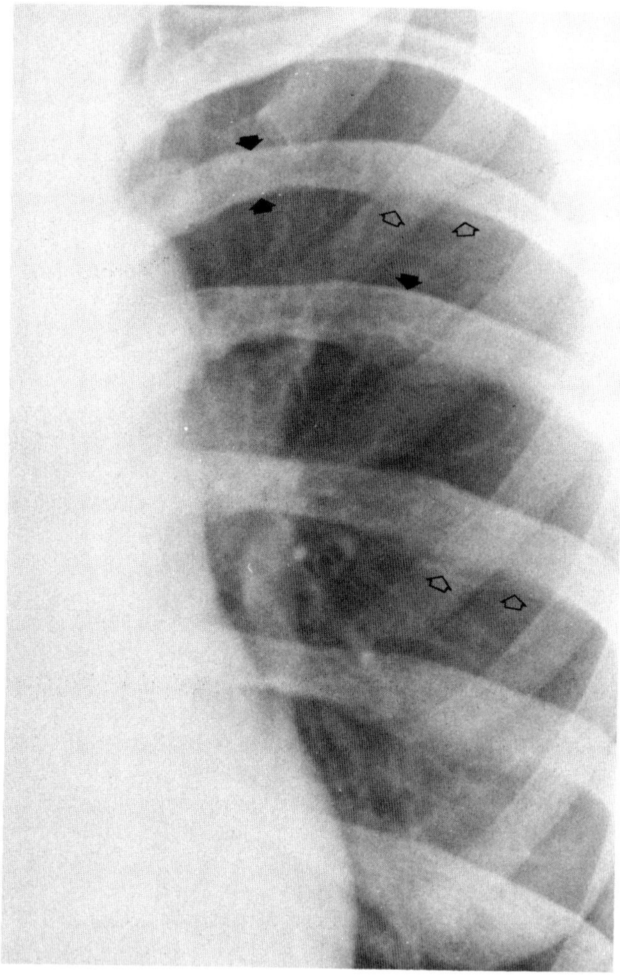

Fig. VIII-11. Chest radiograph in frontal projection. The lower margin of the costal groove (*open arrows*) is indicated at several levels. Each rib has a thin outer layer of compact bone (*black arrows*) which is quite radiopaque; between the cortical layers is the less dense cancellous bone of the rib.

pect of the rib. Between the upper shelf and the inferior sharp margin, a prominent **costal groove** (Fig. VIII-9A) is thus formed to house the intercostal vessels and nerves and to shelter them under the overhanging bony ridge. The accompanying shadow cast by the costal groove *on radiographs* is shown in Figure VIII-11, and its recognition is important. It may erroneously be mistaken for periostitis, or at times even pneumothorax (simulating the visceral pleura).

First Rib: Some ribs, like some people, are more peculiar than others. The first rib (Fig. VIII-12) is short, broad, and flat and the most sharply curved of all the ribs. Its external surface faces upward instead of outward and its internal surface, downward instead of inward. Its head has only a **single facet** for articulation with the first thoracic vertebrae. It has a prominent thick tubercle but *no rib angle* to speak of. **Two grooves** on its upper surface are separated by a ridge, the **scalene ridge,** which ends in a **scalene tubercle** for the Anterior Scalenus. The grooves are for the subclavian artery and vein. The first rib has *no costal groove.* Seen from behind in the articulated skeleton, it lies deep and far anterior to the second rib. Its difficult accessibility and its close relation to the large vessels coming up out of the thorax makes it an important structure to consider anatomically and clinically.

Second Rib: This rib is longer than the first, but it is almost as sharply curved (Fig. VIII-12). It has only a very small tubercle and hardly any non-articulating part to the tubercle. It has a slight rib angle close to the tubercle, but this is not well marked. It does have a short costal groove. Near its middle is a rough eminence or tubercle for the attachment of the second slip of the Serratus Anterior.

Last Three Ribs: The 10th to 12th ribs all have *single facets* at their head ends. The 11th and 12th ribs have *no necks* and *no tubercles.* The 11th has a slight angle line and a shallow costal groove; the 12th lacks both of these. Often the 12th is shorter than the 11th, and it may be shorter on one side than the other. on a standard PA radiograph of the chest (Fig. VIII-13). Of course, the lowermost ribs will tend to be obscured by the Diaphragm and the upper abdominal contents, depending upon the phase of respiration at the instant of filming. The posterior margin of the visible ribs have been numbered for you in Figure VIII-13A, and the anterior aspect of the ribs should be identified by tracing each rib forward from its articulation with the thoracic vertebrae. To aid in this, we have placed arrows along the medial border of the first rib on the patient's left. Note that the first rib is partially hidden anteriorly by the overlapping clavicle. Some individuals show striking ossification of the first costal cartilage which typically has a serrated border inferiorly (Fig. VIII-13 A, B). This typical appearance may be mistaken for a parenchymal pulmonary lesion by the unwary physician. Of course, the first costal cartilages are not the only ones to calcify, as can be seen in Figure VIII-13A. As evident here, this calcification tends to increase in amount lower in the chest.

Costosternal Attachments: Each rib is a cancellous bone with a thin outer layer of compact solid bone covered by periosteum (Fig. VIII-11). The costal cartilages are hyaline surfaced by perichondrium. In their articulation with the sternum (Fig. VIII-14), the first seven or true ribs form diarthrodial gliding joints with the exception of the first costal cartilage which fuses as a synchondrosis with the manubrium. The second costal cartilage, at the angle of Louis, has two small synovial capsules, one with the manubrium, the other with the body of the sternum. **Radiate costosternal ligaments** serve to form stout strengthening strands over these joint capsules. In the lower anterior interspaces where costal cartilages meet, there may be **interchondral synovial joints.** Whether these latter are present

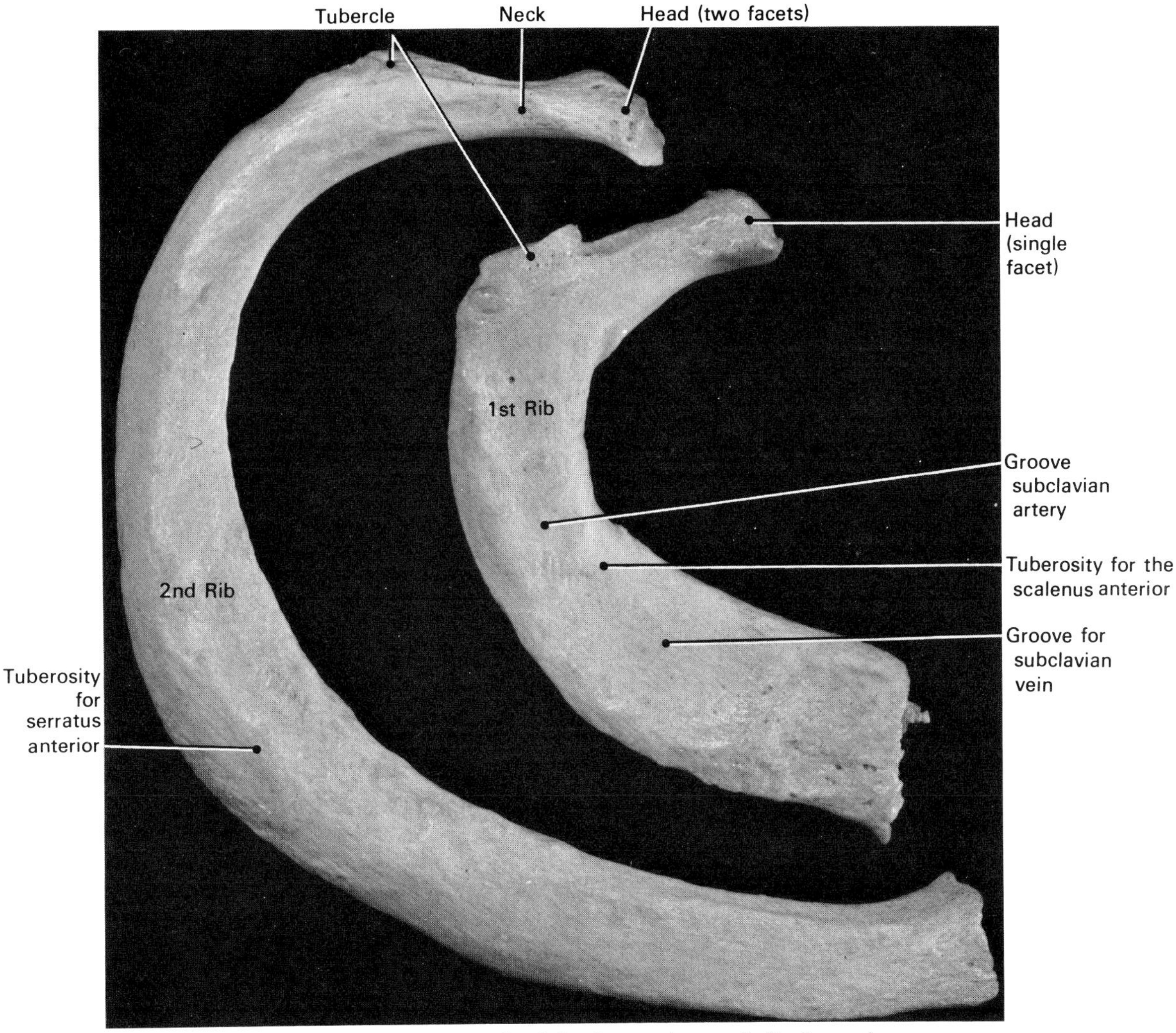

Fig. VIII-12. Photograph of the first and second ribs from above.

or not, and they *are* variable, there are **interchondral ligaments** that stretch between the adjacent borders of the eighth to tenth cartilages to anchor them to each other and to the seventh costal cartilage, the last contact with the sternum.

The first rib is also anchored to the clavicle by means of the **costoclavicular ligament** (Fig. VIII-14B). The insertion of this ligament on the clavicle is usually irregular or notched and, when seen radiographically (Fig. VIII-15), is referred to as the **rhomboid notch** or **fossa.** The clavicle articulates with the manubrium in a synovial joint that contains an interposed articular disk. The costoclavicular ligament, which is quite strong, tends to prevent dislocation at the sternoclavicular joint, while the articular disk cushions shocks transmitted along the length of the clavicle.

Rib Anomalies: The most important rib anomaly is the **cervical rib** which may be either unilateral or bilateral. Cervical ribs are characteristically small and although they frequently articulate with the transverse process of VC_7, they may appear as an elongated transverse process (Fig. VIII-16A). Union of such an odd rib can be with the sternum, the first costal cartilage or with the vertebral end of the first rib. Part of this extra rib may be fibrous, or only its sternal end may be present. Occasionally one may see bony fusion and synostosis of ribs, or even bifid (Fig. VIII-16B) or forked ribs. Only rarely is there a lumbar rib. Figure VIII-16 demonstrates two of the anomalies that may be detected on radiographs.

The Sternum (**Breast Bone**) (Fig. VIII-17 A, B):

This elongated flat bone forms the ventral wall of the thorax in the midline. It has been likened to the broadsword of old because of its shape. The three parts of the sternum, which may eventually fuse into a single bone, are: the **manubrium** (L., that which the hand grasps; handle), which is the superior part; the

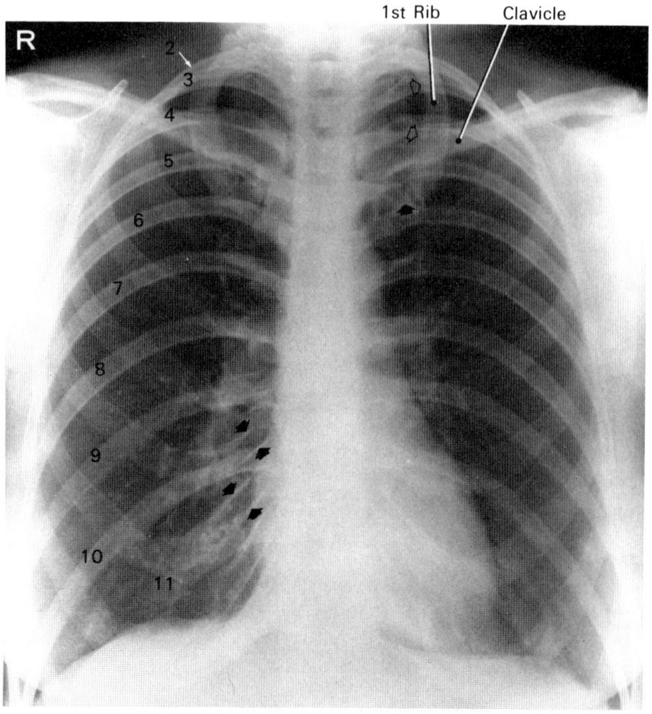

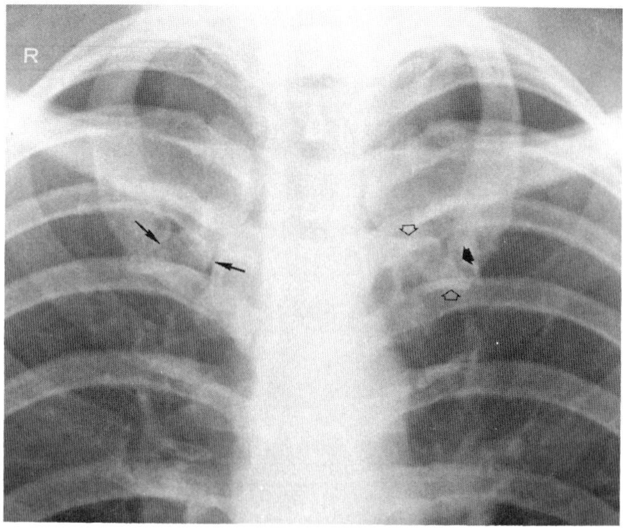

Fig. VIII-13A. PA radiograph of the female chest. The posterior aspect of ribs 2 to 11 have been numbered. The medial margin of the first rib on the left is indicated (*open arrow*). Note the serrated inferior border of the ossified cartilage of the first rib on the patient's left (*black arrow*). On the right the upper and lower borders of the ossified costal cartilages of the fifth and sixth ribs are indicated (*black arrows*). **B.** Close-up of the ossification in the costal cartilages of the first ribs bilaterally. On the patient's right side, *arrows* (→) mark the lucent clefts at the junction of the first rib with the cartilage. Do not mistake these clefts for fractures. On the left the upper and lower margins of the ossified cartilage are outlined (*open arrows*). Again note the vertical cleft (*black arrow*) between the anterior aspect of the first rib and the ossified cartilage.

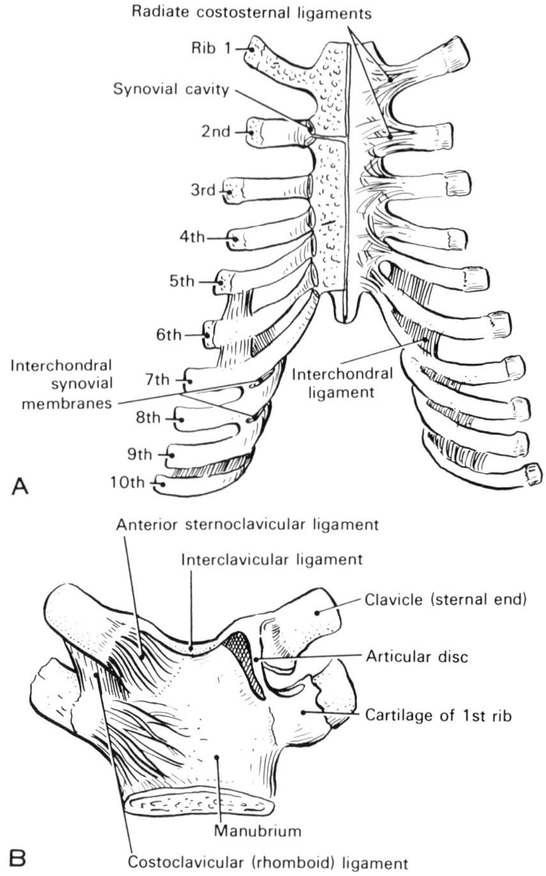

Fig. VIII-14A. Ventral aspect of the interchondral and sternocostal articulations. **B.** Ventral aspect of the sterno-clavicular articulation.

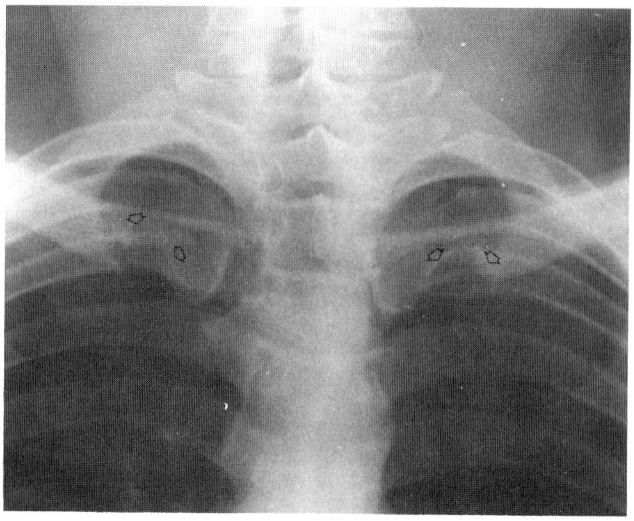

Fig. VIII-15. Radiograph in frontal projection showing a rhomboid fossa or notch bilaterally. Notice the difference in shape and depth on each side. The rhomboid notch is at the site of attachment of the costoclavicular ligament on the clavicle.

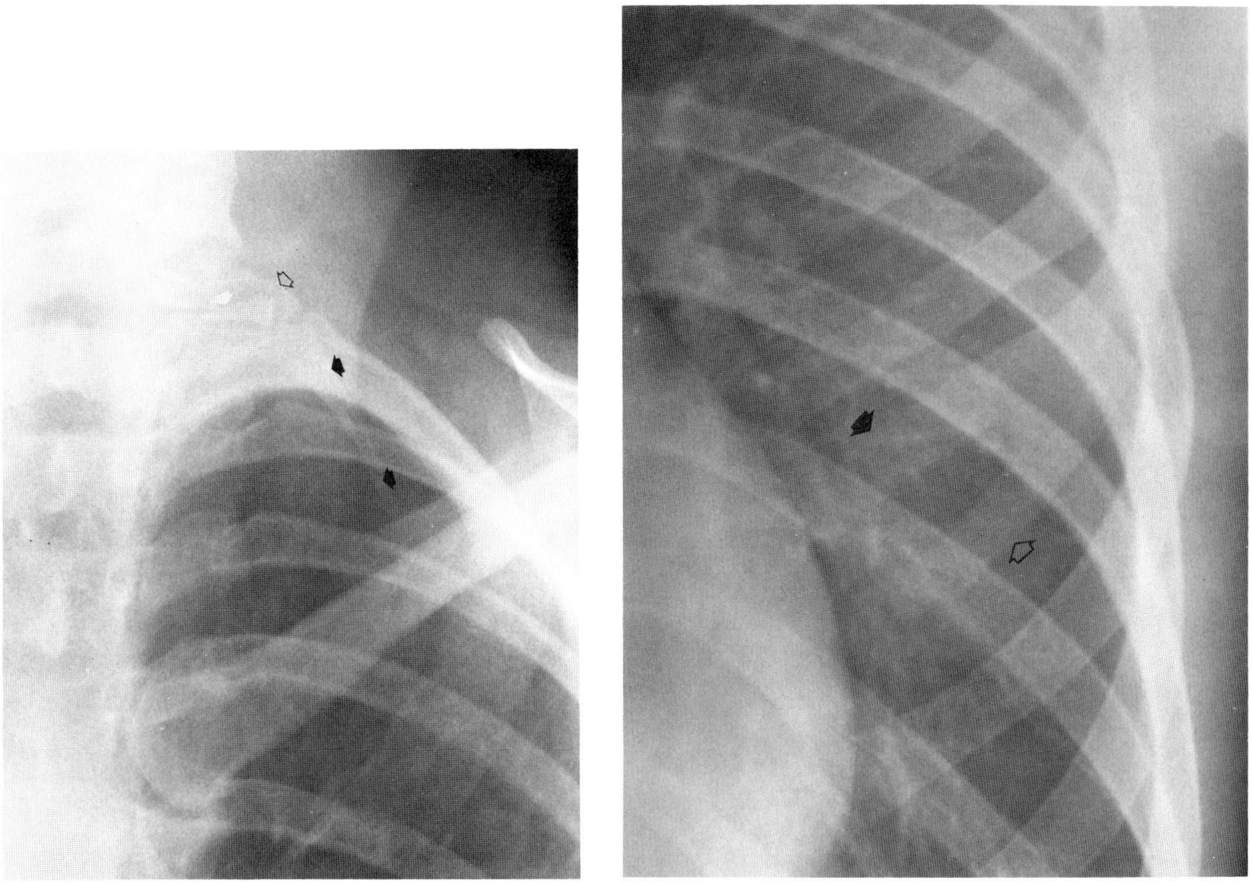

Fig. VIII-16. Rib anomalies. **A.** On the patient's right side the transverse process of VC_7 is elongated. On the left there is a small cervical rib which overlaps the first rib. It may be seen articulating (*arrow*) with the *lateral* and *inferiorly* directed transverse process of VC_7. The transverse process of VT_1 is directed *laterally* and *superiorly*. This difference in direction of these transverse processes is an important diagnostic point to remember for at times the first rib is rudimentary and may be mistaken for a small cervical rib. By noting with which transverse process the rib articulates, the diagnosis becomes apparent. **B.** Bifid rib anteriorly on the patient's left side. Note how broad the rib is anterolaterally as it curves downward to divide into an upper (*black arrow*) and lower part (*open arrow*).

body (corpus; gladiolus; L., small sword) is the middle and largest part; and the xiphoid process (Gr., sword-shaped) is the inferior part.

The **manubrium,** which is about 5 cm. in length, is broader and stouter superiorly than inferiorly. On its median superior edge is a shallow concavity, the **suprasternal** (jugular) **notch** that is readily palpable between the medial heads of the clavicles. At the superolateral margins are the **clavicular notches,** shallow oval articular facets for the articulation of the clavicular heads. Immediately below the clavicular notches, on the lateral borders of the manubrium, are located slight concavities for the attachment of the first costal cartilages. Below the attachment of the first costal cartilage, the manubrium narrows towards its junction with the body of the sternum at the **angle of Louis.** The ridge produced at the angle of Louis is readily palpable and serves as a useful landmark for counting ribs. Laterally at the angle, there is a small demifacet for the cranial portion of the second costal cartilage.

The **body** (gladiolus) of the sternum is about twice as long as the manubrium. It is relatively flat, but its lateral edges are scalloped. At the superior lateral margin of the body is a small demifacet for the attachment of the caudal portion of the second costal cartilage. The costal cartilages of the third to sixth ribs fasten at irregularly spaced facets located along the remainder of the lateral margin of the body of the sternum. The seventh rib finds attachment through its costal cartilage on a facet at the xiphisternal junction laterally. Thus, the inferolateral margin of the body and the superolateral margin of the xiphoid each have a small demifacet. The body of the sternum forms from four ossification centers (sternabrae) which are separated

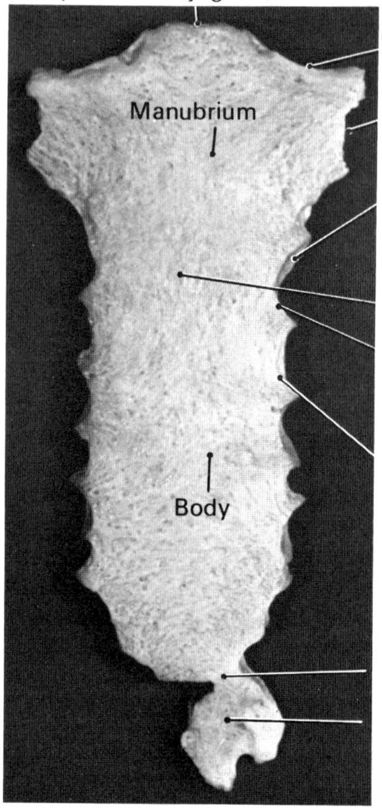

Fig. VIII-17. The sternum. **A.** Frontal view. Note that the xiphoid process is bifid and deviated to the left of midline. **B.** Lateral view.

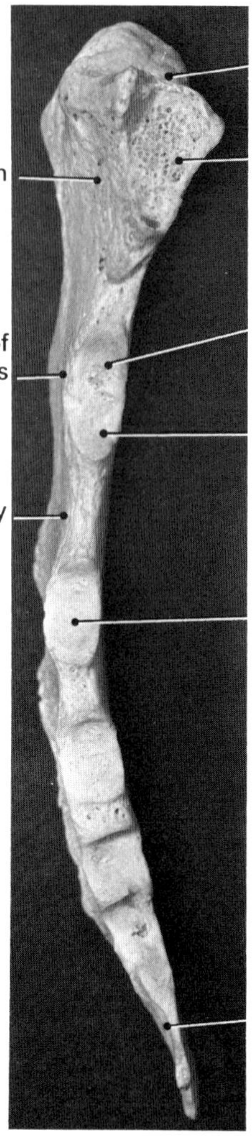

Fig. VIII-17B.

by cartilage, and in some specimens the fine horizontal lines of union of these centers are found etched in the bone.

The **xiphoid process** is quite variable in size, shape, and thickness. In some cases it may not ossify at all. In Figure VIII-17A the xiphoid process is bifid and deviated from the midline.

Clinically the sternum is important to the hematologist because it is rich in marrow and easily accessible. A large bore needle may be pushed through the cortex of the bone into the marrow beneath for aspiration and laboratory evaluation. Thoracic surgeons find that the sternum is readily split lengthwise to gain access to the retrosternal sleeve (anterior mediastinum) as well as the heart and great vessels.

Roentgen Anatomy of the Sternum: This discussion relates to special filming of the sternum, rather than its appearance on "routine" chest radiographs. Whenever abnormalities of the sternum are suspected, such as fracture, osteomyelitis, or tumor, specific techniques must be employed since there is normally so much overlap of this structure by adjacent ribs and mediastinal contents. The standard examination consists of: (1) oblique; and (2) lateral projections (Fig. VIII-18).

1. Oblique Projection: The patient is placed prone on the radiographic table, and the tube is angled 20° to 35° (the larger the patient, the less angulation required). The central X-ray beam is directed toward the midpoint of the body of the sternum. A long exposure time of at least five seconds is employed while the patient is instructed to breathe quietly, thereby blurring out, for the most part, overlying ribs and vascular markings in the lungs.

The three parts of the sternum are readily identified (Fig. VIII-18A). Note the sternoclavicular joint, the suprasternal notch, the facet for attachment of the costal cartilage of the first rib, and the sternomanubrial articulation (angle of Louis). At times there may be horizontal lines of increased density across the body of the sternum, which represent the sites of fusion of the ossification centers.

2. Lateral Projection: The patient is placed in a true lateral position on the radiographic table, and the vertically directed X-ray beam is centered on the body of the sternum. In the infant, the individual sterna-

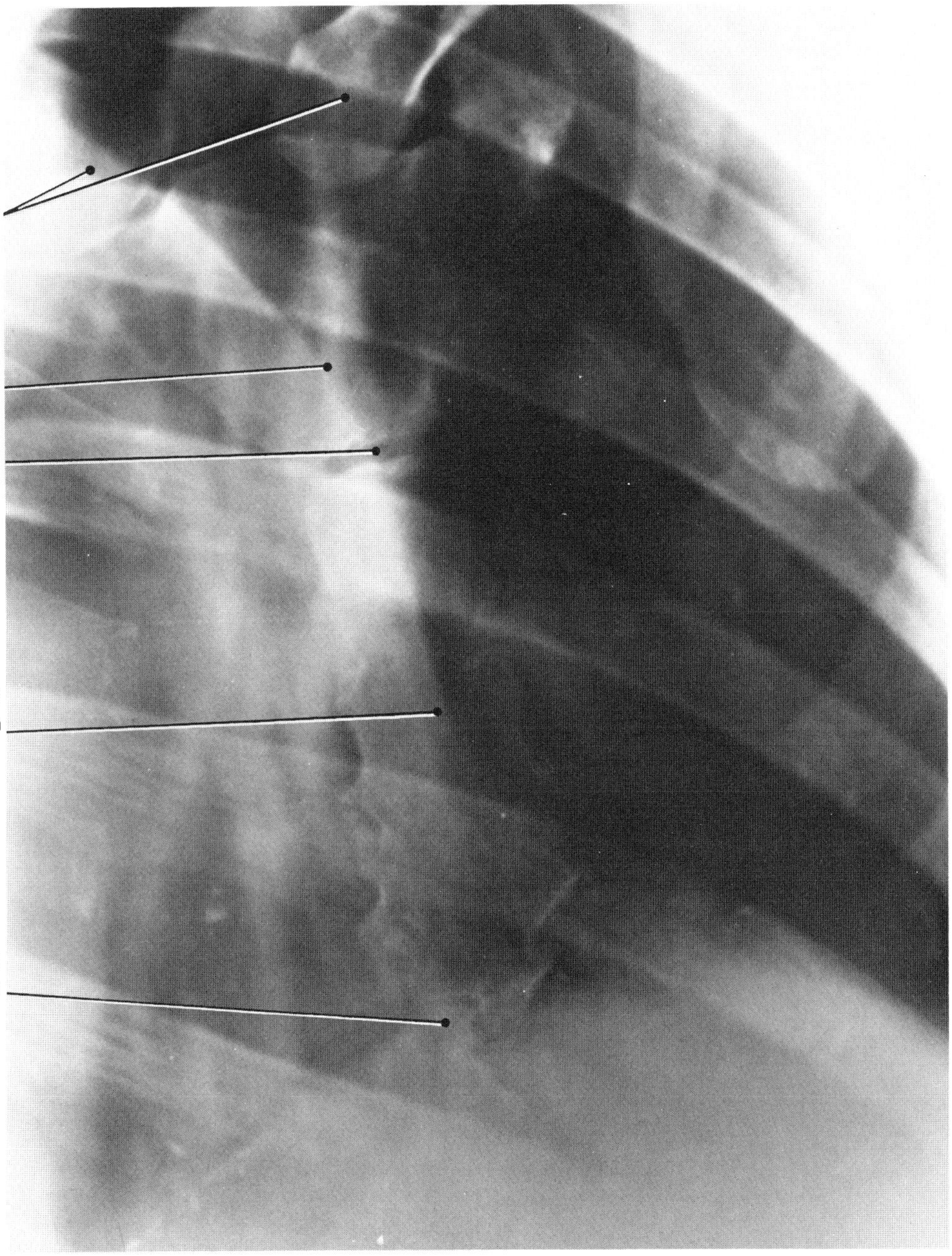

Fig. VIII-18. The sternum as seen on radiographs. **A.** Oblique projection in an adult.

brae may be seen in profile (Fig. VIII-18B). Identify the three parts of the sternum, and note the angle formed at the sternomanubrial junction. The sternoclavicular attachment may be seen, however, both sides are overlapped thereby obscuring fine detail of this area. If the costal cartilages have become ossified much of the anatomy of the sternum will be lost to view.

3. Tomography: If abnormality of the sternoclavicular joints is suspected clinically, it may be necessary to tomogram the joints in the posterior-anterior projection. Figure VIII-18C is an example of such a procedure. Note how shallow the clavicular notches are. The sternoclavicular joint is radiolucent.

The Thoracic Vertebrae

The 12 thoracic vertebrae align in lateral view (Fig. VIII-19) to form a gently-curving, anterior, concave arc. This **primary curvature** was formed *in utero* and it persists after birth. (The sacral anterior concavity is

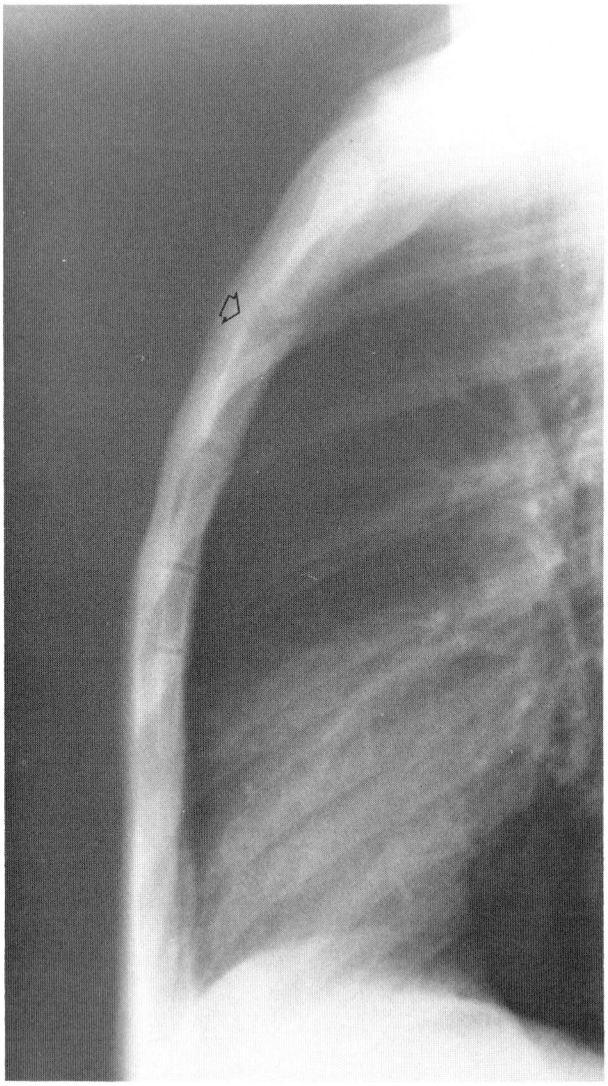

Fig. VIII-18B. The sternum, lateral projection. In this child, five separate osification centers are shown. The number of these centers of ossification is quite variable.

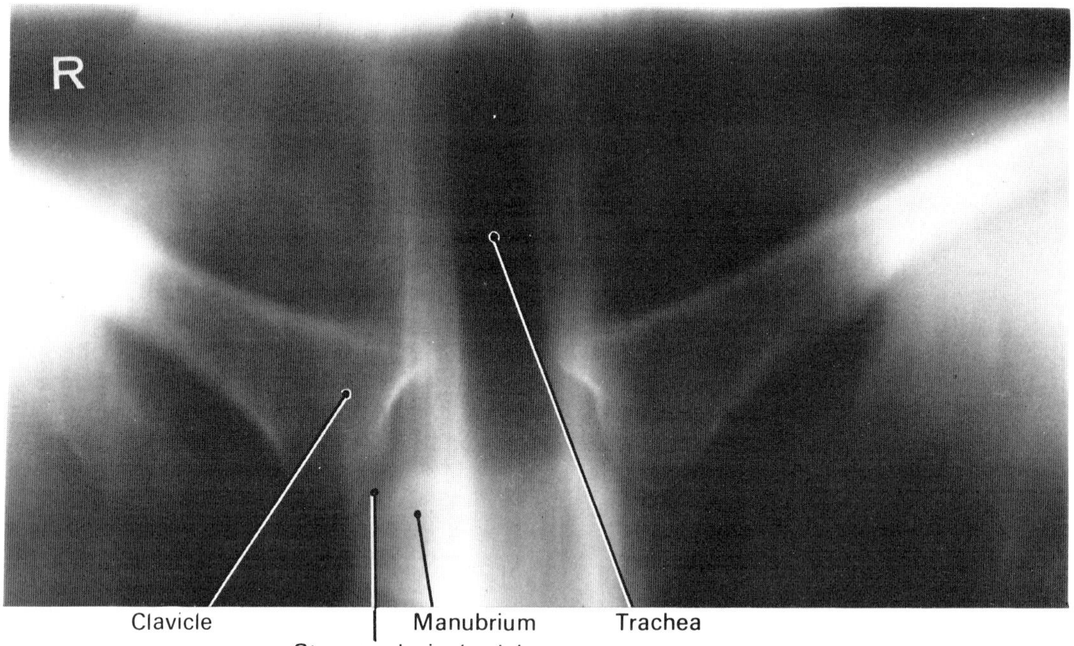

Fig. VIII-18C. The sternum. PA Homogram of the sternoclavicular region.

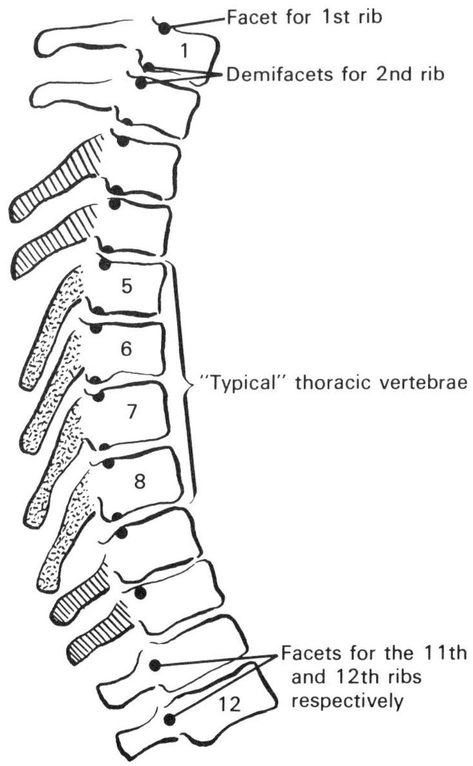

Fig. VIII-19. Schema of the thoracic vertebrae. The facets and demifacets for the rib heads are shown. The extreme inferior obliquity of the spinous processes of the "typical vertebrae" (VT_5 to VT_8) are compared with those above and below. (After Grant.)

also a primary curvature, whereas, the secondary convex curvature of the cervical and lumbar spine form as the infant holds his head erect and begins to walk). Any increase in the anterior concavity of the thoracic spine is referred to as a **kyphosis** (Gr., humpback). An appreciable lateral deviation of the spine from its normal straight line in the frontal view is known as a **scoliosis** (Gr., curvation) (Fig. VIII-24D).

There are, strictly speaking, only four **typical thoracic vertebrae,** namely VT_5 to VT_8. Those vertebrae above (VT_1 to VT_4) and those below (VT_9 to VT_{12}) take on many of the characteristics of the cervical and lumbar vertebrae, respectively. The descriptions to follow will apply to the typical thoracic vertebra, while noting exceptions as they apply to the other vertebrae. You should work carefully with skeletal material, both mounted and unmounted, as you read this section.

Thoracic vertebra differ from their neighbors in the cervical and lumbar area, because they have **facets** on their bodies and their transverse processes (exceptions VT_{11} and VT_{12}) to accommodate the ribs which articulate there. Examine for a moment the thoracic vertebrae of your skeleton, and notice how they gradually

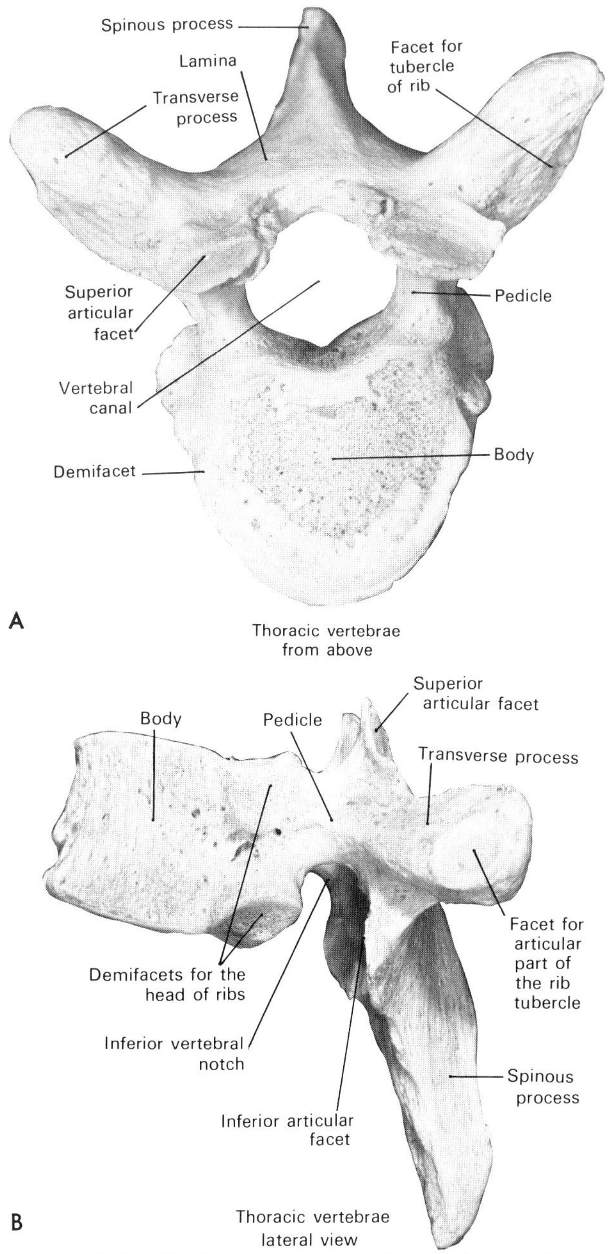

Fig. VIII-20. The features of a "typical thoracic vertebra." **A.** From above. **B.** Lateral view.

enlarge in surface area as you proceed from cranial to caudal end. The bodies are slightly taller dorsally than ventrally which aids in maintaining the normal curvature of the thoracic spine. Realize that the intervertebral disks in the thoracic spine are thinner than those in the cervical and lumbar areas.

The thoracic vertebral **bodies** (Fig. VIII-20A) have slight concave anterior and lateral borders, and their superior and inferior surfaces are also minimally concave. On the superior surface of VT_1 and occasionally

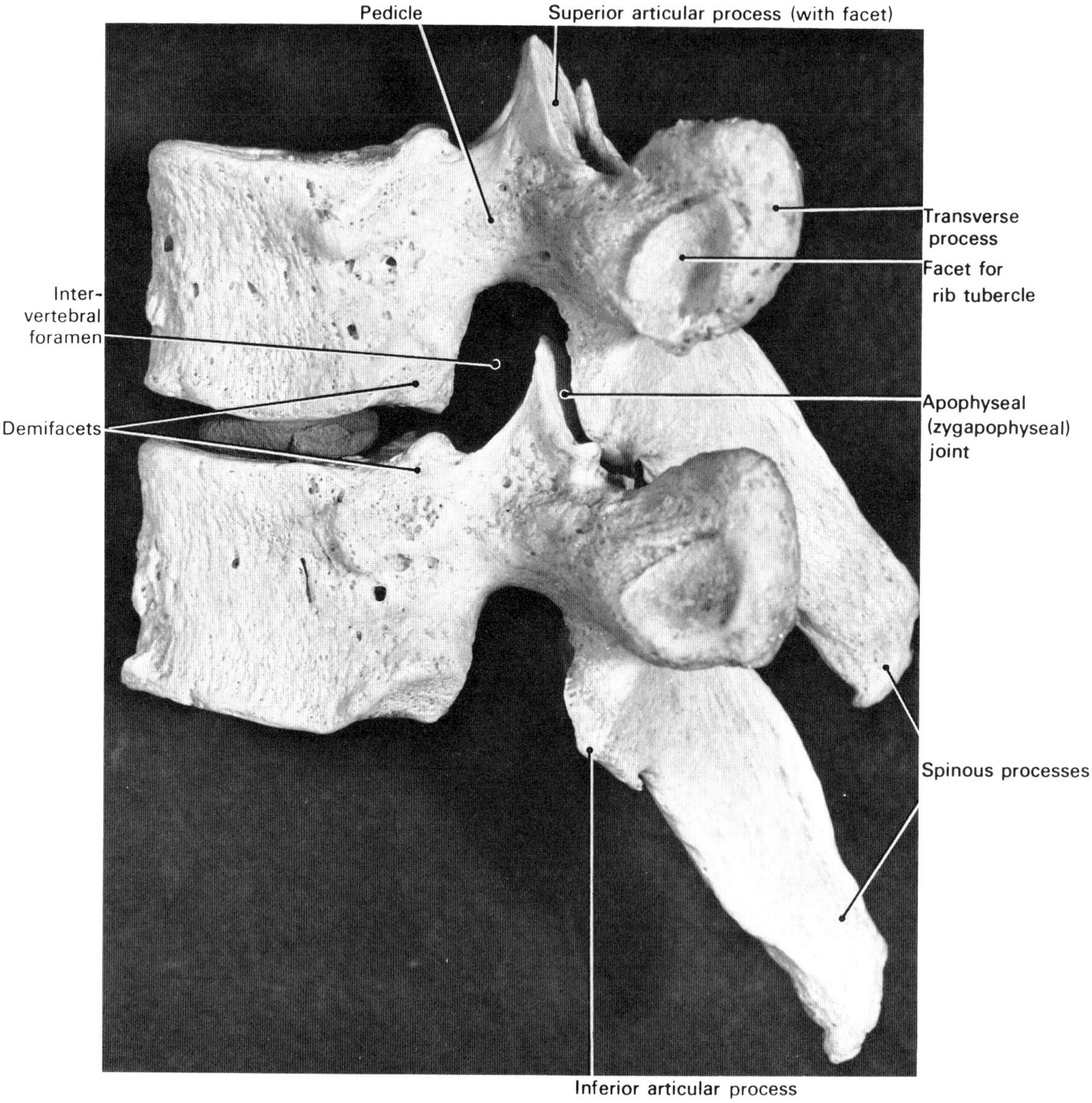

Fig. VII-21. Lateral view of two articulating midthoracic vertebrae.

VT_2, we find raised lips, the **uncinate processes** located posterolaterally. Figure VIII-20A is a midthoracic vertebra, so like other thoracic vertebrae (VT_2 to VT_{12}) it does not have uncinate processes. Two **demifacets** are found laterally on the bodies of VT_2 to VT_9 (Figs. VIII-19, 21). The **superior demifacet** (Fig. VIII-20) at the superior posterolateral margin of the body is for the articulation of the *caudal* portion of the head of its own rib; the **inferior demifacet,** located posterolaterally and inferiorly, accommodates the *cranial* portion of the head of the rib immediately below. The other thoracic vertebrae vary somewhat: VT_{10} has only a single costal facet on its lateral posterior aspect, part of which is on the body and part on the pedicle; VT_{11} and VT_{12} have only a single costal facet located on their pedicles laterally; VT_1 has a single costal facet for the head of the first rib, and a small demifacet posterolaterally and inferiorly for the cranial portion of the second rib.

The **pedicles** are directed posteriorly and slightly cranially. Inferior to the pedicle is the **inferior vertebral notch** which forms a part of the **intervertebral** foramen at VC_7-VT_1. The **superior** and **inferior** forming the inferior boundary of the intervertebral foramen at VC_7 to VT_1. The **superior** and **inferior articular processes** are nearly vertical in position. The superior articular process faces posterolaterally, and the inferior articular process is directed anteromedially. This position of the articular processes permits rotation of the thoracic spine. Between the normal interlocking articular processes are the synovial **apophyseal** (zygapophyseal) **joints** (Fig. VIII-21).

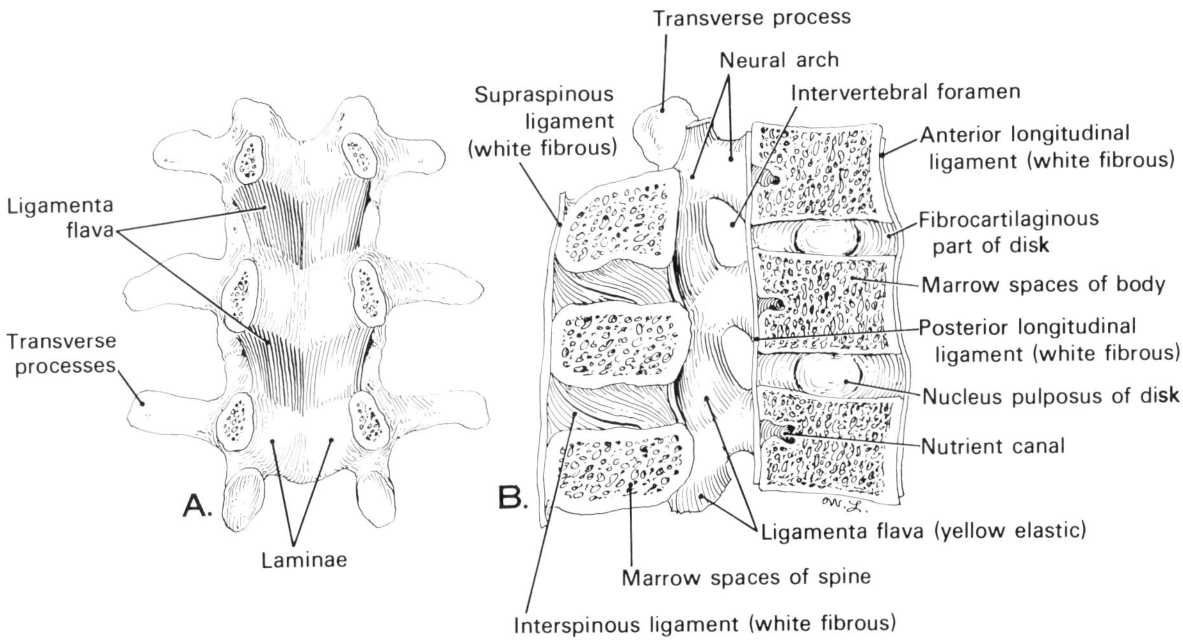

Fig. VIII-22. Ligamentous support of the vertebrae. **A.** The ligamenta flava viewed from the ventral aspect. **B.** Sagittal section.

The **transverse processes** emerge between the articular processes and they extend posterolaterally and slightly cranially. These act as levers for attaching muscles and also as a fulcrum for the ribs. Note how the transverse processes decrease in length from VT_1 through VT_{12}. On the ventral surface of the tips of the transverse processes are the *facets* for articulation of the rib tubercles. These facets are absent from the transverse processes of VT_{11} and VT_{12}. Three **tubercles** have replaced the transverse process of VT_{12}. These tubercles are the **superior,** which is comparable to the mammillary process of the lumbar vertebrae; the **inferior,** which corresponds to the accessory process of the lumbar vertebrae; and the **lateral,** which is comparable to the transverse process.

The **lamina** are quite large in the thoracic spine. They are inclined posteromedially and inferiorly, and overlap the lamina of the body below (imbrication). The **spinous processes** of the thoracic vertebrae (Fig. VIII-19) vary in their posterior direction: those of VT_5 to VT_8 are nearly vertical and overlap the vertebrae below; VT_1, VT_2, and VT_{11}, VT_{12} are horizontal; and VT_3, VT_4 and VT_9, VT_{10} are obliquely directed.

The **vertebral canal** for the spinal cord is rather small and circular throughout most of the thoracic vertebrae. However, where the spinal cord has enlarged for the brachial and lumbosacral plexuses, the vertebral canal also enlarges laterally assuming a triangular shape. The cervical enlargement extends down to the level of VT_2, and the lumbar widening begins at the level of VT_{10}.

The **intervertebral foramen** (Fig. VIII-21) is bounded cranially and caudally by the pedicles of the adjacent vertebrae, anteriorly by the vertebral bodies, and posteriorly by the superior articular process. The intervertebral foramina increase in size from VT_1 to VT_{12}. They are oval in shape and are the exit point for the spinal nerves and vessels.

Ligamentous Support: Figure VIII-22 shows the details of the ligamentous support of the thoracic vertebrae. These details do not differ significantly from prior descriptions in the cervical region. The anterior longitudinal ligament, a narrow band in the neck, thickens and widens as it crosses the thoracic vertebrae and travels toward the sacrum. It attaches firmly to the intervertebral disks and to the margins of the vertebral bodies.

The posterior longitudinal ligament, continuous with the wider tectorial membrane above, narrows in the thoracic region as it covers the bodies and disks in the vertebral canal. Only the pedicles are without ligaments. The lamina are united by short, broad elastic bands, the ligamenta flava. They are pierced by vessels connecting the internal and external vertebral plexuses. Intertransverse ligaments between transverse processes are trivial in the thoracic region, but the interspinous ligaments are stout. The ligamentum nuchae played out as an elastic ligament in the neck, but its position along the tips of the spine is continued by the white, fibrous, supraspinous ligament.

Radiographic Anatomy of the Thoracic Spine and the Costovertebral Articulations: The routine examination of the thoracic spine consists of: (1) anterior-posterior; and (2) lateral projections.

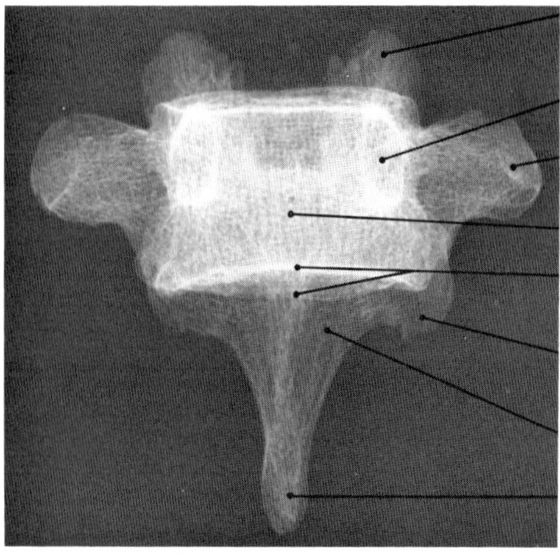

Fig. VIII-23. AP radiograph of an isolated "typical thoracic vertebra." Compare the structures with those of Figure VIII-24.

1. Anterior-posterior Projection: The patient is placed in the supine position on the radiographic table, and the vertically directed X-ray beam is centered on the VT_7 vertebra. The patient is instructed to breathe quietly during the X-ray exposure in order to blur out overlapping structures.

Examine for a moment the AP radiograph (Fig. VIII-23) of a separate "typical" vertebra, and then compare it with the radiograph of the thoracic spine of a patient (Fig. VIII-24). The vertebral body is rectangular in shape, and its lateral margins are concave. Because of the divergence of the X-ray beam upon leaving the tube window and housing, the ventral and dorsal rim of the superior and inferior vertebral surfaces are not precisely superimposed and thus they appear ovoid. In other words, if the ventral and dorsal rims were exactly aligned, there would be a summation effect with resultant increased bone density.

The oval-shaped pedicles are superimposed upon the superior lateral aspects of the body. The inner margins of the pedicles (interpediculate distance) delimits the width of the vertebral canal. Variations from the norm may have pathological significance. These vertebral canal diameters are widest in the thoracic spine at VT_1 and they diminish in size progressively to the level of VT_4 or VT_5 where they remain relatively constant to VT_9, and then gradually increase width to the level of VT_{12}. What have you learned earlier to account for this roentgen observation? Remember that the pedicles should be symmetrical in appearance at the same vertebral level, assuming, of course, that the patient does not have a scoliosis and is not rotated from a true frontal projection (Fig. VIII-24E). Notice in Figure VIII-24 how the pedicles increase in size as the vertebral bodies increase their size from cranial to caudal end of the thoracic spine.

In Figure VIII-24A identify the superior and inferior articular processes, the spinous processes and the lamina. Details of the costovertebral articulations in the

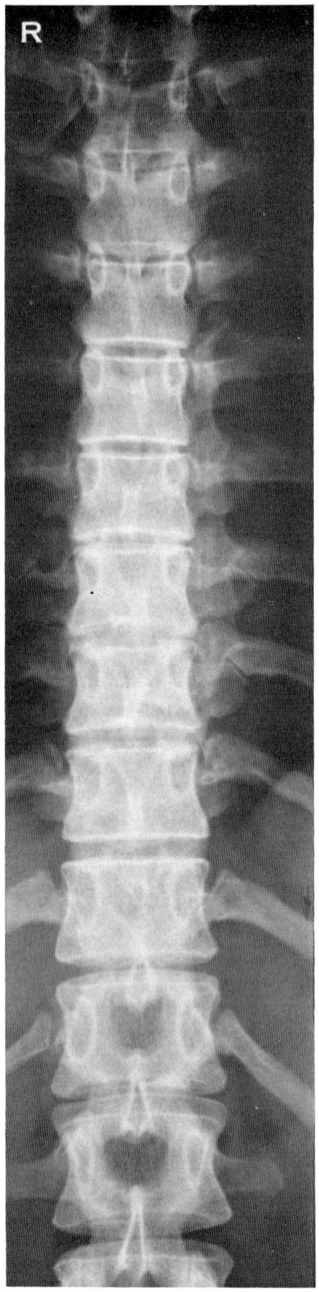

Fig. VIII-24. AP radiographs of the thoracic spine and costovertebral articulations in life. **A.** VT_2 to VL_1 are shown.

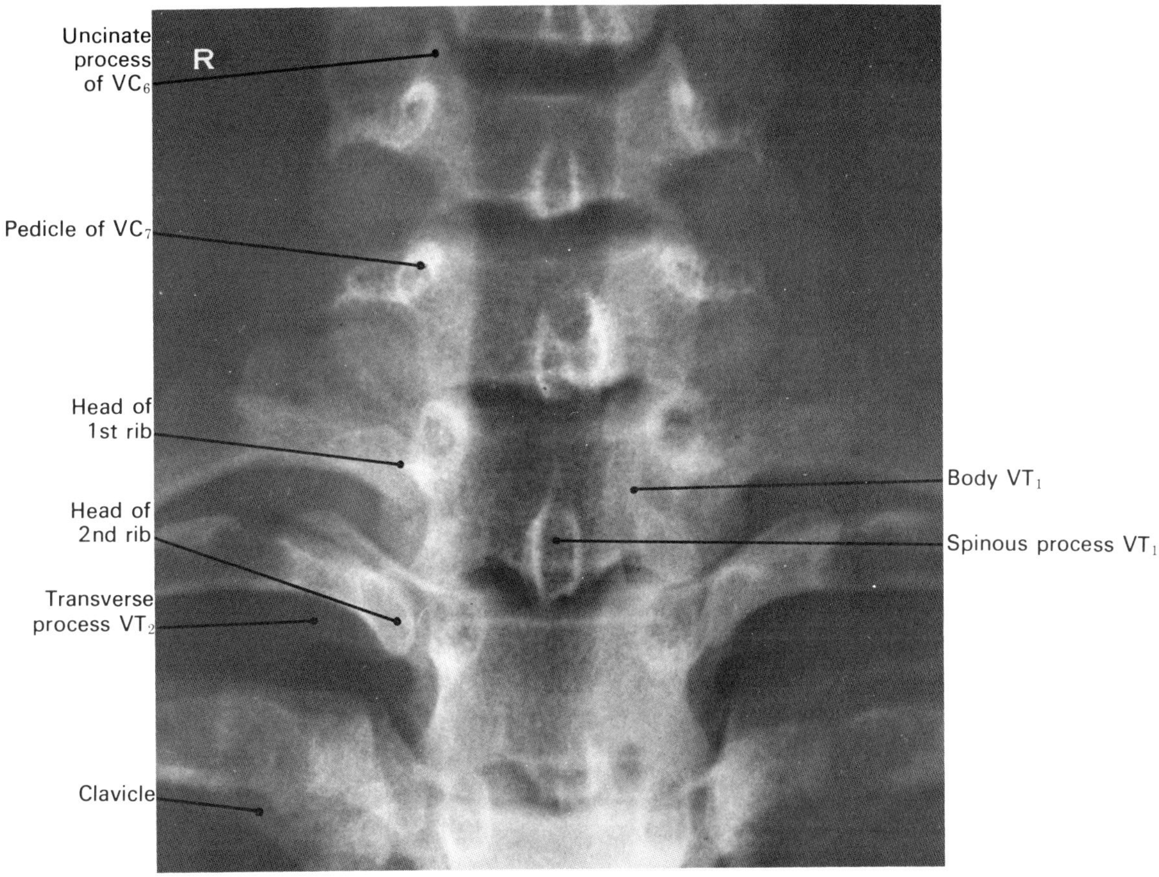

Fig. VIII-24B. Close-up of VC_6 to VT_3. Note the position of the heads of ribs 1 and 2.

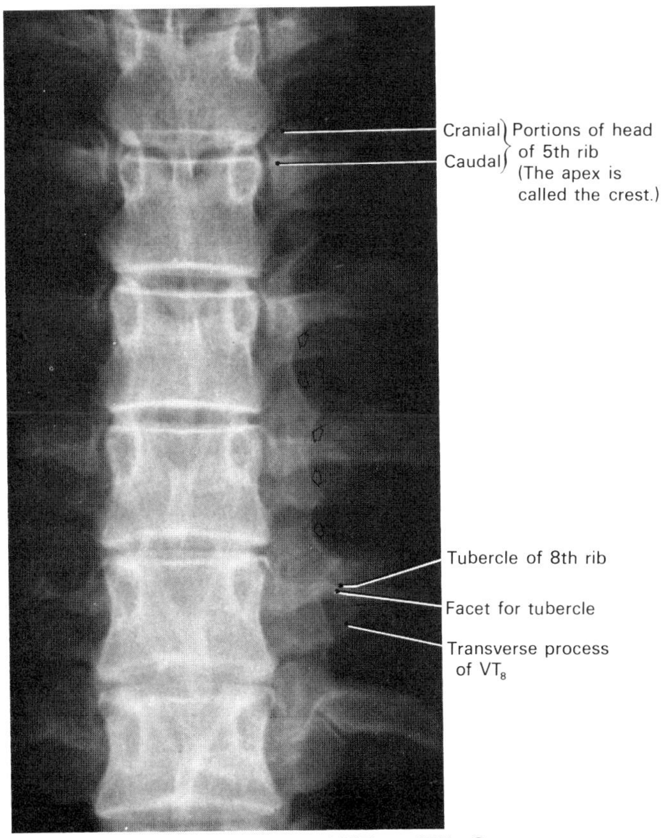

Fig. VIII-24C. Close-up of VT_4 to VT_9. *Open arrows* are on a portion of the scalloped lateral edge of the sternum.

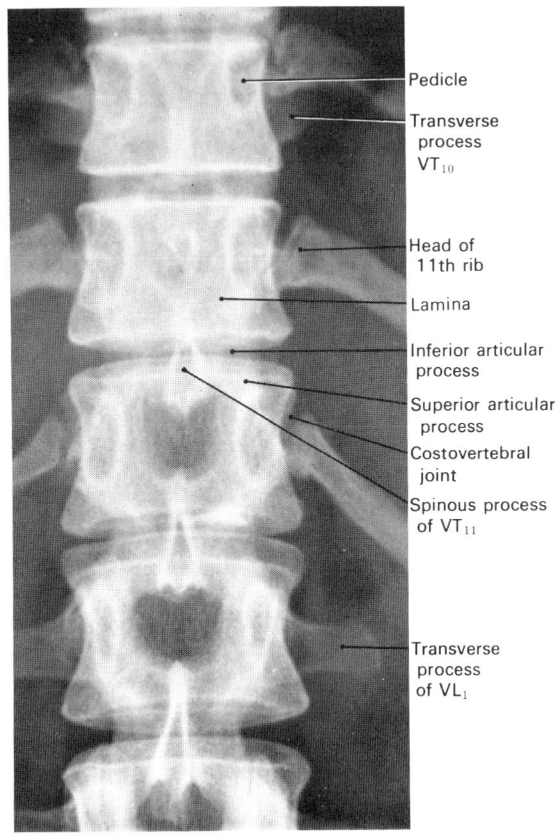

Fig. VIII-24D. Close-up of VT_{10} to VL_1.

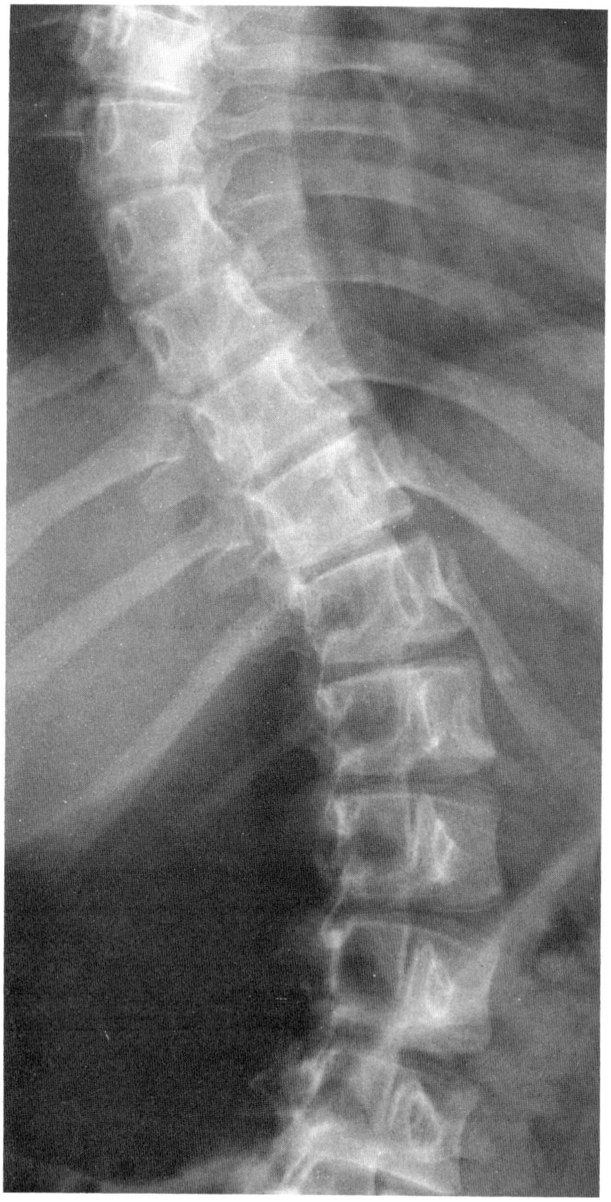

Fig. VIII-24E. A patient with scoliosis of the thoracolumbar spine.

upper, mid, and lower thoracic spine are shown in Figure VIII-24B–D. The rib heads can clearly be seen articulating at the facets and demifacets, and the rib tubercles may be identified articulating with the transverse processes.

2. Lateral Projection: The patient is placed in the true lateral position on the radiographic table, and the central X-ray beam is directed at the VT_7 vertebra. The patient breathes quietly throughout the X-ray exposure in order to blur out the overlapping lateral aspects of the ribs and the vessels within the lungs.

The anatomical details of an isolated vertebra are shown in Figure VIII-25. Compare these details with the lateral radiograph of a patient (Fig. VIII-26), and note immediately that the transverse processes and spinous processes are for the most part obscured by the posterior aspect of the overlapping ribs. The upper two or three vertebrae are usually obscured by the shoulders.

The vertebral bodies are somewhat taller dorsally than ventrally. The superior and inferior vertebral surfaces appear more dense than the inner cancellous bone of the bodies. Identify the pedicles, articular processes, apophyseal joints, and the intervertebral foramina. Between adjacent vertebrae are the radiolucent intervertebral disks. These disks are thicker in the lower thoracic region. As an aid in identifying vertebral levels, look for the 12th rib articulating on the pedicle facet of VT_{12} and count upward to the desired level.

Muscles Attached to the Chest Wall

The thoracic cage is clothed with its own intrinsic musculature which lies deep and is overlaid with the more superficial muscles of the upper limbs. Even some of the abdominal muscles gain attachment to its lower margins. The limb musculature must be dissected away before much of the intrinsic musculature of the back and of the intercostal spaces can be studied.

Superficial Anterior Group: Covering the upper, anterior aspect of the chest wall is the massive **Pectoralis Major,** the lateral extension of which forms the anterior wall of the armpit or **axilla.** Beneath this large muscle will be found the smaller pectoral muscle (**Pectoralis Minor**) and the well-hidden **Subclavius** (Fig. VIII-27).

The Pectoralis Major arises from the medial half of the clavicle by one head and from the sternum and upper six costal cartilages by its broader sternocostal head (Fig. VIII-27A). It inserts upon the humerus by means of a trilaminar tendon. While its chief function is to act as an adductor and a medial rotator of the arm, it can be used to expand the chest when the patient fixes the arms to a solid structure such as the bed rails or the back of a chair. It is not uncommon to find patients with breathing difficulties fixing their arms in this manner to reverse the action of this muscle as an aid to breathing.

The nerve supply to the Pectoralis Major is from the **medial** and **lateral pectoral nerves** which arise from the brachial plexus (C_5 to T_1, anterior primary rami). The predominant supply to the clavicular head comes from C_6 through these nerves, while C_7 supplies the sternocostal portion of the muscle.

The Subclavius (Fig. VIII-27B) arises from the first rib about at the junction of bone and cartilage and does what its name implies: it goes under the clavicle to insert in a groove on the lower surface of the clavicle. The action of this tiny muscle is not really known. From its horizontal alignment it should hold the clavicle steadfast, resisting lateral displacement. But because it attaches to the first rib, some attribute a depressor action to it. Its tiny nerve twig, also from the brachial plexus (C_5 to C_6), enters deep and is hidden in the dissection of the thorax. The muscle is enclosed between the two layers of the **clavipectoral fascia** as they leave the bone. This fascia splits after it leaves the Subclavius to enclose also the Pectoralis Minor below.

The Pectoralis Minor (Fig. VIII-27B) largely hidden by the Major, takes its origin from the anterior ends of the third through the fifth ribs. It narrows markedly as it passes laterally to insert upon the corocoid process of the scapula deep to the Deltoid. Its normal action is to swing the shoulder forward and downward, but with the shoulder fixed it can elevate its ribs of origin and, thus, assist in expanding the upper end of the thoracic cage. Most of us, however, use the muscle in combination with the Serratus Anterior to draw the scapula forward around the chest wall in punching and pushing movements. The muscle is supplied by the medial pectoral nerve which pierces it on its way to the Pectoralis Major.

Laterally the **Serratus Anterior** (Fig. VIII-27B) takes its origin from the first eight ribs and passes backward, tightly bound to the chest wall to terminate on the deep ventromedial border of the scapula. Four of the lower digitations of the Serratus Anterior attach to the inner inferior angle of the scapula. The muscle draws the scapula forward and around the chest wall at the same time. Most of this action is accomplished by a greater pull upon the inferior angle so that this portion of the scapula rotates forward and laterally. We have noted its action in concert with the Pectoralis Minor. It can, however, act as a rib elevator over a considerable stretch of the thoracic cage when the shoulder is fixed. The nerve to the Serratus Anterior is known as the **long thoracic nerve** (of Bell), and it arises from the upper roots of the brachial plexus (C_5 to C_7), descending along the medial wall of the axilla (lateral chest wall) between the muscle and the fascial covering. In the axilla, the nerve runs behind the great vessels and the trunks of the plexus.

Superficial Posterior Group: The more superficial

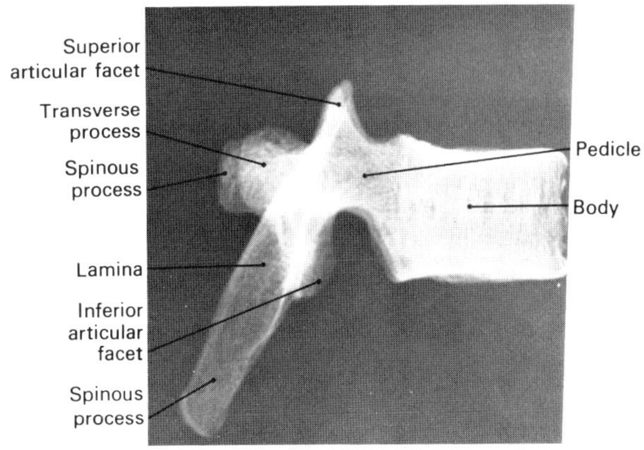

Fig. VIII-25. Lateral radiograph of a dry "typical thoracic vertebra."

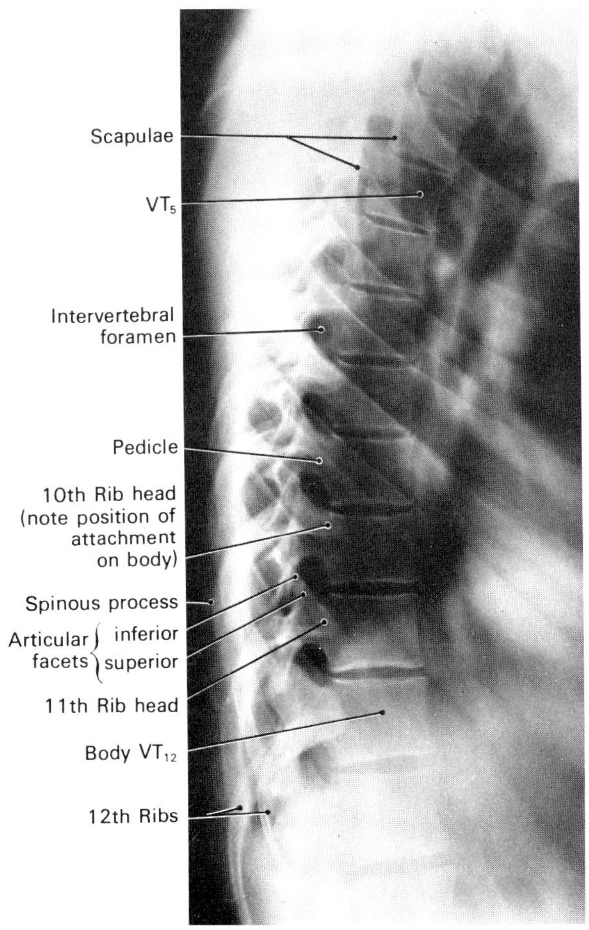

Fig. VIII-26. Lateral radiograph of the thoracic spine in a normal patient. Note the slight rotation of the subject. The magnified ribs are farthest from the film.

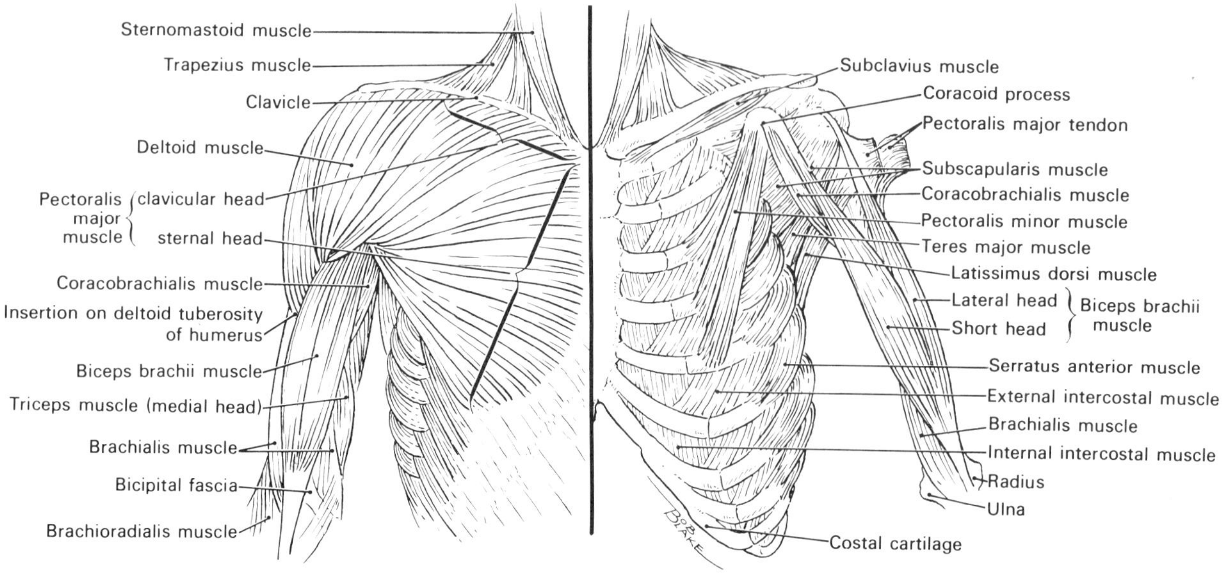

Fig. VIII-27A. Ventral aspect of the superficial muscles of the chest and arm. **B.** Ventral aspect of the deep muscles of the chest and arm.

muscles on the back of the cage also articulate with the shoulder and thorax. They play important roles in the movement of the shoulder, spinal column, and the rib cage as well when they are used as accessory muscles of respiration. Traditionally, we tend to assign function according to the direction of pull by the major muscle bundles. It is well, however, to think of muscles as functional chains and to think of the kinetic action of groups of muscles, since muscles are interdependent and coordinated in their joint endeavors.

We have considered one of these larger superficial back muscles in part before when we examined the posterior triangle of the neck. The **Trapezius** (Fig. VIII-28) arises from the external occipital protuberance, the ligamentum nuchae and all of the vertebral spines down to that of the 12th thoracic vertebra. The muscle fibers pass laterally from this midline origin to approach clavicle and scapula. The upper fibers attach to the lateral aspect of the clavicle, the middle fibers to the acromion along its medial border, and the lower ones to the upper crest of the scapular spine. In this way this large triangular shroud fastens the upper limb girdle to the axial skeleton. The muscle is chiefly supplied by the cranial nerve XI (spinal accessory) but in some individuals its lower portions derive a motor supply from C_3 and C_4. Its sensory supply from muscle and tendon spindles definitely is conveyed over spinal nerves C_3 and C_4. The Trapezius conceals the insertion of several muscles upon the scapula, those of the Levator, Rhomboids, and the Latissimus. The principle action of the Trapezius is upon the shoulder girdle. It retracts the scapula as in squaring the shoulders. In a sense this helps to increase the diameter of the thoracic cage as a result of better posture. The upper and lower fibers working as a unit rotate the scapula in an upward direction, *i.e.*, so that the glenoid cavity points upward, a movement that normally accompanies abduction of the arm. In general, the muscle exercises control over the position and movement of the shoulder during active movements of the arm and, along with the other muscles that insert upon the scapula, it maintains the level and poise of the shoulder. Alone the heavy fibers of the Trapezius which insert on the acromion prevent sagging of the lateral angle, for no other muscles insert this laterad in the shoulder region.

An impressive large **Latissimus Dorsi** (Fig. VIII-28) arises from the lower six thoracic spines and others of its muscular slips come from the broad expanse of the lumbodorsal fascia, the lower four ribs, and the iliac crest and even superior snipets arise from the inferior angle of the scapula. It is a powerful adductor of the arm, but it can assist in violent expiratory efforts such as coughing when its contraction can be felt in the posterior axillary wall to which it contributes as its tendon seeks an insertion upon the humerus. Its nerve, the **thoracodorsal,** comes from the posterior cord of the brachial plexus (C_6–C_8).

A group of three muscles, deep to the Trapezius and Latissimus, all arise from the vertebral column and terminate upon the upper part, medial, and lower parts of the vertebral border of the scapula. These are the **Levator Scapulae,** which insert into the superior angle and medial border as far as the base of the scapular spine, and the **Rhombodii, Major** and **Minor,** which

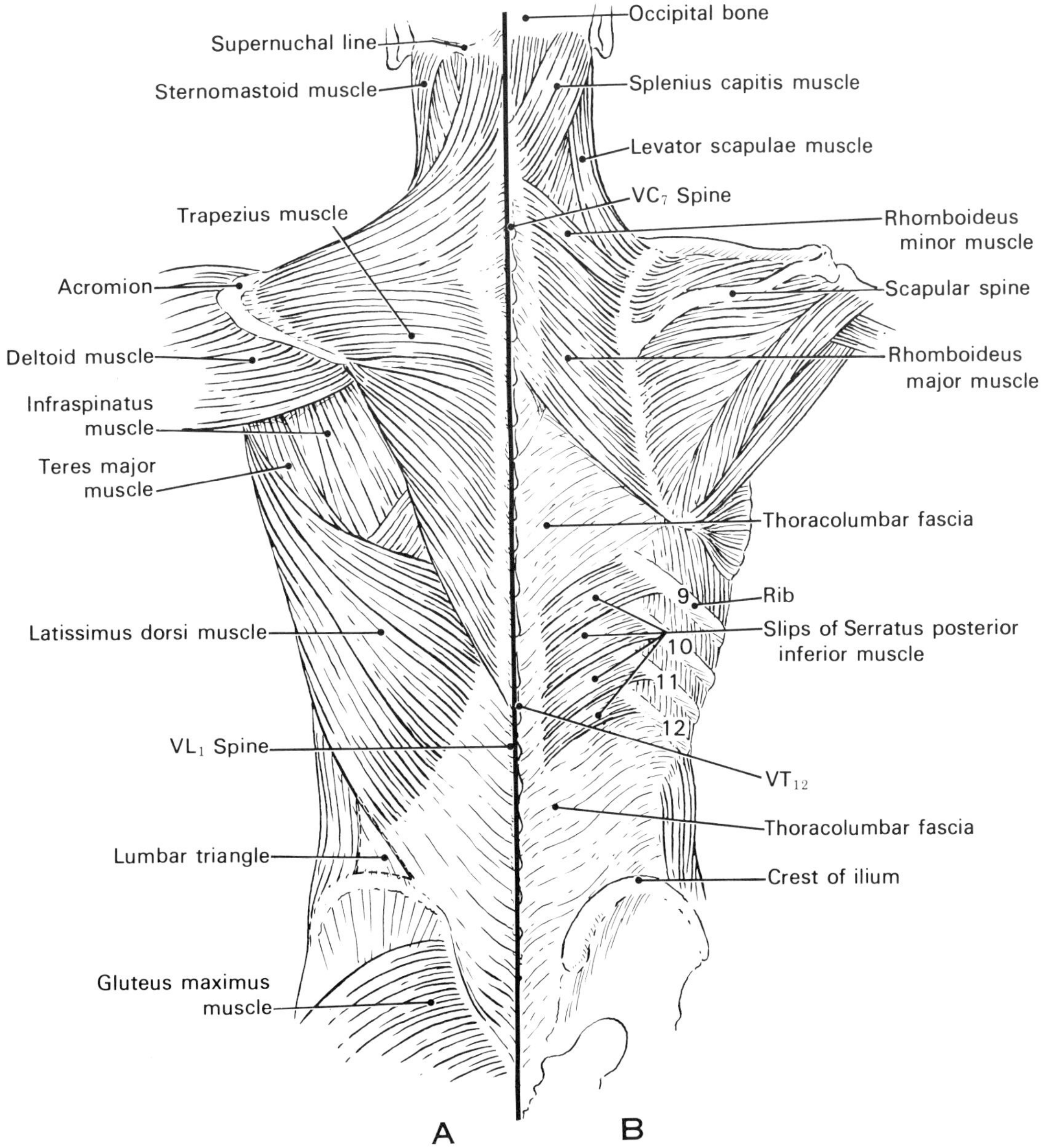

Fig. VIII-28. Superficial (A) and intermediate (B) muscles connecting the vertebrae and upper limb.

insert progressively lower along the medial border of the bone as far as the inferior angle (Fig. VIII-28). These three form a kinetic group of antagonists to the Trapezius and Serratus Anterior in that they rotate the scapula so that the glenoid cavity tips downward. Cooperating with the Trapezius, the Rhomboids pull the scapulae toward each other in the midline. The Levator is innervated by deep branches from the cervical plexus, but the Rhomboids receive fibers from C_5 off the roots of the brachial plexus *via* the **dorsal scapular nerve.**

The **Supraspinatus** arises from the bony supraspinatus fossa of the scapula, passes laterally over the shoulder joint, blends with the joint capsule, and inserts upon the upper of the three facets on the greater tubercle of the humerus. Indeed, the tendons of three muscles (**S**upraspinatus, **I**nfraspinatus, and **T**eres Major) **SIT** on this tubercle. The Infraspinatus, of course, arises from the large fossa inferior to the scapula spine. Like the **Deltoid** in the shoulder region, these muscles are not concerned with the thoracic cage directly, but they act in coordinated ways with some of the muscles just discussed above which do anchor to the cage directly, and they are all innervated from the posterior division of the brachial plexus (suprascapular nerve [C_5 and C_6] and the axillary

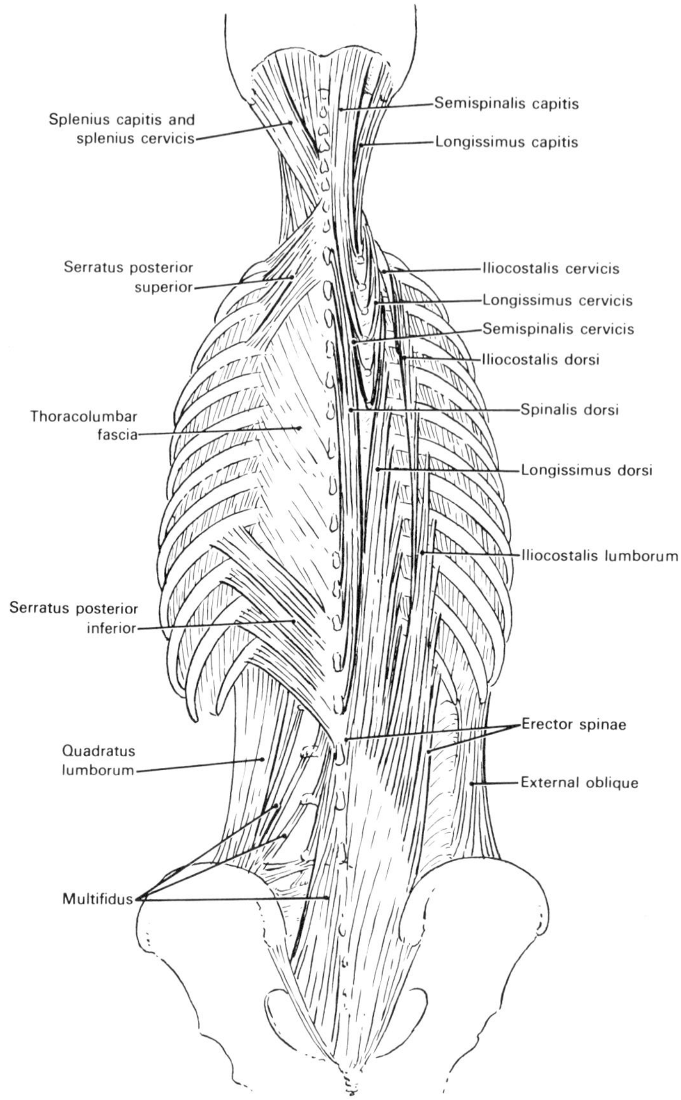

Fig. VIII-29. The deep muscles of the back.

nerve [C_5 to C_6] to the Deltoid, the first from the trunks and the last from the posterior cord of the plexus). The middle portion of the left Deltoid (Fig. VIII-27A), the middle part of the left Trapezius and their counterparts on the right side form a chain or muscle sling which can be used in lifting or holding a load against gravity when both arms are outstretched in the horizontal plane. When the Deltoid abducts the arm alone it can move the scapula from the position of rest through a 90° angle. When the arm is moved over the head, the Deltoid and the Supraspinatus move the scapula while the Trapezius and the Serratus Anterior act upon it as well in a coordinated and synchronized way. The Deltoid arises from the anterior border of the lateral one third of the clavicle, from the acromial lip and from the crest of the spine of the scapula. All of these muscle slips converge to insert upon the Deltoid tuberosity of the humerus. The anterior and posterior fiber bundles of this muscle run as parallel strands but its middle portion (acromial fibers) is a multipennate muscle.

Intermediate Posterior Group: Two quadrilateral plates of muscle lie deep to all of the back and shoulder muscles that we have discussed thus far. They seem to strap down the longitudinal intrinsic muscles of the back. These two are the Serratus Posterior Superior and Inferior (Fig. VIII-29). The Superior arises by a thin aponeurosis that stretches from the lower part of the ligamentum nuchae to the spine of VT_3, thereby overlapping the origin of other neck muscles. The muscle bundles take a downward and lateral course to divide into four muscle digitations which insert on ribs 2 to 5. The Inferior (Fig. VIII-29) is a similar muscle which arises from an aponeurosis attached between the spines of VT_{11} and VL_2. It separates into four, broad, overlapping, thick, muscular straps and these course upward and laterad to insert on the last four ribs. These two muscles act to pull upon the rib cage in opposite directions, the Superior elevating its ribs of attachment, the Inferior depressing its ribs and preventing their being pulled down by the action of the Diaphragm. In this sense the function is respiratory. Intercostal nerves supply the individual muscle slips at the lower border of the last rib which each slip crosses.

Deep Posterior Group (Fig. VIII-29): We have mentioned that the trough between rib angles and spinous processes is largely occupied by the Erector Spinae (formerly called the Sacrospinalis muscle) and by smaller intrinsic muscles of the back. Some of these slips from this complex muscle group were discussed when they appeared in the head and neck (the Splenii, and the Iliocostalis, Longissimus, and Semispinalis Cervicis, for example). It is not always easy to appreciate the complexity of these back muscles because what may have started as a group of segmental muscles running from one vertebra to the next now holds only for the short deep muscles. There has been considerable fusion between the more superficial of the back muscles to form longer ones and tangential splitting has occurred to result in superimposed layers. But the long muscles of the group can be classed into: (a) those that arise from the midline to run laterally upward (the Splenii); (b) those arising at or near the midline to run chiefly longitudinally upward (Erector Spinae); and (c) those that begin laterally and course toward the midline (Semispinalis, Multifidii, and the Rotatores). Finally there is a short segmental group that runs from one vertebrae to the next—the short Rotators, the Interspinales, and the Intertransversarii.

Figure VIII-29 makes it clear that the two Splenii are the only two large muscles of the back that take a markedly lateral course starting from their midline origin. The largest mass among the deep muscles of the back is the **Erector Spinae.** It has a broad and tough tendinous origin from the sacrum, lumbar spines, and the last two thoracic spines and the iliac crest. As it becomes muscular it splits into three longitudinal muscle columns: a lateral **Iliocostalis** muscle, an intermediate **Longissimus,** and a midline, slender **Spinalis** muscle. The first two of these consist of three, regional overlapping parts (lumbar, thoracic, and cervical). Each succeeding upper part arises from the same segments upon which its lower part inserts. The Spinalis has only a thoracic and cervical subdivision.

The inward-slanting or transversospinal group of muscles consists of shorter fiber bundles which vary in length from origin to insertion. The bundles of the Semispinalis span four to six vertebra, the Multifidii, four, and the longest Rotatores two segments at the most. The short Rotatores like the Interspinalis and Intertransversarii limit their span from one vertebra to the next.

Intrinsic Muscles: We now come to the true muscles of the rib cage or the muscles of respiration, proper. These are the Intercostal muscles and the Diaphragm. There are 11 pairs of **External Intercostal** muscles (Fig. VIII-1) which attach from the lower border of the rib above to the upper border of the rib below. Each begins close to the rib tubercle and extends anteriorly as far as the beginning of the costal cartilage, the fibers directed obliquely downward and forward. The gap between costal cartilage and the sternum is completed by a thin **external intercostal membrane** (Fig. VIII-1) which is transparent enough to view the fibers of the **Internal Intercostal** muscle (Fig. VIII-31A, C) deep to the membrane. This muscle forms a second layer from sternum to about the costal angle. Its fiber direction is also oblique but roughly at right angles to the fibers of the external layer. Posteriorly, beyond the rib angle to the vertebral column, the space is again filled by a membrane, the **posterior intercostal membrane.**

The anterior primary rami of the thoracic spinal nerves, as they emerge from their intervertebral foramina, lie against pleura for a short span before they pierce the posterior intercostal membrane. This is true of the intercostal vessels as well. Once they pierce the membrane, they ride forward in the intercostal space until they encounter the Internal muscle. Here the neurovascular bundle dissects the Internal Intercostal into two laminae. The deep layer is then designated as the **Innermost Intercostal** muscle (Fig. VIII-30D) or the **Intercostal Internus.** This deep lamina of the three-plied intercostal wall is not as well defined in the upper interspaces as in the lower ones. Posteriorly in the interspace, the Innermost Intercostal becomes continuous with the slips of the Subcostal muscles that bridge more than one space.

In the internal aspect of the anterior chest wall, deep to the intercostal and internal mammary (internal thoracic) vessels (Fig. VIII-30B), the **Sternocostal** or **Transverse Thoracic** muscle fans from the back of the sternum and xiphoid process to the costal cartilages of ribs two to six. Part of these muscle slips ascend while the lower ones become continuous with abdominal fascia.

Innervation of the Intrinsic Muscles of the Rib Cage: Figure VIII-31 is either a cross-section through a stout, short-legged Toulouse Latrec or through a big-hearted, buxom female leaning over the bar. In either case, this illustration shows the course of the intercostal nerves on the left and the intercostal arteries on the right. It can be seen that the **anterior branch (ramus)** of both spinal nerve and vessel pierces the posterior intercostal membrane to come to lie between Internal and Innermost Intercostal muscles. Nerve twigs supply motor end plates to these muscle fibers, and as the lateral and ventral branches of the nerve pierce the External Intercostal to send a supply to skin, they innervate this outer muscle of the

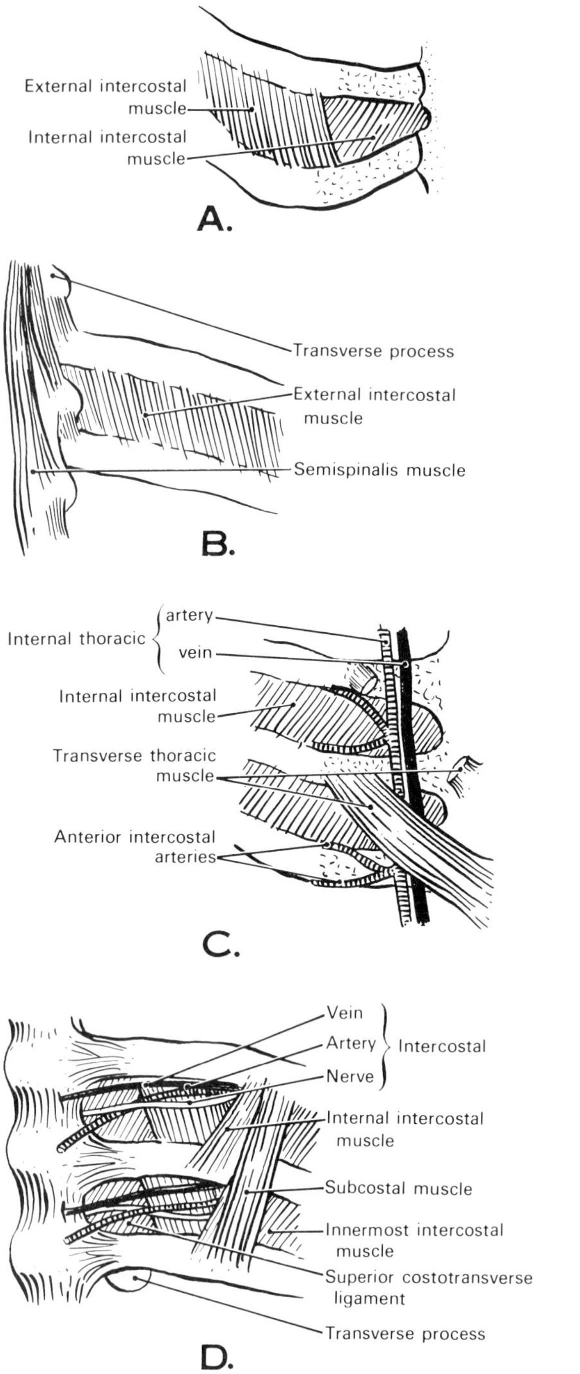

Fig. VIII-30A. Anterior view of the intrinsic muscles of the thoracic cage. **B.** Posterior view of the Internal intercostal muscles at the sternocostal joint. **C.** Posterior aspect of the costovertebral articulation with External intercostal muscles. **D.** Anterior aspect of the costovertebral articulation with the Intercostal muscles.

rib cage. The Subcostals and the Transversus Thoracis are also supplied by intercostal nerves.

The **posterior rami of the spinal nerves** (Fig. VIII-31) do not enter the intercostal space, but pass backward directly to supply skin and the back musculature. Briefly they behave as follows:

1. Like anterior rami, they supply segmental territories in series.
2. They *never* extend into limbs and rarely beyond rib angles.
3. They *never* form plexuses as do the anterior rami.
4. *Generally,* they divide into medial and lateral branches, *but there are exceptions:*

a. C_1 is a single twig to the muscles of the suboccipital triangle only.

b. C_6 to C_8 have no **medial branch** (*i.e.,* provide no cutaneous supply).

c. L_4 to L_5 have no **lateral branch** (*i.e.,* provide no cutaneous supply).

d. Neither do S_4, S_5, and CO_1 branch at all. Rather they unite into one small nerve to the skin over the coccyx (*i.e.,* they lack a muscular branch).

5. So far as the rest are concerned, the medial *or* the lateral branch is the cutaneous one:

a. Above T_6, the *medial* branch is cutaneous.

b. Below T_6, the lateral branch is cutaneous.

6. Some few wander beyond the rib angle:

a. C_2, the greater occipital nerve, reaches the vertex of the skull.

b. T_2 extends to the tip of the acromion.

c. L_1 to L_3 extend down over the buttocks.

Blood Supply: In so far as the **intercostal arteries** (Fig. VIII-31) are concerned, they lie between vein and nerve in the costal spaces. The first and second arise from the costocervical trunk of the subclavian. They are smaller than the others and are often difficult to locate in the process of dissection. The first descends deep to the pleura anterior to the necks of the first two ribs to anastomose with the first aortic intercostal artery. As it passes over the neck of the first rib it lies lateral to the first thoracic chain ganglion of the sympathetic trunk. Here it gives off its branch to the first intercostal space. The second intercostal artery branches off the descending portion of the first as it crosses the neck of the second rib, though this arrangement is not constant and is more prevalent on the right than on the left side. Fortunately, since these vessels are smaller than the other intercostal arteries which spring bilaterally from the posterior aspect of the aorta, they do not bleed as profusely when cut, and this may serve as a clue to the surgeon doing a subpleural dissection. The first through the seventh supply the thorax; eighth through eleventh, the thorax and abdomen; and the twelfth, the abdomen only.

Note in Figure VIII-30D how vein, artery, and nerve, in that order, enter the intercostal space and seek the shelter of the costal groove on the underside of the rib above them. This becomes significant when one enters the space with needle, trocar, or knife. In-

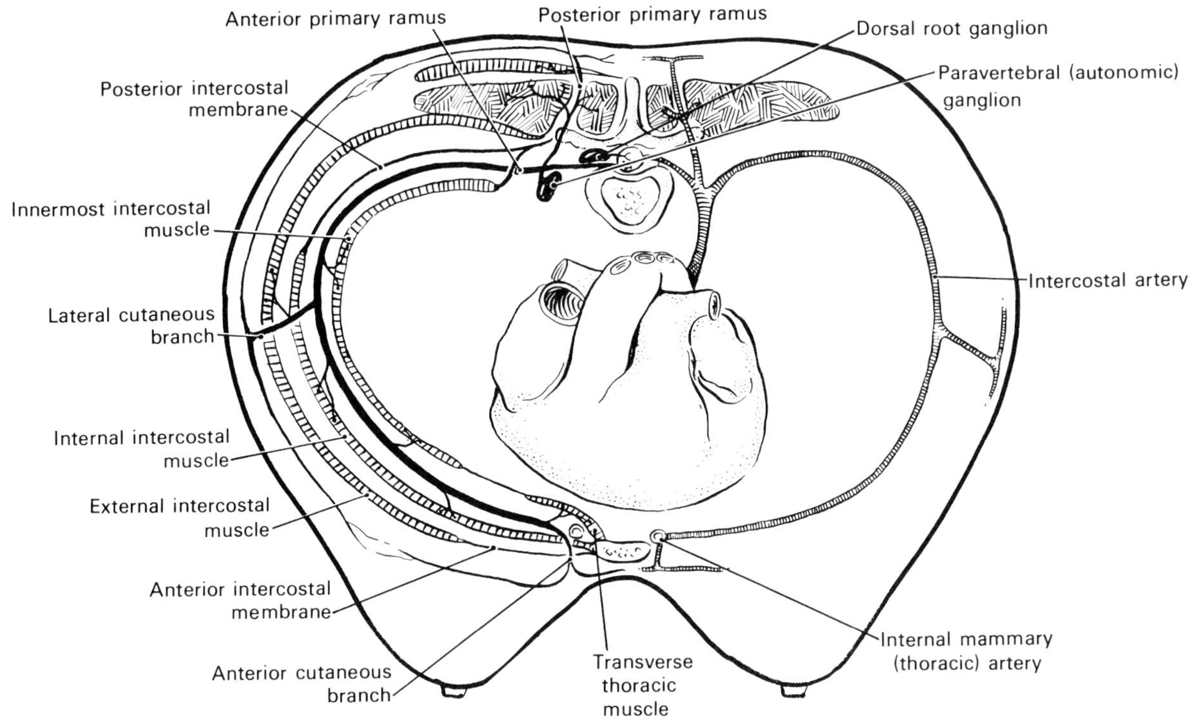

Fig. VIII-31. Innervation and vascular supply of the intrinsic muscles and the rib cage.

struments entering over the *upper border* of a rib will less likely do damage to the elements of the neurovascular bundle in the interspace than when the space is entered just below a rib border (Fig. VIII-32).

The pattern of venous drainage of the chest wall differs considerably from the pattern of arterial supply and it can be quite anomalous in certain individuals. Generally, the scheme is somewhat similar to that sketched in Figure VIII-33 which illustrates a segmental arrangement. The **first intercostal veins** drain directly into their respective brachiocephalic (innominate) veins. Two, three, and four join to form the **superior intercostal vein,** the right one draining into the azygos, the left, into the left brachiocephalic. This left vein is often a good landmark in the chest since it so frequently crosses the arch of the aorta where the left subclavian branches away from the aortic arch. Thus, it tends to lie superficial in the so-called "superior mediastinum" as it approaches the left brachiocephalic vein. The fifth to twelfth drain into the **azygos** vein on the right and into the **hemiazygos vein** on the left.

Below the Diaphragm, the azygos vein communicates with the inferior vena cava and sometimes with the right renal or the lumbar veins. Above the Diaphragm it drains into the superior vena cava. The hemiazygos, on the other hand, empties into the azygos above, and into the left renal or first lumbar vein below.

The internal thoracic (internal mammary) (Fig. VIII-30C) vessels descend in front of the apex of the pleura behind the junction of the subclavian and innominate veins. At the level of the first costal cartilage they are crossed by the phrenic nerve. At this point, the **pericardiophrenic artery** is given off to accompany the nerve to the Diaphragm. This vessel supplies pleura and pericardium as it descends along the side of the pericardial sac. The internal thoracic artery continues to descend behind the ribs about 2 cm. from the lateral margin of the sternum. At the sixth intercostal space it terminates in a **musculophrenic branch** to the Diaphragm and the **superior epigastric artery** which enters the rectus sheath behind the muscle. Moreover, the internal thoracic artery gives off perforating branches in the upper six interspaces which extend laterally to anastomose with branches of the intercostal arteries. In the female, the second and third of these perforating branches take part in supplying the mammary gland. Additional twigs are sent to lymph nodes of the mediastinum and to thymic remnants.

The Diaphragm

This partly muscular partition is dome-shaped. The heart sits upon its convex surface in the floor of the thorax while the liver hangs suspended from the concave undersurface in the abdomen (Fig. VIII-34A, B).

Development: We have noted in Chapter I how the embryo starts life "with its heart in its mouth," so to speak, and that its pericardium eventually rests upon a stout, partial partition, the **septum transversum.**

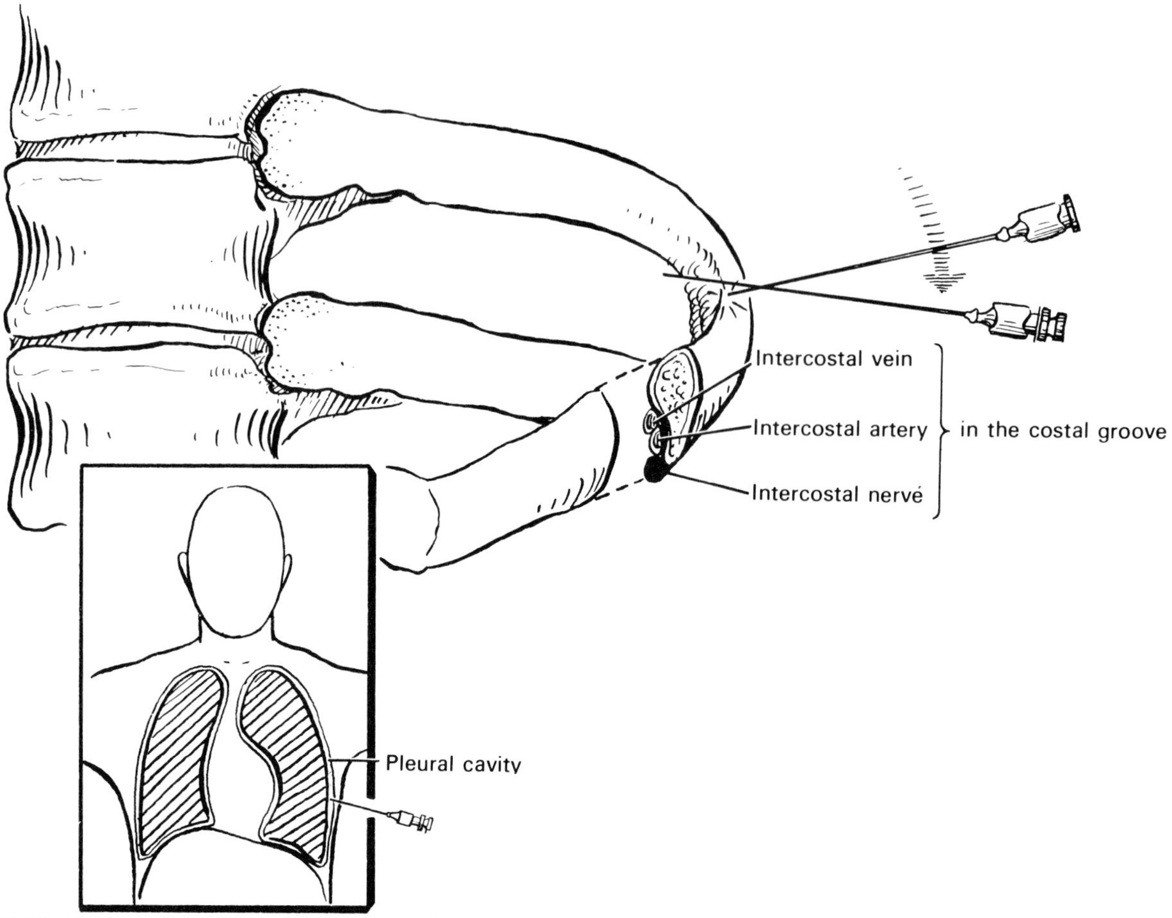

Fig. VIII-32. Contents of the costal groove. The proper needle placement for thoracentesis is shown. Inset shows the needle tip in the pleural cavity.

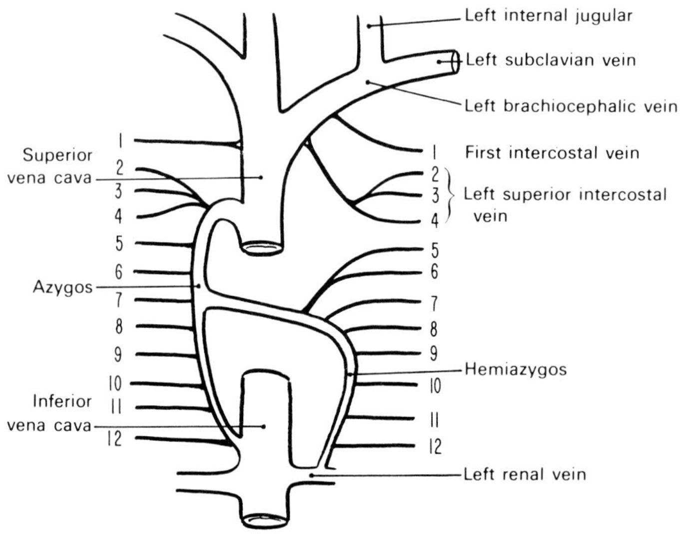

Fig. VIII-33. The azygos system of veins.

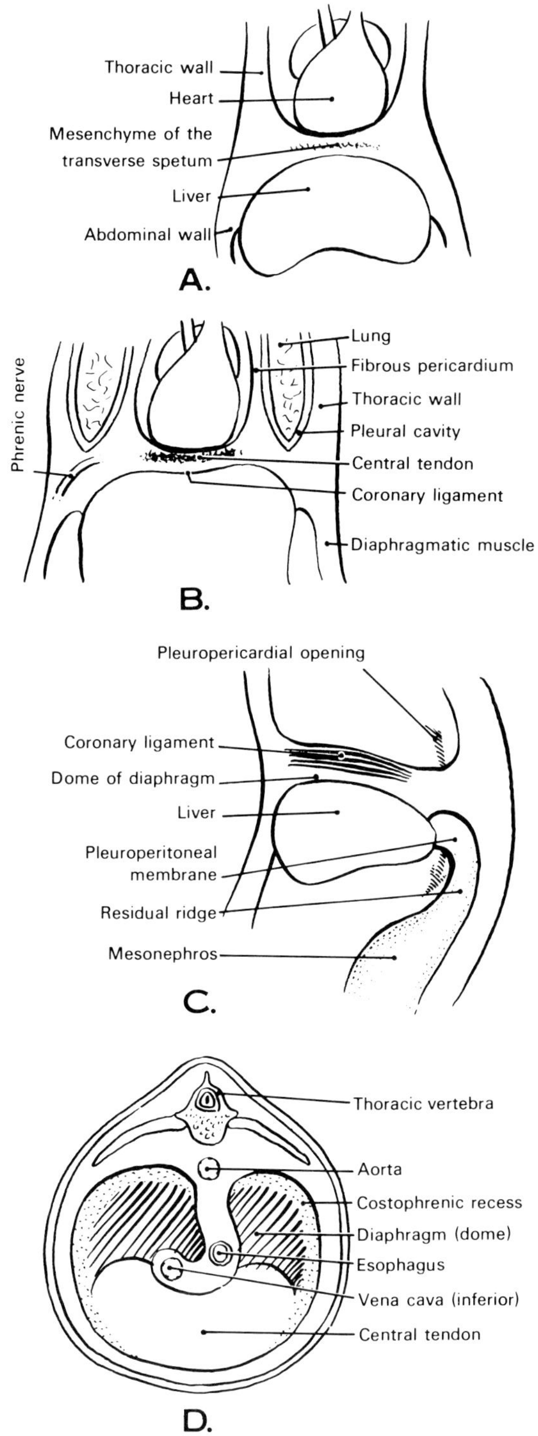

Fig. VIII-34. Formation of the Diaphragm. **A, B.** Anterior view. **C.** Lateral view. **D.** Cross section.

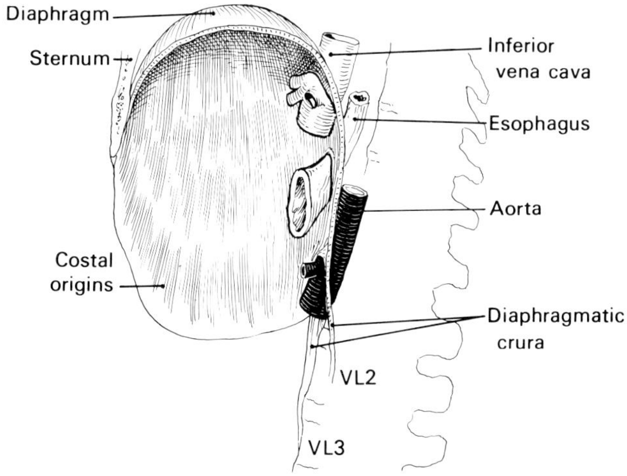

Fig. VIII-35. Diaphragm viewed from the left.

Both were originally located forward in the cervical region, and the septum served to convey venous trunks (common cardinal veins) to the heart. Only after considerable head and trunk bending and caudal migration did this future, partial Diaphragm reach its present level in the trunk. *En route* it picked up its muscle fibers from the **third to fifth cervical myotomes,** and the phrenic nerve to these muscle masses, arising from cord segments C3–C5, followed to the Diaphragm.

As the developing lungs and pleura in the embryonic thorax spread around the sides of the heart, they dissect the lateral body wall into two layers, a thoracic wall lateral to the lung, and a medial fibrous pericardium (Fig. VIII-34B). This widens the thorax at the same time that the liver in its growth from the septum transversum widens the abdominal cavity. In the process, a part of the body wall is drawn in on each side to be added to the septum transversum.

The mesonephric portion of the kidney is now developing in the thoracic wall. Its cranial portion atrophies with time to create a ridge designated as the **pleuroperitoneal membrane.** The membrane lies close to a transient opening between pleural and peritoneal cavities (Fig. VIII-34C). At this early stage of development pleural and pericardial cavities also communicate. The growing bilateral pleuroperitoneal membranes must close off the pleuroperitoneal openings (Fig. VIII-34D) before a diaphragm is complete just as the pleuropericardial membranes must complete the pericardial sac.

The final Diaphragm, then, is a composite of several parts:

1. The connective tissue of the septum transversum which forms the **central tendon** through which the inferior vena cava, esophagus, and aorta pass as major tubes (Fig. VIII-35).

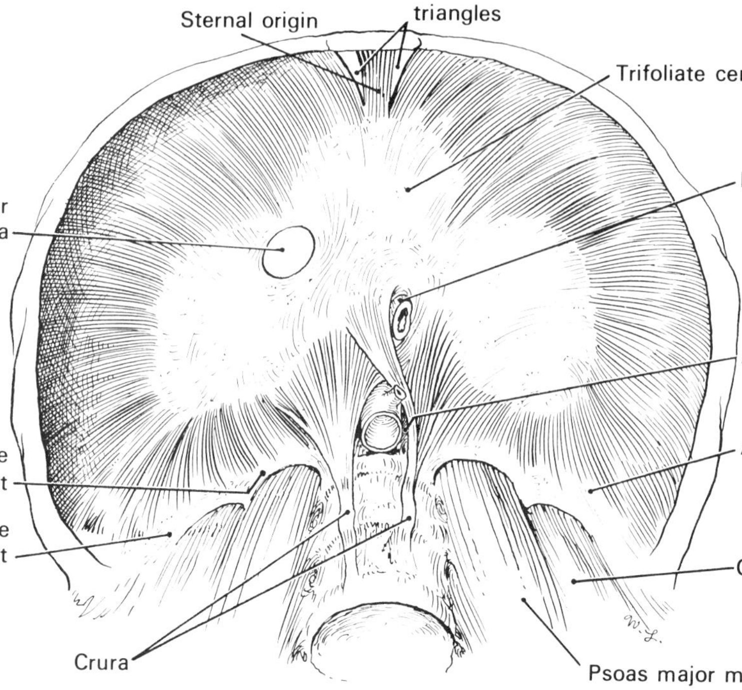

Fig. VIII-36. Diaphragm viewed from below.

2. The **tissue around the esophagus** into which muscle spreads. Portions of this, related to the dorsal mesentery of the abdomen, become the **diaphragmatic crura.**

3. The **pleuroperitoneal folds** make up parts of the **domes** around the central tendon. Muscle infiltrates this region also.

4. **Dissected musculature from the lateral body walls** around the perimeter of the Diaphragm. This has its own lateral wall innervation supplied by intercostal nerves.

The phrenic nerve descends in the pleuropericardial folds and is carried to the Diaphragm and these folds meet to complete the pericardium. The phrenic nerve supplies the muscle described in sections 2 and 3 above.

General Structure: The adult Diaphragm is a sheet of muscle composed of radiating fibers which extend from the lower border of the thorax and from the upper lumbar vertebrae. These fibers converge upon a somewhat trifoliate **central tendon** (Fig. VIII-36, 37). The fibers group into three portions according to their origin:

1. The **sternal portion** consists of two bands which arise from the posterior aspect of the xiphoid process and which are separated from each other by an interval filled with connective tissue (Fig. VIII-36). Laterally these midline bands are separated by a similar interval known as the **sternocostal triangle** through which the **superior epigastric** artery (a branch of the internal mammary artery) enters the sheath of the Rectus Abdominis.

2. The **costal portion** has fibers originating from the cartilages of the lower six ribs. These muscle slips interdigitate with those of the Transversus Abdominis.

3. The **lumbar part,** in continuity with the costal part, arises from two **arcuate ligaments,** medial (internal) and lateral (external), which are heavy fascial thickenings over the Psoas Major and Quadratus Lumborum, respectively (Fig. VIII-36). Above the lateral arcuate ligament, the muscular portion of the Diaphragm may thin at times, especially above the left kidney, to form a weak area in the Diaphragm known as the **lumbocostal trigone.** The crura can be considered as a second lumbar portion, tendinous lower down where they attach to the front of the vertebral bodies but muscular as they approach the central tendon. Between right and left crura is a gap, the **aortic hiatus** where abdominal aorta becomes continuous with thoracic aorta, and through which the **thoracic duct** also passes. The right crus is generally longer than the left (Fig. VIII-36, 37).

Slightly to the left of the midline, enclosed more by the splitting fibers of the upper part of the right crus than the left, lies the **esophageal hiatus.** The vagi pass through here to enter the abdomen with the esophagus.

Farther forward and higher in the tendinous part of the dome is the **foramen for the inferior vena cava** (Fig. VIII-36). This vessel, therefore, undergoes greater excursions during breathing than the other two major tubular organs.

The perforations for smaller structures to pass through the Diaphragm are not as constant in position

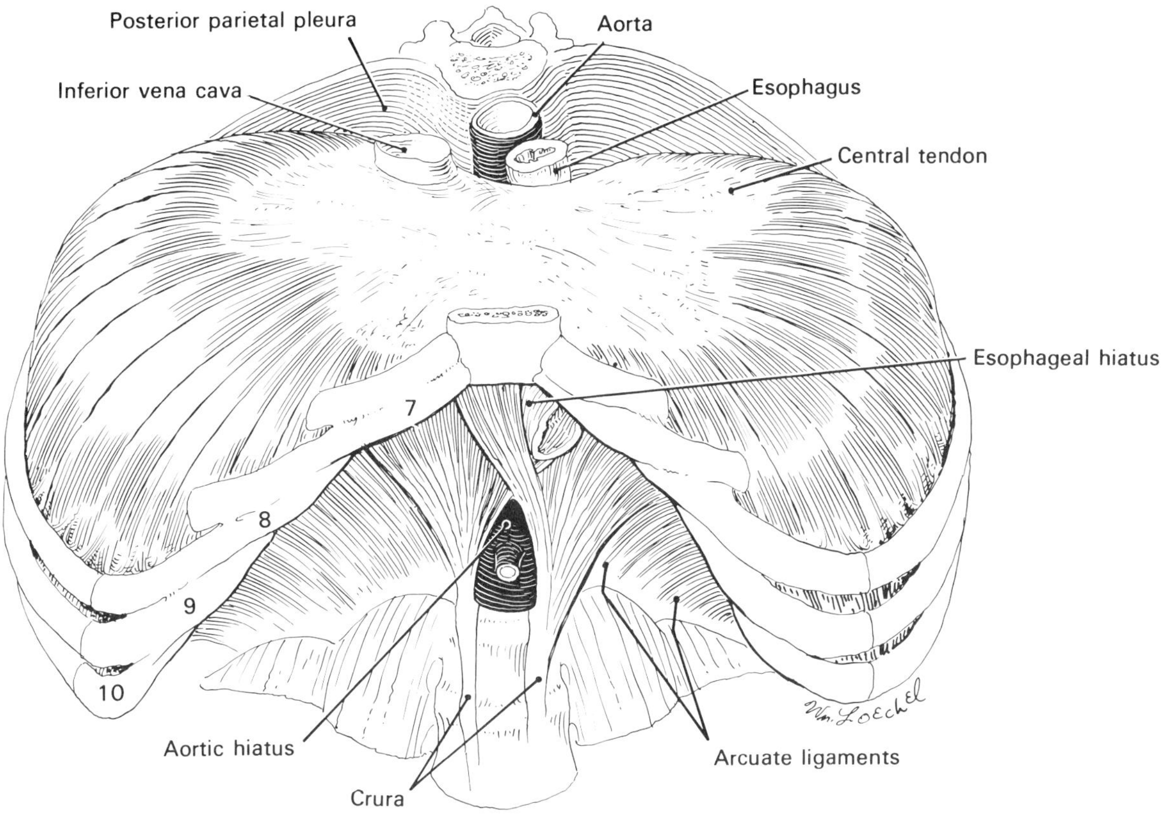

Fig. VIII-37. Diaphragm viewed from in front. Ribs have been cut away for visualization.

as these major openings. Three **splanchnic nerves** pierce each crus. The *azygos* and *hemiazygos veins* leave the abdomen for the thorax behind the crura. The **sympathetic trunk** enters the abdomen behind the medial arcuate ligament. We have already noted the pathways for the superior epigastric artery, thoracic duct, and vagus nerves. Esophageal branches of the gastric arteries and nerves pass through along the esophagus in the reverse direction.

Innervation and Blood Supply of the Diaphragm: The *phrenic nerve* carries the motor supply to the Diaphragm by piercing it and ramifying on its inferior surface. The nerve arises from the ventral rami of the third to fifth cervical spinal segments. The sensory supply to the upper and lower surfaces is also mainly *via* the phrenic, but motor and sensory supply to the peripheral fringes is by the intercostal nerves. Arterial supply is widespread from some intercostal vessels, from the inferior phrenic artery off the aorta, from the musculophrenic artery, and from the pericardiophrenic artery.

Lymphatics of the Diaphragm (Fig. VIII-38): The Diaphragm is richly endowed with lymphatics on both surfaces. These surface vessels unite in a network which is exceedingly fine and mesh-like in the region of the central tendon; however, the network is most prominent in the lateral leaflets. From this network, numerous branches pass out parallel to the muscle fibers and unite with a series of anastomosing stems which drain anteriorly. Ultimately these vessels drain into a single diaphragmatic group of nodes and subsequently pass to the anterior mediastinal lymph nodes. From the network in the lateral leaflets of the central tendon, collecting stems are directed posteriorly and medially toward the aortic hiatus and these empty into the upper coeliac nodes. The lymph nodes on the superior surface of the Diaphragm are situated anteriorly; the coeliac nodes, the principle nodes of the inferior diaphragmatic surface, are located posteriorly. Both of these sets of nodes receive lymph from each surface of the Diaphragm *via* perforating branches which interconnect the upper and lower networks. Other anastomoses of the upper diaphragmatic network are with the lymphatics of the pleura, while the lower diaphragmatic lymphatics connect with the lymphatics of the liver and those of the lateral portions of the peritoneum. Because of the many communicating channels that we have described, it is not difficult to account for the occurrence of pleuritis as a sequel of a subphrenic abscess, or the development of a subphrenic abscess as a consequence of thoracic empyema. The clinical import to understanding the lymphatic system of the Diaphragm is, thus, obvious.

Actions of the Muscles of Respiration

The main muscle of respiration is the Diaphragm, and it is chiefly an inspiratory muscle. The lowering of its dome increases the vertical diameter of the thorax.

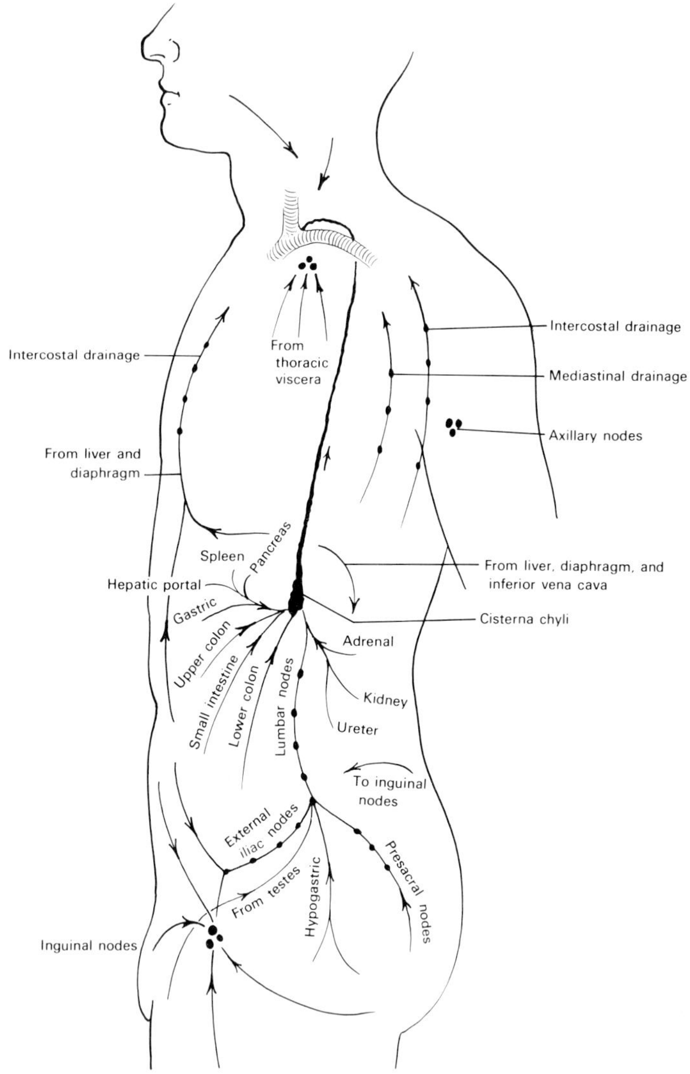

Fig. VIII-38. Pattern and direction of lymphatic drainage (schematic). *Arrows* show direction of flow in lymphatic vessels and the area being drained. Note variability in diaphragmatic region.

The expansion of the 12th rib, to which the Diaphragm is attached, helps increase the transverse diameter.

At rest, when the tidal volume is only a few hundred millimeters, the dome undergoes a vertical movement of about 1.0 to 1.5 cm. During forced expiration, the dome rises high in the thorax compressing the basal pulmonary parenchyma, as well as compressing the heart and widening it in its transverse diameter. Conversely, the Diaphragm sinks low in the abdomen during forced inspiration (Fig. VIII-39A, B). The distance between the location of the diaphragmatic dome in these two phases of respiration is from 6 to 10 cm.

The position of the Diaphragm varies considerably with body position and body habitus. In the *supine* subject, it is displaced several centimeters toward the head, decreasing vital capacity because of the decrease in expiratory reserve volume. With the patient on his *side* (*decubitus*), the Diaphragm is asymmetrical. The highest portion is flattened while the convexity of the lower portion is increased. Respiratory movements are then more apparent in the lower portion. In the erect or the supine position, with every descent of the Diaphragm during inspiration, the abdominal viscera are pushed down, so that the abdominal wall moves slightly forward. This type of "abdominal breathing" is absent when the Diaphragm is paralyzed. With unilateral paralysis, the affected hemidiaphragm pushes several centimeters higher than the intact hemidiaphragm and shows minimal movement in respiration. Indeed, in rapid respiration, the paralyzed side may rise instead of pushing downward even minimally as a result of the drop in intrathoracic pressure on that side as abdominal pressures remain the same or increase slightly. Body habitus too has some bearing upon the position of the Diaphragm. The asthenic individual will commonly have a low lying Diaphragm, while the pyknic patient tends to have a higher positioned dome.

The Subcostal muscles appear capable of rib elevation while the Transversus Thoracis can depress them. A major task of the Intercostal muscles is to keep intercostal tissue taut so that it will not be sucked inward by the decrease in intrathoracic pressure which accompanies inspiration. It has been claimed that one set of these muscles contracts on inspiration, the other on expiration, presumably because of their difference in fiber direction. Perhaps this is the case in quadrupeds, but electromyographic recordings from Intercostal muscles in man demonstrate no differences in function. In quiet respiration there is no recording of electromyographic activity at all and, the Intercostals begin to contract only as respiration deepens forcibly. Even then, External and Internal Intercostals act in unison. It would seem, then, that quiet inspiration is accompanied by a downward depression of the domed Diaphragm and only by some active work on the part of the Subcostals and the Transversus upon the rib cage. Intercostals tense to prevent the interspaces from buckling inward. Expiration is a passive phenomenon. Contracted muscles relax, gravity overcomes any elevation of the rib cage that has occurred, and the lungs recoil as a result of their inherent elasticity in the sense that alveolar and terminal respiratory tube diameters decrease.

In attempts to increase lung volume through active inspiratory efforts, the overall range of muscular movement is not great. The chest expands only a few millimeters more than in normal breathing in increasing its transverse, anteroposterior and longitudinal diameter. Yet, this increment is sufficient to increase lung volume by about three quarters of a liter—a vital factor in patients with labored respiration.

Aside from motions in the joints of the tiny ear ossicles, perhaps no other combinations of joints in the human body are moved more frequently and more continuously than the joints of the thoracic cage.

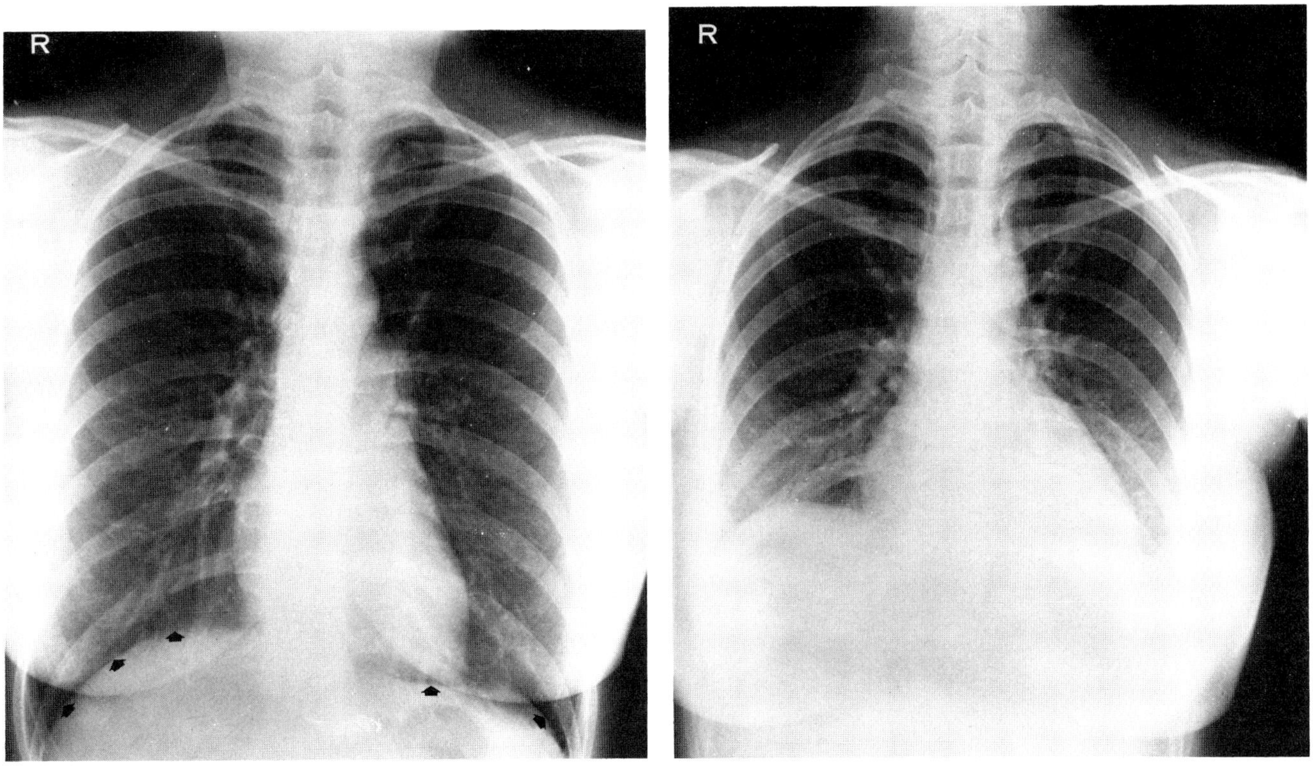

Fig. VIII-39A. Normal adult female chest. PA projection made during deep inspiration. The cranial surface of the Diaphragm is outlined (*arrows*) against the air-filled lung above. The right dome of the Diaphragm is at the level of the sixth rib anteriorly or the 10th posterior intercostal space (between the 10th and 11th ribs). **B.** Same patient as in A. PA projection in expiration. The right dome of the Diaphragm has elevated to the level of the fourth rib anteriorly or the ninth rib posteriorly. Note particularly how large the heart now appears. Extreme *caution* must be exercised in not overinterpreting radiographs made in the expiratory phase of respiration.

While each rib has its own range and variety of movements, they are all combined in a respiratory excursion. Each rib can be regarded as a lever. Its fulcrum lies just outside the costotransverse articulation so that when the rib elevates, its neck depresses. As a result, any small movement at the short neck-end of the lever is magnified disproportionately at the anterior tip of the long end of the lever. The first of the vertebrosternal ribs differs from the rest since it is rigidly fixed to the manubrium, and when its anterior tip elevates, the manubrium moves with it, though in quiet breathing the manubrium is hardly raised. The movement of the second rib, because of its attachment at the manubrium, is also slight in quiet breathing. The costal ends of the third to sixth ribs thrust more forcibly in a ventral direction when they elevate so that they swing the sternum forward and upward through a greater arc than the first two ribs. This is known as the "pump-handle effect" since the sternum is moved up and down, increasing and decreasing the anteroposterior diameter of the chest. This behavior is illustrated in Figure VIII-40. The outward movement of the ribs as they elevate, increases the transverse diameter of the chest as a kind of "bucket-handle effect" (Fig. VIII-40).

The vertebronchondral ribs (including the seventh as well) act as "bucket bails," laterally, to enlarge the thorax, but anteriorly, this outward movement turns out the anterior end of the rib to widen the subcostal angle. This helps widen the anteroposterior diameter of the abdomen and makes room for the displaced abdominal viscera which are depressed by the contracting Diaphragm during inspiration.

The two vertebral ribs are capable of slight movements in all directions, but, generally, as the other ribs elevate, these are depressed and fixed to aid the action of the Diaphragm.

All movements are magnified in forced respiration when accessory respiratory muscles come into play. Probably the most important accessory elevators are the Scalenes and the Sternomastoids. The chief depressors in forced expiration are the oblique and transverse muscles of the anterolateral abdominal wall. Other accessory muscles include many of the superficial ones that clothe and attach to the thorax, the Erector Spinae in particular.

Variations in Shape of Thorax

Certain chests depart from the norm in structure. The *flat chest* has a short AP diameter and a wide transverse diameter. The ribs slope downward and forward

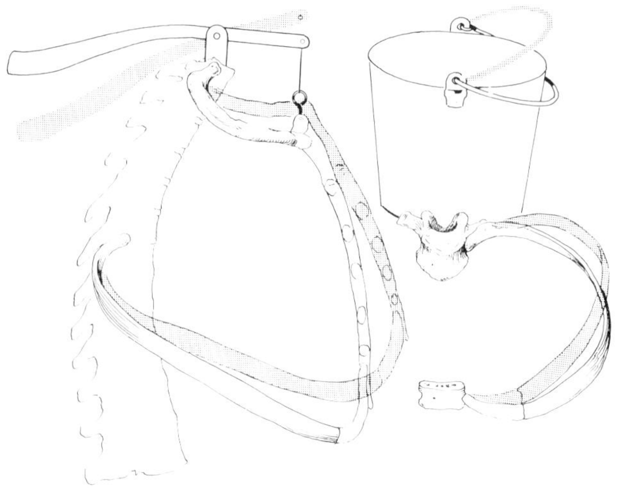

Fig. VIII-40. Schema to show "pump handle" or "bucket handle" action of rib elevation during respiration.

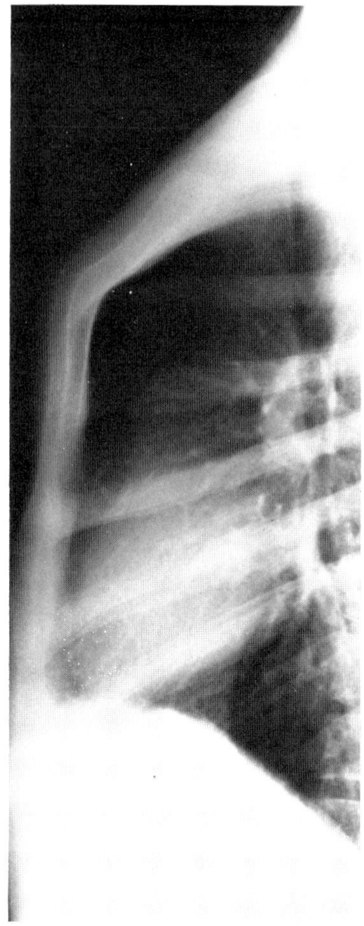

Fig. VIII-41. Pigeon breast or **pectus carinatum.** This diagnosis is usually made on the lateral radiograph where the sternum is seen to protrude farther anteriorly than is normally seen.

with an obliquity that is steeper than normal. The *barrel chest* is short and rounded, the ribs more horizontal and the AP and transverse diameters are almost equal. This type is sometimes seen in chronic, obstructive lung disease such as emphysema. The **pigeon breast (pectus carinatum)** with a prominent sternum need not be associated with rickets alone. It can be congenital (Fig. VIII-41). In the **funnel chest (pectus excavatum)** depicted in Figure VIII-42, there is a marked sinking in of the sternum, especially in its lowest parts. Neither of these last two types necessarily signify structural weakness nor limitations to expansion. The *hope chest* of the female is oblong and rigid, and it usually has a hinged cover. It is a chronic sign of imminent matrimony.

To complete the thorax externally we need only add fascia and skin. It seems a pity to leave the exterior of the thorax without discussing the female breast, an appurtenance far more charming than that of the male, far more useful, and of much greater clinical import.

The Mammae

These structures lie in the superficial fascia, and in the young adult female they possess no true capsule. In the nullipara, each mamma is discoidal, hemispherical or conical in shape. It covers an area extending from the border of the sternum into the axilla and from the second to the sixth or seventh rib (Fig. VIII-43A). Its craniocaudal diameter of about 12 to 15 cm. is somewhat less than its transverse diameter, and the left mamma is often a bit larger than the right. Each mamma weighs 150 to 200 gm, but weight increases to 400 to 500 gm. during lactation.

The deep surface of the gland rests upon the perimysium of the Pectoralis Major (Fig. VIII-43B) and, laterally, on that of the Serratus Anterior. Connective tissue cords, the suspensory ligaments (of Cooper), anchor the mamma to the muscle fascia and dermis.

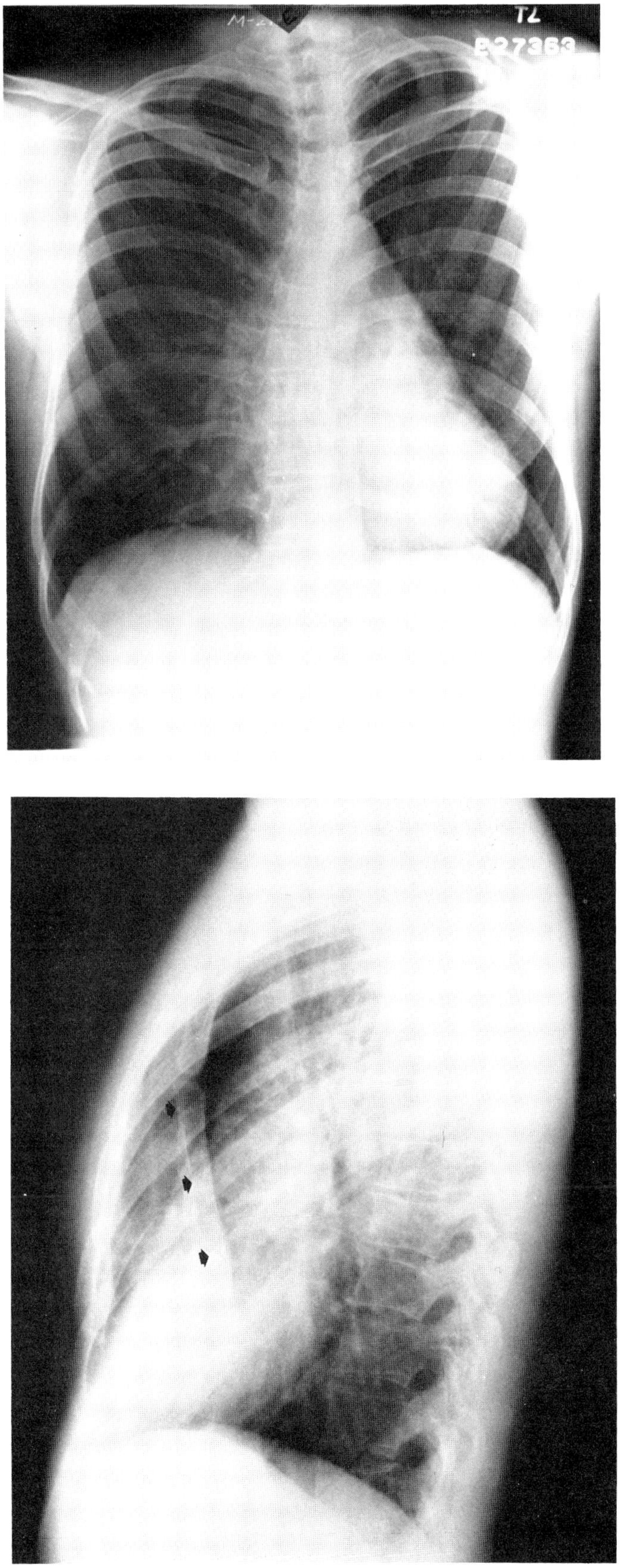

Fig. VIII-42. Funnel chest or **pectus excavatum. A.** In the PA view, the heart is seen to be shifted slightly to the left; the heart is more convex laterally on the left; the right heart border is rather indistinct; there is an apparent denseness in the right lung base and the vascularity here is accentuated due to lung compression by the depressed sternum. **B.** In the lateral radiograph, the sternum is markedly depressed and actually is displacing the heart posteriorly against the vertebral column. In this very severe case, the sternum appears to overlap the heart.

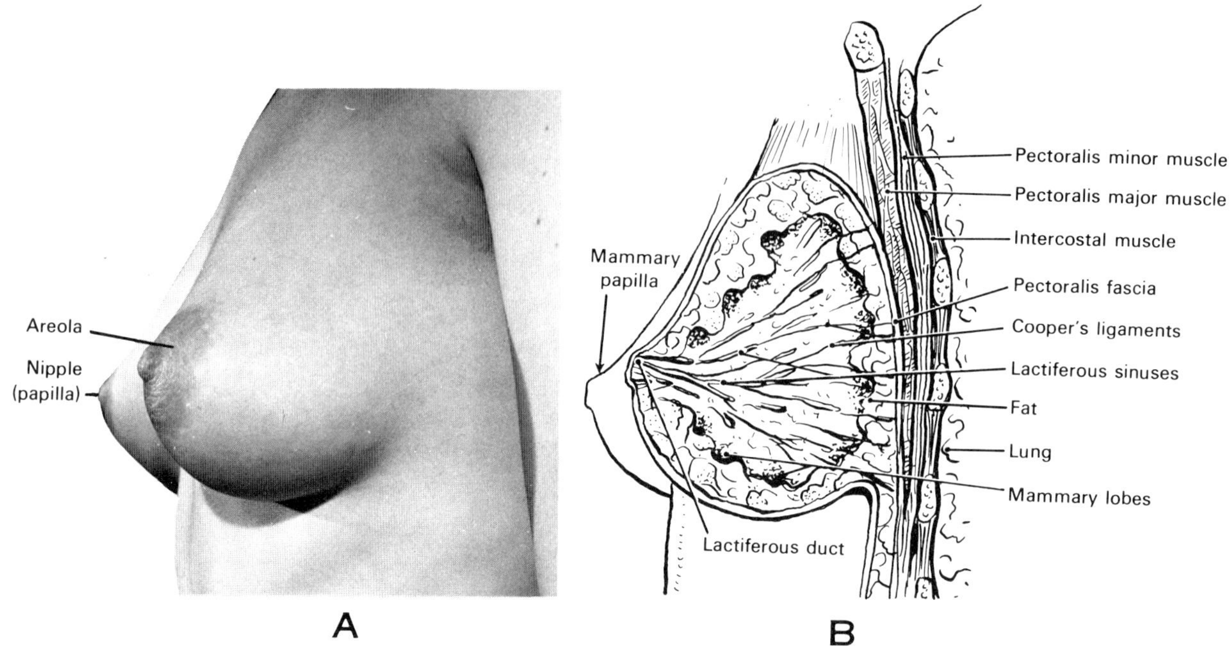

Fig. VIII-43. The adult female breast. **A.** Gross photograph. **B.** Sagittal section (schematic).

The glandular portion of the breast ramifies amongst the fat of the superficial fascia, sending out irregular projections as lobules, one of the longest of these being the axillary tail (of Spence) (Fig. VIII-44A).

The gland develops as solid ectodermal ingrowths which invade the superficial fascia along the *milk line* (Fig. VIII-44B, C) in the 6-week embryo. During the remainder of fetal life, these cords proliferate and gradually form a number of buds. By birth not much more than a main duct system and a few scattered nodules are present in an unimpressive breast. Little change occurs until close to puberty when the breast enlarges as a result of fatty deposition, the proliferation of glandular acini of the non-secretory type, and the enlargement of the ducts. With the onset of the menses there is vascular engorgement and dilatation of acinar lumina in the premenstrual stages only to be followed by regression during postmenstrual intervals. In the early stages of pregnancy the pristine rosy pink of the areola disappears forever once deposition of a brownish pigment increases in this area. The special sweat glands of the areola (glands of Montgomery) proliferate and raise numerous elevations in the pigmented circle about the nipple. The duct system is completed in the first 6 months of pregnancy while the acini reach secretory phase in the last 6 months. The early milk of the 3 days postpartum is thin and yellowish (colostrum), and true milk begins about the fourth day, lasting through weaning. Once nursing terminates the breast gradually returns to prepregnancy size and weight with the loss of considerable glandular tissue and its replacement by fat. The breast becomes more pendulous than before. With menopause there is a gradual reversion to the infantile condition until senile atrophy leaves a sagging fold of skin.

Extra nipples (supernumerary, Fig. XIII-44C) can develop anywhere along the milk line—in axilla, on the abdomen or on inner aspect of the thigh. Ectopic nipples have been described on the back, shoulder, and even on the buttock. All of these are, generally, small and nonfunctional.

The nipple is penetrated by the ducts which have dilatations, the **lactiferous sinuses,** (Fig. VIII-43B), just before they enter the nipple. The level of the nipple varies with the pendulousness of the female organ, and in extreme cases can rest as low as the twelfth thoracic or the first lumbar vertebra.

The **lymphatics of the mammary gland** are exceedingly important to understand since they so frequently serve as channels for the spread of malignancies from one breast to the opposite breast, to the organs of the thorax, and to associated lymph nodes. Drainage of lymphatic channels across the midline has been reported in recent years not to be significant, but some surgeons still hold this to be an important route to consider in the spread of malignant disease.

Superficial lymph vessels lie under the skin of the mammae. The extensive plexus receives afferent channels from the gland and it sends its outgoing or efferent lymphatic channels to various nodes of the chest wall (Fig. VIII-45). Among these are the **pectoral** and **infraclavicular** (subclavicular) nodes. Other efferent channels reach to a deep lymphatic plexus related to the deep fascia on which the gland rests. The deep plexus receives the majority of the direct efferent vessels from the gland. It communicates with pectoral nodes, subscapular nodes, infraclavicular nodes, nodes

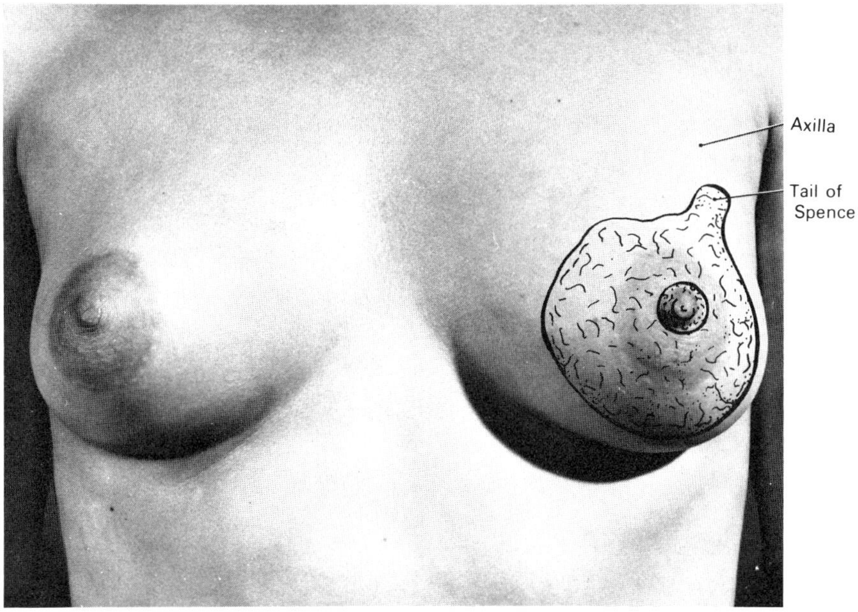

Fig. VIII-44A. The axillary tail of Spence.

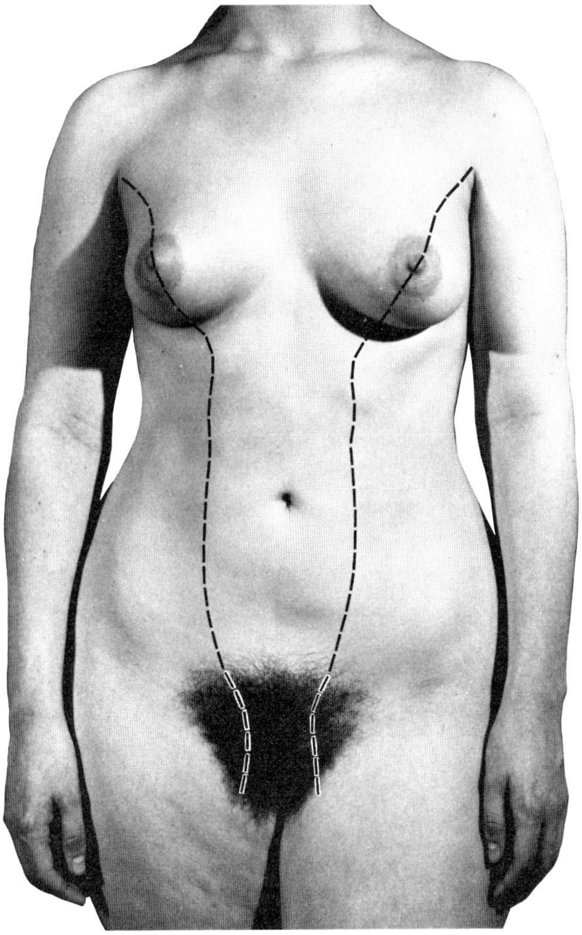

Fig. VIII-44B. Milk lines.

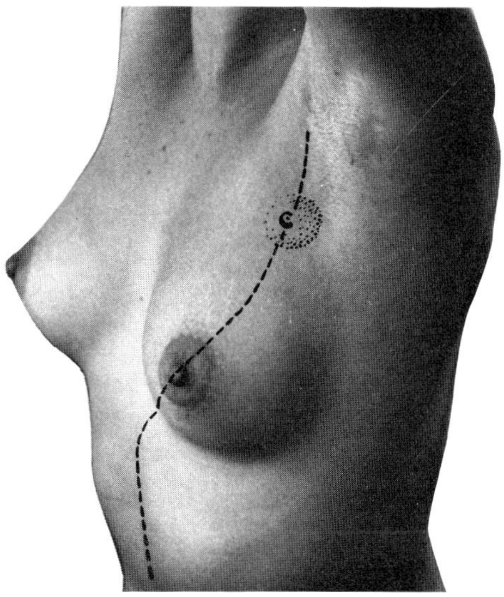

Fig. VIII-44C. Supernumerary nipple on accessory breast.

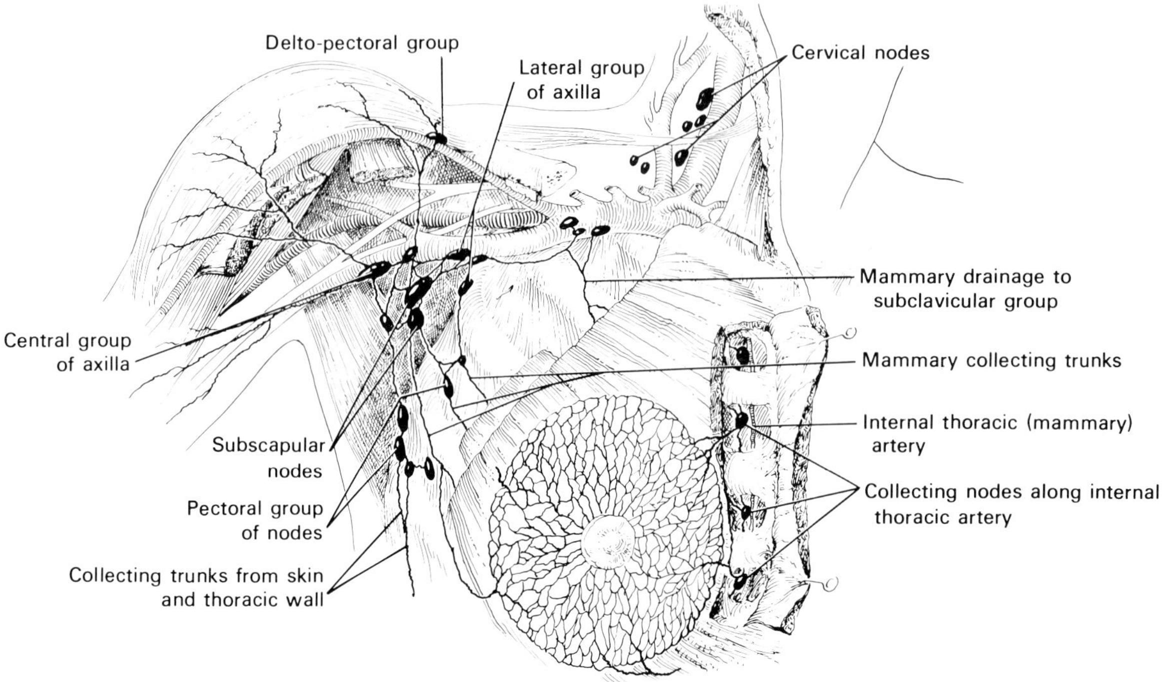

Fig. VIII-45. Lymphatic drainage of the right breast and axilla.

along the course of the internal mammary vessels and even to a subdiaphragmatic plexus of lymph capillaries which crosses the costal margin and penetrates the abdominal wall. All of these specific nodes communicate with other regional nodes that empty either into the right lymphatic duct or the thoracic duct.

In general, lymphatic drainage in the breast and the anterior thoracic wall follows along vascular routes and is primarily to pectoral nodes and secondarily to internal mammary nodes. Individual variation in connections and directions of flow cannot be overlooked, but when axillary nodes become involved in spread of malignancy from the breast, the supraclavicular nodes (lower deep cervical nodes) are also known to be involved in one-sixth to one-half of cases, and the internal mammary group in one-half of the cases. If the supraclavicular and internal mammary nodes are not examined for treatment or excision, but only the axillary nodes are given attention, it can be anticipated that 70% of patients neglected in this way will die of metastatic disease.

The **major source of blood supply to the breast** is from the perforating branches of the internal thoracic (internal mammary) artery and from branches of the lateral thoracic. Intercostal arteries and pectoral branches from the **thoraco-acromial artery** may contribute a partial supply. The thoraco-acromial is a short trunk arising from the axillary artery. Its origin may be partially hidden by the upper portion of the Pectoralis Minor. It pierces the clavipectoral fascia and distributes to the pectoral muscles and mammae, to the Deltoid and the Subclavius, and to the joints of this region. Much of the venous blood passes by way of a superficial plexus, the veins of which show in the mammae of the fair-skinned. Vascular engorgement of these veins during the menses or pregnancy may produce pain and discomfort. Their drainage is easily blocked by clothing straps and binding garments. Even rapidly proliferating glandular tissue can compress these thin-walled veins, and their blockage may contribute to discomfort. The plexus is drained by veins which are companions to the arteries.

PART II. THORAX

Thoracic Cavity Proper

Having constructed and clothed a cage for thoracic structures, our next problem will be that of assembling the occupants within its subcompartments. Recall that the cage has a narrow inlet which communicates freely with the neck. Its boundaries are the sternum in front, the body of the first thoracic vertebra behind, and the first pair of ribs laterally. In fact, the neck structures (trachea, esophagus, veins, and nerves descending from above; arteries and lymphatics ascending from below) serve to completely stuff the inlet. These structures are surrounded by connective tissue sleeves and connective tissue packing material. As all these structures of the inlet continue to the thorax, they will form a central partition which will include the heart to which the great vessels connect. This central partition, or **mediastinum,** will separate two, lateral, pleural chambers

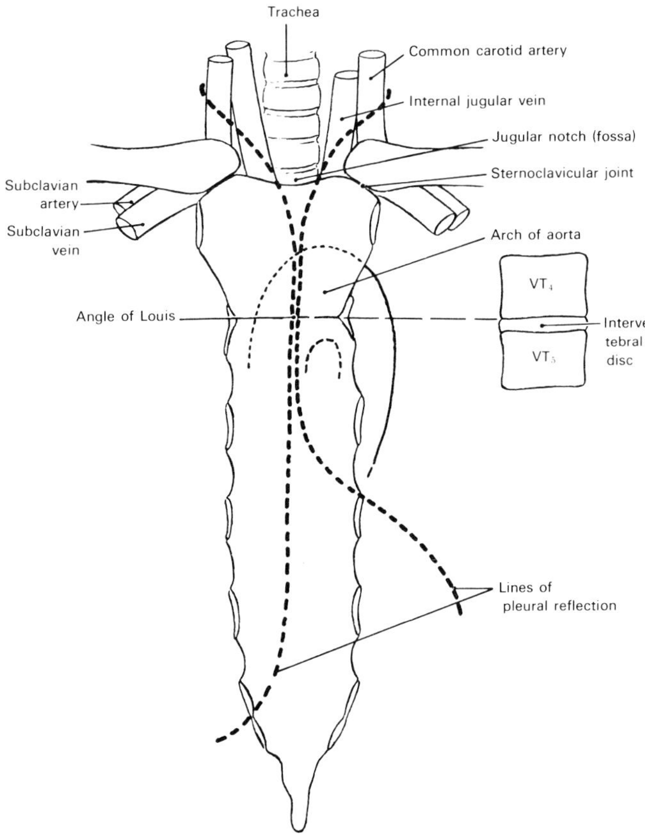

Fig. VIII-46. Scheme of anterior lines of pleural reflection in the mediastinum.

from each other. It is these lateral chambers that the lungs will fill as they bud from the tracheal bifurcation during development.

The outlet of the thorax, bounded by the 12th rib, the costal margins, and the xiphoid process, is effectively stoppered by the arching Diaphragm, a more or less transverse or oblique fibromuscular septum between thorax and abdomen. Any structures leaving the thorax for the abdomen, or entering from it, must perforce punch their way through this septal outlet.

The Mediastinum: The midregion of the thorax, that is the *extrapleural space between the lungs,* has been divided arbitrarily by surgeons and anatomists into a number of subdivisions. But, essentially, the mediastinal compartment contains the heart and pericardium, great vessels, trachea, thoracic duct, and a number of smaller structures including vessels, nerves, and lymphatics.

The intervertebral disk between VT_4 and VT_5 and the sternal angle of Louis are in line with each other anteroposteriorly. This horizontal line marks the level of the bifurcation of the trachea *in the cadaver* and, approximately, the midpoint of the arch of the aorta (between convex and concave borders) *in the living* (Fig. VIII-46).

Through the years, the mediastinum has been described as having *four subdivisions*—a superior, anterior, middle, and posterior mediastinum (Fig. VIII-47A). In our opinion, the "classical" anterior division is hardly a subdivision, and we have called this simply the **retrosternal sleeve,** leaving three divisions of the mediastinum proper: superior, anterior, and posterior—our anterior division containing the heart and pericardium (the middle mediastinum of the older nomenclature). The mediastinum then becomes more meaningful in clinical practice.

The mediastinum, as we visualize it, consists of a **superior** and an **inferior** compartment. The **superior compartment** lies above the horizontal line connecting the sternal angle to the interspace between VT_4 and VT_5 (Fig. VIII-47A). The **inferior compartment** lies below this line. It can be subdivided into a relatively unimportant **retrosternal sleeve** between the mediastinal pleurae laterally and the sternum and pericardium, anteriorly and posteriorly, respectively. The two larger compartments of the inferior compartment are the **anterior mediastinum** containing heart and pericardium; the **posterior mediastinum** between the posterior wall of the pericardium and the bodies of the vertebrae.

Thus:

Mediastinum
- A. Superior
- B. Inferior
  - a. Retrosternal sleeve
  - b. Anterior compartment
  - c. Posterior compartment

The *superior mediastinum* contains the brachiocephalic (innominate) veins, the upper part of the superior vena cava, most of the trachea and the upper portion of the esophagus, termination of the thoracic and right lymphatic ducts, stretches of the vagus and phrenic nerves, part of the sympathetic trunk and the origin of some of the lower sympathetic cardiac nerves, the left recurrent nerve (and occasionally part of the right), the thymus gland and some lymph nodes, and last of all, rather vaguely stated in most texts, the arch of the aorta and its major branches. Now precisely what does one mean by the arch of the aorta being in the superior mediastinum? The *lower concave border* of the aortic arch, arching over the pulmonary vessels rarely rises above the angle of Louis (Fig. VIII-48) in the living. Occasionally, in infants and the elderly, one may find the lower concave border of the aortic arch bending into the superior mediastinum.

The *retrosternal sleeve,* a narrow envelope between sternum and pericardium, lies as an extension of the superior mediastinum downward to the Diaphragm. It is packed with a cobwebby areolar tissue which opposes almost no barrier to blunt dissection with a finger. It may contain a few lymph nodes. Due to the

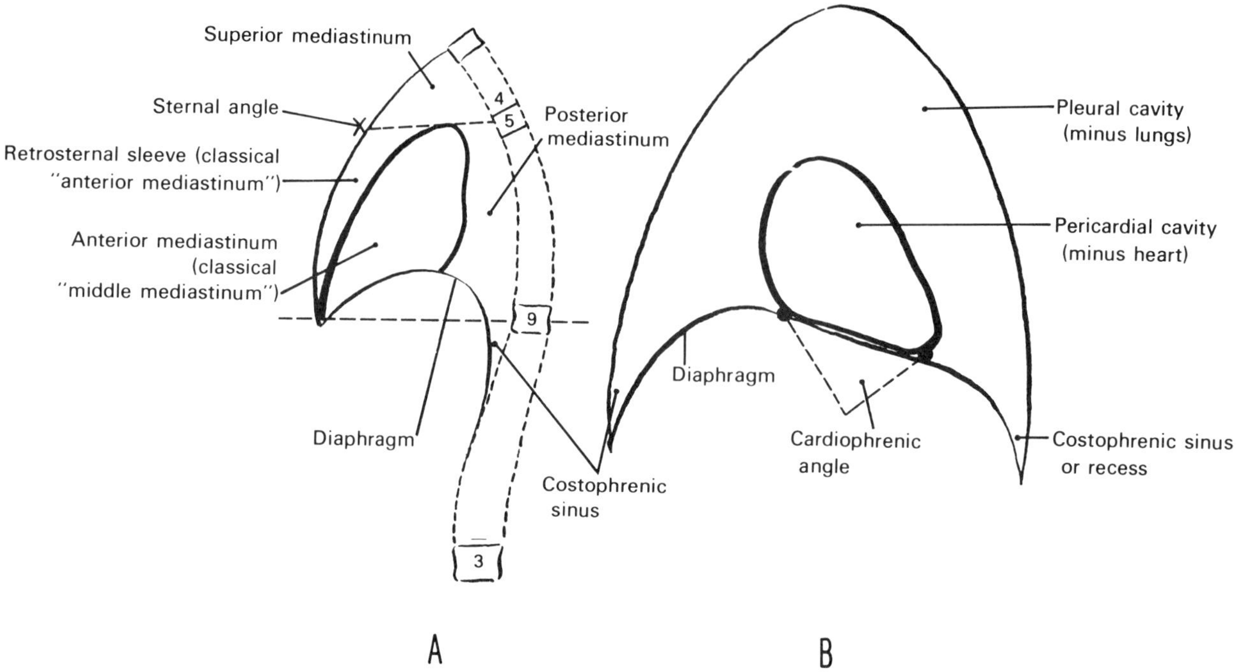

Fig. VIII-47. Divisions of the mediastinum. **A.** Lateral view. **B.** Frontal view.

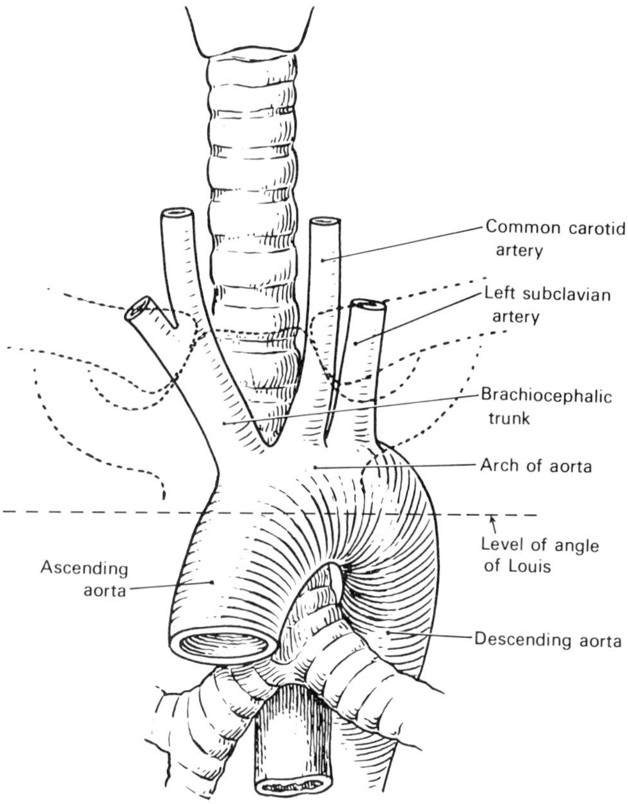

Fig. VIII-48. Relationship of the aortic arch to the Angle of Louis.

distance of the left pleural sac from the midline, the bottom portion of this envelope is more extensive on the left than on the right. Like all mediastinal "spaces," this is a potential space only.

The *anterior mediastinum* (or the classical middle mediastinum) contains the heart and its fibrous pericardial sac (Fig. VIII-49B, C). In close proximity are the phrenic nerve, blood vessels, roots of lungs, the anterior pulmonary plexus, terminal parts of the azygos veins, and the bronchial lymphatics.

The *posterior mediastinum* is bounded anteriorly by the pericardial sac, the roots of the lungs, and by the Diaphragm below. It is continuous, superiorly, with the posterior portion of the superior mediastinum. Within this posterior division are found the thoracic aorta, part of the arch of the aorta, the esophagus and vagus nerves, the thoracic duct, azygos veins, and numerous lymph nodes. The vertebral portions of the aortic intercostal arteries are also located here as well as the arteries to the esophagus, bronchi, pericardium, the greater splanchnic nerves, and the posterior pulmonary plexus.

In the superior mediastinum the esophagus rests against the vertebral column *in the cadaver* (Fig. VIII-50) separated from the trachea in front by loose areolar tissue. This potential space communicates anteriorly with the "pretracheal space," also consisting of loose fibroelastic tissue. *In life* the esophagus does not rest upon the thoracic spine in the superior mediastinum, but rather as shown by transverse tomography, the mediastinal pleurae abut against each other and interpose between the spine and the collapsed esopha-

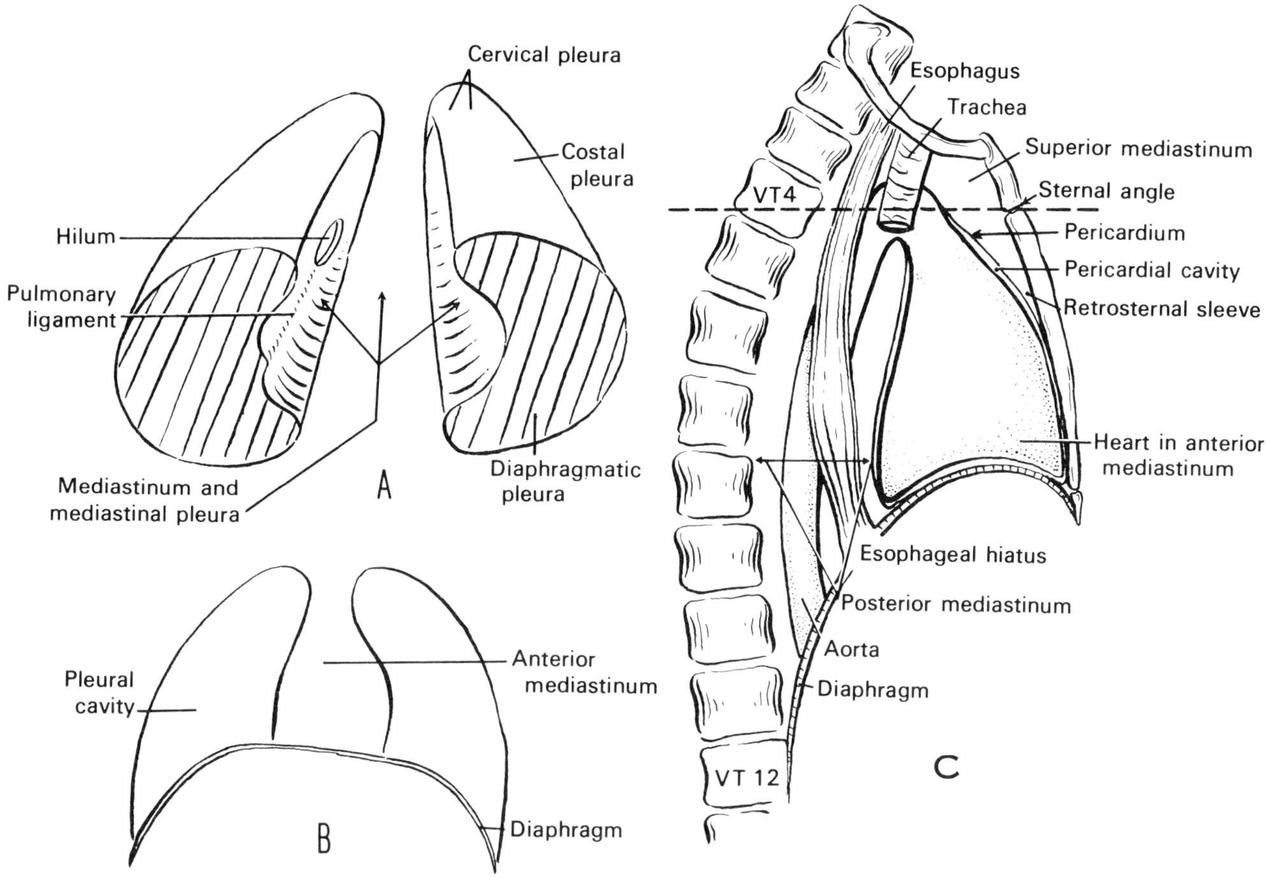

Fig. VIII-49. Schema of the mediastinum. **A.** Posterior-inferior view. **B.** Anterior view. **C.** Lateral view.

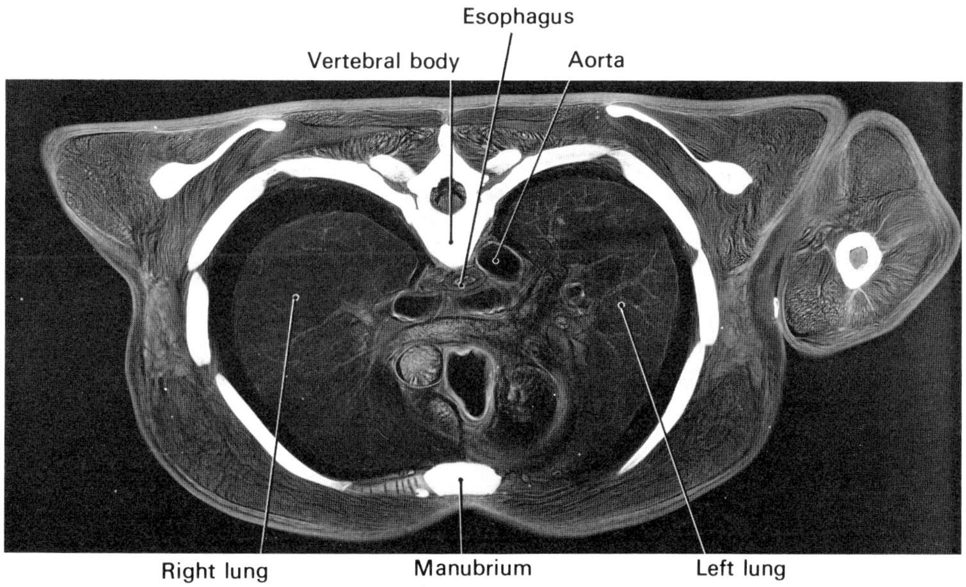

Fig. VIII-50. Radiograph of thick cross section of the thorax in the cadaver showing relation of the esophagus to the vertebra. (Courtesy of B. J. Anson.)

gus.[1, 2] This abuttment of pleurae is termed radiographically the **posterior mediastinal** or **posterior junction line** (Fig. VIII-51A, B).

In the posterior mediastinum, the descending aorta will intervene between vertebral bodies and esophagus. The descending aorta, at first, is slightly to the left of the midline. However, near the Diaphragm it reaches the median plane. The esophagus, meanwhile, angles slightly to the left and ventralward to reach the hiatus of the Diaphragm. This means that anterior to the VT_8 to VT_{10} vertebrae in the cadaver and over a longer distance in the living body, the aorta and esophagus part company leaving a gap between them (Fig. VIII-49C). This gap is again closed by apposition of the right and left mediastinal pleurae to form a double layered **meso-esophagus** between the vascular and digestive tubes. Radiographs (Fig. VIII-51A, B) may show the posterior mediastinal or junction line in the superior mediastinum as a thin white line descending vertically slightly to the left of the midline over the spine. Below the level of the aortic arch, the posterior mediastinal line makes a deep antevertebral incursion to the left of the median plane; however, it is difficult to see on radiographs made with conventional equipment. It is best demonstrated by tomography. We mention the posterior mediastinal line for it, indeed, has clinical importance. Its integrity may be lost by separation of the mediastinal pleurae of each lung by such pathological processes as tumor, infection, and bleeding. Herniation of a portion of lung across the median plane will also alter its normal course.

Throughout its length, the right side of the esophagus is covered with mediastinal pleura except where the azygos crosses the esophagus above the root of the lung. On the left side in the superior mediastinum, the left subclavian artery interposes between the left lung and the esophagus. In the upper portion of the posterior mediastinum, the aorta, which is to the left of the esophagus, prevents mediastinal pleura from contacting this digestive tube; however, in the lower part of the posterior mediastinum, as the esophagus moves toward the left, its left side is in apposition with the left mediastinal pleura. We are talking, of course, of the **parietal pleura** which surfaces the mediastinal walls of the pleural cavities. This interface between the air-containing right lung and the collapsed water density esophagus may be commonly seen on routine frontal chest radiographs (Fig. VIII-51B). This so-called **pleuroesophageal stripe** is usually seen on the right as a gently-curving, medially-directed convex line crossing over the upper thoracic vertebrae and tracheal shadow. It can be traced inferiorly. However, it becomes lost at the level of the azygos vein since as we have mentioned there is no abutment of mediastinal pleura (and underlying lung) against the esophagus at that point. Below this level the line reappears and follows a course toward the left of midline (Fig. VIII-51B). If barium is placed in the esophagus, the thickness of the esophageal wall plus its pleural covering may be detected (Fig. VIII-51C, D). Effacement of the paraesophageal line may signify esophageal distention, tumor, or other mass lesion.

All of the structures in the mediastinum are packed about with loose areolar tissue: trachea, esophagus, vessels, and nerves. This tissue does not offer much of a barrier to the spread of infections coming from within the tubes or the vessels themselves, and a *suppurative mediastinitis* is a dangerous condition no matter where it originates. Wounds of the esophagus, including those made during surgery, can result in a hidden infection detected only by radiographs. Foreign bodies, such as swallowed bone spicules or other sharp objects, may pierce the upper pharynx to enter the "retropharyngeal space" which extends from the skull to approximately the level of VT_4 where it is limited from further descent by the **alar fascia** stretching between the carotid sheaths at this level. Infections of the superior mediastinum can occur in this manner. However, if the infection is through the **prevertebral fascia,** behind the retropharyngeal space, it finds itself in the so-called "danger space." This space lies in front of the vertebral bodies and it extends from the base of the skull to the posterior reaches of the Diaphragm in the posterior mediastinum. Infections in the retropharyngeal space also can carry laterally and anteriorly around the trachea to enter the "pretracheal space" between trachea and the infrahyoid muscles. Figures VIII-52 and 53 illustrate these relationships in part.

The Aorta in the Thorax (Mediastinum): Three parts of this main arterial trunk are found in the mediastinum: its **ascending portion;** its **arch;** and the **thoracic descending aorta.** The latter section of the vessel continues through its own diaphragmatic hiatus to become the abdominal aorta.

The beginning of the ascending aorta, a very short segment of the whole vessel, lies in our anterior mediastinum (classical middle mediastinum) and entirely within the pericardial sac. It is packaged, along with the pulmonary trunk, in a common serous pericardial sheath. The ascending aorta exists from the left ventricle at the level of the third left anterior intercostal space and remains within the pericardial sac until it becomes the aortic arch. As it ascends ventrolaterad toward the right in a slightly oblique plane, the superior vena cava flanks it on the right side and the pulmonary trunk, on the left (Fig. VIII-54). This is because the pulmonary trunk spirals across the root of the ascending aorta as the trunk rises out of the right ventricle. In its short ascent toward the right, the ascending aorta partially overlaps the superior vena cava.

The **aortic arch** begins where the pericardium ceases, at about the level of VT_5. The aorta arcs posterolaterally toward the left side, clearing both trachea and esophagus on their left sides (Fig. VIII-55A). In rising over the right pulmonary artery as well as over the left pulmonary vein and the left main stem bron-

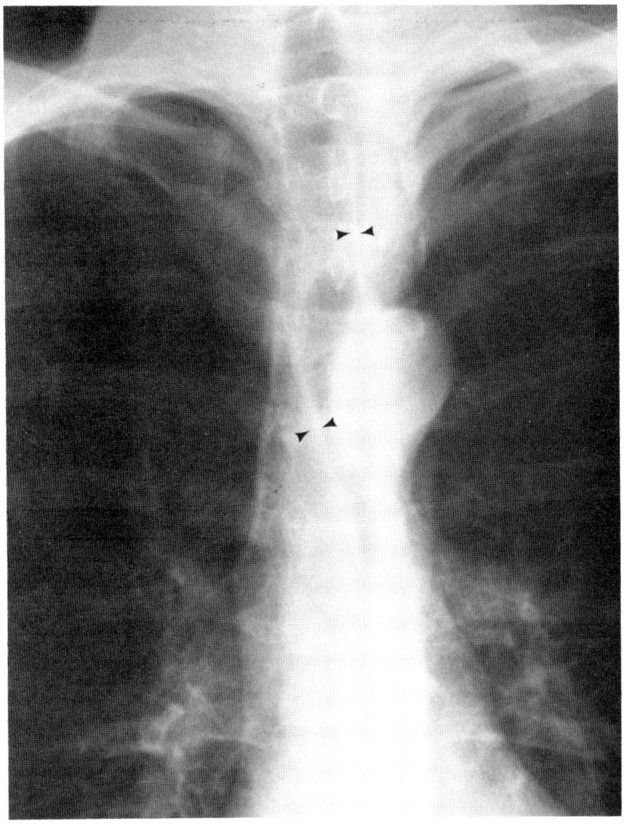

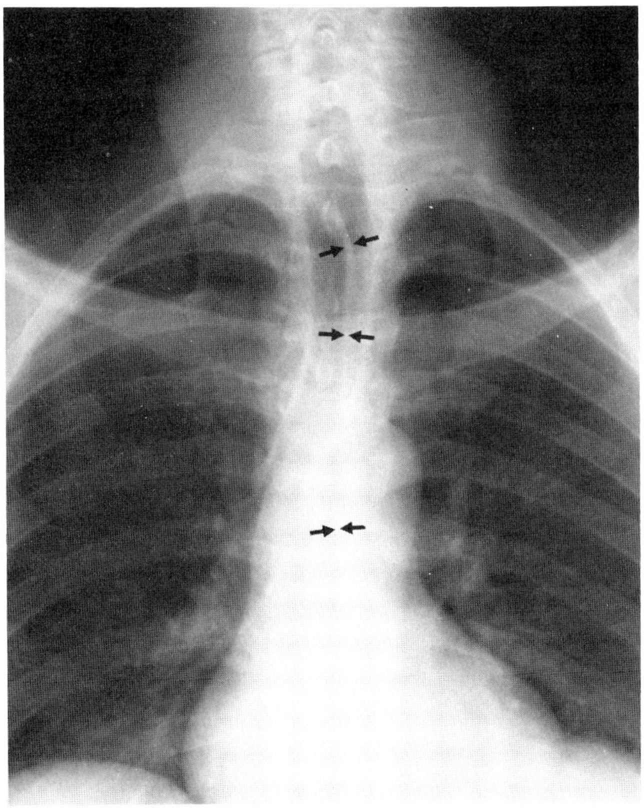

Fig. VIII-51. Frontal radiographic projection demonstrating mediastinal anatomy as seen in life. **A.** The posterior mediastinal line (*upper arrow heads*); the anterior mediastinal line which in this elderly patient is slightly thickened (*lower arrow heads*). **B.** The right pleuroesophageal line (*upper arrows*) blends into the posterior mediastinal line in this young man (*middle arrows*). The anterior mediastinal line is again seen behind the sternum (*lower arrows*). (A and B, Reproduced with permission from Golden's *Diagnostic Roentgenology,* Section 5, the *Gastrointestinal Tract* by R. Golden, C. V. Cimmino, L. C. Collins, J. R. Dreyfus, and M. L. Janower. Williams & Wilkins, Baltimore, 1969.)

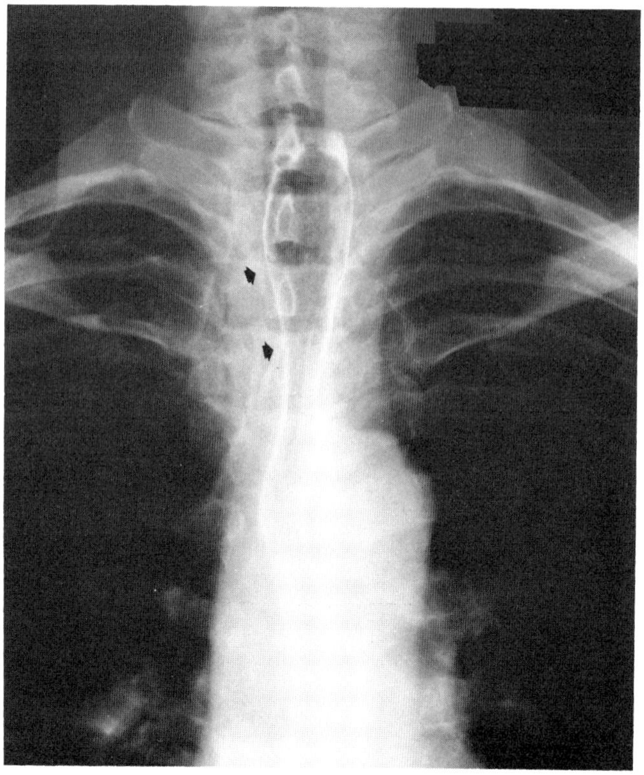

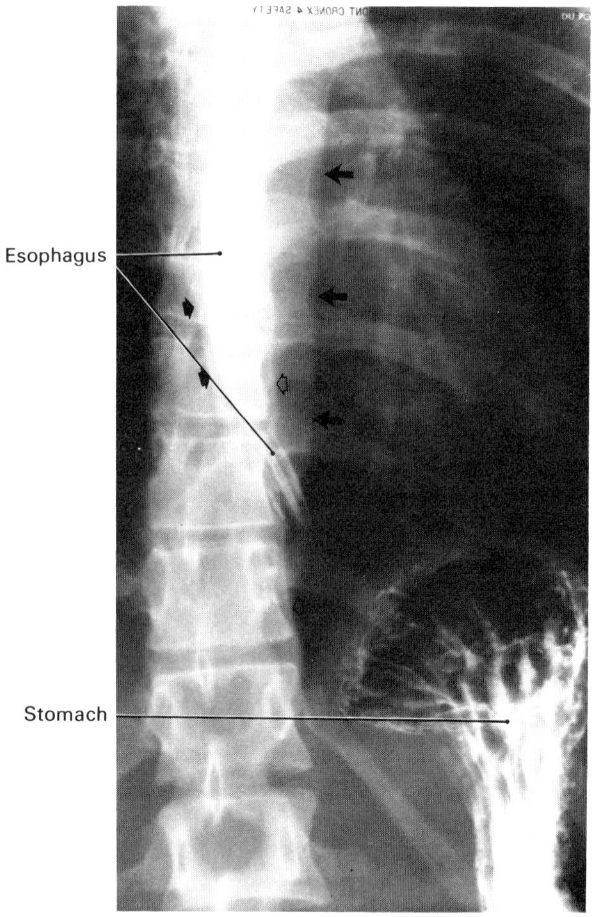

Fig. VIII-51C. The pleuroesophageal line in the upper thorax is outlined by lung laterally (*arrows*) and barium in the esophagus. **D.** The pleuroesophageal line as seen lower in the thorax delineated by the air-filled right lung (*black arrows*) and barium within the esophagus. The paraspinal line on the patient's left (*open arrows*) is seen. Air-containing left lung abutting against the lateral margin of the descending portion of the aorta (*large black arrows*) defines the para-aortic line. Identify the scalloped border of the sternum just medial to the para-aortic line.

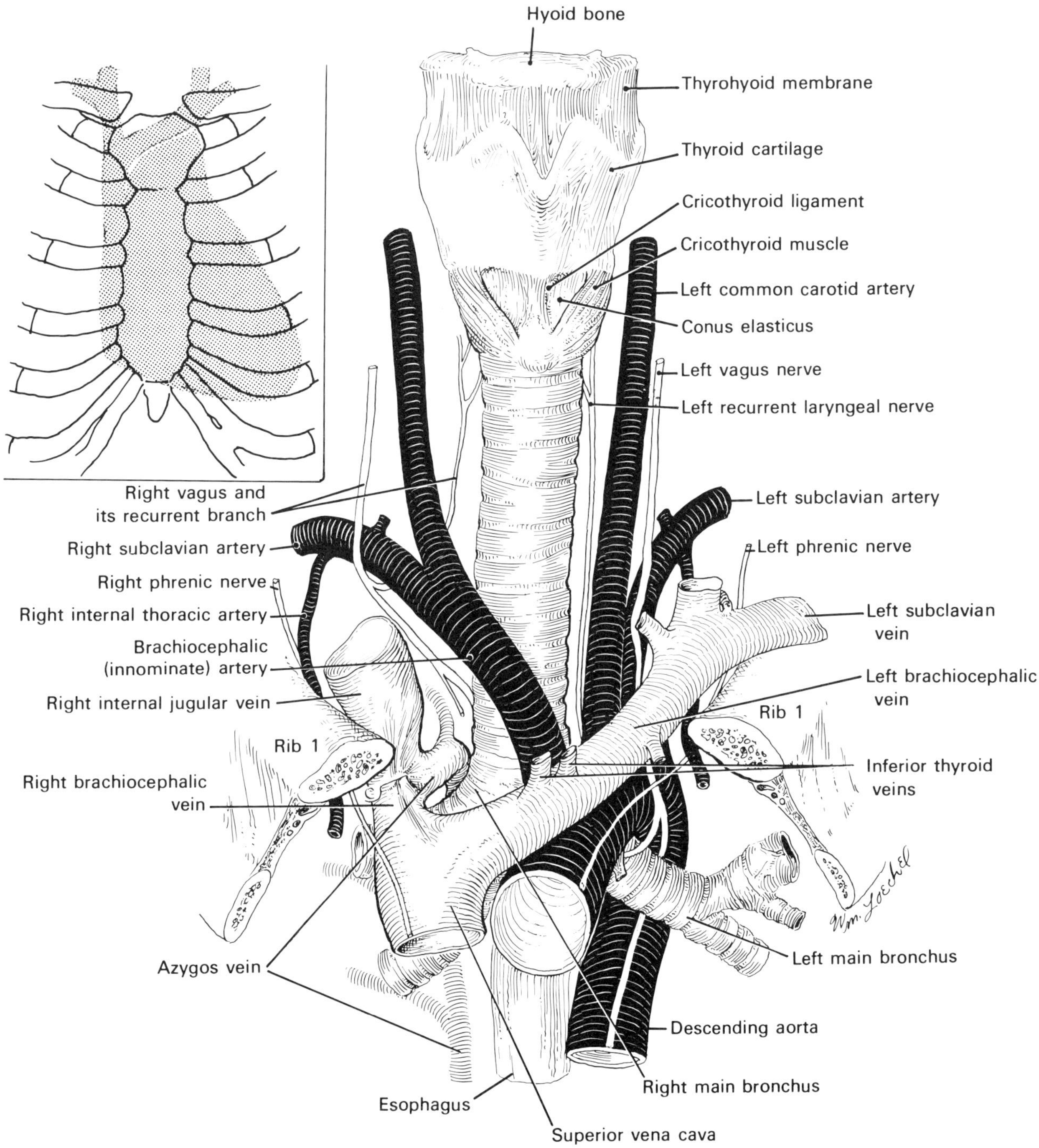

Fig. VIII-52. The trachea and extrapulmonary bronchi seen from in front. (Inset shows heart and great vessels related to anterior thoracic wall.)

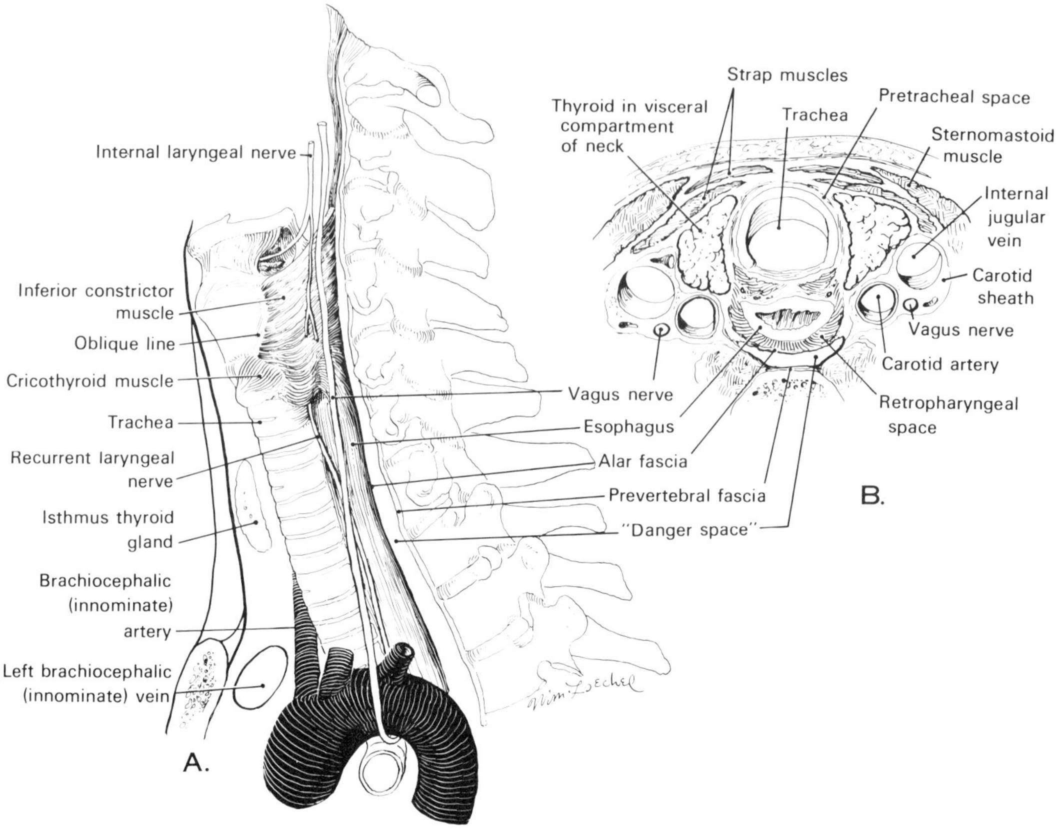

Fig. VIII-53A. Trachea and related structures viewed from left side. **B.** Trachea and related structures in cross-section at level of thyroid gland.

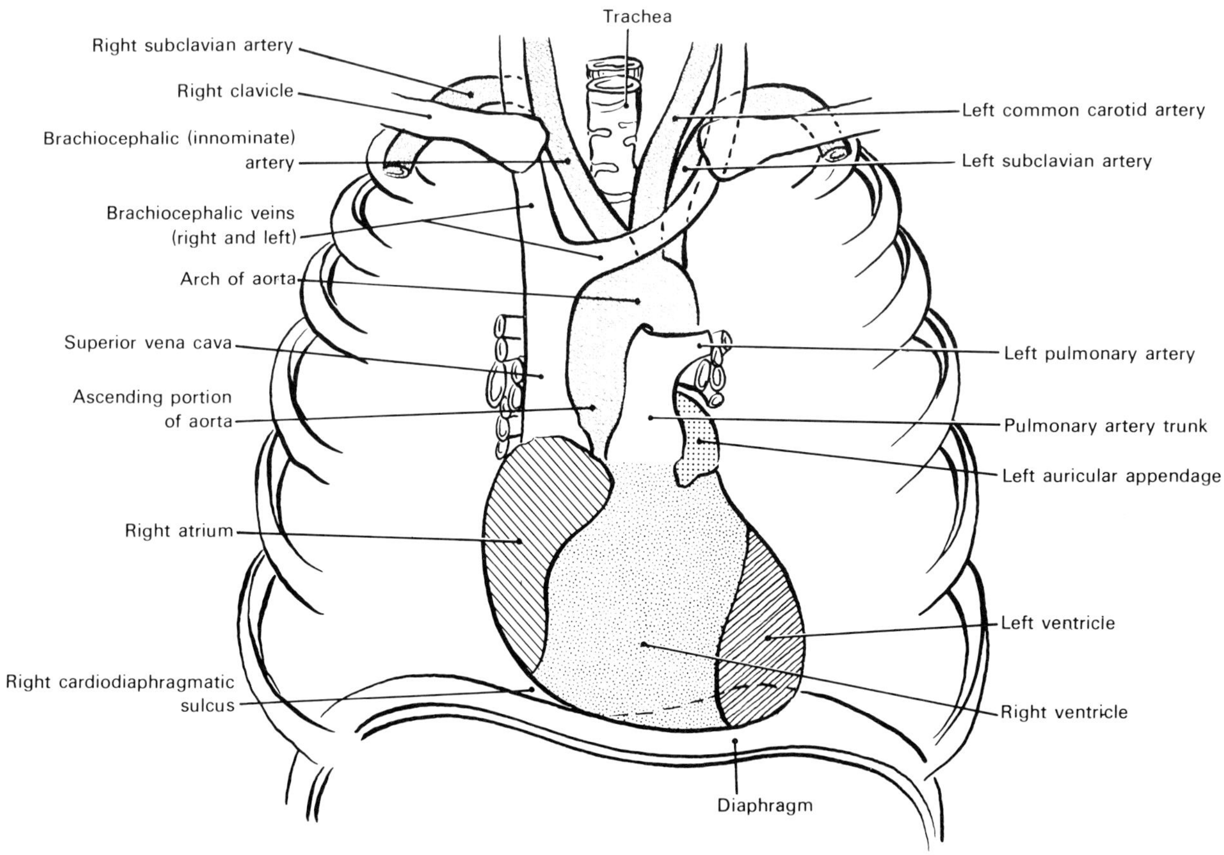

Fig. VIII-54. Schematic drawing showing the relationship of the aorta to the heart and other great vessels in frontal view.

chus, its convex surface in the living may push above into the territory of the superior mediastinum. But, the concave surface of the arch never does so, remaining below this level as the curve is completed and the three main vessels to the head and neck are given off from its convexity. The arch becomes known as the descending thoracic aorta. This occurs at about the same horizontal plane as its origin out of the pericardium, *e.g.*, at the level of VT_5. But now, of course, this portion of the arch lies well posteriorly, virtually upon the vertebral bodies.

The **descending thoracic aorta** (Fig. VIII-55B) is the longest stretch of this main arterial trunk. It passes downward in the posterior mediastinum adjacent to the left anterior borders of the vertebral bodies when it starts its descent through the thorax. By the time it reaches VT_{12} and is ready to burst through the Diaphragm it has reached the exact midline. No fascia or fat pad intervenes between this pulsating vessel and the vertebral bodies.

Branches of the Thoracic Descending Aorta: We can lump the many branches of the thoracic aorta, most of them small ones, into two categories.

A. Visceral branches
   1. Pericardial
   2. Bronchial
   3. Esophageal
   4. Mediastinal
B. Parietal branches
   1. Posterior intercostal
   2. Subcostal
   3. Superior phrenic

The names of the visceral branches tell their own story in respect to their distribution. **Pericardial branches** are tiny and few in number. They supply the dorsal surface of the pericardium.

The **bronchial arteries** (Fig. VIII-55B) are paired, their number varies and even their origin may not be the same. For example, the right bronchial artery generally comes from the uppermost left bronchial artery or from the first intercostal artery to arise from the aorta. In this sense it can hardly be said to be a direct branch of the aorta itself. The two left bronchial arteries (and there may be only one, or more than two) are direct branches of the descending aorta. The bronchial branches seek the posterior aspect of the main stem bronchi and traverse the respiratory tree distally maintaining the same position. Some of their branches do more than serve the bronchial tree. They supply neighboring lymph nodes, areolar tissue of the lung, and are even generous enough to join the vascular plexus upon the esophagus.

The esophagus has its own branches from the descending aorta. They arise as four to five twigs from its ventral wall, the shortest distance between two points since the esophagus overlies the descending aorta over most of its course in the thorax. The **esophageal arteries** (Fig. VIII-55B) tend to form a plexus upon

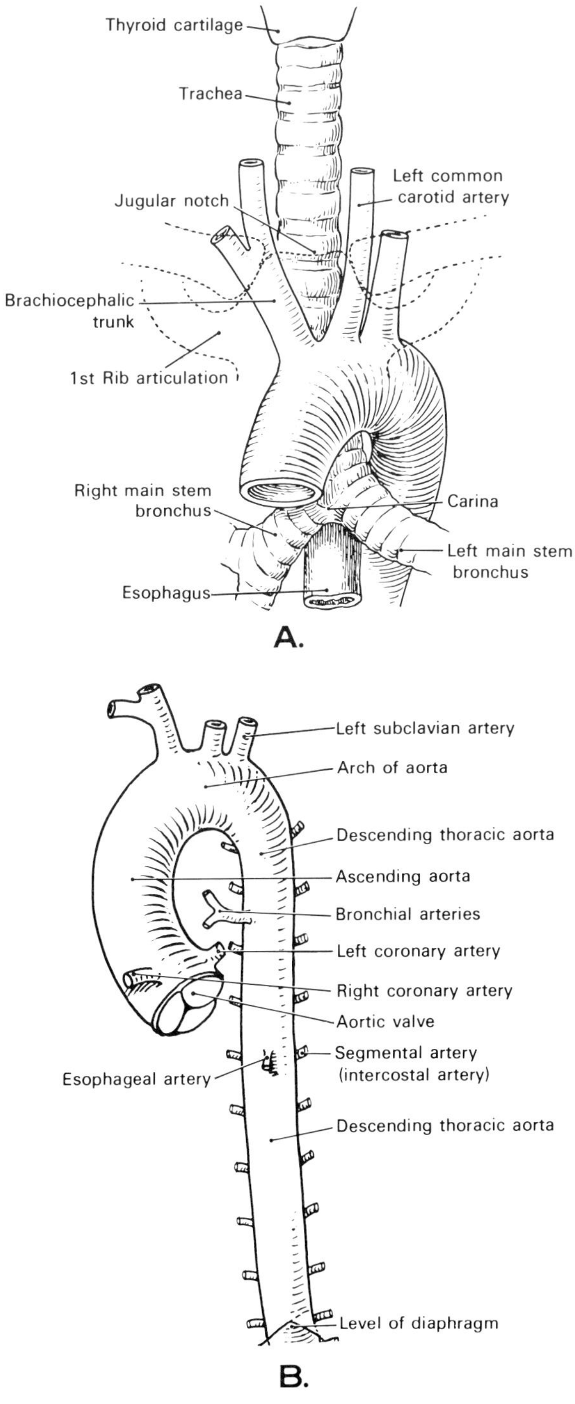

Fig. VIII-55. The aorta in the thorax (mediastinum). **A.** Arch and great vessels. **B.** Ascending, arch, and descending portions of the aorta.

the tube, anastomosing with twigs from the inferior thyroid vessels and those coming up from the left gastric and inferior phrenic arteries.

The **mediastinal branches** supply the lymph nodes and the loose packing tissues and visceral fascias of the mediastinal compartments.

The **posterior intercostal arteries** are those we knew in the intercostal spaces. The aorta does not rise high enough in the thorax to supply them all, but it does give nine pairs of dorsal branches to the rib cage. The first two spaces are serviced by the subclavian artery *via* its costocervical trunk. Because the aorta lies slightly left of the midline in most of its thoracic course the right intercostals must be slightly longer than those on the left side. These vessels cross ventral to the vertebrae but behind the esophagus, thoracic duct, azygos veins, and the parietal pleura. Opposite the heads of the ribs they are also behind the sympathetic trunk and its splanchnic nerves. They split into **dorsal** and **ventral rami.** The dorsal ramus of each vessel runs to the back muscles and the overlying skin and sends twigs through the intervertebral foramina to supply spinal roots, spinal cord, the spinal meninges, and the vertebral bodies. The ventral rami find themselves in the interspaces between Internal and Innermost Intercostal muscles which they pierce to supply, contributing muscular twigs to External Intercostals, the Pectoral muscles, and the Serratus Anterior. Lateral cutaneous branches reach the skin, and in the third to fifth interspaces, twigs are given to the mammae.

The **subcostal arteries** lie below the last rib. These also have their dorsal ramus. The ventral ramus slips below the costal margin between Internal Oblique and Transversus muscles of the abdomen.

The **superior phrenic arteries** are the last to arise at the caudal end of the thoracic aorta, supplying the cranial surface of the Diaphragm.

The Roentgen Divisions of the Mediastinum: For years we have been quite frustrated by the "classical divisions" of the mediastinum as applied to clinical practice. The boundaries and contents of these divisions, based upon cadaver dissections, were found to be meaningless in patients with mediastinal mass lesions. The tumor masses never seemed to be contained within a single division, and it was realized that the contents of the division had changed because our patients were inspiring deeply during radiographic examination.

The proposal for a **"roentgen mediastinum"** by Felson[3] has somewhat alleviated this problem. This is a still different subdivision of the mediastinum from that which we have just described. Felson had observed that mediastinal masses seemed to spread in a craniocaudad direction rather than in an anteroposterior plane, and therefore, he suggested the following division of the mediastinum for clinical use. In Figure VIII-56, a lateral radiograph of the chest, an imaginary line, A-A′, is drawn from the Diaphragm to a point just behind the heart and on upward in front of the trachea into the neck. Line A-A′ then separates the **"roentgen anterior"** from the **"roentgen middle"** mediastinum. A second line, B-B′, is constructed across each of the thoracic vertebrae about 1 cm. posterior to the anterior margin of the bodies. Line B-B′ separates the "roentgen middle" from the **"roentgen posterior"** mediastinum.

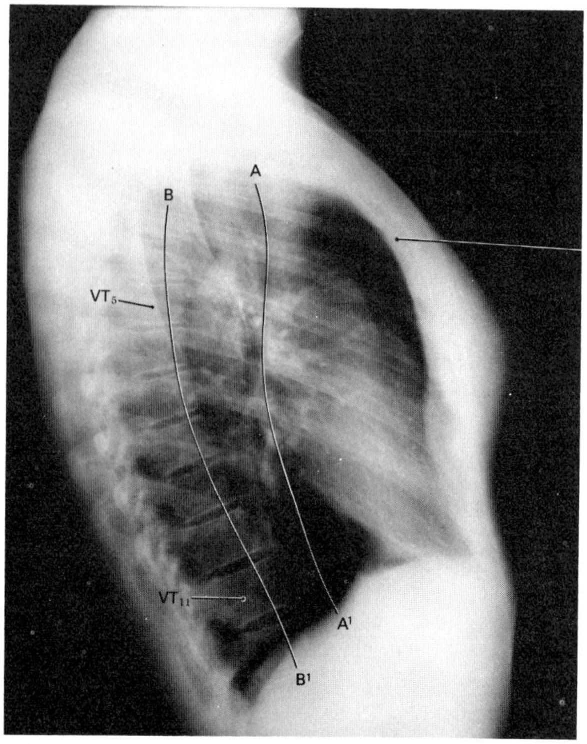

Fig. VIII-56. The "roentgen divisions" of the mediastinum (after Felson). The line A–A′ separates the **"roentgen anterior"** from the **"roentgen middle"** mediastinum. Line B–B′ separates the **"roentgen middle"** and **"roentgen posterior"** mediastinum.

With this delineation of the mediastinum, tumor masses are more easily localized, and by adding such additional information as the size, shape, calcium content, and effect upon adjacent structures, a more accurate determination of the nature of the lesion becomes possible.

The soft tissue structures at the margins of the mediastinum abut against mediastinal pleura and underlying air-containing lung in life. Because of the differences in X-ray absorption between these water density soft tissues and the air density lung, many of the pleural lines become visible on appropriate radiographs. The effacement or loss of these lines is of utmost importance as a means of accurately localizing pathological processes nearby. You should become familiar with the following:

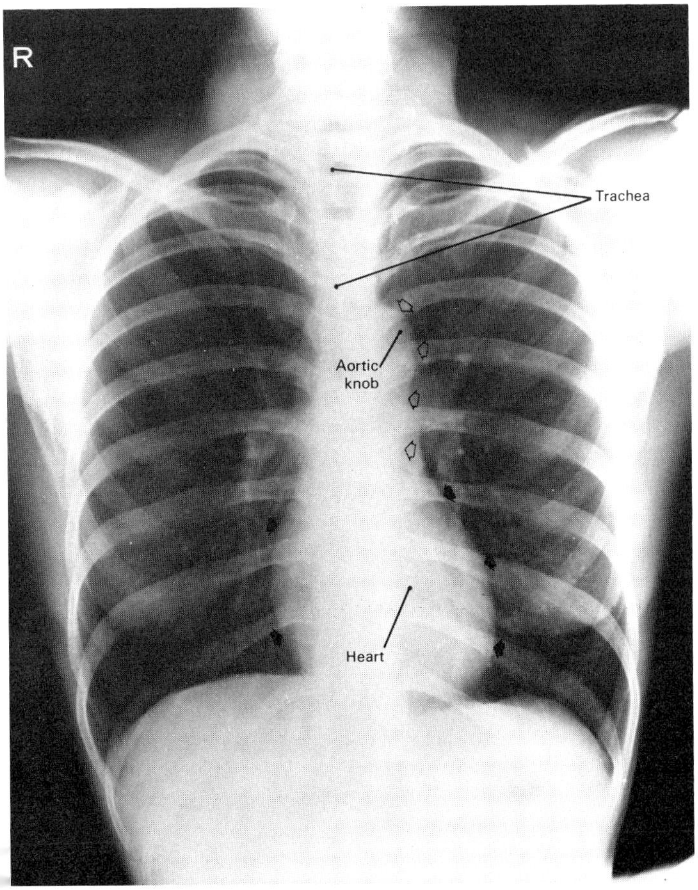

Fig. VIII-57. PA radiograph of the chest. The paracardiac line (*black arrows*); the para-aortic line (*open arrows*) are shown. The trachea is also outlined against air-containing lung.

1. **Anterior mediastinal (junction) line**—page 450.
2. **Posterior mediastinal (junction) line**—page 440.
3. **Paraspinal lines** (right and left)—page 449.
4. **Pleuroesophageal line**—page 440.
5. **Paracardiac line**—outlines the cardiac silhouette against adjacent lung (Fig. VIII-57).
6. **Para-aortic line**—outlines the aortic knob and lateral margin of the descending thoracic aorta (Figs. VIII-51D and 57) in PA view.
7. **Paravenous lines**—portions of the superior and inferior vena cavae may be seen on plain chest radiographs. See page 505.
8. **Paratracheobronchial line**—in deep inspiration, the mediastinal pleura abuts against the right side of the trachea. Thus the thickness of the tracheal wall plus the two layers of pleura may be determined since they are outlined by air within the trachea and adjacent lung (Fig. VIII-57). Rarely is the *medial wall* of the right main bronchus below the carina visualized.

The Esophagus (Figs. VIII-49C and 53A).

This muscular tube does not descend in a straight line through the posterior mediastinum. Neither is it of uniform diameter throughout. Recall that at the lower end of the cricoid cartilage, the esophagus begins. It is *compressed symmetrically* here at the pharyngo-esophageal junction by the Cricopharyngeus muscle. In roentgenograms (Fig. VIII-58) the esophagus is slightly deviated to the left of midline in the neck and superior mediastinum. At the level of the aortic knob, the esophagus is *indented* once more and is displaced to the right. Note how the tube *bulges* at the margins of the aortic arch. These bulges are normal and quite variable in prominence in the individual patient. Their obliteration becomes important in the localization of adjacent mediastinal lesions.

Below the aortic arch the esophagus once again turns to the left and is *indented* by the left main bronchus. The esophagus bulges below this point. Just above the level at which the esophagus joins the cardia of the stomach, there is a final area of physiological *narrowing*. This point is probably due to stricture by the fibers of the Diaphragm, since this occurs *above* the cardia rather than *at* it. There is no true "cardiac sphincter." All four areas of physiological narrowing that we have mentioned are important clinically, since foreign objects in the esophagus may become lodged at these points.

Both vagi reach the esophagus near the **esophageal plexus** upon it, mingling in the plexus with sympathetic contributions. Near the cardia the two vagi disengage from the plexus, the left seeking the anterior aspect of the esophagus to become the anterior "gastric" nerve, the right seeking a posterior position prior to going through the diaphragmatic hiatus.

The tube has the typical plan of a hollow organ (see Chapter IX, p. 559). Its surface lining is stratified squamous epithelium of the non-keratinized type in contrast to the simple columnar epithelium found in the lower digestive tract. Its glands are submucosal rather than in the tunica propria except near the cricoid cartilage and at the junction with the stomach. In these latter two areas typical "gastric cardiac glands" may be present in the tunica propria over short stretches. If you are looking for them in usual histological loan set collections, don't bet heavily on finding them. Another peculiar difference is the presence of longitudinally disposed smooth muscle in its muscularis mucosae. Its main outer muscular coat at its upper end is a continuation of skeletal muscle from the pharynx (hence supplied by branchial motor fibers of the vagus). In its middle reaches, there is a mixture of smooth and skeletal muscle, but its last 8 to 10 cm. will contain no voluntary muscle at all. The smooth

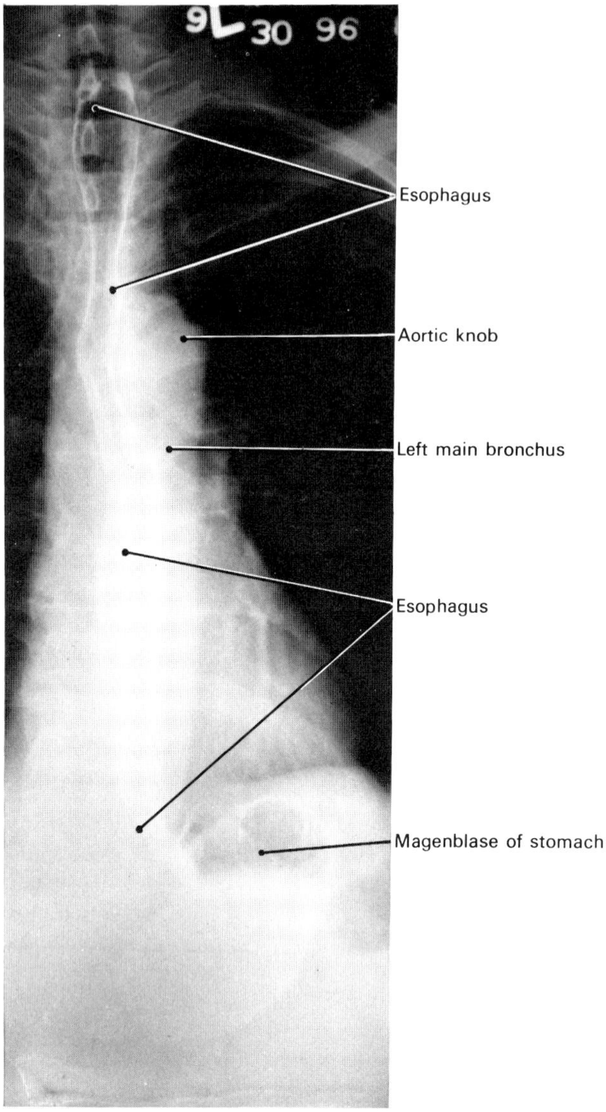

Fig. VIII-58. Radiographs of the normal esophagus. **A.** PA projection with the patient erect.

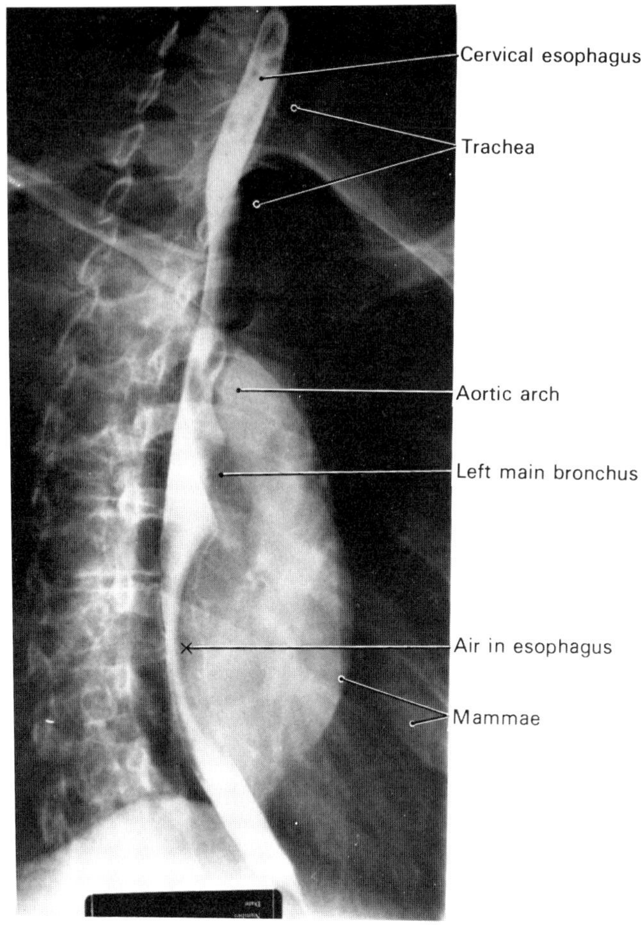

Fig. VIII-58B. Right anterior oblique projection. The right breast casts a dense white shadow when superimposed on the heart.

muscle, of course, is supplied by visceral motor fibers from both vagus and sympathetic nerves.

Blood Supply of the Esophagus: The upper portion receives its blood from the inferior thyroid artery. In its thoracic course, one or more branches of the aorta reach it or twigs come to it from bronchial vessels (which are, of course, aortic branches). It is possible for branches from intercostal vessels to reach it, *e.g.,* third intercostal. At its lower end, branches reach up to it through the hiatus from below the diaphragm either from the coeliac artery directly, from the left gastric artery or even from the inferior phrenic artery.

Veins in the upper portion are small and drain into the inferior thyroid and vertebral veins. In the thoracic region they are larger and empty into the azygos or hemiazygos veins, more generally emptying into the azygos. Near the stomach, venous drainage is downward into the left gastric vein. The veins in this region form an important anastomosis between the abdominal and lower thoracic esophageal veins. They serve as collateral routes of venous drainage when pressure in the portal vein is raised in conditions of portal hypertension or portal obstruction. The difficulty is that internal esophageal varices (hemorrhoids, piles) form which can be troublesome and even fatal when they bleed.

Pleurae

The pleurae are two, distinct, serous membranes which form closed sacs to invest the lungs and which also line the walls of the thoracic space into which the lungs bulge. The thoracic spaces are separated from each other by the mediastinal partition which consti-

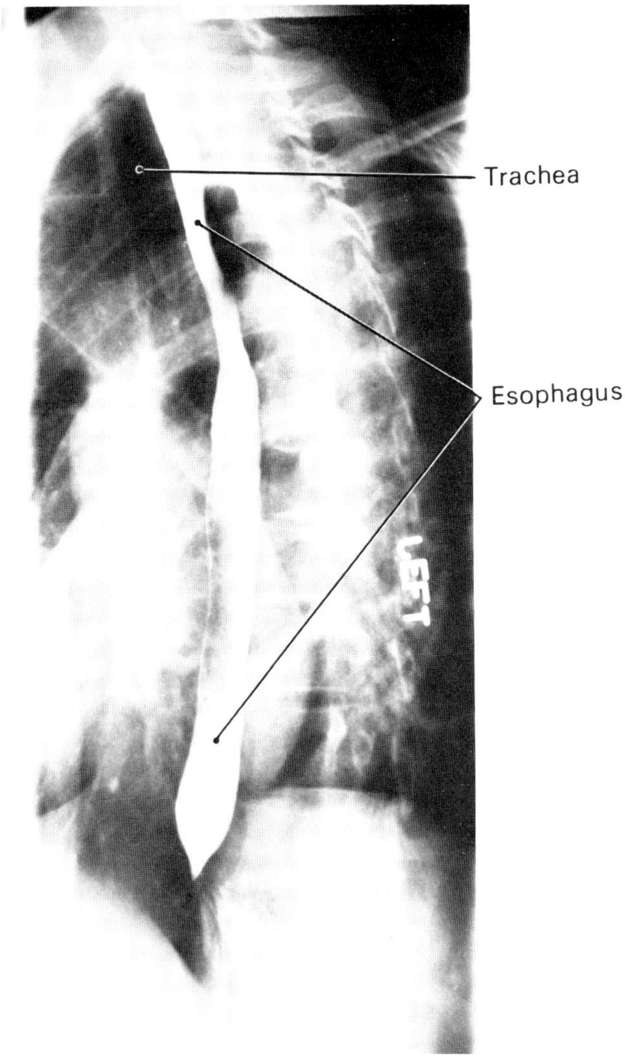

Fig. VIII-58C. Left anterior oblique projection. Note the area of physiological narrowing just above the cardio-esophageal junction.

tutes the medial wall of each thoracic cavity. In each cavity, the pleura consists of a **visceral** and **parietal layer** as a consequence. The visceral layer wraps around the lung as a tight outer garment while the parietal layer lines the inner walls of each cavity. Since each lung fills its thoracic cavity, the two layers are in direct contact. The space between them (which *is* the thoracic cavity) is a potential space only (Fig. VIII-59), and it is never evident unless it is occupied by excess fluid, air, or a foreign body, none of which should be found there except under pathological circumstances. Both layers are capable of secreting a lubricating serous fluid into this space. A minute amount of fluid is normally present to decrease friction as visceral pleura glides by parietal pleura during respiratory excursions of the lung. At the same time, it is this fluid film which provides the surface tension forces to keep these layers applied against each other at all times in the closed chest. The absorption of fluid from this potential space by lymphatics keeps pace with its production. Any inflammatory process which involves the pleural membranes, a condition known as pleurisy, can cause friction, increase serous production, or produce an irritation of the opposing membrane surfaces.

Around the root of the lung at the hilum, visceral and parietal pleura meet in a cuffing fashion to form a narrow fold, the **pulmonary ligament** which descends almost to the Diaphragm in some instances.

The parietal pleura is often subdivided regionally for descriptive purposes into sternocostal pleura, diaphragmatic pleura, mediastinal pleura, and cervical pleura (Fig. VIII-59A, B). It is, of course, all one, uninterrupted, lining sheet. The costal pleura lines the inner aspect of the ribs, the posterior parts of the sternum (except where the connective tissue of the retrosternal sleeve forms a midline partition between the right and left cavities), and the lateral portions of the vertebral bodies.

At the interface between the mediastinal pleura and the thoracic spine, the **paraspinal lines** (right and left) (Fig. VIII-60) become visible on radiographs that have been appropriately exposed (Fig. VIII-51D). On radiographs, the right paraspinal line is usually visible from the pulmonary apex down to the posterior costophrenic recess. The left paraspinal line is only seen from a level below the aortic knob. The distance between this paraspinal line and the vertebral column is quite variable depending upon the underlying soft tissue structures. The hemiazygos vein lies beneath the left paraspinal line, while the azygos vein underlies the right paraspinal line. Clinically, the paraspinal lines are significant since they may be effaced or displaced laterally by paravertebral abscess, bleeding, or tumor.[4-6]

In the upper thorax, the costal pleura continues on as cervical pleura to house the apex of each lung. Below, costal pleura becomes continuous with diaphragmatic pleura, the thoracic covering of the Diaphragm, and, medially, in an anteroposterior plane, it covers the lateral aspects of structures that relate to the mediastinum. In the posterior mediastinum, where aorta and esophagus diverge from each other as they go to their separate diaphragmatic hiatuses, the pleurae of the right and left cavities abut against each other as the *mesoesophagus* (**posterior mediastinal line**) discussed earlier (p. 440). On the left, this pleura runs in an almost straight line along the left lateral margin of the vertebrae, while the right pleura from about the VT_5 to VT_{10} level makes a deep incursion across the midline.[1, 4]

Similarly, in the area of the retrosternal sleeve between sternum and pericardium, from the level of the second to fourth costal cartilages, right and left mediastinal pleurae come into close apposition. This **an-**

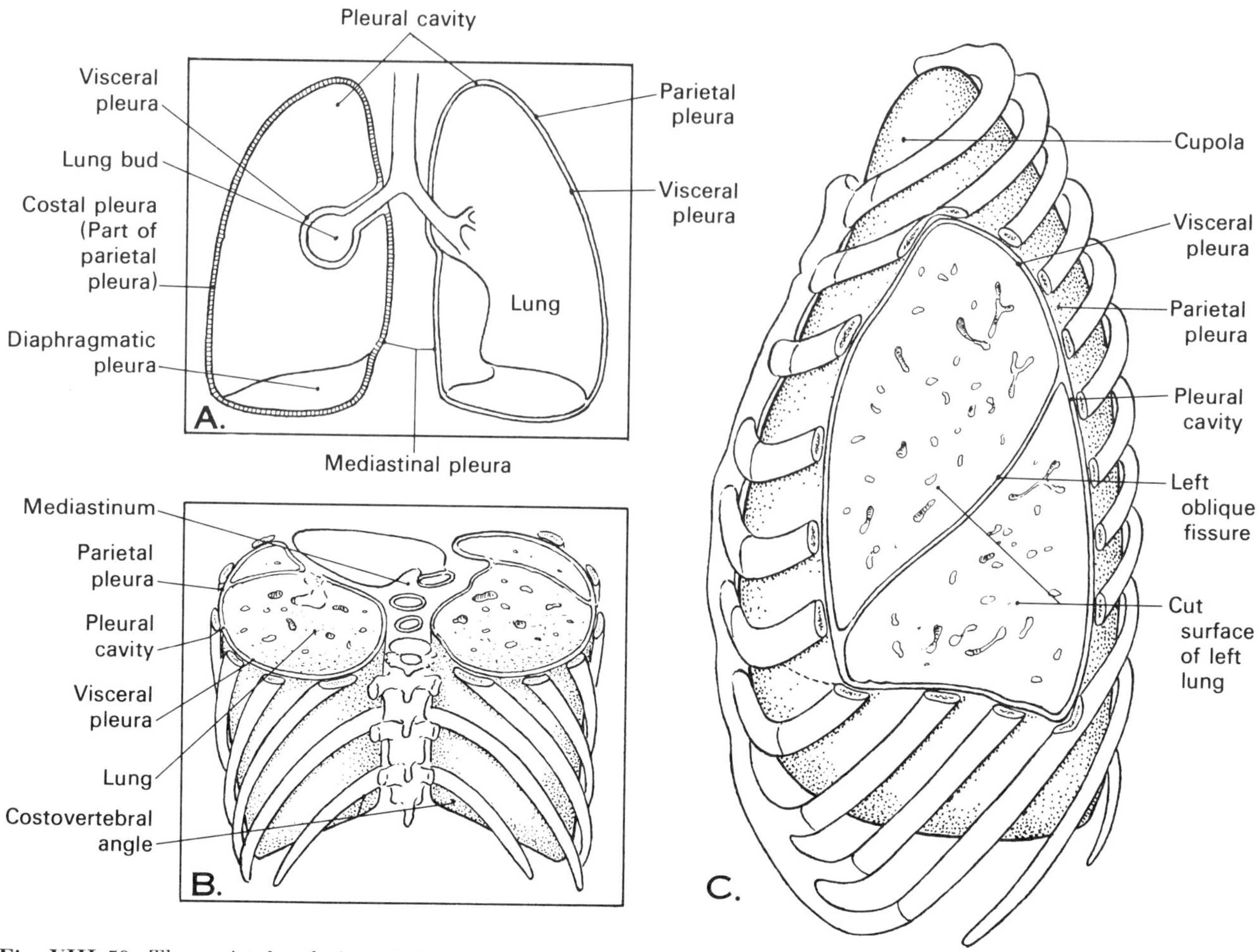

Fig. VIII-59. The parietal and visceral pleurae. **A.** Lung invaginating pleural cavity to gain its visceral pleura and to reduce the space of the cavity. **B.** The costovertebral recesses (angles). Note the position of the mediastinum between pleural cavities in the cross section. **C.** The extent of the parietal pleura and its cupola. Parasagittal section shows lung, visceral pleura, and reduced pleural cavity.

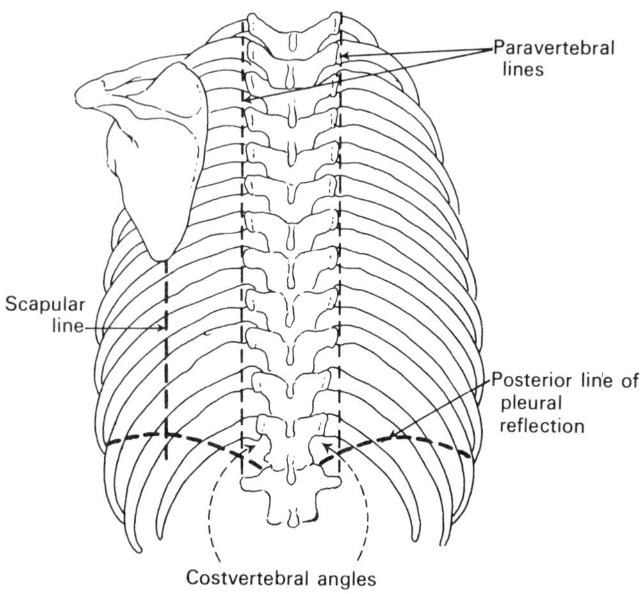

Fig. VIII-60. The paraspinal line.

terior mediastinal or **anterior junction line** may be seen in routine frontal chest radiographs (Fig. VIII-51A, B). The four layers of pleura (visceral and parietal from each lung) stand out sharply against the adjacent lung tissue. In Figure VIII-61 this anterior junction line is depicted schematically. Note that the left anterior line of reflection departs from the right in the region of the **cardiac notch** of the left lung.

While we have said that the "lung fills the thoracic cavity" so that only potential space exists between parietal and visceral pleurae, there are, actually, two regions that are not occupied by lung in the pleural cavities, and the cardiac notch is one of them. Here the costal and mediastinal aspects of the left pleura come into contact with each other over a variable distance depending upon the severity of the cardiac notch and upon how readily this left anterior margin of the lung slips between these pleural layers during inspirational filling. This left anterior pocket of the pleural sac, when devoid of lung, is known as the **costomediastinal recess.** Similarly, along the sides and at the back of the chest, the sharp, thin, lower

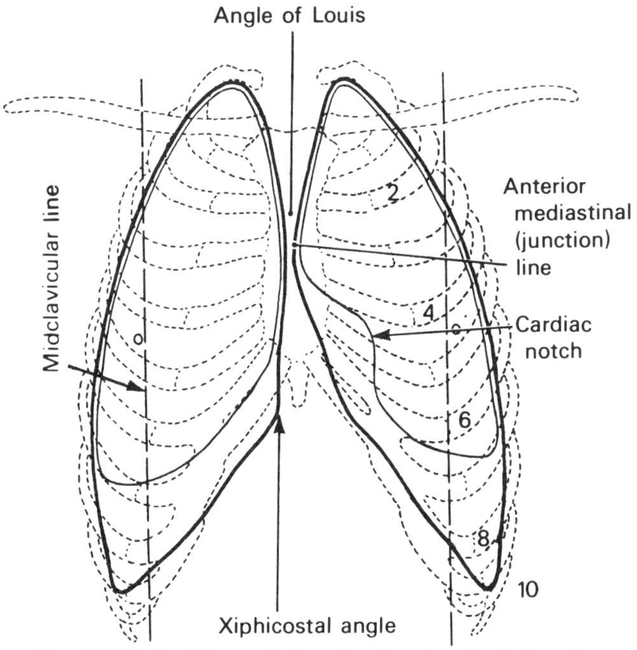

Fig. VIII-61. Lines of pleural reflection. Anterior view—note the cardiac notch on the left and the anterior mediastinal (junction) line.

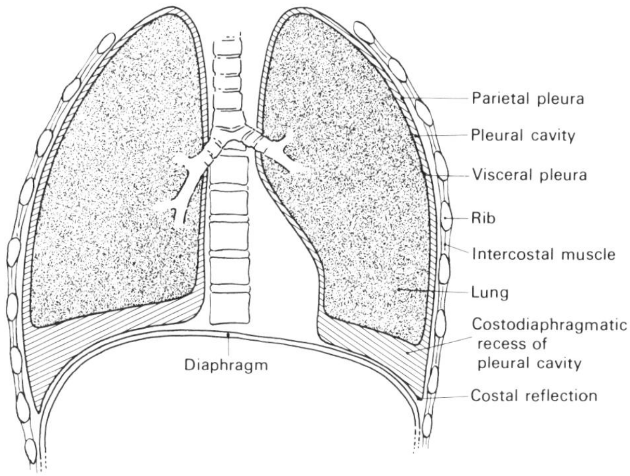

Fig. VIII-62. Coronal section of thorax showing costodiaphragmatic recess.

borders of the lungs do not descend all the way down to the lines of costal reflection of the parietal pleurae. Below the lung margins, costal pleura and diaphragmatic pleura come into apposition, and this potential space, unoccupied by lung, is known as the **costodiaphragmatic recess.** The costodiaphragmatic recess becomes alternately smaller and larger as the lung margins move into and out of it during inspiration and expiration, yet, the lung never completely fills this space, except, perhaps, when full, over-inflationary, active, inspiratory efforts are made as in "taking a deep breath." It should be clear from Figures VIII-59C, and 62 that a knife thrust or bullet entering the *anterior* intercostal spaces 8, 9, 10, and 11, below the line of costal reflection of the pleurae, will enter the peritoneal, not the pleural, cavity. But, at the lower back, between the 12th rib and the vertebral column (**right and left costovertebral angles**), at the **right xiphocostal angle** anteriorly, and just above the clavicle anteriorly, knife and bullet wounds can enter the pleural cavity and result in pneumothorax.

In Figure VIII-63A, the left lung has been removed from its thoracic cavity. It is possible to observe the cut margins of the remaining parietal pleura, and to observe the structures which would impinge upon the mediastinal surface of the lung were it in place (covered by its visceral layer of pleura). The hilar region is occupied by the single left bronchus, the left pulmonary artery, superior and inferior pulmonary veins, all of whose cut surfaces can be identified. Among these structures lie some lymph nodes. Below and anterior to the hilum, it is easy to see why the cardiac notch exists in the left lung in order to compensate for the bulging left ventricle of the heart which occupies considerable space as it pushes toward the left in the middle mediastinum. Above, the mediastinal surface of the lung contacts the pulmonary trunk and the ascending aorta. The arch of the aorta, passing backward above the root of the lung, produces a deep groove on the medial surface of the consolidated lung. The left subclavian artery arises above the aortic arch, contacts the medial aspect of the lung and runs upward with it to turn laterally in front of the apex of the lung in the neck. Posteromedially, the lung is deeply grooved by the descending portion of the thoracic aorta. The vessel keeps the viscus from contacting the bodies of the fourth to ninth thoracic vertebrae.

In the right thoracic cavity (Fig. VIII-63B) other structures leave their imprint on the right lung, particularly if the lung is consolidated. On the right side, the hilum is occupied by the eparterial and hyparterial bronchi, right pulmonary artery, upper and lower pulmonary veins and the ever present lymph nodes, along with branches of the pulmonary plexus of nerves posterior to the bronchus. Below the hilum, there is a hollow, rounded depression on the mediastinal lung surface for right atrium which moves against the lung with its every cardiac contraction. In nestling up tight in front of the hilum, the superior vena cava produces a deep groove along the medial lung aspect. The groove flows toward the cardiac depression, below, and is continuous above with the groove produced by the brachiocephalic vein. Another groove may be discernible as it flows toward the groove for the superior vena cava. This is the impression created by the azygos vein curving anteriorly over the hilum. The depression for the right atrium is limited posteriorly by the hilar structures and the right wall of the esophagus which comes into contact with the medial lung surface here. Most dorsal, the medial surface of the right lung rests upon

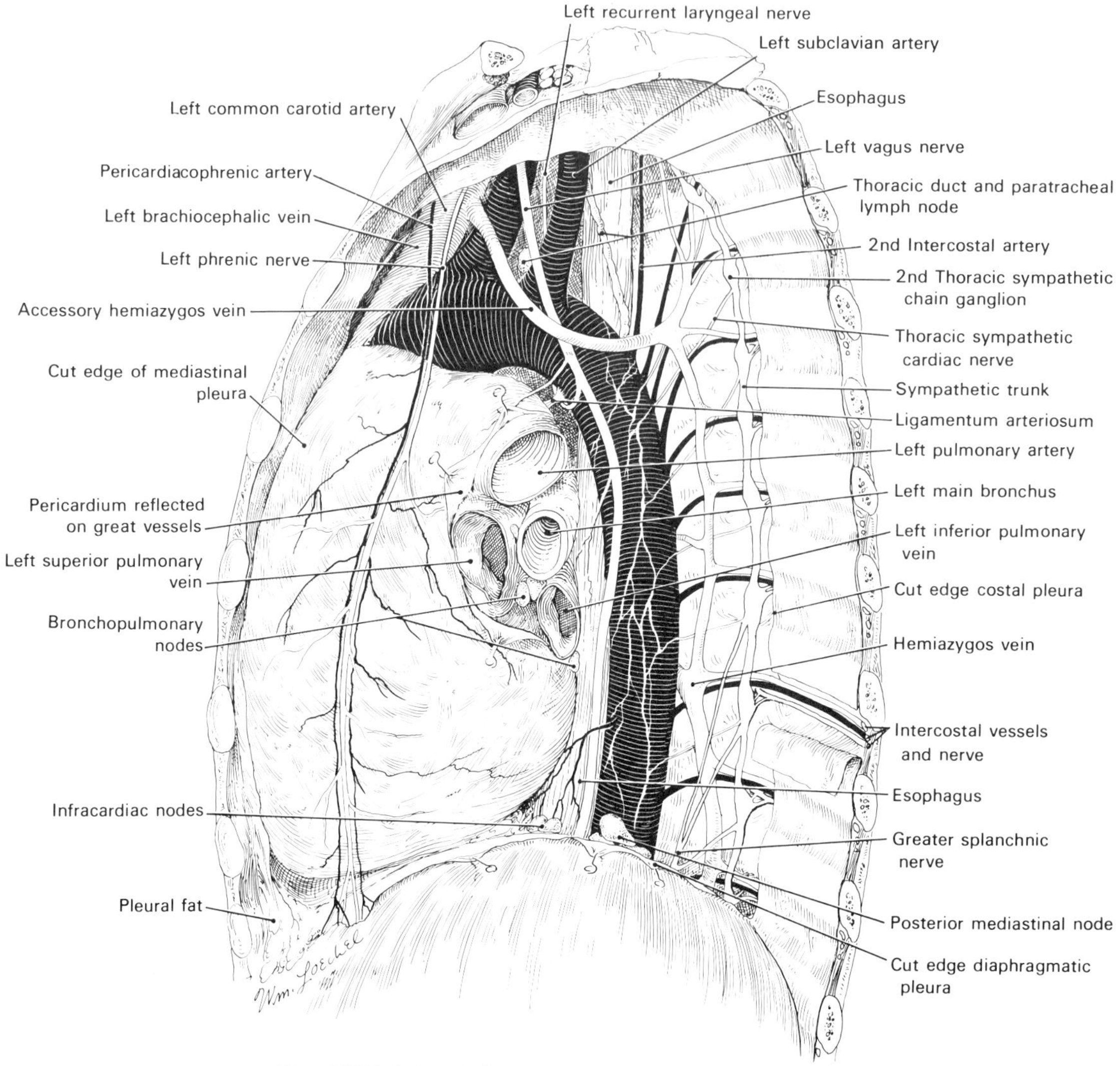

Fig. VIII-63A. Mediastinal relationships on left side.

the rib heads, the sympathetic trunk, the lateral parts of the vertebral bodies, the splanchnic nerves as they arise from the ganglionated trunk, and upon the right posterior intercostal arteries and their accompanying veins.

Blood Supply of the Pleurae: Visceral pleura is vitalized by the bronchial arteries which leave the bronchial tree to reach the pleura *via* interlobar connective tissue septae. Its venous drainage is *via* the pulmonary veins. The parietal pleura, on the other hand, is supplied by local vessels of the body wall, diaphragm, and mediastinum. That is to say, it gets its blood supply from the internal thoracic (mammary), intercostal and pericardiocophrenic vessels and is drained by the accompanying veins.

Nerve Supply of Pleurae: The phrenic nerve (from C_3 to C_5) supplies the mediastinal and central diaphragmatic pleurae. Stimulation in these areas may refer pain along the somatic distribution routes of the supraclavicular nerves since these nerves arise from the same cervical cord segments as the phrenic. Intercostal nerves supply costal and peripheral diaphragmatic pleurae. Stimulation of the costal pleura elicits localized pain while stimulation of the peripheral diaphragmatic pleura refers pain to the lower portions of the thoracic wall or even to the anterior abdominal wall. Sensory endings in visceral pleura are insensitive to tactile stimulation. These are visceral sensory endings supplied by the autonomic innervation of the lung, and they are primarily sensitive to stretch.

The Lung

General Considerations: The lungs are large cone-shaped organs, a right and left lung, each occupying its own thoracic cavity. The right lung is larger, longer, and broader than the left and has three major lobes—

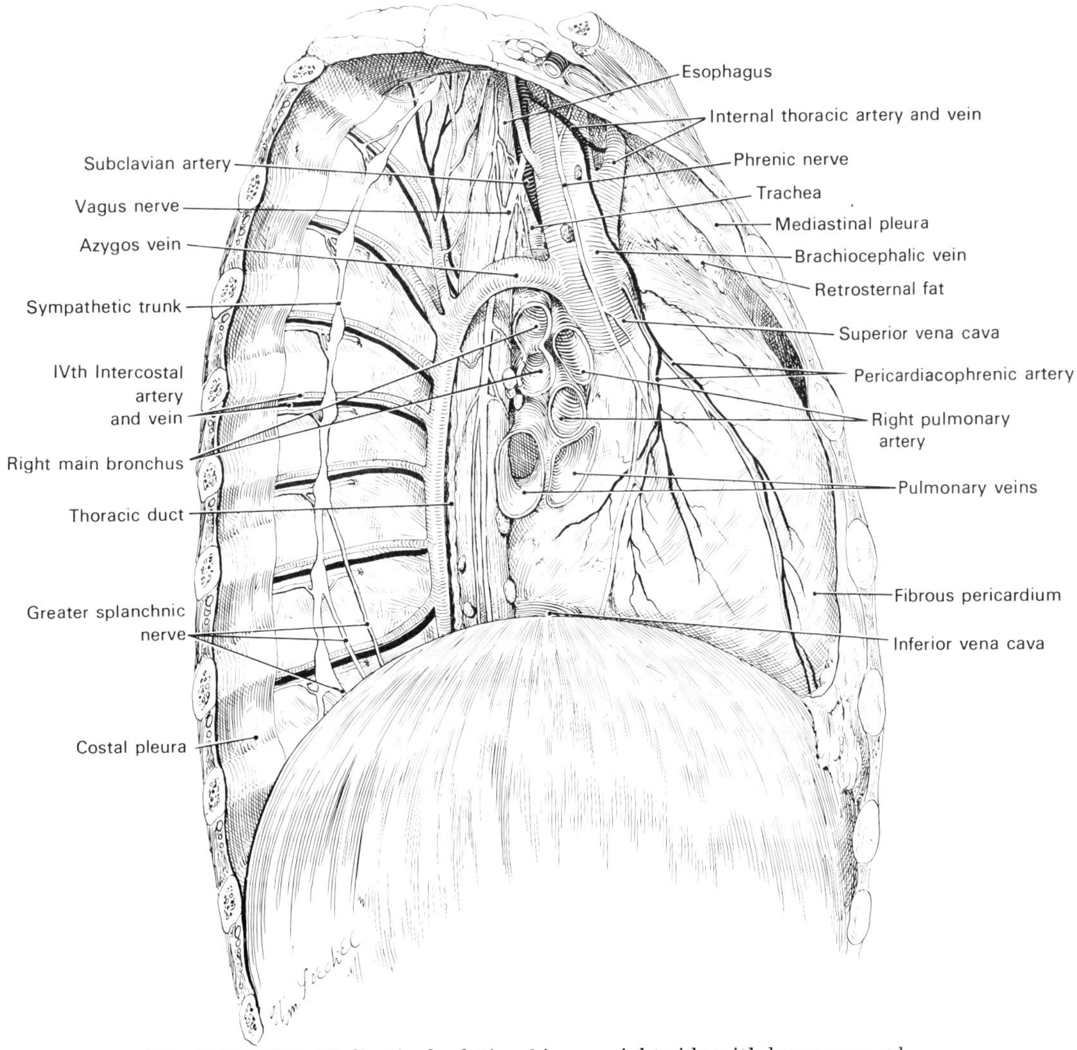

Fig. VIII-63B. Mediastinal relationships on right side with lung removed.

upper, middle, and lower—in contrast to the two lobes, upper and lower, on the left side. (Figs. VII-64 and 71).

Since the left lung is divided into its two lobes by a single **oblique fissure,** it is simpler to describe the left side first. The oblique fissure begins high on the posterior surface of the left lung, about 6.3 cm. from the apex, and traces downward and forward across the medial and costal surfaces so that the inferior border is divided behind and lateral to the cardiac notch. This leaves little of the inferior lobe showing in an anterior view (Fig. VIII-64). The opposed walls of the fissure are lined with visceral pleura. The upper left lobe lies above and in front of the oblique fissure, and it comprises the apex, a substantial portion of the costal and medial surfaces, and the entire anterior border, including the cardiac notch. At the lower margin, the lobe, in the region of the notch, is projected as the **lingula.** This

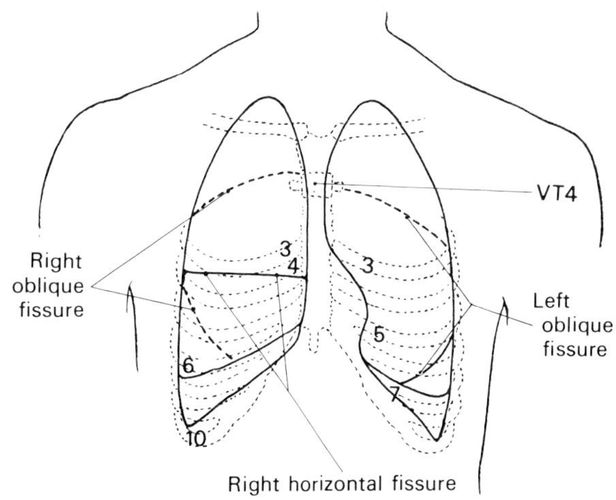

Fig. VIII-64. Location of fissures of lung on surface of trunk.

particular region is difficult to drain and is analagous to the right middle lobe. The lower left lobe lies below and behind the fissure so that most of it is directed posteriorly. Hence, it is not accessible from in front and in the superior portion of the axilla.

One of the fissures of the right lung is comparable to the oblique fissure in the left lung. The other, the **horizontal fissure,** sections a wedge-shaped middle lobe from the upper subdivision of the right lung. This fissure meets the oblique fissure in the midaxillary line and runs, roughly horizontally, across the costal surface on a level with the fourth costal cartilage (Fig. VIII-64). The right lower lobe is the shape and equivalent of the left lower lobe. The upper and middle lobes on the right are the equivalent of the upper lobe on the left. The middle lobe is significant because it is only clinically accessible from in front below the level of the fourth costal cartilage.

Accessory Fissures and Lobes of the Lung: At times extra fissures subdivide the lungs. While they occur in adult lungs, they are not as deeply marked as in infant lungs. They also tend to be more frequent and deeper in the right than the left lung. Some are mere subpleural furrows. As a rule, they seem to mark off one or another bronchopulmonary segment. The most common one has been found on the diaphragmatic surface of the right lung, beginning just anterior to the pulmonary ligament and curving medially toward the vertebral border as if defining the medial basal segment. A second common fissure is sometimes found between superior and basal segments of the lower right lobe as if to cut off the superior bronchopulmonary segment as a separate lobe. The third most common one is in the left lobe (in about 8% adult specimens) seeming to form a small middle lobe the fissure lies between the lingula and the rest of the upper left lobe. Often these are partial and not complete fissures.

So far as lobation is concerned one frequently hears about the **azygos lobe.** But hearing is not seeing since the condition is rare. An azygos lobe is seen in about 0.4% of patients on chest roentgenograms (Fig. VIII-65). It is usually found in the right lung. In the very few instances that it has been seen in the left lung it was most frequently associated there with the persistence of a left azygos vein. It is caused by the vein's taking an abnormal course; usually the vein passes medial to the superior lobe of the lung, arching over the hilum to join the superior vena cava. In the anomolous situation it arches high to indent the parietal pleura over the cupula. It, thus, carries down a fold of the parietal pleura (the meso-azygos) so that fold and vein sink into the apex of the lung dividing the apex into medial and lateral parts.

There have been such things described as accessory or aberrant lobes. There seems to be a fine distinction in terminology here: an "accessory lobe" is or was one that connected with the bronchial tree; an "aberrant lobe" is one originating from some other source. **Tracheal lobes** have been described as arising from supernumerary bronchi, yet Boyden[7] cautions that most of these are probably displaced apical bronchi.

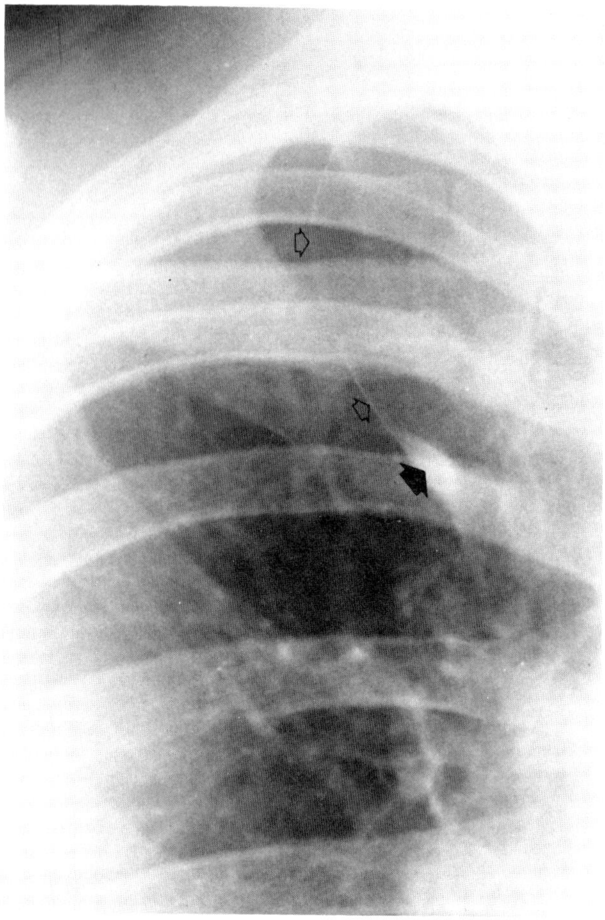

Fig. VIII-65. The azygos lobe. The azygos vein (*black arrow*) appears as an oval white radiodensity. The meso-azygos (two layers of visceral and two of parietal pleura) is seen as a thin white curvilinear line (*open arrows*) extending upward toward the apex of the lung. Lung medial to the meso-azygos is termed the azygos lobe.

The lower **accessory lobe** (Lobe of Rokitansky) is rare but probably occurs more often than tracheal lobes and more often on the left side than on the right. It lies below the normal left lung and sometimes below the diaphragm. It seems to spring from the area of the pulmonary ligament and its blood supply is usually from the aorta. Another rare type has been described as arising from the bronchus near the tracheal bifurcation and opening into the esophagus.

The Development of the Respiratory System

The pulmonary system takes its origin from the early digestive tube as a ventral diverticulum which elongates as the **trachea** (Fig. VIII-66). Its terminal end bifurcates as the **right** and **left lung bud** contain-

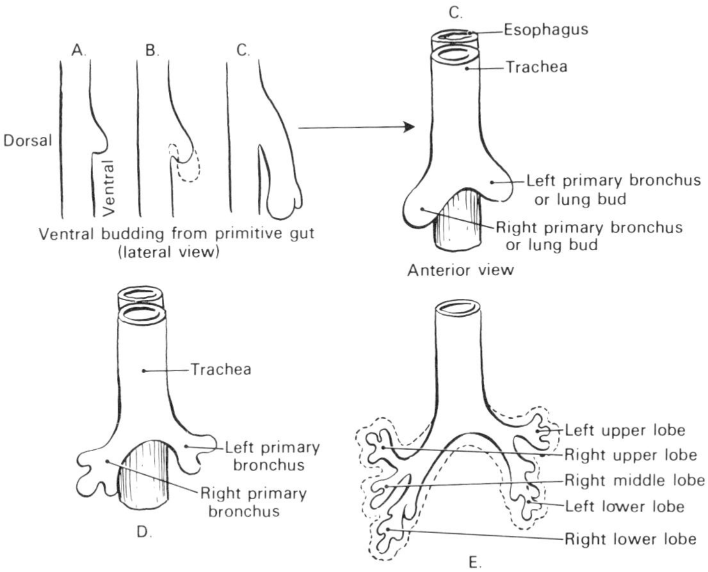

Fig. VIII-66. Stages in development of trachea and lungs (schematic) from 3 weeks to 6 months in human fetus.

ing tubular portions that will represent the **primary bronchi.** The lung buds push mediastinal pleura in front of themselves as they invade the empty pleural cavities. As they expand to fill the cavities, this pleura becomes the visceral pleura of the lungs. Expansion is accompanied by complex subdivision of the primary bronchi until a **primitive bronchial tree is formed.** The younger the fetal specimen, the less likely are the terminal, dichotomous branchings of this tree to be alveoli. Rather, they are apt to be **segmental bronchi** or **terminal bronchioles.** In late fetuses, near term, some of the respiratory elements (as distinct from simple, air-conducting tubes) are present as **respiratory bronchioles, alveolar ducts, sacs,** and terminal, rounded **alveoli.** The first alveoli are apt to be lined by high, columnar, epithelial cells. Near term many of the lining cells are cuboidal and even squamous so that some of the alveoli begin to resemble the air sacs of the air-breathing lung.

But, of course, before birth the lung of the fetus does not contain air. Never the less, its tubular and saccular passageways are filled with the fluids of endogenous secretion, *i.e.*, the secretory products of the lining mucosa. Such a lung is said to be atelectatic in the sense that it has never breathed air. This fetal condition of an underexpanded lung should not be confused with the pathological state of "secondary" atelectasis, or the partial collapse of alveoli in an air-breathing lung that has suffered disease.

A fetal lung can be made to expand near term, whether the fetus is prematurely delivered or still remains within the uterus, for its neurons of the respiratory centers within the brainstem have matured long ago, and its peripheral nerves are in contact with the Diaphragm and intercostal musculature. All that is needed is that the carbon dioxide titer in the circulating blood be high enough to stimulate these nerve cells to discharge rhythmically. Under asphyxial conditions, a late fetus will engage in respiratory-like efforts *in utero* which are rhythmic, at first shallow, and then deep and agonal. Since it cannot take air into the lungs, it will aspirate amniotic fluid. The alveoli will expand beyond normal fetal dimensions, press against each other and become angulated rather than round in appearance. If the asphyxial conditions are not alleviated, the carbon dioxide level in the blood soon reaches depressant levels. The discharge of neurons in the respiratory center is inhibited, the respiratory-like behavior ceases abruptly, and, finally, cardiac muscle also fails to respond. The fetus dies *in utero* and becomes stillborn.

The fact that the fetal, brainstem, respiratory centers are ready to function well before birth means that some very premature fetuses make an effort to breathe at delivery even though lung development has not reached a stage for adequate functioning as an aerating mechanism. Older fetuses, however, will air-breathe and survive, some even after they have aspirated considerable amniotic fluid just prior to premature birth. It is amazing how much fluid can be absorbed by a lung whose alveoli have been expanded by air-breathing.

The lung is not completely developed even after birth since the bronchial tree continues to expand dichotomously for some time. Nor do all alveoli expand with the first breath at birth. It may take hours or days before they all expand. Thus, if death ensues in these early days, parts of the lung still exhibit *fetal atelectasis.* The atelectatic fetal lung or the fluid-filled lung of a stillborn will not float. But, once the lung of a premature or a newborn has breathed air, the air is never completely evacuated from it during expiration, and it will float. This simple test is often used to determine whether an infant was born alive and breathed air for a while before it died, or whether it was stillborn. What would the lung of a victim who had drowned do if subjected to this same test?

From the stand point of its development, the lung can be looked upon as a modified secretory gland. It contains a conducting and a secretory portion. The conducting portion consists of the trachea, bronchi, and bronchioles. These tubes with their mucous secreting lining epithelium are comparable in some ways to the secretory ducts of some of our digestive glands (parotid and submaxillary). The **respiratory units,** namely respiratory bronchioles, with their alveolar ducts and sacs are comparable to the acinar or alveolar portions of such digestive glands. Until recently, it was assumed that the lung alveoli had merely lost their secretory function in transforming into aerating units. But we have just seen that they supplemented the mucous secretions by passing fluids *into* the bronchial tree during embryonic and fetal life. The whole system

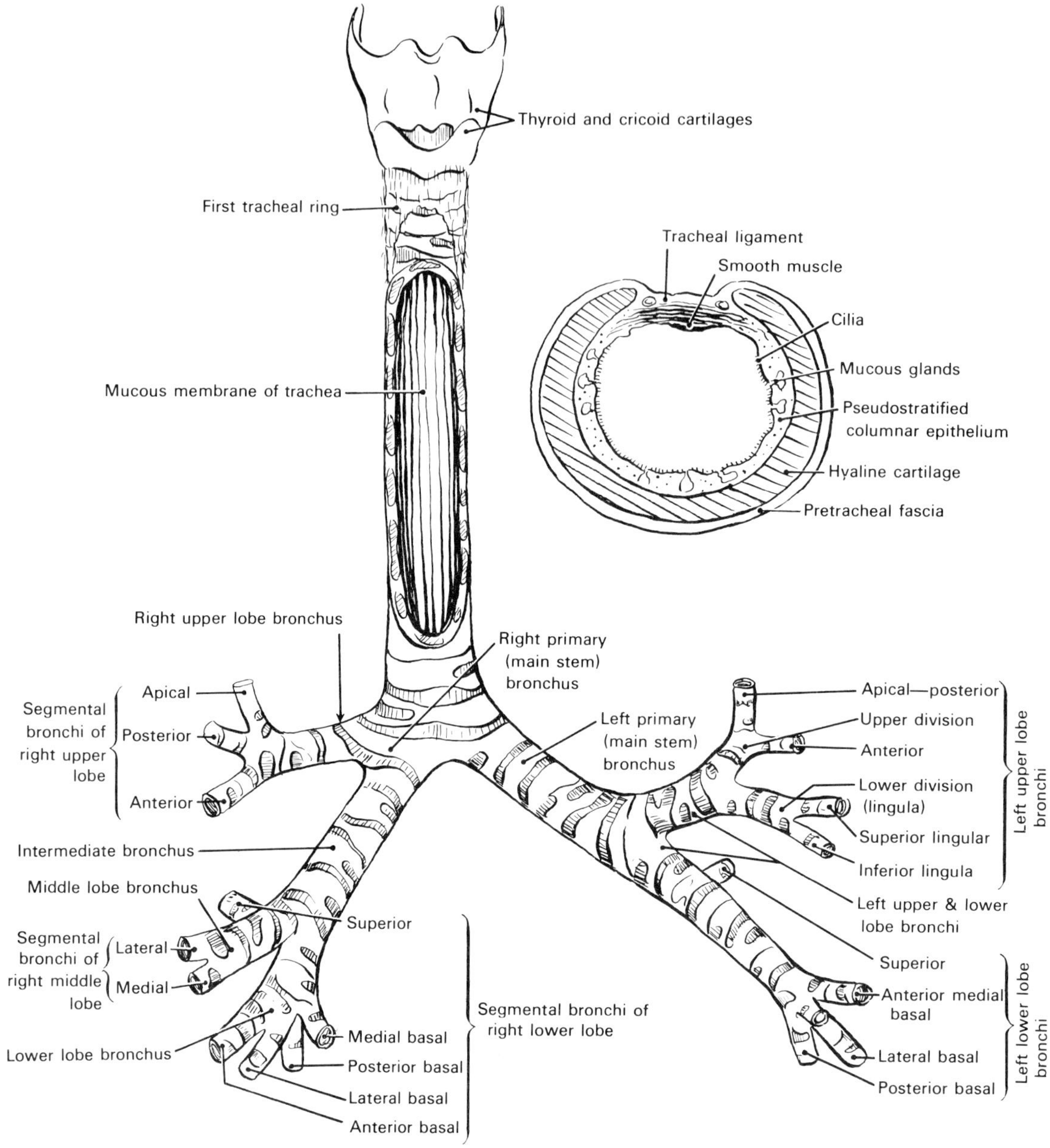

Fig. VIII-67. The conducting tubes of the respiratory tree as far as the segmental bronchi.

is a fluid-filled one, even when the fetus never once engages in amniotic fluid aspiration at any time in the normal course of its development. In adult life certain alveolar lining cells still continue to secrete surfactant (a phospholopid, a mucopolysaccharide, or possibly a combination of both) in to the alveolar lumina.

The Trachea and Bronchial Tree (Fig. VIII-67): While the trachea begins at the lower end of the larynx, it is critical that you realize that its upper two or three rings can be covered by the isthmus of the thyroid, that blood vessels cross the isthmus, and that twigs of the recurrent laryngeal nerve, the all important nerve to the voice box, lie close by. These structures, and the latter in particular, are in danger of severance during *tracheostomy* procedures. While this procedure may have to be executed quickly at times as a life saving measure, there is no call to render the individual voiceless in the process.

Approximately one-half of the 12-cm.-long trachea lies in the neck and the remainder is within the thorax. The trachea terminates by dividing into the right and left main (primary) bronchi. This bifurcation, the cen-

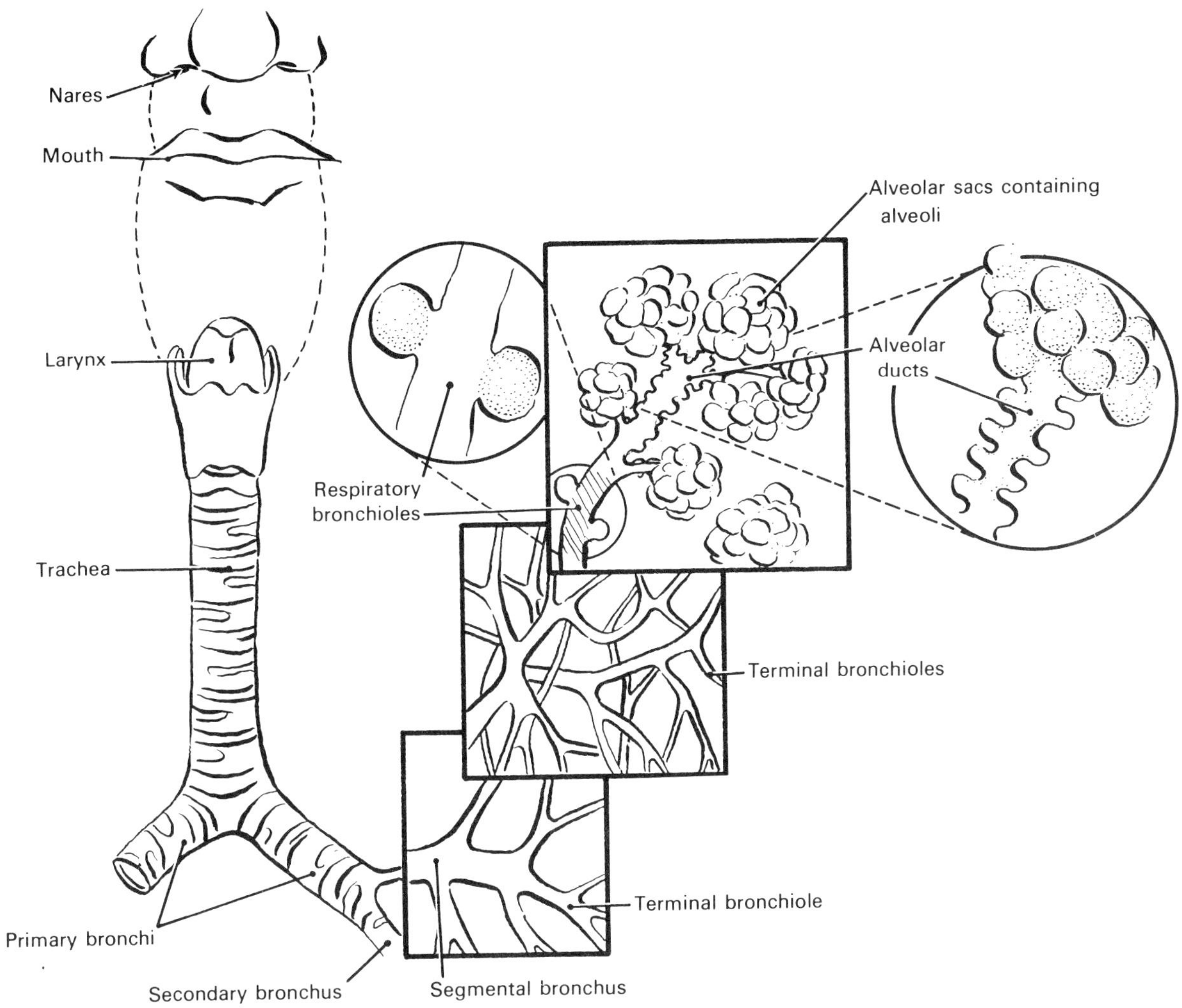

Fig. VIII-68. Parts of the respiratory system from nose to alveolus.

ter of which is termed the **carina,** occurs at the level of VT_4 in the cadaver, and at the VT_6 to VT_7 level in the living subject.

At the bifurcation of the trachea (Fig. VIII-67), the right main (primary) bronchus is wider, shorter (1 to 4 cm.), and runs in a more direct vertical course than the left main bronchus (5 to 7 cm. in length). Perhaps for this reason, the right main bronchus is more apt to collect and direct mis-swallowed objects into the smaller airways of the lung. Large blood vessels accompany each bronchus. The left bronchus lies in closer proximity to the pulmonary trunk, arch, and descending portions of the aorta, than the right main bronchus. Thus, the left main bronchus is more difficult to contend with surgically than the right bronchus.

Each main (primary) bronchus divides into lobar (secondary) branches: three on the right—upper, middle, and lower; two on the left—upper and lower. The division of the primary bronchi occurs nearer the carina on the right than on the left.

Figure VIII-67 shows that the direct continuation of the right main bronchus, after giving off the right upper lobar bronchus, is called the **intermediate** (or **stem**) bronchus. The intermediate bronchus courses for a short distance in the oblique interlobar fissure before it divides into the right middle and lower lobar bronchi. Note also that the left upper lobar bronchus has two divisions: the **upper** and the **lower** (or **lingula**) each of which in turn divides into two segmental (tertiary) bronchi.

Each segmental bronchus gives rise to finer ramifications that are called bronchioles. The smallest of these conducting tubes, the terminal bronchioles, will have an average diameter of 0.5 mm. Tubes of smaller diameter are no longer part of the air-conducting system of the lung but are the beginnings of the respiratory units of the lung (Fig. VIII-68). Respiratory bronchioles, for example, give rise to occasional alveoli which bud from their walls, and then terminate in a number of alveolar ducts. Alveolar ducts are now wider

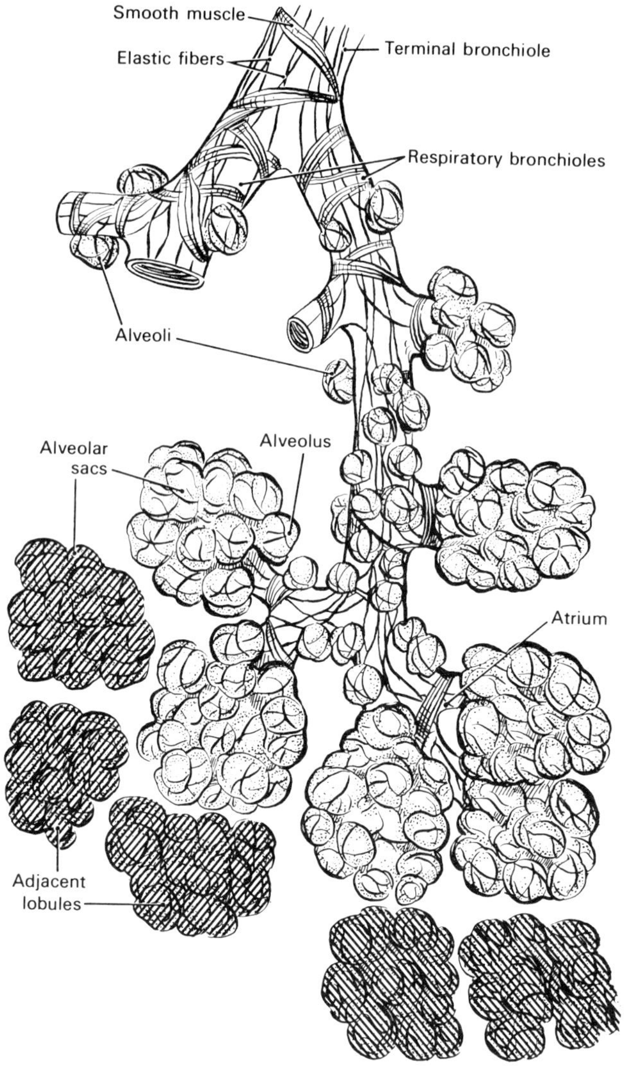

Fig. VIII-69. The respiratory unit of the lung or primary lobule: respiratory bronchiole; its alveolar duct; and alveolar sacs.

channels whose walls are made up of alveoli. The ducts end in one or more large saccular dilatations bearing many alveoli (Fig. VIII-68 and 69). Alveoli are thin-walled air spaces lined with simple squamous epithelium. They tend to be spherical, or have diameters that represent five-sixths of a sphere, if pulmonary pressures or diseases have not been so abnormal as to crowd them and press them out of shape. The connective tissue between them contains considerable elastic tissue, a considerable number of phagocytic cells, and a vast capillary network that is applied closely about the outer walls of the alveoli. While there is very little smooth muscle in the wall of the trachea, it increases in amount in lower parts of the respiratory tree, particularly in the walls of those tubes lacking cartilage (Fig. VIII-69). Cartilage, in the form of irregular plates, is present in bronchi and bronchioles, but disappears in bronchioles of less than 1 mm. in diameter. Smooth muscle continues as far as the atria of the alveolar sacs. It is spirally arranged about the walls of the tubes as far distalward as the respiratory bronchioles. It is circularly disposed about the ostea of alveoli as these bud from alveolar ducts, and again circularly arranged in the atria leading into alveolar sacs. None is found in the alveolar sacs or walls of alveoli (Fig. VIII-68 and 69).

Two factors alter the diameters of bronchi and bronchioles: (1) variations in pressure differences inside and outside the tubes; (2) variations in the contraction of smooth muscle. Variations in intrabronchial and peribronchial pressures, changes in lung volume, and other physiological factors will alter the tension upon smooth muscle of the respiratory tree. The tubular walls are pliable. Although their smooth muscle is, normally, under tonic contraction to some extent, the airways are, generally, held patent because peribronchial pressures are lower than intrabronchial pressures. Contraction of smooth muscle will decrease tubular diameters, and if pressure differences between proximal and distal portions are maintained, a decrease in air flow results.

The smooth muscle of the respiratory tree has a double motor innervation as is true, in general, for most visceral musculature. The parasympathetic, cholinergic fibers of the vagus cause bronchoconstriction when stimulated. The inhalation or injection of certain drugs, such as acetycholine or histamine act upon smooth muscle directly to exert powerful bronchoconstrictor effects. The postganglionic, adrenergic fibers of the sympathetic system, when stimulated, produce relaxation of previously constricted muscle of the bronchial tree. So does injection or aerosal inhalation of epinephrine or norepinephrine. Bronchiospasm, present during asthmatic attack, is relieved by the administration of these drugs.

There are about 550 million alveoli with a mean diameter of 0.2 mm. in the lung. Assuming a spherical shape for them, the entire alveolar surface can be calculated to be about 675 square feet. Consider then, that the blood is collected in the right heart and pumped into the lungs to be spread over an aerating surface about the size of a tennis court. The thin-walled alveoli abut against a vast, fine network of blood capillaries which are supported, in turn, by an elastic network superimposed upon a collagenous supporting framework. This tissue complex is illustrated in Figure VIII-70, an artistic composition suitable for framing and hanging, or for installing as stained glass in chapel windows. The alveolar walls, which separate the capillaries from alveolar gas are thin, never exceeding 0.15 to 0.3 μ in thickness. Some of the alveolar cells are exceedingly tenuous and others are consider-

Fig. VIII-70. The collagenous and elastic framework of the lung. This drawing was made by digesting adjacent histological lung sections with an elastase and a collagenase respectively. The two digested sections were then superimposed upon each other and the resulting pattern traced by the artist. The heavy, dark bands represent elastin while the fine fibrillar network indicates collagen.

ably taller, but all are beyond the level of resolution with the light microscope. Some of the taller cells are thought to secrete a complex surface-acting agent which is part of the fluid film that constantly lines the gas side of the alveolus.

The Bronchopulmonary Segments: The major differences in nomenclature between each lung's bronchopulmonary segments can be seen in Table VIII-1. Following the terminology of Jackson and Huber[8] (Fig. VIII-67 and 71), which is now widely accepted clinically and anatomically, you can see that there are 10 segmental bronchi on the right, and only eight on the left. Some prefer to say that each lung has 10 segments as depicted and numbered in the spacial diagram of Figure VIII-72. However, since the apical (※1) and the posterior (※2) bronchi of the left upper lobe usually arise from a common stem, the two segments are inseparable, and thus the designation apical-posterior segment. The same principle applies to the naming of the anterior-medial basal segment (※7, 8) of the left lower lobe, since these two usually arise from a common stem.

The direction which the bronchi take is important in the development and spread of pulmonary disease and in its diagnosis. Not all physicians must be as versed in these details of the tracheobronchial tree as the radiologist, bronchologist, and thoracic surgeon. These specialists must have a thorough knowledge of the prevailing patterns of the "normal" bronchial tree and, of course, its most common variations. We have depicted for you in Figure VIII-67 the most common pattern of the segmental bronchi, and we relegate discussion of the variations that may be encountered to others.[9, 10]

Through the study of the bronchograms shown in Figures VIII-73 to 78, we can now briefly study the bronchopulmonary segments. The technique of *bronchography* requires the tracheobronchial tree to be anesthetized with a local anesthetic agent, after which a radiopaque contrast medium can be introduced through a catheter to visualize all or selected parts of the segmental anatomy. The indications for bronchography are many, and its usefulness is unquestioned in ruling on many disease entities.

The Right Lung (Figs. VIII-71 and 73 to 75): The right upper lobe bronchus, which arises some 1 to 2 cm. below the carina, leaves the main bronchus at right angles at about the VT_6 level. Within a short span of 1 to 2 cm., it trifurcates, although there may be some normal variation in its pattern of branching from patient to patient. The segmental branches of the right upper lobe bronchus are:

1. Apical Segmental Bronchus: This bronchus ascends steeply upward and slightly laterally to distribute to the medial anterior portion of the upper lobe (Figs. VIII-71 and 73 to 75).

2. Posterior Segmental Bronchus: This bronchus courses posterolaterally and cranially supplying the posterior portion of the upper lobe (Figs. VIII-71 and 73 to 75).

3. Anterior Segmental Bronchus: This bronchus extends anterolaterally and slightly inferiorly to supply the anterimedial region of the upper lobe (Figs. VIII-71 and 73 to 75).

The right middle lobe bronchus arises from the anterior wall of the intermediate bronchus about 3.0 cm. below the right upper lobe bronchus. It courses anterolaterally and inferiorly and divides into two branches:

4. Lateral Segmental Bronchus: This bronchus passes inferolaterally and anteriorly to the lateral-posterior aspect of the middle lobe between the two fissures (Figs. VIII-71 and 73 to 75).

5. Medial Segmental Bronchus: This bronchus arises with the lateral segmental bronchus, however, it takes a course which is more inferolaterally and anteriorly directed. Figures VIII-71 and 73 to 75 show the area of the middle lobe supplied near the right heart border.

The right lower lobe bronchus begins where the intermediate bronchus terminates and the right middle lobe bronchus originates. It divides into five segmental bronchi:

6. Superior Segmental Bronchus: This bronchus arises just below the origin of the right middle lobe

Table VIII-1. The Bronchopulmonary Segments (after Jackson and Huber[8])

Right Lung

Lobe	Segment	No.
Upper	Apical	1
	Posterior	2
	Anterior	3
Middle	Lateral	4
	Medial	5
Lower	Superior	6
	Medial basal	7
	Anterior basal	8
	Lateral basal	9
	Posterior basal	10

Left Lung

Lobe		Segment	No.
Upper	(Upper division)	Apical-posterior	1, 2
		Anterior	3
	(Lower or lingular division)	Superior	4
		Inferior	5
Lower		Superior	6
		Anterior-medial basal	7, 8
		Lateral basal	9
		Posterior basal	10

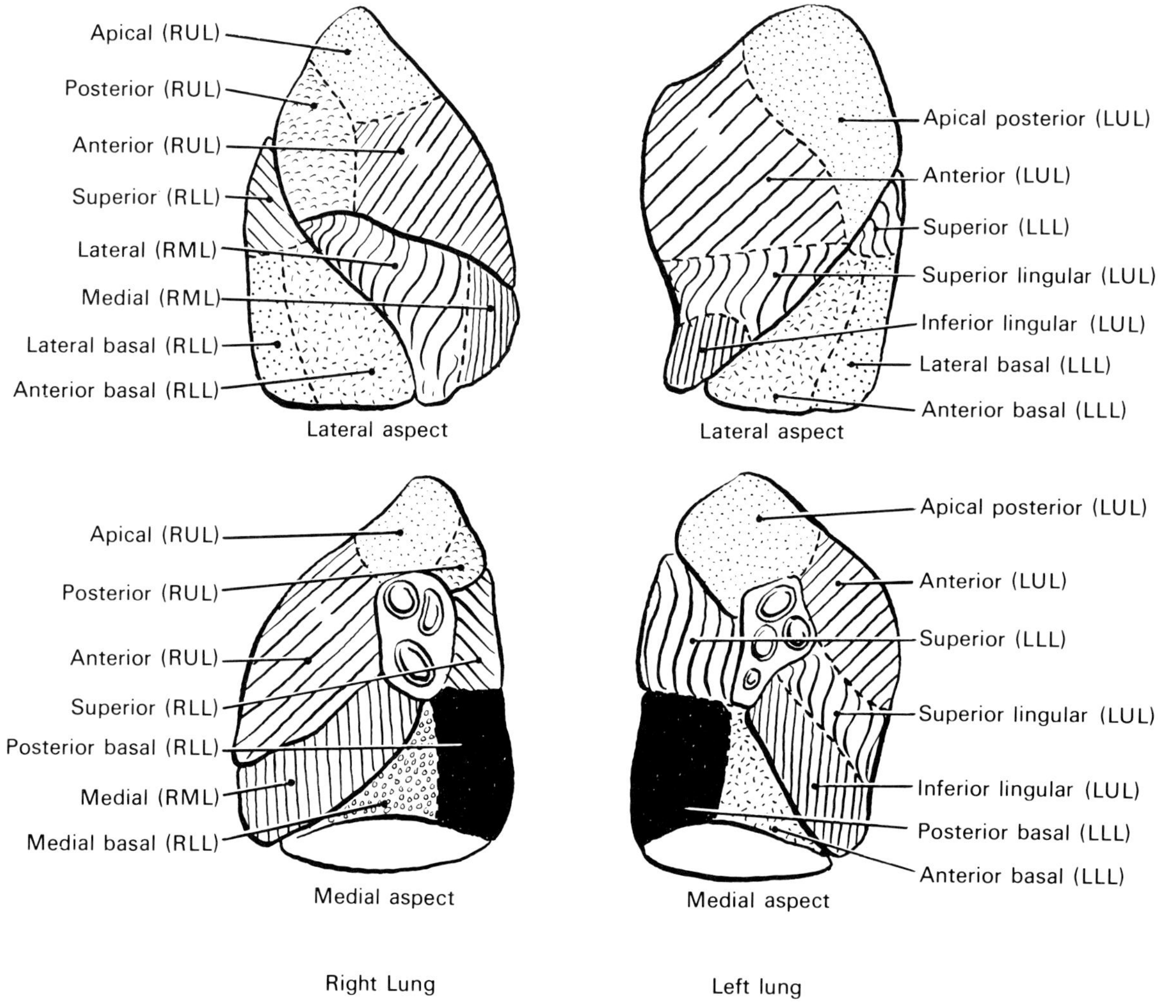

Fig. VIII-71. Bronchopulmonary segments of the right and left lung. (After Jackson and Huber.)

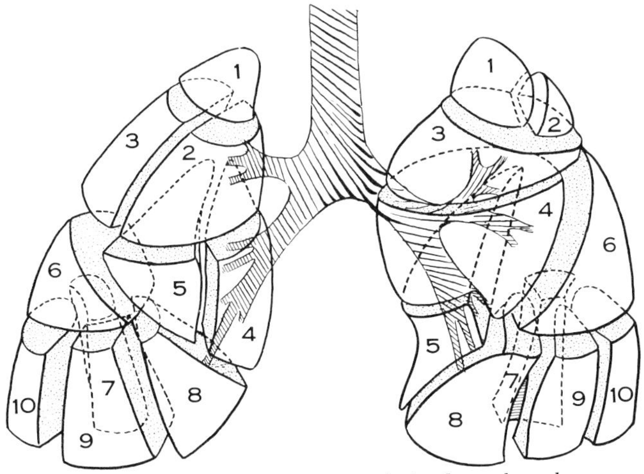

Fig. VIII-72. Spatial concept of the bronchopulmonary segments.

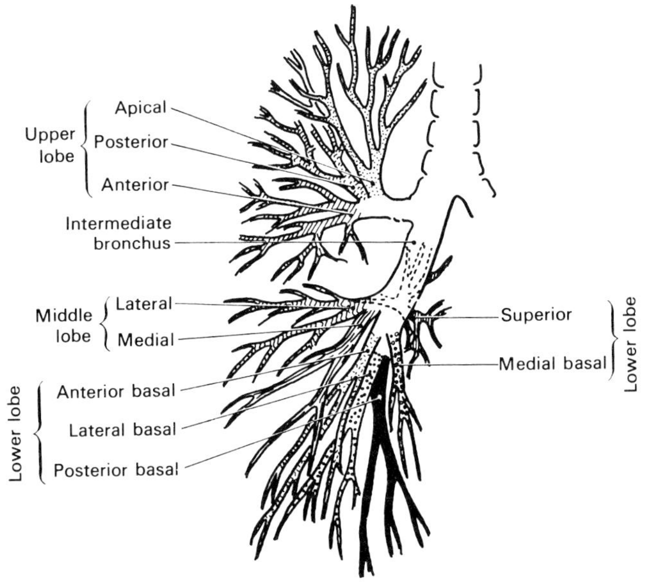

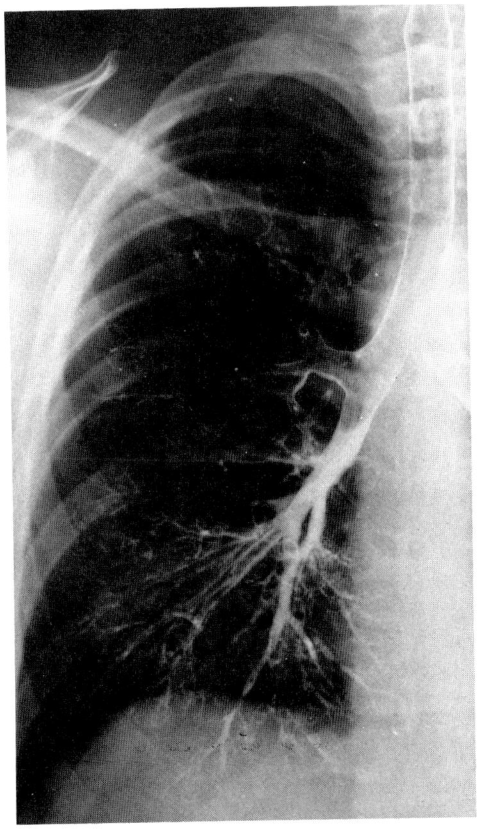

Fig. VIII-73. Bronchopulmonary segments of the right lung—frontal view. **A.** Schematic. **B.** Bronchogram.

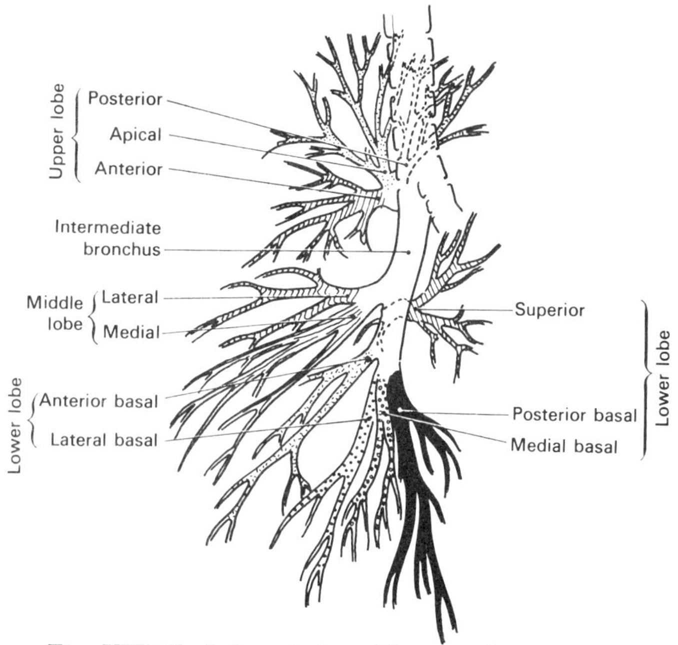

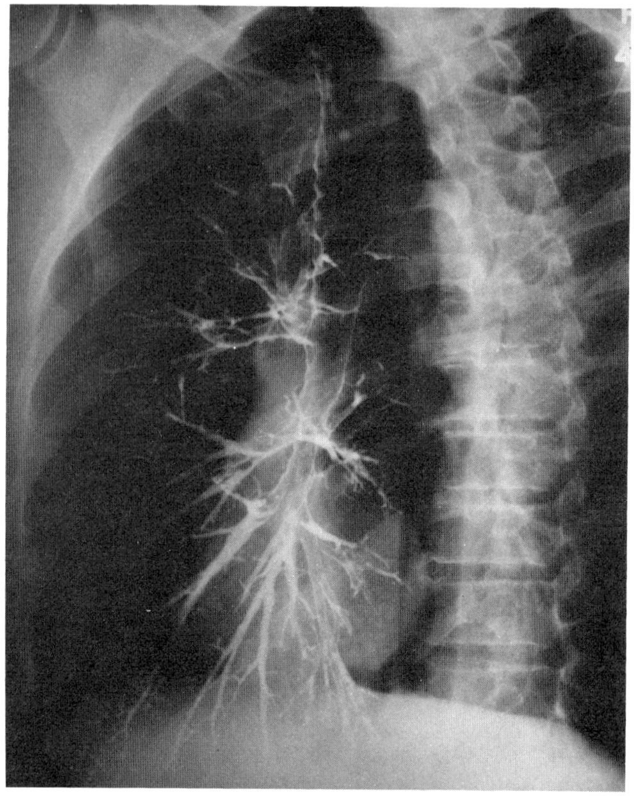

Fig. VIII-74. Left anterior oblique projection of the right lung, showing the bronchopulmonary segments. **A.** Schematic. **B.** Bronchogram.

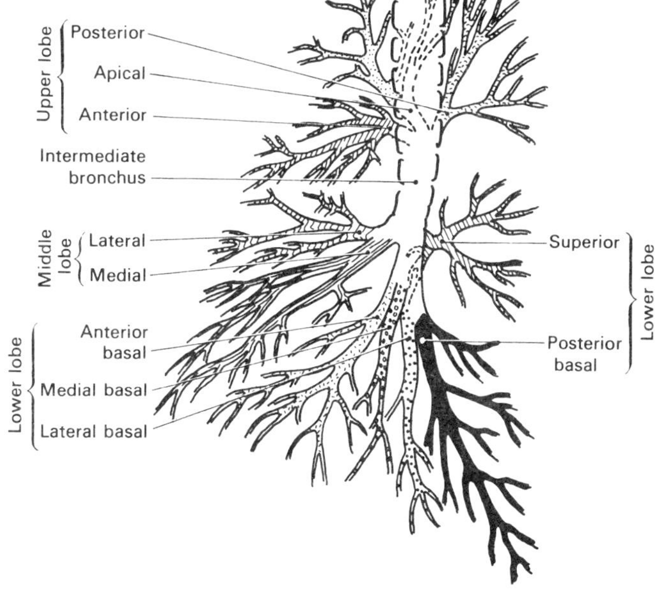

Fig. VIII-75. Schematic drawing of lateral view of right lung to show the bronchopulmonary segments.

bronchus. It passes posterolaterally to supply the apex of the right lower lobe below the oblique fissure (Figs. VIII-71 and 73 to 75).

7. *Medial Basal Segmental Bronchus:* This is the first of the four segmental bronchi to arise from the lower lobe bronchus. It is the most medially placed segment in the frontal projection (Figs. VIII-73 to 75) and supplies the area shown in Figure VIII-71.

8. *Anterior Basal Segmental Bronchus (Figs. VIII-73 to 75):* This bronchus begins on the anterolateral margin of the lower lobe bronchus about 0.5 cm. below the medial basal segment. It extends into the region of the costodiaphragmatic sulcus in its anterolaterally and inferiorly directed course. Its area of distribution is shown in Figure VIII-71.

9. *Lateral Basal Segmental Bronchus (Figs. VIII-73 to 75):* This bronchus comes off the lower lobe bronchus with the posterior basal segmental bronchus, although the latter is considered the direct continuation of the lower lobe bronchus. The lateral basal segmental bronchus supplies an area of the lower lobe lying posterolaterally (Fig. VIII-71).

10. *Posterior Basal Segmental Bronchus:* This bronchus runs posteroinferiorly supplying the area of the lower lobe above the Diaphragm and medial to the lateral basal segment (Figs. VIII-71 and 73 to 75).

As an aid to remembering the relationship of the basal segmental bronchi, anterior, lateral, and posterior (ALP), note that this order holds in frontal,

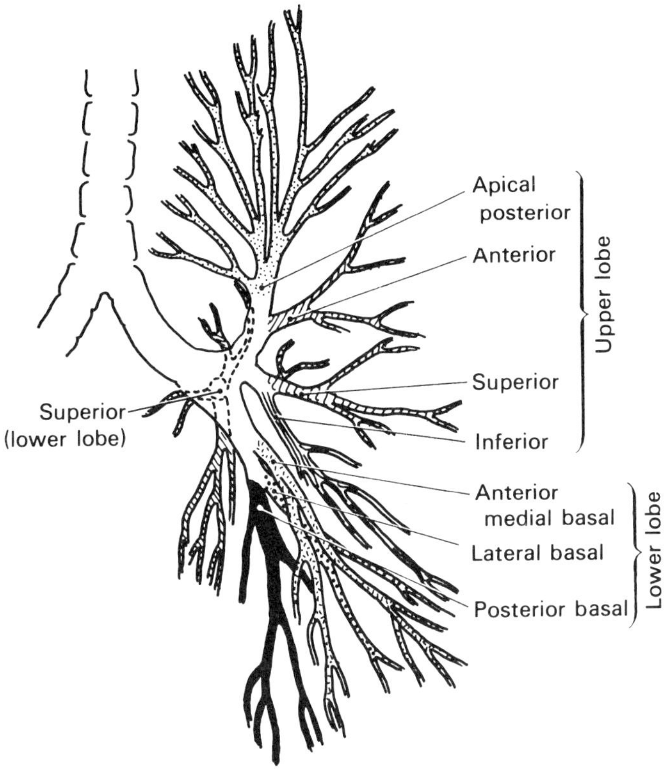

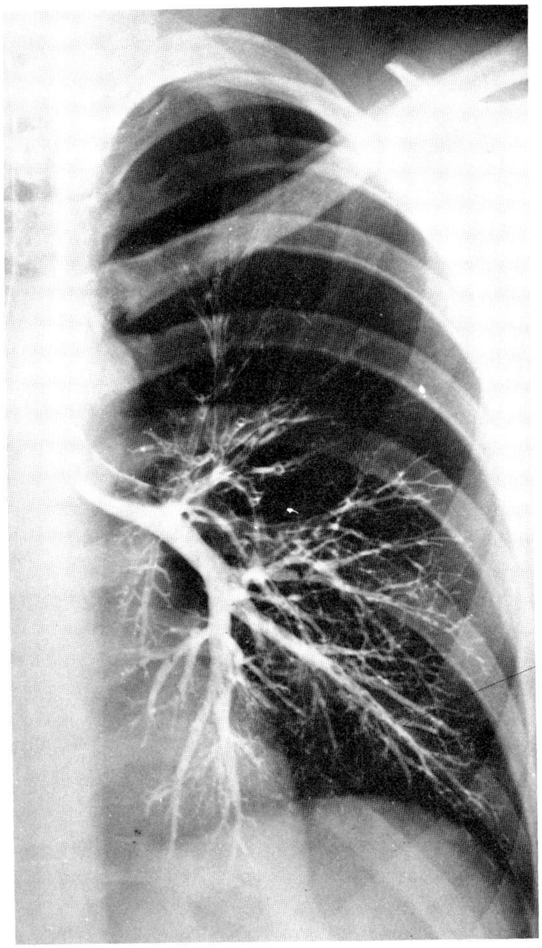

Fig. VIII-76. Frontal view of left lung showing the bronchopulmonary segments. **A.** Schematic. **B.** Bronchogram.

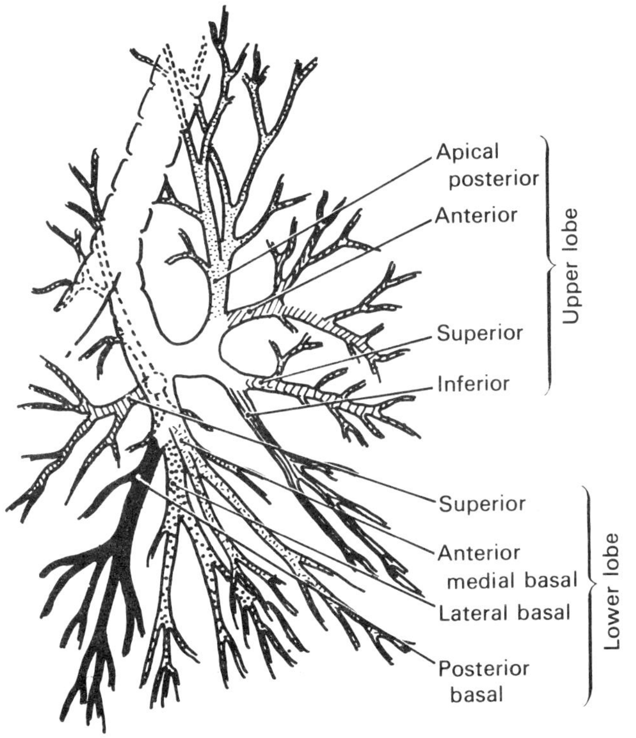

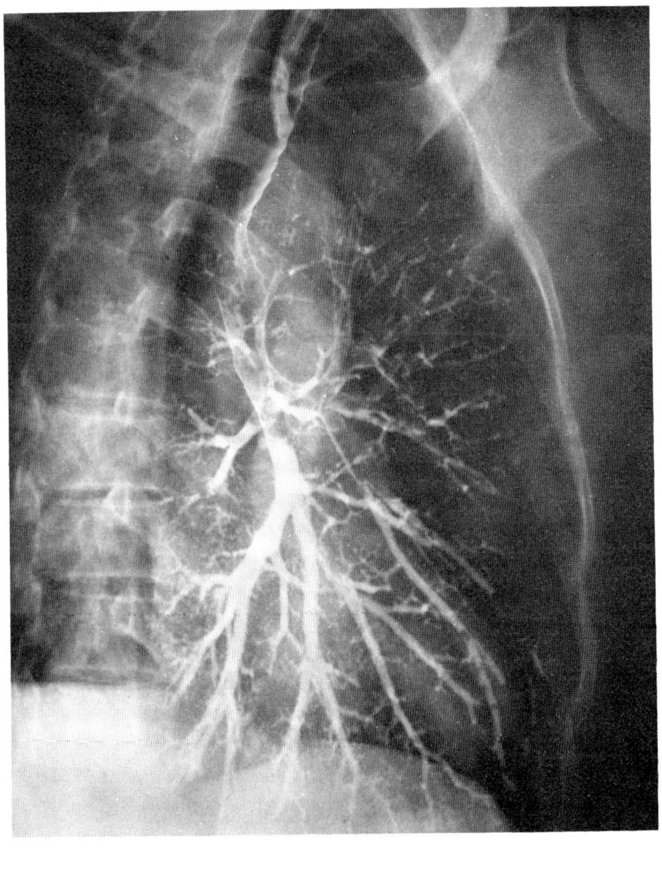

Fig. VIII-77. Right anterior oblique view of left lung. The bronchopulmonary segments are shown. **A.** Schematic. **B.** Radiograph.

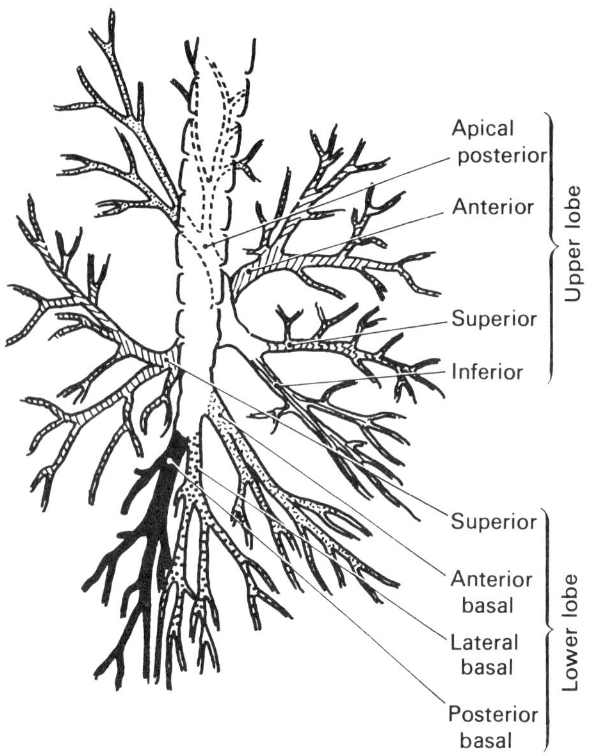

Fig. VIII-78. Schematic. Lateral view of bronchopulmonary segments of the left lung.

oblique, and lateral projections. Only the medial basal segmental bronchus changes its position in each of these three views (Figs. VIII-73 to 75); it is located medial to the posterior segment in the PA projection (ALPm); in the left anterior oblique projection it is interposed between the posterior basal and lateral basal segments (ALmP); and in the lateral projection, it is located between the anterior and lateral basal segments (AmLP).[11] This schema may help you to recall these relationships more easily.

The Left Lung (Figs. VIII-71 and 76 to 78): The left upper lobe bronchus, which is some 1 to 2 cm. in length, courses anterolaterally in a gentle curve before bifurcating into the upper and lower (lingular) divisions. The upper division branches into:

1. Apical-posterior Segmental Bronchus (Figs. VIII-76 to 78): This bronchus is a common stem which gives origin to the segmental bronchi supplying the apical and posterior regions of the upper lobe depicted in Figure VIII-71.

2. Anterior Segmental Bronchus Figs. VIII-76 to 78): This bronchus is the largest of the segmental bronchi of the upper lobe. It passes anterolaterally and upward supplying an area of the upper lobe analagous to its fellow on the right (Fig. VIII-71).

The lower (lingular) division of the left upper lobe bronchus branches into:

3. Superior (Lingula) Segmental Bronchus (Figs. VIII-76 to 78): This bronchus extends obliquely downward and anteriorly to supply the area shown in Figure VIII-71.

4. Inferior (Lingula) Segmental Bronchus (Figs. VIII-76 to 78): This bronchus arises with the superior (lingula) segmental bronchus and follows a similar course but more inferiorly directed. Its area of supply is depicted in Figure VIII-71.

The left lower lobe bronchus begins 4 to 6 cm. distal to the carina. It has a posterocaudal and lateral direction. This bronchus divides into four segmental bronchi:

5. Superior Segmental Bronchus (Figs. VIII-76 to 78): The first branch arises posteriorly about 1 cm. below the origin of the lower lobe bronchus. It passes posteriorly to the area seen in Figure VIII-71.

6. Anterior-medial Basal Segmental Bronchus (Figs. VIII-71 and 76 to 78): This common stem arises from the anterolateral wall of the lower lobe bronchus. It supplies the anterior and medial portion of the lower lobe below the oblique fissure.

7. Lateral Basal Segmental Bronchus (Figs. VIII-76 to 78): This bronchus arises at the terminal bifurcation of the lower lobe bronchus along with the posterior basal segment bronchus. It is directed inferolaterally to the area shown (Fig. VIII-71).

8. Posterior Basal Segmental Bronchus (Figs. VIII-76 to 78): This is the largest of the basal bronchi on the left. It courses postero-inferiorly and slightly laterally to the posterior and medial area of the lower lobe just above the Diaphragm (Fig. VIII-71).

The Blood Supply of the Lung: In devising a separate pulmonary circuit to aerate blood in the lung, the right ventricle of the heart has become the sole pumping chamber for this circuit. The **pulmonary trunk** is a short conduit of some 5 cm. in length and about 2.5 cm. in diameter, which ascends out of the right heart chamber. Like the ascending aorta with which it travels, it is limited pretty much to the pericardium since it is ensheathed in a common serous pericardial covering along with this short segment of the aorta. Because of the position of the heart in the anterior mediastinum, the pulmonary trunk arises ventral to the aorta. And, because it split from the aorta when the spiral bulbo-endocardial cushions formed in development, it twines across the root of the ascending aorta and across the left coronary artery in passing upward out of the right ventricle. The pulmonary trunk curves dorsally to pass between the left auricular appendage and the ascending aorta and finds itself within the concavity of the aortic arch. Here the trunk bifurcates in a T-shaped configuration into its terminal right and left pulmonary arteries. Consult Figure VIII-79.

The Left Pulmonary Artery: This branch of the **"T"** is shorter and narrower than the right one. It runs

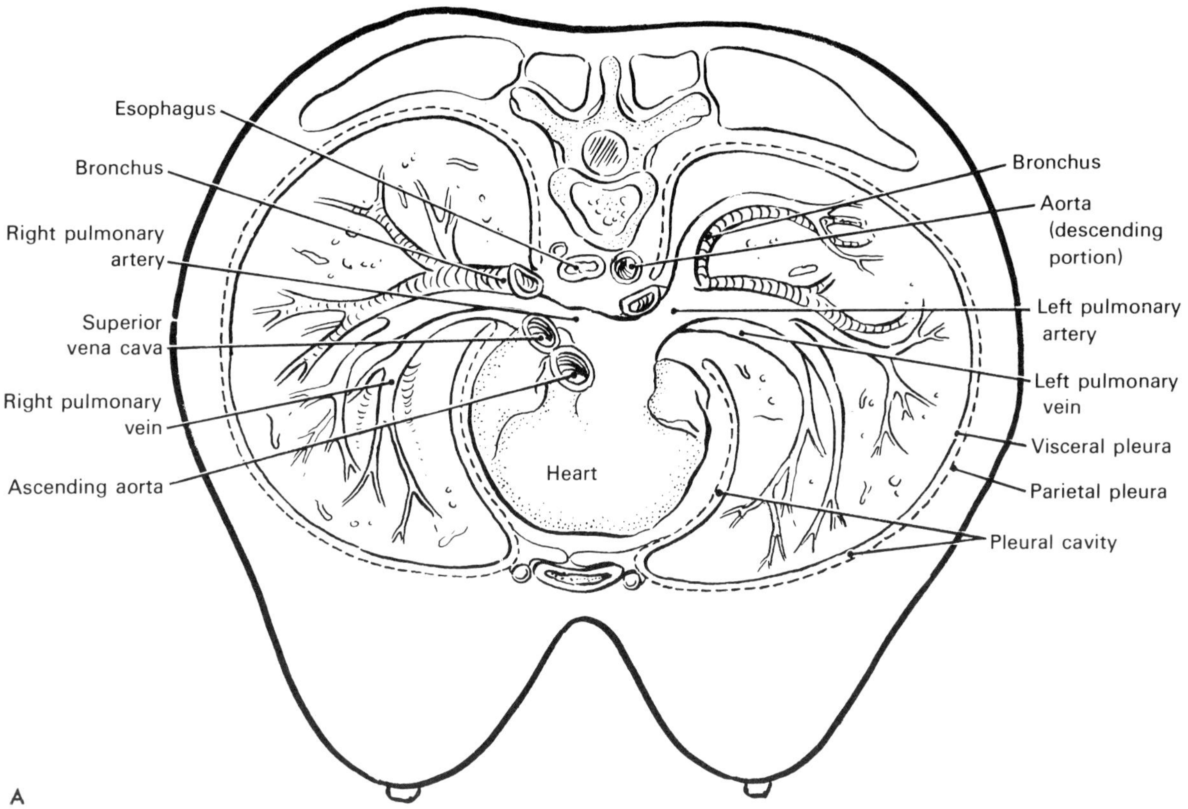

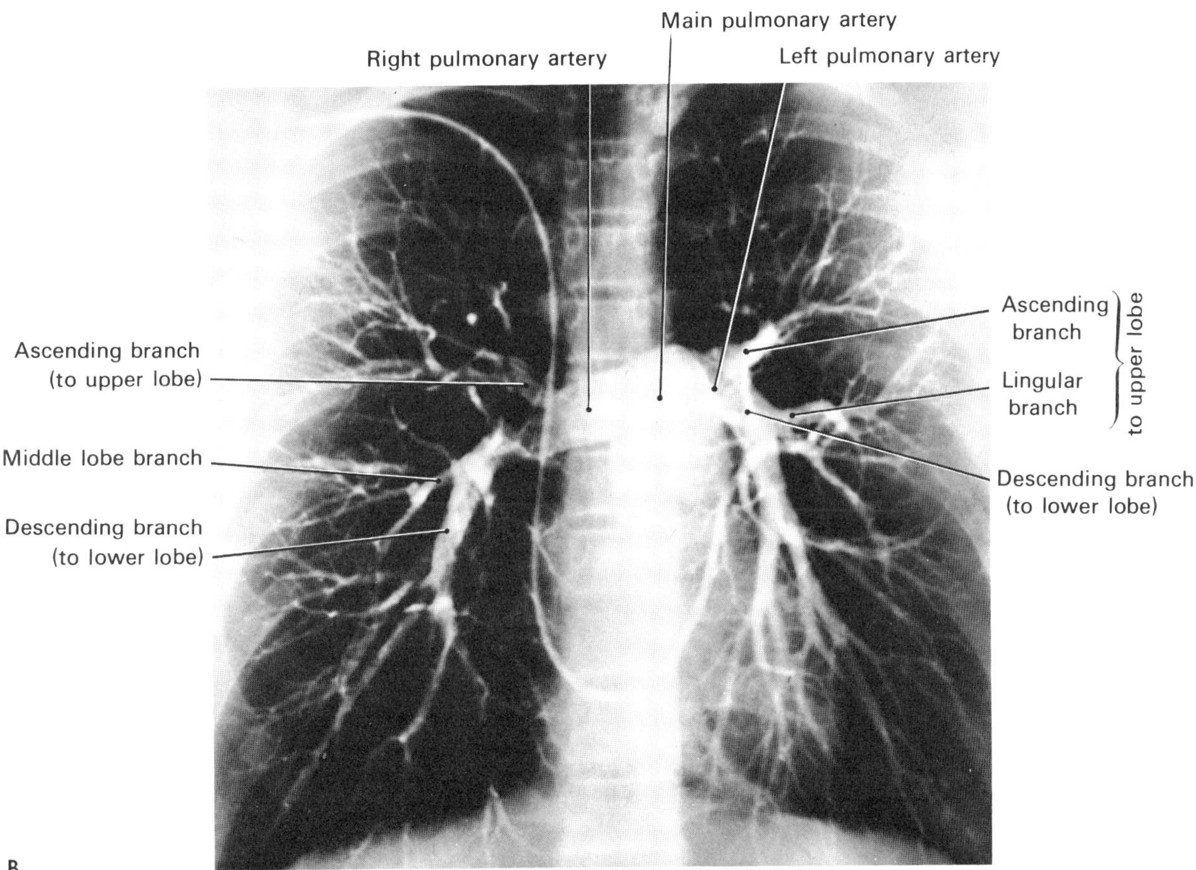

Fig. VIII-79. The pulmonary trunk and its divisions. **A.** Cross section through the thorax. **B.** Frontal radiograph of selective pulmonary arteriogram. Catheter tip is in main pulmonary artery.

horizontally to the left and slightly upward toward the hilus of the lung, crossing ventral to both the descending aorta and the left main stem bronchus on its way. A view of the mediastinal surface of the left lung (Fig. VIII-63) shows the left pulmonary artery cut where it enters the hilus *above* and ventral to the left main stem bronchus. At the hilus, the vessel divides into two large branches, one to each lobe of the left lung (Fig. VIII-79).

Near the origin of the left vessel from the pulmonary trunk a short, stout, fibrous band of connective tissue, the **ligamentum arteriosum,** connects the superior margin of the left pulmonary artery to the concavity of the aortic arch. The left vagus nerve will send its left recurrent laryngeal branch looping back under this fibrous strand. This innocuous, almost invisible bit of connective tissue is all that remains of the once important third shunt in the fetal circulation, the **ductus arteriosus.**

The Right Pulmonary Artery: This is a fairly long and large vessel. As it passes horizontally toward the hilus of the right lung it lies behind the ascending aorta and the superior vena cava but in front of its right main stem bronchus. It will be sandwiched between the bronchus and the superior right pulmonary vein. Within the root of the right lung, the right pulmonary artery divides into a small cranial (ascending) branch to supply the upper lobe and a large caudally directed branch for middle and lower lobes of the right lung (Fig. VIII-79).

While both the pulmonary arteries and veins are closely associated with the primary bronchi, veins tend to pull away from the bronchial tree rather early in its series of branchings so that veins will be found away from it and out in the connective tissue lung septae. The arterial branches, on the other hand, ride piggy-back along the branches of the tree as far distalward as the respiratory bronchioles. This does not mean that each bronchopulmonary segment will have its own arterial supply. It may receive two arteries or it may only be supplied partially from an artery that feeds another segment. Veins, in particular, tend to drain neighboring segments since they elect so quickly to run in intersegmental connective tissue. Thus, while given bronchopulmonary segments can be dissected and removed if disease is localized in them, thus saving the rest of the lung, vessels crossing from one to the other must be identified and tied in such a way as to leave the blood supply to non-diseased segments intact. In this sense, bronchopulmonary segments are *surgical segments* and not true *anatomical segments.*

The terminal pulmonary arterioles at the distal extremities of the tree enter the capillary plexuses (Fig. VIII-80A, B) that envelope the alveolar walls (Fig. VIII-81). The capillary networks drain toward the periphery of respiratory "lobules" so that pulmonary venules start in the interlobular connective tissues. As we noted above, the larger venous channels do not join the bronchial tree until they almost reach the hilar region. Notice (Fig. VIII-81) that the internal veins also drain visceral pleura.

The Pulmonary Veins: Two veins emerge at the hilus of each lung. These are the **superior** and **inferior pulmonary veins.** Each lobule or respiratory unit of the lung generally has one collecting venule. Be certain once more that you understand the term "respiratory unit." It is the pyramidal shaped microscopic lobule which begins with a respiratory bronchiole at its apex and splays out toward the base of the pyramid by branching into a variable number of alveolar sacs, and their component alveoli. If you examine a gross adult lung, its surface will be ringed by a system of irregular interconnecting black rings or hexagons, the whole surface resembling a honey-comb pattern. Each tiny cell (ring or hexagon) of this honey-combed surface represents the base of a pyramid lobule or respiratory unit. The black boundary outlines are due to carbon particles engulfed by the phagocytic cells which dwell in the circumscribing connective tissue. Venules draining such units join to form larger and larger veins until one venous trunk is formed for each lobe of the lung. Since there are three lobes in the right lung, it is customary for the vein of the middle lobe to join with that of the right upper lobe in forming the right superior pulmonary vein.

At the root of each lung (Fig. VIII-82) the **superior pulmonary vein** lies anterior to all other structures that enter the hilus. On the right side it passes behind the superior vena cava to enter the fibrous pericardium and the left atrium. On the left side it passes in front of the descending aorta to reach the pericardial sac and the left atrium.

The inferior pulmonary veins are the lowest hilar structures, situated just above the pulmonary ligaments. The one on the right must cross the right atrium in order to open into the left atrium while the one on the left side, obviously, has easy access to this heart chamber. All of these veins become encased in serous pericardium even before they penetrate the fibrous pericardial sac to pierce the right cranial portion of the left atrium. No valves are associated with these veins or any of their tributaries. The pulmonary veins may be seen entering the left atrium in the angiocardiogram in Figure VIII-83.

Bronchial Vessels: The walls of the respiratory conducting tubes get their own blood supply by way of bronchial arteries. These arise from the aorta or its upper branches as one or more vessels directed toward each side. They accompany the respiratory tree as far as the respiratory bronchioles, breaking up into capillary beds within the walls of bronchi and bronchioles. Arterioles in the vicinity of the respiratory bronchioles empty into the pulmonary capillary bed rather than into bronchiolar beds drained by bronchial veins. Bronchial arteries also supply small vessels to the visceral pleura. Bronchial veins draining local capillary beds

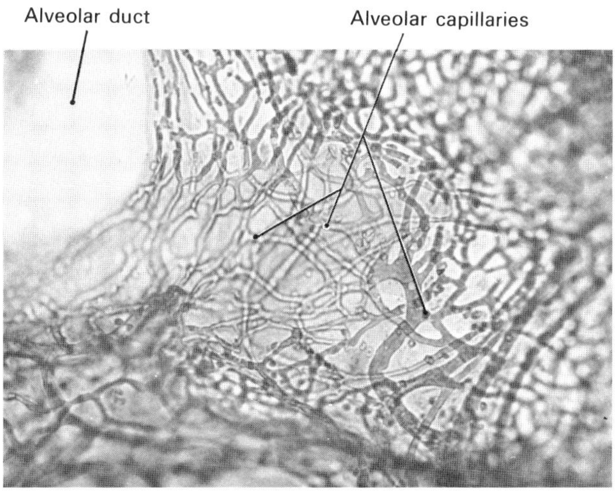

Fig. VIII-80A. Photograph of an alveolar capillary network illustrating the superimposed capillary bed of one alveolus projected on the adjacent alveolar capillary network. The alveolar duct opening is at the upper left. These two systems have been shown to anastomose with each other through the septal walls by way of communicating capillaries. This network allows variations in capillary transit time for the red blood cell as it flows through the lung.

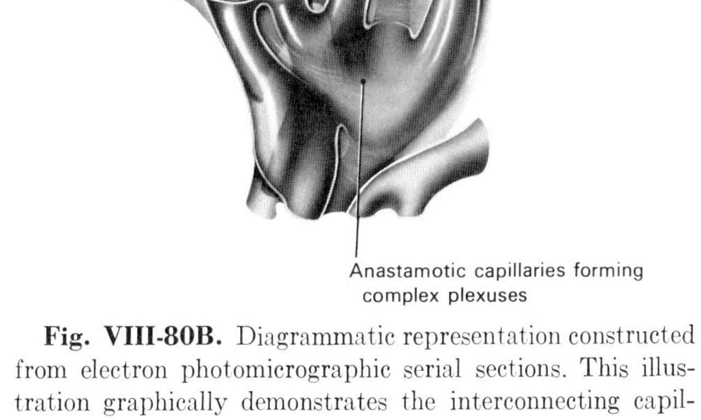

Fig. VIII-80B. Diagrammatic representation constructed from electron photomicrographic serial sections. This illustration graphically demonstrates the interconnecting capillary channels extending from one alveolar surface to the other through the membranous alveolar walls. Interconnecting vessels such as these allow blood flow to proceed from one alveolus to the next for varying distances before returning to the venous side of the pulmonary circulation. (Courtesy Dr. Steve Ryan.)

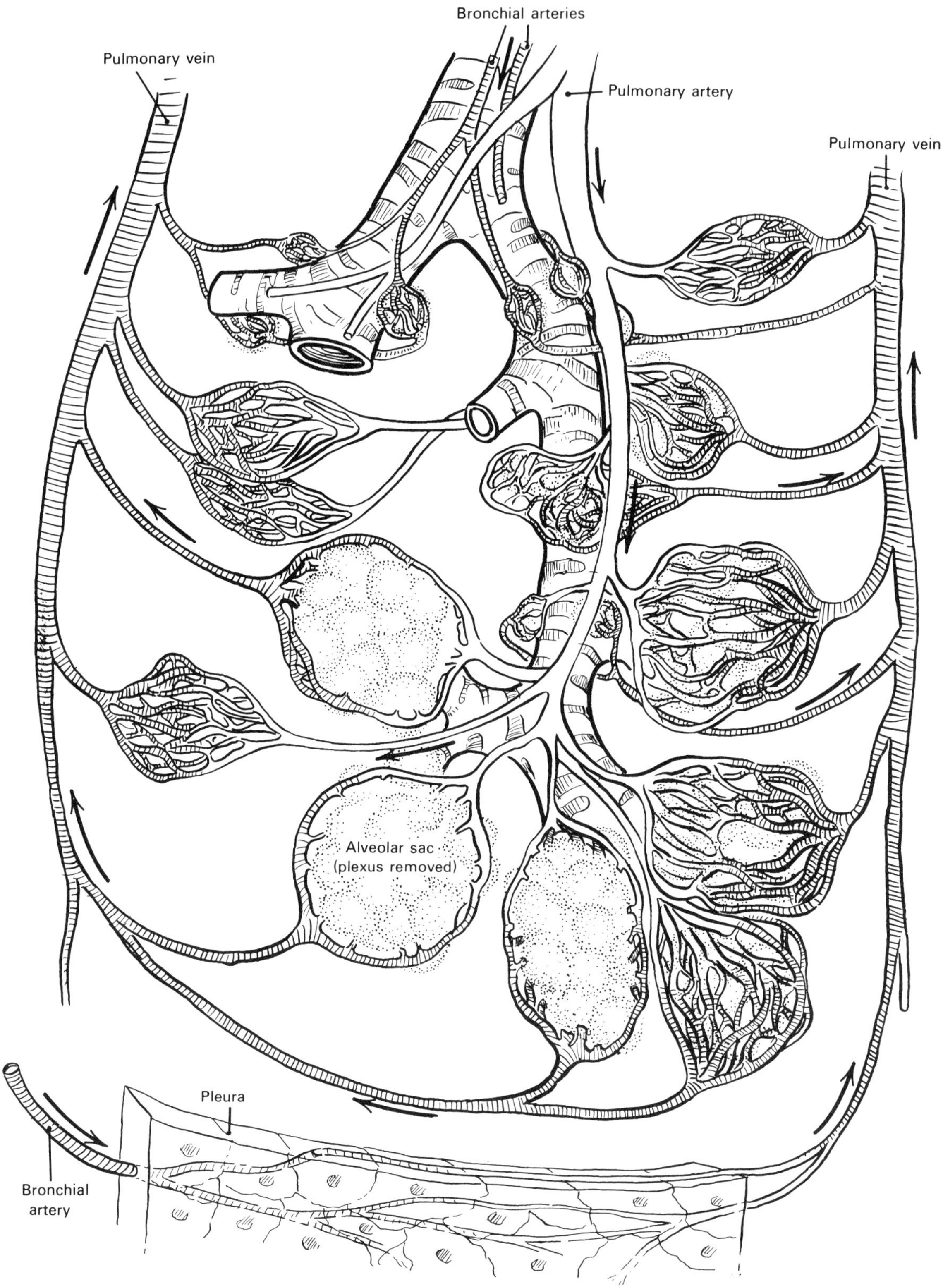

Fig. VIII-81. Schema of the pulmonary microcirculation.

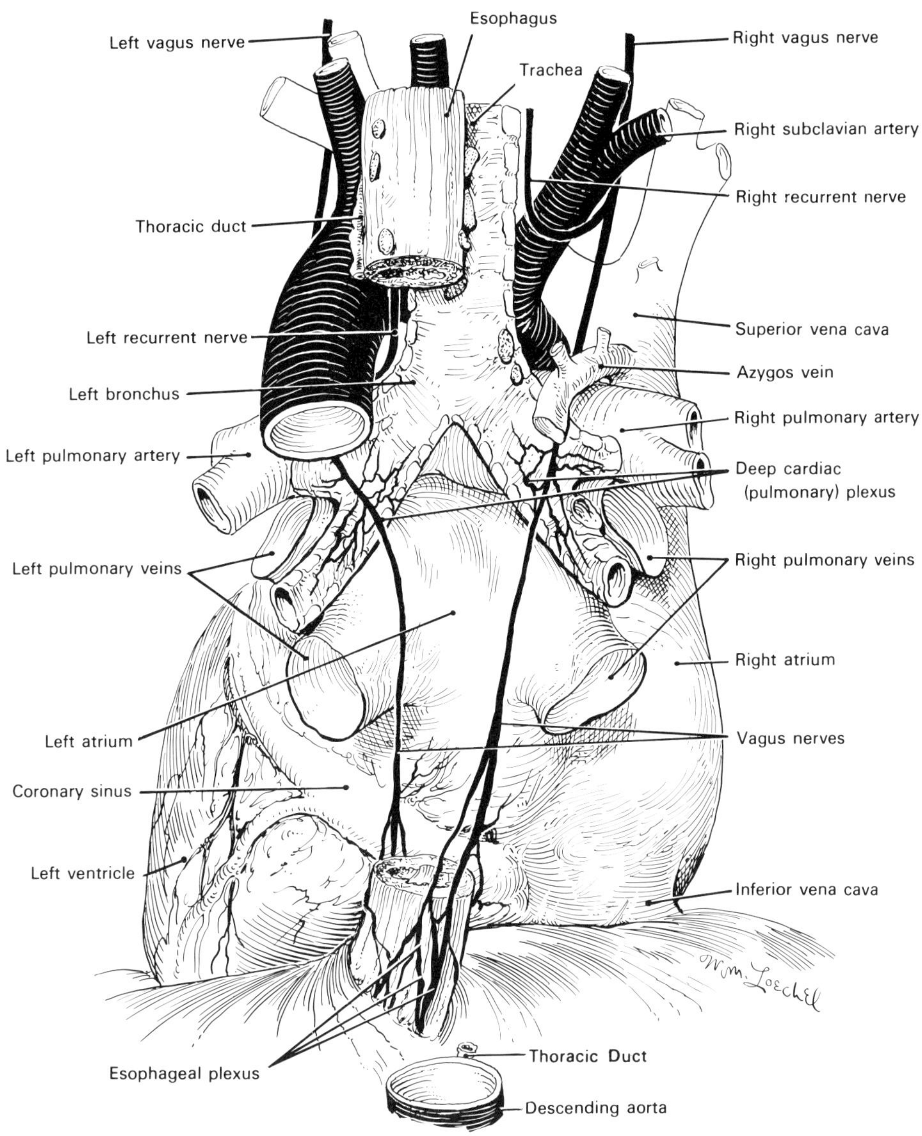

Fig. VIII-82. Structures in superior and posterior mediastinum as seen from behind.

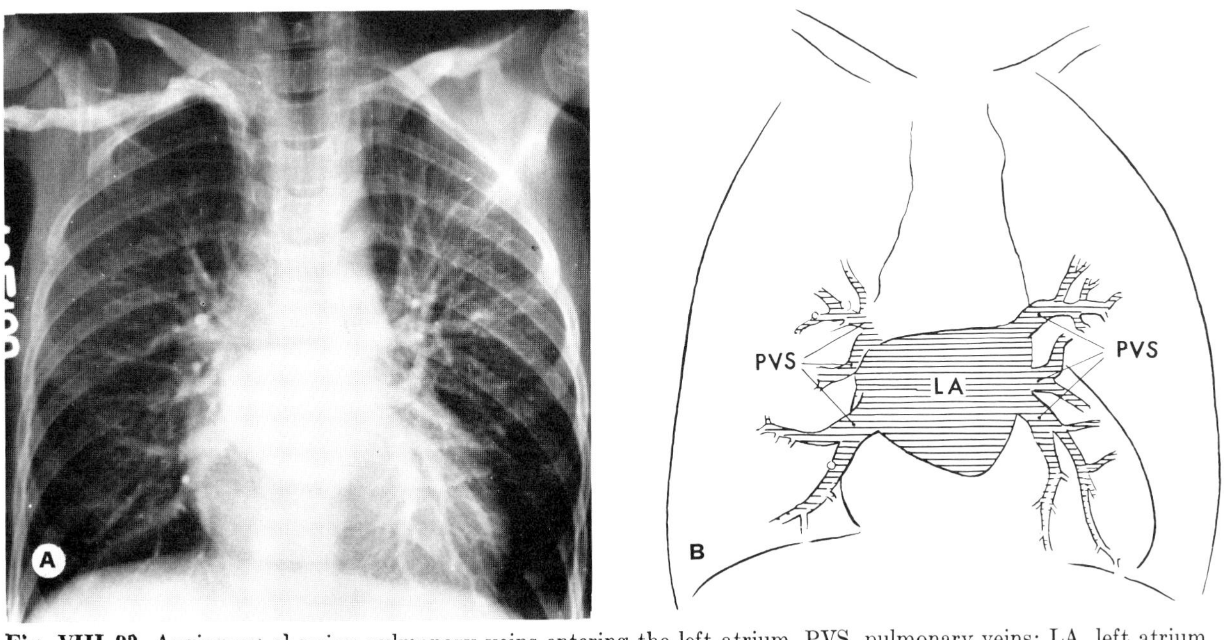

Fig. VIII-83. Angiogram showing pulmonary veins entering the left atrium. PVS, pulmonary veins; LA, left atrium. (From R. N. Cooley and M. H. Schreiber, *Radiology of the Heart and Great Vessels*, 2nd ed., Williams & Wilkins, Baltimore, 1967.)

(Fig. VIII-81) of the respiratory tree return blood to the azygos system of veins.

Pulmonary Shunts: It is well to realize that there are shunting pathways in the lung to return blood to the left heart without circulating it through the pulmonary capillary bed. Such arteriovenous shunts are both short and long depending upon the circuitous route that blood finds to by-pass the alveoli. Pulmonary shunting is not well understood, and may account for the feeling on the part of some investigators that only about 20% of the total surface area for aeration is utilized at any one time as blood is cycled through the lung. Many students also fail to realize that blood is shunted through the pulmonary system before birth, as well as during adult life. Indeed a considerable volume of blood, directed from right ventricle toward the ductus arteriosus, passes through the developing pulmonary system in the fetus even though the distal portions of the developing respiratory tree have not come into close relationship with the vessels in the connective tissue stroma as yet.

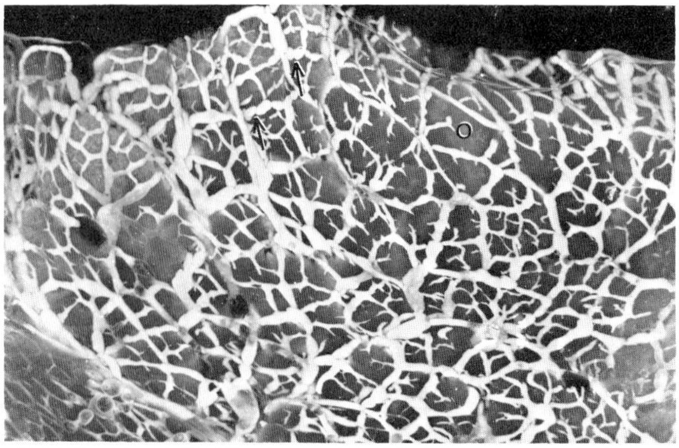

Fig. VIII-84. Well-developed pleural lymphatic vessel plexus, injected with a barium sulphate suspension in a human infant lung; Polyhedral network of large lymphatic vessels outlining more or less the pleural bases of the primary lung lobules; presence of many small side branches within them; occurrence of repeating and limited constrictions usually interpreted as valves (*arrow*) and club-like outpocketings (o) representing blind endings or points of origin. (Courtesy of J. Lauweryns and A. Baert.)

The Lymphatics of the Lung

(*This section is based upon the work and generous contribution of Dr. Joseph M. Lauweryns, Chairman of the Department of Pathology, University of Leuven, Leuven, Belgium.*)

The lymphatics of the lung are one of the most intriguing and challenging anatomical problems of the present day. This is largely because precise data on many aspects of the pulmonary microcirculation are unknown to date. As a general rule, lymphatics are localized in the connective tissue compartments of the lung. Two main sets of vessels can be distinguished: a **superficial** or **pleural** and a **deep** or **parenchymatous** (peribronchovascular) **lymphatic plexus.**

After injection with radiopaque substances or in corrosion casts, the *pleural lymphatics* can be seen to form ramifying, pleomorphic, surface networks composed of large lymphatic vessels which outline, more or less, the pleural bases of the primary lung lobules, those "respiratory units" we mentioned earlier. This is particularly striking in the infant lung (Fig. VIII-84). Within the boundaries outlined by these broader channels, many smaller side branches are present. Frequently, club-like outpocketings, representing blind endings, or perhaps, points of origin, are observed.

In the adult lung, however (Fig. VIII-85), the pleural networks are less numerous. Especially in the upper lobes, the meshwork is not as dense or as closely woven. The drainage of pleural lymph is still a matter of considerable controversy. Since Miller's studies in 1950[12] it has been customary to conceive of a transpleural route for the channeling of superficial lymphatic vessels toward hilar nodes. That is to say, lymphatic channels from costal, diaphragmatic, mediastinal, and interlobar surfaces of visceral pleura gravitate toward the free edges of the lung lobes to end

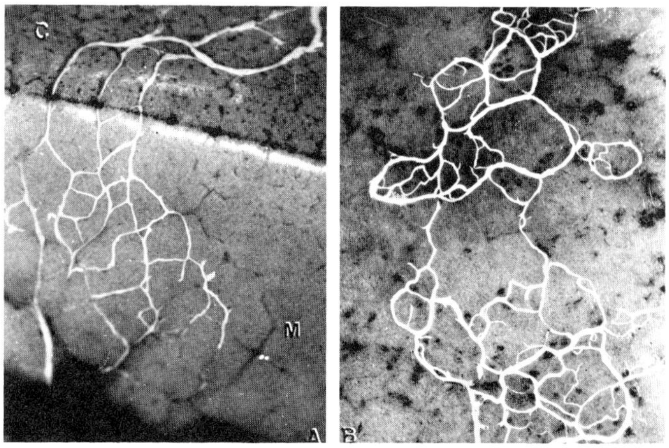

A B

Fig. VIII-85A. Transpleural drainage of pleural lymph in human adult lung; after injection of the pleural lymphatics of the costal surface of the lung (C), the injection mass does go over the free edge of the lung lobe (three vessels) towards the mediastinal surface (M) and the lung hilus. **B.** Classical picture of pleural lymphatics injected in an adult lung. (Courtesy of J. Lauweryns and A. Baert.)

in lymphatic vessels and nodes of the lung hilus (Fig. VIII-86). Recent studies[13-16] reveal that there are some of these connections between pleural and pulmonary lymphatics so that a transpulmonary drainage of pleural lymph does occur in some areas and in some instances. However, these connections in the adult lung are not widespread.

The *plexus of deep or parenchymatous lymphatic vessels* is located in the connective tissue which surrounds the bronchi and blood vessels of the lung. Thus, it might be possible to call these peribronchial, periarterial, perivenous, and septal or interlobular lymphatic channels. Such a distinction, however, is more or less an artificial one since, in many instances, no clear-cut anatomical separation can be made between peribronchial and periarterial lymphatics or, for that matter, between septal and perivenous lymphatics. It is interesting to note, however, that the anatomical location of the pulmonary lymphatics in the interstitial connective tissue spaces around blood vessels and airways corresponds exactly to where fluid first appears in experimentally induced pulmonary edema.[17]

There is still a question as to how deep the "deep" plexus lies. Are there or are there not alveolar lymphatics? Most investigators seem to agree that lymphatics exist at least at the level of respiratory bronchioles and even about alveolar ducts. Recent studies by Lauweryns[18-20] with both the light and electron microscope have *not* shown the presence of lymphatic capillaries at the *level of the air-blood barrier* or *in the interalveolar septa*. These last terms need to be defined rather precisely so that we all know the level at which these observations were made and what is meant by such terms.

1. **The air-blood barrier**[20] (*versus* pathway) consists of:
   a. The alveolar epithelial cell lining.
   b. The interstitium; *i.e.*, basement membrane(s) and any interposed connective tissue elements.
   c. The endothelial lining cells of the blood capillaries.
2. **The interalveolar septum**[20] is the structure between the lumina of two neighboring alveoli. It consists of:
   a. Both outer sides of the epithelial cell linings of adjacent alveoli.
   b. The supporting central framework of connective tissue between these epithelial layers, including the blood capillaries contained therein.

On the other hand, Lauweryns has found **juxta-alveolar** lymphatics, *i.e.*, lymphatic capillaries situated anatomically between *alveolar walls*[20] and the outer border of the interlobular, pleural, perivascular, or peribronchial connective tissue sheets. Here again, it becomes necessary to define an "alveolar wall" so as to distinguish it from an "interalveolar septum."

The *alveolar wall* is the boundary of an alveolus which lies in contact with the connective tissue of either:

a. An *interlobular* septum
b. The pleura
c. The bronchus (bronchiolus)
d. The pulmonary vessels

Such a wall would consist of an alveolar epithelial lining, its subjacent capillaries (when present) and their contiguous connective tissue supports. These last could merge insensibly with the connective tissue surrounds of a, b, c, or d. Thus, we talk here of a different component than the interalveolar septum (which may often be called by some the "interalveolar wall" and so make for a point of confusion).

Between such alveolar walls and the outer borders of the interlobular, pleural, peribronchial, and perivascular connective tissue, juxta-alveolar lymphatics can be demonstrated (Fig. VIII-86). Here they may be separated from the alveolar lumina by a connective tissue layer which can be thin in some areas and may not contain blood capillaries, and, of course, by the alveolar lining cell components. Without being a part of the interalveolar septa or the air-blood barrier, these "juxta-alveolar lymphatic capillaries" do bear a close topographical relationship to alveolar lumina. They may help explain the ready removal of intra-alveolar transudates, exudates, or inhaled materials. Thus, the deep pulmonary lymphatic vessels appear to be disposed in opportune niches.

The Nature of Lung Lymphatics as Seen in the Light Microscope (Fig. VIII-86): The system is composed of lymphatic capillaries and larger lymphatic vessels.[19] The lumina of lymphatic capillaries are relatively large, more or less spherical to oval, especially in the newborn, and they become more elongate and tortuous in the adult. They usually do not contain erythrocytes. The large, flattened endothelial cells rest upon a tenuous connective tissue rich with reticular fibers. Smooth muscle is absent. It is not always easy to differentiate lymphatic from blood capillaries. Characteristics of the latter which help in differentiation are: a lumen of small caliber lined by small epithelial cells whose nuclei are thick and bulging; a lesser concentration of reticular fibers and connective tissue alongside the endothelial tubes; and, of course, the presence of erythrocytes in the lumen where they occur in the tissue section.

The large lymphatic vessels have valves, and smooth muscle is observed occasionally. Veins of similar caliber, of course, lack valves.

Electron Microscopic Characteristics of Lung Lymphatics (Fig. VIII-87): Lauweryns *et al.*[20, 21] show the lymphatic capillaries to be lined with large, flat cells of uneven thickness. Thicker areas, which contain the nucleus or most of the organelles, alternate with thin, lateral, cytoplasmic extensions that have a minimal thickness of 170 Å. The cell outline is irregular since cytoplasmic projections extend both into the ves-

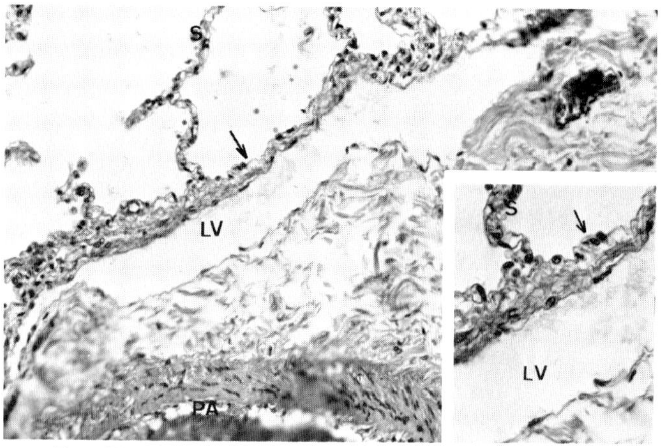

Fig. VIII-86. "Juxta-alveolar lymphatic" capillary (LV) between the outer border of the periarterial (PA) connective tissue and adjacent "alveolar wall" (arrow) of human adult lung. *Insert:* higher magnification of central area of Figure VIII-86; the "alveolar wall" consists mainly of an alveolar epithelial cell lining and connective tissue; it is not to be confused with the "interalveolar septum" (S). (Courtesy of J. Lauweryns.)

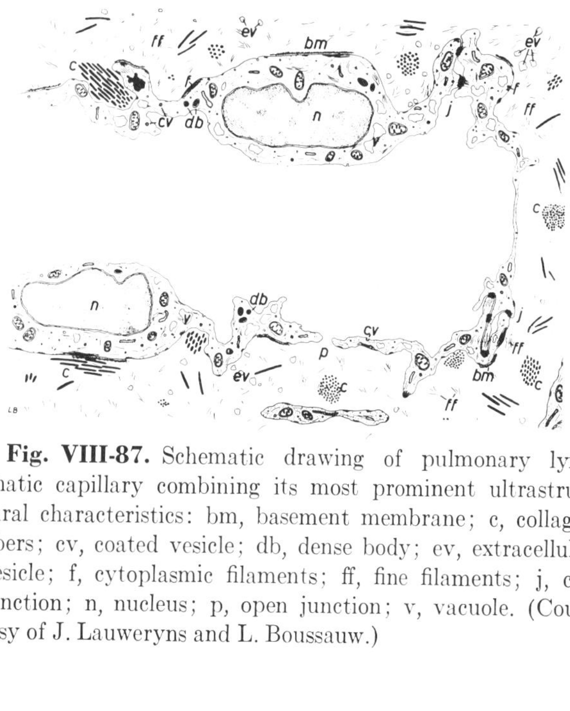

Fig. VIII-87. Schematic drawing of pulmonary lymphatic capillary combining its most prominent ultrastructural characteristics: bm, basement membrane; c, collagen fibers; cv, coated vesicle; db, dense body; ev, extracellular vesicle; f, cytoplasmic filaments; ff, fine filaments; j, cell junction; n, nucleus; p, open junction; v, vacuole. (Courtesy of J. Lauweryns and L. Boussauw.)

sel lumen and out into the neighboring perilymphatic connective tissue spaces. The mitochondria of lymphatics are usually small, and the rough endoplasmic reticulum is poorly developed. Occasionally, a Golgi apparatus, a centriole, multivesicular bodies or lipid droplets are seen. Other cytoplasmic features are round to oval "dense bodies" 70 to 300 $m\mu$ in diameter, "coated vesicles" 70 to 100 $m\mu$ in diameter, micropinocytotic vesicles, and many vacuoles.

As to cell junctions, some may be simple end-to-end unions but most over-lap or interdigitate. The usual attachment devices (zonulae adherentes, zonulae occludentes, maculae adherentes) are noted and occasional intercellular gaps occur. There may be intercellular gaps within these junctions. There are also occasional interruptions or pores along the endothelial cell lining itself.

The endothelium rests upon a thin, interrupted basement membrane, and this can be absent over long distances. The perilymphatic connective tissue contains collagenous fibers and finer filaments, connective tissue cells, and vesicles with a clear content. Frequently, there is close contact, on the one hand, between the collagenous fibers and the fine filaments and, on the other, between the endothelial cell and the basement membrane. These relationships suggest "anchoring mechanisms" which may prevent collapse of the thin-walled structures because of these outside traction forces.

Many of these ultrastructural features serve to differentiate the lymphatic capillaries from the blood capillaries. Their structure also suggests that they play a highly active role in fluid movement across lymphatic walls. Indeed, high permeability is suggested by the absence of a continuous basement membrane and by the occurrence of open junctions, pores, and intercellular gaps. The numerous micropinocytotic vesicles, coated vesicles, and vacuoles suggest active transport mechanisms as well and the possible participation of the endothelial cells in the formation of lymph.

The presence of **valves** in these lymphatic capillaries exists both at the level of the superficial and the deep pulmonary plexuses. Pulmonary lymphatic valves have a structure that is quite unique.[13, 22] A few may be bicuspid like lymphatic and venous valves encountered elsewhere. That is, they may be composed of semilunar cusps attached to opposite sides of vessel walls. But, graphic reconstruction from serial sections show that most of them are entirely different and peculiar to their locus in the lung. The majority of pulmonary lymphatic valves (Fig. VIII-88) are unicuspid, exhibiting the shape of a funnel or "wind tunnel." They are suspended obliquely in the vessel lumen, and are attached to one side of the inner wall of the lymphatic channel by a small triangular structure or septum. This anchoring piece projects into the lumen to connect the funnel neck to the wall of the vessel. These valves can be conceived as single, semilunar cusps attached in such a way that they almost cover the entire interior

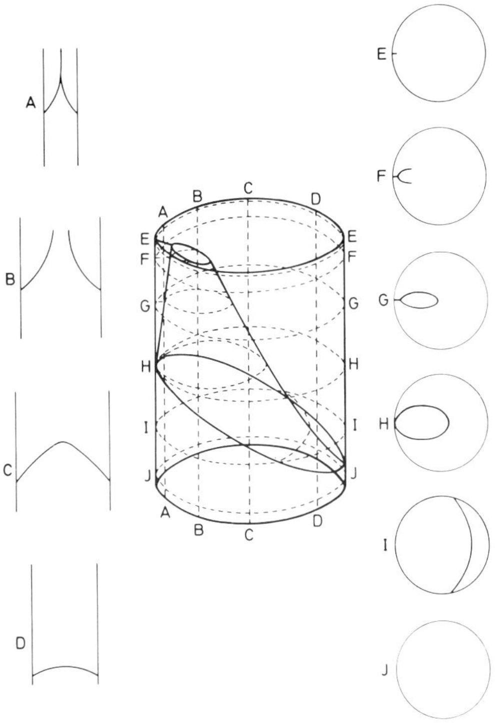

Fig. VIII-88. Schematic drawing (central area) of graphical reconstruction of pulmonary lymphatic valve; unicuspid valve with funnel-like architecture, attached to one side of the lymphatic vessel by a small triangular septum; schematic illustrations of the variable appearance on histological slides of such a valve in vertical section (to the left, A, B, C, D) or in horizontal section (to the right, E, F, G, H, I, J). (Courtesy of J. Lauweryns and L. Boussauw.)

circumference of the vessel and such that the two lateral borders of the cusp may, at times, touch each other on the opposite side of the vessel.

A valvular architecture of this nature is perfectly adapted to insure one way flow of lymph since the valve cannot be reserved or turned inside out. It would be occluded by a reverse direction of flow from that of the normal flow pattern. The valve is also well adapted to assist in lymphatic drainage under the influence of milking or pumping action which respiratory movements and the beating heart impose upon the interstitial connective tissue wherever such pulmonary lymphatic vessels exist.

Innervation of the Lung

In large part, the peripheral innervation is derived from the vagus nerve. As its fibers pass to and around the lung root they give branches to the anterior and posterior pulmonary plexus. Here preganglionic visceral motor fibers of the vagus mingle with postganglionic sympathetic fibers, especially with those from the first four thoracic chain ganglia. Visceral sensory fibers returning *via* the vagus are also found in these plexi. Some of the sympathetic and vagal fibers are motor to the smooth muscle of the respiratory tree. Other sympathetic fibers are motor for the regulation of vasconstriction of the pulmonary arteries, and it has been suggested that some of these may control precapillary sphincters. There are many visceral sensory fibers whose receptor endings are of a proprioceptive nature. These sensory end-organs are distributed in walls of the bronchial tree and, apparently, in the alveolar walls as well. Their reflex responses to inflation and deflation have been identified by physiological measurement even though few, if any, have been identified morphologically as yet. They play a role in reflex regulation of the respiratory centers of the brainstem as alveoli expand during normal respiration, and in altering the rate and sequence of discharge of the nerve cells of the respiratory centers during overinflation or excessive deflation of the lung.

Respiration and the Respiratory System

Among the vertebrates, respiration refers to the exchange of gases that take place between the organism and the atmosphere in which it lives, more specifically, to the intake of oxygen and the discharge of carbon dioxide. Three types of respiration are recognized: external, internal, and true respiration (there, seemingly, being no false kind). *External respiration* involves the passage of oxygen from the air into the alveoli of the lungs and, subsequently, into the blood (and the passage of CO_2 in the reverse direction). The exchange of gases between circulatory fluids (blood, tissue fluids, and lymph) and the cells is *internal respiration. True respiration* refers to the biological oxidation, or utilization of oxygen, by the cells which results in liberation of energy, water, and carbon dioxide.

While its main concern is with respiration, the system can and does carry out other functions such as the production of sound for speech, the elimination of volatile wastes, and, in eliminating water vapor, it helps to regulate body temperature.

Maintenance of life depends upon a continual supply of oxygen by aerobic or anaerobic means, and, to the human being, oxygen can virtually be equated with life. In the process of oxygen acquisition, carbon dioxide must be removed from the tissues continuously. Since the tissues are unable to store any appreciable quantity of oxygen, a cessation of respiration or an interference with the delivery of oxygen to tissues and organs results in malfunction, deterioration, disease, and death. While lack of oxygen is a primary consideration in almost every death, it is ironic that too much oxygen given under high pressures can produce a situation where the body cannot obtain oxygen. Like all medications, the administration of oxygen must be controlled carefully lest an overdose act as a poison. A certain amount of carbon dioxide, of course, is also essential for the maintenance of a respiratory rhythm. An overdose acts as a narcotic to depress and inhibit normal respiration.

External or "pulmonary" respiration takes place at

the level of the primary lung lobule or "respiratory unit." The latter consists of a respiratory bronchiole with a sprinkling of alveolar sacs along its wall and the alveolar ducts and sacs into which it subdivides (Fig. VIII-69). The alveolus is the final microscopic unit of this system, coming, as it does, into intimate contact with the vast capillary bed on the lung. The respiratory unit connects to the exterior through a series of air-conducting passageways, designated, from inside toward the outside, as terminal bronchioles, bronchioles of larger diameter, bronchi, trachea, larynx, the pharynx, and the oral and nasal cavities. (Fig. VIII-68).

Air enters the respiratory system through the nose. We rarely attend to the muscles of the nares which guard the openings into the nasal cavity since they are regarded as minor, unimportant structures in man. They can, however, contract rhythmically during the respiratory cycle, and they account for the flaring of the nostrils which one observes during periods of dyspnea.

The architecture of the nasal cavities in man was not designed so much to serve olfaction as to function as a preconditioning passageway for air going to and from the lungs. It was meant to be a filter that would warm and moisturize air as it trapped dust. When it was built, it was probably not anticipated that carbon particles, nicotine, and other products of tobacco combustion would by-pass it *via* the oral route. Whether the present campaign to close this latter detour will be successful or not remains to be seen. Man tends to be a creature who derives immense satisfaction from all of his vices, and, in this respect, is more spineless than his anatomy belies.

Air, smoke-laden or not, if it passes through the nose, takes the direct route, for the most part, through the inferior meatus to the posterior choanae and nasopharynx. Some portion may take the longer route through the middle meatus, but, rarely, except in sniffing, does much air meander through the superior meatus. But, at times, we are aware of some overpowering olfactory stimuli during quiet breathing, for certain oderiferous substances stimulate the olfactory receptors when they are present in the atmosphere even at very low concentration. The route, in inspiration, continues past the oral pharynx into laryngopharynx and, finally, into the trachea, bronchi, and lungs. In expiration, the most direct reverse route through the nose is *via* the middle meati. The route from nasal cavities traverses nasal, oral, and laryngeal cavities and must pass over the vocal cords before it enters the trachea.

The 12-cm.-long *trachea* tapers in diameter as it progresses distalward. Its connective tissue wall is stiffened by means of C-shaped rings of hyaline cartilage. Their open posterior ends are bridged by a tough membrane (ligament) and by transverse smooth muscle (Fig. VIII-67). The lining epithelium is pseudostratified and surfaced with cilia and mucus-secreting goblet cells. This tube is flexible, and the 20 or so cartilage rings separate when the neck is stretched. The lumen remains open except when crushed by external trauma or when it is obstructed by a foreign body—tumorous or swallowed.

The presence of a mucous film and cilia to move the mucus along is typical of much of the respiratory tract from nose to terminal bronchioles of 0.5 mm. diameter. Although mucous-secreting cells cease before the terminal bronchioles are reached, cilia are still present in these small-diameter air tubes. In the paranasal sinuses, cilia sweep mucus toward the ostia which open into the nasal cavities. Cilia in the nose beat toward the nasopharynx and the cilia there, toward the oropharynx in an upward sweep. In this way, particulate matter trapped by the mucous can be moved and expelled from the airways either by expectoration or by swallowing. Some dust particles reach the final respiratory units of the lung, nevertheless. Their removal is dependent upon the connective tissue phagocytes that dwell among the alveolar septal walls. In this day of smog and automobile exhaust, even the pink lungs of the infant city-dweller soon become blackened by the products engulfed by these phagocytic scavengers. The mobility of cilia is influenced by chemical changes in the blood and by substances applied to them locally. Nicotine has a noticeable effect in inhibiting ciliary action. Even though a neural control over their beating has not been demonstrated, it has been shown that cold depresses ciliary action while increases of temperature in the respiratory tree causes them to beat faster. An increase in the viscosity of mucus will decrease the rate and force of ciliary movement also.

Entrapped foreign material can be expelled in three ways: (1) by the ciliary action just discussed; (2) by an assisting peristaltic movement of the bronchioles; and (3) by the cough reflex. Coughing is caused by an irritation of respiratory passages beyond the level of the nose. Sensory receptors of the glossopharyngeal (IX) or vagus nerve (X) are stimulated by noxious agents. The sensory pathway is to centers in the medulla where motor neurons of the nucleus ambiguus reside whose axons activate branchial skeletal musculature of the pharynx and larynx. Other connections from this brainstem motor center descend to cord levels to bring the Intercostal muscles and the Diaphragm into play.

Sneezing is quite another matter. Here irritation of the sensory receptors of the trigeminal nerve are involved, and the sensory routing through the pons to the medulla takes a slightly different route than that for the cough reflex. Both reflexes initiate a deep inspiration and an abrupt approximation of the vocal cords. The latter remain closed until expiratory pressures build to a high level to force them open. The outward movement of air is, thus, pressurized, explosive and driving—sweeping the foreign material ahead of it like a cork riding on the tide.

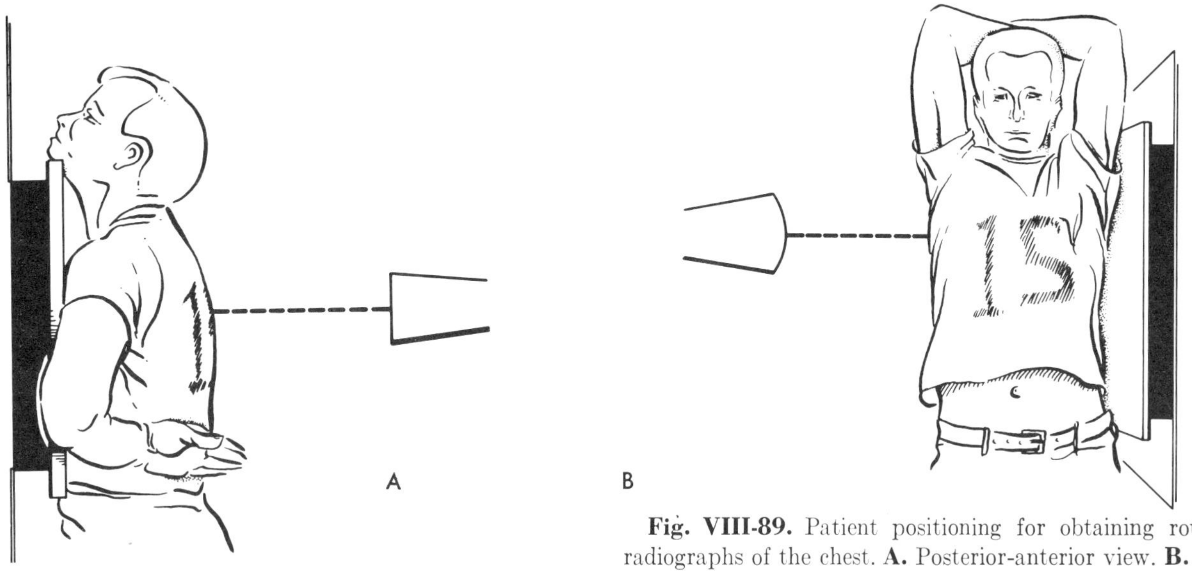

Fig. VIII-89. Patient positioning for obtaining routine radiographs of the chest. **A.** Posterior-anterior view. **B.** Left lateral view.

General Considerations on the Roentgen Anatomy of the Chest

Over the years our students have shown an inordinate interest in chest roentgenograms and their interpretation. Adequate interpretation of chest roentgenograms, however, requires a very thorough knowledge of the normal anatomy, of variations from the norm, and of the manifestations of disease upon those structures. Here, then, we will consider the normal structures that may be observed readily in posterior-anterior and lateral projections of the chest.

1. Positioning: The standard roentgen examination of the chest consists of PA and left lateral views. The left lateral projection places the heart close to the X-ray film and, thereby, reduces the degree of magnification of this important structure. A target to film distance of 6 to 10 feet is employed, and the patient is instructed to inspire deeply for the exposure.

Figure VIII-89A shows the patient positioned for the PA projection. The anterior aspect of the thorax is placed against the film cassette, the head is directed straight ahead, the hands are on the hips with the palms facing backward, and the shoulders have been rotated forward in order to move the scapulae off the upper lung fields. The horizontally-directed X-ray beam is centered on the thorax posteriorly and at right angles to the cassette.

The left lateral radiograph (Fig. VIII-89B) is made with the left side of the thorax against the film cassette. The arms are folded over the head and the horizontally-directed X-ray beam is centered on the right lateral aspect of the thorax, through which it will pass first on its way to the film cassette.

Figure VIII-90 illustrates normal PA and lateral radiographs of the chest.

2. Evaluating the Radiograph: Before commencing a methodical search of the radiographs for abnormalities, first it behooves the physician to evaluate the radiograph before him for:

a) rotation of the patient.
b) position of the patient
c) phase of respiration
d) the effect of b) and c) upon the Diaphragm.

If the patient is rotated from the true PA or lateral position, some distortion of normal structures will occur. Structures which are usually hidden from view may appear, and, of course, those commonly seen may become more pronounced or lost from view, thereby raising the possibility of abnormality. By choosing an anterior and a posterior structure in the sagittal plane as markers, slight degrees of rotation may easily be evaluated in the PA view. It is convenient to guide on the medial (sternal) end of the clavicles, anteriorly, and the spinous process of VT_3 or VT_4, posteriorly. Assuming that the patient is not rotated, then, the distance between the clavicle and the spinous process on each side should be the same (Fig. VIII-91A). A slight degree of rotation (Fig. VIII-91B), however, will result in the claviculospinous distance being increased on the side to which the patient is turned. In the example shown, the patient is rotated to her right side.

In a true lateral projection of the chest, the most posterior aspect of the ribs should overlap exactly. In Figure VIII-92, they do not overlap and, thus, the patient is slightly rotated.

Note that the ribs farthest from the film will, of

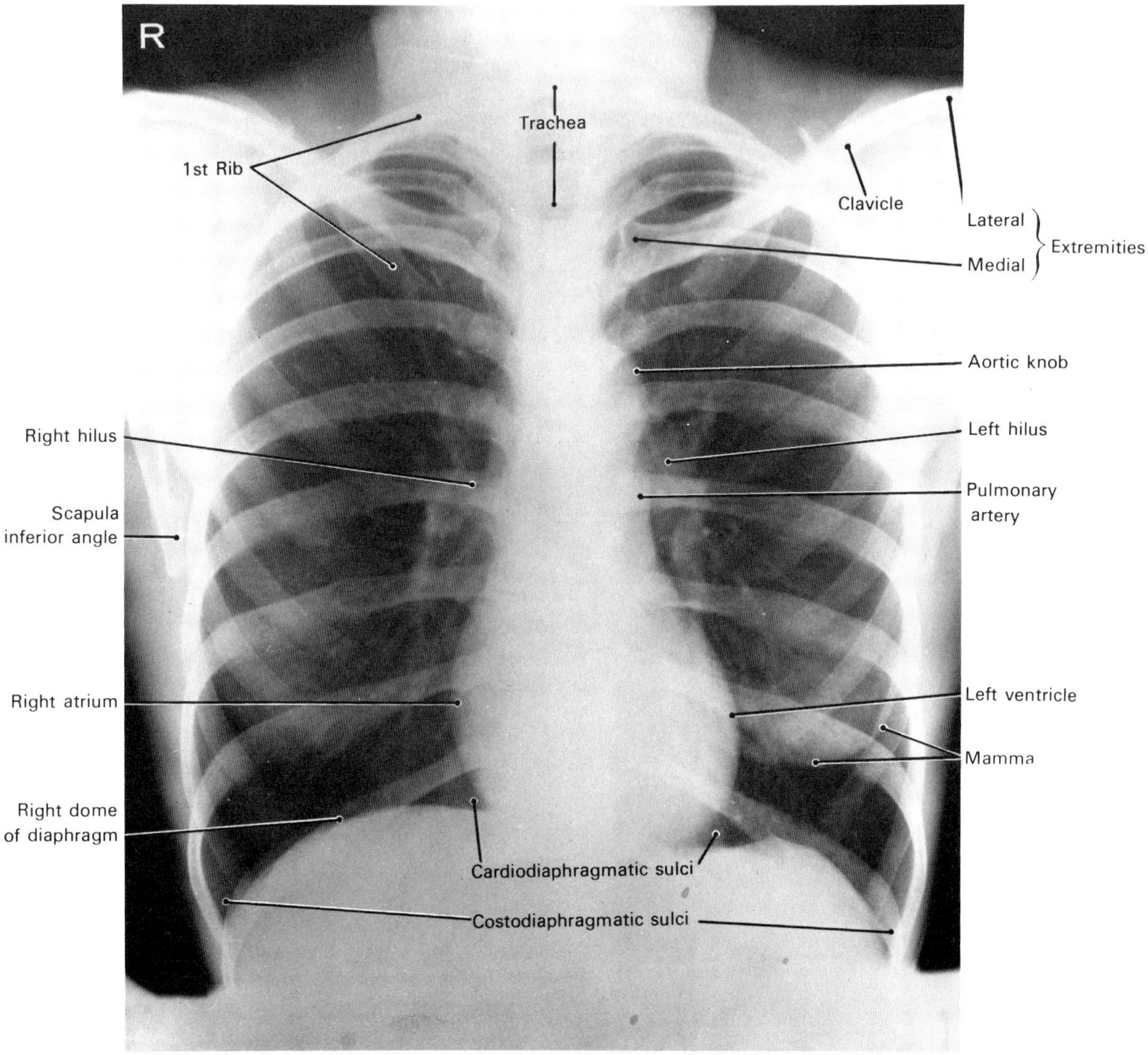

Fig. VIII-90. Normal chest radiographs. **A.** Posterior-anterior.

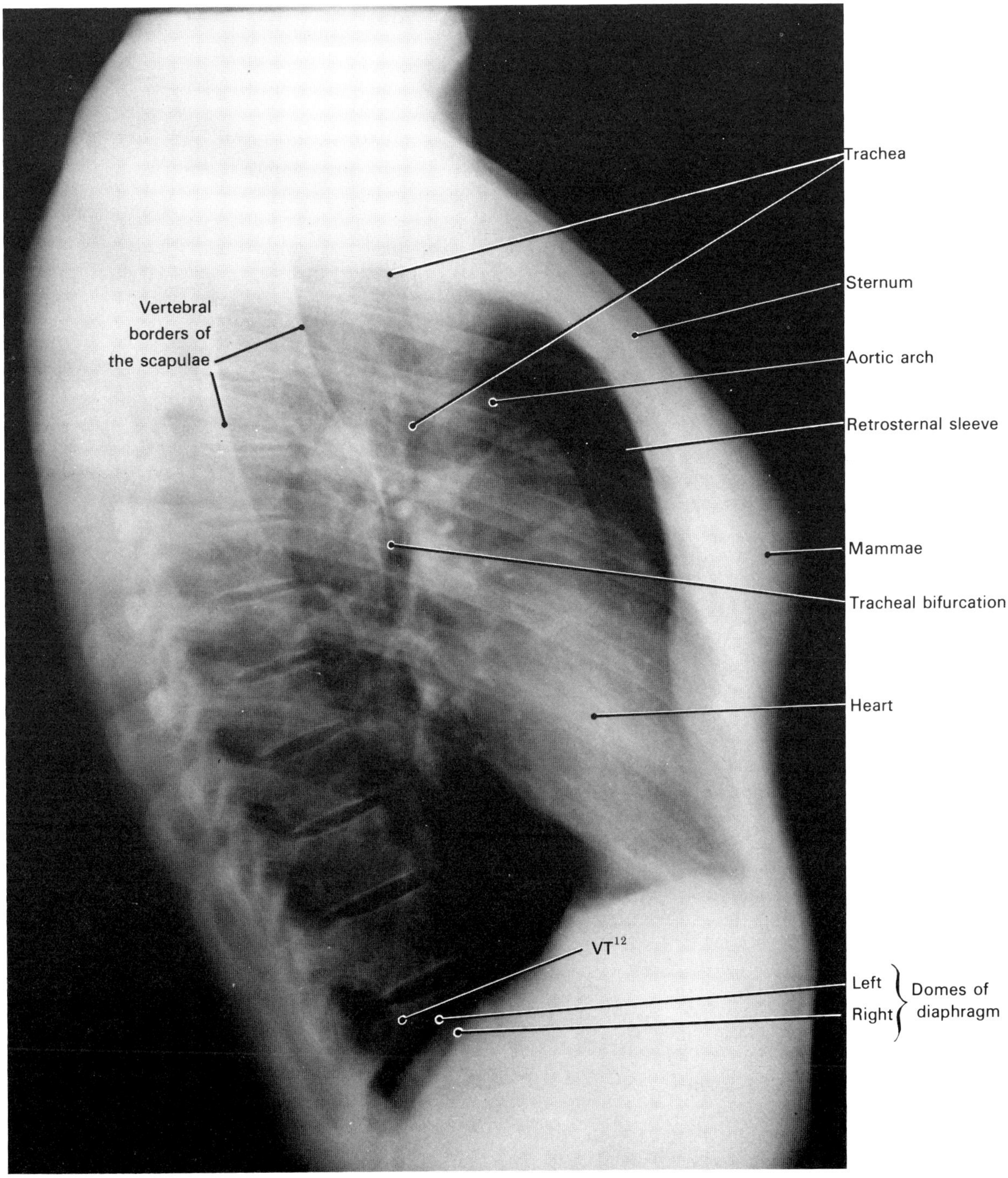

Fig. VIII-90B. Lateral chest radiograph.

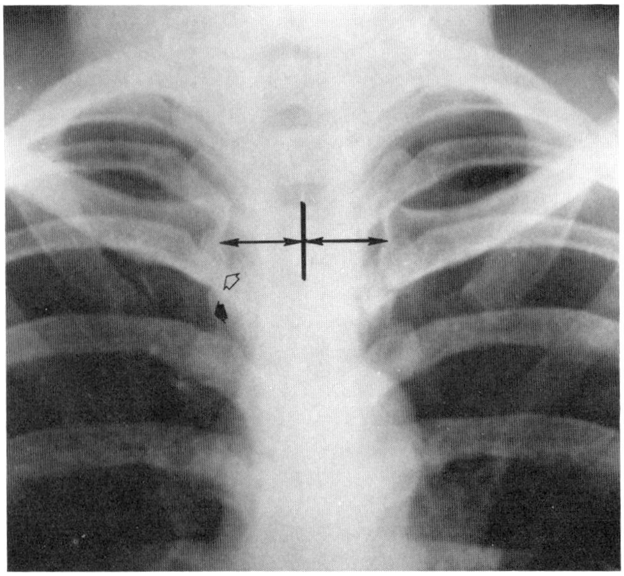

Fig. VIII-91. Evaluating rotation of the patient in frontal radiographs of the chest. Guide on the medial extremity of the clavicle (anterior) and the spinous process of VT_3 or VT_4 (posteriorly). **A.** No rotation.

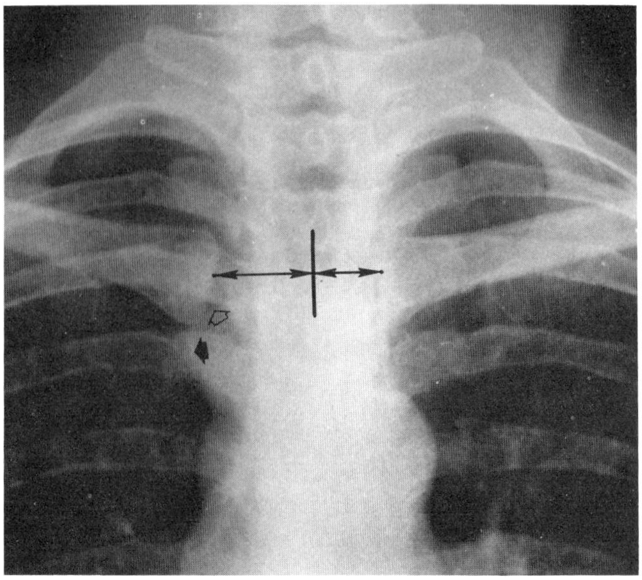

Fig. VIII-91B. Slight rotation of the patient to her right. Clavicular notch on manubrium (*open arrow*); facet on manubrium for the first rib costal cartilage (*black arrow*). Note the way the right lateral margin of the manubrium blends into the shadow of the superior vena cava (*small black arrows*).

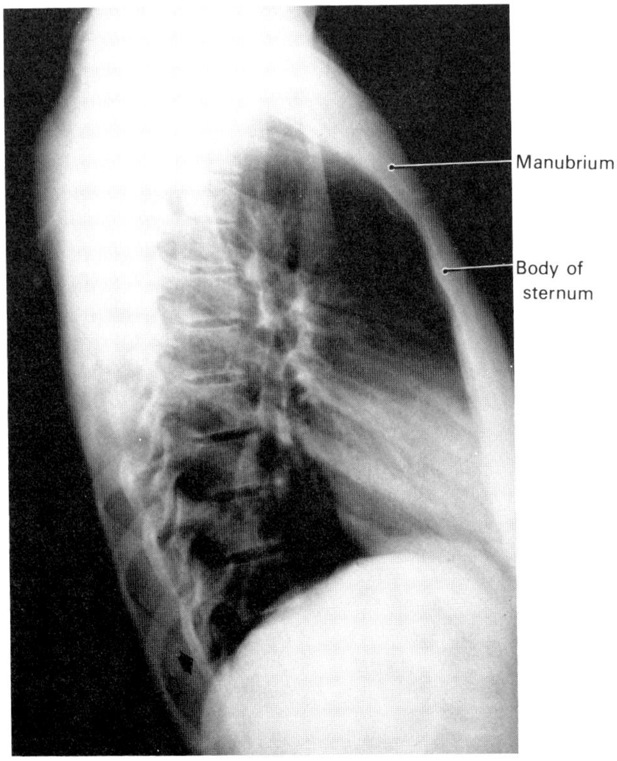

Fig. VIII-92. Slight rotation as seen in the lateral radiograph of the chest. Arrows are on the 11th rib posteriorly. The side closest to the film (*black arrow*) is much more sharply outlined than the side away from the film (*open arrow*).

course, be magnified slightly and they are not as sharply defined as those closer to the cassette.

The position of the patient (supine, erect, or decubitus) and the phase of respiration will influence the level of the Diaphragm and, thereby, the appearance of the heart and basilar lung markings. These factors have been discussed on page 430.

3. Systematic Approach to Evaluating the Chest: Earlier in Chapter IV (p. 182), we told you that most experienced radiologists do not use an orderly search pattern by systems in evaluating any radiograph. For purposes of description, however, it seems reasonable to use artistic license and follow a systematic approach this time. Thus, we will consider the following in order:

a) Soft tissues
b) Bones
c) Diaphragm
d) Trachea
e) Mediastinum
f) Hili and lungs
g) Heart

a) Soft tissues: The soft tissue structures of the chest wall, and those of the breast, cast shadows of varying density depending upon their composition and thickness. The pendulous breasts of the matron are denser and lower in position than the virginal or developing breast (Fig. VIII-90A). The nipples rarely cast a discernible shadow; however, at times they may be seen. When seen they are about 1 cm. in diameter and project over the lower lung fields. The left nipple is slightly lower in position than the right. Occasionally,

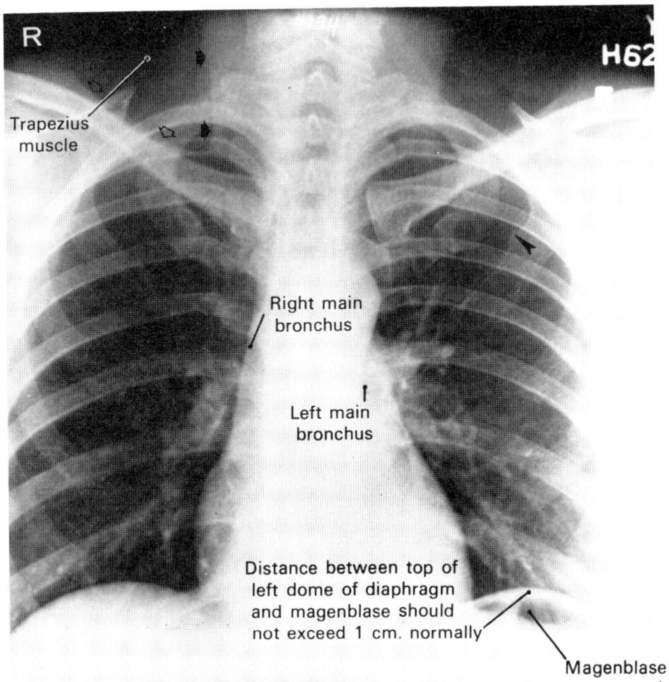

Fig. VIII-93. PA view of the chest. Attend to the vertically directed shadow cast by the Sternomastoid muscle and its overlying skin (*black arrows*). This blends into the companion shadow paralleling the superior border of the clavicle (*open arrows*) cast by overlying skin. A tubercle (*arrowhead*) on the anterolateral aspect of the second rib on the left is for the attachment of the second slip of the Serratus Anterior muscle. This tubercle may not always be discernible on routine radiographs of the chest.

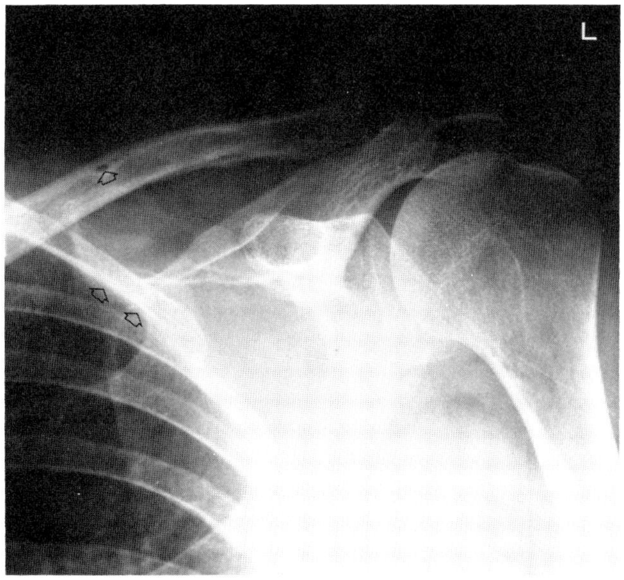

Fig. VIII-94. Frontal projection of the left shoulder showing the clavicular foramen (*arrow*) for the supraclavicular nerve. A second, smaller, foramen is suggested lateral to the one marked. Note the pleural reflection paralleling the inferior border of the second rib posterolaterally (*two arrows*).

the male breast may enlarge unilaterally or bilaterally (gynecomastia) or the male pectoral muscles may hypertrophy with excessive use and these, too, produce striking shadows across the chest wall.

Paralleling the superior margin of the clavicles is an accompanying shadow cast by the skin and subcutaneous tissues which pass over these bones (Fig. VIII-93). Medially, this companion shadow blends into the vertically-descending mass of the Sternomastoid muscle at the latter's attachment to the clavicle. The loss of the clavicular umbra, although not absolute, may be an indication of a subcutaneous, supraclavicular mass. Note also in Figure VIII-93 the density of the Trapezius muscle. High in the apices of the thorax, another companion shadow is seen paralleling the inferior border of the second rib posterolaterally. This is caused by the pleural reflection adjacent to the second rib (Fig. VIII-94).

In the aged patient or the infant, it is not uncommon to visualize folds of loose skin across the lung fields. These folds produce confusing linear densities which, at times, can simulate pneumothorax. Other bewildering shadows may be cast by moles, sebaceous cysts, tumors of the chest wall, braids of hair, and even some creamy, thickly-applied, axillary deodorants. Careful inspection of the patient's skin and the use of fluoroscopy will ferret out these culprits.

b) Bones: (1) The Spine—Chest radiographs are, of course, exposed primarily to evaluate the lungs and heart. Therefore, any information that can be gleaned about the bony structures is a bonus. In the PA projection, the lower cervical and the upper thoracic vertebrae can be seen. The other thoracic vertebrae are, for the most part, obscured by the overlapping sternum and mediastinum. The lateral chest radiograph, however, permits good visualization of the mid and lower thoracic vertebrae, although they are partially overlapped by the ribs. Details of the roentgen anatomy of the thoracic spine has been discussed on page 415.

(2) Clavicles (Figs. VIII-90A and 95).—The clavicles extend laterally from their sternal ends in a lazy "S" toward the acromioclavicular joints. The anterior aspect of the first rib hangs below the clavicular shadow. At times, a foramen or foramina (Fig. VIII-94) may be seen in the superior margin of the clavicle through which the supraclavicular nerve passes. The insertion of the costoclavicular ligament on the medial inferior surface of the clavicle is typically irregular or notched. This rhomboid notch (Fig. VIII-15) should not be mistaken for an osseous abnormality. Identify the sternal end of the clavicle in the lateral radiograph of Figure VIII-95.

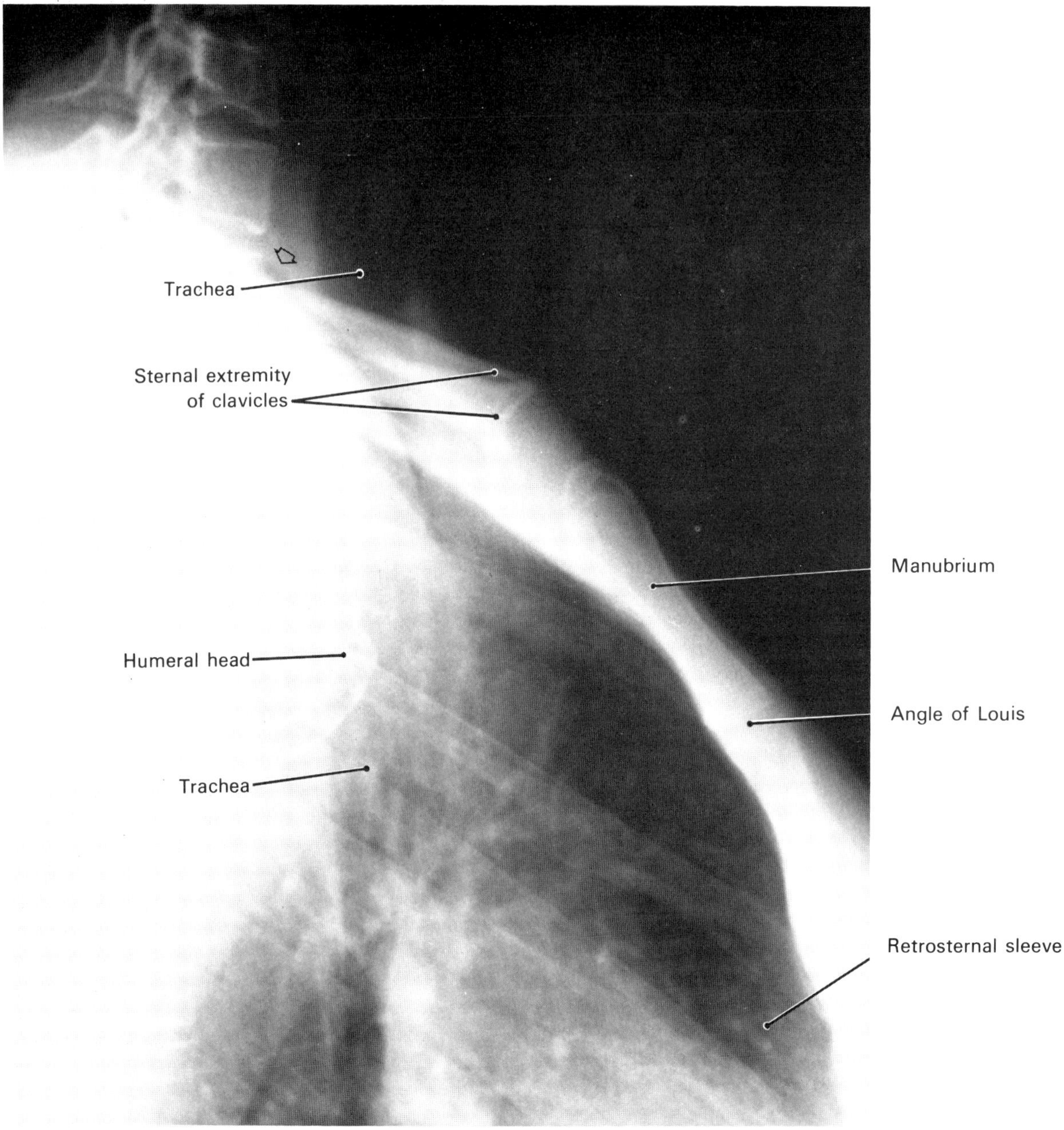

Fig. VIII-95. Lateral radiograph of the chest. For this particular radiograph, the patient's arms have been pulled back and the upper thorax has been thrown forward in a military attitude. The apex of the lung at the cupola is seen (*open arrow*).

(3) Sternum—Usually only the lateral margins of the manubrium are visualized in the PA chest roentgenogram (Fig. VIII-91). Note the flair of the manubrium near the sternoclavicular joints which should not be confused with vascular shadows of the mediastinum. In the lateral view, the three parts of the sternum are seen in profile (Figs. VIII-90B and 95). The angle of Louis serves as a useful reference point, for a line drawn horizontally and posteriorly from the angle intercepts the VT_4 and VT_5 intervertebral disk space.

(4) Shoulder girdles—Usually the vertebral borders of the scapulae, as well as the superior and inferior angles are discernible. In smaller patients, the spines of the scapulae, the glenoid cavities, and the humeral heads are seen. In the lateral projection, the vertebral borders of the scapulae are seen casting vertical densities just behind the trachea and these shadows are superimposed upon the thoracic spine.

(5) Ribs—The ribs stand out sharply against the background of the lucent lung fields. The lowermost ribs will tend to be obscured by the Diaphragm and by the upper abdominal contents, depending upon the phase

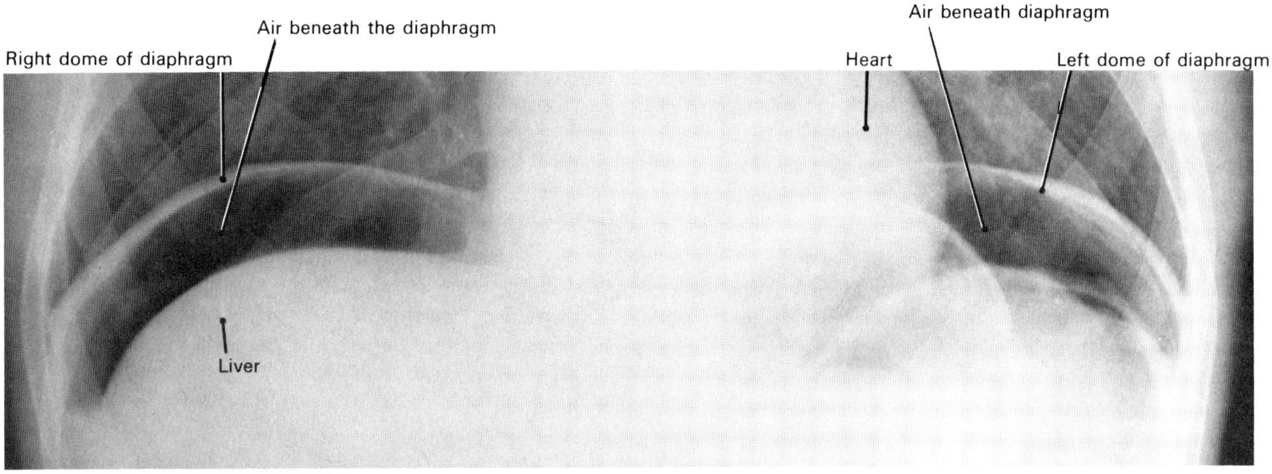

Fig. VIII-96. Pneumoperitoneum. Air has been purposely introduced into the peritoneal cavity to delineate the Diaphragm.

of respiration at the instant of filming. In the PA projection the ribs criss-cross the thorax in a symmetrical pattern. Recall that the bony anterior margins of the first 10 ribs do not reach the sternum directly, but rather they attach by radiolucent costal cartilages. Occasionally, the costal cartilages become calcified, and this calcification tends to increase in amount lower in the chest (Fig. VIII-13). Many individuals have striking calcification of the first costal cartilage only. This has, typically a serrated border inferiorly, and should not be mistaken for a parenchymal pulmonary lesion by the unwary physician. Always compare the ribs on each side of the thorax and there will be less chance of overlooking cervical ribs, absent ribs, or fused ribs.

c) *The Diaphragm:* In the PA projection of the chest (Fig. VIII-90A) the Diaphragm is seen as two domes, a right and a left, separated by the tendinous central portion which is obscured by the heart. Each dome of the Diaphragm arches convexly upward with the highest part of the arc on each side lying at about the middle of the lung fields. Note the position of the cardiodiaphragmatic and costodiaphragmatic sulci in Figure VIII-90A. The right dome of the Diaphragm is usually about one-half interspace higher on the right than the left.

We are, actually, not able to see the Diaphragm *per se* but rather only the superior surface of the highest point of the Diaphragm is seen in this projection as it stands out against the lucent lung above it. Because of the density of the subdiaphragmatic structures, we are unable to define any of the solid organs. It is, however, quite common to see some air within the fundus of the stomach (Fig. VIII-93) immediately beneath the left dome of the Diaphragm. If the distance between the superior margin of the Diaphragm and the top of the fundic air shadow exceeds 1 cm. then an abnormal collection of subpleural or subphrenic fluid should be suspected. On a very rare occasion, air-filled colon may be interposed between the right dome of the Diaphragm and the liver. This finding is usually asymptomatic.

In the lateral projection (Fig. VIII-90B), we may see both domes arching upward from the sternum, attaining their highest position over the posterior aspect of the heart and then arching downward across the vertebral column toward the posterior aspect of the lower ribs. At times, both domes may be superimposed, but usually they are separable. The student should be able to differentiate each dome. The right dome is clearly visible across the thorax, while the left dome becomes obscured anteriorly within the substance of the heart. An additional aid to identifying the left dome of the Diaphragm is, of course, the ever present collection of gas within the fundus of the stomach immediately beneath it.

If air is purposely introduced into the peritoneal cavity (diagnostic pneumoperitoneum), then the actual thickness of the Diaphragm may be seen, as in Figure VIII-96.

A *word of caution* seems justified at this point for those of you who will soon start to interpret chest radiographs of your patients. Do not make the mistake, that is so often done by medical students and beginning residents in radiology, of over-interpreting chest radiographs that have been made in expiration. Many of our students have fallen into this pitfall when they have been shown the radiograph, Figure VIII-97B. All types of pulmonary and cardiac abnormalities have been suggested before it is realized that the Diaphragm was elevated in expiration, and that all of the changes seen in the heart size and pulmonary parenchyma are due to that alone. Figure VIII-97A is the same patient's chest in inspiration; the latter film was made a minute before Figure VIII-97B.

d) *Trachea (Fig. VIII-90A, B):* The air-filled, tubular shadow of the trachea is readily discernible as it descends vertically through the neck into the thorax. The spinous processes of the lower cervical and upper

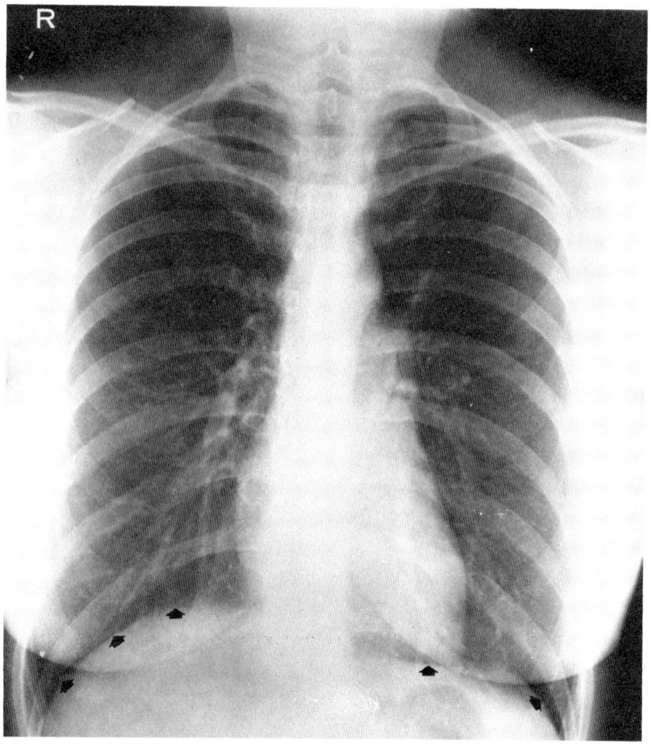

Fig. VIII-97. Phases of respiration as seen in frontal projection of the chest in a normal female patient. **A.** Inspiration—*arrows* outline the cranial surface of the Diaphragm against the adjacent air-containing lung.

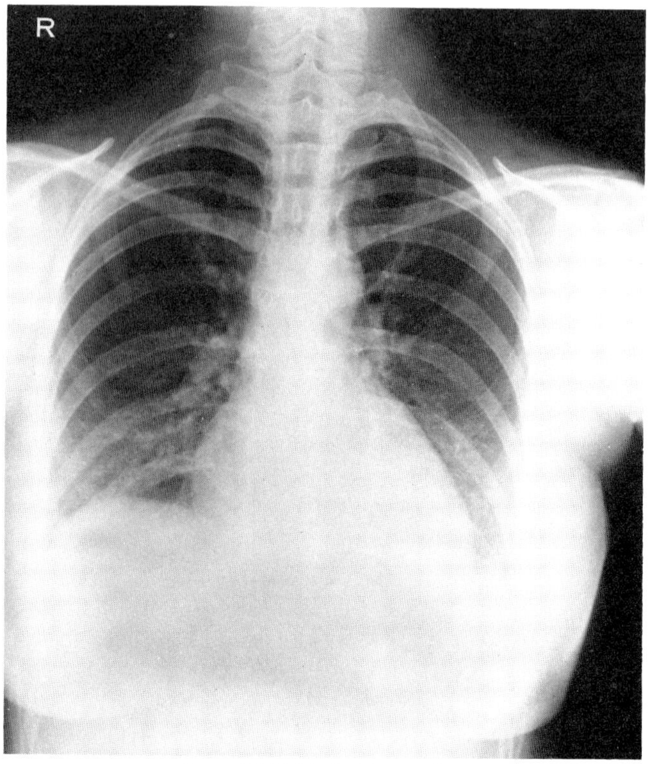

Fig. VIII-97B. Expiration.

thoracic vertebrae can be seen through the lucent air column in the PA view (Fig. VIII-90A). Assuming that the patient is not rotated in the PA projection, then the trachea should remain in its midline position until it reaches the level of the aortic arch where it deviates slightly to the right. This latter point is useful clinically, particularly when examining the chest roentgenograms of a patient where the arch of the aorta is not well visualized. The direction of tracheal displacement at the VT_5 to VT_6 level may tell you whether your patient has a right- or left-sided aortic arch.

In the adult, the trachea bifurcates into right and left main bronchi at the VT_6 or VT_7 level. Note that the right main bronchus descends at a sharper angle than the left main bronchus. Special techniques such as tomography or bronchography are necessary to visualize the secondary and tertiary bronchi.

e) Mediastinum: See page 437.

f) Hili and Lungs: Because of their density, the lungs are relatively lucent compared with other surrounding structures. The two lungs are equal in lucency at the same levels on each side. Both PA and lateral projections are needed to see the full extent of each lung. In the PA view, lung tissue is obscured below the level of the highest point of the domes of the Diaphragm, and, of course, in front of and behind the mediastinum. Thus, those areas obscured in PA projection become visible in the lateral view.

The hilus of each lung (Fig. VIII-90A) is composed primarily of the pulmonary arteries with little contribution from the pulmonary veins and the bronchopulmonary lymph nodes. Superimposed upon the lung fields are branching and rebranching linear densities, the so-called "lung markings." They are produced by the blood-filled pulmonary vessels and not by the intrapulmonary bronchi. The bronchi, which are air-filled, are surrounded by air-filled alveoli, and since their walls are extremely thin, they show no difference in absorption of the X-ray beam any more than do other lung tissues. The pulmonary vessels, on the other hand, will absorb more of the X-ray beam than the surrounding lung, and thereby cast a shadow on the radiograph. Intrapulmonary vessels are slightly larger in caliber in the lower lobes than those in the upper lobes. Within the medial third of each lung, it is common to see rounded densities which represent some of these vascular tubes end-on. These are differentiated readily from the larger bronchi which are also seen end-on occasionally. Cross-sections of bronchi have lucent centers and a thin wall.

The Heart

This large muscular blood vessel occupies a central position in the thorax as our anterior mediastinum (the classic middle mediastinum) which lies between the lung compartments. With it are associated the large conducting arteries and great veins. In close proximity are the parts of the respiratory system within the thorax which we have just discussed, the esophagus, portions of the lymphatic system, and the nerve trunks that either supply thoracic viscera, including the heart, or which are *en route* to and from the neck and abdomen. Packed as closely together as they are in the mediastinum, some of these structures still cast characteristic shadows in roentgenograms. Changes in form of their roentgen outlines may be indications of anomalous anatomy or functional disease.

Pericardium: Because the heart is a rhythmically-pulsating organ, wringing and squeezing as it contracts, it needs a degree of freedom to operate in these crowded quarters. Hence, it is enveloped in a fibroserous, friction-reducing sac, the **parietal pericardium.** A visceral pericardium invests the heart, and this tight coat is continuous at the roots of the great vessels with the parietal pericardium. The situation is similar to the continuity of visceral and parietal pleura at the hilum or root of the lung. Similarly, too, the space between visceral and parietal pericardium is only a potential space, so we find it filled with a modest amount of straw-colored, watery fluid secreted by the surfaces of the pericardium which face each other in the pericardial "cavity." The outer surface of the parietal pericardium is fibrous in contrast to its smooth, inner, mesothelial surface, and this outer layer rests directly upon the Diaphragm, fused to its central tendon and separated from the fundus of the stomach and liver below.

If in Figure VIII-98 you were to pass your right hand under and upward behind the heart as it lies in this pericardial sac, your fingers would end in a cul-de-sac. Your fingers are in the **oblique sinus,** a dead-end pocket. It is an inverted "U," bounded below and on the right by the inferior vena cava and right pulmonary veins, and above and to the left by the left pulmonary veins. Deep to your fingers and to the pericardium upon which they rest in this pocket, separated from the heart by the pericardium which you touch, run the esophagus and the descending aorta with the vagus nerve between them, the aorta to the left of the esophagus and somewhat obliquely behind it.

There is a second sinus related to the heart within its pericardial sac, the **transverse sinus.** If, at the upper end of the pericardial sac, you pass the index finger of your left hand *in front* of the superior vena cava and push it to the left *behind* the aorta and pulmonary artery, it will be in the transverse sinus. It will emerge within the pericardial sac on the left side so that you can reach over to the left with your thumb and touch the tip of your index finger. You have the arterial root of the heart (aorta and pulmonary artery) between thumb and finger. It is easy to see that a ligature could be passed around these vessels *via* the transverse sinus. Notice that you can't pass your finger *between* aorta and pulmonary artery, only *behind* them both, for they are enclosed in a common sheath of visceral pericardium. Notice, too, that the fingers of your right hand in the oblique sinus, do not come in direct contact with your index finger in the transverse sinus even though the index finger can be "felt" there. The fingers of each hand are separated from each other by two layers of serous pericardium which surround the veins entering the left atrium. We have pointed out some significance for the transverse sinus, but there is little of importance we can say for the oblique sinus other than that it is an anatomical cul-de-sac which offers a momentary bit of exploratory entertainment. On the other hand, the geography of its boundaries *are* significant if one wishes to understand the anatomy of the heart *in situ.*

Development of the Heart: The first sign of heart formation in the human is at the end of the third week of gestation.[23] As we learned in Chapter I, at that time the embryonic disk is spread out over the yolk sac. Heart formation is taking place well cephalad in the disk between the splanchnic mesoderm and the entoderm of the yolk sac. Here groups of heart-forming cells appear in clusters and cords. They already exist as solid strands arranged in U-shaped fashion, anteriorly in front of the neural plate and down each side of the embryo, before the first somites appear. As the head end of the embryo lifts off the yolk sac in forward growth, and as the embryo pinches itself off the yolk sac by folding laterally, these cords of cells approach each other ventrally. Some of them form endothelial tubes with a lumen. With continued lateral folding of the body wall, the two, longitudinal, **endocardial tubes** are brought close to each other in the ventral midline where they begin to fuse with each other in a cephalocaudal direction (Fig. VIII-99A–C). Meanwhile, splanchnic mesoderm, which surrounds the fused endothelial tube, consolidates as a **myoepicardial layer,** the inner portion of which is more gelatinous (cardiac jelly) than the outer (Fig. VIII-99D). It is from the loose inner portion that valves of the heart will form. The outer parts will become the cardiac muscle proper and the epicardium (visceral pericardium). By this time, the body wall, in its lateral pinching-in, has folded a part of intra-embryonic coelom around the longitudinal heart tube as a pericardial cavity. The heart is suspended in this cavity, at first, by both a ventral and a dorsal mesentery. The ventral one is very short-lived, but the dorsal one persists until the embryo reaches the 16-somite stage of development when it, too, disappears. This leaves the pericardium attached only at the arterial and venous ends of the heart tube where blood vessels enter and leave. These relationships are shown in Figure VIII-100.

Since the boundaries of the pericardial cavity are fixed, the growing heart, in order to lengthen, must fold upon itself within the cavity, bringing arterial and venous ends close to each other. This is accomplished by the venous end slipping posteriorly and migrating cephalad. The contortions it undergoes in this bending are illustrated in Figure VIII-101.

Almost from the first, the following divisions can be recognized in the longitudinal heart tube. From caudad cephalad: (1) the **sinus venosus,** a small part with thin walls where the veins enter; (2) the **atrium** just beyond, also thin walled; (3) the **ventricle** which is larger and thick walled; and (4) the **bulbus cordis,** a

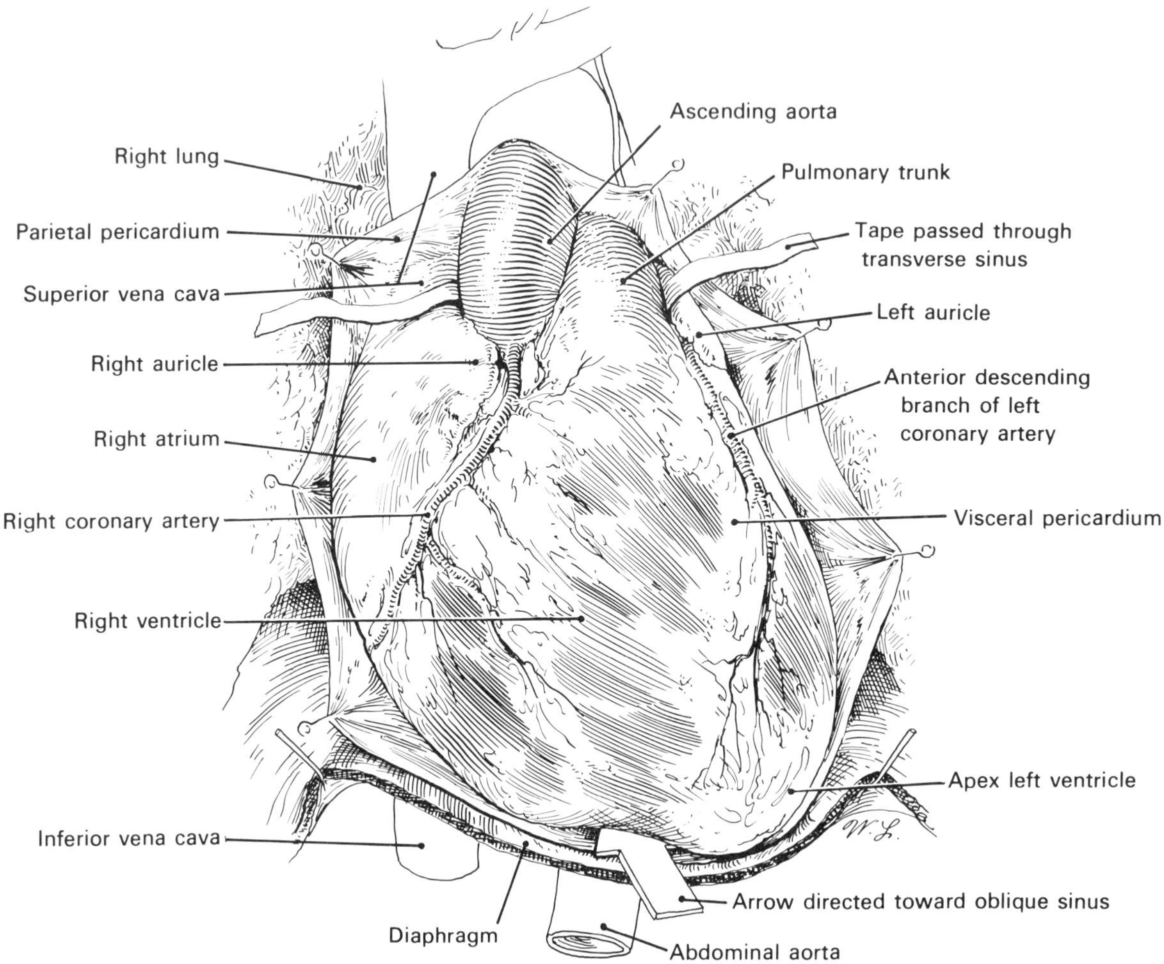

Fig. VIII-98. The heart and pericardium as viewed *in situ.*

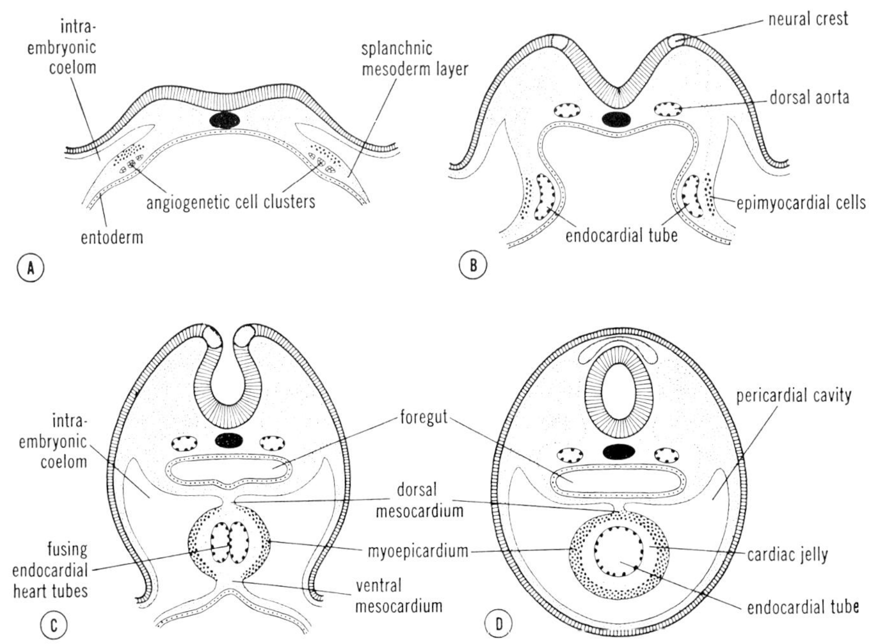

Fig. VIII-99. Schema of fusion of the paired heart tubes. The cross-sections depicted represent embryos of different developmental ages. **A.** Presomite stage at 17 days. **B.** Presomite stage at 18 days. **C.** Four-somite embryo, 21 days. **D.** Eight-somite embryo, 22 days. (From J. Langman, *Medical Embryology,* Williams & Wilkins, Baltimore, 1963.)

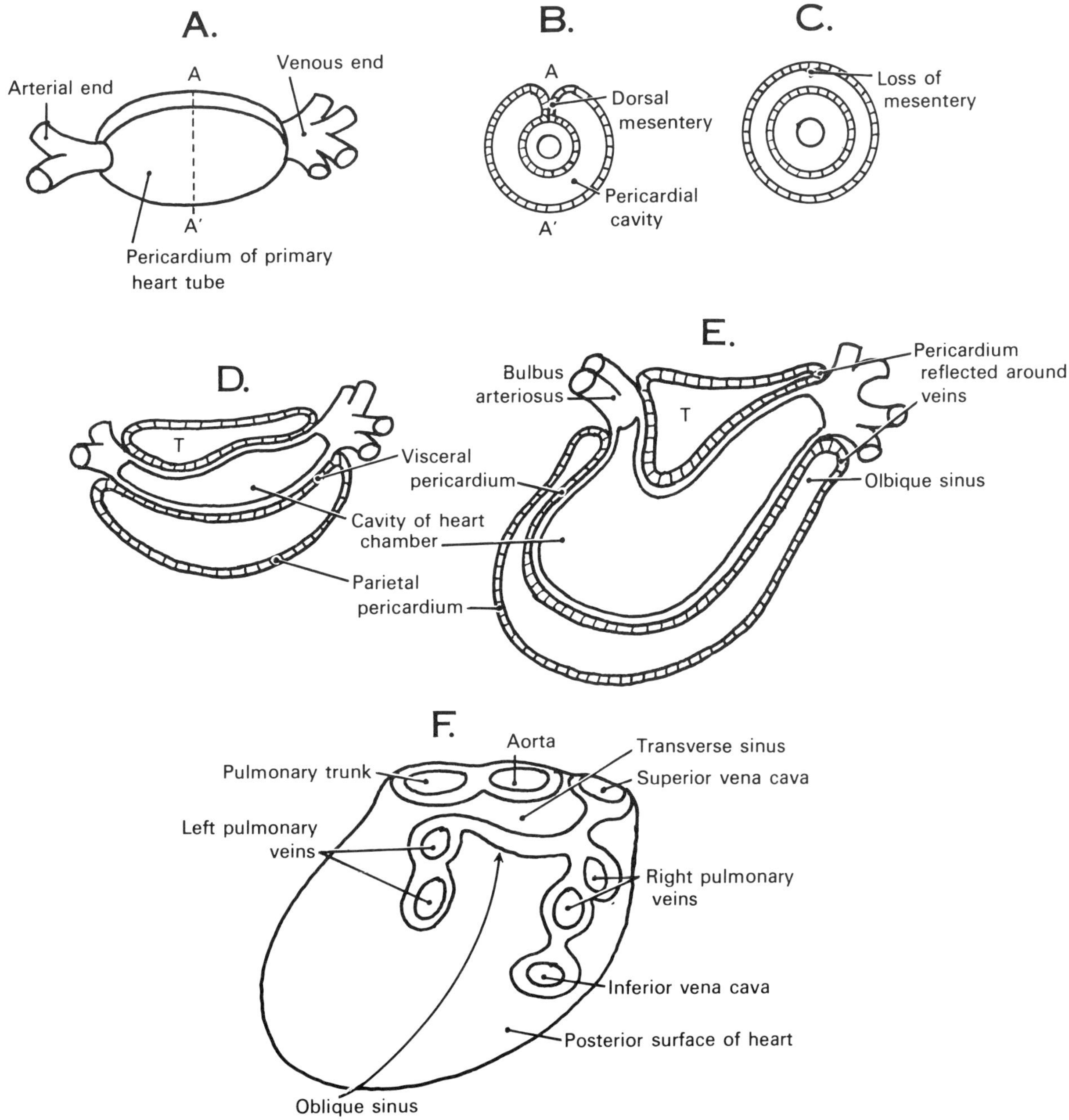

Fig. VIII-100. Schema to illustrate the bending of a longitudinal heart tube within its developing pericardium (pericardial spaces greatly exaggerated). A, arterial and venous ends of the heart tube within its pericardium. B and C, cross sections through a–a′ of A showing gradual loss of a dorsal mesentery and the freeing of the heart within its pericardium. D, early tube flexure, "T" creating a transverse pericardial sinus between venous and arterial ends. E, the approximation and growth of arterial and venous ends; the widening and absorption of veins in the wall of the left atrium; and the reflection of pericardium about the veins in such a way as to create an oblique sinus. F, schema of adult relationships on posterior aspect of heart (visceral pericardium only) to show final locus of transverse and oblique sinuses. (Modified from Williams and Wendell-Smith.)

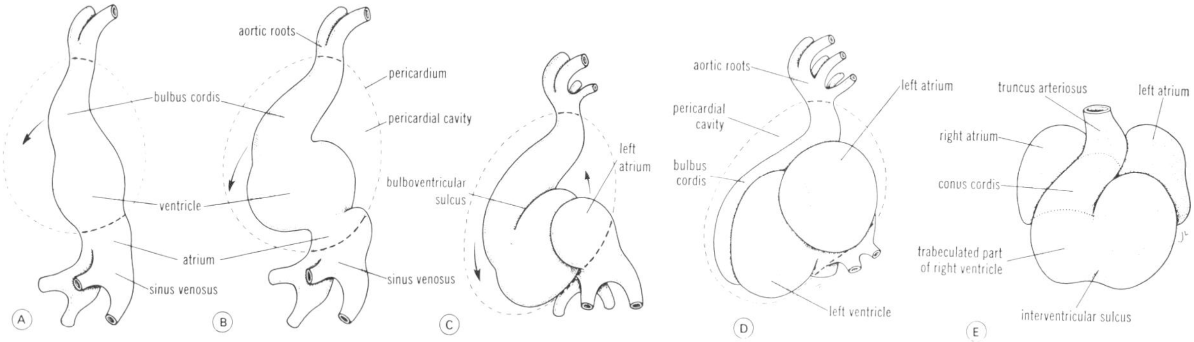

Fig. VIII-101. Formation of the cardiac loop and changing configuration of the heart tube. **A.** In the 8-somite embryo. **B.** In the 11-somite embryo. **C.** In the 16-somite embryo. **D.** At 5 mm. or 28 days, seen from the right side. **E.** Same as D, seen from front. (From J. Langman, *Medical Embryology*, Williams & Wilkins, Baltimore, 1963.)

tapering part of the ventricle which connects in the early embryo with the aortic arches (Fig. VIII-101A). In bending, the tube flexes abruptly at two places, taking on the shape of an "N," approximately. The sinus venosus and the atrium remain in the same caudal position but shift to the midline; the ventricle moves ventrad, caudad, and to the right (its own right); and the bulbus extends cephalad and slightly to the left. The atrium begins to ascend posterior to the ventricle, dragging the sinus venosus with it. The latter gradually moves up posterior to the atrium (Fig. VIII-101B–D). The configuration approaches the adult pattern when the atrium expands into two good-sized pouches, one on either side of the bulbus cordis, the future right and left atria. At the same time, the ventricles expand in size and their muscular walls thicken. An external groove indicates a division into **right** and **left ventricles** with the bulbus connecting with the right side (Fig. VIII-101E).

The bulbus cordis is the part formed by the ascending limb of the cardiac loop and it continues into the roots of the aortic arches. In the 5-mm. embryo of 4½ weeks, three parts can be recognized. The proximal part becomes incorporated in the trabeculated part of the right ventricle. A midsection becomes the **conus cordis** (later the conus arteriosus or infundibulum). An upper portion, the **truncus arteriosus,** will eventually split into pulmonary trunk and aorta. The conus section is designated "cordis" at this stage of development because both ventricular chambers use it as a common outflow tract. The sinus venosus will not remain as a separate chamber in the adult heart but will become incorporated into the right atrium.

In the partitioning of the heart into right and left sides, what actually happens is that a longitudinal septum is formed which runs from end-to-end, through bulbus, ventricle, and atrium. Because this division takes place in the bent heart tube, the process gets difficult to describe and the student often gets confused. We have left the heart in its original longitudinal form in Figure VIII-102 so that you can see, roughly, what has happened.

The original sinus venosus consists of a **transverse part** and a **right** and **left sinus horn.** The transverse part connects with the atrium, and each sinus horn receives three veins in a symmetrical manner. These are the *right* and *left:* (a) **common cardinal;** (b) **umbilical;** and (c) **vitelline** veins (Fig. VIII-102A). The common cardinals receive blood from the anterior and posterior cardinal veins which drain the cephalic and caudal parts of the embryo, respectively. The vitelline veins drain the yolk sac. The umbilical veins are coming from the placenta bringing oxygenated blood to the developing embryo. The right umbilical vein disappears early as the liver forms its portal system. As placental circulation increases and an inferior vena cava develops from a number of veins in the embryo, the left umbilical vein forms a direct connection within the liver substance with the inferior vena cava. It will then lose its connection with the sinus venosus also. The vitelline veins also get interrupted by developing liver cords. Their distal parts get involved in the developing portal system of the liver, while their proximal parts remain attached to the sinus venosus. But, as the left sinus horn regresses, the left vitelline vein also disappears. The proximal stump of the right vitelline becomes the suprahepatic part of the inferior vena cava, the part of that great vessel that pierces the liver to enter the right heart. In this way the sinus venosus loses three of its six veins (the shaded ones in Fig. VIII-102A).

As the walls of the common ventricle and the common atrium thicken, a septum forms to divide the heart into four chambers. The septum develops as several separate parts which gradually become connected end-to-end. In the atrium, two incomplete septa form,

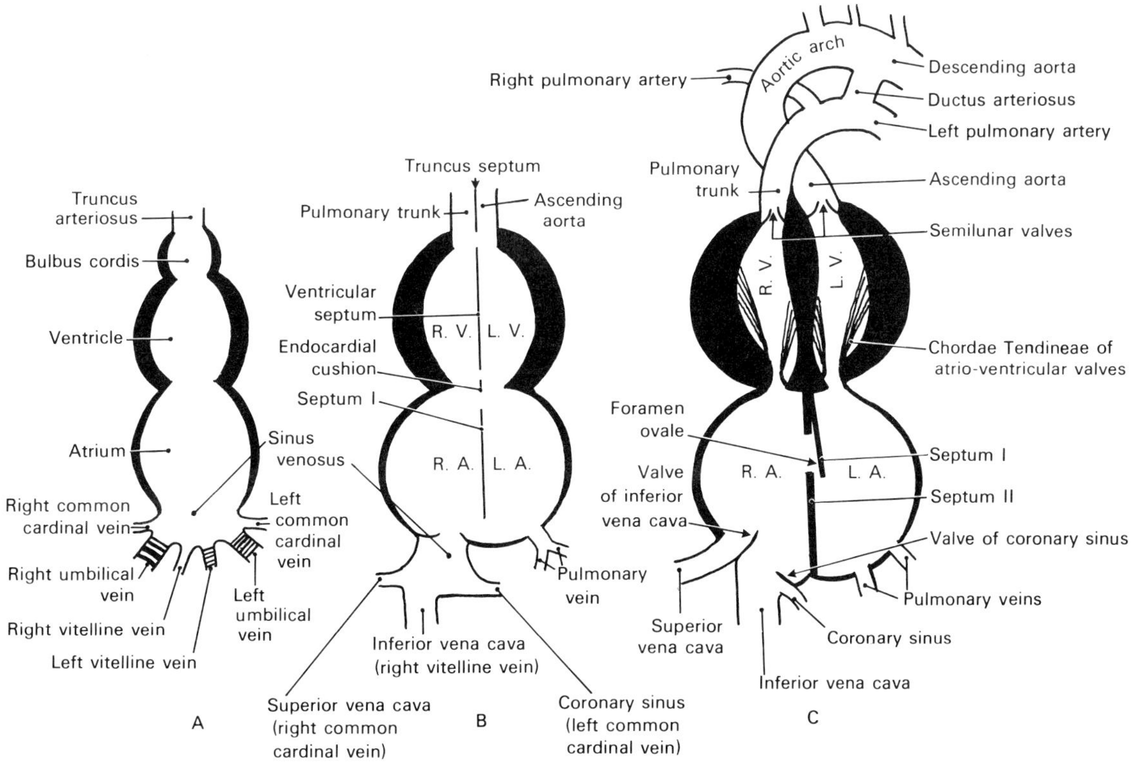

Fig. VIII-102. Changes in the developing heart viewed as if it remained a straight tube (A-C). **A.** Early 2-chambered heart. The shaded veins degenerate and are not shown in B. **B.** Beginning of a septal partitioning; shifting of sinus venosus to the right; degeneration of certain veins and appearance of new pulmonary veins. **C.** Completion of partitioning; absorption of sinus venosus into right atrium; division of bulbus into aortic and pulmonary trunks; the formation of a foramen ovale, ductus arteriosus, and the valves of the heart and great vessels. R.C.C. and L.C.C., Right and left common cardinal veins; R.U.V. and L.U.V., Right and left umbilical veins; R.V.V. and L.V.V., Right and left vitelline veins; R.V. and L.V., Right and ventricles; R.A. and L.A., Right and left atria.

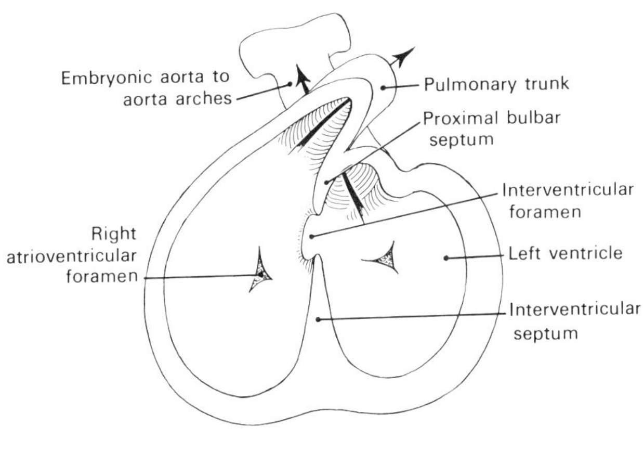

Fig. VIII-102D. A view of C above in the normal position of the heart with a coronal section through the ventricles.

and they both lie to the left of the sinus venosus. The **primary septum (septum I,** Fig. VIII-102B, C) grows out across the atrial cavity from its dorsocephalic wall toward the atrioventricular canal. Before it crosses completely, a good sized hole forms in it. A **second septum (septum II)** arises in a similar manner slightly to the right of septum I. It never becomes complete and its opening is known as the **foramen ovale.** The openings in the two overlapping septa allow blood to flow from right atrium to left but not in a reverse direction, because septum I acts as a valve—the **valve of the foramen ovale.** As the atria contract, septum I is pushed to the right to cover the defect in septum II, preventing back flow into the right atrium. In atrial relaxation, the directed filling stream of the inferior vena cava forces septum I to the left, reopening the foramen ovale.

The **superior vena cava** is the remains of the right common cardinal vein. The proximal part of the right vitelline vein becomes the upper segment of the inferior vena cava. As these vessels form on the right side, all of the blood from the placenta (since the surviving left umbilical vein now empties into the inferior vena cava), and most of the blood from the body of the embryo, enters through the right sinus venosus. The left

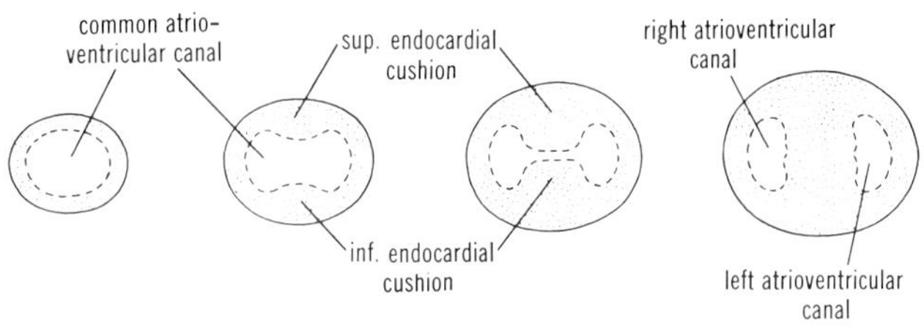

Fig. VIII-103. Role of endocardial cushions in forming the atrioventricular canal. **A–D.** Cross-sectional views at 4-mm., 6-mm., 9-mm., and 12-mm. stages, respectively. (From J. Langman, *Medical Embryology*, Williams & Wilkins, Baltimore, 1963.)

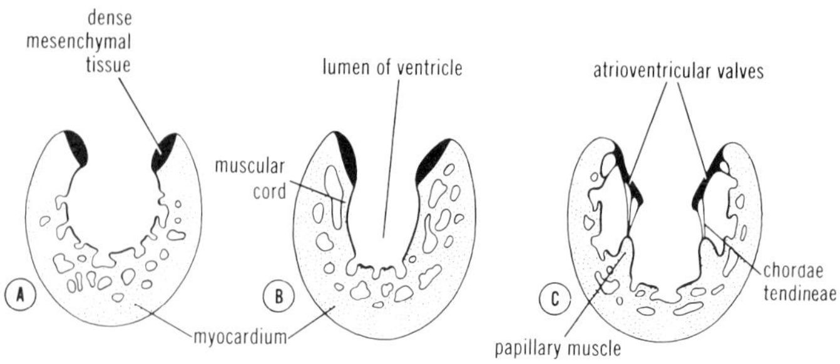

Fig. VIII-104. Formation of the A-V valves and their chordae tendineae as the ventricular wall hollows out. (From J. Langman, *Medical Embryology*, Williams & Wilkins, Baltimore, 1963.)

sinus horn regresses, and only a small volume of blood enters this sinus horn. The proximal portion of the left common cardinal vein dwindles to become the **coronary sinus** and its distal portion regresses to become the **oblique vein of Marshall.** The latter would have been a left superior vena cava if it were not for the gradual loss of the left common cardinal vein which is complete by the 60-mm. stage (10 weeks). Pulmonary veins which later connect with the left atrium are not remnants of these older veins but entirely new ones arising as the lungs develop.

At first the superior and inferior vena cava enter the right atrium through a common sinus venosus, the remnant of the right sinus horn (Fig. VIII-102B). The entrance of this sinus horn is guarded by two valve flaps, the **right and left valves of the sinus venosus.** Near the end of the second month, the right atrium grows rapidly causing the sinus horn to flatten and become a part of the atrial wall. The sinus venosus ceases to exist, having been taken into the right atrium. In the process, the superior and inferior venae cavae are also taken in, and overgrowth of the wall separates them and gives them separate openings . The *left valve of the sinus degenerates,* fusing with septum II. The *right sinus valve regresses* and it becomes, variously, the **crista terminalis,** the **valve of the inferior vena cava,** and the **valve of the coronary sinus,** all of which become useless structures in the postnatal right atrium.

In the early heart, atrium and ventricle communicate by a very narrow passage. This passageway is divided into right and left sides by the formation of two **endocardial cushions.** These are thickenings of the endocardial lining dorsally and ventrally at this narrow opening (Fig. VIII-103). The two meet across the midline and separate off **right and left atrioventricular canals.** The orifices of these canals are surrounded by localized proliferation of mesenchymal tissue which will become the **atrioventricular valves.** On the ventricular side of this piled-up tissue, the muscular walls of the ventricle begin to hollow out so that the valve primordia seem to be anchored to the ventricular walls by relatively thick muscular cords (Fig. VIII-104B). The muscular cords degenerate and are replaced by connective tissue. In this way **chordae tendineae** are formed, and their distal ends attach to conical projections of muscle that is sculptured in the hollowing of the ventricular wall. These we recognize as **papillary muscles** in Figure VIII-104C.

We have seen that the bulk of the **bulbus arteriosus** was absorbed into the right ventricle so that it loses its identity some time during the fifth week of development. Its **truncal portion** develops two longitudinal endocardial ridges, something similar to two elongate endocardial cushions. These thicken into **bulbar ridges** which meet and fuse, but not in the straight-line fashion that our hypothetical sketch in Figure VIII-102 would suggest. Rather these thickenings in the truncus pursue such a spiral course that the **ascending aorta** and **pulmonary trunk,** which they have separated, *intertwine.* The pulmonary trunk crosses ventral to the aorta as a result. The proximal parts of these ridges spiral in such a way that the pulmonary trunk opens from the right ventricle and the aorta from the left ventricle.

We saw the external groove for an **interventricular septum** in the 5-mm. embryo heart (Fig. VIII-101E). This is due to a median projection inward from the base of the common ventricular chamber (Fig. VIII-102D). The septum increases in height as the ventricular sacs become deeper. The ventricular septum is cresentic, reaching up along the anterior and posterior walls to contact the endocardial cushions. Its midsection does not reach this high leaving an **interventricular foramen** which will not be closed off until the seventh week. The foramen is bounded by: (1) the **ventricular septum;** (2) the **proximal bulbar septum,** a continuation of the bulbar ridge; and (3) by the fused midportions of the **endocardial cushions.** The final closure comes from a proliferation of mesenchyme from these three sources and it constitutes the thin, **membranous part of the interventricular septum.** The original, upward-projecting, ventricular septum constitutes the *muscular* part.

The Adult Heart: *Anterior Aspect:* Viewed in the anterior aspect (Fig. VIII-105), the heart has a **sternocostal surface** which is formed, above, by the right and left **auricular appendages** and by the **conus ar-**

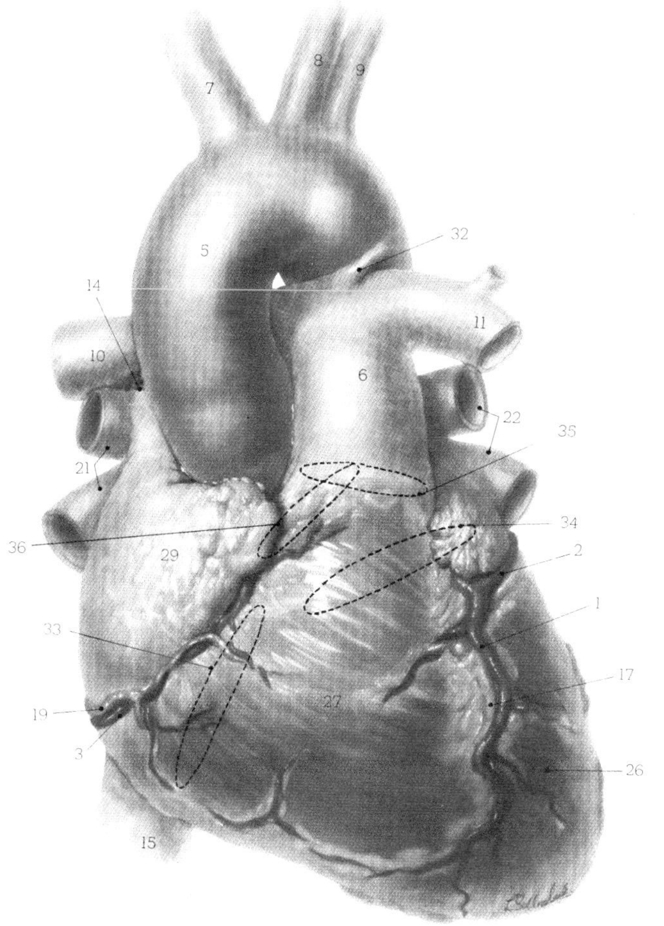

Anterior

Fig. VIII-105. Anterior view of the heart and its vessels. (Courtesy of the American Heart Association.)

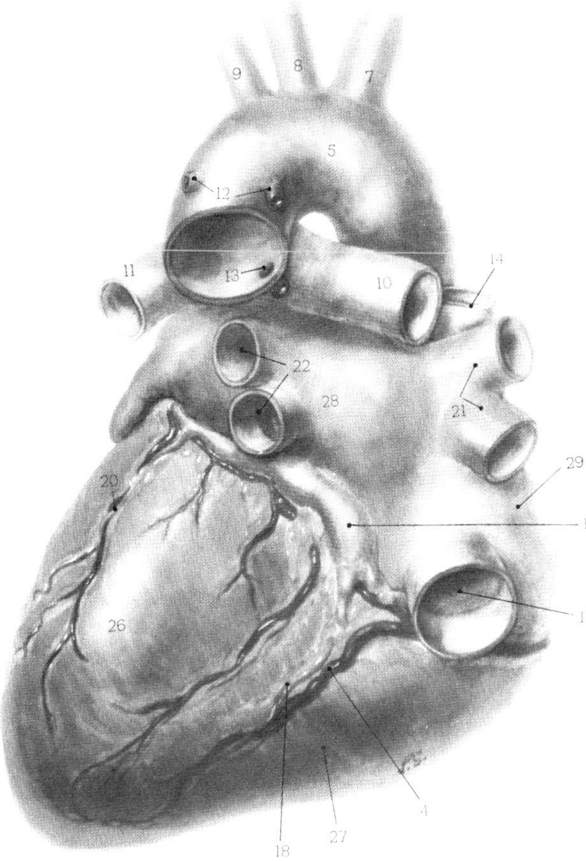

Posterior

Fig. VIII-106. Posterior view of the heart and its vessels. (Courtesy of the American Heart Association.)

Key to Figures VIII-105 and 106. 1. Anterior descending branch of left coronary artery. 2. Circumflex branch of left coronary artery. 3. Right coronary artery. 4. Posterior descending coronary artery. 5. Aorta. 6. Pulmonary artery. 7. Innominate artery. 8. Left carotid artery. 9. Left subclavian artery. 10. Right pulmonary artery. 11. Left pulmonary artery. 12. Intercostal arteries. 13. Left bronchial artery. 14. Superior vena cava. 15. Inferior vena cava. 16. Coronary sinus. 17. Great cardiac vein. 18. Middle cardiac vein. 19. Small cardiac vein. 20. Left posterior ventricular vein. 21. Right pulmonary veins. 22. Left pulmonary veins. 26. Left ventricle. 27. Right ventricle. 28. Left atrium. 29. Right atrium. 32. Ligamentum arteriosum. 33. Tricuspid valve. 34. Mitral valve (bicuspid). 35. Pulmonary valve (semilunar). 36. Aortic valve (semilunar).

teriosus (infundibulum) which leads from right ventricle to **pulmonary trunk.** Inferiorly, two thirds of this aspect of the heart is the province of the **right ventricle,** while only the apical portion of the **left ventricle** presents itself as the remaining one third. The division between the two ventricles is the **interventricular sulcus.** This sulcus contains the **anterior descending branch of the left coronary artery** and the **great cardiac vein.** The right border is formed by the **right atrium,** and this border is continuous, superiorly, with the superior vena cava and, inferiorly, with the inferior vena cava as this vessel comes up through the Diaphragm. The right atrium is separated from the right ventricle anteriorly by the **atrioventricular** (coronary) **sulcus** in which lie the **right coronary artery** and the **anterior cardiac veins.** The convex left border is formed, chiefly, by the left ventricle. The left ventricle is also the main contributor to the **diaphragmatic surface** of the heart. The *base of the heart* is its *posterior surface* since the heart is an inverted triangle with its base uppermost. The base is formed by both atria with the one on the left making the greatest contribution to this surface.

Posterior Aspect: The posterior relationships of the heart are shown in Figure VIII-106 where it becomes quite evident that the left atrium usurps most of the

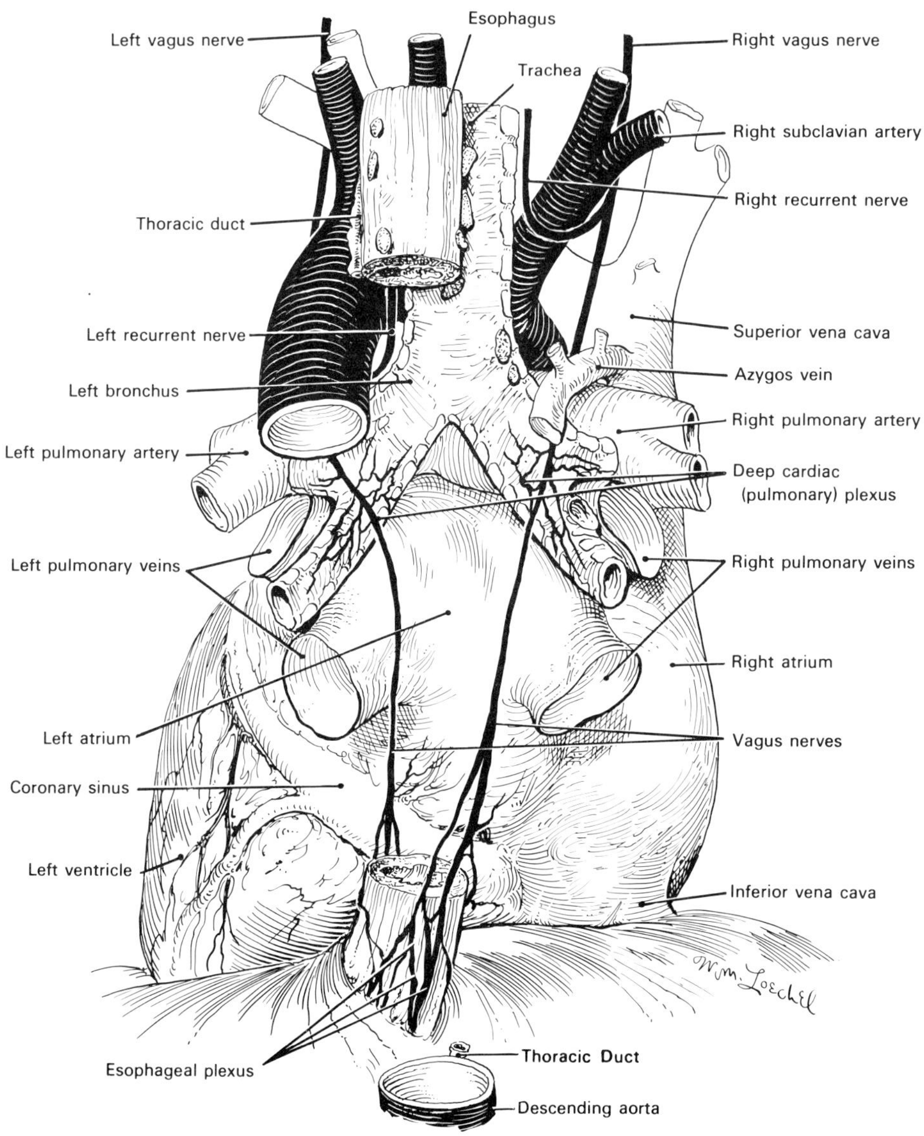

Fig. VIII-107. Structures in superior and posterior mediastinum as seen from behind.

territory. Notice that while there are right and left pulmonary veins, all of them enter the left atrium. The space between their points of entry is the ventral wall of the oblique sinus. The back wall of the sinus has been removed in Figure VIII-107. so that you can view the position of other structures behind the heart. In the atrioventricular sulcus, posteriorly, lies the large collecting sinus for most of the cardiac veins, the **coronary sinus** which empties into the right atrium. The trachea and the primary bronchi lie posterior to the main stems of the bifurcating pulmonary trunk, and also posterior to the pulmonary veins. The aorta arches over the left bronchus while the azygos vein arches over the right main bronchus. Notice the posterior course of the descending aorta, esophagus, and vagi. A further point of note is the relationship of the recurrent laryngeal branches of the vagi. On the right, the recurrent laryngeal nerve makes a loop around the subclavian artery. On the left, it appears that it loops around the aorta, but actually it hooks around a short, ligamentous strand, the **ligamentum arteriosum,** between aorta and pulmonary trunk. This strand is a remnant of a wider, patent channel between these two vessels during fetal life. Actually, it is a remnant of the sixth aortic arch on the left side. Once these laryngeal branches make their loops, they slide in laterally between esophagus and trachea to run cephalad toward the larynx.

The Musculoskeletal System of the Heart: The heart has a kind of skeleton to which its musculature anchors. It is a fibrous system of rings which surround the four openings of the heart that are guarded by valves (Fig. VIII-108). The aortic ring is the strongest of these stout cuffs, and without these cuffs all orifices would stretch and the valves would become useless. In some animals these rings consist of cartilage or even

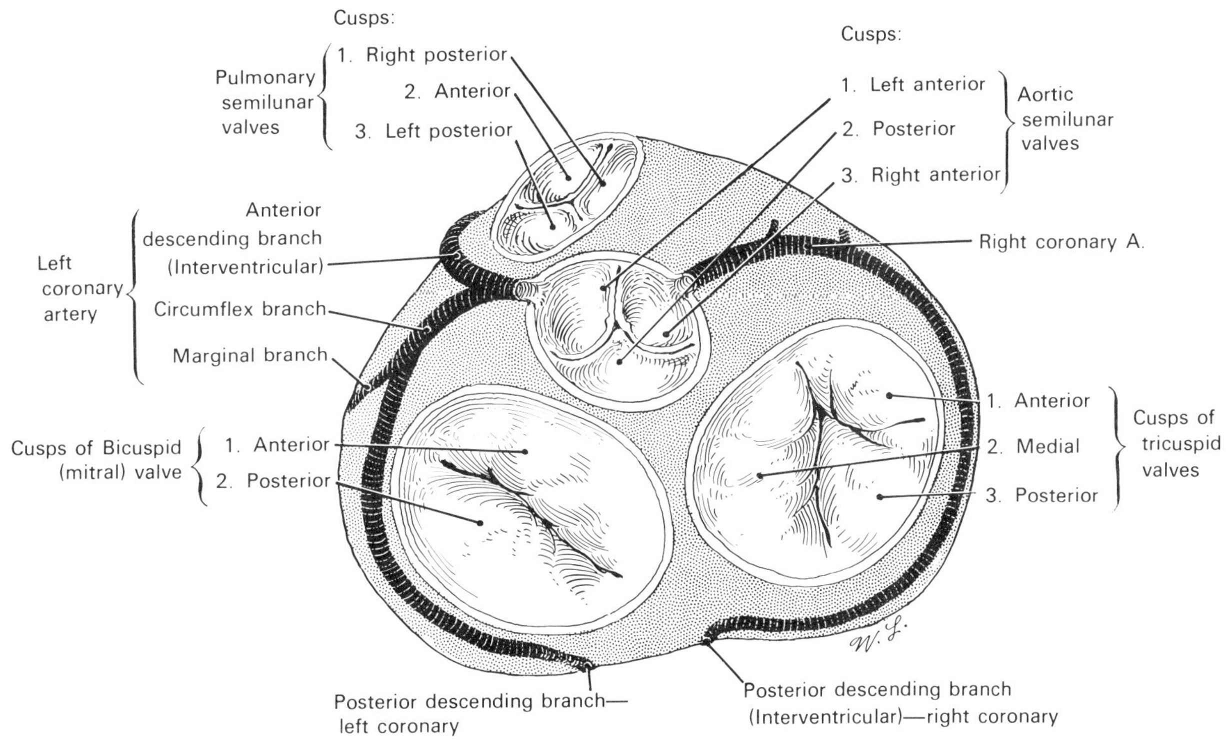

Fig. VIII-108. Heart seen from above with atria removed to show heart valves and origin of coronary arteries. All valves attach to a common atrioventricular connective tissue which is known as the "heart skeleton."

bone. The rest of the skeleton is the usual framework stroma of fibroelastic connective tissue, the basket weave common to many organs of the body.

In the atria, superficial muscle fibers run transversely while the deep ones arch from front to back, both of their ends attaching to the fibrous rings. Circular fibers surround the openings of the great veins.

Even in the ventricles, no matter the course or the length of the fibers, the muscles begin and end upon the rings. There are three layers of these muscles. A superficial layer runs a spiral course to the left, no matter whether the bundles clothe the front, sides or back of the heart. The middle layer forms cylindrical cones in each ventricle. On the left side, these middle-layer fibers arise from the left A-V ring (around the bicuspid valve) and move anteriorly from left to right to insert on the pulmonary and aortic rings and upon the membranous part of the interventricular septum. On the right side, fibers again take their origin from the left A-V ring but circle posteriorly toward the posterior interventricular septum where some run vertically in the septum while others continue around to meet in the anterior part of the septum. The deep layer is really a part of the superficial layer, the deep portion of the spiral which turns around the lower border of the middle cylinder or penetrates it to ascend for its attachment to the rings. This attachment can be direct in some instances, but is often made through papillary muscle, valve cusps or chordae tendineae. It is not necessary to know the exact relationships of these origins and insertions. It is sufficient to realize that whether the muscle runs circularly, spirally, or vertically, it starts and ends by attaching to the ring-skeleton in some fashion.

There is a structural relationship between the amount and thickness of heart muscle in the chambered walls of the heart and the hydrostatic pressures developed in them during systole (Gr., contraction). The atria are small antechambers to the ventricles as their name implies. They only have to move the blood coming from the veins into the relaxed ventricles. The ventricles must be much thicker walled, muscularly, for they have to develop the driving force for the pulmonary and systemic circulations. The resistance through the pulmonary circuit is much less than that through the systemic circuit where pressures of 120 mm. Hg are required normally to propel the blood through the circuit. Hence, you can expect a much more massive musculature in the left ventricle than in the right ventricle. The pressure in the right side is about one-third that in the left and, as a consequence, the walls of the right ventricle are about one-third the thickness of those in the left. Ventricular muscle can *hypertrophy*, especially in cases of high blood pressure, so that the walls grow thicker in some individuals. This is accomplished by an increase in cardiac muscle fiber size, *i.e.*, increase in the number of myofibrils per cell unit and by an increased proliferation of stromal supporting tissues.

The Blood Supply of the Heart: Before we enter the interior chambers of the heart, we should study the coronary vessels which are surface features. The heart

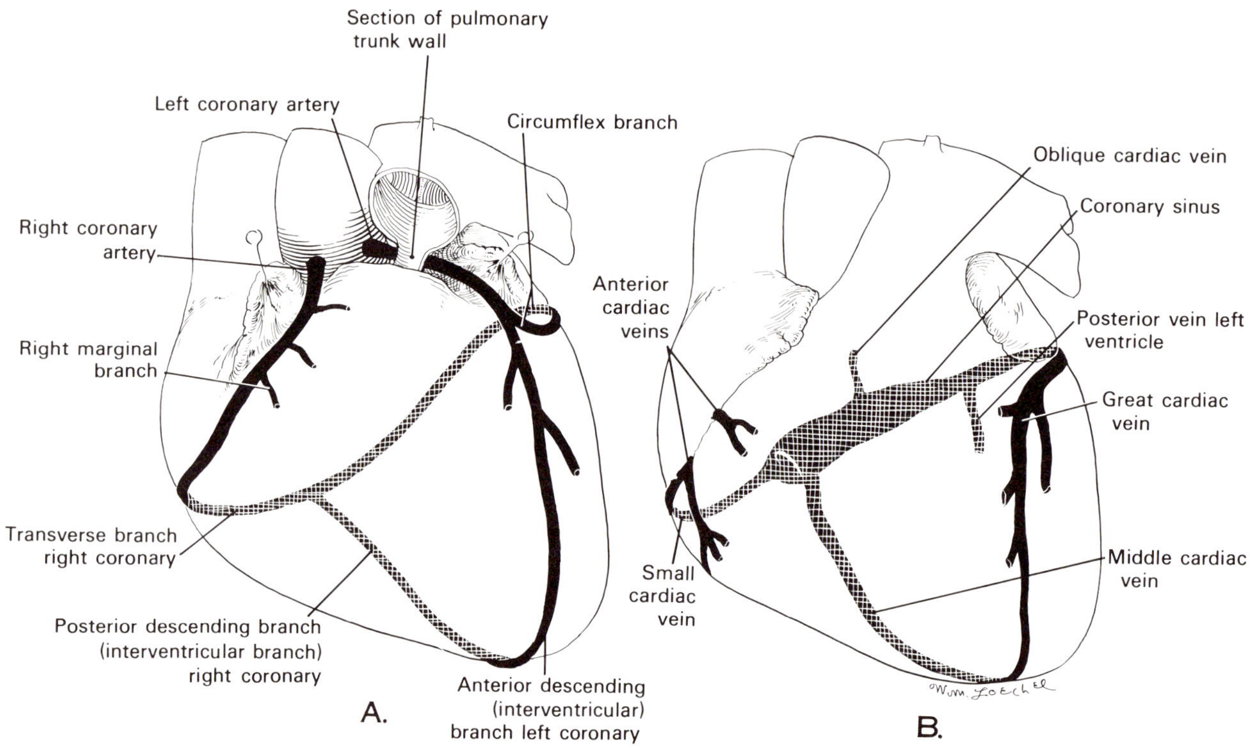

Fig. VIII-109. Schema of usual disposition of (**A**) coronary arteries and (**B**) cardiac veins in the "transparent" heart. Anterior vessels, solid; posterior structures, cross-hatched. In (A) a piece of the pulmonary trunk wall was left in place to show that the left coronary artery crosses behind that vessel.

is never at rest in the true sense except during the brief intervals when atria or ventricles are in diastole (Gr., dilatation), a brief rest period following each of their contractions. Metabolically, the heart's requirement for oxygen is exceeded only by that of the brain, an organ, which, in a sense, seems to be resting all of the time. Thus, the hard-working heart selfishly taps the first blood supply given off by the aorta as this large elastic vessel emerges out of the substance of the left ventricle. Just above the aortic semilunar valves are three slight dilatations in the vessel wall, the **aortic sinuses** (of Valsalva). The **left coronary artery** arises from the **left posterior sinus;** the **right coronary artery** from the **anterior sinus.** No vessel is supplied by the right posterior sinus (Fig. VIII-108).

The Coronary Arteries: Unlike other arteries the flow through coronary vessels occurs in diastole after the aortic semilunar valves are closed. This was first brought to attention by Scaramucci, an Italian of the 17th Century.[24] At first it was thought that this was because the opening of the pocket-like flaps of the aortic valves during systole effectively blocked the orifices of the coronary arteries. Such a simple explanation was not the case for the free margins of the valve rarely reached as far as the openings in the aortic sinuses. Because the coronary vessels lie superficially upon the heart muscle and because their branches penetrate it, they are compressed during systole. The pressure in these arteries is never as high as arterial pressure elsewhere. Neither, however, is the coronary pressure gradient entirely abolished at any time during the cardiac cycle. It is reduced sharply, however, when the heart is in isometric contraction, for then the A-V valves are closed and the pressure in the ventricles is rising in early systole. Coronary pressure is also low late in systole when the semilunar valves are pushed open and the blood is ejected into pulmonary trunk and aorta. During early diastole, when all valves close momentarily as the veins fill the auricles, coronary pressure is highest and rate of flow through the vessels is greatest. Thus it is that the semilunar valves must be closed for much blood to enter the coronary arteries.

The coronary arteries are considered to be specialized vasa vasorum which supply the media of a modified muscular blood vessel. Their branches are end arteries in the sense that they supply regions of cardiac muscle without overlap from other large branches. Even though (as we will show below) there is a rich anastomosis between arterioles, this blood supply is in-

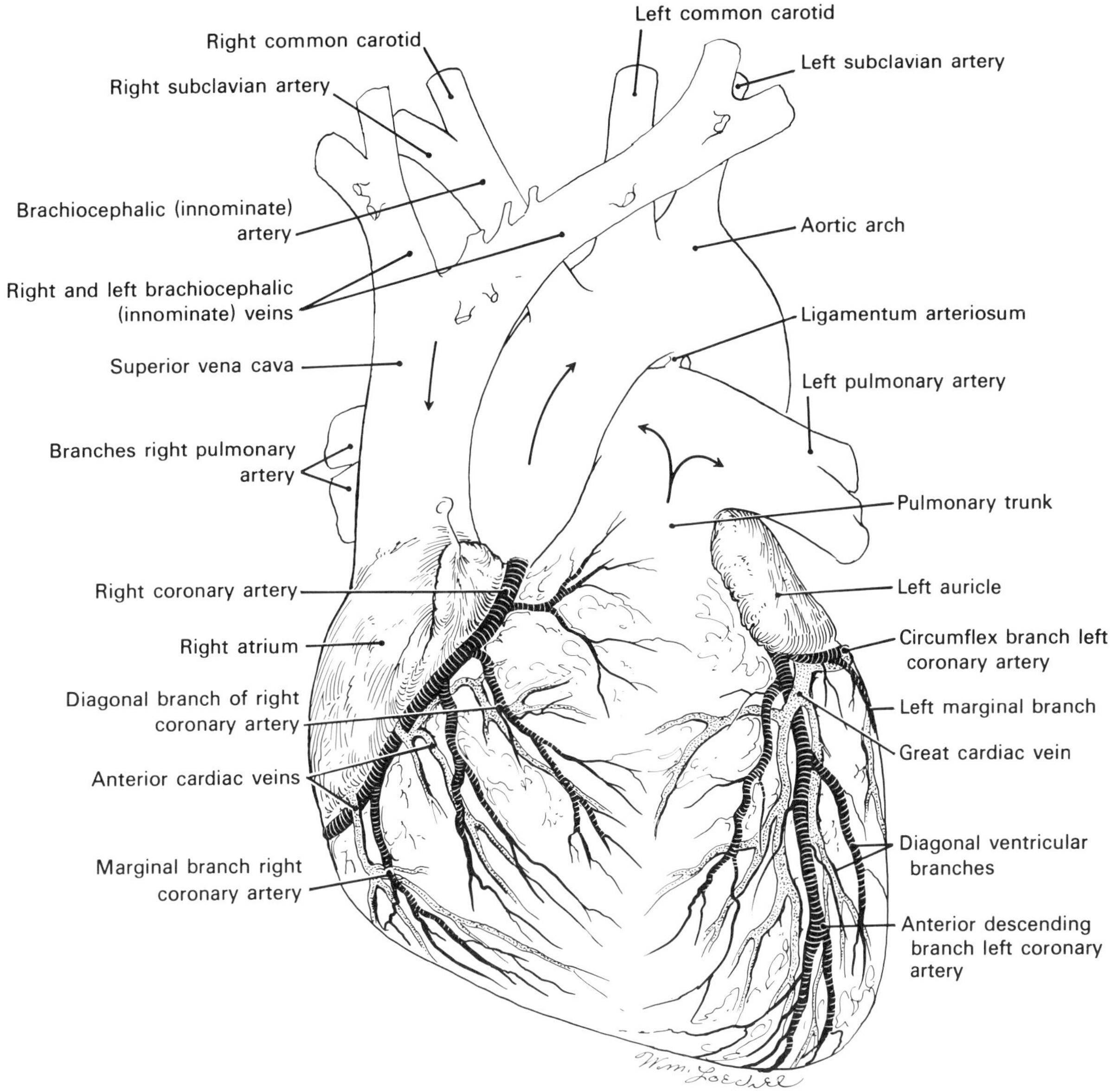

Fig. VIII-109C. Anterior view of the coronary vessels shown in more detail. The arteries are striped; veins, stippled.

sufficient to handle the requirements of cardiac muscle should there be a sudden occlusion of a major branch. As a result, that region supplied by the occluded branch will become infarcted, *i.e.*, become rendered virtually bloodless so that it will die.

The Left Coronary Artery: The ostium, or opening, of the **Left Coronary Artery** at the aortic sinus is about 3 to 5 mm. in diameter. The artery leaves the left posterior sinus to enter the epicardial fat where it is hidden from view (Figs. VIII-105 and 109 A, C, D) between the pulmonary artery and left auricle in the atrioventricular groove. After a course of a centimeter or two in this fat it divides into two branches (and sometimes more than two).

1. The **anterior interventricular** (**anterior descending**) **branch** enters the anterior interventricular sulcus to descend toward the apex of the heart. Near the apex it turns onto the diaphragmatic surface of the heart to ascend in the posterior interventricular sulcus for a few centimeters. Generally, it anastomoses with the **posterior interventricular** (**posterior descending**) **branch of the right coronary** within the sulcus (Figs. VIII-105, 106, and 109). In the anterior interventricular sulcus it may give off two or three diagonal ventricular branches which course in the free wall of the left ventricle (Figs. VIII-105 and 109). These can often be as large in caliber as their parent stem.

2. **The circumflex branch** turns to the left in the coronary (atrioventricular) sulcus between left atrium and left ventricle. It circles around the left (obtuse) margin of the heart to anastomose with the **transverse branch** of the **right coronary.** The circumflex gives off a **left marginal branch** which supplies the left

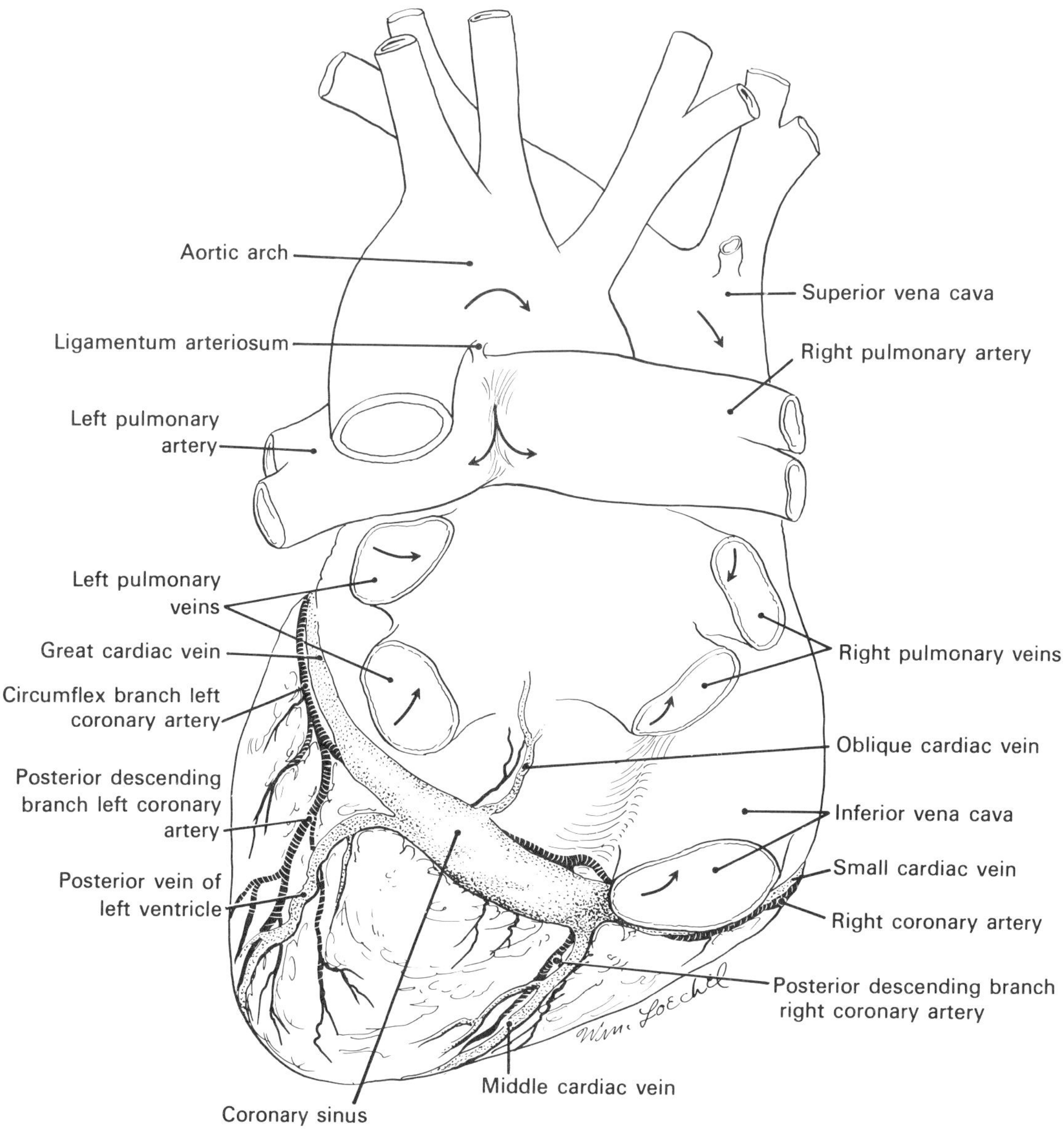

Fig. VIII-109D. Posterior view of the coronary arteries and cardiac veins. Arteries are striped; veins, stippled.

ventricle along its obtuse margin (Figs. VIII-109 and 110).

Thus, the apex and anterior surface of the left ventricle are supplied by the anterior descending artery with its diagonal branches and by the left marginal branch of the circumflex. The diaphragmatic surface of the ventricle is served by the main interventricular stem, in part, and by twigs from the circumflex which descend on its posterior aspect.

The Right Coronary Artery: This vessel arises from the anterior aortic sinus. As it descends obliquely in the right coronary sulcus its origin is covered from view by the pulmonary trunk and the right auricle. Upon emerging from behind the auricle, it continues its descent in the coronary sulcus to the inferior (acute) margin of the heart where it encircles the heart to pass onto its posterior surface. The right coronary has three branches.

1. The **right marginal branch** courses over the anterior surface parallel to the acute right margin and toward the apex (Figs. VIII-105 and 109A). A small twig from this branch frequently supplies the A-V node.

2. The **posterior interventricular branch** (posterior descending) arises on the posterior surface of the heart and descends in the posterior interventricular sulcus to anastomose with the ascending part of the **anterior interventricular branch** of the left coronary artery (Figs. VIII-106, 109A, and 110A).

3. The **transverse branch** is a continuation of the posterior coronary within the coronary sulcus on the posterior aspect of the heart. It anastomoses with the

circumflex branch of the left coronary (Figs. VIII-109A and 110A, B).

There are variations, of course. In some instances the right coronary falls short of meeting the posterior interventricular sulcus. Often the transverse branch gives off several descending branches (Fig. VIII-109D). The right coronary sometimes gives off a large branch diagonally across the anterior surface of the heart (Fig. VIII-109C). Sometimes the main trunk continues as the only atrioventricular sulcal branch clear across to the left side of the heart as a right circumflex branch. In such instances the circumflex branch of the left coronary is absent.

There has been some attempt made to correlate the incidence of coronary disease with the variable distribution of coronary blood supply.[25] In 225 hearts the greater portion of the blood supply came from the right coronary in 48% of the cases. In 34% each vessel tended to supply chiefly its own ventricle, while in 18% the supply of the heart came predominantly from the left coronary. The least vulnerable to arteriosclerotic changes were the hearts in which the blood supply was not predominantly through either artery. The group supplied chiefly by the left coronary was the most susceptible.

Coronary Arterial Supply to Special Heart Structures: There are a number of conical muscular projections in the ventricle known as **papillary muscles.** Their bases attach to the inner ventricular wall while their apices are connected to the cusps of the atrioventricular valves by tendinous strands called **chordae tendinae.** One such papillary muscle is quite constant, attached to the anterior wall of the right ventricle. It is supplied by deep penetrating branches of the left anterior descending artery. The two papillary muscles most frequently present in the left ventricle receive a variable supply. The anterior one receives one or more deep branches of the left anterior ascending artery, or it is supplied by the diagonal branches of the left coronary. The posterior papillary muscle is fed by either the left circumflex branch, or in 20% of the cases where the right coronary covers the territory usually reserved for the left circumflex, it is supplied by the right coronary.

The **interventricular septum** of the heart is usually supplied by the left anterior descending artery. The posterior descending branches of the right coronary rarely penetrate more than 15 mm. This is slightly more than normal thickness of the left ventricular wall, not enough to reach into the septum. The five or six penetrating branches of the anterior descending artery course posteriorly along the right ventricular side of the septum sending smaller twigs into it locally. In this sense, like some other penetrating arterial offshoots, they stem from the high pressure area of the left ventricular endocardium. These branches service the septum inferior to its membranous part. The membraneous portion gets its supply from one long branch of the posterior descending artery. This twig also reaches the region of the **A-V node** and the **bundle of His** (which we will consider below).

Several small arteries are found near the **S-A node** but the primary supply to this localized nodule of specialized conducting tissue is a single, fair-sized artery of about 1 mm. in diameter. In 55% of human hearts this vessel originates from the right coronary 2 to 3 cm. distal to the origin of the parent artery itself. In 45% of the cases this supply comes from the left circumflex branch (Fig. VIII 110A, B). It can, in some instances, come from the main left coronary stem. Either way, it follows along the base of the atria to reach the superior vena cava, circling this vessel in a clockwise or a counter clockwise direction before it penetrates the sinus node which lies between the superior vena cava and the right auricle. It passes through the node, giving off branching twigs, continuing on to supply the myocardium of both atria (Fig. VIII-125 A, C). This is the largest and most constant artery.

We have noted above that the **atrioventricular nodal artery** comes from the posterior descending artery, a branch of the right coronary. It does so about 90% of the time. We also noted that the right marginal branch sometimes helped supply this nodal tissue. The rest of the time the left circumflex provides it. As the right coronary reaches the interventricular sulcus posteriorly it is at a point designated as the **crux.** As the artery crosses the crux, it penetrates in the form of a U-shaped bend, and the artery to the A-V node emerges from the base of the U. The U marks the location of the crux, and it is only a few millimeters from the A-V node and not too far from the ostium of the right coronary artery. A line drawn from the base of this U-turn to the non-coronary sinus of the aorta divides interatrial septum above from interventricular septum below and separates the territories of the right and left atrioventricular valves.

Vascular anastomoses: There are many anastomoses between branches of the coronary arteries in the normal heart. These are all at the arteriolar level and they consist of gently curving vessels of 100 to 300 μ in diameter. These occur in the ventricular epicardium, within the atria and in the interventricular septum. Three large ones, in particular, are often mentioned.

1. *Conus artery:* It may be of right coronary or even aortic origin. It sweeps across the conus arteriosus to meet a branch of the left descending coronary artery. Thus, it serves as a proximal connection between the main right and left coronary supply routes.

2. *Kugel's artery:* It can arise from the right or the left proximal coronary outflows. It courses posteriorly in the base of the interatrial septum to the region of the crux, joining with the vessels there, even with twigs to the A-V node.

3. *The Sinus Node artery:* We have already described this artery as an important cross link between atria.

Many linkages occur between branching vessels in the interventricular septum and the deep vessels in the

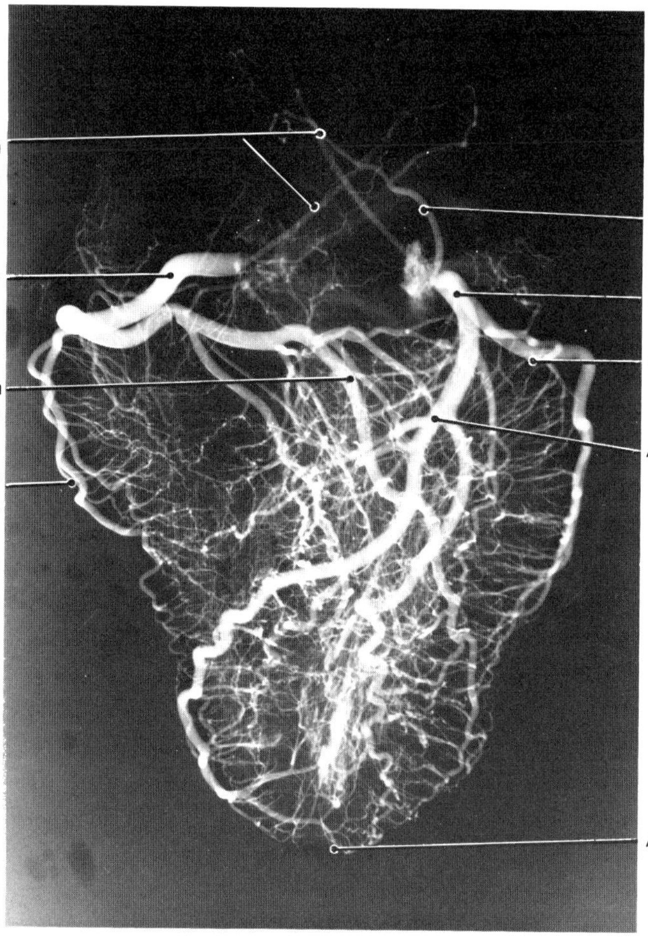

Fig. VIII-110. Micropaque injection of the coronary arterial circulation. **A.** Anterior view showing the right and left main coronary arteries and their respective tributaries.

ventricular walls. Some of these in the neighborhood of 50 to 150 μ have been called **myocardial sinusoids.**

When anastomoses become functional as major flow collaterals, they may enlarge to 1 mm. in diameter and become so tortuous and elongated as to appear in sharp contrast to the normal heart arteries in arteriograms.

The Cardiac Veins (Figs. VIII-105, 106, and 109B, C, D): In most hearts five of the seven cardiac veins empty directly into the **coronary sinus** on the posterior surface of the heart. The others empty directly into the right atrium with the exception of the **Thebesian veins** or **venae cordis minimae** which empty into any of the heart chambers.

Coronary Sinus: This large venous channel measures about 2.5 cm. in length. It is situated within the posterior portion of the coronary (atrioventricular) sulcus. It is partly covered by the muscular fibers of the left atrium, though it opens into the inferomedial aspect of the posterior wall of the right atrium at a point between the opening of the inferior vena cava and the right atrioventricular orifice (Fig. VIII-114). At its right and inferior margins is a rudimentary valve flap, a thin, semilunar fold, the now functionless **valve of the coronary sinus** (or Thebesian valve). The sinus in the embryo was once a fair-sized heart chamber admitting the common cardinal veins and it had functional valves. Even when most of the embryonic sinus was absorbed into the right atrium, a right and left sinus valve of fair size was present early in development. Now, only parts of the right valve show as identifiable remnants in the adult right atrium, *e.g.,* crista terminalis, valve of the coronary sinus, and a larger, but now functionless, valve of the inferior vena cava (Fig. VIII-114). About where the upper arrow appears in the right atrium in Figure VIII-113, the left coronary valve of the embryo became part of the interatrial wall and is no longer a distinguishable feature. The five tributaries of the coronary sinus are:

1. *Great Cardiac Vein* (Fig. VIII-105 and 109B): This vessel begins at the apex of the heart and ascends along side the left descending coronary artery branch. It follows the left coronary circumflex branch to the

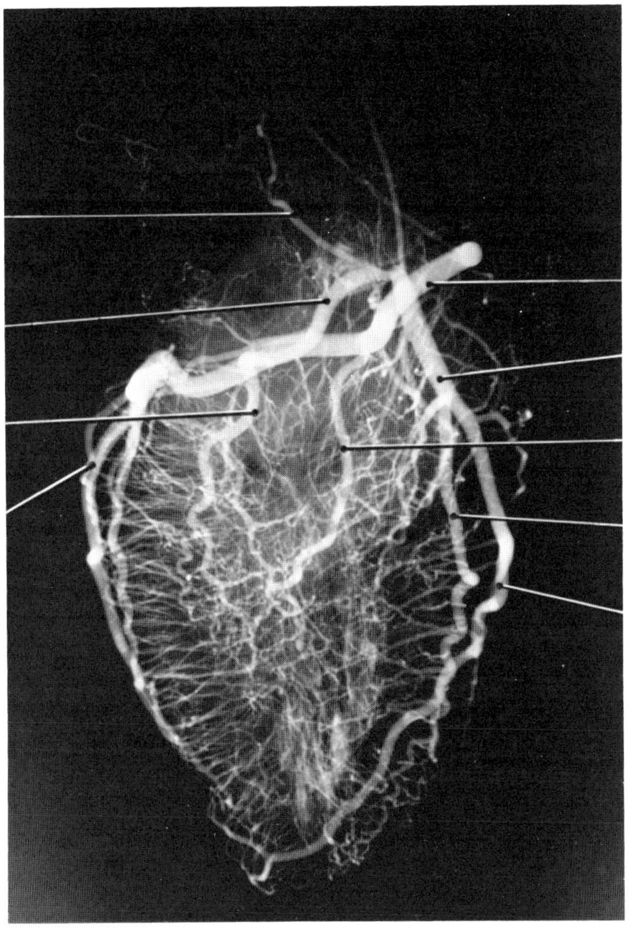

Fig. VIII-110B. Lateral view which demonstrates the penetrating branches of the coronary arterial system and its thorough supply to the anterior and posterior myocardial walls.

posterior aspect of the heart in the coronary sulcus to end in the coronary sinus. Throughout its course it receives tributaries from the ventricle and the left atrium. One large tributary, the **left marginal vein** arises posteriorly near the apex to ascend along the left posterior border of the ventricle.

2. The Small Cardiac Vein (Fig. VIII-105 and 109B): This vessel drains the posterior surfaces of the right atrium and ventricle. It runs in the right posterior portion of the coronary sulcus along with the transverse branch of the right coronary artery and dumps into the right side of the coronary sinus. It receives a large tributary, the **right marginal vein** which ascends along the right cardiac margin. This marginal branch sometimes enters the right atrium directly, by-passing the coronary sinus.

3. The Middle Cardiac Vein (Fig. VIII-106 and 109B): This vessel begins near the apex on the posterior surface of the heart and ascends in the posterior interventricular septum. It receives tributaries from both ventricles along its course, and enters the right side of the coronary sinus.

4. The Left Posterior Ventricular Vein (Fig. VIII-106 and 109B): The origin of this branch is low on the left posterior diaphragmatic surface. It ascends to empty into the left side of the coronary sinus or to become a major tributary of the great cardiac vein.

5. The Oblique Vein of the Left Atrium (Vein of Marshall) (Fig. VIII-109B): This small, almost insignificant vein, begins over the posterior wall of the left atrium and descends obliquely to end in the coronary sinus or into the great vein. It is important embryologically in that it is what is left of the anterior cardinal vein, and what could become, and sometimes does, an anomalous left superior vena cava.

The Anterior Cardiac Veins: These are small and two or three in number. They begin over the anterior surface of the right ventricle, course obliquely upward to the right, cross the coronary sulcus anteriorly and pass directly through the anterior atrial wall into the right atrium.

Aside from some very small **venae cordis minimae** which drain the deep myocardium and empty into all four heart chambers indiscriminately (though most

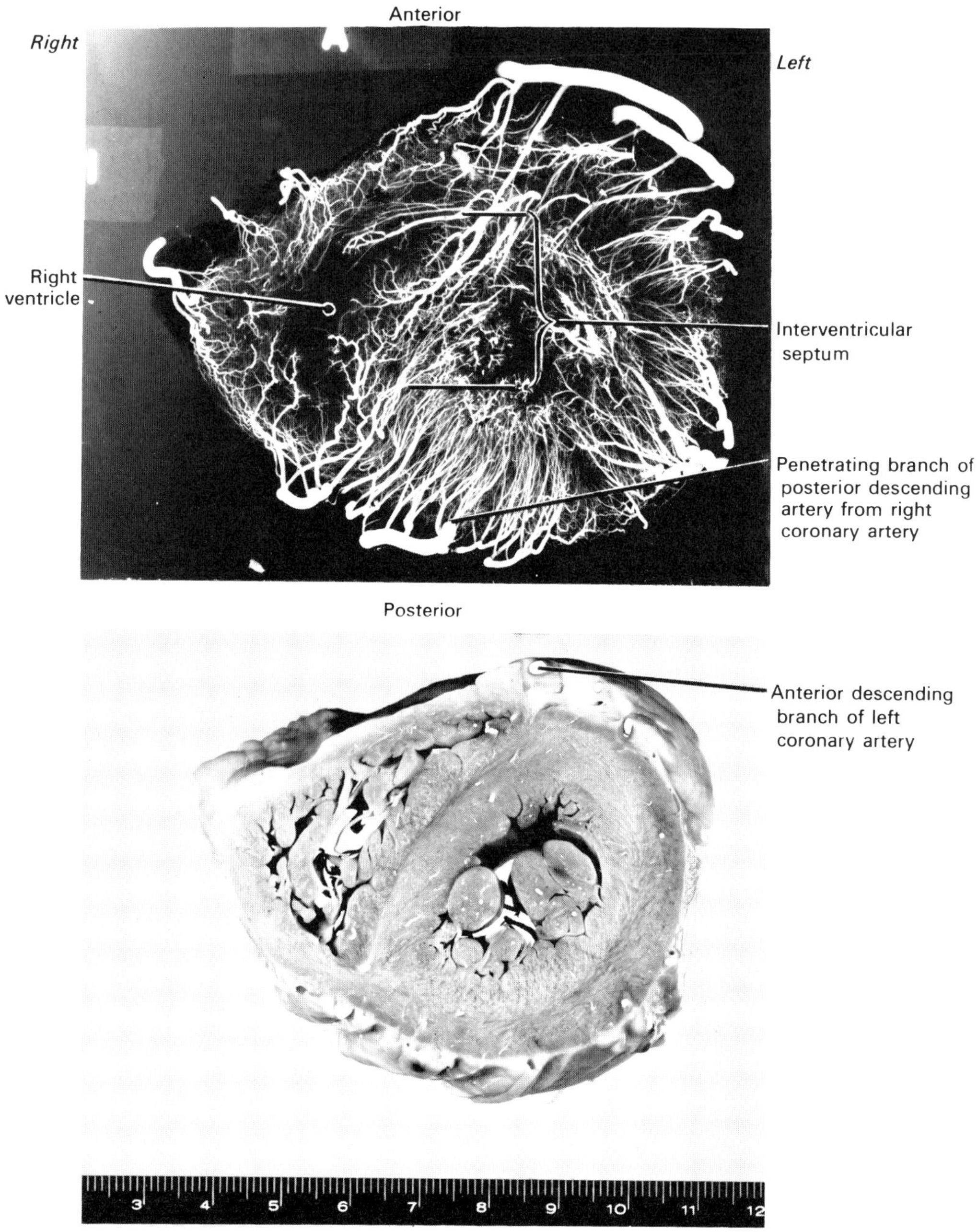

Level I (base)

Fig. VIII-111. Step sections approximately 1 to 1½ inches thick taken from base to apex of the human heart demonstrate the penetration of vessels into the myocardium. Each section is accompanied by its respective gross photograph and the radiographic demonstration of the injected specimen is viewed from dorsal to ventral. **A. Level 1** taken immediately below the atrioventricular valves. The posterior wall of the heart is located at the top of each picture so that the right ventricle is on the reader's left.

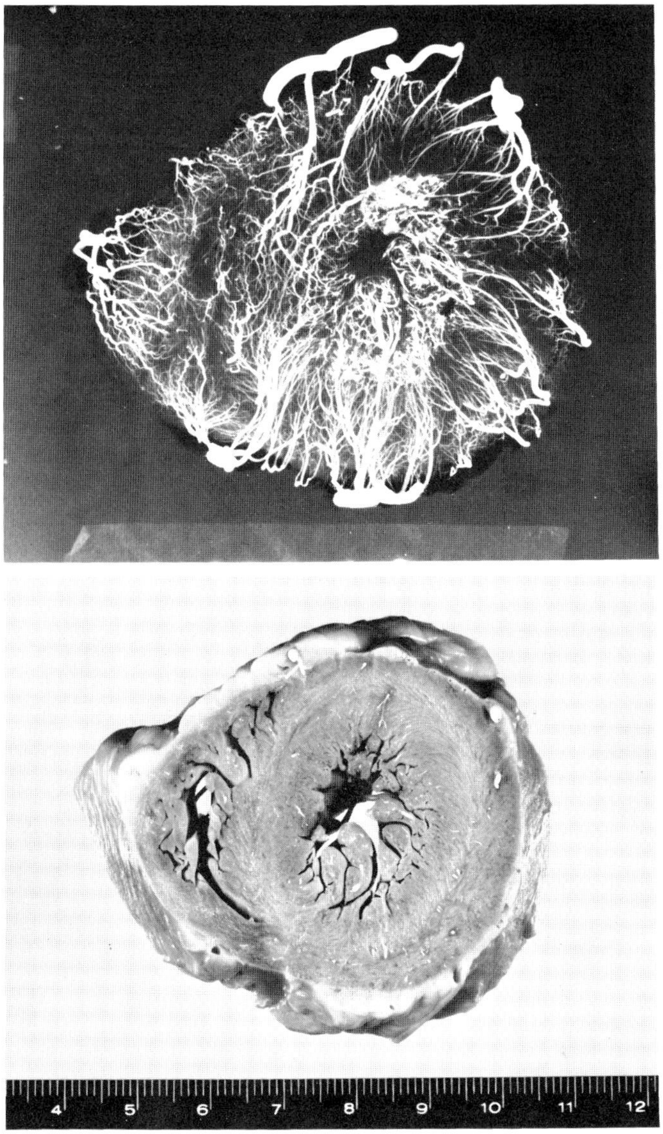

Level II (intermediate)

Fig. VIII-111B. Level 2 is a serial section taken below level 1 as shown in **A.** The same orientation is maintained.

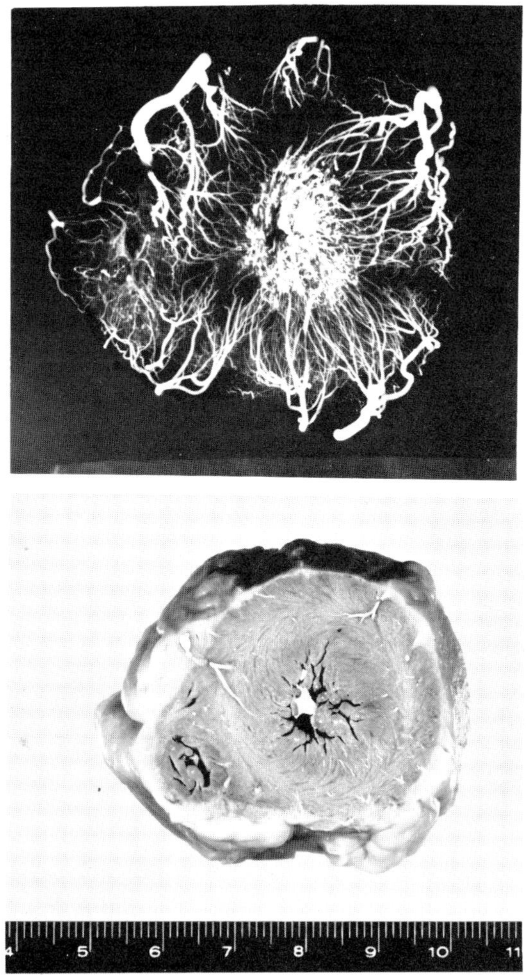

Level III (apex)

Fig. VIII-111C. Level 3 is at the tip of the apex. The left ventricle is contracted and radiopaque material is concentrated in the endocardial surface.

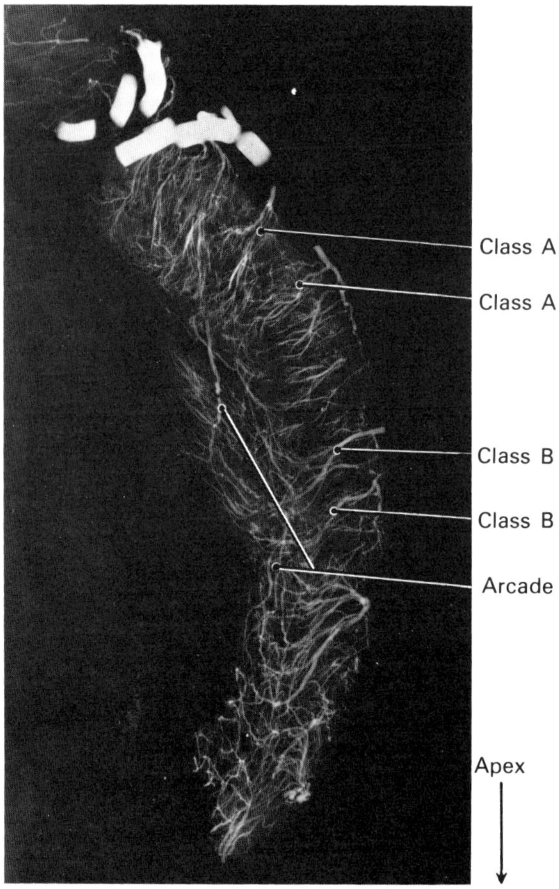

Fig. VIII-112A. This is a normal angiogram of a longitudinal segment of the left ventricular wall. Two types of branches pass into the myocardium at right angles to the main epicardial artery. The class A vessels quickly divide into a fine network, while the class B arteriole demonstrates a larger diameter as it is penetrating the muscle mass and passing to the subendocardial region. Here they form a vascular arcade.

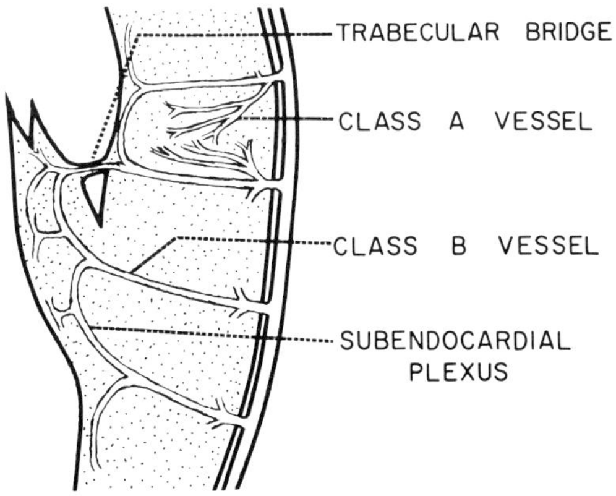

Fig. VIII-112B. A diagrammatic representation on the blood supply to normal papillary muscles. This diagram illustrates the segmental arrangement of the class B vessels and the subendocardial plexus. (Reproduced with permission of The American Heart Journal.)

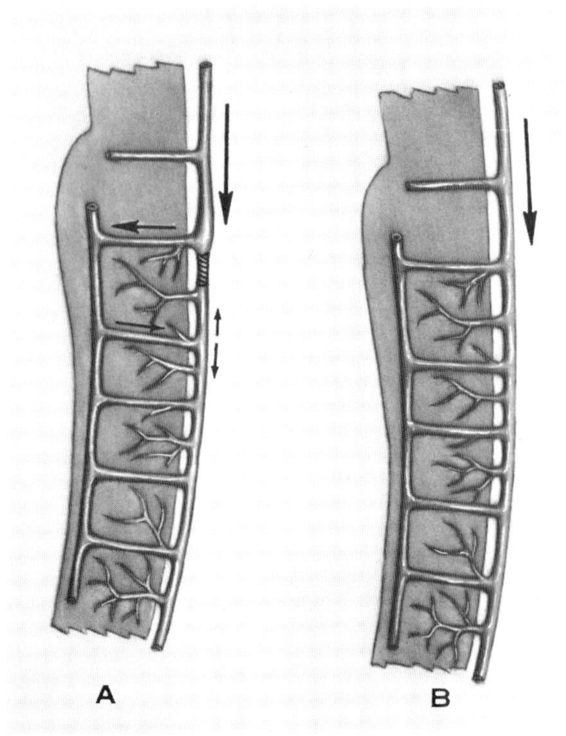

Fig. VIII-112C. A diagrammatic representation of the blood supply to the ventricular wall. Left: This diagram shows the possible effects on blood flow, of occluding an epicardial artery. The obstruction is by-passed but the adequacy of blood supply to muscle located in a more apical position depends on collateral flow through the class B arteries. Right: The distribution of normal vessels is shown in this representation. Comparing this with the left hand figure, note that the arteries which have been classified as class A and class B vessels are similar to the actual injection shown in Figure 112A. The relationship of the class B vessels to the subendocardial arcade is demonstrated in this drawing.

enter the right ventricle), no veins, other than the pulmonary veins returning with oxygenated blood from the lungs, enter the left heart chambers. The pulmonary veins, as we have noted, all empty into the posterior aspect of the left atrium.

Some General Considerations on Coronary Circulation

When we say "coronary circulation" we are not talking about circulation "through the heart chambers" but about the circulation of the blood which supplies the heart, itself, as an organ. Blood entering the coronary arteries may pass through arterioles to enter capillary beds and so perfuse cardiac muscle. The heart muscle possesses an extraordinarily rich capillary bed. Ventricular muscle is more vascular than atrial muscle or than the A-V bundle. As developing cardiac muscle fibers increase in size, they are each supplied by a capillary.[26] In cardiac hypertrophy of the adult, the increase in cardiac fiber size is not followed by an increase in capillarity, so the fiber–capillary ratio remains unchanged. In this sense, the hypertrophied heart suffers a relative reduction of blood supply. The capillary bed is, of course, drained by venules which enter the cardiac veins and, by them, blood is returned to the right atrium.

But upon reaching arterioles, blood flow may be shunted directly to venules. **Arteriovenous anastomoses** are fairly plentiful between small vessels of the myocardium.

We have already indicated the degree of arteriolar anastomoses between fairly large coronary arteries and those effected by smaller myocardial sinusoids. Some of the latter also enter the heart chamber directly, by-passing both capillaries and the venous side of the system. These small arterial channels enter the ventricles when they do not connect with capillary beds. This odd type of circulation was described as early as 1706 by Vieussens.[27]

These tiny direct arterial communications with the ventricles are reminiscent also of the venae cordis minimae first described by Thebesius in 1708.[28] Thus, many of the details of the oddities of coronary circulation have been known for a long while.

On the venous side, too, anastomoses even exist between the coronary sinus and the anterior cardiac veins. This allows blood to flow in both directions or in either direction to reach the right atrium.

All of these provisions for a shifting blood flow make for difficult physiological measurement and interpretation. Studies which demonstrate decreases of resistance in the peripheral parts of the coronary bed or increased rates of flow through some coronary artery do not tell us whether the blood is going through the myocardial capillary bed or through some of the variety of shunts available. Changes in oxygenation based on samplings from the coronary sinus do not tell us whether or how often precapillary shunting occurs. Difficulty of assessing myocardial metabolism from changes in coronary sinus blood are compounded in the presence of focal coronary artery disease because it is impossible to tell how much of the sampled blood comes from the area thought to be ischemic.

Extracardiac anastomoses are also possible for the establishment of collateral circulation through the heart. There are potential communications between atrial twigs of the coronary arteries with branches of the internal thoracic (internal mammary) or even those of the aorta (bronchial, pericardial, and esophageal). The site of anastomosis of these is in the pericardial fat around the great vessels. In the Vineberg surgical procedure, the internal mammary artery, itself, is implanted into the left ventricular myocardium. The question becomes perplexing as to why a hematoma is not formed when a large artery such as this provides systemic pressures that can flow freely into a muscle mass. When you realize that sinusoidal pooling occurs in both epicardium and myocardium, the presence of a multidirectional flow pattern seems to make sense.

After all we have said above, it still does not appear that collateral circulation through anastomoses of one kind or another is terribly significant in the *normal* heart. Nor is it, apparently, significant when a large coronary branch is *suddenly* occluded. Then back flow in the segment of the vessel distal to the occlusion is only about 1 cc. per minute as attempts to establish collateral flow to the area are initiated. Within 1 to 2 minutes, the myocardium supplied by the occluded branch stops contracting.

However, if the occlusion is *gradual,* the situation is different. Pressure in the peripheral section increases and may even approach aortic pressure as collateral channels are forced open. The rate of back flow may be as much as 30 to 100 cc. per minute so that the cardiac tissue is adequately oxygenated. Sometimes, in such situations, when the heart comes to autopsy later on, the main coronary channel is found to be well occluded, yet its occlusion was not the cause of death. The complete blockage of the main stem was a long-standing event, yet the tissue that should have suffered, as a consequence, was not affected to any great degree. In this light, the provision for multiple shunts offer life-prolonging potential for the individual who has suffered a series of "heart attacks" which were gradual in onset.

Arteriole and Capillary Supply of the Myocardium

Human ventricular myocardium has a specific pattern of distributing arterial vessels which is characteristic for this tissue.[29-31] This branching pattern is described in postmortem injected hearts, and the smaller penetrating branches are systematically divided into two classes (Fig. VIII-112A, B).

(1) One group which rapidly divides after entry into the myocardium is classified as **class A.**

(2) The second type forms multiple anastomotic arcades after penetration, to reach the subendocardium. These are **class B.**

These class A (short, rapidly dividing branches) and class B (long, penetrating, plexus forming branches) vessels which form the subendocardial plexus, play a significant role in forming collateral channels in occlusive coronary artery disease. Potentially ischemic foci of cardiac muscle, located distal to a sudden large occlusion of a coronary artery, may produce damaged heart muscle if the blood supply is not re-established. Dynamic factors result in the establishment of collateral blood flow through the subendocardial plexus and this anatomical arrangement is critically significant following acute interruption of coronary artery blood flow (Fig. VIII-112C).

Papillary muscle blood supply is also very cogent to maintenance of myocardial contractility following cardiac insult. This arterial supply is segmentally distributed and reaches the muscle *via* large penetrating branches originating from epicardial arteries radially arranged around the epimyocardium (Fig. VIII-112B).

In general, the arterial supply comes from separate tributaries radiating from a single stem arrangement. The papillary muscle tip, middle portion, and base individually receive a single radiating artery arrangements. Papillary muscle fibrosis most frequently occurs in response to occlusion of a large coronary artery. Secondary alterations in response to fibrosis are of two types:

(1) Fine arborization and outgrowth of **class A** vessels with no **class B** vessel alteration.

(2) Enlargement of subendocardial vessels resulting from interruption of all channels (both class A and class B).

This second type of response points to the augmentation of these vessels utilizing them as collateral channels in moving blood around an area of occlusion.

The Chambers of the Heart. The interior of the heart is divided into right and left atria each with their auricular appendages, and into right and left ventricles. A four-chambered human heart is made possible by the development of interatrial and interventricular septa and by the imposition of atrioventricular valves.

The Right Atrium: This chamber consists of a **sinus venarum** and an "atrial" portion. The latter, because it is ear-shaped, is better described as an **auricular portion.** The large smooth-walled sinus venarum, as its name implies, lies between the two vena cavae, the two large venous channels that empty into the right atrium from above and below (Fig. VIII-113). The auricular portion is merely a small upper pouch or appendage which communicates with the sinus venarum as a kind of rough-walled ante-room. Externally, the auricle tends to overlap the root of the aorta. The boundary line between sinus and "atrial" portions of the right atrium is the **crista terminalis,** a raised, vertical ridge extending from the opening of the superior vena cava to that of the inferior vena cava (Fig. VIII-114). The ridge is all that remains of the upper half of the embryonic right valve of the coronary sinus. Radiating from the crista into the auricle are several muscular ridges, the **musculi pectinati,** resembling the coarse teeth of a comb. They give the auricle its rough-walled appearance (Fig. VIII-114).

The opening of the superior vena cava into the right atrium is devoid of valves. The orifice of the inferior vena cava into this chamber appears to be guarded by a fairly hefty, cresentic valve (*Eustachian valve*), and there is a smaller one just like it at the entrance of the coronary sinus close by (Fig. VIII-114). Both valves are embryonic remnants of the lower half of the right valve of the old coronary sinus. They are non-functional in the adult heart, merely anatomical landmarks of an earlier day in our history.

If you follow the valve of the inferior vena cava to the left you will find it continuous with the **annulus ovalis.** The annulus bounds a wide, shallow depression on the interatrial wall, the **fossa ovalis.** In the fetus the upper end of this fossa could be pushed open under a head of pressure so that a **foramen ovale** existed here as a communication between the right and left ventricle. This was significant for circulation, and its

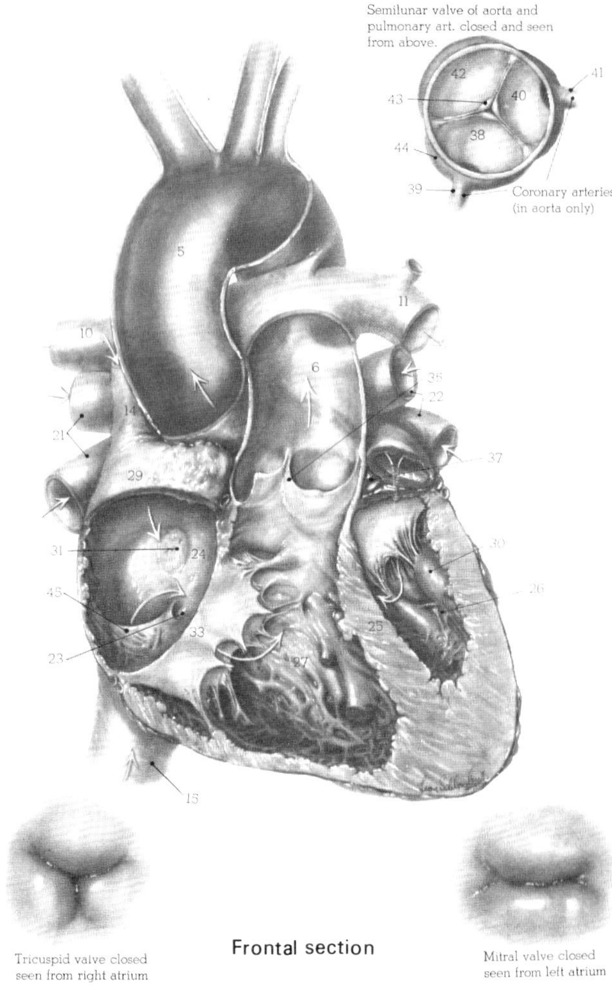

Fig. VIII-113. Section through the heart chambers. *Insets:* Atrioventricular and aortic valves. (Courtesy of the American Heart Association.)

Key. 5. Aorta. 6. Pulmonary artery. 10. Right pulmonary artery. 11. Left pulmonary artery. 14. Superior vena cava. 15. Inferior vena cava. 21. Right pulmonary veins. 22. Left pulmonary veins. 23. Opening of coronary sinus. 24. Interatrial septum. 25. Interventricular septum. 26. Left ventricle. 27. Right ventricle. 29. Right atrium. 30. Anterior papillary muscle. 31. Fossa ovale. 33. Tricuspid valve. 35. Pulmonary valve (semilunar). 37. Left coronary artery. 38. Anterior cusp. 39. Right coronary artery. 40. Left posterior cusp. 41. Left coronary artery. 42. Right posterior cusp. 43. Nodulus semilunaris. 44. Sinus of Valsalva. 45. Eustachian valve.

existence made the valve of the inferior vena cava significant as well. The head of pressure was furnished by the entering stream of blood in the inferior vena cava. This was oxygenated blood coming into the atrium from the placenta. It was not destined for the right heart which was the receiving reservoir for venous blood. Rather, it was destined for the left heart, the pump for the systemic circulation. The valve of the inferior vena cava was so nicely positioned that it could

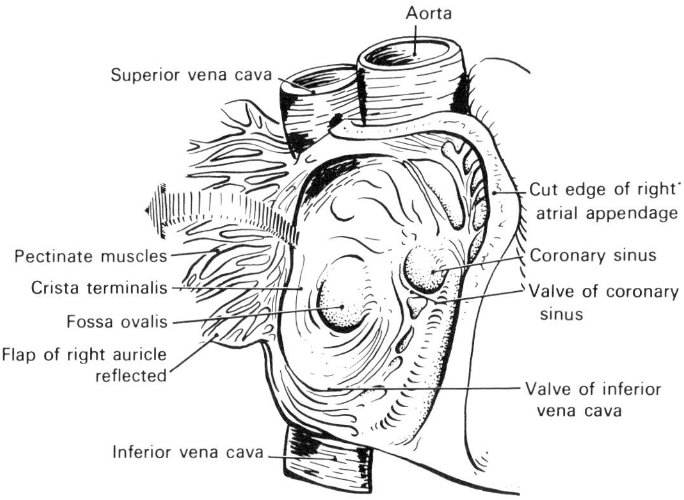

Fig. VIII-114. Details of internal structure of the right atrium.

guide or deflect the inferior caval stream directly against the foramen ovale. In this way the inferior caval stream of oxygenated blood from below was directed past the superior caval stream of venous return from above (destined for the right ventricle). So long as pressures in the chambers were maintained, these caval streams in the right atrium by-passed each other with little to no mixing in this chamber.[32-34] This meant that highly oxygenated blood could leave the left heart *via* the ascending aorta and the vessels of its arch to supply heart and brain; the heart, demanding it for its work load, the brain, because it was an area of rapid growth. Once this oxygenated blood entered the descending aorta it would be greatly diluted with venous blood, for the fetal lung was not yet functioning as an aerating organ. Venous blood from the fetal right atrium was, indeed, pumped over the pulmonary trunk and arteries to the fetal lung, but very little blood, relatively, entered the lung. A large part of it was drained away from this area by a pulmonary shunt, the **ductus arteriosus,** a connection between pulmonary trunk and descending aorta. Hence, parts of the growing body that were supplied by aortic branches below the level of the aortic arch subsisted at levels of reduced oxygenation for a while. It is for these latter reasons that we make a point about these archaic remnants in the adult right atrium. They were once important parts of the fetal life-line. Now, in the adult heart, of course, both caval streams return only venous blood from the entire body, and both streams mix in the right atrium. Neither has any other place to go except into the right ventricle unless there are interatrial or interventricular septal defects as anomalies resulting from faulty embryonic development.

The Right Ventricular Cavity: This muscular chamber is triangular in shape. The superior portion is a funnel (the infundibulum or conus arteriosus) that leads into the pulmonary trunk. Here the walls of the ventricle are relatively smooth. Elsewhere there are many irregular, projecting muscle bundles known as **trabeculae carneae.** Some of these stand up like conical pillars on the ventricular surface. These are the **papillary muscles.** From their apical tips, the threads of the **chordae tendineae** extend to the free margin and ventricular surface of the **tricuspid valve** like the shroud lines of miniature parachutes (Fig. VIII-113). There are usually three of these papillary muscles in the right ventricle but there can be as few as two or as many as five. The posterior one attaches to the inferior wall of the ventricle and sends chordae tendineae to the posterior and septal cusps of the tricuspid valve. A large anterior papillary muscle arises from the anterior wall to anchor to the anterior and posterior cusps by chordae tendineae. One or more septal papillary muscles distribute their chordae to anterior and septal cusps. One well-marked trabeculum forms a bridging, stout band from the interventricular septum to the anterior papillary muscle. This muscle bar is known as the **moderator band** because it was thought to prevent overdistention of the right ventricle. This bridging ribbon does carry conducting tissue across from the septal wall to the anterior wall. Papillary muscles contract when the walls of the chamber contract, and the cusps of the tricuspid valve are drawn tightly together so that they are not forced backward into the atrium as the pressure of the ventricular systole rises. Probably, this is the major role of the moderator band.

Right Atrioventricular Orifice and Valves: This oval aperture between right atrium and ventricle is about an inch in diameter. The cusp-like valves which guard it are attached to a strong, fibrous ring, part of the heart skeleton mentioned above. This **tricuspid valve** of the right heart is composed of roughly three major leaflets which attach around the margins of the orifice and project into the ventricle. Actually, the cusps are not all the same size, and minor leaflets may be intercalated between them. When their edges meet in valve closure (Figs. VIII-108 and 113), it may not always be easy to define the limits of the three cusps which are generally described. The cusps are designated anterior, posterior, and medial or sometimes, more descriptively, as infundibular, marginal, and septal. The **anterior or infundibular cusp** is ventrally attached in the conus region; the **posterior or marginal cusp** curves around the orifice margin from the sternocostal to the diaphragmatic side at the acute margin of the heart; and the **medial or septal cusp,** naturally, attaches to the septal wall. As a rule, the anterior is the largest, the posterior or marginal one, the smallest (not counting accessory intercalated leaflets). Facing the atrium, the surface of the cusps are covered by smooth atrial endocardium. The ventricular surfaces are rough and the margins are ragged, for the chordae tendineae attach on this surface and at the edges. The number of tiny tendinous attachments vary and so do the length of these strands as they descend to papillary muscles.

The **pulmonary orifice** beyond the infundibulum is

also surrounded by a fibrous ring to which the bases of three "vest-pocket" valves attach. These are the **pulmonary semilunar cusps.** The center of the free margin of each of these cusps is thickened as a **nodule** (corpus Arantii), while the **lunulae** on each side of the nodule are much thinner. The nodules come together at the center of the lumen when the valve is closed. As these pocket-flaps are forced open and are flattened against the walls of the pulmonary trunk when blood is ejected from the right ventricle, the elastic recoil in the pulmonary trunk and its arteries forces blood back into the "pockets" at the end of each stroke. This backward rush of blood in early diastole causes the valvules to balloon toward the luminal center so that the three free margins coapt and close off the lumen. The infundibulum, of course, lies below the valves; the **pulmonary trunk,** about 5 cm. long and 2.5 to 3.0 cm. in diameter, above the valves. The trunk winds to the left of the aorta and then dodges beneath the aortic arch where it terminates in right and left pulmonary arteries.

The Left Atrium (Figs. VIII-105 and 106): The upper left chamber forms most of the base of the heart posteriorly. It consists of an atrium proper and a long narrow auricle which projects anteriorly to overlap part of the pulmonary trunk. The inner walls of the atrium are smooth, but the auricular interior is marked by muscular ridges. Superolaterally, four **pulmonary veins,** two on each side, enter the left atrium from behind. The orifices of the pulmonary veins are free of valves.

The Left Atrioventricular Orifice and Valves: The opening from left atrium into left ventricle is smaller than the comparable A-V orifice in the right heart. It is only 2 cm. in diameter, just capable of admitting the tips of two fingers. It is guarded by the **bicuspid or mitral valve** (Figs. VIII-108 and 113). Remember, "The Bishop (with his two-pronged or mitred headgear) is never in the right!"

The **anterior cusp** sits ventrally and laps over onto the right side of the A-V orifice near the aortic orifice. Hence, you will sometimes hear it referred to as the "aortic cusp." The smaller of the two cusps is the **posterior cusp** along the posterior margin of the orifice. In the angles between the two you may encounter lesser leaflets, so that when the two valve flaps close or coapt, their meeting margins appear more scalloped than you would expect them to be (Figs. VIII-108 and 113). As you might expect in the left heart, the chordae tendineae will be quite stout. Strangely, they may be less numerous than those in the right heart. The under surface of the anterior cusp sits so close to the aortic orifice that it forms the posterior wall of the **vestibule of the aorta**. As a consequence, blood will flow over this surface during ventricular systole.

The Left Ventricle (Figs. VIII-105, 106, and 113): The cone-shaped **left ventricle** is longer and narrower than the right. Its wall, we have noted, is much thicker. In systole this ventricle must maintain arterial pressures of 120 mm. Hg and even higher with increasing age. Pulmonary systolic pressures in the right ventricle range from 22 to 30 mm. Hg since the resistance offered by the lungs is so much less. Any condition which increases peripheral vascular resistance in the pulmonary or systemic circulations will produce compensatory hypertrophy of the right or left ventricle, respectively. The trabeculae carneae are finer and more numerous than in the right ventricle, but the papillary muscles, usually two in number, are larger and stronger. The interventricular septum and the upper, anterior parts of the ventricular wall are relatively smooth. The **semilunar aortic valves** (Figs. VIII-108 and 113) are constructed like the pulmonary valves except that they are thicker. The dilatations in the aortic wall above them are more prominent than the shallow sinuses of a similar nature in the pulmonary trunk. We have noted that two of them bear the orifices for the right and left coronary arteries.

The **interventricular septum** (Fig. VIII-113) is a musculomembranous partition. It is muscular and thickest near the apex of the heart and it thins and becomes membranous near the aortic orifice. The membranous part develops separately from the muscular part in the embryo, and it arises from the same tissue that forms the atrioventricular valves. It can be deficient, therefore, leaving communications between the chambers either above or below the valves, or both.

The Roentgen Anatomy of the Heart and Great Vessels

The "normal heart" may be quite variable in shape. At the extremes of normal is the thin (vertical) heart, and the transverse heart; between these is the globular (oblique) shaped heart. The long thin heart may be seen in people with a long thoracic cage and low-lying Diaphragm, whereas the transverse heart is common to the obese or stocky individual. Many variations, then, may be seen depending upon age, body habitus, posture, and phase of respiration.

The standard roentgen examination of the chest is a PA and left lateral projection. When examining the heart for chamber enlargement, we obtain two additional projections: the 45° right anterior oblique and the 60° left anterior oblique. These four projections combined with careful fluoroscopy of the heart comprise the **cardiac series.** It is customary to have the patient swallow barium paste during the procedure to fill out the esophagus which is closely associated with the arch and descending aorta and the left atrium. For our purposes, however, we will omit discussion of the esophagus on the projections to follow.

As we consider the following chest radiographs, it is well to remember that the separate chambers of the heart are not distinguishable. However, you should know the structures that comprise the cardiac silhouette and mediastinum. Once you have learned these it will be easier to consider cardiac abnormalities of a congenital or acquired type. To assist you in more clearly seeing the structures, it would be helpful to

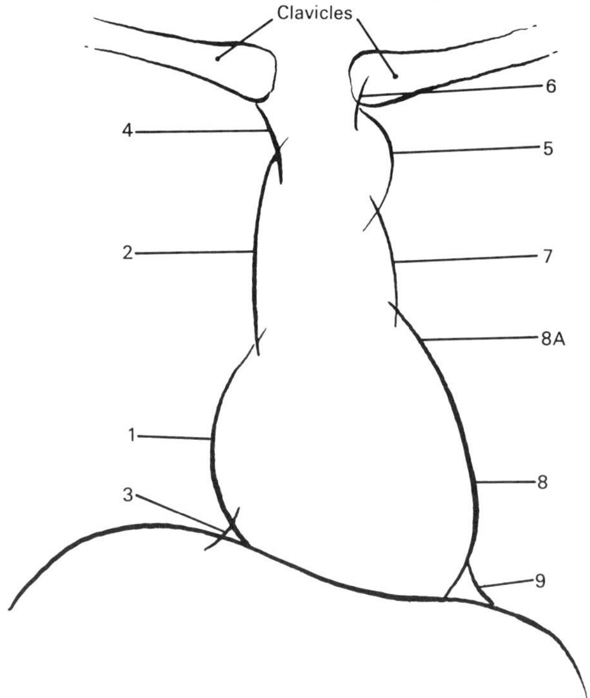

Fig. VIII-115. Roentgen anatomy of the heart in PA view. **A.** Schema of the arcs comprising the cardiac silhouette and great vessels.

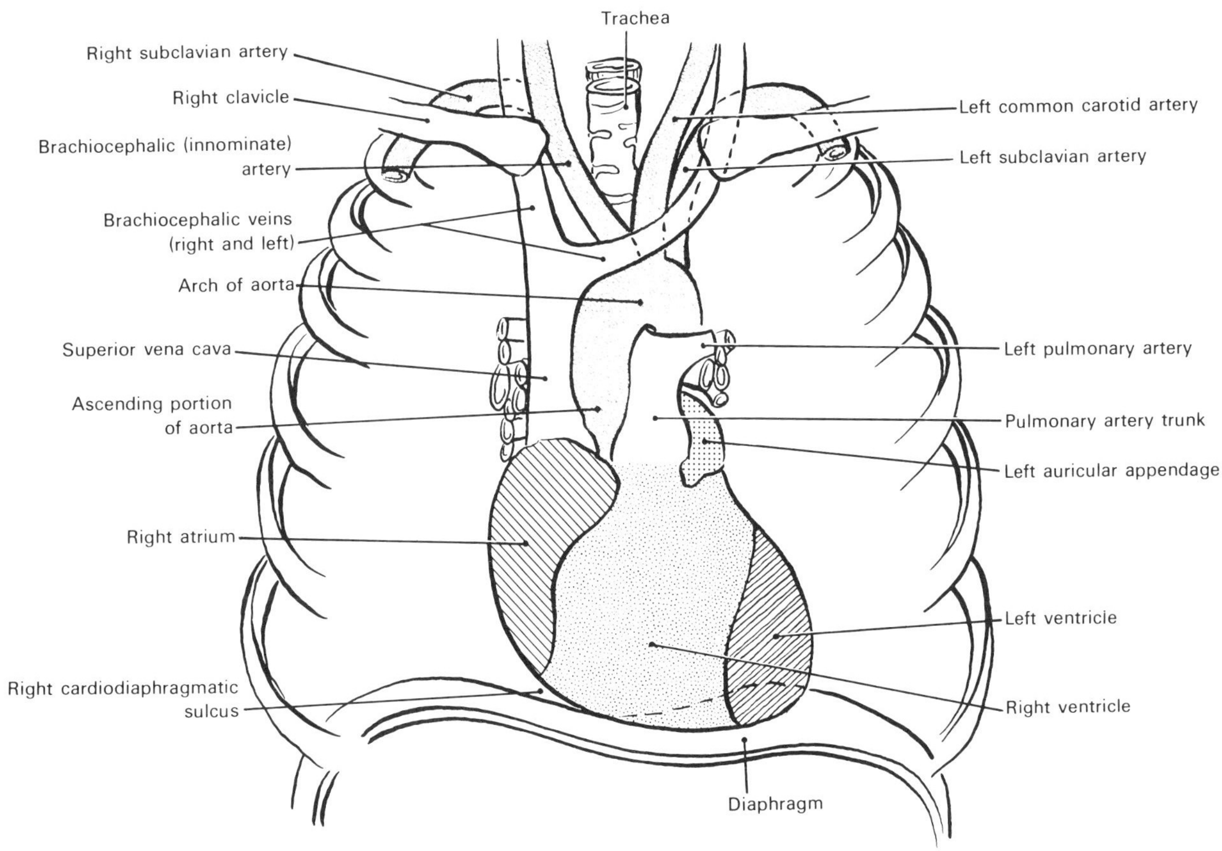

Fig. VIII-115B. Schematic drawing of C.

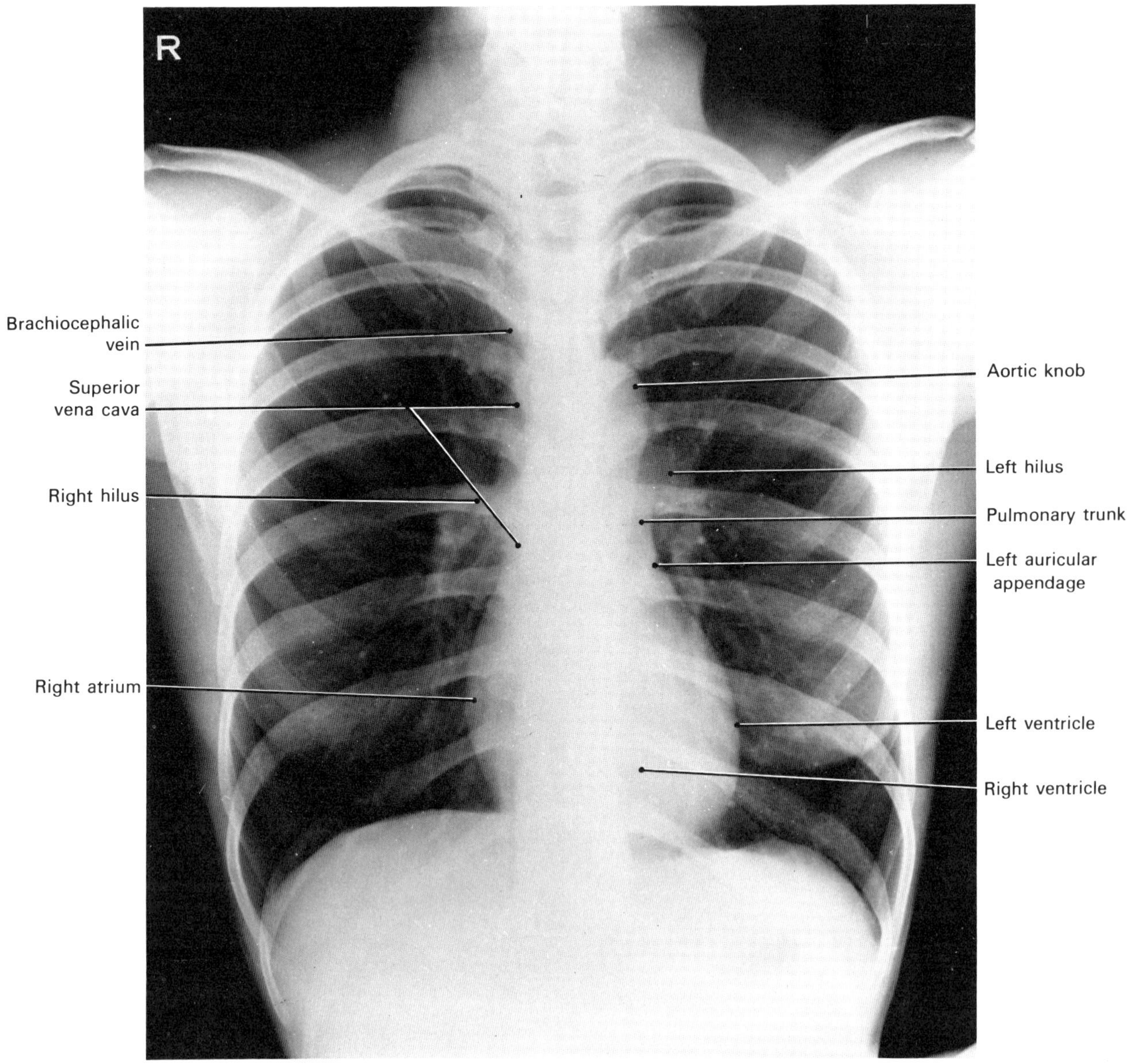

Fig. VIII-115C. PA radiograph of the chest.

have at your disposal a good heart model, which can be rotated to the projection under discussion.

1. Posterior-anterior projection (Fig. VIII-115): Nine separate segmental arcs of variable curvature form the mediastinal and cardiac contour in this projection. Starting from the right dome of the Diaphragm and proceeding cranially are two large arcs, ※1 and ※2. Arc ※1, the lower of the two, is more convex, denser, and extends farther laterally from the right vertebral border than arc ※2. Arc ※1 is formed by the lateral margin of the **right atrium.** Arc ※2, which is flatter, usually is formed by the lateral wall of the **superior vena cava,** although on occasion, the right lateral wall of the ascending aorta may contribute to this shadow. Examine some of these structures in the angiocardiograms in Figure VIII-116.

At times, we see the right cardiodiaphragmatic sulcus bridged by a smooth, faintly outlined, concave arc (※3) which extends caudally and laterally. This is formed by the lateral portion of the **inferior vena cava.**

The last of the four segmental arcs on the right side is formed by the **brachiocephalic (innominate) vein** which ascends cranially toward the medial aspect of the right clavicle (Fig. VIII-115). This concave border should not be confused with the lateral margin of the manubrium which is more laterally located.

On the left side there are five arcs contributing to the silhouette. Below the medial clavicular border, there is a rather prominent convex arc, ※5, which is formed by the foreshortened **aortic arch.** Arc ※5 is commonly referred to as the **aortic knob.** Note, too,

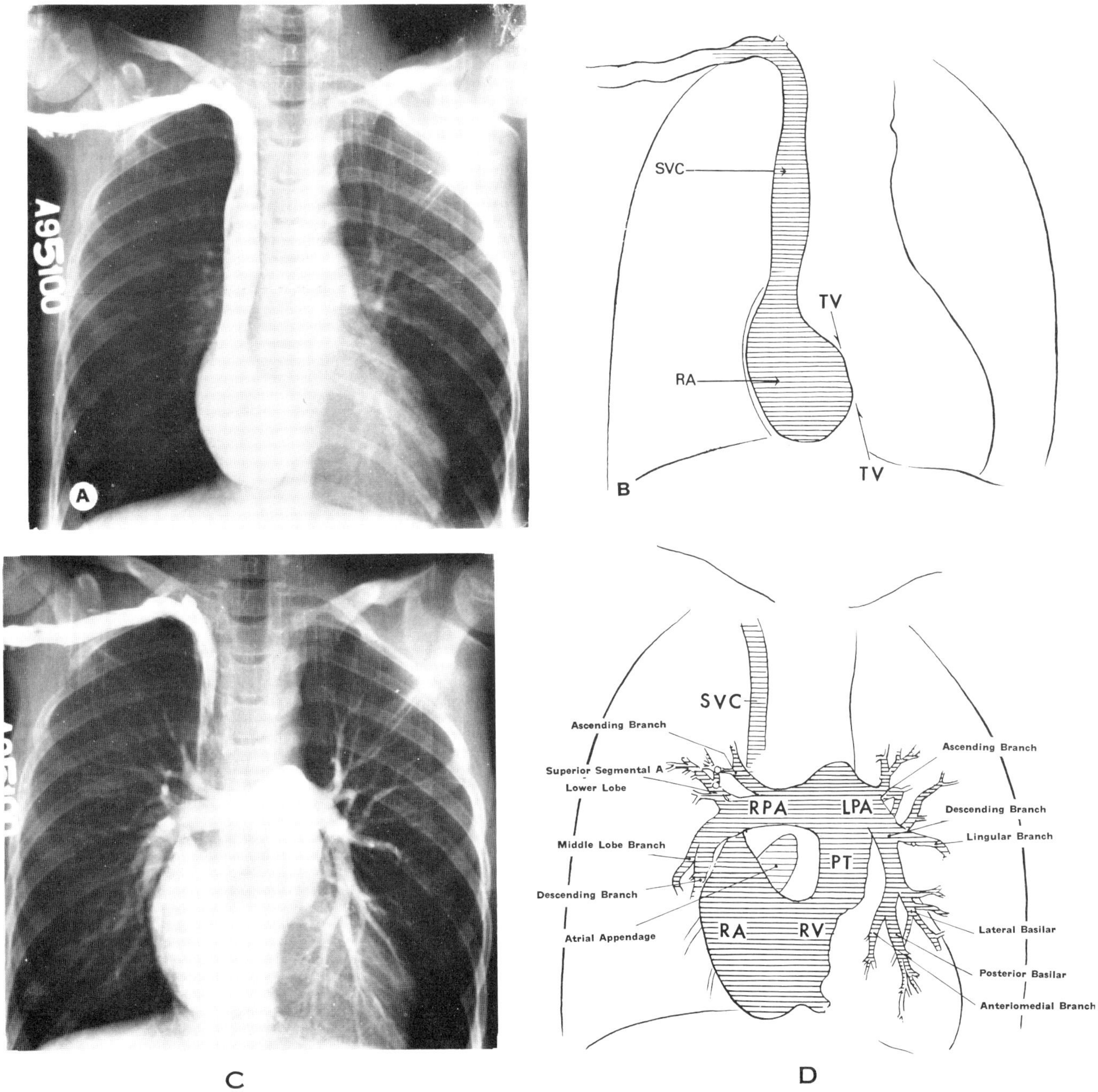

Fig. VIII-116. Intravenous angiocardiogram in frontal view. SVC, superior vena cava; RA, right atrium; TV, region of the tricuspid valve; RV, right ventricle; PT, pulmonary trunk; LPA, left pulmonary artery; RPA, right pulmonary artery. (From R. N. Cooley and M. H. Schreiber, *Radiology of the Heart and Great Vessels*, 2nd ed., Williams & Wilkins, Baltimore, 1967.)

the lateral margin of the descending aorta as it passes toward the Diaphragm. Immediately above the aortic knob, a flattened concave density may be followed all the way to the apex of the left lung. This arc, #6, is produced by the **left subclavian artery.**

The most prominent convex arc, #8, is formed, for the most part, by the anterior wall of the **left ventricle.** A small upper portion of this arc is formed by the **left auricular appendage** which blends imperceptibly into the left ventricular density. The position of the left auricular appendage is designated as arc #8A in Figure VIII-115A. Between the arch of the aorta (aortic knob(and the left auricular appendage is a flattened arc, #7, produced by the **pulmonary trunk.** Note the structures in the angiograms of Figures VIII-116 and 117.

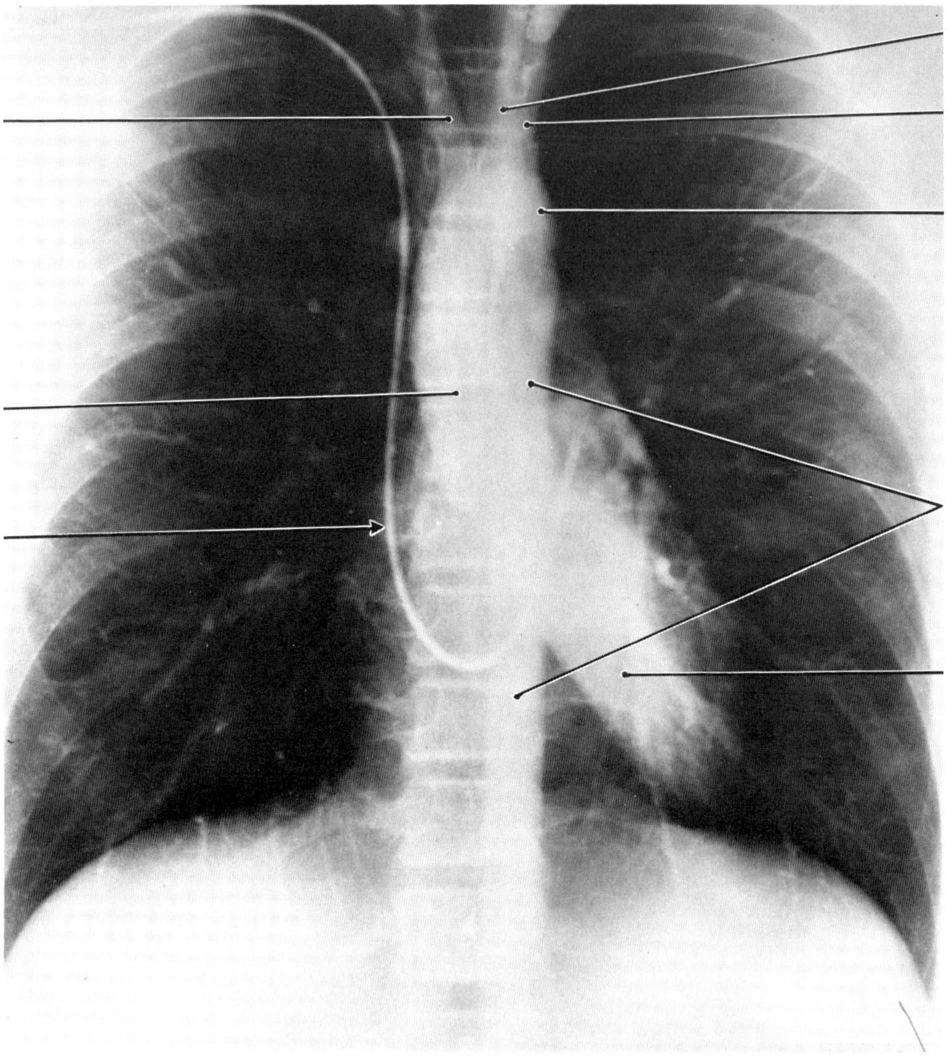

Fig. VIII-117. Selective catheterization of the main pulmonary artery. Some contrast medium is still present in the pulmonary veins and left atrium. Note particularly the thickness of the left ventricular wall.

Within the left cardiodiaphragmatic sulcus, one occasionally sees a concave shadow of variable size. This shadow is of diminished density for it represents the **cardiac fat pad,** a collection of fat and connective tissue situated between the parietal pericardium and the mediastinal pleura. It is located anteriorly overlying the heart in lateral radiographs.

From the foregoing, it is apparent, that with the exception of the left auricular appendage, the left atrium and the right ventricle are obscured by the cardiac silhouette and do not contribute to its borders in this projection. In most patients you will be unable to separate the diaphragmatic border of the heart from the upper abdominal contents and Diaphragm. As you know, the diaphragmatic border of the heart is formed by the right ventricle. Identify all of the structures mentioned in the angiocardiograms of Figures VIII-116 and 117.

Plain radiographs of the chest (teleroentgenograms) will not permit differentiation of the bifurcation of the pulmonary trunk into the left and right pulmonary arteries, nor will you be able to see these pulmonary arteries until they reach the lung roots. The hilus of each lung is formed primarily by the branches of the pulmonary arteries, with little contribution from the bronchopulmonary lymph nodes and pulmonary veins. Examine Figure VIII-116 C, D and note that the usual course of the pulmonary arteries is such that the left is slightly higher than the right. Thus Felson[35] found that 97% of his patients in a large survey showed the left hilus to be higher than the right, while only 3% showed the hili at the same level. This is an extremely important point to remember, for displacement of either hilus up or down may be an important clue to lobar collapse.

2. Right anterior oblique projection (Figs. VIII-118 and 119). This projection is made with the patient rotated 45° so that the right anterior aspect of the chest

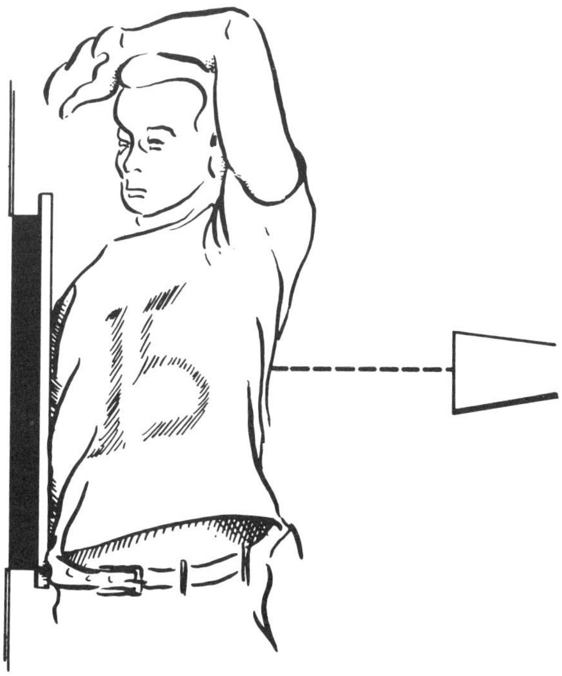

Fig. VIII-118. Positioning of the patient for the right anterior oblique projection of the chest.

touches the film cassette (Fig. VIII-118). Take the heart model in hand and rotate it 45° into the RAO position. Note that as you turn the heart to the left the left ventricle and the left atrial appendage disappear from view, while the left atrium comes into view to contribute the bulk of the right heart border. The heart has cleared the spine; however, the space created is partially obscured by the descending thoracic aorta.

Note in Figure VIII-119 that just above the right dome of the Diaphragm, there is a shallow, concave arc which is produced by the inferior vena cava and its pleural fold. The right heart border is made up by the two atria; the right atrium forming the lower margin and the left atrium above it, a longer portion of this segment. Here they are obscured by the descending aorta.

The bulk of the large convex arc on the left is due to the right ventricle, although there is usually some small contribution from the left ventricle as shown (Fig. VIII-119). Above the right ventricle one may see a slight convex arc formed by the pulmonary conus, and still farther cranially the straight shadow of the ascending aorta and aortic arch is seen in front of the trachea and left main bronchus. The remainder of the arch of the aorta is obscured by the manubrium and other overlapping structures.

Once again, find the left main bronchus and note that just below it, a few centimeters distal to its origin, is the round density cast by the left pulmonary artery seen end-on. Identify some of the branches of the right pulmonary artery crossing the spine to reach the right lung.

3. Left anterior oblique projection: This projection as shown in Figure VIII-120, is made by rotating the patient 60° from lateral so that the left anterior chest wall touches the film cassette. Note that as your heart model is rotated the required amount that the heart clears the spine; however, the descending thoracic aorta is partially superimposed upon the vertebra.

The radiograph schema, Figure VIII-121, shows a smooth convex arc forming the right heart border just above the right dome of the Diaphragm. This is formed by the right ventricle which blends imperceptibly into the density of the right atrium above. Another gently curving convex segmental arc is seen higher yet which is, of course, the ascending aorta. The aortic arch is demonstrated in profile as it courses over the tracheal column until it becomes lost over the spinal vertebral column. The great vessels arising from the aortic arch are usually obscured by other overlapping structures.

The "aortic window" is visible immediately beneath the arch of the aorta and above the left pulmonary artery, which bound it. The aortic window is crossed by the trachea and left main bronchus.

The convex arc forming the left heart border is made up by the posterolateral wall of the left atrium which is situated beneath the left main bronchus. The left atrium comprises most of the heart border but lower down it blends into the left ventricle shadow which curves downward to reach the left dome of the Diaphragm.

The pulmonary artery trunk is for the most part obscured in the LAO projection; however, we can see the end-on view of the right pulmonary artery in the right hilus superimposed over the heart. Below the right pulmonary artery some of the right pulmonary veins can be seen coursing to the left atrium. The left pulmonary artery is demonstrated below the aortic arch and just above the left main bronchus.

4. Lateral projection: A left lateral projection is customarily taken when examining the chest, since this places the heart close to the X-ray film, and thereby, minimizes distortion and magnification. Notice in Figure VIII-122 that the lower anterior convex border of the heart touches the posterior aspect of the sternum. This anterior margin of the heart is formed by the right ventricular wall, and the relatively flat pulmonary conus above. More craniad we may see the separate, anteriorly-directed, convex arc of the ascending aorta. The upper convex border of the aortic arch is seen readily as it crosses over the superior vena caval shadow which is situated just in front of the trachea. The descending thoracic aorta is usually seen just anterior to the vertebra or partially overlapping the spine as it courses toward the Diaphragm below. Inferior to the concave border of the aortic arch a second arch may be seen which is produced by the left pulmonary artery. Some of the branches of the left pulmonary artery overlap the descending aorta. Anterior to the carina, a rounded density cast by the

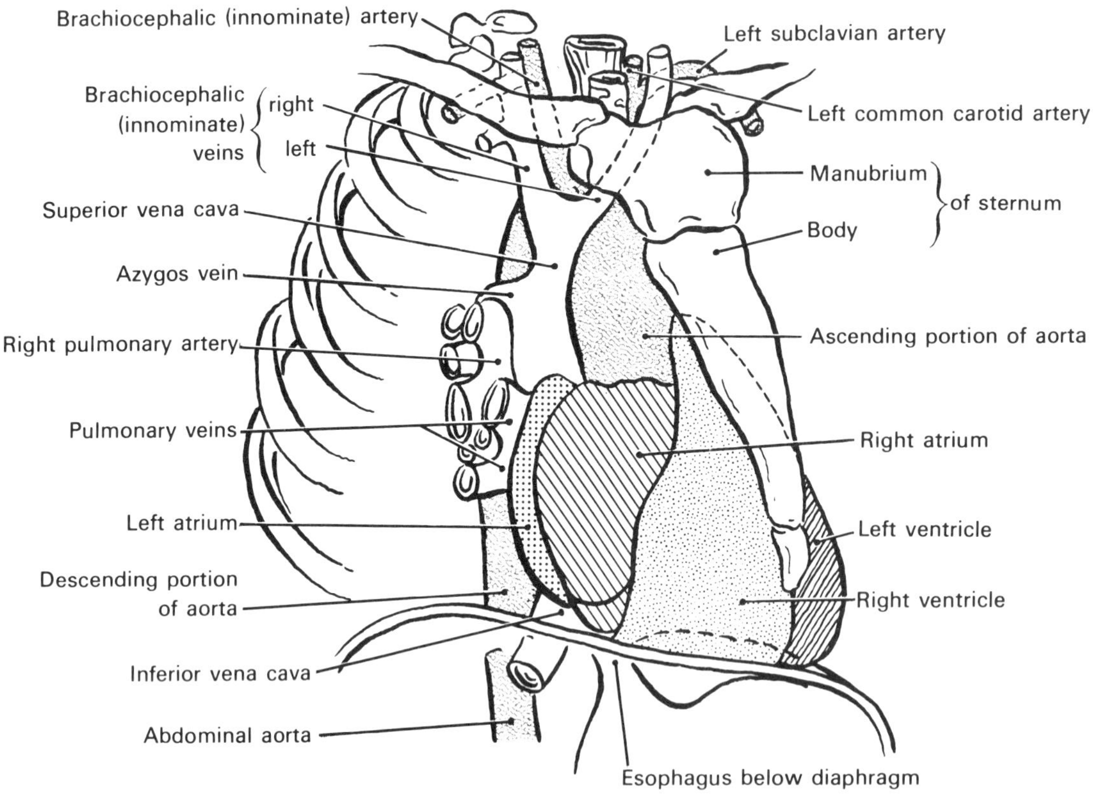

Fig. VIII-119. Roentgen anatomy of the heart and great vessels in the RAO projection. **A.** Schematic drawing of B.

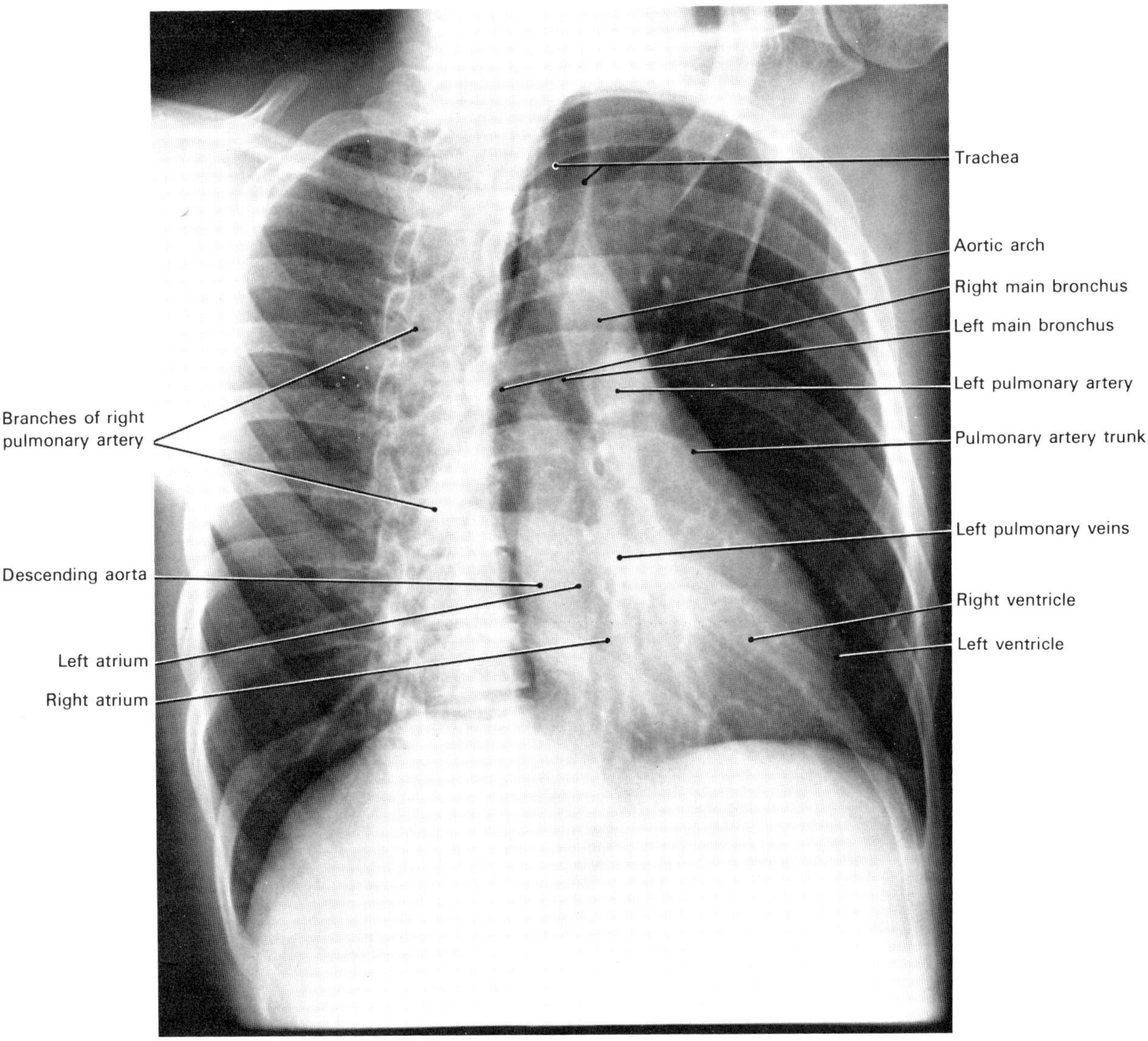

Fig. VIII-119B. Radiograph in RAO projection.

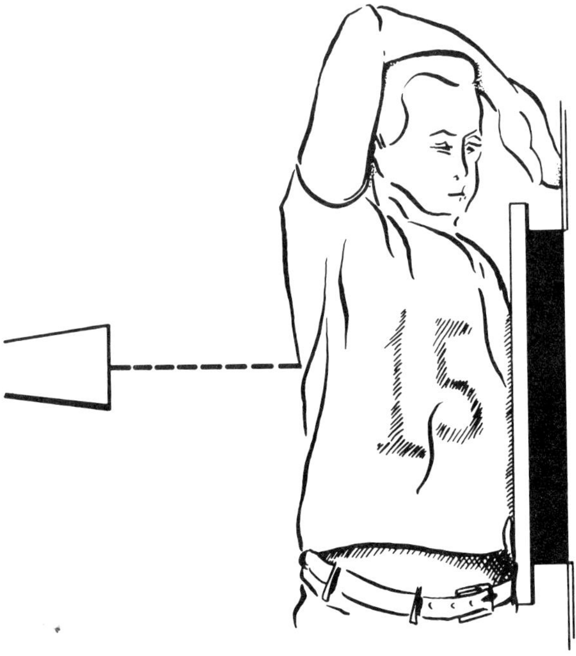

Fig. VIII-120. Positioning of the patient for the left anterior oblique projection of the chest.

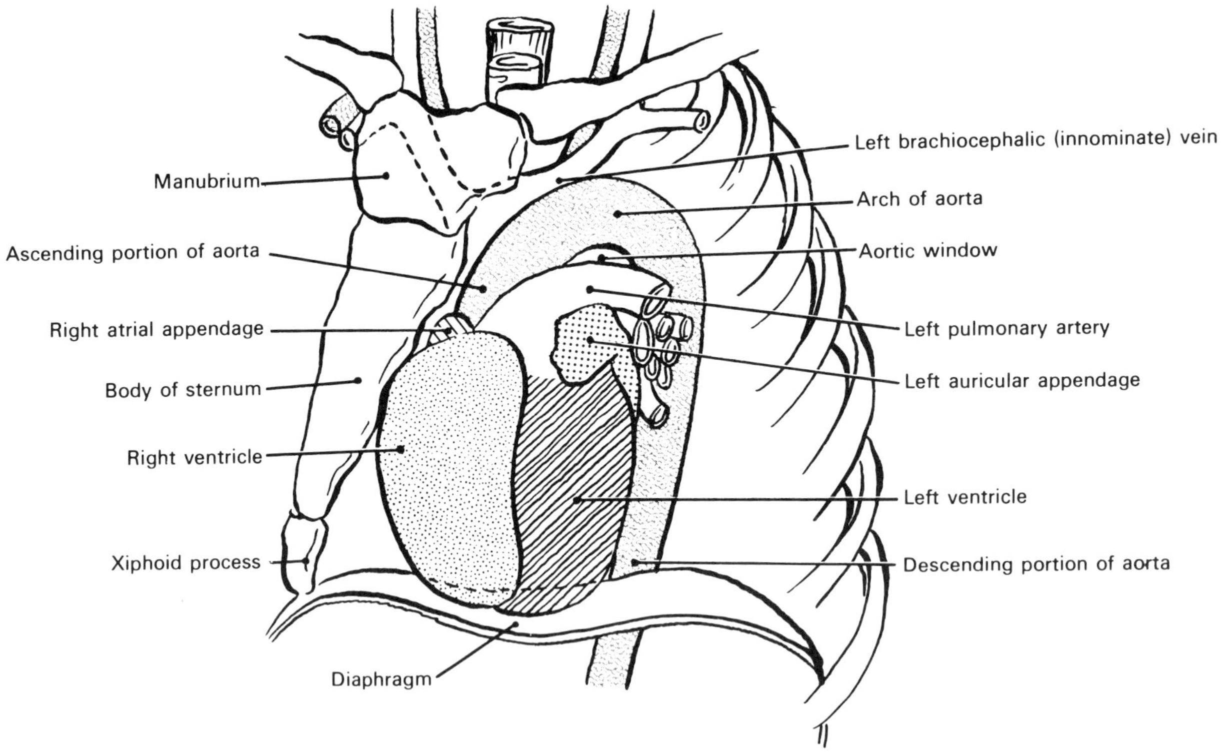

Fig. VIII-121. Roentgen anatomy of the heart and great vessels in the LAO projection. **A.** Schematic drawing of B.

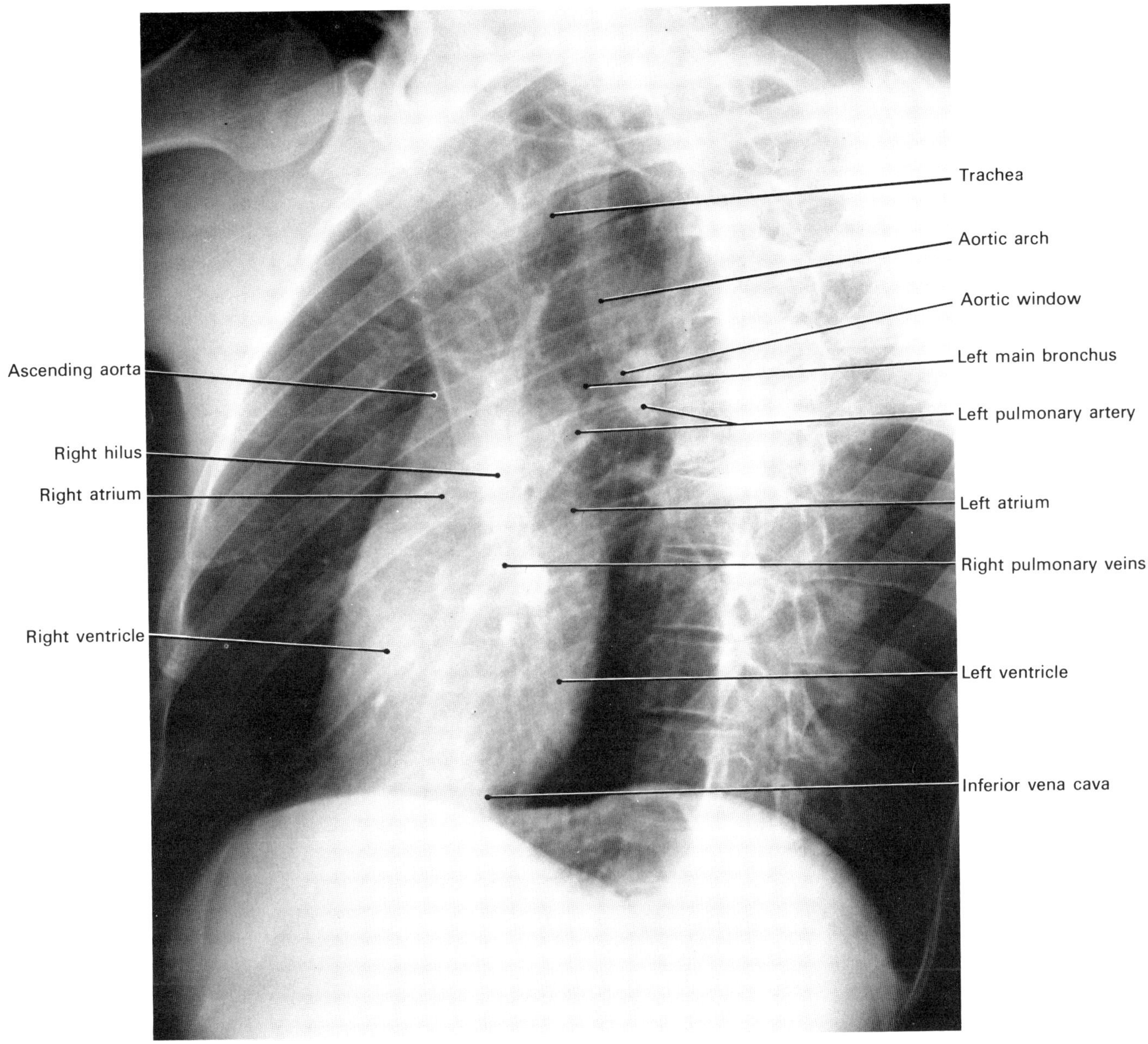

Fig. VIII-121B. Radiographin LAO projection.

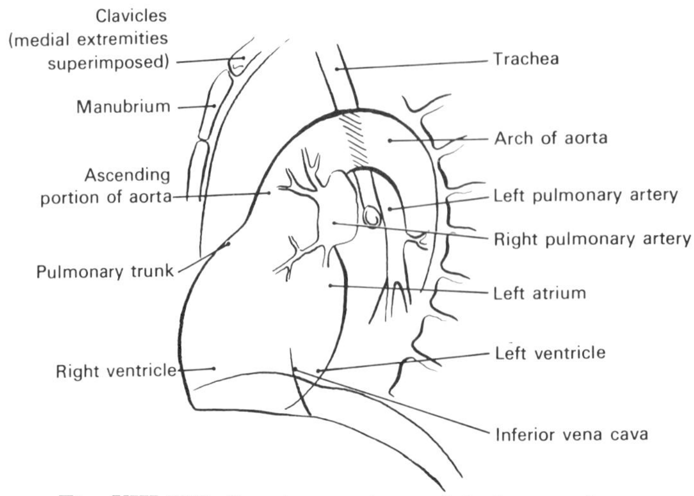

Fig. VIII-122. Roentgen anatomy of the heart and great vessels in the lateral projection. **A.** Schematic drawing of B.

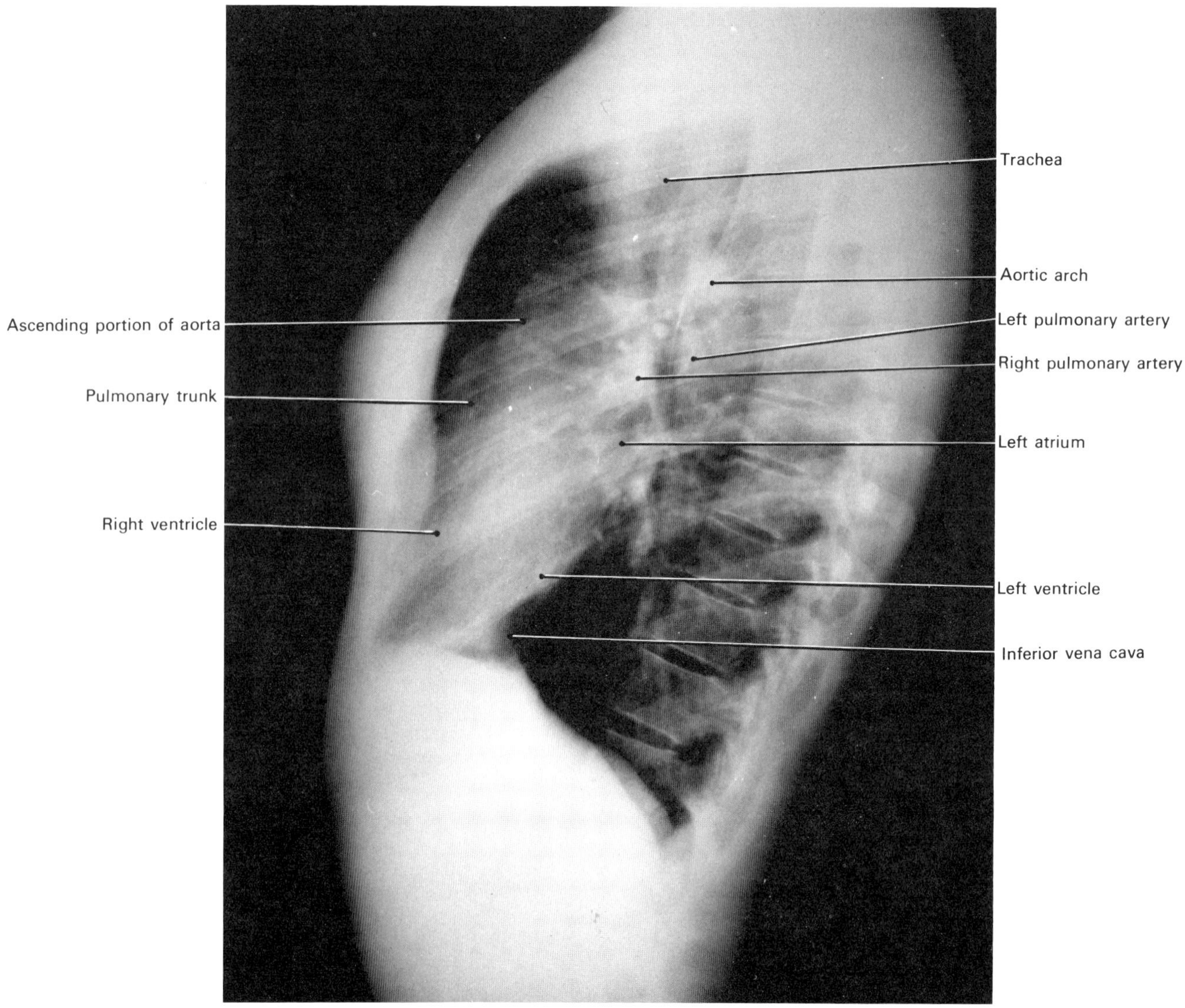

Fig. VIII-122B. Radiograph of a patient in lateral projection.

right pulmonary artery is seen; its branches overlap the aorta and heart.

The posterior convex contour of the heart in the lateral projection is formed for the most part by the left atrium and lowermost by the left ventricle. Just above the Diaphragm we again catch a glimpse of the inferior vena cava passing into the right atrium.

Note all of these structures in the angiocardiograms of Figure VIII-123.

The Conduction System of the Heart

Long before vagal or sympathetic fibers reach the heart to *regulate* its beat, the heart beats rhythmically. In a similar way, long before the vagal and sympathetic fibers reached the fetal gut tube, the fetal intestine was capable of localized constriction. In the case of the intestine this activity was due to an intrinsic autonomic nerve network which migrated there in advance of vagal and sympathetic fibers. In the heart the rhythmic contraction was initiated by specialized cardiac conductile tissue in the form of specialized and localized cardiac muscle cells.

The conductile tissue consists of the sinoauricular or sinoatrial (S-A) node; the atrioventricular (A-V) node; the A-V bundle (or bundle of His); bundle branches and the Purkinje system (Fig. VIII-124). The **S-A node** was discovered by Keith and Flack[36] in 1907 in the adult human heart. It lies embedded in the muscle of the right auricle about at the middle of the crista terminalis. Externally, this corresponds to a shallow groove on the outer surface of the auricle known as the sulcus terminalis. The node would be located deep to a point midway along the sulcus about half way between the entrance of the superior and inferior venae cavae into the right atrium. The node is a minute neuromuscular structure in the adult heart consisting of some slender, spindle-shaped, cardiac muscle cells embedded in connective tissue and intermingled with some autonomic ganglion cells and nerve fibers. The neural elements are late comers to the region. The specialized cardiac muscle fibers of the S-A node are what truly constitute the *pacemaker of the adult heart*. However, the sinus node was also a late comer in the development of the heart and not its original pacemaker as we shall soon see. Ventricles and even atria had their own unique and intrinsic rates of contraction before the sinus region was absorbed into the wall of the right atrium.

Quite separated from the S-A node, the **A-V node** lies in the lower part of the interatrial septum anterior to the opening of the coronary sinus and above the septal leaf of the tricuspid valve. Its discovery by a Japanese pathologist, Tawara, antedates the discovery of the S-A node by a year (1906). The time interval elapse of about 0.19 seconds between the beginnings of atrial and ventricular systoles has been a long known physiological event. It has been estimated that 0.07 seconds is the time required for an impulse initiated at the S-A node to reach the A-V node. The remaining 0.12 seconds has been regarded as delay time at a somewhat slower conducting A-V node. Yet, if the A-V node is depolarized experimentally, there is not this much delay before the ventricles contract. Hence, over the years the bulk of the total time involved between atrial and ventricular systole has been written off as slow conductance from the S-A node through the more refractive, normal, and less specialized cardiac muscle tissue of the atria. This seemed to make sense for two reasons: (1) no specialized tracts of conducting tissue had been discovered between S-A node and A-V node; and (2) once impulses reached the rapidly conducting bundles of His, ventricular contraction was almost instantaneous. Still, there were other odd phenomena noted in certain individuals from time to time and in animal experimentation, namely the business of premature ventricular contractions upon S-A node stimulation.

It has only been within the past 10 years that anatomists, physiologists, and clinicians have begun to define internodal conducting pathways. Even as early as 1914, Kent,[37] in discussing the auriculoventricular junction, described a "by-pass" pathway from auricle to the bundle of His which could avoid the A-V node and could account, perhaps, for premature ventricular contractions. According to him, this appeared to be a fascicle of ordinary atrial myocardium which extended directly from the inferior margin of the right atrium to the upper part of the interventricular septum (Fig. VIII-124). As you might suspect, whether this fascicle is non-specialized cardiac tissue or not, it now bears the name of the "accessory bundle of Kent."

But, beginning with the observations of Bachmann[38] and Wenckebach,[39] and confirmed in human hearts by the anatomical studies of James[39-41] and Holsinger *et al.*,[42] and by electrophysiological studies of Hoffman *et al.*,[43] and Wagner *et al.*,[44] evidence began to accumulate that there might well be several specialized conducting pathways from the S-A node to the A-V node. At least three bundles have been delineated in the atrium (Fig. VIII-124). These pass in various directions from the S-A node and are rich in Purkinje fibers (specialized cardiac muscle cells) as well as in ordinary contractile cardiac muscle. The Purkinje fibers do not seem to be continuous with each other in these tracts, but their presence there could account for the observed speed of conduction and the reproducibility of depolarization patterns in *in vivo* studies with both unipolar and bipolar electrodes.

Of the three bundles, the **anterior internodal tract** (an off-shoot of Bachmann's bundle) and the **middle internodal tract** seem to be of equal length. They begin at the S-A node. The anterior tract spreads to the left around the superior vena cava before the tract splits into two branches. The portion that extends dorsally into the left atrium is Bachmann's old bundle (or old Bachmann's bundle). The other branch crosses the interatrial septum to the A-V node where it be-

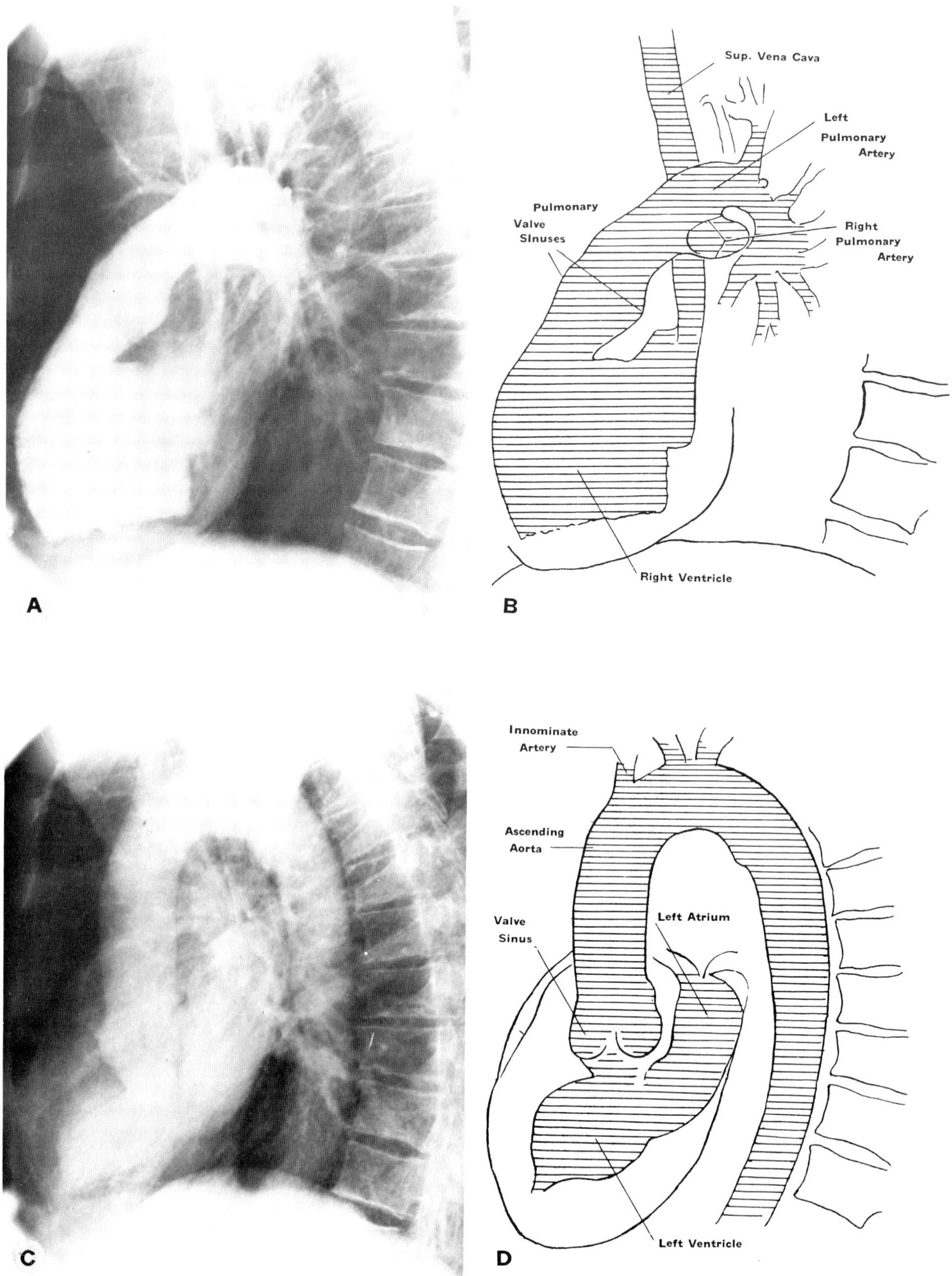

Fig. VIII-123. Angiocardiogram in lateral view. (From R. N. Cooley and M. H. Schreiber, *Radiology of the Heart and Great Vessels,* 2nd ed., Williams & Wilkins, Baltimore, 1967.)

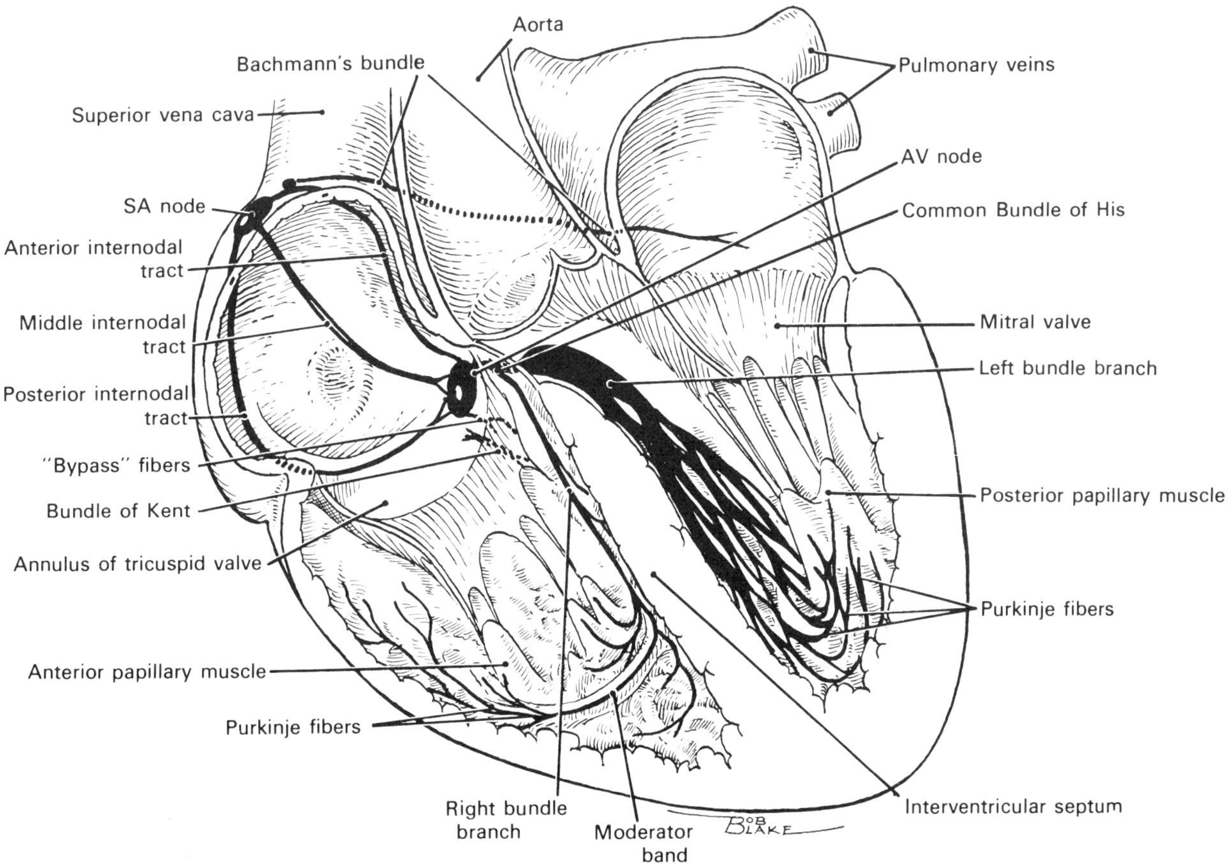

Fig. VIII-124. The conducting system of the heart.

comes enmeshed with other tracts to the node. The middle internodal tract leaves the posterior margin of the S-A node to cross the interatrial septum just ventral to the foramen ovale to end in the A-V node. A third, longer route has been described as the **posterior internodal tract.** It extends from the tail end of the sinus node along the crista terminalis over the Eustachian ridge (the ridge between the valves of the inferior vena cava and coronary sinus) to reach the lower portion of the interatrial wall and A-V node. All of these bundles seem to split off "by-pass" fibers which miss the A-V node to enter portions of the interventricular walls on their own. Some may contribute to the bundle of His. Thus, there may be more "by-pass" opportunities than Kent first imagined. Their exact nature is ill understood, for, day in and day out, in routine heart beats we are faced, as usual, with the normal A-V delay in conduction time. It may be that the "by-pass" fibers serve as a dual innervation to the node.

From the available reports, it would appear that normal depolarization reaches the A-V node *via* the anterior and middle internodal tracts almost simultaneously and there is some delay over the longer posterior route. Severing the first two tracts occasions a 20 to 40 msec. delay, and if all three tracts are cut a complete heart block may result.

To depolarize prematurely a part of the ventricle it would seem to require that atrial impulses circumvent the A-V node. This kind of premature depolarization is the basis for the controversial Wolff-Parkinson-White (W-P-W) syndrome. In this syndrome, impulses reach the right ventricular wall ahead of impulses to the remainder of the ventricles. Patients show EKG changes, frequent episodes of paroxysmal arrhythmias, and premature ventricular contractions. We have had Kent's accessory bundle to account for this for some time, and we have just mentioned other "by-pass" fibers splitting off from the main interatrial conducting routes. Nevertheless, there is the recent report by Cobb *et al.* which indicates that the delta wave and arrhythmias associated with the W-P-W syndrome were eliminated by making a surgical incision in the region of the accessory bundle of Kent, staying well away from the A-V node and the bundle of His. After this incision, conduction was apparently corrected through the normal intact ventricular conducting pathways. They felt that cutting parallel to the crista terminalis (to inter-

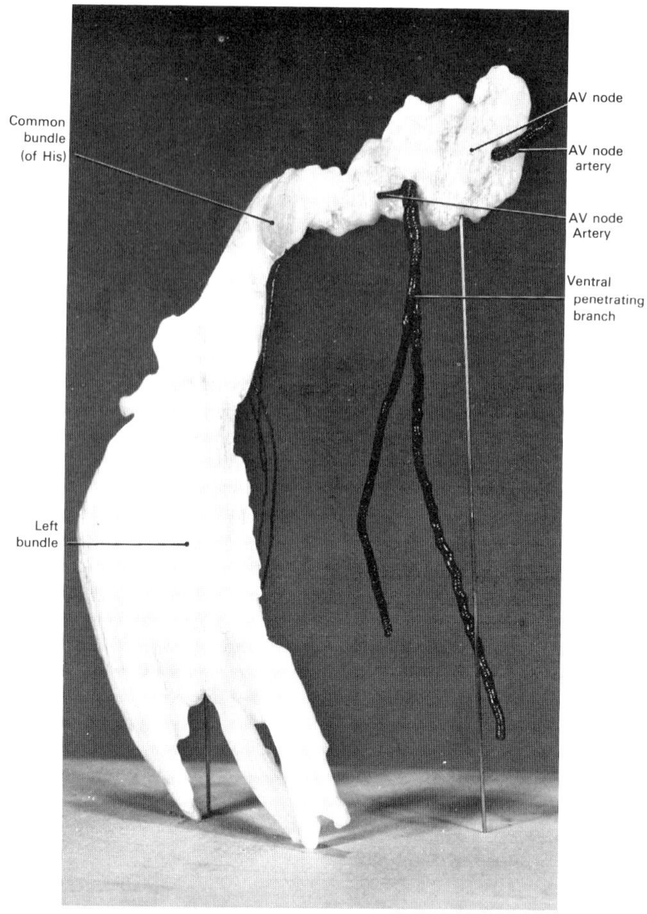

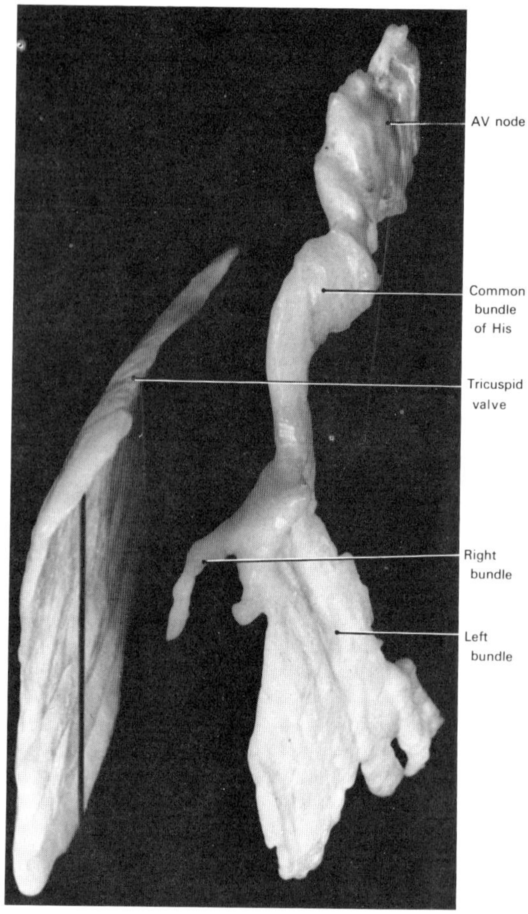

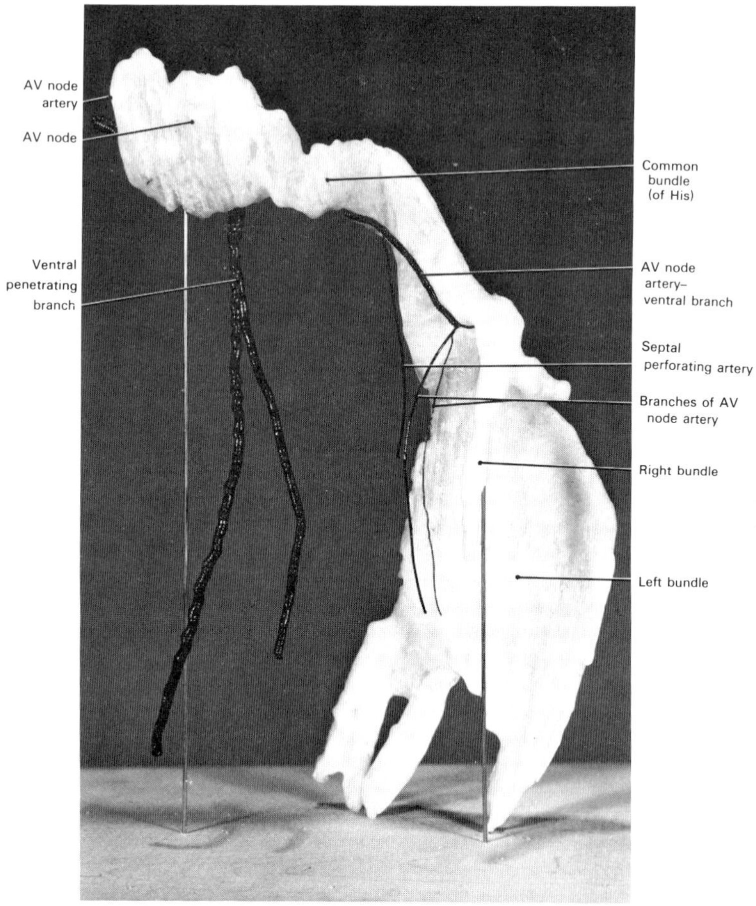

Fig. 125A. A paraffin reconstructed model of the ventricular conducting system of the human heart with accompanying vascular supply related to the anatomical location of this structure. Lateral view (right side). **B.** Anterior view of the paraffin reconstruction of the ventricular conducting system. The free wall of the right ventricle is shown on the left of this photograph. Both the right and left branching bundles into the ventricles originate from the common bundle. **C.** Paraffin reconstruction of the ventricular conducting system. Lateral view (left side). (Courtesy Leona Ayers, M.D.)

fere with the posterior internodal bundles) would have inhibited atrial conduction, but would not have altered the aberrent ventricular behavior of the W-P-W syndrome.

With the identification of these various interatrial bundles and their "by-pass" potentials, we may simply be dealing with anatomical variations from one individual to another, and there may be more than one precise anatomical explanation to account for the clinical significance of the Wolff-Parkinson-White syndrome.

We must now continue with a discussion of the major part of the conducting system, that portion from A-V node to the ventricular walls. Conduction from the A-V node over the A-V bundle and the Purkinje system of cardiac muscle fibers is very rapid. An impulse over this system reaches the apex of the heart and spreads over the entire ventricular musculature in 0.4 seconds. Two individuals have been given credit for the discovery of the A-V bundle and both made their reports on it in the year 1893. Albert Frank Stanley Kent, he of the "accessory bundle" mentioned above, was a British physiologist at Manchester and Bristol. Wilhelm His, Jr., (pronounced "Hiss", like a leaky, steam-radiator valve) taught internal medicine at a number of German Universities. Somehow the eponym of His has survived the test of time as a name for the A-V conducting bundle. This specialized cardiac tissue runs a short horizontal course as a single strand, forward and to the left of its origin from the A-V node. It passes over the septal leaf of the tricuspid valve in the process to reach the upper part of the interventricular septum (Fig. VIII-124). Here it breaks into a right and left branch, each of which follow down opposite sides of the septum in the endocardium. Both give primary branches to papillary muscle and terminate by splaying into the subendothelium of the ventricular walls as innumerable filaments that form a delicate interlacement. If the **bundle of His** and its terminals are held to differ from the specialized **Purkinje fibers** of the heart, we fail to note much difference under the microscope. That they do differ from nodal tissue is quite evident. Otherwise, this tissue appears as a band of hypertrophied cardiac muscle fibers rich in sarcoplasm and glycogen but poor in myofibrils. In this respect, it is similar to what Purkinje described as special cardiac muscle cells of the endocardium. From the gross standpoint, it is not easy to find either of the nodes or the conducting bundle in the dissecting of the human heart. These structures do not stand out as vividly as they do in oxen or sheep hearts. However, that they can be discovered with patience and specialized techniques in the human is attested to by the photographs in Figure VIII-125, a reconstruction of the A-V node and the bundle of His by Dr. Leona Ayers, a former resident in pathology at the Duke University Medical School.

From the viewpoint of early development, the sinoatrial tissue is not the original pacemaker of the embryonic heart. It only assumes this role when the separate parts of the heart tube have completely fused. The mammalian heart begins to beat at the three-somite stage before the paired tubes fuse into one.[45] The initiation of the heart beat in mammals occurs even earlier in development than in the chick[46] or the amphibian.[47] In the mammal, three or four cardiac cells begin to beat in the left ventricle near its junction with a future atrium. About 2 hours later a slower rhythm begins in the unfused right ventricle, retaining its beat until the ventricular parts of the tube fuse. When fusion occurs, the faster left ventricular beat dominates. Not much propulsion of the blood occurs. It just sloshes back and forth in a desultory fashion. Faster contractions begin in the atrial chambers next, but until these fuse with the ventricles, the ventricles continue their slower rhythm. With fusion, the faster atrial beat becomes the pacemaker for the system. Up to this time the sinus region has been quiescent, but it now contracts with the fastest rate of all, and when it becomes incorporated into the right atrium, its beat sets the pace for the developing heart as a whole.

The Fetal Circulation in Review

We have studied the adult heart and have seen how it developed from a longitudinal tube. We have seen some of the embryonic remains within the adult heart, especially in the right atrium. The circulation through the adult heart was outlined in Chapter II, and this may be reviewed in Figure VIII-126. Certain aspects of circulation through the fetal heart were discussed above, and the general concept is presented in Figure VIII-127, which we will now consider.

The digestive organs and the kidneys are all well developed before birth. Throughout the fetal months there has been functional absorption and excretion, but these are carried on through the placenta. As a consequence the umbilical arteries and veins loom far more important than the pulmonary arteries and veins. The lungs remain tardy in development, however, so there is not much return of blood to the left atrium from the pulmonary veins. Certainly the volume returned to the right atrium by the caval veins is far greater. Moreover, the placental circulation through its umbilical vessels is part of the systemic circulation, and the umbilical vein eventually empties into the inferior vena cava.

The valve of the inferior vena cava directs most, if not all, of the well-oxygenated blood returning from the placenta through the foramen ovale in the interatrial wall. Therefore it crosses ventral to the superior caval stream. If it were not for this device, the left atrium would receive only a small amount of the blood returning from the lungs. If superior and inferior caval inflows mixed in the right atrium, the blood in both atria would be equally diluted with venous blood returning from the systemic circulation in general and the entire fetus would be poorly oxygenated.

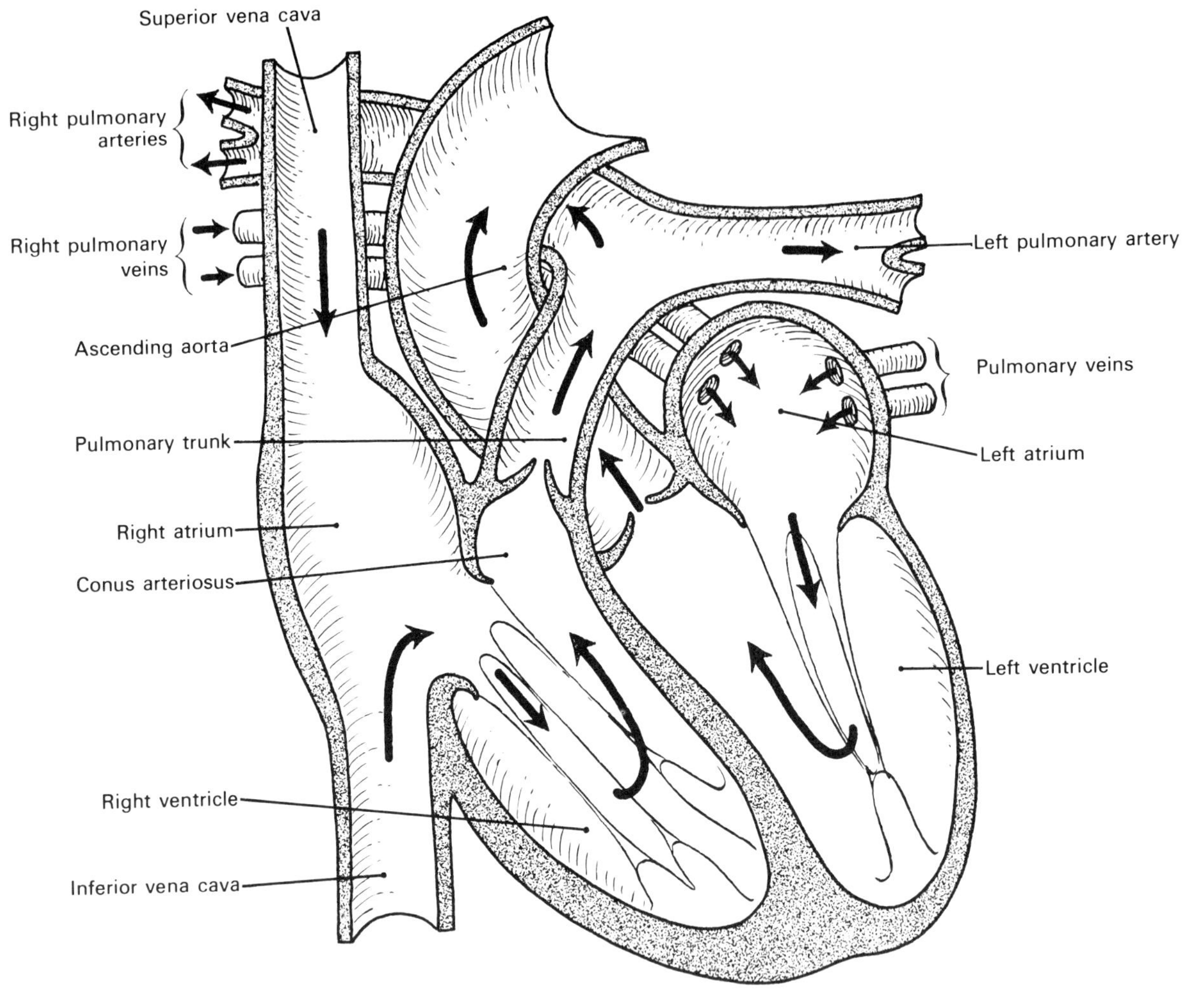

Fig. VIII-126. Plan of circulation through the postnatal or adult heart.

Instead, the superior caval stream is directed through the atrioventricular orifice into the right ventricle where it is pumped over the pulmonary trunk. Some goes to the non-functional lung. However, most passes through a second shunt, the ductus arteriosus. This by-pass leads to the descending aorta and an eventual return to the placenta *via* the umbilical arteries.

The most highly-oxygenated blood, having reached the left heart directly, is pumped over the ascending aorta and its arch to supply the upper parts of the developing embryo. While the developing neurons of the brain can withstand hypoxia far better in the fetus than they can in the infant or the adult, they still have the highest respiratory quotient of any of the tissues and their need for oxygen is even greater than that of cardiac tissue which must also be kept well supplied.

The rest of the body can get along on a reduced supply since metabolic demands are not as great. Once the oxygenated supply rounds the aortic arch into the descending aorta it is greatly diluted by venous blood pouring out of the ductus arteriosus.

Figure VIII-127 shows, by changes in shading, how the oxygenated blood supply from the placenta becomes less "pure" as it is contaminated by channels of venous return at various points along its circulatory route. Even before it reaches the heart it is diluted by some return from the liver and from the inferior vena cava, itself. Actually, the umbilical vein has worn a first shunt through the liver known as the **ductus venosus** to avoid complete contamination from that source. The second shunt through the foramen ovale allows for further contamination of the oxygenated blood by return from the pulmonary veins. The greatest mixing of aerated and venous blood occurs, of course, where ductus arteriosus and descending aorta meet in the third fetal shunt.

What happens at birth? The tying of the umbilical cord stops the output from the placenta through the umbilical vein and its return through the umbilical arteries. The placenta is cut off as an organ of respira-

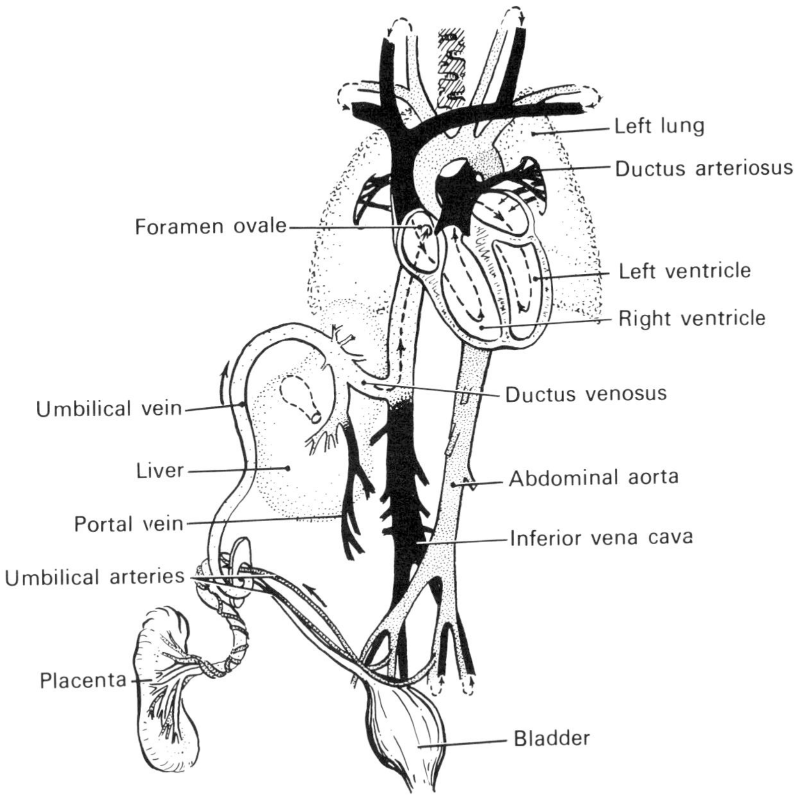

Fig. VIII-127. The fetal circulation. Degrees of reduction in oxygenation are shown by changes in shading at strategic points Least shaded, most highly oxygenated (*e.g.*, umbilical vein); black, most reduced blood (*e.g.*, superior vena cava, pulmonary trunk, ower inferior vena cava, portal return from the gut).

tion, excretion and as a source of food supply. These duties are now foisted, full measure, upon the organs of the new born. Its immediate need is oxygen, without which life is soon extinguished. The need is met by respiratory movements of its own lung as circulating CO_2 builds up in the blood and as the pH of blood changes under sudden hypoxic stress. These factors activate the rhythmic discharge of the neurons of the respiratory center in the hindbrain and pulmonary respiration begins.

There seems to be good reason to expect a reduction in the amount of blood entering the right atrium when umbilical vessels are occluded. It is unlikely that the inferior vena cava has developed to a size larger than is necessary to handle the contingencies of prenatal life. This vessel probably only accommodates gradually to the increased load placed upon it, enlarging in diameter with prenatal growth. What then happens to any extra blood? It seems reasonable to expect new capillary beds (especially those in the lung) to open up to receive more blood as an increase in tonus develops and new muscular activities are attempted.

Thus, filling pressures in the right heart decrease. As negative intrathoracic pressures become stabilized with continued respiration, there should be decreased peripheral resistance within the blood channels of the lung, another factor which should lead to a drop in right atrial pressure. Concomitantly, increased flow through the lungs should increase filling pressure in the left atrium. We can anticipate, then, that septum I (the valve of the foramen ovale) will be pressed against septum II to functionally close the foramen ovale.

In the dog and rabbit, flap-like valves are also active in closing the ductus arteriosus. In the lamb, a muscular sphincter constricts the ductus within a few minutes after breathing starts. The ductus in man lacks valves and a muscular sphincter. Its closure is accomplished gradually over several weeks or months by a growing together of intimal pads. If it remains open, blood should flow from aorta into pulmonary artery. It should be possible to hear the characteristic murmur of a patent ductus in every infant. This is not so. There must, therefore, be some as yet unknown mechanism which brings about functional closure of the ductus in the majority of instances, even though true anatomical closure proceedes at a slower pace. Individuals have lived for several decades with a patent ductus, although the condition predisposes toward bacterial endocarditis, and places additional stress upon the left ventricle which can lead to eventual cardiac decompensation. The condition of patent ductus arteriosus can be corrected surgically.

REFERENCES

1. Knutsson, F.: The mediastinal pleura. Acta Radiol. (Stockholm), *43:* 265–275, 1955.
2. Cimmino, C. V., and Snead, L. O.: The posterior mediastinal line on chest roentgenograms. Radiology, *84:* 516, 1965.
3. Felson, B.: The mediastinum. Seminars Roentgenol., *4:* 41–58, 1969.
4. Gladnikoff, H.: A roentgenographic study of the mediastinum in health and in primary pulmonary carcinoma. Acta Radiol. (Stockholm), Suppl. no. 73, 1948.
5. Berne, A. S., Gerle, R. D., and Mitchell, G. E.: The mediastinum: normal roentgen anatomy and radiologic technics. Seminars Roentgenol., *4:* 3–21, 1969.
6. Figley, M. M.: Mediastinal minutiae. Seminars Roentgenol., *4:* 22–32, 1969.
7. Boyden, E. A.: The distribution of bronchi in gross anomalies of the upper right lobe, particularly lobes subdivided by the azygos vein and those containing pre-eparterial bronchi. Radiology, *58:* 797, 1952.
8. Jackson, C. L., and Huber, J. F.: Correlated applied anatomy of the bronchial tree and lungs with a system of nomenclature. Dis. Chest, *9:* 319, 1943.
9. Brock, R. C.: *The Anatomy of the Bronchial Tree (with Special Reference to the Surgery of Lung Abscess)*, 2nd ed. Oxford, London, 1954.
10. Boyden, E. A.: *Segmental Anatomy of the Lungs.* McGraw-Hill, New York, 1955.
11. Nelson, S. W.: Refresher course on bronchography. American Roentgen Ray Society, New Orleans, Oct. 1968.
12. Miller, W. S.: *The Lung,* 2nd ed. Thomas, Springfield, Ill., 1950.
13. Lauweryns, J. M., Baert, A., and Boussauw, L.: The pulmonary lymphatics: macroscopical and microscopical studies. Proceedings of the 12th Aspen Emphysema Conference, Aspen, Colorado, 1969. Amer. Rev. Resp. Dis., in press.
14. Pennell, T. C.: Anatomical study of the peripheral pulmonary lymphatics. J. Thorac. Cardiov. Surg., *52:* 629–634, 1966.
15. Trapnell: The peripheral lymphatics of the lung. Brit. J. Radiol., *36:* 660–672, 1963.
16. Tobin, C. E.: Pulmonary lymphatics. With reference to emphysema. Amer. Rev. Resp. Dis., *80*(suppl.): 50–57, 1959.
17. Staub, N. C., Nagano, H., and Pearse, M. L.: Pulmonary edema in dogs, especially the sequence of fluid accumulation in lungs. J. Appl. Physiol., *22:* 227–240, 1967.
18. Lauweryns, J. M.: L'angioarchitecture du poupon. Arch. Biol. (Liege), *75* (suppl.): 771–811, 1964.
19. Lauweryns, J. M.: The lymphatic vessels of the neonatal rabbit lung. Acta Anat. (Basel), *63:* 427–433, 1966.
20. Lauweryns, J. M., and Boussauw, L.: The ultrastructure of pulmonary lymphatic capillaries of newborn rabbits and of human infants. Lymphology, *2:* 108–129, 1969.
21. Lauweryns, J. M., and Boussauw, L.: L'ultrastructure des vaisseaux lymphatiques pulmonaires. Comptes Rendus de L'Association Anatomistes, 52e Congres, Paris-Orsay, *138:* 766–775, 1968.
22. Boussauw, L., and Lauweryns, J. M.: Reconstructions graphiques des valvules lymphatiques pulmonaires. Comptes Rendus de L'Association des Anatomists, 54e reunion, Sofia, 30 Mars–3 Avril 1969, in press.
23. Davis, C. L.: The development of the human heart from its first appearance to the stage found in embryos of 20 paired somites. Contrib. Embryol., *19:* 245, 1927.
24. Scaramucci: *Diario Parmeuse,* 1689. Quoted by Porter from Holler's Elementia Physiolog., lib. IV, 459, 1778.
25. Schlessinger, M. J.: Relation of anatomic pattern to pathologic conditions of coronary arteries. Arch. Path. (Chicago), *30:* 403–415, 1940.
26. Wearn, J. T.: The extent of the capillary bed in the heart. J. Exp. Med., *47:* 273, 1928.
27. Vieussens, R.: Nouvelle décourvertes sur le coeur. Toulouse, 1706.
28. Thebesius, A. C.: Dissert. Med.: Inaug. de circulo sanguinis in cordi. Leiden, 1708.
29. Estes, E. H., Entman, M. L., Dixon, H. B., and Hackel, D. B.: The vascular supply of the left ventricular wall. Amer. Heart J., *71:* 58–67, 1966.
30. Estes, E. H., Dalton, F. M., Entman, M. L., Dixon, H. B., and Hackel, D. B.: The anatomy and blood supply of the papillary muscles of the left ventricle. Amer. Heart J., *71:* 356–362, 1966.
31. Hackel, D. B., Estes, E. H., Walston, A., Koff, S., and Day, E.: Some problems concerning coronary artery occlusion and acute myocardial infarction. In press.
32. Windle, W. F., and Becker, R. F.: The course of blood through the fetal heart. Anat. Rec., *77:* 417, 1940.
33. Whitehead, W. H.: A working model of the crossing streams in the fetal right atrium. Anat. Rec., *82:* 277, 1942.
34. Barclay, A. E., Franklin, K. J., and Pritchard, M. M. L.: *The Foetal Circulation.* Blackwell, Oxford, 1944.
35. Felson, B.: *Fundamentals of Chest Roentgenology.* Saunders, Philadelphia, 1960.
36. Keith, A., and Flack, M.: The form and nature of the muscular connections between the primary divisions of the vertebrate heart. J. Anat. Physiol., *41:* 172, 1907.
37. Kent, A. F. S.: Researches on the structure and function of the mammalian heart. J. Physiol. (London), *14:* 233, 1893.
38. Kent, A. F. S.: Observations on the auriculoventricular junction of the mammalian heart. Quart. J. Exp. Physiol., *7:* 193, 1914.
39. James, T. N.: Anatomy of the sinus node of the dog. Anat. Rec., *143:* 251, 1962.
40. James, T. N.: Morphology of the human atrioventricular node with remarks pertinent to its electrophysiology. Am. Heart J., *62:* 756, 1961.
41. James, T. N.: The connecting pathways between the sinus node and A-V node and between the right and left atrium in the human heart. Amer. Heart J., *66:* 498, 1963.
42. Holsinger, J., Wallace, A., and Sealy, W. C.: The identification and surgical significance of atrial internodal conduction tracts. Ann. Surg., *167:* 447, 1968.
43. Hoffman, B. F., Paes de Carvalho, A., de Mello, W. C., and Cranefield, P. F.: Electrical activity of single fibers of the atrioventricular node. Circ. Res., *7:* 11, 1959.
44. Wagner, M. L., Lazzara, Weiss, and Hoffman, B. F.: Specialized conducting fibers in the interatrial band. Circ. Res., *18:* 502, 1966.
45. Goss, C. M.: The first contractions of the heart in rat embryos. Anat. Rec., *70:* 505, 1938.
46. Sabin, Florence R.: Studies on the origin of blood vessels and of red blood corpuscles as seen in the living blastoderm of chicks during the second day of incubation. Contrib. Embryol. (Carnegie Institution of Washington), *9:* (No. 36): 213, 1920.
47. Johnstone, P. N.: Studies on the physiological anatomy of the embryonic heart. II. An inquiry into the development of the heart beat in chick embryos, including the development of irritability to electrical stimulation. Bull. Hopkins Hosp. *36:* 299, 1925.

NINE

The Abdomen

PART I. THE ABDOMINAL WALL

Its Boundaries and Surface Anatomy

Although the major features of surface anatomy were covered in Chapter III, pages 131 to 138, it might be well to reiterate a few boundaries and surface considerations here. The Diaphragm forms a roof for the abdominal cavity (Fig. IX-1A). Where the abdominal cavity continues over into the pelvic cavity there is no floor. To keep structures from tumbling out of the pelvis, a **pelvic diaphragm** lies in a different plane in what is spoken of as the **true** or **lesser pelvis.** This diaphragm forms the floor of the confluent abdominopelvic cavities, and it consists principally of two major muscles with their superior and inferior investing fasciae. These are the **Levatores Ani** and the **Coccygei** (Fig. IX-2). A **false** or **abdominal pelvis** lies *above* the level of the **iliopectineal** or **arcuate line,** a line often called the **pelvic brim.** This line should not be confused with the so-called "white line" or the **arcuate tendon** of the Levator Ani. The latter is merely a thickening of the fascia of the Obturator muscle. This muscle clothes the bony wall of the **true pelvis,** which lies *below* the arcuate line, and the tendon gives rise to the Levator Ani, a chief muscle of the pelvic diaphragm. Figure IX-1 shows that the pelvic cavity is not in direct line with the abdominal cavity. Rather it projects behind as well as below the abdominal cavity proper. The so-called "false pelvis" is truly a part of the abdominal cavity proper.

The manner in which the abdominal cavity is divided into three zones is shown in Figure IX-1B. The first two zones are separated from each other by the *subcostal plane,* an encircling plane drawn around the trunk from the lowest level of the costal margin that you can palpate *from the front.* That would place the plane at the lower level of the 10th costal cartilage. The last two zones of the abdomen are separated by the **intertubercular plane.** This line, in encircling the trunk, passes through the two highest points of the **iliac crests** that you can see or feel *from the front,* that is, the line crosses through the **iliac tubercles** at the level of VL_5. The intertubercular plane separates the **umbilical zone** from the **suprapubic zone** of the "false pelvis" (Fig. IX-1B).

In the plane of the arcuate line, structures common to both abdominal and pelvic cavities cross the pelvic brim in one direction or the other. This is true not only of certain vessels and nerves but of parts of the ureters and the large intestine. By and large, the urinary bladder and the reproductive organs inhabit the pelvis while the other organs shown in Figure IX-3 assume places in the various zones of the abdominal cavity proper. In Figure IX-4B the ventral abdominal wall has been quartered by drawing a **midline** and a **supracristal plane** upon it. This plane passes through the iliac crests at the level of VL_4, and it may correspond closely to the **umbilical plane.** The umbilicus may be more convenient to use to draw this line upon a supine individual since the crests are best viewed from behind. However, the umbilicus varies in position, and it is often lower rather than higher if it is abnormally located whereas the position of the crests are relatively stable points. With these lines in place, organs generally are disposed within the quadrants shown in Figure IX-4A as follows:

A. Upper right quadrant
   1. Right lobe of liver, gall bladder, and biliary ducts
   2. Right kidney and right suprarenal gland
   3. Pyloric and prepyloric parts of stomach
   4. First three portions of the duodenum
   5. Part of ascending and transverse colon

B. Upper left quadrant
   1. Left lobe of liver
   2. Spleen
   3. Left kidney and left suprarenal gland
   4. Lower end of esophagus
   5. Fundus (cardia) and body (corpus) of stomach
   6. The short fourth part of the duodenum and duodenojejunal flexure
   7. Upper part of jejunum
   8. Transverse colon and upper part of descending colon

C. Lower right quadrant
   1. Cecum and appendix
   2. Parts of ascending colon
   3. Parts of ureter (right)
   4. Parts of jejunum and ileum
   5. Right tube and overy of female and part of uterus

D. Lower left quadrant
   1. Parts of descending colon

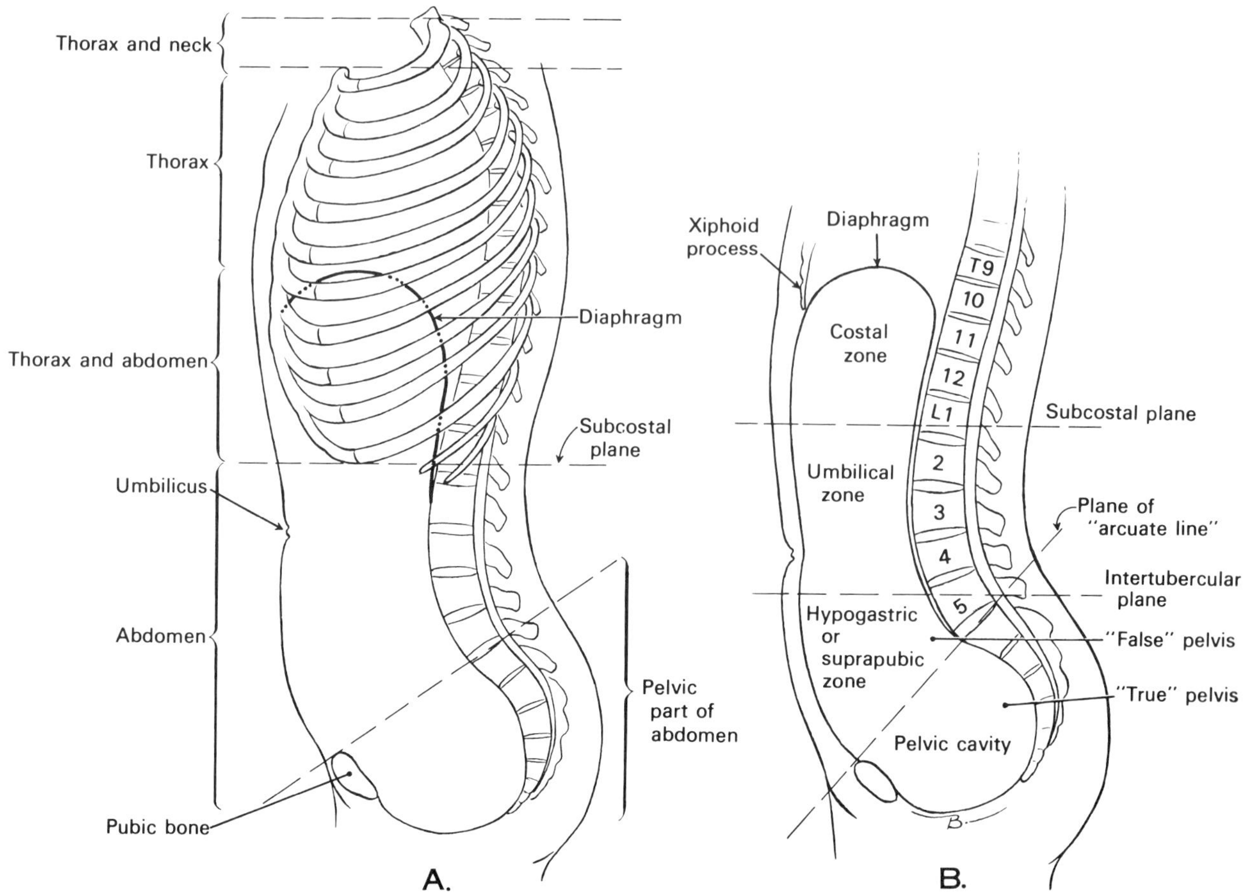

Fig. IX-1. Boundaries (**A**) and subdivisions (**B**) of abdominal cavity seen in lateral and midsagittal view.

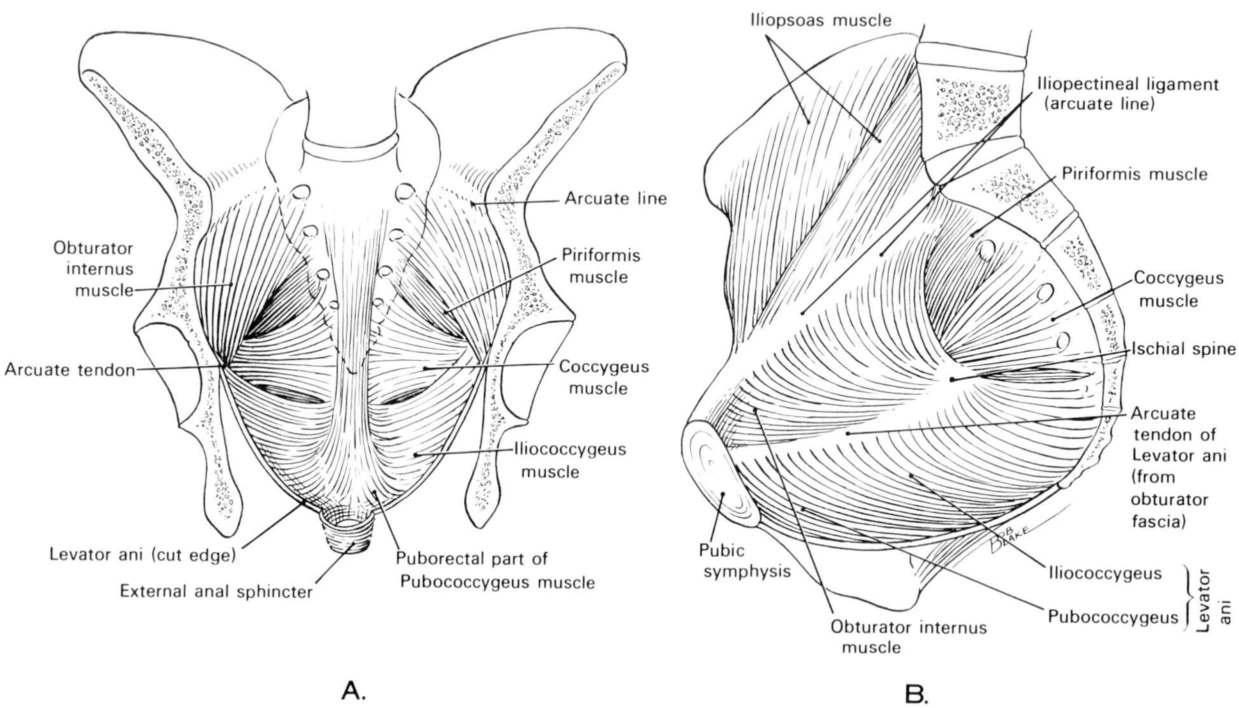

Fig. IX-2. The pelvic diaphragm. **A.** A coronal view of the back part of the diaphragm consisting of Levatores ani and Coccygei. **B.** Lateral view of the right half of the pelvic diaphragm. The *arcuate line* (not tendon) separates false pelvis, above, from true pelvis, below.

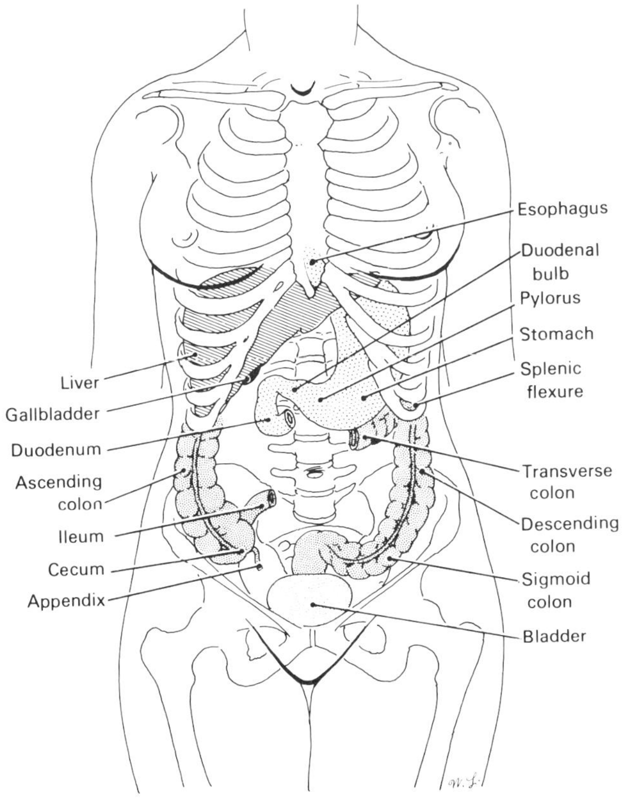

Fig. IX-3. Relation of the organs to skeletal landmarks in the abdominal cavity.

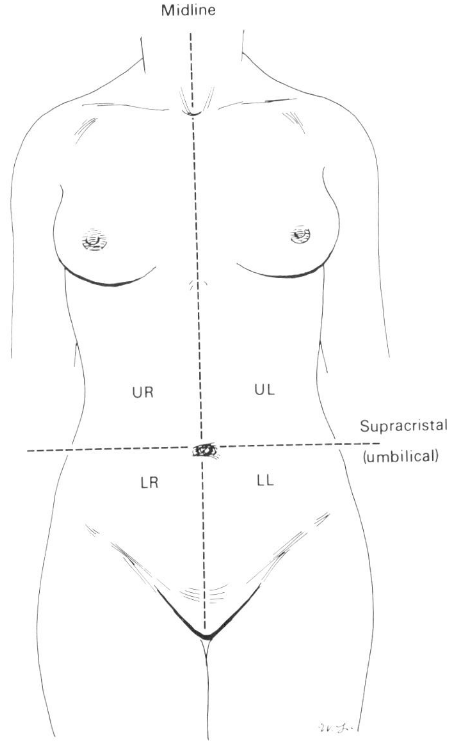

Fig. IX-4. Planes, lines, and zones of the abdomen. **A.** The abdomen in quadrants: UR, upper right quadrant; UL, upper left quadrant; LR, lower right quadrant; LL, lower left quadrant.

2. Sigmoid colon
3. Parts of jejunum and ileum
4. Parts of left ureter
5. Left tube and overy of female and part of uterus

Interestingly, the organs most subject to disease lie in the upper and lower right quadrants so that exploratory incisions tend to be made on that side.

If one wishes a finer subdivision of the abdomen, the ventral abdominal wall can be marked off into nine regions as in Figure IX-4B. An approximate disposition of the organs is shown in Figure IX-3. Only two vertical and two horizontal lines need be drawn. The vertical ones (Fig. IX-4B) are a pair of **lateral (paracentral) lines** which run from the tip of the ninth costal cartilage across the inguinal ligament midway between **anterior superior iliac spine** and **pubic symphysis.** The lower horizontal line is the **intertubercular plane,** through the iliac tubercles, not the crests. The tubercles lie lower and more anterior than the crests. The upper horizontal line is the **transpyloric plane,** a plane midway between the suprasternal notch and the symphysis pubis. The nine regions so delineated between the intersection of these lines and planes are:

1. A pair of right and left **hypochondriac regions** are so named because they lie below, in the sense of "deep to," the ribs. These are anatomical regions and the name does not refer to the neurological or psychiatric individual who complains about vague "pains in the belly" when there seems to be no physical reason for the existence of such complaints.
2. A pair of right and left **lateral regions** were formerly called "lumbar zones," a confusing nomenclature since lumbar vertebrae lie in the midline and extend the length of the abdomen.
3. A pair of right and left **inguinal regions.** These were formerly called "iliac zones," but the iliac crest and the bulk of this bone lie in the lateral region.
4. A midline **epigastric region** lies *"in the neighborhood of* the stomach."
5. A midline **umbilical region** is so called because the umbilicus sits smack-dab in the middle of it (hopefully).
6. A midline, lowermost **pubic region** relates to

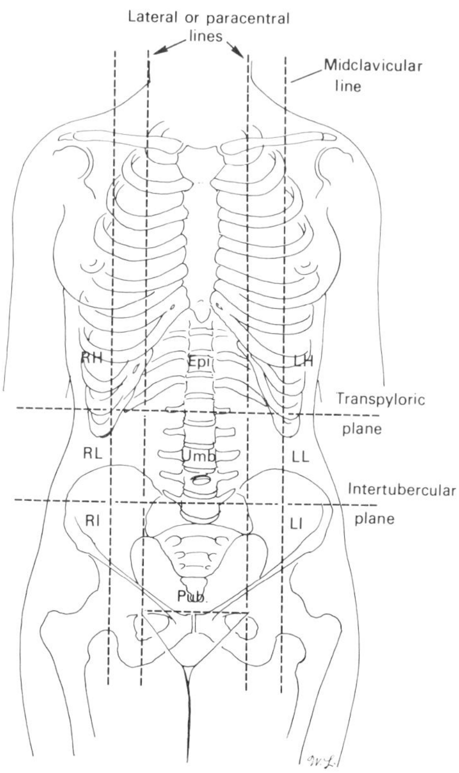

Fig. IX-4B. Conventional zones of the abdomen: RH, right hypochondriac zone; LH, left hypochondriac zone; RL, right lateral (formerly lumbar) zone; LL, left lateral zone; RI, right inguinal (formerly iliac) zone; LI, left inguinal zone; Epi, epigastric region; Umb., umbilical; Pub., pubic (formerly hypogastric) region.

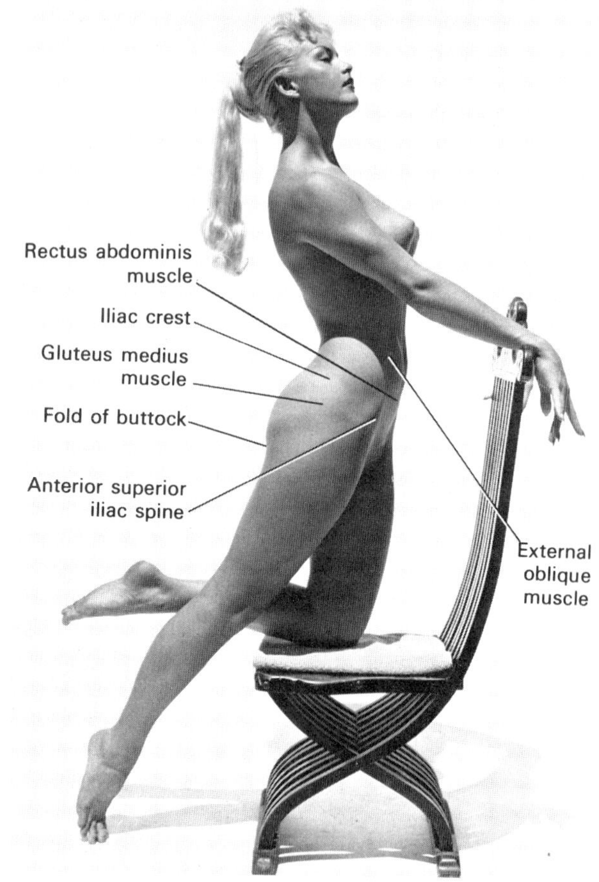

Fig. IX-5. Surface landmarks related to the anterior abdomen and pelvis.

the area of the pubic symphysis. It was formerly called the "hypogastric region," but it certainly was too far south of the border as far as stomachs of most individuals were concerned, and so was a poor choice of terminology.

Nevertheless, variation of organ position among individuals does not always conform so nicely to these regions as Figures IX-4B would have it. This is especially true for the elderly and for the short, stout subject.

The **transpyloric plane** is a useful clinical plane, however, since it approximates the following:

1. The pylorus of the stomach, hence, its name. But this approximation holds better for the supine cadaver than it does for the living subject where the pylorus may lie 1 to 3 inches below this plane.
2. The position of the tip of the 12th rib at the lateral margin of the Sacrospinalis muscle complex.
3. The ninth costal cartilage as it reaches the costal margin.
4. The midpoint of the costal margin.
5. The body of VL_1.
6. The upper end of the semilunar line (of Douglas) which marks the lateral margin of the Rectus Abdominis.
7. Here lie also the pylorogastric junction, the duodeno-jejunal flexure, the neck of the pancreas, and the hilum of each kidney.

It would be well to locate these last landmarks mentioned as well as the following:

1. The **anterior superior spine** of the ilium which is quite prominent in some individuals (Fig. IX-5).
2. The **tubercle of the ilium** marks the highest point along the iliac crest as viewed from the front and, as we noted, it is at the level of VL_5.
3. The **iliac crests,** best seen from behind, lie higher still at VL_4, and a line through both crests marks the supracristal plane as we have noted (Figs. IX-5 and 6).

Fig. IX-6. Surface landmarks of the back and posterior pelvis.

4. Identify the **posterior superior iliac spines** as two dimples in the lower back about 1½ inches from the midline (Fig. IX-6).

5. The **pubic tubercles** lie along the **pubic crest** about an inch from each side of the symphysis. They are hard to feel in unfriendly females, but can be felt in the docile male if a finger is invaginated into the scrotum.

6. The **pubic arch** can be felt in a like manner below the symphysis of the male. In a cooperative female it can be felt below the mons pubis in front of the clitoris.

7. The **xiphoid process** lies at the level of VT_9.

8. The subcostal plane is lower, lower even than the transpyloric plane, at the level of VL_2 or even VL_3.

Figure IX-3 indicates the locus of certain organs in the abdominal cavity; some of these should be related to vertebral levels and other landmarks. The **liver** lies well behind the grill of the costal cartilage. Its inferior border extends from the right 10th costal cartilage obliquely upward and across to almost the left nipple. The **gall bladder** lies about midway along the slanting right costal margin near the tip of the ninth costal cartilage. This is also about at the upper reaches of the lateral margin of the Rectus Abdominis (semilunar line). The transpyloric plane cuts the **neck of the pancreas** at the level of VL_1 in the epigastric region. The tail of this gland lies in the left hypochondriac region. The **hila of the kidneys** also lie along the transpyloric plane. The **spleen** lies posteriorly against the left ninth to 11th ribs. The supracristal plane (VL_4) marks the point where the **aorta terminates** in its bifurcation into the common iliacs.

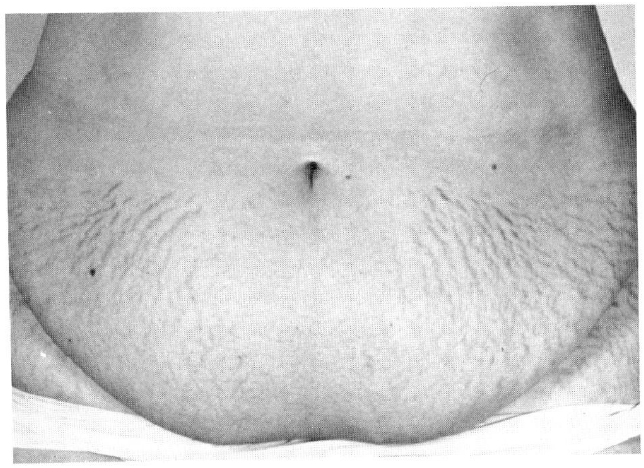

Fig. IX-7. Striae (elongate markings) on skin of a distended abdomen.

In examining **the skin** of the abdomen, striae, *i.e.*, multiple scars, some of which measure as much as 8 cm., are sometimes encountered (Fig. IX-7). These become silvery with age. They probably occur where the skin has been stretched chronically and has been subjected to rupture of the elastic fibers in its reticular layer. Pregnancy, obesity, tumorous growths, or edema were probably contributing factors.

Except where adipose deposition has been slight, veins are not particularly noticeable in the abdominal wall. Above the umbilicus, flow is directed superiorly and below it, inferiorly. When, in certain instances, paraumbilical veins enlarge or engorge they may form a pattern reminiscent of a number of worms crawling out of a pit. This condition has been aptly described as **caput medusae** after the snake-haired goddess, Medusa. The condition generally follows an obstruction of the portal vein, and the paraumbilical veins serve as avenues of collateral venous return. Thus, it is well to contemplate your patients' navels as well as your own. Observe its color and shape, look for everted stumps or protruding nodules. Changes from the norm may signify underlying difficulties. With obesity the abdomen rounds up uniformly. The umbilicus becomes buried in the wall when fat, which accumulates quickly in the abdominal region, deposits superficially around its margins. In pregnancy, of course, the uterus rises out of the pelvic bowl, pushes against the abdominal contents and protrudes the abdominal wall. Any abnormal distention or protrusion of the wall is of clinical import. You can think of the 5 "f's": fetus, flatus, fat, feces, or fluid.

It is also possible at times to detect contractions of the stomach and intestines as you watch the abdomen. They are seen as slow, undulating ripples below the skin and muscle layers. These normal peristaltic waves vary so in amplitude and in distance traveled that many go undetected.

The Ventrolateral Abdominal Wall

The abdomen of the soft-living, beer-imbibing American male adult has often been described facetiously as a pot. As we view it in the examination room with the individual supine, we prefer to describe it as a covered oval-shaped basin, a veritable casserole dish containing a host of interesting goodies. The ends of this dish are the two diaphragms, the thoraco-abdominal, superiorly, and the pelvic one, inferiorly. The floor consists of the lumbar segments of the vertebral column and some deep back muscles (Quadratus Lumborum, and

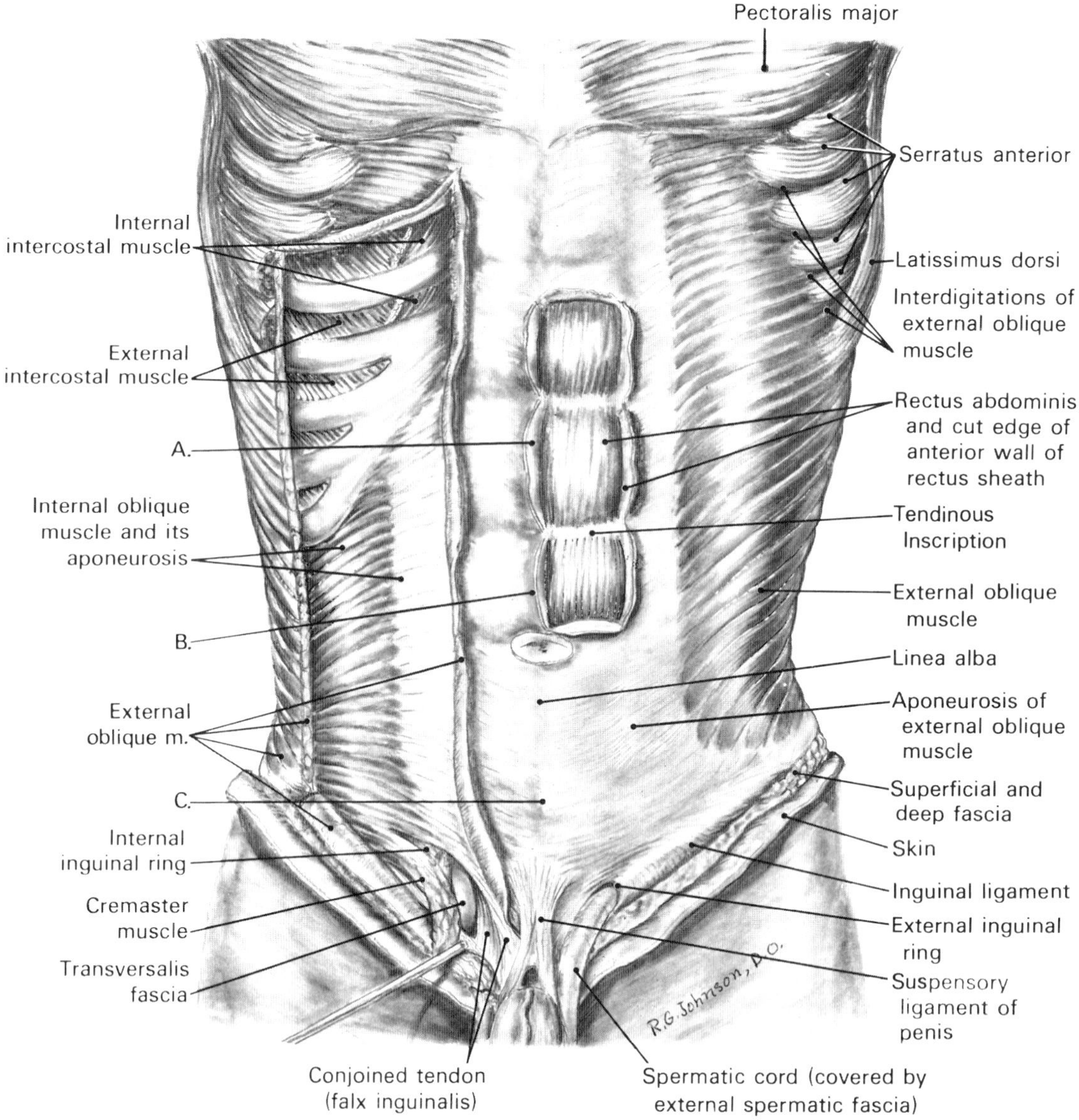

Fig. IX-8. Dissection of ventrolateral body wall. The External oblique muscle of the abdomen has been cut away on your left and structures in the inguinal region have been exposed on that side. The rectus sheath has been opened part way on your right. Cross sections through the rectus sheath at levels A, B, and C are shown in Figure IX-10.

the Psoas muscles). The lower backbone raises such a prominent ridge in the floor that the basin would be a most uncomfortable Sitz-bath. The lips or brims consist of the costal margins, the ilia and pubis. The cover of this basin-like casserole is formed by the flat Rectus Abdominis and the flat aponeurotic tendons of the lateral wall muscles. The lateral walls are the ventrolateral muscles themselves.

The Anterolateral Muscles of the Abdominal Wall: Both Table IX-1 and Figure IX-8 present the salient features about the muscles in this region. The soft ventral contours of the abdomen are formed by the three lateral muscles of the region along with the midline Recti and accumulations of fat (adipose tissue) in the subcutaneous layers of skin and underlying fascia. A faint midline furrow, interrupted as it encircles the umbilicus, separates one **Rectus Abdominus** muscle from the other. This dividing landmark is the **linea alba** (Fig. IX-9), but it is not really seen as a distinct white line until the skin is reflected in the dissection. The linea is a midline aponeurotic junction of the various flattened tendons of the lateral muscles from each side as these tendon sheets pass over or behind the Recti to interdigitate their fibers in the midline. As these aponeurotic tendons of the **External Oblique,** the **Internal Oblique,** and the **Transversus Abdominis** come into relation with the Rectus Abdominis they form the rectus sheath for this longitudinal muscle.

Before the muscles can be uncovered, one has to contend with the **abdominal fascia** to which the dermis of the skin anchors. Most anatomists treat the fascia as two distinct lamina, a **superficial** and a **deep** fascial layer. There is considerable variation in the degree of separation between these two layers, and this often depends upon the degree of obesity of the individual. Classically, at any rate, the **loose superficial fascia** is said to be the fatty layer first described by Pieter **Camper,** a Dutch Anatomist (1722–1789). His description of it comes down to us in his works published in 1803 after his death. The **deep,** tougher, **membranous fascia** was first described in 1809 by Antonio **Scarpa** (1747–1832), an Italian pupil of Morgagni. Even as early as 1854, British writers were altering Scarpa's original description of his layer of fascia. Presumably, they were doing it on the basis of what they saw in the particular bodies they were dissecting and not on the basis of a mistranslation. How much the original description of both these earlier discoverers has changed over the years is evident since their tale no longer fits the story told in current texts. This is because Scarpa and Camper were talking about much the same thing, a connective tissue which could change in quality. But the modernists take them literally, calling loose material Camper's fascia and tough material, Scarpa's fascia. This has probably been influenced by the early dogmatism of our British cousins. Although it's easy to stick to dogma, we ask that you compare these fascial planes in your dissection with those in your neighbors dissection to witness their variable characteristics.

At least we can say the following. Scarpa's fascia is

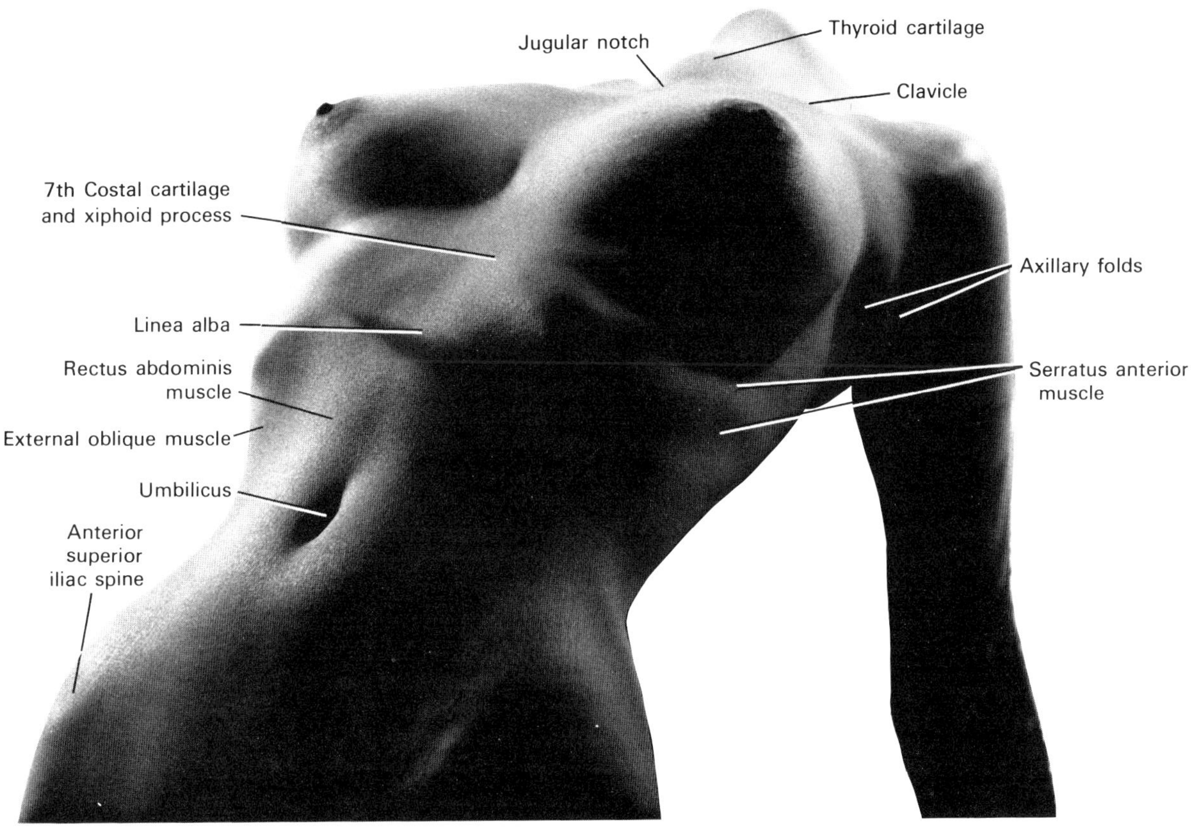

Fig. IX-9. Surface anatomy of the ventral abdominal wall.

not the innominate fascia that invests the muscle below it, though it may be difficult to separate its adherence to the muscle fascia or to the aponeurosis of the External Oblique in places. Scarpa's fascia does form the thick **fundiform,** or **suspensory ligament** of **the penis** by extending to the dorsum and sides of this organ. In the scrotum and in the fat-free penis both fascial layers fuse and lose fat. In the scrotum they gain smooth muscle, a **dartos tunic,** but this muscle is absent in the external penile connective tissue sheath. Thus, the penis is not fat-headed, though its owner may be when he discusses his (or its) prowess. Beyond the inguinal ligament, Scarpa's fascia continues as the **fascia cribrosa** in the upper thigh and as the tough **perineal fascia.** The latter is also called Colles fascia because Colles, an Irishman, got into the act in 1811 by redescribing Scarpa's fascia in the perineal region.

Camper's more classically fatty layer extends into the superficial aspects of the thigh and over the groove between thigh and external genitalia to become the **superficial fascia of the perineum.**

Each **Rectus Abdominis** as seen in cross-section in Figure IX-10 is about 3 inches wide, parallels the midline, and extends from the xiphoid process and the fifth to seventh costal cartilages to the crest of the pubis. When the rectus sheath (to be described below) is opened and the muscle is exposed, it will be seen to be segmented into short stretches by **tendinous inscriptions** (Fig. IX-8). Two such usually lie above the umbilicus, one at it, and, sometimes, one below. These transverse tendon bands attach to the anterior wall of the sheath and are not too evident on the back side of the muscle. Each of them is pierced from behind by twigs from the **superior epigastric arteries,** the con-

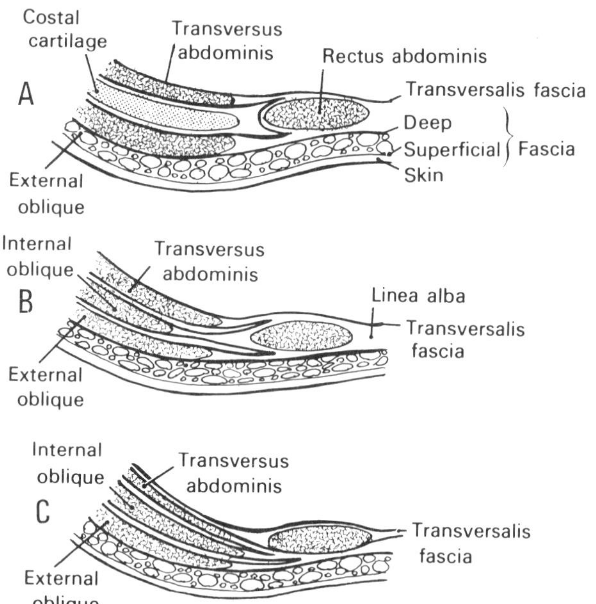

Fig. IX-10. Cross-sections (schematic) through anterior body wall at level A to C of Figure IX-8 to show composition of rectus sheath from outside inward. A, through upper one-third of sheath; B, through middle one-third; C, lower one-third, below arcuate line.

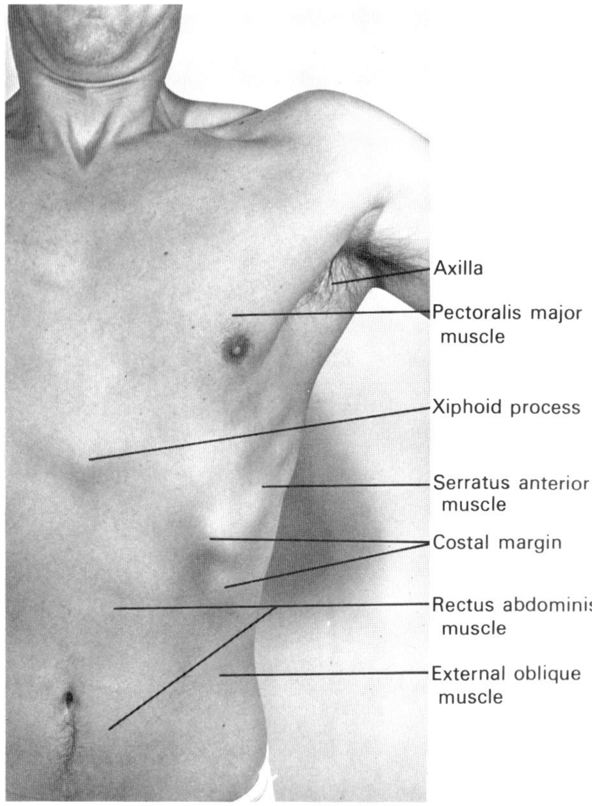

Fig. IX-11. Surface features of the male chest and abdomen.

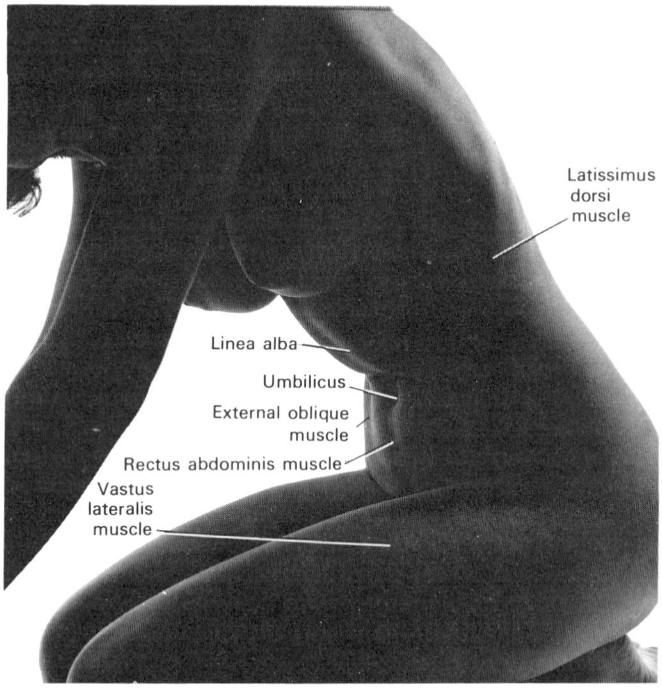

Fig. IX-12. Surface anatomy of the anterior abdominal wall in the female.

tinuations of the internal thoracic (mammary) vessels that run in the posterior compartment of the sheath behind the muscle. In the well-developed male (Fig. IX-11), when the muscle is tensed, the short stretches between inscriptions bulge separately. In the female, with her more softly and smoothly contoured tummy, these contractions are less noticeable (Figs. IX-9 and 12 to 14). In some individuals, the Rectus Abdominis (or its sheath) forms a gentle curve along its lateral margins, the **semilunar line,** as a visible landmark of the muscle borders (Figs. IX-9, and 11 to 13). Do not confuse the term *semilunar line* with the term **arcuate line** (of Douglas) which lies deep in the *posterior wall* of the sheath behind the inferior epigastric vessels (see Fig. IX-15).

The three paired muscles of the lateral walls of the abdomen form a three-plied corset. They are the larger heavier counterparts of the three lamina of muscles we found between the ribs, and their fiber directions tend to take about the same course as the Intercostals. The External Oblique is the only one that can be seen as a feature of surface anatomy, and then only at its lateral origins from the ribs. Its slips of origin are seen only in very muscular or very skinny individuals where its interdigitations or fleshy fingers interlock with slips of the Serratus Anterior (Figs. IX-11 and 14). From these slips of origin (left side of Fig. IX-15), fiber direction is downward and medialward as it was in the case of the External Intercostals of the chest wall. The lower, wide, fleshy belly of the muscle stops short of the semilunar line and does not extend far below the umbilicus. A vast aponeurotic sheet, serving as a flattened, thin tendon, continues medialward to anchor the muscle to the midline, its tendinous fibers interlacing with those from the aponeurosis of the opposite side in the linea alba. Inferiorly, the aponeurosis of the External Oblique bridges from the anterior superior iliac spine to the pubic tubercle. Its lower border is free between these points, thickened, and rolled under. Since it anchors bone to bone at these points, this free edge of the External Oblique aponeurosis is called the **inguinal or Poupart's ligament** (Figs. IX-8 and 15). We will return to a further discussion of the External Oblique aponeurosis when we consider the rectus sheath and the inguinal region (below).

The *deeper* lateral muscles are the Internal Oblique and the Transversus Abdominis, and they complete the three-plied corset. In Figure IX-15, we see them on the side opposite the External Oblique, and much of the Internal Oblique, except its lower portion, has been cut away. Like the Internal Intercostals, the fibers of the Internal Oblique angle just opposite to those of the External Oblique, upward and laterally. If this is not clear in Figure IX-15 it is because we look at these two muscles in the opposite sides of the body so that the fibers of the *left* Internal Oblique have the *same direction* as those of the right External Oblique. Had we laid back the External Oblique of the cadaver's *right* side,

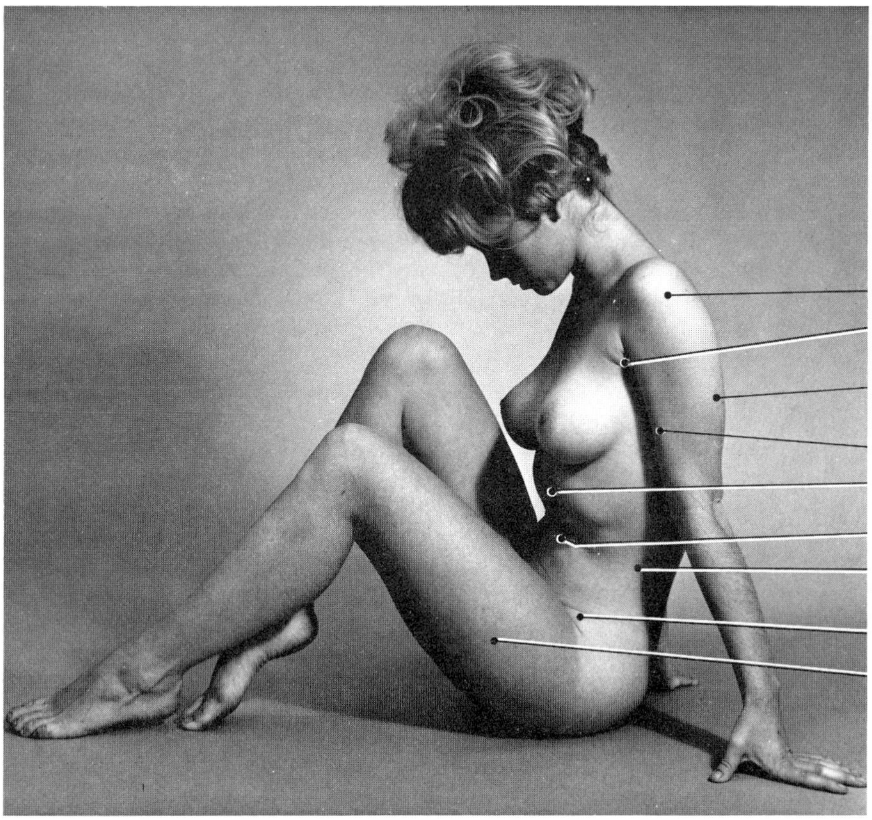

Fig. IX-13. Surface features of the anterior and lateral aspects of the abdominal wall.

then fiber direction in the Internal Oblique of the *right* would have been seen to run in the *opposite direction*. As in the case of the External Oblique, the fleshy fibers do not quite reach the semilunar line, though they do reach farther toward the inguinal region (toward the groin). They also end in an aponeurosis that extends them to the midline.

As you would expect from its name, the deepest of these three muscles, the Transversus, has its fiber direction oriented horizontally. The fleshy fibers extend almost to the semilunar line, and, once again, a tendinous aponeurosis attaches the muscle fibers to the midline. How far toward the inguinal region the muscle fibers of the Transversus extend is a matter of some variability. Probably more often than not, they fall short of reaching the inguinal canal. More often, the muscle fibers of the Internal Oblique are the lowest reaching fibers of all of the three lateral muscle masses in the abdominal wall.

Sometimes, below the level of the arcuate line, a

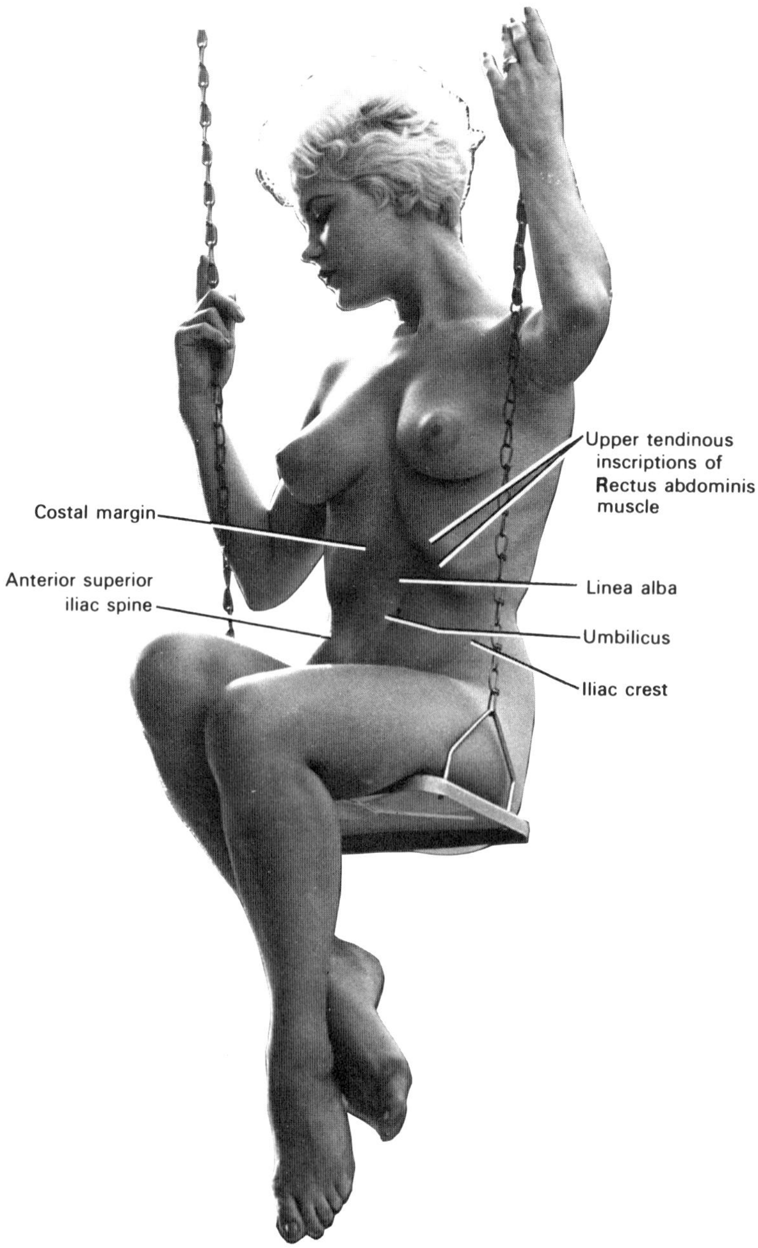

Fig. IX-14. A swing and other surface features.

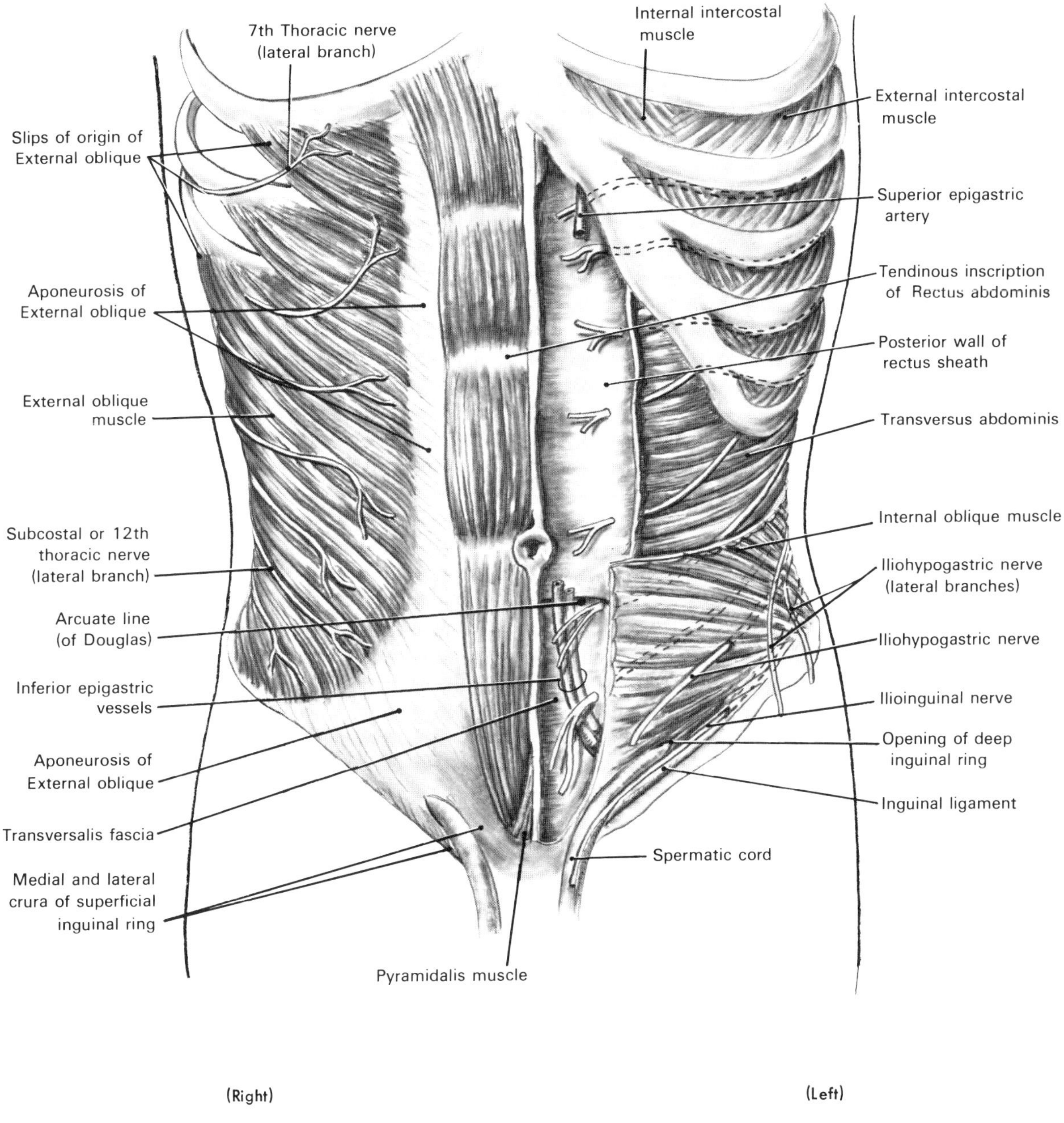

Fig. IX-15. Dissection of anterolateral body wall showing Rectus abdominis and External oblique on your left; the empty rectus sheath, Internal oblique, and Transversus abdominis on your right. Note that the posterior wall of the sheath in vicinity of inferior epigastric vessels and below arcuate line consists of transversalis fascia alone.

Table IX-1. Abdominal Wall Muscles

Muscle	Origin	Insertion	Nerve Supply	Action
External oblique	Outer aspect lower 8 ribs	Anterior part iliac crest; linea alba	Last 5 intercostal nerves; subcostal nerve; ilio-hypogastric nerve	Flex and rotate spine compress and support viscera
Internal oblique	Lumbodorsal fascia; iliac crest; on lateral part of inguinal ligament	Last 4 costal cartilages; linea alba; to pubis *via* conjoined tendon	Same, including ilio-inguinal nerve	Same
Transversus abdominis	Last 6 costal cartilages; lumbodorsal fascia; posterior iliac crest; lateral third inguinal ligament	Linea alba and conjoined tendon	Same as Internal oblique	Compress and support viscera
Rectus abdominis	5th to 7th costal cartilage and xiphoid process	Pubic crest and symphysis	Lower 5 intercostal nerves; subcostal nerve	Flex spine; tense abdomen
Pyramidalis	Pubis and anterior pubic ligament near midline	Linea alba between pubis and umbilicus	Subcostal nerve (when muscle is present)	Tense linea alba
Quadratus lumborum	Lumbodorsal fascia; last 2 inches iliac crest	Medial 2 inches last rib; transverse processes lumbar vertebrae	Subcostal nerve; first 3 lumbar nerves	Flexes spine laterally; fixes and depresses last ribs during inspiration
Psoas major	Bodies and disks of VT_{12}–VL_4	Lesser trochanter femur	2nd and 3rd lumbar nerves	Flexes thigh on hip and *vice versa;* flex spine and pelvis from supine position; laterally or medially rotates femur depending upon whether foot is off the ground or on it
Psoas minor	Bodies and disks of VT_{12}–VL_4	Iliopectineal eminence	1st lumbar nerve (when muscle is present)	Flexes spine (?)

small Pyramidalis muscle sits kitty-corner, superficial to the Rectus within the rectus sheath (Fig. IX-15). It may be present on one side only or entirely absent.

Table IX-1 gives the details of origin, insertion, nerve supply and action of the abdominal muscles. There is no need to memorize such details, but they should be understood. Most of these relationships should be worked out in dissection. Once having displayed them, you are much less apt to forget them than if you rely upon brute memorization. The table is here for use as a guide now, and for quick review should you need to bone up rapidly for special examinations, *e.g.*, State or National Boards.

The **rectus sheath** is formed by the aponeurotic tendons of the three lateral wall muscles as these reach for an insertion upon the linea alba. Indeed, the linea exists because the collagenous fibers from these tendinous sheets of the opposite side interdigitate in the midline. Since the sheath envelopes the Rectus muscle, it consists of anterior and posterior walls or lamellae. Its structure, however, varies regionally. In its upper one-third (Figs. IX-8A and 10A), where its upper fibers contact the ribs, its **anterior lamella** consists of the aponeurosis of the External Oblique only. Its **posterior wall** contains costal cartilage, and, possibly, a bit of the aponeurosis of the Transverse Abdominis. In its middle one-third (Figs. IX-8B and 10B), its *anterior lamella* is the aponeurosis of the External Oblique reinforced by the anterior leaflet of the aponeurosis of the Internal Oblique. This last sheet splits to form an anterior and posterior contribution to the sheath in this region. Thus, the posterior lamella of the sheath in this region consists of the posterior leaflet of the aponeurosis of the Internal Oblique, the aponeurosis of the Transversus Abdominis. Deep to these, the thin **transversalis fascia** (endoabdominal fascia) lines the abdominal wall. Behind the fascia, as always, is the extraperitoneal abdominal fatty tissue and, innermost, the peritoneum itself, a thin, simple squamous lining-layer. These last two coats are not a part of the rectus sheath since, like the transversalis fascia, they are part of the abdominal wall proper. Finally, in its lower one-third below the arcuate line (Figs. IX-8C and 10C), the aponeuroses of each of the lateral muscles cross *in front* of the Rectus to form the anterior wall of the sheath. This leaves the posterior lamella to be made up of the thin transversalis fascia alone. The *arcuate line* at the upper margin of the posterior sheath wall in this region represents the free edge of the fused aponeurotic sheets of the Transversus and posterior leaflet of the Internal Oblique which fail to extend farther caudalward.

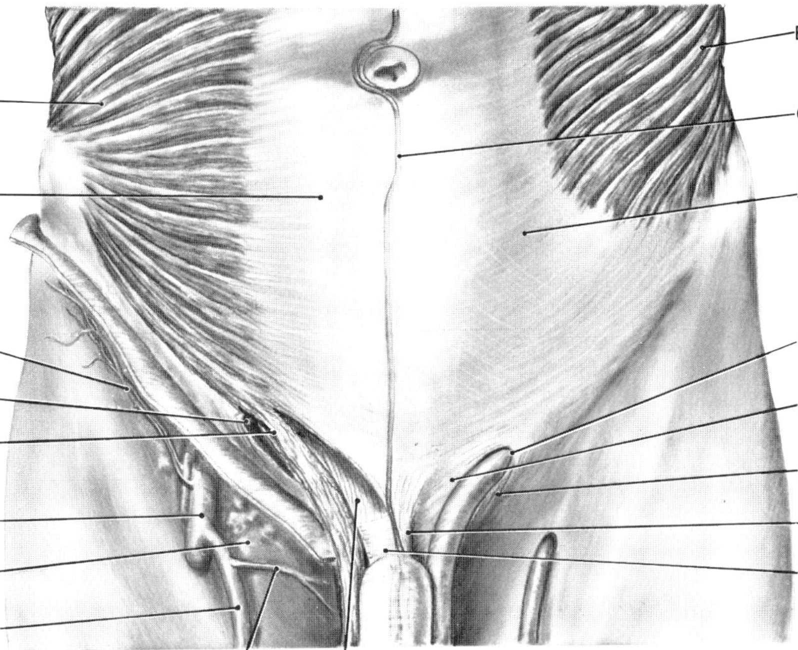

Fig. IX-16. Dissection of the male inguinal region.

The Rectus does not occupy the sheath alone, but has for company within it the **Pyramidalis muscle** when it is present, terminations of the **lower five intercostal nerves,** and the **subcostal nerve** (12th thoracic), as well as the **inferior** and **superior epigastric arteries** and their **venae comitantes** (Fig. IX-15).

The Inguinal Region: Inferiorly, the abdominal muscles (chiefly, their aponeurotic tendons) contribute to the inguinal ligament and inguinal canal (Fig. IX-18). We have mentioned that the inguinal (Poupart's) ligament is nothing more than the lower free edge of the aponeurosis of the External Oblique. Medially, near its attachment to the pubis, the External Oblique aponeurosis splits to form a triangular gap which has a **superior and inferior crus** (Fig. IX-16). **Intercrural fibers** (Figs. IX-17 and 18) interlace at the apex of the triangular area and spin a thin areolar roof over the deficit to convert it into the **superficial inguinal ring.** From the underside of the inguinal ligament (Fig. IX-17), fibers pass to the superior pubic ramus as the **lacunar ligament (or Gimbernat's ligament).** This is a portion of the medial aspect of the inferior crus that rolls under the spermatic cord to attach to the pectineal line, a ridge on the superior ramus of the pubic bone. It is a cresentic triangle with its apex pointing inferomedially toward the pubic tubercle. **Cooper's ligament or the pectineal ligament** can be looked upon as a thickened extension of the lacunar ligament along the pectineal line. Fibers from the opposite aponeurotic sheet of the External Oblique sweep medially beyond the linea alba to dip below the superficial inguinal ring as the **reflected inguinal ligament** (Fig. IX-16). These are the modifications encountered in the aponeurosis of the External Oblique at the inguinal region where the **spermatic cord,** coming up from the testis, makes its entrance into the abdominal cavity. In the embryo, of course, the testis developing in the region of the kidney, made its descent into the scrotum through the inguinal canal to emerge at the superficial inguinal ring. This ring can be easily palpated in the male by pushing a finger into the invaginated scrotal sac. This is done as a routine check to detect possible extension of bowel in the canal or protrusion through adjacent parts of the body wall. In the female, the **round ligament of the uterus** comes through the ring to anchor in the labium majorum. The round ligament in the female corresponds to the **caudal part of the embryonic gubernaculum testis** of the male. The round ligament may or may not be encountered as a well-defined structure in the dissection of the inguinal region of female cadavers.

If there is a superficial inguinal ring, you can be sure it was so named because there is also a **deep inguinal ring.** The latter will be found on the interior of the ventral abdominal wall when the abdomen is opened and the belly muscles are laid back (Figs. IX-15 to 17). The deep inguinal ring is an opening through the transversalis fascia about a half inch above the level of the inguinal ligament just lateral to the origin of the **inferior epigastric artery** from the external iliac artery. The inferior epigastric vessels enter the back of the rectus sheath behind the Rectus Abdominis (Fig. IX-15). The vessels and the ductus deferens do

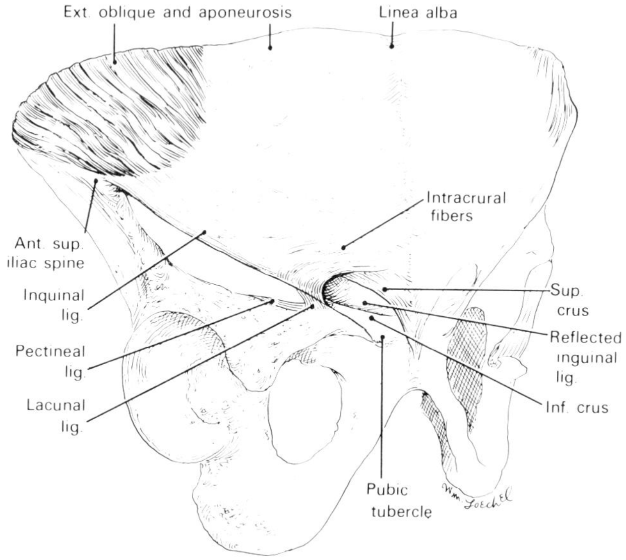

Fig. IX-17. Modifications of the external oblique aponeurosis in the inguinal region.

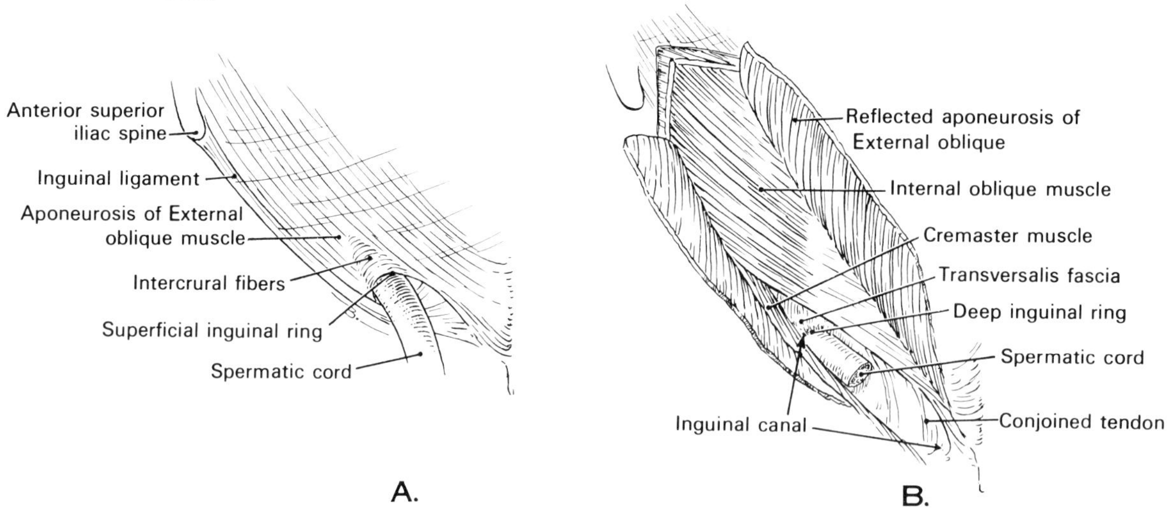

Fig. IX-18. The inguinal canal. **A,** external relationships; **B,** internal relationships.

not pierce the peritoneum but run upon its external surface in the extraperitoneal areolar tissue. This is why we have said that the deep ring is an opening in the *transversalis fascia,* not an opening into the interior of the abdominal cavity (Fig. IX-18). Between superficial and deep rings runs the inch-and-a-half **inguinal canal** (Fig. IX-18).

Obviously, this short, narrow, and rather flat canal has to pierce the body wall structures deep to the skin in this region. The fleshy fibers of neither the External Oblique nor the Rectus encroach upon its territory so they are not pierced by it. We saw its superficial mouth between superior and inferior crura of the External Oblique aponeurosis. The canal must pierce the aponeurosis of the Internal Oblique and Transversus Abdominus as well as the transversalis fascia in its course inward from the external ring. The so-called **extraperitoneal connective tissue** which envelopes the abdominal wall, and in which, in the embryo, the testis as well as certain other visceral organs developed, lies deep to the transversalis fascia. The peritoneum lies deep to all of these and is not involved (in the adult, at least). As a rule, the fleshy fibers of the Transversus do not extend so far inferiorly as to become involved. Some fleshy fibers of the Internal Oblique do, and these contribute loops of skeletal muscle to the spermatic cord (**Cremaster muscle**) as it traverses the canal (Figs. IX-16 and 18). The canal was formed when the **processus vaginalis** (of peritoneum) evaginated the abdominal wall during the early descent of the testis into the scrotum (Fig. IX-19). All that remains of this process is the **tunica vaginalis** in the scrotum of the adult, a displaced piece of peritoneal cavity to house the testis. Besides the structures within the **spermatic cord** in the male, the canal transmits the *cutaneous branch* of the **ilioinguinal nerve,** the genital branch of the **genitofemoral nerve,** and cremasteric blood vessels. The cord, itself, consists of

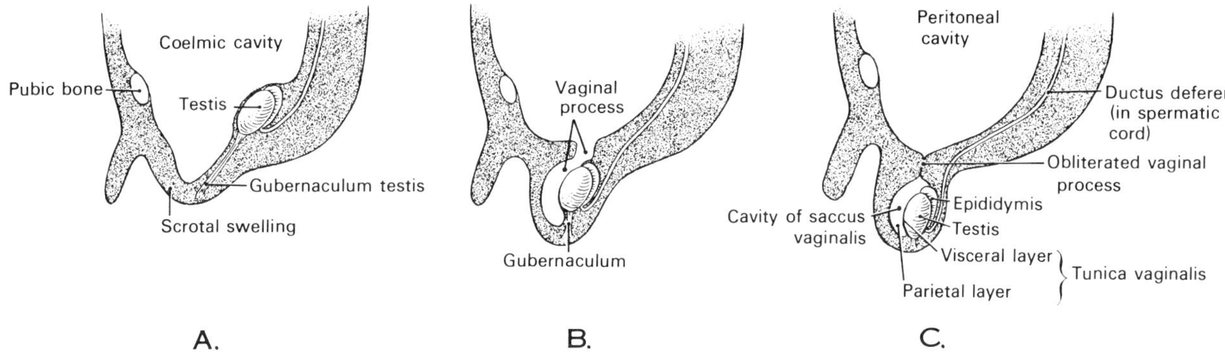

Fig. IX-19. Stages in the descent of the testis. **A.** Early fetus. **B.** Later stage with formation of saccus vaginalis and degeneration of gubernaculum. **C.** Final stage.

Table IX-2. Comparison of the Layers in Abdominal Wall, Scrotum, and Spermatic Cord

Abdominal Wall	Corresponding Layers in Scrotum	
1. Skin	1. Skin	scrotum
2. Superficial fascia A. Camper's (fatty) B. Scarpa's (membranous)	2. (A + B − fat + smooth muscle = Dartos muscle and fascia)	scrotum
3. External oblique aponeurosis	3. External spermatic fascia	coverings of spermatic cord
4. Internal oblique } fibers 5. Transversus abdominis } fibers	4., 5. Cremasteric muscle	coverings of spermatic cord
6. Transversalis fascia	6. Internal spermatic fascia	coverings of spermatic cord
7. Extraperitoneal areolar tissue (between [6] and [8])	7. Areolar tissue surrounding cord structures	spermatic cord constituents
8. Peritoneum	8. Processus vaginalis*	spermatic cord constituents

* May or may not persist in the cord; definitely a part of the scrotal sac.

more than the ductus deferens, namely, the **testicular (internal spermatic) arteries and veins,** the latter forming the complicated network of the **pampiniform plexus;** the **deferential arteries and veins;** the **internal spermatic plexus of nerves** as offshoots of the hypogastric plexus; and the blood and nerve supply mentioned above for the cremasteric muscle. The cord is enveloped in **external spermatic fascia** derived from the areolar tissue of the external ring (an External Oblique contribution from its aponeurosis); **Cremasteric muscle** embedded in cremasteric fascia derived from the Internal Oblique with occasional contribution by the Transversus Abdominis; and finally by a thin **internal spermatic fascia** derived from transversalis fascia. The extraperitoneal areolar tissue of the abdomen becomes continuous as the areolar tissue of the cord which surrounds its vessels, ducts, and nerves. If there is no persistent processus vaginalis then no peritoneal structure is present.

Perhaps the relationships between abdominal wall, cord, and scrotum are best reviewed in Table IX-2 and Figure IX-20. The enumeration used in the table is carried over in the figure so you can find your way through the maze.

The inguinal canal has been described as having a floor, a roof, an anterior, and posterior wall. The floor is the rolled-under, grooved surface of the inguinal ligament and the lacunar ligament inferiorly. The roof is formed by the arching fleshy fibers of the Internal Oblique (and Transversus) which contribute Cremasteric muscle to the cord. The aponeuroses of these two are said to continue over the cord medially and under it to contribute a **conjoint tendon** to the posteromedial wall of the canal. The bulk of this tendon, if it exists outside of the imagination, is of Internal Oblique origin in all probability. At least some thickening into which a surgeon can anchor a suture medially during hernia repair normally passes for a "conjoined tendon." The anterior wall is the External Oblique aponeurosis. The posterior wall is said to be transversalis fascia strengthened inferiorly by the reflected part of the inguinal ligament and the enigmatic conjoined tendon (falx inguinalis) which all good anatomists and surgeons swear by. We show it illustrated in Figure IX-16, but must admit that we are not always lucky in finding it in the course of dissection.

Nerves and Vessels of the Anterior Abdominal Wall (Fig. XI-21): the deep arteries are the musculophrenic, superior epigastric, inferior epigastric, and the deep circumflex iliac.

Musculophrenic Artery: This is a terminal branch of the internal thoracic artery. It slips between fascicles of the Diaphragm from between the ninth and 10th costal cartilages to provide branches to the ab-

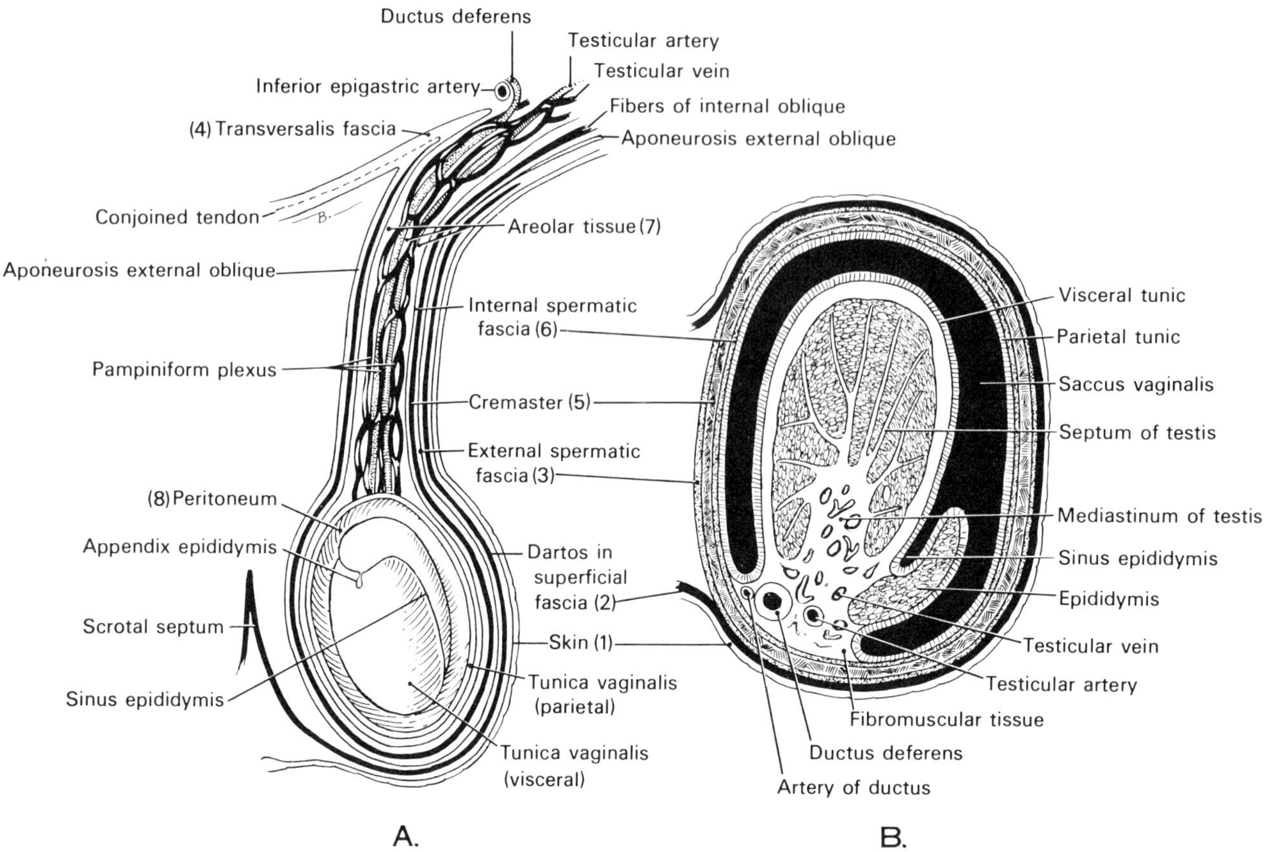

Fig. IX-20. Relation of layers of body wall to spermatic cord and scrotum. **A.** Coronal section through cord and left scrotum. **B.** Cross-section through left testis and scrotum seen from above. Sac is exaggerated in distended condition. (After Cunningham.)

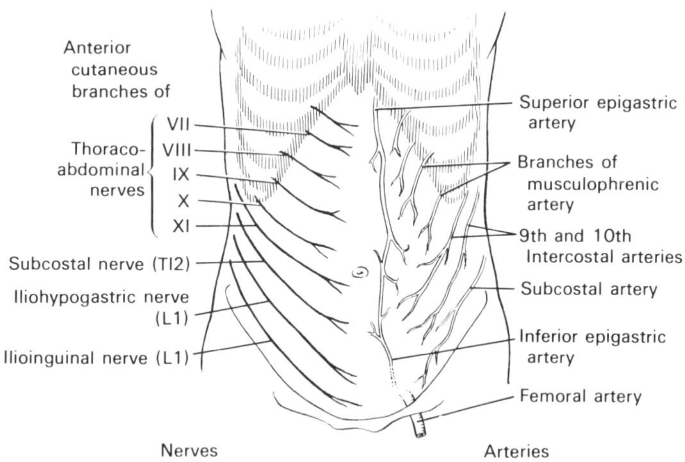

Fig. IX-21. Superficial nerves and arteries of abdomen.

dominal wall. As it traverses the costal angle it also gives side branches laterally to the seventh to ninth intercostal interspaces as anterior intercostal arteries.

Superior Epigastric Artery: As the main, vertically descending, terminal branch of the internal intercostal artery, it pierces between sternal and costal slips of the Diaphragm to plunge into the upper end of the posterior wall of the rectus sheath. It provides the Rectus and overlying skin, and anastomoses with the inferior epigastric artery.

Inferior Epigastric Artery: This branch arises from the external iliac artery. It ascends medialward behind the midpoint of the inguinal ligament in the transversalis fascia. It crosses the arcuate line of Douglas to enter the posterior wall of the rectus sheath and to ascend superiorly behind the Rectus Abdominis which it supplies. It has a **pubic branch** which turns downward behind the inguinal ligament to join with the pubic branch of the obturator artery. It also gives off the **cremasteric branch** which passes through the deep ring to supply Cremasteric muscle and to anastomose with the testicular artery. In females (who for some odd reason lack testicles) it supplies the round ligament.

Not shown in Figure IX-21 are two superficial branches of the femoral artery which ascend from the thigh to cross the inguinal ligament and enter the abdominal fascia close to the inguinal region. These branches are the *superficial epigastric* and the *superficial circumflex iliac* arteries.

Thoraco-abdominal Nerves (T_7 *to* T_{11}): These nerves leave the intercostal spaces at the anterior costal margin by slipping between fascicles of origin of the Diaphragm and Transversus Abdominis. They angle medially and inferiorly across the abdomen, sandwiched between Internal Oblique and Transversus. At the borders of the rectus sheath they pierce ventrally to supply skin. More laterally they supply the muscle fibers of the body wall. Here, too, they give off lateral cutaneous branches, much as do any intercostal nerves of the thorax.

Subcostal Nerve (T_{12}): The last thoracic nerve enters the abdomen behind the lateral arcuate ligament of the Diaphragm. It sends a branch over the iliac crest into the gluteal region and then descends obliquely like any thoraco-abdominal nerve.

Iliohypogasteric Nerve (L_1): This nerve is not always easy to find despite nice pictures of it in your atlas. It runs a course similar to the thoraco-abdominal nerves between Internal Oblique and Transversus. Near the anterior superior iliac spine it gives off a piercing branch through the Internal Oblique which remains below the External Oblique on its way to gluteal muscles. Somewhere close to the superficial ring a branch pierces the aponeurosis of the External Oblique to supply skin. Variability of this lower branch is great. If you can't find it, forget it.

Ilioinguinal Nerve (L_1): It is often easier to locate this nerve than the former one since it emerges through the ring to innervate thigh, scrotum, and the skin in the area of the mons pubis.

The Posterior Abdominal Wall

Recall that the floor of our basin contains the lower part of the vertebral column, *i.e.*, the lumbar vertebrae and their intervening disks. Three sets of muscles clothe the floor or back wall by anchoring, one way or another, to this midline bony column, to the last rib, to the iliac bone and even upon the femur. These muscles are the Psoas Major, the Quadratus Lumborum, and the Iliacus (see Fig. IX-22 and Table IX-1).

All of these muscles are surfaced by a muscle fascia which, indeed, invests them as a perimysium and which is continuous with the transversalis fascia of the ventrolateral abdominal wall (Fig. IX-23). Aptly, when this fascia clothes the ventral aspect of these muscles it is called, in turn, **psoas, quadratus,** or **iliac fascia.**

The Quadratus Fascia and Its Relation to the Thoracolumbar Fascia (Fig. IX-23): The thoracolumbar fascia binds down the deep longitudinal back muscles, *i.e.*, the Erector Spinae group at the sides of the vertebral spines. As a consequence it truly extends from neck to sacrum. In the neck it passes up deep to the Serratus Posterior Superior, and it is quite a thin covering. In the sacral region it spreads from sacral spines to ilium, and it attaches to coccyx and the sacrotuberous ligament. But it is in the lumbar region where it is strongest, and this is why you may see it labeled the "lumbar fascia" in some texts. Also, in the lumbar region the thoracolumbar fascia splits into three laminae (Fig. IX-23).

The *posterior layer* in the lumbar region is very tough and tendinous. It forms the outer perimysium of the Erector Spinae group, anchoring medially to spines and the interspinous ligament. Laterally it gives rise to the Serratus Posterior Inferior and to a goodly portion of the Latissimus Dorsi.

The *middle lamella* clothes the Quadratus Lumborum on its back side and separates this muscle from the Erector Spinae. Laterally, it gives origin to the Inferior Oblique of the abdomen. It attaches to lumbar transverse processes and to the crest of the ilium inferiorly by a strong fibrous band called the **iliolumbar ligament.** The ligament stabilizes the fifth lumbar vertebra, keeping it from rotating excessively at the lumbosacral joint.

The *anterior layer* of thoracolumbar (lumbar) fascia is the **quadratus fascia proper** which we mentioned above as clothing the ventral surface of that muscle. Laterally, this layer gives rise to the Transversus Abdominis. Near the last rib, this fascia forms a tendinous band, the **lateral arcuate ligament,** a partial origin for the Diaphragm.

The **ilio-psoas fascia** is the continuous sheet of connective tissue (dense white fibrous) which covers the ventral aspects of both of these muscles. Above the iliac crest it comes into relationship with the Psoas only. There it is thin. Medially, in that region, it anchors to the vertebral column by a series of arches that bridge the lumbar arteries; laterally, it is continuous with the quadratus fascia; and above, it fuses to form a **medial arcuate ligament.**

Below the iliac crest, this fascia covers both Iliacus and Psoas without separating one from the other. It gets thicker, and attaches to the iliac crest laterally and to the pelvic brim medially. In the thigh, it passes under the inguinal ligament to become the posterior wall of the **femoral sheath** around femoral artery and vein. In the case of a psoas abscess from tubercular disease of the lumbar vertebrae, you can see how the pus could move behind the fascia into the thigh. The spread could not be into the true pelvis for the fascia anchors securely to the pelvic brim, keeping the infection limited to the false pelvis or lower abdominal cavity proper.

The Quadratus Lumborum (Fig. IX-22): This muscle orients obliquely, arising from the last 2 inches of the iliac crest to insert upon the medial 2 inches of the last rib. Along its medial side it gives and receives slips in its anchorage to lumbar transverse processes. The muscle chiefly serves to flex the vertebral column laterad, or, acting with its mate of the opposite side, to extend the spine.

Psoas Major and Minor: These are essentially muscles that move the thigh. They were once residents of the pelvis, but they left their partner, the Iliacus,

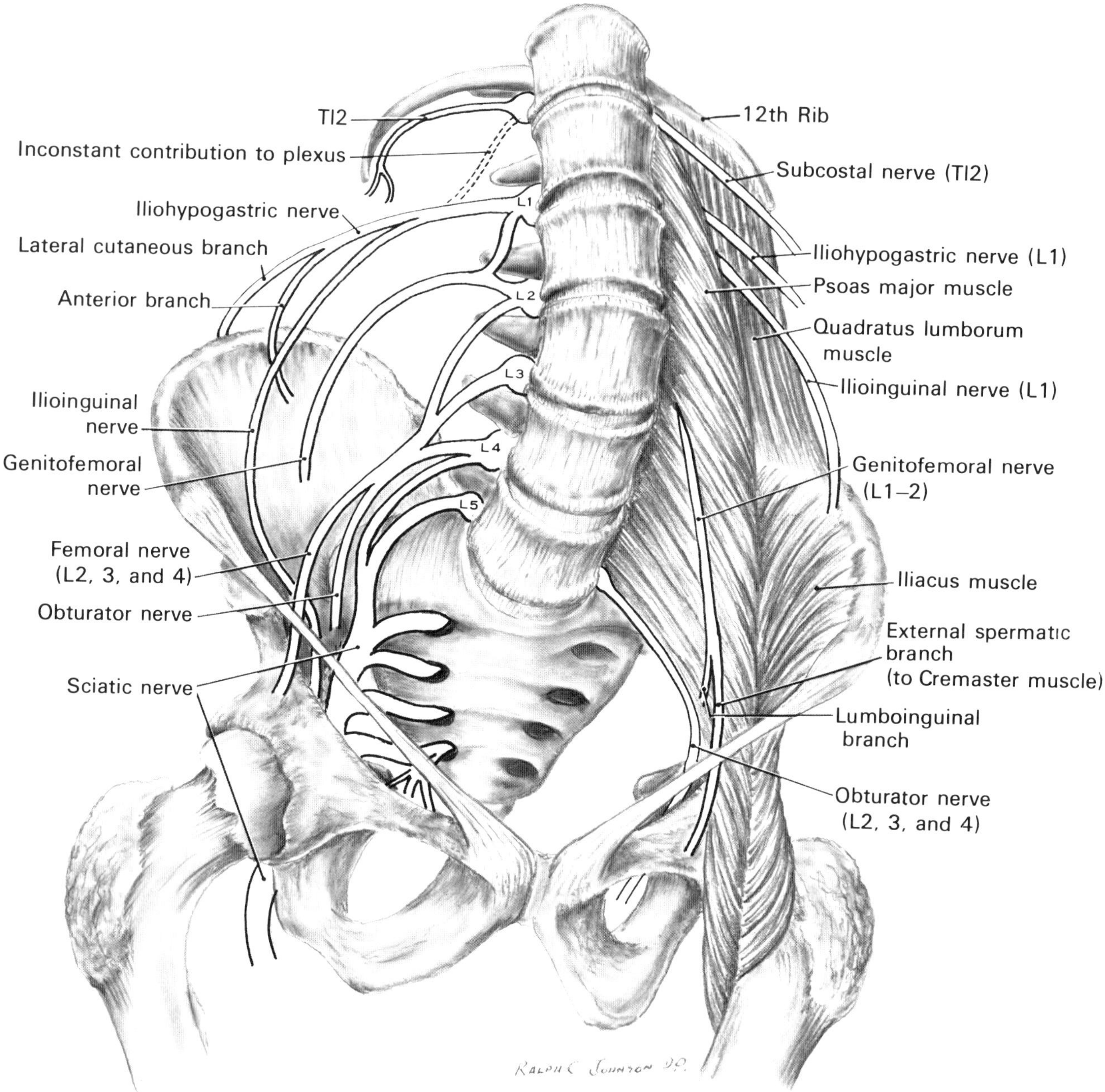

Fig. IX-22. Posterior abdominal muscles (right) and lumbar plexus (left).

behind when they moved north next door to the Quadratus. The Major arises from bodies and disks over a stretch of the vertebral column between VT_{12} and VL_4. Its tendon inserts upon the lesser trochanter of the femur. In combination with the Iliacus, it flexes the thigh on the hip, or in rising from a supine to a sitting position, the hip on the thigh. If the foot is free of the ground, this powerful muscle group can rotate the thigh laterally. If the foot is on the ground with the tibia fixed, the Iliopsoas can rotate the thigh medialward.

The *Psoas Minor*, often absent, is a thin muscle, medially placed. Its long narrow tendon blends with the psoas fascia and ends along the pectineal line.

The **Iliacus** arises from the entire, inner, fan-shaped part of the iliac bone. Its tendon joins that of the Psoas Major on the lesser trochanter of the femur.

Vessels of the Posterior Wall: The deepest vessels of the abdomen arise in pairs from the abdominal aorta, all except one. The exception is the single median sacral artery, a thin, tapering continuation of the aorta itself. All of these deep branches supply the posterior abdominal wall and have nothing to do with the abdominal viscera (Fig. IX-24).

1. The **paired phrenic arteries** are the very first branches of the **abdominal aorta.** They proceed upward to spread over the under surface of the Diaphragm. They often give off a superior suprarenal twig,

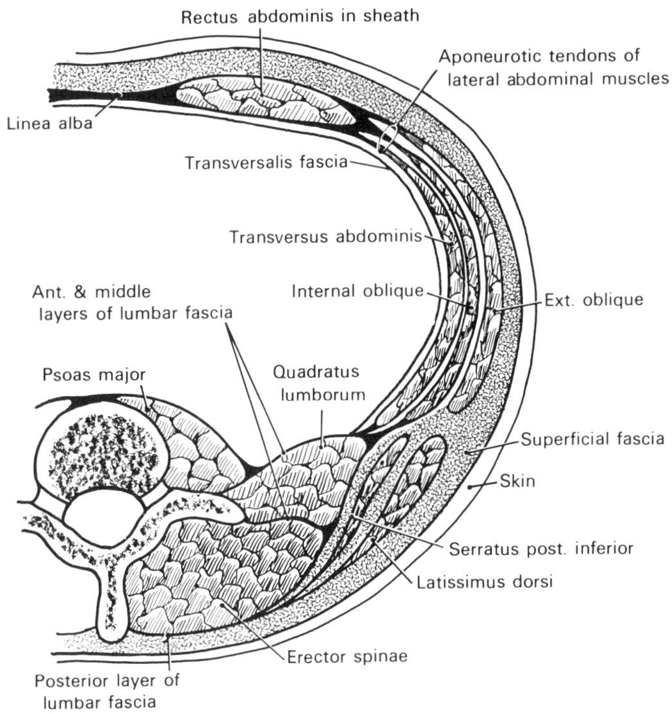

Fig. IX-23. Quadratus fascia and its relation to thoracolumbar fascia.

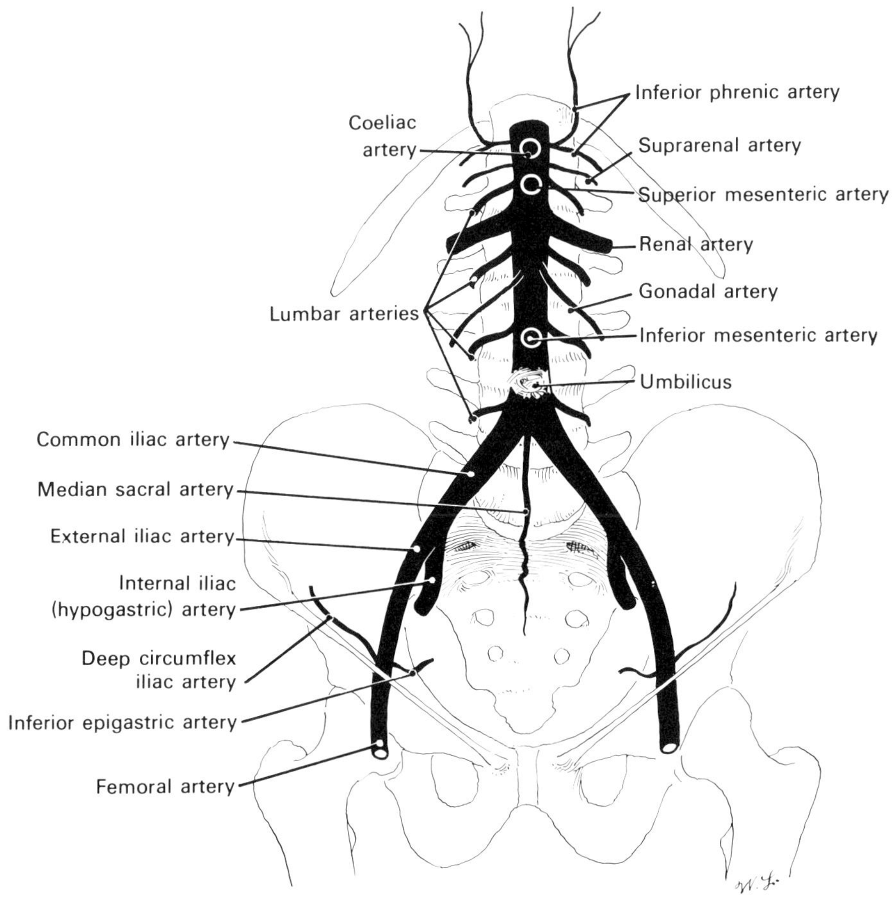

Fig. IX-24. Branches of the abdominal aorta (schematic). (After Grant.)

and in this sense only do they relate to neighboring viscera. Mainly they supply the roof of the abdominal cavity.

Now, the last paired branches of the **thoracic aorta** also enter the abdomen as **subcostal arteries** by passing behind the lateral arcuate ligament and through the Transversus Abdominis. In this sense, they are the highest arterial branches to enter the abdomen posteriorly—but they are interested in supplying its ventrolateral wall.

2. The **lumbar arteries** consist of five pairs—one for each of the lumbar segments—but only four pairs arise from the abdominal aorta because this vessel ends at the level of VL_4. They hug the vertebrae and pass along the back deep to the Psoas and the sympathetic trunk. The fifth pair, associated with VL_5, arise either from the middle sacral artery or from each of the internal iliac arteries. The fifth pair is often called the **iliolumbar arteries.**

3. The **median sacral artery** descends alone over the fifth lumbar vertebra and the sacrum in the midline. It would continue for an indeterminate distance into a tail, if we had one. We have mentioned that it is a direct continuation of the aorta.

Nerves of the Posterior Wall: Among these we would have to list: (1) the **sympathetic trunk** in the lumbar region; (2) the subcostal nerve (since we mentioned its artery); and (3) the ventral rami of the **lumbar nerves** which will form a **lumbar plexus.**

1. The **sympathetic trunk,** on each side, enters the abdomen deep to the medial arcuate ligament, and it follows down in the groove between the backbone and the border of the Psoas. It is, obviously, the continuation of the thoracic sympathetic trunk and, below, it will enter the pelvis behind the common iliac vessels. At the coccyx the paired lateral trunks will join in the midline in the unpaired **ganglion impar.** As in the cervical region, there is considerable variation and fusion of ganglia in the lumbosacral area. There should be a pair for each segmental level, but this rarely holds true as it does in the thoracic region. But you may find at least four pairs of them in most cadavers.

Rami communicantes will communicate from ganglia to lumbar nerves. **White rami** from lumbar spinal nerves to the chain ganglia will exist only at levels L_1 and L_2 of the spinal cord. *All* ganglia will send **gray rami** back to the spinal nerves to be distributed as postganglionic fibers to skin viscera (sweat glands, pilomotor muscle and smooth muscle of arterioles) in the lumbar abdominal skin. Visceral branches from ganglia and trunks will join the plexus formations along the aorta and along the internal iliac arteries. As always these arterial nerve plexuses to abdominal and pelvic viscera contain a mixture of fiber types: (a) unmyelinated postganglionic sympathetics; (b) myelinated parasympathetic preganglionics from the vagus in the upper abdomen and from the pelvic nerve (nervi erigens) in the pelvis and lower abdomen; and (c) myelinated visceral sensory fibers returning from viscera to cord.

2. The **subcostal nerve** follows the same course as its artery (see above) and, after piercing the Transversus, runs between it and the Internal Oblique to give off muscular and cutaneous twigs like any other intercostal nerve, except that it reaches abdominal wall territory and ventrolateral wall territory. Thus, like its artery, it only enters posteriorly and does not remain in that region.

3. The **lumbar nerves** are spinal nerves whose anterior primary divisions leave intervertebral foramina below the level of their corresponding vertebral bodies and dive at once into the substance of the Psoas muscle. These nerves get larger the lower the level of their origin. We have noted their connections to the sympathetic trunk *via* rami communicantes. They give short twigs close to their level of emergence to the Intertransversarii, the Quadratus and to the Psoas. The upper four divide to join one another in plexiform fashion and to give rise to *six named nerves.* The lowest of these dividing branches of the fourth lumbar nerve descends to the fifth as a thick trunk uniting into a **lumbosacral plexus.** All the other branches of the upper four lumbar nerves constitute the **lumbar plexus proper.**

4. The **lumbar plexus** is located within the substance of the Psoas muscle, and is the upper portion of the larger lumbosacral plexus (Figure IX-22). It is formed by the anterior primary divisions of the upper three and a part of the fourth lumbar nerves. Half the time, T_{12} may contribute to it (as shown in Fig. IX-22).

a) L_1, L_2, and L_4 divide into upper and lower branches. The **upper branch of L_1** gives rise (by further splitting) to the **ilioinguinal** and **iliohypogastric nerves.** The **lower branch of L_1** joins the upper branch of L_2 to form the **genitofemoral nerve.** We have already indicated that the lower branch of L_4 joins L_5 to unite lumbar with sacral plexus.

b) The lower branch of L_2, all of L_3 and the upper branch of L_4, all split into **anterior and posterior divisions of the plexus.** The three anterior divisions of these nerves unite to form the **obturator nerve;** the three posterior divisions, the **femoral nerve.** The upper two posterior divisions unite as the **lateral femoral cutaneous nerve.** Sometimes, from the low part of L_3 and the upper part of L_4, a small accessory obturator nerve is formed as an anterior plexus division.

We have already discussed the iliohypogastric and ilioingual nerves when we talked about the ventrolateral body wall. Like the subcostal nerve, this is their territory of distribution. We mentioned the **genitofemoral nerve,** briefly only, in relation to the spermatic cord. It emerges through the anterior surface of the Psoas (Fig. IX-22) and crosses toward the internal inguinal ring on the surface of the Psoas. It divides into the **external spermatic nerve** for the Cremaster muscles and for the skin of scrotum or labia; and into the **lumboinguinal nerve** for the skin of the upper part of the thigh. The **lateral femoral nerve** (not shown), riding upon the Iilacus, crosses under the inguinal ligament to supply the skin of the anterolateral thigh.

Numbness, tingling, and pain over the course of the lateral femoral cutaneous nerve is known as **meralgia paresthetica of Roth.** It occurs in middle aged men after they stand for a long time or after prolonged walking. It could be due to a number of causes, none of them known for certain, *e.g.*, neuritis, the bend in the nerve as it leaves the pelvis, obesity, flat feet, spondylitis, tight clothing, or tight pressure of the iliac fascia upon the nerve bundle.

As a rule, the ilioinguinal, iliohypogastric, and genitofemoral nerves give little trouble. Sometimes there is sensory loss over their area of distribution or referred pain to these regions when the renal pelvis or ureters are diseased, since the nerves lie close to these organs in part of their course. The femoral and obturator nerves supply structures in the thigh and leg and had best be considered with the lower limb.

The Lumbar Vertebrae

General: The five lumbar vertebrae are the largest of the movable segments of the spinal column. They progressively increase in size from VL_1 to VL_5. These vertebrae differ from those in the cervical and thoracic regions. Unlike cervical vertebra, they do not have transverse processes with foramina, and unlike thoracic vertebra, they lack articular facets on their bodies. As the infant begins to ambulate, the **secondary curvature** of the lumbar spine develops. Contributing to the maintenance of this curve are the intervertebral disks which are wedge-shaped and thicker ventrally than dorsally.

In Figures IX-25 to 27, we may study the structures comprising the lumbar vertebra. Matching radiographs have been included; however, these will be elaborated upon later. The vertebral **body** is large and reniform in shape (Fig. IX-28). It is narrower in its anteroposterior diameter than in its transverse diameter. VL_1 and VL_2 are taller dorsally than ventrally, whereas the reverse applies to VL_4 and VL_5. VL_3 is transitional and may resemble its neighbors above or below. The anterior, posterior, and lateral margins are all slightly concave, as are the superior and inferior articular surfaces.

Stout **pedicles** are directed posteriorly from the upper margins of the body (Figs. IX-26 and 28). Immediately below the pedicle is a deep **inferior vertebral notch,** which forms the upper portion of the **intervertebral foramen.** A shallow **superior vertebral notch** etches the upper margin of the pedicle, and this forms the lower part of the invertebral foramen.

The **lamina** are large, sturdy structures which are directed posteromedially and inferiorly. That portion of the lamina, which in Figures IX-25 to 27 is circumscribed by a solder collar, is referred to as the **pars interarticularis.** It is situated between the articular processes.

At the junction of the lamina and the pedicles, the **articular processes** arise. The directional plane of these articular processes varies throughout the lumbar spine. At VL_1 the concave facet of the superior articular process usually faces medially, while the convex facet of the inferior articular process faces laterally. At VL_5 the facet of the superior articular process is now directed dorsally, and the facet of the inferior articular process ventrally. The planes that we have cited are the usual ones to be found, however, it should be realized that there are wide variations, particularly at the VL_4–VL_5 and VL_5–VS_1 levels. If you will survey the skeletal material in your dissecting room, some of these variations will undoubtedly be found. These facet planes may be asymmetrical at the same level. On the dorsolateral tip of each superior articular process a noticeable hummock protrudes as the **mammillary process,** which corresponds to the less conspicuous superior tubercle on the 12th thoracic vertebra.

The **transverse processes** arise from the junction of the pedicles and the lamina. These thin, spatula-shaped processes of the upper three lumbar vertebrae extend laterally, while those of the lower two vertebrae are tipped slightly inferiorly. The longest transvesre process is on VL_3, and they progressively shorten as one moves toward VL_5 or toward VL_1. Dorsally, in the angle between transverse process and mammillary process, there projects a small **accessory process.** If the accessory process becomes elongated it is referred to as a **styloid process.** Extending from the posterior aspect of the lamina is the thick, broad, quadrilateral **spinous process.**

The **vertebral** canal is triangular in shape, and wider than in the thoracic region (Fig. IX-28). Contained within is the tapering end of the spinal cord which ceases at VL_2; its membranes and blood vessels, the filum terminale; and the cauda equina of rootlets that descend from conus to progressively lower intervertebral foramina of exit.

The fifth lumbar vertebra is unique in its massiveness and in its articulation with the sacrum at the **sacrovertebral angle.** While we have characterized transverse processes as slender, those of the fifth are thick and conical, and appear to spring from the body as well as from the roots of the arch. Iliolumbar ligaments attach here as well as do muscles. VL_5 has the smallest spinous process of any of the lumbar vertebrae. Here the inferior articular processes are wider apart than the superior ones, and they articulate with the superior articular processes of the sacrum.

Radiographic Anatomy of the Lumbar Spine: The usual examination of the lumbar spine consists of: (1) anterior-posterior; (2) lateral; and (3) 45° left and right posterior oblique projections. At times other views, such as flexion, extension, and lateral bending films are obtained.

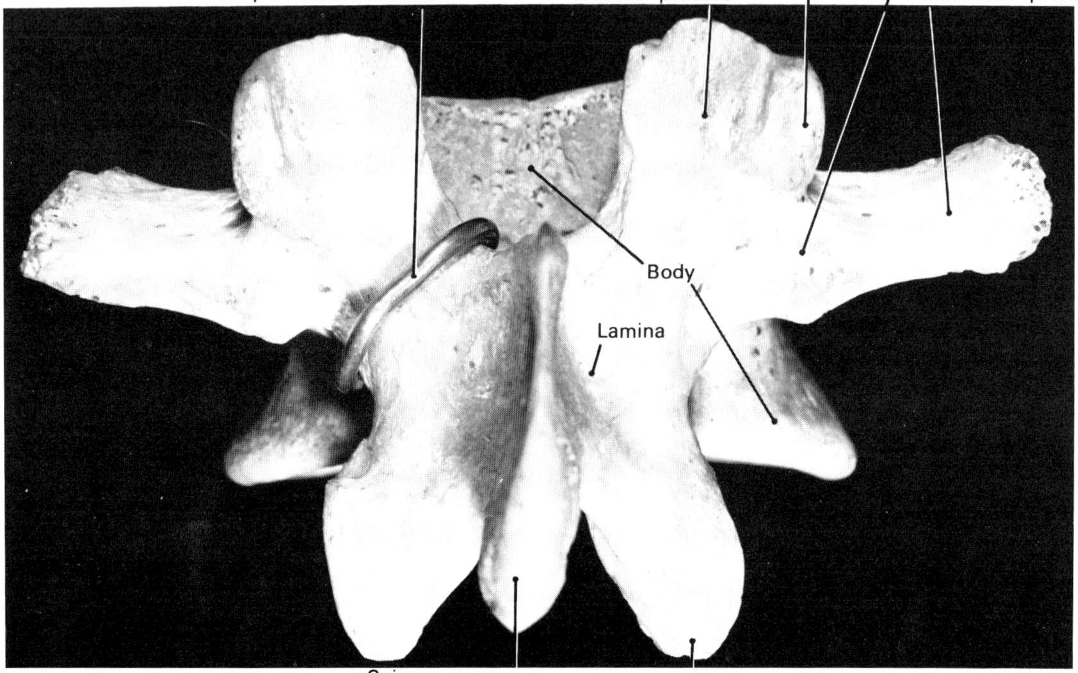

A

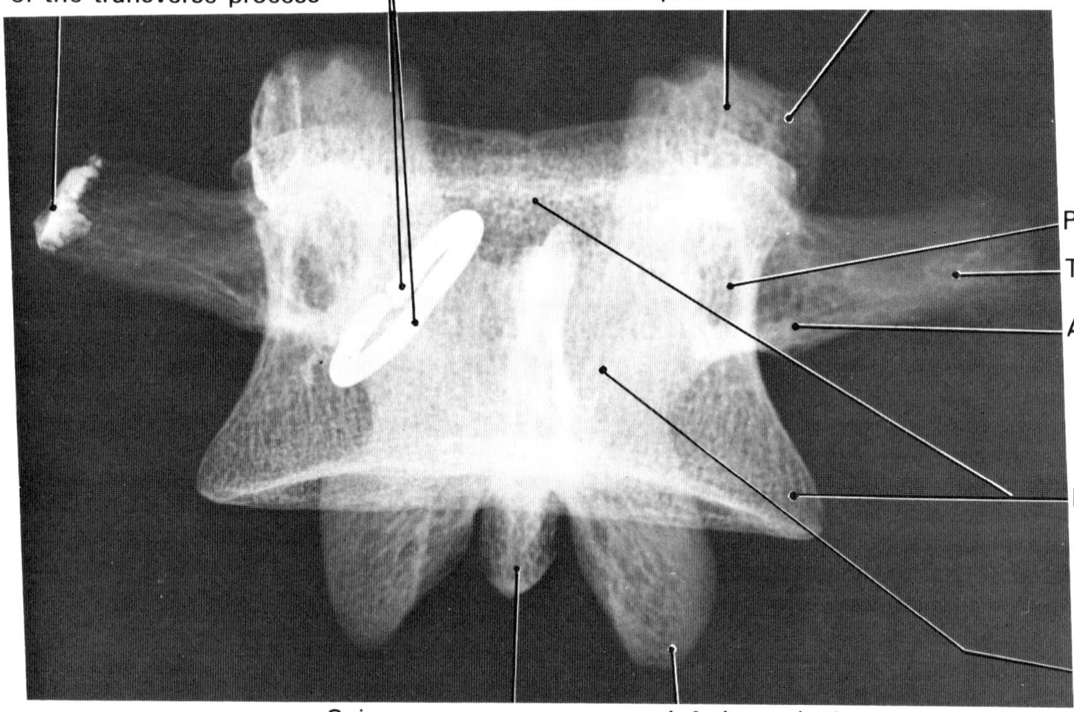

B

Fig. IX-25. Lumbar vertebra, dorsal aspect. **A.** Photograph from position of X-ray film cassette. **B.** Matching radiograph. The solder collar is around the pars interarticularis.

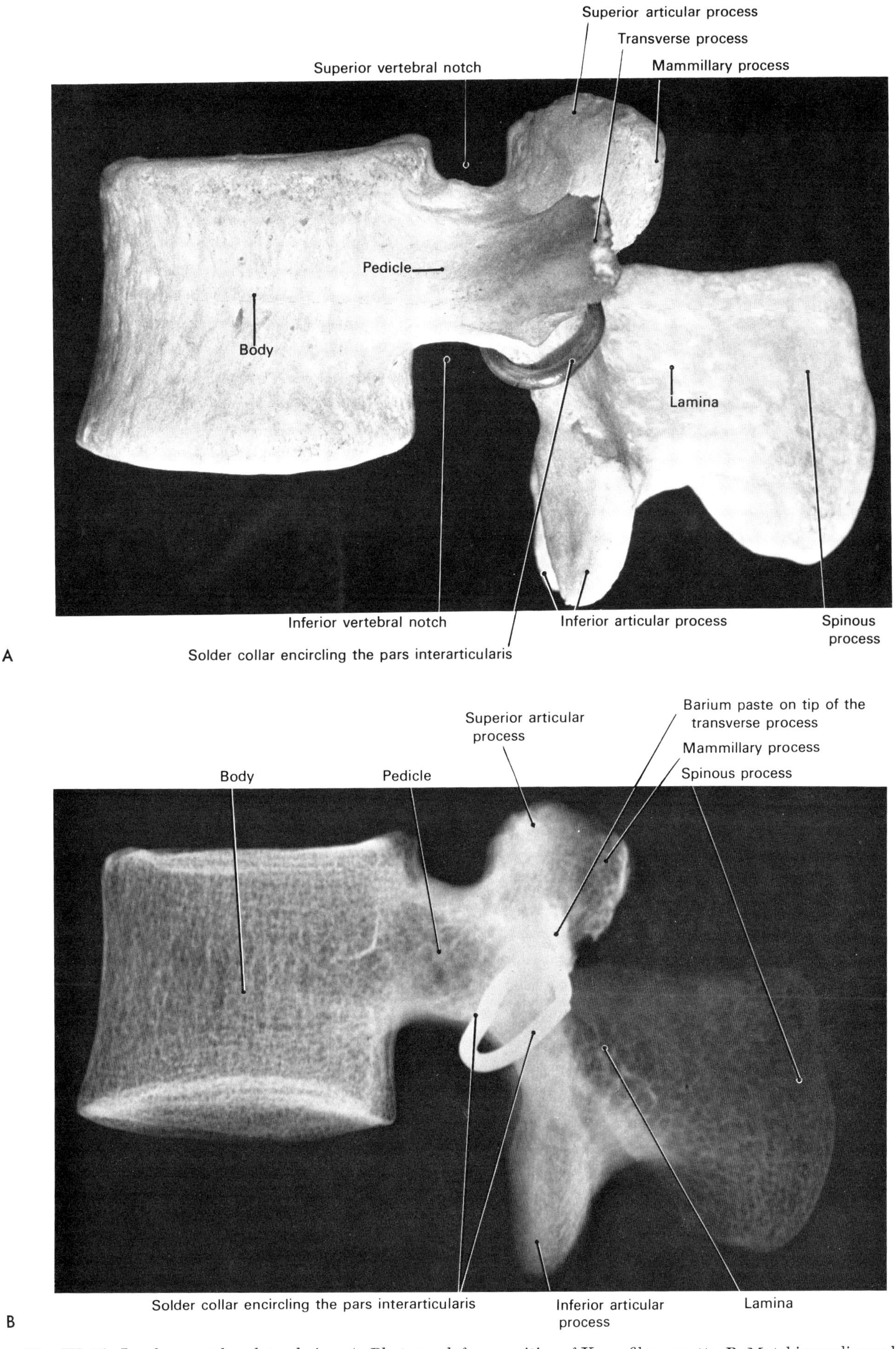

Fig. IX-26. Lumbar vertebra, lateral view. **A.** Photograph from position of X-ray film cassette. **B.** Matching radiograph. Note the position of the solder collar encircling the pars interarticularis.

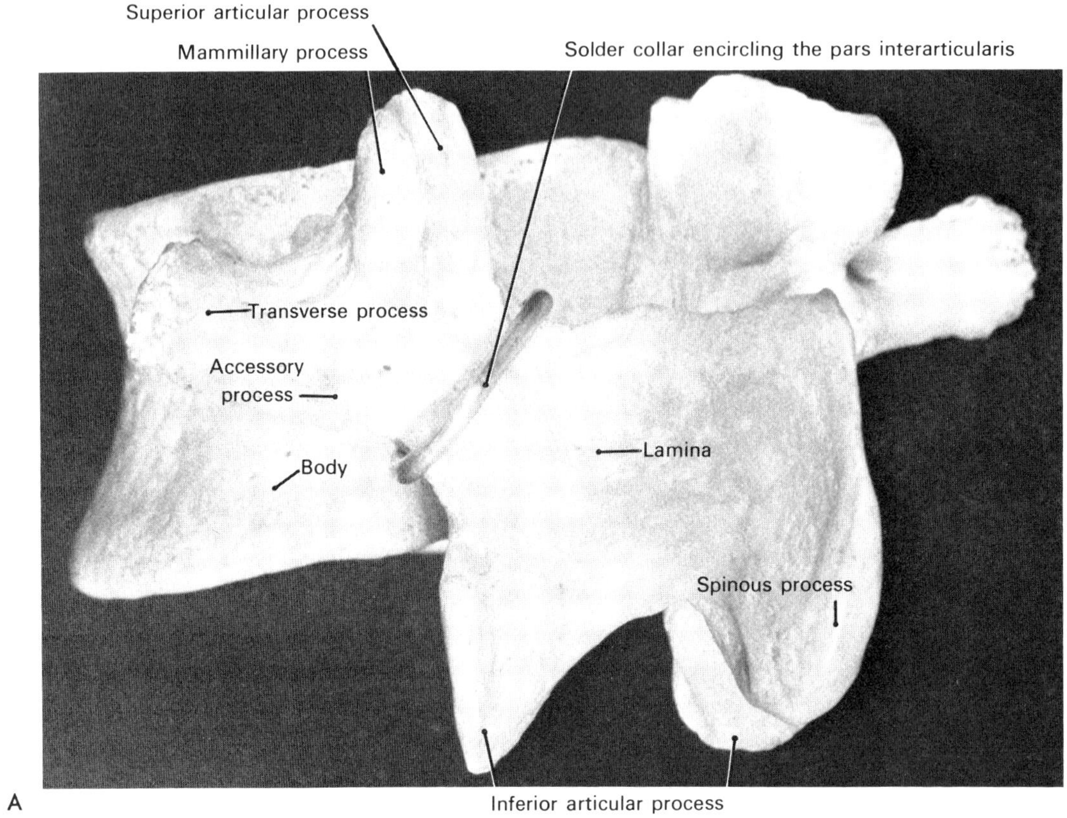

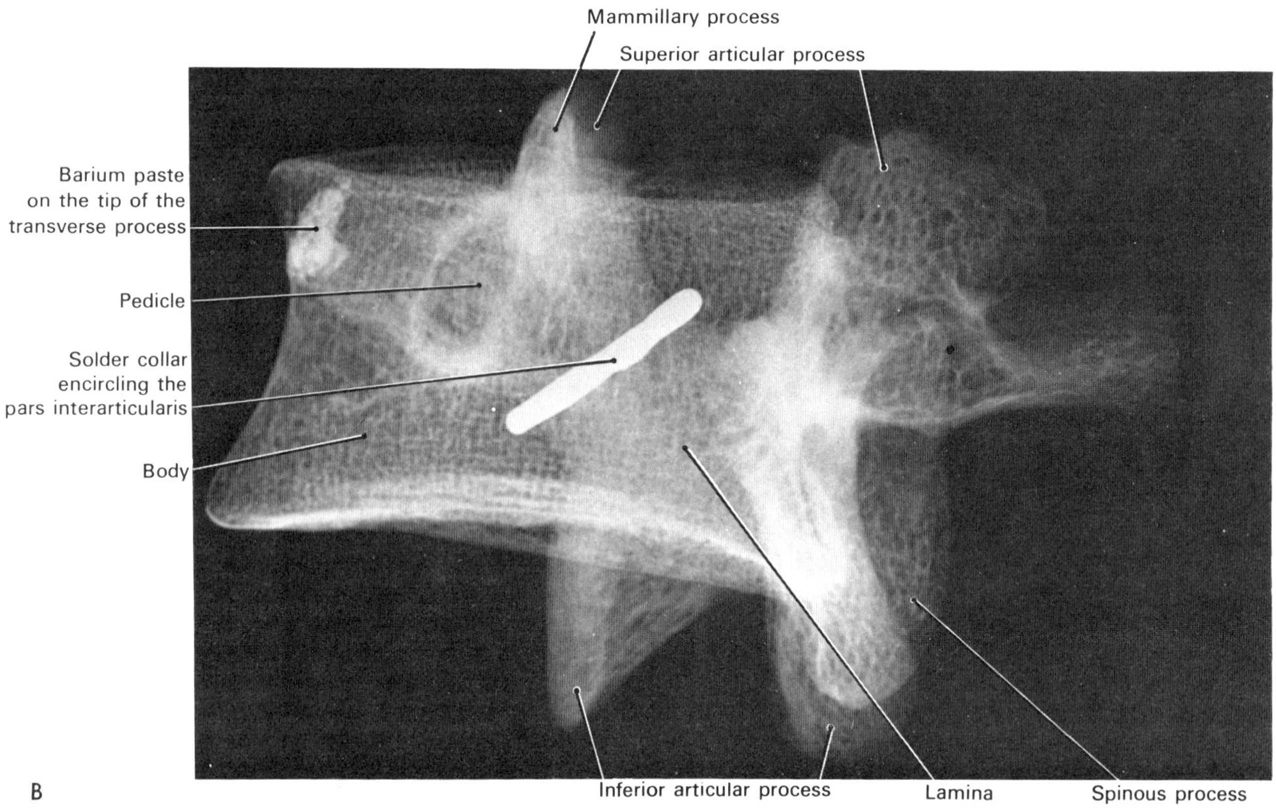

Fig. IX-27. Lumbar vertebra, viewed obliquely from behind. **A.** Photograph from position of X-ray film cassette. **B.** Matching radiograph. The pars interarticularis is encircled by a solder collar.

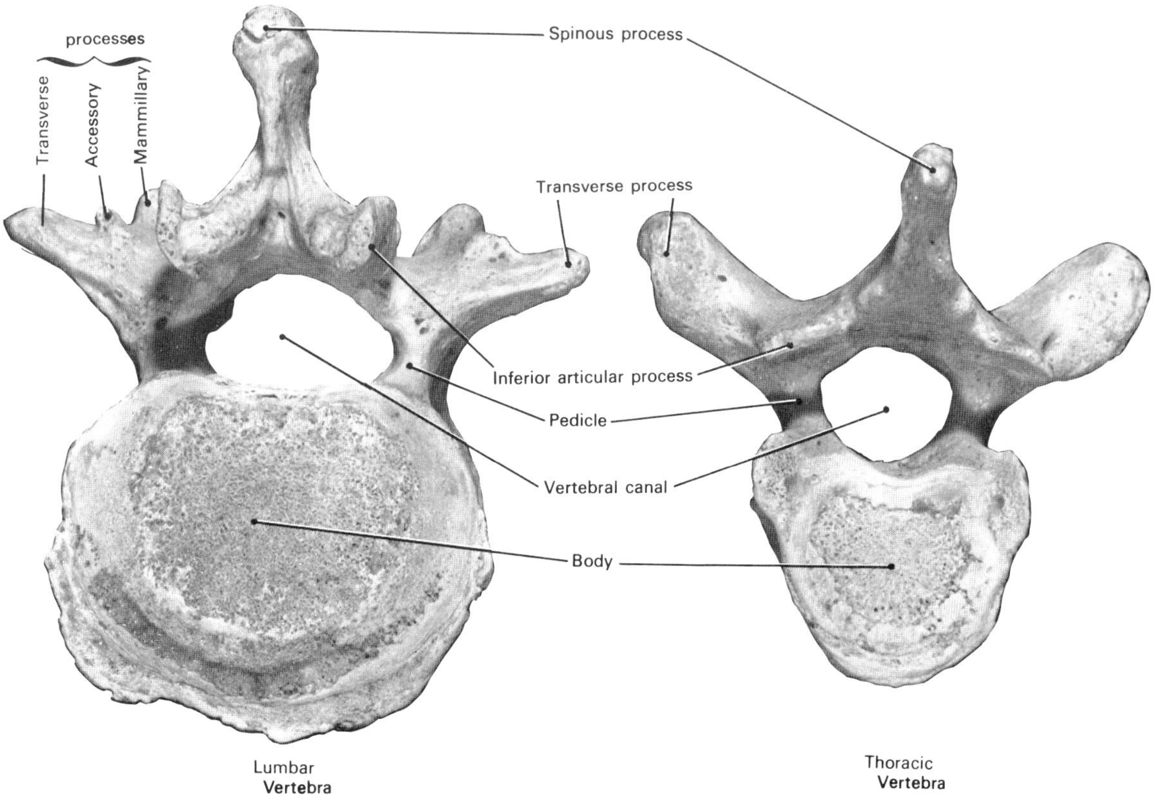

Fig. IX-28. Comparison of lumbar and thoracic vertebrae. Viewed from below.

1. Anterior-posterior Projection: The patient is placed in the supine position on the radiographic table. The legs are flexed at the hips in order to straighten the normal lumbar lordotic curvature, and the vertically-directed X-ray beam is centered on the VL_4 vertebra. The patient is instructed to breathe quietly during the X-ray exposure so that overlapping gas and waste materials within the small intestine will be blurred.

Examine for a moment the schema (Fig. IX-29) of the major component structures of the lumbar vertebra, and the contribution of each to the final roentgen appearance. Compare this with the radiograph of the intact vertebral column (Fig. IX-30). The vertebral body is rectangular in shape with slightly concave borders laterally. The inner cancellous bone is less dense than the cortical bone which rims the superior and inferior vertebral surfaces. Notice in Figure IX-30 that whereas superior and inferior vertebral articular surfaces of VL_4 are seen end-on (summation effect), those of the other vertebrae (especially VL_3) appear oval in shape due to divergence of the X-ray beam.

The stout, oval-shaped pedicles (Figs. IX-25 and 30) are seen at the upper lateral borders of the vertebral bodes. The **interpedicular distance,** the distance between the inner margins of the pedicles, is actually the width of the vertebral canal. Localized widening of the interpedicular distance is abnormal, and the cause should be sought.

Finally, the butterfly-shaped neural arch and associated structures are added in Figures IX-29 and 30. Each structure is labeled: superior and inferior articular processes; lamina; spinous; and transverse processes.

In Figure IX-30, note that the apophyseal joints at the VL_1–VL_2 level on the patient's right are in the sagittal plane, and therefore, are seen end-on. At lower levels of the lumbar spine, these apophyseal joint planes change, as we have described earlier, and are obscured from view.

2. Lateral Projection: The patient is placed in a true lateral position on the radiographic table, and the central X-ray beam is again directed at the VL_4 vertebra. This view affords the best evaluation of the radiolucent intervertebral disk spaces. It is not at all uncommon to find slight narrowing at the VL_5–VS_1 intervertebral disk space in otherwise normal individuals.

The vertebral bodies align in a gentle lordotic curvature (Fig. IX-31), although it should be noted that the greatest curve is in the lower lumbar region. Compare the vertebral height of the five lumbar vertebrae in this

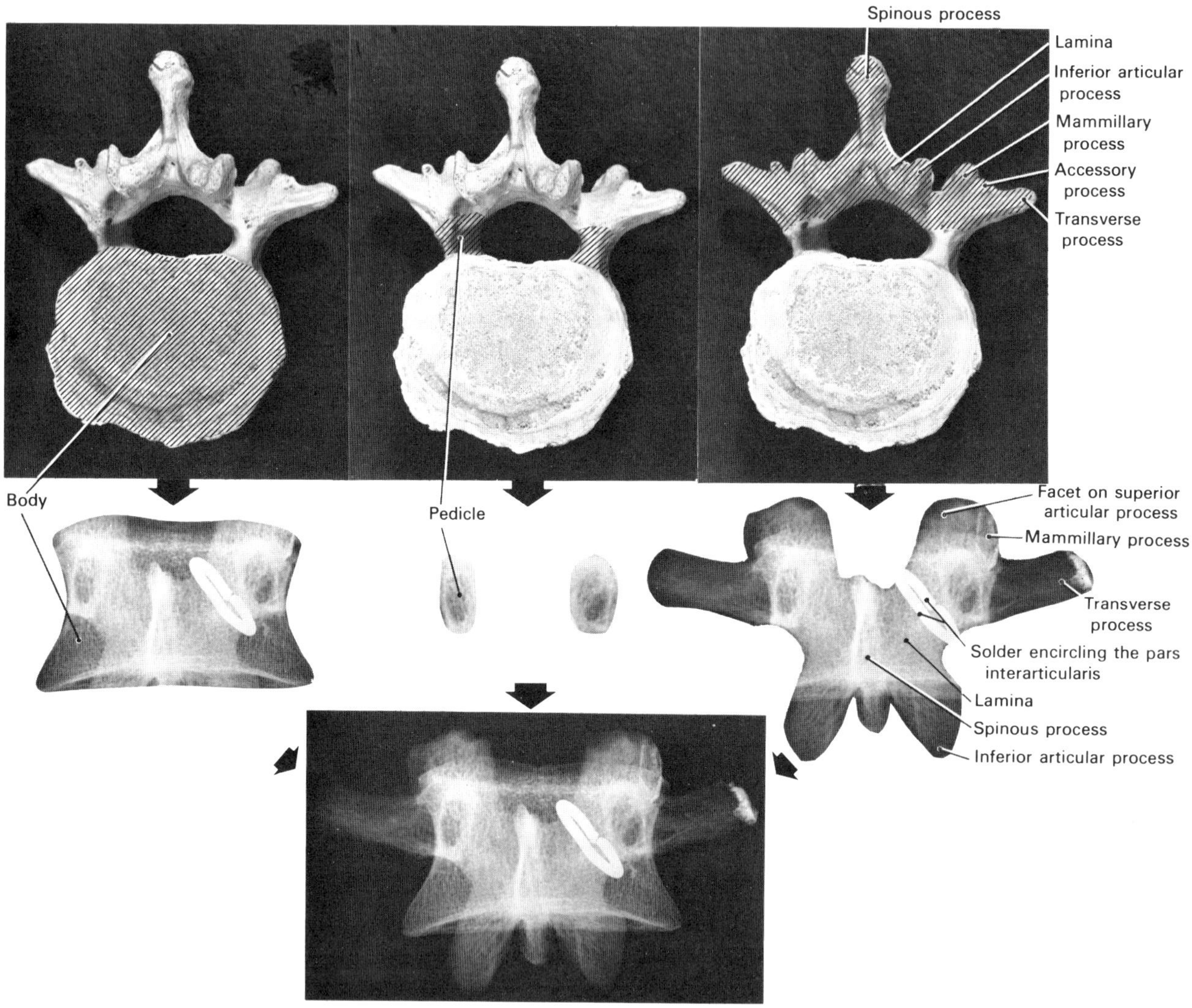

Fig. IX-29. Schema of elements comprising a typical lumbar vertebra.

radiograph in light of the information given you earlier. The intervertebral foramina are, for the most part, oval in shape. These foramina appear to decrease in size at VL_4–VL_5 and VL_5–VS_1. They are bounded by the pedicles of the adjacent vertebrae, their bodies and the superior articular process of the body below.

Identify in Figures IX-26 and 31 the transverse processes seen end-on, the lamina (including the pars interarticularis), and note that the spinous processes are all not well seen. These are usually *burned out*, and can only be seen to good advantage with the aid of a bright light.

3. Forty-five Degree Oblique Projections: The patient is rotated 45° into a right and then a left posterior oblique projection. The vertically-directed central X-ray beam is centered on VL_4.

Examine for a moment the oblique view of the lumbar vertebra and its matching radiograph (Fig. IX-27). If you will use your imagination, you will be able to see a "scotty dog" formed by the vertebral structures. The nose is formed by the transverse process, the eye by the pedicle and proximal portion of the transverse process, the ear by the superior articular and mammillary processes, the front leg by the inferior articular process, the body of the dog is represented by the laminal arch, and, finally, the collar made of solder, surrounds the pars interarticularis. Whenever you examine an oblique view of the lumbar spine (Fig. IX-32), always identify each structure comprising the scotty dog on each side at all levels. This will assure less chance of missing a congenitally or surgically absent structure.

Space does not permit a comprehensive discussion of the myriad of congenital and acquired abnormalites of the lumbar spine. However, we have repeatedly stressed the pars interarticularis to you and for good reason.

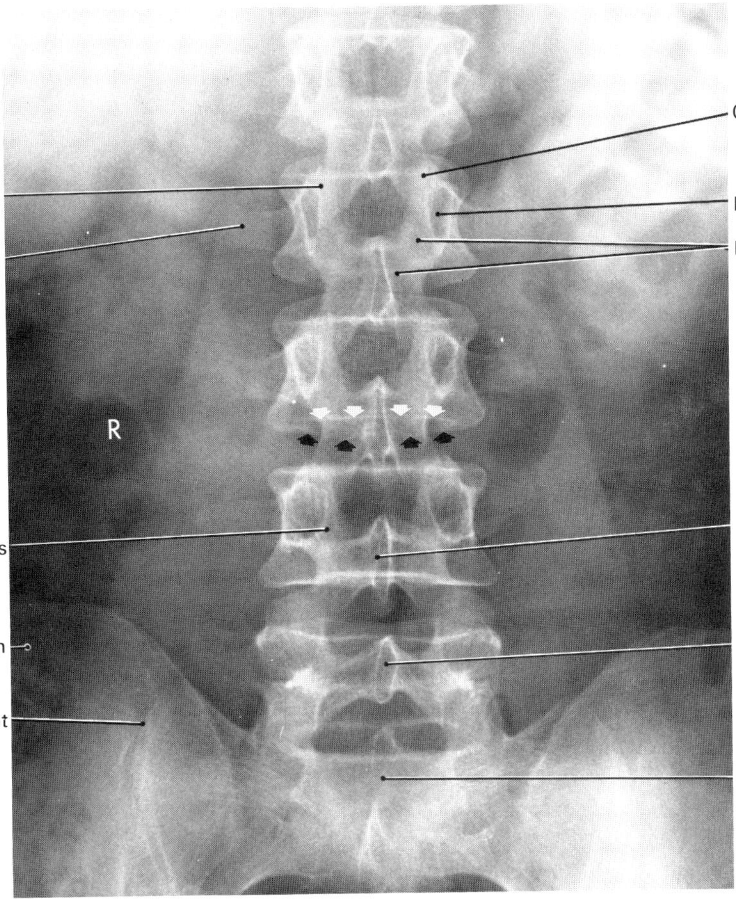

Fig. IX-30. Radiograph of the lumbar spine in the AP projection. Note that the superior and inferior vertebral surfaces of VL_4 are seen end-on (summation effect) while those of the other vertebrae appear oval in shape (black and white arrows) due to divergence of the X-ray beam.

Five per cent of individuals have bony defects of the pars, either on a congenital or acquired traumatic basis. Any level of the lumbar spine may be involved, although the VL_5 level is the most common site. A defect in the pars interarticularis may be either unilateral or bilateral. The latter may afford opportunity for slippage (anterolisthesis) of the body, pedicle, transverse process, and superior articular process of the involved vertebra on the vertebral body below. This is called a **spondylolisthesis.**

PART II. ABDOMINAL CAVITY PROPER

Relation of Organs to the Abdominal Cavity and the Abdominal Wall

It is important that you understand that the viscera and other structures within the abdomen dwell within the extraperitoneal areolar tissue just as the testis did before its descent. This is the loose fibro-elastic connective tissue between transversalis fascia and peritoneum, or, since tranversalis fascia is continuous posteriorly with quadratus and ilio-psoas fascias, between these fascias and peritoneum. Almost all organs dwelling in this tissue relate directly or indirectly to the back walls of the abdominal cavity.

The inner peritoneal lining of the abdominal cavity is a thin sheet of simple squamous epithelium (mesothelium) which rests upon well-vascularized areolar tissue. A serous transudate keeps this **parietal peritoneum** moist. Duplicate peritoneal folds, known as mesenteries or "ligaments," suspend or envelope organs which develop deep to the peritoneum and which bulge into the cavity. Hence, all the structures which we like to think *fill* the abdominal cavity actually lie outside of it beneath a covering of **visceral peritoneum** which is really a reflection of the parietal peritoneum over or about them (Fig. IX-33).

To make this relationship crystal clear, you might take your wife's silk-lined alligator bag. If she'll let you, slit open the outside leather, being careful not to cut through the silk. Next, stuff in through this slit all the junk you usually find inside a woman's purse...

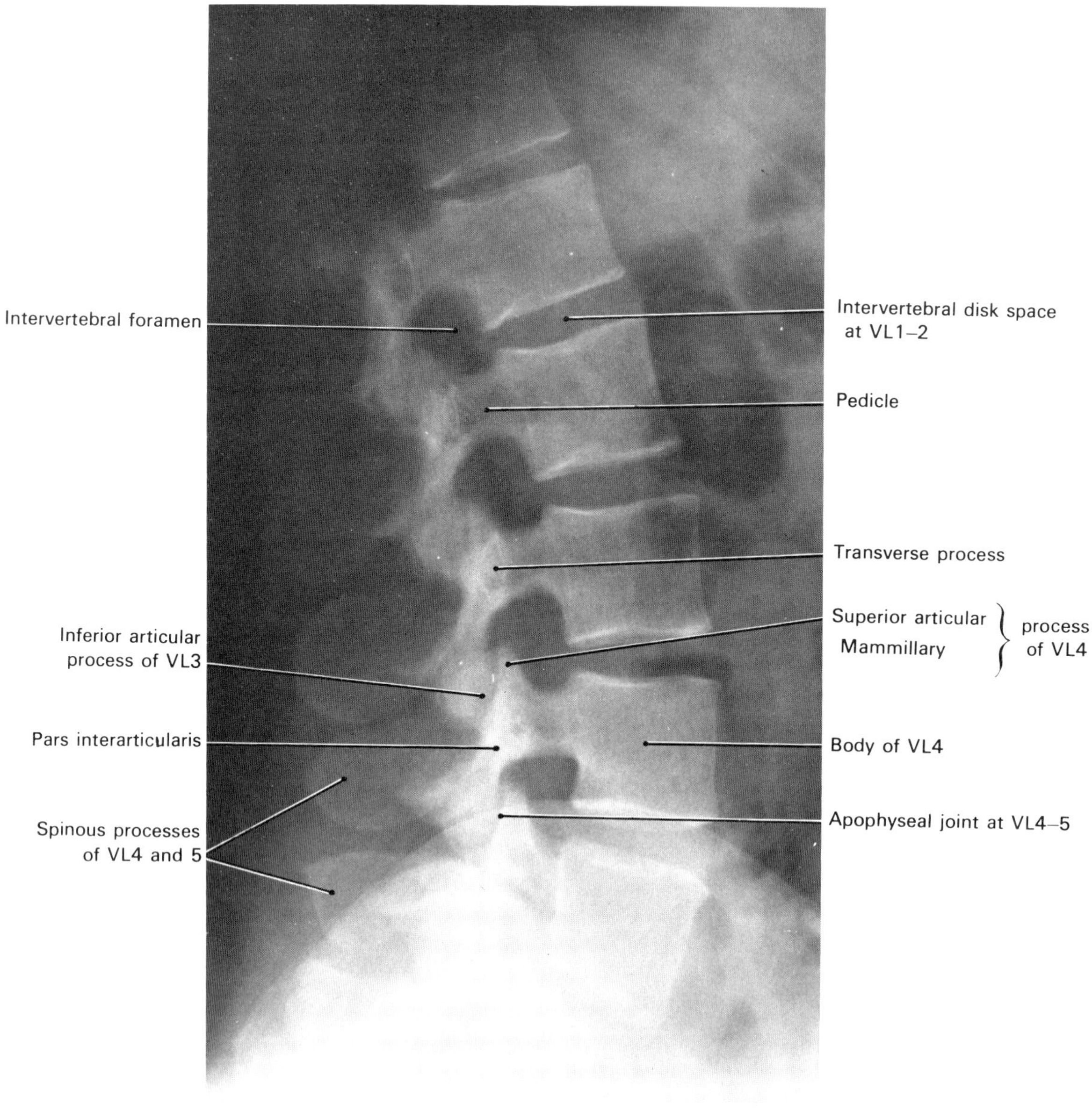

Fig. IX-31. Lateral radiographs of the lumbar spine. **A.** VT_{12} to L_5.

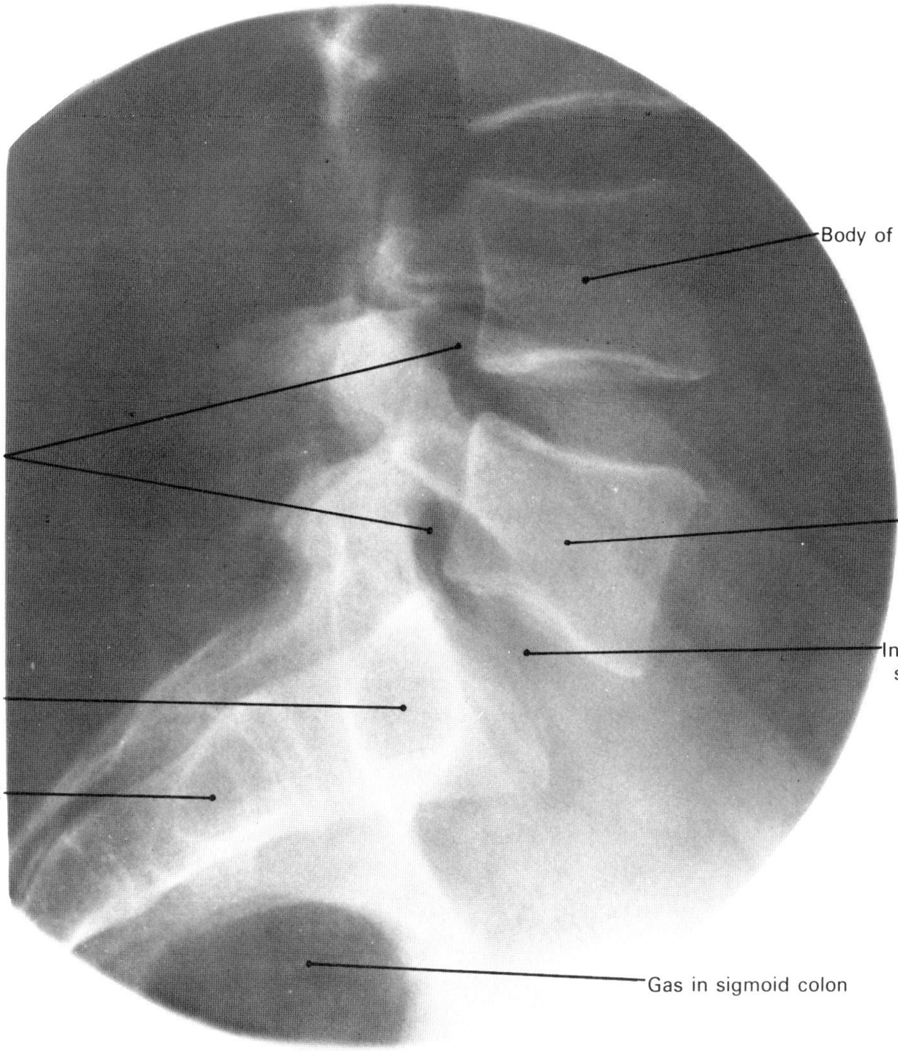

Fig. IX-31B. Coned view of VL_4 to VS_2.

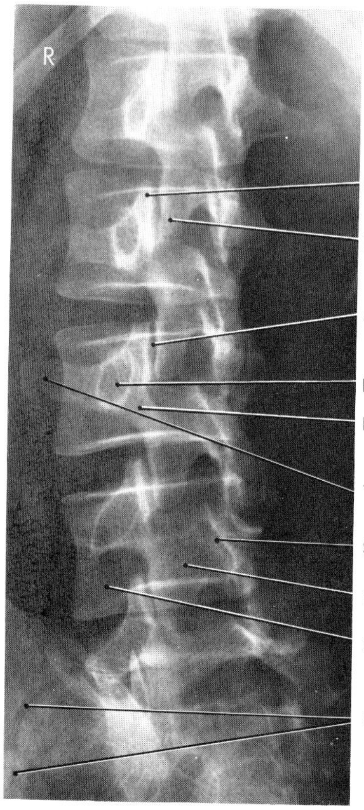

Fig. IX-32. Radiograph of the lumbar spine in the right posterior oblique projection.

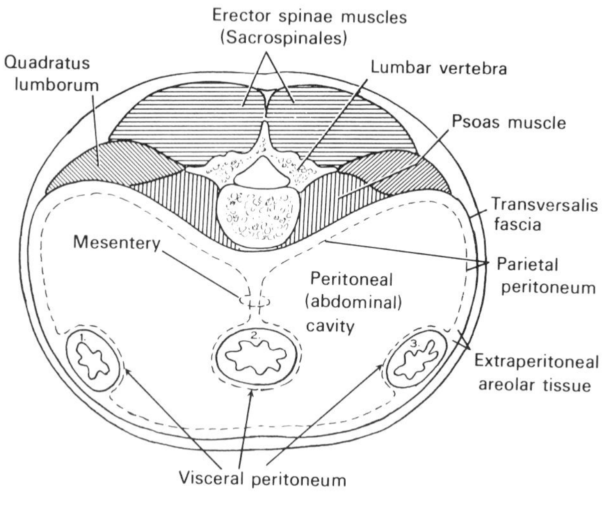

Fig. IX-33. Schema of peritoneal cavity. Organ 1, which developed in extraperitoneal areolar tissue, has now bulged so far into cavity that it is suspended from the posterior body wall by a mesentery. Organs 2 and 3 are said to be "retroperitoneal." See text for explanation.

lipstick, compact, comb, keys, loose change, the baby's bottle, your spare spectacles, the shopping list, and the green stamps. Then open the bag at the catch or zipper and put your hand in it. It is empty, but you can feel the contours of every item as it bulges the silk lining. If you were now to catch an alligator, slit open its abdominal wall (*i.e.*, unzipper it) and reach a hand in through the slit, your hand would be in a very similar empty space, but a bit damper one perhaps. You could only feel the viscera as they bulged the peritoneal lining.

In the human, some organs, like the kidneys or the pancreas, lie so deeply in the subperitoneal connective tissue that only their anterior surface is covered by peritoneum. Others, like the ascending and descending colon, bulge in so far that their sides are also covered. Still others lose all contact with the body wall entirely, dragging before them folds of peritoneum which suspend and cover them, almost completely, *e.g.*, the liver, spleen, and small intestine. Extraperitoneal connective tissue fills in between the lax duplicated folds of peritoneum which suspends these organs so that blood vessels and nerves can reach them. After all, blood vessels and nerves also traverse extraperitoneal connective tissue.

In the above sense, the organs of the abdomen and pelvis relate to their cavities as the heart and lungs do to theirs. There is nothing in the abdominal cavity, anymore than there was anything in the pericardial or pleural cavities, nothing other than the serous transudate between visceral and parietal peritoneum. Surgeons and embryologists both insist upon classifying certain organs as **retroperitoneal.** The term, of course, means "behind the peritoneum." If *all* organs are behind the peritoneum, this seems to be a confusing and redundant term. When the term is used in reference to a given organ, it really means three things: (1) the adult organ no longer has a definable mesentery such as it may have had in embryonic life; (2) it is less surrounded by peritoneum than an organ that has a mesentery; and (3) it is more obviously in association with the parietal peritoneum than an organ with a mesentery, though the surface in contact with the peritoneum is still considered to be covered by *visceral peritoneum.* The sketch in Figure IX-33 may help to make the last statement clear. Organs 1 and 3 are retroperitoneal while organ 2 is not. Embryologically, too, organs 1 and 3, which represent ascending and descending colon, were once as freely suspended by a dorsal mesentery as was organ 1 (the small intestine).

The Development of the Gastro-intestinal Tract and Associated Structures

A primitive, entodermally-lined gut was tucked into the embryo in the process of forming head and tail folds (Fig. IX-34). This tube, in the fourth month, is still in continuity with the yolk sac *via* the **vitelline duct** ventrally (Fig. IX-34C) and with the blind-end-

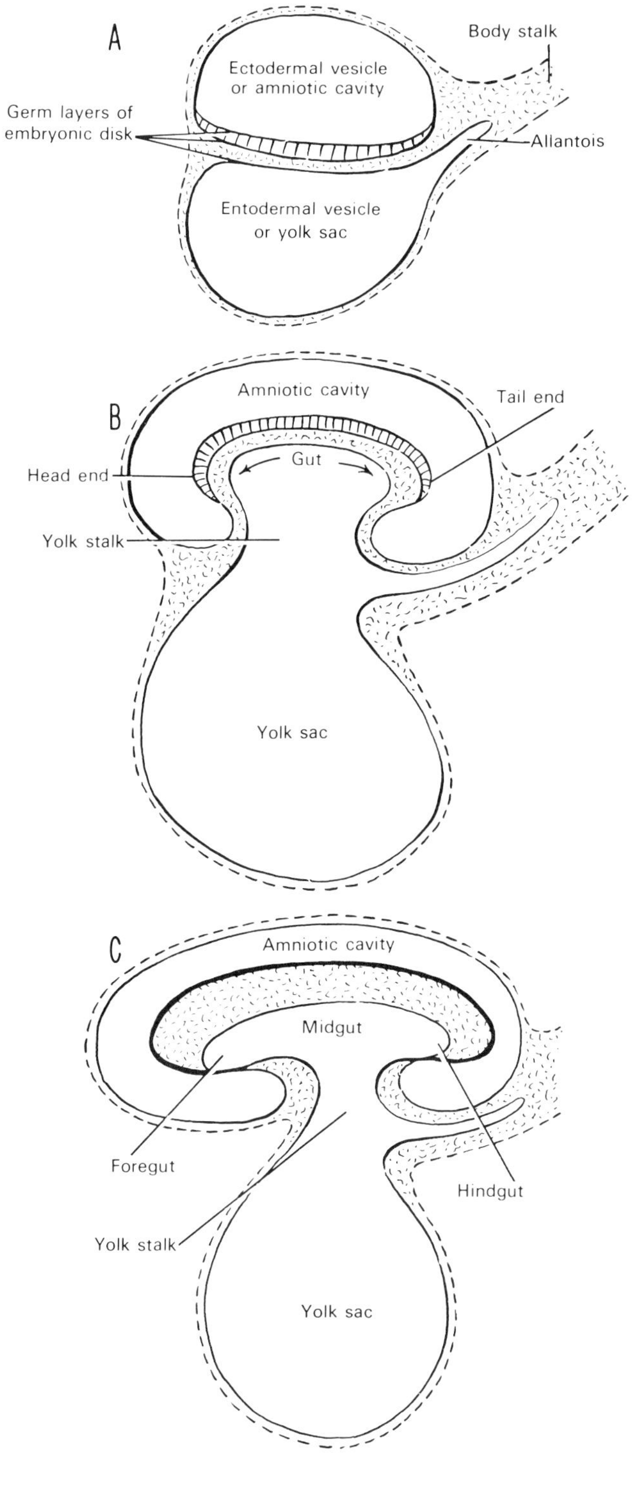

Fig. IX-34. Progressive stages of head and tail folding of embryo to form a gut tube. (After Haines and Mohiuddin.)

ing allantois caudally (Fig. IX-34A, B). Previously, we have discussed changes in the proximal foregut related to tongue, pharynx, pharyngeal pouches, larynx, and the respiratory tree. Here we will be interested mainly in the development of the caudal foregut and midgut regions. The caudal foregut will give rise to esophagus, stomach, duodenum, liver and biliary passages, pancreas, and spleen. It is not easy to define exactly what portion of the developing tube constitutes midgut, but in the adult it is common to regard that stretch of intestine from below the entrance of the common bile duct to a point approximating the proximal two-thirds of the transverse colon as having come from the embryonic midgut. The remainder of the tube from this portion to rectum is considered to have arisen from hindgut territory.

Even in the 32-day embryo, the **esophagus** does not amount to much. It is a short stretch of foregut extending from the respiratory diverticulum to a slight swelling where a stomach is forming below (Fig. IX-35B). But the esophagus will soon elongate to keep pace with the growing neck, heart, and lungs. By 49 days, vacuoles coalesce to form a lumen. In its upper one-third, it acquires a skeletal muscle coat from the surrounding mesenchyme, but the lower two-thirds of its wall is smooth muscle. It has a ciliated epithelial lining on the 70th day, and only by the fifth month does this give way to stratified squamous. Since it is a part of the mediastinum throughout most of its length, it never acquires a real mesentery or serous outer coat. We have mentioned the small tag of mesoesophagus anchoring it to the vertebral wall as the aorta moves away from behind it low in the posterior mediastinum.

The **stomach** appears as a spindle-shaped swelling during the fifth week in the upper part of an otherwise tubular gut of fairly even diameter. The stomach is anchored to the posterior abdominal wall by a dorsal mesentery, while the liver, budding out from the duodenum (Fig. IX-35C) splits a ventral mesentery. Thus, the stomach attaches to the liver by a posterior section of the ventral mesentery which becomes known as the **lesser omentum.** The anterior part of this mesentery is the **falciform ligament.** No ventral mesentery develops below the level of the umbilicus. In Figure IX-35C, note how the rest of the simple gut tube is suspended by the dorsal mesentery, and the way in which the superior mesenteric artery from the aorta continues through the mesentery as the vitelline artery. The artery will become an important landmark during intestinal rotation. Notice also the attachment of the vitelline duct to the intestinal loop. This is the site of a possible **Meckel's diverticulum.**

As the liver, other organs, and the body wall continue to grow at different rates, the stomach is soon forced to make a clockwise rotation of 90° about its long axis. This means that its original left side is brought about to face anteriorly and its right side, posteriorly (Fig. IX-36). This explains why the left vagus

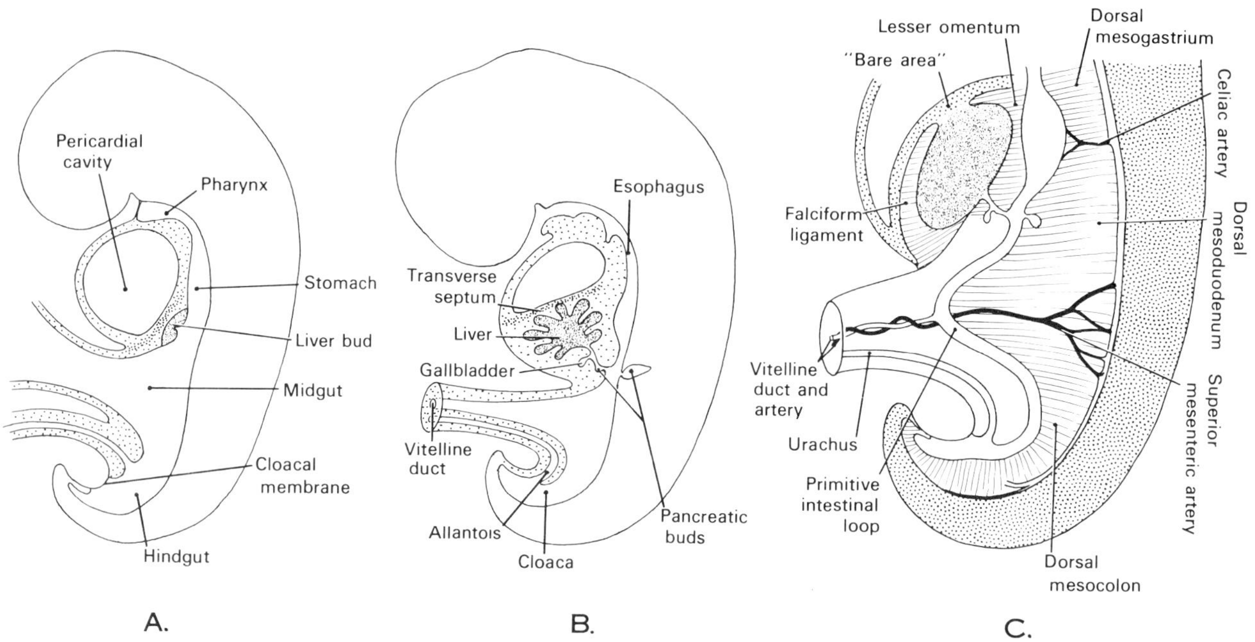

Fig. IX-35. Early stages in the development of a gastrointestinal system (schematic). **A.** Appearance of liver diverticulum at about 3-mm. stage. **B.** Liver cords invading septum transversum and formation of pancreatic diverticuli from gut tube at about 5-mm. stage. **C.** Relation of dorsal and ventral mesentery to digestive tract and its glands. (After Langman.)

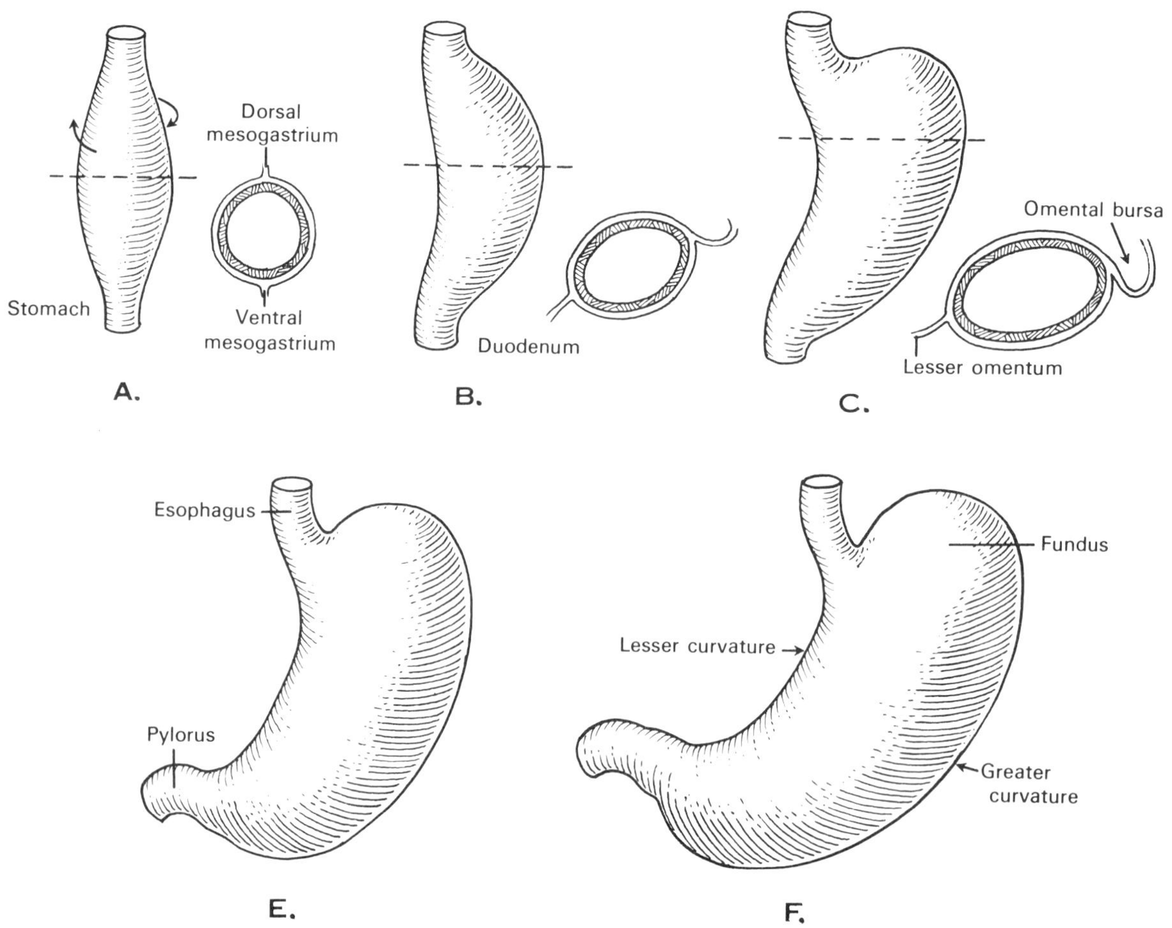

Fig. IX-36. Stages in the rotation and formation of the stomach. Parts **A** to **C** show rotation about longitudinal axis as viewed anteriorly. The adjacent cross-sections show consequent foldings of attached mesenteries. **D** and **E** show a swinging into final position around an anterior-posterior axis. (After Langman.)

will innervate its anterior wall; the right vagus, its posterior wall. During its rotation it also changes shape. It develops **greater and lesser curvatures,** an outpouching **cardia or fundus** near the entrance of the esophagus and a **pyloric outlet** (Fig. IX-36E). However, the **pyloric sphincter** is still weakly developed at birth. **Gastric pits** (foveolae) appear in the gastric mucosa by the seventh week in the embryo and gastric glands bud off by 14 weeks. By the fifth month, enzyme secretion is possible, but production of HCl precursor by the gastric glands may occur considerably later. Peristaltic waves sweep over the stomach from cardia to pylorus in the fifth month, though localized rhythmic constriction may occur earlier. Note that in the rotation of the stomach, the dorsal mesentery is folded on itself to the left creating an **omental bursa,** a blind pocket between two folds of peritoneum (Fig. IX-36C). The opening into this pocket from the right-hand side is the **epiploic foramen.** ("Epiploon" from the Greek was a term for the "intestinal caul." Its Latin counterpart is "omentum," a word whose derivation is most vague. It meant, or came to mean, a fatty membrane, and could have been derived from "opimus," fat or plump, or from "optimus," the best, richest, or fattest. Other etymologists like its more mystic derivation from "omen," since the ancients gravely scrutinized the intestines of birds and animals for prophetic signs. If you don't make it through medical school, the latter occupation might become highly profitable again among the Hollywood clientele now that astrology is back in vogue. All you need is a turban, a mess of chicken gizzards, and a weird imagination. You don't even have to be a psychiatrist or an expert in Yoga.)

The **liver** anlage makes its appearance in the third week (Fig. IX-35A) as a ventral budding from the gut-entoderm in an area that will become the definitive **duodenum.** This sacculation from the gut-tube, the **hepatic diverticulum,** differentiates, cranially, into glandular tissue and bile ducts and into a caudal part which becomes **gall bladder** and **cystic duct.** The cranial part forces its way into the thick splanchnic mesoderm of the septum transversum, the site of the future diaphragm. In a short while, this region that is occupied by the growing liver becomes drawn out as the ventral mesentery (falciform ligament) to which the liver will relate more than it will to the Diaphragm (Fig. IX-37B). The mesentery becomes the **liver capsule** (of visceral peritoneum) except cranially where the liver remains attached to the Diaphragm. Here, if you were to cut the liver away from the Diaphragm, it would show a *"bare area"* uncovered by the mesentery into which it bulged during development. The connective tissue of the original septum transversum invades the liver to form Glisson's capsule around developing lobules and large ducts. As the glandular liver cords proliferate, they come into intimate relationship with the blood sinuses of the vitelline and umbilical veins. From the second to the seventh month, the liver is an active hemopoietic organ as blood cells differentiate between **liver cords** and **sinusoids,** but most of this activity ceases by the time of birth. The liver can, however, be called upon to regenerate blood cells in times of dire emergencies in later life. By the fifth month, bile secretion colors meconium a greenish-yellow.

The main stem of the hepatic diverticulum as it springs from the future duodenum becomes the **main bile duct** (ductus choledochus). Its developmentally distal parts leave the liver as the **right and left bile** ducts. At is developing proximal end, near the duodenum, it buds off a *gall bladder* which, in lengthening, remains connected to the main duct by the *cystic duct* (Fig. IX-38). The total growth of the liver is so rapid (Figs. IX-35 and 37) that by 10 weeks it weighs 10% of the total body weight of the fetus.[1]

Just below the stomach swelling, the first part of the intestinal tube will develop into the **duodenum. Pancreatic diverticula** also bud off from this part of the gut close to the hepatic diverticulum. Here the pancreas has a two-fold origin. Opposite and slightly higher than the hepatic diverticulum, a **dorsal pancreatic bud** pushes from the entodermal tube into the dorsal mesentery. It will form the body and tail of a future pancreas. At the same time (somewhere around the fourth week of embryonic life), a **ventral pancreatic bud** springs from the hepatic stem in the angle between intestine and the gall bladder anlage (Fig. IX-38A). As the stomach rotates, so does the duodenum. The ventral bud is swung around behind the duodenum so it, too, comes to lie in the dorsal mesentery just below and behind the dorsal pancreatic anlage. Part of this rotation involves migration. The ventral bud provides the head for the future gland. Its duct taps the distal portion of the duct of the dorsal pancreas, and it is the combined segments of these two parts that become the **main pancreatic duct** (of Wirsung) which join the common bile duct in the substance of the duodenum at the **ampulla of Vater.** The short proximal stump of the dorsal pancreatic duct either disappears or remains as a very small **accessory pancreatic duct** (of Santorinni) (see Fig. IX-38D). Sometimes (in 10% of the bodies) there may be no union between dorsal and ventral ducts at all so that two, fair-sized, pancreatic ducts enter the duodenum separately.[2] As these ducts branch in the developing pancreatic buds, glandular acini appear in the third fetal month at the terminal ends of the branches. The **islets of Langerhans** (endocrine parts of the pancreas) also differentiate from duct tissue at about the same time. The ducts from which the islets arise lose their lumina. Both insulin and trypsin secretion has been recorded in the fifth fetal month.

The original lumen of the **duodenum** occludes during the sixth to seventh week of development because of the rapid thickening of its walls. The lumen is restored later on by a process of vacuolation. This short stretch of the intestinal tract acquires its characteristic horse-shoe bend quite early, and it cradles the head of the pancreas in its extended arms. The tail of the pancreas reaches to the left to almost tickle the spleen which sucks in its hilum in utter hilarity.

In the fetus, the dorsal mesentery extends from esophagocardiac junction to rectum as Figure IX-37 shows. The part behind the stomach is the **dorsal mesogastrium** or **greater omentum.** Just as the liver developed in the ventral mesentery and subdivided it into falciform ligament and lesser omentum, so the **spleen** develops in the dorsal mesogastrium to split it into a **lienorenal ligament** and a **gastrolienal ligament** (Fig. IX-39). The omental bursa dead-ends between these two folds. In embryos of about 6 weeks, mesenchymal proliferation in the mesentery push nodules of splenic tissue upward to bulge the surface epithelium. The nodules coalesce, and the epithelium becomes the visceral peritoneum of the spleen. The mesenchyme is well vascularized, and it differentiates into **capsule, trabeculae,** and **pulp cords.** Lymphoid tissue, as **splenic corpuscles,** does not appear until the sixth month though the shape of the spleen is well formed by the third month.

Embryologically, all of the small intestine, the ascending colon and a good portion of the transverse colon appear to be derived from the midgut. In the 4-week embryo **the intestine** is a relatively simple tube extending from the stomach to cloaca (Fig. IX-37). It lies in the midplane, bows slightly ventral in the region where the yolk stalk is attached, and it is suspended by the dorsal mesentery. A ventral mesentery relates to the intestine only in the duodenal region. The portion above the yolk stalk (vitelline duct) and above the vitelline artery (superior mesenteric artery) is called the **cranial limb** of the **primitive intestinal loop;** the portion below, the **caudal limb** (Fig. IX-40A). The duodenum would be recognizable because of its association with bile and pancreatic ducts (as we have seen, even if they are not shown in Fig. IX-40). The cranial loop will become the future **jejunum** and most of the **ileum** (notice the spelling of the last word and do not confuse it with "ilium"). The caudal limb is the forerunner of the rest of the intestinal tract, *e.g.*, the *remainder of the ileum,* all parts of the colon (ascending, transverse, and descending), and the **rectum.** By the end of the fifth week, the yolk sac has become smaller and the yolk stalk thinner, and one must mark the caudal limb by a more useful acquisition, the **cecum.** Fortunately, this bulge also marks the boundaries between large and small bowel (Fig. IX-40B). There is rapid growth of the gut during this week so that the primitive tube, in its elongation, finds little room in the restricted abdominal cavity. It is forced to abdicate into extra-embryonic coelom, *i.e.*, into the umbilical cord. Before it herniates into the cord:

1. There has been an anticlockwise rotation *around the superior mesenteric artery as an axis* in such a way that the cranial limb moves to the left and the caudal limb crosses over it to the right (Fig. IX-40B).
2. The cranial limb grows more rapidly than the caudal limb and develops a number of sinuous convolutions. These represent elongation of the jejuno-ileum (Fig. IX-40C).
3. We have already noted the changes in the yolk sac and stalk.

You may find it said that an enlarging left umbilical vein is a major factor in initiating this specific direction of rotation. However, experimental studies indicate that factors much earlier in development have already predetermined that such a rotation will occur. If, in amphibians, the archenteron roof is removed in the gastrula stage of development, when the time comes for gut rotation there is complete transposition of these limbs in mirror image. Even the rotation of the stomach is reverse, and liver and stomach are transposed.

At the 10th week, there now being room in the inn, the intestines return to the abdominal cavity. No one really knows exactly what prompts the return but it is accomplished quickly and, generally, orderly, and the coelom within umbilical cords seals off. The proximal parts of the jejunum lead the parade inward, and the succeeding coils of the small intestine fill the space on the left side. This keeps the future descending colon (which did not herniate) on the left side. Next, the ileal coils progressively file to the right. The large intestine was the last to leave the abdomen and is the last to reenter. In attempting to straighten up, the cecal part is carried slantingly to the right side and the cecum becomes fixed somewhere in the vicinity of the iliac crest. The rest of the entering colon curves sharply to the left, bending at the **splenic flexure,** to become continuous with the descending colon which had remained at home all the while. It is not possible to determine at this time what portion of the tube will become ascending and transverse colon.

The fixed cecum grows a definite blind sac, but growth at its distal end is not as rapid as elsewhere so that a vermiform appendix appears. As the liver decreases in relative size because of overgrowth of body wall, part of the slanting colon pushes to the right to form a **hepatic flexure.** The transverse part has to cross ventral to the duodenum (which also remained at home during all of the filing in and out of the umbilical cord). The duodenum and pancreas have lost their mesentery and have become fixed to the dorsal body wall. Since the transverse colon must cross in front of them, it retains a small part of its mesentery as the **transverse mesocolon,** but ascending and descending colon anchor to the body wall (become retroperitoneal in the surgeon's sense). Perhaps Figure IX-41A to C will clarify the above description. In the figure notice the **greater omentum** hanging down from the greater curvature of the stomach. Most of it has been cut away in Figure IX-41C. It is a double layered apron

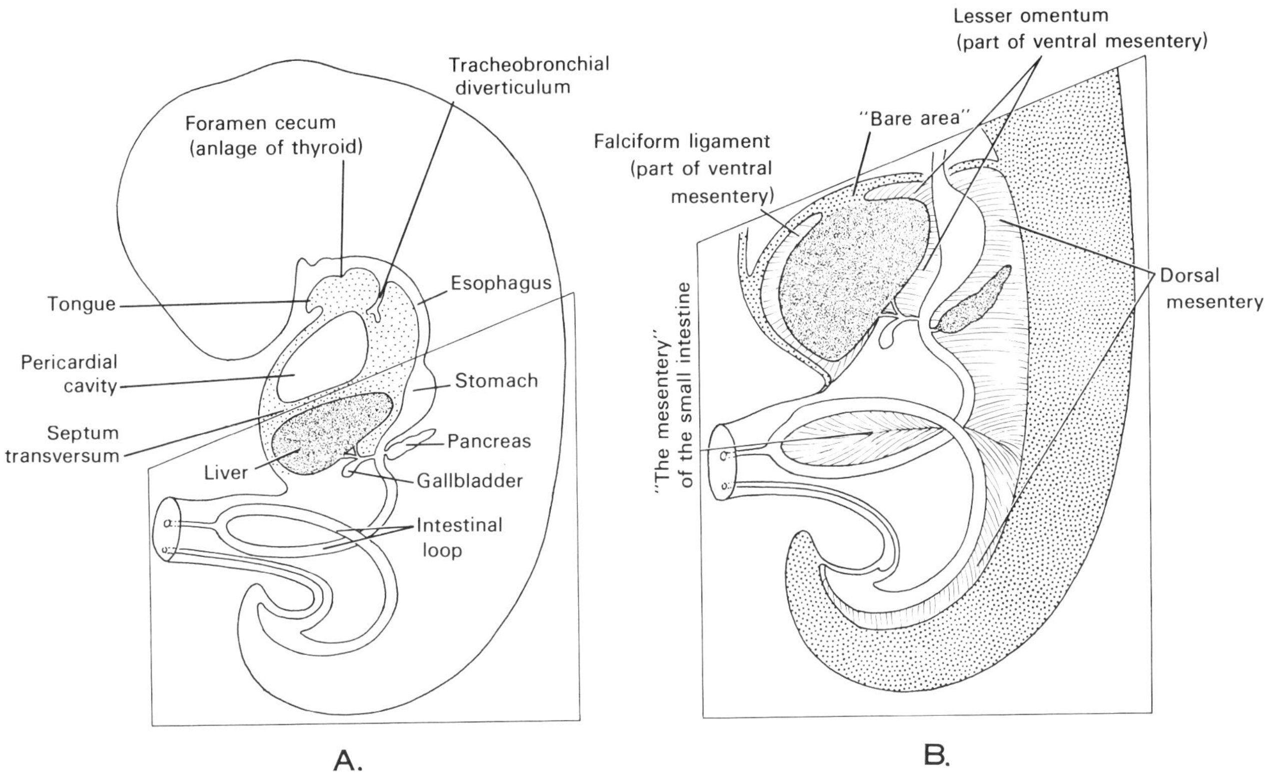

Fig. IX-37. Expansion of liver into abdominal cavity of the growing embryo. **A**, at about 36 days; **B**, in a somewhat older embryo. (After Langman.)

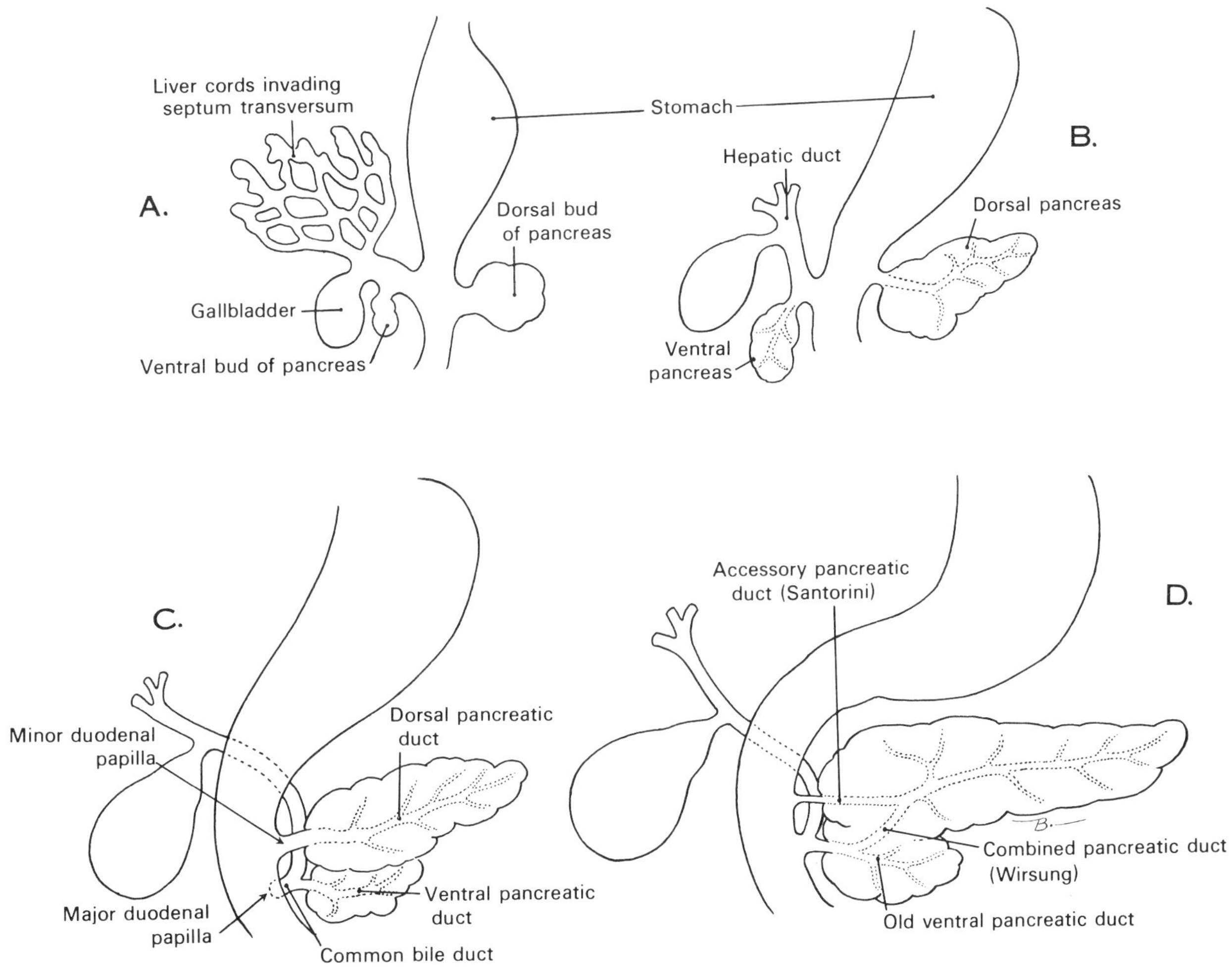

Fig. IX-38. Stages in development of pancreas and biliary duct system. **A,** 30-day-old embryo; **B,** at 35 days; **C,** at 6th week; **D,** fusion of embryonic pancreatic ducts to form the common pancreatic duct of adult stage.

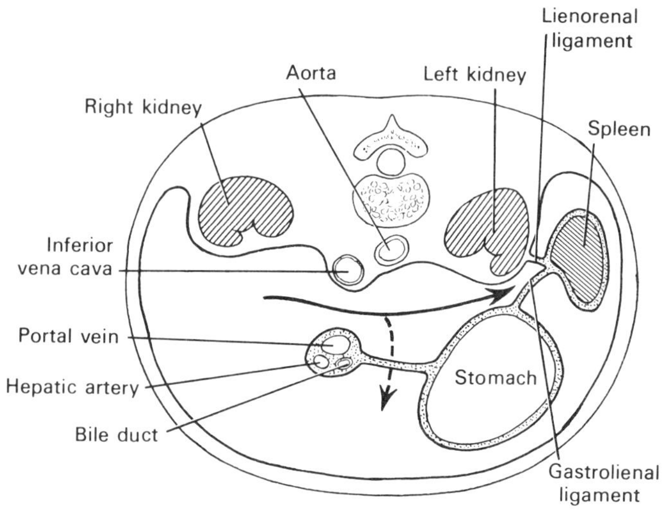

Fig. IX-39. Schematic cross-section through abdomen in region of omental bursa (solid arrow, entering from right through foramen of Winslow). Dotted arrow enters lower recess of bursal sac. (After Grant.)

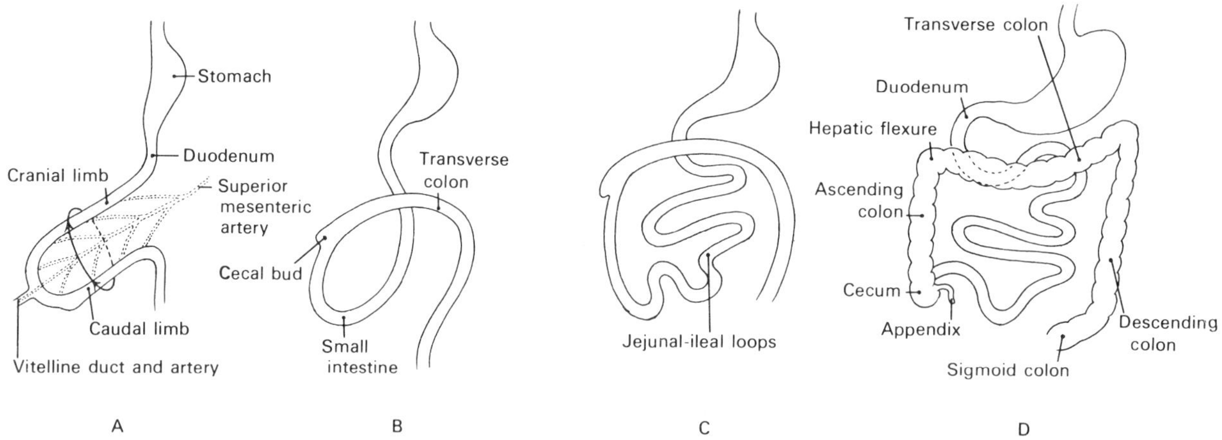

Fig. IX-40. Rotation of the primitive intestinal loop in development of intestinal tract. **A.** Superior mesenteric artery as axis of rotation of primitive loop. **B.** After 180° counterclockwise rotation. **C.** After 270° counterclockwise rotation. **D.** Final position and in advanced stage of development. (After Langman.)

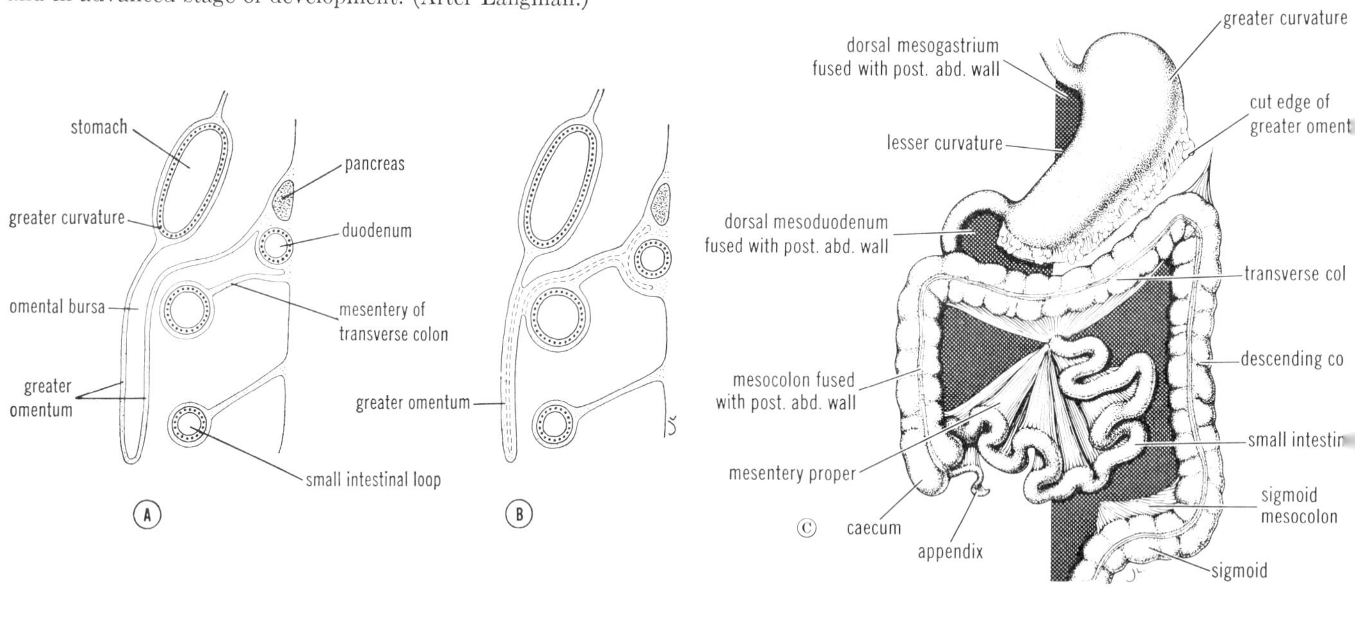

Fig. IX-41. Relationship between greater omentum, stomach, transverse colon, and small intestine. Note retroperitoneal locus of pancreas and duodenum even in early stages. **A.** In the 4-month embryo. **B.** In the newborn. **C.** Mesenteric fusions in the adult. (From Langman, *Medical Embryology*, 2nd ed., Williams & Wilkins, Baltimore, 1969.)

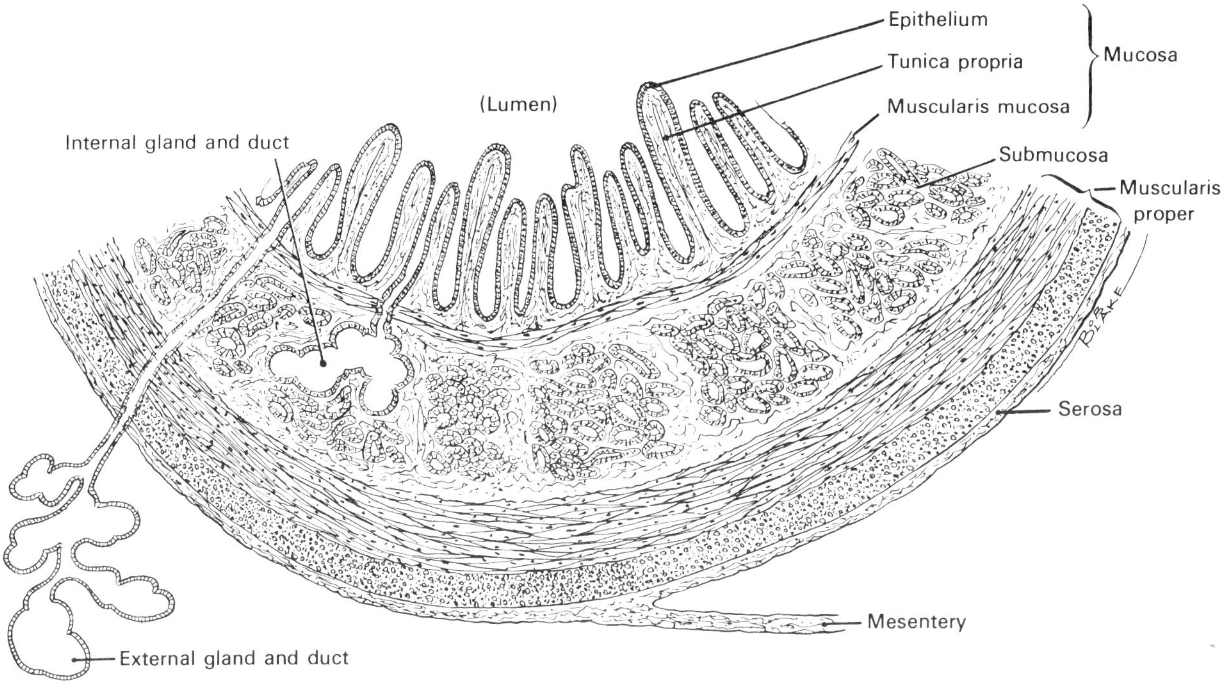

Fig. IX-42. Highly schematized plan of a hollow organ. (See text for details.)

which covers most of the intestines below. The transverse mesocolon will fuse with the underside of this apron (Figure IX-41B). Notice, too, how short is the oblique right to left attachment of the small intestine (Fig. IX-41C), only a matter of 8 inches, yet from this small anchorage to the dorsal body wall it fans out to some 20 feet of small bowel. Embryonic changes in the hindgut can best be discussed when we consider pelvic organs.

The Plan of Hollow Organs

Except for certain large glands, the bulk of the organs of the abdomen are constructed along the lines of hollow tubes. With minor variations, the fundamental design of hollow viscera is much the same (Fig. IX-42).

A. A **mucosa** surrounds a central lumen. Many mucosae consist of three tissue laminae:
   1. **An epithelial lining:**
      a. In the esophagus, it is a multilayered, stratified, squamous epithelium.
      b. Elsewhere in the digestive tract, including the gall bladder, it is a simple columnar epithelium.
      c. In the urinary tract, it is the transitional type.
   2. **A tunica propria:** In general, this is a mixture of loose fibro-elastic and/or reticular and lymphoid tissue upon which the epithelium rests. Not all epithelial, surface-lining cells of the digestive tract are capable of providing, through synthesis, all the secretory enzymes needed for digestion. In fact, many surface cells play more of a protective or absorptive role than a secretory one. Hence, one finds foveolae (epithelial pits leading to glands) and gland cells extending into and occupying the bulk of the tunica propria. Isolated lymph nodules and a scattered diffusion of lymphocytes may infiltrate the connective tissue that surrounds the glands.
   3. **The muscularis mucosae:** This is the last of the three mucosal components. It lies directly under the tunica propria, between it and a submucosa. Except in the esophagus, the smooth muscle of which it is composed is circularly disposed. In the esophagus, the muscularis mucosae is longitudinally oriented in respect to tube length, so in cross sections of the tube the muscle bundles will appear in cross section. This, along with a stratified squamous epithelium, is diagnostic of cross-sections through the esophagus as seen in the light microscope.

B. The **submucosa** lies outermost to the muscularis mucosae. It is a loose fibro-elastic (areolar) tissue containing vessels, nerves, and ganglion cells. The latter are the terminal parasympathetic ganglia of either the vagus nerve or the sacral pelvic nerve (nervi erigens). They are the ganglion cells of **Meisner's submucous plexus.** In the uppermost and lowermost parts of the esophagus, mucous glands may occupy the tunica propria. It should be noted that in most routine sections through the esophagus it is hardly worthwhile spending much time looking for these glands. They are far and few between and are seen in fortuitous cuts only.

However, **Brunner's glands** will *always* be found and in abundance in the submucosa of the duodenum. They are diagnostic of this part of the digestive tract. Nowhere else in the digestive tract will you find glands in the submucosa other than the duodenum and, occasionally, in the esophagus.

C. The **muscularis externa** is the proper muscle coat of the hollow organ. It will vary in thickness in different organs. It is generally two-layered, but in some organs it may be three-layered. Usually it consists of an inner coat of circularly arranged smooth muscle (Fig. IX-42). In the minute amount of connective tissue which separates these layers will lie scattered nests of ganglion cells. These are the autonomic (visceral motor) ganglia of **Auerbach's myenteric plexus.** They, too, are the terminal parasympathetic ganglia of the vagus or the pelvic nerves.

D. Outermost of all is the **serosa** or **serous membrane** as distinct from the inner, **mucous membrane.** Actually, in most organs this is the *visceral peritoneum.* The esophagus, which runs chiefly through the mediastinum lacks a serosa. It has an outer fibro-elastic **adventitia** instead which blends with the connective tissue through which it runs. A true serosa consists of a well-capillarized but thin lamina of loose fibroelastic connective tissue surfaced outermost with simple squamous epithelium (mesothelium). Its surface is kept moist by a *watery* fluid. Mucous membranes, because of glands, keep their surfaces moist with a slimy, thicker, mucus. Mucous membranes, as we have just noted above, are much thicker than serous membranes, and much more distinctly structured.

You will have to get used to the loose usage of the term "membrane" in Anatomy. For some reason these days, the electron microscopist thinks he has a corner on the market in the use of this term since most structures within the cell are "membrane-bound." Here the "unit-membrane," whose thickness is measurable in Angstroms, is regarded as the only proper use of the term. But the electron microscopist is a relative newcomer to biology, hardly dry behind the ears from his recent big splash. Histologists and pathologists have long been bandying around such shocking terms as basement membrane, glassy membrane, hyaline membrane, and the like. To avoid confusion, but mainly to salve his own ego, the electron microscopist would prefer "basal lamina" to "basement membrane." But the term "lamina" has been used in describing many other structural features. Even the gross anatomist has used it for things measurable in millimeters and centimeters. And now you have heard him talk about "serous and mucous membranes" measurable in microns. Electron microscopy to the contrary, these older terms are probably here to stay, as are many other old timers in the medical lexicon.

Every student has to face keeping up with both old and new jargon in every field of rapidly expanding science if he is to understand what he reads. This was as true in 1890 as it is today and as it is likely to be in the future. There is nothing wrong with any of these terms provided the student recognizes the level of organization under discussion and the nature of the "membrane" or "lamina" being discussed. From where we sit, it is the old story of saying: "This is what we sees and this is what we calls it." Your authors can only shrug their shoulders and ask: "Whats all the fuss about?" A marigold by any other name would, probably, stink as much.

Peritoneum, Its Reflections and the Disposition of the Abdominal Organs

The major portion of the *peritoneal cavity* is a most complicated bursal sac, sometimes called **"the greater sac"** because it communicates with a **"lesser sac"** or **omental bursa** as we shall see. We have pointed out that the cavity is lined everywhere with a peritoneum whose surface cells are simple squamous, except over the ovary where they are simple cuboidal. Its surface is kept moist by a serous fluid. The cavity, as we have said, is empty except when opened to air in an operation or when an abnormal amount of fluid increases therein under conditions of abdominal ascites or dropsy. Folds or double layers of visceral peritoneum form to suspend organs, ducts, or blood vessels since all these structures push inward from the areolar tissue layer behind parietal peritoneum, the main continuous lining layer of the cavity. Such folds, suspending the intestines are called **mesenteries.** When they suspend the stomach they are called **omenta,** and when they suspend other structures they are confusingly designated as **ligaments.** So long as the student realizes that these "ligaments" are merely duplications of the peritoneal lining or raised folds and not dense white fibrous connective tissue anchoring bone to bone, the term applies in the sense of a leash or a tie or a binding-structure.

We are now ready to open the abdomen and to explore it in a preliminary way before dissecting or excising organ structures. Hopefully, the preceding discussion on development has prepared you, in part, to read the road maps shown in Figure IX-43A, B. As you cut through the peritoneum of the ventral body wall, the intestines may be more or less hidden by a fatty, omental apron that hangs down over them from the greater curvature of the stomach (Fig. IX-43A). Presumably this **greater omentum** keeps the intestines from forming adhesions with the parietal peritoneum of the ventral body wall. It has some mobility, in that it slips or gravitates from one area to another and can surround an inflamed organ. In some cadavers it may be short and disheveled in appearance resembling a smallish, dirty, moth-eaten dishrag.

It should be easy to spot the large liver hanging from the thoraco-abdominal diaphragm in the right upper quadrant. If you are lucky, the gall bladder may peek out at you from under the sharp, right-hand, inferior border of the liver as a small, shriveled, green-

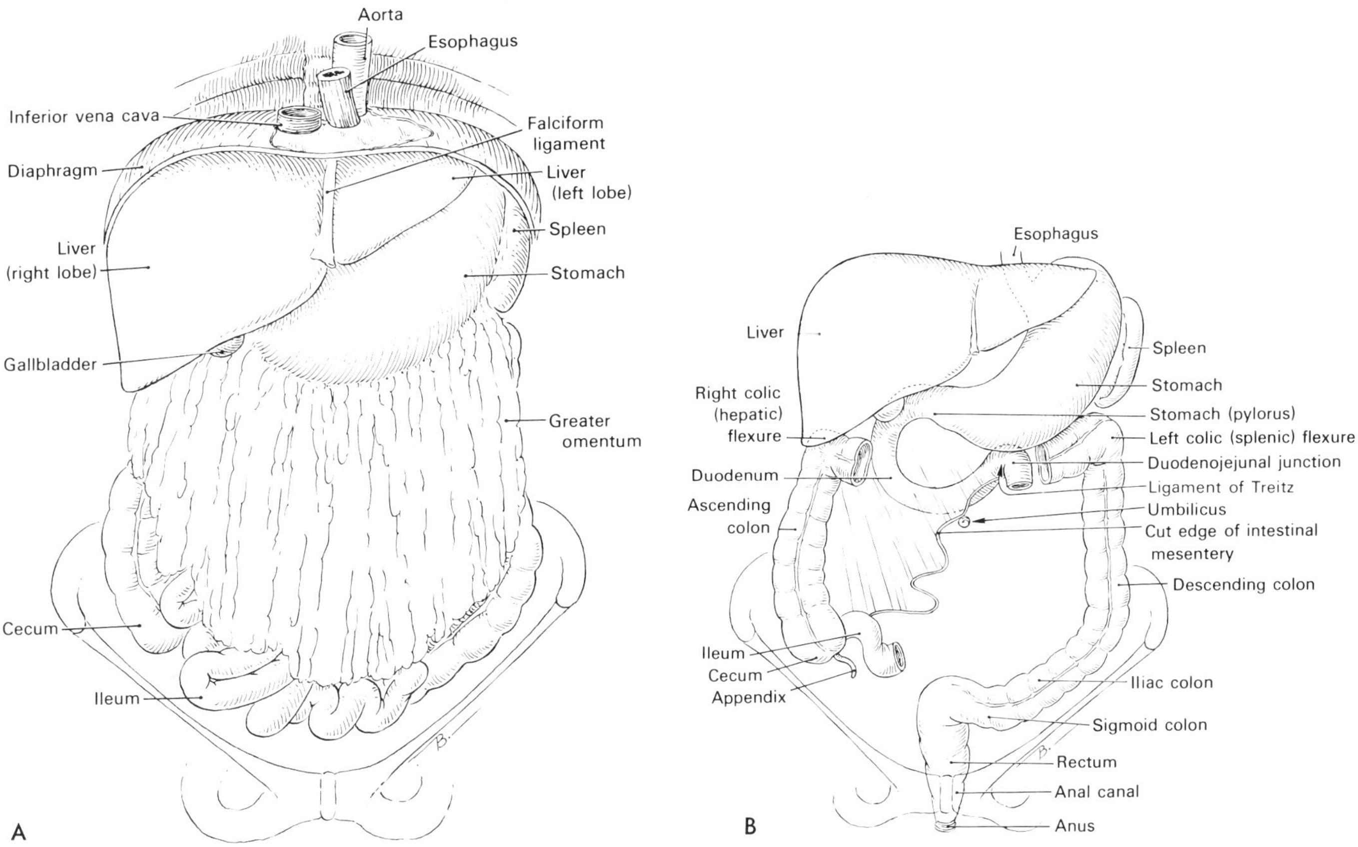

Fig. IX-43. Disposition of the digestive organs in the abdomen. **A.** As seen when abdomen is first opened and with omentum in place. **B.** Relationship with omentum and most of small intestine removed. (After Grant.)

ish sac. Recall that the liver bulged from the septum transversum (diaphragm) into the ventral mesentery in its development, splitting that mesentery into ventral and dorsal portions. You can see the ventral portion as the sickle-shaped **falciform ligament** which anchors to the ventral body wall and seems to cleave the liver into unequal lobes, the right much larger than the left lobe. The ventral mesentery never advanced below the umbilicus, so it has a free inferior margin. Feel the thickening in this free margin. It can be traced from umbilicus to the liver. The thick, non-patent, tubular structure you feel is all that remains of the left umbilical vein, your former lifeline to the maternal placenta. It is now known as the **ligamentum teres hepatis** (round ligament of the liver). Early in embryonic life this vein emptied into the fetal portal vein so that well-oxygenated blood from the placenta was wasted in circulating through the rapidly growing liver before the blood entered the heart. Very quickly, a short-cut channel through the liver, the **ductus venosus,** was worked out to connect the umbilical vein to the hepatic vein and inferior vena cava, and, thus to by-pass the sinusoidal channels of the liver. Figure IX-44 views the liver from behind and from below. We see the obliterated round ligament below, lying in a fissure between left and caudate lobes of the liver. The now obliterated ductus venosus has also become ligamentous (**ligamentum venosum**) and lies above in a fissure between left and quadrate lobes. Note that it connects

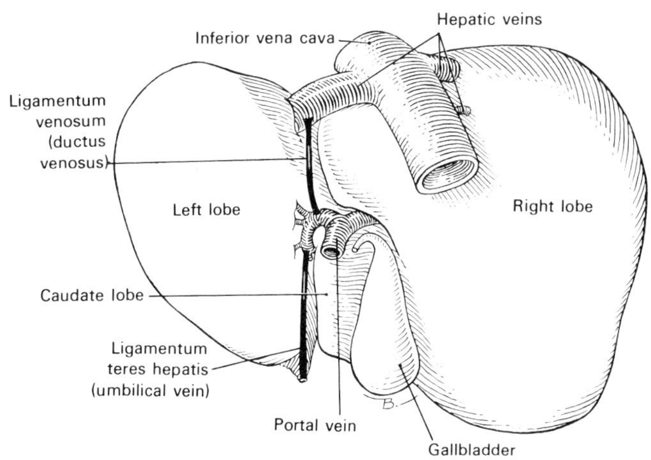

Fig. IX-44. Ligaments and relationships on posterior aspect of liver.

with one of the hepatic veins which leads almost directly to the inferior vena cava. You should try to reconstruct this once important circulatory pathway after you remove the liver in the later dissection of this region. It cannot be visualized well with the liver anchored in place.

The peritoneal investments of the liver fuse on its visceral surface to form a double-layered sheet which extends downward to the lesser curvature of the stomach. This sheet is the **gastrohepatic ligament** (Fig. IX-45) but is better known as the **lesser omentum.** If the lesser omentum anchors the stomach, along its lesser curvature, to liver, the greater omentum, along its greater curvature, anchors stomach to the posterior body wall by a somewhat devious route which we shall point out below.

Locate the stomach somewhat below and to the left of the liver. It may not look like much if it is deflated and flabby in your cadaver (Fig. IX-43A). Reach your hand around its left border to feel the spleen which may be well hidden behind it under the left dome of the Diaphragm (Fig. IX-43B). To see the coils of the small intestine within their picture frame formed by the ascending, descending, and transverse colon (Fig. IX-46), you will have to flip the omental apron and stomach up over the costal margin. You can then trace the *underside* of the omental flap downward toward the transverse colon which appears to be annealed to the backside of the greater omentum (Fig. IX-47). Beyond the transverse colon, the omentum reaches the posterior body wall. This stretch, from transverse colon to body wall is the **transverse mesocolon.** It was once a separate sheet but it has now fused to the greater omentum (see Fig. IX-41C). While you can now identify ascending, descending, and transverse colon and a number of the coils of the small intestine, there are several structures you can't see without some shifting about of structures and some later dissection. You can't see the pancreas, much if any of the duodenum, or the kidneys. The pancreas and duodenum lie behind the dorsal wall of the lesser omental bursa which we haven't located yet. Reach down and laterally around the **hepatic and splenic flexures of the colon** and be certain you can feel and trace the outlines of the deep-lying kidneys. You can't see the urinary bladder either, but you can reach down behind the symphysis pubis and push down anteriorly on the pelvic floor. You will be touching the roof of the bladder, covered over, of course, by visceral peritoneum. Return the omental apron to its original position.

Now examine Figures IX-45 and 48 carefully to find some of the relationships we will now discuss. Figure IX-48 reviews the structures that developed in the ventral and dorsal mesentery when the gut was still a relatively straight, midline tube. Stomach, pancreas, and spleen occupied the dorsal mesentery; the liver occupied the ventral mesentery; and the part of the ventral mesentery between liver and stomach was known as the **gastrohepatic ligament** (lesser omentum). Even at the embryonic stage of life, these structures were supplied by the coeliac artery and the intestines were supplied by the superior and the inferior mesenteric arteries. These relationships will remain true in the cadaver. The kidney did not develop in the mesentery at all, but remained behind the parietal peritoneum in the posterolateral body wall. That is where you will find it in the cadaver, and it will not be supplied by any of the three mesenteric branches of the aorta.

The changes from fetal to adult relationships are shown, in part, in Figure IX-45. The lesser omentum is no longer anterior to the stomach but, as the stomach rotated and developed its lesser curvature, the lesser omentum was dragged posterosuperiorly. It still attaches to the same cleft on the undersurface of the liver in which the ligamentum venosum is harbored. The inferior portion of the lesser omentum has a free border and it attaches to the duodenum as the **hepatoduodenal ligament.** It is simply the lower part of the gastrohepatic ligament. In this free, inferior edge run the cystic and the common bile duct, the latter on its way to pierce the duodenal wall. One can pass a finger below and behind this free border and, when one does so, the finger enters a space behind the stomach. The finger is in the **epiploic foramen** and the space is the *lesser sac* or *omental bursa* (Fig. IX-45). To understand how this sac is formed, study Figures IX-45 and 48, for the posterior wall of the greater omentum is instrumental in its construction. The lesser omentum splits as it clothes anterior and posterior surfaces of the stomach. When these layers meet again at the greater curvature, they fuse once more. They are now part of the old **dorsal mesentery.** In the adult, they drape down as the anterior wall of the omental apron. At the inferior border they turn under and run back to the transverse colon where they split to envelope it. The posterior lamina continues posteriorly to the back body wall, covering duodenum and pancreas on its way. The anterior lamina reflects off the anterior surface of the transverse colon to fuse on the posterior surface of the stomach and lesser omentum. Thus, between the two posterior ascending lamina of the omental apron, above the level of the transverse colon, a space or lesser sac is created (Fig. IX-45). This lesser sac communicates with the "greater sac," which is the peritoneal cavity proper, at the epiploic foramen or foramen of Winslow (see arrows in the diagrams). Find the gall bladder. Follow the cystic duct until your finger is stopped by the duodenum. Slip the finger behind cystic and bile duct, and enter the lesser sac through the foramen of Winslow.

Jacob Benignus Winslow was one of the very famous anatomists of his day (1669–1760). A Dane, he became Professor of Anatomy at the University of Paris in 1743 when he was already 73 years old. It was in 1732 that he struck his thumb in his famous foramen and said: "What a good boy am I." Actually, he was best

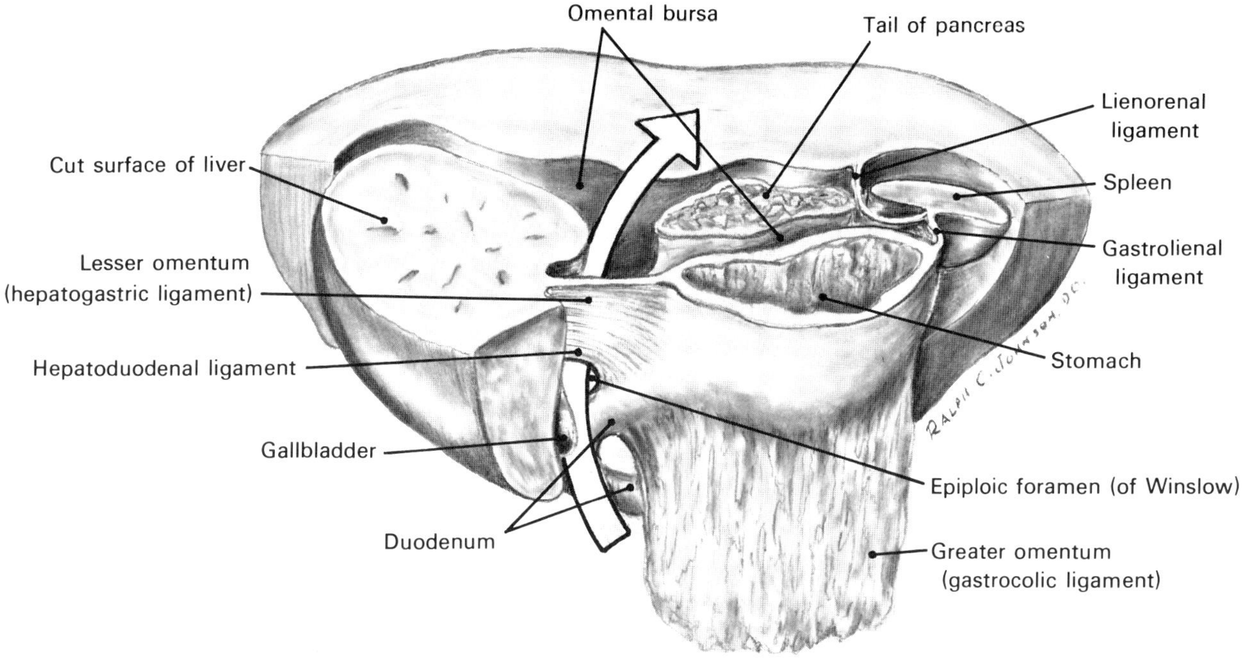

Fig. IX-45. Relationships at the omental bursa at a relatively late fetal stage. **A.** Seen in transverse section. Kidneys are not shown. The pancreas and duodenum are becoming adherent to the posterior body wall. Space behind liver and pancreas is exaggerated.

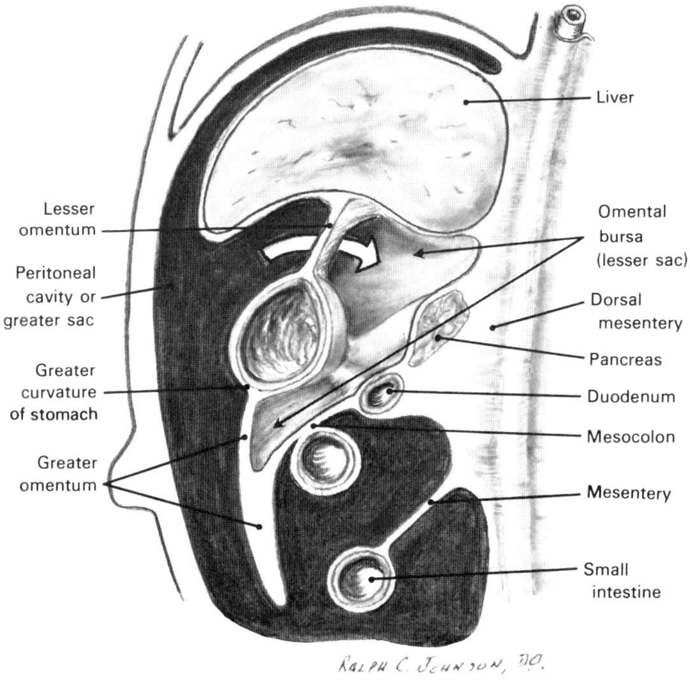

Fig. IX-45B. Same, in sagittal view. Arrow in epiploic foramen.

known for his studies on the eye, ear, and muscles around the knee. His textbook was an authoritative one that remained in vogue for over 50 years. And we remember him for a hole to a space behind the stomach, an entrance to an area that is only good to feel around in to satisfy the whims of pedantic embryologists!

If you do feel around in this space (and what kind of a medical student are you if you don't follow tribal customs?), you will be stopped by the spleen between its **renolienal** and **gastrolienal ligaments** (Figs. IX-39 and 45). If you fish cranially in the lesser sac (and you may have to rip a wider hole in the lesser omentum to get a hand in) you will be stopped by the Diaphragm in an *upper recess* between the inferior vena cava (right) and esophagus (left). If you reach inferiorly in the space, your hand slides behind the stomach and well down between the anterior and posterior walls of the greater omental apron. Depending upon the extent of fusion of these walls, you may even reach down to the very bottom where the anterior wall turns back as the posterior wall. Your hand will be in the space shown in Figure IX-47.

It is easy to see how the lesser omentum became the hepatoduodenal and gastrohepatic ligaments. It is more difficult to figure out what happened to the old dorsal mesentery in the region of the stomach. The bulk of it became the greater omentum. Both spleen and pancreas developed in the dorsal mesentery (Fig. IX-41A). The pancreas and most of the duodenum were pushed against the back body wall and were cov-

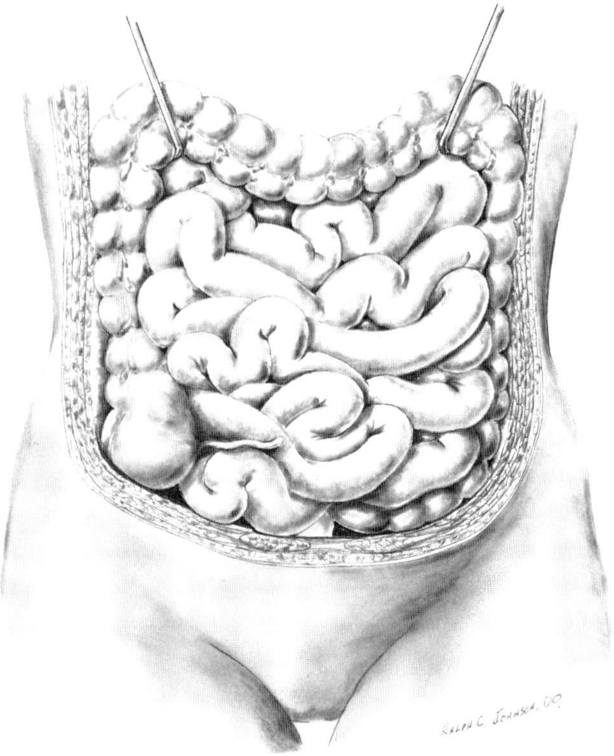

Fig. IX-46. Small intestine framed by large intestine. The stomach and omental apron have been turned up on chest and have been cut away to expose transverse colon.

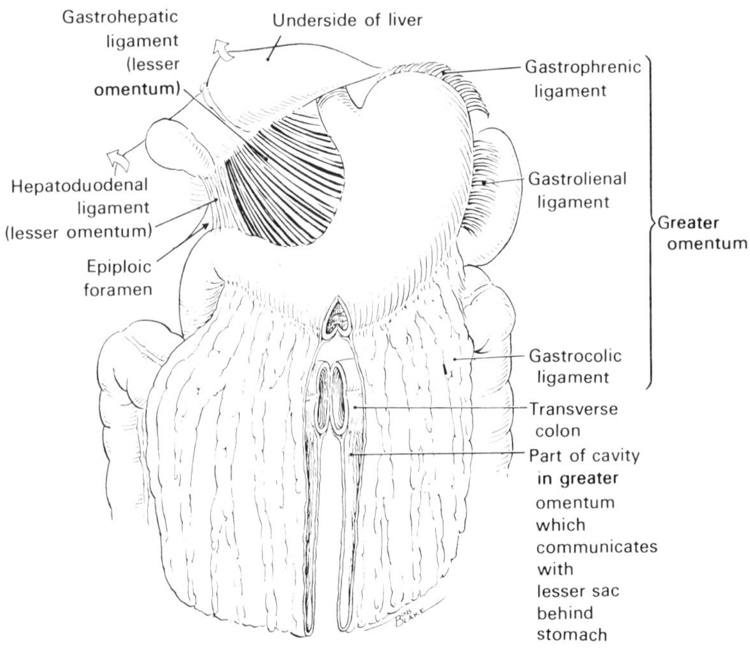

Fig. IX-47. Constituent parts of lesser and greater omentum. The latter consists of the gastrophrenic ligament, gastrolienal ligament and the gastrocolic ligament, all structures which arise from the greater curvature of the stomach.

ered over in turn by the transverse mesocolon and by the posterior lamina of the omental apron (Fig. IX-41B).

The spleen also developed in the dorsal mesentery, but as it was pushed to the left, the anterior lamina remained attached to the stomach near its fundus as the gastroleinal ligament (Fig. IX-45 and 48). The posterior lamina became attached to the posterior body wall over the kidney (Fig. IX-39). Above the spleen, the rest of the dorsal mesentery stretches from stomach to Diaphragm as the gastrophrenic ligament. None of these points are exceedingly important except to account for organ anchorage and for developmental alterations. Attempt to understand them and to look for them as you explore the abdominal cavity. If you don't find all of them, don't weep too much.

Realize that the transverse colon once had its own private mesentery which anchored it alone to the back body wall covering pancreas and duodenum as it did so. But it fused to the posterior lamina of the omental apron in the adult (Fig. IX-41B). The ascending and descending colon have lost their portions of the dorsal mesentery entirely and rest against the posterolateral body walls. The whole 20-odd feet of the small intestine have retained about an 8-inch stretch of dorsal mesentery which fans out from this short base to suspend the entire tube (Fig. 41B). You would have to reel out this entire hose to confirm this, and it is best not to do so until you are ready to excise the small intestine.

We may mention a few other peritoneal folds as we discuss separate organs later. Table IX-3 lists the important ones, tells what structures are contained in them and whether these structures are patent or obliterated. The first five listed are essentially tubes or ducts; the next three are arteries; the next four, arteries and veins; and the last two, veins alone.

Structures Related to the Umbilicus

Whoever attends the accouchement of Florida oranges seems to have considerable expertise in tying off an umbilical cord. We heartily recommend a term of residency with the Florida Citrus Growers Association for all would-be midwives and obstetricians.

A number of fetal, tubular structures meet at the umbilicus (Fig. IX-49). Fortunately, most of them obliterate and some disappear entirely. The **umbilical vessels** maintain their connection. We have seen the **obliterated vein** (round **ligament of the liver**) reach the umbilicus by traversing the inferior margin of the falciform ligament. Both **obliterated umbilical arteries** come up to the umbilicus from the pelvic floor in peritoneal folds as the **lateral umbilical ligaments** (Fig. IX-50). At their lower reaches in the pelvis, they remain patent as the **superior vesical arteries.** The **vitelline vessels** from the yolk sac became the **superior mesenteric artery and vein.** Their persistent connection with the umbilicus is rare, and they normally disappear beyond the intestinal border. Their early presence, and continuity into the cord provided

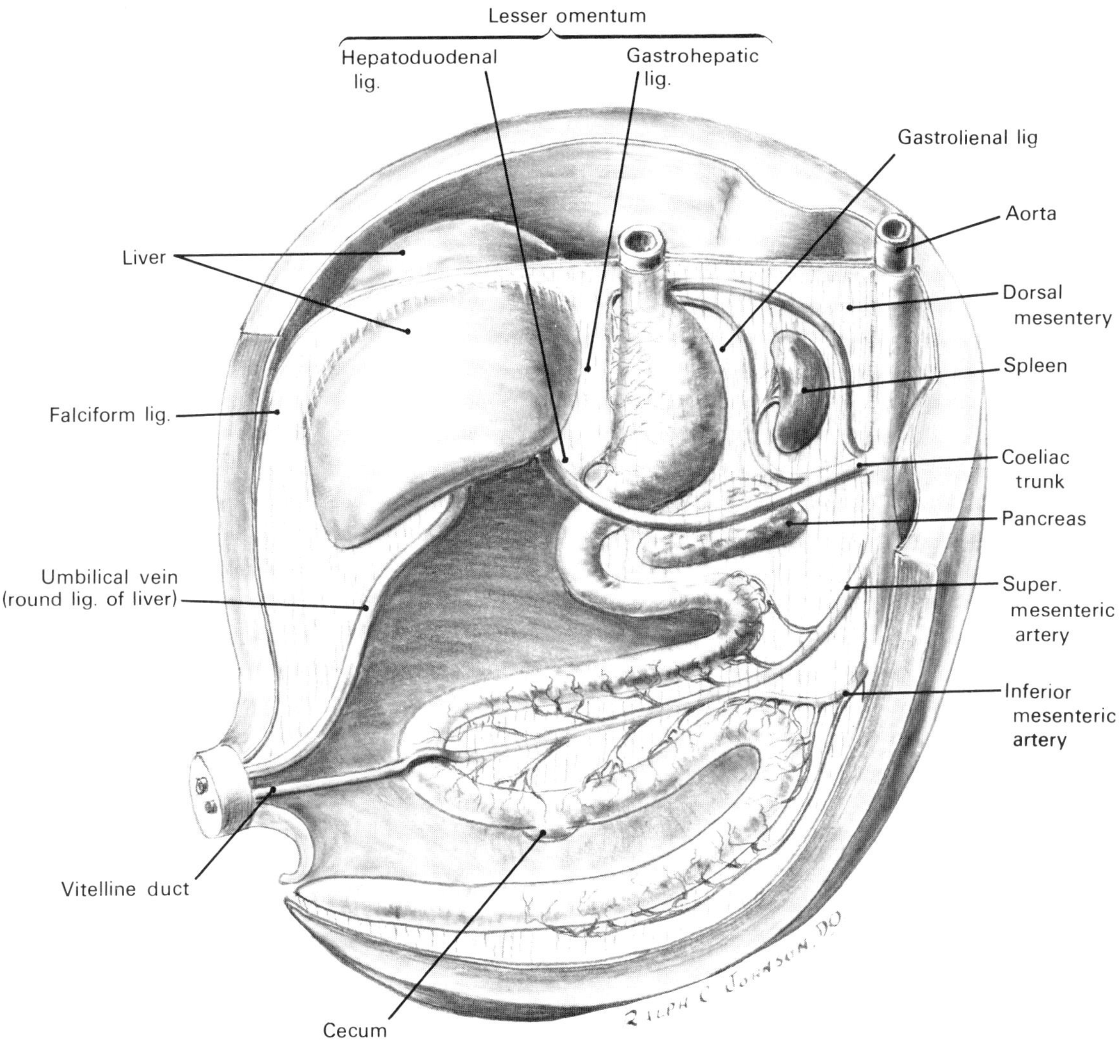

Fig. IX-48. A review of structures related to dorsal and ventral mesentery in early embryo.

Table IX-3. Some Peritoneal Folds and the Structures Contained in Them

Fold	Structure	Patency
1. Lesser omentum	Bile ducts	Patent
2. "The" mesentery	Small intestine	Patent
3. Mesocolon	Transverse colon	Patent
4. Broad ligament of uterus	Uterine tube	Patent
5. Median umbilical ligament	Urachus	Obliterated
6. Right gastropancreatic fold	Hepatic artery	Patent
7. Left gastropancreatic fold	Left gastric artery	Patent
8. Lateral umbilical ligament	Umibilical artery	Obliterated
9. Inferior epigastric fold	Inferior epigastric vessels	Patent
10. Mesentery of appendix	Appendicular vessels	Patent
11. Superior ileocecal fold	Anterior cecal vessels	Patent
12. Suspensory ligament of ovary	Ovarian vessels	Patent
13. Paraduodenal fossae	Inferior mesenteric vein	Patent
14. Falciform ligament	Umbilical vein	Obliterated

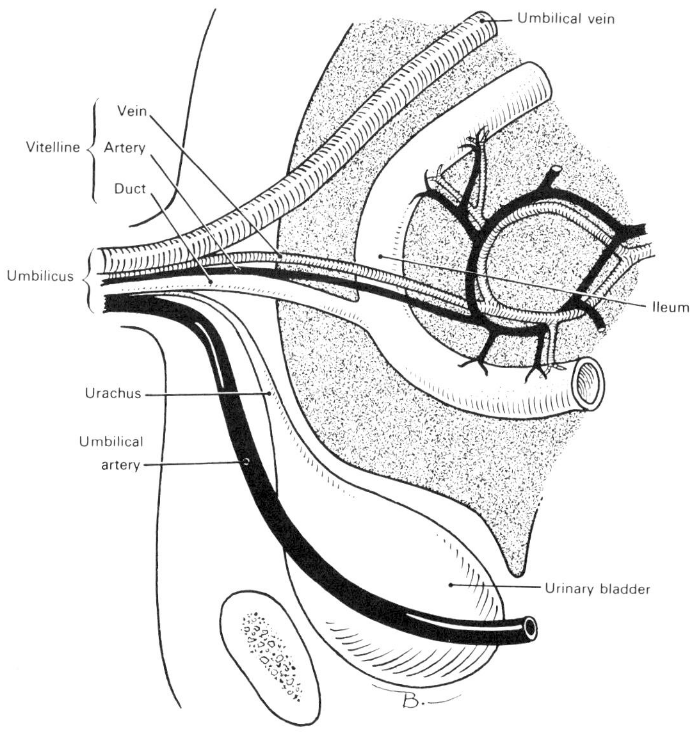

Fig. IX-49. Structures related to the umbilicus. (After Hollinshead.)

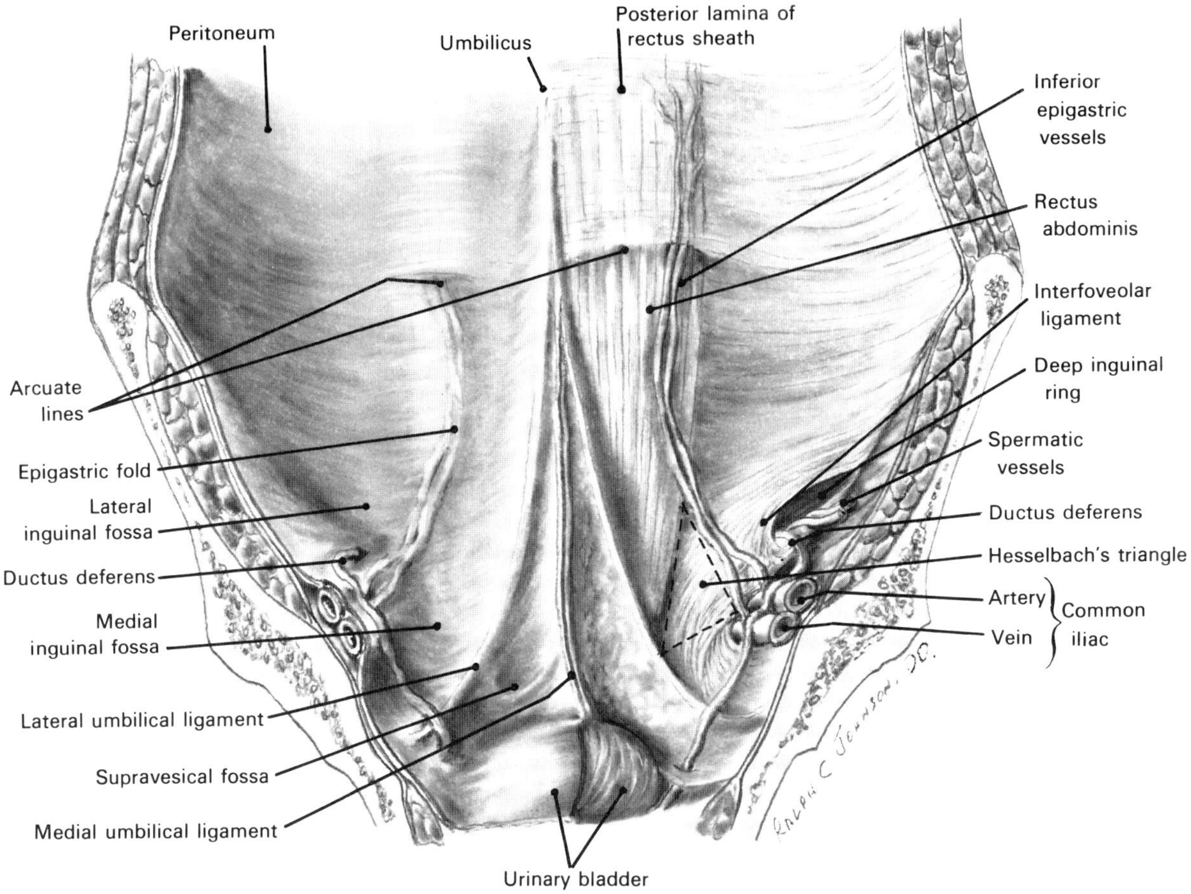

Fig. IX-50. Internal aspect of the ventral body wall. The obliterated urachus is in the medial umbilical ligament; the obliterated umbilical artery, in the lateral umbilical ligament.

the axis around which the primitive intestinal loop rotated. Along with them traveled the **vitelline duct** which attached to the ileal portion of the small intestine. Normally, this once patent connection also dwindles and disappears. But it can persist as: (a) a **vitelline cyst** attached between intestine and umbilicus by short fibrous cords (vitelline ligaments); (b) attached or unattached to the umbilicus as a diverticulum of the ileum (**Meckel's diverticulum**); or (c) as a patent, narrow channel through which a fecal discharge oozes to the surface at the umbilicus (Fig. IX-51). It is possible for the coils of the intestine to twist themselves around such a persistent midline connection to cause a strangulation of the bowel or **volvulus.**

Between the lateral umbilical ligaments on the ventral body wall, a median umbilical ligament is discernible as a midline peritoneal fold reaching up to the umbilicus from the bladder. This fold or plica contains the urachus, the terminal piece of the allantoic duct. Normally, it, too, is obliterated. If patent, urine dribbles forth from the navel. In most of us, fortunately, our navels are water-tight, and they exist as cute but useless "belly-buttons."

Direct your attention now to the inner ventral wall, that is, to the skin and peritoneal flaps you folded back to enter the abdominal cavity. Refer to Figure IX-50, and locate the median and two lateral umbilical folds or plicae. Lateral to the lateral umbilical folds, locate the **plicae epigastrica,** *i.e.,* the peritoneal folds containing the **inferior epigastric vessels.** On the lateral side of this last fold you should be able to find the deep inguinal ring. Trace the epigastric vessels upward until you see them disappear under the arcuate line on the deep side of the rectus sheath.

The triangular depression between the median and lateral umbilical fold is the **supravesical fovea.** The depression between lateral umbilical fold and the plica epigastrica in the inguinal region is the **medial inguinal fovea** or **Hesselbach's triangle,** the area where direct hernias push through the ventral body wall. Try to find the ductus deferens as it loops over the inferior epigastric vessels and courses medialward beneath the peritoneum toward the base of the bladder.

Blood Supply to Abdominal Organs, General Principles

The abdominal organs are situated in such a way that the gastrointestinal tract and its associated partners, the liver, pancreas, and spleen, lie superficial to the genito-urinary tract and its associated partners, the two kidneys, the two suprarenal glands and the gonads (testes or ovaries). The abdominal aorta takes this into account in supplying three, superficially-directed, *single* branches to the "G-I Tract and Asso-

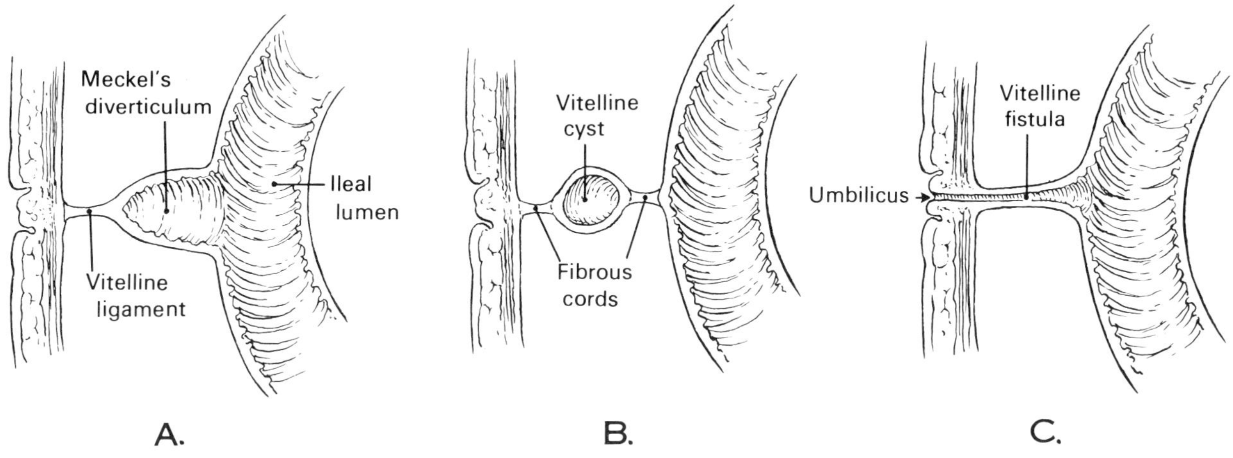

Fig. IX-51. Vitelline duct anomalies. (After Langman.)

ciates" and three sets of deeper, *paired* vessels to the "G-U Tract and Associates." Moreover, it gives off still deeper, chiefly paired branches to the abdominal parieties (body walls). See the simple sketch, Figure IX-52A.

The abdominal aorta, in crossing only four vertebrae, is the shortest segment of the descending aorta. It enters the abdomen at VT_{12} and terminates at VL_4 by dividing into its major branches, the common iliacs to pelvis and extremities. At VL_4, the aorta itself can be considered as tapering off to a fine point as the **median sacral artery** which continues on in the midline over VL_5 and the sacrum (Fig. IX-52B). The paired vessels seen in Figure IX-52E lie deepest of all, and they feed the body walls—all, that is, except the inferior phrenics which give twigs to the suprarenal glands as they pass them *en route* to the Diaphragm. This supply is an accessory one to the glands for these organs receive their own, direct, aortic contribution (Fig. IX-52D).

Aside from the terminal median sacral artery, the only other *single* vessels to spring from the aorta are the **coeliac, superior,** and **inferior mesenteric arteries,** the vessels proper for the G-I tract and its associates. These all arise from the ventral aspect of the main trunk-line (Fig. IX-52 B,C). The **coeliac trunk** is a very short vessel lying at VT_{12} high up along the aorta with the diaphragmatic crura sitting astride it. Embryologically, it arose, believe it or not, at the level of VC_7. As the lungs developed, the aorta lengthened and the Diaphragm was lowered in growth. So also were the stomach and coeliac artery shifted to its final location. As a consequence, the pancreas cannot ascend any higher in the abdomen than it does for it is kept down by both the coeliac stem and the superior mesenteric artery. The latter arises from the aorta not much more than a centimeter below the coeliac stem. The short coeliac trunk breaks up almost at once to send specific branches to the stomach, duodenum, liver and biliary system, pancreas, and spleen.

The **superior mesenteric artery** is a large vessel. It was the old vitelline artery of the embryo. It arises about at the upper part of VL_1, and it crosses the uncinate processes of the pancreas and the inferior part of the duodenum to enter the mesentery of the small intestine. Here it splays out, dividing into several branches to supply all of the small intestine (except the upper part of the duodenum), the cecum and ascending colon, and about half of the transverse colon. In the mesentery it will be accompanied by the superior mesenteric vein which lies to its right. Sometimes the superior mesenteric artery is actually a branch of the coeliac.

The **inferior mesenteric artery** is smaller in diameter than the last vessel. It arises about 3 cm. above the point at which the aorta bifurcates into its common iliacs (about at the middle of VL_3). The vessel will supply the rest of the transverse colon and the descending colon. As it crosses the left common iliac artery, it continues in the sigmoid mesocolon down into the lesser pelvis as the **superior rectal artery.** Thus, it eventually terminates in the upper end of the rectum.

The **renal arteries** are the large, bilateral trunks which arise just below the superior mesenteric artery, about at the disk between VL_1 and VL_2 (Fig. IX-52 B,D). Because of the position of the aorta, the right renal branch is longer than the left one. To reach the right kidney hilus, the right renal artery must pass deep to its own vein, to the inferior vena cava, to the head of the pancreas and to the descending limb of the C-shaped duodenum. The left renal artery reaches its kidney by crossing posterior to its own vein, and posterior to the pancreas and the splenic vein. On its way, it is crossed by the superior mesenteric artery which can clamp down upon it and even constrict it somewhat because of the weight of the small intestine and its mesenteries. The left renal artery is also crossed by the inferior mesenteric vein as this vessel wends its way upward to join the splenic vein. Both renal arteries break into smaller branches as they enter the hila of the kidneys. Like the inferior phrenic arteries, the renals pass close to the suprarenal glands and supply them with **inferior suprarenal twigs.**

The **suprarenal arteries,** proper, had best be called

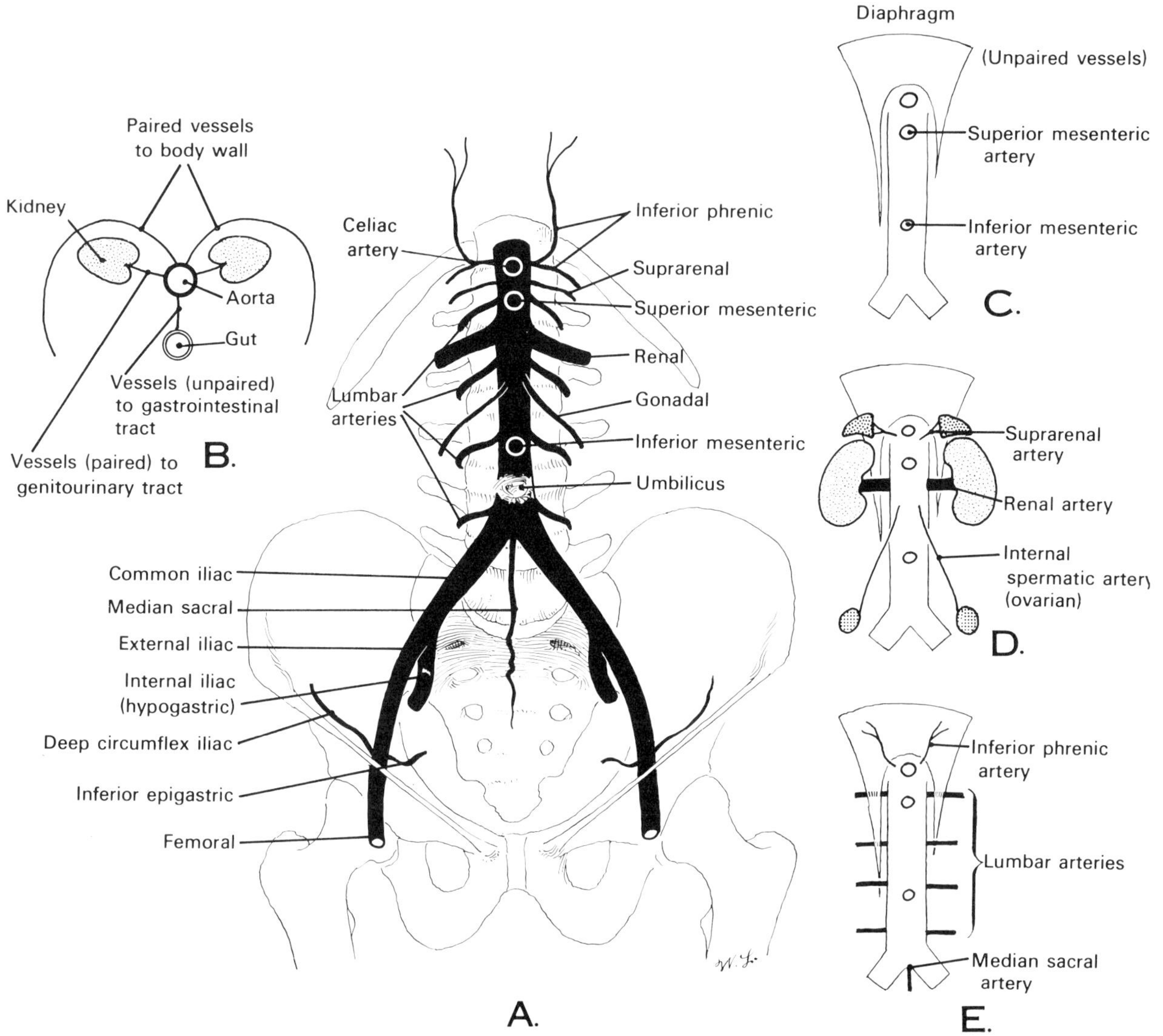

Fig. IX-52. Paired and unpaired branches of the abdominal aorta and their areas of distribution. **A.** Branches related to skeletal landmarks. **B.** Basic plan of supply in the abdomen. **C.** Unpaired vessels for the gastrointestinal tract and its unpaired "glands." **D.** Paired supply to the genitourinary tract and its paired "glands." **E.** Branches to the body wall.

middle suprarenal arteries for the glands receive a superior supply from the inferior phrenic arteries and an inferior supply from the renal arteries. The middle and main vessels to the glands are two small branches of the aorta, which come off, one on each side, about at the level of the superior mesenteric artery (Fig. IX-52 B,D). These were once very respectable branches in the fetus where the gland was much larger. The former size of these organs may account for the earlier need for subsidiary branches from other sources.

The **testicular or internal spermatic arteries** of the male arise as a pair of vessels from the ventral aspect of the aorta between the points of origin of the renal and inferior mesenteric arteries. The testicular arteries diverge from each other, riding caudalward beneath the peritoneum and upon the Psoas. The right-hand artery crosses the inferior vena cava but remains posterior to the middle and iliocolic vessels. The left lies deep to the left colic and sigmoid vessels. Each crosses the ureter and the external iliac artery to reach the deep inguinal ring.

In the female, the counterpart of these gonadal vessels, the **ovarian arteries,** arises in a similar way. At the pelvic brim they enter the leash-like suspensory ligament of the ovary and the broad ligament of the uterus to reach the ovary. In the embryo, testicular arteries were once as short as the ovarian arteries, but they lengthened with the descent of the testes to the scrotum.

General Principles of Nerve Supply to the Abdominal Organs

All organs, except the adrenal (suprarenal and, possibly, kidney and ureter) glands, get a dual sympathetic and parasympathetic innervation. In the bladder, this dual innervation tends to elicit antagonistic responses on the part of the organs. The adrenal gets only a sympathetic supply. The kidney may have no known vagal supply and the ureter appears to have only a visceral sensory supply functionally. The bladder is virtually under control of the parasympathetic system alone, though the sympathetic fibers do regulate its vasculature and its internal sphincter. The sphincter, apparently, functions during sexual orgasm, perhaps in an attempt to keep sperm from backfiring into a watery grave.

The sympathetic fibers emanate from the visceral motor cell columns of the cord at T_6 to L_2. These preganglionic nerve fibers pass in the ventral roots and white rami through their own chain ganglia without synapsing there and continue beyond the sympathetic chain as the splanchnic nerves. Their termination is in the collateral ganglia that cluster about the roots of origin of the coeliac, superior, and inferior mesenteric arteries.

The **greater splanchnic nerve** (T_6 to T_9) consists of preganglionic fibers destined for the **coeliac ganglion.** Some of these pass through the ganglion without synapse and enter the adrenal to terminate among cells in the medulla of this gland.

The **lesser splanchnic** (T_{10} to T_{11}) ends in the superior mesenteric ganglion or in offshoots of it along the renal arteries (aorticorenal ganglia).

The **least splanchnic nerve** (T_{12}), when it is present, runs to the latter also.

Postganglionic neurons of these collateral ganglia send their axons out along coeliac, renal, and superior mesenteric arterial branches to reach the organs supplied by these vessels. The nerve fibers end upon the smooth muscle and glandular cells of the various organs, distributing as far distally as the middle of the transverse colon.

The **lumbar splanchnic nerves** (L_1 to L_2) distribute as preganglionic fibers to the inferior mesenteric ganglion at the root of origin of the inferior mesenteric artery. Postganglionics from this ganglion follow branches of the inferior mesenteric artery to the rest of the colon and rectum. Scattered cell clusters from the inferior mesenteric ganglia extend on over the pelvic brim. Some lumbar splanchnic preganglionics follow into the true pelvis to terminate upon these scattered ganglion cells of the hypogastric plexus. Postganglionic axons follow the branches of the internal iliac (hypogastric) artery to the organs within the true pelvis, *i.e.*, to bladder and parts of the reproductive system.

The parasympathetic preganglionic fibers come from two sources, from the vagi, which follow the esophagus through the Diaphragm, and from the pelvic nerve (nervi erigens). The latter has its origin from visceral motor cells at cord level S_2 to S_4. These parasympathetic fibers end in **terminal ganglia,** *i.e.*, small clusters of parasympathetic neurons lying very close to or within the walls of the organs to be innervated. No parasympathetic fibers go to the adrenal glands. The vagi supply liver, stomach, spleen, pancreas, small intestine, ascending and about half of the transverse colon. The pelvic nerve supplies pelvic organs, and, rising up out of the pelvis along the hypogastric plexus, it supplies descending colon and the rest of the transverse colon. It is always the axons from the terminal ganglia that make the final contact with smooth muscle or gland cells. In the intestine, clusters of these terminal ganglia are located in the submucosa and the muscularis proper. These are the ganglion cells of Meisner and Auerbach, present in these locales in the esophagus, stomach, and the intestinal tract (Fig. IX-63).

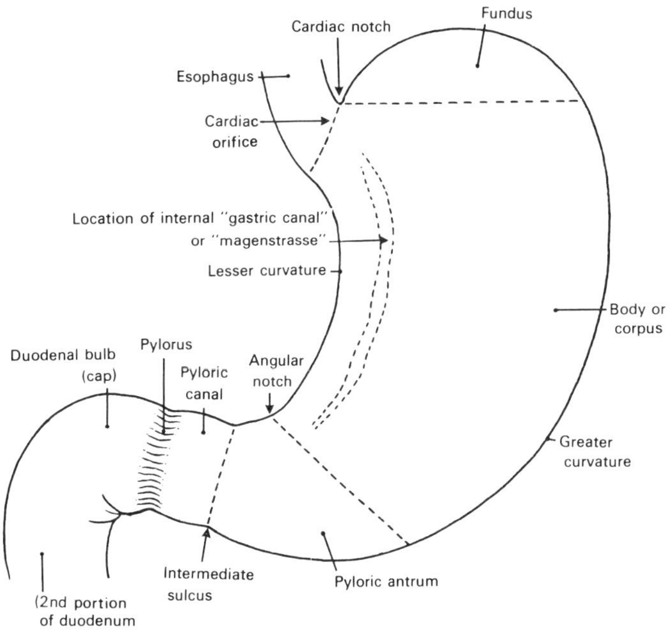

Fig. IX-53. Schema of stomach areas.

The Organs of the Gastro-Intestinal Tract

The Esophagus: *General:* Not much of this portion of the digestive tube will be found in the abdominal cavity. Look for it behind the stubby end of the left lobe of the liver where the liver is grooved as the esophagus punches through the muscular part of the Diaphragm less than an inch from the midline. Most of this gullet resides in the thorax, extending some 25 cm. from pharynx to stomach. It begins at the level of C_6, the level of the cricoid cartilage. In the superior mediastinum, between trachea and the vertebral column, it lies a bit left of the midplane. Trachea and esophagus will be crossed on the left by the aortic arch so that the esophagus runs at first along the right side of the descending aorta. It gradually moves in front and slightly to the left of the aorta to reach its diaphragmatic hiatus. Only a little more than a centimeter of it appears in the abdominal cavity. Here, only its anterior and left sides are covered by peritoneum. It meets the stomach at an oblique angle to form the **cardiac notch** (Fig. IX-53).

The esophagus is a fairly muscular tube which follows the typical plan of a hollow organ. Its exceptional and diagnostic features from a histological point of view are:

1. An inner, mucosal lining-layer of moist, nonkeratinized, stratified squamous epithelium.

2. A muscularis mucosae that is longitudinally rather than circularly disposed.

3. A muscularis proper which varies in composition. In its upper one-third, it is skeletal muscle; its middle one-third consists of a mixture of skeletal and smooth muscle; while its lower two-thirds consists entirely of smooth muscle arranged in the usual way, *i.e.*, an inner lamina of circular smooth muscle and an outer lamina of longitudinally-oriented smooth muscle bundles.

4. It has no true serosa except for the bits of peritoneum on its left and ventral sides as it approaches the stomach. Instead, it is invested outermost by a concentration of fibro-elastic tissue.

The Blood Supply of the Esophagus: This comes from the inferior thyroid, from twigs of the descending aorta, and even from some offshoots of the bronchial arteries, from the left gastric branch of the coeliac artery and from some twigs of the inferior phrenic artery.

The Nerve Supply of the Esophagus: The nerve supply is through a plexus along its outer coat where parasympathetic and sympathetic fibers mingle. The parasympathetic vagi follow down in the lateral grooves between trachea and esophagus. Here, right and left vagus contribute preganglionic twigs to the plexus and, normally, lose themselves in the plexiform network until the lower end of the esophagus is reached. Near the esophageal diaphragmatic hiatus, the vagi reconstitute and emerge as one, two or, rarely, three trunks on each side to continue on over the walls of the stomach as the **anterior** and **posterior gastric nerves.** The right vagal fibers pass posteriorly; the left, anteriorly. There are variations, of course, and the reverse relationship at the stomach has been encountered even without transposition or any other visceral anomaly. Sometimes the two vagi remain discretely separate from the esophageal plexus along its entire extent, simply sending side branches into it. At other times, the plexus fails to form below the level of the roots of the lungs, and the so-called gastric nerves reconstitute themselves some distance cranial to the esophageal-cardiac junction. At any rate, the vagal fibers entering the plexus terminate on the ganglion cells of Meisner and Auerbach in the submucosal and myenteric parts of the esophageal wall. Axons of these ganglion cells reach smooth muscle and glands.

The sympathetic supply differs above and below the Diaphragm (Fig. IX-54). *Below the Diaphragm,* the greater splanchnic nerve serves as the preganglionic sympathetic route. Postganglionic fibers from the coeliac ganglion follow, principally, the left gastric artery. Some, however, follow the hepatic and inferior phrenic arteries to the esophagus. These axons end directly upon smooth muscle or esophageal glands. [The skeletal muscle of the upper esophagus, like that of the larynx, is derived from the vagi, but from its branchial motor nucleus in the medulla (nucleus ambiguous), not from the visceral motor nucleus (dorsal motor nucleus of the vagus).]

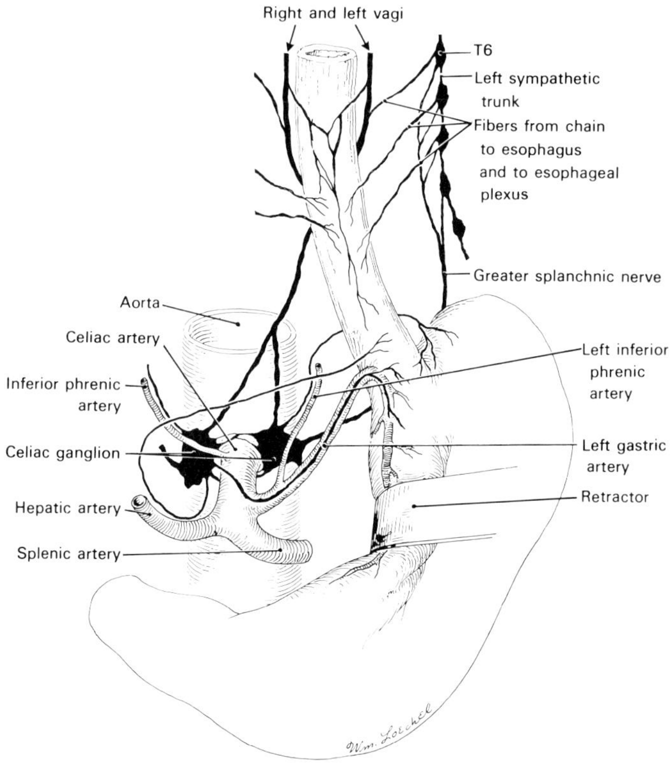

Fig. IX-54. Autonomic innervation of esophagus and cardio-esophageal junction. The vagi are not shown continuing onto the stomach but rather stopped where they *pass through* but do not synapse in the coeliac plexus. Both sympathetic and parasympathetic fibers distribute to plexus on the esophagus and plexuses along the branches of the coeliac artery. Here the inferior phrenic arteries are shown taking origin from the coeliac and left gastric vessels rather than from the aorta directly.

Above the Diaphragm, the splanchnic nerve is not involved. Preganglionics from T_6 to T_9 reach their own ganglia of the sympathetic chain where they synapse. Thoracic sympathetics leave the chain to enter the esophageal plexus where they run to end on smooth muscle and glands.

It would seem safe to say that **visceral sensory fibers** are also found mingled in the esophageal plexus, but their return route is not well known. Probably they return over the vagi. However, frequent reference of esophageal pain to the base of the neck and xiphoid area makes one wonder whether some return is possible *via* the sympathetics and spinal nerves. One could negate this route by arguing that whatever pain is felt might be that involving neighboring structures, not the esophagus itself.

Esophageal Veins: A word about venous return is important largely because of the anastomoses of veins of the portal system with those of the lower part of the esophagus. These serve as collateral routes of venous return when the portal system suffers obstruction. The raised pressures in these veins form internal esophageal varices (esophageal piles) which can be fatal when they bleed. These veins enter the submucosal layers of the esophagus where they have little supporting tissue and hence, they can be quite distensible. The esophageal veins drain into the azygos or hemiazygos veins primarily. At the upper end, drainage is into the inferior thyroid and intercostal veins, some even, into the vertebral veins.

The Stomach: *General Structure:* This muscular organ is, probably, the most distensible of any in the human body. The esophagus joins stomach at the **cardiac orifice** so named because it lies at that part of the Diaphragm near the point at which the heart rests. Histologically, this is also the area where the mucous cardiac glands occupy the tunica propria of both stomach and lower esophagus. Actually, these glands encroach upon the **fundus** of the stomach. The fundus is that ballooned portion to the left of and above the cardiac orifice (Fig. IX-53). The body of the stomach is its major portion, extending along the **lesser curvature** to the **angular incisure or notch.** From the notch to an **intermediate sulcus,** which lies about 2½ cm. from the **pyloric orifice,** the area is known as the pyloric vestibule or antrum. The part from intermediate sulcus to pyloric orifice is the **pyloric canal.** At the orifice, the pylorus ends in a thickened internal muscular ring, the **pyloric sphincter.** The sphincter relaxes at intervals during digestion to allow gastric contents to move along into the duodenum, the first part of the small intestine.

The shape of the stomach varies from individual to individual and from time to time in the same individual depending upon degree of distention, degree of contraction, and the age and body-build of the individual. Its loose omental suspension allows considerable freedom for motility and contractility. In children and husky individuals, it tends to lie transversely, sloping slightly toward the duodenum. Frequently, it is more J-shaped than U-shaped so that its greater curvature can even lie in the greater pelvis when the individual is erect. Thus, the portions of other viscera that can come in contact with the stomach can be quite variable.

Cardiac and fundus are relatively fixed and, hence, tend to move only with the respiratory excursions of the Diaphragm. The cardia lies about 10 cm. behind the seventh left costal cartilage, and 2 to 4 cm. from the midline.

The fundus, lying under the left dome of the Diaphragm, reaches the level of the fifth rib anteriorly and so comes to lie slightly inferior and posterior to the apex of the heart. In the erect individual, the fundus is filled with a bubble of swallowed air known as the "Magenblasse." This is not to be confused with another common German term, the "Magenstrasse" or gastric canal. The canal is formed by two longitudinal, interior, mucosal folds or **rugae** whose edges almost touch to form a tubular passage along the course of the lesser curvature. Swallowed fluids have been purported to take this short-cut channel to the pylorus.

The pylorus, despite its attachment to the duodenum, is also quite mobile by virtue of omental attachment. It comes to lie at any level between the first and third lumbar vertebrae. It may be displaced to the right when the stomach is distended, and, at its highest point, it can hide behind the liver. It lies immediately in front of the pancreas, as a rule, and is separated from this organ by the omental bursa.

The stomach also adheres to the construction plan of a hollow organ (Fig. IX-42). Its surface-lining cells are simple columnar epithelium. These extend into gastric pits or foveolae which sink into the tunica propria to form the necks of gastric glands. Gland types vary regionally. At the cardia and part way into the fundus, the tubular **cardiac glands** have short necks and quickly-branching, short glands which secrete a mucous substance. The fundus and body contain the typical **gastric glands.** Fig. IX-55A, B. Here, the foveolae are moderately long and the glands do not branch much until they reach the bottom of the tunica propria. The relatively straight glands contain innocuous looking chief cells (Fig. IX-55B) which contain many secretory granules, the precursors of the enzyme, pepsin. The cells which attract most attention to themselves are the so-called parietal cells of the gland (Fig. IX-55A, B). They bulge from the glandular wall, and they stain brilliantly pink with eosin. These are the hydrochloric acid-producing cells. The pyloric glands (Fig. IX-55C), on the other hand, have extremely long foveolae which can be confused for intestinal villi, particularly for the villi of the ileum. Look for a striated border at the surface of the cells. If it is present, your specimen is intestine, not pylorus of the stomach. The glands of the pylorus, like those of the cardia, branch profusely almost immediately. Thus, they are most often seen as cross-sections. Like those of the cardia, they lack chief and parietal cells and they are mucous-producing.

Like all parts of the digestive tract, except the esophagus, the muscularis mucosa of the stomach is, typically, circularly-arranged smooth muscle. The muscularis, proper, may consist of three coats of smooth muscle which are rather irregularly arranged along the course of the organ.

Food, mixed with saliva, comes to the stomach where it is rendered into a semifluid **chyme** by the addition of HCl and such enzymes as pepsin, rennin, and lipase. At intervals, the chyme is passed through the pyloris into the duodenum as waves of gastric peristalsis sweep over the stomach. Mucus, added to chyme, serves to lower the pH of the mixture, helping to protect the gastro-intestinal mucosa against autodigestion.

Blood Supply to Stomach (Figure IX-56A, B): The stubby coeliac trunk is the sole arterial supply to the stomach, through its branchings. These branches, however, also supply liver and gall bladder, duodenum and esophagus, pancreas and spleen. The coeliac artery gives rise to three major vessels: **left gastric, hepatic,** and **splenic arteries.** These, in turn, contribute, among themselves, five branches to the stomach.

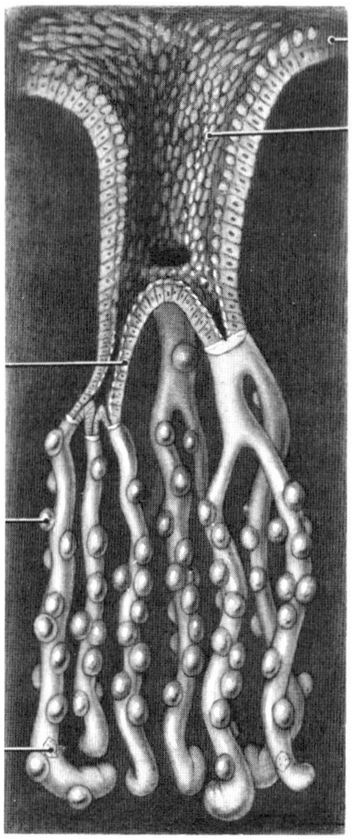

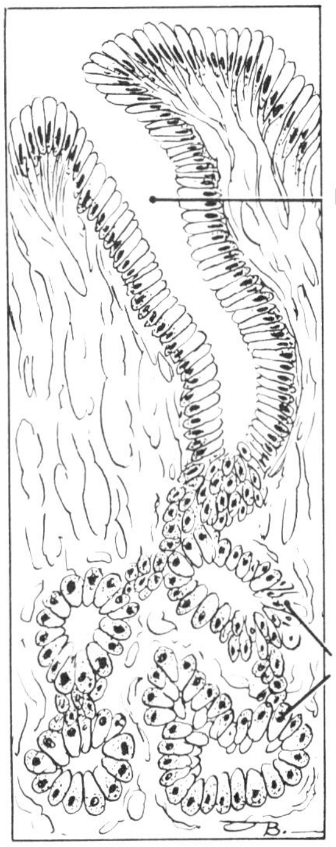

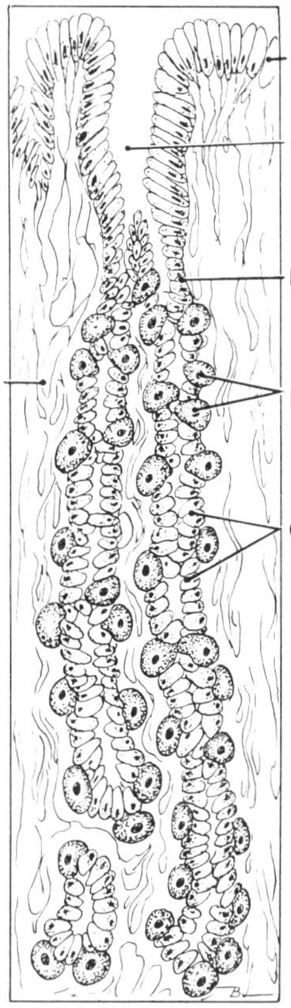

Fig. IX-55. The gastric glands. **A.** A fundic gland in three dimensions. **B.** Sketch of a fundic gland as it appears under the microscope. **C.** Sketch of a pyloric gland as seen microscopically. (After Elias and Pauly.)

A. **Left gastric artery:** This is the smallest of the three branches of the coeliac trunk but the largest of them to give a direct supply to the stomach.
   1. As it courses along the lesser curvature it contributes 4 to 5 **gastric** branches.
   2. It continues through the esophageal diaphragmatic hiatus to supply the esophagus.
   3. It may provide an **aberrant left hepatic artery** which, when present, crosses the lesser omentum to enter the liver *via* the fissure for the ligamentum venosum.

B. **Splenic artery:** This is the largest of the coeliac branches.
   1. Its main course is along the upper border of the pancreas to the spleen.
   2. Near the splenic hilum it gives off 5 to 7 **short gastric branches** which traverse the gastrolienal ligament to the fundus and part of the greater curvature of the stomach.

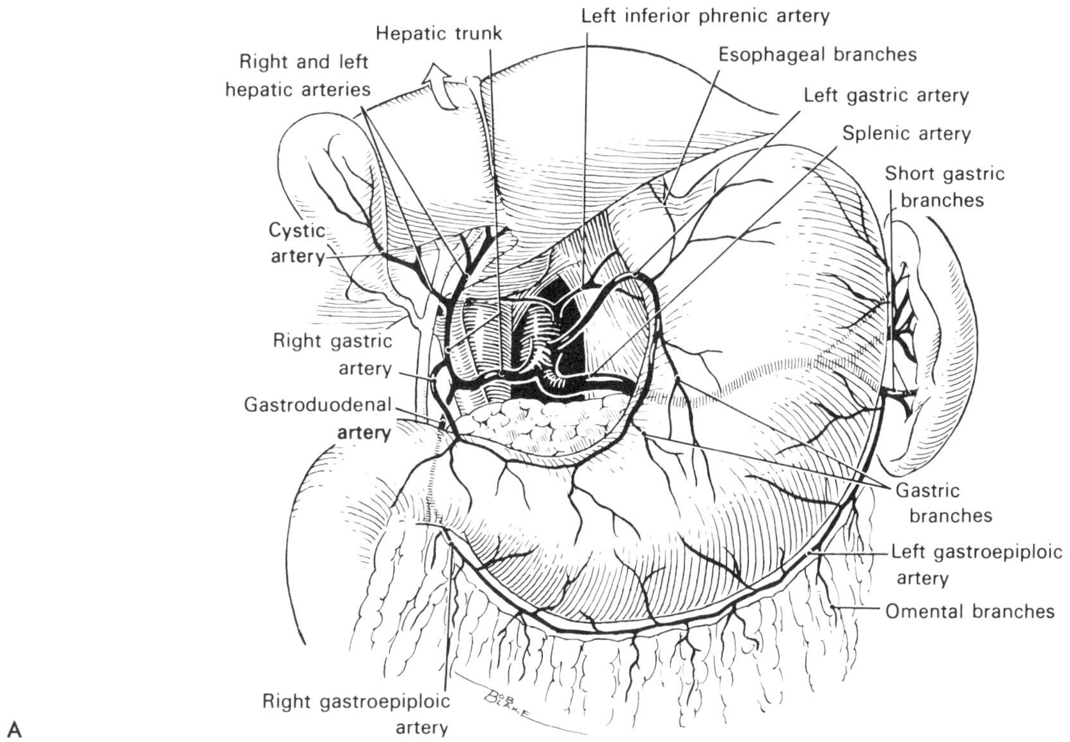

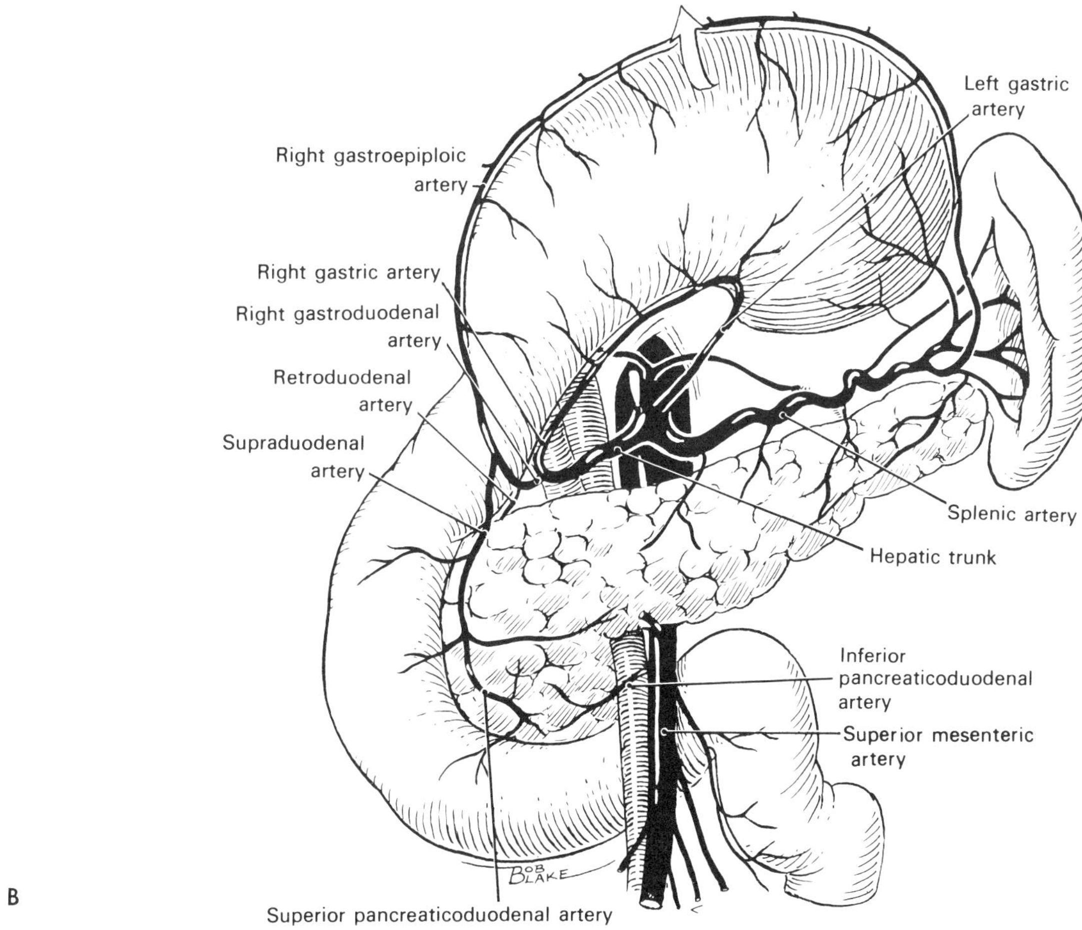

Fig. IX-56. Coeliac arterial supply to stomach and neighboring organs. **A.** Viewed with liver lifted off stomach. **B.** Viewed with stomach freed and turned back off pancreas.

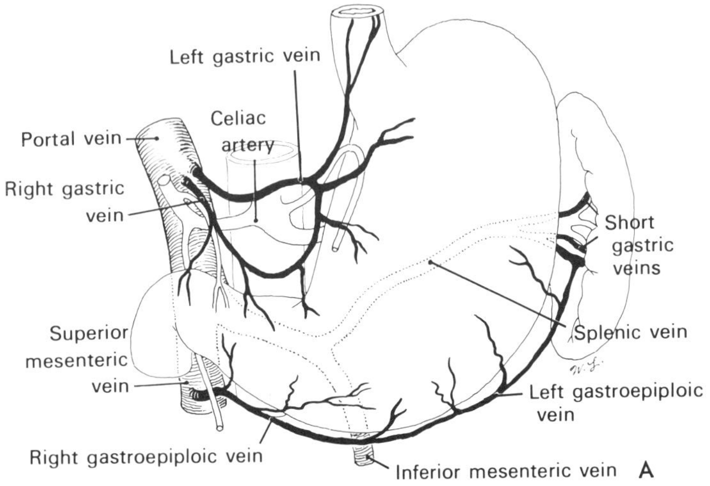

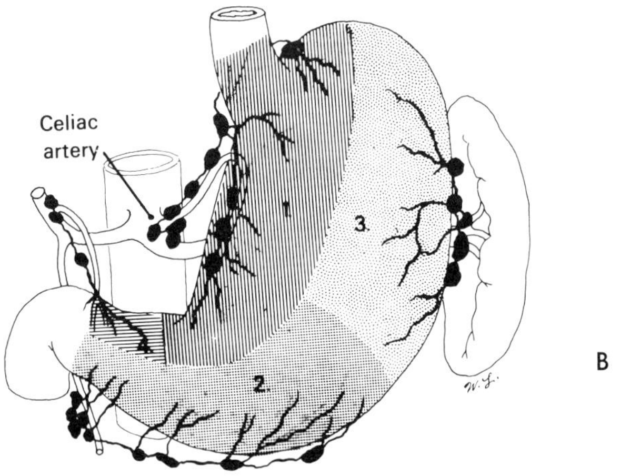

Fig. IX-57. Venous and lymphatic drainage of stomach. **A.** Venous return. **B.** Lymphatic areas of drainage. 1. Lesser curvature to cardiac and left gastric nodes. 2. Greater curvature to right gastroepiploic nodes. 3. Fundus to short gastric nodes. 4. Pyloris to nodes of pancreas and common bile duct.

3. **The gastroepiploic artery** is the largest branch of the splenic. It also enters the gastroleinal ligament, and it is directed to the right along the entire greater curvature of the stomach. It supplies both sides of the stomach, and it sends branches down into the greater omentum.
4. On its way to the spleen it gives pancreatic branches to the body and tail of that gland.

C. **Hepatic artery:** This right-hand branch of the coeliac trunk divides like a tilted "T", *e.g.*, ⊢.

1. The hepatic trunk is the main stem and upper limb of the horizontal "T."
   a. Terminally, the upper limb of the "T" branches into the right and left **hepatic arteries**, one for each major lobe of the liver.
      (1) The right hepatic artery gives rise to the **cystic artery** of the gall bladder.
   b. A **right gastric artery** comes off the base of the upper limb of the "T," or off the main stem, or from the junction of both. It is recurrent to the lesser curvature of the stomach and to the pyloric region.
2. The **gastroduodenal artery** is the descending arm of the horizontal "T." Its task is to supply pylorus and duodenum, chiefly.
   a. **Supraduodenal branch** (Fig. IX-56B). This supplies the superior aspect of the first part of the duodenum. It may arise, as well, from the hepatic trunk or from the retroduodenal artery.
   b. **Retroduodenal artery** (Fig. IX-56B). This arises either above the duodenum and runs dorsal to it to supply bile duct, head of pancreas and the first part of the duodenum; or may form the dorsal side of a C-shaped pancreaticoduodenal arcade.
   c. The **superior pancreaticoduodenal artery** is the smaller terminal branch of the gastroduodenal (Fig. IX-56B). It takes part in the ventral pancreaticoduodenal arcade.
   d. The largest terminal branch is the **right gastroepiploic artery** (Fig. IX-56A, B). This runs in the greater omentum at a variable distance from the greater curvature of the stomach. It may be larger than the left gastroepiploic artery which arose from the splenic artery and with which it anastomoses.

It is obvious that we have discussed more than the specific blood supply to the stomach. The description should help to clarify Figure IX-56, and to give the important blood supply of the neighboring organs.

Veins and Lymphatics of the Stomach: To some extent the veins of the stomach follow the arterial course along the greater and lesser curvatures (Fig. IX-57A). But once they leave these areas, they do not unite into a common trunk as did the arteries. Rather they join the portal vein by routes that are quite variable. We show only a general schema of drainage in Figure IX-57A. Any valves in them are rudimentary ones left over from some fairly robust attempts at valves during embryonic life. As a result of anastomoses between these veins, pressure changes can effect shifts in routes of drainage, a useful arrangement to contend with exigencies should portal hypertension become a problem.

Submucosal lymphatic channels anastomose with each other to form a very dense plexus so that carcinoma can spread easily throughout the stomach wall. Connections exist with the myenteric lymphatic network and the latter, in turn, with a serosal network of channels. The latter drains in a regional kind of way (Figure IX-57B). The major route is by larger lymphatic vessels to the left gastric nodes and to the nodes near the cardia. These nodes drain the major superior portion of the stomach except for the fundus. Another major route is along the greater curvature to inferior gastric nodes along the right gastroepiploic vessels, an important route since carcinoma of the pylorus is common. Several of these nodes lie in the greater omentum some distance from the stomach, and this must be taken into account in surgery for pyloric carcinoma. All these nodes drain to those around the coeliac trunk.

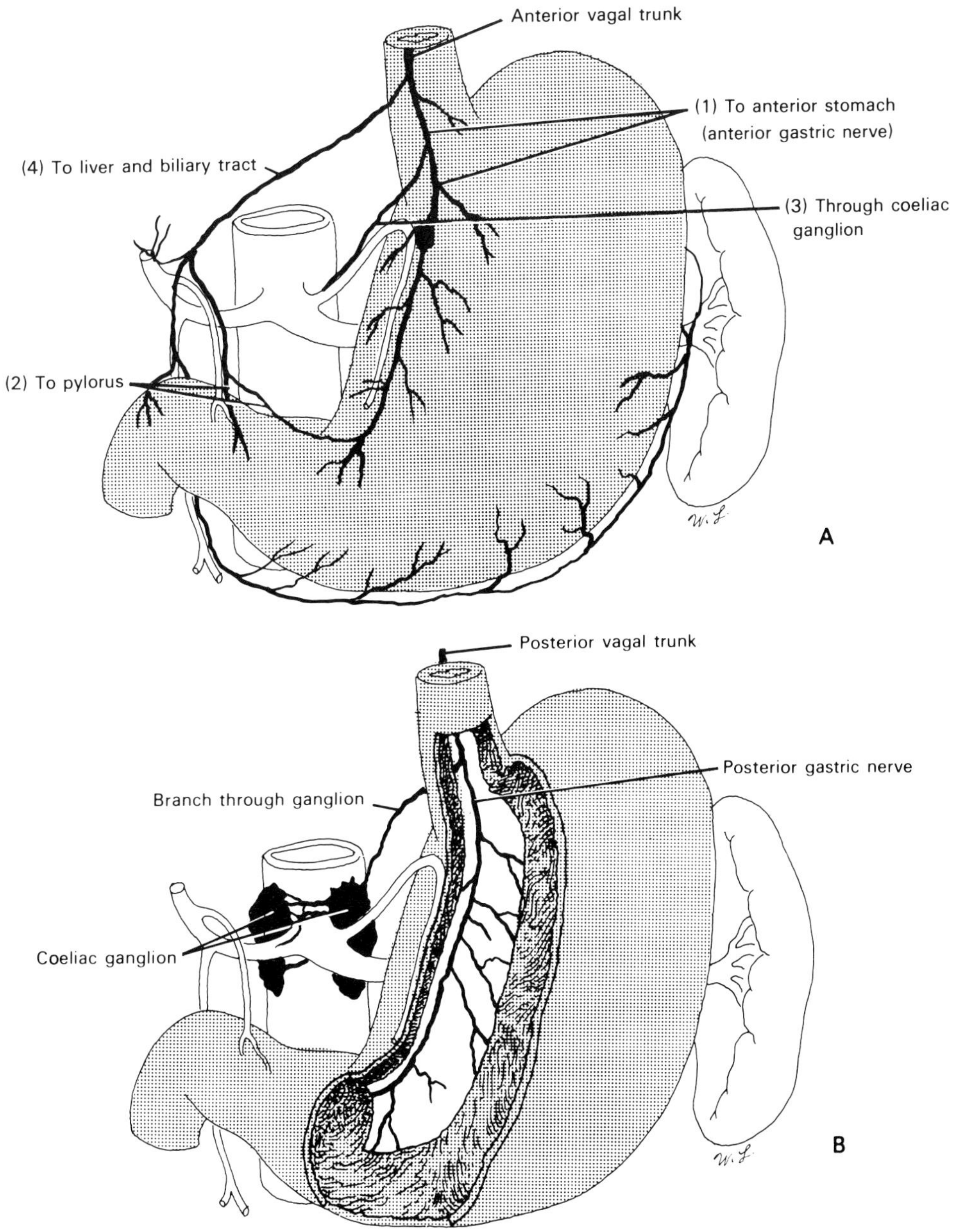

Fig. IX-58. Distribution of anterior and posterior vagal trunks to stomach region. **A,** anterior trunk; **B,** posterior trunk.

Fundus and upper body drain toward nodes which accompany the short gastric and left gastroepiploic vessels, nodes near the splenic hilum. The smallest area of drainage is along the upper border of the pylorus which leads to nodes near the head of the pancreas and along the common bile duct.

Nerve Supply of the Stomach (Fig. IX-58A, B): Recall that the left vagus becomes the **anterior gastric nerve** and the right vagus the **posterior gastric nerve.** The anterior vagal trunk is shown in Figure IX-58A as giving the principal supply (1) to the anterior stomach. It gives off the following subsidiary branches as a rule: (2) a pyloric twig; (3) a branch which runs through the coeliac ganglion without interruption in the ganglion. It follows the left gastric artery to reach the ganglion, and continues on to distribute to stomach, liver, pancreas, and spleen. A branch (4) is shown crossing the lesser omentum to reach the liver and the biliary ducts, pylorus, and duodenum. The posterior vagal trunk (Fig. IX-58B) supplies the dorsal side of the organ. Normally, the posterior trunk also sends a branch *through* the coeliac ganglion which distributes to terminal ganglia in neighboring organs.

The sympathetic supply is the greater splanchnic nerve (T_6 to T_9) which *does* terminate in the coeliac ganglion. Postganglionic axons from the ganglion follow the branches of the coeliac artery to various parts of the stomach and to neighboring viscera supplied by these arterial branches. The vagal fibers breaking away from the main anterior and posterior gastric nerves, of

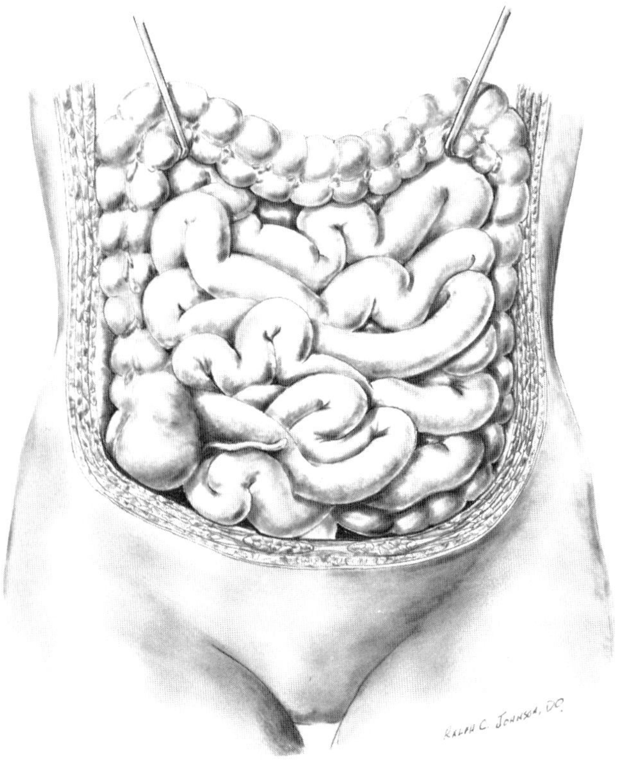

Fig. IX-59. A view of the 20 feet of small intestine crowded in the abdominal cavity.

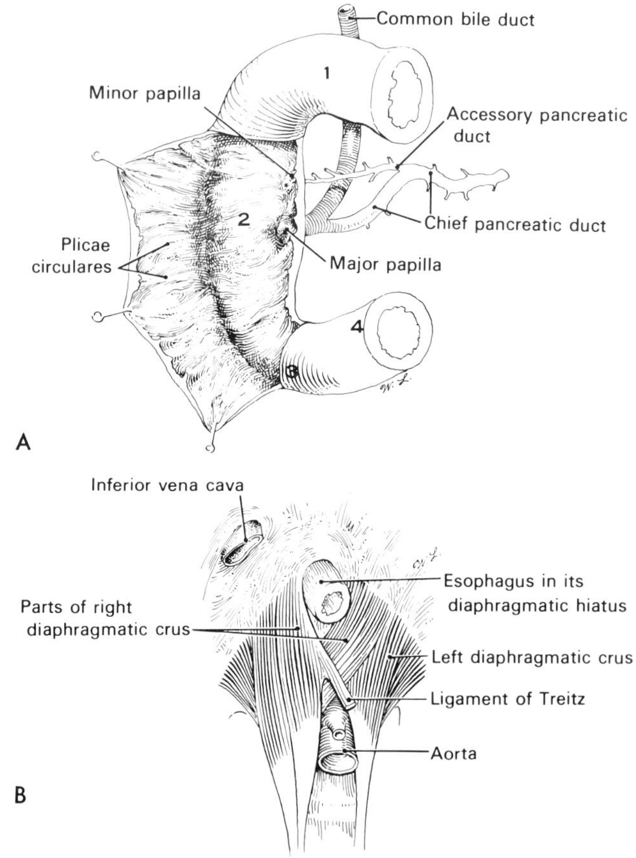

Fig. IX-60. The duodenum. **A.** Its parts, internal structure, and associated bile and pancreatic ducts. 1, Superior part (duodenal cap); 2, descending part; 3, horizontal part (foreshortened in perspective); 4, its short ascending piece at duodenal-jejunal junction. **B.** Its supporting ligament of Treitz.

course, use these same vascular routes to reach neighboring structures.

The Small Intestine (Fig. IX-59): This 20-foot sausage-casing is divided into three parts: **duodenum, jejunum,** and **ileum.** The **duodenum** is the shortest of these portions. About 10 inches long in the human, it was named because it was 12 fingerwidths in length in animal specimens where it was first studied. The duodenum is divided for anatomical orientation into four segments (Fig. IX-60A): a first or superior part, about 2 inches long; a second or descending portion, 3 inches long; a third or horizontal part, 4 inches long; and an ascending piece, about 1 inch in length which ends at the duodenal-jejunal junction suspended by the **ligament of Treitz** (Fig. IX-60B). The duodenum begins at the pylorus about an inch to the right of the midline at the transpyloric plane, and it curves horse-shoe fashion to end just below that plane about 1 inch to the left of the midline. Hence, its beginning and end are no more than 2 inches apart. It has all but lost its mesentery except at its beginning first inch, so it is classed as a retroperitoneal structure along with the pancreas with which it is closely associated. The ligament of Treitz seems to be a part of the right crus of the Diaphragm which fails to swing around the left side of the esophagus at the diaphragmatic hiatus of that organ.

From the histological standpoint, the submucosa of the duodenum is thrown up into circular folds, **plicae circulares.** Somewhat similar mucosal ridges can be seen in the jejunum in Figure IX-61. But the tunica propria is also projected above the surface as broad-fingered, microscopic, intestinal **villi.** These are sur-

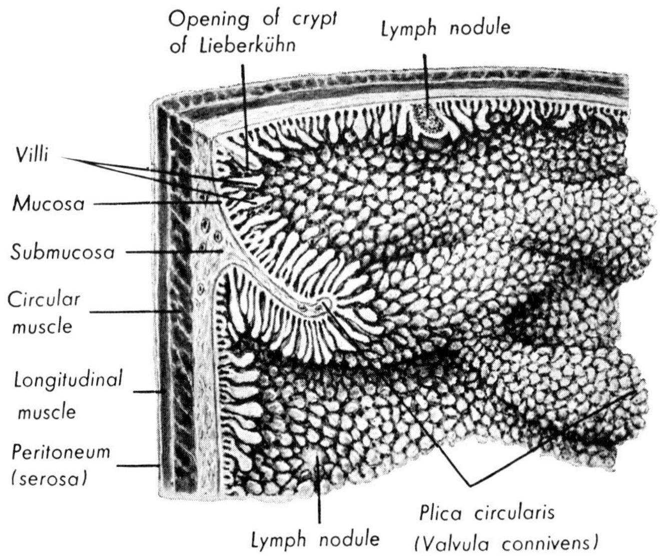

Fig. IX-61. A three-dimensional reconstruction of the jejunal wall to show increase of absorptive surface at microscopic level. (From W. E. Copenhaver, *Bailey's Textbook of Histology,* 15th ed., Williams & Wilkins, Baltimore, 1964.)

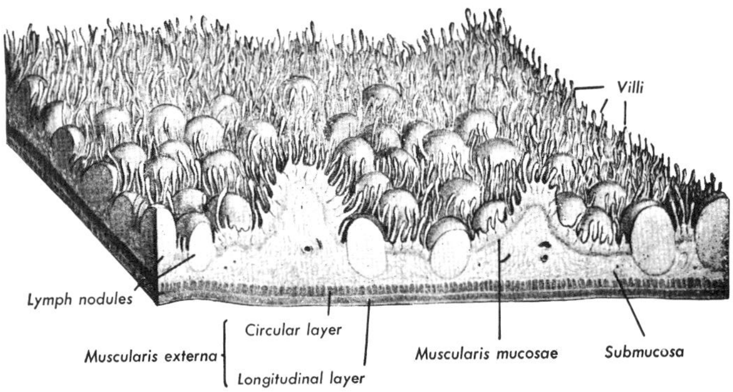

Fig. IX-62. Three-dimensional view of a section of ileum to illustrate tall villi and a portion of a Peyer's patch. (From W. E. Copenhaver, *Bailey's Textbook of Histology*, 15th ed., Williams & Wilkins, Baltimore, 1964.)

faced by columnar cells with a striate border. Electron microscopy shows this border to consist of myriads of **microvilli.** All these modifications—plicae, villi, and microvilli—serve to increase the absorptive surface of the intestinal tract tremendously. As one moves into the jejunum, plicae increase in height and the number of villi superimposed upon them make the inside of the tract appear like velvet. Plicae are shorter in the ileum and less fringed than in the jejunum. They are tall and straight in the ileum (Fig. IX-62). In all of the small intestine, the tunica propria below and between villi is invaginated by tubular glands. The enzyme-secreting glands are known as the **crypts of Lieberkühn.** *Only in the duodenum* do such crypts pierce through muscularis mucosae to enter the submucosa (Fig. IX-63). These submucosal glands of Brunner empty into the bases of the crypts of Lieberkühn where they seem to add a mucous material to the enzyme products of the crypts. The following Table IX-4 details the regional histological differences encountered along the length of the intestinal tract.

The Jejuno-ileum: The length of the small intestine, when excised from the body, is usually quoted as being about 22 feet for the adult, but ranges of 31 feet 10 inches to 15 feet 6 inches have been reported. It may be hard to understand how these measurements come about when one realizes that the average tube one swallows for diagnostic purposes in exploring the digestive tract is only 12 to 14 feet in length, and it usually extends the length of the tract. Apparently, in

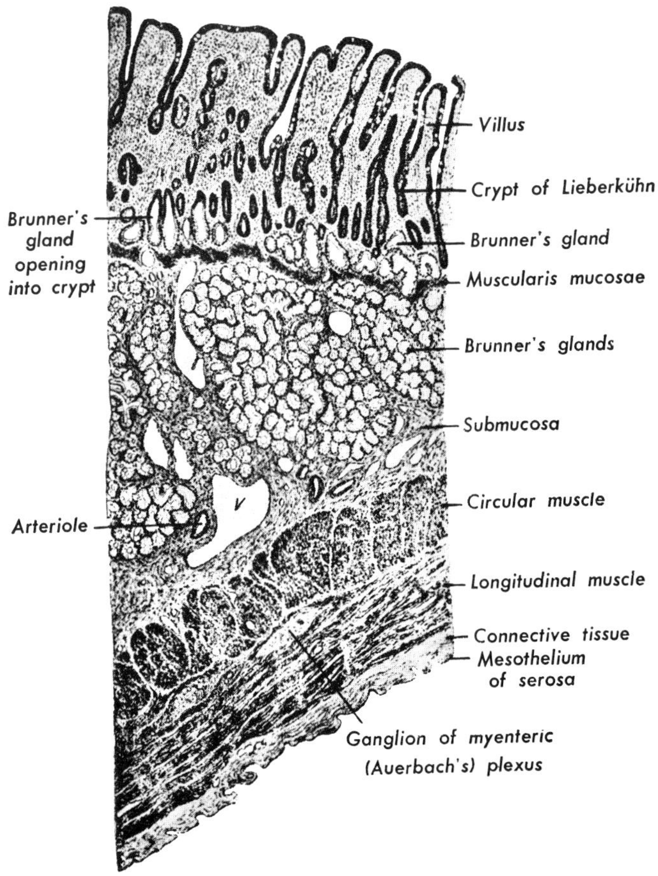

Fig. IX-63. Microscopic longitudinal section through upper duodenum to show crypts of Lieberkühn in tunica propria, Brunner's glands in submucosa. Note: (1) opening of ducts of Brunner's glands into base of crypts by piercing muscularis mucosae; (2) nests of ganglion cells in Auerbach's myenteric plexus. (From W. E. Copenhaver, *Bailey's Textbook of Histology*, 15th ed., Williams & Wilkins, Baltimore, 1964.)

Table IX-4. A Comparison of the Internal Anatomy of Parts of the Intestine

	Duodenum	Jejunum	Ileum	Appendix	Colon
Plicae circulares	Medium height	Very high, thin, and branching	Low	Absent	Absent
Villi	Short and broad	Tall and slender	Fairly tall and broad	Absent	Absent
Crypts of Lieberkühn	Contain zymogenic cells, goblet cells, Paneth's cells, and argentaffin cells			Chiefly goblet cells	Chiefly goblet cells
Brunner's glands	In submucosa	Absent	Absent	Absent	Absent
Lymph nodules	Few, mostly solitary, in tunica propria		In aggregates (Peyer's patches) on antimesenteric border in mucosa	Numerous, generally, solitary, but many aggregate	Solitary when present

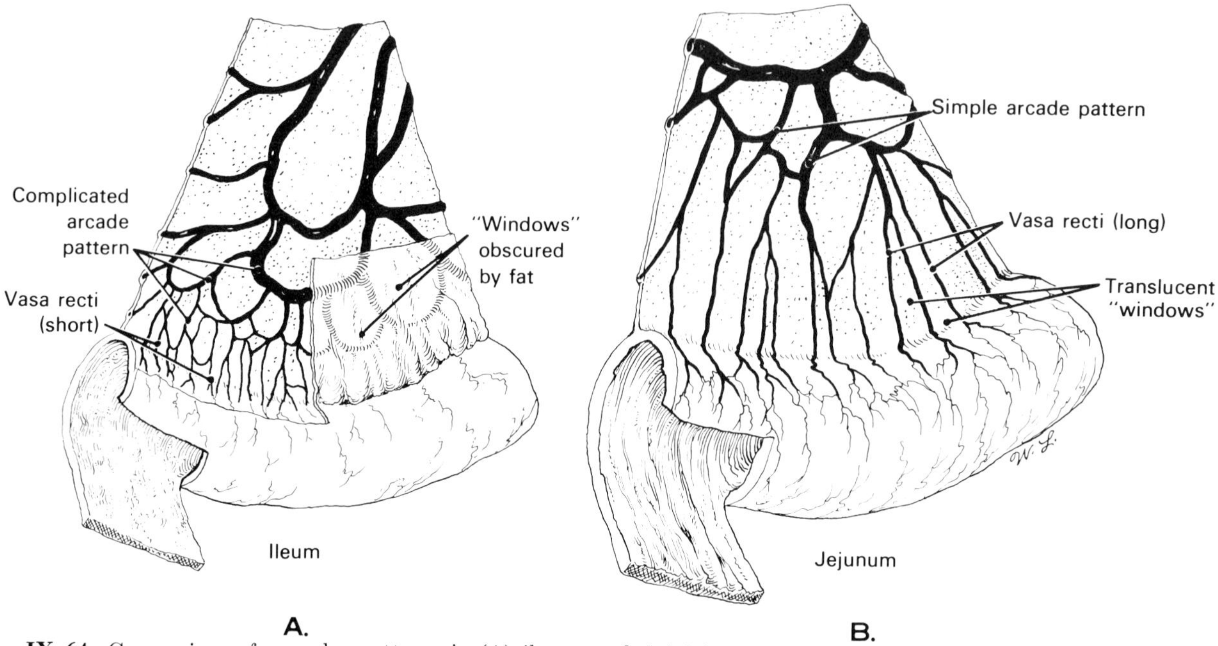

Fig. IX-64. Comparison of vascular patterns in (**A**) ileum, and (**B**) jejunum. Note also differences: (1) in length of arterial recti; (2) amount of mesenteric fat; (3) diameter of tubes.

the living subject, there is a variable length depending upon bowel tone. The whole digestive tube is attached by the mesentery to the back body wall along an oblique line that is only about 8 inches in length. The line of attachment begins at VL_1 or VL_2 on the left side and terminates on the right at about the sacroiliac joint. Its peripheral, enteric border is, however, a massive, folded fan attaching from the duodenal-jejunal junction to the ileocecal junction. About the only blood vessel that runs along the base of attachment is the ileocolic branch of the superior mesenteric artery. The stems of the superior mesenteric vessels occupy the base of the mesentery momentarily as they emerge between duodenum and pancreas before they spread out in the fan portion.

The jejunum will be larger in diameter than the ileum and thicker because of its millions of plushy villi. But it is difficult to say exactly where jejunum becomes ileum, even histologically at times. A rule of thumb is to examine the mesenterial attachment. The **vascular arcades** are less complicated in the jejunal mesentery; the arteriae rectae stemming from the arcades to the gut border are longer; and the mesenteric spaces between the arteriae rectae are more translucent and contain less fat than the ileal mesentery (Fig. IX-64). These criteria can be used to advantage to determine the nature of a loop of bowel when it is pulled up through a small incision, or when you uncoil the small bowel in the dissection laboratory.

Blood Supply of the Small Intestine: **Arterial** blood supply comes entirely from the **superior mesenteric artery,** an unpaired branch of the aorta arising just below the coeliac trunk (Fig. IX-65A). The vessel emerges behind the neck of the pancreas and crosses the duodenum which it clamps down against the posterior body wall like a nut held in the jaws of a nutcracker. The first branches are recurrent ones, the **inferior pancreatico-duodenal arteries** (Fig. IX-56B) contributing to the arcades along the anterior and posterior aspects of the duodenal curvature. The next branches that come off on the right side are the **middle colic, right colic,** and **ileocolic arteries** (Fig. IX-65A). Thus, while the superior mesenteric artery is the sole supply to the small intestine, these three, right-sided, sizable offshoots also supply appendix, cecum, ascending colon, and most (sometimes all) of the transverse colon (Fig. IX-65). The main stem of the superior mesenteric artery continues downward and to the left in the mesentery to provide the arcades and arteriae rectae to the jejuno-ileum. (See the short, cut-off twigs in Fig. IX-65A).

The **venous drainage** is *via* the superior mesenteric vein and gathers chiefly toward one main trunk, the **ileocolic branch.** This latter vessel traverses the base of the mesentery in company with its artery. It receives all the drainage from the small intestine on its left side, and near its upper, right end it receives the right and middle colic veins (Fig. IX-65B). The superior mesenteric vein is the largest contributor to the portal vein.

Lymphatic Drainage of Small Intestine: At the level of capillary supply, the central reticular core of each intestinal villus contains a blood capillary network and a single, blind-ending, lymphatic capillary (Fig. IX-66). The latter is called a "central lacteal," for after the absorption of a fatty meal most fat is returned *via*

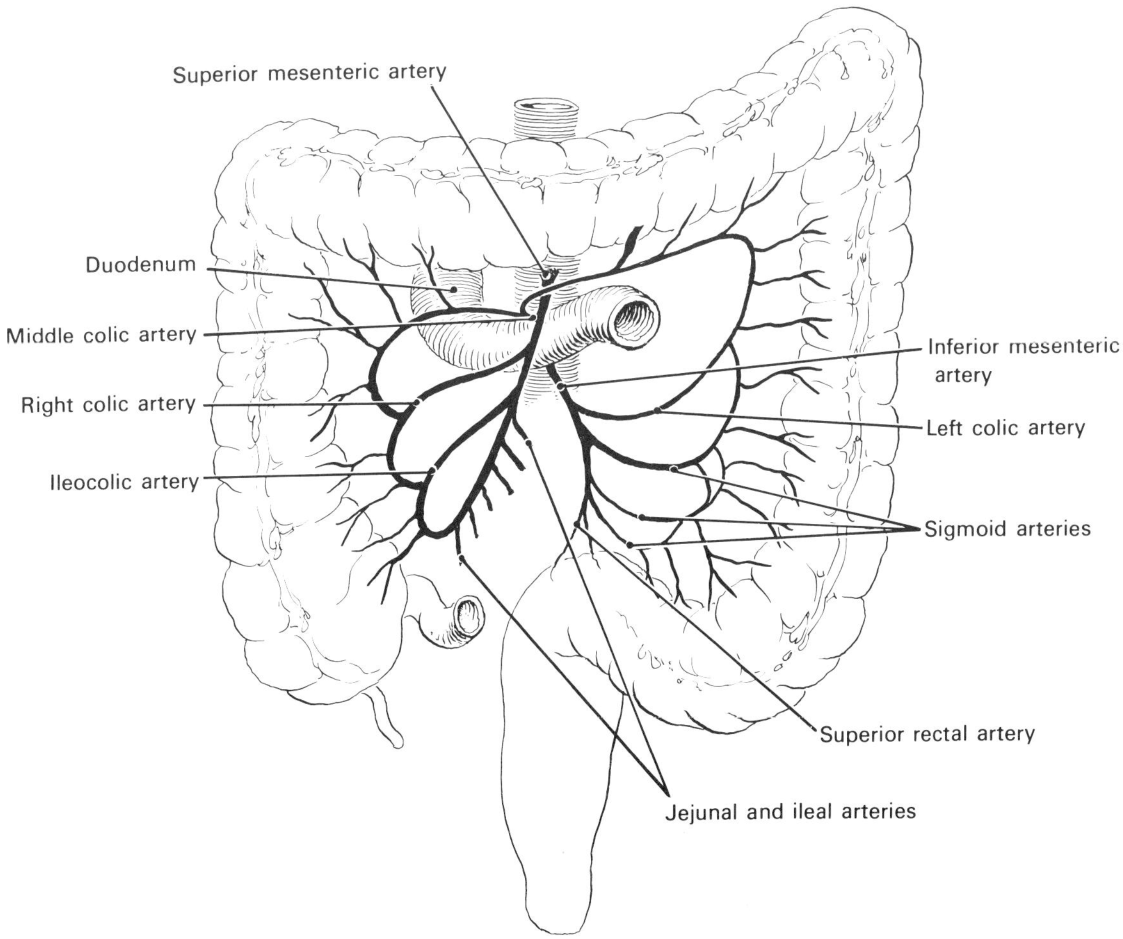

Fig. IX-65. Blood supply of intestinal tract. **A.** Arterial supply to large and small intestine. Besides giving a supply to the jejunum and the ileum, the superior mesenteric artery supplies cecum, appendix, ascending, and most of the transverse colon. The remainder of the tract is shown being supplied by the inferior mesenteric artery.

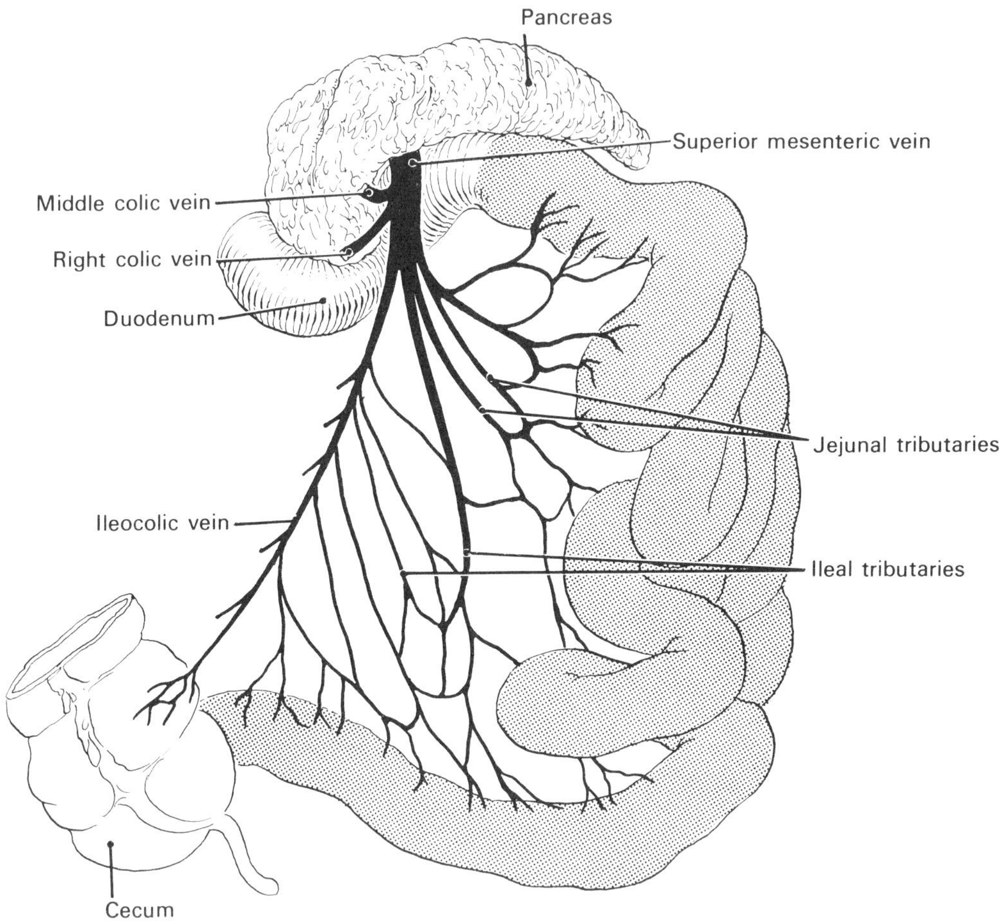

Fig. IX-65. Blood supply of intestinal tract. **B.** Venous drainage of small intestine.

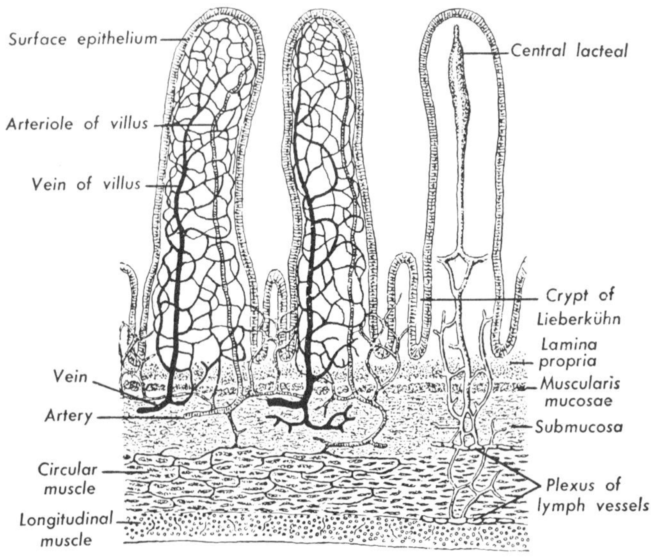

Fig. IX-66. Internal lymphatic drainage of wall of small intestine. Each villus is supplied with a vascular and lymphatic plexus even though they are shown in separate villi here for didactic purposes. Note position of central lacteal in tunica propria of villus and its connection with lymphatic plexus of deeper layers. (From W. E. Copenhaver, *Bailey's Textbook of Histology*, 15th ed., Williams & Wilkins, Baltimore, 1964.)

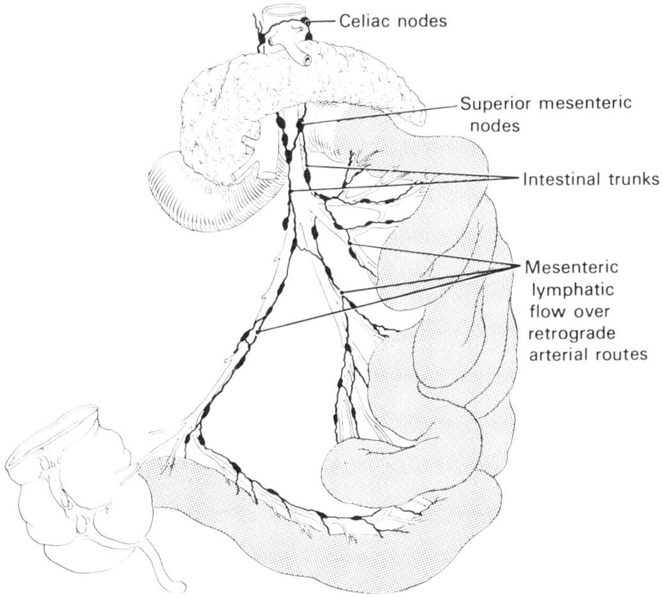

Fig. IX-67. Lymphatic drainage of small intestine toward coeliac nodes.

lymphatic channels which appear milky as a consequence. To appreciate this, you might over-feed the family cat with heavy cream and then examine the intestinal mesentery visually. You had best be adept at surgical embroidery, or you might lose the cat in the process and end up in the dog-house. Nevertheless, **lymphatic drainage** begins in the villous core within central lacteals. Collecting lymphatic channels form as mucosal, submucosal, and serosal plexuses which lead to nodes in the mesentery along the course of the ileocolic vessels, and these, in turn, to nodes at the base of origin of the superior mesenteric artery from the aorta. The latter nodes also receive lymph from the pancreas and the ascending and upper parts of the descending colon. The course is then to coeliac nodes and, eventually, into the cisterna chyli of the thoracic lymph duct (Fig. IX-67).

Innervation of Small Intestine: The vagi and sympathetic postganglionic fibers reach the small intestine as shown in Figure IX-68. The sympathetic postganglionic neurons lie in the superior mesenteric ganglion which surrounds the stump of origin of the artery. This ganglion may be fused, as an overgrowth, to the coeliac ganglion which lies close by. The sympathetic preganglionic supply derives from the lesser splanchnic nerves. The vagal preganglionics mingle with sympathetic fibers along the periarterial routes, but, of course, they are going to their own terminal ganglia of Meisner and Auerbach within the intestinal walls. Visceral sensory fibers return over the plexus *via* the splanchnic route to cell bodies in dorsal root ganglia T_{10} to T_{12}.

The Upper Gastrointestinal and Small Bowel Series—Their Normal Roentgen Anatomy: During the clinical years, students of medicine show a keen interest in gastrointestinal roentgenology. These procedures, in competent hands, are quite accurate in detecting many intrinsic abnormalities of the esophagus, stomach, and small bowel, as well as extrinsic lesions displacing these structures. It is also helpful in diagnosing disorders of motor function which may represent manifestations of systemic disease.

In order to fully appreciate how little detailed anatomy of the gastrointestinal tract may be seen in the cadaver, it would be to your advantage to observe these procedures on patients in the Department of Radiology.

All examinations of the gastrointestinal tract are performed with fluoroscopic guidance, supplemented by spot filming, cineradiography, or television tape recording made during the procedure. Additional radiographs are also obtained after fluoroscopy is completed. Fluoroscopy enables the radiologist to study the dynamics of the GI tract as well as its morphology.

The referring physician and the radiologist should always consult each other about the patient before the examination is performed. The radiologist will then be aware of suspected or known problems (*i.e.,* prior surgery and the results of previous examinations made elsewhere) that might be encountered, which may require modification of the standard procedure. In addition, however, the radiologist is well advised to take his own history, palpate the abdomen of his patient, and review the clinical chart.

1. Procedure: The patient must fast for 12 to 14 hours prior to the examination. If the small bowel is to be examined, some physicians give their patients a cathartic to cleanse the cecum. This greatly speeds the transit time of the *barium meal* (a mixture of chemically pure barium sulfate and water) through the small bowel.

Radiologists differ in their method of performing the GI and small bowel series. One such method is presented. Before introducing any contrast medium into the GI tract, we prefer to obtain a plain (scout, KUB) film of the abdomen (p. 633). The examination is begun with the patient standing erect behind the fluoroscope. Since the X-ray tube is beneath the fluoroscopic table, the positioning of the patient described here will be referable to the fluoroscope which houses the film cassette or film camera. The terminology is identical to that learned earlier in Chapter IV.

If there are no complaints ascribable to the larynx and esophagus, then the radiologist first directs his attention to the cardia. Gravity and peristalsis push the bolus of barium along the esophagus in about 1 second in the normal patient. The contrast medium passes through the cardiac orifice into the body of the stomach where it forms a triangular-shaped shadow whose apex is directed inferiorly (Fig. IX-69A). Gradually, the barium descends between the longitudinal rugal folds near the lesser curvature of the stomach, the so-called gastric (ventricular) canal, toward the pylorus (Fig. IX-69B). This process is termed *canalization* of the stomach.

With his lead-impregnated glove, which is worn to protect the hands against radiation hazard, the radiol-

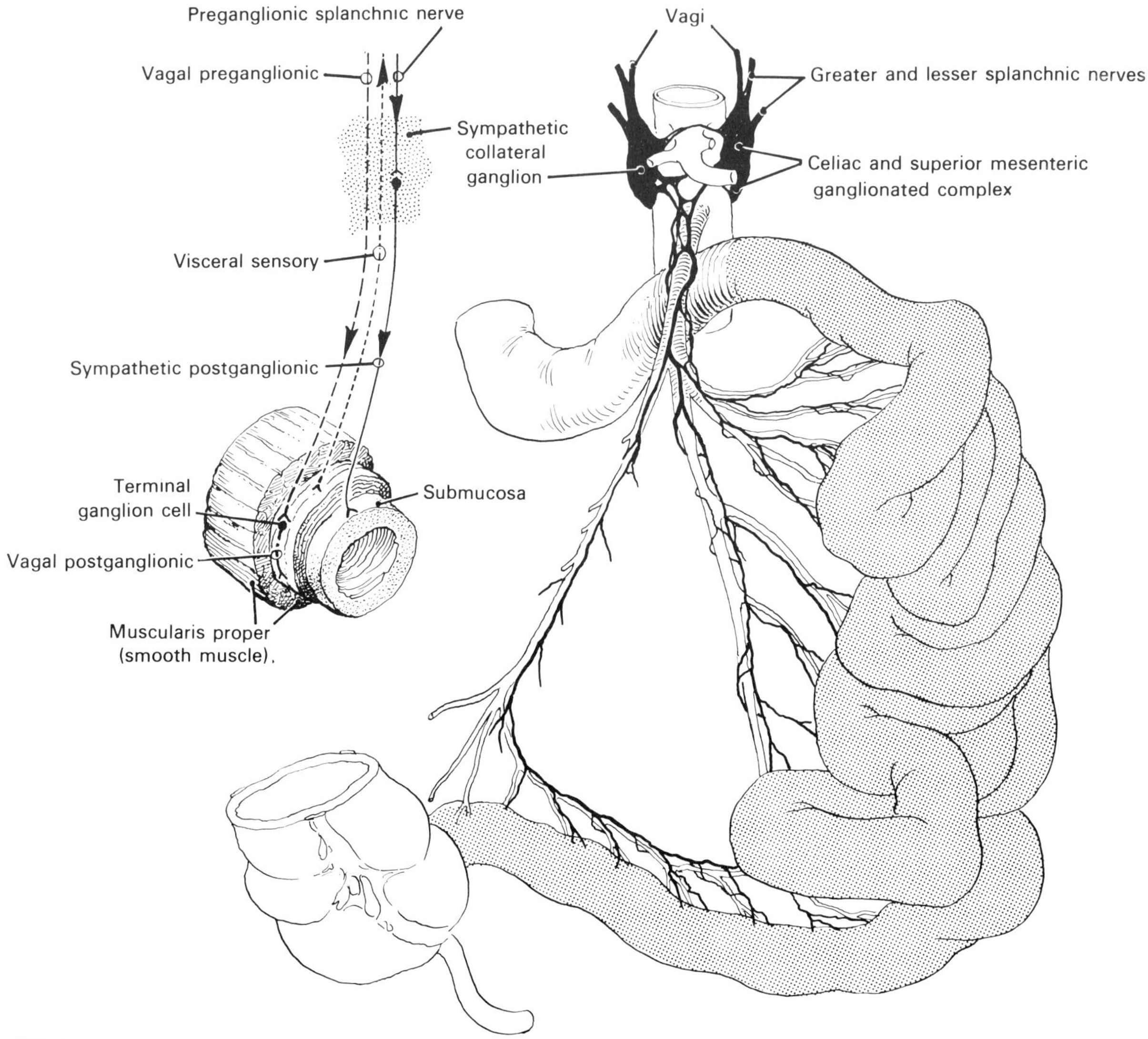

Fig. IX-68. Autonomic innervation of the small intestine. Sympathetic (lesser splanchnic) preganglionics are shown synapsing in superior mesenteric ganglion. Postganglions follow a periarterial route to gut wall. Preganglionic vagal fibers are shown passing through the sympathetic collateral ganglion to reach terminal ganglion cells of Meissner's and Auerbach's submucous and myenteric plexuses.

ogist then massages the small quantity of barium between the elevated **rugal folds** (Fig. IX-69B) of the stomach. Attention is directed to any alteration in the normal pattern of these folds (p. 586), to ulcerations, filling defects, and to the pliability of the gastric wall.

The patient is then directed to drink the remaining barium in his cup (total amount 6 to 8 oz.) in order to study the swallowing function and esophagus. With the stomach now distended with barium, further search is made for abnormalities. By now some peristaltic activity should have pushed some of the barium into the duodenal bulb (cap) (Fig. IX-69B); however, if this has not occurred, then some pressure applied over the antrum will usually force some of the contrast medium into the duodenum. The shape of the patient's stomach will determine which position of the patient—PA, RAO, or LAO—will be best to visualize the bulb. Palpation of the entire duodenum is important, again noting its shape, mucosal pattern, ulcerations, filling defects, and pliability of its walls.

Once the patient has been examined thoroughly in the erect position, the fluoroscopic table is returned to its horizontal position. The systematic search is again performed with the patient in supine, prone, left, and right oblique positions. Invariably some air is swallowed with the barium meal and this is of great help in obtaining a double contrast examination. The air fills the lumen of the stomach or duodenum and the barium particles coat its walls. Some abnormalities are best visualized with such a double contrast technique. Finally, the patient is placed in the left posterior oblique projection. An additional swallow of barium is given,

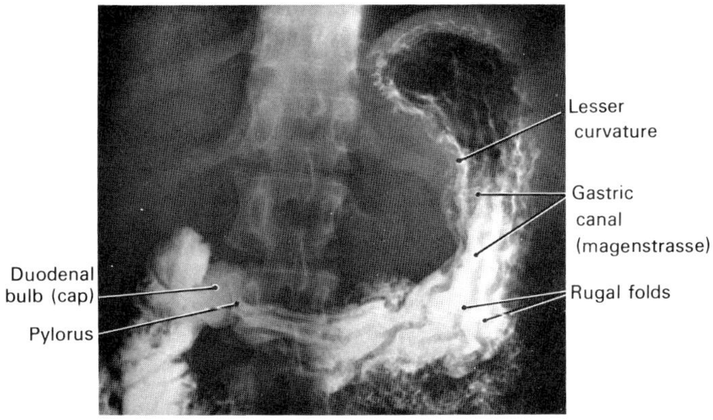

Fig. IX-69. Radiograph of the stomach in frontal projection. **A.** The first few swallows of the barium meal form a triangular shadow in the body of the stomach. **B.** The area of the gastric canal or magenstrasse.

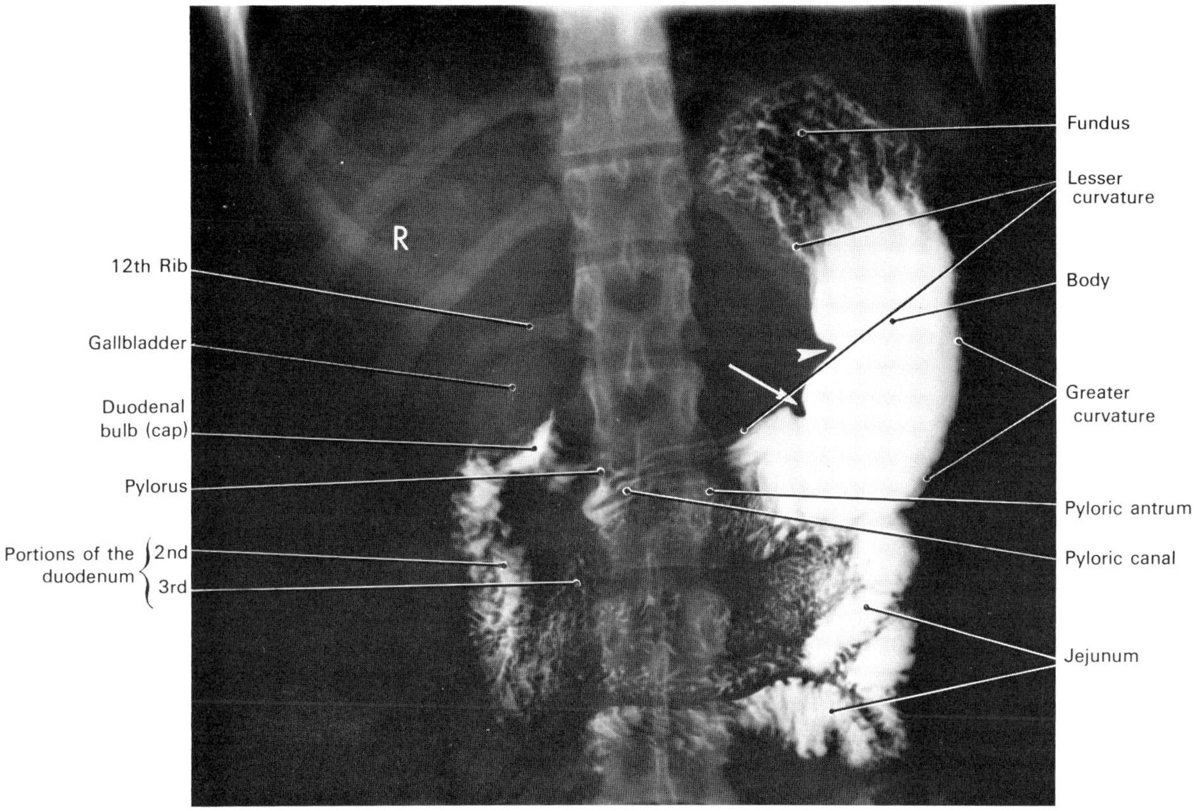

Fig. IX-70. Radiograph of the stomach and proximal small bowel in frontal projection. Arrow indicates the angular notch (incisura angularis). Arrow head indicates a peristaltic wave on the lesser curvature.

and as the bolus nears the cardioesophageal junction, the patient is directed to perform the Müller maneuver (exhaling and then trying to inhale against a closed glottis). After ingesting another swallow of barium, the Valsalva maneuver is performed (after taking a deep breath, the patient forces expiration against a closed glottis). Many other maneuvers and methods, too many to be discussed here, are used to diagnose hiatal hernia, esophageal varices, esophageal reflux, foreign bodies, etc.

At the completion of fluoroscopy, the radiologist directs his technician about other films to be made. The follow-up radiographs are made with the overhead X-ray tube, and therefore, the positions of the patient now refer to the cassette placed beneath the table top. Our routine follow-up examination usually consists of: (1) supine AP (Fig. IX-70); (2) prone PA projections which permit excellent visualization of the greater and lesser curvatures of the stomach, as well as the duodenal loop as it sweeps about the head of the pancreas;

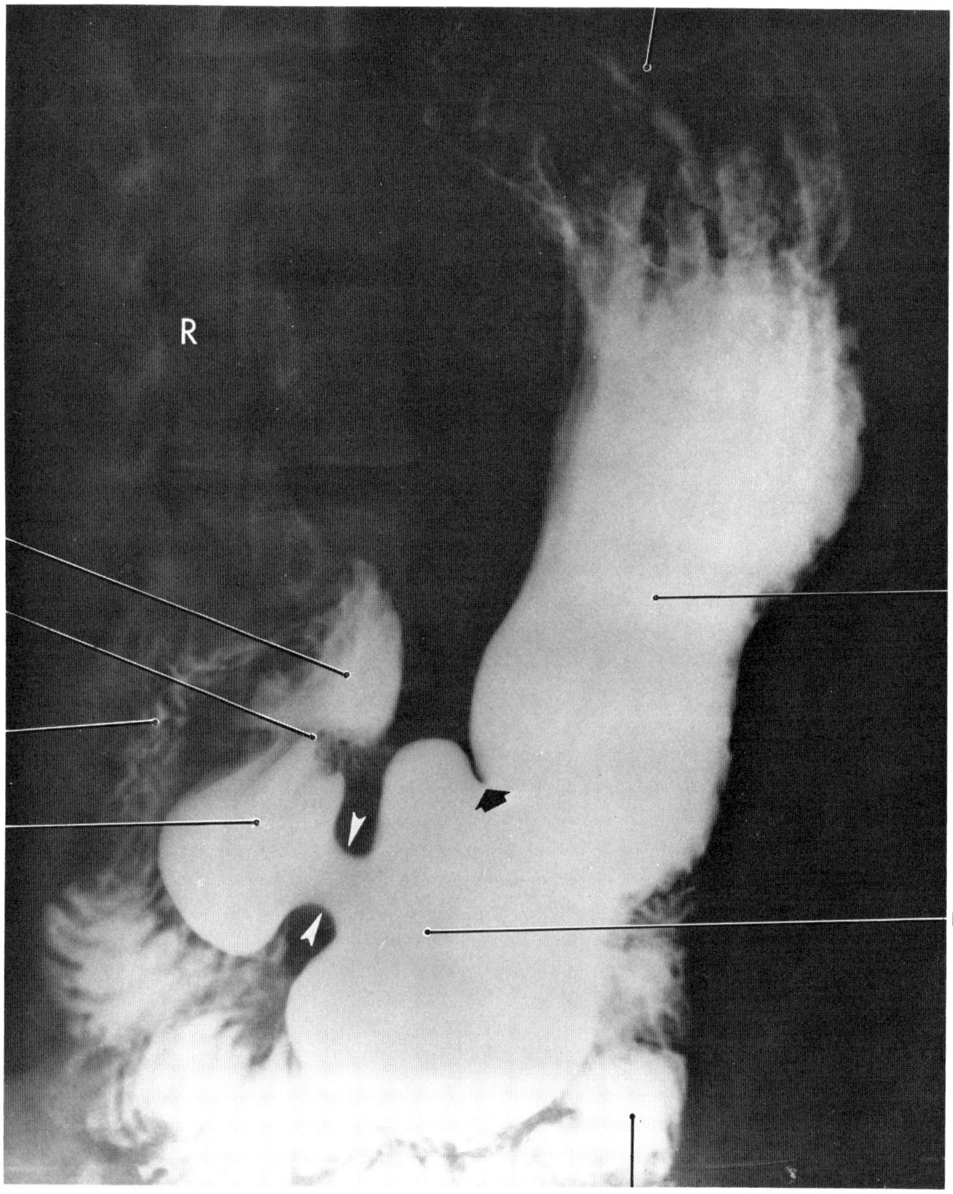

Fig. IX-71. Right anterior oblique projection of the barium-filled stomach and proximal small bowel. Arrow heads (white) show peristaltic wave at approximate junction of the pyloric antrum and canal. Black arrow indicates angular notch.

(3) 45° right anterior oblique which is useful in evaluating the antrum and duodenal bulb (Fig. IX-71); (4) right lateral projection which shows the duodenal bulb, second portion of the duodenum, and the anterior and posterior walls of the stomach (Fig. IX-72).

2. Roentgen Anatomy: A. STOMACH AND DUODENUM: The anatomic divisions of the stomach have been indicated for you in the illustrations (Figs. IX-70 to 72). Realize that the reference points for determining those divisions are all on the lesser curvature of the stomach, and they are rather arbitrarily drawn to the greater curvature (Fig. IX-53).

The fundus lies high in the left upper quadrant beneath the left dome of the Diaphragm. The entrance of the esophagus into the stomach (cardiac opening), which is the junction of the greater and lesser curvatures of the stomach, may at times appear puckered (Figure IX-72). In frontal view, the "radiographic" greater and lesser curvatures more or less parallel each other and at the angular notch (incisura angularis) both bend medially and end at the pylorus. We mention the "radiographic" curvatures, since these differ somewhat from the "anatomic" curvatures. Actually, the greater curvature that we are observing (Fig. IX-70) is more of the ventral wall of the stomach, rather than the actual point of attachment of the greater omentum.[3]

At times, normal structures may be seen to impinge upon the contours of the stomach. An example is shown in Figure IX-70 which is a supine AP projection. In

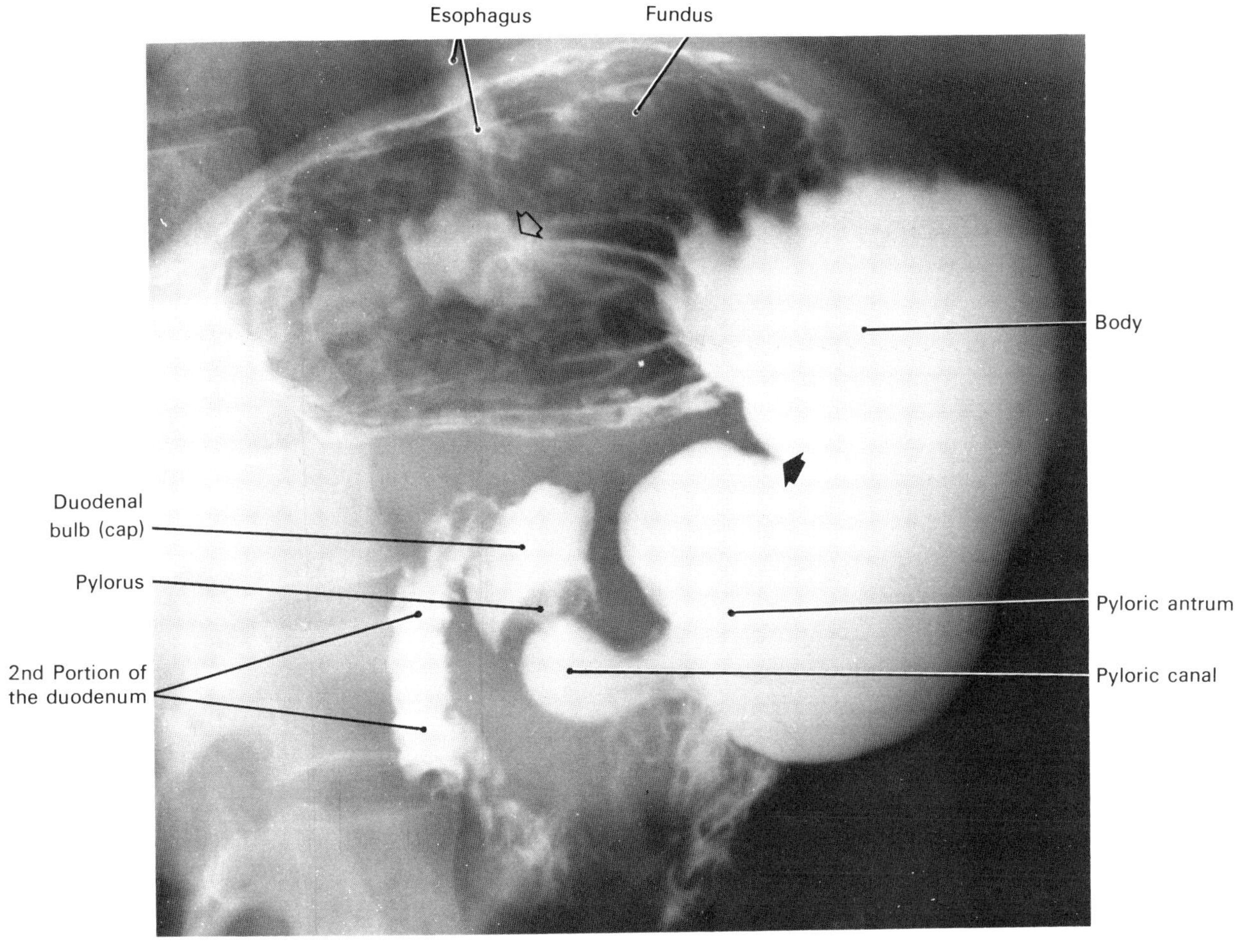

Fig. IX-72. Right lateral projection of stomach and proximal small bowel filled with barium. Black arrow indicates angular notch. Open arrow shows the puckered appearance at the cardio-esophageal junction.

this view, the pyloric portion of the stomach is draped over the vertebral column, and therefore, it is not completely filled with contrast medium. At times, you will find that the spleen or the splenic flexure of the colon leave their imprint upon the fundus and body of the stomach laterally.

The duodenal bulb (cap) is typically triangular-shaped, with a flat or concave base resting upon the pylorus. Its two sides converge toward the second portion of the duodenum. The duodenal cap is best viewed in the right anterior oblique or right lateral projections (Figs. IX-71 and 72). The second stage of the duodenum begins where the herring bone or feathery pattern of the plicae circularis is seen. Note the second part of the duodenum descending vertically to the right of the vertebral column (Fig. IX-70). After some 3 inches, it bends medially and cranially to become the third portion. The most distal inch of the duodenum, before its attachment at the ligament of Treitz, is the fourth portion. All of these divisions are rather arbitrary on the radiographs. Earlier, we mentioned that closely applied structures may imprint the stomach. This is also true for the duodenum. Figure IX-70 shows the gall bladder impressing upon the greater curvature of the duodenum. Enlargement of the head of the pancreas, *i.e.*, tumor, may widen the duodenal loop and in so doing flatten the plicae circularis medially.

Let us now turn our attention to the rugal folds (Fig. IX-73) of the stomach and doudenum. These folds are quite constant in the fundus, body, and pyloric portions of the stomach. They are formed by the mucosa, muscularis mucosa, and the submucosa of the gastric wall, while in the duodenal cap (bulb) they are formed by the mucosa and submucsoa.

In the fundus, these rugae are irregular in size and orientation, thereby forming a mosaic pattern. Along the gastric canal (Magenstrasse), the folds are longitudinally oriented and parallel. Other folds in the body of the stomach nearer the greater curvature are more transverse and wavy in their appearance. Within the pyloric portion of the stomach, the rugal folds become thinner. They course in a direction parallel to the greater and lesser curvatures toward the pylorus. Within the duodenal cap, the folds may be parallel to those of the pyloric portion of the stomach (Fig. IX-71) and almost seem to be a continuation of the latter. At other times, the rugae of the duodenal bulb are perpendicular to the greater and lesser curvature (Fig. IX-

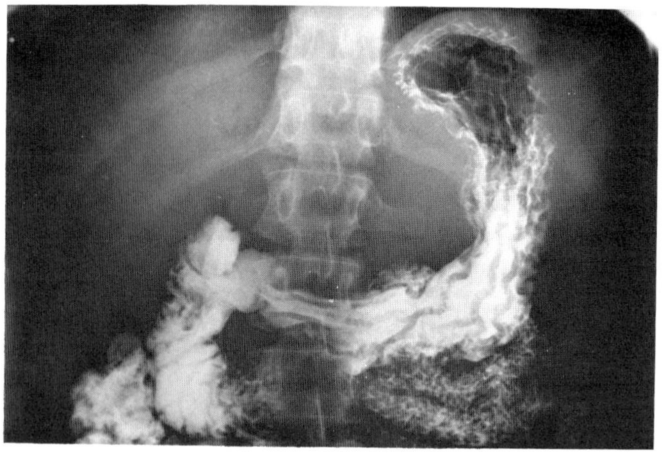

Fig. IX-73. Frontal projection to show the rugal folds of the stomach and proximal small bowel (see text).

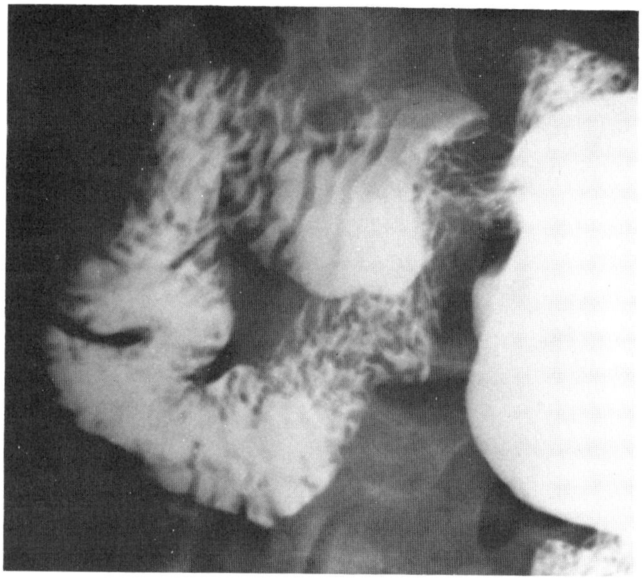

Fig. IX-74. Spot film of the duodenal bulb (cap) and duodenal loop.

74); still other patients show a mixed pattern. At the apex of the duodenal cap, which is the beginning of the second portion of the duodenum, the typical herring bone (feathery) pattern of the plicae circularis is seen (Figs. IX-70 to 72). These circular folds appear radiolucent as they protrude into the barium-filled lumen of the bowel. These circular folds are usually oriented perpendicular to the long axis of the duodenum.

Peristalsis—With thin liquid barium, there is little peristaltic activity perceived in the esophagus in the erect position; however, if a thickened barium paste is given, greater peristalsis is visible. The esophageal peristaltic waves are ring-like contractions which propel the bolus toward the cardia. Once the stomach distends, then circular peristaltic waves (Fig. IX-70) will usually begin in the body of the stomach with some regularity. About three waves per minute sweep toward

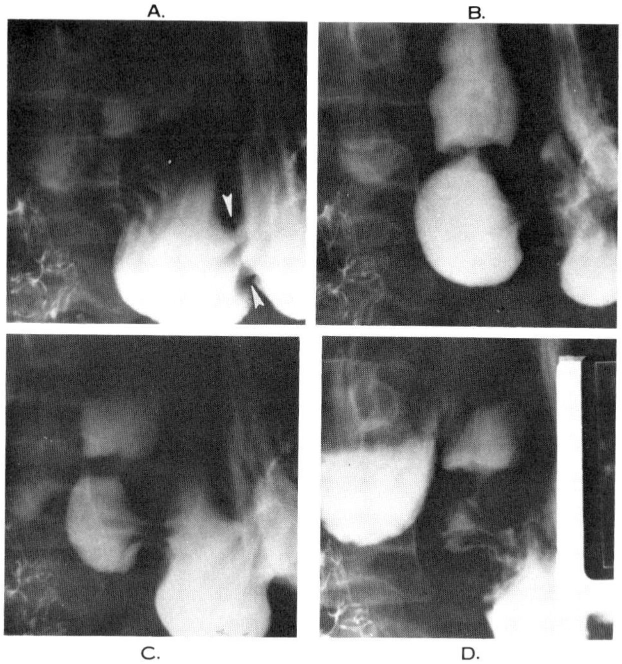

Fig. IX-75. Peristalsis in the pyloric canal and antrum **A.** Peristaltic wave (arrow heads) approaching junction of pyloric antrum and canal. **B.** Muscular flattening of the lesser curvature of the pyloric canal and distention of its greater curvature. **C.** Beginning contraction of the greater curvature of the pyloric canal. **D.** Tubular contraction of the pyloric canal.

the pyloric canal where they cease. In the pyloric canal (Fig. IX-75) emptying begins with muscular contraction along the lesser curvature, thereby flattening this area. Subsequently the greater curvature of the pyloric canal balloons outward and the border zone between the canal and the pylorus constricts down. At this moment, the pylorus is closed. As the pyloric canal then contracts, the pylorus opens and an instant later the contracted area between the canal and antrum relaxes. Most of the contents of the pyloric canal are expelled into the duodenum, although a small amount may be refluxed back into the antrum. With the pylorus again closed, the distended duodenal bulb contracts, propelling its contents into the second portion of the duodenum. Circular contractions within the second part push the contents onward toward the third and fourth portions of the duodenum which have a motor function similar to the mesenteric small bowel.[3]

B. THE MESENTERIC SMALL BOWEL (JEJUNUM AND ILEUM): The mesenteric small bowel is not routinely examined as part of the GI series. Thus, when ordering a GI procedure, the referring physician asks for either an upper GI series, or a GI and small bowel series. Space does not permit consideration of the other available methods of examining the small bowel (retrograde and intubation studies) and therefore only the oral small bowel series will be discussed here.

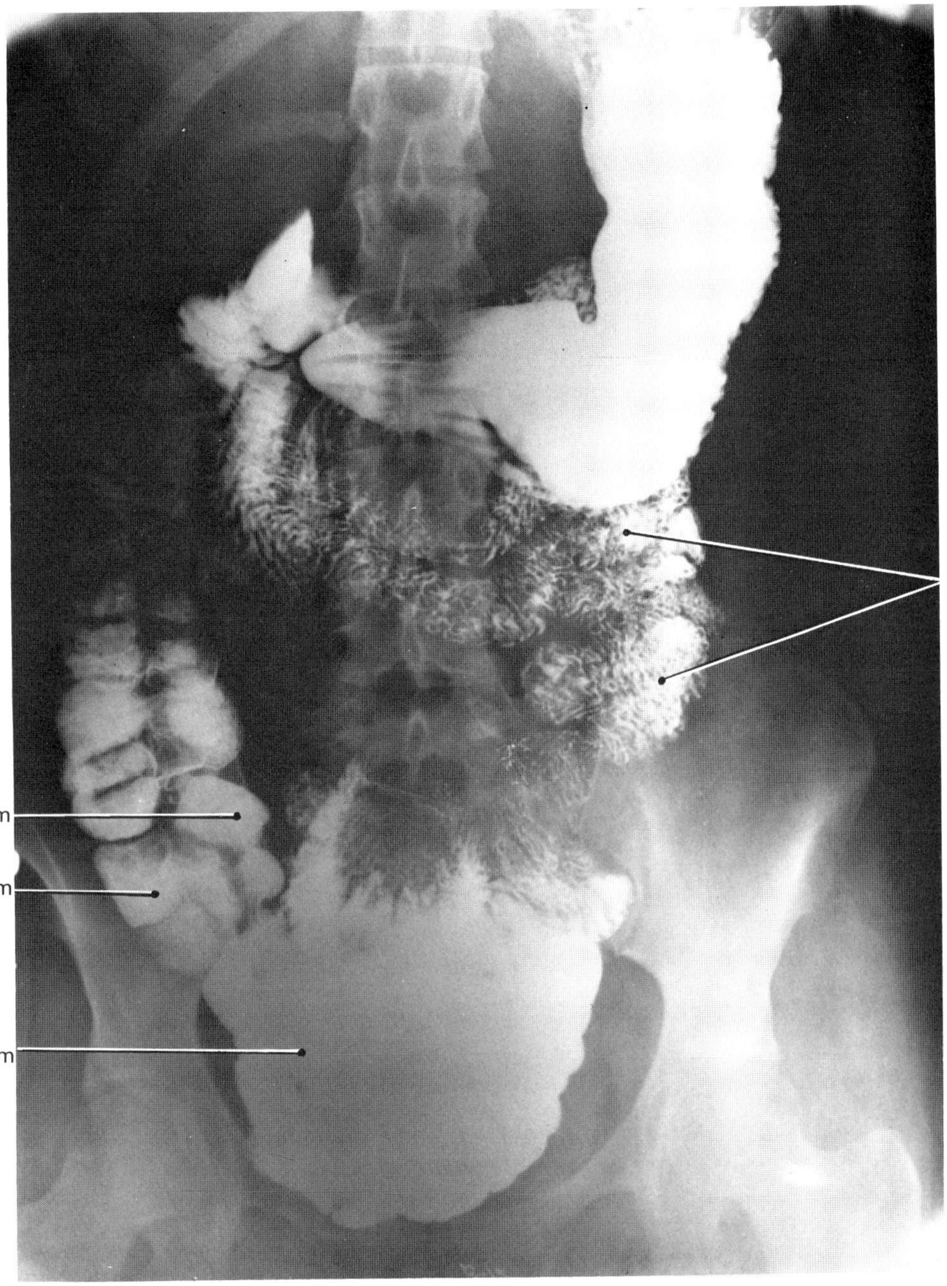

Fig. IX-76. AP projection made during GI and small bowel examination.

Sufficient barium must be given the patient to allow a uniform column to pass through the bowel. We prefer to use 12 to 18 ounces of a colloidal (non-flocculating) barium mixed with water. Such a non-flocculating preparation permits more detailed study of the mucosal pattern of the small bowel. Films may be obtained at 15- to 30-minute intervals in PA or AP projections. As the radiographs are developed, the radiologist views them for abnormalities. If a lesion is detected, then the patient is fluoroscoped and appropriate films made of the abnormality. Some radiologists fluoroscope the small bowel at 30 to 60-minute intervals. It is also advantageous to spot film the terminal ileum and ileocecal valve (p. 597) once the column has entered the cecum, since these areas may not be well visualized on routine films.

Since the position of the individual loops of mesenteric small bowel are so variable, one can only generalize as to their normal anatomic position. In general though, the proximal jejunum is found in the left upper quadrant; the distal jejunum in the left lower quadrant; the proximal ileum in the right midabdomen; the distal ileum in the pelvis; and the ileocecal valve in the right iliac fossa. The terminal ileum rises almost vertically out of the pelvis to enter the cecum (Figs. IX-76 and 78).

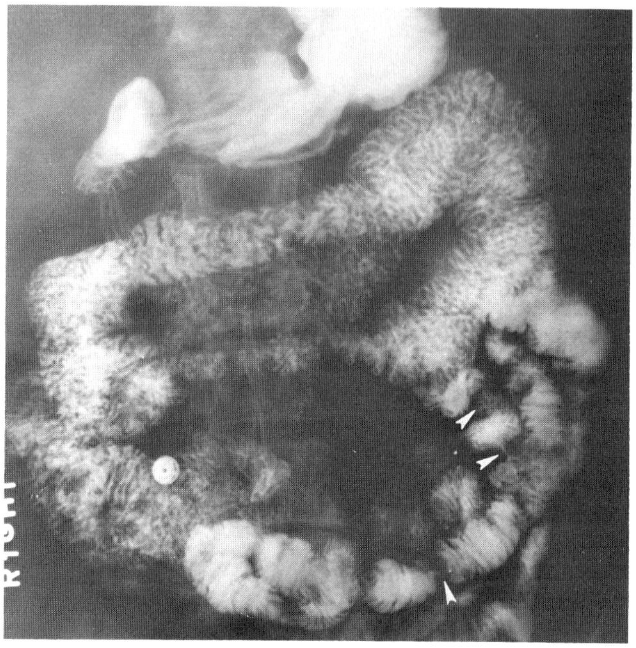

Fig. IX-77. AP projection during small bowel examination. White arrow heads show some areas of peristaltic contraction.

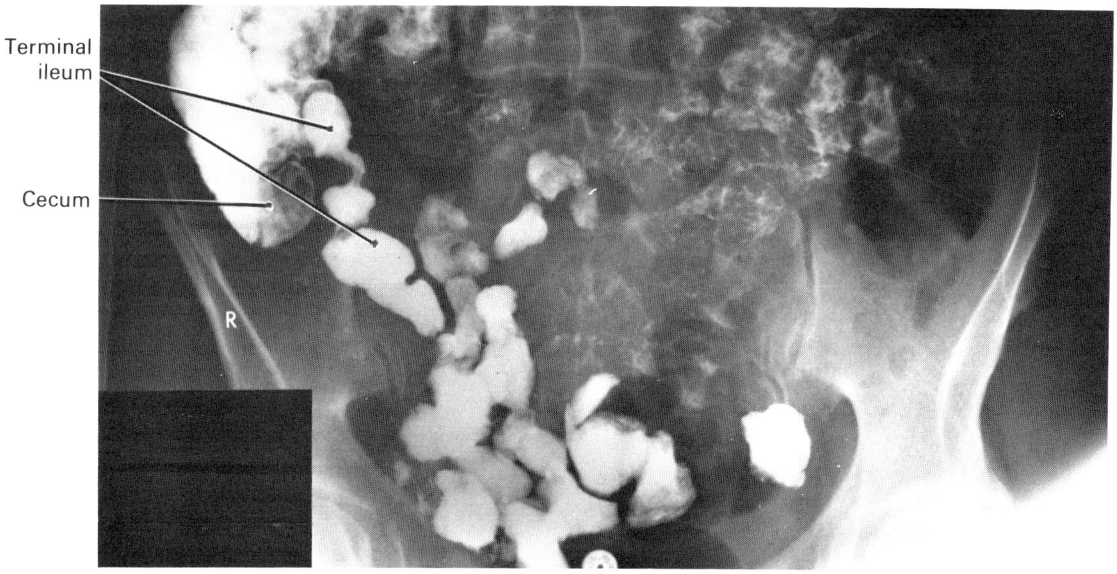

Fig. IX-78. AP projection showing some peristalsis in the terminal ileum which ascends to reach the cecum at the ileocecal valve. Note the sparsity of plicae circularis in the ileum.

Examine closely the radiographs in Figures IX-76 and 77. The jejunal loops have a nearly horizontal course and are joined with sharp, hairpin curves, while the loops of the ileum are arranged more nearly vertically and joined with wider curves. Note that the mucosal pattern of the proximal jejunum closely approximates the pattern seen in the duodenum beyond the bulb. The transverse radiolucent folds (plicae) are made visible by the barium surrounding them. The folds are quite close together and can be seen to transverse the entire width of the lumen of the jejunum. As the column of barium passes into the jejunum, these folds remain sharply visible, whereas, in the ileum the folds are typically obliterated as the lumen is distended. Farther distally in the jejunum and in the ileum, the mucosal folds become smaller, and at times the ileum may appear normally quite smooth. Examine Figure IX-77 and notice that as a wave of peristalsis passes through, the transverse folds become longitudinally oriented.[4]

The caliber of the small bowel decreases from proximal jejunum to terminal ileum normally. Unless there is a marked deposition of fat on the serosa of the bowel, it is not possible to see the thickness of the bowel wall. However, if two distended adjacent loops are visualized, one can roughly judge their combined wall thickness which does not usually exceed 2 mm.

We have considered the ileocecal valve with the colon on page 597.

The Large Intestine: *General Structure:* The average, male, large bowel is slightly longer than in the female, about 4 feet, 8 inches in length. But measurements of 3 feet, 3 inches to 6 feet, 6 inches have been reported. The colon begins at the ileocecal junction. The entire large bowel is divided regionally into cecum and appendix which project downward on the right side as blind-ending sacs; ascending, transverse, and descending colon in the abdomen; sigmoid colon continuing into the pelvis; and rectum and anal canal within the pelvis. Size alone does not differentiate the large from the small intestine for the large bowel can vary greatly in diameter. Its outer coat is, however, distinctively marked. First of all, the outer longitudinal muscle thickens regionally as three, separated strands, the **taeniae.** If you follow the anterior of these, the taenia libre, downward over the cecum it should lead you to the appendix, provided someone hasn't spirited the appendix away already. The intestinal coat is further distinguished by its dilated sacculations or **haustra.** These dilatations probably reflect unequal growth locally in the circular coat of the muscularis. Between the haustra, the colon wall (all four layers) is thrown into folds. These crescent-shaped **semilunar folds (plicae semilunaris)** may be seen on the interior of the colon (Fig. IX-79). The large bowel is also festooned with dripping fatty tags, the **appendices epiploicae.** These are peritoneal saccules stuffed with adipose tissue.

In the cadaver we usually look for the cecum in the iliac fossa but in many living, erect, individuals it can hang down into the true pelvis. Like the rest of the colon, it has its taeniae and haustrae (Fig. IX-79). The mesenteric attachments here may form recesses or fossae which are similar to the fossae around the ligament of Treitz. These fossae were once thought to form internal pockets where loops of bowel could herniate internally and become strangulated. They are no longer considered to be hazardous loci. The superior ileocecal fossa lies above the ileocecal junction, between the lieum and a peritoneal fold containing the **anterior cecal artery.** The fold (**vascular fold of cecum** or **superior ileocecal fold**) is the important landmark, not the fossa, since the fold leads to the vessel. Below the ileum, between the mesentery of the appendix and the **inferior ileocecal fold** (or bloodless fold of Treves), is the ileocecal fossa. The important structure in this area is the meso-appendix since it contains the **appendicular artery.**

At the ileocecal junction in the interior of the cecum will be found two transverse folds that appear to make up an **ileocecal valve** (Fig. IX-79C). They are formed by the protruding lips of the ileum, composed here of mucosa, submucosa, and the circular muscle fibers of the ileum.

The **appendix** is an area of asymmetrical growth of the cecum. It is about 3½ inches long though some appendices have been found twice this length. As a rule, its origin from the cecum is posteromedial. All sorts of diagrams have been drawn about the exact location of the appendix, but it is as mobile as the cecum, so, unless it is bound down by adhesions, its exact locus is of little importance except in the case of the malrotated gut where it may be in the left lower abdominal quadrant rather than in the right. Its congenital absence is rare; so is its duplication which is more often related to a duplication of the entire colon. The general blood supply to cecum and appendix follows that shown in Figure IX-79B, but there are many variations.

The ascending and descending parts of the colon are usually "retroperitoneal," though the ascending colon, especially in its lower segments, can sometimes be more free than the descending colon. There are **paracolic gutters** along the lateral sides of the ascending and descending colon. In the supine individual fluids from the pelvis can traverse these gutters, and, particularly, on the right side above the hepatic flexure of the transverse colon, these fluids can gravitate to a recess, or **pouch of Morrison.** The latter is a large space between Diaphragm and liver above the upper pole of the kidney. Infections from a ruptured appendix, for example, could travel this route to form a subphrenic abscess. There is an old admonition in medicine: if abdominal infection cannot be located readily, look in the pouch of Morrison on the right side.

The ascending colon is about 15 cm. long. As it ascends along the Quadratus Lumborum and above the

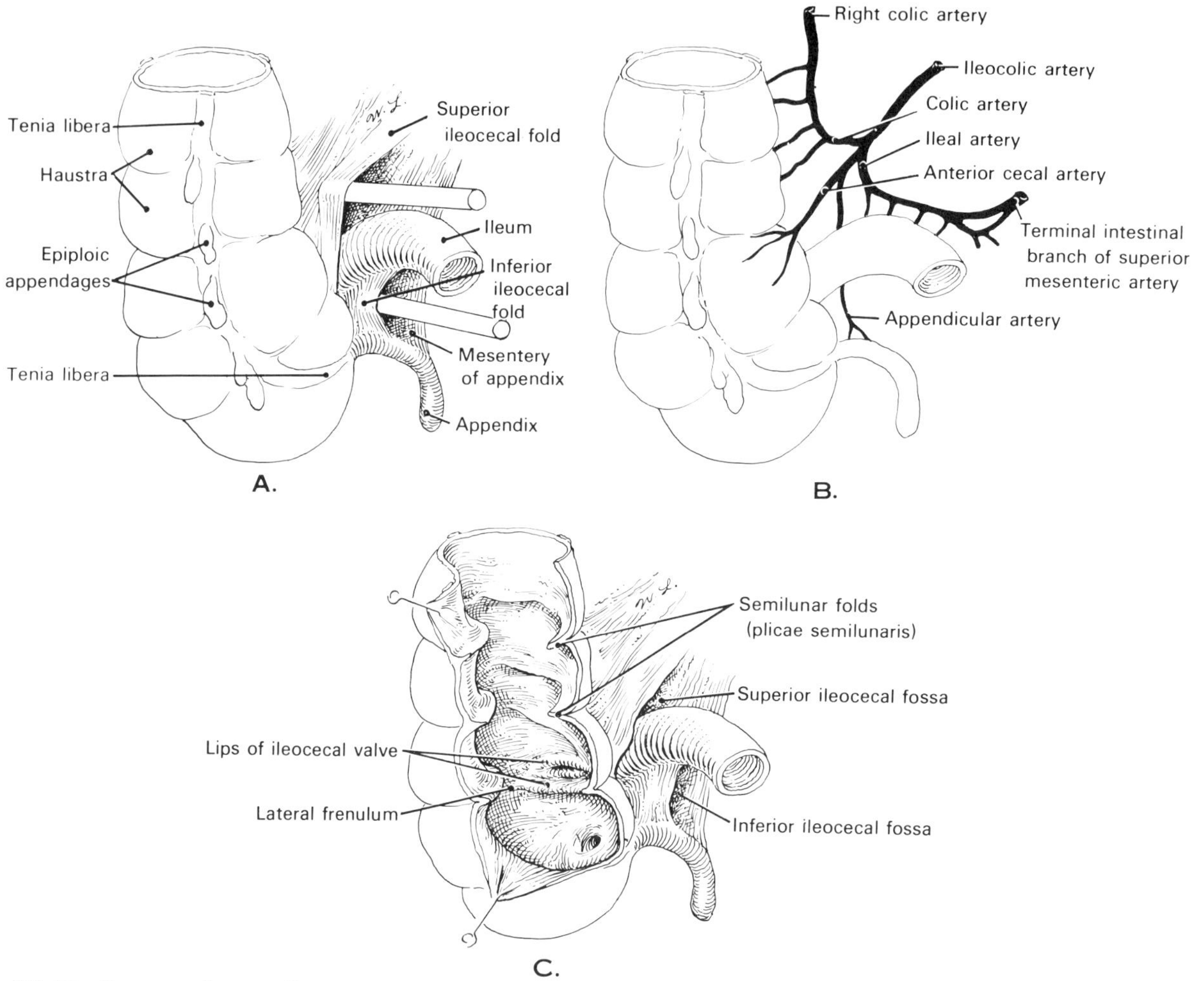

Fig. IX-79. Cecum and appendix. **A.** General anatomy. Dowels in superior and inferior ileocecal fossae. **B.** Blood supply. **C.** Opened to show ileocecal valve.

iliac crest it lies upon the proximal parts of the lateral cutaneous nerve to the thigh and upon the ilioinguinal and iliohypogastric nerves. In pregnancy, the gravid uterus may rise to press the ascending colon against these structures causing pain symptoms along the peripheral distribution of these nerves. In much the same way, the cecum can compress the genitofemoral and femoral nerves and the testicular or ovarian vessels, all of which can become involved if the appendix becomes inflamed acutely.

The transverse colon, all 45 cm. of it, is, of course, the most mobile part of the large bowel, suspended as it is by its mesocolon. It generally arches antero-inferiorly, hammocked between the hepatic and splenic flexures. It may dip below the umbilicus when we sit erect. When distended, it can slip in front or behind the stomach so that percussion of the abdomen produces a tympani that obscures the usual dullness over the liver. This may lead you to suspect "free air" in the peritoneal cavity, but this can be readily checked by means of supine, erect, and decubitus radiographs of the abdomen.

The approximate 30 cm. of descending colon crosses and impinges upon the same structures on the left side that the ascending colon impinged upon the right. It joins the sigmoid colon just anterior to the external iliac vessels.

The **sigmoid colon** varies in length from 15 to 90 cm. but averages about 30 cm. Its extends to the pelvic level of VS_3 where it joins the rectum. Usually it lies relatively free within the lesser pelvis below the level of the small bowel, but if it is long, it may roam about as far as its mesenteric tether will let it.

Blood Supply of Large Intestine: We have seen how the middle, right, and ileocolic branches of the superior mesenteric artery supplied parts of the colon. The ileocolic branch provided smaller cecal and appendicular arteries as well (Figs. IX-65 and 79B). The rest of the colon is supplied from the third of the unpaired branches of the aorta, the **inferior mesenteric artery.**

Its **left colic branch** divides, ascending and descending, to anastomose with the middle colic and ascending sigmoid branches. Two or three sigmoid branches descend to supply the sigmoid colon and the upper part of the rectum (Fig. IX-65). Notice how the branches of the superior and inferior mesenteric arteries, by their anastomoses, appear to form one large, arcing, marginal vessel (marginal artery of Drummond) around the inner border of the large intestine (Fig. IX-65). Contributions to any segment of this arc may be variable from the different vascular stems. The arc may be broken in spots in instances where anastomotic interconnections fail to be made. Thus, the order of supply is not always as nice as that shown in our generalized Figure IX-65. From this marginal artery, both short and long arteriae rectae reach out to the wall of the gut. The short, straight twigs supply the mesocolic taenia and pierce the gut wall to supply muscularis, submucosa, and mucosa. The long arteriae rectae reach around the sides of the large intestine to supply the other two taeniae and all the epiploic tags before the vessels pierce to supply the wall of the colon.

The **venous drainage** of the descending and sigmoid portions of the colon is *via* the **inferior mesenteric vein** which reaches up behind the pancreas to tap the splenic vein or to meet the junction of the splenic and superior mesenteric veins (Figs. IX-57A and 80). These **three unpaired veins** of the "G-I Tract and Its Associates" **form the portal vein** which enters the liver, receiving all returning blood from pancreas, spleen, stomach, small, and large intestine.

Portal-caval Anastomoses: When the portal vein is obstructed, liver disease often being a primary cause, blood from the intestinal tract can still reach the right heart *via* caval return over a number of collateral routes (Fig. IX-80). Some of these are most devious:

1. Where the esophagus joins the stomach, the left gastric vein ascends through the diaphragmatic hiatus to anastomose with the azygos system of veins. Varicosities in these esophageal veins create esophageal piles.

2. The para-umbilical route: Veins from the digestive tract use the falciform ligament to reach umbilicus and anterior body wall. Anastomosis with the deep epigastric veins return blood to the inferior vena cava *via* the external iliacs. Anastomosis with the superficial epigastric veins entails a longer route *via* internal mammary and subclavian veins to the superior vena cava. Distention of these superficial para-umbilical veins results in the configuration of the caput medusae.

3. Superior rectal veins can anastomose with the pelvic venous plexus *via* the middle hemorrhoidal veins. Return to the inferior vena cava is *via* the hypogastric (internal iliac) vein. This is another common area for varices to develop (rectal hemorrhoids or piles).

4. Splenic and pancreatic veins anastomose with the left renal vein to enter the inferior vena cava by this route, once more by-passing the liver or the portal obstruction.

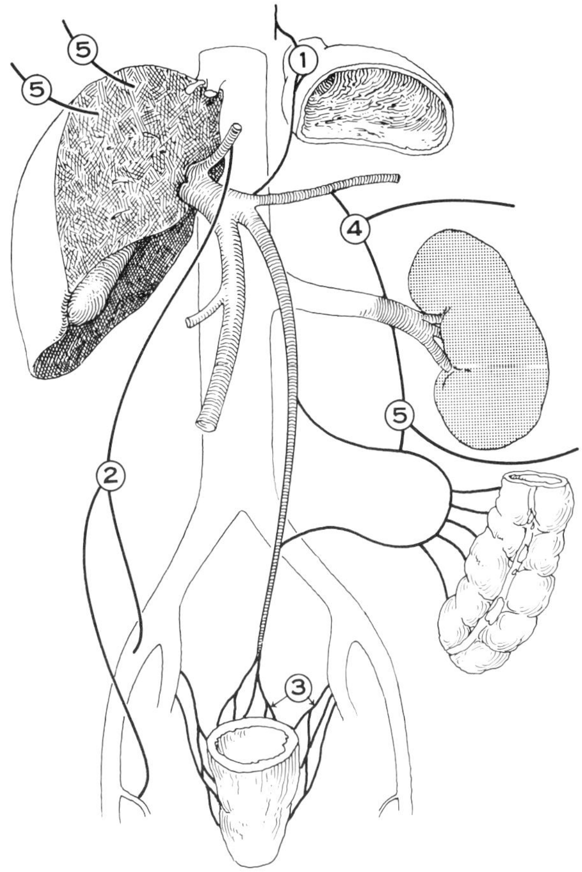

Fig. IX-80. Schema of some principal sites of portacaval anastomoses. The sites, numbered from 1 to 5, correspond with the enumerated paragraphs in the text discussion. (After Grant.)

5. Short veins of Retzius connect splenic and colic veins (and veins of the bare area of the liver) to lumbar veins of the body wall and these enter the inferior vena cava.

Thus, while portal obstruction is a serious matter, there are ways around the road-block. Figure IX-80 reviews these routes.

We have mentioned the *lymphatic drainage* of the stomach and small intestine. Table IX-5 reviews lymphatic drainage for the "G-I Tract and Associates" in general. Figure IX-81 illustrates lymphatic drainage of colon.

Innervation of the Large Intestine: The parasympathetic and sympathetic supply takes a somewhat different routing to the descending and sigmoid portions than it does to the ascending and transverse colon. The latter parts receive their supply in the same way that the small intestine received theirs. As usual all of the supply follows peri-arterial routes.

I. Ascending and transverse colon
   A. Parasympathetic:

Table IX-5. Lymphatic Drainage of the Gastro-intestinal Viscera

Regional Nodes	Draining Nodes	Area Drained
Gastric nodes	Coeliac group of pre-aortic nodes	Stomach
Hepatic nodes		Stomach, duodenum, liver, pancreas
Pancreaticolienal nodes		Pancreas and spleen
Mesenteric nodes	Superior mesenteric group of pre-aortic nodes	Small intestine
Ileocolic nodes		Ileum, cecum, ascending colon
Mesocolic nodes		Transverse colon
Inferior mesenteric nodes	Inferior mesenteric group of pre-aortic nodes	Descending colon
		Sigmoid colon

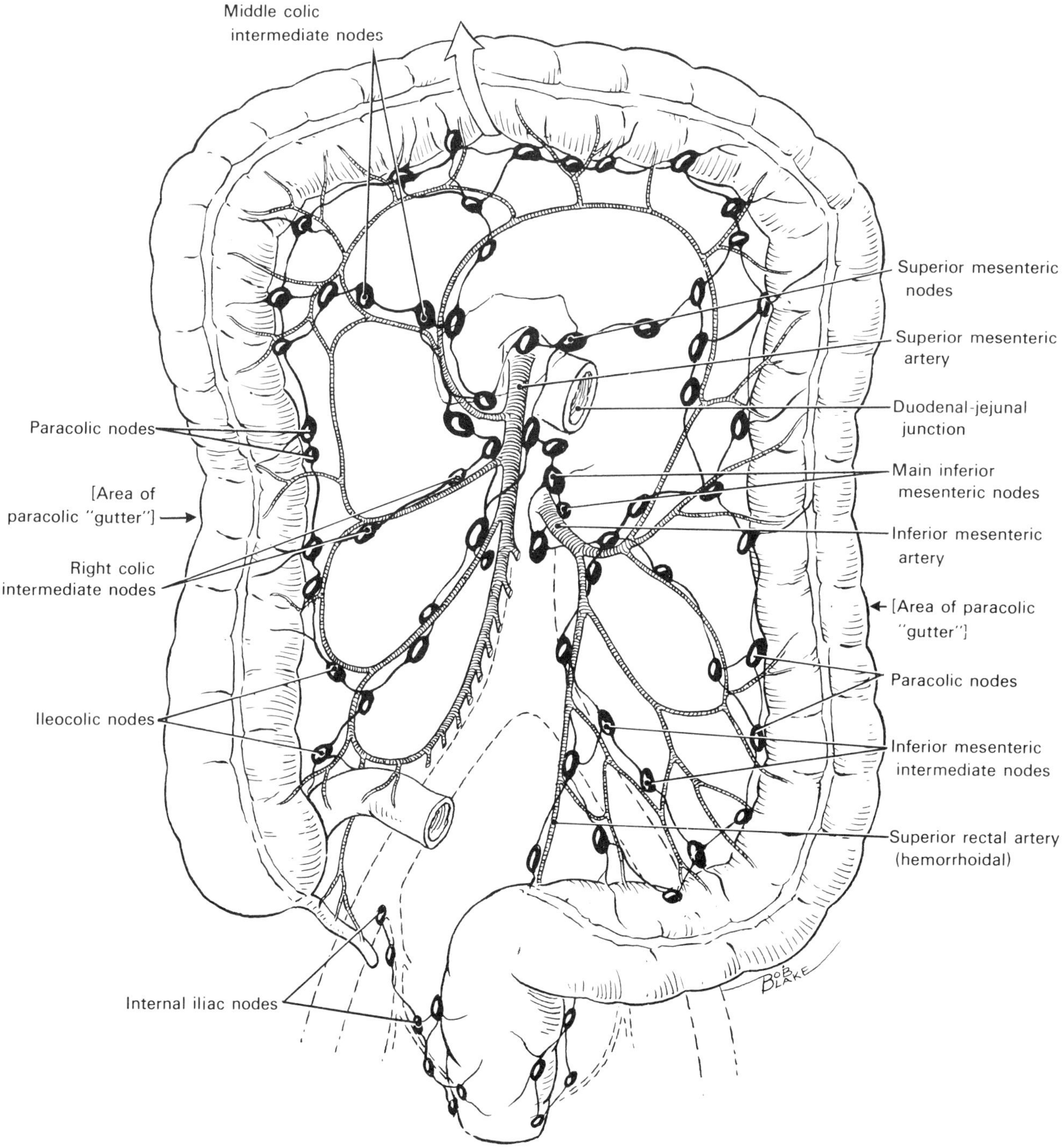

Fig. IX-81. Lymphatic drainage of the colon.

1. Vagus, as preganglionic fibers that have passed *through* coeliac and/or superior mesenteric ganglia to follow middle, right, and ileocolic arteries.
2. Terminal postganglionics in Meissner's and Auerbach's ganglia.

B. Sympathetic:
1. Lesser splanchnic nerve (T_{10} to T_{11}).
2. Postganglionics from superior mesenteric ganglia run peri-arterial course.

II. To descending and sigmoid colon

A. Parasympathetic:
1. Pelvic nerves (S_2 to S_4) *via* pelvic and hypogastric plexus are the preganglionic fibers.
2. Terminal postganglionics lie in Meissner's and Auerbach's plexuses in gut wall.

B. Sympathetic:
1. Lumbar splanchnics (L_1 to L_2)
2. Postganglionics lie in the inferior mesenteric ganglion at the root of the inferior mesenteric artery. Their axons follow the branches of the inferior mesenteric vessels to the descending and sigmoid colon.
   a) From the ganglia, these fibers build up a plexus along the aorta and along the hypogastric arteries to descend over the pelvic brim to reach pelvic viscera. The parasympathetic pelvic nerve contributes to these plexus formations along the large vessels and their branches.

Visceral sensory fibers return to the cord over these various plexuses and through splanchnic and pelvic nerves. There seems to be some unilaterality about the sensory routing. Fibers from the appendix, cecum, ascending colon, and the right half of the transverse colon traverse the superior mesenteric and coeliac plexuses to follow the splanchnic nerves back to cord segments T_9 to T_{11}. The cell bodies of these visceral sensory fibers will be found in the dorsal root ganglia at these levels, and, predominantly, in those on the right side. The fibers from the rest of the transverse and the from the descending colon trace back through the least splanchnic and lumbar splanchnic nerves to dorsal root ganglion cells at cord levels T_{12} to L_2, largely on the left side. The visceral sensory return from the sigmoid, however, appears to be a bilateral affair. It is to dorsal root ganglia S_2 to S_4 over the parasympathetic route *via* the pelvic nerve (nervi erigens).

Additional Remarks on the Histology of the Small and Large Intestine: We have discussed some of the histological changes in the digestive tube already. Typically, the intestinal lining cells display a surface striate membrane which is lacking in the stomach. Electron microscopy reveals this to be composed of microvilli. Thus, it is an adsorptive membrane, and it is found in the large as well as in the small intestine. Plicae circulares reach their greatest height in the jejunum; their lowest, in the ileum; and they disappear almost entirely in the colon. In the ileum and colon the intestinal glands (crypts) display chiefly goblet cells (mucous secretion). These tend to be scattered among the surface cells of villi in the upper parts of the small intestine. Villi are fine and numerous in the jejunum; fewer in number, taller and broader in the ileum, and they disappear in the colon.

Even the lymphoid tissue shows regional changes. Solitary lymph nodules (*not nodes*) are numerous but scattered about singly in the diffuse lymphoid tissue of the tunica propria of stomach, duodenum, jejunum, and colon (Fig. IX-82). But, along the internal antimesenteric border of the ileum, nodules aggregate in patches (of Peyer) (Fig. IX-62). These sometimes cover such an area as to be recognizable by the naked eye when the intestine is slit open lengthwise. They become large enough to encroach upon the submucosa, and frequently erode the luminal mucosa above them wiping out villi locally. The lymph nodules are often so numerous in the appendix as to ring its entire lumen, eroding glands and mucosa in that appendage (Fig. IX-83).

The Rectum (Fig. IX-84): The relationships here will be discussed more fully in the section on the pelvic cavity, but, briefly, the rectum is the continuation of the sigmoid colon to the anal orifice. It passes downward in front of the lower sacrum and coccyx and then takes a posterior bend to its termination. Its terminal part, passing through the pelvic diaphragm, is the **anal canal,** and its opening to the exterior is the **anus.** An area of dilatation, the **ampulla,** occurs just above the anal canal. At its upper end lie three or more **transverse rectal plicae,** while in the anal canal are the **vertical folds or columns of Morgagni.** Each of these longitudinal folds relate to an artery and a vein, the veins being of clinical significance. These are subject to enlargement, rupture, and bleeding, a condition known as **rectal hemorrhoids** or **piles.** These can be internal, inside the canal itself, or external extrusions. Clinically, too, the rectum shows a predilection toward cancer.

The circular muscle, proper, of the rectal tube thickens as an **internal sphincter.** The lower parts of the Levator Ani and the Coccygeal muscles contribute a skeletal muscle, **external sphincter,** quite distinct from the internal, smooth muscle sphincter. The rectum has nothing at all to do with digestion. It acts solely as an excretory organ. The longitudinal muscle contracts in defecation while the internal sphincter relaxes. The external sphincter is, of course, under voluntary control.

The Barium Enema—The Roentgen Anatomy of the Colon: The most valuable method of examining the colon, apart from surgery, is the barium enema. Many disease states, inflammatory, neoplastic, and congenital, may be accurately diagnosed with this technique. In order to understand the examination

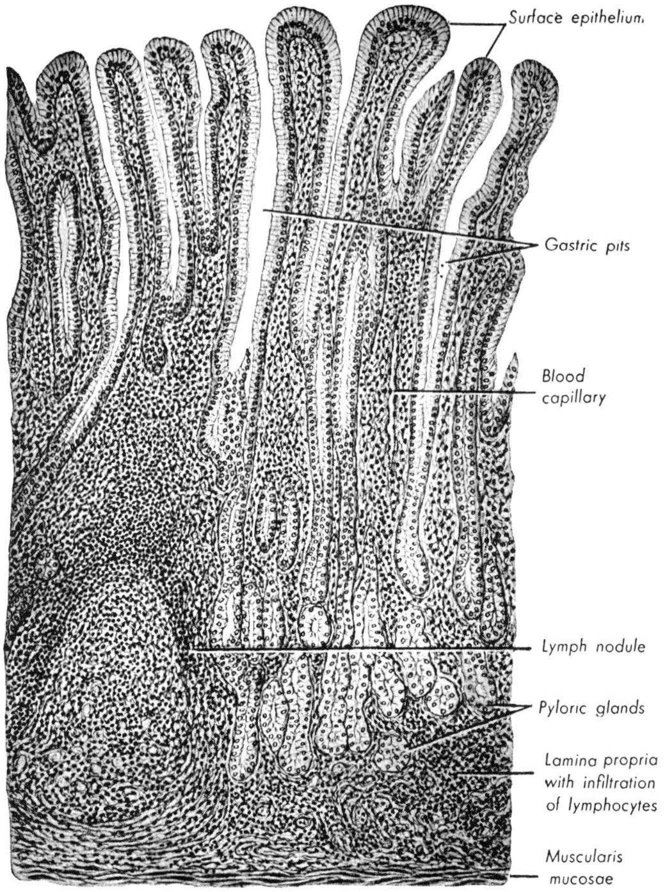

Fig. IX-82. A solitary lymph nodule in the tunica propria of the pyloric canal (From W. E. Copenhaver, *Bailey's Textbook of Histology,* 15th ed., Williams & Wilkins, Baltimore, 1964.)

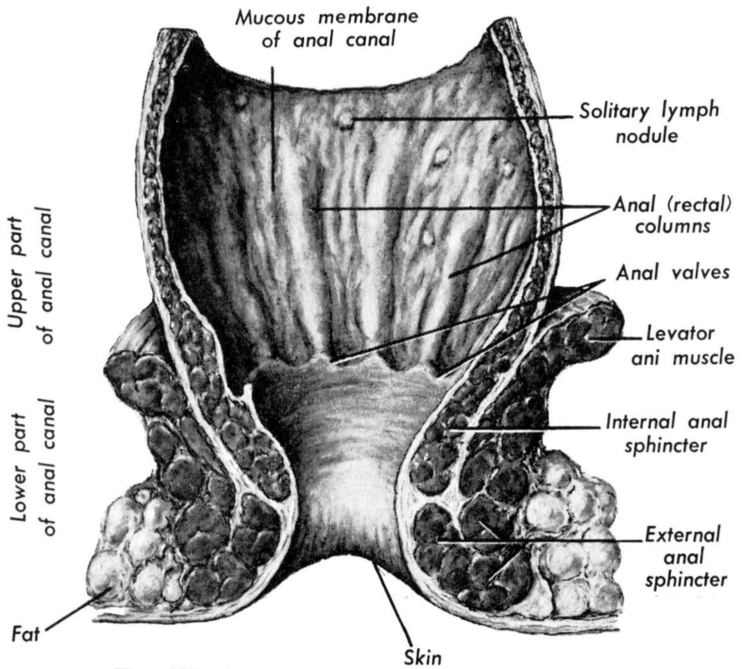

Fig. IX-84. Longitudinal section through anal canal. (From W. E. Copenhaver, *Bailey's Textbook of Histology,* 15th ed., Williams & Wilkins, Baltimore, 1964.)

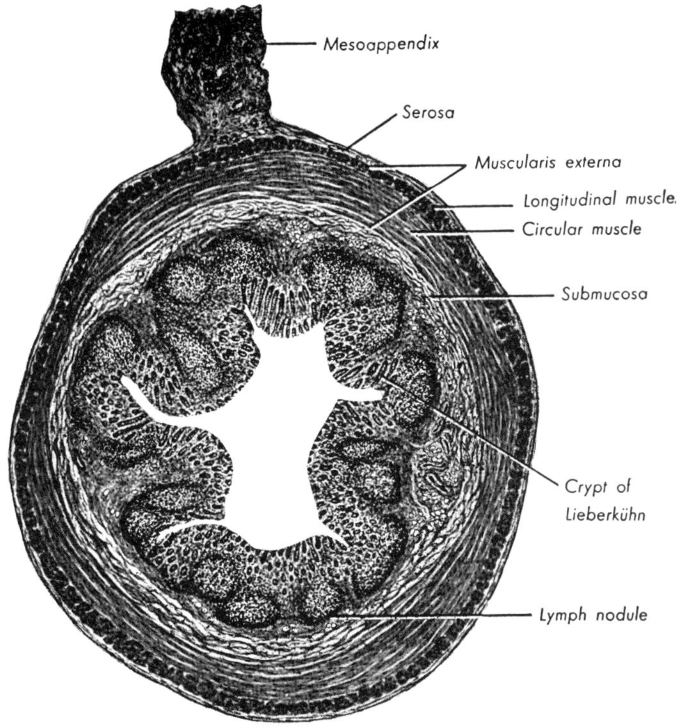

Fig. IX-83. Microscopic cross-sectional view of human appendix. (From W. E. Copenhaver, *Bailey's Textbook of Histology,* 15th ed., Williams & Wilkins, Baltimore, 1964.)

better, we will first discuss briefly the fluoroscopic procedure and then the pertinent anatomic structures that may be seen.

1. PROCEDURE: To examine the colon properly, the bowel must be thoroughly cleansed of all fecal material. If there are no contraindications, this bowel preparation is best accomplished by having the patient eat a low residue diet for several days, and then, on the evening prior to the study, administering a cathartic. Cleansing enemas are given on the morning of the examination. These enemas are done several hours before the patient comes to the radiology department.

It is of the utmost importance for the referring physician to consult with the radiologist about his patient's problem before the examination. In addition the radiologist should examine the patient's abdomen, take a clinical history, and review the patient's medical chart. We also like to obtain a preliminary AP supine KUB (scout) film of the abdomen before actually beginning the contrast examination.

The barium enema is usually started with the patient in a recumbent position. Fluoroscopic guidance is always used with a carefully controlled flow of the barium sulfate-water mixture so that the head of the contrast column may be observed constantly. We prefer not to fill the entire colon and then search for abnormalities, but rather short segments are filled and the enema stopped while these are palpated and compressed with the lead glove protecting the hand. Then, the enema is restarted, and the next proximal portion is filled. Of course, in the search for abnormalities, the patient must be rotated to the best position to show that part of the colon under examination (Fig. IX-85).

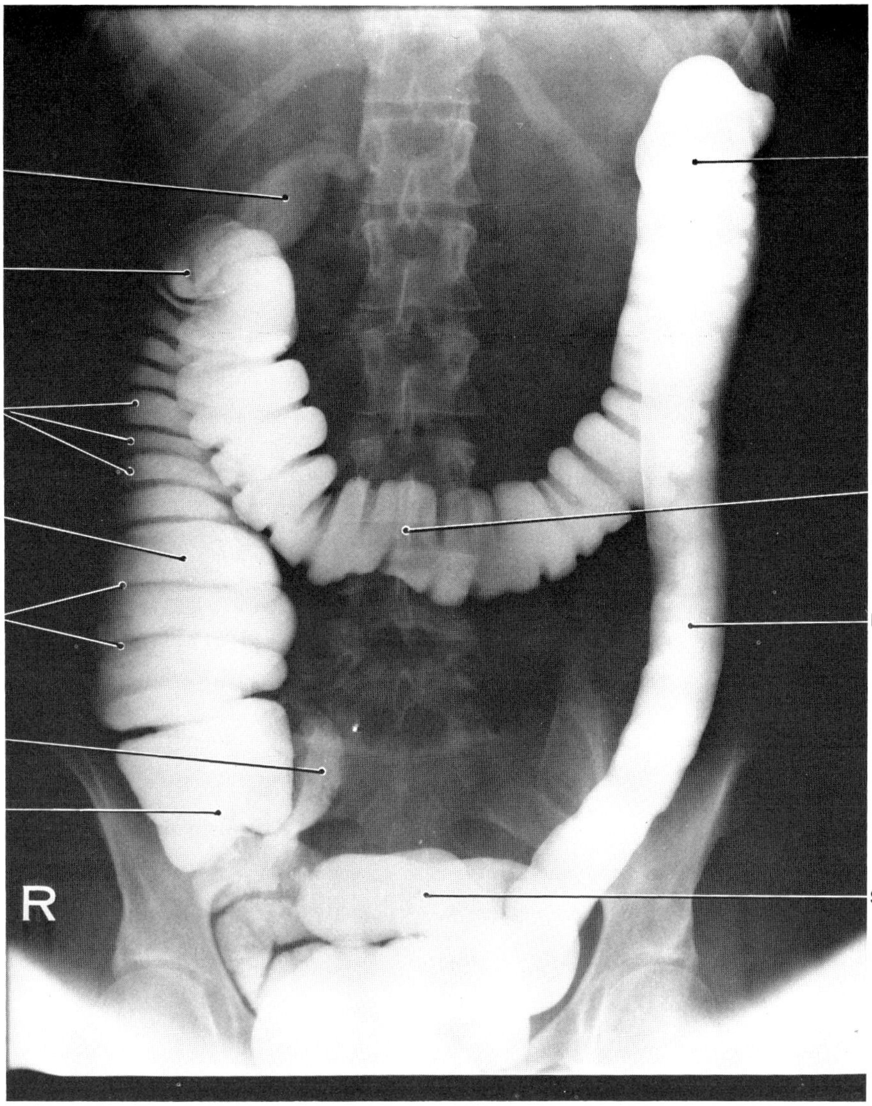

Fig. IX-85. Barium enema—AP projection of the barium-filled colon. Note some reflux of barium into the terminal ileum.

Thus, the sigmoid loop and the hepatic flexure are seen to greatest advantage in the right anterior oblique position, while the splenic flexure is studied in profile in the left anterior oblique position. These positions refer to the patient relative to the fluoroscopic screen. There will be some variation in these projections depending upon redundancy of the colon. To assure complete filling of the cecum, it is important to fill the terminal ileum or appendix or both. The entire procedure may be recorded on television tape for instant replay. In addition, spot films of pathological areas and hard to visualize areas, *i.e.*, the sigmoid loop and the flexures, may be obtained.

Upon completion of fluoroscopy, additional pre-evacuation films are made as specified by the radiologist. The barium is then evacuated by the patient and an additional AP or PA projection of the abdomen obtained in order to study the *mucosal relief*. On occasion, particularly when searching for small intrinsic lesions such as polyps, the *double contrast enema* is useful. The examination may be performed in several ways, but one of the easiest is to perform a regular barium enema and, after filming, siphon off most of the barium mixture. Enough barium must be left to coat the colonic wall adequately. Air is then introduced to distend the bowel lumen. Some radiographs obtained after such an air contrast barium enema are shown in Figures IX-86 and 87. Surprisingly, with so much air in the colon, very few patients become airborne.

2. EVALUATING THE RADIOGRAPHS: Radiologists differ over the radiographic projections that they obtain both during and after fluoroscopy of the colon. In the discussion to follow, we will demonstrate many of the possible views that can be obtained.

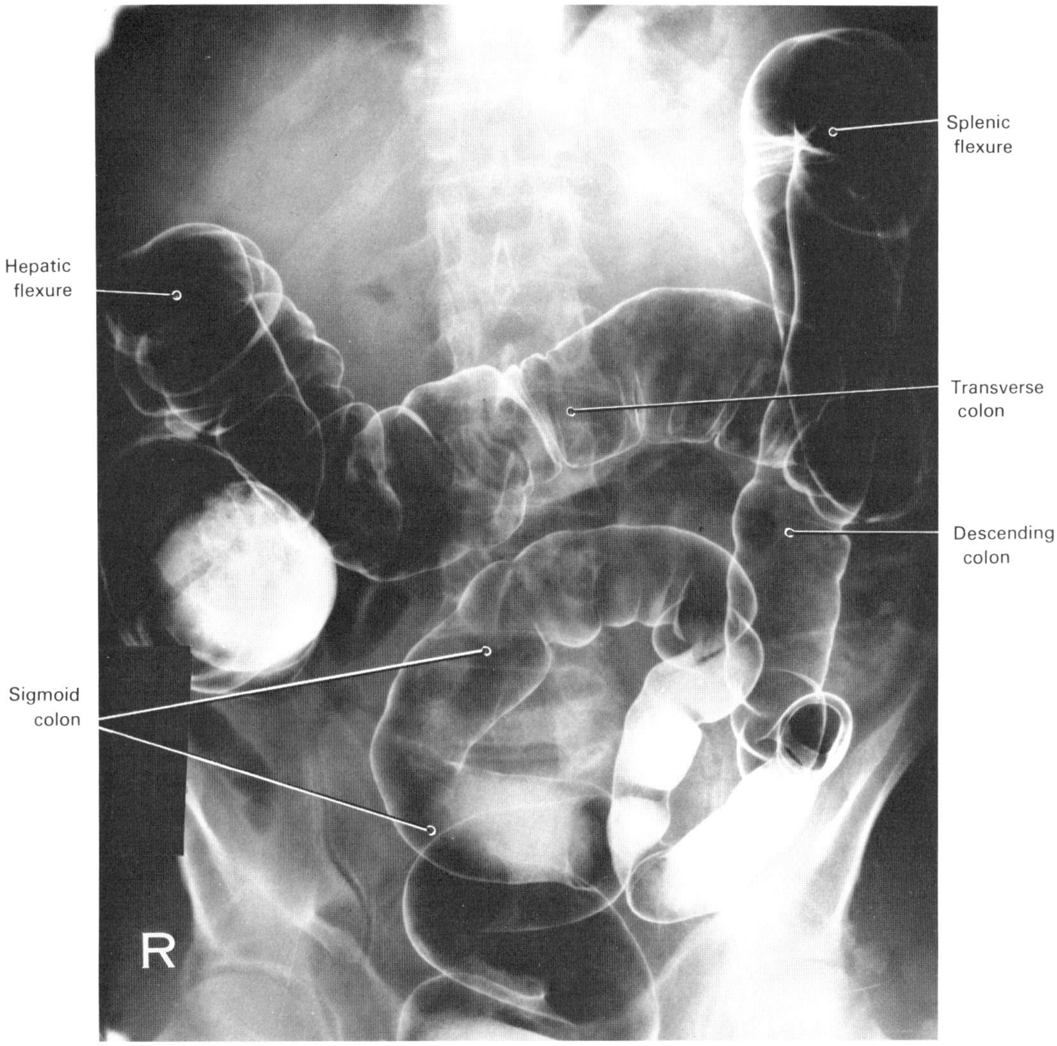

Fig. IX-86. Air contrast barium enema—AP projection. Barium coats the colon wall and air distends its lumen.

The colon, as seemingly everything else that you have studied in anatomy, is inconstant in length, width, and position. The colon may be *short* with very little redundancy, and in this instance the cecum is high in the mid or upper abdomen. More commonly the colon is quite *long* and *redundant*, and the cecum is situated low in the pelvis. Figure IX-85 reveals that the cecum is the widest portion of the colon; however, the large bowel tapers from this point down to the rectosigmoid junction, where is abruptly balloons at the rectal ampulla. Those segments of the colon that are "retroperitoneal," *i.e.*, the ascending and descending colon, are relatively fixed in their location, whereas, the cecum, transverse, and sigmoid colon are quite variable in position depending upon the length of their mesentery.

As you examine the radiographs of Figures IX-85 to 87 take note of the typical sacculations, or haustra, of the colon. They are uneven in length, varying from 1 to 4 cm. Between the haustra, the colonic wall is thrown into folds which are crescent-shaped on the interior of the colon. They are termed the semilunar folds (plicae semilunaris). In Figure IX-85 they appear as lucent bands crossing the long axis of the colon when the latter is filled with barium. The haustra are most pronounced in the ascending and transverse colon, and less so in the other portions. Be certain that you do not mistake this apparent lack of haustral markings in the descending and sigmoid colon for disease states, such as ulcerative colitis. In the chronic stages of colitis the bowel becomes rigid, narrowed, and the mucosa atrophies. The haustral pattern is lost at this point in the disease.

Let us examine in depth the anatomic roentgen features of the colon from proximal to distal end. At the tip of the cecum, the appendix is seen to distend with barium. In this case the appendix is subcecal in position, but in the majority of cases it is retrocecal (Fig. IX-88). The appendix may be mobile depending upon the length of its mesentery. Following appendectomy, the inverted stump of the appendix may be seen protruding into the cecal lumen. Although this appearance may simulate a polypoid tumor, the proper diagnosis

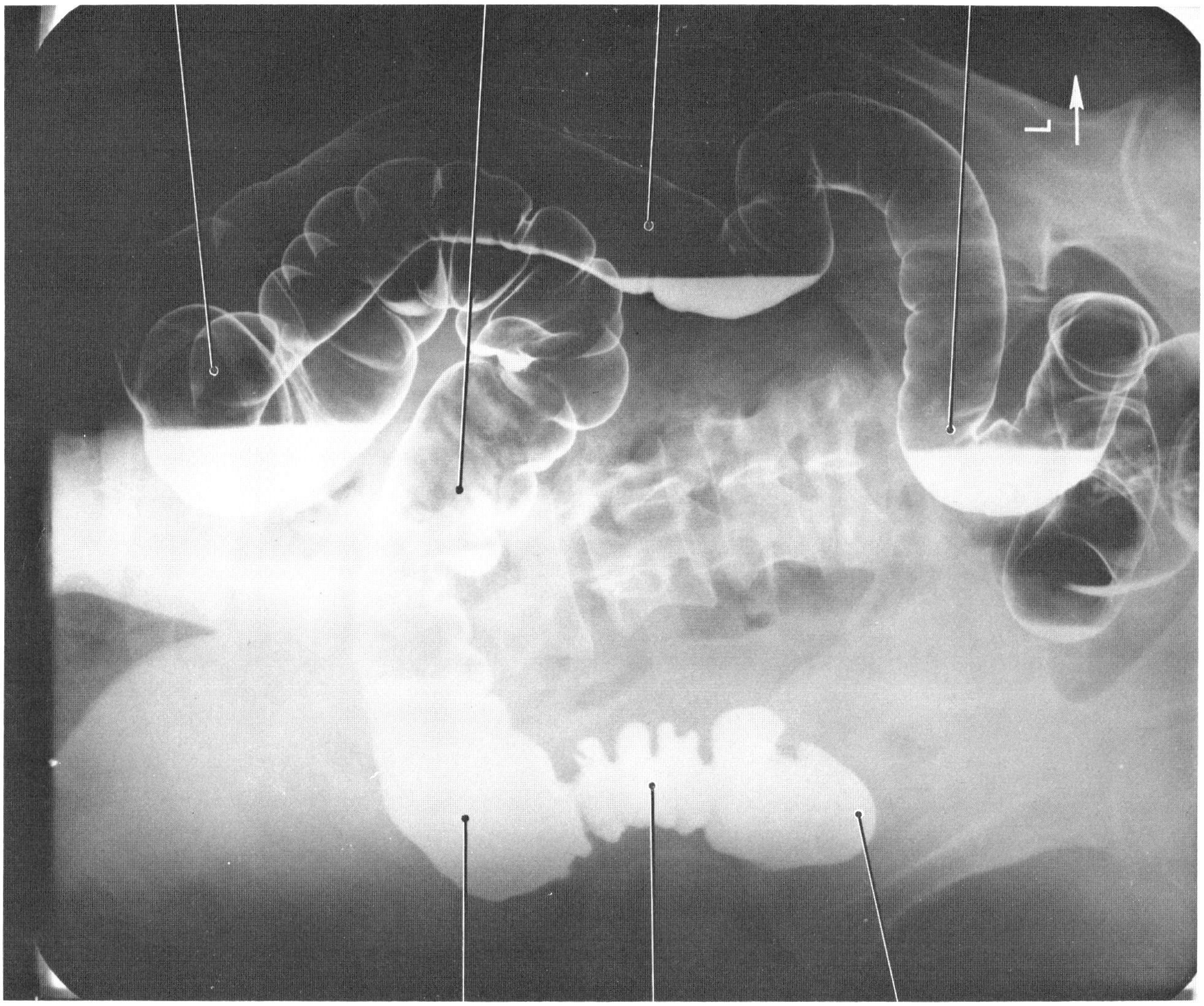

Fig. IX-87. Air contrast barium enema—right lateral decubitus projection. The arrow indicates that the patient's left side was uppermost and his right side on the table.

is usually made based upon the history of prior surgery and the location of the cecal defect.

On the medial or posterior aspect of the cecum there is a shallow indentation caused by the ileocecal valve. In this instance, barium has been refluxed into the terminal ileum. Therefore, we can see the two lucent labial folds of the valve and, sandwiched between them, a thin layer of barium (Fig. IX-88). Occasionally the ileocecal valve becomes hypertrophied or thickened and this may suggest polypoid tumor; however, careful fluoroscopy with compression establishes the true nature of the defect in the barium column.

Above the level of the ileocecal valve, the ascending colon begins. It is situated in the flank and is fixed by its retroperitoneal position in the abdomen. The hepatic flexure is below the liver edge. Here the colon turns toward the midline of the abdomen. By placing the patient in a left posterior oblique projection, this flexure is seen in profile. With hepatomegaly, the hepatic flexure is displaced downward. The transverse colon is on a mesentery and may hang low in the abdomen. It is quite mobile and changes location as the patient changes his position. The splenic flexure lies at a higher lever than the hepatic flexure; however, splenomegaly may alter this by displacing the splenic flexure caudally. In order to visualize the splenic flexure in profile, the patient must be turned to the right posterior oblique position.

Just below the level of the iliac crest, the descending colon emerges from the retroperitoneal space and becomes the sigmoid colon. The latter has a mesentery of variable length, and therefore, as we can see in Figure IX-85 that it may be quite coiled on itself or may even extend across the midline. Marked degrees of mobility and length of the sigmoid and cecum may predispose the patient to **volvulus,** an abnormal twisting

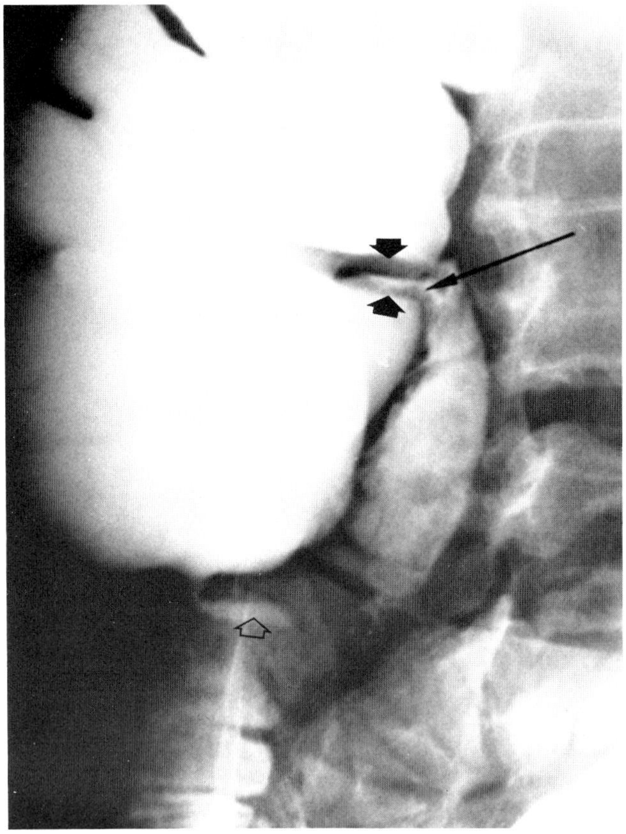

Fig. IX-88. Spot film of the cecum and terminal ileum in the right anterior oblique view (position referable to the fluoroscope). The *open black arrow* indicates the appendix which is located subcecally. The *short black arrows* point to the folds of the ileocecal valve. The *long black arrow* points toward a small quantity of barium between the ileocecal valve lips.

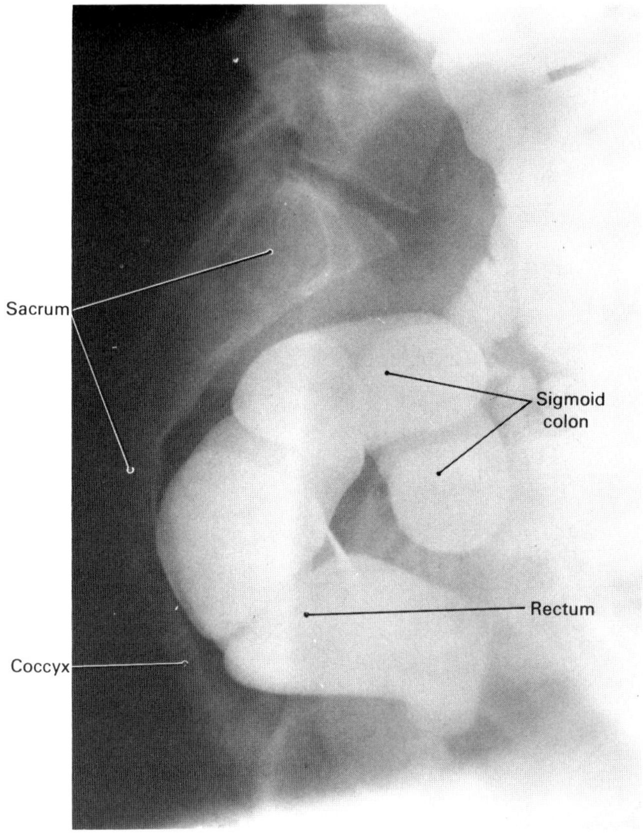

Fig. IX-89. Lateral radiograph of the rectum and sigmoid colon filled with barium.

of the bowel which may result in obstruction of the colonic lumen.

The rectum which is "retroperitoneal" in position, usually begins about the VS_3 level (Fig. IX-89). It is not possible radiographically to see an abrupt transition from sigmoid to rectum. The rectum lies deep in the hollow of the sacrum, and this is best appreciated in the lateral projection.

The postevacuation radiograph of the colon (Fig. IX-90) is an important part of the total examination. It enables the radiologist to view the contracted bowel and the feathery mucosal pattern coated by a thin layer of barium residue. The colonic mucosal relief is typically irregular or transverse to the long axis of the bowel. However, more distally this mucosal pattern may alter and become more parallel to the long axis. On occasion, during evacuation of the barium enema, the ileocecal valve may relax sufficiently to permit marked reflux of the barium into the small bowel (Fig. IX-90B). This permits an excellent view of the lumen of the small bowel as shown.

The Functional Role of the Digestive Tube: Quite obviously, its goal is the digestion of food materials. Predigestion of protein is accomplished by the enzyme, pepsin, secreted by the chief cells of the fundic glands. Since this enzyme functions best in an acid environment, the role of the HCl-secreting parietal cells of the fundic glands is explained. We also have a tendency to swallow small chunks of meat. The acid softens and digests the collagen of its connective framework so that such meat particles can be broken down for easier predigestion. Most of the digestive processes are continued in the small intestine. Indeed, some are just begun there. Here the enzymes require a more salubrious alkaline environment than in the stomach. These protein catalysts are contributed variously by the crypts of Lieberkühn, bile and bile salts from the liver, and a host of enzymes come from the exocrine acini of the pancreas. Polypeptides are converted to amino acids; fats, to fatty acids and glycerol; carbohydrates and complex sugars, to glucose. Only in these simple forms can nutrient constituents be absorbed by the microvillous surface modifications of the cells that clothe interstinal villi. Amino acids and glucose are moved into the blood capillary beds of the villi while fatty acids and glycerol find their way mainly

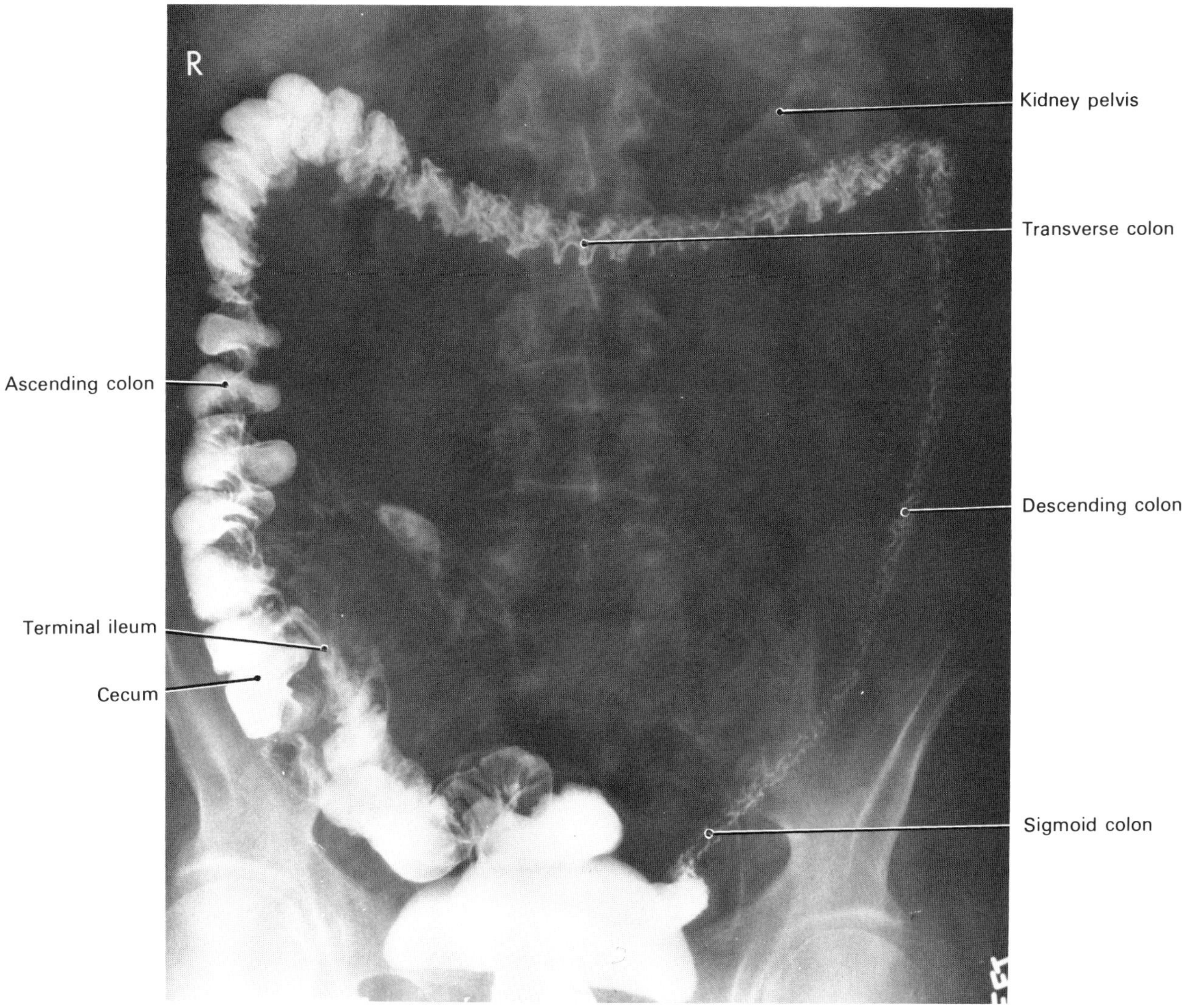

Fig. IX-90. Barium enema. Postevacuation film in frontal projection. **A.** Note the mucosal pattern throughout the various portions of the colon.

into the central lacteals. Some fat absorption does take place through the blood capillaries. Most absorbed products reach the liver first *via* the intestinal veins and the portal system since this is the major return route to the circulatory pump from the intestinal tract. The liver gets the first opportunity to utilize these building blocks, storing glycogen, for example, and reconverting some of the other absorbed constituents into useful by-products of liver manufacture.

Probably no new digestive products are added by the large intestine, for its glands chiefly secrete mucus. Its main job is to absorb water and some remaining nutritive material. In this way the more or less solid feces are formed (L., *faex* = dregs). Bacteria are also incubated here, some of which can synthesize useful products such as certain vitamins. How much food residue remains depends largely upon the amount of cellulose that was ingested. Fecal matter consists mostly of dead bacteria, some living ones and certain substances, some of them putrid, produced by the colon. For these reasons, feces can be produced even during starvation.

Certain matters of clinical relevance may be mentioned here. Ulceration and cancer are common disorders. In gastric ulcers, only the mucosa may be involved, or in severe cases, ulcerative erosion may lead to the complete perforation of the gastric wall. The leakage of gastric contents through such a perforation

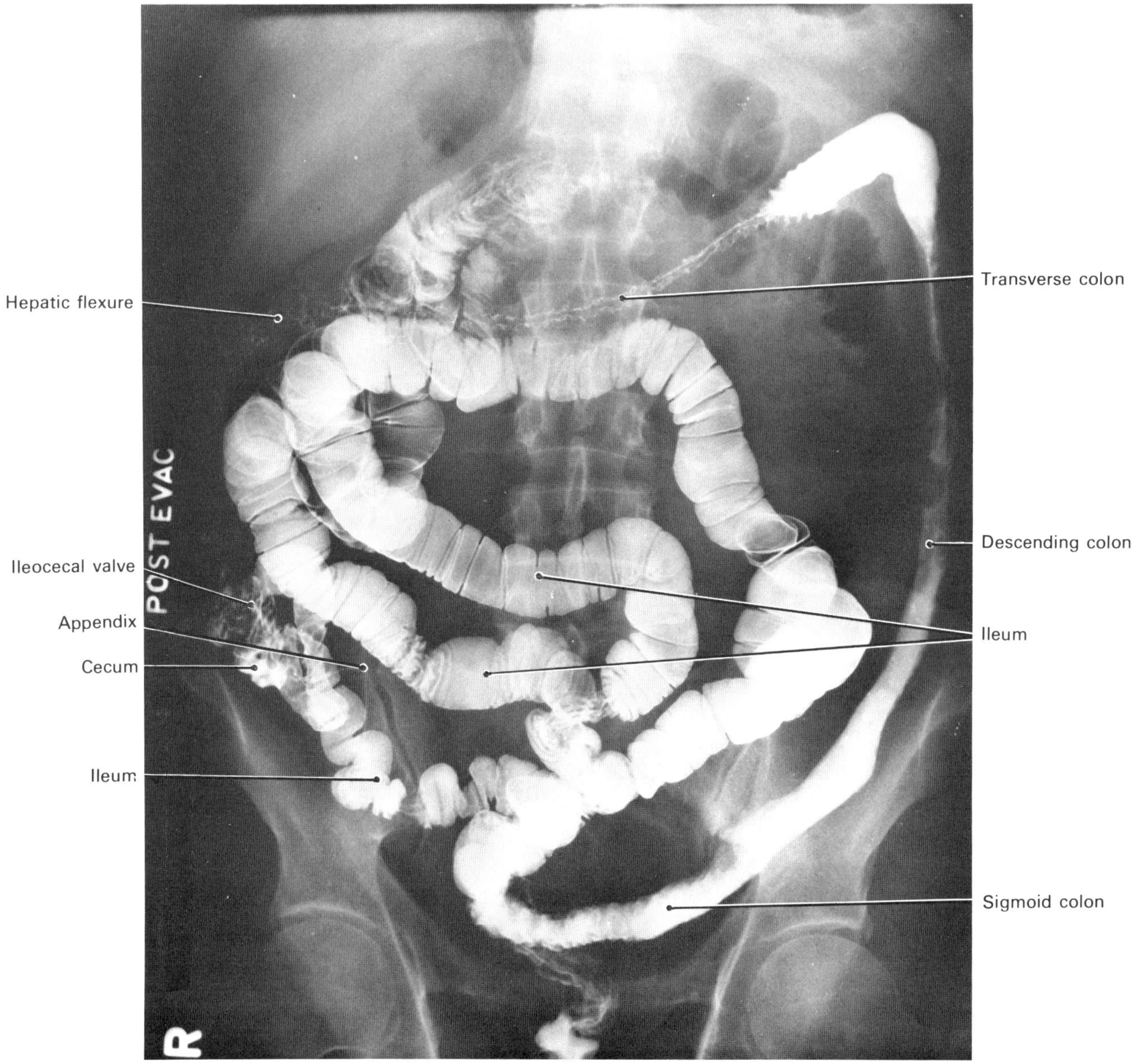

Fig. IX-90B. In this patient, there has been marked reflux of the barium and air into the ileum affording an excellent view of the mucosal pattern.

leads at once to an acute peritonitis which can be rapidly lethal. Posterior perforations can erode the splenic artery, a good-sized vessel, to cause a massive hemorrhage. If the pancreas becomes involved in posterior ulceration, the individual may complain of lumbodorsal pain. But he will also make such complaints if he has a duodenal ulcer, even when this is not entirely perforated. Mild degrees of annoying ulcers can be treated and cured by proper dietary regime. Severe ones require surgery.

The problem of gastric cancer is insidious since it can't be seen or felt, and it may be discovered early enough for corrective surgery only through other complaints. Operative procedures are essential, and massive amounts of stomach may have to be removed, requiring an anastomosis of the esophageal end with the small bowel.

Duodenal ulcer is particularly common in the first part of this organ which receives the acid content from the pylorus. Here again, perforating ulcers are likely to involve the pancreas as well as the biliary tracts and the gastroduodenal vessels.

Another type of disorder, more common in children than adults, is intersusception. This is a telescoping of a segment of the small bowel or colon into the next adjacent lower segment, so as to cause mechanical obstruction. The telescoping can be so severe at times that gut will protrude from the anus.

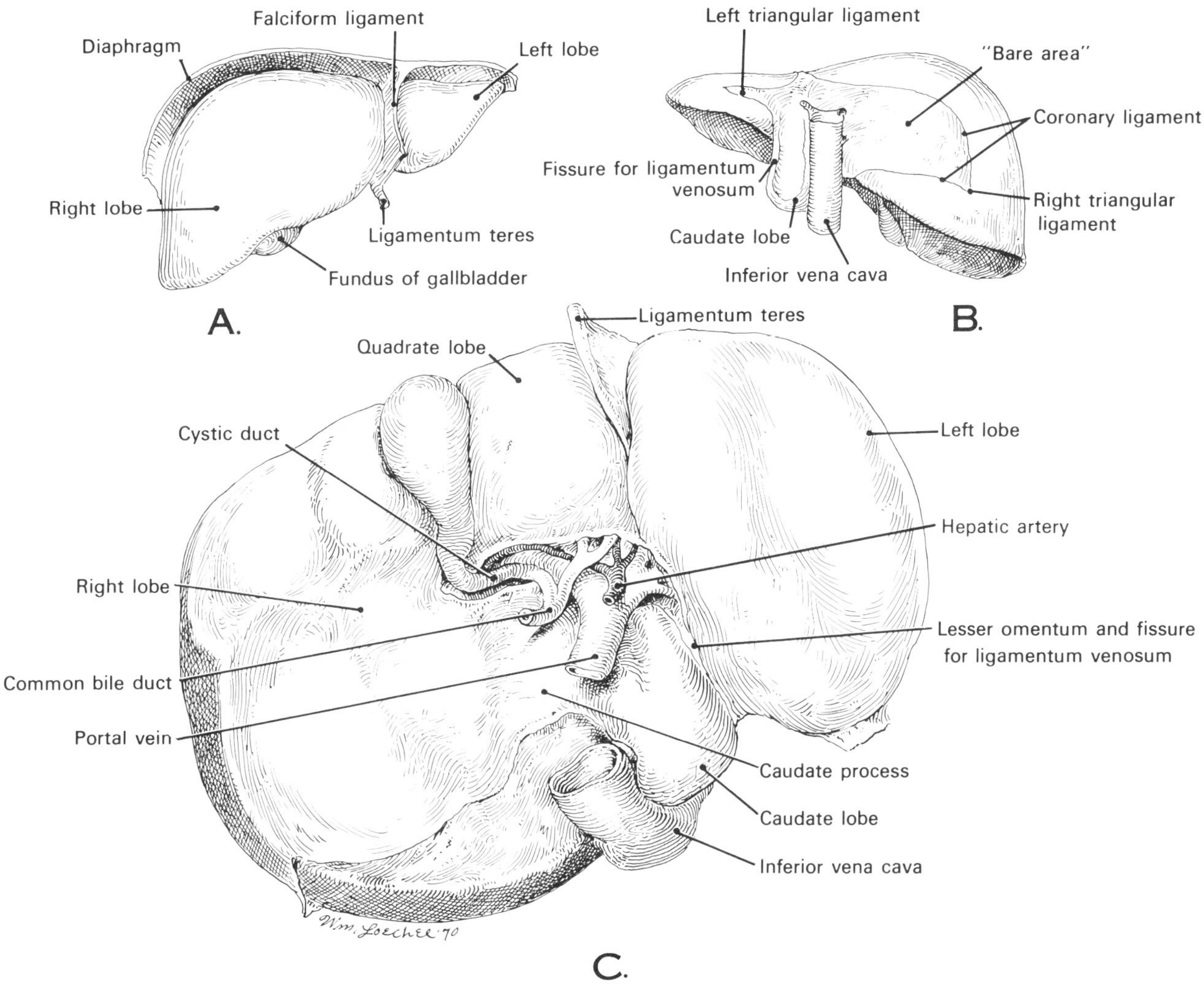

Fig. IX-91. The liver. **A,** anterior view; **B,** posterosuperior view, *i.e.*, view A turned over toward right side; **C,** inferior view *i.e.*, view A turned over cephalically.

Structures Associated with the Gastro-intestinal Tract

The Liver and the Biliary Apparatus: *General Structure:* The liver is the largest, single gland in the body, and it is an apparent Jack-of-all-Trades, despite the fact that the hepatic cells flanking the sinusoids of the microscopic liver lobules all look alike. It contributes no enzymes to the digestive tract, but bile, its chief secretory product, emulsifies fat and helps to provide the alkaline milieu required by the enzymes of the intestinal tract for their optimum function. Bile contains several products of liver manufacture—bile pigment (bilirubin), bile salts, cholesterol, and such crystalloids as are found in tissue juice, even some protein. Bilirubin, itself, is a waste product that is not formed in the liver. It has been passed on to the liver by certain elements of the reticulo-endothelial system which engage in the breaking down of worn-out red cells. The non-iron-containing portion of hemoglobin is bilirubin which the liver has made soluble as one of the bile constituents.

The liver stores glucose by converting it to glycogen, only to restore it to glucose and to release it in the blood stream when the body demands sugar.

The liver also splits amino acids in order to store parts of them as glycogen. It combines the nitrogenous moiety with CO_2 to form urea which is circulated to the kidneys for excretion.

Many of the blood proteins are also formed by the liver, *e.g.*, albumins, globulins (but not the gamma globulins), fibrinogen, and other proteins concerned with blood coagulation. It can even become a hemopoietic organ, a function it once served in the early embryo. We have mentioned here only a few of its 100-odd, purported trades. Let us now examine the anatomy of this important structure (Fig. IX-91).

The liver appears as a four-sided pyramid that lies on one side while its base sits to the right and its apex projects to the left. It has a diaphragmatic and a visceral surface. The visceral surface is cut by an **H-shaped fissure.** The *left arm* of this fissure contains the **ligamentum venosum** and the **round ligament of the liver** (ligamentum teres). The *right arm* contains fossae to hold the **gall bladder** below and the **inferior vena cava** above. The cross arm of the "H" is the **porta hepatis** transmitting the **portal vein, hepatic artery proper, common bile duct, nerves, and lymphatics** (Fig. IX-91C). In an anterior view, it is easy to see that the major **right and left lobes** are separated by the falciform ligament. To find its smaller two lobes, you must examine the inferior aspect of the liver at the site of the H-shaped fissure. The **caudate lobe** lies above the porta hepatis and separates inferior vena cava from the gall bladder. The lobe extends to the left as far as the fissure for the ligamentum venosum. The **quadrate lobe** lies below the porta, between the fissure for ligament teres and the gall bladder (Fig. IX-91C).

It is obvious that the visceral peritoneum of the liver fits it like a glove and covers it except where attachment is to the **"bare area,"** only visualized when the liver is not anchored to the Diaphragm. Below the bare area the liver is relatively free from its diaphragmatic attachment. But the peritoneum is pulled out and reflected as certain other named "ligaments." The falciform ligament is already familiar to us, as is the lesser omentum. Not so familiar is the **coronary ligament,** a kind of perimeter crown around the bare area, its *right* and *left corners* meeting to form the **right** and **left triangular ligaments** (Fig. IX-91B).

Bile canaliculi of the many microscopi liver lobules drain to the periphery of each lobule between double files of hepatic cells to empty into collecting bile ducts all of which eventually are drained by the **right and left hepatic ducts** at the hepatic portal. These latter soon unite as the **common bile duct** in the hepatoduodenal fold in front of the foramen of Winslow. Trace the course of the common duct to the posterior aspect of the duodenum and to its union with the pancreatic duct at the ampulla of Vater. You will have to dissect in the substance of the pancreas, and then you must cut open the duodenum by slitting along the *outer* curvature of this C-shaped tube. You will have to free-up duodenum and pancreas to gain working room. Observe also the origin of the cystic duct and trace it back to the gall bladder.

Blood Supply: The chief vessels carrying blood into the liver are the **portal vein** and the **hepatic artery.** These are seen entering the front door of the liver (hepatic portal), the same door from whence the bile ducts emerged. Portal blood is destined to supply raw materials to the workers along the liver's assembly lines. The workers are the rows of cuboidal liver cells that line up along the liver sinusoids. The raw materials pouring through the sinusoids are the products of digestion and such products as bilirubin from the spleen. The liver assembly-line workers also have to be fed if they are to turn out products of manufacture. The cafeteria line feeds into the assembly line, *i.e.,* both oxygenated blood from the hepatic artery and the venous return of the portal system empty into the same liver sinusoids. Bile secretions are packaged and shipped along the bile canals on the antisinusoidal side of the liver cell. Other products of manufacture are transported by dumping them back into the sinusoids. These latter head for a great sink-hole in the center of each liver lobule, the **central vein,** which is drained by sublobular veins into hepatic veins and then into the inferior vena cava. Stored items are kept in the substance of certain liver cells until demands are made for their release into the sinusoids. The hepatic veins and inferior vena cava can then be regarded as the rear porta or back door of the liver, the way in which all its manufactured products are shipped out from the plant. All, that is, except one special product for one of its best customers. Bile returns to the intestinal tract *via* the front door. Even then, it may be redistilled and aged to a choice vintage by a rerouting into the storage vat, the **gall bladder.**

Clinical Significance: Occasional unwanted byproducts from the storage vat are **gall stones.** These can be formed singly or in great number, and they have been found in many shapes and sizes. Small stones can be passed along the cystic and common bile ducts into the intestine, and they give rise to no trouble. Large ones, which obstruct the passageways, cause pain and must be removed surgically. Usually they are taken from the supraduodenal part of the common bile duct. Since the duct is retroduodenal in its course, often the peritoneum along the duodenal margin must be incised so that it and the head of the pancreas can be freed to gain an approach. If the stone plugs the ampulla of Vater, the duodenum will have to be entered. Any obstruction of the biliary tract system causes bile to back up into the liver where it will be absorbed gradually into the blood stream and give the skin a yellowish to greenish jaundiced tinge. The gall bladder can be removed in gall bladder disease, and a large part of the liver can be sacrificed without undue harm to the individual. Nowadays, parts of a diseased liver can be removed segmentally without sacrificing unnecessarily large portions of it. Healey and Schroy[5, 6] have studied and described **liver segments** on the basis of the branching patterns of the right and left hepatic ducts and vessels. Arteries and veins tend to follow main duct branchings as stems to segments of the liver, much as do the main respiratory ducts to bronchopulmonary segments.

The Gall Bladder: Observe the musculomembraneous sac which constitutes the gall bladder. The organ may be suspended from its fossa of the right lobe of the liver by a mesentery, or it may be anchored down more securely to the liver by a reflection of the peritoneum of the liver. It is about 8 to 9 cm. in length, 3 cm. wide, and it can hold about 30 to 40 cc. Its broadest part, or fundus, may project beyond the inferior margin of the liver. Its body and neck narrow

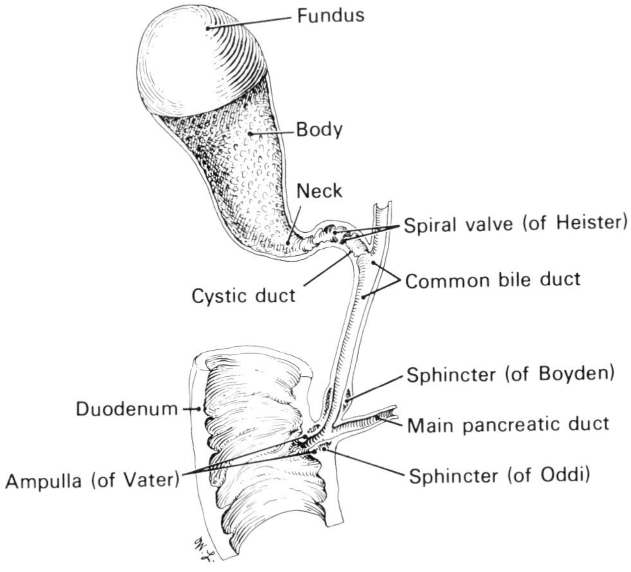

Fig. IX-92. Gall bladder and bile ducts cut open.

conically toward the cystic duct. The latter is a 4-cm.-long tube extending from gall bladder neck to common bile duct. Its internal mucosa is thrown into a series of circular folds like those at the neck of the gall bladder. These folds coil along the inner aspect of the cystic duct creating a kind of spiral valve, the valve of Heister (Fig. IX-92).

Lorenz Heister was a German surgeon and anatomist (1683–1748), a graduate of the University of Leyden. He was an author of several texts in Anatomy and Surgery, and he described the cystic spiral valve in 1717, having also coined the term, "tracheotomy," in 1718. We could fish out several more of these odds and ends of useless nuts and bolts from our anatomical repair kit, and may even do so from time to time. If you try some of them on for size in the recesses of your subconscious, it may make you somewhat erudite but it will do you little harm in the long run. If they don't fit your machinery, they can be discarded with impunity.

Oral Cholecystography—The Roentgen Examination of the Gall Bladder: Cholecystography means simply the radiographic visualization of the gall bladder following ingestion of appropriate radiopaque contrast medium. This procedure is the most commonly employed method for evaluation of the gall bladder. Intravenous and transhepatic cholangiography may be necessary to appraise the status of the bile ducts in patients having pathological conditions such as the postcholecystectomy syndrome, obstructing stone, or when the gall bladder has not been visualized by the oral method. Herein we will concern ourselves only with oral cholecystography.

1. PROCEDURE: We believe it prudent to obtain a plain (scout, KUB) film of the abdomen before administering the oral contrast medium (Telepaque®, Winthrop; Orografin®, Squibb) necessary to visualize the gall bladder. This AP supine film of the abdomen shows many of the anatomical structures within the abdominal cavity. (Details of these structures are described on p. 633.) The real value of the scout film is that abnormalities of the biliary system may be detected before they are obscured by the contrast medium. Other pathological conditions may also be noted, thereby altering the intended plan of radiographic or clinical evaluation.

The oral contrast medium is excreted in the bile and is concentrated in the gall bladder (Fig. IX-93A). Radiologists vary in their methods of studying the gall bladder. We prefer to obtain PA, 20° right posterior oblique, and erect films of the gall bladder, the latter at the time of fluoroscopy when the patient may be rotated into the most desirable position that eliminates overlapping gas shadows within the adjacent bowel. It also permits the radiologist to apply varying amounts of compression to the gall bladder in the search for abnormality. After these initial radiographs have been carefully surveyed and no contraindication found, it is helpful to give the patient a *fatty meal* to study the emptying mechanism of the gall bladder. Every attempt is made to show the cystic duct and the common bile duct in their entirety (Fig. IX-93B) in order to assure completeness of the examination.

2. ANATOMY: Figure IX-93B demonstrates some of the anatomical details that may be seen as the gall bladder contracts. The fundus, body, and neck of the gall bladder—all of which are rather arbitrary divisions—are labeled for you. The spiral valve (of Heister) is evident within the cystic duct. The cystic duct is approximately 4.0 cm. in length and approximately 0.3 cm. in diameter normally. The cystic duct may be seen joining with the common hepatic duct to form the common bile duct which is typically about 0.6 cm. in diameter and about 7.5 cm. in length. Note, too, that some of the contrast medium has refluxed into the hepatic duct as far back as the right and left tributaries. On occasion, the entrance of the common bile duct into the duodenum may be visualized.

The Pancreas (Fig. IX-94): We have noted that this major salivary gland of the abdomen is cradled in the arms of the duodenum, occupies a retroperitoneal position behind the stomach and extends virtually to the spleen. It has a head and, as a consequence, a restricted neck as well as a body which ends in a long tail. The neck is constricted by the superior mesenteric vessels where they emerge from behind it to clamp the third part of the duodenum and left renal vein. Just to the right of this constriction, a projecting part of the pancreatic head is known as the **uncinate** (hooked) **process.** It may tuck itself behind the mesenteric vessels also (Fig. IX-94). A **tuber omentale** is often seen as a prominent process which projects upward from the body toward the lesser omentum.

The pancreas secretes an alkaline digestive juice which is conveyed to the duodenum through the pan-

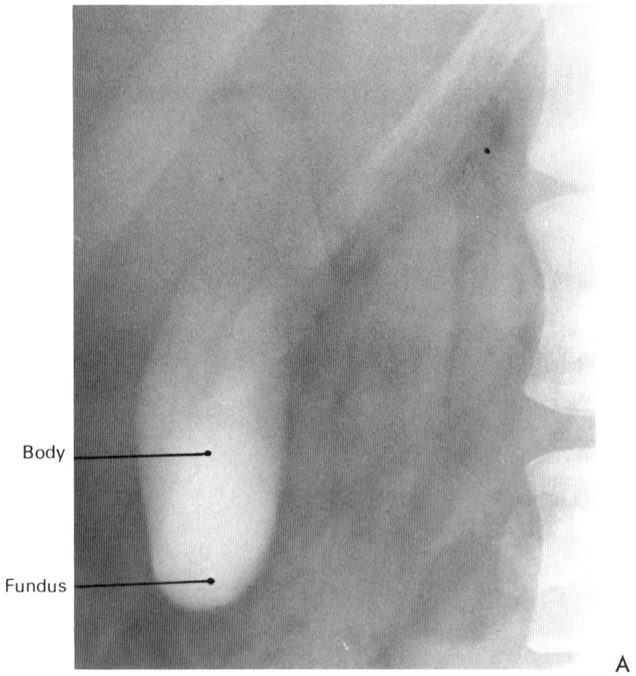

Fig. IX-93. Oral cholecystogram. **A.** 12 hours after ingestion of the contrast medium. Right posterior oblique projection. **B.** Same projection, 10 minutes after ingestion of a fatty meal. Note the fine lucent folds of the spiral valve of Heister in the cystic duct.

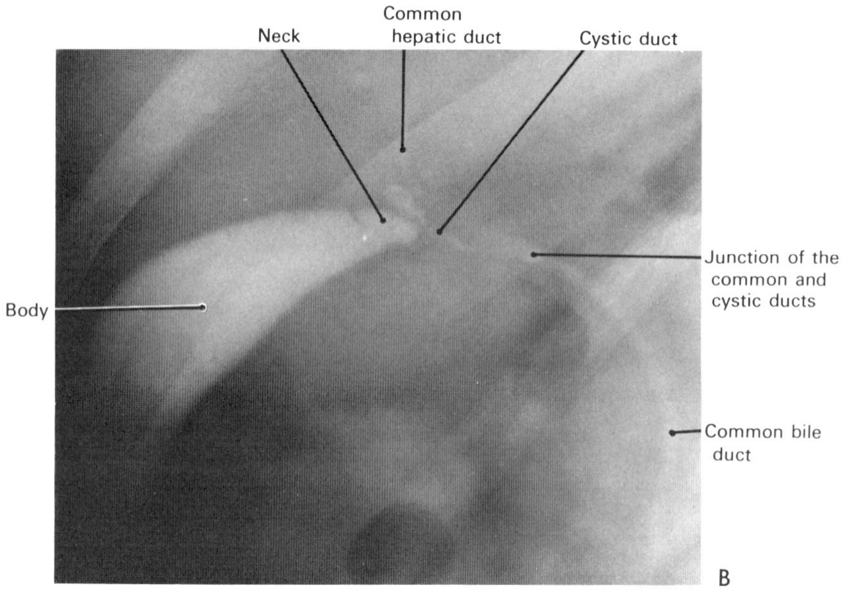

creatic duct (or ducts). At least nine enzymes, along with water, bicarbonate, and salts, make up the 2 liters of juice secreted in 24 hours. There are the proteolytic enzymes trypsinogen, chymotrysinogen, and carboxypeptidase. There is a pancreatic amylase for carbohydrate breakdown and the fat-splitting enzymes, lipase and lecithinase. There are also such hydrolyzing enzymes as ribonuclease and deoxyribonuclease.

From the cells of the islets of Langerhans, the endocrine portion of the gland, insulin and glucagon are secreted into the capillary beds of the gland. The first of these products plays an important role in carbohydrate metabolism. Without the hormone, insulin, the cells of the body are unable to utilize available glucose. The clinical condition resulting from a deficiency of insulin, diabetes mellitus, is characterized by varying degrees of hyperglycemia and glycosuria.

Glucagon has an effect somewhat antagonistic to insulin. It causes an elevation of blood sugar by breaking down glycogen in the liver, hence it is a hyperglycemic-glycogenolytic factor.

The Spleen: This is a lymphoid organ lying to the left of the stomach. It has a heavy fibromuscular capsule of its own, but this is covered by the glistening visceral peritoneum of the dorsal mesentery which the developing organ invaded. Its diaphragmatic surface is smooth and convex (Fig. IX-94) while its visceral surface, facing the stomach and the tail of the pancreas, is triangular in outline. Its perimeter borders face anteriorly, posteriorly, and inferiorly, respec-

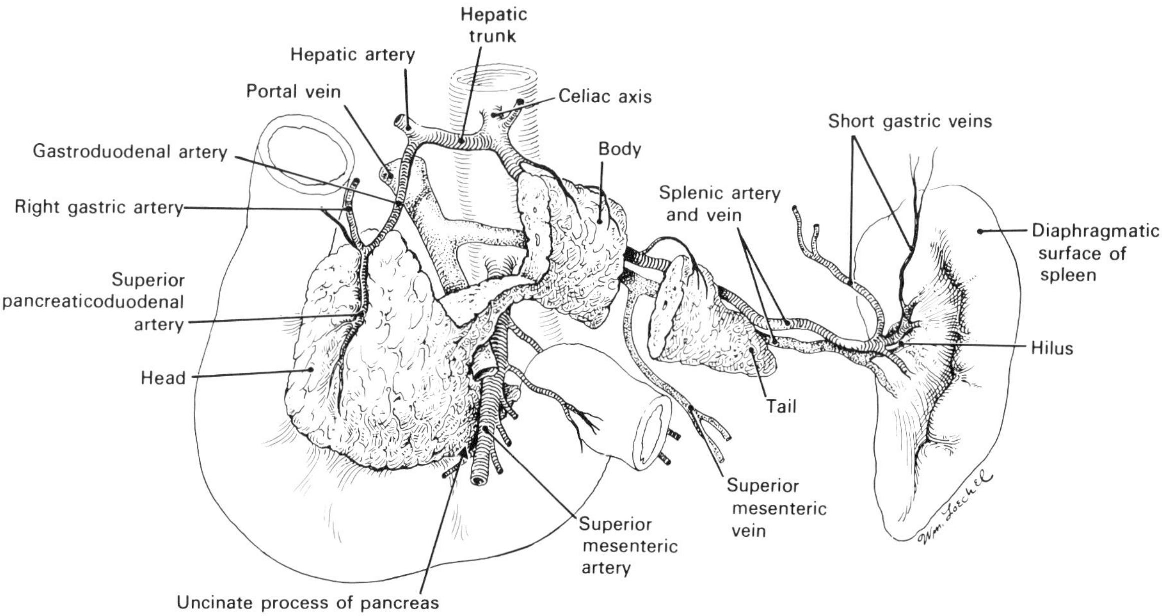

Fig. IX-94. Pancreas and spleen. (In this specimen the right gastric artery arose from the gastroduodenal artery which it does with about an 8% frequency.)

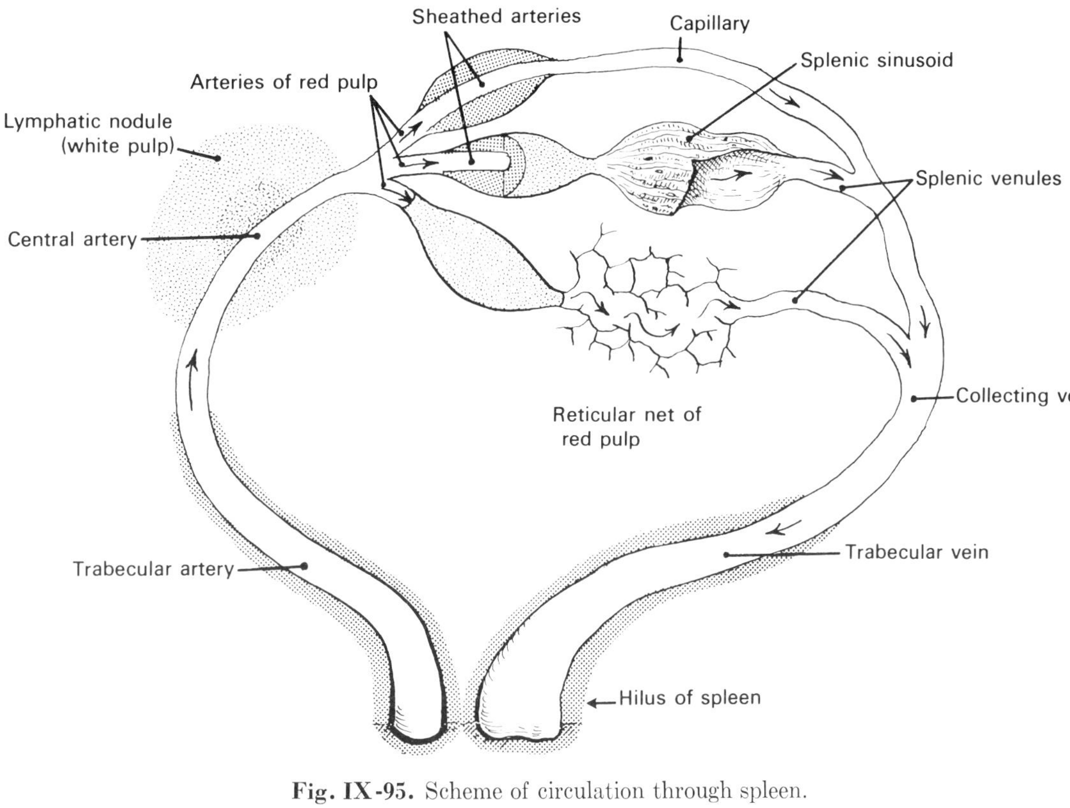

Fig. IX-95. Scheme of circulation through spleen.

tively. The anterior border is notched. The central **hilus** is where splenic vessels enter and leave the organ. It receives the terminal part of the splenic artery, one of the chief branches of the coeliac trunk. Its vein enters the portal vein. Its nerves, to smooth muscle of the capsule and to the trabecular septae that radiate from the capsule, and to the smooth muscle walls of its arteries and arterioles, are derived from the coeliac plexus. It is a highly distensible organ, capable of storing about one-fifth of the volume of circulating red cells. But we can get along without it if we have to.

If lymph nodes act as lymphatic filtering plants where bacteria, debris, and worn-out lymphocytes can be removed from the lymph, then the spleen can be regarded as the filtering plant for circulating blood (Fig. IX-95). Both lymph nodes and the spleen are organs where lymphocytes can be manufactured. In the spleen, this process is relegated to areas of the **"white pulp"** which are seen in histological sections as scattered about among the **"red pulp"** like so many plums in a well-stirred pudding. Actually, "white pulp," or **splenic nodules,** represent modifications of the adventitial coats of the splenic arterioles. As the trabecular branch of the splenic artery enters the "pulp" of the spleen, it loses its normal fibroelastic adventitia and gathers about itself an adventitia of reticular connective tissue. It is now called the **central artery of the splenic nodule,** though the lymphocytes manufactured in its reticular coat make it appear quite eccentric in position within the nodule of white pulp. The branches of the central artery lose their reticular coat as they enter the "red pulp," *i.e.,* the spaces between splenic sinusoids which make up the bulk of the pulp of the spleen. This pulp is, essentially, a reticular connective tissue net supporting specialized sinusoids and the arterioles and venules that connect with these sinusoids. The branches of the central artery (of the node) are then known as **pulp arteries.** They branch away like a number of bristles on a paint brush, hence, the lot of them from a single central artery are termed penicilli. They become **sheathed arterioles** as soon as they loose their muscular coat and gather a tight swirl of reticular tissue about themselves. Finally, they become terminal arterial capillaries. Apparently, these capillaries can end in several ways in the red pulp. They may:

1. Immediately connect with a pulp venous capillary;
2. Empty into splenic sinusoids; or
3. Empty into the reticular tissue network spaces which are drained by venules.

The **splenic sinusoid** is made up of elongated littoral cells. These cells are shaped like barrel staves, and the barrel which they form is the sinusoid, coopered about by rings of reticular fibers. There appear to be physiological sphincters at both ends of these microscopic barrels. In filling, the distal sphincter is closed. Blood rushes in to distend the barrel, and the proximal sphincter closes. The barrel is leaky between the staves so that plasma leaks into the reticular net to be drained away by venules, but cells are retained in storage. An occasional red cell may snap through between the staves much as one snaps a melon seed between the fingers. Some of the littoral cells are phagocytic, as are many cells of the reticular network. No matter whether blood enters sinusoids for storage or whether terminal arterioles dump into the reticular network, particulate matter and worn out red cells can be phagocytized to reclaim hemoglobin and bilirubin. After storage, when the sphincters open, the compaction of cells can be flushed free to return to the circulation through venules and the splenic vein at the hilus. Under conditions of partial hypoxia, as in acclimatizing to high altitudes, the spleen can become hemopoietic. In reverting to its embryonic blood-forming role at a time like this, extra red cells are placed in circulation, thus providing additional hemoglobin for needed oxygen transport.

The Genitourinary System:

General: We have completed the discussion of one large concern within the abdomen, the "G-I Tract and Its Associates," a middle of the road outfit with its unpaired associates, the liver, pancreas, and spleen, and its unpaired blood supply derived from the coeliac, superior, and inferior mesenteric vessels. We now turn our attention to the second major industrial concern in the area, the bicameral "G-U Tract and Its Associates." This firm operates on both sides of the road, has a dual function (genitourinary), and is an amalgamation of paired glands and paired vessels. Even as it started in the business, this company decided quite early to set up shop in such a way that the same equipment could do double duty. Its only concession toward innovation was to turn out two models. There is fundamentally little difference between the two except in exterior streamlining and trim. The urinary plumbing is identical save for the exterior pet-cocks. While the reproductive products differ, there is fundamentally little difference in the urinary machinery which, at times, seems almost interchangeable and quite plastic.

The G-U system is located dorsal to the G-I system so that any part of the latter must cross in front of any part of the former. The gastrointestinal tract, therefore, lies ventral to the suprarenals, kidneys, and gonads; to suprarenal, renal, or gonadal vessels; to the ureters, uterine tubes, and deferent ducts. Among the tubular parts of the G-U system, itself, the genital tubes (ductus deferens and testicular vessels; uterine tubes and ovarian vessels) must cross ventral to the urinary tubes (the ureters). These relationships are made clear in Figure IX-96 which shows how the kidneys and sex glands have migrated to the present locus, and where they finally come to rest relative to the common iliac vessels and their branches. The vessels, of course, lie deepest of all and are crossed by every one of the tubes.

The three paired glands of this system are the suprarenals (adrenals), the kidneys, and the sex glands (testes and ovaries). The suprarenals are endocrine glands which produce an internal secretion and, so, are ductless. The kidneys excrete a filtrate, urine, which is carried to the bladder by ducts, the ureters. The sex glands produce cells and internal secretions. Spermatozoa of the testes reach the male urethra through the deferent ducts; ova of the female reach the uterus over the uterine tubes. Internal gonadal secretions (testosterone, estrongen, and progesterone) enter the blood capillaries directly and require no duct system.

All of these organs developed in the extraperitoneal connective tissue of the abdominal wall like any other abdominal organ. They remain retroperitoneal. The suprarenals developed pretty much at the same level at which they are found in the adult. We have already considered the descent of the testis from its site near the original kidney. The ovaries descended also, but not nearly as drastically. They are not thrust forth into a hostile, external environment to brave the elements and to be banged about by knocked-knees. The

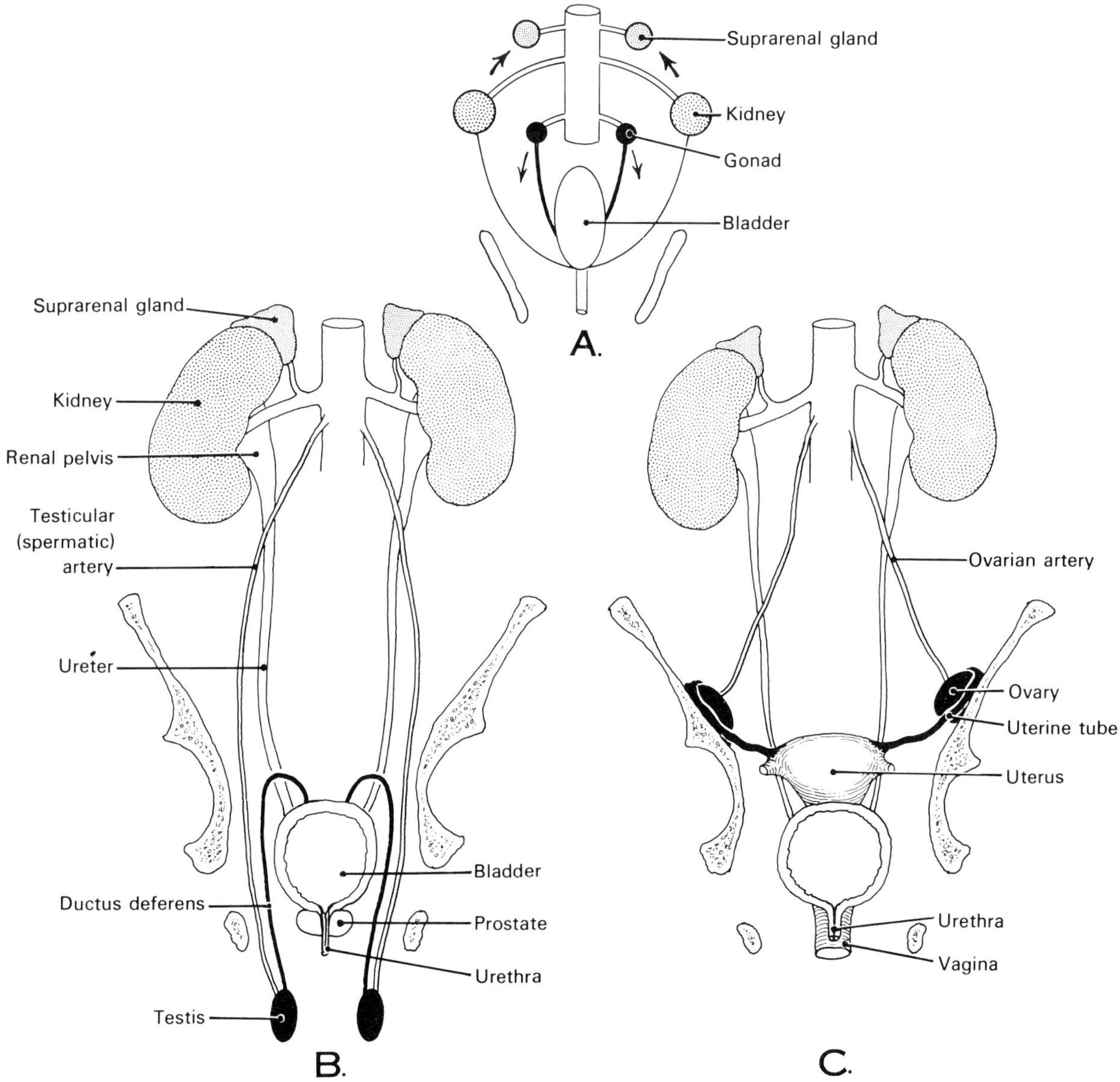

Fig. IX-96. Diagram to show migration and final position of the paired abdominal organs in the adult. **A,** early position in embryo; **B,** final position in the male; and **C,** in the female.

Table IX-6. Summary of the Development of the Human Kidney

Organ	Source of Secretory Tubules	Source of Collecting Tubules	Origin of Excretory Duct	Somite Level of Origin	Stage of Earliest Appearance (Somite Age)	Stage of Maximum Development
A. Pronephros	Segmental	Segmental	None	7–14	13 somites	Middle of 4th week
B. Mesonephros	Unsegmented nephrogenic cord	Unsegmented nephrogenic cord	Old pronephric duct	10–26	18 somites	4th–9th week
C. Metanephros	Caudal nephrogenic mass	End branchings of ureteric bud	Ureteric bud from mesonephric duct	26–28	Start of 5th week	After birth

Organ	Period of Degeneration	No. Tubules in Each Kidney	Characteristics of Tubules	Urinary Function	Permanent Features	Functional Derivatives
A. Pronephros	By end of 4th week	±4	Segmented; short; nephrostome; external glomerulus	None	None	None
B. Mesonephros	5–16 weeks	±4	Non-segmented; S-shaped; internal glomerulus; renal-portal blood system	Only embryonic	Some tubules and ducts are retained	Efferent ductules and ductus deferens of male reproductive system
C. Metanephros	Early order of tubules degenerate 7–20 weeks	±2 million	Complex; long orderly arrangement	Relatively little use prenatally; postnatal use predominantly	Entire organ and duct	The permanent excretory system

kidneys, on the other hand, ascended from their original developmental site, seeming to pull themselves up an arterial ladder as they moved from one set of bilateral branches to the next higher set. Had they not let go of some of these rungs, the kidneys could have more than one renal artery apiece, and sometimes they do. If the paired kidneys fuse low in the pelvis before ascent so as to form a horseshoe kidney, they get caught beneath the inferior mesenteric, midline, arterial stem and can rise no further.

Development of the Kidneys and Their Ducts: For the student who is already familiar with an account of kidney formation in the embryo and fetus, Table IX-6 may be sufficient to aid him in the review of the basic details. What we have to say here is for the uninitiated. The mammalian embryo appears to experiment with three sets of kidneys before it settles upon the definitive one. As it discards one plan for another, it attempts to rescue parts that can serve other uses. Some parts that are not discarded entirely may remain around as useless ornamentation, vestigial remnants in one sex or the other. The three kidneys that are tried in succession are designated: the **pronephros;** the **mesonephros (Wolffian body);** and the **metanephros or permanent kidney.**

Recall from Chapter I that forming mesoderm differentiated into:

1. **Paraxial mesoderm, or somites,** which gave rise to vertebrae, segmental musculature and dermis of the skin in part.

2. **Lateral plate mesoderm** which split into somatic and splanchnic sheets. The former supplied dermal tissue and certain skin muscle; the latter, parts of the walls of viscera, especially their connective tissue and smooth muscle components.

3. An **intermediate mass** (Fig. IX-97A) now is our major concern since it is nephrogenic material. From the very beginning it lay close to the developing, inner body-wall, *i.e.*, to the intra-embryonic coelom.

The Pronephros (*Figs. IX-97 and 98*)*:* This first kidney, functional today only in cyclostomes and a few fish, develops in the 1.7-mm. human embryo and is gone in the 4.25-mm. embryo. That means it starts early in the fourth week and is gone by week-end. In a sense, it is a neck and upper chest kidney, for it develops rostrally in the cervical and upper thoracic levels where the intermediate mass is segmented. Below this level, the nephrogenic mass is not segmented. Specifically, segments 7 to 14 are involved. Each pronephros (since this is a bilateral system) consists of about **7 to 10 small tubules** that connect eventually to a collecting, **pronephric duct.** Anterior tubules develop first and disappear before the posterior ones begin to form (Fig. IX-98). The tubules arise as solid, blind-ending "spigots" from the intermediate mass. As the end of the spigot turns caudally at each segmental level, the solid mass hollows out a lumen, and the terminal end connects with the "spigot" next caudal to it to form the cellecting duct (Fig. IX-99). The other end of the tubule opens into the coelom (Fig. IX-97B). The medial wall of the tubule near the coelom is invaginated by a glomerular capillary tuft, the ending of a short renal twig which comes from the aorta. In man this external pronephric glomerulus is short-lived and apparently non-functional. This arrangement is kept by cyclostomes, some teleosts and dipnoi so that fluids can be filtered out of the coelomic cavity. Other forms add a blood-filtering glomerulus of the metanephric type to their pronephric kidney. But, in man, the pronephric tubules degenerate and only the *pronephric duct* remains to play an important role in the later embryo and even in adult life.

The Mesonephros: This kidney develops from the fourth to the ninth week at the 10th to 26th somite level where the intermediate mass remains unsegmented. It is, thus, a thoracolumbar kidney in the embryo, extending to about the third lumbar vertebrae. It becomes a much larger organ-mass than the pro-

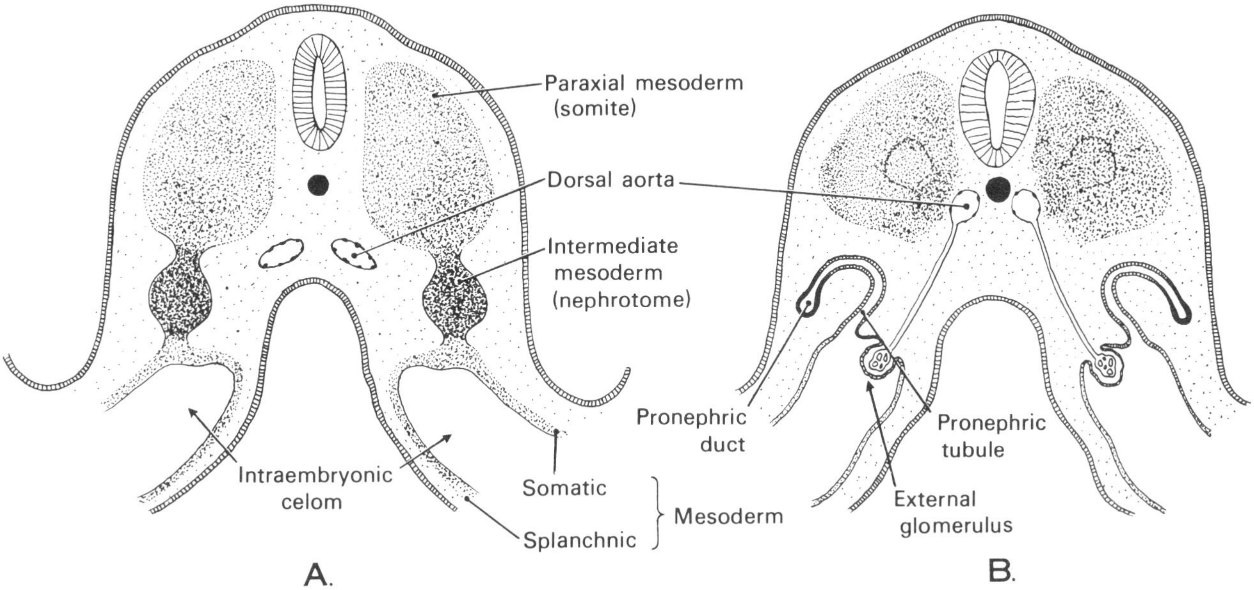

Fig. IX-97. Schematic cross-section through cervical region of early embryos showing formation of pronephric kidney. **A,** at 21 days; **B,** at 25 days. In man the external glomerulus is non-functional and these tubules degenerate rapidly, but the pronephric duct remains. (After Langman.)

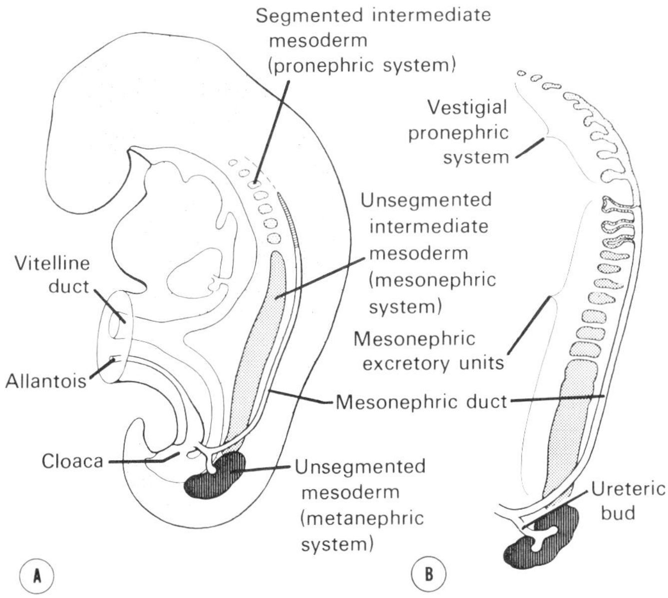

Fig. IX-98. Relation of intermediate mesoderm to the development of the three kidney types. **A.** Showing segmented masses for pronephros and unsegmented masses for mesonephros and metanephros at lower levels. **B.** Early tubule development and partial regression and the development of the ureteric bud. (From Langman, *Medical Embryology,* 2nd ed., Williams & Wilkins, Baltimore, 1969.)

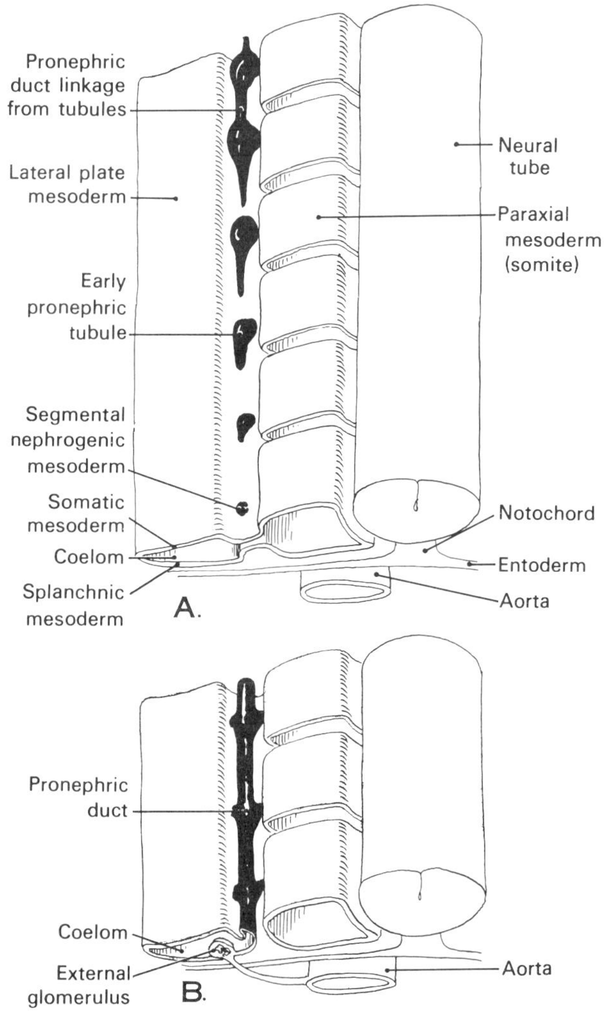

Fig. IX-99. Development of pronephric duct system (3-dimensional model). **A.** Young stage with tubules forming and linking in cephalocaudad progression. **B.** Older stage with pronephric duct formed and serving an external glomerulus to filter wastes from coelom. Such a kidney is functional in lower vertebrates. Only the duct is salvaged in the developing human embryo. (After Arey.)

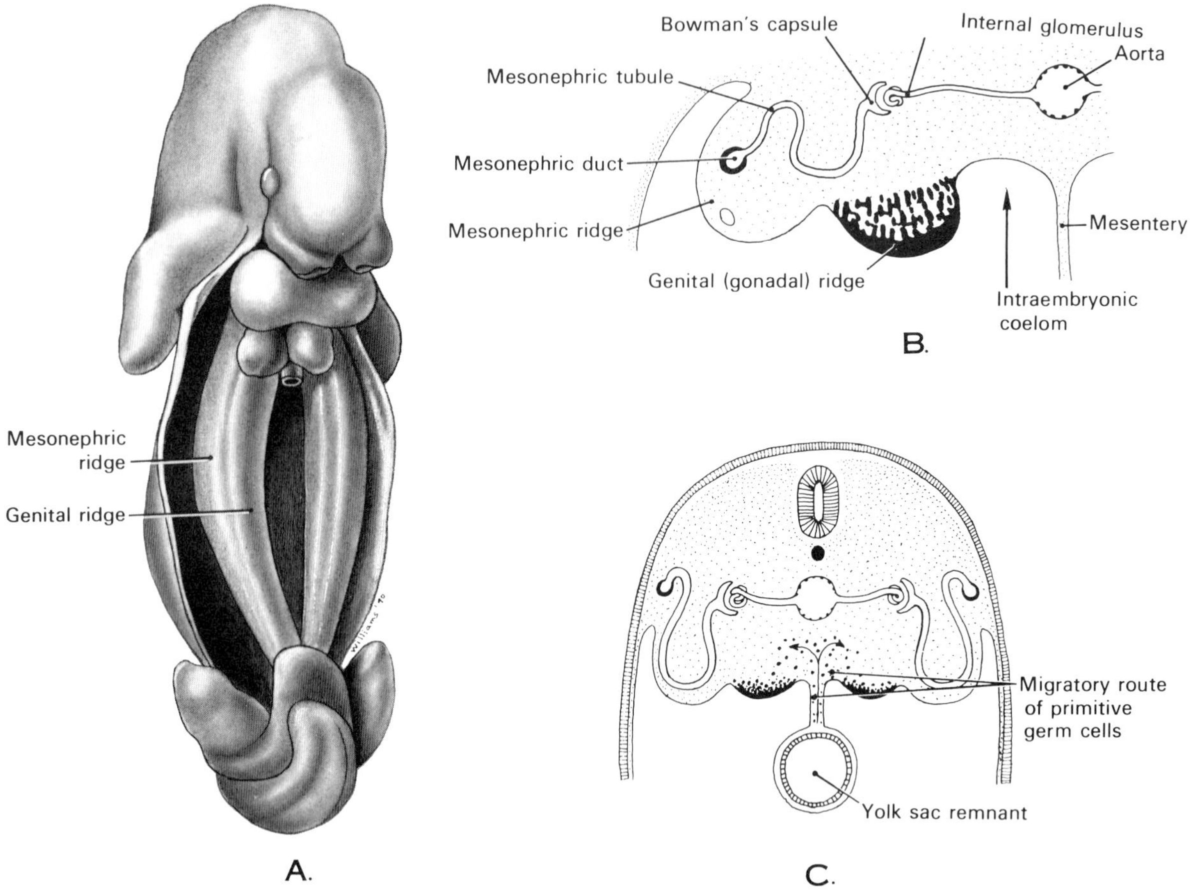

Fig. IX-100. Development of the mesonephric kidney. **A.** About a 5-week embryo showing size and location of mesonephros. **B.** Cross-section midway through mesonephros shown in A. **C.** Diagram to show early origin and migration of primitive germ cells from yolk sac mesoderm to gonadal ridge.

nephros, having about 42 tubules of an advanced style. Not all of these tubules are present at any one time, however. The total mass is prominent enough to create a longitudinal, **mesonephric ridge** which bulges into the coelom (Fig. IX-100A). Note its close proximity to the **gonadal** or **genital ridge.**

Each of the tubules tap the pronephric duct which now conveniently changes its name to the **mesonephric** (or Wolffian) duct. Caspar Friederich Wolff (1733–1794) first described this duct in the chick in his graduation thesis of 1759. He was a German Anatomist who later taught embryology in Russia at St. Petersburg. He probably can be regarded as the father of the germ layer theory in embryology. At any rate, the elongate, S-shaped tubules become invaginated at their distal ends to surround capillary whorls which are the terminal endings of a series of tiny aortic branches at this stage (Fig. IX-100B). Thus, they form **internal glomeruli** as distinct from the external glomeruli of the pronephric system (Fig. IX-100B). The distal end of the tubule and its glomerulus are typical of **renal corpuscles.** The tubule contributes **Bowman's capsule** to the corpuscle. The unit serves, in lower forms, to filter urea and other plasma constituents into the rest of the tubule from whence it is passed into the mesonephric duct. The duct empties caudally into a common reservoir for urinary and digestive tract wastes, **the cloaca** (Fig. IX-101A). This reservoir has not yet subdivided into urinary and rectal compartments. This mesonephric kidney does not function during embryonic life in the human, but a definitive, functional, metanephric kidney, as we know it in the adult, is formed instead.

The Metanephros or Permanent Kidney: This final kidney has a dual origin. Its ureter, pelvis, calyces, and collecting tubules arise from an outgrowth or diverticulum near the caudal end of the mesonephric or Wolffian duct (Fig. IX-101A). The rest is derived from a capping **nephrogenic mass** as the level of the 26th to 28th somite. Hence, this is a low, lumbar or virtual pelvic or, at least, a low abdominal kidney in the embryo. The **ureteric bud** arises from the Wolffian tube during the fifth week. Its growing tip enlarges into a **renal pelvis** (Fig. IX-101B). The unexpanded portion becomes the **ureter,** emptying into the cloaca. The renal pelvis becomes lobulated, expanding into

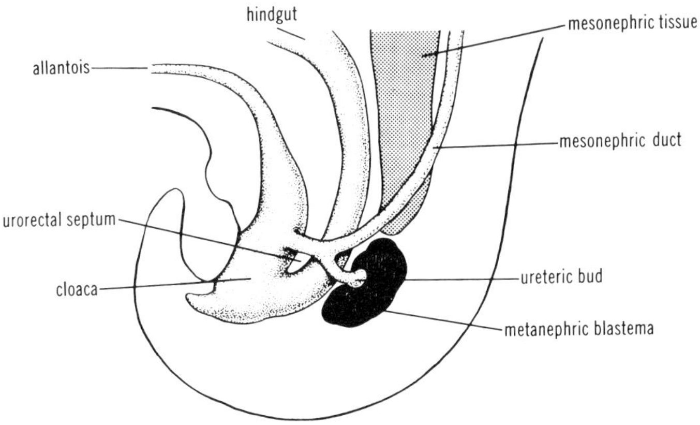

Fig. IX-101. The development of the metanephric kidney. **A.** Ureteric bud arising from metanephric duct to invade metanephric mass in area of hindgut. (From Langman, Medical Embryology, 2nd ed., Williams & Wilkins, Baltimore, 1969.)

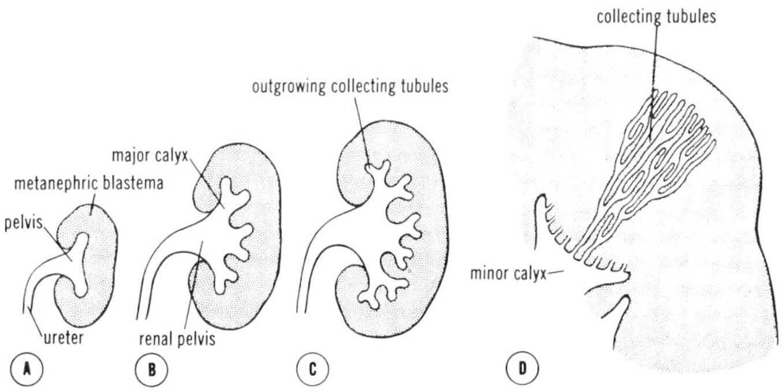

Fig. IX-101B. Stages in development of renal pelvis, calyces, and collecting tubules. (From Langman, Medical Embryology, 2nd ed., Williams & Wilkins, Baltimore, 1969.)

major calyces. Each calyx gives rise to several **minor calyces;** and these, in turn, to collecting tubules. Thus, by the fifth month, the renal pelvis has become a thickly-branched, tree-like structure of about 12 divisions. But before the branching begins, the ureteric bud has invaded the caudal metanephric mass which appears to cap it. With each successive division of the pelvic duct system, the cap also subdivides so that a small mass of nephrogenic tissue lies adjacent to the end of each collecting tubule (Fig. IX-102). These masses will develop into the secretory parts of the **kidney tubule or nephron.** It is the terminal end of this grossly convoluted tubule, derived from the metanephric mass, that will become invaginated by the afferent and efferent glomerular branches of the renal interlobular arteries. Its cup-shaped end, which envelopes the glomerulus, is Bowman's capsule of the adult kidney. The final complex tubule is shown in Figure IX-102C, and here they appear about as they will in the adult. The first renal corpuscles (Bowman's capsules plus glomeruli) appear in the metanephric kidney as early as the second month. New ones continue to develop even after birth. Table IX-6 summarizes the major events in the development of the kidneys.

The Adult Kidneys (Fig. IX-103): These are the organs responsible for removing excess water, salts, toxins, and urea from the blood stream. The pair are of fair size, measuring about 10 cm. in length, 5 cm. in width, and 3 cm. in thickness. The medial and lateral borders are arced, the medial one indenting into a slit-like **hilus.** The hilus is, actually, a hallway into a cavity, the **renal sinus,** which is about 2.5 cm. deep. The sinus is occupied by the renal pelvis, calyces, renal vessels, nerves, and a small amount of adipose tissue. The rest of the organ is solid, moderately elastic, and reddish-brown in color.

These two bean-shaped organs are located retroperitoneally in the upper parts of each of the paravertebral gutters. The right kidney lies about an inch lower than the left because of the position of the liver. The

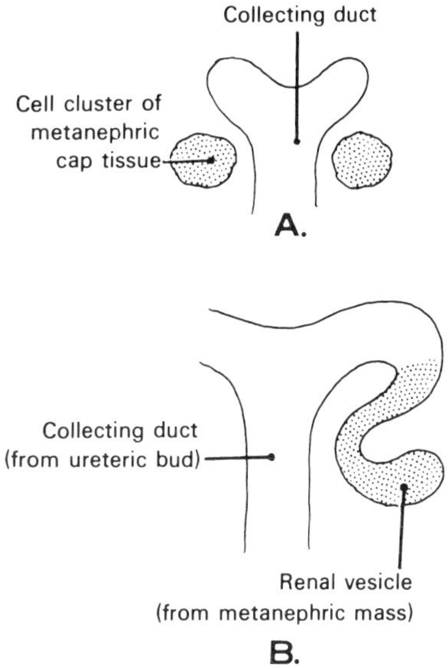

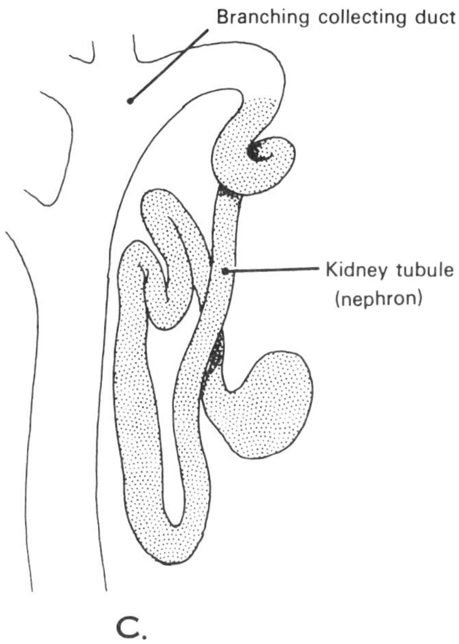

Fig. IX-102. Scheme for the development of a metanephric excretory unit or nephron. A, B, and C show successive stages of union between collecting duct derivative of the ureteric bud and the surrounding metanephric capping tissue.

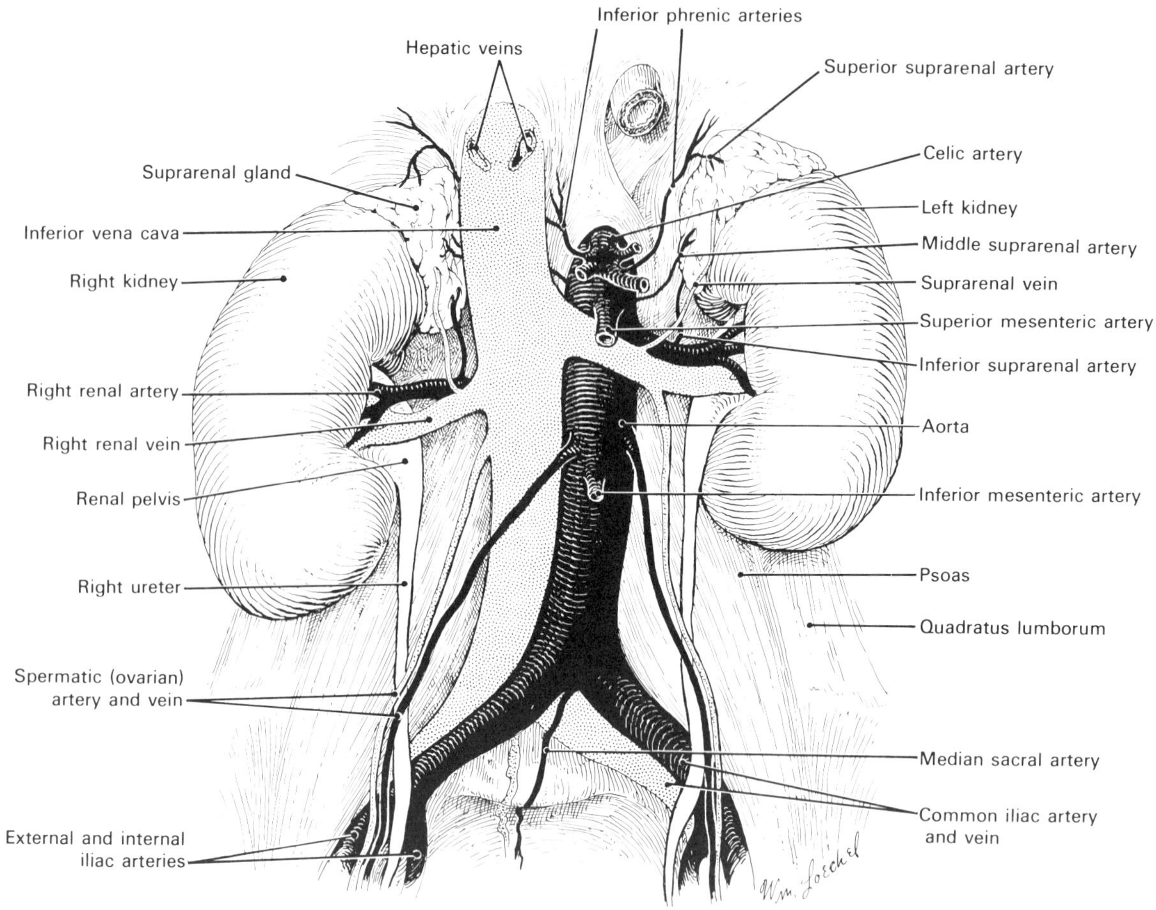

Fig. IX-103. Kidneys and suprarenals in the adult and their blood supply.

posterior surfaces of the kidneys tend to span the 11th and 12th ribs and the transverse processes of VL_1 to VL_3, though the kidneys of the female lie about one-half a vertebral level lower than in the male. The kidneys are somewhat mobile so that they lie a bit lower than normal during expiration and when one is standing than during inspiration and while one is supine. You can consider them as high, lumbar organs in the adult.

The organs have their own fibrous capsule which divides at the hilus into two layers. One layer lines the wall of the renal sinus and the other forms a sheath to cover the vessels before they enter the hilus. Externally, there is a heavy **adipose capsule** which completely invests the organ and dips into the sinus. You might think that the kidney is held in place alone by the peritoneum that covers it. Actually, the subperitoneal connective tissue forms a **renal fascia** passing in front and behind the organ to completely enclose even the adipose capsule. This same fascia passes upward over the suprarenals. To make matters worse, pararenal fat (not to be confused with the adipose capsule or **perirenal fat**) is interposed between renal fascia and the Quadratus Lumborum behind and between renal fascia and the Latissimus and Internal Abdominus posterolaterally (Figure IX-104).

Posteriorly, the kidney relates to the Diaphragm, the left, more so than the right, and to the Psoas Major, Quadratus Lumborum, and the tendon of the Transverse Abdominis. Since a number of nerves (last thoracic, iliohypogastric, and ilioinguinal) and vessels (subcostal and first lumbar) run downward and obliquely across the Psoas and Quadratus, the kidneys lie anterior to them also.

Since the ascending and descending colon, duodenum and pancreas became retroperitoneal in development, portions of them will lie directly upon the kidneys. Where these parts touch kidney, it own peritoneal investment is, of course, raised away from its own immediate anterior surface. It is a safe bet that the upper half of each kidney as well as the lower poles are covered directly by peritoneum and are in close contact with peritoneum.

Laterally, each kidney border extends beyond the lateral limits of the Quadratus Lumborum. Therefore, the kidneys can be reached readily by incising through

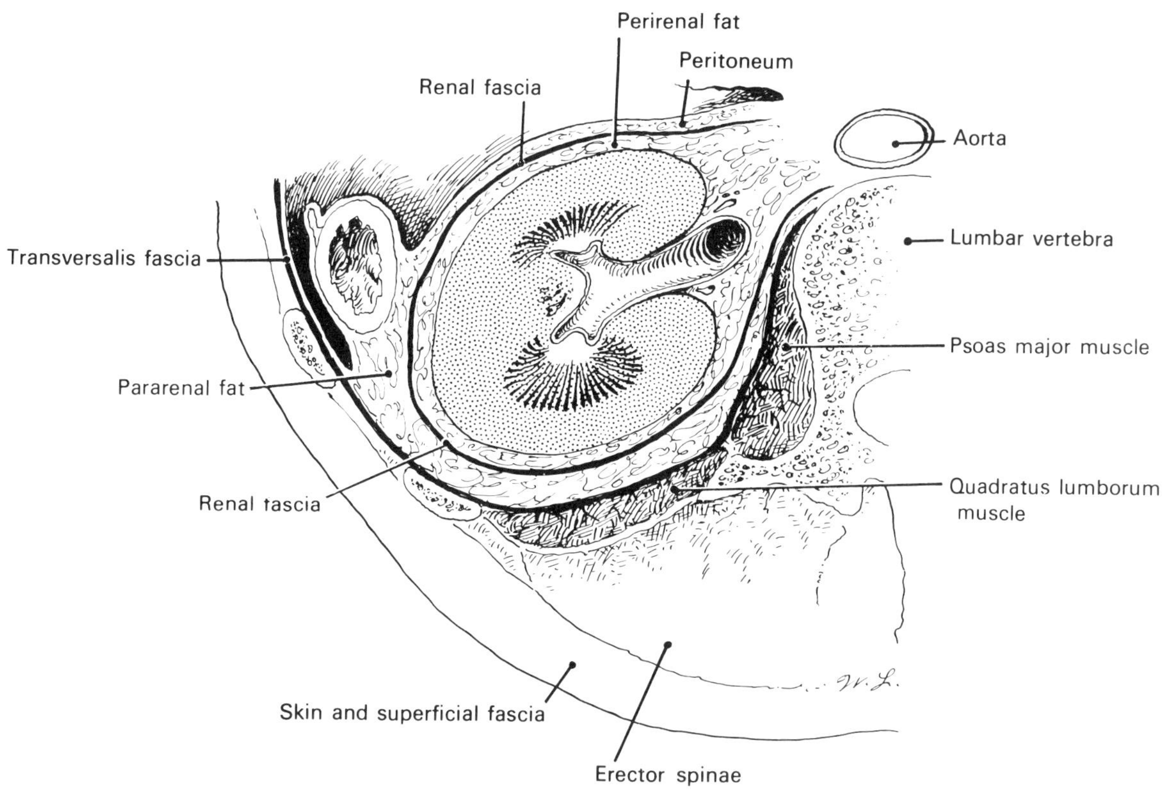

Fig. IX-104. The renal fascia related to posterior abdominal wall.

the posterior body wall at a distance of about 10 cm. from the lumbar spines. The medial border of the right kidney is close to the inferior vena cava, that of the left is about an inch from the aorta. This means that the left renal vein has to be longer than the right renal vein. In its course to the inferior vena cava, it crosses the aorta (Fig. IX-103).

The renal pelvis occupies the bulk of the hilus, and it breaks up into calyces within the sinus. The renal artery enters the hilus anterior to the pelvis, though one of its branches may pass into the hilus posterior to the pelvis. Veins exit ventral to the arteries. Lymphatics hug the pelvis, and autonomic fibers follow in on the adventitial coats of the vessels.

The funnel-shaped renal pelvis collects urine from all parts of the kidney by virtue of its splitting within the sinus into two **major calyces** which are directed toward the upper and lower parts of the kidney, respectively. On occasion, there may be an additional major calyx. All of these intermediate funnels subdivide into still smaller ones, the **minor calyces.** The latter receive one to three **renal papillae** into their cup-shaped ends. There are, therefore, more papillae than calyces. These gather tubular urine from many **collecting ducts** that open at the apices of renal papillae.

If at autopsy a kidney is sectioned sagittally and laid open, these collecting structures can be seen (Fig. IX-105). So can the **pale cortex** and the **darker medulla.** Sometimes even the **renal glomeruli** can be identified, scattered as tiny red dots throughout the cortex. The medulla contains 8 to 18 renal pyramids, striated because of collecting ducts coursing through them. Their bases face the cortex and their apices are the renal papillae. For the location of the various parts of the nephron, in both medulla and cortex, the student should consult Figure IX-106. In places, the cortex dips down between pyramids as **the renal columns.** It should be realized that the cortex is also subdivided into alternate strips of **labyrinths** and **rays.** The labyrinthian segments contain the glomeruli and the proximal and distal parts of the convoluted tubules. The rays contain the straighter parts of the tubule, *e.g.*, descending proximal and ascending distal parts; thick and thin limbs of Henle; and the proximal parts of collecting tubules.

Blood Supply of Kidney (Figs. IX-103 and 105 to 107): As in the case of liver and spleen, to understand the blood supply is to understand organ function. Upon entering the hilus, the renal artery divides into major stems which are directed toward the upper, middle, and lower parts of the kidney. These give rise to **interlobar arteries** that run in the renal columns of the medulla. At the junction of medulla and cortex, interlobar arteries give rise to the horizontally-directed

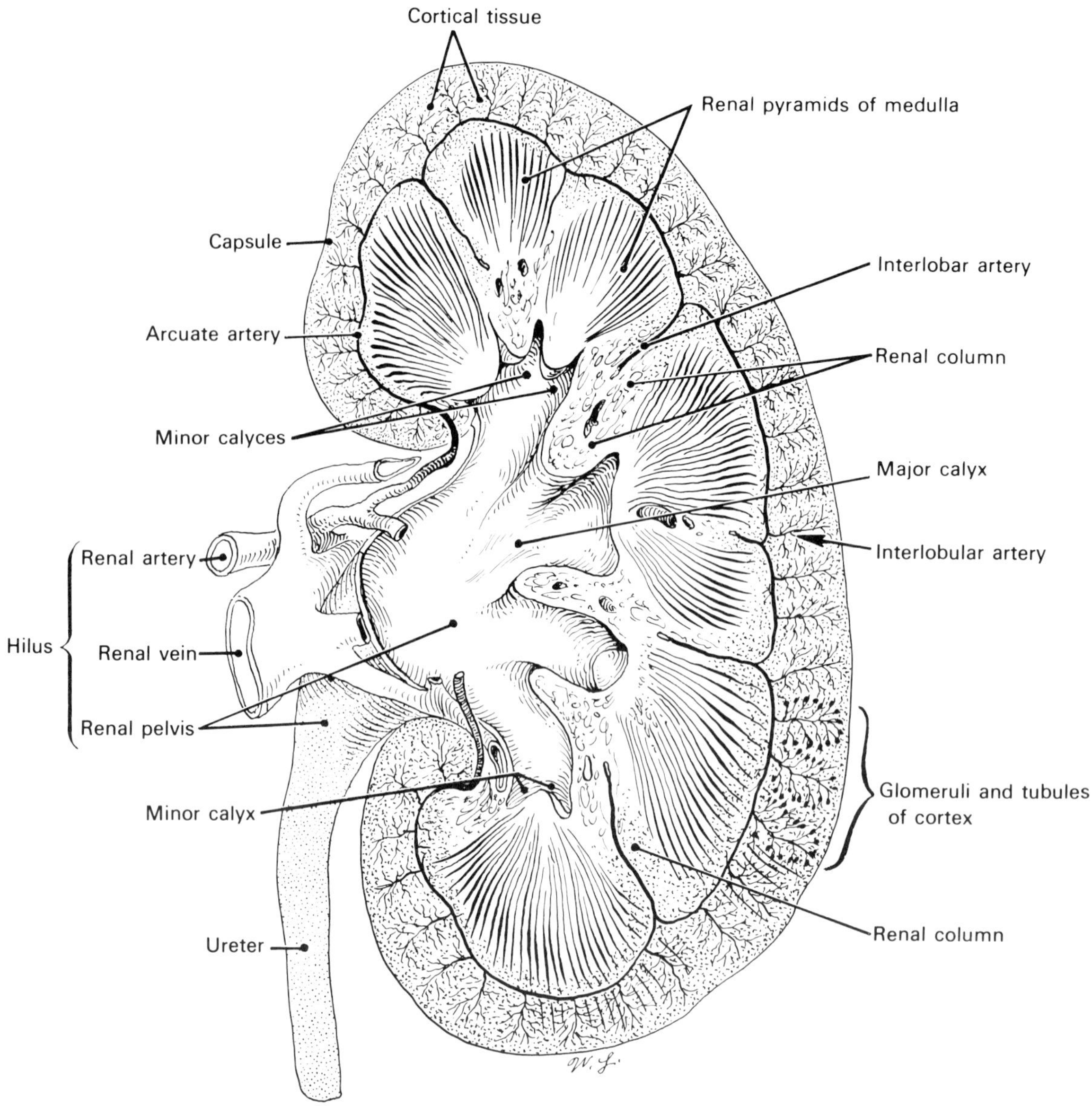

Fig. IX-105. Longitudinal (midsagittal) section through a human kidney.

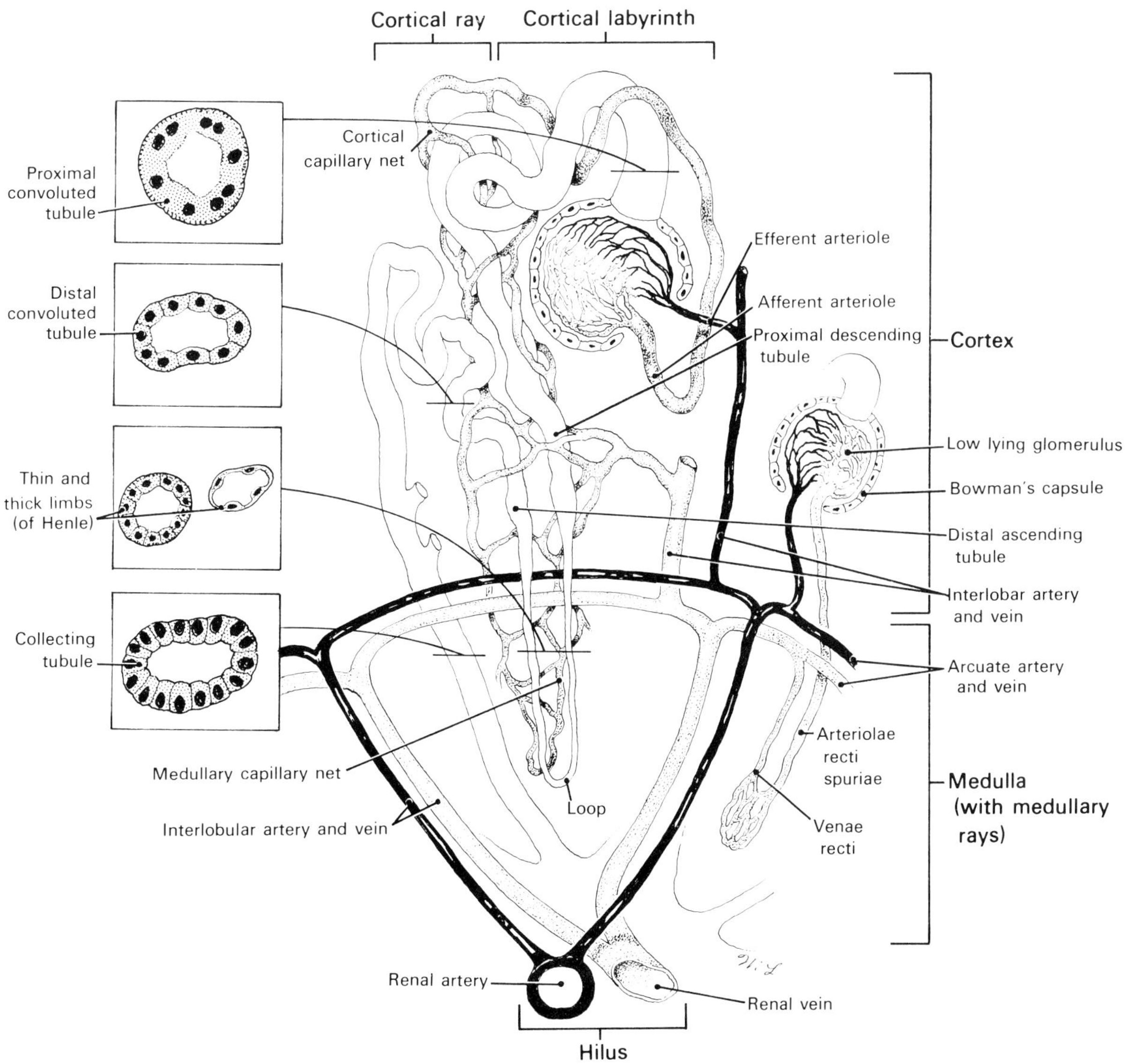

Fig. IX-106. The nephron and the internal blood supply of the kidney. The cortex is divided into alternate ray and labyrinthian portions. Tubular parts of a cortical ray descend into medullary rays. The arcuate arteries mark boundaries of cortex and medulla. To the left, representative cross-sections of tubular parts are shown. Arteriolae rectae spuriae are labeled, but arteriolae verae, which descend into medulla directly from arcuate arteries, have not been shown.

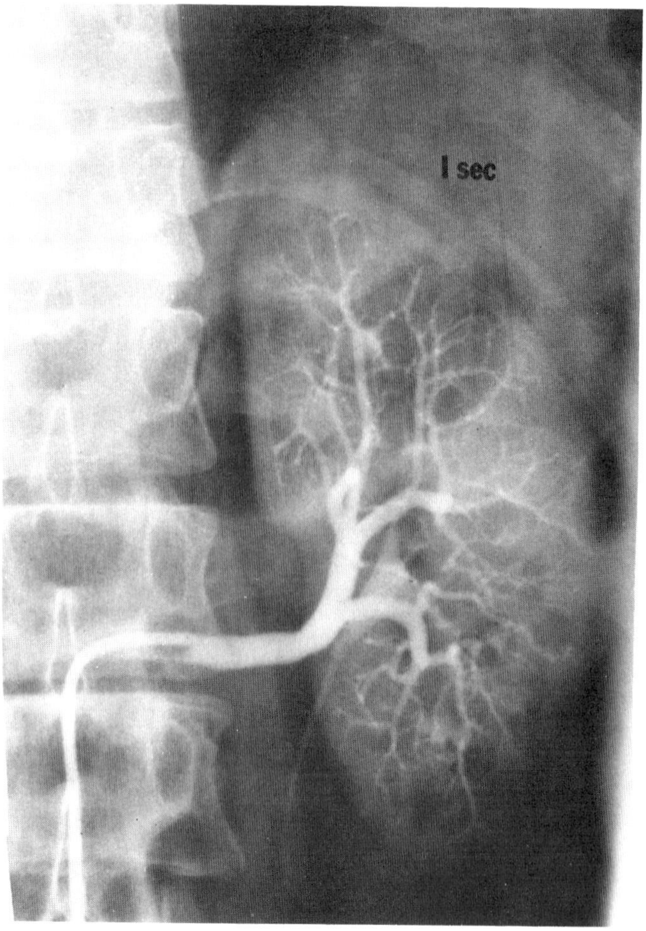

Fig. IX-107. Selective left renal arteriogram. Note the catheter tip in the left renal artery. **A.** One second after injection of opaque contrast medium. The distribution of the segmental renal arteries is shown.

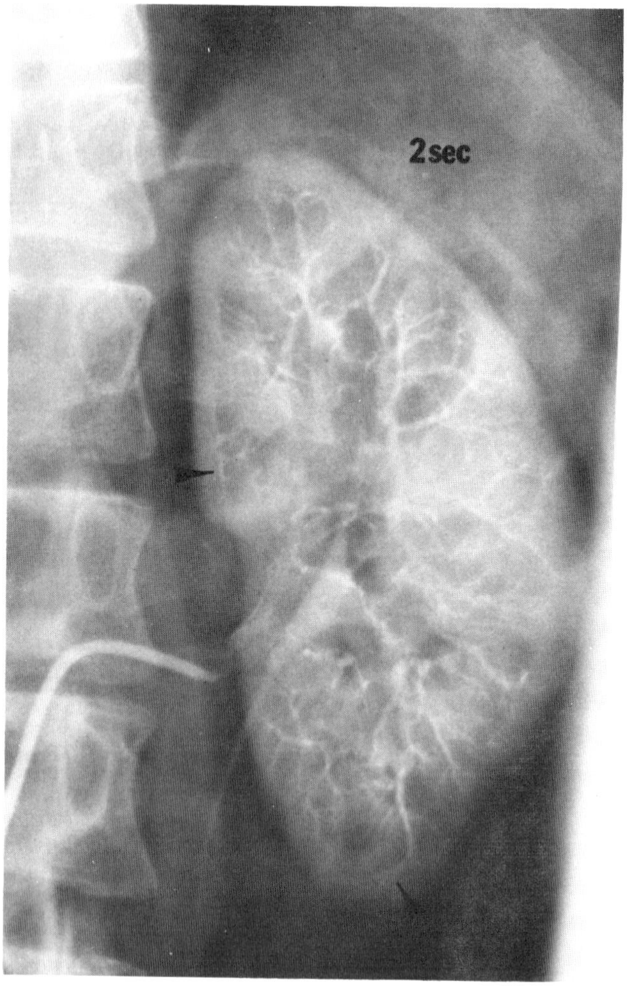

Fig. IX-107B. Two seconds after injection. Arrows show the arcuate artery. Compare with Figure IX-105.

arcuate arteries from which **interlobular arteries** enter cortical rays. Small **afferent arterioles,** which enter glomeruli, arise from the interlobular arteries. **Efferent arterioles** leave the capillary complex of the glomerulus. Here, then, we have a **rete mirabile,** that is a capillary bed supplied and drained by arterioles. As the afferent arteriole approaches the glomerulus some of its muscle cells are replaced by a cuff of myoepithelial cells forming a **juxtaglomerular apparatus.** It is well established that the kidney secretes **renin,** and evidence is accumulating that these cells of the cuff secrete this substance. Renin acts upon another substance in the bloodstream known as **hypertensinogen** to produce **hypertensin** (angiotonin) which causes generalized vasoconstriction. Thus, increased blood pressure produced in this way is known as **renal hypertension.** The emerging efferent arterioles break into a second set of capillaries that surround the labyrinthian and ray segments of the kidney tubules. From arcuate arteries and from glomeruli near the medulla, **arteriolae rectae** extend straight down into the medulla to form capillary networks around nephric tubules and collecting tubules in that region.

Interlobular veins drain the cortical capillary networks and collect from **capsular veins** as well. **Venulae rectae** drain the medulla. Interlobular veins and venulae rectae join **arcuate veins** which enter **interlobar veins.** The latter become the **renal vein** which exits from the hilus.

Nerve Supply and Lymphatic Drainage: Both the vagus and the sympathetic nerves supply the kidney. The role of the vagus is conjectural, but the sympathetics supply the smooth muscle of the vascular tree. Lympatics form both capsular and parenchymatous networks. Both networks drain into upper lumbar nodes.

Function: Urine is excreted from the kidney by means of three processes: *glomerular filtration, tubular reabsorption,* and *tubular secretion.* Electro-

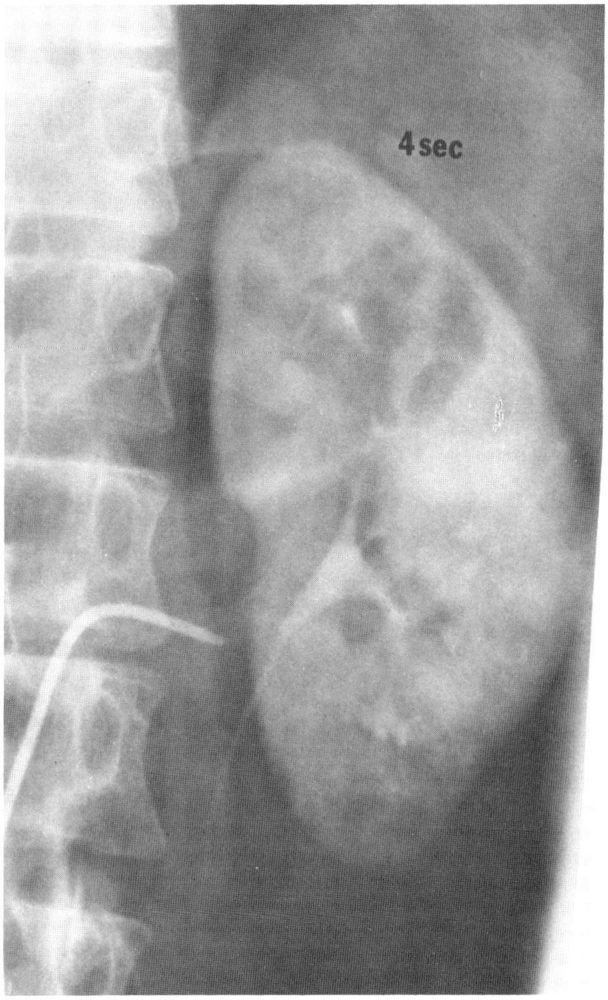

Fig. IX-107C. Four seconds after injection. The so-called nephrogram phase. Contrast medium is opacifying the capillaries as well as the renal excretory system.

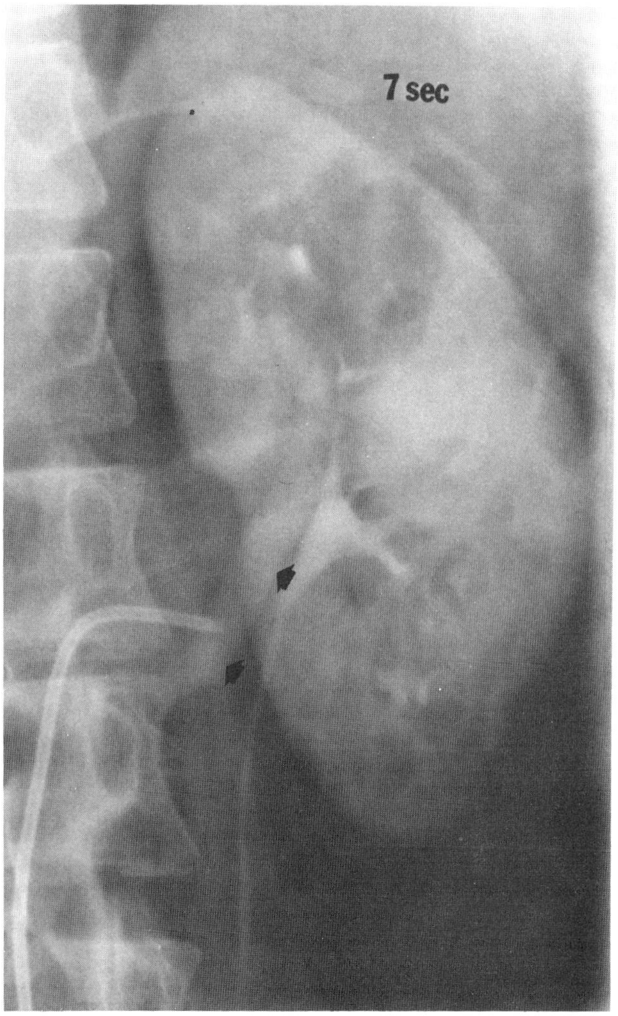

Fig. IX-107D. Seven seconds after injection. Arrow indicates the renal vein. Contrast medium is also seen in the major and minor calyces, renal pelvis, and the left ureter.

lytes, glucose, and water filter from the blood in the glomerular capillaries into Bowman's capsule and the proximal tubule. Filtration is easy and rapid because of glomerular capillary pores and because of the higher hydrostatic pressure. Since the efferent arteriole is smaller than the afferent arteriole, there is a rise in pressure to about 70 mm. Hg as a result of capillary resistance. (Contrast this with a normal capillary-bed pressure of 30 mm. Hg in most places.) The movement is against some resistance, namely, colloidal osmotic presure of plasma and capsular pressure in the tubule, both trying to keep the fluids in the blood stream. Thus, the *effective filtration pressure* amounts to about 20 mm. Hg.

Tubular reabsorption returns most of the water, all of the glucose, and parts of the electrolytes back to the blood, not only through physical diffusion and osmotic mechanisms but also by active transport which requires energy release by the tubule cells.

Tubular secretion means moving substances out of the blood into the tubules by active transport means—a direction which is the reverse of tubular reabsorption. Potassium and hydrogen ions and ammonia are secreted in this way. So are certain drugs, *e.g.*, penicillin, if one can truly speak of kidney secretion. In this latter case we may be dealing simply in semantics.

The Ureters (Fig. IX-108 A, B): These slightly flattened tubes extend from the renal pelvis to the urinary bladder. A few kinks along the 26 to 28 cm. route alter an otherwise fairly uniform diameter. One of these occurs at the ureteropelvic junction where ureter joins renal pelvis; one, where the tube crosses the bony pelvic brim; and another, as it enters the wall of the bladder. The left ureter is slightly longer

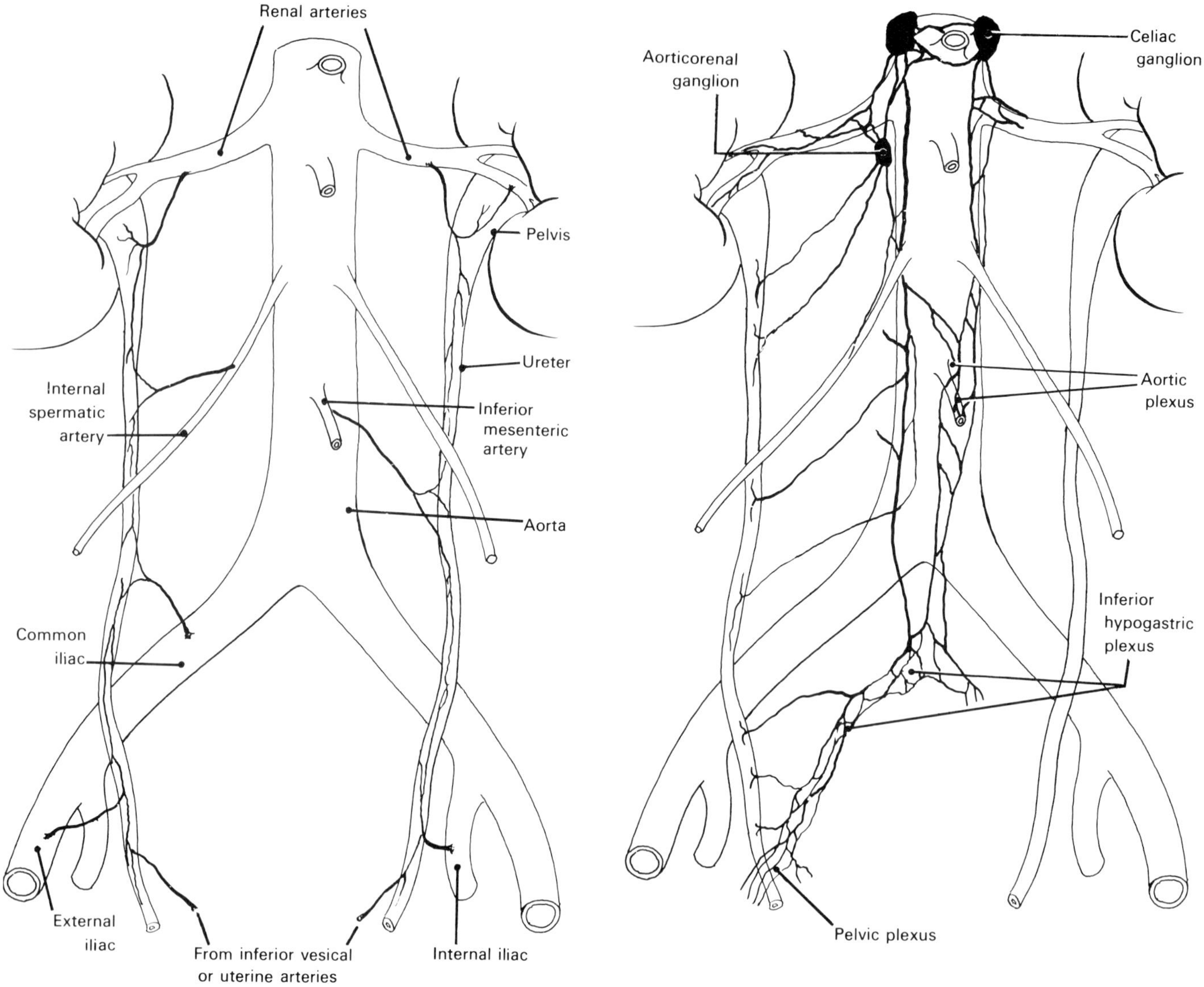

Fig. IX-108. The blood (**A**) and nerve (**B**) supply to the ureters. (After Hollinshead.)

than the right one, and both tend to be longer in males than in females.

As the tubes run their retroperitoneal course to the bladder, they incline slightly medialward. The fat and loose connective tissue around them furnish a kind of sheath. As they make their way over the pelvic brim, they cross the genitofemoral nerve and either the common or the external iliac arteries. Here they can be affected by vascular aneurysms. They themselves, on the right side, are crossed by the right colic and iliocolic vessels; on both sides, by the internal spermatic vessels; and on the left side, by the left colic vessels. The left ureter will be somewhat protected by the sigmoid mesocolon, while the right ureter is not so covered. But, in surgery of the sigmoid colon, the left ureter is in a position of danger.

The Blood Supply: To the ureters blood supply is multiple and variable (Fig. IX-108A). Its proximal end is supplied by twigs from the renal artery; its distal end, from the inferior vesical artery. Along the rest of its route it receives contributions from the internal spermatics (testicular or ovarian vessels); from the iliacs (common, external, and internal); and even aortic twigs.

The Nerve Supply (Fig. IX-108B): This seems to be abundant coming both from above, *via* the aorticorenal plexus, and from below, *via* pelvic and hypogastric plexuses. Wharton[7, 8] seems to have demonstrated that these are mainly visceral sensory fibers, not visceral motor. When cut, there is relief of ureteric pain but no loss of normal peristaltic motility.

The Intravenous (Excretory) Urogram—The Roentgen Anatomy of the Urinary Tract: This radiographic examination permits visualization of the kidneys, ureters, and bladder by means of a radiopaque contrast medium (not a dye). The contrast medium is injected intravenously and it is excreted by glomerular filtration and tubular excretion depending upon the type of medium utilized. *Retrograde pyelography* refers to visualization of the renal pelvis and ureter after injection of contrast medium into a catheter whose tip is placed either in the renal pelvis or ureter. Pyelography may be performed with positive or negative contrast medium.

1. Procedure: Unless the intravenous urogram is performed under emergency conditions, it is advantageous to dehydrate the patient overnight and to prepare the bowel with appropriate cathartics. In the morning, a plain film of the abdomen (KUB) and pelvis is obtained in AP supine projection. This is carefully evaluated by the radiologist prior to administering the contrast medium. The KUB is scrutinized for kidney size, shape and position, as well as for abnormal calcifications overlying the renal outlines, along the course of the ureters, and in the bladder region. In addition, a general evaluation of the other abdominal and pelvic structures is made for clues to suspected or unsuspected disease. The detailed anatomy seen on the plain (scout, KUB) film of the abdomen is considered on page 633.

After the contrast medium is administered, we prefer to obtain radiographs at 5-minute intervals for at least 15 minutes; the 5- and 15-minute films are AP supine projections (Figs. IX-109 and 110A), while the 10-minute films are 25° right and left posterior oblique projections (Fig. IX-110B). The radiologist must evaluate each radiograph as soon as it is developed, since any abnormality encountered may alter the routine study described. Once the physician is satisfied that all possible information about the kidneys, ureters, and bladder has been obtained, then an additional erect AP radiograph is made immediately after the patient has voided. This last projection permits evaluation of renal and ureteral drainage, and the extent of bladder evacuation.

2. Anatomy: Evaluating the Radiographs: If we now proceed to examine some of the radiographs made during an intravenous urogram we will apply some of the detailed anatomy learned earlier. Because the kidneys are surrounded by a generous layer of perirenal fat their bean-shaped outer contour is readily discernible. The left kidney is usually higher than the right kidney (about 2.0 cm. higher on the average) and its midportion is at the level of the VL_1–VL_2 intervertebral disk space. The longitudinal axis of each kidney usually parallels the lateral margin of the Psoas muscle, and, thus, the upper pole of the kidney is closer to the midline than the lower renal pole. Although you will not be able to appreciate it from the radiographs of the living, the kidneys are slightly tilted, such that the anterior surface faces ventrolaterally, and the upper pole lies more dorsal than the lower pole (Fig. IX-111 A, B).

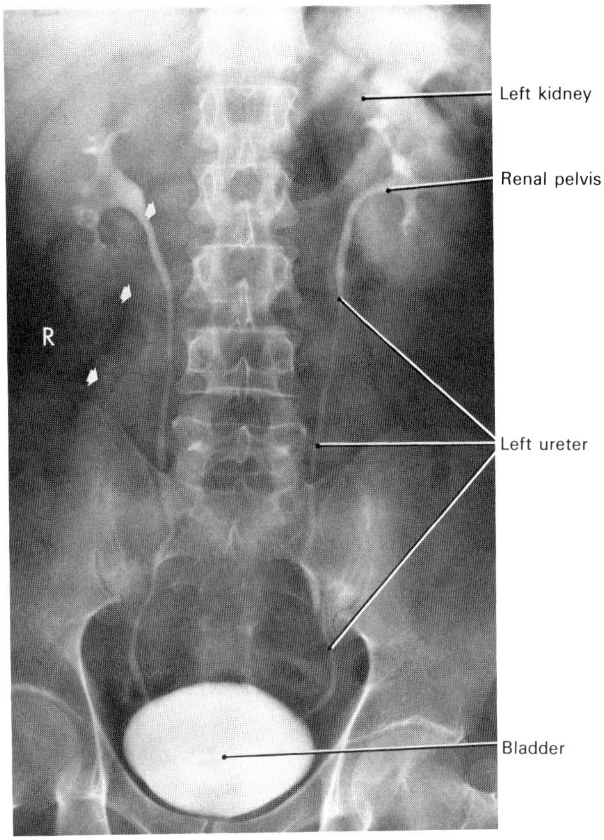

Fig. IX-109. Intravenous urogram. AP projection. The white arrows outline the lateral margin of the Psoas major muscle.

The kidneys normally have some mobility, and therefore they may be seen to shift with the phases of respiration and the position of the patient. In most patients normal renal mobility will be about 3.0 cm. or roughly the height of one vertebral body. Since the kidney may also normally move anteriorly several centimeters, we would caution you that lateral radiograph of the abdomen made with the patient on his side (decubitus) with a vertical X-ray beam should never be used to determine displacement of the kidney by an adjacent soft tissue mass. Such evaluation is best made with a horizontal X-ray beam directed across the abdomen with the patient supine (supine transabdominal projection).

Renal size as determined radiographically is quite variable in the adult, the length ranging from 9 to 14 cm., and the width from 5 to 8 cm.[9-12] Generally the kidneys are of equal size in the same normal individual. However, if the two kidneys differ by more than 1.5 cm. then an abnormality, such as tumor, cyst,

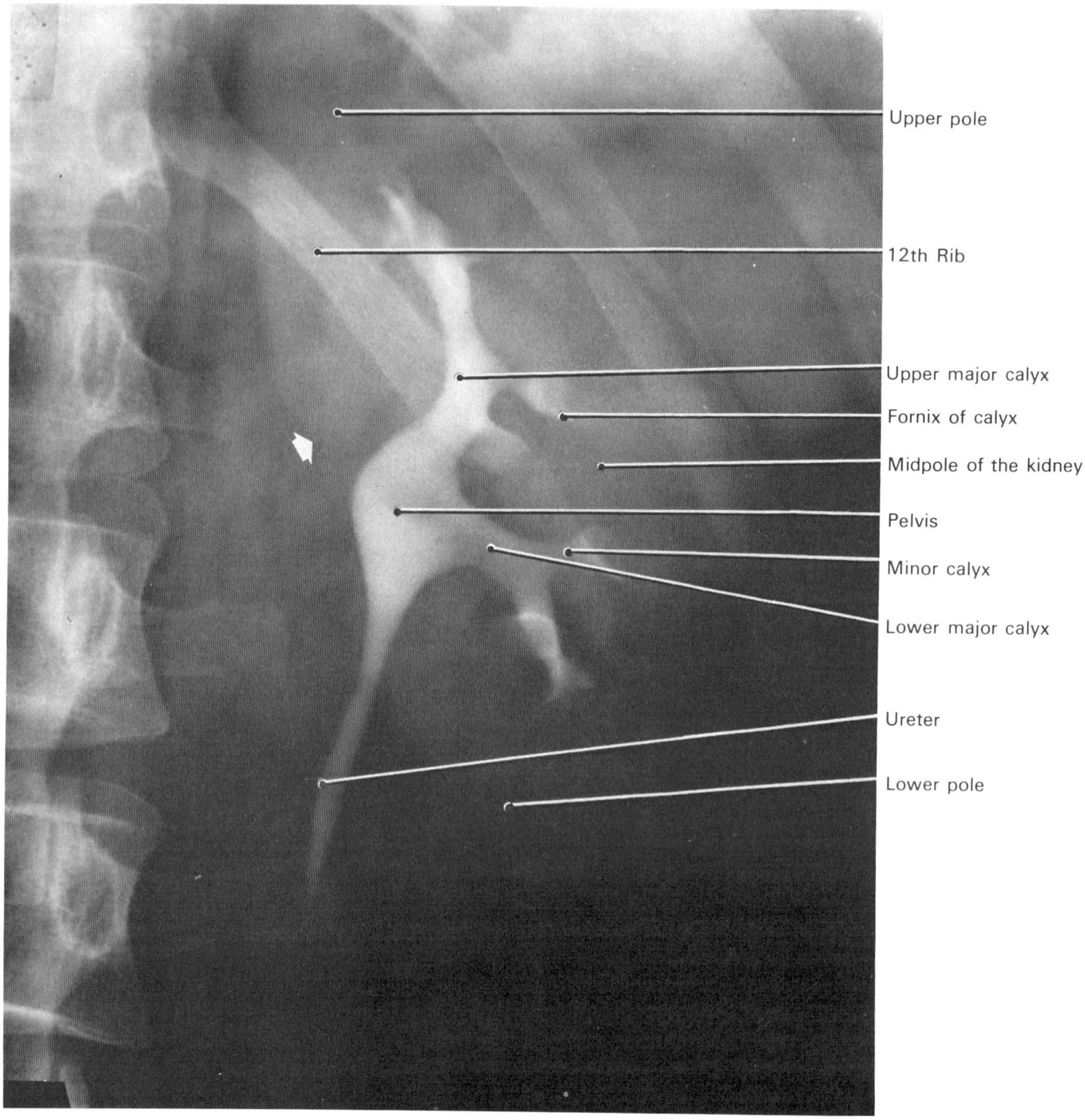

Fig. IX-110. Intravenous urogram—detailed anatomy of the kidneys. **A.** AP projection of the left kidney. (Arrow indicates cranial aspect of hilus).

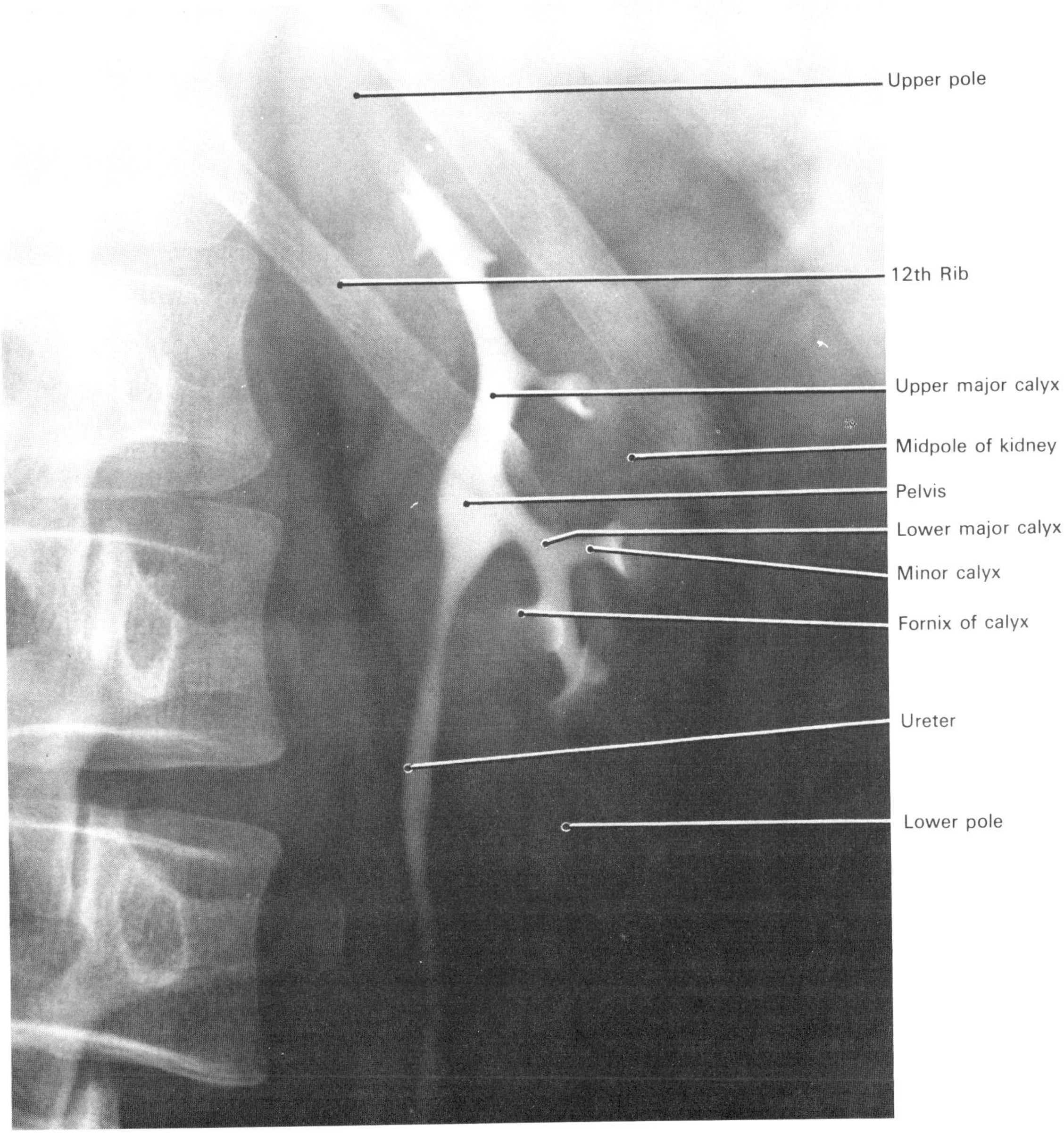

Fig. IX-110B. Left posterior oblique projection of the left kidney.

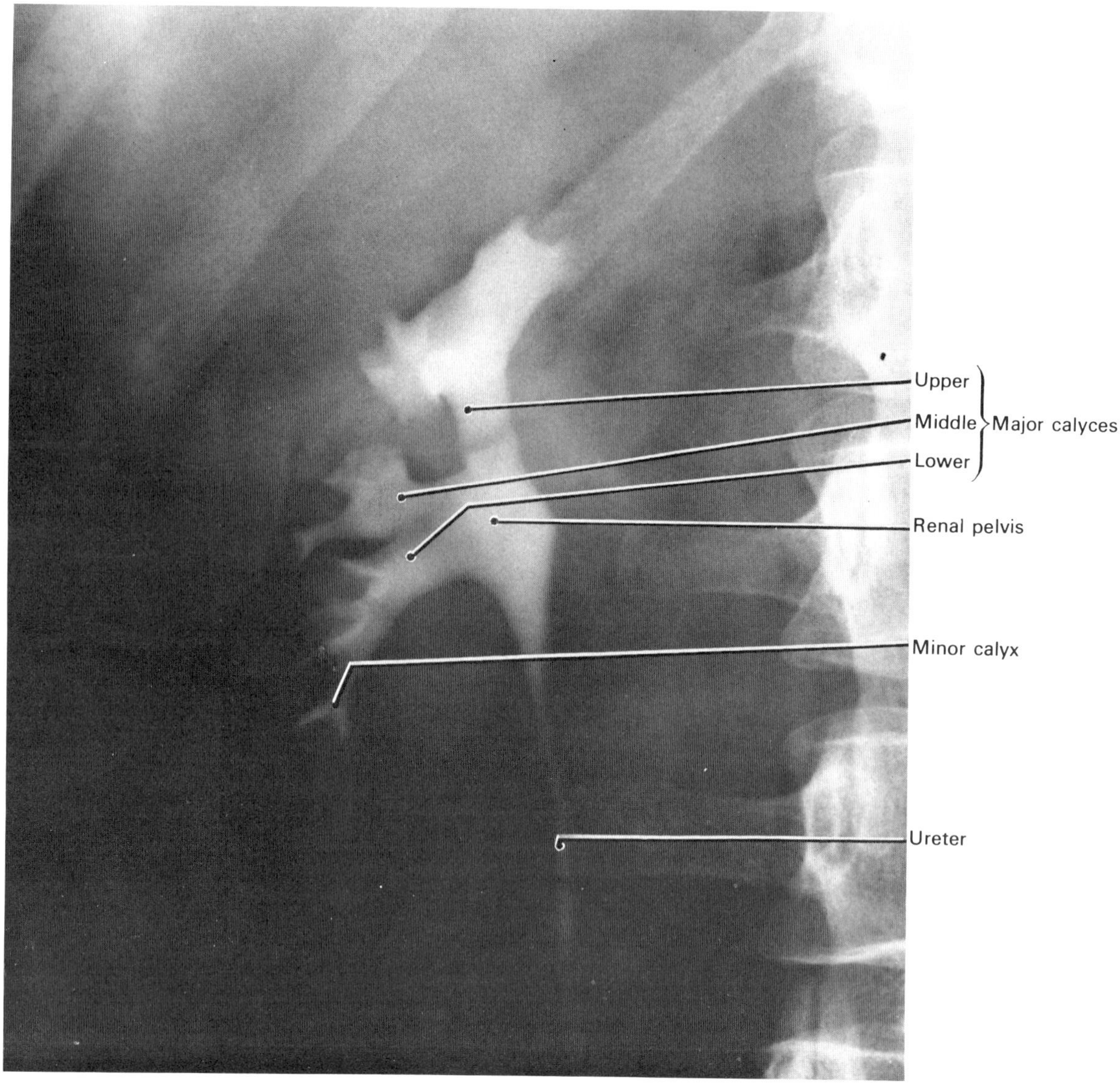

Fig. IX-110C. AP projection of the right kidney.

duplication, or renovascular disease, should be entertained. The width of the normal renal parenchyma, as measured from the outer surface at the midpole to the base of one of the calyces of the middle pole, is usually about 1.5 cm., and this is, of course, greater in the upper and lower poles of the kidney.

Although the renal contour is smooth this outline is normally lost at the hilus (Fig. IX-110A). Any distortion of the renal shadow should raise the possibility of abnormality; however, two variations of normal may unduly alarm the uninformed: fetal lobulation and the dromedary kidney. **Fetal lobulation** may persist into adulthood (Fig. IX-112). It appears as multiple, regular indentations along the cortical margin, such indentations corresponding to the pyramids. This finding is usually bilateral. Occasionally the left kidney may assume a somewhat triangular shape probably due to modeling from the spleen. As a result the upper pole is narrowed while the lower pole is widened. This is called a **dromedary (humpback) kidney.** A similar change may also be seen on the right due to the proximity of the right kidney to the liver.

Figure IX-110A and B are AP supine and left posterior oblique projections, respectively, of the left kidney in the same normal patient. Note the appearance in each projection of the cup-shaped minor calyces. There may be 7 to 14 such minor calyces. The brim of each calyx is termed the **fornix,** and it is normally quite sharp. Loss of sharpness of the fornix or blunting of the calyx is pathological. Recall that from 1 to 3 papillae fit into each calyx; however, the papillae are not normally observed during the intravenous urographic procedure. The minor calyces may be single or clustered on a single neck, and, indeed, wide variations may be seen in the same kidney.

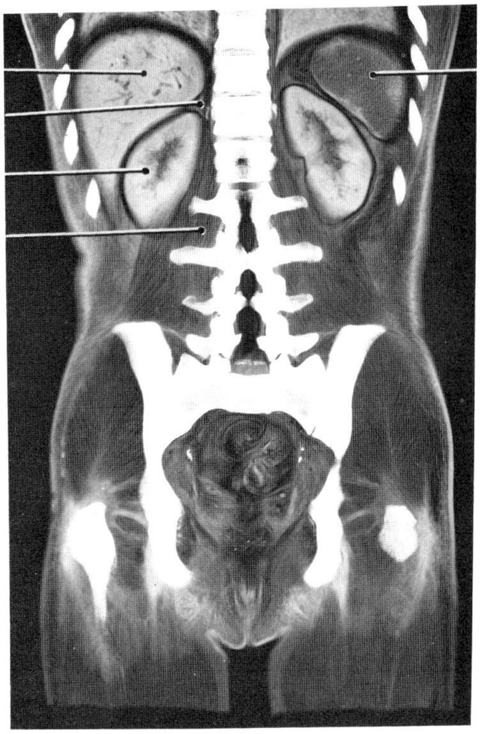

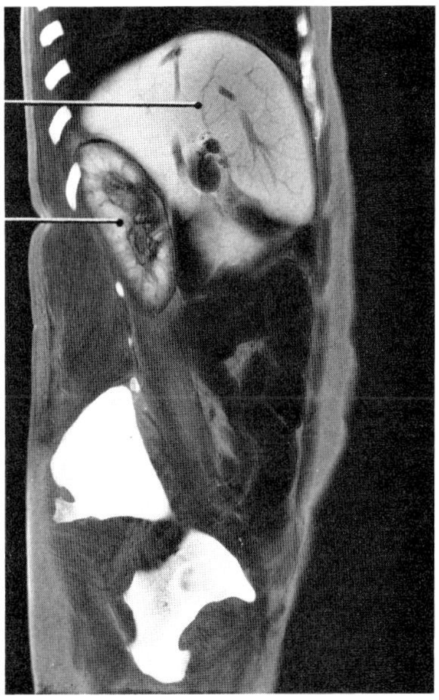

Fig. IX-111. Radiographs of thick section of the cadaver. **A.** Coronal section through the kidneys. **B.** Sagittal section through the right kidney. (Courtesy of Dr. B. Anson.)

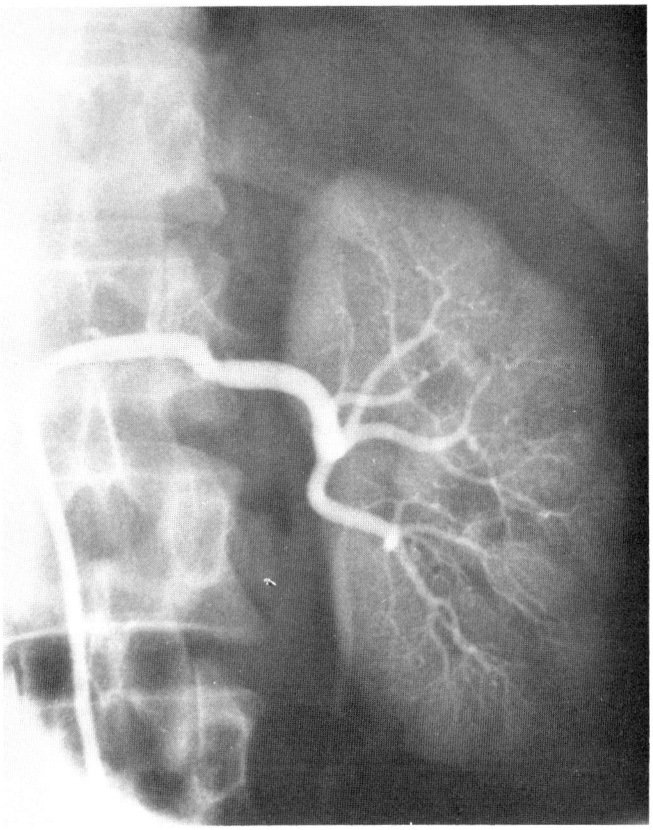

Fig. IX-112. Selective left renal arteriogram showing fetal lobulation. Compare this with Figure IX-107 and note the difference in renal contour between the two illustrations.

There are usually three major calyces, an upper, middle, and lower (Fig. IX-110C); however, the middle major calyx may at times not be clearly defined as such (Fig. IX-110A, B). The major calyces are also inconstant in shape, some being long and thin, and others short and wide.

The renal pelvis, although typically funnel-shaped, may be deep within the renal hilus (intrarenal pelvis), or it may extend beyond the hilus (extrarenal pelvis). The ureteropelvic junction is an indefinite point on the radiographs. The approximate level of this junction is indicated for you in Figure IX-110.

The course of the ureters in their proximal half is not constant. They may descend directly downward from the pelvis and immediately cross over the Psoas muscle, or, for a short course, they may parallel the Psoas before finally crossing over. Then they descend laterally along the spine and, finally, in bending over the iliac artery (Fig. IX-109) they appear to be kinked. Within the pelvis, the ureter curves laterally before coursing medially to enter the bladder at the ureterovesical junction. Usually the ureters are incompletely filled and this is normal. However, if large doses of contrast medium are used it is more common to see more complete ureteral filling (Fig. IX-109).

The bladder is round to oval in shape (Fig. IX-

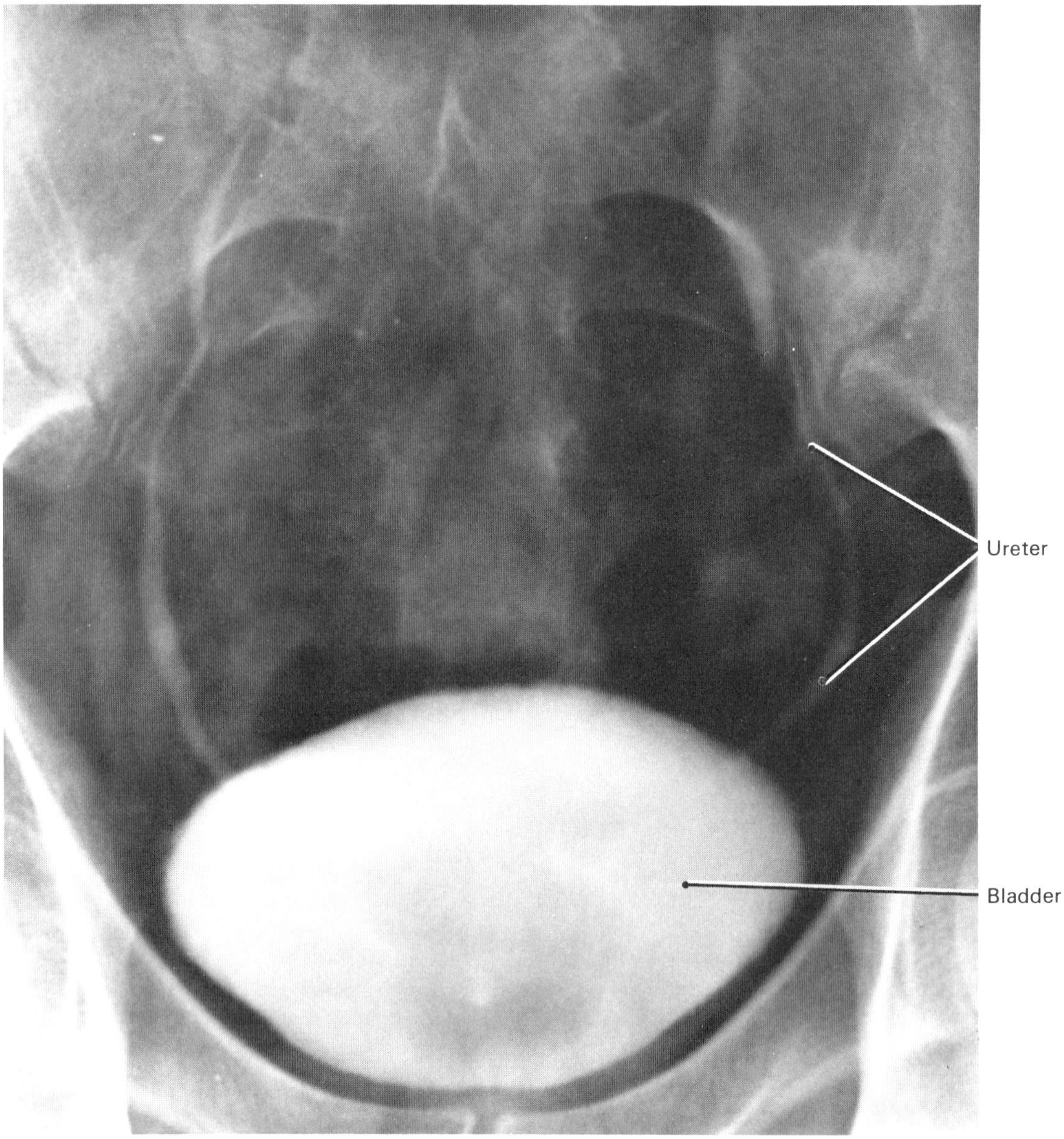

Fig. IX-113. Intravenous urogram showing the bladder in AP projection. **A.** Bladder filled with contrast medium.

113A), and the shape is, in large measure, dependent upon the degree of filling. The base of the bladder appears to rest near or upon the symphysis pubis in frontal views. Its inner wall is smooth, while the outer wall is not normally seen. After voiding (Fig. IX-113B), a very small residium of contrast medium may be evident within the contracted bladder.

The Suprarenal or Adrenal Glands

These organs, as their name implies, lie in the neighborhood of the kidneys, either above them or close to them. They neither develop nor function like the kidneys or the "G-U Tract." Rather, like the spleen, they are a neighborhood concern with a business of their own to run. And, they develop in relation to the autonomic nervous system.

Development: The **cortex** and the **medulla** of these glands have entirely separate developmental origins so that the parts unite into the one organ secondarily. The cortex is of mesodermal origin; the medulla, ectodermal. In the 8-mm. embryo, the cortical cells derive from mesothelium beneath the peritoneal epithelium, well down near the attached base of the dorsal mesentery. There are two waves of cellular proliferation from this source. The first furnishes the large cells of the **fetal cortex** at about the fifth week. At 6½ weeks, a second proliferation forms the smaller cells of the permanent cortex. The growing gland bulges out the dorsal mesentery to a considerable extent, the bulk of large organ having been contributed by the fetal cortex. At birth, this provisional fetal cortex involutes rapidly. Its significance to the fetus is not well understood. Only after birth does the permanent cortex increase in prominence.

The much larger fetal cortex has been called the **X-zone** or the **androgenic zone** on the theory that it elaborates male hormones that could protect the fetus from excess of female hormone reaching it through the maternal placenta. Its size is seen in Figure IX-114.

At about the seventh week, ectodermal medullary

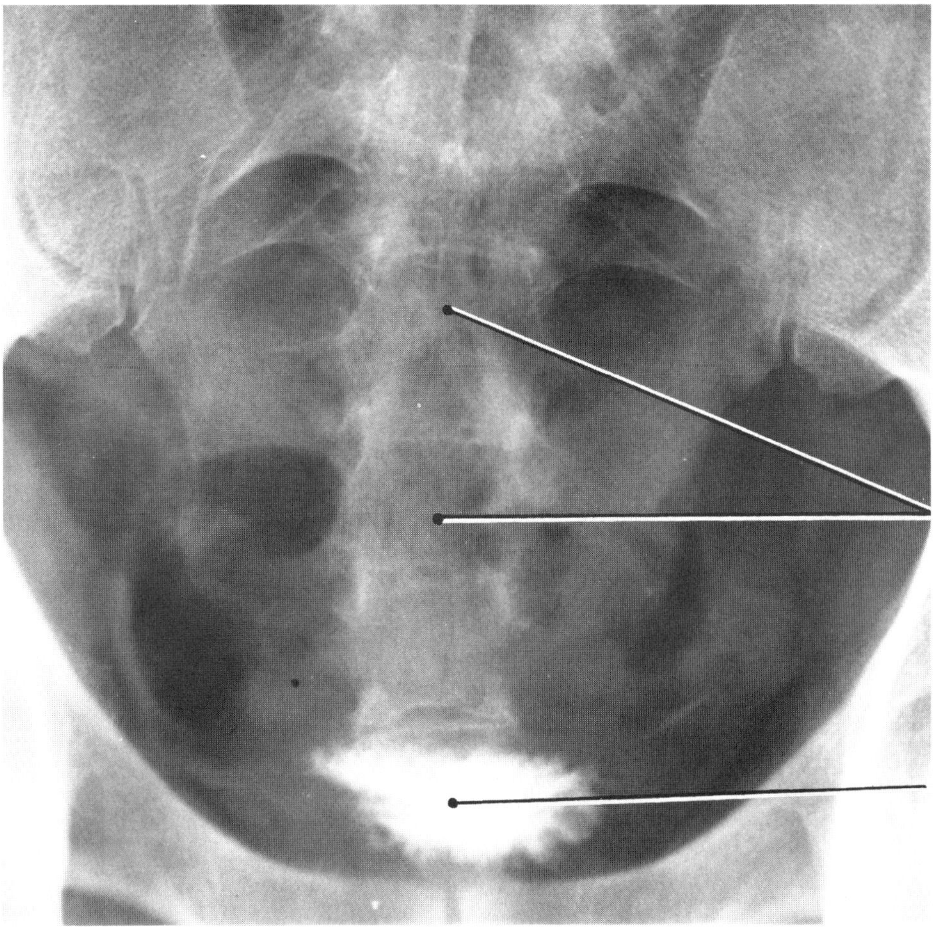

Fig. IX-113B. Bladder after voiding.

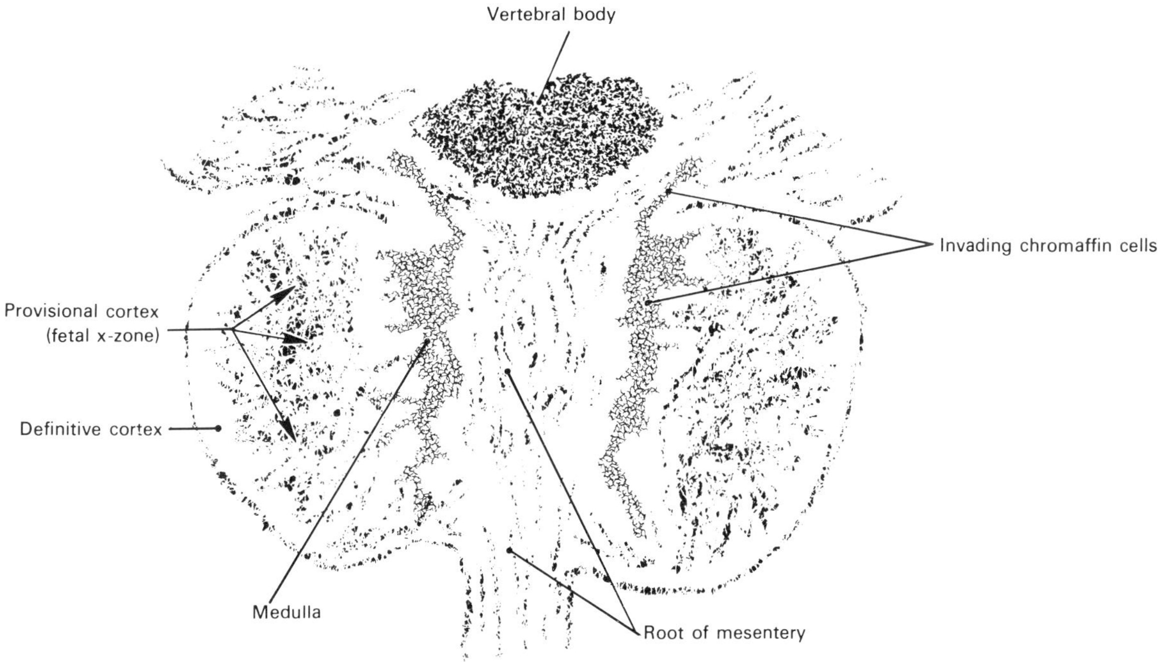

Fig. IX-114. Suprarenal glands in a 17-mm. embryo. The chromaffin cells are just beginning to penetrate the cortical mass.

cells begin to invade the cortical mass. They migrate to the area from two sources: (1) from the **neural crest** of the developing neural tube; and (2) from the **intermediolateral cell columns** of the gray matter of the developing spinal cord. They follow the course of the developing greater splanchnic nerves. In this sense, they are visceral motor sympathetic neurons. But once they reach the gland, they fail to develop neuronal processes, round up, and become secretory cells. These cells give a typical reaction when stained with chrome salts, much as do sympathetic ganglion cells.

Not all "chromaffin cells" that take part in this migration end up in the suprarenal medulla. Some become cells of the **carotid** and **aortic bodies.** Some become **accessory suprarenals,** consisting, chiefly, of medullary cells. These latter come to lie near the main glands, close to the kidneys or below them in relation to the gonads. Others of these migrating cells become, of course, the proper, visceral motor neurons of sympathetic chain and collateral ganglia. Certain mavericks become the **paraganglia,** clusters of chromaffin cells mixed with connective tissue. These are small clumps, as a rule, measuring a millimeter or so in diameter. They parallel the true autonomic ganglia and plexuses. The largest of these, however, are the **aortic paraganglia** (organs of Zuckerkandl). These are bilateral clumps which measure several centimeters in length and lie ventrolateral to the aorta about at the level of the inferior mesenteric artery. They must not be confused in young individuals with the true sympathetic inferior mesenteric ganglion. They probably represent fusions between a number of the larger paraganglia in that particular area. Sometimes as many as 50 or more small nodules constitute this large mass. The paraganglia are thought to be accessory centers for the production of epinephrine since they are so prominent during the infantile period when the suprarenal medulla is attaining full development. They begin to regress about the sixth or seventh year, and by puberty they may be absent entirely.

The Adult Suprarenals (Figs. IX-104 and 115): *General Structure:* These are rather firm, yellowish to brownish, endocrine glands that are much smaller than those seen in the fetus and the infant. They measure about 40 to 60 mm. in length and 20 to 30 mm. in width. Shape varies, though the right one, at times, may have an apical peak making it look like a cocked-hat. Often both glands are semilunar in shape, and both are indented where they appear to cap their respective kidney poles. There is a slight hilus on the ventral surface for the emergence of the single **central vein.** The left gland can be longer and more tongue-like than the right one so that its distal tip may touch the splenic artery, the pancreas or even dip down far enough to touch the renal artery at the kidney hilus.

Suprarenal Nerve and Blood Supply: **The vagus nerve is in no way involved in the innervation of the gland.** As the greater splanchnic nerves pass the gland on their way to the coeliac ganglion, fibers branch from them which join others coming from the coeliac and renal plexuses. This **suprarenal plexus** supplies the medullary portion of the gland. The cortex receives vasomotor fibers only, *i.e.,* no fibers appear to end upon cortical parenchyma. The direct incoming sympathetic fibers from the splanchnics do terminate upon the secretory cells of the medulla. In this sense, the secretory cells of the medulla resemble postganglionic neurons of the sympathetic system. Their role, however, is not one of neural conduction, but, rather, that of elaborating a sympathomimetic, humoral product. The other fibers of the suprarenal plexus are, probably, the vasomotor ones.

If the venous drainage is chiefly single, the arterial supply is multiple. The **inferior phrenic arteries** pass medial to the glands as these vessels make their way to the Diaphragm. They give off the **superior suprarenal arteries** to the gland. Actually, there may be numerous small twigs given off—all of them together being called *the* **superior suprarenal artery** for convenience sake. The aorta, itself, contributes one or more small twigs which constitute *the* **middle suprarenal artery.** The renal or sometimes even the internal spermatic (ovarian) arteries contribute the **inferior suprarenal artery,** again by one or more twigs. Thus, we do have a *multiple* rather than the *triple* supply that is so often mentioned (Fig. IX-115). Indeed, in some instances as many as 50, individual, arterial branches have been counted to a single gland. In contrast, there seems to be only the one large vein leaving the gland. Occasionally, a few, very tiny veins may be seen accompanying some of the arteries. The vein drains into the inferior vena cava on the right side and into the renal vein on the left side. The left vein must pass behind the pancreas to reach the renal vein. *En route* it may be joined by the inferior phrenic vein on the left side.

We shall discuss the other paired associates of the G-U Tract in the next chapter since they relate to the true pelvis. Their paired arteries, we have already observed, arise from the abdominal aorta just below the renal arteries as the internal spermatic or ovarian arteries.

Hernias and Congenital Defects in the G-I and G-U Systems

A **hernia** is an abnormal protrusion of any structure beyond its normal site. In the abdomen the protrusion is often through a weakened point in the wall. For the most part hernias involve a sac of peritoneum. The sac can be empty but it may contain parts of the stomach, omentum, bowel, or even of a solid organ like liver. Hernias occur at several predilected sites. An **epigastric hernia** above the umbilicus occurs along the linea alba. This consists of a bit of fatty tissue pushing through between the medial borders of the Recti and separating collagenous fibers of the linea. As a rule, it does not involve peritoneum, but it can if omentum

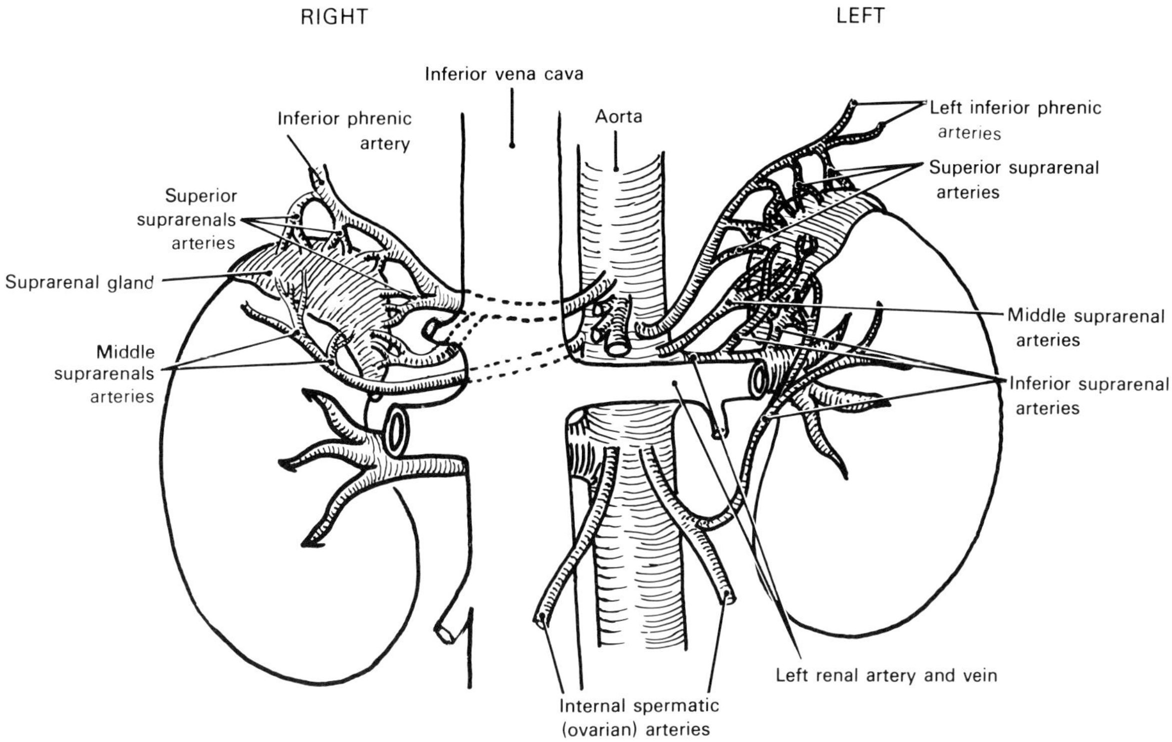

Fig. IX-115. Arterial supply of suprarenal glands. Note that on the left side one of the inferior suprarenals comes from the renal artery; the other, from the internal spermatic artery.

protrudes into it. The incidence of occurrence is about 4.6% of all types of hernias encountered.

There are also minor **lateral line hernias** or **para-umbilical protrusions.** These occur somewhere along the semilunar line of Douglas. There is some uncertainty about ascribing this line to Douglas, a Scotsman who became an anatomist and physician in London (1674–1742). The line was actually described earlier by Spieghel (Adrian van der Spieghel), a Belgian anatomist at the University of Venice (1578–1625). As a consequence, you will sometimes find these lateral line hernias described as **Spieghelian hernias.** Generally, they involve omentum, but sometimes they involve a loop of gut. They can be as small as a pea or as large as an orange. They pass through Transversus and Internal Oblique muscle and spread out under the External Oblique aponeurosis.

The **umbilical hernia** is a congenital one. Recall that, in the early embryo, herniation of the intestinal loops into the umbilical cord was a normal event since there was inadequate room in the small midgut for the rapidly developing digestive tube. As a rule, the intestine returned to the abdominal cavity by the 10th week. This was when the primitive, but large, kidneys degenerated, when the liver stopped increasing in size and when the body wall and abdominal cavity caught up in growth to retain all these contents at the same time. Sometimes this primitive condition of gut protrusion is retained; at other times potential weakness of the wall in this region predisposes toward a secondary

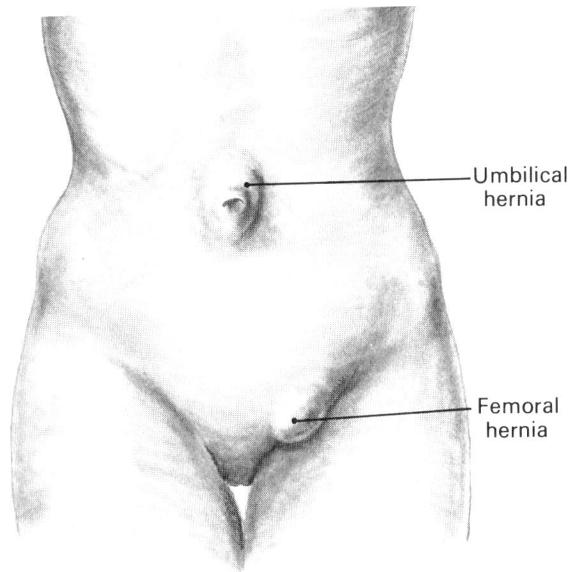

Fig. IX-116. Umbilical and femoral hernia in a female.

protrusion toward the outside. In any event, a condition of umbilical hernia results (Fig. IX-116).

The umbilical hernia has been given a variety of names: gastroschisis, omphalocoele, congenital umbilical hernia, ectopia viscerum, congenital eventration, and amniotic umbilicus. Most people accept these as synonymous and are content with the overall term, umbilical hernia (as are we), but some people like to

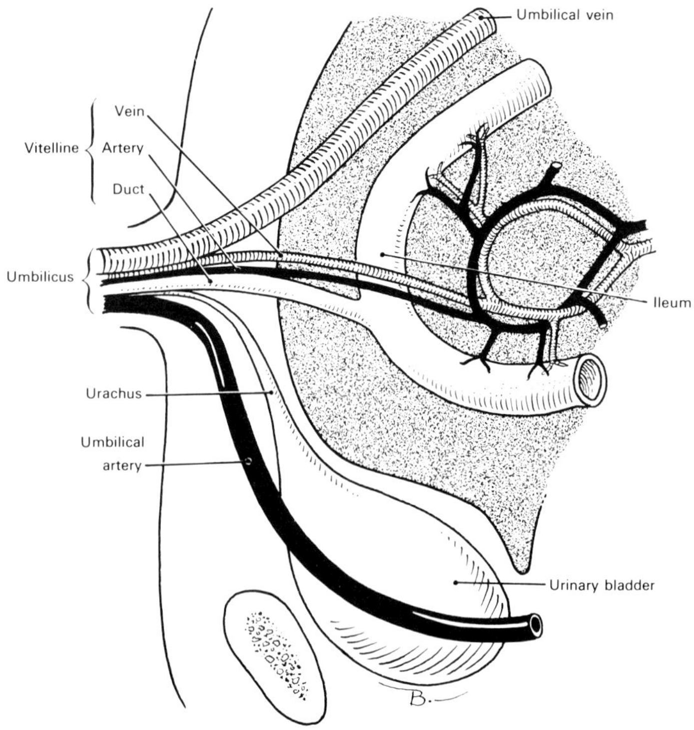

Fig. IX-117. Structures related to the umbilicus. (After Hollinshead.)

make fine distinctions. For example, if the abdominal wall in the newborn has failed to close for some considerable distance in the midline, you may find someone describing the anomaly as a gastroschisis. The word has a peculiar half Latin, half Greek derivation —L. *gaster* = stomach; Gr. *schisis* = fissure. More peculiar still, it is generally the intestine, not the stomach which protrudes through the fissure. If the abdominal gap is small and circular, and if the protruding contents are covered by a thin, transparent membrane, the defect is called an omphalocoele (Gr. *omphos* = navel; *coele* = cavity). But if the protrusion is covered by skin, and the gap is small so that only a small piece of bowel is involved, most people would be content with the garden variety term, umbilical hernia.

Recall that there are a number of other defects of the umbilicus which may or may not be troublesome. All have to do with the persistence of structures that pass through it. Figure IX-117 illustrates the complexity of the plumbing in this small area during fetal life.

The umbilical and vitelline vessels are not involved in anomalous situations, as a rule. The umbilical vein becomes the round ligament of the liver. The umbilical arteries become ligamentous as the lateral umbilical ligaments up to the point where they remain patent as the superior (or middle) vesical arteries. The vitelline vessels obliterate entirely between the umbilicus and the digestive tract. The vitelline duct may maintain its patent connection with the umbilicus, however, so that fecal matter may leak to the exterior. Recall that persistent vitelline duct anomalies can also give rise to a Meckel's diverticulum of the ileum, to vitelline cysts, and to volvulus. The medial umbilical ligament may persist as a patent urachus between the navel and the urinary bladder. The folds and fossae produced on the inner aspect of the ventral body wall by these various remnants of the fetal ducts and vessels are shown in Figure IX-118.

The commonest defects of all in the ventral abdominal wall are the **inguinal** hernias. Of the two types, the **indirect hernia** is far more common than the **direct hernia.** It is paradoxically, the indirect type which enters the deep inguinal ring, traverses the canal, and emerges at the superficial ring. The mistake is to think that since this goes *directly* through the canal, an area of potential patency, it must be a "direct hernia." The term "direct" means *directly through the abdominal wall* (at a point where you would not, at first contemplation, expect an area of weakness). Indirect hernias, in turn, are divided into **congenital** and **acquired types.** Here again one must be careful of meaning, for "congenital" does not necessarily mean the condition was present at birth. The meaning of the term is based upon the condition of the hernial sac. For even if the hernia occurs in later life, it simply means that the tunica vaginalis testis has maintained a patent connection through its funicular process with the peritoneal cavity of the abdomen, and a loop of bowel has invaded it secondarily. The loop will, therefore, lie naked in the tunica vaginalis next to the testes (Fig. IX-119A).

In the acquired type, the herniation is through the deep and superficial inguinal rings and the inguinal canal, and the loop of bowel partially resects the funicular process. The bowel remains wrapped in its own peritoneal investment of the body wall, and does not reach into the cavity of the tunica vaginalis necessarily (Fig. IX-119B). Even if it does extend this far, it is kept from contact with the peritoneum of the tunica by the transversalis fascia and the peritoneal sheath which it has pushed before itself.

Generally, the **direct hernias** protrude *medial* to the inferior epigastric artery and move forward to the **external ring.** Unlike the indirect type, they *do not* cross *lateral* to the vessel to pass through the **deep ring.** Thus, they by-pass the deep ring by coming "directly" through the body wall. The common area of weakness is at the interfoveolar ligament in Hesselbach's triangle. This is an area bounded medially by the lateral umbilical ligament, laterally by the inferior epigastric vessels, and inferiorly by the inguinal ligament (Fig. IX-5). The hernia may be only a general bulging or it may form a true diverticulum, pushing peritoneum and transversalis fascia in front of it. If it emerges from the external ring, more often than not it tends to turn upward rather than to descend toward the testis. But these hernias can also be scrotal.

The other common hernia in this region is the **fem-**

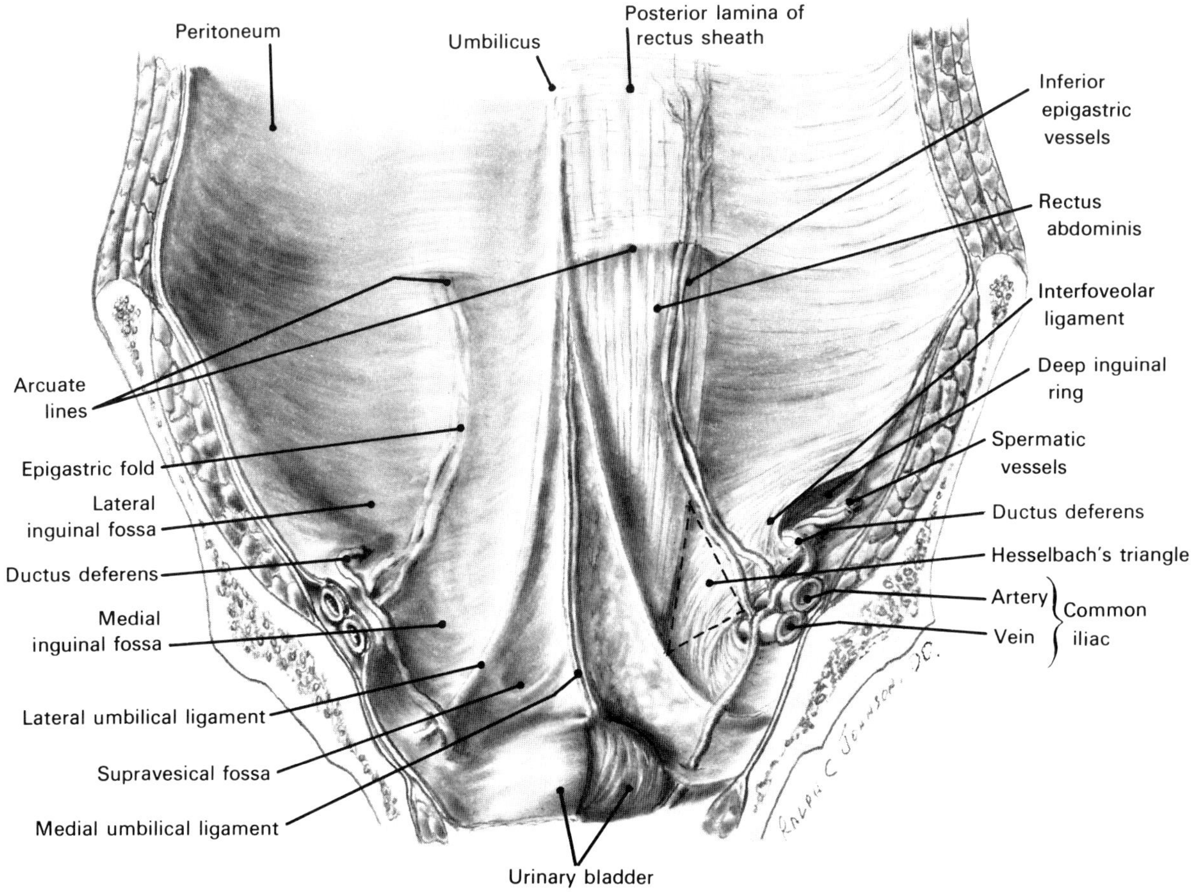

Fig. IX-118. Internal aspect of the ventral body wall. The obliterated urachus is in the medial umbilical ligament; the obliterated umbilical artery, in the lateral umbilical ligament.

oral hernia. But before this can be visualized, you must understand relationships just deep to the inguinal ligament. This structure, you recall, bridged from anterior superior iliac spine to symphysis pubis. If it is a bridge, things must be able to pass under it from the abdomen into the thigh (Fig. IX-120). Laterally, this space is filled by Iliopsoas muscle and fascia. The **femoral nerve** comes through embedded in the muscle. Next, medialward come **femoral artery** and **vein,** in that order, wrapped in the **femoral sheath.** Posteriorly, the sheath consists of iliopsoas fascia of the abdomen. Most medially, the space is plugged by our old friend, the stout **lacunar ligament.** Now between vein and ligament is a potential danger space, the space of the **femoral canal,** its upper end here being known as the **femoral ring.** The ring is closed only by thin peritoneum and transversalis fascia against which rests an unusually large lymph node, the **node of Cloquet** (also called the node of Rosenmueller). Hence, only lymphatic vessels normally traverse this thin area between thigh and leg. Here, loops of bowel can push downward lateral to the lacunar ligament into the thigh creating the typical femoral hernia (Fig. IX-120).

The incidence of inguinal and umbilical hernias is shown in Figure IX-121 based on statistics by Keith[13] from 1000 males and females. Males are predisposed to predominantly more inguinal hernias than females; the female has a far greater incidence of femoral hernias than the male. Overall herniation (of all types) is six times more common in the male than the female. The preponderance of femoral hernia in the female has usually been attributed to the differences in the anatomy of the pelvis between the sexes, or to the shorter lengths of the lacunar ligament in the female. Yet Panton[20] has shown that the female false pelvis is narrower than in the male, and that the lacunar ligament is larger, both factors which ought to protect her against femoral hernias, not render her more susceptible. He did find that the Iliopsoas muscle was much smaller, and that the vascular compartment was larger so that connective tissue filler in the neighborhood of the vessels was more abundant. The shift laterad of the muscles and vessels and the increase in filler tissue seemed to make for an area of greater weakness lateral to the lacunar ligament in the female.

The structures of the gastrointestinal tract are also subject to certain anomalies:

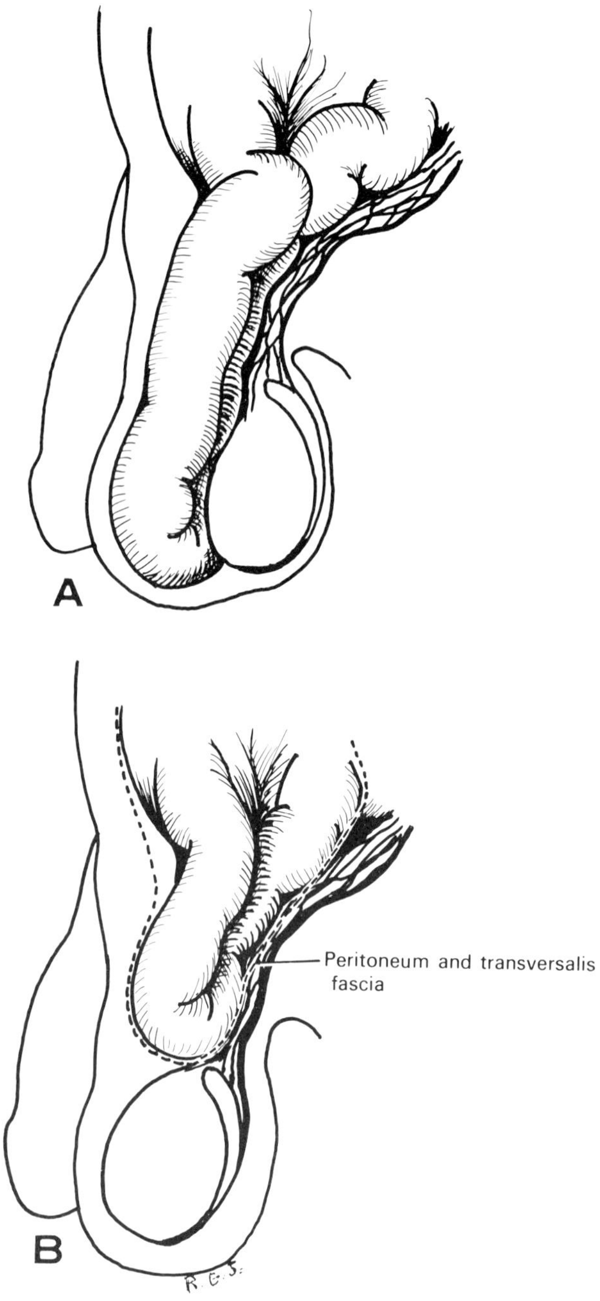

Fig. IX-119. Two types of indirect inguinal hernias. **A.** Congenital type. Bowel has herniated into still open processus vaginalis. **B.** Acquired type. Bowel has redissected a closed processus vaginalis, carrying abdominal peritoneum and transversalis fascia ahead of it.

1. Esophagus: There may be a narrowing or **stenosis** or a **local atresia** (no cavity; a narrowing to a fibrotic cord usually near the point of tracheal bifurcation). Normally, the fetus swallows amniotic fluid and moves it through the digestive tract as soon as fetal peristaltic movements begin.[14-17] This, of course is, impossible with an atretic esophagus. As a consequence, amniotic fluid is not drained away almost as rapidly as it is formed, and the uterus may enlarge greatly (hydramnios). With a normal, patent esophagus, much of the water that is swallowed by the fetus is absorbed in its G-I tract.

2. Stomach: There can be a transposition to the right side, especially when other abdominal organs develop in mirror image. The stomach can be located above the Diaphragm by being attached to a short esophagus. It can become stenotic, the narrowing usually occurring at the pylorus.

3. Duodenum: It can remain atretic if conditions during the sixth and seventh week are not reversed through recanalization.

4. Liver: This gland rarely develops elsewhere than at its usual site. It rarely has more or less than two lobes. The hepatic ducts, the common bile duct, cystic duct, and the gall bladder can duplicate partially or completely. The gall bladder can be absent in man (as it always is in some species, *e.g.*, horse and elephant). Ducts can be narrowed or even completely solid.

5. Pancreas: In the course of migration of the ventral pancreatic bud, it can split and rotate around the duodenum in two directions to encircle it. Sometimes it can constrict the duodenum and the common bile duct completely. There can be accessory bits of pancreatic tissue in various parts of the mesenteries (Fig. IX-122). It can be found anywhere along the intestinal tract from the esophageal-cardiac junction to Meckel's diverticulum, most frequently in the mucosa of the latter and the mucosa of stomach. The ventral bud can migrate with the hepatic diverticulum to form a nodular pancreatic bladder near the gall bladder. We have spoken of the anomalous entrances of its ducts into the duodenum.[15-16]

6. Spleen: This organ can be partially subdivided or appear as multiples. Small accessory spleens are common (Fig. IX-123). They represent the persistence of original uncoalesced nodules, or of the incisures that appear in the developing organ in the third to fourth month.

Maldevelopment accounts for a number of anomalous situations in the "G-U Tract" also.

1. Pelvic and Horseshoe Kidneys: We have noted that the kidneys normally ascend to a high lumbar locus from their pelvic site of origin. Sometimes one fails to "climb the arterial ladder" and remains close to the common iliac artery. This is known as a **pelvic kidney** (Fig. IX-124C). Sometimes both kidneys fuse at the pelvic site to form a **horseshoe kidney** (Fig. IX-124E). Such a kidney gets hung-up at a low lum-

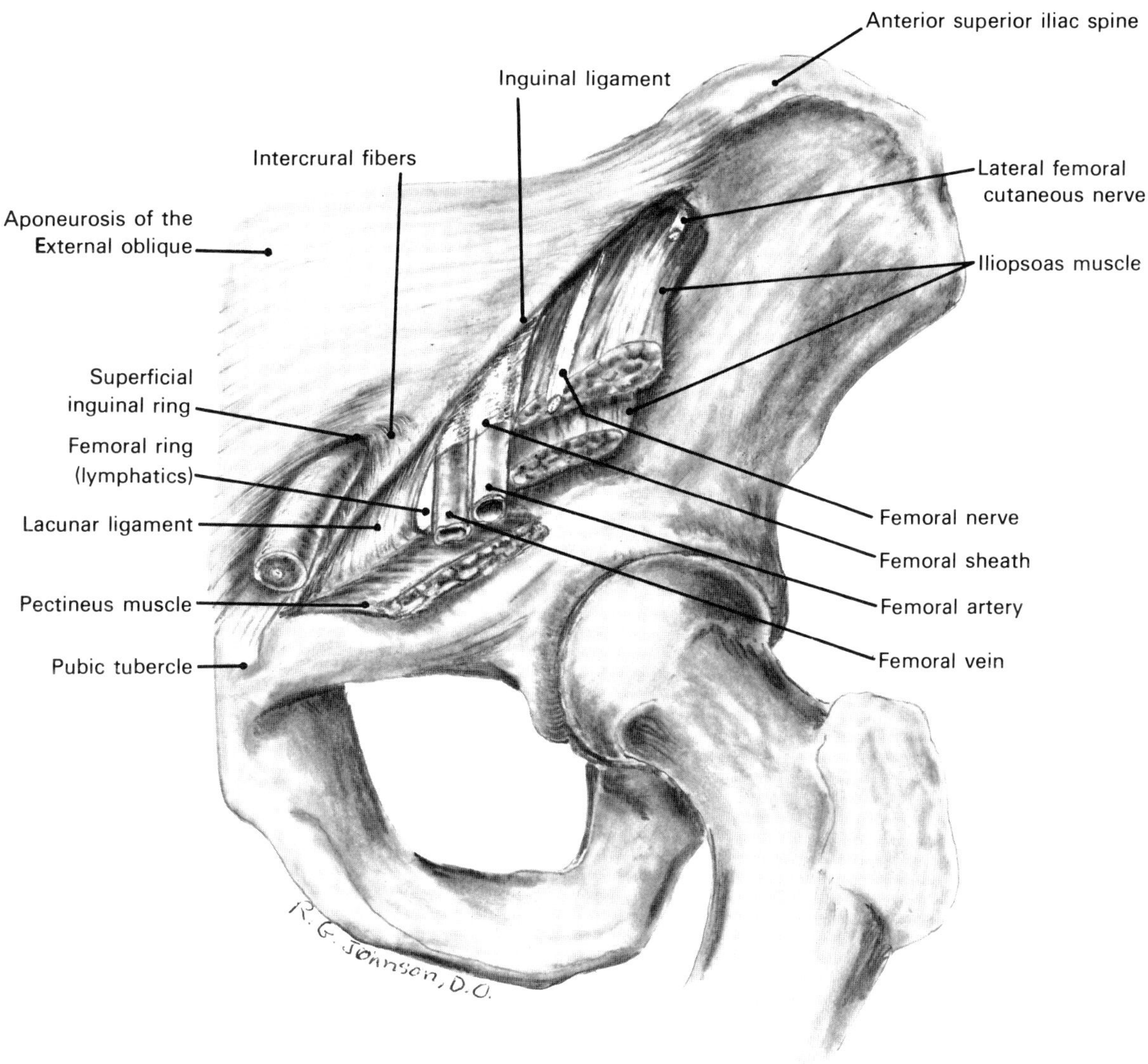

Fig. IX-120. Relationships at the femoral ring.

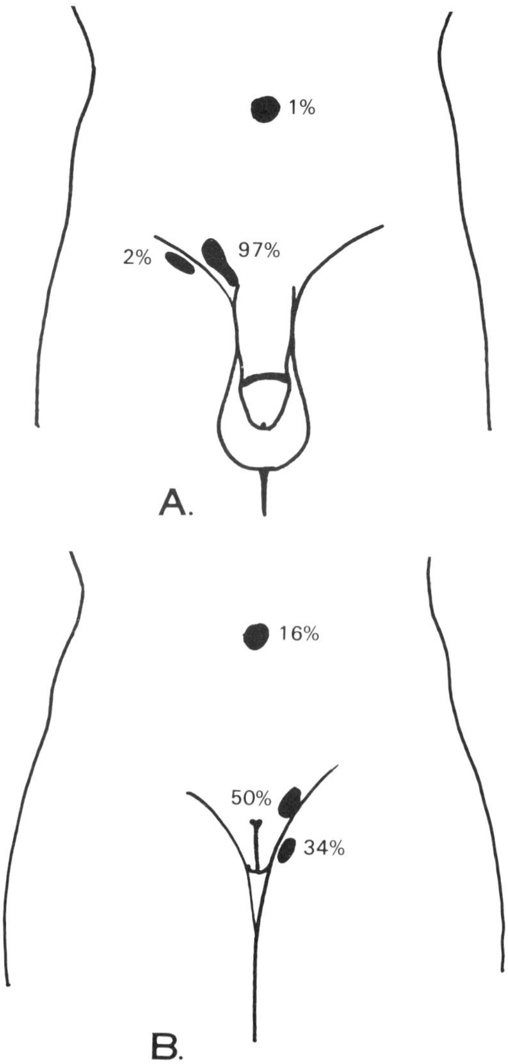

Fig. IX-121. Incidence of umbilical, inguinal and femoral hernias (A) in males and (B) in females. (According to Kieth.)

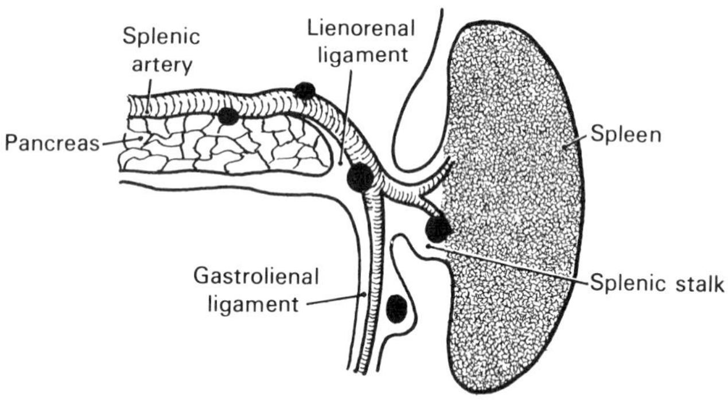

Fig. IX-123. Common sites for accessory splenic tissue. (After Hollinshead.)

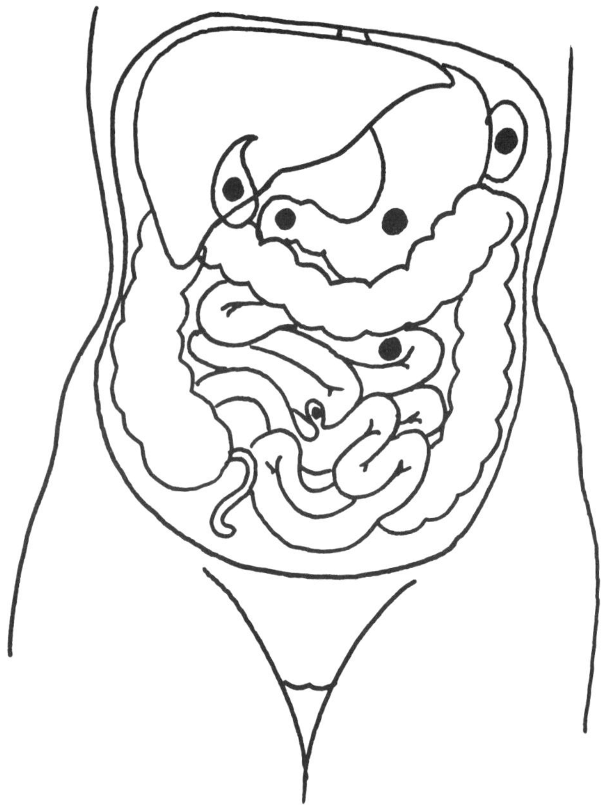

Fig. IX-122. Some locations of accessory pancreatic tissue. (After Hollinshead.)

bar level, hooked, as it were, by the midline, inferior mesenteric artery.

2. Multiple Renal Arteries: In climbing the "rungs" of the arterial ladder, the ascending kidneys sometimes fail to let go of some of the lower arterial branches. Thus, some 3% of kidneys have two or more renal arteries (Fig. IX-124D).

3. The Congenitally Cystic Kidney: If the collecting ducts arising from the ureteric bud failed to unite with some of the excretory tubules formed from the nephrogenic mass, urine could not flow from the kidney. This "non-union" has been the usual explanation for the development of the cystic kidney. More recent evidence[18, 19] points to hyperplasia of the walls of the collecting duct or in blunted portions of the ureteric bud, which constricts in some areas.

4. Other Kidney Anomalies: Some kidneys may retain the fetal lobulated appearance since the nephrogenic mass lobulates in the process of tubule formation. Kidney and ureter can be absent on one side. Kidneys can fuse unilaterally on one side and be absent on the other. In the fetus, the renal pelvis faces ventrally. Either through a kind of "rotation" or a differential growth of the lateral pelvic lip, the renal pelvis normally comes to orient medially. Such "rotation" may

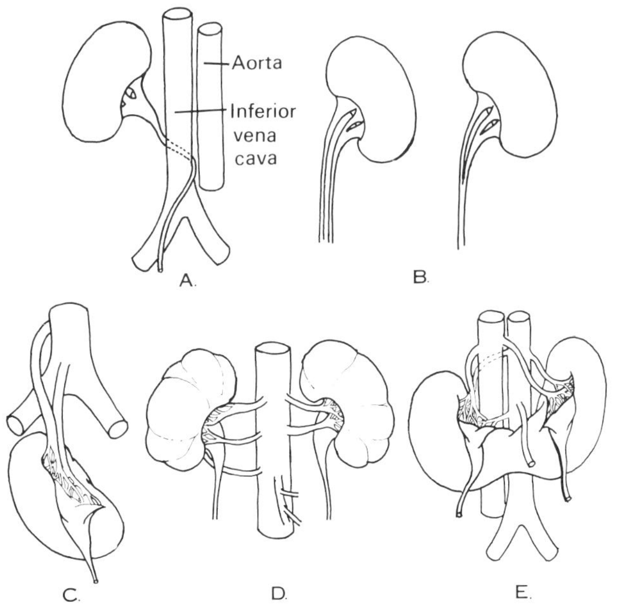

Fig. IX-124. Anomalies of the kidneys and ureters. **A.** Anomalous course of ureter (retrocaval). **B.** Bifid ureters. **C.** Pelvic kidney. **D.** Multiple renal arteries and fetal lobulation. **E.** Horseshoe kidney. (After Grant.)

fail to occur or occur in the opposite direction, so that some renal pelves may be found directed ventrally or laterally.

5. Bifid Ureters: Early splitting of the ureteric bud may give rise to partial or complete duplication of the ureters (Fig. IX-124B).

6. Anomalous Course of Ureter: Because of the complicated ways in which the posterior, supra- and subcardinal veins take part in forming the embryonic inferior vena cava, occasionally a ureter is seen coming out from behind the inferior vena cava, rather than crossing it ventrally (Fig. IX-124A). Even more rare is the ureter which passes behind the external iliac artery.

Plain Film Radiographic Examination of the Abdomen and Pelvis

Now that we have studied the anatomical structures of the abdominal wall, cavity, and pelvis in part, it would be profitable for us to explore the *normal plain* (scout, survey, KUB) *film* of the abdomen and the components that may be visualized on it. The physician should have thorough knowledge of the normal before undertaking the study of the pathological processes affecting these structures.

Plain film studies of the abdomen are somewhat of a problem, since not all of the tissue constituents are as well defined as in other areas of the body. In the chest, for instance, there are wide ranges of differential absorption of the X-rays in the various tissues, whereas, in the abdomen, these gradations are nil. Many organs only become visible because of a handy layer of fat about their borders to set them off from neighboring structures. Still other important details cannot be distinguished at all.

The term "plain film of the abdomen" (scout, survey film, or KUB) means that no contrast media has been used in the examination. A few superannuated physicians prefer the term "flat plate of the abdomen," however, this archaic phrase is best banned from contemporary medical language. It is a hold-over from the time that glass plates rather than celluloid were coated with photographic emulsion for use in radiographic examinations.

Plain film examination of the abdomen is not merely limited to the supine AP projection. Indeed, in many clinical problems, this single view would be an incomplete examination. Often, other projections including erect, prone, oblique, lateral, and lateral decubitus views are made. Herein we will confine our remarks to the supine AP projection, which is the most common of those mentioned.

Consider, for a moment, the radiograph in Figure IX-125 which is the abdomen of a normal patient. Because of the individual's size, not all of the abdominal and pelvic structures are included on the 14- by 17-inch film (standard large size) that has been utilized. Therefore, it is necessary to obtain an additional radiograph to show the entire pelvis.

In Figure IX-125 the lower ribs and thoracic vertebrae, the lumbar spine, and most of the pelvis and sacrum are seen. We will not dwell upon these osseous structures since they have been examined in great detail elsewhere. A word of caution about calcified costal cartilage seems in order though. This all too often overlaps the renal shadows and the upper abdomen where it is frequently mistaken for calculi of the kidneys, gall bladder, pancreas, or vascular calcification. Oblique or lateral radiographs of the abdomen, contrast studies, and even fluoroscopy in these instances are needed to clarify the problem.

In the right upper quadrant, the dense homogeneous shadow of the liver is perceived. The concave inferior liver margin may be seen paralleling the 11th rib (Fig. IX-126). In the asthenic patient, the normal hepatic margin may extend as low as the right iliac crest. Please remember that hepatomegaly and splenomegaly is best evaluated by physical examination and *not* radiographically! The left lobe of the liver is usually not seen as it extends across the midline. The upper margin of the liver is in intimate contact with the right dome of the Diaphragm which outlines it. Laterally, on the right, the liver edge may be seen separated from the abdominal wall by a layer of omental fat (Fig. IX-126).

The spleen may at times be recognized in the left upper quadrant (Fig. IX-127). The lower pole is evi-

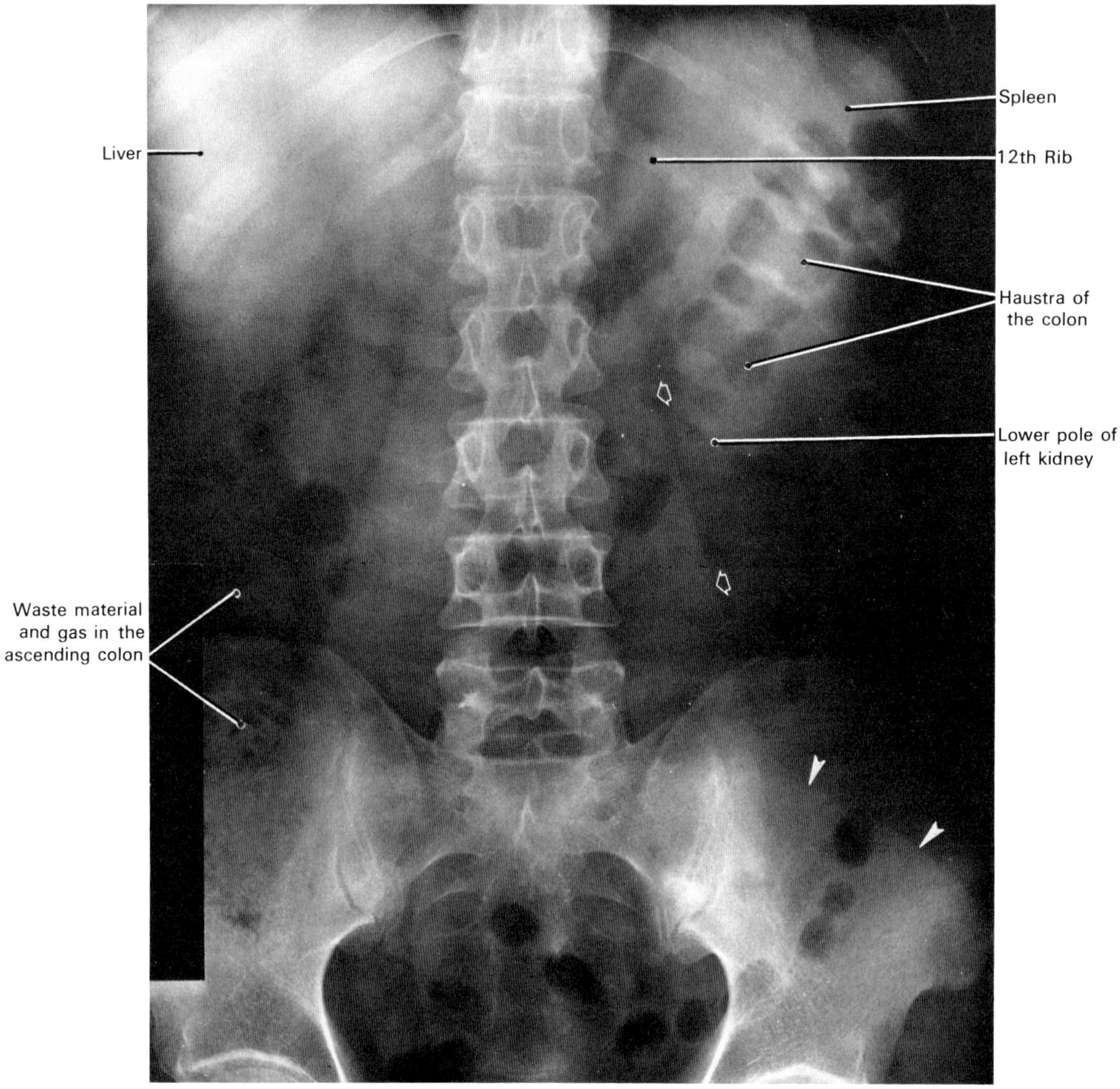

Fig. IX-125. Plain (scout, KUB) film of the abdomen. Open arrows indicate lateral margin of the left Psoas muscle. White arrow heads show the Gluteal muscle crossing the ileum.

dent just below the last rib, and may be outlined by air within the splenic flexure of the colon. Then too, the medial pole may be outlined by gas in the stomach. The definition of the spleen is best whenever a thick fatty layer surrounds it.

The kidneys are of the same water density as the other soft tissues around them, but they are made conspicuous because of their surrounding fat (Fig. IX-128). The ureters, on the other hand, are never visible. The position of the kidneys is quite variable; however, it is fair to say that the right kidney is usually lower than the left kidney. In Figure IX-128 the midpole of the right kidney is about level with the VL_1 vertebra—this level may vary by the height of one vertebra up or down. The long axis of the kidneys usually parallels the lateral Psoas border. Notice how smooth the renal outline is with the exception of the region of the hilum medially.

The lateral margins of the Psoas Major muscles are strikingly visible as they course from the VL_1 level obliquely inferolaterally toward the iliac fossa (Fig. IX-126). In asthenic patients the Psoas may be concave laterally, while in the obese, stocky individual it may be convex laterally. Nearby retroperitoneal inflammatory conditions or neoplasms may cause the Psoas shadow to become indistinct or to be absent, and, therefore, comparison of both sides is important. Occasionally, the lateral margin of the Quadratus Lumborum is evident extending down from the level of the 12th rib toward the iliac spine. A curvilinear line may be seen crossing the ilium that represents the Gluteal muscle border.

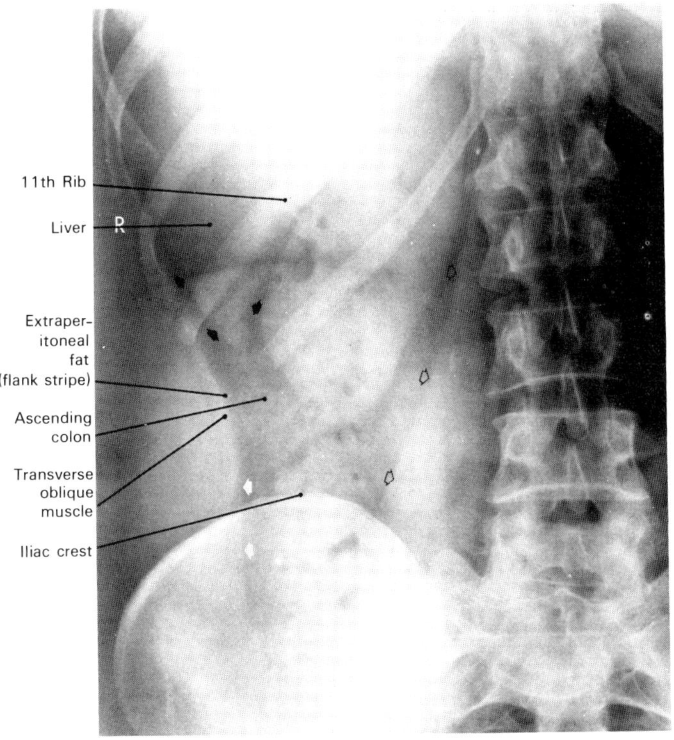

Fig. IX-126. Plain film of the abdomen. Close-up of the right abdomen. Black arrows indicate the tip of the right lobe of the liver outlined against omental fat. Open arrows indicate the lateral margin of the Psoas major muscle. White arrows indicate the parietal peritoneum.

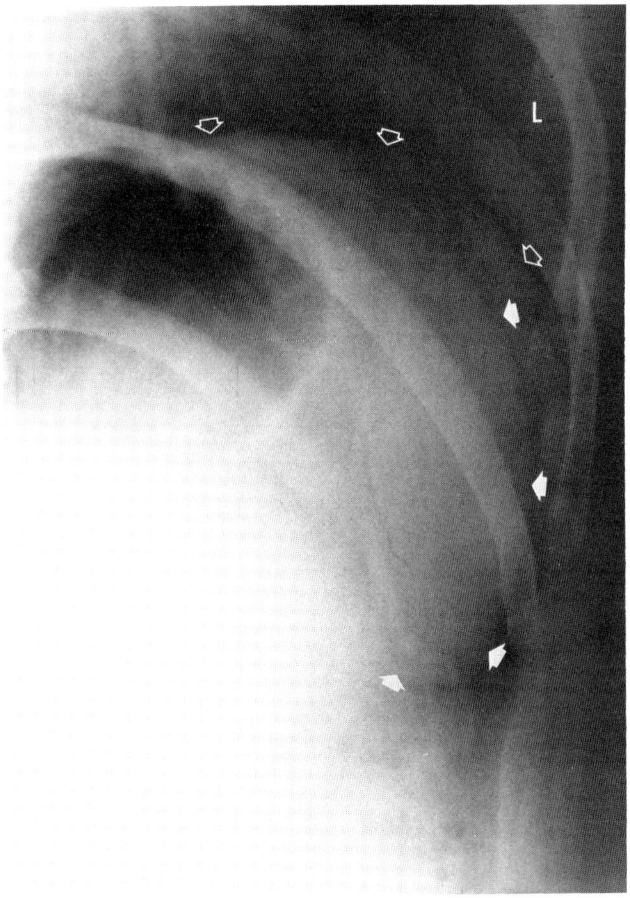

Fig. IX-127. Close-up of the left upper quadrant on KUB film of abdomen. Open arrows indicate top of left dome of Diaphragm. White arrows outline the spleen.

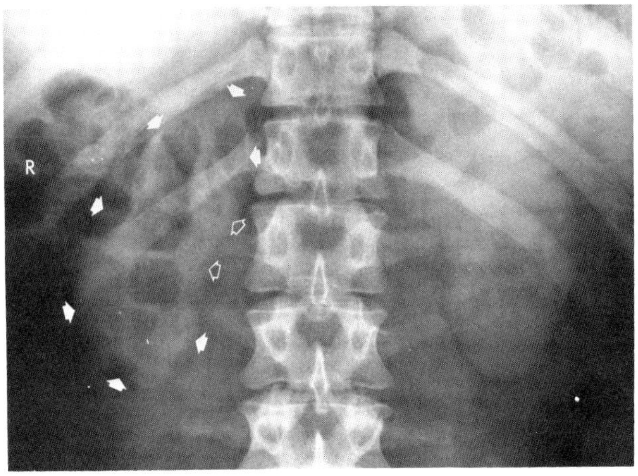

Fig. IX-128. Renal shadows as seen on a plain film of the abdomen. White arrows outline the right kidney. The open arrows show the hilus.

Just behind and above the symphysis pubis, the bladder outline is commonly recognized. It is typically oval in contour, and in the female may be indented superiorly by the uterus.

The intestinal gas pattern is inconstant in the normal adult patient. In Figure IX-125 some small amount of air is visible within the stomach overlying the upper pole of the left kidney and the first two lumbar vertebrae. With the patient erect this gas rises into the fundus of the stomach, and as we have seen on chest radiographs, this so-called "Magenblase" is situated immediately below the left dome of the Diaphragm. For the most part, the small bowel of the adult rarely contains air, whereas small bowel gas is common in infants. The feathery mucous membrane pattern and the location of the air-containing small bowel usually enables one to identify its anatomical segment. The "retroperitoneally"-located ascending colon, descending colon, and rectum are typically located in the flanks and lower pelvis, respectively, whereas the cecum, transverse and sigmoid colon have variable locations depending upon the length of their mesentery and the position of the patient. The colon may contain enough air to reveal its characteristic haustral pattern (Fig. IX-125). Fecal material mixed with gas is evident in the cecum and ascending colon of the patient illustrated. Not uncommonly, dense scybalous waste is evident in the rectum.

The student should take particular note of the normal anatomical structures demonstrable in the flanks. This is very important when evaluating inflammatory processes and nearby accumulated fluids (blood, etc.) in patients with acute abdominal disease. Figure IX-126 is a close-up of the lateral abdominal wall. First, identify the haustrated margin of the ascending colon which contains some waste material. Between the colon and the linear white shadow of the parietal peritoneum is omental fat. The extraperitoneal (properitoneal, flank stripe) fat layer, which is variable in thickness in different patients, is interposed between the parietal peritoneum and the Transverse Oblique. Occasionally fat is seen interposed between the Transverse Oblique, Internal Oblique, and External Oblique muscles. Variable amounts of subcutaneous fat lie lateral to the External Oblique. A discussion of the abnormalities that may be seen on plain films of the abdomen in the patient with an acute abdomen is, of course, beyond the scope of this book.

REFERENCES

1. Langman, J.: *Medical Embryology,* 2nd ed., Williams & Wilkins, Baltimore, 1969.
2. Dawson, W., and Langman, J.: An anatomical-radiological study on the pancreatic duct pattern in man. Anat. Rec., *139:* 59, 1961.
3. Schinz, H. R. *et al.: Roentgen Diagnosis,* 2nd American Ed., Vol. V. Abdomen. Grune & Stratton, New York, 1967.
4. Golden's *Diagnostic Roentgenology,* vol. 2, Williams & Wilkins, Baltimore, 1969.
5. Healey, J. E., Jr., and Schroy, P. C.: Anatomy of the biliary ducts within the human liver: analysis of the prevailing patterns and the major variations of the biliary ducts. A.M.A. Arch. Surg., *66:* 599, 1953.
6. Healey, J. E., Jr., & Schroy, P. C., and Sorenson, R. S.: The intrahepatic distribution of the hepatic artery in man. J. Int. Coll. Surg., *20:* 133, 1953.
7. Wharton, L. R.: The innervation of the ureter, with respect to denervation. J. Urol., *28:* 639, 1932.
8. Wharton, L. R., and Hughson, W.: Denervation of the ureter: a new operation for the cure of persistent painful ureteral spasm. Report of cases. J. Urol., *25:* 145, 1931.
9. Freidenberg, M. J., Walz, B. J., McAlister, W. H., Locksmith, J. P., and Gallagher, T. L.: Roentgen size of normal kidneys. Radiology, *84:* 1022, 1965.
10. Hodson, C. J., Drewe, J. A., Karn, M. N., and King, A.: Renal size in normal children; a radiographic study during life. Arch. Dis. Child., *37:* 616, 1962.
11. Karn, M. N.: Radiographic measurements of kidney section area. Ann. Hum. Genet., *25:* 379, 1962.
12. Moëll, H.: Kidney size and its deviation from normal in acute renal failure. Acta Radiol. (Stockholm), suppl. 206, 5, 1961.
13. Keith, A.: On the origin and nature of hernia. Brit. J. Surg., *11:* 455, 1924.
14. Windle, W. F., Becker, R. F., Barth, E. E., and Schulz, M. D.: Aspiration of amniotic fluid by the fetus. An experimental roentgenological study in the guinea pig. Surg., Gynec. Obstet., *69:* 705, 1939.
15. Becker, R. F., Windle, W. F., Barth, E. E., and Schulz, M. D.: Fetal swallowing, gastrointestinal peristalsis, and defecation in amnio. Surg., Gynec. Obstet., *70:* 603, 1940.
16. Becker, R. F., and Windle, W. F.: Origin and extent of gastrointestinal motility in cat and guinea pig fetuses. Direct observation of fetuses. Amer. J. Physiol., *132:* 297, 1941.
18. Qsathanondh, V., and Potter, E. L.: Pathogenesis of polycystic kidneys. Arch. Path. (Chicago), *77:* 459, 1964.
19. Baxter, T. J.: Cysts arising in renal tubules. Arch. Dis. Child., *40:* 464, 1965.
20. Panton, J. A.: Factors bearing upon the etiology of femoral hernia. J. Anat., *57:* 106, 1923.

TEN

The Perineum and Pelvis

INTRODUCTION

This is a difficult area for the novice to approach and to come away with a reasonable understanding. For this reason we have always felt it was easiest to work from the outside in rather than the other way around. So we will discuss the perineum first and tackle the deeper pelvic structures last. We should, however, introduce the topic by saying a few words about the **pelvic floor,** for the perineum is but a superficial part of this floor.

THE PELVIC FLOOR

You can visualize the female pelvis (or any pelvis, for that matter) as more or less of a cylindrical container into which you might look from the top or which you could examine from the sides. It has to have a floor or bottom else its contents would drop out once this container was supported upright on legs. If you will examine a hat box, you will see that its bottom is attached partway up the sides, and so it will be with the pelvic floor. If this floor is a yielding diaphragm its center will tend to be depressed by the weight of contents placed within it, and it ought to appear slightly funnel-shaped as a consequence. This is the nature of the pelvic floor or **pelvic diaphragm** deep to its internal peritoneal investment.

The *pelvic diaphragm* consists of the **Levatores Ani** and the **Coccygei muscles** and the **fasciae** that clothe their upper and lower surfaces. It also consists of the midline **intercolumnar fascia** into which these paired muscles anchor. Figure X-1 views this pelvic diaphragm in place from below but with the more superficial perineal structures removed that would normally hide the undersurface which you view here. You can see that the anal canal passes through its approximate center, perhaps a bit posterior of center, and that, in the female, the vagina and urethra pierce it anteriorly. The Levatores Ani are the important muscles here, the Coccygei, relatively unimportant.

The *Levator Ani* is most variable in thickness and strength. While its orientation is somewhat that of a shallow funnel (Fig. X-3) it really sits almost horizontally in the pelvic floor. We show it in Figure X-1 divided into its three conventional parts, *i.e.,* **Pubococcygeus, Puborectalis,** and **Iliococcygeus.** It is, however, a more complicated muscle than this by virtue of its relationships with the prostate, vagina, bladder, and rectum.

1. The **Pubococcygeus** arises from the posterior aspect of the pubis and runs back toward the coccyx. It inserts into various structures along the way and, in this sense, it is shown to consist of minor leaflets in Figure X-2. In the male, some medial fibers insert into the prostate as the **Levator Prostatae.** Similar fibers insert as the **Pubovaginalis** into the vagina and urethra of the female and, in concert with similar fibers from the opposite muscle, a **Sphincter Vaginae** is formed around these organs. Posterior to the urethra and vagina, its fibers insert into the intercolumnar fascia. Some fibers continue to the anal canal as **puboanal** fibers. The most posterior fibers of all finally anchor into the **anococcygeal ligament.**

2. The **Puborectalis** joins with its partner of the opposite side to form a sling on the posterior aspect of the anorectal junction. It is really a modified part of the Pubococcygeus. It is responsible for the flexure in the lower bowel at this point. But, as this muscle sling relaxes during defecation, the tube is allowed to straighten. Some of the fibers of this part of the Levator Ani enmesh with those of the **external anal sphincter.**

3. The **Iliococcygeus,** though quite broad, is the least well developed of the whole muscle. It is almost completely fascial and aponeurotic at times. Only a few of its stringy fibers separate the superior and inferior fascial covering of the pelvic diaphragm in this region. The muscle has a wide origin along the thickened arcuate line—a line of obturator fascia from which this portion of the Levator Ani is suspended. It also arises from the ischial spine. The insertion is upon the anococcygeal ligament (Fig. X-1).

The **Coccygeus** contributes to the posterior part of the pelvic diaphragm. It is a tendinous rather than a grossly muscular structure, something of the nature of the Iliococcygeal part of the Levator Ani. It arises from the ischial spine and spreads out over the lower sacrum and coccyx for its insertion (Fig. X-2). Its fibrous portion consolidates to help form the **sacrospinous ligament,** an important landmark in clinical pelvimetry (Fig. X-10).

Besides supporting pelvic viscera, the pelvic dia-

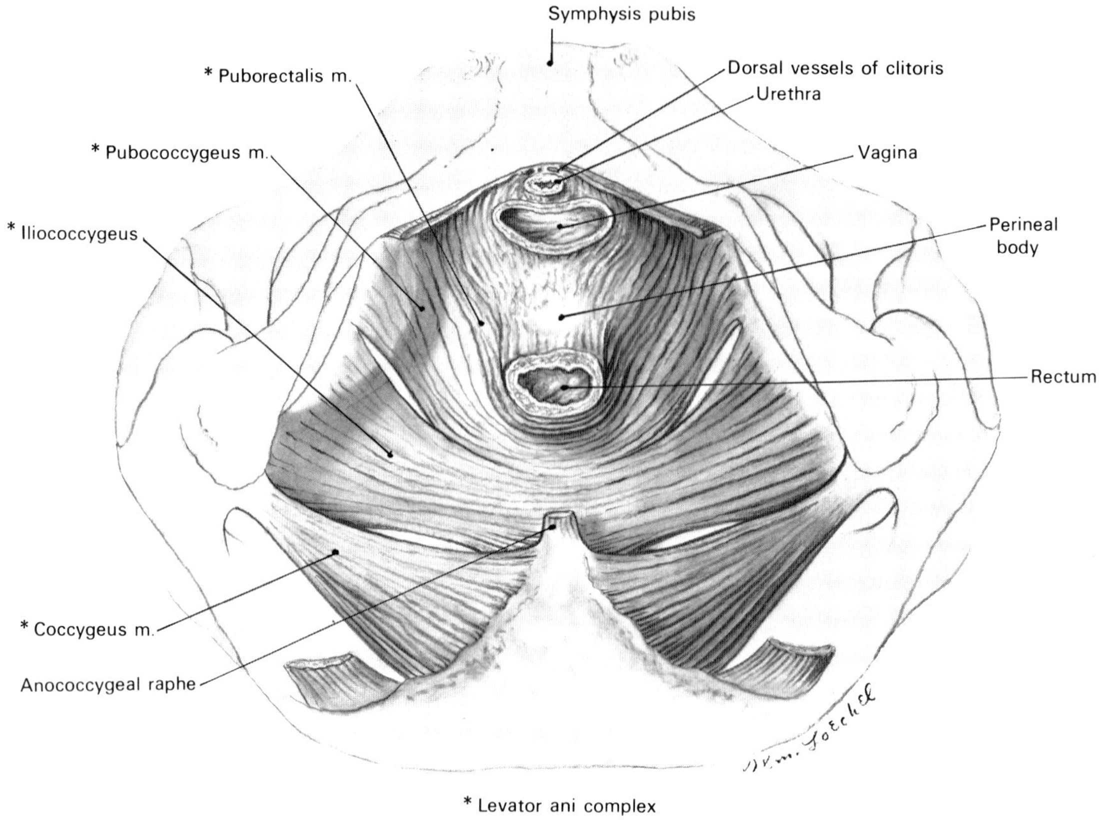

Fig. X-1. The pelvic diaphragm of the female viewed from below (lithotomy position) after perineal compartments have been stripped away.

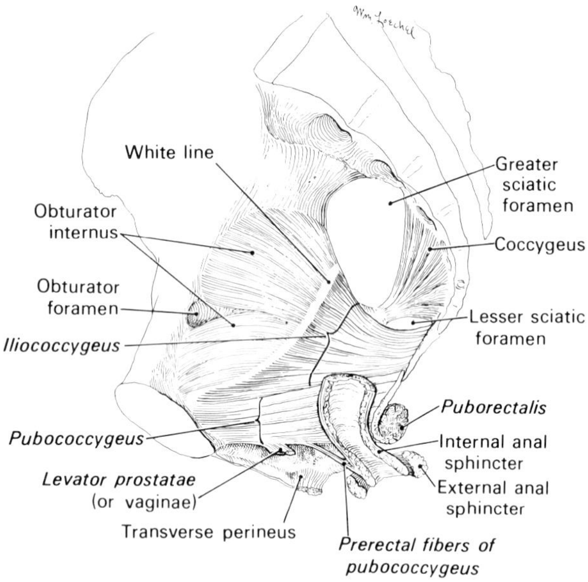

Fig. X-2. Lateral view of the hemisected pelves to show contributions of the Coccygeus and Levator ani to pelvic diaphragm. The subparts of the Levator are underlined. The white line is the tendon of origin of the Levator from the fascia of Obturator internus.

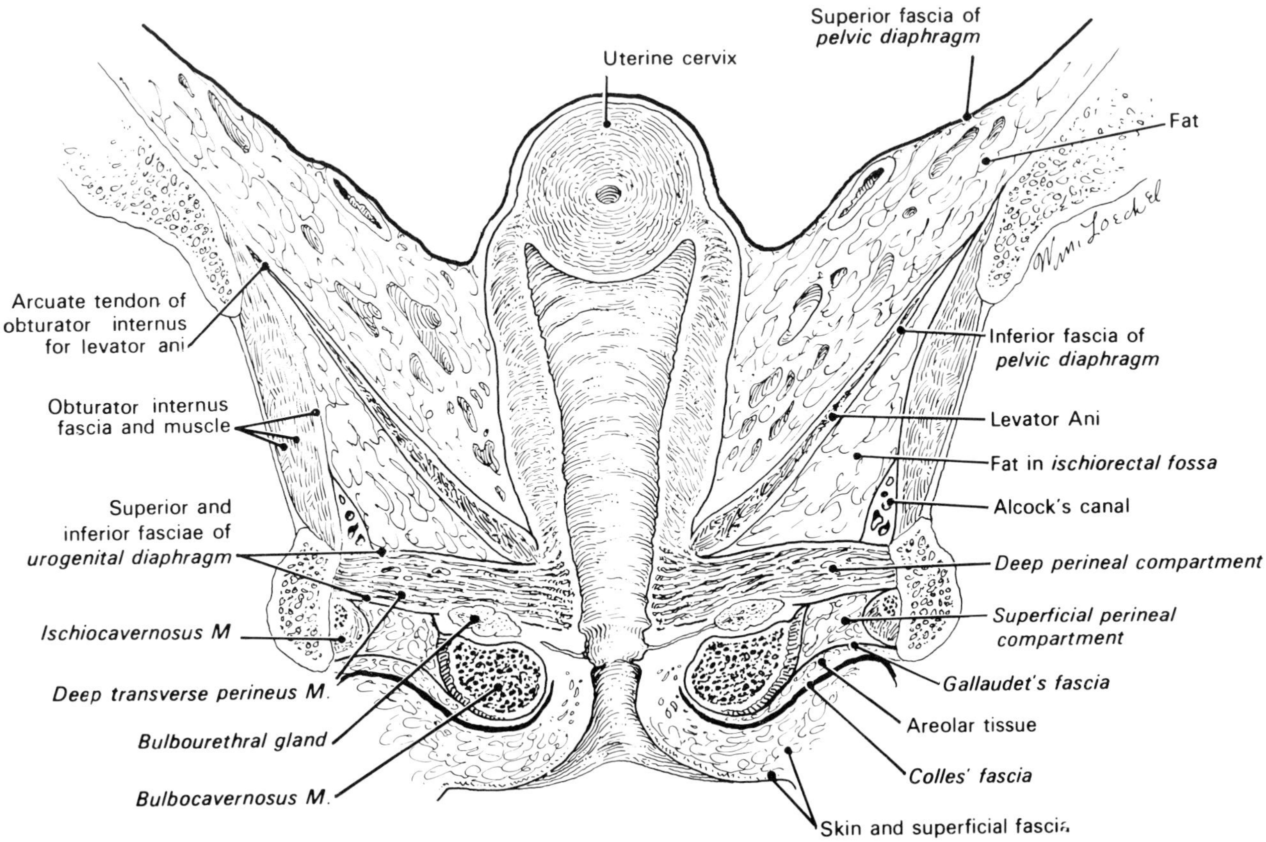

Fig. X-3. Coronal section through pelvic diaphragm and the superficial and deep perineal compartments (pouches) in the female. The deep compartment, with its muscle and fasciae, constitutes the perineal membrane or urogenital diaphragm.

phragm along with the abdominal muscles functions both to increase intra-abdominal pressures and, at the same time, to resist such pressure. Since the Levator Prostatae and Pubovaginalis lie below the bladder, they become functional in the process of urination.

THE PERINEUM

The Female Perineum

We are now ready to consider the perineum which lies superficial to the pelvic diaphragm, and which we view dead on with the body in the lithotomy position. We shall consider the female perineum first. Its boundaries spread laterally to the thighs, posteriorly to the anococcygeal region and anteriorly to the **mons pubis,** a fat pad covered by a triangular mat of pubic hair (Fig. X-4A). The perineum is a diamond-shaped region around the outflow of the bladder, rectum, and reproductive tract (Fig. X-4B). Its superior angle is the **arcuate (inferior) pubic ligament,** and at its inferior angle is the *tip of the* **coccyx.** Its lateral angles are marked by the **ischial tuberosities** upon which we sit. Its upper and lower sides are the **pubic bony arches** and the **sacrotuberous ligaments,** respectively. The ligaments are hidden on each side by the overlying Gluteus Maximus. The upper half of this area is called the **urogenital triangle;** the lower half, the **anal triangle.** The space on each side of the rectum is filled with fat, the fat of the **ischiorectal fossa** (Fig. X-3). This fat covers the posterior parts of the pelvic diaphragm and hides the Levator Ani and Coccygeus from view. The fossa spreads laterally and upward along the slightly sloping sides of the Levator so that it disappears deep to the **perineal membrane** which is the roof of the urogenital triangle. In the center of the perineum is a small fibrous structure, between vagina and rectum, to which the lower border of the perineal membrane anchors at the midline and to which several perineal muscles will attach. This is the **perineal body** (Fig. X-4 A, B). Figure X-4B represents, in a way, the skeletal framework for Figure X-4A.

We shall now identify structures in the perineal region by working progressively deeper, beginning with the superficial skin and the external genitalia.

The external genitalia of the female are related to the **rima pudendus,** the fissure between the **two labia majora.** These latter are the equivalent of the male scrotum. The labia majora unite anteriorly at the

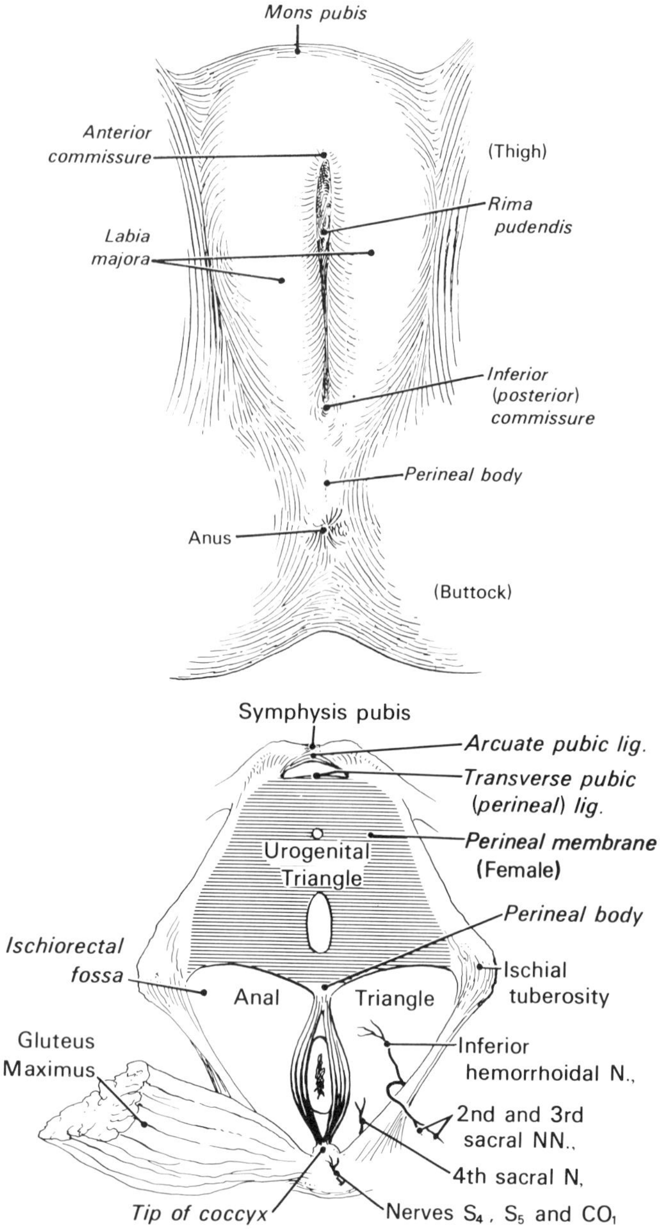

Fig. X-4. Perineal boundaries. **A.** Superficial view (labia closed). **B.** The "perineal skeleton."

mons in the **anterior commissure** and, posteriorly, in the **posterior commissure** (Fig. X-4A). Inferior to the fissure, between vagina and anus, lies the **perineal body.**

In the nullipara, the **labia minora** are hidden by the labia majora. When the latter are spread (Fig. X-5A) it becomes apparent that the **labia minora** bound a space, the **vestibule,** which leads to both the urethral and vaginal orifices. As the labia minora progress toward the mons, they split to form the lateral and medial folds of the **clitoris.** The lateral folds unite to form the **prepuce** of this tiny titilant. The medial folds unite to form a dainty seam or **frenulum** on the inferior aspect of the clitoris. The **urethra** opens in the anterior portion of the vestibule. Flanking it, and slightly caudad, are openings of the **paraurethral ducts of Skene** (Fig. X-5B). The vaginal outlet lies posteriorly in the vestibule with the openings of the vestibular **glands of Bartholin** on either side.

The clitoris itself consists of a **glans,** a **body,** and two **crura,** somewhat as does the penis of the male (Fig. X-5B). Extending downward from the glans is a short **pars intermedia** to which the anterior ends of the **vestibular bulbs** attach. The bulbs lie deep in the lateral floor of the vestibule and would have to be uncovered by dissection to be seen. They remind one of two tiny testes suspended from a minute penis (Fig. X-5B).

Before dissecting in this region, examine the skin of the labia majora and minora. It is different from skin over most of the body. It is hairless on the inner surfaces of these lips. Sebaceous glands here open directly upon the surface, not into the sides of hair follicles, and the sweat glands of the minora are the large "apocrine" type such as one finds in the axilla and around the anus. Both types of glands can be troublesome, giving rise to sweat gland tumors or to sebaceous cysts. The secretions from the ducts of Skene and Bartholin add to the unique odiferous atmosphere of the vestibule. The ducts lend themselves as snug harbors for gonococci.

As in the case of the abdominal wall, the skin rests upon a hypoderm or superficial fascia. In the abdomen, this superficial fascia consisted of an outer fatty lamina, or Camper's fascia and a deeper lamina, more membraneous in nature, Scarpa's fascia. While the fatty portion forms a more or less homogeneous blanket in most abdominal walls, in the pubic region of the female it forms a discrete pad, or eminence known as the mons pubis. In the labial region, fat dips down below the inguinal ligament to stuff the labia majora. Everywhere in the perineal region where some loose, fatty component exists, it rests upon a tougher **Colles' fascia,** the membranous counterpart of Scarpa's fascia. The name change probably indicates that Colles ventured into regions where Scarpa feared to tread. Colles' fascia extends from the inferior border of the pubic rami to the posterior margin of the uro-

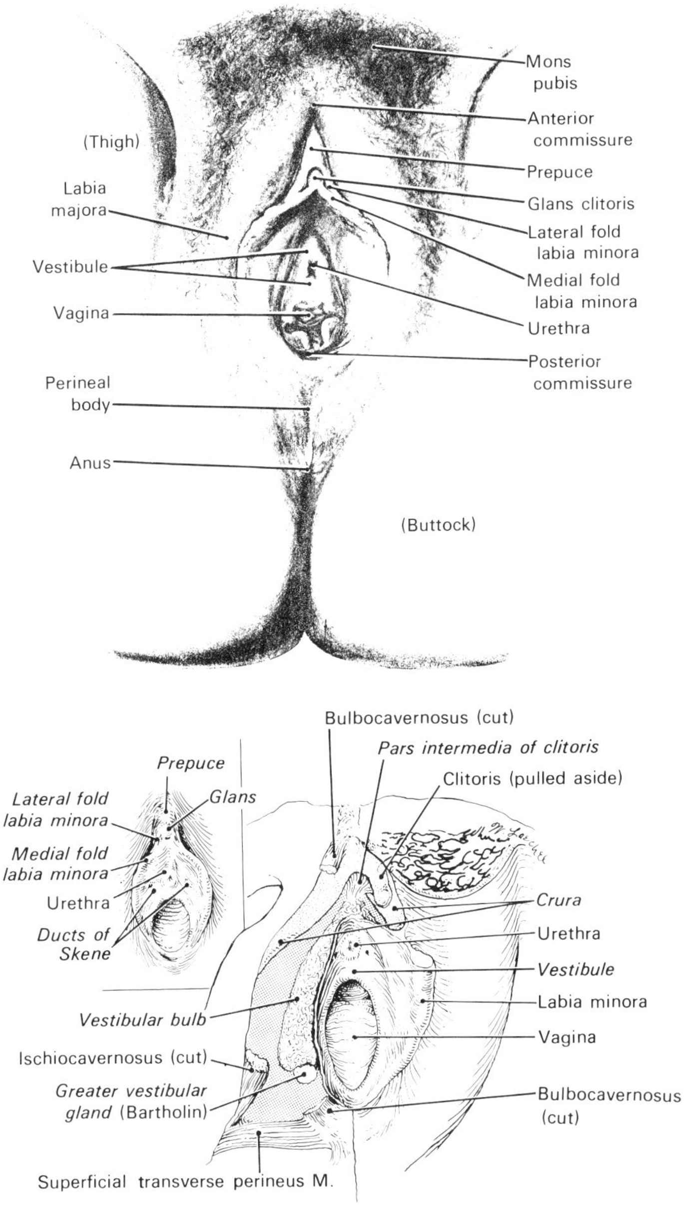

Fig. X-5A. External genitalia of a parous female, labia spread. (From Dickinson, *Atlas of Human Sex Anatomy*, 2nd ed. Williams & Wilkins, Baltimore, 1949) **B.** Superficial compartment of female perineum, partially dissected. *Inset* shows genital relationships prior to dissection.

genital diaphragm. It is backed up by a still deeper but more delicate membranous fascia called Gallaudet's fascia from which it is separated by a miniscule amount of loose areolar tissue (see Fig. X-3). These last three lamina, Colles' fascia, areolar tissue, and Gallaudet's fascia, make up the *superficial wall* of the **superficial perineal compartment.** Since they may appear as one membranous structure, so closely are they applied to each other, you may miss them as a trilaminar tissue as you dissect and go through them all. It is of little consequence if you do. The toughest of them will be Colles' fascia; and if you wish to call all three layers Colles' fascia for simplicity, it is all right with us. It probably means that Colles didn't split hairs as much as Gallaudet did when the latter dissected in this region. The important thing is that their removal gets you into the superficial perineal compartment.

The Superficial Perineal Compartment: The chief contents of this space are three masses of erectile tissue in the female and three muscles associated with these masses. The masses are the vestibular bulbs, the diverging crura of the clitoris, and the tiny pars intermedia of the clitoris. The muscles are the **Bulbocavernosi,** the **Ischiocavernosi,** and the **Superficial Transverse Perinei.** The first pair of these muscles covers the vestibular bulb on each side; the second covers the crura. The last is often a poorly developed muscle. On each side it bridges from the inner surface of the ramus of the ischium to insert into the perineal body. In Figure X-5B the uncovered bulb and crura are seen on the left side after the muscles have been dissected off them. The Ischiocavernosus arises, as did the Superficial Transverse Peronei, from the ischial ramus and it inserts on the lower and medial aspects of the crura. It may play a role in erection of the clitoris by compressing the crus of its side to inhibit flow from this organ. The Bulbocavernosus arises from the perineal body and it passes forward on the side of the vagina. In doing so, it covers and hides the vestibular bulb of its side. It inserts into the pubic arch and into the pars intermedia and dorsum of the clitoris. It acts with its partner as a delightfully weak vaginal constrictor.

The vestibular bulbs are elongate masses of erectile tissue under cover of the Bulbocavernosi, broad behind and tapering in front. They are the homologue of the bulb of the penis and its adjacent spongy tissue.

The remaining structures in the superficial compartment are nerves and vessels which we will discuss below, and the greater vestibular glands (of Bartholin). These last are two tiny bodies inferior to each bulb (Fig. X-5B). Their ducts open into the vagina. They are homologous with the male bulbo-urethral glands.

The Deep Perineal Compartment (Fig. X-7): If the deep fascia of the superficial compartment is removed, we enter the deep perineal compartment. There

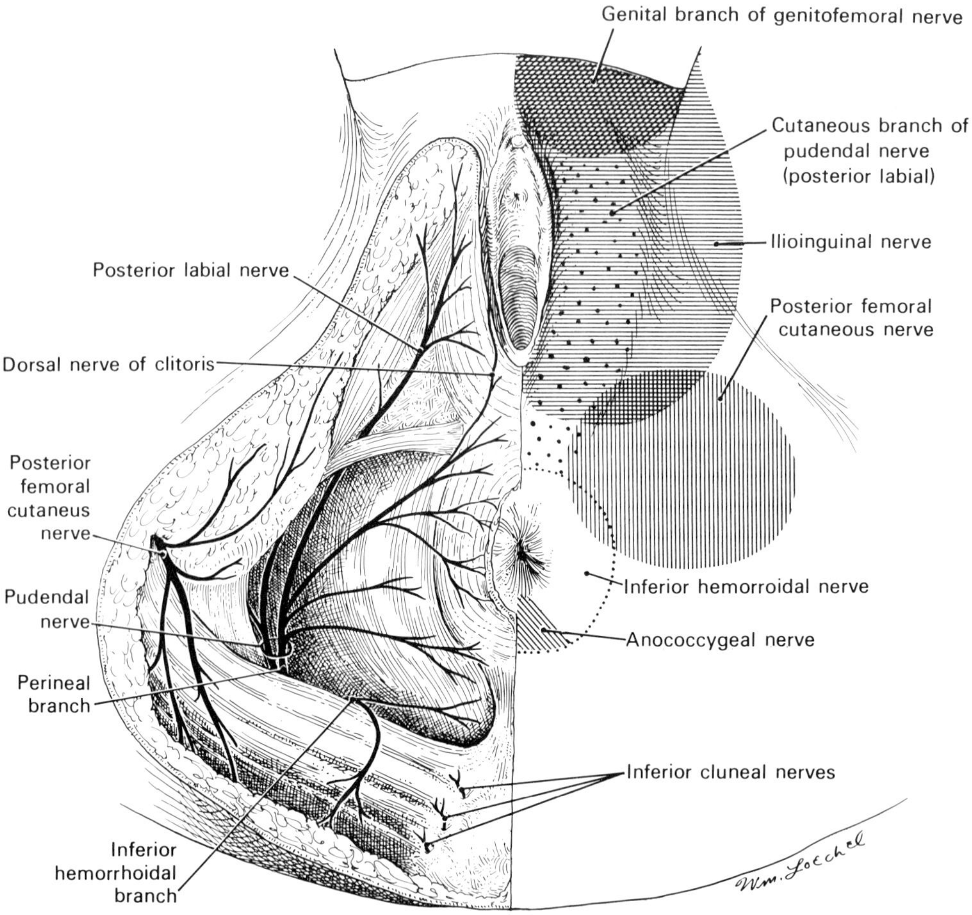

Fig. X-6. The distribution of cutaneous sensory supply to the perineal region. The over-lapping pattern of distribution of five cutaneous nerves are shown on the right side. The left side shows some of the deep nerve supply to the female perineum.

is little exciting in this compartment except for the Deep Transverse Perinei, two muscles of necessity separated by the vagina and urethra in the female. Each muscle arises from the inferior ischial ramus and from the pubis, and both insert into the medial raphe. Each enmeshes with its counterpart of the opposite side to form a kind of sphincter about the vagina and urethra. Some fibers insert anterior to the urethra and a few between the urethra and the vagina. While these latter have been accorded the name **Sphincter Urethrae,** they can hardly act as a true sphincter since the urethra unites with the vagina. Moreover, should the perineal nerve to these fibers be cut, any anticipated incontinence fails to occur.

Actually, the muscle is not entirely limited to the region of the urogenital triangle (at least not in the female). Rather, fibers spread posteriorly to encircle the anus as well and to anchor to the coccyx. Thus, the muscle is not so much triangular as it is quadrangular in shape. At best, however, the fibers of the Deep Transverse Perinei are scanty and well separated as they layer and spread between the superior and inferior fasciae of the deep compartment. Only the inferior fascia, which forms the floor of the deep compartment and adds to the roof of the superficial compartment is tough and dense. The superior fascia is rather indistinct.

It is the muscle and fascia of this deep compartment that is known as the **urogenital or perineal diaphragm (membrane).** Hence, the urethral portion passing through it (and, especially, in the male) is known as the membranous urethra. Remove this diaphragm or fibromuscular membrane and you look down upon the parts of the Levator Ani which contribute to the *pelvic diaphragm* discussed earlier.

The Anal Triangle: Now attend to the posterior aspect of the perineum between ischium and rectum. The Levator Ani, sloping slightly upward in this region, is covered with fat. The fat continues anteriorly into a deep fossa on each side of the rectum. Most of this fossa is hidden from view anteriorly when the structures of the urogenital triangle are in place.

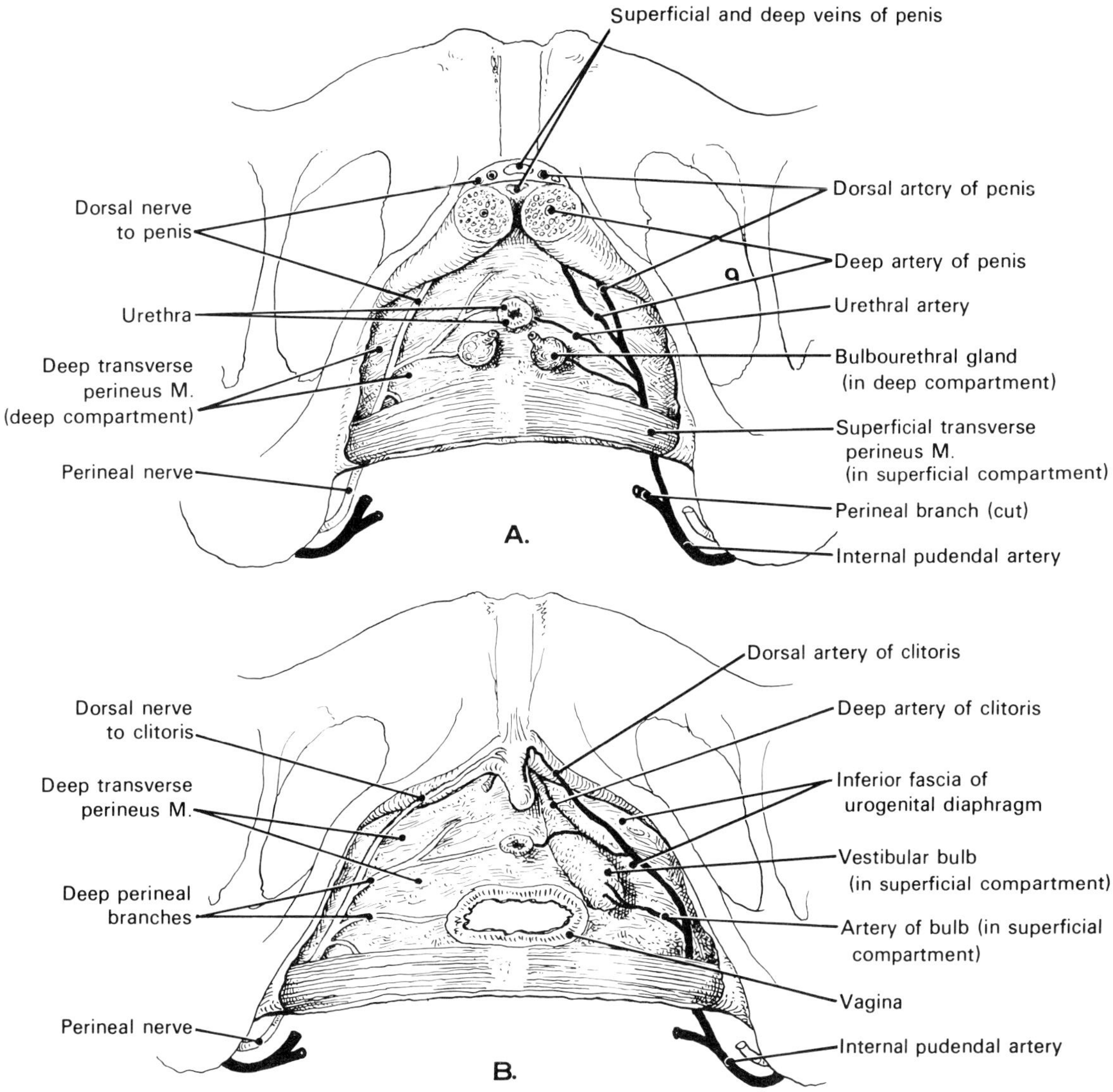

Fig. X-7. The deep perineal compartments of (**A**) the male and (**B**) the female. Except for the Superficial transverse perineus on the right side of B, structures labeled relate to the deep rather than to the superficial compartment.

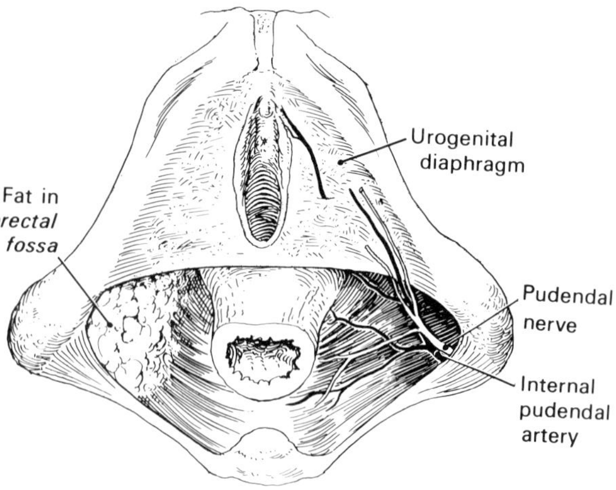

Fig. X-8. The relationship of ischiorectal fossa to the urogenital diaphragm. Some nerves and vessels pass under, some over, and some pierce through the urogenital diaphragm.

Running through the fat of this *ischiorectal fossa* is the pudendal nerve, the major nerve of the area, and the end branches of the internal pudendal artery (Fig. X-8). The fat, of course, is held packed in place by a tough fascia. If the Gluteus Maximus is contracted, it can compact this fatty mass against the rectum as a handy, extra, emergency sphincter until a handy evacuation site can be discovered.

Development and Innervation: A word about the development of this region may help to clarify its nerve supply. The hindgut of the embryo terminates in the early stages at a common **cloacal membrane** (Fig. X-9A). The hindgut gives rise to the distal third of the transverse colon, descending colon, the sigmoid, rectum and upper end of the anal canal. The terminal part of the hindgut opens into a common urinary-fecal reservoir, the **cloaca** along with the allantois, the terminal end of which will become a urinary bladder. As development progresses, a tissue ridge or septum arises in the junction angle between allantois and rectum. This **urorectal septum** divides the cloaca into an anterior part or **urogenital sinus** (the area of the urogenital triangle in the adult) and an **anorectal canal** (the area of the anal triangle and future ischiorectal fossa). The septum reaches the cloacal membrane in the seventh week (Fig. X-9B) at which point a **primitive perineum** is born. The anterior closing plate of the primitive perineum is the urogenital membrane or diaphragm which will consist later of the Deep Transverse Peronei and their fasciae, pierced by urethra and vagina in the female.

Meanwhile, the posterior portion of the cloacal membrane becomes surrounded by mesenchymal swellings so that it lies in an ectodermal depression or **proctodeum.** The mesodermal swellings covered by skin are known as the **anal folds.** The membrane here

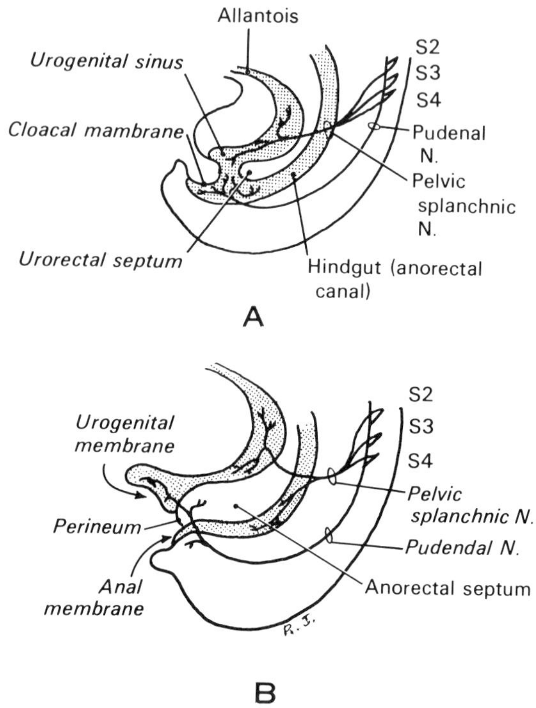

Fig. X-9. Division of cloaca into urogenital sinus and anorectal canal and the course of pudendal and pelvic splanchnic nerves to these areas. **A.** In the 5-week embryo. **B.** At 8 weeks.

ruptures to open the anal canal to the exterior. Thus, the lining of the upper end of the anal canal is endodermal in origin from the hindgut and will be vascularized by the hindgut artery, the **inferior mesenteric.** The lower third of the canal lining comes from the indented ectoderm and is supplied by the systemic rectal arteries, the **inferior hemorrhoidals** which are branches of the internal iliacs within the pelvis.

The region of the pelvis and perineum will be innervated by the anterior divisions of spinal nerves S_2 to S_4 (Fig. X-9). These nerves are somatic motor to skeletal muscle; visceral motor (parasympathetic) to smooth muscle and glands of the area; and both somatic and visceral sensory. They reach the region either as the **pudendal nerves** or as the **pelvic splanchnic nerves** (nervi erigens). The pudendal nerves supply the perineal muscles. The pelvic splanchnics supply pelvic visceral contents.

In the adult the *pudendal nerve* takes a somewhat devious course (Fig. X-10B). But in the embryo, it went merrily on its unimpeded way to reach perineal structures (Fig. X-10A). As a bony pelvis organized, certain ligaments began to push the nerve around. The **sacrotuberous** and **sacrospinous ligaments** delineated two spaces, one large and one small, as the **greater** and **lesser sciatic foramina** (Fig. X-10B). Thus, the nerve was forced first to leave the pelvis *via* the greater foramen and to return to it *via* the lesser.

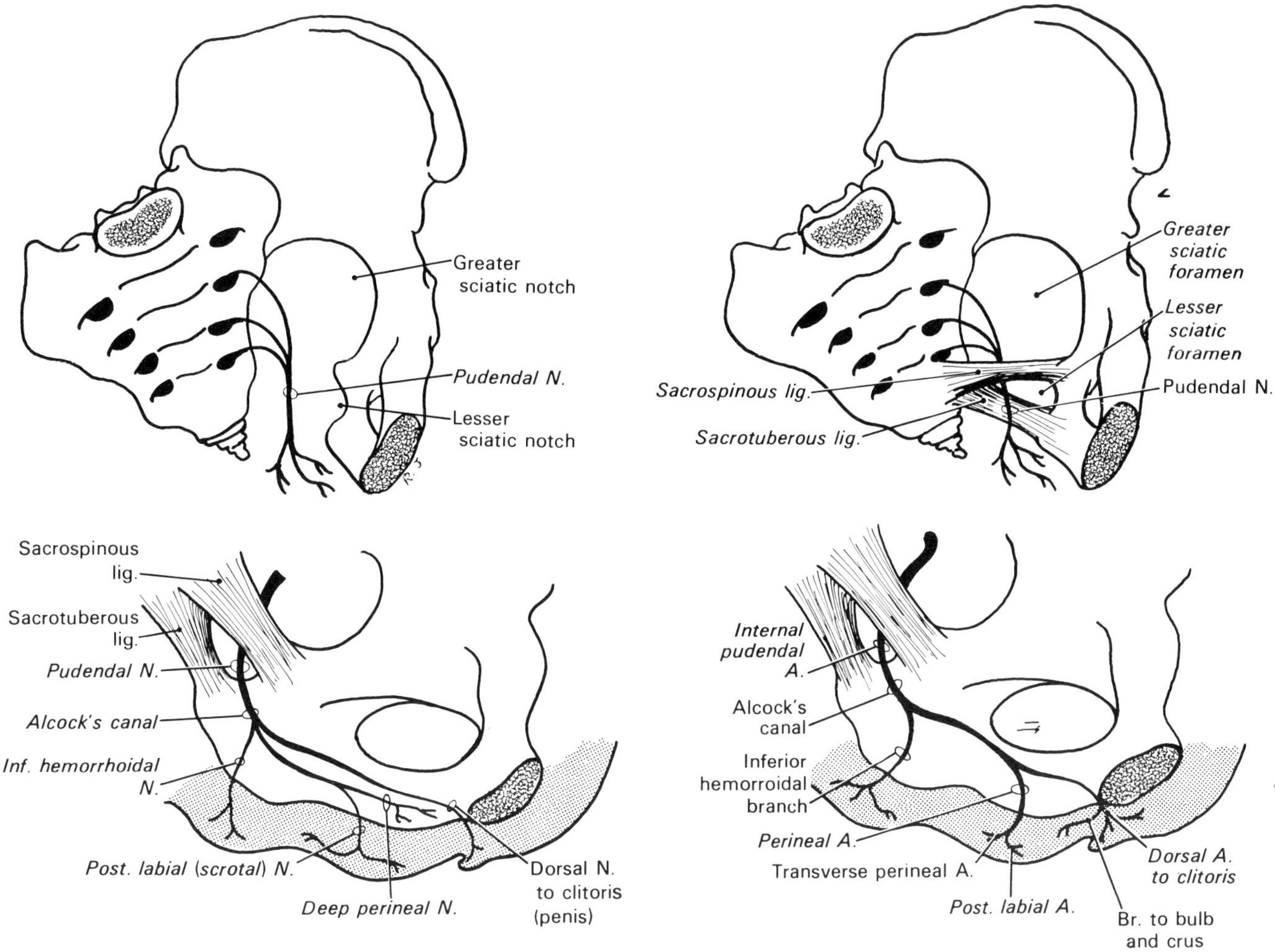

Fig. X-10. The course of the pudendal nerve and the internal pudendal artery. **A.** In the early embryo. **B.** After sacrotuberous and sacrospinous ligaments form. **C.** Chief branches of the nerve. **D.** Chief branches of the artery.

Upon its return, it encountered a tunnel on the lateral wall of the ischiorectal fossa formed by a splitting of the obturator fascia, and so it found itself in **Alcock's canal** (Fig. X-10C).

Upon entering the canal, the pudendal nerve splits into three main branches: (1) **the inferior hemorrhoidal;** (2) **the perineal nerve;** and (3), in the female, the **dorsal nerve to the clitoris.** The perineal nerve, itself, divides into the **posterior labial nerve** and the **deep perineal nerve.** The posterior labial supplies labia majora and then sends a branch caudal to the region of the perineal body. The deep perineal supplies the Levator Ani in part, the Deep and Superficial Peronei, the Bulbocavernosus, the Ischiocavernosus. The nerve to the bulb is given off from the branch to the Bulbocavernosus. The dorsal nerve supplies crura, prepuce, and glans of the clitoris. Thus, the *chief role* of the pudendal nerve is to afford a somatic motor supply to the area. The nerve differs from the artery, essentially, in that all its branches enter the deep perineal compartment and all twigs must pierce the inferior fascia of that compartment to reach structures in the superficial compartment. The artery has a superficial branch which enters the compartment by crossing over (or under) the Superficial Transverse Perineus at the posterior margin of the urogenital diaphragm.

The pudendal nerve is also a main sensory supply to the skin of the perineal region. But it gets an assist in this regard from three other nerves which overlap the region, namely: the **posterior femoral cutaneous;** the **ilioinguinal;** and the **genitofemoral** (see Fig. X-6). The posterior femoral cutaneous nerves are derived from the anterior divisions of S_1 to S_3, and they supply the skin along the posterior parts of the labia majora and adjacent lateral skin. The ilioinguinal nerves (T_{12} to L_1) supply skin over the mons, the anterior parts of the labia majora and upper medial thigh. The genitofemoral nerves (L_1 to L_2), *via* its genital branch which accompanies the round ligament of the uterus into the labia, supply anterior portions of the labia majora also. In so far as the pudendal nerve is concerned, only its deep perineal branch fails to be sensory to a skin area in the region. Even its hemorrhoidal twig sends branches to the skin (Fig. X-6). There are anococcygeal and inferior cluneal branches as sensory fibers in the area, but these are interlopers from the region of the buttocks proper.

If you mark a point midway between anus and ischial tuberosity you can enter with a hypodermic needle to block any of the four major nerve supplies to the perineum. It is a matter of directing the point and shaft of the needle from this midway site. If you direct the needle obliquely toward the ischial spine you block the *pudendal nerve.* If you direct the needle horizontally and laterally toward the ischial tuberosity and even a bit posterior of the tuberosity, you block the *posterior femoral cutaneous.* To block the *ilio-inguinal and genitofemoral supplies,* pass the needle anteriorly beneath the skin lateral to the labia and toward the superficial inguinal ring.

Since the pelvic splanchnic nerves are motor and sensory to the viscera of the pelvis (though some of their twigs have to reach glands of the perineum) we will delay discussing them until we enter the pelvis.

Blood Supply of the Perineum: Fortunately, the **internal pudendal artery** is the main vessel in the region. It follows the course of the pudendal nerve. The vessel is a branch of the anterior division of the hypogastric (internal iliac) artery. The artery and nerve to the perineum almost have the same order of branching but not quite (Figs. X-10D and 11). As the vessel leaves Alcock's pudendal canal, it gives its branch to the anal triangle, the **inferior hemorrhoidal (rectal) artery** which runs through the fat of the ischiorectal fossa. It breaks up into many twigs to supply fascia, Levator Ani, external and internal anal sphincter, and skin. Deep twigs enter the anal canal to anastomose with the middle and superior hemorrhoidal branches of the inferior mesenteric arteries.

The artery then divides into deep and superficial branches to enter the perineal pouches or compartments. The superficial branch enters the superficial pouch as the **perineal artery** to supply the muscles of the pouch. Its terminal twigs move forward through the superficial compartment as the **posterior labial arteries** (or **posterior scrotal** arteries of the male) to supply the skin.

The deep branches pierce the free posterior edge of the urogenital diaphragm to run between its superficial and deep fascia in the deep compartment among the strands of the Deep Transverse Perineus, supplying it and its urethral and vaginal sphincteric portions.

A twig, the **artery of the bulb** (of the vestibule or the penis) must pierce the inferior fascia to enter the superficial compartment where this structure lies.

In a like manner, the **dorsal and deep arteries of the clitoris** (penis), as terminal branches of the deep division of the perineal artery, must leave the deep compartment by piercing its inferior fascial floor to enter these structures which lie in the superficial compartment. In the male, since the bulbo-urethral glands are found in the deep compartment, they are supplied there by the deep arterial branches. The greater vestibular glands (Bartholin glands) of the female, lying in the superficial compartment, get their supply from the superficial arterial branches.

Coming into the region also will be the **artery and vein of the round ligament.** These are branches of the inferior epigastric vessels which follow the round ligament of the uterus through the superficial inguinal ring. Coming through the ring with these vessels will be the ilio-inguinal nerve and the genital branch of the genitofemoral nerve and, of course, the round ligament itself. Do not expect to see much of a structure when you look for the round ligament in most female cadavers. It usually continues around the pubic tubercle and into the labium majus on each side. But it may end somewhere within the inguinal canal as an indistinct strand.

If there is an internal pudendal artery in the area, we can expect an *external one.* This vessel arises from the femoral artery below the inguinal ligament at the fossa ovalis where the femoral emerges into the thigh. The external pudendal is small in caliber as a rule. It courses medialward to end in the prepuce of the clitoris. Small twigs also enter the labium majus as the **anterior labial arteries.**

Episiotomy: Sometimes, during obstetrical deliveries, it becomes necessary to enlarge a vaginal outlet; or perineal tears occur in the process of childbirth. Such natural lacerations are frequent obstetrical occurrences which involve the perineal body and the muscles attaching thereto. They have to be repaired and it is well to know that the urogenital diaphragm (perineal membrane) is the toughest tissue in the area into which to anchor a suture. When the posterior portion of the vagina is purposely cut to enlarge the outlet, the episiotomy procedure is intended to spare the tearing of the perineal body. The incision can be a midline or a mediolateral one. The mediolateral incision has the advantage of extending the opening a further distance than the midline incision, but it will encounter the Bulbocavernosus and Pubococcygeus muscles with possible bleeding from the bulb. Its main disadvantage is that it is not as easy to repair as the midline incision which heals readily and is less painful. In a midline incision the cut involves skin, subcutaneous fascia, the Superficial Transverse Perineus, deep fascia and Deep Transverse Perineus (*i.e.,* the urogenital diaphragm). The Puborectalis can be inadvertently severed if the cut is made too deep, but the Bulbocavernosus and Pubococcygeus will not lie in the path of the cut as they do in the case of a medio-

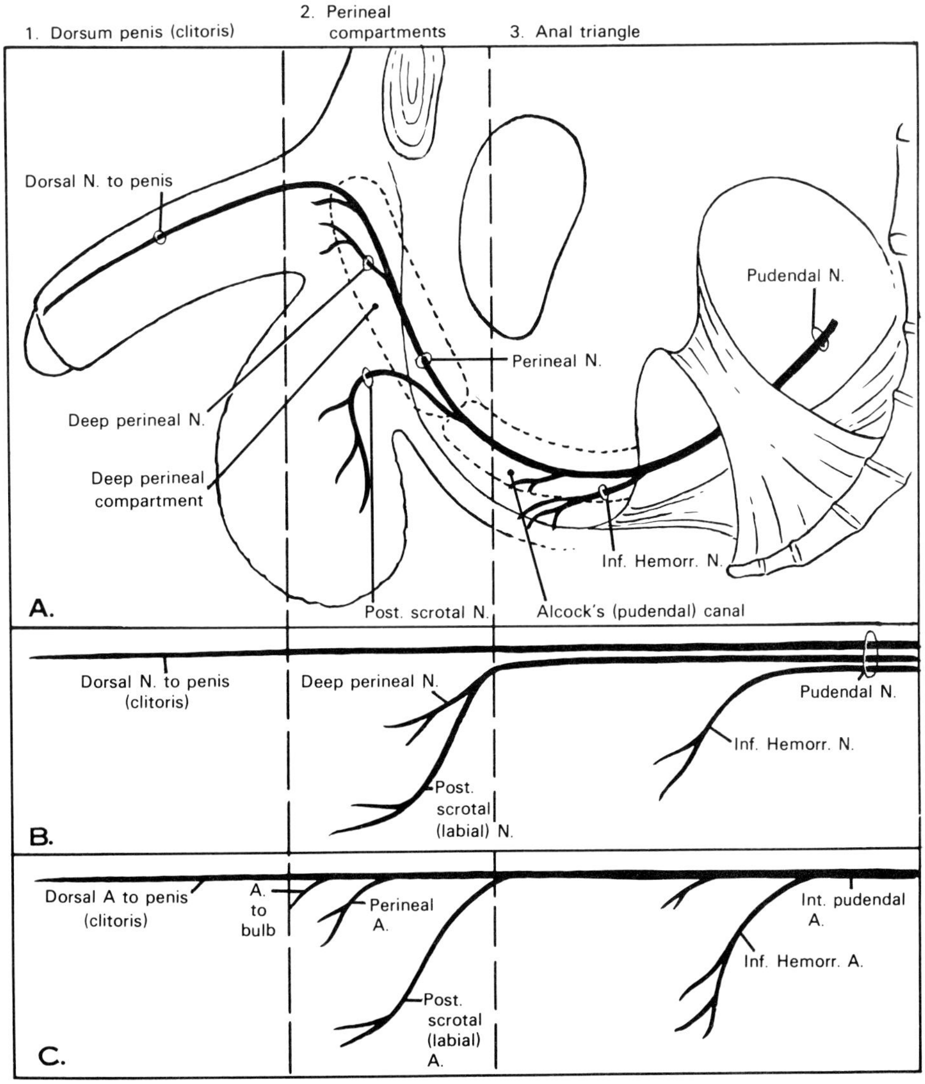

Fig. X-11. Comparison of course and distribution of pudendal nerve and internal pudendal artery in male. **A.** Course of nerve (artery follows same course). **B.** Distribution of nerve branches to anal triangle, perineal pouches, and to dorsum penis. **C.** Distribution of arterial branches to same areas.

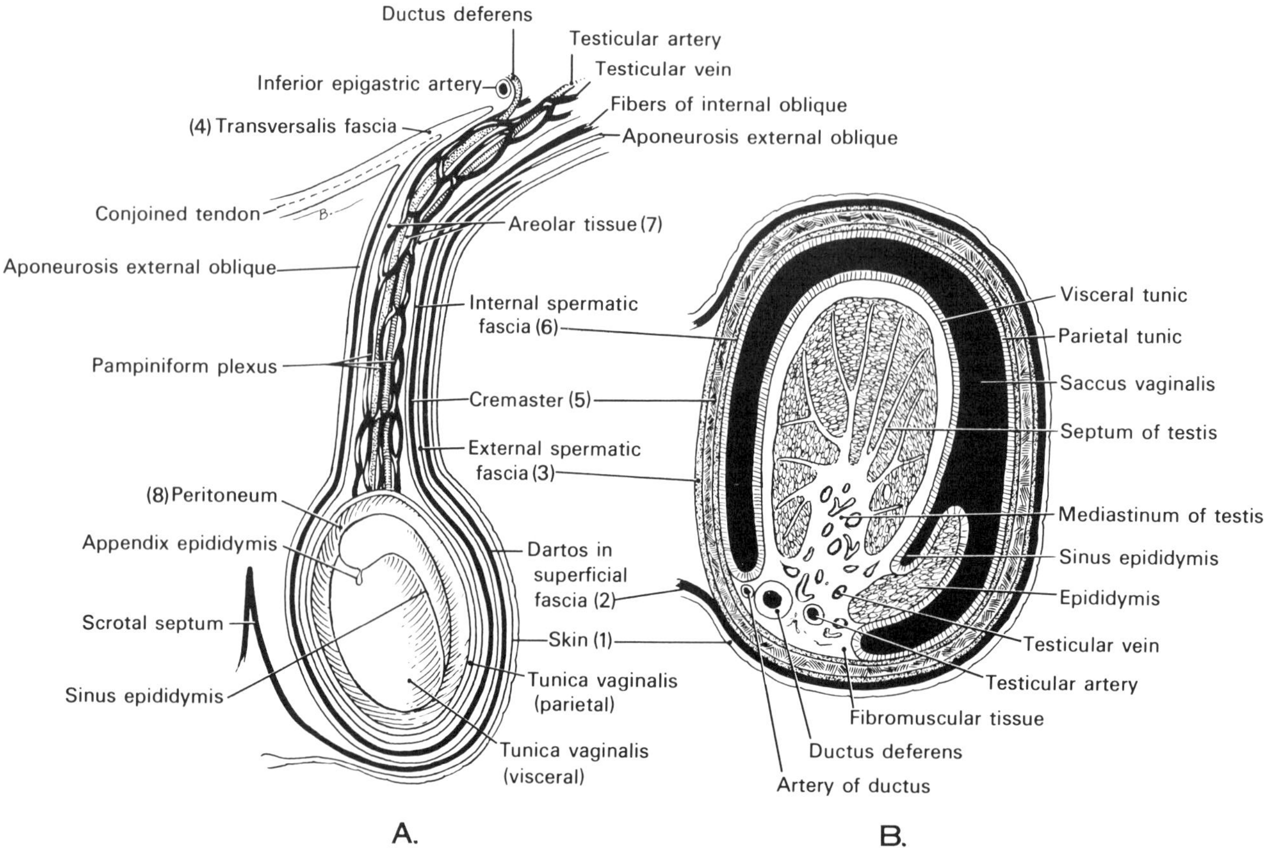

Fig. X-12. Relation of layers of body wall to spermatic cord and scrotum. **A.** Coronal section through cord and left scrotum. **B.** Cross-section through left testis and scrotum seen from above. Sac is exaggerated in distended condition. (After Cunningham.

lateral incision. Otherwise, the tissues encountered are the same in both procedures.

The Male Perineum

The External Genitalia of the Male: The scrotum and penis differ little from the labia and clitoris of the female, nor, conveniently, does the perineum differ too markedly between the sexes (Fig. X-16). But whatever slight differences there are, they have served to make an otherwise dull existence entertaining. We deal largely with a functional difference in the case of the scrotum, a dual urinary and reproductive duct in the case of the penis, and a matter of robustness in the case of perineal muscle and erectile tissue. Lacking such a fat pad as the mons pubis, the pubic hair distribution in the male is diamond-shaped rather than triangular, and fat does not invade the scrotum as it does the labia majora.

The **scrotal pouch,** situated below the penis and the pubic symphysis, is partitioned into two sacs each containing a male gonad or **testis** and its associated **epididymis** as well as the lower portion of the spermatic cord with its coverings (Fig. X-12). The sac consists of: (1) skin; (2) Dartos muscle, which largely replaces superficial fat and Colles' fascia here; (3) a thin lamina of external spermatic fascia; (4) slips of) Cremaster muscle; (5) external spermatic fascia; and (6) an inner lining of parietal peritoneum, the tunica vaginalis. The **Dartos muscle** is smooth muscle whose strands are embedded in a bit of loose areolar tissue. When this muscle contracts, as it does in the cold or during sexual stimulation, the scrotum presents the appearance of a sad, worried, wrinkled prune. It is much more relaxed in a calm, warm environment. The other structures mentioned are direct carry-overs from the body wall as it is organized in the inguinal region. External spermatic fascia is a continuation of the External Oblique aponeurosis, spun thin at the diverging crura of the superficial inguinal ring. Some few scanty Cremasteric fibers may add to the scrotal wall in its upper part. These slips of skeletal muscle are derived from the Internal Oblique. The internal spermatic fascia is a continuation of the transversalis fascia. The Transversus Abdominis contributes neither to cord coverings or scrotum since it fails to extend to the inguinal region except as an aponeurotic slip behind the superficial ring where it may (or may not) take part in the formation of a conjoined tendon. The peritoneal tunica vaginalis, once continuous with the peritoneal cavity, but cut off from it by the obliteration of

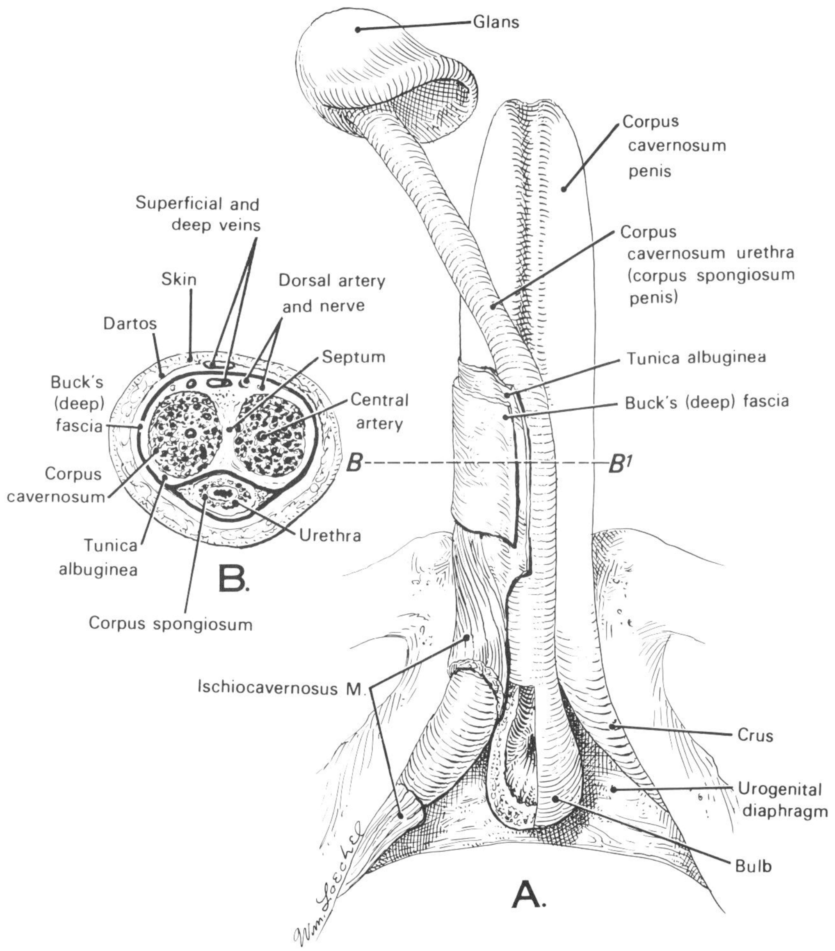

Fig. X-13A. Crural relationships and tunics of the penis. **B.** Cross-section taken through A at B-B′.

the processus vaginalis, provides a home away from home for the testis (Fig. X-12). It reflects as a visceral tunic over testis and epididymis.

Scrotal Blood and Nerve Supply: Anteriorly, the scrotum derives its blood from the external pudendal artery, a branch of the femoral artery. Posteriorly, branches of the internal pudendal artery reach the scrotum. An additional supply comes to it from cremasteric and testicular arteries which traverse the spermatic cord.

The ilio-inguinal and genitofemoral nerves supply cutaneous twigs to its anterior wall. The posterior scrotal nerves, branches of the perineal division of the pudendal nerve, supply the posterior wall. The posterior femoral cutaneous nerves help to innervate scrotal skin much as they do in the case of the female labia.

It should be realized that lymphatics of the scrotum drain into superficial inguinal nodes.

The Penis: The male copulatory organ consists of a **root** and a **body.** The root is anchored to the roof of the superficial perineal pouch. It forms the erectile tissue of the **penile bulb** and its two crura. The crura attach to the body of the organ as do those of the clitoral crura of the female, but they do not compose the entire erectile organ nor do they contribute to the glans as is true of the clitoris. In the superficial perineal compartment the crura are covered by the Ischiocavernosi muscles as they were in the female (Fig. X-16), but the crura pass forward into the penile shaft as the conjoined **corpora cavernosa** (Fig. X-13A).

The bulb is located between the crura. It is not split into paired structures as in the female, but is single and is tunneled through by the male urethra. Rounded inferiorly and laterally, it flattens against the roof of the superficial perineal compartment (Fig. X-13A). The urethra pierces the perineal membrane (urogenital diaphragm) of the deep compartment as the **membraneous urethra** to enter the bulb, and its elongate, penile extension as the **cavernous urethra** (Fig. X-13B). Within the pelvis, the first part of the urethra has pierced the prostate which rests about the neck of the bladder (Fig. X-14).

Hence, the male urethra, much longer than that of

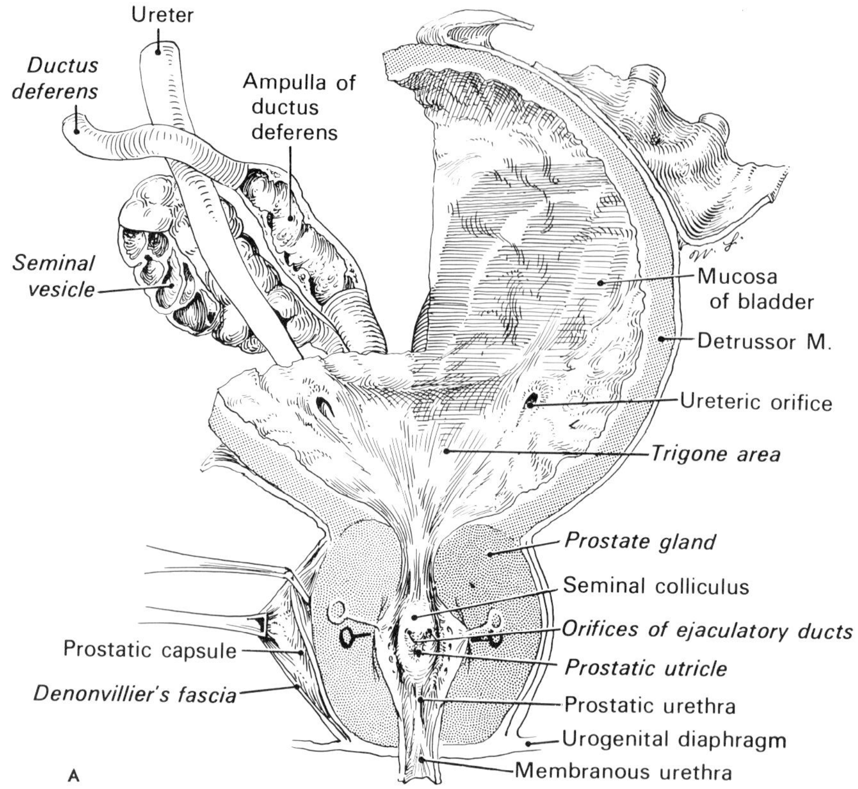

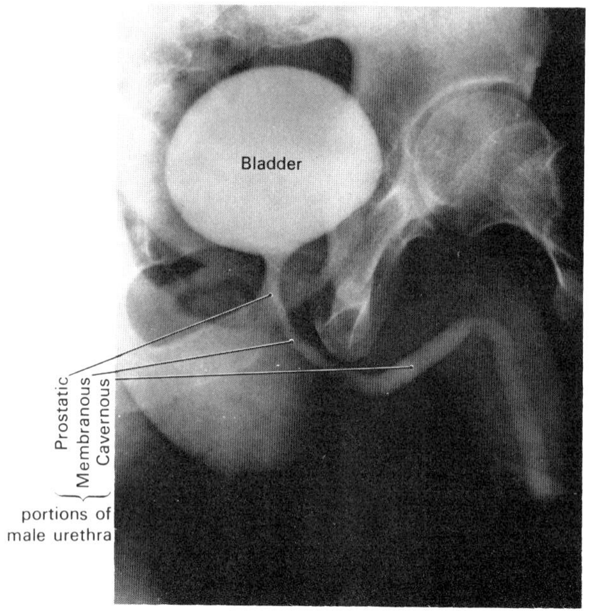

Fig. X-14A. Relationship of male reproductive ducts and glands to bladder and ureters. **B.** Voiding cystogram (radiograph) in the male. Note the three portions of the male urethra. Patient is in the left posterior oblique position.

the female, consists of **prostatic, membraneous,** and **cavernous** parts—not, as one of our nurses stated: "prostatic, membraneous, and carnivorous!" Obviously, her last date had been with a "wolf."

Both the crura and the elongate extension of the bulb contribute to the *body* of the penis, and it is the free end of the bulb which contributes to the terminal **glans penis** (Fig. X-13). The crura continue into the pendulous organ as the erectile corpora cavernosa. They form the dorsum and sides of the organ, and their blunt tips insert into the capping glans. The continuation of the bulb into the body constitutes the **corpus spongiosum** surrounding the urethra (Fig. X-13B). The latter fits into a groove formed on the underside of the merging crura, and, as indicated, provides the bulbous glans into which the free ends of the corpora cavernosa fit. When we speak of the dorsum of the penis we mean the part which faces forward when the organ is in its flaccid state, but which faces the belly wall when the organ is erect. Naturally, we are talking about the belly wall of its owner.

The glans, as in the case of the clitoris, is covered by a **prepuce** (or foreskin) unless the patient has been circumcised. The prepuce is separated from the body of the organ by a constriction at the neck of the glans. The **external urethral orifice** opens as a slit at the tip of the glans. The prepuce envelopes the glans for a variable distance as a double fold of skin.

Like scrotal skin, that of the penis is dark, but it is thin, elastic, and smooth. Some few hairs are present at its base. Except at the glans, the skin is loosely attached to a fat-free subcutaneous fascia which is continuous with that of the scrotal Dartos and with the superficial perineal fascia. Even a few slips of Dartos muscle may be present. The deep fascia is inordinately tough and is rather appropriately designated as **Buck's fascia.** Buck seems to have extended himself farther afield than either Scarpa or Colles, but, fundamentally, all were describing the same stuff while grabbing onto a different part of the proverbial blind-man's elephant. If Buck thought he had taken hold of the trunk, he was mistaken.

The **tunica albuginea** of the corpus cavernosum lies deep to Buck's fascia (Fig. X-13B), and it forms a dense fibrous sheath about the erectile component. Its superficial collagenous fibers orient longitudinally while its deep fibers encircle each corpus to meet in the midplane as the **septum.** The septum is thick and complete near the root of the penis but it becomes irregularly interrupted distally so that erectile tissue communicates between the gaps. A similar, but less rigid, tunica albuginea surrounds the spongiosum which contains the urethra. Extensions of these sheaths reach into cavernosum and spongiosum to form the walls of the blood-filled cavernous spaces in the erectile portions of the organ. The fibrous trabeculae contain a bit of smooth muscle as well, and they conduct vessels and nerves into the deep parts of the organ.

Aside from snakes and lizards which have hemipenes and crocodiles and turtles which have something approaching the nature of a penis, this copulatory organ is most highly developed in mammals. Mere possession of such an organ of insertion does not mean it is adequate for copulation. It has to be made that way through suitable nervous stimulation, whether directly physical or mental, in which case it increases in length, girth, and rigidity in less than 5 seconds. Fortunately for clothing manufacturers, this state is unnecessary at times other than copulation so that the penis is normally more neatly reduced in size and flaccid. In various mammals its reduction to a non-active state is variously solved. It can be tucked away from view completely, can be folded conveniently S-shaped, can retreat into a special sheath, or remain external, punily shrunken. In man, along with tree shrews and bats, it is truly external and pendulous.

Man has made much of the glorious size of this organ in the erect state; but, in most instances, when *erect*, it is not much more than a mere 6¼ inches (15.6 cm.) in length and about 1½ inches (3.75 cm.) in diameter. This is indeed puny, compared to a good 3 feet in the bull, 5 feet in the elephant, and 8 feet in some whales! Many mammals, but not man, possess a bony or cartilagenous penis. While a femur generally looks like a femur in all plantigrade animals, the os penis is most unique in its variety of styles. It can be a rod or a plate. It can be straight or bent into a double curve. It comes spoon-shaped or pronged, and it can be square, triangular, or round in cross section.[1]

Blood and Nerve Supply of Penis: The hydraulics of erection is the hydraulics of a hose whose nozzle has been closed down to restrict flow under pressure. The bulk of the blood to the penis is supplied by the internal pudendal artery. The proximal portion of its skin gets its supply from the external pudendal, a branch of the femoral. We have seen how the internal pudendal supplies the perineal compartments. Its **superficial branch** supplies the three muscles of the superficial pouch or compartment and the cutaneous **posterior scrotal branches.** Its **deep branch** enters the deep compartment by piercing the posterior edge of the urogenital diaphragm. It rides amongst the fibers of the Deep Transverse Perineus which it supplies. It sends an **arterial twig** through the inferior fascia of the diaphragm into the superficial compartment *to the bulb,* the erectile tissue of which it supplies. A **urethral branch** is also given off in the deep compartment. It pierces the fascia also to enter the sulcus between corpora cavernosum and corpus spongiosum to reach the urethra within the latter (Fig. X-15).

The deep artery of the penis (Fig. X-15), obviously larger than its clitoral counterpart in the female, also leaves the deep perineal compartment for the superficial compartment, piercing near the ischiopubic ramus. It enters the crus of the penis distributing along the length of the corporus cavernosum as a **central artery** supplying arterioles to the cavernous spaces. These arterioles spiral somewhat and have a heavy smooth muscle media which chiefly remains constricted in the relaxed state of the organ. These **helicine arteries** are most abundant in the proximal parts of the corpora cavernosum.

The **dorsal artery of the penis** (Fig. X-15) runs the longest course in the deep compartment before piercing into the superficial compartment. It does so near the **transverse pubic ligament** to pass forward on the dorsum of the penis between the layers of the **suspensory ligament.** The latter is a condensation of the aponeurosis of the External Oblique Muscle of the abdomen, you will recall. The artery will come to lie between the dorsal vein and dorsal nerve of the penis (Fig. X-13B) below the penile (Buck's) fascia. The artery also continues on to supply the glans. It supplies fascia and skin of penis in general.

At times a dorsal artery comes from an **accessory pudendal artery** which rides on the superior side of the urogenital diaphragm (superior when the body is in a standing position, deep as viewed from the lithotomy position). This vessel passes forward and comes through to the penis between the arcuate pubic ligament and the anterior border of the urogenital diaphragm.

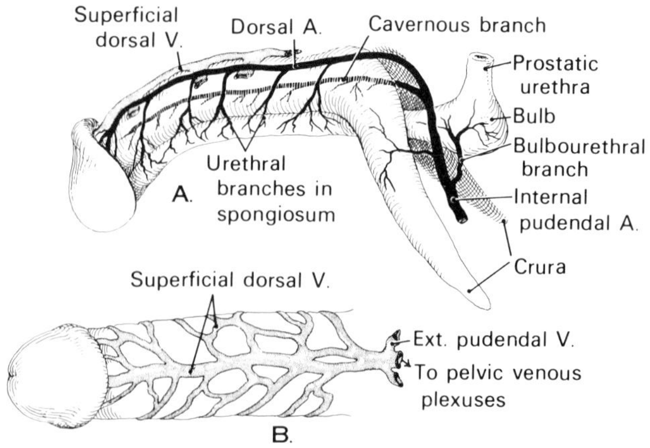

Fig. X-15. Blood supply of penis. **A.** Arterial. **B.** Venous.

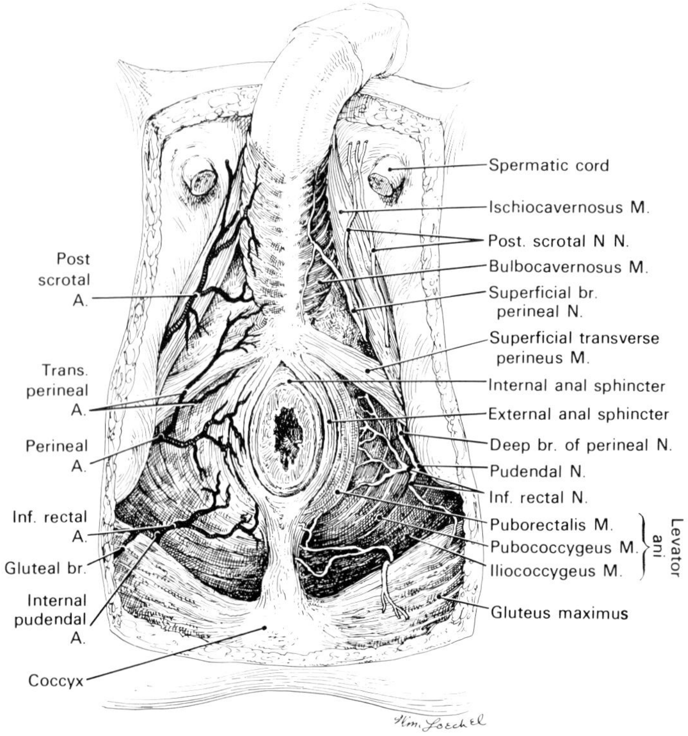

Fig. X-16. Male perineum and anal triangle.

Upon entering cavernous areas, arteriolar twigs traverse trabeculae which consist of collagenous tissue heavily laced with elastic fibers and some smooth muscle strands. Some arterioles end in capillary plexuses, others in the "helicine arteries" of which we spoke. The latter open into the cavernous spaces. It is thought that some capillaries also open into these spaces. The spaces empty by one route or another into the **dorsal vein of the penis** (Figs. X-13 and 15). This vein consists of two parts: The **superficial** and the **deep vein.** The superficial part of the dorsal vein runs in the subcutaneous tissue external to Buck's fascia. It drains prepuce and skin of the penis and empties into the external pudendal vein (sometimes into the obturator vein). The deep vein lies below Buck's fascia. It drains the glans and the cavernous spaces. It passes between the diverging arms of the suspensory ligament of the penis and pierces the urogenital diaphragm high between arcuate and transverse pubic ligaments. It empties into the prostatic plexus of veins within the pelvis. Some of its side branches join the superficial vein near the base of the penis.

During erection, the cavernous spaces are filled with blood from the helicine arteries. As the spaces expand, the thinner walled venules of the deep vein which drains them are compressed. The turgidity of the organ pushes the tough Buck's fascia against the superficial vein compressing it also, and so flow out of the organ is restricted until the media of the helicine arteries constrict after orgasm or until disturbing erotic stimuli have been put aside mentally. Then gradual venous drainage restores the organ to its flaccid state. Undoubtedly, clitoral erection functions similarly but on a smaller scale. During sexual arousal, other erogenous zones respond in the female, *e.g.*, erection of nipples; and in both sexes the cavernous spaces of the nasal mucosa fill with blood, warming air which enters the nose. Warm air aids in olfactory acuity so that sexual odors (and certain artificial perfumes) have a strong sexual arousal value. Such odors play even a keener role among lower forms of animal life. In man, obviously, sight and touch appear to have higher arousal values along with fanciful, vicarious dreaming and imagination.

Structures in the Superficial Perineal Compartment of the Male (Fig. X-16): These differ little from those structures seen in this same region in the female. They are, as partially discussed already, a *pair* of **crura** and a **bulb.** The **bulb of the penis** remains in the median plane, undivided in the male. It corresponds to the bulb of the vestibule in the female perineum except that the vestibular bulb, like other perineal structures in the female superficial compartment, is divided into right and left halves by the passage of the vagina. The crura of the penis correspond to the crura of the clitoris.

There are **three small perineal muscles:** (1) an Ischiocavernosus upon each crus; the Bulbocavernosus upon the bulb, united by a median raphe in the male but divided in the female; and the Superficial Transverse Peronei meeting at the peroneal body in each sex.

In addition, there are the perineal and scrotal branches of the pudendal nerve (labial branches in the female) and the perineal and scrotal branches of the internal pudendal arteries with their accompanying veins. As we have noted, *all* branches of the pudendal nerve enter the deep compartment first and must pierce the inferior fascia of the urogenital diaphragm to reach the bulb and other structures in the superficial compartment.

Nothing comparable to the *greater vestibular glands* (of Bartholin) are present in the superficial compartment of the male.

Structures in the Deep Perineal Compartment of the Male

1. The membranous part of the male urethra. This corresponds to the total urethra of the female.
2. The muscle of the compartment is thin and similar to that in the female. It forms a more definite **sphincter urethrae** in the male which appears somewhat more separate from the Deep Transverse Perinei muscles proper than it does in the female. In the female, the sphincteric portion of this muscle surrounds the vagina as well.
3. A pair of **bulbo-urethral glands** (of Cowper). These lie on each side of the urethra. Their ducts do not open into the membranous urethra, however, but pass forward through the perineal membrane into the caverous urethra 1 inch beyond the membrane. They secrete a non-viscous material into the urethra at the time of ejaculation. This combines with the fluids of the seminal vesicle and of the prostate, and, along with sperm, the mixture constitutes **semen.**
4. The end branches of the internal pudendal artery which are: a) artery to the bulb (retrograde in both sexes); b) deep artery to the penis (clitoris) is retrograde back through the perineal membrane (urogenital diaphragm) to crura in both sexes.
5. The dorsal nerve to penis (clitoris).
6. The deep dorsal vein of the penis corresponds to the dorsal vein of the clitoris. In both cases this vessel would have to pass through both compartments and pierce the urogenital diaphragm between transverse and arcuate pubic ligaments to enter the pelvic cavity. The counterpart of a superficial dorsal vein of the penis is also found in the female clitoris. This latter vein remains entirely superficial within the fascia, emptying into the greater saphenous vein *via* the external pudendal vein.

The Contents of the Ischiorectal Fossa: These contents within the anal triangle are the same in both sexes:

1. Fat.
2. The inferior rectal nerve which runs latero-medialward through the fat.
3. The inferior rectal vessels which accompany the nerve.
4. Anteriorly in the fossa are parts of the scrotal (labial) nerves and vessels and the transverse perineal vessels.
5. The perineal branch of S_4 going from the posterior angle of the fossa to the external anal sphincter.

In Figure X-17 the superficial compartment, the ischiorectal fossa and certain branches of the pudendal artery are shown on the left side. A deeper dissection on the right side shows the distribution of the deep arterial branch of the pudendal to bulb, crus, and penis, though the terminal shaft of the penis is not shown. The bulbo-urethral gland has been removed from the deep compartment. The point where the urethra goes through the urogenital diaphragm from prostatic to cavernous urethra is shown and labeled "membranous urethra." The Levator Ani has been removed; that is, the *pelvic diaphragm* has been removed. This allows us to look down upon the posterior surface of the bladder, prostate, seminal vesicle, and ductus deferens.

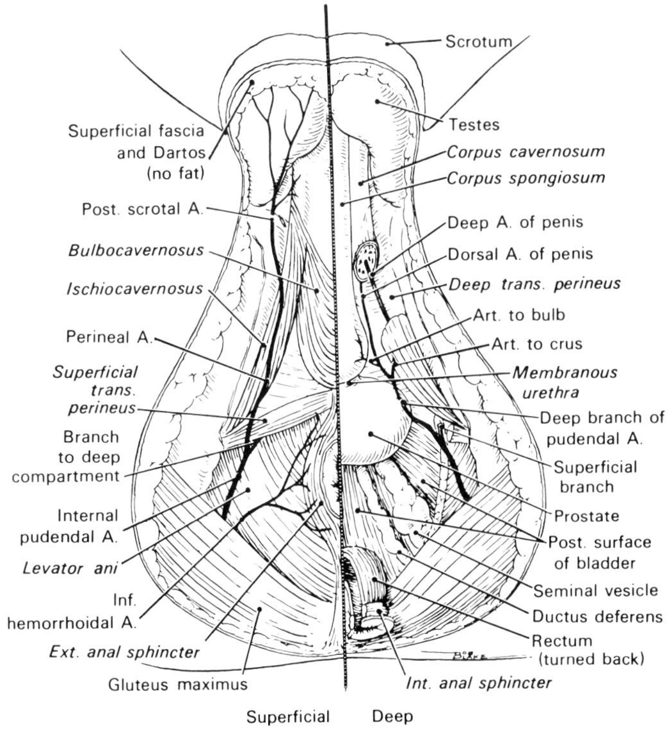

Fig. X-17. Superficial (left) and deep (right) compartments of male perineum.

THE PELVIS

Bony Parts

The framework of the pelvis consists of the two innominate bones (os coxae), the sacrum and the coccyx. These are held togther by stout white fibrous ligaments and they articulate at a few slightly movable joints. The large, irregularly shaped **innominate,** or **hip** bones (Fig. X-18 A–E) articulate with each other at the **symphysis pubis** as the radiograph (Fig. X-18E) shows. Each hip bone articulates with the broad **auricular surface** of the **sacrum.** The **coccyx** is just a slightly movable tail piece at the end of the sacrum. Joined in this manner, the two hip bones form anterior and lateral walls of the pelvis while sacrum and coccyx make up the bony posterior wall.

If you examine the large depression about midway along the outer aspect of each hip bone, *i.e.*, its **acetabulum** ("vinegar cup"), you will see where an early fusion of three bones contributed to make up this one, large, odd-shaped bone. The once separate bones are the **ilium,** the **ischium,** and the **pubis.**

The Ilium: This portion of the hip bone consists of an upper, large, wing-shape portion and a smaller **body.** The body contributes to the superior two-fifths of the acetabulum.

The external lateral view of the ilium (Fig. X-18A) presents a wide plate with a superior, free crest. It is marked by some faint lines and it has some spinous projections fore and aft. You have palpated the length of the **iliac crest** at your own waist line and have located the **iliac tubercle** when you were considering the **intertubercular plane.** You know that the lateral abdominal muscles and the Latissimus anchor to the inner, intermediate, and outer lips of the crest. Very early, you located the **anterior superior** and **posterior superior iliac spines** and learned that the posterior of these placed you at the level of the second sacral vertebral spine. If there are named spines that are superiorly located, you can be certain no one tacked "superior" to them unless there were also **anterior inferior** and **posterior inferior iliac spines.** Locate these now (Fig. 18 A–C, E). The Sartorius muscle, running from hip to knee, takes origin from the anterior superior spine; the Rectus Femoris muscle originates from the anterior inferior spine.

The faint lines we mentioned are the gluteal lines (anterior, inferior, and posterior) (Fig. X-18A) and they help to localize the areas of origin of the gluteal muscles. The largest of these, the Maximus, attaches to the smallest area of the broad, winged ilium. It attaches between crest and **posterior gluteal line.** The Medius originate between the much more extensive **anterior gluteal line** and the crest. Between anterior and inferior lines, the Minimus originates (except for a small anterior spot for the Tensor Fasciae Latae). The area between the inferior gluteal line and the acetabular rim is occupied by the capsule of the hip joint and the capsule reflects upon the neck of the femur whose head articulates in the acetabulum.

Observe that the inner aspect of the iliac wing (Fig. X-18B) forms as smooth, shallow dish, the **iliac fossa.** The area is almost completely occupied by the Iliacus muscle which originates there. On the bare bone, an **arcuate line** internally separates fossa from the body of the ilium. It also marks the boundary of the **false pelvis** (an abdominal region) from the **true pelvis** below the line. Actually, the arcuate line is but a part of a longer **terminal line** which involves part of the pubis and sacrum in the delineation of the true pelvic brim.

The surface of the ilium facing the sacrum is quite rough, and it has an auricular surface to match the ear-shaped articular surface of the sacrum.

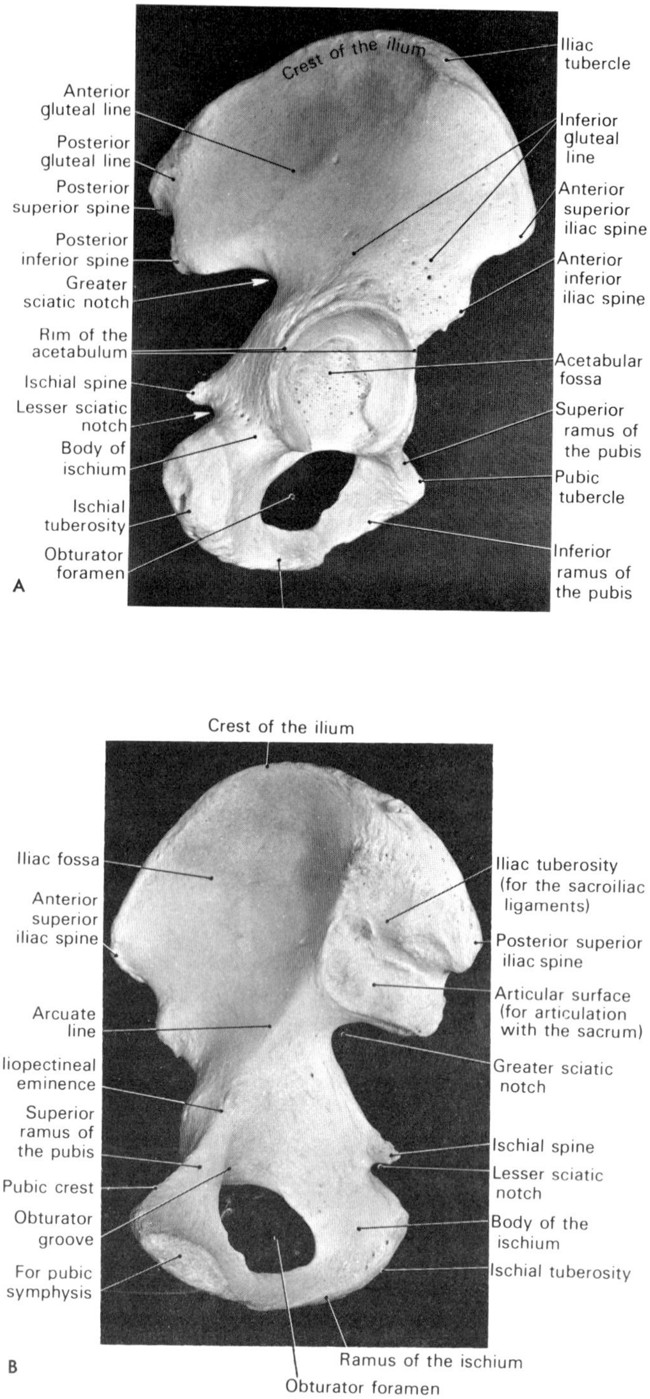

Fig. X-18. The hip bone (os coxae; innominate bone). **A.** External surface. **B.** Internal surface.

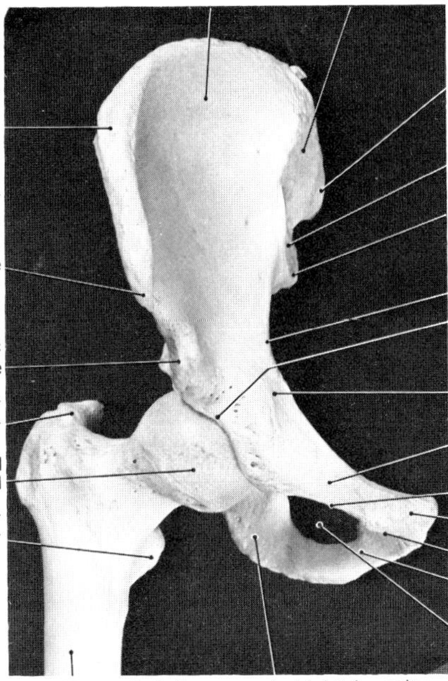

C

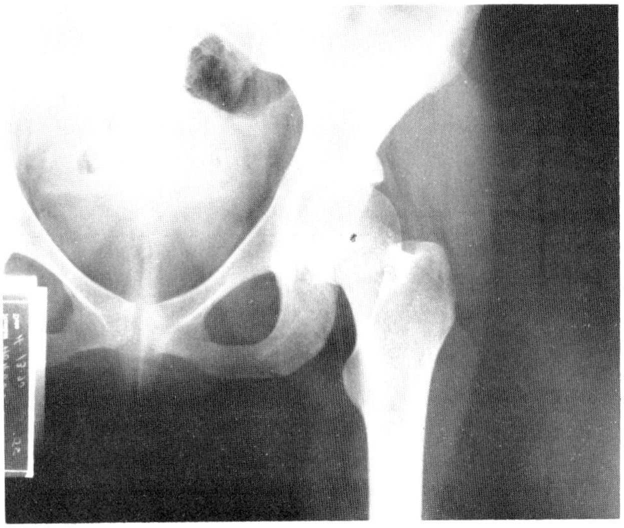

D

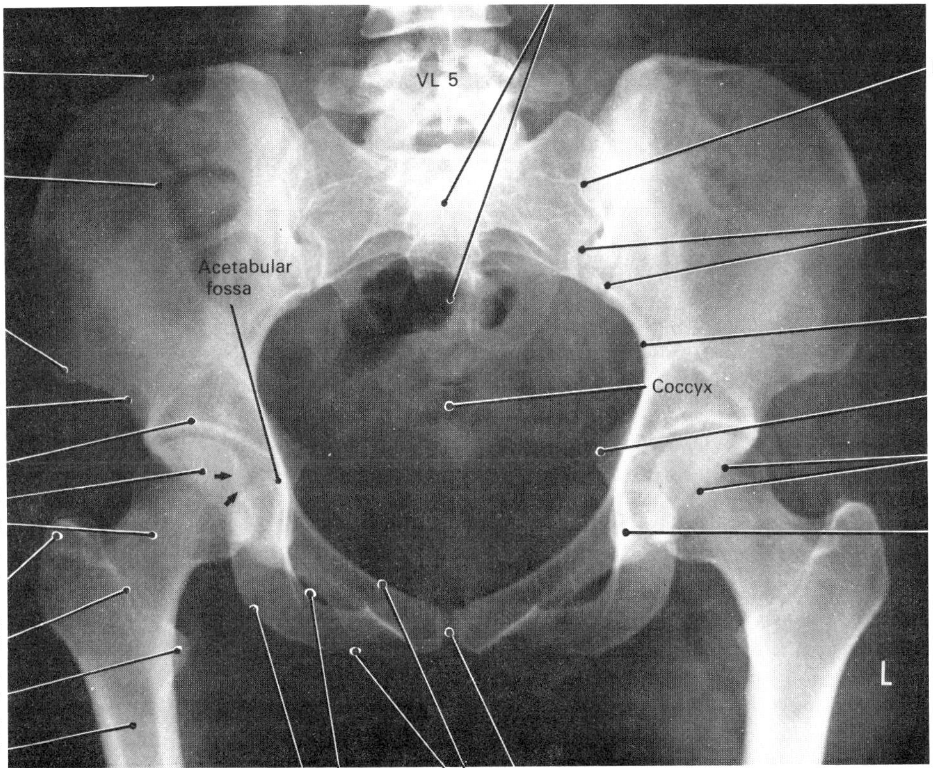

E

Fig. X-18. The hip bone (os coxae; innominate bone). **C.** In articulation with the femur. **D.** AP projection of the left hip with the femur externally rotated. **E.** AP projection of the female pelvis. The *arrows* on the patient's right indicate the fovea capitis in the femoral head. Note the left femur is in internal rotation; the right femur is in neutral position.

The Ischium (Fig. X-18A and B): This portion of the innominate bone makes up its postero-inferior part. It has a body and one ramus. The body also contributes about two-fifths to the acetabulum. Its smooth inner surface contributes to the wall of the true pelvis and there it gives origin to the Obturator Internus. Its **ischial tuberosity** provides anchorage for the hamstrings. The **ischial spine** is an important landmark. Between it and the posterior inferior iliac spine lies the **greater sciatic notch.** The latter becomes a foramen when sacrotuberous and sacrospinous ligaments are in place. The following structures pass through the **greater sciatic foramen:**

1. The Piriformis muscle
2. The sciatic nerve
3. The posterior femoral cutaneous nerve
4. The inferior gluteal artery
5. The nerve to the Quadratus Femoris
6. The nerve to the Obturator Internus
7. The internal pudendal vessels
8. The pudendal nerve

The last three do not remain outside the pelvis for long, for the **lesser sciatic notch** below the ischial spine is also converted to a foramen by the same ligaments. The three structures return to the pelvis by the **lesser sciatic foramen** while the tendon of the Obturator Internus emerges from it.

The **ischial ramus** helps to form the lower boundary of the obturator foramen and the boundary of the pelvic outlet which we found closed by the pelvic and urogenital diaghragms. In the latter regard, it gives support to the penile (clitoral) crus and offers attachment for the Ischiocavernosus muscle. On the outside, the Obturator Externus and Adductor Magnus (in part) attach to the ramus.

The Pubis: This part of the innominate bone has a body and sports two rami, one directed superiorly and the other inferiorly (Fig. X-18C). The ends of each pubic bone, as they meet at the fibrocartilaginous symphysial joint, appear as a rough oval in cross-section (Fig. X-18B). The superior border is sharp forming the **pubic crest.** Medially along the crest, the **pubic tubercle** forms a prominent bulge. The Pyramidalis and the Rectus Abdominis muscles attach to the crest while the inguinal ligament attaches to the tubercle. Other muscles attach to the internal and external aspects of the bone below the level of the crest, the Adductor Longus and Obturator Externus on its outer aspect; the Obturator Internus and Levator Ani, in part, on the inner aspect.

The **superior ramus** contributes the last one-fifth to the acetabulum, and, in its laterad progression, it carries the sharp superior border of the bone to the ilium, meeting the latter at the **iliopubic** or **iliopectineal eminence** (Fig. X-18C). To this sharp border or **pubic pecten** a number of smaller ligaments gain attachment, *e.g.*, the lacunar ligament, the pectineal ligament, the conjoined tendon, and the reflected part of the inguinal ligament. The iliopectineal eminence blends into the arcuate line of the ilium as part of the longer terminal line. The lateral, inner border of the superior ramus contributes to the **obturator groove** (Fig. X-18B) which transforms into a canal when the **obturator membrane** covers over the wide **obturator foramen.** Thus, the massive hole we see in the skeleton between the superior and inferior pubic ramus does not exist in the living.

The **inferior ramus** meets with the ramus of the ischium. The Adductor Brevis, a part of the Adductor Magnus and the Obturator Externus take origin from its outer aspect while the urogenital diaphragm, the Sphincter Urethrae, crus penis, and Obturator Internus attach, in part, to the inner aspect.

The Acetabulum: This deep hemispheric cavity accepts the head of the femur (Fig. X-18C). It is divided into an articular and non-articular area. The smooth articular part is peripheral and bears the acetabular rim. It is semilunar in shape and deficient inferiorly (Fig. X-18A). The non-articular portion is somewhat circular in shape, lies at a deeper level, is rough, and it completes the inferior margin at the **acetabular notch.** A transverse ligament bridges the notch to convert it into a foramen through which vessels and nerves can enter the joint. The synovial membrane of the joint is applied to the inner margin of the articular portion thus leaving the central, rough area outside of the joint cavity. In this way the short ligament of the head of the femur, arising from a pit, the **fovea capitis** in the femoral head, does not have to enter the joint cavity to anchor upon the central, non-articular area.

The articulations of the innominate bone with the sacrum are fully displayed in the radiograph (Fig. X-18E).

Roentgen Anatomy of the Pelvis (Fig. X-18E): This examination consists of an AP projection which is made with the patient supine on the radiographic table. The central X-ray beam is perpendicular to the cassette, and directed at a point about 1½ inches above the symphysis pubis. In order to evaluate the entire length of the femoral neck, the foot is internally rotated about 15°. Notice in Figure X-18E, that the left femur was properly positioned, and therefore the lesser trochanter is seen only in part. In the same radiograph, the right femur was maintained in neutral position, and therefore the lesser trochanter is seen better. Note, however, that less of the femoral neck is seen. In Figure X-18D, the leg has been allowed to rotate externally, thus bringing the lesser trochanter into profile, and causing extreme foreshortening of the femoral neck.

Identify the posterior superior iliac spine which is overlapped in this projection (Fig. X-18E) by the ala of the sacrum. Trace the crest of the ilium as it curves around ending above the anterior superior iliac spine. Just above the roof of the acetabulum is the anterior inferior iliac spine. Recall that the sacroiliac joint takes an oblique course from front to back, and therefore, it cannot be seen *en face*. However, we can see its anterior and posterior margins which form separate lines.

Examine the acetabulum carefully in Figure X-18D. The roof of the acetabulum normally has a dense, radiopaque bony cortex which is easily followed medially until it becomes interrupted by the deeper acetabular fossa. At the lower margin of the acetabulum is a "tear-drop"-shaped shadow (Köhler's teardrop[2]) formed by overlapping structures. Now trace the oblique line cast by the posterior rim of the acetabulum which lies lateral to the more faintly visible anterior rim. Within the acetabulum, the femoral head is seen. Note the concavity in the femoral head cast by the fovea capitis—this is the point of attachment of the round ligament. The femur will be discussed further in Chapter XIII.

Below the acetabulum is the ischial tuberosity with its characteristically-roughened lower margin. The pubic bones articulate at the symphysis pubis; however, the cartilagenous disk separating them is radiolucent. The obturator foramina show differing appearances in the various types of pelves. In Figure X-18E it appears as a slit, while in Figure X-18D it shows a more oval configuration. It is formed by the ischial and pubic bones. Identify the very prominent ischial spines which appear to project medially into the pelvis (Fig. X-18E).

The Sacrum (Fig. X-19A–D): The original five sacral vertebrae generally fuse into one wedge-shaped or shovel-shaped mass which articulates firmly with the hip bones as we have observed. Along with the coccyx, the sacrum forms the posterior wall of the true pelvis.

Its ventral or pelvic surface (Fig. X-19B) is smooth and concave. The transverse lines which cross it delineate narrow disks between the original separate vertebrae. Thus, between the ridges, we look at the ventral aspects of sacral vertebral bodies. The first two, about equal in size, are larger than the lower three all of which are also equal to each other in size. The anterior sacral foramina line up laterally to transmit the first four sacral arteries, veins, and nerves. The Iliacus muscle takes a partial origin from the tip of the alar wing of the first segment, and the Piriformis arises from the deep grooves lateral to the second, third, and fourth foramina.

The dorsal surface (Fig. X-19A) is very rough and it is raised in the midline as a **median sacral crest.** The crest is formed by the vestigial spinous processes of VS_1 to VS_4. The first process is large, the next two tend to fuse, and the last is rudimentary. The sloping laminae to the fourth and fifth of these spines fail to fuse and they leave an open **sacral hiatus,** the termination of the sacral canal. The rough area lateral to the median crest gives origin to the Multifidis and the very strong Erector Spinae. The more **lateral intermediate crest** would normally have become the articular processes of the separate unfused vertebrae. Now there is only one large pair of articular facets cranially for the fifth lumbar vertebra and a smaller pair caudally on each side of the sacral hiatus, the **sacral cornua,** for articulation with the coccygeal cornua. Lateral to the intermediate crest lie the **posterior sacral foramina,** smaller in diameter than the anterior ones since they need only transmit the small posterior rami of the sacral nerves. Most of all, what would have been transverse processes form the **lateral sacral crest.** (Fig. X-19A–D).

The lateral surface (Fig. X-19C) is broad cranially and narrows markedly caudally. It is formed by what would be the costal elements in other vertebrae. The extent of the **auricular surface** here varies but most often includes the first three segments. It is covered with fibrocartilage for articulation with the ilium. The bumpy area posterior to the articular surface affords ligamentous attachments, and is called the **sacral tuberosity.** Caudal to each sacral cornu is a notch that converts to a foramen when the sacrum articulates with the coccyx by virtue of the presence of the transverse process of the first coccygeal vertebra and its associated sacrococcygeal ligament. Here emerges the posterior ramus of the fifth sacral nerve. There is a true intervertebral disk between sacrum and coccyx but it may transform into a synostosis late in life.

If you look down upon the disarticulated sacrum from above, you view its **base** (Fig. X-19D). Centrally is the articular surface of the body of the first segment and its arch. These boundaries form the cranial opening into the sacral canal. Laterally, on each side of the arch laminae, are superior articular processes, each with a prominent mammillary process. Beyond these you view the broad alar expanse of the lateral part of the sacrum.

The Coccyx: This remnant of our ancestral tail has regressed to the extent that we are left with usually four tiny, terminal vertebrae. There may be five and sometimes only three. The first piece is usually separate while the last three pieces fuse together.

The first piece is the largest and broadest of all. It has a body, a pair of blunt transverse processes and two processes which should have met in a spine to form a vertebral arch. Instead, the latter processes extend almost vertically upward as cornua to unite with the sacral cornua at the open sacral hiatus. These coccygeal cornua represent pedicles and superior articular processes. In joining with the sacrum, they help to form the foramina of exit for the fifth sacral nerves, *i.e.*, for the posterior rami of these nerves.

The transverse processes connect to the sacrum by ligaments and the ligaments form foramina for the exit of the anterior rami of the fifth sacral nerves.

The second piece, much smaller than the first, has a hint of transverse processes and its vertebral arch has regressed to a pair of hardly noticeable tubercles. The third and fourth segments are mere nubbins represent-

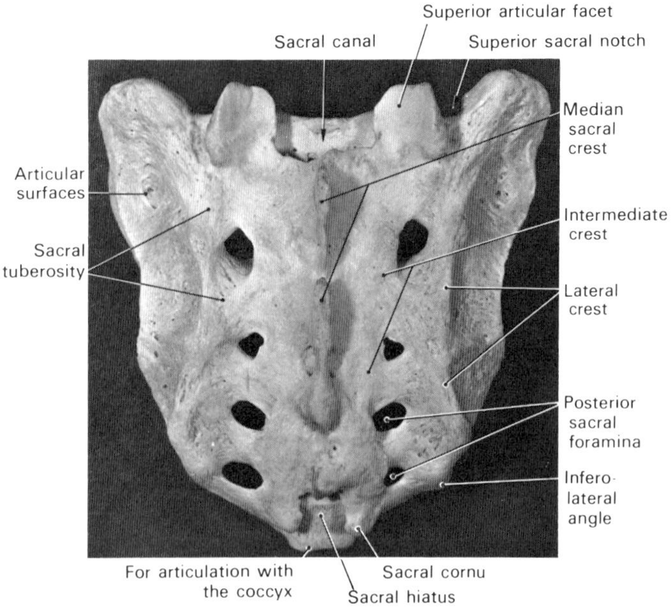

A

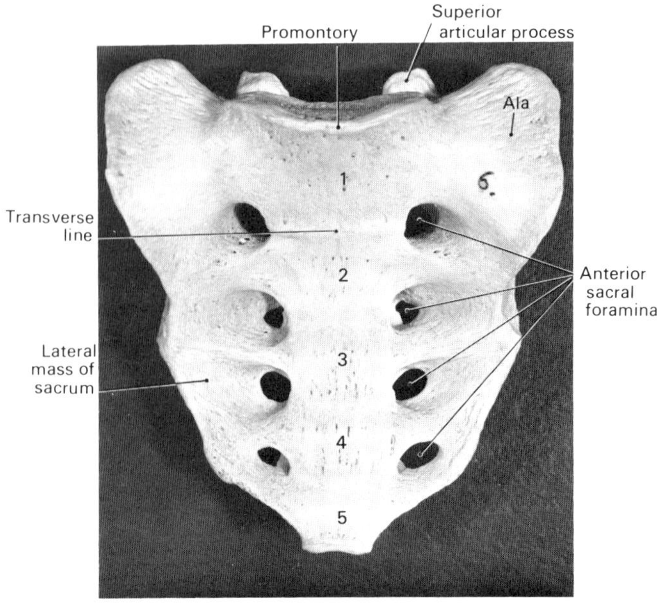

B

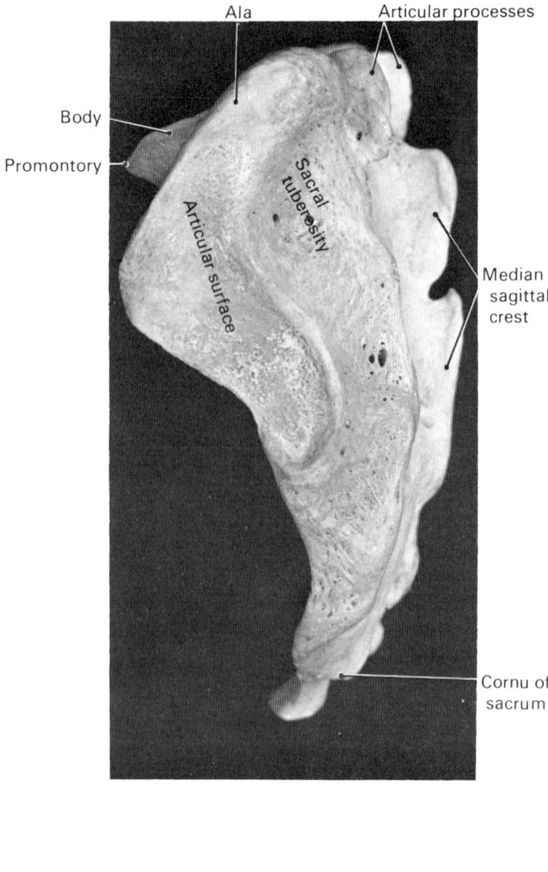

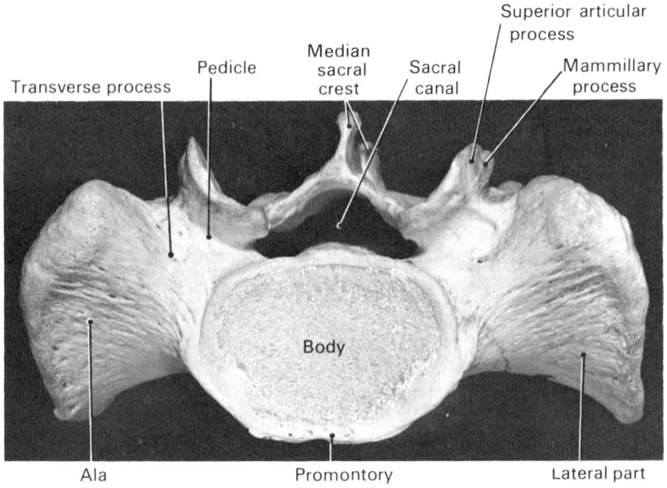

D

Fig. X-19. The sacrum. **A.** Dorsal aspect. **B.** Ventral aspect. **C.** Lateral surfaces. **D.** The base of the sacrum as seen from above.

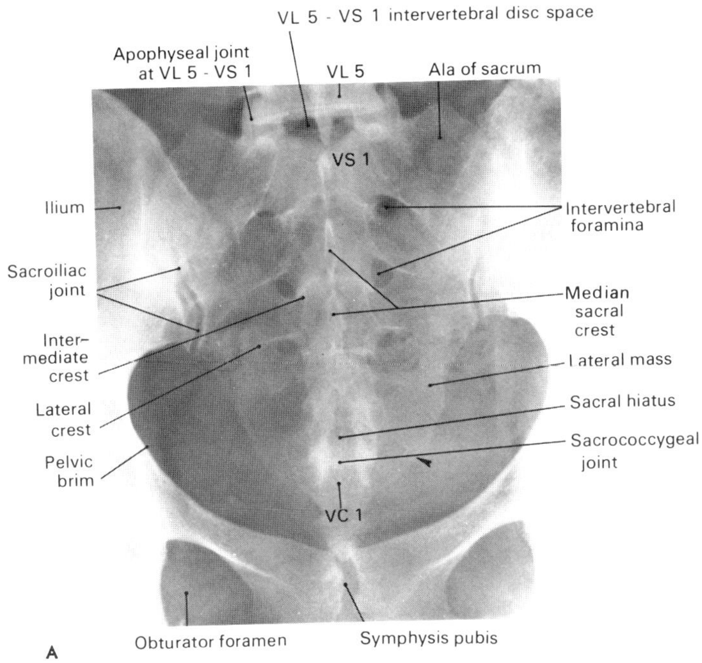

A

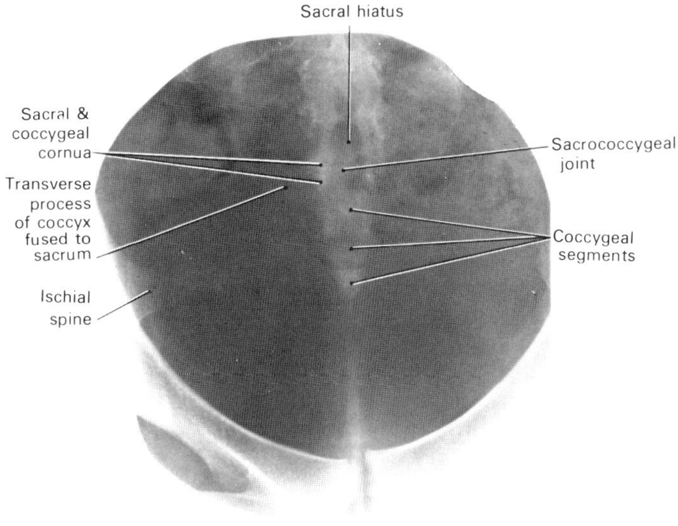

B

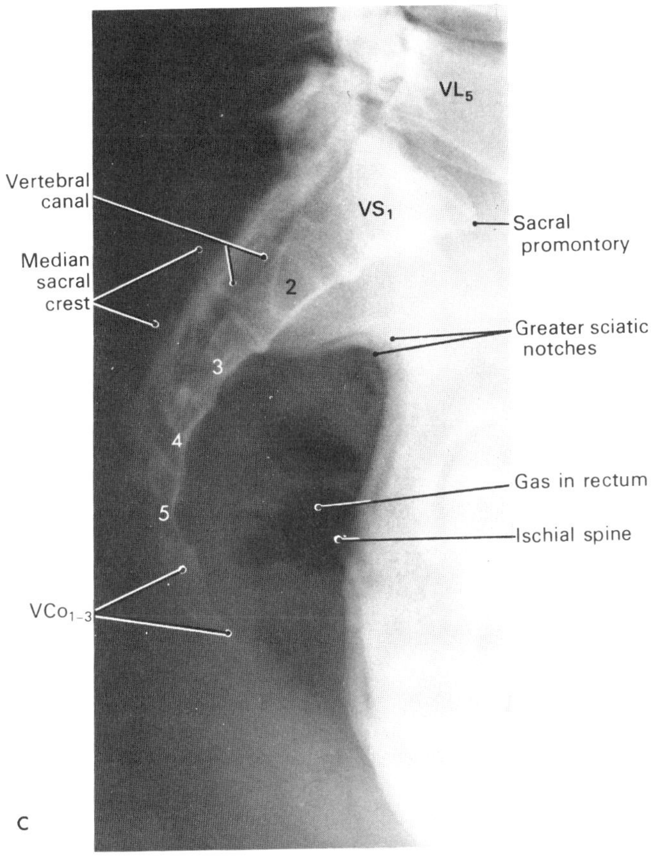

C

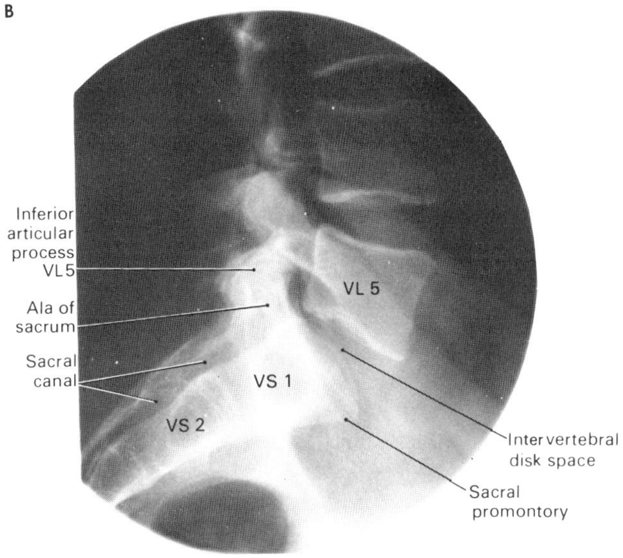

D

Fig. X-20. The sacrum and coccyx. **A.** AP projection of the sacrum. **B.** AP projection of the coccyx. **C.** Lateral projection of the sacrum. **D.** Lateral spot film of VL_5 to VS_2.

ing vertebral bodies only. The Gluteus Maximus attaches along the dorsolateral aspects of the first three segment; the external anal sphincter, to the dorsal tip of the fourth. Ventrally, these small bones offer some slight attachment for the Coccygeus and Iliococcygeus muscles and for the sacrotuberous and sacrospinous ligaments.

The anterior curvature of this ancestral tailpiece curves more sharply toward the pubis in males than in females where the pelvic outlet needs to be wider in fashioning a birth canal.

Roentgen Anatomy of the Sacrum and Coccyx: To visualize these osseous structures adequately, it is desirable to have the rectum and colon free of gas and all waste material. The patient is placed in the supine position and the central X-ray beam is directed 15° toward the head to view the sacrum (Fig. X-20A). The beam is angled 10° toward the feet to see the coccyx (Fig. X-20B). A true lateral view is also obtained as part of the routine examination (Fig. X-20C, D).

In Figure X-20A, identify the apophyseal joints formed by the superior articular facet of VS_1 and the inferior articular facet of VL_5. As we have mentioned earlier, when discussing the lumbar spine, these apophyseal joints are usually in a coronal plane at this

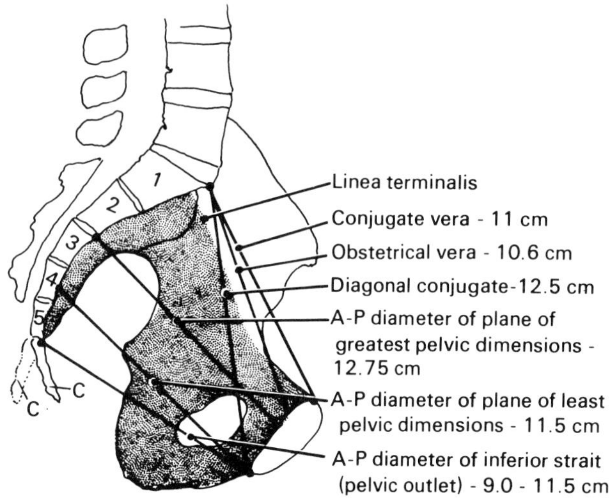

Fig. X-21. Diameters of planes in false pelvis (white) and true pelvis (shaded). The borderline between the two pelves marks the linea terminalis. Lateral view in hemisected pelvis.

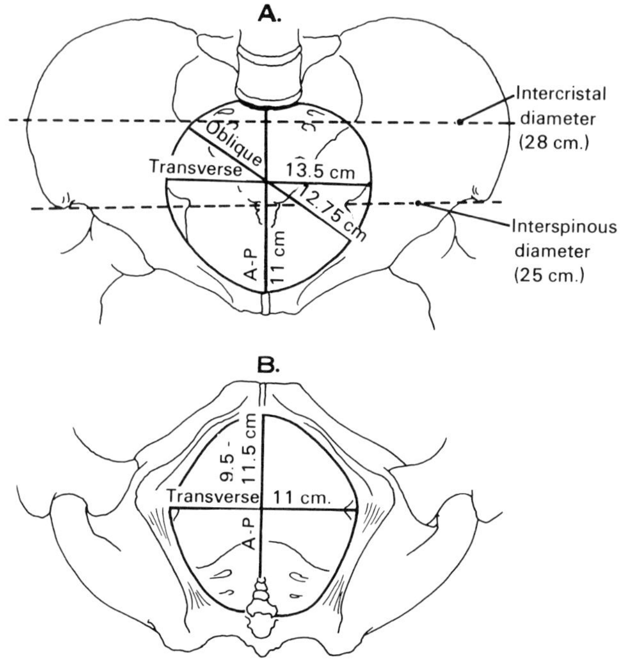

Fig. X-22. The dimensions of (**A**) pelvic inlet and (**B**) pelvic outlet; normal values in centimeters.

level of the spine; however, as you can see, this patient is an exception to the general rule. In most patients one would not expect to look through these joints at this level.

Identify the five segments of the sacrum, which may be more easily recognized in the lateral projection (Fig. X-20C). Note the dense shadows cast by the median sacral crest (Fig. X-20A). Immediately below the last tubercle is the sacral hiatus which appears relatively radiolucent. Just lateral to the median sacral crest, the intermediate crest may be seen on each side. Lateral to the articular tubercles are the intervertebral foramina. In this patient there appears to be a fifth intervertebral foramen (Fig. X-20A, B). Actually, the transverse process of the coccyx (*arrow*) has fused with the sacrum, thereby producing the foramen. Many such variations are encountered daily. The coccyx, typically, has four segments which decrease in size caudally, but this is quite variable. Identify the dense sacral and coccygeal cornua.

In Figure X-20C, we can appreciate the anterior concavity of the sacrum. Each of the five sacral segments is identifiable, separated by a dense white line representing the sites of fusion between them. The superior anterior margin of VS_1 is called the sacral promontory. The narrow sacral canal can be seen separating the posterior surfaces of the sacral segments from the laminae. The coccygeal segments may be seen individually.

The Terminal Line:

The division of the pelvis into **major (greater or false)** and **minor (lesser or true)** parts is effected by the **terminal line** (Fig. X-21). The abdominal cavity lies above this line and the true pelvic cavity below it. Of the pelvic organs, only the distended bladder and the pregnant uterus protrude above the line at times. This boundary line is compound in nature, contributed to by the **pubic pecten** (pectineal line), the *arcuate line* of the ilium; and a sacral part, the rounded angle between the base and pelvic surface of the sacrum. The pubic pecten is a well-defined ridge passing laterally from the pubic tubercle. The inner surface of the ilial wing is limited by the arcuate line, a rounded ridge which separates wing from internal surface of the body of the ilium.

The Pelvic Planes Useful as Obstetrical Measurements[3]

Two measurements of the *false pelvis* are always described; however, they are rarely employed clinically. They relate to the size of the false pelvis. The first of these, the **interspinous diameter,** is the distance between the anterior superior iliac spines, and measures about 25 cm. The second is the **intercristal diameter,** the widest diameter between the external lips of the iliac crests, and measures about 28 cm. (Fig. X-22).

Three planes of the *true pelvis* are important: the planes of the **inlet** and the **outlet,** and the **midpelvic plane.** The terminal lines, meeting at pubis and sacrum, form the inlet, the first of the passageways to be encountered by the infant's head during labor (Fig. X-22). Three diameters are commonly given for the *inlet:* (1) The **anteroposterior** or **true conjugate (conjugate vera) diameter** (Fig. X-21) is the distance from the sacrovertebral angle (sacral promontory) to the upper inner margin of the symphysis pubis. It usually measures 11.0 cm. or more in the normal female pelvis. (2) The **transverse diameter** (Fig.

Table X-1. Pelvic Measurements

	True Conjugate or Anterior-Posterior		Oblique		Transverse	
	Inches*	Centimeters	Inches*	Centimeters	Inches*	Centimeters
Inlet	4	11	4½	12.75	5	13.5
Midpelvis	4½	11.5	†	†	4 or greater	10.0 or greater
Outlet	5	9–11.5	†	†	4¼	11.0

* Inches are useful as rough average, not as exact as centimeter measurements.
† Meaningless clinically.

X-22A, B) from one terminal line across to the other is, obviously, the greatest width of the inlet. It is measured at right angles to the true conjugate about 5.0 cm. anterior to the sacral promontory. This cannot be measured clinically! Normally, the AP and transverse diameter should sum to 23 cm. (or about 9 inches) to insure an adequate pelvic inlet. (3) The third diameter, often measured by X-ray, is the **oblique** from the sacroiliac articulation of one side to the iliopectineal eminence of the other (Fig. X-22A). Table X-1 gives the measurements of all these diameters for all three pelvic planes in terms of both centimeters and inches as average figures. The measurements in inches are easiest to commit to memory, but you will, no doubt, record measurements in centimeters, so both are given.

In obstetrics, the **diagonal conjugate** (AP diameter) is always measured, since it is an important clinical measurement of the female pelvis from which the true conjugate can be estimated (Fig. X-21). The gloved index and long finger are placed in the vagina and their tips directed along the inner surface of the sacrum until the tip of the middle finger touches the sacral promontory. The distance between the finger tip and the lower margin of the pubis represents the diagonal conjugate. This measures about 12.5 cm. With this figure and subtracting 1.5 to 2.0 cm., the **true conjugate** (conjugate vera) can be estimated. Normally the conjugate vera measures 11.0 cm. or more. Realize that the true conjugate is the *most important diameter* in clinical obstetrics. Unfortunately, it cannot be measured in the living patient directly. The true conjugate does *not* represent the shortest AP diameter of the inlet. The **obstetrical conjugate,** a line drawn from the sacral promontory to the inner surface of the pubis a few millimeters below the upper margin, is the shortest. It is only a few millimeters shorter than the true conjugate (Fig. X-21) or about 10.6 cm.

If we consider only the true conjugate (conjugate vera) at the **pelvic inlet,** one can arrive at a useful rule of thumb:

1. If this diameter is less than 11.0 cm., the inlet is contracted.
2. If it is from 9.0 to 10.0 cm., forceps delivery may be required.
3. If it is 8.0 to 8.5 cm., a trial with forceps may be made. It may not prove successful.

Midpelvis: None of the following clinically-important planes can be measured in the patient, except by radiographic means. (1) The **plane of the least AP diameter,** the line from the postero-inferior aspect of the pubis through the ischial spines to the sacrum, measures approximately 11.5 cm. (Fig. X-21). (2) The **transverse (interspinous) midplane diameter** between the ischial spines measures about 10.0 cm. or more. In depth, this lies at about the level of the fourth sacral segment. Here the spines offer the next critical barrier to the passage of the infant head in the birth canal. If this diameter is less than 9.5 cm., one can expect operative intervention during labor about 50% of the time. This plane would be the transverse measurement at the plane of the least AP diameter mentioned above. (3) One final plane, although of no obstetrical significance, is frequently described in standard anatomy texts. This is the **AP diameter of the greatest pelvic** dimension (12.75 cm.) (Fig. X-21). This line is drawn from the center of the pubic symphysis to the junction of the second and third sacral segments. Do not confuse it with the oblique plane mentioned earlier which also measured 12.75 cm.

Outlet: The second most important clinical measurement of the pelvis is the estimation of the (1) **bituberous (intertuberous) distance.** This transverse distance between the ischial tuberosities can be measured by placing the clenched fist between the tuberosities. Discounting the thumb, the distance across the knuckles is about 9 cm. If the bituberous distance is less than 7.5 cm., you can usually discount vaginal delivery. A measurement greater than 8.0 cm. is normal. (2) If you add to this measurement the AP distance from the bituberous line to the tip of the sacrum (the **posterior sagittal diameter**) which is approximately 6 to 7.5 cm., the sum, about 15 cm., is taken as a sign that the pelvic outlet is adequate. (3) A third **AP diameter** of the **inferior strait** (Fig. X-21 and 22B) is also used, extending from the lower margin of the pubis to the sacral tip (11.5 cm.).

In observing radiographs of the pelvis, sex can be determined by the following observations:

1. A heart-shaped pelvis signifies male; a circular to oval one, female.
2. The angle between the pubic rami will be narrow in the male (see Table X-2) and wide for the female.
3. Soft shadows of penis and scrotum are a "give-away."

Table X-2 lists a number of ways in which the bony pelvis of male and female can be differentiated.

In moving through the birth canal from inlet to pel-

Table X-2. Comparison of the Male and Female Gynecoid Pelvis

Item	Male	Female
Bones	Heavy, rough	Light and thin
Joint surfaces	Large	Small
Muscle markings	Fairly prominent	Indistinct
Sacrum	Long, narrow, curved	Short, wide, flat, curving only in lower part
Ilia	Rather vertical	Less vertical
Iliac fossa	Deep	Shallow
Greater sciatic notch	Narrow	Wide
False pelvis	Narrow, deep	Wide, shallow
True pelvis	Narrow, deep	Wide, shallow
Pelvic inlet	Heart-shaped	Circular to oval
Pelvic canal	Long segment of a short cone (long and tapering)	Short segment of a long cone (short with parallel sides)
Sacroiliac articular facet	Extends well down to 3rd segment	Extends only to upper border of 3rd segment
Sub-pubic angle	Narrow and pointed (the angle of the middle and index finger)	Wide and rounded (the angle of the thumb and index finger)
Symphysis pubis	Deep	Narrow
Anterior superior spine	Inverted	Everted
Inferior pubic ramus	Marked everted surface for crus of penis	Not present as a bony marking
Obturator foramen	Round to oval	Oval to triangular
Pelvis	Tilted backward	Tilted forward
Acetabulum	Large	Small
Coccyx	Curved forward to reduce outlet	Does not descend as low or curve in as much to encroach on outlet
Ischial tuberosities	Inturned	Everted

vic outlet, the infant's head is forced to make about a 90° turn (Fig. X-23) in following the path of the **pelvic axis.** As we pointed out, the head, in turn, must contend with the possibilities of a contracted inlet, proximity of ischial spines at midpelvis and of ischial tuberosities at the outlet. More than that, it must contend with the vagaries of pelvic configurations in the female, which will now be considered.

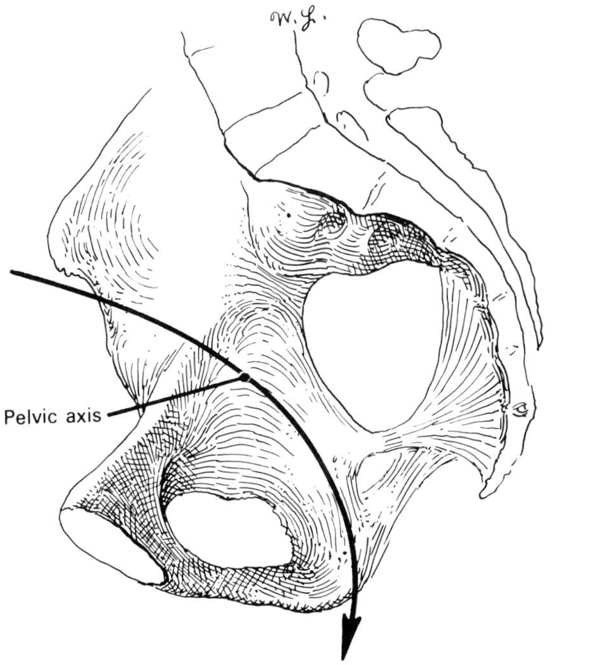

Fig. X-23. The pelvic axis marks the path taken by the infant's head in its passage through the birth canal. The axis is not a straight line but one of almost 90°.

Pelvic Shape

Four basic shapes are generally described: **gynecoid, android, anthropoid,** and **platypelloid** (Fig. X-24). Most pelves fall somewhere between these in basic architecture.

The Gynecoid Pelvis: This ought to be the archtype of the female pelvis. It has a good, wide, circular inlet; a wide subpubic arch; insignificant ischial spines and a neatly-rounded, sacrosciatic notch. Moreover, it is shallow with fairly straight walls. If any structure should assure a reasonably uneventful delivery, it is this type of pelvis. However, it only occurs in about 40% of cases. A comparison between male and female gynecoid pelves are set forth in Table X-2.

The Android Pelvis: This type has a heart-shaped inlet and is more typical of the male pelvis. The ischial spines are quite prominent and the subpubic arch is narrowed. Females with extreme examples of this type pelvis frequently encounter difficult forceps deliveries, and dystocia is common. It occurs in about 32% of white females and 16% in non-whites.

The Ape-like or Anthropoid Pelvis: This is a com-

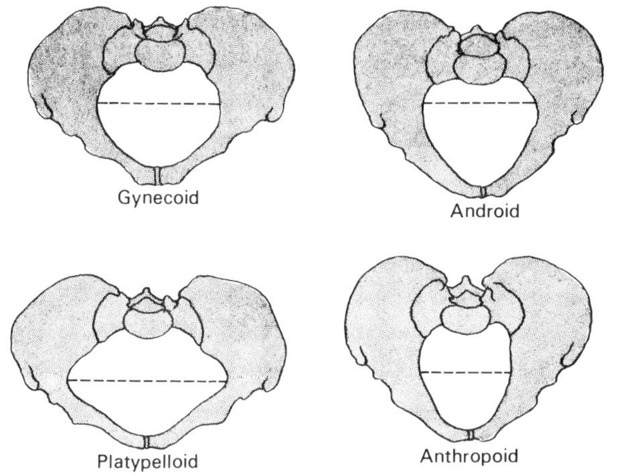

Fig. X-24. The four main types of female pelves. Notice a real difference in hind and fore pelvis as delineated by the transverse plane in the four types.

mon female type, more common among non-whites (40%) than whites (20%). Its sides are long and narrow and its AP diameter of the inlet is greater than the transverse diameter. The sacrum is long, often running to six segments; hence, the pelvic cavity is deep. The subpubic angle is narrow and the ischial spines are prominent while the sacrosciatic notch is wide. The fetal head commonly presents with the occiput posterior, which requires forceps rotation. Dystocia is frequently encountered.

The Platypelloid Pelvis: This is a flattened gynecoid type. Its occurrence rate is low, of the order of 1 to 3% in females. The AP diameter is shorter than usual, the transverse being the greater diameter. It is a very shallow pelvis with a short sacrum.

X-ray Pelvimetry

In the management of his obstetrical patient, the obstetrician has the primary responsibility. He uses roentgen pelvimetry based upon his training and experience. Some physicians feel that the procedure should never be used, while others feel it is to be utilized "routinely" in all primiparae and, of course, in those multipara who have, in the past, experienced difficulty in labor. Between these two extremes are those physicians who realize that in certain circumstances, X-ray pelvimetry is a valuble adjunct to their clinical observations. Pelvimetry is performed to provide the obstetrician with certain basic information: the shape and size of the maternal pelvis; presentation, position and size of the fetus; and the fetal attitude *in utero*. It may also prove valuable in localization of the placenta.

Pelvimetry is usually performed close to term, but generally not earlier than 36 weeks of gestation. Frequently, it is performed when the patient has gone into labor, and this may, actually, be the most optimal time. Remember that roentgen pelvimetry cannot be used as the sole factor in determining whether the patient will deliver from below.

There are, literally, hundreds of methods of performing X-ray pelvimetry. The ideal method should be simple and provide the most accurate measurements of the pelvic diameters. One such method, that of Colcher and Sussman,[4] which we have used for more than a decade, will be presented here. We realize fully that this technique may not be used at the medical school in which you are now training; however, it will demonstrate some of the basic principles, many of which are used in the other methods employed throughout the world.

In 1944, Colcher and Sussman[4] described a new and simplified method of roentgen pelvimetry. It required no fancy and costly equipment, and the method of measuring the pertinent pelvic diameters could be taught readily to the novice. In order to correct for magnification, an isometric ruler is used which permits direct measurement in each of the two projections used: (1) The *AP projection* is made with the patient in supine position, with the knees and hips semiflexed and the knees separated. The authors state that in this position the transverse diameter of the pelvic inlet, the bispinous (midpelvis), and the bituberous (outlet) diameters are all within the same plane and parallel to the table top. The female radiological technician carefully palpates the ischial tuberosities of the patient and places the isometric ruler at this level parallel to the table top. (2) In the *lateral projection,* the patient is placed in a true lateral position, and the isometric ruler is placed in the gluteal fold parallel to the table. The central X-ray beam is directed at the greater trochanter.

Figure X-25A is an AP projection showing the three transverse diameters to be measured: (1) the actual inlet (A-A′), at its widest point; (2) the midpelvic diameter (B-B′) measured between the spines of the ischium; and (3) the outlet (C-C′) or bituberous diameter. The points C and C′ are found easily by constructing a line from the lateral margin of the pelvic brim down to the lateral border of the obturator foramen and extending this through the ilium.

In the lateral projection (Fig. X-25B), the anteroposterior diameters are determined, again with the isometric ruler for correction of magnification. (1) Colcher and Sussman feel that the true inlet intersects the widest diameter of the pelvis. Therefore, the actual inlet is measured by drawing a line from the posterior superior surface of the symphysis pubis (point G) to the sacrum (point I) through a point E. Point E is halfway between the pelvic brim and the sciatic notches, and roughly one third of the distance from I to G as measured from the sacrum. Thus the AP diameter (G to I) intersects the transverse diameter A-A′ at the level of point E. (2) The midpelvic AP diameter (P to M) is measured from the posterior inferior aspect of the

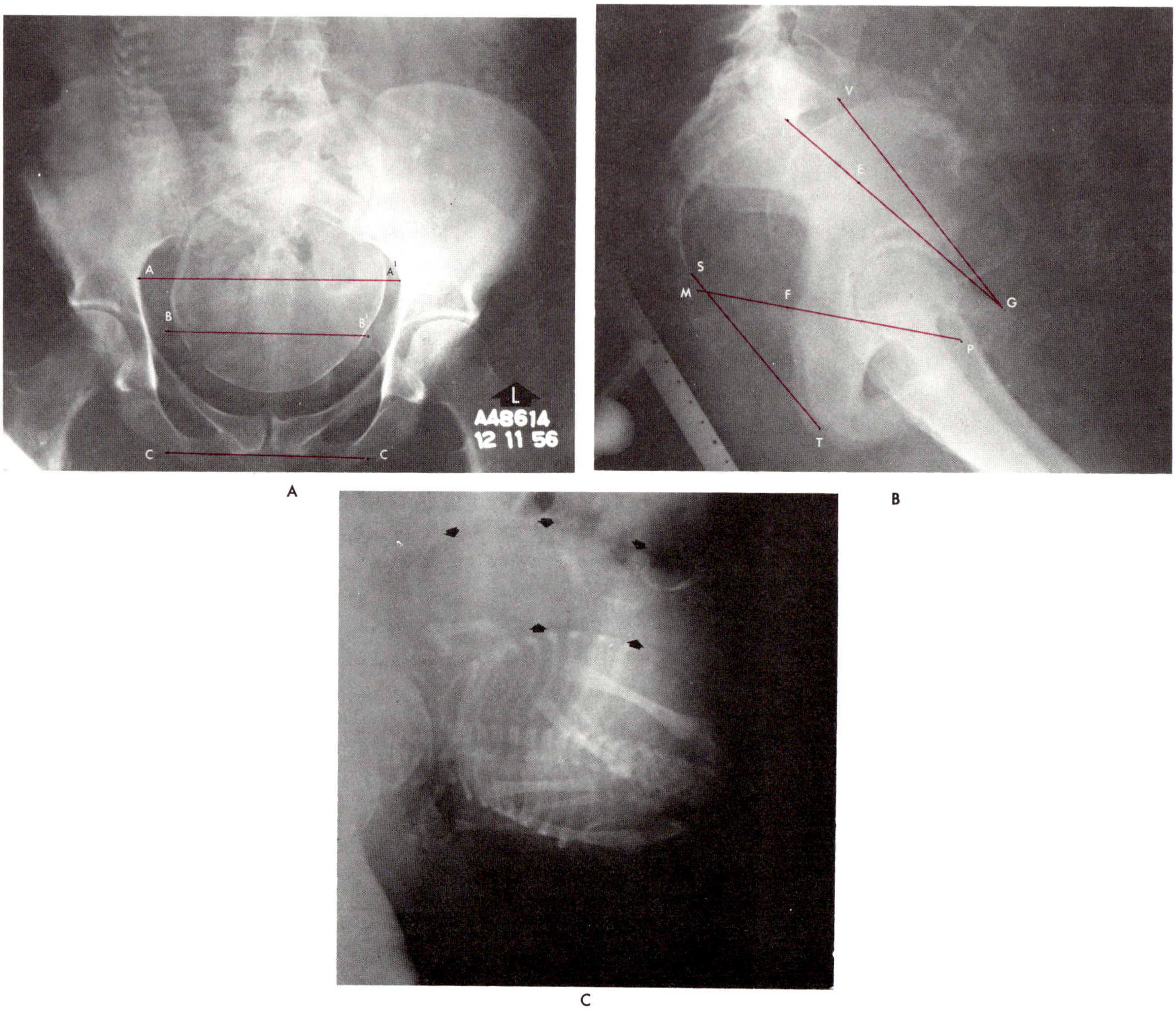

Fig. X-25. X-ray pelvimetry using the Colcher-Sussman method. **A.** AP projection. **B.** Lateral projection. Note the isometric ruler on your left. **C.** Placental localization. *Arrows* outline the placental shadow.

symphysis pubis through the midpoint of the ischial spines (point F) to the sacrum. Thus, the AP diameter M to P should intersect the lateral diameter at the point F. (3) The outlet AP diameter (S to T) is drawn from the midpoint of the intertuberous diameter to the sacrococcygeal junction. The AP diameter (S to T) and the transverse diameter (C-C′) intersect at point T, the midtuberal point.

By adding the AP and transverse diameters together, the following range of normal values is given: (1) actual inlet = 22.0 to 25.5 cm.; (2) midpelvis = 20.0 to 22.0 cm.; (3) outlet = 16.0 to 18.5 cm. Occasionally, it is deemed advisable to measure the true conjugate, points G to V also (Fig. X-25B). The normal range of measurement is 11.0 to 11.5 cm.

Other pertinent information is also provided to the obstetrician such as the size of the fetal head and any sign of molding. The size of the head is determined by measuring the longest and shortest diameters in each projection, correcting for distortion with the isometric ruler. Accuracy will be greater if the fetal head is midline.

In addition to the above projections, we make an additional radiograph in AP or lateral projection to view the entire fetus, and attempt to localize the placenta. With proper technique, the latter is quite easy to see (Fig. X-25C). At the site of placental attachment to the inner uterine wall, the wall, which is of uniform thickness (about 1.0 to 1.5 cm.), bulges out to several times its normal thickness. The placenta is typically sickle-shaped.

There are many methods to estimate the age of the fetus *in utero;* however, where possible the simplest method of determining a term pregnancy is to visualize the distal femoral epiphysis which is present in approximately 90 to 95% of mature fetuses.

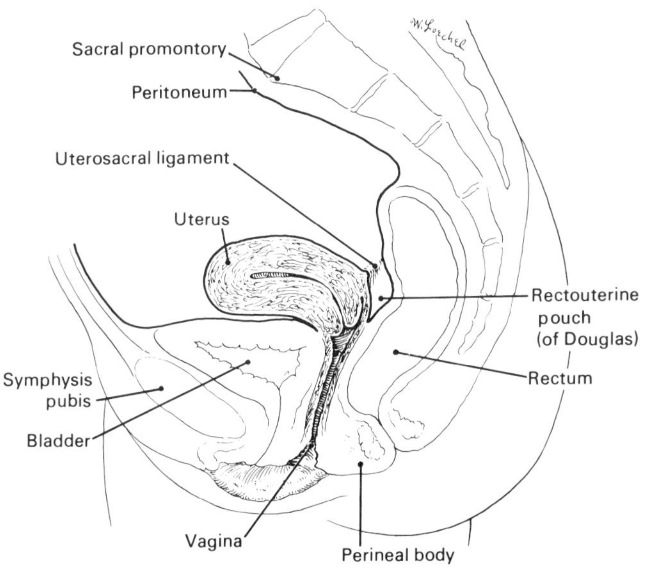

Fig. X-26. Midsagittal section through uterus and pelvis, viewed from the left.

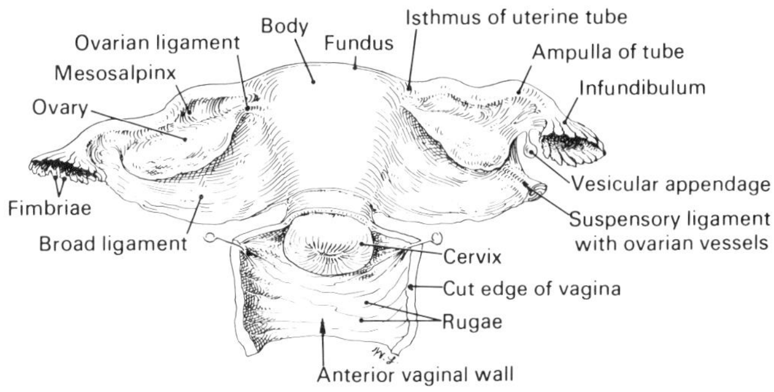

Fig. X-27. Uterus, ovaries, and fallopian tubes. Posterior view. The back wall of the vagina has been cut and spread open.

PELVIC ORGANS IN THE FEMALE

The Vagina (Fig. X-26)

This organ, some 6 to 8 cm. in length on its anterior wall, and 7 to 10 cm. on its posterior wall[3] is actually only a potential space, since its walls are in apposition. The vagina opens into a vestibule below, and communicates with the uterus above. Since the vagina serves as the lower end of the birth canal it may be markedly distended, particularly from front to back, and less so from side to side because of attachments to the pelvic diaphragm and the lateral cervical ligaments.

The vaginal opening is separated from the vestibule, in most virgins, by a fold or **hymen,** usually having one or more holes within it. At times, the hymen will be imperforate, and, therefore, lacks an opening.

The inner mucosa of the vagina bears longitudinal ridges on the anterior and posterior walls. Numerous transverse ridges with valleys between extend laterad from these vaginal columns. This rugose nature of the vaginal mucosa is more apparent before pregnancy than afterward, as most husbands soon discover.

It should be realized that the base of the female bladder rests upon the ventral vaginal wall and that the the urethra is embedded in this wall. Posteriorly, the vaginal wall is separated from the anal canal, from outermost inward, by the perineal body, the rectum and the depths of the peritoneal uterorectal pouch. The latter continues off the back wall of the uterus, dips down part way over the posterior vaginal wall and reflects upward upon the rectum (Fig. X-26). Obviously, if an instrument is directed posteriorly into the vagina in a crude attempt to enter the cervix, it will be pushed through the posterior vaginal wall and enter the peritoneal cavity. This relationship follows from your awareness of the positioning of the uterus and cervix in respect to the vagina. You can be certain that peritonitis will follow if this is done. It has occurred sometimes during amateur attempts at abortion with a hatpin or some other equally-lethal device.

Laterally, you can expect the uterus, the anchoring Levator Ani and the pelvic fascia to lie near the vaginal fornices. If the vagina has any sphincters, they are not particularly strong ones. A weak sphincter surrounds the outlet into the vestibule. A slightly heftier one lies deep to the vestibular bulb, consisting of muscular slips from the Deep Transverse Perineus and from the Levator Ani.

The **vaginal artery** (Fig. X-31) can arise from several possible sources. It can be a branch of the uterine artery, spring separately from the internal iliac, arise in common with the inferior vesical branch to the bladder, or be an independent branch of the internal pudendal artery. It supplies the cervix in part, the vaginal mucous membrane, the vestibular bulb, the fundus of the urinary bladder. and even adjacent parts of the rectum.

The **vaginal venous plexus** and the uterine plexus join with the vesical plexus and communicate with the internal iliac vein.

The Uterus (Fig. X-26 to 29)

The uterus of the non-pregnant female is shaped like a Bartlett pear. The stem-end. which projects into the vagina is its **cervix.** The rounded, butt-end, is its **fundus,** and the main portion. is its **body** (Fig. X-27). The uterine tubes spring laterad below the level of the bulging fundus. The walls of the organ are thick with smooth muscle, and it is covered with the peritoneum of the broad ligament. Its collapsed interior cavity is lined with a mucosa which varies in height during the period of the ovarian cycle. The mucosal lining cells are simple columnar.

The length of the uterus, from **cervical os** to fundus, is 5.5 to 8.0 cm. in the nulliparous female and 9.0 to 9.5 cm. in the multipara,[3] as the "womb with a view" will show (Fig. X-28). The external os is rounded in the

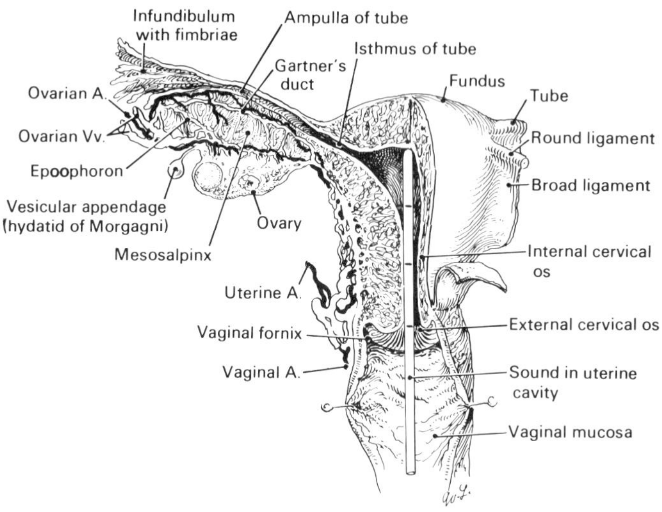

Fig. X-28. Section through vagina, uterus, tube, and ovary. A uterine sound has been placed in the uterine cavity.

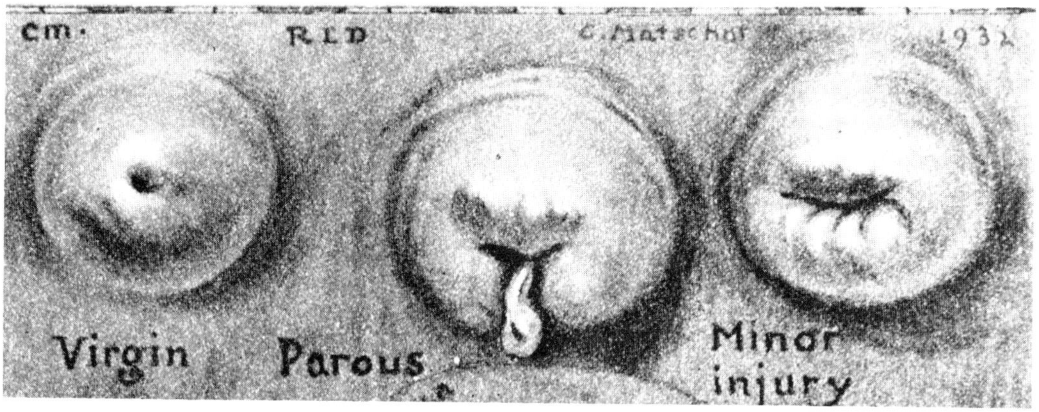

Fig. X-29. The cervix in the nulliparous and multiparous female. (From R. L. Dickinson, *Atlas of Human Sex Anatomy*, 2nd ed., Williams & Wilkins, Baltimore, 1949.)

non-pregnant female, but it becomes slit-like after parturition (Fig. X-29). The mucosa of the cervical cavity folds into **plica palmata,** and its coloration is darker than that of the body of the uterus so that its junction with the body can be determined in sagittal sections of the organ. Thus, the cervical canal has an **internal os** as well before it opens up into the wider, flaring, uterine cavity proper.

Since most of the cervix projects into the vagina, it is surrounded by a circular vaginal trough, the **fornix,** deeper posteriorly by about 2 cm. than it is anteriorly. In the sectioned vagina and cervix (Fig. X-28), obviously, we only see the sides of the fornix. In a specimen such as this, however, a finger could be slipped behind the cervix into the deeper posterior fornix.

You may be amazed at the disparity in size between uterus and cervix at different ages. In the fetus the cervix is larger than the body of the uterus. It is still twice the size of the body during childhood, but at puberty the body of the uterus finally enlarges as continued growth of the entire body of the female takes place. Interestingly, in old age, it reverts back to the childhood ratio.

Normally, the uterus does not sit straight upright in the pelvis, but is *anteflexed.* That is, it bends ventralward at the internal cervical os to form a 170° angle with the cervix at this point. At the same time, the uterus is *anteverted* (turned so that the cervical axis is at an angle of 90° to the *axis of the vagina*) which means that the body of the uterus lies in almost a horizontal plane within the pelvis. Its anterior wall rests upon the bladder, while, posteriorly, beyond the **recto-uterine pouch** (of Douglas), is the large bowel or coils of small bowel. Any forward bending of the uterus (anteflexion) which decreases the 170° angle at the internal os places the uterus in *malposition.* Thus, while a normal uterus is *both* anteflexed and ante-

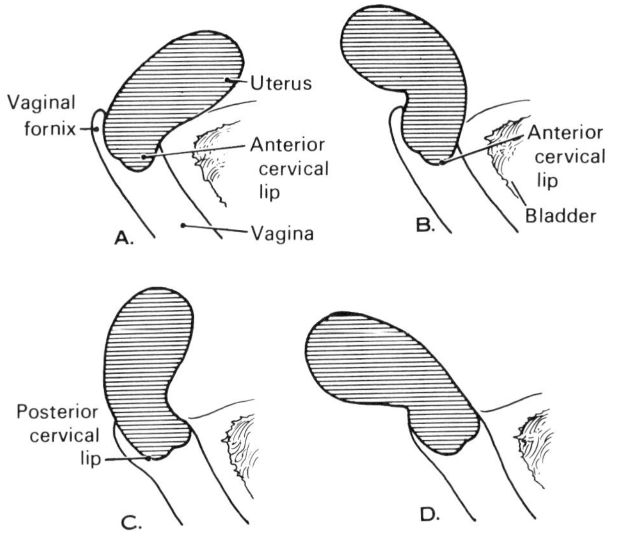

Fig. X-30. Positioning of uteri in lateral view as seen from right side. **A.** Normal posture in respect to vagina and bladder; anteflexed and anteverted. **B.** Retroflexion, uterus still anteverted (note anterior cervical lip). **C.** Retroversion, uterus still anteflexed. **D.** Retroversion and retroflexion. B to D are abnormal positionings.

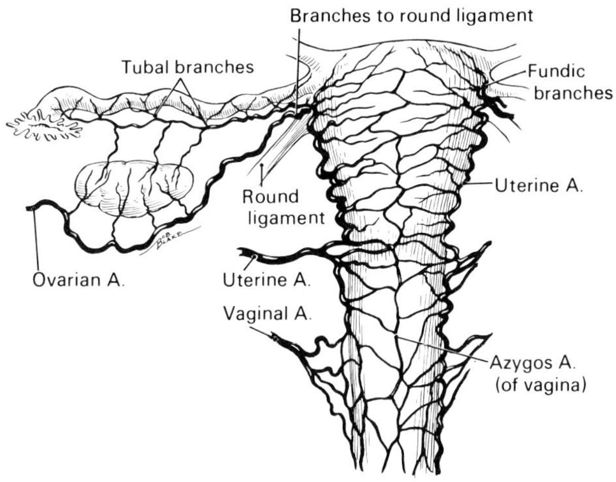

Fig. X-31. Arterial supply of uterus and vagina.

verted to the extent indicated above, anteflexion in respect to cervical axis can proceed beyond the normal state. If the uterus is *retroflexed* or *retroverted* it is also in malposition.

When the uterus is in its normal position, vaginal examination allows one to feel the *anterior lip of the cervix* (Fig. X-30A). *Retroversion of the uterus* tips the long axis of the cervix upward and backward. It does not interfere with the angle between cervix and body of uterus, but it changes the angle between cervix and vagina. Thus, in retroversion, one feels the *posterior lip of the cervix.* In a sense, the *whole organ* (cervix and body) has *tipped posteriorly* by 10° to 30°. Note, we say the uterus is *retroverted in respect to position of cervix with vagina,* but it is still *anteflexed* in respect to position of cervix and body of uterus.

In *retroflexion,* the body of the uterus bends upward and backward in relation to the long axis of the cervix. The relation of cervix to vagina remains unchanged (the anterior cervical lip still presents to a finger in the vagina) (Fig. X-30B). Thus, in retroflexion, the uterus is still *anteverted* in that cervix and vagina maintain a 90° relationship. It is, however, quite possible to find the uterus in a position of both retroversion and retroflexion (Fig. X-30D). In the latter case, it will again be easier to feel the posterior lip of the cervix than its anterior lip in a vaginal examination.

There are also other features which involve malpositioning of the uterus. The uterus can lie in normal anteflexion and anteversion but be in a position much closer to the sacrum than normal. It is then said to be in *retrocession,* probably because the ligaments anchoring the uterus to the sacrum foreshorten while those anchoring it to the bladder lengthen. This movement does not seem to alter the length of the round ligament.

The reverse, *anteposition,* moves the uterus closer to the symphysis pubis. The uterosacral ligaments have lengthened while both the round and uterovesical ligaments have shortened.

At times, because of weakening of the urogenital and pelvic diaphragms, and the cardinal and uterosacral supporting ligaments, **prolapse** of the cervix may occur. In first degree prolapse, the cervix pushes down into the vagina, but remains within its confines. In second degree prolapse, the cervix protrudes into the introitus. In third degree prolapse or **procidentia,** the cervix protrudes through the vaginal introitus and the body of the uterus is pulled down into the vaginal canal.

A most important relationship with the cervix is the close proximity of the **ureters** as they wend their way from the kidneys to the bladder. At the lateral fornices, they lie lateral to the supravaginal cervix and just beneath the uterine blood vessels which cross them to supply the vagina. As a result, kidney stones which have made their way down in the ureter can be palpated here vaginally. Ureters are easily placed in danger here of being inadvertently clamped or severed during hysterectomy, especially if the anatomy of the cervical area has been distorted by disease.

Uterine Vessels: The **uterine artery** springs from the medial aspect of the internal iliac (hypogastric) artery. It rides medialward upon the Levator Ani toward the cervix. About 2 cm. from the supravaginal cervix it either crosses the ureter itself or sends a vaginal branch across the ureter. Either or both supply twigs to the ureter at this point. Once the vessel reaches the side of the uterus, it sends a very tortuous main trunk up along the side of the uterus (Fig. X-31). Here, of course, it runs between the peritoneal layers of the broad ligament. It sends descending branches downward as well toward the cervix and vagina and these anastomose with the branches of the **vaginal artery.** With the latter it forms a median channel both on the dorsal and ventral aspects of the vagina, the **azygos artery of the vagina.** The ascending portion supplies body and fundus, branches to the uterine tube and the round ligament. It finally anastomoses superiorly with the **ovarian artery.**

The **uterine veins** accompany the arteries and drain mainly into the iliac veins, but some follow vaginal and vesical veins to the pelvic plexus of veins.

The Uterine (Fallopian) Tubes

The egg-transport ducts of the human female are not much more than 4 inches long. They lie in the free edge of the broad ligament (Figs. X-27 and 28). Their proximal ends open into the uterus near its fundus while their fimbriated, distal ends lie close to the ovary with some of the finger-like processes in contact with the ovary. It is customary to consider four parts of this short tube, simply because of the regional changes of its internal mucosa. The flaring **infundibulum** with its fringed folds constitutes the trumpet-shaped ovarian segment of the tube. While the whole of the tubal mucosa is thrown into longitudinal folds, the folds at the infundibulum are thin, high, and intricate. They subdivide the lumen into a labyrinth of such complicated, narrow chinks that one wonders how the egg ever makes its way down the tube at all. Fortunately, the egg is small when it enters this part of the tube and does not increase much in size during cleavage following fertilization. Still, it must be sought out and fertilized here before any useful progress toward the uterus becomes worthwhile. If anything, perhaps, the complicated folds prevent too much of a gang-attack on the part of single-purposed sperm. Many sperm lose their way, entrapped in the mucosal folds.

The next segment beyond is the **ampulla** which gradually narrows into a terminal **isthmus** before the tube pierces the uterine mucosa. The ampulla is still a relatively dilated part of the tube, making up about two-thirds of its length. The tubal lumen contains more clear passage than at the infundibulum, but the mucosal folds are still complicated in their branching. The isthmus is slender and narrow. Here the mucosa is oriented into low, unbranched, longitudinal folds.

Finally, the last segment is its **intramural portion,** the segment that continues through the uterine wall to attain the uterine cavity. It is quite possible for an ectopic tubal pregnancy to implant in any of these parts of the tube, though the common site is somewhere along the ampulla. The danger, of course, is internal hemorrhage as the tube ruptures or as placental villi begin to erode its vascular supply. The ampullary and intramural portions of the tube are its most vascularized parts.

Epithelial cells of the tube are of two types: some ciliated and some mucoid-secreting. Egg reception is aided by cilia which beat inward toward the uterus. The fimbria help at ovulation when they become turgid and sweep over the ovary, mothering it tenderly. Downward transportation is effected chiefly through peristaltic activity of the muscularis of the tube, and it is probably tubal contraction that helps push sperm upward, rather than the Australian crawl of the sperm themselves. The mucoid secretion reduces friction in general, and it may serve as tubal nourishment for the egg. The fact that the female peritoneal cavity opens to the exterior *via* the fimbriated tubal orifice, uterus and vagina, creates a potential channel for peritoneal infection from the exterior which is not present in the male.

Hysterosalpingography—The Radiographic Anatomy of the Normal Uterus and Fallopian Tubes

Hysterosalpingography is a useful and accurate method of evaluating the uterine cavity and the oviducts. Its principle value in clinical practice is in appraising problems of abnormal uterine bleeding and infertility. Pathological conditions that are detected may or may not have been suspected clinically. The procedure may be of immense value to the gynecological surgeon in his therapy.

1. Procedure: Hysterosalpingography is usually performed about a week after the menses is completed. In most instances, no premedication is required unless the patient is extremely apprehensive. Our technique begins with a plain film of the pelvis in order to search for abnormal calcifications, soft tissue masses and osseous lesions. With the patient in the lithotomy position, a self retaining cannula (Fig. X-32A) is placed into the external cervical os. The cannula tip reaches to the level of the internal os. Thus the os is plugged, thereby preventing reflux of the radiopaque contrast medium back into the vagina, An oily or water soluble contrast medium is slowly instilled with constant pressure on a syringe which is attached directly to the patent cannula. The radiologist monitors the procedure with a fluoroscope, and spot films are obtained after each 1 to 2 cc. of medium is introduced. Such filming, after small increments of contrast medium enter the uterine cavity, minimizes the chance of missing small lesions. On the average about 5 cc. of contrast medium will be used to fill the uterine cavity and the fallopian tubes. In cases of suspected sterility, it is of the utmost importance to observe spillage of the medium into the peritoneal cavity to prove patency of one or both oviducts (Fig. X-32A). In addition to the spot films, the procedure may be recorded by means of cineradiography and television tape recording.

2. Anatomy: In Figure X-32A, we can see the tip of the opaque cannula in the uterine isthmus. The uterine cavity is typically triangular in shape, with the apex of the triangle directed toward the symphysis pubis. The inner surface of the cavity is smooth and the lateral walls of the body are either straight or slightly concave, as in this case (Fig. X-32A). The wall of the fundic portion of the cavity may also be concave or straight (Fig. X-32A, B). Notice that the uterus is

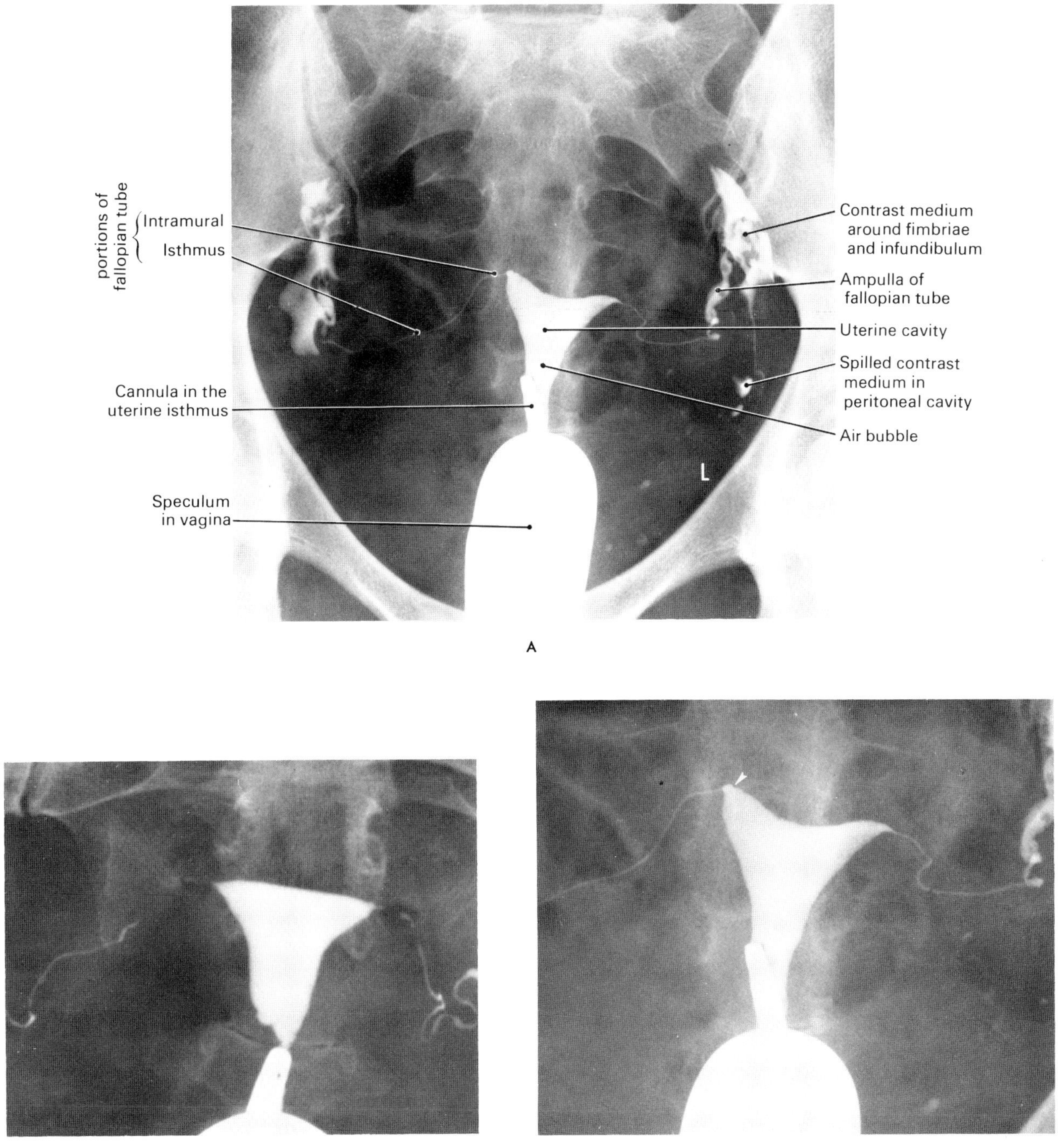

Fig. X-32. Hysterosalpingography. **A.** AP view of the filled uterine cavity and fallopian tubes. There has been spillage of the contrast medium into the peritoneal cavity bilaterally. In this case, the interior wall of the fundic portion of the uterine cavity is concave. **B.** Uterine cavity filled with contrast medium. The fundic portion in this case is straight. **C.** Closeup of (A) to show the radiolucent mucosal fold (*arrow*) which is occasionally observed at the transition point between uterine cornu and the fallopian tube. This is a normal finding.

tipped slightly to the patient's left. At times the uterus may be retroverted, thereby appearing foreshortened on the radiographs. Since there is normally no fat or air surrounding the uterus, its outer wall may not be seen.

The fallopian tubes (oviducts, uterine tubes) are about 10 cm. in length. On occasion, near the transition between the uterine cornu and the fallopian tube, a tiny radiolucent fold of mucous membrane may be observed (Fig. X-32C). Just lateral to the fold the intramural (uterine) portion of the oviduct may bulge slightly. Although not seen in all patients, the mucous membrane fold and the triangular-shaped, pretubal bulge are normal.[5, 6]

The intramural (uterine) portion of the fallopian tube is about 2.0 cm. in length (Fig. X-32A). The isthmus portion of the oviduct is approximately 5.0 cm. long with a thick wall and tiny lumen. The transition from isthmus to the ampulla of the oviduct is not absolute; however, the ampula is characteristically tortuous, thin-walled, and has a larger lumen. The infundibulum is funnel-shaped, and, in Figure X-32A, it is seen surrounded by contrast medium that has spilled out through the abdominal ostium into the peritoneal cavity.

The Ovary

Students generally look for something of fair size when they search for ovaries in the female cadaver. Don't expect to see anything much larger than a shelled almond, about 1½ inches long. You will also have to look for the ovaries on the back side of the broad ligament before you give them up as lost entirely. Their position along the lateral pelvic wall is by no means as fixed as textbooks lead you to think. They may be prolapsed down in the rectouterine pouch (cul-de-sac of Douglas). Their normal position is said to be opposite the **ovarian fossa.** This is a depression which has the external ilac vessels in front and the ureter and internal iliac vessels behind. The obturator nerve also crosses the fossa. Review the information presented on the ovary in Chapter I.

The **ovarian artery** (Fig. X-28 and 31) is, of course, derived directly from the aorta as the lower of the paired series of vessels related to the genitourinary tract and its associates. It supplies the ovary, principally, but it gives twigs to the uterine tube and even, perhaps, to the uterine fundus near its site of anastomosis with the uterine artery (Fig. X-31).

Pelvic Ligaments

As in the case of the abdominal viscera, certain peritoneal folds, masquerading under the name of "ligaments," anchor uterus, uterine tubes, and ovaries to the pelvic body walls.

The largest of these ligaments in the female has aptly been called the **broad ligament.** It passes from the sides of the uterus to the lateral pelvic wall (Fig. X-33A, B). Three other ligaments curve away from the superior wall of the broad ligament. Realize, in

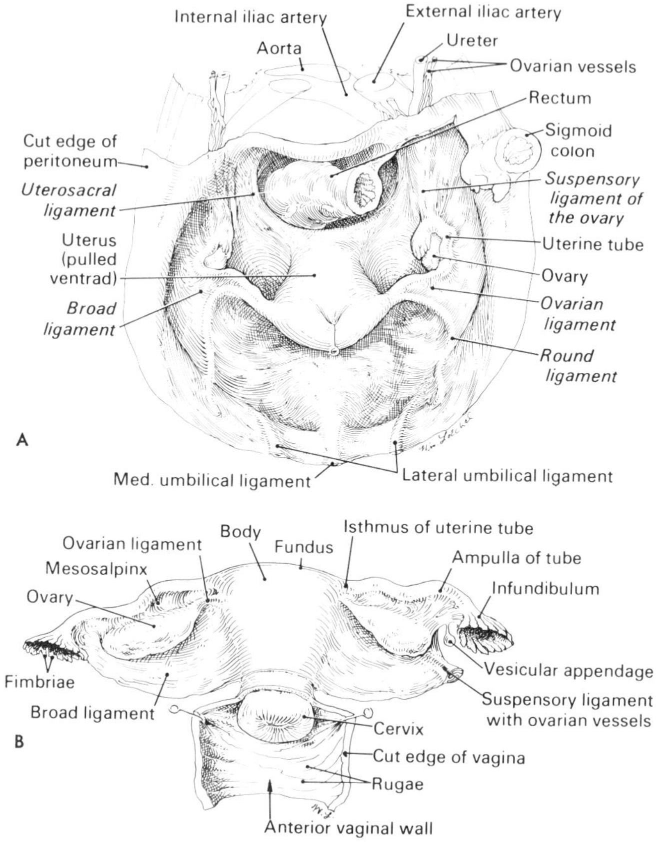

Fig. X-33A. Anterior view of the female pelvic organs. The uterus has been pulled further ventrad than its normal positioning. **B.** Uterus, ovaries, and fallopian tubes. Posterior view. The back wall of the vagina has been cut and spread open.

Figure X-33A, that the uterus has been retracted ventrad from its normal position so that these could be viewed. The most anterior of the three is the **round ligament** (ligamentum teres of the uterus). It courses laterally and ventrally in the margin of the broad ligament to gain the deep inguinal ring. It traverses the inguinal canal of the female to anchor in the labium majus. Traveling with it will be the funicular branch of the ovarian artery, some veins of the ovarian plexus, a twig from the inferior epigastric artery and the genital branch of the genitofemoral nerve. Often there is considerable fat which makes it difficult to find the ligament and most of these other structures during dissection. The round ligament can be fairly robust sometimes, but mostly it doesn't amount to more than a fibrous strand, for all the fuss dissection guides make of it.

The most posterior of the three structures, curving from the superior margin of the broad ligament, is the **ovarian ligament.** It lies between mesosalpinx (the mesentery of the uterine tube) and the mesometrium of the uterus itself. It is short, extending from uterus to uterine pole of the ovary, anchoring ovary to uterus. At the uterine end, it is in direct continuation with the round ligament which courses ventrally away from it.

Embryologically, the ovarian ligament and the round ligament are the counterparts of the gubernaculum testis in the male, except that the ovary stays put and is rarely moved into the labium majus (the homologue of half the male scrotum).

We must distinguish the ovarian ligament from the **suspensory ligament** (clinically called the **infundibulo-pelvic ligament**) of the ovary, even though the former also suspends the ovary in a way. The suspensory ligament lies at the other pole of the ovary, *i.e.*, at its tubal end (Fig. X-33B). It runs laterad between the layers of the broad ligament to the pelvic wall, and it is part of a leash which conducts ovarian artery and vein.

Uterine Ligaments

We have talked of peritoneal folds which suspend ovaries, tube, and the uterus in part, but we have not said much about the main supporting structures of the uterus, itself. The main support is the pelvic diaphragm (chiefly, Levator Ani) and the urogenital diaphragm. The following "ligaments" are also involved: cardinal ligaments (the most important clinically of the supporting structures); uterosacral; uterovesical; and round ligaments. Some of these have been mentioned briefly before.

1. The **cardinal ligaments** (transverse cervical or Mackenrodt's ligaments) pass to the lateral pelvic walls along the course of the Levator from cervix and upper vagina. They are tough, white fibrous, connective tissue strands mixed with some smooth muscle. The ureters pierce through their upper part and the uterine vessels pierce them as these vessels pass through the base of the broad ligament.
2. The **uterosacral ligaments** (Fig. X-33A) attach to the cervix near the lateral fornices and pass posterolaterally around the rectouterine pouch. The pouch is a peritoneal cul-de-sac between uterus and rectum. This peritoneum dips down between uterus and rectum far enough to cover about a fourth of the posterior aspect of the vagina. Beyond the pouch, the uterosacral ligaments join the perirectal fascia and attach to the sacrum.
3. The **uterovesical ligaments** run from cervix forward in a direction opposite that of the uterosacral ligaments. They join the perivesical fascia (or ligaments) where they act as a sling for the bladder and anchor to the inferior pubic rami.

It is changes in length of these ligaments, or weakening of them and the diaphragmatic supports, that lead to malpositioning of the uterus for the most part. The stoutest of them all, obviously, are the cardinal ligaments since they are more than peritoneal folds. Strange as it may seem, for its size, the broad ligament only gives slight support to the uterus.

4. The **round ligament** (Fig. X-33A) already described, runs in the upper part of the broad ligament. All the rest run through the base of the broad ligament to the base of the uterus or cervix.

Lymphatic Drainage of the Female Reproductive Tract

A number of routes can be used. The body of the uterus drains into the external iliac nodes and so does the cervix, in part. The cervix also drains posterolaterally, following the uterine artery toward the internal iliac nodes. Posteriorly, the cervical lymph channels follow the rectouterine fold toward the sacral nodes.

The uterine fundus, ovary and tube have a route of their own toward the aortic nodes. Still, one must be careful to check inguinal nodes in uterine carcinoma, for some lymphatic drainage is along the course of the round ligament to this region.

Cyclic Changes in Uterus and Ovary

Figure X-34 illustrates how changes in the ovarian cycle relate to build-up and sloughing off of the uterine mucosa periodically during the child-bearing years. Most of the lining of the uterine cavity except that at the cervix is involved. Build-up is followed by breakdown and repair over a variable interval of 21 to 35 days, the average menstrual cycle consuming 28 days. Day 1, the first day of menstrual flow, can be considered as the start of a cycle that progresses insensibly in five stages to, let us presume, day 28.

At its height, the uterine mucosa consists of three layers:

1. A surface layer, the **compact layer,** containing the necks of uterine glands and some large stromal cells of connective tissue origin which contain considerable glycogen.
2. A next deeper **spongy layer** containing the dilated parts of cervical glands. These two layers are called the **functional layers.**
3. Deepest of all is the **basal layer** containing the blind ends of the glands and the basal parts of spiral arterioles which make their way up through the other layers.

During menstruation, the basal layer and perhaps a bit of the overlying spongy layer are retained. The rest is sloughed off in menstrual discharge on days 1 to 5 of the menstrual cycle.

A. Stage I could be called the **stage of resurfacing** after menstruation. It starts before bleeding stops as epithelial cells begin to glide over the remains of the basal layer. This repair is completed by day 5 or 6.

B. Days 7 to 14 constitute the **proliferative stage:** Back at the ranch in the ovary, a fresh crop of primary ova are each aspiring to become, under the influence of estrogen, the "Egg of the Month." One or two will make it to the Graafian follicle stage and rarely more than one to the fulfillment of ovulation. Meanwhile, the uterus is busy preparing a home for an expected bride. The mucosa thickens from a millimeter or less to a height of 2 mm. or more. Glands increase in number and lengthen in wavy columns. Connective tissue

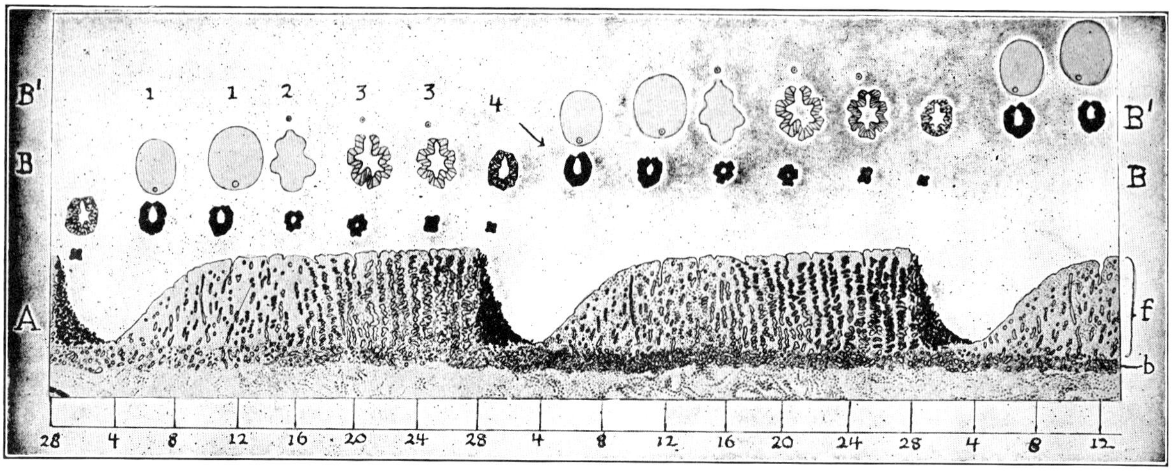

Fig. X-34. Diagram illustrating relation of menstruation to ovulation. **A.** Cyclic changes in uterine mucosa; **B, B'.** Ovarian cycles; *b*, basal layer; and *f*, functional layer of mucosa; *1, 1*, maturing follicle; *2*, rupture of follicle and discharge of ovum (ovulation); *3, 3*, corpus luteum in full function; *4* and remaining figures, degenerating corpus luteum. Numbers at base indicate days of menstrual cycle. (After Schroeder.) (From W. E. Copenhaver: *Bailey's Textbook of Histology* 15th ed., Williams & Wilkins, Baltimore, 1964.)

cells multiply rapidly and some fill with glycogen. But the growth of coiled arteries from the basalis upward is still quite laggard.

C. Days 15 to 27 constitute the **secretory stage:** The egg has ovulated and is hopefully waiting at the church to consumate union on its very doorstep, the ampulla of the tube, with a mate it has never seen and to which it has had no introduction until initial contact is made. It may be overwhelmed with ardent suitors, or it may languish away if none ever arrive. Occasionally, some arrive in a state too decrepit to carry out their natural function. It is then that the egg dies a disdainful death. The uterus seems unmindful of this and carries on to refurbish the home and to prepare the wedding feast. The glands swell and secrete in abundance and take on a baggy, saccular appearance. Their secretion is also rich in glycogen as is that of certain connective tissue cells which aspire to become the succulent, **decidual cells** of pregnancy. All this is governed by progesterone secreted by the corpus luteum of the ovary. Now, too, the coiled arteries spiral tightly and reach for the surface which is moist and swollen with tissue juice and secretion.

D. Day 28 is a fateful day if implantation of the fertilized ovum has not occurred. It is the day of decision or the **ischemic stage.** In this period of a day or two the coiled arteries begin to constrict intermittently. Blood flow to the area is depleted at these intervals and this continues for several hours. The functional layers become pale, lose fluids and secretions, and begin to shrink but there is yet no bleeding.

E. Days 1 to 5 bring us full circle to the **menstrual stage.** The ischemic functional lamina die and slough. There is capillary leakage, and the alternate constricting and dilating of coiled arteries pour forth blood. Other vessels rupture through necrosis. Patches of dead tissue tear away so that the compactum and much of the spongiosum is lost. The unfertilized egg has long since met its demise, fragmenting within the tube.

In the case of a blessed event, the picture is quite different. The fertilized egg, undergoing cleavage stages, reaches the uterus and seeks its nidation site, usually on a lateral wall, before the end of the endometrial secretory stage. The egg and its adnexa imbed and proceed to fashion a placenta to which the uterus contributes the decidua basalis, the portion below the implant, *i.e.*, the spongiosa containing the glycogen-rife decidual cells. The fetus contributes the chorionic plate at the site of implant, and here the chorionic villi grow longest and flourish while the rest of the villi of the chorionic frondosum regress. Review Chapter I to refresh yourself on the nature of these villi, placental circulation, and the relation of the fetal membranes to the wall of the uterine cavity.

MALE ORGANS OF REPRODUCTION IN THE ADULT PELVIS

We noted previously that the male reproductive system consisted of: (1) the **scrotum** which contains (2) two testes; an excretory duct system consisting of (3) the **epididymides,** (4) the **ductus** or **vas deferens,** (5) the **ejaculatory ducts** and (6) **seminal vesicles;** (7) the **spermatic cord;** (8) the **penis** traversed by (9) the **urethra,** a common conveyer of urine and seminal

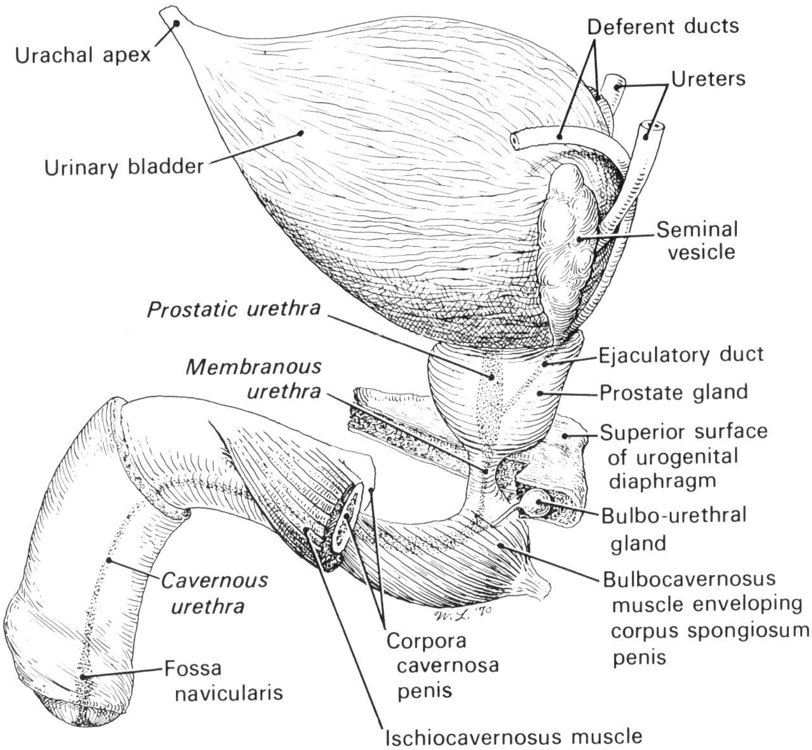

Fig. X-35. The male urethra and its relation to the urogenital organs.

fluids; and, finally, (10) the **prostate** and **bulbourethral glands** as accessory glands to the system.

We have discussed scrotum, penis, and spermatic cord in sufficient detail here and elsewhere in the text. Some few details should now be added in relating the other parts of the adult system to pelvic anatomy. Except for the scrotal contents and the spermatic cord, the principle parts of the system are shown in Figure X-35.

The Testes (Fig. X-36)

Ordinarily, these ovoid, flattened, male reproductive organs averaging 4 to 5 cm. in length and 2 to 3 cm. in thickness are found outside of the pelvis within the scrotal sac. In development, however, we saw these organs arise, internally, at the level of the fetal kidney. The definitive metanephric kidney was a pelvic organ which climbed an "arterial ladder" to reach an intra-abdominal site while the testes descended to their exterior site. Should the testes fail to descend, they could remain within the pelvis or within the inguinal canal close to the external ring. One or both might do so. In an *ectopic migration,* a testis can be found at odd external locations as well, *e.g.,* in the thigh, in front of the pubis, in the perineum, or superficial to the inguinal ligament. In children it is not uncommon to find that cremasteric contraction has pulled the testis out of the scrotum up close to the external ring. Such *retracted testes* can be returned to their proper harbor by a hot bath and gentle massaging from above.

The tunica vaginalis, as visceral peritoneum, covers the testis everywhere except where it attaches to the epididymis and the spermatic cord. This tunic is not shown in the simple sketch, Figure X-36, but a heavy, white fibrous, inelastic capsule, the tunica albuginea, is delineated there. This capsule sends in septae which converge toward the upper pole at the mediastinal region which contains the rete testis. The septae compartmentalize the testis into as many as 400 **lobules** each of which is occupied by two or more highly convoluted, seminiferous tubules. Stretched out to full length, one of these sex-cell manufacturing canals would extend to as much as 2 feet in length. It is from the epithelium of the tubules that spermatozoa are formed (see Chapter I). The contorted tubules converge toward the rete network to which they connect by **straight tubules** (**tubuli recti**) (Fig. X-36). Most of the lining cells of the straight tubules consist of Sertoli cell elements so that little spermatogenesis occurs here. Straight tubules can be regarded as the first segments of the duct system. The endocrine **interstitial cells** lie in the connective tissue of each lobule outside the seminiferous tubules.

We have previously discussed herniation into the scrotal sac. Problems can arise within the closed-off sac itself as fluids distend the tunica vaginalis to form

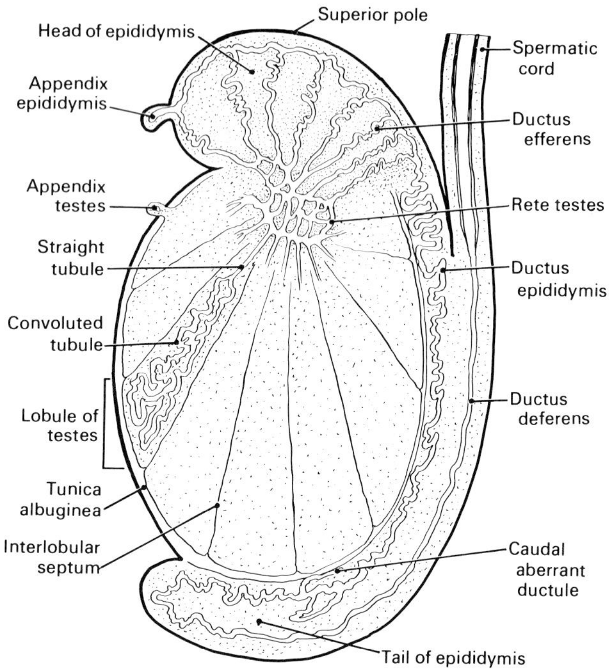

Fig. X-36. Schema of the tubular channels in testis and epididymis.

a **hydrocoele.** Such fluid cysts can be primary manifestations or they can arise secondarily through testicular disease. They come in four anatomical types:

A. The **vaginal cyst,** which simply distends the tunica vaginalis and, so, is limited to the scrotum. The fluid must surround ventral and lateral sides of the testis as a consequence.

B. The **congenital hydrocoele,** which obviously communicates with the peritoneal cavity of the abdomen *via* inguinal canal.

C. The **infantile hydrocoele,** which extends as a finger of the unobliterated processus vaginalis often as far as the external ring but no farther.

D. The **hydrocoele of the spermatic cord,** where a partial, unobliterated segment of the processus vaginalis becomes distended with fluid.

These cysts are quite different from **epididymal cysts** lying *above* the testis. Here the tubules of the epididymis distend with secretory fluids. The latter cysts lie within, not outside, the reproductive channels.

The Epididymis

This C-shaped structure (Fig. X-36) has a head, situated at the upper pole of a testis; a narrowing **body,** clasping the posterior body of the testis, and a **tail** attaching at the inferior extremity of the testis. This terminal part has its medial border facing the terminal part of the spermatic cord from which it derives its blood, lymphatic, and nerve supply. The lateral and posterior surfaces of the entire organ lie free, covered by visceral peritoneum of the tunica vaginalis which dips into a cleft between testis and epididymis, the **sinus epididymis.**

Entering the head of the organ are about a dozen **efferent ductules** (Fig. X-36). They leave the rete area of the testis as fairly straight tubes but they coil considerably as they enter the epididymis. Each coiled part forms an epididymal lobule. The epithelial characteristics change as the straight portions become convoluted; and, histologically, it is the convoluted parts that we identify as epididymis with **stereocilia** (microvilli) under the microscope. The coiled epididymal ducts all empty into a single tube, the **ductus epididymis,** which itself becomes more and more convoluted as it forms the body and tail of the organ. Without a sharp transition, it becomes continuous with the **ductus deferens** of the spermatic cord. The epididymal duct is only about 0.4 mm. in diameter but may extend, when straightened, for some 21 feet.

The Ductus (Vas) Deferens

This much thicker duct (2 to 3 mm. in diameter) is also much shorter. Its 18-inch length seems to be almost a magic anatomical number for that is roughly the length of the spinal cord, the femur and the thoracic lymph duct. The "window-cord" toughness of this muscular duct can be palpated through the scrotum and through the soft tissues of the spermatic cord. It travels in the latter through the inguinal canal in company with the fascias and Cremaster muscle of the cord, the testicular artery and pampiniform system of veins, and the lymph and nerve supply of the testis and epididymis. It has its own arterial supply, the **deferential artery,** which can be either a branch of the umbilical, internal iliac, or middle hemorrhoidal (rectal) artery. Its own vein empties into a plexus of veins upon the seminal vesicle which in turn join vesical plexuses on the urinary bladder or pudendal plexuses to reach the internal iliac veins.

Leaving the cord, the ductus deferens loops over the origin of the inferior epigastric artery and passes, subperitoneally, caudad and laterad in the pelvic wall. It passes medial to the distal end of the ureter (Fig. X-35), bends caudad in an effort to reach the midplane, and lies on the posterior wall of the bladder just medial to the seminal vesicle. It terminates adjacent to its fellow of the opposite side in the midplane just dorsal to the prostate. Terminally, it dilates into an irregular, spindle-shaped **ampulla.** Here its lumen is larger but its mucous lining is more folded. These folds branch and anastomose and produce recesses which may invade the muscularis of the organ. The epithelium here is somewhat akin to that in the seminal vesicles.

The Ejaculatory Duct

This is the short (0.8 inch), slender termination of the ductus deferens. It seems to be formed by union of ampulla and seminal vesicle. It pierces the glandu-

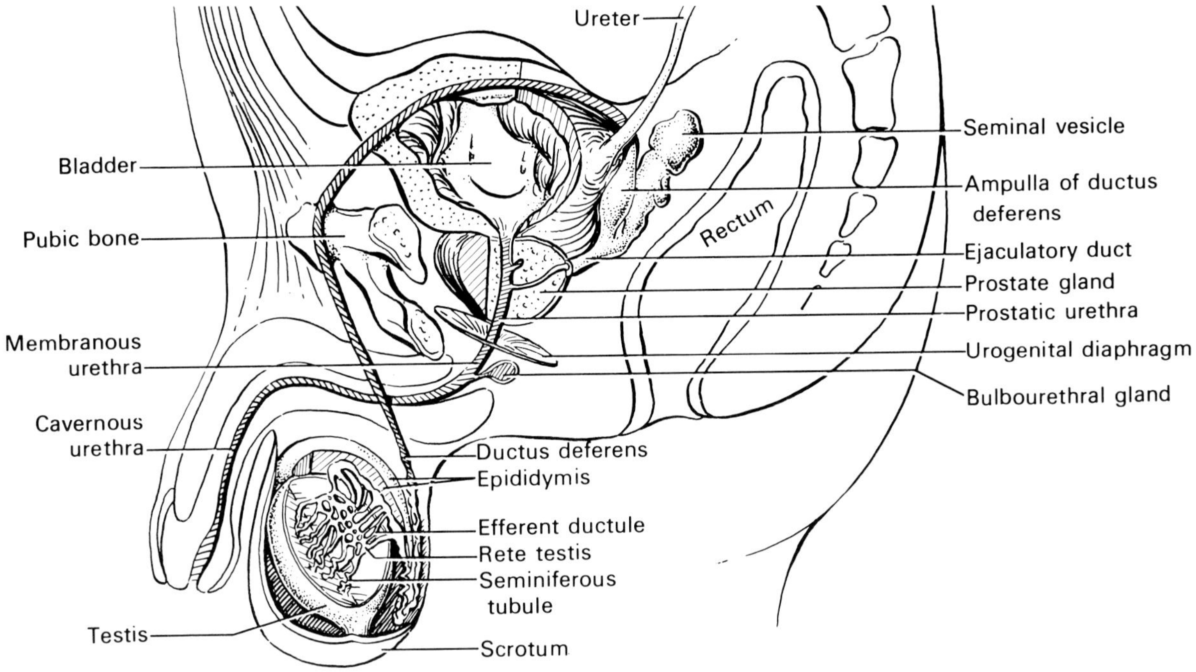

Fig. X-37. Midsagittal schema of parts of the male reproductive system. (The spermatic cord associated with the ductus deferens is not shown nor is a second lamina of the urogenital diaphragm below the bulbourethral gland. These were deleted to show schema of duct system to advantage.)

lar tissue of the prostate to open into the **prostatic urethra** at the **urethral crest.** The routing from testes to urethra is depicted in Figure X-37.

Up to this point we have dealt entirely with those parts of the system concerned with the formation and transport of spermatozoa. Sperm develop motility only gradually as they pass through the male reproductive tract as is shown quite clearly in a cine film produced by Blandau and his co-workers at the University of Washington.[7] In the seminiferous tubules and efferent ductules sperm show only weak vibratory motion *in vitro;* whereas those from the distal part of the head of the epididymis move in swift circular fashion. Rapid forward movement with longitudinal rotation characterize fully mature sperm. This kind of motion is achieved in the caudal epididymis.

Thus, sperm mature in their passage through the epididymis. They move through the straight tubules and rete washed along by the fluids secreted there. They must pass through these regions rather quickly for few sperm are hardly ever seen here. They pass through the epididymis slowly as they ripen and mature. Its thick, viscid secretion supplies sperm with nutritive substances. Passage may take as long as 6 weeks. The epididymis is, then, the real storage reservoir for maturing spermatozoa.

The ductus and its ampullary and ejaculatory terminus serve for rapid transit. Powerful contractions of it and of the urethra forcibly expel spermatozoa along their merry way under the stimulus of the climax during orgasm. And so, like lemmings in their suicidal surge, they either spill on barren soil or plunge into the acid environment of some vagina to perish by the million except on those rare occasions when fertilization is accomplished by a scant few of their number.

The Auxiliary Genital Glands of the Male

These are the seminal vesicle, the prostate, and the bulbourethral glands whose secretory products contribute to seminal fluid.

The Seminal Vesicle (Fig. X-35): This is a diverticulum off the terminal end of its own ductus deferens. It is a highly convoluted, 2-inch long, glandular sac of no more than 0.7 inches in width. It can be stretched straight to about 4 inches in length. The mucosa bears intricate folds, elevations and recesses. While the gland may be lacking in such species as carnivores and insectivores, where it is present it is highly dependent upon hormonal control for its very development and for its secretory functioning. Testosterone seems to exert a prime influence, for castration before puberty results in underdevelopment of the gland and, after puberty, in its regression and lack of secretory behavior. The seminal vesicle has considerable luminal storage capacity (4 ml.). The sugar, fructose, is a chief product. It is possible to palpate the seminal vesicle by rectal examination. When it becomes tuberculous, it enlarges considerably. We have indicated previously that it is false to consider it as a storehouse for spermatozoa. Whatever few sperm may be encountered there by histological examination probably got there quite by chance either after death or as a result of sexual excitement without ejaculation. The fact that other animals lack such a storage reservoir also makes it unlikely that storage is a prominent role in man.

The Prostate: This is a pyramidal-shaped, fibromuscular gland of about 1½ inches in diameter found wrapped about the urethra at the base of the urinary bladder (Fig. X-35). It is about the shape and size of a chestnut—if you can still remember chestnuts shined to a high gloss to be carried in the pocket as a "lucky" piece or roasted by some Italian vendor. Superiorly, the prostate sits at the bladder neck so that the urethra enters near its ventral border. Inferiorly, the gland rests directly upon the external sphincter of the bladder (urethra) in the deep perineal pouch. In front of it (ventral) lies the **retropubic space** filled with extraperitoneal fat which separates prostate and bladder from the pubic bone. Here is the prostatic plexus of veins. Behind (posteriorly), it is kept free of the rectum by a double layer of fascia (Denonvillier's fascia). This fascia continues over the seminal vesicles and laterad to the pelvic fascia. Fortunately, it serves as a good buffer between rectum and prostate so that prostatic carcinoma rarely penetrates into the rectum. In rectal surgery care should be taken to see that the fascia is left clothing the prostate and urethra. The Levator Ani lies laterally.

Since the urethra runs almost vertically through the ventral part of the gland (Fig. X-35), its posterior aspects constitute its **lateral lobes,** separated from each other by a slight **median groove.** The groove terminates at the posterior base in a more distinct median notch anterior to which is a funnel-shaped depression for the reception of the ejaculatory ducts. At the depression, two grooves course ventrad and laterad to cut off a median elevation. When this area becomes enlarged, it is called the **middle lobe** of the gland. Under normal circumstances, when seen in median section between urethra and ejaculatory ducts, it looks more like a simple commissure between the lateral lobes than a separate lobe. It can hypertrophy with age, however, and lead to prostatic obstruction without being detected by rectal examination. In this sense, it is significant to know that a middle lobe exists.

The gland has a fibro-elastic capsule. But when we speak of the capsule of the prostatic gland, we must be certain of what we speak for there is a condensation of extraperitoneal fascia outermost around the true capsule, and between these two lies the prostatic venous plexus. The pathologist will also use the term, capsule, in another sense. When the gland hypertrophies, the adenomatous tissue which is to be removed by prostatectomy compresses normal glandular tissue peripheralward. This compressed "capsule" of the pathologist is left behind when the central mass is enucleated.

The whole gland is really an aggregate of 30 to 50 tubulo-alveolar glands grouped into five lobes by fibro-elastic stromal septae which are also rich in smooth muscle. The separate glandular units empty into the prostatic urethra, not by one, but by 15 to 30 ducts.

The blood supply of the organ comes from the inferior vesical artery and from a branch of the middle rectal (hemorrhoidal) artery. The venous plexus is part of the pudendal plexus which surrounds the bladder and communicates behind with the rectal plexus and above with the plexus on the bladder. Its lymphatics drain into the internal iliac nodes. Its nerve supply reaches it from the pelvic plexus—sympathetics *via* hypogastric plexus and parasympathetics *via* nervi erigens.

How does one approach a prostate in the event of a benign hyperplasia or carcinoma? A transvesical approach across the bladder is possible, or one could enter the retropubic space to reach it directly, or it could be approached from the perineum anterior to the rectum. In the case of carcinoma, planes of cleavage between prostate and surrounding fascias may be obliterated so that the only avenue may be a transurethral one in which chunks of tissue are fished out with a cystoscope. The gland can be removed with no detrimental effect upon reproductive prowess.

In every day life, the gland is best approached per rectum with a rubber-gloved finger. Its lateral lobes are easily palpable against the anterior rectal wall and they can be massaged to collect prostatic fluid. This secretion, with its distinct odor, is a thin, mildly acid milk which makes up the bulk of seminal fluid. It contains lipid, acid phosphatase and a goodly amount of citric acid.

In the aged, particularly, prostatic concretions or corpora amylacea may abound within the gland. Variable in size, should they increase to 2 mm. in diameter, they are likely to be retained to calcify as calculi. These lamellated structures consist of protein and nucleic acids.

The bulbourethral glands (of Cowper): We observed these glands in the deep perineal compartment. They lie on each side of the membraneous urethra (Fig. X-35) between the fascial layers of the urogenital diaphragm. Their ducts run forward some 3 to 4 mm. in the corpus spongiosum of the penile bulb before entering the urethra. They get their blood supply in the perineum from the artery of the bulb. They add a mucoid secretion to the seminal fluid.

DEVELOPMENT OF THE REPRODUCTIVE GLANDS AND THEIR DUCTS IN MALE AND FEMALE

While sex is determined at the time of fertilization, the sex glands or gonads do not declare themselves as male or female structures until the seventh week of embryonic development. They appear as an indifferent gonad as early as the fifth week, however. Thus, the

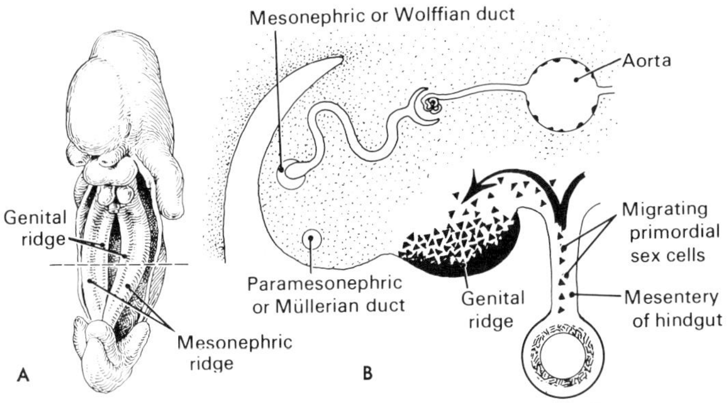

Fig. X-38. The genital ridge. **A.** The genital ridge associated with the mesonephros of a 9-mm. human embryo. **B.** Transverse section at level shown in A to diagram migration of primordial sex cells from yolk sac to ridge. (Modified from Arey and Langman.)

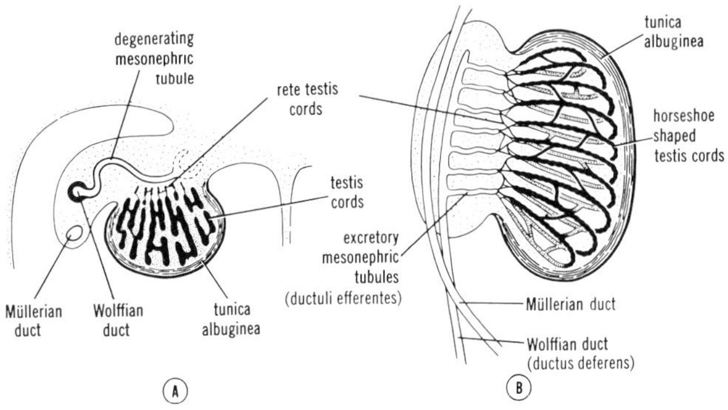

Fig. X-39. Development of a testis from the genital ridge. **A.** In the eighth week. **B.** At the fourth month. (From J. Langman, *Medical Embryology*, 2nd ed., Williams & Wilkins, Baltimore, 1969.)

two sexes are very similar at their beginning and any uniqueness between them arises by differential development of seemingly indifferent rudiments. Development goes at a pace that is much slower than even that of the kidneys, and function is postponed since reproductive prowess is not witnessed much before the time of puberty, several years after birth. Endocrine function exerts itself much earlier in these organs, even in embryonic life, for the sex hormones are important in influencing the development of certain other bodily parts.

The Genital Ridge

This is not the ridge over which the battle of the sexes is fought nightly. It is a strip of peritoneal epithelium along the ventromedial surface of each mesonephros. Growth here has caused a raised expansion into the coelomic cavity (Fig. X-38 A,B). The ridge makes its appearance in the 4-week embryo, and at that time there are no sex cells in it. They do not appear until the sixth week. The inner aspect of this epithelial mass becomes subdivided into many **sex cords** connected to the superficial **germinal epithelium** from which they grow (Figs. X-39A and 40A). Recall from Chapter I that the primordial germ cells arise initially in the wall of the yolk sac. They migrate along the dorsal mesentery of the hindgut to enter the area of the genital ridge at 6 weeks. They come to lie between the proliferating cords and partly in the underlying mesenchyme (Fig. X-38B). We have indicated previously (Chapter I) that current evidence favors the view that all definitive sex cells of the genital glands are descended from these early migrants. A current minority view still holds that all later functional sex cells originate locally from the germinal epithelium as they are needed in later life.

The genital ridges *are* the **indifferent gonads,** and at first they are as long as the mesonephros. But as the gonads grow, they shorten and thicken and pull away from the mesonephros by an undercutting of lateral grooves until they are suspended by a double-layered, peritoneal mesentery, the **mesorchium** or **mesovarium,** depending upon sex.

The Testis

Two features, by means of which a male gonad can be recognized as a testis, make a sudden appearance in the indifferent gonad. These are the **tunica albuginea** and the **branching and anastomosing testis cords** (Fig. X-39). The tunica albuginea is a heavy concentration of connective tissue between surface epithelium of the gonad and the sex cords. The surface germinal layer in the male gonad thins to a flat, visceral peritoneum.

The sex cords, themselves, first lacking a lumen, are of three types: (1) those near the mesorchium form a network, **the rete testis;** (2) short tubules radiate from the rete as **tubuli recti** and connect with; (3) the longest aspects of the cords, the contorted or convoluted parts, which when these latter canalize, become the **convoluted or seminiferous tubules.** Lumina are not acquired until puberty. In these cords are found: (a) the primary germ cells (**the spermatogonia**) and (b) supporting cells which become **Sertoli cells** when the testis matures at puberty. In the mesenchyme between the cords or future tubules, **interstitial cells** (or cells of Leydig), the endocrine secretors of testosterone, make their appearance as early as the fourth or sixth month of development *in utero.*

The Ovary

The female gonad differentiates more slowly than the male gonad. It retains its surface germinal epithelium as a layer of **cuboidal cells** and it does *not* differentiate a very distinct tunica albuginea below this surface layer. Rather than lengthening and convoluting, the primitive sex cords of the female gonad are broken up into clusters by invading mesenchyme. Early clusters of an irregular nature, containing primitive germ cells, form near the mesovarium, an area

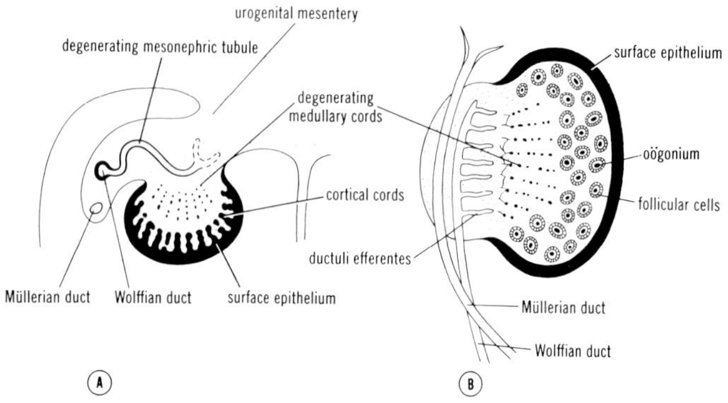

Fig. X-40. Development of an ovary from the genital ridge. **A.** At the seventh week. **B.** At the fifth month. (From J. Langman, *Medical Embryology*, 2nd ed., Williams & Wilkins, Baltimore, 1969.)

which will become the **medulla** of the ovary as it becomes invaded by blood vessels (Fig. X-39A).

As the medullary cords degenerate, a secondary growth of **cortical cords** sprout from the germinal epithelium. These also break into clusters of cells which appear to surround primitive germ cells located nearer the outer surfaces of the gonad. Such cells are called **follicular cells** (Fig. X-39B). It is the germ cells of these secondary cords near the surface which are said to become the definitive **oogonia.**

The Genital Ducts

In the 6-week embryo two duct systems begin to declare themselves as the future transport ducts for the holocrine products of these sex glands, *i.e.*, for ova and sperm. Both ducts are paired organs: (1) some old friends, already on the scene, *e.g.*, the **mesonephric or Wolffian ducts** (Fig. X-40); (2) a new paired system which runs a parallel course with the first as the **paramesonephric or Müllerian ducts.** The first will serve as the male genital ducts, the second as female genital ducts. If the terms are confusing, think of the male as a "wolf" and the female as a "mule" who requires gentle handling and some coaxing. This should help you keep the two duct systems related to the proper sexes.

The Müllerian duct arises in the mesonephric (not the genital) ridge by an invagination and closing-off of the coelomic epithelium. Notice its location in Figure X-38B, close to the Wolffian duct. Both pairs of ducts extend the length of the mesonephros to the cloaca (Fig. X-41). Thus, while the male and female gonads have identical embryonic origins, their ducts have different origins.

Cranially, the Müllerian ducts are open, as you might expect them to be if they are to become the fimbriated ostia of the uterine tubes. The Müllerian ducts lie lateral to the mesonephric ducts until they cross the latter ventrally to fuse with each other in the midline. At first, at the site of fusion, each duct is separated from the other by a septum. The partition even-

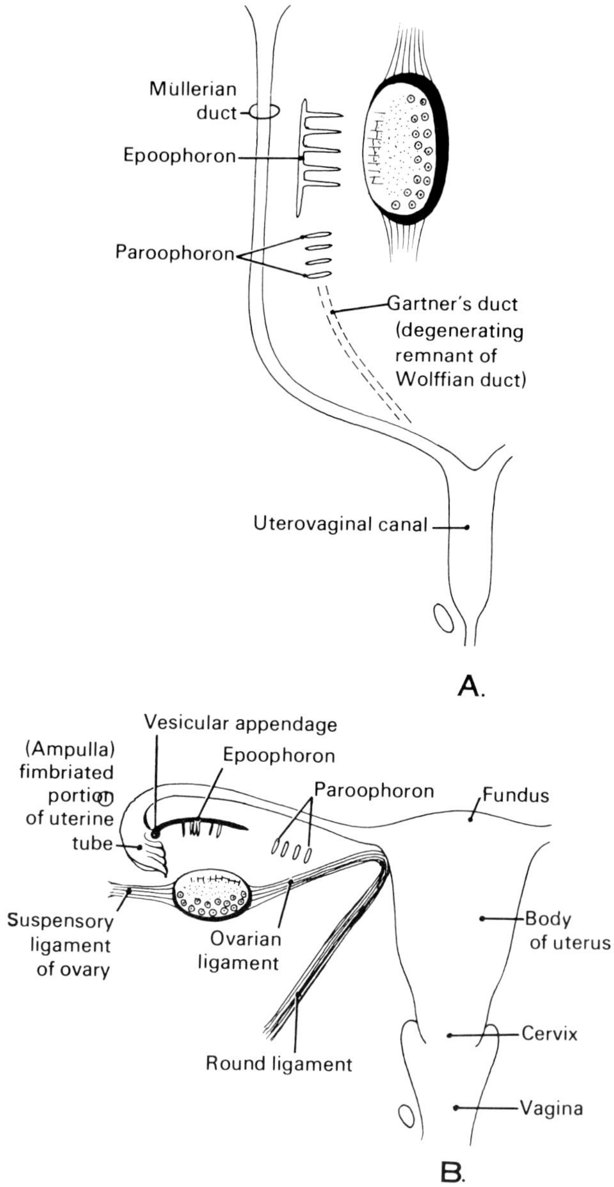

Fig. X-41. Schema of the genital ducts at the end of the second month (**A**) and after the descent of the ovary (**B**). (Modified from Langman.)

tually breaks down so that a common **uterovaginal canal** is formed. The solid tip of this common duct moves caudally until it contacts the inside wall of the urogenital sinus. Its tiny swelling there is the **Müllerian tubercle** (Figure X-42).

Fate of the Wolffian (Mesonephric) Duct: In both males and females, its cranial portions are used as the excretory duct of the first two, short-lived, embryonic kidneys. But only its caudal stretch, associated with the developing metanephric or permanent kidney, is retained as the ureter in both sexes. When the middle kidney or **mesonephros** degenerates in the male, its duct, above the level of the metanephros, persists along with some 26-odd, mesonephric kidney tubules. In both sexes, the rete portions of the gonadal

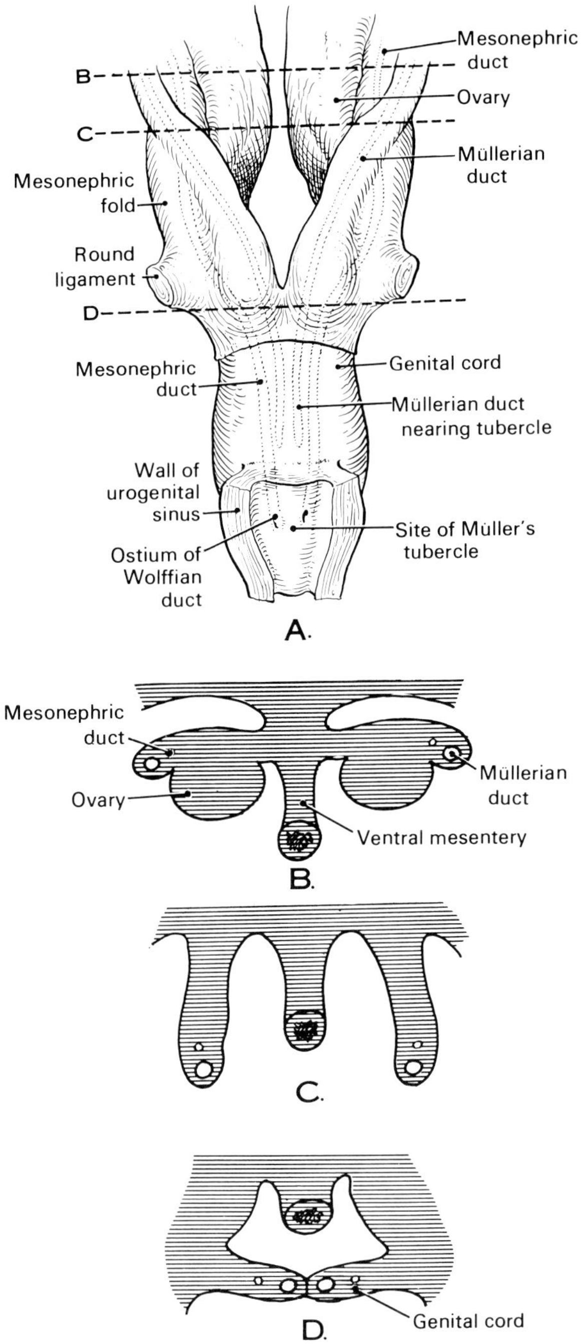

Fig. X-42. Course of genital ducts and their relationship at the urogenital sinus in the formation of the genital cord. **A,** a 3-dimensional construct at approximately 2 months. **B–D,** cross-sections through A where indicated.

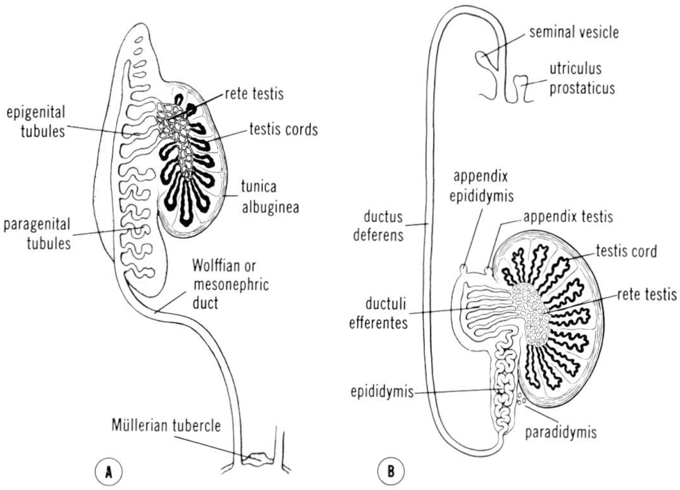

Fig. X-43. Development of the genital ducts in the male (**A**) at 4 months and (**B**) after the descent of the testes. (From J. Langman, *Medical Embryology*, 2nd ed., Williams & Wilkins, Baltimore, 1969.)

sex cords unite with some of these left-over mesonephric tubules. The ovary does have a rete near its medulla, just as there is a rete testis in the male gonad.

In the male these tubules of the old second kidney become the **efferent ducts (ductuli efferentes) of the epididymis** (Fig. X-43B). The part of the mesonephric duct into which such tubules emptied becomes the **duct of the epididymis,** and its more caudal part becomes the **ductus deferens.** The latter will lose its contact with the ureteric bud to empty alone into a urethra below the level of the bladder. The terminal part becomes the **ejaculatory duct,** giving rise to a **seminal vesicle** as a blind diverticulum, much in the manner that the ureteric bud, itself, grew from the mesonephric duct earlier in development. Thus, upper, older parts of a urinary duct system, no longer useful to the organism, are revamped for use as genital ducts in the male.

Some of the old, 26-odd, tubules which failed to become efferent ductules in the male, persist as vestigial structures: (1) the **paradidymis;** (2) the **cranial;** and (3) the **caudal aberrant ductules.** The blind, cranial end of the upper part of the mesonephric duct also remains attached to the epididymis as the **appendix epididymis** (Fig. X-43). The appendix testis is a mesonephric or Wolffian duct derivative.

Now, in the female, there are *no functional* remains of the Wolffian ducts. There are only *vestigial remains.* These are: (1) remnants of the **rete ovarii** in the ovarian medulla; (2) the **epoophoron;** (3) the **paroophoron;** (4) parts of **Gartner's duct;** and (5) the **vesicular appendix.**

The **epoophron** (organ of Rosenmüller) is a remnant of a longitudinal piece of the mesonephric duct which lies in the mesosalpinx. It parallels the uterine tube and it is closed at both ends. It has a number of aberrant mesonephric tubules attached to it, so it is the homologue of the efferent ductules and part of the epididymal duct of the male.

Near the proximal end, or mouth, of the uterine tube there may be one or more vesicular blebs, the **vesicular appendices** (hydatids of Morgagni). They are about the size of a pea and, like the appendix epididymis, they are pieces of the blind end of the Wolffian duct.

The **paroophoron** represents some of the unattached, caudal, aberrant mesonephric tubules left behind in the mesosalpinx. They usually disappear entirely by adult life.

Gartner's duct is nothing more than chunks and pieces of the old Wolffian duct found distributed at any level from epoophoron to the vaginal epithelium (Fig. X-44).

Fate of the Müllerian Duct: In the female, **uterus** and **vagina** develop from the joined caudal ends of these ducts while the **uterine (Fallopian) tubes** develop from their upper portions.

In the male, only two remnants of this duct appear to persist. At the site of the Müllerian tubercle in the **prostatic urethra** of the male is a raised hillock with an opening that leads into a tiny pouch of varying depth. This is the **seminal colliculus** with its **prostatic utricle** (vagina masculina or uterus masculina). This is all that the male inherits as a vestigial uterovaginal canal.

Capping the testis, is another appendage, the **appendix testis,** thought, generally to be part of the funneled end of the Müllerian duct. However, some also tend to regard it as an aberrant mesonephric tubule (Figs. X-43 and 44).

The External Genitalia in Development

So alike are the external genitalia of the two sexes at the beginning that it is difficult to determine by superficial scrutiny (Fig. X-45) whether a very young embryo would have become a boy or a girl. Once a few homologues are realized, even the finished product no longer appears to be all that different. Still, we all know what the difference is. Even the little girl who asked the clerk in the candy store for a male chocolate Easter bunny knew the difference.

"Young lady," the clerk replied, "They are all alike. In fact, I can't see *that* (snapping his fingers) much difference between any of them."

"Oh, yes there is!", cried the little girl. "The male Easter bunnies have *that* (snapping her fingers) much more chocolate!"

In embryos of about 5 weeks, at a time when the urogenital sinus has already been divided from the cloaca, a conical elevation, the **genital tubercle,** arises at the urogenital diaphragm. On the underside of the tubercle a **urogenital groove** appears, flanked by two ridges, the **urogenital folds** (Fig. X-45A). The tubercle soon assumes a cylindrical shape in both sexes.

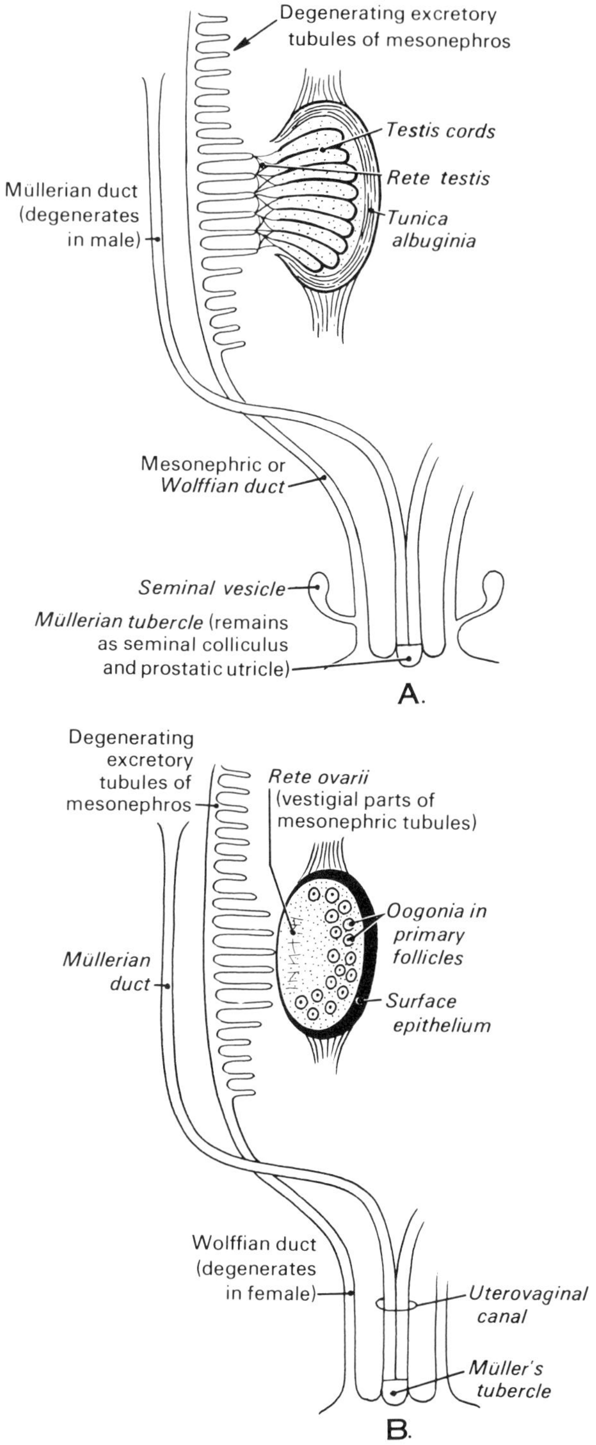

Fig. X-44. Diagram of the development of the genital ducts in (**A**) the male and (**B**) the female. (Modified from Langman.)

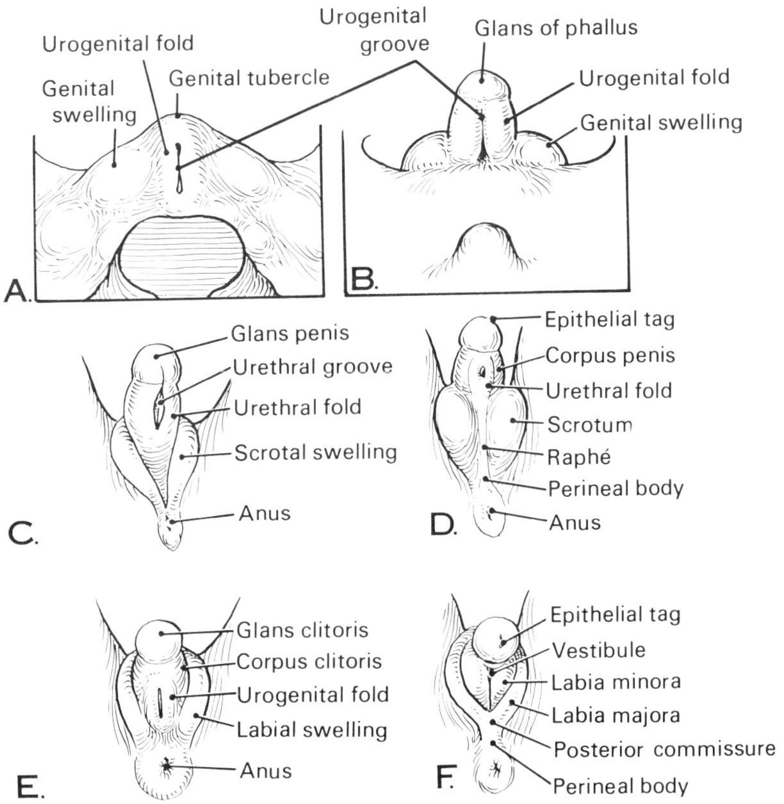

Fig. X-45. Differentiation of external genitalia in the male and female. **A–B,** indifferent stage at almost 7 and 8 weeks; **C–D,** the male at 10 and 12 weeks; **E–F,** the female at 10 and 12 weeks.

Very early this **phallus** develops a **glans** at its tip. The urethral groove extends to the glans in the male, but not quite to the glans in the female. Still, it is difficult to recognize sex with a certainty in embryos under 50 mm. in crown-rump length. The indifferent rudiment will soon develop a pair of **labioscrotal swellings,** but this feature does not help to make it any easier to tell the sex of the early embryo.

Differentiation in the Male: Elongation of the phallus produces a **penis.** The **phallic urethra** is represented by the urethral groove which begins to roll in as a **urethral canal** at about the end of the second month. Closure of the groove moves from base to tip and is completed at about the third month. The pelvic part of the urogenital sinus forms the connecting part of the urethra from the penile base to the site of the ejaculatory duct entrance. Epidermal folds appear at the base of the glans as a **prepuce** by the middle of the fourth month.

Scrotal swellings migrate caudally to the base of the penis where they fuse in the midline to form the **scrotum.** Two extensions from the peritoneal cavity move through the inguinal canal into these sacs. The testis and epididymis, on each side, descend into the sacs behind these peritoneal extensions. They drag behind them the ductus deferens, vessels, and nerves through the inguinal canal. Thus, the testis bulges into the scrotal sac behind its investment of visceral peritoneum. The rest of the cavity of the sac consists of parietal peritoneum derived from the processus vaginalis. The vaginal process separates from the peritoneal cavity as the inguinal canal closes.

Differentiation in the Female: The labial swellings correspond to the scrotal swellings of the male. They remain in their primitive condition, failing to fuse in the midplane, and develop into the fleshy *labia majora.* The edges of the urethral groove do not fuse to form a phallic urethra but remain separated as the *labia minora.* The urogenital sinus, from the entrance of the Müllerian tubercle to the exterior, shortens to become the **vaginal vestibule** (Fig. X-46). This is a shallow area into which the adult vagina and urethra both empty (Fig. X-45F). The cavernous portion of the phallus develops only slightly as the *clitoris.* The latter develops a glans and prepuce. The whole tiny organ projects freely for a time but it soon becomes wholly embedded in the tissues of the *vulva.* The term, vulva, includes labia, vestibule and clitoris but not the deeper vaginal sheath. The prepuce normally hides the glans which usually protrudes only upon stimulation. The female phallus lacks a corpus spongiosum and, hence, does not contain the urethra.

The female urethra, then, includes only that portion of the male counterpart that would normally lie be-

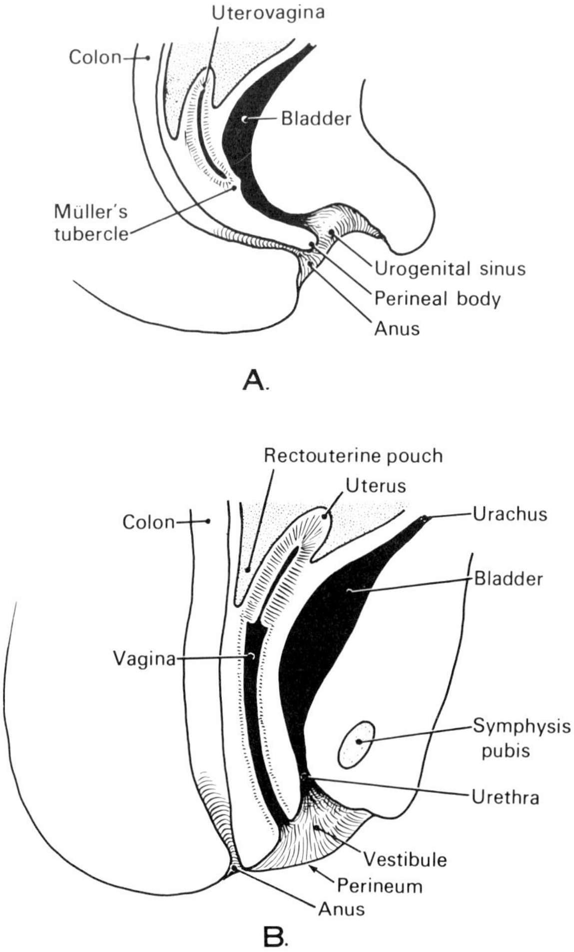

Fig. X-46. Shortening of a urogenital sinus into a shallow vestibule. Midsagittal views in female fetuses. **A.** At 10 weeks. **B.** At 5 months.

tween bladder and ejaculatory duct. In the female, the labia must be spread to see the urethral orifice which lies well below the clitoris, between the latter and the vaginal orifice, as most intrepid male explorers have, no doubt, discovered. Strange as it may seem, the common female complaint is that these same explorers too often seem to be unable to discover the clitoris when it counts most. Surely, this can't be true of true anatomists!

The genital swellings, which constitute the labia majora, fuse in front of the anus as a **posterior commissure** (Fig. X-45F) so that these parts form a horseshoe-shaped rim, open toward the umbilicus. Cranially, this gap is later closed by the *mons pubis* which develops independently of the primitive genital swellings.

Development of the Urinary Bladder and Urethra

The **urorectal septum** divides the cloaca into a **rectal** and **urogenital part** (Fig. X-47) in the seventh week of embryonic life. The urethral membrane ruptures to the exterior shortly thereafter. Now, the allantois and the urogenital ducts are tributaries of the urogenital sinus, not of the anorectal portion of the cloaca. These ducts lie close together on the dorsal aspect of Müller's tubercle in the 9-week fetus (Fig. X-48A).

The urogenital sinus in its early stages shows three distinct regions: (1) its cephalic extention from Müller's tubercle to the allantois is its **vesicourethral portion;** (2) its middle section is a narrow isthmus beginning caudal to Müller's tubercle and is known as the **pelvic part;** and its somewhat expanded caudal and terminal part is the **phallic portion** (Fig. X-49). It is really the last two parts which are known as the **urogenital sinus,** proper.

The first portion, or vesicourethral region, will produce the bladder. Growth in its apical part lags so that the allantois dwindles to a fibrotic cord. We have long ago identified this in the cadaver as the **median umbilical ligament,** from umbilicus to bladder on the internal aspect of the ventral body wall. A very short segment of this area just above Müller's tubercle does not expand into bladder but remains narrow and short. It becomes the *upper prostatic urethra* of the male, and *all* of the *urethra* in the female. The rest of the longer male urethra, *e.g.,* its lower prostatic, its membranous, and its cavernous portions, derive from the pelvic and phallic parts of the urogenital sinus.

In their primitive relationship, the ureters did not open directly into the bladder in either sex, but rather into the Wolffian ducts (Fig. X-49A). But, in later development (Fig. X-49B), as the terminal parts of the Wolffian ducts became absorbed into the bladder wall, the ureters attained independent entrances into the bladder. The Wolffian ducts, as the sperm ducts of the male, retained their original entrances into the *urethra* rather than into the bladder, proper. Hence, they enter a short distance below the bladder (Figs. X-48B and 49B).

When you open the bladder in a dissected cadaver to explore the **trigone area,** you will see that the ureters open superolaterally at the trigone region. The urethra leaves the region inferiorly. This trigone area is not of mesodermal origin, as was once thought. Rather, the mesodermal duct tissue degenerates as the ureters are absorbed in the bladder wall so that the trigonal epithelium is of entodermal origin as is the rest of the bladder.

A review of the urogenital homologues is presented in Table X-3. The table is so arranged that the indifferent structures are in the central column while the male and female corresponding parts, arising from the indifferent fetal part, are contrasted in the right and left hand columns. The important persisting parts are underlined. If they are vestigial, they are placed in parentheses. If a homologous structures does not develop in one sex or the other, this fact is indicated by the word, "none."

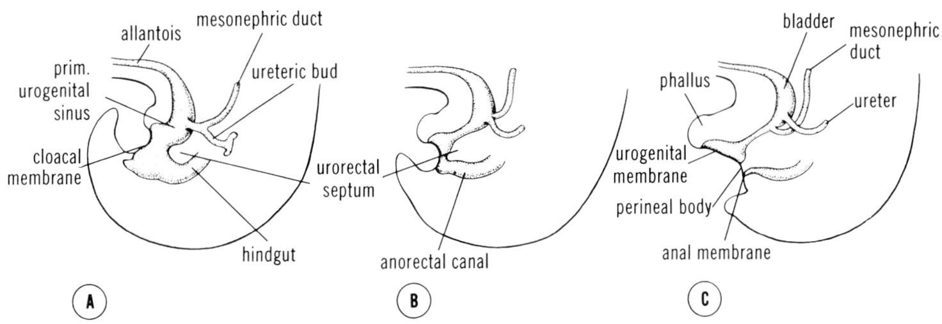

Fig. X-47. The division of cloaca into urogenital sinus and anorectal canal. **A.** At end of 5 weeks. **B.** At 7 weeks. **C.** At 8 weeks. (From J. Langman, *Medical Embryology* 2nd. ed., Williams & Wilkins, Baltimore, 1969.)

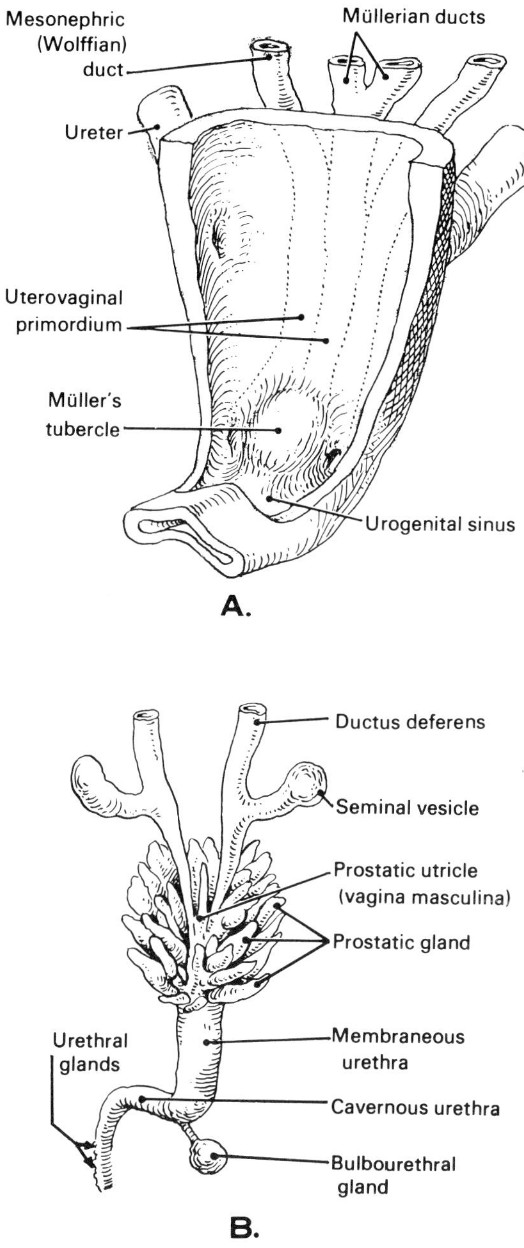

Fig. X-48. Three-dimensional constructs to illustrate differentiation of urogenital sinus. **A.** Anterior view in 9-week female fetus. **B.** Posterior view in 4-month male fetus.

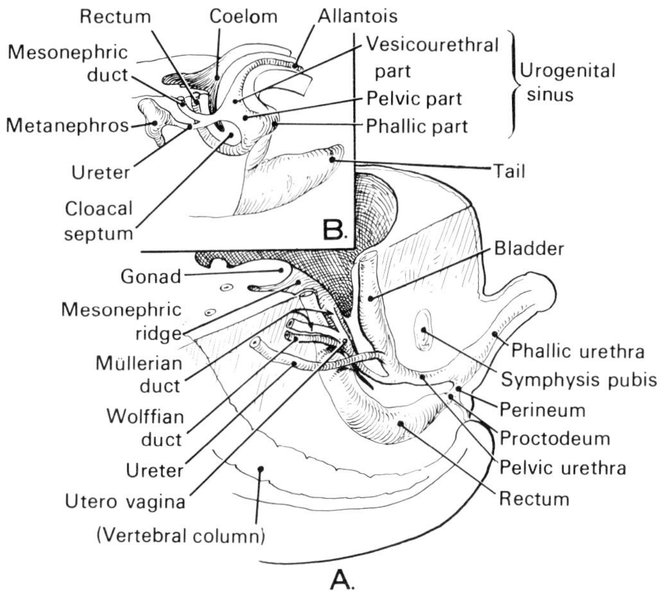

Fig. X-49. Division of the cloaca. **A,** late stage at 9 weeks; **B,** as an inset, an earlier stage at about 5½ weeks.

Table X-3. Urogenital Homologues

(Persisting vestigial parts are in parentheses; none = no counterpart)

Male	Indifferent Stage	Female
Testis	Gonad	Ovary
1. None		1. **Cortex**
2. **Seminiferous tubules**		2. **Primary medulla**
3. **Rete testis**		3. (Rete ovarii)
1. (Mesorchium)	Genital ligaments	1. **Mesovarium**
2. None		2. **Suspensory ligament of ovary**
3. (Ligamentum testis)		3. **Proper ligament of ovary**
4. (Gubernaculum—caudal part)		4. **Round ligament of uterus**
5. (Gubernaculum—as a whole)		5. None
6. None		6. **Broad ligament of uterus**
	Mesonephric collecting tubules	
1. **Efferent ductules;** (cranial aberrant ductule)	1. Cranial group	1. (Epoophoron; aberrant ductules)
2. (Paradidymis; caudal aberrant ductules)	2. Caudal group	2. (Paroophoron)
1. (Appendix epididymis)	Mesonephric (Wolffian) duct	1. (Vesicular appendage)
2. **Ductus epididymis**		2. (Duct of epoophoron)
3. **Ductus deferens** and **seminal vesicle**		3. } (Gartner's duct)
4. **Ejaculatory duct**		4. } (Gartner's duct)
5. **Ureter, renal pelvis** and **collecting tubules**		5. **Ureter, renal pelvis** and **collecting tubules**
1. (Appendix testis)	Müllerian duct	1. **Uterine tube**
2. None		2. **Uterus**
3. None		3. **Vagina**—upper part?
1. **Bladder**	Vesicourethral primordium	1. **Bladder**
2. **Upper prostatic urethra**		2. **Urethra**—**entire**
	Urogenital sinus	
1. **Lower prostatic urethra**	1. } Pelvic portion	1. **Vestibule**—nearest vagina
a. (Prostatic utricle, *i.e.*, vagina masculina)		a. **Vagina**—lower part
b. **Prostate gland**		b. **Urethral glands;** (para-urethral ducts)
2. **Membranous urethra**	2. } Pelvic portion	2. **Vestibule**—middle part
3. **Cavernous urethra**	3. Phallic portion	3. **Vestibule**—between labia minora
a. **Bulbo-urethral glands**		a. **Vestibular glands** of Bartholin
b. **Urethral glands** of Littre		b. **Lesser vestibular glands**
1. **Penis**	1. Phallus	1. **Clitoris**
a. **Glans penis**	a. Glans	a. **Glans clitoridis**
b. **Urethral surface of penis**	b. Lips of urogenital groove	b. **Labia minora**
c. **Corpora cavernosa penis**	c. } Shaft	c. **Corpora cavernosa clitoridis**
d. **Corpus cavernosum urethrae**	d. } Shaft	d. **Vestibular bulbs**
2. **Scrotum**	2. } Genital swellings	2. **Labia majora**
3. **Scrotal raphe**	3. } Genital swellings	3. **Posterior commissure**
4. None	4. Medial cranial swelling	4. **Mons pubis**

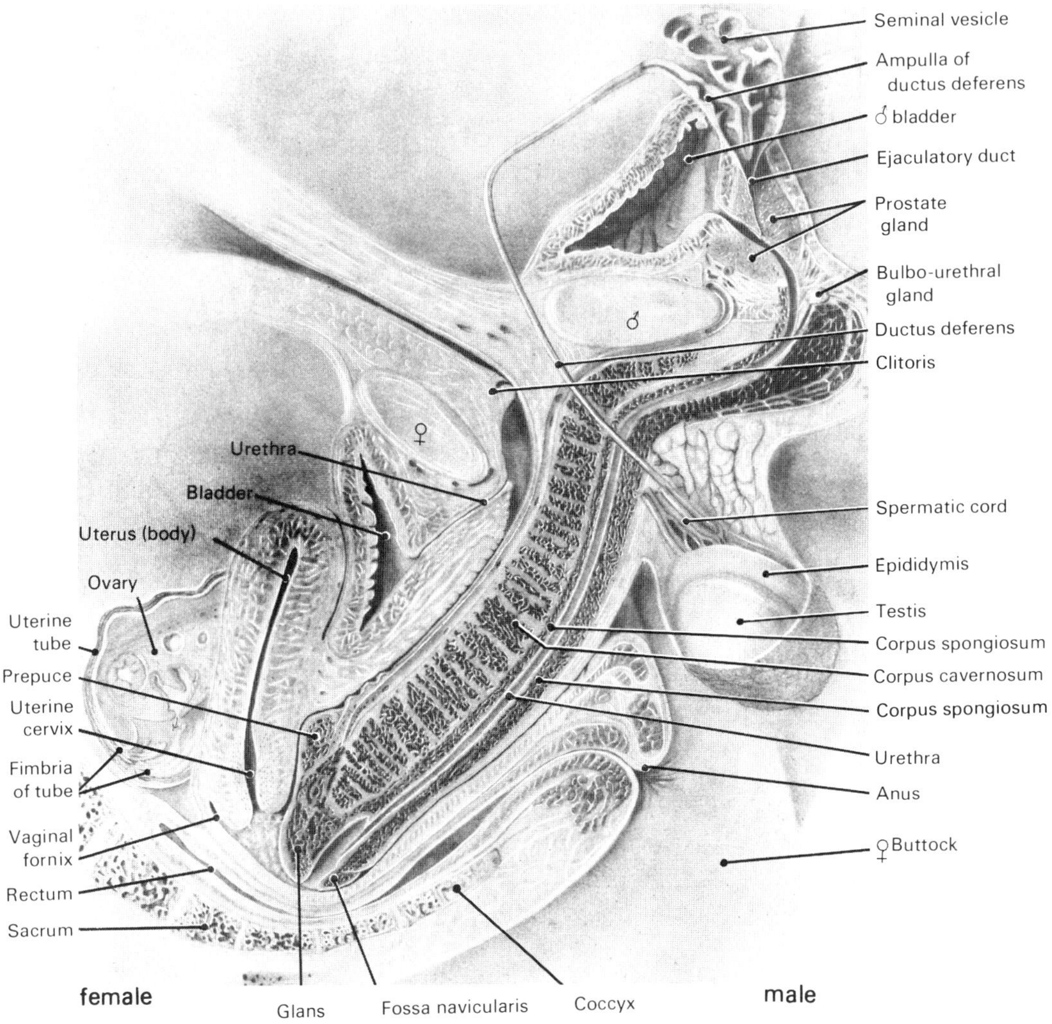

Fig. X-50. Midsagittal sections of male and female pelves showing structures of the urogenital systems that persist into adulthood. The respective pubic bones are marked ♀ and ♂. (From R. W. Dickinson, *Atlas of Human Sex Anatomy*, 2nd. ed. Williams & Wilkins, Baltimore, 1949.)

Figure X-50 provides an excellent comparative demonstration of those structures of the male and female urogenital systems that persist into adulthood.

OTHER PELVIC ORGANS

The Rectum

The rectum is a 5-inch segment of large bowel between sigmoid colon and the anal canal. Its name would signify that it is a straight segment but this is true in some lower forms only. It is not true in man. The rectum begins at the third segment of the sacrum, completely lacking in mesentery and taeniae. It is bent concave as it follows the slope of the sacrum to the urogenital diaphragm. At its caudal end, however, it turns sharply posteriorly and caudalward at the **perineal flexure.** This is caused by the sling placed around it by the Puborectalis muscle. It may dilate into an ampulla just caudal to this flexure and feces may impact at this site. As a rule, the rectum does not retain feces. Feces generally await evacuation in the caudal part of the sigmoid colon.

Peritoneum reflects off the rectum over the bladder in the male to form a slight **retrovesical pouch;** whereas in the female, the ventral peritoneal investment reaches the level of the vaginal fornices before ascending up over the bladder. Thus a **rectouterine** or **rectovaginal pouch** is formed. The peritoneum virtually surrounds the rectum completely at its cranial pole but, caudally, only the sides are covered. After the peritoneum reflects upon the bladder, the rectum is left without further peritoneal coverage.

Obviously, the middle sacral artery will lie behind the rectum. As the lower sacral nerves emerge from their anterior foramina they are in danger of becoming irritated and involved in cases of rectal carcinoma.

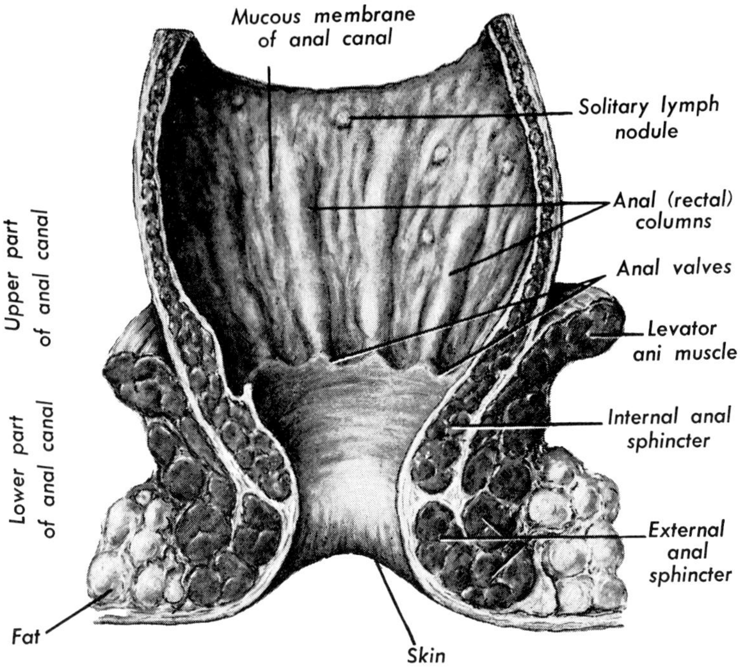

Fig. 51. Mucous surface of anal canal. Redrawn. (Courtesy of Wyeth, Philadelphia, Pa.) (From W. E. Copenhaver, *Bailey's Textbook of Histology,* 15th ed., Williams & Wilkins, Baltimore, 1964.)

Recall that Denonvillier's fascia (not peritoneum) separates the rectum, below bladder level, from prostate and seminal vesicles. To the sides, as always in the pelvic region, is the Levator Ani which will take part in anal sphincter formation.

The Anal Canal (Fig. X-51): This is the last inch and a half of the digestive tube leading to the exterior. The mucosa here is thrown up into eight magnificent columns, the columns of Morgagni. These have been celebrated since 1751 when Morgagni made the first trip through the alimentary canal in a kayak and photographed them in passing.[8] The ends of these folds join one another near the anal orifice. They create anal valves with anal sinuses between them. Above the valves, simple columnar epithelium gives way to a stratified cuboidal type; below, it is stratified squamous epithelium to the anal orifice. Here hair, sweat glands and some large apocrine glands make their appearance. The submucosa contains a wealth of hemorrhoidal vessels and the circular smooth muscle forms an internal sphincter. Outermost is the skeletal muscle sphincter derived from the Levator Ani.

The Urinary Bladder

This organ is the most anterior one in the pelvis. It would seem strange if it were not described as spherical, but this would only be true when it was full of urine. It holds about a pint, ordinarily, when it is not overdistended. But when it is empty, it is a four- or five-sided structure. Its superior surface lies beneath the peritoneum and it is at the oblique plane of the pelvic inlet just behind the pubic body. The *superior surface* rounds and pushes upward into the abdomen (false pelvis) only when the bladder fills. As it empties, the weight of the bowel presses down upon it to render this surface concave. There are two, sloping, *inferolateral sides* which rest on the pelvic diaphragm and which meet anteriorly. For this reason one often finds an *anterior surface* described. It is the sometimes slightly rounded *posterior surface,* not the superior one, which is called its **fundus.** Superior and lateral surfaces meet at an anterior peak which becomes continuous with the **middle umbilical ligament** on the ventral body wall. The ligament is what is left of the urachus. It is generally a non-patent peritoneal fold stuffed with a fibrotic cord. The superior surface, from vertex to fundus, is designated the **body** of the bladder. The region surrounding the urethral ostium is sometimes called the bladder *cervix* or neck.

The bladder does not rest directly against the pubic bone but is separated from it by the **prevesical** or **retropubic space** (Cave of Retzius). This consists of adipose tissue disposed between two layers of transversalis fascia. Laterally, there are paravesical spaces. In the male (Fig. X-37), we have already noted the posterior relations of the bladder to ductus deferens, seminal vesicles, and rectum. Note once more how the ureters reach its posterolateral corners by passing laterad of the ductus deferens. Note its relation again to the prostate. The urethral ostium of the bladder opens directly into the prostatic urethra.

In the female, the posterior surface of the bladder adheres to the cervix of the uterus and to the upper portion of the vagina. In both sexes some coils of the small intestine and part of the sigmoid colon rest upon its uppermost surface.

The bladder anchors in place by false and true "ligaments," neither of which are really ligamentous. The false ones are peritoneal folds over the bladder surface which run toward anterior, lateral, and posterior abdominal walls. The true ligaments are thickenings of endopelvic fascia. Two of these extend, in the male, from the anterior surface of the prostate to the pubis. In the female they come off the inferior aspect of the bladder itself. These are the **puboprostatic or pubovesical ligaments** and they are medially placed so that, in the male, the dorsal vein of the penis passes into the pelvis under the arcuate pubic ligament and passes between them on its way to the pelvic plexus of veins. A pair of **lateral true ligaments** reach each lateral pelvic wall from the side of the prostate as well. The sacrogenital folds of peritoneum also extend from the bladder base to posterior pelvic wall as anchoring structures. They contain some muscle fibers within them. In the female, these latter are the recto-uterine folds.

The renal pelvis, ureteric mucosa and the internal surface of the bladder alike are surfaced by a distensible **transitional epithelium** whose surface cells bulge in folds when the cavities of these organs are empty but which flatten as they distend. A darker colored triangular region exists internally at the bladder base. At this **trigone** there is much less mucosal folding and a transverse ridge extends between the openings of the two ureters. A median elevation extends toward the urethral ostium below.

There is a trilaminar arrangement of smooth muscle in the bladder wall. The outermost layer is longitudinally disposed and forms a loop at the bladder outlet. This is the layer designated as the **Detrussor muscle.** The middle layer is circularly arranged and is thicker than the outer one. The inner layer is best developed in the trigone region where it tends to be longitudinal. It does form a kind of sphincteric loop about the urethral orifice, however. Along with the loop formed by the outer layer, whose longitudinal fibers pull upward and caudalward upon contraction, an effective internal sphincter is formed at the bladder neck.

The main vessels to the bladder are the **superior vesical arteries** from the umbilical artery and the **middle and inferior vesical arteries** directly from the internal iliacs (hypogastrics). It gets additional twigs from vessels which supply rectum, uterus, and vagina.

While both sympathetic and parasympathetic fibers supply this organ, as they do most all viscera, functionally, micturition is under parasympathetic control. The sympathetics seem mainly involved in regulating the vascular supply of the bladder and in closing its internal sphincter during sexual excitement.

The Phenomenon of Micturition: Distention stimulates sensory stretch receptors within the bladder wall. These messages are relayed to the sacral center for reflex bladder contraction in the second to fourth sacral segments of the cord *via* the nervi erigens. Only babies and certain neurologically inflicted individuals wet their pants when these messages arrive at cord level. Most of us have been trained to exert cerebral inhibitory control over these cord centers until more propitious surroundings allow us to obey these signals of frantic distention. We undertake an ambulatory search for cover or for the rewarding symbols "Men" or "Ladies." Thus, we exert voluntary control through somatic motor nerves (pudendal) to the skeletal muscle of the **external urethral sphincter** which remains contracted until a proper haven for release has been found. It is then that we allow the parasympathetic motor reflex mechanism full play. Detrussor muscles contract and internal and external sphincters open under the head of pressure. When the bladder is empty, the external sphincter is contracted under voluntary control; but we may continue to milk the urethra of that one last drop by contracting the bulbocavernosus muscles. Urination can be started in a partially full bladder by exerting intra-abdominal pressure in order to increase intravesical pressure, but contracting of abdominal muscle alone is not responsible for bladder emptying. Urine has to be forced into the urethra by events which normally stimulate the sensory receptors within the bladder wall and within the urethra as urine begins to flow through it. The parasympathetic viscerovisceral cord reflex has to be triggered over its normal route.

The diseased bladder can be made to empty more or less frequently or to contain more or less urine than normal. Frequency of urination increases with inflammatory conditons. Bladder walls become permanently contracted in tuberculosis so that the organ holds less. Obstructive diseases (benign tumors of the prostate) cause strictures and double or triple the bladder's retentive capacity. Certainly neurological lesions of the cord interfere with bladder functioning. Destruction of visceral motor centers at S_2 to S_4 eliminates action of the viscerovisceral reflex so that, if any contraction is to occur at all, it will depend upon the isolated effort of the Detrussor muscle itself. If the cord is severed above S_2, voluntary control over the external sphincter is lost and we have an **automatic or cord bladder** which empties when it is full, a kind of infantile reversion. Destruction of sacral rootlets removes the sensory limb of the viscerovisceral reflex arc. With no sensations of awareness of distention able to register in consciousness, the bladder tends to overdistend while it awaits its central inhibitory message for release.

Having considered the pelvis and perineum as bodily regions and having accounted for the pertinent structures in these areas, we are left to make a quick survey of the blood and nerve supply, particularly that to the pelvic organs. We have already mentioned this in part so that it becomes a matter of tying major trunk lines together.

Blood Supply to Pelvis

Four arteries deliver blood to the pelvis:

1. The **internal iliac,** or **hypogastric** which arises from the common iliac;
2. the **ovarian** and
3. the **middle sacral.** Both of these latter are direct branches of the aorta;
4. and, the **superior hemorrhoidal,** a branch of the inferior mesenteric.

These four vessels and the branches of the large internal iliac trunk are shown in Figure X-52.

The **ovarian arteries** are two vessels of small diameter which originate from the front of the aorta just distal to the renal arteries. They descend over the pelvic brim lateral to the ureters to enter the vascular leash of the suspensory ligament of the ovary (infundibulopelvic ligament). We have seen them lying between the layers of the broad ligament, forming a vascular arch from which ovary, tube, and uterine fundus derive a blood supply (Fig. X-31).

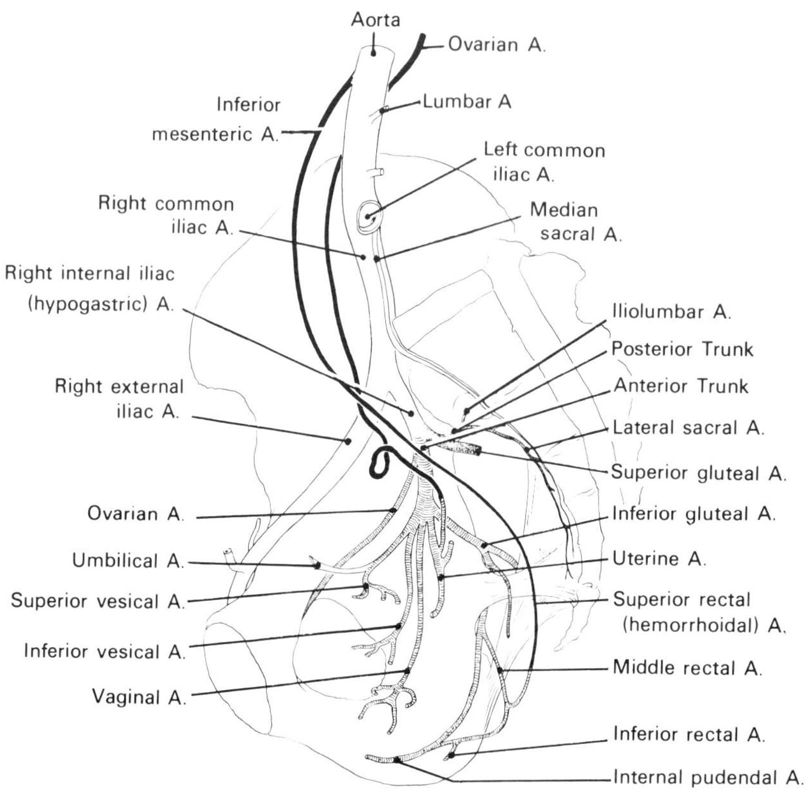

Fig. X-52. Blood supply of the female pelvis. Black vessels indicate abdominal sources of supply to pelvis; those that are stippled form the posterior trunk of the local pelvic supply from the internal iliac (hypogastric) artery; the striped vessels form the anterior trunk of local pelvic supply.

The main pelvic supply is, of course, the much larger **hypogastric artery.** It not only supplies pelvic viscera, but the pelvic walls, buttock, external genitalia, and medial aspect of the thigh. This is quite a task for a 2-inch vascular spigot which springs from the common iliac at the lumbosacral joint. It supplies the pelvic viscera mainly from its anterior trunk. Its posterior trunk is primarily concerned with pelvic wall structures.

The **uterine artery** is often the largest branch of the anterior division of the hypogastric. We have noted its course in Figure X-31; but, perhaps, did not realize that it serviced all of the organs in the female pelvis except the rectum. It forms a loop with the ovarian artery after it supplies part of the vagina, cervix, and body of the uterus. It often gives a **middle vesical artery** to the bladder; has tubal and ovarian branches, and sends a supply to the round ligament of the uterus. The **upper vaginal arteries** may arise from the cervical branch of the uterine artery or from the main stem of the uterine artery. **Lower vaginal branches** come from another source, most often from the **inferior vesical branch of the hypogastric.**

All of the **arteries to the bladder** arise **indirectly** from the anterior division of the hypogastric. We have seen the **middle vesical artery** come indirectly from the uterine artery. A **vesicle branch** to the superior aspect of the bladder also arises indirectly from the umbilical artery, a branch of the hypogastric. The distal ends of the umbilical arteries obliterate in the ventral body wall as the paired lateral umbilical ligaments. It is possible for the middle vesical artery to be a branch of either the umbilical or of the superior vesical arteries. The only vesical branch to come directly from the hypogastric is the **inferior vesical artery;** and, as you might suspect, it is directed toward the bladder base.

Three vessels supply the **rectum:**

1. a superior rectal (hemorrhoidal) artery which is a terminal branch of the inferior mesenteric artery (Fig. X-52);
2. a middle rectal which is a direct branch of the anterior division of the hypogastric (Fig. X-52);
3. and the inferior rectal in the perineal region as a branch of the internal pudendal as we have seen (Figs. X-9 to 11).

The **parietal branches to the pelvic wall** are also three in number and all come from the **posterior division** of the hypogastric artery (Fig. X-52):

1. the **iliolumbar,** as the most proximal branch of the posterior division, supplying lumbar and iliac twigs to the walls;

2. the **lateral sacral artery;**
3. and the largest branch of the division, the **superior gluteal artery.**

The lateral sacral arteries may be paired or may arise as a single trunk. Either way, the vessels split into superior and posterior divisions. The superior division enters the first anterior sacral foramen; the inferior one, the second to fourth anterior foramina after they cross the Piriformis muscle on the way. The posterior division may anastomose with the middle sacral artery.

The superior gluteal artery leaves the upper part of the greater sciatic notch to pass behind the Piriformis muscle after it passes between the first sacral nerve and the lumbosacral trunk.

An inferior gluteal artery is generally a branch of the **anterior division of the hypogastric** which also leaves the greater sciatic foramen, but lower down between Piriformis and Coccygeus. It is important since the **internal pudendal artery** may spring from it to leave the pelvis by the same route, only to return to the pelvis *via* the lesser sciatic foramen. We have seen this latter artery supply the inferior rectal artery to the pelvis while it, itself, makes its way to the perineum and external genitalia.

Nerve Supply to the Pelvis

Pelvic nerves are concerned partly with pelvic contents and partly with the innervation of the lower extremity.

Parasympathetic nerves to pelvic viscera enter as the nervi erigens or pelvic nerve. They distribute with the second to fourth sacral nerves and mingle in the pelvic plexuses. Such plexuses consist of parasympathetic preganglionic fibers; sympathetic postganglionic fibers; and visceral sensory fibers which arise from sensory receptors within the viscera. The plexuses form a skein-like entanglement around the pelvic organs. Parasympathetic postganglionics arise from ganglion cells *within* the organ walls as a rule.

Sympathetic contributions to the plexuses arise as postganglionic fibers from the inferior mesenteric ganglion of the abdomen. Their preganglionic fibers lay at cord levels L_1 and L_2 and reached the ganglion as the lumbar splanchnic nerves. Some fibers may have followed the sympathetic chain into the plexus area, but this is not the rule. Postganglionic sympathetics prefer arterial routes in the abdomen and head, so most enter plexuses along the aorta to leave at its termination as the hypogastric plexus. The latter plexus crosses the pelvic brim to enter the pelvis along vessels as right and left pelvic plexuses. Even so, the pelvic plexuses distribute as an anterior division to bladder and reproductive organs and a posterior division to the rectum.

The **ganglionic sympathetic trunk** also enters the pelvis on each side of the sacrum anteriorly. We have indicated that some small contribution might come from these paired paravertebral chain ganglia which meet, incidentally, in the midline at the coccyx in a single **ganglion impar.** But the main task of the lower chain ganglia is to supply the lower extremity through **gray rami** communicates to the lower lumbar and sacral spinal nerves (there are no **white rami** below L_2).

In addition to parts of the **lumbar** and **sacral plexuses** in the pelvis, there are the **coccygeal** and the **pudendal plexuses.** Now we must be careful not to confuse plexuses formed by the **anterior divisions of spinal nerves** (which these are) with **autonomic plexuses** on the organs of the pelvis, which the hypogastric and pelvic plexuses represent. Because spinal nerves receive visceral motor postganglionic components to be distributed to peripheral skin does not make spinal nerves parts of autonomic plexuses. Students often intermix the two kinds of plexuses in their minds by failing to realize that spinal nerves are merely being used as **visceral motor conveyors** to peripheral parts (something in the manner in which the cranial nerve V gets used in the head). Spinal nerves (except S_2–S_4) do not supply internal viscera. Neither does the cranial nerve V.

So now we find parts of spinal nerve plexuses in the pelvis also. Only one branch of the lumbar plexus invades the pelvis as the **obturator nerve.** This is only natural since the nerve has to exit from the obturator canal and obturator foramen which are located in the pelvic region. The nerve travels with the obturator artery and vein; and like all structures in **the pelvis,** these lie *beneath* the peritoneum of the pelvic wall. How else could they leave foramina without piercing the inner parietal covering? Nothing ever pierces body cavity linings except at diaphragmatic partitions between cavities (between thorax and neck, the mediastium is a thick midline partition). Again, the point needs to be made, since we all use terms so loosely at times—*nothing* is *in* the pelvic cavity, neither bladder, rectum nor any organ; neither vessels nor nerves. All these structures are in the body wall in the pelvic region, except that some protrude more obviously into the cavity *behind* their **visceral** coverings. When we say nothing is in the pelvic cavity, we except a bit of serous fluid; and in the female, occasionally some air coming in *via* vagina, uterus, and tube. In disease states, we consider other matters.

All the branches of the sacral, coccygeal, and pudendal plexuses are found in the pelvic region, running through the pelvis, so to speak. The **sacral plexus** is formed by the first lumbar and the first four sacral spinal nerves. The **pudendal plexus** is formed by parts of the second to fourth sacral nerves. The much less significant **coccygeal plexus** is formed by parts of the fourth and fifth sacral and the coccygeal nerves. What are all these plexuses doing here? They are forming named nerves to be distributed, chiefly, outside the pelvis to the gluteal region, the leg, the perineum and to the neighborhood of the buttock. They will use the greater sciatic foramen to leave the pelvis; some will sneak back in the lesser sciatic foramen to reach the perineum; others will pierce through the sacrotuberous ligament to emerge from the pelvis.

Eight named nerves will form from the sacral plexus:

1. the nerve to Quadratus Femoris and Inferior Gemellus which enters the gluteal region *via* the greater sciatic foramen in company with the internal pudendal artery;
2. the nerve to the Obturator Internus and Superior Gemellus which also accompanies the internal pudendal artery in its exit;
3. the superior gluteal nerve;
4. the inferior gluteal nerve, these latter two to reach gluteal muscles and the Tensor Fasciae Latae;
5. the common peroneal (note spelling);
6. the tibial nerve, these latter two to supply the lower extremity;
7. a sensory nerve, the posterior femoral cutaneous;
8. and, finally, two twigs which may or may not leave the pelvis in delivering their supply to the Piriformis muscle.

The **pudendal plexus** gives rise to a **perineal nerve** as a motor branch to the Levator Ani and Coccygeus muscle and to the out-again, in-again, **pudendal nerve** whose course in Alcock's canal and whose perineal distribution we have considered previously.

The **coccygeal plexus** has small branches which combine as the anococcygeal nerve, pierce the sacrotuberous ligament and supply skin in the coccygeal region.

Thus, there are autonomic plexuses and plexuses formed by anterior divisions of the spinal nerves in the pelvis. In respect to a supply to pelvic organs, it is the automonic plexuses in which we are interested.

We have mentioned the **nervi erigens** as the **parasympathetic component** of intermingled nerve plexuses upon pelvic viscera and the **hypogastric plexus** as the main routing of a **sympathetic supply** to pelvic organs. Sometimes you will see the term, "presacral nerve" used for the hypogastric plexus. It is neither a nerve nor is it necessarily presacral in position. It is a direct continuation of the aortic plexus into the pelvis and it follows the bifurcation of the internal iliacs (hypogastric arteries) into the pelvis in plexiform strands. It is composed of nerve fibers and some scattered ganglion cells. Its fiber components are:

1. postganglionic sympathetics coming *down* from the inferior mesenteric ganglia located at the root of the inferior mesenteric artery, the last of the three unpaired arteries of the aorta.

2. preganglionic sympathetic fibers from **lower lumbar chain ganglia** which will end in the scattered ganglia of the plexus itself.

3. postganglionic sympathetics from these local ganglia of the plexus.

4. preganglionic fibers of the nervi erigens (parasympathetic) reaching *up* to sigmoid colon, descending colon and to about the middle of the transverse colon to complete the parasympathetic supply of the digestive tube.

5. There may be some sensory visceral fibers in it.

The Pelvic Autonomic Plexuses

A right and a left plexus embed in the loose connective tissue between the visceral organs and the lateral pelvic wall. These form an irregularly disposed network of fibers of the same kinds listed under the hypogastric plexus above and with which they are continuous. There may even be a few isolated, scatered, **sympathetic ganglion cells** as **collateral ganglia** distributed this deep in the pelvic region. All of this plexiform neural tissue tends to follow along the vascular branchings of the hypogastric artery, using the vessels to reach the viscera. In the male, the pelvic plexuses flank the rectum, nearby parts of the bladder, and seminal vesicle. In the female, they lie posterior to the rectum at first and then extend forward beside the vagina. Extensions from these mixed parasympathetic and sympathetic pelvic plexuses reach various pelvic organs, and we mentioned the extension of a parasympathetic component upward toward parts of the colon where it must become a part of the hypogastric plexus in its course. This latter has to be so since parasympathetic contributions to the nervi erigens arise low in the pelvis from sacral cord segments S_2 to S_4.

The extensions of the pelvic plexuses to visceral organs tend to bear specific names. Just be sure you know that they are a mixture of parasympathetic and sympathetic fibers as outlined above. Thus we find:

1. a **rectal plexus,** accompanying vessels to the wall of the rectum and anal canal;
2. a **vesical plexus,** disposed around the wall of the bladder. Offshoots of this innervate smooth muscle and glands of the ureters, seminal vesicles, and vas deferens (as far as the epididymis, actually);
3. a **prostatic plexus** of considerable size which supplies seminal vesicle, in part, and prostatic and membranous urethra;
4. a **uterine plexus** which passes up in the broad ligament to service smooth muscle and glands of cervix, uterus and tubes;
5. a **vaginal plexus,** largely parasympathetic in nature, to vaginal glands and female urethra;
6. a **cavernous plexus,** probably an offshoot of the prostatic plexus in the male, to carry autonomic fibers to erectile tissues of the penis. There is a similar supply to the clitoris in the female. It is possible that the pudendal nerve also conveys some parasympathetic and, certainly, sympathetic fibers to penis and clitoris.

You must realize also that **visceral sensory fibers** emanating from receptor endings in pelvic organ walls return in these various plexiform routes to cell bodies in dorsal root ganglia. Those responsive to pain caused by over distension of organ walls (overly-full bladder; distension of rectum or anal canal in difficult evacuation) probably travel sympathetic routes and find their ganglia at lumbar cord levels. But proprioceptive stretch receptors associated with normal mechanisms of filling and emptying; *e.g.*, bladder contraction reflexes, travel the parasympathetic route to dorsal root ganglia S_2 to S_4.

As a rule, parasympathetic fibers in the pelvic region are concerned with (1) contraction of expulsion musculature (bladder and rectum) and (2) with inhibition of the smooth muscle sphincters of bladder and rectum. We have pointed out above that micturition is, fundamentally, a parasympathetic phenomenon and that opening of the internal sphincter is by a head of pressure upon bladder contraction. Keeping the internal sphincter closed is an involuntary sympathetic task. Closing the skeletal muscle, external sphincter is voluntary.

Sympathetic function in the pelvis is more related to regulating the walls of blood vessels, contracting seminal vesicles and male reproductive ducts, and to contracting the internal urethral sphincter, not during any phase of normal micturition, but during sexual excitement.

Opinions vary about the relative parasympathetic and sympathetic control of the uterus. The picture is more confused than that which concerns urinary bladder and rectum because so many hormonal influences are involved and because physiological changes in state during pregnancy and parturition are, rather, abnormal instead of normal events relative to the uterus. Perhaps, sympathetic activity produces uterine contraction and vasoconstriction; parasympathetic activity, uterine inhibition and vasoconstriction. We do know this, however; visceral sensory impulses from the cervix reach the sacral dorsal root ganglia *via* the pelvic sacral nerve (nervi erigens portion of pelvic plexus). Visceral sensory impulses from the rest of the uterus traverse sympathetic routes to lower thoracic and upper lumbar dorsal root ganglia.

REFERENCES

1. Smith, A.: *The Body,* Ch. 5. Walker & Co., New York, 1968.
2. Köhler, A., and Zimmer, E. A.: *Borderlands of the Normal and Early Pathologic in Skeletal Roentgenology.* Translated and edited by S. P. Wilk. Grune & Stratton, New York and London, 1968.
3. Eastman, N. J., and Hellman, L. M.: *Williams' Obstetrics.* Appleton-Century-Crofts, New York, 1966.
4. Colcher, A. E., and Sussman, W.: A practical technique for roentgen pelvimetry with a new positioning. Amer. J. Roentgen., *51:* 207, 1944.
5. Schneider, P.: The problem of the "tubal sphincter" of the intramural portion of the fallopian tube. Amer. J. Roentgen., *48:* 527, 1942.
6. Siegler, A. M.: *Hysterosalpingography.* Hoeber (Harper-Row), New York, 1967.
7. Gaddum, P., Blandau, R. J., and Hayashi, R.: *Sperm Maturation in the Male Reproductive Tract: Development of Motility.* Film division of the Univ. of Washington Press, Seattle, 1968.
8. Morgagni, G. B.: *Memoiri: Per Canalo Alimento cum Rodo et Reelo.* Columbo Publishers, Bologna, 1751.

ELEVEN

The Circulatory System

INTRODUCTION

We have talked about various parts of the circulatory system from time to time so that regional parts of it are scattered throughout the book. This has not allowed you to grasp an overall view of the system, and undoubtedly it is difficult for you to recall from whence certain important vessels arise. We shall try to put the system in order here, but, even so, certain topics such as the details on the coronary vessels, the unpaired branches of the abdominal aorta and the like are best picked up in the regional chapters. Thus, while we will mention all the important vessels in a systemic fashion, we will not bother to give the complete ramifications of some of them if they have been reasonably covered elsewhere in the text. One main purpose is to do a little better job on some of those that were left dangling in the breeze.

ARTERIAL SYSTEM

Divisions of the Aorta

After all, the aorta and its major branches, and their subsequent branchings, constitute the entire arterial supply of all parts of the body except for the lungs. It constitutes the **systemic circulation** emanating from the left heart as distinct from the **pulmonary circulation** emanating from the right heart. The best way to get a clear picture of the systemic circulation is to follow blood first into the head and then, in order, into the upper limb, chest, abdomen, pelvis and, finally, to the tips of the toes. It is well, of course, to follow it back to the heart again, thus, completing the round trip. But we will set the return route off in a special section called the venous system. It will be only fair to add accessory venous drainage under the category of the lymphatic system.

The primary artery of the body is the **aorta.** It is shaped like a tapering cane which arches backward to descend upon the bodies of the vertebrae from the level of VT_5 to VL_4. Here it terminates by dividing into its major trunks, the **common iliac vessels.** Actually, the pointed terminal end of the cane continues straight on as a narrowing **median sacral artery** upon the body of VL_5 and the midline of the sacrum, and, if we had a tail, it would continue into this balancing or prehensile organ. Its divisions and major branches can be outlined as follows (see Fig. XI-1A, B):

The Ascending Aorta: This short 5.0 cm. segment lies within the pericardium wrapped in a serous pericardial sheath along with the pulmonary trunk—its embryological conjoined twin of the right heart. It leaves the left ventricle at the level of the third intercostal space, and ascends, tilted ventralward and obliquely toward the right. It courses upward a short distance between the pulmonary trunk on its left side and the superior vena cava on its right. Its chief branches are the right and left coronary arteries (Fig. XI-1B).

The Aortic Arch (Fig. XI-1A, C): This arched handle to the cane begins where the aorta leaves the pericardium. It arches posteriorly and to the left to clear both the trachea and esophagus on their left side. The arch loops over the right pulmonary artery, and the left pulmonary veins and left main stem bronchus. It does not change its name to the descending aorta until it reaches VT_5 on the left or about the level from whence the aorta left the pericardium. Three major branches arise from the arch:

1) The innominate or brachiocephalic artery which gives off: a) right common carotid artery; b) right subclavian artery.
2) Left common carotid artery.
3) Left subclavian artery.

Descending Thoracic Aorta (Fig. XI-1A, B): From VT_5 this great trunk descends through the thorax upon and along the left anterior bodies and disks of the vertebral column. It progressively seeks the midline and certainly reaches it at VT_{12} where it leaves the thorax through the aortic hiatus in the diaphragm. Its chief branches are:

1) The intercostal arteries, most but not all of them.
2) The superior phrenic arteries.
3) Esophageal arteries.
4) Certain bronchial arteries.
5) Pericardial branches.
6) Mediastinal branches.

The Descending Abdominal Aorta (Fig. XI-1A, B):

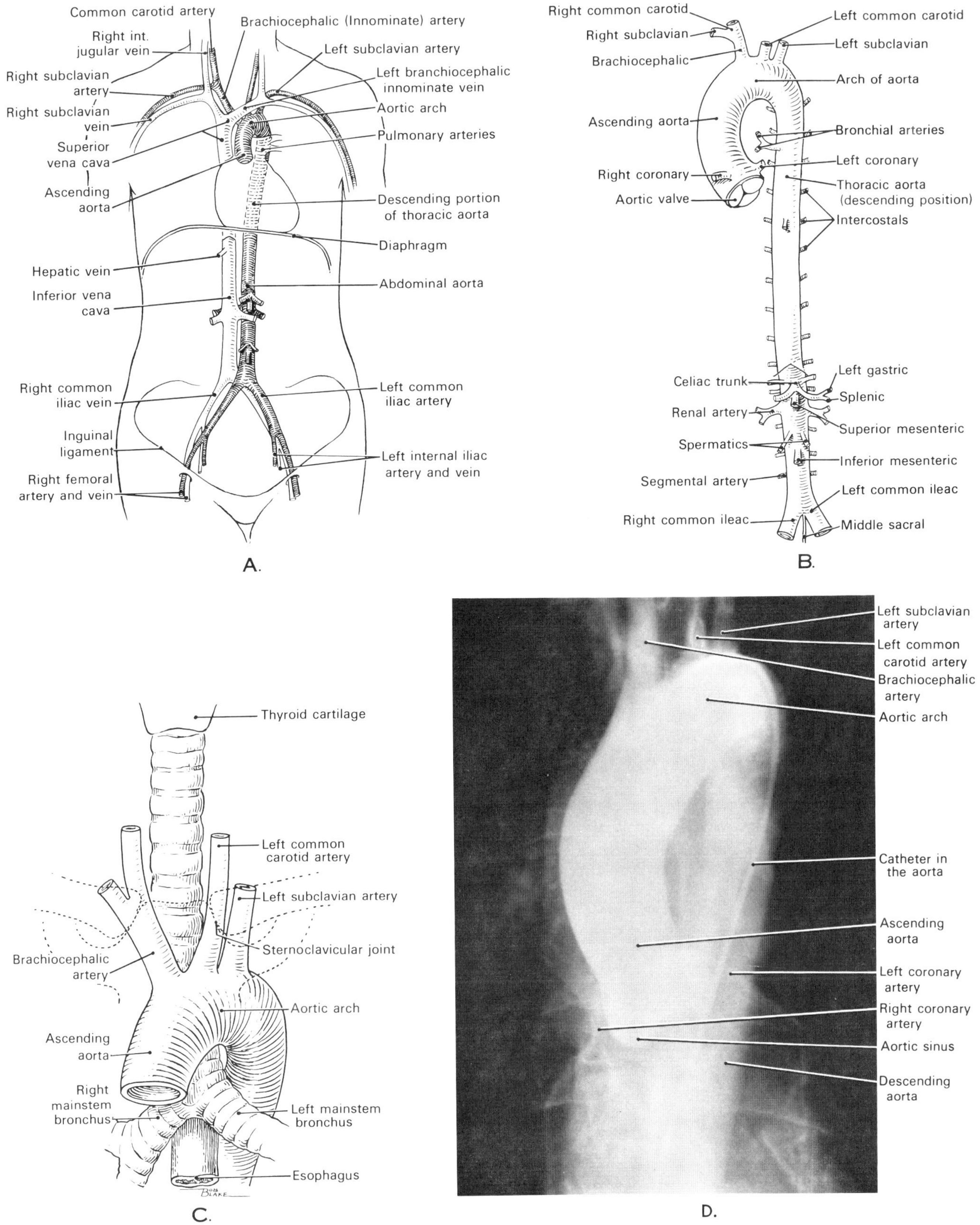

Fig. XI-1. The aorta. **A**. Diagram of the course of the aorta and greater vessels. **B**. A sketch of the chief branches of the aorta. **C**. The aortic arch. **D**. Retrograde thoracic aortogram, frontal projection.

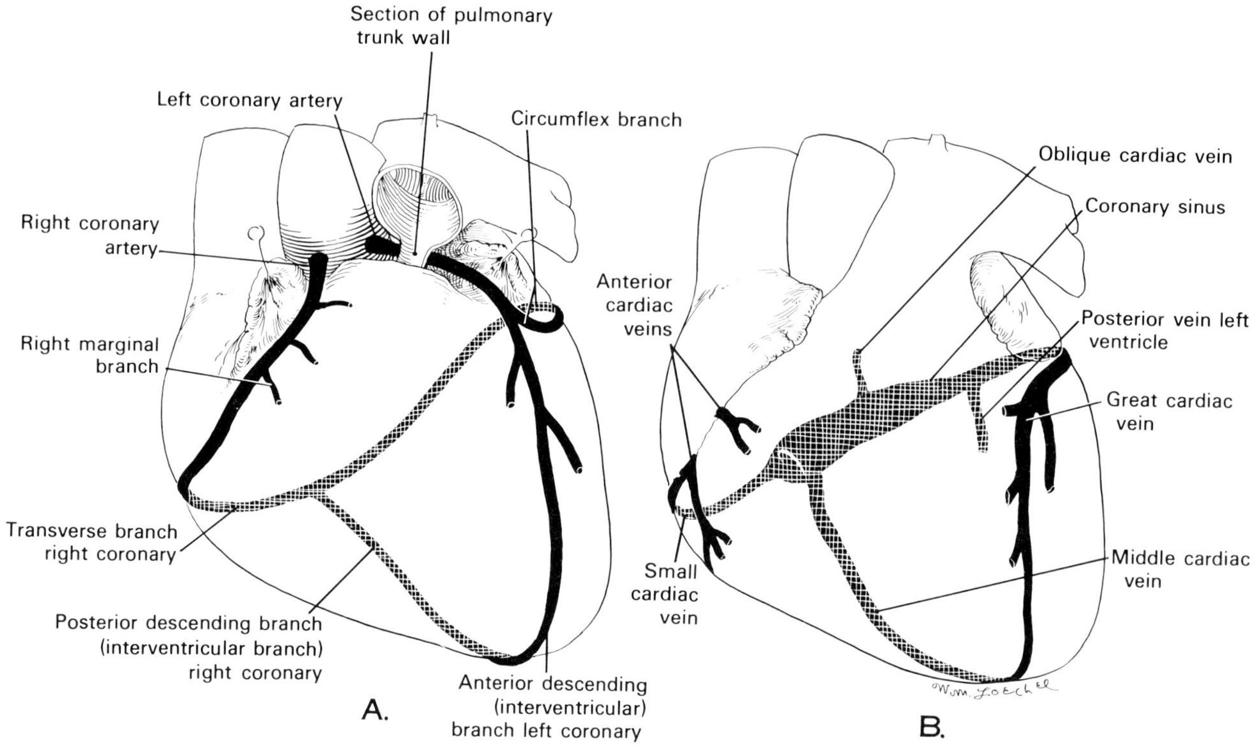

Fig. XI-2. Schema of usual disposition of (**A**) coronary arteries and (**B**) cardiac veins in the "transparent" heart. Anterior vessels, solid; posterior vessels, cross-hatched. In (**A**) a piece of the pulmonary trunk wall was left in place to show that the left coronary artery crosses behind that vessel.

This continuation of the main trunk-line below the diaphragm holds the midline upon the bodies of the first four lumbar vertebrae. The aorta has been decreasing in diameter along its course from its widest width of 3.0 cm. in its ascending segment to about 1.75 cm. at its point of bifurcation into the common iliacs at VL_4. This shortest segment, the descending abdominal aorta, gives off the most branches:

For G.I. tract and its partners	1. Coeliac artery 2. Superior mesenteric artery 3. Inferior mesenteric artery	Single, unpaired branches
For body wall	4. Inferior phrenic arteries	Paired branches
For G.U. tract and its partners	5. Suprarenal arteries 6. Renal arteries 7. Spermatic (ovarian) arteries	Paired branches
For body wall	8. Lumbar arteries	Paired branches
	9. Middle sacral artery—*unpaired terminal*	
For pelvis and lower extremities	10. Common iliac arteries—*paired terminals* a. Internal iliac arteries b. External iliac arteries	

We can now consider some of the sub-branchings of the aortic divisions and their major stems.

Ascending Aorta: The only branches we are concerned with here are the two coronary arteries arising from two of the three aortic sinuses. We have discussed them fully under the blood supply of the heart in the Chapter VIII on the thorax, page **492**. We present on the chart on the next page a simple outline of the major branchings only and request that you review this material in Chapter VIII (see also Fig. XI-2A–D).

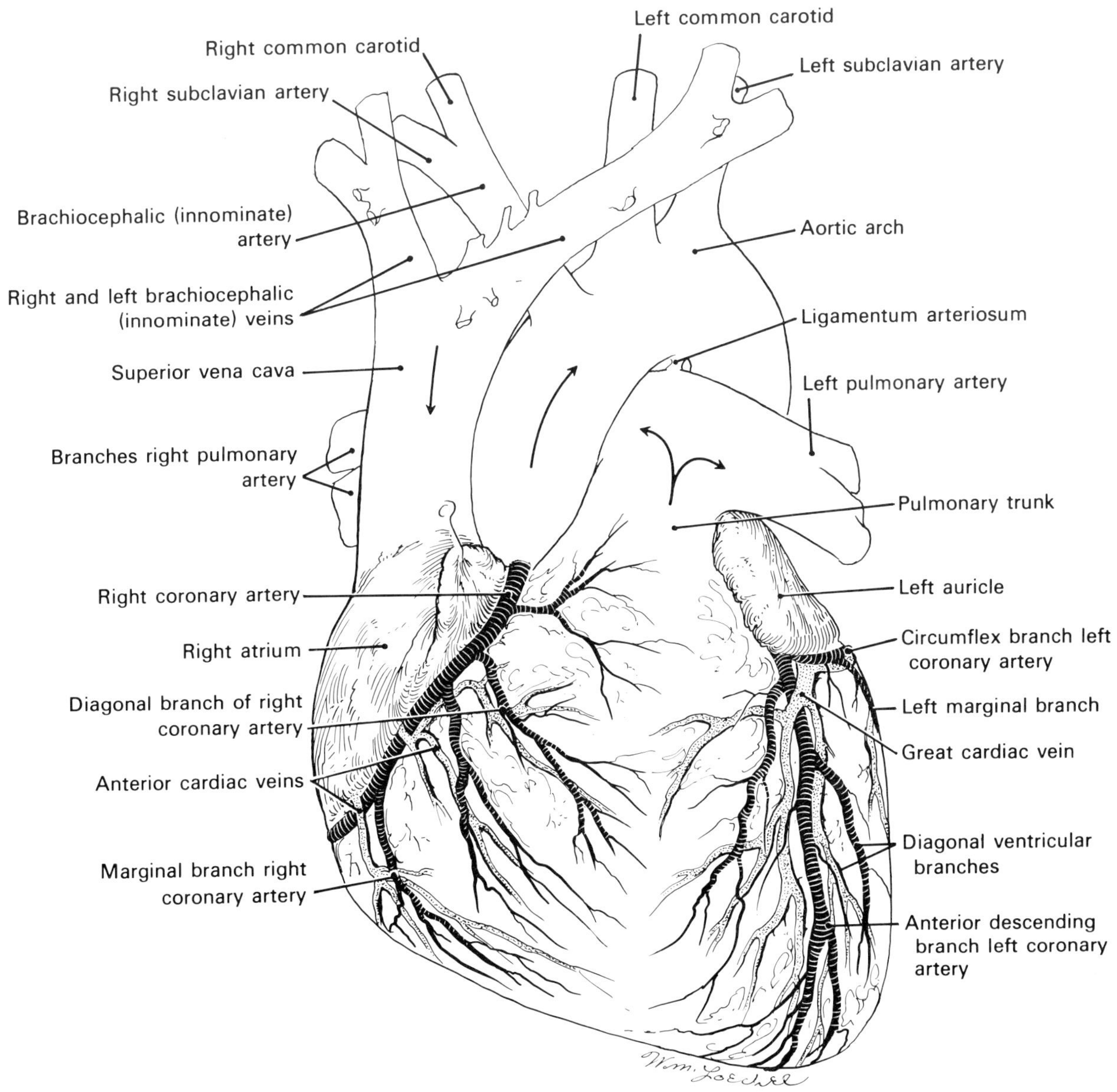

Fig. XI-2C. Anterior view of the heart to show the coronary arteries and cardiac veins in greater detail. Arteries are striped; veins, stippled.

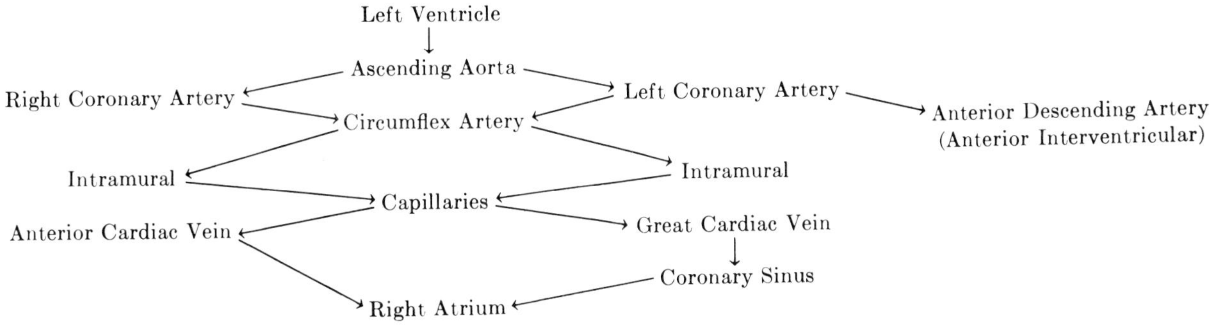

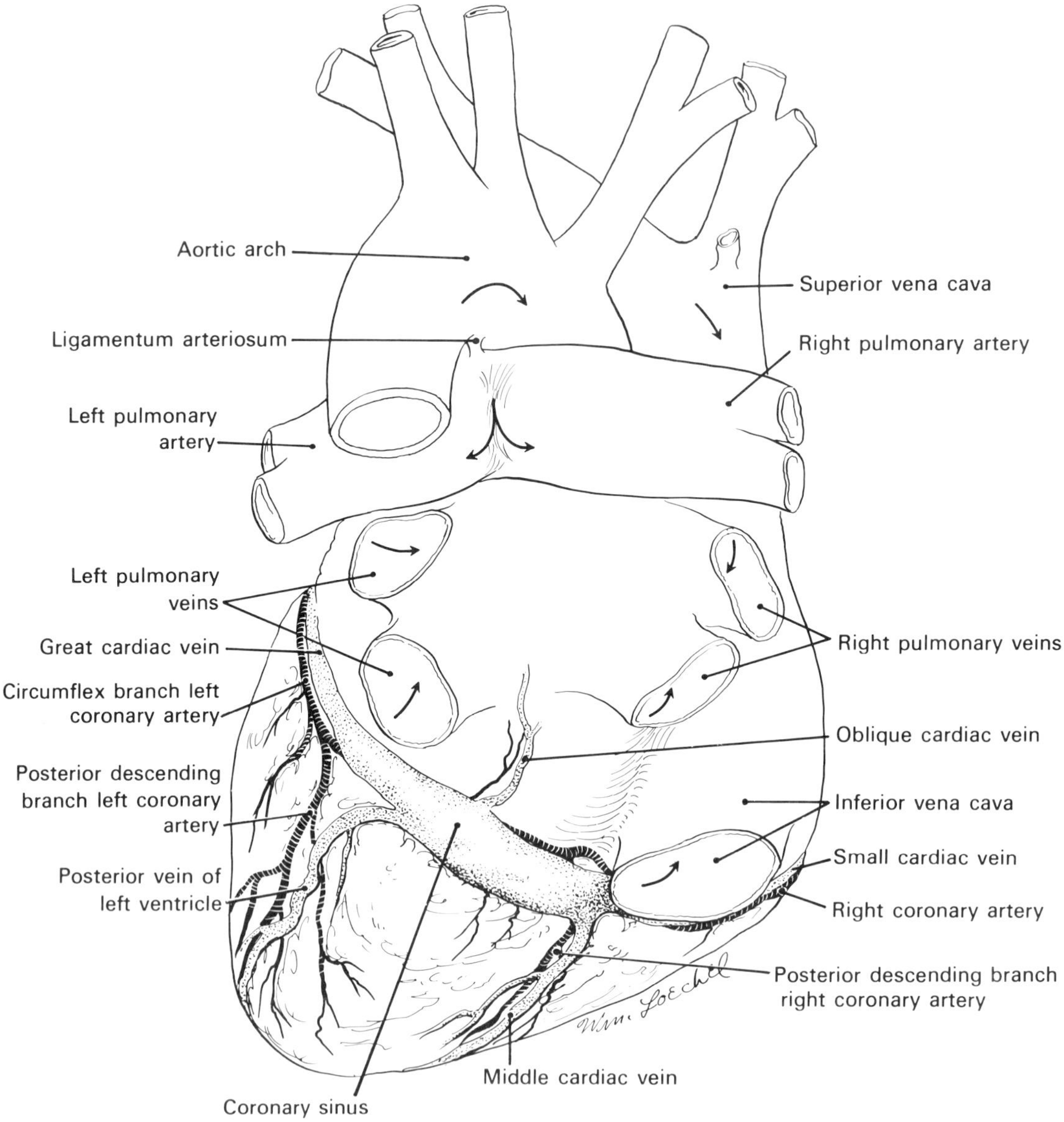

Fig. XI-2D. Posterior aspect of the heart showing the coronary vessels. Arteries are striped; veins, stippled.

The Aortic Arch: The **brachiocephalic** or **innominate artery,** while the largest in diameter of the three vessels to spring from the arch, is its first and shortest stem. It passes upward through the superior mediastinum behind and along the right side of the **left brachiocephalic vein.** Recall that there is only *one* brachiocephalic artery, but *two* brachiocephalic veins. The vessel lies ventral to the trachea. Behind the right sternoclavicular joint, the brachiocephalic artery bifurcates into its terminal vessels, the **right subclavian** and **right common carotid** arteries (Fig. XI-1C). It is easy to see why this once "no-name" artery has come to be called the brachiocephalic since its terminal branches supply the right upper arm or brachium and the right side of the head and neck.

The next two branches of the arch, in quick succession, are the **left common carotid** and the **left subclavian arteries** as separate stems (Fig. XI-1C).

These, two remain deep to the left brachiocephalic vein and ascend vertically to the left of the esophagus and trachea until they come to lie behind the left sternoclavicular joint. From here, they wend their separate ways as do their partners on the right side.

The Common Carotids and Their Branches: These vessels have no branches as they ascend vertically in the neck on either side of the trachea and esophagus until they reach the level of the thyroid cartilage. There they divide into **internal** and **external carotid arteries.** In their course they are bound in the heavy connective tissue of the carotid sheath along with the internal jugular vein (behind them) and the vagus nerve (between vein and artery). Once they emerge from under cover of the Omohyoids and the Sternomastoids, they are fairly superficially placed in the carotid triangle of the neck, and they have diverged somewhat. Where they give off their terminal branches, they have a slight dilatation in their walls, the **carotid sinus.** Here the wall contains specialized receptor organs that respond to changes in blood pressure and that alter the heart rate reflexly through connections within the central nervous system.

THE EXTERNAL CAROTID ARTERIES (FIGS. XI-3 AND 4): These vessels supply almost everything in the face and neck. They do not enter the cranial or orbital cavities nor do they quite take care of the lower neck, leaving that task for some of the early branches of the subclavian arteries. Since they lie superficial to the internal carotids, their pulse can be felt in the carotid triangle. They ascend upon the pharyngeal constrictors and, at the jaw angle, they pierce the parotid glands. Just below the zygomatic arch they give off their terminal branches (superficial temporal and maxillary arteries). But they give off several branches before this. The names of these branches and their subsequent branchings tell, in general, the area supplied so that further description beyond mere listing is hardly warranted.

1) **Superior Thyroid Artery** (Figs. XI-3 and 4). Its branches are: a) infrahyoid artery; b) sternomastoid artery; c) superior laryngeal artery; d) cricothyroid artery.

2) **Ascending Pharyngeal Artery** (Fig. XI-4). This is a deep vessel between internal carotid and the pharynx: a) pharyngeal arteries; b) inferior tympanic artery; c) meningeal arteries, *via* jugular foramen and foramen lacerum, supply the posterior aspects of dura.

3) **Lingual Artery** (Figs. XI-3 and 4). Its branches are: a) suprahyoid artery; b) dorsal lingual arteries which even reach soft palate and epiglottis; c) sublingual artery; d) deep lingual artery.

4) **Facial (External Maxillary) Artery** (Figs. XI-3 and 4). Obviously, it supplies facial musculature. Its other branches are: a) ascending palatine artery; b) tonsillar artery; c) submandibular artery; d) submental artery; e) superior and inferior labial arteries; f) lateral nasal artery; g) angular artery, a terminal branch along side of the nose to the medial angle of eye.

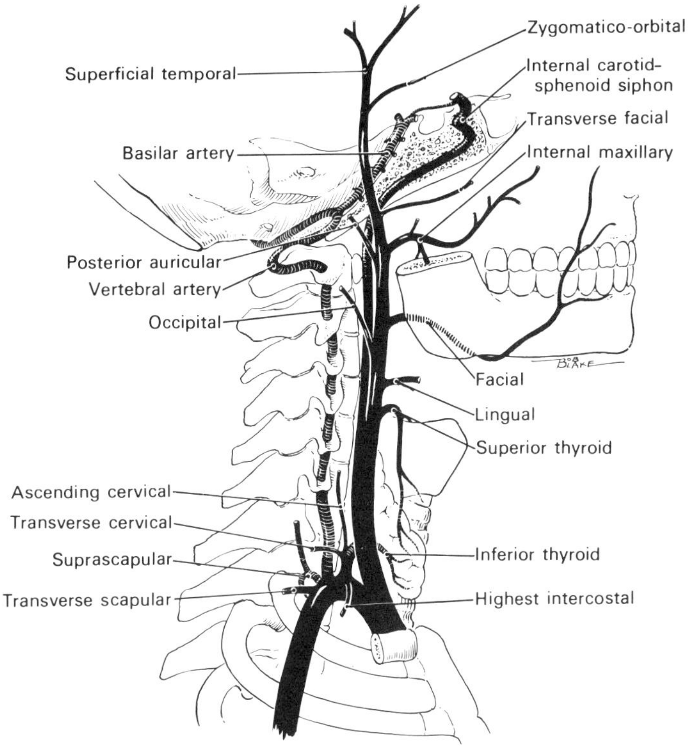

Fig. XI-3. Diagram of the subclavian, vertebral, and carotid arteries and their distribution to the head and neck.

Through the ophthalmic artery, it effects an anastomosis with the internal carotid.

5) **Occipital Artery** (Figs. XI-3 and 4). This vessel disappears under the upper portion of the Sternomastoid and mastoid process to emerge once more in the suboccipital triangle from whence it courses over the scalp. a) It gives twigs to the ear, mastoid air cells, and dura; b) it has short muscular branches to the Digastric, Stylohyoid, Sternomastoid, and to the Splenius and Longisimus Capitis.

6) **Posterior Auricular Artery** (Fig. XI-3). It supplies the auricular muscles as well as the Sternomastoid and the back of the scalp.

7) **The Superficial Temporal Artery** (Figs. XI-3 and 4) is a mainline, terminal branch heading north in front of the ear to spread out in a profusion of branches below, over, and above the zygomatic arch. It supplies parotid, zygomatic, and temporal regions, and even the Orbicularis Oculi.

8) **The Maxillary (Internal Maxillary) Artery** (Fig. XI-3) is the larger of the two terminal branches of the external carotid. Unfortunately, it quickly disappears from sight at the mandibular neck, heading for the infratemporal fossa. The Lateral Pterygoid, which it crosses, divides the vessel into three regional parts called, simply, Part I, Part II, and Part III. Very conveniently, the first two parts run with the branches of V_3 (mandibular division of the trigeminal) and the last with V_2 (maxillary division of the trigeminal nerve).

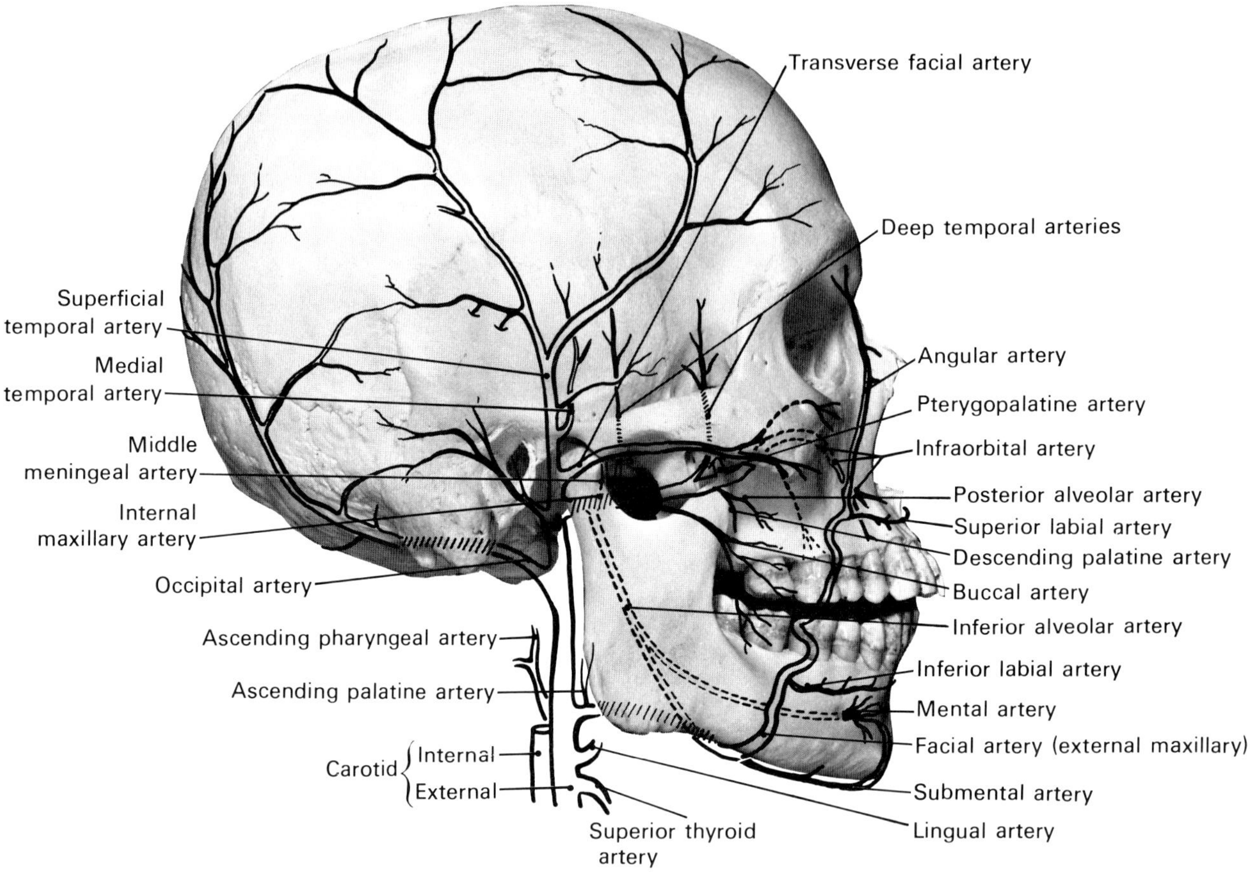

Fig. XI-4. Arteries of the face and scalp.

a) Part I courses from the parent stem to the posterior border of the Lateral Pterygoid. The main task of this portion of the vessel is to supply: temporomandibular joint, external auditory meatus, outer surface of the tympanic membrane, lower teeth (**inferior alveolar or dental artery**), and the important **middle meningeal artery** which ascends through the foramen ovale to supply the dura and trigeminal ganglion.

b) Part II, the part lateral to the Lateral Pterygoid, is essentially concerned with muscles: the two Pterygoids (Medial and Lateral), the Temporalis and to the Masseter, all muscles of mastication supplied by V_3. To make matters confusing, it gives off a **buccal branch** to the Buccinator, a facial muscle in the area which is normally supplied by cranial nerve VII. It is not always easy to keep nerve branches and vascular branches straight. You have to make a little effort.

c) Part III consists of a short stretch in the pterygopalatine fossa where it breaks up into several vessels rapidly and terminates as the sphenopalatine artery. It supplies small arteries to: the upper teeth, the palatine canals, an infraorbital twig, a small artery to the vaginopharyngeal canal, an artery of the pterygoid canal, and branches to the conchae, septum, and paranasal sinuses.

THE INTERNAL CAROTID ARTERY (FIG. XI-3): This large vessel is devoid of branches in the neck. Its main task is to supply all the brain (except occipital lobe and inferior portion of the temporal lobe) and to supply the orbital cavity. Its branches in the cranial cavity are:

1) Caroticotympanic artery.
2) Ophthalmic artery.
3) Posterior communicating artery.
4) Choroid artery to lateral ventricle.
5) Middle cerebral artery.
6) Anterior cerebral artery.

The internal carotid ascends through the pharynx in the carotid sheath along with the internal jugular vein and vagus nerve. Alone, it enters the carotid canal in the petrous part of the temporal bone. It snakes through the bone to its termination by crossing the foramen lacerum and making a hairpin turn along the medial side of the anterior clinoid process. After this arduous passage through "fat-man's squeeze" in hard bone, it is quite willing to give off its terminal cerebral arteries. But it gives off two prior branches in the cranium:

The **caroticotympanic branch** leaves through a small hole in the posterior wall of the carotid canal to supply the tympanic region of the middle ear cavity.

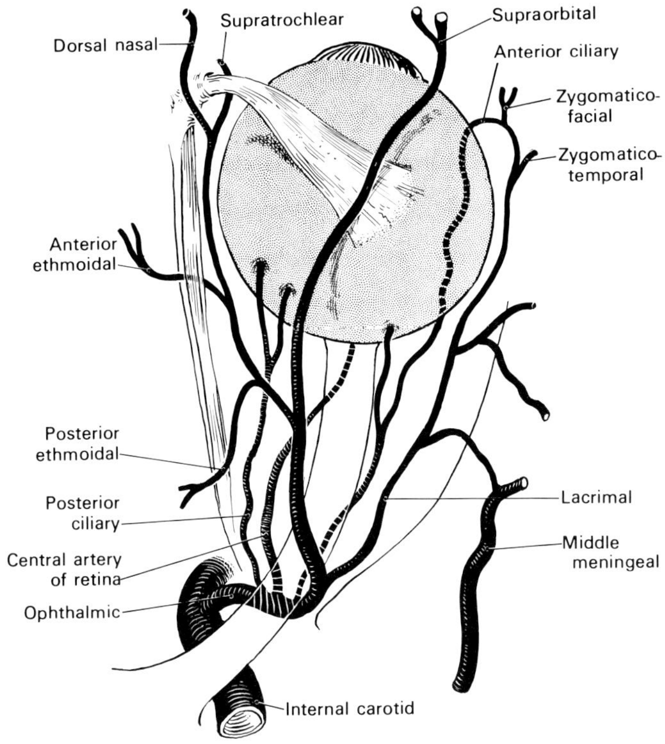

Fig. XI-5. Branches of the ophthalmic artery.

Its other branch is the *busy* **ophthalmic artery** which arises immediately below the anterior clinoid process just as the internal carotid first pierces the dura mater. Recall that the ophthalmic artery enters the royal gate or optic canal to reach the structures in the orbital cavity. Recall, too, its most important branch, is the **central artery to the retina** which enters the optic nerve. Besides this, the ophthalmic artery has its hands full doing a host of jobs through its following branches: a) lacrimal artery to tear gland, generally its first large branch, b) middle meningeal recurrent branch to dura, c) zygomatic branches to cheek and temple through zygomaticotemporal and zygomaticofacial foramina, d) supraorbital branch to scalp, e) frontal branch—sometimes as an extra branch to scalp, f) palpebral branches to upper and lower lids and conjunctiva, g) diploic branches, h) periosteal branches to orbital roof, i) muscular branches to extraocular muscles, j) trochlear branch to pulley of Superior Oblique, k) ciliary branches—short, long, and anterior to eyeball, l) posterior ethmoidal branch to ethmoid air cells, m) meningeal branches to cribriform plate, n) anterior ethmoidal branches to air cells, o) branch to frontal sinus, p) nasal branches to middle meatus, and, q) cutaneous external nasal branch. We do not make this list for you to memorize, but to make you aware of the heroic efforts of the ophthalmic artery (Fig. XI-5).

Sometimes the internal carotid supplies a **posterior communicating branch** to the circle of Willis (see below). Mostly this vessel comes from the middle cerebral (Figs. XI-6 and 7).

As a rule the internal carotid gives off the **choroid artery,** not to the choroid coat of the eyeball, but to the **choroid plexus** of the **lateral ventricle** in the **cerebral hemisphere,** an important vessel in the formation of cerebrospinal fluid (Figs. XI-6 and 7). Note the following vessels in a lateral view of a carotid arteriogram: internal carotid artery; ophthalamic artery; posterior communicating artery; anterior choroidal artery, and the anterior cerebral artery (Fig. XI-7A, B).

The caroticotympanic branch, the ophthalmic artery, the posterior communicating branch and the choroid artery all leave before the internal carotid breaks into its terminal branches:

The middle cerebral artery (Figs. XI-6 to 8), supplies the dorsolateral surfaces of frontal, parietal, and temporal lobes. It also supplies the **basal ganglia,** those nuclear gray masses of nerve cells deep within the cerebral hemispheres.

The anterior cerebral artery (Figs. XI-6 to 8), supplies the medial and basal aspects of the frontal and parietal lobes. A **short anterior communicating artery** connects the two anterior cerebrals across the midline to complete the circle of Willis anteriorly (Figs. XI-6 and 8).

Note the location of the following vessels in the AP half-axial carotid arteriogram: internal carotid artery branching into the middle and anterior cerebral arteries; and the anterior communicating artery (Fig. XI-8A, B).

Subclavian Artery and Its Branches: While this vessel becomes the **axillary artery** and the main route of supply to the upper limbs, it also completes the supply to the head and neck through four of its branches and the supply of the anterior thoracic wall through a fifth branch. Like the maxillary artery, it is conveniently divided into three parts by a muscle. In this case the muscle is the Anterior Scalene which crosses ventral to the vessel. Part I lies medial to the muscle border; Part II, behind the muscle; Part III beyond the lateral muscle border to the outer border of the first rib. The origins of the Subclavian from the aortic arch differ slightly on each side of the body. These relationships are shown in Figure XI-1C. The branches to neck and anterior trunk virtually all arise from the Part I of the subclavian, though the costocervical trunk may barely qualify as coming off Part II (just barely behind the Anterior Scalene) (Fig. XI-9).

VERTEBRAL ARTERIES: Not only are these the first branches of the subclavian, but it is important that we consider them at once since they relate directly to the supply of blood to the brain and enter into the completion of the circle of Willis. They enter the foramina in the transverse processes of cervical vertebrae at the level of VC_6 and follow up through them to enter the cranium through the foramen magnum. Once inside the cranium, they find themselves along the ventrolateral sides of the medulla, angling toward the midline so that the vertebral arteries of opposite sides unite

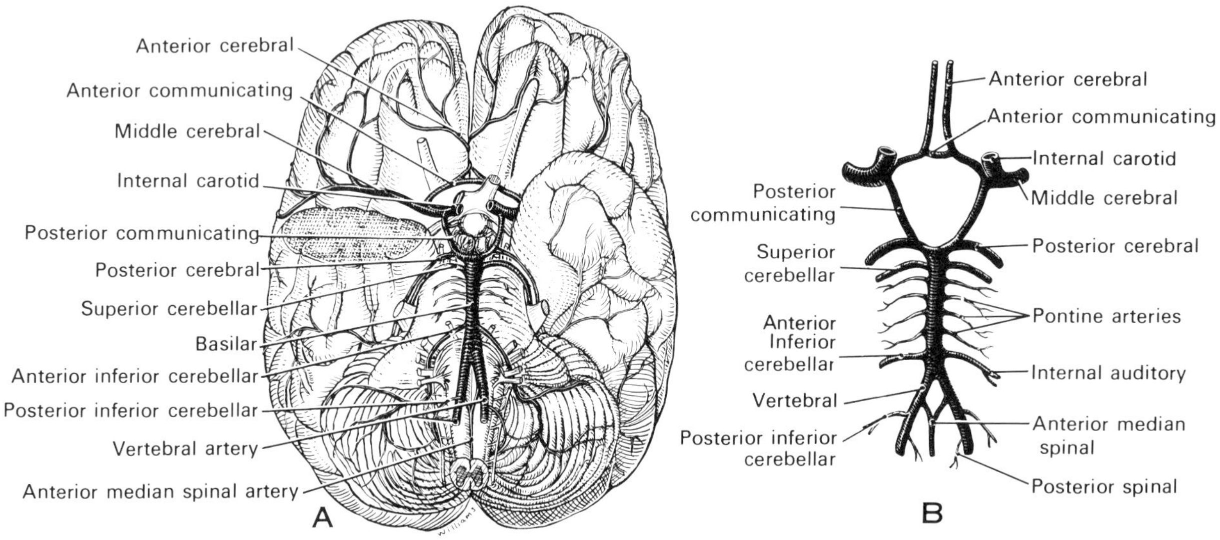

Fig. XI-6. Arteries at the base of the brain. *Inset:* The circle of Willis.

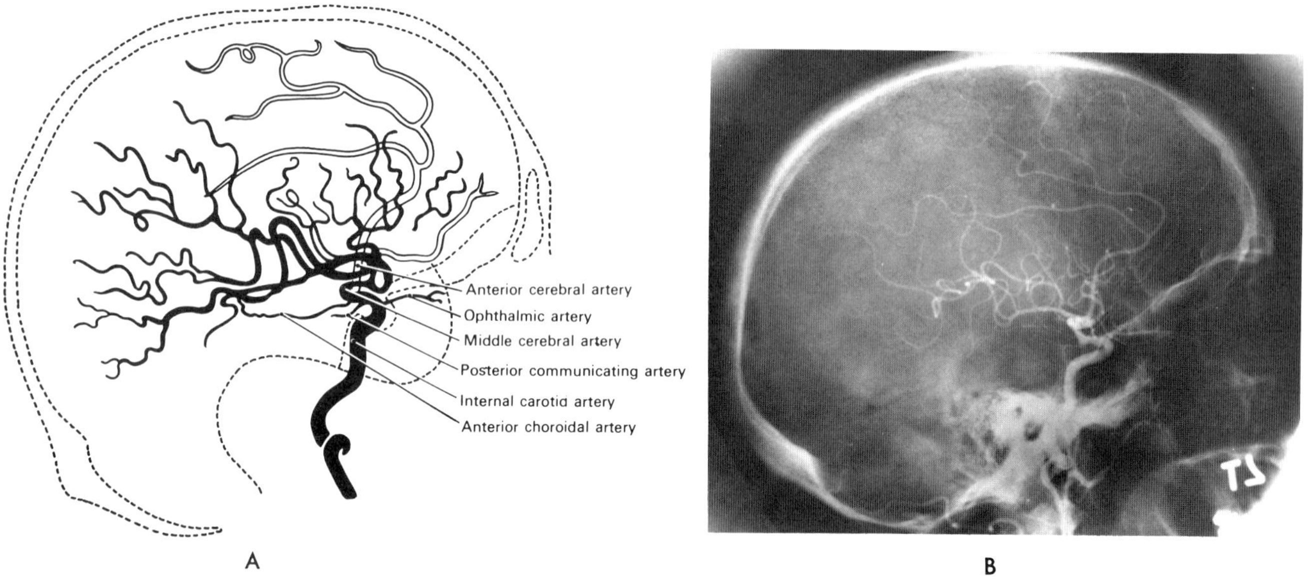

Fig. XI-7. Carotid arteriogram—lateral projection. **A.** Diagram of middle cerebral artery and branches (black). Anterior cerebral artery and branches (white). Not all branches are labeled here. **B.** Matching radiograph.

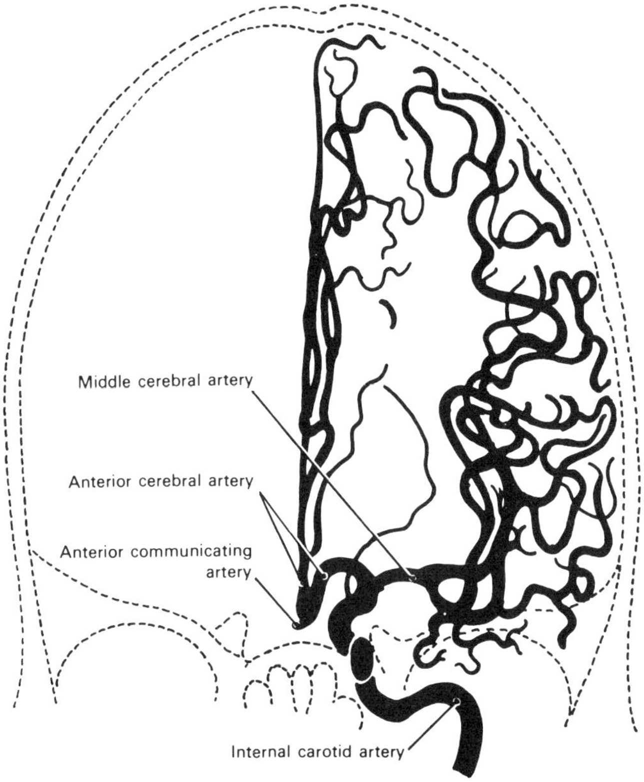

A

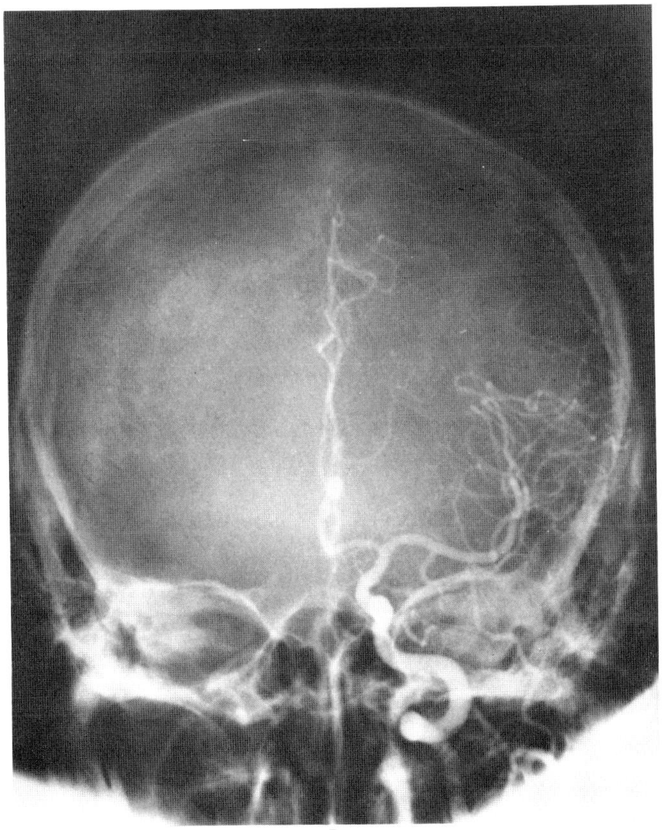

B

Fig. XI-8. Carotid arteriogram—AP half-axial projection. **A.** Shows distribution of middle cerebral vessels laterally and anterior cerebral vessels in the midline. (Not all the branches are labeled here.) **B.** Matching radiograph.

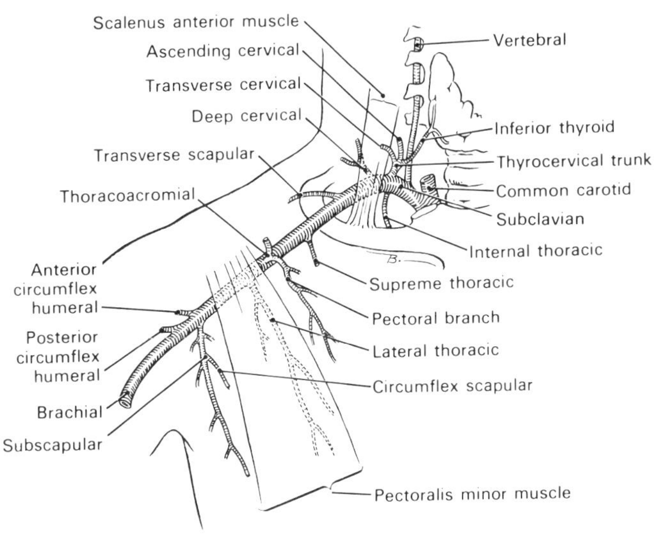

A

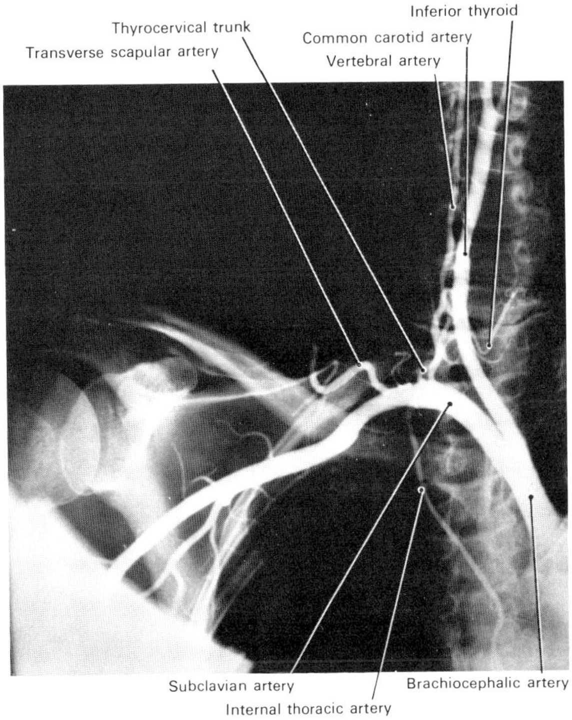

B

Fig. XI-9. The subclavian and axillary arteries and their respective branches. **A.** Schematic drawing of arterial branches. **B.** Aortic injection in the living patient demonstrating the major branches.

as the single **basilar artery** at the pontine-medullary junction. The branches of the vertebral arteries are (see Fig. XI-6):

1. **Medullary twigs** to the medulla in the caudal part of the brainstem
2. Posterior inferior cerebellar arteries
3. Anterior spinal arteries
4. Posterior spinal arteries
5. The fusion of the vertebrals to form the basilar artery

The anterior spinal arteries run a recurrent course to supply medulla on it ventral aspect, and re-exit from the foramen magnum to supply the cord. Before leaving the cranium, the anterior spinal arteries of each side fuse as the single **artery of the ventral sulcus** of the cord.

The **posterior spinals** also supply medulla and are recurrent to the cord through the foramen magnum. They follow down the cord in the posterior intermediate sulci. (We shall discuss these anterior and posterior spinal arteries more fully below in a section on the *blood supply of the spinal cord and vertebral column.*)

The **posterior inferior cerebellar artery** supplies the back inferior aspect of the cerebellum, that large motor regulating center attached to the brainstem. An important branch of this vessel is the **choroid artery of the fourth ventricle,** another area, in dorsal pons and medulla, where cerebrospinal fluid is formed.

The Basilar artery, (Figs. XI-6, 10, and 11) gives off a number of named branches to the brainstem, cerebellum, and, finally, through its terminal branch, to the cerebral hemisphere. a) The **anterior inferior cerebellar artery** supplies (as you might suspect) the anterior aspect of the underside of the cerebellum. b) The **internal auditory artery** enters the internal auditory meatus to supply inner ear, both the cochlea and labyrinth. c) There are a number of **branches to the pons.** d) The **superior cerebellar** artery for the superior aspect of the cerebellum. e) Its all important branches are the terminal **posterior cerebrals** for the completion of the supply to the hemispheres, namely to the occipital lobes and inferior part of temporal lobe. The posterior cerebrals have two important small branches of their own. a) The **choroid artery to the third ventricle** between the thalami, a diencephalic region where cerebrospinal fluid is also formed. b) **The posterior communicating artery,** or at least it connects with this vessel to complete the posterior link in the circle of Willis.

Locate certain branches of the vertebral and basilar arteries on the vertebral arteriograms shown in Figures 10A and B and 11A and B.

The inset, Figure XI-6 shows what we mean by the anastomotic **circle of Willis** between the cerebral vessels at the base of the brain. It consists of: a) the posterior cerebrals, b) the posterior communicating link between them and the middle cerebrals (or internal carotids), c) the middle cerebral arteries, themselves,

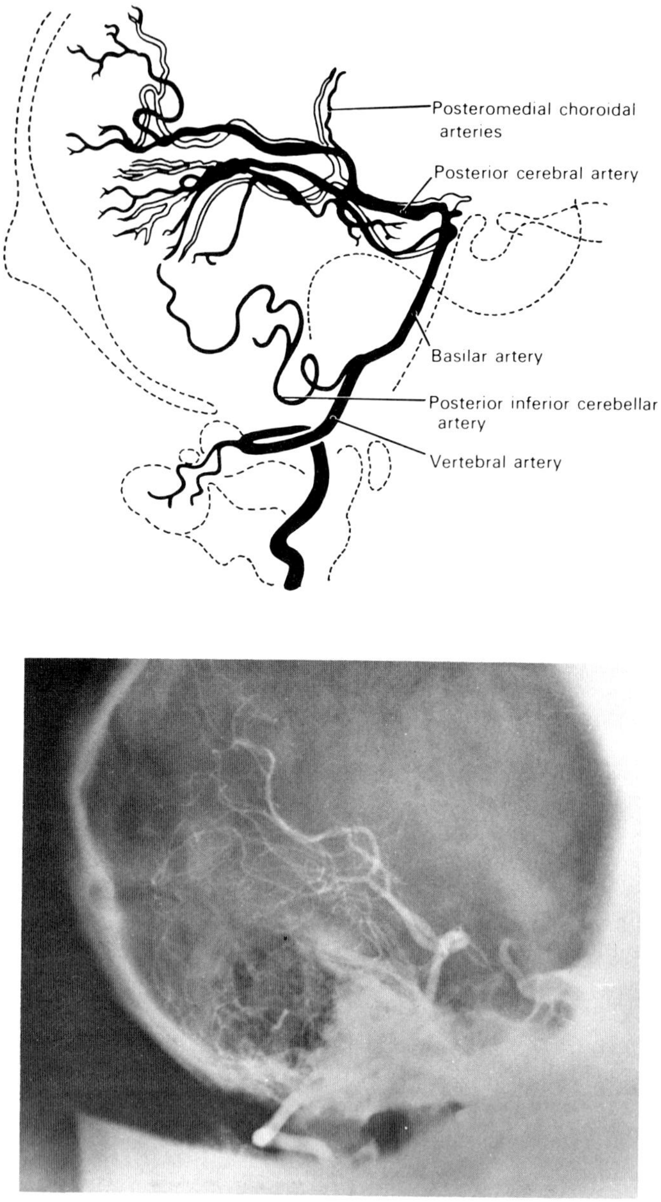

Fig. XI-10. Vertebral arteriogram—lateral projection. **A.** Diagram to show basilar and posterior cerebral arteries. (Not all branches are labeled.) **B.** Matching radiograph.

d) the anterior cerebral arteries, and e) the anterior communicating crosslink.

Under abnormal circumstances, such as blockage of a major cerebral vessel, blood can be shunted around the circle to reach both hemispheres. Probably such shunting does not take place under normal circumstances while the major vessels stay patent and healthy.

Through the basilar artery, the vertebrals and the internal carotids are the only suppliers of blood to the entire central nervous system except for parts of the cord. The basilar supplies thalamus, midbrain, pons, the parts of the cerebrum not served by the internal

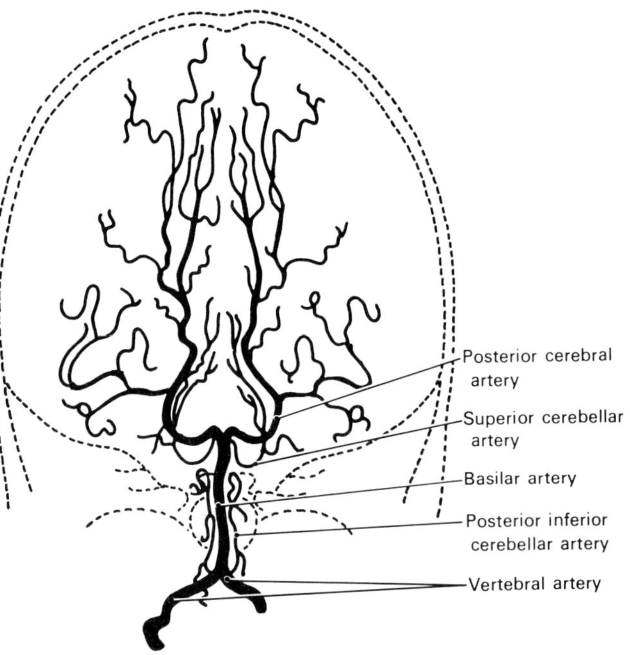

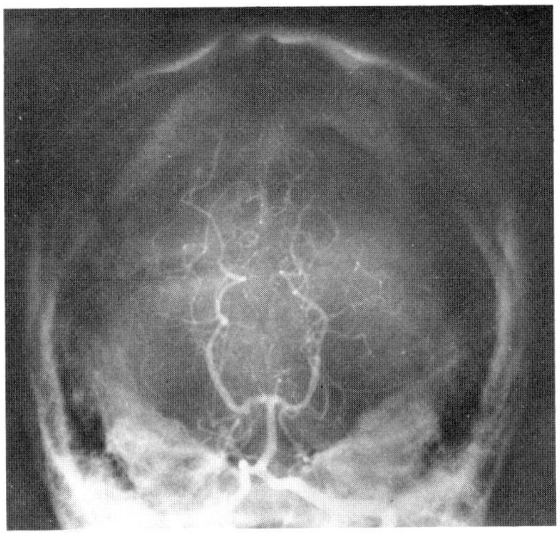

Fig. XI-11. Vertebral arteriogram—AP half-axial projection. **A.** Diagram to show basilar and posterior cerebral arteries. (Not all branches are labeled.) **B.** Matching radiograph.

carotids, and it supplies most of the cerebellum to boot. The vertebrals, themselves, take care of the rest of the cerebellum, the medulla, and a goodly stretch of the spinal cord.

Having discussed only the first branch of the subclavian artery, we are ready to consider its remaining branches (Fig. XI-9). Two of these complete the supply of the lower neck and the last contributes to the supply of the ventral thoraco-abdominal wall:

THYROCERVICAL TRUNK is a short stem which branches at once into:

a) Inferior thyroid
b) Transverse cervical artery
c) Suprascapular artery

COSTOCERVICAL TRUNK, another short, quickly-branching stem: a) **first and second intercostals**—these arise here because the aorta did not ascend high enough to provide branches to the first two costal interspaces, and b) **ascending deep cervical artery** which passes up behind the cervical transverse processes as the highest artery of all of the subclavian derivatives to help in the supply of posterior neck muscles.

Before we consider the internal thoracic artery, and the continuation of the subclavian into the axilla as the axillary artery, it should now be realized that we have supplied the entire head and neck from the arch of the aorta as follows:

1. **External Carotid:** All the neck from the upper part of the thyroid gland cranialward; all the face externally and internally; and all the head except for its cranial and orbital cavities. The external nasal artery may be the only exception in the supply to the face since it is derived from the internal carotid through the ophthalmic artery.
2. **Thyrocervical and Costocervical Trunks:** Rest of lower neck including lower part of thyroid gland.
3. **Internal Carotid:** Most of the brain and all of the orbital cavity.
4. **Vertebrals and Basilar:** Rest of brain, brainstem, cerebellum, and the upper spinal cord; by additions from intercostal and lumbar arteries, the vertebrals can be regarded as supplying the cord in its entirety.

THE INTERNAL THORACIC (INTERNAL MAMMARY ARTERY): This is the one branch of the subclavian to arise from its undersurface. It descends vertically behind the clavicle to enter and course through the thorax one finger's breadth lateral to the sternal margins. Its branches and divisions are: a) anterior intercostals, one for each interspace, b) musculophrenic—a terminal branch to the Diaphragm, and c) superior epigastric—a terminal continuation which enters the rectus sheath from above and behind the muscle to supply it.

THE AXILLARY ARTERY AND ITS BRANCHES (FIG. XI-9): In a sense, this is just a name change for the subclavian as it continues from the neck into the axilla, just as the **brachial artery** into the upper arm is a continuation of this subclavian trunk. In its confines of the axilla, the axillary artery is bound together with the nerves of the brachial plexus and with its companion vein in the tough fibroelastic axillary sheath, and is well embedded in fat. More often the term "axillary vein" in this region means "veins" and their numerous, entangling tributaries. In this small space, this short axillary artery serves two purposes: a) to supply the walls of the axilla; b) to supply shoulder, upper arm muscles, and the shoulder joint. It does these jobs as follows (see Fig. XI-9):

1. A **thoracoacromial artery** arises first as a rule. It is a short-stemmed vessel which branches at once just below the clavicle to supply, principally, the pectoral muscles of the anterior axillary wall, the Deltoid in part, and the Subclavius.

2. A small **highest (supreme) thoracic artery** may actually be the first branch of the axillary, but it is often absent or becomes one of the branches of the thoracoacromial. It may supply the anterior pectoral muscles, but, chiefly, it anastomoses with the internal thoracic or the intercostals.

3. The **lateral thoracic** runs along the medial wall of the axilla to supply the Serratus Anterior.

4. The **subscapular artery** has the major task of supplying the posterior axillary wall, *i.e.*, Latissimus and Subscapularis. This it does through its longest branch: a) The **thoracodorsal;** b) Its **circumflex scapular branch** curls around the lateral edge of scapula supplying in its course Subscapularis, Teres Major and Minor, the long head of the Triceps, and the Deltoid.

The final branches of the axillary are the **posterior** and **anterior humeral circumflex** arteries. The former follows the axillary nerve through the quadrangular space in the back wall of the axilla to distribute to the Deltoid and the shoulder joint. The anterior circumflex humeral rounds the humerus in the opposite direction. It supplies Corachobrachialis, Biceps Brachii, and Deltoid, and anastomoses with the posterior humeral circumflex.

The Brachial Artery: This is the chief continuation of the subclavio-axillary trunk into the upper arm. If we were to outline at this point the complete *deep circulation* in the *right upper limb*, the order of events would be as follows:

Left Ventricle
↓
Ascending Aorta
↓
Brachiocephalic Artery (not present on left side)
↓
Subclavian Artery
↓
Axillary Artery
↓
Brachial Artery
↙ ↘
Radial Artery Ulnar Artery
↘ ↙
Superficial and Deep Volar Arches
↓
Digital Arteries
↓
Arterioles
↓
Capillaries
↓
Venules
↓
Superficial and Deep Volar Venous Arches
↙ ↘
Radial Vein Ulnar Vein
↘ ↙
Brachial Vein
↓
Axillary Vein
↓
Axillary Vein
↓
Innominate Vein
↓
Brachiocephalic Vein (one on each side)
↓
Superior Vena Cava
↓
Right Atrium

But, in the limbs there is, in addition, a system of superficial veins which returns blood from distal parts independent of this main arteriovenous line. So we would have to add this alternative venous return route:

Dorsal Venous Network of the Hand
↙ ↓ ↘
Cephalic Vein Basilic Vein
Median Cubital Vein
↓
Brachial Vein
↓
Axillary Vein

The deep and superficial routes for both dorsal and ventral aspects of the arm are shown in Figures XI-12 and 13. We can now return to a consideration of the **brachial artery,** *per se.* It enters the arm on the medial side of the humerus, and the median nerve is its companion in its course along the upper arm. About midway it moves to the midline of the humerus where it lies upon the Brachialis muscle. At the elbow, the artery is sandwiched between the Biceps tendon laterally and its old friend, the median nerve, medially. Once it crosses the elbow joint it breaks up into the radial and ulnar arteries, arteries of the forearm. Where the vessel enters the upper arm, it relates to all three major nerves. The radial nerve lies behind it, the medial nerve next to it and the ulnar farther medialward than the median. The median, we have said, stays with the artery. As the radial and ulnar nerves diverge from the artery, the vessel sends a branch along with each:

THE PROFUNDA BRACHII (FIG. XI-12 AND 13): This largest of the branches of the brachial takes off high in the arm to follow the radial nerve behind the humerus. It is the blood supply of the Triceps. It runs to the elbow to anastomose with vessels in the forearm.

SUPERIOR ULNAR COLLATERAL ARTERY (FIG. XI-12): It arises close to the profunda but follows the ulnar nerve.

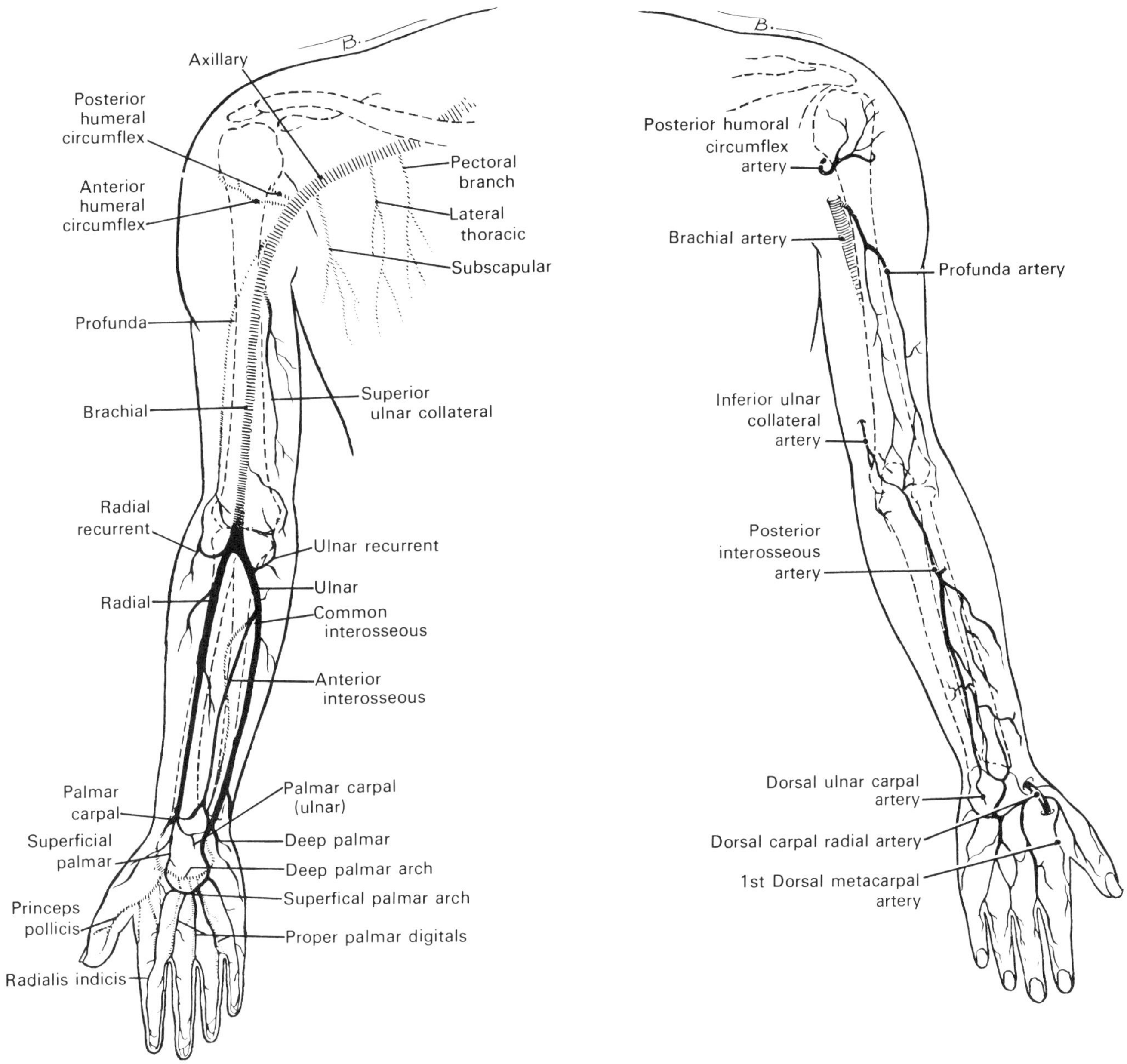

Fig. XI-12. Arterial supply of upper limb. **A.** Volar surface. **B.** Dorsal surface. Superficial vessels are black; deep vessels shaded.

Hence it remains medial and runs to the medial epicondyle of the humerus to supply flexor muscles that arise there.

INFERIOR ULNAR COLLATERAL (FIG. XI-12 AND 13): This is a smaller medial branch arising near the elbow joint on the ulnar side of the arm to supply elbow joint and flexor muscles. As the radial and ulnar arteries bifurcate from the brachial in the forearm they send **recurrent arteries** back to the elbow to anastomose with the last three branches of the brachial artery.

Arteries of the Forearm and Hand: (Figs. IX-12 and 13)

RADIAL ARTERY: Of the two major forearm arteries, this appears to be more of a direct continuation of the brachial even though it often somewhat smaller than the ulnar. In the forearm, it runs under the partial shelter of the medial border of the Brachioradialis all the way to the wrist. It then winds dorsolaterally around the carpus deep to the tendons of the Abductor Pollicis Longus and the Extensors Pollicis Longus and Brevis, to the space between the proximal metacarpals of the thumb and index finger. Finally, it passes between the two heads of the First Dorsal Interosseous to form the deep volar or palmar arch. Its branches are numerous but need not all be recalled. We underline the important ones below:

In forearm

1. **Recurrent radial collateral** *at elbow* (Fig. XI-13)

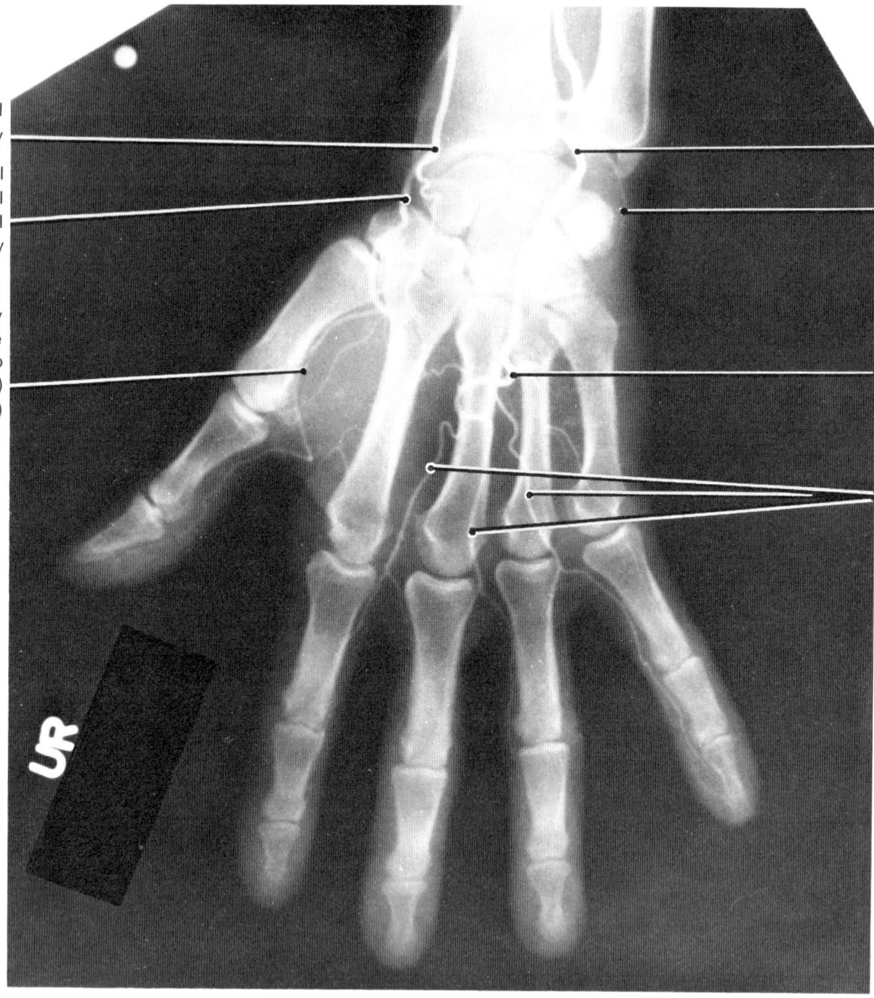

Fig. XI. 12C. Radiograph of the right hand and wrist showing the arterial supply as demonstrated by arteriography.

2. Many **muscular branches** to muscles on radial aspect of forearm
3. Palmar carpal branch

At wrist

4. Superficial palmar artery
5. Dorsal carpal
6. First dorsal metacarpal
7. **Principal artery** *to thumb* (Princeps Pollicis)
8. **Radial artery** *to index finger* (Radialis Indicis)
9. **Deep palmar arch**

At the hand

10. Palmar metacarpals
11. Perforating branches to clefts of fingers to anastomose with superficial palmar arch
12. Recurrent branch from palmar arch to ascend on palmar aspect of wrist

Thus, the **radial** at the hand is principally concerned with the *blood supply of thumb and index finger* and in forming the **deep palmar arch.**

THE ULNAR ARTERY: This artery runs deep beneath several flexor muscles to reach the medial side of the forearm. In the lower half of its course, it appears lateral to the Flexor Carpi Ulnaris and is partly sheltered by its tendon. It enters the hand lateral to the pisiform bone quite near the skin surface, uncovered by muscle. It ends in the formation of the **superficial palmar arch.** Its branches are as follows:

1. **Anterior and posterior ulnar recurrent arteries** to the elbow joint.
2. The **common interosseous**—a short vessel arising just distal to the radial tuberosity. It immediately splits into: a) anterior interosseous, and b) posterior interosseous—These run down on the front and back of the **interosseous membrane** between radius and ulna to give off muscular twigs.
3. Routine **muscular branches** directly from the ulnar artery, itself
4. Palmar carpal branches
5. Dorsal carpal branch
6. Deep palmar branch
7. **Superficial palmar arch**
8. **Common palmar digital** artery. a) **proper palmar digitals** to contiguous sides of index, middle, ring, and little fingers.

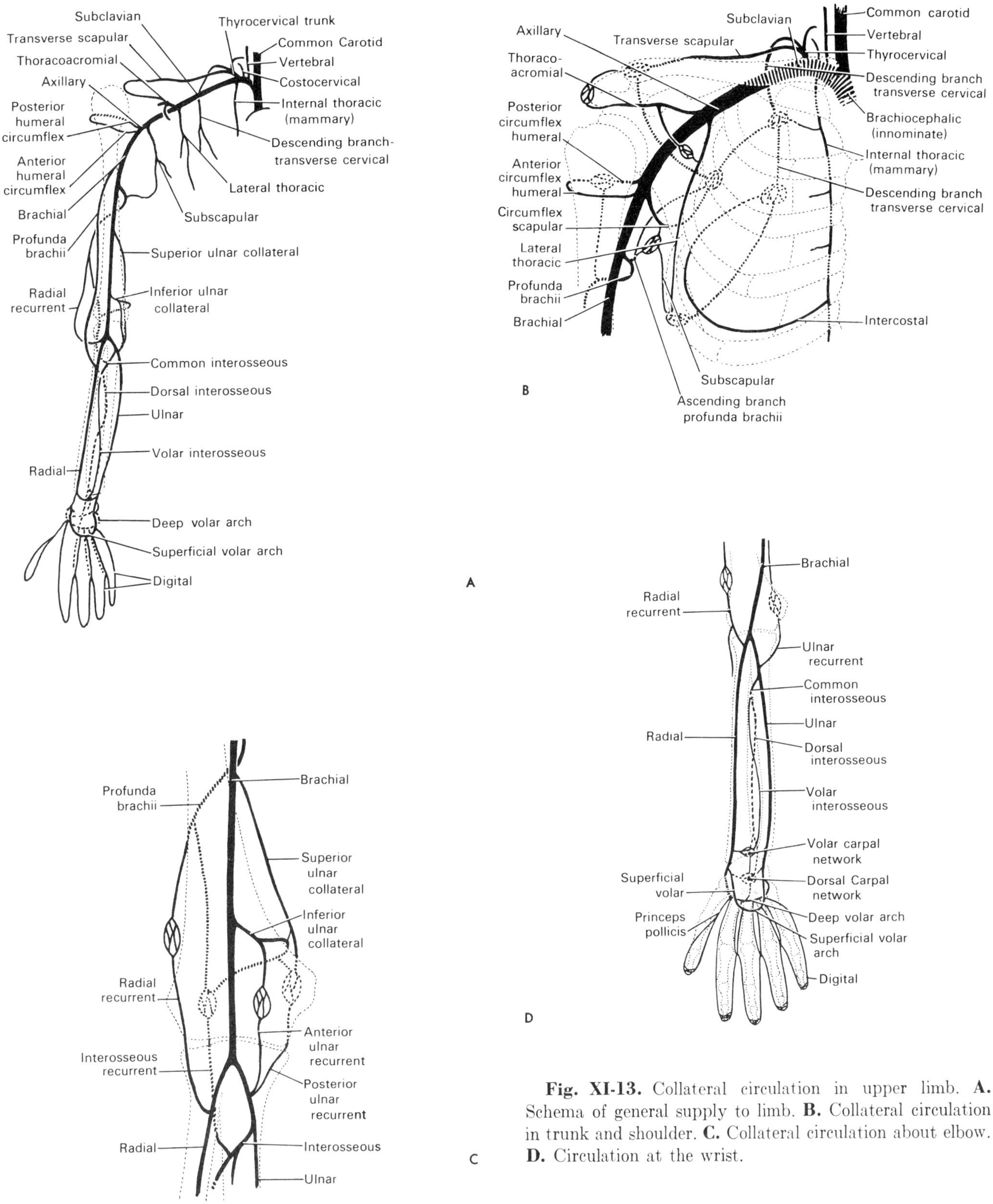

Fig. XI-13. Collateral circulation in upper limb. **A.** Schema of general supply to limb. **B.** Collateral circulation in trunk and shoulder. **C.** Collateral circulation about elbow. **D.** Circulation at the wrist.

The **superficial palmar arch** is the larger of the two in the hand. It traverses the palm from pisiform bone to about the level of the outstretched thumb. The ulnar artery is its main contributor, and it is completed laterally by a branch of the radial artery. It is covered only by the palmar aponeurosis and skin. The **deep palmar arch** lies just proximal to the superficial arch but on a much deeper plane. It runs distal to the bases of the metacarpals. The radial artery is its chief contributor, but a deep branch of the ulnar completes it medially. Both arches give off **palmar branches** that meet at the clefts between the fingers. Each such union provides **digital branches** that run down on the contiguous sides the fingers along side the flexor tendon sheaths. These vessels end in networks at the finger tips.

Up to this point, blood has only been moved through the short stretch of the ascending aorta and its arch to fill all of the arteries mentioned above. We are now ready to consider the branches of its longest segment, the **descending aorta.** Still, this long segment of the vessel is probably the easiest portion to understand and the shortest portion to discuss.

The Thoracic Aorta: This segment is longer than the abdominal aorta with which it is continuous. It covers eight vertebral bodies and disks (VT_4 to VT_{12}) compared to 3½ in the abdomen. Again, this longest portion of the aorta is easier to comprehend than its shortest. As in the abdomen, it has visceral and parietal branches. The area to which most of these distribute is self-evident because of the names of the vessels, and all are, chiefly, bilaterally distributed.

A. Parietal Branches
1. **Posterior intercostal arteries.** The first nine of these are old friends found in the costal interspaces 3 to 11. Number 1 and 2, recall, came from the costocervical trunk.
2. The 12th pair, since they travel below the last rib, are the **subcostal arteries.**
3. The **superior phrenic arteries** are the last small twigs of the thoracic aorta before it enters the abdomen. They are distributed to the vertebral portion of the Diaphragm along its upper surface.

B. Visceral Branches
1. **Pericardial.** A few small branches to the dorsal surface of the pericardial sac.
2. **Bronchial arteries.** These vary in size, number, and origin. The *right one* does not arise from the aorta as a rule, but from the first aortic intercostal artery in the third interspace, or from the uppermost left bronchial artery. *There are usually a pair of left bronchial arteries.* These do arise from the thoracic aorta, the upper one at the level of the tracheal bifurcation, the lower one close to the level of the left bronchus. Both spring from the front of the aorta.
3. **Esophageal arteries.** These are four or five in number and also arise from the front of the aorta. These pass slantingly downward to form anastomoses on the esophagus with branches from the inferior thyroid, and ascending branches of the left inferior phrenic and left gastric arteries.
4. **Mediastinal branches.** Many small aortic twigs to lymph glands and loose areolar tissue of the posterior mediastinum.

The Abdominal Aorta (Fig. XI-14): Having described this vessel and its branches fully in the previous chapter on the abdomen, we shall list branches only, and then catagorize some of the branches of the three unpaired vessels of the aorta to the digestive system since it is these details that are often forgotten.

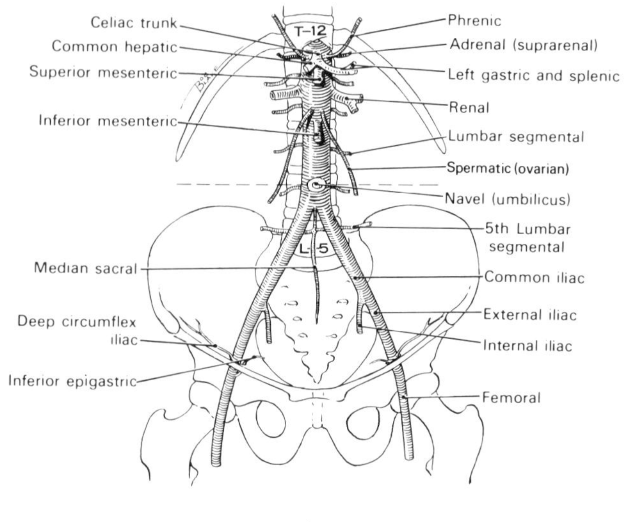

A

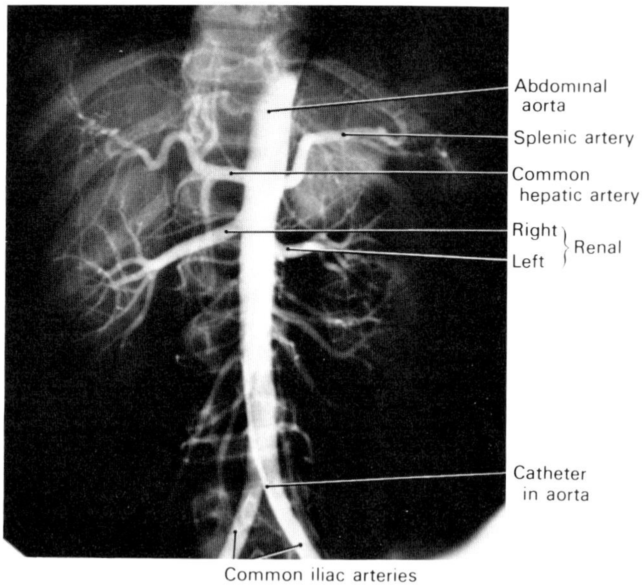

B

Fig. XI-14. The abdominal aorta and its branches. **A.** Schematic representation of branching pattern. **B.** Retrograde aortic injection in the living patient; major branches are identified.

A. Visceral Branches
1. Coeliac artery
2. Superior mesenteric artery

3. Inferior mesenteric artery
4. A pair of middle suprarenal arteries
5. A pair of renal arteries
6. A pair of testicular or ovarian arteries

B. Parietal Branches

7. Paired inferior phrenic arteries
8. Five pairs of lumbar arteries
9. The middle sacral artery

C. Terminal Branches

10. The common iliac arteries

The Coeliac Trunk (Fig: XI-15):

1. **Left Gastric Artery:** it is the smallest of the three main branches of the trunk but larger than the right gastric. It runs along the lesser curvature of the stomach. a) Its cardio-esophageal branch ascends into the thorax. b) Sometimes it gives rise to the inferior phrenic artery c) It can arise from the aorta, from the splenic artery or the hepatic artery.

2. **The Splenic Artery:** only its terminal branches reach the spleen. It snakes along the upper border of the pancreas giving off: a) Numerous pancreatic branches. b) Left gastroepiploic to greater curvature of stomach and to the omentum. c) Short gastric arteries. These are 5 to 7 in the gastrolienal ligament as a supply to the stomach.

3. **The Common Hepatic Artery.** It is smaller than the splenic. It runs first horizontally to the right along the cranial head of the pancreas to pylorus and duodenum. At the duodenum it gives off its descending gastroduodenal artery and then ascends vertically to the porta hepatis. Its various branches are: a) Gastroduodenal artery, which branches in turn into: retroduodenal artery, right gastroepiploic artery, and superior pancreaticoduodenal artery. b) The right gastric artery arises between hepatic arteries proper and the gastroduodenal. It supplies pylorus and lesser curvature and anastomoses with left gastric. c) Right hepatic artery. From this, as a rule, arises the cystic artery. d) Left hepatic artery.

Superior Mesenteric Artery (Fig. IX-16):

1. The inferior pancreaticoduodenal artery anastomoses with superior pancreaticoduodenal branch of gastroduodenal artery to form the anterior and posterior pancreatic arcades.

2. Intestinal arteries: 12 to 15 branches are given to the jejunum and ileum. All arise from the left side of the superior mesenteric.

3. Iliocolic artery. This is the most caudal right hand branch. Its branches are a) colic to ascending colon, b) anterior and posterior cecal branches, c) appendicular artery, d) ileal artery—to caudal portion of ileum as it approaches cecum.

4. The right colic artery arises midway on the right side of superior mesenteric to supply ascending colon.

5. The middle colic artery arises on the right just caudal to pancreas to supply transverse and part of descending colon.

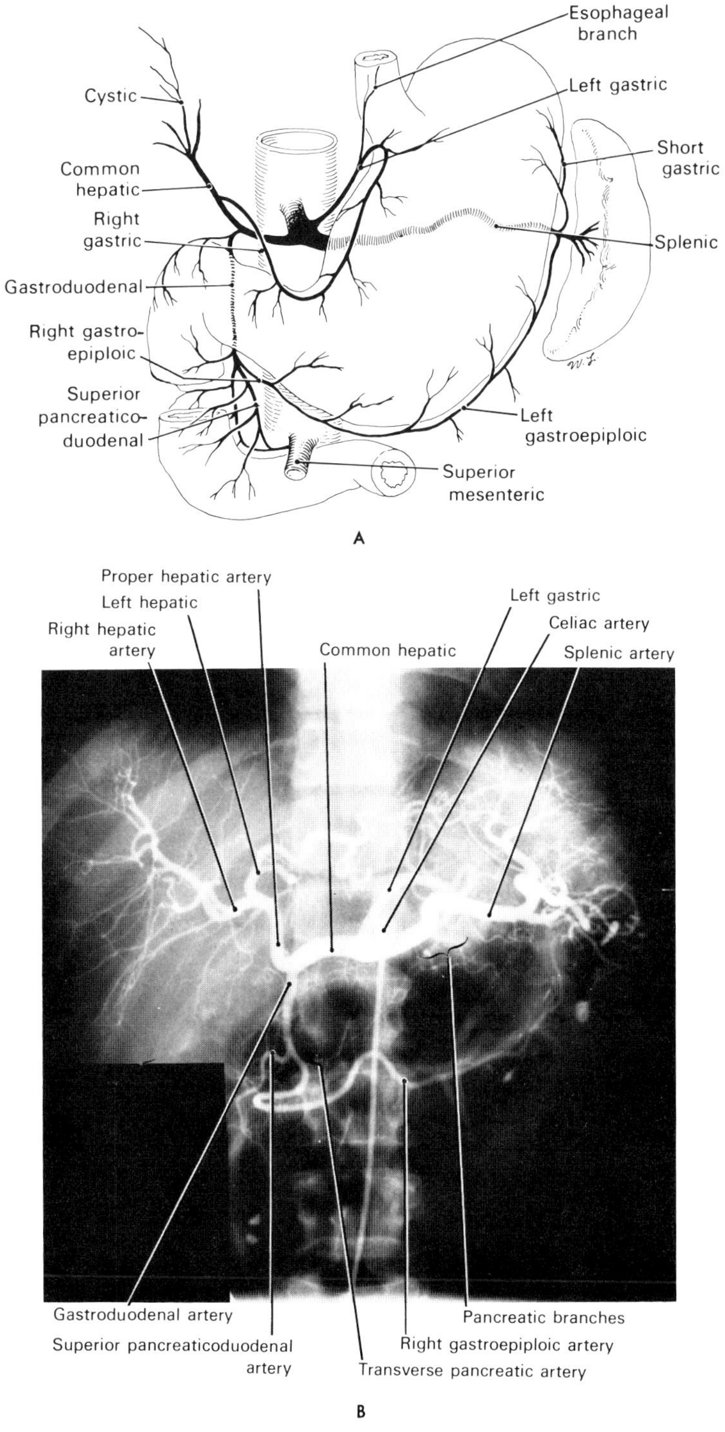

Fig. IX-15. The coeliac artery and its branches. **A.** Schematic drawing of the arterial branches. **B.** Selective catheterization of the coeliac artery in a living patient.

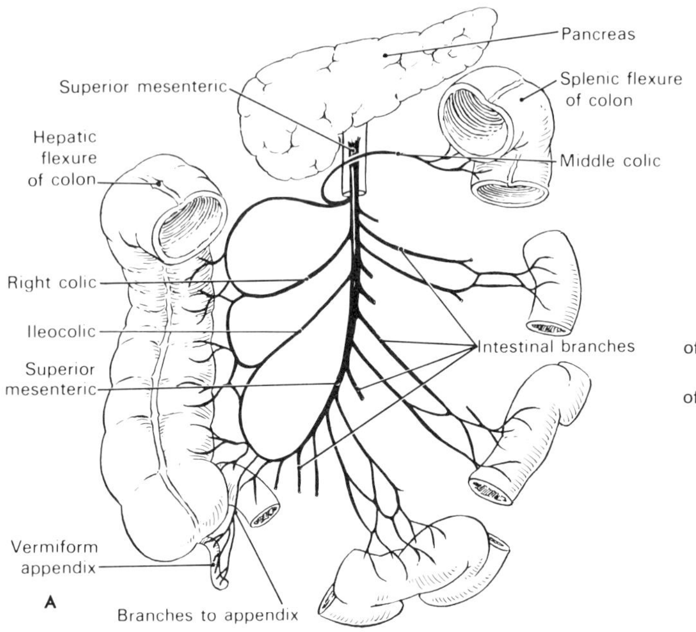

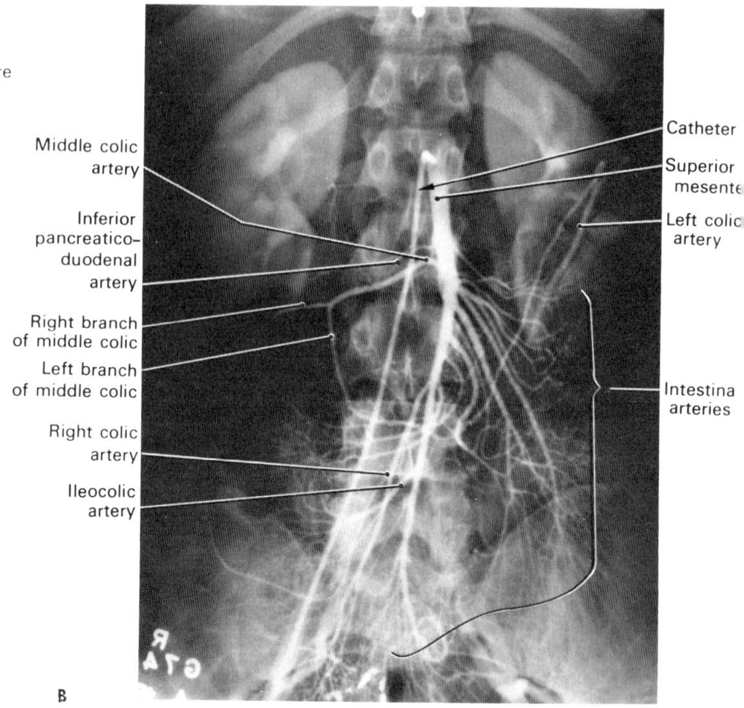

Fig. XI-16. The superior mesenteric artery and its branches. **A.** Schematic diagram of the arterial tree. **B.** Selective catheterization of the superior mesenteric artery in a patient.

Inferior Mesenteric Artery (Fig. XI-17):

1. **Left Colic Artery:** distributes to descending colon and, at times, to left part of transverse colon.
2. **Sigmoid Arteries:** These are two or three deep vessels, deep in that they lie under Psoas, ureters, and the testicular arteries of the left side. They supply lower part of descending, pelvic, and sigmoid colon.
3. **Superior Rectal (Hemorrhoidal) Artery:** opposite the third sacral vertebra this vessel breaks into two branches to descend on each side of rectum almost to anus. Several branches leave it to enter rectum and course in submucosa to level of internal sphincter. Here they anastomose with branches of middle and inferior rectals derived from the internal iliac and its branches.

The Common Iliac Arteries: These terminal branches of the aorta must, through their own branching, supply pelvic organs and the lower limbs. They are short trunks of about 5 cm. in length which arise at VL_4. They diverge from each other, and at VL_5 they break into their main terminal branches, the **internal iliac (or hypogastric)** arteries for the supply of pelvic structures and the **external iliacs** for each of the limbs. See Figure IX-18A for this division and its relation to the companion veins. Aside from the large terminal trunks, the common iliacs give off small branches to:

Psoas Major—lower part
Peritoneum and areolar tissue in the area
Sometimes the iliolumbar artery
Sometimes an accessory renal artery

We shall consider first the terminal supply to the pelvis, buttocks, and genitalia.

THE INTERNAL ILIAC (HYPOGASTRIC) ARTERY: It is said to divide into a **posterior division** whose branches are all destined for the parieties and buttock and an **anterior division** whose branches are visceral, parietal, and genital (Fig. XI-18B and C). But, its branching is most variable so that sometimes even these major divisions are difficult to delineate, and any two branches anywhere may arise from one common trunk. Still, once the individual branches themselves are discovered, they are usually doing what textbooks say they should be doing. The hypothetical situation can be outlined as follows:

A. Posterior division
  1. Iliolumbar artery
  2. Lateral sacral arteries
  3. Superior gluteal arteries
  4. Inferior gluteal artery.

B. Anterior division
  1. Parietal and genital branches: a) obturator, b) internal pudendal.
  2. Visceral branches: a) umbilical, b) inferior vesical, c) middle rectal (hemorrhoidal), d) uterine.

Iliolumbar artery: This vessel runs upward to the medial border of Psoas behind the obturator nerve on the external iliac vessels. Upon reaching the muscle, it divides: a) **lumbar branch** goes to Psoas and Quadratus Lumborum. It anastomoses with last lumbar artery from the aorta and sends the **spinal branch** to exit through intervertebral foramen between VL_4 and VS_1 to supply cauda equina. b) **iliac branch:** This branch descends to supply the Iliacus, iliac bone, and gluteal and abdominal muscles through anastomotic branches with neighboring vessels.

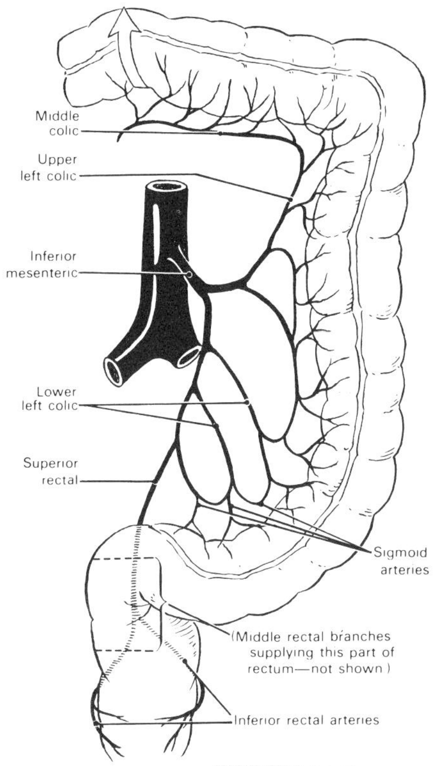

Fig. XI-17. The inferior mesenteric artery and its branches. **A.** Schematic artistic representation of the arborizing pattern. **B.** Selective catheterization of the inferior mesenteric artery with contrast medium in a living patient.

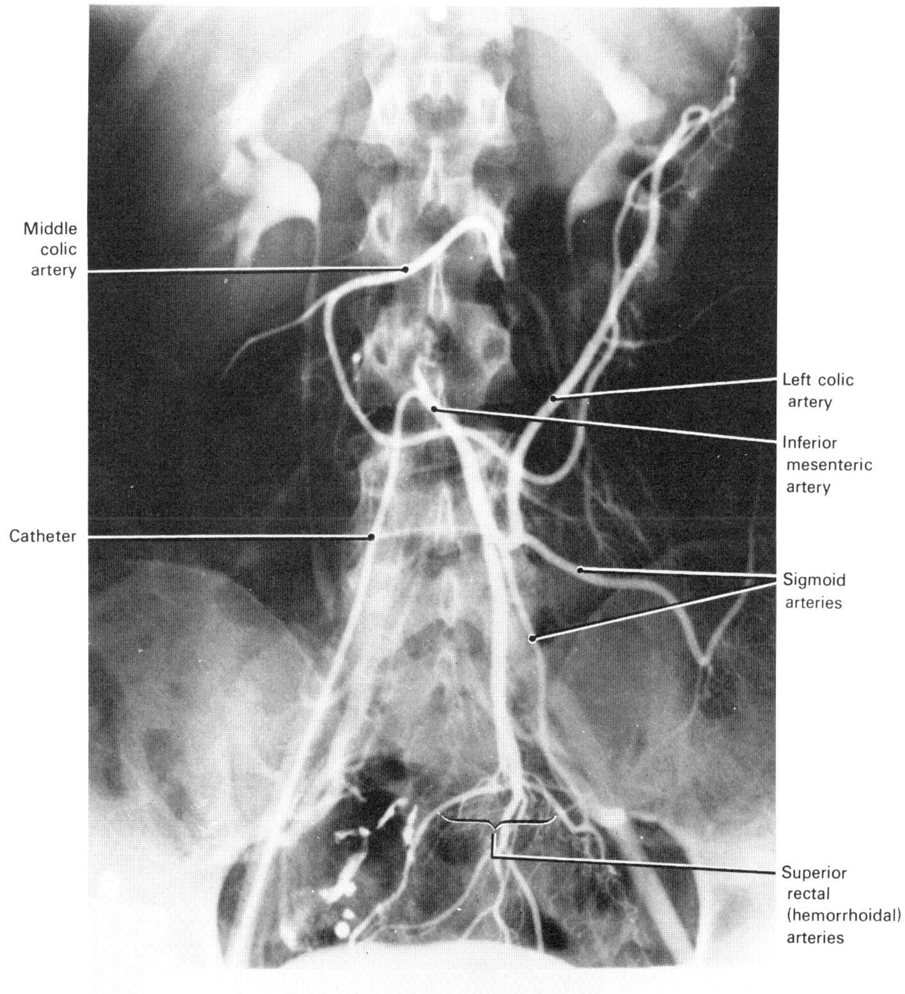

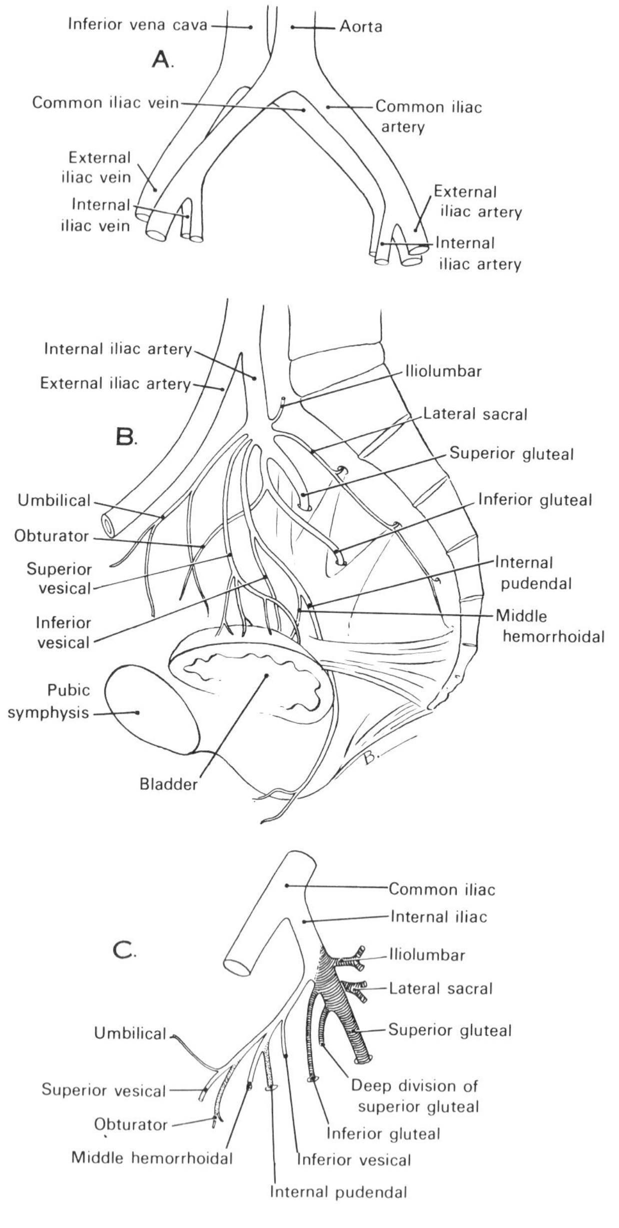

Fig. XI-18. Common and internal iliac arteries. **A.** Relation of iliacs to inferior vena cava. **B.** Diagrammatic distribution of the internal iliac or hypogastric artery in the pelvis. **C.** The anterior (white) and posterior divisions (hatched) of the internal iliac artery. The posterior division, along with the obturator and internal arteries, can be said to supply the pelvic walls. The rest supply the pelvic viscera. In the female, uterine and vaginal arteries arise from the anterior division.

Lateral sacral arteries: There are usually two. These anastomose with the median sacral artery (the continuation of the aorta), enter anterior sacral foramina to supply structures in the sacral canal and emerge in posterior sacral foramina as cutaneous branches to supply skin of the sacral region.

Superior gluteal artery: This is by far the largest branch of the internal iliac and, when followed, it appears as if it were its direct continuation. Despite this, it is a very short vessel which branches quickly. It supplies the Gluteus Medius and Minimus chiefly, but gives branches to Maximus, Iliacus, Piriformis, and Obturator Internus as well. It leaves the pelvis by passing *above* the superior border of the Piriformis muscle.

Inferior gluteal artery: The vessel leaves the pelvis through the lower part of the sciatic foramen *between* the Piriformis and Coccygeus. *Inside the pelvis,* it supplies these muscles and the Levator Ani. *Outside,* it supplies the gluteal muscles and the Maximus in particular: a) **arteria comitans nervi ischiadici** is a special branch to the large sciatic nerve, b) cutaneous branches distribute to buttock and posterior thigh, c) branches supply the hip joint.

The Obturator Artery: The artery runs ventralward on the lateral wall of the pelvis to the obturator canal and foramen. (The obturator foramen is not the large hole you see in the os coxae of the skeleton. This is mostly filled by a membrane and the hole through which nerve and vessel pass is small.)

Inside pelvis: a) muscular branch supplies Iliacus, b) vesical branch is given to the bladder, and c) pubic branch—anastomoses with inferior epigastric artery.

Outside the pelvis, after exit from the foramen: a) muscular anterior branches to Adductor muscles of thigh, Obturator Externus, Gracilis, and Pectineus, b) muscular posterior branch distributes to muscles attached to ischial tuberosity (hamstrings), c) acetabular branch reaches hip joint.

Internal Pudenal Artery: This artery leaves the pelvis through the **greater sciatic foramen** using the same exit route as the inferior gluteal artery, but it re-enters pelvic cavity through the **lesser sciatic foramen** (Fig. XI-19) to run in the lateral wall of the ischiorectal fossa, *i.e.,* in Alcock's canal formed by a splitting of the obturator fascia. Its branches are as follows: a) muscular: supply Levator Ani, Obturator Internus, Piriformis, and Coccygeus. b) **inferior rectal arteries.** (Superior rectal artery arises from inferior mesenteric; middle rectal will be a separate branch of the internal iliac.) The inferior rectal pierces Alcock's canal to supply muscle and skin in the anal region. It anastomoses with superior and middle rectals (hemorrhoidals). c) **Perineal artery** (Fig. XI-20): **superficial transverse perineal artery** is given to the muscle of the same name, **muscular branches** reach Bulbocavernosus and Ischiocavernosus muscles of penis, **posterior scrotal branches** distribute to Dartos muscle and skin of scrotal sac, and branches go to **perineal**

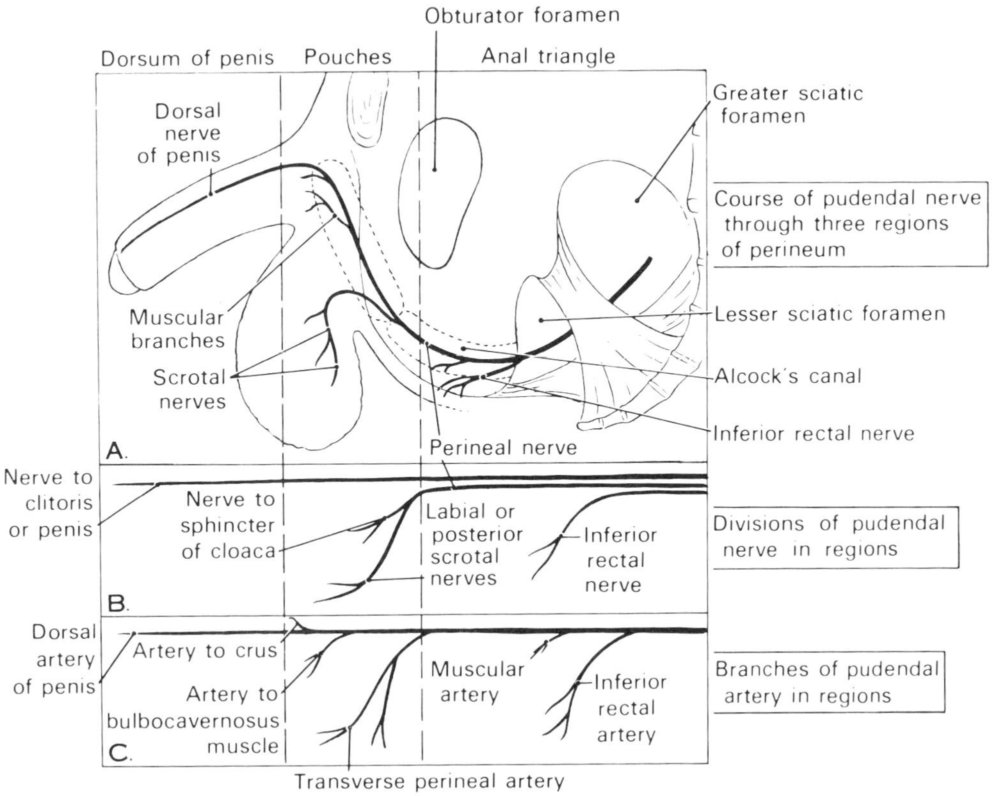

Fig. XI-19. The course of the pudendal artery. The division and regions supplied by the pudendal nerve (B) and artery (C) are presented in tabular form.

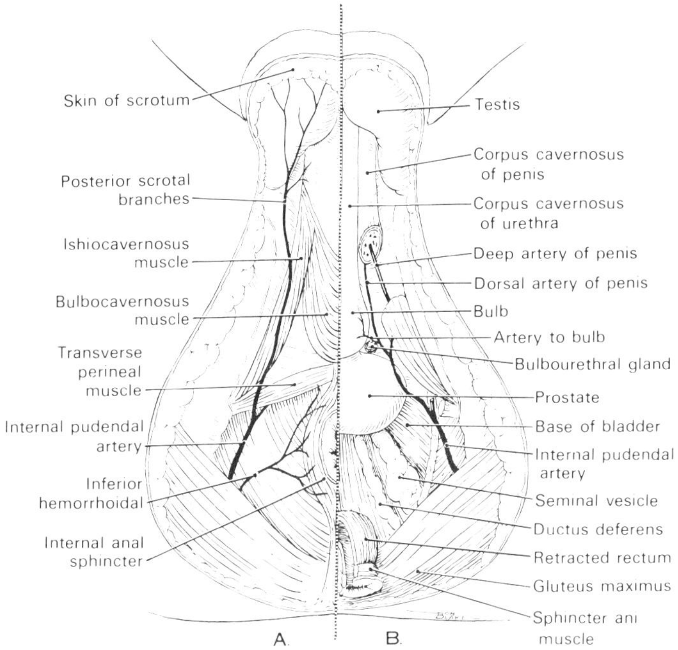

Fig. XI-20. The pudendal artery. **A.** Superficial branches (left). **B.** Deep branches (right).

body between penile bulb and anus. d) **Artery of the penile bulb**—this short but stout branch of the internal pudendal supplies bulb and substance of corpus spongiosum (Fig. XI-16B). e) **Urethral artery**—supplies corpus caverosum and glans. f) **Deep artery of penis**—is a terminal branch (Fig. XI-20) given off in urogenital diaphragm to crus of penis for a supply to corpus cavernosum. g) **Dorsal artery of penis** (Fig. XI-20) arises deep in perineal compartment as a terminal branch also. It pierces the urogenital diaphragm between the crus of the penis and the pubic arch, passing between the two layers of the suspensory ligament to the dorsum of the penis. It lies between the dorsal nerve and the deep dorsal vein. It supplies the skin of the penis and glans and the fibrous sheath around the corpus cavernosum.

[**In the female,** the **pudendal artery** is smaller but runs a similar course and gives off similar branches. A perineal branch supplies the labia instead of the scrotum; the artery of bulb supplies the bulbus vestibuli and vaginal erectile tissue (of which there is a minimal amount); deep artery of the clitoris (instead of penis) supplies the corpus cavernosum clitoridis, while the dorsal artery supplies dorsum of that organ, its tiny glans and its prepuce.]

This leaves us to account for the **visceral branches of the internal iliac** (hypogastric) now that we have considered its parietal, gluteal, and genital branches.

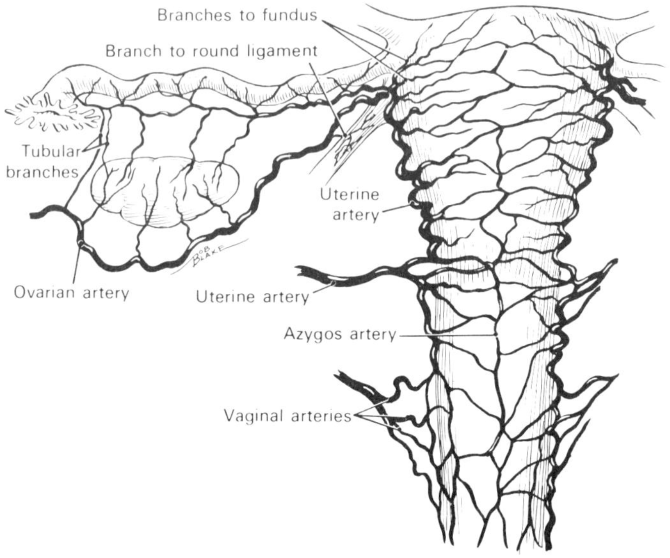

Fig. XI-21. Uterine, vaginal, and ovarian arteries as seen from behind.

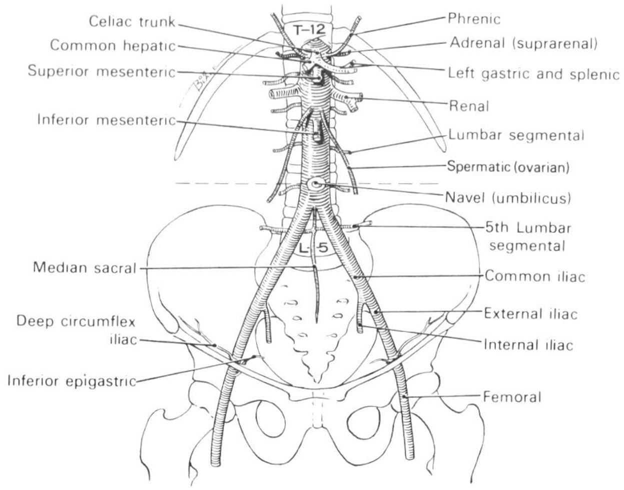

Fig. XI-22. Common and external iliac arteries.

Actually, the iliolumbar, lateral sacral, superior, and inferior gluteals can be considered to be branches to the posterolateral parieties, while the obturator and internal pudendal are branches to the anterior parieties of the pelvis. True, some viscera were innervated by some of the named offshoots of these branches, *e.g.*, the vesical artery to the bladder, the inferior rectal artery, the artery to bulbourethral glands and the like, but the main branches were *not* predominately ones to the *organs within the pelvis*, in the sense that the following visceral branches claim to be:

The umbilical artery (Fig. XI-18B and C): This is not the umbilical artery of the fetus which we know in the adult as the lateral umbilical ligament, but, rather, its proximal *patent* portion in the adult which still connects with the internal iliac. This branch gives off the following smaller branches to the bladder: a) **superior vesical artery**—in turn, branching into the artery to ductus deferens and b) *middle vesical*—branching into a supply to the ureter and to the seminal vesicle.

The **inferior vesical artery** is a direct branch *to the bladder* from the internal iliac artery (Fig. XI-18B, C): and has branches to the prostate and to the seminal vesicles.

Middle rectal (hemorrhoidal) artery (Fig. XI-18B, C): This is the *direct* branch of the internal iliac to the rectum. It singly sends branches to the prostate and seminal vesicles.

Uterine artery of female (Fig. XI-21): This is a medial branch of the internal iliac which runs on the Levator Ani toward the uterine cervix. On its way it sends a small twig to the ureter. When it reaches the uterus it ascends along its side in a serpentine fashion between the layers of the broad ligament. In its ascending course, it anastomoses with the ovarian artery (direct branches from the aorta). In its descending course it forms the **azygos artery of the vagina.** Medial **vaginal branches** are sent to anastomose with this latter vessel either as direct branches from the internal iliac, but chiefly, as branches from the internal pudendal, or, sometimes, from the inferior vesical artery.

THE EXTERNAL ILIAC ARTERY (FIGS. XI–22 AND 23): This is the larger branch of the common iliac artery destined for the lower limb. It passes obliquely distad and laterad along the medial border of the Psoas Major as it leaves the bifurcation of the internal iliac. It crosses *under* the inguinal ligament midway between the superior iliac spine and symphysis pubis where it changes its name to the **femoral artery** of the thigh. Thus, we have a situation comparable to the subclavian of the upper limb. Before passing under the inguinal ligament it gives off its side branches:

1. Twigs to the Psoas and neighboring lymph nodes.
2. The **inferior epigastric** (Fig. XI-23) is a landmark at the margin of the deep inguinal ring and a boundary (lateral one) of Hesselbachs triangle. The ductus deferens, on entering the ring, winds laterally and dorsally around the vessel to turn toward the midline and reach the prostate, seminal vesicle, and bladder. You might say that both the ductus and the round ligament of the female hook around this vessel. The vessel, itself, enters the posterior aspect of the rectus sheath at the arcuate line to ride behind the muscle, anastomosing with the superior epigastric artery from above. [Terminal branch of internal thoracic (mammary)]. The branches of the inferior epigastric are: a) cremasteric (external spermatic) branch. It runs with the spermatic cord to supply the Cremaster muscle and other cord coverings. It is much smaller when it accompanies round ligament of female. b) Pu-

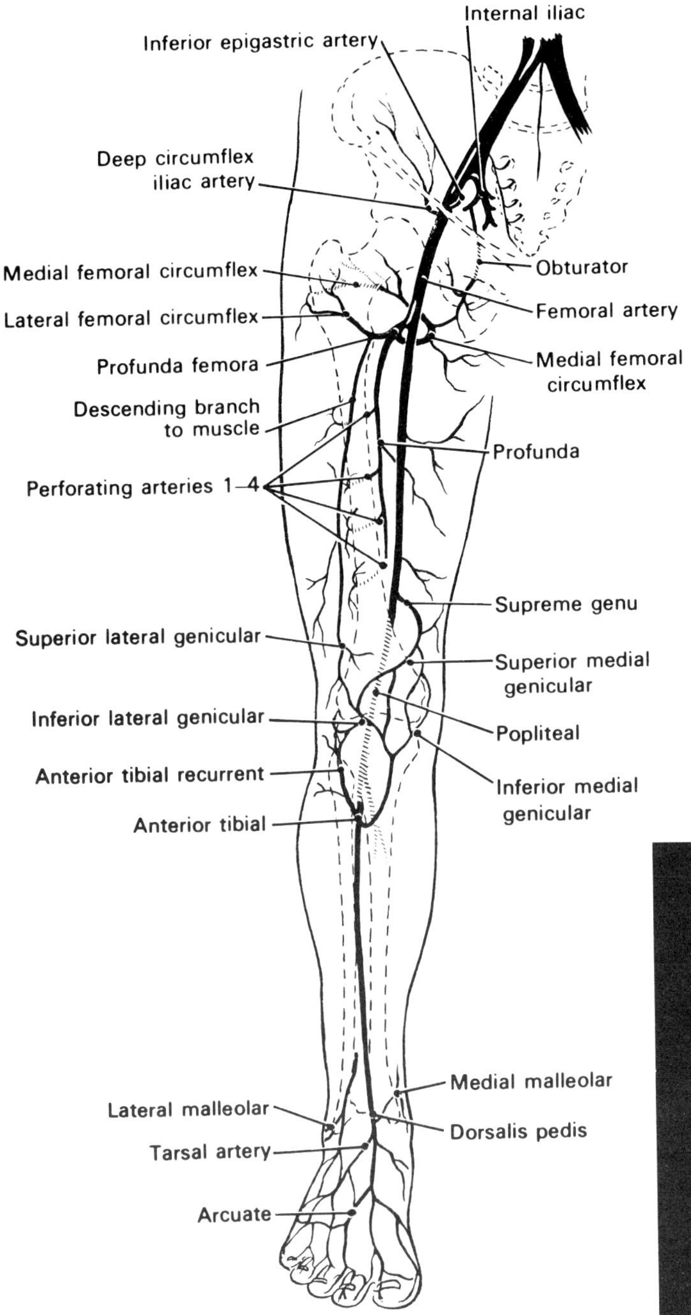

A

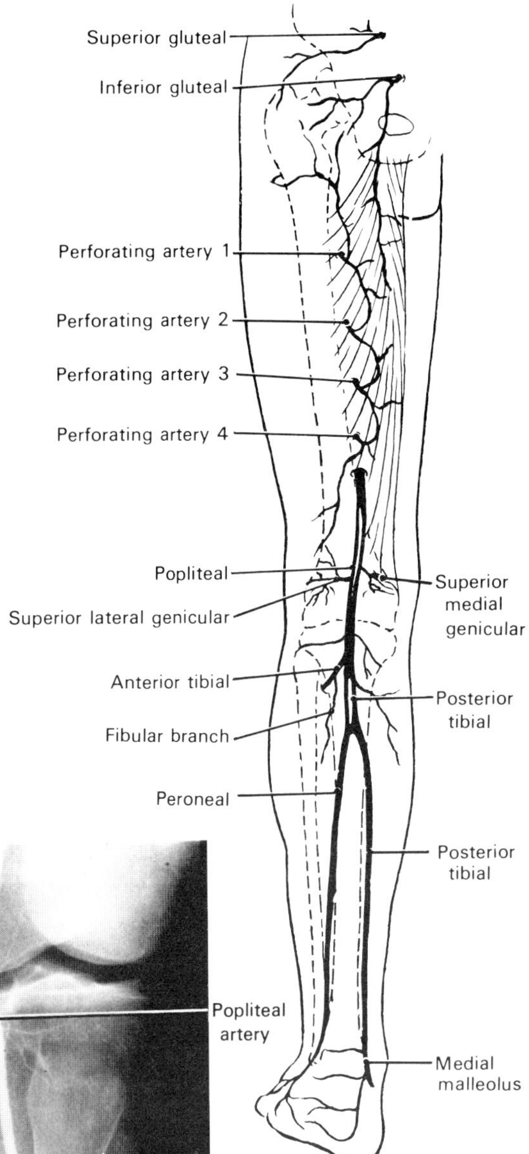

B

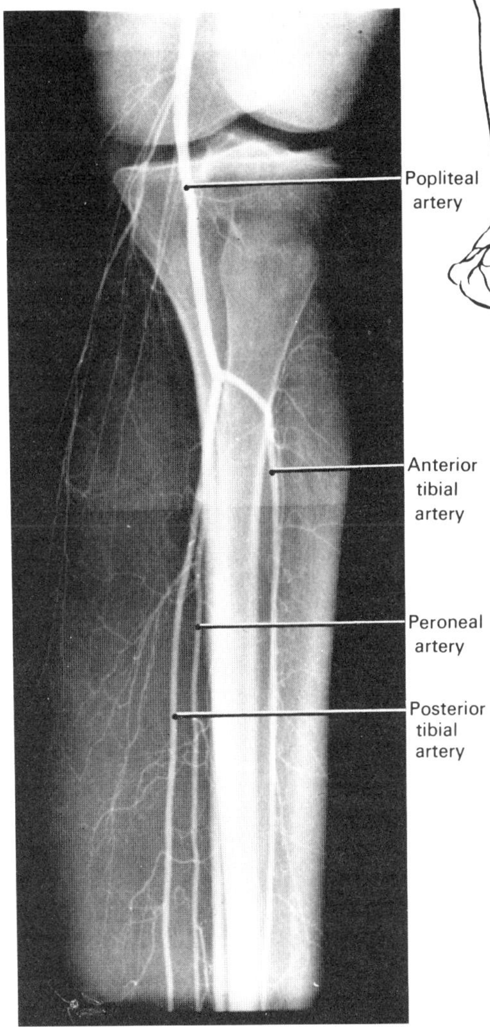

C

Fig. XI-23. Arteries of the lower limb. **A.** Anterior view. **B.** Posterior view. Note that deeper vessels have been cross-hatched. **C.** Femoral arteriogram of the left leg as seen in frontal view.

bic branch. It courses along inguinal ligament and medial margin of internal ring to pubis. c) Muscular branches to abdominal muscles.

3. **Deep circumflex iliac artery** (Fig. XI-23): This vessel arises from the external iliac near the inferior epigastric but takes a lateral course toward the anterior superior iliac spine, crest of ilium and over the posterior crest margin to finally anastomose with the iliolumbar artery. a) Its **muscular branches** supply Internal Oblique and Transversus, Psoas, Iliacus, Sartorius, and Tensor Fasciae Latae. b) It also gives off cutaneous branches along its course.

The Femoral Artery (*Fig. XI-22*): As the direct continuation of the external iliac into the thigh, it holds to the midline of the anterior compartment of the thigh. It descends vertically to meet the obliquely placed femur about 4 inches above the knee. Here the vessel passes through a hole in Adductor Magnus to run behind the distal portion of the femur into the popliteal fossa. It then changes its name to the **popliteal artery.** The femoral artery has to supply both the front and the back thigh. Thus, its branches, like the parent trunk, will have to perforate the adductor muscles to reach posterior ones, for the parent trunk lies upon adductor muscles in its course. The branches of the femoral artery are:

1. Superficial epigastric artery
2. Superficial circumflex iliac artery
3. Superficial external pudendal artery
4. Deep external pudendal artery
5. Muscular branches artery
6. Profunda femoris artery (Fig. XI-23)
7. Descending genicular artery (Fig. XI-23)

The first two, superficial epigastric and superficial circumflex iliac, are ones you flung aside, buried in the subcutaneous fascia, when you reflected the muscles of the abdominal wall. They arise close together about a centimeter below the inguinal ligament. **The superficial epigastric** crosses the inguinal ligament medially in its ascent toward the umbilicus. It distributes to superficial fascia, lymph nodes, and skin along its course. The **superficial circumflex iliac** runs laterad, parallel to the inguinal ligament, to the iliac crest. It lies in the superficial fascia and dermis of the abdominal skin. It parallels the course of the deep circumflex iliac artery with which it anastomoses.

The next two branches, superficial external pudendal and deep external pudendal, are also small and frequently overlooked. They arise close to the preceding two vessels, remain deeper, and supply the skin of the lower abdominal wall, penis, and scrotum (labium majus in female). If you see all four of these arteries in your dissection, that is the mark of a careful dissector. If you don't see them, don't weep unduly. Some days, you can't make a nickel with all that fat and all those veins.

Muscular branches to the Sartorius, Vastus Medialis, and Adductors should be found easily.

The profunda femoris (Fig. XI-23 and 24) should give you no trouble. It is a large vessel arising posterior to the femoral trunk itself, 2 to 5 cm. below the inguinal ligament. It may lie lateral to the femoral and then cross behind it to the medial side. It terminates deep to the Adductor Longus in the distal third of the thigh in a small branch (fourth perforating artery) that pierces the Adductor Magnus. Its branches are: a) medial femoral circumflex artery (Fig. XI-23), b) lateral femoral circumflex artery (Fig. XI-23), c) perforating arteries (Figs. XI-23 and 24), and d) muscular arteries.

The **circumflex arteries** can arise directly from the femoral artery on their own. Both are important arteries, and, obviously, one goes one way, and the other, the other way. **The medial femoral circumflex** passes backward above the adductors and reaches the back of the thigh high enough up to anastomose with the inferior gluteal artery. It supplies the hip joint, the Adductor muscles, the Gracilis, and the Obturator Externus. Small branches run to the head of the femur also. If these are damaged in a fracture of the neck of the femur, the femoral head can necrose.

The **lateral femoral circumflex artery** is the important lateral branch of the femoral (or profunda) artery, indeed its only major lateral branch. Thus, it must take care of the massive extensor muscle of the knee, the Quadraceps Femoris in all its parts (Rectus and Vasti Lateralis, Medialis and Intermedius) as this vessel ascends and descends along the lateral aspect of the thigh. It also supplies the Tensor Fasiae Latae.

There are usually three named **perforating branches** of the profunda, conveniently, the first, second, and third. And, for good measure, its termination is called the fourth. They all pass posteriorly to reach the hamstring muscles which arise from the ischial tuberosity. The first of these perforates proximally or above the Adductor Brevis; the second anterior to this muscle and the third just distal to it. They all punch holes through the Adductor Magnus in the process.

The **muscular branches** of the profunda, itself, end in the Adductor muscles, but some of these also punch through to the hamstrings.

The last branch of the femoral artery (before it becomes the popliteal) is the **descending genicular artery.** a) Its **articular branch** obviously descends, and on the medial side, to provide the knee joint, anastomosing there with the recurrent tibial artery. It sends a branch transversely across the patella to anastomose with the lateral superior genicular branch of the popliteal artery. b) It gives off a medially direct **saphenous branch** which follows the saphenous nerve to the medial side of the knee and proximal leg.

The Popliteal Artery (*Fig. XI-23*): This is a direct extention of the femoral in the popliteal fossa. It extends from the aponeurotic hiatus in the Adductor Magnus to the lower border of the Popliteus muscle

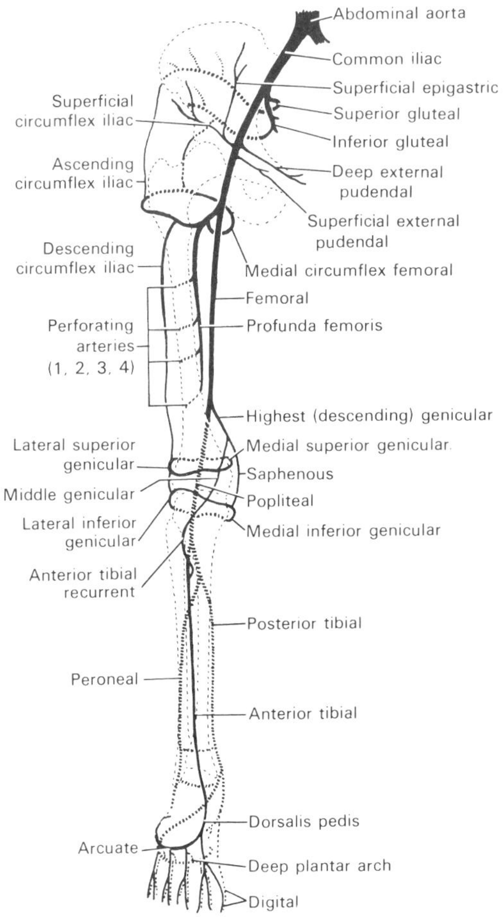

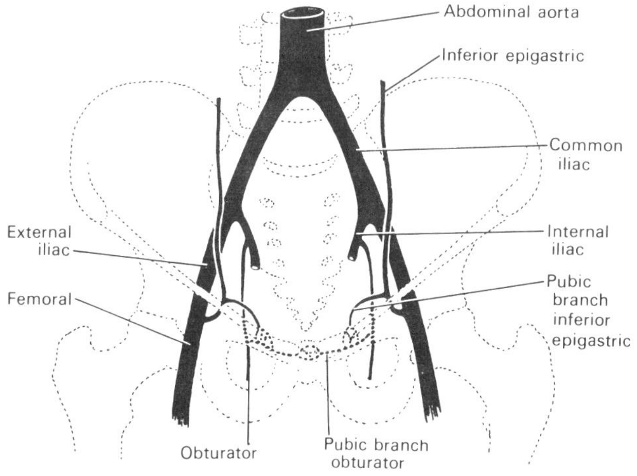

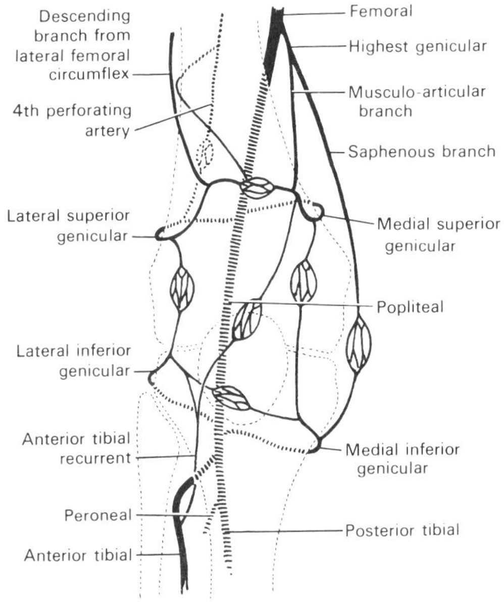

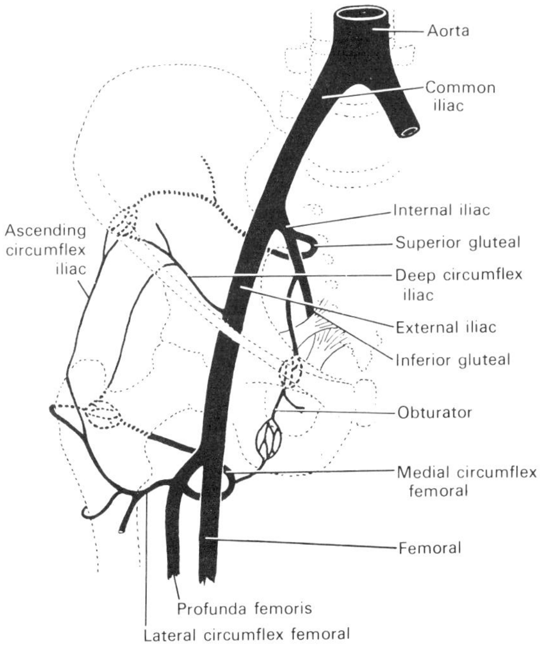

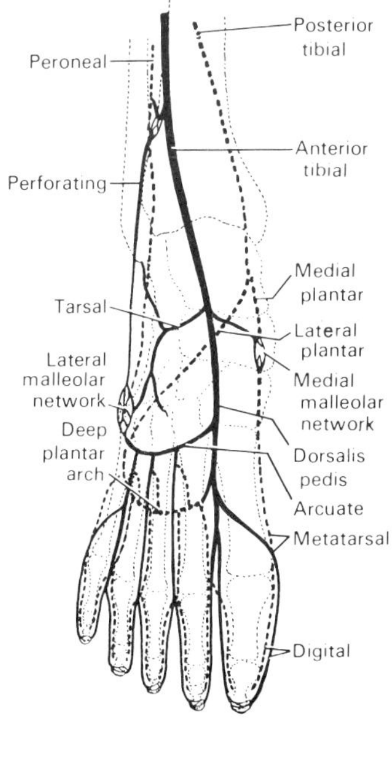

Fig. XI-24. Collateral circulation in lower limb. **A.** General schema of supply. **B.** Collateral anastomoses in the pubic region. **C.** Collateral circulation at pelvic girdle and hip. **D.** Collateral circulation about the knee. **E.** Collateral circulation of the ankle and foot.

where it divides into **anterior** and **posterior tibial arteries.** Its branches can be described briefly:

1. Two or three **superior muscular arteries** at its proximal end supply the Adductor Magnus and the hamstring group.
2. Two branches of fair size, the **medial** and **lateral sural arteries** distribute to Gastrocemius, Soleus, and Plantaris. They arise opposite the knee joint.
3. A host of small arteries supply the knee joint (medial and lateral superior genicular; middle genicular; medial and lateral inferior genicular). These all anastomose with the descending genicular from the femoral and the anterior and posterior tibial recurrent arteries to form complex superficial and deep arterial plexuses at the knee around the patella and the contiguous ends of the femur and tibia (Fig. XI-23 and 24).
4. Its terminal anterior and posterior tibial arteries.

The Anterior Tibial Artery (Figs. XI-23 and 24): This artery reaches the front of the leg by passing forward above the interosseous membrane that stretches between fibula and tibia. It runs down the anterior surfaces of this membrane much as did the volar interosseous branch of the ulnar artery in the arm. Before it terminates in the foot, the branches of the anterior tibial are:

1. **Anterior and posterior tibial recurrent arteries** to the knee joint and tibiofibular joint (Fig. XI-23)
2. The fibular artery (Fig. XI-23) though this can come from the posterior tibial artery as well at times. It wraps around the neck of the fibular, pierces the Soleus which it supplies, and ends in the Peroneus Longus.
3. Its musclar branches are many to the ankle extensors lying on either side of the vessel.
4. Anterior medial and lateral malleolar arteries to the ankle joint. The medial vessel on the big toe side passes under the tendons of the Extensor Hallucis Longus and Tibialis Anterior. The lateral malleolar artery passes deep to the tendons of the Extensor Digitorum Longus and Peroneus Tertius. These vessels, anastomosing with others in the area, forming complex medial and lateral malleolar networks.
5. The **dorsalis pedis artery** is the direct continuation of the anterior tibial onto the dorsum of the foot (Figs. XI-23 and 25). It passes to the great toe side of the foot to reach the proximal part of the first intermetatarsal spaces where it terminates. Its branches are: a) Lateral and medial tarsal arteries to the malleolar networks. The lateral tarsal supplies the Extensor Digitorum Brevis (Fig. XI-25). b) An **arcuate artery:** This passes laterally over the bases of the metatarsals under the Extensor Digitorum Brevis and it gives off the **second to fourth dorsal metatarsal arteries** which run to the webs between the toes. Here they branch to run down the continguous sides of adjacent toes as the **dorsal digital branches** (Fig. XI-25). c) The **first dorsal metatarsal artery** continues to the cleft between the first and second toes. A medial branch passes deep to the External Hallucis Longus to supply the medial side of the great toe. The lateral branch acts like a digital branch supplying contiguous sides of the first and second toes (Fig. XI-25). d) The **deep plantar artery:** This vessel penetrates deep into the sole of the foot between the two heads of the first Dorsal Interosseous muscle (Fig. XI-25). It unites below with the lateral plantar artery to complete the plantar arch (Fig. XI-25) and sends a branch, the **first plantar metatarsal artery** to act on the sole of the foot as b. and c. above acted on the dosrum of the foot.

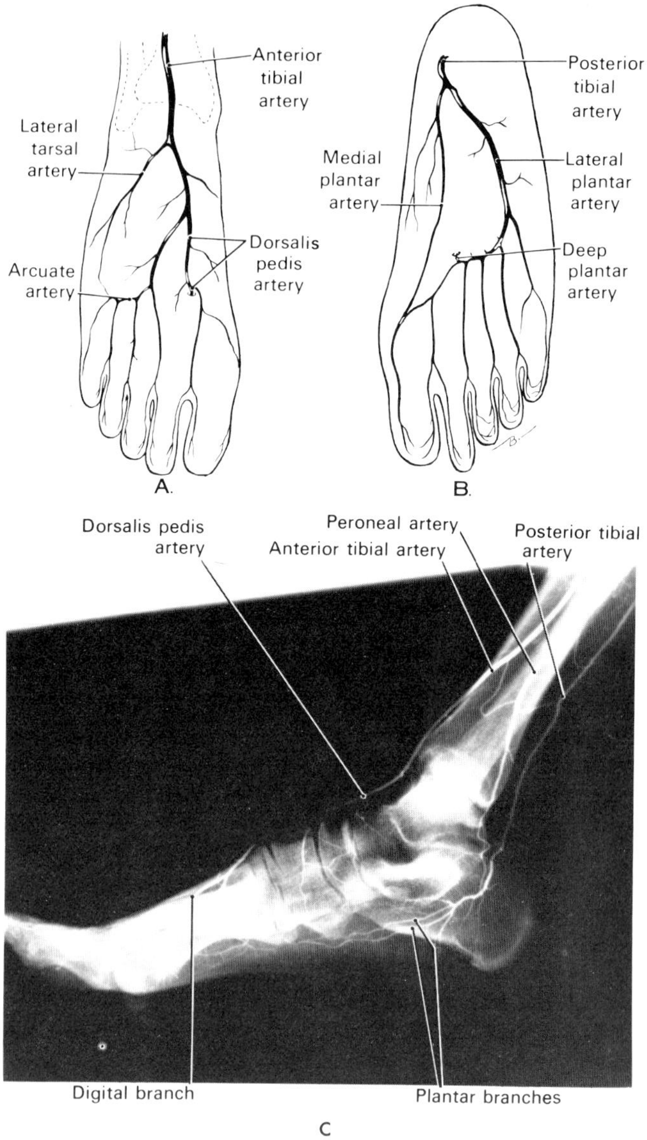

Fig. XI-25. Arteries of the foot. **A.** Anterior tibial becoming the dorsal pedis artery. **B.** Posterior tibial becoming medial and lateral plantar arteries. **C.** Lateral radiograph of the major vessels of the lower leg and foot injected with contrast medium.

The Posterior Tibial Artery (Fig. XI-23): This is, actually, the larger of the two terminal branches of the politeal artery and it remains on the posterior aspect of the leg as its name implies. It descends deep to the Soleus and the Gastrocnemius, but superficial to the deep flexor muscles. Under cover of the Abductor Hallucis, it divides into its **terminal medial and lateral plantar arteries** just behind the medial malleolus.

1. Its major branch in the leg is the **peroneal artery.** This artery arises from the posterior tibial about 2.5 cm. distal to the lower border of the Popliteus muscle. Since it rides deep upon the fibula, it has been known as the fibular artery. It supplies that bone with a nutrient vessel. Its muscular branches go to Soleus, Tibialis Posterior, Flexor Hallucis Longus, and the Peronei. It gives lateral branches to the ankle and heel.

2. The **tibial branch** of the posterior tibial artery is a nutrient artery for that bone. a) Its many muscular branches supply Soleus and the deep flexors of the ankle. b) Like its peroneal branch, it has tarsal and calcaneal branches to the medial aspect of the ankle and heel. c) The **medial plantar artery** is the smaller of its terminal branches in the sole of the foot, and is devoted principally to the muscles of the great toe. d) The **lateral plantar artery,** entering into the formation of the plantar arch, supplies the rest (Fig. XI-25).

The Pulmonary Circulation

Refer to pages 464 to 470 in the Chapter on the Thorax. The concepts are simple to grasp. We have expounded sufficiently there, and feel no need to extend this explanation further.

VENOUS SYSTEM

The Systemic Veins

The veins of the heart (venae cordis) can be quickly reviewed in Figure XI-2. All, except the **anterior cardiac veins** which empty directly into the right atrium, and the **venae cordis minimae** (Thebesian veins) which open indiscriminately into all chambers—all empty into the coronary sinus. To name the others will be sufficient here:

Great cardiac vein (left coronary)
Small cardiac vein (right coronary)
Middle cardiac vein (of the posterior intraventricular sulcus)
Posterior vein of left ventricle
Oblique vein (of Marshall) of left atrium.

The Veins of the Head and Neck

The Cranial Venous Sinuses (Fig. XI-26): These are wide-mouthed, endothelially-lined channels that course between the endosteal and periosteal layers of the dura, the outermost of the meninges covering the brain. Two of these sinuses run in the fold of dura that separates the medial surfaces of the cerebral hemispheres:

1. **The superior sagittal or longitudinal sinus** lies in the upper convex portion of the **falx cerebri** (Fig. XI-26A). The vessel begins at the crista galli and foramen cecum where it may communicate with nasal veins, and it proceeds posteriorly in the midline, grooving the under surfaces of the bones of the calvaria. By receiving blood from the superior cerebral veins of the dorsolateral hemispheric surfaces, it enlarges as it progresses. It receives diploic veins and emissary veins as well. Finally, it terminates, generally on the right side, in the horizontally-oriented **transverse sinus.**

2. In the bottom part of the flap-like falx cerebri runs the **inferior sagittal sinus,** (Fig. XI-26A). It has a short course to traverse in this sickle-shaped flap, and it terminates in the short **straight sinus.** It collects blood from the medial surface of the hemispheres.

3. The right and left **transverse sinuses** (Figs. XI-26A, B) are among the largest of the sinuses. They begin at the internal occipital protuberance and follow laterad along the margins of the horizontally disposed **tentorium cerebelli,** the tent-like roof separating cerebellum from the occipital lobes. They end at the base of the petrous bone where the tentorium anchors, and, from that point, the transverse sinuses on each side empty into the **sigmoid sinuses** (Fig. XI-26A, B).

4. The **sigmoid sinuses** (Fig. XI-26A, B) follow a short, sinuous course from the occipital squama to the mastoid angle of the parietal bone, to the mastoid part of the temporal bone and to their termination at the jugular process of the occipital bone. They are the great cess pools that drain all but one of the dural sinuses into the **internal jugular vein** as the latter makes its exit from the jugular foramen (Fig. XI-26A).

5. The **petrosquamous sinus,** when present, is a skinny channel traveling along the squamopetrous junction of the temporal bone. It empties into the transverse sinus.

6. The **occipital sinus** is also a small one and, perhaps, the smallest. It may be double. It rides in the **falx cerebelli** between the cerebellar hemispheres. It starts at the foramen magnum where it communicates with the internal posterior vertebral plexus of veins, and it ascends along the occipital bone to the confluence of sinuses (Fig. XI-26A, B).

7. The **straight sinus** (Fig. IX-26A) begins as a fairly robust triangular channel at the posterior end of the inferior sagittal sinus. As a rule, it terminates slightly left of the midline in the left transverse sinus. It receives the **great cerebral vein of Galen** which is formed by the two **internal cerebral veins.**

8. The **confluence of sinuses (Torcular Herophili)** (Fig. XI-26A, B). You may have noticed how many of the above (superior sagittal, straight, and occipital sinuses) all flowed to one collecting point near the internal occipital protuberance to be drained laterally to the sigmoid sinus by the transverse sinuses. This area is the confluence. The right transverse sinus usually receives the currents of flow from the superior sagittal sinus, the left, from the straight sinus.

There are, in addition, a series of dural sinuses more

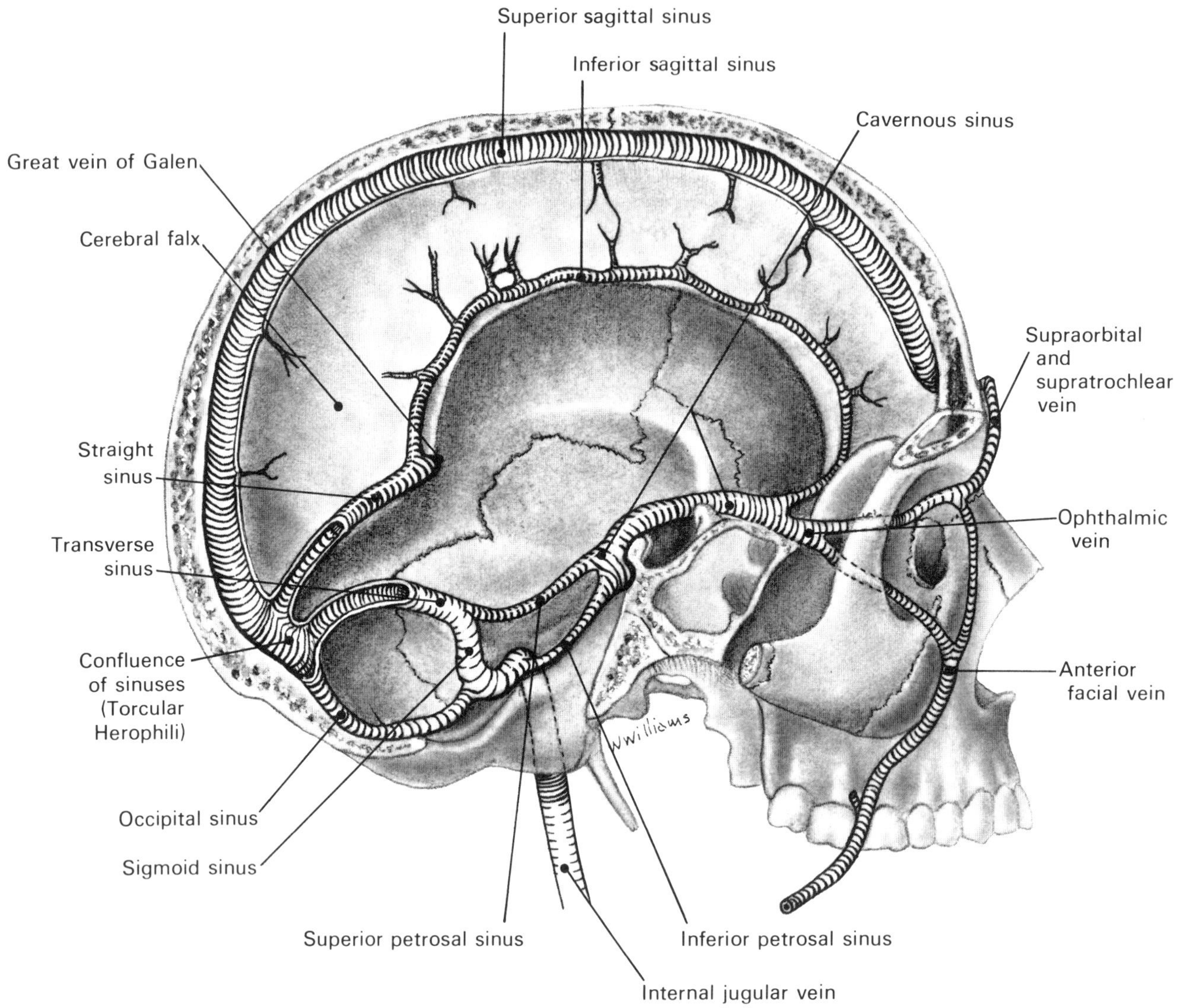

Fig. XI-26. Venous sinuses of the dura. **A.** Lateral view.

anteriorly placed in the floor of the cranium. The **cavernous sinus** is so named not because it is a massive cave or a ballooned structure but because it is a honeycomb of many little interconnecting caves. Its internal organization is more akin to the erectile tissue of the corpus cavernosum penis. It surrounds the sella turcica, and in its walls, or through it, will plow: **the internal carotid artery, cranial nerves III, IV, VI,** and the **ophthalmic and maxillary divisions of V** (Fig. XI-26A, B). It receives: a) superior and inferior ophthalmic veins, b) the sphenoparietal dural sinus, c) the pterygoid plexus of veins (through foramina ovale and lacerum), and d) the angular and facial veins (through superior opthalmic vein). It drains by two routes: a) through superior petrosal dural sinuses to transverse sinuses and b) through inferior petrosal sinuses *directly* into internal jugular vein.

9. **The sphenoparietal sinuses** (Fig. XI-26B) are narrow channels along the free margins of the lesser wings of the sphenoid. They receive temporal scalp veins through diploic veins.

10. The **superior petrosal sinus** (Fig. XI-26A, B) runs along the ridge of the petrous part of the temporal bone to join the transverse sinus just as the latter empties into the sigmoid sinus.

11. If the superior petrosal sinus takes the high road from the cavernous sinus, the **inferior petrosal sinus** takes the low road and gets to the internal jugular vein before it (Fig. XI-26A). It follows the lateral margins of the clivus between petrous part of temporal bone and the basilar part of the occipital bone. It is its own boss. It makes direct contact with the internal jugular vein, employing no intermediary to reach it.

12. The **basilar plexus** consists of several small interlacing dural venous channels over the basilar part of the occipital bone. They drain the anterior vertebral plexus of veins and dump into the inferior petrosal sinuses.

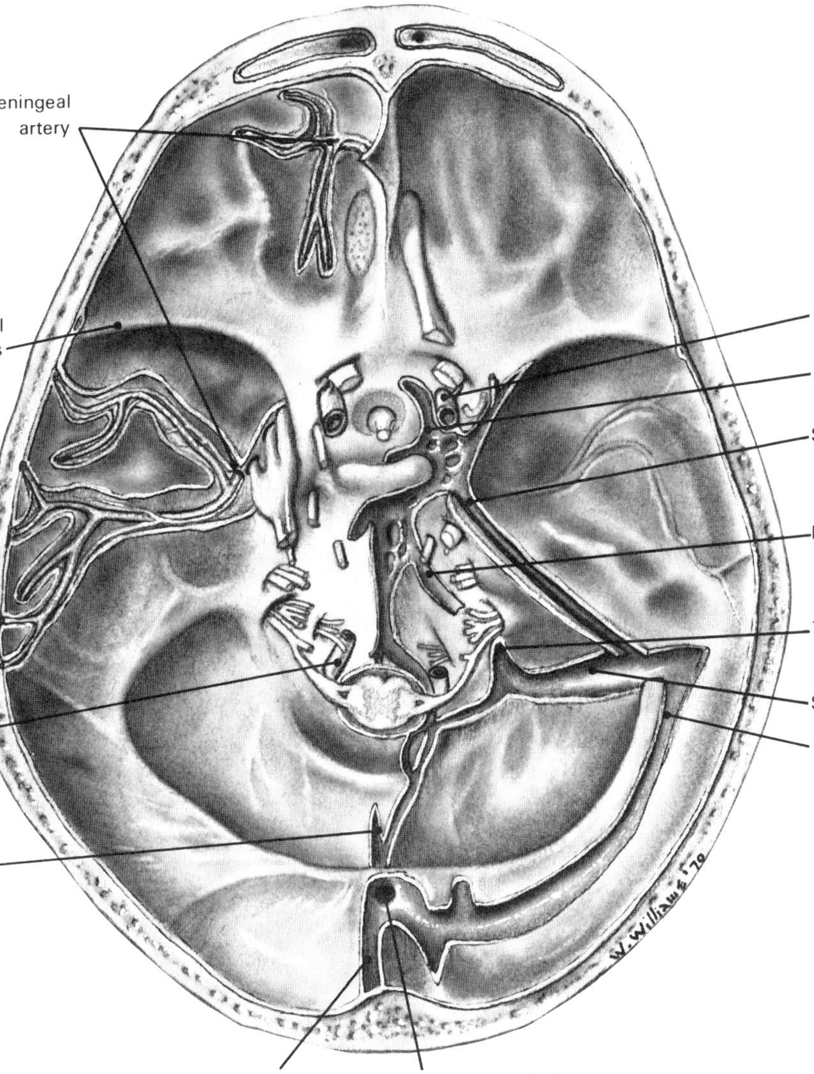

Fig. XI-26B. View of dural sinuses from above after removal of calvaria and brain.

Diploic Veins of the Skull: We have mentioned these channels before in the substance of the cranial bones. As sutures ossify, the channels may form linkages between bones where they were once confined to separate bones. Four are well known because of their connections with the dural sinuses:

1. Frontal Diploic: Connects supraorbital vein to superior sagittal sinus.

2. Anterior Temporal Diploic: This vein, strangely, is confined mostly to the frontal bone, but it connects one of the deep temporal veins by way of a hole in the greater wing of the sphenoid to the sphenoparietal sinus.

3. Posterior Temporal Diploic: It is truly in the parietal bone and ends in the transverse sinus either through an opening at the mastoid angle or through the mastoid foramen.

4. The Occipital Diploic Vein: It is the largest of these internal bony channels and it opens into the confluence or the transverse sinus. It connects externally with the occipital vein.

Emissary Veins: There are a number of these, (some are most inconstant) which connect scalp and other parts of the head and neck to dural sinuses. Some of these interlink through diploic channels to achieve this end. Flow is, generally, away from the brain. A few of the more constant ones (some of which we may have already named in passing) are:

1. Mastoid Emissary: This connects the posterior auricular or the occipital vein of the scalp to the transverse sinus through the mastoid foramen. It is apt to employ diploic channels.

2. Parietal Emissary: This also employs diploic veins to connect scalp veins to the superior sagittal sinus.

3. Rete of the Hypoglossal Canal: This network of fine channels joins vertebral veins and deep veins of neck with the transverse sinus.

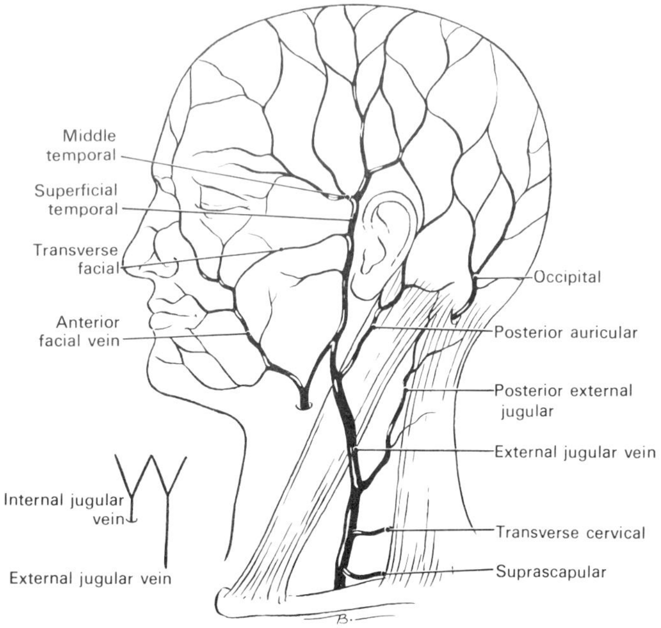

Fig. XI-27. Superficial veins of face and scalp. *Inset* shows a "W" formed as some vessels enter the internal jugular vein high in the neck while others join the external jugular vein superficial to the Sternomastoid muscle.

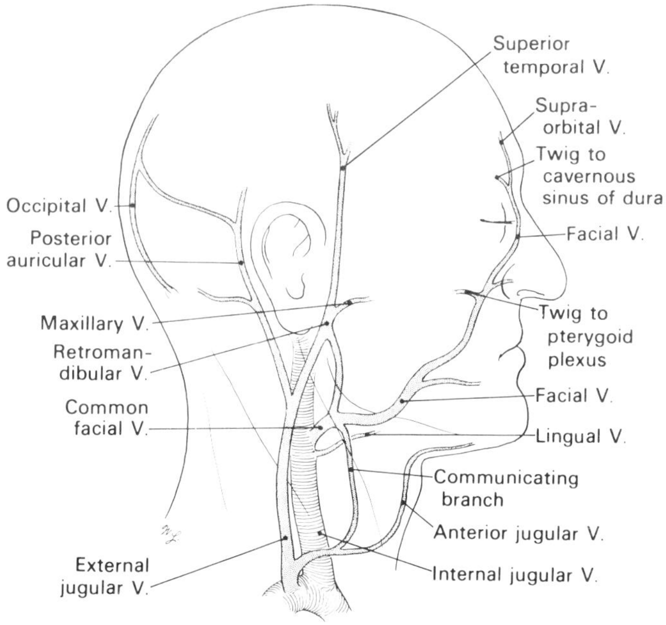

Fig. XI-28. Venous drainage of the head.

4. Condyloid Emissary: This vein connects deep veins of neck with transverse sinus through a canal in the occipital condyle.

5. Rete of Foramina Ovale and Lacerum: These tiny network emissaries connect the pterygoid plexus of veins with the cavernous sinus.

6. Vein of the Foramen Cecum: We have already mentioned this vein as a route between the nose and the superior sagittal sinus.

The Great Veins of the Neck: These return blood from the head, face, and cervical regions. The large collecting trunks are three in number (Fig. XI-27 and 28):

1. Internal jugular veins
2. External jugular veins
3. Vertebral veins

The simple arrangement of the **superficial** drainage of the face and head is shown in Figure XI-27. A pattern in the form of "W" is formed (but don't always bet on it) with some veins draining into the internal jugular, others into the external jugular.

Internal Jugular Vein (Figs. XI-27 and 28): Needless to say, but we say it for emphasis anyway, the **internal jugular** alone drains the cranial and orbital cavities. At the same time, this great trunk drains a considerable portion of the face and neck. Thus, as it passes down through these regions it receives many tributaries. In this respect it differs from its companion artery (internal carotid) which runs through the area without branching until it enters the cranium. We saw the internal jugular vein exit from the skull at the jugular foramen after receiving the flow from the dural sinuses. Here it is dilated into the **superior bulb.** It takes a vertical course through the neck within the carotid sheath where it lies somewhat posterolateral to the internal and common carotid vessels. At the root of the neck its junction with the subclavian vein forms the **brachiocephalic vein** on its own side. There are **right** and **left branchiocephalic veins** and these lie anterior to the single, right brachiocephalic artery and the left common carotid artery. Just before it ends, the internal jugular vein has a valve, and the vessel expands into an **inferior bulb.** Its tributaries are:

1. The **lingual veins** which join the internal jugular directly about at the level of the hypoglossal nerve. A branch from the tip of the tongue may first enter the **common facial vein** which joins the internal jugular lower down (Fig. XI-28).
2. The **pharyngeal** veins: These may enter the internal jugular directly or through the lingual veins, the superior thyroid vein or the common facial vein.
3. The **superior and middle thyroid veins.**
4. The **common facial vein** (Fig. XI-28) enters somewhere below the level of the hyoid bone. Since it is formed by a junction of the **anterior** and **posterior facial veins,** blood from the following sources is also returned to the internal jugular vein: a) frontal and supraorbital veins of forehead and scalp (Fig. XI-28), b) angular vein of side of nose, c) superior and inferior palpebral veins, d) superior and inferior labial veins, e) buccinator and masseteric veins, f) blood from pterygoid plexus through the deep facial vein to the posterior facial vein, g) below the mandible, from the submental, palatine, and submandibular veins, and h) part of the drainage from in front and behind the

auricle, and from the temporomandibular joint and auricle *via* the **posterior facial vein.** An anterior branch of this latter vein also connects with the external jugular vein.

External Jugular Vein (Fig. XI-27): This more superficial venous trunk drains the rest of the external face and neck, leaving some deep parts of the neck to be drained by the vertebral veins. The external jugular begins in the substance of the parotid gland and courses superficial to the Sternomastoid muscle to end in the subclavian vein. Near its termination, it has two sets of valves, one as it enters the subclavian, and one set a bit above the level of the clavicle. The tributaries of the external jugular vein are:

1. POSTERIOR FACIAL VEIN (OR RETROMANDIBULAR VEIN) (FIG. XI-28): This vein is formed by a union of: a) superficial temporal vein (Fig. XI-28), b) internal maxillary vein (Fig. XI-28), and c) posterior auricular vein which enters the **posterior division** of the posterior facial vein (Fig. XI-28).

Thus, the posterior facial vein can drain similar territories (posterior auricular region, the temple, and the deep pterygoid plexus) into the external jugular vein as we see it do here; or, by its anterior division, into the internal jugular vein as mentioned above.

2. POSTERIOR EXTERNAL JUGULAR VEIN: This vein lies between the Splenius and Trapezius muscles and returns blood from the skin and muscles of the upper posterior neck region. It joins the external jugular about midway in the neck.

3. TRANSVERSE CERVICAL AND TRANSVERSE SCAPULAR VEINS: These tributaries drain the lower neck and shoulder region and reach the external jugular near its termination.

4. ANTERIOR JUGULAR VEIN (FIG. XI-28): This also reaches the external jugular near its termination. It begins at the hyoid bone and lies on the Sternohyoid muscle, draining submandibular region and the supra- and infra-strap muscles of the anterior neck. It forms a **transverse venous arch** with its fellow across the midline just above the sternum, and, in this region, it drains the larynx and thyroid gland. It is also a vein that sends side branches into the internal jugular so that venous drainage from the anterior neck can be returned by both great venous trunks.

5. OCCIPITAL VEIN (FIG. XI-28): This vein, draining the back of the scalp and neck, is another one which can't seem to make up its mind about which major route to take. If it follows its companion artery, it ends in the internal jugular vein. If it joins the posterior auricular vein, it becomes a tributary of the external jugular.

The Vertebral Veins: These veins form *outside* of the cranium in the suboccipital region from many small tributaries in the area, and from this plexus they leave the vertebral canal by crossing above the posterior arch of the atlas. They then enter a foramen in the transverse process of the atlas and descend through the processes of the other cervical vertebrae as a dense plexus of vessels that surround the ascending vertebral artery. They leave the foramen in the transverse process of VC_6 to descend and end in the innominate (brachiocephalic) vein of their own side. Tributaries of this system of veins are:

1. Connections at the upper end with the transverse dural sinus *via* the condyloid canal of the occipital bone.

2. Anterior vertebral vein. This leaves the foramina of upper cervical transverse processes and follows the course of the ascending cervical artery in the neck only to rejoin the vertebral system near its termination.

3. Deep cervical vein. This is another deviate that leaves the parent vertebral plexus to follow the artery of the same name through the neck. But it rejoins the vertebral vein before the latter ends in the brachiocephalic vein.

BLOOD SUPPLY OF SPINAL CORD AND VERTEBRAL COLUMN

At this point we ought to realize that much of the blood supply, both arterial and venous, for the spinal cord and vertebral column is contributed by the vertebral vessels with assists from a number of others along the length of the segmental back bone.

Arterial Supply

We have noted that the **anterior** and **posterior spinal arteries** (Fig. XI-29) were recurrent branches arising from the vertebral arteries *after* the latter had entered the foramen magnum.

1. As the **anterior spinal artery** (formed by a union of medially directed branch from each vertebral artery on the ventral aspect of the medulla) descends in the median ventral sulcus of medulla and cord, it is reinforced by a series of branches which enter certain of the intervertebral foramina along the length of the vertebral column. Such contributing twigs come from: a) the ascending cervical, inferior thyroid and the vertebral arteries, themselves, in the neck region, b) internal intercostal arteries at thoracic levels, and c) lumbar, iliolumbar, and lateral sacral arteries in the abdominal and pelvic regions.

These entering branches not only complete the median anastomotic chain, which is the **anterior spinal artery** begun by the vertebrals cranially, but they supply radicular twigs to spinal rootlets (the lower ones to filaments of the cauda equina); a supply to the meninges of the cord; and lateral, circumferential branches around the periphery of the cord to anastomose with the descending chains of the **posterior spinal arteries** (Fig. XI-30).

The anterior (median) spinal artery, itself, sends deep twigs into the cord to supply ventral and ventrolateral horns of gray matter and the ventral and lateral funiculi of white matter. It may even reach the basal portions of the dorsal horn of gray matter.

2. The **posterior spinal arteries,** unlike the anterior spinals do not fuse into a single anastomotic chain but remain as two separate channels (Fig. XI-29) de-

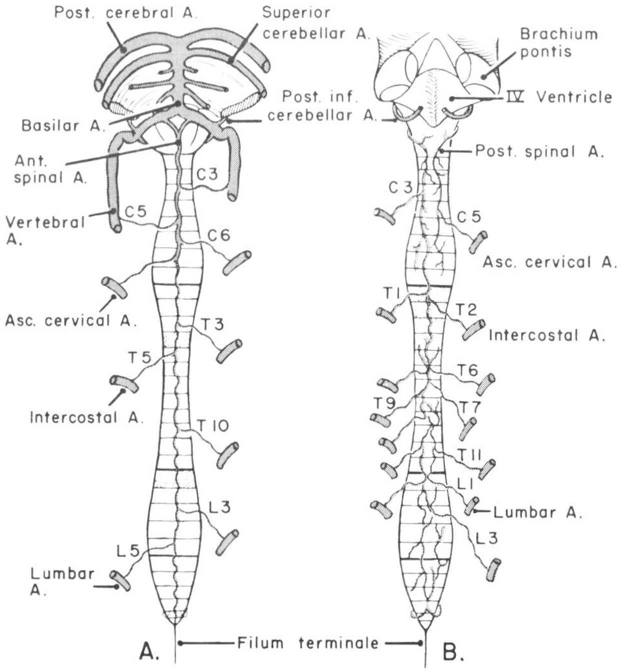

Fig. XI-29. Arteries of the spinal cord. **A.** ventral aspect; **B.** dorsal aspect. The stippled areas of the cord are those most vulnerable to vascular deprivation when contributing stem vessels are injured. The levels of entry of common radicular branches are given (C_5, T_5, etc.). (From Truex and Carpenter *Human Neuroanatomy*, 6th ed., Williams & Wilkins, 1969.)

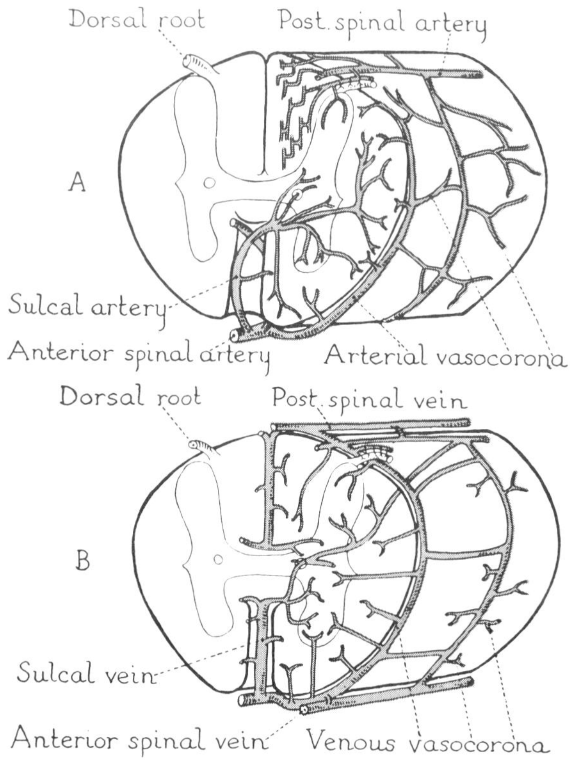

Fig. XI-30. Schema of blood supply to a section of the spinal cord. **A.** Arterial. **B.** Venous. (From Truex and Carpenter, *Human Neuroanatomy*, 6th ed., Williams & Wilkins, 1969.)

rived as a branch from each vertebral artery at the level of the caudal medulla. The chains descend along the dorsal intermediate sulci of the cord reinforced by radicular arteries from the dorsal rootlets. These latter arteries are contributed by the same incoming branches mentioned above. The posterior spinal chains are not as robust as the anterior spinal chain. They supply the dorsal gray matter and the posterior funiculi of the cord by deep penetrating twigs (Fig. XI-30); the meninges; and anastomose with the lateral circumferential branches mentioned above.

3. The same incoming twigs, arriving through intervertebral foramina serve as **spinal branches** to the **vertebral bones.** As each enters the **vertebral canal** it forms ascending and descending branches that anastomose with vessels that enter intervertebral foramina at successively lower levels. It should be noted that not every intervertebral foramina serves to conduct such branches to spinal cord and vertebrae. The number and locus of foramina and branches involved is quite variable in different individuals, but are adequate, for the most part, to build anastomotic chains. Spinal anastomotic channels are built up laterally along the *posterior surfaces of the bodies* of each vertebra within the vertebral canal close to the pedicles. Deep branches penetrate bodies and vertebral processes.

Venous Drainage

Drainage of the bony column, meninges, the cord substance, and even neighboring muscles attached to the column is by way of very intricate venous plexuses disposed along the length of the column. There is an outer, **external venous plexus** around the outer margins of the vertebrae and there are **anterior and posterior deep venous plexuses** within the vertebral canal itself in the epidural fatty tissue between dura and bony canal.

1. External Plexus (Fig. XI-31): This plexus around bodies, spines, and transverse processes of vertebrae serves to connect with tributaries of the diploic vessels emanating from within the bony parts of the vertebrae. The external plexus drains chiefly into vertebral, occipital, and deep cervical veins because it is best developed in the neck. Elsewhere **intervertebral veins** connect this plexus (as well as the deep plexus) with intercostal, lumbar, and lateral sacral veins.

2. Deep Vertebral Plexuses (Fig. XI-31): These lie along the anterior and posterior aspects of the internal surfaces of the vertebral canal. At each level anterior and posterior veins communicate with each other through **venous rings** which encircle the canal. Drainage is through bone *via* **basivertebral veins** which

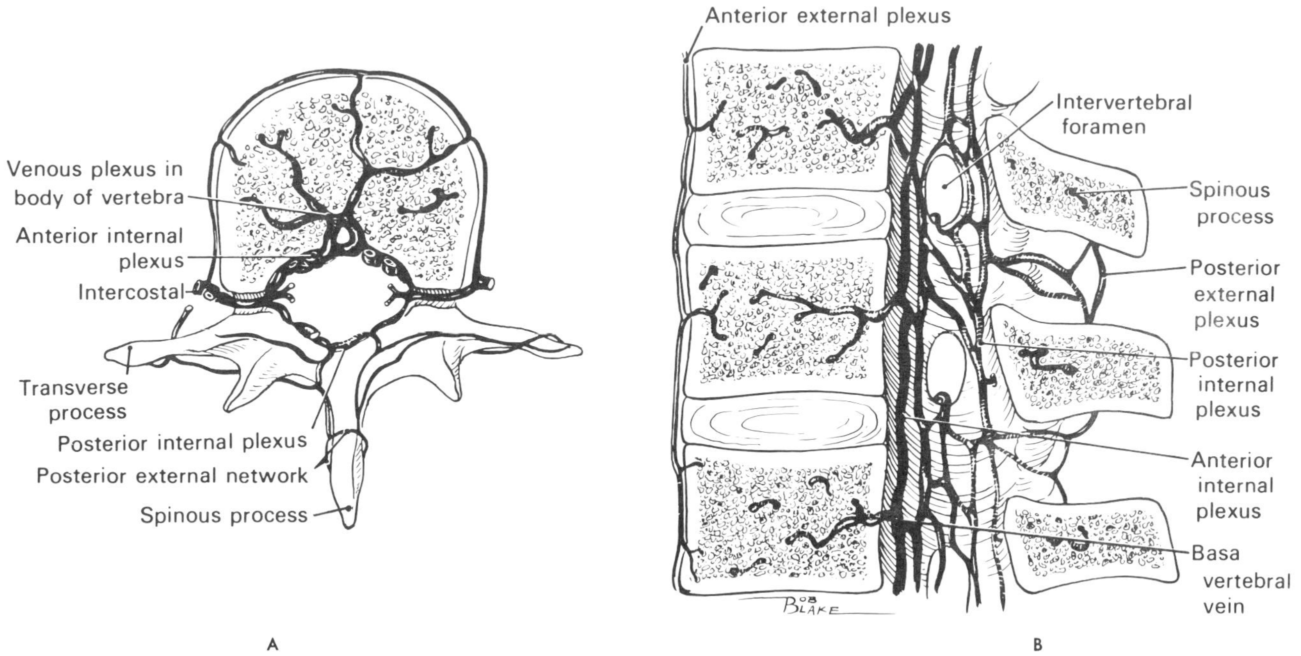

Fig. XI-31. Vertebral venous plexuses. **A.** Viewed in transverse section. **B.** Viewed in midsagittal section.

reach the external plexus by burrowing through vertebral bodies; or by the **intervertebral veins** which leave certain of the intervertebral foramina. (See 1 just above.)

3. Longitudinal Plexuses: There are a series of finer, **longitudinal plexuses** of veins *in the pia mater of the cord itself.* These form longitudinal channels in front of the anterior fissure of the cord; behind the posterior sulcus; and behind the nerve rootlets. They drain into the deep vertebral plexuses mentioned above.

Batson's Venous Channels

In the "40's" Batson[1, 2] discovered that the veins of the vertebral column constituted a collateral circulation of a most unique and, perhaps, unexpected route. When he injected radiopaque materials in the dorsal vein of the penis in cadavers, the material was visualized by X-ray in the veins of the vertebral column, skull, and interior of the cranium. The routing was *via* the prostatic plexus of veins into the veins of the sacrum, ilium, and lumbar vertebrae, by-passing the inferior vena cava entirely.

When Batson injected small veins of the breast, again in cadavers, the injection medium found its way into the clavicle, intercostal veins, the head of the humerus, the cervical vertebrae and the dural sinuses. Up to this point, one might have written off this entire routing by pointing out that pressures used to move injection materials through dead tissues have little bearing upon flow patterns in the living.

So Batson injected thorium dioxide into the dorsal vein of the penis in a living monkey. It is not reported who held the monkey. The material went directly into the inferior vena cava in the unrestrained monkey seeming to prove our last point. However, when intra-abdominal pressure was increased by application of a belly-band, a kind of monkey girdle, the material reached the vertebral veins readily. This simulated the pressure increases that occur in coughing or straining (as in defecation). Batson pointed out that these routes were commonly used for the spread of metastases from tumors, routes taken from prostate and breast, to reach bony structures of the pelvis, shoulder girdle, vertebral column and, eventually, the venous channels in the head.

The Superior Vena Cava (Fig. XI-32)

This 7 cm.-long great vessel is formed by the meeting of the two brachiocephalic (innominate) veins. It should be obvious that since these latter vessels are themselves formed from the junction of internal jugular and subclavian veins, that the superior vena cava drains the entire head and upper limb.

We have just discussed the venous return from the head and neck. The return of deep veins to the upper limb was outlined earlier in this chapter (p. 704). It should be evident that were you to begin at the superficial and deep palmer arches, and were you to enumerate the names of the arteries in reverse order, you would have the names of the deep veins. You would also locate them approximately since they are companion veins to the arteries, and they would all lead, eventually, to the subclavian vein. We have outlined the exceptional superficial veins of the upper extremity in this chapter (p. 704). We found them emptying into the brachial or the axillary veins, both of which are direct routes to the subclavian trunk. Figures XI-33A and B should provide sufficient review of these structures.

The superior vena cava, of course, also returns blood

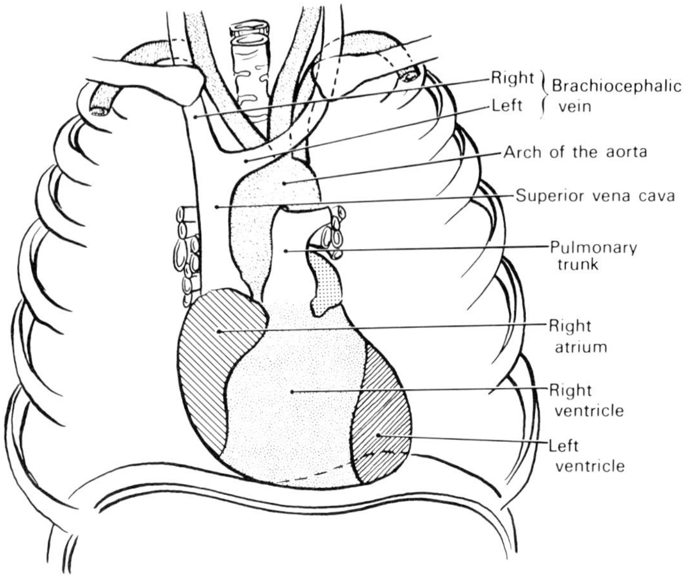

Fig. XI-32. Superior vena cava and brachiocephalic veins.

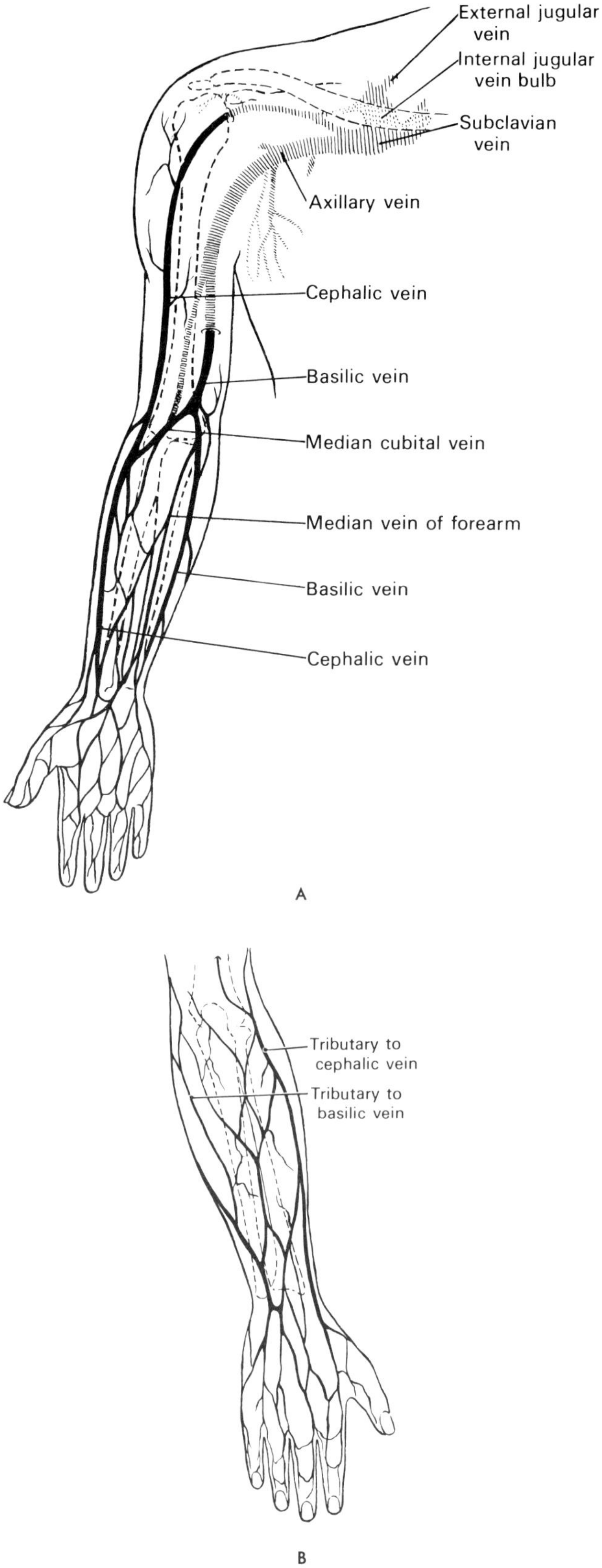

Fig. XI-33. Superficial veins of the upper extremity. **A.** Flexor surface. **B.** Extensor surface.

to the right atrium from the entire thorax except for the return from the lung which enters the left atrium. This thoracic drainage reaches the superior vena cava through the **azygos system of veins.** Indeed, because this system of vessels anastomoses with lumbar veins, parts of the abdominal and pelvic walls are serviced by the superior vena cava in a circuitous way.

This main caval conduit, whose chief task is to drain most of the entire body above the Diaphragm, begins just below the costal cartilage of the first rib and it ends in the right atrium at the upper border of the the third costal cartilage. Like aorta and pulmonary trunk, where it relates to the pericardial sac it is invested with pericardium.

The Azygos System of Veins in the Thorax (Fig. XI-34)

1. The Azygos Vein: This begins opposite the first or second lumbar vertebra. It can start as an offshoot of the right renal vein or the inferior vena cava; or of the ascending lumbar vein. It travels into the thorax through the aortic hiatus and lies under the parietal pleura along the right side of the vertebral column. It continues upward in this position to the level of VT_4 before it arcs anteriorly over the hilus of the right lung to empty into the superior vena cava. Its tributaries are: a) right subcostal vein near the Diaphragm, b) upper three or four intercostal veins, including the superior (supreme) intercostal vein, c) some esophageal, mediastinal, and pericardial veins, d) the right bronchial vein, e) the hemiazygos vein of the left side.

2. The Hemiazygos Vein: It starts in the abdomen at the same level as the azygos but on the left side either from the left renal or as a branch of the ascending lumbar vein. It pierces the left diaphragmatic crus

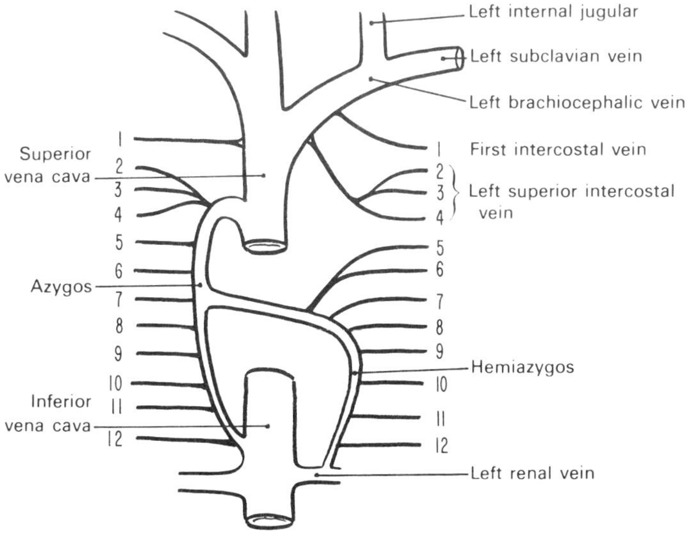

Fig. XI-34. The azygos system of veins.

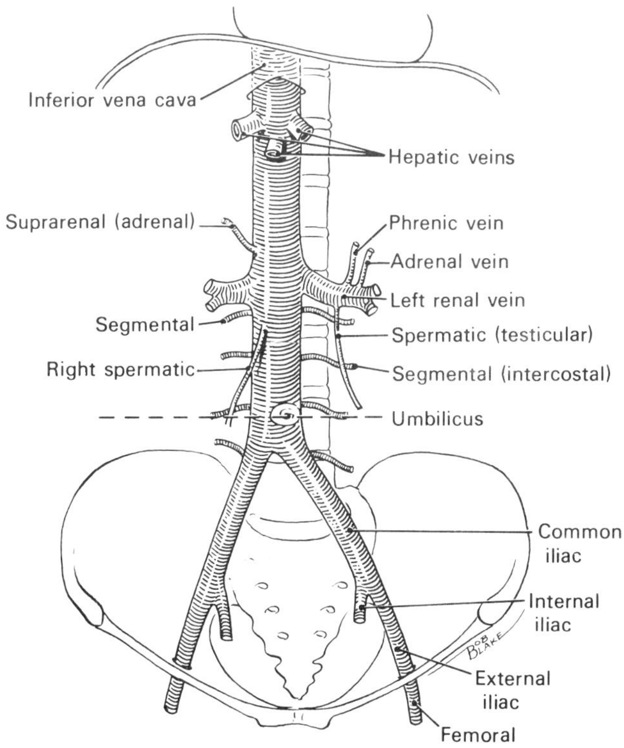

Fig. XI-35. Abdominal branches of the inferior vena cava.

and ascends for a variable distance to the left of the vertebral column, often no higher than VT_9. At its highest point it crosses the bodies of the vertebral column behind parietal pleura to end in the azygos vein on the right side. Its tributaries are in general: a) lower intercostal veins, b) left subcostal veins, c) mediastinal and esophageal veins.

3. The Accessory Hemiazygos: Sometimes this vein is in an integral part of the hemiazygos vein which would then have reached its highest level in the thorax. If it is a vein in its own right, it *descends* in the left thorax and crosses the vertebral column to join the azygos on the right side at about the level of VT_8. Its tributaries are: a) the veins of the upper intercostal spaces on the left not covered by the hemiazygos, b) the left bronchial vein (sometimes this vein, the left supreme intercostal vein and the left superior phrenic vein all empty into the left brachiocephalic veins).

The **brachiocephalic veins** (Fig. XI-32) besides receiving the vertebrals, the subclavians and the internal jugulars, receive the **inferior thyroid veins,** the **internal thoracics** (mammaries) and the **right** and **left thoracic lymph channels** about where subclavian and internal jugular veins meet.

Venous Drainage of the Abdomen

In the last analysis, abdomen, pelvis, and lower limb drain into the right atrium through the other major caval trunk, the inferior vena cava.

The Inferior Vena Cava

This great vessel is formed to the right of VL_5 where the common iliac veins join. The large trunk ascends along the vertebral bodies, keeping to the right of the aorta. It grooves the posterior surface of the liver as it moves to its hiatus in the Diaphragm. In piercing this partition it angles slightly medialward to pierce the fibrous pericardial sac. It enters the lower portion of the right atrium almost in direct line with the superior vena cava which enters the same chamber from above. Its thoracic course is less than a half inch, but, in its entirety, the inferior vena cava is the largest blood vessel in the body.

Abdominal Branches of the Inferior Vena Cava (Fig. XI-35)

1. Common iliac veins
2. Lumbar veins (four pair)
3. A pair of testicular veins emerging from the pampiniform plexus along the right and left spermatic cords (ovarian veins in the female)
4. A pair of renal veins
5. A pair of suprarenal veins
6. A variable number of hepatic veins emerging from the substance of the liver close to the Diaphragm

Since the inferior vena cava lies so far to the right side, both lumbar veins and renal veins are longer on the left side of the body than they are on the right, in order to reach this large vessel. As a consequence, too, the left suprarenal and the left testicular vein drain into the left renal vein and do not enter the inferior vena cava directly as do their partners on the right side. Moreover, none of the return from the veins of the digestive tract reaches the inferior vena cava directly. The liver and a portal system intervene so that the return is eventually effected by the hepatic veins.

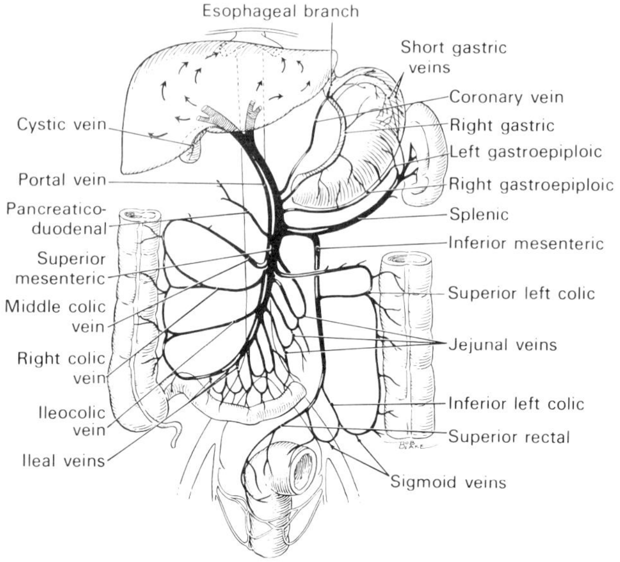

Fig. XI-36. The portal vein and its system of tributaries. Anastomoses at the proximal and distal ends of the gastro-intestinal tract are shown between the portal system and the esophageal veins on the one hand, and the portal system and the inferior hemorrhoidals on the other. There are other connections (not shown here) between the portal and systemic circulation.

The Portal Vein and Its Tributaries

Blood sent to the G-I tract, liver, pancreas, and spleen by the three unpaired arteries of the aorta is returned *via* the hepatic portal system over the portal vein (Fig. XI-36). A portal system is not called such because the vessels enter a "gateway" or portal crevice in some specific organ. Its definition is broader than this, and signifies that blood collected from one capillary bed is passed through a secondary capillary bed. In the abdomen this means that blood, collected from the capillary beds of the intestinal villi, from the walls of the gut and the mucosa of the colon and stomach, and from pancreas and spleen is returned by veins to empty into the special, wide, sinusoidal capillaries between the cords of the liver parenchyma.

The **portal vein** of some 8 cm. in length is formed near the pancreatic head by three major vessels: **superior mesenteric, splenic,** and **inferior mesenteric veins,** though the third vessel, just named, generally enters the second before the second joins with the first.

The portal vein ascends in the right border of the lesser omentum near its free edge at the foramen of Winslow. At the hepatic portal it divides into its right and left branches to enter the liver along with accompanying branches of the hepatic artery. a) Its **right branch** receives the **cystic vein** before it empties into the right lobe. b) Its small distal **left branch** enters caudate and left lobes of the liver. Its two chief proximal tributaries are: a) **splenic vein,** which receives: short gastric veins, left gastro-epiploic vein, pancreatic veins, and **inferior mesenteric vein** (superior hemorrhoidal vein, sigmoid vein, and left colic vein). b) **superior mesenteric vein** which receives intestinal veins, iliocolic vein, right colic vein, middle colic vein, right gastro-epiploic vein, pancreaticoduodenal veins, coronary (vein of lesser curvature of stomach), and pyloric vein.

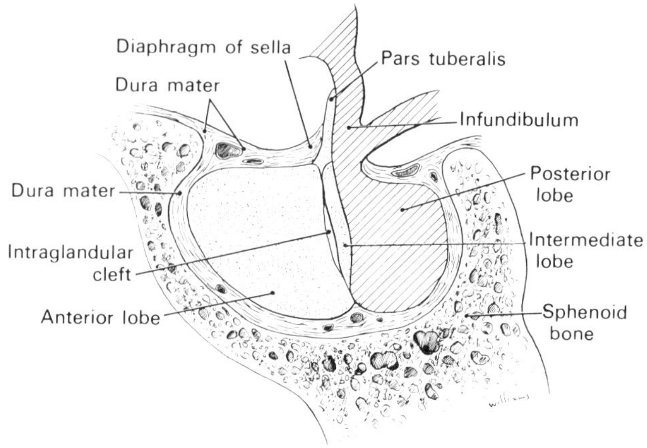

Fig. XI-37. The principal parts of the pituitary gland (Hypophysis).

We have discussed the important channels of the portal collateral circulation previously. Recall quickly, that these were:

1. Paraumbilical veins
2. Connections of gastric with esophageal veins
3. Connections between inferior mesenteric vein with the pelvic inferior hemorrhoidal veins
4. Diaphragmatic veins with azygos system
5. The veins of Retzius (intestinal branches to body wall)

Liver sinusoids drain into:

1. Central veins of liver lobules
2. Sublobular veins
3. Hepatic veins. These valveless veins, three in an upper group and a number of small veins constituting a low group, leave the liver substance to enter the inferior vena cava where that vessel grooves the posterior surface of the liver.

The student should be aware that this is not the only portal system in the body. Another smaller, but equally important, portal system is that related to the **pituitary gland** or **hypophysis** (Fig. XI-37).

The Hypophyseal Portal System: To understand the system, realize that the gland consists of: (1) an **anterior lobe** (adenohypophysis) derived from the entoderm of Rathke's pouch (see Chapter I, page 21), and that this part has no direct connection with the basal surface of the brain; (2) and a **posterior lobe,** or **neural lobe,** derived from the neural tube and connected to the diencephalon by a stalk or **infundibulum.** The anterior lobe of the pituitary is further

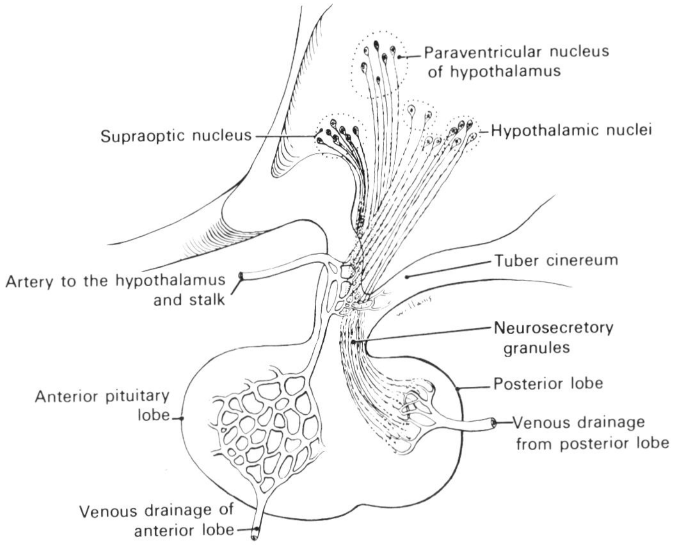

Fig. XI-38. The hypophyseal portal system. Neurosecretions from the supra-optic and paraventricular regions are conveyed *via* the supra-opticohypophyseal tract of nerve fibers to capillary beds in the posterior or neural lobe. Other types of hypothalamic neural humors are conveyed by the hypophyseal portal system to the anterior lobe. See text for more detailed discussion.

subdivided into three parts: (1) the **pars distalis;** (2) the **pars intermedia;** (3) and the **pars tuberalis.** The pars distalis is what we ordinarily think of when we speak of the anterior lobe in a histological or functional sense. It is the bulk of that glandular lobe. The pars intermedia is a small portion lying between pars distalis and the posterior lobe. It may be somewhat cystic at times or contain thyroid-like follicles. The pars tuberalis is simply an upward extention of the pars intermedia along the pituitary stalk or infundibulum (Fig. XI-37).

With this orientation, Figure XI-38 may begin to make some sense. Small arterial twigs from the internal carotid artery or from the posterior communicating artery of the circle of Willis form a capillary plexus in the pars tuberalis. The venous side of the plexus drains into a portal system of veins which empty into the capillary sinuses of the anterior lobe (pars distalis). The infundibulum leading to the posterior lobe does not appear to share this portal supply. It has its own arterial supply and its own venous drainage. If the anterior lobe gets an additional arterial supply of its own (like the liver from the hepatic artery), it is small in comparison to the portal venous supply which reaches it.

Axons of nerve cells in the hypothalamus in the region of the median eminence liberate humoral substances into the capillary plexus of the pars tuberalis. When these substances are carried to the anterior lobe *via* the hypopsyseal portal system, they either facilitate or inhibit glandular secretion. On the other hand, fibers from the supra-optic region of the hypothalamus by way of the supra-opticohypophyseal tract traverse the infundibulum to release ADH. (antidiuretic hormone) into the area of the capillary beds of the posterior lobe.

Venous Return Over Common Iliacs: We are left finally to consider the venous drainage of the pelvis and lower limbs.

The **middle sacral veins** parallel the median sacral artery and join to form a single vein which empties generally into the **left common iliac,** or into the angle between the two iliacs.

Tributaries of the Internal Iliac: Except for the fetal umbilical vein which passes cranialward toward the liver, and for the iliolumbar veins which join the common iliacs on each side as a fifth pair of lumbar veins in the region of VL_5, all the other tributaries of the internal iliac or hypogastric vein correspond with the arterial branches in the region. These are:

1. Superior gluteal veins
2. Inferior gluteal veins
3. Internal pudendals
4. Obturators
5. Lateral sacrals
6. Middle rectals
7. Dorsal veins of penis
8. Vesical veins
9. Uterine veins
10. Vaginal veins

So far as the tributaries of the **external iliacs** are concerned you need only note that they have **inferior epigastric, deep circumflex iliac,** and **pubic** veins which correspond to similar arteries in the inguinal and pubic area. Name the deep arteries of the lower limb in reverse from the plantar arches upward, and you have the names and general locus and course of the deep veins of the lower extremity. Figure IX-39 reviews the superficial veins of the lower limb. If the inferior vena cava is the *largest* blood vessel in the body, the **great saphenous veins** are the *longest,* beginning at the medial margin of the foot and ending in the femoral vein about 3 cm. below the inguinal ligament. The **small saphenous** starts out as a lateral vein posterior to the lateral malleolus but it reaches the back of the leg to end in the popliteal vein between the heads of the Gastrocnemeii.

The Lymphatic System

The lymphatic system begins in the meshwork of loose connective tissues as blind-ended capillaries, wider and more irregular in shape and diameter than blood capillary networks. The unit lymphatic capillaries anastomose into rich plexuses. The plexuses are drained by collecting trunks of **afferent vessels** which course toward *regional lymph nodes.* **Efferent lymphatic vessels** leave nodes to become larger drainage trunks which eventually reach the two large **terminal collecting trunks,** the large **thoracic duct** on the left and the smaller **right lymphatic duct.** The latter collects lymph from the entire right side of the body above

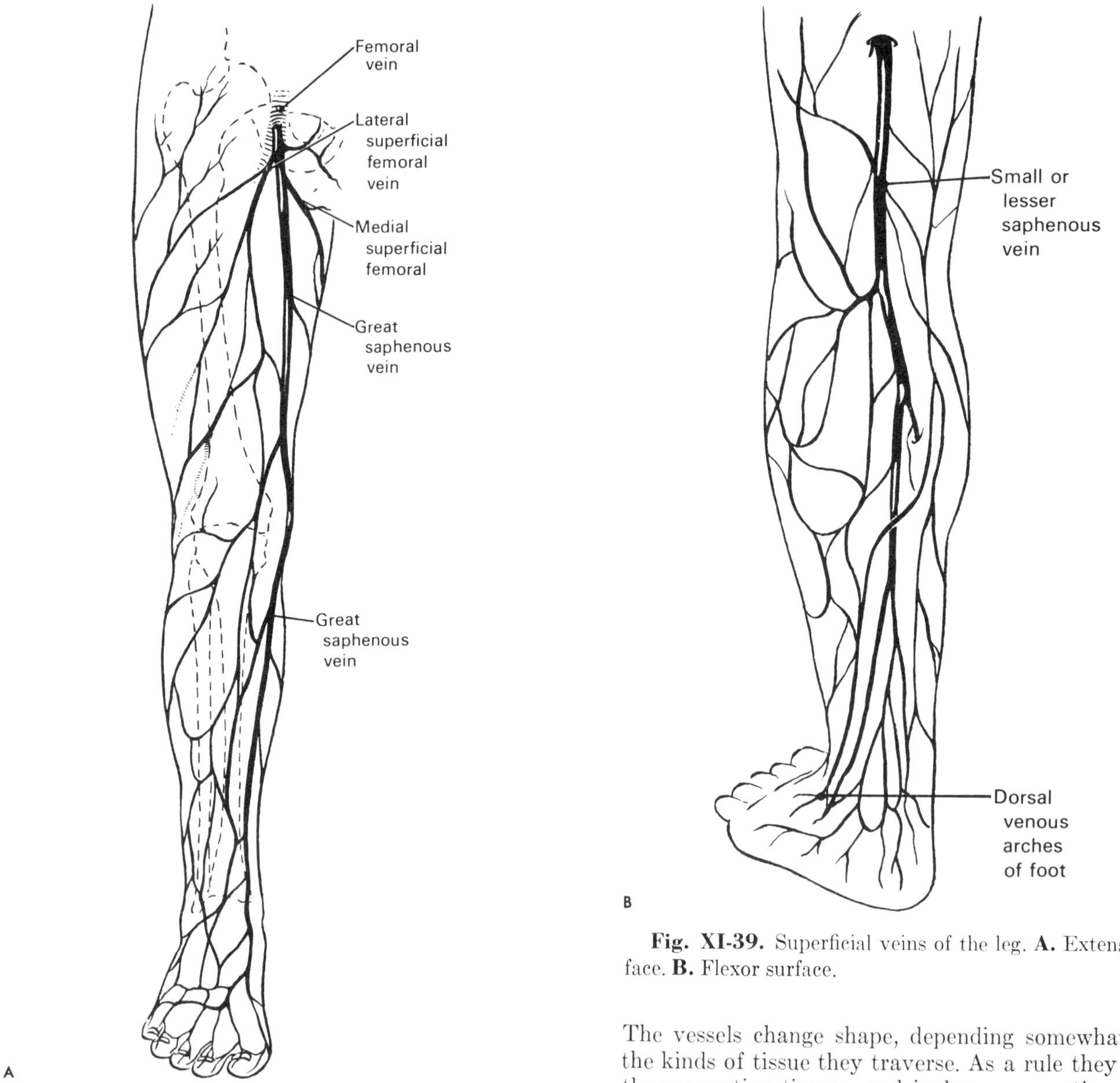

Fig. XI-39. Superficial veins of the leg. **A.** Extensor surface. **B.** Flexor surface.

the thoraco-abdominal diaphragm while the thoracic duct receives lymph from the entire left half of the body and the right half below the diaphragm. Each of the terminal trunks finally empties into the junction of the internal jugular and subclavian veins on their respective sides of the body.

Lymphatic capillaries, like ordinary capillaries have endothelial walls. Unlike endothelial cells of blood capillaries which are diamond-shaped with serrated interlocking cell borders, lymphatic endothelial cells are jig-saw puzzle-shaped, interlocking with irregular, but larger and smoother, protruding border parts. The walls of collecting trunks are thickened by the addition of adventitial coats that resemble those of veins. The number of valves in lymphatics far exceeds those in veins, and valves are present even at the capillary level.

The vessels change shape, depending somewhat upon the kinds of tissue they traverse. As a rule they prefer the connective tissues, and in loose connective tissues their lumen diameter is apt to be fairly wide. As they course through muscle, the diameter may be constricted to a slit. Alternate contraction and relaxation of muscle, as a consequence, constricts and dilates the vessels, exerting a milking action to help propel lymph on its way. Lymphatics will be found both on the surface of organs and deep within them. Collecting trunks generally group near blood vessels and tend to lie superficial to them.

Lymph Nodes

These bean-shaped structures vary in size and distribution, and are always intermediate between the capillary plexuses and the larger collecting trunks that lead to terminal trunks. These little organs have a capsule of fibroelastic connective tissue from which stout connective tissue trabeculae penetrate into the

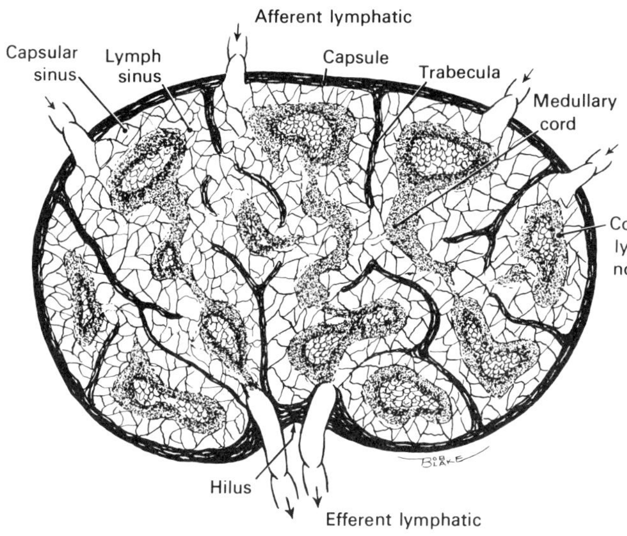

Fig. XI-40. Structure of a lymph node.

organ, branching and anastomosing as they do so (Fig. XI-40). From this stouter framework, reticular tissue stretches from capsule to trabeculae and from trabeculae to trabecula, much like fishermens' nets hung out to dry, except that this is, of course, a three-dimensional network. Reticular tissue consists of mesenchymal-like reticular cells "holding hands" by means of their cytoplasmic processes. Reticular fibers (young collagenous fibers) reinforce the net by hugging the boundaries of their parent cells and only here and there extend a cautious foot out into the surrounding matrix as if to test its temperature. The matrix in this case is the fluid lymph which trickles through the interstices of the reticular network.

In the perimeter of the node below its capsule, some of the multipotential reticular cells grew tired of holding hands. They had maternal aspirations so they retracted their processes, rounded up and became great with child. These are the large **lymphocytoblasts** whose mitotic divisions give rise to the myriades of **large** and **small lymphocytes** in nodular groups which form the **cortex** of this organ. Generally, the offspring around mother's knee form the paler centers of cortical nodules. The cells of the so-called **germinal centers** are younger and larger than the smaller, darker, more mature lymphocytes in the perimeter of the nodules.

Lymph enters the node through a series of perforating *afferent lymphatics* around the outer perimeter of the capsule. They wash a marginal sinus, directly under the capsule, free of lymphocytes. The lymph, as it percolates through the reticular net toward the efferent lymphatics which exit at the hilus, carries along mature lymphocytes from the edges of the cortical nodules. These pin-ball down through the reticular net in irregular fashion, some of them getting piled up as **medullary cords** of lymphocytes deeper in the node. Lymph channels surround them and the trabecular borders as the lymph pours through the net. Such channels in the gland are called **lymph sinuses** or **medullary sinuses.** These are hardly structured channels, but, rather, fortuitous ones. Their shape, and that of the medullary lymph cords, changes from day-to-day, hour-to-hour, and minute-to-minute as newly formed lymphocytes from the cortex are washed inward and downward to enter the efferent lymphatic channels, seeking their way into the world of the systemic circulation.

In the medulla, other restless cells of the net have also retracted their processes, left the net, and have become ameboid. These are the wandering phagocytes of the node, the scavenger cells, which feast on bacteria and particulate debris transported in lymph as it seeps through the nodes.

Shape, size, number, color, and consistency of nodes vary. They are more numerous, large and obvious in children. They increase or regress in size depending upon their freedom from bacterial or tumor invasion. In the lung they blacken as phagocytes engulf dust particles. But in the abdomen they remain a healthy pink. In acute regional infections, certain nodes swell and become tender and painful. They may even become abscessed in a losing battle with bacteria. In chronic illnesses they increase in number and, while they may enlarge, they tend to remain firm.

We have noted that lymphatics not only drain the tissue spaces, but that in intestinal villi the central lacteals engage in fat absorption from the intestinal tract. After a particularly fatty meal, the blood stream may contain many microscopic fat globules or **chylomicrons.**

The Lymphatics of the Head and Neck

An overall scheme of the lymphatic system is shown in Figure XI-41. That for the head and neck, in particular, is seen in Figure XI-42. There are no nodes or lymphatics within the cranial cavity. A collar of superficial nodes ring the neck from the lower border of the mandible, about the ear to the external occipital protuberance and around to the other lower mandibular border. Deep nodes lie along the deep vessels of the neck and head. Lymphatic channels drain to nodes of the superficial ring and these, in turn, drain toward the deep nodes. The pharyngeal and oral tonsils are also drained by lymphatic channels. The important nodes are as follows:

A. Superficial Nodes:

1. An occipital group
2. Retroauricular or mastoid group
3. Parotid group
4. Submandibular nodes below and along the mandible
5. Submental nodes under the chin

This whole group is palpable.

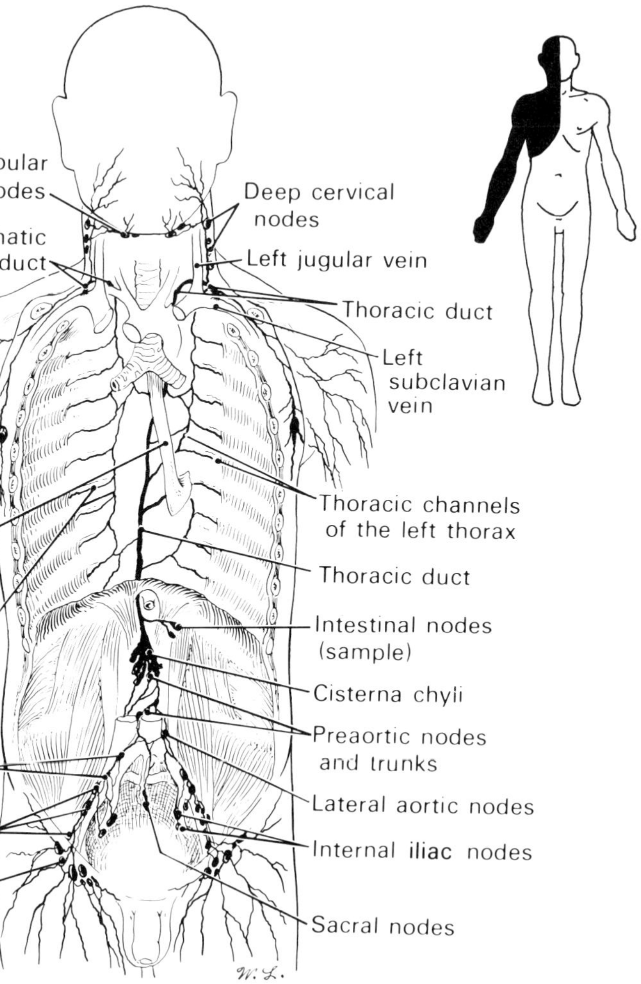

FIG. XI-41. General plan of lymphatic drainage. Inset shows drainage areas of the right lymphatic duct (black) and of the thoracic duct (white).

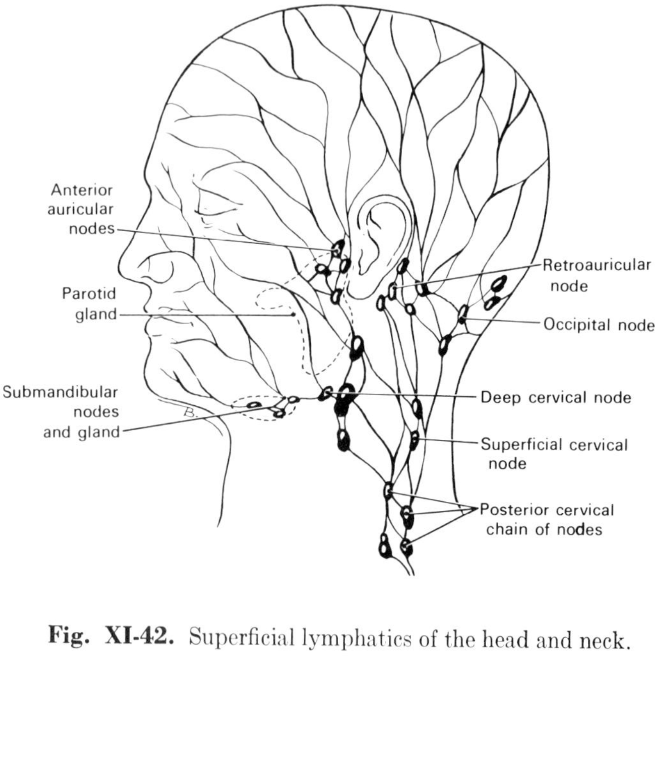

Fig. XI-42. Superficial lymphatics of the head and neck.

B. Deep Nodes:

1. Retropharyngeal group between pharynx and vertebral column
2. Deep cervical nodes along internal jugular
3. Supraclavicular and subclavian groups

It is easy to imagine scalp drainage following these courses: frontal lymphatics to the parotid nodes; parietal channels to the mastoid group; and occipital channels to the occipital group of the back of the head. Those from the lids reach the parotid group; those from the cheeks, gums and skin of the nose reach the submandibular group. All of these nodes drain into the deep cervical chain of nodes. The nasal cavity, air sinuses, and pharyngeal channels drain into retropharyngeal and cervical nodes. The submandibular nodes can become tender and swollen in infections of the oral tonsils, and retropharyngeal nodes can become quite large with nose and throat infections, especially in children. The lymphatic channels of the tongue take several routes: those from the tip to the submental nodes; from the sides, to submandibular nodes; and,

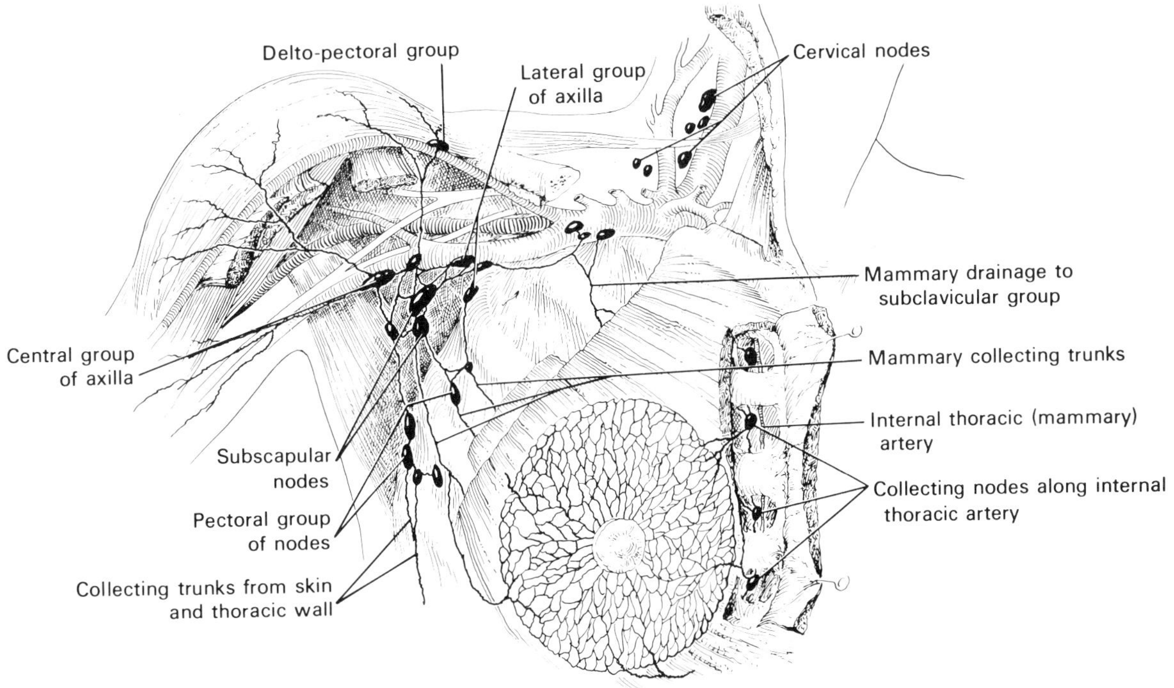

Fig. XI-43. Lymphatic drainage of the mammary gland. The axillary lymph nodes.

from the back of the tongue, to cervical nodes. All in all, the patterns of drainage are quite logical ones, crossing toward the nearest nodes in the vicinity.

Lymphatics of the Upper Limb

All of the superficial and deep lymphatic drainage from the limb is directed over long lymphatic channels to the **axillary nodes.** There may be a few scattered nodes in other parts of the arm. One common group, in particular, is the epitrochlear nodes in the deep fascia just proximal to the medial epicondyle of the humerus. Into these, drains lymph from the ulnar aspect of the arm, little and ring fingers, and medial aspect of the hand.

Axillary Nodes (Fig. XI-43): These drain the thorax from the umbilical level upward as well as the upper limb. They are distributed in the axilla as follows:

1. A humeral or lateral chain along the axillary vein which receives nearly all the lymph from the arm.
2. A pectoral group lying along the lateral thoracic artery which drains anterior chest and breast.
3. A posterior subscapular chain along the subscapular vessels. These nodes drain lower neck and shoulder.
4. An anterior infraclavicular group upon the Pectoralis Minor into which efferent channels from other groups drain.
5. A central axillary group at the axillary apex which also collects from the other group of nodes.

These nodes drain the back as well, from the midline and above the iliac crests. The lymphatic channels are particularly fine and numerous in the fingers and palms. The collecting trunks are about 30 in number in the forearm and coalesce into half that number over the upper arm.

Lymphatics of the Thorax (Fig. XI-44): There are several small superficial groups of nodes and the deeper bronchial groups.

A. Superficial:

1. Sternal nodes behind the sternum receive lymph from the inner aspects of the thoracic wall and Diaphragm.
2. Intercostal nodes drain the chest wall. These occupy the dorsal parts of the intercostal spaces.
3. Anterior mediastinal nodes. These lie ventrally in the superior mediastinum ventral to the aortic arch and in close proximity to the brachiocephalic veins. They drain the thymus and pericardium and receive efferent vessels from the sternal nodes. Their own efferent vessels form the right and left bronchomediastinal trunks.
4. Posterior mediastinal nodes lie dorsal to the pericardium related to the esophagus and thoracic aorta. They drain esophagus, liver, diaphragm and dorsal pericardium and their efferent vessels enter the thoracic duct.

B. Deep Bronchial Group: These relate to the trachea and bronchi and break into the following groups:

1. Tracheobronchial nodes in the angle between trachea and primary bronchi and along side of the trachea.

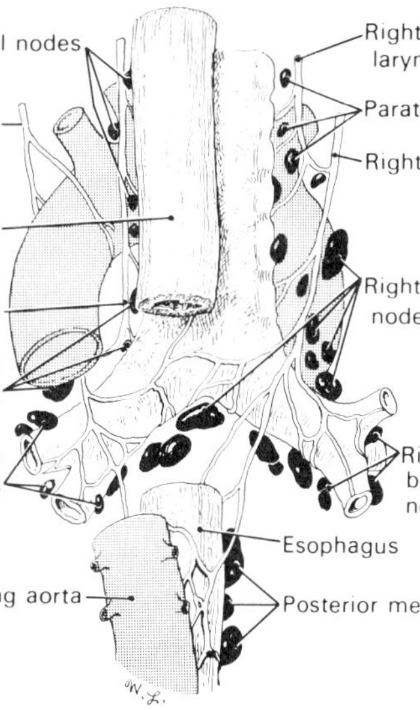

Fig. XI-44. Tracheobronchial nodes (from behind).

2. Bronchopulmonary nodes at the lung hilum and the larger bronchial branches.

Review the concepts of the lymphatic drainage of the lungs by returning to the chapter on the thorax, pages 470 to 473.

Lymphatics of the Abdomen

One could subdivide the lymph nodes of the abdomen into any number of arbitrary groupings. Since so many of them intercommunicate with each other, we'll keep it simple. The major groupings are distributed along the aorta and particularly in relation to the three unpaired vessels.

1. Aortic Nodes: They tend to be more numerous along the left side of the vessel than along the right in the stretch between left renal artery and left common iliac. Above the level of the renal arteries, the aortic group intermingles almost insensibly with the coeliac and superior mesenteric groups. Below, they blend into the inferior mesenteric group. A few nodes, the lateral lumbar nodes in particular, chase along some of the aortic branches to the parieties.

2. Coeliac Nodes (Fig. XI-45): These cluster about the coeliac trunk and follow out along its branches. Many are bunched along the upper pancreatic border and some scatter in the omenta.

3. Superior Mesenteric Group (Fig. XI-45): These number in the hundreds, and they are scattered irregularly within the mesentery of the small intestine. In this sense, like the others, they follow roughly the distribution of the arterial branchings.

4. Inferior Mesenteric Group (Fig. XI-45): These align themselves more neatly along the major vascular branches of this artery than do the last group. They tail out along the sigmoid colon as scattered superior hemorrhoidal, sacral, and internal iliac clusters.

The lymphatic drainage of the stomach appears to be regional (Fig. XI-46), though boundaries between the regions are indistinct since there are so many internodal connections. Yet, lower esophagus and fundus drain to nodes along the left side; the pyloric region to coeliac nodes to the right, and the greater curvature to pancreatic nodes. All flow seems to be toward the nodes clustered about the coeliac stem eventually, and from this into the thoracic duct. The wall of the small intestine is drained into mesenteric nodes, and the efferent vessels of these enter the nodes about the base of the superior mesenteric artery. The relay to this last station before channels reach the thoracic duct is through a variable number of nodes. The cranial part of the large intestine also reaches superior mesenteric nodes while its more caudal parts seek the inferior mesenteric group. It is easy to visualize the pancreas as draining into the nodes along its upper border and, hence, to the coeliac group. This is also the course taken by lymph from the gall bladder and the bulk of the liver, but the diaphragmatic surface of the latter drains upward into thoracic nodes. The lymph channels of the spleen drain to the left group of coeliac nodes. The kidneys and suprarenals tend to drain into aortic nodes. There are few nodes along the renal arteries, however. Suprarenals may also be drained partially by the coeliac nodes and even upward into the thoracic nodes. All final channels to Rome are *via* the Aqueductus Thoracis not the Cloaca Maxima. The latter leads to Bangkok with side branches to Haiti as any world traveler who follows his nose knows.

Lymphatics of the Pelvis (Fig. XI-47): Once more the grouping is an arbitrary one and, chiefly, vascular in distribution.

1. Common iliac nodes. These interconnect the aortic group of the abdominal cavity and the pelvic chains along the course of the internal iliac artery with those from the inguinal region.
2. The external iliac group. These tend to drain the lower anterior abdominal wall and inguinal nodes.
3. The internal iliac group drain pelvic viscera, perineum, and buttocks. If these are, in reality, the true pelvic nodes those of the external iliac group are the nodes of the pelvic brim (or false pelvis). The lymph channels of pelvic viscera follow the vessels of the given viscus to nodes along the vascular course. Efferent lymphatics from these nodes move lymph in bucket-brigade fashion to those of the pelvic brim, to those of the common iliac group, to aortic nodes, to thoracic duct. Lymph channels from the testis, upper uterus, ovary, and upper rectum flow, as a rule, directly into aortic nodes over larger collecting routes to the abdominal cavity proper. Thus, they evade much of the bucket-brigade.

In abdomen and pelvis, lymphatic networks within organs lie at superficial levels in outer coats or capsules and at deep levels within the connective tissue

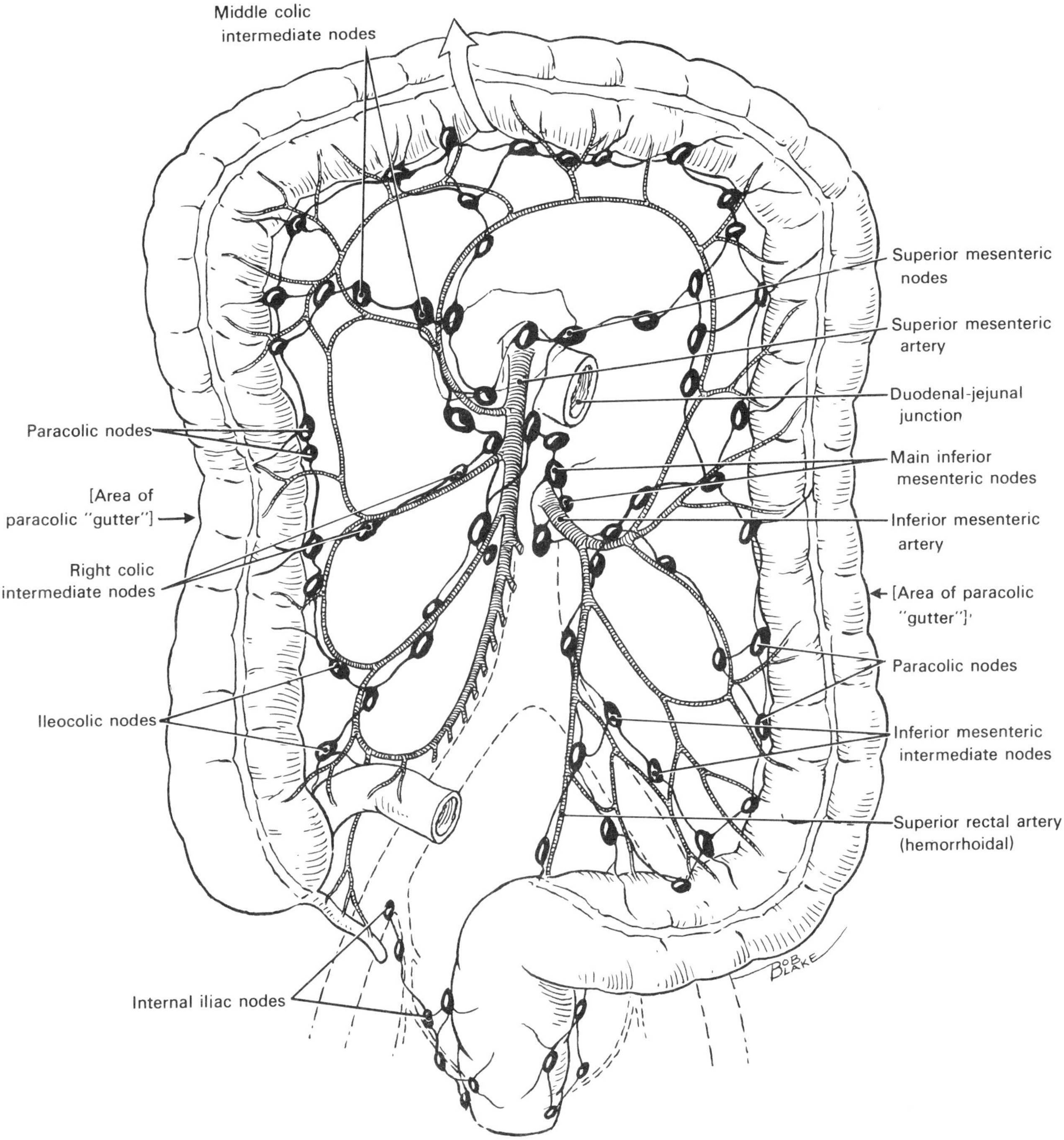

Fig. XI-45. Lymphatic drainage of the large bowel.

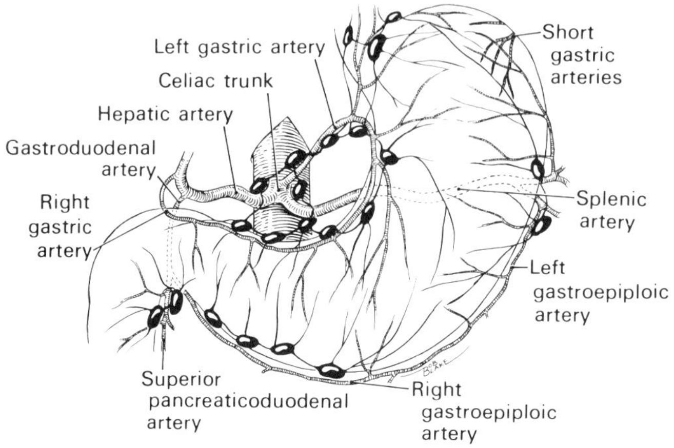

Fig. XI-46. Lymphatic drainage of the stomach.

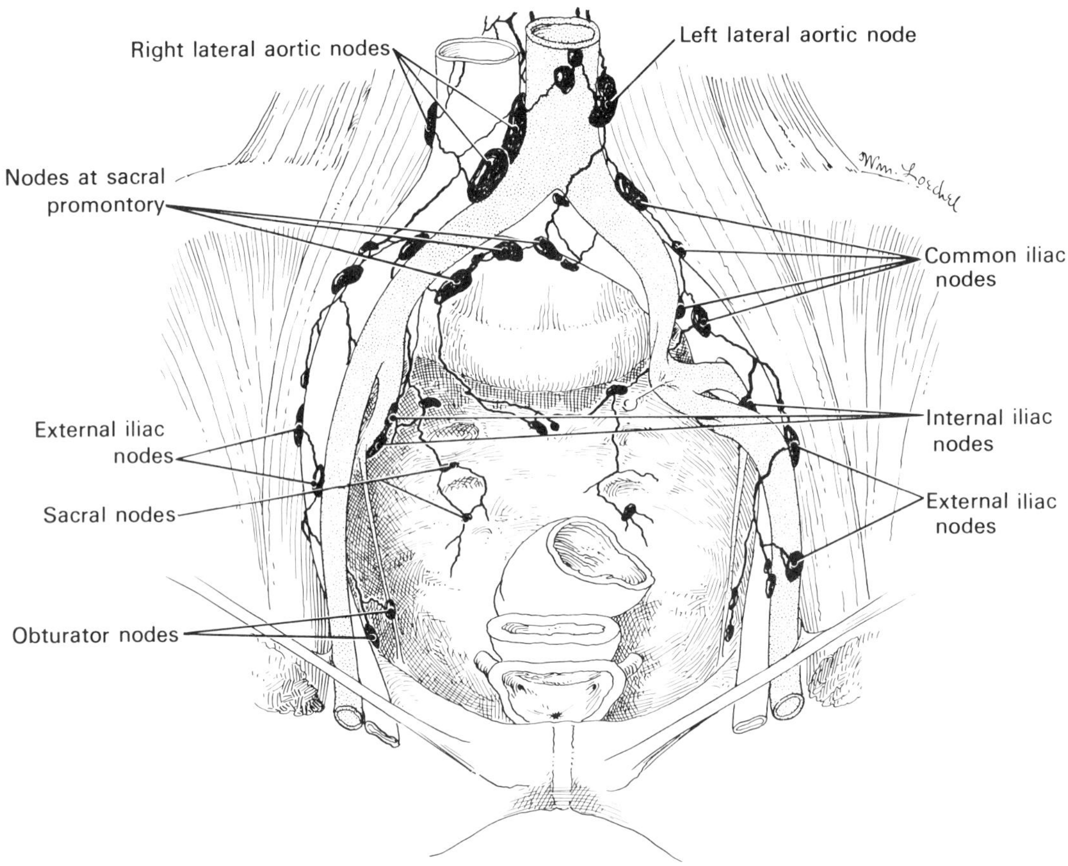

Fig. XI-47. Lymph nodes of the pelvic wall. Left common iliac retracted to show chain of nodes behind vessel as well.

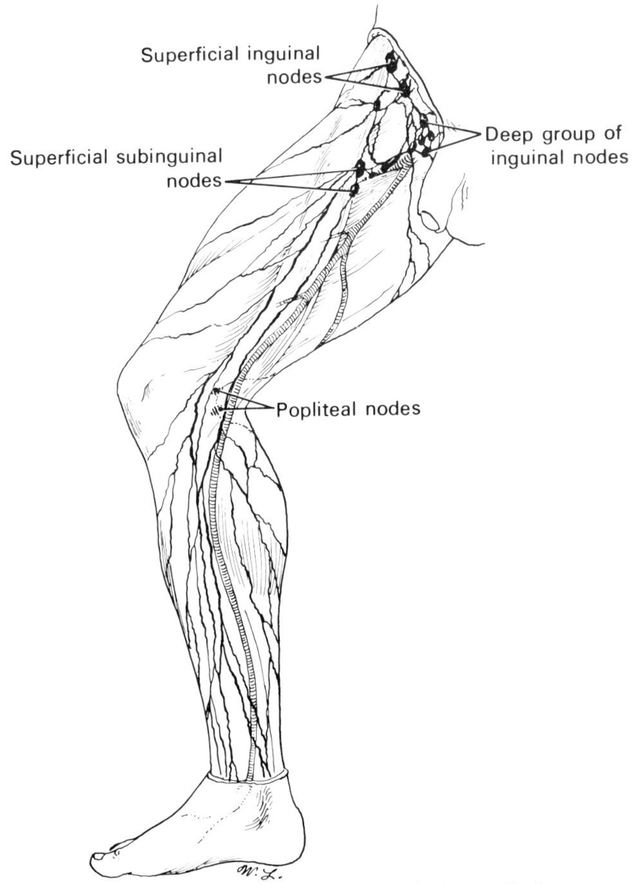

Fig. XI-48. Lymphatics of the lower limb.

framework. Drainage from the organ is toward hilar sites or to loci where the chief blood supply enters the organ.

Lymphatics of the Lower Limb (Fig. XI-48): The lymphatic channels of the leg, like those in the arm, are long and run to end in nodes clustered at the proximal part of the member. In this case the terminal depot is the **inguinal nodes.** Near the knee, as near the elbow, there may be a few scattered intermediate nodes along this long drainage route. These tend to lie behind the knee near the popliteal vein. They collect from lymph channels in the calf that have followed the small saphenous vein, or have tagged along with the posterior tibial and peroneal arteries. There may be a small but inconstant anterior tibial node which lies on the anterior interosseous membrane proximally.

The Inguinal Nodes (Fig. XI-48): These vary from 10 to 20 in number, and they all cluster in the upper part of the femoral triangle. If you draw a horizontal line where the greater saphenous vein joins the femoral vein, those above this line constitute a superficial group; those just below it, a deep group of nodes.

1. Superficial Group: These receive superficial (or skin) lymphatics from the lower limb; from the skin of the penis and its prepuce (or prepuce of clitoris); from scrotum (or labia majora); skin of anal and perineal region and buttocks; and skin of the lower abdominal region. (Note that drainage of the testes is not involved here.) The testis drains directly to aortic nodes in the lower abdomen. Nor is drainage of the glans of the penis or the clitoris involved. These drain to deep inguinal nodes.

A medial group of superficial lymphatic channels follows the course of the great saphenous vein from the first, second, and third toe and dorsum of the foot and from the anterior and posterior aspects of the medial malleolus up the leg. They pass with the vein behind the medial condyle of the femur and on to the superficial nodes of the groin. A lateral group of lymphatic channels, lesser in number than the medial group, climb the fibular aspect of the leg, crossing the tibia below the knee to join with the lymphatic channels on the medial side of the thigh. Others pass posterior to the lateral malleolus, ascend posteriorly on the leg along the small saphenous vein to enter popliteal nodes. Collecting channels from the gluteal region pass laterally around the thigh while those from the inner aspect of the buttocks wind inferomedially around the thigh to reach superficial inguinal nodes.

2. The Deep Inguinal Group: These nodes are fewer in number than in the superficial group, as few as one to three. They lie deep to the fascia lata on the medial side of the femoral vein. The most distal lies just below junction of saphenous with femoral vein. A middle one is the one most frequently absent, and the highest is generally the large **node of Cloquet or Rosenmüller.** It must be nice to have a lymph node named after you. In the next edition of this text your authors have decided to pick one specific node each in the thorax, abdomen, and pelvis and to call them the nodes of Becker, Wilson, and Gehweiler. This is probably our only chance to go down in anatomical history.

The deep nodes drain lymphatic channels which follow the deeper vessels of the limb. These channels accompany the pedal and anterior tibial vessels and the plantar and posterior tibial vessels to course along, finally, with the femoral vein. However, lymph channels which follow the obturator, superior, and inferior gluteal vessels end in internal iliac nodes. The deep nodes receive afferent vessels from the superficial nodes, from glans penis and clitoris, and from the channels coming to them along the femoral vessels as mentioned above.

REFERENCES

1. Batson, O. V.: The function of the vertebral veins and their role in the spread of metasteses. Ann. Surg., *112:* 138–149, 1940.
2. Batson, O. V.: The role of the vertebral veins in metastatic processes. Ann. Intern. Med., *16:* 38–45, 1942.

TWELVE

The Endocrine System

GENERAL

A group of ductless, internally-secreting glands comprise the endocrine system. Those usually included are the pituitary or hypophysis, the thyroid, thymus, suprarenals, pancreas, testes, and ovaries. One could include, perhaps, paraganglia, pineal body, and placenta; and, in one sense, the mucosa of stomach and small intestine. All of these mentioned, or at least specific cells in each of them, have the ability to secrete specific products called **hormones.** These products enter the blood stream to be carried to certain **target areas** which may consist of a tissue, another organ or the body at large. The hormones, as a rule, act some distance from their site of origin, though this may not be true in some instances. For example, stomach and intestinal mucosae produce hormones for home consumption. Even though the products secreted there are circulated through the entire body, they return to exert their effect upon local structures related to the digestive tract. Most hormones, however, go to work some distance away from home.

Hormones may goad the target organ into action so that it, in turn, produces a specific hormone or product; or they may be more concerned with the maintenance of some steady state of behavior. But, through the principle of negative-feedback, they can also inhibit specific actions of the target organ. The endocrine organs are capable not only of influencing and integrating the behavior of other bodily parts, but their products interact to regulate themselves.

If such organs lack ducts, then they must have a rich and intimate vascular supply to carry away their secretions. In general, their secretory cells are arranged in clumps, cords, or plates in such a way that sinusoids and capillaries either abut against these cells or interweave among them. Sometimes, as in the ovary or thyroid, a single layer of cells forms a ball or a sac with a hollow luminal cavity, much like a blastocyst. Such **follicles** have a closely-woven capillary plexus about their exterior surface.

Endocrine tissue doesn't have to be epithelial in nature. It can be neural, *e.g.*, neurohypophysis or the cells of the adrenal medulla and paraganglia. It can even be connective tissue, *e.g.*, the interstitial cells of the testis. Of course, in many instances, besides the secreting parenchyma, a substantial amount of stromal tissue enters into shaping up the final endocrine organ.

The simple diagram, Figure XII-1, localizes the important endocrine organs in a rough way. At least it places them in the proper areas—head, neck, mediastinum, abdomen, or pelvis. The thymus, perhaps, is "off base" and too low in this crude sketch.

We could begin with the pituitary, but since it secretes hormones for the personal attention of many other endocrine organs, it might be best to treat it later. Do not forget, however, that the very secretions of the target organs which it affects may, themselves, exert a regulatory effect upon the pituitary.

THE THYROID GLAND

The gland arose as an epithelial bud which detached from the floor of the embryonic pharynx in the neighborhood of the base of the tongue. It migrated down around the upper tracheal rings where it sits "like a shield." It extends its two lateral lobes, each about half the size of a golf ball, downward from the caudal border of the thyroid cartilage to the sixth tracheal ring. A narrow isthmus of thyroid tissue interconnects the lobes by passing in front of the third tracheal ring.

A coronal section through an entire gland is shown in Figure XII-2. Sometimes the gland has a pyramidal lobe which marks part of the course of its former migratory pathway. You can see that branches of the superior and inferior thyroid arteries have penetrated the tissue at all margins to form capillary plexuses about the tiny, rounded, white areas which are the thyroid follicles. We would need a microscope to see them well since they measure only 50 to 500 μ in diameter and their total number is around 20 million. But, in the lobe is a clearly visible, black, colloid cyst. Whether this represents tremendous expansion of one follicle, or a coalescence of many, is difficult to say; but at least it is pathologic. Actually, all we see here is a hole in the tissue; the cystic area having fallen out in sectioning.

Follicle cells approach cuboidal epithelium in height but their height varies with the activity state of the gland. The cells are low when the gland is underactive, *i.e.*, when follicles are storing colloid (Fig. XII-3C),

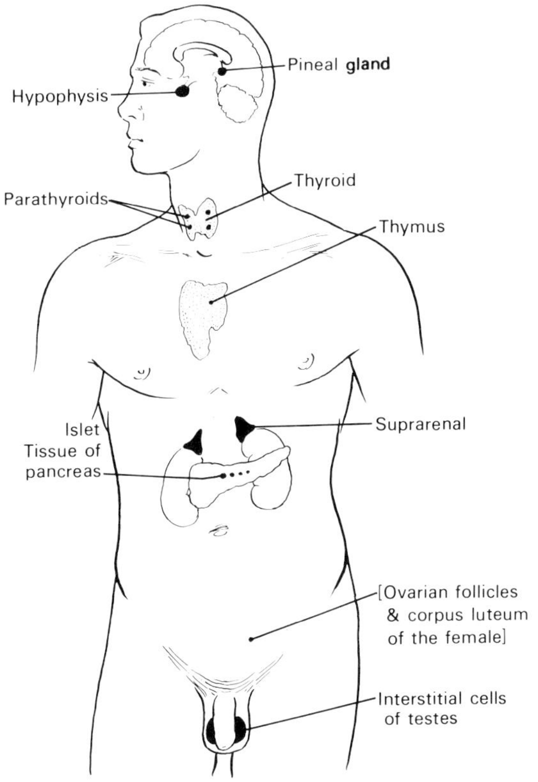

Fig. XII-1. Diagrammatic locus of endocrine organs.

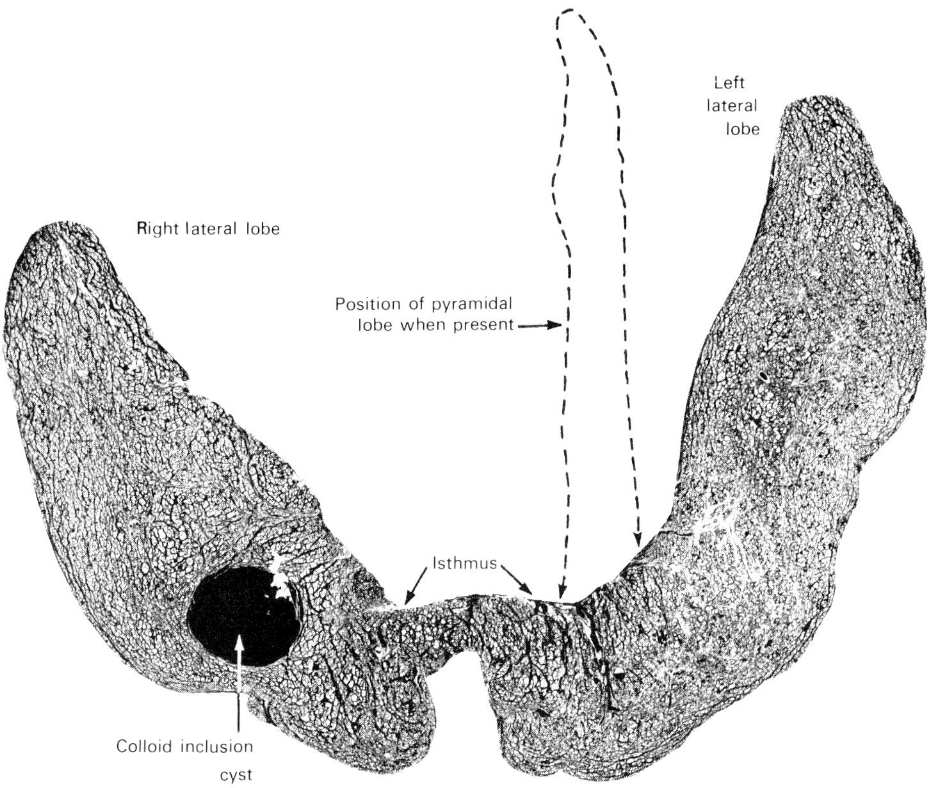

Fig. XII-2. Coronal section of entire adult thyroid gland. Note unequal size of lateral lobes and the colloid inclusion cyst (benign) in the right lobe.

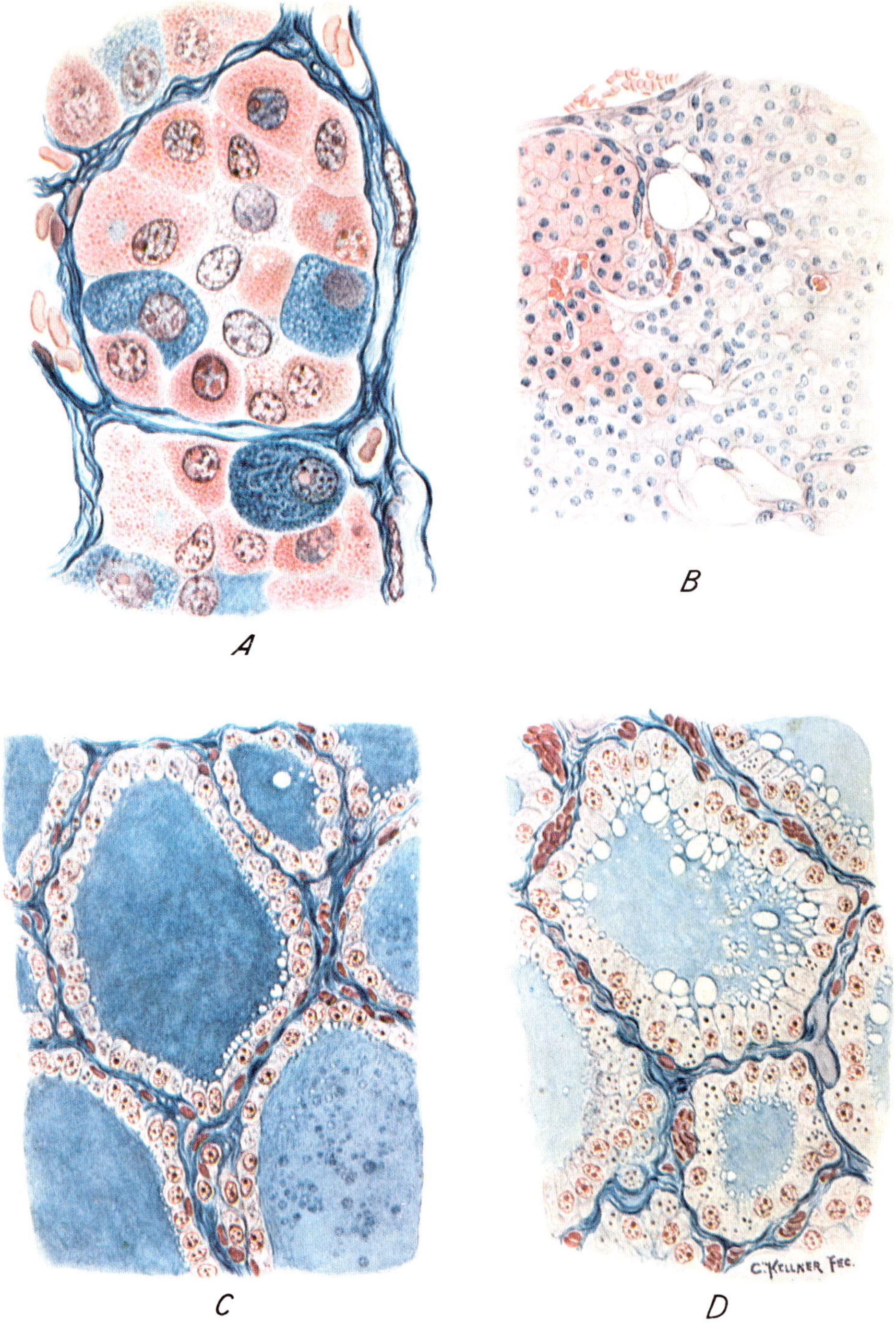

Fig. XII-3. Drawings of the normal anterior hypophysis and parathyroid of man, and the normal and activated thyroid of the rhesus monkey.

A. Anterior hypophysis of an executed man, 38 years old. Cords of cells are from the central part of the gland. In 3 of the acidophiles (pink), the Golgi area is stained bluish. In the basophile in the lower cord, a negative picture of the Golgi apparatus is shown as clear canals. Chromophobes are present in the middle of the central cord. The rather large amount of connective tissue shown in the drawing is characteristic of the human anterior hypophysis. A colloid mass (blue) is shown at the lower part of the inferior cell cord. Modified Masson stain. × 1250

B. Parathyroid of normal human adult. The oxyphils are pink and the chief cells bluish purple. The spaces shown were filled with fat. Hematoxylin and eosin stain. × 350

C. Thyroid of a normal adult rhesus monkey. A few peripheral and central colloid vacuoles are shown. Modified Masson stain. × 570

D. Activated thyroid of a normal adult rhesus monkey. The thyroid was activated by injections of an anterior hypophysis extract. The epithelium is high and many absorption vacuoles are present. Colloid droplets and vacuoles are present in the thyroid epithelium. Modified Masson stain. × 570. (From W. M. Copenhaver, *Bailey's Textbook of Histology*, 15th ed., Williams & Wilkens, Baltimore, 1964.)

but taller when follicles engage in releasing stored hormone into the capillary plexus (Fig. XII-3D). Colloid is rich in nucleoproteins and it contains thyroglobulin and enzymes. Thyroglobulin is, itself, a glycoprotein containing iodinated amino acids.

The cuboidal follicle cells extract inorganic iodide from the adjacent blood stream. They concentrate it and link it with tyrosine to form monoiodo and diiodotyrosine. Linked with thyroglobulin, these substances become the two thyroid hormones stored in the colloid, namely, **triiodothyronine** and **tetraiodothyronine.** The latter is generally known as **thyroxine.** When these compounds are secreted back into circulation, the ratio of thyroxine to the triiodo compound is 10:1. They react much the same physiologically though the triiodothyronine is many times more potent than thyroxine.

As these hormones enter the blood stream, they attach rapidly to serum proteins. Since there are no other important iodinated compounds in the body, the amount of iodine bound to protein (PBI) is used as a clinical measure of the amount of thyroid hormone in the blood. A normal PBI range is 4 to 8.5 μg. per 100 ml. of serum.

Through feedback to the pituitary, when there is a *decline* of thyroid hormone in blood, certain cells in the pituitary release **Thyroid Stimulating Hormone (TSH).** This causes the thyroid to release its hormones into the general circulation to reach *body tissues in general.* Here they serve as catalysts to hasten oxidation in tissue cells, *i.e.,* to regulate cellular enzymes which control energy metabolism. In this way, thyroid hormone influences all metabolic processes including growth.

Obviously, if *too much* thyroid hormone enters the blood, it exerts a depressant effect upon the pituitary and no more TSH is released from that gland. The thyroid follicles enter a manufacturing and storage phase until the scales are tipped once more for active release of thyroid hormone.

Certain pale cells in the thyroid are suspected of secreting another hormone, **calcitonin.** This reducer of blood-calcium levels offsets parathyroid influences.

What happens when the thyroid gland atrophies or when there is insufficient production of TSH by the pituitary? Obviously, less thyroid hormone reaches the blood stream. The use of energy by cells falls and so will the basal metabolic rate. The individual complains of being cold. He increases his glycogen and fat storage instead of utilizing these substances for oxidation.

Thyroid insufficiency *in the adult* leads to **myxedema.** Thyroid follicles begin to store non-iodinated colloid. The patient gains weight but, paradoxically, complains of loss of appetite. His blood cholesterol levels rise. He has lowered blood pressure and low heart and respiratory rates. His gastrointestinal tract is sluggish and so we find him resorting to enemas and suppositories. As a result, he is most apathetic and lethargic. We can recognize his problem, in part, by examining his skin which we find puffy, thick, and leathery. His hair is brittle and breaks, and he worries about the amount of it in the teeth of his comb. All of these symptoms can be overcome by treatment with thyroid extracts.

In the *newborn,* congenital absence of the gland results in the pitiful **cretin dwarf.** A complete slowing of metabolic processes at this stage in life results in the failure of sexual, skeletal, and mental growth and development.

The opposite side of the coin can turn up as well. It is possible for the thyroid to enlarge and become hyperactive. The cells of the body use up energy, exhausting glycogen and fat stores. As the metabolic rate increases, body temperatures rises. To compensate, the individual sweats in order to cool himself. Instead of lowering— blood pressure, heart, and respiratory rates now rise. Again paradoxically, while appetite is healthy, weight falls off. Bowels are loose and the patient tries everything in the Pharmacopeia to solidify stools. He is excitable and apprehensive and may show muscular tremors. If he exhibits "pop-eyes" (exophthalmos), it may not be due to hyperthyroidism, but to some other secondary effects of hormonal imbalance. Treatment with I^{131} or surgical removal of part of the gland may help alleviate his symptoms.

The picture of thyroid malfunctioning which we have just painted may not be this pure and simple. We are picturing symptomatic highlights; not individual naunces, for, as in every other circumstance, individuals react differently to the same set of conditions.

THE PARATHYROIDS

Most of us have four, tiny, brownish parathyroid glands attached to the back of the capsule of each lateral thyroid lobe. Some of us have less and in some cases there are accessory ones scattered in the mediastinum or in the fascial planes of the neck. They are ovoid in shape, and they remind one of apple seeds in size.

The two small ones shown as sagittal sections in Figure XII-4 are from the human. Those shown from the dog are obviously larger.

The epithelial cells of the glands are irregular, connecting masses. Each mass is totally channeled by capillaries. There are two cell types: chief cells and oxyphil cells. The first are the most numerous and the only ones present until middle childhood. Chief cells are identified as *light* and *dark* cells. Secretory granules are rare in the light cells and they are regarded as inactive or resting cells. The smaller dark cells are full of secretory granules and are felt to be in an active, secretory state.

Oxyphil cells are characteristic in man but absent in most mammals. They are larger than chief cells, abundant in granules, acidophilic, and are scattered between chief cells, in large or small clumps (Fig. XII-3B).

So far as can be determined, the chief cells appear to

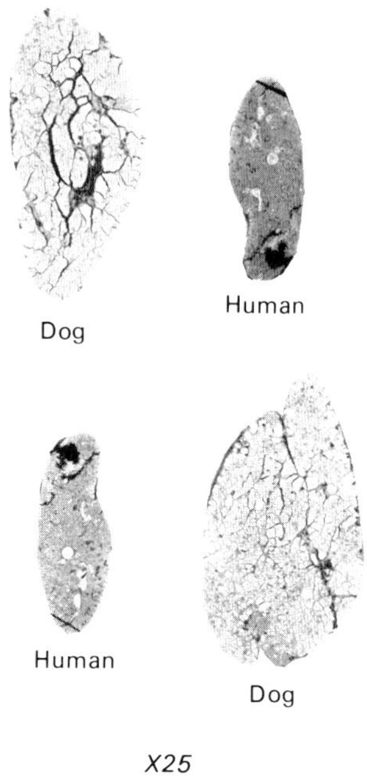

Fig. XII-4. The parathyroid glands of man and dog in sagittal section. Enlarged 25 ×.

be the hormone secretors (**Parathormone or PTH**). Hormone is released into the adjacent capillaries and circulated to all bodily tissues. However, not all tissues are sensitive to it. The hormone is important in calcium and phosphorus metabolism. Its action is primarily upon kidney tubules, bone, and intestine to maintain the blood calcium level at 11 mg. per 100 ml. of plasma. This level is critical for normal neuromuscular excitability. The hormone releases calcium and phosphorus from bone by increasing their solubility. It increases tubular reabsorption of calcium and tubular excretion of phosphorus. Thus, calcium content in the blood is raised while phosphorus is excreted in urine. The hormone goads the intestinal tract to absorb calcium and phosphorus from food stuffs. When calcium ion concentration decreases in the blood stream, parathormone is secreted; conversely, as blood calcium concentration rises, hormone secretion is depressed. An elevated phosphorus level in blood also may, indirectly, stimulate the gland to release hormone.

There seems to be no regulatory control exerted over these glands by the pituitary. The size of the parathyroids does not seem to change after hypophysectomy and neither does the Ca:P ratio change from its normal serum ratio of 3:1.

If the glands atrophy or are lost accidentally in the removal of thyroid tissue, the result is a severe convulsive disorder called **tetany.** Bones increase their calcium and phosphorus content, the gut fails to absorb these ions from food, and more calcium is lost in urine. The concentration of calcium ions in blood falls. When it falls below 6 mg per 100 ml. of plasma, tetany follows. There is nervousness, muscular twitching, and spasms in facial and limb musculature. These alarming symptoms can be relieved by injecting calcium or extract of parathyroid. It obviously wouldn't help much to ingest these substances since the gut is now insensitive to their absorption.

Parathyroid tumors may lead to increased secretion of parathormone. Obviously, this is going to pull calcium from bone and gut and increase its absorption by the kidney tubule. Bones soften and deform when calcium plasma levels reach 16 mg. per 100 ml., and calcium begins to deposit in odd places, *e.g.*, the kidney to produce kidney stones. Surgical removal of the parathyroid tumor helps restore balance, or calcitonin can be administered to accomplish the same result.

THE SUPRARENAL (ADRENAL) GLANDS

These somewhat triangular cocked-hats perch upon the cranial pole of each kidney. If you cut a fresh one open transversely, you can recognize certain regions by color. The **cortex** is yellow externally and it is a reddish-brown internally while the **medulla** has a softer gray cast. Figure XII-5 is a photograph of such a sliced organ. In black and white it is not as easy to recognize these regions grossly, but they are labeled for you.

Developmentally, cortex and medulla, you will recall, are separate components. The former grew out from peritoneal mesothelium and the latter, like sympathetic ganglia, migrated from neural crest origin. Phylogenetically, the two parts remain separated in fishes and intermingle in amphibians and reptiles. Only in mammals does the medulla become concentrated as a centralized mass.

The blood supply is very rich. You will recall the number of twiglets coming to these organs from inferior phrenic and renal arteries and from the aorta itself. The cortex gets its supply from a number of arterioles in the capsule which break into wide capillary sinusoids that filter downward between the zonular cell columns of the cortex. A main supply of blood for the medulla is the venous blood from the cortical capillaries, but some arterial branches reach the medulla before they break up in a terminal network within its parenchyma. All the blood of the gland drains into collecting venules which empty into one, large, central, suprarenal vein (Fig. XII-6). This leaves at the hilus to empty into the inferior vena cava on the right side and into the renal vein on the left side of the body.

The cortical cells are arranged radially in long, slender cords with often ill-defined zonation into three regions: **glomerulosa, fasciculata,** and **reticularis** (Fig. XII-7). The cells in the outer glomerular zone tend to dispose themselves in slanting arches or whorled balls; those of the fasciculata form long, parallel cords. Cholesterol is abundant in these cells, also fatty acids and neutral fat. The cells of the reticularis are most ir-

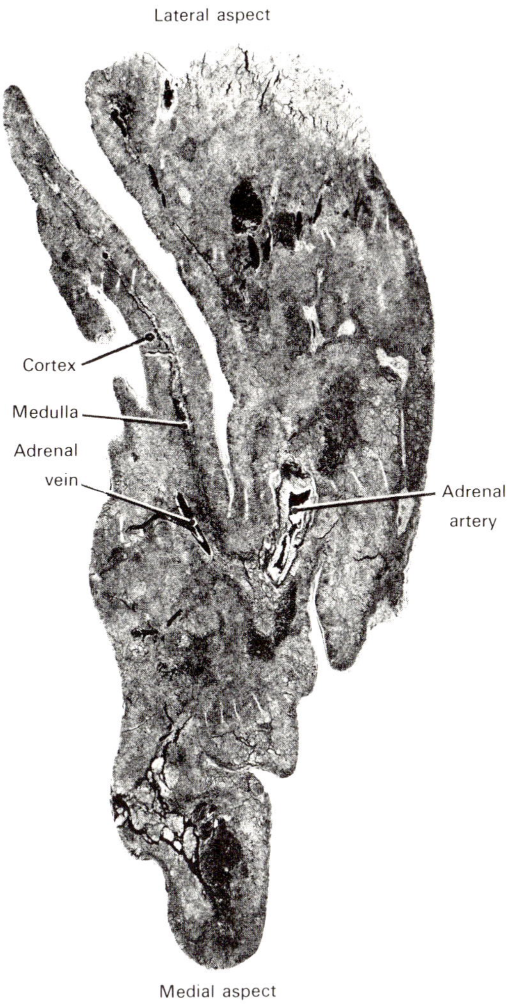

Fig. XII-5. A photograph of a transverse section through the human suprarenal after removal of the perirenal fat. Cortical and medullary regions and adrenal vein and artery are indicated.

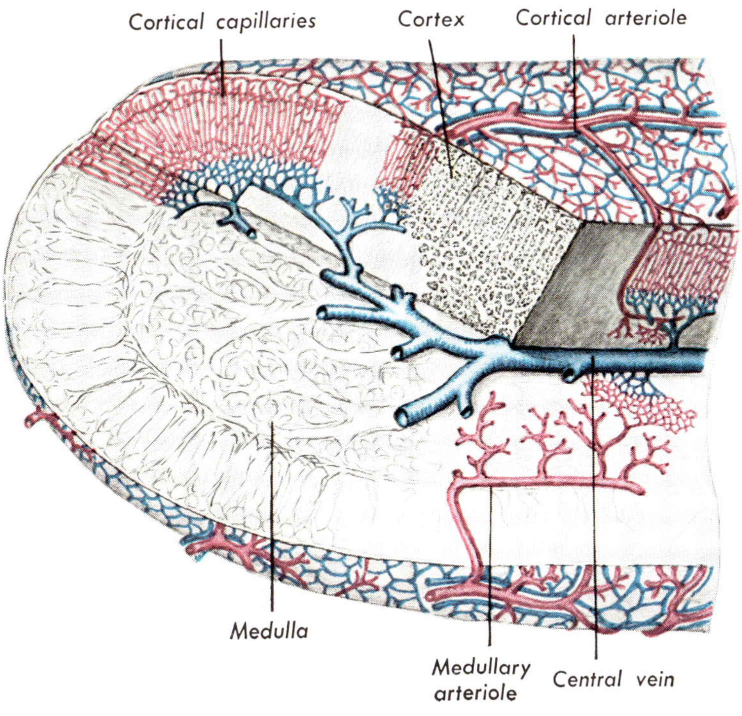

Fig. XII-6. Blood supply of adrenal of dog. One end, only, of the adrenal is shown. (After Flint.) (From W. M. Copenhaver, *Bailey's Textbook of Histology*, 15th ed., Williams & Wilkins, Baltimore, 1964.)

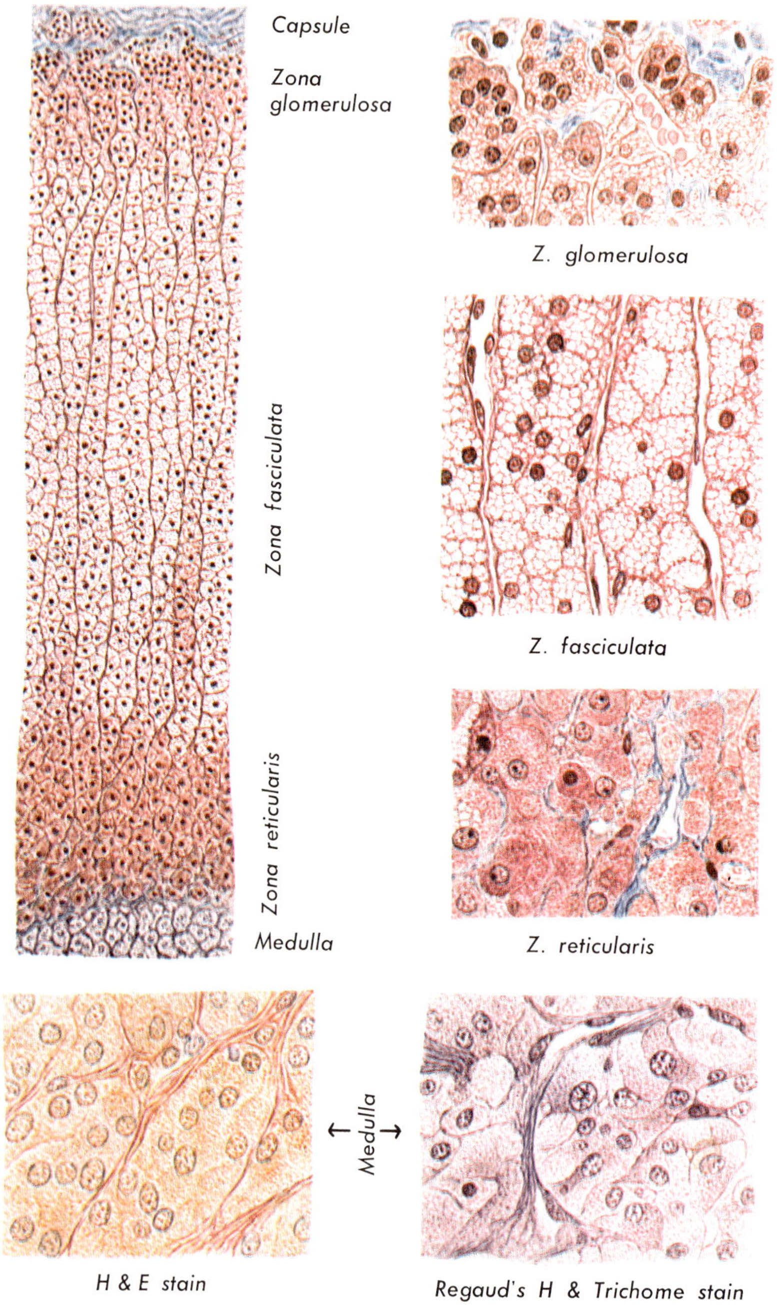

Fig. XII-7. Sections through the right adrenal of a man 35 years old. The adrenal was removed in an attempt to give relief from a testicular tumor with metastases. The tissue was fixed in Helly's fluid (formol Zenker). The section used for all the drawings, except the lower left one, was stained with Regaud's hematoxylin and Masson's trichrome. The low power (upper left) is × 205, the others × 435. (From W. M. Copenhaver, *Bailey's Textbook of Histology*, 15th ed., Williams & Wilkins, Baltimore, 1964.)

regular in their arrangement, forming anastomosing cords.

There is some evidence of functional specialization in these zones. The glomerulosa appears to produce mineral corticoids, hormones which regulate water and salt balance. The cells of the fasciculata appear to secrete glucocorticoids which regulate carbohydrate metabolism; the reticularis produces hormonal steroids of male and female types. Thus, the suprarenal cortex is essential for life, even though some nine-tenths of it probably can be removed without untoward effects. Forty steroid compounds have been extracted from the cortex. Seven of them have mineral and glucocorticoid propensities, while the rest act somewhat like sex hormones.

Mineral corticoids promote retention of sodium and water, favoring the excretion of potassium. Thus, deoxycorticosterone and aldosterone act upon the kidney tubules. Such glucocorticoids as cortisone and hydrocortisone act upon the liver to form sugar from protein, increasing the blood sugar level and increasing glycogen storage. The cortical androsteroids are akin to the androgens of the testis and to estrogen and progesterone of the ovary. They influence sex functions and the development of secondary sexual characteristics.

The principal of negative feedback to the pituitary operates in the suprarenal as it does in the case of the thyroid. As blood corticoid levels fall, adrenocorticotrophic hormone (ACTH) is secreted by certain cells of the anterior pituitary to stimulate the release of corticoids from suprarenal cortex. Aldosterone production in the glomerulosa can apparently be stimulated by an alteration in volume of extracellular tissue fluid.

Cortical destruction by tuberculosis has a fatal expectancy. There is insufficient resorption of sodium and water by the kidney tubule. Blood sodium levels decrease so that body fluids show serious electrolyte imbalance. Potassium builds up in tissue fluids. As normal osmotic relationships change, blood volume falls and blood pressure falls. Blood urea increases. Glucocorticoids fail to produce glucose from protein so that blood sugar levels also fall. A condition known as Addison's disease results, in which there is muscular wasting and weakness. Weight loss is severe and fluids are lost through diarrhea and vomiting. The patient becomes anemic and skin pigmentation increases.

Suprarenal cortical tumors may lead to excessive production of all corticoids. As blood sodium level rises, so does the level of water. As a result, crystalloids increase in the tissue spaces. Besides edema production, excess glucose production leads to obesity of face and trunk. Overproduction of suprarenal sex hormones leads to virilism with body hair growth in adult women; precocious development of secondary sexual characteristics in children; and muscular hypertrophy in very small boys so that they appear herculean in proportions.

The medullary region of the suprarenal gland comprises about 10% of the organ in man. Its cells are organized in tubular formations around postcapillary venules. One surface of the cell abuts on the capillary; the other, on the venule so that the cell demonstrates definite polarization. Sympathetic fibers are directed toward the capillary side where the nucleus resides. Golgi apparatus and secretory precursor granules face and discharge into the venule. Two hormones, epinephrine and norepinephrine, are produced. Each affects tissues differently and norepinephrine may be a precursor of epinephrine. Normally, these products are stored and so little of them is released into the blood stream as to have little significant effect. These products are released in greater quantity under conditions of emotional stress, however; and then they excite activity at sympathetic nerve endings in tissue, especially smooth muscle, cardiac muscle, and glands.

PARAGANGLIA

In addition to the suprarenal medulla, there are bits of chromaffin tissue distributed along the paravertebral sympathetic chain and in areas near sympathetic collateral ganglia. Like the adrenal medulla, these tissues stain yellow to dark brown when exposed to chromate salts. They contain epinephrine or 5-hydroxytryptamine which may account for the color reaction. Like sympathetic ganglia and medullary cells, they are neural crest derivatives. The largest of these clumps (organ of Zuckerkandl) measures about 1 cm. The rest measure between 1 to 3 mm. in diameter. After the second year of life many regress so that not too many are found in the adult as accessory epinephrine secretory centers.

Carotid and aortic bodies were once classed with paraganglia, but they are chemoreceptor centers for reflex regulation of respiration and vasomotor control. They are not secretory endocrine tissues and do not react with chormate salts.

THE HYPOPHYSIS (PITUITARY GLAND)

This most complex of the endocrine glands consists of anterior and posterior portions which differ in origin, structure, and function. The anterior part or **adenohypophysis** is glandular and was derived from an upward invagination of ectoderm in the roof of the embryonic mouth cavity (Fig. XII-8A). This upward pouching pinched off, vesicle fashion, to form **Rathke's pouch.** We can consider the pouch to have an anterior and posterior wall, an upward extension and a central cavity (Fig. XII-8 B,C). The vesicular pouch, as we indicated, freed itself from any connection with the mouth cavity.

The posterior part or **neurohypophysis** arose as a downward evagination of the floor of the primitive brain in the region of the future hypothalamus (Fig. XII-8B). It became a solid stalk with a dilated end. The stalk remains attached to the hypothalamus at

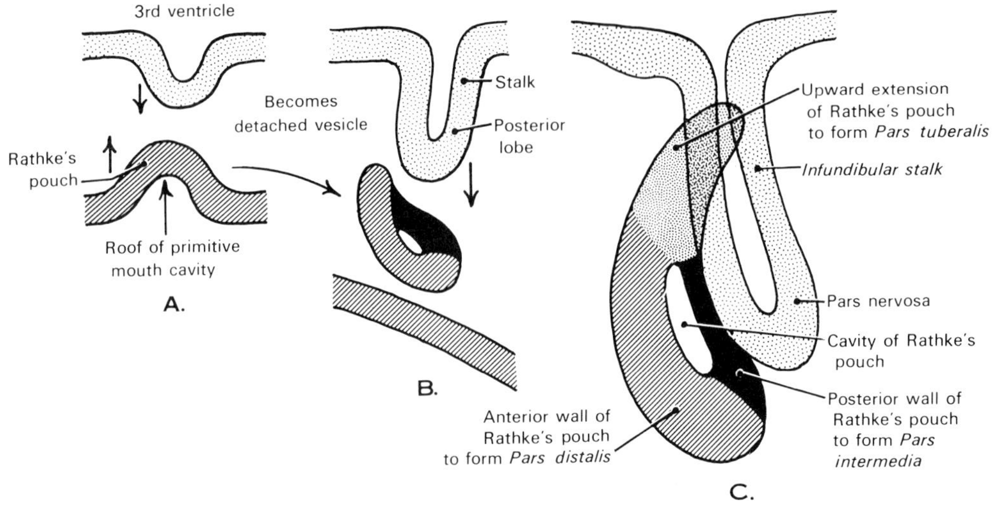

Fig. XII-8. Schema of the development of the hypophysis. **A.** Diverticula from primitive mouth cavity and third ventricle converge. **B.** Rathke's diverticulum detaches as a vesicle. **C.** Confluence of diverticula and named parts of each which give rise to regions of adult gland.

the region of the **median eminence** in the floor of the third ventricle (Fig. XII-10). The vesicle of Rathke's pouch migrated upward to meet the evaginating infundibular stalk of the posterior part of the forming pituitary. Its upward extension wrapped itself about the stalk as the **pars tuberalis** of the gland (Fig. XII-10). Its anterior wall became the **pars distalis** of the gland. Its cavity reduced to chink-like spaces, some of which may be fluid-filled cysts in man. Its posterior wall became a narrow strip of tissue between anterior and posterior lobes, the **pars intermedia** of the gland. The dilated termination of the infundibular stalk of the neurohypophysis became the **pars nervosa** or posterior lobe of the gland. The gland, the size of a small grape, lies hidden in the sella turcica of the body of the sphenoid bone under a diaphragmatic covering of dura mater which surrounds the stalk. While it normally weighs about a half a gram, it enlarges during pregnancy, and it may weigh 1.0 gm. in women who have had children.

Figure XII-9 is a photograph, considerably enlarged, of a coronal or frontal section through the gland, third ventricle and adjacent hypothalamus. It would be difficult to see much, if any, of the posterior lobe in such a cut this far forward. We do see pars tuberalis wrapped around the stalk, while the pars distalis of the anterior lobe makes up the bulk of the gland. In the stalk can be discerned the nerve fibers of the **hypothalamo-hypophyseal tract** and veins of the **hypophyseal portal system.**

The adenohypophysis or anterior lobe of the gland then consists of a number of regions in which cell types vary in the way they react to stains and in their characteristics under the electron microscope. This glandular part of the pituitary consists of: pars distalis

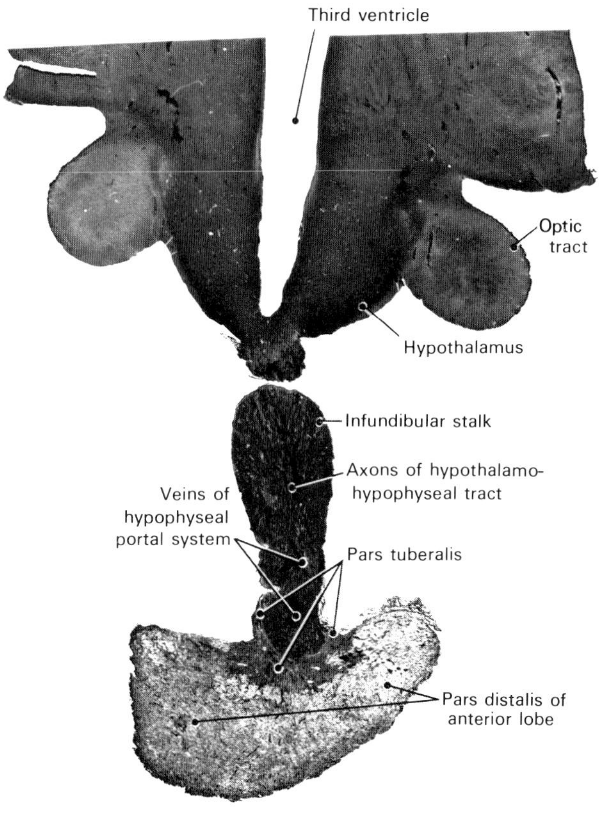

Fig. XII-9. Photograph of a coronal section through the hypothalamus and pituitary gland. In such a section the cut does not include the posterior lobe. If any part of the posterior lobe can be seen at all, it is only where it blends to the base of the stalk near pars tuberalis.

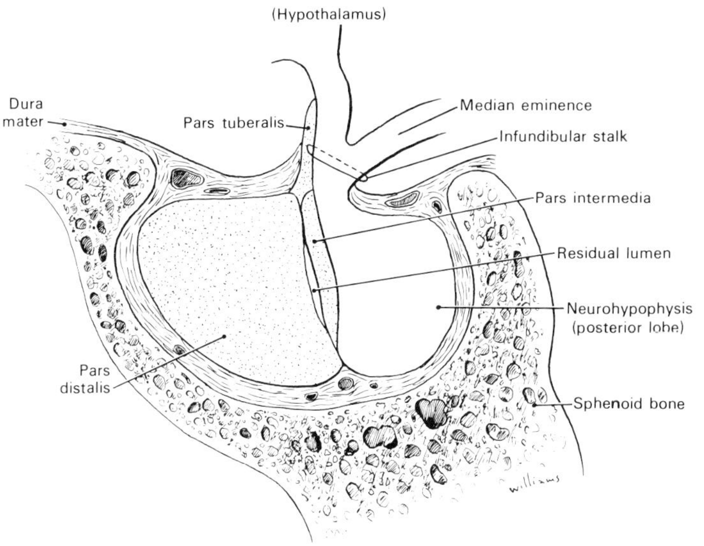

Fig. XII-10. The pituitary gland (hypophysis) in the sella turcica of the sphenoid bone. The parts of the adenohypophysis (anterior lobe) are stippled while parts of the neurohypophysis (posterior lobe) as well as the hypothalamus above appear in white.

(the major portion of the anterior lobe), pars tuberalis, pars intermedia, and the residual lumen. These are, as we learned, all derivatives of Rathke's pouch.

The pars distalis contains colorless **chromophobe** cells which, so far as can be determined, are not active hormone secretors (Fig. XII-3A). They may be inactive phases of the **chromophils** which do form granules and secrete specific hormones; or they may be unspecialized stem cells capable of giving rise to chromophil types.

The Hormone Producing Cells (Fig. XII-3A)

Some chromophilic cells of the pars distalis are acidophilic, reacting to acid dyes. Because of the way their granules stain, some are called orangeophils (*alpha* cells) and others, carminophils (*epsilon cells*). These acidophils can react with basic dyes such as safranin, however. Then there is also a group of basophilic chromophils in the pars distalis: the *beta, gamma,* and *delta* cells. These react with aniline blue, which, paradoxically, is an acid dye.

Functionally, the pars distalis is responsible for producing the following types of hormones:

1. Growth Hormone or Somatotropin (STH): The orangeophils appear to be its source. While it stimulates growth in general, the hormone is especially active at epiphyses. As you might suspect, undersecretion on the part of these cells leads to **dwarfism;** oversecretion to **giantism;** but in the full grown adult to **acromegaly,** *i.e.*, bony thickening in face, jaw, nose, hands, and feet.

A reduction in somatotropin secretion alone produces a **Lorain dwarf.** While there is delayed skeletal and sexual development, the child is well proportioned and mentally alert.

Widespread destruction of the anterior lobe, hypothalamus, and even the posterior lobe, leads to a **Fröhlich's dwarf.** This is a stunted, obese child with a great craving for sugar. Sexual development is arrested. The child sleeps much of the time, is lethargic and mentally subnormal. There are deficiencies in other endocrine glands and their hormonal products.

2. Gonadotropins: These are three hormones of the pars distalis which influence sex cells:

a. Follicular Stimulating Hormone (FSH): The *delta* basophil seems responsible for its production. It promotes growth of ovarian follicles and stimulates the testis to produce spermatozoa.

b. Luteinizing Hormone (LH): The *gamma* basophil is implicated. It produces a corpus luteum after follicular rupture at ovulation, and it stimulates interstitial cells of the male testes to produce testosterone.

c. Luteotrophic Hormone (Lactogenic Hormone; Prolactin) (LTH): The carminophil is said to be responsible for its production. It causes the maturely developed breast to secrete milk.

3. The Thyrotropic Hormone (TSH): *Beta* basophils are responsible for a product which stimulates and maintains thyroid activity.

4. Adrenocorticotropic Hormone (ACTH): This substance goads the suprarenal cortex to activity and seems to be produced by a subtype of *beta* basophil.

No specific hormones are known to be produced by the **pars tuberalis** even though the cells there are finely granular and faintly basophilic in their reaction to stains. Colloid cysts seem to form here.

In the **pars intermedia,** cells produce a **Melanocyte Stimulating Hormone (MSH)** the actual role of which is somewhat uncertain in man.

The posterior lobe consists of pars nervosa, the median eminence and the infundibular stalk. There are several types of odd cells in all three of these regions but none of them are hormone producing. All three areas receive the same blood supply and contain fibers of the hypothalamo-hypophyseal tract. Cell bodies of neurons in the hypothalamus contribute their axons to this tract and pass neurosecretory products along the axons. The terminal endings of these fibers are swollen with secretory granules and are known as Herring bodies. The swollen blebs end in close proximity to the capillary networks of the posterior lobe.

Two active factors of neurosecretory origin have been extracted from the posterior lobe:

1. Antidiuretic Hormone (ADH or Vasopressin): It stimulates the kidney tubule to resorb water. Lack of this hormone leads to polyuria (excessive passage of urine) and polydypsia (excessive water intake). These are symptoms of **diabetes insipidus.** Normal urinary output is about 1 to 1½ liters per day of a straw-colored fluid with a specific gravity of 1.020 to 1.032. Individuals with diabetes insipidus put out 20 to 30 liters per day of a pale, colorless urine with a low specific gravity of 1.001 to 1.002. Harvey Cushing summed it up aptly when he said, "And all this activity

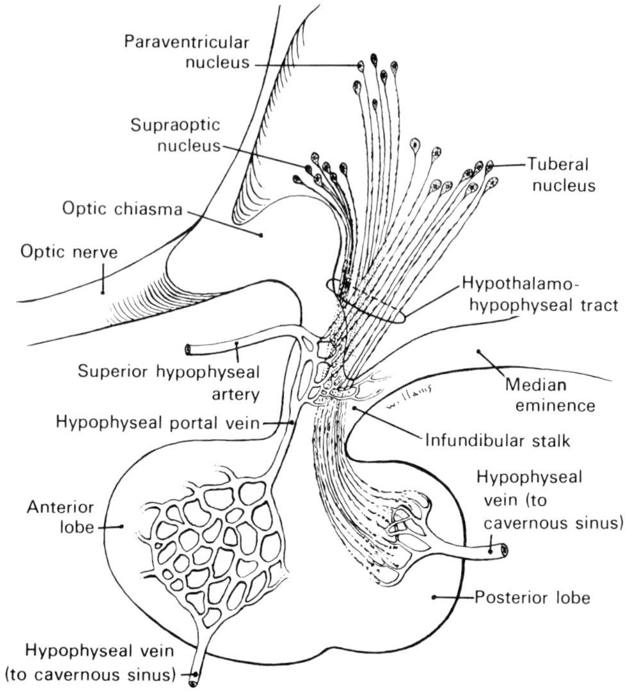

Fig. XII-11. The hypophyseal portal system and the hypothalamo-hypophyseal tract.

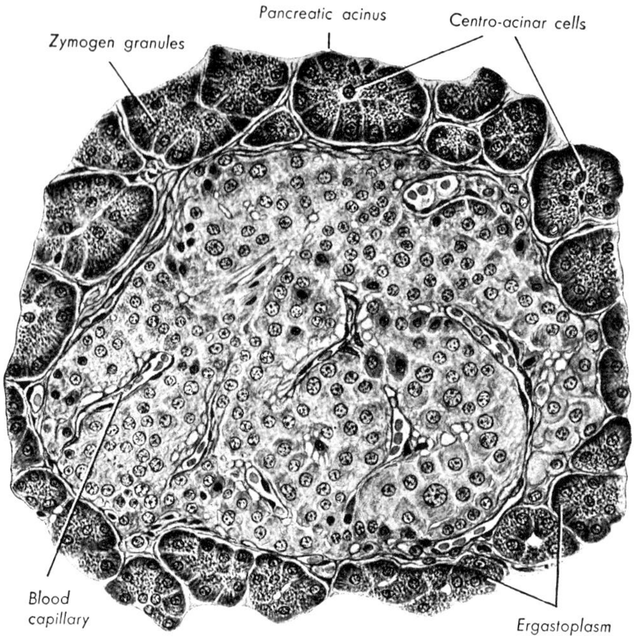

Fig. XII-12. A section through an island of Langerhans and surrounding acini. Human pancreas. Hematoxylin and eosin × 450. (From W. M. Copenhaver, *Bailey's Textbook of Histology*, 15th ed., Williams & Wilkins, Baltimore, 1964.)

is under the control of a tiny gland which is, itself, no larger than an ordinary pea."

In high concentrations, ADH also raises blood pressure by causing contraction of smooth muscle in arterioles.

2. Oxytocin: This hormone also tends to act upon smooth muscle, but upon smooth muscle that is concerned with infant-raising. It contracts the distended uterine wall during labor and acts upon the sacs and ducts of the fully lactating breast to let down milk during suckling.

The blood supply of the pituitary, like that of the liver, is based upon a portal system (Fig. XII-11). Blood to the anterior lobe comes from several **superior hypophyseal arteries** as branches off the circle of Willis. Some of these go directly to the sinusoidal capillary beds between the secretory cords and clusters of cells in the anterior lobe. Other branches of these arteries anastomose in capillary networks of the median eminence and infundibular stalk. Veins from these latter networks enter the capillary beds of the anterior lobe as **hypophyseal portal veins.** In this way, neurohumoral substances from the region of the median eminence can influence the adenohypophysis.

The **inferior hypophyseal artery,** a branch of the internal carotid, feeds the smaller-diametered capillaries of the posterior lobe. These capillaries are peculiar in that their walls are fenestrated, presumably to allow for passage of the neurosecretions reaching them over the **hypothalamo-hypophyseal tract** (Fig. XII-11). All capillary beds, whether in the anterior or posterior lobes, are drained by veins into the cavernous sinus which rings the sella turcica in which the gland rests.

THE PANCREATIC ISLETS

The pancreas is essentially a digestive gland modeled somewhat on the order of a misplaced salivary gland. Digestive enzymes are carried over its duct system into the duodenum. However, it contains within its lobules about a million scattered **islets of Langerhans** as isolated bits of endocrine tissue whose cells secrete directly into capillaries adjacent to them. Islet tissue varies from clusters of a few cells to masses as much as 3 mm. in diameter (Fig. XII-12).

At least three types of granule-containing, secretory cells are described: *Alpha* cells whose granules are water soluble; *beta* cells, which contain alcohol soluble granules; and *delta* cells, much fewer in number and which may represent a cyclic variant of the other two. The *beta* cells produce the **insulin** essential for carbohydrate metabolism. It renders carbohydrates oxidizable and muscle-storable. Without it, sugar spills over in the urine in diabetes mellitus. In *beta*-cell tumors blood sugars decrease as insulin is oversecreted. The *alpha* cells secrete **glucogen,** a hormone which offsets insulin action so that the level of blood sugar can increase to a useful utilizable level in the body.

THE THYMUS

It is difficult to classify this organ as an endocrine gland in a strict sense. It plays a role in insuring *antibody formation* and the development of *immune responses* to foreign proteins. Yet the thymus itself, does not appear to produce antibody.

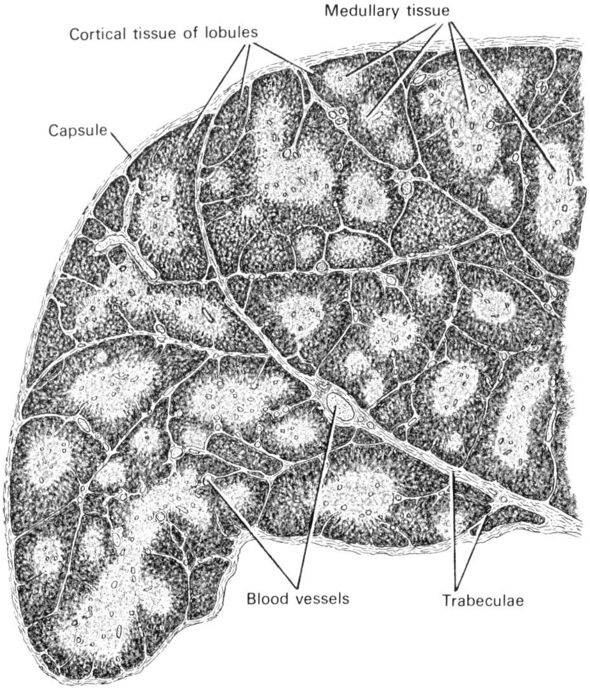

Fig. XII-13. Section through a part of the human thymus at birth. × 12. (From W. M. Copenhaver, *Bailey's Textbook of Histology,* 15th ed., Williams & Wilkins, Baltimore, 1964.)

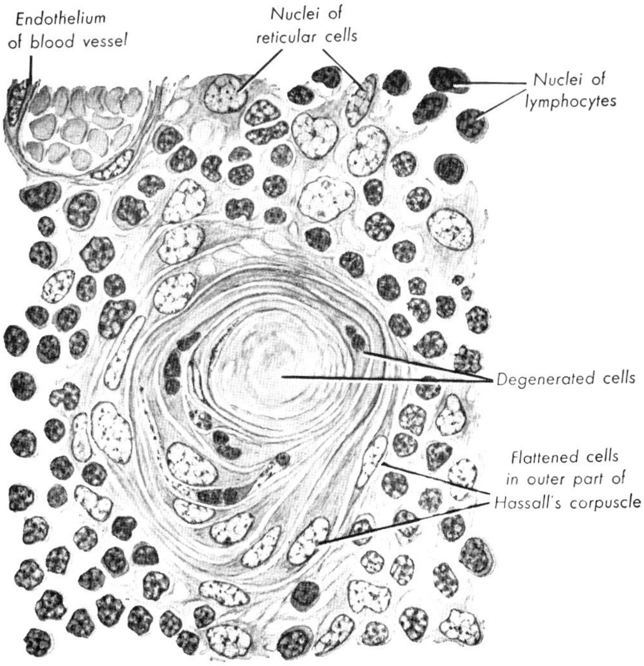

Fig. XII-14. Portion of medulla of thymus at birth showing Hassall's corpuscle. × 1134. (From W. M. Copenhaver, *Bailey's Textbook of Histology,* 15th ed., Williams & Wilkins, Baltimore, 1964.)

It is, in reality, a flattish, bilobed mass of lymphoid tissue beneath the sternum, large in the newborn but regressing after puberty. It has a cortex containing massive lymphoid nodules with germinal centers (Fig. XII-13). The lymphocytes and some plasma cells produced in these lymphopoietic centers have been confusingly and, probably, erroneously designated as "thymocytes" over the years. This came about because they were thought to be derived from cells of the peculiar epithelio-reticular network on which they lie. The connective tissue framework of thymus is unlike that of spleen or lymph node. It is not regular reticular connective tissue derived from mesoderm. Recall that the thymus, like tonsil and parathyroids, was of pharyngeal wall origin—entodermal in its ancestry.

The medulla consists entirely of this framework material and harbors over a million large-size thymic corpuscles or Hassall's corpuscles whose function is unknown. They seem to be degenerative areas in the epithelio-reticulum (Fig. XII-14).

Involution of the thymus would make it appear that its usefulness is relegated to the formative years of life. Removal of the organ at birth results in a severe lymphocytic shortage and a lack of antibody. The young organism does not repel invading microorganisms nor reject foreign bodies of other kinds. Perhaps the peculiar epithelial reticulum does secrete a humoral factor into the blood which effects the production of lymphocytes in other organs and imparts immunological competence. This tissue does not seem to give rise to the very lymphocytes present within its cortex, however, for thymic transplants never give rise to a new crop of a so-called "thymocytes."

THE INTERSTITIAL CELLS OF THE TESTES (FIG. XII-15)

We have indicated that luteinizing hormone (LH) of the pituitary influences nests of cells in the interlobular connective tissue of the testes to produce testosterone. Testosterone reacts with tissues of the body to produce the secondary sexual characteristics of the male. Hair grows on the face, in the axilla, and on abdomen and pubis. Sometimes it spreads almost ubiquitously except upon the pate, as your senior author can attest. The larynx enlarges and the voice deepens. Penis, scrotum, and prostate enlarge, and, with this emergence, the psychological outlook also changes as sexual instincts develop or warp, depending upon the social milieu. None of these things occur if testicular atrophy ensues before puberty. If atrophy occurs after puberty, spermatogenesis ceases and sterility ensues. Testosterone injections can overcome delayed puberty. Interstitial cells, apparently, also secrete a small amount of estrogen.

THE HORMONES OF THE OVARY

Follicular-stimulating hormone of the anterior lobe of the pituitary stimulates maturation of ovarian follicles, urging them to attain the Graafian stage. The cells of the internal theca of the follicle produce estrogen under this influence (Fig. XII-16). **Estrogen** is stored in the liquor folliculi in the antrum of the growing follicle and it is also released into the capillaries of the surrounding ovarian stroma. Estrogen, as the female sex hormone, brings about changes in secondary sexual characteristics at puberty. It assists in increasing height

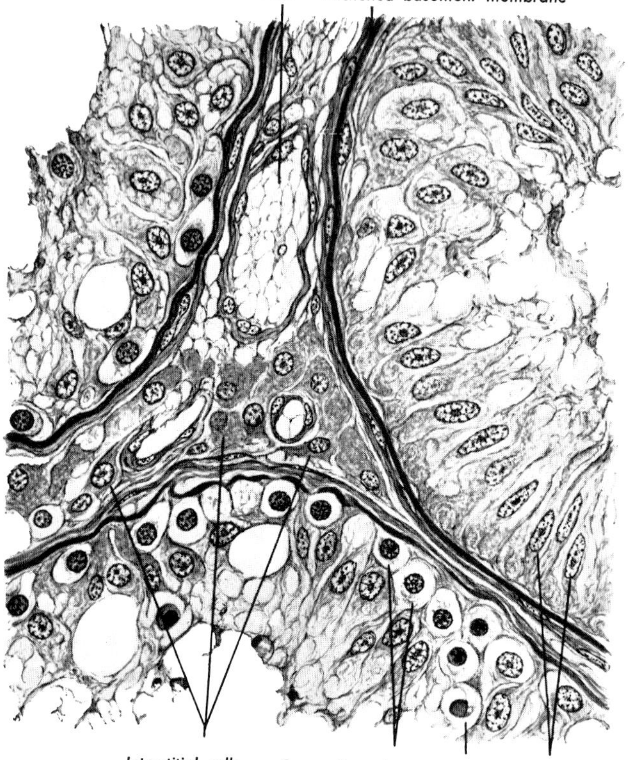

Fig. XII-15. Parts of three seminiferous tubules showing pronounced degeneration. The tissue was from a healthy middle-aged man. Camera lucida drawing. × 565. (From W. M. Copenhaver, *Bailey's Textbook of Histology*, 15th ed., Williams & Wilkins, Baltimore, 1964.)

through bone growth and it broadens the pelvis. Fat is deposited in nice places to round out the contours of shoulders, hips, and thighs. Mammary glands enlarge. Hair appears in the axilla and in its characteristic pubic pattern. The Fallopian tubes lengthen, the uterus enlarges and its endometrium proliferates in readiness for cyclic changes. Vaginal epithelium thickens. Concomitant psychological and personality alterations, gradually or quite suddenly, transform little girls into little women. Thus the "little bit of difference" in a woman's built-in biology urges her to ensnare a man. Such is the curse of estrogen.

Once an ovary has ovulated, luteinizing hormone stimulates the granulosa lutein cells of the forming corpus luteum to produce **progesterone** (Fig. XII-17). This second female sex hormone pushes endometrium growth to its succulent secretory stage. The uterus is preparing a proper nest for the egg it hopes will become fertilized. Should fertilization fail, the endometrial functional layers slough in menstruation, and the monthly cycle inevitably begins anew. If, one day, a wayward egg is waylaid and pregnancy ensues, progesterone secretion continues in the corpus luteum to maintain the uterus in its secretory state and to preserve the embryo.

Obviously, at the end of a cycle or after pregnancy, the estrogen content of a new monthly shower of developing follicles must influence the pituitary to momentarily decrease LH production and increase FSH production. Similarly, in the event of a pregnancy, a high titer of progesterone must be able to influence the pituitary to decrease FSH production. Still, even during pregnancy, the blood levels of both female hormones are on the rise.

The latter may be due in part to the fact that the developing **placenta** also synthesizes a gonadotrophic hormone similar to the luteinizing hormone. Its role is to assure the preservation of the corpus luteum. The cytotrophoblastic layer of the chorion and chorionic villi seem to be responsible for its production. Cytotrophoblast gradually becomes expended in the production of syncytial trophoblast. During pregnancy the syntrophoblastic cells of the placental villi secrete large amounts of estrogen and progesterone and so tend to take over the ovarian hormone-producing functions during gestation. In this sense, we must consider the placenta to be an endocrine gland. Frequently, we regard it merely as an organ for fetal-maternal interchange of nutrients and blood-gases.

THE PINEAL BODY

We mention this structure last since its endocrine function has been variously postulated and has been as vigorously denied over all the years. Descartes pegged it as the seat of the soul since it sat so centrally in the cranium. It dangles on a stalked outgrowth from the roof of the epithalamus between the habenular nuclei (Fig. XII-18). It is only 7 mm. in length, lobulated, and thinly encapsulated. Septae of reticular fibers penetrate from the capsule to incompletely surround masses of epithelial cells. The cells have long branched processes with bulbous ends (Fig. XII-19). Some of the cells appear homogeneous; others, granular or lipoidal. Neuroglial elements, chiefly astrocytes, form a support framework for the cells (Fig. XII-20). The blood vessels are limited to the capsule and septae. Nerve fibers (unmyelinated) enter to supply the vessels. Other nerve fibers (myelinated) appear to be entering bundles from the brain substance itself.

The function of the gland is obscure and uncertain, even though much has been written about it of late. It was once regarded as a vestigial third eye since the epiphysis of mammals is a vestige of a photoreceptive system in lower vertebrates. Its cells do seem to secrete melatonin and serotonin, both in the rat and man. These products in the rat undergo marked circadian rhythms. Melatonin, injected into rats, slows the estrus cycle and the ovaries lose weight. There is some evidence that pineal tumors in children cause delayed puberty. Hence, there is a current "feeling," but not much more, that the gland may have something to do with the timing of the estrus and menstrual cycles. Certainly, serotonin acts as an effective vasoconstrictor, and melatonin causes amphibian melanophores to retract their pigmented processes. It is, however, difficult as yet to find adequate experimental and clinical proof that this gland is a functioning endocrine gland.

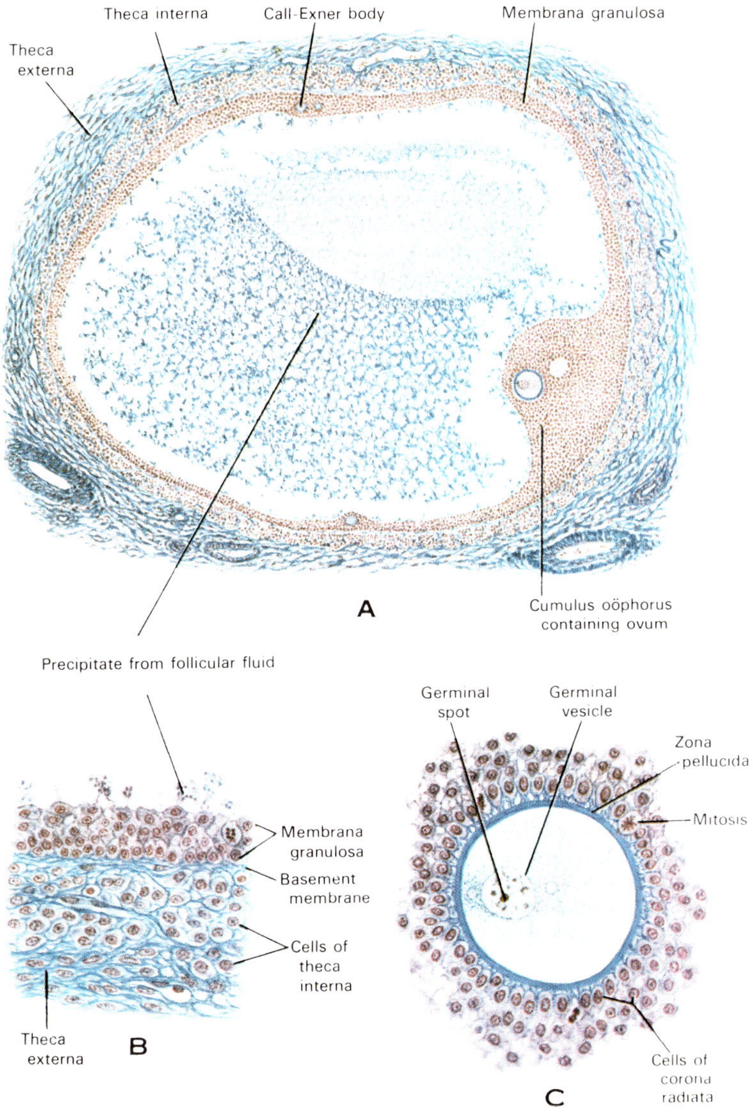

Fig. XII-16. On this plate is illustrated a follicle with a maximum dimension of 4.5 mm. Masson's trichrome stain. **A.** A low power view. × 53. **B.** A portion of the wall of this follicle. × 287. **C.** The ovum of this follicle. × 287. (From W. M. Copenhaver, *Bailey's Textbook of Histology*, 15th ed., Williams & Wilkins, Baltimore, 1964.)

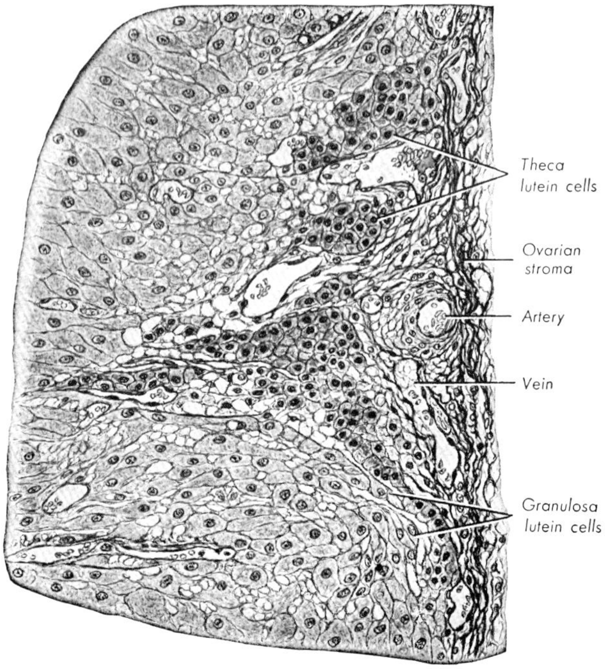

Fig. XII-17. Section through the peripheral part of corpus luteum removed during the progravid stage of the sexual cycle. Masson's trichrome stain. × 175. (From W. M. Copenhaver, *Bailey's Textbook of Histology*, 15th ed., Williams & Wilkins, Baltimore, 1964.)

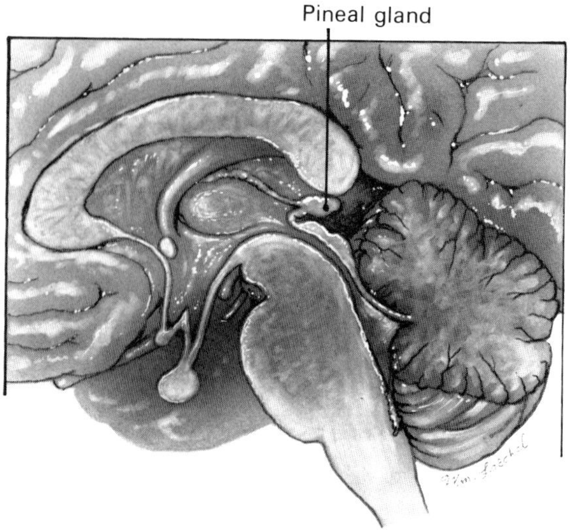

Fig. XII-18. Midsagittal section through brain and brainstem to show locus of pineal gland.

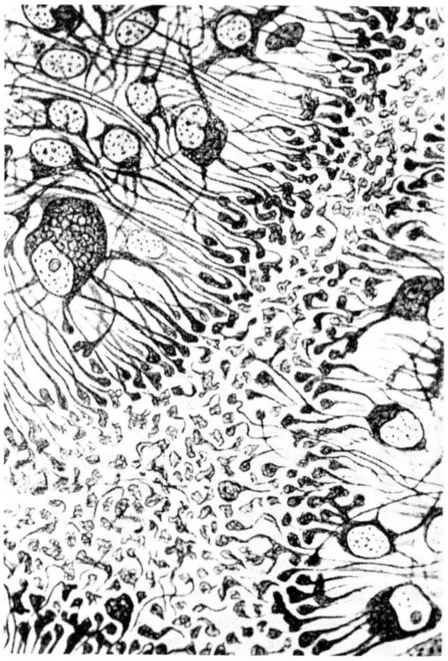

Fig. XII-19. Section of a pineal gland of a boy. The parenchymal cells and their processes were impregnated with silver. The processes are shown extending into an interlobular septum. Collagenous tissue and glia are not shown with this technique. (After del Rio-Hortega.) Semidiagrammatic. (From W. M. Copenhaver, *Bailey's Textbook of Histology*, 15th ed., Williams & Wilkins, Baltimore, 1964.)

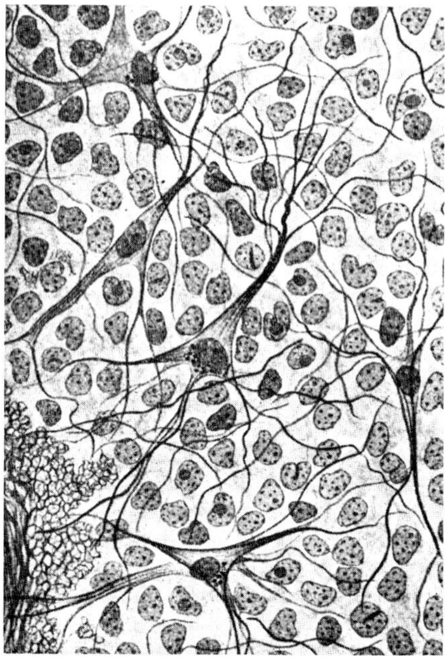

Fig. XII-20. Pineal gland showing the glial cells and their processes. The nuclei of the parenchymal cells are also shown. Silver preparation. (After del Rio-Hortega.) (From W. M. Copenhaver, *Bailey's Textbook of Histology*, 15th ed., Williams & Wilkins, Baltimore, 1964.)

THIRTEEN

The Upper and Lower Limbs

INTRODUCTION

While most other textbooks of anatomy have continued to devote large sections to the detailed anatomy of the limbs, we have elected to change our approach to the subject and limit the facts to the minimum. Although all medical students dissect the limbs, in actuality few remember enough of the complex anatomy to put it to practical use. For those of you who plan specialty training, there is no question that additional dissections in depth will be required of you during medical school and residency training. In order to serve the needs of those "superspecialists," we have included a bibliography at the end of the chapter which should prove useful in the future. To the vast majority of students of anatomy, the facts gathered and tabulated here will certainly suffice for the moment.

At this time, it would be to your advantage to review the surface anatomy of the limbs previously explained in Chapter III.

BONES OF THE SHOULDER GIRDLE AND UPPER LIMB

Two bones form the shoulder girdle: **clavicle** and **scapula** (Fig. XIII-1). The clavicle attaches medially to the manubrium of the sternum and laterally to the acromion of the scapula. Both points of articulation are by means of synovial joints; hence, the clavicular attachment is not an entirely rigid, immovable strut. The sternomanubrial attachment is the more complex one (Fig. XIII-2) since here two synovial cavities separated by an articular disk form a double gliding joint. The acromioclavicular joint is also a gliding joint but of simpler construction, consisting, generally, of a single synovial-lined cavity. Sometimes, however, an articular disk may be present in whole or in part. Quite in contrast, the scapula in no way articulates with the trunk.

The Scapula (shoulder blade)

This large, flat bone stands away from the upper dorsal part of the rib cage, held close to it by muscular attachments only. It has a spade-shaped **body** which expands superolaterally into a **head** bearing the **glenoid cavity.** The head is supported by a stout but somewhat constricted neck. The **costal or ventral**

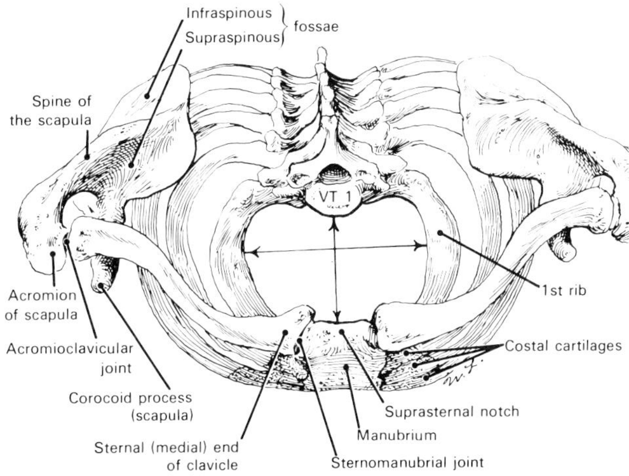

Fig. XIII-1. The shoulder girdles as viewed from the thoracic inlet.

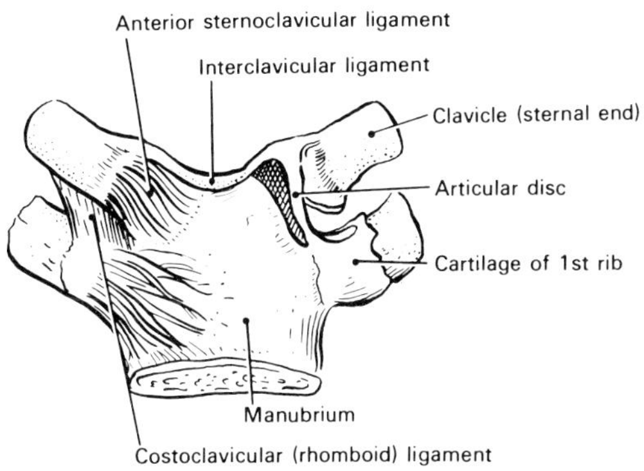

Fig. XIII-2. The sternoclavicular joint and associated ligaments.

surface of the bone (Fig. XIII-3A) faces the rib cage but is kept from the cage by the Serratus Anterior, a muscle which inserts upon the **medial or vertebral border** of the scapula and then continues laterally to its origin from the 4th to 9th ribs. The bulk of the costal surface of the bone forms the markedly-ridged **subscapular fossa** to give origin to the Subscapularis muscle. Laterally on this surface, a stout, smooth ridge descends from the neck to the *inferior angle* of the bone marking the lateral border of the triangular scapula.

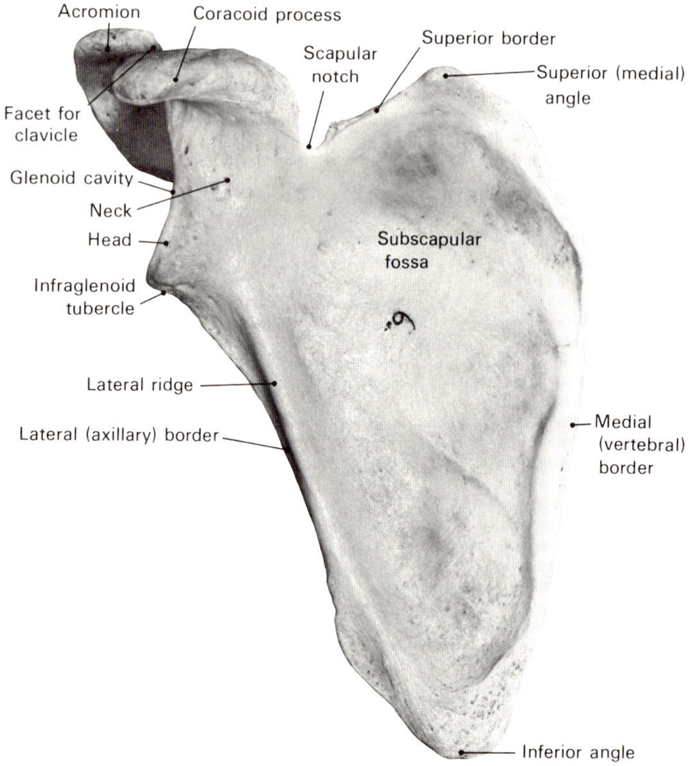

Fig. XIII-3. The right scapula. **A.** Costal surface.

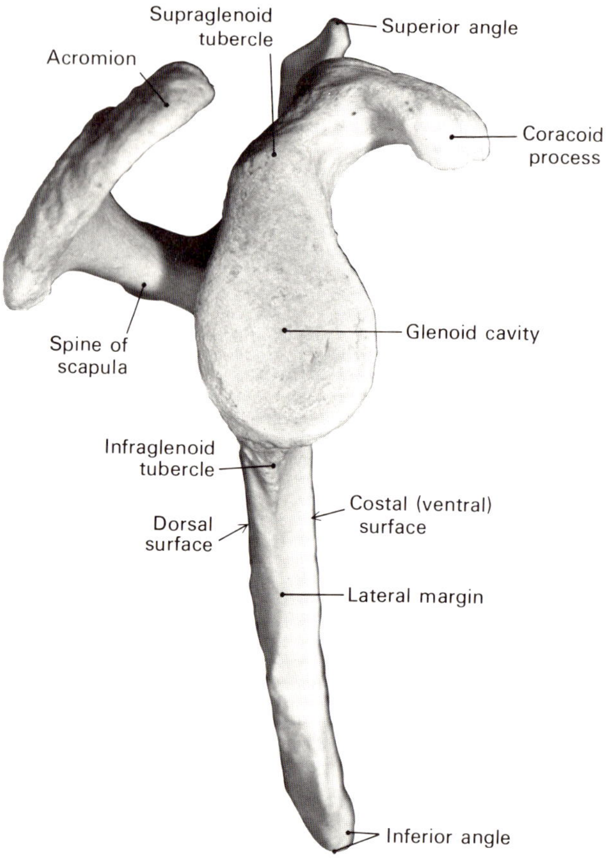

C. Lateral aspect.

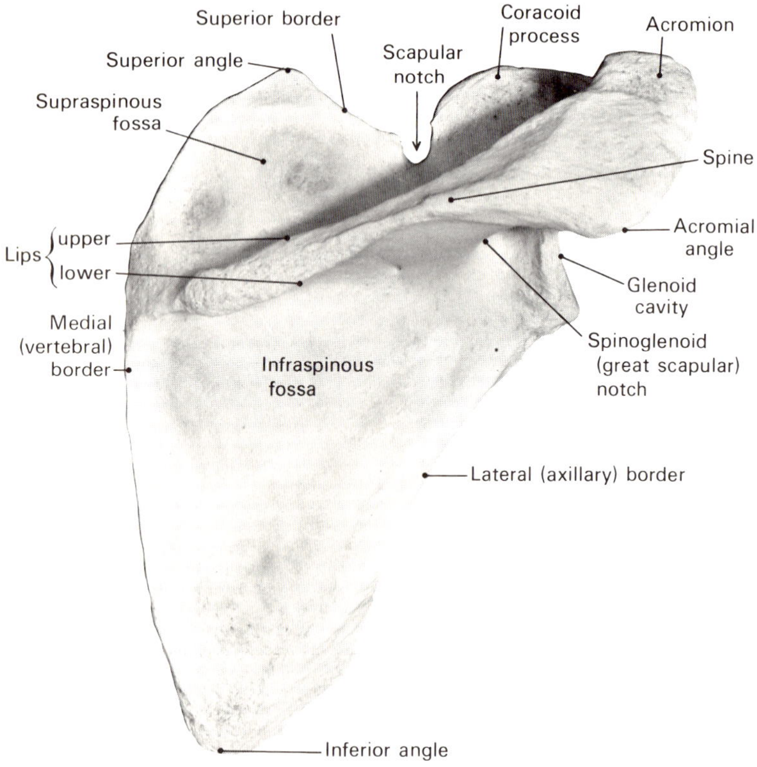

B. Dorsal surface.

A beak-like **coracoid process** projects upward from the costal aspect of the neck of the bone, pointing, in the manner of a bent finger, forward and laterally. Since the rib cage tapers markedly in its approach to its cervical outlet; the coracoid process projects forward around the lateral side of the cage, and it can be felt by pushing a finger inward and laterally just inferior to the clavicle in the region of the deltopectoral triangle. Its free extremity will give origin to the conjoined tendons of the Coracobrachialis and the short head of the Biceps Brachii. Certain ligamentous structures also attach to the process, e.g. the coracohumeral ligament, the conoid ligament and the costocoracoid ligament. The irregular upper surfaces of the process allow for the insertion of the Pectoralis Minor and for the anchorage of the trapezoid ligament. Except for this process and for the large subscapularis fossa, there are no other prominent features of note on the ventral aspect of this bone.

The *dorsal surface* of the scapula is a bit more complex (Fig. XIII-3B). A prominent **spine** divides this surface into two fossae, an upper, small, **supraspinatous fossa** and a lower, greater, **infraspinatous fossa.** These depressions give rise to muscles bearing the same names as the fossae. These fossae communicate laterally, at the neck of the bone and below the spine, in the **spinoglenoid** (**greater scapular**) **notch.** Superolaterally, the concave articular surface of the head of the scapula forms the glenoid cavity for reception of the head of the humerus (Fig. XIII-3C). Immediately above and below the lips of the glenoid cavity are the roughened **supra** and **infraglenoid** tubercles. The upper-most tubercle gives origin to the long head of the Biceps; the lower one, to the long head of the Triceps muscle of the upper arm.

The strong spine is a somewhat triangular plate of bone whose apex lies at the vertebral border of the scapula (Fig. XIII-3B). From this point the spine is directed obliquely upward to its base which is concave and becomes free of the main bone about at the middle of the neck of the scapula. Here an **inferior transverse ligament** will arch over the suprascapular (or transverse scapular) artery and the suprascapular nerve as these structures course from the suprascapular to the infrascapular fossa. The anterior border of the spine continues laterally as a subcutaneous **crest** to end in the free tip of the shoulder or **acromion.** The broad **acromial process** overhangs the glenoid cavity. The broad, roughened crest allows for the insertion of the large Trapezius muscle of the superficial back and neck cranially and, caudally, for the origin of the massive shoulder muscle, the Deltoid. A small, oval, articular facet on the medial border of the acromion is for articulation with the lateral or acromial end of the clavicle.

The **vertebral border** of the scapula forms the medial side of its triangular outline. Three muscles insert in descending order upon the dorsal aspect of this border: Levator Scapulae, Rhomboideus Minor and Rhomboideus Major. The Latissimus Dorsi may or may not attach at the inferior angle. The **upper border** of the triangular bone is gently curved, and just as it approaches the coracoid process laterally, there is the **scapular (lesser) notch** (Fig. XIII-3B). The notch is bridged by the **superior transverse scapular ligament.** The suprascapular nerve passes below the ligament within the notch. Usually, the suprascapular vessels course above the ligament to enter the suprascapular fossa. The Omohyoid muscle (inferior belly) arises from the ligament and from the bone just medial to the notch. Two other muscles arise from the scapulae from the dorsal aspect of its **lateral (axillary) border** which is roughened on their account. The Teres Minor starts its origin just below the neck of the bone. This point is marked by a groove over the lateral border of the scapula for the circumflex scapular artery. The Teres Major arises lower down along the lateral border nearer to the inferior angle.

The Clavicle (collar-bone)

We have mentioned previously that this is a unique bone in several respects: (1) it is the first bone of the body to ossify; (2) it is usually described as a membrane bone but since it forms epiphyses at each end, it ossifies in part as a cartilaginous bone; and (3) on occasion, it may be pierced by one or more of the supraclavicular nerves and, at times, even by the phrenic nerve or strands of the phrenic.

The clavicle is classed as a long bone of irregular shape. Thus, it has two articulating ends, **sternal** and **acromial,** and a shaft of double curvature (Fig. XIII-4A and B). Its medial two-thirds is convex anteriorly in the horizontal plane and its distal one-third concave anteriorly. As it sits in the body, it displays an anterior surface and is thick and prismatic in its medial two-thirds but flattened along its lateral one-third. At its thickened sternal end, the clavicular head of the Pectoralis Major originates along its lower part and the Sternomastoid along its upper part. Midway the anterior surface is smooth and is covered only by the skin and Platysma. Its flat, lateral end looks directly upward so that the Deltoid arises in front and the Trapezius inserts behind on this superiorly directed surface (Fig. XIII-4B).

Turn the bone over as if you were to pluck it from the articulated skeleton in order to examine its inferior or deep surface (Fig. XIII-4A). Look for an irregular concave depression (the rhomboid fossa or notch) at the thickened sternal end. Here the costoclavicular ligament attaches and the Sternohyoid muscle has its origin. This area, seen radiographically (p. 764), is frequently scalloped and has a faint sclerotic margin. Lateral to the area of attachment for the ligament is a long, shallow groove to house the Subclavius muscle. This groove often contains the nutrient foramen into which a branch of the suprascapular (transverse scapular) artery will enter.

At the flat end posteriorly, there is a rough elevation, the **coracoid tubercle.** Its medial portion is the **conoid tubercle** while laterally it forms an oblique line, the **trapezoid line.** These two subareas give attachment to the medial and lateral parts of the **coracoclavicular ligament** anchoring clavicle to coracoid process of the scapula. The **costoclavicular ligament** at the sternal end anchors clavicle to the first rib. In addition, the clavicle is held in place by the synovial capsules at its two ends. The sternal end has an articular facet which extends to articulate inferiorly with the first rib (Fig. XIII-4A). The remainder of the facet articulates in the clavicular notch of the manubrium. At the acromial end there is a small, oval, beveled facet for articulation with the scapula.

The Humerus (Fig. XIII-5A–D)

This largest and longest bone of the upper limb extends from the scapula above to elbow below where it articulates with the radius and ulna, bones of the forearm. The smooth **head,** covered by cartilage (Fig. XIII-5B and D), forms the proximal, spherical (one-third of a sphere) end of the bone which articulates in the glenoid cavity. When the humerus is in a neutral position, the head is directed postero-medially and upward (Fig. XIII-5A). Just distal to the articular cartilage of the head is a slight constriction, the **anatomical neck** of the humerus, to which the fibrous capsule of the glenohumeral joint attaches. Anteromedially, beneath the anatomical neck, projects the **lesser tubercle (tuberosity)** which receives the tendon of the Subscapularis muscle. A larger projection on the lateral side, proximally, forms the **greater tubercle (tuberosity)**. Its three facets accommodate the tendons of the

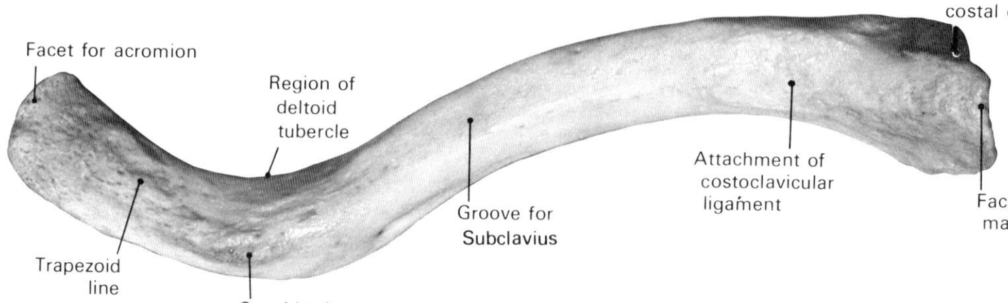

A. Viewed from below.

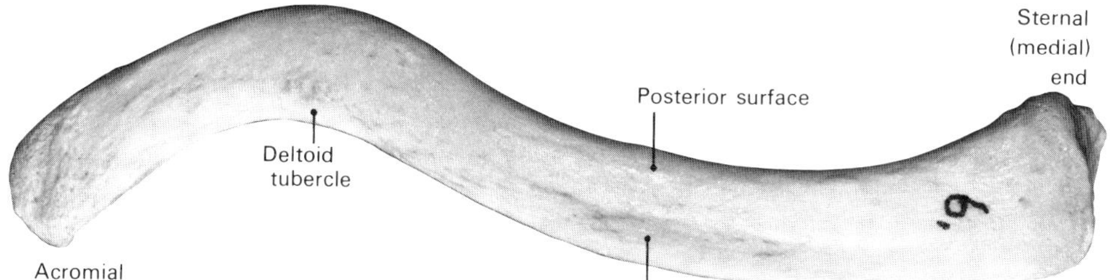

B. Viewed from above

Fig. XIII-4. The right clavicle.

Supraspinatus, **I**nfraspinatus and **T**eres Minor which **SIT** here in a row from front to back. Between greater and lesser tubercles anteriorly runs a groove, the **intertubercular (bicipital) groove.** Between the tubercles, a transverse ligament converts the upper end of the groove into a canal. In addition to housing the tendon of the long head of the Biceps and the synovial sheath of this tendon, the groove affords the site of insertion of the Latissimus Dorsi and it transmits a branch of the anterior (humeral) circumflex artery. Downward continuations of the tubercles actually form the side **crests** of the groove. The Pectoralis Major inserts along the lateral crest; the Teres Major, on or near the medial crest in the upper end of the humerus.

Distinct from the anatomical head is a narrowed zone about an inch below the tubercles, the **surgical neck,** often involved in fracture. On the posterior aspect of the bone (Fig. XIII-5B), the lateral head of the Triceps takes origin just below the surgical neck and extends to about midlateral shaft where it meets the **spiral groove** for the radial nerve as this nerve makes its way around the back of the bone to gain its lateral margin. The medial head of the Triceps clothes the rest of the posterior shaft below the level of the spiral groove. We have already observed that the third head or long head of the Triceps arose from the infraglenoid tubercle of the scapula. All extensor muscles occupy the posterior aspect of the upper limb. The Triceps is the extensor group related to the posterior compartment of the upper arm. The radial nerve supplies all the extensor-supinator muscles of the entire limb. In supplying the Triceps, it is only logical to look for the groove of this nerve on the posterior aspect of the humerus.

The **shaft or body** of the humerus is roughly cylindrical in contour but it widens and flattens distally where it flares out into the **medial** and **lateral epicondyles** (Fig. XIII-5A and B). Midway along the lateral margin of the anterior aspect of the shaft is the more or less prominent **deltoid tuberosity,** the site of insertion of the Deltoid muscle. Almost opposite, on the medial margin, the Coracobrachialis inserts on the shaft. Below these points, the shaft is clothed anteriorly by the Brachialis muscle, a large flexor of the forearm. Distally, along the **lateral epicondylar ridge** (the flared part of the bone) two other muscles, the Brachioradialis and Extensor Carpi Radialis Longus, arise. From the medial epicondyle itself, the Pronator Teres and four *flexor* muscles of the wrist and hand take origin. The lateral epicondyle, in contrast, gives rise to the Supinator and four *extensor* muscles. Thus, we see that the anterior and medial parts of the humerus favor a flexor-pronator group of muscles; the posterior and lateral parts, an extensor-supinator group in so far as their action upon forearm, wrist and hand is concerned.

Between and beneath the epicondyles, the humerus terminates in its articular **condyle.** A bony ridge divides this condyle into two parts: the **capitulum** and the **trochlea** (Fig. XIII-5A). The capitulum lies laterally and is directed anteriorly and inferiorly for articulation with the head of the radius. It is, thus, spheroidal in shape. The trochlea is a larger, pulley-shaped articular surface lying medially and lapping

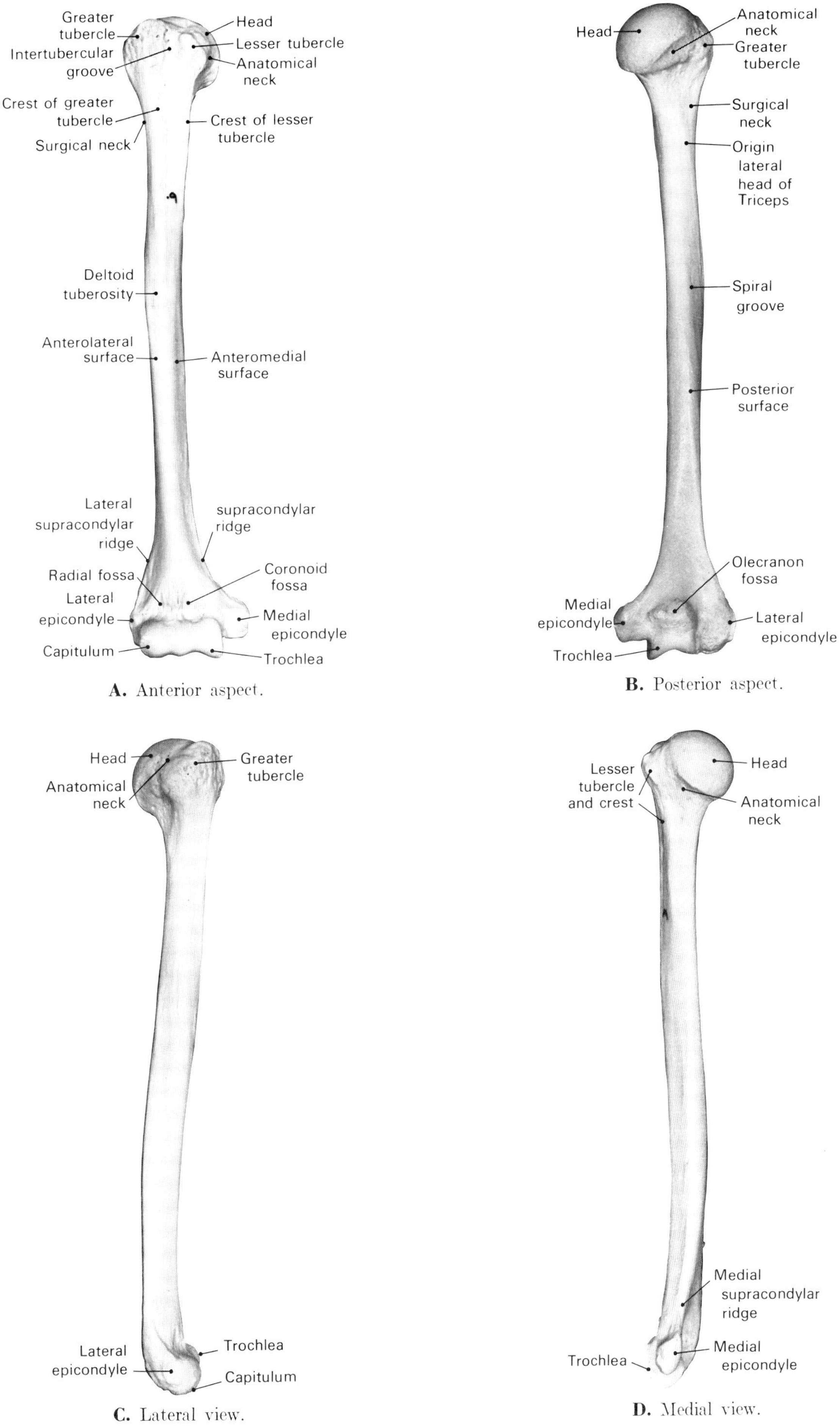

A. Anterior aspect.

B. Posterior aspect.

C. Lateral view.

D. Medial view.

Fig. XIII-5. The right humerus.

over on both the anterior and posterior aspects of the bone. It articulates with the **semilunar notch** of the ulna. Note three depressions above the condyle (Fig. XIII-5A and B). Posteriorly, there is the large, triangular-shaped **olecranon fossa** to receive the olecranon process of the ulna when the arm is fully *extended* at the elbow joint. Anteriorly, a small **radial fossa** is found just above the capitulum and a larger, deeper, **coronoid fossa** exists above the trochlea. These last two fossae accept the head of the radius and the coronoid process of the ulna when the arm is fully *flexed* at the elbow joint.

The Ulna (Fig. XIII-6A–D)

This prismatic bone is the longest and the most medial of the two bones of the forearm when the body is in the anatomical position. It is largest at its proximal (elbow) end while its **head** is small and distal. Observe the deep **trochlear (semilunar) notch** at the proximal end (Fig. XIII-6A). A bony ridge divides the notch into two articular surfaces. The notch belongs partly to a thick, upward projection of the shaft, the **olecranon,** and to the **coronoid process** which projects as a transverse shelf at the antero-inferior lip of the notch. The superior surface of the olecranon is rough to receive the extensor tendon of the Triceps (Fig. XIII-6B). Its posterior surface is smooth and palpable as the "crazy bone" beneath the skin of the elbow (Fig. XIII-6C). The inferior aspect of the coronoid process is rough to receive the flexor tendon of the Brachialis. Between this rough area and the main shaft is the **ulnar tuberosity.** An oblique fibrous cord passes from it onto the radius. There is a still smaller tubercle, more proximal on the medial margin of the coronoid process which gives origin to the superficial muscle which flexes the fingers and to the proximal ulnar collateral ligament. Between these tubercles runs an oblique crest, the **pronator ridge,** for the origin of the ulnar head of the Pronator Teres (Fig. XIII-6B). At the inferior tip of the ridge the long flexor muscle of the thumb arises.

The lateral margin of the coronoid process drops off into a depression, the **radial notch,** an articular surface for the disk-shaped radial head (Fig. XIII-6D). An **anular ligament,** attaching at the anterior and posterior margins of this notch, holds the radial head to the ulna and allows the head to rotate when the radius crosses the ulna in supination.

The upper four-fifths of the ulnar shaft anteriorly is clothed by the deep flexor muscle of the fingers; the lower one-fifth, by the Pronator Quadratus. On the posterior surface (Fig. XIII-6C), the shaft also is marked by an **oblique line** which runs from the posterior extremity of the radial notch to the postero-lateral margin of the bone. Some fasicles of the Supinator attach to this line; but more effectively, it divides the posterior shaft into a small, proximal triangular area for the insertion of the Anconeus and into the large remaining distal area of the posterior shaft proper. The medial portion below the oblique line is covered by the deep flexor muscle of the fingers and by the ulnar wrist extensor, both of which attach here by an aponeurosis. The posterolateral aspect of this lower shaft segment (facing the radius) gives origin to three muscles, two other extensors and an abductor, e.g. long extensor of thumb and extensor of the index finger; long abductor of the thumb. The exact role of the small Anconeus is not well known though, generally, it is described as aiding the Triceps in extension at the elbow. The truth of this is debatable. It would appear to be more of a weak abductor of the ulna when the forearm moves from extreme supination to extreme pronation.[1]

An **interosseous membrane,** attaching to both bones, fills the gap between lateral border of ulna and medial border of radius. This interosseous border along the ulna is rather ill-defined at its distal aspect. Proximally, it forms the outlines for the **supinator fossa** (Fig. XIII-6D).

The distal articular end of the ulna is small and is composed of the **head** and a posteromedial, downward-projecting **styloid process** (Fig. XIII-6D). The rounded, lateral circumference of the head articulates with the distal ulnar notch of the radius. An articular disk is interposed inferiorly between the head and the carpal bones. To the rounded summit of the styloid process, the distal ulnar collateral ligament attaches and its posterior surface (Fig. XIII-6C) is grooved by the tendon of the ulnar wrist extensor.

The Radius (Fig. XIII-7A–D)

This shorter bone of the forearm is lateral in position. Unlike the ulna, its small cylindrical **head** lies proximally and its distal end is expanded. The proximal end bears the disk-shaped **head,** a **neck** and the **radial tuberosity** to which the tendon of the Biceps attaches (Fig. XIII-7A). We have seen how the head is held in the radial notch of the ulna. This kind of articulation (the **superior radio-ulnar joint**) is a pivot joint. The surface of the wafer-shaped disk, the size of a Tums or a Rollaid, is concave for articulation with the capitulum of the humerus.

The shaft or body of the bone increases in size from upper to lower end. It is concave anteriorly where it faces the ulna in its upper three-fourths and flattened at its distal one-fourth. Posteriorly, its upper and lower thirds are convex and smooth; its middle one-third, concave (Fig. XIII-7C).

Anteriorly, an oblique line runs from the radial tuberosity to midlateral shaft (Fig. XIII-7A). It gives attachment to the oblique fibrous cord between radius and ulna and to the radial head of the superficial flexor of the digits. Above the line laterally, the Supinator inserts; at the lateral extreme of the line the Pronator Teres inserts; while below the line, medially, the origin of the long flexor of the thumb occupies the concavity of the shaft. The widely flared, anterior, distal end of the shaft provides a broad insertion for the Pronator Quadratus.

Posteriorly (Fig. XIII-7C), the smooth, rounded, proximal shaft still provides space for insertion of the Supinator. Its middle third is covered by the fleshy parts of the long abductor of the thumb and by its

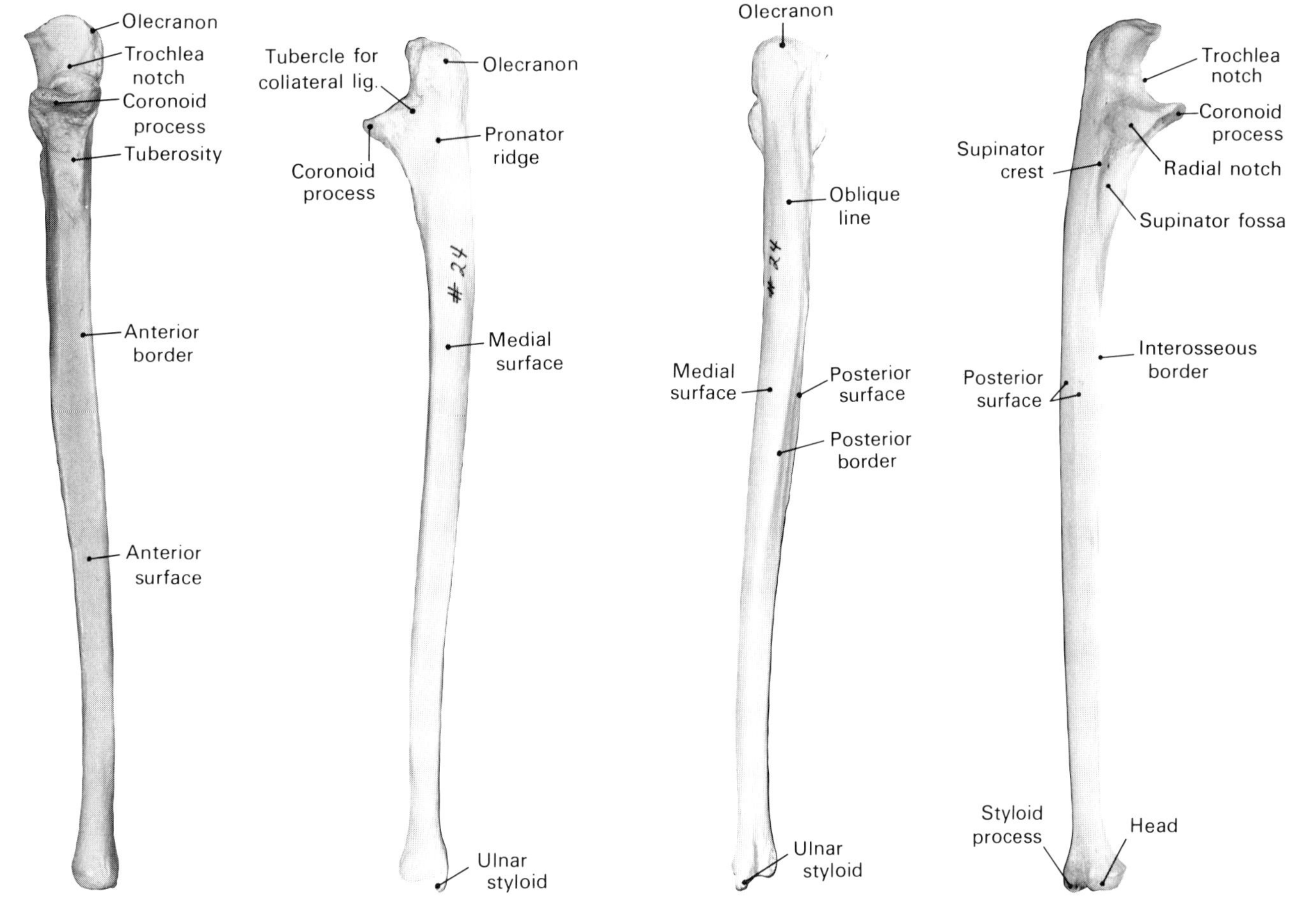

A. Anterior view. **B.** Medial view. **C.** Posterior view. **D.** Lateral view.

Fig. XIII-6. The right ulna.

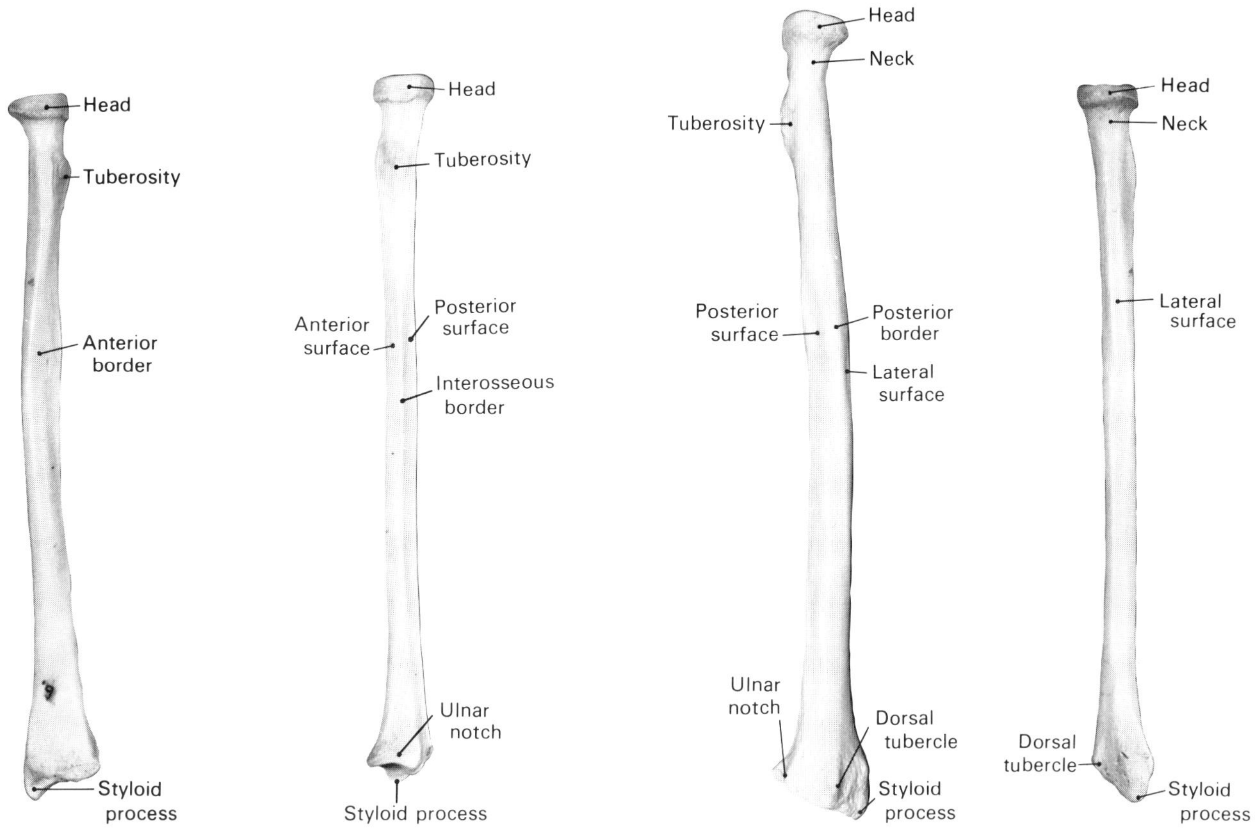

A. Anterior view. **B.** Medial aspect. **C.** Posterior view. **D.** Lateral aspect.

Fig. XIII-7. The right radius.

short extensor muscle. The lower third of the posterior shaft is covered, in part, by the tendons of these two muscles as they groove the medial **styloid process** of the radius. Other tendons wear grooves into the distal, posterior, flared part of the bone, e.g. those of the long and short radial wrist extensors; the long extensor of the thumb, the common extensor of the digits and the special extensor for the index finger. The styloid process also, at its tip, provides anchorage for the radial collateral ligament of the wrist and the Brachioradialis muscle inserts into its base.

The bone bears a medial **ulnar notch** at its distal end where the head of the ulnar articulates in the **distal radio-ulnar joint** (Fig. XIII-7B and C). Its inferior articular surface is concave and bears two facets for articulation with two carpal bones—a medial, oval one for the lunate bone and a broader, triangular one for the scaphoid (navicular). The radius alone of the two forearm bones articulates with the wrist and so it carries the wrist about the ulnar in movements of pronation and supination.

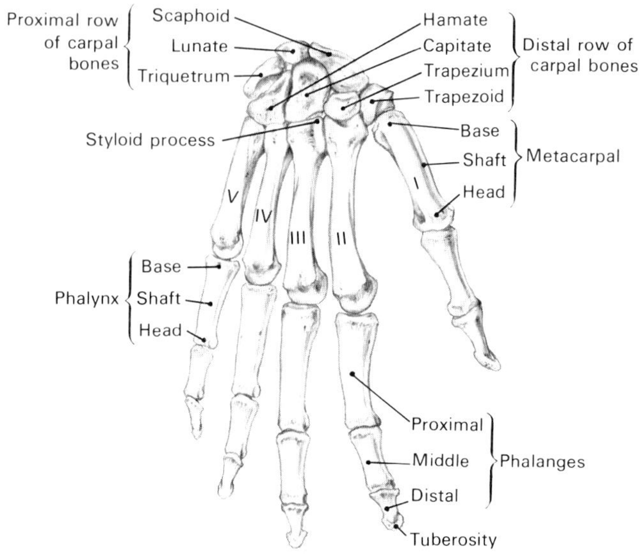

A. Dorsal view.

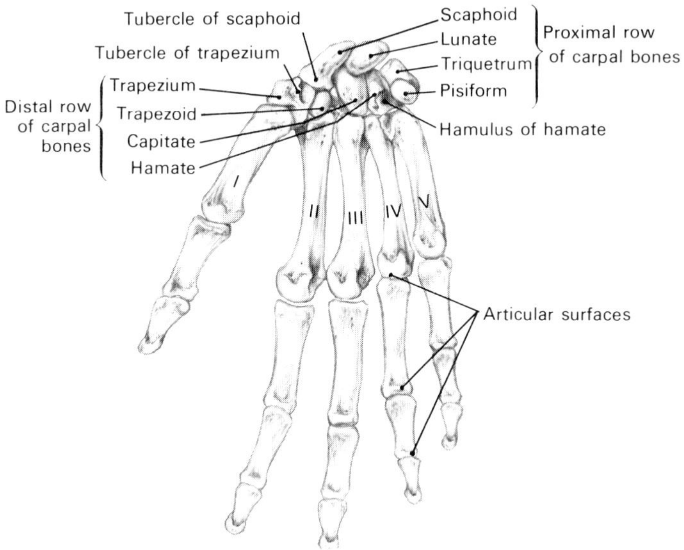

B. Palmar view.

Fig. XIII-8. The right hand and wrist.

The Hand (Fig. XIII-8A and B)

The bones comprising the skeleton of the hand are: (1) eight short **carpal bones** known collectively as the **carpus** (bones of the wrist); (2) five **metacarpals** (bones of the **metacarpus** which form the palm of the hand); and (3) 14 **phalanges** in all which are the bones of the **digits.** Recall that metacarpals and phalanges are miniature long bones and that digits (fingers) are numbered I to V viewed from radial to ulnar side of the hand, the thumb counting as number I. Fingers can also be designated as index, middle, ring and little fingers if you wish to be certain that people understand perfectly which one you mean.

The Carpal Bones: These bones are conveniently arranged into two rows of four each, a proximal and a distal row to make up the wrist. Again, from radial to ulnar side, the proximal row is composed of: the **scaphoid** (navicular), **lunate, triquetrum** (triangular), and **pisiform** bones. The distal row consists of: the **trapezium** (greater multangular) **trapezoid** (lesser multangular), **capitate,** and **hamate** bones (Fig. XIII-8A and B). If you learn their names at this time, that is probably all you will recall about them two months hence and even then, this may depend upon what dirty jingle you elect as a memory device. Perhaps, it were better to forget the names entirely and attend to some generalizations about them:

1. They are short, irregular bones and tend to be six-sided.
2. The anterior and posterior sides are rough for ligamentous attachments which will knit them into a structural unit.
3. Distal and proximal surfaces are smooth and articular.
4. The lateral surfaces are also smooth if they articulate laterally with their neighbors; rough if they do not.
5. Except for the lunate, they are broader on their dorsal surface than on their ventral aspect.
6. They all have a cancellous center surrounded by a thin layer of compact bone and there is hyaline cartilage at their articular surfaces.

You will notice that the palm of your hand is concave and that the back of it is convex. This is because a connective tissue **flexor retinaculum** on the palmar surface bridges from scaphoid and trapezium laterally to pisiform and lunate medially arching the bones into a **carpal sulcus** which the retinaculum converts to a **carpal tunnel** or canal so that flexor tendons can pass into the hand. An excellent view of this sulcus is seen in the radiograph of Figure XIII-9.

Three of the four bones in the proximal row articulate either directly with the radius, or through the articular disk, with the ulna to form the radiocarpal (wrist joint). Scaphoid (navicular) and lunate engage facets on the radius; the triquetrum articulates with

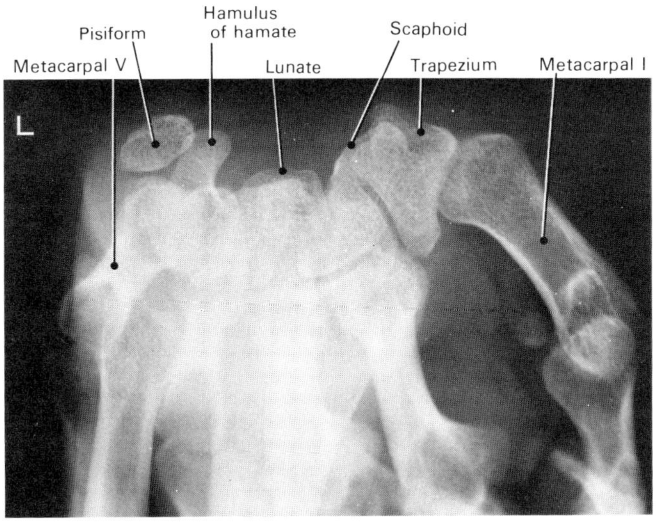

Fig. XIII-9. Radiograph of the carpal tunnel (canal) in axial projection.

the disk interposed between it and the ulna. The tiny, pea-shaped pisiform in this proximal row articulates solely with the triquetrum. There is a sinous **midcarpal** joint between the proximal and distal rows which permits some flexor action on its own account at the wrist.

Figure XIII-8B shows how the distal row articulates with the proximal row and with metacarpals. The trapezium articulates with the first metacarpal, the trapezoid with the second; the capitate, largest of the carpals in the distal row, with the second, third, and fourth; and hamate with the fourth and fifth metacarpal.

The Metacarpals: The **heads** of these miniature long bones are at their distal ends where they articulate with their own opposing proximal phalanges. These articular heads extend further on the palmar or volar surface of the hand than on the dorsal surface. Each head has a small tubercle on each side on its dorsal surface and anterior to these tubercles are notches for the attachment of collateral ligaments and joint capsules.

The **shafts** of these bones tend to be concave on their medial and lateral sides where Interossei attach. Each have a prominent volar ridge at midshaft. The distal two-thirds of the dorsal shaft is smooth, triangular and flat because this area is covered by extensor tendons.

The proximal **base** of typical metacarpals is broader dorsally than on the volar surface. The number of facets varies depending upon how many carpal bones are contacted in articulation at the wrist. There are also articular facets on the sides of metacarpals II to V where they articulate with each other.

Metacarpal I is the thickest and shortest of all and its palmar surface is flat and rotated so that it faces anteromedially. Its head (distal end) is flatter and wider than the others, and it sports two tubercles on the **palmar surface** for two sesamoid bones in the heads of the Flexor Pollicis Brevis (short thumb flexor). Its base forms a saddle-shaped articulation with the trapezium which enhances its mobility. There is a tubercle on its radial side for the Abductor Pollicis Longus. Hence, this particular metacarpal varies more than the others from a generalized pattern.

Metacarpal III has a styloid process on the radial side of its dorsal base. Distal to it the tendon of the Extensor Carpi Radialis Brevis attaches.

Metacarpal V has a small tubercle on its ulnar side for the insertion of the tendon of the Extensor Carpi Ulnaris.

The Phalanges: The thumb has two phalanges—a proximal and a distal; the other four digits have three: proximal, middle, and distal. The proximal row, obviously, articulate with the metacarpals. The phalanges are also considered to be miniature long bones and so have a shaft and two ends.

The **proximal phalanges** are unique in that their proximal **bases** have round, concave, articular surfaces. Their distal **heads** are pulley-shaped with two articular surfaces separated by a depression. Like the metacarpals, these articular surfaces extend further on the palmar than on the dorsal surfaces of the hand. Palmar surfaces of the **shafts** are flat; dorsal surfaces are rounded; and the sides are rough for the attachment of flexor tendon sheaths.

The middle phalanges have two articular surfaces at the base separated by a bony ridge. These facets fit nicely in the condylar heads of the proximal phalanges. They themselves also have distal condylar (pulley-shaped) heads, and their shafts are flat on the palmar side and rounded on the dorsal side.

The distal phalanges are the shortest of all. They, too, have two facets at their proximal bases for articulation with the middle phalanges. Their shafts are similar to those of other phalanges in terms of flatness and convexity. They taper distally to end in horseshoe-shaped **tuberosities** (tufts) where they are rough and expand into the pulp of the finger tips.

ROENTGEN ANATOMY OF THE SHOULDER GIRDLE AND UPPER LIMB

Introduction

Since there are 126 bones comprising the appendicular skeleton, it is not practicable to examine nor to describe all of the roentgen anatomical features of these bones. Only the principle features will be reviewed. If the student will use a mounted skeleton while reading this section, the information will be more meaningful.

Most radiologists have a "routine" series of radiographs which they utilize to study the osseous system. These projections are used to provide the most information with the least radiation exposure or discomfort to the patient. In the seriously ill or injured patient it is not always possible to follow such a routine series, and, indeed, it may become necessary to use the physician's knowledge of anatomy to devise new projections in order to acquire the necessary radiographic diagnosis.

As in other chapters of this book, we will very briefly describe the method of obtaining the projection under consideration in order to understand the anatomical structures visualized better.

1. The Shoulder Girdle

As you now know, the shoulder girdle consists of the clavicle and scapula, and of course, it connects the upper limb to the trunk. Since the proximal humerus is included in the shoulder joint, we will elect to consider its anatomy with the girdle.

a. AP Projection—External Rotation Position (Fig. XIII-10A): The patient may be in either the erect or supine position with the posterior aspect of the scapula resting upon the cassette. The humerus is rotated so that its epicondyles are parallel with the film. The arm is slightly abducted. Centering of the central X-ray beam is perpendicular to the cassette and directly at the corocoid process.

Thus, in Figure XIII-10A, we are actually seeing the humerus in its normal anatomical position relative to the glenohumeral joint. Since there is normally some obliquity of the glenoid cavity of the scapula, it appears oval in shape. The posterior labrum of the glenoid is partly overlapped by the humeral head. Identify the densely radiopaque corocoid process which in part obscures the spine of the scapula. Follow the scapular spine along its entire course from medial end to its continuation as the acromion. Also observe the suprascapular notch and the spinoglenoid notch. Trace the lateral, medial, and superior borders of the scapula. Most of the body of the scapula is obscured by ribs lying in front of it.

In the same illustration (Fig. XIII-10A), the distal two-thirds of the clavicle is seen. Its curvature, which is in an anteroposterior plane, cannot be appreciated. Examine carefully the normal apposition of the distal clavicle and the acromion. This is a frequent site of trauma with subluxation or dislocation of the acromioclavicular joint, as well as a site of fractures of the clavicle and scapula.

We mentioned earlier that in this position, the humerus is in the normal anatomical position. The greater tubercle of the humerus is now in profile such that you may clearly see the facet to which the tendon of the Supraspinatus muscle attaches. Note the dense curvilinear line cast by the lesser tubercle. Between the two tubercles is the intertubercular groove. Curving horizontally across the proximal humerus is the remnant of the epiphyseal line. Approximately 1 inch below the tubercles is the surgical neck of the humerus. Examine for a moment the shaft of the humerus with its dense compact cortical bone. Notice that this normally thins out proximally in the metaphysis and it becomes extremely thin over the humeral head.

When studying radiographs of the osseous system, it is of the utmost importance to view each of the aforementioned structures with the aid of a *magnifying glass*. Look carefully for distortions of the trabecular pattern, bone destruction, new bone formation, fractures, etc. Always look at the soft tissues for signs of swelling, loss of muscle mass, and, of course, for foreign bodies and calcifications. The latter are common findings in the tendons of the rotator cuff (Supraspinatus, Infraspinatus, Teres Minor, all of which attach on the greater tubercle; and the Subscapularis which inserts on the lesser tubercle). Such calcifications may also burst out of the tendons into bursal and sub-bursal areas of the region.

b. AP Projection—"Neutral" Position (Fig. XIII-10B): The patient remains in the same position as described in (a); however, the epicondyles of the humerus are internally rotated 45°. Again the central X-ray beam is directed at the corocoid process.

This projection affords a view of the two other facets of the greater tubercle, namely those giving attachment to the Infraspinatus and Teres minor tendons. The lesser tubercle is obscured in this view.

c. AP Projection—Internal Rotation Position (Fig. XIII-10C): In this projection, the humerus is more internally rotated, such that the coronal plane of the epicondyles of the distal humerus are perpendicular with the film plane.

We may now see a profile view of the lesser tubercle upon which the tendon of the Subscapularis attaches. The intertubercular groove may be seen also.

d. Transthoracic Projection (Fig. XIII-10D): At times the patient is unable to abduct the arm, and this projection permits us to view the proximal humerus and its relation to the glenoid cavity. Place the injured shoulder laterally against the erect radiographic table. The epicondyles of the humerus should be perpendicular to the film plane. Raise the normal arm above the head and rest the forearm atop the head. The central X-ray beam is directed at the glenohumeral joint, and, of course, the beam is perpendicular to the table top. When possible we prefer to utilize a breathing technique—the patient breathes quietly, and a long exposure time (5 to 6 seconds with low milliamperage) is used. In this manner the rib cage and lung markings will be blurred out thus affording greater osseous detail of the shoulder girdle.

Examine in Figure XIII-10D the normal relationship of the humeral head to the glenoid cavity. Dislocations of the humerus out of the glenoid as well as displacement and angulation of humeral fractures may be well demonstrated in this view. For the most part, other structures are obscured.

e. Superior-inferior Axial Projection (Fig. XIII-10E): The cassette is placed upon the table top, and the patient abducts his arm and extends the shoulder over the cassette. The central X-ray beam is directed 5° to 15° toward the elbow while centering upon the shoulder joint.

This projection shows you yet another view of the acromioclavicular articulation, and the relation of the head of the humerus to the glenoid cavity. Also identify the corocoid process, the scapular notch, the spine of the scapula, and the superior and lateral borders of the scapula. The lesser tubercle and intertubercular groove may also be seen.

f. Clavicle: (*1*) *AP projection* (*Fig. XIII-11A*): With the patient supine upon the table top, the central X-ray beam is directed perpendicularly at the film holder centering on the midpoint of the clavicle.

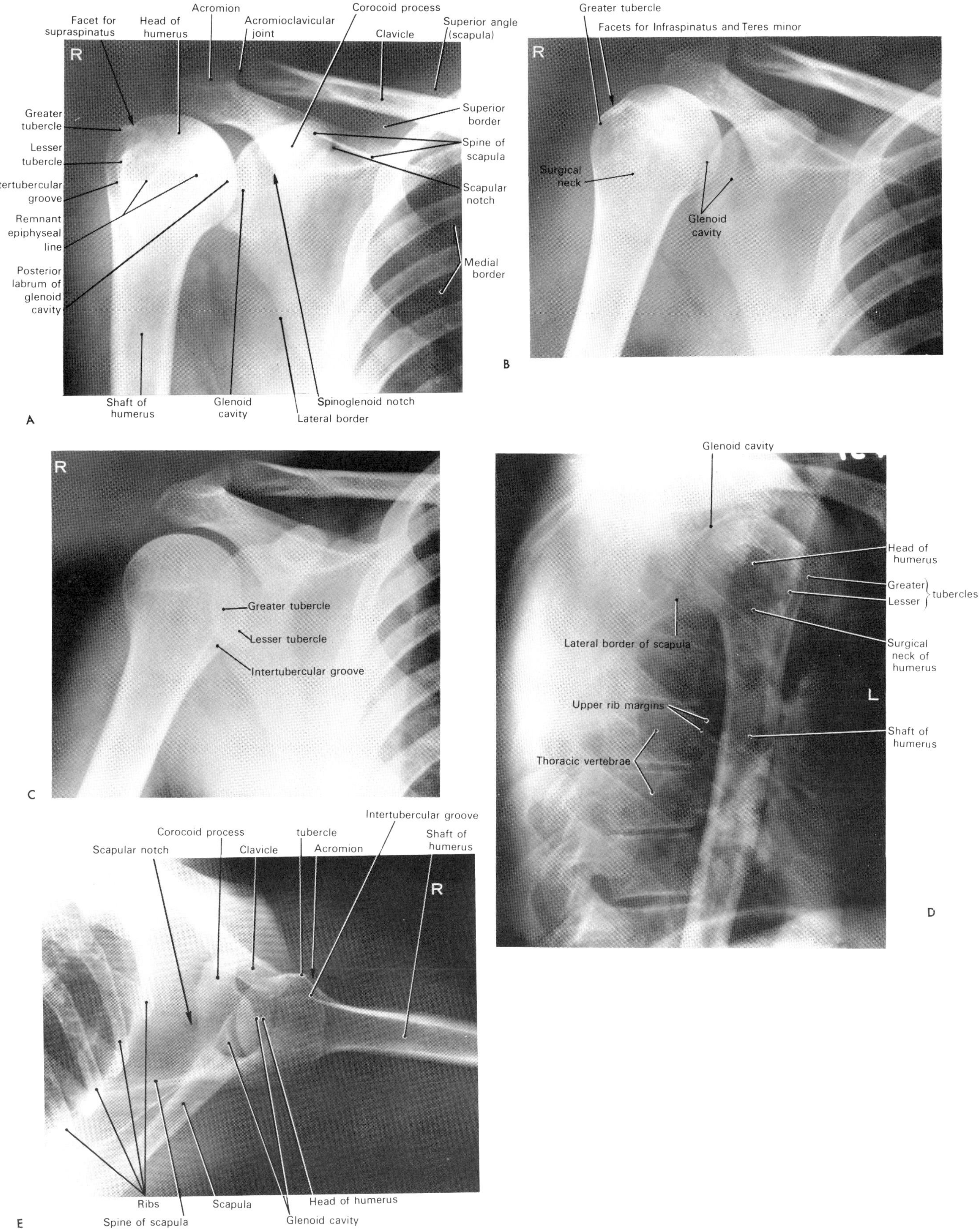

Fig. XIII-10. Radiographs of the shoulder girdle and proximal humerus. **A.** AP projection of the right side—external rotation position. **B.** AP projection same side—"neutral" position. **C.** AP projection right side—internal rotation position. **D.** Transthoracic projection left shoulder girdle and proximal humerus. **E.** Superior-inferior axial projection on the patient's right side.

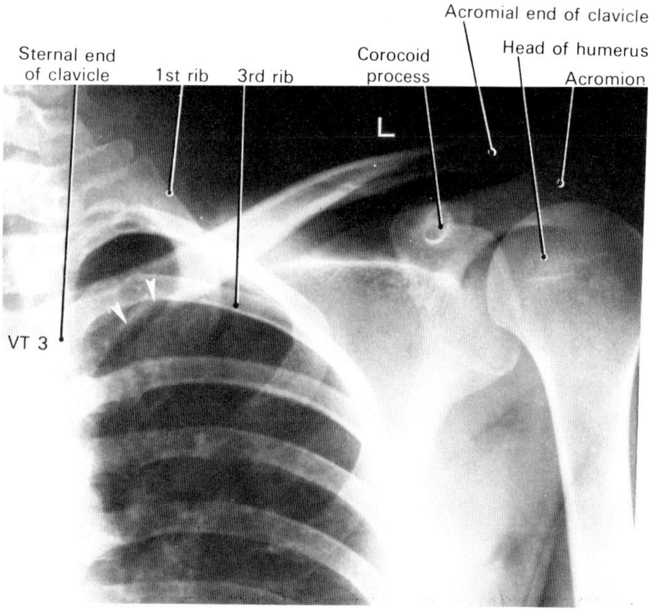

A. AP projection. *White arrows* indicate the rhomboid notch.

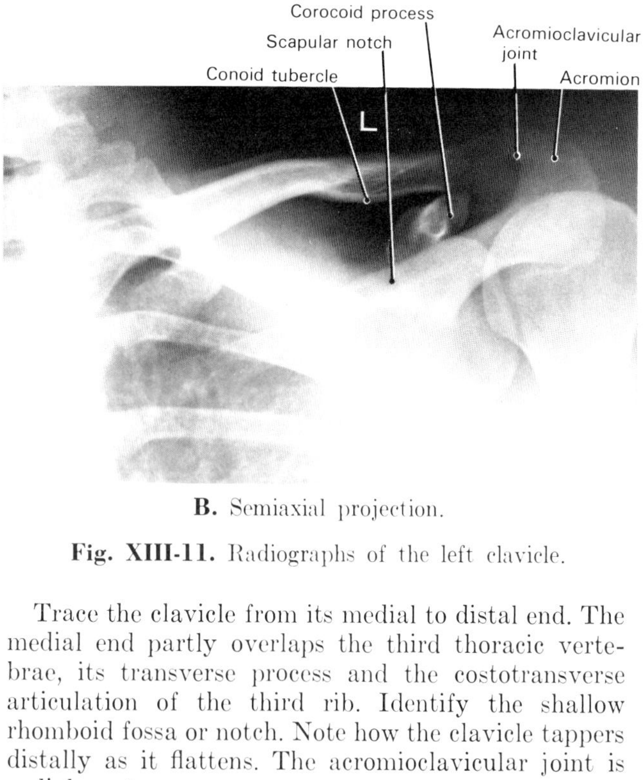

B. Semiaxial projection.

Fig. XIII-11. Radiographs of the left clavicle.

Trace the clavicle from its medial to distal end. The medial end partly overlaps the third thoracic vertebrae, its transverse process and the costotransverse articulation of the third rib. Identify the shallow rhomboid fossa or notch. Note how the clavicle tappers distally as it flattens. The acromioclavicular joint is radiolucent.

(*2*) *Semi-axial projection* (*Fig. XIII-11B*): With the patient supine, the central X-ray beam is directed at the supraclavicular fossa with the tube rotated 25° to 30° cranially.

This projection is useful in detecting fractures which when not displaced may be missed in the AP view. Note the prominent conoid tubercle to which the conoid ligament attaches. This view affords us some appreciation of the double curve of the clavicle.

g. The Sternoclavicular Joints. (*1*) *Right and left anterior projections* (*Fig. XIII-12A*): The patient is placed in the prone position and rotated to throw the sternum off the vertebral column. The exact degree of rotation is best determined by palpation of the sternum and the vertebral column. The central X-ray beam is perpendicuar to the table top and directed at the sternoclavicular joints.

In Figure XIII-12A, the fifth rib partially obscures the joint; however, it is still readily identifiable. The articular facet of the clavicle can be seen articulating with the small, shallow clavicular facet of the manubrium. Recall that there is an articular disk separating two synovial joints at this articulation. By comparing both oblique projections, abnormalities are more easily detected.

(*2*) *PA Projection* (*Fig. XIII-12B and C*): The patient is placed in the prone position, and the central X-ray beam is directed perpendicular to the table top at the sternoclavicular joints. Identify the manubrium, the joints, and the contribution of the medial ends of the clavicles to the suprasternal notch.

h. Scapula: (*1*) *AP Projection* (*Fig. XIII-13A*): With the patient in the supine position on the X-ray table top, the upper limb is abducted 90°. The elbow is flexed and the forearm is placed on the table top. This manuever rotates the scapula laterally. The central X-ray beam is directed perpendicular to the film holder at the midscapula.

Identify the glenoid cavity and the abducted humeral head articulating within it. The scapular spine partly overlaps the upper portion of the glenoid cavity. Trace out the three borders of the scapula, its angles, tubercles, the corocoid process, acromion, and the scapular notch.

(*2*) *Posterior Oblique Lateral Projection* (*Fig. XIII-13B*): Rotate the patient from the prone position on the table top until the scapula is perpendicular to the film holder. The central X-ray beam is directed at the center of the scapula, and, of course, perpendicular to the table top.

The scapula has now been projected off the rib cage. The glenoid cavity is visualized *en face*. Note the normal position of the head of the humerus to the glenoid. Extending inferiorly from the glenoid cavity is the heavy bony ridge which lies anterior to the axillary border of the scapula in this view. Identify the corocoid process, the spine of the scapula and the acromion. Notice that the distal end of the clavicle appears to overlap the acromion—a phenomenon of projection.

2. The Elbow

a. AP projection (Fig. XIII-14A): The extended elbow is placed on the film cassette with the forearm in supination. The central X-ray beam is directed perpendicular to the cassette and at the elbow joint.

Figure XIII-14A shows the distal humerus as it flairs out as the supracondylar ridges laterally ending on the prominent medial epicondyle and the more flattened lateral epicondyle. Distal to the epicondyles

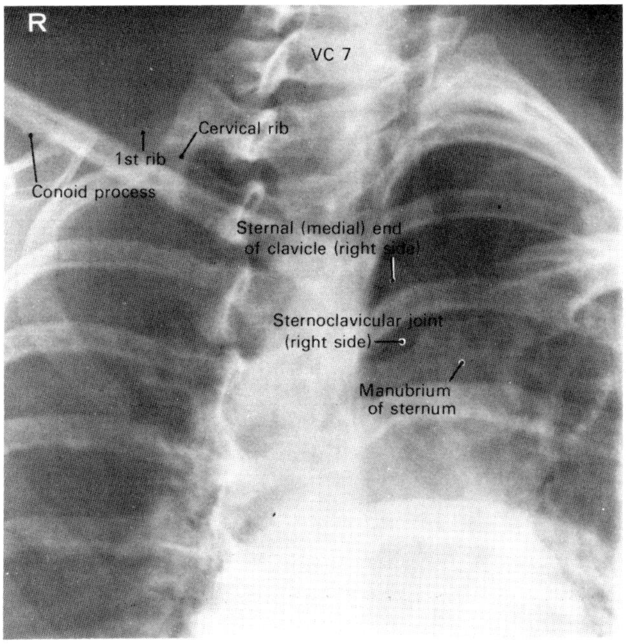

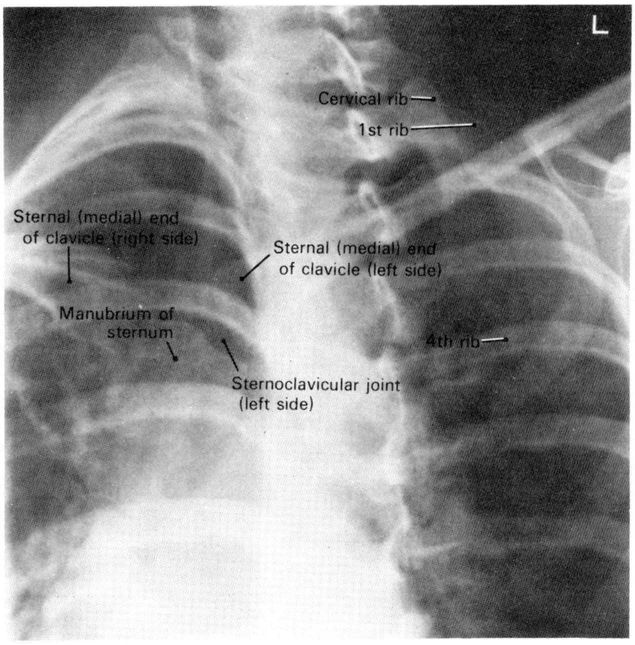

A. Right and left anterior oblique projections.

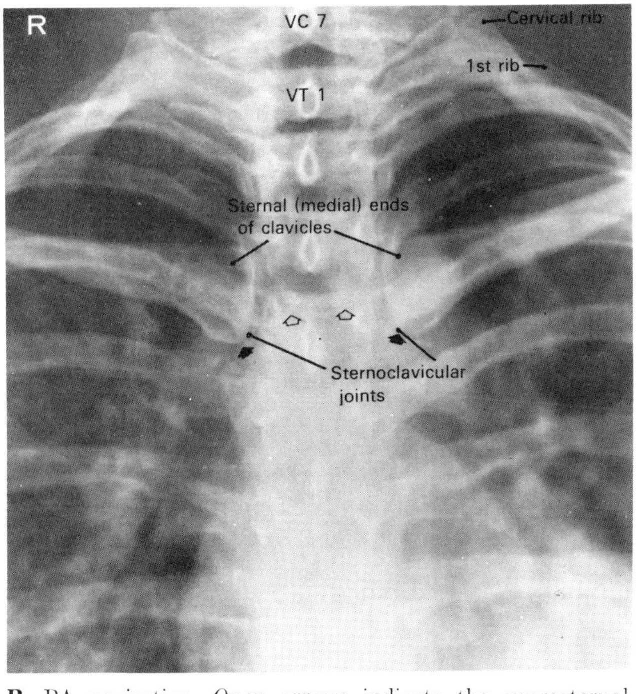

B. PA projection. *Open arrows* indicate the suprasternal (jugular) notch. Black arrows show the clavicular notches.

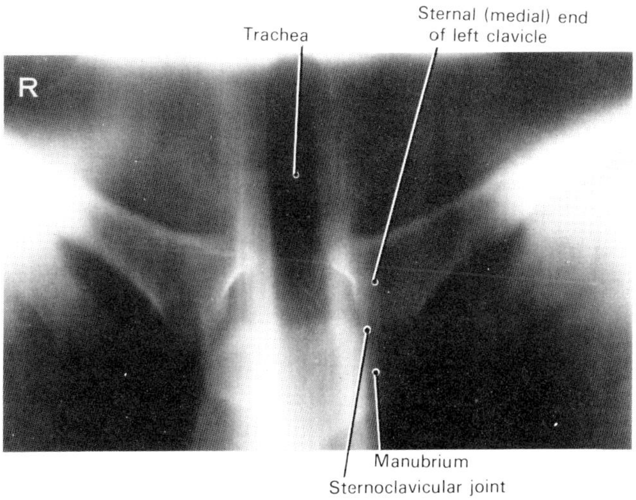

C. PA tomogram of the sternoclavicular joints.

Fig. XIII-12. Radiographs of the sternoclavicular joints.

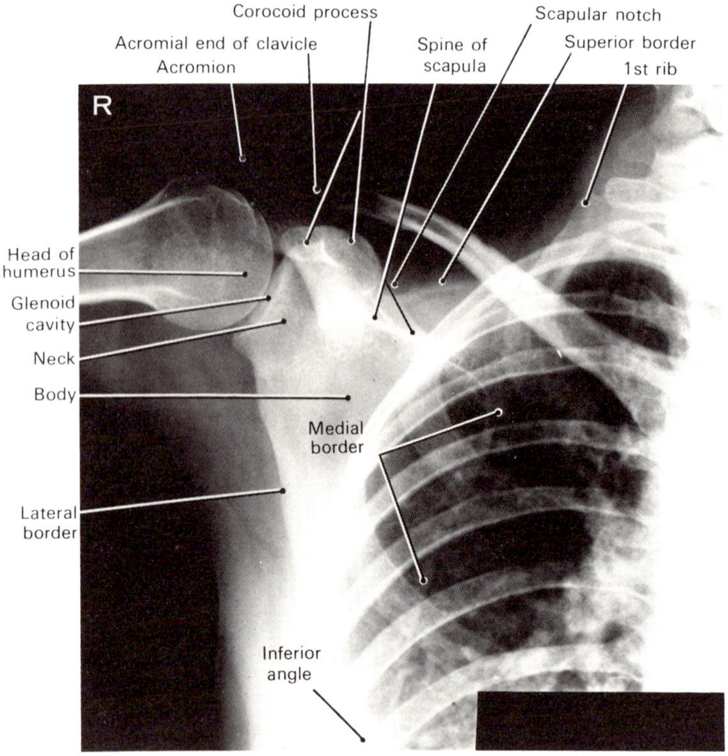

A. AP projection.

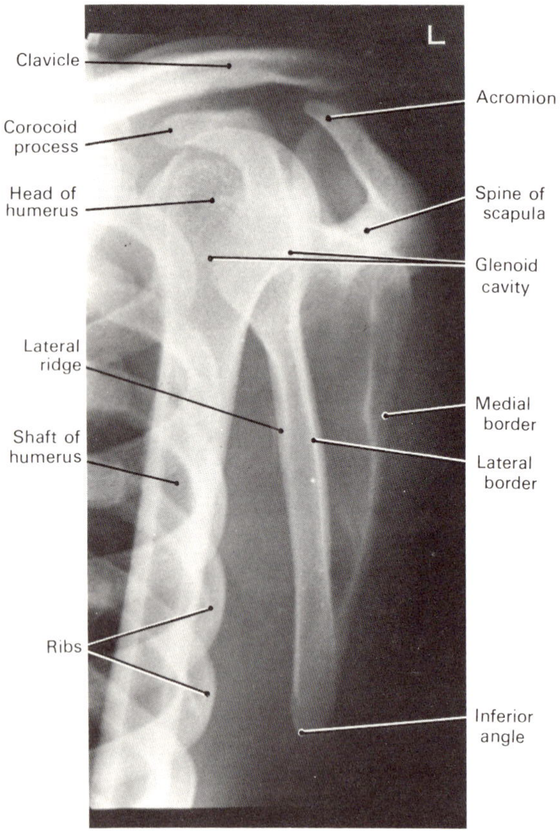

B. Posterior oblique lateral projection.

Fig. XIII-13. Radiographs of the scapula.

are the convex-shaped capitulum and the pulley-shaped trochlea. The medial lip of the trochlea is much more prominent than its lateral lip which results in the "carrying angle" at the elbow. Thus, a line drawn through the central axis of the humerus forms an angle of 165° to 170° with the axis of the forearm. The radius and ulna are angled laterally (valgus).

Between the epicondyles is a radiolucent area formed by the superimposed olecranon and coronoid fossae. Just lateral to these fossae is the less apparent radial fossa (Fig. XIII-14A).

The head of the radius, situated opposite the capitulum, is disk-shaped and has a concave articular surface. The neck is well visualized; however, the radial tuberosity overlaps the ulna. The olecranon of the ulna is superimposed upon the distal humerus and can be seen in the olecranon fossa with the arm fully extended. Immediately below the trochlea is the coronoid process of the ulna, and just distal to it the rough ulnar tuberosity may be seen. Note, too, the radial notch of the ulna in which the radial head rotates.

b. Lateral Projection (Fig. XIII-14B): The patient is positioned with the elbow flexed 90°, and the humerus and forearm are placed on the cassette in the true lateral position. The thumb should point up at the ceiling. The central X-ray beam is directed perpendicular to the cassette and at the elbow joint.

Before examining the osseous details, the first thing to observe are the soft tissues! There are three fat pads about the distal humerus which are intracapsular and extrasynovial in position. One, the largest, is situated over the olecranon fossa *posteriorly,* and it is not normally seen. The other two, collectively known as the **anterior fat pad,** are situated in the coronoid and radial fossae. As you can see in Figure XIII-14B the anterior fat pad is rather triangular in shape and it sits tightly against the bone. Elevation of these fat pads, the so-called *"positive fat pad sign,"* indicates synovial thickening or fluid (blood in the case of a fracture) in the joint. This should immediately alert the physician to search harder for the etiology. This sign will not be seen if a fracture about the elbow does not communicate with the synovial cavity, or if the capsule of the joint ruptures permitting blood to ooze into the nearby tissues.[2]

In Figure XIII-14B, the epicondyles as well as the capitulum and trochlea are superimposed. A dense radiopaque shadow is seen which represents the superimposed supracondylar ridges.

Observe the normal appearance of the trochlear notch, olecranon, and coronoid process which are all seen in profile. The coronoid process is partly obscured by the head of the radius. Also identify the radial tuberosity which is commonly called pathological by the tyro.

3. The Forearm

a. AP Projection (Fig. XIII-15A): The patient is positioned so that the forearm is in supination and the elbow is fully extended. The central X-ray beam is

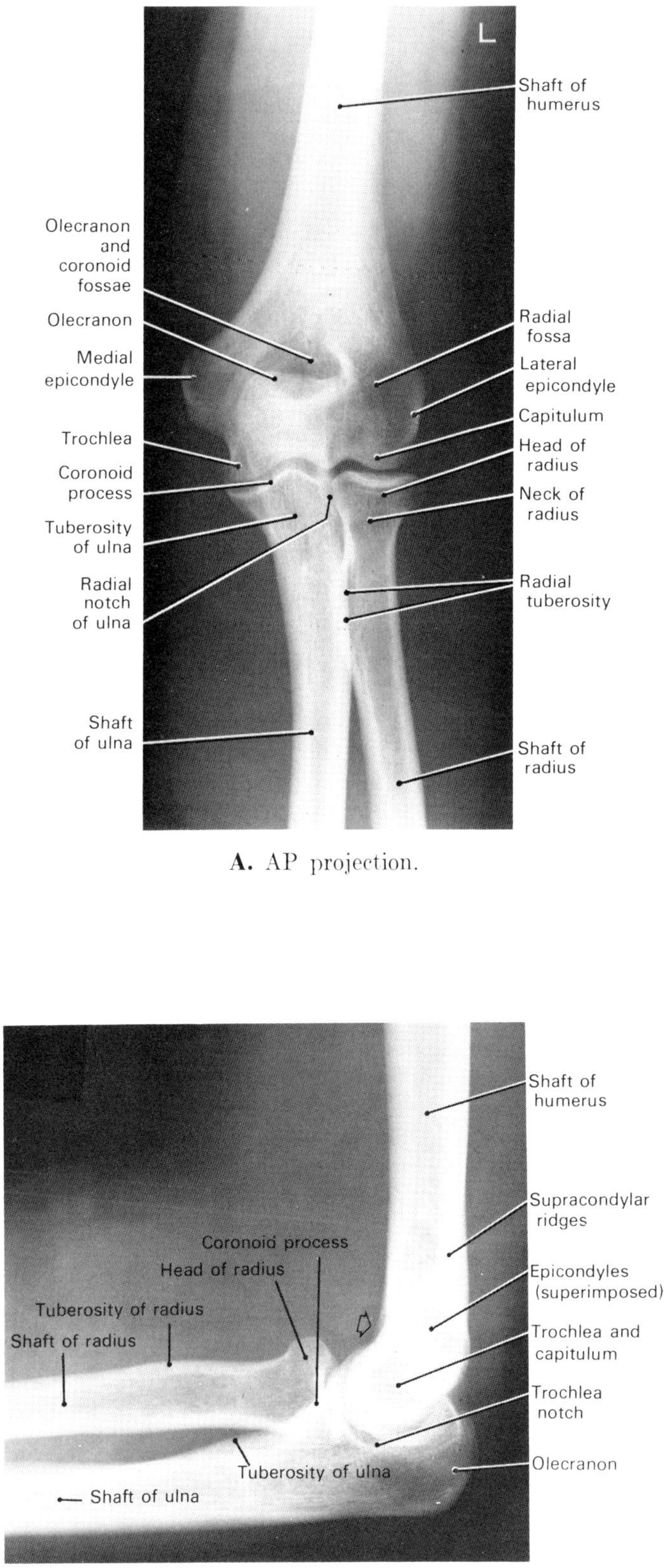

A. AP projection.

B. Lateral projection. The *arrow* indicates the anterior fat pad tucked tightly against the distal humerus. This is actually composed of two fat pads which are superimposed in the radiograph. The two fat pads are situated in the radial and coronoid fossae.

Fig. XIII-14. Radiographs of the elbow.

perpendicular to the film holder and passes through the midpoint of the forearm. The elbow and wrist joints must be on the radiograph.

Note how the radius and ulna bow as they pass distally from the elbow with some divergence of the two bones in their distal two-thirds. They articulate distally at the inferior radio-ulnar joint. We cannot see the interosseous membrane which is radiolucent.

b. Lateral Projection (Fig. XIII-15B): The elbow is flexed 90° and the humerus placed in the true lateral position on the table top. The ulnar surface of the forearm rests upon the table with the thumb extended upward, thus maintaining the forearm in a position of supination. This positioning assures a true lateral view of both the wrist and the elbow.

At the elbow the radius is seen to lie anterior in position to the ulna, while at the wrist the two bones are superimposed. Let us now examine the wrist in detail.

4. The Wrist

a. PA Projection (Fig. XIII-16A and B): The palmar surface of the hand and wrist are placed on the film holder. The central X-ray beam is directed at the scaphoid, and the beam is perpendicular to the film plane. In actual practice, we prefer to have the wrist in ulnar deviation (Fig. XIII-16A) rather than straight. The reason for this is that the scaphoid is thereby elongated and seen to better advantage. Since this is the most commonly fractured carpal bone, the break is thus more readily diagnosed.

The radial styloid process normally extends 1.0 to 1.5 cm. beyond the ulnar styloid process. Notice that the wrist joint appears wider distal to the ulna (Fig. XIII-16A and B). This radiolucent space is occupied by the articular disk and two synovial joints. Observe the normal articulation of the radius and ulna distally at the inferior radio-ulnar joint.

You should have no problem in identifying the two rows of carpal bones in this projection. Figure XIII-16A and B shows the change in shape of the scaphoid with radial and ulnar deviation; it becomes foreshortened in the former and elongated in the latter. Examine the lunate which at its base is convex to fit into the radial articular surface. The entire bone has a somewhat rhomboid shape. It is of the utmost importance to always bear this in mind particularly in cases of trauma. The rhomboid shape changes in this projection to a triangle when the lunate is dislocated anteriorly. Of course, the diagnosis of lunate dislocation is quite easy in the lateral projection. The distal aspect of the lunate normally overlaps the capitate. Observe the pea-shaped pisiform superimposed upon the triquetrum which has an oblong shape.

Of the four bones comprising the distal row of the carpus, the capitate is the largest. Lateral to the capitate, the trapezoid and trapezium are partly overlapped. The hamate has a triangular shape. The oval density upon its distal aspect is its hamulus.

b. Lateral Projection (Fig. XIII-16C): The wrist

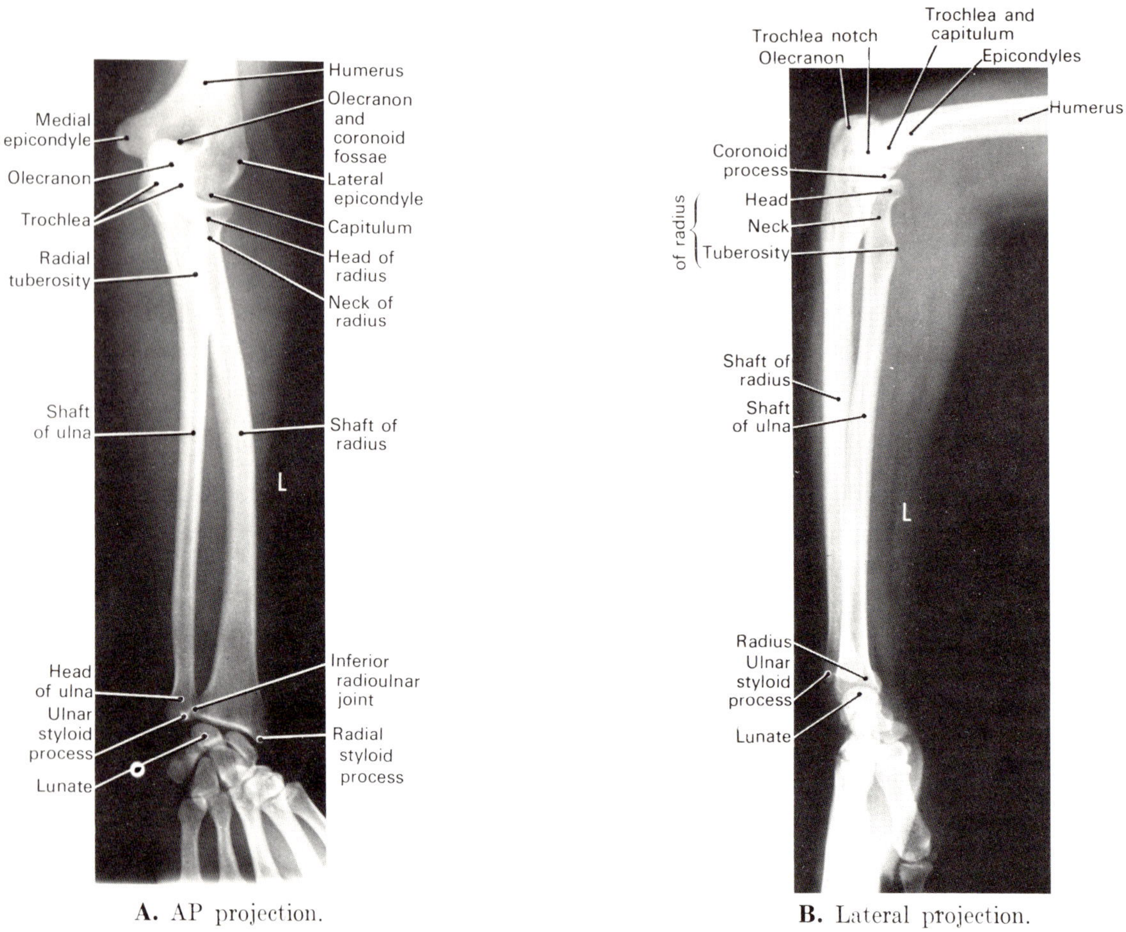

Fig. XIII-15. Radiographs of the left forearm.

is placed in the true lateral position so that its ulnar side is on the film holder. Once again the perpendicularly-directed central X-ray beam is centered on the carpus.

Before examining the carpal bones let us look at another important soft tissue sign which will unquestionably prove useful to you in clinical practice. In Figure XIII-16C we have pointed to a gently-curved, anteriorly-convex, fat line. This fat lies between the Pronator Quadratus and the tendons of the Flexor Digitorum Profundus.[3] The fat line is located anterior to the lower one-sixth of the radius and ulna. In many instances the detection of elevation or loss of this fat line will tip off to the presence of pathology in the region, whether it be fracture or infection. It is helpful to have the normal side for comparison until you have learned more about the normal and its variations. If there is a difference in the two sides of the body, suspicion of abnormality should be high, and additional views including both obliques of the wrist are called for. Remember, however, that fractures distal or proximal to the attachment of the Pronator Quadratus muscle are not likely to produce changes in this fat line.

Take a mounted specimen of the hand and wrist and carefully compare it with the radiograph in Figure XIII-16C. As you can see from the radiograph, most of the carpal bones are hopelessly superimposed upon each other; however, with some practice, you will be able to pick out most of them easily. Begin with the most obvious carpal bone, the semilunar-shaped lunate with its proximal convexity resting in the concave distal articular surface of the radius. Now identify the concave distal articular surface of the lunate into which the capitate fits. Next find the scaphoid which is inclined slightly anteriorly. Immediately behind the lunate a small portion of the triquetrum may be observed; however, the pisiform is almost totally obscured. The trapezium is readily identified, while the trapezoid is superimposed upon the trapezium and capitate. The hamulus of the hamate is seen although the remainder of the bone is indistinct.

In spite of the frustrations that you will experience

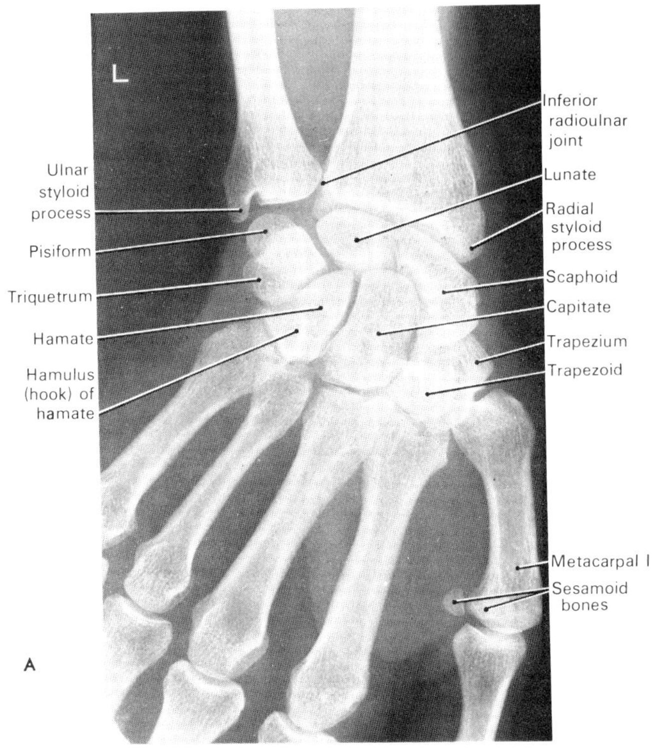

A

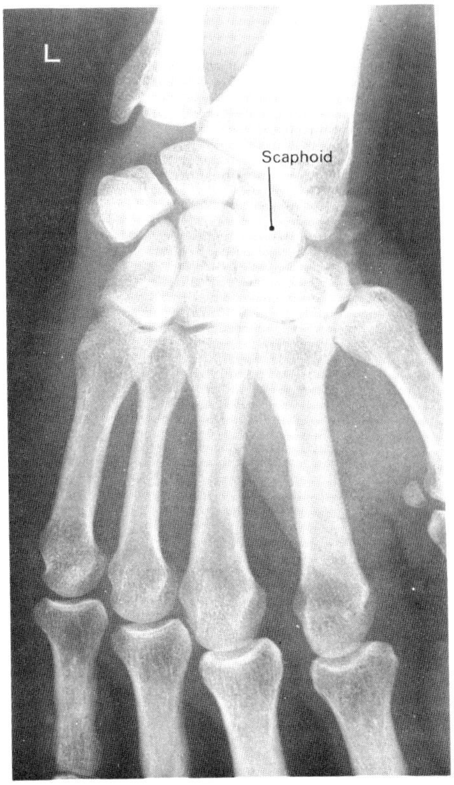

B

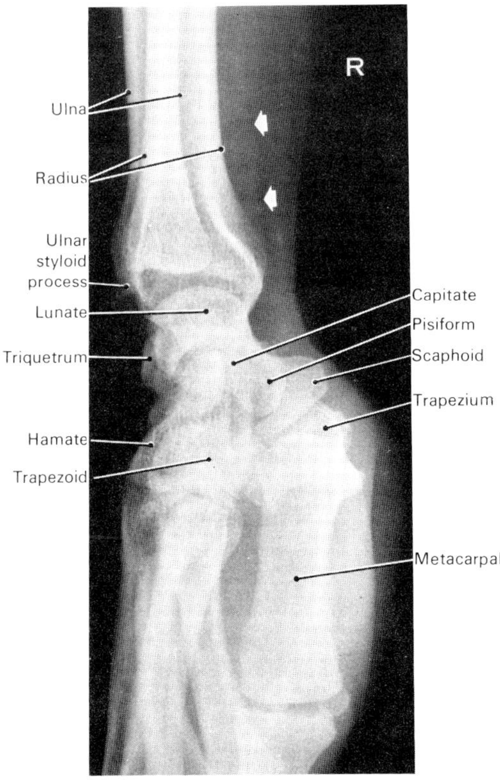

C

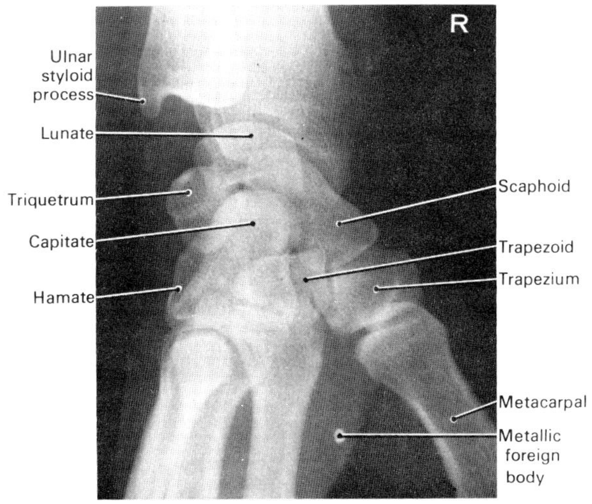

D

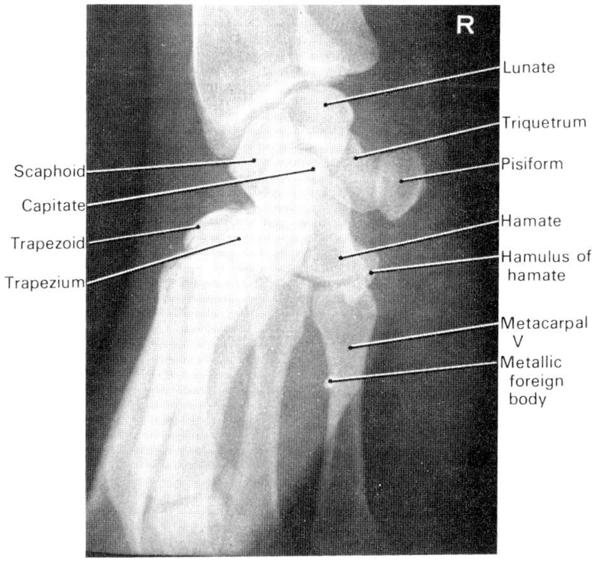

E

Fig. XIII-16. Radiographs of the wrist. **A.** PA projection in ulnar deviation. **B.** PA projection in radial deviation. Note the scaphoid and its change in shape when compared to **A.** **C.** Lateral projection. *White arrows* indicate the fat line over the Pronator quadratus and beneath the tendons of the Flexor digitorum profundus. **D.** Medial anterior oblique (semipronation oblique) projection of the patient's right wrist. **E.** Medial posterior oblique (semisupination oblique) projection of the right wrist.

for a time in identifying the individual carpal bones, this projection is of great importance. Indeed in some instances, it will be the only projection in which certain fractures may be accurately diagnosed.

c. Medial anterior oblique (**semipronation oblique**) (Fig. XIII-16D): The palm of the hand is down, but the wrist is placed at a 45° angle with the film holder. Thus, the medial anterior surface of the wrist is against the film. This view is excellent for viewing the scaphoid, trapezoid, and trapezium.

d. Medial posterior oblique (**semisupination oblique**) (Fig. XIII-16E): By rotating the hand so that the palm is now up, the medial posterior aspect of the wrist is against the film holder. It forms an angle of 45° with the film, and the central X-ray beam is directed perpendicularly at the carpus. This projection affords an excellent view of the pisiform, triquetrum, and hamate.

5. The Hand

a. PA Projection (Fig. XIII-17A): The hand is placed palm down on the film holder, and the fingers are extended fully and separated slightly. The perpendicular central X-ray beam is directed at the metacarpals.

We have already considered the carpal bones and need dwell there no longer. Observe the relative size and shape of the phalanges and metacarpals. In this view, the thumb is in an oblique position normally. Along the sides of the shafts of the middle and proximal rows of phalanges rough, raised margins are occasionally observed. These are for the attachments of the Interossei muscles.

Two sesamoid bones are typically seen opposite the head of the first metacarpal, and one opposite the head of the fifth metacarpal. These lie within the flexor tendons on the palmar surface.

b. Lateral Projection (Fig. XIII-17B): The hand is positioned in a true lateral position perpendicular to the film holder. By abducting the thumb slightly it may be evaluated in a PA position. As you can see, the other metacarpals and phalanges of the fingers are superimposed, and little information can be gleaned about them. Therefore, if a particular finger is suspected of having an injury, it should be viewed laterally with all of the other fingers flexed (Fig. XIII-17C).

c. Medial Anterior Oblique (Fig. XIII-17D): The hand is placed palm down such that the medial anterior aspect of the hand forms a 45° angle with the

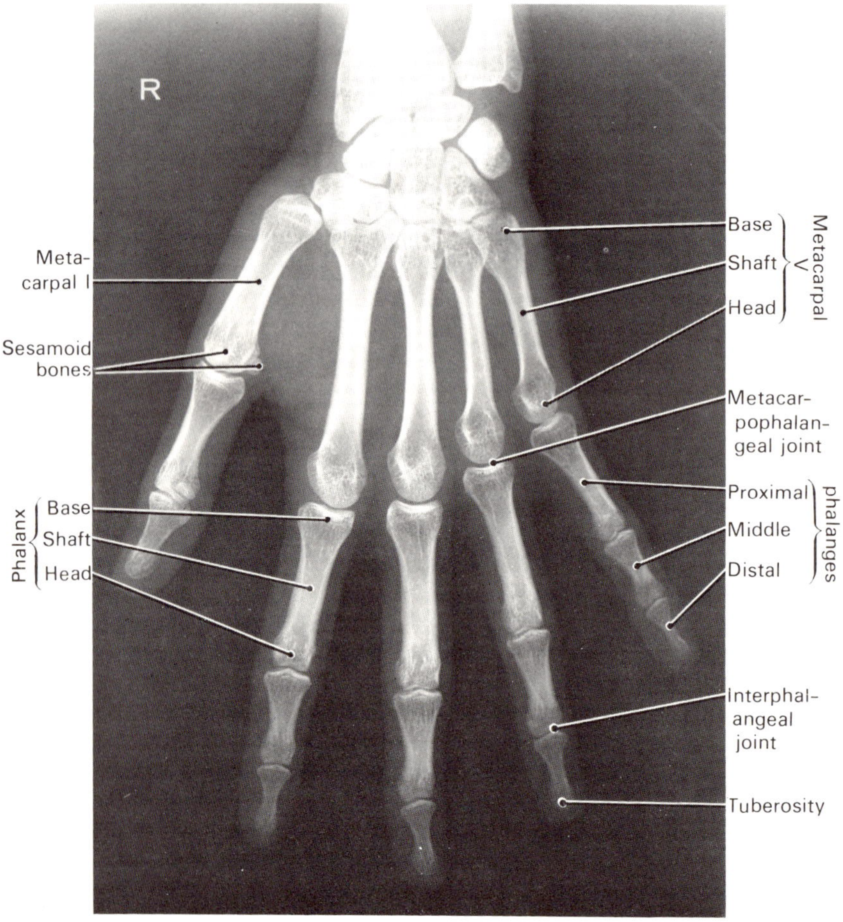

A. PA projection of the right hand. **Fig. XIII-17.** Radiographs of the hand.

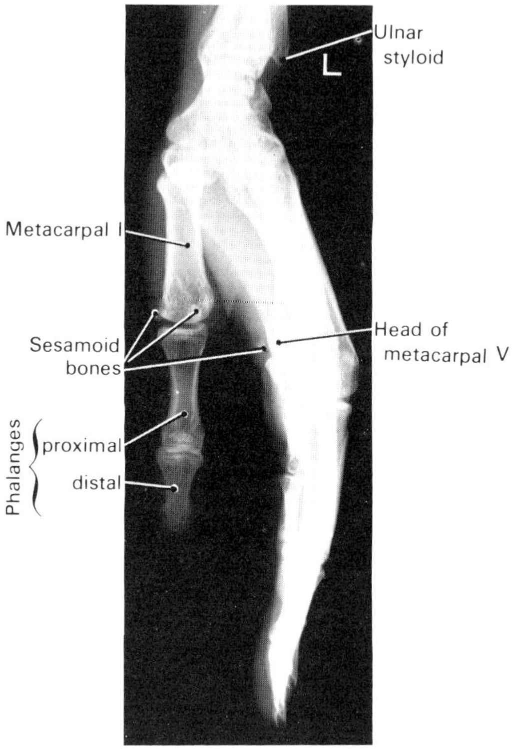

Fig. XIII-17B. Lateral projection of the left hand.

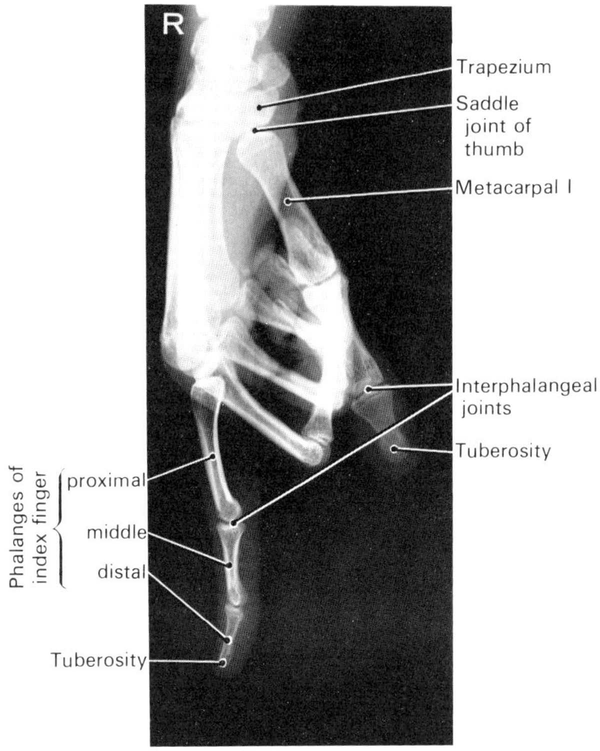

Fig. XIII-17C. Lateral projection of the phalanges of the index finger of the right hand.

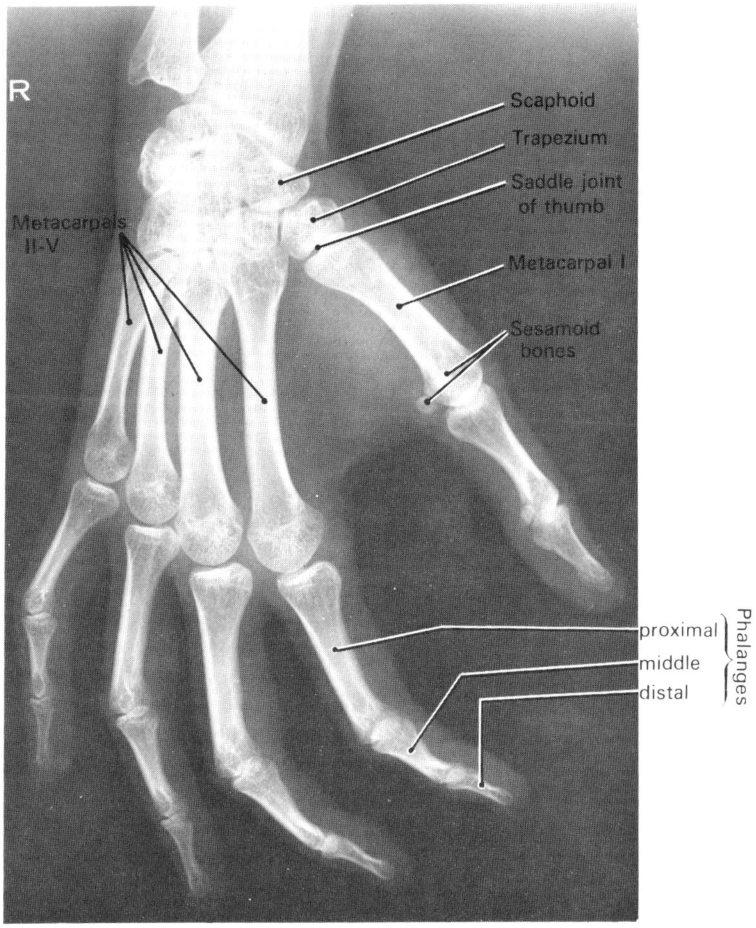

Fig. XIII-17D. Medial anterior oblique projection of the right hand.

film holder. The fingers are slightly flexed. The direction of the X-ray beam is the same as before, perpendicular to the film plane and at the midpoint of the hand. Notice that there is some superimposition of the metacarpals at both ends, while their shafts appear separated. Because of the flexion of the fingers the phalanges appear foreshortened.

THE BONES OF THE LOWER LIMB

The Hip Bone (os coxae; innominate bone)

As you know, the pelvis is composed of four bones: laterally the two hip bones, and dorsally the sacrum and coccyx. These bones are held together by stout white fibrous ligaments and they articulate at the sacroiliac joint and the symphysis pubis.

Figure XIII-18A–C shows the large acetabulum on the lateral aspect of the hip bone. This is the point at which three, separate, primary bones fused to contribute to the hip bone of the adult viewed here. The three parts, the ilium, ischium, and the pubis, are distinct from each other in children (Fig. XIII-18D).

The Ilium

This portion of the hip bone consists of an upper, large, wing-shaped **ala** and a smaller **body.** The body contributes to the superior two-fifths of the acetabulum.

The external view of the ilium (Fig. XIII-18A) shows a wide plate with a superior **crest.** The lateral abdominal muscles and Latissimus Dorsi attach to the inner, intermediate, and outer lips of the iliac crest. Recall that the iliac crest ends fore and aft at the anterior and posterior superior iliac spines. Locate the anterior inferior and posterior iliac spines in Fig. XIII-18A–C. The Sartorius muscle, which runs from hip to knee, takes origin from the anterior superior iliac spine; the Rectus Femoris muscle originates from the anterior inferior iliac spine.

The ala has three faint **gluteal lines:** anterior, inferior, and posterior, on its lateral aspect. These lines help in localizing the areas of origin of the gluteal muscles. The largest of these, the Gluteus Maximus, attaches to the small area between the iliac crest and the posterior gluteal line. The Gluteus Minimus originates (except for a small anterior spot for the Tensor Fasciae Latae) between the anterior and inferior gluteal lines. An area between the inferior gluteal line and the acetabular rim is occupied by the capsule of the hip joint. The capsule reflects upon the neck of the femur whose head articulates in the acetabulum.

The inner aspect of the ala forms a smooth, shallow dish, the iliac fossa. The area is almost completely occupied by the Iliacus muscle which originates there. On the bare bone, an arcuate line (Fig. XIII-18B) internally separates the fossa from the body of the ilium. The arcuate line is but a part of the longer terminal line which involves portions of the pubis and sacrum in delineating the true pelvic brim.

That part of the surface of the ilium facing the sacrum is quite rough, and bears an auricular surface for articulation with the ear-shaped articular surface of the sacrum. Immediately above the auricular surface is the roughened, elevated, iliac tuberosity to which the dorsal sacroiliac ligament attaches. The Sacrospinalis and Multifidus muscles originate on this tuberosity.

The Ischium (Figs. XIII-18A and B)

This portion of the hip bone makes up its posteroinferior part. It has a body and one ramus. The body contributes about two-fifths to the acetabulum. Its smooth inner surface contributes the wall of the true pelvis and there it gives origin to the Obturator Internus. Its ischial tuberosity provides anchorage for the hamstrings. Identify the ischial spine, and note that the greater sciatic notch lies between it and the posterior inferior iliac spine. The greater sciatic notch becomes a foramen when the sacrotuberous and sacrospinous ligaments are in place. The following structures pass through the foramen: the Piriformis muscle; the sciatic and posterior femoral cutaneous nerves; the inferior gluteal artery; the nerve to the Quadratus Femoris; the nerve to the Obturator Internus; internal pudendal vessels; and the pudendal nerve.

The last three structures return to the pelvis through the lesser sciatic foramen which is situated between the ischial spine and tuberosity. The foramen is created when the sacrospinous and sacrotuberous ligaments are in place and bridge the lesser sciatic notch. Recall that the tendon of the Obturator Internus emerges from the lesser sciatic foramen.

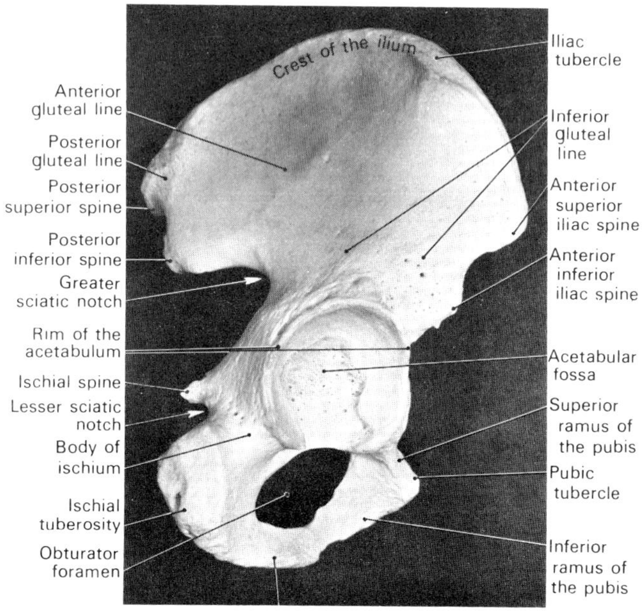

Fig. XIII-18. The hip bone (os coxae; innominate bone). **A.** External surface.

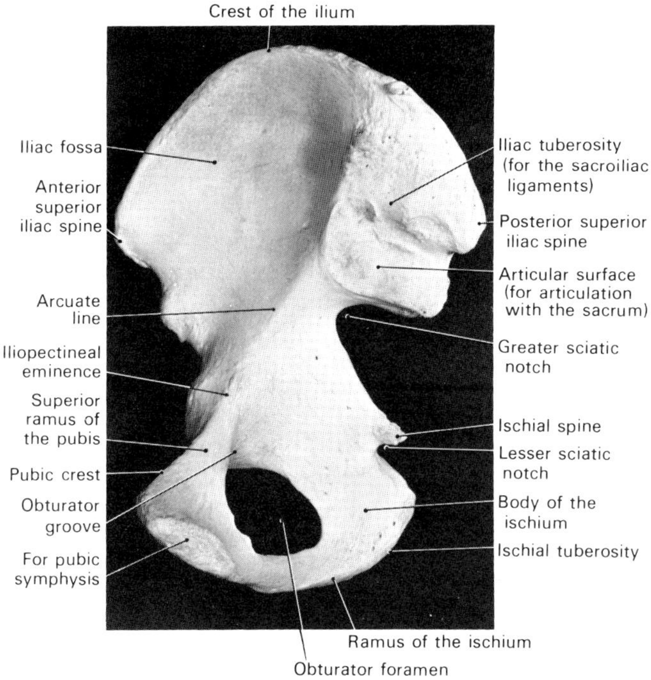

XIII-18B. Internal surface.

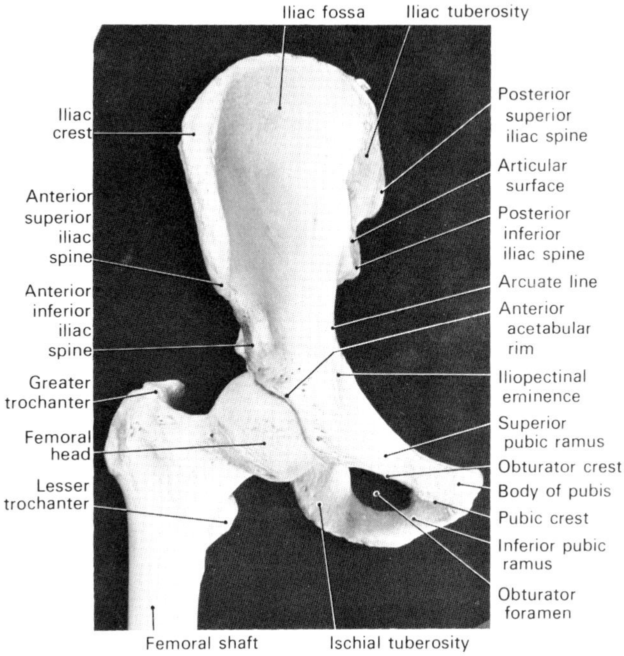

XIII-18C. The hip bone. In articulation with the femur.

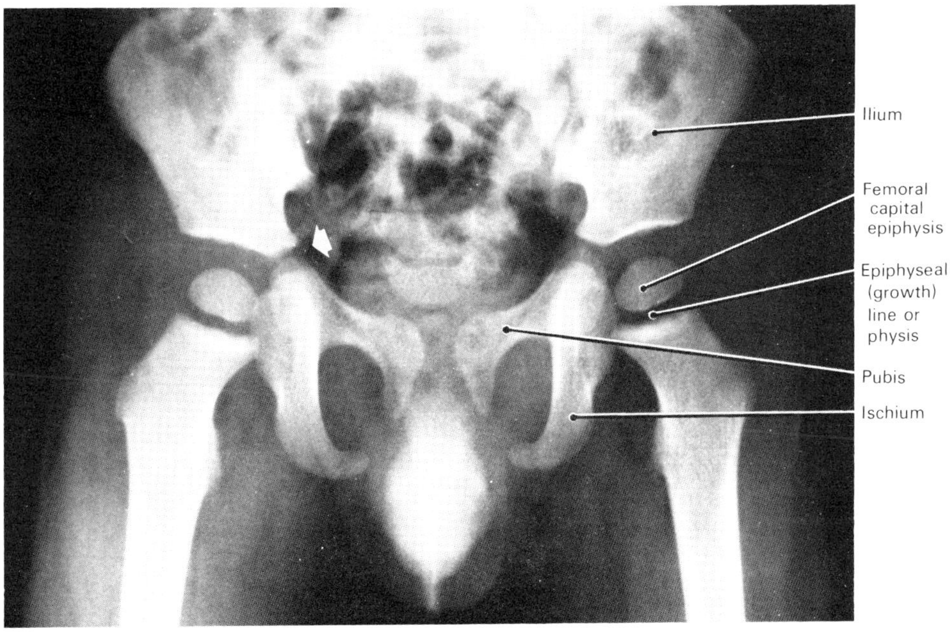

XIII-18D. The hips of an infant. The arrow indicates the site of the triradiate cartilage. Also note the epiphysis of the femoral head.

The ischial ramus helps to form the lower boundary of the obturator foramen, and the boundary of the pelvic outlet which is closed by the pelvic and urogenital diaphragms. In the latter regard, it gives support to the penile (clitoral) crus and offers attachment for the Ischiocavernosus muscle. On the outside, the Obturator Externus and Adductor Magnus (in part) attach to the ramus.

The Pubis

This portion of the hip bone has a body and sports two rami, one superiorly and the other inferiorly directed (Fig. XIII-18C). The ends of each pubic bone, as they meet at the fibrocartilagenous symphysial joint, apear as a rough oval on cross-section (Fig. XIII-18B). The superior border is quite sharp forming the pubic crest. Medially along the crest the pubic tubercle forms a prominent bulge. The Pyramidalis and Rectus abdominis muscles attach to the crest while the inguinal ligament attaches to the tubercle. Other muscles attach to the internal and external aspects of the bone below the level of the crest; the Adductor Longus and Obturator Externus on its outer aspect; the Obturator Internus and Levator Ani, in part, on the inner aspect.

The superior ramus contributes the last one-fifth to the acetabulum, and in its laterad progression, it carries the sharp superior border of the bone to the ilium meeting the latter at the iliopubic or iliopectineal eminence (Fig. XIII-18C). To this sharp border or pubic pecten, a number of smaller ligaments gain attachment, *e.g.*, the lacunar ligament, the pectineal ligament, the conjoined tendon and the reflected part of the inguinal ligament. The iliopectineal eminence blends into the arcuate line of the ilium as part of the longer terminal line. The lateral, inner border of the superior ramus contributes to the obturator groove (Fig. XIII-18B) which transforms into a canal when the obturator membrane covers over the wide obturator foramen. Thus, the massive hole we see in the skeleton between the superior and inferior pubic ramus does not exist in the living.

The inferior ramus meets with the ramus of the ischium. The Adductor Brevis, a part of the Adductor Magnus and the Obturator Externus take origin from its outer aspect while the urogenital diaphragm, the Sphincter Urethrae, crus penis, and Obturator Internus attach in part to the inner aspect.

The Acetabulum

This deep, hemispheric cavity accepts the head of the femur (Fig. XIII-18C). It is divided into an articular and non-articular area. The smooth articular part is peripheral and bears the acetabular rim. It is semilunar in shape and deficient inferiorly (Fig. XIII-18A). The non-articular portion, the acetabular fossa, is somewhat circular in shape, lies at a deeper level, is rough, and it completes the inferior margin of the acetabular notch. A transverse ligament bridges the notch to convert it into a foramen through which vessels and nerves can enter the joint. The synovial membrane of the joint is applied to the inner margin of the articular portion thus leaving the central acetabular fossa outside of the joint cavity. In this way, the short ligament of the head of the femur, arising from a pit, the fovea capitis in the femoral head, does not have to enter the joint cavity to anchor upon fossa.

The Femur (Fig. XIII-19A–D)

This is the longest, heaviest bone of the body, and like all long bones consists of a **shaft** (body) and two ends. Its proximal end is comprised of a head, neck, and two trochanters. The femoral **head** is spherical, actually forming about two-thirds of a sphere. It is directed medially, upward and anteriorly. Except for a small pit, the **fovea capitis,** the head is quite smooth and in life is covered by hyaline cartilage. The fovea capitis (Fig. XIII-19B and D) is located posteriorly and inferiorly of the center of the head. The ligament of the femoral head attaches at this point.

The **neck** of the femurs connects the head to the shaft in the vicinity of the trochanters. The neck is a stout bar of bone which is roughly pyramidal in shape, and is flattened anteriorly. The long axis of the neck makes an angle of 120° to 130° with the long axis of the shaft. This has been termed the **angle of inclination.** If this angle is decreased we describe this as a *coxa vara* clinically, while if the angle is increased it is termed a *coxa valga*. Consider for a moment the superior and inferior borders of the neck of the femur. The superior border is shorter (Figs. XIII-19A), and slightly concave. It is pitted by many small nutrient foramina. The inferior border is longer and more deeply concave.

Anteriorly, at the junction of the shaft and neck is a rough bony ridge, the **intertrochanteric line** (Fig. XIII-19A). It begins superiorly at the **cervical tubercle** and it is directed inferomedially. At its lowest point it joins the less prominent **spiral line,** the latter turns onto the posterior aspect of the femur to become continuous with the medial lip of the **linea aspera** (Fig. XIII-19B).

Posteriorly a prominent ridge of bone, the **intertrochanteric crest** joins the posterior aspect of the greater trochanter to the lesser trochanter. On the crest, at about the junction of its proximal and middle thirds, is a small, rounded protuberance which is called the **quadrate tubercle** (Fig. XIII-19B). At the junction of the neck and shaft of the femur, there is a large, laterally-positioned, irregular **greater trochanter.** It is directed upward and medially, and laterally it is convex. Its upper posterior margin overhangs the **trochanteric fossa** (Fig. XIII-19D).

The conically-shaped **lesser trochanter** projects medially off the posteromedial surface of the femur at the junction of the neck and shaft.

The **shaft** (body) of the femur, which is circular on cross-section, is bowed anteriorly, and it is inclined inferomedially. The **anterior surface** is convex and smooth. In the midthird of the posterior aspect of the shaft is a raised, rough area, the linea aspera (Fig. XIII-19B). Medial and lateral lips border it. Observe

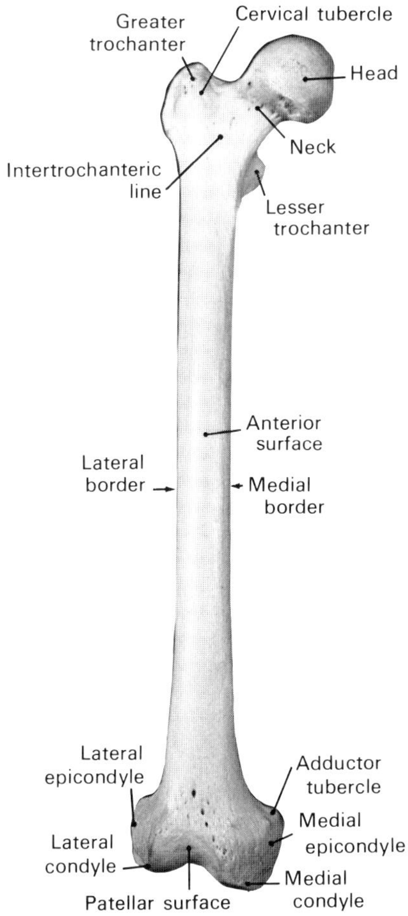

A. Anterior view.

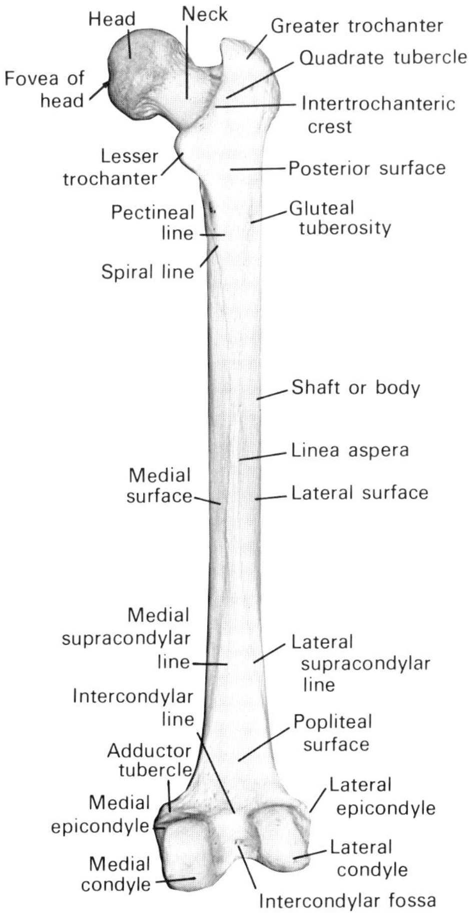

B. Posterior view.

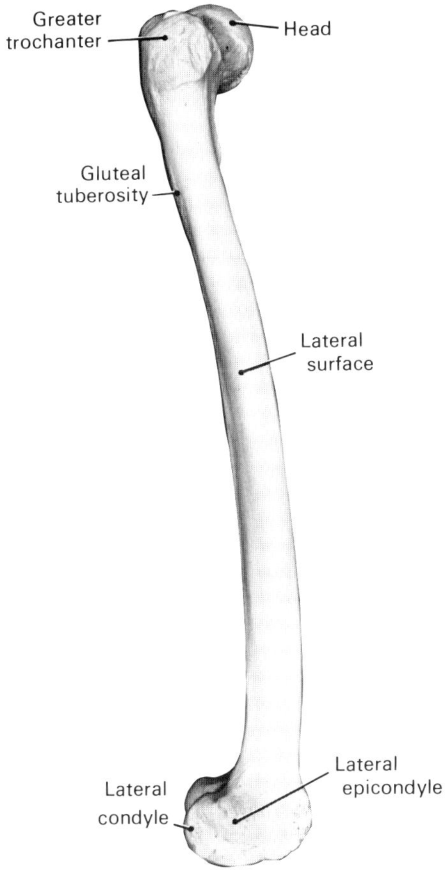

C. Lateral view.

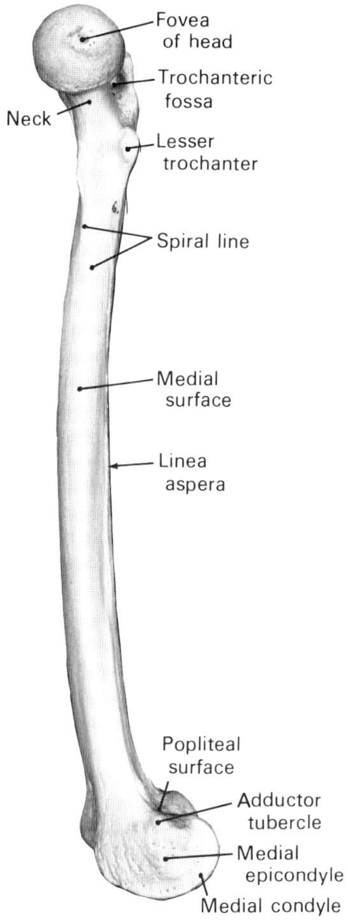

D. Medial view.

Fig. XIII-19. The right femur.

carefully that the linea aspera bifurcates proximally and distally, thereby forming two triangular regions. Proximally, the lateral lip of the linea aspera continues as the **gluteal tuberosity,** while the medial lip continues as the **spiral line.** Within this proximal triangle formed by the gluteal tuberosity and the spiral line, is a slightly raised line, the **pectineal line,** which runs from the posterior margin of the lesser trochanter to the linea aspera (Fig. XIII-19B).

Distally (Fig. XIII-19B) and posteriorly the lateral lip of the linea aspera is continued as the **lateral supracondylar line** which terminates on the **lateral epicondyle.** The medial lip of the linea aspera continues distally as the **medial supracondylar line** ending at the **adductor tubercle.** Between the two supracondylar lines is a flattened **popliteal surface,** bounded inferiorly by the raised **intercondylar line** which separates the popliteal surface from the **intercondylar fossa** inferiorly.

The distal end of the femur is expanded into two large, smooth **condyles, medial** and **lateral** (Fig. XIII-19A–D). Both condyles join anteriorly; however, inferiorly and posteriorly they are separated by the **intercondylar fossa.** At the anterior junction of the condyles is a depression, the **patellar surface** (Fig. XIII-19A). A groove divides the patellar surface into two parts, the lateral being the larger and its lateral lip is more prominent and extends farther anteriorly than its counterpart medially. The medial and lateral condyles are convex inferiorly from front to back, and from side to side. Note that the medial condyle is thinner than the lateral condyle, while the latter is stouter. A large bony prominence projects laterally off the medial condyle, and it is called the **medial epicondyle.** A smaller **lateral epicondyle** projects to the side from the lateral condyle (Fig. XIII-19B–D).

The Patella (Kneecap) (Fig. XIII-20A and B)

The largest sesamoid bone of the body is embedded in the tendon of insertion of the Quadraceps Femoris muscle. This bony prominence on the front of the knee, demonstrated best when the knee is extended, sinks into the intercondylar notch during flexion of the knee joint. It protects the joint and offers great leverage for the action of the muscles.

The patella, as we view it in Figure XIII-20A and B, is roughly triangular in shape. It has two **surfaces:** anterior and articular. The **anterior surface** (Fig. XIII-20A) is convex from superior to inferior and from side to side. It is typically roughened and has many holes for nutrient vessels. The **articular surface** (Fig. XIII-20B) is covered by hyaline cartilage and articulates with the femur on the patellar surface of the femoral condyles. Observe that a raised ridge divides the articular surface into two facets, the smaller medial and the larger lateral facet. Immediately below the facets is a **non-articular portion.** This semicircular area is sharply set off from the articular facets by a groove. The area is depressed and has an irregular surface (Fig. XIII-20B). It gives attachment to the patellar ligament.

Three **borders,** superior, medial, and lateral, and an inferior **apex** may also be observed. The **superior border** is rather flat and has antero-inferior inclination. The **medial** and **lateral borders** converge at the apex of the patella.

The Tibia (Shin Bone) (Fig. XIII-21A–D)

This long bone is located on the anteromedial side of the leg. It is the second largest bone of the skeleton. As with all long bones, it has two **ends** joined by a **shaft** (body).

The proximal portion of the tibia ends in two **condyles, medial** and **lateral** (Fig. XIII-21A). The **superior articular surfaces (facets)** are smooth and oval in shape. Their central portions serve for articulation with the condyles of the distal femur. The **medial facet** is slightly concave, while the **lateral facet** is not only concave from side to side, but slightly convex from front to back. The two articular surfaces are separated by the **intercondylar eminence** (spine of the tibia) from the sides of which project the **medial** and **lateral intercondylar tubercles** (Fig. XIII-21B). The surfaces of the articular facets are prolonged onto these tubercles. Anterior and posterior to the intercondylar eminence are two intercondylar depressions for attachment of the cruciate ligaments and menisci. Anteriorly the condyles of the tibia are joined by a roughened, triangular-shaped surface. The apex of this area ends in the prominent elevated **tibial tuberosity** (Fig. XIII-21A and C). The medial condyle has an irregular convex medial surface. On its posterior surface is a shallow groove. Note that the lateral condyle is more prominent than the medial condyle (Fig. XIII-21B). On the posterior margin of the lateral condyle, there is a circular, flat facet for the articulation of the fibular head (Fig. XIII-21A).

The shaft of the tibia has no anterior surface, only medial, lateral and posterior surfaces. Anteriorly a sharp, palpable **crest** descends from the tibial tuberosity almost to the distal end of the bone. The lateral margin or **interosseous border** is a distinct line of bone from which an **interosseous membrane** bridges to fill the gap between tibia and fibula (Fig. XIII-21D). As in the forearm between radius and ulna, blood vessels will run down on either side of this membrane and muscles will use it as well as they use the bones for their origins. There is a less distinct line marking the **medial border** which begins at the back of the medial condyle and ends posteriorly on the medial malleolus distally (Fig. XIII-21B). In the upper part, a portion of the Popliteus inserts on this border and some fibers of the Flexor Digatorum Longus originate in part from the middle third of it. However, the bulk of these muscles attach to the posterior surface of the bone. Nearly all of the **medial surface** is also palpable and subcutaneous (Fig. XIII-21A). Only the tendons of the Gracilis, Sartorius, and Semitendinosus insert high up on this surface near the anterior crest.

The **lateral surface** (Fig. XIII-21A) gives origin to the Tibialis Anterior while its lower one-third is bare, crossed only by two extensor tendons (Extensor

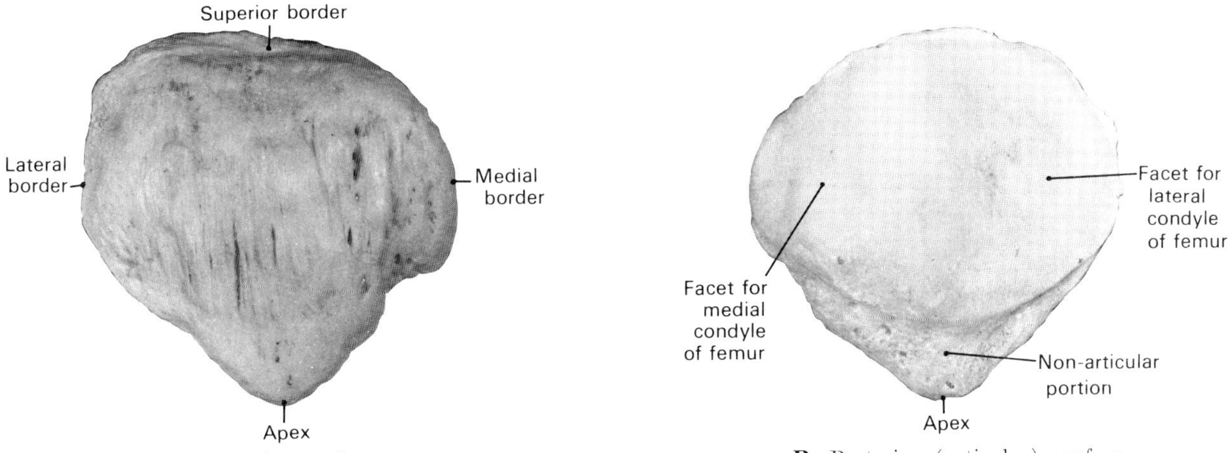

A. Anterior surface. **B.** Posterior (articular) surface.

Fig. XIII-20. The right patella.

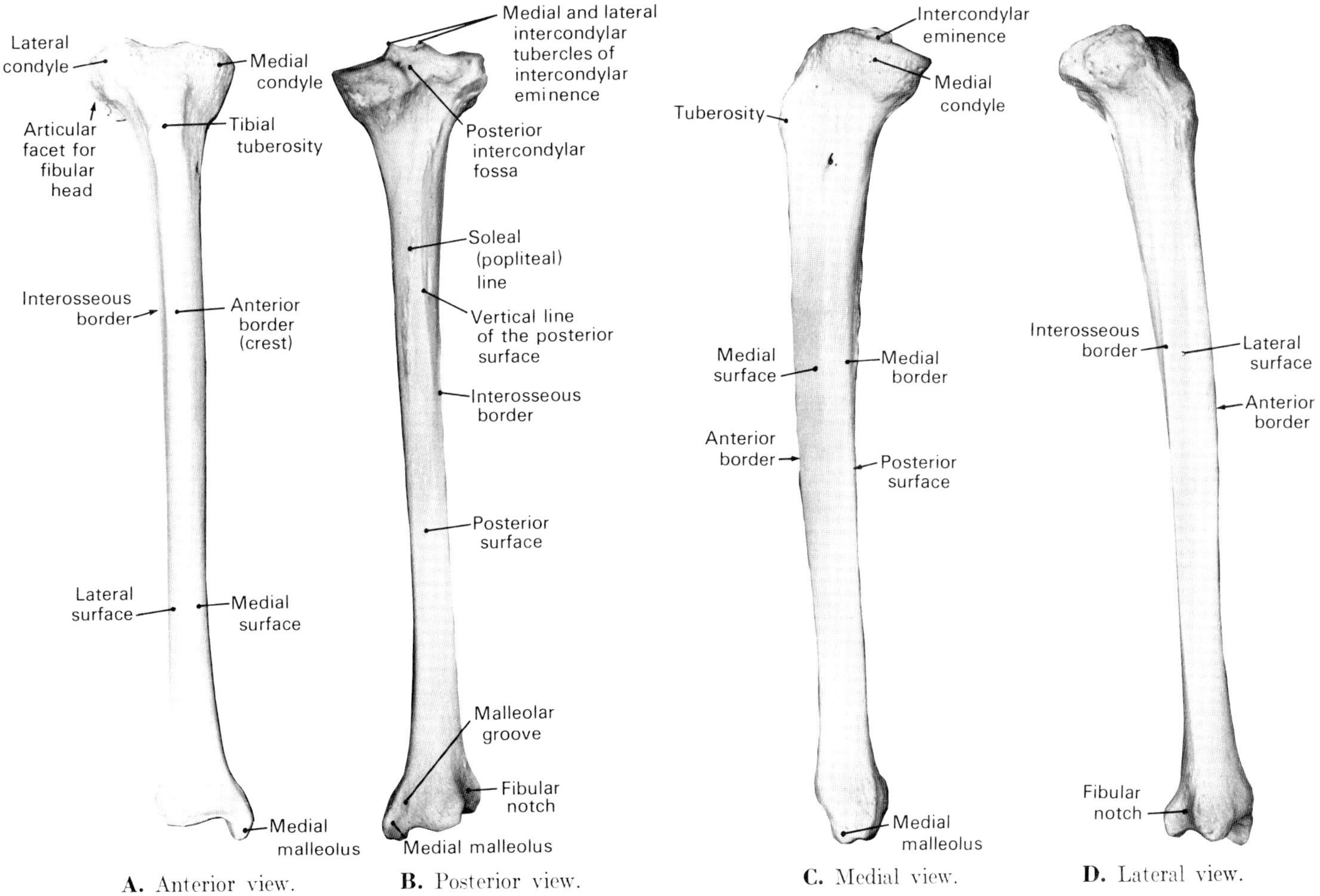

A. Anterior view. **B.** Posterior view. **C.** Medial view. **D.** Lateral view.

Fig. XIII-21. Right tibia.

Hallucis Longus (long extensor of great toe) and Extensor Digitorium Longus). In the lower limb the extensor muscle groups are anterior; the flexor groups, posterior.

The **posterior surface** bears a prominent oblique ridge (the popliteal (soleal) line) in its upper third (Fig. XIII-21B). The line marks the lower boundary of insertion for the Pollitius and the point of origin for part of the Soleus. Below the line in the middle third of the bone, the bulk of the Flexor Digitorum Longus originates. The lower end of the bone is again crossed by tendons, this time flexor tendons, which groove the bone at its flared end. These are flexor tendons for toes and ankle.

The prominent features at the distal end of the bone are the **medial malleolus,** that bony projection often called the ankle bone; a **lateral fibular notch** into which the fibular inserts; and two inferior articulating areas, the **inferior facet** and the **malleolar facet,** both of which articulate with one of the tarsal bones, namely the **talus** (Fig. XIII-21B and D).

The Fibula or Calf Bone (Fig. XIII-22)

This long, slender bone lying lateral to the tibia is not weight bearing as is the tibia and it is there primarily for muscle attachments. While it is shorter than the tibia its distal end projects beyond the tibia to also articulate with the talus. Its proximal end or **head** is irregular in shape and expanded (Fig. XIII-22B). An anterio-medial facet on the head articulates with the tibia in the **superior tibiofibular joint.** A superior eminence of the head projects posterolaterally as the **apex** (styloid process).

Anteriorly, most of the shaft except the distal one-fifth gives origin laterally to the Peroneii (Longus, Brevis, and Tertius) and medially to toe extensors. Similarly, except for the posterolateral half, the back side of the bone gives origin to muscles (Soleus and the flexor of the great toe). Obviously, the **medial border** has a crest for the interosseous membrane. The anterior border terminates in a downward projecting knob, the **lateral malleolus** or ankle bone. It has a triangular facet for the talus. Posteriorly there is a **malleolar fossa** (sulcus) to transmit the tendons of the Perineus Longus and Brevis.

The Foot (Fig. XIII-23A–D)

The foot contains 25 to 26 six bones: seven tarsal bones, classified as short bones; five metatarsals; and 13 to 14 phalanges. Variability in the latter number is due to the possible fusion of the distal two phalanges of the little toe in some instances.

The Tarsus (Fig. XIII-23A and B): The seven tarsal bones consist of the talus, calcaneus, cuboid, navicular, and three cuneiform bones. The largest of these is the calcaneus with the talus as the runner-up. An old ditty has it that "the leg bone attaches to the ankle bone and the ankle bone attaches to the foot bone." Would it were that simple. Only one ankle bone, the talus, articulates with two of the leg bones,

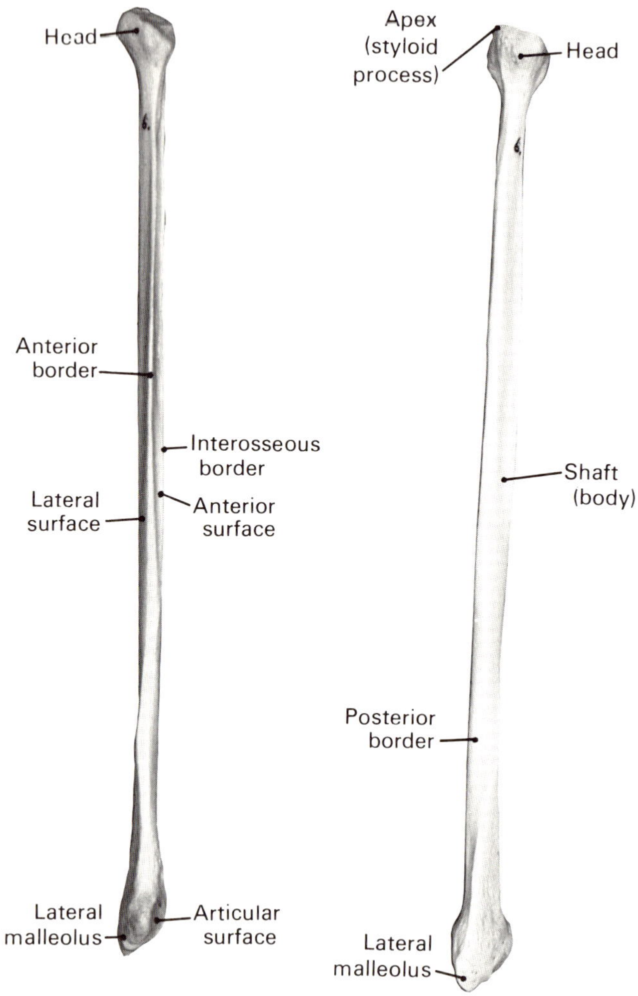

A. Anterior view. **B.** Posterior view.

Fig. XIII-22. The right fibula.

tibia and fibula, at the ankle joint. The anteromedially-directed talus rests upon the distal two-thirds of the calcaneus and, in front, it articulates with the navicular. The cuboid, in turn, lies in front of the calcaneus. These four—talus with navicular and calcaneous with cuboid—form the **transverse tarsal joint** anteriorly. The posterior aspect of the calcaneous forms the "heel" of the foot. Still further anteriorly, the three **cuneiforms—medial, intermediate** and **lateral**—find themselves wedged between the navicular and the bases of the first three metatarsals. This leaves the fourth and fifth metatarsals to articulate with the cuboid. Medially the cuboid must also articulate with the lateral cuneiform. How these bones take part in forming the arches of the foot will be considered later in this chapter.

The Talus (Fig. XIII-23A–D): This bone has a definite, anteromedially-directed expansion or **head** which faces the navicular bone; a constricted **neck** behind the head, rough and pitted superomedially for ligamentous attachments and nutrient foramina; and a large posterior portion or **body.** The body is squarish in cross section, and its proximal collection of

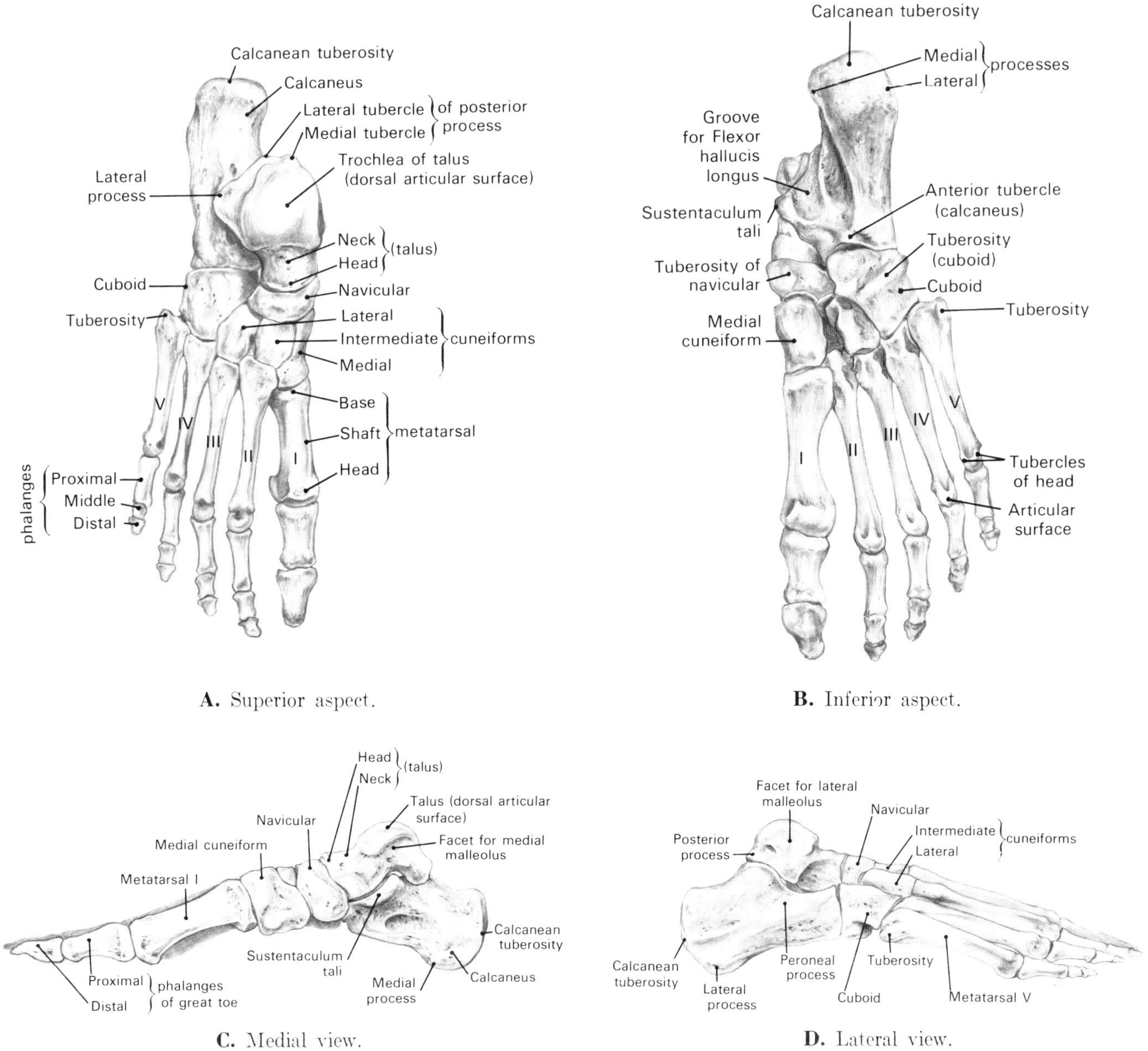

Fig. XIII-23. The right foot.

articular surfaces are known as the **trochlea** (Fig. XIII-23A). This expansive articular area is convex from front to back and concave from side to side, forming a large pulley-shaped arc and a comma-shaped medial surface for the tibia and medial malleolus and a lateral surface and lateral triangular facet for the fibula and lateral malleolus (Fig. XIII-23C and D). Posteriorly, the body narrows to its **posterior process** (Fig. XIII-23D) which is grooved for the tendon of the Flexor Hallucis Longus (Fig. XIII-23B). The posterior surface also projects as **medial** and **lateral tubercles** that flank the groove (Fig. XIII-23A). Upon occasion the whole posterior process ossifies separately, and it is then called the **os trigonum.**

The inferior or distal part of the bone has three **articular facets,** anterior, middle, and posterior, for articulation with the calcaneous. Of these, the **posterior** one is by far the largest to articulate with a corresponding facet on the calcaneus. The **sulcus tali** runs between the posterior and middle facets. It overlies a corresponding groove on the proximal surface of the calcaneus. The channel so formed harbors the interosseous talocalcaneal ligament. The **middle facet** is actually situated on the head and neck of the under surface of the bone so that it articulates with the **sustentaculum tali** of the calcaneus. The **anterior facet** is on the head, laterally placed, to articulate with the superoanterior surface of the calcaneous (Fig. XIII-23C).

The anterior surface of the **head** is large and round for articulation with the navicular. It bears two inferior facets—the anterior one just mentioned above and a semioval facet which rests upon the plantar calcaneonavicular ligament. As one looks down upon the proximal end of the talus, the neck can be seen distinctly to separate head from body (Fig. XIII-23A).

The Calcaneus (Fig. XIII-23A–D): This largest of the tarsal bones sits below and slightly lateral to the talus with its long axis directed anteriorly. It extends most posteriorly in the foot with its massive tuberosity forming the heel. The posterior part is, thus, nonarticular in contrast to the anterior portion. As you look down upon the proximal anterior surface you are struck by the three articular facets, posterior, middle, and anterior, which match those we have noted on the inferior surface of the talus. Between the posterior and middle facet is the matching **sulcus calcanei** for the talocalcaneal ligament. The middle facet sits on the shelf-like **sustentaculum tali** which the tendon of the Flexor Hallucis Longus grooves on the plantar aspect (Fig. XIII-23B and C).

The lateral surface of the calcaneus is relatively flat except for two small projections, the **retrotrochlear eminence** and the **peroneal trochlea.** The first gives attachment to the calcaneofibular ligament and the second has a grooved inferior surface to serve as a pulley for the Peroneus Longus tendon. These two bumps or projections together are called simply, the **peroneal process** (Fig. XIII-23D).

Inferiorly the plantar surface of the massive **calcaneal tuberosity** (heel) is grooved so that it divides into a large **medial** and a small **lateral process** (Fig. XIII-23B). The small, rounded lateral process gives origin in part to the Abductor Digiti Minimi. To the medial process attach the Flexor Digitorium Brevis, the Abductor Hallucis, and the plantar aponeurosis. The rest of the Abductor Digiti Minimi arises from the groove between the processes. The rough area above the processes gives attachment to various ligaments and to the lateral head of the Quadratus Plantae. Anteriorly, the plantar surface of the bone terminates in the **articular facet** for the **cuboid bone** (Fig. XIII-23A, B, and D).

Cuboid (Fig. XIII-23A, B, and D): This six-sided short bone lies laterally and it articulates with the calcaneus, the fourth and fifth metatarsals and the lateral (third) cuneiform. Its **posterior articular surface** is saddle-shaped for articulation with the calcaneus. Anteriorly it has two **articular facets** divided by a ridge into a small medial facet for articulation with the fourth metatarsal, and a larger lateral facet for the fifth metatarsal. Medially, on its upper midportion, is a small oval facet to articulate with the lateral cuneiform. Beginning on the lateral surface of the cuboid is a deep tendinous groove which extends onto the plantar surface. The posterolateral limit of this groove has a prominent ridge, the **tuberosity of the cuboid** (Fig. XIII-23B).

Navicular (Fig. XIII-23A, B, and D): Lying between the talus and the cuneiforms is the crescent-shaped navicular. Its **proximal articular surface** is concave to accommodate the head of the talus. Distally, it presents three distinct **facets** for articulation with the three cuneiforms. Both upper and lower surfaces are roughened. Medially there is a small bony projection which is directed downward and back, the **tuberosity of the navicular** (Fig. XIII-23B).

The Cuneiforms (Fig. XIII-23A–D): As their name implies, they are wedge-shaped. Situated between the navicular, the cuboid and the first three metatarsals, they are named: the **medial** (first); **intermediate** (second); and **lateral** (third) cuneiforms. The medial and lateral may be seen (Fig. XIII-23A) to project farther forward than the intermediate cuneiform. The base of the second metatarsal fills the resulting gap. The **proximal articular facets** of the cuneiforms are slightly concave for articulation with the navicular. Their **distal articular facets** are more flattened.

The Metatarsals (Fig. XIII-23A–D): Five of these miniature long bones comprise the metatarsus, joining the tarsus with the phalanges. As with every long bone, they characteristically have two ends joined by a shaft (body). They are numbered I to V from the medial side.

Each metatarsal possesses certain features common to all. The proximal end or **base** is enlarged and wedge-shaped with the base of that wedge dorsal in position. Each articulates proximally with the tarsal bones and laterally with the adjacent metatarsals. Both dorsal and plantar surfaces of the base are rough for ligamentous attachments. The **shaft** (body) is convex dorsally and concave on the plantar surface. Beginning at the base of the metatarsal, the plantar surface extends

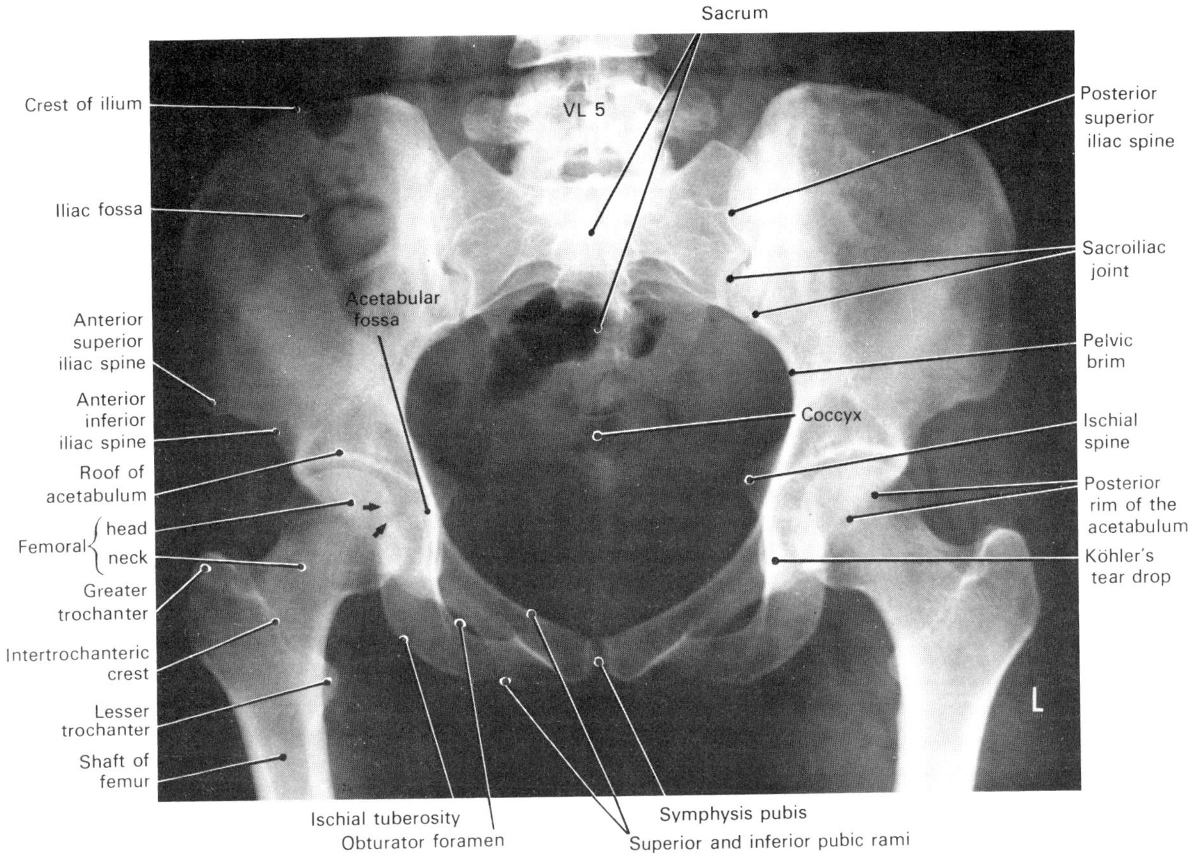

Fig. XIII-24. AP projection of the female pelvis. The *arrows* on the patient's right indicate the fovea capitis in the femoral head. Note the left femur is in internal rotation; the right femur is in neutral position.

distally only to bifurcate near the head ending in a **tubercule** on either side of the head (Fig. XIII-23B). The distal end or **head** has a convex, smooth articular surface which extends onto the plantar surface. The plantar surface is grooved for the passage of flexor tendons. The sides of the head are flat with central depressions. Each side dorsally has small tubercles for ligamentous attachments.

The Phalanges (Fig. XIII-23A–D): With the exception of the great toe, which has only two (proximal and distal), every toe has three phalanges. Like the hand, they are arranged into proximal, middle, and distal rows. Those of the great toe are the largest and strongest.

1. The Proximal Phalanges: All of the phalanges are miniature long bones, and as you should certainly know by now, they have two ends and a shaft. The proximal end or **base** has an oval articular surface (facet) to accept the corresponding metatarsal. The **shaft** is concave laterally and medially, convex dorsally, and flattened on the plantar surface. The **head** is pulley-shaped for articulation with the base of the middle phalanges.

2. The Middle Phalanges: These bones decrease in size progressively from the second to the fifth toe. In clinical practice, it is not at all uncommon to find the middle and distal phalanges of the fifth toe fused. The proximal end or **base** has two articular facets separated by a raised ridge of bone for articulation with the proximal phalanges. The **shaft** is similar to that of the proximal phalanges. The **head** is also pulley-shaped to accommodate the base of the distal phalanges.

3. The Distal Phalanges: These bones have a broad articular **base** proximally to articulate with the middle phalanges. The **shaft** is concave laterally and medially, convex dorsally, and flat on the plantar surface. The distal ends are expanded and roughened to support the nail and its pulp.

ROENTGEN ANATOMY OF THE PELVIC GIRDLE AND LOWER LIMB

The Pelvic Girdle and Upper Femur (Figs. XIII-24 and 25):

1. AP Projection with the Femur in Neutral Position (Fig. XIII-25A): The patient is in the supine position on the radiographic table with the foot pointing straight up. The central X-ray beam is centered over the hip joint, perpendicular to the cassette.

We have already considered the pelvis in some detail elsewhere (Chapter X, p. 656). Review those anatomic features in Figure XIII-24. Identify the spherical femoral head articulating in the acetabulum. The smooth, thin cortical bone of the head of the femur is interrupted by a notch, the fovea capitis. Superimposed upon the head of the femur is the posterior rim of the acetabulum (Figs. XIII-24 and 25A). The anterior rim of the acetabulum lies more medially. At the junction of the femoral neck and shaft, the greater and lesser trochanters may be seen. Between the trochanters is an oblique linear line cast by the superimposed intertrochanteric line and crest.

Note the angle of inclination between the central axis of the femoral neck and shaft. Recall that this normally measures 120 to 130°. You will undoubtedly hear of two other "lines" about the hip. The first is called **Shenton's line** and it is a gently curving line which in the normal patient follows the upper margin of the obturator foramen and becomes continuous with the inferior surface of the femoral neck (Figure XIII-25A). The second line is referred to as **Skinner's line** and it is a horizontal line which is drawn perpendicular to the long axis of the shaft of the femur. It commences on the upper margin of the greater trochanter and must pass through or below the fovea capitis. Disruption of these lines may be an important clue to some disease process about the hip joint.

2. AP Projection with the Femur Internally Rotated (Fig. XIII-25B): The patient remains in the same position described above; however, in this position the foot is internally rotated 15°. This overcomes the normal anteversion of the femoral neck placing it parallel to the film.

The femoral neck is now seen in its entirety. Thus any fractures of the neck will be more readily detectable. The lesser trochanter is almost lost from view

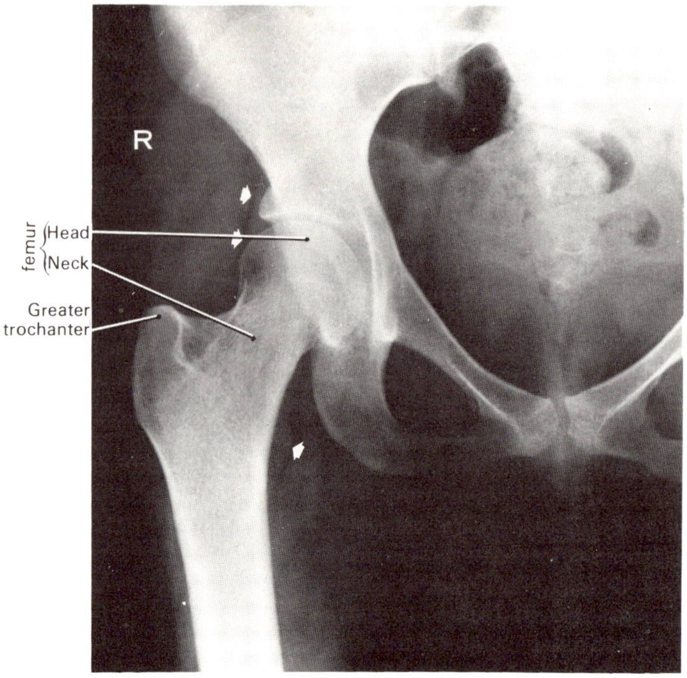

B. AP projection with the femur internally rotated. *White arrows* outline the capsule of the hip joint.

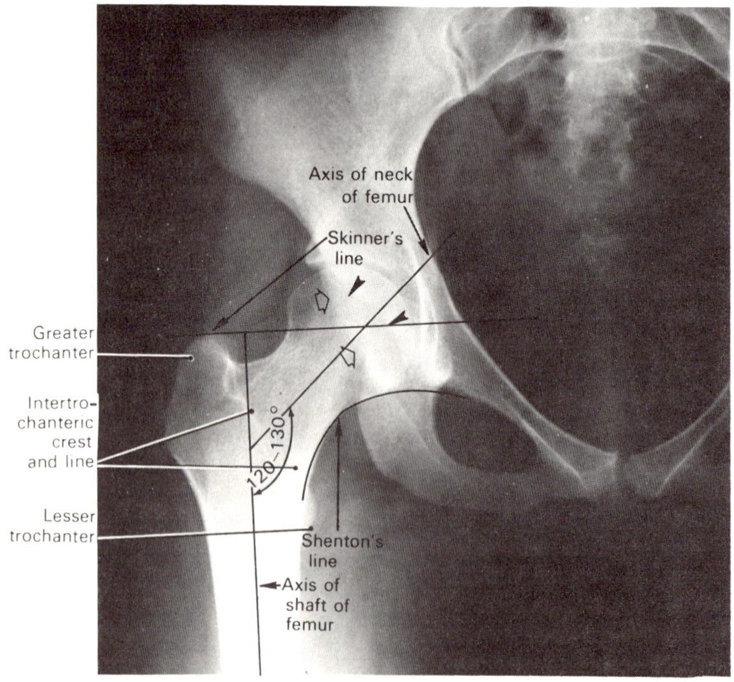

A. AP projection with the femur in neutral position.

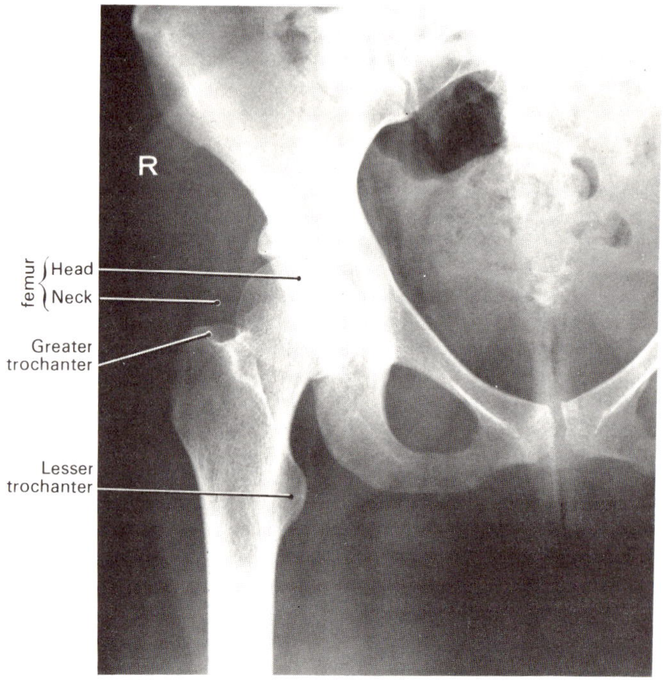

C. AP projection with the femur externally rotated.

Fig. XIII-25. Radiographs of the pelvic girdle and proximal right femur. *Black arrows* indicate the anterior rim of the acetabulum; *open arrows*, the posterior rim.

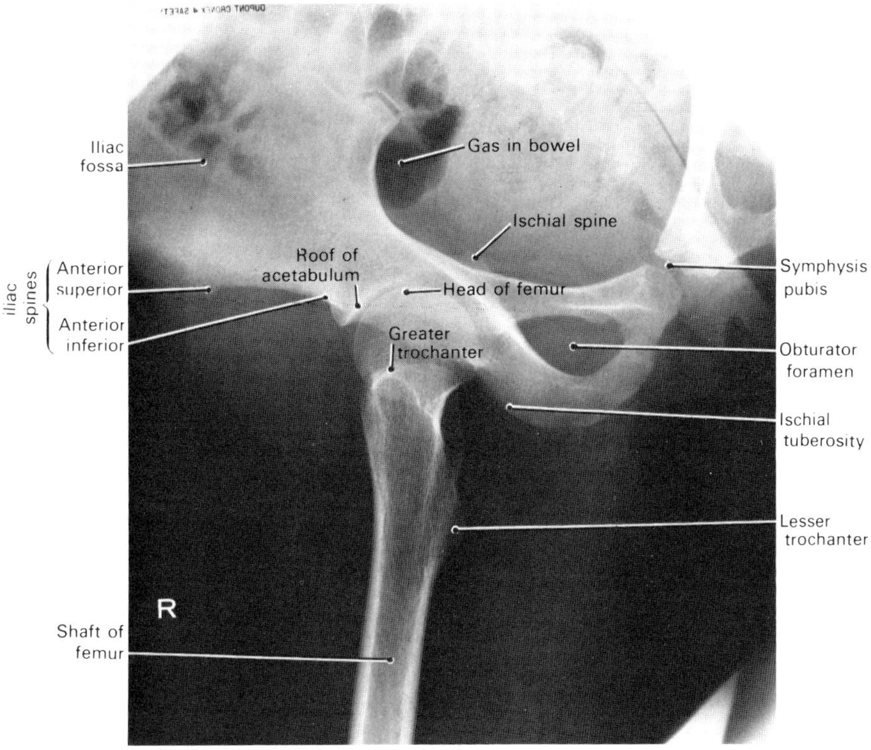

Fig. XIII-25D. Frog-leg lateral of the right hip.

in this projection; however, the greater trochanter is well visualized.

As in the shoulder girdle, we must always examine the soft tissues carefully for signs of disease. In Figure XIII-25B, the capsule of the hip joint has been outlined for you. Whenever possible have both sides of the body for comparison of the soft tissues, since this will improve your chances for detecting minor abnormalities. Bulging of the hip capsule or loss of its shadow may represent the earliest changes in some disease states. Calcifications about the greater trochanter and ischial tuberosity are not infrequent findings. These occur in nearby tendons or bursae, and may be the source of your patient's pain. Look for them with a "hot light" and a magnifying glass.

3. AP Projection with the Femur Externally Rotated (Fig. XIII-25C): This projection is made exactly like those mentioned above, except that the foot is allowed to rotate externally.

The neck of the femur is now *foreshortened,* while the lesser trochanter is thrown into profile. *Caution:* in patients with trauma to the hip, this view is fraught with danger in evaluating fractures of the femoral neck. In those cases you must have an AP projection with the femur internally rotated.

4. Frog-leg Lateral of the Hip (Fig. XIII-25D): With the patient supine on the X-ray table top, the heels are brought together with the knees flexed and the thighs everted and placed nearly flat on the table. The central X-ray beam is directed at the hip joint and is, of course, perpendicular to the film holder.

In this view the femur is seen in true lateral view. Note that the greater trochanter overlaps the femoral neck which is foreshortened. The lesser trochanter is well visualized.

The Knee

1. AP Projection (Fig. XIII-26A): With the patient in supine position, the knee is extended. The central X-ray beam is directed through the joint space.

The patella is projected over the distal end of the femur. The apex of the patella lies above the knee joint space.

The femur is covered with cortical bone which is much thicker along its shaft than over its ends. Observe the prominent adductor tubercle just above the medial epicondyle. The lateral femoral epicondyle appears more prominent than its medial counterpart. The condyles of the femur are convex distally. Their posterior parts are visible and are somewhat rectangular-shaped. Between the condyles is the intercondylar notch, the most distal aspect of which is opposite the tubercles of the intercondylar eminence of the proximal tibia. The articular surfaces of the tibial condyles are concave. Normally the medial portion of the fibular head and its apex (styloid process) are obscured by the superimposed lateral condyle of the femur.

2. Lateral Projection (Fig. XIII-26B): The knee is slightly flexed, and the lateral aspect placed atop the table. The beam passes through the knee joint.

Ideally one would like to have the femoral condyles superimposed, but alas, this is not the case in Figure

XIII-26B. Because of this minor degree of rotation it is not possible to see the retropatellar space clearly. Just above the outlines of the condyles is a dense oblique line cast by the intercondylar notch.

One cannot appreciate the concavities of the articular surface of the tibia in a lateral view. However, the intercondylar eminence may be seen rising up to overlap the femoral condyles.

Again observe that the fibular head is partially obscured by the tibial condyles.

3. Intercondyloid Fossa Projection (Fig. XIII-26C): The patient is placed in prone position, the knee is flexed 40°, and the foot supported on sandbags. The central X-ray beam is angled about 40° caudally, centering on the joint. This permits an excellent view of the knee joint space, and gives us greater appreciation of the depth of the intercondyloid notch.

4. Tangential View of Patella (Fig. XIII-26D): With the patient in prone position, the knee is flexed fully and the central ray directed along the articular surface of the patella. Note the prominent bony ridge separating the small medial articular facet, and the larger lateral facet. The anterior surface of the patella is characteristically roughened. Fine fissues at tendinous insertions should not be mistaken for fracture.

The Ankle

1. AP Projection (Fig. XIII-27A): With the ankle dorsiflexed, the central X-ray beam is directed vertically through the ankle joint. The trochlea of the talus is surrounded by the inferior surface of the tibia from above, and medially by the medial malleolus of the tibia. Laterally, the lateral malleolus of the fibula borders it. The ankle joint is relatively uniform in width throughout. The inferior tibiofibular joint is not seen in profile in this particular projection.

2. Medial Oblique Projection (Fig. XIII-27B): With the ankle dorsiflexed, the leg is rotated 45° medially, and the vertically-directed, central X-ray beam is centered on the ankle joint. This view brings the inferior tibiofibular joint into profile. It also shows the malleolar fossa as an oval lucency with a faint sclerotic margin. This projection also permits some visualization of the talocalcaneal and talonavicular joints. The cuboid bone may also be evaluated as well as the base of the 5th metatarsal.

3. Lateral Projection: This will be considered with the foot.

4. 45° View of the Subtalar Joint (Fig. XIII-27C): We usually perform this projection with the patient standing on the film cassette. The central X-ray beam is directed 45° through the posterior surface of the dorsi-flexed ankle. This projection is excellent in evaluating the presence of talocalcaneal coalition, an abnormal bar of bone joining the talus and calcaneus. It is also useful in evaluating fractures of the calcaneus and for viewing the posterior aspect of the talocalcaneal joint.

The Foot:

1. Lateral Projection (Weight Bearing) (Fig. XIII-28A): This projection is made with the patient standing in order to demonstrate the longitudinal arch of the foot. The film holder is parallel to the foot. A horizontally-directed, central x-ray beam is projected at the navicular-cuboid articulation.

Observe the convex surface (trochlea) of the talus articulating with the concave tibial and fibular surfaces. The medial malleolus of the tibia is not seen; however, the distal fibula may be seen superimposed upon the body of the talus. The talus narrows at its neck before broadening out at its head. The distally placed head of the talus is convex and it articulates with the disc-shaped navicular. Posteriorly the body of the talus narrows to a small posterior process. Immediately beneath it we may see the posterior facet for the calcaneus. There are actually three facets on the talus for articulation with the calcaneus: posterior, middle, and anterior. The posterior facet is the largest and it comprises the entire inferior surface of the body of the talus (Fig. XIII-28A). Between the posterior and middle articular surfaces is a groove, the sulcus (sinus) tali. The middle and anterior facets are not seen to advantage here.

Since the talus has three articular facets for articulation with the calcaneus, then of course, the calcaneus has three similar facets: posterior, middle, and anterior. The posterior facet is immediately beneath its corresponding member on the talus. The other two are not seen to advantage. Observe the dense shadow of

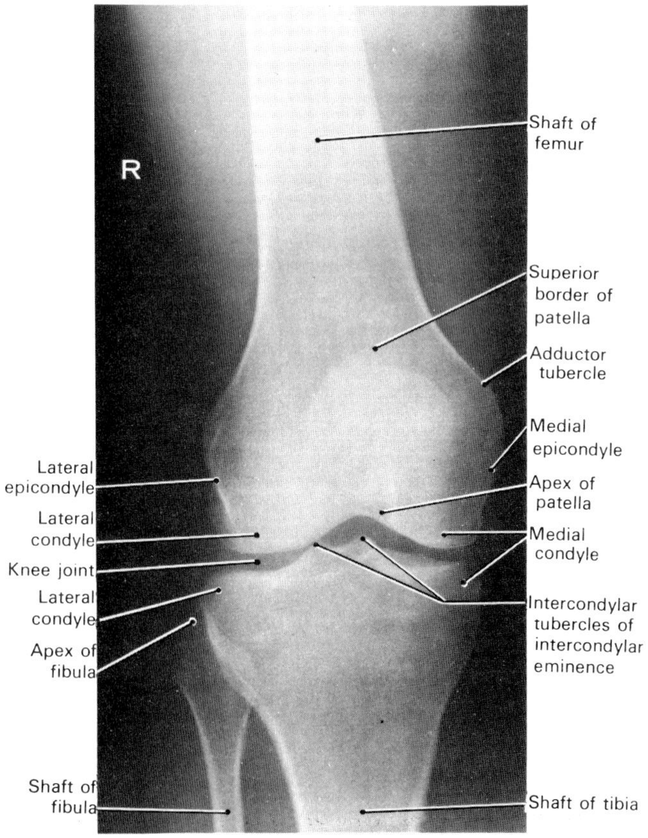

A. AP projection.

Fig. XIII-26. Radiographs of the knee.

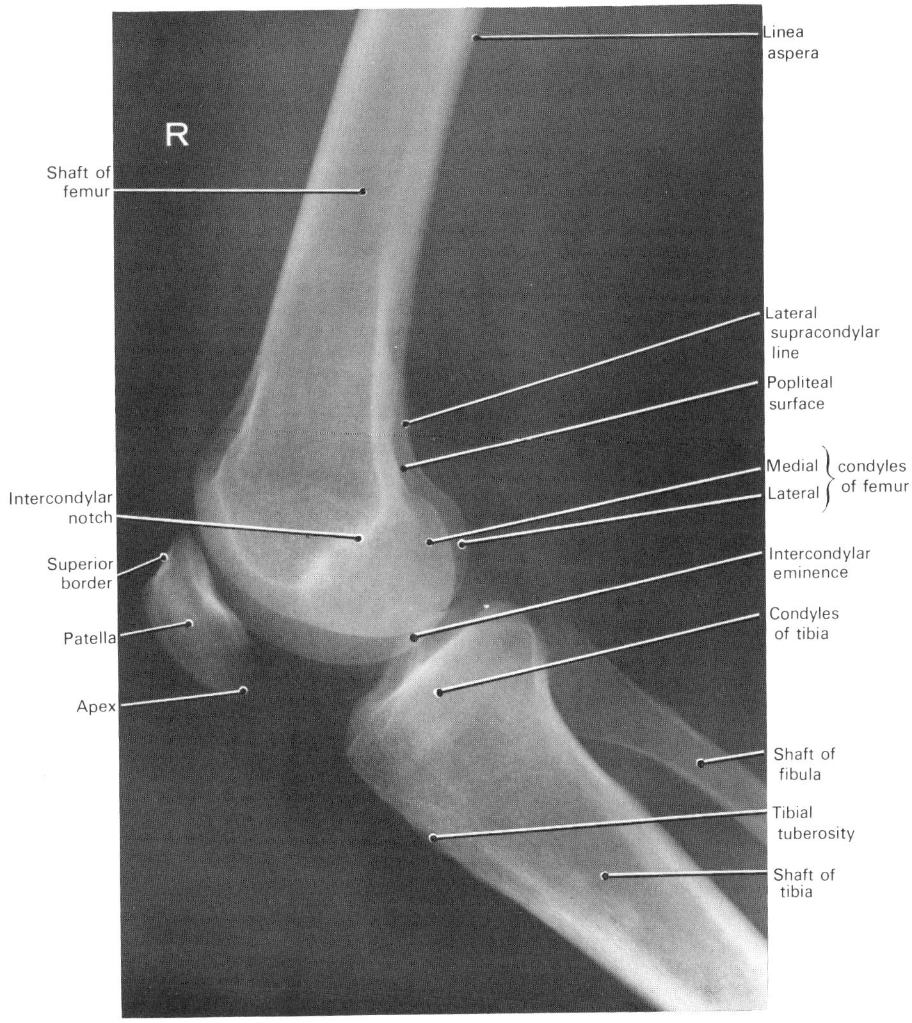

B. Lateral projection.

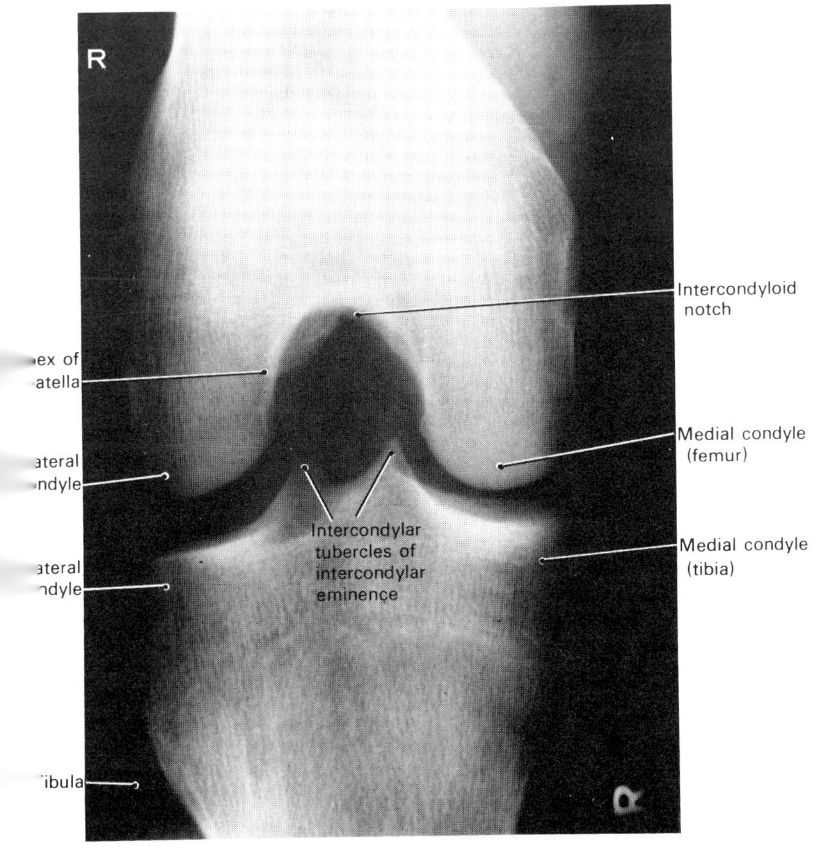

C. Intercondyloid fossa projection.

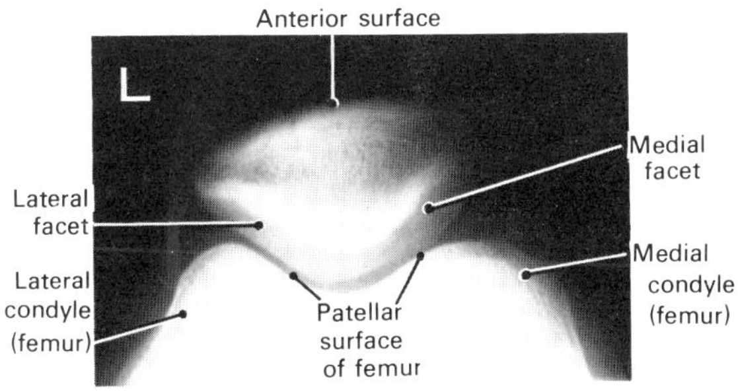

D. Tangential view of the left patella.

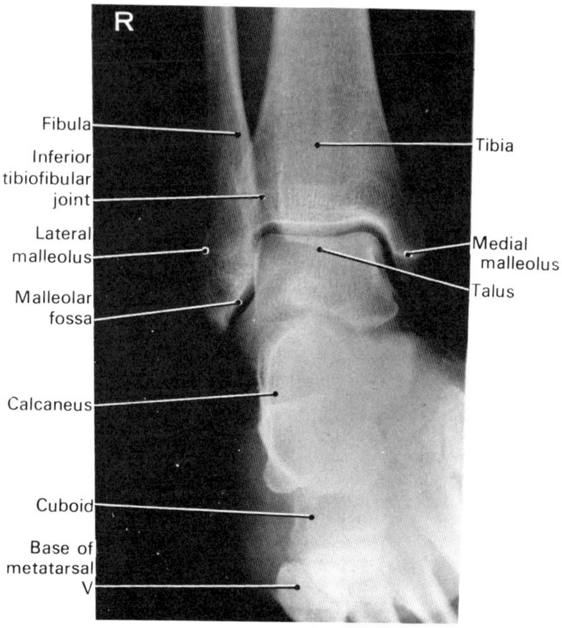

A. AP projection.

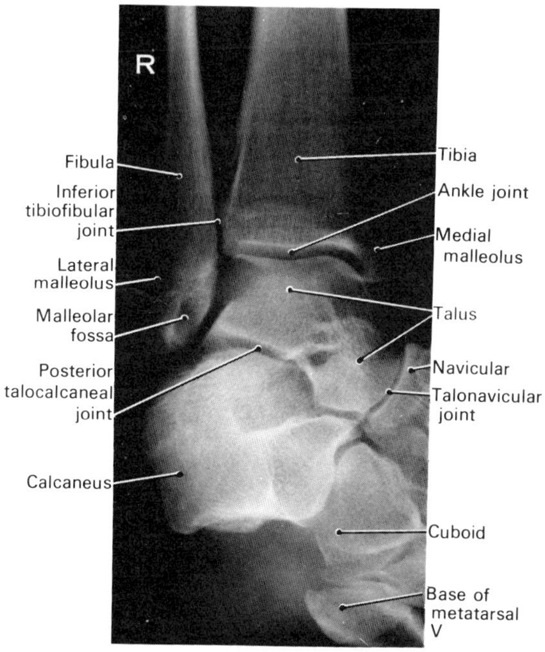

B. Medial oblique projection.

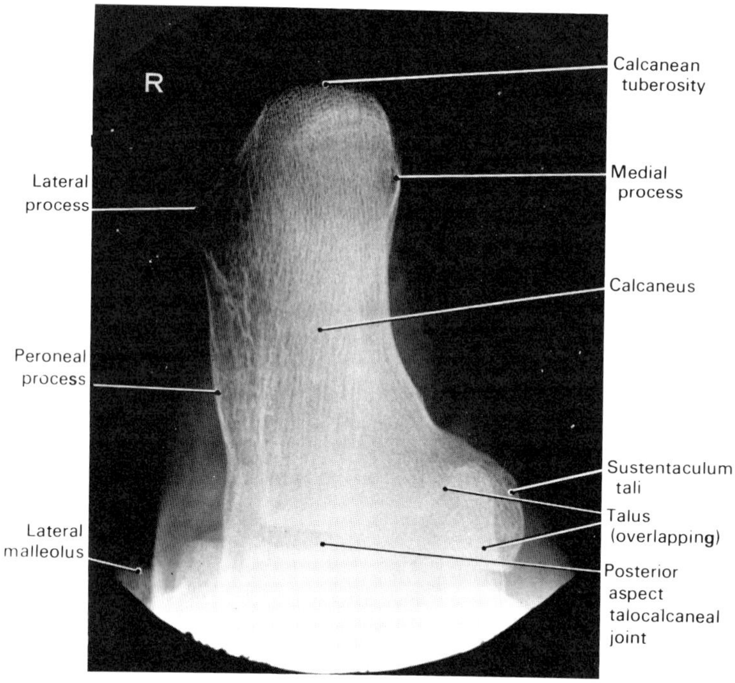

C. 45° view of the subtalar joint.

Fig. XIII-27. Radiographs of the right ankle.

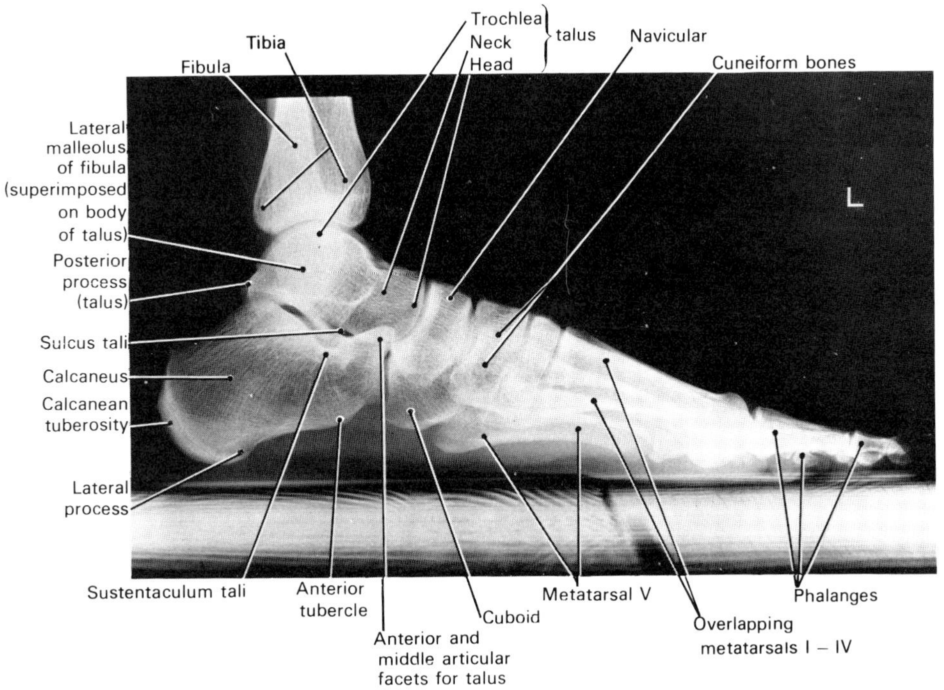

A. Lateral projection (weight bearing).

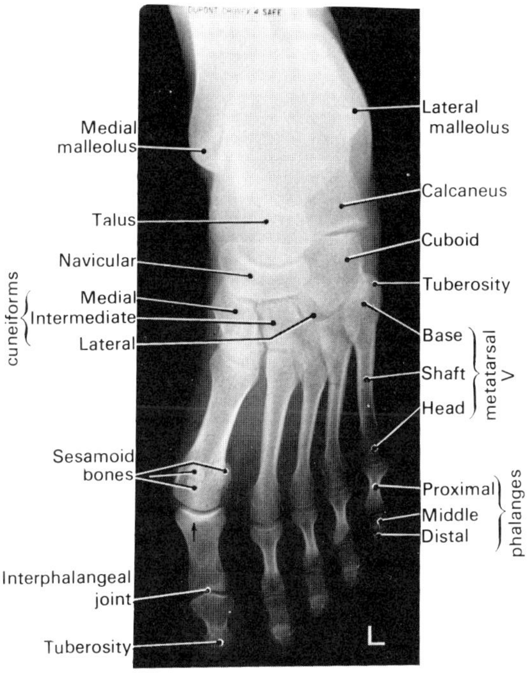

B. AP (dorsoplantar) projection.

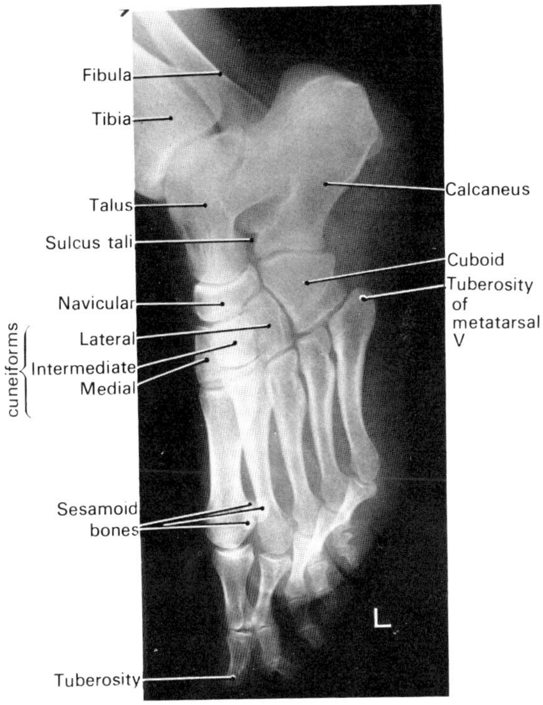

C. Lateroanterior oblique projection.

Fig. XIII-28. Radiographs of the left foot.

the sustentaculum tali which lies on the medial surface of the calcaneus immediately beneath its middle facet. This structure supports the talar head. Other structures on the calcaneus that may be seen are: the anterior tubercle, lateral process, inferior, posterior, and superior surfaces. Anteriorly the calcaneus articulates with the cuboid.

The disk-shaped navicular is concave posteriorly. Anteriorly it is convex, and sports three facets for articulation with the three cuneiforms, all of which are superimposed in this projection. Note how they partly overlap the cuboid.

We have seen how the cuboid articulates with the calcaneus, but this view does not permit us to see its anterior articulations with the fourth and fifth metatarsals, nor its medial articulation with the third cuneiform.

The fifth metatarsal is seen in part, while the other four metatarsals overlap as do the phalanges.

2. AP (Dorsoplantar) Projection (Fig. XIII-28B): With the patient supine on the X-ray table, the knee is flexed and the plantar surface of the foot is placed on the film holder. The central X-ray beam is directed vertically at the film centering on the base of the third metatarsal.

Observe the phalanges, and note their relative size and shape. The interphalangeal joints of the second through fifth toes are not clearly seen since the toes are in slight flexion. The dense white horizontal line at the base of the proximal phalanx of the great toe is the remnant of the epiphyseal line.

The first metatarsal is much larger in width but shorter than the other metatarsals. Immediately beneath the head of the first metatarsal on the plantar surface are three sesamoid bones. The base of the first metatarsal articulates with the medial cuneiform and the base of the second metatarsal. The second through the fifth metatarsals are narrow and long. Their bases overlap such that we cannot apreciate the intermetatarsal joints.

Identify the individual tarsal bones, but notice that there is some overlapping of them. Because of the normal curvature of the foot, we are not able to see clearly the individual tarsal joints. Only the anterior portions of the talus and calcaneus are visualized because of overlap from the tibia and fibula.

3. Lateroanterior Oblique Projection (Fig. XIII-28C): The dorsolateral aspect of the foot is placed upon the film holder, and the central X-ray beam is directed vertically at the film centering upon the base of the third metatarsal.

Clearly demonstrated are the talus, calcaneus, navicular, and cuboid bones. All three cuneiforms are overlapped, as are the bases of the first three metatarsals. Note the articulation between the cuboid and the bases of the fourth and fifth metatarsals. The midtarsal joint, that is between the talus and navicular, and the calcaneus and cuboid, should also be observed for it is at this point that the movements of eversion and inversion occur.

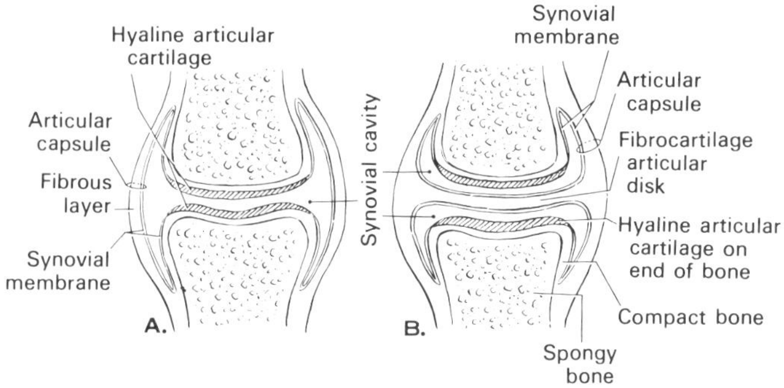

Fig. XIII-29A. Diagram of usual type of synovial (diarthrotic) joint. **B.** Synovial joint with an articular disk. (Schematic).

JOINTS

A joint is the union between one bone and another. The joints were classified according to the method of union between the participating bones in Chapter II. They may be fibrous, cartilaginous, or synovial, and they may be completely immovable, slightly movable, or freely movable. Refer to Chapter II, page 58, for a review of these joints and the subclassifications in each group.

All the highly movable joints of the body are synovial (Fig. XIII-29). These diarthroses (a joint in which there is a separation, or articular cavity) have at least one axis of movement with the exception of the gliding joints (arthrodial). The gliding joints, with their irregular shapes, have no axis of movement at all. Such joints are found between the carpal and tarsal bones. The hinge and pivot joints are uniaxial, a pivot joint allowing rotation about a vertical axis. The hinge joint permits movement about a frontal horizontal axis, producing flexion or extension. The condyloid and saddle joints are bi-axial, the axes being in the frontal horizontal and the sagittal horizontal planes. The ball and socket joints are multi-axial, permitting flexion and extension (about a frontal horizontal axis), abduction and adduction (about a sagittal horizontal axis), and rotation (about a vertical axis). Circumduction is the result of moving through a sequential combination of flexion, extension, abduction, and adduction and is, therefore, performed about two axes. Circumduction may occur in condyloid, saddle, and ball and socket joints.

LIGAMENTS

Composed of dense collagenous tissue with varying amounts of elastic fibers, these structures occur around joints, joining bone to bone, and restricting the movements of the joint. Ligaments not only prohibit movement in an undesirable plane, but they also limit the extent of normal movement. When movement in a normal plane is slowly arrested, the ligaments involved become increasingly taut. However, if an abnormal movement occurs, the ligaments protecting the joint become tense immediately and will tear if the force bearing on the joint is too great.[1] Examples of limited movement can be seen in the leg where extension cannot progress beyond a certain point in either the hip or the knee. In both instances, it is due to the presence of strong ligaments that the motion is stopped.

Every synovial joint has ligaments. A **capsular ligament** is merely a localized thickening of the fibrous capsule of the joint and is not a separate entity. An **accessory ligament,** on the other hand, is an independent structure located near a joint but not as an integral part of the capsule.

SKELETAL MUSCLES OF THE LIMBS

Types of Attachment

Review Chapter II, page 59, for general myology, types of muscle tissue, structure and organization of muscle fibers, and attachments. The type of attachment of muscle to bone will often leave its mark upon the bone. As you read about the bones of the limbs, you learned of various tubercles, well-defined, roughened areas, and lines. Attachments by tendons are at limited sites, either in well-delineated areas on the bone or upon a tubercle. Attachments by aponeurosis cover a length of bone and are linear. They often produce raised lines or ridges on the bone. Attachments by "fleshy fibers" rather than a distinct tendon usually cover a large area of the bone and do not leave many, if any, signs of their having been there.

Origins and Insertions

The *origin* of a muscle is usually thought of as the proximal attachment. It is often characterized by stability (the "fixed end") and by the lack of a long tendon. The *insertion* is, by convention, the more distal attachment. It frequently involves a long tendon and is considered to be the "moving end." Such a functional definition of origin and insertion is highly relative, since the action of a given muscle may be different under various circumstances. There are many movements in which the insertion is stationary and the origin is the end that moves. For example, when picking up an object from a table, the elbow is flexed, and the forearm (point of insertion) moves toward the upper arm (point of origin). When the more athletic students in your class chin themselves, the situation is reversed. The movement at the elbow joint is still flexion, but it is the upper arm (formerly the "origin" or "fixed end" of the muscle attachment) that now moves toward the stabilized forearm.

Generally speaking, the origin of a muscle covers a wider area than its insertion which is usually tendinous. When considering the intrinsic muscles of the head, the neck, and the trunk, it is often preferable to use terms like "upper end" or "pelvic attachment" rather than "origin" and "insertion."

CHARACTERISTICS OF SKELETAL MUSCLE

Striated muscle tissue has the properties of **elasticity, extensibility,** and **contractility.** Elasticity and extensibility allow a muscle to be stretched like a rubber band and then to return to its normal resting length. The average muscle fiber can be stretched approximately half again as long as its resting length. Contractility is the unique characteristic of muscle tissue. The degree of a muscle's total shortening depends upon its internal structure as well as the length of its individual fibers. As you will recall, the *all or none law* for muscle contraction states that each muscle *fiber* (not the entire muscle) contracts maximally or not at all. Those muscles whose fibers run parallel to each other and to the long axis of the muscle will shorten as a whole in proportion to the shortening of the individual fibers. Pennate and fan-shaped muscles whose fibers are arranged obliquely will shorten as a whole to a lesser degree. Experimenters variously report that a muscle can shorten as much as one-half to two-thirds of its normal resting length.

Types of Contraction

Thus far, we have assumed that a muscle shortens when it contracts; however, muscles are capable of three distinct types of contraction. These are based upon whether the muscle shortens, lengthens, or maintains the same length during contraction. An **isotonic contraction** is one in which the muscle actually shortens while performing its work. One end is stabilized while the other end pulls the bone to which it is attached and moves it around the axis of the joint. This is the type of contraction seen in most physical activities.

Isometric contractions have become popularized on the news-stands over the last few years as the desk-bound worker's way to good muscle tone through surreptitious exercise on company time. An isometric contraction is a static contraction in which the length of the muscle does not change. By pressing the palms of your hands together, you can make the muscles of the arms and shoulder girdle contract, or become tense. If you press both hands together with equal force, no movement occurs, and the muscles remain at the same length during the entire period of the effort.

Finally, there is the **eccentric contraction,** a seemingly paradoxical term. It is frequently referred to as a "lengthening" contraction, although in actuality it does not really lengthen; rather, it returns from a shortened condition to its normal resting length. Such eccentric contractions occur when a student puts down a heavy arm-load of anatomy books. There is a gradual releasing of the contracted muscles against the force of gravity. This produces a smooth, controlled movement.

Actions and Functions of Muscles

The **action** of a muscle is the way in which it moves a part, *i.e.,* the motion that it imparts to the joint it crosses. These are flexion, extension, abduction, adduction, and rotation. Muscles also have **functions.** A purposeful movement of any part of the body requires a considerable amount of muscular activity beyond that directly responsible for the movement itself. Since the muscles causing the movement must have a stable base from which to work, the bone, or bones, providing attachment for the origins of the muscles must be stabilized (fixed) by the other muscles. Unwanted movement in the joint must be prevented by still other muscles, and so it becomes apparent that the muscles involved in a single well-coordinated movement have many and various functions. These roles are categorized as agonists, antagonists, stabilizers, and neutralizers. The category of agonists may be further broken down into prime movers, secondary movers, and synergists. The term "synergist" is mentioned here only because you may be exposed to it by one of your instructors or may find it in another book. Synergistic muscles are those that contract at the same time as the prime mover, performing a function to facilitate or to potentiate the effect of the prime mover. Hence, it can be argued that a secondary mover, a stabilizer, a neutralizer, and even an antagonist may all be synergists. The term is ill-defined in common usage, and we shall avoid its use in this text.

The **agonists** are the muscles which are directly responsible for bringing about the movement with which we are concerned. The chief muscle producing the desired movement is the **prime mover.** Other muscles aiding in the desired movement are the assistants, or **secondary movers.** These secondary movers may contract at the same time as the primary mover or they may come into play as the movement progresses through its range of motion.

Antagonists are muscles which oppose the action of the prime mover and thus aid in producing a smooth motion by controlling the speed and power of the agonists. They also act as brakes in bringing a motion to a stop.

The **stabilizers** most commonly function to steady, or fix, the bone to which the contracting muscle is attached so that the prime mover may cause an effective movement of the bone on which it inserts. The grouping also includes those muscles which contract statically (isometrically) to steady or support some part of the body against the pull of gravity, such as the raised arm when the person is working with his hands at shoulder level.

The last category, **neutralizers,** are those muscles which act to prevent an undesired action of one of the movers. If a prime mover, or agonist, both flexes and abducts, and abduction is the desired movement, then an extensor must contract to counteract, or neutralize, the undesirable flexion. Occasionally, two movers may have one action in common and secondary actions which oppose each other. In such a situation, the muscles work together to produce the desired movement and neutralize each other's secondary actions.

In summary, then, there are four basic categories of muscle function: agonist, antagonist, stabilizer, and neutralizer. As with the origin and insertion of a muscle, these roles are highly relative. A muscle may function as a prime mover in one situation and as a neutralizer, stabilizer, or antagonist in other situations.

Gravity

A great deal of misunderstanding concerning the functions of muscles could be avoided if the student would bear in mind the influence of gravity in aiding, replacing, or hindering muscles. For example, when a man, standing upright, flexes his elbow, the Brachialis must overcome gravity in raising the forearm. When this same person extends his flexed elbow, gravity functions as the extensor, aiding extension and replacing the Triceps. The Brachialis must work eccentrically to control the movement. In short, whether flexing or extending his elbow, the man is using the Brachialis muscle in the movement. He is not using his Triceps as one might suppose.

Nomenclature

The names given to muscles are descriptive of their various characteristics. Geometrical *shape* is reflected in such names as Triangularis, Trapezius, or Quadratus Femoris. General *form* is expressed in such names as Serratus Anterior and Latissimus Dorsi (both of these names also give the location of the muscle, *i.e.*, a serrated muscle on the anterior part of the body and the broadest muscle of the back). Other examples of naming by *location* are the Temporalis, the Iliacus, and the Tibialis Anterior. *Contrasting features* are noted in muscle names like Peroneus Longus and Peroneus Brevis. Finally, *actions* are described in the names of the muscles such as Levator Scapulae and Pronator Teres. Since there is a fairly obvious reason for the name of each muscle, remembering them should not be that difficult.

JOINTS AND MUSCLES OF THE SHOULDER GIRDLE AND UPPER LIMB

Since man walks erect, his upper limbs are free and are adapted for use in prehension. Each arm articulates with the trunk through one joint, the sternoclavicular joint, on its respective side. This general lack of bony attachment to the trunk allows the upper limb a great range of movement, but sacrifices its security. However, man is apparently willing to grant his arms their freedom for the advantages that prehensile upper limbs give him over all other mammals.

Articulations of the Shoulder Girdle

The shoulder girdle consists of the two clavicles and the two scapulae. The acromioclavicular articulation between the acromion process of the scapula and the outer end of the clavicle is an arthrodial joint allowing only small, gliding movements. It is strengthened and stabilized by the **acromioclavicular ligament** and the **coracoclavicular ligament.** The coracoclavicular ligament is actually two fibrous bands, the **conoid,** passing from the base of the coracoid process of the scapula to the conoid tubercle on the clavicle; and the **trapezoid,** extending from the top of the coracoid process to the trapezoid ridge of the clavicle.

The second articulation between the bones of each side of the shoulder girdle is the sternoclavicular joint.

The movements of this joint are greater than those of the acromioclavicular joint, allowing limited motion of the clavicle in all three planes. It is bound together and protected by four ligaments: the **anterior sternoclavicular,** the **posterior clavicular,** the **interclavicular,** and the **costoclavicular.** The costoclavicular ligament connects the clavicle to the upper border of the first costal cartilage. Movements of the shoulder girdle are defined in terms of the movements of the scapulae; however, the student must remember that every movement of the scapula involves motion in both the acromioclavicular and the sternoclavicular joints.

Muscles of the Shoulder Girdle: There are six muscles associated with the movements of the scapula: three anterior muscles—the **Subclavius,** the **Pectoralis Minor,** and the **Serratus Anterior**; and three posterior muscles, the **Levator Scapulae,** the **Trapezius,** and the **Rhomboids (Major** and **Minor**). These produce the movements of elevation, depression, abduction (protraction), adduction (retraction), upward tilt and reduction of upward tilt, and upward and downward rotation. Table XIII-1 summarizes the information needed about these muscles and their actions.

The Shoulder Joint (Scapulohumeral Joint)

Just as every movement of the shoulder girdle involves both the acromioclavicular and sternoclavicular articulations, every movement of the arm on the trunk is shared by the scapula. As the arm travels through its wide range of motions, the scapula moves to place the glenoid cavity in the most favorable position for the head of the humerus.

The shoulder joint is a ball and socket joint formed by the articulation of the spherical head of the humerus with the shallow glenoid cavity of the scapula. A strip of fibrocartilage, the **glenoid labrum,** is attached to the periphery of the cavity, deepening it somewhat and providing a cushion against which the humeral head rolls. Surrounding the joint is a loose, sleevelike, **fibrous capsule** attached proximally to the margin of the glenoid cavity and distally to the anatomical neck of the humerus. The capsule is reinforced both by ligaments and by muscle tendons. The most important of the ligaments is the **coracohumeral ligament,** a broad band passing from the lateral border of the coracoid process to the anterior border of the greater tubercle of the humerus. It blends with and thickens the upper anterior portion of the capsule. Three glenohumeral ligaments (superior, middle, and inferior) are found blending into the anterior and lower portions of the fibrous capsule as they pass obliquely from the front and lower part of the anatomical neck of the humerus medially and upwards to attach on the supraglenoid tubercle (Fig. XIII-32).

The strength of a joint depends upon its bony formation, its ligaments, and the muscles crossing it. The shoulder joint has a tenuous bony apposition, relatively insignificant ligaments, and a multitude of muscles. The **rotator (musculotendinous) cuff** of the shoulder is formed by four short muscles placed closely about the humeral head. This cuff provides the chief strength of the joint, retaining the head of the humerus in its socket, and, with the assistance of the overhanging coracoacromial arch, protecting the shoulder joint anteriorly, superiorly, and posteriorly. There is no support below save for the loose articular capsule. The coracoacromial arch is a union of the coracoid and acromion processes of the scapula by the coracoacromial ligament, thus forming a bridge, or arch, over the top of the shoulder joint. The four short muscles of the rotator cuff are the **Supraspinatus** above, the **Infraspinatus** and **Teres Minor** behind, and the **Subscapularis** in front. All the muscles involved in the movements of the shoulder joint are tabulated in Table XIII-2. Although it is not always clear-cut, these muscles may be grouped according to their position in relation to the joint as an aid in remembering them. Thus, the muscles are:

Superior—middle portion of Deltoid
  Supraspinatus
Inferior—Latissimus Dorsi
  Teres Major
  long head of the Triceps (primarily a muscle of the elbow joint)
Anterior—Pectoralis Major
  Coracobrachialis
  anterior portion of Deltoid
  Subscapularis
  Biceps (again, primarily a muscle of the elbow joint)
Posterior—posterior portion of Deltoid
  Infraspinatus
  Teres Minor

The Elbow Joint and Proximal Radio-ulnar Joint

The elbow and radio-ulnar joints work in concert to make possible the many positions of the hand, its dexterity and usefulness. The humerus joins with the radius and ulna to form a hinge joint whose only motions are flexion and extension. It may therefore be anticipated that the muscles of the arm are massed in front of and behind the humerus to flex and extend the elbow. There are three articulations, the humero-ulnar, the humero-radial, and the proximal radio-ulnar, which share a single synovial capsule and a continuous joint space. The upper end of the radius touches the ulna at the radial notch of the ulna, and it is held in place there by the **annular ligament.** There are four ligaments of the elbow proper, the **anterior and posterior ligaments** being thin layers of fibers which blend with and strengthen the anterior and posterior portions of the capsule. The collateral ligaments are strong, thick, fan-shaped bands which prevent abduction and adduction of the elbow. These are the **ulnar collateral (medial) ligament** and the **radial collateral (lateral) ligament** (Fig. XIII-33).

Muscles of the Arm

With the exception of the Coracobrachialis (Table XIII-2), all the muscles of the arm act principally on the elbow joint. The **anterior (flexor) compartment**

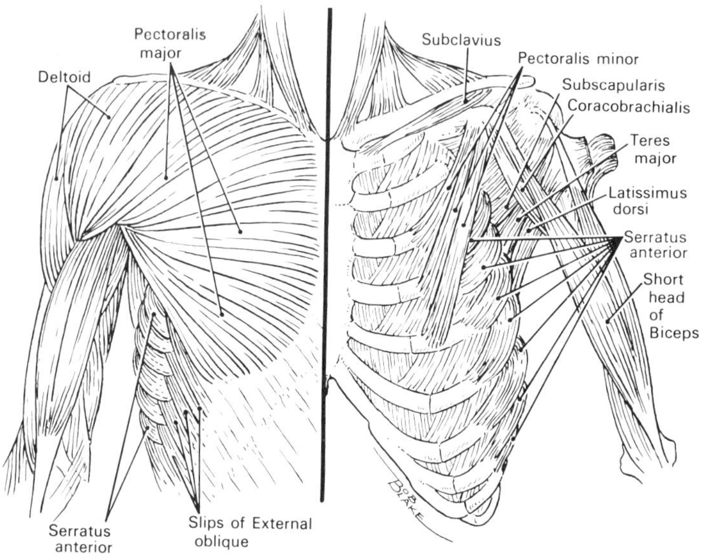

Fig. XIII-30. Muscles of the chest and shoulder region. In addition, the origin of the Coracobrachialis is shown from the coracoid process in common with the short head of the Biceps. Slips of the External oblique are shown interdigitating with those of the Serratus anterior.

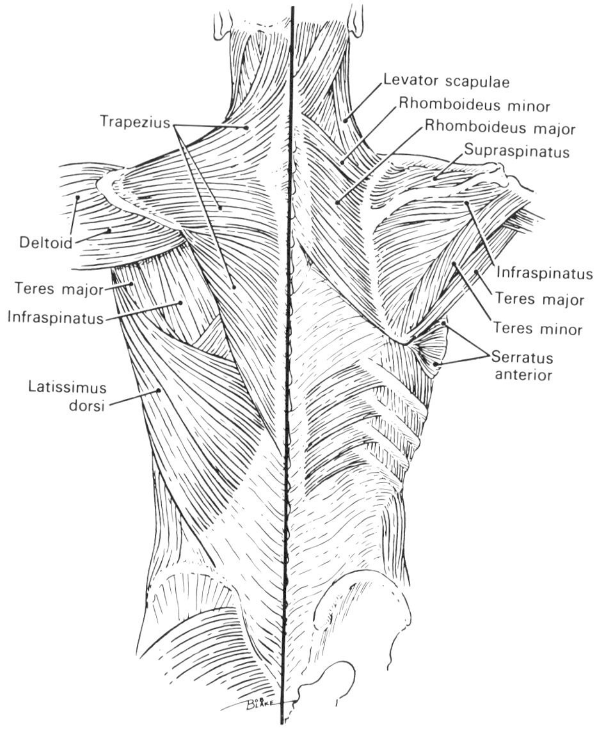

Fig. XIII-31. Muscles of the shoulder. The more superficial muscles are seen on the left; deeper ones, on the right.

Table XIII-1. Muscles of the Shoulder Girdle (Figs. XIII-30 and 31)

Muscle	Nerve	Origin and Insertion	Primary Action
Subclavius	$C_{5,6}$ Nerve to Subclavius	(O) Upper surface of first rib at junction with cartilage (I) Underside of middle half of clavicle	Depression of clavicle? Stabilization of sternoclavicular joint
Pectoralis minor	C_8, T_1 Medial pectoral nerve	(O) Anterior surface of third, fourth, and fifth ribs near their cartilages (I) Tip of coracoid process of scapula	Adduction and extension of arm
Serratus anterior	$C_{5,6,7}$, Long thoracic nerve	(O) Outer surface of upper nine ribs at side of chest (I) Anterior surface of vertebral border and inferior angle of scapula	Upward rotation of scapula
Levator scapulae	$C_{3,4}$ Nerves to Levator scapulae	(O) Transverse processes of first four cervical vertebrae (I) Vertebral border of scapula between medial angle and scapular spine	Elevation of scapula
Trapezius	$C_{3,4}$ and spinal accessory nerve	(O) Occipital bone; ligamentum nuchae; spinous processes of seventh cervical and all thoracic vertebrae (I) Part I: Posterior border of lateral one-third of clavicle Part II: Top of acromial process Part III: Upper border of spine of scapula Part IV: Root of spine of scapula	Part I: Elevation Part II: Elevation and upward rotation Part III: Adduction Part IV: Upward rotation, depression, assists in adduction
Rhomboids major and minor	C_5 Dorsal scapular nerve	(O) Spinous processes of seventh cervical and first five thoracic vertebrae (I) Vertebral border of scapula from spine to inferior angle	Retraction of scapula

Table XIII-2. Muscles of the Shoulder (Figs. XIII-30 and 31)

Muscle	Nerve	Origin and Insertion	Primary Action
Supraspinatus	$C_{5,6}$ Suprascapular nerve	(O) Medial two-thirds of supraspinatus fossa above spine of scapula (I) Top of greater tubercle of humerus	Abduction of arm (part of rotator cuff)
Deltoid	$C_{5,6}$ Axillary nerve	(O) Acromial process, outer one-third of clavicle, and lower margin of spine of scapula (I) Lateral aspect of humerus near midpoint of shaft	Abduction of arm (anterior and posterior portions also aid in flexion and inward rotation, and extension and outward rotation respectively)
Teres minor	$C_{5,6}$ Axillary nerve	(O) Axillary border and posterior surface of scapula below scapular spine (I) Posterior aspect of greater tubercle of humerus	External rotation of arm (part of rotator cuff)
Subscapularis	$C_{5,6}$ Upper and lower subscapular nerves	(O) Anterior surface of scapula (I) Lesser tubercle of humerus	Internal rotation of arm (part of rotator cuff)
Teres major	$C_{5,6}$ Lower subscapular nerve	(O) Posterior surface of inferior angle of scapula (I) Anterior surface of humerus below head, just medial to tendon of Latissimus dorsi	Extension and internal rotation of arm
Infraspinatus	$C_{5,6}$ Suprascapular nerve	(O) Axillary border & posterior surface of scapula below scapular spine (with Teres minor) (I) Posterior aspect of greater tubercle of humerus	External rotation of arm (part of rotator cuff)
Pectoralis major (upper portion)	$C_{5,6,7}$ Lateral pectoral nerve	(O) Medial two-thirds of clavicle (I) Lateral surface of humerus just below head by a wide, flat tendon	Adduction and flexion of arm
Pectoralis major (lower portion)	C_8, T_1 Medial pectoral nerve	(O) Anterior surface of sternum, cartilages of first six ribs, slip from aponeurosis of External oblique abdominal muscle (I) Lateral surface of humerus just below head by a wide, flat tendon	Adduction and extension of arm
Latissimus dorsi	$C_{6,7,8}$ Thoracodorsal nerve	(O) Spinous processes of lower six thoracic and all lumbar vertebrae, posterior surface of sacrum, crest of ilium, and lower three ribs (I) By tendon on anterior surface of humerus below head, anterior to and parallel with tendon of Pectoralis major	Extension and adduction of arm
Coracobrachialis	C_5-C_7 Musculocutaneous nerve	(O) Coracoid process of scapula (I) Inner surface of humerus opposite Deltoid attachment	Adduction and flexion of arm

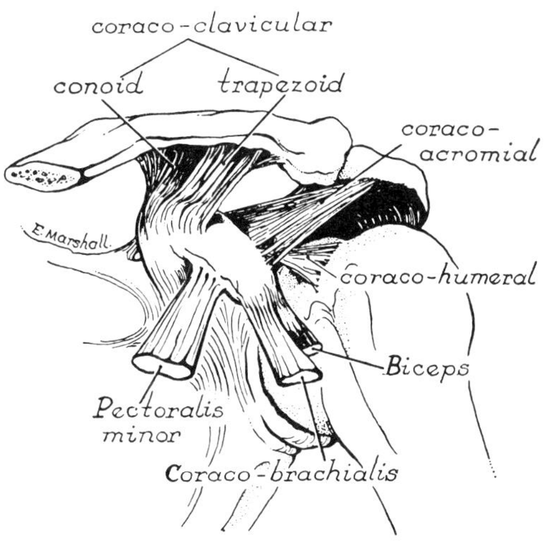

Fig. XIII-32. Ligaments and tendons of left coracoid process from the front. (From J. V. Basmajian, *Primary Anatomy*, 6th ed., Williams & Wilkins, Baltimore, 1970.)

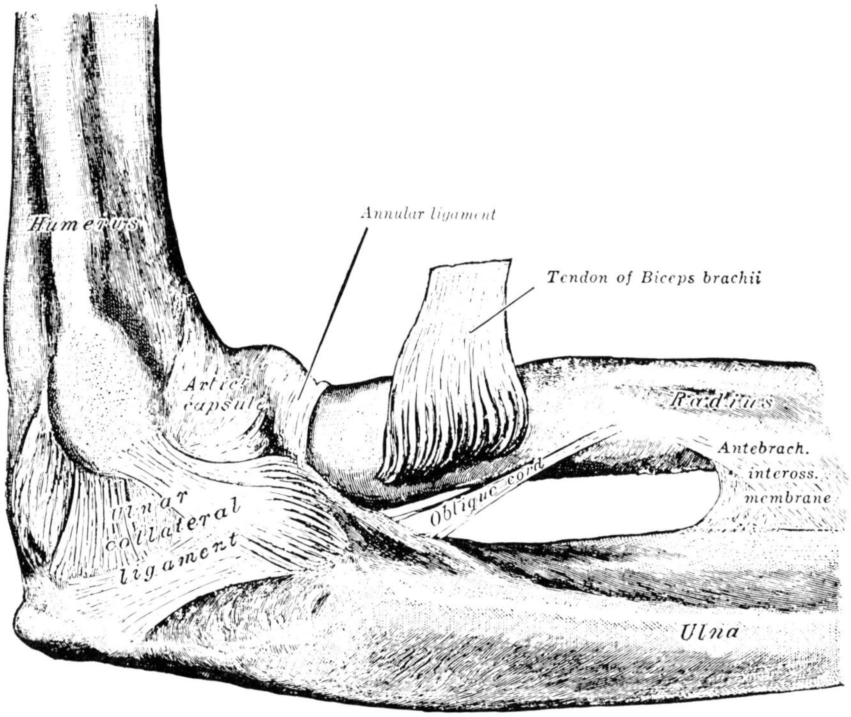

Fig. XIII-33. Medial aspect, left elbow joint. (From C. M. Goss, *Gray's Anatomy of the Human Body*, 28th ed., Lea & Febiger, Philadelphia, 1966).

contains the **Biceps Brachii, Brachialis, Brachioradialis,** and the **Pronator Teres.** The **posterior (extensor) compartment** contains the **Triceps Brachii, Anconeus,** and **Supinator.** In order that the flexor muscles may act to advantage, the "rotator cuff" muscles must stabilize the head of the humerus in the glenoid cavity. The "cuff" muscles, in turn, cause movements of the scapula which must also be fixed by other muscles acting upon it. Thus, what appears to be a simple motion, such as flexion of the elbow, is supported by a complex chain reaction to stabilize the more proximal portions of the axial skeleton. The muscles of the arm are listed in Table XIII-3. Note that while all the muscles listed attach to the upper arm they move the forearm.

Of peculiar interest in this group of muscles is the **Biceps Brachii,** which as its name implies, has two heads. The **long head** arises high from the supraglenoid tuberosity. Its tendon, wrapped in a special synovial covering, crosses through the joint cavity, over the head of the humerus, emerges from the capsule and finds itself restrained in the intertubercular groove by the transverse humeral ligament. Halfway down the arm it fuses with the **short head** of the muscle which has sprung from the coracoid process of the scapula. The conjoined fleshy bellies unite upon a single tendon to insert upon the posterior part of the radial tuberosity. The tendon gives off a broad fibrous band from its medial aspect as it crosses the bend of the elbow. This is the **lacertus fibrosis** or **bicipital aponeurosis** which passes medially to become coextensive with the fascia of the flexor muscles of the forearm, holding down the brachial artery in the process. The band allows the muscle to act as a supinator as well as a flexor of the elbow joint. The **Triceps,** with its three heads, is really three muscles in one and covers the entire posterior surface of the humerus. Its long head also crosses the shoulder joint and assists the rotator cuff in holding the humeral head in place.

The Radio-ulnar Joints

The proximal articulation of the radius and ulna, allowing the radius to rotate and cross the ulna, was discussed with the elbow joint because of its intimate relationship with the elbow and the muscles crossing it. There are two more points at which these bones articulate. The **intermediate radio-ulnar joint** and the distal articulation, the **inferior radio-ulnar joint.** The mid radio-ulnar union (intermediate radio-ulnar joint), Figure XIII-33, is effected by the oblique cord, the interosseous membrane, and the posterior layer of the fascia covering the Pronator Quadratus. These membranes together fill the space between the radius and ulna almost entirely and, with the bones themselves, divide the forearm into an anterior flexor compartment and a posterior extensor compartment. The joint is a syndesmosis.

At the lower end of the two long bones of the forearm, the small disk-like head of the ulna fits into the ulnar notch on the medial side of the radius. There is a loose joint capsule to permit the movement of pronation. The end of the ulna does not participate in the wrist joint proper, but rather, it is spanned by a **triangular fibrocartilaginous disk** (Fig. XIII-41) attached to the styloid process of the ulna at its apex

Table XIII-3. Muscles of the Arm

Muscle	Nerve	Origin and Insertion	Primary Action
Biceps brachii (Fig. XIII-34)	$C_{5,6}$ Musculocutaneous nerve	(O) Long head: upper margin of glenoid cavity Short head: tip of coracoid process of scapula (I) Radial tuberosity	Flexion and supination of forearm
Brachialis (Fig. XIII-35)	$C_{5,6}$ Musculocutaneous nerve	(O) Anterior surface of lower one-half of humerus (I) Anterior surface of coronoid process of ulna	Flexion of forearm
Brachioradialis (Fig. XIII-36)	$C_{5,6}$ Radial nerve	(O) Upper two-thirds of lateral supracondylar ridge of humerus (I) Lateral side of base of styloid process of radius	Flexion of forearm
Pronator teres (Fig. XIII-37)	C_5-C_7 Median nerve	(O) Medial epicondyle of humerus and medial side of coronoid process of ulna (I) Lateral surface of radius near middle	Pronation of forearm; assists in flexion
Triceps brachii (Fig. XIII-38)	C_5-C_8 Radial nerve	(O) Long head: infraglenoid tubercle Lateral head: posterior surface of upper half of humerus Medial head: posterior surface of lower two-thirds of humerus (I) Olecranon process of ulna	Extension of forearm
Anconeus (Fig. XIII-39)	$C_{6,7}$ Radial nerve	(O) Posterior surface of lateral epicondyle of humerus (I) Lateral side of olecranon process and posterior surface of upper part of ulna	Extension of forearm (?); abducts ulna during extreme pronation

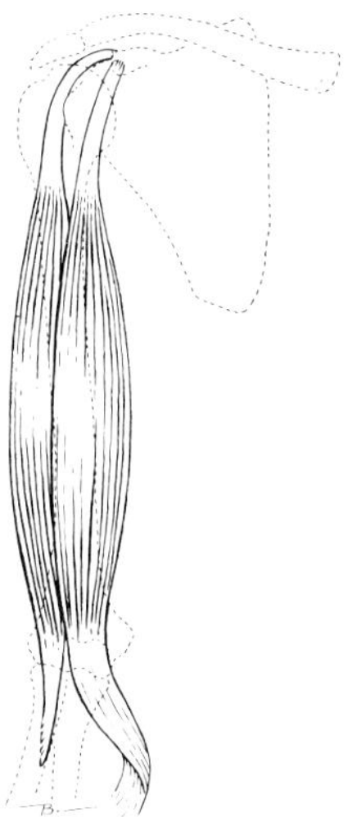

Fig. XII-34. Biceps brachii.

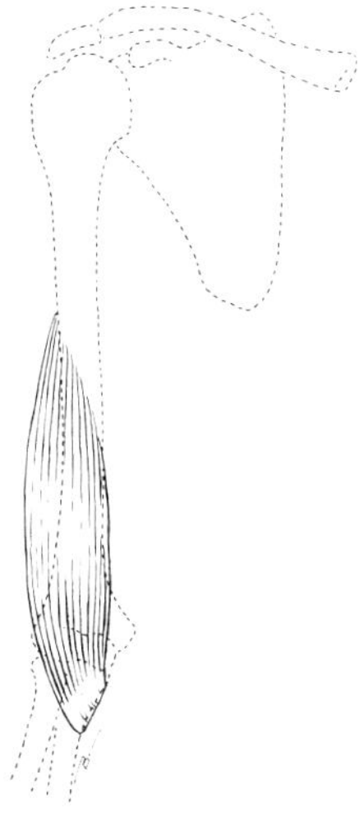

Fig. XIII-35. Brachialis.

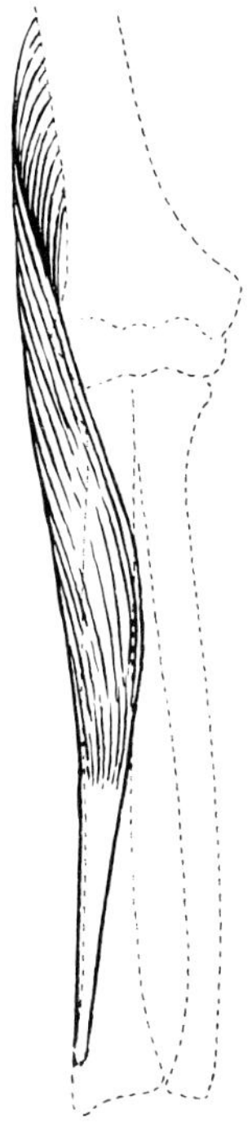

Fig. XIII-36. Brachioradialis.

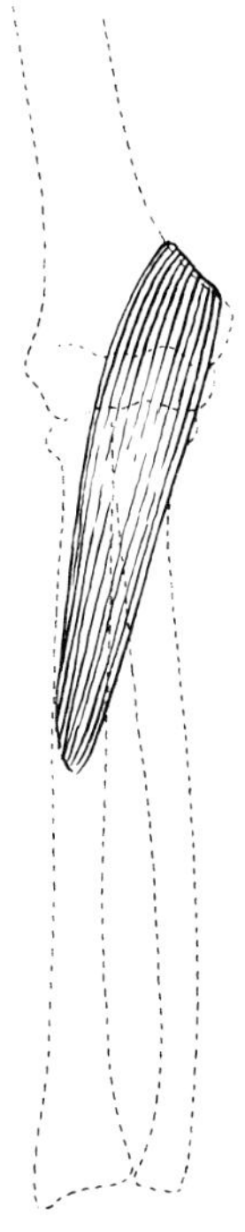

Fig. XIII-37. Pronator teres.

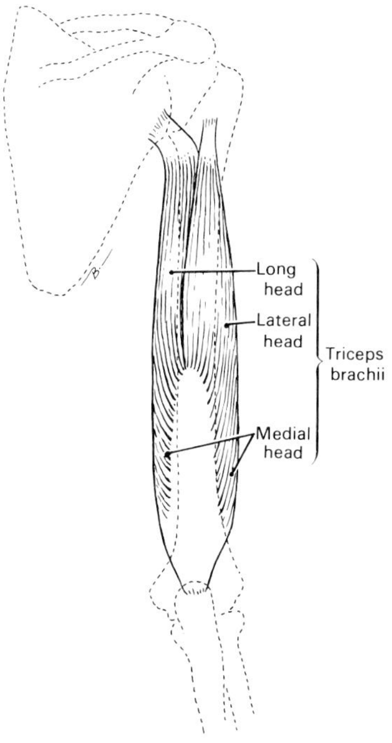

Fig. XIII-38. Posterior compartment of the upper arm: the Triceps.

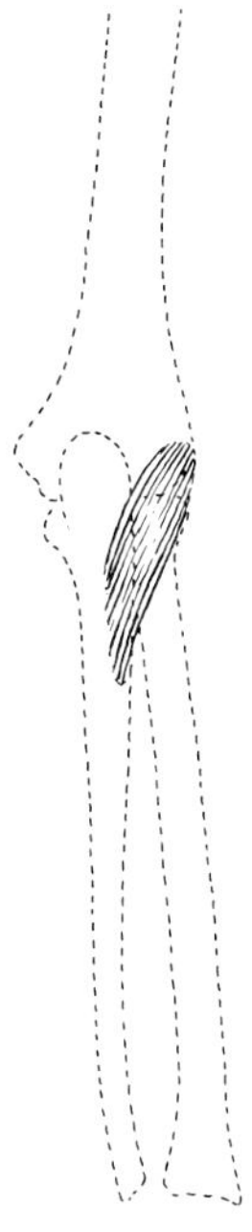

Fig. XIII-39. Anconeus.

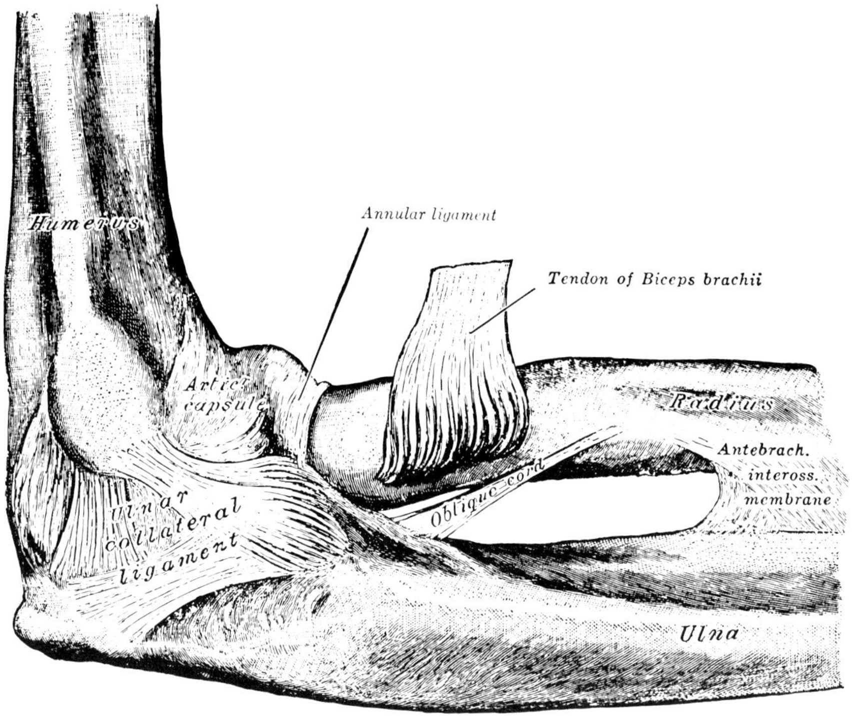

Fig. XIII-40. Medial aspect, left elbow joint. (From C. M. Goss, *Gray's Anatomy of the Human Body*, 28th ed., Lea & Febiger, Philadelphia, 1966.)

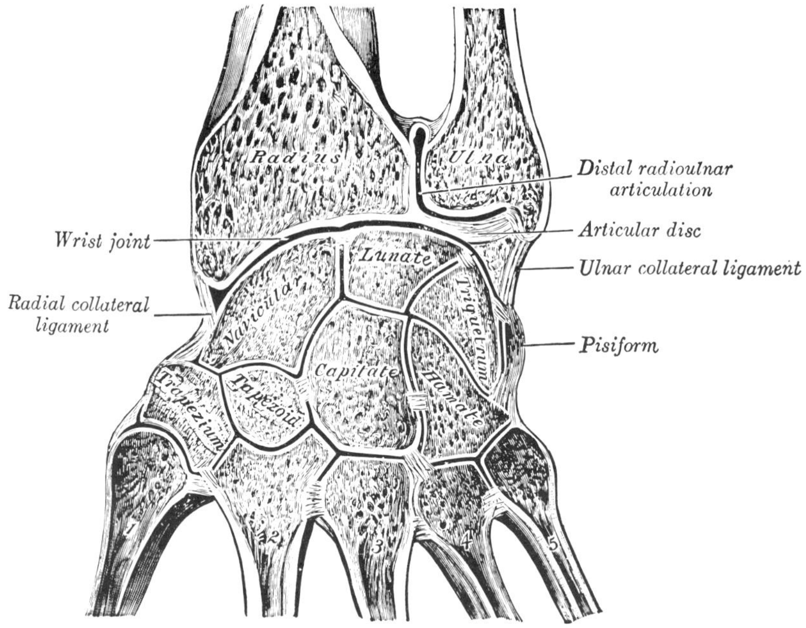

Fig. XIII-41. Vertical section through the articulations at the wrist showing the synovial joints. (From C. M. Goss, *Gray's Anatomy of the Human Body*, 28th ed., Lea & Febiger, Philadelphia, 1966.)

and to the medial margin of the lower end of the radius at its base. The **palmar** (anterior) and **dorsal** (posterior) **radio-ulnar ligaments** are lax to allow the radius to rotate on the ulna.

Structure of the Wrist or Radiocarpal Joint

The union of the slightly concave, oval-shaped surface of the radius and the articular disk spanning the lower end of the ulna with the slightly convex, oval-shaped surface of the proximal row of carpal bones forms a condyloid joint. As you may remember, a condyloid (ovoid) joint allows movement in two planes, forward and backward, and from side to side. Together these movements constitute circumduction. The **ulnar** and **radial collateral ligaments** on either side of the wrist joint descend from the styloid processes of the radius and ulna to the outer carpal bones and serve to reinforce the joint capsule. The fibers of the **anterior** (palmar) and **posterior** (dorsal) **radiocarpal ligaments** pass obliquely inferomedially from the lower end of the radius to the proximal row of carpal bones and to the capitate. This arrangement forces the carpal bones to follow the radius as a unit during pronation and supination (Fig. XIII-42A and B).

Joints of the Carpus and Hand

The articular surfaces of each of the carpals, metacarpals, and phalanges have been described earlier in this chapter. Adding to the mobility of the wrist is the gliding movement permitted between the proximal and distal rows of carpal bones, the **midcarpal joint** (ginglymus or hinge joint). The carpal bones each have their own arthrodial joints; however, they share a continuous, single joint cavity among them and their movement upon each other is slight. The carpometacarpal joints, except that of the thumb, also share a continuous joint cavity with the intermetacarpal joints and with the distal parts of the intercarpal joints of the distal row (Fig. XIII-41). The pisotriquetral joint and the very important carpometacarpal joint of the thumb each have an independent joint space. The carpometacarpal joints, II through V, have limited

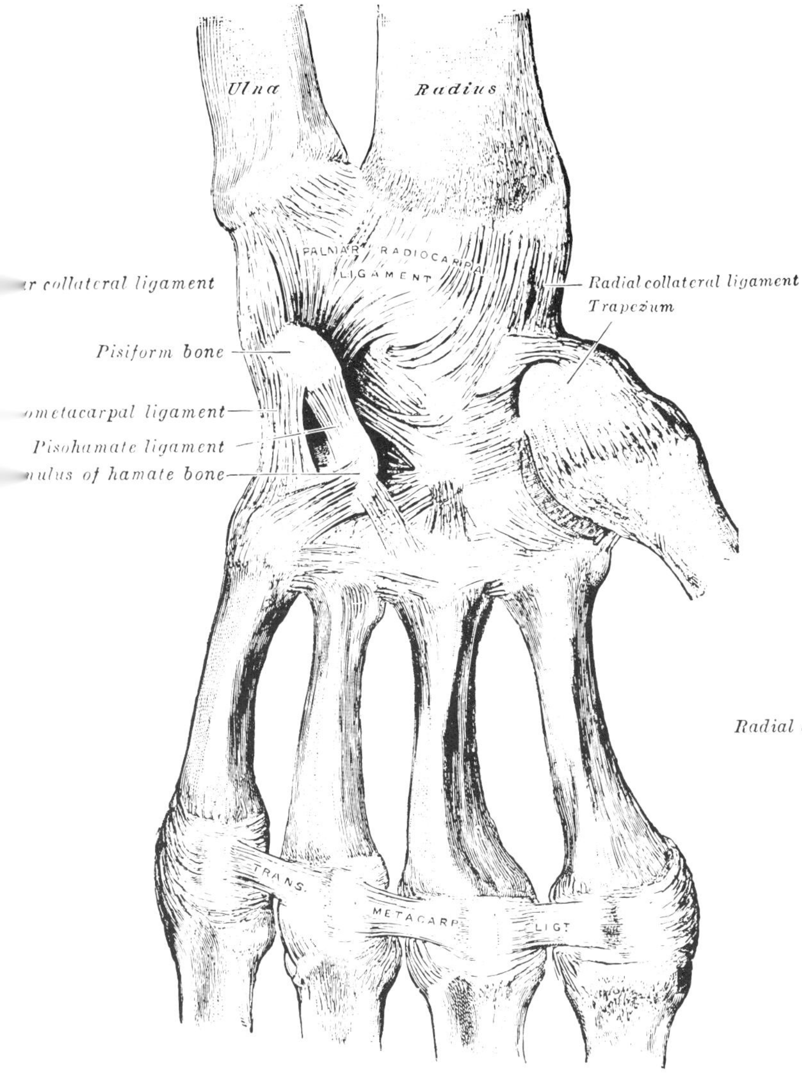

A

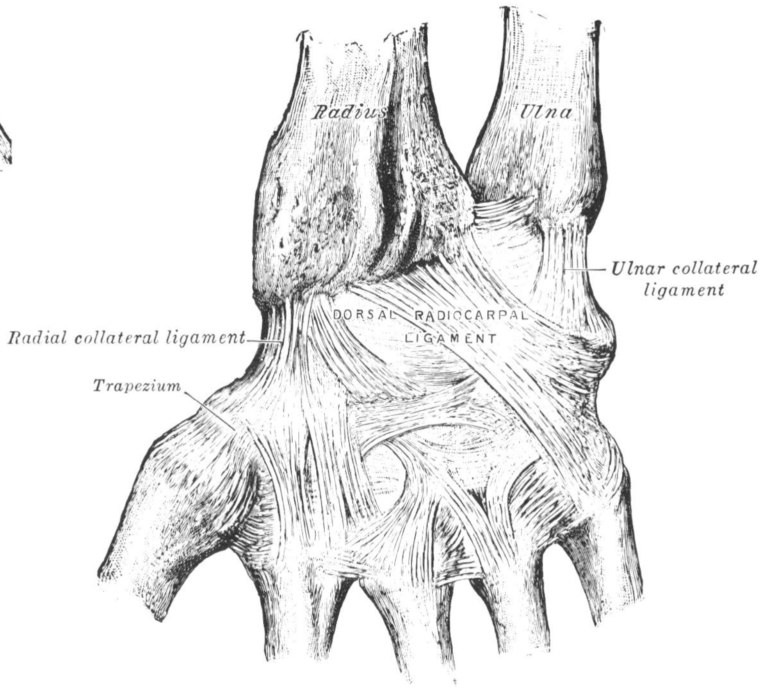

B

Fig. XIII-42A. Palmar aspect of the left wrist and metacarpus showing the ligaments. **B.** Dorsal aspect of the ligaments of the left wrist. (From C. M. Goss, *Gray's Anatomy of the Human Body*, 28th ed., Lea & Febiger, Philadelphia, 1966.)

movement, permitting a man to "cup" the palm of his hand. The first metacarpal (of the thumb) articulates with the trapezium by means of a saddle joint which allows angular movements in any plane and some rotation. The usefulness of the hand is achieved by the great mobility of the thumb joint. It should be noted that the first metacarpal is "placed" differently from the others. It is turned so that, while the fronts and backs of the other metacarpals face anteriorly and posteriorly, the corresponding surfaces of the first metacarpal face medially and laterally. Therefore, the motion that carries the thumb laterally away from the fingers is extension and the movement of the thumb forward away from the palm of the hand is abduction.

There are many ligaments binding the joints between the carpal bones and the metacarpals. The strongest of the carpal ligaments, the **radiate ligament,** is on the anterior surface of the carpus. Its tough fibers radiate from the capitate to the scaphoid, lunate, and triquetrum bones.

Metacarpophalangeal and Interphalangeal Joints: The metacarpophalangeal joints, uniting the proximal phalanx with its corresponding metacarpal for each of the four fingers, is a condyloid joint permitting flexion, extension, abduction, and adduction. The point of reference for abduction and adduction of the fingers is the middle finger. The middle finger, its metacarpal, and the capitate with which the metacarpal articulates, form the **axial line** for the movements of the hand. Abduction and adduction for the fourth, fifth, and index fingers are lateral movements away from and back toward the middle finger. One does not speak of abduction of the third finger but rather of lateral deviation (radial and ulnar deviation). The student will find, if he experiments briefly, that he cannot abduct his fingers and flex them at the same time. As flexion at the metacarpophalangeal joints progresses, the fingers come together. This is due both to the shape of the heads of the metacarpals, which are flattened in front, and to the eccentrically attached, fan-shaped **collateral ligaments** which are slack when the joint is extended and taut when it is flexed (Fig. XIII-43). The articulation can be palpated proximal to the web of the finger but distal to the metacarpal head.

There are no ligaments across the posterior aspect of the metacarpophalangeal joints. Instead, they are protected by the "extensor hood" produced by an expansion of the long extensor tendons passing along the dorsum of each finger. The collateral ligaments have already been mentioned. Of greater interest is the dense fibrocartilaginous pad on the flexor surface of each joint called the **palmar ligament** or **plate.** These plates protect the flexor tendons from friction as they cross the joint. Each plate is connected to its neighbor by deep transverse ligaments of the palm resulting in a continuous fibrous band stretching from one side of the palm to the other and holding the metacarpal heads together (Fig. XIII-43A).

Once again, the thumb stands alone. Although it has great freedom of movement in the carpometacarpal

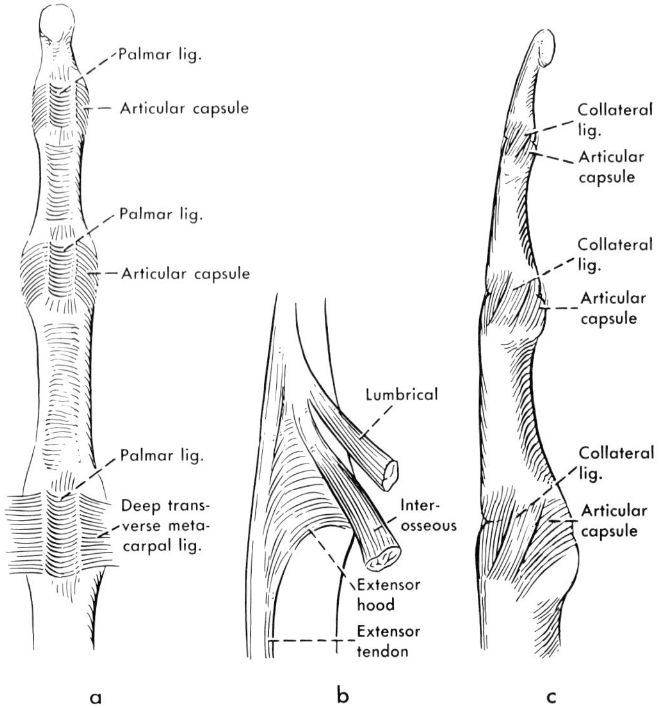

Fig. XIII-43. Metacarpophalangeal and interphalangeal joints. **A.** Palmar view. **B.** Lateral view of the metacarpophalangeal joint with the extensor tendon in place. **C.** Lateral aspect of a digit. (From H. Hollinshead, *Textbook of Anatomy*, 2nd ed., Harper & Row, New York, 1967.)

joint, its metacarpophalangeal joint is more limited than those of the fingers. It is more like an interphalangeal joint, allowing only flexion and extension and some slight rotation. It possesses a palmar ligament, but it is not connected to the neighboring index finger.

The nine interphalangeal joints are hinge joints; hence, their only movements are flexion and extension. Each joint possesses a capsule which is protected dorsally by expansions of the extensor tendons, laterally by collateral ligaments, and anteriorly by small palmar plates (ligaments) (Fig. XIII-43). These palmar ligaments prevent hyperextension of the interphalangeal joints. If one of them should be ruptured, the associated phalanx may become locked in hyperextension.

Muscles of the Forearm and Hand

The muscles of the forearm have many talents, acting upon the elbow, wrist, and hand. The intrinsic muscles of the hand, however, have only limited opportunities and act upon the digits alone. The forearm muscles form fleshy masses below the medial and lateral epicondyles of the humerus where they have their proximal attachments. They taper off rapidly as they approach the wide base of the radius and continue into the hand as long, strong tendons. The flexor tendons all cross the wrist joint on the anterior side of the forearm, most of them passing under the **flexor retinaculum (transverse carpal ligament).** The retinaculum is a strong broad ligament attached to the marginal bones of the carpus. It helps to maintain the anterior concavity of the carpus while forming a kind of bridge. The space under the bridge has been called the **carpal tunnel,** lined with synovial tissue and divided into three compartments by septa for the passage of the flexor tendons (Fig. XIII-44A).

There are two masses of muscles in the palm of the hand, the **thenar eminence** at the base of the thumb and the **hypothenar eminence** at the base of the little finger. Deep to these masses, behind the long tendons and associated with the metacarpals, are the other muscles of the palm.

The forearm muscles can be readily divided into two groups. The **flexor muscles** arise from the medial epicondyle and occupy the medial border and anterior surface of the forearm. The **extensors** attach proximally upon the lateral epicondyle and occupy the lateral border and posterior surface of the forearm. The posterior border of the ulna, from olecranon to wrist, is the divider between the flexors and the extensors. The hollow in front of the elbow between the lateral and medial muscle masses is the cubital fossa.

The flexor-pronator muscle group is listed in Table XIII-4. The six muscles in this group are disposed in three layers. In general, the deeper the layer, the more distal the attachments of the muscles. The **superficial flexors** are: the **Flexor Carpi Radialis, Palmaris Longus,** and **Flexor Carpi Ulnaris.** Of these, the Palmaris Longus is of passing interest. It is apparently disappearing and many people may be missing it in one or both arms. When present, the tendon, which lies superficial to the flexor retinaculum, can be seen standing out prominently in the midline while the wrist is flexed against resistance. The **intermediate layer** claims one muscle, the **Flexor Digitorum Superficialis** (Flexor Digitorum Sublimus), and the deep layer has two, the **Flexor Digitorum Profundus** and the **Flexor Pollicis Longus.** Also present in the anterior compartment of the forearm are the **Pronator Teres** (Table XIII-3) and the **Pronator Quadratus** (Table XIII-4).

The extensor-supinator muscle group is summarized in Table XIII-5. There are only two layers of these muscles, the **superficial** and the **deep.** Of these, the

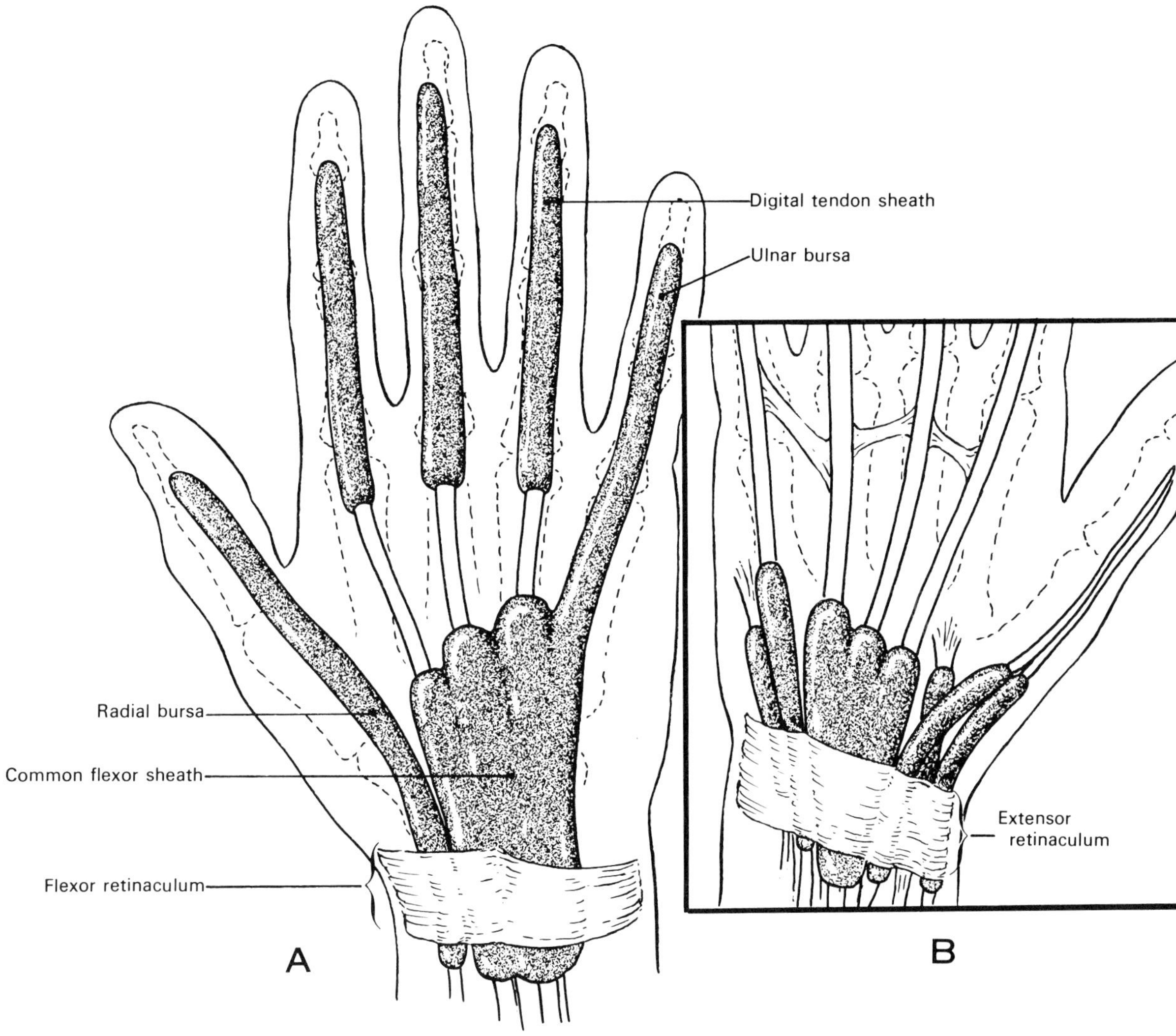

Fig. XIII-44. Retinaculi and tendon sheaths. **A.** Flexor retinaculum. **B.** Extensor retinaculum.

Table XIII-4. Flexor-Pronator Muscles of the Forearm

Muscle	Nerve	Origin and Insertion	Primary Action
Flexor carpi radialis (Fig. XIII-45)	$C_{6,\ 7}$ Median nerve	(O) Medial epicondyle of humerus (I) Anterior surface of base of second metacarpal	Flexion at wrist
Palmaris longus (Fig. XIII-46)	$C_{7,\ 8}$ Median nerve	(O) Medial epicondyle of humerus (I) Transverse carpal ligament and palmar aponeurosis	Flexion at wrist
Flexor carpi ulnaris (Fig. XIII-47)	C_8, T_1 Ulnar nerve	(O) By two heads, from medial epicondyle of humerus and medial border of olecranon process of ulna (I) Palmar surface of pisiform and hamate, and base of fifth metacarpal	Flexion and adduction at wrist
Flexor digitorum superficialis (Fig. XIII-48)	C_7-T_1 Median nerve	(O) Humeroulnar head: medial epicondyle of humerus, ulnar collateral ligament, medial margin of coronoid process Radial head: oblique line on anterior surface of shaft of radius (I) By four tendons to the four fingers (each tendon splits and attaches to the sides of the base of the middle phalanx)	Flexion of middle phalanges of fingers
Flexor pollicis longus (Fig. XIII-49)	C_7-T_1 Median nerve	(O) Anterior surface of midshaft of radius (I) Anterior surface of base of distal phalanx of thumb	Flexion of distal phalanx of thumb
Flexor digitorum profundus (Fig. XIII-50)	C_8, T_1 Radial part of muscle supplied by median nerve, ulnar part by ulnar nerve	(O) Upper two-thirds of anterior and medial surfaces of ulna (I) By four tendons to the four fingers (each tendon passes through the corresponding tendon of the Flexor digitorum superficialis and attaches to the base of the distal phalanx)	Flexion of distal phalanges of fingers
Pronator teres (Fig. XIII-51)	C_5-C_7 Median nerve	(O) Medial epicondyle of humerus and medial side of coronoid process of ulna (I) Lateral surface of radius near middle	Pronation of forearm
Pronator quadratus (Fig. XIII-51)	C_7-T_1 Median nerve	(O) Anterior surface of lower one-fourth of ulna (medial attachment) (I) Anterior surface of lower one-fourth of radius (lateral attachment)	Pronation of forearm

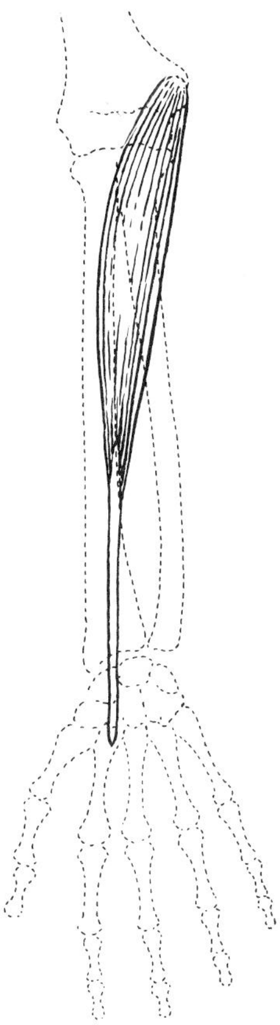

Fig. XIII-45.

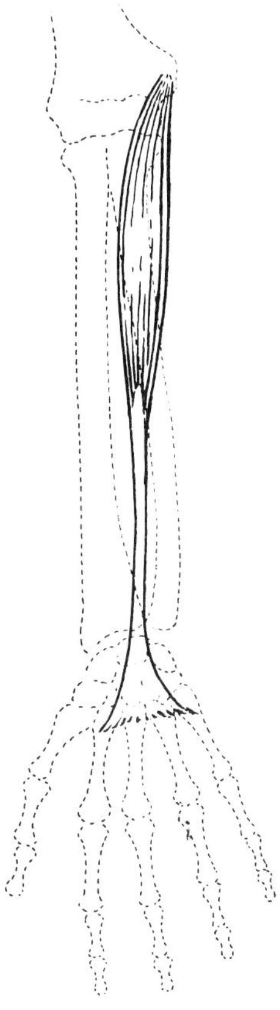

Fig. XIII-46.

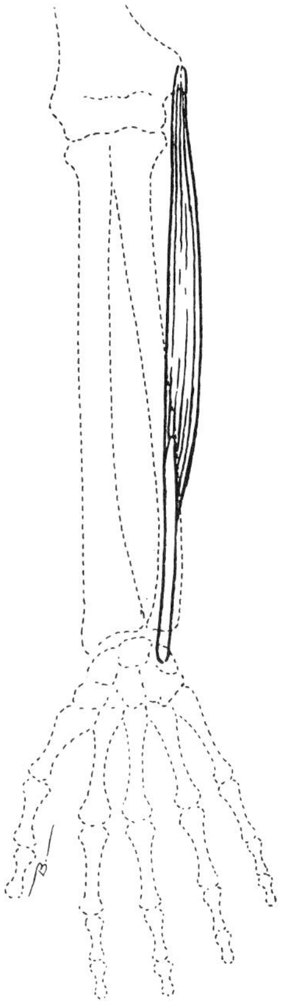

Fig. XIII-47.

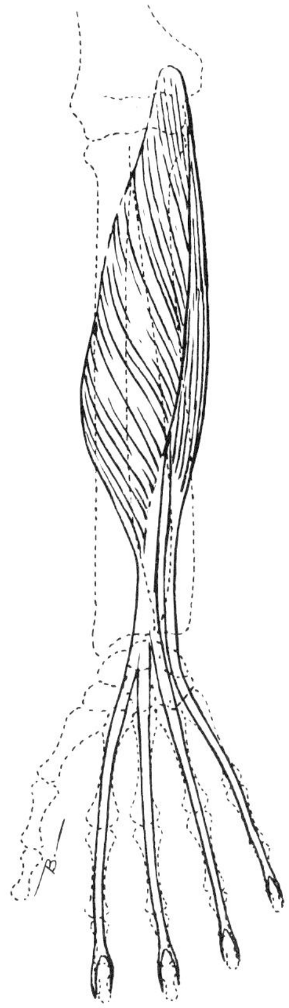

Fig. XIII-48.

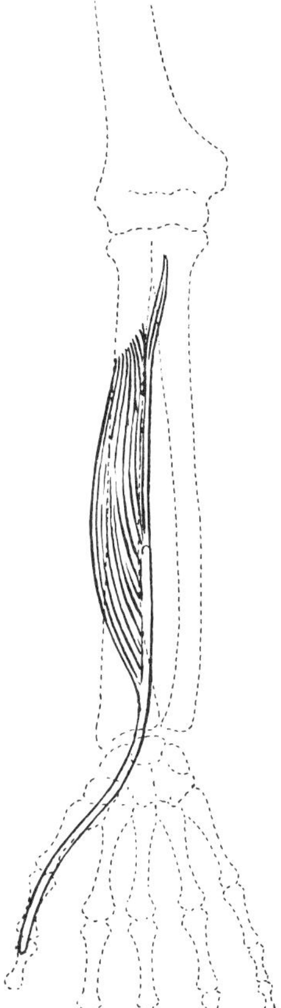

Fig. XIII-49.

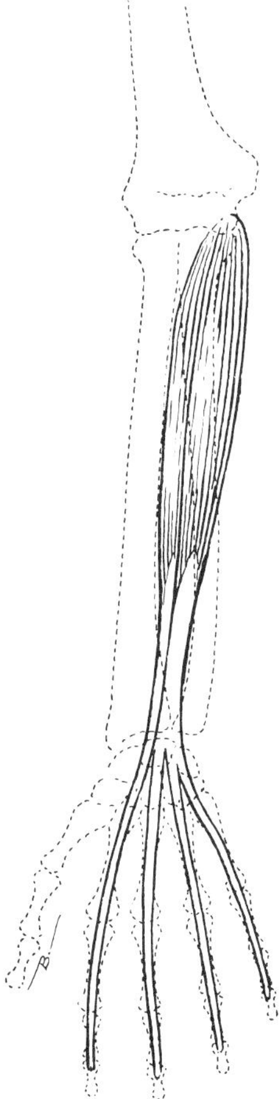

Fig. XIII-50.

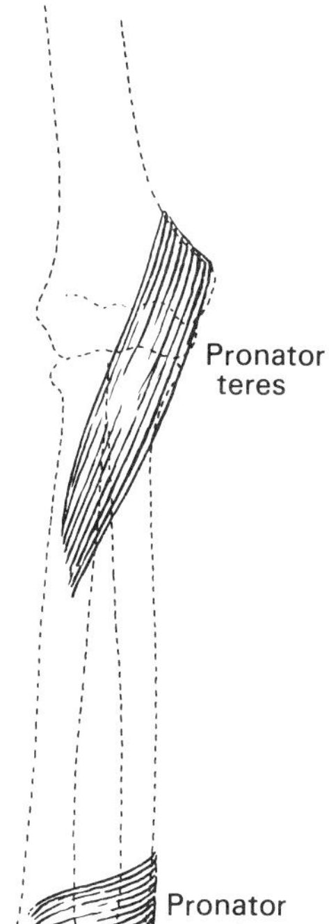

Fig. XIII-51.

Muscles of the flexor-pronator compartment of the forearm:

Fig. XIII-45. Flexor carpi radialis.
Fig. XIII-46. Palmaris longus.
Fig. XIII-47. Flexor carpi ulnaris.
Fig. XIII-48. Flexor digitorum superficialis.
Fig. XIII-49. Flexor pollicis longus.
Fig. XIII-50. Flexor digitorum profundus.
Fig. XIII-51. Pronators of the forearm.

Table XIII-5. Forearm Muscles of the Extensor-Supinator Group

Muscle	Nerve	Origin and Insertion	Primary Action
Extensor carpi radialis longus (Fig. XIII-52)	$C_{6, 7}$ Radial nerve	(O) Lateral epicondyle of humerus and supracondyloid ridge above it (I) Posterior surface of base of second metacarpal	Extension and abduction at wrist
Extensor carpi radialis brevis (Fig. XIII-53)	$C_{6, 7}$ Radial nerve	(O) Lateral epicondyle of humerus (I) Posterior surface of third metacarpal	Extension and abduction at wrist
Extensor carpi ulnaris (Fig. XIII-54)	C_6-C_8 Radial nerve	(O) By two heads, from lateral epicondyle of humerus and middle third of posterior ridge of ulna (I) Posterior surface of base of fifth metacarpal	Extension and adduction at wrist
Supinator (Fig. XIII-55)	$C_{5, 6}$ Radial nerve	(O) Lateral epicondyle of humerus, adjacent part of ulna, radial collateral and annular ligaments (I) Lateral surface of upper one-third of radius	Supination of forearm
Extensor digitorum (Fig. XIII-56)	C_6-C_8 Radial nerve	(O) Lateral epicondyle of humerus (I) By four tendons to the four fingers (each tendon divides into three slips; the middle slip inserts on the dorsal surface of the second phalanx; the other two reunite and attach to the dorsal surface of the base of the distal phalanx (see Fig. XIII-62))	Extension of all joints of digits II to V
Extensor digiti minimi (Fig. XIII-57)	C_6-C_8 Radial nerve	(O) Proximal tendon of Extensor digitorum (I) Unites with little finger tendon of Extensor digitorum	Extension of all joints of digit V
Extensor indicis (Fig. XIII-58)	$C_{7, 8}$ Radial nerve	(O) Dorsal surface of lower half of ulna (I) Unites with index finger tendon of Extensor digitorum	Extension of all joints of digit II
Extensor pollicis longus (Fig. XIII-59)	$C_{7, 8}$ Radial nerve	(O) Dorsal surface of middle one-third of ulna (I) Dorsal surface of base of distal phalanx of thumb	Extension of both phalanges of thumb; extension and adduction of metacarpal of thumb
Extensor pollicis brevis (Fig. XIII-60)	$C_{6, 7}$ Radial nerve	(O) Dorsal surface of radius below Abductor pollicis longus (I) Dorsal surface of base of first phalanx	Extension of proximal phalanx and metacarpal of thumb; abduction at wrist
Abductor pollicis longus (Fig. XIII-61)	$C_{6, 7}$ Radial nerve	(O) Dorsolateral surface of ulna below Anconeus, dorsal surface of radius near middle, interosseus membrane (I) Lateral surface of base of first metacarpal	Extension and abduction of thumb; flexion and abduction at wrist

The following muscles of the forearm extensor-supinator group are seen on the opposite page:

Fig. XIII-52. Extensor carpi radialis longus.
Fig. XIII-53. Extensor carpi radialis brevis.
Fig. XIII-54. Extensor carpi ulnaris.
Fig. XIII-55. Supinator.
Fig. XIII-56. Extensor digitorum.
Fig. XIII-57. Extensor digiti minimi.
Fig. XIII-58. Extensor indicis.
Fig. XIII-59. Extensor pollicis longus.
Fig. XIII-60. Extensor pollicis brevis.
Fig. XIII-61. Abductor pollicis longus.

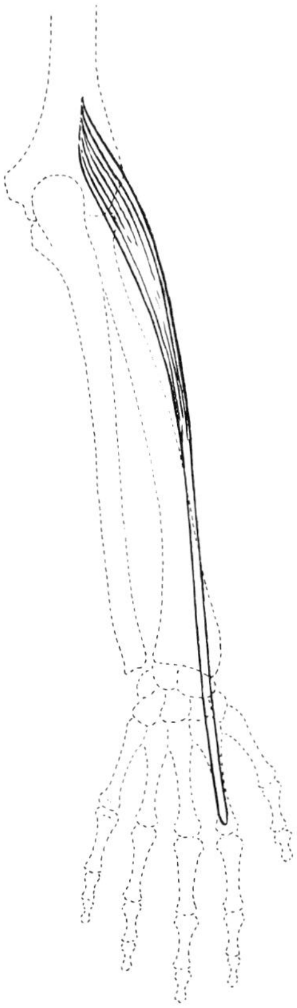

Fig. XIII-52.

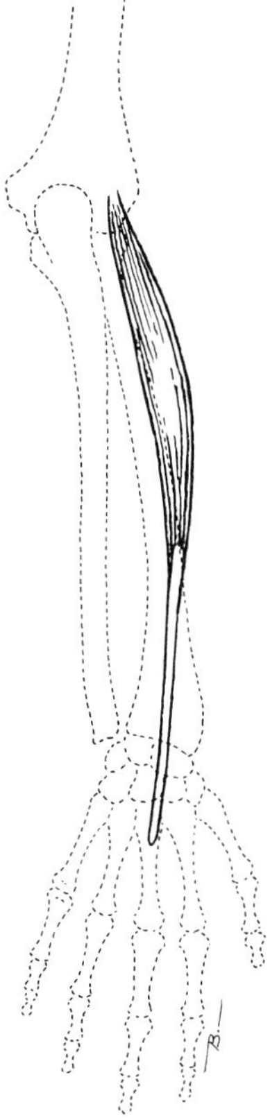

Fig. XIII-53.

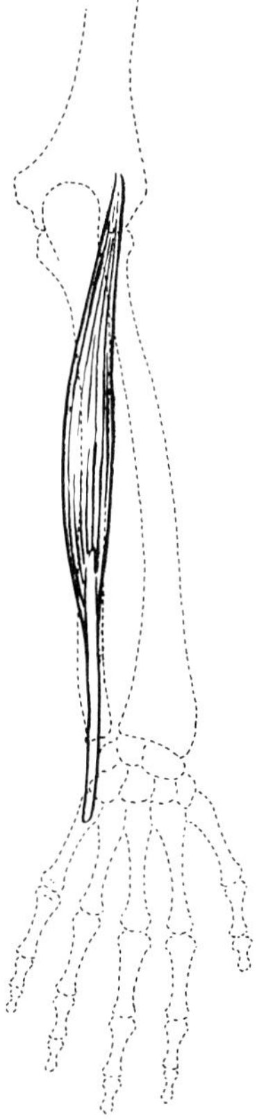

Fig. XIII-54.

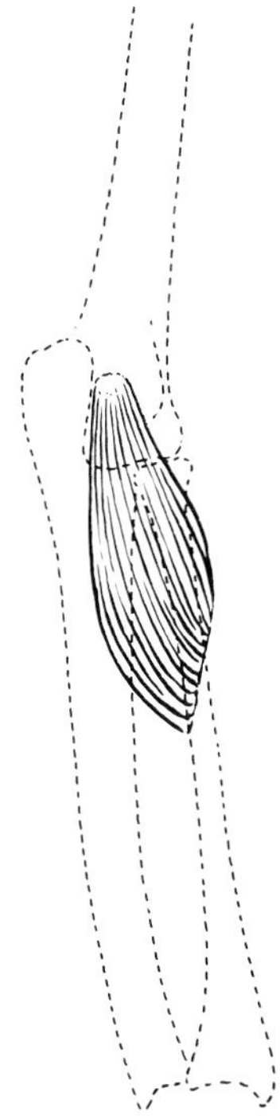

Fig. XIII-55.

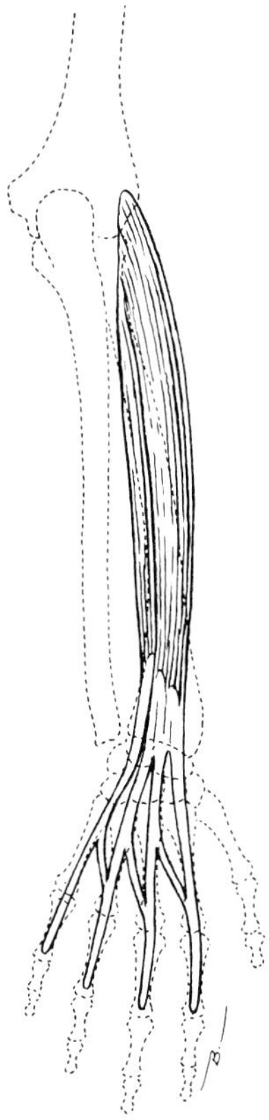

Fig. XIII-56.

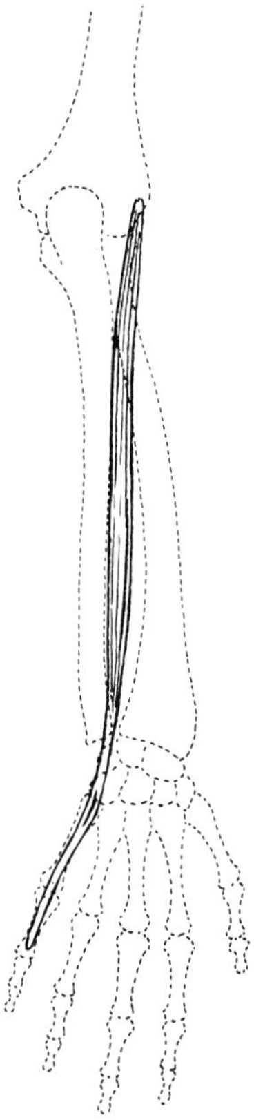

Fig. XIII-57.

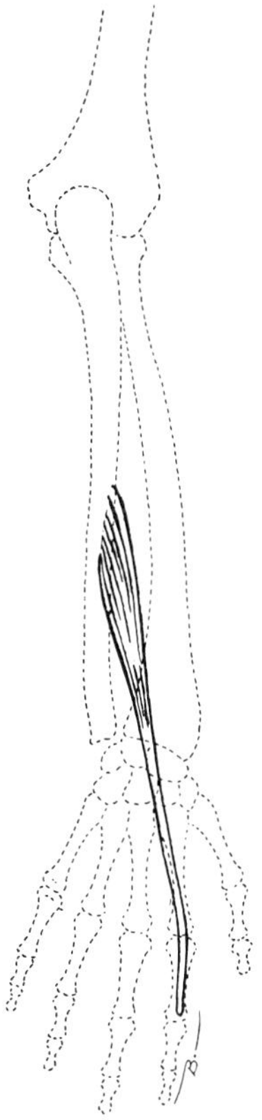

Fig. XIII-58.

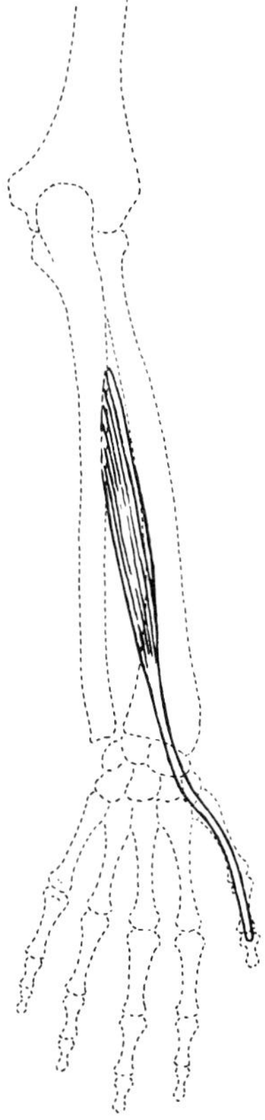

Fig. XIII-59.

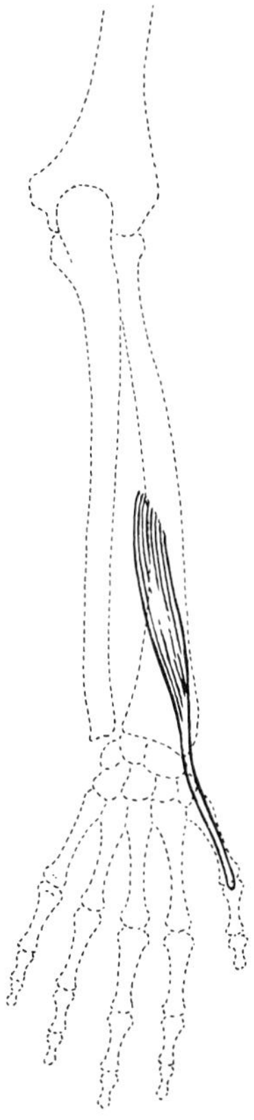

Fig. XIII-60.

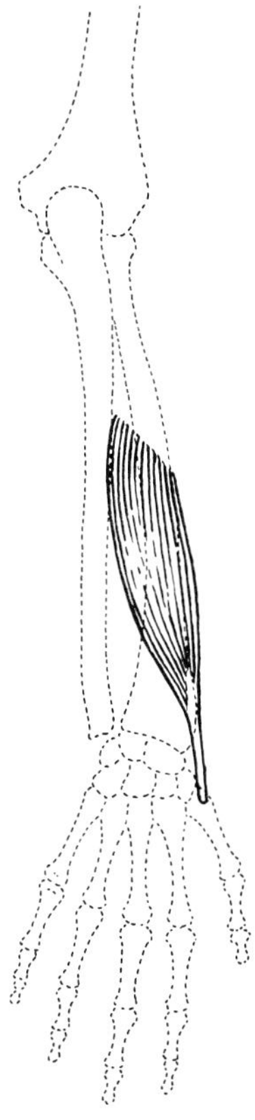

Fig. XIII-61.

superficial muscles may be further subdivided into **lateral** and **posterior groups** by the masses of the three deep thumb muscles that also become superficial about halfway down the forearm. All of the superficial muscles originate from a broad **common extensor tendon** which is attached to the anterior portion of the lateral epicondyle of the humerus, the investing deep fascia, and intermuscular septa. The lateral superficial extensors reach their origins as far up the lateral supracondylar ridge as the spiral groove while the posterior superficial muscles confine their origins entirely to the common extensor tendon and the slightly more distal intermuscular septa and deep fascia of the adjacent muscles. The common extensor tendon passes behind the humeroradial joint and the head of the radius. The superficial extensors are: **Extensor Carpi Radialis Longus, Extensor Carpi Radialis Brevis** (the lateral group); and the **Extensor Digitorum** (Communis), **Extensor Digiti Minimi,** and **Extensor Carpi Ulnaris** (the posterior group). The **deep extensors** surface between the Extensor Carpi Radialis Brevis and the Extensor Digitorum. These are the **Extensor Pollicis Longus, Extensor Pollicis Brevis, Abductor Pollicis Longus,** and the **Extensor Indicis.** Their blended proximal attachments are from the posterior surface of the interosseous membrane and the adjacent radius and ulna below the level of the **Supinator.** Although the Supinator is not an extensor muscle, it will appear in Table XIII-5 because it is found in the extensor compartment of the forearm. Recall that the Biceps Brachii is also an important supinator even though it is in the flexor compartment of the upper arm.

Just as there is a flexor retinaculum bridging most of the flexor tendons on the anterior surface of the wrist, there is an **extensor retinaculum,** a transverse band of fibrous tissue holding the extensor tendons in place in their bony grooves on the posterior and lateral surfaces of the wrist. There are a series of partitions extending from the deep surface of the extensor retinaculum to the edges of the bony grooves on the radius to form six compartments (tunnels) through which the extensor tendons pass (Fig. XIII-44). These tunnels are lined with synovial sheaths. The retinaculum has the further use of binding the carpal bones together so that they move as a unit to follow the radius during pronation.

Length of the Long Digital Tendons: Before leaving the muscles of the forearm to discuss the small muscles of the hand, a comment should be made on the flexion and extension of the wrist and fingers. A characteristic of all muscles crossing two or more joints is that they are not long enough to permit a full range of motion in more than one joint at a time. This is true of the hamstrings which flex the leg at the knee and extend the thigh at the hip. It is true of the flexors and extensors of the wrist and fingers where the muscles move all the joints in the same direction, *i.e.*, a flexor causes flexion at all the joints it crosses rather than opposite movements as in the leg. Try to make a fist and simultaneously flex the wrist as far as it will go. You will find that either you must extend your fingers somewhat (relax the fist) or fail to reach complete flexion of the wrist. Feel the pull in the long extensor muscles. The Extensor Digitorum will not stretch across the flexed joints sufficiently to permit full flexion at each joint. Similarly, it is impossible to achieve complete extension of the fingers and hyperextension at the wrist because the Flexor Digitorum Superficialis and Profundus will not stretch enough to allow it. In short, strong finger action requires a rigid wrist in a near neutral position. Strong wrist action requires relaxed fingers.

Intrinsic Muscles of the Hand (See Table XIII-6)

There are 18 small muscles intrinsic to the hand. These group conveniently into the **thenar muscles,** the mass at the base of the thumb, the **hypothenar muscles,** massed at the ball of the little finger, and, in addition, the **Lumbricals,** the **Adductor Pollicis,** and the **Interossei.** The thenar muscles are: the **Abductor Pollicis Brevis, Opponens Pollicis,** and **Flexor Pollicis Brevis.** Correspondingly, the muscles of the hypothenar eminence are: the **Abductor Digiti Minimi, Opponens Digiti Minimi,** and the **Flexor Digiti Minimi Brevis.** See Table XIII-6 for the summary of these muscles.

The four *Lumbricals* are, by virtue of their tendinous insertions, flexors of the metacarpophalangeal joints and extensors of the interphalangeal joints. It is due to the Lumbricals that you can curl your hand into a "writing" position. The tendons of insertion attach to the radial side of the tendons of the Extensor Digitorum (Fig. XIII-62). From there they pass backwards to join the expansion of the extensor tendon causing them to act as extensors of the interphalangeal joints.

The *Interossei* are the abductors and adductors of the fingers. Since the thumb has its own abductor and adductor and the fifth finger has its own abductor, only seven more muscles are required. Four of these seven are abductors (one for the index finger, two for the middle finger to move it either way from the axial line, and one for the fourth finger) and these are known as the **Dorsal Interossei** because they are best seen there filling the spaces between the metacarpals. The remaining three are adductors (one for the index finger, one for the fourth finger, and one for the fifth finger). These are called the **Palmar Interossei.** Besides acting primarily as adductors and abductors of the fingers, these muscles have the same action as the Lumbricals due to the union of their tendons with the extensor tendon expansions on the backs of the fingers.

The **Palmaris Brevis** is a superficial sheet of muscle that passes from the flexor retinaculum to the skin at the medial border of the hand. Like the Palmaris Longus, if it is missing, it is not missed by the owner.

INNERVATION AND BLOOD SUPPLY OF THE UPPER LIMB

In the tables of muscles we have noted the common segmental origin of the nerves and named the major peripheral nerves. All motor and sensory nerves of the shoulder and arm muscles are supplied by the *brachial*

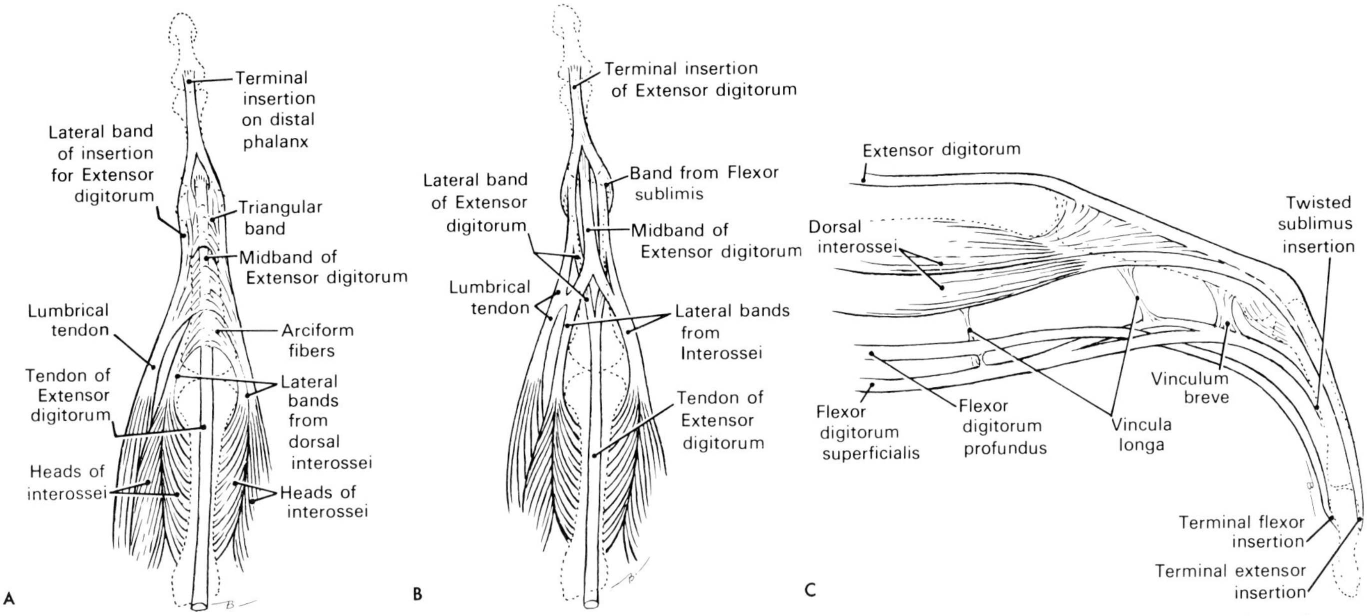

Fig. XIII-62A. Extensor tendon apparatus covered by triangular band and arciform fibers. Note how tendinous portions of dorsal interossei and the lumbrical cooperate with the common extensor tendon to form the apparatus. Anchorage is on distal and middle interphalangeal joints and on the metacarpophalangeal joint. **B.** The same extensor tendon apparatus seen at **A** but with fibrous sheath removed. Note now how the band of the Flexor digitorum superficialis also twists upward on the sides to join in the extensor apparatus. **C.** A lateral view of a finger to show relationship of flexor and extensor tendons as they near their termination. Note tendinous bands or vincula which connect the flexor tendons to the phalanges.

plexus. Traditionally, students have had to memorize this plexus and be able to draw it. We suggest you do this to be safe only under emergency situations: just before state and national board examinations and prior to the appropriate examinations in medical school if you realize that your particular instructor is a hard-nose. Otherwise, as far as the upper limb is concerned, the important point to grasp is the specific roles of the five major nerves which emanate from the plexus to enter the arm and forearm. The plexus gathers fibers from five roots, C_5 to T_1, and delivers them—after reorganizing into three trunks, splitting into six divisions, reuniting into three cords, and splitting again into five major peripheral nerves—to the upper limb. It combines the spinal nerves into three divisions: medial, lateral, and posterior. The **medial** and **lateral divisions** form three nerves, the **musculocutaneous,** the **median,** and the **ulnar** which supply the flexor and pronator muscles of the anterior aspect of the limb. The **posterior division** gives off the **radial nerve** to supply the extensor and supinator muscles on the dorsal aspect of the limb. The posterior division also provides the **axillary nerve** to the Deltoid. Refer to Chapter III for discussion and illustrations of the nerve supply of the upper limb.

The most important job of the peripheral arteries is to deliver blood to the muscles. As the main arteries pass along the length of the limb, they give off numerous branches and twigs to the individual muscles, most of which are simply referred to as "muscular branches." Anastomoses between muscular branches and collateral branches are frequent and serve to insure a constant supply of blood to distal segments regardless of "kinking" or excessive pressure on the vessels during movement of the joints. The general circulatory plan to the arm is simple. All the named arteries run deep and are accompanied by veins of the same name to return blood to the heart. There are superficial veins but no comparable superficial arteries. Refer to Chapter III for a more complete discussion of the superficial veins.

The **subclavian artery** lies behind the clavicle and gives off branches to the thoracic wall, neck, and shoulders before becoming the **axillary artery.** The latter also gives branches to the shoulder, clavicular region, and thoracic region before entering the arm as the **brachial artery.** The brachial artery gives off a deep branch that follows the radial nerve behind the shaft of the humerus. It then bifurcates to form terminal branches, a lateral **radial artery** and a medial **ulnar artery,** which pass along their respective bones. Both give **superior** and **inferior collateral branches** that anastomose around the elbow joint. The ulnar artery sends a **deep branch** to pierce the intermuscular connective tissue septum between the bones. This branch forks to run on the anterior and posterior aspects of the septum, supplying the flexor and extensor muscle groups of the forearm. **Deep** and **superficial palmar arches** are formed between the radial and ulnar arteries at the styloid processes of each bone. Digital branches to the fingers leave these arches. The deep veins follow the same route back up the arm to empty into the subclavian veins. For the details of the circulation and lymphatic drainage of the upper limb, refer to Chapter XI, page 733.

Table XIII-6. Intrinsic Muscles of the Hand

Muscle	Nerve	Origin and Insertion	Primary Action
Abductor pollicis brevis (Fig. XIII-63)	C_5-T_1 Median nerve	(O) Anterior surface of flexor retinaculum, trapezoid and scaphoid bones (I) Radial side of base of first phalanx of thumb	Abduction of thumb
Flexor pollicis brevis (Fig. XIII-64)	C_5-T_1 Median nerve to superficial head C_8-T_1 Ulnar nerve to deep head	(O) Superficial head: trapezoid and adjacent part of transverse carpal ligament (flexor retinaculum) Deep head: ulnar side of first metacarpal (I) Superficial head: radial side of base of first phalanx of thumb Deep head: ulnar side of base of first phalanx of thumb	Flexion of metacarpophalangeal joint of thumb
Opponens pollicis (Fig. XIII-65)	C_5-T_1 Median nerve	(O) Anterior surface of trapezoid and transverse carpal ligament (I) Entire radial border of anterior surface of first metacarpal	Opposition of thumb
Abductor digiti minimi (Fig. XIII-66)	C_8, T_1 Ulnar nerve	(O) Pisiform bone and tendon of Flexor carpi ulnaris (I) Ulnar side of base of first phalanx of little finger and ulnar border of aponeurosis of Extensor digiti minimi	Abduction of fifth digit
Flexor digiti minimi brevis (Fig. XIII-66)	C_8, T_1 Ulnar nerve	(O) Hook of hamate and adjacent transverse carpal ligament (I) Ulnar side of base of first phalanx of little finger	Flexion at metacarpophalangeal joint of fifth digit
Opponens digiti minimi (Fig. XIII-67)	C_8, T_1 Ulnar nerve	(O) Hook of hamate and adjacent transverse carpal ligament (I) Entire ulnar border of fifth metacarpal bone	Opposition of fifth metacarpal, "cupping" of hand
Lumbricals I and II III and IV (Fig. XIII-68)	C_5-T_1 Median nerve C_8, T_1 Ulnar nerve	(O) Tendons of the Flexor digitorum profundus in the middle of the palm (I) Radial side of the tendons of the Extensor digitorum	Flexion at metacarpophalangeal joints, extension at interphalangeal joints of digits II, III, IV, and V
Adductor pollicis (Fig. XIII-69)	C_8, T_1 Ulnar nerve	(O) Entire shaft of third metacarpal bone & carpals adjacent to its base (I) Ulnar side of base of proximal phalanx of thumb	Adduction of metacarpal and flexion at metacarpophalangeal joint of thumb
Palmar interossei (3) (Fig. XIII-70)	C_8, T_1 Ulnar nerve	(O) 1st: Ulnar side of second metacarpal 2nd: Radial side of fourth metacarpal 3rd: Radial side of fifth metacarpal (I) 1st: Ulnar side of base of first phalanx of index finger and expansion of Extensor digitorum tendon 2nd: Radial side of base of first phalanx of fourth finger and Extensor expansion 3rd: Radial side of base of first phalanx of fifth finger and Extensor expansion	Adduction of digits II, IV, and V; flexion at their metacarpophalangeal joints; extension at their interphalangeal joints
Dorsal interossei (4) (Fig. XIII-71)	C_8, T_1 Ulnar nerve	(O) Opposite surfaces of adjoining metacarpals, filling the interspaces. (I) Base of proximal phalanx and aponeurosis of extensor muscles on the thumb side of the index finger, the ulnar side of the ring finger, and each side of the middle finger	Abduction of digits II, III, and IV; extension of interphalangeal joints of digits III and IV
Palmaris brevis	C_8, T_1 Ulnar nerve	(O) Palmar aponeurosis (I) Skin of the ulnar border of the hand	Wrinkling of skin of hypothenar eminence

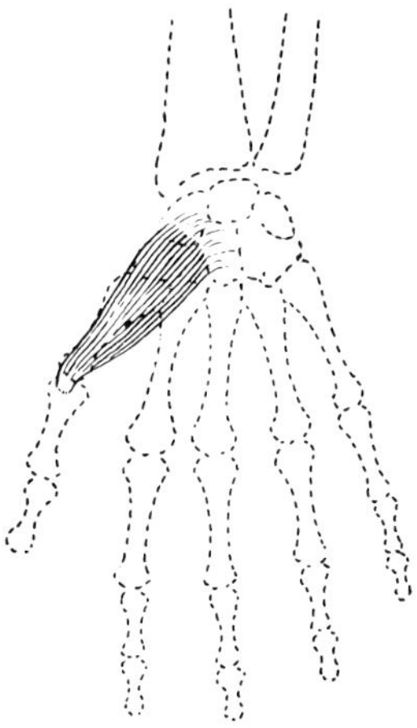

Fig. XIII-63.

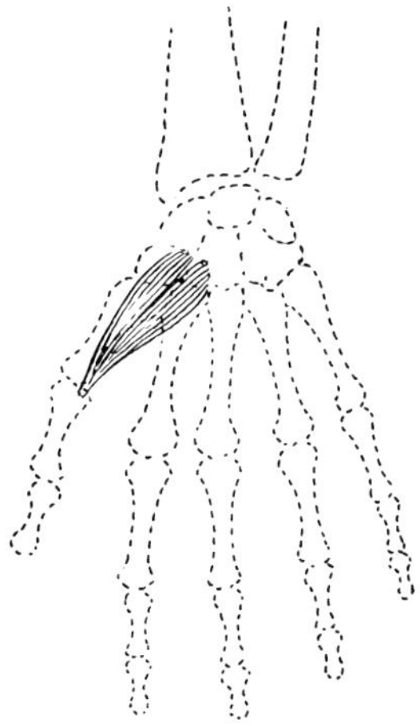

Fig. XIII-64.

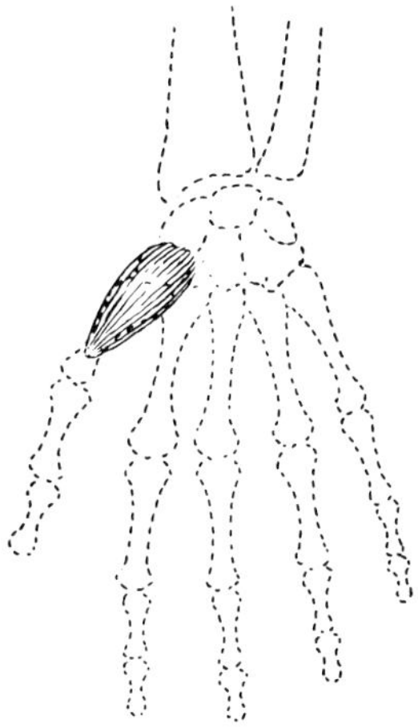

Fig. XIII-65.

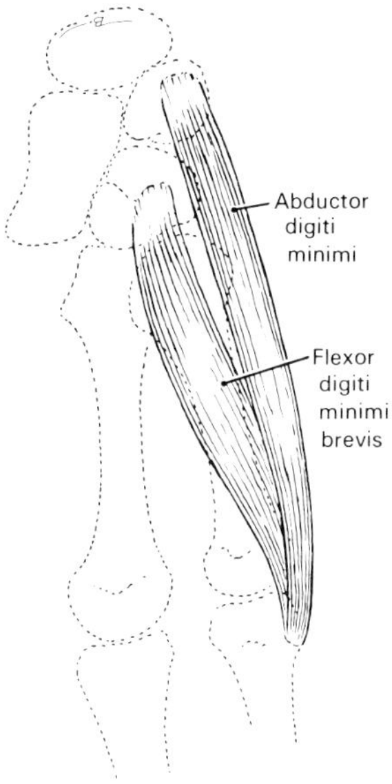

Fig. XIII-66.

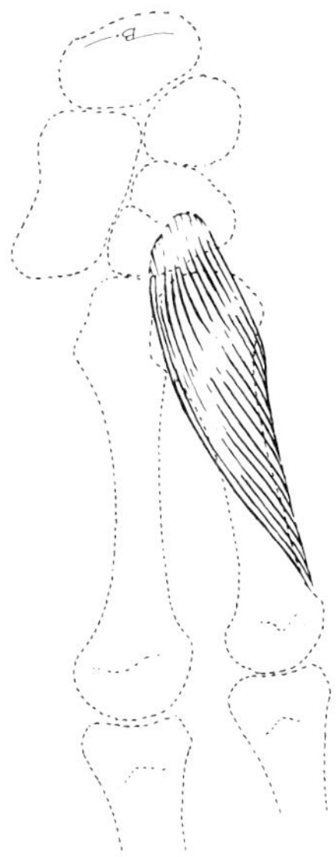

Fig. XIII-67.

Fig. XIII-63. Abductor pollicis longus.
Fig. XIII-64. Flexor pollicis brevis.
Fig. XIII-65. Opponens pollicis.

Fig. XIII-66. The abductor and small flexor of the little finger.

Fig. XIII-67. Opponens digiti minimi.

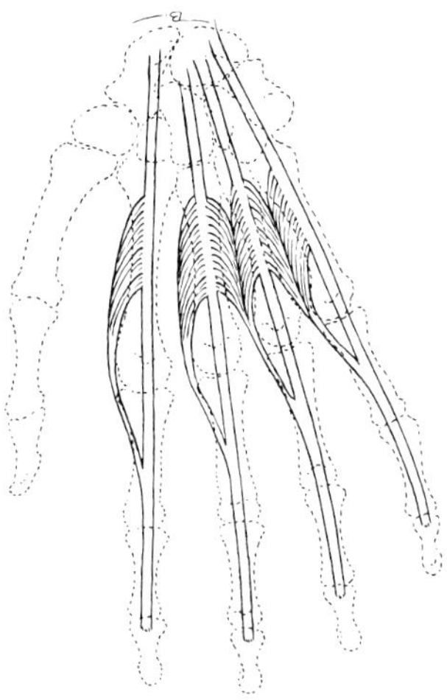

Fig. XIII-68.

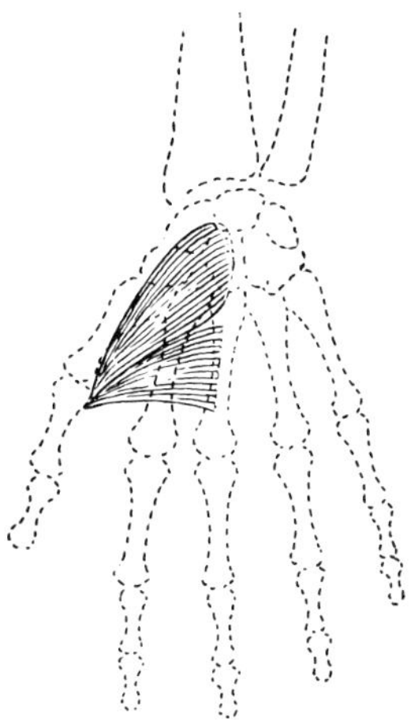

Fig. XIII-69.

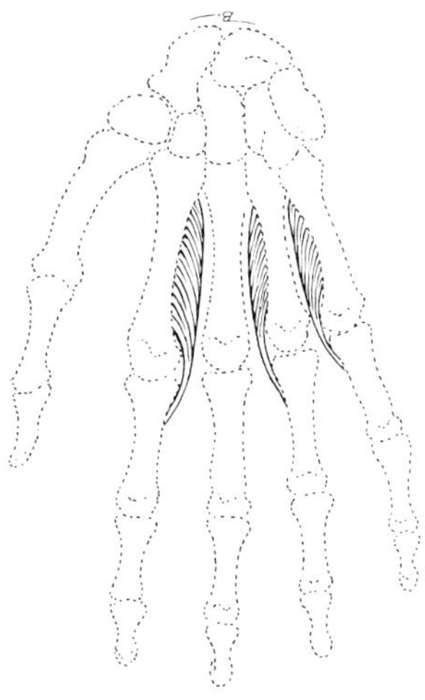

Fig. XIII-70.

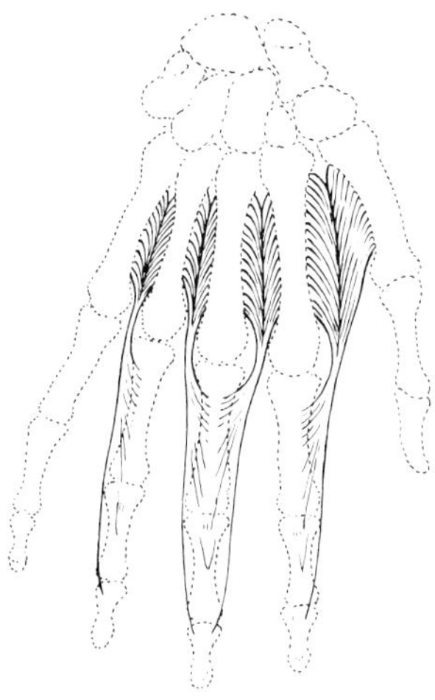

Fig. XIII-71.

Fig. XIII-68. Lumbricals.
Fig. XIII-69. Adductor pollicis.

Fig. XIII-70. Palmar interossei.
Fig. XIII-71. Dorsal interossei.

JOINTS AND MUSCLES OF THE PELVIC GIRDLE AND LOWER LIMB

The primary function of the lower limb is locomotion. It follows that the joints must be stable, their movements more limited and not as delicate as those of the upper limb. It is easiest to learn the muscles if you think in terms of the actions necessary to produce and maintain the standing position and to walk. For instance, the great muscle masses of the lower limb are found in the buttock, at the front of the thigh, and at the back of the leg (calf). These are the muscles which must work against gravity to bring the body from the sitting to the standing position. There are several muscles in the thigh which cross both the hip joint and the knee. Thus, they can perform two movements at the same time with a minimum of effort.

Pelvic Girdle

The pelvic girdle is more rigid and less mobile than the shoulder girdle. Whereas the right and left sides of the shoulder girdle can move independently of one another, the pelvic girdle moves as a single unit. The joint between the sacrum and the ilia permits only enough movement to give the pelvis resilience as it takes weight. The symphysis pubis, on the other hand, tends to try to spread apart when a person is sitting and to be compressed when the person stands. Again, movement is slight. What little movement the girdle as a whole does possess occurs at articulations between the last lumbar vertebra and the sacrum, and the femur and pelvis. Thus, "flexing" the pelvis is mostly a matter of limited movement at the lumbosacral joint enhanced by flexion of the adjacent lumbar vertebrae. For all intents and purposes, the pelvic girdle is practically fixed and the muscles arising from it and from the lumbar vertebrae to cross the hip joint produce movement at that joint. Movements of the pelvis are: upward rotation ("flexing"), downward rotation ("extending"), lateral rotation (to the same or the opposite side), and lateral tilting, or raising, of one side of the pelvis higher than the other.

Hip Joint

The head of the femur fits into the acetabulum of the pelvis to form a typical ball and socket joint. As in the shoulder, there is a rim of fibrocartilage attached to the rim of the acetabulum, the **glenoid labrum,** which again serves to deepen the socket. The entire joint is enclosed in a strong, extensive fibrous capsule which is attached proximally to the margin of the acetabulum and below to the distal margin of the neck of the femur, thus including the neck as well as the head within its synovial membrane. The capsule itself is reinforced by three ligaments named according to the bones that they unite.

One of the strongest capsular ligaments in the body is the **iliofemoral ligament** (Fig. XIII-72A). It is actually triangular, but is often called the "Y" ligament because of its appearance. The base of the Y attaches to the acetabular margin and the anterior inferior iliac spine. The limbs of the Y are very thick and attach to the upper and lower ends of the trochanteric line. In this position it reinforces the anterior portion of the joint capsule and prevents the body from rotating backward on the head of the femur, thus maintaining an erect posture with little or no muscular involvement at this point.

The **pubofemoral ligament,** attached above to the pubic portion of the acetabular brim and blending below with the fibers of the capsule, protects the anteroinferior aspect of the joint. It helps to prevent excessive abduction of the femur and assists the iliofemoral ligament in checking extension at the hip. Between the pubofemoral and iliofemoral ligaments there is an exposed area of weak capsule. This triangular area is protected by the tendon of the iliopsoas muscle.

Reinforcing the lower posterior aspect of the joint is the **ischiofemoral ligament** (Fig. XIII-72B). Attaching to the ischium behind and below the acetabular margin, it passes to the trochanteric fossa, the fibers blending with the capsule. This ligament relaxes during hip flexion and tightens during extension, helping to make the extended position of the joint the most stable one.

All three ligaments tend to limit internal rotation of the femur because this movement wraps them more tightly around the joint. Conversely, they unwind during external rotation so that this movement must be checked solely by muscles. Since all parts of the capsule are relaxed during flexion and external rotation of the thigh, dislocation of the hip is most likely to occur in this position.

There are two more ligaments associated with the hip: the **transverse acetabular ligament** which bridges the acetabular notch to complete the acetabular ring; and the **ligament of the head** (ligamentum teres, a misnomer since the band is not round by any stretch of the imagination) (Fig. XIII-73). This flat band, sheathed in synovial membrane, lies within the hip joint. It is attached to the non-articular surface in the acetabulum and to the fovea capitis, the small pit near the center of the head of the femur. By its position, it should help to check abduction of the femur, but apparently does not become tense enough to be of any use.

Movements of the hip joint are flexion, extension, abduction, adduction, circumduction (a combination of the first four movements), and internal and external rotation. Internal (medial) rotation of the hip causes the kneecaps to face each other. External (lateral) rotation causes the kneecaps to turn outward.

Knee Joint

The largest joint in the body, the knee is a masterful work of engineering. It must meet the requirements of stability and mobility, and in the other joints which we have considered, one virtue has been sacrificed for the other. With its massive condyles for bearing weight, strong ligaments to resist stress, a wide range of motion,

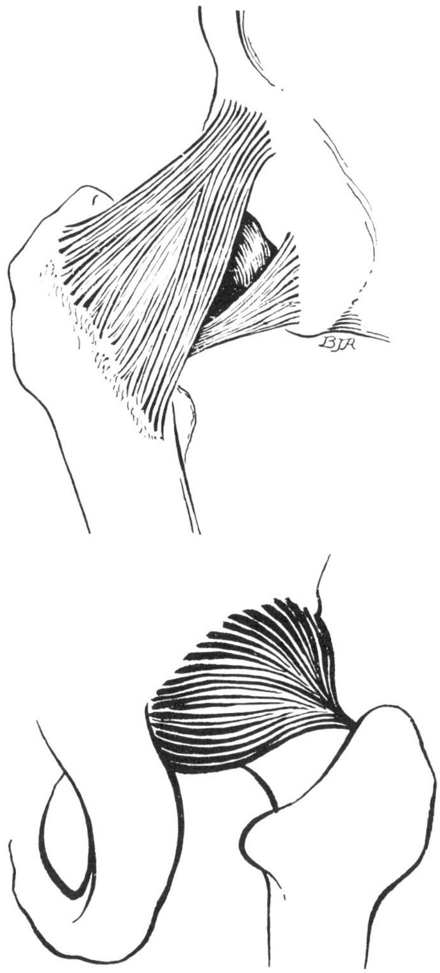

Fig. XIII-72A. Right iliofemoral ligament from in front. It forms the important part of the fibrous capsule. **B.** Posterior part of the capsule of the hip is the ischiofemoral (and pubofemoral) ligament. (From J. V. Basmajian, *Primary Anatomy,* 6th ed., Williams & Wilkins, Baltimore, 1970.)

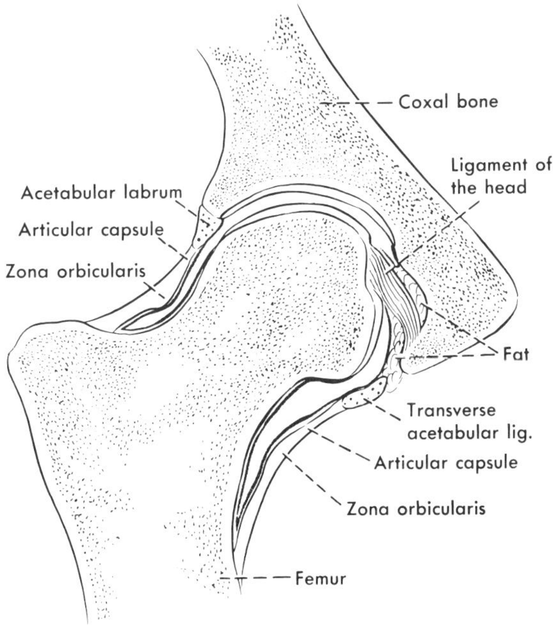

Fig. XIII-73. Frontal section through the hip joint. (From H. Hollinshead, *Textbook of Anatomy,* 2nd ed., Harper & Row, New York, 1967.)

and powerful musculature, the knee joint is admirably suited to its tasks. It is a modified hinge joint, allowing only flexion or extension, but its structure appears more like two condyloid joints lying side by side.

Both the femur and the tibia have been well described earlier in this chapter. It is important to note that, because of the femoral obliquity, the lateral condyle of the femur receives most of the body weight. It is the larger of the two condyles. The memory aid, "The outer is stouter; the inner is thinner," was around when we were students, and it is still around today. The medial condyle is not to be outdone, however, and makes up for its smaller bulk by being half an inch longer from front to back than the lateral condyle. When the tibia moves across the femoral condyles during extension of the knee, it traverses all of the articular surface of the lateral condyle before extension is completed. In order to finish the movement, the tibia must rotate laterally on its long axis to take advantage of the extra half-inch of articular surface still available on the medial condyle. This brings us to the **anterior cruciate ligament** which brakes the motion. It also acts as the pivot around which the tibia rotates.

The Articular Capsule and Ligaments of the Knee: The articular capsule is a thin, strong fibrous membrane which is strengthened nearly all around the knee by various bands and extensions of the muscle tendons crossing the joint. These bands blend with, and become inseparable from, the capsule. Anteriorly the knee joint is well protected by the **insertion of the Quadriceps on the patella,** the **patella** itself, and the **patellar ligament to the tibia.** On either side of the Quadriceps tendon, the articular capsule is formed by fibers from the **fascia lata** and expansions from the Quadriceps tendon which are called the **medial** and **lateral patellar retinacula.** See Figure XIII-74. Posteriorly the capsule is composed of interlacing fibers which arise from the condyles and the sides of the intercondylar fossa of the femur. The posterior part of the capsule, therefore, is anterior to the cruciate ligaments and excludes them from the joint cavity. Posterior to the cruciate ligaments and protecting them is the **oblique popliteal ligament** which derives its tissue from the Semimembranosus tendon. There is an opening in the capsule at the back of the knee, reinforced by the arching fibers of the **arcuate popliteal ligament,** to permit passage of the tendon of origin of the Popliteus muscle. See Figure XIII-75.

Laterally, as is typical of a hinge joint, there are collateral ligaments. The **collateral tibial ligament** is a broad, flat band on the medial side of the joint. Attached proximally to the medial epicondyle of the

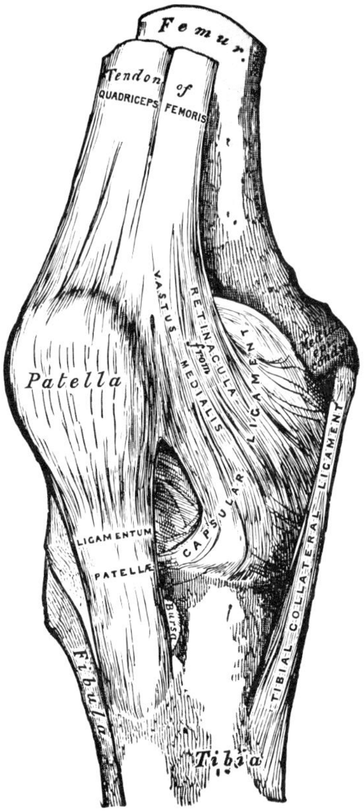

Fig. XIII-74. Anterior view of the right knee joint. (From C. M. Goss, *Gray's Anatomy of the Human Body*, 28th ed., Lea & Febiger, Philadelphia, 1966.)

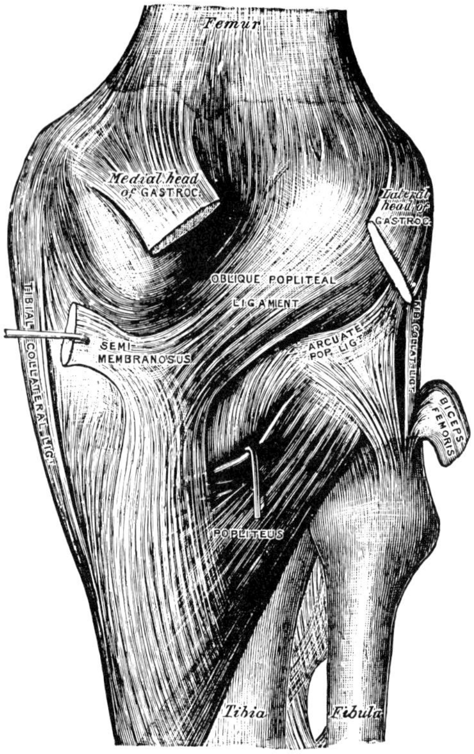

Fig. XIII-75. Posterior aspect of the right knee joint. (From C. M. Goss, *Gray's Anatomy of the Human Body*, 28th ed., Lea & Febiger, Philadelphia, 1966.)

femur below the adductor tubercle and distally to the medial condyle of the tibia, it checks extension of the knee and prevents lateral motion. This ligament is also firmly attached to the medial meniscus. It is the collateral tibial ligament and the medial meniscus which are most often involved in athletic injuries to the knee. There are two of these cartilaginous layers, the **lateral** and **medial menisci,** (semilunar cartilages) which serve to reduce friction between the bones and to adjust the unmatched surfaces of the two femoral condyles with the relatively flat articular surfaces of the tibial head (Fig. XIII-76).

On the lateral side of the knee joint is the **collateral fibular ligament,** which, like its opposite number, checks extension and prevents medial motion. It is a strong, rounded cord running from the lateral epicondyle of the femur to the lateral surface of the fibular head. Reinforcing the collateral ligaments are a prolongation of the iliotibial band on the lateral side and expansions from the Sartorius and Semimembranosus medially.

Another pair of ligaments, the **anterior** and **posterior cruciate** ("crossed") **ligaments,** are named according to their attachments to the tibia and the fact that one crosses over the other (Fig. XIII-76). They remain fairly taut no matter what the position of the knee, but become even more tense at the extreme ends of the range of motion. They are considered to be extremely important to the stability of the joint, preventing anteroposterior displacement of the tibia, limiting extension of the knee, and preventing rotation of the femur on the tibia in the extended position. They lie outside the joint capsule but are covered anteriorly and laterally by reflections of the synovial membrane. The *anterior cruciate ligament* ascends upward and backward from the *anterior* intercondyloid fossa to attach on the back part of the medial surface of the lateral condyle. The *posterior cruciate ligament* is shorter and stronger than the anterior. It passes upward and forward from the *posterior* intercondylar area of the tibia to the lateral side of the medial femoral condyle.

Deep Fascia of the Thigh: The *fascia lata* (broad fascia) forms a tough, fibrous layer that completely invests the muscles of the thigh. Additional longitudinal fibers join it on the lateral aspect of the thigh to form the **iliotibial tract.** Into this tract run the attachments of the Tensor Fasciae Latae muscle and a large part of the insertion of the Gluteus Maximus. Distally, the fibers of the iliotibial tract reinforce the capsule of the knee joint and attach anterolaterally to the lateral femoral condyle and to the lateral tibial condyle. About the knee, its fibers blend with the medial and lateral

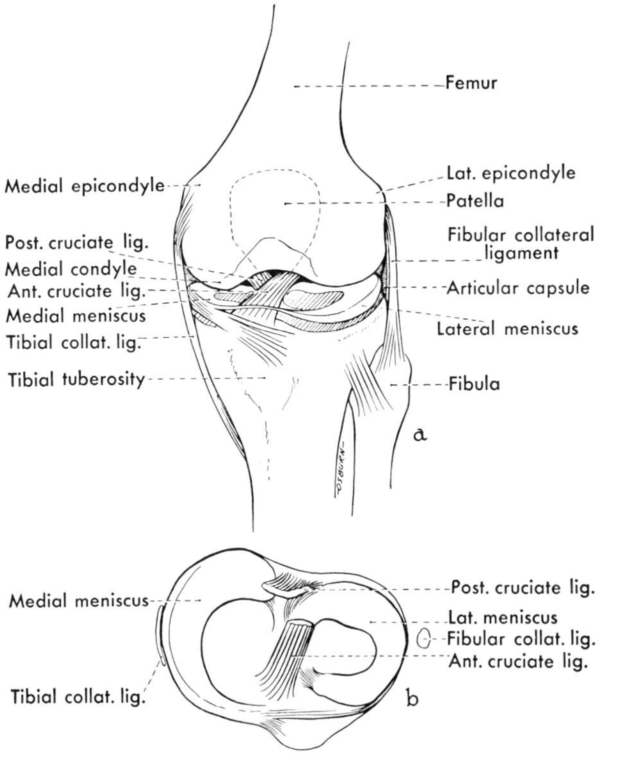

Fig. XIII-76. Primary structures of the left knee joint. **A.** Anterior aspect. **B.** The tibia from above with the femur removed. (From H. Hollinshead, *Functional Anatomy of the Limbs and Back,* 3rd ed., W. B. Saunders, Philadelphia, 1969.)

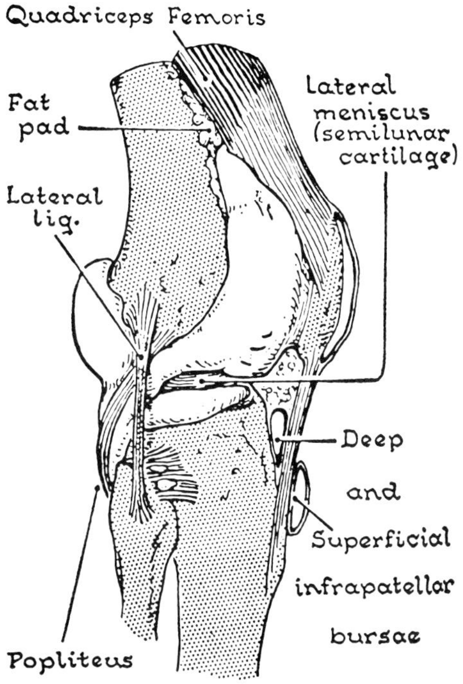

Fig. XIII-77. Knee joint filled with latex to show the extent of the synovial cavity; lateral view. (From J. C. B. Grant and J. V. Basmajian, *Grant's Method of Anatomy,* 7th ed., Williams & Wilkins, Baltimore, 1965.)

patellar retinacula. Proximally, the fibers of the fascia lata extend upward to attain attachment to the iliac crest, the sacrotuberous ligament, and the ischial tuberosity posterolaterally. Toward the front of the body, it attaches to the inguinal ligament and the pubis.

Synovial Membrane and Bursae (Fig. XIII-77): The synovial membrane of the knee is the most extensive in the body, loose, and much folded. It begins at the proximal border of the patella and runs beneath it and the Quadriceps Femoris, supported by a small muscle, the **Articularis Genu,** which is inserted into the membrane. It is separated from the patellar ligament by the **infrapatellar fat pad.** From the lateral borders of the articular surface of the patella, the synovium extends beneath the aponeuroses of the Vasti (part of the Quadriceps Femoris) and sends two fringe-like **alar folds** into the interior of the joint. These join distally, deep within the joint, to form a single continuous band, the **patellar fold.** On either side of the joint, the synovial membrane lines the articular capsule to its point of attachment to the menisci (the collateral ligaments). The membrane covers the proximal surfaces of the menisci, turns under, and crosses their distal surfaces. Behind the joint the membrane is reflected anteriorly across the cruciate ligaments which, you will recall, are outside the joint capsule and are also outside the synovial cavity.

There are numerous bursae about the knee, the most important being the pre-, infra-, and suprapatellar bursae. The patellar bursae may communicate with the synovial cavity. Briefly, there are four bursae anterior to the joint, four bursae laterally, and five medially. These insure the friction-free movement of the many ligaments and tendons over each other and across the bones of the joint.

Muscles of the Buttock and Thigh

These muscles divide themselves into three groupings according to their anatomic position in the thigh. The **anterior group** (two-joint muscles) is concerned with flexion at the hip and extension at the knee. The **adductor** or **anteromedial group** are "one-joint" muscles concerned with adduction and flexion of the thigh (Fig. XIII-78); and there is a **posterior group** (two-joint muscles, again) which causes extension at the hip and flexion at the knee (Fig. XIII-79). There is some overlap of functions and innervations of these groups, but a safe rule of thumb is that the anterior muscles are innervated by the **femoral nerve,** the anteromedial ones by the **obturator nerve,** and the muscles of the posterior group by the **sciatic nerve.** Refer to Charts XIII-1 and 2 for the muscles in these groups.

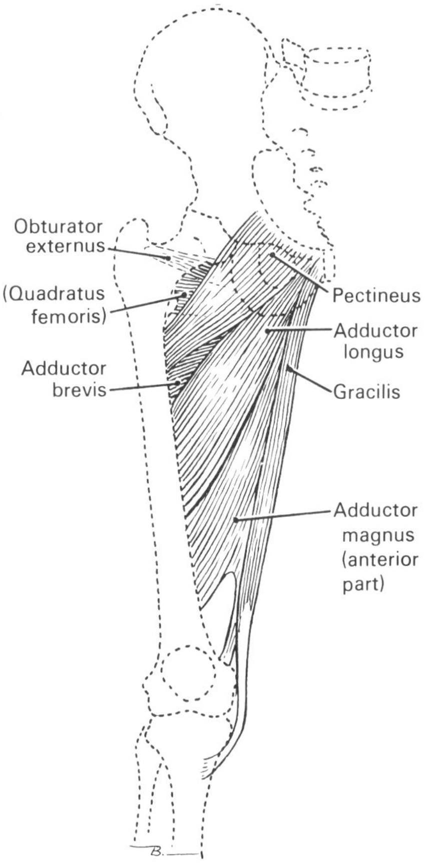

Fig. XIII-78. The muscles of the anteromedial thigh or adductor compartment. A part of the Quadratus femoris, not a union member of this fraternity, also shows.

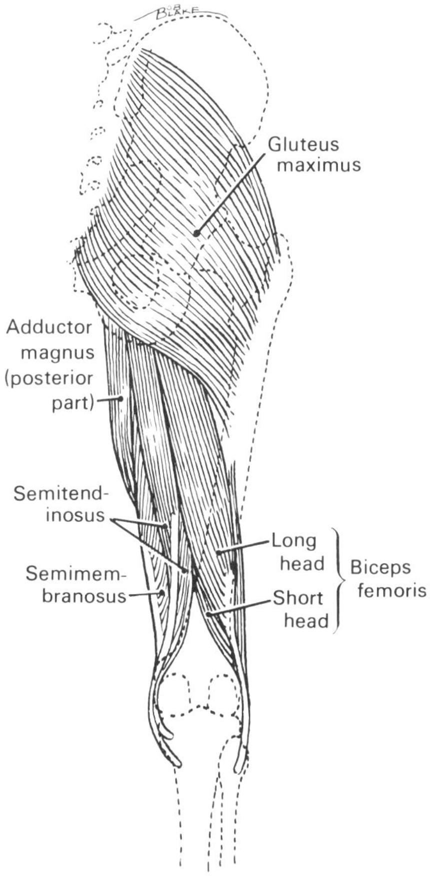

Fig. XIII-79. The hamstring group of the posterior compartment of the thigh.

Chart XIII-1. Muscle Groups of the Thigh

Anterior	Anteromedial	Posterior
Sartorius	Adductor longus (anterior part)	Semitendinosus
Quadriceps femoris	Gracilis	Semimembranosus
Rectus femoris	Adductor brevis	Biceps femoris
Vastus medialis	Adductor magnus (anterior part)	Adductor magnus (posterior part)
Vastus lateralis	Obturator externus	
Vastus intermedius		
Articularis genu		
Iliopsoas		
Psoas major		
Iliacus		
Psoas minor		
Pectineus		

Chart XIII-2. Muscles of the Buttock

Gluteus maximus	Obturator internus
Gluteus medius	Superior gemellus
Gluteus minimus	Inferior gemellus
Tensor fasciae latae	Quadratus femoris
Piriformis	

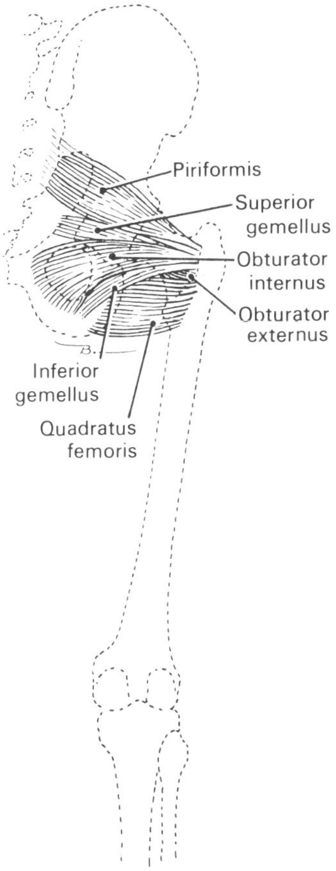

Fig. XIII-80. Deep gluteal region: the external rotators of the hip.

There are several comments to be made on the muscles in Tables XIII-7 and 8. The Sartorius has popularly been called the "tailor's" muscle by students. It is the longest muscle in the body and relatively weak. Because it winds obliquely across the surface of the thigh from the anterior superior iliac spine to the medial side of the knee, it is supposed to help fold the leg "tailor-fashion." In this position the hip will be flexed, abducted, and laterally rotated and the knee will be flexed and medially rotated. Actually, the Sartorius cannot do any of this alone.

Of the four heads of the Quadriceps, the Rectus Femoris is the only one to cross the hip joint as well as the knee. At the hip it acts as a flexor of the thigh. The Psoas Minor, although not really a muscle of the thigh, was included in Table XIII-7 because of its intimate association with the Psoas Major. Indeed, some texts no longer discuss the Minor as a separate muscle.

The six deep external rotators (Obturator Externus and Internus, Gemellus Superior and Inferior, Quadratus Femoris, and Piriformis) are listed as being outward rotators of the hip. Some of these muscles also have a secondary function such as abduction or adduction, but, more important, they are well located to help hold the head of the femur in the acetabulum (Fig. XIII-80).

Everyone agrees that the Adductor Magnus adducts the thigh, but that is all that they agree upon. Some kinesiologists claim that the upper portion of the muscle acts as a flexor of the thigh and the lower portion is an extensor. Others assert that its primary action is extension at the hip and that it acts as an adductor only when the thigh is already extended. Apparently, it also has a very slight internal rotatory action about which everyone agrees once again.

In Table XIII-8, the hamstring muscles (Semitendinosus, Semimembranosus, and Biceps Femoris) and the Adductor Magnus are listed as being innervated by the sciatic, the tibial, or the common peroneal nerves. Each of these muscles is supplied by multiple branches of the sciatic nerve. Occasionally, the sciatic nerve will divide into its tibial and common peroneal parts unusually high in the thigh. When this occurs, the Semitendinosus, Semimembranosus, long head of the Biceps, and part of the Adductor Magnus receive their nerve supply from the tibial nerve. The short head of the Biceps is supplied by the common peroneal nerve.

Tibiofibular Joints (Fig. XIII-100)

Just as the radius and ulna of the forearm have three joints, so also do the tibia and fibula. Proceeding down the leg from the knee, the first joint encountered is, predictably, the **superior** (proximal) **tibiofibular joint.** It is arthrodial, permitting gliding movement between the head of the fibula and the tibia. This slight play in the joint protects the fibula from being broken by forces pushing upward from the heel. A small, flat, round facet on the head of the fibula articulates with a small facet on the posterolateral part of the lateral condyle of the tibia. The joint capsule is very strong anteriorly. Posteriorly, it is crossed by the Popliteus tendon.

The **middle** and **inferior** (distal) **tibiofibular joints** are both syndesmoses. The interosseous membrane (forming the middle joint) fills the space between the two bones for their entire lengths with only a small open space at each end for the passage of blood vessels and nerves. The *inferior tibiofibular joint* is a very strong, fibrous union. The lower end of the fibula, just above the malleolus, fits into a hollow on the lateral side of the lower end of the shaft of the tibia where it is snugly kept in place by the **interosseous tibiofibular ligament.** The **anterior** and **posterior tibiofibular ligaments** run obliquely downward and laterally between the adjacent margins of the bones and behind the joint. The inferior **transverse tibiofibular ligament** lies anterior to the posterior ligament and passes across the back of the joint from the lateral malleolus to the posterior border of the articular surface of the tibia, nearly to the medial malleolus.

All of the ligaments between the tibia and fibula run in the same oblique direction. They prevent the fibula from being pulled downward by the muscles attached to it while still allowing it to move slightly upward and giving the ankle joint suppleness. Although the

Table XIII-7. Anterior and Anteromedial Muscles of the Thigh

Muscle	Nerve	Origin and Insertion	Primary Action
Iliopsoas (Fig. XIII-81)	L_2-L_4 Femoral nerve	(O) Psoas major: anterolateral aspect of the lumbar vertebrae and their transverse processes Iliacus: inner surface of the ilium (I) Together on, and below, the lesser trochanter of the femur	Flexion at hip
Sartorius (Fig. XIII-82)	$L_{2,3}$ Femoral nerve	(O) Anterior superior iliac spine (I) Tibia, anterior and medial surface just below condyle	Flexion and rotation at hip and knee
Quadriceps (Fig. XIII-83A and B)	L_2-L_4 Femoral nerve	(O) Rectus femoris: anterior inferior iliac spine and ilium just above acetabulum Vastus medialis: medial lip of linea aspera, lower half of intertrochanteric line, and upper part of medial supracondylar line Vastus lateralis: lateral lip of linea aspera, upper part of intertrochanteric line, greater trochanter, lateral lip of gluteal tuberosity Vastus intermedius: anterior and lateral surfaces of upper two-thirds of shaft of femur (I) By a common tendon upon the patella and, through the patellar ligament, upon the tibial tuberosity	Extension at knee
Articularis genu	$L_{3,4}$ Femoral nerve	(O) Front of femur underneath Vastus intermedius (I) Upper part of synovial membrane of knee joint	Pulls synovial membrane of knee joint upward
Pectineus (Fig. XIII-84)	$L_{2,3}$ Femoral nerve (occasionally innervated by obturator nerve instead)	(O) Superior ramus of pubis along pectineal line (I) Pectineal line just below the lesser trochanter	Flexion and adduction at hip
Psoas minor (Fig. XIII-81)	L_2-L_4 Femoral nerve	(O) Anterolateral surfaces of vertebrae, usually T_{12} and L_1 (I) Superior ramus of pubis	Assists in upward rotation of pelvis
Adductor longus (Fig. XIII-85)	$L_{2,3}$ Obturator nerve	(O) Anterior surface of pubis (I) Medial lip of middle half of linea aspera	Adduction and flexion at hip
Adductor brevis (Fig. XIII-86)	$L_{3,4}$ Obturator nerve	(O) Body and inferior ramus of pubis (I) Line from lesser trochanter to linea aspera and upper one-fourth of linea aspera	Adduction and flexion at hip
Gracilis (Fig. XIII-87)	$L_{3,4}$ Obturator nerve	(O) Anterior aspect of lower half of symphysis pubis and upper half of pubic arch (I) Medial surface of tibia just below condyle	Adduction at hip, flexion at knee
Adductor magnus (anterior part) (Fig. XIII-88)	$L_{3,4}$ Obturator nerve	(O) Inferior ramus of pubis and ischium, lateral border of inferior surface of ischial tuberosity (I) Linea aspera, medial supracondylar line, and adductor tubercle on medial condyle of femur	Adduction and flexion at hip
Obturator externus (Fig. XIII-89)	$L_{3,4}$ Obturator nerve	(O) Outer surface of pelvis, rim of obturator foramen and obturator membrane (I) Trochanteric fossa on medial side of greater trochanter	External rotation at hip

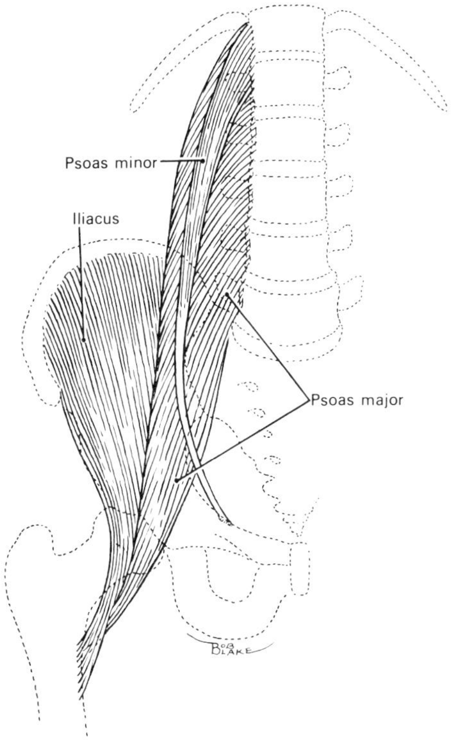

Fig. XIII-81.

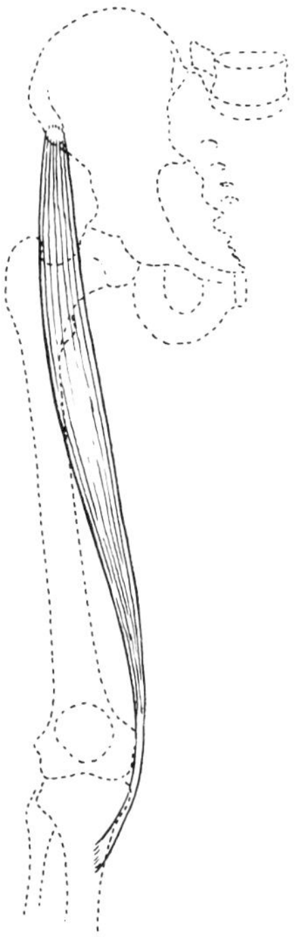

Fig. XIII-82.

Fig. XIII-81. The hip flexors.
Fig. XIII-82. Sartorius.

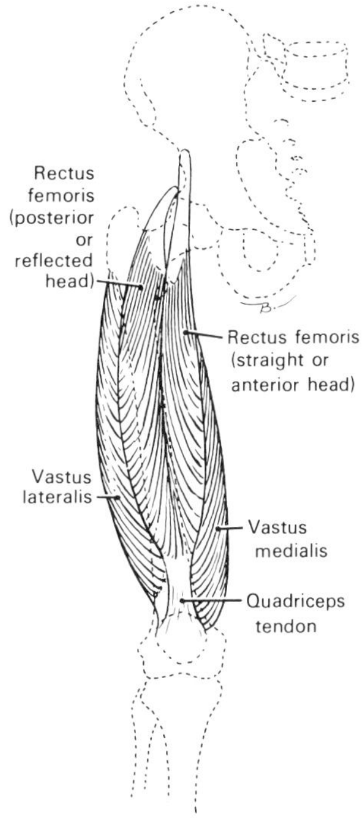

Fig. XIII-83A.

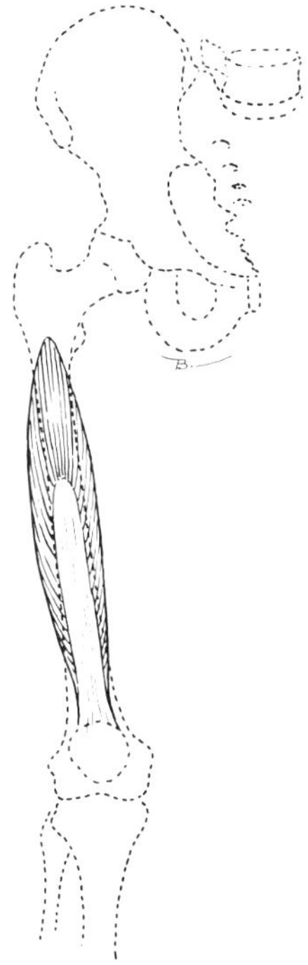

Fig. XIII-83B.

Fig. XIII-83A. The extensor compartment of the thigh: the Quadriceps femoris. The deeper Vastus intermedius cannot be seen.

Fig. XIII-83B. Vastus intermedius portion of Quadriceps muscle.

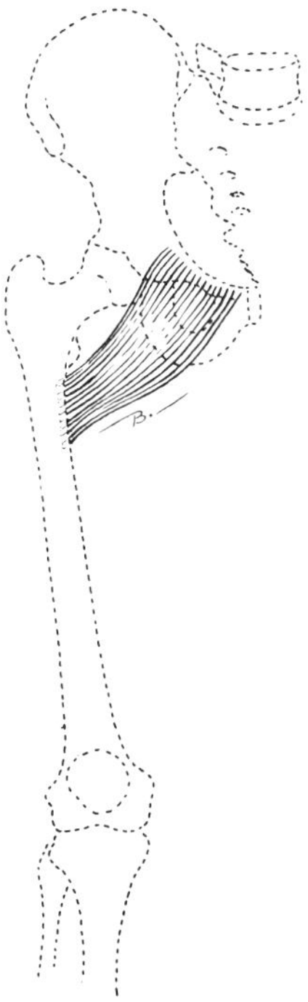

Fig. XIII-84.

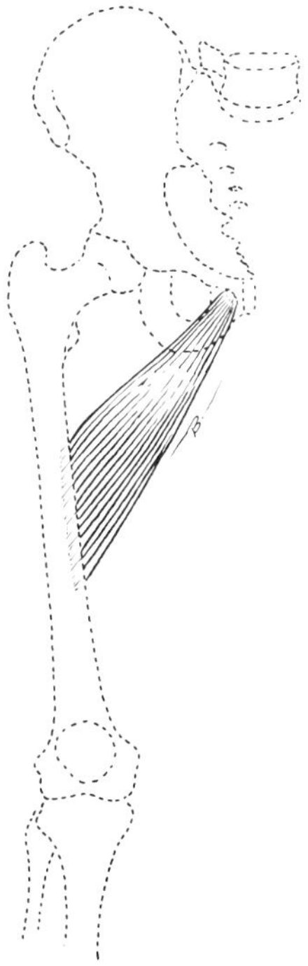

Fig. XIII-85.

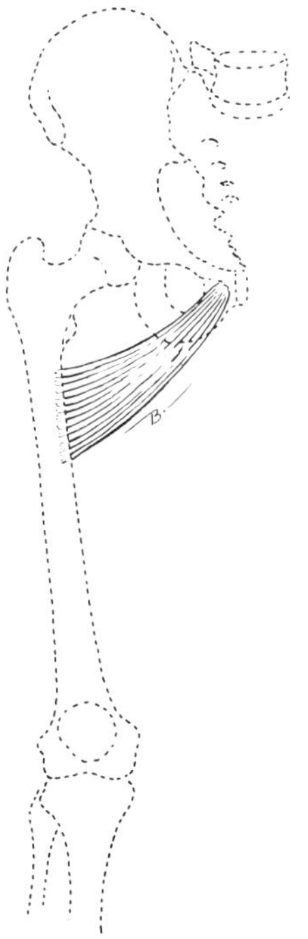

Fig. XIII-86.

Fig. XIII-84. Pectineus.
Fig. XIII-85. Adductor longus.
Fig. XIII-86. Adductor brevis.

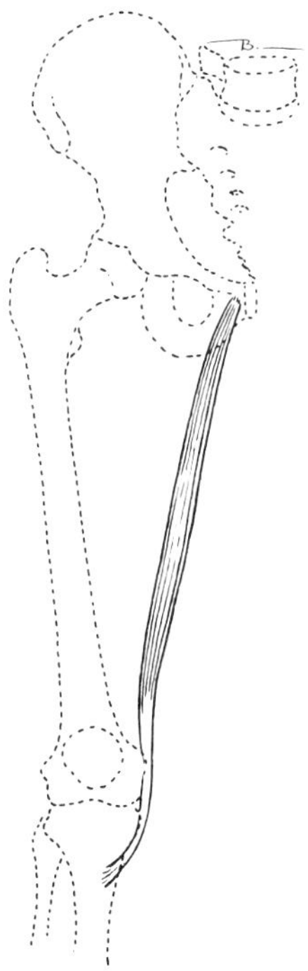

Fig. XIII-87.

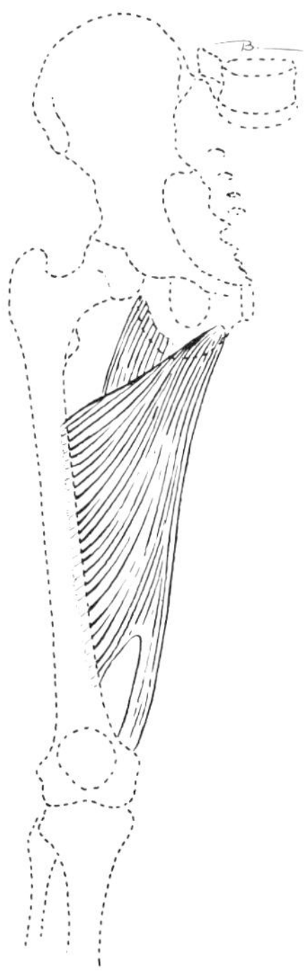

Fig. XIII-88.

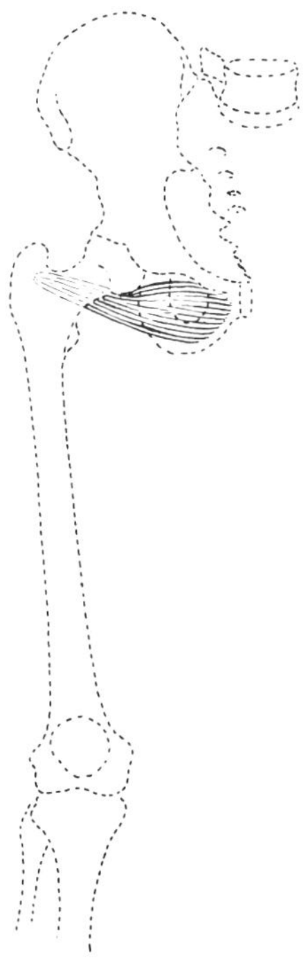

Fig. XIII-89.

Fig. XIII-87. Gracilis.
Fig. XIII-88. Adductor magnus.
Fig. XIII-89. Obturator externus.

Table XIII-8. Muscles of the Buttock and Posterior Thigh

Muscle	Nerve	Origin and Insertion	Primary Action
Gluteus medius (Fig. XIII-90)	L_4-S_1 Superior gluteal nerve	(O) Posterior surface of ilium between crest, posterior gluteal line, and anterior gluteal line (I) Oblique ridge on lateral surface of greater trochanter	Abduction and external rotation at hip
Gluteus minimus (Fig. XIII-91)	L_4-S_1 Superior gluteal nerve	(O) Posterior surface of ilium between the anterior and inferior gluteal lines (I) Anterior border of greater trochanter	Abduction and internal rotation at hip
Tensor fasciae latae (Fig. XIII-92)	L_4-S_1 Superior gluteal nerve	(O) Anterior part of outer lip of iliac crest and outer surface of anterior superior iliac spine (I) Iliotibial tract of fascia lata on anterolateral aspect of thigh about one-third of the way down	Flexion and internal rotation at hip
Gluteus maximus (Fig. XIII-93)	L_5-S_2 Inferior gluteal nerve	(O) Posterior gluteal line of ilium and adjacent part of crest; posterior surface of lower part of sacrum and side of coccyx (I) Posterior surface of femur on ridge below greater trochanter; iliotibial tract of fascia lata	Extension and adduction at hip
Piriformis (Fig. XIII-94)	$S_{1,2}$ Nerve to Piriformis	(O) Anterior surface of sacrum (I) Inner surface of upper part of greater trochanter	External rotation at hip
Obturator internus (Fig. XIII-95)	L_5-S_2 Nerve to Obturator internus and Superior gemellus	(O) Inner surface of obturator membrane and rim of obturator foramen (I) Greater trochanter of femur just above the trochanteric fossa	External rotation at hip
Superior gemellus (Fig. XIII-96)	L_5-S_2 Nerve to Obturator internus and Superior gemellus	(O) Ischial spine (I) Upper border of Obturator internus tendon	External rotation at hip
Inferior gemellus (Fig. XIII-96)	L_4-S_1 Nerve to Quadratus femoris and Inferior gemellus	(O) Ischial tuberosity (I) Lower border of Obturator internus tendon	External rotation at hip
Quadratus femoris (Fig. XIII-97)	L_4-S_1 Nerve to Quadratus femoris and Inferior gemellus	(O) Ischial tuberosity (I) Posterior surface of femur about half way between the lesser and greater trochanters	External rotation at hip
Semitendinosus (Fig. XIII-98)	L_5-S_1 Sciatic nerve, (tibial nerve)	(O) Ischial tuberosity blended with the long head of the Biceps femoris (I) Upper part of medial surface of shaft of tibia	Flexion at knee, extension at hip
Semimembranosus (Fig. XIII-98)	L_5-S_2 Sciatic nerve (tibial nerve)	(O) Lateral aspect of ischial tuberosity (I) Posterior surface of medial tibial condyle	Flexion at knee, extension at hip
Biceps femoris long head (Fig. XIII-99)	L_5-S_2 Sciatic nerve (tibial nerve)	(O) Ischial tuberosity blended with Semitendinosus (I) Lateral side of head of fibula and lateral condyle of tibia	Flexion at knee, extension at hip
Adductor magnus (posterior part) (Fig. XIII-88)	$L_{4,5}$ Sciatic nerve (tibial nerve)	(O) Inferior ramus of pubis, ramus of ischium, and ischial tuberosity (I) Linea aspera, medial supracondylar line, and adductor tubercle on medial condyle of femur	Extension at hip
Biceps femoris short head (Fig. XIII-99)	L_5-S_1 Sciatic nerve (common peroneal nerve)	(O) Lateral lip of linea aspera (I) Lateral side of head of fibula and lateral condyle of tibia	Flexion at knee

The following muscles of buttocks and thigh are shown in place on the opposite page:

Fig. XIII-90. Gluteus medius.
Fig. XIII-91. Gluteus minimus.
Fig. XIII-92. Tensor fasciae latae.
Fig. XIII-93. Gluteus maximus.
Fig. XIII-94. Piriformis.

Fig. XIII-95. Obturator internus.
Fig. XIII-96. Superior and inferior gemelli.
Fig. XIII-97. Quadratus femoris.
Fig. XIII-98. The medial hamstrings.
Fig. XIII-99. Biceps femoris.

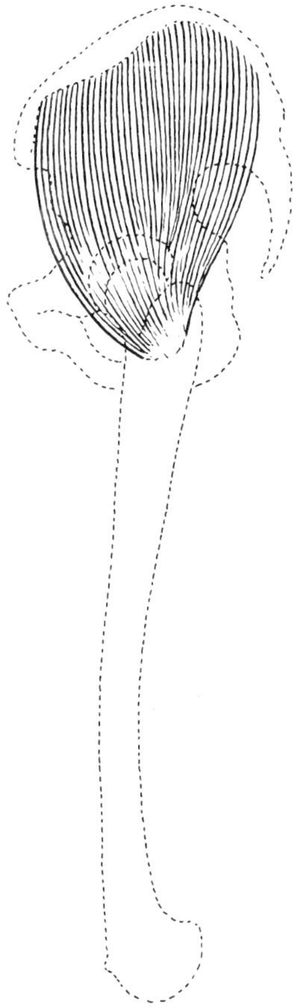

Fig. XIII-90.

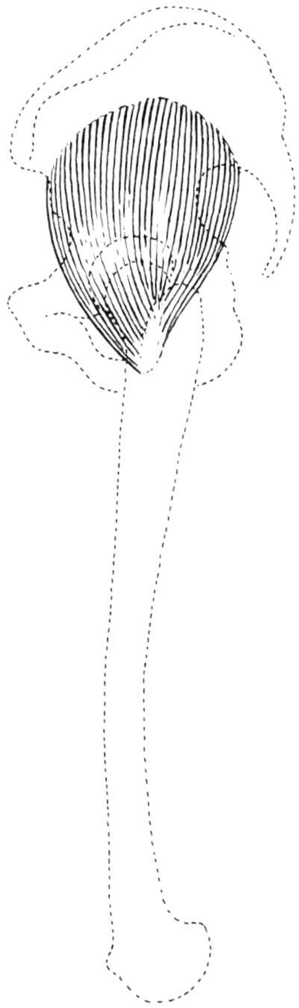

Fig. XIII-91.

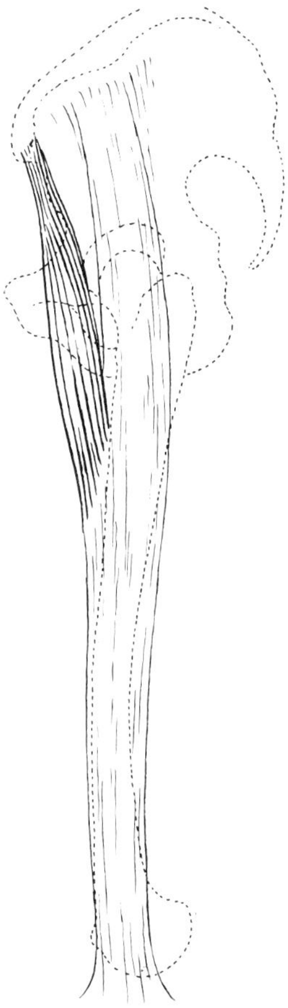

Fig. XIII-92.

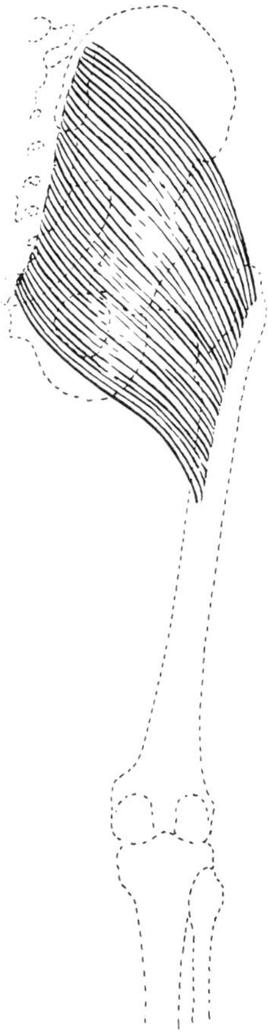

Fig. XIII-93.

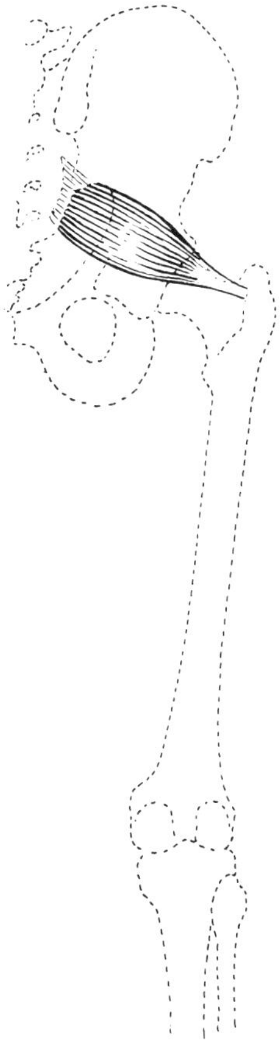

Fig. XIII-94.

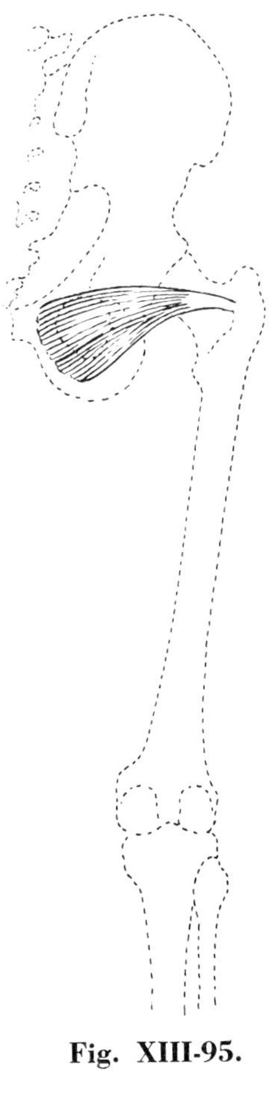

Fig. XIII-95.

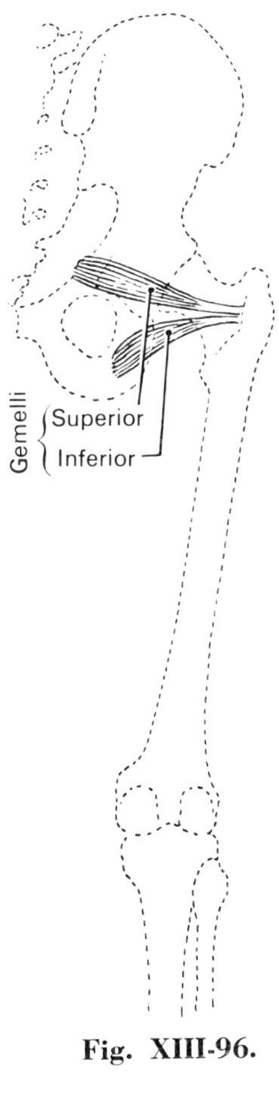

Fig. XIII-96.

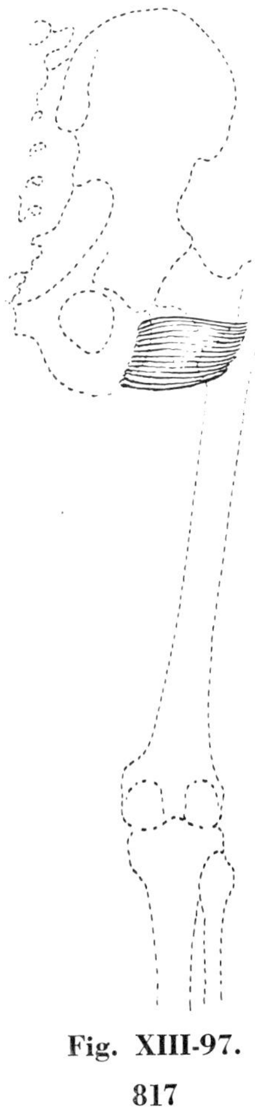

Fig. XIII-97.

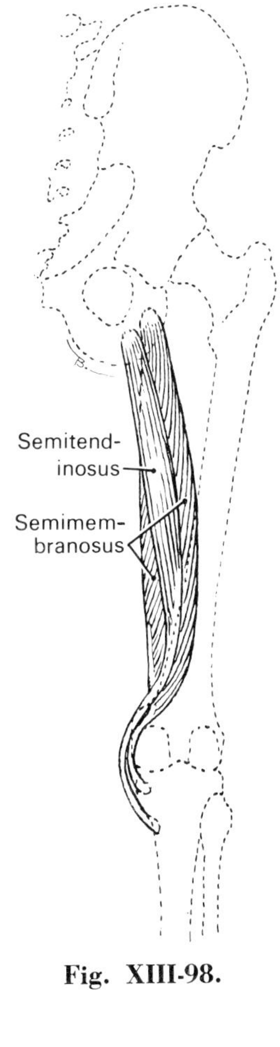

Fig. XIII-98.

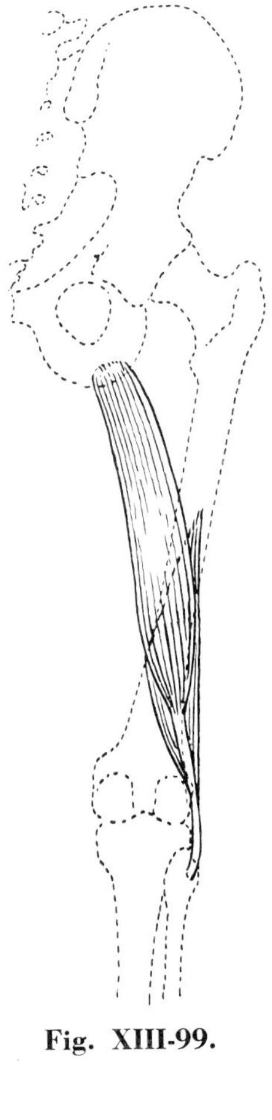

Fig. XIII-99.

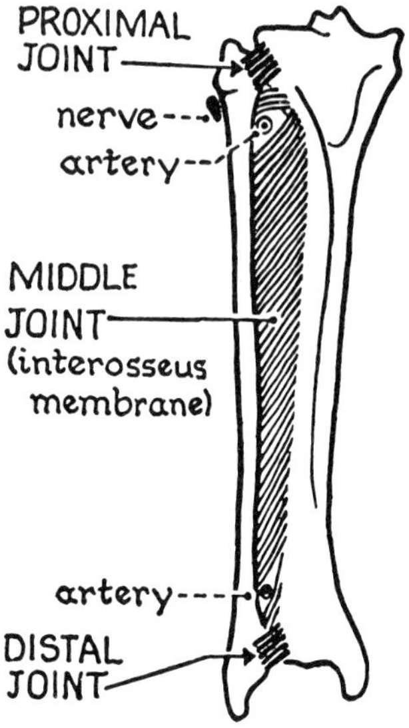

Fig. XIII-100. The tibiofibular articulations. Note the unity of direction of the ligamentous fibers. (From J. C. B. Grant and J. V. Basmajian, *Grant's Method of Anatomy,* 7th ed., Williams & Wilkins, Baltimore, 1965.)

fibula does not articulate with the femur and therefore does not transmit any of the weight from the thigh, it is a very important bone in the leg. Not only does it give attachment to many muscles, it also stabilizes the ankle joint. Normally, the talus, the uppermost bone of the carpus, is held firmly between the two malleoli so that only flexion and extension can occur. If the lateral malleolus is fractured or part of the fibula is missing, the ankle may be badly twisted.

The Ankle Joint

The hinge joint between the foot and the leg is made up of three bones, the tibia, the fibula, and the talus. As we noted above, the talus is gripped firmly between the two extensions of the tibia and fibula, the medial and lateral malleoli respectively. The talus itself is slightly wider in front than behind. Thus, when the foot is dorsiflexed (upper surface of foot is brought toward leg), the talus is more firmly wedged between the malleoli and the joint is very secure. Conversely, when the foot is plantar-flexed, as when one stands on tiptoes, the narrower part of the talus lies between the malleoli, and the ankle joint is less secure.

Ligaments of the Ankle: The capsule of the talocrural (ankle) joint is thin anteriorly and posteriorly but well reinforced laterally by strong ligaments. The most interesting of the ligaments lies on the medial side of the joint where it must make up for the shortness of the tibial extension. This is the **deltoid ligament,** so called because it fans out from its attachment on the medial malleolus much as the Deltoid muscle fans out over the shoulder. It has four parts: the **anterior** and **posterior tibiotalar,** the **tibiocalcaneal,** and the **tibionavicular ligaments,** all attaching to the medial as-

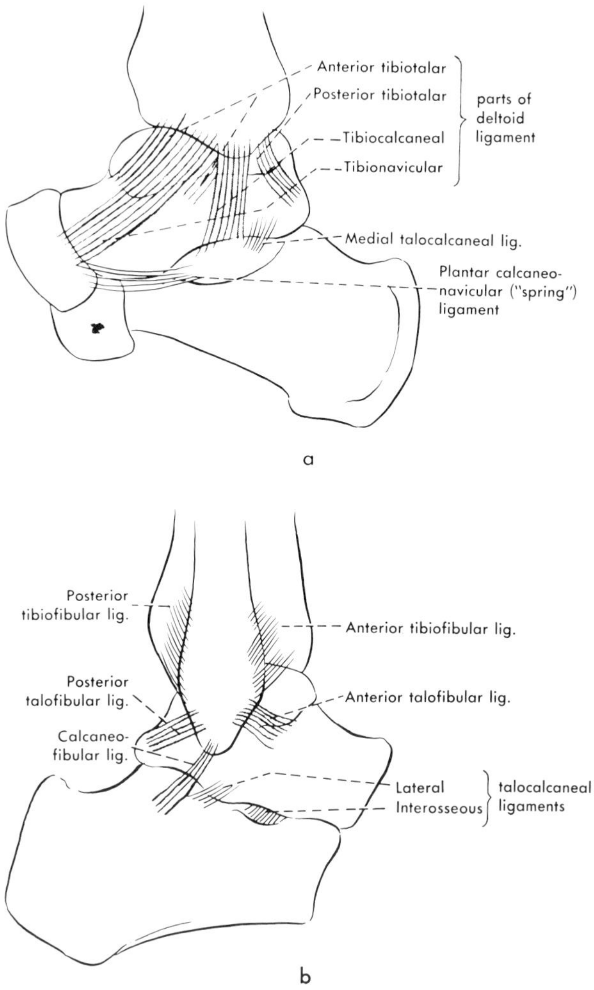

Fig. XIII-101. The ligaments of the ankle and subtalar joints. **A.** Medial view. **B.** Lateral view. (From H. Hollinshead, *Functional Anatomy of the Limbs and Back,* 3rd ed., W. B. Saunders, Philadelphia, 1969.)

pects of their respective bones (Fig. XIII-101A). Together they resist eversion of the foot, a sideward movement causing the sole of the foot to face away from the other foot. Weakness of this ligament may predispose the owner to flatfootedness.

There are three ligaments on the lateral aspect of the ankle. Attached to the lateral malleolus of the fibula, they are the **anterior talofibular ligament,** the **posterior talofibular ligament,** and the **calcaneofibular ligament** (Fig. XIII-101B). These ligaments prevent excess inversion of the foot (turning the soles of the feet so that they face each other). When the ankle is sprained, some of these ligaments are torn. Most commonly it is the lateral ligaments which are injured by forced inversion of the foot.

The external investing layer of deep fascia around the ankle has embedded in it several distinct fibrous bands which bind down the tendons situated anterior and posterior to the ankle on their way to the foot.

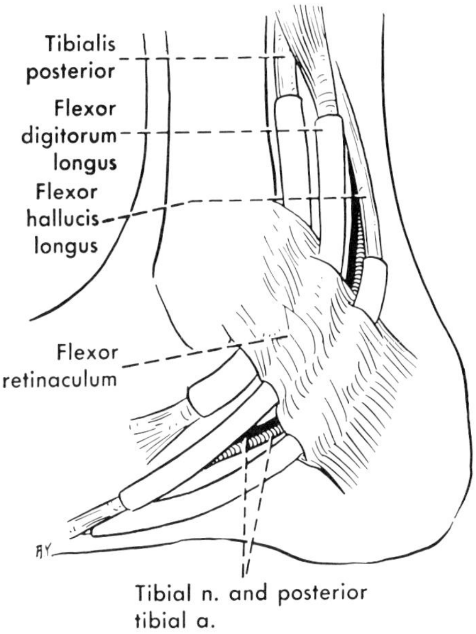

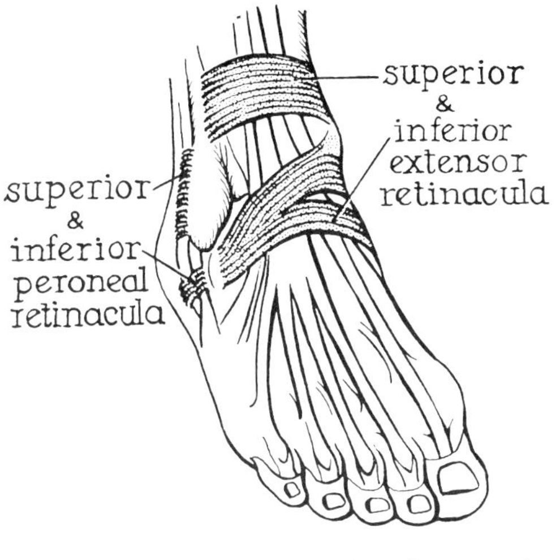

Fig. XIII-102A. The flexor retinaculum at the ankle. **B.** Extensor and peroneal retinacula. (**A.** From H. Hollinshead, *Textbook of Anatomy*, 2nd ed., Harper & Row, New York, 1967. **B.** From J. V. Basmajian, *Primary Anatomy*, 6th ed., Williams & Wilkins, Baltimore, 1970.)

There are three sets of these retinacula to match the three muscle groups of the leg. The **flexor retinaculum** extends from the medial malleolus posteromedially to the calcaneus (Fig. XIII-102A). In sequence from the medial side, it binds down the tendons of the Tibialis Posterior and Flexor Digitorum Longus, the posterior tibial vessels and tibial nerve, and last, the tendon of the Flexor Hallucis Longus. The tendo calcaneus lies superficial to the retinaculum. The **peroneal retinacula** bind down the tendons of the Peroneus Longus and Brevis. The **superior peroneal retinaculum** runs from the lateral malleolus distally to the lateral surface of the calcaneus. The **inferior peroneal retinaculum** is attached to the lateral surface of the calcaneus posteriorly but becomes continuous with the fibers of the inferior extensor retinaculum. A fibrous septum to the peroneal trochlea separates the tendons of the Peroneus Longus and Brevis at that point. The **extensor retinacula** are also paired (Fig. XIII-102B). The **superior extensor retinaculum** passes transversely across the fibula and tibia. Deep to it are the tendons of the Extensor Digitorum Longus, Extensor Hallucis Longus, Peroneus Tertius, and Tibialis Anterior. The anterior tibial vessels and deep peroneal nerve also run underneath it. The **inferior extensor retinaculum,** once called the cruciate ligament, is a Y-shaped band situated anterior to the ankle joint. The stem of the Y is attached laterally to the proximal surface of the calcaneus. It passes medially as a double layer, enclosing the tendons of the Peroneus Tertius and Extensor Digitorum Longus. At this point the two limbs of the Y diverge. The upper limb blends with the periosteum on the medial malleolus. The lower limb blends with the plantar fascia on the medial side of the foot. Having passed deep to the superior extensor retinaculum, the Extensor Hallucis Longus now passes deep to the limb of the inferior one, but the Tibialis Anterior passes through its fibers. The more distal limb of the Y passes over the tendons of the Extensor Hallucis Longus and Tibialis Anterior, the vessels, and the nerves. All the tendons making the trip across the ankle joint, whether under or through their various retinacula, are encased in synovial sheaths to prevent friction of tendon against bone.

Muscles of the Leg

Of the muscles of the leg, just one, the Popliteus, crosses the knee joint only. Two more cross both the knee and ankle joints, and the rest act at the ankle and sometimes on more distal parts of the foot. They can be separated into three groups: *anterior* muscles, *lateral* muscles, and *posterior* (*calf*) muscles. The **anterior muscles** are generally **dorsiflexors** at the **ankle** and **extensors** of the **toes.** The **lateral group everts the foot,** and the **posterior muscles** are **plantar flexors** at the **ankle** and **flexors** of the **toes. Inversion** of the foot is caused by some of the anterior and posterior muscles working together.

The nerve supply comes from three major branches of the sciatic nerve. The **tibial nerve** supplies the posterior muscles; the **superficial peroneal nerve** goes to the lateral group; and the **deep peroneal nerve** innervates the anterior group. The student would be correct if he suspected that the superficial and deep peroneals were the terminal branches of the common peroneal nerve, itself a branch of the sciatic which we met in the thigh. Refer to Chart XIII-3 for the muscles in these groups (page 822).

The posterior muscles, the largest of the three groups, are subdivided into superficial and deep layers, each invested by its own fascial covering. The **Gastrocnemius,** the **Soleus,** and the **Plantaris** all insert on the back end of the calcaneus. The most superficial of this **upper layer** is the Gastrocnemius, which entirely cov-

Table XIII-9. Muscles of the Leg

Muscle	Nerve	Origin and Insertion	Primary Action
Gastrocnemius (Fig. XIII-103)	$S_{1,2}$ Tibial nerve	(O) By lateral and medial heads from posterior surface of each femoral condyle (I) Posterior surface of calcaneus by calcaneal tendon (tendon of Achilles)	Plantar flexion at ankle
Soleus (Fig. XIII-104)	$S_{1,2}$ Tibial nerve	(O) Posterior surface of head of fibula and upper two-thirds of shaft; Popliteal line and medial border of middle one-third of tibia (I) With Gastrocnemius on posterior surface of calcaneus by calcaneal tendon	Plantar flexion at ankle
Plantaris (Fig. XIII-104)	L_4-S_1 Tibial nerve	(O) Lateral epicondyle of femur just above attachment of lateral head of Gastrocnemius (I) On calcaneus anteromedial to calcaneal tendon	Plantar flexion at ankle
Popliteus (Fig. XIII-105)	L_5, S_1 Tibial nerve	(O) Lateral surface of lateral condyle of femur (I) Posterior surface of tibia, above popliteal line	Rotation and flexion at knee
Tibialis posterior (Fig. XIII-106)	L_5, S_1 Tibial nerve	(O) Posterior surface of upper two-thirds of tibia beginning at popliteal line; medial surface of upper two-thirds of fibula and posterior surface of interosseus membrane (I) Tuberosity of navicular bone with fibers to all the other tarsals *except* the talus; bases of second, third, and fourth metatarsals	Adduction and inversion of foot
Flexor digitorum longus (Fig. XIII-107)	L_5, S_1 Tibial nerve	(O) Posterior surface of middle three-fifths of tibia (I) Plantar surface of base of distal phalanx of each of the four lateral toes	Flexion of lateral four toes; plantar flexion of foot and ankle
Flexor hallucis longus (Fig. XIII-105)	L_5-S_2 Tibial nerve	(O) Posterior surface of lower two-thirds of fibula (I) Plantar surface of base of distal phalanx of great toe	Flexion of big toe; plantar flexion of foot
Peroneus longus (Fig. XIII-108)	L_4-S_1 Superficial peroneal nerve	(O) Lateral condyle of tibia; lateral surface of head and upper two-thirds of fibula (I) Lateral margin of plantar surface of medial cuneiform and base of first metatarsal	Eversion of foot; plantar flexion of ankle and foot
Peroneus brevis (Fig. XIII-109)	L_4-S_1 Superficial peroneal nerve	(O) Lateral surface of lower two-thirds of fibula (I) Tuberosity on lateral side of base of fifth metatarsal	Eversion of foot; plantar flexion of foot
Tibialis anterior (Fig. XIII-110)	L_4-S_1 Deep peroneal nerve	(O) Lateral condyle and upper two-thirds of lateral surface of tibia (I) Plantar surface of base of first metatarsal and medial surface of medial cuneiform	Inversion and dorsiflexion of foot
Extensor digitorum longus (Fig. XIII-111)	L_4-S_1 Deep peroneal nerve	(O) Lateral condyle of tibia and upper three-fourths of anterior surface of fibula (I) By four tendons on dorsal surface of second and third phalanges of four lesser toes	Extension of four lesser toes, dorsiflexion of foot
Peroneus tertius (Fig. XIII-112)	L_4-S_1 Deep peroneal nerve	(O) Anterior surface of lower one-third of fibula (I) Dorsal surface of base of fifth metatarsal	Eversion and dorsiflexion of foot
Extensor hallucis longus (Fig. XIII-112)	L_4-S_1 Deep peroneal nerve	(O) Middle half of anterior surface of fibula (I) Dorsal surface of base of distal phalanx of great toe	Extension of great toe

The following muscles which move the leg, ankle, and foot are shown in place on the opposite page.

Fig. XIII-103. Gastrocnemius.
Fig. XIII-104. Triceps surae and Plantaris.
Fig. XIII-105. Popliteus and Flexor hallucis longus.
Fig. XIII-106. Tibialis posterior.
Fig. XIII-107. Flexor digitorum longus and Quadratus plantae.
Fig. XIII-108. Peroneus longus.
Fig. XIII-109. Peroneus brevis. Note that the tendon of the muscle runs underneath the superior and inferior peroneal retinacula.
Fig. XIII-110. Tibialis anterior, Extensor digitorium brevis and Extensor hallucis brevis.
Fig. XIII-111. Extensor digitorum longus.
Fig. XIII-12. Peroneus tertius, Extensor hallucis longus and Dorsal interossei.

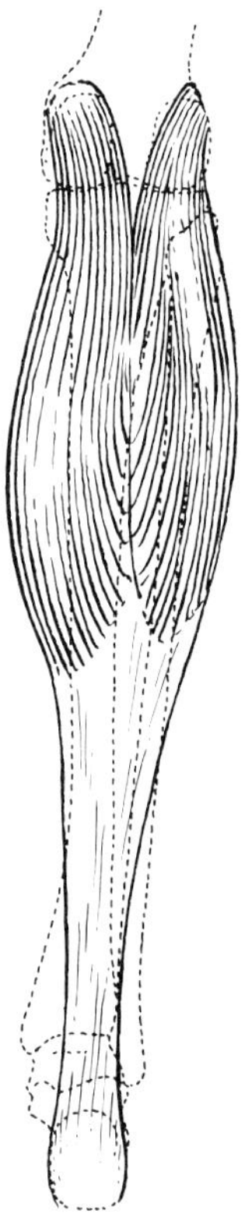

Fig. XIII-103.

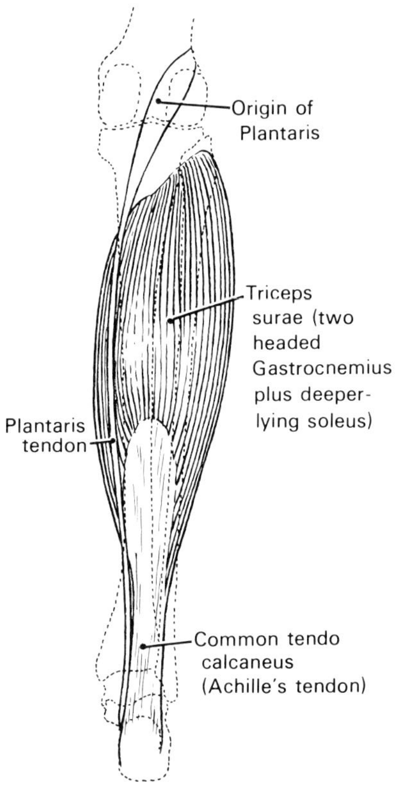

Fig. XIII-104.

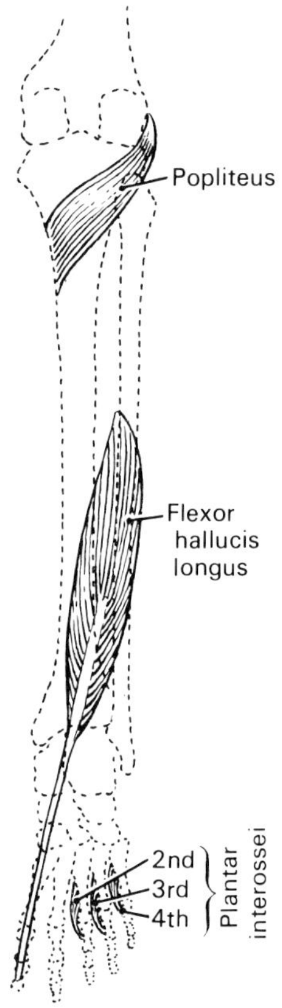

Fig. XIII-105.

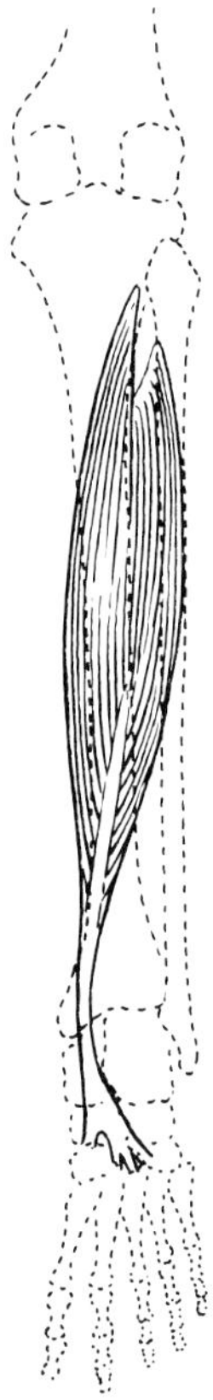

Fig. XIII-106.

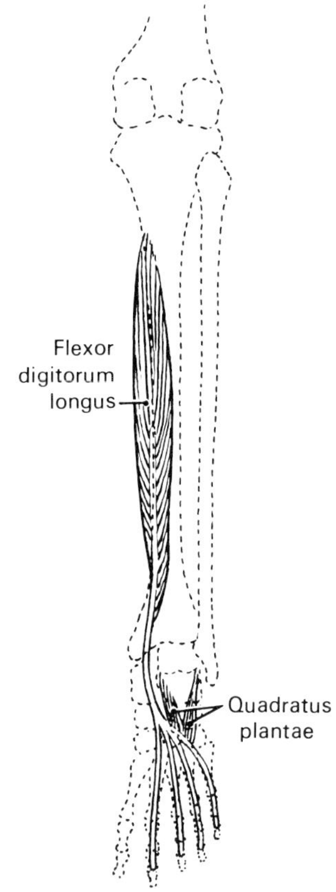

Fig. XIII-107.

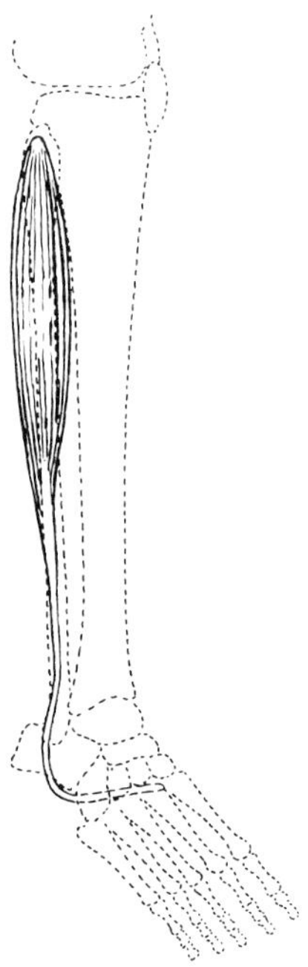

Fig. XIII-108.

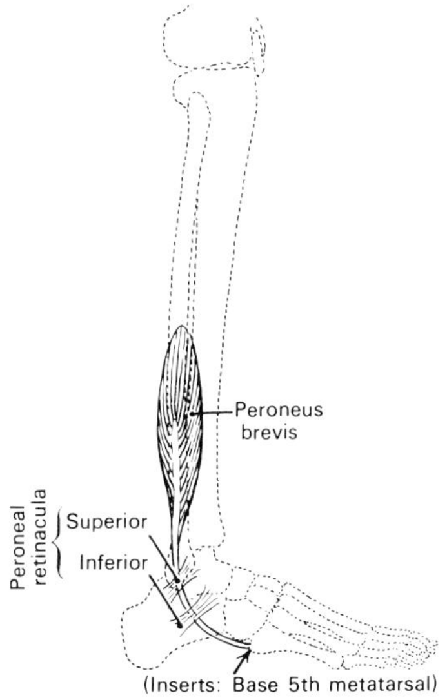

Fig. XIII-109.

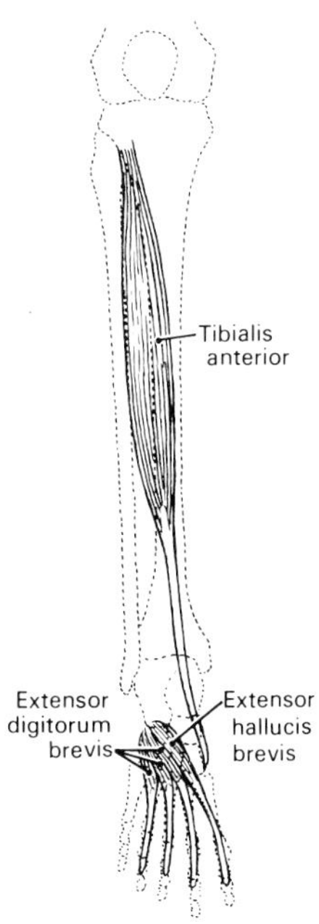

Fig. XIII-110.

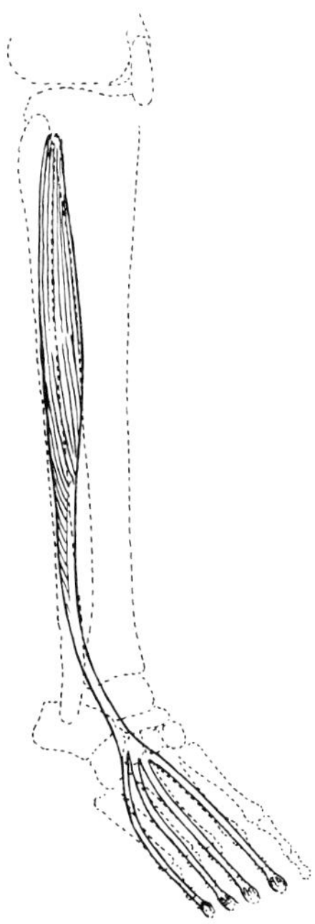

Fig. XIII-111.

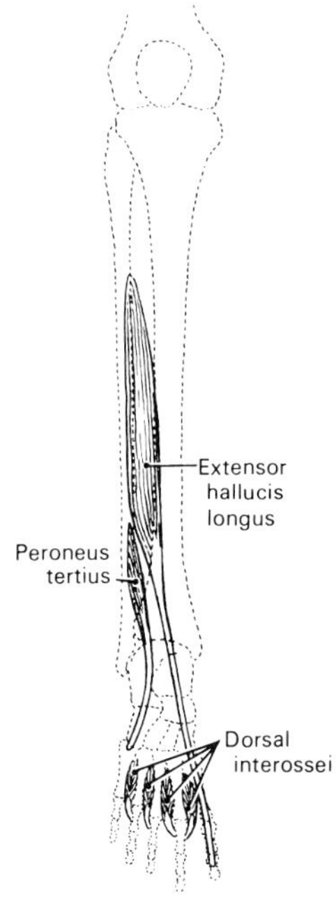

Fig. XIII-112.

Chart XIII-3. Muscle Groups of the Leg

Anterior	Lateral	Posterior
Extensor digitorum longus	Peroneus longus	Gastrocnemius
Peroneus tertius	Peroneus brevis	Soleus
Extensor hallucis longus		Plantaris
Tibialis anterior		Popliteus
		Flexor hallucis longus
		Flexor digitorum longus
		Tibialis posterior

ers the calf and is the only muscle mass readily palpated there. The Soleus shares a common tendon of insertion with the Gastrocnemius and together they are called the **Triceps Surae,** the Gastrocnemius having two heads and the Soleus one.

The **deep muscles** of the **calf** are the **Popliteus,** the **Flexor Hallucis Longus,** the **Flexor Digitorum Longus,** and the **Tibialis Posterior.** Hallucis is to great toe as Pollicis is to thumb. The last three muscles cross the ankle joint and continue into the foot deep to the flexor retinaculum. The Tibialis Posterior is the deepest of the deep muscles and, due to its multiple attachments on the plantar surface of the tarsal bones and its direction of pull, it is of great importance to the maintenance of the longitudinal arch of the foot. The Flexor Hallucis Longus, a strong pennate muscle, provides the "push-off" from the foot in walking, running, and jumping. Table XIII-9 tabulates the information about the muscles of the leg.

Articulations of the Foot

Just as the articulations of the carpal bones enhanced the movements of the wrist, so the joints between the tarsal bones provide greater movement of the foot. As you will recall, there are seven tarsal bones, and only the *talus* articulates with the bones of the leg. The talus sits upon the calcaneus (the "heel" bone) and joins with it at the important **subtalar joint.** Between the front ends of both the talus and calcaneus is the **transverse tarsal joint (midtarsal joint)** uniting them with the more distal tarsals. These are the joints which allow *supination* and *pronation* of the foot. You have already learned what the movements of eversion and inversion are. Supination and pronation, like circumduction, are combination movements. Supination is inversion and adduction. Pronation is eversion and abduction.

Because of the longitudinal arch of the foot, the weight of the body, while standing, is transmitted to the ground only through the posterior end of the calcaneus and the distal ends (heads) of the metatarsals. Both the front ends of the calcaneus and the talus are normally off the ground and form the lateral and medial parts of the arch. The toes provide the thrust in walking when the body's weight is shifted forward onto the ball of the foot.

The foot can be divided transversely into three units by the transverse tarsal joint and the tarsometatarsal articulations. The posterior unit contains the talus and calcaneus. The middle unit is comprised of the five small tarsal bones: navicular, cuboid, and the medial, intermediate, and lateral cuneiforms. This leaves the metatarsals and phalanges to fill the anterior unit.

The *subtalar joint,* in the posterior unit, is a synovial joint of the arthrodial variety between the inferior surface of the talus and the anterior and superior surfaces of the calcaneus. Because there is a cartilaginous articular facet lined with synovial membrane, the contact between the talus and the **plantar calcaneonavicular ligament** is also considered part of the joint. Besides being important as a contributor to supination and pronation, this joint also possesses the "spring ligament." The plantar calcaneonavicular ligament passes from the lower surface of the calcaneus to the lower surface of the navicular, forming a sling on which the head of the talus rests. In this position it provides support for the highest part of the longitudinal arch, and supposedly gives the arch some of its "spring."

The *transverse tarsal joint* (midtarsal joint) consists of two articulations. Laterally the calcaneus articulates with the cuboid (calcaneocuboid joint) in a non-axial, gliding joint. The medial articulation, the **talonavicular joint,** is tri-axial, and resembles a shallow ball and socket joint (Fig. XIII-113). Together they supplement dorsal and plantar flexion at the ankle, allow slight adduction and abduction of the anterior unit of the foot on the posterior part, and very slight inversion and eversion.

The articulations between the navicular, the cuboid, the cuneiforms and the bases of the second and third metatarsals (intertarsal joints) are of the arthrodial type, permitting only slight movement. Although each joint is a separate entity, they all share a single synovial cavity. Thus far we have said little about the ligaments of these joints with the exception of the plantar calcaneonavicular ligament. The tarsal bones are bound together by three sets of ligaments. Collectively, they are the **dorsal tarsal ligaments** which are thin and found on the dorsum of the foot; the **plantar tarsal ligaments,** on the sole of the foot and very heavy, acting as supports to the bony arch; and the numerous ***interosseous tarsal ligaments*** between adjacent surfaces of the bones. Generally, they are named according to the bones on which they attach, such as the plantar intercuneiform ligament or the dorsal cuboideonavicular ligament.

Besides the calcaneonavicular ligament which has already been discussed, there is one other plantar ligament of import. The **long plantar ligament** (Fig. XIII-114) extends from the calcaneus posteriorly to the oblique ridge on the plantar surface of the cuboid and anteriorly to the bases of the second, third, fourth, and fifth metatarsals. Thus, it passes along most of

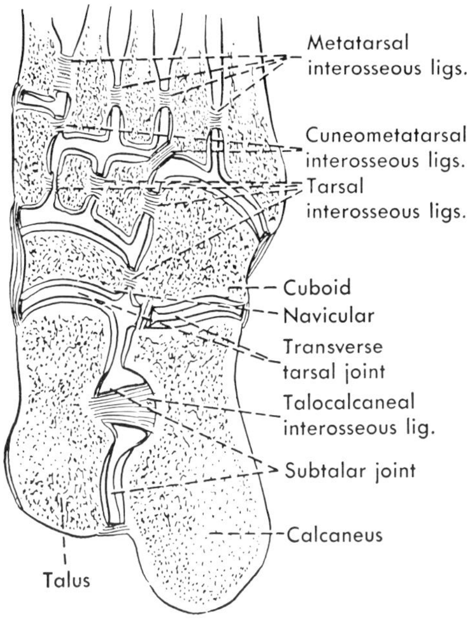

Fig. XIII-113. The intertarsal joints of the foot as shown in an oblique section. (From H. Hollinshead, *Functional Anatomy of the Limbs and Back*, 3rd ed., W. B. Saunders, Philadelphia, 1969.)

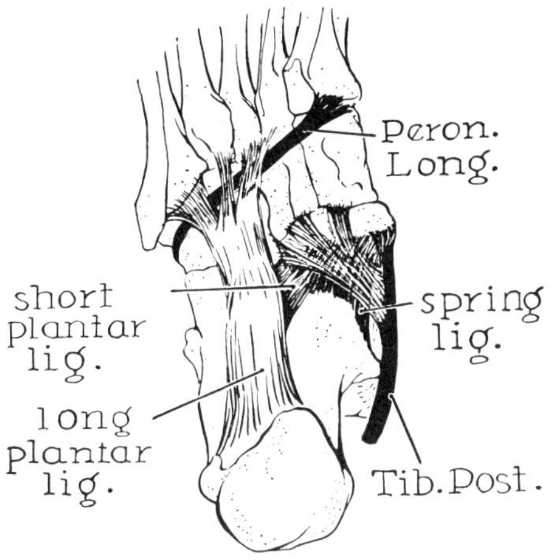

Fig. XIII-114. Important ligaments and tendons in the sole of the (right) foot help maintain arches. (From J. V. Basmajian, *Primary Anatomy*, 6th ed., Williams & Wilkins, Baltimore, 1970.)

the length of the lateral side of the longitudinal arch and is the chief support for it.

The **tarsometatarsal joints** are non-axial. The articulation between the first metatarsal and the medial cuneiform has a separate joint space. The second and third tarsometatarsal joints are incorporated in the common synovial cavity of the lesser intertarsal joints mentioned above. The corresponding joints between the fourth and fifth metatarsals and the cuboid are confluent. They are strengthened by dorsal and plantar tarsometatarsal ligaments and interosseous ligaments.

There are two sets of **intermetatarsal articulations,** those at the bases and those at the heads of the metatarsals. Usually there is no articulation between the bases of the first and second metatarsals. The rest, however, are tightly bound together so that only minimal motion is allowed. There are the usual dorsal, plantar, and interosseous ligaments between the bones. The heads of the metatarsals are connected by transverse metatarsal ligaments, including the head of the first metatarsal as the great toe does not enjoy the freedom that the thumb does.

The **metatarsophalangeal joints** are condyloid so that the toes may flex, extend, and move slightly laterally (abduction and adduction). The joint capsules are reinforced dorsally by the individual slips of the **extensor tendon.** Laterally there are fan-shaped **collateral ligaments** similar to those of the metacarpophalangeal joints of the hand. On the plantar surface, the joint capsule is completed by the heavy **plantar ligaments** (**plates**) over which the flexor tendons glide. Since the metatarsophalangeal joint of the big toe takes the weight of the body when moving, it is equipped with two sesamoid bones in the tendons of the Flexor Hallucis Brevis to cope with the burden. These bones also protect the tendon of the Flexor Hallucis Longus.

The **interphalangeal joints** of the foot are like those of the hand. They are hinge joints reinforced by collateral ligaments on the sides, the extensor tendon dorsally, and a plantar ligament on the flexor surface.

The *fascia* of the dorsal surface of the foot is thin and unremarkable. The **plantar aponeurosis** (plantar fascia), however, is extremely strong and well developed. The central portion is the thickest. Attached to the calcaneus, it advances forward, becoming broader and thinner and dividing near the heads of the metatarsals into five processes, one for each toe. These processes help to reinforce the flexor digital tendon sheaths and form arches around the tendons to attach to the distal ends of the metatarsals and the bases of the proximal phalanges. The central portion of the aponeurosis is continuous with the thinner lateral and medial portions. However, at the lines of junction, it sends deep intermuscular septa into the foot, dividing the short muscles of the foot into compartments. The tendons of the Flexor Digitorum and the Lumbrical muscles lie in the central compartment. The muscles of the big and little toes lie in the medial and lateral compartments respectively.

The Arches of the Foot

All the weight of the body transmitted to the talus from the tibia and fibula, is, in turn, transmitted backward and down to the back end of the calcaneus, or forward and down to the heads of the metatarsals. The weight is divided approximately equally between the calcaneus and the metatarsal heads. Between these weight-bearing points are the **arches** of the foot.

There are two arches, the longitudinal and the transverse. The **longitudinal arch,** anatomically, comes in two parts, a higher medial arch and a lower lateral one.

The medial part is made up of the calcaneus, talus, navicular, three cuneiforms, and the three medial metatarsals. The lateral component also begins with the calcaneus and then proceeds forward through the cuboid to the heads of the two lateral metatarsals. The inner part of the arch, with its flexibility and its curving arch, is well designed to absorb shock, as the foot must do during locomotion. The lateral part is nearly flat and lacks mobility so that it is better designed for weight support. *Functionally,* studies indicate that both parts of the longitudinal arch and the transverse arch act as a single, complex arch through which stress spreads out in all directions. The stress at any particular point is proportional to the height of the point on the curve, whichever curve that may be. The keystone of the arch is the talus.

The **transverse arch** runs from side to side of the foot. The head of the talus and the navicular bone form the highest part of the arch on the medial side, the calcaneus and cuboid the highest point on the outer side. From these points forward, the transverse arch gradually flattens out until it reaches the heads of the metatarsals. Some people say there is a transverse metatarsal arch and some deny it. The point of controversy is that the "transverse metatarsal arch" flattens out completely when weight-bearing. The concavity called the metatarsal arch exists only when in a non-weight-bearing position. Both the longitudinal arch and the transverse arch in the region of the midtarsal joint maintain an arched configuration. They act more like springs, giving a little, but not completely.

Muscles of the Foot

There are only two intrinsic muscles of the *dorsum* of the foot: the **Extensor Digitorum Brevis** and the **Extensor Hallucis Brevis.** The Extensor Digitorum Brevis sends three tendons toward the second, third, and fourth toes. However, the tendons never quite make it to their destinations and unite with those of the long extensor going to the same toes instead. The Extensor Hallucis Brevis is often considered a part of the Extensor Digitorum Brevis. It sends a single tendon to the base of the proximal phalanx of the big toe. These two short muscles assist the long extensors in extension of the toes.

So much for the top of the foot. The plantar musculature, including the long tendons of the toes, form four layers. The short plantar muscles act as elastic ties for the arches of the foot. All their attachments are within the foot itself. Posteriorly, they arise from the under surface of the calcaneus. Anteriorly, they attach near the bases and along the shafts of the metatarsals.

Layer I, the most superficial, contains the **Abductor Hallucis,** the **Flexor Digitorum Brevis,** and the **Abductor Digiti Minimi.** Layer II consists of the **tendons** of the Flexor Hallucis Longus and Flexor Digitorum Longus. The intrinsic muscles are the **Quadratus Plantae** and the four **Lumbricals.** The third layer of muscles is composed of the **Flexor Hallucis Brevis, Flexor Digiti Minimi Brevis,** and the **Adductor Hallucis.** The deepest muscles, in Layer IV, are the seven **Interossei.** These are kept company by the *tendons* of Peroneus Longus and Tibialis Posterior. When discussing the actions of the interossei, remember that in the *foot,* it is the *second toe* and its *metatarsal* which are the axial line, not the third. For a summary of these muscles, see Table XIII-10.

INNERVATION AND BLOOD SUPPLY OF THE LOWER LIMB

In the tables of muscles we have noted the usual segmental origin of the nerves and named the major peripheral branches. Just as there is a brachial plexus for the arm, there is a lumbosacral plexus for the pelvic girdle and lower limb. The plexus is formed from the anterior rami of the spinal nerves L_1 to S_3. Most anatomists prefer to separate this complex mass of nerves into the **lumbar plexus,** L_1 to L_4 inclusive, and the **sacral plexus,** L_4 to S_3. Note that the anterior ramus of L_4 is involved with *both* plexuses. Of the six named branches of the lumbar plexus issuing from the interior of the Psoas muscle where they formed, only the lowest two, large branches make their way into the thigh to supply the muscles there. There are many small cutaneous branches splitting off the **femoral** and **obturator nerves,** but we are concerned here with innervation of the muscles. The femoral nerve passes under the inguinal ligament and immediately divides into a series of smaller nerves to supply the anterior group of thigh muscles and the hip and knee joints. The obturator nerve escapes from the pelvis through the obturator foramen and divides at once into anterior and posterior branches to supply the anteromedial thigh muscles, the capsules of the hip and knee joints, and a variable patch of skin on the medial aspect of the thigh.

The *sacral plexus* gathers almost all its fibers into one huge branch—the **sciatic nerve**—which is the largest nerve in the body. Passing through the greater sciatic foramen and lying deep to the Gluteus Maximus, it runs down the middle of the posterior compartment of the thigh, supplying the muscle there. Approximately at the upper limit of the popliteal space, the sciatic nerve divides into two branches: the **tibial nerve** and the **common peroneal nerve.**

The *tibial nerve* continues across the knee and down the posterior aspect of the leg (calf). It gives twigs to the muscles of the back of the knee and the leg. Passing behind the medial malleolus, it divides into its terminal branches, the **medial** and **lateral plantar nerves,** which supply the intrinsic muscles of the foot and the skin of the sole.

The *common peroneal* (lateral popliteal) *nerve* does not wait as long as the tibial to divide. As soon as it comes forward around the neck of the fibula, it divides into the **superficial peroneal nerve,** supplying the Peroneus Longus and Brevis and a large area of skin, and the **deep peroneal nerve.** The deep peroneal nerve passes along the anterior surface of the interosseous membrane and supplies the anterior muscles of the leg and foot.

There really are branches of the sacral plexus other than the sciatic nerve. These supply the muscles of the buttock. The **superior** and **inferior gluteal nerves** supply the three Glutei and the Tensor Fasciae Latae. *Small branches* of the plexus run to the rotators of the hip and are named according to the muscles they supply.

Circulation: The abdominal aorta, as you know, bifurcates into the common iliac arteries. Each common iliac artery in turn bifurcates to form the internal and external iliac arteries. It is at this point that we must pay attention to the general plan of the blood supply to the lower limb. Having begun the discussion in this chapter with the pelvic girdle and the muscles of the buttock, we must follow the route of the **internal iliac artery** at least far enough to see that it supplies two vessels to the gluteal region. The **superior gluteal artery** leaves the interior of the pelvis through the greater sciatic foramen and, as its name implies, supplies the superior part of the region. The second vessel also passes through the greater sciatic foramen and emerges to supply the muscles of the lower part of the gluteal region and upper part of the posterior thigh as the **inferior gluteal artery.**

The **external iliac artery** continues the general line of direction of the common iliac. It passes under the inguinal ligament to enter the thigh anteriorly and in the midline as the **femoral artery.** Very shortly after entering the thigh, the femoral artery gives rise to the **medial femoral circumflex artery.** This passes posteriorly above the adductor muscles and anastomoses with the inferior gluteal artery. It carries blood to the hip joint and upper part of the thigh in back. The **lateral femoral circumflex artery** may arise from a common stem with the medial branch or may arise nearby. It supplies the entire length of the lateral side of the thigh and gives off numerous small branches to the Quadriceps muscle which is its *raison d'etre.* The third branch of the femoral artery in the thigh is the **profunda femoris artery** which runs in the midline deep among the anterior muscles. As it goes, it gives off a series of **perforating branches** that pass through the insertions of the adductor muscles to reach the posterior muscles which it supplies richly. The profunda femoris terminates in the thigh as a final perforating branch itself.

As the femoral artery approaches the popliteal space, it turns around the medial aspect of the femur to run obliquely behind the knee as the **popliteal artery.** This gives off branches called the **genicular arteries** (L. *genu* = knee) to the knee joint. They form an encircling network of vessels about the knee, encompassing the lower end of the femur and the upper ends of the tibia and fibula and supplying twigs to the joint and its adjacent muscles.

As the popliteal artery approaches the superior tibiofibular articulation, it divides into the **posterior** and **anterior tibial arteries.** The posterior tibial artery gives off a large branch, the **peroneal artery,** which runs along behind the fibula primarily to supply the Flexor Hallucis Longus. The posterior tibial artery, meanwhile, is descending vertically beneath the Gastrocnemius and the Soleus and with numerous muscular branches to the posterior muscles of the leg and to the ankle joint.

As it passes behind the medial malleolus, the posterior tibial artery bifurcates into its terminal branches, the **medial** and **lateral plantar arteries.** These run among the muscles of the sole of the foot. The medial plantar artery provides blood to the medial muscles of the foot, primarily those of the great toe. The lateral plantar artery supplies all the other plantar muscles and the skin of the sole of the foot. The plantar arteries anastomose with each other as well as with the dorsalis pedis artery.

The *anterior tibial artery* comes forward between the tibia and fibula above the interosseous membrane. It runs down the front of the leg deep to the muscles, giving branches to them until it reaches the ankle. There it becomes the **dorsalis pedis artery,** supplying the dorsum of the foot and toes. It dives through the foot in the space between the first and second metatarsals and distal to their bases to anastomose with the medial and lateral plantar arteries.

The venous return of the blood in the lower limb is accomplished by deep veins accompanying the arteries of the same name. For details of the circulation and lymphatic drainage of the limb, refer to Chapter XI, page 737. The superficial vessels are noted in Chapter III.

REFERENCES

1. Ray, R. D., Johnson, R. J. and Jameson, R. M., Rotation of the forearm: an experimental study of pronation and supination. J. Bone & Joint Surg. *33-A:* 993, 1951.
2. Bledsoe, R. C., and Izenstark, J. L.: Displacement of fat pads in disease and injury of the elbow—a new radiographic sign. Radiology, *73:* 717–724, 1959.
3. MacEvan, D. W.: Changes due to trauma in the fat plane overlying the pronator quadratus muscle: a radiologic sign. Radiology, *82:* 879–886, 1964.

GENERAL REFERENCES

1. Gardner, Gray, and O'Rahilly: *Anatomy,* ed. 3 Saunders, Philadelphia, 1969.
2. Grant, J. C. B., and Basmajian, J. V.: *Grant's Method of Anatomy,* ed. 7. Williams & Wilkins, Baltimore, 1965.
3. Gray, Henry: *Anatomy of the Human Body,* ed. 28, edited by C. M. Goss. Lea & Febiger, Philadelphia, 1966.
4. Hollinshead, H.: *Functional Anatomy of Limbs and Back,* ed. 3. Saunders, Philadelphia, 1969.
5. Hollinshead, H.: *Textbook of Anatomy,* ed. 2. Hoeber (Harper-Row), New York, 1967.
6. Morris, Sir Henry: *Morris' Human Anatomy,* ed. 12, edited by B. J. Anson. McGraw-Hill, New York, 1966.
7. Wells, Katharine F.: *Kinesiology,* ed. 5. Saunders, Philadelphia, 1967.
8. Basmajian, J. V.: *Muscles Alive: Their Function Revealed by Electromyography,* ed. 2. Williams & Wilkins, Baltimore, 1967.
9. Basmajian, J. V.: *Primary Anatomy,* ed. 5. Williams & Wilkins, Baltimore, 1964.

Table XIII-10. Muscles of the Foot

Muscle	Nerve	Origin and Insertion	Primary Action
Layer I			
Abductor hallucis (Fig. XIII-115)	L_5, S_1 Medial plantar nerve	(O) Medial process of tuberosity of calcaneus (I) Base of proximal phalanx of great toe	Abduction and flexion of great toe
Flexor digitorum brevis (Fig. XIII-116)	L_5, S_1 Medial plantar nerve	(O) Medial process of calcaneal tuberosity; deep surface of center portion of plantar aponeurosis and intermuscular septa (I) By four tendons to the sides of the second phalanx of the four lateral toes	Flexion of four lateral toes
Abductor digiti minimi (Fig. XIII-115)	$S_{1,2}$ Lateral plantar nerve	(O) Lateral process of tuberosity of calcaneus; plantar surface of calcaneus, and distal part of medial process; plantar aponeurosis (I) Lateral side of base of first phalanx of little toe	Abduction of little toe
Layer II			
Quadratus plantae (Fig. XIII-117)	$S_{1,2}$ Lateral plantar nerve	(O) Medial head: medial concave surface of calcaneus Lateral head: lateral border of inferior surface of calcaneus; from long plantar ligament (I) Lateral margin and dorsal and plantar surfaces of the tendon of the Flexor digitorum longus	Assists flexion of four lesser toes
Lumbricals (Fig. XIII-118)	First lumbrical L_5, S_1–Medial plantar nerve All others $S_{1,2}$—Lateral plantar nerve	(O) From tendons of Flexor digitorum longus (I) Dorsal expansions of tendons of Extensor digitorum longus of proximal phalanges	Flexion of the four small toes
Layer III			
Flexor hallucis brevis (Fig. XIII-117)	L_5, S_1 Medial plantar nerve	(O) The cuneiform bones (I) By common tendon with Abductor hallucis on medial side of proximal phalanx of big toe; by second tendon in common with Adductor hallucis on lateral side of first phalanx	Flexion of great toe
Flexor digiti minimi brevis (Fig. XIII-116)	$S_{1,2}$ Lateral plantar nerve	(O) Cuboid and base of fifth metatarsal (I) Towards plantar surface on lateral side of proximal phalanx of little toe	Flexion of little toe
Adductor hallucis (Fig. XIII-117)	$S_{1,2}$ Lateral plantar nerve	(O) Oblique head: long plantar ligament and bases of second, third, and fourth metatarsals Transverse head: capsules and deep transverse metatarsal ligaments of third, fourth, and fifth metatarsophalangeal joints (I) Lateral side of base of first phalanx of great toe, together with Flexor hallucis brevis	Adduction of great toe
Layer IV			
Plantar interossei (3) (Fig. XIII-119)	$S_{1,2}$ Lateral plantar nerve	(O) Third, fourth, and fifth metatarsals, respectively (I) Medial side of proximal phalanges of third, fourth and fifth toes	Adduction and flexion of three lateral toes
Dorsal interossei (4) (Fig. XIII-115)	$S_{1,2}$ Lateral plantar nerve	(O) From adjacent metatarsals between which they lie (I) First interosseous: medial side of proximal phalanx of second toe Second, third, and fourth interossei: lateral side of second, third and fourth toes	Abduction and flexion of toes II, III, and IV
Dorsal group			
Extensor digitorum brevis and Extensor hallucis brevis (Fig. XIII-110)	L_5, S_1 Deep peroneal nerve	(O) Upper and lateral surfaces of calcaneus (I) Hallucis: by single tendon to dorsal surface of base of proximal phalanx of big toe Digitorum: by three tendons into lateral sides of tendons of Extensor digitorum longus of second, third and fourth toes	Extension of toes

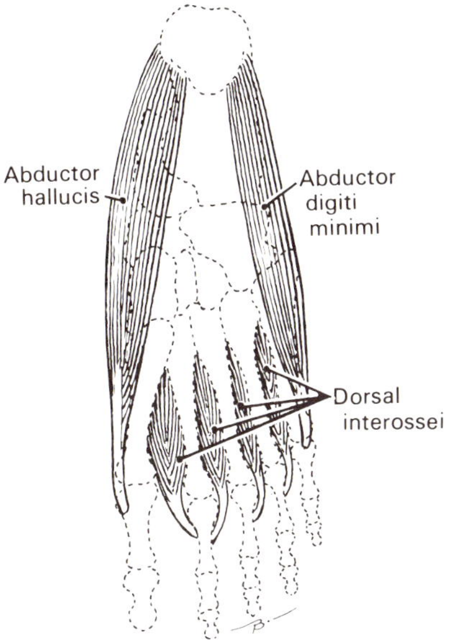

Fig. XIII-115.

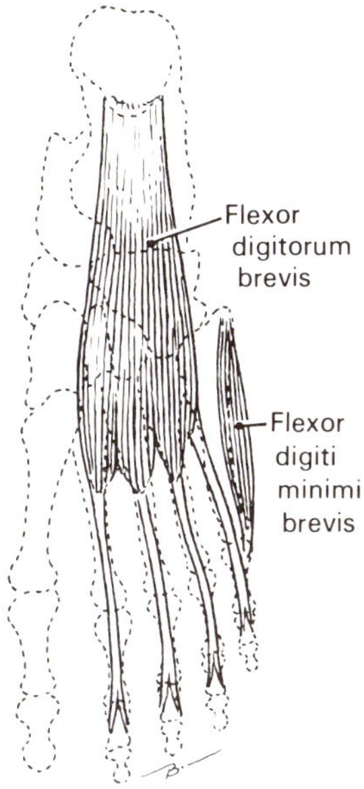

Fig. XIII-116.

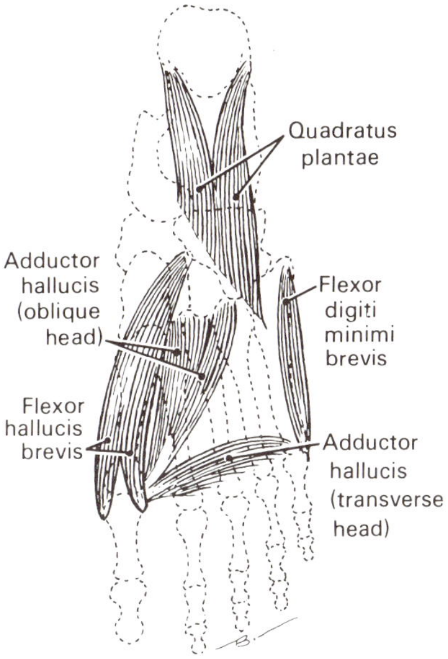

Fig. XIII-117.

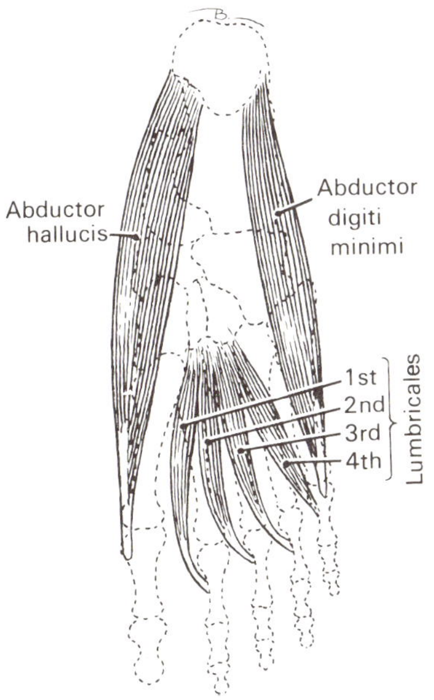

Fig. XIII-118.

Fig. XIII-115. The Dorsal interossei. The abductors of the great and little toes have been left in place.

Fig. XIII-116. Flexor digitorum brevis and Flexor digiti minimi brevis.

Fig. XIII-117. The muscles of the third layer of the foot. The Quadratus plantae (a second layer muscle) has been left in place.

Fig. XIII-118. Lumbricals, Abductor hallucis and Abductor digiti minimi.

Fig. XIII-119. Flexor hallucis brevis and Plantar interossei.

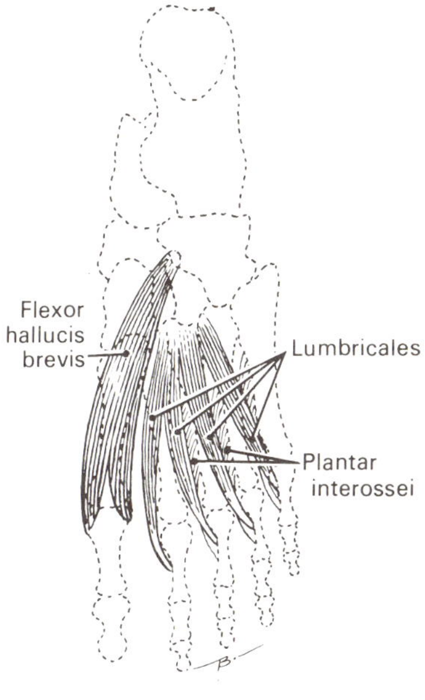

Fig. XIII-119.

FOURTEEN

The Central Nervous System

PART I—THE CENTRAL NEURAXIS

Introduction

Throughout the text we have considered various parts of the **peripheral nervous system,** *e.g.,* the 31 pairs of spinal nerves and their sensory root ganglia; the 12 pairs of cranial nerves and their sensory ganglia; and the autonomic or visceral nervous system with its outlying motor ganglia. It is time now to present a very basic view of the **central nervous system.** Very schematically, Figure XIV-1 indicates that it consists of the paired cerebral hemispheres; the brainstem and its associated cerebellum; and the spinal cord. In such a simple view as in this Figure, the upper diencephalic part of the brainstem, the thalamus, is not in sight since it lies buried in the midline between hemispheres.

All of these structure which constitute the central nervous system (CNS) comprise the **central neuraxis** which is encased in bone. The brain and brainstem lie within the bony cranium and the spinal cord lies within the vertebral canal. Essentially, we inherited a tubular central neuraxis from our chordate ancestors. Recall, from Chapter I, how it rolled up from a flat, ectodermal plate in the midline above the forming notochord. You should not be surprised to find it canalized in certain areas and even quite cavitous in others. Indeed, all of us have larger "holes in our head" than we might care to admit. We get around the problem nicely by calling them **ventricles.** We shall discover odd-shaped **lateral ventricles** in each hemisphere; a V-shaped narrow cleft roofed over by ependyma and pia, the **third ventricle,** in the thalamus; and a shallow trough, the **fourth ventricle** in pons and medulla. The latter is partly roofed in the manner of the third ventricle and partly by solid neural tissue, the **superior medullary velum.** The cerebellum lies above this roof (Fig. XIV-2). The midbrain is pierced by a narrow channel, the **cerebral aqueduct** (of Sylvius). The spinal cord has a pin-point canal, hardly patent in the adult, plugged as it is by ependymal debris. Most of these hollow channels were much wider in the embryo. As the walls of the developing neural tube thickened and reshaped, the bore of the tube was altered to conform to local configurations. All of these hollow channels communicate with each other and they are filled with cerebrospinal fluid—all except

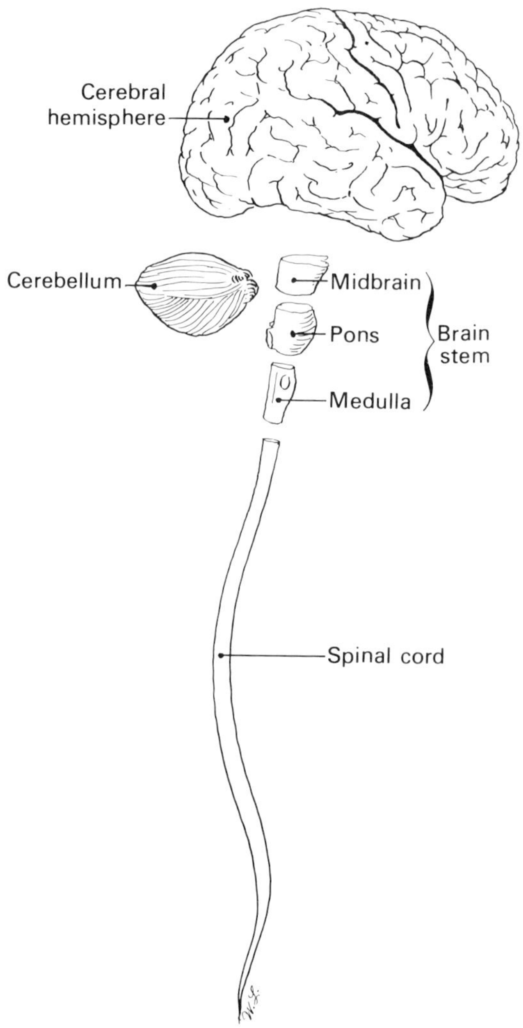

Fig. XIV-1. Schema of parts of the central nervous system.

the central canal of the cord where there can be only a mere trickle or film of fluid.

Indeed, we shall see that cerebrospinal fluid is formed from the **choroid plexuses** within the ventricles. These plexuses are formed by capillary tufts which invaginate walls or roofs of the tube in places where these walls are only paper-thin endymal. The fluid eventually escapes from openings in the fourth ventricle to enter

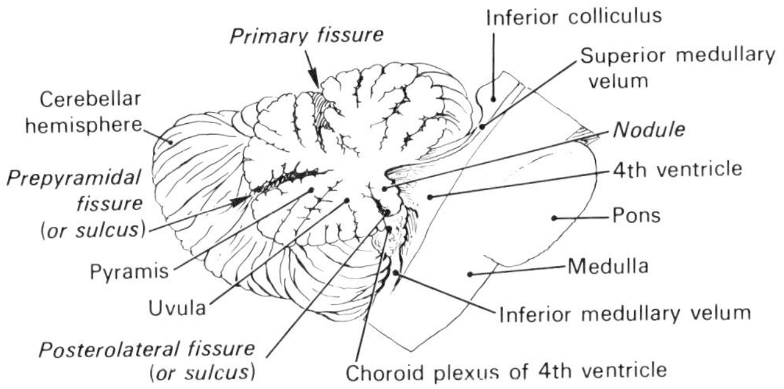

Fig. XIV-2. Midsagittal section through vermis of cerebellum, looking toward left cerebellar hemisphere. Note relationship to hindbrain and fourth ventricle and attend to important cerebellar fissures (sulci).

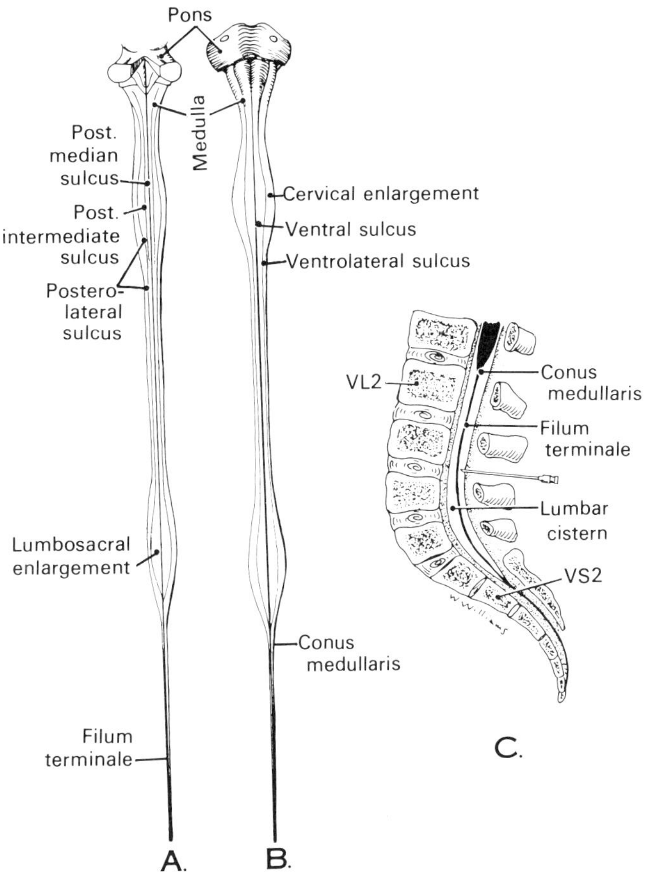

Fig. XIV-3. The spinal cord (**A**) posteriorly and (**B**) anteriorly viewed. (**C**) Indicates its termination at the upper part of VL_2 in the vertebral canal. The sac of dura-arachnoid continues beyond to terminate at VS_2. Cerebrospinal fluid can be withdrawn from this cistern.

the subarachnoid space between pia and arachnoid, membranes which envelope cord and brain. In this manner, a water cushion is placed between the central neuraxis and the bone which confines the neural tissue. Tougher dura mater, of course, is interposed between bone and arachnoid.

General Anatomy: It would be well to review the data relative to the spinal cord in Chapter II. Here we shall enlarge a bit upon that discussion and may not cover all of the basic points mentioned there. Some of the same illustrations will be offered for review purposes and we may add some new ones.

The spinal cord, about 18 inches long in men and 2 inches shorter in women, extends upward in the vertebral canal to become continuous with the medulla at the foramen magnum. It is shorter than the vertebral canal itself, occupying only the upper two-thirds of it. It terminates caudally in the canal at the tapered conus in the region of the small of the back (approximately at the upper border of VL_2). It is a slightly flattened cylinder, wider than it is thick, and wider in the cervical and lumbosacral segments than elsewhere (Fig. XIV-3). Its close pial coat spins out as a thread-like internal filum which runs from the conus to VS_2. Here, the sac of dura-arachnoid, which forms the lumbar cistern, terminates. Dura and pia now continue as an external filum through the sacral hiatus to anchor upon coccyx as the **coccygeal ligament.** Thus, the cord is anchored in the vertebral canal by fusion with the medulla superiorly and by the ligament inferiorly. Its lateral anchorings are the pial **denticulate ligaments** (Fig. XIV-4) and the attachments of dural sleeves over emerging nerve rootlets at each intervertebral formina.

A water-cushion of cerebrospinal fluid occupies the space between arachnoid and pia and so surrounds the cord within the canal. Below the conus, this subarachnoid space expands into the lumbar cistern, making cerebrospinal fluid available for sampling without the risk of cord injury. The relationship of cord and rootlets to the meningeal coats is reviewed in Figure XIV-4.

Since 11 cord segments (L_1 to coccygeal) are packed closely near the conus (Fig. XIV-5), the long nerve rootlets in this part of the vertebral canal must run a considerable distance in the canal before they reach their respective lower intervertebral foramina for exit as spinal nerves. Thus, they form a cauda equina about the filum terminale.

An almost continuous series of nerve rootlets emerge from the **posterolateral sulcus** of the cord and another set from the **venterolateral sulcus.** These **sensory** and **motor roots,** respectively, join at the intervertebral foramina to become spinal nerves (31 pairs of them). The sensory root has a swelling upon it which contains the unipolar, dorsal root, ganglion cells. The exception, possibly, is C_1 which, more often than not, lacks a ganglion. Its major role is to supply the tiny muscles of the suboccipital triangle and it rarely sends sensory twigs to overlying skin. Each motor (ventral) root emerges from the cord and each sensory (dorsal) root enters the cord by means of a series of fine, root filaments (Figs. XIV-4 and 6B). It is the variable distance between the cranial and caudal extent of these rootlets at any given level that comprises a spinal cord segment. Thus, cord segments are not necessarily all of the same length.

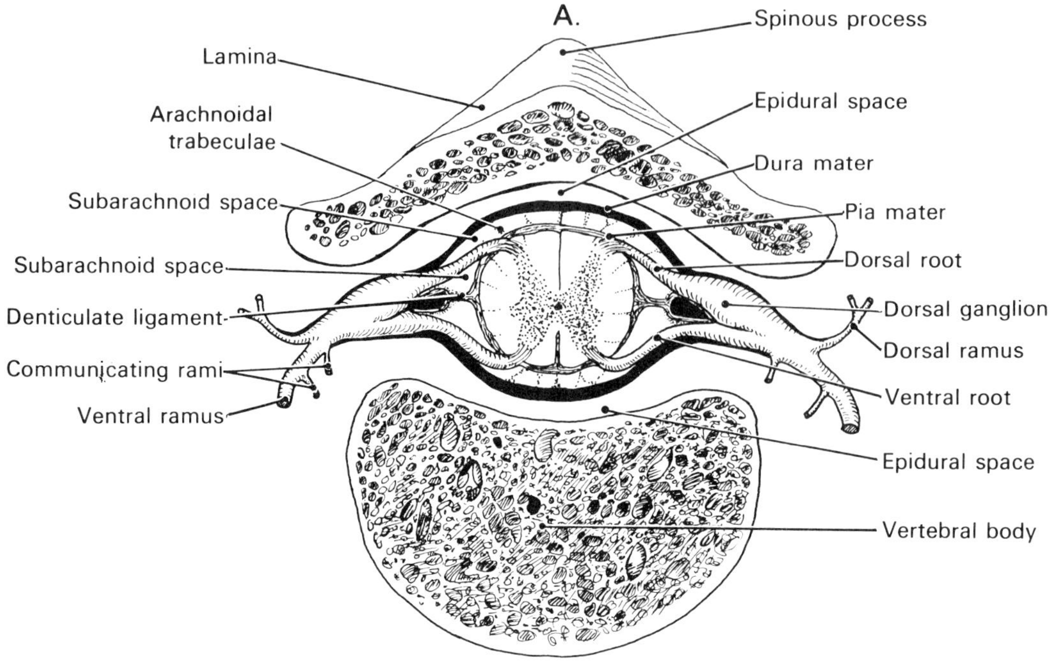

Fig. XIV-4. Relation of cord and rootlets to their meningeal coverings. **A.** Cross-sectional view at midthoracic cord level.

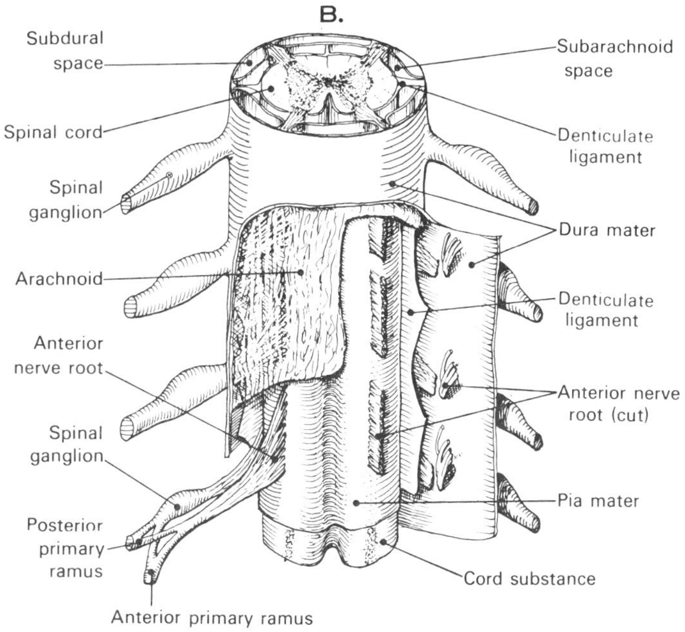

Fig. XIV-4B. Longitudinal view with meninges cut away in stages to expose filamentous rootlets and the pial denticulate ligament.

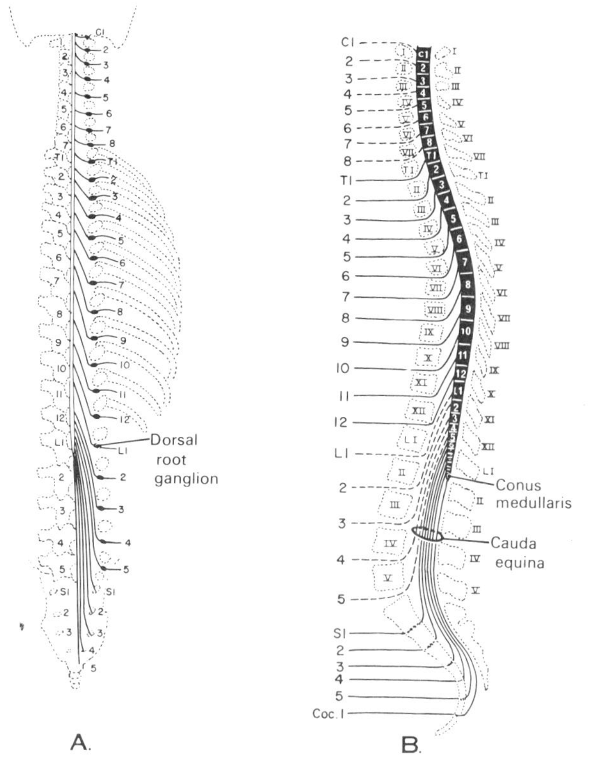

Fig. XIV-5. Relation of vertebral segments to emergence of cord rootlets. **A,** dorsal view; **B,** lateral view.

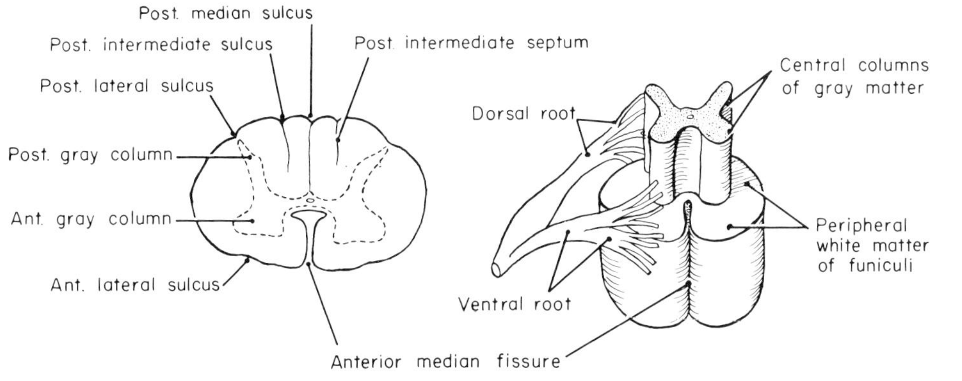

Fig. XIV-6. Topography of cervical spinal cord. **A.** Sulci dividing cord section into funiculi. **B.** Root filaments, gray and white matter. (From R. C. Truex and M. B. Carpenter, *Human Neuroanatomy*, 6th ed. Williams & Wilkins, Baltimore, 1969.)

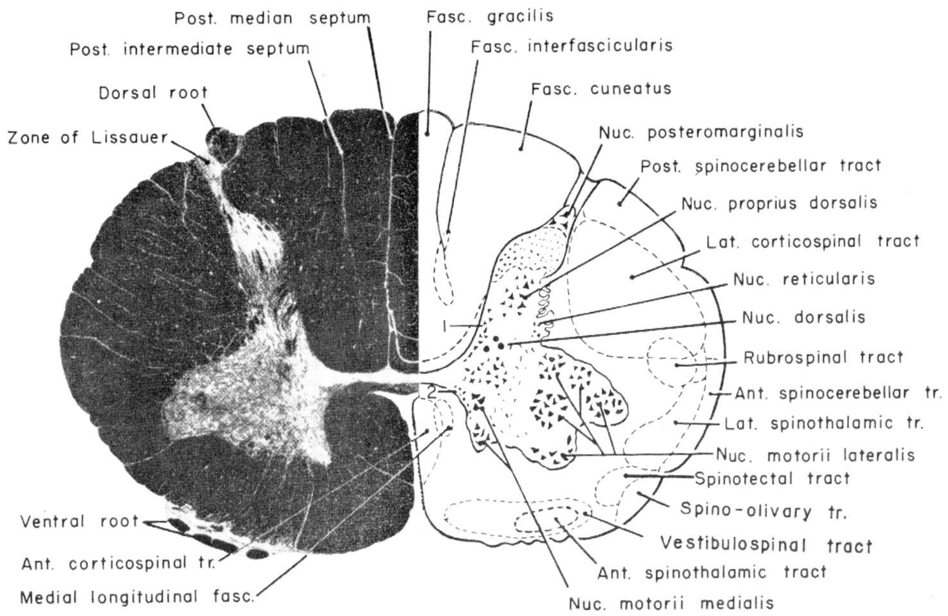

Fig. XIV-7. A section through eighth cervical section of the cord; Weigert myelin stain on left side. On the right, the section illustrates the locus of certain named fiber tracts in funiculi and named nuclei or cell columns in the gray matter at this level. 1 and 2 represent the posterior and anterior cornucommissural nuclei, respectively. (From R. C. Truex and M. B. Carpenter, *Human Neuroanatomy*, 6th ed. Williams & Wilkins, Baltimore, 1969.)

Cord Sulci and Funiculi: There are other sulci besides those from which cord rootlets emerge (Figs. XIV-3 and 6A). The **ventral sulcus** is the deepest of all and it helps to orient cords cut in cross-section. The posterior median sulcus is not much more than a tight seam in the dorsal midline, much less of an infolding than the ventral sulcus. Also, from cord segments C_1 to T_6, there are faint **posterior intermediate sulci,** one on each of the posterior median sulcus (Fig. XIV-6). Below T_6 this sulcus disappears.

On each side between posterior median sulcus and posterolateral sulcus lies the **posterior funiculus** of white matter (Fig. XIV-7). It contains fiber tracts, the **fasciculus cuneatus** lateral to the posterior intermediate septum and the **fasciculus gracilis** medial to that septum. Below T_6 the posterior funiculus consists only of fasciculus gracilis. These fasciculi or fiber tracts are axons of sensory neurons conducting touch, pressure, proprioception and vibratory sensibility from the homolateral side of the body. The tracts do not cross to the opposite side until these axons synapse upon cells in the medulla. After this synapse, axons in the **nucleus gracilis** and **nucleus cuneatus** of the medulla cross to ascend to the thalamus in a fiber tract known as the **medial lemniscus.** After a second synapse in **thalamic nuclei,** axons from nerve cells in the thalamic nuclei conduct to the cerebral cortex. In this way these sensory modalities originating on the *right* side of the body register in consciousness in the **postcentral gyrus** of the **parietal lobe** in the *left* hemisphere, and *vice versa*.

The **lateral funiculus** lies between dorsolateral and

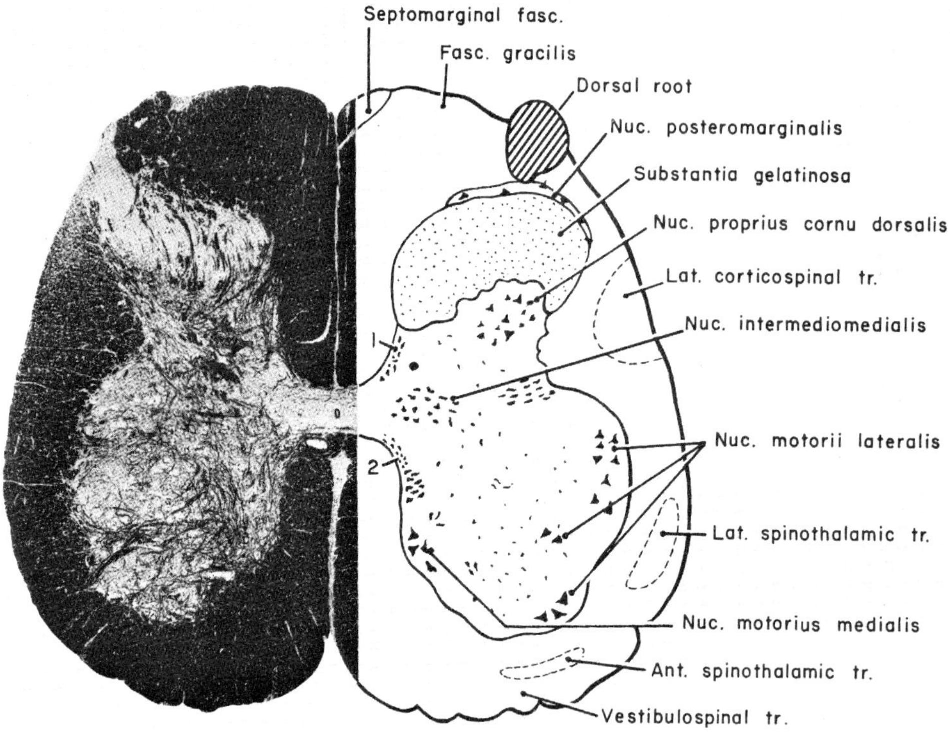

Fig. XIV-8. Spinal cord section similar to that of Figure XIV-7, but at third sacral level. (From R. C. Truex and M. B. Carpenter, *Human Neuroanatomy*, 6th ed. Williams & Wilkins, 1969.)

ventrolateral sulci. In its white matter are ascending and descending tracts of various kinds (Figs. XIV-7 and 8). Some of the ascending tracts never reach conscious levels but only as high as the cerebellum for reflex control of motor regulation. One of the ascending tracts, the **lateral spinothalamic tract** conducts pain sensations from one side of the body to thalamic nuclei of the opposite side. Crossing occurs after the entering sensory fibers synapse in a cell column on the side of entry, the **substantia gelatinosa** (Fig. XIV-8). Axons of these cells cross to the opposite side almost at once to build up the lateral spinothalamic tract. While some pain may register in consciousness at the level of the thalamic nuclei, again axons from these nuclei carry impulses onto the postcentral gyrus of the contralateral hemisphere.

The tract shown labeled **anterior spinothalamic** in Figures XIV-7 and 8 is classically illustrated and discussed in textbook and clinical literature as a special tract conducting a specific modality (light touch) to conscious level. There is no real proof of this. Indeed, it is probably unlikely that it conducts to a thalamic level much less to a conscious level or that it conducts the specific modality ascribed to it.

The tract was postulated long ago by clinicians who, in a wool gathering mood, were attempting to explain certain tactile phenomena in patients with cord lesions which they were unable to explain without dreaming this tract into existence. There may have been some poorly contrived experiments in cats which tended to suggest that the tract might exist in cats. And so we, as men, inherited it. Galenic reporting perpetuated it in our texts. God knows how many other named fiber tracts, named sensory receptor end-organs, and named nuclear cell columns are as equally revered, or dogmatically classified as to function and as equally uncertain as to fact.

An important *descending* tract in the lateral funiculus is the **lateral corticospinal tract** (Figs. XIV-7 to 9). This tract arises from contiguous motor neurons in the **sensorimotor cortex** (pre- and postcentral gyri) of a hemisphere and descends into the contralateral lateral funiculus of the cord. The bulk of its crossing occurs in the **pyramids** of the ventral **medulla.** It reaches motor neurons in the ventral horn of gray matter after one or more synapses on interneurons of the intermediate gray matter of the cord. Axons of ventral horn cells leave *via* ventral roots and spinal nerves for skeletal muscle. The tract is responsible for the initiation of voluntary movement in localized muscle masses.

Other important descending tracts in the lateral funiculus arise at brainstem levels. They descend as **rubrospinal** tracts, **lateral reticulospinal** and **vestibulospinal** tracts (Fig. XIV-9). Some of these may not be named in the figures shown, but they have important

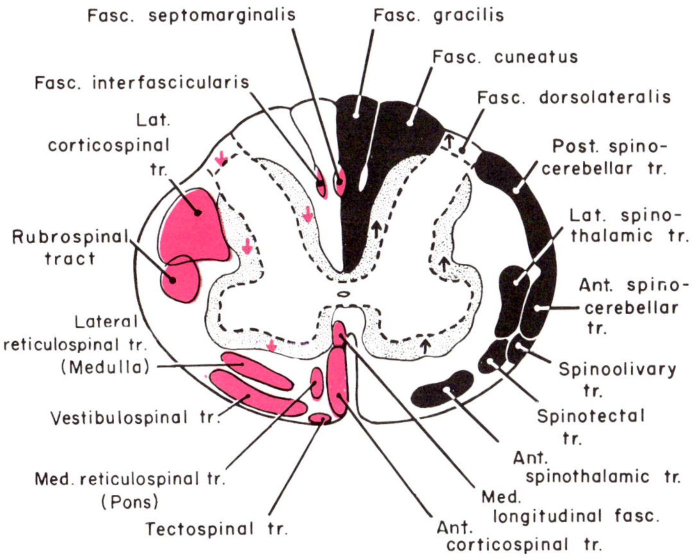

Fig. XIV-9. Diagram of ascending (*black*) and descending (*red*) pathways of the spinal cord. The fasciculus proprius system (*stippled*) and dorsolateral fasciculus contain both ascending and descending nerve fibers. (From R. C. Truex and M. B. Carpenter, *Human Neuroanatomy*, 6th ed., Williams & Wilkins, Baltimore, 1969.)

influences upon the ventral horn cells also, either as facilitators or inhibitors of its neuromotor activities at cord level. The ventral horn cell is the final common pathway to skeletal muscle. Its behavior is, therefore, influenced by many converging tract systems upon it, including its reflex connections with incoming sensory fibers, either directly or indirectly through interneurons.

The **ventral funiculus** of white matter lies between ventrolateral and ventral sulci. This smaller region still contains a number of ascending and descending fiber tracts. There are two of significance along the margin of the ventral sulcus, the **medial longitudinal fasciculus** (sulcomarginal tract) and the **ventral (anterior) corticospinal tract** (Fig. XIV-9). The former is an old tract phylogenetically and the first tract to myelinate ontogenetically. It is a descending portion of a brainstem tract related to vestibular function. It coordinates neck and trunk movements with head and eye movements. The second of these sulcal tracts is a continuation of the corticospinal system of fibers which failed to cross in the medulla. They enter the cord ventrally and cross late by ending on cord interneurons whose axons are commissural, to reach ventral horn cells of the opposite side.

Among other tracts in the ventral funiculus are the **ascending spinotectal** (from cord to roof of midbrain) and **spino-olivary** tracts (Fig. XIV-9), the latter going from spinal cord to a large odd-shaped group of cells in the medulla known as the **inferior olive.** Probably, the so-called ventral spinothalamic tract, mentioned above as a questionable tract, is part of the spino-olivary complex carrying touch to the cerebellum and no farther. An important group of *descending* fibers are the **ventral** (medial) **reticulospinals** from rostral brainstem levels.

The approximate locus of some of the tracts we have just mentioned are shown in the various funiculi of white matter in Figure XIV-9. It is not really important that you know their names until you make an in-depth study of neuroanatomy. You can, however, get a rough idea of their locus now.

As a rule, ascending, sensory-conducting tracts occupy the perimeter of the funiculi except in the case of the posterior funiculus where such sensory tracts fill it (Fig. XIV-9). We can regard them as **suprasegmental tracts** which project to cerebellum, brainstem or, finally, to the cortex as they ascend. Descending motor tracts tend to lie deep to these ascending fiber bundles or **fasciculi.** Mixed in among the motor tracts at this depth in the white matter are certain **intersegmental** tracts such as reticulo-spinal and spino-reticular bundles. These latter traverse several segments of the cord or run between cord and the brain stem. Deepest of all, next to the central gray, are short **intrasegmental tracts** which run only from one cord segment to the next above or below. These have been called the **ground bundles** or the **fasciculi proprii.**

The fundamental point to grasp is that white matter of the cord consists of these tracts, certain specialized connective tissue elements of the CNS, and blood vessels. These fiber tracts are axons of cells which lie either in the dorsal root ganglia or in the gray matter of the CNS at any level, cord, brainstem or cortex.

The Gray Matter: For some time you have been aware of the H-shaped central gray matter. Its dorsal horn contains second-order, sensory neurons upon which axons or collaterals of dorsal root ganglion cells terminate. The axons of these second-order, sensory neurons enter the white matter to build up some of the ascending fiber tracts which we mentioned above. Other neurons in the dorsal horn of gray and in the intermediate gray, between dorsal and ventral horns, are more or less

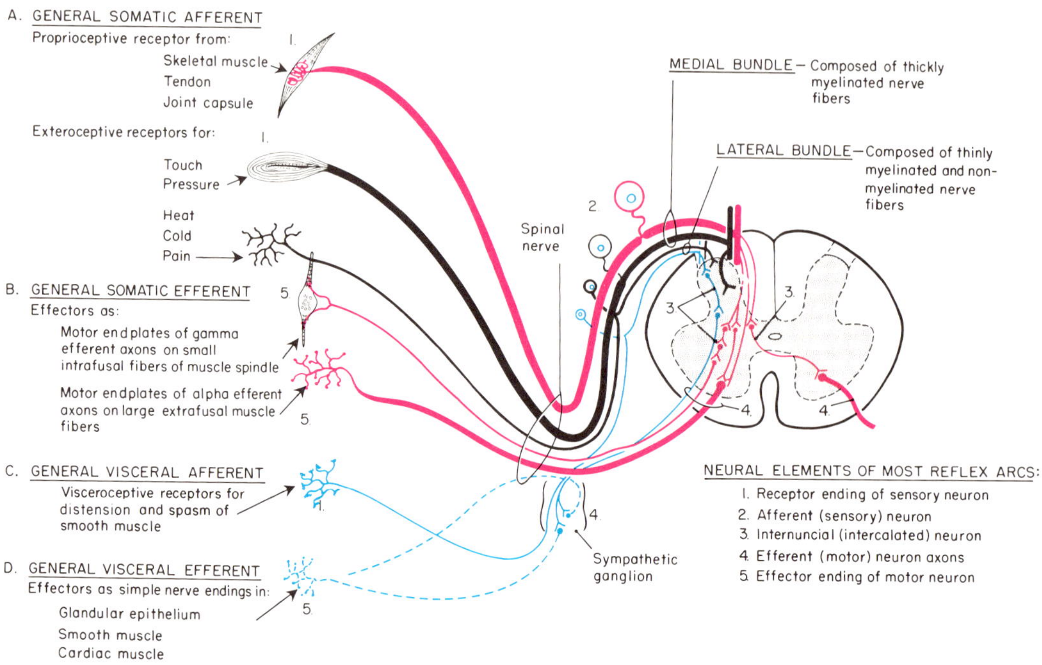

Fig. XIV-10. Diagram of functional components of a thoracic spinal nerve and the arrangement of dorsal root fibers as they enter the spinal cord. Skeletal muscle afferent fibers are indicated in *red.* Visceral afferent and efferent fibers are shown in *blue.* An afferent fiber from a Pacinian corpuscle (*black*) and a thin pain fiber (*black*) also are shown. Numbers in the diagram correspond to neural elements that form reflex arcs. (From R. C. Truex and M. B. Carpenter, *Human Neuroanatomy,* 6th ed., Williams & Wilkins, Baltimore, 1969.)

neutral interneurons. They link incoming sensory axons to motor cells in the ventral horn at the level of entry to a given segment, or they send their axons out into the ground bundles to ascend or descend to motor cells of the ventral horn at higher or lower levels.

The ventral horn cells themselves are **somatic motor neurons** whose axons supply skeletal muscle derived from somites.

In only a few places are these intermediate horns or roots of gray matter which break up a generalized H-shaped configuration. There is a projecting bit of intermediate gray at cord segments C_1 to C_4 for **branchiomotor neurons,** the spinal part of the spinal-accessory nerve. These become part of cranial nerve XI to serve Trapezius and Sternomastoid derived from branchial arch mesoderm.

Again, from T_1 to L_2, there is an intermediate projection of gray matter, the **intermediolateral cell column,** which contains the **visceral motor preganglionic neurons** of the sympathetic nervous system. The parasympathetic visceral motor neurons at S_2 to S_4 hardly change the configuration of gray matter. Both of these localized cell clusters in the cord supply smooth muscle, cardiac muscle, and glands.

Thus, Figures XIV-7 and 8 show named cell clusters in various parts of the gray matter. We call such functionally related groups of neurons (really the cell bodies and dendrites of neurons) **nuclei.** Since, however, they may not be limited to one cord segment, but to several contiguous cord segments, we often call them **cell columns.** A functional cell column may run the entire length of the spinal cord, *e.g.,* the sensory cells of the substantia gelatinosa, or the somatic motor neurons in the ventral horn. Other cell columns are discrete, running for a prescribed number of segments only, *e.g.,* the sensory cells of Clarke's column; the branchial motor cell column at C_1 to C_4; or the visceral motor cell columns at T_1 to L_2, or S_2 to S_4. Axons of motor cell columns tend to leave the central neuraxis. Similar functional nuclei or cell columns will be found in the gray matter throughout the brainstem. In the cortex, gray matter is distributed over the surface of folds of white matter to form gyri. We know the functional significance of cells in certain of the cortical gyri but not in all of them.

We can now ask what does an incoming dorsal root fiber do when it enters the cord? First, realize that it is an axon of a cell in the dorsal root ganglion. Its peripheral process came from sensory receptors in the skin or from proprioceptive endings in muscles, tendons, or joint capsules (Fig. XIV-10). The entering axon is not a single terminal twig all of the time. It gives off a number of collateral terminal branches as it enters the cord; hence, it can begin to engage in a number of coordinated activities at the same time.

1. It can build up projectional ascending pathways in the white matter at once before it ever begins to

synapse. This it does in the case of the posterior funiculus (fasciculus gracilis and cuneatus). These axons ascend all the way to the medulla before a synapse is made.

2. The incoming axon or one of its collateral branches can end upon a second-order, sensory neuron in the dorsal or intermediate gray. It would be the axon of this second neuron, then, which would build up a named projectional ascending tract in the white matter. We found that some of these tracts could go all the way to the thalamus before a second synapse occurred (lateral spinothalamic tract).

3. The axon or its collateral could end on interneurons in the dorsal, intermediate, or ventral gray. The axon of this cell could enter the ground bundles: or other fasciculi for intersegmental connections.

4. The axon or one of its collaterals could synapse directly upon a motor neuron at the segment of entry (or indirectly upon it through an interneuron at that segment). The only direct connections known are from primary afferents of muscle spindles. By means of such sensory-motor connections, a sensory stimulus can elicit a motor response which is virtually instantaneous and involuntary, long before the reflex act registers in our cortex. In many acts, if it registers at all, we fail to heed or attend to it for it becomes part of the automatic movement pattern of our limbs in particular. Many of these reflex connections are subliminal anyway, especially those arising as stimuli from muscles, tendons, and joints which tend to maintain **muscular tonicity** even when the incoming stimuli are too weak to effect a total muscular contraction. **Muscle tone** is also facilitated by descending fiber tracts from the labyrinths of the inner ear which interconnect with brainstem **vestibular nuclei.** We met such vestibulospinal tracts in the lateral and ventral funiculi.

Clinical Correlates: Without going into the effects of injury to tracts in the white matter at this time, there are a few clinical correlates which we can make about spinal cord function.

What would be the consequence, for example, of injury or severance of all dorsal roots involved in the brachial plexus, *i.e.*, dorsal roots C_5 to T_1? Since dorsal roots are sensory roots, would you anticipate any motor dysfunction? An obvious consequence would be complete **anesthesia** in the limb involved. No sensory messages of any kind from it are reaching the central neuraxis. We have just learned that some of this information comes from sensory proprioceptors in muscle, tendon, and joints. This kind of sensory input directed toward ventral horn cells (motor neurons) contributes in a large measure to muscular tonicity. We have now denied a large measure of this reflex stimulation to flow out to muscle. We would find the muscles of that limb in a state of **hypotonicity.** While ventral horn cells would react to impulses coming to them over the lateral and ventral corticospinal tracts, muscle in the affected limb would not be too willing to be moved to action.

Severance or injury to one single dorsal root produces hardly any noticeable effect in so far as appreciation of sensory loss is concerned. To anesthetize a localized skin area (or dermatome) at least three contiguous dorsal rootlets must be involved in injury. The upper and lower of these three overlap, by one half each, the same entire skin area supplied by the middle one of the three. Thus, cutting the single middle one (or any one of the three, for that matter) would elicit no detectable sensory loss in that segmental skin area or dermatome.

We constantly localize painful or tactile impulses applied to any particular spot of our hides in a fairly precise manner. We mentioned that the fasciculus gracilis and cuneatus conducted touch impulses to specific parts of our cortex and pain was likewise conducted over the lateral spinothalamic tract to conscious centers of awareness in our cerebral cortex. We pay far less attention to sensations from our viscera, aside from feelings of bladder distention or bowel fullness (or emptiness before mealtimes). It is only when certain of our viscera become diseased (heart disease, gall bladder disease, or appendicitis are common examples) that we try to localize pain coming from them. If we are laymen with only vague ideas about anatomy, we are often not quite exact in localizing such pain. We tend to refer the visceral pain to some skin area where we know we have experienced pain before. Thus, patients with coronary disease may complain of pain along the medial border of the left arm. Patients with gall bladder disease may localize it at the inferior angle of the scapula, and patients with appendicitis, to the epigastrium rather than to the lower right quadrant. This is because the visceral sensory fibers which conduct pain enter over dorsal rootlets that normally are involved in conducting pain from these odd skin areas. The entering axons end upon a common pool of second-order neurons that normally responded to pain from these skin areas. There is no pain in these skin areas at the moment, but incoming pain from a viscus which we have hardly ever attended to before is stimulating the common pool of second-order sensory neurons in the dorsal gray matter. We "refer" or interpret the pain as coming from a somatic sensory origin. This is, at least, one concept to explain the odd clinical phenomenon of **referred visceral pain**. You can see that unless you know something about central nervous system mechanisms, you could be misled in a diagnosis of the patient's problems. Your best bet is always to work with good anatomists as patients. At least, they will be more reliable in localizing visceral pain.

Many times an early amputee wishes he could put his lost foot in his mouth and bite down upon it to make the pain he feels in that **phantom limb** real. Even though his leg is lost below the knee, for a long time he will complain about "pins and needles" and other worrisome sensations in a foot which he "feels" somewhere out in space. Sometimes he swears the "foot" is right at the end of his amputated stump. He is much more aware of sensation in the distal part of the amputated limb than in its proximal parts. In an amputation at the ankle, it is the lost toes that are most bothersome, for example. With a lost hand, he can still "feel" the ring he wore on his finger. It appears that those parts of the sensory cortex with the greatest representation for limb parts (thumb and great toe area; finger areas) are hypersensitive after limb loss, and any stimulation of the peripheral stump activates these cortical centers and brings them into sharp awareness in the patient. His lost parts become real to him again. He can even feel these phantom parts move through objects in outer space, and this aggravates him no end. We are dealing here with a kind of central or cortical referred pain.

When ventral roots are cut or ventral horn cells are damaged (as in poliomyelitis), there is no longer a common motor pathway to muscle. There is a flaccid paralysis. Muscles lose tone, of course (atonia); reflexes are lost (areflexia); and muscles degenerate (atrophy).

When descending tracts to ventral horn cells are injured in the cord, more often than not, there are exaggerated tendon reflexes (hyperreflexia), and muscle is maintained in a state of hypercontraction (hypertonia). The explanation of this is quite complicated and would require that you have much more knowledge of cortical and brainstem pathways than we can impart to you at this time.

Myelography—The Roentgen Anatomy of the Spinal Subarachnoid Space: The myelographic examination is of value in the diagnosis of many abnormalities of the vertebral canal and its contents. These abnormalities include herniated intervertebral disk, as well as many neoplastic, inflammatory, traumatic, and vascular lesions. A brief comment on the examination procedure itself will precede consideration of the normal anatomy to be seen.

1. Procedure: Positive (opaque) or negative (gas) contrast media may be used alone or in combination for studying the spinal subarachnoid space. The choice of contrast medium is largely dependent upon the clinical problem encountered. However, the most common method employs the use of oily, iodized, positive medium, usually Pantopaque.

Before contemplating a myelogram, radiographs of the spine should be obtained and evaluated thoroughly. These radiographs may disclose abnormalities hitherto unsuspected.

The spinal level at which the lumber puncture is performed is in large measure dependent upon the spinal level of clinical interest. Thus, if the cervical or thoracic region is to be examined, then the lumbar puncture should be performed at the VL_4–VL_5 or VL_5–VS_1 level (Fig. XIV-3B). However, if interest is primarily in the lumbar region, then the lumbar puncture should be at the VL_2–VL_3 or VL_3–VL_4 level. It is *not* advisable to insert the spinal needle at the exact level of suspected abnormality, since the needle tip usually produces a defect in the opaque contrast column which may simulate a lesion.

Meticulous attention to technique is required to assure an aseptic procedure and accurate placement of the spinal needle in the midline. Once the dura has been punctured by the beveled needle tip, then the needle is advanced an additional 1 to 2 mm. The stylet is removed from the needle, and if the tip is, indeed, within the subarachnoid space, then spinal fluid should flow freely from the site. To be absolutely certain of this, however, it is advisable in some instances to instill 0.5 cc. of the oily contrast medium through the needle into the subarachnoid space. With the patient in a prone position, the free and rapid flow of Pantopaque may be observed under the fluoroscope by tilting the table up and down. Contrast medium within the *subdural space* moves rather *slowly*. Once the physician is certain of a satisfactory *subarachnoid* injection, then additional Pantopaque is instilled. The total quantity of contrast medium used varies between 2 and 26 cc. depending upon the clinical problem, the region to be examined, as well as the size of the vertebral canal. Pantopaque is an iodized oil which readily fragments unless care is exercised in moving the patient slowly.

For the procedure, a footrest and shoulder brace are affixed to the fluoroscopic tilting table to assure that the patient will not slide off the table. Once the Pantopaque has been placed into the subarachnoid space, the stylet is replaced in the needle and the area covered with a sterile towel. Subsequent fluoroscopic examination is done with the needle remaining in place, although on occasion, it must be removed in order to examine the patient in the supine position.

Initially, the entire column of Pantopaque is pooled in the caudal sac by standing the patient in a semi-erect position. By gradually lowering the patient's head from the semi-erect position, the contrast medium may be positioned over each spinal segment. When examining the thoracic and cervical regions, the patient is placed in Trendelenberg position (head down); however, care must be exercised to keep the patient's head hyperextended in order to avoid having the contrast medium enter the cranial subarachnoid space.

Sites of suspected or real abnormality are recorded on spot films in frontal and 45° oblique projections. In addition, lateral projections utilizing a horizontal X-ray beam directed against a grid cassette are obtained. Many times it proves useful to obtain both decubitus films of the spine when examining the thoracic and lumbar regions. Once filming is completed, then the Pantopaque is pooled and centered over the needle tip so that it may be aspirated from the subarachnoid space.

2. The Roentgen Anatomy of the Subarachnoid Space as Demonstrated by Myelography: Figure XIV-4A and B will briefly review for you the anatomic relationships between the spinal cord and its three investing membranes. Immediately adjacent to the spinal cord is the delicate pia mater. The next outermost layer is the arachnoid mater. This is separated from the pia by the subarachnoid space which is filled by

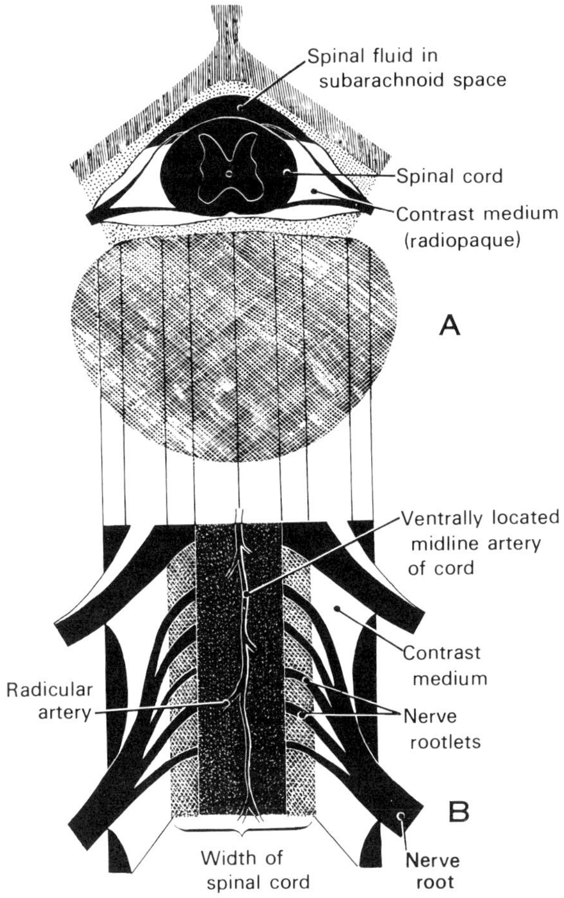

Fig. XIV-11. Schematic illustration of structures to be visualized in a cervical myelogram utilizing Pantopaque. **A.** Cross-section with the patient in prone position. Note the Pantopaque in the most dependent portion of the subarachnoid space. The entire subarachnoid space is not filled with contrast medium. **B.** Frontal view showing Pantopaque outlining the spinal cord, the nerve roots, and anterior spinal artery. Compare this schema with the actual myelographic picture in Figure XIV-13A.

spinal fluid. The outermost membrane is the tough and fibrous dura mater. In Figure XIV-4B, the distance between the dura and the arachnoid has been exaggerated. The subdural space is actually only a potential space, for, in reality, the spinal fluid in the subarachnoid space pushes the arachnoid mater tightly against the dura mater.

In myelography, the entire subarachnoid space is usually not filled with contrast medium. This is particularly important to realize, especially when viewing Pantopaque myelograms. As we can see in the schematic drawing (Fig. XIV-11), with the patient in the prone position for the examination, the Pantopaque has settled to the most dependent portion of the subarachnoid space, since it has a higher specific gravity than spinal fluid. Exactly how much of the subarachnoid space is filled is, of course, dependent upon the quantity of the medium utilized, the volume of the vertebral canal, and the length of the region it occupies. Obviously, then, if one is suspicious of a dorsally-placed lesion, the patient must be examined in the supine position after the spinal needle has been removed.

Recall that the enlargement of the spinal cord for the brachial plexus begins at VC_3, gradually attains its largest transverse diameter at VC_5 to VC_6, and then tapers to the level of VT_2 (Fig. XIV-3B). The narrowest portion of the thoracic cord is at the VT_8 to VT_9 level. At VT_{10}, the lumbar enlargement of the cord begins (Fig. XIV-3A) and the spinal cord terminates in the adult at the upper margin of VL_2. The cord is attached to the coccyx by the filum terminale which is largely pia mater. It is relatively easy to measure the transverse diameter of the spinal cord in the cervical and upper thoracic regions and to compare these figures against tables of normal values. This becomes a difficult problem lower in the spine. The spinal cord in the cervical and upper thoracic regions lies in close apposition to the posterior margins of the vertebral bodies with the patient in the prone position. More caudally, the spinal cord occupies a more posterior position in the vertebral canal (Fig. XIV-5B).

It is difficult to estimate the sagittal diameter of the spinal cord in the Pantopaque myelogram because of inadequate filling of the subarachnoid space. Therefore, it is best delineated by means of pneumomyelography (Fig. XIV-12). Charts are available for determining the range of normal values at various ages.

A. THE CERVICAL MYELOGRAM: Identify the relatively radiolucent spinal cord in the frontal radiograph (Fig. XIV-13A). The cord is not intensely radiolucent, since there is a thin layer of contrast medium beneath its ventral surface (Fig. XIV-11). Lateral to the spinal cord is the densely radiopaque contrast medium (Pantopaque). On each side in the cervical region, this Pantopaque column presents a slightly concave smooth border laterally. Note that just cranial to the intervertebral foramina, there are bilaterally symmetrical outpouchings of the arachnoid and dura which are termed the **root** or **axillary pouches.** This is the site from which the nerve roots exit from the thecal sac. Careful study of Figure XIV-13A and B will reveal the faintly radiolucent nerve roots. At times we may identify more than two lucent lines, since it is not at all uncommon in the cervical cord to find nerve rootlets forming several bundles which course inferolaterally before joining to form their respective dorsal and ventral root. Observe in Figure XIV-5A the course of the nerve roots from their origin in the spinal cord to their point of exit through the intervertebral foramina. Contrast their relatively short and almost horizontal course with the nerve roots in the thoracic, lumbar and sacral regions.

In the frontal projections (Fig. XIV-13A), the anterior spinal and radicular arteries may be observed. The radicular arteries accompany the nerve roots and provide their vascular supply. How can you tell the radicular arteries from the nerve roots? Simply stated: the nerve roots do not reach the midline of the cord!

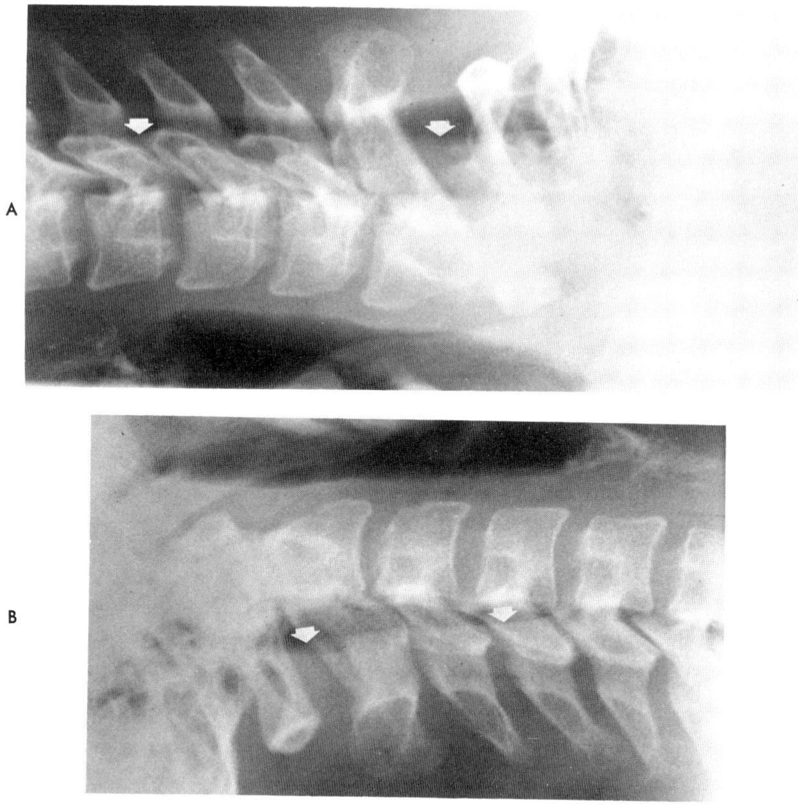

Fig. XIV-12. Pneumomyelogram in the cervical region. **A.** Lateral projection made with a horizontally-directed X-ray beam. The patient is in the prone position. The *arrows* indicate the sagittal diameter of the cervical cord as outlined by air. **B.** Same normal patient in the supine position. The cervical cord has fallen away from the anterior margin of the vertebral canal.

Figure XIV-13C is a lateral projection of a cervical myelogram in a normal patient. The patient was in the prone position, and the horizontally-directed central X-ray beam was directed against a grid cassette placed beside the patient's neck. Notice that at the VC_3 to VC_7 levels, there is about 1 mm. separating the posterior surface of the vertebral bodies from the anterior margin of the Pantopaque column. This space is filled by the posterior longitudinal ligament and the dura mater of the spinal cord. At the VC_1 to VC_2 level, the anterior margin of the contrast column is farther from the posterior margins of the odontoid and the body of VC_2. A review of Figure XIV-13D will make the reason for this appearance obvious to you. Notice that the posterior longitudinal ligament at this point must rise up to cross posterior to the transverse ligament behind the odontoid process.

Approximately 4 mm. from the posterior border of the midcervical vertebral bodies (Fig. XIV-13C), the ventral surface of the spinal cord is seen. Since a large quantity of contrast medium was used in this case, it is also possible to delineate the dorsal surface of the cord. Paralleling the midportion of the spinal cord is a faint radiolucent line which is the denticulate (dentate) ligament (Fig. XIV-13C). Just above the denticulate (dentate) ligament you may perceive many tiny obliquely oriented nerve rootlets. At the VC_5–VC_6 and

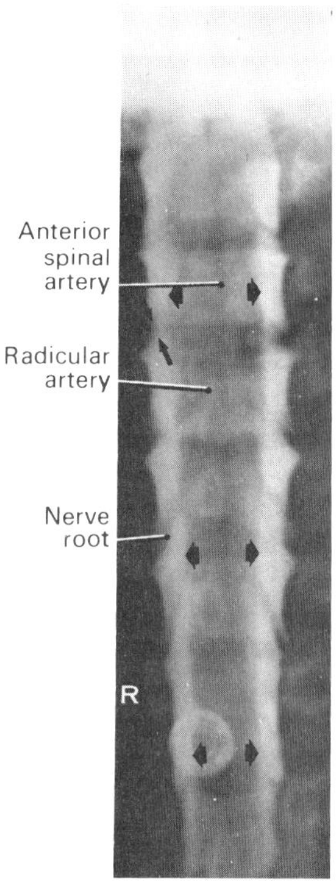

Fig. XIV-13A.

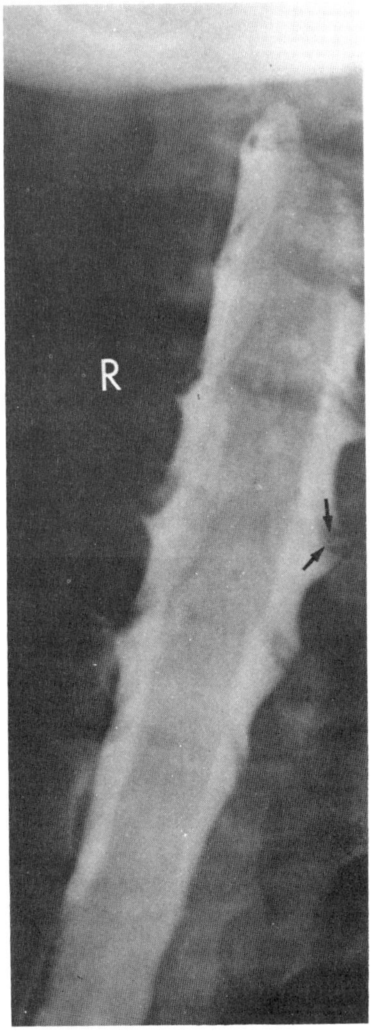

B

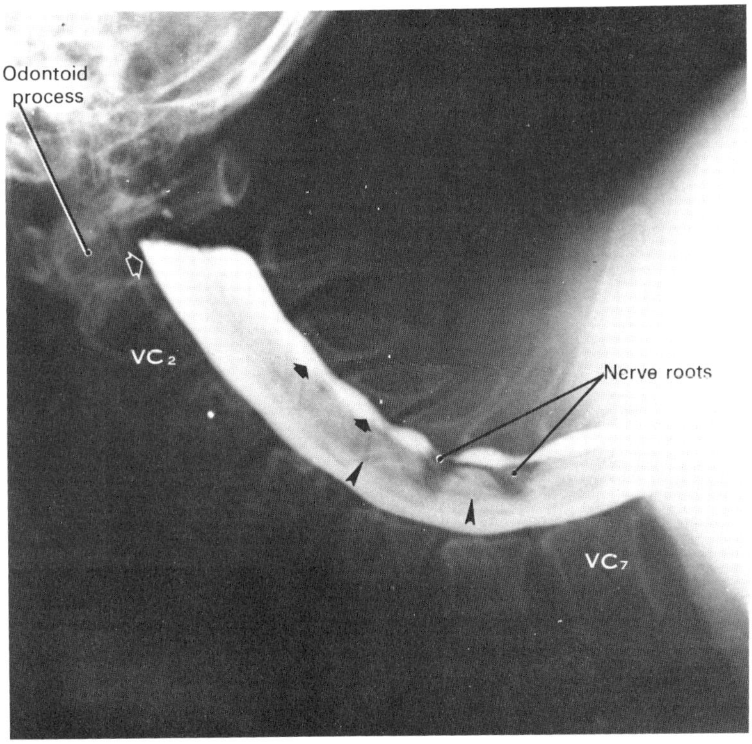

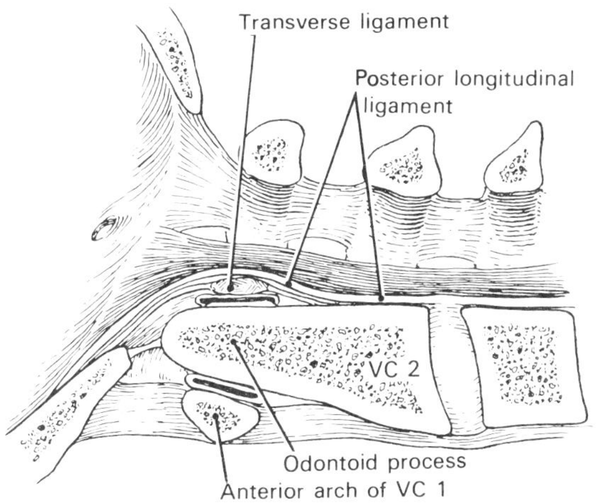

D

Fig. XIV-13. Cervical myelogram. **A.** AP spot film made at fluoroscopy with the patient prone. The *large black arrows* delineate the lateral boundaries of the cervical and upper thoracic cord. *Small black arrows* indicate the radiolucent nerve roots at their point of emergence from the subarachnoid space. **B.** Oblique projection made at the time of fluoroscopy. *Arrows* indicate the nerve roots near their exit point from the subarachnoid space. **C.** Lateral projection made with a horizontally directed X-ray beam. The patient is in the prone position with the neck hyperextended. *Black arrows* delineate the dorsal surface of the cervical cord. *Black arrow heads* show the linear radiolucency produced by the denticulate (dentate) ligament. The *white arrow* is at the point where the column of contrast medium courses over the posterior longitudinal ligament, which elevates to pass over the transverse ligament (see part D). **D.** Schematic illustration of a sagittal section through the upper cervical spine. Note that the posterior longitudinal ligament elevated off the posterior border of VC_2 to pass over the transverse ligament behind the odontoid process.

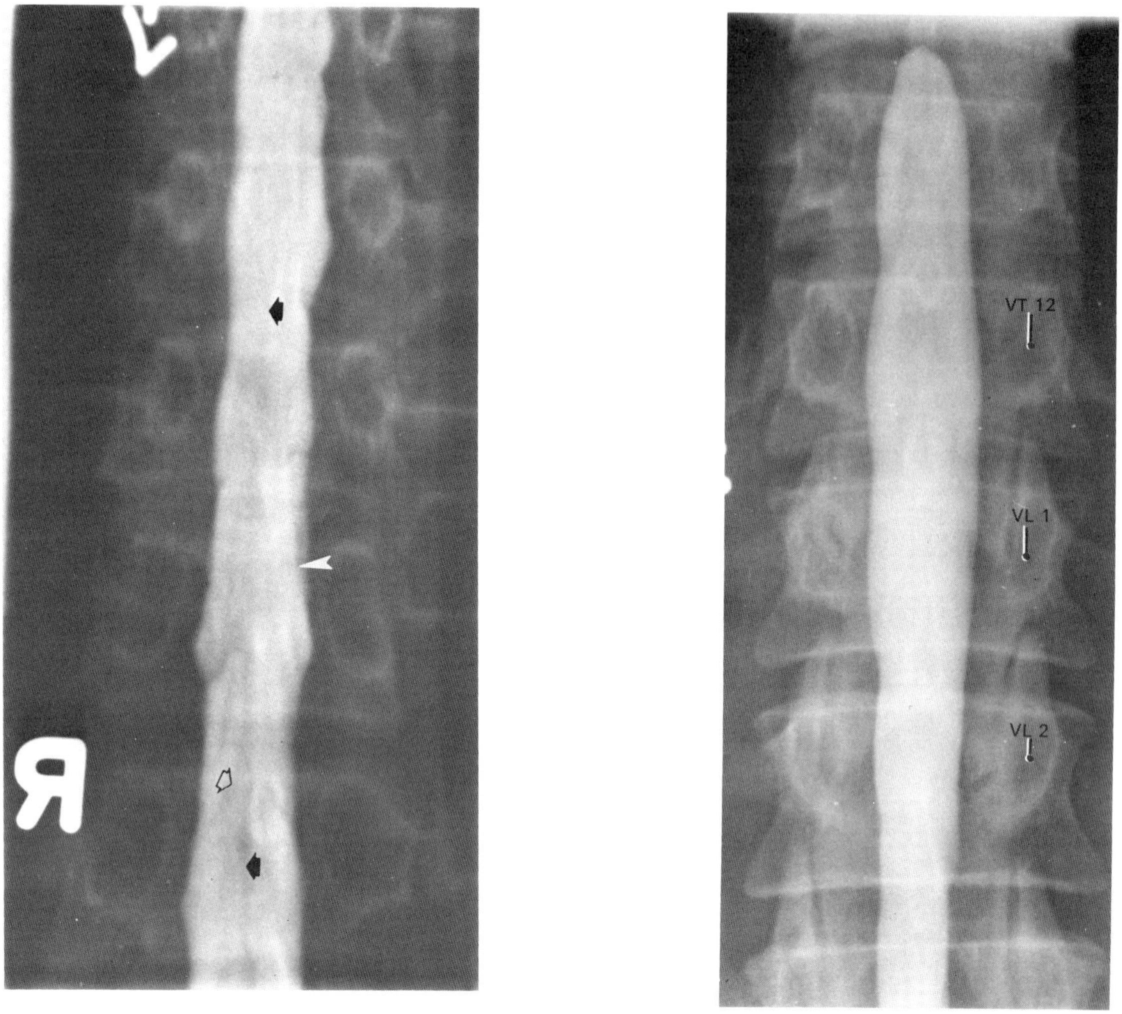

Fig. XIV-14. Thoracic and upper lumbar myelogram. **A.** Frontal view through the midthoracic region. *Black arrows* show the anterior spinal artery. *Open black arrow* shows a radicular artery. The *white arrow* indicates the radiolucent nerve root. **B.** Frontal view of the lower thoracic and upper lumbar region. The spinal cord terminates at the upper margin of VL_2.

VC_6–VC_7 levels there are posterior radiolucent defects in the Pantopaque column produced by the nerve roots at those levels.

B. THE THORACIC MYELOGRAM: In frontal projections, the upper thoracic spinal cord is more readily identifiable than lower in the thoracic region (Fig. XIV-14A, B). It is possible to see the obliquely-directed nerve roots, but the axillary pouches in the mid and lower thoracic spine are usually not well defined in this projection, since the contrast medium does not often rise up high enough. Remember that here the spinal cord is posteriorly positioned preventing the contrast medium from outlining these features well. Decubitus projections can, however, be most helpful in comparing symmetry of the axillary pouches in this region.

C. THE LUMBAR MYELOGRAM: Since the spinal cord terminates in the adult at the upper margin of VL_2, only the dense radiopaque contrast medium can be seen below that level (Fig. XIV-15). The appearance of the Pantopaque column is influenced by several factors including: the size and shape of the subarachnoid space; the point of termination of the caudal sac; the effect of structures adjacent to the subarachnoid space; and the order of the nerve roots of the cauda equina. Shapiro[1] describes two basic patterns to be found in the configuration of the opaque oil column: (1) a cylindrical column; and (2) a biconcave column. In the first example (Fig. XIV-15A), the nerve roots are aligned across the subarachnoid space. Their order places the more cephalic root lateral to the next lowest root, thus the root L_3 is lateral to L_4, and L_4 is lateral to L_5, etc. The biconcave configuration occurs when the roots of the cauda equina are arranged into two large bundles. Here the nerve roots are organized such that each root lies posterior to the one cephalic to it (Fig. XIV-15B).

By comparing the caudal sacs of the cadavers at your disposal, you will find that there is quite a bit of

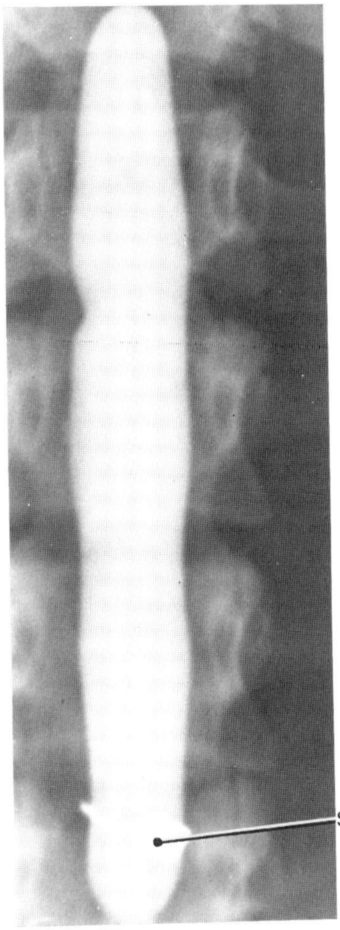

Fig. XIV-15. Lumbar myelogram. **A.** Example of the cylindrical-type of pattern. AP projection.

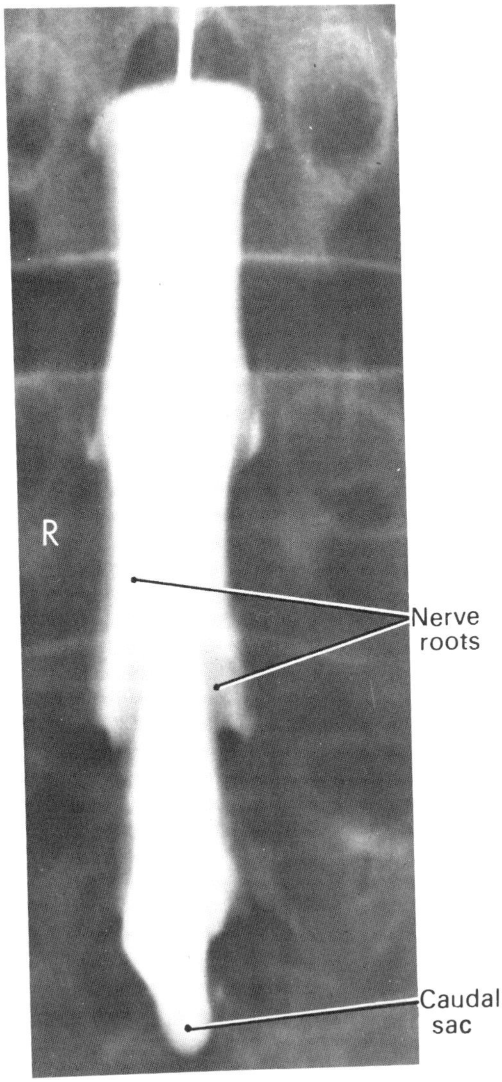

Fig. XIV-15B. Biconcave-type column of contrast medium. AP projection.

variation. It may terminate at VL_5–VS_1 or extend downward to VS_4, and it is variable in shape and size.

In Figure XIV-15B note that the axillary (root) pouches are quite symmetrical. Identify the nerve root as it exits from the arachnoid and dura. Realize that the nerve root crossing the VL_4–VL_5 intervertebral disk space is the L_5 nerve, and, thus, it would be affected by a herniation of the disk at VL_4–VL_5. Similarly a herniation of the disk at the VL_5–VS_1 level would encroach upon the nerve root of S_1 (Fig. XIV-5A).

In the lateral projection which is obtained with the patient prone, the contrast medium is centered over the area of clinical interest. Notice that there is quite a distance in this case (Fig. XIV-15D) between the posterior margin of the lumbar vertebra and the anterior margin of the oil. This space is variable in patients, and in some the oil may be much closer to the vertebral bodies. Normally, this space is occupied by the posterior longitudinal ligament, the dura, and epidural soft tissue (principally areolar tissue, and the epidural venous plexus). If this space is large, it is possible to have a large herniated disk and not see it in the lateral projection.

The Brain and Associated Brainstem

General Concepts: One almost has to consider cerebral hemispheres and brainstem together since they can only be separated crudely by dissection. Together they constitute the brain, an organ weighing about 3 pounds or about 2% of the total body weight. In life, it is a rather quivery substance; more like a mold of Jello than the solid head-cheese it appears to be when fixed in formalin. As a consequence, the neurosurgeon draws it out through an aspirator rather than attacking it with a scalpel when he wishes to excise a portion of it. He uses cautery rather than suture to control bleeding in the area, for there is little substance into which to

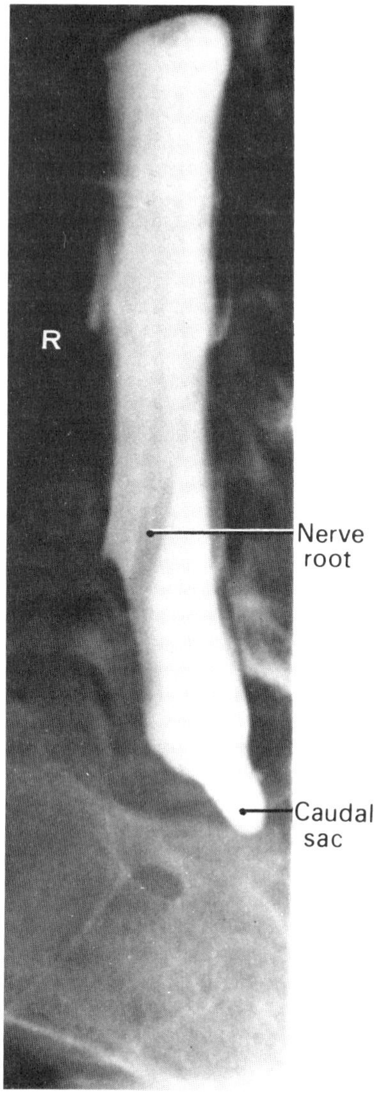

Fig. XIV-15C. Oblique projection made at fluoroscopy.

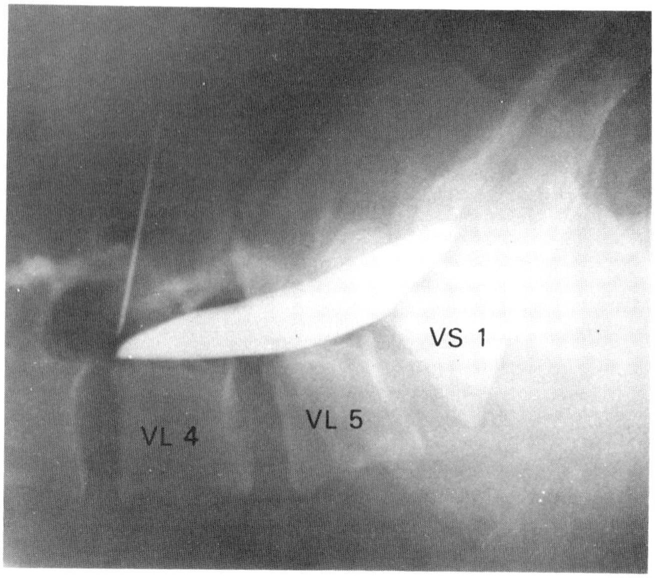

Fig. XIV-15D. Lateral projection made with the patient prone, utilizing a horizontally directed X-ray beam. Note the distance between the anterior margin of the oily contrast column and the posterior margins of VL_5 and VS_1.

anchor a suture. Like the spinal cord, the mass is held together by the network of blood vessels which pervade it and associated specialized connective tissue (astrocytes, oligodendroglia, and microglial cells). There is little extracellular space between nervous tissue and its vascular and connective tissue framework, far less than formerly suspected, but also, probably, far more than the 4% recently proposed by electron microscopists. It is probably closer to 25%. That there would be *no* extracellular space, as concluded by some astute scientists using the tools of modern sophistication, borders upon the realm of stupidity.

Yet, such a concept was in vogue a short few years ago, and many an astute scientist looked at the evidence and nodded his head in aquiescence. If there were no extracellular space, there would be no need for connective tissues. If there is anything fundamental about anatomy and physiology, it is that substances pass into and out of cells from extracellular space, no matter how minimal, through living cell membranes. Extracellular space is the milieu for gas and fluid interchange. Such space is invariably of a watery substance in one form or another. Blood cells would be free (and useless) in capillaries except for the plasma in which they float. It is transudates of plasma which enter the extracellular spaces of connective tissues. When blood cells escape from capillaries, they find themselves in extracellular space and in tissue fluid. So it seems normal to find nerve cells and their processes receiving nutrients and emptying their metabolic wastes or their products of cellular manufacture into extracellular spaces like any other cell in the body.

Since the brain substance has, itself, such a gelatinous consistency, we can understand why it is enveloped in delicate membranes, the meninges, and why, between the inner two of these, there is a watery cushion of cerebrospinal fluid.

We divide the brain (including its brainstem) into five structural regions:

1. The telencephalon (end brain) or the **cerebral hemispheres (cerebrum)**
2. The diencephalon (between brain) or **thalamus**
3. The mesencephalon or **midbrain**
4. The metencephalon (after brain) or **pons**
5. And the myelencephalon (marrow brain or the **medulla.** The spinal cord is part of the myelencephalon in development.)

Four and five, above, are often called the hindbrain or bulb. The **cerebellum** attaches to 3, 4, and 5 above by separate brachia (**brachium conjunctivum** to midbrain; **brachium pontis** to pons; and inferior brachium or **restiform body** to medulla).

Two through five, above, constitute the **brainstem**

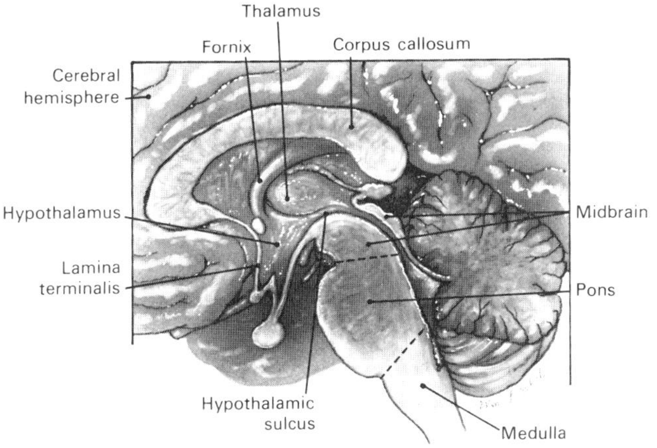

Fig. XIV-16. Median sagittal section of human brain stem. For identification of other structures see Figure XIV-21.

which stops rostrally at the **lamina terminalis** where cerebrum or hemispheres begin (Fig. XIV-16).

Since fiber tracts flow from hemispheres into thalamic regions and from thalamus to hemispheric portions, one can't simply pull the brainstem out and leave cerebral hemispheres behind. Moreover, the thalamus developed from a common brain vesicle along with the telencephalon. Both derive from the prosencephalon of the embryonic neural tube. This may be why the thalamus is considered by some to be a part of *the* brain while the parts caudal to it are designated *the* brainstem. We regard brainstem to include thalamus for arbitrary, but useful, conceptualization. The foramen magnum marks the caudal reaches of the brainstem every bit as much as the lamina terminalis marks its rostral limits.

The **cerebral hemispheres,** considered as if they were not integrally interconnected with the thalamic part of the brainstem, consist of: (1) an outer **cortical lamina** of gray matter folded over gyri and dipping into sulci and fissures; (2) a massive underlying area of **white matter** in which deeper, gray nuclei—the **basal ganglia**—lie; and (3) a stout, broad, commissural, connecting piece, the **corpus callosum,** bridging between the two hemispheres (Fig. XIV-17).

Topography of the cerebral hemispheres: We divide each hemisphere into five or six lobes more for convenience and for the way it relates to brain case and cranial fossae than for the sake of significance. Each lobe, certainly, relates to more than one kind of brain function. No lobe can act independently of the others. If lobes do not function well, one without the other, it might be asked how well would hemispheres function, cleft from one another? Patients have been reported who have had surgical separation of the hemispheres.[2, 3] Not only the corpus callosum but other cerebral commissures have been cut through, *e.g.,* anterior and hippocampal commissures. No noticeable aberrations in everyday behavior, temperament or intellect seems to have followed. Yet, in the intact brain, damage to certain areas in either hemisphere often leads to permanent dysfunction while damage in other areas in one hemisphere may, in time, be "taken over" by the other hemisphere.

Since it seems to have been deemed good for students to divide the brain into lobes for whatever reason, we had best get at it. In a dorsolateral view of the hemisphere removed from the brain case (Fig. XIV-18A), find a deep, somewhat horizontal fissure, the **lateral** or **Sylvian fissure,** which separates away the thumb of the mitten-shaped hemisphere. The thumb is the **temporal lobe** and it ordinarily sits snugly in the middle cranial fossa. A **fissure** is just a very deep **sulcus.**

About in the midline of the hemisphere, two parallel gyri should be found which orient almost vertically with the lateral fissure. They will be separated by the **central sulcus** which may, or may not, reach down to the lateral fissure. The sulcus also may, or may not, be continuous between the two gyri. But the two gyri are the only ones oriented in this plane among the wormy convolutions that are displayed over the brain's surface. You should be able to find them readily enough even if they take more of an oblique rather than a vertical course in your specimen. That portion of the brain above the lateral fissure and anterior to the central sulcus is the **frontal lobe.** Its normal habitat is the anterior cranial fossa upon which it rests in the brain case.

The central sulcus (of Rolando) (Fig. XIV-19) separates the **precentral gyrus** of the frontal lobe from the **postcentral sulcus** of the parietal lobe. It is not going to be easy to limit the **parietal lobe** without, first, rather arbitrarily cutting off the **occipital lobe** on the dorsolateral aspect of the brain. To do this, one must first find the **parieto-occipital sulcus** on the *medial aspect* of the hemisphere (Fig. XIV-18C). This sulcus is more prominent on the medial surface and laps over onto the dorsolateral surface for only a very short distance. Trace it around to the dorsolateral surface. Draw an imaginary line from the parieto-occipital sulcus superiorly to the **preoccipital notch** inferiorly (Fig. XIV-18A). You have cut off the *occipital lobe* on the dorsolateral surface of the hemisphere.

Now you can extend the line of the lateral fissure to meet the midpoint of the imaginary line you have just drawn. This maneuver will place the *parietal lobe* above the fissure and between central sulcus and your imaginary line (Fig. XIV-18A).

You have just delineated the four major lobes of the brain. To see a fifth hidden lobe, you will have to pull the temporal lobe down and outward or cut a portion of it away. This exposes the **gyri of the insula** (island of Reil) which lie buried deep to the lateral fissure (Fig. XIV-18B). The important gyri to locate in this region are some short oblique ones of the temporal lobe which run toward the posterior part of the insula, the **transverse temporal gyri.** The anterior one of these is known as Heschl's convolution.

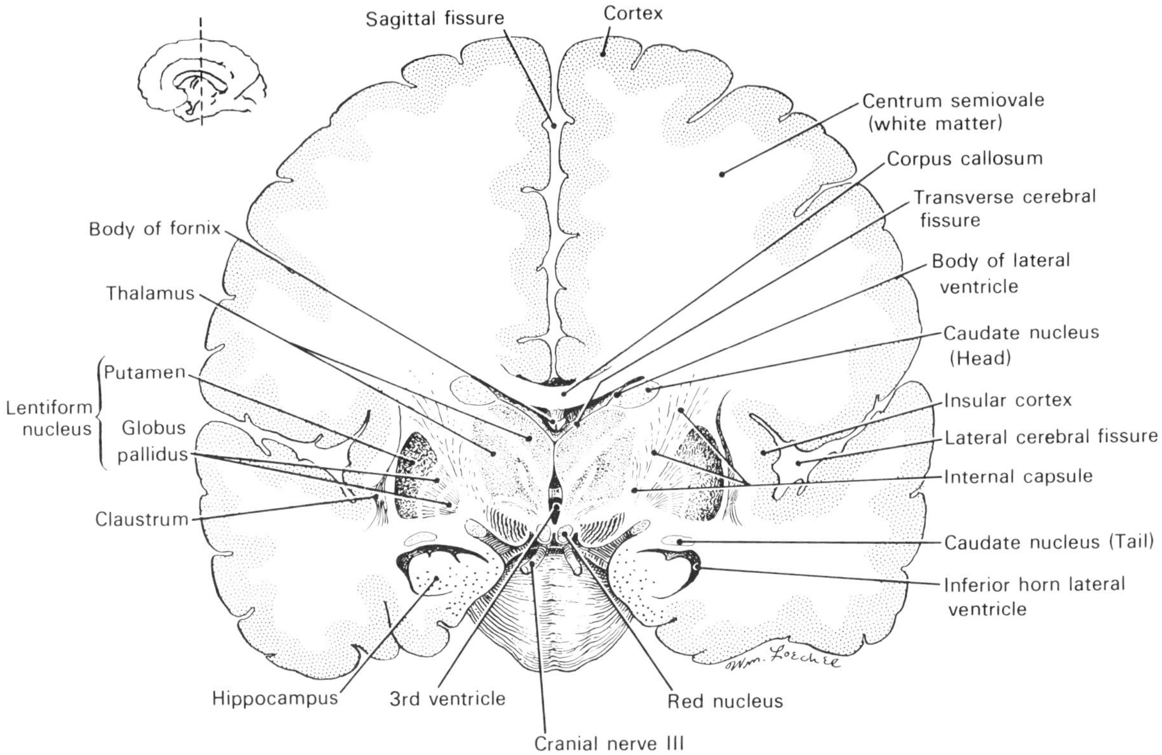

Fig. XIV-17. Coronal or frontal section through hemispheres as shown in inset. Note in particular the basal ganglia (caudate and lentiform nuclei) and their relation to thalamus and the ventricles. Note also proximity of putamen to insular cortex. The nternal capsule, a white fiber tract, lies between caudate and lentiform superiorly and between globus pallidus and thalamus nferiorly.

Look once more upon the medial aspect of the hemisphere (Fig. XIV-18C). It is not as easy to find parts of the major four lobes here. Still, the central sulcus partially divides the **paracentral lobule** which continues over onto the dorsolateral surface as precentral and postcentral gyri. Anything anterior to central sulcus, all the way to the **lamina terminalis,** is frontal lobe.

The parieto-occipital sulcus forms the boundary between parietal and occipital lobes on the medial aspect of the hemisphere. Thus, only a short stretch of the parietal lobe shows here between central sulcus and parieto-occipital sulcus.

In the occipital lobe, the **calcarine fissure or sulcus** meets with the parieto-occipital fissure in a Y-shaped configuration to hold the wedge-shaped **cuneus** between its upper arms. Below the calcarine fissure lies the **lingual gyrus** of the occipital lobe; but under the stem of the "Y," the lingual gyrus becomes part of the temporal lobe.

Uncus, hippocampal gyrus, and **fusiform gyrus** make up the inferomedial parts of the temporal lobe. The *fusiform gyrus* lies between **collateral sulcus** and **inferior temporal sulcus,** just below the lingual gyrus, in the temporal lobe.

Sometimes you will hear of a **limbic lobe** on the medial aspect of the hemisphere as if it were a separate sixth lobe. It will be said to consist of the **subcallosal gyrus** and **parolfactory region;** the **gyrus cingulus** which lies just above the corpus callosum and follows its arched course; the *isthmus of the hippocampal gyrus, the hippocampal gyrus and uncus.* These parts are older parts of the brain phylogenetically, constituting the archicortex, paleocortex and mesocortex (gyrus cingulus) as distinct from all the rest or newer parts of the hemispheres—the neocortex. Obviously, the "limbic lobe" takes in parts of frontal, parietal, and temporal lobes because it swings through all these territories. It was once strongly felt that it had special functions all of its own which had to do with olfaction, affect, and vegetative mechanisms. But these functions so interrelate with centers in frontal and temporal lobes that it can no longer really be regarded as a separate lobe either on the basis of function or morphology.

You may find also that some texts call the gyrus cingulus, the isthmus of the hippocampus and the hippocampal gyrus by one name, *e.g.,* the **gyrus fornicatus.** This also has no relation to the function it may connote. Fornix, from which its name was derived, means "bent" or "arched." Your authors still insist that the gyrus must have been so named because it screws around the brainstem.

Functional localization: Figures XIV-19 and 20 name many other gyri and sulci on the dorsolateral and medial surfaces of the hemispheres and there are still more on the inferior aspects of the hemisphere. It is hardly worth learning all of them and certainly not

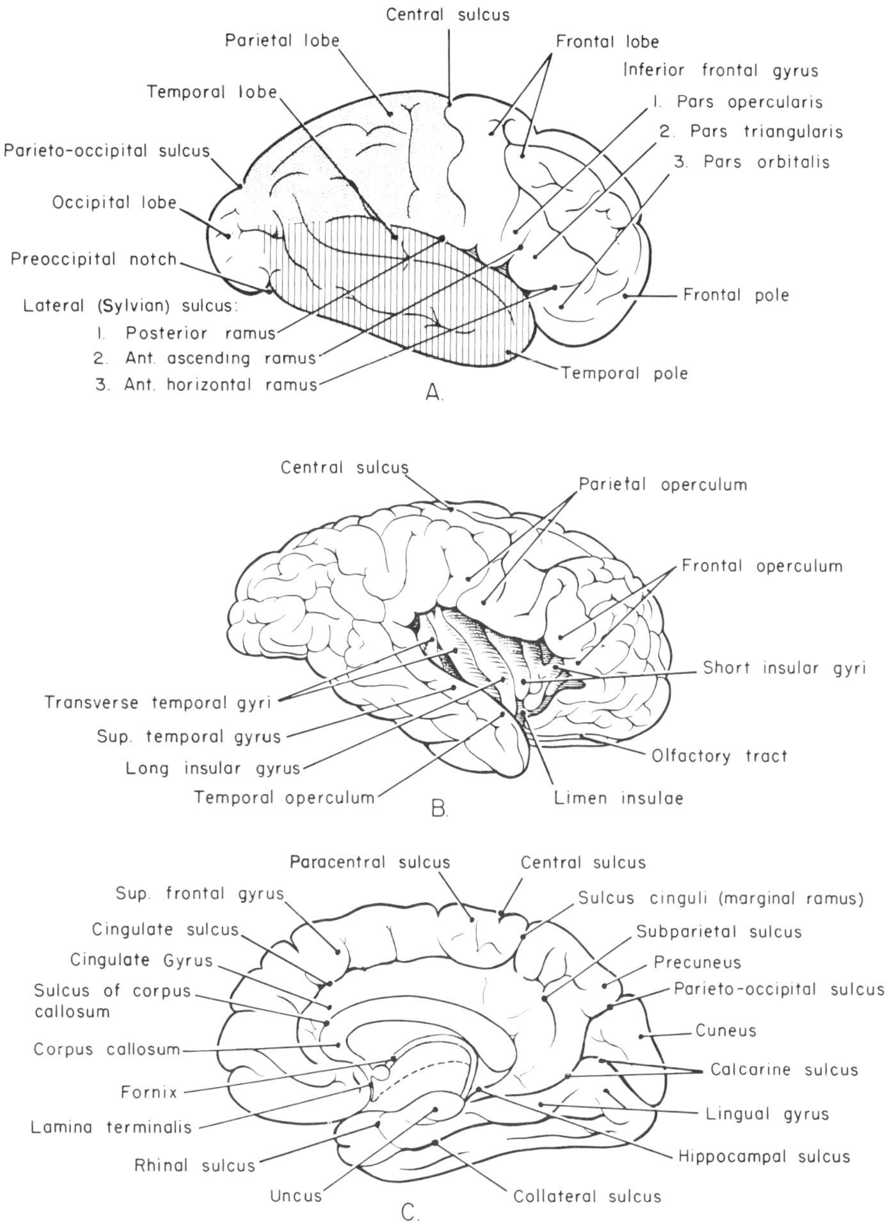

Fig. XIV-18. The right cerebral hemisphere. **A.** Division into lobes. **B.** Retraction of opercular lips to expose fifth lobe or insula. **C.** Gyri and sulci on medial surface of the hemisphere. (From R. C. Truex and M. B. Carpenter, *Human Neuroanatomy*, Williams & Wilkins, Baltimore, 1969.)

worth even naming some of the gyri unless you can relate them to some useful function.

There are a basic, irreducible, minimal number of them that we should learn to locate right now. The rest can wait until you study neuroanatomy in depth later.

1. Precentral Gyrus: The *precentral gyrus* anterior to the central sulcus in the frontal lobe has been located (Fig. XIV-19). It has been called the **voluntary motor cortex.** If it is stimulated electrically, precisely localized movements can be elicited in the lightly anesthetized subject. Stimulating medially along the paracentral lobule will produce movements in the toes and lower leg. Near the dorsolateral margin, thigh muscles can be brought into play. As one moves inferiorly along the gyrus toward the lateral fissure, muscles of the trunk, shoulder, arm, hand, and particularly thumb and fingers are activated; then neck and facial muscles, particularly tongue; and lowest of all, larynx with some gutteral vocalization. It is as if we had laid out a mannequin on the gyrus with its head at the lateral fissure and with its knees bent over the dorsolateral rim so that its legs dangled medially between the hemispheres.

Oddly, some of this motor activity can be elicited from the postcentral gyrus as well. Hence, the two in association are known as the **sensorimotor cortex.**

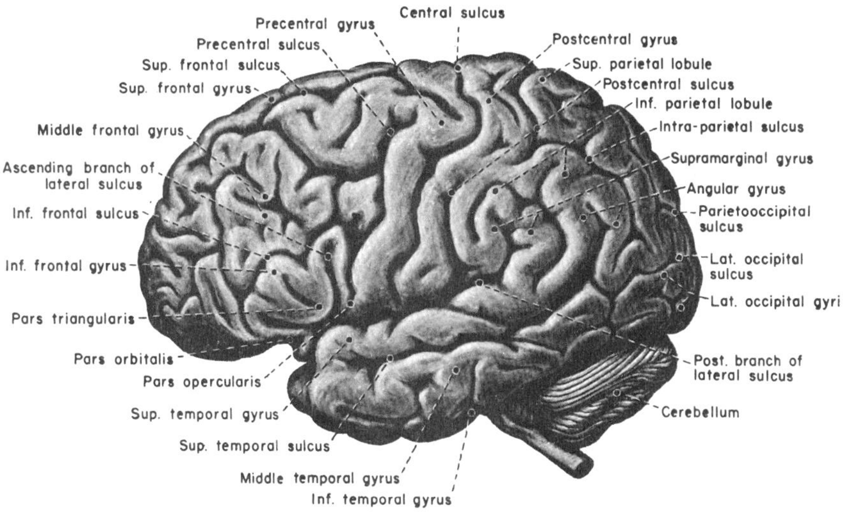

Fig. XIV-19. Lateral surface of left hemisphere showing principal gyri and sulci. (From R. C. Truex and M. B. Carpenter, *Human Neuroanatomy*, 6th ed., Williams & Wilkins, Baltimore, 1969.)

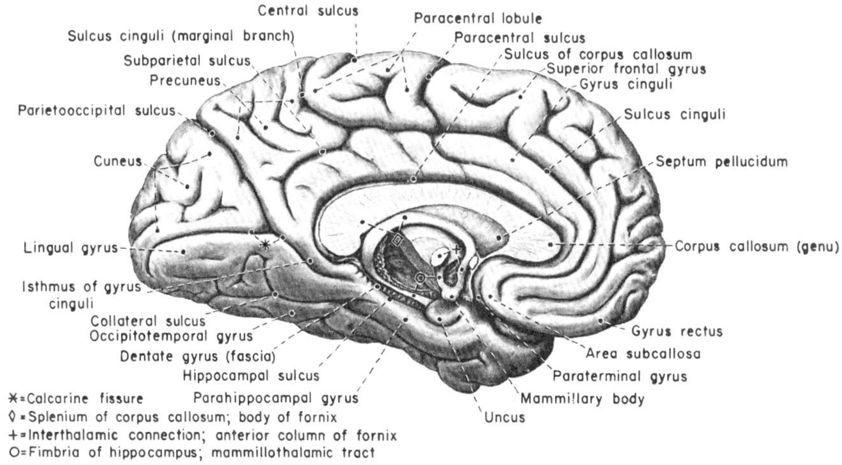

Fig. XIV-20. Medial surface of left hemisphere showing principal gyri and sulci. The ependyma and part of the thalamic nuclei were removed to expose the relations of the anterior column of the fornix and the mammillothalamic tract. (From R. C Truex and M. B. Carpenter, *Human Neuroanatomy*, 6th ed., Williams & Wilkins, Baltimore, 1969.)

2. Postcentral Gyrus: The **postcentral gyrus** posterior to the central sulcus lies in the parietal lobe (Fig. XIV-19). It is the general **somesthetic receptive center** of the brain, registering pain, temperature, touch and proprioception from the opposite side of the body. Thus, the right side of the body projects to the left postcentral gyrus and *vice versa,* just as the left precentral gyrus initiates muscular movements in the right side of the body and *vice versa.* Again, body topography is projected upon the postcentral gyrus as it was upon the precentral gyrus. Stimulation of the postcentral gyrus in the semiconscious patient will not, however, elicit any of the sensory modalities of pain, touch, pressure, or proprioception; only vague tingling sensations in very localized regions of the body depending, of course, upon where along the **gyrus** the electrode is placed.

3. Anterior Transverse Temporal Gyrus: The **anterior transverse temporal gyrus** (Heschl's convalution), almost buried in the insula, is the **auditory receptive center** (Fig. XIV-18B). The cochlea of the inner ear projects topographically upon it so that its coils for progressive pitch perception are laid off along it in an anteroposterior direction.

4. Cuneus and Lingual Gyrus: **Cuneus and lingual gyri,** separated by the calcarine fissure in the occipital lobe, constitute the **primary visual receptive center** of the brain (Fig. XIV-20). Upper quadrants of the retina project along the upper banks of the calcarine fissure, *i.e.,* on cuneus; while lower quadrants project along its lower bank, *i.e.,* the lingula. The reverse, of course, is true of *visual fields* out in space. Upper quadrants of the visual field project upon lingula; lower quadrants, on cuneus. The macula or area of central vision spills out over the occipital pole.

5. Hippocampal Gyrus and Uncus: The **olfactory receptive center** in man is restricted to a very small region at the **anterior tip of the hippocampal gyrus** and the **anterodorsal surface of the uncus** (Fig. XIV-18C). These regions in the temporal lobe, known as the prepiriform and periamygdaloid regions, constitute the primary cortical center for perceiving smells.

The sensory ganglion cells for olfaction lie in the nasal mucosa in the region of the superior conchae and on the median nasal septum high in the nose. Axons of these cells make up the olfactory nerve filaments which pass through the cribriform plates of the ethmoid to synapse in the **olfactory bulbs** which rest upon the plates. Axons of the bulb form the **olfactory tract** which follows the **lateral olfactory stria** to the primary olfactory receptive center. Of all the senses of man, this one makes no thalamic connections before it reaches its cortical receptive center.

6. Gyri Related to Language Function: There are, in addition to the five gyri just mentioned, three other gyri in the *dominant hemisphere* to which you should attend. In right handed individuals the left hemisphere is the dominant one and *vice versa.* Three gyri for the symbolic formulation of language reside in the dominant hemisphere only, or at least function there normally. One of these lies in the frontal lobe and two in the parietal lobe.

A. MOTOR SPEECH CENTER (BROCA'S AREA). At its anterior end, the lateral fissure continues into the frontal lobe as two small sulci or rami, an **ascending branch or ramus** and a **horizontal branch or ramus** (Fig. XIV-19). These small extensions of the fissure end among three, small curlique-like gyri just anterior to the bottom of the precentral gyrus. These are known as the **pars opercularis, pars triangularis,** and **pars orbitalis** or the **inferior frontal gyrus.** It is the opercularis and triangularis that constitute Broca's area or the motor speech center. It is here that words are formulated into meaningful speech. The area lies next to the area in the precentral gyrus which is capable of moving the vocal cords, lips, mouth, and tongue.

B. AUDITORY SPEECH CENTER OR SUPRAMARGINAL GYRUS (FIG. XIV-19). Now chase the posterior end of the lateral fissure back as far as it will go until its ends are capped by a horseshoe-shaped gyrus of the parietal lobe. This is the supramarginal gyrus. Here speech symbols are interpreted as meaningful sounds. We understand the spoken word. Note that it lies close to where the anterior transverse temporal gyrus disappears deep into the lateral fissure toward the insula. This latter gyrus, of course, is the auditory receptive center.

C. VISUAL SPEECH CENTER OR ANGULAR GYRUS (FIG. XIV-19). Find the **superior temporal sulcus.** It lies just below the superior temporal gyrus. Trace the superior temporal sulcus posteriorly as far as it will go. It too is capped by a horseshoe-shaped gyrus in the parietal lobe, the angular gyrus. Here we are able to interpret the written word as we see it. Note that it lies in proximity with the visual cortex of the occipital lobe.

Lesions in these three areas *of the dominant hemisphere* lead to **aphasias** of one type or another which interfere with language function. Either the individual cannot name objects which he sees (even if he knows how they are used), doesn't understand what he hears, or can't read the printed page. None of these aphasias are apt to be pure and distinct since these areas interconnect with each other by associational fiber tracts. Hence, symptoms tend more often to be of a mixed type when one area or the other is damaged. Loss may persist for a long time and may require tedious speech and language retraining. Eventually, the same areas in the non-dominant hemisphere may "learn" to handle symbolic language formulation reasonably well. Thus, an aphasia is a memory loss of a language function which may be relearned in time.

Other Functional Areas in the Cortex: There are other *motor areas* in the brain besides the precentral gyrus which are capable of initiating movement. We will find them in the frontal lobe just anterior to the precentral gyrus, in the superior part of the parietal lobe posterior to the postcentral gyrus, in the occipital lobe anterior to the occipital pole and in the superior temporal gyrus. None of these motor areas will be able to control the very precise movements of the distal extremities. Rather, they are concerned with postural regulation of the large proximal muscle masses of the limb, the so-called **antigravity muscles.** These muscles hold us upright on legs so that gravity does not pull us to the floor or they raise the arm up against gravity which always wants to pull it down. Only the precentral gyrus can move the small distal muscles which we use for precise purposeful manipulation of our external environment.

Some cortical motor areas were thought capable of inhibiting or suppressing movement once it was initiated either voluntarily (by precentral gyrus) or reflexly at cord or brainstem levels. At the present time, we are no longer certain that these experimentally discovered **suppressor areas** really exist. They may have been artifacts of experimentation as the exposed cortex dried out.

No cortical area can be singled out as essential for learning. There is some evidence that lesions in the temperoparietal region result in memory losses of one form or another.

If we try to assess the purpose of the prefrontal lobe of man, that portion anterior to its motor centers, we are somewhat at a loss. People have had both frontal lobes shot away and have lived perfectly normal lives after recovery from the original insult. These findings run quite counter to the odd behavior, loss of judgement, and personality changes reported after frontal lobotomies in man. It probably depends upon whether certain connections between frontal lobe and the hy-

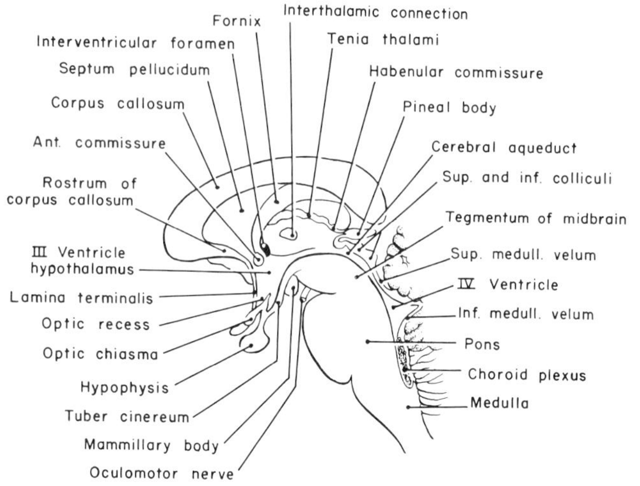

Fig. XIV-21. Sketch of a midsagittal section of brainstem. Compare with Figure XIV-11. (From R. C. Truex and M. B. Carpenter, *Human Neuroanatomy,* Williams & Wilkins, Baltimore, 1969.)

pothalamus and the "limbic lobe" regions have been kept intact or not.

Anatomy of the Brainstem in Midsagittal View: A cut between hemispheres along the sagittal fissures will slice through corpus callosum and the attached brainstem (Figs. XIV-16 and 21). It will cut midway through the third and fourth ventricles and will allow us to delineate certain brainstem areas.

The corpus callosum, septum pellucidum, anterior commissure, and fornix are all parts of the cerebrum in this plane of section.

The diencephalon or *thalamic region* begins caudal to the fornix and ends at the *midbrain.* A line drawn from the **oculomotor nerve** transversely to the top of the **superior colliculus** marks the rostral limit of the midbrain (Fig. XIV-16). The **pons** is seen to bulge markedly to the left forming rostral and caudal indentations on this, its ventral, margin. The bulge is the **basis pontis.** Transverse lines drawn at the rostral and caudal ends of this bulge cut the pons from the midbrain above and from the rest of the brainstem, the *medulla,* below (Fig. XIV-16).

Thalamus and hypothalamus form the medial walls of a narrow third ventricle. This ventricle is bridged sometimes by an **interthalamic connection** or **massa intermedia** shown cut in cross-section here (Fig. XIV-21). The midbrain is tunneled by the **cerebral aqueduct** (of Sylvius) so that its bulky **tegmentum** lies ventral to the aqueduct (to the left in Fig. XIV-21) and its **tectum** or roof, consisting of **superior** and **inferior colliculi** lies dorsal to the aqueduct. The aqueduct connects the third with the fourth ventricle. The latter is a shallow trough lying partly on the dorsal aspects of pons and medulla. It is roofed over by the **superior** and **inferior medullary vela** and by a thin, ependymal roof caudally which contains a choroid plexus. Dorsal to this roof of the fourth ventricle sits the cerebellum, barely shown in Figure XIV-21 but almost in full view in Figure XIV-16.

Let us now consider all of these regions from above downwards.

The Cerebral Commissures: Regard the C-shaped **corpus callosum.** This main interhemispheric commissure consists of a broad, arched band of white matter whose fibers (axons) cross through it transversely so that they would be cut in cross-section in this plane. Its roof lies in the bottom of the sagittal fissure. The roof has a thin mantle of gray matter over it, the **indusium griseum** or **hippocampal rudiment** which is a vestige of the hippocampus of lower forms. Its **rostrum** or anterior portion bends to thin out as the **lamina terminalis.** The lamina represents the caudal end of a forward growing telencephalon in the embryo. But, since a forming brain case soon restricted forward growth of the cerebrum, it bent or arched back over the diencephalon so that now the thalamus lies between and below the hemispheres instead of caudal to them as it was, strictly, in the young, embryonic longitudinal neural tube. The back end of the corpus callosum broadens as its blunt **splenium.**

The **anterior commissure** and **fornix** are two additional interhemispheric connections. Between them and the corpus callosum, stretches a thin curtain or partition, the **septum pellucidum.** This forms part of the medial wall of each *lateral ventricle.* Were these cerebral ventricles numbered, they would be the first and the second ventricles; but they are hardly ever called thus. They communicate with the third ventricle through a hole in their medial walls just posterior to the column of each fornix. This is the **foramen of Munro** or the **interventricular foramen.** The septum pellucidum does not allow us to look into the lateral ventricle of this right hemisphere, but we can see the foramen through which it communicates with the third ventricle (Figs. XIV-21 and 16).

The **anterior cerebral commissure** lies between lamina terminalis and the column of the fornix. It is a small bundle of axons cut in cross-section in the midsagittal plane. Such a cut through it belies its odd shape, for laterally in the hemispheres it breaks into two pairs of arms; one set is anteriorly directed into each olfactory bulb and one set posterolaterally directed toward the basal ganglia, the deep gray masses of the hemisphere (Fig. XIV-22). The dotted line in Figure XIV-22 shows where this commissure was cut in cross-section in Figure XIV-21.

The third cerebral commissure, the **fornix,** is also odd-shaped and its contours cannot be fully appreciated in midsagittal view either. In Figure XIV-21 we see the **column of the fornix** anterior to the foramen of Munro after this column has come up from the **mammillary bodies** of the hypothalamus. The **body** of the fornix arches as a flattened ribbon to disappear from sight beneath the splenium of the corpus callosum. This is because it is moving quite laterally at this point, behind the plane of the page, to reach the roof of the inferior horn of the lateral ventricle in the temporal lobe of this right hemisphere. There, its fim-

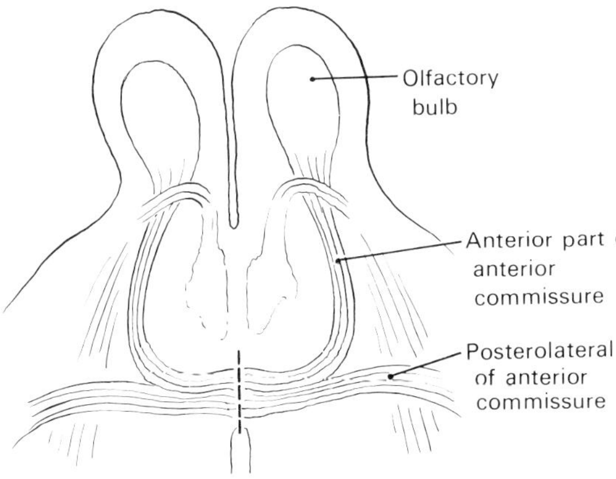

Fig. XIV-22. The anterior commissure as seen in the mouse brain. (After Cajal.) (The dotted line cuts the commissure in the plane in which it is seen in Figs. XIV-16 and 17.)

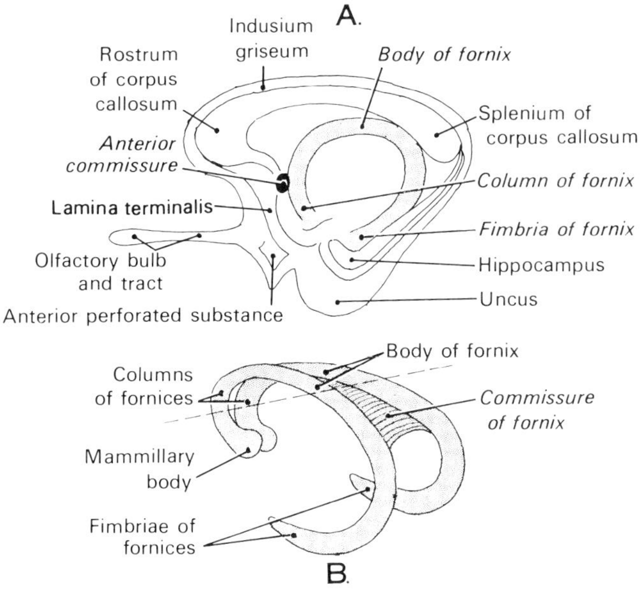

Fig. XIV-23. The fornix. **A.** In relation to rhinencephalon or small brain. **B.** Its configuration seen as if it were dissected free. The portion of the fornix above the dotted line in B is all of the fornix that can be seen in a midsagittal section through corpus callosum and brainstem.

briated end will contact the rolled-in portion of the hippocampal gyrus near the uncus (Fig. XIV-23A). Were we to dissect it free on both sides, we would see that the fornices of each hemisphere met in a body as their columns came up out of each mammillary body (Fig. XIV-23B). Posteriorly, under the splenium, the fornices would diverge again to reach each temporal lobe. A triangular connecting piece beweeen them posteriorly is the **hippocampal commissure** or **commissure of the fornix.** We learned that fornix meant arched so that the fornix is a kind of arched commissure which interconnects cerebral hemispheres and parts of the hypothalamus.

There is an apocryphal tale about an American student in Dr. Edinger's laboratory in Germany. When he came across this commissure in one of his dissections of the brain, he inquired in broken German: "Wass ist dass for?" To this, the good Doctor replied: "Dass ist for nix!" And so, it is said, the structure got its name. Staten Island was named in a similar fashion. As the Dutch sailed into New York, a sailor tapped Hendrik Hudson on the shoulder, pointed, and asked: "Iss dat an island?"

The Diencephalon: The thalamic region forms the medial walls of a V-shaped third ventricle. The **tenia thalami** (Fig. XIV-21) is the cut edge of a paper-thin ependymal roof which normally bridges across the ventricle from one dorsal thalamus to the other. The thalamus proper is the large ovoid mass below the fornix; and, in this specimen, there is a **thalamic connection** below the roof which also bridges across the ventricle. The anterior part of this ovoid thalamus stops as the foramen of Munro and posteriorly it becomes the **epithalamus** which includes **pineal body;** the **habenular commissure** above the pineal body; and **posterior commissure** below it. The latter commissure, unnamed in Figure XIV-21, is the small ventral bulge between pineal body and superior colliculus of the midbrain. Habenular and posterior commissures interconnect the thalami of opposite walls of the third ventricle. They are diencephalic, not cerebral, commissures, just as is the **massa intermedia** when it is present, as it is in this specimen.

Below the ovoid dorsal thalamus proper is a slight S-shaped sulcus which runs from the foramen of Munro posteriorly to the entrance of the cerebral aqueduct. This is the **hypothalamic sulcus.** It separates thalamus proper above from hypothalamus below. Caudally, below the sulcus, a bit of **subthalamus** intervenes between sulcus and midbrain tegmentum.

The **hypothalamus** is the medial wall of the anteroinferior part of the third ventricle where it narrows into a funnel directed toward the **optic chiasma** and the **infundibulum of the hypophysis.** It is limited anteriorly by the lamina terminalis and posteriorly by the **tuber cinereum** and *mammillary body.*

The Midbrain: This short stretch of the brainstem is divisible into *basis, tegmentum,* and *tectum.* It is not as easy to visualize these parts in midsagittal view (Fig. XIV-21) as it is in cross-section (Fig. XIV-24). However, in Figure XIV-21, the oculomotor or cranial nerve III can be seen emerging from the **interpeduncular fossa** on the *ventral aspect* of the brainstem. Figure XIV-26 shows this fossa flanked by two, stout, diverging fiber bundles—the **cerebral peduncles** or **basis pedunculi.** These bundles bring descending fiber tracts, which arise at cortical levels, down to brainstem and cord levels. The corticospinal fibers which become lateral and ventral corticospinal tracts in the cord, occupy the middle three-fifths of these cerebral peduncles (Fig. XIV-24). The peduncles make up the *basis of the midbrain.*

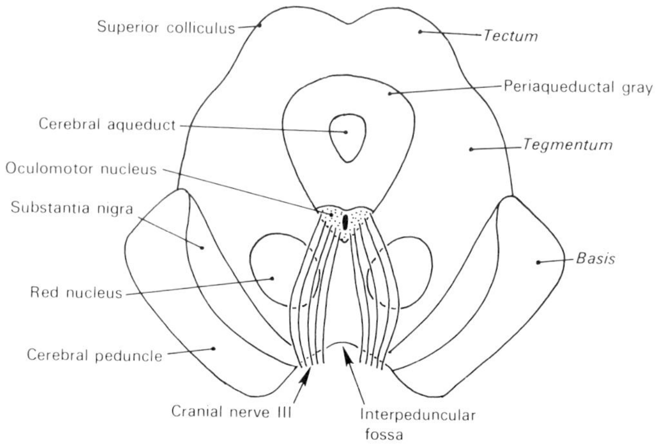

Fig. XIV-24. Schematic cross-section through midbrain to show basic regions: tectum, tegmentum, and basis.

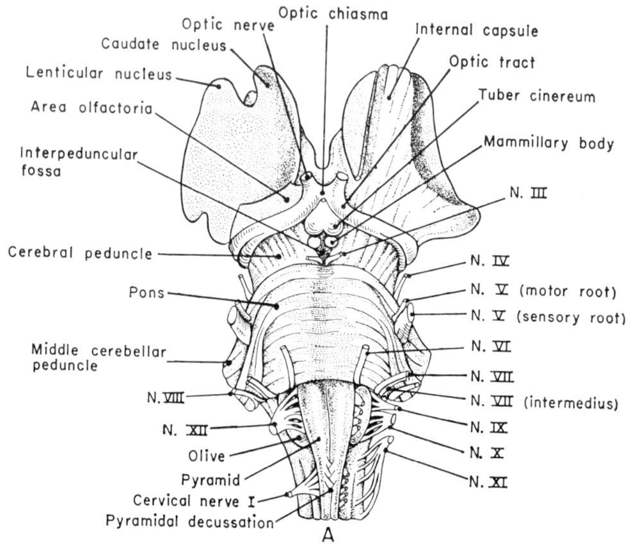

Fig. XIV-25. Anatomy of ventral aspect of brainstem. (From R. C. Truex and M. B. Carpenter, *Human Neuroanatomy*, 6th ed., Williams & Wilkins, Baltimore, 1969.)

Between peduncles and aqueduct, other ascending and descending fiber tracts and various patches of gray matter (named nuclei) make up the *midbrain tegmentum.* In a freshly cut brainstem, a large, rounded, pink area may be identified in the tegmentum. This is the prominent **red nucleus** which plays a role, along with the cerebellum, in giving rise to motor integrative pathways concerned in postural control and regulation. Motor nuclei which give rise to the axons of cranial nerves III and IV lie in the tegmentum in the midline below the aqueduct.

The *roof or tectum* of the midbrain lies above the aqueduct and consists of two pairs of hillocks, the **superior and inferior colliculi.** Since they lie dorsally, they are best seen in a full dorsal view of the brainstem (Fig. XIV-26). Here the pineal body will overhang the superior pair. The four colliculi, together, are known as the **corpora quadrigemina** of the midbrain. The *superior colliculi* are concerned with **visual reflexes;** the *inferior colliculi* with **auditory reflexes,** *e.g.,* turning head, eyes, neck, and even trunk toward the direction of a sound; or moving the whole body away from an approaching sound (automobile horn). Such reflex messages can be routed to cord level to motivate anterior horn cells over **tectospinal pathways,** or they can be routed to the cerebellum *via* central tegmental bundles.

Midsagittal Views of Pons and Medulla (Fig. XIV-21): These are not very illuminating on the whole. Both areas have a basis and a tegmentum. The latter may sometimes be referred to as the **reticular formation** because the gray matter is cut up into many gray nuclear patches by criss-crossing fiber bundles as to appear to be a virtual jungle impervious, at first glance, to mental penetration by beginning students. One nucleus, for example, is so difficult to find in this crazy-quilt patterning that it has been called the **nucleus ambiguus.**

If the hindbrain has any true tectum at all, it probably is the roof structure over the fourth ventricle (the medullary vela and the choroid plexus). One seldom hears the word tectum spoken in reference to pons and medulla, but it does crop up in reference to the **tectal or roof nuclei of the cerebellum** (globose and fastigial nuclei).

The *basis pontis* is rather prominent in ventral view (Fig. XIV-25) where it is seen as a heavy collection of transversely running fasciculi grooved in the midline by a sulcus. This is the **basilar sulcus** which is occupied by the **basilar artery** formed by union of right and left vertebral arteries after the latter enter the cranium through the foramen magnum.

If the medulla has a true basis, it is the columnar **pyramids** on its ventral aspect (Fig. XIV-25). These fiber bundles now contain the corticospinal tracts alone before they decussate to enter the cord. They have come down grouped in the middle three-fifths of the peduncles only to scatter somewhat in the basis of the pons. But they have gathered neatly together once more in the pyramids of the medulla. All other descending tracts at this level lie in the tegmentum or reticular formation of the medulla distinctly separated from the pyramids. Close to the foramen magnum, fascicular bundles in the pyramids begin, alternately, to cross to the right and to the left to become the lateral corticospinal tract of the cord. This area of **pyramidal decussation** tends to wipe out a short stretch of the ventral sulcus of the medulla (Fig. XIV-26). Some fibers of the pyramids fail to decussate and these descend into the cord as the ventral corticospinal tracts.

The Ventral Aspect of the Brainstem (Fig. XIV-25 and 27): While we have already considered several structures on the ventral side of the brainstem, this was done merely to help you understand its features in midsagittal view. We need now to locate cranial

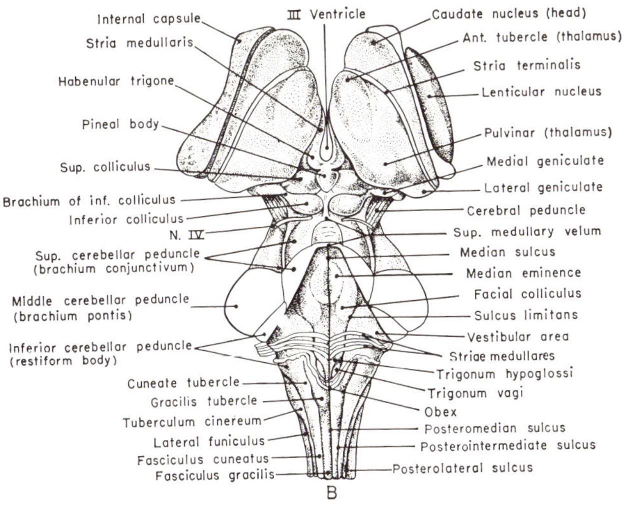

Fig. XIV-26. Anatomy of dorsal aspect of brainstem. (From R. C. Truex and M. B. Carpenter, *Human Neuroanatomy*, 6th ed., Williams & Wilkins, Baltimore, 1969.)

nerves and to gain a fuller concept of the configuration of the brainstem.

We are not going to be concerned with the upper parts of Figure XIV-25 beyond the optic nerves. These are parts of the central gray nuclear masses of the cerebral hemispheres which constitute the basal ganglia. A glance at Figure XIV-27, makes it apparent at once that when the brainstem and hemispheres are intact, very little of the thalamus and midbrain can be seen in a ventral view of the brainstem. These parts lie deep and if they diverge laterally, even a little bit, they are masked by the poles of the temporal lobes. Pons and medulla do, however, come into reasonable view.

Cranial nerve I or the **olfactory** nerve is not at all related to the brainstem. It is a part of the cerebral hemisphere (Fig. XIV-27). The *olfactory bulb* and *tract* lie on the inferior aspect of the brain in a sulcus just lateral to two, small, straight gyri of the frontal lobe, the **gyri recti.** Posteriorly, the *olfactory tracts* diverge into **medial** and **lateral olfactory striae.** The

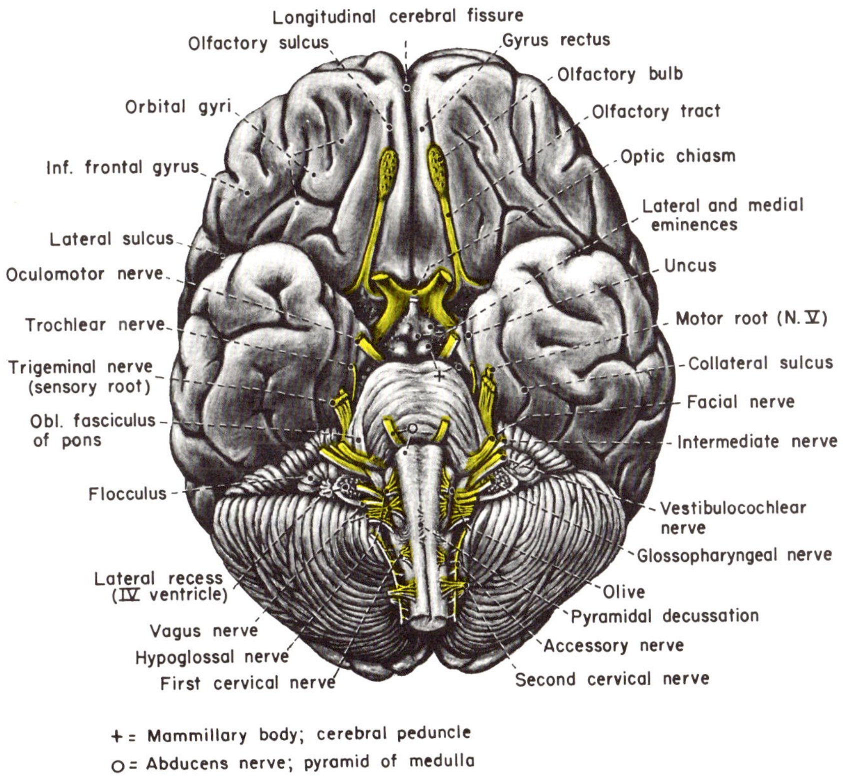

Fig. XIV-27. Inferior surface of the brain with attached cranial nerves. (From R. C. Truex and M. B. Carpenter, *Human Neuroanatomy*, 6th ed., Williams & Wilkins, Baltimore, 1969.)

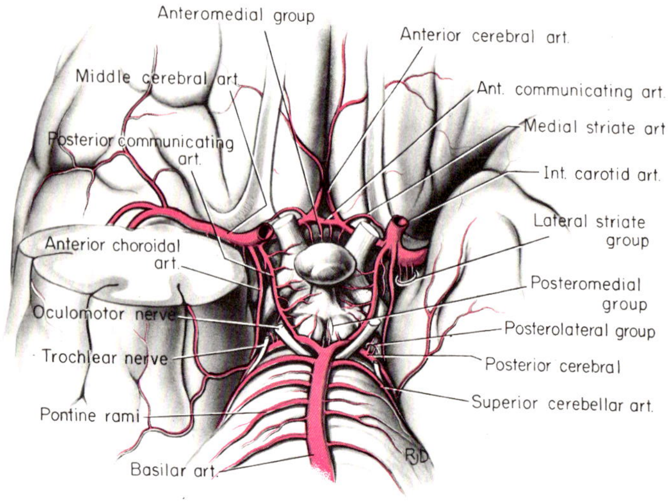

Fig. XIV-28. Arterial branches from circle of Willis to the basal ganglia. Ventral aspect of brain. These small perforating vessels form anteromedial and posteromedial, posterolateral and lateral striate groups. Also shown are the medial striate and anterior choroidal arteries. (From R. C. Truex and M. B. Carpenter, *Human Neuroanatomy,* 6th ed. Williams & Wilkins, Baltimore, 1969.)

lateral one will course to its cortical receptive center, the prepiriform and periamygdaloid regions of the temporal lobe. The medial stria and a very short intermediate stria make many complicated reflex connections with structures in the frontal and temporal lobes and with the diencephalon and, eventually, with other parts of the brainstem.

Just posteriorly between the diverging olfactory striae, the base of the brain is perforated. This is the region of the **anterior perforated substance.** Here many fine twigs from the arterial circle of Willis penetrate to the deep underlying gray matter of hemispheres, their basal ganglia (Fig. XIV-28). One of these small vessels, the lenticulostriate artery (of Charcot) is known as the artery of cerebral hemorrhage because of its frequency of involvement in paralytic strokes. It feeds descending as well as ascending pathways between basal ganglia and thalamus in a region called the **internal capsule** (Fig. XIV-17). The internal capsule is, in a sense, an upward continuation of the cerebral peduncles of the midbrain. However, the internal capsule not only contains the motor fibers which will enter the peduncles from above, it contains as well sensory ascending fibers from thalamus to cortex and fibers from all the cortical sensory receptive centers returning to the thalamus for reflex connections in parts of the brainstem. In light of the above, the small patch of anterior perforated substance on the inferior base of the brain bears some significance.

The **optic nerve** is a diencephalic nerve; and if you do not care to consider the thalamus as a part of the brainstem, then, like the olfactory nerve, you can regard the optic nerve as a part of the brain proper. Indeed, retina and optic nerve were outgrowths of the embryonic, diencephalic, brain vesicle. For that reason, you should not find it strange that the **optic tracts** wend their way back home to the thalamus to one of its major nuclear masses, the **lateral geniculate body.** In Figure XIV-25 we see the two optic nerves partially decussating in the **optic chiasma.** Recall that the chiasma lies in a sulcus between the anterior clinoid processes above the sella turcica. The sulcus is directed toward each optic canal in the lesser wings of the sphenoid bone.

It should not be strange, then, to find the **stalk of the pituitary** and the **tuber cinereum** of the *hypothalamus* lying just caudal to the optic chiasma at the base of the brain. Just caudal to these structures are the *mammillary bodies* of the hypothalamus. These are all that can be seen of the hypothalamic (or diencephalic) structures in a ventral view of the brainstem. They lie in the interpeduncular fossa created by the diverging cerebral peduncles which quickly disappear from sight under the temporal lobes (Figs. XIV-25 and 27).

The optic tracts emerge from the chiasma to encircle the peduncles in their lateral course to reach the lateral geniculate body on the lateral aspect of the brainstem (Fig. XIV-29). Actually, the lateral geniculate body and the medial geniculate body just beyond and below it, constitute the parts of the **met-thalamus.** They are partially hidden under a bulging posterior part of the dorsal thalamus properly known as the pulvinar (Fig. XIV-26).

We need say little more about the *midbrain* in ventral view other than to call your attention once more to the **oculomotor** or **cranial nerve III** as it emerges from the interpeduncular fossa. This and the **trochlear nerve** (**IV**) are midbrain nerves. *All cranial nerves, except the trochlear, have a ventral emergence from the central neuraxis. Cranial nerve IV emerges from the brainstem dorsally* (Fig. XIV-22) by piercing the superior medullary velum of the fourth ventricle just below the inferior colliculus. Then, it moves laterad to encircle the cerebral peduncle, to reach the ventral aspect of the brainstem. We see its fine filament coming into view ventrally in Figures XIV-25 and 27. It will join III and VI to enter the superior orbital fissure in order to reach the extra-ocular muscles of the globe in the orbital cavity.

In the **ventral pontile region,** fibers of the basis pontis stream laterad to gather into stout **middle cerebellar peduncles** or to form a **brachium pontis** on each side. **Trigeminal nerve rootlets** break through these arms or brachia which connect cerebellum to pons. There is a stout **sensory root** and a finer **motor root** of the **trigeminal** or **cranial nerve V** in the lateral angle between pons and cerebellum (Fig. XIV-27). The ophthalmic division of the sensory root enters the superior orbital fissure along with III, IV, and VI; the maxillary sensory root enters the foramen rotundum; and the mandibular sensory root and all of the motor root drop through the foramen ovale to reach the infratemporal fossa.

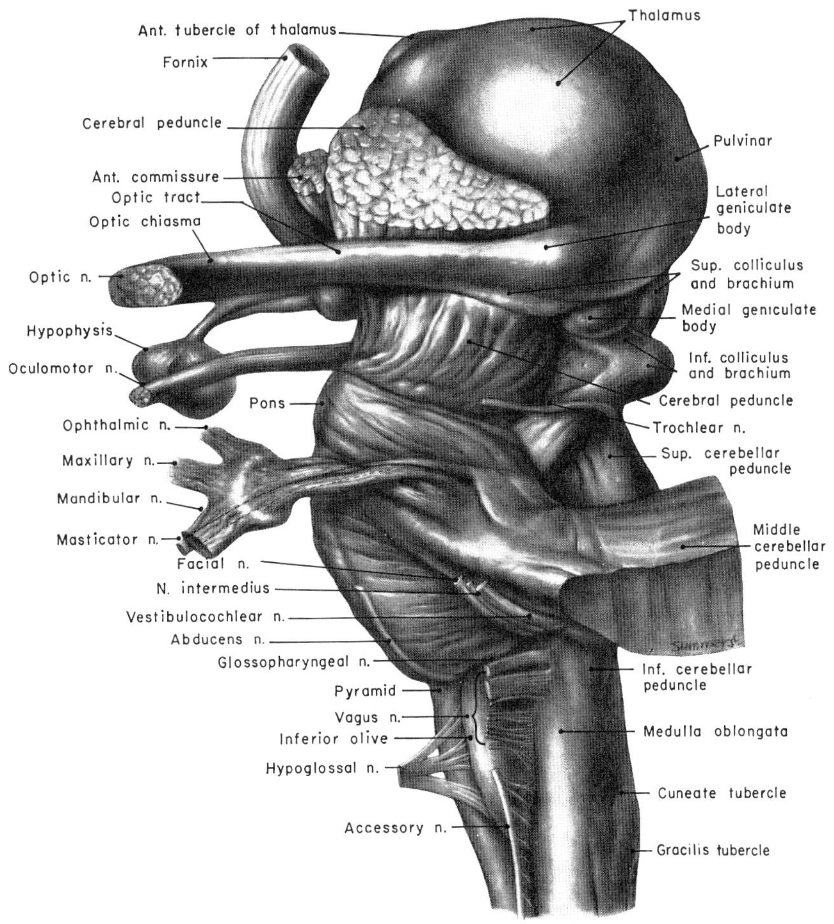

Fig. XIV-29. Lateral view of the brain stem. (From R. C. Truex and M. B. Carpenter, *Human Neuroanatomy*, 6th ed., Williams & Wilkins, Baltimore, 1969.)

Cranial nerve VI or the **abducens** arises out of a ventral transverse sulcus between the pons and medulla close to the midline.

Cranial nerves VII and VIII arise in the cerebellar pontomedullary angle (Fig. XIV-27). They lie near the entrance to the internal auditory meatus of the petrous part of the temporal bone when the brain is placed in its brain case. Cranial nerves V through VIII are pontile nerves.

The **ventrolateral sulcus** runs along the lateral margins of the pyramids of the **medulla.** Out of the sulcus emerges **cranial nerve XII,** or the **hypoglossal,** which will leave the skull through the hypoglossal canal in the occipital condyle.

In a ventral view, a fair-sized bulge can be seen on the brainstem behind the rootlets of XII (Fig. XIV-25). This is the **olive.** Internally in the medulla it appears as an odd-shaped corrugated mass of gray matter known as the **inferior olive.** In size it is much superior to a smaller olive in the tegmentum of the pons, known as the **superior olive.** As far as olives go, your authors would settle for the inferior ones in their martinis over the superior ones. The olive lies between dorsolateral and ventrolateral sulci. **Cranial nerves IX, X, and XI** emerge in order, in rostrocaudal progression, from the **dorsolateral sulcus** behind the bulging inferior olive. Since they can be seen easily in ventral view, they can be lumped with ventrally emerging fibers though, quite accurately, they have a dorsolateral emergence (Fig. XIV-29). All exit in the jugular foramen, that cleft between the petrous part of the temporal bone and the occipital bone in the posterior cranial fossa.

The Brainstem in Dorsal View (Fig. XIV-26): The brainstem with cerebellum removed looks much more complicated on its dorsal aspect than it does on its ventral aspect. It is not too difficult to understand if you examine it systematically.

The big, lateral, ovoid masses rostrally are the gray masses of the thalamus proper. They bulge up into an **anterior nuclear mass** rostrally and pull out into an overhanging **pulvinar** posteriorly. Their most lateral parts make up the floor of the bodies of the lateral ventricles.

Their **dorsomedial surfaces,** as we look down on them from above, lie in no ventricle at all but in the **transverse cerebral fissure.** In Figure XIV-26 their *medial sides* are hidden from view for they make up the walls of the third ventricle to which the medial aspect of the hypothalamus contributes as well. We

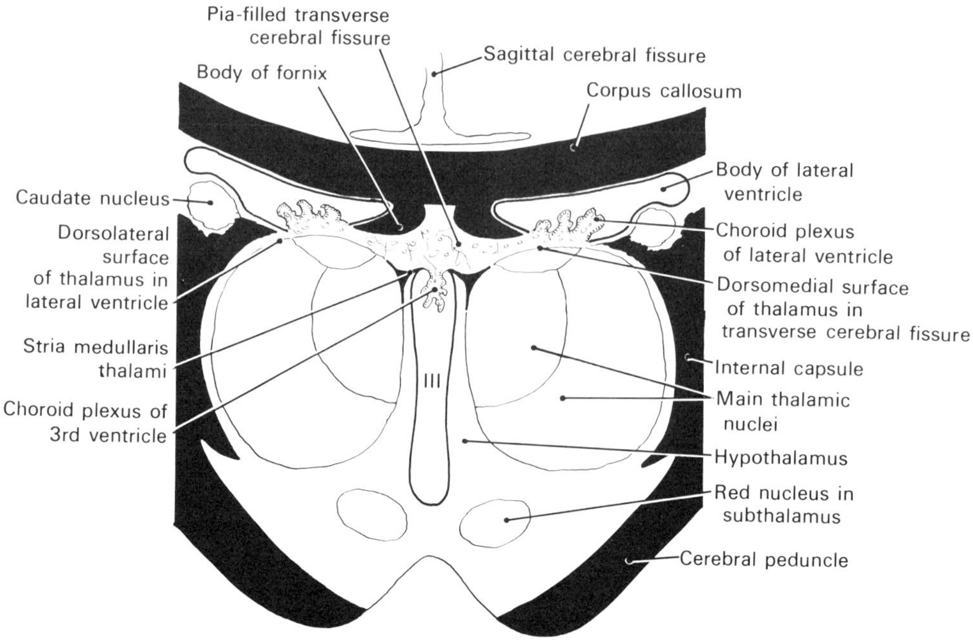

Fig. XIV-30. Schematic coronal section through thalamic mass to show relation of thalamus, caudate nucelus, and fornix to transverse cerebral fissure, lateral and third ventricles.

see the cleft of the third ventricle separating those parts of the thalamus which we examined in midsagittal view (Figs. XIV-16 and 21). Normally, from one **stria medullares thalami** to the other, this roof is bridged with a thin ependyma which blood vessels of the pia mater invaginate as a choroid plexus. We would have to take a coronal section through the two thalami proper to conceive of the ways in which they are involved in both the lateral ventricles and the third ventricle (Fig. XIV-30).

In such a coronal section, we see that the corpus callosum forms a roof for both lateral ventricles and also for the **transverse cerebral fissure,** the area that appears under corpus callosum and fornix in Fig. XIV-30. The **caudate nucleus,** one of the basal ganglia of the hemispheres, makes up the lateral wall of the body of each lateral ventricle. The medial wall consists of the lateral margin of the ribbon-like fornix and a thin piece of ependyma which bridges from fornix to dorsolateral thalamus below. This thin-walled cleft between lateral fornix and thalamus is known as the **choroid fissure.** It is nothing more than the lateral extent of the transverse cerebral fissure which is prevented from entering the lateral ventricles by the thin, ependymal, bridging tissue. So blood vessels of pia mater push laterally against this thin ependyma to shove a choroid plexus into each lateral ventricle.

We can see (Fig. XIV-30) that the *dorsomedial surface* of the thalamus lies in the **floor of the transverse cerebral fissure** while the *dorsolateral surface* makes up the **floor of each lateral ventricle,** at least of the **body** of the peculiar-shaped lateral ventricle. We will discuss its total configuration later. Ependyma bridges across the third ventricle from stria medullaris thalami to stria medullaris thalami of the opposite side. Again pial vessels push down on this thin roof to hang a choroid plexus into the third ventricle. The medially-lined ependymal walls of thalamus and hypothalamus complete the trough of the third ventricle which is sandwiched between the opposing thalamic territories on each side.

In dorsal view (Fig. XIV-26), the dorsal thalamic masses appear as diverging buttocks, and they were once called "the anchae" or "buttocks" by Mondino, an Italian anatomist. Galen called them thalami or "the bed or chambers of the brain." He had them confused, in part, with the ventricles. If the thalamus is the bed of the brain, then the hypothalamus is probably the proverbial chamber that is found under the bed.

Peeking out from under the pulvinar laterally are the medial and lateral geniculate bodies. These make up the *met-thalamus.*

In the midline, both stria medullares thalami, the habenulae and their commissure, the pineal body and the posterior commissure under it, all make up the *epithalamus.* The nebulous *subthalamus* cannot be clearly indicated in gross view anywhere but is best seen in oblique sections cut through thalamus and midbrain regions. We need not worry about it at this time. It would lie ventrolateral to the posterior commissure.

In the dorsal midbrain we see the colliculi of the corpora quadrigemina quite clearly. Each connects by means of a **quadrigeminal branchium** to the geniculate bodies. Superior colliculi, related to visual reflexes, connect by a narrow strand the **superior quadrigeminal brachium** with, as you might suspect, the lateral geniculate body since it is associated with the optic tract (Fig. XIV-29). The medial geniculate body is involved in relaying auditory messages to the anterior transverse temporal gyrus so the **inferior quadrigeminal brachium connects** it to the inferior colliculus (Fig. XIV-29).

The trochlear nerve is seen making its dorsal emergence in the superior medullary velum (Fig. XIV-26). The cerebral peduncles reach toward the area of the basal ganglia lateral to the ovoid thalamic masses.

The entire dorsal aspect of the pons and the rostral dorsal half of the medulla are roofed over by the medullary vela and caudally by a thin ependymal roof. The sloping sides of this roof is formed in part by the three paired brachia which attach cerebellum to the midbrain, pons, and medulla, respectively. These are designated as: **superior cerebellar peduncle** or **brachium conjunctivum, middle cerebellar peduncle** or **brachium** pontis, and **inferior cerebellar peduncle** or **restiform body** (Figs. XIV-26 and 31). The dual names of each pair may be used indiscriminately by different authors. It is well to understand that to which they refer.

The superior medullary velum stretches between two fiber tracts which come laterally out of the cerebellum and converge toward each other as they move toward the midbrain. In the midbrain they will decussate to reach opposite red nuclei. Hence, the superior cerebellar peduncle is known as the brachium conjunctivum since it will meet to cross its mate of the opposite side.

The middle cerebellar peduncles are the arms of basis pontis coming around laterally from the ventral aspect of the pons to enter the cerebellum. They bring motor fibers from frontal, parietal, occipital, and temporal cortices into the cerebellum as **extrapyramidal tract fibers**—fiber bundles that never enter the pyramids of the medulla.

The inferior brachium or restiform body comes up from the medulla carrying principally sensory proprioceptive tracts from cord and medullary levels into the cerebellum. Some tactile and vestibular information also traverses this tract.

Of the three brachia, basically, the middle and inferior ones act mainly as cerebellar input bundles while the superior one is a main cerebellar outflow bundle. These three brachia are seen cut across in Figure XIV-26 after the cerebellum has been detached. This allows you to look down upon the rhomboid fossa which is the floor of the fourth ventricle.

The upper half of the **rhomboid fossa** is in the pons; the lower half in the rostral part of the medulla. The division between pons and medulla on this floor is marked by the **striae medullares** (**acousticae**). This is a poor name for these laterally directed fibers which leave faint marks on the floor of the fossa. It was once thought they were related in some way with the cochlear nuclei in the area. It is now known that they are axons of **arcuate nuclei of the medulla** which carry extrapyramidal motor information into the medulla by joining the restiform body (inferior brachium).

A **median sulcus** divides the fossa longitudinally. There is a curving sulcus lateral to the median sulcus. This is the **sulcus limitans** of the embryonic neural tube. Medial to it, between it and the median sulcus, motor nuclei lie under its ependymal mantle while lateral to it lie second order sensory nuclei of the cranial nerves—nuclei of termination for entering axons of cranial sensory ganglion cells. The axons of these second-order neurons build up ascending and descending fiber tracts in the tegmentum of the brainstem.

One of the motor nuclei bulges the pontile floor of the rhomboid fossa as the **facial colliculus** (Fig. XIV-26). This is where the *motor fibers* of the facial nerve bend around the *motor nucleus* of the *abducens* as the facial nerve courses toward its exit from the brainstem. Lateral to the sulcus limitans, far laterad near the lateral recesses of the pontile part of the fossa, are the **sensory vestibular nuclei** and the **dorsal** and **ventral cochlear nuclei.**

In the midline of the fossa in the medulla is a triangular area, the **hypoglossal trigone** (Fig. XIV-26). Deep to this tiny raised area lies the motor nucleus of the hypoglossal nerve. Just lateral to it is a tinier triangular area, the **vagal trigone** or the **ala cinerea** ("gray wing"). It marks the locus of the **dorsal motor nucleus of the vagus** (or the **visceral motor nucleus** of the vagus) which contributes to the cranial parasympathetic outflow.

This vagal nucleus lies more dorsal and superficial in the medulla than does the **branchial motor nucleus** or **nucleus ambiguous** which contributes vagal fibers to the striated muscle of the pharynx, larynx, and upper esophagus. Motor fibers of IX and XI as well as X arise from the nucleus ambiguous, and all are branchial motor in function.

In the closed part of the medulla, dorsally, two fiber bundles from the posterior funiculus of the cord course upward through the foramen magnum separated by the posterior median sulcus and the intermediate septum. These are the **fasciculus gracilis and cuneatus** and they terminate in two swellings just caudal to the rhomboid fossa, the **clava** or **gracile tubercle** and the **cuneate tubercle** (Fig. XIV-26). Second-order sensory nuclei for the terminations of these sensory tracts lie under these superficial medullary swellings. Lateral to the cuneate tubercle (Fig. XIV-26), is another swelling—the **tuberculum cinereum.** Do not confuse this with the **tuber cinereum** of the hypothalamus. The tuberculum cinereum is where the **spinal nucleus of the trigeminal nerve** gets large in the interior of the medulla and raises a dorsolateral swelling. It is a second-order, sensory neuron for entering fibers from the trigeminal semilunar ganglion which descend to synapse in the medulla. These conduct pain and temperature sensations from the face.

Some of these same features can be viewed in the lateral aspect of the brainstem in Figure XIV-29.

To appreciate the nature of the choroid plexus over the posterior half of the fourth ventricle, see Figure XIV-31. The plexus is a T-shaped invagination of the ependyma by pial capillaries. At the bottom of the stem of the "T" is a hole in the ependymal roof, the midline *foramen of Magendie*. There are ependymal

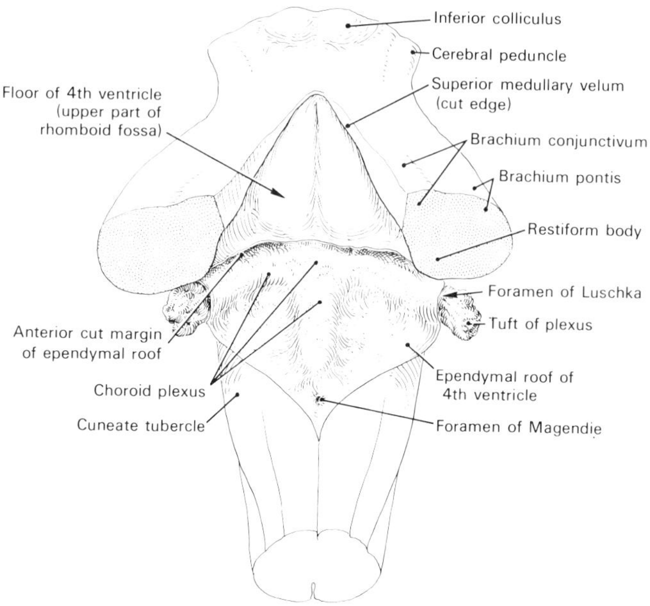

Fig. XIV-31. The choroid plexus of the fourth ventricle.

openings at each lateral recess of the medullary part of the rhomboid fossa, the *foramina of Luschka.* The choroid plexus is seen protruding through the foramina of Luschka in Figure XIV-31. Through these three apertures, cerebrospinal fluid manufactured by the choroid plexuses of the ventricles leaves for the subarachnoid space between pia and arachnoid as these meningeal coverings invest all of the brain and spinal cord externally.

The Anatomy of the Basal Ganglia and the Cerebellum: The basal ganglia and the cerebellum represent two portions of the central nervous system that are dedicated to the regulation of postural control mechanisms. One lies buried deep within the white matter of the hemispheres; the other hangs as an external appendage to the brainstem.

Students always have trouble understanding the configuration of the basal ganglia and in relating them to neighboring structures in the brain. Perhaps, if you will use your Superman X-ray vision to peer at them through the lateral aspect of the hemisphere (Fig. XIV-32), you will realize their peculiar shape and get the feel that these gray nuclear masses lie deep in the hemispheric white matter. In (A) of the figure we have lifted them out of the hemisphere; in (B) you view them as if you were looking at them through the frontal pole of the hemisphere; and at (C), from the occipital pole. In (D) and (E) we have taken horizontal slices through the hemisphere at two depths to let you see their relation to each other and to the thalamus through which the horizontal section will also cut.

Figure XIV-33 is a more realistic horizontal representation than our schema of Figure XIV-32. This is a good place to make some initial observations before returning to study the three-dimensional views. Observe the following in order in Figure XIV-33:

1. Thin **mantle of cortical gray** in this horizontal slice through the hemisphere. Frontal pole is up; occipital pole is at the lower part of figure with cerebellum showing between occipital lobes.
2. Find the buried cortical gray mantle of **insular cortex** in the folds of the lateral fissure.
3. Note the massive amount of **white matter** everywhere beneath the gray cortex.
4. Find the third ventricle flanked by the gray, ovoid masses of **thalamus proper.**
5. Above the thalamus and lateral to it are other large, gray masses. The ovoid mass above, making up the lateral wall of the lateral ventricle, is one of the basal ganglia, the *caudate nucleus.* Here we cut through the **head** of this tailed structure.
6. Find the **inferior horn** of the lateral ventricle as a hole (not labeled) on the left side of Figure XIV-33. It will have a piece of the **crus of the fornix** in it. It will also have a piece of the **tail** of the *caudate nucleus* in its dorsolateral wall. Thus, because the *caudate* is an arching, C-shaped structure, like the fornix, it will be cut twice; once in the anterior horn and once in the inferior horn into which it swings upon reaching the temporal lobe (see (D) as well as the main diagram in Fig. XIV-32).
7. Now notice the pie or wedge-shaped gray masses lateral to the thalamus in Figure XIV-33. Their lateral, somewhat curved outer surfaces lie close to the insular cortex. If we scraped away the gray matter of the insulae, its white matter, and a thin, detached part of the insula (the claustrum)—we would soon be down upon the outer surface of the **lentiform** (lens-shaped) **nucleus,** the second of the basal ganglia.

   The lentiform nucleus consists of two named parts. Its outer, large, convex portion is the **putamen;** its lighter-colored, inner, wedge-shaped part closest to the thalamus is the **globus pallidus.** Anteriorly, on the base of the frontal lobe, putamen (but not globus pallidus) and the head of the caudate nucleus fuse (Fig. XIV-27). As we proceed dorsalward in the frontal lobe, a cleft opens up between putamen and the caudate. These large nuclei touch only at the base of the brain (see slice (E) in Fig. XIV-32).
8. Return to Figure XIV-33. Notice that fiber tracts run in the cleft between lentiform nucleus and caudate anteriorly and between lentiform and thalamus posteriorly. This bundle of fibers, which appears V-shaped in horizontal section, is known as the **internal capsule.** Fibers are coursing through it in a longitudinal direction from gray cortex of the hemisphere above the level of this slice, down through it to enter the cerebral peduncles of the midbrain with which it is continuous.

We would have to slice lower in Figure XIV-32 to cut through the third basal ganglia, the **amygdaloid nucleus.** It lies at the end of the

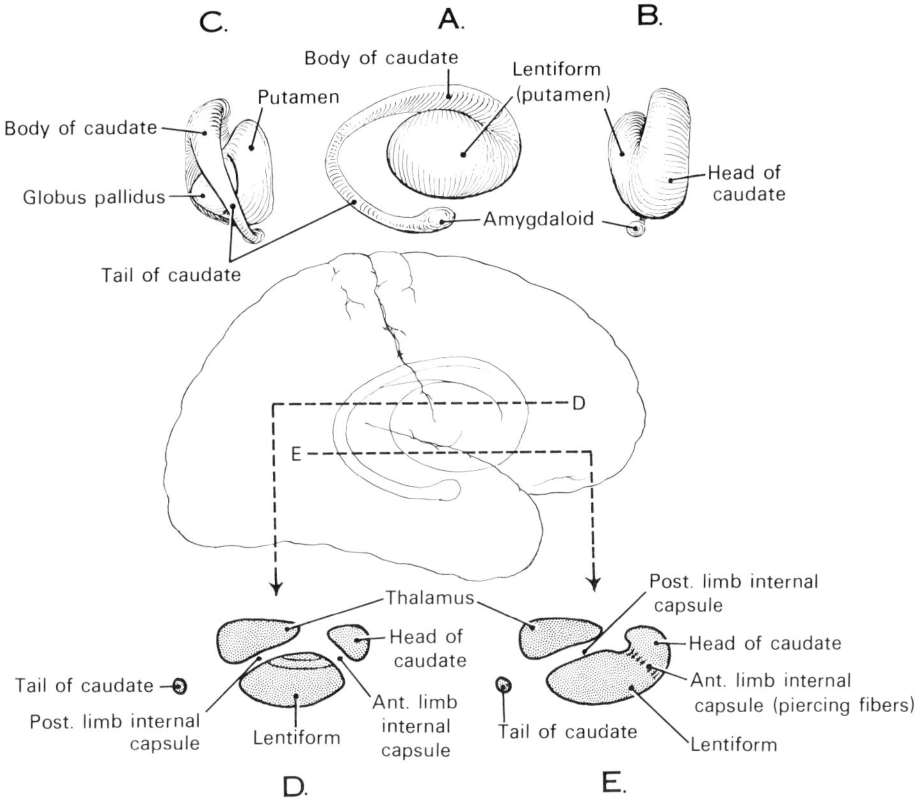

Fig. XIV-32. The basal ganglia. **A.** As if removed from depth of central figure. **B.** As viewed through frontal pole of hemisphere. **C.** As viewed through occipital pole. **D.** In horizontal section as at level shown. **E.** In horizontal view at lower level shown.

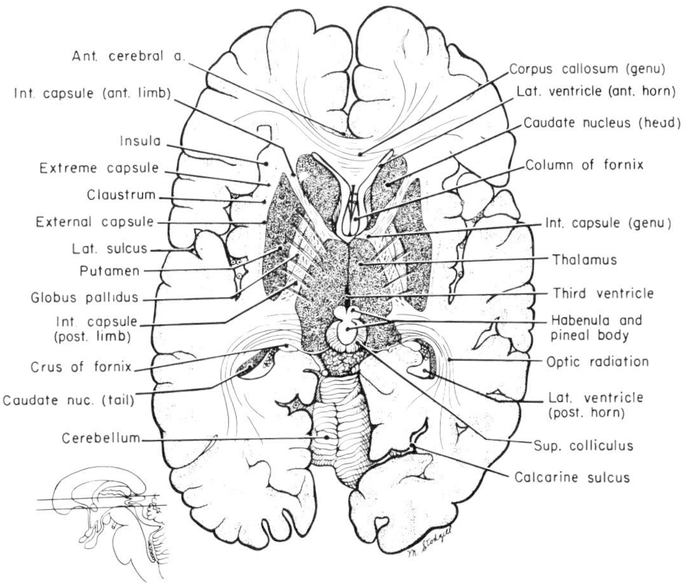

Fig. XIV-33. Sketch of horizontal section through hemisphere cut at level shown by top line of inset. (From R. C. Truex and M. B. Carpenter, *Human Neuroanatomy*, 6th ed. Williams & Wilkins, Baltimore, 1969.)

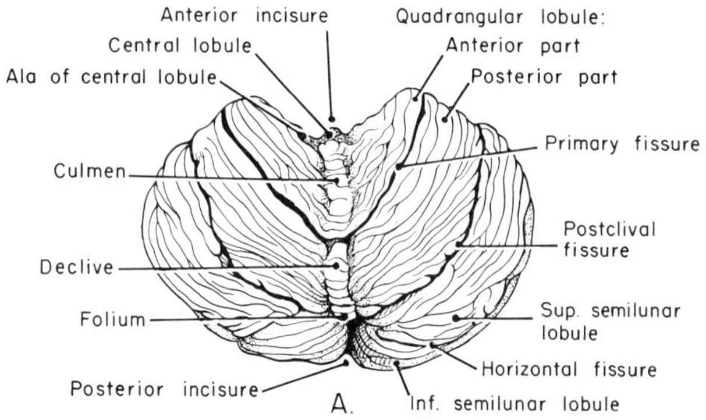

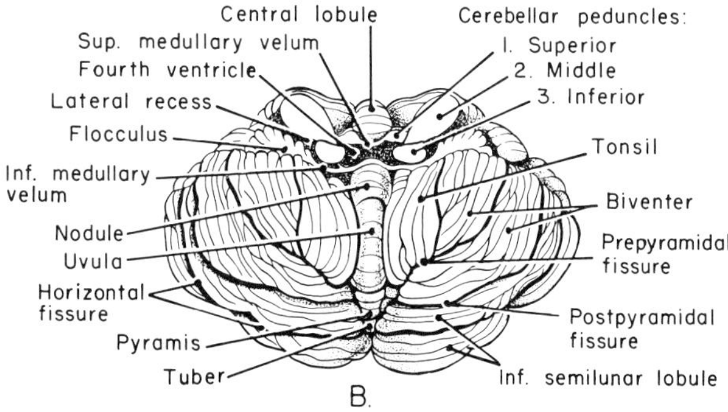

Fig. XIV-34. Topography of human cerebellum. **A.** Superior surface. **B.** inferior surface. (From R. C. Truex and M. B. Carpenter, *Human Neuroanatomy*, 6th ed., Williams & Wilkins, Baltimore, 1969.)

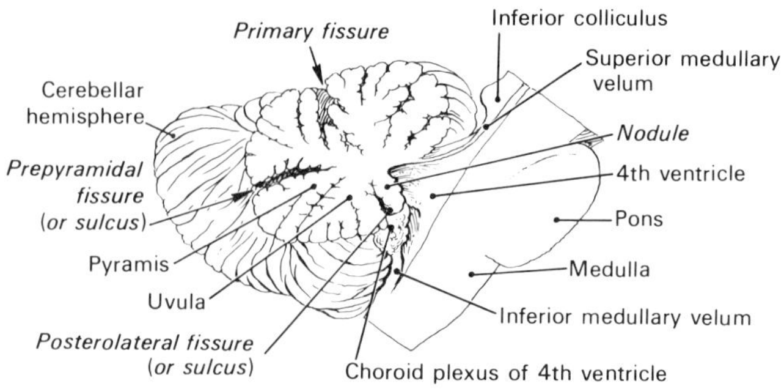

Fig. XIV-35. Midsagittal section through vermis of cerebellum, looking toward left cerebellar hemisphere. Note relationship to hindbrain and fourth ventricle and attend to important cerebellar fissures (sulci).

tail of the caudate in the interior of the temporal lobe close to the uncus. Perhaps it should not be regarded as a part of the basal ganglia at all.

We hope that these two figures and a little mental gymnastics will give you a "feel" for the shape and approximate locus of these gray nuclear masses of the hemisphere. Aside from this exercise in cerebral geometry and geography, we will say no more about them here.

The cerebellum (Fig. XIV-34) consists of **two hemispheres** and a central C-shaped, worm-like process, the **vermis.** Like the cerebral hemispheres, the cerebellum has an outer gray mantle or cortex draped over a central core of white matter which projects into its many **folia.** The folia are formed by numerous sulci and fissures. Also, like the cerebral hemispheres, it contains central gray masses, the **cerebellar nuclei.**

The many names applied to regional parts of the cerebellum (Fig. XIV-34) are frightening, to say the least. They are also meaningless and useless to learn unless you wish to become a comparative anatomist who specializes in cerebelli throughout the phyla. They are handles which specialists use in comparing cerebellar parts in different species. The names are of no earthly, or even heavenly, use to practitioners of the healing arts. The important anatomical landmarks are *three fissures.* These divide the cerebellum not only into phylogenetic lobes, but into functional ones which make the organ understandable clinically.

The first of the fissures, the **posterolateral fissure** lies caudal to a tiny portion of each hemisphere known as the **flocculus.** The fissure (unlabeled) runs at the leader line which is labeled "flocculus" in Figure XIV-34B. The fissure follows medialward along the line of the inferior medullary velum to cut off the paired flocculi of each hemisphere as well as a part of the vermis known as the **nodule.** The nodule is shown in a midsagittal section through the vermis in Figure XIV-35 and the posterolateral fissure is shown just posterior to the folium which makes up the tiny vermal nodule. Thus, the posterolateral fissure cuts off a **flocculonodular lobe.** This represents the archeocerebellum and it is *principally vestibular in function.* That is, it receives sensory input from the vestibular ganglia of the semicircular canals of the inner ear and from vestibular nuclei of the hind brain. It sends axons both directly and indirectly (*via* synapses with certain of the central nuclei of the cerebellum) to the **vestibular nuclei** of pons and medulla. Vestibular nuclei build fiber tracts for coordination of eyes, head and trunk movement; for increase in extensor tone in antigravity muscles; and for regulation of equilibrium and balance.

The importance of this small lobe of the cerebellum is demonstrated when it becomes damaged. Then the body shows signs of oscillation. The individual walks on a broad base as if drunk. There is oscillation of the head and oscillatory movements of the eyes (nystagmus).

A second deep fissure, the **primary fissure,** cuts off an **anterior lobe** of the cerebellum (Fig. XIV-34A). This is also phylogenetically an old part, the paleocerebellum. It consists of all parts of the hemisphere anterior to the fissure (toward the superior medullary velum) including those folia of the vermis anterior to the fissure (Figs. XIV-34 and 35). There is no need to learn the names of the parts involved so long as you realize that this portion of the cerebellum is its anterior lobe.

The anterior lobe is *chiefly spinal in function,* though it does get some vestibular input and some input from the cerebral cortex over corticopontine tracts. Mainly reflex proprioceptive pathways from spinal cord and medulla come to it over the inferior cerebellar peduncle or restiform body. It receives reflex tactile information as well, for the cerebellum needs to know not only where the limbs are located in the environment if it is to effect useful adjustments in postural contact, it needs to know also the nature of contact of these limbs with the environment. Is the foot on ice or cinders? Is it planted firmly on the stair tread to allow the body to be raised up? Has the hand a good grasp on an object so it will not drop the object? You can imagine the cerebellum asking questions of this sort as reflex sensory information reaches it.

When the anterior lobes suffers injury, the organism tends to go into rigidity. That is, hypertonicity increases in the extremities and movement is difficult under such spastic states. The anterior lobe may be more important in the above respects, in quadrupeds, such as cats, than in man. In man, many of the functions of the middle lobe lap over into the anterior lobe.

The third fissure of note, the **prepyramidal fissure,** cuts off another part of the paleocerebellum, the **posterior lobe.** In man the posterior lobe is vermal only; that is, it has no bilateral parts in the hemispheres as have all the other cerebellar lobes. The fissure is seen in Figure XIV-35 lying posterior to a folium called the pyramis for its apparent pyramidal shape. The posterior lobe then consists entirely of two vermal folia (pyramis and uvula, though their names are unimportant) which lie between the posterolateral and the prepyramidal fissures. This small posterior lobe is also *chiefly spinal in function.* It does receive some vestibular and corticopontine input and also some auditory and visual input. The inferior olive receives this latter information from superior and inferior colliculi of the midbrain by way of the central tegmental bundle. The inferior cerebellar brachium conveys it to posterior lobe.

Posterior lobe lesions are rare. They lead to oscillation, *i.e.,* disturbances of equilibrium and to inability to stop forward progression easily once it starts. The individual tends to crash into obstacles as if he is unable visually to gauge distances between himself and objects in his path.

We have named the three important fissures, *posterolateral, primary,* and *prepyramidal,* and alliterative series of clefts which cut off *three* of the *four* lobes that make up the cerebellum. Certainly, the flocculonodular lobe and the posterior lobe are tiny and the anterior lobe is not large. This leaves all the rest of the great bulk of the hemispheres and that part of the vermis between primary and prepyramidal fissure unaccounted (Figs. XIV-34 and 35). It is the newest part of the cerebellum, the neocerebellum, or great **middle lobe** which came into being with the flowering of the cerebral hemispheres of higher vertebrates. As a consequence, its greatest input *via* the middle cerebellar peduncle or brachium pontis will come from corticopontile pathways. The cortex is informing the cerebellum of motor acts in progress and is requesting that the cerebellum, as an error-detecting device, keep an eye on them. This the cerebellar does through visual, auditory, tactile, vestibular and proprioceptive feedback as the body moves about in its environment. The middle lobe, *via* the superior cerebellar brachium or brachium conjunctivum keeps motor cortex, basal ganglia, red nucleus, and subthalamus informed. It allows the motor cortex, through basal ganglia and cerebellum, to alter and make fine adjustments of movements in progress. Moreover, through fastigiobulbar tracts and other brainstem pathways, the cerebellum exerts a direct influence upon the lower motor neurons of the cord, the ventral horn cells.

Middle lobe lesions result principally in loss of muscular tone (hypotonicity) and many odd effects. There is **dysmetria,** or excessive range of movement. There are errors in direction or rate of movements and loss of force of movement. There is decomposition of movement so that acts become jerky as if the individual were going through "the manual of arms" a step at a time. Finally, there is true oscillation. While a movement of intention, *e.g.,* reaching for an object, might start off smoothly enough, the closer to the termination of the act the movement progresses, the more grossly tremorous and shaky the movement becomes. This oscillatory behavior has been called an **intention tremor.** The individual, for example, would have great difficulty in picking up a glass of water and drinking it. He might even knock over the glass to begin with, or shake out most of its contents if he were successful in grasping it.

Both basal ganglia and the cerebellum act as negative feed-back, servomechanisms in a system designed for motor control and regulation. In this sense, like mechanical and electronic feed-back systems, the two cardinal *signs of malfunction* they exhibit are **oscillations and rigidities.** This seems to be every bit as true of basal ganglia malfunction as it does of cerebellar malfunction.

Hence, lesions in the *globus pallidus* and in the associated **substantia nigra** of the midbrain lead to the hypokinetic-rigid syndrome of **Parkinson's disease** or **paralysis agitans.** The disease is characterized by **rigidity and tremor.** Voluntary movements are retarded but force of movement is well preserved. The facies are relatively immobile and mask-like. Movements are hard to initiate. The voice may have a monotonous quality. The "pill-rolling" tremors of the fingers and wrist are **tremors of rest.** They disappear once an actual useful movement is underway.

Lesions in the **corpus striatum** leading to atrophy in the *putamen* and *caudate* nucleus give rise to choreatic, jerky movements; again a type of oscillation, as in **Huntington's chorea.** At first limited to small areas of face or the limbs, they soon involve all the bodily musculature so that in an advanced stage, the patient is never at rest. They can interfere with speech and swallowing.

Another type of abnormal movement is encountered in **athetoid** behavior. This consists of slow, bizarre movements of forceful extension of the metacarpophalangeal or interphalangeal joints with abduction of the wrist and extension of the arm and hand. As in all

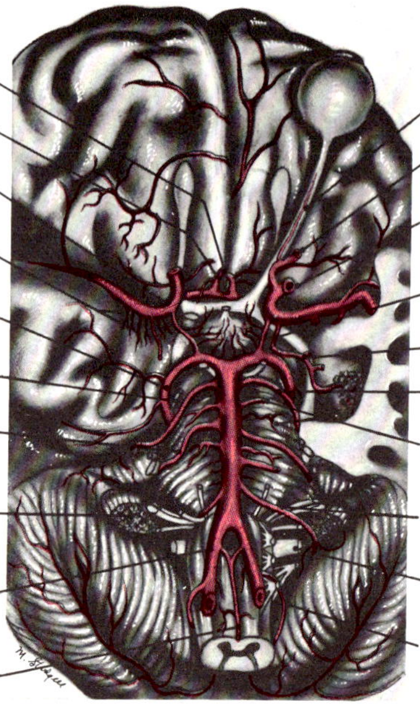

Fig. XIV-36. Blood supply and cranial nerves. Ventral aspect of brain. (From R. C. Truex and M. B. Carpenter, *Human Neuroanatomy*, 6th ed., Williams & Wilkins, Baltimore, 1969.)

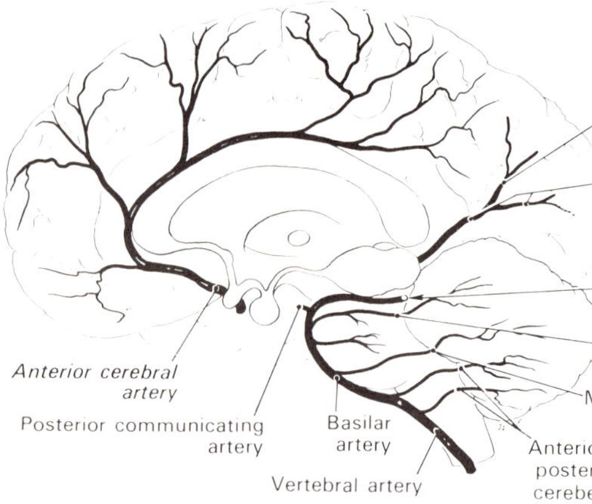

Fig. XIV-37. Major arterial supply of medial hemispheric surface.

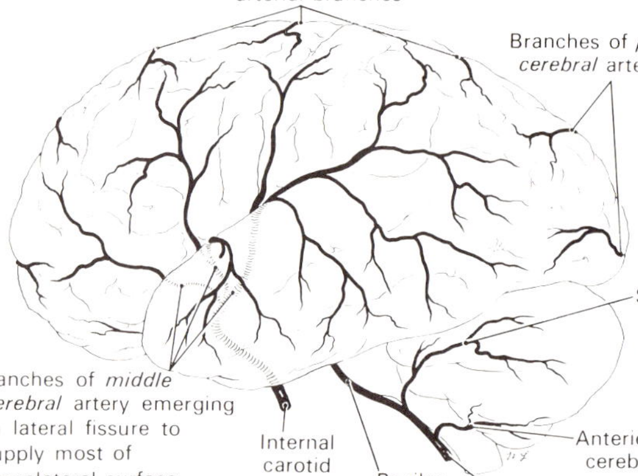

Fig. XIV-38. Major arterial supply to dorsolateral surface of hemisphere and cerebellum.

of these basal ganglia disturbances, frequency increases under emotional stress.

Sometimes when **subthalamic nuclei** are involved (and basal ganglia make connections with this area), **hemiballismus** is observed. This involves large-scale movements of arms, legs, and trunk on one side.

It is still an unsettled matter as to whether basal ganglia are involved in tracts which project downward to the cord directly as does the cerebellum. Most evidence points to interconnections with the thalamus and a relay back to the motor cortices and to areas of the motor cortex concerned with affect or emotional control. Yet, it does not seem wise to rule out descending projections entirely at this time.

Blood Supply of the Brain

The entire arterial supply of the brain is derived from two paired arteries—the **vertebrals** and the **internal carotids.** The vertebrals traverse the foramina transversaria of the cervical vertebrae to enter the cranium *via* the foramen magnum. They find themselves coursing along the ventrolateral aspect of the medulla, converging toward each other (Fig. XIV-36). Their union at pontomedullary junction forms the **basilar artery.** Through the terminal branches of the basilar (the **posterior cerebral arteries**), the vertebrals contribute to the supply of occipital lobe (Fig. XIV-37) and to the inferior parts of the temporal lobe (Fig. XIV-38).

The internal carotids enter the base of the skull *via* the carotid canal, find themselves crossing over the foramen lacerum and wending their way up along side the anterior clinoid processes. Here they give off their main two large branches, the **anterior,** and **middle cerebral arteries.**

The anterior cerebral arteries (Fig. XIV-37) supply the medial aspects of each hemisphere. Their uppermost twigs lap over the dorsal margin of the hemisphere to supply the dorsolateral surface of the brain as well—a territory not quite reached by the middle cerebral artery (Fig. XIV-39). The latter vessels cut deep to the temporal poles (Fig. XIV-38) to reach the lateral fissure and to surface out of it onto most of the dorsolateral hemisphere. Neither anterior nor middle cerebrals encroach beyond the parieto-occipital sulcus or reach the occipital pole, so they leave the supply of the occipital lobe to the **posterior cerebral artery** entirely. Neither do they bother with the basal or inferior surface of the temporal lobe.

Vertebral and internal carotid systems interconnect *via* a **posterior communicating artery** which interlinks the carotid with the posterior cerebral around the sides of the optic chiasma and pituitary in the hypo-

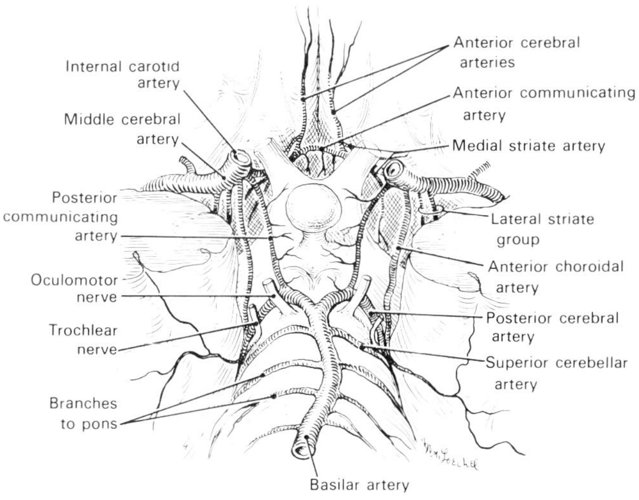

Fig. XIV-39. The arterial circle of Willis, striate and choroidal arteries in relation to hypophysis and the third and fourth cranial nerves.

thalamic region of the brainstem (Fig. XIV-38). Sometimes the connection is with the middle cerebral rather than with the internal carotid.

Areas Supplied by Basilar and Vertebrals Together (Fig. XIV-36):

1. The hindbrain (pons, medulla, and cerebellum).
2. A few twigs to ventral midbrain.
3. A few twigs to caudal diencephalon, especially in hypothalamic regions.
4. *Via* the posterior cerebral artery, they serve the occipital lobe and inferobasal parts of temporal lobe.

Area of Supply of the Vertebrals Only (Fig. XIV-36):

1. **Recurrent anterior spinal artery** through the foramen magnum to cord.
2. **Recurrent posterior spinal arteries** to cord.
3. Small twigs to medulla.
4. **The posterior inferior cerebellar artery** to postero-inferior surfaces of cerebellum and medulla.
   (a) An important branch of this artery is the **choroidal artery for the fourth ventricle.**

Area Supplied by Basilar Alone (Fig. XIV-36):

1. **Anterior inferior cerebellar artery** for antero-inferior surface of cerebellum and pons.
2. An auditory branch to internal auditory meatus. Twigs from this branch also supply pons.
3. Several pontine branches.
4. The **superior cerebellar artery** to superior aspects on cerebellum and pons.
5. **The posterior cerebral artery:**
   (a) Besides the parts of the brain noted above, this vessel sends twigs to: (1) ventral midbrain; (2) posterior thalamus; (3) and gives off the important **choroid artery to the third ventricle.**

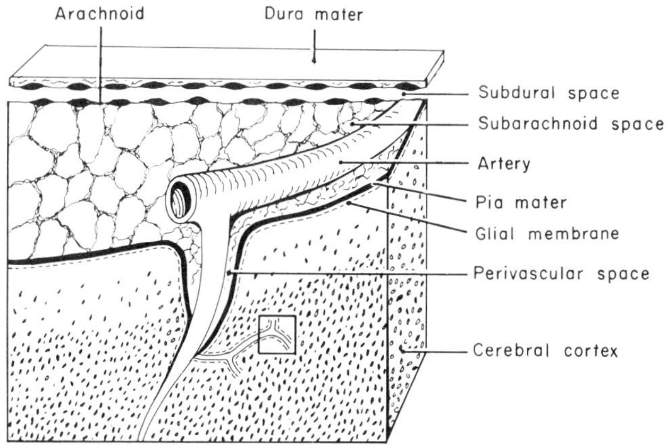

Fig. XIV-40. Sketch of subarachnoid and perivascular space on surface of hemisphere. (From R. C. Truex and M. B. Carpenter, *Human Neuroanatomy,* 6th ed., Williams & Wilkins, Baltimore, 1969.)

General Territory of Supply of Internal Carotids:

1. The hemispheric parts not supplied by the posterior cerebral branch of the basilar.
2. Anterior parts of the thalamus.
3. Choroidal arteries to lateral ventricles.
4. Deep structures of the hemispheres; *e.g.,* corpus striatum and internal capsule (the areas related to the so-called basal ganglia).
5. Twigs to pituitary gland which may come from various parts of the circle of Willis.

The circle of Willis is an anastomotic arterial circle at the base of the brain (Fig. XIV-38), linking together the vertebral and carotid systems. Its components are:

1. Posterior cerebrals.
2. Posterior communicating arteries.
3. Internal carotids.
4. Anterior cerebrals.
5. **Anterior communciating artery,** a short twig bridging across the midline to link the anterior cerebrals of each hemisphere.

Under ordinary circumstances, vertebral and carotid supplies deliver to their own specific territories and, probably, do not mix. But let there be an occlusion of a *major* trunk line to the brain, and, then, shunts through the circle can be opened up so that the brain can continue to be supplied fairly adequately with one of its prime requisites—oxygen (and its other prime requisite, glucose).

The Deep Arteries (Fig. XIV-40): Surface branches send twigs deep into the white matter below the gray cortical mantle and to the central gray nuclei (basal ganglia) lying still deeper.

There are no anastomotic connections between these deeply penetrating branches until they reach precapillary size. In this sense they are virtually *end arteries.* That is, they can become occluded easily by thrombi to shut off some local area of supply. Should this occur, brain softening and localized neuronal death ensues. We need not go into details about the distribution of deep arterial branches at this time since our main aim all along in this chapter has been to outline fundamental relations that will help you better understand neuroanatomy which is a rather specialized topic.

Carotid and Vertebral Angiography: This highly complex technique is now being performed by neuroradiologists who have subspecialized in this challenging field. Close cooperation is needed between the neuroradiologist and the neurosurgeon or neurologist in order to assure greater accuracy of diagnosis, and, of course, to perform the proper examination. By studying the arterial, capillary, and venous systems of the cerebrum and cerebellum, localization of space occupying mass lesions (tumor, abscess, hematoma, etc.) or vascular lesions (arteriovenous malformations, aneurysms, and occluded vessels) may be achieved.

The examination is performed by rapidly injecting an iodine-containing contrast medium, *i.e.*, Hypaque or Renografin directly into the internal carotid or vertebral arteries. Special radiographic equipment permits the taking of films rapidly as the contrast medium courses through the circulatory system. Usually, frontal and lateral projections are utilized, although, on occasion, other projections are needed to achieve the correct diagnosis.

The complexities of the subject are best left to your course in neuroanatomy. We will show you here both carotid and vertebral arteriograms, schematically, with matching radiographs from normal patients (Figs. XIV-41 and 42). The venous drainage will similarly be demonstrated later (Fig. XIV-48). For those of you who wish to explore the subject further, additional references are given at the end of this chapter.

General Pattern of Venous Drainage: All the venous return from brain capillaries eventually reaches the sinuses of the dura mater (Fig. XIV-43). In the last analysis, by one channel of these *wide dural veins* or another, all blood reaches the internal jugular vein either *via* transverse and sigmoid sinuses which form a common channel, or *via* the inferior petrosal sinus which makes its way to the jugular vein alone. More than that, veins other than cerebral veins empty into the dural sinuses:

1. Some facial veins like the ophthalamic and internal maxillary.
2. Some emissary veins from the scalp.
3. Diploic veins of cranial bone.
4. Cerebrospinal fluid (in a sense, venous return from arterial capillary beds of the choroid plexuses) *via* arachnoidal villi into the superior sagittal sinus mainly.

Recall that **surface arteries** lie on pia mater within the subarachnoid space over the brain before they send branches deep to reach the capillary beds within the brain substance. As a consequence, **surface cerebral veins** also will lie on pia in the subarachnoid space, *i.e.*, between arachnoid above and pia below. Figure XIV-45 shows some of these surface veins, though you must imagine that you are looking through arachnoid at them. They will have to pierce arachnoid to enter the *subdural space* before they can empty into dural sinuses.

There are, essentially, *four chief routes of venous drainage:*

1. By way of the **superior cerebral veins.** Some eight to 10 of these drain dorsally (upward) into the superior sagittal sinus.
2. By way of the **middle cerebral veins.** These drain downward over the lateral and inferior surfaces of the hemispheres toward the cavernous sinus (Fig. XIV-45).
   a. But, by way of two great anastomotic venous channels on the dorsolateral hemispheric aspect, the **vein of Trolard** and the **vein of Labbé,** these middle cerebral veins can drain upward into the superior sagittal sinus, or posterolaterally into the transverse sinus directly (Fig. XIV-45).
3. There are a group of **inferior cerebral veins** on the basal portion of the brain. Those related to the basal part of the frontal lobe empty either into the inferior sagittal sinus, cavernous sinus (Fig. XIV-44), or superior petrosal sinus. Those on the basal aspect of temporal and occipital lobes usually seek out the transverse sinus.
4. Some small veins on the anteromedial surface of of the hemisphere; a small branch of the middle cerebral coming deep from the insula; and a deep vein from the striate portion of the basal ganglia all join forces to form a **basal vein** (Figs. XIV-46 and 47). It contributes to the **great cerebral vein** (**of Galen**) at the splenium of the corpus callosum.
   a. The vein of Galen receives, besides, two **internal cerebral veins** and the **superior median vein of the cerebellum.** Through the great vein of Galen much of the deep structures of the area are drained into the straight sinus and thence into the transverse.

The internal cerebral veins are formed: (1) by the **terminal veins** which lie in the angle between dorsal thalamus and caudate nucleus on the floor of the lateral ventricle; and (2) by the **choroid vein of the lateral ventricle.** This system drains fornix, hippocampus, and corpus callosum and parts of the thalamus and midbrain—all into the great cerebral vein.

The **cerebellum** is drained by three veins: (1) superior median vein into the vein of Galen; inferior median vein into the straight sinus directly; and the superior and inferior lateral veins into the transverse sinus.

In the **pons,** two large venous trunks on each side of the basilar artery empty into the superior petrosal sinus. The **veins of the medulla** are not well known. They parallel the vertebral arteries and may drain

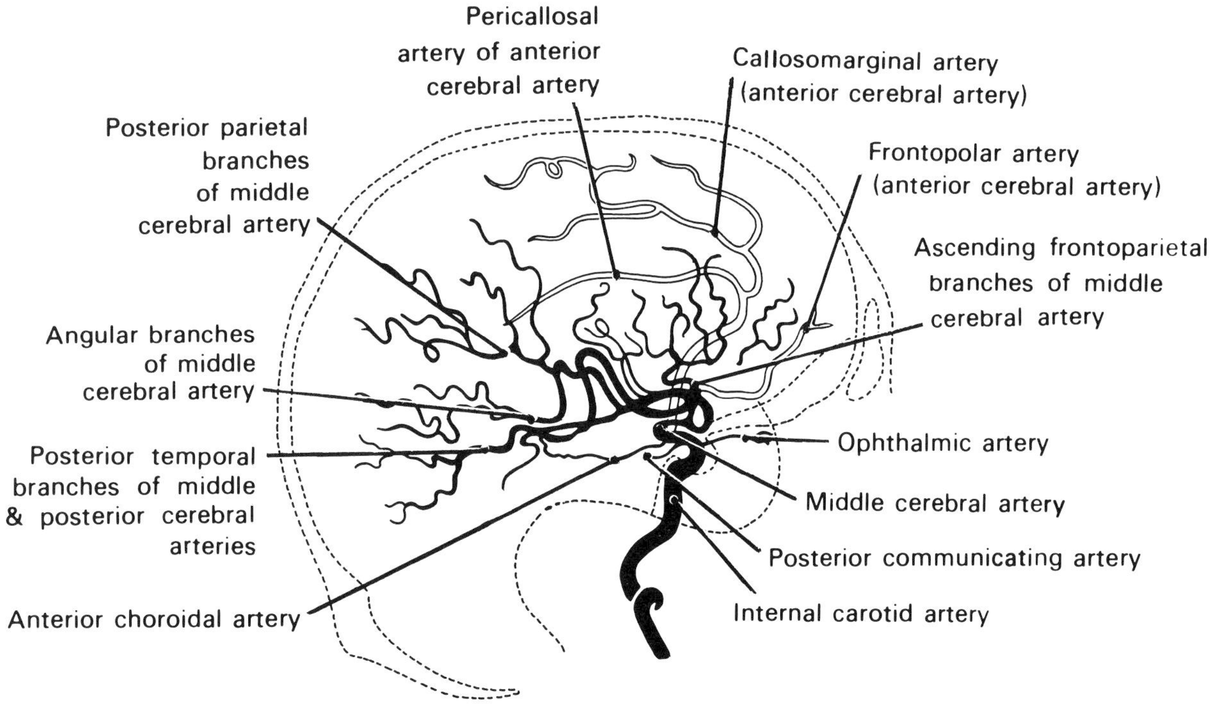

Fig. XIV-41. Internal carotid angiogram. **A.** Lateral projection-schematic.

either into pontine veins or the plexus of veins about the spinal cord.

Ventricular System

The cavitous ventricular system is shown in phantom views in Figure XIV-49 A and B. Viewed laterally in the hemispheres, each lateral ventricle is an arch-shaped cavity which projects into all lobes. It has a short-wide, blunt, **anterior horn** in the frontal lobe. Its **body** narrows and arches slightly in the parietal lobe and it projects a short, pointed, **posterior horn** into the occipital lobe. Laterally, it swings a curving horn into the temporal lobe. This is known at the **inferior horn** of the lateral ventricle. Near the junction of body and anterior horn, a connection is made through the downward and laterally projecting interventricular foramina (of Monro) with the midline third ventricle. The latter is a narrow V-shaped cleft between the thalami and hypothalami. It is bridged, at times, by a massa intermedia. Just below its posterior recess it communicates with the cerebral aqueduct of the midbrain. The latter, in turn, empties in the shallow rhomboid fossa over the dorsal surface of the pons and the rostral part of the medulla.

There is a choroid plexus fed by arterial capillaries in each lateral ventricle but it is limited to body and inferior horn only. *No choroid plexus is found in the anterior and posterior horns.* This is because the plexus follows the course of the curving fornix which

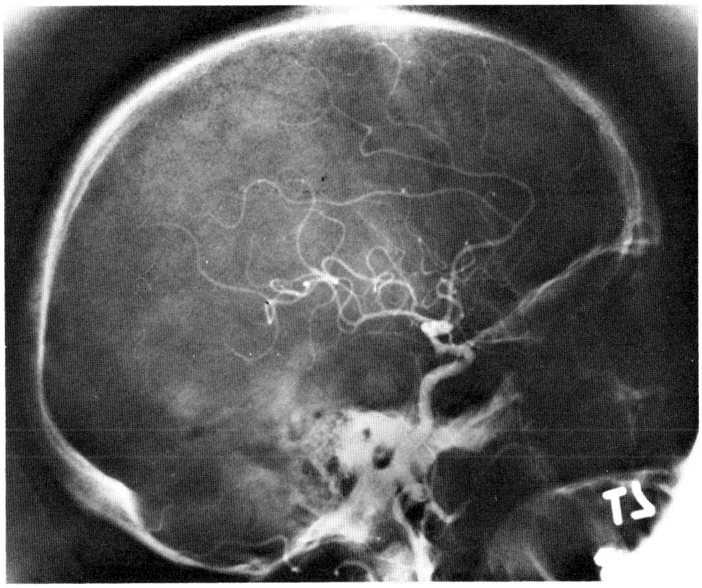

Fig. XIV-41B. Lateral projection in a patient.

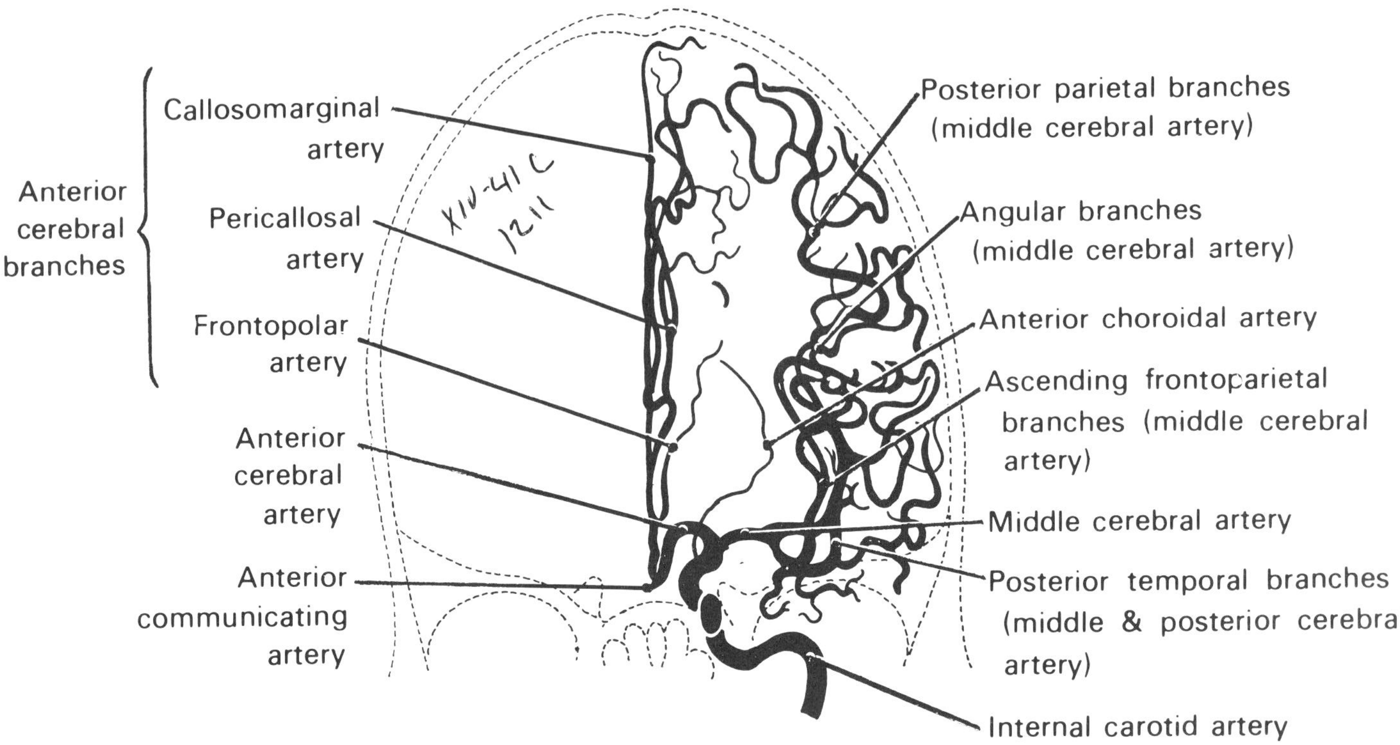

Fig. XIV-41C. AP half-axial projection-schematic.

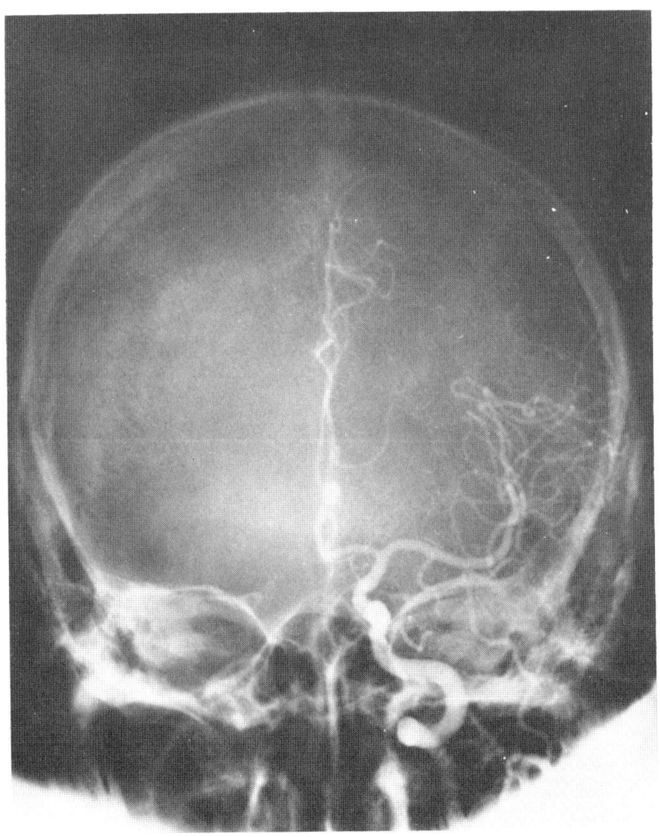

Fig. XIV-41D. Same projection as **C** in a patient.

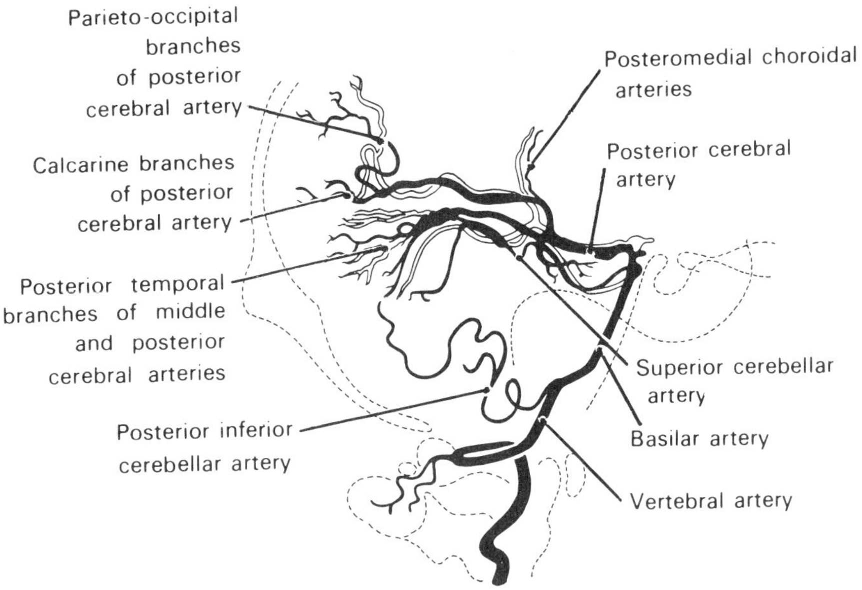

Fig. XIV-42. Vertebral angiogram. **A.** Lateral projection-schematic.

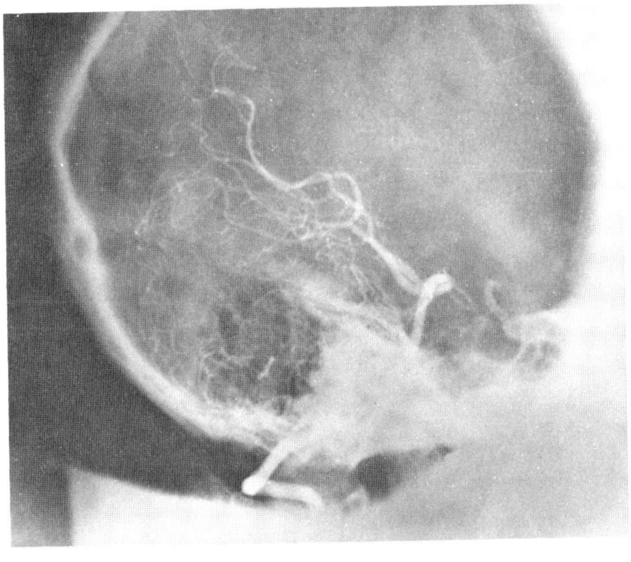

Fig. XIV-42B. Lateral projection in a patient.

bears the curving choroid fissure (lateral extent of the transverse cerebral fissure) all along its edge. Ependymal lining of the ventricle is all that exists along this line between transverse cerebral fissure and ventricle so that capillary tufts can be tucked into the ventricular lumen only along this curved course. We have noted the loci of choroid plexuses in the third and fourth ventricles already. Figure XIV-49B shows the marked lateral bend of the lateral ventricle as it moves anteriorly into the temporal lobe.

Meninges and Cerebrospinal Fluid: We have referred to the **meninges** here and there in the text for one brief reason or another. It is time to consolidate our concepts of them. The meninges are membranes which invest the brain and spinal cord. Because the outer covering or dura mater is so tough, it has been called the **pachymenix.** The middle and inner investments of arachnoid and pia mater are fine and thin and together are known as the **leptomeninges.**

The Spinal Dura: This is a tubular coat of tough, fibroelastic connective tissue lined with smooth mesothelium. Since it extends from the foramen magnum to VS_2, it is a continuation of the **inner layer** of cranial dura (Fig. XIV-50). Its *external attachments* are as follows:

1. Superiorly, to the periosteum of the posterior parts of the bodies of VC_2 and VC_3.

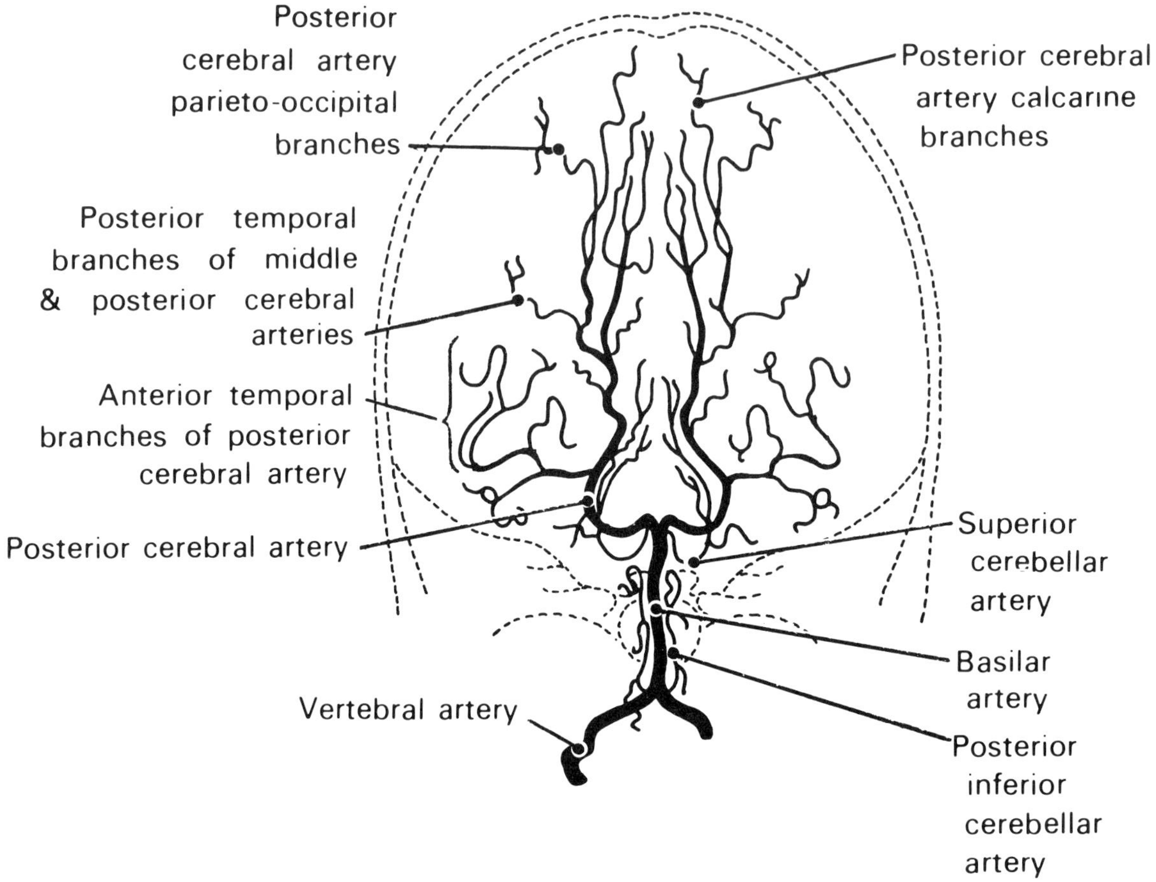

Fig. XIV-42C. AP half-axial projection-schematic.

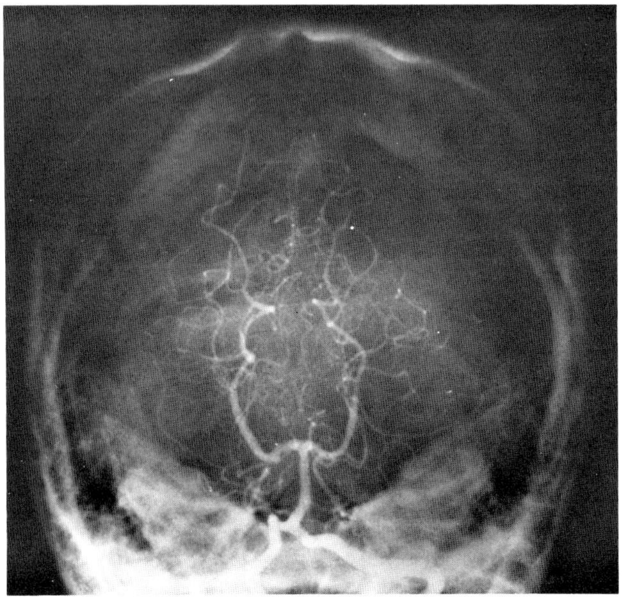

Fig. XIV-42D. Same projection as **C** in a patient.

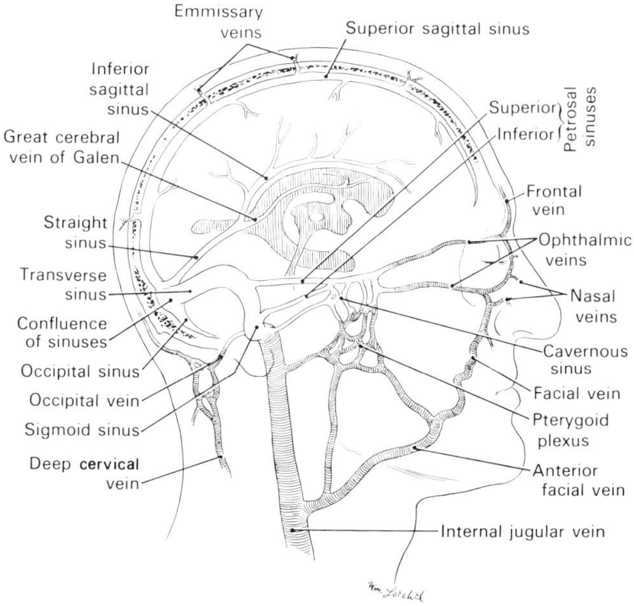

Fig. XIV-43. The dural sinuses and extracranial connections. Intracranial channels (except ventricles) are clear; extracranial vessels are shaded.

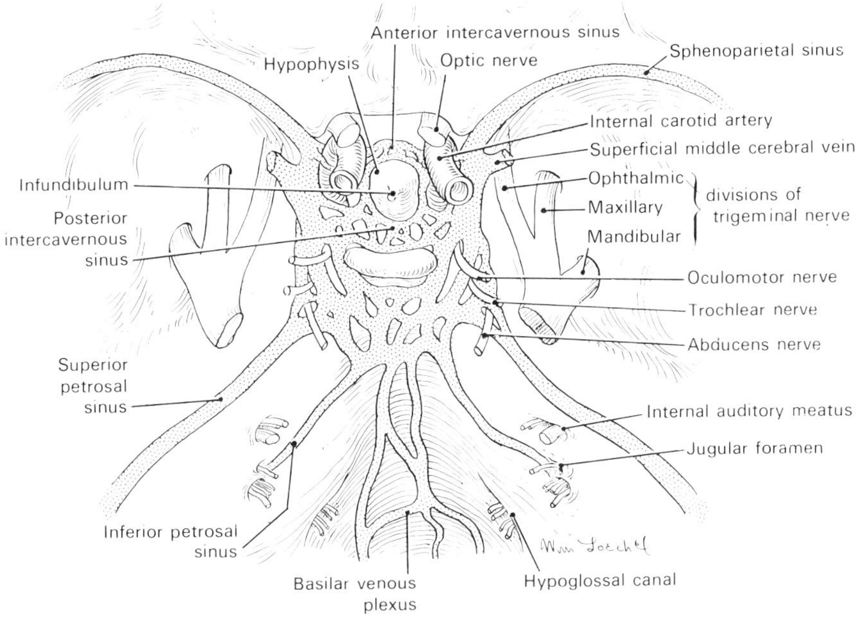

Fig. XIV-44. The cavernous sinus and its relationships.

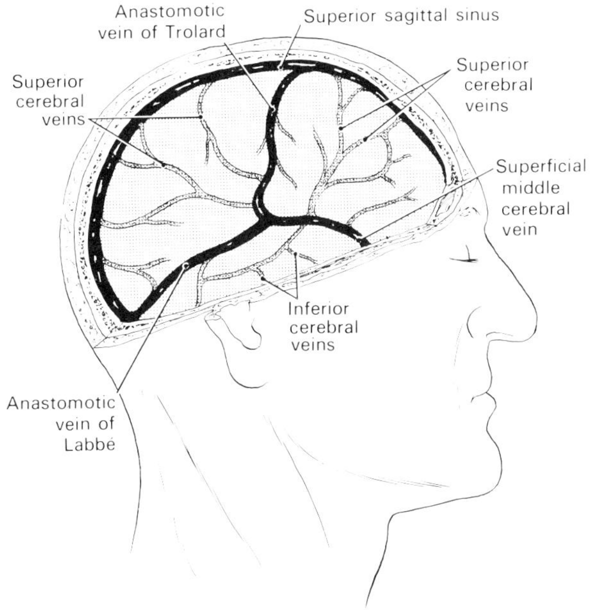

Fig. XIV-45. Venous drainage of the dorsolateral surface of the hemisphere by anastomotic veins. (See text for description.)

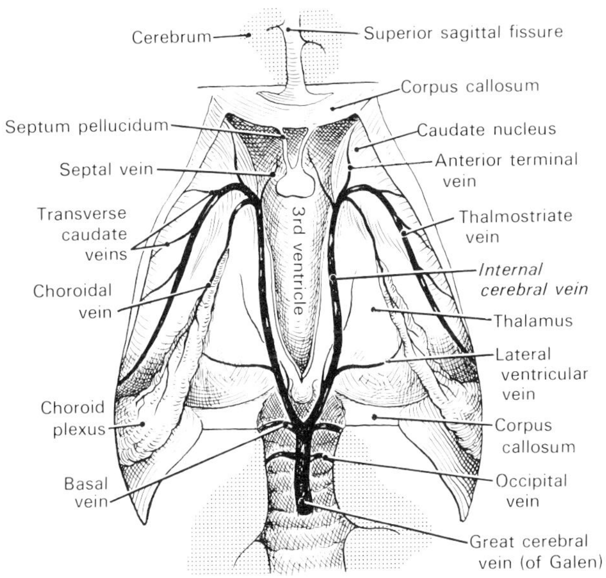

Fig. XIV-46. Internal cerebral veins and their tributaries.

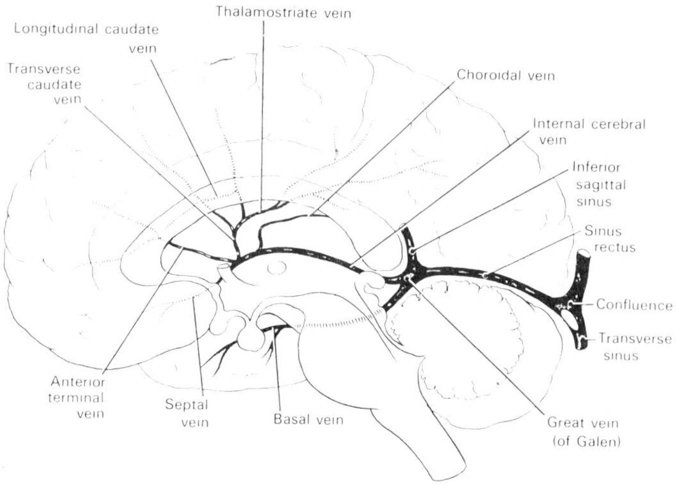

Fig. XIV-47. Midsagittal view of the internal cerebral veins and their relation to great vein and dural sinuses.

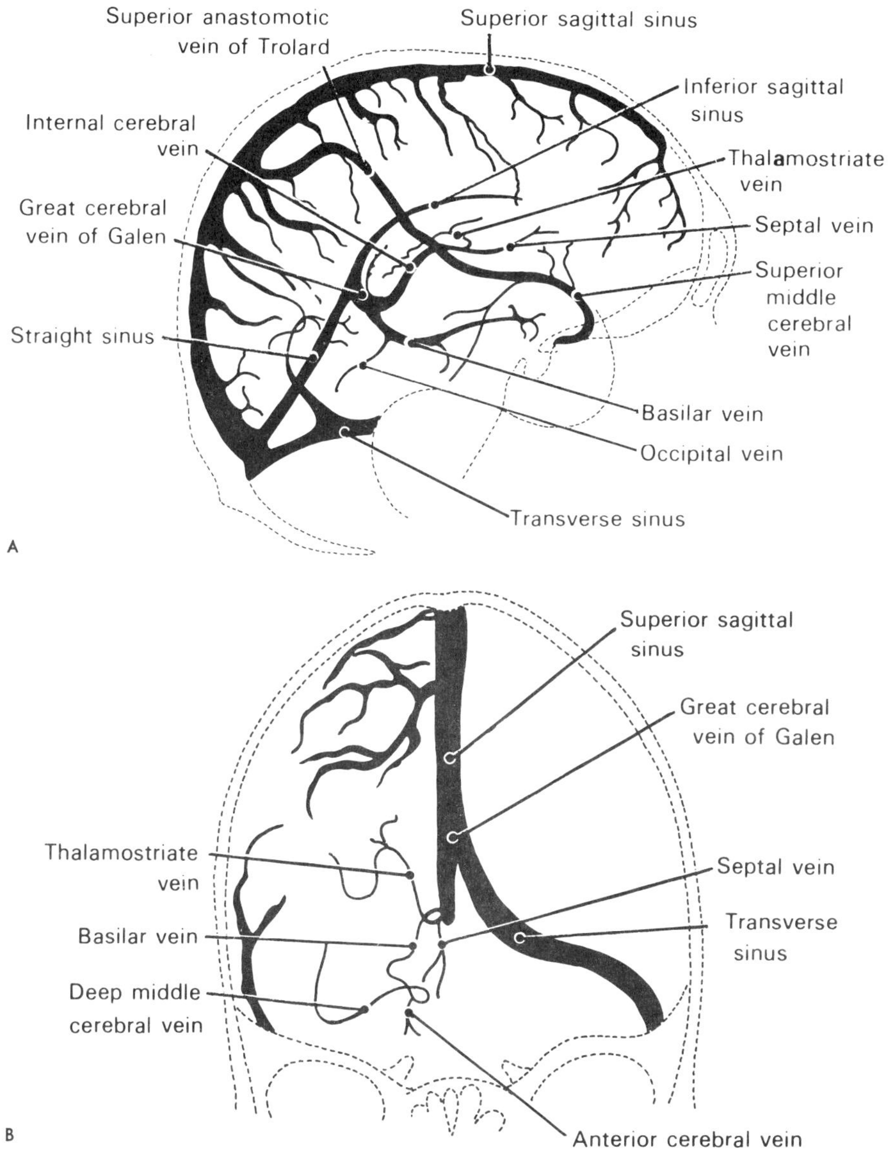

Fig. XIV-48. Internal carotid venogram (schematic). **A.** Lateral projection. **B.** AP half-axial projection.

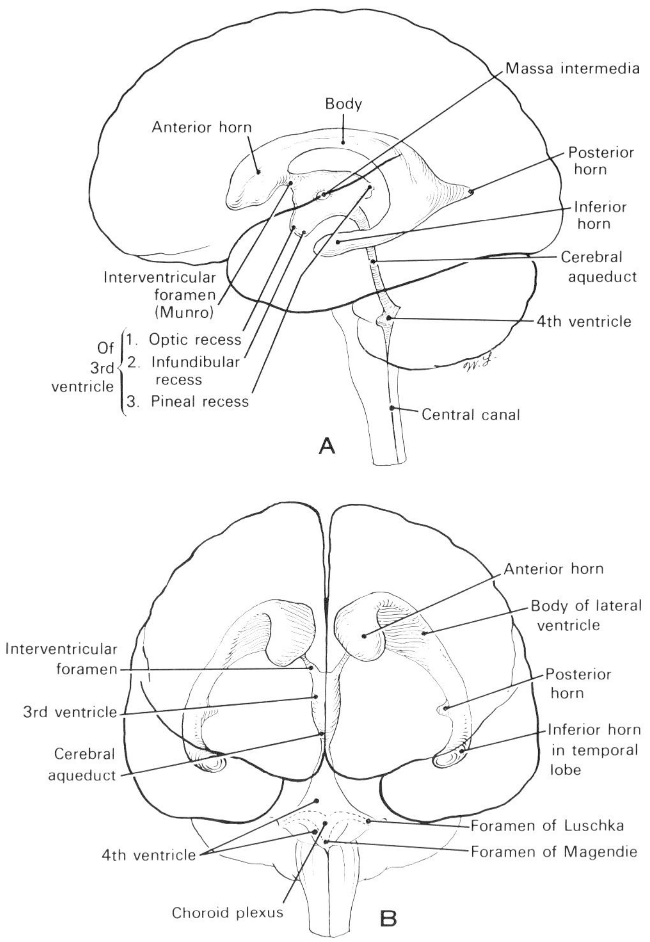

Fig. XIV-49. The ventricular system. **A.** Viewed in lateral aspect. **B.** As seen in frontal view.

2. Inferiorly, to the posterior surface of the coccyx as the **external coccygeal ligament.** This is by way of the **external filum terminale,** a pial thread to which dura fuses below the level of VS_2.
3. Laterally, as dural sleeves of nerve rootlets which extend to the intervertebral foramina.
4. Ventrally in the vertebral canal, to the **posterior longitudinal ligament** of the vertebral column. This ligament runs along the posterior aspects of the vertebral bodies.
5. Posteriorly in the canal, spinal dura is free. Its *internal attachments* are:
   a. With arachnoid by way of **arachnoidal trabeculae.**
   b. With pia at the serrations of the pial **denticulate ligament.**

There is a spinal **epidural space** between dura and most of the periosteum of vertebral canal (Fig. XIV-51). The space contains an internal vertebral plexus of veins, areolar, and adipose tissue. At the sacrum, this space is used to administer caudal anesthesia.

The Cranial Dura: This is, essentially, a double-layered structure. The **inner layer** constitutes periosteum of the cranial cavity; the **outer layer** is surfaced by mesothelium and it duplicates itself as:

1. Falx cerebri.

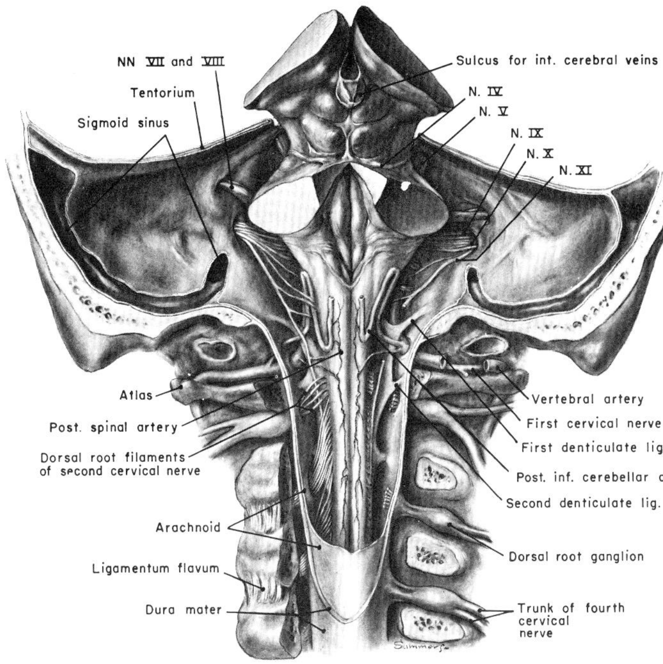

Fig. XIV-50. Posterior spinal view of cord, brain stem, and meninges. (From R. C. Truex and M. B. Carpenter, *Human Neuroanatomy,* 6th ed., Williams & Wilkins, Baltimore, 1969.)

2. Falx cerebelli.
3. Tentorium cerebelli between occipital lobes and the cerebellar hemispheres.
4. The diaphragma sella.

Clefts between inner and outer layers form:

1. Mechel's cave at the apex of the petrous bone to house the semilunar ganglion of V.
2. The endolymphatic sac.
3. The venous dural sinuses.
4. Venous lacunae—lateral extensions of some of the sinuses.

The Arachnoid: This "spider-webby" tissue forms a delicate, avascular, fibrous membrane which is lined and covered by mesothelial cells (Fig. XIV-40). Even its thin extensions, the arachnoidal trabeculae, are so covered. It clothes the brain and brainstem and extends downward in the vertebral canal as far as VS_2 where it terminates. Its trabeculae cross the **subdural space** to anchor to dura. This potential space may contain a film of serous fluid. Its **subarachnoidal trabeculae** anchor to pia. Between the interstices of this bridging network of trabeculae is the **subarachnoid space.** The arachnoid is not a close investment of the neural tissues and it bridges across sulci and fissures and from one elevated contour to that next adjacent (Fig. XIV-52).

The Pia Mater: This "tender" membrane, while thin, is thicker than arachnoid. It is fibrous and highly vascularized. It applies itself closely to the neural tissue, dipping into all crevices. It is surfaced by mesothelium, and cerebral vessels run upon it in the subarachnoid space between pia and arachnoid. As vascular branches penetrate neural tissue, pia-arachnoid follows in only a relatively short distance.

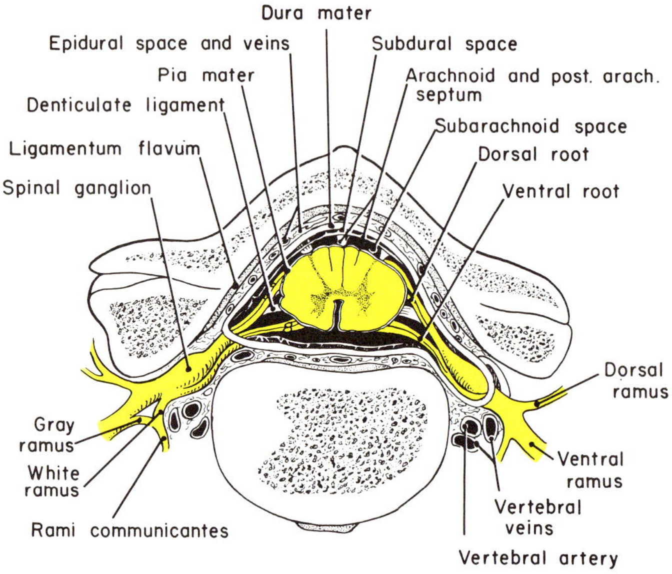

Fig. XIV-51. Meningeal coverings of spinal cord seen in cross-section (From R. C. Truex and M. B. Carpenter, *Human Neuroanatomy,* 6th ed., Williams & Wilkins, Baltimore, 1969.)

The **spinal pia** dips into the ventromedian fissure of the cord to form a longitudinally-thickened band called the **linea splendens.** Laterally and longitudinally along the cord, pia duplicates into linear folds; one on each side. Between root filaments, some 21, toothed, serrations of this fold, the *denticulate (dentate) ligament,* extend to dura as anchorage (Figs. XIV-50 and 51). The uppermost serration attaches on the inner margin of the foramen magnum. As the cervical portion of the *spinal accessory* nerve (XI) emerges from cord segments C_1 to C_4, its rootlets ride upward and through the foramen magnum on the dorsal aspect of the denticulate ligament (Fig. XIV-50). The anterior aspect of this ligament is used to guide the surgeon's scalpel to the locus of the lateral spinothalamic tract in cord tractotomies for the elimination of pain.

The Cranial Pia: As a close investment of brain and stem, it dips into all sulci and fissures carrying blood vessels with it. It enters the transverse cerebral fissure to impinge upon the ependymal walls along the choroid fissure, and it rides over the ependymal roof of the third ventricle while in the transverse fissure. It slips between medulla and cerebellum to ride over the ependyma of the fourth ventricle. In these places, the blood vessels invaginate ependyma to form **choroid plexuses.**

Practical Considerations Relative to Meninges: **Meningitis** is an inflammation of these investments of the central neuraxis. It can be a pachymeningitis or a leptomeningitis. Neoplastic growth of meningeal tissues can compress and destroy underlying neural tissue. The meninges are sometimes involved in congenital defects, as follows:

A. **Spina bifida** is a failure of the vertebral neural arches to close. If the protruding sacculation:

1. contains meninges only, it is called a **meningeocoele;**
2. contains neural tissue and meninges, it is called a **meningomyelocoele.**
3. The most severe type of defect of this kind is when the neural tube is immature, unclosed, open to air, and fails to have even skin as a covering. Since no meninges are present in this instance, this defect is called a **myelocoele.**

B. **Cranial bifida,** due to a failure of closure of the cranial bones, is usually occipital in locus. If the protrusion contains neural tissue and meninges, it is called a **meningo-encephalocoele,** if only neural tissue protrudes, it is an **encephalocoele.**

The Subarachnoid Space: This space, between pia and arachnoid, filled with *cerebrospinal fluid,* is seen enveloping brain, brainstem, and cord in Figure XIV-53. The relation of this space to all three meninges is evident in the diagram. In the spinal region, the space is narrow at best except below cord termination. Between VL_1 and VS_2 it widens into the **lumbar cistern.**

Over most of the brain and stem, the space is, perhaps, even narrower than around the cord, except where arachnoid bridges between structures widely. The widest of these areas are also called **cisterns.** Some of the largest of these (Fig. XIV-52) are:

1. The cisterna chiasmatica anterior to the optic chiasma on the floor of the diencephalon.
2. The cisterna interpeduncularis between the cerebral peduncles of the midbrain.
3. The cisterna ambiens along the lateral sides of the midbrain.
4. The cisterna pontis (Hilton's water bed) on the basilar surface and along the sides of the pons.
5. The cisterna cerebellomedullaris or the cisterna magna. This great cistern lies between the inferior surface of cerebellum and the superior surface of the medulla. Here cerebrospinal fluid samples are sometimes extracted from infants.

Cerebrospinal Fluid: Cerebrospinal fluid is formed by ependymal cells which engage in active secretion. There is active transport of Na^+ and Cl^- across ependymal cell membranes into the ventricles. Water follows passively. The bulk of the formation is carried out by the ependymal cells of the choroid plexuses. These cells, whose basal surfaces face the invaginated tufts of blood capillaries, have free surfaces facing the ventricle. The free surface is beset with microvilli and the lateral and basal surfaces also show marked foldings of the plasma membrane. The basal surface rests upon a delicate connective tissue matting of pia-arachnoid rich in thin-walled capillaries.

Some fluid is thought to be added by ependymal and glial cells elsewhere in the ventricular walls where these walls are devoid of choroid plexus; and also by capillaries of the pia-arachnoid which run over the external surfaces of the brain. Fluid from this last source would, obviously, be added directly to the subarachnoid space. That is, it would not have to traverse ventricular chan-

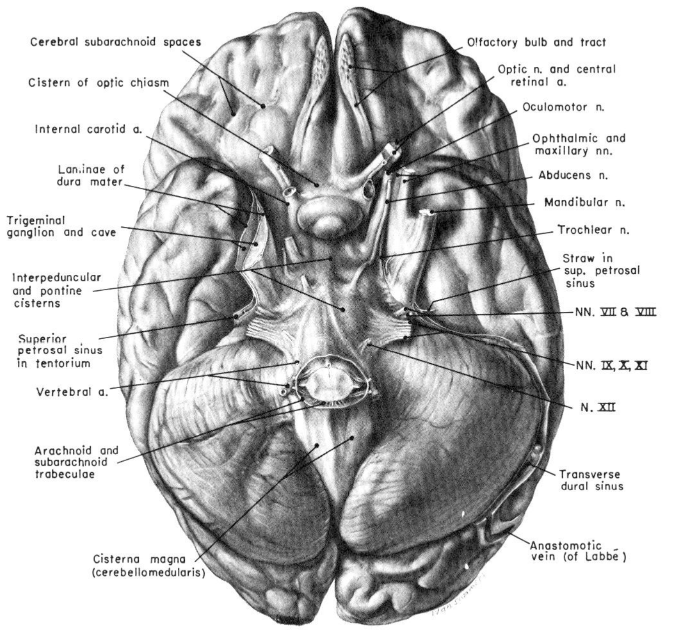

Fig. XIV-52. Inferior view of brain, cranial nerves, and meninges showing locations of subarachnoid cisterns. (From R. C. Truex and M. B. Carpenter, *Human Neuroanatomy*, 6th ed., Williams & Wilkins, Baltimore, 1969.)

nels to reach subarachnoid space *via* the foramina of Magendie and Luschka.

The total cerebrospinal fluid system shown in Figure XIV-53 holds about 130 ml. Of this, approximately 80 ml. lie in the ventricles and 50 ml. lie out in subarachnoid space. Gravity flow carries it down around the cord into the lumbar cistern. There is hardly any fluid in the central canal of caudal medulla and cord, the lumen being of tiny diameter and often full of ependymal debris.

While the system holds about 130 ml. at any one time, daily turnover through formation and absorption varies considerably; figures quoted range from 45 to 300 ml. per day. Thus, it is not a static water cushion. We shall discuss absorption below.

The fluid itself, a transudate of plasma, is normally clear, colorless and slightly alkaline. It has a specific gravity ranging from 1.004 to 1.007. While it is often said to be acellular, a lymphocyte count of 3 to 5 per cu. mm. is normal. In meningitis, tuberculosis, and syphilis the count goes up to 100 to 1000 per cu. mm. It contains about 10 mg per 100 ml. protein (globulins and albumins). A reading of as much as 30 to 40 mg per 100 ml. in a lumbar puncture sample is considered a high normal value. Glucose is present at 50 to 80 mg per 100 ml. and inorganic salts and water in the same proportion as in blood plasma. The chemical constitution varies in pathological states. For example, in purulent meningitis the sugar decreases early; NaCl, late. In acute syphilis, NaCl content drops before sugar content is decreased.

Obviously, arterial pressures in the choroid plexuses have to be higher than cerebrospinal fluid pressures in the ventricles; and the latter have to be lower than venous pressures if there is to be a gradient of flow of cerebrospinal fluid toward absorption sites. In the incumbent individual, cerebrospinal fluid pressure varies between 60 to 180 mm. H_2O (not mercury), but it can rise as high as 300 mm. H_2O in the lumbar cistern while sitting. In the latter posture, pressure will be low at the cisterna magna. You can expect pressures to change with alterations in heart rate and respiration.

Absorption of Cerebrospinal Fluid: Certainly, fluid can't continue to be forced out of the fourth ventricle into the subarachnoid spaces without there being some mechanism for absorption if neural damage is to be avoided as a consequence of compression in a tight cranium. Absorption around the cord by capillary plexuses is said to be slow. Most absorption occurs through **arachnoidal villi** disposed along the superior sagittal sinus of the dura (Figs. XIV-53 and 54) and, perhaps, along other dural sinuses as well. Villi have been described in spinal arachnoid,[4] but they are not often mentioned.

In the cranium at least, arachnoidal tufts are pushed through dura into dural sinuses. The sinuses are wide venous channels between periosteal and endosteal layers of cranial dura. In this way cerebrospinal fluid returns to the systemic circulation. It is possible for certain materials to pass from cerebrospinal fluid into the extracellular spaces of neural tissues where, presumably, they would be absorbed in local capillary beds.

Function of Cerebrospinal Fluid: The watery contents of the subarachnoid space provides a cushion for the central neuraxis. Its important function is to maintain a constant intracranial pressure. Any change in intracranial volume will be reflected by a change in cerebrospinal fluid volume. If a brain lesion is destructive, fluid volume increases. If the lesion is neoplastic, fluid volume decreases. Thus, intracranial pressures are maintained and normal brain tissue, not involved in the lesion, will not be subject to damage. Rapid changes in intracranial volume must be compensated through reflex activation of vasomotor centers of the brainstem to either increase rate of production or rate of absorption of fluid. Any increase at all of intracranial pressure reflects inadequate compensation on the part of the cerebrospinal fluid system.

Practical Considerations Relative to Cerebrospinal Fluid: The fluid can be sampled by puncture of the lumbar cistern in the clinic or by puncture of the lateral ventricle in the operating room. Other reasons for entering the fluid system are for manometric purposes; to inject anesthetic agents; to inject antibiotic agents to combat meningitis; or to obtain roentgenograms.

A **pneumo-encephalogram** is an extensive series of roentgenograms of the skull following partial replacement of cerebrospinal fluid with air or oxygen by means of a *lumbar puncture*. The gases are less dense than fluid for the passage of X-rays. The procedure partially drains cisterns as well as ventricles—air entering the latter through the foramina in the roof of the fourth

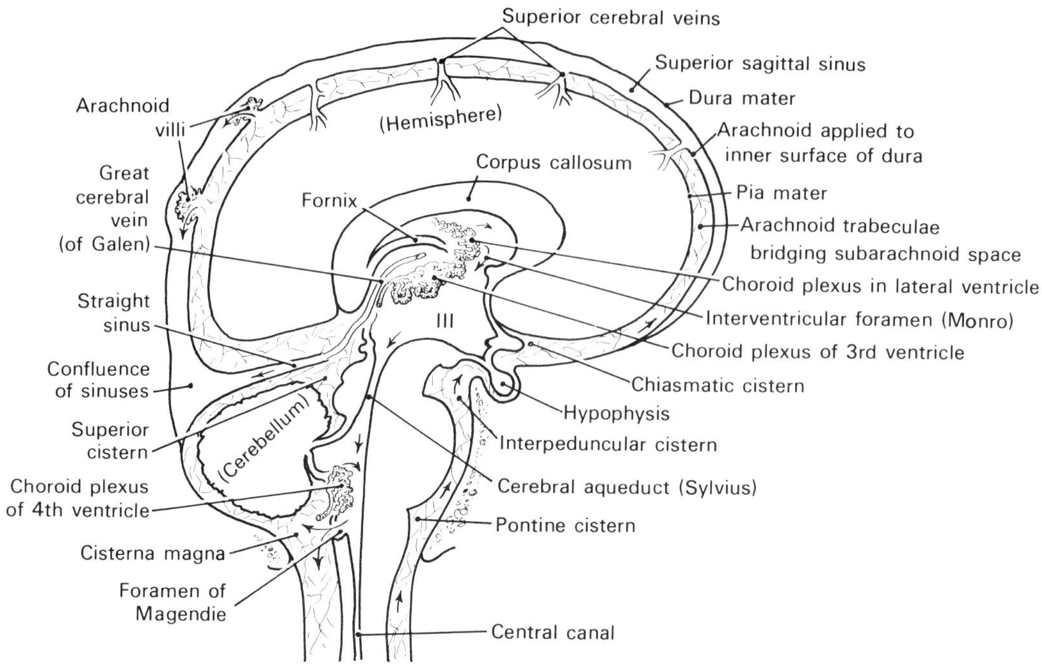

Fig. XIV-53. Meninges, ventricles, and subarachnoid spaces. Arrows indicate route of flow of cerebrospinal fluid from formation to cerebral absorption sites.

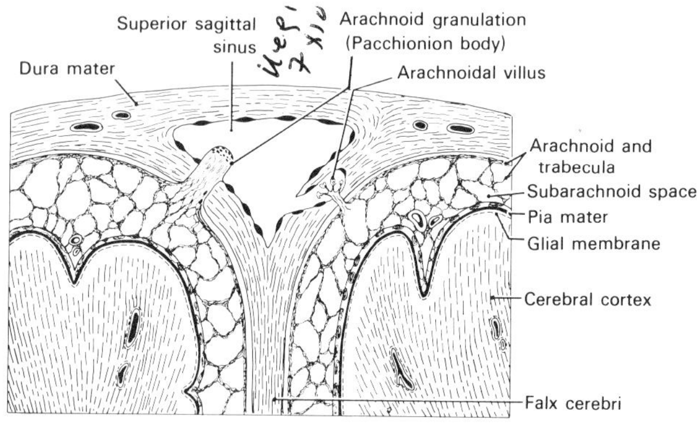

Fig. XIV-54. Absorption of cerebrospinal fluid *via* arachnoidal villi. (From R. C. Truex and M. B. Carpenter, *Human Neuroanatomy,* 6th ed., Williams & Wilkins, Baltimore, 1969.)

ventricle. The procedure is contraindicated when papilledema or cerebellar lesions are evident.

A **ventriculogram** is essentially the same thing except that gases are introduced directly into the ventricles after a burr hole has been placed in the calvaria. Only the ventricular system is drained. It is a safer procedure than encephalography when expanding lesions are expected.

Hydrocephalus is any condition which prevents flow or absorption of cerebrospinal fluid. Sites of obstruction (Fig. XIV-53) would be: (1) the cerebral aqueduct; (2) the foramina of the fourth ventricle; (3) plugged arachnoidal villi; (4) or congenital displacement of the cerebellum and part of the hindbrain through the foramen magnum (Arnold-Chiari malformation). Hydrocephalus can, thus, be of two types:

1. **Internal or obstructive,** involving structures related to the ventricular part of the system, as in the case of (1) and (2) above.
2. Or **external (communicating)**, involving the subarachnoid channels and absorption sites as in (3) and (4) above.

Internal hydrocephalus will compress brain substance against bony cranial walls. External hydrocephalus collapses the brain inward away from the brain case. Both may produce large heads especially in infants where cranial sutures have not fused.

The Autonomic Nervous System

General Plan:

I. The ANS is basically a visceral motor system innervating all smooth muscle, cardiac muscle, and glands in the body (except sebaceous glands which have no nerve supply). There is, of course, a visceral sensory counterpart just as there is a somatic sensory counterpart for the somatic motor system. For didactic purposes, however, we shall consider the ANS as if it were a **motor system** only. It consists of **two** principal divisions.

The Key to the ANS is the number 2:

A. A thoracolumbar outflow, or **sympathetic** division arising from:
   1. The intermediolateral cell column of cord segments T_1 to L_2.

B. A craniosacral, or **parasympathetic** division:
   1. Cranial outflow *via* visceral motor nuclei of cranial nerves, 3, 7, 9, 10.
   2. Sacral outflow from visceral motor cell column of cord segments S_2 to S_4 (Fig. XIV-55).

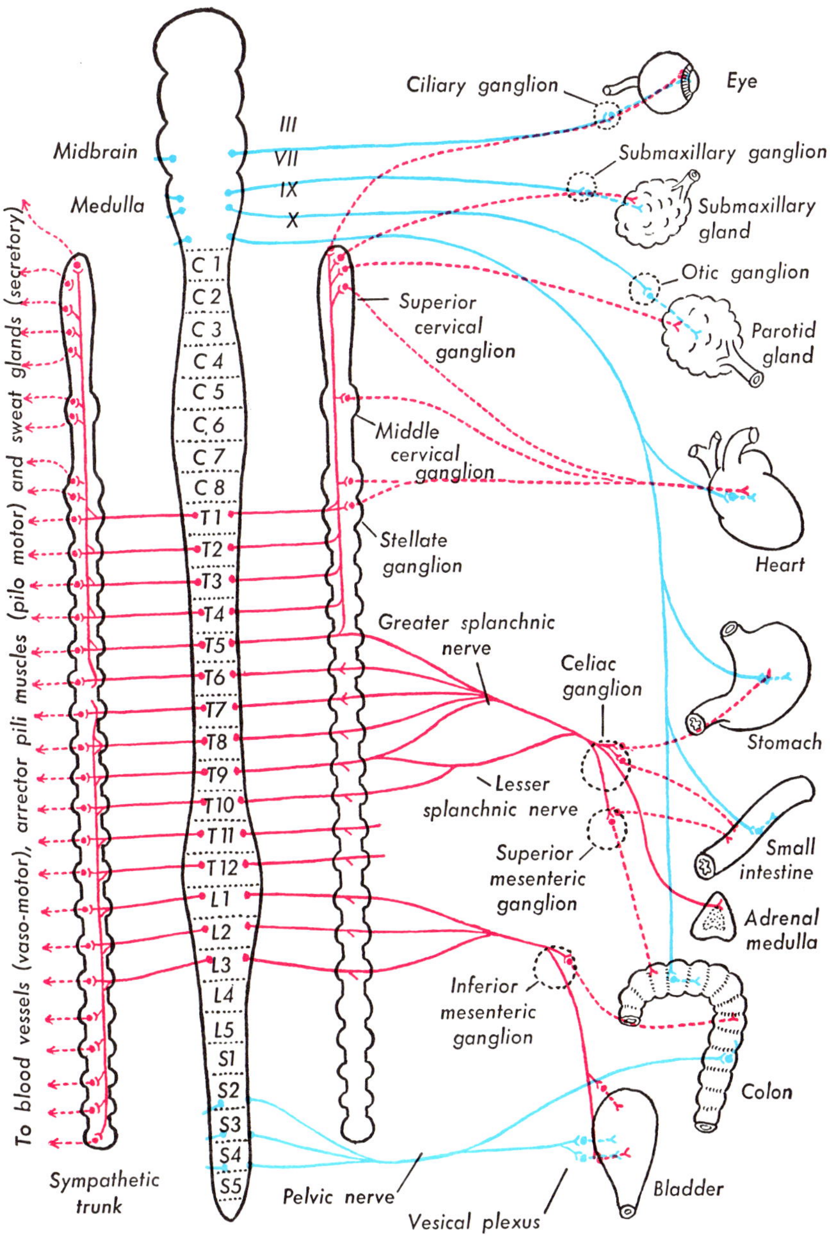

Fig. XIV-55. Diagram showing general arrangement of the autonomic system. The sympathetic components as shown in *red*; while the parasympathetic components are in *blue*. *Solid lines* represent preganglionic fibers; *broken lines*, postganglionic fibers. For clearness the sympathetic fibers to the blood vessels, hair, and sweat glands are shown separately on the *left*. (From R. C. Truex and M. B. Carpenter, *Human Neuroanatomy*, 6th ed., Williams & Wilkins, Baltimore, 1969.)

II. **Basic plan of sympathetic division** (Fig. XIV-56)

A. Lateral horn cells from *two* **continuous** regions of the cord supply the preganglionic fibers of this system.
   1. Thoracic cord (T_1 to T_{12}).
   2. Lumber cord (L_1 to L_2).

B. There are always *two*, and never more than *two*, neurons concerned with the innervation of any structure containing smooth muscle, cardiac muscle or glands (adrenal, an exception).
   1. A preganglionic neuron in the cord.
   2. A postganglionic neuron in the periphery.

C. Postganglionic neurons lie in one of *two* places: (Fig. XIV-55)
   1. In the *two* ganglionic chains of the sympathetic trunk.
      a. These ganglia are known as **chain ganglia** (or paravertebral ganglia, since they parallel the vertebral column).
   2. In outlying **collateral ganglia** (or prevertebral ganglia); for didactic purposes, we can say collateral ganglia are found in *two* places:
      a. *In the neck,* some distance forward in the main sympathetic trunk. Such ganglia supply postganglionic fibers to the visceral structures in the head and thorax.
         (1) Superior, middle, and inferior cervical sympathetic ganglia.
      b. *In the abdomen* at the base of the three unpaired vessels of the aorta.
         (1) Coeliac, superior, and inferior mesenteric ganglia which supply visceral structures in the abdomen and pelvis.

D. Chain ganglia lie in *two* parallel longitudinal cords of nervous tissue, the **sympathetic trunks.**
   1. Right and left trunks meet in a central ganglion, the **ganglion impar,** in the coccygeal region.
   2. As a rule, there is a chain ganglion for each cord segment.
      a. But variation is common, particularly at lower levels where some may be missing.
      b. In cervical region considerable fusion has occurred in that the theoretical eight becomes three.
         (1) These, the **superior, middle,** and **inferior cervical sympathetic ganglia,** lie considerably "upstream" in the trunk from the rest of the ganglia, enough so that we can arbitrarily class them as the collateral ganglia for the head and thorax.
         (2) In some forms, *e.g.*, cat, the inferior cervical sympathetic ganglia fuses still further with the 1st thoracic chain ganglia to form the **stellate ganglion.** This occurs in man, but not as frequently as is generally taught.
   3. Chain ganglia connect to spinal nerves by *two* rami (Fig. XIV-56):
      a. **White** ramus—by which preganglionic fibers leave the spinal nerves to reach the sympathetic trunk.
      b. **Gray** ramus—by which postganglionic fibers from chain ganglia reach spinal nerves recurrently to be distributed with spinal nerve to smooth muscle of peripheral blood vessels, sweat glands, and pilo-arrector muscle.
   4. *Only chain ganglia T_1 to L_2 are served by white rami.*
      a. This is a point students often fail to realize, but it should be obvious, since cord segments T_1 to L_2 only, supply the preganglionic outflow *via* their spinal nerves.
      b. All other ganglia receive their preganglionic supply through the sympathetic trunk, the preganglionic fibers traveling rostally or caudally in the trunk to reach them.
   5. All axons from cells in chain ganglia return as gray rami to their (Fig. XIV-56) respective spinal nerves.
      a. Superior cervical sympathetic ganglia returns 4 gray rami—one each to spinal nerves C_1 to C_4.
      b. Middle cervical—2 gray rami, C_5 and C_6.
      c. Inferior cervical—2 gray rami, C_7 and C_8.

E. The preganglionic fiber synapses in one of *two* places (*never* first in one and then the other, but *only* in one *or* the other) (Fig. XIV-55):
   1. A chain ganglia, or
   2. A collateral ganglia.
      a. To reach collateral (*i.e.*, relatively outlying) ganglia in the neck, preganglionic axons use the upper part of the sympathetic trunk.
      b. To reach collateral (prevertebral) ganglia in the abdomen, preganglionic axons traverse the trunk in part to enter the **splanchnic nerves** emanating from the sympathetic trunk.

F. Splanchnic nerves are chiefly bundles of preganglionic fibers (there are some visceral sensory fibers in them also) (Fig. XIV-55)
   1. **Thoracic splanchnics:**
      a. **Greater splanchnic** from cord segments (T_5 to T_{10}).
         (1) Termination—coeliac ganglion.
      b. **Lesser splanchnic** (T_9 to T_{11}).
         (1) Termination—superior mesenteric ganglion.
      c. **Least or lowest splanchnic** (T_{12}).
         (1) Termination—**superior mesenteric ganglion** or an aorticorenal ganglion.
   2. **Lumbar splanchnic** (L_1 and L_2).
      a. Termination—**inferior mesenteric ganglion.**

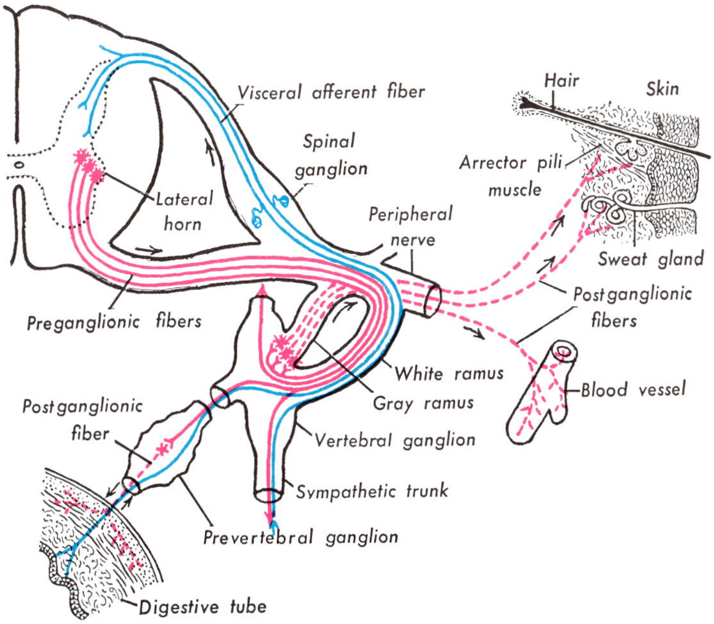

Fig. XIV-56. Diagram of neural reflex arcs of the sympathetic system. (From R. C. Truex and M. B. Carpenter, *Human Neuroanatomy*, 6th ed., Williams & Wilkins, Baltimore, 1969.)

G. The prevertebral plexuses are formed by the splanchnic nerves and by postganglionic fibers from the coeliac, superior, and inferior mesenteric ganglia. These plexuses run along the aorta and its branches. Joining these fibers in the plexuses are preganglionic fibers from the vagus and pelvic nerves. These latter fibers merely run *through* but never synapse *in* the coeliac, superior, or inferior mesenteric ganglia. Visceral sensory fibers are present also.
   1. Subdivisions of coeliac plexus:
      a. Phrenic plexus
      b. Hepatic plexus
      c. Splenic plexus
      d. Gastric plexus
      e. Adrenal plexus
      f. Renal plexus
      g. Spermatic (ovarian) plexus
      h. Superior mesenteric plexus
      i. Inferior mesenteric plexus
      j. Abdominal aortic plexus
   2. Hypogastric plexus:
      a. This forms the connecting link between the inferior mesenteric plexus and the pelvic plexuses.
   3. Pelvic plexus and subsidiaries:
      a. Hemorrhoidal
      b. Vesical
      c. Prostatic (uterine)
      d. Cavernous (vaginal)

III. **Basic Plan of Parasympathetic system** (Fig. XIV-55)
   A. Preganglionic fibers arise from neurons located in *two widely separated* regions.
      1. From brainstem *via* cranial nerves III, VII, IX and X
         a. The vagus is the only one of these nerves concerned with innervation of thoracic and abdominal viscera.
      2. From sacral cord segments S_2 to S_4 *via* pelvic nerves.
   B. Postganglionic neurons lie in terminal ganglia *within* or *close to* the structure innervated.
      1. Terminal ganglia of the head
         a. **Ciliary**—postganglionics over third nerve to smooth muscle of pupilloconstrictor and to ciliary muscle of iris angle.
         b. **Sphenopalatine**—postganglionics over nerve VII to lacrimal glands, nasal glands, and palatine glands.
         c. **Submaxillary**—postganglionics over nerve VII to submaxillary and sublingual gland.

d. **Otic**—postganglionics over nerve IX to parotid gland.
2. The terminal ganglia of the vagus and pelvic nerves lie in or near the thoracic, abdominal, and pelvic viscera.
   a. One example would be the ganglion cells of the **myenteric** (**Auerbach's**) and **submucous** (**Meisner's**) plexuses of the gastro-intestinal tract about which vagal and pelvic nerve preganglionics terminate.

C. Contrast with sympathetic system:
1. In both systems, there are *always two and never more than two* neurons for the innervation of any structure.
2. The parasympathetic system has *nothing to do* with the sympathetic trunk or its chain ganglia.
   a. Hence, it is not involved with the spinal nerves or their visceral rami.
   b. Hence, there is *no* parasympathetic innervation for smooth muscle of peripheral blood vessels, sweat glands, or piloarrector muscle.
3. While parasympathetic fibers traverse the abdominal and pelvic plexuses, they synapse in *terminal ganglia only*, never in the collateral ganglia along the plexuses.
4. Parasympathetic postganglionic fibers therefore are *quite short* since their cell bodies *lie in* or *close to* the structure innervated. Sympathetic postganglionic fibers are *relatively long* since chain and collateral ganglia lie *some distance from* the structures innervated.
5. There is no *parasympathetic supply to the adrenal.* Nor is there a *postganglionic* sympathetic supply since the cells of the adrenal medulla are, themselves, modified sympathetic ganglionic cells.
6. Most visceral structures receive *both* a sympathetic and parasympathetic supply, but functionally the systems act antagonistically. There is little consistency about this antagonism, however. In one structure the sympathetic excites and the parasympathetic inhibits action, yet their roles may be diametrically reversed for another structure, the sympathetic inhibiting action and the parasympathetic exciting it. In general:
   a. Parasympathetic system regulates visceral motor activity commensurate with the calm, normal, physiological state.
   b. Sympathetic system usurps control during stress, situations which require total bodily mobilization (fright, flight, or fight).

IV. **Higher centers for regulating autonomic activity.**

A. Visceral motor centers of cerebral cortex.

B. Hypothalamic nuclei:
1. **Anterior hypothalamic** nuclei control **parasympathetic** activity in general.
2. **Posterior hypothalamic** nuclei control **sympathetic** activity.

C. Visceral motor centers of brainstem.
1. Pneumotaxic center of pons.
2. Respiratory center of medulla.
3. Vasomotor centers of medulla for carotid and aortic arch reflexes.
4. Center for swallowing reflex in medulla.
5. Center for vomiting reflex in medulla.
6. Center for cough reflex in medulla.
7. All of these brainstem motor centers are activated through secondary visceral sensory neurons of the solitary nucleus. Sensory impulses from cranial nerves VII, IX, and X over the tractus solitarius play into this nucleus.

V. **Visceral sensory neurons**

A. Receptor endings of various types in viscera.
1. Free branching endings with bulbs.
2. Free skeins.
3. Lightly encapsulated tangles.
4. Lamellated corpuscles.

B. Locus of cell body.
1. **Dorsal root ganglia.**
   a. Peripheral fiber traverse splanchnics, sympathetic trunk, and white ramus to reach dorsal root.
2. Other visceral sensory fibers traverse *facial*, *glossopharyngeal*, and *vagus* to reach cell bodies in:
   a. Nodose ganglia (X).
   b. Petrous ganglia (IX).
   c. Geniculate ganglia (VII).

PART II—THE ORGANS OF SPECIAL SENSE

The Eyeball

Introduction: The eye (Fig. XIV-57) has been likened to a camera. While the comparison may not always be a good one, some similarities exist. In keeping with the present era of plastics, the **sclera** or outer coat can be looked upon as a plastic case which gives shape to the structure and contains the working parts. The case is made mobile by muscles which insert upon it to point its lens at appropriate objects in the visual field. The eye has a shutter mechanism of sorts in the eyelids. At least when they are closed, not too much light passes the lens. It has a complex lens system consisting of the **cornea,** the **aqueous humor,** the **crystalline lens,** and the **vitreous body.** The crystalline lens has the ability to bulge or to flatten in focusing upon near and far objects respectively. This ability is lacking in lenses of fixed focal lengths in cameras. Either their lenses have to be changed or they have to be moved back and forth to bring things into focus on the film. The eye, like the better cameras, has an **automatic iris diaphragm** which stops down or opens up, depending upon the light intensity. The *pigmented*

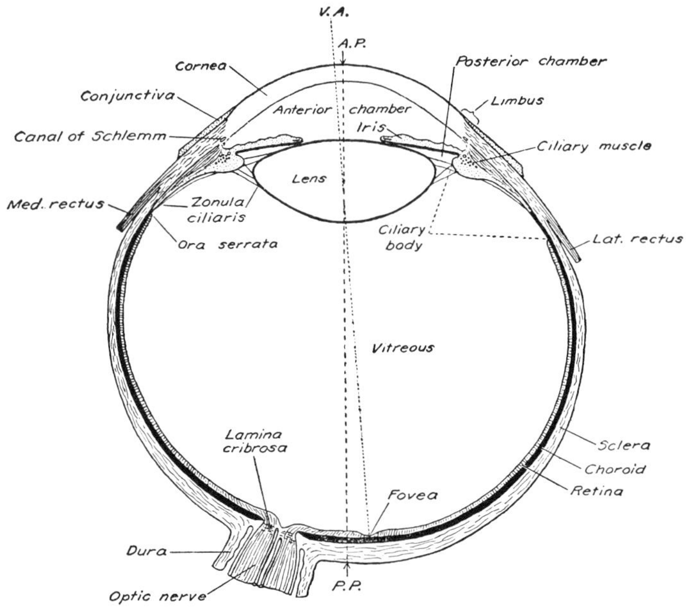

Fig. XIV-57. Schematic horizontal meridional section of right eye. × 3. A.P., anterior pole; P.P., posterior pole; V.A., visual axis. (From W. M. Copenhaver, *Bailey's Textbook of Histology*, 15th ed., Williams & Wilkins, Baltimore, 1964.)

parts of the **choroid coat** and of the *non-photosensitive parts* of the **retina** line the insides to convert the eye into a black box. The photosensitive retina is the film which must be developed in the dark room of the brain. Film processing is almost instantaneous, much faster even than the Polaroid Corporation has been able to achieve. There is no waiting for a little bell to ring. Most of the time, there is no loss of true color quality; and at night black and white snapshots can be taken under dim light. Still, old photographs fade with time even though safely stored in memory banks.

The Sclera: This hard outer coat of the eyeball consists of dense, white, fibrous connective tissue mixed with a few elastic fibers. It is a modification and a continuation of the dura mater which comes forward as a sleeve over the optic nerve. The other meninges invest the nerve up to its point of contact with the globe, but they continue no farther. The coat is tough enough to hold suture and to withstand considerable intraocular pressure. While pierced by vessels, it is poorly vascular. It is broken through posteriorly at the **lamina cribrosa** for passage of optic nerve fibers and the central artery and vein of the retina. Other blood vessels fail to reach the retina directly by this route. The extraocular muscles peg into the sclera. Other nerves and vessels, other than retinal vessels, pierce sclera in several places.

The Cornea: The bulging, transparent, anterior curvature of the globe is a modification of sclera known as the cornea. It is about 0.5 mm. thick at its central point. Its anterior epithelium is non-keratinized, stratified squamous in type but it lacks dermal papillae so that its blood supply is remote at best. Carbon dioxide is eliminated directly at its surface. This lowers the pH of fluids that bathe the anterior surface of the eye and damages the back side of contact lenses. Hence, wearers who feel they must utilize such devices for purely esthetic reasons must place a drop of buffering solution between the artificial lens and the cornea. The cornea is richly supplied by free-ending nerve fibers as true grit often demonstrates.

The corneal epithelium rests upon a homogeneous basal lamina (**Bowman's membrane**) which is highly protective against trauma and bacterial invasion. It does not degenerate when injured. It terminates at the corneoscleral junction (at the **limbus**).

About 90% of the corneal thickness constitutes its **substantia propria.** It consists of some 60 or more collagenous lamellae with flattened fibroblasts between the layers. The cementing substance of these layers is pure hyaluronic acid sulphate responsible for corneal transparency. This layer is avascular, deriving tissue fluid from vessels at the limbus.

The substantia propria rests upon an inner basal lamina (**Descemet's membrane**) which reacts with resorcin fuchsin as if it contained elastin, but no elastic fibers are present in it.

The lining of the cornea which faces the aqueous humor is called "corneal endothelium," a poor name since endothelium is the simple squamous epithelium which lines blood vessels. The inner corneal surface doesn't seem to be epithelial in nature in the true sense, but consists of flattened mesenchymal cells.

The Choroid Coat: This middle coat of the eyeball lies between sclera and retina. It is a thin coat, 0.1 to 0.2 mm., yet it is three-layered. Its **supra-** or **epichoroid lamina,** next to sclera, contains elastic fibers and large pigmented cells with branching processes. These are dopa-positive chromatophores whose granules contain melanogen. They give the choroid coat a blue-purple cast, like the skin of a Concord grape. Hence, the choroid coat sometimes has been called the uveal coat (uva = grape). Pigment cells of the retina, in contrast, are dopa-negative since they contain fuchsin, not melanogen (DOPA, **d**i-**o**xy-**p**henyl-**a**lanine). One will find a scattering of longitudinal smooth muscle in the epichoroid. These fibers are making their way forward to enter the **ciliary body.** In addition, the long posterior ciliary arteries run in the epichoroid.

The middle lamina of the choroid is its **vascular layer proper.** It contains the short posterior ciliary vessels and nerves and the four vortex veins. These veins are peculiar in that a number of venous channels become confluent to form a whorl-like patterning from which a single vortex vein arises. There are four of them, two superiorly and two inferiorly; and they leave the eyeball through emissaria in the sclera.

The deep lamina of the choroid, nearest the retina, contains some of the largest capillaries in the body. These feed the outermost laminae of the retina and, at the ciliary body, contribute to the formation of aqueous humor. This lamina is surfaced by a membrane (Bruck's membrane) which is partly elastic and partly cuticular. The cuticle may be contributed by the pigment cells of the retina which lie next adjacent.

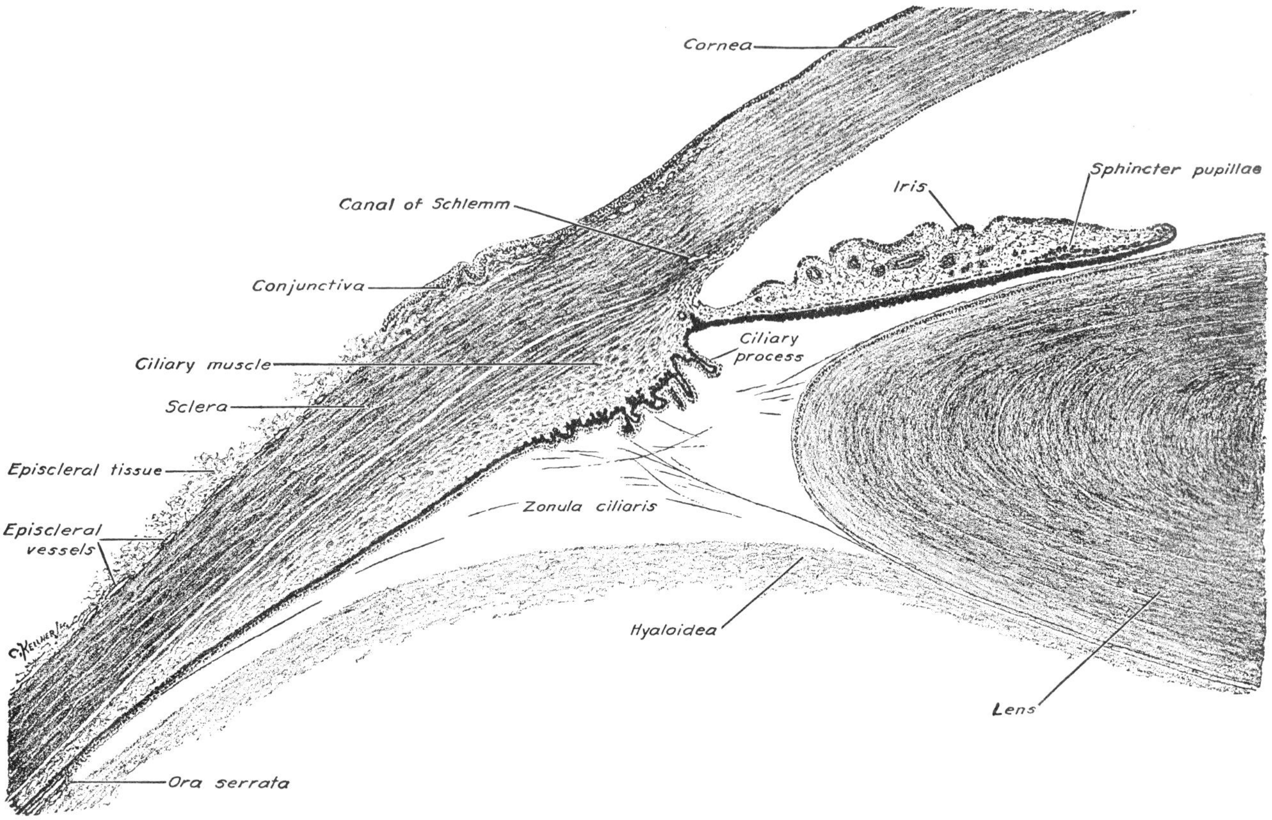

Fig. XIV-58. Horizontal meridional section through anterior portion of human eye. (From W. M. Copenhaver, *Bailey's Textbook of Histology*, 15th ed., Williams & Wilkins, Baltimore, 1964.)

Anteriorly, the choroid coat modifies to form the ciliary body and the iris diaphragm.

The Ciliary Body: This body is a thickening of the choroid coat behind the **iris angle** (Figs. XIV-58 and 59). Some three lamina of smooth muscle contribute to its bulk. The outermost fibers are those coming up from the choroid below to peg into the **sclera spur** (Fig. XIV-59). The middle group of smooth muscle bundles are radially arranged to anchor into the scleral network. Innermost are the circular bundles which ring the globe at the iris angle. From the surface of the ciliary body, **ciliary processes** project inward toward the lens. You must conceive of the body and its processes as arranged ring-fashion around the lens (Fig. XIV-60). Here, the back of the eyeball has been cut off so that you can look forward to view the back of the lens. It is not easy to appreciate this arrangement of ciliary processes by looking at horizontal, meridional sections through the eyeball, such as Figures XIV-57 through 59.

Some 70 to 80 ciliary processes extend outward from the circular ciliary body. The processes are pale in contrast to the dark pigment in the valleys between them (Fig. XIV-61). The *pigment* is contributed by the **non-photosensitive layer of the retina** which continues beyond its photosensitive portion to line ciliary processes and the underside of the iris diaphragm. The **photosensitive layer of the retina** ends anteriorly . . . well posterior to the ciliary body at the **ora serrata.** The ora serrata forms a scalloped anterior margin which encircles the globe also. (Find the ora serrata in Figs. XIV-57, 58, and 60.)

The **Zonula Ciliaris** (Fig. XIV-58) is a system of fine, rather homogeneous fibers. These gather to form a thickened band which radiates from the lens capsule to the ciliary body. The fibers fix themselves to the valleys between the ciliary processes and to the ciliary body substance itself.

Ciliary muscle is innervated by the parasympathetic visceral motor fibers of cranial nerve III (oculomotor). Normally, when ciliary muscle is relaxed, the weight of the choroid coat draws the zonula ciliaris taut, so the lens flattens in its elastic capsule. The eye is then in its normal position of rest and is focused for far distance. To accommodate for near vision, the lens must bulge in its elastic capsule. Ciliary muscle is contracted, the choroid coat is pulled forward, zonula ciliaris is relaxed and the elastic lens capsule bulges as the lens thickens. This phenomenon is called **accommodation.** It is a reflex which involves occipital cortex as well as the superior colliculus and the Edinger-Westphal component of the oculomotor nucleus. The cortex is involved because we consciously shift focus from far to near; but we have no control over

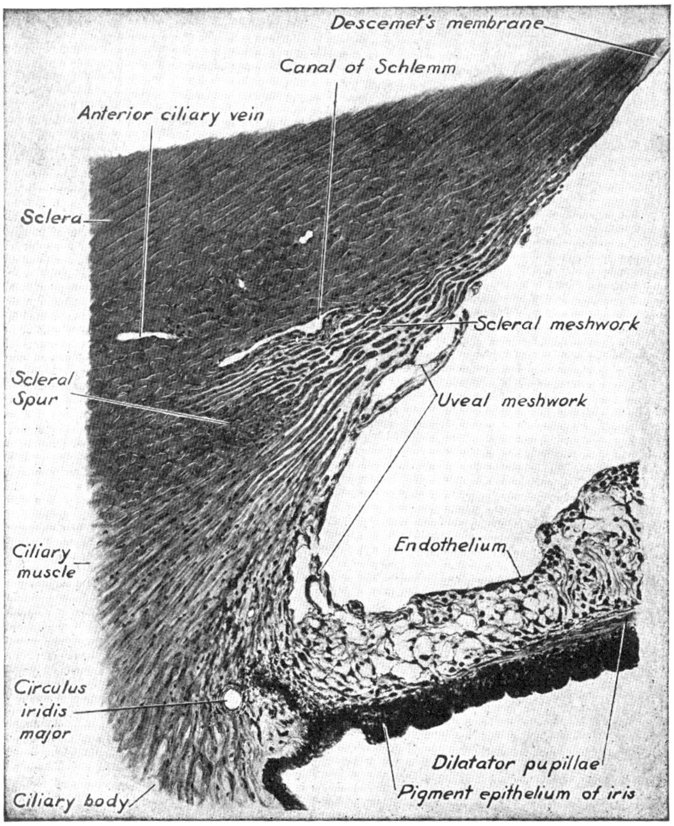

Fig. XIV-59. The iris angle. (From W. M. Copenhaver, *Bailey's Textbook of Histology,* 15th ed., Williams & Wilkins, Baltimore, 1964.)

the reflex lens action or the contraction of the ciliary muscle.

The surface cells of the ciliary processes themselves are not pigmented. They are large and cuboidal to columnar in shape. They play an active role in secreting aqueous humor into the **posterior chamber** of the eye. They add ascorbic acid, salts, and bicarbonate to the aqueous fluid. The posterior chamber sits behind the lens and iris diaphragm, between these structures and the vitreous (Fig. XIV-57). Fluid passes from *posterior chamber* through the pupil (hole) of the iris diaphragm into the **anterior chamber** which lies between cornea and iris diaphragm. Absorption occurs in the margins of the anterior chamber in the region of the scleral meshwork (Fig. XIV-59). We will discuss absorption below when we consider the nature of the *iris angle.*

The Iris Diaphragm (Figs. XIV-57 and 58): This structure is an extension of the choroid coat also. It extends in front of the lens as a truncated cone with a central aperature, the **pupil.** It is continuous with the ciliary body at its attached borders. Its surface lining, facing the anterior chamber, consists of flattened mesenchymal cells similar to those which line the inner aspect of the cornea. Its connective tissue stroma contains branching, pigmented chromatophores which contain yellow-brown granules. When granules are plentiful, the coloration of the iris is brown. When granules are sparse, the iris is blue. Thus, "blue eyes" are caused by a reflection of light from a sparse supply of yellow-brown pigment granules. There are no blood vessels in the anterior surface of the iris. Rather, vessels lie more posteriorly. Its posterior surface is covered by the pigment epithelium of the non-photosensitive retina. The inner layer of the retinal cells modify into myoepithelial elements giving rise to the **Dilator Pupillae,** radially arranged smooth muscle of the diaphragm capable of pulling open the margins of the pupil; and to the **Sphincter Pupillae,** circularly arranged smooth muscle at the pupil margin.

Postganglionic axons of cranial nerve III from the ciliary ganglion innervate the Sphincter causing it to constrict the pupil reflexly in the face of bright light. Postganglionic axons from the superior cervical sympathetic ganglion innervate the Dilator as well as smooth muscle of blood vessels in the eyeball. The connections for the pupillolight reflex take place in the **pretectal region** just rostral to the superior colliculus of the midbrain and involve neither occipital cortex nor the lateral geniculate body.

The Iris Angle (Fig. XIV-59): This is the angle of the anterior chamber. It contains a meshwork of scleral tissue. It extends posteriorly to the ciliary body and it is continuous anteriorly with Descemet's membrane of the cornea. The lamellated spaces of the meshwork are known as **the spaces of Fontana.** They communicate by holes in the lamellated plate with each other, with the anterior chamber and with the **canal of Schlemm** which lies in the scleral furrow (Fig. XIV-59). The canal is an endothelially-lined space which encircles the globe in this region. It communicates with 20 to 30 small branches of the anterior ciliary veins.

Thus, the iris angle is organized to absorb aqueous humor from the anterior chamber. The fluid, itself, was produced by secretory cells in the posterior chamber—cells lining ciliary body and its processes. The normal rate of production is slow, 2 cu. mm. per minute. Since the inelastic sclera keep the eye under an intra-ocular pressure of 20 to 25 mm Hg, pressure in the capillary layers of the ciliary body must be higher than this to be able to form aqueous humor.

The cells, we have noted, secrete a considerable amount of hyaluronic acid. In other connective tissues this tends to exist as a gel. In the eye it is 95% depolymerized because of the enzyme hyaluronidase. Hence, aqueous humor is watery. Other constituents are 0.77% NaCl, traces of urea and glucose and hardly any protein. The circulation is forward into the anterior chamber through the pupil and toward the iris angle. There it is absorbed by the spaces of Fontana, the canal of Schlemm and the anterior ciliary veins which flow together to form vortex veins.

Blockage of this drainage route leads to **glaucoma.** There is hardening of the globe as intra-ocular pressure increases, opacity of the lens, corneal anesthesia, restricted visual field, lessened visual power, colored halos around artificial lights, excavation at the optic papilla, and gradual blindness.

The **true limbus** is the exact site of corneoscleral

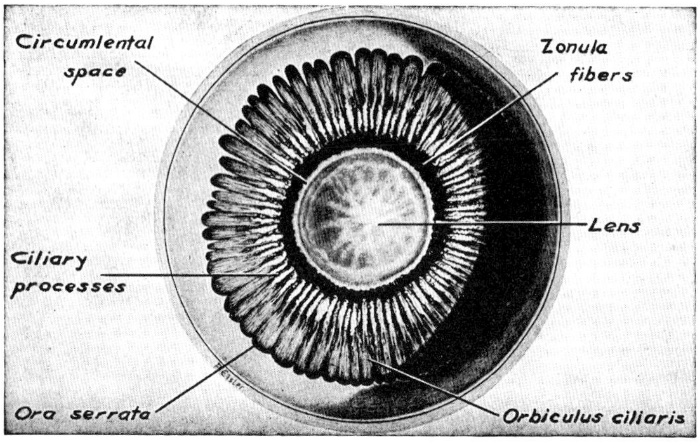

Fig. XIV-60. Ora serrata, ciliary body, zonula ciliaris and lens as viewed from behind after removal of the vitreous body. (From W. M. Copenhaver, *Bailey's Textbook of Histology*, 15th ed., Williams & Wilkins, Baltimore, 1964.)

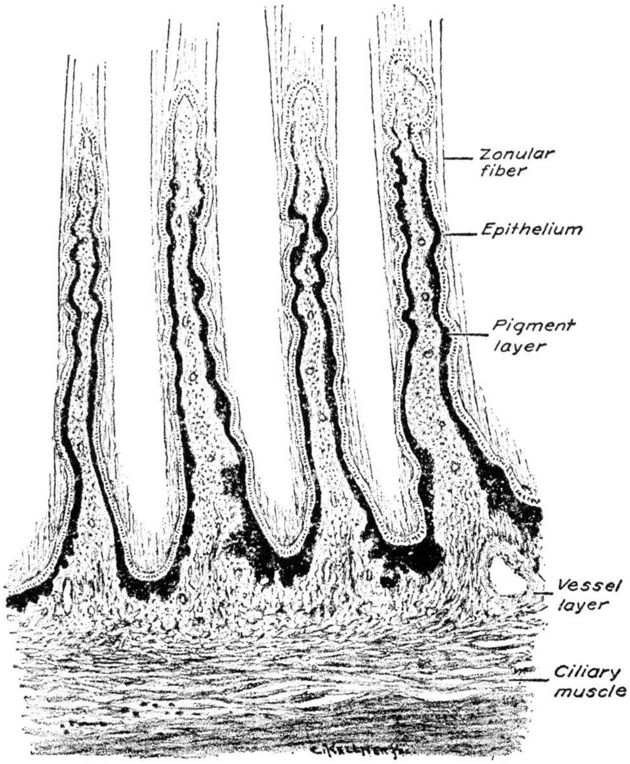

Fig. XIV-61. Enlarged view of ciliary processes. (After W. M. Copenhaver, *Bailey's Textbook of Histology*, 15th ed., Williams & Wilkins, Baltimore, 1964.)

junction. The **limbus of Tenon's capsule,** the anterior anchorage of the fascia bulbi to the globe, is 1.5 mm. posterior to the true limbus. The **conjunctival limbus,** where the inner lining of the eyelid fuses with cornea, is 1.0 mm. anterior to the true limbus. It is in the interval between conjunctival limbus and limbus of Tenon's capsule that corneoscleral trephine procedures are performed for relief of glaucoma (decreasing the intraocular pressure).

The Retina: This internal tunic is the pars optica of the eyeball, the part which transforms the physical stimulus of a light source into nervous impulses.

The retina arose embryologically from ectoderm as an outgrowth in the form of an **optic vesicle** from the diencephalon. The vesicle became invaginated into a double-layered **optic cup** by a developing **lens placode** which pushed inward from surface epithelium of the embryo. The inner, invaginated layer of the optic cup developed as the photosensitive retina; the outer layer became its pigmented, non-nervous layer. In histological sections these two layers may separate leaving the pigmented layer attached to the choroid. The photosensitive retina, after all, anchors to the globe only at the ora serrata anteriorly, and to the **optic disk** posteriorly where its optic nerve fibers leave to pierce the other coats of the eyeball. Held up to light, the detached retina is paper thin, delicate, and transparent. Yet, in terms of histology, it is a 10-layered structure. Compare it to the choroid in Figure XIV-62 where the full thickness of the sclera is not completely shown.

The retina begins at layer 5 in Figure XIV-62. This is its pigmented layer applied to the choroid. Indeed, the cuticle of the lamina vitrea (4) in the Figure may even be a part of the retina. Strangely, its photosensitive elements, the **rods** and **cones,** lie deepest of all in the photosensitive retina close to the pigmented layer. An **external** (in the sense of lying closer to the outer wall of the eyeball than to its inner wall) **limiting membrane** separates the photosensitive or dendritic parts of the rods and cones from their cell nuclei in the **outer nuclear layer.** The axons of rods and cones make up an **outer plexiform layer** to synapse with the dendrites of **bipolar cells** of the retina whose nuclei make up the **inner nuclear layer.** Bipolar cell axons form an **inner plexiform layer** to synapse with dendrites of the **ganglion cells** of the retina. Axons of the ganglion cells give rise to the fibers of the optic nerve. These fibers lie at the inner surface of the retina, directly under its internal **limiting hyaline membrane** which abuts against the gelatinous vitreous substance of the globe. Thus, light has to pierce all these retinal layers to effect rod and cone cells.

The nature of interconnections between cellular elements in these layers is depicted in Figure XIV-63. Notice that there are **horizontal cells** (11 and 12) and **amacrine cells** (13 to 15) scattered among various layers. These are for intraretinal connections.

The human retina has been calculated to contain

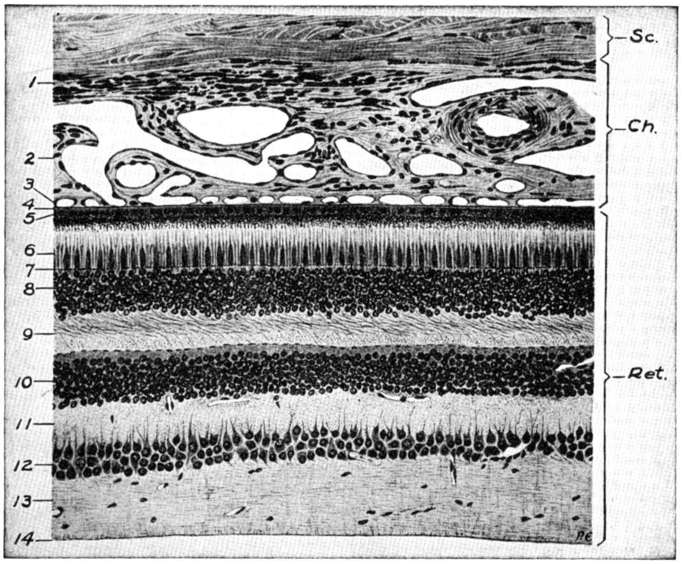

Fig. XIV-62. Section through retina, choroid and sclera. Sc, Sclera; Ch, choroid, Ret, retina; 1, suprachoroid; 2, vessel layer; 3, choriocapillaris; 4, lamina vitrea; 5, pigment epithelium; 6, layer of rods and cones; 7, external limiting membrane; 8, outer nuclear layer; 9, outer plexiform layer; 10, inner nuclear layer; 11, inner plexiform layer; 12, ganglion cell layer; 13, nerve fiber layer; 14, internal limiting membrane. (From W. M. Copenhaver, *Bailey's Textbook of Histology,* 15th ed., Williams & Wilkins, Baltimore, 1964.)

over 100 million rod cells, the most photosensitive of the visual elements, capable of responding to a single quantum of radiant energy to produce electrogenesis. Still, it takes the excitation of 6 rods to produce a detectable visual response.[5] There are about 6 million cones, the elements responsive to color vision. Rods are cells of twilight vision; cones operate in bright light and require more intense radiant energy to become excitable. Despite the plethora of photosensitive elements, only 1 million optic nerve fibers leave the retina. It is obvious that there are 100 times as many rods and 6 times as many cones as there are optic nerve fibers. Unless rods and cones grouped into clusters of a certain number of each necessary to fire a single ganglion cell whose axon is one optic nerve fiber, the concept of "photographic grain" in the retina would largely disappear. There does not appear to be this kind of order. Interconnections are most variable and elaborate between the elements. Only 2.5 mm. lateral to the optic disk, at the **fovea centralis,** do cone cells synapse with single rod cells to set up isolated transmission lines from visual receptor cells to brain. There are no rods at the fovea centralis. Elsewhere, rods and cones intermix. Toward the periphery (ora serrata) rods predominate over cones. The fovea is a depression in a yellow pigmented spot of the retina, the **macula lutea** (yellow spot). Elsewhere the retina has a reddish cast (visual purple) due to the

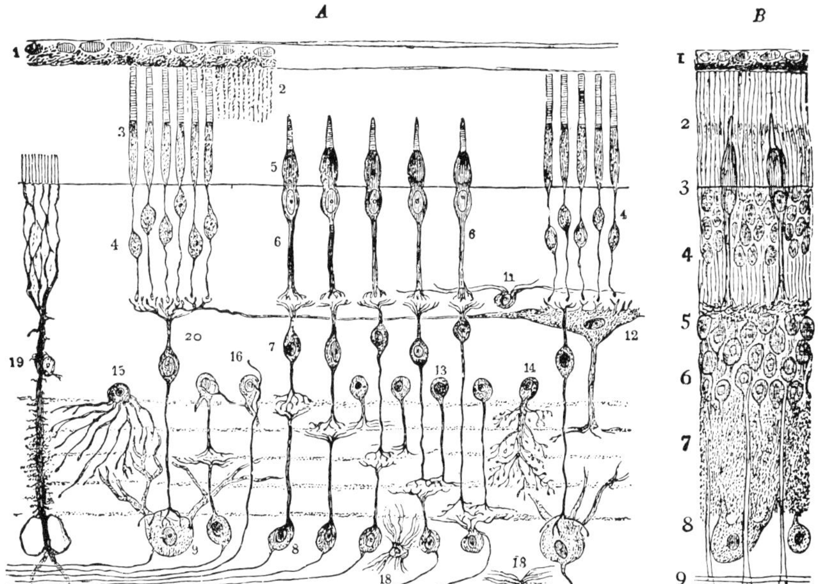

Fig. XIV-63A. Scheme of retina as shown by Golgi method. **B.** Section of retina to show layers as demonstrated by hematoxylin-eosin method.

A. 1, Pigment epithelium; 2, processes of pigment epithelium cells; 3, rods; 4 rod nuclei and fibers; 5, cones; 6, cone fibers; 7 cone bipolar cells of inner nuclear layer; 8, smaller cells of ganglion cell layer; 9, larger ganglion cells; 10, ganglion cell axons forming nerve fiber layer; 11 and 12, horizontal cells; 13, 14, 15 and 16, amacrine cells; 18, neuroglia cells; 19, Müller's fiber; 20, rod bipolar cell of inner nuclear layer.

B. 1, Pigment epithelium; 2, layer of rods and cones; 3, external limiting membrane; 4, outer nuclear layer; 5, outer plexiform layer; 6, inner nuclear layer; 7, inner plexiform layer; 8, layer of ganglion cells; 9, nerve fiber layer; 10, internal limiting membrane. (From W. M. Copenhaver, *Bailey's Textbook of Histology*, 15th ed., Williams & Wilkins, Baltimore, 1964.)

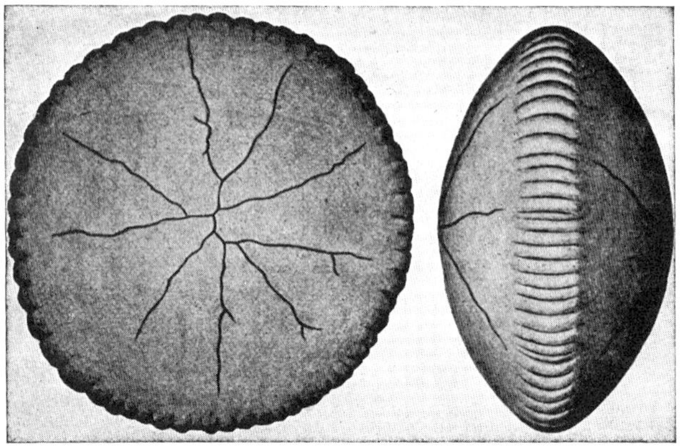

Fig. XIV-64. The lens, viewed from behind and from the side. (From W. M. Copenhaver, *Bailey's Textbook of Histology,* 15th ed., Williams & Wilkins, Baltimore, 1964.)

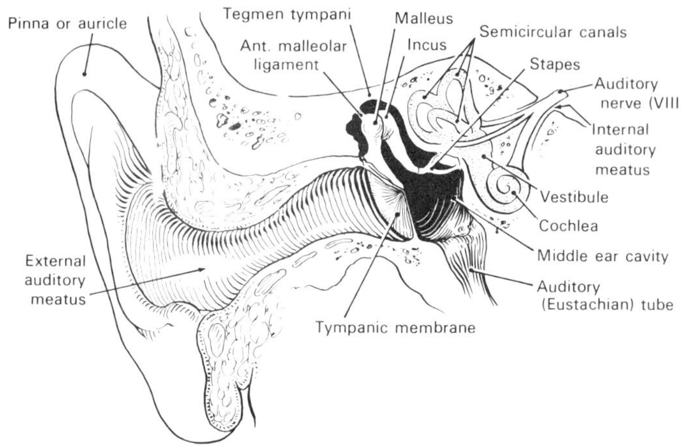

Fig. XIV-65. The basic parts of the ear; external, middle, and inner.

rhodopsin of the rod cells. The fovea centralis is the area of sharpest vision in bright light, and the extraocular muscles turn the eyeball so that the lens system will be brought to best focus here for keenest vision.

There are no rods and cones at the optic disk where optic nerve fibers from all retinal parts converge to leave the eyeball. Hence, light falling here falls on a blind spot. By focusing on an object and by pushing laterally against the eyeball, you can get the lens to focus on this blind spot and see the focused object disappear from view.

The Lens (Fig. XIV-64): The crystalline lens contains a capsule, anterior epithelium, and the lens substance. The capsule is a highly elastic membrane which sheaths the lens but is always thinnest on the posterior aspect. On the sides there is a thickening for insertion of the zonular fibers as a suspensory ligament. The posterior aspect bulges most and is devoid of epithelial cells. Hence, the epithelium, as cuboidal cells, covers the anterior surface and laps the lens' equator where elongate cells are arranged in meridional rows. The lens fibers make up the lens substance. These long, flat, prismatic elements are derivatives of the embryonic posterior epithelium of the lens placode which got used up in the making of lens substance. The older and deeper fibers have lost their nuclei. An amorphous cementing substance binds the fibers together.

The cornea, anterior chamber, lens, and vitreous constitute the *refracting media of the eye.* The cornea, itself, has a refractive index of 1.376 and a radius of curvature of 7.8 mm. Both of these values are greater than that of the lens itself. Hence, much of the refraction of light takes place at the cornea. The anterior chamber has a refractive index of 1.330 while that of the lens is 1.360 and the vitreous, 1.334. The radius of refraction of the lens is 6.9 mm.

The Vitreous Body: This structure is not a part of the posterior chamber. It lies between lens and retina as a firm but transparent jelly, forming the main stuffing of the interior of the globe. There is a broad, shallow depression on its anterior surface, the **patellar fossa,** which has nothing to do with the knee joint. It simply accommodates the posterior convexity of the lens. A **hyaloid canal** leads from the fossa posteriorly through the vitreous. The ophthalmic artery once continued forward in the canal as the fetal hyaloid artery.

To review coverings of the optic nerve; the retina as seen in the ophthalmoscope; the lids and lacrimal apparatus; and other details of the eye, see the appropriate sections in the chapter on "Surface Anatomy," pages 97–103.

The Ear

Introduction: The ear, concerned with the functions of equilibrium and hearing, is divided into three parts: **external, middle,** and **inner ear** (Fig. XIV-65). The first two parts serve to conduct sound waves to receptor cells located in the inner ear.

The external ear consists of the **pinna or auricle;** and **external auditory meatus** and an ear drum or **tympanic membrane.** The middle ear contains a cavity, the **tympanum proper;** the **tiny ossicles** or ear bones; an **auditory** (Eustachian) **tube** communicating with the pharynx; and the **mastoid air cells.** The inner ear, embedded in the petrous part of the temporal bone, contains **osseous** and **membranous labyrinths** of the **cochlear** or **auditory system** and the **semicircular canals** of the **vestibular system.** Opening into the petrous bone in the middle cranial fossa is the **internal auditory meatus** bringing into the bone the fibers of cranial nerves VII and VIII (facial and auditory or stato-acoustic nerves).

The External Ear: The *pinna* serves as a poor, relatively immobile funnel to converge sound waves in man. It is put to better use for this purpose in many animals. Its nerve supply is interesting since it lies between branchial arch and dermatome areas. Dermatomal mesoderm is of somite origin. Hence, the skin

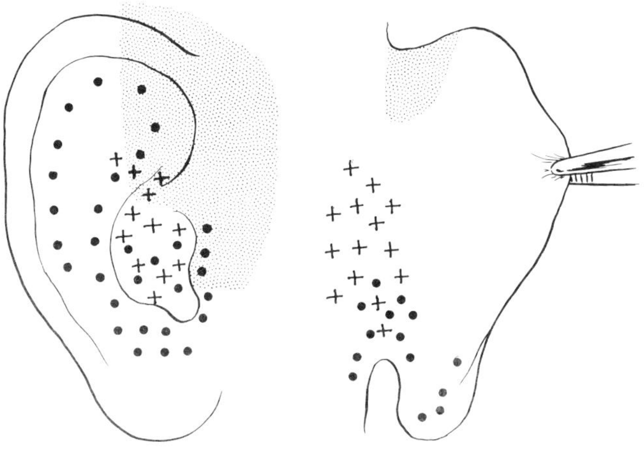

Fig. XIV-66. Sensory supply of external ear. **A.** Lateral aspect. **B.** Posterior aspect. Shaded area, auriculotemporal branch of V; unshaded parts, C_2 to C_3; crosses, probable distribution of IX and X; circles, probable distribution of VII. Region of distributions and zones of overlap vary in different individuals.

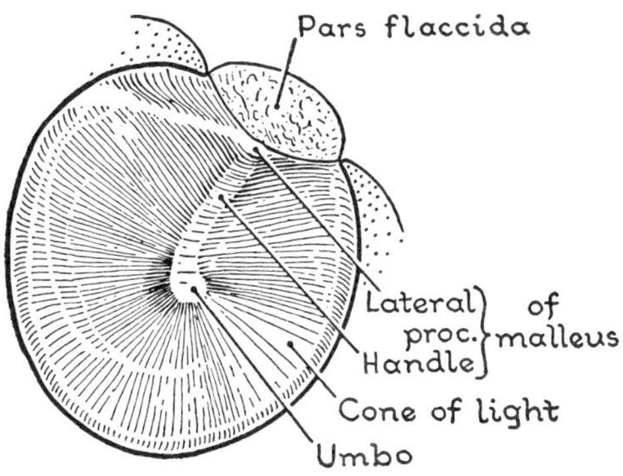

Fig. XIV-67. The tympanic membrane, on lateral view. (From J. C. B. Grant and J. V. Basmajian, *Grant's Method of Anatomy*, 7th ed., Williams & Wilkins, Baltimore, 1965.)

of the pinna or auricle is supplied by both cranial and cervical nerves (Fig. XIV-66). The auriculotemporal branch of V supplies the supero-anterior aspect and the **central concavity** of the **concha.** Posterior and inferior aspects are innervated by sensory branches of C_2 and C_3. Most of this latter area is the territory of the great auricular nerve; the mastoid tip is by the lesser occipital nerve; and the posteromedial tip, by the small occipital.

The greatest part of the concavity of the concha is innervated by cranial nerve VIII. Areas probably supplied by cranial nerves IX and X in the concha are variable and overlap the territory of VII and V. But the lower back part of the ear (Fig. XIV-66B) is an exclusive skin patch for X.

The *external auditory meatus* is tortuous and almost S-shaped. The skin of its upper surface is supplied by sensory fibers of V while its lower parts are innervated by the sensory fibers of X. The stratified squamous epithelium of this canal stops where it surfaces the inner aspect of the tympanic membrane or ear drum. The purpose of the meatus is to conduct sound waves against the membrane, and to keep ears dirty by an accumulation of ear "wax" (which isn't wax).

The Tympanic Membrane (Fig. XIV-67): Review the details on this conical structure in the chapter on "Surface Anatomy," page 103. Pay attention there to its appearance in the otoscope. The membrane is obliquely placed and attached by a cartilaginous ring. It is pulled inward as a slight cone by its attachment to the manubrium of the malleus at the umbo. It consists of a connective tissue core in which elastic fibers are arranged in double layers, in concentric circular arrangement crossed by radial fibers from a central hub, much in the form of a woven spider web. Its outer surface, facing the external auditory meatus, we found covered with a stratified squamous mucosa. Its inner surface, facing the middle ear cavity, is lined with a pseudostratified mucosa continuous, through the Eustachian tube, with that of the nasopharynx. Its upper segment, the pars flaccida, lacks elastic fibers.

The Middle Ear Cavity: The cavity of the middle ear lies directly behind the ear drum. It is narrow, somewhat rounded, compressed at its center and wider peripherally than at the center. It is approximately 15 mm. deep along a vertical axis; 2 mm. across at its center; 4 mm. across at its floor and 6 mm. across at its roof. Imagine looking with X-ray vision into the left ear of your colleague. Be sure no one is looking in his right ear or you may only see another eye. The general shape of the cavity, in bold schematics, is like the sketch in Figure XIV-68. The tympanic membrane blocks a hole in the lateral wall of a sump pit. The roof of this pit slants and is made of thin bone, the **tegmen tympani.** Since the roof extends higher than the upper border of the ear drum, this space—the **epitympanic recess** will house the head of the **malleus** and the **incus,** two of the ear ossicles. A narrow channel is directed posteriorly at the **aditus.** It will widen into an **antrum** which leads into the maze of **mastoid air cells** within the mastoid process of the temporal bone. The epitympanic recess opens anteriorly into the auditory or Eustachian tube which is directed medially toward the nasopharynx. The floor of the sump pit rests upon the jugular bulb. Its anterior wall is a common one for the carotid canal. In its posterior wall, the facial nerve (VII) descends toward the stylomastoid foramen. Running through a bony wedge between artery and vein is the tiny tympanic canal or Jacobson's canal for the tympanic branch of IX to enter the middle ear cavity (Fig. XIV-70).

If we were to stand on the *medial wall* of this cavity so that we looked across it at the inner aspect of the ear drum (Fig. XIV-69), we would see the relationship of malleus to incus in the epitympanic recess and the fusion of the manubrium of the malleus

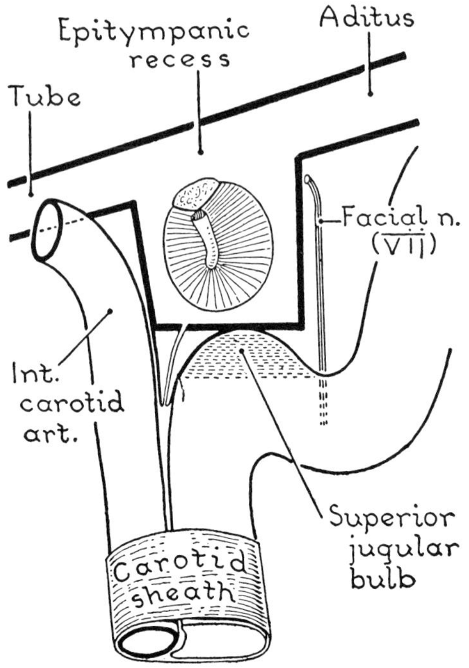

Fig. XIV-68. The tympanic cavity likened to a sump pit on the course of the auditory tube, epitympanic recess, and aditus. (From J. C. B. Grant and J. V. Basmajian, *Grant's Method of Anatomy*, 7th ed., Williams & Wilkins, Baltimore, 1965.)

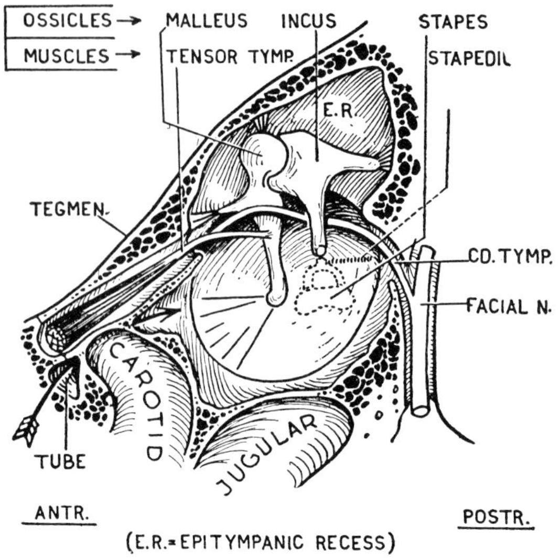

Fig. XIV-69. Medial wall of the tympanic cavity. (From J. C. B. Grant and J. V. Basmajian, *Grant's Method of Anatomy*, 7th ed., Willliams & Wilkins, Baltimore, 1965.)

to the drum at the umbo. Bridging *across* the cavity, toward us, to the medial wall, is the third little ear ossicle, the **stapes.** It is shaped like a tiny stirrup with its foot plate filling a hole in the medial wall at the **oval window.** The malleus is anchored to the bony wall of the epitympanic recess by a **superior and anterior malleolar ligament.** A posterior ligament anchors the incus to the posterior wall.

Anteriorly, roof and floor converge to form the auditory tube. Above the auditory tube is a semicanal in the roof housing the **Tensor Tympani muscle** whose tendon inserts upon the manubrium of the malleus to tense the ear drum. This muscle, along with the Tensor Palatini, is of first branchial arch origin. Hence, it will be innervated by the trigeminal nerve (V). It contracts reflexly to sounds of high frequency (high pitch) which would set the ear drum vibrating at a painful level. Thus, it takes part in a protective reflex which increases the rigidity of the ossicular chain and, as a consequence, it dampens conduction rate. Posteriorly, we see the facial nerve, in its facial canal, heading for the stylomastoid foramen. In its course it gives off two twigs which enter the middle ear cavity from behind. The highest of these is its **chorda tympani** branch (Figs. XIV-69 and 70A). The chorda crosses between the stapedial process of the incus and the manubrium of the malleus to reach the petrotympanic fissure anteriorly on its way to the infratemporal fossa where it will join the lingual branch of V. It is visceral motor to sublingual and submaxillary glands and special sensory from taste buds of anterior two-thirds of the tongue.

The second short branch is a branchial motor twig to the **Stapedius muscle.** This tiny muscular derivative arises from a small pyramidal eminence in the posterior wall of the middle ear cavity and inserts upon the stapes (Fig. XIV-69). Since it is a second branchial arch muscle, its supply, naturally, is from VII. It contracts reflexly to loud sounds (sound waves of high amplitude) to draw back on the stapes. This prevents the stapes from setting up high amplitude waves in the perilymph of the inner ear against which the stapes acts. When the facial nerve is compressed in the facial canal, patients with Bell's palsy (facial nerve paralysis) may complain of **hyperacusia.** The ordinary voice sounds overly loud to them since the stapes now swings freely without check by the stapedial reflex mechanism. Note how deeply the jugular bulb and carotid canal indent the thin floor of the middle ear cavity (Fig. XIV-69).

There are other tiny, bony channels in the petrous bone related to nerves and vessels of the middle ear cavity (Fig. XIV-70). The geniculate ganglion of VII lies in the **genu of the facial canal.** The posterior limb of this canal, we have seen, heads for the stylomastoid foramen, giving off the chorda tympani and stapedial branches to the middle ear cavity. The **hiatus of the facial canal** is directed anteriorly and pierces near the apex of the petrous bone. In it runs the **greater superficial petrosal branch of VII** to reach the pterygopalatine ganglion in the fossa of the same name. It is visceral motor to lacrimal, nasal, and palatine glands.

The tympanic branch of IX pierces the floor of the cavity between jugular bulb and carotid canal in the **lower segment of Jacobson's (tympanic) canal.** The

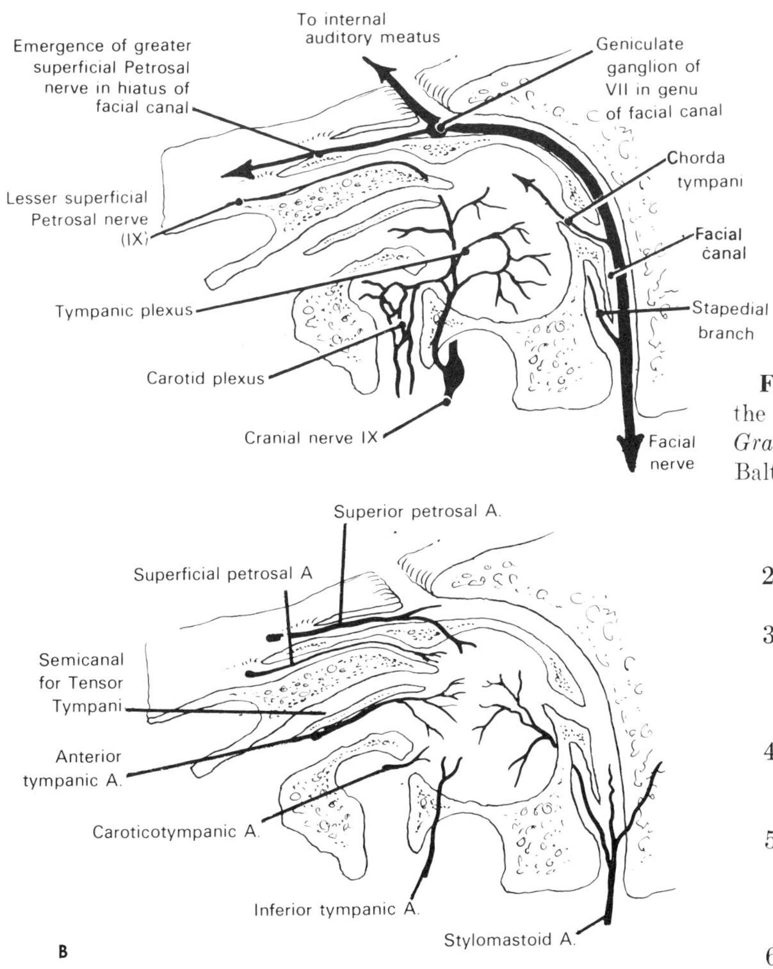

Fig. XIV-70. Nerves (**A**) and arteries (**B**) of the tympanic cavity. (Modified from Hollinshead.)

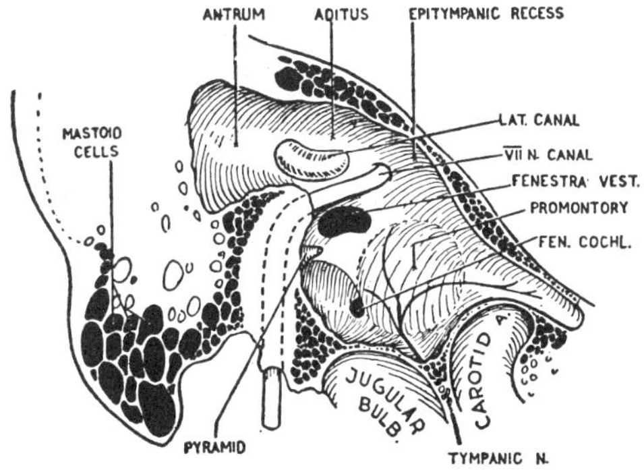

Fig. XIV-71. Lateral wall of the tympanic cavity and the ossicles. (From J. C. B. Grant and J. V. Basmajian, *Grant's Method of Anatomy,* 7th ed., Williams & Wilkins, Baltimore, 1965.)

nerve gives a sensory supply to the mucosa of the middle ear and then sends its visceral motor branch through the **upper segment of Jacobson's canal,** just below the anterior limb of the facial canal. This branch is the **lesser superficial petrosal nerve of IX** heading for the foramen ovale, the infratemporal fossa, the otic ganglion, the auriculotemporal branch of V and, finally, the parotid gland. The sensory twigs of IX join with sympathetic fibers which have entered the carotid canal with the artery. They leave the artery through a **caroticotympanic canal** to form a carotid plexus on the medial wall of the cavity. Figure XIV-70B shows the arteries which utilize these same bony channels to gain access to the middle ear.

Now we need to stand on the *lateral wall* of the cavity with our backs against the ear drum as we look across at its medial wall. You may have to duck your head a bit if you are tall, to keep from bumping the tegmen tympani. You should observe the features depicted in Figure XIV-71:

1. The position of the posterior limb of the facial canal between jugular bulb and mastoid process.
2. The aditus widening posteriorly into an antrum which communicates with mastoid air cells.
3. Below and anterior to the bend of the facial nerve lies the *oval window* or fenestra vestibulis in the medial wall of the cavity. Into this, the stapes will fit.
4. The pyramidal process of bone from which the Stapedius muscle arises is seen in the posterior wall.
5. Tucked under a curving bulge of the medial wall postero-inferiorly is the *round window* or fenestra cochleae. This is covered by the mucous membrane of the cavity.
6. The big bulge or promontory of the medial wall lies anterior to both windows and the carotid plexus breaks up under its mucosal covering.
7. A prominent ridge above the nerve is formed by the lateral semicircular canal as it pushes inward against the medial wall.

The mucosa of this wall covers the medial aspect of the petrous part of the temporal bone. The windows in it open into a series of interconnected bone channels, the **osseous labyrinths** of the inner ear. The oval window leads into a wide part of the labyrinth known as the vestibule. The round window taps the **scala tympani** of the basal part of the coiled cochlea.

The foot-plate of the stapes is held in the oval window by a semicartilaginous ligament. As sound waves impinge upon the ear drum to set it in vibratory motion, the manubrium will be moved. Since the head of the malleus interdigitates cog-tooth fashion with the incus, it will move the incus. The stapedial process of the latter ossicle will move the foot plate of the stapes against **perilymph,** the fluid which fills the osseous labyrinths. The stapes is not pushed back and forth like a plunger or piston. Rather, it rocks on a hinged joint at its base so that only its superior portion moves in and out like a hinged flour bin of a kitchen cabinet. For every inward movement of the stapes against perilymph, there is a compensatory bulging of the membrane over the round window into the middle ear cavity and *vice versa.* If fluid is compressed in rigid bony channels, something has to give somewhere. The round window serves to absorb the thrust of the stapes against the fluid.

Conduction of Sound through the Middle Ear: The tympanic membrane is a universal resonator. It is set in motion easily by sound waves of any frequency and it dampens as readily. In this respect it is like the magnetic diaphragm of a telephone receiver. The vibrations of the membrane are best transmitted to the inner ear over the hydraulic mechanism of the ossicular chain. This route can energize the heavy fluid perilymph 1000 times more effectively than can air waves, set up by the ear drum, which cross the middle ear cavity to impinge against the membranous round window. Even so, about 55% of the vibrations act against the round window as air-conducted waves across the middle ear cavity. If incus and stapes are disarticulated, hearing is still possible over the round window route, but a 40 decibel deafness ensues over the middle ranges of tonal frequencies. There is less loss at high and low sound frequencies. This loss can be decreased to 28 decibels if ossicles and tympanic membrane are removed completely.

While the major action of the round window is to serve as a compensatory take-up mechanism for the stapes, it does seem to aid auditory acuity. If it is sealed, deafness results. If the stapes becomes ankylosed (fused rigidly in the oval window), it is still possible to hear *via* air conduction across the middle ear cavity. Bone conduction is partly responsible for this maintenance of hearing. The round window seems to fractionate its behavior under such circumstances, the upper part pushing inward against the scala vestibuli of the cochlea; the lower part, bulging away from the scala typmani. The significance of these scala or galleries will be made clear below.

Under the latter circumstances of ankylosis, the otolaryngologist achieves better sound conduction by constructing a *new round window* in the bend of the lateral semicircular canal above (**fenestration operation**). Now, as air waves cross the middle ear cavity, they can bulge the *old* round window inward while the *new* window compensates by bulging out, and *vice versa.* The *new* window communicates with the vestibule, just as the ankylosed stapes did before it became immobilized. The *old* round window acts against the scala tympani as it always did.

Thus, **air conduction** through the middle ear cavity can take two routes, normally:

A. From external auditory meatus → tympanic membrane → ossicular chain → intralabyrinthian fluids.

B. From external auditory meatus → tympanic membrane → air in middle ear cavity → round window → intralabyrinthian fluids.

Bone conduction of sound waves is another possibility, for the sound of a tuning fork placed against the mastoid process or a tooth can be heard clearly when conduction through the middle ear cavity is impaired. Again, there is a choice of routing:

A.

Skin → periosteum → a skull bone ↘
Tooth → peridontal membrane → a skull bone ↙
→ other skull bones
↓
parts of the petrous pyramid
↓
otic capsule of the osseous labyrinth
↓
intralabyrinthian fluids

Or, B.

Skin → periosteum ↘
Tooth → peridontal membrane ↙
→ a skull bone
↓
other skull bones
↓
walls of middle ear cavity
↓
air of middle ear cavity
↓
round window
↓
intralabyrinthian fluids

The Inner Ear: Let us first examine once more some important *external landmarks on the petrous bone* (Fig. XIV-72):

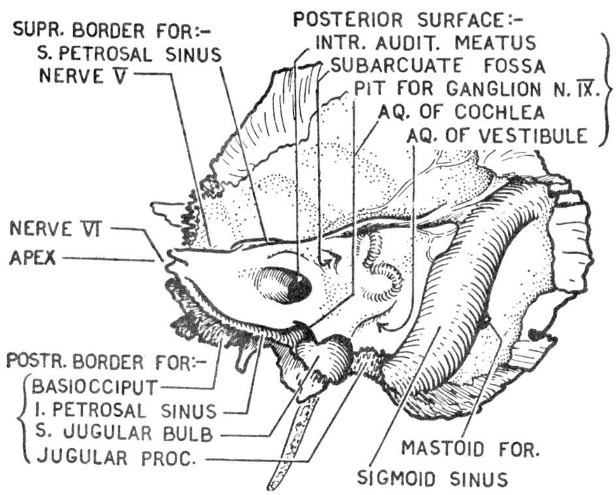

Fig. XIV-72. Pertinent landmarks related to petrous part of temporal bone. (From J. C. B. Grant and J. V. Basmajian, *Grant's Method of Anatomy,* 7th ed., Williams & Wilkins, Baltimore, 1965.)

1. Find the orifice of the internal auditory meatus. Sensory fibers of VII and VIII emerge here to enter the brainstem near the lateral recess of the rhomboid fossa. Visceral motor and branchial motor axons of VII enter from superior salivatory and facial motor nuclei of the pons.
2. The **subarcuate fossa** marks the outline of the superior semicircular canal in the bone.
3. There is a depression for the **endolymphatic sac.** The canal opening into the depression is the **endolymphatic canal** or **aqueduct of the vestibule.** It is the opening of the **endolymphatic duct.** The sac is just a dural covering over the depressed area. The duct connects sac to the **membranous saccule** and **utricle** in the vestibule of the osseous labyrinth. The sac is a storage reservoir for excess **endolymph** manufactured from blood capillaries within the **membranous labyrinth.**
4. On the medial edge of the petrous bone near the midpoint of the jugular foramen is another tiny canal, the **perilymphatic canal** or **aqueduct of the cochlea.** It has been said to connect with the subarachnoid space around the cranial nerve IX as this nerve makes its exit through the jugular foramen. Through it, it has been purported, **perilymph** (really cerebrospinal fluid) gains entrance to the **osseous labyrinth** near the base of the cochlea. We have pointed out in Chapter VI that Wharton claims that the membrane over this opening is impervious to cerebrospinal fluid. The source of perilymph remains a point of confusion that has not been satisfactorily explained by experimental efforts to date.
5. The jugular bulb, the superior and inferior petrosal sinuses, and the transverse and sigmoid sinuses leave their depressions marked on the bone.
6. At the apex, there is a depression for the semilunar ganglion of V, under cover of dura in Meckel's cave. The greater and lesser superficial petrosal nerves pierce to the surface near the apex.

Osseous versus Membranous Labyrinth: The **osseous labyrinth,** were it dissectible from hard petrous surroundings, would have the shape shown in Figure XIV-73 with its oval and round windows facing the middle ear cavity. The section containing the windows is its **vestibule.** The three **semicircular canals,** in three planes of space, orient posterolaterally in the petrous bone. The 2½ turns of its snail-shell **cochlea** sit anteromedially. The bony labyrinths or channels interconnect and they are filled with **perilymph,** or they *would* be completely filled with perilymph if a **membranous labyrinth** were not suspended *within* the bony labyrinth.

The tubular membranous labyrinth is also fluid filled, but with *endolymph* (Fig. XIV-74). In this Figure, the relationship of osseous labyrinth to middle ear cavity is shown as well as the relationship of the membranous labyrinth to osseous labyrinth. The stapes can be seen communicating with the vestibule

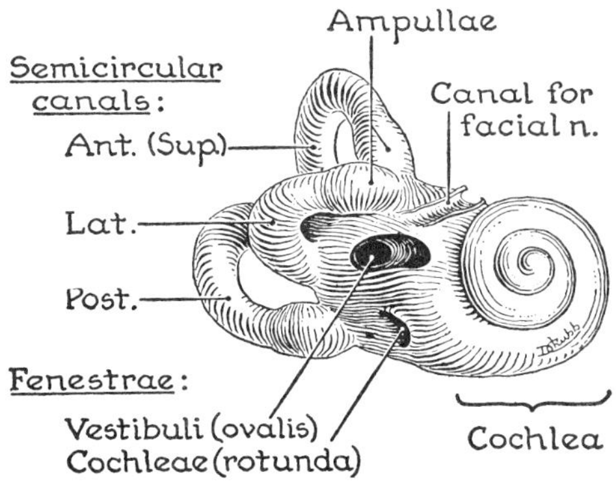

Fig. XIV-73. The bony labyrinth. (From J. C. B. Grant and J. V. Basmajian, *Grant's Method of Anatomy,* 7th ed., Williams & Wilkins, Baltimore, 1965.)

and the round window communicates with the lower gallery or scala tympani at the base of the cochlea.

The regional parts of the membranous labyrinth also interconnect with each other. The semicircular parts of the membranous labyrinth connect at their dilated **ampullary ends** with a swelling, the **utricle.** Utricle and saccule of the vestibule join by way of the branching endolymphatic duct which is seen ascending to its storage reservoir under the dural covering of the petrous bone. Saccule joins the membranous cochlear *via* a short duct, the **ductus reunions.** It is in the membranous labyrinth that the **neuro-epithelium** of the cochlea and that of the semicircular canals and vestibule are housed.

The Bony Cochlea: Were we to take a longitudinal cut through the coiled cochlea (Fig. XIV-75), we would see a central pyramidal core of bone (petrous bone) with a tunnel leading to its base. The tunnel transmits the cochlea or auditory nerve to the internal auditory meatus. This central core of bone is the **modiolus** of the cochlea. Around it winds a shelf of bone for 2½ turns, *the spiral bony lamina,* like the threads of a screw or the stairs of a lighthouse. This thread is the medial wall of the osseous labyrinth of the cochlea. The lamina projects out into space laterally, further in its basal coils than in its apical ones, but it never reaches across to the lateral walls of the spiraling osseous labyrinth. Since the tubular bony channels wind around the modiolus, they will be cut in several sections, wider at the base than at the narrowing apex (Fig. XIV-75). Thus, the bony thread or spiral bony lamina only *incompletely* separates the tubular bony passageways of the osseous labyrinth into **upper and lower galleries** or scala (Fig. XIV-75–1). The separation is completed by placing a triangular-shaped, winding, membranous labyrinth between the lateral wall the bony spiral shelf (Fig. XIV-75–5). The upper scala in the osseous labyrinth is now separated from the lower scala by a membranous tube filled with endolymph. The upper scala is called the scala vestibuli and it communicates with the oval

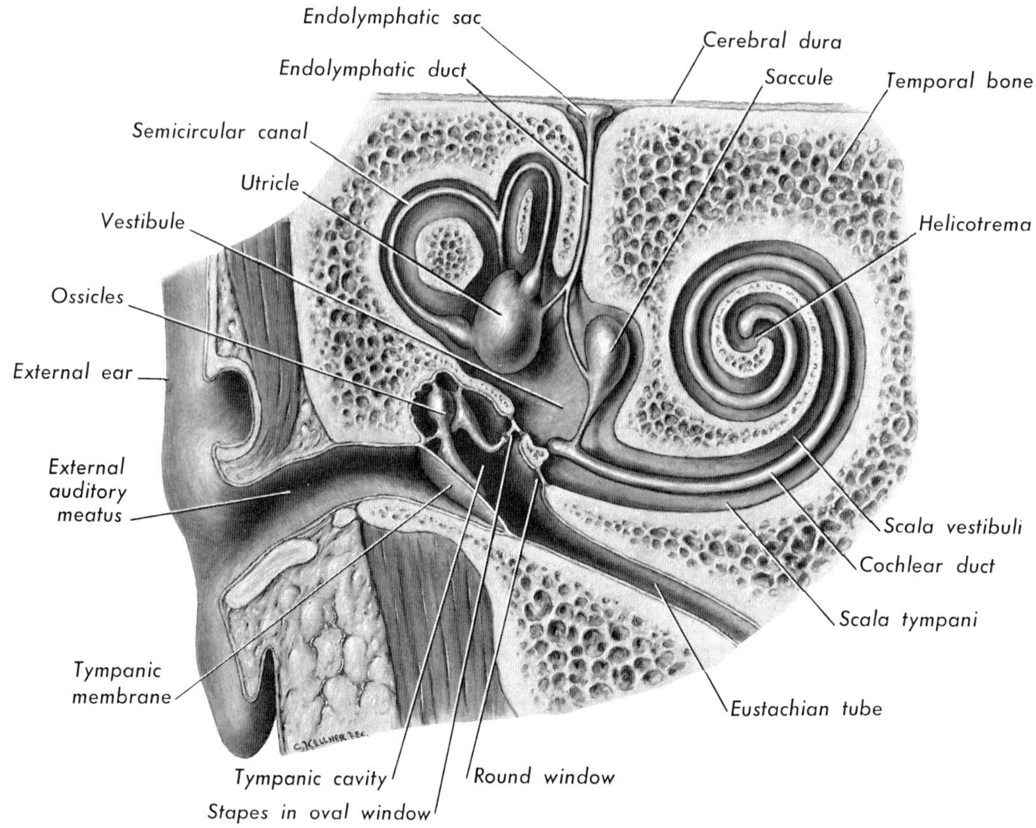

Fig. XIV-74. Diagrammatic drawing to show the relationship of the structures of the external ear, middle ear and internal ear. No attempt has been made to maintain correct proportions. (From W. M. Copenhaver, *Bailey's Textbook of Histology*, 15th ed., Williams and Wilkins, Baltimore, 1964.)

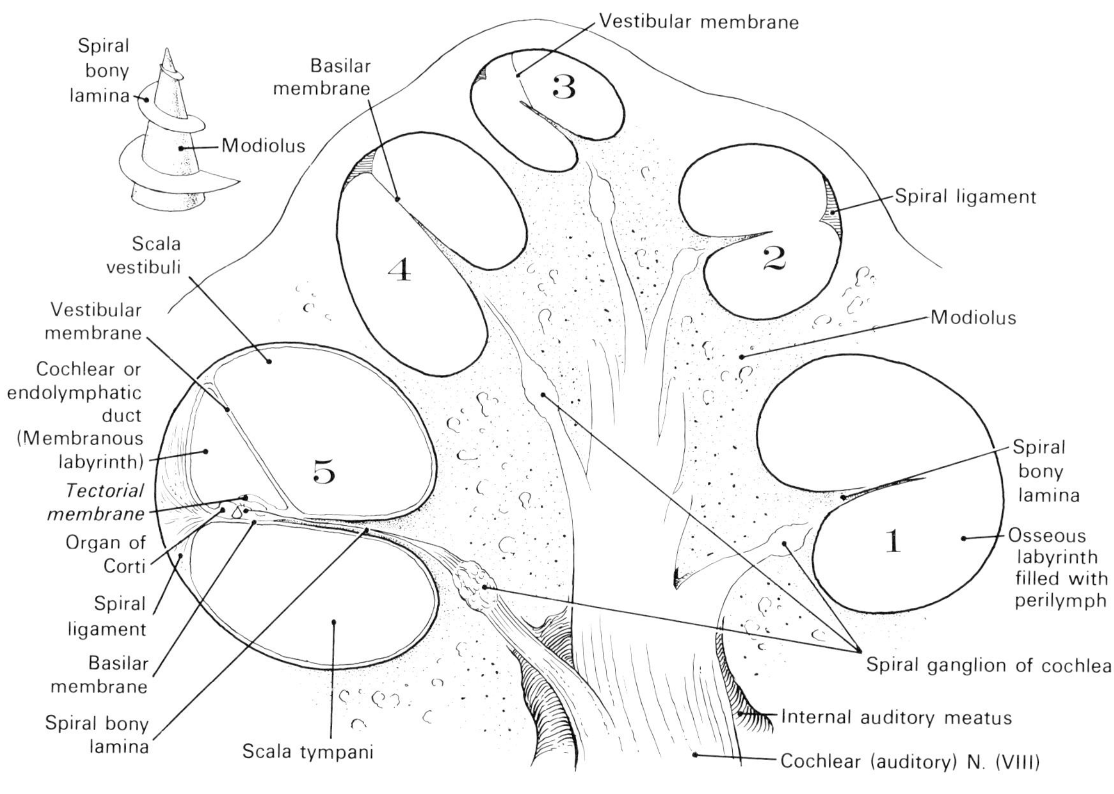

Fig. XIV-75. Sketch of longitudinal section through cochlea. The cone-like nature of modiolus and spiral bony lamina is indicated at upper left. 1 to 5 show stages in which the osseous labyrinth is partitioned into upper and lower galleries by interposition of a membranous labyrinth.

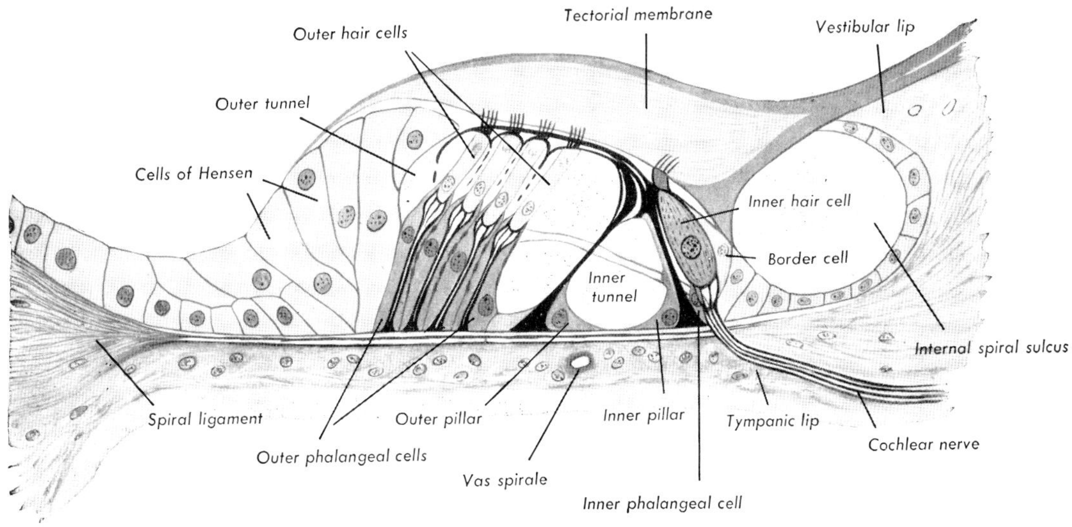

Fig. XIV-76. Radial section of human organ of Corti, from the upper end of the first coil. (From W. M. Copenhaver, *Bailey's Textbook of Histology*, 15th ed., Williams & Wilkins, Baltimore, 1964.)

window of the vestibule. The lower scala is the scala tympani and communicates with the round window. At the apex of the modiolus, **the helicotrema**—scala tympani and scala vestibuli become confluent (Fig. XIV-74).

In the proximal margins of the spiral bony lamina lie the bipolar ganglion cells of the cohlea (Fig. XIV-74). They make up the **spiral ganglion.** The dendrites of these cells reach out through the lamina to contact sensory receptor cells within the membranous labyrinth or **cochlear duct.** Their axons traverse the central core of the modiolus to exit from the internal auditory meatus as the cochlear nerve (cranial nerve VIII, or the auditory nerve).

The Membranous Labyrinth (Cochlear Duct): The walls of this tube are triangular in shape. The lateral wall of the tube is formed by the **spiral ligament** which anchors to the lateral wall of the osseous labyrinth. This is a modification of periosteum. It contains blood vessels, the larger of which is a vein. Its capillary bed is held responsible for the formation of endolymph which fills the cochlear duct. The ligament is stout in the basal turns of the cochlear duct (Fig. XIV-75–5), but it gets greatly reduced in size and strength in the apical coils (Fig. XIV-75–3). Thus, both ligament and spiral bony lamina increase in size as one progresses from apical to basal coils. This means that the open distance between them (Fig. XIV-75) is, actually, longer in the apical coils than in the basal coils.

Bridging between spiral ligament and spiral bony lamina, to form a floor for the triangular membranous labyrinth is the **basilar membrane** (Fig. XIV-75). The chief components of this membrane are the basilar fibers or "auditory strings." Once thought to be elastic connective tissue fibers, their elasticity is now held in doubt. The fibers are embedded in a homogeneous matrix. The basilar membrane, then, is longer in the apical coils near the helicotrema than in the basal coils of the cochlear duct near the vestibule. Resting upon the membrane is the complicated **spiral organ of Corti** made up of hair cells and supporting cells. This organ is the sensory neuro-epithelium of the cochlea. Supporting cells are fewer and shorter in basal coils than in apical coils where they are taller and many. The basilar membrane is, thus, more heavily weighted in its apical coils.

To complete the triangular walls of the cochlear duct, there is a thin, slanting roof, the **vestibular** (Reisner's) **membrane.** It separates perilymph in the osseous scala tympani from the endolymph of the cochlear duct, just as the basilar membrane separates endolymph from the perilymph in the osseous scala tympani.

The Organ of Corti (Fig. XIV-76): Supporting cells of the organ vary from cuboidal to tall columnar. Those on the medial and lateral margins of the cellular organ serve primarily as additional weights for the basilar membrane. Some near the medial margin form a curved structure, the **spiral** sulcus, to help support a **tectorial membrane.** The names given groups of supporting cells are not worth learning. The important supporting elements are near the center of the organ. These are the **phalangeal cells** which hold the **hair cells.** The latter are the sensory receptors. Another group of supporting cells are the **pillar cells** which form the **spiral tunnel.**

Hair cells are found on each side of the tunnel, a single row on the medial side (bony lamina side) and a row of four on the lateral side (spiral ligament side). The dendrites of the bipolar cells in the spiral ganglion (embedded in the base of the bony lamina) innervate the hair cells, even crossing the tunnel to reach the lateral ones.

Hanging over the hair cells is a gelatinous canopy, the **tectorial membrane.** It is a rather stiff immobile structure. When the hair cells are at rest (not excited to transmit a neural impulse), they are bent against the membrane and embedded in it. They send impulses over the cochlea nerve when they are made to straighten, which they will do when the basilar membrane swings down away from the tectorial membrane.

Functioning of the Inner Ear: Is it possible to understand the basic functioning of the inner ear in its role as an analyser of incoming sound waves? The mechanism is complicated and only partially understood at best. In trying to simplify concepts, we trust we will not lead you too far astray.

Basic Concepts: First of all, as to the order of simple events:

1. As the stapes is caused to rock in sympathy with sound wave frequencies impinging upon the tympanic membrane, it acts upon the fluid which fills the osseous vestibule.
2. Fluid waves are set in motion throughout the perilymph, but those of significance for hearing are those which traverse the perilymph in the osseous scale vestibuli of the wide basal coils of the cochlea.
3. All that separates this wave motion from endolymph is the thin vestibular roof of the cochlear duct all along its length from base to helicotrema. It is set into vibration.
4. The vibration of the vestibular membrane sets fluid waves in motion in the endolymph.
5. This, in turn, acts upon the basilar membrane which transmits the motion to the perilymph in the lower osseous gallery or scala tympani.
6. The fluid wave is now directed against the round window.
7. Thus, for every inward swing of the stapes, the basilar membrane is ultimately depressed, straightening the hair cells, and the round window is bowed outward into the middle ear cavity. As the stapes swings out of the oval window, the basilar membrane returns to its normal position, the hairs bend against the tectorial membrane in a resting position, and the round window relaxes.

Resonance Theory: How, then, does the basilar membrane analyze incoming pitch frequencies? An old, neat, theory which appeared to fit the anatomical facts was proposed by von Helmholtz.[6] This **resonance theory** was once thought to provide the ideal answer. It is not adequate nor consistent with recent findings. Yet, it is well to consider it momentarily.

1. Not all parts of the basilar membrane were thought to vibrate at once, to a single sound frequency or to a combination of frequencies.
2. Only those "auditory strings" of the basilar membrane which were pulled by the spiral ligament to the right tautness and were weighted appropriately along the length of the cochlea by the overlying supporting cells of the organ of Corti—only those "strings" *in selected places* along the membrane vibrated in sympathy with the fluid waves impinging upon them. Thus, only certain hair cells would be stimulated and only certain auditory nerve fibers would be discharged. In the end, conduction over the auditory tract would excite certain parts of the anterior transverse temporal gyrus in the cortex.
3. This was predicated upon the elasticity of the "auditory strings." Those in the basal coils were tightly stretched by a taut, spiral ligament; they were short, and lightly weighted. In the apical coils, they were long, lax, and heavily weighted by cells—like the long strings of a harp weighted by coils of copper wire. Auditory strings in between basal and apical coils were stretched less and less taut and weighted more and more heavily. Thus, the basilar membrane was conceived as an organ of strings that could be set to vibrate in selective spots in sympathetic resonance.
4. Indeed, basal coils do respond to high pitch (high frequency sound waves) and apical coils to low frequency sound waves. Intermediate coils respond to graded frequencies in between. The range in man is from 16,000 cycles (double vibrations) per second in the basal coils to 16 cycles per second in the apical coils. Experiments on guinea pigs in insulated chambers, exposed day in and day out to a tone of one frequency showed death of hair cells in the proper locus of the basilar membrane with a graded series of tone frequencies used in different animals. Boiler makers tend to show degeneration of hair cells in the apical coils after constant exposure to overstimulation with low frequency sounds.

What, then, is wrong with the old von Helmholtz resonance theory?

1. As von Békésey[7] has shown, it can only account for 4½ of the 10½ octaves available to the human ear.
2. The von Helmholtz theory is based upon a "standing wave," and this does not appear to be the form of the activating fluid wave.
3. The old theory is based upon the elasticity of the basilar membrane which proves to be non-elastic and not under tension.
4. The new theory proposed by von Békésey does *not* argue against "place" or spot stimulation of hair cells for the tonal response of the membrane.
5. A "traveling" fluid wave, von Békésey showed, will stimulate an inelastic membrane at points where that wave reaches its maximum peak. Waves of different frequencies reach peak value at different distances of travel.
6. High frequency waves travel a short distance before they peak and would activate the lower basal coils of the cochlea.
7. Low frequency waves travel through the fluid for greater distances before they peak and so would reach apical coils and even turn round the helicotrema.
8. The traveling wave theory along an inelastic basilar membrane explains matters of pitch analysis better than the von Helmholtz theory. It covers all octaves of human hearing.

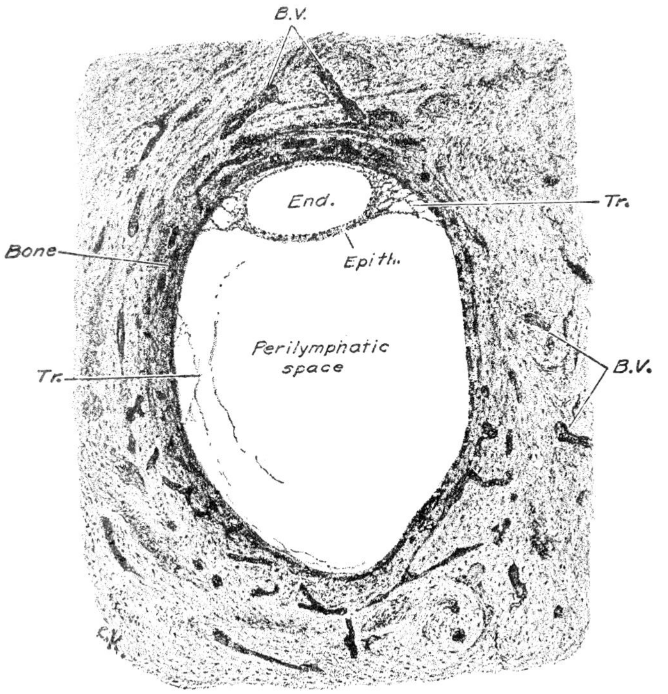

Fig. XIV-77. Cross-section of superior semicircular canal of 15-year-old girl. BV., Blood vessel; End., endolymphatic space; Epith., epithelium; Tr., connective tissue trabeculae in perilymphatic space. (From W. M. Copenhaver, *Bailey's Textbook of Histology,* 15th ed., Williams & Wilkins, Baltimore, 1964.)

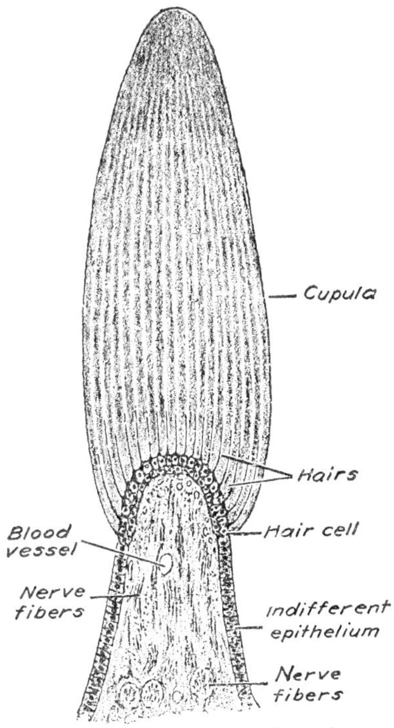

Fig. XIV-78. Crista ampullaris of newborn child. (From W. M. Copenhaver, *Bailey's Textbook of Histology,* 15th ed., Williams & Wilkins, Baltimore, 1964.)

9. But, it rests upon experiments with cadaver tissues in tests for elasticity and upon ingeniously constructed mechanical models. If it does not exactly reflect what may be happening in the living cochlea, it is a thought-provoking piece of research. It does not necessarily help, more than other theories, to explain differences in the loudness perception of sounds.

The Vestibular Organ of the Inner Ear: The membranous parts of the semicircular canals are small-diameter tubes filled with endolymph. They lie along the periosteal borders of their osseous canals which, of course, contain perilymph (Fig. XIV-77). Except for regions where the neuro-epithelium heightens as **cristae** in the ampullae or dilatations of each canal (Fig. XIV-78), the canals consist of a thin tunic of fibrous connective tissue lined with simple squamous epithelium.

Cristae are connective tissue ridges which bear taller supporting cells and hair cells in the form of a cupula. The hair cells, again, are the sensory neuro-epithelium innervated by dendrites of the vestibular division of cranial nerve VIII. The hairs are embedded in a conical gelatinous mass (Fig. XIV-78). Fluid movement (if indeed such can occur in such fine capillary tubes) tilts the gelatinous mass and so stimulates the hair cells. Since each canal is oriented in a different plane of space, cristae are activated differentially depending upon bodily orientation. Cristae respond by tilting under conditions of angular acceleration, *i.e.,* rotation of the body about some vertical axis. They help to orient us in space when we are turned rapidly, help to maintain balance and make reflex connections with the eyes through brainstem connections (*e.g., via* the medial longitudinal fasciculus).

Otolith organs form the sensory receptors in the **utricle** and **saccule** of the vestibule. Their arrangement is somewhat similar to the cristae. The **maculae** (neurosensory regions) of the utricle and saccule sit at right angles to each other (Fig. XIV-79). The mass of cells in each macula is much broader in area than in cristae and the cells are multilayered rather than single-layered. The projecting hair cells at the surface are short. The broad gelatinous mass which envelopes the hair cells contains tiny crystalline bodies, *the otoliths,* consisting of calcium carbonate combined with a protein. Depending, again, upon the orientation of the head, the gel can be tipped to stimulate hair cells which excite dendrites of the vestibular division of VIII. The maculae seem more concerned with static balance and with altering posture during linear acceleration.

The vestibular division of VIII (vestibular nerve) has its bipolar ganglion cells located in **superior** and **inferior vestibular ganglia** just inside the mouth of the internal auditory meatus. The dendritic or peripheral processes of the superior ganglion innervate hair cells of the macula of the utricle and the cristae of the superior and horizontal (lateral) semicircular canals. The dendrites of the inferior ganglion run to

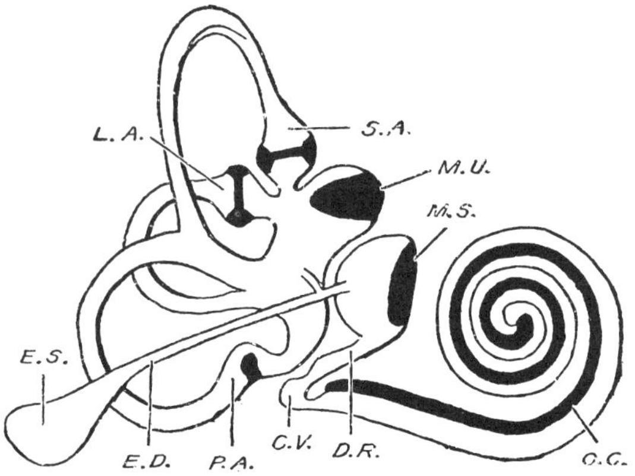

Fig. XIV-79. Diagram of left membranous labyrinth viewed from the medial aspect. The neuroepithelial areas are shown in *black*. C.V., Caecum vestibulare; D.R., ductus reuniens; E.D., endolymphatic duct; E.S., endolymphatic sac; L.A., ampulla of lateral semicircular canal; M.S., macula sacculi; M.U., macula utriculi; O.C., organ of Corti in cochlear duct; P.A., ampulla of posterior semicircular canal; S.A., ampulla of superior semicircular canal. (From W. M. Copenhaver, *Bailey's Textbook of Histology*, 15th ed., Williams & Wilkins, Baltimore, 1964.)

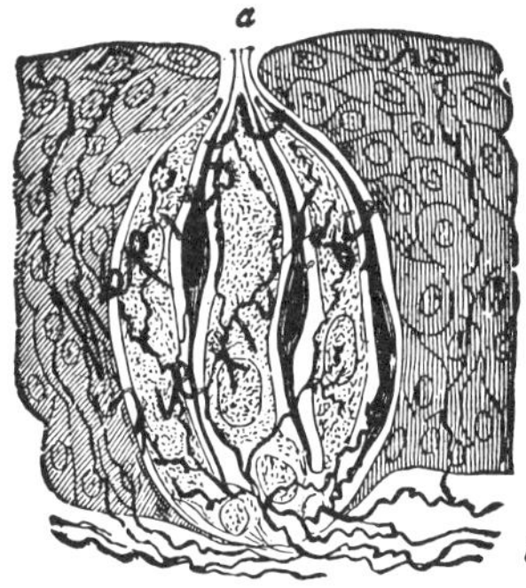

Fig. XIV-80. Taste-bud from side wall of circumvallate papilla. (Merkel-Henle.) a, Taste pore; b, nerve fibers, some of which enter the taste-bud—intragemmal fibers—while others end freely in the surrounding epithelium—intergemmal fibers. (From W. M. Copenhaver, *Bailey's Textbook of Histology*, 15th ed., Williams & Wilkins, Baltimore, 1964.)

the macula of the saccule and the crista of the posterior semicircular canal. A connecting twig, the nerve of Oort, is directed into the cochlea.

The Organ of Smell

The olfactory organ, as we have indicated at other times, consists of the olfactory portion of the nasal mucosa. It is brownish-yellow in color in contrast to the ordinary reddish respiratory mucose which lines the bulk of the nose. The olfactory mucosa sits high up in the nose over superior concha and laps over on the upper medial walls of the nasal septum. Once more, sustentacular cells are more numerous than neuro-epithelial cells. These cells contain a pigment. The olfactory cells lie between the sustentacular or supporting cells and their nuclei tend to be more deeply placed. A delicate process extends to the surface where it terminates in several minute hair-like processes. This actually is its dendrite. Basally, a longer process becomes its axon which courses through the pores of cribriform plate of the ethmoid bone to synapse in the olfactory bulb. These cells are, therefore, neurons as distinct from the epithelial hair cells of the cochlea and semicircular canals. We have discussed the course of axons of cells in the olfactory bulb to prepiriform and periamygdaloid cortex.

The Organ of Taste

The **taste buds** form the organs of taste. They are found in ovoid patches (Fig. XIV-80) in the epithelium (stratified squamous) which lines the troughs of the circumvallate papillae of the tongue; a few in the epithelium of fungiform papillae; in the lateral folds of folliate papillae on the sides of the tongue; and in the mucosa of the epiglottis. The broadest cells of a taste bud are sustentacular elements. **Taste cells** are slender and fusiform in shape. A peripheral hair-like process extends to the surface through a **taste pore.** The basal processes are tapering and may end in a forked branch.

Dendritic fibers of cranial nerves VII, IX, and X end about the basal process of the taste cell. In this sense, taste cells are neuro-epithelial elements like the hair cells of the cochlea and semicircular canals, not neurons like the olfactory receptor cells.

The facial nerve serves the taste buds of the anterior two-thirds of the tongue approximately; the glossopharyngeal nerve, the buds of the posterior one-third; and the vagus serves the taste buds of the epiglottis.

REFERENCES

1. Shapiro, R.: *Myelography,* 2nd ed. Yearbook, Chicago, 1968.
2. Gazzaniga, M. S., Bogan, J. E., and Sperry, R. W.: Observations on visual perception after disconnection of the cerebral hemispheres in man. Brain, *88:* 221–236, 1965.
3. Gazzaniga, M. S., and Sperry, R. W.: Language after section of the cerebral commissures. Brain, *90:* 131–148, 1967.
4. Elman, R.: Spinal arachnoidal granulations with special reference to cerebrospinal fluid. Bull. Johns Hopkins Hosp., *34:* 99–104, 1923.
5. Hecht, S., Schlaer, S., and Pirenne, M. H.: Energy, quanta and vision. J. Gen. Physiol., *25:* 819–840, 1942.
6. Helmholtz, H. L.: *Sensations of Tone.* Longmans, New York, 1930.
7. Békésey, G. von, and Rosenblith, W. A.: In *Handbook of Experimental Psychology,* Ch. 27, edited by S. S. Stevens. Wiley, New York, 1951.

General References:

1. Truex, R. C., and Carpenter, M. B.: *Human Neuroanatomy,* 6th ed. Williams & Wilkins, Baltimore, 1969.
2. Copenhaver, W. M.: *Bailey's Textbook of Histology,* 15th ed. Williams & Wilkins, Baltimore, 1964.

Neuroradiology References—General

1. Tavaras, J. M., and Wood, E. H.: *Diagnostic Neuroradiology.* Williams & Wilkins, Baltimore, 1964.
2. Wilson, M.: *The Anatomical Foundation of Neuroradiology of the Brain.* Little, Brown, Boston, 1963.

*Index**

* Italicized page numbers denote illustrations.